P9-BZM-325

Handbook of
Fluids in Motion

Handbook of Fluids in Motion

Edited by

Nicholas P. Cheremisinoff
Ramesh Gupta

230 Collingwood, P.O. Box 1425, Ann Arbor, Michigan 48106

Library of Congress Catalog Card Number 82-70706
ISBN 0-250-40458-3

Manufactured in the United States of America

Butterworths, Ltd., Borough Green
Sevenoaks, Kent TN15 8PH, England

PREFACE

This book is intended to supply both the practicing engineer and the researcher with an authoritative reference work that comprehensively covers the field of applied fluid mechanics. Sound engineering design cannot be performed without proper interpretation of the physics controlling the flow phenomenon. That is, proper understanding of physical laws, which engineers refer to as fluid mechanics, is necessary to rationally design for heat transfer and mass transfer, and to achieve proper chemical kinetics. To implement these rational design principles requires a departure from the traditional brute force approach of empiricism in obtaining process operating parameters, and requires that you first establish criteria for modeling the fluid dynamics of the system. To this end, many chapters emphasize modeling approaches based on first principles. The work goes beyond this endeavor in that many chapters also give direction to apply fluid transport analogies to addressing heat transfer, mass transfer, and kinetic issues.

The book is organized in five sections. Section I deals with Single-Phase Flows and gives concepts applied in later sections to describe multiphase flow systems. Emphasis is placed on more advanced topics and some subjects, although important, have been excluded from this section because of their wide availability in standard textbooks. Section II deals with Gas-Liquid Flows. This section covers from empirical methodology to rigorous modeling techniques for flow analysis. Section III covers Gas-Solid Flows with emphasis on design-related principles, and contains several chapters with detailed design procedures in pneumatic transport, fluidized beds and solids feeding applications. A point of clarification for this section is that solid/gas systems for both dense and dilute phase applications are analogous to gas-liquid flow systems. Section IV covers Liquid-Solid Flows, with several chapters giving design-related guidelines for transporting coarse and fine suspensions. Finally, Section V is entitled Special Topics Flow Phenomena. The bulk of this section contains information on liquid-liquid flow systems. In addition, an extensive treatise on the state-of-the-art of charge generation and transport of low conductivity flows is included.

To ensure the highest degree of reliability, the services of a large number of specialists were enlisted. This book presents the efforts of these 50 experts. Additionally, it brings you the experience and opinions of scores of engineers and researchers who aided with advice and suggestions in reviewing and refereeing the material presented. Each contributor is to be regarded as responsible for the statements in his chapter. Extensive bibliographies are given with each chapter.

Nicholas P. Cheremisinoff
Ramesh Gupta

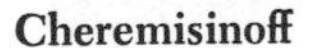

Cheremisinoff **Gupta**

Nicholas P. Cheremisinoff heads the Reactor and Fluid Dynamics Modelling Group in the Chemical Engineering Technology Division of Exxon Research and Engineering Company, Florham Park, New Jersey. He is involved in research and engineering on atomization, fluid bed reactor design, shale oil development, coal gasification/liquefaction, instrumentation and equipment design, and numerical modeling of complex flow systems. He is also a visiting professor in chemical engineering at the University of Missouri–Rolla. Dr. Cheremisinoff received his BS, MS and PhD in Chemical Engineering from Clarkson College of Technology. He is the author of numerous papers and books, including Ann Arbor Science Publishers' *Fluid Flow–Pumps, Pipes and Channels; Cooling Towers–Selection, Design and Practice; Gasohol for Energy Production; Chemical and Nonchemical Disinfection.* He is a member of a number of professional and honor societies, including AIChE, Tau Beta Pi and Sigma Xi.

Ramesh Gupta is with Exxon Research and Engineering Company, Florham Park, New Jersey, where he does research on multiphase chemical reactors. Prior to Exxon, he was at Air Products and Chemicals Inc. Dr. Gupta obtained his PhD in Chemical Engineering from Princeton University (1973), and was a Research Fellow at the California Institute of Technology (1973–1975). Dr. Gupta holds several patents and has authored papers on separations, adsorption, biomedical engineering, polymers and chemical reactors. He is a member of AIChE, ACS and Sigma Xi.

CONTRIBUTORS

M. B. Ajinkya, Exxon Research & Engineering Company, Florham Park, New Jersey

D. Azbel, Department of Chemical Engineering, University of Missouri–Rolla, Rolla, Missouri

J. P. Batycky, Petroleum Recovery Institute, Calgary, Alberta, Canada

M. Bohnet, Institute for Process Technology, Technical University of Braunschweig, Braunschweig, Federal Republic of Germany

J. Bridgwater, Department of Chemical Engineering, University of Birmingham, Birmingham, England

L. H. Chen, Department of Chemical Engineering, West Virginia University, Morgantown, West Virginia

T.-Y. Chen, UOP Process Division, Des Plaines, Illinois

N. P. Cheremisinoff, Exxon Research & Engineering Company, Florham Park, New Jersey

P. N. Cheremisinoff, Department of Civil and Environmental Engineering, New Jersey Institute of Technology, Newark, New Jersey

D. Chisholm, Glasgow College of Technology, Glasgow, Scotland

S. W. Churchill, Department of Chemical Engineering, University of Pennsylvania, Philadelphia, Pennsylvania

W.-D. Deckwer, Universität Oldenburg, Oldenburg, Federal Republic of Germany

D. E. Dietrich, JAYCOR, San Diego, California

N. Epstein, Department of Chemical Engineering, University of British Columbia, Vancouver, British Columbia, Canada

L. T. Fan, Department of Chemical Engineering, Kansas State University, Manhattan, Kansas

P. Freymuth, Department of Aerospace Engineering Sciences, University of Colorado, Boulder, Colorado

R. L. Gandhi, Bechtel Petroleum, Inc., San Francisco, California

S. R. Goldman, Los Alamos Scientific Laboratory, Los Alamos, New Mexico

J. R. Grace, Department of Chemical Engineering, University of British Columbia, Vancouver, British Columbia, Canada

R. A. Greenkorn, School of Chemical Engineering, Purdue University, West Lafayette, Indiana

R. Gupta, Exxon Research & Engineering Company, Florham Park, New Jersey

P. Harriott, School of Chemical Engineering, Cornell University, Ithaca, New York

S. Hartland, Swiss Federal Institute of Technology, Zürich, Switzerland

H. Hofmann, Institut für Technische Chemi I, Universität Erlangen-Nürnberg, Erlangen, Federal Republic of Germany

D. T. Y. Kao, Department of Civil Engineering, University of Kentucky, Lexington, Kentucky

H. H. Klein, JAYCOR, San Diego, California

D. H. Laird, JAYCOR, San Diego, California

A. I. Liapis, Department of Chemical Engineering, University of Missouri–Rolla, Rolla, Missouri

F. G. McCaffery, Occidental Research Corporation, Irvine, California

R. D. Patel, Exxon Research & Engineering Company, Florham Park, New Jersey

G. K. Patterson, Department of Chemical Engineering, University of Missouri–Rolla, Rolla, Missouri

N. Rajaratnam, Department of Civil Engineering, University of Alberta, Edmonton, Alberta

S. J. Rossetti, Exxon Research & Engineering Company, Florham Park, New Jersey

M. F. Scharff, JAYCOR, San Diego, California

A. M. Scott, Koninklijke/Shell-Laboratorium, Amsterdam, The Netherlands

Y. T. Shah, Department of Chemical and Petroleum Engineering, University of Pittsburgh, Pittsburgh, Pennsylvania

C. A. Shook, Department of Chemistry and Chemical Engineering, University of Saskatchewan, Saskatoon, Saskatchewan, Canada

S. Simone, Refineria de Amuay, Lagoven S.A., Falcon, Venezuela

A. H. P. Skelland, School of Chemical Engineering, The Georgia Institute of Technology, Atlanta, Georgia

P. E. Snoek, Bechtel Petroleum, Inc., San Francisco, California

S. L. Soo, Department of Mechanical and Industrial Engineering, University of Illinois at Urbana–Champaign, Urbana, Illinois

T. Sridhar, Department of Chemical Engineering, State University of New York at Buffalo, Buffalo, New York

B. Srinivas, JAYCOR, San Diego, California

L. Steiner, Swiss Federal Institute of Technology, Zürich, Switzerland

C. Tien, Department of Chemical Engineering and Materials Science, Syracuse University, Syracuse, New York

J. P. Wagner, Exxon Research & Engineering Company, Florham Park, New Jersey

W. P. Walawender, Department of Chemical Engineering, Kansas State University, Manhattan, Kansas

J. Weisman, Department of Chemical and Nuclear Engineering, University of Cincinnati, Cincinnati, Ohio

C. Y. Wen, Department of Chemical Engineering, West Virginia University, Morgantown, West Virginia

F. A. Zenz, Chemical Engineering Department, Manhattan College, New York, New York

CONTENTS

Section I: Single-Phase Flows

Section II: Gas-Liquid Flows

Section III: Gas-Solids Flows

Section IV: Liquid-Solids Flows

Section V
Special Topics: Flow Phenomena

SECTION I

SINGLE-PHASE FLOWS

CONTENTS

CHAPTER 1
TRANSPORT COEFFICIENTS OF LIQUIDS

T. Sridhar*
Department of Chemical Engineering
State University of New York at Buffalo
Buffalo, New York 14260

CONTENTS

INTRODUCTION

In almost every branch of engineering, reliable values of material properties are necessary in the design and operation of processes. An enormous amount of data has been collected and correlated over the years, but a significant gap exists between demand and availability. The engineer still must rely on common sense and experience to evaluate physical properties. In chemical engineering calculations it is often necessary to evaluate liquid-phase properties. Among these, the importance of transport coefficients is well appreciated. Rarely is a problem encountered that does not involve the transport of heat, mass or momentum. It is hardly possible to solve problems involving flowing fluids or heat transfer without the knowledge of liquid viscosity or thermal conductivity. There are numerous examples of situations involving mass transfer with or without chemical reaction. In applied kinetics the rate of reaction frequently is controlled by the diffusion of reactants through a film or within a catalyst pellet. In separation

*Present address: Department of Chemical Engineering, Monash University, Clayton, Australia.

processes diffusion and mass transfer play key roles. In all these instances, there is a need to be able to calculate the diffusion coefficient of the solute through the solvent.

From a theoretical viewpoint, our understanding of the liquid state is still in a preliminary stage of development, and data on transport coefficients can provide important checks on a proposed model. A good model of the liquid state should provide reliable equations for predicting transport and other properties. At present, we are far from reaching this goal, even though some progress has been made. In this chapter, methods for predicting viscosity, diffusion coefficients and thermal conductivity of Newtonian liquids are presented. A brief overview of the various models of the liquid state is given. Comparison with experimental data indicates the accuracy of the empirical equations and thus suggests which equations should be used.

Definitions

In general, transport coefficients are defined as proportionality constants between the flux and the corresponding driving force. Diffusion is the flux of mass across a concentration gradient, $\partial c/\partial x$, and is embodied as Fick's law. The flux of momentum as a result of a velocity gradient, $\partial v/\partial x$, gives viscosity and is described by Newton's law. Fourier's law relates thermal conductivity to heat flux and a temperature gradient:

$$J_H = -k_L \frac{\partial T}{\partial x} \text{ (Fourier's Law)} \tag{1}$$

$$J_D = -D_{AB} \frac{\partial c}{\partial x} \text{ (Fick's Law)} \tag{2}$$

$$J_M = -\eta \frac{\partial v}{\partial x} \text{ (Newton's Law)} \tag{3}$$

In each case, the flux is in the direction of decreasing gradient; however, the coefficients themselves are always positive.

Most organic liquids have viscosities below 1 cP (10^{-3}N-sec/m^2) and thermal conductivity between 1×10^{-4} and 3×10^{-4} kW/m-°K. Diffusion coefficients in liquids are of the order of 10^{-9}m^2/sec.

THEORIES OF THE LIQUID STATE

At present, liquid state theories are still in a preliminary stage of development. There does not exist, for the liquid state, a theory that can rival the kinetic theory of gases. Unlike the mechanism of momentum transfer in gases, in which molecules move from one layer to another, momentum transfer in liquids has to take interatomic forces into account. The density of liquids is higher than gases so that the average distance between atoms is not much greater than the range of these forces. In general, the prevailing theories of the liquid state can be divided into gas-like and solid-like models. Such a distinction is quite arbitrary as many variations of such classifications also can be postulated. Brush [1] has written an excellent review of the various models. They are also examined in the classic text by Hirschfelder et al. [2].

Hard Sphere Models

The rigorous kinetic theory of gases assumes that molecules undergo two-body collisions and that the molecular diameter is small compared to the mean free path. While these assumptions may be valid for dilute gases, it breaks down when a dense medium is considered. Enskog [3] modified the theory to take into account the finite size of molecules. The closer spacing of molecules causes energy and momentum transfer to occur when two molecules collide. The finite size of molecules increases the frequency of collisions, this increase being counteracted by the shielding effect of more closely spaced molecules. Enskog showed that the difference in frequency of collision from that for dilute gases depends only on the local thermodynamic state and molecular diameter [4]. Hirschfelder et al. [2] give more details on the Enskog model. Longuet-Higgins and Pople [5] considered a model consisting of nonattractive hard spheres undergoing a random walk motion. Using the model they were able to derive expressions for the transport coefficients. Comparison with experimental data is not good and is attributed to the neglect of attractive forces. Several modifications have been proposed. Longuet-Higgins and Valleau [6] included attractive forces through a square well potential function, producing much better agreement with experimental data [7,8]. Dymond [9] achieved good accuracy in comparison with experimental data by using the method of molecular dynamics to obtain correction factors for the hard sphere model of Enskog.

Rate Theory

Eyring [10] devised the absolute rate theory to explain the rate of a chemical reaction. He assumed that during a chemical reaction there is one slow step that can be identified with an activated step. The first application of the reaction rate theory to transport phenomena was given by Eyring [11]. The Eyring rate theory attempts to explain the diffusion process on the basis of a cell model for the liquid, which is thought to contain holes or vacancies. In this theory, the coefficient of diffusion is expressed as

$$D_{AB} = \frac{\lambda^2}{\xi} k_1 \tag{4}$$

where λ = the distance between successive equilibrium positions of the diffusing molecule,
ξ = a parameter that describes the geometric configuration of the diffusing molecule and its nearest neighbors,
k_1 = the rate constant for a unimolecular rate process.

This parameter, k_1, may be predicted by means of the absolute rate theory [12]. The rate theory equation for diffusion is given by

$$D_{AB} = \frac{kT}{\eta} \left(\frac{N}{V'_B}\right)^{1/3} \exp \frac{[E_\mu - E_D]}{RT} \tag{5}$$

The above equation was derived by Eyring and March [13] based on the significant structure theory. Eyring and co-workers [14,15] have assumed that activation for diffusion and viscosity are equal. While this is true for self-diffusion, it is only an approximation in the case of mutual diffusion.

Gainer and Metzner [16], based on the application of the above equation to available experimental data, arrived at a value of 6 for ξ. The geometric factor was found to equal 8 for ethanol and methanol. The suggested significance of ξ is that it is the number of bonds formed between the diffusing molecule and its cage of neighboring molecules. This remains one of the weakest links of the absolute rate theory approach to diffusion. The Gainer and Metzner [16] modification still did not predict low-viscosity systems with any confidence. The general prediction procedure is cumbersome because of the need to evaluate the various terms of the equation. Furthermore, the Gainer-Metzner equation is valid only for liquid-liquid systems.

The fundamental rate process for viscous flow is considered to be the movement of a molecule to a vacant site by squeezing through the bottleneck between its nearest neighbors. The external force lessens the free energy of activation for a jump in the direction of force. This leads to an expression for the net velocity of flow of one layer with respect to the one next to it [12]:

$$V_1 = 2ak_o^1 \sinh[F/2nkT] \tag{6}$$

This equation predicts viscosity to be dependent on external force (non-Newtonian behavior). Assuming that this force is small and that the free energy for forming a hole can be assumed to be proportional to the energy of vaporization,

$$\eta = hn\, e^{3.8T_b/T} \tag{7}$$

This expression is useful for obtaining rough estimates.

The rate theory is applied to thermal conductivity by considering that the speed of sound in liquids is much greater than the speed of the molecules themselves. Assuming that the energy is transported in a similar manner, Eyring [11] derived

$$k_L = n^{2/3}k\sqrt{8/\pi\gamma}\, C^{(liq)} \tag{8}$$

This equation is valid for monoatomic liquids. The Eucken correction can be applied to make the result valid for polyatomic molecules. Bridgman [17] has shown that the above equation does not predict the pressure dependence of thermal conductivity.

Lattice Models

In this class of models, each molecule is confined to a cage by its nearest neighbors. The molecules remain in their lattice sites while some lattice sites are empty. Viscous flow takes place by the jumping of molecules from one lattice position to another under the influence of the applied force. Variations of this model can be found in the literature. In some models, the lattice as a whole vibrates and thus transfers momentum. In other models, the molecule vibrates around an equilibrium position and occasionally transfers momentum to its nearest neighbors.

Frenkel [18] considered the mobility of a molecule as it moved through a liquid. He considered the mobility to be the ratio of the molecule's average velocity and the resistance to the molecule. This resistance was evaluated using Stokes' law. The final expression for viscosity, assuming that the mean time spent by the molecule at an equilibrium position is exponentially related to temperature, is

$$\mu = [kT\,\tau_o/\pi\bar{a}\delta_1^2]\, e^{W/kT} \tag{9}$$

Absolute values predicted by this equation are 2 to 3 orders of magnitude higher than experimental data.

Furth [19] considered momentum transfer as taking place by the Brownian motion of holes within the liquid. Andrade [20] attempted to look at liquids from the solid state point of view. In this model the molecules are vibrating at a certain position with frequencies equal to that at the melting point and the following equation is obtained:

$$\mu = 5.1 \times 10^{-4}\, (AT_m)^{1/2}\, (V_A)^{-2/3} \tag{10}$$

Andrade [21] later extended this theory to take into account the temperature dependence of viscosity:

$$\mu = A_1 \exp\,(C_1/T) \tag{11}$$

Although the above equation is rarely used to predict viscosities, it has formed the basis for various empirical equations.

A lattice model was also used by Horrocks and McLaughlin [22]. These investigators calculate the change in potential energy of a molecule displaced from its cell due to its neighbors at the centers of their own cell. The excess energy due to the temperature gradient is assumed to be transferred with a frequency dependent on molecular mass and intermolecular forces. By differentiating their equation with respect to temperature at constant pressure, Horrocks and McLaughlin [22] were able to show that the temperature coefficient of thermal conductivity is negative and is a linear function of the coefficient of expansion.

A number of investigators have used variations of the cell model to study the process of diffusion [23]. Houghton developed a cell model that treats diffusion and viscous flow as a stochastic process. The model assumes a cubic cell and restricts consideration to the nearest neighbors of a central molecule. The diffusion coefficient is then given as

$$D = \frac{RT\rho}{6\eta M} \left(\frac{V}{N}\right)^{2/3} \tag{12}$$

This equation is similar to the Stokes-Einstein relationship.

Cohen and Turnbull [24] describe a "free volume" model. Adjacent to each molecule is a free volume of varying size. The continuous motion of molecules gives rise to density fluctuations. When a void of critical size is formed adjacent to a molecule, a diffusive displacement takes place. The final expression for diffusion coefficient is

$$D = g\sigma\,(3kT/M)^{1/2} \exp\,(-\gamma_1 V_c/V_f) \tag{13}$$

where γ_1 is a correction factor. This equation has been applied to liquid metals and nonelectrolytes [8,25-27]. Since Equation 13 contains two adjustable parameters, the agreement with experimental data is usually good.

Macedo and Litovitz [28] have combined the Eyring rate theory and Cohen-Turnbull free volume model. They specified that, in addition to a critical free volume adjacent to it, the molecule also must acquire an energy greater than some critical value for diffusion to occur. Self-diffusion is given by

$$D = A^{*} \exp\left[\frac{-E}{RT} - \frac{\gamma_1 V_c}{V_f}\right] \tag{14}$$

Again adjustable parameters allow good agreement with theory.

Statistical Mechanical Theories

Statistical mechanical theories envisage molecular motion taking place in small steps in a random walk fashion. In these theories, the friction coefficient in Brownian motion is related to the autocorrelation function of the force acting on the molecule. It provides a sophisticated approach to the problem of relating molecular friction coefficients to the properties of the liquid. In a review, Bearman [29] shows the similarities between statistical mechanical theories and rate theories in relating mutual and self-diffusion coefficients. The Lamm-Dullien approach [30-32] is based on very similar equations. Dullien [32], using Lamm's theory, evaluated a molar average friction coefficient. Observing that the molar average friction coefficients and viscosity are two different ways of averaging the friction coefficient, Dullien [32] derived the following equation for self-diffusion coefficients:

$$\delta = \left[\frac{2\eta VD}{RT}\right]^{1/2} \tag{15}$$

The above equation is not based on any particular model of the liquid state and is expected to be applicable to Newtonian liquids. It has been shown [32] that if elementary kinetic theory expressions are introduced in the above equation, it reduces to $\delta = 2\lambda/3$, which is a well known result from kinetic theory. Most empirical equations for diffusivities can be obtained from the above equation by postulating various relationships for the dependence of δ on liquid-phase properties.

SOURCES OF EXPERIMENTAL DATA

To verify the reliability of both new methods of measuring transport properties and new equations for predicting these properties, it is essential to be able to locate sources of reliable experimental data. Various data compilations have appeared from time to time, and a few are mentioned here.

Diffusion Coefficients

Reid and Sherwood [33] and Reid et al. [34] give an extensive compilation of diffusion coefficients in aqueous solutions and organic liquids. Sovova [35] presents a comprehensive compilation of most diffusion data along with original references. Similar compilations, although somewhat less comprehensive, are contained in papers by Akgerman and Gainer [36] and Hayduk and Buckley [37].

Viscosity

The Thermophysical Properties Research Institute has published a monograph containing a large amount of data on liquid viscosities [38]. Reid and Sherwood [33] and Reid et al. [34] also tabulated a large amount of data. Bridgman's [39] book contains data on viscosities at high pressures. Dullien [40] also has published some accurate data. International critical tables and Batschinski's [41] paper also could be consulted.

Thermal Conductivity

The *Thermophysical Properties Research Literature Retrieval Guide* [42] contains a compilation of papers published prior to 1964. Touloukian's monograph [43] contains a more up-to-date compilation of data on thermal conductivity. Tsederberg's book [44] tabulates thermal conductivity data for organic liquids at 30°C. Extensive compilations are also available in papers by Missenard [45] and Reid et al. [34]. Also useful are the papers by McLaughlin [46], Robbins and Kingrea [47] and Sakiades and Coates [48,49].

EMPIRICAL EQUATIONS

The theoretical developments already outlined do not provide a reliable basis for predicting transport coefficients. However, a large number of empirical relationships have appeared in the literature, some of which allow reasonable predictions. This section outlines such methods for each transport coefficient, followed by a comparison of the accuracy of the various equations.

Thermal Conductivity

There is no scarcity of estimating techniques for calculating thermal conductivities of liquids. The small variation in the thermal conductivity of different liquids allows various schemes for estimating purposes. One of the earliest correlations published was due to the work of Weber [50]:

$$k_L = 359\ (10^{-3}) C_p\ \rho^{4/3}/M^{1/3} \tag{16}$$

Several modifications of this equation have appeared. The most well known is the Robbins and Kingrea correlation [47]:

$$k_L = \frac{(88.0 - 4.94H)\,10^{-3}}{\Delta S^*} \left(\frac{0.55}{T_r}\right)^{N_1} C_p\ \rho^{4/3} \tag{17}$$

where

$$\Delta S^* = \Delta H_v/T_b + R\ln(273/T_b) \tag{18}$$

The parameter H depends on molecular structure, and N_1 depends on the liquid density. These parameters can be estimated using Table I. A much simpler method was suggested using the equations of Sato and Riedel, described by Reid et al. [34]:

$$k_L = \frac{2.64\ (10^{-3})}{M^{1/2}}\ \frac{3 + 20\ (1 - T_r)^{2/3}}{3 + 20\ (1 - T_{rb})^{2/3}} \tag{19}$$

This method is not recommended for branched-chain and light hydrocarbons.

Chhabra et al. [51] applied Hildebrand's fluidity theory to thermal conductivity of liquid. They showed that the reciprocal of thermal conductivity is a linear function of molal volume. Figure 1 shows that this is true for some normal alkanes and toluene. They postulate that thermal conductivity starts at its value at the freezing point and varies linearly with molal volume:

$$\frac{1}{k_L} = \frac{1}{k_O} + \alpha_1 \left(\frac{V - V_o}{V_o}\right) \tag{20}$$

The constant α was shown to be proportional to MB/C_p, and k_o was estimated from a modified version of the Scheffy and Johnson correlation [33].

The final expression is

$$\frac{1}{k_L} = 2.09 \frac{MB}{C_p} \left(\frac{V - V_o}{V_o}\right) + 442.16 (T_m)^{0.216} M^{0.3} \tag{21}$$

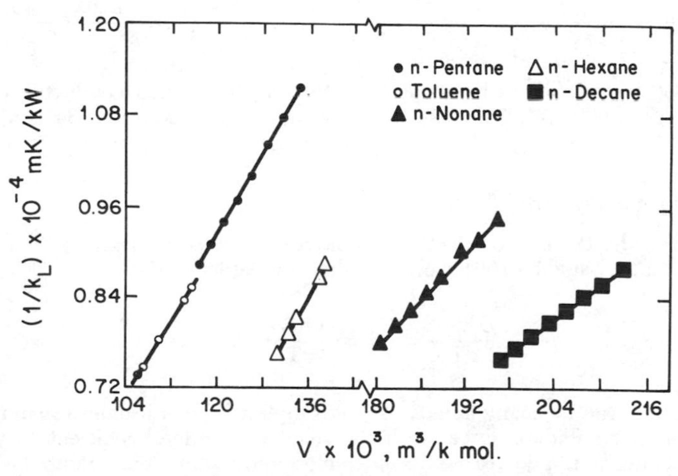

Figure 1. Dependence of thermal conductivity on molar volume.

Table I. Predictions of Equations For Thermal Conductivity

Compound	Temperature (°K)	Thermal Conductivity × 10⁴kW/m-°K			Data Source
		Experimental	Chhabra et al.	Robbins and Kingrea [47]	
Propane	323	0.783	0.783	0.787	[33]
n-Butane	277	1.096	1.148	0.745	[52]
n-Pentane	303	1.135	1.181	1.320	[33,53]
	313	1.100, 1.197	1.140	1.247	
	323	1.066	1.100	1.184	
n-Hexane	293	1.308	1.286	1.392	[33]
	303	1.267	1.247	1.321	
	311	1.234	1.216	1.176	
	333	1.134	1.132	1.133	
n-Heptane	303	1.255, 1.280	1.234	1.268	[33,53]
n-Octane	293	1.323, 1.363	1.410	1.994	[53,54]
	311	1.224, 1.297	1.357	1.826	
	350	1.160	1.167	1.146	
n-Nonane	323	1.220	1.251	1.155	[53]
n-Decane	313	1.300	1.333	1.370	[53]
n-Tetradecane	316	1.370	1.264	1.330	[52]
n-Hexadecane	316	1.414	1.246	0.993	[52]
Toluene	293	1.289, 1.360	1.294	1.410	[33,55-57]
Chloroform	293	1.030, 1.190	0.977	1.178	[33,55]
2-Methylpentane	305	1.084	1.215	1.100	[33]
Cyclopentane	293	1.322	1.375	1.255	[33]
Cyclohexane	293	1.243	1.427	1.393	[53]
Carbon Tetrachloride	273	1.100	1.213	1.093	[55]
	293	0.951, 1.033	1.184	1.086	[55,58]
	303	0.931, 1.017	1.050	1.079	[55,58]
	308	0.983	1.034	1.075	[33]
Chlorobenzene	293	1.368	1.356	1.337	[55]
Bromobenzene	293	1.117	1.268	1.096	[55]
Iodobenzene	293	0.987, 1.021	1.276	0.967	[33,55]
Benzene	293	1.359, 1.477	1.342	1.543	[4,6,8,33,57,58]
	303	1.440	1.295	1.464	[58]
	313	1.352, 1.400	1.252	1.418	[57,58]
	323	1.368	1.211	1.402	[33]
	343	1.31	1.137	1.292	[57]

Several other correlations have appeared that relate thermal conductivity to various fluid properties [45,49,59]. These have been reviewed by Reid et al. [34] and McLaughlin [46].

Temperature and Pressure Effects

In general, the thermal conductivity of liquids decreases linearly with temperature. Horrocks and McLaughlin [60] show that their model leads to

$$\frac{d\ln k_L}{dT} = -2.75\frac{d\ln V}{dT} + 0.0015 \qquad (22)$$

This relation is true for many liquids. As the simplest liquids usually expand more than complex ones, this finding agrees with the smaller negative coefficients usually found for complex liquids. Liquid viscosity is much more sensitive to temperature and molecular structure than thermal conductivity.

Figure 2 shows the variation of thermal conductivity with temperature for a few liquids. Predictions using Equations 17 and 21 also are shown. Special attention is drawn to the data on toluene, which has been recommended as a standard by McLaughlin [46]. Equation 21 is a better predictor of temperature dependence than the Robbins and Kingrea correlation. Thermal conductivity is independent of pressure for pressures below about 50 atm. Bridgman's [39] book contains most of the available high-pressure thermal conductivity data. Kamal and McLaughlin [61] show that the variation of thermal conductivity with pressure at constant temperature is proportional to isothermal compressibility. Lenoir [62] has published a graphic method of calculating the effect of pressure on thermal conductivity of liquids. These charts have been devised using Bridgman's data and are expected to be very accurate.

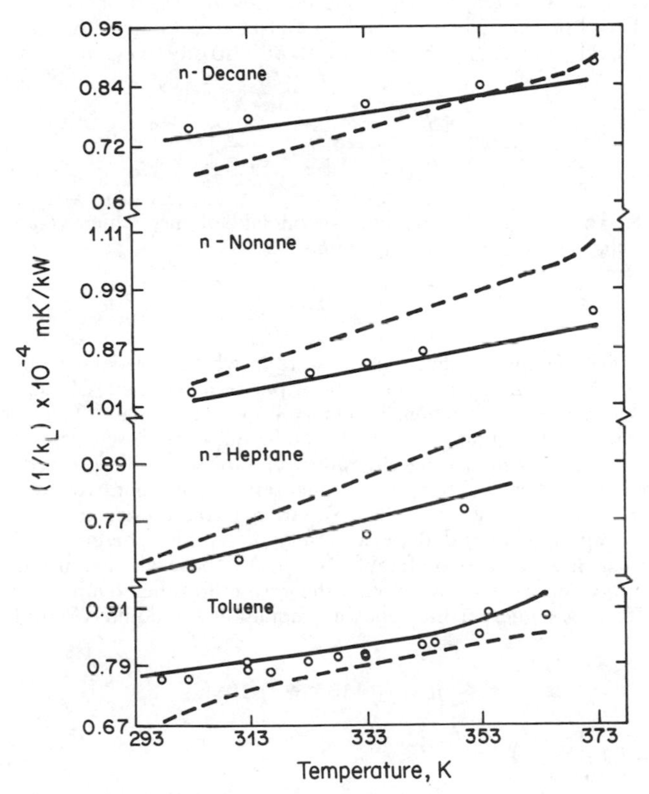

Figure 2. Temperature dependence of thermal conductivity. The prediction by Robbins-Kingrea equation (broken line) and the equation of Chhabra et al. [51] (Continuous line) also are shown.

Comparison of Equations

Based on a comprehensive survey, Reid and Sherwood [33] recommended the Robbins and Kingrea correlation. Errors are usually less than 5%. The Sato and Riedel equation is much simpler to use, but the errors often reach 15–20%. Table I presents a comparison between the equations of Chhabra et al. and the Robbins and Kingrea correlation. In general, both equations often show the same accuracy. More precise comparisons are of little use because of the wide variation in experimental data of different investigators [34]. However, because the equation of Chhabra et al. is not based on a group contribution technique, it is more valuable for liquids containing rare structural features.

Viscosity

Although gas viscosities can be estimated based on sound theory, there is no comparable basis for estimating liquid viscosities. In general, it is preferable to use experimental data if these are available. In this section, some of the empirical relationships for estimating liquid viscosities will be discussed. Only the more widely used equations will be discussed. Gambill [63] has surveyed the relevant literature in two review articles. Other reviews have appeared in the book by Reid and Sherwood [33].

Hildebrand [64] has extended the work of Batschinski to arrive at an elegant representation of liquid viscosities. The reciprocal of viscosity was shown to be directly related to molar expansion:

$$\frac{1}{\mu} = \phi = B\left[\frac{V - V_0}{V_0}\right] \tag{23}$$

V is the molal volume and V_0 the intrinsic molal volume, which is an approximate corresponding states ratio of the critical volume:

$$V_0 = 0.3V^* \tag{24}$$

The intrinsic molal volume is defined as the state in which molecular motion ceases and fluidity is zero. The parameter B denotes the capacity of molecules to absorb externally imposed momentum due to rotation, bending or molecular mass. The above equation has been shown to be valid also for liquid metals and compressed gases [65]. The use of the above equation for predicting viscosities relies on our ability to predict V_0 and B. While Hildebrand has shown B to be a function of molecular weight, the exact dependence has yet to be established. The equation is rather sensitive to values of V_0. Table II gives values of V_0 and B for a variety of liquids. For other substances, it is suggested that at least B, and preferably B and V_0, be evaluated using experimental data and the equation used to extrapolate the results to other temperatures.

Thomas [66] has suggested the following equation for liquid viscosities below the boiling point:

$$\mu = (0.1167\,\rho^{0.5})\,10^{\alpha} \tag{25}$$

where $$\alpha = \theta\left(\frac{1}{T_r} - 1\right) \tag{26}$$

The viscosity is in centipoise, density, ρ, in g/cm^3, and T_r is the reduced temperature, T/T^*. The parameter θ is a constant, which depends on the structure of the liquid and is obtained by adding the contribution of atoms. The values are given in Table III. In many cases, the resulting errors are large. This method should not be used for alcohols, acids, heterocyclics, amines and aldehydes. Comparison with experimental data is shown in a later section.

Morris [67] presented another group contribution method:

$$\log\frac{\mu_L}{\eta_L^*} = J\left(\frac{1}{T_r} - 1\right) \tag{27}$$

Table II. Parameters of Hildebrand's Equation [65]

Compound	B, cP^{-1}	V_0 cm^3/mol
Methane	35.5	32.0
Ethane	25.5	49.6
Propane	19.8	61.0
Butane	19.2	77.7
Pentane	18.7	94.0
Hexane	18.0	111.0
Heptane	17.7	129.1
Octane	17.0	146.4
Nonane	16.9	165.0
Decane	16.6	183.0
Dodecane	15.5	217.0
Pentadecane	14.0	273.0
Cyclopentane	17.6	83.3
Cyclohexane	17.2	103.2
i-Butane	21.3	82.2
2,2-Dimethyl Propane	23.9	104.4
i-Pentane	20.0	98.2
2-Methyl Pentane	20.7	113.5
2-Methyl Hexane	20.7	131.0
2,2,4-Trimcthyl Pentane	19.9	150.5

Table III. Structural Contributions for θ [66]

C	−0.462
H	0.249
O	0.054
Cl	0.340
Br	0.326
I	0.335
C_6H_5	0.385
S	0.043
CO (ketones and csters)	0.105
CN (nitriles)	0.381
Double Bond	0.478

Table IV. Pseudocritical Viscosity [67]

Compound	η^*_L, cP
Hydrocarbons	0.0875
Halogenated Hydrocarbon	0.148
Benzene Derivatives	0.0895
Halogenated Benzene Derivatives	0.123
Alcohols	0.0819
Organic Acids	0.117
Ethers, Ketones, Aldehydes, Acetates	0.096
Phenols	0.0126
Miscellaneous	0.10

$\eta_L{}^*$ is a constant for each type of liquid (Table IV), and J is obtained from structural contributions given in Table V:

$$J = [0.0577 + \sum_i (b_i n_i)]^{1/2} \tag{28}$$

As is generally found with such group contribution techniques, large errors are to be expected for the first few members of a homologous series.

Orrick and Erbar [72] have suggested another group contribution method:

$$\ln\left(\frac{\mu}{\rho_L M}\right) = A_2 + B_2/T \tag{29}$$

A_2 and B_2 are obtained from group contributions given in Table VI.

Table V. Structural Contributions for the Morris Equation [67]
$$J = (0.0577 + \sum_i b_i n_i)^{1/2}$$

Group	b_i
CH_3, CH_2, CH	0.0825
Halogen-Substituted CH_3	0.0
Halogen-Substituted CH_2	0.0893
Halogen-Substituted CH	0.0667
Halogen-Substituted C	0.0
Bromine	0.2058
Chlorine	0.1470
Fluorine	0.1344
Iodine	0.1908
Double Bond	−0.0742
C_6H_4 Benzene Ring	0.3558
Additional H in Ring	0.1446
CH_2 as Saturated Ring Member	0.1707
CH_3, CH_2, CH Adjoining Ring	0.0520
NO_2 Adjoining Ring	0.4170
NH_2 Adjoining Ring	0.7645
F, Cl Adjoining Ring	0.0
OH for Alcohols	2.0446
COOH for Acids	0.8896
C=O for Ketones	0.3217
O=C–O for Acetates	0.4369
OH for Phenols	3.4420
–O– for Ethers	0.1090

Other correlations include the corresponding states method of Stiel and Thodos [33]. Various empirical approaches are reviewed by Bondi [68] and Gambill [63]. More recently, Van Velzen et al. [69], based on an extensive study of the effect of molecular structure on viscosity, devised another group contribution method using the Andrade equation. The method is somewhat complicated to use and does not result in a much greater accuracy than the equations discussed above.

Temperature and Pressure Effects

Liquid viscosities are very sensitive to, and decrease with, temperature. The behavior is generally represented by the Andrade equation:

$$\mu = A_3 \exp. (B_3/T) \tag{30}$$

Table VI. Orrick and Erbar Group Contributions for A and B [72]

Group	A_2	B_2
Carbon Atoms[a]	$-(6.95 + 0.21\,n)$	$275 + 99\,n$
R—C(R)—R (C bonded to three R, with one free bond)	−0.15	35
R—C(R)(R)—R	−1.20	400
Double Bond	0.24	−90
Five-Membered Ring	0.10	32
Six-Membered Ring	−0.45	250
Aromatic Ring	0	20
Ortho Substitution	−0.12	100
Meta Substitution	0.05	−34
Para Substitution	−0.01	−5
Chlorine	−0.61	220
Bromine	−1.25	365
Iodine	−1.75	400
—OH	−3.00	1600
—COO—	−1.00	420
—O—	−0.38	140
—C(—)=O	−0.50	350
—COOH	−0.90	770

[a]n=number, not including those in groups shown above.

Barlow et al. [70] also recommended

$$\mu = A_4 \exp \frac{B_4}{(T - C_4)} \tag{31}$$

A form of the Arrhenius equation also has been used. To estimate viscosities at high temperatures, Reid et al. [34] recommended the Letsou and Stiel [71] method:

$$\mu_L\,\xi_1 = (\mu_L\xi)^\circ + \omega\,(\mu_L\xi)' \tag{32}$$

where

$$(\mu_L\xi)^\circ = 0.01574 - 0.02135T_r + 0.0075T_r^2 \tag{33}$$
$$(\mu_L\xi)' = 0.042552 - 0.07674T_r + 0.034T_r^2 \tag{34}$$
$$\omega = \text{Pitzer's acentric factor.}$$

If two or more viscosity values are available, the Hildebrand equation (Equation 23) based on molal volumes can be used to extrapolate data.

Moderate pressures seem to have very little effect on liquid viscosities; however, high pressures result in an increase in viscosity. The magnitude of the increase seems to be related to molecular complexity [39]. There are no reliable methods of estimating this effect. Hildebrand [65] has shown that his equation is also valid at high pressures, and pressure influences viscosity through changes in molal volume. This hypothesis has yet to undergo comprehensive testing. Bridgman's results show that over a wide range viscosity is linear with pressure. This could be used to extrapolate data up to a few thousand atmospheres.

Comparison of Equations

Extensive comparison of some of these equations is given by Reid and Sherwood [33] and Reid et al. [34]. A less comprehensive comparison is shown in Table VII. The errors are quite large in many cases, so it is preferable to use experimental data whenever these are available. Molecular structure has a strong effect on liquid viscosities. At present, there seems to be no way of quantifying the effects of molecular weight, branching, association and flexibility on liquid viscosities. All the more well known methods are based on group contribution techniques and, hence, extrapolation to liquids containing rare structural features invariably will lead to large errors. There is little to choose between these equations, although it does appear that the Morris method gives the least amount of error.

Table VII. Comparison of Calculated and Experimental Viscosities

Compound	T °C	Experimental μ(cp)	Percentage Error		
			Thomas [66]	Orrick and Erbar [72]	Morris [67]
Propane	−190	13.8	12.0	7.3	49
	−40	0.205	2.9	25	10
Pentane	−120	2.35	−8.6	1.0	3.0
	−40	0.428	−10	3.3	−1.2
	30	0.216	0	11	3.7
Hexane	−60	0.888	−8.5	−2.9	−4.5
	40	0.262	0.8	5.3	3.5
	70	0.205	2.0	4.9	3.2
Heptane	−90	3.77	26	21	17
	20	0.418	1.2	1.9	1.7
	100	0.209	1.4	3.3	4.8
Carbon Tetrachloride	0	1.369	0.1	−20	26
	70	0.534	−1.9	−20	12
Chloroform	0	0.7	−24	−40	−12
	30	0.502	−18	−34	−12
Acetone	−90	2.075	15	25	5
	0	0.389	−5.8	6.7	−14
	60	0.226	−0.5	8.3	−4.1
Chlorobenzene	0	1.054	−15	−1.4	−12
	40	0.639	−4.4	0.6	−11
Acetic Acid	10	1.45	51	22	−5.5
	40	0.901	49	15	6.1
	110	0.416	36	5.3	−2.2
1-Butene	−110	0.79	−19	22	16
	−40	0.26	−15	18	5.7
2-Methyl Butane	−50	0.55	7.3	13	15
	30	0.205	0	10	3.2
i-Butane	−80	0.628	22	23	19
	−10	0.239	15	24	13
Phenol	50	3.02	76	0	17
	100	0.783	42	−37	15
Toluene	−20	1.07	18	19	7.2
	60	0.38	6.1	10	4.8
	110	0.249	3.2	6.8	4.2
O-Xylene	0	1.108	16	−3.1	29
	40	0.625	9.8	−5.0	23
	100	0.345	6.1	−3.7	18

[a]Experimental values are listed by Reid et al. [34].

Diffusion Coefficients

The various theoretical approaches to diffusion in liquids have helped provide increasing insight into the diffusional process. Disregarding certain relatively minor differences, there are basically four different approaches in use at present: (1) the hydrodynamic theories, (2) the kinetic theory of liquids, (3) the absolute rate theory, and (4) various friction coefficient approaches, including statistical mechanical theories. Empirical approaches based on some of the above theories have had limited success, but their fundamental use is in question.

Detailed reviews have been presented by Himmelblau [73], Akgerman and Gainer [74] and Ghai et al. [75].

Hydrodynamic theories usually start with the Nernst-Einstein equation, which describes the diffusivity of a single particle of A diffusing through a liquid medium, B, by the equation

$$D_{AB} = kT \frac{U_A}{F_A} \tag{35}$$

in which D_{AB} denotes the diffusivity, and F_A is the force opposing the motion of particle A. When the opposing force used is the Stokes drag on a particle moving through a liquid,

$$F_A = 6\pi\mu_B U_A R_A \tag{36}$$

The diffusivity is given by the familiar Stokes-Einstein equation:

$$D_{AB} = \frac{kT}{6\pi \mu_B R_A} \tag{37}$$

This equation was independently proposed by Einstein [76]. This equation has been shown to predict diffusivities close to experimentally found values for systems in which a large spherical molecule is diffusing into a solvent that appears as a continuum to the diffusing species, but it is not applicable when the size of the solute molecules approaches that of the solvent.

The Eyring rate theory attempts to explain the diffusion process on the basis of a cell model for the liquid, which contains vacancies or holes. In this theory, the coefficient of diffusion is expressed as

$$D_{AB} = \frac{\lambda^2}{\xi} k_1 \tag{38}$$

Using the rate theory, Akgerman and Gainer [74] extended the approach of Gainer and Metzner [16] to gas-liquid systems:

$$D_{AB} = \frac{kT}{\xi\eta}\left[\frac{N}{V'_B}\right]^{1/3}\left[\frac{M_B}{M_A}\right]^{1/2} \exp\left[\frac{E_\mu - E_D}{RT}\right] \tag{39}$$

They distinguished, for both viscous and diffusive flow, a hole-forming and hole-jumping contribution. Assuming a hole-forming mechanism identical for both processes,

$$[E_\mu - E_D] = E^j_{BB}\left\{1 - \left[\frac{E^j_{AA}}{E^j_{BB}}\right]^{1/(\xi_A + 1)}\right\} \tag{40}$$

where E^j_{BB} and E^j_{AA} are the jumping energies for solute A and solvent B molecules, respectively. To calculate E^j_{BB}, the equation of Laidler and Glasstone [12] was used:

$$E^{j}_{BB} = -\frac{R \ln \frac{\mu_2}{\mu_1} + \frac{R}{2} \ln \frac{T_2}{T_1}}{\frac{1}{T_1} - \frac{1}{T_2}} \tag{41}$$

where μ_1 and μ_2 are the viscosities of B at temperature T_1 and T_2. For E^{j}_{AA}, Akgerman and Gainer [74] proposed an empirical relationship:

$$E^{j}_{AA} = 24.6 \times 10^{6} \, M_A{}^{-0.186} \tag{42}$$

This is by far the most comprehensive work to be based on the rate theory. Akgerman and Gainer derived their equations for the diffusion of gases in liquids, but with the assumptions made it should be valid for liquid-liquid systems when the diffusing component is at infinite dilution. Akgerman and Gainer [74] tested the equation for 49 gas-liquid systems and, in most cases, found better agreement with the data from the literature than for previously developed equations.

Akgerman [77] extended the above equations to liquid-liquid systems by defining a new relationship for the geometric factor. Comparison with experimental data showed a marginal improvement over other available equations.

Arnold [78] derived an expression for gaseous diffusivity based on the classical kinetic theory for gases, and applied this to the liquid state. Three assumptions relative to the collision rate have been made: all collisions are assumed to be binary; the collision rate is unaffected by the volume occupied by the molecules; and intermolecular attractions are negligible. To correct for these effects he introduced association factors in his equation:

$$D_{AB} = \frac{B_1 \left[\frac{1}{M_A} + \frac{1}{M_B}\right]^{0.5}}{A_A \, A_B \, \mu^{0.5} [V_A{}^{1/3} + V_B{}^{1/3}]^2} \tag{43}$$

B_1 was found to be 0.01 at 293°K. For associated systems, the association factors A_A and A_B were determined from diffusivity data, and no way has been shown in the literature to predict them. Thus, only for ideal nonassociating systems can the Arnold equation be used to predict diffusivities.

Several empirical equations have appeared from time to time, but have had rather limited success in predicting diffusion coefficients. The earliest of these and still the most widely used is the Wilke-Chang equation. Wilke's [79] correlation was based on relations suggested by the Eyring theory of absolute reaction rates and the Stokes-Einstein equation. The correlation was made through the group $F = T/D_{AB}\mu$. Wilke and Chang [80] further tested the above approach and found that F was a function of the solute molal volume. To determine the effect of solvent properties on diffusivity, a wide range of variables, such as solvent molal volume, etc., were tested. Of these, the solvent molecular weight appeared to correlate the data most successfully. This led to the Wilke-Chang correlation:

$$D_{AB} = 7.4\,(10^{-8}) \, \frac{\bar{\alpha}\,(M_B)^{0.5}}{\mu_B\,(V'_A)^{0.6}} \tag{44}$$

No general method for determining the association factor, $\bar{\alpha}$, is available. The above equation is not dimensionally consistent, and viscosity is expressed in centipoise. The introduction of an association parameter is one of the main difficulties in attempts to develop correlations of diffusion coefficients for liquids. Associated molecules behave as large molecules and diffuse more slowly; the degree of association varies with the mixture composition and molecular type in a manner not well understood. Even values of the association factor reported by several investigators vary [35,81].

Another difficulty encountered in attempts to develop a general correlation is that D_{AB} evidently depends in part on the shape of the molecules to be, in general, some 30% greater than for substances composed of essentially spherical molecules having the same molal volume at the normal boiling point. The other parameter on which there is no consensus of opinion is the molar volume that appears in most of the equations. The molar volume of a gas, at its normal boiling point, is computed in an additive fashion from the contributions of individual atoms in the molecule of the gas [33] or is determined experimentally. As can be seen from comparison of published tables [73,80, 82], these values for some gases differ considerably.

Scheibel [83], also using the hydrodynamic theory, confirmed the approach of Wilke [79]. Neglecting any association, he correlated the data by

$$D_{AB} = 8.2 \times 10^{-2} \frac{T}{\mu V'^{1/3}_A} \left[1 + \left(\frac{3V'_A}{V'_B}\right)^{2/3}\right] \tag{45}$$

The above modification to the Wilke-Chang equation is not universal, and changes in the form of the equation for various solvents is recommended [33].

The Othmer and Thakar correlation [84] followed the rate theory. A plot of log D_{AB} against the vapor pressure of a reference substance should give a slope from which the activation energy could be determined in accordance with the Clausius-Clapeyron equation. Similar plots were obtained for the viscosity of water. This suggested a linear relationship between log D_{AB} and log μ, which led to the following correlation:

$$D_{AB} = 14 \times 10^{-5}[V'^{0.6}_A \; \mu_B \; \mu_W{}^{1.1\Delta H_B/\Delta H_W}]^{-1} \tag{46}$$

where ΔH_B and ΔH_W are the latent heat of vaporization of the solvent and water, respectively.

Othmer and Thakar [84] report that this agrees with a large number of data with an average deviation of about 5%. Agreement with nonaqueous solvents is considerably poorer.

Sitaraman et al. [85], following the earlier development of limited correlations by Ibrahim and Kuloor [82], developed a general empirical correlation, which, unlike that of Wilke and Chang [80], is said to hold for diffusion of water at low concentrations in organic solvents:

$$D_{AB} = 5.4 \times 10^{-8} \left[\frac{M_B{}^{1/2} \, \Delta H_A{}^{1/3} \, T}{\mu_B \, V'^{1/2}_A \, \Delta H_B{}^{0.3}}\right]^{0.93} \tag{47}$$

Again, η_B is the solvent viscosity in centipoise. The average reported error was about 13%.

Hayduk and Laudie [81] tested the reliability of the Wilke-Chang equation. They concluded that a revised form of the equation related the diffusion data better:

$$D_{AB} = 1.33 \times 10^{-4} \, V'^{0.6}_A \, \mu_B{}^{-1.4} \tag{48}$$

However, this equation does not take into account any association of the solute with the solvent; therefore, large deviations can be expected. Hayduk and Cheng [86] suggested a relationship,

$$D_{AB} = \phi_1 \, \mu \, \phi_2 \tag{49}$$

for the diffusion, at infinite dilution, under isothermal conditions, the constants ϕ_1 and ϕ_2 depending only on the solute properties. Simons and Ponter [87] conclude that these types of equations have marginal utility. Sovova [35] tested the Hayduk and Cheng [86] hypothesis of a simple viscosity-diffusion coefficient relationship using a large number of data. The data were correlated by three different sets of constants—ϕ_1 and ϕ_2—one for diffusion in water, the second for diffusion in aromatic hydrocarbons and the third set for aliphatic hydrocarbons. The aromatic hydrocarbons are approximately spherical, and the aliphatic hydrocarbons are mainly linear molecules. A close look at this type of classification using some molecular asymmetry factor, say the Pitzer's acentric factor, shows little trend. Hence, it is hard to justify such approaches. Tyn and Calus [88] have suggested a group contribution technique. However, the analytical expressions discussed above are more convenient to use.

Sridhar and Potter [89,90] derived an equation for predicting diffusion coefficients in liquids. Starting with Dullien's [40] equation for self-diffusion coefficients and assuming that mutual diffusion coefficients are dependent on molecular cross section, they derived the following equation:

$$D_{AB} = 0.088 \frac{V_B^{*4/3}}{N^{2/3}} \frac{RT}{\eta V_o} \frac{1}{V_A^{*2/3}} \tag{50}$$

For diffusing molecules that are large (liquid-liquid system), better accuracy is obtained if the molal volume, V_B, is used instead of V_o in the above equation [91].

Umesi and Danner [92] have presented a simple equation for diffusivities in liquids. They chose the radius of gyration [34] to define the size-shape relationship of the solvent and solute and its effect on diffusivities:

$$D_{AB} = 2.75 \times 10^{-8} \frac{T}{\mu} (\bar{R}_B/\bar{R}_A)^{2/3} \tag{51}$$

This equation gives large errors when permanent gases are the diffusing species. It seems to be particularly suitable for the diffusion of hydrocarbon gases in nonpolar solvents.

Temperature and Pressure Effects

At present, the effect of temperature on diffusivity is not known precisely. The correlations presented previously do not indicate the temperature dependence of diffusivity, as some of the parameters are themselves temperature dependent. The equations do not indicate the correct form of temperature dependency for all systems. This is illustrated [36] by the xenon-water system over the temperature range of 283°K to 333°K, where the experimental diffusivity increases far more rapidly with temperature than indicated by either the Akgerman-Gainer equation or the Wilke-Chang correlation. However, for many systems the temperature dependence is closely approximated by both these expressions, including cases in which the actual magnitude of the predicted diffusivity might be in error by a constant factor.

Unver and Himmelblau [93] correlated their data within 11% by a simple quadratic-type expression in temperature. Nir and Stein [94] propose a nonlinear relationship between log D_{AB} and 1/T. This occurs because they envisage two possible diffusion mechanisms: a flow mechanism and a lattice movement mechanism. The flow mechanism results from the movement of solvent molecules and is the only available mode of diffusion for large molecular-weight diffusants, at least an order of magnitude larger than that of the solvent. If one mode of diffusion operates, a single activation energy is obtained.

The widely used and accepted form of equation to describe the temperature dependence is the Arhennius-type equation:

$$D_{AB} = A_5 \exp\left(-\frac{E_{D_{AB}}}{RT}\right) \tag{52}$$

All work derived from the rate theory is based on

$$D_{AB} = A_6 T \exp\left(-\frac{E_{D_{AB}}}{RT}\right) \tag{53}$$

Both expressions suggest a plot of log D_{AB} against $\frac{1}{T}$ to represent the temperature dependence. The rate theory expression will show a positive deviation from this plot at higher temperatures. Simons and Ponter [87] show, using order of magnitude values, that based on available diffusion data it is difficult to decide between the above two equations. It is concluded that both forms described the diffusion process well and that the concept of a constant activation energy is a good approximation for small temperature changes.

Table VIII. Experimental and Calculated Gas-Liquid Diffusivities

			Predicted (m^2/sec) 10^9			
System	Temperature (°K)	Experimental[a] (m^2/sec) 10^9	Akgerman-Gainer [36]	Wilke-Chang [80]	Umesi et al. [96]	Sridhar-Potter [89,90]
CO_2-Water	293	1.62-1.77	1.79	1.79	0.5	1.80
	298	1.85-1.98	2.03	2.05	0.57	2.05
	313	2.75-2.80	2.83	2.91	0.81	2.94
O_2-Water	293	2.01-2.30	2.11	2.14	0.7	2.10
	298	2.07-2.60	2.40	2.41	0.79	2.40
	313	3.33-3.80	3.36	3.49	1.14	3.44
	333	5.7	4.95	5.19	1.69	5.10
N_2-Water	293	1.66-2.60	2.11	1.88	0.74	1.85
	298	1.80-2.25	2.37	2.14	0.84	2.11
	313	2.83-4.30	3.40	3.06	1.21	3.02
	333	6.50	4.94	4.62	1.79	4.50
Ne-Water	293	3.00	2.70	2.82	0.54	3.08
	313	5.39	4.30	4.58	0.89	5.05
	333	8.08	6.30	6.79	1.32	7.49
Cl_2-Water	293	1.22	1.40	1.45	0.5	1.49
	298	1.40-1.51	1.60	1.65	0.57	1.70
CO-Water	293	2.03	2.11	1.91	0.73	1.81
	313	3.62	3.40	3.11	1.19	2.96
	333	5.68	5.00	4.66	1.77	4.39
SO_2-Water	293	1.40-1.66	1.48	1.51	0.35	1.51
	298	1.83-2.04	1.69	1.75	0.40	1.72
	313	2.59	2.38	2.49	0.57	2.47
CO_2-Ethanol	279.4	2.45	1.22	1.35	1.11	1.59
	298	3.42	1.81	2.05	1.72	2.45
N_2-Benzene	298	6.93	3.74	4.09	6.14	5.38
O_2-Cyclohexane	302.6	5.31	3.88	3.5	4.22	4.74
CH_4-Carbon Tetrachloride	298	2.89	5.89	3.96	2.9	3.31
O_2-Carbon Tetrachloride	298	3.82	3.6	3.17	4.4	3.88
Ar-Carbon Tetrachloride	298	3.63	4.54	4.1	2.99	3.86
N_2-Carbon Tetrachloride	298	3.42	5.0	3.84	4.7	3.51
CO_2-Carbon Tetrachloride	298	2.95	3.32	2.70	3.16	3.05

[a]Experimental values are listed in the paper by Akgerman and Gainer [36]. Where more than one experimental value is available, the range is shown.

Pressure dependence of diffusion coefficients has received very little attention. Lynch and Potter [95] measured the diffusion of nitrogen in cyclohexane at pressures up to 170 psi. At these pressures, diffusion coefficients seem little affected by pressure. At higher pressure, because molecular spacing is forcibly reduced, diffusion coefficients would be expected to decrease with pressure. It is not possible to predict the extent of this decrease.

Comparison of Equations

Reid et al. [34] present an excellent comparison of the various equations for estimating diffusion coefficients. For diffusion of large solutes in aqueous systems, the Wilke-Chang equation is recommended. The errors are of the order of 10%. Similar errors are obtained also by using the Schiebel correlation.

Akgerman and Gainer [36] compare their equation with some of the correlations presented here. They concluded that their equation was superior to the other correlations. Umesi et al. [96] have performed an extensive analysis of the various equations. Their data base is much more restrictive than that used by Reid et al. [34]. For diffusion of hydrocarbon gases, the equation of Umesi et al. [96] seems to be preferable with its average error about 15%. For the diffusion of nonpolar gases in nonpolar liquid, the Sridhar-Potter equation and the Tyn-Calus method yield similar errors. For polar solutes in nonpolar solvents, the Umesi correlation is definitely superior, with an average error around 12%.

To illustrate the errors involved, a smaller selection is shown in Table VIII for gas-liquid systems. The Umesi correlation gives large errors when the diffusing species is a permanent gas. Among others, the Sridhar-Potter equation is marginally better; it has the advantage of being simpler to use. Figure 3 compares the Akgerman-Gainer and Sridhar-Potter equations for two typical systems. While both equations give similar predictions for aqueous systems, the Sridhar-Potter equation is superior for diffusion in organic liquids. One of the problems encountered when comparing different equations is the lack of data covering a wide temperature range. Lynch and Potter [95] published such data for the nitrogen-cyclohexane system. Figure 4 shows that, for this system, the Sridhar-Potter equation accurately predicts both the absolute values and the temperature dependence. Table VIII also shows the wide range of experimental data re-

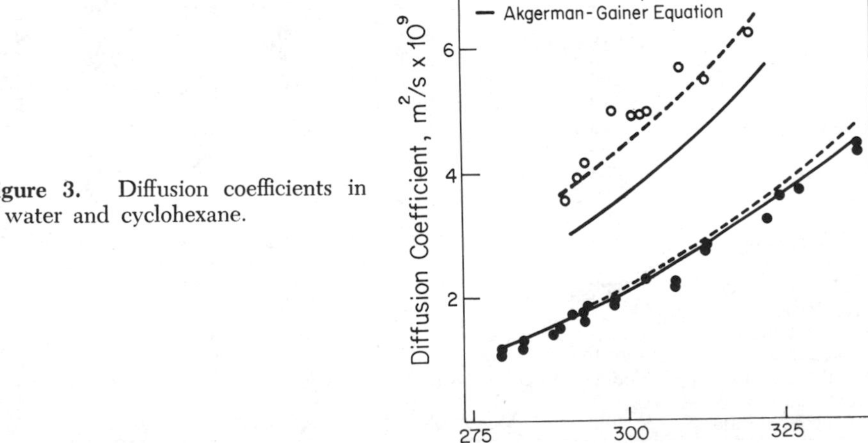

Figure 3. Diffusion coefficients in water and cyclohexane.

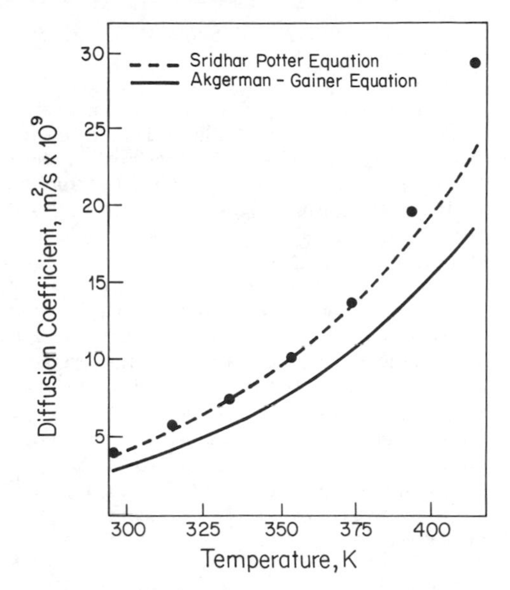

Figure 4. Temperature dependence of diffusion coefficients for nitrogen in cyclohexane [95].

Table IX. Diffusion in Liquid-Liquid Systems

			$m^2/sec \times 10^9$		
Solute	**Solvent**	**Temperature (°K)**	**Experimental Data**[a]	**Sridhar-Potter [89,90]**	**Wilke-Chang [80]**
Hexane	Benzene	288	1.78	1.64	1.42
Methanol	Benzene	288	1.28	1.25	1.36
Methanol	CCl_4	298	2.61	2.65	3.20
Ethanol	CCl_4	298	1.95	2.10	2.54
Acetone	CCl_4	298	1.70	1.80	2.23
Benzene	CCl_4	298	1.54	1.57	1.96
Benzene	Hexane	298	4.64	5.15	4.49
Cyclohexane	Hexane	298	3.77	4.60	3.98
Decane	Hexane	298	3.02	2.94	2.68
2-Methylbutane	Hexane	298	4.40	4.60	3.98
Phenol	CCl_4	298	1.37	1.55	1.82
Pentane	CCl_4	298	1.57	1.39	1.73
2-Methylbutane	CCl_4	298	1.49	1.40	1.73
Toluene	CCl_4	298	1.40	1.38	1.74
Cyclohexane	CCl_4	298	1.27	1.40	1.73
Eicosane	CCl_4	298	0.664	0.57	0.78
Decane	CCl_4	298	1.09	0.89	1.16
Methanol	Benzene	298	3.82	3.63	3.43

[a]Experimental data are listed by Hayduk and Buckley [37].

ported. This factor, more than anything else, makes precise statistical comparison difficult. Table IX presents a comparison for liquid-liquid systems. It illustrates that the Sridhar-Potter correlation is preferable to the Wilke-Chang correlation. In conclusion, the Sridhar-Potter equation is recommended for diffusion of nonpolar solutes in nonpolar solvents. The Umesi correlation is recommended for the diffusion of polar solutes and hydrocarbon solutes in nonpolar solvents.

CONCLUSIONS

It is obvious from the foregoing analysis that much needs to be done to improve our capability of predicting transport coefficients. More attention needs to be paid to the prediction of liquid viscosities. At the moment, the more successful methods are based on structural contributions, but such methods do not throw much light on the actual mechanism of momentum transfer. Their use in the overall development of the field is questionable. Viscosity may hold the key to unravelling the mysteries of the liquid state.

There exists a reasonable data base to allow new approaches to be tested adequately. There are, however, few data that can explain fully the effect of pressure and temperature, especially on diffusion coefficients. This results partly because, unlike thermal conductivity and viscosity, the measurement of diffusivity is more difficult. This is especially true when the diffusing solute is a gas. The literature contains data from experimental work spanning nearly a century. There is considerable disagreement between data from various sources. Some time needs to be taken to sort these discrepancies to build a data base containing reliable data.

Finally, we are far from achieving the ability to predict transport coefficients from a rigorous theoretical basis without introducing adjustable parameters of dubious physical significance.

NOMENCLATURE

A	atomic weight
A_1	constant, Equation 11
A_2	parameter, Equation 29
A_3	constant, Equation 30
A_4	constant, Equation 31
A_5	Preexponential factor, Equation 52
A_6	Preexponential factor, Equation 53
A^*	Preexponential factor, Equation 14
A_A, A_B	association factors
$\bar{a}$	distance between lattice points
a	molecular radius
B	Hildebrand parameter, Equation 23
B_1	constant, Equation 43
C	concentration
C_1	constant, Equation 11
C^{liq}	speed of sound in liquid phase
C_p	specific heat at constant pressure
D	self-diffusion coefficient
D_{AB}	mutual diffusion of solute A in solvent B
E	activation energy, Equation 14
E^j_{AA}	activation energy for solute molecules to jump into prepared holes
E^j_{BB}	activation energy for solvent molecules to jump into prepared holes
E_D	activation energy for diffusion
E_μ	activation energy for viscosity
F	applied force
F_A	force opposing motion of solute
g	geometric parameter, Equation 13
h	Planck's constant
H	parameter, Equation 17
J_D	diffusive flux
J_H	heat flux
J_M	momentum flux
k	Boltzmann's constant
k'_o	frequency of movement of molecule in the absence of external force
k_o	thermal conductivity at V_o
k_1	rate constant
k_L	thermal conductivity
M, M_A, M_B	molecular weight
n	number density
n_i	number of particular groups in a molecule
N	Avogadro's number
N_1	parameter, Equation 17
P_C	critical pressure
R	gas constant
R_A	radius of diffusing molecule
T	temperature
T_b	boiling point
T_c	critical temperature
T_m	melting point
T_r	reduced temperature
T_{rb}	reduced boiling point, T_b/T_c

U_A relative velocity
V molal volume
V_o intrinsic molal volume
V_1 velocity of one fluid layer with respect to another
V_A', V_B' molal volume at normal boiling point
V_C critical void volume
V_f free volume
V^* critical volume
W activation energy
x distance

Greek Symbols

α parameter, Equation 26
α_1 constant, Equation 20
$\overline{\alpha}$ association factor, Equation 44
γ ratio of specific heats
γ_1 geometric parameter to correct for free volume overlap
δ momentum transfer distance, Equation 15
δ_1 potential parameter
ΔH_B latent heat of vaporization
ΔH_V molar heat of vaporization at normal boiling point
ΔH_W latent heat of vaporization of water
ΔS^* modified Everett entropy of vaporization
θ parameter, Equation 26
η viscosity
η_L^* pseudocritical viscosity
λ distance between equilibrium position of molecules
$\overline{\lambda}$ mean free path
μ viscosity in centipoise
μ_B solvent viscosity in centipoise
$(\mu\,\xi)^\circ$ defined by Equation 33
$(\mu_L\xi)'$ defined by Equation 34
ξ geometric parameter
ξ_1 $T_c^{1/6}\,M^{-1/2}\,P_C^{-2/3}$
ρ density
σ molecular diameter, Equation 13
ϕ_1, ϕ_2 constants, Equation 49

REFERENCES

1. Brush, S. G. *Chem. Rev.* 62:513 (1962).
2. Hirschfelder, J. O., C. F. Curtiss and R. B. Bird. *Molecular Theory of Gases and Liquids* (New York: John Wiley & Sons, Inc., 1954).
3. Chapman, S., and T. G. Cowling. *The Mathematical Theory of Non-Uniform Gases*, 2nd ed. (New York: Cambridge University Press, 1952).
4. Bird, R. B. *Adv. Chem. Eng.* 1:156 (1956).
5. Longuet-Higgins, H. C., and J. A. Pople. *J. Chem. Phys.* 25:884 (1956).
6. Longuet-Higgins, H. C., and J. P. Valleau. *Mol. Phys.* 1:284 (1958).
7. Ertl, H., R. K. Ghai and F. A. L. Dullien. *AIChE J.* 20:1 (1974).
8. Naghizadeh, J., and S. A. Rice. *J. Chem. Phys.* 36:2710 (1962).
9. Dymond, J. H. *J. Chem. Phys.* 60:969 (1974).
10. Eyring, H. *J. Chem. Phys.* 3:107 (1935).
11. Eyring, H. *J. Chem. Phys.* 4:283 (1936).
12. Glasstone, S., K. J. Laidler and H. Eyring. *The Theory of Rate Processes* (New York: McGraw-Hill Book Co., 1941).
13. Eyring, H., and R. P. March. *J. Chem. Ed.* 40:562 (1963).
14. Ree, T., and H. Eyring. In: *Rheology—Theory and Application*, F. R. Eirich, Ed. (New York: Academic Press, Inc., 1958).
15. Ree, F. H., T. Ree and H. Eyring. *Ind. Eng. Chem.* 50:1036 (1958).
16. Gainer, J. L., and A. B. Metzner. *AIChE-IChE Symp. Series No. 6* (1965).
17. Bridgman, P. W. *Proc. Am. Acad. Arts Sci.* 59:109 (1923).
18. Frenkel, J. *Kinetic Theory of Liquids* (New York: Dover Publications, Inc., 1955).
19. Furth, R. *Proc. Camb. Phil. Soc.* 37:252 (1941).
20. Andrade, E. N. *Phil. Mag.* 17:497 (1934).
21. Andrade, E. N. *Phil. Mag.* 17:698 (1934).
22. Horrocks, J. K., and E. McLaughlin. *Trans. Farad. Soc.* 56:206 (1960).

23. Houghton, G. *J. Chem. Phys.* 40:1628 (1964).
24. Cohen, M. H., and D. Turnbull. *J. Chem. Phys.* 31:1164 (1959).
25. Robinson, R. C., and W. E. Stewart. *Ind. Eng. Chem.* 90 (1968).
26. Kessler, D., A. Weiss and H. Witte. *Ber Bunsenges* 71:3 (1967).
27. Collings, A. F., and R. Mills. *Trans. Farad. Soc.* 66:2761 (1970).
28. Macedo, P. B., and T. A. Litovitz. *J. Chem. Phys.* 42:245 (1965).
29. Bearman, R. J. *J. Phys. Chem.* 65 (1961).
30. Lamm, O. *Acta. Chem. Scand.* 6:1331 (1952).
31. Lamm, O. *Acta. Chem. Scand.* 8:1120 (1954).
32. Dullien, F. A. L. *Trans. Farad. Soc.* 59:856 (1963).
33. Reid, R. C., and T. K. Sherwood. *Properties of Gases and Liquids,* 2nd ed. (New York: McGraw-Hill Book Co., 1966).
34. Reid, R. C., J. M. Prausnitz and T. K. Sherwood. *Properties of Gases and Liquids,* 3rd. ed., (New York: McGraw-Hill Book Co., 1977).
35. Sovova, H. *Res. Rep. 29, Inst. Chem. Proc. Fund.,* Prague (1975).
36. Akgerman, A., and J. L. Gainer. *J. Chem. Eng. Data* 17:372 (1972).
37. Hayduk, W., and W. D. Buckley. *Chem. Eng. Sci.* 27:1997 (1972).
38. Touloukian, Y. S., S. C. Saxena and P. Hersterman. *Thermophysical Properties of Matter,* Vol. 11 (New York: Plenum Publishing Corp., 1975).
39. Bridgman, P. W. *The Physics of High Pressure* (London: Bell, 1958).
40. Dullien, F. A. A. *AIChE J.* 18:62 (1972).
41. Batschinski, A. J. *Z. Phys. Chem.* 84:643 (1913).
42. Touloukian, Y. S., J. K. Gerritsen and N. Y. Moore. *Thermophysical Properties Research Literature Retrieval Guide* (New York: Plenum Publishing Corp., 1967).
43. Touloukian, Y. S., P. E. Liley and S. C. Saxena. *Thermophysical Properties of Matter,* Vol. 3 (New York: Plenum Publishing Corp. 1970).
44. Tsederberg, N. V. *Thermal Conductivity of Gases and Liquids* (Cambridge, MA: The MIT Press, 1965).
45. Missenard, A. *Comp. Rend.* 260:5521 (1965).
46. McLaughlin, E. *Chem. Rev.* 64:380 (1964).
47. Robbins, L. A., and C. L. Kingrea. *Hydrocarb. Proc. Pet. Ref.* 41:133 (1962).
48. Sakiades, B. C., and J. Coates. *AIChE J.* 1:275 (1955).
49. Sakiades, B. C., and J. Coates. *AIChE J.* 3:121 (1957).
50. Weber, H. F. *Wiedmann's Ann. Phys. Chem.* 10:103 (1880).
51. Chhabra, R. P., T. Sridhar, P. H. T. Uhlherr and O. E. Potter. *AIChE J.* 26:522 (1980).
52. Mathur, V. K., J. D. Singh and W. M. Fitzgerald. *J. Chem. Eng. Japan* 11:67 (1978).
53. Mallan, G. M., M. S. Michaelin and F. J. Lockhart. *J. Chem. Eng. Data* 17:412 (1972).
54. Kandiyoti, R., E. McLaughlin and J. F. T. Pittman. *J. Chem. Soc. Farad. Trans.* 68:860 (1972).
55. Pachaiyappan, V., and K. R. Vaidyanathan. *Proc. 6th Symp. on Thermophysical Properties* (1973).
56. Ziebland, H. *Int. J. Heat Mass Transfer* 2:273 (1961).
57. Venart, J. E. S. *J. Chem. Eng. Data* 10:239 (1965).
58. Poltz, H., and R. Jugel. *Int. J. Heat Mass Transfer* 10:1075 (1967).
59. Viswanath, D. S. *AIChE J.* 13:850 (1967).
60. Horrocks, J. K., and E. McLaughlin. *Trans. Farad. Soc.* 59:1709 (1963).
61. Kamal, J., and E. McLaughlin. *Trans. Farad. Soc.* 60:809 (1964).
62. Lenoir, J. M. *Pet. Ref.* 36:162 (1957).
63. Gambill, W. R. *Chem. Eng.* 66:129 (1959).
64. Hildebrand, J. H. *Science* 174:490 (1971).
65. Hildebrand, J. H. *Viscosity and Diffusivity–A Predictive Treatment* (New York: John Wiley & Sons, Inc., 1977).
66. Thomas, L. H. *J. Chem. Soc.* 573 (1946).
67. Morris, P. S. M. S. Thesis, Polytechnic Institute of Brooklyn, Brooklyn, NY (1964).
68. Bondi, A. *Ann. N. Y. Acad. Sci.* 53:870 (1951).

69. Van Velzen, D., R. L. Cardoza and H. Langenkamp. *Ind. Eng. Chem. Fund.* 11:20 (1972).
70. Barlow, A. J., J. Lamb and A. J. Matheson. *Proc. Roy. Soc.* A292:322 (1966).
71. Letsou, A., and L. I. Stiehl. *AIChE J.* 19:409 (1973).
72. Orrick, C., and J. H. Erbar. Quoted in Reference 34.
73. Himmelblau, D. M. *Chem. Rev.* 64:527 (1964).
74. Akgerman, A., and J. L. Gainer. *Ind. Eng. Chem. Fund.* 11:373 (1972).
75. Ghai, R. K., H. Ertl and F. A. L. Dullien. *AIChE J.* 19:881 (1973).
76. Einstein, A. *Investigation on the Motion of Brownian Movement* (New York: Dover Publications, Inc., 1956).
77. Akgerman, A. *Ind. Eng. Chem. Fund.* 15:78 (1976).
78. Arnold, J. H. *J. Am. Chem. Soc.* 52:3937 (1930).
79. Wilke, C. R. *Chem. Eng. Prog.* 45:218 (1949).
80. Wilke, C. R., and P. Chang. *AIChE J.* 1:264 (1955).
81. Hayduk, W., and H. Laudie. *AIChE J.* 20:611 (1974).
82. Ibrahim, S. H., and N. R. Kuloor. *Brit. Chem. Eng.* 5:795 (1960).
83. Scheibel, E. G. *Ind. Eng. Chem.* 46:2007 (1954).
84. Othmer, D. F., and M. S. Thakar. *Ind. Eng. Chem.* 45:589 (1953).
85. Sitaraman, R., S. H. Ibrahim and N. R. Kuloor. *J. Chem. Eng. Data* 8:198 (1963).
86. Hayduk, W., and S. C. Cheng. *Chem. Eng. Sci.* 26:635 (1971).
87. Simons, J., and A. B. Ponter. *Can. J. Chem. Eng.* 53:541 (1975).
88. Tyn, M. T., and W. F. Calus. *J. Chem. Eng. Data* 20:106 (1975).
89. Sridhar, T., and O. E. Potter. *AIChE J.* 23:590 (1977).
90. Sridhar, T., and O. E. Potter. *AIChE J.* 23:946 (1977).
91. Sridhar, T. PhD Thesis, Monash University, Australia (1978).
92. Umesi, N. O., and R. P. Danner. *Ind. Eng. Chem. Process Des. Dev.* (1982).
93. Unver, A. A., and D. M. Himmelblau. *J. Chem. Eng. Data* 9:428 (1964).
94. Nir, S., and W. D. Stein. *J. Chem. Phys.* 55:1598 (1971).
95. Lynch, D. W., and O. E. Potter. *Chem. Eng. J.* 15:197 (1978).
96. Umesi, N. O., R. P. Danner and T. E. Daubert. Documentation Report 13-80, American Petroleum Institute (1980).

CHAPTER 2
GOVERNING EQUATIONS FOR SINGLE-PHASE FLOWS

N. P. Cheremisinoff
Exxon Research & Engineering Company
Florham Park, New Jersey 07932

CONTENTS

INTRODUCTION

The ancient Romans employed lead pipes to collect water, while the early Chinese conveyed natural gas through bamboo tubes. Despite these early engineering feats, centuries passed before fluid mechanics became a science. Gradually, fundamental laws describing the physics of flow phenomena began to be understood and applied to engineering problems. The science has gained impetus since the early 1800s when the necessity for sanitary conditions began creating a demand for water distribution systems. These have now become ubiquitous throughout the developed nations. Discoveries of uses for oil and natural gas gave further impetus to the design and construction of millions of kilometers of additional pipelines, reaction vessels and processing operations.

Our knowledge of fluid mechanics has expanded greatly with principles that are applied not only to fluid transport, but to an overwhelming number of process operations. Because it is not possible to rationally design a piece of equipment or process for whatever purpose without an understanding of the fluid mechanics, this book was devised. An understanding of the physics of flow phenomena is required, regardless of the intent of design, whether simply to transport fluids; process them via mixing; for separating, drying or coating; to transfer heat; remove certain constituents via adsorption or absorption; or promote chemical reactions.

This chapter presents a review of fundamental principles and theorems that have evolved from studies of single fluid flow. These principles underlie the analysis of virtually all problems of fluid mechanics and, in fact, are applied to understanding and modeling multicomponent flows, which are discussed in subsequent chapters. Specifically, this chapter reviews the general differential equations of conservation of mass and momentum, and the law of conservation of energy. The first is often referred to as the principle of continuity, which accounts for the mass of fluid flowing through any system. The last is the mechanical energy equation, which accounts for the energy interconversions that occur in the flowing fluid. Reviewed also in this chapter are the classical theories of turbulent flow, along with the semi-empirical models describing shear and velocity distribution for simple flow configurations. More recent advances in turbulence theory are presented in Chapter 4.

CONTINUITY AND EQUATIONS OF MOTION

The principle of *continuity* is a statement of the law of conservation of mass. Consider a cubical element of fluid flowing in space with respect to the coordinates shown in Figure 1. A simple material balance for this fluid element is as follows:

$$\left\{\begin{matrix}\text{Mass}\\ \text{accumulation}\end{matrix}\right\} = \left\{\begin{matrix}\text{Mass into cubical}\\ \text{element}\end{matrix}\right\} - \left\{\begin{matrix}\text{Mass out of cubical}\\ \text{element}\end{matrix}\right\} \tag{1}$$

The mass input into the x face of the element and the mass out at the x + Δx face can be obtained over some differential time element, dt; denoting small increments as derivatives, the material balance becomes

$$\frac{\partial \rho}{\partial t}\mathrm{dxdydzdt} = \rho\mathrm{udydzdt} - \left(\rho u + u\,\frac{\partial \rho}{\partial t}\,dx + \rho\,\frac{\partial u}{\partial x}\,dx\right)\mathrm{dydzdt} \tag{2}$$

Rearranging terms,

$$-u\frac{\partial \rho}{\partial x} - \rho\,\frac{\partial u}{\partial x} = \frac{\partial \rho}{\partial t} \tag{3a}$$

or

$$-\frac{\partial(\rho u)}{\partial x} = \frac{\partial \rho}{\partial t} \tag{3b}$$

where u refers to the fluid velocity in the x-direction.

This last differential equation is the law of conservation of mass, where both u and ρ are functions of x and t. However, Equation 3b only describes flow in the x-direction, and the more general statement is

$$\frac{\partial \rho}{\partial t} = -\left(\frac{\partial}{\partial x}\rho u + \frac{\partial}{\partial y}\rho v + \frac{\partial}{\partial z}\rho w\right) \tag{3c}$$

Table I summarizes the various forms of the continuity equation for both compressible and incompressible fluids under steady- and unsteady-state flow conditions. Table II gives the general statement of continuity in rectangular, cylindrical and spherical coordinates.

Table I. Various Forms of the Continuity Equation[a]

	Compressible Fluids		Incompressible Fluids	
	Unsteady State	Steady State	Unsteady State	Steady State
One-Dimensional Flow (x-direction)	$\frac{\partial(\rho u)}{\partial x} = -\frac{\partial \rho}{\partial t}$	$\frac{\partial(\rho u)}{\partial x} = 0$	$\frac{\partial u}{\partial x} = 0$	$\frac{\partial u}{\partial x} = 0$
Two-Dimensional Flow (x and y direction)	$\frac{\partial(\rho u)}{\partial x} + \frac{\partial(\rho v)}{\partial y} = \frac{-\partial \rho}{\partial t}$	$\frac{\partial(\rho u)}{\partial x} + \frac{\partial(\rho v)}{\partial y} = 0$	$\frac{\partial u}{\partial x} + \frac{\partial v}{\partial y} = 0$	$\frac{\partial u}{\partial x} + \frac{\partial v}{\partial y} = 0$
Three-Dimensional Flow	$\frac{\partial(\rho u)}{\partial x} + \frac{\partial(\rho v)}{\partial y} + \frac{\partial(\rho w)}{\partial z} = -\frac{\partial \rho}{\partial t}$	$\frac{\partial(\rho u)}{\partial x} + \frac{\partial(\rho v)}{\partial y} + \frac{\partial(\rho w)}{\partial z} = 0$	$\frac{\partial u}{\partial x} + \frac{\partial v}{\partial y} + \frac{\partial w}{\partial z} = 0$	$\frac{\partial u}{\partial x} + \frac{\partial v}{\partial y} + \frac{\partial w}{\partial z} = 0$

[a] v represents fluid velocity in the y-direction; w represents fluid velocity in the z-direction.

Table II. Equation of Continuity in Different Coordinates

Coordinate System	Axes	Equation
Rectanguar	x-, y-, z-	$\frac{\partial \rho}{\partial t} + \frac{\partial}{\partial x}(\rho u) + \frac{\partial}{\partial y}(\rho v) + \frac{\partial}{\partial z}(\rho w) = 0$
Cylindrical	r-, θ-, y-	$\frac{\partial \rho}{\partial t} + \frac{1}{r}\frac{\partial}{\partial r}(\rho r u_r) + \frac{1}{r}\frac{\partial}{\partial \theta}(\rho u_\theta) + \frac{\partial}{\partial y}(\rho w) = 0$
Spherical	r-, θ-, ϕ-	$\frac{\partial \rho}{\partial t} + \frac{1}{r^2}\frac{\partial}{\partial r}(\rho r^2 u_r) + \frac{1}{r \sin\theta}\frac{\partial}{\partial \theta}(\rho u_\theta \sin\theta) + \frac{1}{r \sin\theta}\frac{\partial}{\partial \phi}(\rho u_\phi) = 0$

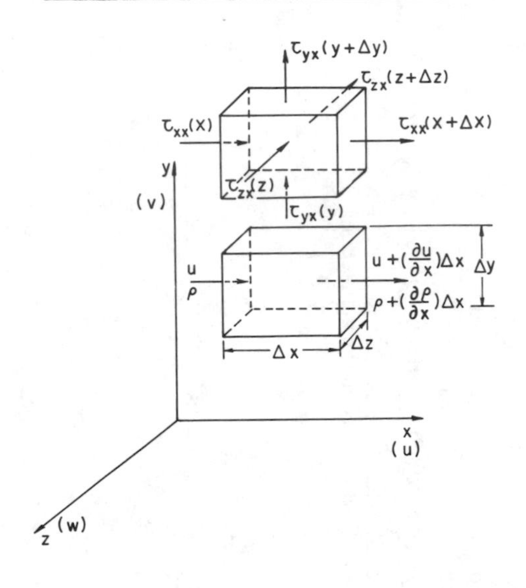

Figure 1. One-dimensional flow of a cubical element of fluid. The top element shows the directions in which the x-component of momentum is transported through the surfaces.

A series of equations also may be derived describing the rate of momentum transport through the surfaces of the cubical fluid element in Figure 1. The general statement of a momentum balance for laminar, rectilinear flow is

$$\left\{\begin{matrix}\text{Rate of momentum}\\ \text{into a fluid}\\ \text{element}\end{matrix}\right\} - \left\{\begin{matrix}\text{Rate of momentum}\\ \text{out of a fluid}\\ \text{element}\end{matrix}\right\} + \left\{\begin{matrix}\text{Pressure forces and}\\ \text{gravity forces}\end{matrix}\right\}$$
$$= \left\{\begin{matrix}\text{Rate of momentum}\\ \text{accumulation}\end{matrix}\right\} = 0 \text{ at steady state} \tag{4}$$

The arrows on the upper fluid element in Figure 1 indicate the direction in which the x-component of momentum is transported through the fluid volume surfaces. Momentum transport into and out of the fluid element occurs because of (1) the bulk fluid motion, and (2) the velocity gradients. The first mechanism is strictly convection, whereas the latter is attributed to molecular transfer.

Following the general momentum balance statement, an expression describing the unsteady-state momentum transport for the x-component is developed:

$$\underbrace{\frac{\partial}{\partial t}\rho u}_{①} = \underbrace{-\left(\frac{\partial}{\partial x}\rho u u + \frac{\partial}{\partial z}\rho w u + \frac{\partial}{\partial y}\rho v u\right)}_{②}$$

$$\underbrace{-\left(\frac{\partial}{\partial x}\tau_{xx}+\frac{\partial}{\partial z}\tau_{zx}+\frac{\partial}{\partial y}\tau_{yx}\right)}_{③}\underbrace{-\frac{\partial P}{\partial x}+\rho g_x}_{④} \tag{5}$$

where term (1) = rate of momentum accumulation
term (2) = rate of momentum in
term (3) = rate of momentum out
term (4) = sum of forces acting on fluid element

Equation 5 is the equation of motion for the x-component of momentum.
For the y- and z-components, the equations of motion are

$$\frac{\partial}{\partial t}\rho v = -\left(\frac{\partial}{\partial x}\rho uv+\frac{\partial}{\partial y}\rho vv+\frac{\partial}{\partial z}\rho wv\right)$$
$$-\left(\frac{\partial}{\partial x}\tau_{xy}+\frac{\partial}{\partial y}\tau_{yy}+\frac{\partial}{\partial z}\tau_{zy}\right)-\frac{\partial P}{\partial y}+\rho g_y \tag{6}$$

$$\frac{\partial}{\partial t}\rho w = -\left(\frac{\partial}{\partial x}\rho uw+\frac{\partial}{\partial y}\rho vw+\frac{\partial}{\partial z}\rho ww\right)$$
$$-\left(\frac{\partial}{\partial x}\tau_{xz}+\frac{\partial}{\partial y}\tau_{yx}+\frac{\partial}{\partial z}\tau_{zz}\right)-\frac{\partial P}{\partial z}+\rho g_z \tag{7}$$

Table III gives the general equations of motion in three different coordinate systems.
For incompressible, Newtonian fluids, the above equations are the Navier-Stokes equations, summarized in rectangular coordinates as follows.

x-component:

$$\rho\left(\frac{\partial u}{\partial t}+u\frac{\partial u}{\partial x}+v\frac{\partial u}{\partial y}+w\frac{\partial u}{\partial z}\right)=-\frac{\partial P}{\partial x}+\mu\left(\frac{\partial^2 u}{\partial x^2}+\frac{\partial^2 u}{\partial y^2}+\frac{\partial^2 u}{\partial z^2}\right)+\rho g_x \tag{8}$$

y-component:

$$\rho\left(\frac{\partial v}{\partial t}+u\frac{\partial v}{\partial x}+v\frac{\partial v}{\partial y}+w\frac{\partial v}{\partial z}\right)=-\frac{\partial P}{\partial x}+\mu\left(\frac{\partial^2 v}{\partial x^2}+\frac{\partial^2 v}{\partial y^2}+\frac{\partial^2 v}{\partial z^2}\right)+\rho g_y \tag{9}$$

z-component:

$$\rho\left(\frac{\partial w}{\partial t}+u\frac{\partial w}{\partial x}+v\frac{\partial w}{\partial y}+w\frac{\partial w}{\partial z}\right)=-\frac{\partial P}{\partial z}+\mu\left(\frac{\partial^2 w}{\partial x^2}+\frac{\partial^2 w}{\partial y^2}+\frac{\partial^2 w}{\partial z^2}\right)+\rho g_z \tag{10}$$

The equations of motion and continuity form the basis for describing any viscous flow problem. By applying proper boundary conditions based on the flow system geometry and the physics of the flow phenomenon, the equation of motion can be solved for the appropriate flow parameter of interest. For example, consider the simple case of an incompressible fluid flowing down a vertical tube, as shown in Figure 2. To derive an expression for the velocity profile for laminar flow in cylindrical coordinates one may ignore end effects for a steady-flow solution. Hence, by inspection of the equation in cylindrical coordinates in Table III, noting that $u_\theta = u_r = 0$, the following is obtained:

$$\rho v\frac{\partial v}{\partial y}=-\frac{\partial P'}{\partial y}+\mu\left[\frac{1}{r}\frac{\partial}{\partial r}\left(r\frac{\partial v}{\partial r}\right)+\frac{\partial^2 v}{\partial y^2}\right] \tag{11}$$

Table III. General Equation of Motion in Different Coordinates

Coordinate System	Component	Equation
Rectangular (x, y, z)	x-	$\rho\left(\frac{\partial u}{\partial t} + u\frac{\partial u}{\partial x} + v\frac{\partial u}{\partial y} + w\frac{\partial u}{\partial z}\right) = -\frac{\partial P}{\partial x} - \left(\frac{\partial \tau_{xx}}{\partial x} + \frac{\partial \tau_{yx}}{\partial y} + \frac{\partial \tau_{zx}}{\partial z}\right) + \rho g_x$
	y-	$\rho\left(\frac{\partial v}{\partial t} + u\frac{\partial v}{\partial x} + v\frac{\partial v}{\partial y} + w\frac{\partial v}{\partial z}\right) = -\frac{\partial P}{\partial y} - \left(\frac{\partial \tau_{xy}}{\partial x} + \frac{\partial \tau_{yy}}{\partial_y} + \frac{\partial \tau_{zy}}{\partial z}\right) + \rho g_y$
	z-	$\rho\left(\frac{\partial w}{\partial t} + u\frac{\partial w}{\partial x} + v\frac{\partial w}{\partial y} + w\frac{\partial w}{\partial z}\right) = -\frac{\partial P}{\partial z} - \left(\frac{\partial \tau_{xz}}{\partial x} + \frac{\partial \tau_{yz}}{\partial y} + \frac{\partial \tau_{zz}}{\partial z}\right) + \rho g_z$
Cylindrical (r, θ, y)	r-	$\rho\left(\frac{\partial u_r}{\partial t} + u_r\frac{\partial u_r}{\partial r} + \frac{u_\theta}{r}\frac{\partial u_r}{\partial \theta} - \frac{u_\theta^2}{r} + u_y\frac{\partial u_r}{\partial y}\right) = -\frac{\partial P}{\partial r} - \left(\frac{1}{r}\frac{\partial}{\partial r}(r\tau_{rr}) + \frac{1}{r}\frac{\partial \tau_{r\theta}}{\partial \theta} - \frac{\tau_{\theta\theta}}{r} + \frac{\partial \tau_{ry}}{\partial y}\right) + \rho g_r$
	θ-	$\rho\left(\frac{\partial u_\theta}{\partial t} + u_r\frac{\partial u_\theta}{\partial r} + \frac{u_\theta}{r}\frac{\partial u_\theta}{\partial \theta} + \frac{u_r u_\theta}{r} + u_y\frac{\partial u_\theta}{\partial y}\right) = -\frac{1}{r}\frac{\partial P}{\partial \theta} - \left(\frac{1}{r^2}\frac{\partial}{\partial r}(r^2\tau_{r\theta}) + \frac{1}{r}\frac{\partial \tau_{\theta\theta}}{\partial \theta} + \frac{\partial \tau_{\theta y}}{\partial y}\right) + \rho g_\theta$
	y-	$\rho\left(\frac{\partial u_y}{\partial t} + u_r\frac{\partial u_y}{\partial r} + \frac{u_\theta}{r}\frac{\partial u_y}{\partial \theta} + u_y\frac{\partial u_y}{\partial y}\right) = -\frac{\partial P}{\partial y} - \left(\frac{1}{r}\frac{\partial}{\partial r}(r\tau_{ry}) + \frac{1}{r}\frac{\partial \tau_{\theta y}}{\partial \theta} + \frac{\partial \tau_{yy}}{\partial y}\right) + \rho g_y$
Spherical (r, θ, φ)	r-	$\rho\left(\frac{\partial u_r}{\partial t} + u_r\frac{\partial u_r}{\partial r} + \frac{u_\theta}{r}\frac{\partial u_r}{\partial \theta} + \frac{u_\phi}{r\sin\theta}\frac{\partial u_r}{\partial \phi} - \frac{u_\theta^2 + u_\phi^2}{r}\right) = -\frac{\partial P}{\partial r} - \left(\frac{1}{r^2}\frac{\partial}{\partial r}(r^2\tau_{rr}) + \frac{1}{r\sin\theta}\frac{\partial}{\partial \theta}(\tau_{r\theta}\sin\theta)\right.$
		$\left. + \frac{1}{r\sin\theta}\frac{\partial \tau_{r\phi}}{\partial \phi} - \frac{\tau_{\theta\theta} + \tau_{\phi\phi}}{r}\right) + \rho g_r$
	θ-	$\rho\left(\frac{\partial u_\theta}{\partial t} + u_r\frac{\partial u_\theta}{\partial r} + \frac{u_\theta}{r}\frac{\partial u_\theta}{\partial \theta} + \frac{u_\phi}{r\sin\theta}\frac{\partial u_\theta}{\partial \phi} + \frac{u_r u_\theta}{r} - \frac{u_\phi^2\cot\theta}{r}\right)$
		$= -\frac{1}{r}\frac{\partial P}{\partial \theta} - \left(\frac{1}{r^2}\frac{\partial}{\partial r}(r^2\tau_{r\theta}) + \frac{1}{r\sin\theta}\frac{\partial}{\partial \theta}(\tau_{\theta\theta}\sin\theta) + \frac{1}{r\sin\theta}\frac{\partial \tau_{\theta\phi}}{\partial \phi} + \frac{\tau_{r\theta}}{r} - \frac{\cot\theta}{r}\tau_{\phi\phi}\right) + \rho g_\theta$
	φ-	$\rho\left(\frac{\partial u_\phi}{\partial t} + u_r\frac{\partial u_\phi}{\partial r} + \frac{u_\theta}{r}\frac{\partial u_\phi}{\partial \theta} + \frac{u_\phi}{r\sin\theta}\frac{\partial u_\phi}{\partial \phi} + \frac{u_\phi u_r}{r} + \frac{u_\theta u_\phi}{r}\cot\theta\right)$
		$= -\frac{1}{r\sin\theta}\frac{\partial P}{\partial \phi} - \left(\frac{1}{r^2}\frac{\partial}{\partial r}(r^2\tau_{r\phi}) + \frac{1}{r}\frac{\partial \tau_{\theta\phi}}{\partial \phi} + \frac{1}{r\sin\theta}\frac{\partial \tau_{\phi\phi}}{\partial \phi} + \frac{\tau_{r\phi}}{r} + \frac{2\cot\theta}{r}\tau_{\theta\phi}\right) + \rho g_\phi$

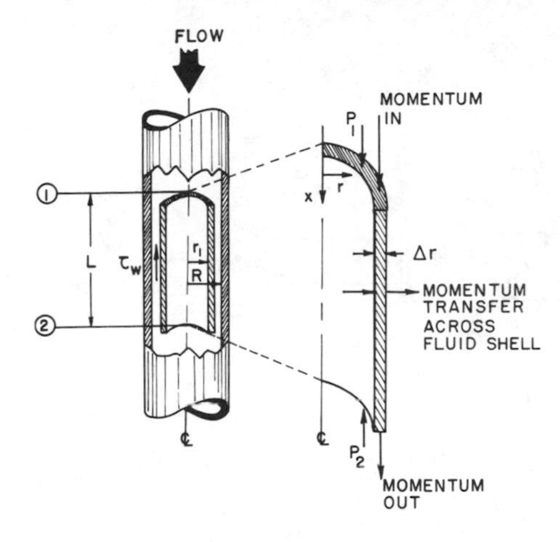

Figure 2. Incompressible fluid flowing downward in a vertical tube. A cylindrical shell of fluid can be imagined if one chooses to derive the volocity profile of the fluid from a momentum balance.

The continuity equation reduces to

$$\frac{\partial v}{\partial y} = 0 \tag{12}$$

This also means that $\frac{\partial^2 v}{\partial y^2} = 0$. Hence, Equation 11 becomes

$$0 = -\frac{dP'}{dy} + \mu \frac{1}{r}\frac{d}{dr}\left(r\frac{dv}{dr}\right) \tag{13}$$

Applying a "no-slip" boundary condition: $v = 0$ at $r = R$, and, after integrating,

$$v = \frac{\Delta P' R^2}{4\mu L}\left[1 - (r/R)^2\right] \tag{14}$$

Equation 14 is the expression for a parabolic velocity profile describing laminar flow in a tube.

Flow problems involving the determination of fluid velocity as a function of time and space require simultaneous solution of the momentum and continuity equations. For many situations the flow is such that these equations may be greatly simplified. In the one-dimensional flow problem treated above, the two equations in the y- and z-directions (or the θ- and y-directions in cylindrical coordinates) were eliminated. In two-dimensional flow problems, two momentum equations are required. Obviously, terms vanish when describing the coordinate direction in which no flow exists.

THE ENERGY BALANCE

Total Energy Balance

The mechanical energy equation is best described by an example such as the system illustrated in Figure 3. The flow system shown has fluid entering at point (1) and exiting at point (2), with the same mass flow rate, W. That is, the flow system is at steady state and, hence, there is no progressive accumulation or depletion of fluid at any point.

The fluid entering at point (1) has a certain amount of energy associated with it. This energy exists in several forms. First, since the fluid has a velocity, U, it brings to the system *kinetic energy* defined as $U_1{}^2/2g$. Second, because the incoming flow is

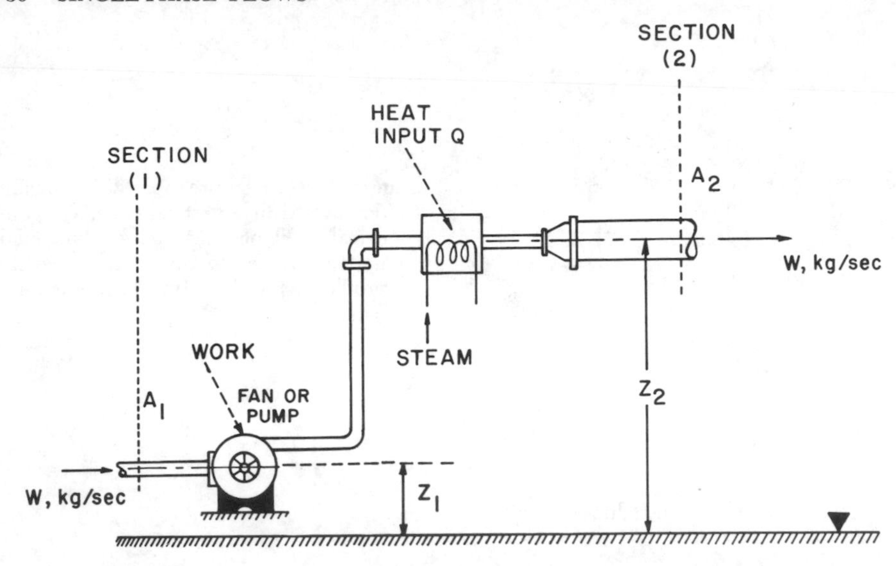

Figure 3. Example system for describing the total energy balance.

elevated at a distance, Z_1, above the specified horizontal datum plane, it possesses potential energy of Z_1-meters of fluid. It also brings *internal energy,* E_1, which consists of thermal, chemical or combinations of these and other sources. Finally, there is mechanical energy associated with the incoming fluid. That is, the incoming fluid is forced into the system under pressure. This pressure is exerted by the upstream fluid forcing the flow past point (1). Mechanical energy is defined as the product of the force exerted by the flowing fluid and the distance over which it acts. Hence, it is the fluid pressure per unit area times the cross-sectional area of flow, P_1A_1. If, for calculation purposes, the mass of fluid flowing per unit time past point (1) is 1 kg, then the distance through which the force acts is the specific volume of the fluid divided by the cross section, $\dot{\nu}_1/A_1$. Furthermore, since the work performed is the product of force and disance, then the energy expended is equal to pressure times the specific volume:

$$W_M = P_1A_1\ (\dot{\nu}_1/A_1) = P_1\dot{\nu}_1 \tag{15}$$

Between points (1) and (2) of the system diagram occurs an exchange of energy between the surroundings and system. This interchange of energy is in the form of heat, Q, and work, W, which can be expressed in terms of the mass of flowing fluid. In Figure 3, both energies are added to the flow from the surroundings and, thus, are denoted by positive signs.

At point (2) the fluid departs from the system with energy in the same four forms that entered at point (1): kinetic, potential, internal and mechanical energies. To maintain conservation of energy requires equality between the sum of all energies entering and the sum of those leaving. This statement is, in fact, the total energy balance:

$$Z_1 + P_1\dot{\nu}_1 + \frac{U_1^2}{2g} + E_1 + Q + W' = Z_2 + P_2\dot{\nu}_2 + \frac{U_2^2}{2g} + E_2 \tag{16}$$

Evaluation of the individual terms in Equation 16 often poses difficulties simply because of the lack of experimental data at a system's intended operating conditions. Hence, basic assumptions are often imposed to Equation 16 when making engineering computations. In applying assumptions, it is important that the meaning of each term be clearly understood to justify approximations.

Kinetic Energy Term

The kinetic energy term, $U^2/2g$ is based on a small stream of fluid having a local velocity or flow in which the velocity profile across the cross section is flat. In reality, there is a velocity gradient across the passage, as was shown by the derivation for laminar flow in a tube (Equation 14). For the turbulent flow situation, the velocity profile is more flat, and it is customary to use an average velocity, $\overline{U}$, in the kinetic energy term, i.e., $\overline{U}^2/2g$. The error introduced by using $\overline{U}^2/2g$ is generally not serious, but its magnitude depends on the shape of the conduit's cross section. Also, since the $\overline{U}^2/2g$ terms appear on each side of the energy balance equation, the error tends to cancel out. In the case of a parabolic velocity profile, the error introduced by using $\overline{U}^2/2g$ can be significant. We can derive an exact expression for the kinetic energy term for this case.

Considering the case of laminar flow, we can use Equation 14 to describe the velocity profile, and an expression for the volumetric flowrate in a tube is

$$q = \int_0^R u2\pi r dr = \frac{\pi R^4 \Delta P'}{8\mu L} \tag{17}$$

Defining the average velocity as the volumetric flowrate divided by the cross-sectional area of flow for a single cylindrical tube, then

$$\overline{U} = -\Delta P' R^2/8\mu L \tag{18}$$

The local kinetic energy through the cylindrical shell is

$$d(KE) = dw\, u^2/2g \tag{19}$$

where dw is the mass rate of flow through an annulus, assuming the fluid flowing in a section of tube may be visualized as an infinite number of cylindrical shells of thickness r (refer to Figure 2). The mass of fluid flowing per unit time through the annulus is $2\pi r dr\, u\rho$.

From Equation 14 it can be shown that

$$u = 2\overline{U}\,(R^2 - r^2)/R^2$$

Then,

$$d(KE) = \frac{dw\, u^2}{2g} = \frac{(2\pi r dr\, u\rho) u^2}{2g} \tag{20}$$

The average kinetic energy per unit mass of fluid flowing is the integral of this expression divided by the fluid mass flowing through the entire cross section of pipe.

$$\begin{aligned}\hat{KE} &= \frac{\int d(KE)}{\pi r_1^2 u\rho} = \frac{8\overline{U}^2}{g r_1^8}\int_0^{r_1} (r_1^6 r dr - 3r_1^4 r^3 dr + 3r_1^2 r^5 dr - r^7 dr) \\ &= \frac{8\overline{U}^2}{g r_1^8}\left(\frac{r_1^8}{2} - \frac{3r_1^8}{4} + \frac{3r_1^8}{6} - \frac{r_1^8}{8}\right) \\ &\simeq \frac{\overline{U}^2}{g}\end{aligned} \tag{21}$$

Hence, for the parabolic velocity profile case, $\overline{U}^2/g$ provides a reasonable estimate of the kinetic energy term in the total energy equation.

Variations in point velocity over each section of flow in the system of Figure 3 must be evaluated when applying the energy equation. For computation purposes, it is sometimes easier to apply the energy equation to a differential portion of the pipe flow, although this may pose additional problems in evaluating other terms.

Heat and External Work

The term Q accounts for the total heat that is added to the system per unit mass of flowing fluid. This definition is restricted, however, only to the heat that enters the fluid through the appropriate heat transfer area supplied by the system or vessel walls. That is, it is only heat that is supplied by the surroundings or some external source, such as steam to a heat exchanger. Excluded from this term is any heat generated by friction. This latter energy form is already dissipated in the flow and is accounted for in the internal energy term. The addition of heat to the system is denoted by a positive sign on Q in Equation 16. Note that the positive addition of heat does not always involve a rise in temperature for the system and, in fact, it is not uncommon for a temperature drop to occur for this case.

External work energy, W′, is most commonly supplied by a mechanical device such as a pump or compressor. Work energy also may be removed from the fluid system by such devices as engines, prime movers, etc. If no work is either added or removed from the system, the term W′ vanishes in Equation 16.

Internal Energy

The internal energy term, E, accounts for thermodynamic properties. Note that both E and P are properties of the fluid that are uniquely determined by point conditions. The sum of these two terms is most readily treated as a single function, enthalpy:

$$H = E + P\dot{\nu} \tag{22}$$

Enthalpy itself is uniquely determined by the fluid's point conditions. Unfortunately, like E, the absolute value of H cannot be determined; however, differences in value above an arbitrary reference can be evaluated. Evaluation of thermodynamic properties is treated in the literature [1-5].

The Bernoulli Equation

The volume of a liquid decreases linearly with pressure. For gases, volume is inversely proportional to pressure. For example, if a sample of water is subjected to a pressure change from 1 to 2 atm, less than 10^{-3}% reduction in volume occurs. If a sample of air were subjected to the same pressure change, its volume would be reduced by more than a factor of two. Because of this characteristic of insignificant volume changes from moderate variations in pressure, liquids are referred to as incompressible fluids.

When treating problems involving steady flow of incompressible fluids under conditions in which friction is small and in the absence of external effects, the energy balance based on mechanical energy forms may be applied with confidence. For situations involving compressible fluid flow, the total mechanical energy at the system discharge is often significantly greater than the upstream section. This is especially true when large pressure drops between sections occur.

Any quantity of flowing fluid undergoes an expansion when it performs mechanical work. This work is then expended on the fluid immediately ahead of it. Realize, however, that the fluid element in question also acquires an equivalent quantity of mechanical energy from the fluid behind it. This results in an overall increase in the mechanical energy at the expense of the internal energy of the fluid or of externally added heat energy [6,7].

To account for self-expansion work and for total friction due to fluid flow, the mechanical energy balance is modified to the following form:

$$Z_1 + P_1\dot{\nu}_1 + \frac{u_1^2}{2g} + \int_1^2 Pd\dot{\nu} = Z_2 + P_2\dot{\nu}_2 + \frac{u_2^2}{2g} + \Sigma F \quad (23)$$

In applying the correction term, $\int_1^2 Pd\dot{\nu}$, it should be noted that the mechanical energy output is almost always less than the input. Unlike $P\dot{\nu}$, E and H, the integral is not a point function of the conditions. Instead, it is a function of the path of expansion or compression.

ΣF represents the total friction produced by fluid flow. Walker et al. [7] define fluid friction as the mechanical energy rendered nonavailable due to irreversibilities in the flow process. For example, assume the pump between points (1) and (2) in Figure 3 delivers in reality $\hat{W}'$ joules of mechanical work to the fluid. Note that $\hat{W}'$ is less than the work term W′ in the total energy equation since it includes friction in the pump. That is,

$$\hat{W}' = W' - F_p \quad (24)$$

where F_p is friction energy in the pump.

Hence, the most general form of Equation 23 is

$$Z_1 + P_1\dot{\nu}_1 + \frac{u_1^2}{2g} + \int_1^2 Pd\dot{\nu} + \hat{W}' = Z_z + P_2\dot{\nu}_2 + \frac{u_2^2}{2g} + \Sigma F \quad (25)$$

Equation 25 is the familiar Bernoulli equation. In more compact form, it is written as follows:

$$\hat{W}' - \int_1^2 \dot{\nu}dP = \Delta Z + \Delta u^2/2g + \Sigma F \quad (26)$$

where

$$\Delta u = u_2^2 - u_1^2$$

and noting that

$$\int_1^2 Pd\dot{\nu} + P_1\dot{\nu}_1 - P_2\dot{\nu}_2 = -\int_1^2 \dot{\nu}dP \quad (27)$$

Both the Bernoulli equation and the total energy equation represent the fundamental basis for solving problems in fluid mechanics. Note that application of Equation 27 requires not only the thermodynamic properties of the flowing system to be defined, but also the term ΣF. Although ΣF may be evaluated from studies of the mechanisms of flow, the traditional engineering approach has been based on empiricism.

LAMINAR FLOW

Wall Shear Stress

Shearing stresses exist in the boundary layer of fluid flowing within a system of fixed boundaries (e.g., when flowing through a pipe or duct or over a surface). These stresses are exerted in the opposite direction to the flow and, therefore, may be thought of as the forces of resistance to flow. By performing a force balance over a length of tube, L, the following expression is obtained:

$$\pi R^2 P_1 - \pi R^2 P_2 + \pi R^2 L \rho g - 2\pi R L \tau_w = 0 \tag{28}$$

or

$$\frac{\Delta P}{\rho} + Lg = \frac{2L\tau_w}{\rho R} \tag{29}$$

Defining the sum of all the friction forces to be $\Delta P/\rho$, then

$$\Sigma F = \frac{\Delta P}{\rho} = \frac{P_1 - P_2}{\rho} - Lg \tag{30}$$

Solving Equation 29,

$$\tau_w = \frac{\Delta P}{\rho}\left(\frac{\rho R}{2L}\right) = \frac{R\Delta P}{2L} \tag{31}$$

Note that τ_w is actually τ_{rx} at $r = R$.
Hence, for any r, where $r \leqq R$, the fluid shear stress is

$$\tau_{rx} = \frac{r\Delta P}{2L} \tag{32}$$

Combining Equations 31 and 32, τ_{rx} is expressed in terms of the shearing force (shear stress) at the wall:

$$\tau_{rx} = \tau_w (r/R) \tag{33}$$

This last expression states that the shear stress distribution across the tube is linear and varies from zero at the tube centerline to τ_w at the tube wall. The shear stress distribution is illustrated in Figure 4 along with the parabolic velocity profile for laminar flow.

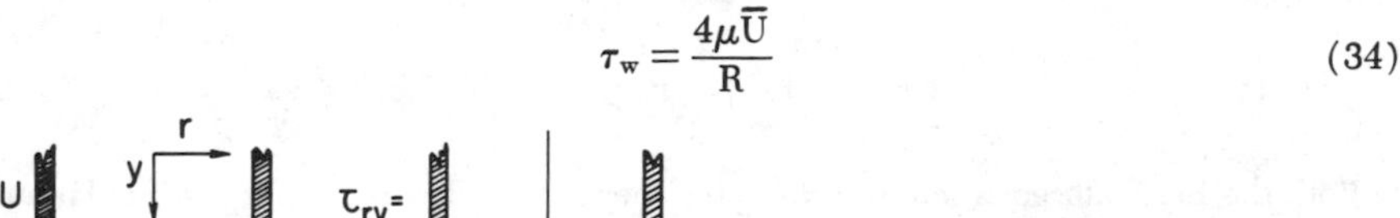

Since shear stress is the force of resistance to flow, it is ultimately a measure of the friction losses. For laminar flow, a relatively simple relation can be obtained. By combining the expression for the velocity profile (Equation 14) and Newton's law of viscosity ($d\tau = d(\mu A du/dr)$), an alternative expression for the wall shear is obtained:

$$\tau_w = \frac{4\mu \overline{U}}{R} \tag{34}$$

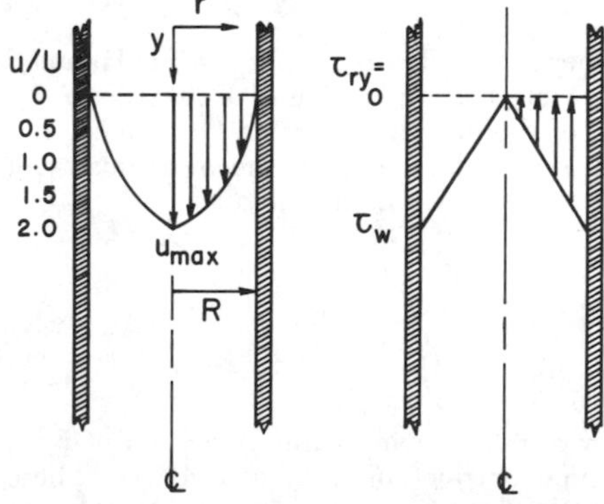

Figure 4. Velocity profile and shear stress distribution for laminar flow in a tube.

This expression defines wall shear stress in terms of fluid viscosity, average velocity and the tube radius. Combining Equations 31 and 34 and expressing the tube radius, R, in terms of the inside diameter of the tube, the following expression written in differential form is obtained:

$$\frac{-dP}{dx} = \frac{32\mu\bar{U}}{g_c D^2} \tag{35}$$

This equation is Poiseuille's law and defines the pressure loss due to friction in terms of fluid viscosity, average fluid velocity and tube diameter. The lost work due to friction is conventionally correlated by the parameter "friction factor." The friction factor is a dimensionless parameter, which enables a compact expression to be formulated to include tube diameter, fluid properties, fluid velocity and the pressure gradient equivalent to the frictional resistance per unit length of tube.

Friction Factor

Since shear stresses represent the force of resistance to flow, they are ultimately a measure of the frictional losses. Friction losses in turbulent flow are known to be proportional to the kinetic energy of the fluid per unit volume and the area of the solid surface contacting the flow. The force per unit area, A_w, acting on a solid surface such as a pipe wall is

$$F/A_w \propto \frac{\rho U^2}{2g_c} = f\,\frac{\rho U^2}{2g_c} \tag{36}$$

where f is a proportionality constant.

We note that F/A_w is really the wall shear, τ_w; hence, $\tau_w = f(\rho U^2/2g_c)$.

By substituting for τ_w in Equation 32, the proportionality constant, f, is solved for. Factor f, the friction factor, is referred to as the "Fanning friction factor" for pipe flow:

$$f = \frac{g_c R}{\rho U^2}\,\frac{\Delta P}{L} \tag{37a}$$

or

$$f = \frac{g_c D}{2\rho U^2}\,\frac{\Delta P}{L} \tag{37b}$$

where D is the pipe diameter.

For laminar flow in a tube, Poiseuille's equation may be used to derive the following relation:

$$f = \frac{8\mu}{RU\rho} = \frac{16}{N_{Re}} \tag{37c}$$

N_{Re} is the Reynolds number, ($N_{Re} = DU\rho/\mu$).

The friction factor definition in Equation 37c is good for Reynolds numbers up to 2000.

TURBULENT FLOW

Turbulence Structure

Laminar flow described above consists of a steady advance of fluid layers such that a streamline flow is maintained. If the fluid's flowrate is increased sufficiently, this pattern is no longer maintained and the flow becomes unsteady. This latter flow situation is characterized by chaotic movements of portions of the fluid in different directions superimposed on the main flow of the fluid; the phenomenon is referred to as turbulence. The complex movement of fluid elements within the flow can only be described in terms of time-averaged values.

Momentum transfers between adjacent regions of flowing fluid play a dominant role in establishing turbulence. These momentum transfers are inertial effects, which cause velocity and fluid density to take on important roles. By contrast, in laminar flow only purely viscous effects determine the nature of the flow. Turbulent flow dominates when inertial effects (denoted by ρu^2) become large in comparison to viscous forces (denoted by $\mu u/D$).

The transition from laminar to turbulent flow for a small portion of flowing fluid can be illustrated by visualizing the flow to consist of distinct parallel streamlines of fluid. By introducing a random disturbance, density and velocity are in part dampened out by viscous forces. As more momentum is transferred into the disturbance, inertial forces eventually become much greater than the viscous term, and eddy currents are formed. The Reynolds number best describes this flow transition since it is the ratio of inertial to viscous forces, i.e., $N_{Re} = (\rho u^2)/(\mu u/D)$.

Numerous investigators [8-14] have studied the flow of fluids in pipes, confirming the flow regime dependency on Reynolds number. For flow in smooth pipes, the flow is always laminar for Reynolds numbers up to about 2000. Between Reynolds numbers of 2000 and 4000, the flow undergoes a transitional change from laminar to turbulent flow. Above Reynolds numbers of 4000, the flow is generally fully established turbulent flow.

The formation of turbulent eddy currents is best described by imagining ourselves to be moving within the fluid at the same velocity as the fluid's mean flowrate. If the flow is in a pipe, what one sees in the vicinity of the wall are patches of fluid torn into small eddies by the strong shearing forces within the fluid. These eddies tend to migrate toward the pipe centerline, some combining to form larger eddies, while many smaller ones simply diffusing into the "core" fluid with decaying intensity. One would further observe that these eddies are in fact superimposed on a much faster overall mean flow.

There are several situations that can lead to turbulent flow. One situation just described is that of rapid flow of a fluid past a solid surface. This situation leads to unstable, self-amplifying velocity fluctuations, which form in the fluid in the vicinity of the wall and spread outward into the main fluid stream. In a similar manner, turbulent eddies are formed from the velocity gradients established between a fast-moving fluid and a slower-moving fluid. A third general way in which turbulence is induced is by the relative movement of an object through the fluid streams. Examples of this last case are an impeller blade on an agitator and a sphere or cylinder falling through a fluid medium. These situations cause eddies to form in the wake, resulting in an increase in the resistance of the movement of the object (called "form drag").

In the case of stirred vessels, turbulence can be very intense near the tips of the rotor blades. Most of the turbulence throughout the vessel arises from velocity gradients, whereby portions of high-velocity fluid are thrown from impeller blades into slower-moving fluid. Some of this turbulence is attributed, however, to the high shearing over the blades themselves, which creates separation behind each blade or neighboring baffles.

Another mixing-related turbulence arises from submerged jets. In submerged jets (also referred to as "free turbulent jets"), the fluid is expelled from a nozzle into a mass of miscible fluid, which is essentially at rest with respect to the discharging stream. The free jet expands outward in the form of a cone, entraining large amounts of the surrounding fluid at the periphery of the jet. The sizes of eddies formed are relatively uniform across any given section of such a jet. When a submerged jet is introduced to an immiscible fluid, the stream is referred to as a *restrained turbulent jet*. Turbulent eddies protrude from the sides of the free jet but are restrained from breaking away by surface tension forces. This type of turbulence decays quickly, as fluctuations are damped by the elastic forces associated with surface tension effects. Chapter 9 describes the theory of turbulent jets.

It should be noted that the turbulent condition set by $N_{Re} > 4000$ is a general rule and one based mainly on experience with pipe flow. In fact, turbulent conditions can be generated at much lower Reynolds numbers. One example is flow in a corrugated conduit, where vortex shedding at the angularities can induce strong eddying at Reynolds numbers of only a few hundred. This principle is also applied to inline static mixers used for mixing two liquid streams or a chemical additive and a liquid. By means of fins, baffles or an arrangement of packing material through which fluids flow, vortex shedding occurs, promoting eddies assisting in the necessary mass transfer. Another example of this application is in heat exchangers used for viscous liquids. Here, only moderate flowrates are needed to promote eddy formation in narrow tubes. This promotes good heat transfer even though Reynolds numbers are only as high as perhaps 500–600. In these applications, the sharpness of the angularities is important since eddies shed much more readily from sharp corners than from smooth surfaces.

Turbulence may be described quantitatively in terms of the instantaneous velocity fluctuations occurring in local patches of flowing fluid. The instantaneous velocity in the x-direction of flow is

$$u_x = \bar{u}_x \pm u'_x \tag{38}$$

where $\bar{u}_x$ is the time-averaged velocity at any point in the flowing fluid and u'_x is the instantaneous velocity fluctuation. Fluctuation velocities may be positive or negative at times, with their time average being zero. That is,

$$\bar{u}'_x = 0 \tag{39}$$

The amplitude of fluctuation velocities in the x-direction can be expressed in terms of the mean of the squares of the fluctuating velocities, obviously making them positive. This is denoted by $\overline{(u'_x)^2}$. A root mean square fluctuation velocity would be

$$u'_{x_{rms}} = [\overline{(u'_x)^2}]^{1/2} \tag{40a}$$

where $u'_{x_{rms}}$ is always positive. $\bar{u}x_{rms}$ provides an indication of the *intensity of turbulence* in the x-direction.

$$u_y = \bar{u}_y \pm u_y' \tag{40b}$$

$$\bar{u}_y' = 0$$

and

$$u'_{y_{rms}} = \sqrt{\overline{(u'_y)^2}} \tag{40c}$$

The product $\overline{u'_x u'_y}$ for any point in the fluid is not necessarily zero. The product of u'_x and u'_y has been time-averaged, and these fluctuations are correlated with a particular eddy. These fluctuating velocities are responsible for establishing the degree of dissipation of energy, mass transfer and heat transfer in turbulent flows.

Visualize the fluid as consisting of individual parcels. In turbulent flow there is a continuous interchange of these parcels between adjacent regions of the fluid. This exchange of mass involves both gains and losses of momentum. That is, each parcel of fluid leaving a region with some velocity takes momentum with it to the next fluid region, which, in turn, may be moving at a different mean velocity. Fluid parcels transposed laterally from a faster-moving region tend to accelerate the slower-moving fluid region; conversely, a slow-moving parcel transposed to a faster-moving region has a retarding effect.

The rate of momentum transfer per unit area is $\rho u'_y u'_x$ and can be expressed in terms of a stress, i.e., equivalent force per unit area:

$$\tau = -\rho u_x' u_y' \tag{41a}$$

where the time average of this product is the Reynolds stress:

$$\tau = -\overline{\rho u_x' u_y'} \tag{41b}$$

Continuing with this model of parcel exchanges, assume that a particular parcel is carried a distance, y_1, from the wall to a new distance, y_2. Further assume that the parcel is moved from a slower-moving section to a faster-moving one. The instantaneous deficit in momentum occurring at distance y_2 may be expressed as follows:

$$\text{Momentum deficit} = -(\Delta y)d(\rho\bar{u}_x)/dy \tag{42}$$

where $\Delta y = y_2 - y_1$.

If momentum is conserved, then

$$\rho u_x' = -(\Delta y)d(\rho\bar{u}_x)/dy \tag{42a}$$

or

$$\tau = -\rho u_x' u_y' = u_y'(\Delta y)d(\rho\bar{u}_x)/dy \tag{42b}$$

Alternately, this last expression may be written as follows:

$$\tau = -\overline{\rho u_x' u_y'} = \rho l u'_{y_{rms}} d\bar{u}_x/dy \tag{43}$$

Equation 43 is obtained by representing the root mean squares of the displacement, Δy, and the velocity fluctuation, u'_y, by l and $u'_{y_{rms}}$, respectively The parameter l is defined as the mean distance of travel of the parcels of fluid before they are mixed into the new fluid region. We can think of l as a measure of the "scale of turbulence eddies" (commonly referred to as the *mixing length* or *Prandtl eddy length*).

By analogy to Newton's law,

$$\epsilon = \tau/(d\bar{u}_x/dy) = \rho l u'_{y_{rms}} \tag{44}$$

where ϵ is the eddy viscosity. Equation 44 can be written in the alternative form:

$$\tau = \rho\epsilon_M d\bar{u}_x/dy \tag{45}$$

where ϵ_M is the eddy kinematic viscosity (or eddy viscosity of momentum), defined by $\epsilon_M = l u'_{y_{rms}}$.

Prandtl [15] hypothesized that ϵ_M is proportional to the velocity fluctuations and to their distances (this is analogous to the product of the average molecular velocity and the mean free path in the kinetic theory of gases).

Since $u'_{y_{rms}}$ is difficult to obtain directly, Prandtl assumed u'_y to be related simply to the velocity fluctuation ($u'_y \propto l\, d\bar{u}_x/dy$) and wrote the Reynolds stress in the following form:

$$\tau = \rho l^2 (d\bar{u}_x/dy)^2 \tag{46a}$$

or, alternately,

$$\tau = (\rho l^2 d\bar{u}_x/dy)(d\bar{u}_x/dy) \tag{46a}$$

Hence, Equation 44 may be expressed as

$$\epsilon = \rho l^2 (d\bar{u}_x/dy) \tag{47}$$

With this definition, the eddy kinematic viscosity may be written as $\epsilon_M = l^2(d\bar{u}_x/dy)$.

Finally, we can account for both viscous and turbulent contributions with an equation analogous to Newton's law of viscosity:

$$\tau = -(\nu + \epsilon_M)d\bar{u}_x/dy \tag{48}$$

In the last series of equations, only time-averaged values for the velocity components are necessary. From this point on we will drop the bar (—), realizing that values are time-averaged. Equation 48 describes the shear distribution for turbulent flow in terms of the velocity gradient.

The parabolic velocity distribution characteristic of laminar flow is due to the effects of viscous forces between adjacent fluid layers. In turbulent flow, the shape of the velocity profile is very different. The majority of the velocity gradient occurs in the fluid region close to the wall where some streamline flow continues to persist; however, as we enter into the fluid core, the velocity profile is more blunt than in the laminar case and is nearly flat at the pipe centerline. The reason for this flatness is that in turbulent flow inertial forces are high and have influence over a large portion of the tube. Figure 5 compares the velocity profiles for laminar and turbulent flow in a pipe. It is of interest that although the two profiles differ greatly, both flows have the same average velocity.

Many early investigations were directed at studying the velocity profiles of turbulent flows. Nikuradse [9] extensively studied turbulent flow velocity distributions in smooth circular tubes, showing that the profile always takes on the same general shape shown in Figure 5. From these investigations, a number of attempts have been made to develop from first principles an equation for describing the velocity distributions in tubes. In the discussions to follow, we shall review the principal models developed for describing

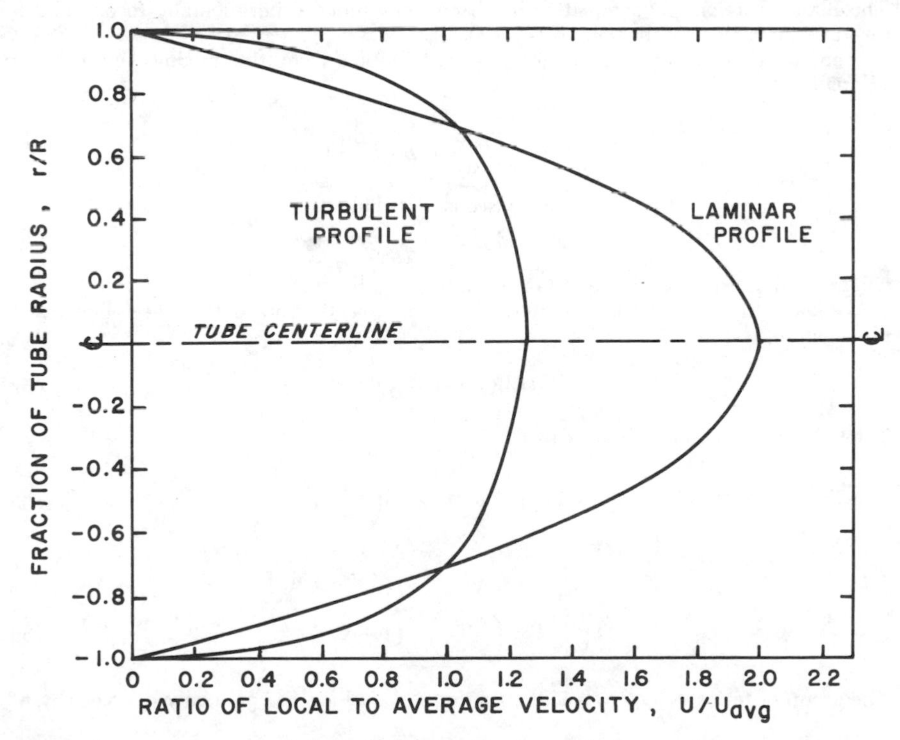

Figure 5. Comparison of the velocity profiles for turbulent and laminar flows in a tube. Note that the average fluid velocity is the same in each case.

turbulent flow in pipes. Much of the early work between the period 1920–1955 forms the basis for our current understanding of turbulence and has enabled sound engineering principles to be established.

Prandtl's Mixing Models

The most extensive study of turbulent velocity profiles was that of Prandtl [13,15]. Prandtl's, as well as Nikuradse's [9], experiments showed evidence that the velocity of the fluid at the wall is zero for smooth pipes. Prandtl further postulated that there is a layer of fluid very close to the wall that is in laminar motion, and that only viscous forces contribute to shearing in this region. That is, at the wall $u \simeq 0$ for y close to 0. Hence, from τ_w the velocity profile in the laminar layer can be computed. The earliest attempt to describe the flow in the turbulent region resulted in a power law relation:

$$\frac{u}{u_{max}} = C(y/R)^{1/7} \tag{49}$$

where C is a constant, y is the distance from the pipe wall, and R is the tube radius. Equation 49 is the 1/7th power law velocity distribution and is related to the Blasius [16] friction factor as follows:

$$\tau_w = \frac{f\rho\overline{U}^2}{2g_c} \tag{50}$$

The Blasius friction factor equation has been determined experimentally to be valid for turbulent Reynolds numbers up to 100,000.

Prandtl further derived a logarithmic distribution from the mixing length theory (Equation 48):

$$\tau_w g_c = \underbrace{\mu\frac{du}{dy}}_{\text{viscous shear}} + \underbrace{\rho\left(l\frac{du}{dy}\right)^2}_{\text{turbulent shear}} \tag{51}$$

where y is defined as being the distance normal from the wall.

We know from laminar flow analysis that $\tau = \tau_w(r/R)$ and, noting the following approximation,

$$(r/R) \simeq 1 - y/R \tag{52}$$

The following relation is obtained:

$$\tau_w g_c(1 - y/R) = \rho\left(l\frac{du}{dy}\right)^2 \tag{53}$$

or

$$l\left(\frac{du}{dy}\right) = \left(\frac{\tau_w g_c}{\rho}\right)^{1/2}(1 - y/R)^{1/2} \tag{54}$$

The group of terms $(\tau_w g_c/\rho)^{1/2}$ is the *friction velocity* and is commonly denoted by u^*:

$$u^* = \sqrt{\frac{\tau_w g_c}{\rho}} \tag{55}$$

For conditions at the wall, $1 - y/R$ reduces to 1 and

$$u^* = l\frac{du}{dy} \tag{56}$$

Prandtl hypothesized that the eddy mixing length is proportional to the distance from the wall by some constant that he hoped to be universal. That is,

$$l = ky \tag{57}$$

Equation 56 may be restated as

$$\frac{du}{dy} = u^*/ky \tag{58}$$

Or, on integration,

$$u = \frac{u^*}{k}\, l\mathrm{n}\, y + C \tag{59}$$

Using the eddy viscosity, Equation 47, which also may be written as $\epsilon = k^2y^2(du/dy)$ from the above considerations, and solving for the constant of integration, C, in Equation 59, Prandtl developed the following velocity distribution equation:

$$u = u_{max} + 2.5u^* l\mathrm{n}(y/R) \tag{60}$$

von Kárman Analogy

von Kárman [17,18] proposed the velocity distribution through similarity theory, developing the following series of equations:

$$\epsilon_M = \frac{k^2(du/dy)^3}{(d^2u/dy^2)^2} \tag{61}$$

$$l = \frac{k(du/dy)}{d^2u/dy^2} \tag{62}$$

$$\tau = \frac{\rho k^2(du/dy)^4}{g_c(d^2u/dy^2)^2} \tag{63}$$

Applying the following boundary conditions, for

$$y = 0, \quad du/dy = \infty$$

and for

$$y = R, \quad u = u_{max}$$

the velocity distribution expression was derived:

$$u = u_{max} + \frac{1}{k}u^*[l\mathrm{n}(1 - \sqrt{1 - y/R}) + \sqrt{1 - y/R}] \tag{64}$$

where the constant, k, must be evaluated empirically from the velocity profile data.

From Prandtl's theory and the data of Nikuradse [9], Reichardt [19] and others, von Kárman was able to construct a universal velocity profile through three equations. Each equation describes a different region in the flow: a laminar boundary layer, a buffer layer and the turbulent fluid core.

From the friction velocity (Equation 55), the following expression was obtained:

$$\tau_w = \frac{u^* \rho}{g_c} = \frac{\mu}{g_c} \frac{u_{\delta_1}}{\delta_1} \tag{65}$$

Equation 65 defines conditions at the edge of the boundary layer, where $y = \delta_1$ (i.e., the boundary layer thickness) and du/dy is approximated by u_{δ_1}/δ_1. u_{δ_1} is the velocity of the fluid at the edge of the boundary layer.

For conditions at the edge of the boundary layer, Prandtl's velocity distribution equation (Equation 60) may be written as

$$\frac{u_{\delta_1}}{u^*} = \frac{u_{max}}{u^*} - 2.5 ln(R/\delta_1) \tag{66}$$

The following manipulation can be made with the log term $R/\delta_1(u^*/u^*)$ and since $u_{\delta_1}/u^* =$ constant, $= C_1$, then $u^* = C_1\nu/\delta_1$.
This term then becomes

$$\frac{Ru^*}{\delta_1\left(\frac{C_1\nu}{\delta_1}\right)}$$

and Equation 66 now becomes

$$\frac{u_{max}}{u^*} = \frac{u_{\delta_1}}{u^*} + 2.5\left[ln\left(\frac{Ru^*}{\nu}\right) + ln\left(\frac{1}{C_1}\right)\right] \tag{67}$$

The form of Equation 66 enables all the constants to be collected into one term:

$$\frac{u_{max}}{u^*} = C_2 + 2.5\, ln\left(\frac{Ru^*}{\nu}\right) \tag{68}$$

where

$$C_2 = \frac{u_{\delta_1}}{u^*} + 2.5\, ln\,(1/C_1)$$

By substituting Equation 68 for u^* back into Prandtl's equation (Equation 60), the following is obtained:

$$\frac{u}{u^*} = C_2 + 2.5\, ln\left(\frac{yu^*}{\nu}\right) \tag{69}$$

Equation 69 may be written in its most general form by introducing the following dimensionless terms:

$$u^+ = u/u^*, \text{ velocity parameter} \tag{70}$$

$$y^+ = yu^*/\nu, \text{ friction distance parameter} \tag{71}$$

Where y^+ is the dimensionless distance from the wall and u^* is the dimensionless velocity. Hence, the universal velocity distribution equation for turbulent flow in circular pipes is

$$u^+ = C_2 + 2.5\, ln y^+ \tag{72}$$

Further details of the derivation are given by Cheremisinoff [20].

von Karman evaluated the constant C_2 in terms of three equations that empirically fit different portions of the dimensionless velocity profile u^+ versus y^+:

$$\text{Laminar Sublayer} \ldots \; u^+ = y^+, \text{ for } 0 < y^+ < 5 \tag{73}$$

$$\text{Buffer Region} \ldots \; u^+ = -3.05 + 5.0 \ln y^+, \text{ for } 5 < y^+ < 30 \tag{74}$$

$$\text{Turbulent Core} \ldots \; u^+ = 5.5 + 2.5 \ln y^+, \text{ for } y^+ > 30 \tag{75}$$

The universal velocity profile has been used extensively in developing useful analogies between momentum and heat transfer. In Chapter 14, the above principles are applied to analyzing flows involving heat transfer and two-phase (gas-liquid) flows.

Deissler's Model

Deissler [11,21-25] proposed a universal velocity profile model based on a description of the laminar and buffer layers in terms of the exponential decay of eddies penetrating these layers.

Using von Karman's model, $\tau = -(\mu + \epsilon\rho)(du/dy)$, in which the eddy viscosity could be defined for $y^+ > 30$ as

$$\epsilon = k \frac{(du/dy)^3}{(d^2u/dy^2)^2} \tag{76}$$

For the region $0 < y^+ < 30$, the following relationship was reveloped:

$$\epsilon = n^2uy\,(1 - e^{-\rho n^2 uy/\mu}) \tag{77}$$

where n is a constant equal to 0.10.

Equation 77 predicts that the eddy viscosity approaches zero as the wall is neared. In dimensionless form the velocity profile takes the following form:

$$y^+ = \frac{1}{n} \frac{\frac{1}{\sqrt{2\pi}} \int_0^{nu^+} e^{-[(nu^+)^2/2]} d(nu^+)}{(1/\sqrt{2\pi}) e^{-[(nu^+)^2/2]}} \tag{78}$$

This last expression applies to the laminar and buffer regions for $0 < y^+ < 26$. Equation 78 must be numerically integrated to obtain values for u^+.

For the turbulent core, Deissler obtained the following expression:

$$u^+ = 3.8 + 2.78 \ln y^+ \tag{79}$$

TURBULENT FRICTION FACTORS AND DRAG COEFFICIENTS

Flow Through Pipes and Ducts

In smooth pipe flow, the viscous sublayer near the wall is very thin at high Reynolds numbers. The introduction of roughness elements limits the thinning of the viscous sublayer. Roughness elements are protrusions or interstices on the pipe walls. Standard pipe, even when new, has many imperfections and is considered to be quite rough. Drawn tubing, when new, and plastic pipe generally are smooth-walled. As fluid flows past these roughness elements, eddies are formed behind them, causing so-called "form drag."

The height of a single unit of roughness is denoted by ϵ' and is referred to as the *roughness parameter*. From dimensional analysis, the friction factor, f, is both a function of the Reynolds and the *relative roughness*, ϵ'/D, where D is the pipe diameter. A different relationship between the friction factor, f, and N_{Re} is obtained for each magnitude of the relative roughness.

Studies by Nikuradse [26,27], Rouse [28] and Moody [29] have led to the development of generalized friction factor charts.

Moody [29] has given a complete friction factor plot for smooth and rough pipes (a log-log plot summarizing experimental data on friction factor as a function of Reynolds number and relative roughness), which is given in Figure 6. Moody also developed a chart giving the relative roughness of commercial pipe as a function of the diameter for different pipe materials. A portion of this plot is given in Figure 7.

For partially turbulent flows (transition region), the Colebrook [30] equation gives reasonable estimates of the friction factor for pipe flow:

$$\frac{1}{\sqrt{f}} = 2 \log_{10} \left(\frac{\epsilon'/D}{3.7} + \frac{2.51}{N_{Re}\sqrt{f}} \right) \tag{80}$$

Table IV lists various friction factor correlations for pipe flow and their range of applicability.

Thus far, this chapter has reviewed the simple case of flow through circular tubes. Unfortunately, few studies have examined in detail frictional losses and velocity profiles for flows in irregularly shaped conduits. Early studies [31,32] resulted in obtaining velocity profiles for Newtonian fluids flowing in various noncircular conduits; however, no generalized equation for them has been obtained. Typical examples from these studies are given in Chapter 13.

Recently, considerable interest in gas flow patterns in large-diameter ducts and irregularly shaped conduits has materialized in the area of synthetic fuels production. In oil shale retorting, for example, large vertical ducts called life pipes are being proposed for contacting hot flue gas with crushed shale. In these devices, the crushed shale

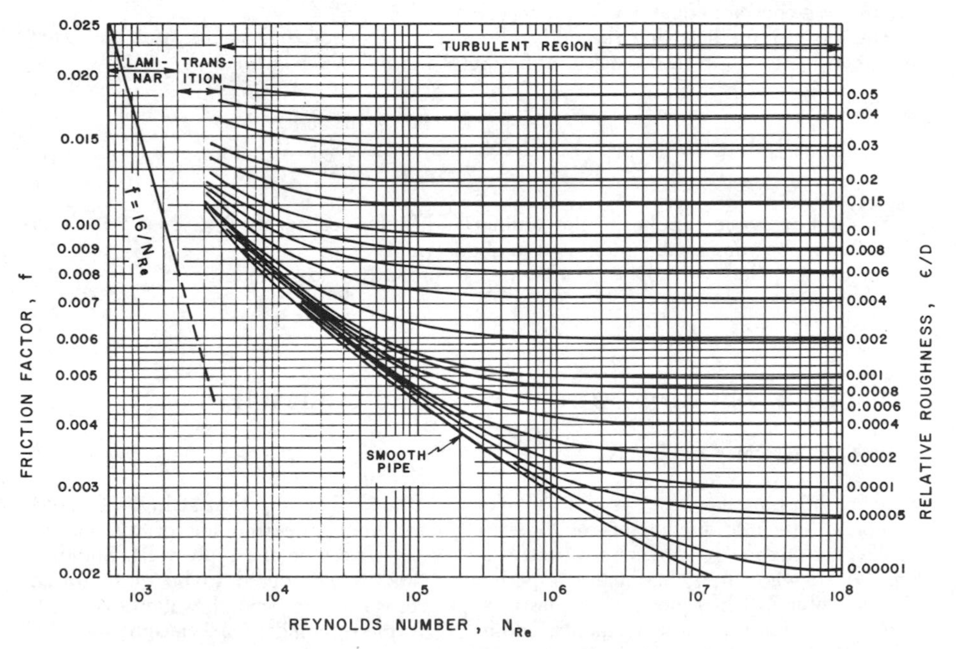

Figure 6. Moody chart for obtaining friction factors [29].

is introduced and transported vertically through the lift pipe to the next stage of the operation (usually the retort). Since efficient heat transfer depends on the degree of contact between shale particles (which have a wide particle-size distribution) and the gas, the specific velocity profiles encountered are of keen interest.

Of particular interest are the secondary flow patterns that develop and their impact on the overall heat transfer process. Figure 4 in Chapter 13 shows typical secondary flow patterns reported by Prandtl [8] for water flow in noncircular conduits. This type of motion was observed to be superimposed over the longitudinal motion of the fluid streams. The effect of secondary flow on the lines of constant velocity causes these lines to transpose in the direction of secondary flow. This results in the corners of these lines being pushed towards the corner. In the vicinity of the wall, they are pushed away from the wall.

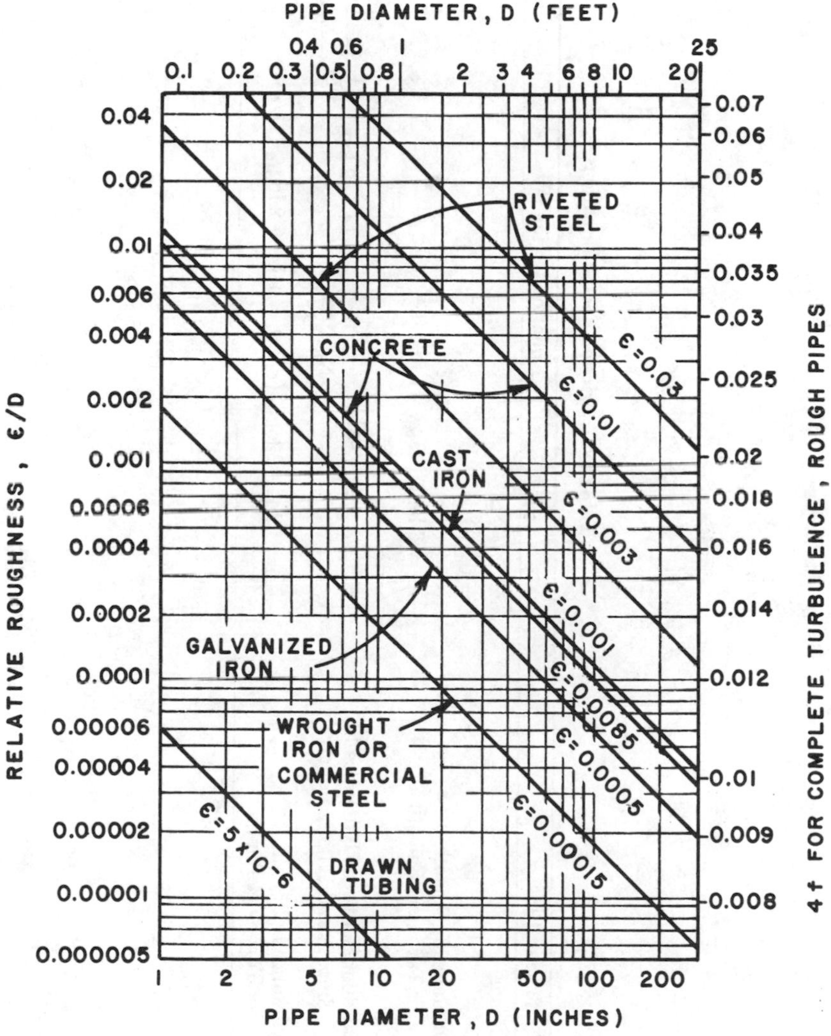

Figure 7. Moody plot of relative roughness of different pipe materials [29].

Table IV. Friction Factor Equations

Type of Flow	Equation Giving f	Range of Application
Laminar	$f = 64/N_{Re}$	$N_{Re} < 2100$
	$f = 0.316/N_{Re}^{0.25}$	$4000 < N_{Re} < 10^5$
Hydraulically Smooth or Turbulent Smooth	$\frac{1}{\sqrt{f}} = 2 \log_{10}(N_{Re}\sqrt{f}) - 0.8$	$N_{Re} > 4000$
Transition between Hydraulically Smooth and Wholly Rough	$\frac{1}{\sqrt{f}} = 2 \log_{10}\left(\frac{\epsilon/D}{3.7} + \frac{2.52}{N_{Re}\sqrt{f}}\right) = 1.14 - 2 \log_{10}\left(\frac{\epsilon}{D} + \frac{9.35}{N_{Re}\sqrt{f}}\right)$	$N_{Re} > 4000$
Hydraulically Rough or Turbulent Rough	$\frac{1}{\sqrt{f}} = 1.14 - 2 \log_{10}(\epsilon/D) = 1.14 + 2 \log_{10}(D/\epsilon)$	$N_{Re} > 4000$

Unfortunately, no extensive studies have been aimed at friction factors for flows in noncircular conduits. Knudsen et al. [1] note that the Blasius friction factor is applicable for turbulent flow for Reynolds numbers up to 100,000:

$$f_B = 0.079\, N_{Re}^{-1/4} \tag{81}$$

In lieu of more exact correlations, the Moody plot is used to estimate f from a Reynolds number based on a characteristic length term. For circular tubes this characteristic length is the diameter; however, for conduits having cross sections other than circular, an equivalent diameter is used. The equivalent diameter is defined as four times the hydraulic radius, which is the ratio of the cross-sectional area of flow to the wetted perimeter of the conduit:

$$D_E = 4R_H = 4\,\frac{\text{cross section of flow}}{\text{wetted perimeter}} \tag{82}$$

The above definition is suitable for conduits of constant cross section. There are, however, situations in which the conduit cross section varies periodically over the system. For these cases, Walker et al. [2] proposed the following definitions for the hydraulic radius:

$$(R_H)_{max} = \frac{\text{maximum cross section}}{\text{wetted perimeter}} \tag{83}$$

$$(R_H)_{min} = \frac{\text{minimum cross section}}{\text{wetted perimeter}} \tag{84}$$

$$(R_H)_v = \frac{\text{volume of free space per unit length}}{\text{area of wetted surface per unit length}} \tag{85}$$

$(R_H)_v$ is defined as the volumetric hydraulic radius, and the equivalent diameter is defined as before.

When a fluid flows through conduit of changing cross section or shape, its velocity is changed either in direction or magnitude. This results in additional friction over that formed in straight pipe flow. This additional friction includes form friction or drag, which produces vortices that develop during boundary layer separation. Figure 8 illustrates several situations in which this occurs.

Figure 8(A) shows that when the cross section of a conduit is enlarged suddenly, the fluid stream separates from the wall and flows into the enlarged section as a jet. This fluid jet expands, filling the entire cross section of the larger conduit. The region between the expanding jet and the wall is filled with fluid in vortex motion, which is a characteristic of boundary layer separation. There is considerable friction generated in this turbulent region.

Figure 8(B) shows the effect of suddenly reducing the conduit cross section. The fluid stream cannot flow around the sharp edge, causing it to break contact with the conduit wall. This situation also results in a jet that flows into the stagnant fluid in the smaller section. The jet first undergoes contraction and then expands, filling the smaller cross section. Downstream from the point of contraction, the normal velocity distribution is reestablished. The cross section of minimum flow area where the fluid jet changes from a contracted to an expanded stream is referred to as the *vena contracta.*

The last example shown in Figure 8 (sketch C) shows boundary separation in a gradually diverging channel. In this situation, because of the increase of cross section in the flow direction the velocity of the fluid decreases. In accordance with the Bernoulli equation, a decrease in fluid velocity must be accompanied by an increase in pressure. Visualize two fluid streamlines as shown in this sketch; *a* close to the wall, and a′a′ at some short distance from the wall. The rise in pressure over a specified length of

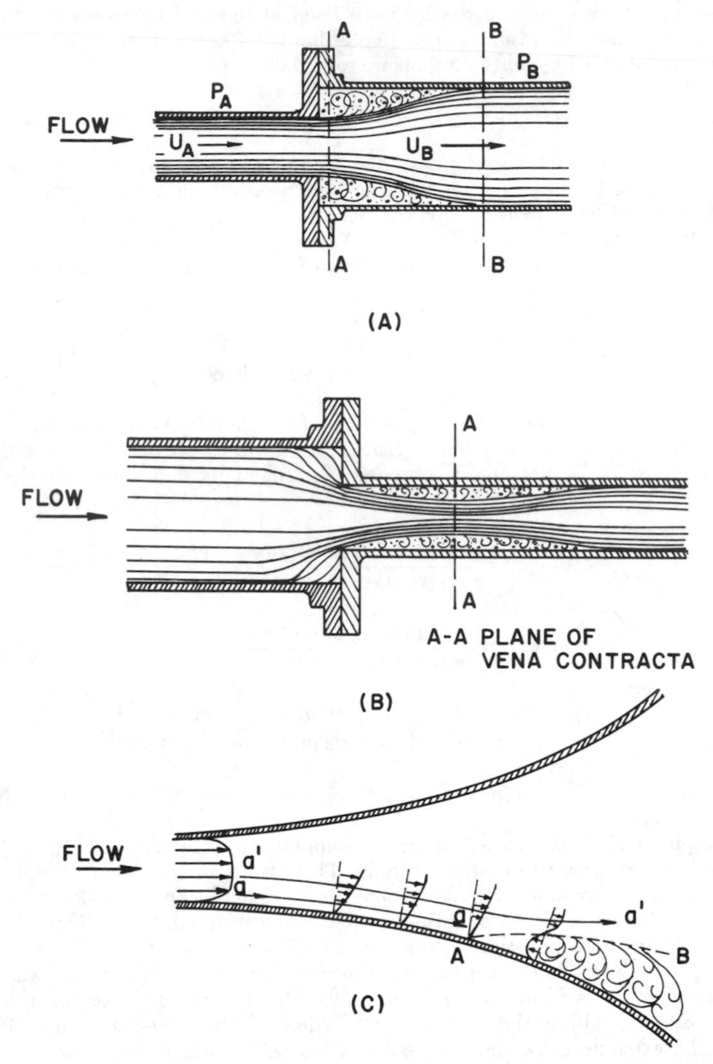

Figure 8. Flow situations resulting in boundary layer separation: (A) sudden enlargement of conduit cross section; (B) sudden contraction of cross section; (C) gradual enlargement of cross section.

channel must be the same for both streamlines (the pressure across any flow area is uniform). This means that the loss in velocity head is the same for both filaments. The initial velocity head of a′a′ is greater than that of aa, since aa is closer to the wall. At a specific point downstream, the velocity of streamline aa becomes zero; however, the velocities of other streamlines farther from the wall still remain positive. This point is denoted by A in Figure 8(C). As shown, beyond point A the velocity nearest the wall will change sign, and a backflow situation is created between aa and the wall. In this region, we then see a separation of the boundary layer. Point A is referred to as the *separation point*. The dashed line AB denotes the edge of this fluid backflow region where the fluid velocity changes sign and is referred to as the *line of zero tangential velocity*. Again, the vortices formed between the wall and the separated fluid stream beyond the separation point result in large form-frictional losses.

Frictional losses from sudden expansions and contractions as illustrated in Figure 8 are not readily estimated and require the use of empirical correlations. Losses in sudden expansions are attributed mainly to turbulence generated by the impact of the high-velocity fluid stream into a region of slowly moving fluid downstream. This means that the energy dissipated depends on the difference in initial and final values of fluid velocity. Although difficult to justify from a theoretical basis, frictional pressure losses for sudden contractions and expansions have been shown to follow an equation of the following form:

$$\Delta P_f/\rho = \frac{K\overline{U}^2}{2g_c} \tag{86}$$

Equation 86 is an approximation of the friction loss due to sudden enlargements and contractions (in English units $\Delta P_f/\rho$ is ft-lb$_f$/lb$_m$). Parameter K describes the fraction of the final kinetic energy content that is dissipated and is a function of the cross-section change. K, called the resistance coefficient, is shown plotted in Figure 9 as a function of the ratio of cross-section changes for sudden enlargements and contractions. If sections are noncircular in shape, equivalent diameters may be used in obtaining values of K. Note that the general relation given by Equation 86 is largely empirical and is known to be most reliable when conditions are decidedly turbulent. In actual use, Equation 86 is combined with the Bernoulli equation:

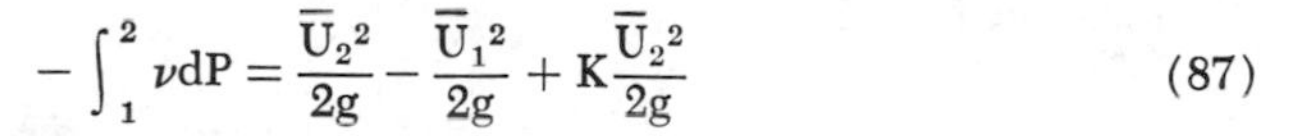

$$-\int_1^2 \nu dP = \frac{\overline{U}_2^2}{2g} - \frac{\overline{U}_1^2}{2g} + K\frac{\overline{U}_2^2}{2g} \tag{87}$$

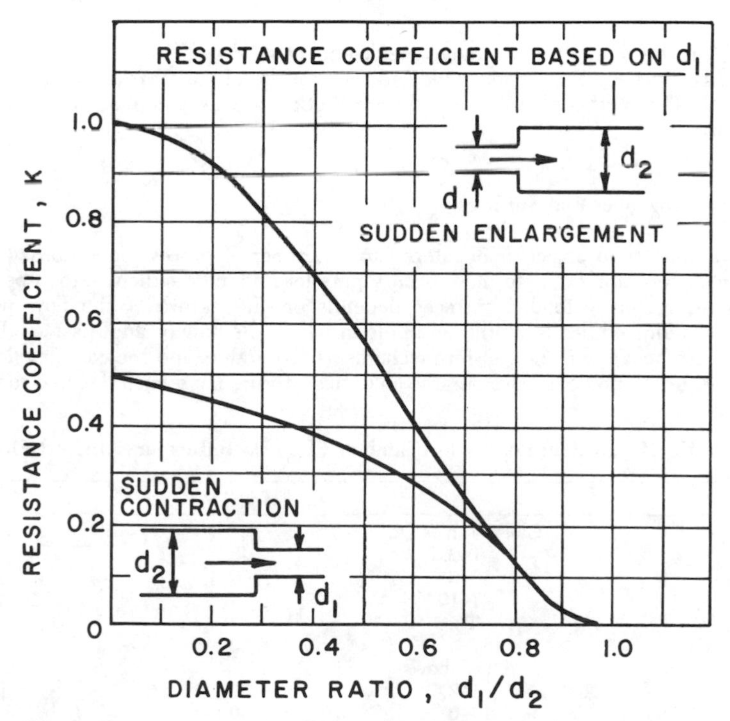

Figure 9. Resistance coefficients for cross-sectional changes [33].

Flow Through Bends and Fittings

The effects of bends and fittings such as elbows, valves, tees, etc., are evaluated empirically through a fictitious *equivalent length* of straight pipe having the same diameter and that would develop the same frictional pressure loss. This equivalent length is added to the length of actual straight pipe whence the total friction loss can be computed from the Fanning pressure loss equation. That is,

$$L = \text{length of straight pipe} + L_{eq} \tag{88}$$

where L_{eq} is the equivalent length of all fittings whence the total frictional pressure loss may be computed from

$$(\Delta P)_f = \frac{1}{144}\left(\frac{4fL}{D_E}\right)\left(\frac{\rho \bar{U}^2}{2g_c}\right) \tag{89}$$

where $(\Delta P)_f$ = frictional pressure loss (psi)
L = total pipe length (ft.)
D_E = equivalent diameter (ft.)

Typical values for the ratio of equivalent length to nominal diameter for standard fittings are given in Table V.

Equivalent lengths also have been correlated in terms of resistance coefficients with the following formula:

$$L_{eq} = \left(\frac{D_E}{4f}\right)\Sigma K \tag{90}$$

where ΣK is the sum of the resistance coefficients of all fittings.

Typical values for K for bends, ells and tees can be obtained from the plots given in Figure 10. The methods outlined in the subsection are best applied to turbulent flow conditions.

Drag from Flow over Flat Surfaces

Flow around solid objects is of interest to a number of process operations. In tubular heat exchangers, for example, flow occurs parallel and perpendicular to tube bundles. Heat transfer from extended surfaces depends on the nature of the flow past these surfaces. A fluidized bed is another example in which the flow of fluid past solid particles governs both heat and mass transfer mechanisms. To understand the complex flow mechanisms for these situations, considerable analytical studies have been devoted to modeling

Table V. Equivalent Length to Diameter (L_{eq}/D) Ratios for Standard Fittings

	L_{eq}/D
Globe Valves (in).	
1–2.5	45
3–6	60
7–10	75
Tees	
1–4	60
90° Elbows	
1–2.5	30
3–6	40
7–10	50

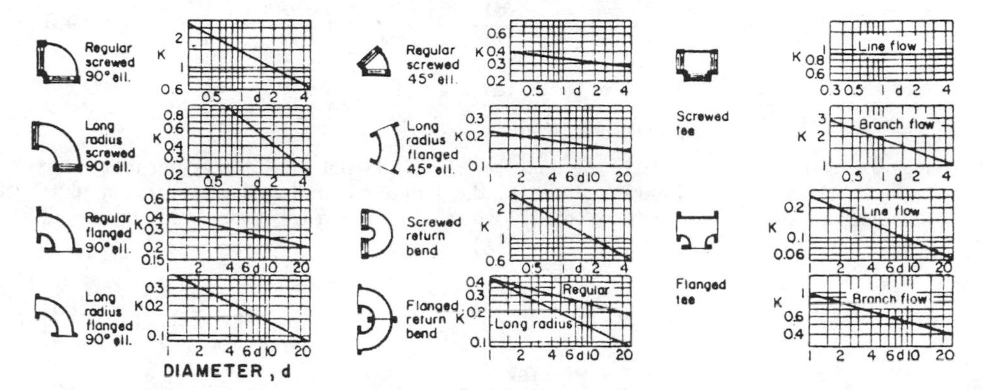

Figure 10. Resistance coefficients for bends, ells and tees [34].

and describing simple flow situations. Some of the conclusions arrived at in examining simple flow situations, such as flow over a flat plate, may be extended to more complex flow situations.

Drag is the resistance to movement of a solid in a fluid. It is caused by the shear stresses exerted in the boundary layer of the fluid adjacent to the solid surface. For a completely laminar boundary layer, shear forces are viscous only. If the boundary layer is turbulent, then the resistance results from velocity fluctuations of the fluid within the boundary layer. There are basically two types of drag. First, if the resistance is caused by stresses in the boundary layer alone, as in the case of flow over a flat surface, the resistance is referred to as surface drag. However, if the flow is around a two-dimensional object, such as a cylinder, separation of the boundary layer occurs and the fluid immediately behind the solid body becomes turbulent. This type of drag caused by the turbulent wake is called form drag and depends on the geometry of the body past which the fluid flows.

The drag coefficient is defined as follows:

$$F = \frac{f\rho\bar{U}^2A_w}{2g_c} \tag{91}$$

or

$$F = \frac{f_D\rho\bar{U}^2A_p}{2g_c} \tag{92}$$

F is the force of resistance exerted on the solid by the flowing fluid. In Equation 91, A_w is the wetted area of the immersed body. A_p in Equation 92 is the projected area of the body on the plane normal to the flow direction. Since the drag coefficient f in Equation 91 is based on the wetted area, it is mainly applicable to flat plates or thin streamlined struts. The coefficient f_D (Equation 92) is based on the projected area of the body and therefore applies to cylinders and bodies of revolution. Hence, when Equation 92 is used to compute the force, it includes both surface drag and form drag.

Consider an incompressible fluid flowing past a thin, flat plate, which is oriented parallel to the flow direction. The length of the laminar boundary layer will depend on the main stream turbulence and mean velocity, the distance from the leading edge and the fluid's kinematic viscosity.

The momentum equations of flow in two dimensions for an incompressible fluid with constant viscosity are

$$\frac{\partial u}{\partial t} + u\frac{\partial u}{\partial x} + u\frac{\partial u}{\partial y} = -g_c\frac{\partial \Omega}{\partial x} - \frac{g_c}{\rho}\frac{\partial P}{\partial x} + \nu\left(\frac{\partial^2 u}{\partial x^2} + \frac{\partial^2 u}{\partial y^2}\right) \tag{93}$$

$$\frac{\partial v}{\partial t} + u\frac{\partial v}{\partial x} + v\frac{\partial v}{\partial y} = -g_c\frac{\partial \Omega}{\partial y} - \frac{g_c}{\rho}\frac{\partial P}{\partial y} + \nu\left(\frac{\partial^2 u}{\partial x^2} + \frac{\partial^2 v}{\partial y^2}\right) \tag{94}$$

where Ω is the force potential of the fluid, i.e., the fluid's potential energy per unit mass.

An exact solution to these equations for the laminar boundary layer on a flat plate immersed in a fluid with uniform velocity was first obtained by Blasius [35] by use of the stream function, Ψ:

$$\frac{\partial \Psi}{\partial y} = u \tag{95}$$

$$\frac{\partial \Psi}{\partial x} = -v \tag{96}$$

Equations 95 and 96 satisfy the two-dimensional continuity equation $\left(\frac{\partial u}{\partial x} + \frac{\partial v}{\partial y} = 0\right)$.

Defining U as the velocity at the edge of the boundary layer and noting that $\frac{\partial U}{\partial x} = 0$, the momentum expression over the entire thickness of the boundary layer becomes

$$u\frac{\partial u}{\partial x} + v\frac{\partial u}{\partial y} = \nu\frac{\partial^2 u}{\partial y^2} \tag{97}$$

Blasius reduced Equation 97 to an ordinary differential equation by introducing the variables η and ϕ:

$$\eta = y/2\left(\frac{U\rho}{\mu x}\right)^{1/2} \tag{98}$$

$$\Psi = \left(\frac{\mu U x}{\rho}\right)^{1/2}\phi \tag{99}$$

The solution obtained by Blasius is in the form of a Taylor series solution and is expressed in terms of one of the constants of integration:

$$\phi = \frac{C_2\eta^2}{2!} - \frac{C_2^2\eta^5}{5!} + \frac{11\,C_2^3\eta^8}{8!} - \frac{375\,C_2^4\eta''}{11!} + \ldots \tag{100}$$

where $C_2 = 1.32824$. The derivation is given in detail by Cheremisinoff [20].

Howarth [36] computed values for ϕ, ϕ' and ϕ''' as functions of η. Table VI gives selected values computed by Howarth along with values of u/U. Using these values, flow conditions at any point in the x-y coordinate system may be obtained.

von Kárman [37] analyzed the flow in the boundary layer using Newton's second law to develop an integral relationship for the velocity distribution. His analysis considered a two-dimensional region of fluid, which included the boundary layer of differential length, dx. It can be shown that in this region the total rate of momentum increase is equal in both magnitude and direction to the forces acting on the boundaries of the region.

For the constant U, von Kármán's integral momentum equation for the boundary layer on a flat plate is

$$\frac{\tau_w g_c}{\rho} = \frac{\partial}{\partial x}\int_0^\delta u(U - u)\,dy \tag{101}$$

Table VI. Computed Values for Blasius Parameters [36]

η	ϕ	ϕ'	ϕ''	u/U
0	0	0	1.32824	0
0.4	0.1061	0.5294	1.3096	0.2647
0.8	0.4203	1.0336	1.1867	0.5168
1.2	0.9223	1.4580	0.9124	0.7290
1.6	1.5691	1.7522	0.5565	0.8761
2.0	2.3058	1.9110	0.2570	0.9555
2.4	3.0853	1.9756	0.0875	0.9878
2.8	3.8803	1.9950	0.0217	0.9915
3.2	4.6794	1.9992	0.0039	0.9996
3.6	5.4793	2.0000	0.0005	1.0000
3.8	5.8792	2.0000	0.0002	1.0000

The boundary layer thickness, δ, is the distance normal from the wall where the point velocity is within 1% of the mainstream velocity. That is, δ is the y value where $u/U = 0.99$. Using the values given in Table VI, it can be shown that

$$\frac{\delta}{x} = \frac{4.96}{\sqrt{N_{Re_x}}} \tag{102}$$

where N_{Re_x} is the local Reynolds number.

Pohlhausen [38] modified von Kármán's integral equation (Equation 101) to obtain the following expressions for the laminar boundary layer thickness and velocity profile:

$$\delta = 4.64\ (\nu x/U)^{1/2} \tag{103}$$

$$\frac{u}{U} = \frac{1.5}{4.64}\frac{y}{\sqrt{\nu x/U}} - \frac{y^3}{199.8\ (\sqrt{\nu x/U})^3} \tag{104}$$

For laminar flow, the Blasius solution predicts the following relation for the total drag coefficient for a plate of length, L:

$$f = \frac{1.328}{\sqrt{N_{Re_L}}} \text{ for } 2 \times 10^4 < N_{Re_L} < 5 \times 10^5 \tag{105}$$

where N_{Re_L} is the total Reynolds number for flow over the flat plate ($=UL/\nu$).

At low Reynolds numbers ($10 < N_{Re_L} < 3000$), the following empirical equation may be used [39]:

$$f = 2.90\ (N_{Re_L})^{-0.60} \tag{106}$$

Using the analogy for flow through circular tubes, von Kármán [40] derived the form of the velocity-profile equation describing the turbulent portion of the boundary layer. For any point, x, on the plate, the relation between point velocity, u, and distance normal from the plate, y, is

$$\frac{u}{U} = (y/\delta)^{1/7} \tag{107}$$

Equation 107 is the power law for the velocity distribution over flat plates.

The turbulent boundary layer thickness may be computed from the Blasius friction factor equation for circular tubes. Following von Karman [40], the friction factor equation is used in expressing the shear stress at the plate's boundary:

$$\tau_w = 0.0228 \frac{\rho U^2}{g_c} \left(\frac{\nu}{U\delta}\right)^{1/4} \tag{108}$$

Combining this expression with von Kármán's integral equation (Equation 101),

$$\frac{\partial}{\partial x} \int_0^\delta u(U - u)dy = 0.0228U^2 \left(\frac{\nu}{U\delta}\right)^{1/4} \tag{109}$$

Using the 1/7th power law relation (Equation 107) for obtaining a value of u and substituting into Equation 109, integration can be performed:

$$\frac{\delta}{x} = 0.376 \, (N_{Re_x})^{-1/5} \text{ for } 5 \times 10^5 < N_{Re_x} < 10^7 \tag{110}$$

For local Reynolds numbers above 6.5×10^5, the following relation may be used [41]:

$$\frac{\delta}{x} = 0.1285 \, (N_{Re_x})^{-1/7} \tag{111}$$

By substituting Equation 110 in Equation 108, it can also be shown that

$$\frac{2\tau_w g_c}{\rho U^2} = 0.0585 \, (N_{Re_x})^{-1/5} \tag{112}$$

$$f' = 0.0585 \, (N_{Re_x})^{-1/5} \tag{113}$$

The last expression gives the local drag coefficient for the turbulent boundary layer. The total friction factor is obtained as outlined earlier:

$$f = 0.074 \, (N_{Re_L})^{-1/5} \text{ for } 5 \times 10^5 < N_{Re_L} < 10^7 \tag{114}$$

Several other expressions for local and total drag coefficients have been derived for turbulent boundary layers on flat plates. These have been summarized by Cheremisinoff [20].

All drag coefficients given thus far in this chapter describe flow over a smooth plate. Surface roughness for flat plates is described in terms of the *admissible roughness.* This is the roughness above which an increase in the drag coefficient occurs over that for a smooth plate. The following expression can be used to determine admissible roughness:

$$\frac{U\epsilon_{ad}}{\nu} = 10^2 \tag{115}$$

where ϵ_{ad} is the admissible height of roughness projections.

For flat plates having a roughness below the admissible roughness, any of the smooth plate expressions are acceptable for estimating the drag coefficient. The following equation can be used for computing f for $U\epsilon_{ad}/\nu > 2000$:

$$f = (1.89 + 1.62 \log_{10} L/\epsilon_{ad})^{-2.5} \tag{116a}$$

where $10^2 < L/\epsilon_{ad} < 10^6$.

The thickness of the turbulent boundary layer on the surface of a rough plate can be computed from the following expression [43].

$$\delta = 0.259 \left(\frac{x}{\epsilon_{ad}}\right)^{2/3} \epsilon_{ad} \, (1 - 0.00059x) \tag{116b}$$

where quantities x, ϵ_{ad} and δ are in cm.

Flow Past Two- and Three-Dimensional Objects

Now consider fluid flowing past a two-dimensional object such as a cylinder. The fluid is accelerated as it passes over the front portion of the object and then decelerated after it passes the thickest portion of the object. This sequence of events causes a separation of the boundary layer.

The boundary layer undergoes separation at a point on the object where the pressure gradient is zero. The reason for this is as follows: The boundary layer thickness increases with distance in the flow direction. The portion of fluid in the mainstream accelerates because of its movement around the object. This acceleration is actually an increase in the fluid's kinetic energy, which must be accompanied by a decrease in the pressure, i.e., $\partial P/\partial x$ becomes negative. As the mainstream fluid travels past the object, the expanding cross section of flow requires a deceleration of the fluid and, hence, a corresponding increase in the pressure ($\partial P/\partial x$ becomes positive). Hence, the boundary layer actually flows against an adverse pressure gradient as it moves around the object, resulting in a distortion in the velocity profile in the boundary layer. To maintain flow in the direction of the adverse pressure gradient, the boundary layer must separate from the solid surface, and this point of separation is actually a stagnation point (that is, the velocity gradient $\partial u/\partial y$ at the surface becomes zero). Hence, at any point on the surface of the object prior to separation u = 0 and v = 0. Prandtl's [44] momentum equations for the boundary layer are

$$u \frac{\partial u}{\partial x} + v \frac{\partial u}{\partial y} = - \frac{g_c}{\rho} \frac{\partial P}{\partial x} + \nu \frac{\partial^2 u}{\partial y^2} \tag{117}$$

$$\frac{\partial u}{\partial x} + \frac{\partial v}{\partial y} = 0 \tag{118}$$

whence it follows that

$$\frac{\partial P}{\partial x} = \frac{\mu}{g_c} \frac{\partial^2 u}{\partial y^2} \text{ at } y = 0 \tag{119}$$

The significance of Equation 119 is that it predicts a change in sign of $\partial P/\partial x$, the term $\partial^2 u/\partial y^2$ changes sign; thus, the velocity-profile curve exhibits a point of inflection.

Defining δ as the boundary layer thickness for fluid flowing over a circular cylinder, and U and P as the velocity and pressure for the undisturbed portion of the flowing stream, respectively, we can develop relationships for the velocity and pressure distributions. The outer edge of the boundary layer represents the point at which the velocity gradient becomes zero. The velocity at the edge of the boundary layer, u_{max} differs from the mainstream velocity and is a function of positions on the surface of the cylinder in relation to the leading edge:

$$u_{max} = 2 \, U \sin \theta \tag{120}$$

where θ represents the angle of departure of the boundary layer from the cylinder.

Bernoulli's steady-flow equation may be applied between the mainstream and the edge of the boundary layer (the implication being that the main fluid stream is irrotational); thus,

$$P_\theta + \frac{\rho u^2_{max}}{2g_c} = P + \frac{\rho U^2}{2g_c} \qquad (121)$$

where P_θ is the static pressure in the boundary layer and is constant across thickness δ.

Rearranging to solve for u_{max},

$$u_{max} = U\sqrt{\frac{P - P_\theta}{\rho U^2/2g_c} + 1} \qquad (122)$$

Equation 122 can be used to compute the velocity at the edge of the boundary layer from the knowledge of the pressure distribution over the cylinder. If the effect of fluid viscosity is negligible, the pressure distribution over the surface of the cylinder can be estimated from

$$\frac{P_\theta - P}{\rho U^2/2g_c} = 1 - 4\sin^2\theta \qquad (123)$$

When a two-dimensional cylinder is immersed in a flowing stream, it has a considerably greater drag on it than the case of a flat plate. This is due to the separation of the boundary layers on two-dimensional objects and the subsequent formation of a turbulent wake behind the body.

Several investigators [45-48] have measured total drag coefficients past cylinders. Equation 124 is just one expression for the drag coefficient for flow past circular cylinders when no boundary layer separation occurs [45].

$$f_D = \frac{8\pi}{N_{Re_o}(2.002 - \ln N_{Re_o})} \qquad (124)$$

Equation 124 applies only for low Reynolds numbers ($N_{Re_o} < 0.5$). Note that for fluids flowing past immersed cylinders, Reynolds numbers are expressed in terms of the cylinders outside diameter, D_o. One of the most extensive studies on the drag on cylinders, including noncircular cylinders, is that of Delany and Sorenson [48]. Their investigation covered cylindrical, elliptical, square, rectangular and triangular cylinders.

Bodies of revolution of interest are spheres, ellipsoids and disks. We define the radius in the plane normal to the flow direction as r_o and the Reynolds number for the flow as D_oU/ν, where $D_o = 2r_o$. The projected area of the body is πr_o^2, and the overall length of the body in the direction of flow is L.

When flow occurs past any of these objects, the fluid motion near the object's surface is three-dimensional with respect to a rectangular coordinate system. The general boundary layer equations for flow over bodies of revolution are

$$u\frac{\partial u}{\partial x_1} + v\frac{\partial u}{\partial y} = -\frac{g_c}{\rho}\frac{\partial P}{\partial x_1} + \nu\left(\frac{\partial^2 u}{\partial y^2} + \frac{1}{r}\frac{\partial r}{\partial y}\frac{\partial u}{\partial y}\right) \qquad (125)$$

$$\frac{\partial (ru)}{\partial x_1} + \frac{\partial (rv)}{\partial y} = 0 \qquad (126)$$

Equations 125 and 126 are the momentum and continuity equations, respectively. The terms in these equations are as follows:

x_1 = distance from the leading edge of the body
y = distance normal to the body surface
r = radius of transverse cross section at point (x_1, y)

The definitions of distance terms used in Equations 125 and 126 for flow past spheres are given in Figure 11. The radius of curvature at any point is r_c, and at the sphere's

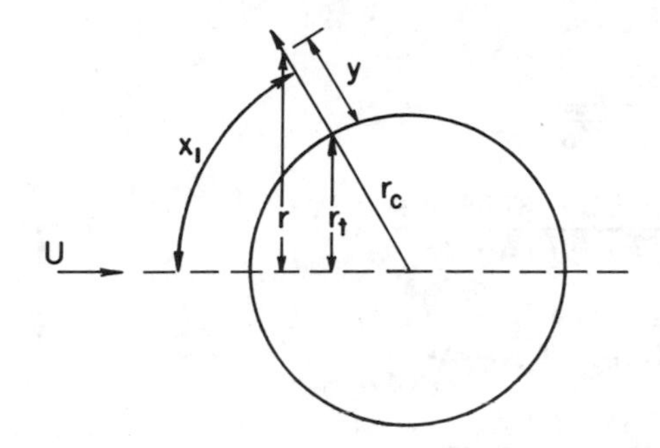

Figure 11. Defines distance terms for three-dimensional equations describing flow over a sphere.

surface (i.e., $y = o$) the transverse cross section of the body has a radius r_t. These definitions assume $y \ll r_c$ and that δ is small in comparison to r. These assumptions tend to break down in the region of the leading edge as r goes to 0.

Solutions to Equations 125 and 126 provide expressions for the boundary layer velocity profile, the boundary thickness and local drag coefficients. The following three-dimensional integral momentum equations were developed from Equations 125 and 126 [49]:

$$\frac{\partial}{\partial x_1}\int_o^\delta u^2 dy - u_{max}\frac{\partial}{\partial x_1}\int_o^\delta u dy + \frac{1}{r_t}\frac{\partial r_t}{\partial x_1}\left(\int_o^\delta u^2 dy - u_{max}\int_o^\delta u dy\right) = \delta u_{max}\frac{\partial u_{max}}{\partial x_1} - \frac{\tau_w g_c}{\rho} \tag{127}$$

To solve this expression, information is needed on the pressure distribution over the immersed sphere. The pressure distribution is given by the following expression:

$$\frac{P_\theta - P}{\rho U^2/2g_c} = 1 - \frac{9}{4}\sin^2\theta \tag{128}$$

(Note the resemblance to Equation 123.)

where θ = angle measured from the forward stagnation point of the sphere
P_θ = static pressure at angle θ
P = static pressure of the undisturbed stream

It can be shown from Bernoulli's equation that

$$\frac{u_{max}}{U} = 1.5\sin\theta \tag{129}$$

A number of empirical correlations are available for estimating u_{max} (the velocity at the edge of the boundary layer).

A simple approach used in solving Equation 127, and one that also has been used in solving the two-dimensional equation, is to assume a velocity profile. A simplified approach is to use a velocity profile empirically evaluated to a polynomial fit:

$$u = c_1 y + c_2 y^2 + c_3 y^3 + c_4 y^4 \tag{130}$$

Using this approach, Tomotika [50] obtained the following expression for the laminar boundary layer thickness:

$$\frac{U}{\nu r_o}\frac{\partial(\delta^2)}{\partial\theta} = \phi_1(\Delta)\frac{U}{u_{max}} - \phi_2(\Delta)\cot\theta\frac{U}{\partial u_{max}/\partial\theta} + \phi_3(\Delta)\frac{\delta^4}{U\nu^2}\frac{\partial^2 u_{max}}{\partial\theta^2} \tag{131}$$

where

$$\Delta = \frac{\delta^2}{r_o \nu} \frac{\partial u_{max}}{\partial \theta} \quad (132)$$

$$\phi_1(\Delta) = \frac{7{,}257.6 - 1{,}336.2\,\Delta + 37.92\,\Delta^2 + 0.8\,\Delta^3}{213.12 - 5.76\,\Delta - \Delta^2} \quad (133)$$

$$\phi_2(\Delta) = \frac{426.24\,\Delta - 3.84\,\Delta^2 - 0.4\,\Delta^3}{213.12 - 5.76\,\Delta - \Delta^2} \quad (134)$$

$$\phi_3(\Delta) = \frac{3.84 + 0.8\,\Delta}{213.12 - 5.76\,\Delta - \Delta^2} \quad (135)$$

A parabolic velocity profile also may be assumed:

$$\delta^2 = \frac{30\nu}{u_{max}^{9} r_t^2} \int_o^{x_1} u_{max}^{8} r_t^2 \, dx_1 \quad (136)$$

To develop an expression for the drag coefficient of a sphere moving slowly through a viscous fluid, Stokes' law can be used to define the drag force:

$$F = \frac{3\pi D_o \mu U}{g_c} \quad (137)$$

Equation 138 is based on the condition that the boundary layer does not separate from the sphere.

For the flow of a viscous fluid past a sphere, the following relation is applicable in computing the drag force:

$$F = \frac{3\pi D_o \mu U}{g_c} \left(1 + \frac{3}{16} N_{Re_o} \right) \quad (138)$$

For $N_{Re_o} < 1.0$, the drag coefficient from Stokes' law is

$$f = \frac{24}{N_{Re_o}} \quad (139)$$

Combining the above relations, the total drag coefficient describing the motion of spheres (neglecting neighboring wall effects) is

$$f_D = \frac{24}{N_{Re_o}} \left(1 + \frac{3}{16} N_{Re_o} \right) \quad (140)$$

Equation 140 is applicable in the range $N_{Re} < 2$. For drag coefficients for flow past a sphere contained in a cylinder of diameter D_w, use the following relation:

$$f_D = \frac{24}{N_{Re_o}} (1 + 2.4\,(D_o/D_w)) \quad (141)$$

Equation 141 can be used for rough estimates of the total drag coefficient up to Reynolds numbers of about 100. To estimate drag coefficients at Reynolds numbers outside the range of Equations 140 and 141, the data of Wieselsberger [46] for spheres,

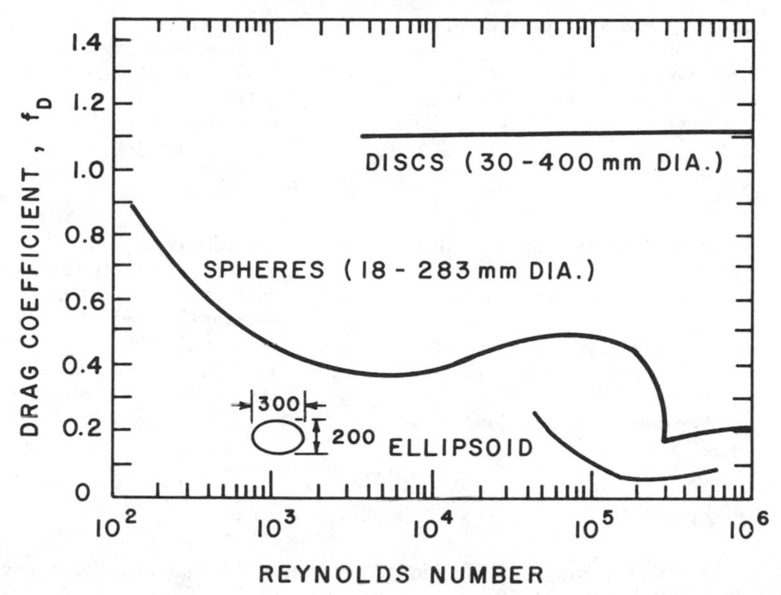

Figure 12. Plot of drag coefficient for flow past bodies of revolution at high Reynolds numbers (curves based on data of Wieselberger [46]).

disks and ellipsoids may be used. A summary of these data is given in Figure 12. It should be noted that there is a distinct effect of aspect ratio on the drag coefficient, where the aspect ratio is defined as the diameter:length ratio of the body. For ellipsoids, for example, the drag coefficient tends to decrease with values of the aspect ratio. As $2r_o/L$ decreases, the object becomes more streamlined and, hence, the resistance to the flow becomes less.

NOMENCLATURE

A	cross-sectional area (m^2)
A_w, A_p	wall surface area and projected area (m^2)
c	constant in Equation 59
C_2	constant in Equation 72
D	diameter (m)
D_E	equivalent diameter, see Equation 82 (m)
E	internal energy (J/kg)
F	frictional forces (N)
F_p	friction energy (J)
f, f_D	friction factor and drag coefficient
g	gravitational acceleration term (m/s^2)
H	enthalpy (J/kg-mol)
K	friction loss coefficient, see Equation 86
KE	kinetic energy (J/kg)
k	universal constant, see Equation 64
L	length (m)
L_{eq}	equivalent length of piping components (m)
l	Prandtl mixing length (m)
N_{Re}	Reynolds number
n	constant equal to 0.10
P	pressure (Pa)
$\Delta P'$	pressure drop including static and head ($= P + fgz$) (Pa)
Q	heat (J/kg)
q	volumetric flowrate (m^3/s)
R	tube radius (m)
R_H	hydraulic radius (m)
r	radius (m)
t	time (s)
$\overline{U}$, U	average velocities (m/s)
u	velocity in x-direction (m/s)
u_x, u_y	instantaneous velocity components (m/s)
u_{rms}	root mean square velocity (m/s)

u^+ dimensionless velocity, see Equation 70
u^* friction velocity (m/s)
v, w velocities in y- and z-directions (m/s)
W′ work (J)
w mass flowrate (kg/s)
y^+ dimensionless distance, see Equation 71
Z elevation (m)

Greek Symbols

Δ parameter defined in Equation 132
δ boundary layer thickness (mm or cm)
ϵ' roughness height (mm or in.)
ϵ_M eddy viscosity of momentum (mm^2/s)
η variable defined in Equation 98
θ angle (°)
μ viscosity (cp)
ν kinematic viscosity (mm^2/s)
$\dot{\nu}$ specific volume (kg/m^3)
ρ density (kg/m^3)
τ, τ_{rx} shear stress (Pa)
τ_w wall shear stress (Pa)
Ψ stream function, refer to Equations 95 and 96

REFERENCES

1. Reynolds, W. C., and H. C. Perkins. *Engineering Thermodynamics* (New York: McGraw-Hill Book Co., 1970.
2. Lee, J. F., F. W. Sears and D. L. Turcotte. *Statistical Thermodynamics* (Reading, MA: Addison-Wesley Publishing Co., Inc., 1963).
3. Weber, H. C. *Thermodynamics for Chemical Engineers* (New York: John Wiley & Sons, Inc., 1939).
4. McAdams, W. H. *Heat Transmission* (New York: McGraw-Hill Book Co., 1942).
5. Badger, W. L., and W. L. McCabe. *Elements of Chemical Engineering* (New York: McGraw-Hill Book Co., 1936).
6. Knudsen, J. G., and D. L. Katz. *Fluid Dynamics and Heat Transfer* (New York: McGraw-Hill Book Co., 1958).
7. Walker, W. H., W. K. Lewis, W. H. McAdams and E. R. Gilliland. *Principles of Chemical Engineering*, 3rd ed. (New York: McGraw-Hill Book Co., 1937).
8. Stanton, T. E., D. Marshall and C. N. Bryant. *Proc. Roy. Soc. (London)* 97A:413 (1920).
9. Nikuradse, J. *VDI-Forschungsheft* 356 (1932).
10. Reichardt, H. *NACA TN* 1047 (1943).
11. Deissler, R. G. *NACA TN* 2138 (1950).
12. Moody, L. F. *Trans. ASME* 66:671 (1944).
13. Prandtl, L. Z. *Ver. Deut. Ing.* 77:105(1933); *NACA TN* 720 (1933).
14. von Karman, T. *NACA TN* 611 (1931).
15. Prandtl, L. Z. *Angew. Math. Mech.* 5:136 (1925); *NACA TN* 1231 (1949).
16. Blasius, H. *Mitt. Forschungsarb.* 131:1-40(1913).
17. von Karman, T. *NACA TN* 611 (1931).
18. von Karman, T. *J. Aeronaut. Sci.* 1:1 (1934).
19. Reichardt, H. *NACA TN* 1047 (1943).
20. Cheremisinoff, N. P. *Fluid Flow: Pumps, Pipes and Channels* (Ann Arbor, MI: Ann Arbor Science Publishers, Inc., 1981).
21. Deissler, R. G. *NACA TN* 2242 (1950).
22. Deissler, R. G. *Trans. ASME* 73:101 (1951).
23. Deissler, R. G., and C. S. Eian. *NACA TN* 2629 (1952).
24. Deissler, R. G. *NACA TN* 3145 (1954).
25. Deissler, R. G., and M. F. Taylor. *NACA TN* 3451 (1955).
26. Nikuradse, J. *Proc. Int. Cong. Appl. Mech.* (3rd Congress, Stockholm) 1.239 (1933).

27. Nikuradse, J. *VDI Forschungsheft* 361 (1933).
28. Rouse, H. *Elementary Mechanics of Fluids* (New York: John Wiley & Sons, Inc., 1948).
29. Moody, L. F. *Trans. ASME* 66:671 (1944).
30. Colebrook, C. F., and D. White. *J. Inst. Civil Eng., London* 11:133 (1938-1939).
31. Prandtl, L. *Proc. Int. Cong. Appl. Mech.* (2nd Congr., Zurich) 62 (1927).
32. Nikuradse, J. *Ing-Arch.* 1:306 (1930).
33. Simpson, L. L. "Process Piping: Functional Design," *Chem Eng.* 76(8): 167-181 (1969).
34. *Chem. Eng.* 75(13):198-199 (1968).
35. Blasius, H. Z. *Math. Phys.* 56:1 (1908).
36. Howarth, L. *Proc. Roy. Soc.* (*London*) 164A:547 (1938).
37. von Karman, T. Z. *Angew, Math. Mech.* 1:233 (1921).
38. Pohlhausen, K. Z. *Angew, Math. Mech.* 1:252 (1921).
39. Janour, Z. *NACA TN* 1316 (1951).
40. von Karman, T. *NACA TN* 1092 (1946).
41. Falkner, V. M. *Aircraft Eng.* 15:65 (1943).
42. Schlichting, H. *NACA TN* 1218 (1949).
43. von der Hegge-Zijnen, B. G. *Verhandel. Koninkl. Akad. Wetenschap. Amsterdam Afdeel. Natuurk.* 31:499 (1928).
44. Prandtl, L. *NACA TN* 542 (1928).
45. Lamb, H. *Phil. Mag.* 21:112 (1911).
46. Weiselsberger, C. *Ergeb. Aerodyn. Versuchsanstalt Göttingen* 2:22 (1923).
47. Wieselsberger, C. *Ergeb. Aerodyn. Versuchsanstalt Göttingen* 1:120 (1923).
48. Delany, K., and N. E. Sorenson. *NACA TN* 3038 (1953).
49. Millikan, C. B. *Trans. ASME* 54:APM-3 (1932).
50. Tomotika, S. *Brit. Aeronaut. Res. Comm., R&M* 1678 (1935).

CHAPTER 3
PRINCIPLES OF DIMENSIONAL ANALYSIS

R. Gupta
Exxon Research & Engineering Company
Florham Park, New Jersey 07932

CONTENTS

INTRODUCTION

For many engineering problems, theoretical equations describing a phenomenon are either not known or are in a form that cannot be solved conveniently or accurately. Often, because theoretical solutions are not possible, engineering practice must resort to empirical correlations for design and analysis. For problems involving fluid flow, these correlations often are developed on the basis of model studies in which flow conditions, fluid properties and size of the test system may be different. This chapter provides an introduction to the theory of dimensional analysis. Covered are the use of dimensional analysis in the development of empirical correlations and its application to the design and analysis of engineering systems. The applications of dimensional analysis extend to several branches of science and engineering and are not limited to fluid dynamics. An important bibliography on the subject is provided in the literature [1-10]. This chapter is based in part on the work of Langhaar [1].

PRINCIPLES OF DIMENSIONAL ANALYSIS

Dimensional analysis is based on the premise that the equation describing a phenomenon must be dimensionally consistent, even when the equation is unknown. That is, different variables must occur in the equation in a manner such that each term of the equation has the same dimensions. By dividing each term of the equation by any one term, the equation describing the phenomenon can be transformed into the following generalized form:

$$f(\pi_1, \pi_2, \pi_3, \cdots \pi_p) = 0 \tag{1}$$

where $\pi_1, \pi_2 \cdots \pi_p$ are dimensionless groups of variables.

The advantages of dimensional analysis are illustrated in the following example. Chapter 2 examines the drag force on a smooth spherical body submerged in a stream of incompressible fluid. Let us determine the form of drag force equation here. From experimental data or experience, it is known that the drag force, F, depends on the sphere diameter, D, fluid viscosity, μ, fluid density, ρ, and fluid velocity, V. That is,

$$F = f(D, \mu, \rho, V) \tag{2}$$

If Equation 2 is dimensionally consistent, then by using the procedure to be described later in the chapter, it must be reducible to the following form:

$$\frac{F}{\rho V^2 D^2} = f\left(\frac{\rho VD}{\mu}\right) \tag{3}$$

or

$$\pi_1 = f(\pi_2)$$

That the groups $\pi_1 = \left(\frac{F}{\rho V^2 D^2}\right)$ and $\pi_2 = \left(\frac{\rho VD}{\mu}\right)$ are dimensionless is readily checked by observing that F has the dimensions of MLT^{-2}; similarly, $\rho \equiv ML^{-3}$, $V \equiv LT^{-1}$, $\mu \equiv ML^{-1}T^{-1}$ and $D \equiv L$, where M, L and T denote mass, length and time, respectively.

Dimensional analysis has thus greatly simplified the drag force problem. Instead of determining the drag force from a relation between five different variables in Equation 2, it can be determined from a relation between only two dimensionless groups in Equation 3. For example, a plot of π_1 vs π_2 may be constructed from experimental data. From such a plot, F is readily calculated for any sphere if D, μ, ρ and V are known.

Dimensional analysis provides some insight into how different variables in a problem may be related to each other. By lumping the variables into groups, the number of parameters in the analysis of a problem is reduced. However, dimensional analysis neither promises a complete solution to a problem, nor provides any insight into the inner mechanism of a phenomenon. The dimensionless groups, which are always fewer in number than the original variables, still must be related to each other on the basis of additional theoretical knowledge or experimental data.

Application of dimensional analysis is based on the premise that the starting equation is dimensionally consistent. It may not always be assumed that any arbitrary equation is dimensionally consistent unless the equation is fundamental; that is, the equation contains all the variables required in its analytical derivation. For example, in the problem of drag force on a sphere, if the fluid is water, it can be argued that ρ and μ may be excluded, since they are constant for water. Then

$$F = f(D, V) \tag{4}$$

Dimensional analysis cannot be applied to Equation 4 because it is neither fundamental nor dimensionally consistent. The variable F contains the dimensions of mass, whereas neither of the variables D and V does. Dimensional analysis alone cannot tell which variables influence the phenomenon and should be included in the formulation of the problem. It also fails to exclude variables that are irrelevant to the phenomenon. For example, in the case of drag force on a sphere, experimental or theoretical knowledge is necessary to state that F, D, V, μ and ρ form a complete set of variables. Extraneous dimensionless groups would appear in the solution if the fluid surface tension, which is known not to influence the drag force, was included in the analysis by mistake.

SYSTEMATIC DETERMINATION OF DIMENSIONLESS GROUPS

Once all variables affecting a phenomenon are known, the pertinent dimensionless groups must be decided on. Innumerable dimensionless groups can be formed by combining the variables in many different ways. However, only a few dimensionless groups, which must form a complete set, are necessary to describe the phenomenon. Two requirements for a set of dimensionless groups to be complete are as follows:

1. The groups within the set should be independent of each other such that no group within the set is a product of powers of the other groups within the set.
2. All possible dimensionless groups outside the complete set can be obtained as product of powers of the groups within the complete set.

Thus, many complete, but different, sets of dimensionless groups may be constructed from the large number of possible dimensionless groups. The phenomenon may be described by any one of these complete sets. The selection of specific dimensionless groups to form a complete set depends on the ease with which the selected groups can be regulated or varied in the experimental investigation of the phenomenon. Very often the chosen groups have a physical significance.

A rigorous method of computing the number of dimensionless groups in a complete set and determining a complete set of dimensionless groups is illustrated below by way of an example. The method was originally described by Langhaar [1]. A basic knowledge of determinants and matrices is needed. Proof of the theorems involved in obtaining a solution may be found in books on advanced algebra.

Illustration: As noted in Chapter 2, the pressure drop of an incompressible fluid flowing in a straight section of pipe has the following functional dependence:

$$\Delta P = f(l, \mu, \epsilon, \rho, D, V)$$

where
$l =$ pipe length (L)
$\mu =$ fluid viscosity ($ML^{-1}T^{-1}$)
$\epsilon =$ pipe roughness (L)
$\rho =$ fluid density (ML^{-3})
$D =$ pipe diameter (L)
$V =$ fluid velocity (LT^{-1})

We wish to determine the number of dimensionless groups required to describe the phenomenon

$$f(\Delta P, l, \mu, \epsilon, \rho, D, V) = 0$$

and determine a complete set of dimensionless groups.

Any dimensionless group formed from the variables has the form

$$\pi = (\Delta P)^{k1} (l)^{k2} (\mu)^{k3} (\epsilon)^{k4} (\rho)^{k5} (D)^{k6} (V)^{k7}$$

The dimensions of π are, therefore,

$$(ML^{-1}T^{-2})^{k1} (L)^{k2} (ML^{-1}T^{-1})^{k3} (L)^{k4} (ML^{-3})^{k5} (L)^{k6} (LT^{-1})^{k7}$$

The only restriction on the exponents k1 - - - - k7 is that their numerical values be such that π is dimensionless.
Thus,

$$\left.\begin{aligned} k1 \quad\quad + k3 \quad\quad + k5 \quad\quad\quad\quad &= 0 \\ -k1 + k2 - k3 + k4 - 3k5 + k6 + k7 &= 0 \\ -2k1 \quad\quad - k3 \quad\quad\quad\quad\quad\quad - k7 &= 0 \end{aligned}\right\} \quad (5)$$

The coefficients of the various k's in Equation 5 are the same as the elements of the following matrix created by writing the dimensions of the seven variables in different columns.

	k1	k2	k3	k4	k5	k6	k7
	ΔP	l	μ	ϵ	ρ	D	V
M	1	0	1	0	1	0	0
L	−1	1	−1	1	−3	1	1
T	−2	0	−1	0	0	0	−1

The largest nonzero determinant of the above dimensional matrix is of order three and, therefore, the matrix has a rank 3. That a nonzero determinant of order three exists may be seen by constructing from the last three columns of the matrix the following determinant:

$$\begin{vmatrix} 1 & 0 & 0 \\ -3 & 1 & 1 \\ 0 & 0 & -1 \end{vmatrix}$$

which has a nonzero value.

A rank 3 implies that all three equations represented by Equation 5 are linearly independent. That is, none of the three equations is redundant in the sense that it can be obtained by a linear combination of the other two. A rank 2 would have implied that one of the equations was dependent and, therefore, should be ignored.

The number of dimensionless groups in a complete set is equal to the number of variables minus the rank of their dimensional matrix. In the above illustration, since there are seven variables and the rank of the matrix is 3, the number of dimensionless groups in the complete set is four.

The above conclusion could be reached without resorting to matrix algebra by the use of Buckingham's theorem [7], which states that the number of dimensionless groups is equal to the number of variables minus the number of dimensions in the problem. Normally, the rank of the matrix is the same as the number of dimensions in the problem. Buckingham's theorem is, however, only a convenient rule of thumb and should be used with caution.

Equation 5 comprises three independent equations in seven unknowns. Since the number of unknowns is larger than the number of equations, the system is undetermined. The best that can be done is to solve for three unknowns in terms of the remaining four unknowns. Solving for k5, k6 and k7 in terms of k1, k2, k3 and k4 from Equation 5,

$$\left.\begin{aligned} k5 &= -k1 \qquad\quad -k3 \\ k6 &= \qquad\quad -k2 - k3 - k4 \\ k7 &= -2k1 \qquad -k3 \end{aligned}\right\} \tag{6}$$

Obviously, many values of k5, k6 and k7 are possible, depending on the values of k1, k2, k3 and k4. If we assign the value k1 = 1 and k2 = k3 = k4 = 0, then the solution is k5 = −1, k6 = 0, and k7 = −2. If k2 = 1 and k1 = k3 = k4 = 0, then the solution is k5 = 0, k6 = −1 and k7 = 0. If k3 = 1 and k1 = k2 = k4 = 0, then k5 = −1, k6 = −1 and k7 = −1. Similarly, if k4 = 1 and k1 = k2 = k3 = 0, then the solution is k5 = 0, k6 = −1 and k7 = 0.

The four different solutions obtained above can be arranged in a matrix:

	ΔP	l	μ	ϵ	ρ	D	V
	k1	k2	k3	k4	k5	k6	k7
π_1	1	0	0	0	−1	0	−2
π_2	0	1	0	0	0	−1	0
π_3	0	0	1	0	−1	−1	−1
π_4	0	0	0	1	0	−1	0

Each row of the matrix contains values of the seven exponents in a dimensionless group. Hence,

$$\pi_1 = \frac{\Delta P}{\rho V^2}, \pi_2 = \frac{l}{D}, \pi_3 = \frac{\mu}{\rho D V}, \text{ and } \pi_4 = \frac{\epsilon}{D}$$

These four dimensionless groups form a complete set. Since ΔP, l, μ and ϵ occur in only one of the groups, none of the groups can be a product of powers of the other three. The relation between the four dimensionless groups may be written as

$$\frac{\Delta P}{\rho V^2} = f\left(\frac{l}{D}, \frac{\mu}{\rho D V}, \frac{\epsilon}{D}\right)$$

Furthermore, since ΔP is expected to be a linear function of l,

$$\frac{\Delta P}{\rho V^2} = \frac{l}{D} f\left(\frac{\mu}{\rho D V}, \frac{\epsilon}{D}\right)$$

It should be noted that the solutions to Equation 6 were obtained by assigning numerical values to k1, k2, k3 and k4 in a manner that only one of the k's had a nonzero value at a time. The result is that ΔP, l, μ and ϵ occur in only one group and the four dimensionless groups are independent. This is an important characteristic of the method. The seven variables were arranged deliberately in a particular order, and the solution of k5, k6 and k7 was obtained in terms of k1, k2, k3 and k4 to obtain the four familiar dimensionless groups—$\frac{\Delta P}{\rho V^2}, \frac{l}{D}, \frac{\epsilon}{D}$ and $\frac{\rho D V}{\mu}$. It was not essential that a solution be obtained in terms of k1 to k4. Any three unknowns could have been solved for in terms of the remaining four. The only requirement is that the coefficients of the three k's being solved for constitute a nonzero determinant in the dimensional matrix. Thus, a solution could have been obtained for k1, k2 and k3 in terms of k4, k5, k6 and k7 since the coefficients of k1, k2 and k3 in the dimensional matrix also yield a nonzero determinant. This way, an alternate set of dimensionless groups would have been obtained. As previously noted, a phenomenon may be desecribed by alternative sets of dimensionless groups.

MODEL TESTING

Very often, information about the performance of an as yet unbuilt engineering system is obtained by studying the performance of a small-scale replica or model of the system. Model tests are of use when the necessary information about a process or phenomenon is not known or cannot be predicted from theory, and when tests on the actual system are either too expensive or not possible. An example is the design of a new airplane wing on the basis of tests on a model wing in a wind tunnel. Another example is the scaleup of a chemical process unit such as a slurry reactor. Many times the purpose of model tests is limited to determining only certain important aspects of the phenomenon, since it may be impractical to build a model that would represent the phenomenon in its entirety. For example, in the design of a new airplane wing, the key objective of a model study could be to predict the conditions at which flutter occurs.

For a model to be completely similar with the prototype, it is essential that dimensionless groups forming a complete set have the same numerical values for both the model and prototype. This is possible only if geometric similarity exists, i.e., the ratio of all corresponding dimensions between the model and the prototype is the same. To achieve complete similarity, physical properties and experimental conditions for the model test must be such that identical values of dimensionless groups are obtained between the model and the prototype. A very often encountered problem in model testing is that physical properties and test conditions that give identical dimensionless groups cannot be chosen. In such cases, the dimensionless groups that have a greater influence on the phenomenon are maintained identical between the model and the prototype, while the less important groups are allowed to differ and an approximate solution is obtained. If the model simulation is not for the entire phenomenon but only for some specific aspects of the phenomenon, then only those dimensionless groups that influence these specific aspects may be maintained identical.

INSPECTION ANALYSIS OF DIFFERENTIAL EQUATIONS AND ANALOGIES

In the foregoing, a procedure was described for obtaining dimensionless groups from the variables involved in a phenomenon. If the differential equations governing the phenomenon are also known, the dimensionless groups and the necessary conditions for similarity can be obtained more conveniently through inspection analysis.

Inspection analysis is a method of determining what conditions are necessary for similarity to exist in geometrically similar systems. It involves converting the governing differential equations into dimensionless form by the use of characteristic length, time, velocity, temperature, etc., of the system and determining the conditions under which the dimensionless differential equations will yield identical solutions for two geometrically similar systems.

Again, consider flow around a sphere. The flow is governed by the equations of continuity and motion. If constant density and viscosity are assumed, these equations are

$$\frac{\partial V_x}{\partial x} + \frac{\partial V_y}{\partial y} + \frac{\partial V_z}{\partial z} = 0 \tag{7}$$

and

$$\rho\left(\frac{\partial V_x}{\partial t} + V_x\frac{\partial V_x}{\partial x} + V_y\frac{\partial V_x}{\partial y} + V_z\frac{\partial V_x}{\partial z}\right) = -\frac{\partial p}{\partial x} + \mu\left(\frac{\partial^2 V_x}{\partial x^2} + \frac{\partial^2 V_x}{\partial y^2} + \frac{\partial^2 V_x}{\partial z^2}\right) + \rho g_x \tag{8a}$$

plus two additional Equations 8b and 8c for y- and z-components of velocity. Equations 8b and 8c are similar to Equation 8a and are omitted for brevity. Here, V_x, V_y and V_z are the x-, y- and z-components of fluid velocity, and ρ and μ are fluid density and viscosity, respectively. p represents the pressure and g_x is the x-component of acceleration due to gravity. Equations 7 and 8 can be made dimensionless by using quantities that are characteristic of the system. For flow around a sphere, convenient characteristic quantities are the sphere diameter, D, and the approach velocity of the fluid, V_o. The choice of characteristic quantities is quite arbitrary but should be specified carefully. Using D and V_o, we may define the following dimensionless variables:

$$V_x^* = \frac{V_x}{V_o},\ V_y^* = \frac{V_y}{V_o},\ V_z^* = \frac{V_z}{V_o}$$

$$x^* = \frac{x}{D},\ y^* = \frac{y}{D},\ z^* = \frac{z}{D},\ t^* = \frac{tVo}{D} \text{ and}$$

$$p^* = \frac{p\text{-}po}{\rho V_o^2} \tag{9}$$

where po is a reference pressure. Substituting for V_x, V_y, V_z, x, y, z, t and p from Equation 9 into Equations 7 and 8, the following equations are obtained:

$$\frac{\partial V_x^*}{\partial x^*} + \frac{\partial V_y^*}{\partial y^*} + \frac{\partial V_z^*}{\partial z^*} = 0 \tag{10}$$

and

$$\frac{\rho V_0^2}{D}\left(\frac{\partial V_x^*}{\partial t^*} + V_x^* \frac{\partial V_x^*}{\partial x^*} + V_y^* \frac{\partial V_x^*}{\partial y^*} + V_z^* \frac{\partial V_x^*}{\partial z^*}\right)$$

$$= -\frac{\rho V_0^2}{D}\frac{\partial p^*}{\partial x^*} + \frac{\mu V_0}{D^2}\left(\frac{\partial^2 V_x^*}{\partial x^{*2}} + \frac{\partial^2 V_x^*}{\partial y^{*2}} + \frac{\partial^2 V_x^*}{\partial z^{*2}}\right) + \rho g_x \quad (11)$$

Multiplication of the last equation by $\frac{D}{\rho V_0^2}$ yields

$$\left(\frac{\partial V_x^*}{\partial t^*} + V_x^* \frac{\partial V_x^*}{\partial x^*} + V_y^* \frac{\partial V_x^*}{\partial y^*} + V_z^* \frac{\partial V_x^*}{\partial z^*}\right)$$

$$= -\frac{\partial p^*}{\partial x^*} + \frac{\mu}{\rho V_0 D}\left(\frac{\partial^2 V_x^*}{\partial x^{*2}} + \frac{\partial^2 V_x^*}{\partial y^{*2}} + \frac{\partial^2 V_x^*}{\partial z^{*2}}\right) + \frac{D g_x}{V_0^2} \quad (12)$$

Observe that each term in Equations 10 and 12 is dimensionless and that the variables are lumped together into dimensionless groups, $\frac{\rho V_0 D}{\mu}$ and $\frac{V_0^2}{g_x D}$. The dimensionless group $\frac{\rho V_0 D}{\mu}$, called the Reynolds number, is a ratio of the variables in the first and third terms of Equation 11, which represent the inertial and viscous forces, respectively, in the overall momentum balance. Thus, physically, the Reynolds number signifies the relative importance of inertial force to viscous force. Reynolds number of less than unity, for example, means that the viscous forces are large in comparison to the inertial forces. The second dimensionless group, $\frac{V_0^2}{D g_x}$, is called the Froude number and represents the relative importance of the inertial to gravity forces.

Now, if the variables ρ, V_0, D, μ and g in two systems (the model and prototype) are such that the Reynolds and the Froude numbers for the two systems are the same, then both systems are described by identical dimensionless equations (10 and 12). If the dimensionless initial and boundary conditions for the two systems are also the same, which is feasible only if geometric similarity exists, then the solution of the two systems also must be the same. In other words, similarity exists. Inspection analysis thus shows that specification of the two dimensionless groups, the Reynolds and Froude numbers, completely defines the system. These two dimensionless groups therefore form a complete set. The procedures described above can be used to obtain a complete set of dimensionless groups for any phenomenon if the governing differential equations are known.

A knowledge of the governing differential equations also helps to reveal analogies between different phenomena and design correlations. The dimensionless differential equations describing a large number of problems in heat, mass and momentum transport are exactly of the same form. For example, the differential equations and the initial and boundary conditions for a specific heat transfer process may be obtainable by changing the notation and certain dimensionless groups in the equations for an analogous problem in mass or momentum transfer. If, for two different problems, the dimensionless differential equations and the initial and boundary conditions are identical in form, the solutions also must be identical. For this reason, many analogies exist between heat, mass and momentum transfer correlations. If the correlation for one process is known, the correlation for an analogous process may be obtained by an appropriate change in notation.

IMPORTANT DIMENSIONLESS GROUPS AND THEIR SIGNIFICANCE

The Reynolds and Froude numbers obtained in the foregoing example of flow around a sphere were shown to represent the ratios of different forces. Other physically meaningful dimensionless groups are obtained, depending on the phenomenon under investigation. A list of the dimensionless groups that occur in the engineering problems is found in the literature [5,10]. The most common of these groups and their significance are listed in Table I.

Table I. Dimensionless Groups and their Significance

Dimensionless Group	Symbol	Definition	Significance (ratio)
Reynolds Number	N_{Re}	$\frac{\rho VL}{\mu}$ ρ = fluid density V = fluid velocity μ = fluid viscosity L = characteristic dimension	$\frac{\text{Inertial force}}{\text{Viscous force}}$
Froude Number	N_{Fr}	$\frac{V^2}{Lg}$	$\frac{\text{Inertial force}}{\text{Gravitational force}}$
Euler Number	N_{Eu}	$\frac{p}{\rho V^2}$ p = pressure	$\frac{\text{Pressure}}{2 \times \text{velocity head}}$
Mach Number	N_{Ma}	$\frac{V}{V_c}$	$\frac{\text{Fluid velocity}}{\text{Velocity of sound}}$
Weber Number	N_{We}	$\frac{\rho LV^2}{\sigma}$ σ = surface tension	$\frac{\text{Inertial force}}{\text{Surface tension force}}$
Drag Coefficient	C_D	$\frac{(\rho - \rho^1)\, Lg}{\rho V^2}$ ρ = density of object ρ^1 = density of surrounding fluid	$\frac{\text{Gravitational force}}{\text{Inertial force}}$
Fanning Friction Factor	f	$\frac{D}{L}\,\frac{\Delta P}{2\rho V^2}$ D = pipe diameter L = pipe length	$\frac{\text{Wall shear stress}}{\text{Velocity head}}$
Nusselt Number (heat transfer)	N_{Nu}	$\frac{hL}{k}$ h = heat transfer coefficient k = thermal conductivity	$\frac{\text{Total heat transfer}}{\text{Conductive heat transfer}}$

Table I—continued

Prandtl Number	N_{Pr}	$\frac{C_p\mu}{k}$ C_p = heat capacity	$\frac{\text{Momentum diffusivity}}{\text{Thermal diffusivity}}$
Peclet Number (heat transfer)	N_{Pe}	$\frac{C_p \rho VL}{k}$ $= N_{Re} N_{Pr}$	$\frac{\text{Bulk heat transport}}{\text{Conductive heat transfer}}$
Grashof Number	N_{Gr}	$\frac{gb^3\rho^2\beta\Delta T}{\mu^2}$ β = coefficient of expansion ΔT = temperature difference b = height of surface	$N_{Re} \times \frac{\text{Buoyancy force}}{\text{Viscous force}}$
Stanton Number	N_{St}	$\frac{h}{\rho VC_p}$ $= N_{Nu}N_{Re}^{-1}N_{Pr}^{-1}$	$\frac{\text{Heat transferred}}{\text{Thermal capacity of fluid}}$
J Factor for Heat Transfer	j_H	$\frac{h}{\rho VC_p}\left(\frac{C_p\mu}{k}\right)^{2/3}$	Proportional to $N_{Nu}N_{Re}^{-1}N_{Pr}^{-1/3}$
Nusselt Number (mass transfer)	N_{Nu}	$\frac{k_cL}{D}$ k_c = mass transfer coefficient D = molecular diffusivity	$\frac{\text{Total mass transfer}}{\text{Diffusive mass transfer}}$
Schmidt Number	N_{Sc}	$\frac{\mu}{\rho D}$	$\frac{\text{Momentum diffusivity}}{\text{Molecular diffusivity}}$
Peclet Number (mass transfer)	N_{Pe}	$\frac{LV}{D} = N_{Re}N_{Sc}$	$\frac{\text{Bulk mass transport}}{\text{Diffusive mass transport}}$
J Factor for Mass Transfer	j_D	$\frac{k_c}{V}\left(\frac{\mu}{\rho D}\right)^{2/3}$	Proportional to $N_{Nu}N_{Re}^{-1}N_{Sc}^{-1/3}$

REFERENCES

1. Langhaar, H. L. *Dimensional Analysis and Theory of Models* (New York: John Wiley & Sons, Inc., 1951).
2. Johnstone, R. E., and M. W. Thring. *Pilot Plants, Models, and Scale-Up Methods in Chemical Engineering* (New York: McGraw Hill Book Co., 1957).
3. Birkhoff, G. *Hydrodynamics* (Princeton, NJ: Princeton University Press, 1960).
4. Klinkerberg, A., and H. H. Mooy. "Dimensionless Groups in Fluid Friction, Heat, and Material Transfer," *Chem. Eng. Prog.* 44 (1):17 (1948).
5. Catchpole, J. P., and G. Fluford. "Dimensionless Groups," *Ind. Eng. Chem.* 58 (3): 46 (1966).
6. Catchpole, J. P., and G. Fluford. "Dimensionless Groups," *Ind. Eng. Chem.* 60 (3): 71 (1968).
7. Buchingham, E. "On Physically Similar Systems; Illustration of the Use of Dimensional Equations," *Phys. Rev.* 4 (4): 345 (1914).

8. Lord Rayleigh. "The Principle of Similitude," *Nature* 95: 66 (1915).
9. Bird, R. B., W. E. Stewart, and E. N. Lightfoot. *Transport Phenomenon* (New York: John Wiley & Sons, Inc., 1960).
10. Boucher, D. F., and G. E. Alves. "Dimensionless Numbers," *Chem. Eng. Prog.* 55 (9): 55 (1959).

CHAPTER 4
HISTORY OF THERMAL ANEMOMETRY

Peter Freymuth
Department of Aerospace Engineering Sciences
University of Colorado
Boulder, Colorado 80309

CONTENTS

INTRODUCTION

Based on the existing body of literature on thermal anemometry, its historical development is sketched in this chapter. Three periods of development are distinguished and important developments in these periods are traced to their origins.

There are several routes to gain access to the historical development of thermal anemometry. Recollections of scientists actively involved in its development are some of the most valuable sources. Unfortunately, at this point recollections seem to be confined to those by Burgers [1], Schubauer [2] and Lowell [3]. Lowell, in his recollections of work at NACA-Lewis for the period 1944-1952, states: "Hopefully, someone will soon undertake to write a treatise on the entire field, and these early histories will be more definitively recorded, for many institutions, than has here been done." It seems unlikely that his hope will come true. In the absence of a wealth of recollections, authors interested in the historical development of thermal anemometry are forced to rely on the published literature in their assessments. The result is a more generic picture of the historical development. Recently, a comprehensive bibliography of thermal anemometry has been developed by Freymuth [4], which features chronological order of references and, in conjunction with the referenced original literature, represents an entry route to the historical development.

A more limited but valuable route to the historical development is represented by the numerous reviews of thermal anemometry, which contain materials of historical interest, notably the reviews by Burgers [5], Corrsin [6], Sandborn [7] and Comte-Bellot [8], as well as the paper by Benson and Brundrett [9], who admirably reviewed the development of temperature measurement in flows.

Historical essays can have quite different objectives. Recollections highlight particular events in a historical development. One may try to show the influence of the general scientific and technological progress on the development of thermal anemometry. A related objective has recently been pursued by Weske [10], who tried to show the interaction between the development of hot-wire anemometry and the statistical theory of turbulence (Table I).

Table I. Overview of the Development of Hot-Wire Anemometry in Correlation to the Development of the Statistical Theory of Turbulence

Motivation Problem	Name(s) Group	Hot-Wire Anemometry Contribution	Year	Year	Name(s) Group	Theory of Turbulence Contribution
Thermal Transport	L. V. King, Montreal	Convection of heat of small cylinders	1914	1807-1822	J. B. Fourier	Fourier functions (see Kampé de Fériet, 1948)
				1921	G. I. Taylor	Theory of turbulent motions
Wind Speed and Direction	Huguenard, Magnan, Planiol, Paris	Measurement of unsteady flows	1924			
Boundary Layers, Velocity, Fluctuations	J. M. Burgers, B. G. von der Hegge-Zijnen Delft	Measurement in laminar and turbulent boundary layers	1924			
Wind Tunnel Turbulence Structure	H. L. Dryden, R. H. Heald, U.S.-NBS	Investigation of turbulence in wind tunnels	1926			
Thermal Inertia of the Hot-Wire	H. L. Dryden, A. M. Kuethe, NBS	Measurement of velocity fluctuations, L-R and C-R compensation	1929 1930			
Directional Characteristics; Time Constant	M. Ziegler, Delft	C-R compensation, directional effect, turbulence oscillograms	1931 1932			
Hot-Wire Measurements	J. M. Burgers, Delft	Review	1931			
Hot-Wire Measurements	M. Ziegler, Delft	Proposal of the constant-temperature schematic	1934			
				1935	H. Gebelein	Turbulence (physical statistics and hydrodynamics)
				1935	G. I. Taylor	Statistical theory of turbulence

				1938	Th. v. Kármán, L. Howarth	Statistical theory of isotropic turbulence
Turbulent Stresses	H. K. Scramstad, NBS	Shear stresses in turbulent flow	1939			
				1941	A. N. Kolmogoroff, M. G. U., Moscow	Local isotropy
Extension of the Frequency Range for High Velocities	J. R. Weske, Cleveland	Use of tungsten wire, constant-temperature anemometer, electronics by E. R. Jervis	1943			
Turbulent Boundary Layers	G. O. Schubauer, P. S. Klebanoff, NBS	Theory and application of hot-wire anemometers	1946			
				1948	J. Kampé de Fériet, Lille	Spectral tensors
Supersonic Flows (f < 70 kHz)	L. S. G. Kovasznay, Johns Hopkins University	Hot-wire measurements in supersonic flows	1950			
		Development of turbulence measuring equipment	1953	1953	G. K. Batchelor	Theory of homogeneous turbulence
Correlations	A. Favre, Marseille	Correlation in space and time	1953			
				1956	A. A. Townsend	Structure of turbulent shear flow
Signal Evaluation and Analysis	P. S. Klebanoff, F. N. Frenkiel, NBS	Tape recording and evaluation of three-dimensional fluctuations	1970			
				1972	J. C. Rotta	Turbulent flows

This chapter has limited objectives. It tries to identify periods of historical development and to trace important concepts and developments to their origin in the literature. Showing the maturing of concepts and developments and aiming for comprehensiveness must be left to the future.

PERIODS OF THE HISTORICAL DEVELOPMENT OF THERMAL ANEMOMETRY

Thermal anemometry seems to be one of the most intensely covered fields of instrumentation science. This became apparent during the collection of references for a comprehensive bibliography of this specialty [4]. The bibliography contains more than 1600 references up to 1980 in chronological order, and from this to a view of the historical development is a small step. If one plots the number of papers published each year over the years, a publication rate curve of thermal anemometry results as shown in Figure 1. There is considerable scatter in the data, but it seems obvious that the curve peaks around 1970. The field was still very vigorous in 1980, but the curve is nevertheless now declining. It seems that thermal anemometry is rapidly reaching maturity, and laser anemometry is entering areas where thermal anemometry exhibits marginal performance. By considering more details of the publication rate curve and in view of the content of published papers, it seems reasonable to divide the historical development of thermal anemometry into three periods.

1. The **incipient period** goes back to the nineteenth century and ends in 1908. It is characterized by few publications, often spaced years apart. Subjects of relevance to thermal anemometry, like heat transfer from fluids at rest or moving, were considered for the first time and provide reference points for subsequent researchers. Furthermore, the invention of the constant-current hot-wire anemometer falls into this period, but the usefulness of the instrument for practical applications was not demonstrated within this period.

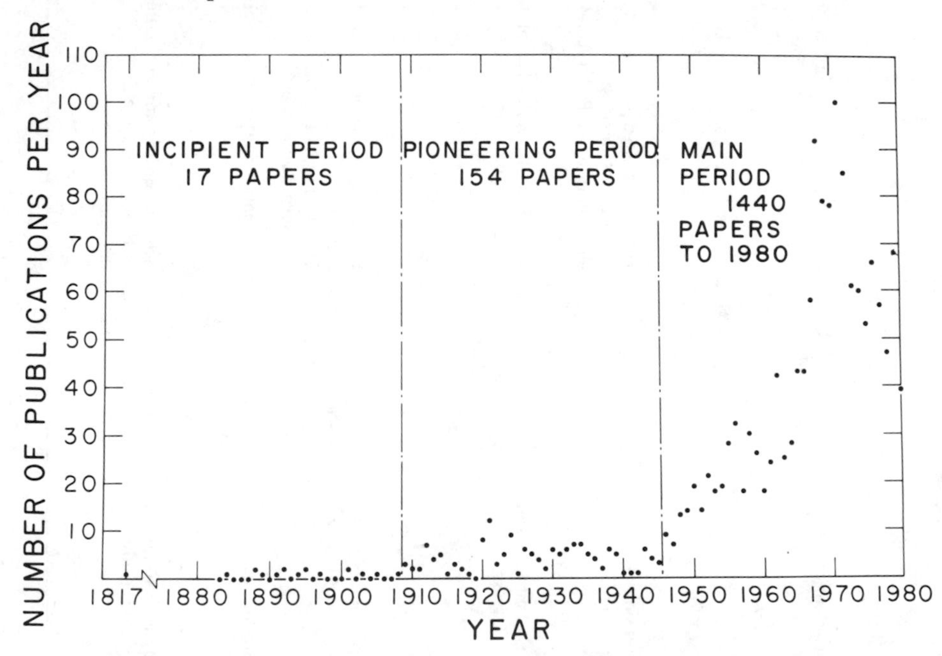

Figure 1. Publication rate curve of thermal anemometry, 1817-1980.

2. The **pioneering period** starts in 1909 and ends in 1945. This period is characterized by a low and somewhat erratic, but continuous, publication rate, in which basic concepts were formulated and proven by practical application. The constant-temperature concept, the concepts of feedback control and automatic recording, the measurement of flow direction, static and dynamic calibration, electronic amplification and frequency compensation for rapid fluctuations, specialties like protective wire coatings and measuring of blood flow all were developed and applied successfully on an individual basis. Hot-wire anemometry became established as an "art." This period discovered, but hardly tapped, the vast potential of thermal anemometry for flow research.
3. The **main period** of thermal anemometry starts in 1946 and is still continuing. In this period, thermal anemometry was developed from an art into a standard experimental method. Instruments became available routinely on a commercial basis. An appropriate opening statement for the main period was provided by the title of the early paper "Extended Applications of the Hot-Wire Anemometer" [11]. The application of thermal anemometry was extended indeed to all phases of experimental flow research, including stability, transition, turbulence, heat transfer, compressible flow, unsteady aerodynamics, stratified and two-phase flows. It seems likely that we are now in the late phases of the main period, which may expire near the end of the 1980s.

This author offers the speculation that a final period may follow the main period, which then would be characterized by a low publication rate. In this period, overlooked gaps in the knowledge of thermal anemometry are likely to be filled in, the domain of application should become well established and adjustments to new technologies should take place rendering a fully developed measuring tool of experimental fluid mechanics.

ORIGIN OF CONCEPTS AND DEVELOPMENTS IN THE EARLY PERIODS

After the rather generic overview of the previous section, let us ask some questions surrounding the incipient and pioneering periods and find the answers.

When was the constant-current hot-wire anemometer invented?

The constant-current anemometer was proposed by Oberbeck in 1895 [12]. He cites, as perceived benefits, its compactness and relatively fast response in view of the thin wires that can be used. He proceeds to test a platinum wire with a diameter of 0.05 mm and a length of 7.5 cm in the flow in front of an electrically driven fan. He erroneously finds a linear relation between heat transfer and velocity and does not demonstrate the viability of his concept in a practical application.

In the same year, Weber [13] detected that a glowing platinum wire lost in luminosity when exposed to a weak air current. He failed to suggest using this effect for measurement purposes. This omission was made up for by Dau [14], who was a graduate student of Weber's.

The constant-current concept was proposed again by Shakespeare [15] in 1902, Koepsel [16] in 1908, Kennelly et al. [17] in 1909, Bordoni [18] and Morris [19] in 1912, and Schrodt [20] in 1914.

The first one to actually use a constant-current anemometer was its inventor, Riabouchinsky[21], in 1909, and this marks the beginning of the pioneering period of thermal anemometry. It should be mentioned that Kennelly et al. [17] used a whirling arm for wire testing, whereas Morris [19] used a wind tunnel.

When was the constant-temperature hot-wire anemometer invented?

The constant-temperature anemometer was invented by Morris in 1912, complete with Wheatstone bridge, and he proceeded to its immediate application in 1913 [22].

Automatic feedback control to constant temperature by means of an electromechanical controller was achieved by Hill, Hargood-Ash and Griffith [23] in 1921. Electronic feedback was first proposed by Ziegler [24] in 1934, who showed the reduction in thermal inertia associated with this method. The first anemometer with electronic feedback which worked was built by Weske [25,26] in 1943, with the help of Dr. E. R. Jervis.

Which other concepts were used in early anemometry?

Thomas [27] introduced the first automatic feedback control circuit into thermal anemometry in 1909. It was an electromechanical controller to keep the temperature difference in front of and behind a flow heater constant. The Thomas meter represents, together with the anemometer by Riabouchinsky [21], the earliest applications. It measured natural gas flow and featured an automatic recording system. It was the first "wake sensing probe." Kennelly et al. [17] proposed a constant-voltage anemometer. Dau [14] used a constant-voltage over a Wheatstone bridge anemometer to measure the in- and outflow of gases from a box, which developed as a consequence of changing winds moving over the box. Automatic recordings showed signals that look as turbulent as any (Figure 2). Dau does not mark a time scale on his recordings, but it appears to be low frequency. Dau's arrangement seems to be a crude predecessor of an orifice hot-wire anemometer. Dau also used the glow from a wire for flow measurement.

A constant-power anemometer was considered by Huguenard, Magnan and Planiol [28,29] in 1924. Ziegler [30] developed the first pulsed-wire anemometer to measure low air speeds.

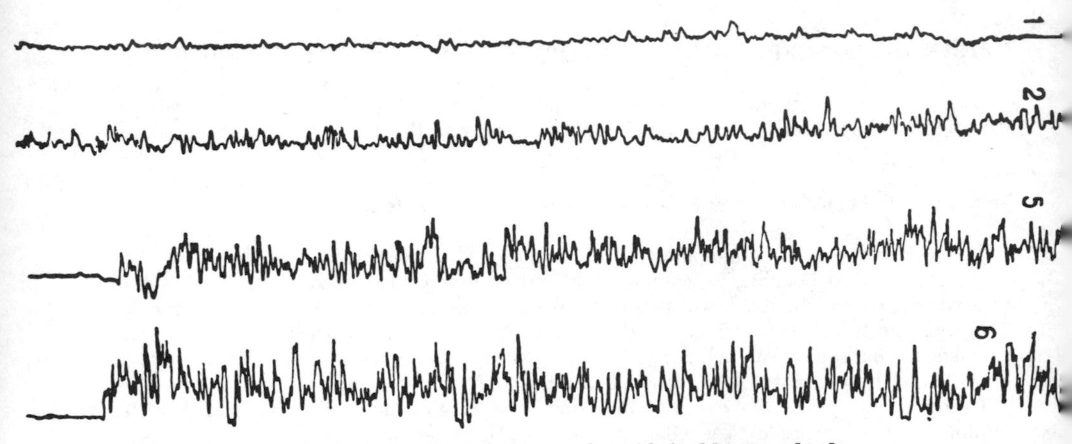

Figure 2. Turbulent records published by Dau [14].

What background information was available to early investigators?

Early investigators, notably King [15] and Kennelly et al. [17], cited the papers by Dulong and Petit [31] in 1817, Ayrton and Kilgour [32] in 1892, Petavel [33] in 1901, and by Langmuir [34] in 1912 on heat transfer from wires in gases at rest. Kennelly et al. [17] were attracted to this topic in connection with the problem of overheating of electrical power transmission lines and the heat transfer in light bulbs, and give some additional references on this and on radiative heat transfer.

The earliest sensible experimental results on forced convection from wires were those by Ser [35] in 1888 for air flow and by Rogovsky [36] in 1903 for water flow. Both experimenters found a square root dependence of convection loss on velocity and approximately linear dependence on temperature difference. A convective velocity was added by Kennelly et al. [17] to accommodate the free convection influences at low velocities, and King [15] added a more practical heat loss term instead. On a theoretical basis, Boussinesque [37] also suggested a square root dependence, and King [15] confirmed this. Unfortunately, all the theoretical work was flawed by the assumption of potential flow around the wire.

When was dimensional analysis of heat transfer from wires first considered?

Such an analysis was first considered by King [38] (Figure 3) in 1916, who suggested Reynolds number dependence, as well as what we now call Prandtl number dependence. He obtained his inspiration from a 1915 paper by Rayleigh [39] entitled "The Principle of Similitude." An actual dimensionless presentation of data goes back to Davis [40] in 1922.

When was the directional sensitivity of the hot-wire recognized and explored?

The directional sensitivity of hot-wires was recognized by King [15] but explored by Thomas [41,42] in 1920 and by Bailey [43] in 1921. A relatively high perfection was reached by Simmons and Bailey [44] in 1927. In 1924, Huguenard, Magnan and Planiol [28,29] utilized a device similar to a split film probe. In the following decade numerous directional determinations were made. In 1943 Weske [25,26] utilized the directional sensitivity of a rotating hot wire with axis of rotation slanted to the flow direction for a dynamic calibration of the frequency response of an anemometer.

Which other corrections of the basic heat transfer relation for hot wires were considered?

Dependence on humidity was considered by Kennelly et al. [17] and, in more detail, by Schubauer [45] and Paeschke [46]. The influence was small. King [15] considered heat conduction from the wires to the prongs. Kennelly and Sanborn [47] investigated

Figure 3. Louis Vessot King around 1920 at McGill University. (Photograph obtained from Professor S. Corrsin, Johns Hopkins, who obtained it from Professor B. Newman at McGill.)

the influence of atmospheric pressure. In 1917 Faye [48] invented a compensation circuit to compensate for the influence of varying flow temperature on anemometer output. Van der Hegge Zijnen [49] detected the influence of wall proximity on heat transfer from the wire and was the first to notice intermittency in the turbulent boundary layer, an effect noticed again by Ziegler [50].

Effects like radiation, buoyancy, wire evaporation, vibrations of the wire, generation of vortex streets behind the wire and dust deposition were considered by early investigators; for details, the reader is referred to the literature [5].

What types of wires were used in the early periods?

Platinum wires remained the mainstay of the art; however, Kennelly et al. [17] used copper wires and proposed platinized copper wires. Morris [19] experimented with tinned iron and copper wires and with tantalum and nickel wires. He finally used platinum wires. Tungsten wires were mentioned as a good possibility if good electrical joints could be made. Tungsten wires were first used by Weske [25]. Petersen [51] was interested in the measurement of rapidly changing high temperatures and used 0.0008-in.-diameter (0.02 mm)-Pt-Ir wires. He found that tungsten wires oxidized at 900° C and thus he did not use them. For a thermocouple he used platinum and a platinum-rhodium alloy, an alloy later used also by Davies and David [52] for the measurement of flame temperatures. In 1920 Thomas [41,42] used a glass-coated platinum wire in a natural gas flow to avoid catalytic effects. The first experimenter to use a Wollaston wire was Ziegler [30] in 1931. It had a platinum core of 0.00012 in. (0.003 mm) in diameter and a maximum length of 0.04 in. (1 mm).

How long have rapid velocity fluctuations been measured?

The earliest measurements of this kind were made by Huguenard et al. [28,29] in 1923 (turbulence), followed by Relf and Simmons [53] in 1924 and Tyler [54] in 1926 (vortex streets). Burgers [55] was the first to make correlation measurements. Simmons and Salter [56] measured turbulent spectra, and Skramstad [57] measured shear stresses.

How did electronics enter thermal anemometry?

A first attempt of electronic amplification of a hot-wire signal goes back to Teegan [58] in 1926 and, in the same year, Tyler [54] used a stepup transformer to increase voltage. Successful signal amplification was established by Gupta [59] in 1927. Two years later, Dryden and Kuethe [60] presented their compensation method for the accurate measurement of rapid turbulent fluctuations. A similar method has been described by Abraham [61] in 1897 for improving the response of an oscillograph, as pointed out by Ziegler [30a].

The introduction of electronic feedback control by Ziegler [24] and Weske [25,26] has already been mentioned. It is remarkable that Weske incorporated bridge unbalance and trim into his design, features that are prominent in modern anemometers. In 1924 King [62] introduced a squaring circuit into the constant-current anemometer for partially linearizing its output, and more complete linearization was achieved by Huguenard, Magnan and Planiol [28,29]. Unfortunately, linearizers are of little use in connection with constant-current anemometers in a dynamic situation. A more lasting achievement was the invention by Ziegler [24] of a linearizer for the constant-temperature anemometer. Electronic square wave testing of the frequency response of the constant-current anemometer was pioneered by Ziegler [30] in 1931. Many early investigators preferred direct dynamic testing methods, like probe shaking, as did Dryden and Kuethe [60], Salter and Raymer [63] and Simmons [64]. As already mentioned, Weske [25,26] used the slanted rotation of a wire for this purpose.

What other flow properties were measured in the early periods?

Steam temperature was measured in 1891 by Hall [65], using a thermoelement, and by Callendar and Nicholson [66] in 1897, using thin (0.001-in.-diameter) platinum wires. It is interesting that a discussion attached to the 1891 paper contains the earliest worry about the influence of thermal lag of a sensor. Sensor lag was considered more thoroughly by Retchy [67] in 1912 for a velocity probe and by Petersen [50] in 1913 for a temperature probe. Even though velocity and temperature measurements in flows developed in parallel, there was little information exchange between these two techniques, to the detriment of both.

Concentration measurements in gases were introduced by Koepsel [16]. A hot-wire microphone was invented by Tucker and Paris [68] in 1921. Biomedical anemometry started with the measurement of blood flow by Gibbs [69] in 1933.

In conclusion, it seems that many developments of the main period had their roots in the early history of thermal anemometry. It must be stated, however, that developments in thermal anemometry often occurred in ad hoc fashion (with the result of "reinventing the wheel" several times over), rather than based on any historical development.

THE MAIN PERIOD OF THERMAL ANEMOMETRY

The development of the main period is difficult to assess because of the enormous body of literature that has accrued. Another difficulty consists in the fact that the main period is not yet complete, so an impartial assessment is unlikely. This author leaves a detailed assessment to the future but offers a brief and preliminary chronology that tries to trace important developments to their origin within the main period. Many ideas within the mainstream of development of thermal anemometry (the measurement of local velocity and its fluctuations) have been reinvented several times over, and many concepts needed a long time to approach maturity. The chronology that follows does not give an account of this but only traces developments to their origin.

CHRONOLOGY OF THE MAIN PERIOD

1947 Corrsin [11] pushed for the extended use of thermal anemometers.

1948 Frenkiel [70] considered the influence of hot-wire length on the small-scale measurement of turbulence. The first stability theory of the constant-temperature anemometer was proposed by Ossowsky [71]. Second-order dynamic theory of the constant-temperature anemometer by Kovasznay [72].

1949 Nonlinear dynamic theory for constant-current anemometers by Betchov [73]. Flush-mounted probe for shear stress measurements introduced by Ludwieg [74].

1950 Investigation of heat loss from wires in supersonic flows by Kovasznay and Toernmark [75] and by Lowell [76].

1953 Theory of hot film probes by Lowell [77].

1955 Hot-film anemometer introduced by Ling [78].

1960 Hot-wire arrays introduced by Kresa and Molloe-Christensen [79]. Electronic square-wave testing of constant-temperature anemometers introduced by Staritz [80].

1961 First transistorized constant-temperature anemometer built by Wehrmann [81].

1963 Third-order dynamic theory of constant-temperature anemometers by Berger et al. [82].

1964 Measurement in liquid mercury by Sajben [83].

1965 Two-phase flow measurement by Goldschmidt [84]. Measurement in polymer solutions by Zakin and Patterson [85].

1966 Sine wave testing for constant-temperature hot-wire anemometers, by Freymuth [86].
1967 Bellhouse-Schultz [87] effect for noncylindrical hot films. Influence of probe stem by Hoole and Calvert [88]. Data evaluation using digital techniques by Frenkiel and Klebanoff [70].
1968 Synchronized sampling, later called "conditional sampling," introduced by Kovasznay [89]. Nonlinear theory for constant-temperature anemometers by Freymuth [90].
1970 Influence of wire support on dynamics of the resistance thermometer, by Maye [91].
1971 Dynamic theory of "short wire" in constant-temperature operation, showing influence of conduction to the wire supports, by De Haan [92].
1975 Feedback control theory for constant-temperature hot-film anemometers, by Weidmann and Browand [93].
1978 First bibliography of thermal anemometry aimed at comprehensiveness, by Freymuth [4].

REFERENCES

1. Burgers, J. M. "Early Developments of Hot-Wire Anemometry in the Netherlands," *Adv. in Hot-Wire Anemometry, Proc.* (1968), pp. 25-28.
2. Schubauer, G. B. "Early Developments in Hot-Wire Anemometry at NBS and a Look at Some Elsewhere," *Adv. in Hot-Wire Anemometry, Proc.* (1968), pp. 13-24.
3. Lowell, H. H. "Early (1944-1952) Hot-Wire Anemometer Developments at the NACA Lewis Research Center," *Adv. in Hot-Wire Anemometry, Proc.* (1968), pp. 29-37.
4. Freymuth, P. "A Bibliography of Thermal Anemometry," *TSI Quart.* (1978, 1982).
5. Burgers, J. M. "Hitzdrahtmessungen," *Handb. Exptl.-Physik* 4 (Part 1): 636-667 (1931).
6. Corrsin, S. "Turbulence: Experimental Methods," *Handb. Physik* 8 (Part 2): 523-590 (1963).
7. Sandborn, V. A. *Resistance Temperature Transducers* (Fort Collins, CO: Metrology Press, 1972).
8. Comte-Bellot, G. "Hot-Wire Anemometry," *Annual Rev. Fluid Mech.* 8: 209-231 (1976).
9. Benson, R. S., and G. W. Brundrett. "Development of a Resistance Wire Thermometer for Measuring Transient Temperatures in Exhaust Systems of Internal Combustion Engines," *Temperature, its Measurement and Control in Science and Industry* 3 (Part 2): 631-653 (1962).
10. Weske, R. "Seminar Talk on the History of Hot-Wire Anemometry" at University Karlsruhe, 1979.
11. Corrsin, S. "Extended Application of the Hot-Wire Anemometer," *Rev. Sci. Instrum.* 18: 469-471 (1947).
12. Oberbeck, A. "Ueber die Abkuehlende Wirkung von Lufstroemen, *Annalen Physik Chemie* 56: 397-411 (1895).
13. Weber, L. "Ueber die Waermeleitung der Gase," *Schriften Naturwissenschaftlichen Vereins zu Schleswig Holstein* 10 (Part 2): 313 (1895).
14. Dau, R. "Erdatmung und Messung Schwacher Luftstroeme," *Inaugural-Dissertation,* Universitaet Kiel (1912).
15. King, L. V. "On the Convection of Heat from Small Cylinders in a Stream of Fluid," *Phil. Trans. Roy. Soc., London* A214: 374-432 (1914).
16. Koepsel, A. "A Continuous Gas Testing Apparatus," *Deutsche Phys. Gesellschaft, Verh.* 10.20: 814-827 (1908).
17. Kennelly, A. E., C. A. Wright and J. S. Bylevelt. "The Convection of Heat from Small Copper Wires," *Trans. Am. Inst. Elect. Eng.* 28: 363-396 (1909).

18. Bordoni, U. "The Electrical Measurement of Wind Velocity," *Electrician* 70: 275-276 (1912).
19. Morris, J. T. "The Electrical Measurement of Wind Velocity," *Engineering* 94: 892-894 (1912).
20. Schrodt, J. "Experimental Investigation of Linear Gas Motion by a New Method," Dissertation, Budapest, Hungary (1914).
21. Riabouchinsky, D. "Appareil pour l'Etude du Frottement de L'Air Contre un Plan," *Bull. Inst. Aerodyn. Koutchino* 2: 115-120 (1909).
22. Morris, J. T. "The Distribution of Wind Velocity in the Space Surrounding a Circular Rod in a Uniform Current of Air," *Engineering* 96: 178-181 (1913).
23. Hill, L., and H. Hargood-Ash. "The Caleometer," *London Phys. Soc., Proc.* 33: 169-171 (1921).
24. Ziegler, M. "The Construction of a Hot-Wire Anemometer with Linear Scale and Negligible Lag," *Verh. Koninkl. Akad Wetenschappen, Amsterdam*, 15: 3-22 (1934).
25. Weske, J. R. "A Hot-Wire Circuit With Very Small Time Lag," *NACA TN* 881 (1943).
26. Weske, J. R. "Methods of Measurement in High Air Velocities by the Hot-Wire Method," *NACA TN* 880 (1943).
27. Thomas, C. C. *"An Electric Gas Meter," ASME J.* 31: 1325-1341 (1909).
28. Huguenard, E., A. Magnan and A. Planiol. "Sur un Anemometre a fil Chaud a Compensation," *Compt. Rend. Acad. Sci. Paris* 176: 287-289 (1924).
29. Huguenard, E., A. Magnan and A. Planiol. "Sur une Methode de Measure de la Vitesse et la Direction Instantanées du Vent," *La Technique Aeronautique* 14: 798-806; 854-862 (1924).
30. Ziegler, M. "A Complete Arrangement for the Investigation, the Measurement and Recording of Rapid Airspeed Fluctuations with Very Thin and Short Hot Wires," *Proc. Koninkl. Akad. Van Wetenschappen, Amsterdam*, 34: 663-673 (1931).
30a. Ziegler, M. "A Direct Method of the Measurement of Low Air Speeds," *Proc. Roy Acad. Van Weterschappen, Amsterdam* 36: 426-431 (1933).
31. Dulong, M., and M. Petit. "Du Refroidissement dans l'air et Dans Le Gaz," *Annales Chemie Physique* 7: 337 (1817).
32. Ayrton, W. E., and H. Kilgour. "The Thermal Emissivity of Thin Wires in Air," *Phil. Trans. Royal Soc.* A183: 371-405 (1892).
33. Petavel, J. E. "On the Heat Dissipated by a Platinum Surface at High Temperatures," *Phil. Trans. Roy. Soc.* A197: 229-254 (1901).
34. Langmuir, I. "Conduction and Convection of Heat in Gases," *Phys. Rev.* 34: 401-422 (1912).
35. Ser, L. "Convection, Traite de Physique Industrielle," 1: 142-162 (1888).
36. Rogovsky, E. "Sur La Conductibilité Exterieure de Fils D'argent Plongés dans L'eau," *Compt. Rend. Acad. Sci., Paris* 136: 1391-1393 (1903).
37. Boussinesque, J. "Conductibilité Exterieure ou Superficielle Representative, pour un Corps Donné, du Poivoir Refroidissant d'une Courant Fluide," *Compt. Rend. Acad. Sci., Paris* 140: 64-70 (1905).
38. King, L. V. "The Linear Hot-Wire Anemometer and its Application in Technical Physics," *J. Franklin Inst.* 181: 1-25 (1916).
39. Lord Rayleigh. "The Principle of Similitude," *Nature* 95: 66 (1915).
40. Davis, A. H. "The Cooling Power of a Stream of Viscous Fluid," *Phil. Mag., Ser. 6* 44: 940-944 (1922).
41. Thomas, J. S. G. "The Hot-Wire Anemometer: its Application to the Investigation of the Velocity of Gases in Pipes," *Phil. Mag., Ser. 6* 39: 505-534 (1920).
42. Thomas, J. S. G. "The Directional Hot-Wire Anemometer: its Sensitivity and Range of Application," *Phil. Mag., Ser. 6* 40: 440-665 (1920).
43. Bailey, A. "Directional Hot-Wire Anemometer," *Techn. Rep. Adv. Comm. Aeron., London* R & M No. 777 (1921).
44. Simmons, L. F. G., and A. Bailey. "A Hot-Wire Instrument for Measuring Speed and Direction of Airflow," *Phil. Mag., Ser. 7* 3: 81-96 (1927).

45. Schubauer, G. B. "The Effect of Humidity in Hot-Wire Anemometry," *J. Res. Nat. Bur. Stds.* 15: 575-578 (1935).
46. Paeschke, W. "Feuchtigkeitseffekt bei Hitzdrahtmessungen," *Physik. Zeitschrift* 36: 564-565 (1935).
47. Kennelly, A. E., and H. S. Sanborn. "The Influence of Atmospheric Pressure upon Forced Thermal Convection from Electrically Heated Platinum Wires," *Proc. Am. Phil. Soc.* 8: 55-77 (1914).
48. Faye, R. D. "Improvement in Hot-Wire Anemometers," *J. Franklin Inst.* 183, 783 (1917).
49. Van der Hegge Zijnen, B. G. "Measurements of the Velocity Distribution in the Boundary Layer Along a Plane Surface," Dissertation, Delft, The Netherlands (1924).
50. Ziegler, M. "Oscillographic Records of Turbulent Motion Developing in a Boundary Layer from a Sheet of Discontinuity," *Proc. Koninkl. Akad, Van Wetenschappen, Amsterdam*, 35: 419-426 (1932).
51. Petersen, A. "Verfaben zur Messung Schnell Wechselnder Temperaturen, Mitteilungen ueber Forschungsarbeiten," 143; also, in shorter form: *Zeitschr. Ver. Deutsch. Ing.* 58: 602-610 (1914).
52. Davies, W., and W. T. David. "Flame Temperatures," *Phil. Mag.* 12: 1043-1057 (1931).
53. Relf, E. F., and L. F. G. Simmons. "The Frequency of Eddies Generated by the Motion of Circular Cylinders Through a Fluid," *Techn. Rep. Aeron. Res. Comm., London* R&M No. 917 (1924).
54. Tyler, E. "The Use of the Hot-Wire Detector in Determining the Path of Vortices," *J. Sci. Instrum.* 3: 398-400 (1926).
55. Burgers, J. M. "Experiments on the Fluctuations of the Velocity in a Current of Air," *Proc. Roy. Acad., Amsterdam* 29: 547-558 (1926).
56. Simmons, L. F. G., and C. Salter. "An Experimental Determination of the Spectrum of Turbulence," *Proc. Roy. Soc. London* A165: 73 (1938).
57. Skramstad, H. K. "An Experimental Method for Measurement of Shearing Stresses in Turbulent Flow," *Phys. Rev.* 55: 1141-1142 (1939).
58. Teegan, J. A. C. "A Thermionic Valve Method of Measuring the Velocity of Air Currents of Low Velocity in Pipes," *Phil. Mag., Ser.* 7 1: 1117-1120 (1926).
59. Gupta, B. L. "Application of Thermionic Valves to Hot-Wire Anemometry," *J. Sci. Instrum.* 4: 202-205 (1927).
60. Dryden, H. L., and A. M. Kuethe. "The Measurement of Fluctuations of Air Speed by the Hot-Wire Anemometer," *NACA Tech. Rep.* 320 (1929).
61. Abraham, H. "Sur le Rheograph a Induction Abraham-Carpentier," *Jo. Physique, Ser. 3* VI: 356; *Compt Rend. Acad. Sci.* 124: 758 (1897).
62. King, R. O. "The Measurement of Air Flow," *Engineering* 117: 249-251 (1924).
63. Salter, C., and W. G. Raymer. "Direct Calibration of Compensated Hot-Wire Recording Anemometer," A.R.C. R&M No. 1628 (1934).
64. Simmons, L. F. G. "Note on Errors Arising in Measurements of Turbulence," A.R.C. R&M No. 1919 (1939).
65. Hall, E. H. "A Thermo-Electric Method of Studying Cylinder Condensation in Steam Engine Cylinders," *Trans. AIEE* 8: 236-245 (1891).
66. Callendar, M. C., and J. T. Nicholson. "On the Law of Condensation of Steam Deduced from Measurements of Cylinder Temperature Cycles of a Steam Engine," *Proc. Inst. Civil Eng.* 130 (Part 1): paper 3024 (1897).
67. Retchy, C. "Beitraege zur Untersuchung Annaehernd Geordneter Lufstroeme, der Motorwagen," 15: 438-468; 524-528 (1912).
68. Tucker, W. S., and E. T. Paris. "A Selective Hot-Wire Microphone," *Phil. Trans. Roy. Soc., London* A221: 389-430 (1921).
69. Gibbs, F. A. "A Thermoelectric Bloodflow Recorder in the Form of a Needle," *Proc. Soc. Exp. Biol., NY* 31: 141-146 (1933).

70. Frenkiel, F. N. "Etude Statistique de la Turbulence, Part 2, Influence de la Longueur d'un fil Chaud Compense sur La Measure de la Turbulence," O.N.E.R.A. Rapp. Tech. No. 37 (1948).
71. Ossowsky, E. "Constant Temperature Operation of the Hot-Wire Anemometer at High Frequency," *Rev. Sci. Instrum.* 19: 881-889 (1948).
72. Kovasznay, L. S. G. "Simple Analysis of the Constant Temperature Feedback Hot-Wire Anemometer," Johns Hopkins University Report Aero/JHU CM-478 (1948).
73. Betchov, R. "Theorie Non-Lineaire de L'anemometre a Fil Chaud," *Koninkl. Acad. Van Wetenschappen Proc.* 52: 195-207 (1949).
74. Ludwieg, H. "Ein Geraet zur Messung der Wandschubspannung Turbulenter Reibungsschichten," *Ingenieurarchiv* 17: 207 (1949).
75. Kovasznay, L. S. G., and S. I. A. Toernmark. "Heat Loss of Wires in Supersonic Flow," Bumble Bee Series Report No. 127 (1950).
76. Lowell, H. H. "Design of Hot-Wire Anemometers for Steady-State Measurements at Transonic and Supersonic Air Speeds," *NACA TN* 2117 (1950).
77. Lowell, H. H. "Response of Two-Material Laminated Cylinder to Simple Harmonic Environment Temperature Change," *J. Applied Phys.* 24: 1473-1478; *NACA TN* 3514 (1955).
78. Ling, S. C. "Measurement of Flow Characteristics by the Hot-Film Technique," PhD thesis, University of Iowa (1955).
79. Kresa, K., and E. Molloe-Christensen. "A Multiple Hot-Wire Array for Measurement of Downwash Integrals," *J. Aerospace Sci.* 27: 141-142 (1960).
80. Staritz, R. F. "Die Elektronische Messung der Stroemungsgeschwindigkeit und der Turbulenz," *VDI Zeitschrift* 102: 94-97 (1960).
81. Wehrmann, O. "Weiterentwicklung und Neuartige Anwendung der Hitzdrahtmesstechnik," *Konstruktion* 13: 183-186 (1961).
82. Berger, E., P. Freymuth and E. Froebel. "Anwendung der Regeltechnik bei der Entwicklung eines Konstant-Temperatur-Hitzdrahtanemometers," *DVL Reports* 282/283 (1963).
83. Sajben, M. "Hot-Wire Anemometer for Measurement in Liquid Mercury," *Sc. D. Thesis,* M.I.T., Mech. Eng. Dept. (1964).
84. Goldschmidt, V. W. "Measurement of Aerosol Concentration with a Hot-Wire Anemometer," *J. Colloid Sci.* 20: 617-634 (1965).
85. Zakin, J. L., and G. K. Patterson. "Measurement of Intensity of Turbulence in Drag Reducing Organic Polymer Solutions," Report on NASA contract NGR-26-0003, Chem. Eng. Dept., University Missouri-Rolla (1965).
86. Freymuth, P. "Ueber Einige Spezielle Probleme der Hitzdrahtmesstechnik," Report DLR FB 66-03 (1966).
87. Bellhouse, B. J., and D. L. Schultz. "The Determination of Fluctuating Velocity in Air With Heated Thin Film Gauges," *J. Fluid Mech.* 29: 289-295 (1967).
88. Hoole, B. J., and J. R. Calvert. "The Use of a Hot-Wire Anemometer in Turbulent Flow," *J. Roy. Aeron. Soc.* 71: 511-513 (1967).
89. Kovasznay, L. S. G. "Should We Still Use Hot-Wires?" *Adv. in Hot-Wire Anemometry, Proc.* 1-12 (1968).
90. Freymuth, P. "Nonlinear Control Theory for Constant Temperature Hot-Wire Anemometers," *Adv. in Hot-Wire Anemometry, Proc.* 203-218 (1968).
91. Maye, J. P. "Error Due to Thermal Conduction Between the Sensing Wire and its Supports When Measuring Temperature," *DISA Information* 9: 27-29 (1970).
92. De Haan, R. E. "Dynamic Theory of a Short Wire Normal to an Incompressible Air Flow, Constant Resistance Operation," *Appl. Sci. Res.* 24: 231-260 (1971).
93. Weidmann, P. D., and F. K. Browand. "Analysis of a Simple Circuit for Constant Temperature Anemometry," *J. Phys. E. Sci. Inst.* 8: 553-560 (1975).

CHAPTER 5

TURBULENT MIXING AND ITS MEASUREMENT

Gary K. Patterson
Department of Chemical Engineering
University of Missouri-Rolla
Rolla, Missouri 65401

CONTENTS

MIXING AND TURBULENT DIFFUSION

The degree of unmixedness of two fluids in contact is called the segregation. If two miscible fluids are mixing in a turbulent flow with an initial segregation of a relatively small scale, two processes combine to cause the fluids to mix. One is the stretching of fluid elements by the turbulence, causing the concentration gradients of the two fluids to increase and also drawing the regions of the two fluids closer together. The other process is molecular diffusion, which ultimately mixes the two fluids on a molecular level and which is greatly enhanced by the turbulent stretching. The process may be modeled one-dimensionally, either timewise [1] or spatially [2].

If the two fluids enter the mixing region in such a way that they fill two separate but adjacent regions, so that fluid dispersion over relatively large distances must occur to bring about mixing, then a third process—turbulent diffusion—must take place. Turbulent diffusion may be treated as a large-scale phenomenon. The process must now be modeled as a two-dimensional [3,4] or a three-dimensional system. It must be emphasized that the difference between this case and the former is the size of the mixing region (pipe, jet, stirred-mixer, etc.) compared to the initial scale of the segregation. A stirred tank, for instance, with fluid injection by sparge rings with tiny jets would fit the case of relatively small initial segregation scale, while most mixing jets or diffusion-flame combustors require substantial turbulent diffusion for mixing and would fit the second case.

The mixing and diffusion phenomena described above are purely physical processes. If the fluids involved react chemically under the conditions of mixing, considerable interaction between the two processes may occur. This is particularly true if the reaction or reactions occur at about the same rate as the mixing, so that concentration gradients are increased by reaction, *and* reaction rate is increased by mixing rate. Very important interactions occur if more than one reaction occurs simultaneously, the selectivity to specific products being a strong function of mixing rate.

MODELS OF MIXING

Mixing Model Complexity

The modeling of mixing processes serves as motivation for the necessary measurements. The first half of this chapter is, therefore, about how mixing is modeled.

Some mixing processes simply blend ingredients. However, most mixing is carried out to bring about some chemical reaction, interaction or combination of the mixed fluids. Therefore, most mixing may be characterized as being done in a reactor. The least complex model one might conceive for a mixing vessel with chemical reaction would be a *lumped system,* in which the effect of the degree of mixing on chemical conversion is correlated with externally observable variables. Those variables could be the Damköhler number based on chemical rate constant and average residence time; some measure of mixing intensity, such as impeller power per unit volume or mass; and geometric factors, such as impeller:tank diameter ratio, tank diameter:height ratio or static mixer element size:pipe size ratio. Corrsin [1] has described such an approach. Such an approach may give successful narrow-range correlations for single-parameter variations for a specific chemical and mixer system. Any chance of a very general result, however, is slim.

The treatment of the mixed reactor using residence time distribution (RTD) information is essentially a Lagrangian approach—the history of each fluid element is analyzed. If the degree to which each fluid element mixed with other fluid elements throughout its history in the mixing vessel were determined, it then would be possible to model the mixing effects on any chemical reaction. This is generally not possible, but an approximation may be made using a model based on the movement of fluid elements within the mixer. The RTD approach has been reviewed completely by Nauman [5].

In the coalescence-dispersion (c-d) approach, packets of fluid entering the vessel are moved through it in a way to approximate the convection pattern in the real case. In contrast to the segregated model based on RTD concepts, the packets are allowed to mix with one another according to a set of rules designed to simulate the mixing rates that actually would occur. Since each packet maintains its identity throughout its passage through the reaction vessel, the model is Lagrangian, but Eulerian information on the distribution of concentrations may be obtained at each time interval. Coalescence-dispersion is easily used to simulate unsteady-state processes and works well for complex chemistry. Nonpremixed feed streams are no problem. Descriptions of the c-d method are given in the literature [6-9] as are related methods by Spalding (ESCIMO) [10] and Rhodes et al. [11].

Coalescence-dispersion modeling may be done by actually simulating the events for each fluid packet or by using mathematical functions describing probabilities of coalescence directly. In either case, such modeling consumes computer time. Much more efficient computation is achieved if an Eulerian model is conceived in which balance equations about macroscopic elements or cells of the mixer are written. If the segregation, as well as molar concentration, is regarded as a property of the fluid, then balance equations may be written for all cells in the vessel for each of these two properties, plus any others of importance that may vary with reaction conversion, such as temperature, viscosity or density. If the equations are solved in the unsteady-state mode, a good computer code for ordinary differential equations is necessary to solve the set. Stiff sets may result that require special methods to handle them.

If the balance equations are solved for a steady-state solution and they are linear, then simple matrix inversion suffices. If, however, any equation set is nonlinear, a Newton method for solution is necessary. Even so, the cell-balance approach is much faster than the coalescence-dispersion simulation method.

The coalescence-dispersion and cell-balance models may be used where convection patterns in the vessel and local mixing rates are known, whence flowrates into and out of cells can be calculated. Where this is not known, where greater generality is desired or where substantial turbulent diffusion exists, a *finite-difference* model (or finite-element model, in some cases) might be best. The cell-balance model is, of course, an approximation to the finite-difference model and gives the same results when the cell size is small enough to have the same resolution. An advantage of the finite-difference technique, however, is that momentum balance equations and mass balance equations with significant turbulent diffusion may be solved. The computer time required depends on the number of balance equations and the number of grid points in the simulation. Methods for unsteady-state problems have been formulated and used in finite-difference modeling, but steady-state problems are most often solved using this approach.

There are various levels of complexity that may be used for finite-difference modeling to obtain the distributions of fluid velocities, turbulence intensities, rates of mixing (or turbulence energy dissipation rates and turbulence scales), segregations, thermal energies and concentrations for each component in the mixing vessel. The hierarchy of such models for the hydrodynamic part (velocities, turbulence, energy dissipation and scales) has been discussed in detail by Launder and Spalding [12]. To model mixing of chemical components, it is necessary to use a model of sufficient complexity that the local mixing rate information can be derived. Usually this means a hydrodynamic model, which includes balance equations for mass, momentum, turbulence energy and turbulence energy dissipation rate, even though simpler finite-difference models exist. Formulations based on the use of turbulence energy and turbulence energy dissipation rate balances are called *two-equation models* because they involve two equations beyond the mass and momentum equations. These formulations are the most widely used and accepted for turbulence modeling.

Transport Equations for Hydrodynamics and Mixing

The most fundamental approach to modeling the interactions of mixing, turbulent diffusion and convection is through the formulation and solution of the differential balance equations that describe the process. In this chapter, the general formulation will be followed by various simplifications and variations of the modeling method.

The modeling of mixing without large-scale turbulent diffusion, both with and without chemical reaction, has been studied extensively. Both hydrodynamic [1,2] and coalescence-dispersion (c-d) [6,13] methods have been used to model mixing successfully in such one-dimensional systems.

Two-dimensional mixing systems are considerably more difficult to model because of the turbulent diffusion that affects all the dependent variables—concentration, velocity, turbulence energy, turbulence energy dissipation rate and segregation if a full finite-difference model is used. Such two-dimensional systems have been modeled successfully without reaction for the round free jet [14], the confined round jet [15,16] and the coaxial free jet [4], as well as some other more complex geometries. The round free jet [16] and the coaxial free jet [4] also have been modeled for the second-order reaction occurring between the two mixing fluids. Lockwood and Naquib assumed chemical equilibrium [16]; Patterson et al. [4] did not. All the modeling efforts above were finite-difference models based on the use of the two-equation turbulence model for the hydrodynamics.

The differential balance equations that must be included in a general formulation are those for overall mass, momentum, component mass (concentration) and thermal energy. To facilitate the solution of the momentum balance, the balances for the turbulence energy and turbulence energy dissipation rate or turbulence length scale are frequently added to complete a two-equation model. More balance equations may be necessary if a full Reynolds stress model for the hydrodynamics is used: three Reynolds stress balance equations replace the turbulence energy equation.

To improve the mixing model, it is necessary to add balance equations for the segregation of each component. An alternate scheme with equations for the velocity-concentration correlations also has been employed. Hill [17] has given a useful review of the various modeling methods commonly used. If temperature fluctuations are important, it also may be necessary to include balance equations for temperature-velocity and temperature-concentration interactions (correlations).

Modeling of Turbulence

Here, the two-equation approach using equations for turbulence energy and dissipation rate will be described. The overall mass balance equation, frequently called the continuity equation, is as follows:

$$\frac{D\rho}{Dt} = -\rho\left(\frac{\partial U_i}{\partial x_i}\right) \tag{1}$$

where $D/Dt = \partial/\partial t + U_j\partial/\partial x_j$.

The momentum balance equation for instantaneous variables (constant density and viscosity) is

$$\frac{DU_i}{Dt} = -\frac{1}{\rho}\frac{\partial p}{\partial x_i} + \nu\left(\frac{\partial^2 U_j}{\partial x_i \partial x_j}\right) + g_i \tag{2}$$

After Reynolds-averaging and rearrangement,

$$\frac{D\bar{U}_i}{Dt} = -\frac{1}{\rho}\frac{\partial \bar{p}}{\partial x_i} + \nu\left(\frac{\partial^2 \bar{U}_i}{\partial x_j \partial x_j}\right) - \overline{u_i u_j} + \bar{g}_i \tag{3}$$

To model the momentum valence equation based on velocity gradients and turbulent kinematic viscosities, Equation 3 is rewritten in two dimensions as follows:

$$\frac{D\bar{U}_1}{Dt} = -\frac{1}{\rho}\frac{\partial \bar{p}}{\partial x_1} + \frac{\partial}{\partial x_1}\left(\frac{\nu_e + \nu}{1}\frac{\partial \bar{U}_1}{\partial x_1}\right) + \frac{\partial}{\partial x_2}\left(\frac{\nu_e + \nu}{1}\frac{\partial \bar{U}_1}{\partial x_2}\right) \tag{4a}$$

$$\frac{D\bar{U}_2}{Dt} = -\frac{1}{\rho}\frac{\partial \bar{p}}{\partial x_2} + \frac{\partial}{\partial x_1}\left(\frac{\nu_e + \nu}{1}\frac{\partial \bar{U}_2}{\partial x_1}\right) + \frac{\partial}{\partial x_2}\left(\frac{\nu_e + \nu}{1}\frac{\partial \bar{U}_2}{\partial x_2}\right) \tag{4b}$$

The turbulent kinematic viscosity, ν_e, is a momentum diffusivity and the body force term, g_i, has been dropped.

If Equation 2 is multiplied by u_i, the fluctuating velocity in the i-direction, each of the resulting three equations is Reynolds-averaged, then the corresponding terms are summed; the turbulence energy equation results:

$$\frac{Dk}{Dt} = \underbrace{-\frac{\partial}{\partial x_i}\left(\overline{u_i k} + \frac{1}{\rho}\overline{u_i p}\right)}_{\text{(diffusion)}} \underbrace{- \overline{u_i u_j}\frac{\partial \bar{U}_i}{\partial x_j}}_{\text{(production)}} \underbrace{- \nu\overline{\left(\frac{\partial u_i}{\partial x_j}\right)^2}}_{\text{(dissipation)}} \tag{5}$$

k represents the turbulent kinetic energy, $1/2\ (\overline{u_i^2})$. Again, to model the turbulence energy equation based on turbulence energy gradients, the equation is rewritten as follows:

$$\frac{Dk}{Dt} = \underbrace{\frac{\partial}{\partial x_i}\left(\frac{\nu_e + \nu}{\sigma_k}\frac{\partial k}{\partial x_1}\right) + \frac{\partial}{\partial x_2}\left(\frac{\nu_e + \nu}{\sigma_k}\frac{\partial k}{\partial x_2}\right)}_{\text{(diffusion)}}$$

$$+ \underbrace{\nu_e\left\{\left(\frac{\partial \bar{U}_1}{\partial x_2} + \frac{\partial \bar{U}_2}{\partial x_1}\right)^2 + 2\left(\frac{\partial \bar{U}_i}{\partial x_i}\right)^2\right\}}_{\text{(production)}} \underbrace{- \epsilon}_{\text{(dissipation)}} \tag{6}$$

The diffusivity for the turbulence energy is given by the ratio of the momentum diffusivity and a constant σ_k of the order one. The term for turbulence energy production rate is frequently reduced to $\nu_e(\partial\bar{U}_1/\partial x_2 + \partial\bar{U}_2/\partial x_1)^2$, and the term for dissipation rate is reduced to one dependent variable, ϵ, which must be modeled by a balance equation.

The balance equation for the turbulence energy dissipation rate may be obtained by taking the derivative of Equation 2 with respect to x_1, multiplying through by $2\nu \, \partial u_i/\partial x_1$, then Reynolds averaging. After some rearrangement the dissipation term as in Equation 5 appears as the dependent variable. For simplicity the complete equation is not shown. The model balance equation is given in the same form as the model equation for turbulence energy:

$$\frac{D\epsilon}{Dt} = \frac{\partial}{\partial x_1}\left(\frac{\nu_e + \nu}{\sigma_\epsilon}\frac{\partial \epsilon}{\partial x_1}\right) + \frac{\partial}{\partial x_2}\left(\frac{\nu_e + \nu}{\sigma_\epsilon}\frac{\partial \epsilon}{\partial x_2}\right) + \underbrace{C_1\nu_e\left\{\left(\frac{\partial \bar{U}_1}{\partial x_2} + \frac{\partial \bar{U}_2}{\partial x_1}\right)^2 + 2\left(\frac{\partial \bar{U}_i}{\partial x_i}\right)^2\right\}\epsilon/k}_{\text{(production)}} - \underbrace{C_2\epsilon^2/k}_{\text{(dissipation)}} \quad (7)$$

The production and dissipation terms "model" the actual collection of terms in the complete equation. Such an approximation is necessary to force closure of the set of differential balance equations so that a solution can be found. Again, the term $2(\partial\bar{U}_i/\partial x_i)^2$ is frequently dropped.

In Equations 1-7 the values of the "constants" are somewhat dependent on the flow geometry, but for most cases the following are recommended by Jones and Launder [18]: $C_1 = 1.44$, $C_2 = 1.92$, $\sigma_k = 1.0$, $\sigma_\epsilon = 1.3$. The turbulent kinematic viscosity, $\nu_e = C_D k^2/\epsilon$, where $C_D = 0.09$.

These equations constitute the $k - \epsilon$ two-equation model of turbulence. This model is only one of several methods of modeling the hydrodynamics of turbulent flows. The well-known Prandtl mixing-length and von Karman similarity models for the turbulent (eddy) viscosity lead to zero-equation models, that is, zero differential equations beyond the continuity equation and the momentum balance. There are several one-equation models that are useful for boundary layer flows [12]. They are usually not useful for the mixing problems considered in this chapter because most mixing is done in complex geometries, for which two-equation models are more effective. Other two-equation models are the $k - l$ model, the $k/l^2 - k$ model, the $k^{3/2}/l - k$ model and the $k^{1/2}/l - k$ model. Launder and Spalding [12] state that the three most thoroughly tested ($k - \epsilon$, $k - l$, and $k/l^2 - k$) are essentially equivalent. The relationship for the length scale $l = k^{3/2}/\epsilon$ is assumed to hold in each of these models.

More complex modeling methods, such as the Reynolds stress models, are available for hydrodynamic modeling of turbulent flows. Unfortunately, the number of equations to solve is larger, which is a great disadvantage since the mixing model also requires several equations. Also, as shown below, the $k - \epsilon$ model for hydrodynamics fits perfectly with the model for mixing.

Modeling Turbulent Mixing with Chemical Reaction

The differential balance equation for a component in a turbulent mixing flow may be written as

$$\frac{DC_i}{Dt} = D_{i?}\left(\frac{\partial^2 C_i}{\partial x_n^2}\right) + R_i \quad (8)$$

$D_{i?}$ is the diffusivity of component-i in a mixture. If the source term R_i is modeled by an irreversible second-order reaction between components A_i and A_j, then

$$R_i = -K\,C_iC_j \quad \text{for } i \neq j \quad (9)$$

K is the rate constant for $A_i + A_j \rightarrow A_k$.

If C_i, U_n and K are divided into average and fluctuating quantities, such as $\overline{C}_i$ and c_i, then the resulting equation is Reynolds-averaged to yield

$$\frac{D\overline{C}_i}{Dt} = D_{i?}\left(\frac{\partial^2 \overline{C}_1}{\partial x^2}\right) - \frac{\partial}{\partial x_n}(\overline{u_n c_i}) - \overline{K}(\overline{C}_i\overline{C}_j + \overline{c_i c_j}) - \overline{kc_i}\overline{C}_j - \overline{kc_j}\overline{C}_i - \overline{kc_i c_j} \tag{10}$$

(diffusion) (reaction)

If the temperature is constant or if temperature fluctuations are not highly correlated with concentration fluctuations, the last three terms of Equation 10 involving k may be dropped. That will be done throughout the remainder of this chapter because the modeling of such temperature fluctuation effects is not yet well developed.

The simplified form of Equation 10 may be rewritten in the two-dimensional form of Equations 4, 6 and 7, as follows:

$$\frac{D\overline{C}_i}{Dt} = \frac{\partial}{\partial x_1}\left(\frac{\nu_e + \nu}{\sigma_c}\frac{\partial \overline{C}_i}{\partial x_1}\right) + \frac{\partial}{\partial x_2}\left(\frac{\nu_e + \nu}{\sigma_c}\frac{\partial \overline{C}_i}{\partial x_2}\right) - \overline{K}(\overline{C}_i\overline{C}_j + \overline{c_i c_j}) \tag{11}$$

This is the model component mass balance equation. The last term is later given as $P_{\overline{c}_1}$.

If Equation 8 is multiplied by c_i then Reynolds-averaged, the following transport (balance) equation for segregation c_i^2 occurs:

$$\frac{D\overline{c_i^2}}{Dt} = -\frac{\partial}{\partial x_n}(\overline{u_n c_i^2}) - 2\overline{c_i u_n}\frac{\partial \overline{C}_i}{\partial x_n} - 2D_{i?}\overline{\left(\frac{\partial c_i}{\partial x_n}\right)^2} - \overline{K}(\overline{C}_i\overline{c_i c_j} + \overline{C}_j\overline{c_i^2} + \overline{c_i^2 c_j}) \tag{12}$$

(turbulent diffusion) (production) (dissipation) (reaction effect)

Just as in the hydrodynamic part, a number of terms must be modeled to produce closure of the set of differential balance equations. Other equations may be generated, such as a balance equation for $\overline{u_n c_i}$, but closure always must be approximated at some level. Here, closure with the segregation equation will be shown.

It is necessary to model the dissipation term and the production term to account for mixing and diffusing without reaction. If second-order reaction occurs, the terms involving $\overline{c_i c_j}$ and $\overline{c_i^2 c_j}$ also must be modeled all in terms of $\overline{c_i^2}$ and/or $\overline{C}_i$. Corrsin [1] showed that the rate of mixing of miscible components, hence the rate of dissipation of segregation, may be modeled as a function of a scalar length scale and the rate of turbulence energy dissipation in isotropic turbulence. Because both the final rate of mixing and the rate of turbulence energy dissipation are dependent primarily on the smallest scales of turbulence, which tend to be isotropic, Corrsin's model has proven useful even for shear flow turbulence [2,4,19]. His relationship may be written as

$$2D_{i?}\overline{\left(\frac{\partial c_i}{\partial x_n}\right)^2} = 2\,\overline{c_i^2}/[4(k/\epsilon) + (\nu/\epsilon)^{1/2}\ln N_{Sc}] \tag{13}$$

Here the scalar length scale does not appear because it has been assumed to be approximately equal to $k^{3/2}/\epsilon$, a seemingly crude assumption, but one that seems to work. The usefulness of Equation 13 is that it involves k and ϵ as hydrodynamic variables, which are provided by solutions to Equations 6 and 7. Spalding [3] modeled the segregation dissipation rate as $Cg_2\epsilon\overline{c_i^2}/k$, where Cg_2 was about 0.2. If in Equation 13 the Schmidt number, N_{Sc}, is small, the segregation dissipation rate becomes $0.5\epsilon\overline{c_i^2}/k$, which is a higher rate of dissipation than that predicted by Spalding's term.

The turbulent diffusion term is modeled as in all other cases and shown in the final equation.

The most difficult term to handle is the segregation production term. Spalding [3] modeled the segregation production term analogously to the turbulent energy production term (see Equation 6). Therefore, he arrived at the following:

$$2\overline{c_i u_n}\frac{\partial \overline{C}_i}{\partial x_n} = -Cg_1 \nu_e \left(\frac{\partial \overline{C}_i}{\partial x_n}\right)^2 \tag{14}$$

where Cg_1 has a value of approximately 3.

Another approach to the problem is to make use of two ideas: (1) new fluid diffusing into a region by turbulent transport contributes to segregation, regardless whether the fluid is perfectly mixed; and (2) the segregation for unmixed components is given by the product of their average concentrations. A balance equation for imaginary total segregation may be written as follows:

$$-\underset{\text{(convection)}}{\frac{D(\overline{C}_i(C_T-\overline{C}_i))}{Dt}} = \underset{\text{(production)}}{2\overline{c_i u_n}\frac{\partial \overline{C}_i}{\partial x_n}} \tag{15}$$

It is assumed that the turbulent diffusion of component-i is responsible for the production rate of segregation, and that the rate of imaginary total segregation production is the same. The imaginary total segregation does not diffuse; therefore, no diffusion term is involved.

The two-dimensional model equation for segregation with second-order reaction is now

$$\frac{D\overline{c_i^2}}{Dt} = \frac{\partial}{\partial x_1}\left(\frac{\nu_e+\nu}{\sigma_c^2}\frac{\partial \overline{c_i^2}}{\partial x_1}\right) + \frac{\partial}{\partial x_2}\left(\frac{\nu_e+\nu}{\sigma_c^2}\frac{\partial \overline{c_i^2}}{\partial x_2}\right) + \frac{D(\overline{C}_i(C_T-\overline{C}_i))}{Dt}$$
$$- 2\overline{c_i^2}/[4(k/\epsilon) + (\nu/\epsilon)^{1/2}\ln N_{Sc}] - K_{ijk}(\overline{C}_i\overline{c_i c_j} + \overline{C}_j\overline{c_i^2} + \overline{c_i^2 c_j}) \tag{16}$$

The combination of the last three terms is later given as $P_{c_i^2}$.

Closure Approximations for Second-Order Chemical Reactions

Two terms have not yet been modeled: $\overline{c_i c_j}$ and $\overline{c_i^2 c_j}$. Modeling of these terms is really the crux of modeling mixing with second-order chemical reactions. Closure of the component balance and segregation balance equations may be accomplished by a variety of methods: assumption of chemical equilibrium or near chemical equilibrium, instantaneous local mixing and reaction, higher-order moment closure, use of probability density functions of the component concentrations, and the use of simple physical models such as coalescence–dispersion, fluid-strand diffusion, eddy mixing and "exchange-with-the-mean." These latter methods actually are substitutes for the segregation balance equation and not closure methods in the usual sense. Details of each of the closure methods mentioned will be elaborated below.

Equilibrium Assumption

If the chemical reaction is very fast compared to the mixing rate, it may be assumed that any mixed reactants are immediately converted. No rate expression is necessary, therefore, and the last term of Equation 12 is related to the rate of mixing. Brodkey [20] showed that the segregation, $\overline{c^2}$, of two totally segregated fluids is the product of their average concentrations, $C_1C_2 (= \overline{c_1^2} = \overline{c_2^2})$. The rate of decrease of $\overline{C_1C_2}$ due only to reaction may be expressed as follows:

$$-\left(\frac{\partial(\overline{C}_1\overline{C}_2)}{\partial t}\right) = -\overline{C}_1\left(1+\frac{\overline{C}_2}{\overline{C}_1}\right)\left(\frac{\partial \overline{C}_1}{\partial t}\right)_r = \left(\frac{\partial \overline{c_1^2}}{\partial t}\right)_m \tag{17}$$

because $\left(\frac{\partial \overline{C}_1}{\partial t}\right)_r = \left(\frac{\partial \overline{C}_2}{\partial t}\right)_r$ when they are due only to reaction. Therefore, the rate of reaction decrease of the component concentration becomes

$$-\left(\frac{\partial \overline{C}_1}{\partial t}\right)_r = -\left(\frac{\partial \overline{C}_2}{\partial t}\right)_r = -\left(\frac{\overline{C}_1}{\overline{C}_1{}^2 + \overline{C}_1\overline{C}_2}\right)\left(\frac{\partial \overline{c_1{}^2}}{\partial t}\right)_m \tag{18}$$

With the rate of segregation decrease, $-\left(\frac{\partial \overline{c_i{}^2}}{\partial t}\right)_m$, given by the Corrsin relation, Equations 11 and 12 become

$$\frac{D\overline{C}_1}{Dt} = \frac{\partial}{\partial x_1}\left(\frac{\nu_e + \nu}{\sigma_{\bar{c}}}\frac{\partial \overline{C}_i}{\partial x_1}\right) + \frac{\partial}{\partial x_2}\left(\frac{\nu_e + \nu}{\sigma_{\bar{c}}}\frac{\partial \overline{C}_i}{\partial x_2}\right) - \frac{\overline{C}_i\overline{c_i{}^2}}{\overline{C}_i{}^2 + \overline{C}_i\overline{C}_j}\,\frac{1}{[4(k/\epsilon) + (\nu/\epsilon)^{1/2}\ln N_s]} \tag{19}$$

$$\frac{D\overline{c_i{}^2}}{Dt} = \frac{\partial}{\partial x_1}\left(\frac{\nu_e + \nu}{\sigma_c{}^2}\frac{\partial \overline{c_i{}^2}}{\partial x_1}\right) + \frac{\partial}{\partial x_2}\left(\frac{\nu_e + \nu}{\sigma_c{}^2}\frac{\partial \overline{c_i{}^2}}{\partial x_2}\right) + \frac{D(\overline{C}_i(C_T - \overline{C}_i))}{Dt} - 2\overline{c_i{}^2}/[4(k/\epsilon) + (\nu/\epsilon)^{1/2}\ln N_S] \tag{20}$$

The last term of Equation 12 does not appear in Equation 20 because with total segregation it becomes zero. Equations 19 and 20 include no effect of reaction on mixing rate, a possible source of error. Because no other moments besides $\overline{c_i{}^2}$ appear in the equations, they are closed.

The condition under which the chemical equilibrium approach is valid may be expressed in terms of the ratio of the time constants of homogeneous reaction and mixing. The time constant for the second-order reaction is $\tau_R = (KC_i)^{-1}$, and the time constant for mixing is $\tau_M = [4(k/\epsilon) + (\nu/\epsilon)^{1/2} \ln N_S]$. To apply the equilibrium approximation, it is necessary that $\tau_M >> \tau_R$. If $\tau_R >> \tau_M$, then the mixing rate has no effect on the reaction and it may be treated as a homogeneous reaction (see next section). If τ_R and τ_M are of the same order, it is necessary to use a closure method, which accounts for the interaction of mixing and reaction.

The foregoing may be illustrated by reference to Figures 1 and 2. Figure 1 illustrates the real case in which the concentration profiles for mixing components are essentially

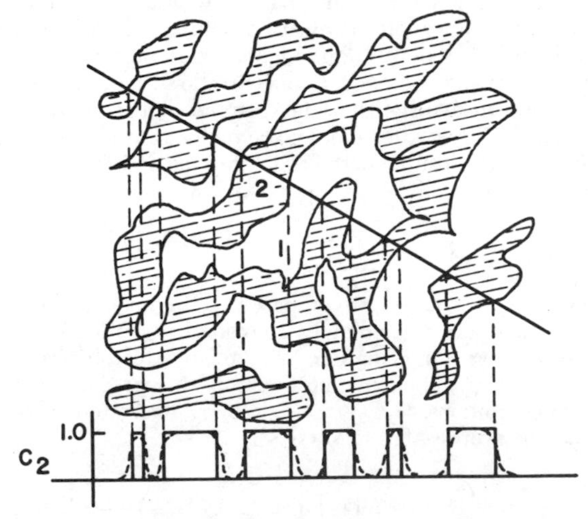

Figure 1. Mixing process for two fluids.

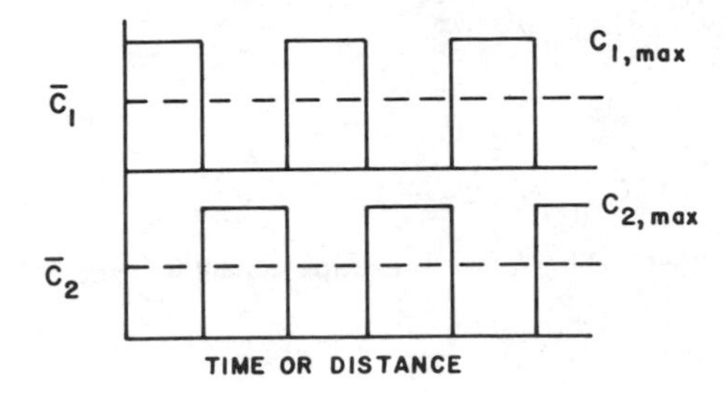

Figure 2. Concentration profiles for completely segregated fluids.

continuous and interdiffusion is significant. That occurs with slow or even rapid reactions. Figure 2 shows the complete segregation maintained when an infinitely fast reaction occurs. No significant interdiffusion results.

Instantaneous Local Mixing and Reaction Assumption

In many cases, a "conserved scalar" approach is used in the formulation of balance equations for fast reactions and mixing. The method has been reviewed thoroughly by Bilger [21]. Gosman et al. [22] have used the mixture fraction, a conserved scalar, to formulate a model for combustion, which is useful in many types of problems. The mixture fraction is defined as

$$\xi = \frac{\overline{C_k^o} - \overline{C_{k2}}}{\overline{C_{k1}} - \overline{C_{k2}}} \tag{21}$$

where 1 and 2 refer to feeds 1 and 2. $\overline{C}_k$ is the unreacted concentration of component $-k$. For a given value of ξ, which corresponds to a given fraction of feed 1 in the mixture, it is assumed that the fluids are totally *mixed* instantaneously and reaction is instantaneously *complete*. ξ is a conserved scalar because it is not a function of reaction extent.

For a second-order, irreversible reaction that is very fast, as follows—$A_1 + nA_2 \rightarrow A_3$—if the mixing is also very fast the local concentrations of A_1, A_2 and A_3 are functions of ξ and the feed concentrations. If $\overline{C}_{12}^{f}$ and $\overline{C}_{21}^{f}$ are zero (only component 1 in feed 1 and only component 2 in feed 2), the local concentrations with instantaneous reaction are as follows:

$$\overline{C}_1 = \xi\overline{C}_{11}^{f} + n(\xi - 1)\overline{C}_{22}^{f} \text{ for } \xi > \xi_s; \overline{C}_1 = 0 \text{ for } \xi \leq \xi_s$$

$$\overline{C}_2 = (1 - \xi)\overline{C}_{22}^{f} - \frac{1}{n}\overline{C}_{11}^{f} \text{ for } \xi < \xi_s; \overline{C}_2 = 0 \text{ for } \xi < \xi_s$$

Such a formulation does not require the computation of the degree of mixing, so along with equations for the hydrodynamics, the following conserved mass equation may be used to model the chemical reaction in two dimensions:

$$\frac{D\xi}{Dt} = \frac{\partial}{\partial x_1}\left(\frac{\nu_e + \nu}{\sigma_\xi}\frac{\partial \xi}{\partial x_1}\right) + \frac{\partial}{\partial x_2}\left(\frac{\nu_e + \nu}{\sigma_\xi}\frac{\partial \xi}{\partial x_2}\right) \tag{22}$$

There is no generation term.

Spalding [23] has proposed a method for modeling the effect of incomplete mixing when the reaction is much faster than the mixing. In addition to the hydrodynamic equations and Equation 22 (with a source term $\overline{S}_\xi$), Equation 20 (with the Spalding term for $\overline{c_k^2}$ — production) is used. The level of $\overline{c_k^2}$ is used to determine the values ξ_+ and ξ_-, the upper and lower extents of concentration fluctuations. The source term was given as follows:

$$\overline{S}_\xi = -\{\alpha r_- + (1-\alpha) r_+\} C_{EBU} |\partial \overline{U}/\partial x_2| \tag{23}$$

where $\alpha = (\xi_+ - \xi)/(\xi_+ - \xi_-)$

$$r_+ = (\xi_+ \overline{C}_{1k} - \overline{C}_{k+})(\overline{C}_{k+} - C^*_{k+})/(\xi_+ \overline{C}_{1k} - C^*_{k+})$$

$$r_- = (\xi_- \overline{C}_{1k} - \overline{C}_{k-})(\overline{C}_{k-} - C^*_{k+})(\xi_- \overline{C}_{1k} - C^*_{k-})$$

$\overline{C}_{k+}$ and $\overline{C}_{k-}$ must lie on a straight line with $\overline{C}_i$. The "eddy breakup" constant $C_{EBU} \approx$ 0.5.

Higher-Order Moment Closure

If other balance equations for correlations such as $\overline{u_i c_k}$ and for higher-order moments such as $\overline{c_k{}^3}$ are formed, closure may take place at a higher order than in Equations 12 and 13. Typically, at some level the moments are assigned to be zero so that closure is accomplished.

Borghi [24] has investigated the possibilities for higher-order moment closure. He included in the source term for the component mass balance all the terms generated in Equation 10. By expressing the instantaneous rate constant as an Arrhenius term, $R_i = KC_1C_2T_{ref}\exp(-T_a/T)$, the source term was eventually put into a series expansion form for which the higher-order terms approached zero under certain circumstances. Model equations were necessary for the terms $\overline{h^2}$ (mean square enthalpy fluctuation), $\overline{c_1{}^2}$, $\overline{c_1 c_2}$, $\overline{c_2 h}$ and $\overline{c_2 \xi}$ for a two-component reacting mixture, in addition to the basic hydrodynamic equations, a mass balance equation for component-2, a balance equation for ξ and a balance equation for $\overline{H}$, the average enthalpy. The complete model was a 13-equation model for only two reacting components, an illustration of the complexity of complete modeling of mixing with reaction.

Probability Density Functions

Since the source terms in the mass balance equations generate all of the difficult, but important, higher-order terms that in some way must be modeled, attempts have been made to represent the statistical distribution of the basic variables, which are concentrations of each component and temperature (or enthalpy). At small scales, where most of the important interactions occur in mixing with a chemical reaction, the distributions of those quantities become more random than at the large scales, where great coherency exists. This has led many investigators to use approximations to the probability density distributions or functions (pdf) of each of the important quantities to close the equations. Figure 3 illustrates the typical pdf in a real mixing system.

Although very few measurements of pdf for mixing components have been made, a number of modelers have proposed model pdf for use in modeling chemical reactions. Lockwood and Naquib [16] have proposed a clipped-normal pdf for the function $p(\xi)$, where $\overline{\phi_1 \phi_2} = \int_0^1 \phi_1(\xi)\phi_2(\xi)p(\xi)d\xi$ represents the averaged value of products

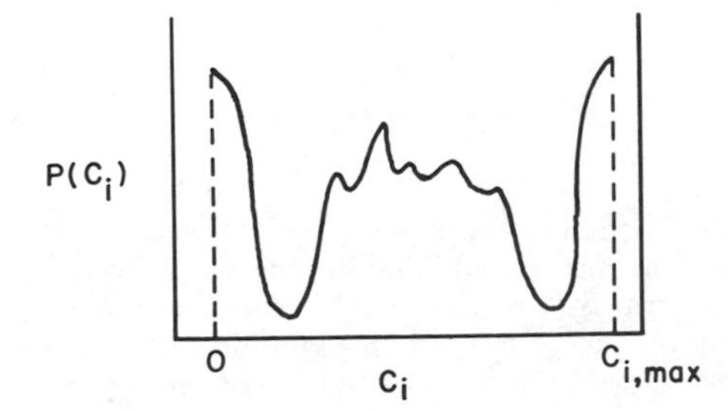

Figure 3. Probability density functions for mixing—probability of finding a concentration, c_i.

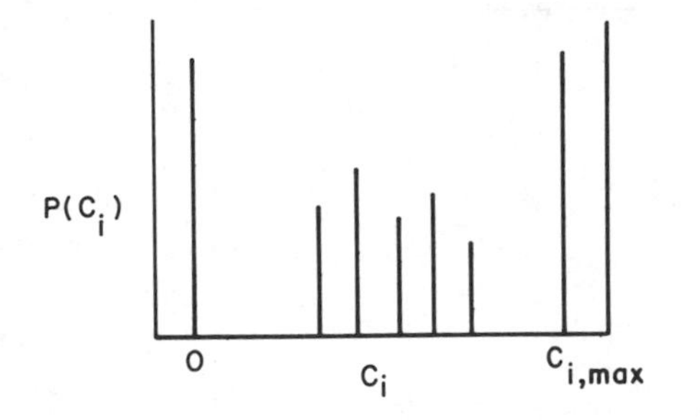

Figure 4. Spiked pdf representative of the "most typical eddy" proposal.

of functions of ξ, such as c_1c_2. Any correlation quantity used in the closure of the model equations is thus determined. Another proposed pdf is the spiked-distribution generated by the "most typical eddy," described by Donaldson [25]. The type of pdf produced by the "most typical eddy" idea is shown in Figure 4.

A much simpler version of that approach was formulated previously by Patterson [2] and may be called the "interdiffusion" model. Application of the interdiffusion model results in the following relationships:

$$\begin{aligned} \overline{c_1c_2} &= -\overline{c_1{}^2}(1-\gamma)/(\beta(1+\gamma)) \\ \overline{c_1{}^2c_2} &= 2(\overline{c_1{}^2})^{3/2}\gamma(1-\gamma)^{1/2}/(\beta(1+\gamma)^{3/2}) \end{aligned} \tag{24}$$

where $\beta = C_{10}/C_{20}$ and $\gamma = (\beta\overline{C_1C_2} - \overline{c_1{}^2})/(\beta\overline{C_1C_2} + \overline{c_1{}^2})$. Comparisons with experiments showed that best results were obtained when $\overline{c_1{}^2c_2}$ was set equal to zero. The interdiffusion model is a three-spike pdf: one at zero, one at pure component and one at the average concentration.

A relatively new approach to closure called the "three-equation model" has been tested by Panagopoulos [26]. In the three-equation model, segregation balance equations for components 1 and 2 and a balance equation for $\overline{c_1c_2}$ are solved with the assumption that $\overline{c_1{}^2c_2}$ and $\overline{c_1c_2{}^2}$ are zero. This approach actually gives a faster computer solution than the interdiffusion model described above.

Application to Complex Geometries

An advantage of the fundamental approach to modeling the hydrodynamics and mixing described so far is its ability to describe even quite complex two- and three-dimensional geometries. Examples of such applications of modeling of mixing and reaction are given by Spalding for a bunsen burner [23], by Ellail et al. [27] for a three-dimensional combustion chamber, by Lockwood and Nagib [16] in a round free jet diffusion flame and by Patterson et al. [4] in a coaxial mixing jet. Figures 5 and 6 show resulting velocity and turbulence profiles for a coaxial jet when a two-equation model (k-ϵ) is used [4]. Figures 7, 8 and 9 show corresponding profiles of concentration and segregation without reaction, and concentration profiles with reaction, when the interdiffusion model is applied. Not enough is known about the similarity rules that may exist in such complex, reacting systems needed to scale them up or down without resorting to full modeling. Therefore, the recommended scaleup method in such a case would be to confirm the modeling at the pilot scale with experimental measurements, then compute results at larger scales. Since the modeling method above depends on physics as fundamental as possible, some confidence should be possible in the scaleup based on it.

For less complex geometries, which are essentially one-dimensional or for which turbulent diffusion is not important, models based on interconnected tanks or cells may be applied. The finite-difference solution of the hydrodynamic equations is not sought then, but flowrates between the tanks usually are determined from experimentally

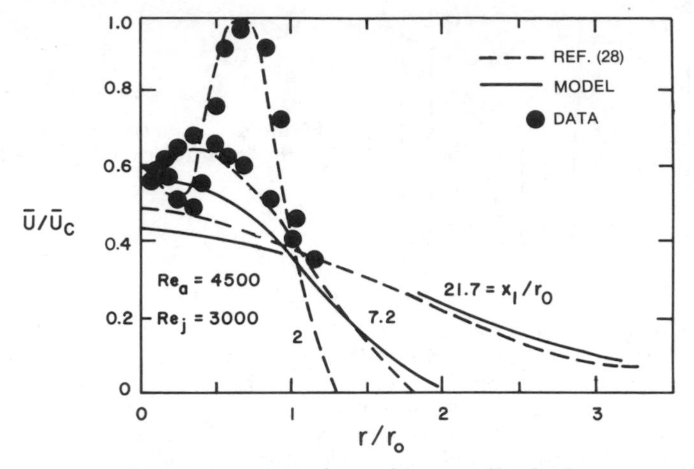

Figure 5. Coaxial jet velocity profiles [28].

measured flow patterns. Levels and scales of turbulence for each of the tanks also are usually determined from experiments. Only the component mass balance and segregation balance equations are solved (Equations 12 and 13). Such an approach has been applied to the turbulent tubular reactor [2] and the stirred-tank reactor [29]. Figure 10 shows results of such modeling in mixing tanks with no reaction. Scaleup of the mixing [30] and reaction in such geometries seems possible based on the use of the proper dimensionless groups and absolute geometric similarity. Figures 11 and 12 show dimensionless correlations of modeling results for second-order reaction in stirred tanks (turbine and propeller). Patterson [31] reviewed the results of the application of this approach.

Modeling Complex Reaction Chemistry with Mixing

Conversion, yield and selectivity in complex (multiple) reactions in a homogeneous mixing environment are determined essentially by the relative rates of reaction and

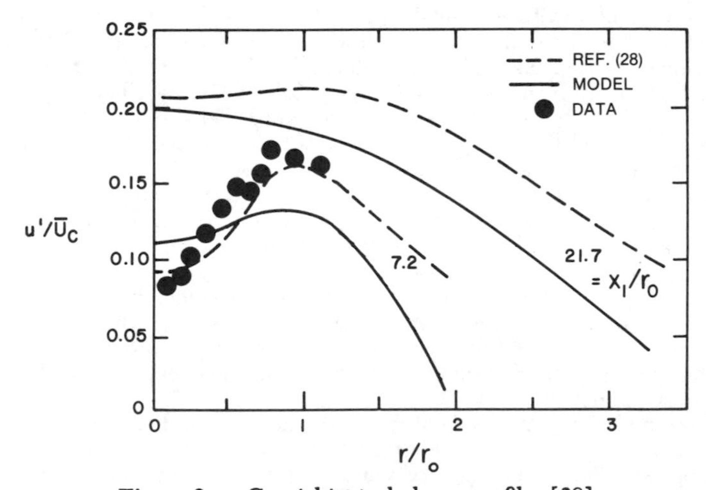

Figure 6. Coaxial jet turbulence profiles [28].

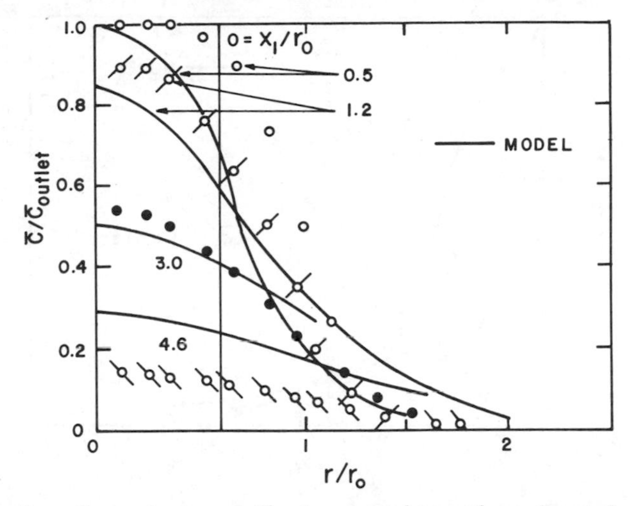

Figure 7. Concentration profiles in a coaxial jet without chemical reaction.

diffusion, just as in heterogeneous reactions. In a homogeneous mixing environment, however, there does not exist a relatively unchanging solid surface to and from which diffusion occurs and on which the reaction occurs. In the heterogeneous case, separation of the diffusion and reaction steps is a classic and very successful procedure for most reactions. In the homogeneous mixing case, diffusion and reaction are inseparably intertwined, and the concentration gradients in a given fluid locality change constantly, making the application of a quasi-steady-state model questionable at best.

A number of investigators have attempted to apply simple, one-parameter (non-distributed) models of mixing to the complex reaction problem by considering the relative rates of the final mixing process and chemical reactions. A review of the one-parameter models—spherical eddy by Bourne et al. [32] and Nabholz et al. [33], fluid strand by Truong and Methot [34], interaction by exchange with the mean by David and Villermaux [35] and the various versions of coalescence-dispersion [6-9] have been given by Patterson [31].

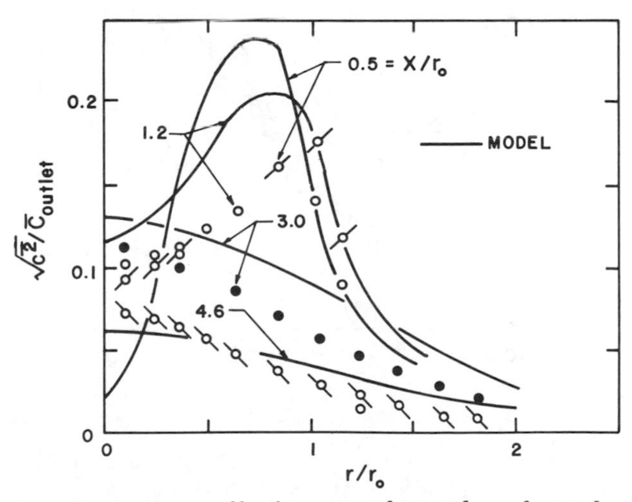

Figure 8. Segregation profiles for a coaxial jet without chemical reaction.

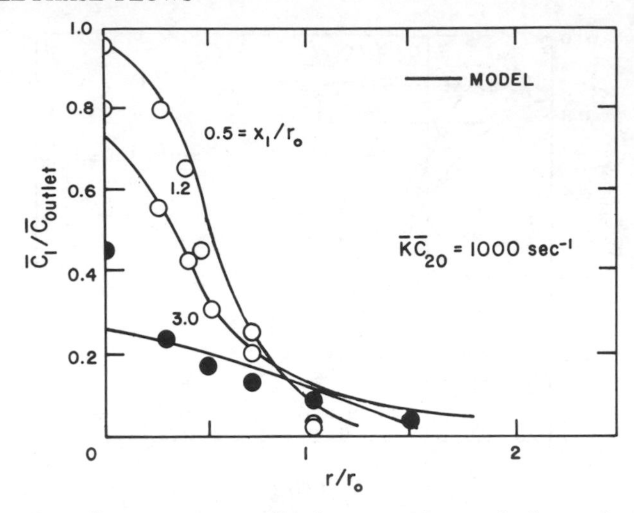

Figure 9. Concentration profiles for a coaxial jet with chemical reaction.

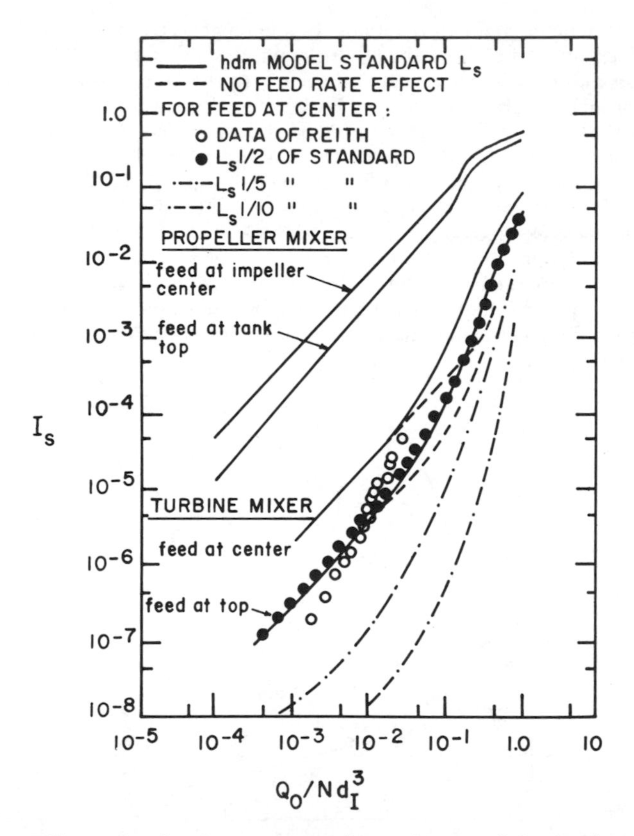

Figure 10. Effluent levels of segregation for tanks stirred by Rushton turbines and propellers. Turbines and propellers centered in tank; impeller diameter, d_I, one-third of tank diameter, D; $Q_O/Nd^3{}_I$ is a measure of feedrate:pumping rate ratio.

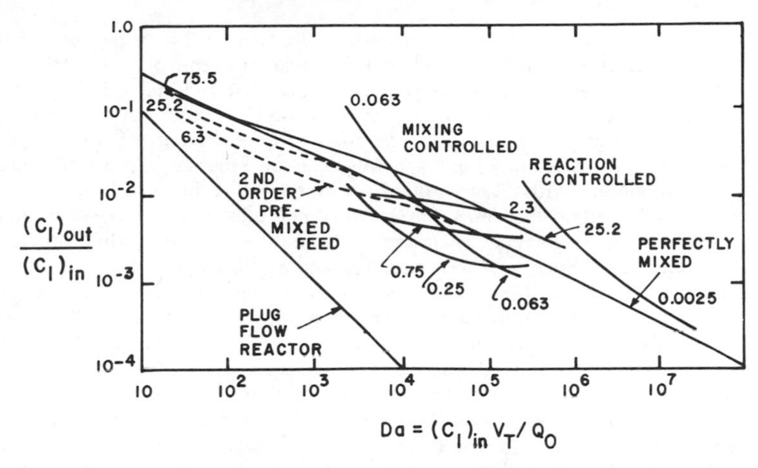

Figure 11. Conversion in the effluent for second-order reactions in a Rushton turbine-stirred tank. Turbine centered in tank; $d_I = D/3$; parameters on curves are mixing intensity given by $(\epsilon/L^2_s)^{1/3}{}_{center}K(C_1)_{in}$.

None of the nondistributed, one-parameter models can account for the relative effects of bulk circulation, large-scale diffusion and local mixing that lead to spatial distributions of segregation, conversion and yield in reactors. They also do not include interactions between reactants and turbulent mixing. It is necessary to make use of distributed models and closures with adequate interactions to include such effects. Patterson [29] described a model for second-order reactions applied to a segmented reactor that was successful in describing effects of circulation and local mixing in propeller- and turbine-stirred reactors. By using a finite-difference solution method, he and co-workers applied the same model to a coaxial jet reactor [4]. As formulated, however, the model is not adequate for complex reactions. In the next two sections a coalescence-dispersion (c-d) model and a closure approximation with interactions between reactants and the turbulent mixing process will be described.

Coalescence-Dispersion Modeling of Complex Reactions

The coalescence-dispersion (c-d) model has been shown to do an adequate job in simulating complex reactions in a mixing tank [6]. In the cases simulated, the tanks

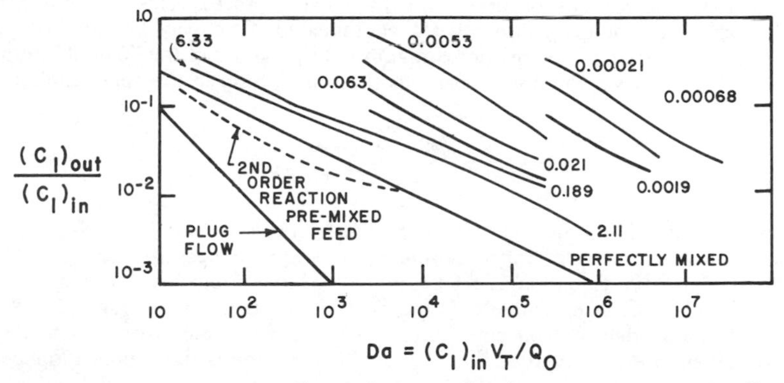

Figure 12. Conversion in the effluent for second-order reactions in a propeller-stirred tank. Conditions are as in Figure 11.

were divided into volume segments, each with 100 c-d sites. By using the relationship below to determine the c-d rate in each volume segment, the proper rates of mixing were simulated. Apparently, the scale size and rate of individual mixing occurrences were adequate to account, in some way, for important interactions between reactants in complex reactions and with the physical mixing process. A great advantage of the c-d method is the easy use of any homogeneous reaction kinetics (almost any number of reactions and any order) without any added computational difficulty.

Use of the c-d method becomes particularly complex when large-scale turbulent diffusion (such as in jet mixers) is a dominant feature of the mixing process (two-dimensional fluid mechanics). When such turbulent diffusion occurs, not only must c-d sites be moved within the mixer to simulate convection (flow), but coalescence must occur at a frequency and distance (to simulate the turbulent mixing scale) to properly model the diffusion. In the tank model described by Canon et al. [6], the sites were moved to simulate the fluid circulation, but coalescence occurred only between adjacent sites. Large-scale diffusion therefore was not simulated in the tank.

To simulate both local mixing and large-scale turbulent diffusion, two c-d parameters were made available for adjustment to match the physical process [36]. The c-d rate was again determined by the following equation [6]:

$$I = 1333(\epsilon/k)(\bar{\tau}/N) \tag{25}$$

where $\bar{\tau}$ = the average residence time in a volume element
N = the number of coalescence sites in each volume element
k = the turbulence kinetic energy
ϵ = the rate of k-dissipation.

The diffusion scale was determined by the distance to which coalescence may occur. The model equation used was

$$L_{c\text{-}d} = 60(\nu_e/k^{1/2})(N/V) \tag{26}$$

where V is the volume of a segment. Computations based on Equations 25 and 26 were made for a complex reaction to determine the yield behavior for a mixing jet, but no comparisons with data have yet been made.

Pratt [9] has provided an alternate treatment of the c-d method for two-dimensional fluid mechanics. His approach makes use of a statistical analog to the c-d process, so Monte Carlo computations are not done.

Complex Reaction Closure with Interactions

If more than one reaction occurs during the mixing process, closure approximations for finite difference modeling are greatly complicated. One way to simplify the approach to generating such closure approximations is by the use of the "paired-interaction" assumption [37]. For such multiple reactions the source term in the concentration equation is

$$P_{\bar{c}_i} = \sum_{k=1}^{m} \sum_{j=1}^{n} \overline{K_{ijk} C_j C_k} \tag{27}$$

where $m \times n$ = number of reaction, K_{ijk} is positive when $k \neq i$ and $j \neq i$; K_{ijk} is negative when $k = i$ or $j = i$; C_j is one for first-order-producing reactions; and $K_{ijk} = 0$ for non-reacting combinations. Equation 27 results in a very large number of terms if completely expanded for $C_i = \bar{C}_i + c_i$, where c_i is the fluctuation about the average $\bar{C}_i$. Assuming that K_{ijk} is independent of concentration and that each pair of reactants mixes independently of the presence of the others, to avoid many terms the source term becomes

$$P_{\bar{C}_i} = \sum_k \sum_j \bar{K}_{ijk}\{\bar{C}_k\bar{C}_j - \sqrt{\overline{c_k^2}\,\overline{c_j^2}}\,(1-\gamma_{kj})/(1+\gamma_{kj})\} \tag{28}$$

where the degree of interdiffusion is

$$\gamma_{kj} = [\bar{C}_k\bar{C}_j - \sqrt{\overline{c_k^2}\,\overline{c_j^2}}]/[\bar{C}_k\bar{C}_j + \sqrt{\overline{c_k^2}\,\overline{c_j^2}}]$$

A degenerate case of Equation 28 is one in which only one second-order reaction occurs, i.e., $i = k = 1$ and $j = 2$. Therefore,

$$P_{\bar{C}_1} = \bar{K}_{121}\{\bar{C}_1\bar{C}_2 - \sqrt{\overline{c_1^2}\,\overline{c_2^2}}\,(1-\gamma_{12})/(1+\gamma_{12})\} \quad (29)$$

which is nearly equal to the result of using Equation 24 if β $(=\bar{C}_{10}/\bar{C}_{20})$ is close to one.

As mentioned following Equation 24, the best results were obtained when the term $\overline{c_1^2c_2}$ was set to zero. This is again true when the paired-interaction approach is used. As above, the segregation source term now becomes

$$P_{\overline{c_i^2}} = -2\sum_k\sum_j \bar{K}_{ijk}\{\overline{c_k^2}\,\bar{C}_j\sqrt{\overline{c_k^2}\,\overline{c_j^2}}\,(1-\gamma_{kj})/(1+\gamma_{kj})\} - D_{Ti} + P_{Di} \quad (30)$$

where γ_{kj} is defined as above. Again, the reaction term is the same as when Equation 24 is used, if β is close to one.

Equations 28 and 30 are simplified if the degree of interdiffusion is substituted into them. They become

$$P_{\bar{C}_i} = \sum_k\sum_j \bar{K}_{ijk}\,\{\bar{C}_k\bar{C}_j - \overline{c_k^2}\,\overline{c_j^2}/\bar{C}_k^2\} \quad (28a)$$

$$P_{\overline{c_i^2}} = \sum_k\sum_j \bar{K}_{ijk}\,\{\overline{c_k^2}\bar{C}_k - \overline{c_k^2}\,\overline{c_j^2}/\bar{C}_j\} - D_{Ti} + P_{Di} \quad (30a)$$

The terms D_{Ti} and P_{Di} are the same for complex reactions as for single reactions, since they do not depend directly on reaction kinetics. They are as follows:

$$D_{Ti} = 2\,\overline{c_k^2}/[4.1(k/\epsilon) + (\mu/\rho\epsilon)^{1/2}\ln N_{Sc}] \quad (31)$$

$$P_{Di} = \frac{\partial[\bar{C}_i(C_T - \bar{C}_i)]}{\partial t} + \underline{\nabla}\cdot[\bar{C}_i(C_T - \bar{C}_i)\underline{U}] \quad (32)$$

As explained by Patterson [2], Equation 31 results from the use of Corrsin's derivation of the rate of segregation decrease in turbulence with a high Schmidt number, N_{Sc}. Equation 32 allows computation of the segregation increase that would occur due to large-scale diffusion if reaction and turbulent mixing were not present. In a numerical computation process, Equation 32 is solved simultaneously with the basic transport equations for mass, momentum, turbulence energy, turbulence energy dissipation, concentrations and segregations. The full set of transport equations for a two-component (single reaction) system was shown by Pratt [9] and Truong and Methot [34]. The first four equations that model the flow and turbulence are basically the k-ϵ method of Jones and Launder [18].

The paired-interaction method was applied to the case of the turbine-agitated tank, and the results shown in Figure 13 were obtained. Preliminary comparisons with limited data show successful modeling of the yield of R in the reactions A+B→R, A+R→S.

If the three-equation model were applied to the complex reaction case, a balance equation for the concentration of each reacting component, the segregation of each reacting component and the $\overline{c_ic_j}$ - correlation for each reacting combination would be solved simultaneously. As in the one-reaction case, triple correlations of concentration fluctuations would be assumed to be zero. For a two-reaction case (both second-order), this would lead to 14 equations, including hydrodynamic equations. This approach has not yet been tested.

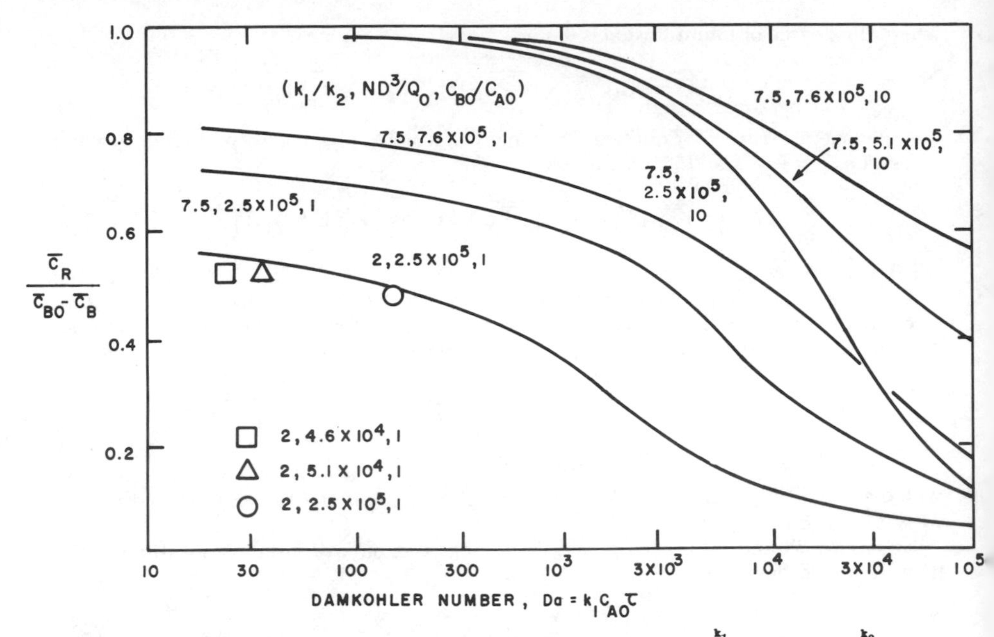

Figure 13. Prediction of yield of R in the reactions $A + B \xrightarrow{k_1} R$; $R + B \xrightarrow{k_2} S$ in a Rushton turbine-stirred tank. $N D^3/Q$ is a measure of the pumping:feedrate ratio [34].

Tests of the interdiffusion, paired-interaction and three-equation models against the experimental data from the turbulently mixed tubular reactor of Toor and co-workers [38,39] have been presented in the literature [7,9,16,18,27,34].

Concluding Remarks on Modeling Mixing and Mixed Reactors

Modeling of turbulently mixed reactors is not a simple and straightforward problem. The single-parameter scaling rules for stirred tanks, based on power dissipation per unit volume to some exponent, have some basis, since local energy dissipation per unit volume is involved in most mixing rate models; however, the complex interactions of convection, turbulent diffusion, mixing rate and chemistry make scaleup a many-parameter problem. To improve our ability to predict accurately the behavior of large reactors involving turbulent mixing, fundamental methods of modeling are a necessity.

When exothermic reactions are involved, as in combustion or other gas-phase reactions with large heat release, the turbulent mixing characteristics are greatly affected by the rise in temperature, and the resulting rise in viscosity can be great enough in the regions of mixing and reaction to greatly reduce the mixing rate because of reduced turbulence. This is a kind of interaction that has not yet been modeled adequately, although those interested in combustion are working hard on it.

A problem of great interest to the chemical industry is the two-phase reaction system, which depends on turbulent transport on one or both sides of the interface. Such problems arise often in stirred-tank systems because of the mechanical convenience of suspending solids and/or gases in a stirred liquid. In such cases, the suspended phase may be reactant or catalyst. With the increased understanding of turbulence in stirred tanks, and other mixers, and new knowledge of effects of turbulence on interfacial transport, the modeling of chemical reactions in such systems is becoming possible. Intense research efforts in this area are necessary and likely.

The greatest problem facing industrial reactor designers is how to predict or scale up mixing effects of the product yield from multiple chemical reactions, some consecutive and some competitive. Work is now being done in this area by a few university groups and by some in industry. The greater knowledge now available on the details of turbulence characteristics in various types of reactors and the effects of such turbulence and the resulting mixing on simple second-order reactions make progress on the yield problem easier. More work needs to be done, however, both in obtaining new and detailed data on product yields in reactors with well characterized turbulent mixing and in developing new modeling and scaleup approaches for testing against such data.

MEASUREMENT OF MIXING

The modeling efforts for mixing described above provide the motivation for the experimental work necessary to test the models and confirm theoretical ideas. The main measurement problems relate to velocities and concentrations.

Velocity and Turbulence Measuring Techniques

A number of techniques have been developed for measuring the velocity and the level of turbulence in a flowing fluid, some of which may be applied to mixers. These techniques may be classified into two groups. In the first group, a tracer or other indicator is introduced into the fluid to make the flow pattern observable by some detecting apparatus outside of the flow field. In the second group, a detecting element (probe) is introduced into the fluid, and the velocity is measured as a function of some property change of the element. The nonprobe techniques are emphasized for mixing because of interference problems discussed below.

Laser-Doppler anemometry and flow visualization studies belong to the first group; hot-wire and hot-film anemometry belong to the second group.

Flow Visualization

One of the first methods of examining the velocity in stirred tanks was through flow visualization studies. Tracer particles are introduced into the fluid, and photographs of the moving particles are taken. The particles appear as streaks on the film, with the length of a streak proportional to the particle velocity and the exposure time. This method has been used by Sachs and Rushton [40], Cutter [41], Schwartzberg and Treybal [42] and Nienow [43]. Recently, Tatterson et al. [44] have developed a three-dimensional stereoscopic method of particle visualization from which velocities in all three directions may be determined.

Hot-Wire and Hot-Film Anemometry

The detecting element of the hot-wire anemometer is a very fine, short metal wire. Hot-film anemometry uses a small metal film plated onto an inert substrate, such as glass. Anemometer measurements are based on the convective heat loss from the electrically heated element caused by the flow of the fluid past the sensor. The resistance of the metal is a function of temperature; as the temperature of the metal decreases, so does its electrical resistance. The rate of transfer to the fluid is a function of fluid velocity; by measuring the power dissipated by the element, the velocity can be determined.

Two different methods can be applied to measure turbulent velocities: the constant-current method and the constant-temperature method. Figure 14(a) shows the schematic of the constant-current system. The current through the sensor is maintained constant by using a large resistor in series with the element. Since the current is constant, Ohm's law can be used to determine the resistance of the element from the voltage across nodes a and b. Constant-current operation works best at low turbulence levels. The constant-temperature system is shown in Figure 14(b). The servoamplifier is controlled by the bridge between nodes a and b, which indicates a change in the resistance of the probe. The heating current output of the servoamplifier is controlled by the bridge imbalance; in this way, the probe is maintained at constant temperature.

A hot-wire sensor must have two important characteristics: a high temperature coefficient of resistance, and an electrical resistance such that it can be heated at a reasonable voltage and current. The most common materials of construction are tungsten, platinum and a platinum-iridium alloy.

The shape most commonly used for hot-wire or film sensors is the cylinder, usually normal to the general flow direction, but sometimes at an angle to allow sensing of turbulence in more than one direction. The wire sensors are usually in the range of 0.0025 to 0.025 mm, and the film sensors, which are generally platinum plated on glass, are in the range of 0.025 to 0.15 mm. Both types are usually about 1 mm in length.

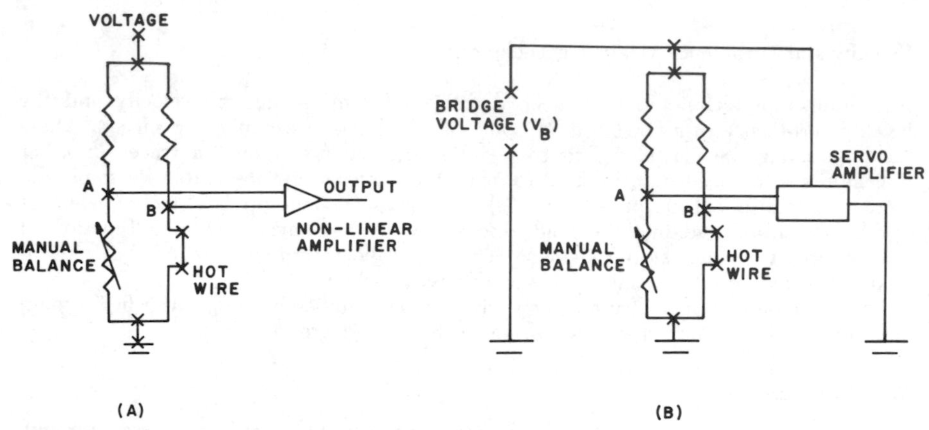

Figure 14. Hot-wire and hot-film anemometry: (a) constant-current system; (b) constant-temperature system.

Another common shape for film probes is the 30° wedge. In most cases, the tip of the wedge is normal to the flow and about 1 mm in length. The wedge has the advantage of no vortex shedding but cannot be used as close to a wall or obstruction as can the cylinder or wire probe.

The response of hot-wire and hot-film probes is, unfortunately, very nonlinear. The calibration curve for either probe type can usually be fit over workable velocity ranges with a modification of King's law [45] as follows:

$$e = A\,u^n + B \tag{33}$$

The anemometer voltage (bridge voltage) is given by e, and velocity by u. The exponent n can have values of 0.3–0.8 in usual applications. Because of the difficulties associated with the nonlinearity of the hot-wire/-film probe calibration, linearizers are available that give an output voltage proportional to velocity over a narrow range, as follows:

$$e = A'u \tag{34}$$

This form is more convenient for cross-correlation measurements in which more than one probe is used.

A major problem with hot-film and hot-wire anemometry is that the probe must be large enough to obtain the needed sensitivity (heat transfer rate), which disrupts the flow patterns near the probe and limits the spatial resolution. In mixing flows, a major problem is that the flow usually has variable directionality, so the probe cannot be oriented properly. Only mixing flows with a strong directionality, such as jet mixers without recirculation, have been measured extensively by hot wires or films. The subject of hot-wire and hot-film anemometry has been reviewed in detail by Davies et al. [46], Durst [47], Rodi [48], Kanevce and Oka [49] and Sandborn [50]. A history of the development of this widely used technique is given in Chapter 3.

Laser-Doppler Anemometry

General Principles. The laser-Doppler anemometer (LDA) uses a laser to provide incident light, which is scattered by moving particles in the fluid. The laser is an ideal light source because it is coherent and monochromatic. The LDA is ideal for measurements of velocities and turbulence in mixing flows because no probe disturbs the flow and the response of the anemometer is linear. The relationship between the frequency of the scattered light and the frequency of the incident light is

$$f_s = f_i + \frac{n}{\lambda}\vec{U} \cdot (\vec{e_s} - \vec{e_i}) \tag{35}$$

where f_s = frequency of scattered light
f_i = frequency of incident light
n = index of refraction of the medium surrounding the particle (≈ 1 in air)
λ = wavelength of incident light
$\vec{U}$ = velocity vector
$\vec{e_s}$ = unit vector in the scattering direction
$\vec{e_i}$ = unit vector in the incident direction

The frequency shift of the light is called the Doppler frequency (f_d):

$$f_d = f_s - f_i \tag{36}$$

Therefore,

$$f_d = \frac{n}{\lambda}\vec{U} \cdot (\vec{e_s} - \vec{c_i}) \tag{37}$$

This equation shows that the Doppler frequency is proportional to the fluid velocity. Since $\vec{U} \cdot (\vec{e_s} - \vec{e_i}) = U\ (2 \sin \theta/2)$, where θ is the angle between beams, Equation 35 may be rewritten for velocity as follows:

$$U = \lambda f_d/(2 \sin \theta/2) \tag{38}$$

The laser-Doppler anemometer consists of the laser, the beam splitter, the lenses, the photomultiplier and the electronics. They can be arranged in a number of different ways, as described in the following section.

Modes of Operation. The laser, beam splitter, lenses and photodetector may be arranged for several different modes of operation. The most important of these, illustrated by Figure 15, are the reference-beam mode, the dual-beam mode and the differential-Doppler mode. Because the differential-Doppler mode is the most commonly used, only it will be described in detail.

The differential-Doppler mode uses two incident beams of equal intensity focused to create the measuring volume, as shown in Figure 16. The photomultiplier picks up

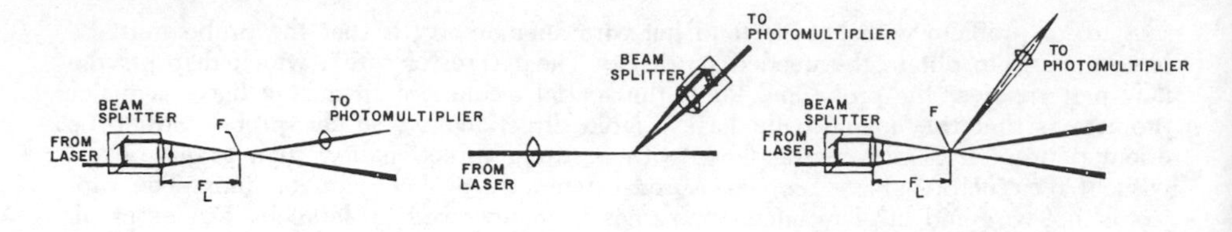

Figure 15. Modes of operation.

two scattered signals, one from each beam. Each scattered beam has a frequency shift relative to the beam from which it originated:

$$f_{s1} = f_{i1} + \frac{n}{\lambda}\vec{U} \cdot (\vec{e}_{s1} - \vec{e}_{i1}) \tag{39}$$

$$f_{s2} = f_{i2} + \frac{n}{\lambda}\vec{U} \cdot (\vec{e}_{s2} - \vec{e}_{i2}) \tag{40}$$

The beam frequency is

$$f_d = f_{i2} - f_{s2} \tag{41}$$

But

$$f_{i1} = f_{s2} \tag{42}$$

$$\vec{e}_{s1} = \vec{e}_{s2} \tag{43}$$

Therefore,

$$f_d = \frac{n}{\lambda}\vec{U} \cdot (\vec{e}_{i2} - \vec{e}_{i1}) \tag{44}$$

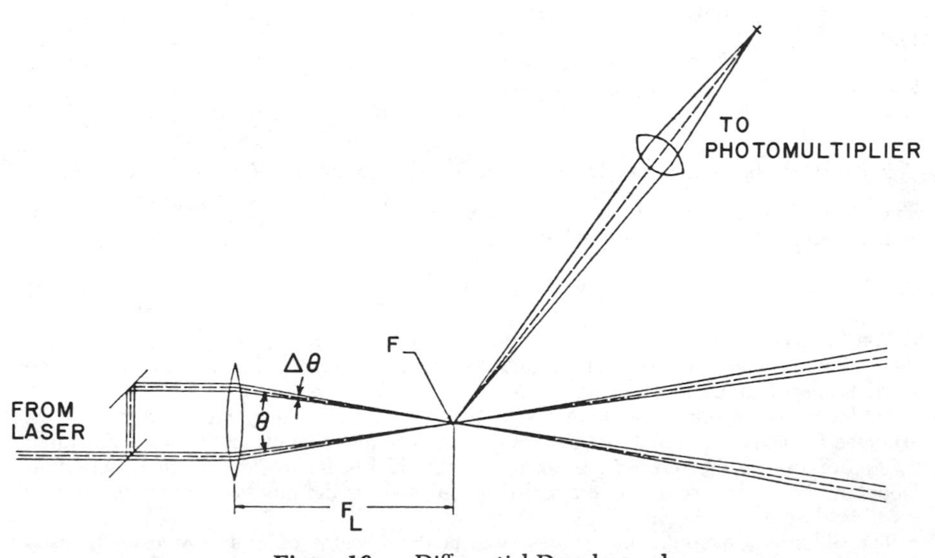

Figure 16. Differential-Doppler mode.

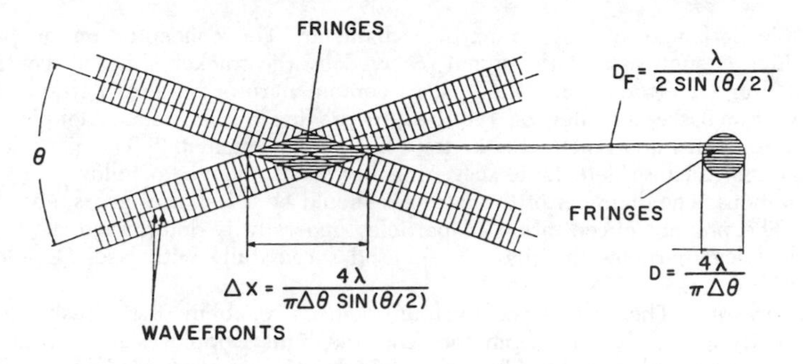

Figure 17. Beam intersection for differential-Doppler mode.

Again, the flow is measured normal to the plane of the beam intersection, parallel to the vector $(\vec{e}_{i2} - \vec{e}_{i1})$. Therefore, Equation 38 can be used to calculate the velocity.

The differential-Doppler mode also is called the "fringe" mode. Near the focal point, the two plane-wave radiation fronts interfere constructively and destructively to produce parallel planes of bright and dim illumination, called interference fringes. Figure 17 shows the beam intersection with the fringe. Adjacent planes of maximum illumination are separated by a distance, D_f:

$$D_f = \frac{\lambda}{2 \sin (\theta/2)} \tag{45}$$

As the particule moves through the fringes, it scatters light at a rate proportional to the particle velocity, as follows:

$$f = \frac{U}{D_f} = \frac{2 U \sin (\theta/2)}{\lambda} \tag{46}$$

where U is the velocity component normal to the fringes. Each beam tapers to a minimum diameter due to the focusing effect of the lens at the beam splitter. The minimum beam diameter (D_v) is

$$D_v = \frac{4 \lambda}{\pi \Delta\theta} \tag{47}$$

where $\Delta\theta$ is the angle of beam convergence shown in Figure 16. The intersection of the beam forms an ellipsoid, which can be described by the equation

$$z^2 + y^2 \cos^2 (\theta/2) + x^2 \sin (\theta/2) = (D_v/2)^2 \tag{48}$$

The volume contained within this ellipsoid is

$$V_v = \frac{\pi (D_v{}^3)}{3 \sin \theta} \tag{49}$$

A complete explanation of the fringe mode has been made by Brayton et al. [51]. Laser-Doppler anemometry has been discussed thoroughly by Durst et al. [52].

Seeding. For the laser-Doppler anemometer to function properly, light must be scattered by particles in the fluid. Signal quality depends on a number of factors: concentration of particles, size distribution of the particles, laser power, mode of operation and the ratio of fringe spacing to particle diameter. The concentration of particles should be high enough so that the signal received by the tracker is nearly continuous, minimizing dropout errors. Some fluids may contain natural seeding particles or impurities; additional seeding may be required to provide a continuous Doppler signal.

Particle size is an important factor in laser-Doppler anemometry. The particles must be large enough to cause detectable scattering but small enough to follow the smallest turbulent motions. The diameter of the particles should be 0.1–1 μ in gases, and 1–10 μ in liquids. The proper concentration of particles generally is determined by trial and error. Table I lists particles that have been used successfully with laser-Doppler anemometry [53].

Signal Processing. There are basically four methods to obtain the measurements of average velocity and turbulence from the scattered light Doppler signal produced by the photomultiplier or photodiode. The first method is the simplest and, historically, also the first. In that method, the spectrum of the Doppler signal is measured. The breadth of the spectral peak is a measure of the turbulence intensity and the average frequency corresponds to the average velocity. This method is rarely used except for very basic studies of new methods and to monitor LDA performance.

The second method makes use of a signal tracker, which is a servooscillator that produces an output voltage proportional to the oscillator frequency. The circuit continuously matches the oscillator frequency to the Doppler frequency, usually holding the last measured frequency when there is no measurable Doppler signal (dropout) because no particle is in the measuring volume. Figure 18 illustrates the instrumentation typically used with a tracker.

The third method is called the counter technique. Instead of measuring Doppler frequency, the number of light fringe crossings (as indicated by variations in scattered intensity) is counted for each particle moving through the volume. This method is useful for much lower particle concentrations because it does not depend on a continuous signal, as does the tracker. Each particle contributes statistically to the computation of the average velocity and to the turbulence intensity (root-mean-square velocity variation). The counter method is probably the most flexible in its application. The fourth method for extraction of velocity information from the LDA signal is direct digital recording of the converted analog Doppler signal from the photomultiplier

Table I. Methods of Seeding

Flow Medium	Method Generation	Seeding Material
Water		Polystyrene
		Milk, cream powder
Air	Atomization	Water vapor
		Di-octyl phthalate (LOOP)
		Teflon®[a] dust
		Silicon oil droplets
	Fluidization	Titanium dioxide
		Geon (PVC)
		Aluminum
		Aluminum oxide
	Chemical reaction	Ammonium chloride
		Stannic chloride
	Combustion	Tobacco smoke
		Magnesium oxide
		Smoke bombs
		Smoke pellets
Burning Gas	Combustion	Silicon dioxide
		From burning of hexamethyledisiloxane
		Magnesium oxide
	Atomization	Silicon oil droplets

[a]Registered trademark of E. I. du Pont de Nemours and Company, Inc., Wilmington, Delaware.

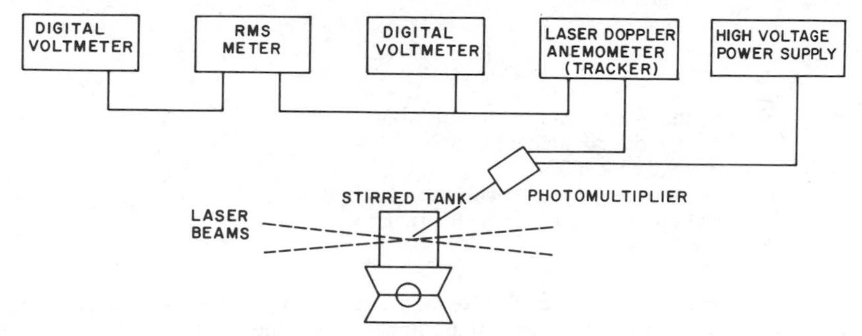

Figure 18. Laser-Doppler anemometer.

receiver. Durst and Tropea [54] described this method in which a fast analog-digital converter (10-100 MHz conversion rate) feeds short bursts of the signal into a mini- or microcomputer for storage on disk and later analysis. The analysis of the digitized signals involves counting of zero crossings during the middle portion of each burst for a predetermined time period, then calculation of the particle velocity, making use of the known fringe spacing. The method is similar to the use of a counter, but has the advantage of storage of the burst signals for a variety of later treatments.

Concentration and Segregation Measuring Techniques

Several methods can be used to measure concentration at a point. Early studies assumed that the diffusion of heat is equivalent to that of mass; temperature measurements were used to determine diffusion rates. This assumption, can produce significant errors, however, particularly for low Prandtl numbers. Four methods of measuring concentration of a dissolved, but not totally mixed, component at a point—electrical conductivity, light absorption, light scattering, fluorescence measurements—will be discussed.

Conductivity

Electrical conductivity measurements are based on detection of the current flowing between two electrodes. One requirement is that the liquid solution be electrically conductive; highly ionic solutions, such as salt solutions, meet this requirement. A small probe containing the two electrodes is introduced into the flow. It is fed by a constant voltage supply, and concentration is measured as a function of the current flowing through the gap between the electrodes. The circuitry consists of the probe, a signal oscillator, an ac amplifier, a rectifier, a wave amplifier and squaring and averaging circuits. Typical measuring volume size is on the order of 0.3 mm^3. Mean values, root-mean-square values and spectral distribution functions can be obtained. Measurements of this type were made by Manning and Wilhelm [55] and Reith [56], as well as by many others.

Light Absorption

Another method to measure concentration is based on the amount of light absorbed by the solution. A light-absorbing dye is placed in the flow. The probe consists of fiber optic conductors, which transmit the light into the measuring volume and receive the incident light. It is generally assumed that Beer's absorption law is valid, so that light absorption is simply related to the point concentration, as shown in Equation 50:

$$\overline{V} = \overline{V}_O \exp(-B\,\overline{C}) \tag{50}$$

and

$$c' = v' / (B\overline{V}) \tag{51}$$

where $\overline{V}$ = mean voltage from photomultiplier
$\overline{V_O}$ = voltage due to incident light
$\overline{C}$ = average concentration
c' = root-mean-square concentration fluctuation
v' = root-mean-square voltage fluctuation
B = Beer's constant

Measurements made by Lee and Brodkey [57] suffered from limitations in sample volume size, frequency response and signal:noise ratio; the sample volume size was 5.6×10^{-1} mm^3. Nye and Brodkey [58] devised an improved probe with a measuring volume of 1.2×10^{-2} mm^3. Mean and fluctuating concentrations can be measured in this way.

Light Scattering

The light scattering technique was first suggested by Becker et al. [59]. The apparatus used in this technique is illustrated in Figure 19. An intense beam of light is focused in the flow field. Seeding particles in the fluid scatter the incident light; this scattered light is then collected by a lens outside of the flow field and focused on a photomultiplier. Ideally, the signal produced by the photomultiplier is proportional to the concentration of particles in the measuring volume and, when that volume is small enough, to the local concentration. Since the response of the phototube is linear with light intensity within the normal range of operation, problems of nonlinearities that complicate most other methods are absent. Average concentration is measured with an averaging voltmeter, and fluctuating concentration is measured with a root-mean-square voltmeter. This method has certain limitations and drawbacks, however. The most significant limitation is that many sources of noise may enter the signal. Becker et al. [59] have analyzed the noise that is inherent in this technique: optical background noise, source fluctuation noise, dark noise, particle shot noise and electronic shot noise.

Optical attenuation noise is caused by the signal-attenuating effects of absorption and scatter along the optical path through the mixing field. If the marker particles are present in low concentration in the tank, optical attenuation noise level is generally negligible. Optical background noise can be caused by: (1) light from extraneous sources; (2) stray light from the excitation lamp, the incident beams and the scattered light, all reflected by the surroundings; (3) rescatter of scattered light in the mixing field; and (4) emission from the mixing field, such as that encountered with flames. In laboratory conditions, sources of extraneous light can be blocked, minimizing

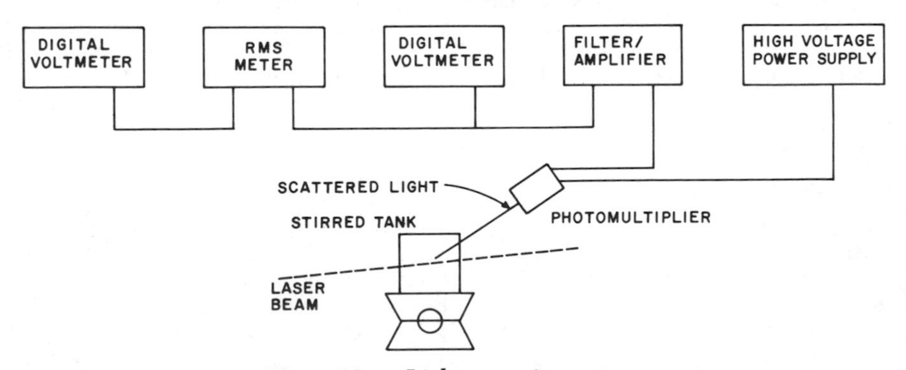

Figure 19. Light scattering system.

causes (1) and (2); rescatter noise can be neglected if optical attenuation noise level is minimized. Emission by the mixing field can be coped with by using a lamp that radiates strongly in a different part of the spectrum from that of the mixing field.

For light scattering measurements to be meaningful, the source of light must be of constant intensity. Variation in the intensity of the light source is called source fluctuation noise. If a good light source is used, source fluctuation is minimal. Dark noise, which is caused by thermal effects within the photomultiplier, is independent of the light flux and generally can be neglected. If the number of particles passing through the control volume is not great enough to validate the assumption of marker continuum, or the size of the particles varies significantly, marker shot noise is present. Electronic shot noise is caused by the particulate nature of every electric current. For proper operation of the light scatter technique, both sources of shot noise must be minimized. Experimental measurements in jets using the light scattering technique have been made by Shaughnessy and Morton [60] and Lee [53].

The optical system for measuring concentration is basically the same as used in laser-Doppler anemometry, as shown in Figure 19. A calibration curve demonstrates the linear relationship between concentration and voltage:

$$\overline{C} = a\,\overline{V} + b \tag{52}$$

where a = slope of the line
b = y-intercept of the line
$\overline{C}$ = concentration at the measuring point
$\overline{V}$ = average voltage produced by the photomultiplier

The value of b is usually small enough to be negligible. Once the value of a is determined, concentration can be measured.

The quantities of interest are the ratio of the local concentration in the tank, $\overline{C}/\overline{C}_m$, and the segregation, $\overline{c^2}/\overline{c_o^2}$. As has been explained previously, the concentration is a linear function of the voltage:

$$\overline{C} = a\,\overline{V} + b \tag{53}$$

and b is negligible:

$$\overline{C} = a\,\overline{V} \tag{54}$$

Therefore,

$$\overline{C}/\overline{C}_m = a\,\overline{V}/\,a\,\overline{V}_m = \overline{V}\,/\,\overline{V}_m \tag{55}$$

Equation 55 was used to obtain the concentration ratio results.

To define the value of $\overline{c^2}\,/\,\overline{c_o^2}$, the value of the initial segregation, $\overline{c_o^2}$, must be established. The initial segregation in a mixer is usually defined as the segregation between the two streams entering. By examining a point at the entrance of the mixer, complete segregation would be evidenced by a pulse of stream 1 passing through the measuring point, followed by a pulse of stream 2. The mean-square fluctuation of concentration then can be defined by integrating the concentration fluctuation over the total flow through the point:

$$\overline{c_i^2} = \frac{1}{Q}\int_0^Q \overline{c_i^2}\,dq = \frac{1}{Q}\int_0^Q (C_i - \overline{C}_i)^2\,dq \tag{56}$$

where C_i = instantaneous concentration of component i
$\overline{C}_i$ = average concentration of component i
c_i = fluctuation from the average concentration

The average mixed concentration of component 1 in the tank is proportional to the flowrate of stream 1 entering the tank.

$$\overline{C}_1 = \frac{Q_1}{Q_1 + Q_2} C_{1,\,\mathrm{enter}} \tag{57}$$

where $C_{1,\,\mathrm{enter}}$ is the concentration of component 1 entering the reactor in stream 1. Equation 56 can be integrated to obtain

$$(\overline{c_1^2})_O = \frac{1}{Q_1 + Q_2} (Q_1 (\overline{c_1^2})_{\mathrm{stream}\,1} + Q_2 (\overline{c_1^2})_{\mathrm{stream}\,2}) \tag{58}$$

$$= \frac{1}{Q_1 + Q_2} (Q_1 (C_1 - \overline{C}_1)^2{}_{\mathrm{stream}\,1} + Q_2 (C_1 - \overline{C}_1)^2{}_{\mathrm{stream}\,2}) \tag{59}$$

Since the flow in this experiment was nonreacting, component 1 can be considered to be stream 1, and component 2 can be considered to be stream 2.
Therefore,

$$C_1 = C_{1,\,\mathrm{enter}} \qquad \text{(in stream 1)} \tag{60}$$

$$C_1 = 0 \qquad \text{(in stream 2)} \tag{61}$$

Substituting Equations 60 and 61 into Equation 59 gives

$$(\overline{c_1^2})_O = \frac{1}{Q_1 + Q_2} (Q_1 (C_{1,\,\mathrm{enter}} - \overline{C}_1)^2 + Q_2 (0 - \overline{C}_1)^2) \tag{62}$$

From Equation 57,

$$C_{1,\,\mathrm{enter}} = \frac{Q_1 + Q_2}{Q_1} \overline{C}_1 \tag{63}$$

Substituting Equation 63 into Equation 62 and solving,

$$(\overline{c_1^2})_o = \frac{Q_2}{Q_1} \overline{C}_1^2 \tag{64}$$

If R is defined as the ratio of the feedrates, Q_2/Q_1, then

$$\overline{c_O^2} = (\overline{c_1^2})_O = R\,\overline{C}_1^2 \tag{65}$$

If the signal and noise are uncorrelated [53], then

$$\overline{c^2} = (\overline{v^2} - \overline{v^2}_{\mathrm{noise}})\, a^2 \tag{66}$$

and

$$\begin{aligned} \overline{c^2}/\overline{c_O^2} &= (\overline{v^2} - \overline{v^2}_{\mathrm{noise}})\, a^2/R\,\overline{V}_m^2 a^2 \\ &= (\overline{v^2} - \overline{v^2}_{\mathrm{noise}})/R\,\overline{V}_m^2 \end{aligned} \tag{67}$$

Equation 67 is used to calculate the segregation results.

Fluorescence

Measurement of local concentration to obtain concentration profiles and segregation by use of fluorescence is very similar to the light scattering method. A beam of light

Table II. Fluorescent Systems Useful for Concentration Measurements

Mixing without Reaction	
Acridine	e 400 nm; f 440 nm
Fluoroscein	e 485 nm; f 520 nm
Eosin (basic solution)	e 517 nm; f 540 nm
Rhodamine 6 g	e 526 nm; f 555 nm
Rhodamine b	e 550 nm; f 605 nm
Mixing with Reaction	
Fluorescamine (Reacts with certain amino acids, becoming nonfluorescent)	e 390 nm; f 474 nm
Acridine (Reacts with alkyl halide, becoming nonfluorescent)	e 400 nm; f 440 nm
Cerium(III) Sulfate (May be oxidized by $Fe_2(SO_4)_3$ in acid solution, becoming nonfluorescent, or vice-versa)	e 260 nm; f 350 nm

is focused on the point of measurement, and the fluorescence induced in a compound in solution is measured by a photomultiplier. The experiment is controlled so that concentrations and light intensities are in a range in which fluorescence intensity is proportional to local concentration.

For measurements in nonreacting systems, fluorescent dyes may be used. Table II shows various dyes and other fluorescent compounds that may be used along with their excitation range and fluorescence range. For measurements of concentrations in reacting systems, reactions must be found that involve either reactants or products that fluoresce. Table II also shows some systems that may be used for such experiments.

As Table II indicates, one of the difficulties with the fluorescence method is that the use of ultraviolet light is frequently necessary, both for excitation and as the fluorescent light. For open systems, such as gas jets, that is no particular problem, but for measurements within closed vessels, it is necessary to use quartz windows. Few glasses and no plastics transmit any portion of the ultraviolet spectrum.

To maintain linearity of the fluorescence intensity-concentration relationship, the concentration must be maintained low enough to avoid secondary scattering and significant absorption of the incident beam. Details of the method for use in liquid systems have been given by Patterson [61]. Figure 20 shows apparatus used to make measurements of reactant product concentrations within a stirred reactor.

Measurements of Turbulence and Mixing in Jets and Pipes

Turbulence and momentum transport effects have been studied for many years in jet flows and injection into pipes. Local concentration and segregation, however, have been measured only recently in jet and pipe mixers. Results are presented here for pipe injection mixing, circular jets mixing with ambient fluid and coaxial jets. These configurations are typical of the many designs used in commercial apparatus to promote mixing.

Pipe Injection

Figures 21 and 22 show comparisons of measurements of concentration and segregation profiles for injection into a pipe made by Becker et al. [62], with modeling of the same phenomenon by Patterson et al. [4]. The jet of dispersing fluid was introduced into the pipe at the same velocity as the pipe fluid and was small enough to be con-

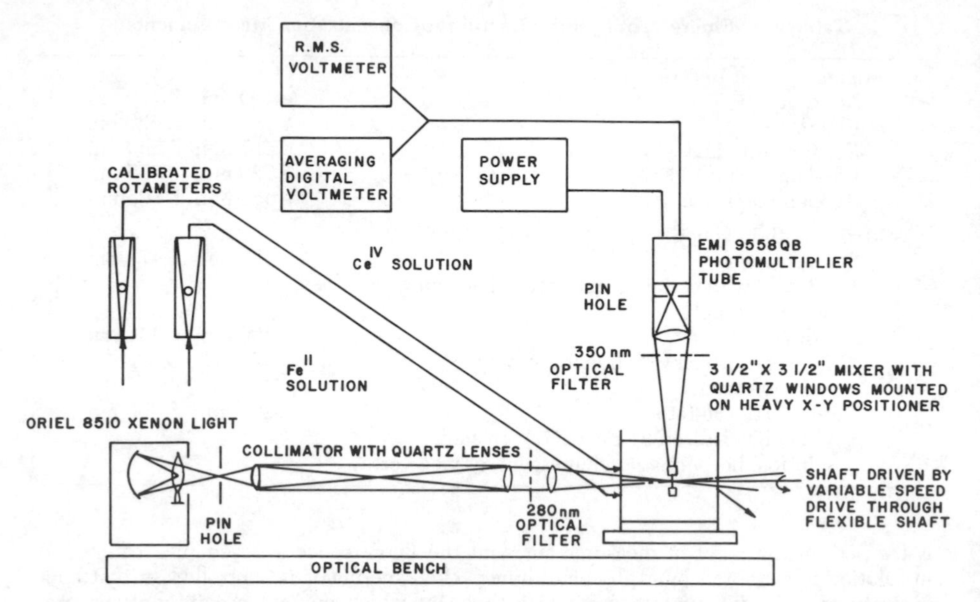

Figure 20. Apparatus to measure concentration of Ce^{III} produced by oxidation of Fe^{II} to Fe^{III} by Ce^{IV}.

sidered a point source. The measurements of concentration and segregation (mean square of concentration fluctuations) were made using the light scattering technique with smoke particles in air as the scattering medium, a method used again later by Becker et al. [59] and Becker and Smith [63]. The modeling was done using a finite difference solution of Equations 11 and 16, along with equations for mass, momentum, turbulence energy and energy dissipation rate.

There is generally good agreement between modeling and experiment, except when the injection nozzle diameter, dj, is too large. The injection nozzle in the experiment was small enough to be considered a point source. Other point source dispersion measurements in pipe flow have been made by Lee and Brodkey [19] using the fiber optic light absorption probe.

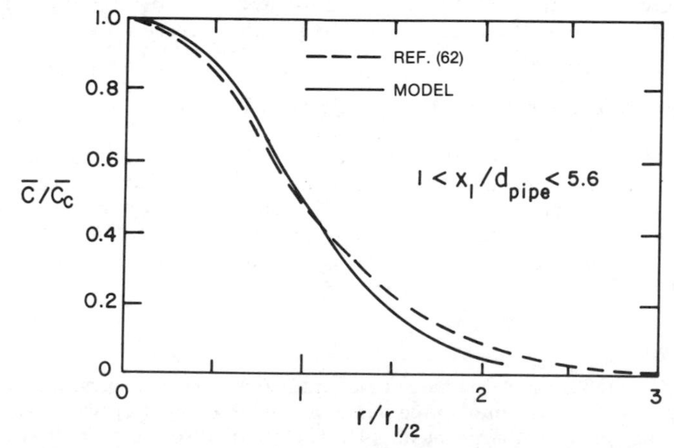

Figure 21. Concentration profile for pipe dispersion [62].

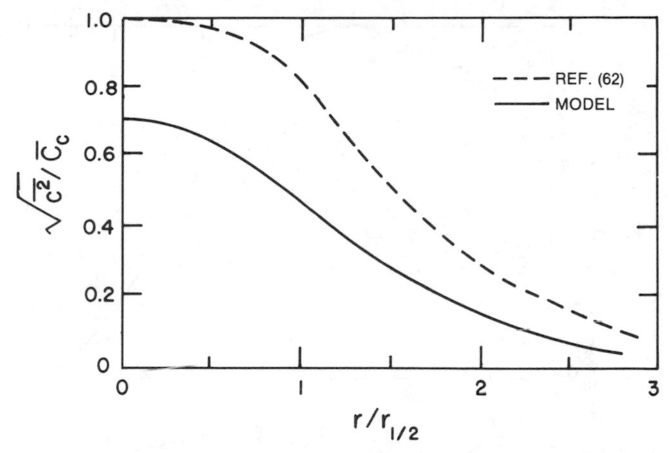

Figure 22. Segregation profile for pipe dispersion [62].

Circular Jets

Mixing of the fluid issuing from round, circular nozzles with the surrounding entrained fluid has been measured by many. Representative of such measurements are those by Becker et al. [59]. They made use of the same light scattering technique with smoke particles dispersed in air that Becker et al. [62] used in the pipe injection experiments. Figures 23 and 24 show comparisons of those measurements with modeling results of Patterson et al. [4]. The modeling was done as mentioned above, using Equations 11 and 16. Spalding [3] also has reported comparisons of measurements and modeling of concentration and segregation in round jets.

Coaxial Jets

Typical measurements of velocity, turbulence, concentration and segregation in a coaxial jet without reaction are shown in Figures 5-8. Figure 9 shows the concentration profiles measured and modeled for a very fast chemical reaction between an inner and an outer gas [4]. Durao and Whitelaw [28] also made measurements of velocity and turbulence in coaxial jets. Some of their results are shown in Figures 5 and 6.

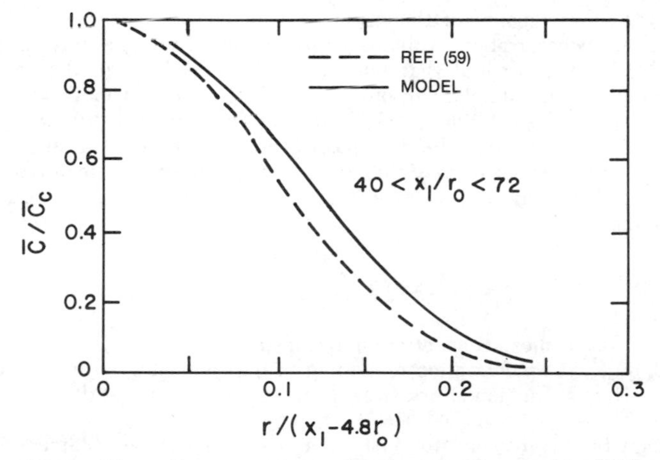

Figure 23. Concentration profile for a single round jet [59].

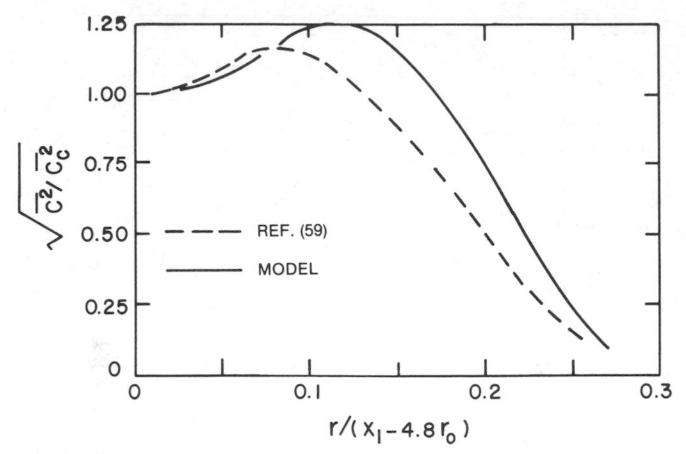

Figure 24. Segregation profile for a single round jet [59].

Studies of Turbine-Agitated Vessels

Many approaches have been taken in an effort to explain the turbulent mixing occurring in stirred tanks. The simplest approach is to use stimulus-response methods. This technique measures the response of the concentration (of a component or of a tracer) in the exit stream to a change in the entrance stream concentration, and relates the response to process variables such as tank size, impeller speed, impeller design or tank design. A result of the stimulus-response method is the mixing time, defined as that time after tracer injection required to reach some deviation from the final concentration, as measured in the effluent. The approach that contains the most information about turbulent flow patterns is one in which the concentration and velocity are measured at various points within the reactor. These data, while less useful in solving daily engineering problems, provide a basis for the theoretical interpretation of the turbulent mixing problem and the formulation of advanced models for more accurate scaleup.

Residence Time Distribution Measurements

By measuring the response of the exiting stream of the reactor to a change in the entrance stream concentration, valuable information about the mixing process can be obtained. Elements of fluid that take different paths through the reactor require different lengths of time to pass through the vessel. The distribution of these times is called the exit age distribution, $E(t)$, or the residence time distribution of the fluid. If the assumption is made that the fluid passes through the vessel as elements, with no intermixing of adjacent elements, the age distribution of the material in the exit stream indicates how long each element was within the reactor. The average exit concentration is, therefore,

$$\overline{C}_a = \int_0^\infty C_{a,\,element} E(t)\, dt \tag{68}$$

where $\overline{C}_a$ = mean concentration in exit stream
$C_{a,element}$ = concentration remaining in the element aged between t and t+dt
$E(t)$ = fraction of exit stream aged between t and t+dt

Using this equation, conversion for first-order reactions can be obtained directly from concentration data. Levenspiel [64] provides a thorough review of this technique and residence time distribution theory. Hubbard and Patel [65,66] used dimensional analysis to relate residence time distributions to dimensionless variables. While residence time studies can be useful in predicting concentrations in the exit stream, a quantity very important to the engineer, they give little insight into the behavior of the fluid while it is in the tank.

Mixing Time Studies

A similar method is to measure the concentration at some point in the batch reactor and to define a mixing time as that required to reach some deviation from the final mixed concentration at that point. The first mixing time correlations were made by Fox and Gex [67] for propeller and jet-mixed tanks by making visual observations based on the amount of time to neutralize the fluid using an acid-base reaction. Norwood and Metzner [68] made similar measurements in turbine-agitated vessels. Mixing time studies are particularly useful in estimating the performance of batch-mixing tanks. For large vessels, Middleton [69] has used a similar technique. A radio pill flow follower was introduced into a batch mixer. Recirculation time, which can be related to the mixing time, was measured by determining the frequency of pill passage through the impeller region.

Detailed Velocity and Concentration Studies

Velocity Studies. Early attempts to measure the velocity of the fluid leaving a turbine were made by Sachs and Rushton [40] by making photographs of tracer particles suspended in the fluid. The particles appeared as streaks, the lengths of which were directly proportional to the velocity. Tangential, radial and axial velocities in the discharge stream of a turbine impeller were measured. They discovered that the maximum radial velocity does not occur at the impeller tip but rather at a distance from the tip, due to the axially moving fluid being entrained by the discharge stream. It also was determined that the radial velocity was pulsatile, with the maximum velocity occurring about 50° ahead of the moving blade.

Cutter [41] used a similar photographic technique to make detailed studies of the fluid velocity throughout the stirred tank. The assumption made by Sachs and Rushton [40] that the mean velocity at a point is proportional to the impeller speed was confirmed and extended to the root-mean-square values of the turbulent fluctuations. It was also discovered that the scale of turbulence increased with distance from the impeller, a probable result of the breakup of the impeller stream. Turbulence intensities were on the order of 60%.

A similar study using the streak photography technique to measure velocities throughout the tank was made by Schwartzberg and Treybal [42]. Their study concentrated on the vertical plane velocity of the fluid. They found that with the exception of the vertical plane, including the baffles, where the vertical plane velocity is 17% higher than average, the vertical plane velocity did not vary significantly with angular position. They found that the average vertical plane velocity, $\overline{U}_v$, could be described as

$$\overline{U}_v = 1.387 \frac{ND_I^2}{(D_T^2H)^{1/3}} \tag{69}$$

where N = impeller speed
D_I = impeller diameter
D_T = tank diameter
H = height of fluid

By analysis of the fluctuating velocities, it was concluded that the turbulence in the tank is distinctly nonisotropic. The average intensity of turbulence was 35%.

An improved photographic technique was used by Komasawa et al. [70] to measure the turbulence quantities of fluid flow. Previous methods suffered from the limitations of shutter speed, wide sampling region and the period over which a continuous measurement was run. A high-speed cinephotography technique was used to overcome these limitations. Their goal was to prove the validity of this technique. Their results compared favorably with those of Mujumdar et al. [71] and of Rao and Brodkey [72] which will be described later.

Tatterson et al. [44] have developed a stereoscopic film technique to examine particle motion in stirred tanks. Their studies have concentrated so far on the development of coherent structures (vortices, etc.) in the mixer, but eventually could contribute to the overall study of three-dimensional velocity fields in mixers.

Desouza and Pike [73] measured velocities in a stirred tank using a directional pitot tube in a high-velocity region of the impeller and hot-wire anemometer, where velocities are low and the turbulence is less intense. They described the flow from the impeller using a tangential jet model and concluded that for geometrically similar impellers, jet width was proportional to distance from the center.

Nagata et al. [74] made similar studies in the impeller stream, using a pitot tube to measure velocity at high Reynolds numbers. The photographic technique of Sachs and Rushton [40] was used at lower Reynolds numbers. Both baffled and unbaffled tanks were used. The flow pattern was more erratic with baffles, due to increased turbulence caused by the baffles.

Hot-wire and hot-film anemometry also can be used to measure velocities in stirred tanks. Mujumdar et al. [71] measured mean and fluctuating radial velocities in the impeller stream. They used air as the working fluid, based on the assumption that small-scale turbulence characteristics would be similar at equal Reynolds numbers, regardless of the fluid in the tank. Their results indicated that the intensity of the turbulence shows a rapid increase from a level of 5% near the impeller tip to 35% at a radial position of one-third of the distance from the impeller tip to the wall.

Günkel and Weber [75] made a similar study using a shielded hot-wire anemometer to measure the radial, tangential and axial velocities in the impeller stream. They discovered the presence of four vortices between each blade of the impeller, two above the disk and two below, by placing hot-film probes on the surface of the impeller. By making an energy balance over the impeller stream, they showed that the energy put into the tank via the impeller is dissipated in the bulk of the tank, which does not include the impeller stream.

All the studies mentioned in this section have involved the mixing of fluid in batch-tank reactors. Rao and Brodkey [72] studied the continuous-flow tank stirred by a flat-bladed open impeller. They measured the mean and fluctuating radial, axial and tangential velocities in the impeller stream using a hot-film anemometer. They confirmed the pulsatile velocity of the fluid that leaves the impeller.

The development of the laser-Doppler anemometer provided a valuable tool for studying turbulent parameters; since no physical probe is used, there is no interference with the flowing fluid. Reed et al. [76] investigated the turbulent flow in the impeller stream using laser-Doppler anemometry. The average and root-mean-square tangential, radial and axial velocities were measured. An impeller speed of 1000 rpm was used. The measurements revealed the absence of a symmetry plane in the impeller stream of a flat-blade, disk-type turbine located one-third the way up from the base of the tank. Turbulence intensities of 100% were not unusual; near the baffles, turbulence levels of several hundred percent were observed. The periodic motion of the fluid exiting the impeller observed by Rushton was confirmed.

A new technique for flow study is being developed that uses a self-scanning photodiode array. Kaniwamo et al. [77] are examining flow patterns of a liquid in a stirring vessel using this method.

Very recently, Zipp [78] carried out detailed measurements in a Rushton turbine-stirred baffled tank, in which radial and tangential velocities and turbulence were measured at many locations within the tank. Laser-Doppler anemometry using tracker signal analysis was used to make the measurements, with milk as the scattering medium.

Comparisons were made with previous velocity and turbulence data. Figure 25 shows a comparison of radial average velocities in the impeller stream at distances from the

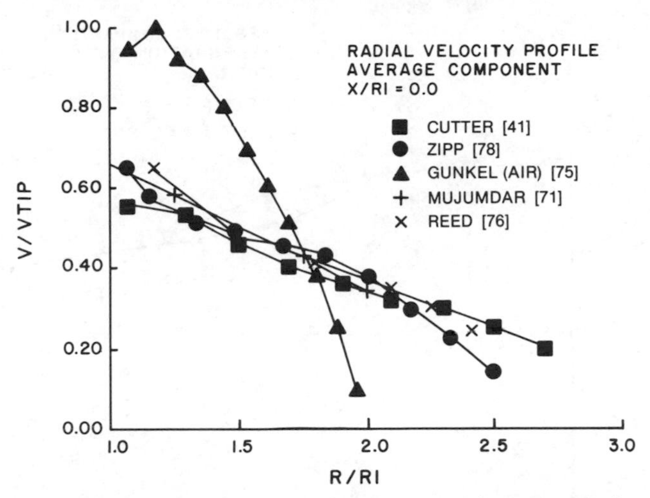

Figure 25. Comparison with previous radial average velocities. All data in impeller stream at mid-plane between baffles [41,71,75,76,78].

tank axis given in impeller radii. All data are similar, even though many tank sizes and Reynolds numbers are represented. Figure 26 shows measurements made by Zipp [78] of the radial turbulence component in the impeller stream.

Concentration Studies. Various methods have been used to make concentration measurements. Manning and Wilhelm [55] have made steady-state measurements in a continuous flow-stirred tank using conductivity methods. The traced solution was injected directly under the impeller, and water was fed into the bottom of the tank. Measurements were made at various radial, axial and angular positions. The ratio of ionic solution to water flowrate was 130; probe volume was approximately 0.3 mm^3. Rotation speed and impeller size were varied. As was expected, the highest segregation was obtained with the smallest impeller and lowest rotation rate.

Manning and Wilhelm's [55] study concentrated on measurements at 1500 rpm, a speed at which the average concentration is nearly uniform and the concentration

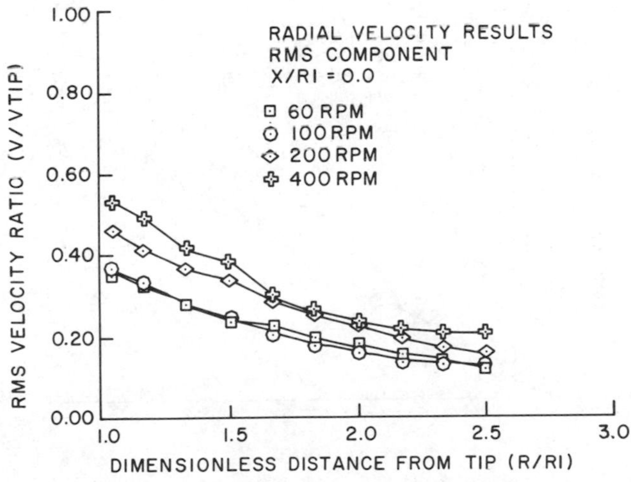

Figure 26. Comparison of radial turbulence results.

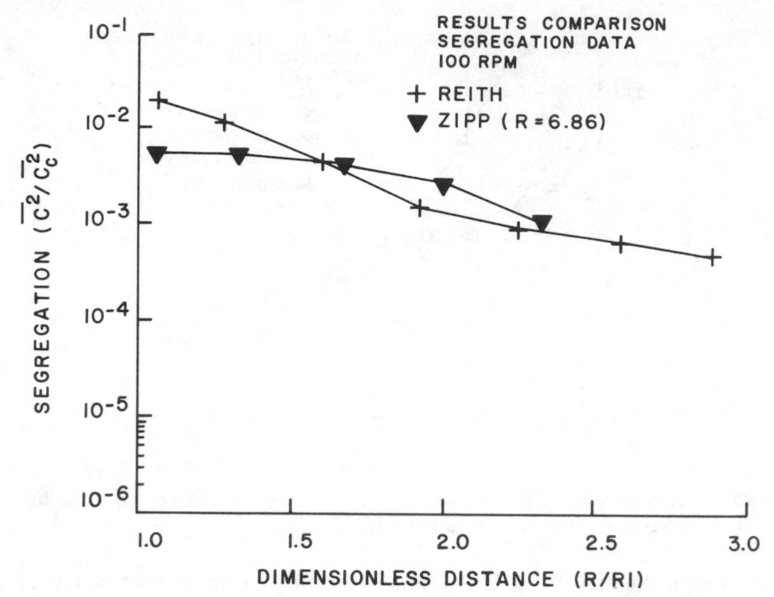

Figure 27. Comparison with previous segregation results–impeller stream, 100 rpm [56,78].

fluctuations are near zero at all points except those near the injection point, i.e., the impeller. The root-mean-square concentration fluctuations varied little with respect to angular position, as did the average concentration, except near the baffles. At 1500 rpm, concentration fluctuations on the order of 11% were observed near the impeller, decaying to 1% near the wall. The maximum fluctuation was 40%, with the smallest impeller at the lowest speed.

A similar study was made by Reith [56], with the exception that both the water stream and the ion-solution stream were injected directly on either side of the impeller. The probe was an improved version of the one used by Manning and Wilhelm [55].

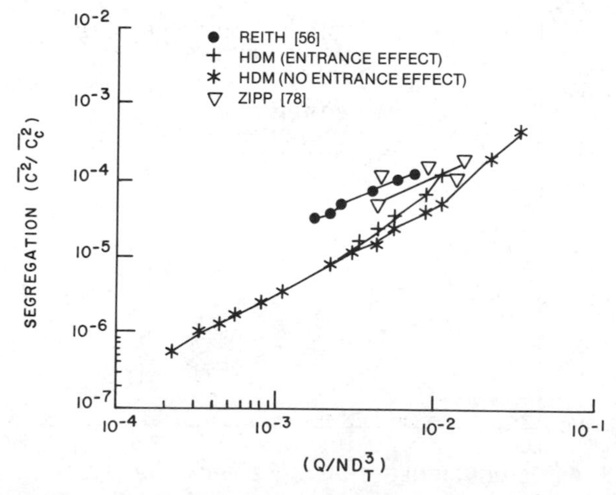

Figure 28. Comparison with previous segregation results [56,78].

It was more sensitive and had better frequency response characteristics. Intensity of segregation was a maximum at 60 rpm, nearest to the tip of the impeller.

Zipp [78] measured concentration and segregation using a light scattering method, as described above. The instrumentation necessary for the measurements is shown in Figure 19. Measurements of segregation in the impeller stream with a Rushton turbine are shown in Figure 27. Zipp's data are compared with data measured by Reith [56] using the conductivity method. Figure 28 shows comparisons of values of segregation in the effluent at various flowrates for Zipp's data, Reith's data and modeling results using the hydrodynamic model (HDM) described above [29,30].

CONCLUSION

Measurements of velocity, turbulence, local concentration and segregation in mixing fluids have progressed to the point that application can be extended beyond the research laboratory into the industrial process development laboratory. Laser-Doppler anemometry has reached a level of standardization of method and equipment that its use is reliable. Light scattering and fluorescence for measurement of local concentration are still under development but soon should offer viable methods to study mixing and reactor problems in the industrial environment. Fluorescence, in particular, offers the possibility of studying mixing and other transport effects on complex multiple reactions where yields are the critical problem.

NOMENCLATURE

a	constant in Equation 52
A, A'	constants in King's law
b	constant in Equation 52
B	Beer's constant
$\overline{C}$	average concentration
c'	root-mean-square concentration
C_1, C_2	constants in Equation 7
C_{EBU}	eddy breakup constant
C_D	constant in momentum diffusivity term ($C_D = \nu_e \epsilon / k^2 = 0.09$)
$C_{i,j}$	concentration
D_f	separation distance of planes of maximum illumination
$D_{i?}$	diffusivity of component i in a mixture
D_I	impeller diameter
D_T	tank diameter
D_V	minimum beam diameter
e	anemometer voltage
$\vec{e}_i, \vec{e}_s$	unit vectors in incident and scattering directions, respectively
$E(t)$	exit age distribution times
f_d	Doppler frequency
f_i	frequency of incident light
f_s	frequency of scattered light
h'	mean square enthalpy fluctuations
H	height
$\overline{H}$	average enthalpy
I	coalescence-dispersion reaction rate
k	turbulent kinetic energy term $1/2\ (\overline{u_i^2})$; universal constant in Prandtl mixing length equation
K	reaction rate constant
l	Prandtl mixing length
n	exponent in Equation 33

N	number of coalescence sites; impeller speed
N_{Sc}	Schmidt number
pdf	probability density function
P	pressure
P_{ci}	source term in concentration equation
$P_{c_i^2}$	source term in segregation equation
Q	total volumetric flow
R	ratio of feedrates (Q_2/Q_1)
R_i	instantaneous rate
R_i	source term $= -K\, C_i\, C_j$
S_ξ	source term, Equation 23
t	time
T	temperature
T_a	ambient temperature
T_{ref}	reference temperature
u	fluid velocity
u_i	fluctuating velocity in i-direction
$\overline{U}_v$	average vertical plane velocity of mixed fluid
V	volume
$\overline{V}$	mean voltage from photomultiplier
$\overline{V}_o$	voltage due to incident light
ν'	root-mean-square voltage fluctuation

Greek Symbols

α	parameter in Equation 23
β	parameter defined in Equation 24
γ	parameter defined in Equation 24
γ_{kj}	degree of interdiffusion
ξ	mixture fraction, Equation 21
ϵ	rate of k-dissipation
λ	wavelength of light
θ	angle between incident and reflective beam
$\overline{\tau}$	average residence time
τ_M	time constant for mixing
τ_R	reaction rate time constant
σ_K	constant in turbulent energy definition
ν	kinematic viscosity
ν_e	momentum diffusivity

REFERENCES

1. Corrsin, S. *AIChE J.* 10: 870 (1964).
2. Patterson, G. K. In: *Fluid Mechanics of Mixing*, Proc. ASME meeting, Fluids Eng. Div., Atlanta, GA (1973).
3. Spalding, D. B. "Concentration Fluctuations in a Round Turbulent Free Jet," *Chem. Eng. Sci.* 26: 95 (1971).
4. Patterson, G. K., W. C. Lee and S. J. Calvin. Paper presented at Symp. on Turbulent Shear Flows, Pennsylvania State University, April, 1977.
5. Nauman, E. B. *Chem. Eng. Comm.* 8: (1981).
6. Canon, R. M., K. W. Wall, A. W. Smith and G. K. Patterson. *Chem. Eng. Sci.* 32: 1349 (1977).
7. Rao, D. P., and A. R. Rao. *Chem. Eng. Sci.* 29: 1809 (1974).
8. Evangelista, J. J. PhD Thesis, City College of New York, New York, NY (1970).

9. Pratt, D. T. "Mixing and Chemical Reaction in Continuous Combustion," *Prog. Energy Combustion Sci.* 1: 73 (1976).
10. Spalding, D. B. Imperial College, London, Mech. Eng. Dept., Rep. No. HTS/76/13.
11. Rhodes, R. P., P. T. Harsha and C. E. Peters. "Turbulent Kinetic Energy Analyses of Hydrogen-Air Diffusion Flames," *Acta Astron.* 1:443 (1974).
12. Launder, B. E., and D. B. Spalding. *Mathematical Models of Turbulence* (New York: Academic Press, Inc., 1972).
13. Rao, D. P., and I. J. Dunn. *Chem. Eng. Sci.* 25: 1275 (1970).
14. McKelvey, K. N., H. N. Yieh, S. Zakanycz and R. S. Brodkey. *AIChE J.* 21: 1165 (1975).
15. Elghobashi, S. E., W. Run and D. B. Spalding. *Chem. Eng. Sci.* 32: 161 (1977).
16. Lockwood, F. C., and A. S. Naquib. *Combustion and Flame* 24: 109 (1975).
17. Hill, J. C. *Chem. Eng. Educ.* Winder, 34 (1979).
18. Jones, W. P., and B. E. Launder. *Int. J. Heat Mass Transf.* 16: 119 (1973).
19. Lee, J., and R. S. Brodkey. *AIChE J.* 10: 187 (1964).
20. Brodkey, R. S. *Phenomena of Fluid Motions* (Reading, MA: Addison-Wesley Publishing Co., Inc., 1967).
21. Bilger, R. W. *Turbulent Reacting Flows*, P. A. Libby and F. A. Williams, Eds. (Heidelberg: Springer Publishers, 1977).
22. Gosman, A. D., W. M. Run, A. K. Runchal, D. B. Spalding and M. Wolfshtein. *Heat and Mass Transfer in Recirculating Flows* (New York: Academic Press, Inc., 1969).
23. Spalding, D. B. In: *Turbulent Combustion*, L. A. Kennedy, Ed. (Washington: American Institute of Aeronautics and Aerospace, 1978).
24. Borgi, R. In: *Turbulent Mixing in Nonreactive and Reactive Flows*, S. N. B. Murthy, Ed. (New York: Plenum Publishing Corp., 1975), p. 131.
25. Donaldson, C. duP. In: *Turbulent Mixing in Nonreactive and Reactive Flows*, S. N. B. Murthy, Ed. (New York: Plenum Publishing Corp., 1975), p. 131.
26. Panagopoulos, J. "Turbulent Mixing in a Chemically Reactive Flow with Second-Order Chemical Reaction," M. S. Thesis, University of Missouri-Rolla (1981).
27. Ellail, M. M. M. A., A. D. Gosman, F. C. Lockwood and I. E. A. Megahud. In: *Turbulent Combustion*, L. A. Kennedy, Ed. AIAA, 1978, p. 163.
28. Durao, D., and J. H. Whitelaw. "Turbulent Mixing in the Development Region of Coaxial Jets," *J. Fluids Eng., Trans. ASME* (1973).
29. Patterson, G. K. In: *Application of Turbulence Theory to Mixing Operations*, R. S. Brodkey, Ed. (New York: Academic Press, Inc., 1975).
30. Patterson, G. K. *Proc. 1st Eur. Conf. on Mixing and Centrifugal Separation*, paper A4, BHRA Fluid Eng., Cambridge, England (1974).
31. Patterson, G. K. *Chem. Eng. Comm.* 8: 25 (1981).
32. Bourne, J. R., U. Moergeli and P. Rys. paper presented at the 2nd European Conference on Mixing, Cambridge, England, April, 1977.
33. Nabholz, F. R. J. Ott and P. Rys. paper presented at the 2nd European Conference on Mixing, Cambridge, England, April, 1977.
34. Truong, K. T., and J. C. Methot. *Can. J. Chem. Eng.* 54: 572 (1976).
35. David, R., and J. Villermaux. *Chem. Eng. Sci.* 30: 1309 (1975).
36. Patterson, G. K. "Modelling Complex Chemical Reactions in Flows with Turbulent Diffusive Mixing," paper presented at the 70th Ann. AIChE Meeting, New York, 1977.
37. Patterson, G. K. *Proc. 2nd Symp. on Turbulent Shear Flow*, Imperial College (1979).
38. Vassilatos, G., and H. L. Toor. *AIChE J.* 11: 666 (1965).
39. Mao, K. W., and H. L. Toor. *AIChE J.* 16: 49 (1970).
40. Sachs, J. P. and J. H. Rushton. "Discharge Flow from Turbine-Type Impellers," *Chem. Eng. Prog.* 50: 597 (1954).
41. Cutter, L. "Flow and Turbulence in a Stirred Tank," *AIChE J.* 12: 35 (1966).
42. Schwartzberg, H. G., and R. E. Treybal. "Fluid and Particle Motion in Turbulence Stirred Tanks," *Ind. Eng. Chem. Fund.* 7: 1 (1968).
43. Nienow, A. paper presented at the 73rd annual AIChE meeting, Chicago, IL, 1980.

44. Tatterson, G. B., H. C. Yuan and R. S. Brodkey. "Stereoscopic Visualization of the Flows for Pitched Blade Turbines," *Chem. Eng. Sci.* 35: 1369 (1980).
45. Hinze, J. O. *Turbulence* (McGraw-Hill Book Co., 1959).
46. Davies, P. O., M. J. Fisher and M. J. Baratt. "The Characteristics of the Turbulence in the Mixing Region of a Round Jet." *J. Fluid Mech.* 15: 337 (1963).
47. Durst, F. "Evaluation of Hot-Wire Anemometer Measurements in Turbulent Flows," Imperial College, Report of Mechanical Engineering Department, ET/TN/A/9 (1971).
48. Rodi, W. "A New Method of Analyzing Hot-Wire Signals in Highly Turbulent Flow and its Evaluation in a Round Jet," *DISA Information* 17: 9 (February 1975).
49. Kanevce, G., and S. Oka. "Correcting Hot-Wire Readings for Influence of Fluid Temperature Variations," *DISA Information* 15: 21 (October 1973).
50. Sandborn, V. A. *Resistance Temperature Transducers* (Fort Collins, CO: Metrology Press, 1972).
51. Brayton, D., H. Kalb and F. Crossway. "A Two Component Dual Scatter Laser Dopper Velocimeter with Frequency Burst Signal Readout," in *The Use of the Laser Doppler Velocimeter for Flow Measurements,* Purdue University Conference (1972).
52. Durst, F., A Melling and J. H. Whitelaw. *Principles and Practice of Laser-Doppler Anemometry* (New York: Academic Press, Inc., 1976).
53. Lee, W. "Light-Scattering Measurements of Turbulence and Concentration in an Annular Mixing Jet," PhD Thesis, University of Missouri-Rolla, 1976).
54. Durst, F., and C. Tropea. "Processing of Laser-Doppler Signals by Means of a Transient Recorder and a Digital Computer," Rpt. SFB80/E/118, University of Karlsruhe, West Germany (1977).
55. Manning, F. S., and R. Wilhelm. "Concentration Fluctuations in a Stirred Baffled Vessel," *AIChE J.* 9: 12 (1963).
56. Reith, I. R. "Generation and Decay of Concentration Fluctuations in a Stirred Baffled Vessel," *AIChE-IChE Symp. Series* 10: 14 (1965).
57. Lee, J., and R. Brodkey. "Light Probe for the Measurement of Turbulent Concentration Fluctuations," *Rev. Scientific Instrum.* 34: 1068 (1963).
58. Nye, J., and R. Brodkey. "Light Probe for Measurement of Turbulent Concentration Fluctuations," *Rev. Scientific Instrum.* 38: 26 (1966).
59. Becker, H. A., H. C. Hottel and G. C. Williams. "On the Light Scatter Technique for the Study of Turbulence and Mixing," *J. Fluid Mech.* 30: 259, 285 (1967).
60. Shaughnessy, E. J., and J. B. Morton. "Laser Light-Scattering Measurements of Particle Concentration in a Turbulent Jet," *J. Fluid Mech.* 80: 129 (1977).
61. Patterson, G. K. "Development of a Fluorescence Method for Reaction Conversion Measurement in a Turbulent Mixing Tank," paper presented at the AIChE national meeting, Detroit, MI, August, 1981, also, *Physiochem. Hydro.*, in press.
62. Becker, H. A., R. E. Rosensweig and J. R. Gwozdz. "Turbulent Dispersion in a Pipe Flow," *AIChE J.* 12: 964 (1966).
63. Becker, H. A., and K. I. M. Smith. "Mixing between Diametrically Opposed Jets in a Transverse Pipe Flow," *Proc. Symposium on Turbulence,* 5th ed., G. K. Patterson and J. L. Zakin, Eds. (Princeton, NJ: Science Press, 1979).
64. Levenspiel, O. *Chemical Reaction Engineering* (New York: John Wiley & Sons, Inc., 1972).
65. Hubbard, D. W., and H. N. Patel. "Dimensional Analysis for Imperfect Mixing Processes, paper presented at the 62nd annual meeting of AIChE, 1969.
66. Hubbard, D. W., and H. N. Patel. "Hydrodynamic Measurements for Imperfect Mixing Processes, *AIChE J.* 17: 1387 (1971).
67. Fox, E., and V. Gex. "Single Phase Blending of Liquids," *AIChE J.* 2: 539 (1956).
68. Norwood, K., and A. Metzner. "Flow Patterns and Mixing Rates in Agitated Vessels," *AIChE J.* 6: 432 (1960).
69. Middleton, J. "Measurement of Circulation Within Large Mixing Vessels," *Proc. Third European Conf. on Mixing,* paper A2, Cambridge, England (1979).

70. Komasawa, I., R. Kuboi and T. O. Otake. "Measurement of Turbulence in Liquid by Continual Pursuit of Tracer Particle Motion," *Chem. Eng. Sci.* 29: 641 (1974).
71. Mujamdar, A. S., B. Huang, D. Wolf, M. E. Weber and W. J. M. Douglar. "Turbulence Parameters in a Stirred Tank," *Can. J. Chem. Eng.* 48: 475 (1970).
72. Rao, M. A., and R. S. Brodkey. "Continuous Flow Stirred Tank Turbulence Parameters in the Impelled Stream," *Chem. Eng. Sci.* 27: 137 (1972).
73. Desouza, A., and J. Pike. "Fluid Dynamics and Flow Patterns in Stirred Tanks with a Turbine Impeller," *Can. J. Chem. Eng.* 50: 15 (1972).
74. Nagata, S., K. Yamamoto, K. Hashimoto and Y. Naruse. *Mem. Fac. Eng. Kyoto Univ.* 21: 260 (1959).
75. Günkel, A. A., and M. E. Weber. "Flow Phenomena in Stirred Tanks," *AIChE J.* 21: 931 (1975).
76. Reed, X. B., M. Princz and S. Hartland. "Laser Doppler Measurements of Turbulence in a Stirred Tank," *Proc. Second European Conf. on Mixing*, Cambridge, England (1977).
77. Kaniwano, M., N. Ohshima, T. Koga and T. Motoyoshi. "Measurement Method of Flow Pattern of a Liquid in a Stirring Vessel Using an Image Sensor," Department of Chemical Engineering, National University, Yokohama, Japan (1980).
78. Zipp, R. P. "Light Scattering Measurements of Turbulence and Concentration in a Stirred Tank Reactor," MS Thesis, University of Missouri-Rolla (1981).

CHAPTER 6
NON-NEWTONIAN FLOWS

R. D. Patel
Exxon Research & Engineering Company
Florham Park, New Jersey 07932

CONTENTS

INTRODUCTION

The writer of a chapter on non-Newtonian flow is faced with a difficult task indeed. The field is vast. Entire books have been written on the subject, some focusing only on specialized aspects. In addition, it is a very active field with several specialized journals devoted to publishing the latest results.

Faced with this task, we have elected to focus attention on results that the practicing engineer would find useful in the design of equipment to handle non-Newtonian fluids. Necessarily, this emphasizes flow in ducts and pipes. For our purposes, the fluid is considered to be homogeneous (although it may not be in actuality), with a given "rheology" or flow behavior. Many multiphase heterogeneous fluids such as slurries and liquid-liquid dispersions can be treated this way from the fluid mechanics viewpoint. Specialized aspects of multiphase flows are considered elsewhere in this book. We have also elected not to elaborate on flows of viscoelastic fluids, except to mention them in passing. Viscoelasticity is of major interest perhaps only in polymer manufacture and polymer processing, and the interested reader should consult one or more of the recent texts on this subject [1-6].

The major sections of this chapter are divided as follows. As an introduction, the methodology of fluid dynamic analysis to be followed is discussed briefly. This sets the stage for a discussion of the flow behavior of non-Newtonian fluids and their classification. Laminar and turbulent flow are then treated. To conclude, some other aspects of non-Newtonian flows are pointed out that are of engineering interest, and sources of further information are given.

CONSERVATION LAWS AND FLUID MECHANICS ANALYSIS

The objective of fluid mechanics analysis is to determine one or more of the following for a given flow situation: (1) the velocity distribution, (2) the flowrate as a function of pressure drop, and (3) the mechanical forces acting on solid surfaces wetted by the fluid. To take a simple example, for fluid flowing in a pipe we may be interested in the fluid velocity as a function of position in the pipe, the flowrate for a given pressure difference, or vice versa, and the force acting on the pipe because of the fluid motion.

Fluid flow analysis is founded on three governing precepts: (1) the laws of conservation of mass, momentum and energy; (2) the "rate expression" or constitutive equation of the particular fluid that expresses how the fluid behaves under the action of a shear stress; and (3) the external constraints or boundary conditions that act at the boundaries of the fluid and thereby affect its motion. This methodology is covered in Chapter 2 of this volume, and a detailed discussion will not be given here. However, a brief review is germane to our discussion of non-Newtonian flows.

Distributed Parameter and Lumped Parameter Systems

For our purposes, a system is a portion of fluid or a piece of equipment to which we wish to devote our attention. Returning to our example for flow in a pipe, one system that could be chosen would be the pipe itself. This system would have an entrance and an exit, where fluid enters and leaves, respectively, and fixed walls that are wetted by the fluid. By choosing such a macroscopic system, we are confining our attention to macroscopic effects only, such as the relation between the fluid flowrate and the imposed pressure difference between the entrance and the exit. An analysis using this system cannot tell us about microscopic effects, such as the distribution of velocity with position. Such a macroscopic system is termed a *lumped parameter* system because we are, in effect, lumping the microscopic effects to determine the macroscopic behavior.

Another possible system we could choose would be an infinitesimal volume at some point within the pipe. This is a microscopic, or *distributed parameter,* system and allows for the determination of flow properties such as velocity at each point within the pipe. Once such distributed properties are obtained, we may readily derive the lumped parameter results by integrating over the volume of the lumped system. For example, once the velocity distribution in the pipe is known, the flowrate is easily determined by integrating the velocity over the pipe cross-section. It is not possible to reverse the procedure to obtain a distributed result (the velocity distribution) from a lumped result (the flowrate). For engineering purposes, the lumped parameter approach is favored because it yields results of engineering interest and also because the distributed parameter approach is often too complex, and solutions are unobtainable.

The three governing precepts mentioned above therefore may be applied to either lumped or distributed systems to yield a set of equations that must be solved to obtain the desired results. Under steady-state conditions, the lumped approach yields algebraic equations and the distributed approach yields partial or ordinary differential equations, depending on the simplifying assumptions used.

Conservation Equations and Methodology

We confine ourselves to fluids of constant density and steady-state conditions. For distributed systems, the important conservation equations are the conservation of mass (equation of continuity) and the conservation of momentum (equation of motion). The energy equation does not contribute useful information for distributed parameter analysis of incompressible isothermal flow. The equations of continuity and motion are given in Table I as Equations 1 and 2. Here, v is the velocity vector, p is the pressure, g is the acceleration due to gravity, and τ is the stress. The equation of motion is essentially Newton's second law of motion and equates the mass times the acceleration of the fluid to the sum of the forces acting on the fluid. The forces acting on the fluid are the pressure and gravity forces, and the viscous forces (related to τ) that are set up when the fluid is in motion. This equation cannot be solved for v without further knowledge of these viscous forces. Thus, a "rate expression" is necessary that gives the relation between τ, the viscous stresses and the velocity, v (or, strictly speaking, the gradient of velocity). The nature of this relation is what distinguishes Newtonian from non-Newtonian fluid mechanics, and is the subject of the section in this chapter entitled Classification of Fluid Behavior. Once τ is expressed correctly in terms of v, then the problem can be solved in principle after the boundary conditions are applied. A typical boundary condition is the "no-slip" condition wherein the velocity of the fluid at a solid surface is taken to be the same as the velocity of the surface.

For lumped systems, the important conservation equations are those of mass and mechanical energy. The conservation of momentum is used to obtain forces acting on solid surfaces wetted by the fluid and is not needed for the purposes of this chapter. The mechanical energy equation often is called the Bernoulli equation, and its form, assuming constant density, is Equation 4. Here, z is the vertical coordinate, $\hat{W}$ the rate of work done by the fluid, and $\hat{E}_v$ is the rate of dissipation (irreversible conversion) of mechanical to internal energy, both per unit mass of fluid flowing. This equation relates the change in kinetic, potential and pressure energy over the exit and entrance of the system to the useful work done, and the irreversible conversion of mechanical forms of energy to internal energy through friction.

The mechanical energy equation finds frequent use in engineering calculations. For example, the pressure drop over a system may be calculated from Equation 4 if the other terms are known. All the other terms except $\hat{E}_v$ usually are known, and some way of obtaining $\hat{E}_v$ is necessary. This is usually done through the friction factor, which is determined for a flow by analytical means or via a correlation. For turbulent flow, analytical methods fail and a correlation is necessary (see the section entitled Turbulent Flow); for laminar flow, analytical expressions are often available (see the section entitled Laminar Flow).

To consider non-Newtonian flows, we must first elaborate on the "rate expressions" or constitutive relations for such fluids. With such relations in hand, we may, in principle, obtain velocity distributions for non-Newtonian flows. (In practice, only a limited number of cases with laminar flow are amenable to analytical solutions.) Using these velocity distributions, other results may be obtained readily. Equally important, and perhaps more so for engineering purposes, we need to develop correlations for friction factors for laminar or turbulent non-Newtonian flows including flow in straight sections of pipe and in fittings. These are used in the Bernoulli equation to make engineering calculations.

Table I. Governing Equations and Conservation Laws for Distributed and Lumped Parameter Systems (steady-state and constant density, notation as in Bird et al. [7])

Quantity		Distributed Parameter	Lumped Parameter[a]
Conservation	Mass	$\nabla \cdot \mathbf{v} = 0$ (1)	$\Delta w = 0$ (3)
	Momentum	$\rho \mathbf{v} \cdot \nabla \mathbf{v} = -\nabla \mathbf{p} + \rho \mathbf{g} + \nabla \cdot \boldsymbol{\tau}$ (2) Mass × acceleration = pressure + gravity + viscous forces (per unit volume of fluid)	[b]
	Mechanical Energy	[b]	$\Delta \left[\frac{1}{2} \frac{\langle v^3 \rangle}{\langle v \rangle} + gz + \frac{p}{\rho} \right] = -\hat{W} - \hat{E}_v$ (4)[c,d] Change of [kinetic + potential + pressure energy] = work done + conversion of mechanical to internal energy (per unit mass of fluid flowing)
Rate Expression		$\boldsymbol{\tau}$ as function of velocity gradient	$\hat{E}_v$ as function of friction factor, f, or drag coefficient
Boundary Conditions		e.g., $\mathbf{v}$ specified at solid surfaces	—

[a] Δ denotes output − input.
[b] Not useful for our purposes.
[c] The symbol $< \, >$ denotes an average over the cross section. $\langle v^3 \rangle / 2 \langle v \rangle$ is the kinetic energy of the fluid per unit mass and accounts for the velocity distribution over the cross section.
[d] Also called the Bernoulli Equation.

CLASSIFICATION OF FLUID BEHAVIOR

This section discusses the behavior or deformation of a fluid under the action of a shearing stress. In the purest sense of the word, a *fluid* is that form of matter that deforms continuously under the action of any shearing stress. When the stress is removed deformation ceases, but the fluid does not return to its original configuration. This behavior is in contrast to that of a *solid*, which does not deform continuously under a shearing stress, but instead attains a definite equilibrium-deformed state for a particular stress. When the stress is removed, the solid returns to its original configuration. (Here we are assuming that the solid is being stressed below its yield value.) In practice we do not insist on such an absolute definition of fluid.

There are many materials that exhibit both fluid- and solid-like properties and are still treated as fluids. For example, certain materials will not deform continuously under the action of any non-zero shearing stress, but only if a certain "yield" stress is exceeded. Other materials exist that deform continuously but when the stress is removed will partially recover their original configuration. For our purposes, we will consider a fluid to be a form of matter that exhibits continuous deformation under some range of shearing stress and may partially recover its original configuration when the stress is removed.

The deformation of a fluid under the action of a stress is the subject of the science of *rheology*, the study of the deformation and flow of materials. This is a complex and difficult subject that is still being developed and has been treated extensively elsewhere [1,2,5,6,8]. Rheologists are concerned with both the measurement of the deformation of a fluid with stress and the formulation of mathematical relations between deformation rates (or velocity gradients) and stress (τ). These relations, called rheological models, or "constitutive equations," can be used in the momentum balance equation to compute for a given laminar flow situation, velocity profiles, volumetric flowrates, etc. While a great number of such equations have been formulated, their use often entails the evaluation of multiple parameters from data, which is difficult in practice. Further, the complexity of many of these equations precludes their use for engineering design. Unfortunately, parameters evaluated from measurements under steady shear often cannot represent data from, say, oscillatory shear. For these reasons, only very simple constitutive equations are in use for engineering purposes, and only these will be discussed in some detail. Naturally, these simple rheological models are limited in their ability to represent all aspects of fluid behavior and are suitable only under restricted conditions. Details of the more complex models are given in the above references.

Deformation of a Fluid

Consider a fluid held between two large parallel plates separated by a small gap, Y, as shown in Figure 1. The lower plate is held stationary while the upper plate is moved at a constant velocity, V, through the action of a force, F. A thin layer of fluid adjacent to each plate will move at the same velocity as the plate. (This is the "no slip" assumption, and holds true for all except a few fluids.) Molecules in fluid layers between these two extremes will move at intermediate velocities. For example, a fluid layer, B, immediately below A will experience a force in the x direction from layer A and a smaller retarding force from layer C. Layer B then will flow at a velocity lower than V. This progression will continue to the layer adjacent to the lower plate, whose velocity will be zero. Under steady-state conditions, the force, F, required to produce the motion becomes constant and will be related to the velocity as

$$\frac{F}{A} = f\left(\frac{V}{Y}\right) \tag{5}$$

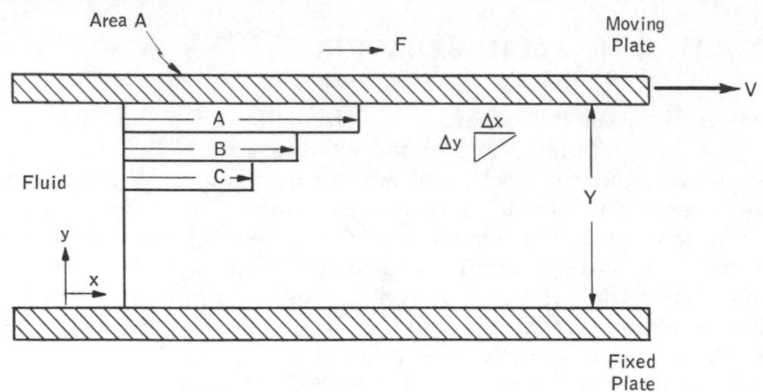

Figure 1. Deformation of a fluid.

where A is the area of each plate, and f( ) is a function depending on the fluid in question and the temperature and pressure.

The ratio F/A is the shear stress on the fluid at the upper plate, and V/Y is the velocity gradient. On a local basis, that is, at any point within a fluid, we may generalize the above expression as

$$\tau_{yx} = f\left(\frac{dv_x}{dy}\right) \tag{6}$$

Here, τ_{yx} is the local shear stress (force per unit area) acting in the x direction on a plane perpendicular to the y axis, and v_x is the velocity in the x direction, $\frac{dv_x}{dy}$ being the velocity gradient on a local basis. Equation 6 is the constitutive equation for the fluid in question. In graphic form it is referred to as the flow curve, or rheogram for the fluid.

It is instructive to interpret the velocity gradient as follows. We may write

$$\frac{dv_x}{dy} = \frac{d}{dy}\left(\frac{dx}{dt}\right) = \frac{d}{dt}\left(\frac{dx}{dy}\right) \approx \frac{d}{dt}\left(\frac{\Delta x}{\Delta y}\right) \tag{7}$$

The term $\frac{\Delta x}{\Delta y}$ is the shear strain on the fluid, and we therefore see that $\frac{dv_x}{dy}$ is the rate of shear strain, or simply shear rate. Thus, for fluids, the shear stress is a function of the rate of shear strain, or simply shear rate. (For solids, on the other hand, the shear stress would be a function of strain rather than the rate of strain.) The shear rate is sometimes called rate of deformation and is given the symbol $\dot{\gamma}$.

For a homogeneous fluid containing small molecules, the constitutive equation (Equation 6) is usually fairly simple, but for a multiphase mixture, a solution or a liquid containing large molecules, quite complex relations result. In the first case, the shear stress-shear rate relation is often linear through the origin, and such fluids are termed Newtonian. For such simple fluids, the internal structure is unaffected by the magnitude of the imposed shear rate. For complex fluids, an imposed shear results in changes in the internal structure of the fluid so that the constitutive relation becomes much more complicated. For example, in a liquid containing large molecules (a molten polymer or a polymer solution) at low shear rates, the molecules would remain randomly coiled, much as they are in the fluid at rest. The fluid structure would remain unchanged in this range, and the shear stress-shear rate relation would be linear, as for a Newtonian fluid.

As the shear rate is increased, the randomly coiled molecules would tend to line up in the flow direction, changing the structure of the fluid and, hence, the nature of the constitutive relation. The progressive lining-up or disentangling of the molecules results in the fluid becoming less viscous, or "thinner," in its flow properties. Increasing the shear further would continue to cause more and more molecules to line up, so that the stress-shear rate relation would remain nonlinear, until a limiting shear is reached when all the molecules have lined up. Further shear rate increases would not result in structural change, and the constitutive relation would return to a Newtonian relation, albeit a different one from that for low shear rates. Thus, the progression is from Newtonian to non-Newtonian and back to Newtonian as the shear rate is increased. Structural changes also occur for suspensions of solid particles in a liquid, or for liquid-liquid emulsions, resulting again in complex flow curves for such fluids. We will discuss these relations further below.

Before we examine fluid behavior in detail, we must recognize that our simple experiment will not detect certain unusual fluid characteristics. In a steady flow such as exists in our experiment, time-dependent properties and solid-like or elastic behavior cannot be exhibited. However, consider what would occur if the force moving the upper plate were removed. In most cases, the upper plate would continue to move, but with decreasing velocity until it and the fluid eventually came to a stop. If the fluid were *viscoelastic* (that is, possessing both viscous and elastic properties), the plate and the fluid would first slow down as before, but after coming to a stop some motion in the negative x direction would occur as the fluid sought to recover its original configuration. Only partial recovery would be attained.

Next consider what happens if the shear rate on a fluid were changed instantaneously. (This is difficult to achieve experimentally, since in our experiment the inertia of the plate and the fluid itself will result in only a gradual change.) For many fluids, any resulting changes in internal structure occur very rapidly, and a new stress level corresponding to the new shear rate is reached instantaneously following Equation 6. Such fluids are termed *time-independent*. For *time-dependent* fluids structural changes are much slower, and the shear stress changes slowly until ultimately a steady value is reached corresponding to the new shear rate. Finally, some fluids do not deform until a certain "yield" stress value is exceeded. In our experiment, F/A would have to be greater than the yield stress for flow to occur for such a fluid.

It is convenient to divide fluids into three classes in terms of their flow behavior: time-independent purely viscous fluids, time-dependent purely viscous fluids, and viscoelastic fluids (Table II). Of these three classes, the time-independent purely viscous fluids are best understood. A great deal of work has been done on viscoelastic fluids, but few results of use in engineering design are available. Time-dependent fluids are less common, and perhaps the least is known about these fluids.

Time-Independent Fluids

Newtonian Fluids

As mentioned briefly above, the simplest class of real fluids is comprised of the Newtonian fluids, whose constitutive equation is given by

$$\tau = \mu\dot{\gamma} \tag{8}$$

Table II. Classification of Fluid Behavior

Fluids			
Purely Viscous			
Time-Independent		Time-Dependent	
No Yield Stress	Yield Stress	Thixotropic	Viscoelastic
Newtonian	Bingham	Rheopectic	
Pseudoplastic	Yield-pseudoplastic		
Dilatant	Yield-dilatant		

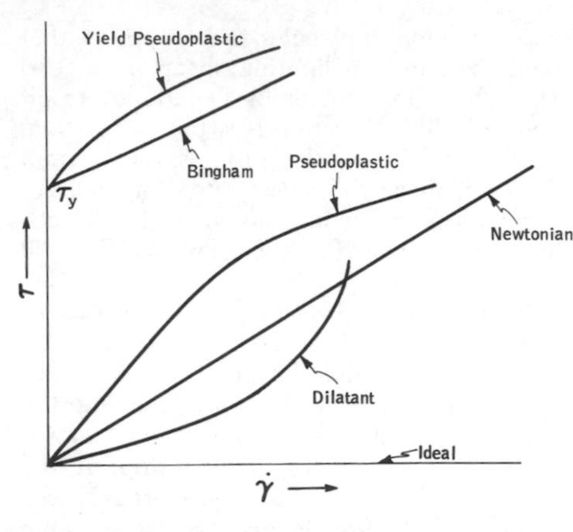

Figure 2. Rheograms on arithmetic coordinates.

where for convenience we have dropped the yx subscripts on τ. Equation 8 is known as Newton's law of viscosity. The viscosity, μ, is a fluid property and is a function of temperature and pressure only (for a single-phase, single-component system). The viscosity of liquids decreases, and that of gases increases, with temperature. An *ideal* or perfect fluid is one whose viscosity is zero, and thus can be sheared without the application of a shear stress. It is an artifact in that no such fluids exist, but finds use in the theory of potential flows.

The dimensions of μ are readily developed. The stress τ has dimensions $ML^{-1}t^{-2}$ and $\dot{\gamma}$, t^{-1}. Thus, μ has dimensions $ML^{-1}t{-}1$. In the CGS units system (M in g, L in cm, t in sec) the viscosity has units g cm^{-1} s^{-1}, known as poise. For most Newtonian fluids this is a rather large unit, and the centipoise (cP), which is 0.01 poise, is more convenient. The viscosity of water at 20°C is 1.00 cP and that of air is 0.0181 cP.

The flow curve or rheogram of a Newtonian fluid is a straight line through the origin in arithmetic coordinates. The slope of the line is the viscosity so that the entire class of Newtonian fluids can be represented by a family of straight lines through the origin. It is often convenient to plot the rheogram using logarithmic coordinates, as it allows a greater range of data and also permits easy comparison of Newtonian versus non-Newtonian behavior. On such coordinates, a Newtonian fluid has a rheogram that is a straight line of unit slope and whose intercept at a shear rate of unity is the viscosity. Figures 2 and 3 illustrate these alternative methods of representation.

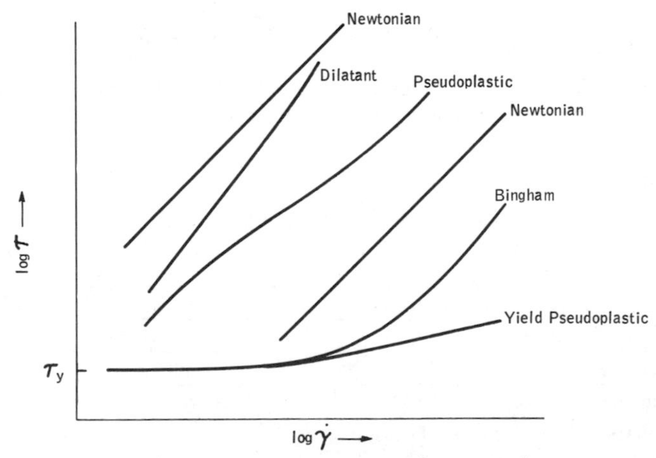

Figure 3. Rheograms on logarithmic coordinates.

Most low-molecular-weight liquids and solutions, and all gases, are Newtonian. Homogeneous slurries of small spherical particles in gases or liquids at low solids concentration are also frequently Newtonian. There is thus a large class of fluids for which a single property, the viscosity, characterizes the flow behavior.

A far larger class of fluids does not follow Newton's law of viscosity and, furthermore, may have other characteristics such as time-dependence, viscoelasticity and yield. A fluid with any of these characteristics is termed non-Newtonian. In general, there are no completely satisfactory constitutive equations for non-Newtonian fluids, in contrast to the situation for Newtonian fluids.

For non-Newtonian fluids it is convenient to define an "apparent viscosity," η_a, as

$$\eta_a = \tau/\dot{\gamma} \tag{9}$$

The apparent viscosity is a function of $\dot{\gamma}$ for non-Newtonian materials and is analogous to the Newtonian viscosity, μ, for Newtonian fluids. However, whereas the Newtonian viscosity does not vary with shear rate, the non-Newtonian apparent viscosity does.

Pseudoplastic or Shear-Thinning Fluids

Pseudoplastic fluids often also are called shear-thinning because their apparent viscosity decreases with shear rate. In other words, the rate of increase of shear stress for such fluids decreases with increased shear rate. Increased shear breaks down the internal structure within the fluid very rapidly and reversibly, and no time-dependence is manifested. Examples of fluids that exhibit shear-thinning are polymer melts and solutions, mayonnaise, suspensions such as paint and paper pulp and some dilute suspensions of inert particles. Many of these fluids also exhibit other non-Newtonian characteristics, such as viscoelasticity in the case of polymer solutions and melts, and time-dependence in the case of paints. Thus pseudoplasticity is but one important characteristic of such a non-Newtonian fluid and does not necessarily describe all of its non-Newtonian features. Many pseudoplastics are shear-thinning at intermediate shear rates and are Newtonian at low and high shear rates. A mechanism for this phenomenon for fluids containing long molecules such as polymer solutions has been mentioned above. Rheograms for pseudoplastics are illustrated in Figures 2 and 3.

Many constitutive equations of varying complexity have been proposed for these fluids. Because of its simplicity, the power law is the most widely used, even though it does not describe the Newtonian extremes found for most pseudoplastics.

Power Law

The power law, or Ostwald-de Waele equation, is

$$\tau = K\dot{\gamma}^n \tag{10}$$

and is a two-parameter equation. The parameter K is the *consistency index* and n is the *flow behavior index*. Both are functions of temperature and pressure, but K is more sensitive to temperature than n. Pressure dependence of these parameters has not been investigated.

It is clear that if $n = 1$, Equation 10 reduces to Newton's law of viscosity. For $n < 1$, we have pseudoplastic or shear-thinning behavior, whereas if $n > 1$, dilatant or shear-thickening behavior is shown as will be detailed below. The apparent viscosity for a power law fluid is

$$\eta_a \equiv \tau/\dot{\gamma} = K\dot{\gamma}^{n-1} \tag{11a}$$

showing that if $n < 1$, η_a decreases with $\dot{\gamma}$. The value of n is the slope of the rheogram on logarithmic coordinates.

The fit of the power law equation to experimental data is usually good over several orders of magnitude in $\dot{\gamma}$. However, at low and high $\dot{\gamma}$ Newtonian behavior is obtained and agreement is poor. In spite of these limitations, use of the power law gives good predictions of pipeline pressure drop if the parameters are fitted over the same range of shear encountered in the wall region of the pipe.

It should be pointed out that Equation 10 is valid only for $\dot{\gamma} > 0$. Depending on the coordinate system used to solve a flow problem, $\dot{\gamma}$ may be negative over some portion of the flow field, in which case the correct form of Equation 10 is

$$\tau = K \mid \dot{\gamma} \mid^{n-1} \dot{\gamma} \tag{11b}$$

For our purposes, we will always assume $\dot{\gamma} > 0$ and "simple shear" with only one nonzero component of the velocity gradient. For extensions of Equation 11 to more complicated flow fields and where $\dot{\gamma}$ may be negative, the reader is referred to the work of Bird et al. [7]. These authors place a negative sign in the constitutive equations because of their interpretation of τ as momentum flux rather than shear stress. This sign should be ignored to conform to our notation.

Other Equations

Many other equations have been proposed for pseudoplastics, but they have not been used as widely as the power law. Also, they are generally more complex, and some involve three or more parameters, rather than only two. The principal advantage of these more complex models is that they can predict shear-thinning behavior as well as the tendency to Newtonian behavior at one or both extremes of shear rate. Since these models are not in wide use, we will not discuss them in detail, but rather refer the reader to Table III, where a few are listed along with their main features. Further details may be found in the literature [2,7,9].

Dilatant or Shear-Thickening Fluids

Dilatant fluids often are called shear-thickening because their apparent viscosity increases with shear rate. The rate of increase of shear stress for such fluids increases with shear rate. By using appropriate values of the parameters, equations developed for pseudoplastics can be applied to dilatant fluids. For example, the power law equation may be used with n greater than unity for dilatant fluids. Rheograms for these fluids are illustrated in Figures 2 and 3. Dilatancy is not as common as pseudoplasticity and generally is observed for fairly concentrated suspensions of irregular particles in liquids. A commonly accepted mechanism [10] for dilatancy is that at low shear rates the particles in such fluids are densely packed but still surrounded by liquid, which lubricates the motion of adjacent particles. At higher shear rates the dense packing breaks up progressively, forcing liquid out of more and more of the interstices between particles. There is now insufficient liquid to lubricate the motion of such adjacent "dried out" particles, and the shear stress increases with shear rate at a higher rate than before. Many dilatant fluids also exhibit volumetric dilatancy, that is, an increase in volume with shear rate, as well as viscous dilatancy, the increase in apparent viscosity with shear rate. Examples of dilatant fluids are aqueous suspensions of titanium oxide, suspensions of starch and quicksand.

Plastic Fluids—Fluids with Yield

Certain fluids do not deform continuously unless a limiting or "yield" stress is exceeded. They are sometimes called plastic fluids or Bingham fluids, although the latter implies that they follow a particular constitutive equation. In the simplest case, such fluids behave exactly like Newtonian fluids once the yield stress is exceeded, in that their rheograms are Newtonian rheograms shifted upward (Figure 2). These are

Table III. Some Constitutive Equations for Pseudoplastic (or dilatant) Fluids

Model Name	Constitutive Equation	Apparent Viscosity	Limiting Newtonian Prediction: Low Shear	Limiting Newtonian Prediction: High Shear	Remarks[a]
Power law (Ostwald-de Waele)	$\tau = K\dot{\gamma}^{n}$	$K\dot{\gamma}^{n-1}$	–	–	Two parameters: K, n Newtonian limits not predicted.
Prandtl-Eyring	$\tau = A \sinh^{-1}(\dot{\gamma}/E)$	$A\dot{\gamma} \sinh^{-1}(\dot{\gamma}/B)$	A/B	–	Two parameters: A, B Based on Eyring's kinetic theory of liquids
Ellis	$\dot{\gamma} = (\phi_o + \phi_1 \tau^{a-1})\tau$	$(\phi_o + \phi_1 \tau^{a-1})^{-1}$	ϕ_o^{-1} for $\alpha > 1$ only	ϕ_o^{-1} for $\alpha < 1$ only	Three parameters: ϕ_o, ϕ_1, α If $\phi_1 = 0$ gives Newtonian equation If $\phi_o = 0$ gives power law equation
Reiner-Phillippoff	$\dot{\gamma} = \left[\mu_\infty + \frac{\mu_o - \mu_\infty}{1 + (\tau/\tau_s)^2} \right]^{-1} \tau$	$\mu_\infty + \frac{\mu_o - \mu_\infty}{1 + (\tau/\tau_s)^2}$	μ_o	μ_∞	Three parameters: μ_o, μ_∞, τ_s μ_o and μ_∞ are the limiting viscosities
Sisko	$\tau = a\dot{\gamma} + b\dot{\gamma}^{c}$	$a + b\dot{\gamma}^{c-1}$	a	–	Three parameters: a, b, c. Combination of Newtonian and power law

[a] All these equations predict shear-dependent viscosity.

called Bingham fluids. Pure Bingham behavior is rare. In other cases, the flow curve is that of a pseudoplastic shifted upward, and such fluids are termed yield pseudoplastics. Yield dilatant behavior also may be encountered. The commonly accepted mechanism for the behavior of plastic fluids is that the fluid at rest contains a structure sufficiently rigid to resist shear stresses smaller than the yield stress, τ_y. When this stress is exceeded, the structure collapses and deformation is continuous, as for nonplastic fluids. Examples of fluids exhibiting plastic behavior are paint systems, suspensions of finely divided minerals such as chalk in water, and some asphalts. The yield values can be quite small for water suspensions and very large for materials such as asphalt. Plastic behavior is necessary in a paint to prevent flow when it is applied as a vertical film. Constitutive equations for such fluids are modifications of the Newtonian or the power law expressions.

Bingham Model

The Bingham equation is given by

$$\begin{aligned} \tau &= \tau_y + \eta\dot{\gamma} \qquad \tau > \tau_y \\ \dot{\gamma} &= 0 \qquad\qquad\quad \tau < \tau_y \end{aligned} \tag{12}$$

where τ_y is the yield stress, and η is the "plastic viscosity" or "coefficient of rigidity," the slope of the flow curve in arithmetic coordinates. On logarithmic coordinates the curve is asymptotic to τ_y at low $\dot{\gamma}$ and approaches a slope of unity for high $\dot{\gamma}$ (Figure 3). Note that for $\tau < \tau_y$, $\dot{\gamma} = 0$, implying that there is no deformation in this region. The apparent viscosity for Bingham fluids is given by

$$\eta_a \equiv \tau/\dot{\gamma} = \eta + \tau_y\dot{\gamma}^{-1} \tag{13}$$

showing that the apparent viscosity decreases with shear rate. For very high shear rates the effect of the yield stress becomes negligible, the apparent viscosity levels off to the plastic viscosity, η, and the fluid behaves as a Newtonian fluid.

Yield-Power Law Model

This model is given by

$$\begin{aligned} \tau &= \tau_y + K\dot{\gamma}^n \qquad \tau > \tau_y \\ \dot{\gamma} &= 0 \qquad\qquad\quad \tau < \tau_y \end{aligned} \tag{14}$$

and is simply a combination of the Bingham and power law equations (Figures 2 and 3). As before, pseudoplastic or dilatant behavior obtains, depending on whether n is less than or greater than unity. An example of yield-pseudoplastic liquids would be a clay-water suspension. Yield-dilatant behavior is less common.

Time-Dependent Fluids

The fluids discussed so far have the property that their flow behavior responds instantaneously to sudden changes in shear. For example, if a pseudoplastic fluid described above were subjected to a sudden change in shear rate, the new stress given by its constitutive equation would be attained instantaneously. Such fluids are termed time-independent. (Actually, a finite nonzero time is required for the new stress to be achieved, but this time is very short in terms of our normal time frame.) However, there exists a class of fluids termed time-dependent fluids for which the response time is appreciable. For such fluids, the sudden application of a change in shear rate would result in the shear stress changing slowly with time until the new equilibrium shear stress corresponding to the changed shear rate is reached. The postulated mechanism for time-dependence is that the time scale for structural changes within the fluid due to shear is large compared to the time scale of shear and the very short time scales for time-independent fluids. Since the flow behavior depends on the fluid structure, the shear stress responds slowly to an imposed change in shear rate. Thus, the shear stress becomes a function of shear rate and time until steady conditions are reached. Time-dependence may be exhibited by fluids that otherwise may be termed pseudoplastic, dilatant or plastic. Time-dependent fluids generally are classified as thixotropic or rheopectic.

Thixotropic Fluids

Thixotropic fluids break down under shear. At a given shear rate, the shear stress slowly decreases until an equilibrium value is reached. Such fluids behave as time-dependent pseudoplastics. An illustration of thixotropic behavior is shown in Figure 4, which shows the change in apparent viscosity of the fluid when a higher shear rate, $\dot{\gamma}_2$, is suddenly imposed at time t_1. At times less than t_1 the fluid has been sheared at $\dot{\gamma}_1$ for a long time, so that an equilibrium apparent viscosity, η_{a1}, is manifested. For a time-independent pseudoplastic, the apparent viscosity will decay to η_{a2} immediately. For a thixotropic fluid, there will be only a slow decay over some measurable time $(t_2 - t_1)$ until the equilibrium viscosity, η_{a2}, is reached. If the new shear rate is a *decrease* over the original value, $\dot{\gamma}_1$, the apparent viscosity will *increase* slowly with time. Such instantaneous changes in $\dot{\gamma}$ cannot be achieved in practice because of fluid and equipment inertia.

A more practical way to detect thixotropy (or any time-dependent behavior) is to subject the fluid to a programmed change in shear rate with time, increasing from zero shear rate to some peak value and then decreasing back to zero. A thixotropic fluid under such a program would produce a hysteresis loop for τ versus $\dot{\gamma}$, as shown in Figure 5.

For a pseudoplastic, no hysteresis would be shown, and the same curve would be traced out for increasing and decreasing shear. We should note, of course, that the area within the loop depends both on the degree of thixotropy as well as the time scale of change of the shear rate. If the shear rate is changed slowly enough, even a highly thixotropic fluid would produce a curve with no hysteresis. Also to be noted is the position of the curve for increasing shear relative to that for decreasing shear, which follows from the fact that the fluid is basically pseudoplastic with an apparent viscosity that decreases with increasing shear rate. Examples of thixotropic fluids are paints, ketchup and food materials, oil well drilling muds and some crude oils. Thixotropy

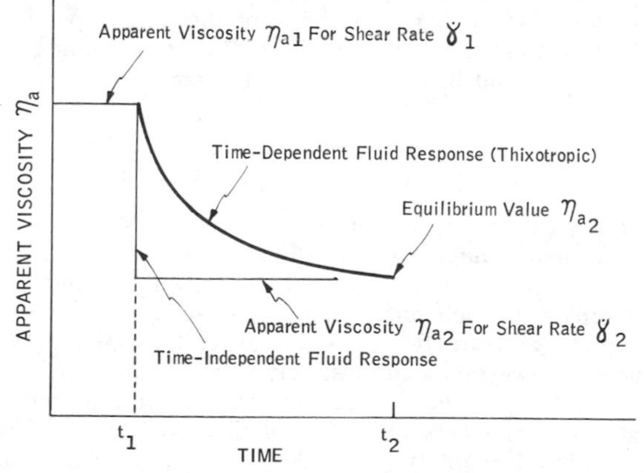

Figure 4. Response of time-dependent fluid to change in shear rate (thixotropic case).

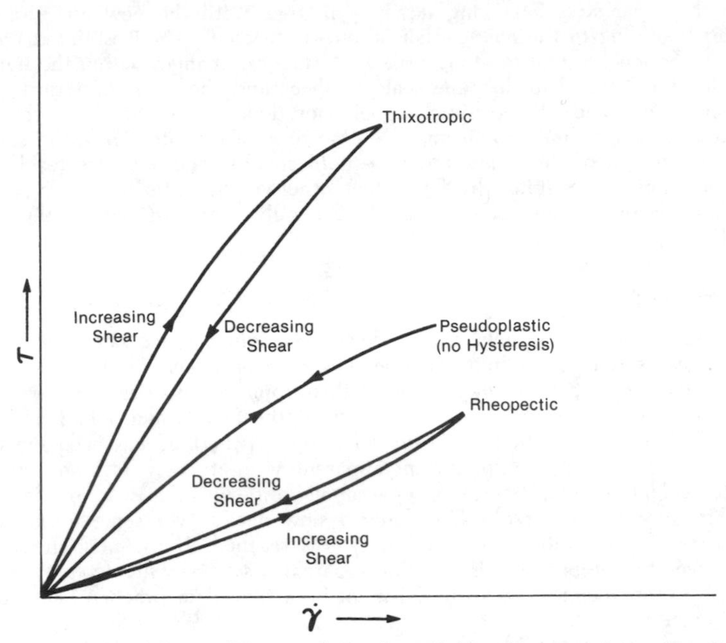

Figure 5. Hysteresis loops for time-dependent fluids. (Arrows show the chronology of the imposed shear rate.)

is necessary in paint to allow it both to flow easily after the large shear imposed by brushing and then to recover a more viscous character after a short period of standing. Govier and Aziz [11] have summarized some of the rheological models proposed for thixotropic fluids. A review of the subject has been given by Bauer and Collins [12].

Rheopectic Fluids

For rheopectic fluids, shear stress at a constant shear rate increases slowly with time until an equilibrium value is reached. Such fluids behave as time-dependent dilatant fluids. Under the programmed change in shear rate described above, the stress-shear rate curve forms a hysteresis loop, but of a different shape than a thixotropic fluid (Figure 5). As before, the location and shape of the loop depend on the shear rate program, as well as on the degree of rheopexy. Rheopectic fluids are rare, examples being gypsum suspensions and bentonite clay suspensions.

Viscoelastic Fluids

For the fluids discussed so far, if a previously imposed shear stress is removed, deformation ceases, but there is no tendency for the fluid to recover its original undeformed state. Certain fluids do have the property of partially recovering their original state after the stress is removed. Such fluids thus have properties akin to elastic solids, as well as viscous liquids, and are termed viscoelastic. Examples of viscoelastic liquids are molten polymers and polymer solutions, egg white, dough and bitumens.

The elastic property of such fluids leads to some interesting and unusual behavior. The classic example is the phenomenon of rod-climbing, or the "Weissenberg effect" exhibited by these fluids. If a rotating cylinder or rod is immersed in a purely viscous liquid, the liquid surface is depressed near the rod because of centrifugal forces. In a viscoelastic liquid, on the other hand, liquid climbs up the rod because of normal stresses generated by its elastic properties. This can be observed during mixing of flour dough and in stirred polymerization reactors. Another phenomenon is the marked swelling in a jet of viscoelastic liquid issuing from a die. As a result, extrusion dies must be designed properly to produce the desired product cross section. These and other phenomena have been strikingly demonstrated in a film by Markovitz [13].

For these viscoelastic liquids, some of the work done on the fluid in, say, forcing it through a tube, is stored in the fluid as normal stresses as opposed to being dissipated into heat in the case of purely viscous liquids. This stored energy is released when the fluid emerges from the tube and results in swelling of the emerging fluid jet. The normal stresses generated within the tube relax when the fluid emerges from the tube, and such fluids are said to exhibit stress relaxation. Some liquid-liquid mixtures consisting of droplets of one liquid dispersed in the other also exhibit viscoelasticity. Elastic energy is stored when the spherical droplets are distorted by shear and is released through the action of interfacial tension when the shear is removed.

Elastic effects come into play mainly during the storage or release of elastic energy. Thus, such effects are important in the entrance and exit sections of tubes and during flow accelerations and decelerations caused by changes in cross section, or by imposed oscillations or by turbulence. For steady laminar flow in a tube or channel of constant cross section, elastic effects are not important except near the entrance and exit. On the other hand, for flow in fittings or in turbulent flow these effects may become important. Marked viscoelasticity is observed mainly in molten polymers and concentrated polymer solutions and generally is not considered important for pipeline flow of other non-Newtonians, such as slurries.

It is clear that for viscoelastic fluids the flow behavior cannot be represented as a relation between shear stress and shear rate alone. Rather, it will depend on the recent history of these quantities, as well as their current values. Constitutive equations for such fluids therefore must involve shear stress, shear rate and their time derivatives. It is clear also that the time derivatives involved must be those applicable to a particular blob of fluid as it moves through the system, and not those from the viewpoint of, for example, a stationary observer. The latter has no direct impact on fluid behavior; the former represents time rates of change experienced by the fluid itself. A large number of constitutive equations of this general type have been proposed involving, among other features, various types of time derivatives. It is beyond the scope of this work to discuss these in any detail, particularly because their use in engineering calculations is not widespread. For illustrative purposes we may mention the simple three-constant Oldroyd model, which is applicable for low shear rates:

$$\tau + \lambda_1 \frac{d\tau}{dt} = \mu \left(\dot{\gamma} + \lambda_2 \frac{d\dot{\gamma}}{dt} \right) \tag{15}$$

Here, λ_1 and λ_2 are two relaxation times and μ is the viscosity. We may first observe that if λ_1 and λ_2 are both zero, the equation reverts to that for a Newtonian fluid. If only λ_2 is zero, the equation reduces to that for a Maxwell fluid. The Maxwell fluid is an early and simple model for a viscoelastic fluid based on a mechanical analogy. The fluid is represented as a spring and dashpot in series, the spring representing the elastic part and the dashpot the viscous part. Both the Maxwell fluid and the fluid represented by Equation 15 show stress relaxation. If flow is stopped,

$$\dot{\gamma} = \frac{d\dot{\gamma}}{dt} = 0$$

the stress decays or relaxes as e^{-t/λ_1}. If stress is removed, the shear rate in a Maxwell fluid becomes zero immediately, while for the fluid of Equation 15 the shear rate decays as e^{-t/λ_2}. This simple three-constant model has been shown to represent the behavior of certain viscoelastic fluids at low shear rates.

Rheological Measurements—Viscometry

Determination of the rheology of a fluid requires an apparatus that permits the measurement of shear stress and shear rate at the same location in the fluid and, if viscoelastic properties are required, also the measurement of normal stresses. The parallel plate apparatus of Figure 1 is obviously impractical, and some alternative scheme must be devised. A large number of viscometers of different design have been developed, and it is not our purpose to explore the various types in any detail. For the purposes of this chapter, it will be sufficient to assume that the shear stress-shear rate relation is described by one of the models discussed above and that the model parameters are supplied. However, to give the reader a flavor of what is involved, a very brief discussion of viscometry is in order. All viscometers are constructed to give a flow field that is well described using the equations of conservation of mass and momentum. Usually this means there is but one nonzero velocity component, which varies in but one coordinate direction. Under these conditions, the kinematics of the motion are known exactly and the shear stress and shear rate can be calculated from easily measured quantities. Table IV compares three types of viscometers that have been used and indicates the general way they operate. For example, the concentric cylinder-type instrument consists of a stationary bob concentric with a rotating cup. Fluid is held between the two. Laminar flow must prevail. The torque, T, on the bob and the angular velocity, Ω, of the cup are measured. Shear stress on the bob surface is readily obtained directly from T. Shear rate at the bob surface cannot be calculated directly from Ω unless the gap between the bob and cylinder is very small. Otherwise, a correction that turns out to be dependent on the variation of T with Ω must be applied. Normal stress measurements can be made from the difference in pressure at the bob and cup surfaces. The cone and plate instrument has the advantage that shear rate is independent of fluid behavior and may be directly calculated from Ω. Normal stresses also are measured more easily. The capillary instrument is simple in construction and is capable of providing data at high shear rates. For further details, the reader is referred to the book by van Wazer et al. [14] and the references they cite.

LAMINAR FLOW

This section discusses some of the laminar flows of non-Newtonian fluids, but will be confined to power law fluids and Bingham-type or yield-power law fluids as these models are in much wider use in industry than any of the other models. Many solutions for laminar flow of these and other non-Newtonians are available for various geometries. The books by Skelland [9], Govier and Aziz [11] and Cheremisinoff [15] list some of these. The fluid mechanics literature should be consulted for more recent solutions. Flow anomalies caused by time-dependent behavior or viscoelasticity will not be discussed.

The usual methodology for distributed parameter analysis of laminar flows, discussed at length by Bird et al. [7], may be divided into five steps:

1. Simplify the velocity field by judicious assumptions.
2. Simplify the equations of conservation of mass and momentum based on 1, above.
3. Use the velocity field of (1) in the constitutive equation for the fluid and substitute for τ in the momentum conservation equation.
4. Apply the boundary conditions on velocity.
5. Solve for the velocity profile and other derived quantities such as flowrate, force on solid surfaces, etc.

This approach can be taken for a number of simple geometries such as tubes, annuli, falling films and parallel plates. For all these cases, only one velocity com-

Table IV. Summary Of Common Viscometer Types

Viscometer Type	Capillary	Concentric Cylinder	Cone and Plate
Description	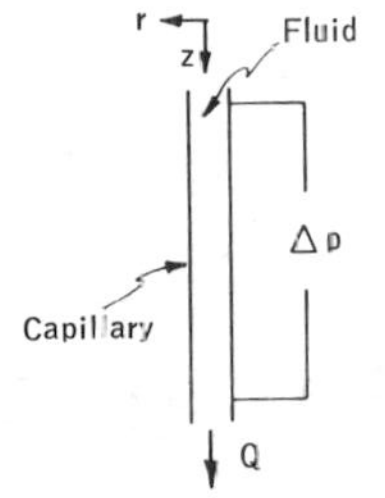 r z Fluid Capillary Δp Q	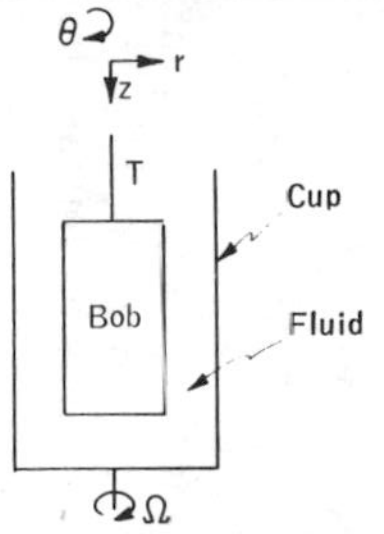 θ r z T Cup Bob Fluid Ω	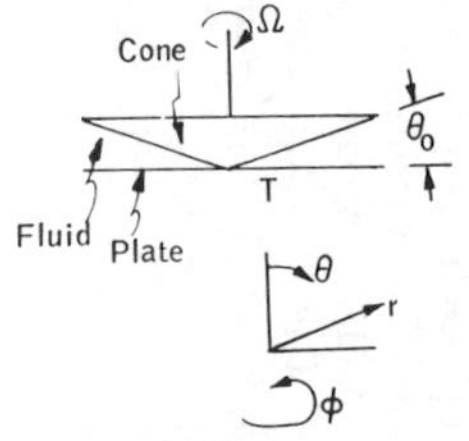 Cone Ω θ_0 T Fluid Plate θ r ϕ
Measured Variables	Δp—Pressure gradient Q—Fluid flowrate	T—torque on stationary bob Ω—angular velocity of cup	T—torque on stationary plate Ω—angular velocity of cone
Velocity Field	$v_z = v_z(r)$	$v_\theta = v_\theta\ (r)$	$v_\phi = v_\phi\ (r, \theta)$
Shear Stress	At tube wall; related to Δp	At bob surface; related to T	At plate; related to T
Shear Rate	At tube wall; related to Q	At bob surface; related to Ω	At plate; related to Ω
Comments	Shear rate varies with r; relation of $\dot{\gamma}$ to Q depends on fluid behavior—how Δp varies with Q	Shear rate varies with r; relation of $\dot{\gamma}$ to Ω depends on fluid behavior—how T varies with Ω	If angle θ_0 is small ($< 1°$), shear rate independent of r; relation of $\dot{\gamma}$ to Ω is independent of fluid behavior
Normal Stress can be obtained from	Axial thrust on capillary; swelling of jet	Difference in pressure at surface of bob and cup	Normal force on cone or plate; pressure distribution across diameter of cone or plate

ponent is nonzero and, further, it depends on only one coordinate direction. Such problems yield to rather simple methods of solution. At the same time, the results provide valuable insight into the physical phenomena underlying non-Newtonian flows in general. We will confine our discussions to some of these simple cases.

It is often simpler to take a more indirect route than the steps outlined above to arrive at the flowrate-pressure drop relation. The classical example of this is the Rabinowitsch-Mooney equation for flow in a pipe, which will be discussed first.

Rabinowitsch-Mooney Equation for Circular Tube

For steady, fully developed flow in a pipe or tube (Figure 6), the velocity field is simply $v_z(r)$, and all other velocity components are zero. If the fluid is inelastic and time-independent, the only nonzero component of stress is τ_{rz}, the shear stress in the z direction acting on a surface normal to the r direction, and τ_{rz} depends only on r. For this situation, the equation of motion for the z direction reduces to

$$\frac{1}{r}\frac{d}{dr}(r\tau_{rz}) = \frac{dp}{dz} \tag{16}$$

where gravity has been neglected, and dp/dz is the pressure gradient. (Gravity can be included by defining a dynamic pressure $P = p - \rho gz$; dp/dz may be shown to be a constant and is equal to $\Delta p/L$.) Equation 16 also may be obtained from a force balance on a differential volume within the tube.

Using the symmetry condition that $\tau_{rz} = 0$ at $r = 0$, we may integrate the above to obtain

$$\tau_{rz} = \frac{r}{2}\frac{dp}{dz} \tag{17}$$

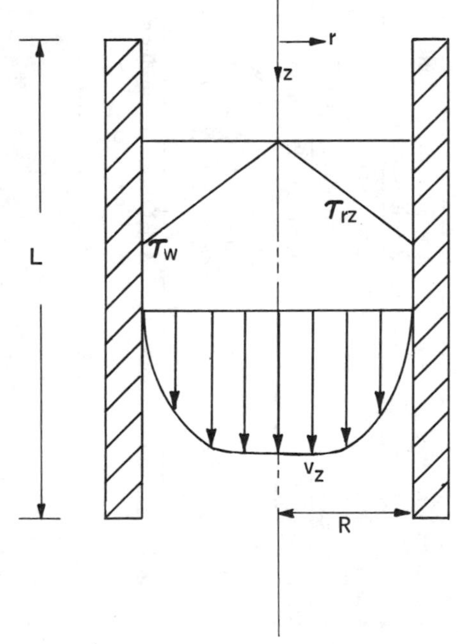

Figure 6. Flow in a tube.

or

$$\frac{\tau_{rz}}{\tau_w} = \frac{r}{R} \tag{18}$$

where τ_w is the shear stress at the wall and, using Equation 17, is given by

$$\tau_w = \frac{R}{2}\frac{dp}{dz} = \frac{R}{2}\frac{\Delta p}{L} \tag{19}$$

To obtain further details of the flow, two approaches are possible. In the standard approach, the constitutive equation for the fluid would be used in Equation 17 to obtain an equation for v_z, which then could be solved with the help of the "no-slip" boundary condition at the wall. The second approach is of advantage when the constitutive equation is unknown, and is of special value for non-Newtonian fluids. Using this approach we proceed as follows:

The volumetric flowrate, Q, is given by

$$Q = \int_0^R v_z \, 2\pi r dr \tag{20}$$

which may be integrated by parts to give

$$Q = \pi[v_z r^2 - \int r^2 \, d \, v_z]_{r=0}^{r=R} \tag{21}$$

Using the no-slip assumption we have the boundary condition

$$r = R, \quad v_z = 0 \tag{22}$$

which eliminates the first term in Equation 21. We now assume an arbitrary constitutive equation as

$$-\frac{dv_z}{dr} = f(\tau_{rz}) \tag{23}$$

where the negative sign is used because dv_z/dr is negative here. Changing variables in Equation 21 using Equation 23 yields the Rabinowitsch-Mooney equation:

$$Q = \frac{\pi R^3}{\tau^3_w} \int_0^{\tau_w} \tau^2 f(\tau) \, d\tau \tag{24}$$

which relates the volumetric flowrate to the pressure gradient (since τ_w depends on $\Delta p/L$ only) and the constitutive relation (Equation 23). Equation 24 also can be used to measure the function, f, from capillary viscometer data.

The flowrate-pressure gradient relations for various non-Newtonian fluids flowing in a tube can be constructed from Equation 24 by substituting the appropriate forms of f. Results are shown in Table V. For yield-type fluids, $f(\tau)$ is discontinuous because for $0 \leq r \leq r_c$ the shear stress is less than τ_y, the yield stress. This means that dv_z/dr (or f) is zero here, or that the fluid in this region moves in plug flow. The integral in Equation 24 is readily evaluated, however. Note further that once Equation 23 is known (say as data), the integral in Equation 24 can be evaluated numerically, if necessary. A rheological model is not required. If τ_w (or Δp) is known and Q is to be determined, an iterative procedure must be followed. If the velocity distribution $v_z(r)$ is sought, we must revert to the standard solution method.

Table V. Flowrate for Non-Newtonian Fluids in a Circular Tube (laminar flow)

Fluid	Constitutive Relation (Equation 23), f	Flowrate, Q
Newtonian	τ/μ	$\frac{\pi R^3 \tau_w}{4\mu}$
Power Law	$(\tau/K)^{1/n}$	$\frac{\pi R^3 n}{3n+1}\left(\frac{\tau_w}{K}\right)^{1/n}$
Bingham Plastic	$0 \leqslant r \leqslant r_c \quad f = 0$ $r_c \leqslant r \leqslant R \quad f = (\tau - \tau_y)/\eta$ $r_c = 2L\tau_y/\Delta p$	$\frac{\pi R^3 \tau_w}{4\eta}\left[1 - \frac{4}{3}\left(\frac{\tau_y}{\tau_w}\right) + \frac{1}{3}\left(\frac{\tau_y}{\tau_w}\right)^4\right]$
Yield-Pseudoplastic	$0 \leqslant r \leqslant r_c \quad f = 0$ $r_c \leqslant r \leqslant R \quad f = \left(\frac{\tau - \tau_y}{K}\right)^{1/n}$ $r_c = 2L\tau_y/\Delta p$	$\frac{\pi R^3 n(\tau_w - \tau_y)^{\frac{n+1}{n}}}{(K)^{1/n}\tau_w^3}\left[\frac{(\tau_w - \tau_y)^2}{1+3n} + \frac{2\tau_y(\tau_w - \tau_y)}{1+2n} + \frac{\tau_y^2}{1+n}\right]$

Velocity Profiles in a Circular Tube

Following the standard method outlined previously, we illustrate the procedure for obtaining the velocity distribution of a power law fluid. In doing so, we will repeat portions of the above derivation, but in more detail.

1. We assume that the flow is steady and fully developed and that the fluid has constant density. The simplified velocity field has been discussed already, i.e., $v_z = v_z(r)$; other components are zero.
2. The equation of continuity (conservation of mass) shows that if v_r and v_θ are both zero, then $\partial v_z/\partial z = 0$, confirming that $v_z = v_z\,(r)$ only. The components of the equations of motion* (conservation of momentum) simplify as follows if we assume τ_{rz} is the only nonzero stress term and τ_{rz} is independent of z (fully developed flow)

r-equation
$$\frac{\partial p}{\partial r} = 0 \tag{25}$$

θ-equation
$$\frac{1}{r}\frac{\partial p}{\partial \theta} = 0 \tag{26}$$

z-equation
$$\frac{\partial p}{\partial z} = \frac{1}{r}\frac{d}{dr}(r\tau_{rz}) \tag{27}$$

Equations 25 and 26 show that $p = p\,(z)$ only and, therefore, that

$$\frac{dp}{dz} = \frac{\Delta p}{L} \tag{28}$$

We therefore obtain Equation 16 as before, but the assumptions behind this equation are now more clear.

3. The constitutive equation for a power law fluid in our coordinate system reduces to

$$\tau_{rz} = K\left(-\frac{dv_z}{dr}\right)^n \tag{29}$$

*Chapter 3 of Bird et al. [7] contains tables of these equations in various coordinate systems.

where the negative sign is required because dv_z/dr is negative here. Substituting Equation 29 into Equation 16, we obtain

$$\frac{1}{r}\frac{d}{dr}\left[rK\left(-\frac{dv_z}{dr}\right)^n\right]=\frac{dp}{dz}=\frac{\Delta p}{L} \tag{30}$$

which is a second-order equation for v_z.

4. The boundary conditions on velocity are

$$r = R \qquad v_z = 0$$

$$r = 0 \qquad v_z = \text{finite, or } \frac{dv_z}{dr} = 0 \tag{31}$$

(The second condition is equivalent to the condition $r = 0$, $\tau_{rz} = 0$.)

5. The solution of Equation 30 with boundary conditions (Equation 31) is

$$v_z = \left[\frac{\Delta p}{2KL}\right]^{1/n} \frac{nR}{n+1}^{\frac{n+1}{n}} \left[1 - \left(\frac{r}{R}\right)^{\frac{n+1}{n}}\right] \tag{32}$$

or

$$\frac{v_z}{\bar{v}} = \left(\frac{3n+1}{n+1}\right)\left[1 - \left(\frac{r}{R}\right)^{\frac{n+1}{n}}\right] \tag{33}$$

where $\bar{v}$ is the cross-sectional average velocity in the tube and is related to the flowrate, Q, by

$$Q = \pi R^2 \bar{v} \tag{34}$$

Plots of the velocity distribution are given in Figure 7 and show the change in shape as the flow index, n, is varied. For pseudoplastic fluids ($n < 1$) the profiles are flatter than for Newtonian fluids ($n = 1$), and, as n approaches zero, plug flow results. For dilatant fluids ($n > 1$) the profiles are less flat, and, as n becomes very large, a triangular profile is approached. This variation in the velocity profiles has a significant effect on heat transfer rates for such fluids.

The *velocity gradient*, or *shear rate* in the tube, may be obtained from Equation 32 as

$$\dot{\gamma} = -\frac{dv_z}{dr} = \left(-\frac{\Delta p}{2KL}\right)^{1/n} r^{1/n} \tag{35}$$

At $r = 0$, $\dot{\gamma}$ is zero, and for small r, $\dot{\gamma}$ is also small. We have discussed in the section entitled Classification of Fluid Behavior that the power law is invalid for small $\dot{\gamma}$, so we may expect inaccuracies to result. However, in practice errors are slight, and the equations derived above agree with data quite well.

The *volumetric flowrate*, Q, has been obtained previously (Table V), and can also be derived by inserting Equation 32 into Equation 20. The flowrate-pressure drop relation is useful in sizing pipelines and pumps and, for this purpose, is frequently represented in the form of the friction factor or drag coefficient. The friction factor, f, may be defined as

$$\hat{E}_v = 4f\frac{L}{D}\frac{\bar{v}^2}{2} \tag{36}$$

(The more common definition of f is in terms of the force exerted on the pipe walls [7].) $\hat{E}_v$ represents the frictional "losses" per unit mass of fluid flowing. For our pipe

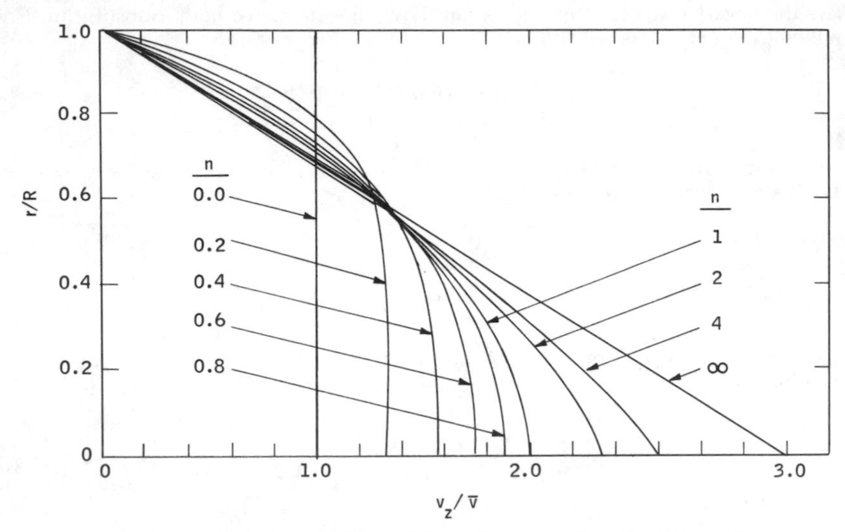

Figure 7. Velocity profiles for flow of a power law fluid in a tube.

flow situation, if gravity is neglected as before, the lumped parameter mechanical energy balance, or Bernoulli equation (4), reduces to

$$\frac{\Delta p}{\rho} = \hat{E}_v$$

so that

$$\Delta p = 4\,f\,\frac{L}{D}\,\frac{\bar{v}^2}{2} \tag{37}$$

(This is sometimes called the Fanning equation and also may be used as a definition of f.) The friction factor, f, now can be derived for laminar flow of a power law fluid using Table V and Equations 34 and 37. From Table V,

$$\Delta p = \frac{2L\tau_w}{R} = 2KL\left(\frac{3n+1}{n}\right)^n \frac{\bar{v}^n}{R^{n+1}} \tag{38}$$

Using Equation 37,

$$f = 16/\left[\frac{D^n\bar{v}^{2-n}\rho}{K}\,8\left(\frac{n}{6n+2}\right)^n\right] = 16/Re_{n,2} \tag{39}$$

where the power law-Reynolds number, $Re_{n,2}$ is defined as

$$Re_{n,2} = \frac{D^n\bar{v}^{2-n}\rho}{K}\,8\left(\frac{n}{6n+2}\right)^n \tag{40}$$

Another definition is

$$Re_{n,1} = \frac{D^n\bar{v}^{2-n}\rho}{K} \tag{41}$$

Both of these Reynolds numbers are in use and both reduce to the Newtonian Reynolds number when $n = 1$. However, use of $Re_{n,2}$ results in Equation 39 being identical to the Newtonian friction factor equation, $f = 16/Re$. If $Re_{n,1}$ is used instead, the friction factor expression varies with n. A plot of f versus $Re_{n,2}$ is a single straight line on logarithmic coordinates. If f is plotted against $Re_{n,1}$, a line is produced for each value of n and clearly shows the effect of n on the transition from laminar to turbulent flow. In fact, the transition Reynolds number increases with decreasing n. The friction factor-Reynolds number relation will be discussed further below under scaleup (Figure 9).

Equation 38 shows that if n is much less than 1 (a highly pseudoplastic liquid), Δp is insensitive to $\bar{v}$, implying that use of the pressure drop over a straight length of pipe to measure flowrate will not give accurate results. Conversely, increasing the flowrate of such a fluid will not incur a substantial pressure drop penalty.

Velocity distributions for other non-Newtonian fluids flowing in tubes may be derived following the same procedure as outlined above. Rather than repeat the derivations, we list the results in Table VI and discuss them briefly below.

For *Bingham plastics,* the velocity profile is a plug flow at the center of the tube surrounded by an annular region where the velocity decreases from the plug velocity to zero at the wall (Figure 8A). The edge of the plug is given by r_c, and if $r_c = R$ or

$$\Delta p = 2L\tau_y/R \tag{42}$$

there will be no flow. Equation 42 therefore gives the minimum pressure drop required for flow to ensue. It also shows that if a Bingham fluid is held in a vertical tube, it may or may not flow out under the force of gravity, depending on the tube radius, R, and the yield stress, τ_y.

The friction factor for a Bingham fluid depends on two dimensionless groups. One form for the relation is shown in Table VI and uses the Bingham Reynolds number $Re_B = D\bar{v}\rho/\eta$, and the Hedstrom number, $He = \tau_y D^2 \rho/\eta^2$. Another representation, using the yield number,

$$Y = D\tau_y/\bar{v}\eta \tag{43}$$

is given by Govier and Aziz [11]. The Hedstrom number is independent of $\bar{v}$ and depends only on tube diameter and fluid properties, which is an advantage. The relation for f must be evaluated numerically and plots of f versus Re with He as parameter may be constructed as in Figure 8B. If $\tau_y = 0$ ($He = 0$), the usual Newtonian relation

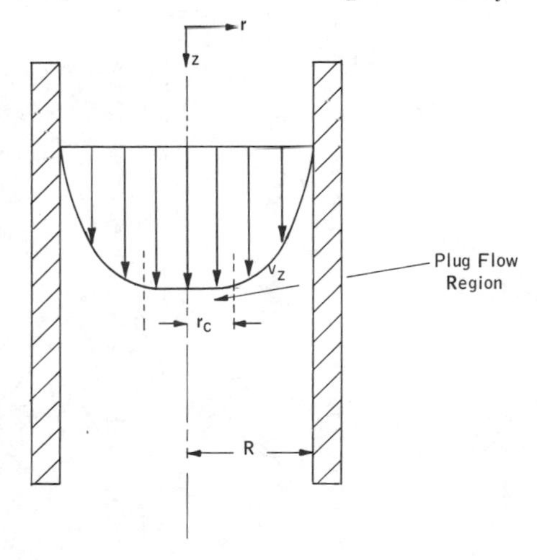

Figure 8A. Typical velocity distribution for Bingham fluid in a tube.

Table VI. Velocity Distributions and Friction Factors for Non-Newtonian Fluids in Circular Tubes (laminar flow)

Fluid	Velocity Distribution	Friction Factor
Newtonian	$\frac{\Delta p R^2}{4\mu L}\left[1-\left(\frac{r}{R}\right)^2\right]$	$f = 16/\mathrm{Re} = 16/(D\bar{v}\rho/\mu)$
Power Law	$\left(\frac{\Delta p}{2KL}\right)^{1/n}\frac{nR^{\frac{n+1}{n}}}{n+1}\left[1-\left(\frac{r}{R}\right)^{\frac{n+1}{n}}\right]$	$f = 16/\mathrm{Re}_{n,2} = 16/\left[\frac{D^n\bar{v}^{2-n}\rho}{K}\,8\left(\frac{n}{6n+2}\right)^n\right]$ $= \left(\frac{16}{\mathrm{Re}_{n,1}}\right)\left(\frac{1}{8}\right)\left(\frac{2+6n}{n}\right)^n$
Bingham Plastic[a]	$0 \leqslant r \leqslant r_c \quad v_z = \frac{\Delta p R^2}{4\eta L}\left(1-\frac{r_c}{R}\right)^2$ $r_c \leqslant r \leqslant R \quad v_z = \frac{\Delta p R^2}{4\eta L}\left[1-\left(\frac{r}{R}\right)^2\right] - \frac{\tau_y R}{\eta}\left[1-\left(\frac{r}{R}\right)\right]$	$\frac{1}{\mathrm{Re}_B} = \frac{f}{16} - \frac{1}{6}\frac{\mathrm{He}}{\mathrm{Re}^2_B} + \frac{1}{3}\frac{\mathrm{He}^4}{f^3\mathrm{Re}^8_B}$ $\mathrm{He} = \frac{\tau_y D^2\rho}{\eta^2},\ \mathrm{Re}_B = \frac{D\bar{v}\rho}{\eta}$
Yield-Power Law[b]	$r_c \leqslant r \leqslant R \quad v_z = \frac{n}{n+1}\frac{2L}{\Delta p K^{1/n}}\left[\left(\frac{R\Delta p}{2L}-\tau_y\right)^{\frac{n+1}{n}} - \left(\frac{r\Delta p}{2L}-\tau_y\right)^{\frac{n+1}{n}}\right]$ $0 \leqslant r \leqslant r_c \quad$ as above with $r = r_c$	$f = f\left(\frac{D^n\bar{v}^{2-n}\rho}{K}, \frac{\tau_y D^2\rho}{K^2}, n\right)$

[a] Here, $r_c = 2L\tau_y/\Delta_p$.
[b] Friction factor relation is complex; best approach is direct Q vs Δp relation.

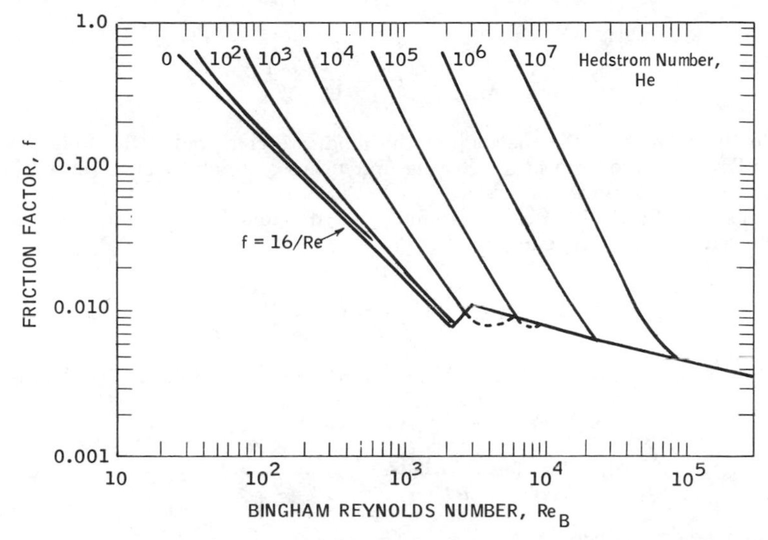

Figure 8B. Friction factor-Reynolds number relation for Bingham plastics [11].

is obtained. The laminar-turbulent transition depends on He, as may be seen in Figure 8B. This is discussed further in the next section.

For *yield-power law* fluids, the velocity distribution is as shown in Table VI. Here again there is a plug flow region in the central part of the tube and an annular region where the velocity varies. This is the most general velocity profile for the fluids considered in Table VI, and the other profiles can be derived from this one. For $n = 1$, the profile reverts to that of a Bingham fluid, and if $\tau_y = 0$ the power law profile is obtained. The Newtonian result obtains for $\tau_y = 0$ and $n = 1$. The friction factor-Reynolds number relation is somewhat unwieldy, and it is more convenient to directly use the flow rate-pressure drop relation from Table V. In dimensionless form, the friction factor depends on the power law-Reynolds number, the Hedstrom number and n.

Generalized Correlation and ScaleUp: Laminar Flow

Metzner and Reed [16] have developed a generalized scaleup and correlation procedure for laminar flow of time-independent non-Newtonians flowing in a tube. In this approach, the Rabinowitsch equation (24) is rearranged as

$$\frac{Q}{\pi R^3} = \frac{1}{4}\left(\frac{8\bar{v}}{D}\right) = \frac{1}{\tau_w{}^3}\int_0^{\tau_w} \tau^2 f(\tau)\,dt \tag{44}$$

which shows that for any fluid, τ_w is a function of only $(8\bar{v}/D)$:

$$\tau_w = \frac{D}{4}\frac{\Delta p}{L} = \phi(8\bar{v}/D) \tag{45}$$

This equation implies that we may scale up the pressure drop-flowrate relation from a model to full scale by running the two at the same value of $\bar{v}/D$ using the same fluid. Then Equation 45 shows that

$$\tau_{W_{model}} = \tau_{W_{full\ scale}}$$

or

$$\frac{(\Delta p/L)_{\text{full scale}}}{(\Delta p/L)_{\text{model}}} = \frac{D_{\text{model}}}{D_{\text{full scale}}} \tag{46}$$

Thus, without knowing the rheology of the fluid, we can predict the fully developed laminar flow pressure drop in a full-scale unit using one data point from a laboratory scale unit at the same $\bar{v}/D$.

Metzner and Reed [16] further assumed that the function ϕ in Equation 45 could be expressed as a power function:

$$\tau_w = \frac{D}{4}\frac{\Delta p}{L} = K'(8\bar{v}/D)^{n'} \tag{47}$$

Using this form along with Equation 37, which defines the friction factor, one may write

$$f = \frac{\left(\frac{D}{4}\frac{\Delta p}{L}\right)}{\rho\bar{v}^2/2} = \frac{K'(8\bar{v}/D)^{n'}}{\rho\bar{v}^2/2} \tag{48}$$

which may be rewritten as

$$f = 16/\left(\frac{D^{n'}\bar{v}^{2-n'}\rho}{\gamma}\right) \tag{49}$$

where

$$\gamma = K'\,8^{n'-1}$$

Equation 49 is analogous to the friction factor relation for laminar Newtonian flow and indeed reverts to this relation when $n' = 1$ and $K' = \mu$. In analogy with the Newtonian relation, we may define a generalized non-Newtonian Reynolds number as

$$Re_{n'} = \frac{D^{n'}\bar{v}^{2-n'}\rho}{\gamma} \tag{50}$$

Equivalently, the *effective viscosity*, μ_{eff}, of a non-Newtonian fluid may be defined by writing

$$Re_{n'} = Re = D\bar{v}\rho/\mu_{eff} \tag{51}$$

where

$$\mu_{eff} = K'\,8^{n'-1}\,\bar{v}^{n'-1}\,D^{1-n'} \tag{52}$$

is the viscosity that makes the non-Newtonian friction factor relation identical to the Newtonian relation:

$$f = 16/Re \tag{53}$$

Metzner and Reed [16] were able to correlate data for a large number of non-Newtonian fluids (including plastic fluids) over a wide range of Reynolds numbers by plotting f versus $Re_{n'}$. They found that the curves for Newtonian fluids fit the data very well on such a plot (Figure 9). Govier and Aziz [11] report that other workers have confirmed these results. To use the Metzner-Reed approach, a plot of τ_w versus $8\bar{v}/D$ must be constructed from data over the same range of $8\bar{v}/D$ (or τ_w) as will be used in the design. Values of K′ and n′ are then calculated at the appropriate value of $8\bar{v}/D$, and the friction factor obtained from Figure 9. No particular form of constitutive equation is needed, thus avoiding problems such as the inadmissability of the power law for low and high shear rates. On a double logarithmic plot of the $D\Delta p/4L$ versus $8\bar{v}/D$, n′ is the slope of the tangent to the curve at some $8\bar{v}/D$ and K′ is the intercept made by the tangent at $8\bar{v}/D$ equal to unity. For a general non-Newtonian fluid, n′ and K′ vary with $8\bar{v}/D$. Using the flow equations already developed for power law and Bingham fluids, the relations between n′ and K′ and the constants characterizing these fluids may be constructed (Table VII). It may be seen that n′ and K′ are independent of $8\bar{v}/D$ for power law fluids. $Re_{n'}$ for power law fluids is identical to $Re_{n,2}$ (Equation 40), and the laminar curve of Figure 9 is identical to Equation 39.

Flow Between Parallel Plates

Another simple, yet important, flow geometry is flow between two large stationary parallel plates. The practical case of this flow is flow in a rectangular duct of large aspect ratio so that the duct edges can be neglected. The solution is valid away from the duct edges. Solutions that account for all the duct walls are available [9] but do not provide much more physical insight and are necessarily quite complex.

The flow situation is sketched in Figure 10. Following the same procedures outlined above for circular tubes, it may be shown that the shear stress, τ_{yz}, acting in the z direction on a surface normal to the y direction is given by

$$\tau_{yz} = y\Delta p/L \tag{54}$$

and the shear stress at the wall, τ_w, is, therefore,

$$\tau_w = H\Delta p/2L \tag{55}$$

where H is the distance between the plates and $\Delta p/L$ is the pressure gradient.

An equation analogous to be the Rabinowitsch-Mooney equation for pipes can be derived as

$$Q = \frac{WH^2}{2\tau_w{}^2}\int_0^{\tau_w} \tau f(\tau)\,d\tau \tag{56}$$

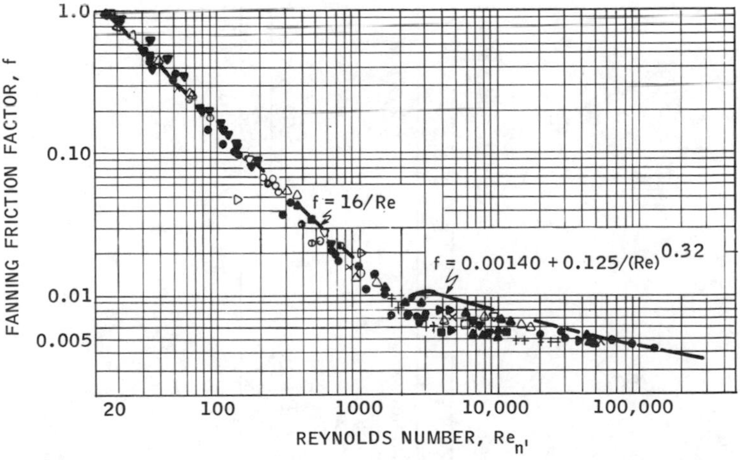

Figure 9. Friction factor vs generalized Reynolds number: time-independent non-Newtonian flow in a tube (curves are Newtonian correlations) [16].

Table VII. The Metzner-Reed Constants, n′, K′, for Various Non-Newtonian Fluids [17]

	n′	K′	γ
Newtonian	1	μ	μ
Power Law	n	$K\left(\frac{1+3n}{4n}\right)^{n}$	$K\left(\frac{1+3n}{4n}\right)^{n} 8^{n-1}$
Bingham Plastic ($x = \tau_y/\tau_w$)	$\frac{1 - 4x/3 + x^4/3}{1 - x^4}$	$\tau_w \left[\frac{\eta}{\tau_w(1 - 4x/3 + x^4/3)}\right]^{n'}$	$8^{n'-1}\tau_w \left[\frac{\eta}{\tau_w(1 - 4x/3 + x^4/3)}\right]^{n'}$

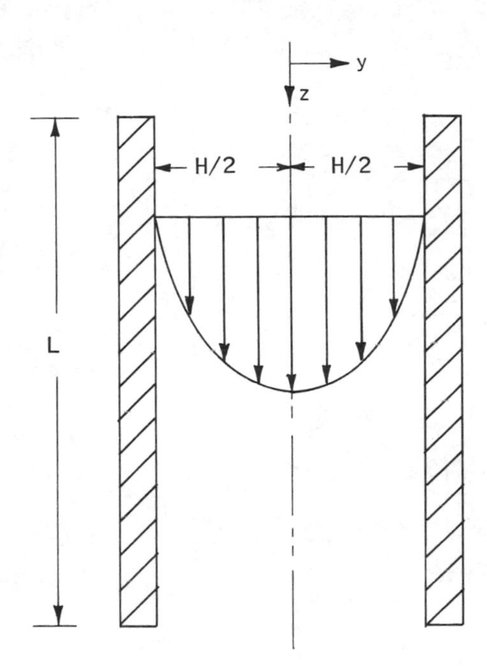

Figure 10. Flow between parallel plates.

where W is the width of the plates. Knowing the constitutive relation $f(\tau)$, Equation 23, the integral may be evaluated numerically or by using a constitutive equation. Velocity profiles may be calculated as for flow in tubes. Some results for different flow models are given in Table VIII.

Flow in Other Geometries

Flow in *concentric annuli* for Bingham plastics [18-20] and power law fluids [19,21] have been considered in the literature. Skelland [9] gives a detailed discussion for both of these cases. For the Bingham plastic an annular plug flow develops as shown in Figure 11. Expressions for the velocity profile and the volumetric flowrate are quite complex but can be readily computed.

Charts are also available for quick calculations [19,20]. In an annulus the stress becomes zero at the point of maximum velocity, as sketched in Figure 11. As discussed earlier, the power law equation is a poor representation for fluid rheology at low stress levels and predicts infinite apparent viscosity at zero stress. Real non-Newtonians exhibit Newtonian behavior in this region. For flow in tubes where zero stress occurs at the centerline, the volumetric flowrate is not greatly affected because r is small near the centerline (see Equation 20). In an annulus, appreciable error may result because r is large at the point of zero stress. Thus, flowrate predictions for power law fluids in annuli give erroneous results, as has been pointed out by Vaughn and Bergman [21]. These authors provide a method of using tube flow results to predict flow in annuli.

Solutions also may be developed for flow between *rotating cylinders* and for plane *Couette flow* or flow between flat plates, one of which is moving. These are discussed by Bird [7] and Skelland [9]. Solutions of this type have application in rotational viscometers.

A variety of other problems have been considered, including flow in a falling film, in cone and plate viscometers, over submerged objects, radially between disks, etc. Some of these are discussed in the books cited in the references. Boundary layer flow

Table VIII. Velocity Distributions and Flowrates for Non-Newtonian Flow Between Parallel Plates (laminar flow)

Fluid	Velocity Distribution, v	Flowrate, Q
Newtonian	$\frac{\tau_w H}{4\mu}\left[1-\left(\frac{2y}{H}\right)^2\right]$	$\frac{\tau_w H^2 W}{6\mu}$
Power Law	$\left(\frac{2\tau_w}{HK}\right)^{1/n}\frac{n}{n+1}\left(\frac{H}{2}\right)^{\frac{n+1}{n}}\left[1-\left(\frac{2y}{H}\right)^{\frac{n+1}{n}}\right]$	$\frac{nWH^2}{2(2n+1)}\left(\frac{\tau_w}{K}\right)^{1/n}$
Bingham Plastic	$y_c \leqslant y \leqslant \frac{H}{2},\quad \frac{\tau_w H}{4\eta}\left[1-\left(\frac{2y}{H}\right)^2 - 2\frac{\tau_y}{\tau_w}\left(1-\frac{2y}{H}\right)\right]$ $0 \leqslant y \leqslant y_c$ as above with $y = y_c = H\tau_y/2\tau_w$	$\frac{WH^2\tau_w}{6\eta}\left[1-\frac{3}{2}\left(\frac{\tau_y}{\tau_w}\right)+\frac{1}{2}\left(\frac{\tau_y}{\tau_w}\right)^3\right]$
Yield-Power Law	$y_c \leqslant y \leqslant \frac{H}{2},\ \frac{Hn}{2(n+1)}\left(\frac{\tau_w}{K}\right)^{1/n} \times$ $\left[\left(1-\frac{\tau_y}{\tau_w}\right)^{\frac{n+1}{n}} - \left(\frac{2y}{H}-\frac{\tau_y}{\tau_w}\right)^{\frac{n+1}{n}}\right]$ $0 \leqslant y \leqslant y_c$ as above with $y = y_c = H\tau_y/2\tau_w$	$\frac{WH^2 n}{2(n+1)}\left(\frac{\tau_w}{K}\right)^{1/n}\left(1-\frac{\tau_y}{\tau_w}\right)^{\frac{n+1}{n}}\left[1-\left(\frac{n}{2n+1}\right)\left(1-\frac{\tau_y}{\tau_w}\right)\right]$

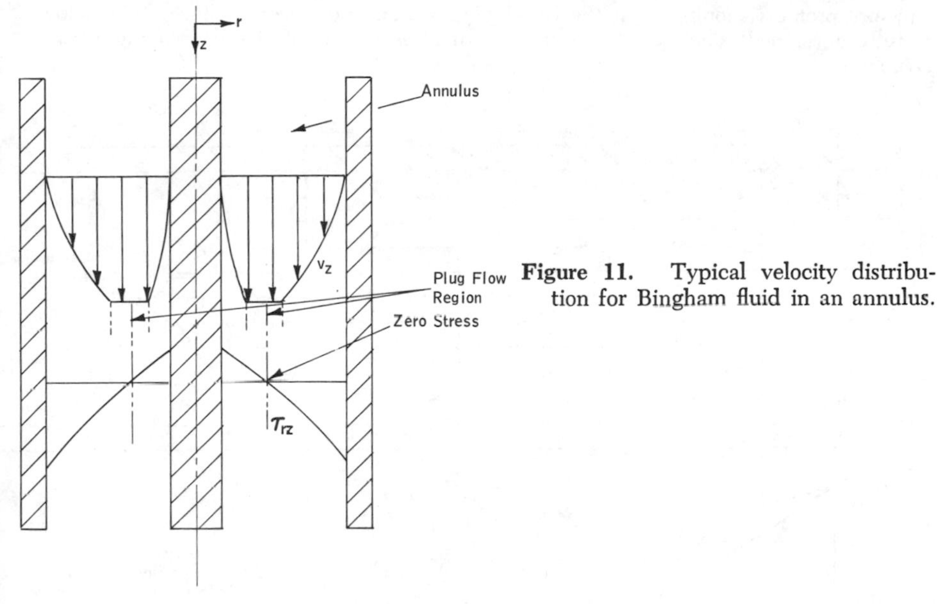

Figure 11. Typical velocity distribution for Bingham fluid in an annulus.

of power law liquids has been treated by Acrivos et al. [22]; other references have been given by Schowalter [6].

Entrance Effects

In our discussion of flow in ducts of constant cross section (tubes, annuli, etc.), we have assumed the flow to be fully developed. This implies that the axial velocity is independent of axial position, which is only true away from the entrance and exit of the duct. Exit effects are much less significant than entrance effects, and not much is known about them. Entrance effects have been investigated to a limited extent for non-Newtonian fluids.

Figure 12 depicts the velocity and pressure distribution at the entrance of a tube or duct. Fluid enters the duct from a large reservoir and the entering velocity distribution is flat, as shown. The velocity of the fluid adjacent to the wall becomes zero as soon as the fluid enters the duct. As the fluid penetrates farther into the duct, adjacent layers are subjected to increasing drag from the wall layer and they, in turn, exert drag on fluid layers adjacent to them. A progressively thicker zone is set up where the velocity varies with distance from the wall. This zone is the well known boundary layer. Fluid in the boundary layer therefore flows at velocities lower than the entrance velocity. To satisfy continuity, the fluid in the core must have a higher velocity than the entrance velocity, so the core region accelerates. The boundary layer grows in thickness as we go farther into the duct and ultimately grows to fill the entire cross section. From this point on we have fully developed flow, and equations derived earlier apply as long as flow remains laminar. The entrance length, L_e, is usually defined as the distance from the entrance at which the centerline velocity becomes very close (within 98-99%) to the fully developed value.

Figure 12 also shows the pressure change with axial distance. The upper straight line shows the distribution if the flow were fully developed throughout the duct. The lower curve is the actual distribution. In the entrance region the pressure decreases more rapidly than in developed flow because of the acceleration of the core fluid (pressure energy converted to kinetic energy) and because of the greater velocity gradients in the wall boundary layer, which result in greater wall shear stresses. The pressure profile is nonlinear in the developing region and becomes linear when flow is fully developed. The overall pressure gradient is greater if the developing region is included.

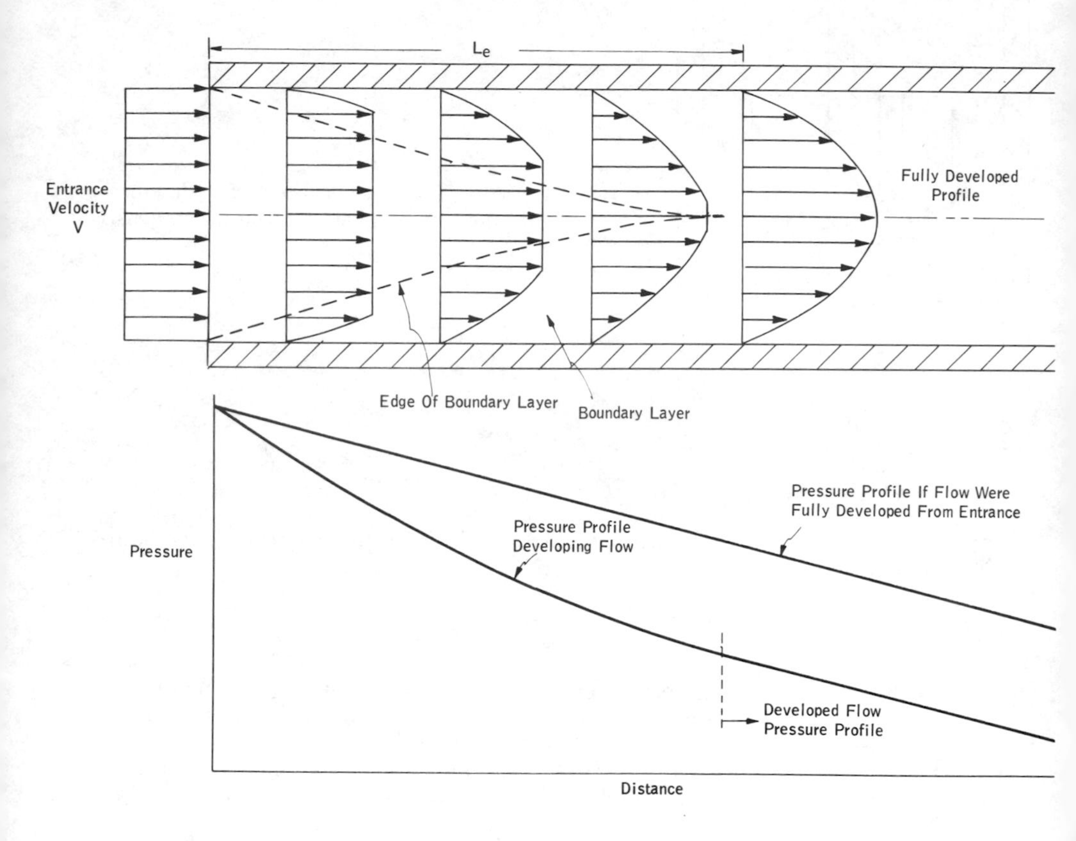

Figure 12. Velocity and pressure distributions in entrance region of a tube.

Since the solutions we developed earlier do not apply in the developing region, it is of interest to know how significant this region is and how the pressure drop is affected. Collins and Schowalter [23,24] have developed entrance region solutions for power law fluids flowing in tubes and in channels. Their solutions utilize a boundary layer analysis in the near-entrance region and perturbation of the fully developed flow near the fully developed region, where boundary layer theory does not apply. Good agreement was obtained with previous results for Newtonian fluids and with data for both Newtonian and non-Newtonian fluids. Dimensionless entrance lengths, $L_e/RRe_{n,1}$ (98% of centerline velocity attained), as a function of the flow behavior index, n, are shown in Figure 13 from Collins and Schowalter [23]. It may be seen that for the same Reynolds number entrance lengths increase as n decreases up to $n \approx 0.1$. No results were provided for $n > 1$. The pressure drop in the entrance region was given by these authors as

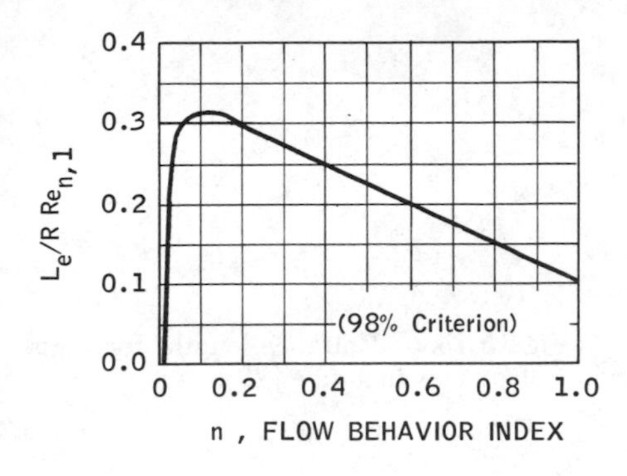

Figure 13. Entrance length for power law flow in a tube [23].

$$\frac{\Delta p}{\rho \bar{v}^2/2} = \frac{2^{n+2}\left(\frac{1+3n}{n}\right)^n}{Re_{n,1}} \frac{x}{R} + c \tag{57}$$

where x is the distance from the entrance. The factor c depends on n only and is given in Figure 14.

Entrance lengths for Bingham plastics flowing in tubes have been considered by Michiyoshi et al. [25] and by Chen et al. [26]. Entrance lengths from these two sources are shown in Figure 15 from Chen et al., which gives their results and those of Michiyoshi. The dimensionless entrance length, $L_e/R\ Re_B$, is given as a function of τ_y/τ_w and shows that entrance lengths decrease as the yield stress, τ_y, increases. This agrees with physical intuition in that as τ_y increases, the plug flow core of the fully developed profile becomes larger and the profile becomes more and more similar to the flat profile at the tube entrance. The distance required to attain the developed profile therefore decreases. Chen et al. [26] claim that their results are more accurate than those of Michiyoshi. In any case, their entrance lengths are more conservative for design purposes. Pressure drops in the entrance region also have been given by these authors. We will not discuss these here.

When these non-Newtonian fluids approach Newtonian behavior ($n \rightarrow 1$, or $\tau_y \rightarrow 0$) the results cited above reduce approximately to the generally accepted entrance length relation for Newtonian fluids:

$$L_e = 0.125\ R\ Re \tag{58}$$

Flow Through Fittings

No analytical treatment of flow through fittings is available. Skelland [9] cites the work of Weltmann and Keller [27] who did experimental work on such flows for

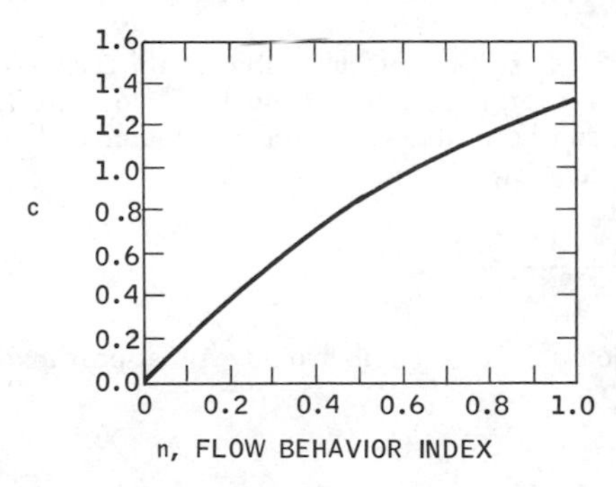

Figure 14. Entrance region pressure drop correction for power law flow in tubes [23].

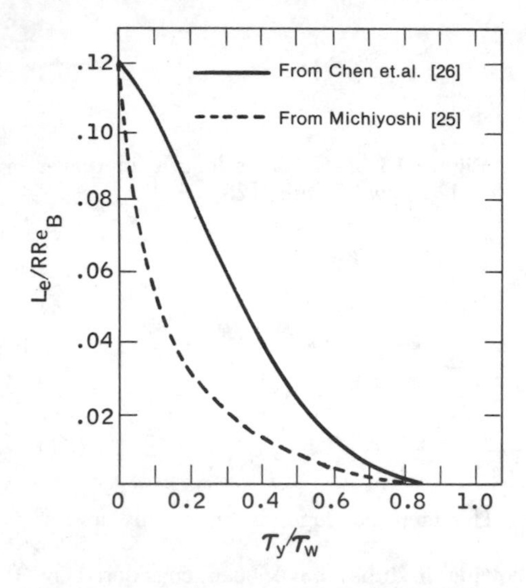

Figure 15. Entrance lengths for Bingham flow in a tube [26].

Bingham plastics and for pseudoplastic fluids. Pressure losses were found to be comparable to those for Newtonian fluids. Therefore, the conventional friction loss factors in fittings that are available for Newtonian fluids may be used for design of piping systems for non-Newtonian fluids. However, Heywood [28] has stated that at low Reynolds number, losses for non-Newtonian fluids may be as much as ten times greater than for Newtonian fluids. If the fluid is viscoelastic, greater losses will occur because of the elastic effects that come into play in such flows.

Kinetic Energy

The kinetic energy term, $\langle v^3 \rangle / 2 \langle v \rangle$, in the mechanical energy equation (4) is conveniently written as

$$\frac{1}{2}\frac{\langle v \rangle^2}{\alpha} \text{ or } \frac{1}{2}\frac{\bar{v}^2}{\alpha}$$

where α is a factor that accounts for the distribution of velocity in the duct, and is given by

$$\alpha = \bar{v}^3 / \langle v^3 \rangle \tag{59}$$

(Note that $\langle v^3 \rangle$ is the average over the cross section of the cube of the velocity, while $\bar{v}^3$ is the cube of the average velocity over the cross section.) If the velocity profile is flat, $\alpha = 1$, and for other profiles α can be evaluated readily. For laminar flow of power law fluids in tubes, we have [10], using Equation 32,

$$\alpha = \frac{(1 + 2n)\ (3 + 5n)}{3(1 + 3n)^2} \tag{60}$$

For Bingham plastics, the exact expression for α is cumbersome. An approximate relation, accurate to within 2.5%, is

$$\alpha = \frac{1}{2 - \tau_y/\tau_w} \tag{61}$$

TURBULENT FLOW

When the Reynolds number becomes sufficiently large, the ordered layered structure of laminar flow becomes unstable and turbulent flow ensues. Transport of momentum in turbulent flow is through the motion of macroscopic eddies rather than through molecular action, as for laminar flow. Under turbulent flow all of the laminar flow results do not apply, and different design relations must be developed. We will first consider the transition from laminar to turbulent flow, followed by prediction methods for friction factors and velocity profiles in tubes. This discussion will be confined to only a few of the available methods. The literature [9,11] should be consulted for other methods.

Transition From Laminar to Turbulent Flow in Tubes

A number of workers have presented correlations for transition for non-Newtonian fluids. These correlations are usually in the form of a critical Reynolds number, Re_c, above which laminar conditions do not prevail. As is well known, the critical Reynolds number in pipe flow of Newtonian fluids is generally accepted as 2100, although if external disturbances and vibrations are minimized and a very smooth entrance is provided, laminar flow may be observed for Reynolds numbers as high as 40,000 [29]. For practical cases, $Re_c = 2100$ should be used. For non-Newtonian fluids, the critical Reynolds number varies with the type of non-Newtonian and the degree of departure from Newtonian behavior, as measured by the flow index, n, or the yield stress, τ_y.

For *power law* fluids, Ryan and Johnson [30] give the critical Reynolds number as

$$(Re_{n'})_c = (Re_{n,2})_c = \frac{6464n}{(1+3n)^2\,(2+n)^{-(2+n)/(1+n)}} \tag{62}$$

which reduces to the Newtonian critical Reynolds number of 2100 when $n = 1$. As n decreases, $(Re_{n,2})_c$ from Equation 62 increases until $n \approx 0.4$, after which it decreases rapidly. This is at variance with data of friction factor versus generalized Reynolds number of Dodge and Metzner [31] (see below), which show that the critical Reynolds number increases monotonically as n decreases from unity. Further, values from Equation 62 are not in complete agreement with these data, even for $n > 0.4$. In any case, the largest critical Reynolds number reported for power law fluids is in the region of 3000, which from the design point of view is not significantly different from the Newtonian value of 2100. If the fluid is viscoelastic, the transition to turbulent flow is substantially affected, with $(Re_{n,2})_c$ as large as 10,000 [32].

For *Bingham plastics,* on the other hand, the critical Bingham Reynolds number, $(Re_B)_c$, can be substantially different from 2100 (see Figure 8B). Hanks [33] has given the following expression:

$$(Re_B)_c = \frac{He}{8x_c}\left[1 - \frac{4}{3}x_c + \frac{1}{3}x_c^4\right] \tag{63}$$

where x_c is the critical value of x $(x = \tau_y/\tau_w)$ and is given by

$$\frac{x_c}{(1-x_c)^3} = \frac{He}{16800} \tag{64}$$

Thus, for Bingham fluids the critical Bingham Reynolds number depends on the degree of plastic behavior, as represented by the Hedstrom number. This may be seen in Figure 16, which is a plot of the above equation. When He becomes zero the Newtonian value applies, and as He increases, the critical Reynolds number increases mono-

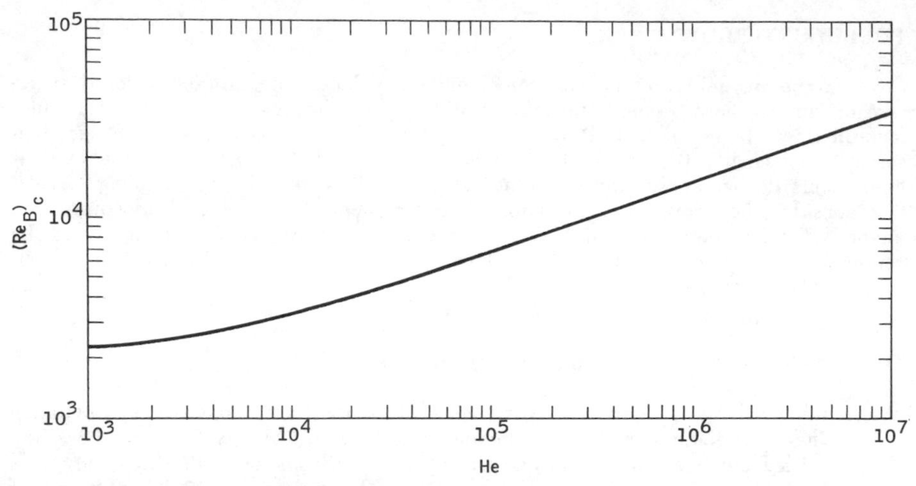

Figure 16. Critical Bingham Reynolds number for flow in a pipe [33].

tonically. For He $= 10^7$, transition to turbulent flow occurs at Reynolds numbers on the order of 50,000, whereas Equation 63 predicts $(Re_B)_c = 34{,}000$ when He $= 10^7$, and underpredicts $(Re_B)_c$ for He greater than 10^5.

More recently, Froishteter and Vinogradov [34] have given the following relation for the critical Reynolds number for a yield-power law fluid:

$$\left.\begin{aligned}
(Re_{n'})_c &= 2100/C \\
\text{where}\quad C &= \frac{1}{w^3}\left[\frac{x^2}{2} + (1-x^2)\,A + x(1-x)B\right] \\
w &= x^2 + 2x(1-x)\,\frac{m+1}{m+2} + (1-x^2)\,\frac{m+1}{m+3} \\
A &= \frac{3(m+1)^3}{2(m+3)(m+2)(3m+5)} \\
B &= \frac{6(m+1)^3}{(m+2)(2m+3)(3m+4)} \\
m &= \frac{1}{n} \\
x &= \tau_y/\tau_w
\end{aligned}\right\} \tag{65}$$

Values of $(Re_{n'})_c$ from this equation are plotted as a function of n and x in Figure 17. It may be seen that for constant x, the critical Reynolds number decreases continuously as n increases (fluid becomes less pseudoplastic) and levels off for n approximately equal to five. The critical Reynolds number increases as x increases (fluid becomes more plastic) for any value of n. These results are in qualitative agreement with the data. For pseudoplastic fluids, Dodge and Metzner [31] found experimentally that $(Re_{n'})_c = 2700$ for $n' = 0.726$ and 3100 for $n' = 0.38$. Equation 65 predicts $(Re_{n'})_c$ of 2253 and 2636, respectively, showing only fair agreement.

Friction Factors in Turbulent Flow

A convenient way of calculating pressure drops because of friction losses is through the Fanning friction factor. The friction factor was defined above for laminar flow

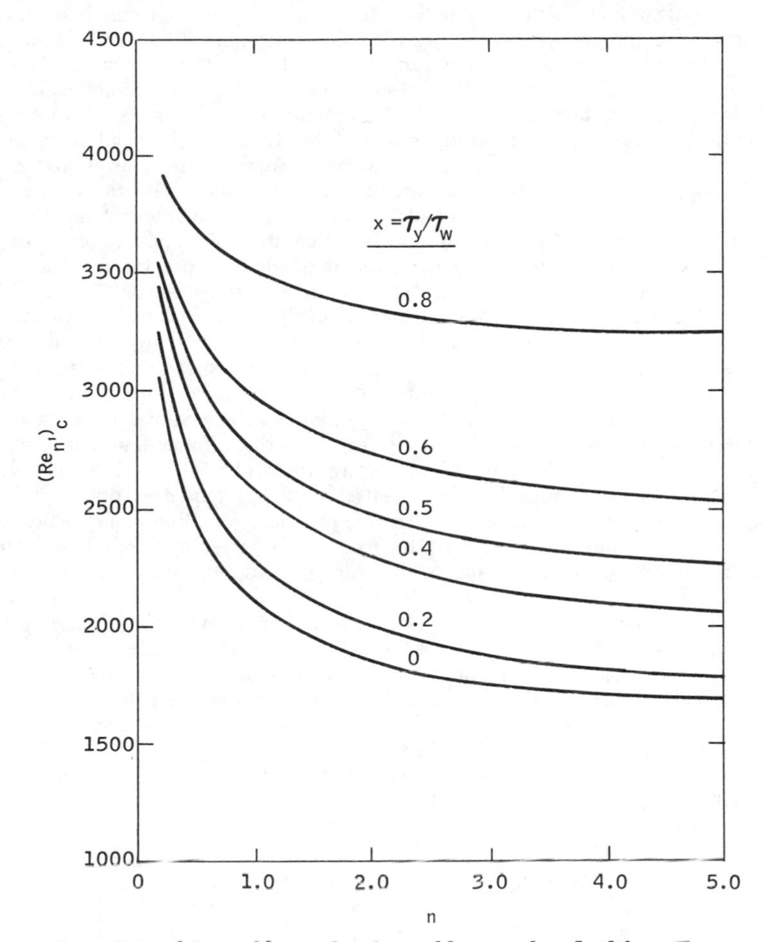

Figure 17. Critical Reynolds number for yield-power law fluid from Equation 65.

(Equation 37) and was shown to depend only on the Reynolds number and, for Bingham fluids, also on the Hedstrom number. Dimensional analysis may be used to show that the same situation prevails in fully developed turbulent flow. Many expressions have been proposed for the friction factor in turbulent flow. These are usually based on semitheoretical arguments that lead to equations with constants that must be determined using experimental data. We do not wish to develop these expressions here, but rather will confine ourselves to one suggested by Dodge and Metzner [31]. The interested reader will find more details and other correlations in the books by Skelland [9] and Govier and Aziz [11].

Extending the laminar relation of Metzner and Reed [16] to turbulent flow, Dodge and Metzner [31] obtained the friction factor as a function of the generalized Reynolds number, $Re_{n'}$, as

$$\frac{1}{\sqrt{f}} = \frac{4.0}{(n')^{0.75}} \log_{10}\left[Re_{n'}\, f^{(1-n'/2)}\right] - \frac{0.40}{(n')^{1.2}} \tag{66}$$

This relation is implicit in f, and a chart of f versus $Re_{n'}$ (Figure 18) is somewhat more convenient to use. The chart shows clearly the effect of the exponent n′ on the friction factor in turbulent flow. Values of n′ less than unity (pseudoplastic behavior) give lower friction factors (at the same $Re_{n'}$) than do Newtonian fluids, and for n′ greater than unity the friction factor becomes progressively larger. When $n' = 1$ (Newtonian fluid), Equation 66 reduces to the classical equation of Nikuradse presented in standard texts [29]. For highly dilatant fluids, $n' > 2$, Equation 66 and Figure 18 must be used with caution because certain assumptions underlying their development may no longer be valid. Further, no confirming data are available in this range, as is also the case for highly pseudoplastic liquids. (These regions are represented by the dashed curves in Figure 18.) Extrapolation to $Re_{n'}$ greater than 10^5 is permissible but not for $n' > 2$.

Although the form of Equation 66 suggests that its main application is to power law fluids, Dodge and Metzner [31] have recommended its use for Bingham plastics as well, and have verified its applicability with experimental data. For Bingham fluids, as with any other fluid, n′ and K′ must be evaluated at the wall shear stress prevailing in the flow. For power law fluids, this does not present any difficulty since n′ and K′ are constants. For other fluids, n′ and K′ vary with shear rate and shear stress, so that an iterative calculation is necessary. First the pressure drop must be assumed and the wall shear stress calculated from Equation 19. The values of K′ and n′ are then obtained at the calculated wall shear stress from the laminar flow curve τ_w versus $8\bar{v}/D$. These are used in Equation 66 or Figure 18 to obtain f, and Δp is calculated using Equation 37. The procedure is repeated until agreement is obtained between the calculated and assumed values of Δp. In practice, for Bingham plastics flowing in smooth tubes for Re_B greater than $(Re_B)_c$ the friction factor depends only on Re_B for fully turbulent flow. The relation in this range is essentially that for a Newtonian fluid.

Heywood [28] has pointed out some practical difficulties with the Dodge and Metzner approach. The parameters n′ and K′ must be estimated at the wall shear stress prevailing in the turbulent flow, but from laminar data. The problem is that often these laminar data are not at high enough shear rate, usually because of the onset of turbulence in the experiments. Thus n′, K′ are estimated at too low a shear rate, which means that n′ is underestimated. This, in turn, leads to values of f from the Dodge and Metzner correlation that will be too low. As long as this is kept in mind, the correlation can be used with confidence.

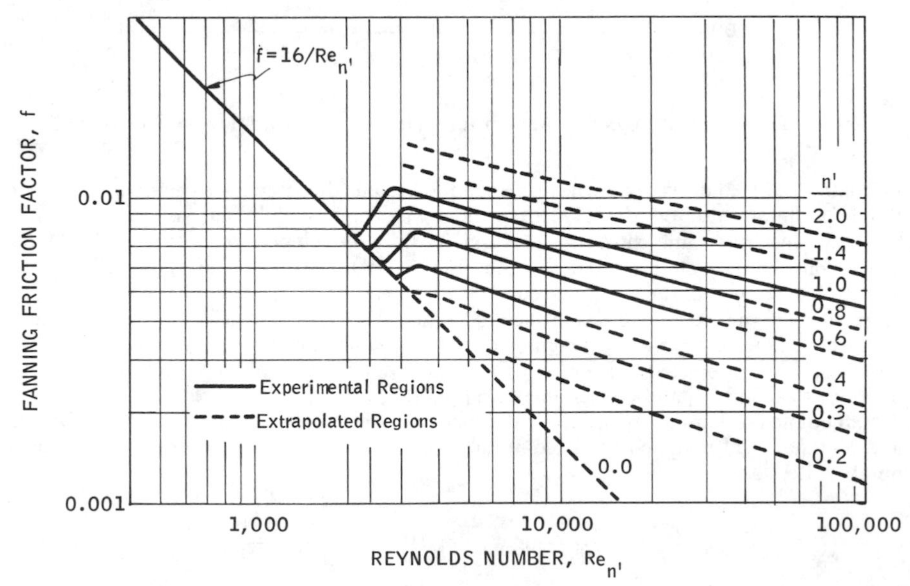

Figure 18. Friction factor-Reynolds number relation for non-Newtonian fluids [31].

Other approaches to obtain the friction factor have been suggested by Tomita, by Clapp and by Torrance (see Grovier and Aziz [11]). A recent paper by Masuyama and Kawashima [35] provides design charts for friction factors for pseudoplastic liquids that are claimed to be more accurate than previous methods. Correlations for rough-walled pipes have been reviewed by Govier and Aziz [11]. In most cases, the Newtonian values for rough pipes are suitable for design for pseudoplastic fluids. Friction factor relations for annuli and for rectangular ducts may be found in the book by Skelland [9]. For Newtonian fluids, turbulent pressure drop in noncircular cross-sectional ducts may be estimated using the hydraulic radius of the cross section [7]. The literature does not report such a practice for non-Newtonian flows, and hence this approach cannot be recommended for such flows.

For viscoelastic fluids, friction factors are found to be much lower than predicted by the Dodge and Metzner relation. Govier and Aziz [11] have reviewed some available correlations.

Velocity Distribution in Turbulent Flow

As in the case of Newtonian fluids, time-averaged velocity profiles of non-Newtonians in turbulent flow are much flatter than in laminar flow. These profiles have been developed in conjunction with the derivation of the friction factor relations by the workers cited above. Although they have little utility for design purposes, we mention one such profile to convey the general form of the prevailing results. Dodge and Metzner [31] have derived a profile for power law fluids analogous to the universal velocity distribution for Newtonian fluids. Some errors in the original equations have been pointed out by Skelland. The Dodge-Metzner profile, as corrected by Skelland [9] is

Laminar Sublayer

$$u^+ = (y^+)^{\frac{1}{n}} \tag{67}$$

Turbulent Core

$$u^+ = \frac{5.66}{n^{.75}} \log_{10} y^+ - \frac{0.566}{n^{1.2}} + \frac{3.475}{n^{.75}} \left[1.960 + 0.815n - 1.628n \log_{10} \left(3 + \frac{1}{n} \right) \right] \tag{68}$$

where

$$u^+ = u/u^*$$
$$y^+ = y^n u^{*2-n} \rho/K \tag{69}$$
$$u^* = \sqrt{\tau_w/\rho}$$

and $y = R - r$ is the distance from the pipe wall into the fluid. By contrast to the Newtonian universal velocity distribution discussed in Chapter 2, the constants in the above were obtained from friction factor measurements rather than experimental velocity data. As a result, the thickness of the sublayer was not obtained.

For engineering calculations, the *kinetic energy correction factor*, α, defined in Equation 59, may be taken as unity for turbulent flow of non-Newtonian fluids. This is a reasonable approximation because the velocity distribution is close to flat.

Drag Reduction

Dilute solutions of polymers under turbulent flow conditions sometimes exhibit a considerable decrease in pressure drop compared to that for the solvent alone (say water) flowing at the same flowrate. For example, solutions of polyethylene oxide in water at concentrations of less than 50 ppm (wt) of the polymer show a drag reduction of more than 30% [6]. This effect, called the Tom's phenomenon, after its discoverer, has attracted considerable interest because of its obvious economic importance. Drag-reducing additives have found use in such diverse applications as increasing the flowrate of water in fire hoses and of crude oil in long distance pipelines for a given pressure drop. No completely satisfactory explanation for the phenomenon is available.

Some of the qualitative features of drag reduction are that often less than 200 ppm of the additive are needed, and no measurable change in the solution rheology results compared to the pure solvent. Increasing additive concentration increases drag reduction only up to a point, beyond which drag increases, very likely because of increased viscosity [6]. Some evidence exists that less drag reduction is experienced in large-diameter pipes than in small-diameter pipes. The interested reader can find more details on this interesting phenomenon in the reviews that appear periodically, for example, that by Virk [36].

Entrance Effects

Few data are available on the entrance length necessary to achieve fully developed turbulent flow or on the pressure drop in this region. As in the case of laminar flow, the pressure gradient in the entrance region is greater than that in fully developed flow. Dodge and Metzner [31] found that

$$L_e/D < 15$$

for turbulent flow of non-Newtonians with $n' < 1$. This value is considerably less than that for a Newtonian fluid, for which some 50 diameters are usually needed to obtain a fully developed flow [37]. Limited data indicate that the pressure gradient in the entrance region is best estimated from Newtonian correlations.

Substantial deviations from Newtonian behavior occur for viscoelastic fluids. Much greater entry lengths are found in drag-reducing systems [11].

Flow Through Fittings

As for laminar flow, little information is available on frictional losses in fittings. Current practice is to use the loss coefficients developed for Newtonian fluids as long as viscoelastic effects are negligible.

Metering and Control

Not much information exists on the design of control valves for non-Newtonian fluids or for their metering. Skelland [9] cites some relevant papers, and Cheremisinoff [15,38] provides some discussion. We have noted already that for laminar flow of pseudoplastic liquids the pressure drop in a straight length of pipe is insensitive to the flowrate. Therefore, use of the frictional pressure drop as a measure of flowrate is not to be recommended. In general, with proper calibration, variable area meters such as rotameters have been found to be suitable for non-Newtonian fluids. No information is available on the use of variable head meters such as orifice or venturi meters. The high viscosity of most non-Newtonian fluids would result in orifice Reynolds numbers smaller than the recommended values and would make such meters unsuitable without calibration. A magnetic flowmeter was used by Dodge and Metzner [31], but such devices are limited to fluids showing some degree of electrical conductivity. Skelland [9] cites rotary volumetric meters as suitable where abrasion is not a problem. Elbow meters also may be used with proper calibration.

SUMMARY AND CONCLUSION

In this chapter we have presented some of the available methods for engineering calculations of flow of non-Newtonian fluids. The major types of non-Newtonian behavior have been covered, and design equations for laminar and turbulent flow of time-independent non-Newtonian fluids have been developed. We also have considered the laminar-turbulent transition in tube flow, flow in the entrance region and flow in fittings. Insufficient information is available on flow through fittings and in control valves, and on metering of non-Newtonian fluids. However, sufficient information exists, and has been presented here, for the calculation of pressure drops in piping systems or in other ducts of constant cross section. A very detailed treatment of such calculation methods for pipe flow has been given by Govier and Aziz [11], along with methods for time-dependent fluids and viscoelastic fluids. Recent reviews on this topic by Heywood [28] and Cheremisinoff [15,38] also are useful. It is important to recognize that the non-Newtonian flow parameters, e.g., n and K, used in these design methods should be measured for the fluid in question. Extrapolation of rheological data from one non-Newtonian fluid to another apparently similar one can lead to large errors [9].

Because of the limited objectives of this chapter it has been necessary to omit many topics, such as boundary layer flow and flow over submerged objects. In addition, we have but touched on the subject of viscoelastic fluids. The interested reader will find many of these items covered in the books by Schowalter [6] and, to a lesser extent, Skelland [9]. An early review article on non-Newtonian technology by Metzner [10] also is to be recommended. Skelland has given an interesting and useful discussion on optimization of piping systems, including the optimum choice of pumping temperature and also whether laminar or turbulent flow is to be preferred. The books by Bird et al. [1], Middleman [5], Han [3] and McKelvey [4] will be of interest to those involved in the flow of polymeric materials. A collection of reviews under the title *Rheology: Theory and Applications*, edited by Eirich [39] is an excellent source of information. Many other texts not mentioned here also are available.

NOMENCLATURE

a	constant in Sisko model (Table III)
A	area; constant in Prandtl-Eyring model (Table III); constant in Froishteter-Vinogradov equation (65)
b	constant in Sisko model (Table III)
B	constant in Prandtl-Eyring model (Table III); constant in Froishteter-Vinogradov equation (65)
c	constant in Sisko model (Table III); factor in Equation 57
C	constant in Froishteter-Vinogradov equation (65)
D	tube diameter
$\hat{E}_v$	rate of irreversible conversion of mechanical to internal energy per unit mass of fluid
f	function in Equation 5; friction factor
F	force
g	acceleration of gravity
H	distance
He	Hedstrom number, $\tau_y D^2 \rho / \eta^2$
K	power law consistency index
K′	factor in Metzner-Reed equation (47)
L	length
L_e	entrance length to obtain fully developed flow
m	reciprocal of flow behavior index, Froishteter-Vinogradov equation (65)
n	power law flow behavior index
n′	factor in Metzner-Reed equation (47)
p	pressure
P	dynamic pressure
Q	volumetric flowrate

r	radial coordinate
r_c	radius of plug flow core of Bingham flow in tube (Table VI)
R	tube radius
Re	Reynolds number for Newtonian fluid, $D\bar{v}\rho/\mu$
Re_B	Reynolds number for Bingham fluid, $D\bar{v}\rho/\eta$
$Re_{n,1}$	power law-Reynolds number, $D^n\bar{v}^{2-n}\rho/K$, Equation 41
$Re_{n,2}$	power law-Reynolds number, $\frac{D^n\bar{v}^{2-n}\rho}{K} 8 \left(\frac{n}{6n+2}\right)^n$, Equation 40
$Re_{n'}$	generalized non-Newtonian Reynolds number of Metzner and Reed, $D^{n'}\bar{v}^{2-n'}\rho/\gamma$, Equation 50
$(Re_i)_c$	critical Reynolds number (of type i) for onset of turbulent flow
t	time
T	torque
u^+	dimensionless turbulent velocity, Equation 69
u^*	friction velocity, $\sqrt{\tau_w/\rho}$, Equation 69
v	velocity
v_x, v_y, v_z	components of velocity in x, y and z directions
$\bar{v}$	cross-sectional average velocity
V	velocity
w	mass flowrate
W	width
$\hat{W}$	work done by fluid per unit mass of fluid
x	coordinate; ratio of yield stress to wall stress, τ_y/τ_w, for Bingham fluid
x_c	critical τ_y/τ_w at onset of turbulent flow
y	coordinate
y^+	dimensionless distance for turbulent velocity profile, Equation 69
Y	distance; yield number for Bingham fluid, $D\tau_y/\bar{v}\eta$, Equation 43
z	coordinate
α	kinetic energy correction factor, Equation 59
γ	factor in Metzner-Reed equation (49)
$\dot{\gamma}$	shear rate
Δ	difference between value at outlet and inlet of a system, (Table I)
Δp	difference between inlet and outlet pressure
η	plastic viscosity for Bingham fluid
η_a	apparent viscosity, Equation 9
θ	coordinate
λ_1, λ_2	relaxation times, Equation 15
μ	viscosity
μ_o, μ_∞	constants in Reiner-Philippoff model (Table III)
μ_{eff}	effective viscosity, Equation 52
ρ	density
τ	deviatoric stress tensor
τ_w	wall shear stress
τ_y	yield stress for Bingham or yield-power law fluid
$\tau_{yx}, \tau_{yz}, \tau_{rz}$	shear stress components
ϕ	coordinate
ϕ_o, ϕ_1	constants in Ellis model (Table III)
Ω	angular velocity
$< >$	average over cross section

REFERENCES

1. Bird, R. B., R. C. Armstrong and O. Hassager. *Dynamics of Polymeric Liquids*, Vols. I & II (New York: John Wiley & Sons, Inc., 1977).
2. Fredrickson, A. G. *Principles and Applications of Rheology* (Englewood Cliffs, NJ: Prentice-Hall, Inc., 1964).
3. Han, C. D. *Rheology in Polymer Processing* (New York: Academic Press, Inc., 1976).
4. McKelvey, J. M. *Polymer Processing* (New York: John Wiley & Sons, Inc., 1962).
5. Middleman, S. *The Flow of High Polymers: Continuum and Molecular Rheology* (New York: Interscience Publishers, 1968).
6. Schowalter, W. T. *Mechanics of Non-Newtonian Fluids* (Oxford, England: Pergamon Press, Inc., 1978).

7. Bird, R. B., W. E. Stewart and E. N. Lightfoot. *Transport Phenomena* (New York: John Wiley & Sons, Inc., 1960).
8. Lodge, A. S. *Elastic Liquids* (New York: Academic Press, Inc., 1964).
9. Skelland, A. H. P. *Non-Newtonian Flow and Heat Transfer* (New York: John Wiley & Sons, Inc., 1967).
10. Metzner, A. B. In *Advances in Chemical Engineering*, Vol. 1, T. B. Drew and J. W. Hoopes, Eds. (1956), pp. 79-150.
11. Govier, G. W., and K. A. Aziz. *The Flow of Complex Mixtures in Pipes* (New York: Krieger Publishing Co., 1977).
12. Bauer, W. H., and E. A. Collins. In *Rheology: Theory and Applications*, F. R. Eirich, Ed., Vol. 4, (New York: Academic Press, Inc., 1967), pp. 423-459.
13. Markovitz, H. "Rheological Behavior of Fluids," National Committee for Fluid Mechanics Films (Encyclopedia Brittanica Educational Corp., Chicago, IL, 1972); see also "Illustrated Experiments in Fluid Mechanics," *NCFMF Book of Film Notes* (Cambridge, MA: The MIT Press, 1972).
14. Van Wazer, J. R., J. W. Lyons, K. Y. Kim and R. E. Colwell. *Viscosity and Flow Measurement* (New York: Interscience Publishers, 1963).
15. Cheremisinoff, N. P. *Fluid Flow: Pumps, Pipes and Channels*, (Ann Arbor, MI: Ann Arbor Science Publishers, Inc., 1981).
16. Metzner, A. B., and J. C. Reed. *AIChE J.* 1:434 (1955).
17. Metzner, A. B. *Ind. Eng. Chem.* 49:1429 (1957).
18. Laird, W. M. *Ind. Eng. Chem.* 49:138 (1957).
19. Fredrickson, A. G., and R. B. Bird, *Ind. Eng. Chem.* 50:347 (1958).
20. Fredrickson, A. G., and R. B. Bird. *Ind. Eng. Chem. Fund.* 3:383 (1964).
21. Vaughn, R. D., and P. D. Bergman. *Ind. Eng. Chem. Proc. Des. Dev.* 5:44 (1966).
22. Acrivos, A., M. J. Shah and E. E. Petersen. *AIChE J.* 6:312 (1960).
23. Collins, M., and W. R. Schowalter. *AIChE J.* 9:804 (1963).
24. Collins, M., and W. R. Schowalter. *AIChE J.* 9:98 (1963).
25. Michiyoshi, L., K. Mizuno and Y. Hoshiai. *Int. Chem. Eng.* 6:373 (1966).
26. Chen, S. S., L. T. Fan and C. L. Hwang. *AIChE J.* 16:293 (1970).
27. Weltmann, R. N., and T. A. Keller. *NACA TN* 3389 (1957).
28. Heywood, N. I. *I. Chem. E. Symposium Series* No. 60:33 (1980).
29. Schlichting, H. *Boundary Layer Theory*, 6th Ed. (New York: McGraw-Hill Book Co., 1968).
30. Ryan, N. W., and M. M. Johnson. *AIChE J.* 5:433 (1959).
31. Dodge, D. W., and A. B. Metzner. *AIChE J.* 5:189 (1959).
32. Metzner, A. B., and M. G. Park. *J. Fluid Mech.* 20:291 (1964).
33. Hanks, R. W. *AIChE J.* 9:306 (1963).
34. Froishteter, G. B., and G. V. Vinogradov. *Rheol. Acta.* 16:620 (1977).
35. Masuyama, T., and T. Kawashima. *Bull. JSME* 22:48 (1979).
36. Virk, P. S. *AIChE J.* 21:625 (1975).
37. Knudsen, J. G., and D. L. Katz. *Fluid Dynamics and Heat Transfer* (New York: McGraw-Hill Book Co., 1958).
38. Cheremisinoff, N. P. *Applied Fluid Flow Measurement: Fundamentals and Technology* (New York: Marcel Dekker, Inc., 1979).
39. Eirich, F. R., Ed. *Rheology: Theory and Applications, Vols. I-IV* (New York: Academic Press, Inc.).

CHAPTER 7
MIXING AND AGITATION OF NON-NEWTONIAN FLUIDS

A. H. P. Skelland
School of Chemical Engineering
The Georgia Institute of Technology
Atlanta, Georgia 30332

CONTENTS

INTRODUCTION

In contrast to the position that prevailed 15–20 years ago, it is presumably no longer necessary to begin an article on non-Newtonian fluids with an outline of their characteristics. Several texts [1-4] have dealt with these materials in some detail.

Although mixing is one of the oldest processing operations, its theoretical foundations are among the least developed. This arises partly from the lack of a generally accepted definition of either mixing or agitation and partly because of the complexity of the relevant hydrodynamic equations. As a consequence, most studies to date have been based on analysis of the integral physical quantities involved, such as power consumption, circulation capacity and mixing time. This leads to a substantial dependence on dimensional analysis in the correlation of experimental data and the interpretation of results from models.

The great breadth of the subject is indicated by the many reviews that have appeared on various subsections within the field, a few of which are noted below.

- criteria for mixing [5]
- batch mixing in stirred vessels [1,5,6]
- mixing in continuous flow through stirred vessels [5]
- mixing of highly viscous fermentation broths [7]
- static mixers [6,8]

- mixing of viscoelastic fluids [9]
- mixing in extruders [6,10-12]
- heat transfer to non-Newtonian fluids in agitated vessels [13]

Among books devoted to the general subject of mixing may be cited the two-volume work edited by Uhl and Gray [14] and the still more recent text by Nagata [15].

It is evident, therefore, that comprehensive coverage of the subject cannot be undertaken in a single chapter. Carreau et al. [16] note that the primary objectives of the design of a stirred vessel are the attainment of a specified degree of homogeneity and the enhancement of the rate of heat and/or mass transfer. The following treatment will deal with topics relating to these objectives.

SIMILARITY CRITERIA

The scaleup of results obtained from experiments on models and the reproduction of flow patterns achieved on two scales of operation require an awareness of geometric, kinematic and dynamic similarity between the model and the large-scale prototype.

Geometric similarity prevails between two bodies of different size if all counterpart length dimensions bear a constant ratio.

Kinematic similarity exists in two geometrically similar units of different size when all velocities at counterpart positions bear a constant ratio.

Dynamic similarity occurs in two geometrically similar units of different size if all corresponding forces at counterpart positions bear a constant ratio. "Corresponding" forces distinguish between inertial, gravitational, viscous, surface tension, elasticity and other forces, all of which may occur in a fluid system. Consider counterpart positions in systems 1 and 2, where the different types of force occurring are F_a, F_b, F_c, F_d, etc. For dynamic similarity we have

$$\frac{F_{b_1}}{F_{b_2}} = \frac{F_{c_1}}{F_{c_2}} = \frac{F_{d_1}}{F_{d_2}} = \ldots \frac{F_{a_1}}{F_{a_2}} = \text{constant} \tag{1}$$

or

$$\frac{F_{a_1}}{F_{b_1}} = \frac{F_{a_2}}{F_{b_2}}, \quad \frac{F_{a_1}}{F_{c_1}} = \frac{F_{a_2}}{F_{c_2}}, \quad \frac{F_{a_1}}{F_{d_1}} = \frac{F_{a_2}}{F_{d_2}} \tag{2}$$

Kinematic and dynamic similarity both require geometric similarity so that counterpart positions can be identified in the two systems. Some of the various types of force that may arise during fluid motion will now be formulated algebraically.

Inertial force is associated with the "reluctance" of a body to change its state of rest or motion [17]. Consider the mass flow of fluid with linear velocity V through area A during time dt, where A is normal to V. Then $dm = \rho V A\, dt$, in which ρ is the fluid density. The inertial force F_i = mass (acceleration), or

$$dF_i = dm \frac{dV}{dt} = \rho V A\, dt \frac{dV}{dt}$$

$$\int_0^{F_i} dF_i = \int_0^V \rho A V\, dV$$

$$F_i = \frac{1}{2} \rho A V^2$$

But A = (constant) L^2, where L is the length dimension chosen to characterize the geometry of the system. Therefore,

$$F_i \propto \rho L^2 V^2$$

For agitated vessels L is customarily chosen as the impeller diameter, D, and the representative velocity V is taken to be at the impeller tip. Then $V = \pi D N$, or $V \propto D N$. The proportionality expression for inertial force thus becomes

$$F_i \propto \rho D^4 N^2 \tag{3}$$

Now the change in V due to F_i (namely dV/dt) may be countered by the change in V due to the *viscous force,* F_v, where

$$F_v = \mu \frac{du}{dy} A'$$

for a Newtonian fluid.

Let

$$\frac{du}{dy} = (\text{const}') \frac{\Delta u}{\Delta y} = (\text{const}^x) \frac{(u_1 = V) - (u_2 = 0)}{(y_2 - y_1) \propto L} = (\text{const}^*) \frac{V}{L}$$

$$A' = (\text{const}'')L^2$$

Then

$$F_v \propto \mu VL$$

and for an agitated vessel

$$F_v \propto \mu D^2N \tag{4}$$

also

$$\text{the } \textit{gravitational force} = F_g \propto \rho D^3 g \tag{5}$$

and

$$\text{the } \textit{surface tension force} = F_{S.T.} \propto \sigma D \tag{6}$$

The use of the proportionality sign $\propto$ above arises from the substitution of the impeller diameter D for the appropriate length dimension in each case.

Criteria for Purely Viscous Non-Newtonian Fluids

These fluids are free from elastic effects. Suppose that, in Equation 2, F_a, F_b, F_c, and F_d are replaced by F_i, F_v, F_g, and $F_{S.T.}$, respectively, for systems 1 and 2. If, in addition, the above proportionalities are used in place of F_i, F_v, F_g, and $F_{S.T.}$, the following results are obtained:

$$\left(\frac{D^2N\rho}{\mu_A}\right)_1 = \left(\frac{D^2N\rho}{\mu_A}\right)_2, \text{ or } N_{Re_1} = N_{Re_2} \tag{7}$$

$$\left(\frac{DN^2}{g}\right)_1 = \left(\frac{DN^2}{g}\right)_2, \text{ or } N_{Fr_1} = N_{Fr_2} \tag{8}$$

$$\left(\frac{D^3N^2\rho}{\sigma}\right)_1 = \left(\frac{D^3N^2\rho}{\sigma}\right)_2, \text{ or } N_{We_1} = N_{We_2} \tag{9}$$

In Equation 7 an apparent viscosity μ_A has been introduced; this term, which will be defined quantitatively later, is chosen to "characterize" the rheological behavior of the non-Newtonian fluid under the particular flow conditions involved.

Equations 2, 7, 8 and 9 show that the attainment of dynamic similarity requires respective equality of certain well known dimensionless groups on scales 1 and 2. The presence of geometric and dynamic similarity ensures that kinematic similarity is also obtained. Thus, from Equation 7,

$$\frac{V_1}{V_2} = \frac{(DN)_1}{(DN)_2} = \frac{(D\rho)_2 \mu_{A1}}{(D\rho)_1 \mu_{A2}} = \text{constant}$$

Criteria for Viscoelastic Fluids

Consider a viscoelastic fluid undergoing laminar couette flow, as shown in Figure 1, where the inner cylinder is rotating and the outer cylinder is stationary.

Stresses τ_{11}, τ_{22} and τ_{33} normal to the surface of shearing are generated, in addition to the shearing stress τ_{21}. (Note that the first subscript identifies the direction normal to the surface on which the stress acts, while the second subscript shows the direction in which the stress acts. This convention conforms to that used by some [1,18] and is the reverse of that employed by others [6, p. 15; 9].)

If the fluid were Newtonian the shearing stress would be written as

$$\tau_{21} = \mu \frac{dv_1}{dr} = \mu \dot{\gamma} \tag{10}$$

where dv_1/dr is negative, accounting for the direction of τ_{21} in Figure 1, and μ is constant. For a non-Newtonian fluid, however, we have

$$\tau_{21} = \mu_a(\dot{\gamma})\dot{\gamma} \tag{11}$$

where the apparent viscosity $\mu_a(\dot{\gamma})$ is a function of shear rate, $\dot{\gamma}$. If the fluid is a power law material we may write

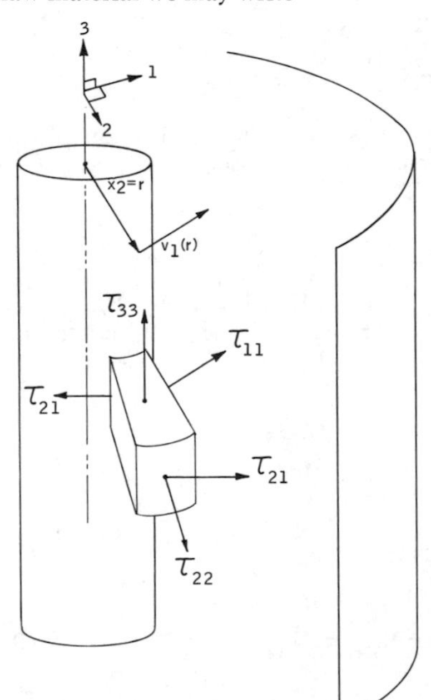

Figure 1. Surface stresses on a differential element of viscoelastic fluid in couette flow.

$$\tau_{21} = K(\dot{\gamma})^n \tag{12}$$

so that

$$\mu_a = K(\dot{\gamma})^{n-1} \tag{13}$$

This quantity is readily measured, for example, in a cone-and-plate viscometer by measuring the torque T required to rotate the cone of radius R at a speed of N rps, giving [1, p. 48-49]

$$\mu_a = 3T/2\pi R^3\dot{\gamma}; \quad \dot{\gamma} = \frac{2\pi N}{\phi} \tag{14}$$

ϕ is the angle between the cone and plate surfaces, in radians.

For Newtonian fluids,

$$\tau_{11} = \tau_{22} = \tau_{33} = -p \tag{15}$$

where p is the hydrostatic pressure. In this case, therefore, the primary and secondary normal stress differences, $\tau_{11} - \tau_{22}$ and $\tau_{22} - \tau_{33}$, respectively, are both zero. For viscoelastic fluids, however,

$$\tau_{11} - \tau_{22} = f_1(\dot{\gamma}) = \sigma_1(\dot{\gamma})\dot{\gamma}^2 \tag{16}$$

where the primary normal stress coefficient, $\sigma_1(\dot{\gamma})$, increases toward a constant value as $\dot{\gamma}$ decreases toward zero.

The coefficient σ_1 can be measured at a given shear rate $\dot{\gamma}$ in a cone-and-plate viscometer. This is done by determining the axial force F tending to separate the cone from the plate, measured by a transducer. Then [19,20]

$$\sigma_1 = 2F/\pi R^2\dot{\gamma}^2 \tag{17}$$

Ulbrecht and co-workers [9,19] found experimentally that

$$\sigma_1 = h(\dot{\gamma})^{m-2} \tag{18}$$

where h and m are empirical constants somewhat analogous to K and n in Equation 13.

The secondary normal stress difference is believed by some to be only about 10% of the primary normal stress difference [9]; others support the Weissenberg hypothesis [1, p. 51], which indicates that $\tau_{22} - \tau_{33}$ may be neglected. Provisionally, we have

$$\tau_{22} - \tau_{33} = f_2(\dot{\gamma}) = \sigma_2(\dot{\gamma})\dot{\gamma}^2 \tag{19}$$

Equations 13–19 provide components for the formulation of the following dimensionless criteria relevant to the mixing of viscoelastic fluids:

$$\text{Reynolds number} = N_{Re} = \frac{D^2N\rho}{\mu_a(\dot{\gamma})} \tag{20}$$

$$\text{Weissenberg number} = N_{Wi} = \frac{\sigma_1(\dot{\gamma})}{\mu_a(\dot{\gamma})}N \tag{21}$$

$$\text{Viscoelastic ratio} = N_{VR} = \frac{\sigma_2(\dot{\gamma})}{\sigma_1(\dot{\gamma})} \tag{22}$$

The Reynolds number may alternatively take the form used in Equation 7, as will be shown later. The Weissenberg number has also been called the Deborah number [9,21]; its physical significance is as follows:

$$N_{Wi} = N_{De} = \frac{\text{characteristic material time}}{\text{characteristic process time}} \tag{23}$$

The denominator is customarily represented by $1/N$ in agitated vessels [9], while the numerator is defined as

$$\frac{\sigma_1(\dot{\gamma})}{\mu_a(\dot{\gamma})} = \frac{\tau_{11} - \tau_{22}}{(\dot{\gamma}')^2} \bigg/ \frac{\tau_{12}}{\dot{\gamma}'} \tag{24}$$

where $\dot{\gamma}'$ is described as the characteristic shear rate [9,22,23]. If the Weissenberg hypothesis prevails, the group N_{VR} in Equation 22 becomes zero.

POWER CONSUMPTION

The power P needed to rotate an impeller in a vessel filled with a Newtonian fluid may be expressed as a general function of the variables perceived as relevant in the following way:

$$P = f_3\ (D, D_T, H, C, S, L, W, J, \rho, \mu, N, g, B, R) \tag{25}$$

where D and D_T = impeller and vessel diameters, respectively
H = the liquid depth
C = the height of the impeller above the vessel floor
S = the impeller pitch
L and W = length and width of the impeller blades
J = the width of the baffles
ρ and μ = density and viscosity of the fluid
N = the rotational speed of the impeller
g = the acceleration due to gravity (included because of the possible presence of waves and vortices)
B = the number of impeller blades
R = the number of baffles.

Dimensional analysis gives

$$\frac{P}{D^5N^3\rho} = N_P = f_4\left[\left(\frac{D^2N\rho}{\mu}\right), \left(\frac{DN^2}{g}\right), \left(\frac{D_T}{D}\right), \left(\frac{H}{D}\right), \left(\frac{C}{D}\right), \left(\frac{S}{D}\right), \left(\frac{L}{D}\right), \left(\frac{W}{D}\right), \left(\frac{J}{D}\right), B, R\right] \tag{26}$$

The first three groups, including that on the left side, are the Power number, the Reynolds number and the Froude number. For a fixed geometry the last nine terms are constant, giving

$$N_P = f_5[N_{Re}, N_{Fr}] \tag{27}$$

Surface waves and vortices are absent when baffling is adequate or the impeller is properly offset from center. The Froude number is accordingly deleted:

$$N_P = f_6[N_{Re}] \tag{28}$$

Rushton et al. [24] evaluated f_4-f_6 experimentally for many types of impellers and expressed their results as plots of N_P vs N_{Re}. Three flow regimes were found.

In laminar flow ($N_{Re} < 10$),

$$N_P = K_L (N_{Re})^{-1} \text{ or } P = K_L D^3 N^2 \mu \qquad (29)$$

where $K_L = f$ (impeller, geometry)

For transitional flow, $10 \leq N_{Re} \leq 1000$ to 10,000.

In turbulent flow ($N_{Re} > 1000$ to 10,000, depending on the system), N_P is not a function of N_{Re} when the vessel is fully baffled. Consequently

$$P = K_T D^5 N^3 \rho \qquad (30)$$

where $K_T = f$(impeller, geometry)

For unbaffled vessels, however, N_P continues to decrease with increasing N_{Re}, rendering Equation 30 inapplicable.

When one fluid is dispersed in another, in which it is immiscible, the interfacial tension effects are accommodated by introducing the Weber number, $D^3N^2\rho_d/\sigma$, into the right-hand side of Equations 26-28.

When the fluid in the vessel is non-Newtonian, however, the variations in shear rate with distance from the impeller give rise to corresponding variations in fluid consistency. To allow for this, Metzner and Otto [25] and Magnusson [26] identified an average shear rate $(du/dr)_A$ in the vessel. The apparent viscosity μ_A corresponding to $(du/dr)_A$ equals the viscosity of that Newtonian fluid that would show exactly the same power consumption for agitation under identical conditions, at least in laminar flow. It was found empirically that

$$\dot{\gamma}_A = \left(\frac{du}{dr}\right)_A = k_s N \qquad (31)$$

This finding was confirmed by Metzner and Taylor [27]. The experimental evaluation of k_s for a given geometry proceeds as follows:

1. N_P is measured for a particular N.
2. The corresponding N_{Re} is read from the appropriate plot for Newtonian fluids on Rushton et al.'s chart [24].
3. The apparent viscosity μ_A for these conditions is computed from this N_{Re}. (Laminar flow is used for maximum sensitivity to changes in μ_A.)
4. Next, $(du/dr)_A$ is obtained from the flow curve for the particular fluid under consideration [1, Ch. 1-2], because $\mu_A = \tau/(du/dr)_A$.
5. The constant k_s is then evaluated from Equation 31. The procedure is repeated at various N to give an average k_s.

A compilation of experimental k_s values for a variety of geometries appears in Table I taken from [1]. Skelland [1] condensed the published data on N_P vs $D^2N\rho/\mu_A$ for non-Newtonian fluids with turbine, propeller, paddle and anchor impellers into a single chart, as shown in Figure 2, for use with Table I.

The prediction of power consumption in a particular non-Newtonian agitation system for a given impeller speed proceeds in the following manner:

1. Compute $(du/dr)_A$ from Equation 31, with the appropriate k_s from Table I.
2. Evaluate the corresponding μ_A from the flow curve as $\tau/(du/dr)_A$.
3. Calculate the Reynolds number, $D^2N\rho/\mu_A$, and read the Power number, $P/D^5N^3\rho$, from the appropriate curve in Figure 2. P is then found from N_P.

Table I. Key to Figure 2 [1]

Curve	Impeller	Baffles	D (ft)	D_T/D	n	$k_s(n < 1)$
A-A	Single turbine with 6 flat blades	4, $J/D_T = 0.1$	0.167–0.67	1.3–5.5	0.05–1.5	11.5±1.5
A-A_1	Single turbine with 6 flat blades	None	0.167–0.67	1.3–5.5	0.18–0.54	11.5±1.4
B-B	Two turbines, each with 6 flat blades and $D_T/2$ apart	4, $J/D_T = 0.1$	–	3.5	0.14–0.72	11.5±1.4
B-B_1	Two turbines, each with 6 flat blades and $D_T/2$ apart	4, $J/D_T = 0.1$ Or none	–	1.023–1.18	0.14–0.72	11.5±1.4
C-C	Fan turbine with 6 blades at 45°	4, $J/D_T = 0.1$ Or none	0.33–0.67	1.33–3.0	0.21–0.26	13±2
C-C_1	Fan turbine with 6 blades at 45°	4, $J/D_T = 0.1$ Or none	0.33–0.67	1.33–3.0	1.0–1.42	13±2
D-D	Square-pitch marine propellers with 3 blades (downthrusting)	None, (i) shaft vertical at vessel axis, (ii) shaft 10° from vertical, displaced R/3 from center	0.417	2.2–4.8	0.16–0.40	10±0.9
D-D_1	Same as for D-D but upthrusting	None, (i) shaft vertical at vessel axis, (ii) shaft 10° from vertical, displaced R/3 from center	0.417	2.2–4.8	0.16–0.40	10±0.9

$D\text{-}D_2$	Same as for D-D	None, position (ii)	1.0	1.9–2.0	0.16–0.40	10±0.9
$D\text{-}D_3$	Same as for D-D	None, position (i)	1.0	1.9–2.0	0.16–0.40	10±0.9
E-E	Square-pitch marine propeller with 3 blades	4, $J/D_T = 0.1$	0.5	1.67	0.16–0.60	10
F-F	Double-pitch marine propeller with 3 blades (downthrusting)	None, position (ii)	–	1.4–3.0	0.16–0.40	10±0.9
$F\text{-}F_1$	Double-pitch marine propeller with 3 blades (downthrusting)	None, position (i)	–	1.4–3.0	0.16–0.40	10±0.9
G-G	Square-pitch marine propeller with 4 blades	4, $J/D_T = 0.1$	0.392	2.13	0.05–0.61	10
$G\text{-}G_1$	Square-pitch marine propeller with 4 blades	4, $J/D_T = 0.1$	0.392	2.13	1.28–1.68	–
H-H	2-bladed paddle	4, $J/D_T = 0.1$	0.283–0.416	2–3	0.16–1.68	10 $(n < 1)$
–	Anchor	None	0.927	1.02	0.34–1.0	11±5
–	Cone impellers	0 or 4, $J/D_T = 0.08$	0.33–0.5	1.92–2.88	0.34–1.0	11±5

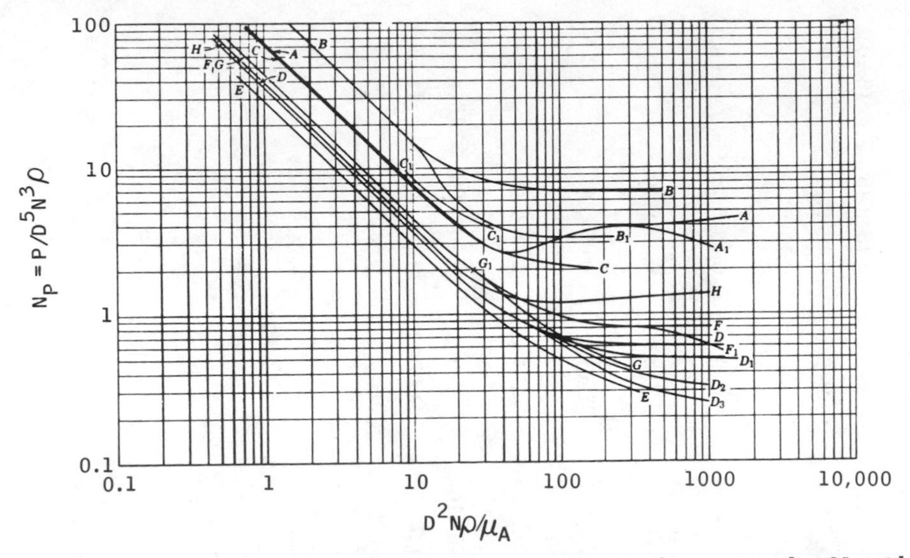

Figure 2. A condensation of published data on correlation of power number N_p with N_{Re} ($= D^2N\rho/\mu_A$) in the agitation of non-Newtonian fluids (keyed to Table I) [1].

One consequence of an alternative approach by Calderbank and Moo-Young [28,29], which is confined to power law fluids, is an extension of k_s to incorporate a dependence on n, the flow behavior index of the fluid. They obtain

$$k_s = B\left(\frac{4n}{3n+1}\right)^{\frac{n}{1-n}} \tag{32}$$

which may again be used with Equation 31 to compute $(du/dr)_A$. The following empirical relationships were found for the constant B:

1. For high-speed impellers,

$$\text{(i)}\quad n < 1;\ \frac{D_T}{D} > 1.5;\ R = 4;\ \frac{J}{D_T} = 0.1$$

$$B = 11 \pm 10\% \tag{33}$$

$$\text{(ii)}\quad n > 1;\ \frac{D_T}{D} < 3.0;\ R = 4;\ \frac{J}{D_T} = 0.1$$

$$B = \frac{22(D_T/D)^2}{(D_T/D)^2 - 1} \pm 15\% \tag{34}$$

2. For anchor agitators, $n < 1$; $D_T/D < 1.4$:

$$B = 9.5 + \frac{9(D_T/D)^2}{(D_T/D)^2 - 1} \pm 10\% \tag{35}$$

It may be noted that the term involving n in Equation 32 has the value 0.84±8.5% for $0.05 \leqq n \leqq 1.68$, and is thus essentially a constant. Evidently, therefore, a linear relationship between $(du/dr)_A$ and N is predicted by both Metzner and Otto [25] and Calderbank and Moo-Young [28,29], with only minor dependence on n.

The viscous behavior of many non-Newtonian materials is often represented by the power law of Equation 12 (with τ conventionally positive) over limited ranges of shear rate. If this is applied to the viscoelastic criteria in Equation 20, with the aid of Equations 13 and 31, we have

$$N_{Re} = \frac{D^2 N \rho}{K(\dot{\gamma}_A)^{n-1}} = \frac{D^2 N \rho}{K(k_s N)^{n-1}} = \frac{D^2 N^{2-n} \rho}{k_s^{n-1} K} \tag{36}$$

and from Equations 13, 18, 21 and 31,

$$N_{Wi} = \frac{h(\dot{\gamma}_A)^{m-2}}{K(\dot{\gamma}_A)^{n-1}} N = \frac{h}{K} (k_s N)^{m-n-1} N = \frac{h}{k_s K} (k_s N)^{m-n} \tag{37}$$

Helical ribbon agitators are particularly effective for mixing highly viscous fluids in which the flow is necessarily laminar [15, p. 54-55; 30-33]. The use of turbulent conditions for mixing is normally precluded, for instance, by the high consistency of polymer melts and concentrated polymer solutions [9].

A sketch of a typical double helical ribbon mixer appears in Figure 3, which shows some of the terminology to be used.

The power consumed by a helical ribbon impeller is substantially dependent on the clearance between the outer edge of the ribbon and the vessel walls ($D_T - D$), and on the pitch of the impeller, S. For pseudoplastic non-Newtonian fluids (i.e., time-independent fluids without a yield stress), Nagata [15, pp. 54-56, 76-80] obtained the correlation of experimental data shown in Figure 4. The measurements were made for the following geometry: $D/D_T = 0.95$, $W/D_T = 0.1$, $S/D = 1$; it appeared also that $H/D_T \doteq 1$.

In the laminar region ($D^2 N \rho / \mu_A \leq 100$) Nagata presents the correlation below for double helical ribbon impellers with $W/D_T = 0.1$:

$$N_P = \frac{P}{D^5 N^3 \rho} = 74.3 \left(\frac{D_T - D}{D}\right)^{-0.5} \left(\frac{D}{S}\right)^{0.5} \left(\frac{D^2 N \rho}{\mu_A}\right)^{-1} \tag{38}$$

and for the case of $D = 0.95\ D_T$ he gives

$$N_P = \frac{P}{D^5 N^3 \rho} = 340 \left(\frac{S}{D}\right)^{-0.5} \left(\frac{D^2 N \rho}{\mu_A}\right)^{-1} \tag{39}$$

Curiously, the coefficient 340 is nearly 5% greater than that obtained by substituting $D = 0.95\ D_T$ in Equation 38. For square-pitched ribbons ($S = D$), Nagata then obtains $N_P = 340/N_{Re}$. It is important to note that Equations 38 and 39 contain the apparent

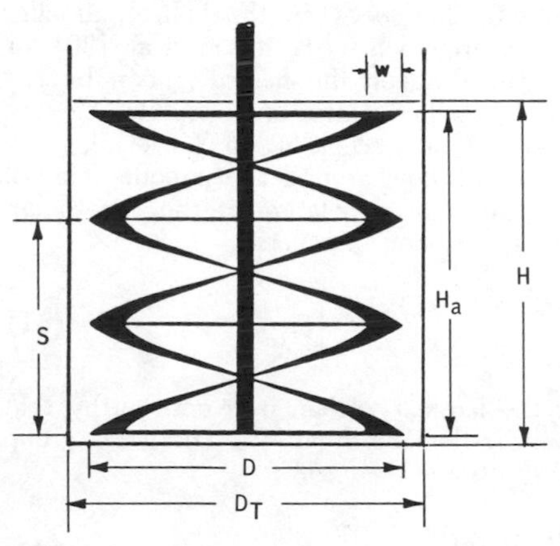

Figure 3. Sketch of a double helical ribbon agitator system: B = number of impeller blades or ribbons; L = length of one blade or ribbon.

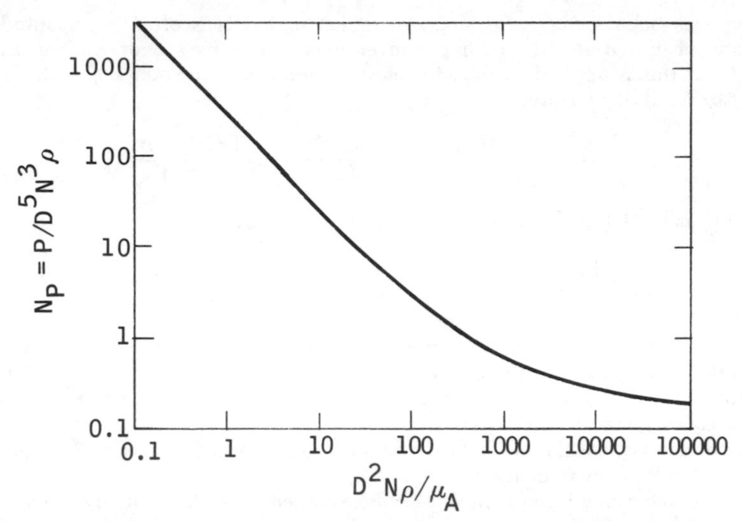

Figure 4. Power correlation for double-helical ribbon agitators in pseudoplastic liquids with $D/D_T = 0.95$, $W/D_T = 0.1$, $S/D = 1$ and $H/D_T \doteq 1$ [15].

viscosity μ_A, obtained from the flow curve for the fluid, using $(du/dr)_A$ computed from Equation 31 at the prescribed N, as described earlier. For this purpose Nagata [15, p. 78] reports a value of 30 for k_s when $D/D_T = 0.95$ and $S/D = 1$. For a single helical ribbon with $D/D_T = 0.94$ and $S/D = 0.5$, Coyle et al. [34] found k_s to be 25.

Bingham plastic fluids are characterized by a yield stress, τ_y, and the equation to the flow curve is

$$\tau_{yx} = \tau_y + \eta\ (du/dy) \tag{40}$$

where η is the "plastic viscosity." Highly concentrated slurries of fine particles suspended in a liquid often show this type of behavior. For such materials Nagata [15, pp. 66-72] found the effect of $D_T - D$ in Equations 38 and 39 to be slight; he recommends replacing the group $(D_T - D)/D$ by unity in such cases, except when the viscosity of the suspending fluid is high.

Attempts at a more fundamental understanding of the mechanism of power consumption were made by Bourne and Butler [35] and Chavan and Ulbrecht [36,37]. They assumed that the power is consumed in couette flow between the helical impeller and the vessel wall. This contrasted with the later approach by Patterson et al. [30] for Newtonian fluids, whose model assumes that the drag on the helical ribbon by the fluid causes the torque exerted on the rotating impeller. Unfortunately, their elegant analysis was not very adaptable to non-Newtonian fluids, according to Yap et al. [38], because of increased complexities in analysis of the flow around the impeller blades. These authors [38] nevertheless attempted a partial extrapolation of the Newtonian model to obtain the following expression for helical ribbon impellers:

$$N_P = 24\ B\left[(N_{Reg})^{0.93}\left(\frac{D_T}{D}\right)^{0.91}\left(\frac{D}{L}\right)^{1.23}\right]^{-1} \tag{41}$$

where B is number of ribbons. Although for the laminar region, it is noteworthy that the exponent on the generalized Reynolds number deviates from −1. The latter group is defined as

$$N_{Reg} = \frac{D^2\ N\rho(t_1\ \dot{\gamma}_e)^{2s_1}}{\mu_o} \tag{42}$$

where μ_o is the viscosity at zero shear rate, t_1 is a time constant for the fluid associated with its non-Newtonian behavior, and $\dot{\gamma}_e$ is the effective shear rate in the vessel. Guidance in evaluating these quantities is provided in the literature [16,38,39]. Equation 41 was said to be most successful with fluids that do not have a high degree of elasticity.

A helical screw impeller in a coaxial draught tube within a cylindrical vessel is often used for mixing highly viscous fluids. The device is sketched in Figure 5a. Nagata [15, p. 56] says that the power consumption for such mixers may be approximated by that for a double helical ribbon impeller of the same pitch, provided that D_s/D_r for the screw is the same as D/D_T for the helical ribbon.

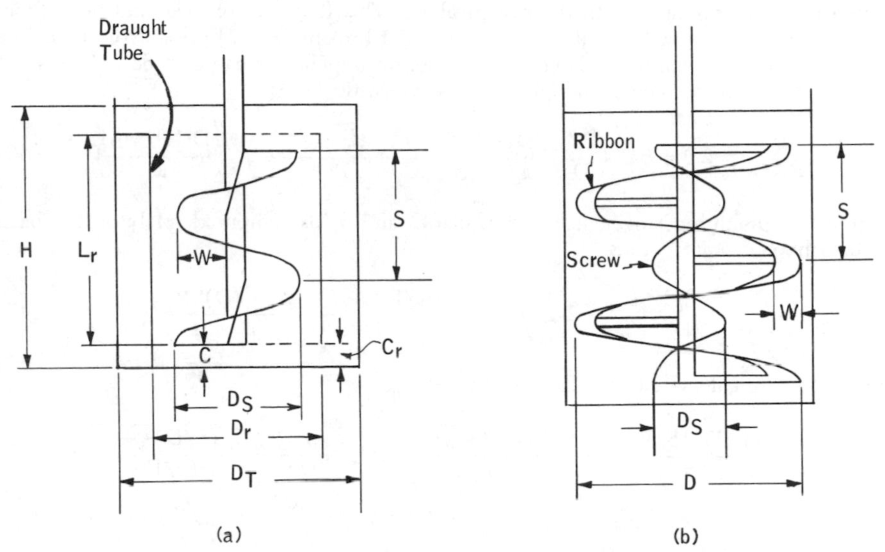

Figure 5. (a) A helical screw impeller in a draught tube; (b) a combined ribbon-screw impeller. Symbols D_T, C and H are common to (a) and (b). L is the length of the screw flight in (a) and of the helical ribbon in (b).

An alternative estimate of power consumption by the screw impeller in a draught tube may be made from the relation given by Chavan and Ulbrecht [36,37], based on the couette, coaxial-cylinder flow model referred to earlier. (As in all cases in which alternative procedures are offered, the designer presumably will take that giving the most conservative indication for the problem at hand.) For either pseudoplastic or viscoelastic fluids to which the power law applies, and in laminar flow ($D_s^2 N^{2-n} \rho/K \leq 10$), these authors [36,37] give

$$\frac{P}{D_s^5 N^3 \rho} = \frac{\pi a_s}{2}\left(\frac{d_e}{D_s}\right)\left(\frac{D_r}{d_e}\right)^2 \left\{\frac{4\pi}{n[(D_r/d_e)^{2/n} - 1]}\right\}^n \left(1 + \frac{D_s}{C}\right)^{0.37}$$
$$\left(\frac{D_T - D_r}{L_r}\right)^{-0.046} \left(\frac{C_r}{D_s}\right)^{-0.036} \left(\frac{D_s^2 N^{2-n} \rho}{K}\right)^{-1} \tag{43}$$

where

$$\frac{d_e}{D_s} = \frac{D_r}{D_s} - 2\frac{W}{D_s} \Big/ \ln\left\{\frac{(D_r/D_s) - 1 + 2(W/D_s)}{(D_r/D_s) - 1}\right\} \tag{44}$$

and

$$a_s = \frac{(L/D_s)(S/D_s)}{3\pi}\left[\frac{\pi\sqrt{(S/D_s)^2+\pi^2}}{(S/D_s)^2} + \ln\left(\frac{\pi}{S/D_s} + \frac{\sqrt{(S/D_s)^2+\pi^2}}{(S/D_s)}\right)\right]$$

$$\{1-[1-2(W/D_s)]^2\} + \pi[1-2(W/D_s)](L/D_s) \qquad (45)$$

Equation 43 was found to apply for impeller rotation in either direction and to be unaffected by elastic effects for $N_{Re} < 10$. The authors [36,37] explain why the equation does not apply to screw extruders.

Finally we will consider combined ribbon-screw impellers of the type sketched in Figure 5b. Coyle et al. [34] report that the combination of the inner screw with the helical ribbon significantly decreased the blending time for pseudoplastic fluids, compared to results obtained with the ribbon alone. This finding, however, did not extend to the blending of Newtonian fluids. Chavan and Ulbrecht [36,37] present the following expression for the combined helical ribbon-screw impeller in laminar flow ($D^2 N^{2-n} \rho/K \leq 10$) of power law pseudoplastic or viscoelastic fluids:

$$\frac{P}{D^5N^3\rho} = 2.5\,\pi\, Ba_s'\left(\frac{d_e'}{D}\right)\left(\frac{D_T}{d_e'}\right)^2\left\{\frac{4\pi}{n[(D_T/d_e')^{2/n}-1]}\right\}^n\left(\frac{D^2N^{2-n}\rho}{K}\right)^{-1} \qquad (46)$$

where B is the number of helical ribbon blades and d_e' is computed solely on the basis of the ribbon geometry, thus,

$$\frac{d_e'}{D} = \frac{D_T}{D} - 2\frac{W}{D}\Big/\ln\left\{\frac{(D_T/D) - 1 + 2(W/D)}{(D_T/D) - 1}\right\} \qquad (47)$$

The term a_s' is calculated on the same basis, as

$$a_s' = \frac{(L/D)(S/D)}{3\pi}\left[\frac{\pi\sqrt{(S/D)^2+\pi^2}}{(S/D)^2} + \ln\left(\frac{\pi}{(S/D)} + \frac{\sqrt{(S/D)^2+\pi^2}}{(S/D)}\right)\right]$$

$$\{1-[1-2(W/D)]^2\} \qquad (48)$$

It is interesting that although Equations 46-48 were first developed for helical ribbon impellers alone, they were found [36,37] also to be valid for combined helical ribbon-screw impellers, provided that the surface area of the screw does not exceed that of the ribbon. When this condition is fulfilled, the power consumption for the combined unit may be safely predicted on the basis of the ribbon geometry only (see also Coyle et al. [34]). The expressions apply for rotation of the impeller in either direction.

The reader may note that Equations 44 and 47 above are taken from the corrigenda [37] to the original paper [36], and that Equations 45 and 48 are in the corrected form supplied by J. Ulbrecht [40].

The foregoing material shows that the influence of purely viscous non-Newtonian fluid characteristics on power consumption in mixing is fairly well accommodated. Furthermore, viscoelastic properties exert no influence in the laminar region ($N_{Re} < 10$ to 100, depending on the system) [9]. This has been confirmed for helical screw and ribbon impellers in polymer solutions [36,37,41], for paddles [42] and for turbines [43,44].

In the transition region ($100 < N_{Re} < 2000$), however, viscoelastic effects require introduction of the Weissenberg number [45,46], giving

$$N_P = f(N_{Re}, N_{Wi}) \qquad (49)$$

Torque on the impeller may be suppressed by viscoelastic effects in the transition region [9].

CIRCULATION CAPACITY, CIRCULATION TIME AND MIXING TIME

The pumping capacity of an impeller is clearly responsible for the overall flow around a vessel, generating fresh interface even within a single-phase liquid. This promotes the action of mixing by molecular and eddy diffusion. The pumping or circulation capacity may be measured, with some accuracy, by velocity traverses across the stream ejected by the impeller. More often, and with less accuracy, it is estimated by determining the average time for some suitable small-sized particle of neutral buoyancy to complete one circulation loop within the vessel, as established by visual observation and stopwatch measurement.

In either case the results are correlated as the dimensionless circulation number, N_{Ci}, defined as

$$N_{Ci} = \frac{q}{ND^3} \doteq \frac{V}{\theta_c ND^3} \tag{50}$$

where q = the volumetric discharge rate of the impeller
V = the volume of liquid in the vessel
θ_c = the time for one circulation loop to be completed.

Early studies on circulation capacities of propeller and turbine agitators in liquids of moderate viscosity are well summarized by Gray [47].

An alternative way of correlating circulation times is in the form of a dimensionless circulation time, $N\theta_c$, and this method was adopted by Holmes et al. [48] for Newtonian fluids stirred by six-flat-bladed, centrally mounted disk turbines in baffled vessels, as follows:

$$N\theta_c = 0.85\left(\frac{D_T}{D}\right)^2; \ N_{Re} > 2(10^4) \tag{51}$$

The mechanism just outlined indicates why N_{Ci} and $N\theta_c$ have been considered as measures of performance for mixers; in this regard, a recent paper by McManamey [49] extends earlier work relating circulation models to the prediction of mixing time under turbulent conditions. The latter quantity has been variously defined as the time interval after the addition of some tracer for some arbitrary fraction of the final uniform concentration to be attained. McManamey's model is in terms of an alternative mixing rate constant for the decrease of concentration fluctuations; this is regarded as a less arbitrary measure of mixing performance, although the mixing constant and the customary mixing time are nevertheless shown to be related.

McManamey's analysis shows that radial flow impellers, typified by turbines, produce more rapid mixing than axial flow impellers, such as propellers and fan turbines. In line with this observation, Gray [47] notes that the circulation number, q/ND^3, ranges from 0.5 to 2.9 for turbines, whereas it is only 0.4 to 0.61 for propellers. McManamey [49] explains this in terms of the longer single recirculation loops for axial flow impellers, compared with the two shorter recirculation loops for radial flow impellers. Early correlations of mixing time for turbine and propeller agitators with fluids of moderate consistency are reviewed in the literature [1, Ch. 9].

In contrast, however, to the turbulent Newtonian conditions of McManamey's [49] treatment, many non-Newtonian mixing systems are confined to laminar or transitional flow regions, for which forms of helical ribbon or helical screw and draught tube devices are usually employed.

Carreau et al. [16] measured both circulation times (θ_c) and mixing times (θ_m) in high-consistency Newtonian, pseudoplastic and viscoelastic fluids stirred with helical ribbon impellers. The ratio θ_m/θ_c was about 3.5 for the Newtonian fluids, increasing two- or threefold for the slightly elastic Carboxy Methyl Cellulose (CMC) solutions, and increasing markedly for the highly elastic fluids (up to sevenfold).

Nagata [15, pp. 80,199] shows the dimensionless mixing time number, $N\theta_m$, to be constant in the laminar region, regardless of fluid consistency, for helical ribbon mixers with purely viscous fluids in geometrically similar units of different size. This observation is qualified somewhat by Carreau et al. [16], who cite conflicting observations by Johnson [50] to the effect that θ_m decreased with increasing viscosity due to more efficient drag flow. However, they [16] note the later work of Coyle et al. [34], indicating that θ_m was unaffected by viscosities greater than 15 Poises. This was confirmed experimentally by Carreau et al. [16], who also examined the effects of blade width W, wall clearance ($D_T - D$), and fluid rheology on θ_m.

They [16] showed that the mixing time, θ_m, increased by three- to sevenfold as the elastic characteristic time (Equation 24) increased. The pumping capacity of the helical ribbon was correspondingly reduced by increasing elasticity. Furthermore, in the absence of significant stagnant zones, it was found that the mixing effectiveness (defined as $V/ND^3\theta_m = 1/\text{dimensionless mixing time}$) is directly related to the axial pumping capacity of the ribbon impeller.

Turning to the configuration of a helical screw inside a draught tube, Chavan and Ulbrecht [51] examined axial discharge rates, circulation times and axial velocities for Newtonian, pseudoplastic and viscoelastic fluids.

Once again the circulation number N_{Ci} was independent of impeller speed, Reynolds number and rheology in the laminar region for Newtonian and pseudoplastic fluids ($D^2N^{2-n}\rho/K \leq 10$ or 100, depending on geometry), where power law behavior was shown to be occurring. Elastic effects, however, reduced the circulation number by up to 50% in a way that also depended on geometry.

A straightforward correlation is provided [51] for the circulation number in terms of the geometry of the screw and draught tube unit. The empirical nature of the expression requires its application within the tabulated ranges of a large number of geometrical variables used in the study.

In a subsequent paper, devoted to blending approximately equal amounts of polymer solutions with different rheological properties, Ford and Ulbrecht [52] obtained an empirical correlation for the dimensionless blending time, $N\theta_B$, using a helical screw in a draught tube. The end point for blending was assessed by optical means.

The fluids included Newtonian, pseudoplastic and viscoelastic materials, and the correlation takes the form of a plot of

$$N\theta_B(\mu_{OM}/\mu_{OL})^{-z}(1 + 0.52\,N_{Wi})^{-1.25}$$

vs

$$N_{Re} = D^2N\rho/\mu_a' \tag{52}$$

with

$$\mu_a' = \mu_o - mN^{b-1} \tag{53}$$

where μ_o, m and b are parameters of the flow curve of the *final* mix. Furthermore, the exponent z on the ratio (μ_{OM}/μ_{OL}) in the ordinate is related to the initial orientation of the two components to be blended. Thus, $z = 0.059$ when the more viscous liquid is initially in the upstream position in the draught tube, whereas $z = 0.17$ when the more viscous liquid is initially in the downstream position in the draught tube. Also μ_{OM} and μ_{OL} are the zero-shear viscosity of the more and less viscous components, respectively [e.g., 1, Ch. 1]. The correlation appears in Figure 6.

It may be noted that the form of correction for viscoelastic effects embodying the Weissenberg number, N_{Wi}, in the ordinate of Figure 6 resembles that outlined by Ulbrecht [9], although differing in detail.

Attention will finally be given to the combined ribbon-screw impellers shown schematically in Figure 5(b). Chavan et al. [19] examined the influence of pseudoplasticity and viscoelasticity on mixing in such devices; in the laminar region ($D^2N^{2-n}\rho/K < 10$)

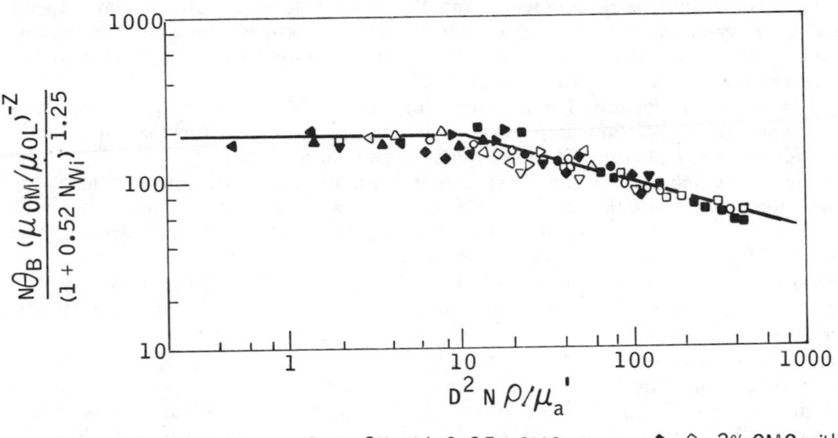

Figure 6. Dimensionless blending time correlation. The plain points refer to the upstream position and the full points to the downstream position of the more viscous fluid in the draught tube. Concentrations are given by approximate figures and serve for orientation only [52].

they found that $N\theta_c$ and $N\theta_m$ were constants for a given geometry, unaffected by the shear thinning viscosity of the fluid. However, the mixing time increased by a factor of two for viscoelastic materials. The interaction between shearing and normal stresses in viscoelastic fluids caused substantial reduction in the axial velocity component, with substantial increase in the angular component of velocity.

Coyle et al. [34], using a ribbon-screw combination, found the mixing time of nonelastic liquids to be independent of apparent viscosity, μ_A, above 15 Poises, with $N\theta_m$ being constant for a given geometry in laminar flow. Measurements of both circulation and mixing times showed that $\theta_m \doteq 3\theta_C$.

In some cases the range of applicability of fan turbines and of helical impellers overlap, depending on the fluid consistency involved and the degree of mixing regarded as satisfactory. Coyle et al. [34] have prepared an interesting comparison of the merits of the two alternative forms of agitation when satisfactory results are obtainable from either source. This is shown in Table II. The relative cost data are for 1970.

Clear differences are seen between the power, speed, torque and cost for the two types of impeller to achieve a given degree of mixing. The heat generated by the mixer

Table II. Comparison of an Axial Flow Turbine and a Helical Impeller for Equal Blend Time and Equal Heat Transfer Characteristics [34][a]

	Relative Values			Initial
Impeller Type	Horsepower	Speed	Torque	Cost
Axial Flow Turbine	1	1	1	1
Helical Impeller	1/3 to 1/10	1/15 to 1/6	1.5 to 3.0	2.5 to 3.5

[a]Ranges for a helical impeller compared to an axial flow turbine are from field observations with both Newtonian and non-Newtonian fluids from 10,000 to 50,000 cP at 5 sec.$^{-1}$

itself, compared to that to be removed from the charge, is frequently a major factor in selecting between the two forms of impeller. Other features, such as the relative scales of mixing produced and the possible use of a high-speed line blender in the outlet streams, are discussed by Coyle et al. [34].

The flow patterns generated in Newtonian and purely viscous non-Newtonian fluids by axial and radial flow impellers, such as propellers and turbines, are qualitatively well known, consisting of the familiar recirculation loops.

Secondary flows unknown in purely viscous fluids may arise, however, in viscoelastic materials containing a rotating turbine (simulated, for experimental simplicity, by a rotating sphere [53]). The competing influence of shearing and normal stresses may cause a torroidal vortex to form above and below the rotating body, the vortices becoming segregated at a particular numerical combination of Weissenberg and Reynolds numbers [5, pp. 225-229; 9]. Material trapped in these vortices fails to mix with the rest of the material in the vessel, thereby jeopardizing the success of the operation.

In contrast, no such segregated vortices have been observed with propeller agitators, the screw propulsion process apparently being enhanced by the viscoelasticity to accentuate the usual axial recirculation pattern [9]. This constitutes an advantage for axially thrusting propellers over turbines and paddles when agitating viscoelastic liquids.

HEAT TRANSFER

Metzner and Otto's [25] concept of an apparent viscosity μ_A corresponding to an average shear rate $(du/dr)_A$ in the vessel was described earlier in the section located between Equations 30 and 35, together with the modifications introduced by Calderbank and Moo Young [28,29]. The success of μ_A in correlating power consumption in agitated vessels prompted Skelland [1, p. 347] in 1967 to suggest its use in correlating agitated-side heat transfer coefficients for purely viscous non-Newtonian fluids in the following general way:

$$\frac{hD_S}{k} = f\left[\left(\frac{D^2N\rho}{\mu_A}\right), \left(\frac{c_p\mu_A}{k}\right), \left(\frac{\mu_A}{\mu_{AW}}\right)\right] \tag{54}$$

where D_S characterizes the heat transfer surface.

A large number of such correlations, mostly in terms of μ_A, have subsequently appeared; many have been reviewed and classified by Edwards and Wilkinson [13] for paddles, turbines, propellers and anchors in both jacketed vessels and vessels with heating or cooling coils.

As an example, consider the relation obtained experimentally by Skelland and Dimmick [54] in 1969 for heat transfer between coils and pseudoplastic fluids with propeller agitation in baffled vessels. These authors used about 22 pseudoplastic fluids, which turned out to obey the power law, with $0.528 \leq n \leq 0.910$. Various sizes of axially mounted propellers were used; they were of marine type, each with three blades and square pitch. Various-sized coils were also tried, and both downthrusting and upthrusting agitation were employed. Additional details appear in the original paper [54]; the resulting correlation is

$$\frac{h_cD_e}{k} = 0.258\left(\frac{D^2N\rho}{\mu_A}\right)^{0.62}\left(\frac{c_p\mu_A}{k}\right)^{0.32}\left(\frac{\mu_A}{\mu_{AW}}\right)^{0.2}\left(\frac{D}{D_T}\right)^{0.1}\left(\frac{D_c}{D_T}\right)^{0.5} \tag{55}$$

where h_c is the individual heat transfer coefficient on the agitated side of the coil, and D_e is the external diameter of the coil tube. In the case of a power law fluid,

$$\tau = K\left(\frac{du}{dr}\right)_A^n$$

$$\mu_A = \frac{\tau}{(du/dr)_A} = \frac{K(du/dr)_A^n}{(du/dr)_A} = \frac{K}{(du/dr)_A^{1-n}}$$

and substituting Equation 31 for $(du/dr)_A$,

$$\mu_A = \frac{K}{(k_sN)^{1-n}} \tag{56}$$

combining Equations 55 and 56 for power law fluids,

$$\frac{h_cD_e}{k} = 0.258\left(\frac{D^2N^{2-n}\rho}{k_s^{n-1}K}\right)^{0.62}\left(\frac{c_pK}{k(k_sN)^{1-n}}\right)^{0.32}\left(\frac{K}{K_W}\right)^{0.2}\left(\frac{D}{D_T}\right)^{0.1}\left(\frac{D_c}{D_T}\right)^{0.5} \tag{57}$$

The mean deviation between Equation 55 or 57 and the results from 123 runs in both heating and cooling operation was 35.5%. In the form of Equation 55 the expression is not restricted to power law materials, but applies in general to any time-independent non-Newtonian fluid for $332 \leq D^2N\rho/\mu_A \leq 2.6(10^5)$. Furthermore, it reduces exactly to the recommended Newtonian form [14, Vol. 1, pp. 294-295] with regard to the effects of six key variables (D, N, D_T, D_e, μ, μ/μ_W).

Because the temperature of the fluid varies during the heating or cooling process, a question arises concerning the temperature at which to evaluate μ_A in Equation 55 (or K in Equation 57). Skelland and Dimmick [54] showed that evaluation of μ_A at the mean run temperature, $(T_1 + T_2)/2$, was adequate throughout their work.

As a result of the non-Newtonian nature of the liquids used, the variation in shear rate from point to point in the vessel causes corresponding variations between the local apparent viscosity, μ_a, and the overall average value, μ_A, used to characterize the run. The extent of this deviation between local μ_a and average μ_A depends on the coil and propeller diameters and the shaft speed. Heat transfer rates, which depend on local μ_a, therefore show corresponding deviations from run to run. This source of deviation is not present in Newtonian systems and might be expected to increase with increasingly non-Newtonian behavior, e.g., decreasing n. This is confirmed by the data of Hagedorn and Salamone [55] for heat transfer in the simpler geometry of a jacketed vessel. Thus their mildly non-Newtonian results ($0.69 \leq n \leq 1.0$) were correlated with an average deviation of ± 14%, whereas for their highly non-Newtonian data ($0.36 \leq n \leq 0.69$) the average deviation from their correlation increased to ±20%.

Scaleup in Coil-Heated, Baffled Vessels with Propeller Agitation

Suppose that one or both of the following criteria are required to duplicate some product characteristic when scaling up from pilot plant to industrial scale, using power law fluids.

Criterion A. Equal rates of heat transfer per unit surface of coil in the two vessels.

Criterion B. Equal rates of heat transfer per unit volume of vessel contents. (Either of these criteria might previously have been noted as being fulfilled when product duplication was obtained on the two scales.)

Confining attention to power law fluids, Equation 57 expands to

$$h_c = \phi\, D_e^{-0.5}\, D^{1.34}\, D_T^{-0.6}\, N^{0.92-0.3n} \tag{58}$$

where

$$\phi = 0.258\, k^{0.68}\, \rho^{0.62}\, c_p^{0.32}\, (K/K_W)^{0.2} k_s^{0.3(1-n)} K^{-0.3} \tag{59}$$

It will be assumed that the same fluid is heated or cooled over temperature ranges with the same mid-point and that the tube wall temperatures have the same average value in the two different-sized vessels, denoted by subscripts 1 and 2. Therefore

$$\phi_{av,1} = \phi_{av,2}$$

and for Criterion A,

$$h_{c1}\,\Delta T_{1m1} = h_{c2}\,\Delta T_{1m2} \tag{60}$$

Combining Equations 58 and 60 and rearranging,

$$\frac{N_1}{N_2} = \left[\left(\frac{D_{e_1}}{D_{e_2}}\right)^{0.5}\left(\frac{D_2}{D_1}\right)^{1.34}\left(\frac{D_{T_1}}{D_{T_2}}\right)^{0.6}\frac{\Delta T_{1m_2}}{\Delta T_{1m_1}}\right]^{\frac{1}{0.92-0.3n}} \tag{61}$$

This defines the ratio of impeller shaft speeds in the two vessels to obtain Criterion A.
To achieve Criterion B,

$$\frac{q_1}{V_1} = \frac{q_2}{V_2} \tag{62}$$

or

$$\frac{h_{c_1}\pi D_{e_1} L_{c_1}\Delta T_{1m_1}}{H_1(\pi D^2{}_{T_1}/4)} = \frac{h_{c_2}\pi D_{e_2} L_{c_2}\Delta T_{1m_2}}{H_2(\pi D^2{}_{T_2}/4)} \tag{63}$$

where L_c is the length of the submerged coil and H is the height of liquid in the vessel. Equation 58 is substituted for h_{c_1} and h_{c_2}, and the result solved for the ratio of shaft speeds needed to achieve Criterion B on the two scales as

$$\frac{N_1}{N_2} = \left[\frac{H_1}{H_2}\left(\frac{D_{T_1}}{D_{T_2}}\right)^{2.6}\left(\frac{D_{e_2}}{D_{e_1}}\right)^{0.5}\frac{L_{c_1}}{L_{c_2}}\left(\frac{D_2}{D_1}\right)^{1.34}\frac{\Delta T_{1m_2}}{\Delta T_{1m_1}}\right]^{\frac{1}{0.92-0.3n}} \tag{64}$$

For simultaneous duplication of Criteria A and B, Equations 61 and 64 give

$$\frac{H_1}{H_2} = \frac{D_{e_1}}{D_{e_2}}\left(\frac{D_{T_2}}{D_{T_1}}\right)^2\frac{L_{c_1}}{L_{c_2}} \tag{65}$$

so that Equation 65 and either Equation 61 or 64 must both be satisfied. Equation 65, however, cannot be satisfied when strict geometrical similarity prevails on the two scales, so that criteria A and B cannot both be achieved in such a case.

Jackets are of course frequently used instead of coils for heating or cooling; as a matter of design interest one may note a correlation, recently published by Baker and Walter [56], for predicting the heat transfer coefficients inside jackets fitted to provide jetting action of the heating or cooling medium.

Helical ribbon impellers are widely used in promoting heat transfer to non-Newtonian fluids of high consistency when operation is confined to the laminar and transitional flow regimes. For heat transfer in jacketed vessels agitated by double helical ribbon impellers, and when $D^2N\rho/\mu_A$ is below 1000, Nagata [15, pp. 99-102] gives

$$\frac{h_j D_T}{k} = 1.75\left(\frac{D^2N\rho}{\mu_A}\right)^{1/3}\left(\frac{c_p\mu_A}{k}\right)^{1/3}\left(\frac{\mu_A}{\mu_{AW}}\right)^{0.2}\left(\frac{D_T - D}{D_T}\right)^{-1/3} \tag{66}$$

where μ_A has the meaning it had earlier in considerations of non-Newtonian power consumption and μ_{AW} is the value of μ_A at the wall temperature. h_j is the individual agitated-side coefficient of heat transfer to or from the jacket, and dimensional symbols are as defined in Figure 3.

Equation 56 could be used to convert Equation 66 into a form suitable for power law fluids, in a manner similar to that which led to Equation 57. Scaleup relationships comparable to Equations 58–65 could then be developed for the attainment of Criteria A (written for unit surface of the jacket) and B on two different scales of operation.

Scraped Surface Heat Exchangers

Scraped surface heat exchangers have found extensive application for non-Newtonian fluids of high consistency that may tend to adhere to the heat transfer surface. Such devices feature rotating blades that continually scrape material from the surface; they are exemplified by the Votator [57], the Dopp Kettle [58] and the helical ribbon impeller fitted with rubber scrapers [15, pp. 101-103].

The theory of heat transfer in such units will be developed in some detail because it is found to apply well in some cases and badly in others; furthermore, the difference between the two categories is instructive.

Consider the fluid in the vicinity of the heat transfer surface between scrapings by two consecutive blades. It is assumed that

1. plug flow prevails in the fluid over the depth of heat penetration;
2. the fluid thickness exceeds the depth of heat penetration; and
3. there is complete mixing of the fluid displaced by the scraper blade with the bulk fluid after each displacement from the vicinity of the heat transfer surface.

The general equation for unsteady-state three-dimensional conduction will first be developed, and then reduced to that for one-dimensional thermal transport for the case at hand.

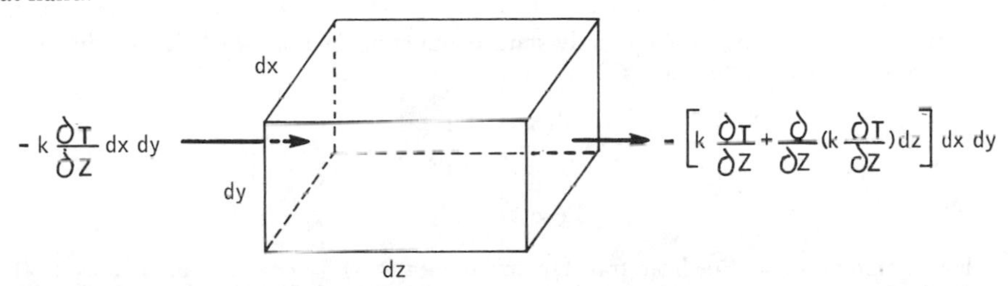

Figure 7. Unsteady-state conduction into and out of the volume element dx dy dz in the z direction.

Figure 7 shows the fluid volume element dx dy dz in the vicinity of the heat transfer surface, awaiting removal by the next scraper blade. Unsteady-state conduction may occur through all six faces of the element, depending on the symmetry of the temperature field.

The rate at which heat enters through the left face of the element by conduction is given by Fourier's law as

$$-k \frac{\partial T}{\partial z} dx\, dy$$

The rate at which heat leaves by conduction through the right face is

$$-\left[k\frac{\partial T}{\partial z} + \frac{\partial}{\partial z}\left(k\frac{\partial T}{\partial z}\right) dz \right] dx\, dy$$

The rate of accumulation of heat within the volume element from these two sources is therefore

$$\frac{\partial}{\partial z}\left(k\frac{\partial T}{\partial z}\right) dx\, dy\, dz$$

Similarly, from the top and bottom faces the accumulation rate is

$$\frac{\partial}{\partial y}\left(k\frac{\partial T}{\partial y}\right) dx\, dy\, dz$$

and from the front and rear faces it is

$$\frac{\partial}{\partial x}\left(k\frac{\partial T}{\partial x}\right) dx\, dy\, dz$$

Another expression for the rate of accumulation of heat in the volume element is

$$dx\, dy\, dz\, c_p\rho\,\frac{\partial T}{\partial t}$$

so that for constant thermal conductivity k,

$$\frac{\partial T}{\partial t} = \alpha\left(\frac{\partial^2 T}{\partial x^2} + \frac{\partial^2 T}{\partial y^2} + \frac{\partial^2 T}{\partial z^2}\right) \tag{67}$$

where α is the thermal diffusivity, $k/\rho c_p$.

For one-dimensional conduction, say in the z direction, Equation 67 reduces to

$$\frac{\partial T}{\partial t} = \alpha\frac{\partial^2 T}{\partial z^2} \tag{68}$$

This is the basic equation for unsteady-state conduction in the radial (or z) direction.

The boundary conditions are

$$T(z,o) = T_\infty$$
$$T(o,t) = T^*$$
$$T(\infty,t) = T_\infty$$

The temperature at the heat transfer surface $(z = o)$ is constant at T^*, as is the temperature T_∞ at z effectively equal to infinity. If T' is defined as $T - T_\infty$ and $T^{*\prime}$ as $T^* - T_\infty$, Equation 68 may be written as

$$\frac{\partial T'}{\partial t} = \alpha\,\frac{\partial^2 T'}{\partial z^2}$$

and the boundary conditions become

$$T'(z,o) = o$$
$$T'(o,t) = T^{*\prime}$$
$$T'(\infty,t) = o$$

The Laplace transform of both sides is taken with respect to time t:

$$\int_o^\infty e^{-pt}\frac{\partial^2 T'}{\partial z^2} dt = \frac{1}{\alpha}\int_o^\infty e^{-pt}\frac{\partial T'}{\partial t}\, dt$$

or

$$\frac{d^2F}{dz^2} = \frac{1}{\alpha} pF - T'(z,o) = \frac{p}{\alpha} F$$

where

$$F(z,p) = L_t[T'(z,t)]$$

Hence

$$\frac{d^2F}{dz^2} - \frac{p}{\alpha} F = 0$$

The second boundary condition transforms to $F(0,p) = T^{*\prime}/p$. The solution of the ordinary differential equation for F, subject to $F(0,p) = T^{*\prime}/p$ and F remains finite as $z \rightarrow \infty$, is

$$F(z,p) = \frac{T^{*\prime}}{p} e^{-\sqrt{pz^2/\alpha}}$$

The function having this as its Laplace transform is found from tables; for example, as item 8 in the table of Laplace transforms given by Crank [59]:

$$T'(z,t) = T^{*\prime} \left[1 - \frac{2}{\sqrt{\pi}} \int_0^{z/2\sqrt{\alpha t}} e^{-\theta^2} d\theta \right] \tag{69}$$

and inserting the definitions of T' and $T^{*\prime}$,

$$\frac{T^* - T}{T^* - T_\infty} = \operatorname{erf} \frac{z}{2\sqrt{\alpha t}} \tag{70}$$

where erf is the error function, defined generally as

$$\operatorname{erf} \zeta = \frac{2}{\sqrt{\pi}} \int_0^{\zeta} e^{-\theta^2} d\theta$$

in which θ is any dummy variable, used merely to describe the function to be integrated; θ is eliminated by the limits of integration. Equation 70 gives the temperature in the fluid near the heat transfer surface as a function of position z and time t.

Now the instantaneous heat flux from the surface is

$$\frac{q}{A} = -\alpha c_p \rho \left(\frac{\partial T}{\partial z} \right)_{z=0} \tag{71}$$

and evaluating $(\partial T/\partial z)_{z=0}$ from Equation 70,

$$\frac{q}{A} = (T^* - T_\infty) c_p \rho \left(\frac{\alpha}{\pi t} \right)^{1/2}$$

The total thermal energy penetrating the fluid in an exposure time t_e is

$$\int_0^{t_e} \frac{q}{A} dt = (T^* - T_\infty) c_p \rho \left(\frac{\alpha}{\pi} \right)^{1/2} \int_0^{t_e} t^{-1/2} dt$$

$$= 2(T^* - T_\infty) c_p \rho \left(\frac{\alpha t_e}{\pi} \right)^{1/2}$$

and the average rate of transfer during exposure is obtained by dividing the last expression by t_e:

$$\left(\frac{q}{A}\right)_{av} = 2(T^* - T_\infty)c_p\rho\left(\frac{\alpha}{\pi t_e}\right)^{1/2}$$

or

$$\frac{(q/A)_{av}}{T^* - T_\infty} = h = 2\left(\frac{c_p\rho k}{\pi t_e}\right)^{1/2} = 1.1284\left(\frac{c_p\rho k}{t_e}\right)^{1/2} \tag{72}$$

The time of exposure, t_e, is the time between passage of two consecutive blades. For n_b scraper blades rotating at N rps, $t_e = 1/n_bN$ sec, where the blades are uniformly distributed around the shaft. Then the scraped surface heat transfer coefficient h is given by

$$h = 1.1284(c_p\rho k n_b N)^{1/2} \tag{73}$$

Rearranging into dimensionless form,

$$\frac{hD}{k} = 1.1284\left(\frac{D^2N\rho}{\mu_A} \cdot \frac{c_p\mu_A}{k} \cdot n_b\right)^{1/2} \tag{74}$$

It is important to note that Equation 74 predicts that the scraped surface heat transfer coefficient h varies with rotary shaft speed N raised to the power 0.5.

Skelland [57] used Houlton's [60] data on water in the Votator to show that h did indeed vary with the square root of N. This exponent, or one close to it, was subsequently confirmed by several workers, but only for processed fluids of low to moderate consistency or viscosity [61-63].

In contrast, with fluids of high consistency or viscosity the exponent on N is found to be much lower. Thus, Skelland [57] found an exponent of 0.17; Bott and Azoory [63] found 0.18; and recently Ramdas et al. [64] found the exponent on N to be 0.113. Similar departure from the indications of the penetration theory for high-viscosity fluids has been reported by Mizushina et al. [65]. It was concluded by Ramdas et al. [64] that "the penetration theory model is unsuitable for predicting the coefficients for viscous fluids in laminar flow."

It is believed [63,64] that the third assumption underlying the development is not fulfilled in such cases; complete mixing between the bulk fluid and that supposedly displaced from the wall by the scraper does not occur for high-consistency materials. This, of course, contrasts with the performance as predicted on low-viscosity fluids. Correlations of coefficients for such scraped surface heat exchangers may be found in the literature [57,61-64].

Turning to helical ribbon impellers fitted with two rubber scrapers, Nagata [15, p. 101-103] gives

$$\frac{h_jD_T}{k} = 5.4\left(\frac{D^2N\rho}{\mu_A}\right)^{1/3}\left(\frac{c_p\mu_A}{k}\right)^{1/3}\left(\frac{\mu_A}{\mu_{AW}}\right)^{0.2} \tag{75}$$

This was developed for a vessel with D_T of 30 cm. Nagata comments that the departure of the exponent on the Reynolds number—and hence on N—from 1/2 shows that "the experimental correlation does not agree with theory." This is presumably for the reasons associated with high-consistency fluids described above.

MASS TRANSFER

The success of the term μ_A in correlating power consumption and heat transfer in agitated non-Newtonian fluids suggests that it may be effective in mass transfer correlations also. Some evidence for this will now be given.

Gas-Liquid Systems

Yagi and Yoshida [75] measured the rate of desorption of oxygen from two Newtonian and two non-Newtonian fluids, the latter exhibiting both shear thinning and viscoelastic behavior. The operation was performed in sparged, baffled vessels, agitated by a six-flat-blade turbine.

All their data, from both the Newtonian and non-Newtonian liquids, were "fairly well" correlated by the following single empirical equation:

$$\frac{k_L a D^2}{D_L} = 0.060 \left(\frac{D^2 N \rho}{\mu_A}\right)^{1.5} \left(\frac{DN^2}{g}\right)^{0.19} \left(\frac{\mu_A}{\rho D_L}\right)^{0.5} \left(\frac{\mu_A V_s}{\sigma}\right)^{0.6} \left(\frac{ND}{V_s}\right)^{0.32} \left[1 + 2(N_{Wi}')^{0.5}\right]^{-0.67} \quad (76)$$

where $k_L a$ is the liquid-phase mass transfer volumetric or capacity coefficient, in which a is the interfacial area per unit volume, V_s is the superficial gas velocity (length/time), and μ_A is the average apparent viscosity, obtained from the flow curve in the way described earlier as

$$\mu_A = \tau/(du/dr)_A = \tau/k_s N \quad (77)$$

In Yagi and Yoshida's study k_s was taken to be 11.5.

For Newtonian fluids μ_A becomes μ and the Weissenberg number of course equals zero. The Weissenberg number, N_{Wi}', used in Equation 76 is conventional in definition, being

$$N_{Wi}' = \frac{\text{characteristic material time}}{\text{characteristic process time}}$$

but the numerator is defined in accordance with Prest et al. [66]. It is the reciprocal of the shear rate at which the reduced complex viscosity, i.e., the ratio of the apparent viscosity μ_A' to the zero-shear viscosity μ_o, is 0.67.

An earlier study of CO_2 absorption by aqueous Carbopol solutions in turbine-agitated, sparged vessels was reported by Perez and Sandall [67]. However, they found it necessary to discard data from two of their three non-Newtonian solutions, due to excessive foaming. In addition, one of the allegedly dimensionless groups used in their correlation, DV_s/σ, is in fact not dimensionless. Details will therefore not be given here.

Next, the question of dispersing one immiscible liquid in another in an agitated vessel will be considered. Such problems arise, for instance, in liquid-liquid extraction operations.

Liquid-Liquid Dispersion

Skelland and Seksaria [68] dispersed a variety of pairs of immiscible Newtonian liquids in each other, using equal volumes of each liquid. The operation was carried out in baffled cylindrical vessels, successively using four types of impellers, namely propellers, pitched-blade turbines, flat-blade turbines and curved-blade turbines. Three different sizes of each impeller were used, axially mounted in turn at H/4, H/2 and 3H/4. Two impellers on the shaft were also studied.

Visual observation showed the minimum impeller speed N_m at which one phase was completely dispersed as droplets in the other, leading to the following empirical correlation:

$$\frac{D^{1/2}N_m}{g^{1/2}} = C_1\left(\frac{D_T}{D}\right)^{a_1}\left(\frac{\mu_c}{\mu_d}\right)^{1/9}\left(\frac{\Delta\rho}{\rho_c}\right)^{0.25}\left(\frac{\sigma}{D^2\rho_c g}\right)^{0.3} \tag{78}$$

The values of C_1 and a_1 are tabulated by Skelland and Seksaria [68] for each impeller type and location. Depending on impeller and position, the average percentage deviation between predicted and observed N_m ranged from 4.28% to 18.15%. However, almost all average deviations were below 12% on a total of 195 measurements.

It was found that impeller speeds up to 1000 rpm are in some cases insufficient to ensure complete dispersion, thereby confirming the need for the study. In a subsequent paper Skelland and Lee [69] showed that gross uniformity of dispersion was obtained using N values that were about 8% greater than N_m, on average.

An intriguing question that now arises is whether Equation 78 could be extended to accommodate a time-independent, non-Newtonian continuous phase, perhaps by the use of an appropriate μ_{cA} in place of μ_c. For agitation of two-phase Newtonian fluids, Treybal [70] shows that the Power number-Reynolds number plot coincides with that for single-phase liquids when the Power and Reynolds numbers are formulated using

$$\rho_M = \rho_c\phi_c + \rho_d\phi_D \tag{79}$$

$$\mu_M = \frac{\mu_c}{\phi_c}\left(1 + \frac{1.5\,\mu_d\phi_D}{\mu_d + \mu_c}\right) \tag{80}$$

where ϕ_c and ϕ_D are volume fractions of the continuous and disperse phases.

Whether μ_{cA} would be that value that fitted Equation 80 when used in place of μ_c is a matter for speculation. Alternatively, some other formulation of μ_{cA}, such as Equation 56, may prove adequate.

The effects of a non-Newtonian disperse phase are also unknown, although if the drops behaved as rigid particles (high σ, low $\Delta\rho$, small diameter), then the effects of μ_d might not be important. In this regard, both Harriott [71] and Calderbank [72] state that small drops and bubbles behave as rigid spheres when suspended in agitated liquids. In contrast, both Treybal [70, p. 416] and Glen [73] say that the surface of such liquid drops is mobile. Clearly further study of this question of dispersion in non-Newtonian liquid-liquid systems is indicated.

Liquid-Liquid Systems

Skelland and Lee [74] measured drop size and continuous-phase mass transfer coefficients in Newtonian liquid-liquid extraction systems in agitated vessels. Five liquid-liquid systems were used, with one phase dispersed as droplets in the other, and with heptanoic acid transferring from the drops to the continuous phase. Centrally mounted six-flat-blade turbines provided the agitation, and dispersed-phase volume fractions ϕ_D between 0.03 and 0.09 were used in baffled cylindrical vessels with $D_T = H$.

The continuous-phase mass transfer coefficient was correlated as

$$\frac{k_c}{\sqrt{ND_c}} = 2.932\,(10^{-7})\,\phi_D{}^{-0.508}\left(\frac{D}{D_T}\right)^{0.548}\left(\frac{D^2N\rho_c}{\mu_c}\right)^{1.371} \tag{81}$$

The Sauter-mean droplet diameter when about 50% of the possible mass transfer had occurred was chosen for correlation as an average value during batch operation, denoted by $\bar{d}_{32}$. The result was

$$\frac{\bar{d}_{32}}{D} = 6.713(10^{-4})\,\phi_D{}^{0.188}\left(\frac{D}{D_T}\right)^{-1.034}\left(\frac{D^2N\rho_c}{\mu_c}\right)^{-0.558}\left(\frac{\mu_c}{\sqrt{\rho_c D\sigma}}\right)^{-1.025} \tag{82}$$

The average absolute deviations between predicted and experimental values for Equations 81 and 82 are 23.8% and 10.9%, respectively. The corresponding capacity coefficient, $k_c a$, is readily obtained, since $a = 6\,\phi_D/\bar{d}_{32}$.

These expressions were combined with a cost analysis to obtain relationships for the optimum values of impeller speed, vessel volume and power consumption, at which the total cost of the extraction process is a minimum. Scaleup relationships were also developed.

As in the treatment of liquid-liquid dispersions presented in the previous section, it is intriguing to speculate whether these Newtonian expressions could be extended to the case of a time-independent non-Newtonian continuous phase, perhaps by using an appropriate average apparent viscosity, μ_{cA}, in place of μ_c. Once more, some formulation such as Equation 56 could prove adequate in the case of a power law fluid. Such an approach seemed effective for Yagi and Yoshida [75] in the case of gas-liquid systems, but the effects of the greater viscosity and density of the dispersed phase in liquid-liquid systems are not known.

If some such extension to a purely viscous non-Newtonian continuous phase is feasible, then the reader may find it an interesting exercise to rework the optimization and scaleup relationships from the original paper [74] into corresponding non-Newtonian terms. Evidently more study is needed on this question.

NOMENCLATURE

A, A'	areas
a	interfacial area per unit volume
a_s, a_s'	dimensionless surface areas, Equations 45 and 48
B	number of impeller blades
B	constant, see Equations 32-35
C	height of impeller above vessel floor
C_r	height of bottom of draught tube above vessel floor
c_p	specific heat (energy/(mass)(temp))
D	impeller diameter
D_c	diffusivity in continuous phase ((length)^2/time)
D_e	external diameter of coil tubing
D_L	diffusivity in liquid phase ((length)^2/time)
D_r	diameter of draught tube
D_s	diameter of screw impeller
D_T	vessel diameter
d_e, d_e'	equivalent diameter given by Equations 44 and 47
$(du/dr)_A$	average shear rate in an agitated vessel (see Equation 31) 1/time
$\bar{d}_{32}$	Sauter mean drop diameter when about 50% of the possible mass transfer has occurred
F, F_a, F_b, F_c, F_d	force, types of force
$F_g, F_i, F_{S.T.}, F_v$	gravitational, inertial, surface tension and viscous forces
g	acceleration due to gravity
H	liquid height in vessel
h	empirical constant in Equation 18
h	individual coefficient of heat transfer (energy/(time)(area)(temp))
h_c, h_j	individual heat transfer coefficients on the agitated side of a coil and a jacket, respectively (energy/(time)(area)(temp))
J	width of baffles
K	fluid consistency index in the power law (mass time^{n-2} length^{-1})
K_w	value of K at the wall temperature
k	thermal conductivity (energy/(time)(length)(temp))
k_c	continuous-phase mass transfer coefficient (length/time)
k_L	liquid phase mass transfer coefficient (length/time)
k_s	constant in Equation 31

L	length dimension chosen to characterize the geometry of the system
L	length of impeller blade
L_c	length of submerged coil
L_r	height of draught tube
m	mass
m	empirical constant in Equation 18
N	rotational speed of impeller (rev/time)
N_{Ci}	Circulation number, Equation 50
N_{De}	Deborah number, Equation 23
N_{Fr}	Froude number, Equation 8
N_m	minimum impeller speed for complete liquid-liquid dispersion, without regard to uniformity (rev/time)
N_P	Power number, Equation 26
N_{Re}	Reynolds number, Equation 7
N_{Reg}	a "generalized" Reynolds number, defined by Equation 42
N_{VR}	viscoelastic ratio, Equation 22
N_{We}	Weber number, Equation 9
N_{Wi}	Weissenberg number, Equation 21
N_{Wi}'	modified form of N_{Wi}, defined in text below Equation 77
n	flow behavior index in the power law, dimensionless
n_b	number of scraper blades
P	power (energy/time)
p	hydrostatic pressure
q	pumping capacity or volumetric discharge rate of impeller (vol/time)
q	rate of heat transfer (energy/time)
R	radius of cone
R	number of baffles
r	radial coordinate
S	pitch of impeller
T	torque, (force) (length)
T	temperature
T^*	temperature of heat transfer surface
T_∞	temperature at z = "infinity"
T'	$T - T_\infty$
$T^{*\prime}$	$T^* - T_\infty$
ΔT_{1m}	logarithmic mean of the temperature differences causing heat transfer at the beginning and end of a run
t	time
t_e	time of exposure to heat transfer
t_1	defined in text below Equation 42
u	velocity (length/time)
V	velocity (length/time *or* volume of liquid in vessel)
V_s	superficial gas velocity (length/time)
$v_1(r)$	velocity parallel to annulus walls at distance r from axis (Figure 1) (length/time)
W	width of impeller blade
x,y,z	coordinate directions
$x_2 = r$	distance in radial direction (Figure 1)
z	an exponent in the ordinate of Figure 6, equaling 0.059 when the more viscous liquid is initially in the *upstream* position in the draught tube, and 0.17 when it is in the reverse location
α	thermal diffusivity, $k/\rho c_p$ $((\text{length})^2/\text{time})$
$\dot{\gamma}$	shear rate, du/dr, dv/dr, etc. (1/time)
$\gamma_A = (du/dr)_A$	an average shear rate in an agitated vessel, described between Equations 30 and 31, (1/time)
γ_e	the effective shear rate in the vessel—see Equation 42 and following text (1/time)
$\dot{\gamma}'$	the characteristic shear rate, see Equation 24 and following text, (1/time)
η	plastic viscosity for a Bingham plastic fluid, Equation 40 (mass/(length) (time))
$\theta_B, \theta_C, \theta_m$	blending, circulation and mixing times
μ_A	average apparent viscosity for agitation corresponding to $(du/dr)_A$ (mass/(length) (time))
μ_{AW}	value of μ_A at the wall- or heat transfer surface-temperature
$\mu_a = \mu_a(\dot{\gamma})$	apparent viscosity, defined by Equations 11-13 (mass/(length) (time))
μ_a'	a form of apparent viscosity defined in Equations 52 and 53 and surrounding

text (mass/(length)(time))

μ_c continuous phase viscosity (mass/(length)(time))

μ_{cA} a value of μ_A (defined above) for the continuous phase (mass/(length)(time))

μ_d disperse-phase viscosity (mass/(length)(time))

μ_M viscosity of a mixture of two liquid phases, Equation 80 (mass/(length)(time))

μ_O viscosity at zero shear rate (mass/(length)(time))

μ_{OL}, μ_{OM} zero shear rate viscosities of the less and more viscous components, respectively (mass/(length)(time))

ρ, ρ_c, ρ_d density, of the continuous and disperse phases, respectively (mass/vol)

ρ_M density of a mixture of two liquid phases, Equation 79 (mass/vol)

$\Delta\rho$ $|\rho_c - \rho_d|$ (mass/vol)

σ surface or interfacial tension (force/length)

$\sigma_1(\dot{\gamma})$ the primary normal stress coefficient, Equation 16

$\sigma_2(\dot{\gamma})$ the secondary normal stress coefficient, Equation 19

τ shear stress (force/area)

τ_y yield shear stress (force/area)

τ_{yx} shear stress in x direction on surface normal to y (force/area)

τ_{21} shear stress in direction 1 on surface normal to direction 2 (force/area)

$\tau_{11}, \tau_{22}, \tau_{33}$ normal stresses in the respective directions 1, 2 and 3, on surfaces respectively normal to the directions 1, 2, and 3 (force/area)

ϕ angle between cone and plate surfaces in a cone-and-plate viscometer (radians)

ϕ_c, ϕ_D volume fractions of the continuous and disperse phases

REFERENCES

1. Skelland, A. H. P. *Non-Newtonian Flow and Heat Transfer* (New York: John Wiley & Sons, Inc., 1967).
2. Wilkinson, W. L. *Non-Newtonian Fluids* (Oxford, England: Pergamon Press, Inc., 1960).
3. Middleman, S. *The Flow of High Polymers* (New York: John Wiley & Sons, Inc., 1968).
4. Schowalter, W. R. *Mechanics of Non-Newtonian Fluids* (Oxford, England: Pergamon Press, Inc., 1978).
5. Chavan, V. V., and R. A. Mashelkar. In: *Advances in Transport Processes*, Vol. 1, A. S. Mujumdar, Ed. (New York: John Wiley & Sons, Inc., (1980), pp. 210-252.
6. Middleman, S. *Fundamentals of Polymer Processing* (New York: McGraw-Hill Book Co., 1977), Chapt. 12.
7. Pace, G. W. *The Chem. Engr. (London)* 833-837 (1978).
8. Chen, S. J., and A. R. MacDonald. *Chem. Eng.* 105 (March 19, 1973).
9. Ulbrecht, J. *The Chem. Engr. (London)* 347-353, 367 (1974).
10. Pinto, G., and Z. Tadmor. *Polymer Eng. Sci.* 10: 279 (1970).
11. Tadmor, Z., and I. Klein. *Engineering Principles of Plasticating Extrusion* (New York: Van Nostrand Reinhold Co., 1970), Chapt. 7.
12. Bigg, D., and S. Middleman. *Ind. Eng. Chem. Fund.* 13: 66 (1974).
13. Edwards, M. F., and W. L. Wilkinson. *The Chem. Engr. (London)* 328-335 (1972).
14. Uhl, V. W., and J. B. Gray, Eds. *Mixing: Theory and Practice*, Volumes I and II (New York: Academic Press, 1966).
15. Nagata, S. *Mixing—Principles and Applications* (New York: John Wiley & Sons, Inc., 1975).
16. Carreau, P. J., I. Patterson and C. Y. Yap. *Can. J. Chem. Eng.* 54: 135 (1976).
17. Lindsay, R. B. *Basic Concepts of Physics* (New York: Van Nostrand Reinhold Co., 1971), p. 92.

18. Metzner, A. B. In: *Handbook of Fluid Dynamics,* Section 7, V. L. Streeter, Ed., (New York: McGraw-Hill Book Co., 1961).
19. Chavan, V. V., M. Arumugam and J. Ulbrecht. *AIChE J.* 21: 613 (1975).
20. Ford, D. E., and J. Ulbrecht. *AIChE J.* 21: 1230 (1975).
21. Metzner, A. B., J. L. White and M. M. Denn. *AIChE J.* 12: 863 (1966).
22. Chavan, V. V., D. E. Ford and M. Arumugam. *Can. J. Chem. Eng.* 53: 628 (1975).
23. White, J. L., and A. B. Metzner. *J. Appl. Polymer Sci.* 7: 1867 (1963).
24. Rushton, J. H., E. W. Costich and H. J. Everett. *Chem. Eng. Prog.* 46: 395, 467 (1950).
25. Metzner, A. B., and R. E. Otto. *AIChE J.* 3: 3 (1957).
26. Magnusson, K. *IVA* 23 (2): 86-99 (1952).
27. Metzner, A. B., and J. S. Taylor. *AIChE J.* 6: 109 (1960).
28. Calderbank, P. H., and M. B. Moo-Young. *Trans. Inst. Chem. Eng.* (*London*) 39: 337-47 (1961).
29. Corrigenda, to Ref. 28 *Trans. Inst. Chem. Eng.* (*London*) 40: facing p. 1 (1962). (Further correction appears in Equation 9.61 of Ref. 1.)
30. Patterson, W. I., P. J. Carreau and C. Y. Yap. *AIChE J.* 508 (1979).
31. Parker, N. H. *Chem. Eng.* 71 (12): 165 (1964).
32. Gray, J. B. *Chem. Eng. Prog.* 59 (3): 55 (1963).
33. Johnson, R. T. *Ind. Eng. Chem. Process Des. Dev.* 6: 340 (1967).
34. Coyle, C. K., H. E. Hirschland, B. J. Michel and J. Y. Oldshue. *AIChE J.* 16:903 (1970).
35. Bourne, J. R., and H. Butler. *The Chem. Engr.* (*London*) 180: CE202 (1964).
36. Chavan, V. V., and J. Ulbrecht. *Ind. Eng. Chem. Process Des. Dev.* 12: 472 (1973).
37. Corrigenda to Ref. 36, *I/EC Process Des. Dev.* 13:309 (1974).
38. Yap, C. Y., W. I. Patterson and P. J. Carreau. *AIChE J.* 25: 516 (1979).
39. Carreau, P. J. *Trans. Soc. Rheol.* 16 (1): 99 (1972).
40. Ulbrecht, J. Private communication (August 28, 1981).
41. Chavan, V. V., and J. Ulbrecht. *Trans. Inst. Chem. Eng.* (*London*) 51: 349 (1973).
42. Schümmer, P. *Chem.-Ing.-Tech.* 41: 1156 (1969).
43. Kelkar, J. V., R. A. Mashelkar and J. Ulbrecht. *Trans. Inst. Chem. Eng.* (*London*) 50: 343 (1972).
44. Kelkar, J. V., R. A. Mashelkar and J. Ulbrecht. *Chem. Eng. Sci.* 28: 664 (1973).
45. Kale, D. D., R. A. Mashelkar and J. Ulbrecht. *Nature* (*London, Phys. Sci.*) 242 (115): 29 (1973).
46. Kale, D. D., R. A. Mashelkar and J. Ulbrecht. *Chem.-Ing.-Tech.* 46: 69 (1974).
47. Gray, J. B. *Mixing,* Vol. I, V. W. Uhl and J. B. Gray, Eds. (New York: Academic Press, Inc., 1966), Chapt. 4.
48. Holmes, D. B., R. M. Voncken and J. A. Dekker. *Chem. Eng. Sci.* 19: 201 (1964).
49. McManamey, W. J. *Trans. Inst. Chem. Eng.* (*London*) 58: 271 (1980).
50. Johnson, R. T. *Ind. Eng. Chem. Process Des. Dev.* 6: 340 (1967).
51. Chavan, V. V., and J. Ulbrecht. *Chem. Eng. J.* 6: 213-223 (1973).
52. Ford, D. E., and J. Ulbrecht. *AIChE J.* 21: 1230 (1975).
53. Kelkar, J. V., R. A. Mashelkar and J. Ulbrecht. *Trans. Inst. Chem. Eng.* (*London*) 50: 343 (1972).
54. Skelland, A. H. P., and G. R. Dimmick. *Ind. Eng. Chem. Process Des. Dev.* 8: 267 (1969).
55. Hagedorn, D., and J. J. Salamone. *Ind. Eng. Chem. Process Des. Dev.* 6: 469 (1967).
56. Baker, C. K., and G. H. Walter. *Heat Transfer Eng.* 1 (2): 28 (1979).
57. Skelland, A. H. P. *Chem. Eng. Sci.* 7: 166 (1958).
58. Huggins, R. E. *Ind. Eng. Chem.* 23: 749 (1931).
59. Crank, J. *The Mathematics of Diffusion* (Oxford, England: Clarenden Press, 1956).
60. Houlton, H. G. *Ind. Eng. Chem.* 36: 522 (1944).
61. Harriott, P. *Chem. Eng. Prog. Symp. Ser.* 55 (29): 137 (1959).
62. Skelland, A. H. P., D. R. Oliver and S. Tooke. *Brit. Chem. Eng.* 7: 346 (1962).
63. Bott, T. R., and S. Azoory. *Chem. and Process Eng.* 50 (1): 85-90 (1969).

64. Ramdas, V., V. W. Uhl, M. W. Osborne and J. R. Ortt. *Heat Transfer Eng.* 1 (4): 38-46 (1980).
65. Mizushina, J., et al. *Kagaku Kogaku* 31 (12): 1208 (1967).
66. Prest, W. M., R. S. Porter and J. M. O'Reilly. *J. Appl. Polymer Sci.* 14: 2697 (1970).
67. Perez, J. F., and O. C. Sandall. *AIChE J.* 20: 770 (1974).
68. Skelland, A. H. P., and R. Seksaria. *Ind. Eng. Chem. Process Des. Dev.* 17: 56 (1978).
69. Skelland, A. H. P., and J. M. Lee. *Ind. Eng. Chem. Process Des. Dev.* 17: 473 (1978).
70. Treybal, R. E. "Liquid Extraction," 2nd ed. (New York: McGraw-Hill Book Co., 1963), p. 415.
71. Harriott, P. *AIChE J.* 8: 93 (1962).
72. Calderbank, P. H. In: *Mixing,* Vol. II, V. W. Uhl and J. B. Gray, Eds. (New York: Academic Press, Inc., 1967).
73. Glen, J. B. "Mass Transfer in Disperse Systems," Ph.D. dissertation, University of Canterbury, Christchurch, New Zealand (1965).
74. Skelland, A. H. P., and J. M. Lee. *AIChE J.* 27: 99 (1981); also 28(6): 1043.
75. Yagi, H., and F. Yoshida. Ind. Eng. Chem. Process Des. Dev. 14:488 (1975).

CHAPTER 8
COMPRESSIBLE FLOWS

Stuart W. Churchill
The Carl V. S. Patterson Professor of Chemical Engineering
The University of Pennsylvania
Philadelphia, Pennsylvania 19174

CONTENTS

INTRODUCTION

The postulate of constant density greatly reduces the equations governing fluid motion and thereby simplifies their solution. With minor exceptions, this assumption is justified for liquids. For most applications involving gases (as illustrated in other chapters of this book) the use of a constant, mean value for the density provides a reasonable and practical approximation.

However, in those processes that involve gases at high velocities or induce large pressure ranges, compressibility may be a significant, or even a controlling, factor. Examples are depressuring operations, pressure relief systems, mechanical compressions, shock compressions, detonations, thermocouple measurements, pitot-tube measurements, flow through pipelines and flow through porous media. Such flows will be examined in this chapter. One example is given of a significant effect of compressibility in flowing liquids—hammer due to the sudden closing of a valve.

Particular attention is given to the criteria for significant effects of compressibility. For example, in high-velocity flows the effect of compressibility is characterized by the ratio of the velocity to the acoustic velocity, i.e., by the Mach number, whereas in flow through porous media the Reynolds number may be a sufficient criterion.

High velocities generally imply large Reynolds numbers. As the Reynolds number increases, the dependence on the viscosity decreases. The postulate of negligible viscous effects also greatly simplifies the equations of motion. The consequent solutions provide useful approximations for fluid-mechanical behavior at large velocities.

Attention in this brief treatment is restricted to steady, quasi-one-dimensional flows. Further details of the derivations outlined herein are provided by Churchill [1]. Use of books on gas dynamics, such as that by Emmons [2], is suggested for treatments of two-dimensionality and other such complexities.

STEADY, REVERSIBLE (INVISCID), QUASI-ONE-DIMENSIONAL FLOW

If viscous effects are neglected, most internal flows and many external flows can be approximated as quasi-one-dimensional in the local direction of flow. In steady flow, the path of a particle of fluid follows a *streamline*. The streamlines passing through a closed curve form a *stream tube*, as illustrated in Figure 1. This stream tube represents the boundaries of a filament of fluid during changes in direction and velocity.

According to Newton's second law, the rate of change of momentum is equal to the net sum of the applied forces. For the filament of fluid in Figure 1, the applied forces in the direction of flow are the pressure times the cross-sectional area, and the component of gravity in the direction of flow times the mass of fluid. The momentum equals the velocity times the mass rate of flow. The resulting balance on a differential length of fluid is

$$-AdP - g\left(\frac{dz}{ds}\right)Ads = wdu \tag{1}$$

where
A = cross-sectional area of filament (m^2)
P = mean pressure across A (Pa)
g = acceleration due to gravity (m/s^2)
z = elevation (upward distance) (m)
s = distance along filament (m)
w = mass rate of flow in stream tube (kg/s)
u = mean velocity across A (m/s)

Within the stream tube the mass rate of flow is constant and

$$w = u\rho A \tag{2}$$

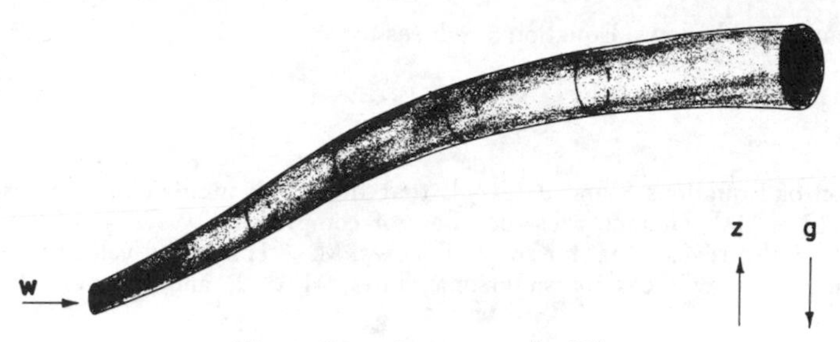

Figure 1. A stream tube [1].

where here, ρ = density, kg/m^3. Replacing w in Equation 1 by uρA, cancelling out ds, dividing through by ρ and rearranging, results in

$$udu + gdz + \frac{dP}{\rho} = 0 \tag{3}$$

Equation 3 is the differential form of *Bernoulli's equation* for compressible flow. Although it was obtained above from a force-momentum balance, Equation 3 can also be interpreted as an energy balance since energy and momentum are not independent for reversible flows.

For all gas flows in which the effect of compressibility is significant, the gravitational term in Equation 3 is negligible. Hence, the reduced expression

$$udu + \frac{dP}{\rho} = 0 \tag{4}$$

is used hereafter.

Criterion for Significant Effect of Compressibility

Differentiating Equation 2 and dividing through by uρA yields

$$\frac{du}{u} + \frac{d\rho}{\rho} + \frac{dA}{A} = 0 \tag{5}$$

which can be rewritten as

$$\frac{du}{u} + \frac{d\rho}{dP}\left(\frac{dP}{\rho}\right) + \frac{dA}{A} = 0 \tag{6}$$

The *acoustic (sonic) velocity* for any material is

$$a = \left(\frac{dP}{d\rho}\right)^{1/2} \tag{7}$$

Substituting for dP/ρ from Equation 4 and for dP/dρ from Equation 7 permits re-expression of Equation 6 as

$$\frac{du}{u}\left(1 - M^2\right) = -\frac{dA}{A} \tag{8}$$

where $M = u/a =$ *Mach number*

For incompressible flows, Equation 5 reduces to

$$\frac{du}{u} = -\frac{dA}{A} \tag{9}$$

Comparison of Equations 8 and 9 reveals that the Mach number or, more specifically, the factor $(1 - M^2)$, characterizes the effect of compressibility.

Equation 8 also reveals that for subsonic flows, $M < 1$, and the velocity decreases as the area increases, whereas for supersonic flows, $M > 1$, and the velocity increases.

Integral Form

Equation 4 can be integrated formally from any location 1 to any location 2 to obtain

$$\frac{u_2^2 - u_1^2}{2} = \int_{P_1}^{P_2} \frac{dP}{\rho} \tag{10}$$

Equation 10 is an integral form of Bernoulli's equation for horizontal, compressible flow.

u_2 can be eliminated from Equation 10 per Equation 2 to obtain

$$u_1^2 = \frac{2\int_{P_2}^{P_1} \frac{dP}{\rho}}{\left(\frac{\rho_1 A_1}{\rho_2 A_2}\right)^2 - 1} \tag{11}$$

Also,

$$u_2^2 = \frac{2\int_{P_2}^{P_1} \frac{dP}{\rho}}{1 - \left(\frac{\rho_2 A_2}{\rho_1 A_1}\right)^2} \tag{12}$$

and

$$w^2 = \frac{2(\rho_2 A_2)^2 \int_{P_2}^{P_1} \frac{dP}{\rho}}{1 - (\rho_2 A_2/\rho_1 A_1)^2} \tag{13}$$

For many expansions $(\rho_2 A_2/\rho_1 A_1)^2 \ll 1$, permitting the corresponding simplification of Equations 11–13.

Equations 11–13 provide a basis for the prediction and correlation of flow through venturi tubes, nozzles and orifices. It is evident that a relationship between P and ρ is required to evaluate the integral.

EQUATIONS OF STATE FOR GASES

In general,

$$\rho = \phi\{P, T, y_i\} \tag{14}$$

where T = absolute temperature (K)
y_i = mole fraction of components of gas

For an ideal gas,

$$\rho = Pm/RT \tag{15}$$

where m = molar mass (kg/mol)
R = universal gas law constant = 8.3143 J/mol-K

Nonideal behavior generally can be represented by

$$\rho = Pm/RTZ \tag{16}$$

where Z is called the compressibility factor.

The compressibility factor for most gases has been correlated successfully in terms of the reduced pressure and temperature. Many other equations of state of varying complexity and range of applicability have been developed. An alternative approach is through the fugacity, defined by the expression

$$\frac{dP}{\rho} = \frac{RT}{m} d\,ln\,(f) \tag{17}$$

The fugacity coefficient $\eta = f/P$ also has been correlated in terms of the reduced pressure and temperature.

Correlations for Z and η, as well as a discussion of various equations of state, can be found in most books on thermodynamics (see, for example, that by Balzhiser and Samuels [3]), and hence will not be reproduced here. Subsequent derivations will be for an ideal gas only. However, the equivalent developments can be carried out, at least numerically, using Z, η or other equations of state.

INTEGRAL OF EXPANSION

In general, some restraint (such as constant temperature) is necessary to permit evaluation of the integral in Equations 10–13, even when the thermodynamic behavior of the fluid is known. Attention here will be limited to three special cases, but one or the other of these is generally an acceptable approximation in applications.

Mean Value

In general,

$$\int_{P_2}^{P_1} \frac{dP}{\rho} = \frac{P_1 - P_2}{\rho_m} \tag{18}$$

where ρ_m is some mean value. If the variation of ρ is slight, the error in choosing a mean value cannot be great. If ρ varies monotonically with P, which is the usual case, ρ_m lies between ρ_1 and ρ_2, and the fractional error in using the arithmetic average cannot be greater than $(\rho_1 - \rho_2)/(\rho_1 + \rho_2)$.

Isothermal Processes

For the isothermal expansion or compression of an ideal gas,

$$\int_{P_2}^{P_1} \frac{dP}{\rho} = \int_{P_2}^{P_1} \frac{RT}{Pm} dP = \frac{RT}{m} ln\left(\frac{P_1}{P_2}\right) \tag{19}$$

$$= \frac{RT(P_1 - P_2)}{m}\left(\frac{ln\,(P_1/P_2)}{P_1 - P_2}\right) = \frac{P_1 - P_2}{\rho_1 - \rho_2}\,ln\left(\frac{\rho_1}{\rho_2}\right) = \frac{P_1 - P_2}{\rho_{lm}} \tag{20}$$

where ρ_{lm} = log-mean value of ρ.

Comparison of Equations 18 and 20 indicates that the correct mean density for isothermal flow of an ideal gas is the log-mean.

For steady flow, the first law of thermodynamics can be written in the integral form:

$$q - \tilde{w} = \Delta H + \frac{\Delta u^2}{2} + g\Delta z \tag{21}$$

where q = heat entering fluid stream from surroundings (J/kg)
$\tilde{w}$ = work done by fluid stream on surroundings (J/kg)
H = specific enthalpy of fluid (J/kg)

For free flow, $\tilde{w} = 0$, and for an isothermal expansion of an ideal gas, $\Delta H = 0$ (see, for example, Churchill [1] or Balzhiser and Samuels [3] for a proof). Hence, if the change in potential energy is neglected, the heat required to maintain a constant temperature is

$$q_T = \frac{u_2{}^2 - u_1{}^2}{2} \tag{22}$$

From Equations 10 and 19 it follows that

$$q_T = \frac{RT}{m} \ln(P_1/P_2) \tag{23}$$

Also, as [3]

$$dH = TdS + \frac{dP}{\rho} \tag{24}$$

it follows that

$$q_T = T(S_2 - S_1) \tag{25}$$

Here, S = specific entropy of fluid, J/kg-K.
Thus, the required heat can be seen to be equal to the increase in the kinetic energy, also to $(RT/m)\ln(P_1/P_2)$ and $T(S_2 - S_1)$.

Adiabatic Processes

For free adiabatic reversible flow, $\tilde{w}$, q and ΔS are zero. For an ideal gas with a constant heat capacity, it can be shown (see, for example, Churchill [1] or Balzhiser and Samuels [3] that

$$\frac{\rho}{\rho_1} = \left(\frac{T}{T_1}\right)^{\frac{1}{\gamma-1}} = \left(\frac{P}{P_1}\right)^{\frac{1}{\gamma}} \tag{26}$$

where $\gamma = C_p/C_v$ = heat capacity ratio
C_p = specific heat capacity at constant pressure (J/kg-K)
C_v = specific heat capacity at constant volume (J/kg-K)

Then,

$$\int_{P_2}^{P_1} \frac{dP}{\rho} = \int_{P_2}^{P_1} \left(\frac{P_1}{P}\right)^{1/\gamma} \frac{dP}{\rho_1} = \frac{\gamma P_1{}^{1/\gamma}}{(\gamma-1)\rho_1}\left(P_1{}^{\frac{\gamma-1}{\gamma}} - P_2{}^{\frac{\gamma-1}{\gamma}}\right) \tag{27}$$

$$= \left(\frac{\gamma}{\gamma - 1}\right)\frac{P_1}{\rho_1}\left[1 - \left(\frac{P_2}{P_1}\right)^{\frac{\gamma-1}{\gamma}}\right] \tag{28}$$

$$= \frac{\gamma}{\gamma - 1}\left(\frac{P_1}{\rho_1} - \frac{P_2}{\rho_2}\right) \tag{29}$$

$$= C_p(T_1 - T_2) \tag{30}$$

The latter expression follows from Equation 15, and

$$(C_p - C_v)m = R \tag{31}$$

is derived in most books on thermodynamics, such as in Balzhiser and Samuels [3].

For an adiabatic process without work and with a negligible change in potential energy, Equation 21 reduces to

$$\frac{u_2^2 - u_1^2}{2} = H_1 - H_2 \tag{32}$$

From Equations 10 and 30, or directly from Equation 32, it follows that

$$\frac{u_2^2 - u_1^2}{2} = C_p(T_1 - T_2) \tag{33}$$

Thus, in an adiabatic reversible (isentropic) expansion of an ideal gas, the temperature decreases in direct proportion to the increase in the kinetic energy.

The heat capacity ratio, γ, is relatively invariant with pressure and varies only moderately with temperature, as illustrated in Table I for air and methane. Values at ambient conditions are given in Table II for a number of common gases.

Nonadiabatic, Nonisothermal Processes

Most high-velocity processes are essentially adiabatic. The effect of heat exchange with the surroundings is to shift the behavior toward that of an isothermal process. These two idealized conditions therefore generally indicate the limiting behavior of real processes. It may be noted that letting $\gamma = 1$ (at the appropriate point to avoid a

Table I. Variation of γ of Air and Methane with Temperature

T, °C	0	100	200	300	400	500	600	700	800	900	1000	2000	3000
Air	1.410	1.397	1.398	1.378	1.367	1.356	1.347	1.338	1.331	1.325	1.319	1.243	1.283
Methane	1.318	1.274	1.228	1.196	1.174	1.158	1.145	1.136	1.129	1.122	1.117		

Table II. Approximate, Representative Values of γ at 1 atm and 298°K

Gas	γ
C_2H_6, C_2H_4, C_2H_2	~ 1.2
CH_4, H_2O[a], H_2S, NH_3, Cl_2, CO_2	~ 1.3
Air, H_2, N_2, O_2, CO, NO and HCl	~ 1.4
Monatomics	~ 1.67

[a]At saturation pressure.

singularity) in the above expressions for adiabatic reversible processes yields the corresponding expressions for isothermal behavior. Hence, a possible approximation for intermediate behavior is to replace γ with an empirical value, k, intermediate between 1 and 8.

FLOW IN NOZZLES

Combining Equation 28 with Equation 10 for negligible u_1 (or with Equation 12 for large A_1) gives the following limiting expression for adiabatic reversible flow from a large vessel:

$$u^2 = \left(\frac{2\gamma}{\gamma - 1}\right)\frac{P_1}{\rho_1}\left[1 - \left(\frac{P}{P_1}\right)^{\frac{\gamma-1}{\gamma}}\right] \tag{34}$$

The subscript 2 has been dropped to emphasize that this expression gives the velocity for any pressure. The corresponding expression for the mass rate of flow is

$$w^2 = \left(\frac{2\gamma}{\gamma - 1}\right)A^2P_1\rho_1\left[1 - \left(\frac{P}{P_1}\right)^{\frac{\gamma-1}{\gamma}}\right]\left(\frac{P}{P_1}\right)^{2/\gamma} \tag{35}$$

Since the mass rate of flow is invariant in a nozzle, Equation 35 provides a relationship for the variation of the pressure with the cross-sectional area. Equation 35 indicates that the mass flux density, $G = w/A = u\rho$, approaches zero in the limits of $P = 0$ and $P = P_1$. Hence, a maximum possible mass flux density exists at some intermediate pressure. Equating the derivative dG/dP to zero gives the following *critical (or throat) pressure* corresponding to this maximum value of G:

$$\left(\frac{P_c}{P_1}\right) = \left(\frac{2}{\gamma + 1}\right)^{\frac{\gamma}{\gamma-1}} \tag{36}$$

It follows that

$$\frac{T_c}{T_1} = \left(\frac{\rho_c}{\rho_1}\right)^{\gamma-1} = \frac{2}{\gamma + 1} \tag{37}$$

$$G_c^2 = \gamma P_1\rho_1\left(\frac{2}{\gamma + 1}\right)^{\frac{\gamma+1}{\gamma-1}} \tag{38}$$

and

$$u_c^2 = \frac{2\gamma P_1}{(\gamma + 1)\rho_1} = \frac{\gamma P_c}{\rho_c} = \left(\frac{dP}{d\rho}\right)_{cs} \tag{39}$$

The latter expression indicates that the velocity corresponding to the critical conditions is sonic.

For air and other diatomic gases (with $\gamma \cong 1.4$),

$$\frac{P_c}{P_1} \cong \left(\frac{2}{1.4 + 1}\right)^{\frac{1.4}{1.4-1}} = 0.528 \tag{40}$$

Converging Nozzles

If the external pressure, P_∞, is greater than P_c, as given by Equation 36, substitution of the exit area in Equation 35 gives the rate of flow. For any external pressure, P_∞

less than P_c, the rate of flow is given by Equation 38 (which is equivalent to Equation 35 in terms of P_c) and the throat area. The gas then expands irreversibly outside the nozzle from P_c to P_∞.

Converging-Diverging Nozzles

Equation 35 has two real positive roots, i.e., two values of P for each A and w. The greater root corresponds to subsonic reexpansion in the diverging section, as indicated by curve 3 in Figure 2. The lesser root corresponds to supersonic expansion in the diverging section, as indicated by curve 5 in Figure 2. If the external pressure is less than P_2, the flow is obtained from Equation 35 in terms of P_∞ and the exit area. If the external pressure is decreased below P_5, as indicated by P_6 in Figure 2, the flow within the nozzle will be unchanged, and an irreversible expansion from P_5 to P_∞ will occur outside the nozzle. This is called *underexpansion.* If the external pressure is between P_2 and P_5, as illustrated by P_4, completely reversible flow cannot be attained. An irreversible, shock recompression will occur within, or just outside, the nozzle. This undesirable condition is called *overexpansion.* The same mass flux density, as given by Equation 38 and the area of the throat, is obtained for all external pressures less than P_2.

Although $P_c/P_1 \cong 0.5$, it should be noted that critical flow can be attained even with a ratio of external pressures, P_2/P_1, approaching unity. The minimum external pressure, P_2, which will produce sonic flow depends on the ratio of the area of the exit to the area of the throat, according to the expression

$$\left(\frac{A_t}{A_e}\right)^2 = \frac{2}{\gamma - 1}\left(\frac{\gamma + 1}{2}\right)^{\frac{\gamma+1}{\gamma-1}}\left[1 - \left(\frac{P_2}{P_1}\right)^{\frac{\gamma-1}{\gamma}}\left(\frac{P_2}{P_1}\right)^{\gamma/2}\right. \tag{41}$$

Real Nozzles

The shape of the nozzle, i.e., the relationship between area and length, is not determined by the above expression. The change in the area must be gradual enough to prevent separation. The converging section may be quite steep, but the total internal angle of the diverging section should not exceed 0.122 rad (7°). Deviations from ideality due to viscosity, two-dimensionality and separation are discussed by Emmons [2]. Such deviations are sometimes correlated for a particular nozzle by multiplying the right side of Equations 35 and 38 by C_n^2 and then correlating the *nozzle coefficient,* C_n, with the Reynolds number at the throat. Such a correlation is illustrated in Figure 3 for an ASME long-radius flow nozzle [4], with a converging section of length D_2 and a straight following section of length $0.6D_2$, where D_2 is the throat diameter.

Venturi Tubes

A Venturi tube is a converging-diverging nozzle used to measure the rate of flow. Ordinarily, P_1 is measured upstream and P_2 at the throat. Equation 12 is applicable, and the value of ρ_2 from Equation 26 should be used in the integral and in the denominator. However, a mean value of ρ is usually used as an approximation. The correction coefficient for Venturi tubes ranges from 0.9 to 0.99 and approaches the latter value for $Re_1 > 100{,}000$.

Orifices

The fluid stream through the hole in an orifice plate necks down to a lesser diameter, called the *vena contracta,* somewhat downstream from the plate, and then reexpands.

The flow is quite reversible down to this minimum jet diameter; hence, the above equations for isentropic expansions are applicable. However, the diameter of the orifice plate usually is used in the correlating equation, rather than the actual diameter

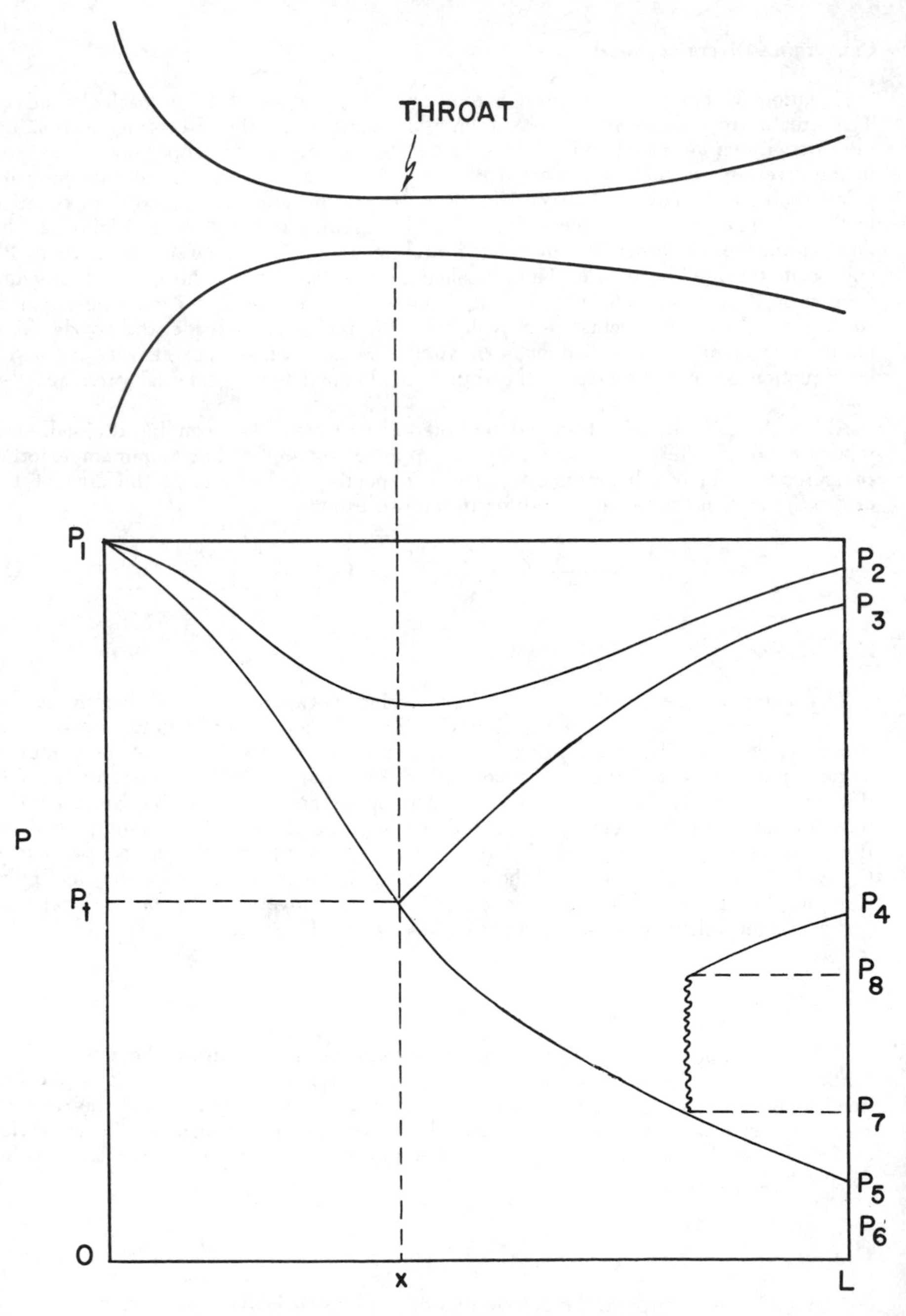

Figure 2. Pressure profiles in a nozzle [1].

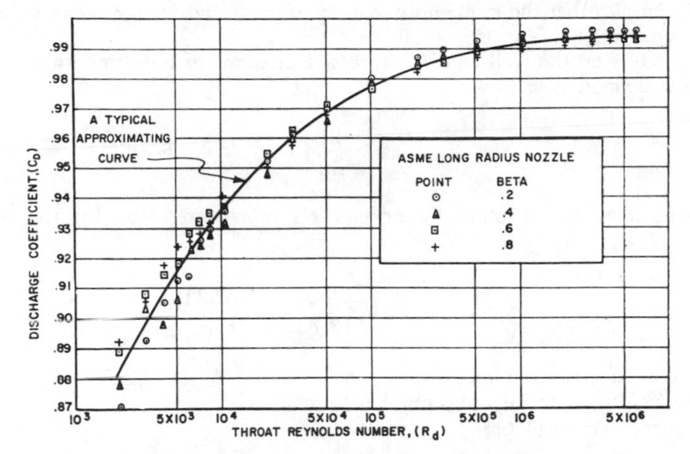

Figure 3. Discharge coefficient for ASME long-radius flow-nozzle [4].

at the downstream point of pressure measurement; hence, the coefficient of correction for the orifice differs significantly from unity. For subsonic flow, a mean density, rather than that given by Equation 26, usually is used as an approximation. The orifice coefficient for subsonic flow approaches the theoretical value of $\pi/(\pi + 2) \cong 0.611$ for the *vena contracta* for larger Re (see, for example, the book by Churchill [1]). For an orifice rounded off on the upstream side, Grace and Lapple [5] recommend a coefficient of 0.91, and for rounding off on the downstream side, a coefficient of 0.82–0.86.

Depressuring

Equations 35 and 38 are applicable for transient depressuring of a vessel, with P_1 determined from a material balance on the remaining gas.

Propulsion

Equation 34 is applicable to jet and rocket engines (see, for example, the books by Churchill [1] and Balzhiser and Samuels [2].

FLOW IN PIPES

Irreversibilities cannot be neglected in flow through a channel of constant cross section. A force-momentum balance taking into account such irreversibilities can be written for a differential length, dx, as follows:

$$\tau p dx + A dP + w du = 0 \tag{42}$$

where x = distance down channel (m)
τ = mean shear stress over perimeter (Pa)
p = perimeter of channel (m)
A = cross-sectional area of channel (m^2)
P = pressure (Pa)
w = mass rate of flow (kg/s)
u = mean velocity (m/s)

Equation 42 implies that the momentum can be represented by the mean velocity, and that the net gravitational force is negligible.

The shear stress on the wall can be expressed in terms of the dimensionless *Fanning friction factor,* defined as

$$f \equiv \frac{2\tau}{\rho u^2} \tag{43}$$

For turbulent flow, this friction factor can be represented [6] by the correlating equation

$$\frac{1}{\sqrt{f}} = -1.74\, ln\left\{\frac{e}{3.7\, D} + \frac{1.252}{Re f^{1/2}}\right\} \tag{44}$$

where e = effective roughness (m)
$Re = Du\rho/\mu$ = Reynolds number for channel
μ = viscosity (Pa-s)

As an approximation for nonround channels,

$$D = 4A/p \tag{45}$$

In a channel of constant cross section, the mass flux density, G, is invariant and, hence, a more convenient variable than the linear velocity, u. Substituting in Equation 42 for τ from Equation 43 and for u from Equation 46 gives, on rearrangement,

$$f dx = -\frac{D\rho dP}{2G^2} + \frac{D}{2}\frac{d\rho}{\rho} \tag{46}$$

Equation 33 holds for irreversible, as well as reversible, adiabatic flows. Dropping the subscript 2, substituting G/ρ for u, for C_p from Equation 31 and for ρ from the ideal gas law (Equation 15) gives, on rearrangement,

$$\frac{P}{P_1} = \frac{\rho}{\rho_1}\left[1 - \left(\frac{\gamma-1}{2\gamma}\right)\frac{G^2}{P_1\rho_1}\left(\left(\frac{\rho_1}{\rho}\right)^2 - 1\right)\right] \tag{47}$$

Differentiating Equation 47 with respect to ρ and substituting for dP in Equation 46 gives

$$\frac{4f}{D}dx = \left(\frac{\gamma+1}{\gamma}\right)\frac{d\rho}{\rho} - 2\left(\frac{\gamma-1}{2\gamma} + \frac{P_1\rho_1}{\gamma^2}\right)\left(\frac{\rho}{\rho_1}\right)d\left(\frac{\rho}{\rho_1}\right) \tag{48}$$

Finally, assuming constant f and integrating from ρ_1 at $z = 0$ to ρ_2 at $x = L$, gives

$$\frac{4fL}{D} = \frac{\gamma+1}{\gamma} ln\left(\frac{\rho_2}{\rho_1}\right) + \left[1 - \left(\frac{\rho_2}{\rho_1}\right)^2\right]\left(\frac{\gamma-1}{2\gamma} + \frac{P_1\rho_1}{G^2}\right) \tag{49}$$

It is apparent from Equation 49 that $G \to 0$ as $\rho_2 \to \rho_1$ and, also, as $\rho_2 \to 0$. Hence, a maximum value exists for some intermediate, ρ_2, and can be found by equating $dG/d\rho_2$ to zero. If f again is assumed to change negligibly with G, this procedure gives

$$\frac{P_1\rho_1}{G_c^2} = \frac{\gamma+1}{2\gamma}\left(\frac{\rho_1}{\rho_{2c}}\right)^2 - \frac{\gamma-1}{2\gamma} \tag{50}$$

Elimination of P_1 between Equations 47 and 50 reveals that

$$G_{2c}^2 = \gamma P_{2c}\rho_{2c} \tag{51}$$

hence, that

$$u_{2c}^2 = \gamma P_{2c}/\rho_{2c} \tag{52}$$

Thus, despite the irreversibilities, the maximum flow in a channel occurs when the velocity at the exit becomes sonic; the irreversibilities simply increase the required external pressure ratio, P_1/P_2.

Eliminating ρ_2/ρ_1 between Equations 49 and 50 produces

$$\frac{P_1\rho_1}{G_c^2} = \frac{4fL}{D} + \frac{1}{\gamma} + \frac{\gamma+1}{2\gamma} ln\left\{\frac{\gamma-1}{\gamma+1} + \left(\frac{2\gamma}{\gamma+1}\right)\frac{P_1\rho_1}{G_c^2}\right\} \tag{53}$$

If the external pressure is less than P_{2c}, the rate of flow is given by Equation 53. ρ_1/ρ_{2c} then can be calculated from Equation 50 and P_{2c}/P_1 from Equation 52, thereby testing the postulate of $P_{2c} > P_\infty$. If the external pressure is greater than P_{2c}, the rate of flow can be calculated from Equation 49 for a given ρ_2/ρ_1, and the corresponding P_2/P_1 from Equation 47. Trial and error is necessary to determine the appropriate ρ_2/ρ_1 for the specified P_∞. In either case, an additional, but rapidly converging, trial-and-error procedure is necessary to evaluate f from Equation 44.

The above equations all can be reduced for the other limiting case of isothermal flow by setting $\gamma = 1$. Coulson and Richardson [7] assert that the adiabatic rate of flow exceeds that for isothermal flow by less than 20% for all practical cases, and by less than 5% for $L/D > 1000$.

SHOCK WAVES

Explosions, electrical discharges and the movement of bodies at high velocities create compression waves. Under some conditions these waves coalesce into a single strong wave traveling at supersonic velocity. Such a wave is called a shock wave and exhibits a near discontinuity in pressure density and temperature. Such waves are important because of their potential for damage, but also because of their use to produce high temperatures in the laboratory.

Here, the wave will be considered to create a finite discontinuity in pressure, temperature, density and velocity. Eddy and molecular transport are neglected. Ideal gas behavior and adiabatic conditions are postulated. The symbols used to describe the wave are shown in Figure 4. The equations of conservation are simpler if written with the velocities referred to the wave, as indicated in Figure 5. Mass and momentum force balances can be written as follows:

$$u_1'\rho_1 = u_2'\rho_2 \tag{54}$$

$$P_1 + u_1'^2\rho_1 = P_2 + u_2'^2\rho_2 \tag{55}$$

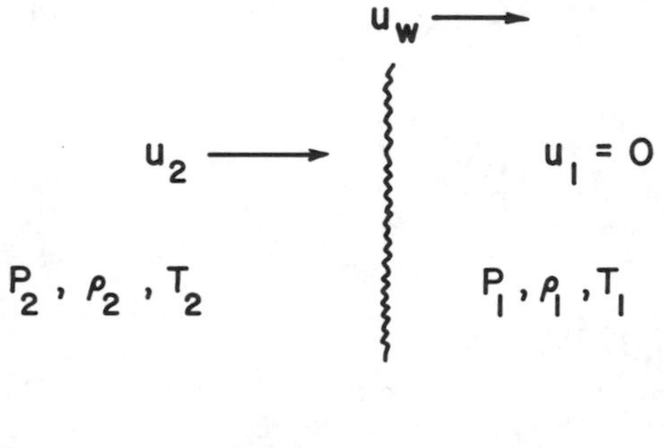

Figure 4. Notation for a shock wave moving through a stagnant gas [1].

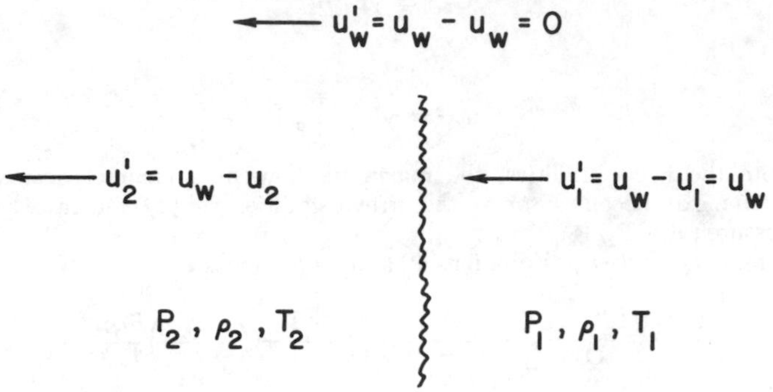

Figure 5. Stationary frame of reference for a shock wave [1].

Equation 33 is applicable as an energy balance. With Equations 15 and 31 it can be converted to

$$\frac{u_1^2 - u_2^2}{2} = \frac{\gamma}{\gamma - 1}\left(\frac{P_2}{\rho_2} - \frac{P_1}{\rho_1}\right) \tag{56}$$

By combining Equations 54 and 55,

$$u_1'\rho_1 = u_2'\rho_2 = \left(\frac{P_2 - P_1}{\frac{1}{\rho_1} - \frac{1}{\rho_2}}\right)^{1/2} \tag{57}$$

Eliminating u_1' and u_2' between Equations 54 and 57 then yields

$$\frac{P_2}{P_1} = \frac{\beta(\rho_2/\rho_1) - 1}{\beta - \rho_2/\rho_1} \tag{58}$$

It follows that

$$\frac{u_1'}{u_2'} = \frac{\rho_2}{\rho_1} = \frac{1 + \beta(P_2/P_1)}{\beta + P_2/P_1} \tag{59}$$

and

$$\frac{T_2}{T_1} = \frac{\beta + P_2/P_1}{\beta + P_1/P_2} \tag{60}$$

Also, that

$$M_1'^2 = \frac{\gamma - 1}{2\gamma}\left(1 + \beta\frac{P_2}{P_1}\right) \tag{61}$$

and

$$M_2'^2 = \frac{\gamma - 1}{2\gamma}\left(1 + \beta\frac{P_1}{P_2}\right) \tag{62}$$

where here,

$$M_1' \equiv u_1'\left(\frac{\gamma P_1}{\rho_1}\right)^{1/2} \tag{63}$$

$$M_2' \equiv u_2 \left(\frac{\gamma P_2}{\rho_2}\right)^{1/2} \tag{64}$$

From Equation 24,

$$ds = \frac{dH}{T} + \frac{dP}{\rho T} = \frac{C_p dT}{T} - \frac{R}{m}\frac{dP}{P} \tag{65}$$

Integrating for constant C_p gives

$$S_2 - S_1 = C_p ln\left(\frac{T_2}{T_1}\right) - (C_p - C_v) ln\left(\frac{P_2}{P_1}\right) \tag{66}$$

or, in terms of M_1',

$$\frac{S_2 - S_1}{C_p} = ln\left\{1 - \frac{2}{\gamma + 1}\left(1 - \frac{1}{M_1'^2}\right)\right\} + \frac{1}{\gamma} ln\left\{1 + \frac{2\gamma}{\gamma + 1}(M_1'^2 - 1)\right\} \tag{67}$$

It is apparent that for

$$\begin{array}{ll} M_1' > 1 & S_2 - S_1 > 0 \\ M_1' = 1 & S_2 - S_1 = 0 \\ M_1' < 1 & S_2 - S_1 < 0 \end{array}$$

Therefore, according to the second law of thermodynamics, M_1' must be greater than unity. It follows that a shock wave is necessarily supersonic with respect to the gas ahead, and that P_2/P_1, ρ_2/ρ_1 and T_2/T_1 are all greater than unity. Thus, a shock wave is necessarily a compression wave; a shock expansion is excluded.

Reflected Waves

Normal reflection off a solid body produces a further increase in pressure, temperature and density. As illustrated in Figure 6, the velocity behind the reflected wave is zero. The frame of reference of the reflected wave, as illustrated in Figure 7, produces a situation completely analogous to that of the initial wave in the standing frame of reference (as illustrated in Figure 5). Hence, Equations 54–60 are applicable, with u_2'', u_3'', ρ_2 and ρ_3 replacing u_1', u_2', ρ_1 and ρ_2, respectively, etc.

From Equation 57,

$$u_2 = u_1' - u_2' = (P_2 - P_1)^{1/2}\left(\frac{1}{\rho_1} - \frac{1}{\rho_2}\right)^{1/2} \tag{68}$$

It follows that

$$u_2 = u_2'' - u_3'' = (P_3 - P_1)^{1/2}\left(\frac{1}{\rho_2} - \frac{1}{\rho_3}\right)^{1/2} \tag{69}$$

Equating these two expressions for u_2 gives

$$\frac{P_3 - P_2}{P_2 - P_1} = \frac{\dfrac{\rho_2}{\rho_1} - 1}{1 - \rho_2/\rho_3} \tag{70}$$

Substituting for ρ_2/ρ_1 from Equation 59 and, correspondingly, for ρ_3/ρ_2, yields an expression that can be solved to obtain

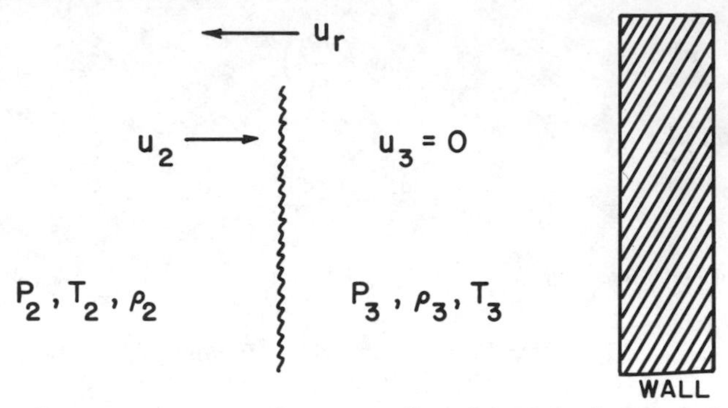

Figure 6. Notation for a normally reflected shock wave [1].

$$\frac{P_3}{P_2} = \frac{(\beta + 2)\frac{P_2}{P_1} - 1}{\frac{P_2}{P_1} + \beta} \tag{71}$$

Hence,

$$\frac{P_3}{P_1} = \frac{(\beta + 2)\frac{P_2}{P_1} - 1}{1 + \beta(P_1/P_2)} \tag{72}$$

The other conditions behind the reflected wave are readily obtained by substituting P_3/P_2 for P_2/P_1 in the expressions for the initial wave.

Generation of Shock Waves

A shock wave can be generated by a moving piston. In this case the velocity of the piston is represented by u_2. It follows from Equations 68 and 59 that

$$\frac{P_2}{P_1} = 1 + \frac{\gamma(\gamma + 1)}{4}\left(\frac{u_2}{a_1}\right)^2 \left\{1 + \left[1 + \left(\frac{4a_1}{(\gamma + 1)u_2}\right)^2\right]^{1/2}\right\} \tag{73}$$

where

$$a_1 = (\gamma P_1/\rho_1)^{1/2}$$

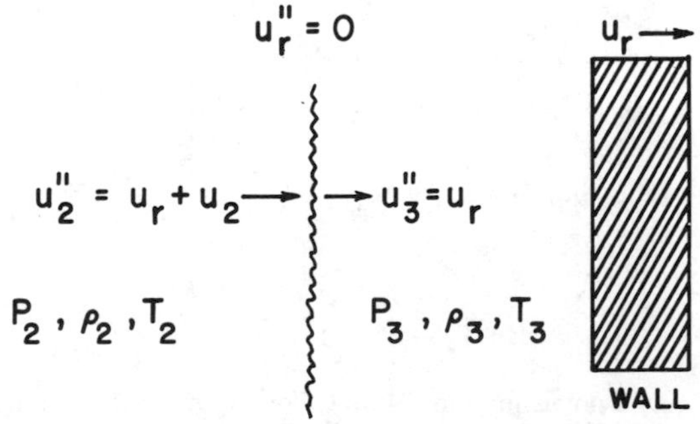

Figure 7. Stationary frame of reference for a normally reflected shock wave [1].

A shock wave can be generated in the laboratory by rupturing a diaphragm between the high- and low-pressure section of a *shock tube,* as illustrated in Figure 8. The generating equation is

$$\frac{P_0}{P_1} = \frac{P_2}{P_1}\left[1 - \left(\frac{P_2}{P_1} - 1\right)\left(\frac{\beta_1 - 1}{\beta_0 - 1}\right)\left(\frac{\gamma_1 T_1 m_0}{\gamma_0 T_0 m_1 (\beta_1 + 1)\left(1 + \beta_1 \frac{P_2}{P_1}\right)}\right)^{1/2}\right]^{-(1+\beta_0)} \quad (74)$$

where the subscript 0 indicates the conditions in the driver section of the shock tube. This expression is based on the assumption that an isentropic rarefaction wave propagates upstream.

Real Waves

Predictions of the wave velocities from the above expressions are in good agreement with experimental data. The predictions of temperature and pressure are in fair agreement on-the-mean with experimental values. The short-term deviations observed are presumed to be associated with the time of relaxation of the rapidly heated molecules of gas. Decomposition and ionization of the heated gas also may cause deviations, particularly for the reflected wave.

Additional experimental and theoretical considerations are presented by Emmons [2] and others.

DETONATION WAVES

A detonation wave is a shock wave driven by a chemical reaction. Detonation waves have greater destructive power than explosions (isentropic expansions). Hence, their behavior is of direct interest in processing and storing explosive materials [8,9].

Equations 32, 54 and 55 are applicable with

$$H_2 - H_1 = \Delta H_{R_1} + C_{P_2}(T_2 - T_1) \quad (75)$$

where ΔH_{R_1} = enthalpy change due to reaction at T_1 (J/kg)
C_{P_2} = mean heat capacity of reacted mixture between T_1 and T_2 (J/kg-K)

Eliminating u_1' and u_2' and assuming the ideal gas law then leads to

$$\frac{\Delta H_{R_1}}{C_{P_2} T_1} + \frac{T_2}{T_1} - 1 = \left(\frac{\gamma_2 - 1}{2\gamma_2}\right)\left(\frac{P_2}{P_1} - 1\right)\left(\frac{m_2}{m_1} + \frac{T_2 P_1}{T_1 P_2}\right) \quad (76)$$

where subscripts 1 and 2 designate the unreacted and reacted mixtures, respectively. From Equations 32, 54 and 55,

$$H_2 - H_1 = \left(\frac{P_2 - P_1}{2}\right)\left(\frac{1}{\rho_2} + \frac{1}{\rho_1}\right) \quad (77)$$

Differentiating Equation 77 for fixed P_1, ρ_1 and H_1 gives

$$dH_2 = \left(\frac{1}{\rho_1} + \frac{1}{\rho_2}\right)\frac{dP_2}{2} - \left(\frac{P_2 - P_1}{2}\right)\frac{d\rho_2}{\rho_2^2} \quad (78)$$

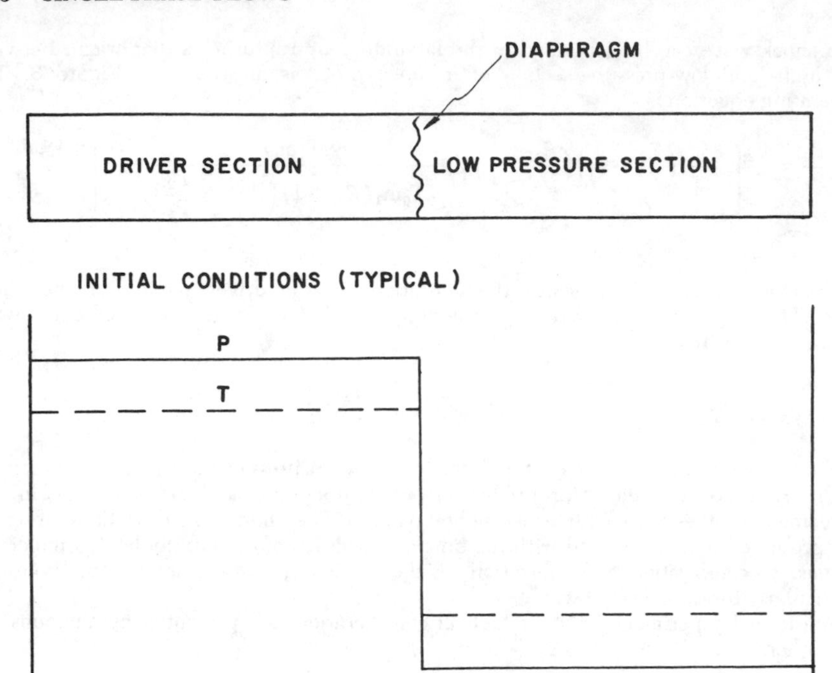

CONDITIONS AT t>0

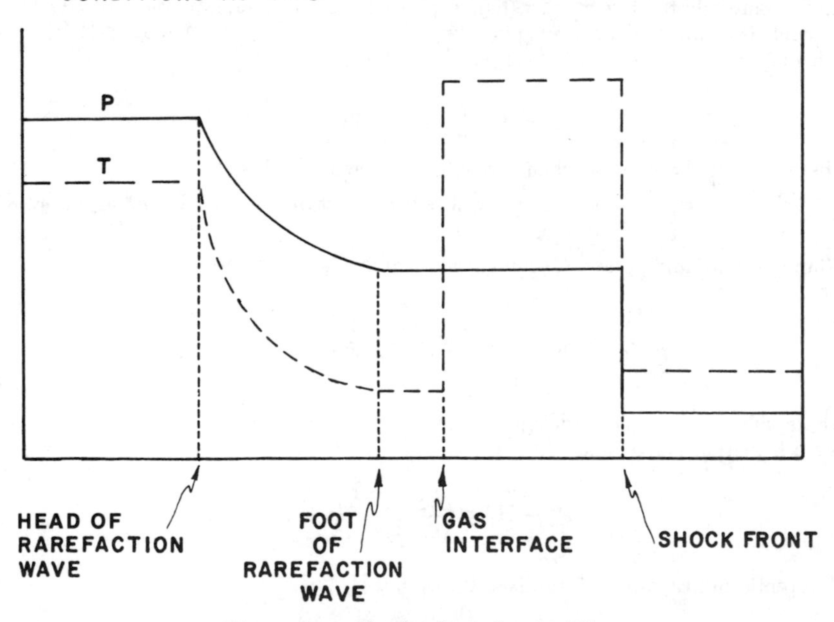

Figure 8. Shock tube behavior [1].

From Equation 24 for condition 2 with the same restrictions,

$$dH_2 = T_2 dS_2 + \frac{dP_2}{\rho_2} \tag{79}$$

Equating these two expressions for dH_2 and setting $dS_2 = 0$, thus giving a minimum (or maximum) in entropy, leads to

$$u_2' = \left(\frac{dP_2}{d\rho_2}\right)_S^{1/2} \tag{80}$$

This expression, which indicates that the burned gas moves at acoustic velocity relative to the wave, is known as the *Chapman-Jouguet condition.* It has been confirmed by countless experiments.

Substitution of

$$u_2' = (\gamma_2 P_2/\rho_2)^{1/2} \tag{81}$$

in Equation 57 and rearrangement leads to

$$\frac{T_1}{T_2} = \frac{m_1}{\gamma_2 m_2}\left(\gamma_2 + 1 - \frac{P_1}{P_2}\right)\frac{P_1}{P_2} \tag{82}$$

Equations 76 and 82 can be combined to obtain

$$\left(\frac{P_2}{P_1}\right)^2 + 2\left[\frac{m_1}{m_2}\left(\frac{\Delta H_{R_1}}{C_{P_2} T_1} - 1\right)\gamma_2 + \frac{\gamma_2 - 1}{\gamma_2 + 1}\left(\frac{3\gamma_2 + 1}{2}\right)\right]\frac{P_2}{P_1} - \frac{m_1}{m_2}\left(\frac{\Delta H_R}{C_{P_2} T_1} - 1\right)\frac{2\gamma_2}{\gamma_2 + 1} - \gamma_2\left(\frac{\gamma_2 - 1}{\gamma_2 + 1}\right) = 0 \tag{83}$$

The positive root of this quadratic equation gives P_2/P_1 as a function of m_2, ΔH_{R_1}, C_{P_2} and γ_2.

A possible method of solution with a computer is as follows:

1. Assuming complete reaction, compute P_2/P_1 from Equation 83 and T_2/T_1 from 82.
2. Calculate the equilibrium composition for this P_2 and T_2, and then m_2, ΔH_{R1}, C_{p2} and γ_2.
3. Using these values, recompute P_2/P_1 from Equation 83 and T_2/T_1 from 82.
4. Repeat steps 2 and 3 to convergence.
5. Calculate ρ_2, u_1', u_2', etc.

The behavior resulting from reflection of the detonation wave can be computed similarly. Further details concerning this procedure are given by Churchill [1] and others.

Experimental Confirmation

Detonation velocities computed from the above model are in almost exact agreement with experimental values, as illustrated in Figure 9 for hydrogen and oxygen. Computed pressures for detonation waves and their reflection are in reasonable agreement with

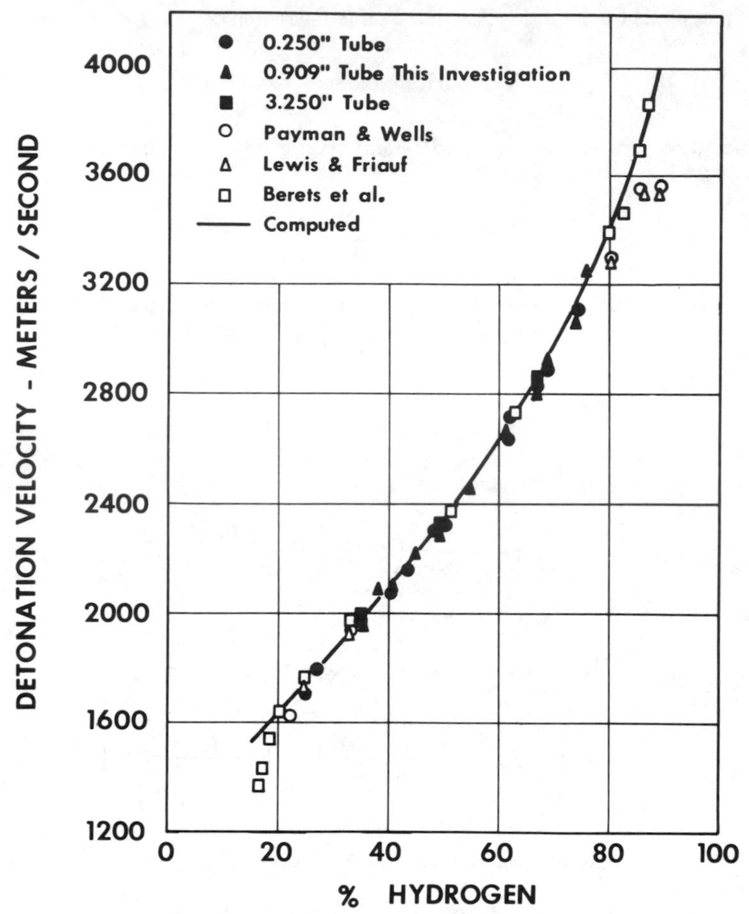

Figure 9. A comparison of experimental and computed detonation velocities for $H_2 - O_2$ mixtures [1].

experimental measurements for a wide range of conditions (see, for example, the book by Churchill [1]). The destruction of equipment has been found to be correlated with the reflected (impact) pressure.

Approximate Model for Detonation and Shock Waves

Making the assumption of a negligible change in the molecular weight and heat capacity ratio, Adamson and Morrison [10] derived the following approximate expression for the Mach number of a detonation wave:

$$M_1' = \left[\left(\frac{\gamma+1}{2}\right)\left(\frac{-\Delta H_R}{C_p T_1}\right)\right]^{1/2} + \left[\left(\frac{\gamma+1}{2}\right)\left(\frac{-\Delta H_R}{C_p T_1}\right) + 1\right]^{1/2} \tag{84}$$

Morrison [11] proposed that $(-\Delta H_R)$ in Equation 84 be estimated as two-thirds of the value for complete reaction, and that γ and m be evaluated at the initial conditions.

Adamson and Morrison [10] further generalized the corresponding solution for the other characteristics for both detonation and shock waves as

$$\frac{P_2}{P_1} = 1 + \frac{\gamma F}{\gamma + 1}(M_1'^2 - 1) \tag{85}$$

$$\frac{u_2'}{u_1'} = \frac{\rho_1}{\rho_2} = 1 - \frac{F(M_1'^2 - 1)}{(\gamma + 1)M_1'^2} \tag{86}$$

$$\frac{T_2}{T_1} = \left[1 + \frac{\gamma F}{\gamma + 1}\left(M_1'^2 - 1\right)\right]\left[1 - \frac{F(M_1'^2 - 1)}{(\gamma + 1)M_1'^2}\right] \tag{87}$$

and

$$M_2'^2 = \frac{(\gamma + 1 - F)(M_1'^2 - 1) + \gamma + 1}{F\gamma(M_1'^2 - 1) + \gamma + 1} \tag{88}$$

where F = 1 for detonation waves and 2 for shock waves.

If the further reaction on reflection is neglected, the following expression is applicable for P_3/P_2 for a detonation:

$$\frac{P_3}{P_2} = 1 + \frac{\left(\gamma + 1\right)\left(1 - \frac{P_1}{P_2}\right)^2}{4\gamma}\left[1 + \left(1 + \left[\frac{4\gamma}{\left(1 - \frac{P_1}{P_2}\right)\left(\gamma + 1\right)}\right]^2\right)^{1/2}\right] \tag{89}$$

Further details are given by Churchill [1], and complexities are examined by Emmons [2].

STAGNATION PROPERTIES

If a gas stream is slowed down by flowing over or impacting on a surface, or by expanding into a large reservoir, the temperature increases due to the conversion of the kinetic energy to thermal energy. For free, adiabatic flow, Equation 33 is applicable. Setting u_2 to zero and designating the corresponding *stagnation temperature* by the subscript s gives

$$\frac{T_s}{T_1} = 1 + \frac{u_1^2}{2C_pT_1} \tag{90}$$

which can be rewritten as

$$\frac{T_s}{T_1} = 1 + \left(\frac{\gamma - 1}{2}\right)M_1^2 \tag{91}$$

It follows that T_s is invariant in any adiabatic process, whether reversible or not. Equation 91 is applicable, therefore, to flows with and without friction, e.g., to pipes as well as nozzles.

For adiabatic and reversible flows (such as approximated in nozzles) of an ideal gas with constant heat capacity, Equation 26 is applicable. Combination with Equation 91 gives

$$\frac{P_s}{P_1} = \left[1 + \left(\frac{\gamma - 1}{2}\right)M_1^2\right]^{\frac{\gamma}{\gamma - 1}} \tag{92}$$

and

$$\frac{\rho_s}{\rho_1} = \left[1 + \left(\frac{\gamma - 1}{2}\right) M_1^2\right]^{\frac{1}{\gamma - 1}} \tag{93}$$

It is apparent that the *stagnation pressure*, P_s, and the stagnation density, ρ_s, also are invariant in such flows.

Expanding Equation 92 for small M_1,

$$\frac{P_s}{P_1} = 1 + \frac{\gamma}{2} M_1^2 \left(1 + \frac{M_1^2}{4} + \ldots\right) \tag{94}$$

$$= 1 + \frac{u_1^2 \rho_1}{2P_1}\left(1 + \frac{u_1^2 \rho_1}{4\gamma P_1} + \ldots\right) \tag{95}$$

For incompressible flow,

$$\frac{P_s}{P_1} = 1 + \frac{u_1^2 \rho}{2P_1} \tag{96}$$

Therefore, the second term inside the brackets of Equations 94 and 95 represents the first-order effect of compressibility.

Similarly, expanding Equation 93,

$$\frac{\rho_s}{\rho_1} = 1 + \frac{M_1^2}{2} + \frac{M_1^4}{4} + \ldots \tag{97}$$

The first-order effect of compressibility on the density, $(M_1^2/2)$, is seen to be twice that for the pressure, $(M_1^2/4)$.

From Equations 94 and 97 it is apparent that for incompressible, adiabatic flow,

$$\frac{T_s}{T_1} = 1 + \frac{u_1^2 \rho}{2P_1} \tag{98}$$

Comparison with Equation 91 indicates that the coefficient of u_1^2 for compressible flow is only $(\gamma - 1)/\gamma$ times that for incompressible flow, whereas ρ_s and P_s are greater.

The stagnation density for adiabatic flow of an ideal gas in a pipe can be determined by setting $u_2 = 0$ in Equation 33 and simplifying to obtain

$$\frac{u_1^2}{2} = \frac{\gamma}{\gamma - 1}\left(\frac{P_s}{\rho_s} - \frac{P_1}{\rho_1}\right) \tag{99}$$

The same process as used to obtain Equation 49 then yields

$$\frac{4fL}{D} = ln\left(\frac{\rho_s}{\rho_1}\right)^2 + \frac{1}{\gamma}\left(\frac{1}{M_1^2} + \frac{\gamma - 1}{2}\right)\left[1 - \left(\frac{\rho_s}{\rho_1}\right)^2\right] \tag{100}$$

The stagnation pressure can, in turn, be calculated using the ideal gas law and T_s from Equation 91.

Applications

The stagnation temperature is important in thermometry, but viscous dissipation and convective and radiative heat transfer must be taken into account [1]. It is evident that compressibility also must be taken into account when dealing with gases.

The impact side of a Pitot tube reads the stagnation pressure insofar as viscous dissipation and heat transfer are negligible.

COMPRESSIBLE FLOW THROUGH POROUS MEDIA

Flow through porous media provides an example of the possible significant effect of compressibility at a very low Mach number—a large change in pressure resulting from viscous rather than inertial effects. (This situation also was implied above for channel flows with $P_\infty > P_{c^*}$.)

The viscous drag can be represented by Ergun's equation [12]:

$$f_p = \frac{150}{Re_p} + 1.75 \tag{101}$$

where
$$f_p = \frac{d_p\epsilon^3}{\rho u_o^2(1-\epsilon)}\left(-\frac{dP}{dx}\right)$$
$$Re_p = \frac{d_p u_o \rho}{\mu(1-\epsilon)}$$
ϵ = void fraction
u_o = superficial velocity (in absence of packing) (m/s)
$d_p = 4\epsilon/a_v$
a_v = specific aerodynamic surface of packing per unit volume of bed (solid and fluid) (m^{-1})

The above definition of f_p implies that gravitational and inertial effects are negligible. The term aerodynamic surface implies the neglect of internal voids in the particles composing the packed bed.

Combination of Equation 101 with the definitions of f_p and Re_p gives

$$\frac{150(1-\epsilon)\mu}{d_p u_o \rho} + 1.75 = \frac{d_p\epsilon^3}{\rho_o u_o^2(1-\epsilon)}\left(-\frac{dP}{dx}\right) \tag{102}$$

Owing to the large heat capacity and thermal conductivity of the solid phase, the assumption of isothermal conditions is a reasonable approximation. The assumption of ideal gas behavior then allows integration to obtain

$$L = \frac{d_p\epsilon^3 RT}{1.75\, m u_o^2(1-\epsilon)}\, ln\left\{\frac{\dfrac{85.7(1-\epsilon)\mu RT}{d_p u_o m} + P_1}{\dfrac{85.7(1-\epsilon)\mu RT}{d_p u_o m} + P_2}\right\} \tag{103}$$

Expanding the logarithmic term for small $P_1 - P_2$ gives

$$L = \frac{d_p\epsilon^3 RT}{1.75\, m u_o^2(1-\epsilon)}\left\{\frac{P_1/P_2 - 1}{\dfrac{85.7(1-\epsilon)\mu RT}{d_p u_o m P_2} + 1}\right\}\left[1 - \frac{\frac{1}{2}\left(\left(\frac{P_1}{P_2}\right) - 1\right)}{\dfrac{85.7(1-\epsilon)\mu RT}{d_p u_o m P_2} + 1} + \ldots\right] \tag{104}$$

The second and higher-order terms in the square bracket represent the effect of compressibility. The denominator inside both brackets reduces to unity as u_0 increases.

HAMMER

Compressibility has one noteworthy effect in the flow of liquids. If a valve is suddenly closed in a pipe carrying a moving liquid, a pressure wave will propagate upstream, stopping the flow by compressing the liquid. The amplitude of this pressure wave may be sufficient to cause failure of the pipe or fittings.

Figures 6 and 7 are applicable for this behavior if subscripts 2 and 3 are replaced by 1 and 2, respectively, and u_r by u_w. The mass and momentum balances corresponding to Equations 54 and 55 then take the form

$$\rho_1(u_1 + u_w) = \rho_2 u_w \tag{105}$$

and

$$P_1 + \rho_1(u_1 + u_w)^2 = P_2 + \rho_2 u_w^2 \tag{106}$$

Elimination of ρ_1 between Equations 105 and 106 and simplification then gives

$$P_2 - P_1 = u_1 u_w \rho \tag{107}$$

For a perfectly rigid pipe, it can be shown that u_w is necessarily the acoustic velocity of the liquid. However, most pipes have bulk moduli of the same order of magnitude as that of liquids. In this case, the wave velocity is given by the expression

$$u_w = a\left(1 + \frac{E_l D}{E_s \delta}\right)^{-1/2} \tag{108}$$

where
$a = (E_l/\rho)^{1/2}$ = acoustic velocity (m/s)
E_l = bulk modulus of liquid (Pa)
E_s = bulk modulus of solid (pipe) (Pa)
D = diameter of pipe (m)
δ = thickness of pipe wall (m)

Values of E_l and/or a for many liquids and E_s for many solids can be found in handbooks. A more detailed treatment of hammer and related behavior can be found in books such as that by Streeter and Wylie [13].

NOMENCLATURE

A area (m^2)
a acoustic velocity (m/s)
a_v specific aerodynamic surface of porous material per unit volume (m^{-1})
C coefficient
C_p specific heat capacity of constant pressure (J/kg-K)
C_v specific heat capacity at constant volume (J/kg-K)
D 4A/p = hydraulic diameter (m)
d_p $4\epsilon/a_v$ = effective diameter of porous material (m)
E bulk modulus (Pa)
e effective roughness (m)
F coefficient (= 1 for detonation; = 2 for shocks)
f fugacity (in Equation 17) (Pa); Fanning friction factor (in Equation 43)
f_p friction factor for porous media
G $u\rho$ = mass flux density (kg/m^2-s)
g acceleration due to gravity (m^2/s)
H specific enthalpy (J/kg)
k empirical exponent, replacing γ in Equation 26
L length (m)
M u/a = Mach number
m molar mass (kg/mol)
P absolute pressure (Pa)

p perimeter (m)
q heat transferred to the system (J/kg)
R universal gas constant = 8.3143 J/mol-K)
Re $Du\rho/\mu$ = Reynolds number
Re_p $d_p u_o \rho/\mu$ = effective Reynolds number for porous media
S specific entropy (J/kg-K)
s distance in direction of flow (m)
T absolute temperature (K)
u velocity (m/s)
u_o superficial velocity (m/s)
w mass rate of flow (kg/s)
$\tilde{w}$ specific work done by system on surroundings (J/kg)
x distance down channel (m)
y_i mole fraction of ith component
z elevation (distance in upward direction) (m)
Z Pm/RT = compressibility factor
β $(\gamma+1)(\gamma-1)$
γ C_p/C_v = heat capacity ratio
δ thickness of pipe wall (m)
ϵ void fraction
η f/P = fugacity coefficient
μ viscosity (Pa-s)
ρ specific density (kg/m³)
τ mean shear stress on wall (Pa)
Q function of

Superscript

′ indicates velocities relative to wavefront

Subscripts

0,1,2 specific condition or location
c critical
e exit
l liquid
lm logarithmic mean
m mean
n nozzle
R due to reaction
S isentropic
s stagnation (for $u_2 = 0$) in Equations 90–100; solid (pipe wall) in Equation 108
T isothermal
t throat
∞ external

REFERENCES

1. Churchill, S. W. *The Practical Use of Theory in Fluid Flow, Book I, Inertial Flows,* (Thornton, PA: Etaner Press, 1980).
2. Emmons, H. W., Ed. *Fundamentals of Gas Dynamics, Vol. III, High Speed Aerodynamics and Jet Propulsion* (Princeton, NJ: Princeton University Press, 1958).
3. Balzhiser, R. E., and M. R. Samuels, *Engineering Thermodynamics* (Englewood Cliffs, NJ: Prentice-Hall, Inc., 1977).
4. Benedict, R. P. "Most Probable Discharge Coefficients for ASME Flow Nozzles," *J. Basic Eng.* 88: 734-744 (1966).
5. Grace, H. P., and C. E. Lapple. "Discharge Coefficients of Small Diameter Orifices and Flow Nozzles," *Trans. ASME* 73: 639-647 (1951).
6. Churchill, S. W. "Empirical Expressions for the Shear Stress in Turbulent Flow in Commercial Pipes," *AIChE J.* 19: 375-376 (1973).
7. Coulson, J. M., and J. F. Richardson. *Chemical Engineering, Vol. I, Fluid Flow, Heat Transfer and Mass Transfer* (New York: McGraw-Hill Book Co., 1954), p. 57.
8. Ginsburgh, I., and W. L. Bulkley. "Hydrocarbon-Air Detonations-Industrial Aspects," *Chem. Eng. Prog.* 59 (2): 82-86 (1963).
9. Randall, R. N., J. Bland, W. M. Dudley and R. B. Jakobs. "The Effects of Gaseous Detonations Upon Vessels and Piping," *Chem. Eng. Prog.* 53 (12): 574-580 (1957).
10. Adamson, T. C., Jr., and R. B. Morrison. "On the Classification of Normal Detonation Waves," *Jet Propulsion* 25: 400, 403 (1955).
11. Morrison, R. B. "A Shock Tube Investigation of Detonative Combustion," Report AF33 (038) 12657, UMM-97, Willow Run Research Center, University of Michigan, Ann Arbor, MI (1952).
12. Ergun, S. "Fluid Flow Through Packed Columns," *Chem. Eng. Prog.* 48: 89-94 (1952).
13. Streeter, V. L., and E. B. Wylie. *Hydraulic Transients* (New York: McGraw-Hill Book Co., 1967).

CHAPTER 9
MIXING OF GASES

Milind B. Ajinkya
Exxon Research and Engineering Company
Florham Park, New Jersey 07932

INTRODUCTION

Mixing of two or more gases is encountered frequently in the process industry. In most cases mixing is a precursor to carrying out gas-phase reactions, the intention being the rapid mixing of gaseous reactants so that the desired reactions will take place in a more or less homogeneous gas phase. The purpose of this chapter is to outline some of the current practices in industry used to achieve mixing of gases as efficiently as possible. The emphasis will be on pipeline and jet mixing, with different configurations thereof. How the extent of mixedness affects chemical reactions has been a popular research topic for several years but is not discussed here.

Before moving on to the main subject matter, a few examples of industrial importance would be in order.

1. **Chlorination of gaseous hydrocarbons:** Allyl chloride is an intermediate formed in the process of making epichlorohydrin and synthetic glycerine [1,2]. It is made by reacting propylene and chlorine in the gas phase at elevated temperatures by premixing the two gases. Several parallel and series reactions are possible [3]; hence, the yields of the desired products depend highly on how well propylene and chlorine are mixed before reacting.

2. **Oxidation of ethylene:** Ethylene oxide is formed by reacting air or oxygen and ethylene over a catalyst. The two gases are premixed before entering the reactor. In addition to the product selectivity and yield of the process, safety is a major consideration since ethylene and oxygen form explosive mixtures in certain proportions. It is extremely important, therefore, that the mixture of ethylene and air/oxygen be as homogeneous as possible.

3. **Air oxidation of ammonia:** In the oxidation of ammonia to produce nitric acid, mixing of air with ammonia in the shortest possible time is desirable. This is essential to avoid product loss from side reactions.

4. **Air pollution:** Gas-gas mixing finds application in the abatement of air pollution, for example, where stack gases disperse into the crosswinds in the form of jets.

The above are but a few of a number of gas-mixing problems encountered in the process industry. Although some research articles have appeared on modeling turbulent fluid mixing in recent years, mixer design is still a highly empirical procedure and much engineering judgment is called for. The state-of-the-art has not progressed to the point where a practicing design engineer can use reliable correlations to design mixers. If the engineer has access to laboratory facilities and the time, he can design scaled-down models of the commercial design and study the critical scaleup issues, e.g., the dynamic similarities or the geometric similarities. Even for troubleshooting a malfunctioning commercial mixer design, the latter is the approach followed by many industrial labs.

PIPELINE MIXING

Pipeline mixing usually is the simplest means of mixing two or more gases. The process is slow, but if enough pipelength is available, as in most interconnecting process piping between vessels, adequate mixing can take place. However, the efficiency of the mixing process can be increased significantly, and the mixing length, that is, the amount of pipe required, can be reduced to roughly 2–10 pipe diameters from 30–100, if a tee is used to inject one of the fluids perpendicular to the other at sufficiently high velocities [4].

One of the first studies on pipeline tee mixing was carried out by Chilton and Genereaux [5] 50 years ago. Using smoke tracers, they studied the mixing of CO_2, SO_2, air etc., into an air stream. Mainstream air flowed through a 1.75-in.-diameter pipe, whereas secondary air flow was introduced perpendicular to the mainstream through a tee whose diameter could be varied between 0.25 in. and 1.25 in. They found an optimum mass velocity of injection at two to three times the mainstream mass velocity for achieving good mixing in two to three pipe diameters of the main piping. Chilton and Genereaux [5] also tried other injection methods, but none was more efficient than the simple tee mixer.

Narayan [6] studied pipeline mixing with the three configurations shown in Figure 1. He used air and air/CO_2 with a constant velocity ratio of 1:3.5 to study the mixedness

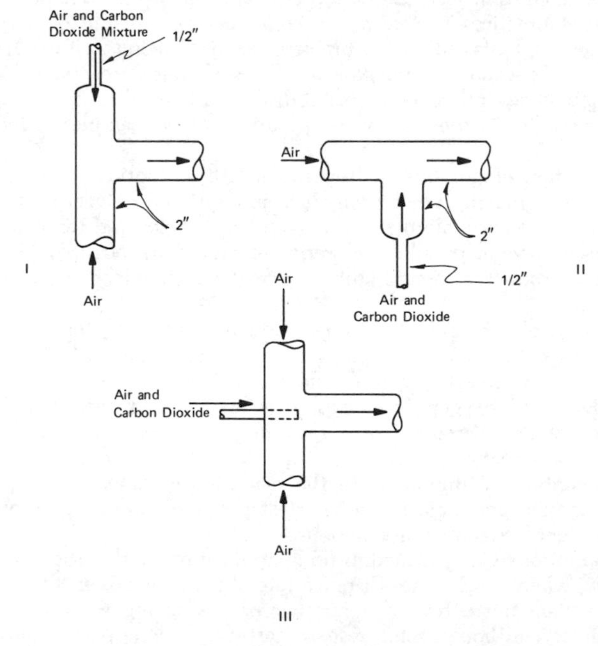

Figure 1. Pipeline mixers [6].

by measuring the specific gravity of the sampled stream. He found that good mixing was achieved for the first arrangement in 4–5 pipe diameters, whereas the same was achieved in about 8 diameters with the second arrangement. In the case of the third configuration, about 250 diameters were required, indicating the poor mixing efficiency of this design. Narayan's results confirmed those of Chilton and Genereaux and, thus, further demonstrate the effectiveness of simple tee mixers in gas-gas mixing. Some more critical comments on these two studies can be found in the work by Simpson [4,7]. Simpson [7] carried out a very simple analysis of pipeline turbulence combining Corrsin's work [8,9] with Brodkey's [10] and Laufer's [11] data on turbulence. He showed that (Figure 2) for gases having a Schmidt number, (N_{Sc}) ~ 1, the mixing length to diameter ratio for reducing the root-mean-square concentration fluctuations (a') to 1% of their initial value (a'_o) is 30–40 at a Reynolds number (N_{Re}) of 10,000, and increases to 40–70 at $N_{Re} \simeq 10^7$.

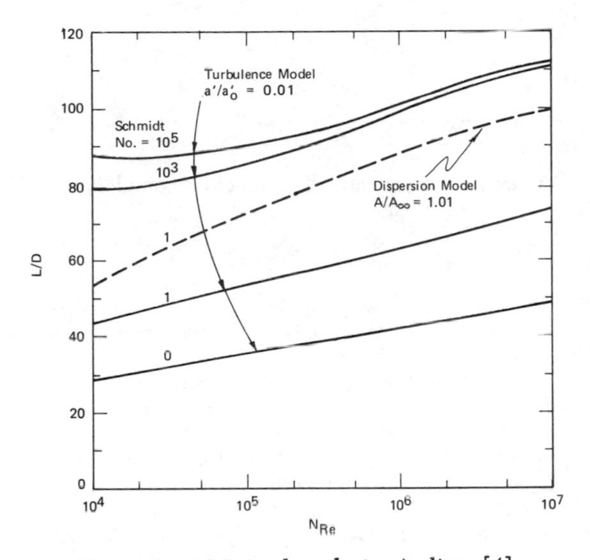

Figure 2. Mixing lengths in pipelines [4].

The above results roughly agree with Beek and Miller's [12] analysis of turbulent pipeflow. Simpson [7], however, has shown that for liquids (N_{Sc} ~ 1000), the sensitivity to Reynolds number is not great. He reported that 85–88 pipe diameters are needed at $N_{Re} \simeq$ 10,000, compared to ~120 as estimated by Beek and Miller. Simpson also used the dispersion model to estimate the length required for the centerline tracer concentration to reach 1.01 times the equilibrium value. As shown in Figure 2, results using the dispersion model fall within the results obtained using the turbulence models.

JET MIXING

Another important means of achieving gas mixing is through jet formation. Jets are formed by forcing fluids into a surrounding medium through a device like a nozzle or slot. We can attempt to formally define a jet as follows [13]: When there is motion of a liquid or gas, a tangential separation of surfaces can occur due to velocity or temperature differences or gradients in species concentrations. Both sides of this surface are termed jets. Jets may be moving in the same or opposite directions, in crossflow or in a swirling motion. The eddies on the tangential surfaces cause an exchange of momentum, heat and mass transfer across the jet.

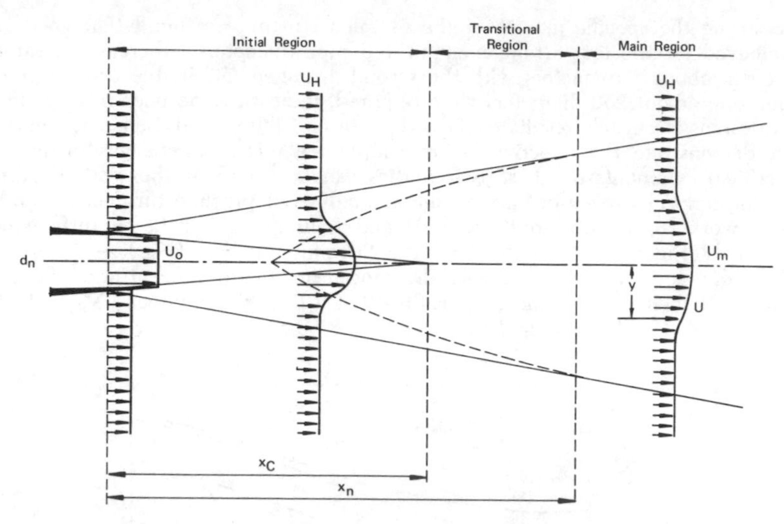

Figure 3. Schematic diagram of jet boundary layer.

Jets in Coflow

A jet in coflow is shown schematically in Figure 3. The fluid issuing through a nozzle is assumed to have a uniform velocity, U_o. The surrounding medium is assumed to have a uniform velocity, U_H. The jet expands by drawing some of the surrounding medium into itself and also by slowing down. Until the end of the jet core is the initial region (also known as the potential region). The velocity in the potential region is constant at U_o. The jet Reynolds number remains constant throughout the jet structure and is equal to its value at the nozzle exit. When the jet Reynolds number is less than 300, the jet is considered to be in laminar flow [7]. Between 300 and 1000 there is unsustained turbulence in the jet boundary layer. Simpson [7] has analyzed a round turbulent jet, assuming a simple plug flow model to describe the centerline velocity decay as

$$\frac{U_m}{U_o} = \sqrt{\frac{\rho_o}{\rho_m}} \Big/ \left[1 + 2\frac{X}{d_n} \tan\left(\frac{\theta}{2}\right) \right] \tag{1}$$

If $\rho_o = \rho_m$, $\theta = 14°$ and X is large, the mass flowrate, W_m, at X is related to that at the nozzle exit as

$$\frac{W_m}{W_n} \cong 0.25 \left(\frac{X}{d_n}\right) \tag{2}$$

This is an approximate measure of the entrainment of the surrounding medium into the jet itself.

A more realistic approach would be to assume that the jet velocity varies radially, and if U_o is the velocity in the potential region,

$$U_m = U_o \text{ for } \frac{X}{d_n} \leq \frac{X_C}{d_n} \tag{3}$$

and

$$U_m/U_o = \frac{X_m}{d_n} \cdot \frac{d_n}{X} \text{ for } \frac{X}{d_n} > \frac{X_C}{d_n} \tag{4}$$

where the core length is correlated as follows [14]:

$$\frac{X_C}{d_n} = 2.13\, N_{Re}^{0.097}, \qquad 10^4 < N_{Re} < 10^5 \tag{5}$$

Jets in Crossflow

For a designer, the use of crossflowing jets for mixing usually will represent the most interesting configuration. A schematic diagram of a jet in a lateral deflecting flow is shown in Figure 4. Abramovich [13] presents a simple empirical relation for the jet path, based on experiments with air jets from a circular nozzle discharging into a crossflow:

$$\frac{X}{d_n} = \frac{q_m}{q_o}\left(\frac{Y}{d_n}\right)^{2.55} + \frac{Y}{d_n}\left(1 + \frac{q_m}{q_o}\right)\cot\alpha' \tag{6}$$

where X and Y are the coordinates of any point on the deflecting jet axis, d_n is the diameter of the nozzle, and α' is the angle between the direction of the axis of the nozzle and the lateral flow direction. The quantities q_m and q_o are proportional to the kinetic energies of the respective streams, e.g.,

$$q_m = \rho_m \frac{U_m{}^2}{2} \text{ and } q_o = \rho_o \frac{U_o^2}{2}$$

where U_m is the centerline crossflow velocity and U_o is the discharge velocity in the nozzle. Abramovich [13] states that Equation 6 is valid over the following ranges of q_o/q_m and α':

$$2 \leq q_o/q_m \leq 22 \text{ and } 45° \leq \alpha' \leq 90°$$

It is also indicated that the same correlation can handle nonisothermal ($T_m \neq T_o$) jets through the quantities q_m and q_o. Abramovich has also derived a simple semitheoretical relation to describe the path of a deflecting jet, which is given below:

$$\frac{Y}{d_n} = 14.4\sqrt{\frac{\rho_o U_o{}^2}{C_n \rho_m U_m{}^2}}\log\left[1 + 0.1\frac{X}{d_n}\left(1 + \sqrt{1 + 20\frac{d_n}{X}}\right)\right] \tag{7}$$

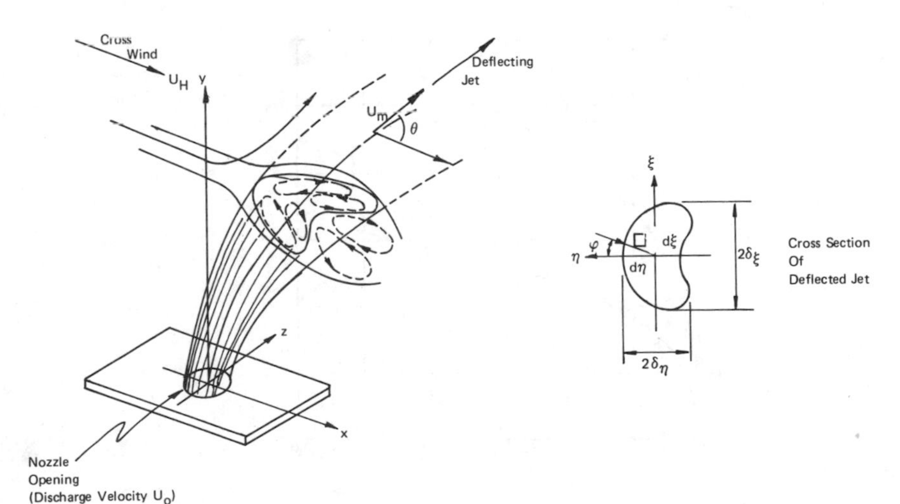

Figure 4. Schematic of deflecting jet in crossflow.

where α' is $\pi/2$, and C_n is a coefficient whose value lies between 1 and 3. For $C_n = 3$, some comparisons between the experimental data and the predictions of Equation 7 are shown in Figure 5. Relatively more recent work by Patrick [15] shows that

$$\frac{Y}{d_n} = 1.91\left[\frac{\rho_o}{\rho_m}\left(\frac{U_o}{U_m}\right)\right]^{0.606}\left(\frac{X}{d_n}\right)^{0.303} \tag{8}$$

When recast into the other form, we get

$$\left(\frac{X}{d_n}\right) = 0.12\,\frac{\rho_m}{\rho_o}\cdot\frac{U_m^2}{U_o^2}\left(\frac{Y}{d_n}\right)^{3.3} \tag{9}$$

This form of the correlation is consistent with that of Abramovich's above, except for the exponent on Y/d_n and the proportionality constant. Patrick [15] has summarized most of the available experimental work on jet penetration, which includes the early work done at NASA by Callaghan and Ruggeri [16] at the Lewis Flight Propulsion Research Lab. Callaghan and Ruggeri experimented with various crossflow and jet velocities and densities through square and elliptical orifices and concluded that the jet penetration can be expressed as

$$\frac{Y^{1.65}}{\sqrt{C_n d_n}} = 2.91\,\frac{\rho_o U_o}{\rho_m U_m}\,\frac{X^{1/2}}{C_n d_n} \tag{10}$$

Norster and Chapman [17] correlated their data to express the maximum jet penetration as

$$\left(\frac{Y}{d_n}\right)_{max} = 1.224\left(\frac{\rho_o U_o^2}{\rho_m U_m^2}\right)^{1/2} \tag{11}$$

Keffer and Baines [18] have plotted their data on jet penetration as a function of various jet to mainstream velocity ratios, $R = U_o/U_m$ (Figure 6). The data more or less fit the correlation

$$\frac{Y}{d_n} = 1.0\left[\sqrt{\frac{\rho_o}{\rho_m}}\cdot\frac{U_o}{U_m}\right]^{0.85}\left(\frac{X}{d_n}\right)^{0.38} \tag{12}$$

if we assume $\rho_o = \rho_m$.

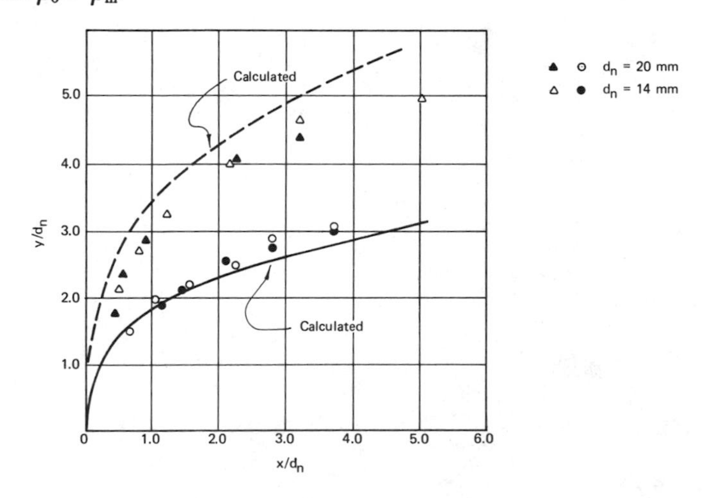

Figure 5. Jet trajectory in crossflow [13].

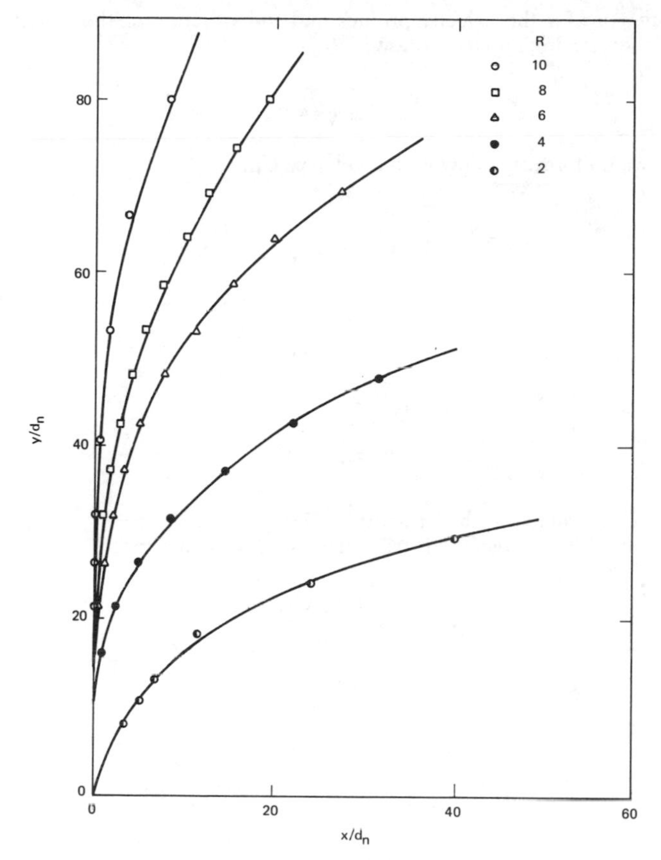

Figure 6. Jet centerline trajectory [18].

A number of theoretical studies have emerged recently that characterize single- and multiple-jet discharge into crossflowing streams. Sucec and Bowley [19] investigated a rectangular jet ejecting at various angles into a crossflow of arbitrary velocity distribution (as opposed to uniform). They solved the momentum integral equation, which takes into account the entrainment by the finite difference technique, and compared their results with the available experimental data. Campbell and Schetz [20] assumed an elliptical cross section of the jet (axes ratio 5:1). They also solved the momentum integral equation by the same technique. Stoy and Ben-Haim [21] used an arbitrary distribution for the crossflow velocity and solved the equation governing the conservation of momentum by a predictor corrector method, comparing their results with their own experimental data. Patankar et al. [22] approached the problem differently, treating the injected jet flow in three dimensions.

Few attempts have been made to describe the behavior of multiple jets in cross flow. Kamotani and Greber [23] reported some experimental findings on the trajectory of crossflow multiple jets for different nozzle spacings and velocity ratios. Makihata and Miyai [24] investigated the penetration of multiple jets into a crossflow by spacing two nozzles next to each other such that there was interaction between the issuing jets. The spacing was 0.8 d_n. Using a momentum integral equation, which took into account entrainment, buoyant flux and the drag force, they were able to derive theoretical relations to describe the jet trajectories and compare them with their own experimental data and that of others. The agreement was fairly good. They also developed the

following expressions for the velocity profiles and the variation of the centerline jet velocity for single jets issuing into a crossflow:
Potential region:

$$f = e^{-\alpha'\sigma^4} \tag{13}$$

where $\nu = 0$ for a uniform crossflow with a velocity U_H.
For

$$\frac{\sqrt{n} \cdot \zeta}{d_n} (1 + 10/K_U) \leq 6.15 \tag{14}$$

and

$$\alpha' = \ln 0.5$$
$$\frac{U_m}{U_o} = 1.0$$
$$\sigma = \frac{\xi}{\delta} = \frac{\eta}{\delta}$$

assuming symmetry, where δ is the jet width at $U/U_m = 0.5$.
For a crossflow of arbitrary velocity distribution, i.e., $\nu = \nu$, and for

$$\frac{\sqrt{n} \cdot \zeta}{d_n} \left(1 + \frac{10N^{\nu}}{K_U}\right) \leq 6.15$$

$$f = e^{-\alpha'\sigma^4} \tag{15}$$

where $N = \frac{y}{y}$

and

$$\frac{U_m}{U_o} = 1.0$$

The jet trajectory was given by the following relations:
For

$$\nu = 0 \text{ and } \frac{\sqrt{n}\zeta}{d_n} \left(1 + \frac{10}{K_U}\right) > 6.15$$

$$\nu = e^{-\alpha'\sigma^2}$$

where $\alpha' = -\ln 0.5$, $\sigma = \frac{\xi}{\delta} = \frac{\eta}{\delta}$

and

$$\frac{U_m}{U_o} = \frac{7.8}{\frac{\sqrt{n} \cdot \zeta}{d_n} \left(1 + \frac{10}{K_U}\right) + 1.65} \tag{16}$$

with

$$\frac{U_m}{U_H \cos^{\theta}} \geq 1.0$$

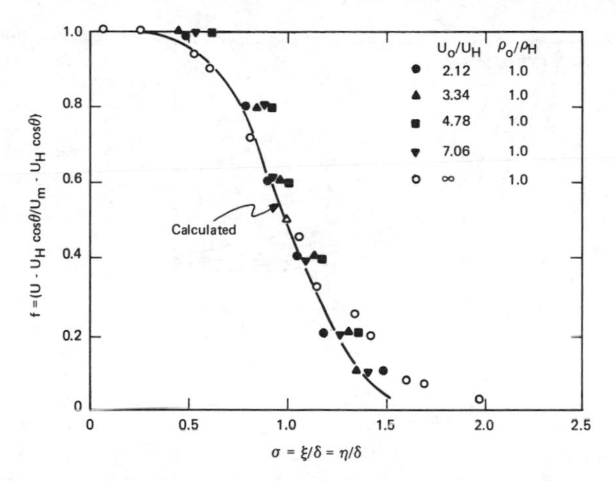

Figure 7. Velocity distribution in the deflecting jet cross section for the potential flow (i.e., in the initial region of core length) [24].

when $\nu = \nu$ and $\frac{\sqrt{n}\zeta}{d_n}\left(1 + \frac{10N^{\nu}}{K_U}\right) > 6.15.$

$$f = e^{-\alpha'\sigma^2}$$

and

$$\frac{U_m}{U_o} = \frac{7.8}{\frac{\sqrt{n}\zeta}{d_n}\left(1 + \frac{10N^{\nu}}{K_U}\right) + 1.65} \tag{17}$$

with

$$\frac{U_m}{G\ (Y)\ \cos\theta} \geq 1$$

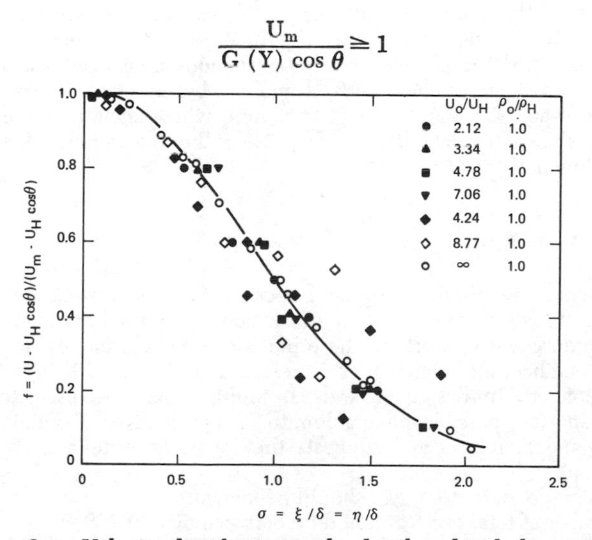

Figure 8. Velocity distribution in the developed turbulent jet [24].

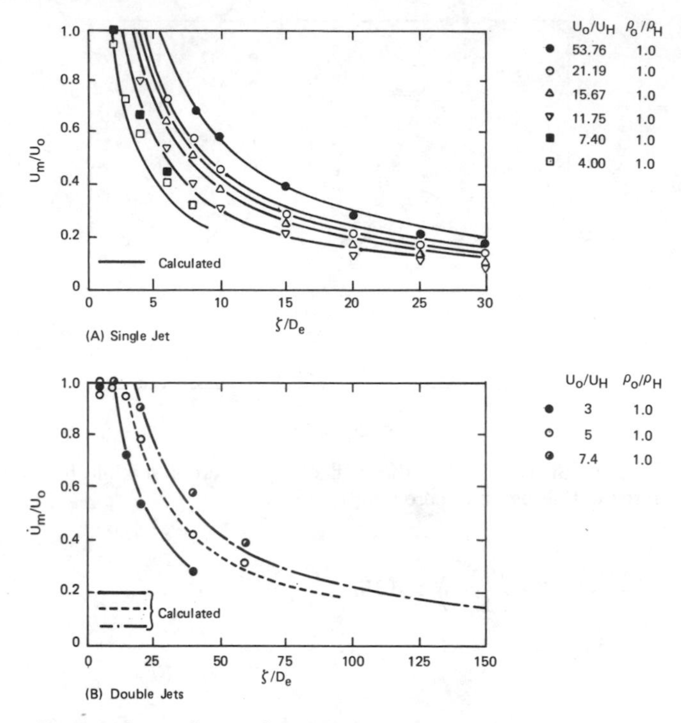

Figure 9. Decay of jet velocity along the jet trajectory for single and double jets [24].

Makihata and Miyai [24] showed that their data, and also the data of Daniel et al. [25], Keffer and Baines [18] and Patrick [15] were described adequately by the above correlations (Figures 7–9).

Recently, Crabb et al. [26] have measured the velocity characteristics of a jet in crossflow encompassing the entire mixing region. Laser-Doppler anemometry was used upstream, where the turbulence intensity is high, and hot-wire anemometry downstream. Their results confirmed the general trends of previous investigations. They concluded that the jet in crossflow is characterized mainly by pressure forces that cause the bending of the jet and the double vortex structure, which should simplify the flow-field calculations, as shown by Patankar et al. [22], without a detailed knowledge of the structure of turbulence.

INDUSTRIAL MIXERS

Having gained some insight into the basics of pipeline mixing and jet mixing, it would seem appropriate to consider some of the commonly used mixer designs. Simpson's [4] excellent comprehensive work on the topic is of practical interest for those involved in design and engineering. The reader is referred to Figure 10, in which four basic designs are depicted. In designs I and II, multiple jets are used to achieve good mixing in the shortest possible mixing length. Simpson gives a detailed algorithm to design types I and II mixers and suggests the following criteria as being important:

1. The design pressure drop, ΔP, should be known.
2. The number of total orifices should be between 50 and 200.
3. The orifices should be placed so they do not interact with each other and, thus,

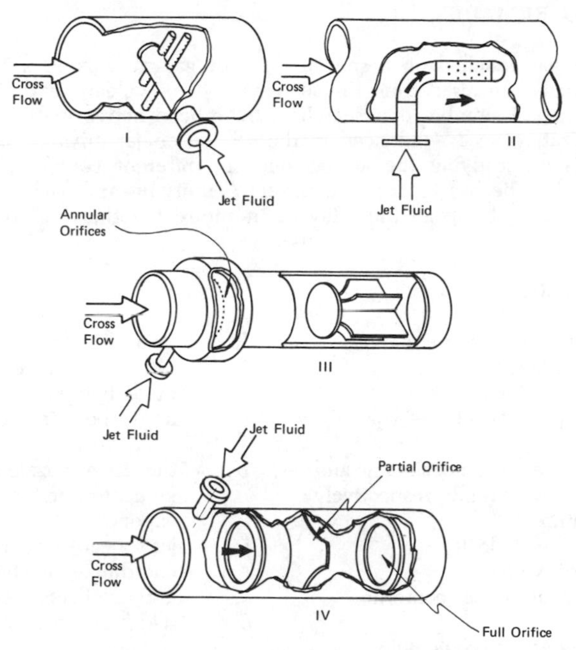

Figure 10. Some industrial mixers [4].

the jet behavior can be analyzed as though they are independent (noninteracting).

4. The orifice Reynold's number should be ~5000.
5. In the first mixer, a good flow distribution along the length of each finger will be ensured if the finger cross-sectional area is a minimum 1.5 times the total orifice area in that finger.
6. There should be sufficient lengths of main piping upstream and downstream of the mixer to avoid pressure and velocity variations across the main pipe.

Type III mixer is used primarily for rapid mixing of reactants for fast gas-phase reactions. Type IV mixers involve installation of full or partial orifices. Details of these are found in Simpson [7]. These mixers are especially suitable for installation in existing process piping systems.

Hartung and Hiby [27] have experimented with the mixer design shown in Figure 11. Each baffle occupied half the pipe cross section and was placed one pipe diameter apart from the next baffle. Though Hartung and Hiby used this design to investigate liquid mixing, the design seems equally suitable for gas mixing.

Like coflowing and crossflowing jets, jets in swirling flow also are used for mixing gases efficiently. Many combustors use swirling flows [28]. It is an extensively developed separate field and will not be dealt with here.

The motionless mixers commercially available from manufacturers will not be discussed, as they have been widely used for liquid-liquid mixing, especially for relatively viscous fluids. Nevertheless, it is felt that they may be used successfully in gas mixing applications.

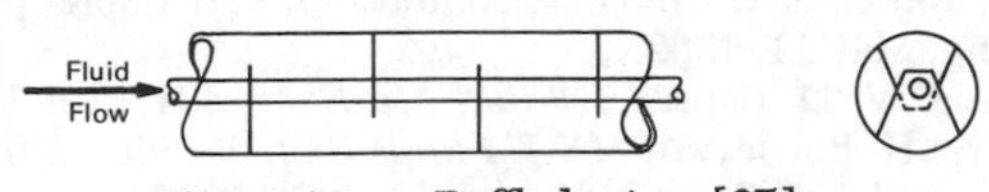

Figure 11. Baffled mixer [27].

CONCLUDING REMARKS

This chapter has touched on a special topic of gas-gas mixing as practiced in the industry. As can be surmised from the sections above, a clear-cut design methodology has yet to emerge. A number of researchers have been active in the field of jets and their characterization, as is evidenced in the section on jet mixing. In spite of such a heavy emphasis on studying the jet behavior in different configurations, a designer is bewildered when he has to design a mixer. A unifying approach with the designer in mind is needed to bring all the relevant literature together and simplify his work.

NOMENCLATURE

a', a'_o	RMS concentration fluctuations
C_d	orifice coefficient
d_n	nozzle diameter
f	$(U - U_H \cos\theta)/(U_m - U_H \cos\theta)$
q_o, q_m	kinetic energies of nozzle flow and jet centerline fluids, respectively
T_o, T_m	temperatures
U_m	jet centerline velocity
U_o	nozzle exit velocity
W_m	mass flowrate of jet centerline fluid
W_n	mass flowrate of nozzle fluid
X	axial distance from nozzle
Y	vertical distance from nozzle
α'	angle between nozzle and X-axis
θ	angle between jet trajectory and X-axis
ρ_o	density of nozzle fluid
ρ_m	jet centerline fluid density
ν	exponent
ζ	jet trajectory coordinate
ξ, η	coordinates in the cross section of the deflected jet
δ	width of the deflected jet

REFERENCES

1. Groll, H. P. A., and G. Hearne. *Ind. Eng. Chem.* 31:1530 (1939).
2. Fairburn, A. W., H. A. Cheney and A. J. Cherniavsky. *Chem. Eng. Prog.* 43:280 (1947).
3. Smith, J. M. *Chemical Engineering Kinetics* (New York: McGraw-Hill Book Co., 1981).
4. Simpson, L. L. *Chem. Eng. Prog.* 70:77 (1974).
5. Chilton, T. H., and R. P. Genereaux. *AIChE Trans.* 25(102) (1930).
6. Narayan, B. C. MS Thesis, University of Tulsa, Tulsa, OK (1971).
7. Simpson, L. L. *Turbulence in Mixing Operations,* R. S. Brodkey, Ed. (New York: Academic Press, Inc., 1975).
8. Corrsin, S. *AIChE J.* 3:329 (1957).
9. Corrsin, S. *AIChE J.* 10:870 (1964).
10. Brodkey, R. S. *AIChE J.* 12:403 (1966).
11. Laufer, J. *NACA TN* 1174 (1954).
12. Beek, J., Jr., and R. S. Miller. *Chem. Eng. Prog. Symp. Series 110,* 25(55):23 (1959).
13. Abramovich, G. N. *The Theory of Turbulent Jets* (Cambridge, MA: The MIT Press, 1963).
14. Harsha, P. T. Arnold Engineering Development Center, Tech. Report AEDC-TR-71-36 (1971).
15. Patrick, M. A. *J. Inst. Fuel* 425 (1967).
16. Callaghan, E. E., and R. S. Ruggeri. *NACA TN* 61615 (1948).
17. Norster, E. R., and C. S. Chapman. "Combustion and Fuels" paper presented at the A. R. C. meeting, May 11, 1962.
18. Keffer, J. F., and W. D. Baines. *J. Fluid Mech.* 15:481 (1963).
19. Sucec, J., and W. H. Bowley. *ASME J. Fluids Eng.* 98:656 (1976).
20. Campbell, J. F., and J. A. Schetz. *AIAA J.* 11:242 (1973).

21. Stoy, R. L., and Y. Ben-Haim. *ASME J. Fluids Eng.* 95:551 (1973).
22. Patankar, S. V., D. K. Basu and S. A. Alpay. *AMSE J. Fluids Eng.* 99:758 (1977).
23. Kamotani, Y., and I. Greber. NASA CR-2392 (March 1974).
24. Makihata, T., and Y. Miyai. *ASME J. Fluids Eng.* 101:217 (1979).
25. Daniel, T. L., S. Chan and J. F. Kennedy. IIHR Report No. 140 (August 1972).
26. Crabb, D., D. F. G. Durão and J. H. Whitelaw. *ASME J. Fluids Eng.* 103:142 (1981).
27. Hartung, K. H., and J. W. Hiby. *Chem. Ing. Tech.* 44:1051 (1972).
28. Béer, J. M., and N. A. Chigier. *Combustion Aerodynamics* (New York: Halstead Press, 1972).

CHAPTER 10
THEORY OF TURBULENT JETS

N. Rajaratnam
Department of Civil Engineering
University of Alberta
Edmonton, Alberta T6G 2G7
Canada

CONTENTS

INTRODUCTION

This chapter discusses the turbulent mixing and diffusion of jets of a Newtonian fluid discharging into the same fluid. These jets sometimes are referred to as submerged jets to differentiate them from other classes of jets. Because in engineering practice one deals mostly with turbulent jets only this type will be discussed here. For more detailed treatment of the topics in this chapter, refer to the book entitled *Turbulent Jets* [1].

PLANE JET

Consider a rectangular jet of water (or air) discharging from a nozzle with a thickness of $2b_0$ and length, B, much larger than its thickness, with an almost uniform velocity of U_0 into a large body of water (or air) at rest. Observation of this flow indicates that the jet mixes with the surroundings forming turbulent shear layers all around. Because the jet is very long, let us neglect the end regions and consider only the central part of the jet for the present. The turbulence created by the jet penetrates outward into the ambient fluid and inward into the jet, thereby reducing the thickness of the central region of potential flow, generally known as the potential core. After the jet has traveled an axial distance of approximately $12b_0$, the turbulence has penetrated to the jet axis itself and the potential core has vanished. This initial part of the jet, with the potential core surrounded by shear layers, is known as the flow development region, whereas the region downstream of the end of the potential core is referred to as the region of fully developed flow. We will first study this region of fully developed flow, and consider the initial region after discussing shear layers.

Let us look at some basic characteristics of the plane jet as observed from experiments before attempting to develop a theoretical structure. Simple observations show that as the jet travels farther from the nozzle, it grows in thickness; secondly, it entrains the surrounding fluid; thirdly, the velocity in the jet decreases continuously. With reference to the definition sketch in Figure 1, if u is the time-averaged velocity in the axial or x direction at any point (x,y), the typical u(y), or velocity profiles at different x stations, are shown in Figure 2(a) from the early observations of Foerthman [2], where $\bar{x}$ is the axial distance from the nozzle and x is that measured from a suitable virtual origin (to be discussed later). The velocity profiles at different sections appear to have the same shape; let us replot these profiles in a dimensionless form with u/u_m against y/b, wherein u_m is the maximum value of u at any section occurring on the axis and b = y where u is equal to (say) 0.5 u_m. These two quantities can be referred to, respectively, as the velocity and the length scales. We find from Figure 2b that all these profiles are described by one general curve, and the velocity profiles possessing this property are said to be "similar." This similarity property is possessed by many turbulent jet flows and is of great importance in developing general methods of solution.

The second interesting property is that the jet grows rather slowly, that is, its thickness at any section is small compared to the distance of this section from the nozzle. As a result, such jet flows belong to the class of "slender flows" or boundary layer type of flows, and for such flows some powerful approximations to the full Reynolds equations of motion can be made. It is useful to point out here that in such slender flows the (time-averaged) transverse velocity component, v, is much smaller than the axial component, u.

For a plane turbulent jet in which the primary flow (u,v) is steady, if the flow field does not vary in the z direction (which is along the length of the nozzle) and the piezometric pressure does not vary in the axial direction (as in the case of a jet issuing into a large expanse of the surrounding fluid), the relevant Reynolds equations can be reduced to [1]

$$u \frac{\partial u}{\partial x} + v \frac{\partial u}{\partial y} = \frac{1}{\rho} \frac{\partial \tau}{\partial y} \tag{1}$$

wherein most of the viscous stress terms also have been neglected and ρ is the mass density of the fluid. The shear stress, τ, is equal to the sum of τ_t, the turbulent shear

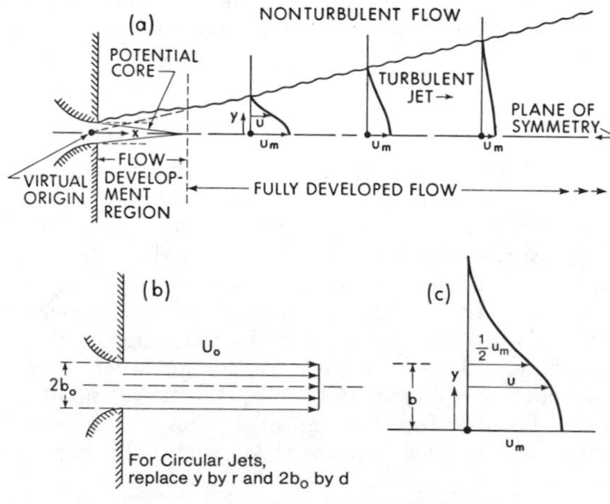

Figure 1. Plane and circular jets.

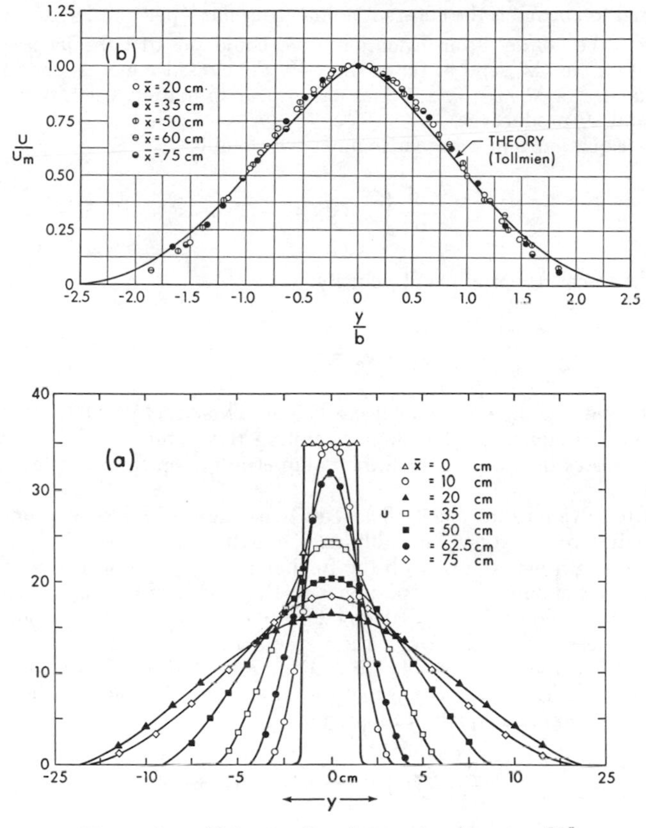

Figure 2. Velocity distribution in plane jets [2].

stress ($-\rho\overline{u'v'}$, the primes denote fluctuating quantities) and the laminar shear stress, $\tau_l = \mu\ \partial u/\partial y$ (μ being the coefficient of dynamic viscosity). For most turbulent jet problems, $\tau_t >> \tau_l$ and, hence, τ in Equation 1 can be considered as the turbulent shear stress itself. An exception to this rule will arise later when we consider turbulent wall jets, where the viscous shear stress will dominate as the wall is approached instead of the turbulent shear stress. The continuity equation reduces to

$$\frac{\partial u}{\partial x} + \frac{\partial v}{\partial y} = 0 \tag{2}$$

Equations 1 and 2 could be referred to as the equations of motion for the plane turbulent jet.

Integrating Equation 1, using Equation 2 with respect to y from y = 0 to infinity, and using the Liebnitz rule [1], leads to

$$\frac{d}{dx}\int_0^\infty \rho u^2 dy = 0 \tag{3}$$

Equation 3 says that the axial momentum flux of the jet is invariant or remains constant for all values of x. Integrating Equation 3 with respect to x,

$$2\int \rho u^2 dy = M_o \tag{4}$$

where M_o equal to $2b_o\rho U_o^2$ is the constant momentum flux (per unit length of the nozzle) that issues from the nozzle. From Equation 4 we could say that the jet is characterized by its momentum flux, M_o, and as far as the fully developed region is concerned all that is important is M_o (or M_o/ρ), and it does not appear to matter which particular combination of b_o and U_o produces it.

Integrating Equation 2 in a similar manner we obtain

$$\frac{d}{dx}\int_o^\infty u\,dy = -v_e \tag{5}$$

where v_e is the so-called entrainment velocity.
We will now write

$$-v_e = \alpha_e\, u_m \tag{6}$$

wherein α_e is known as the entrainment coefficient. The structure of Equation 6 can be developed from the equations of motion, assuming the similarity of v velocity profiles. Equation 5 expresses the rate at which the volumetric flux in the jet is increasing in the axial direction.

If we want to solve for the velocity field (u,v), because we have three unknowns, u, v and τ, and only two equations, an additional equation is needed. Tollmien used the Prandtl mixing length hypothesis, with the further assumption that the mixing length is constant at any x station and is proportional to the width of the jet, whereas Goertler assumed a constant eddy diffusivity at any x station. Details of these solutions can be studied in the literature [1,3]. There are a number of other closure models and the reader is referred to the literature [4] for a discussion of these. In this analysis we are going to assume an empirical equation that describes the experimental observations very satisfactorily. The empirical equation adopted is

$$\frac{u}{u_m} = f(\eta) = \exp\left[-0.693\,\eta^2\right] \tag{7}$$

where $\eta = y/b$
b = value of y
$u = u_m/2$
u_m = value of u on the axis of the jet.

Having assumed the dimensionless velocity profile, $f(\eta)$, to predict the velocity field completely, all we need is to predict the relations $u_m(x)$ and $b(x)$, the decay of the velocity scale and the growth of the length scale, respectively.

Let us further assume

$$\left.\begin{aligned} u_m &= k_1\, x^p \\ b &= k_2\, x^q \end{aligned}\right\} \tag{8}$$

where p and q are unknown exponents and k_1 and k_2 are the characteristic coefficients of the jet which do not depend on the variable x.

Considering the integral momentum equation, using the similarity property of the velocity profiles we can show that

$$\frac{d}{dx}\left\{u_m^2\, b \int_o^\infty f^2 d\eta\right\} = 0 \tag{9}$$

from which it follows that

$$2p + q = 0 \tag{10}$$

From the integral continuity equation,

$$\frac{d}{dx}\left\{u_m b \int_0^\infty f d\eta\right\} = \alpha_e u_m \tag{11}$$

from which it follows that

$$p + q - 1 = p \tag{12}$$

Solving Equations 10 and 12: $q = 1$ and $p = -1/2$. Thus, for a plane turbulent jet,

$$\left.\begin{aligned} u_m &= \alpha \frac{1}{\sqrt{x}} \\ b &= \alpha x \end{aligned}\right\} \tag{13}$$

Using the integral momentum and continuity equations, we can show that

$$k_1 = \sqrt{\frac{b_o}{k_2 F_2}}\, U_o \tag{14}$$

$$k_2 = \frac{2\alpha_e}{F_1} \tag{15}$$

where $F_1 = \int_0^\infty f d\eta$ and $F_2 = \int_0^\infty f^2\, d\eta$.

For the exponential profile suggested in Equation 7, $F_1 = 1.065$ and $F_2 = 0.753$.

Among the three coefficients k_1, k_2 and α_e, at least one will have to be determined experimentally. It is convenient to evaluate k_2 from experimental observations; then k_1 and α_e can be evaluated easily.

If the jet carried a passive pollutant or tracer in it, we could study the mixing and spread of the tracer as shown below. If c is the time averaged concentration of this pollutant at any point in the jet in some suitable units (say mg/l), then from the pollutant conservation equation and with the slender flow approximations we can obtain the simplified equation

$$u\frac{\partial c}{\partial x} + v\frac{\partial c}{\partial y} = \epsilon_y \frac{\partial^2 c}{\partial y^2} \tag{16}$$

wherein the eddy diffusivity, ϵ_y, for the pollutant concentration has been assumed to be constant at any x station in the jet. Integrating Equation 16 across the jet from $y = 0$ to a large value, we can reduce Equation 16 to

$$\frac{d}{dx}\int_0^\infty ucdy = 0 \tag{17}$$

Integrating Equation 17 with respect to x,

$$2\int_0^\infty ucdy = P_o \tag{18}$$

where P_o is the pollutant flux in the jet. If c_o is the pollutant concentration at the nozzle,

$$P_o = 2b_o U_o c_o \tag{19}$$

Experimental observations indicate that the pollutant concentration profiles at the different x stations are similar. That is,

$$\frac{c}{c_m} = h(\eta) \tag{20}$$

where c_m is the concentration on the axis for a particular value of x. If we assume that

$$c_m = k_3 x^s \tag{21}$$

where s is the unknown exponent and k_3 is the x-independent coefficient, from Equation 17 we can show that $s = -1/2$. Because the pollutant is known to spread slightly faster than the momentum in the jet, if $b_c = kb$, where $b_c = y$ and where $c = 1/2\ c_m$ and k is a constant to be evaluated from experiments, from Equation 18 one obtains the result

$$u_m c_m b \int_0^\infty f\, h\, d\eta = b_o U_o c_o \tag{22}$$

and

$$k_3 = c_o \sqrt{\frac{b_o F_2}{k_2 F_3^2}} \tag{23}$$

where

$$F_3 = \int_0^\infty f\, h\, d\eta$$

The earliest experiments on the plane jet were performed by Foerthmann [2]. Since then, many other experiments on the plane jet have been done by Albertson et al. [5], Zijnen [6,7], Heskestad [8] and others. These measurements have shown conclusively that the u velocity profiles at different x sections are indeed similar in the fully developed region, and that the exponential equation (Equation 7) is indeed a good approximation to this similarity curve. The length scale b grows linearly with x, but the coefficient k_2 has been found to vary to some extent with the turbulence level and the degree of uniformity of the velocity profile at the nozzle. For practical purposes, k_2 could be given a value of 0.097, or even 0.1. For the concentration profiles, it appears that $k \simeq 1.47$ [6,7]. Hence, taking k_2 as equal to 0.097, $\alpha_e = 0.052$, $k_1 = 3.70\ U_o\sqrt{b_o}$ and $k_3 = 3.17\ c_o\sqrt{b_o}$. We can recast the equations for the scales of the plane jet as

$$b = 0.097\, x \tag{24}$$

$$\frac{u_m}{U_o} = \frac{3.70}{\sqrt{x/b_o}} \tag{25}$$

$$\frac{c_m}{c_o} = \frac{3.17}{\sqrt{x/b_o}} \tag{26}$$

In these equations, x is the axial distance measured from a virtual origin, from which the ideal jet of momentum flux of M_o and zero diameter originates, and this virtual origin appears to be located slightly behind the nozzle [9]. For practical purposes, unless $x/2b_o$ is small, we could consider $\bar{x}$ the distance from the nozzle essentially the same as x itself.

If Q is the volumetric flux at any x station, and if Q_o is the flux from the nozzle (per unit length), we can show that

$$\frac{Q}{Q_o} = 0.44\sqrt{x/b_o} \tag{27}$$

Thus, at $x = 100b_o$ and $Q = 4.4Q_o$, and $3.4Q_o$ is the flux entrained from the surroundings. If E and E_o are, respectively, the kinetic energy flux at any section and at the nozzle, we can show that

$$\frac{E}{E_o} = \frac{2.64}{\sqrt{x/b_o}} \tag{28}$$

Thus at $x = 100b_o$, $E = 0.26E_o$, and one can calculate from Equation 28 the decay of the kinetic energy along the jet. Experimental observations [6,7] show that the eddy diffusivity $\epsilon \simeq 0.003x\, u_m$. Experimental observations [8] also have shown that the profiles of $\overline{u'^2}$, $\overline{v'^2}$, $\overline{w'^2}$ and $\overline{u'v'}$ in terms of $u_m{}^2$ at different sections are similar, with the maximum values being, respectively, about 8%, 3.5%, 4.2% and 2% approximately.

CIRCULAR JET

Let us next consider a circular jet of water (or air) issuing from a nozzle of diameter d with a uniform velocity of U_o into a large body of water (or air) at rest (Figure 1). The velocity difference produces a turbulent axisymmetrical shear layer surrounding the potential core, which disappears at an axial distance equal to about 6d. This region is known as the flow development region, which is followed by a fully developed flow region in which the maximum axial velocity, u_m (occurring on the axis of the jet), decays continuously as the axial distance, $\bar{x}$, from the nozzle increases. The jet grows in thickness but still behaves like a slender flow. In addition, the jet entrains the surrounding fluid, thereby diluting any passive pollutant that it might transport.

If u and v are the time-averaged velocities in the axial (x) and radial (r) directions at any point (x,r) in the jet and if the primary flow in the jet is steady, the relevant Reynolds equations can be simplified—using the conditions of axisymmetry and slender flow and the observation that laminar stresses are negligible compared to the corresponding turbulent stresses—to obtain [1]

$$u\frac{\partial u}{\partial x} + v\frac{\partial u}{\partial r} = \frac{1}{\rho r}\frac{\partial r\tau}{\partial r} \tag{29}$$

where τ is the total shear stress at any point in the axial direction equal to the sum of the laminar and the turbulent components. For turbulent jets, if the nozzle Reynolds number is at least a few thousand, the viscous shear stress can be neglected and τ becomes equal to the turbulent shear stress itself. The continuity equation reduces to

$$\frac{\partial ru}{\partial x} + \frac{\partial rv}{\partial r} = 0 \tag{30}$$

Equations 29 and 30 are referred to as the equations of motion for the circular turbulent jet.

Let us multiply Equation 29 by ρr and integrate with respect to r from $r = 0$ to a large value (for convenience, infinity) of r outside the jet. With the use of Equation 30, this operation reduces Equation 29 to

$$\frac{d}{dx}\int_o^\infty 2\pi r\, dr\, \rho u^2 = 0 \tag{31}$$

which says that the flux of the axial (x) momentum in the jet is preserved as the jet travels away from the nozzle. Integrating Equation 31 with respect to x gives

$$\int_{o}^{\infty} 2\pi r dr\, \rho u^2 = M_o = \frac{\pi d^2}{4}\rho U_o^2 \tag{32}$$

wherein M_o is the momentum flux of the jet as it leaves the nozzle.

If we next integrate the continuity equation from $r = 0$ to $r = \bar{b}$, where $u \simeq 0$, using the Liebnitz rule, we obtain

$$\frac{d}{dx}\int_{o}^{\bar{b}} 2\pi r\, dr\, u = -v_e\, 2\pi\bar{b} \tag{33}$$

where v_e is the entrainment velocity. Let

$$-v_e = \alpha_e\, u_m \tag{34}$$

where α_e is the entrainment coefficient. With Equation 34, Equation 33 can be reduced to

$$\frac{d}{dx}\int_{o}^{\bar{b}} r dr\, u = \alpha_e\, \bar{b}\, u_m \tag{35}$$

Experimental observations on the u velocity profiles at different x sections have shown that these velocity profiles possess the similarity property, and we can write

$$\frac{u}{u_m} = f\left(\frac{r}{b}\right) = f(\eta) \tag{36}$$

where for any x section, u is the axial (x) velocity at a radial distance of r, u_m is the value of u on the axis, b is a length scale, generally taken as equal to r where $u = 1/2\, u_m$. The velocity profile similarity function $f(\eta)$ can be found by solving Equations 29 and 30 along with an auxiliary equation. For solutions of this kind using the mixing length idea and the eddy diffusivity concept, see the book by Rajaratnam [1], and for solutions using other kinds of closure models, see the book by Launder and Spalding [4]. In this chapter, we are going to assume the convenient exponential function described by Equation 7, which is very satisfactory for most purposes.

Having introduced the similarity concept and Equation 7, to solve for the u velocity at any point in the jet all we have to do is to predict the variation of the velocity and length scales u_m and b, respectively. Then, using the continuity equation, the v velocity profile can be predicted [1]. The turbulent shear stress profile also can be predicted using, in addition, Equation 29.

Let us now assume that $u_m = k_1\, x^p$ and $b = k_2\, x^q$, where k_1 and k_2 are x-independent parameters and p and q are the unknown exponents.

To evaluate the unknown exponents and parameters, consider Equation 35, which can be rewritten as

$$\frac{d}{dx} b^2 u_m \int_{o}^{\beta} \eta f d\eta = \beta\alpha_e\, b u_m \tag{37}$$

where β (equal to $\bar{b}/b$) is approximately equal to 2.5 for the exponential function. Letting $F_4 = \int_{o}^{\beta}\eta f d\eta$, wherein F_4 is a numerical constant; for both sides of Equation 40 to be of the same power in x, $2q + p - 1 = q + p$, which reduces to $q = 1$. That means the circular jet grows linearly in the x direction. Considering Equation 31,

$$\frac{d}{dx} u_m^2 b^2 \int_0^\infty \eta f^2 \, d\eta = 0 \qquad (38)$$

where $F_5 = \int_0^\infty \eta f^2 \, d\eta$ is a constant. Then we can write $2p + 2q = 0$. Since $q = 1$, $p = -1$. Thus, for a circular turbulent jet, $u_m \propto 1/x$ and $b \propto x$. To evaluate k_1 and k_2 from Equation 37,

$$\frac{d}{dx}(F_4 k_1 k_2^2 x) = \beta \alpha_e k_1 k_2 \qquad (39)$$

or

$$k_2 = \frac{\beta \alpha_e}{F_4} \qquad (40)$$

From Equation 32 and Equation 38,

$$8F_2 k_1^2 k_2^2 = U_o^2 d^2$$

or

$$k_1^2 k_2^2 = \frac{M_o}{2\pi\rho F_5} \qquad (41)$$

We thus find that among the three unknowns, α_e, k_1 and k_2, if we evaluate one experimentally the other two can be calculated.

If the jet carries a passive pollutant, we could study the distribution of the pollutant in the jet using the pollutant conservation equation. For the circular jet, the pollutant conservation equation can be reduced to

$$u \frac{\partial c}{\partial x} + v \frac{\partial c}{\partial r} = \frac{1}{r} \frac{\partial}{\partial r}\left(r \epsilon_r \frac{\partial c}{\partial r}\right) \qquad (42)$$

where c is the time-averaged pollutant concentration at any point, and ϵ_r is the eddy diffusivity for pollutant transport. Integrating Equation 42 with respect to r from $r = 0$ to $r = \infty$, we obtain

$$\frac{d}{dx} \int_0^\infty r\,dr\, uc = 0 \qquad (43)$$

which says that the pollutant flux is conserved as the jet travels away from the nozzle. Integrating Equation 43 with respect to x, we get

$$\int_0^\infty r\,dr\, uc = \frac{P_o}{2\pi} \qquad (44)$$

where P_o is the pollutant flux from the nozzle, equal to $\frac{\pi d^2}{4} U_o c_o$, c_o being the pollutant concentration at the nozzle.

Experimental observations on the concentration profiles at different x sections indicate that

$$\frac{c}{c_m} = h(r/b) = h(\eta) \qquad (45)$$

where c_m is the maximum value of c at any section occurring at $r = 0$, and

$$\frac{c}{c_m} = \exp\left[-.693\left(\frac{r}{kb}\right)^2\right] \tag{46}$$

where k appears to be a constant, equal to about 1.17 [10]. If we assume that $c_m = k_3 x^s$ from Equation 43,

$$\frac{d}{dx} b^2 u_m c_m \int_o^\infty \eta fh \, d\eta = 0 \tag{47}$$

From Equation 47 we can write $2q + p + s = 0$, with $q = 1$, $p = -1$, $s = -1$; that is, $c_m \propto \frac{1}{x}$. To evaluate k_3 from Equation 44,

$$b^2 u_m c_m F_6 = \frac{P_o}{2\pi} \tag{48}$$

where $F_6 = \int_o^\infty \eta fh \, d\eta$

Equation 48 can be reduced to

$$k_3 = \frac{P_o}{2\pi} \frac{1}{F_3 k_1 k_2^2} \tag{49}$$

where $F_6 = 0.722 \dfrac{k^2}{1 + k^2}$

If k is known and knowing k_2 and k_1, k_3 can be calculated.

Experiments on circular jets have been performed by Trupel [11], Corrsin [12], Hinze and Zijnen [10] and Albertson et al. [5], among others. From these numerous studies, the u velocity profiles have been shown to be similar [1], and the exponential equation is a good approximation to this similarity profile. The length scale has been found to grow linearly with the axial distance, x, even though the value of k_2 has been found to vary to some extent with the turbulence level and nonuniformity of the velocity distribution in the jet at the nozzle. For practical purposes, a value of 0.097 has been suggested by Abramovich [3]. Strictly speaking, x is measured from a virtual origin, located behind the nozzle, but for reasonably large values of x/d, x could conveniently be measured from the nozzle itself.

With $k_2 = 0.097$ and with $F_5 = 0.361$ from Equation 41, k_1 is equal to $dU_o/0.164$. Then the velocity scale u_m in terms of the jet velocity at the nozzle is given by the expression

$$\frac{u_m}{U_o} = \frac{6.13}{x/d} \tag{50}$$

wherein the numerical factor is slightly smaller than the value suggested by Rajaratnam [1]. The entrainment coefficient, α_e, can be evaluated from Equation 40. With $F_4 = 0.722$, $\beta \simeq 2.5$ and $k_2 \simeq 0.097$, $\alpha_e = 0.028$. Regarding k_3 for the pollutant concentration, from Equation 57 with $k = 1.17$

$$k_3 = \frac{21.34}{\pi} \frac{P_o}{U_o d} \tag{51}$$

and the scale for pollutant concentration, in terms of the concentration at the nozzle c_o, reduces to

$$\frac{c_m}{c_o} = \frac{5.34}{x/d} \tag{52}$$

If Q is the volumetric flux at any x section, we could show that

$$\frac{Q}{Q_o} = 0.33\,\frac{x}{d} \tag{53}$$

where Q_o is the volumetric flux at the nozzle. At $x/d = 100$, $Q = 32Q_o$, where $31Q_o$ is the contribution from the entrainment. If E is the kinetic energy flux at any section and E_o is that at the nozzle,

$$\frac{E}{E_o} = \frac{4.1}{x/d} \tag{54}$$

Turbulence measurements in the circular jet have been made by Corrsin [12] and Wygnanski and Fielder [13]. Profiles of $\overline{u'^2}$, $\overline{v'^2}$, $\overline{w'^2}$, and $\overline{u'v'}$ in terms of $u_m{}^2$ have been found to be similar, with the maximum values of the respective ratios equal to 8%, 6%, 6% and 1.7%.

SHEAR LAYERS

Let us consider a plane jet of large (or semiinfinite) width and uniform velocity of U_0 flowing over a stagnant mass of the same fluid, as shown in Figure 3. The intense shear at the surface of velocity discontinuity induces turbulent mixing and, as a result, the stagnant fluid is accelerated and a portion of the jet loses some momentum. The thickness of the layer affected by this momentum exchange is known as the shear layer, and its thickness at any x station can be denoted by $\overline{b}$. In Figure 3, the ϕ_1 and ϕ_2 lines denote the edges of the shear layer.

The velocity profile measurements by Liepmann and Laufer [14] have shown that the distribution of u/U_o across the shear layer is similar and, here again, the exponential equation describes the similarity profile well.

The equations of motion for the plane shear layer are the same as Equations 1 and 2. Assuming that $u/U_o = f(\eta)$ where $\eta = y/b$ (b being a suitable length scale), and assuming that $\tau/\rho U_o{}^2 = g(\eta)$ and using Equation 1 to obtain the relation for v as

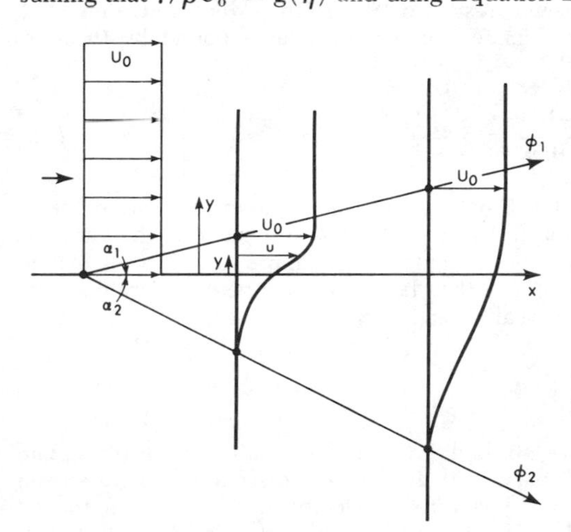

Figure 3. Plane shear layers.

$$v = \int_{y_*}^{y} \frac{\partial v}{\partial y}\, dy \tag{55}$$

where y_* = y at the location of v being equal to zero, Equation 1 can be reduced to

$$g' = b' [\eta f f' - f' F_7] - b' h(x) f' \tag{56}$$

where the primes on g and f denote differentiation with respect to η, and the prime on b denotes differentiation with respect to x, $F_7 = \int \eta f' \, d\eta$ and $h(x) = F_7(\eta_*)$. For Equation 56 to be satisfied, $b \propto x$ and $y_* \propto x$.

Tollmien [1] used the Prandtl mixing length hypothesis with the assumption that the mixing length $l \propto b$ and solved Equation 1 using Equation 2 to obtain the u velocity distribution in the shear layer. This indicates that on the y = 0 plane, $u = 0.68\, U_o$. For the solution to be complete, we need the experimental determination of just one parameter, $a = [2(l/x)^2]^{1/3}$. Goertler [1] employed the constant eddy viscosity model and solved the general shear layer problem, wherein a uniform stream with velocity of U_o flows over a slower stream with a velocity of U_1.

Experiments on the simple shear layer have been made by Liepmann and Laufer [14], Albertson et al. [5] and Wygnansky and Fiedler [15], among others. Using these results, we find that Tollmien's a = 0.084. With this value of a, for the plane turbulent shear layer $b = 0.115x$, $\bar{b} = 0.263x$, and the angles α_1 and α_2 of the outer limits of the shear layer (Figure 3) are equal to 4.80° and 9.50°, respectively.

With the knowledge of plane shear layers, we can easily find the length of the potential core, x_o, of plane jets as

$$x_o = \frac{b_o}{\tan 4.8°} = 11.91\, b_o \tag{57}$$

Combining the uniform velocity in the potential core and the velocity distribution in the shear layer, the characteristics of the flow development region of plane jets can be found [1].

If now we consider the general (or compound) shear layer, which forms between a deep stream with uniform velocity of U_o flowing tangentially over another deep stream with a smaller uniform velocity of U_1, the solutions of Tollmien and Goertler can be extended [1]. Also, using the experimental results of Watt [16], Miles and Shih [17], Yule [18,19] and others, if Δb is the thickness of the shear layer bound by the lines of $(u - U_1)/(U_o - U_1) = 0.1$ and 0.9,

$$\frac{\Delta b}{0.165x} = \frac{\alpha - 1}{\alpha + 1} \tag{58}$$

where α is equal to the ratio of U_o to U_1. Further, if α_1 and α_2 are the angles of penetration of the shear layer into the faster and slower streams, respectively, the variation of α_1 and α_2 with $1/\alpha = U_1/U_o$ is shown in Figure 4. In Figure 4, it is seen that as $1/\alpha$ approaches unity, α_2 approaches α_1 and they both approach zero. Rajaratnam [1] provides a similar analysis of axisymmetrical shear layers.

JETS IN COFLOWING STREAMS

In the first three sections, we have studied primarily the diffusion of plane and circular jets in a large expanse of stationary fluid. If the jet is discharged into a stream flowing in the same direction as the jet, but with a velocity of U_1 less than the jet velocity of U_o, then we have the problem of a jet in a coflowing stream. We will discuss

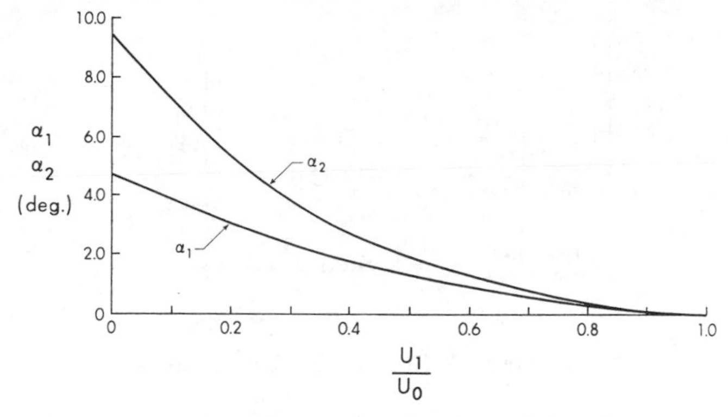

Figure 4. Spreading angles of compound shear layers.

in some detail the diffusion of a plane jet in a coflowing stream; for a similar treatment of circular jets refer to the literature [1].

With reference to Figure 5, in the flow development region we have a potential core with a compound shear layer on the top and bottom. Using the results presented in the previous section, this region can be analyzed; this section will deal primarily with the fully developed flow region. In this region, experimental observations have shown that if the flow in the jet is looked at with respect to the coflowing stream, then the u velocity profiles possess the similarity property. That is, if $U = (u - U_1)$ and U_m is the maximum value of U equal to $(u_m - U_1)$, then $U/U_m = f(\eta)$, where $\eta = y/b$, b being equal to y where $U = 1/2\ U_m$. It also has been observed that $f(\eta)$ is well described by the exponential relation discussed earlier. Then for predicting the structure of the mean flow in the jet immersed in a large coflowing stream we will have to predict the manner of variation of U_m and b with the axial distance.

Equations 1 and 2 are the equations of motion, and integrating Equation 1 with Equation 2 one can obtain the integral momentum equation as

$$\frac{d}{dx}\int_0^\infty \rho u\,(u - U_1)\,dy = 0 \tag{59}$$

Equation 59 states that the excess momentum of the jet (with respect to the freestream) is preserved. We can reduce Equation 59 to the form

$$\frac{d}{dx}\left[b\frac{U_m}{U_1}\int_0^\infty f d\eta + b\left(\frac{U_m}{U_1}\right)^2\int_0^\infty f^2\,d\eta\right] = 0 \tag{60}$$

Now consider the so-called "strong jet case," where $U_m/U_1 >> 1$. For this case, the first term in Equation 60 can be dropped as an approximation, whereas for the "weak jet case," for which $U_m/U_1 << 1$, the second term in Equation 60 can be neglected, leaving only the first term. If we assume that $U_m \propto x^p$ and $b \propto x^q$, then with these simplifications $q + 2p = 0$ for the strong jet case, whereas $p + q = 0$ for the weak case. To derive another equation for each one of these two cases, we can do a similarity analysis of the equations of motion. From such an analysis [1] it can be seen that for the strong jet case $q = 1$, and this leads to $p = -1/2$. For the weak jet case, $q - p = 1$. As a result, $p = -1/2$ and $q = 1/2$. For the intermediate case, in which $U_m \sim U_1$, it does not appear to be possible to have a simple power law type of variation.

Using dimensional arguments, we can show that for the jet in a coflowing stream

$$\frac{U_m}{U_1} = F(\sqrt{x/\theta}) \tag{61}$$

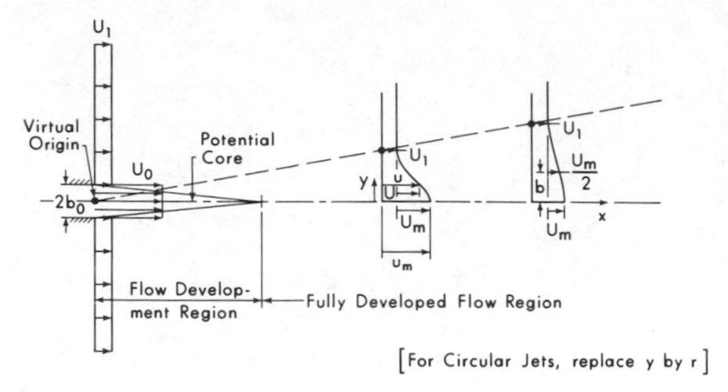

Figure 5. Jet in a coflowing stream.

where F denotes a function and θ is the momentum thickness defined by the equation

$$M_o = 2\theta \, \rho U_1^2 \tag{62}$$

where M_o is the excess momentum in the jet. A number of methods have been proposed to predict the function F in Equation 61 [1]. We will simply present a correlation developed by Pande and Rajaratnam [20], using the integral momentum and energy equation, along with a hypothesis regarding the turbulence production in the transition region from the strong jet to the weak jet flows (Figure 6). This correlation appears to describe the experimental data better than the other computational schemes. Once the velocity scale has been determined for a given problem, the growth of the length scale can be obtained using the integral momentum equation. Turbulence characteristics of the plane jet in a coflowing stream have been studied by Bradbury [21].

CIRCULAR JETS WITH SWIRL

In the section entitled Circular Jet we considered the diffusion of a circular jet. If we give this jet a certain amount of swirl before it leaves the nozzle, we would find that the jet spreads more rapidly than the corresponding nonswirling jet. The maximum velocity, u_m, decays more rapidly, and if the swirl is relatively strong, we could produce reverse flow. Swirling jets have been used in combustion chambers and burners and could be used effectively in situations in which large dilutions are needed in relatively short distances. In this section, we are going to study the characteristics of the swirling circular jet in the region of fully developed flow.

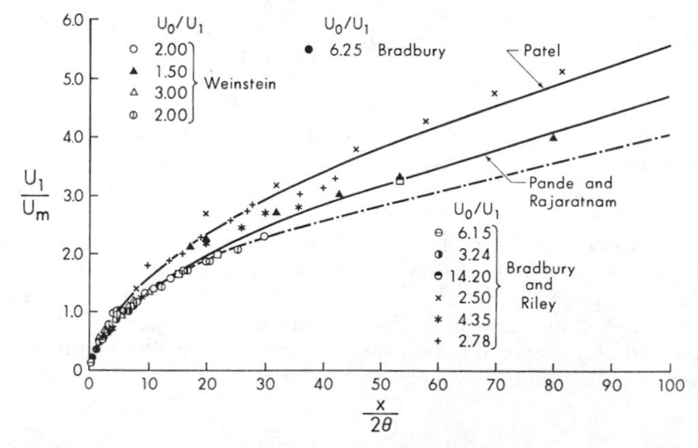

Figure 6. Velocity scale for plane jets in coflowing stream [20].

For the circular jet with swirl, the Reynolds equations can be reduced to the form [1]

$$u\frac{\partial u}{\partial x} + v\frac{\partial u}{\partial r} = -\frac{1}{\rho}\frac{\partial p}{\partial x} + \frac{1}{\rho r}\frac{\partial}{\partial r}\tau_{rx} \tag{63}$$

$$\frac{w^2}{r} = \frac{1}{\rho}\frac{\partial p}{\partial r} \tag{64}$$

$$u\frac{\partial w}{\partial x} + v\frac{\partial w}{\partial r} + \frac{vw}{r} = \frac{1}{\rho}\frac{\partial}{\partial r}\tau_{r\phi} + \frac{2}{\rho r}\tau_{r\phi} \tag{65}$$

where u, v and w are the time-averaged velocities in the axial (x), radial (r) and peripheral (ϕ) directions, and τ_{rx} and $\tau_{r\phi}$ are the turbulent shear stresses. The continuity equation is Equation 30. Integrating Equation 63 with respect to r from r = 0 to r = ∞, using Equation 30, we can obtain the integral expression

$$\frac{d}{dx}\int_0^\infty (p + \rho u^2)\, rdr = 0 \tag{66}$$

which says that the pressure plus axial momentum flux, W, is conserved. If we perform a similar operation on Equation 65, we obtain

$$\frac{d}{dx}\int_0^\infty rdr\, \rho uw\, r = 0 \tag{67}$$

which says that the moment of the peripheral momentum flux or the angular momentum, T, is preserved. Using W, T and the radius of the jet, r_o, we can formulate the swirl number, $S = T/Wr_o$. We will see later that the behavior of the swirling jet depends very much on its swirl number.

Experimental observations of Chigier and Chervinsky [22,23] have indicated that, at least for moderate values of S, the u and w velocity profiles are similar (see Figure 7). That is,

$$\frac{u}{u_m} = f(\eta) \text{ and } \frac{w}{w_m} = g(\eta)$$

where $\eta = r/b$ and u_m and w_m are the scales for u and w velocities, respectively. For the case of weak swirl, such that $u_m >> w_m$, using the integral Equations 66 and 67 and the integral continuity equation with the entrainment hypothesis, it can be shown [1] that $b \propto x$; $u_m \propto 1/x$ and $w_m \propto 1/x^2$. For the case with $u_m \sim w_m$, performing a similarity analysis on the x direction equation and considering the two integral momentum equations, it can be shown that $b \propto x$; $u_m \propto 1/x\sqrt{x}$ and $w_m \propto 1/x\sqrt{x}$.

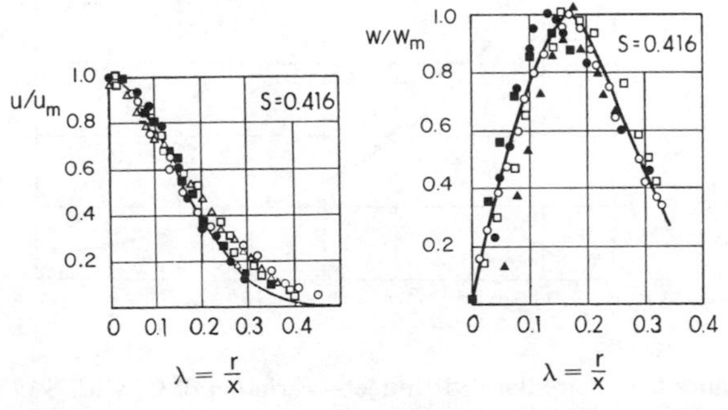

Figure 7. Velocity profiles in circular swirling jets [23].

The experimental observations of Chigier and Chervinsky [22,23] for S up to 0.64 agree with the exponents derived above for the case of $u_m >> w_m$. If x is the angle defined by the relation $\tan \alpha = b/x$, α is connected with S by the relation

$$\alpha\ (\text{deg}) = 4.8 + 14\ S \tag{68}$$

The u velocity profiles were described satisfactorily by the exponential equation for S up to about 0.42. The w velocity profiles were described by the equation

$$\frac{w}{w_m} = C\lambda + D\lambda^2 + E\lambda^3 \tag{69}$$

where $\lambda = r/x = \eta\, b/x$ and C, D and E are constants that depend on the swirl number, S.

Using the similarity property of the u and w velocity profiles and the two integral momentum equations, Chigier and Chervinsky [22,23] developed an equation for u_m/U_o, which could be approximated by the simpler expression

$$\frac{u_m}{U_o} = \frac{C_1}{x/d} \tag{70}$$

where the coefficient C_1 decreases continuously with S, as shown in Figure 8. The corresponding relation for w_m is

$$\frac{w_m}{w_{mo}} = \frac{\hat{C}_1}{(x/d)^2} \tag{71}$$

where w_{mo} is value of w_m at the nozzle, and the variation of the coefficient $\hat{C}_1$ with S is not known satisfactorily at this time. Some turbulence measurements on swirling jets have been made by Pratt and Keffer [24].

JETS IN DUCTS

In the earlier sections on circular jets, we studied the diffusion of circular jets in a large mass of the same fluid. If the jet enters a circular duct of diameter D $(= 2R_o)$,

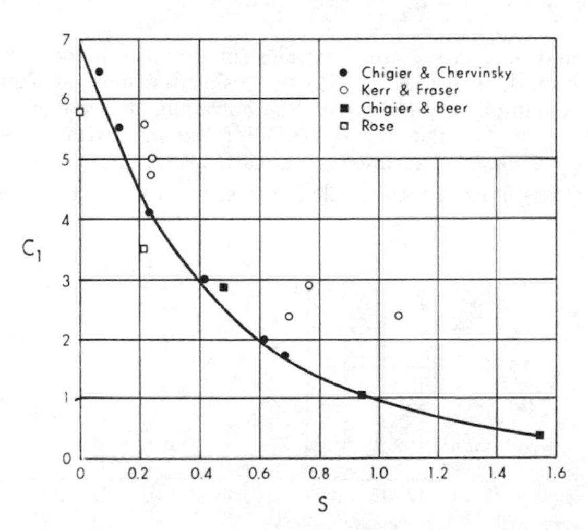

Figure 8. Circular swirling jet—Variation of C_1 with S [23].

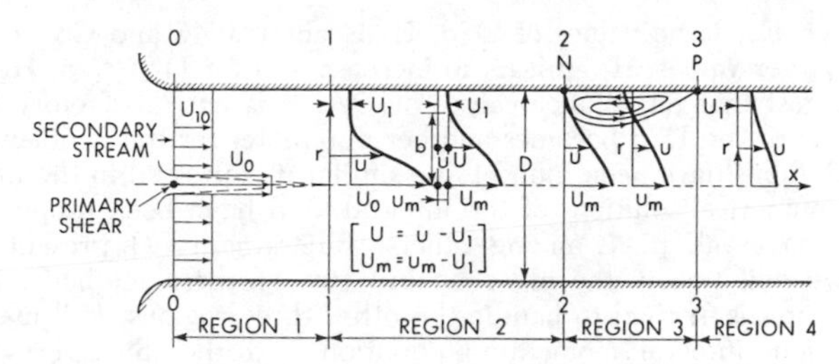

Figure 9. Circular jet in a duct.

which is only a few times the diameter of the jet, which carries a flow with mean velocity of U_1, then depending on the magnitude of U_o/U_1 and D/d the flow assumes different configurations. These will be discussed in this section. Rajaratnam [1] provides a similar study of plane jets in plane ducts.

Figure 9 shows a circular jet of diameter d and velocity U_o issuing axially into a circular duct, which carries what is known as the secondary stream. The jet itself is referred to as the primary stream, and let U_{10} be the velocity of the secondary stream at the plane of the nozzle. The resulting flow can be divided into a number of characteristic regions. At the end of region 1 the potential core of the jet is consumed. In regions 1 and 2, due to the entrainment by the jet, the secondary stream velocity decreases. As a result, the pressure increases in the longitudinal direction. If the jet velocity is large enough, the whole secondary stream is consumed by the jet at the end of region 2, and region 3 is characterized by an annular recirculation region. In region 4, the flow gradually degenerates to fully developed pipe flow.

The equations of motion for this problem are Equations 29 and 30, considered earlier, with the addition of $-1/\rho\ dp/dx$ to the right-hand side of Equation 29. If we integrate the continuity equation from $r = 0$ to $r = R_o$, the radius of the duct, we get

$$\frac{d}{dx}\int_o^{R_o} rdru = 0 \tag{72}$$

which says that the volumetric flux, Q, is invariant in the x direction. Next let us multiply Equation 29 by ρr and integrate from $r = 0$ to $r = R_o$ to obtain

$$\frac{d}{dx}\int_o^{R_o} (p + \rho u^2)\ rdr = 0 \tag{73}$$

if the wall shear is neglected. Equation 73 says that the pressure plus momentum flux, W, is also invariant in the x direction. Combining Q, W, ρ and R_o, the following dimensionless parameter H, is formulated:

$$H = \frac{1}{\sqrt{2\pi}}\frac{Q}{\sqrt{W/\rho}\, R_o} \tag{74}$$

This parameter is connected with the Craya-Curtet number, C_t, by the relation

$$H = \frac{1}{\sqrt{(1 + 2/C_t^2)}} \tag{75}$$

Experiments on circular jets in ducts have been performed by Mikhail [25], Becker et al. [26], Curtet and Ricou [27], Barchilon and Curtet [28], Exley and Brighton [29] and Razinsky and Brighton [30], among others. Based on these observations, we find that for H (or C_t), less than a critical value of H_c (or C_{tc}), a recirculation region or eddy forms in the duct. For large values of D/d, H_c is about 0.46 and $C_{tc} \simeq 0.73$. As D/d decreases from larger values, H_c appears to increase and for $D/d \simeq 5$, $H_c \simeq 0.6$. These observations suggest that for smaller values of D/d it is not satisfactory to replace the jet by a point source, and D/d becomes another parameter for this problem. Further, the u velocity distributions have been found to be similar if expressed in the proper form.

A number of numerical solutions of the integral kind have been proposed by Mikhail [25] and Brighton et al. [31], among others. Rajaratnam [1] presents a discussion of these, and we will review the ideas behind two of these methods. Hill neglected region 1 and set up his method to handle the other three regions. Hill used the integral continuity equation, integral momentum equation and the integral moment of momentum equation. For the secondary stream, he used the Bernoulli equation and the boundary layer on the duct wall was neglected, but a partial wall shear correction was made. Brighton et al. [31] used the integral continuity equation and integrated the momentum equation to three different outer levels. They used the Bernoulli equation for the secondary stream and introduced the wall shear stress. The results of these two methods have been found to be satisfactory.

CIRCULAR JETS IN CROSSFLOW

Circular jets injected into crossflows have been studied because of their many applications in industry and environmental problems, and also because it is a rather interesting and complex form of shear flow. Consider a circular jet of diameter, d, issuing with a uniform velocity of U_0 into a freestream or crossflow of large extent with an (almost) uniform velocity of U_1 as shown in Figure 10. Experimental observations show that close to the nozzle, the jet gets deflected due to the pressure field imposed by the crossflow. Due to the differential bending of the different parts of the diffusing jet, the jet acquired a characteristic kidney shape [3,32]. The deformed section of the jet houses on its sides a pair of attached vortices [33]. Entrainment of the freestream fluid and momentum causes further deflection of the jet and, after some more distance, the jet is moving only at a small differential angle from the crossflow and the cross section of the jet is almost completely occupied by the bound vortex pair. In addition to this bound vortex system, the crossflow creates another vortex system that is shed from the deformed jet [34].

It is known from experimental observations that the length of the potential core gets reduced because of the presence of the crossflow. In the fully developed region, in the ζ-η plane (Figure 10), the u velocity profiles have been found to be similar across the jet, with the exponential equation being a good approximation [35,36]. If η_b and ζ_b are the corresponding length scales, they are described approximately by

$$\frac{\eta_b}{d} = 0.25\,\xi/d \text{ and } \frac{\zeta_b}{d} = 0.05\,\xi/d \tag{76}$$

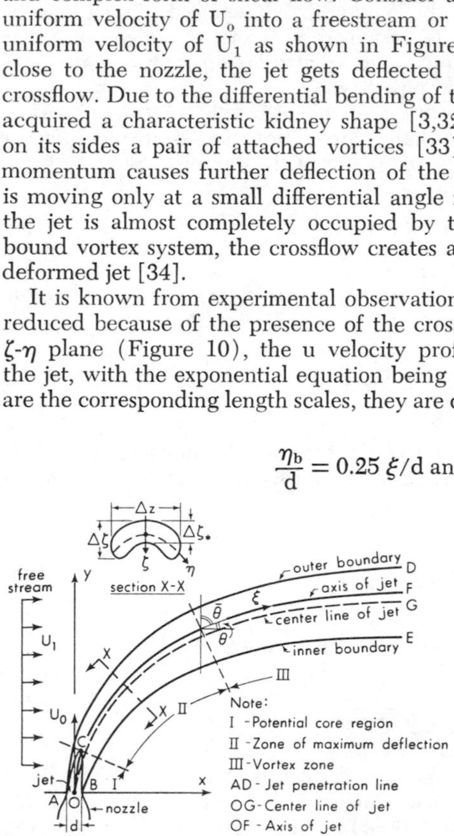

Figure 10. Circular jet in a crossflow.

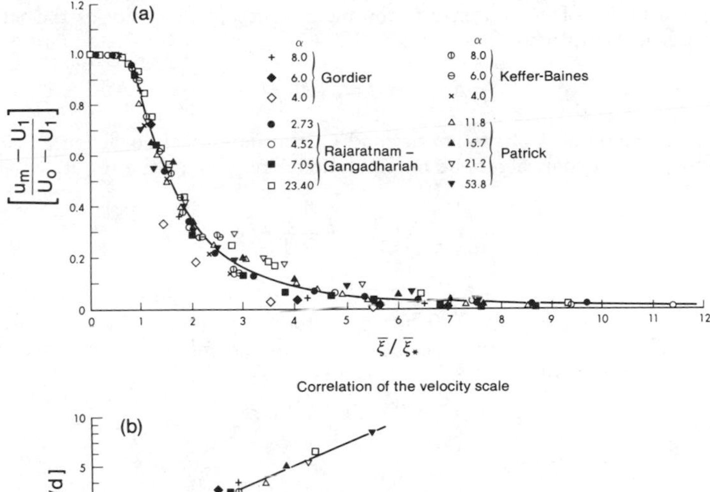

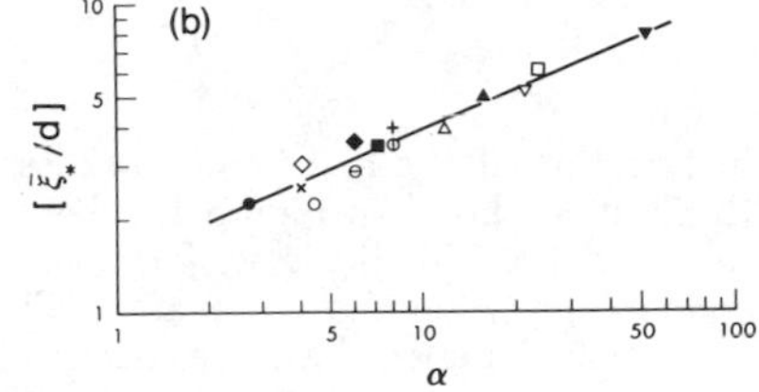

Figure 11. Correlation of velocity scale for circular jet in crossflow [36].

with ξ measured from virtual origins located behind the nozzle by distances of 1.3 d and 2.6 d, respectively, (for α at least in the range of 3 to 7 where $\alpha = U_o/U_1$). A useful correlation for the velocity scale is shown in Figure 11, where ξ_* is a length specified by $u_m - U_1 = (U_o - U_1)$ on a double-log plot [38].

If α_e is the entrainment coefficient defined by the relations

$$v_e = \alpha_e \,(u_m - U_1 \cos \theta)$$

$$v_e = \frac{1}{P}\frac{dQ}{d\xi}$$

where v_e is the entrainment velocity, θ is the angle of the deflected jet with the cross stream, P is the perimeter, and Q is the approximate volumetric flux in the deflected jet, α_e has been found to increase with $\bar{\xi}/\alpha d$, with $\bar{\xi}$ being measured from the nozzle. Further, the entrainment coefficient is significantly larger than that for the simple jet. For a range of the velocity ratio up to about 7, the ratio of the volumetric flux can be described by the equation

$$\frac{Q}{Q_o} = 0.54\ (\xi/d)^{1.22} \tag{77}$$

Considering the axis of the deflected jet, a number of methods have been proposed [1]. We will present here the outline of a method recently suggested [36]. For the deflected jet,

$$C_d \frac{\rho U_1^2}{2} \sin^2\theta \, d\xi \, \Delta Z = A \frac{d\xi \, \rho \, \overline{u}^2}{r} \tag{78}$$

where, with reference to Figure 10, A is the area of cross section, ΔZ is the transverse width, r is the radius of curvature, $\bar{u}$ is the mean (section) velocity of the jet, and C_d is a drag coefficient. Further,

$$\rho A \bar{u}^2 \sin\theta = C A_o \rho U_o^2 \tag{79}$$

where A_o is the value of A at the nozzle, and C is an empirical coefficient. Using Equations 78 and 79, after some algebraic manipulations one obtains the result

$$\frac{x}{d} = \frac{C_d}{\pi\alpha^2}\left(\frac{y}{d}\right)^2\left(\frac{\bar{\beta}}{3}\frac{y}{d} + 1\right) \tag{80}$$

in which C_d can be given a value of 1.5 and $\bar{\beta}$ is an empirical coefficient which is mainly a function of α. Equation 80 has been found to predict well the experimental jet profiles and Equation 80 also has been extended to oblique jets. Rajaratnam [1] discusses plane jets in crossflow.

WALL JETS

In the plane jet considered in the first section, if the half-jet of thickness b_o is made to discharge tangentially onto a smooth plate, as shown in Figure 12, then the resulting flow configuration is known as the plane wall jet. This plane wall jet will have a flow development region in which the boundary layer on the plate also will have to be considered. Considering now the fully developed region downstream of the end of the potential core, experimental observations have shown that (Figure 13a) this is a slender flow. Further, at any x section, as y increases from zero the u velocity increases to a maximum of u_m at $y = \delta$, where δ is the thickness of the boundary layer; and for y greater than δ, u decreases continuously from u_m to zero. If these profiles are plotted in a dimensionless form with u_m and b where b = y at u = 1/2 u_m as the scales, as shown in Figure 13b, they are found to be similar.

The equations of motion are Equations 1 and 2 discussed before, with the difference that τ now must be interpreted as the sum of the turbulent shear stress and the laminar shear stress, which dominate over the former as the boundary (or wall) is approached. Integrating Equation 1 with respect to y from y = 0 to y = ∞, we get

$$\frac{d}{dx}\int_0^\infty \rho u^2 dy = -\tau_o \tag{81}$$

where τ_o is the wall shear stress at any distance, x. If τ_o is neglected as a first approximation,

$$\frac{d}{dx}\int_0^\infty \rho u^2 dy = 0 \tag{82}$$

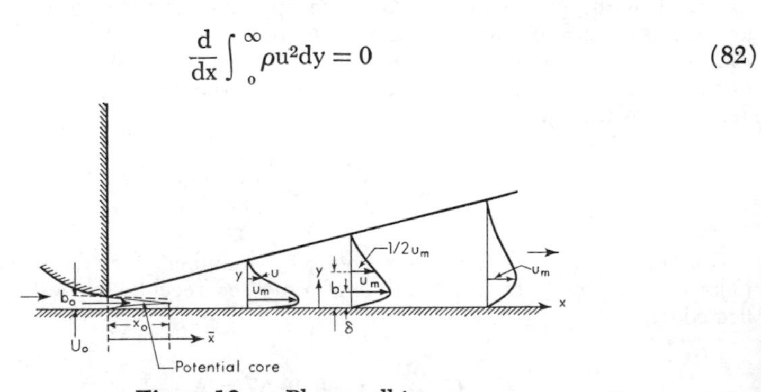

Figure 12. Plane wall jet.

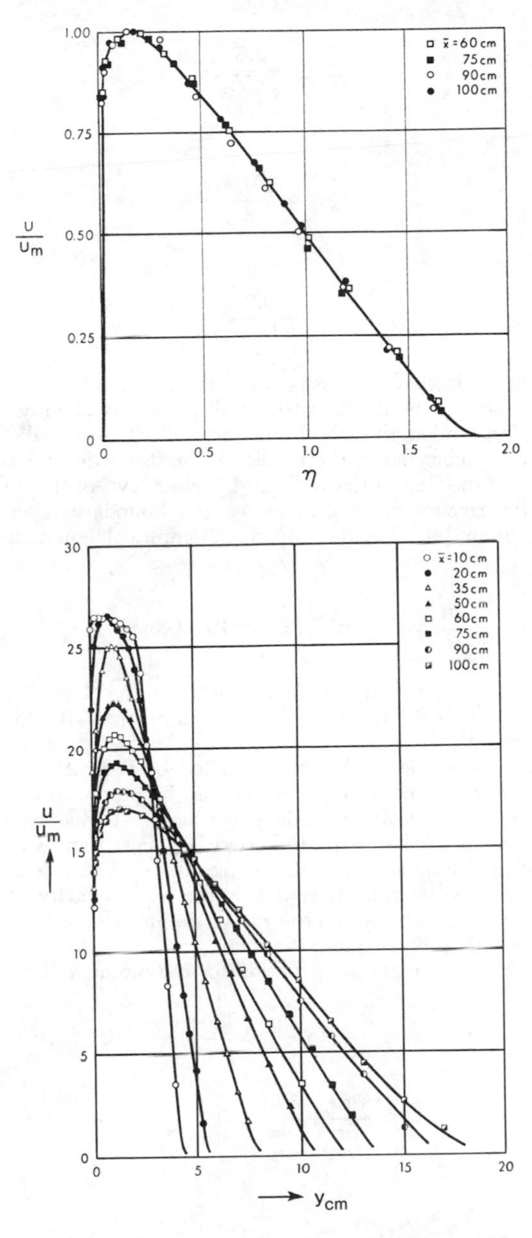

Figure 13. Velocity profiles in plane wall jets [2].

If $u_m \propto x^p$ and $b \propto x^q$ from Equation 82, after rewriting it in the form of Equation 9, $2p + q = 0$. A similarity analysis of Equation 1 with the neglect of the viscous term gives $q = 1$. The entrainment hypothesis also would give $q = 1$. Then $p = -1/2$. Then the variations will be the same as those for the plane jet. Dimensional considerations indicate that $\tau_o \propto 1/x$.

Experiments on plane wall jets have been made by Foerthmann [2], Sigalla [37], Myers et al. [38] and Schwarz and Cosart [39], among others. Based on these results, we could write

$$b = 0.068x \tag{83}$$

$$\frac{u_m}{U_o} = \frac{3.5}{\sqrt{x/b_o}} \tag{84}$$

and

$$\tau_o = c_f \frac{\rho U_o^2}{2} \tag{85}$$

where

$$c_f = \frac{0.2}{(x/b_o)(U_o b_o/\nu)^{1/12}} \tag{86}$$

and ν is the coefficient of kinematic viscosity. Further, $\delta \simeq 0.16b$ and $\alpha_e = 0.035$. It can be noticed that the velocity scale for the plane wall jet is only slightly less than that for the plane (free) jet, but the length scale, b, for the wall jet is only about 0.7 times that of the free jet. In the boundary layer, the u velocity profiles can be described by power laws or a combination of the "law of the wall" and "defect law" of the turbulent boundary layers, whereas in the free-mixing region above the boundary layer the exponential equation is valid. Verhoff [40] has developed an empirical equation to describe the complete profile as

$$\frac{u}{u_m} = 1.48\, \eta^{1/7} [1 - \text{erf}\,(0.68\eta)] \tag{87}$$

where erf represents the error function. The length of the flow development region has been found to vary from 6.1 b_o to 6.7 b_o for $U_o b_o/\nu$ from 10^4 to 10^5. Turbulence observations on plane wall jets have been made by Mathieu and Tailland [41]. The effect of wall roughness on plane jets has been studied by Rajaratnam [42].

We will next consider the radial wall jet, which can be produced by the impingement of a circular jet on a large circular boundary (or wall). Radial wall jets on smooth walls have been studied by Bakke [43], Heskestad [8], Poreh and Cermak [44], Poreh et al. [45] and Beltaos and Rajaratnam [46], among others. With reference to Figure 14, if u is the time-averaged radial velocity and v is the axial velocity at any point, then experiments indicate that the u velocity profiles are similar and that the similarity curve is essentially the same as that for the plane wall jet.

We can show [1] that the equations of motion for the radial wall jet are

$$u\frac{\partial u}{\partial r} + v\frac{\partial u}{\partial z} = \frac{1}{\rho}\frac{\partial \tau}{\partial z} \tag{88}$$

$$\frac{\partial ru}{\partial r} + \frac{\partial rv}{\partial z} = 0 \tag{89}$$

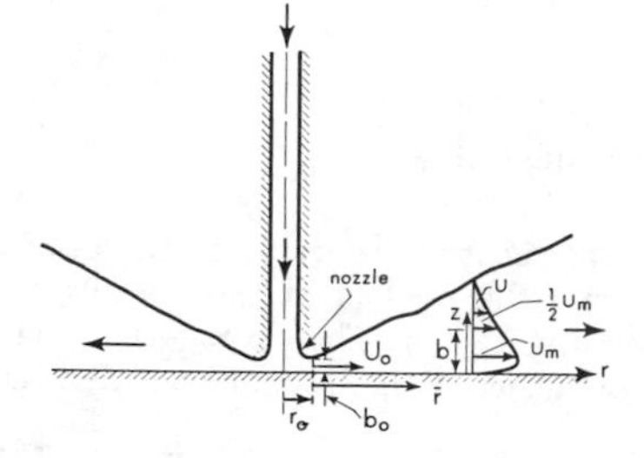

Figure 14. Radial wall jets.

where τ is the sum of the turbulent and laminar shear stresses. Integrating Equation 88 with respect to z for z from 0 to ∞,

$$\frac{d}{dr}\int_{0}^{\infty} \rho u^2 \, r dz = -r\tau_o \tag{90}$$

where τ_o is the wall shear stress. If the right-hand side is neglected, then if $u_m \propto r^p$ and $b \propto r^q$, we get $2p + q + 1 = 0$. Integrating the continuity equation,

$$\frac{d}{dr} r \, u_m \, b \int_{0}^{\infty} f \, d\eta = r\alpha_e \, u_m \tag{91}$$

which gives $q = 1$, then $p = -1$. Going back to Equation 90, or from dimensional considerations, it follows that $\tau_o \propto 1/r^2$.

Based on Bakke's [43] results, $b = 0.078r$. For the velocity scale, using Heskestad's [8] experiment we could suggest the following approximate relation:

$$\frac{u_m}{U_o} = \frac{3.5}{\sqrt{\left(\frac{r}{r_o}\right)\left(\frac{r}{b_o}\right)}} \tag{92}$$

where b_o and r_o are, respectively, the thickness and radius of the wall jet at the source.

For radial wall jets produced by impinging circular jets (issuing from a nozzle of diameter, d, located at a height, H, above the wall) the velocity scale for the wall region downstream of the impinging jet is given by the equation

$$\frac{u_m}{U_o} = \frac{1.03}{r/d} \tag{93}$$

where U_o is the velocity at the nozzle, and r is the radial distance from the impingement point. In this region, $b = 0.087r$. Beltaos and Rajaratnam [46] give a more detailed treatment of circular impinging jets. Rajaratnam [1] discusses cylindrical wall jets.

THREE-DIMENSIONAL JETS

Earlier we discussed the diffusion of plane and circular jets. Jets that issue from nozzles that are neither axisymmetrical nor rectangular, with a very large aspect ratio or width to height ratio (B/h) (Figure 15) are referred to here as three-dimensional jets. Consider the decay of the centerline or maximum velocity with x for a jet issuing from a rectangular nozzle with an aspect ratio of, say, 20. If U_o and u_{mo} are, respectively, the efflux and centerline velocities, the variation of u_{mo}/U_o with x/h will be as shown schematically in Figure 15. In region I, $u_{mo}/U_o = 1$ and this is the potential core region. In region II, the velocity decay is the same as that of plane jets, and in region III u_{mo} varies inversely with x, which is the decay rate of axisymmetrical jets. Trentacoste and Sforza [47,48] referred to regions II and III, respectively, as the characteristic decay region and axisymmetrical-type decay region. If we consider, as a second case, the jet form of a square or equilateral triangular nozzle, we will observe that the potential core region is followed by the axisymmetrical-type decay region. Three-dimensional jets with the three distinct regions are referred to as (three-dimensional) slender jets, and those that have only the potential core and axisymmetrical-type regions are referred to as the (three-dimensional) bluff jets.

For slender jets from rectangular nozzles, Figure 16 shows the decay of u_{mo}/U_o with x/h. For any given aspect ratio, after the end of the potential core the velocity decay is along the plane jet curve up to some value of x/h, beyond which it falls along

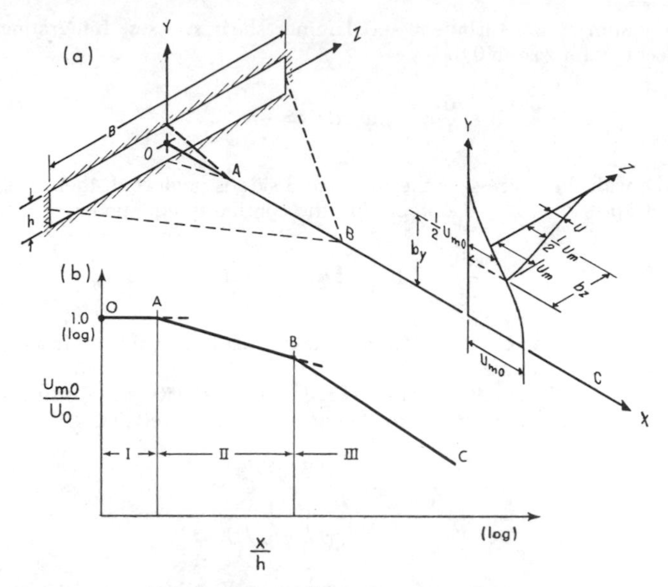

Figure 15. Three-dimensional jets.

the circular jet curve. In the second region, Trentacoste and Sforza [47] found that the u velocity distribution was similar along the Y axis, but was nonsimilar along the Z axis. But in the third region, similarity existed in both directions.

Bluff jets originating from square nozzles have been studied by Yevdjevich [49], Trentacoste and Sforza [47,48] and Pani [50]. Trentacoste and Sforza [47] also studied the bluff jet from a equilateral triangular nozzle. For the jet from a square nozzle, Pani [50] found that in the axisymmetrical-type decay region the u velocity distribution was similar in the Y and Z directions, and the respective length scales b_y and b_z were equal to 0.097 x.

The equations of motion for a bluff jet are

$$u\frac{\partial u}{\partial x} + v\frac{\partial u}{\partial y} + w\frac{\partial u}{\partial z} = -\left[\frac{\partial \overline{u'v'}}{\partial y} + \frac{\partial \overline{u'w'}}{\partial z}\right] \tag{94}$$

and

$$\frac{\partial u}{\partial x} + \frac{\partial v}{\partial y} + \frac{\partial w}{\partial z} = 0 \tag{95}$$

Integrating Equation 94 using Equation 95, we can show that

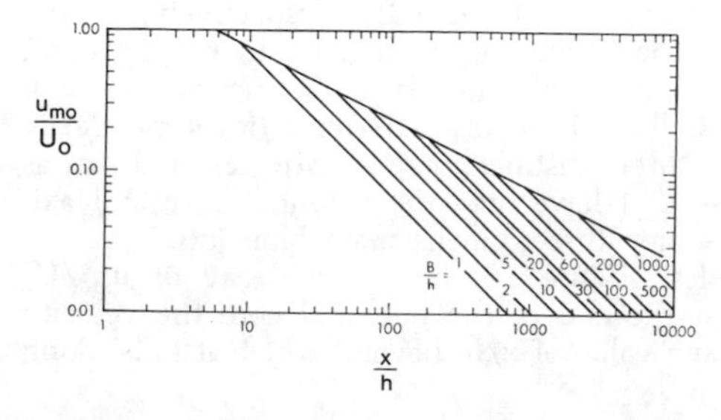

Figure 16. Velocity scales for slender jets [49].

$$\frac{d}{dx}\int_{o}^{\infty}\int_{o}^{\infty}\rho\, u^2\, dydz = 0 \tag{96}$$

which simply says that the momentum flux is conserved. If $u_{mo} \propto x^p$, and b_y and b_z are proportioned to x^q, from Equation 96 we can write $2p + 2q = 0$. If we assume that u_{mo} is a function of only the momentum flux, ρ and x, then from dimensional considerations we can show that

$$\frac{u_{mo}}{U_o} = \frac{C}{(x/\sqrt{A})} \tag{97}$$

where A is the area of the nozzle with $p = -1$, $q = 1$. Pani [50] has been able to correlate the results for square, triangular, rectangular (aspect ratio of 10) and elliptical jets on the basis of Equation 96.

Three-dimensional wall jets also could be divided into two groups as slender and bluff wall jets, for which Rajaratnam [1] gives a preliminary discussion.

NOMENCLATURE

- A — area of cross section of nozzle, also of jet in crossflow
- a — Tollmien coefficient
- B — length of jet
- b — length scale, also suffix for length scale
- b_o — half-thickness (width) of nozzle
- C — coefficient
- C_t — parameter
- c — time-averaged pollutant concentration
- D — diameter of duct
- d — diameter of nozzle
- e — suffix
- F — definite integrals (with suffixes)
- f — function (with suffixes), also suffix
- g — function
- H — parameter
- h — function, also height of nozzle
- k — ratio of length scales, with suffixes to denote characteristic coefficients
- l — suffix to denote laminar flow
- M_o — momentum flux at nozzle, also excess momentum flux
- m — suffix to denote maximum value
- p — pressure, also exponent
- Q — volumetric flux
- q — exponent
- R — radius of duct
- r — radial distance
- r_o — radius of nozzle
- S — swirl number
- s — exponent
- T — angular momentum
- t — suffix to denote turbulent flow
- U_o — jet velocity at nozzle
- U_1 — velocity of free stream
- U — velocity relative to freestream
- u — time-averaged axial velocity at any point
- v — time-averaged velocity in y, r or z direction
- v_e — entrainment velocity
- W — pressure plus momentum flux
- w — time-averaged velocity in peripheral direction
- x — axial distance from virtual origin
- $\bar{x}$ — axial distance from nozzle
- y — coordinate distance
- z — coordinate distance

Greek Symbols

- α — velocity ratio
- α_e — entrainment coefficient
- $\underline{\beta}$ — ratio of length scales
- $\overline{\beta}$ — coefficient
- δ — boundary layer thickness
- ϵ — eddy diffusivity
- ζ — coordinate
- η — coordinate, also dimensionless distance
- θ — angle of deflected jet with crossflow
- λ — dimensionless distance
- μ — coefficient of dynamic viscosity

ν coefficient of kinematic viscosity
ξ coordinate
ρ mass density
τ shear stress
ϕ angular coordinate

Primes denote velocity fluctuations and differentiation; bars denote (generally) time averages.

REFERENCES

1. Rajaratnam, N. *Turbulent Jets,* (Amsterdam, The Netherlands: Elsevier Scientific Publishing Co., 1976).
2. Foerthmann, E. "Turbulent Jet Expansion," English translation *NACA TN*-789 (1936).
3. Abramovich, G. N. *The Theory of Turbulent Jets,* English translation (Cambridge, MA: The MIT Press, 1963).
4. Launder, B. E., and D. B. Spalding. *Mathematical Models of Turbulence* (New York: Academic Press, Inc., 1972).
5. Albertson, M. L., Y. B. Dai, R. A. Jensen and H. Rouse. "Diffusion of Submerged Jets" *Trans. ASCE* 115:639–697 (1950).
6. Zijnen, B. G. van der Hegge. "Measurements of the Velocity Distribution in a Plane Turbulent Jet of Air," *App. Sci. Res., Sec.* A 7:256–276 (1958).
7. Zijnen, B. G. van der Hegge. "Measurements at Turbulence in a Plane Jet of Air by the Diffusion Method and by the Hot Wire Method," *Appl. Sci. Res. Sec.* A 7: 293–313 (1958).
8. Heskestad, G. "Hot-Wire Measurements in a Plane Turbulent Jet," *Trans. ASME, J. Appl. Mech.* 1–14 (1965).
9. Flora, Jr., J. J., and V. W. Goldschmidt. "Virtual Origins of a Free Plane Turbulent Jet," *J. AIAA* 7:2344–2346 (1969).
10. Hinze, J. O., and B. G. van der Hegge Zijner. "Transfer of Heat and Matter in the Turbulent Mixing Zone of an Axially Symmetrical Jet," *J. Appl. Sci. Res.* A1:435–461 (1949).
11. Schlichting, H. *Boundary Layer Theory,* 6th ed., (New York: McGraw-Hill Book Co., 1968).
12. Corrsin, S. "Investigation of Flow in an Axially Symmetric Heated Jet of Air," *NACA Wartime Report* W-94 (1946).
13. Wygnanski, I., and H. Fiedler. "Some Measurements in the Self Preserving Jet," *J. Fluid Mech.* 38:577–612 (1969).
14. Liepmann, H. W., and J. Laufer. "Investigation of Free Turbulent Mixing," *NACA TN*-1257 (1947).
15. Wygnanski, I., and H. Fiedler. "The Two Dimensional Mixing Region," *J. Fluid Mech.* 41:327–361 (1970).
16. Watt, W. E. "The Velocity Temperature Mixing Layer," Report 6705, Department of Mechanical Engineering, University of Toronto, Toronto, Ontario (1967).
17. Miles, J. B., and J. S. Shih. "Similarity Parameter for Two-Stream Turbulent Jet Mixing Region," *J. AIAA* 6:1429–1430 (1968).
18. Yule, A. J. "Two Dimensional Self-Preserving Turbulent Mixing Layers at Different Free Steam Velocity Ratios," Aeronautical Research Council, England (1971).
19. Yule, A. J. "Spreading of Turbulent Mixing Layers," *J. AIAA* 10:686–687 (1972).
20. Pande, B. B. L., and N. Rajaratnam. "Turbulent Jets in Coflowing Streams," *J. Eng. Mech. Div., ASCE* 105 (EM6):1025–1038 (1979).
21. Bradbury, L. J. S. "The Structure of a Self-Preserving Turbulent Jet," *J. Fluid Mech.* 23:31–64 (1965).
22. Chigier, N. A., and A. Chervinsky. "Experimental and Theoretical Study of Turbulent Swirling Jets Issuing from a Round Orifice," *Israel J. Technol.* 4:44–54 (1965).

23. Chigier, N. A., and A. Chervinsky. "Experimental Investigation of Swirling Vortex Motion in Jets," *Trans. ASME, J. Appl. Mech.* 443–451 (1967).
24. Pratt, B. D., and J. F. Keffer. "Swirling Turbulent Jet Flows. I, The Single Swirling Jet," Rept. 6901, Department of Mechanical Engineering, University of Toronto, Toronto, Ontario (1969).
25. Mikhail, S. "Mixing of Coaxial Streams Inside a Closed Conduit," *J. Mech. Eng. Sci.*, 2:59–68 (1960).
26. Becker, H. A., H. C. Hottel and G. C. Williams. "Mixing and Flow in Ducted Turbulent Jets," in *Proc. 9th Int. Symp. on Combustion* (1962), pp. 7–20.
27. Curtet, R., and F. P. Ricou. "On the Tendency to Self-Preservation in Axisymmetric Ducted Jets," paper presented at the Fluids Engr. Conference, ASME, 64-FE-20: 1–11, (1964).
28. Barchilon, M., and R. Curtet. "Some Details of the Structure of an Axisymmetric Confined Jet with Backflow," paper presented at the Fluids Engr. Conference, ASME, paper 64-FE-3:1–17, 1964.
29. Exley, J. T., and J. A. Brighton. "Flow Separation and Reattachment in Confined Jet Mixing," paper presented at ASME, paper 70FE-B:1–11 1970.
30. Razinsky, E., and J. A. Brighton. "Confined Jet Mixing for Nonseparating Conditions," *Trans. ASME, J. Basic Eng.* 333–349 (1971).
31. Brighton, J. A., E. Razinsky and D. A. Bowlus. "Turbulent Mixing of a Confined Axisymmetric Jet," *Proc. Fluidics Internal Flows*, 2:57–90 (1969).
32. Rajaratnam, N., and T. Gangadharaiah. "Circular Jets in Crossflow," Technical Report, Department of Civil Engineering, University of Alberta, Edmonton, Alberta (1980).
33. Moussa, Z. M., J. W. Trischka and S. Eskinazi. "The Near Field in the Mixing of a Round Jet with a Cross-stream," *J. Fluid Mech.* 80 (1):49–80 (1977).
34. McMahon, H. M., D. D. Hester and J. G. Palfery. "Vortex Shedding from a Turbulent Jet in a Crosswind," *J. Fluid Mech.* 48:73–80 (1971).
35. Keffer, J. F., and W. D. Baines. "The Round Turbulent Jet in a Cross-Wind," *J. Fluid Mech.* 8:481–496 (1963).
36. Rajaratnam, N., and T. Gangadharaiah. "Axis of a Circular Jet in Crossflow," *Water, Air, Soil Poll.* 15:317–321 (1981).
37. Sigalla, A. "Measurements of Skin Friction in a Plane Turbulent Wall Jet," *J. Roy. Aeronaut. Soc.* 62:873–877 (1958).
38. Myers, G. E., J. J. Schauer and R. H. Eustis. "The Plane Turbulent Wall Jet—Part I, Jet Development and Friction Factor," *Trans. ASME, J. Basic Eng.* (1963).
39. Schwarz, W. H., and W. P. Cosart. "The Two Dimensional Turbulent Wall Jet," *J. Fluid Mech.* 10:481–495 (1961).
40. Verhoff, A. "The Two Dimensional Turbulent Wall Jet with and without an External Stream," Report 626, Princeton University (1963).
41. Mathieu, J., and A. Tailland. "Jet Parietal," *Comp. Rend. Acad. Sci. Paris* 261: 2282–2286 (1965).
42. Rajaratnam, N. "Plane Turbulent Wall Jets on Rough Boundaries," Technical Report, Department of Civil Engineering, University of Alberta, Edmonton, Alberta (1965).
43. Bakke, P. "An Experimental Investigation of a Wall Jet," *J. Fluid Mech.* 2:467–472 (1957).
44. Poreh, M., and J. E. Cermak. "Flow Characteristics of a Circular Submerged Jet Impinging Normally on a Smooth Boundary," *Proc. 6th Mid-Western Conference on Fluid Mechanics*, (1959), pp. 198–212.
45. Poreh, M., Y. G. Tsuei and J. E. Cermak. "Investigation of a Turbulent Radial Wall Jet," *Trans. ASME, J. Appl. Mech.* 457–463 (1967).
46. Beltaos, S., and N. Rajaratnam. "Impinging Circular Turbulent Jets," *Proc. ASCE J. Hydraul. Div.* 100:1313–1328 (1974).
47. Trentacoste, N., and P. M. Sforza. "An Experimental Investigation of Three Dimensional Free Mixing in Incompressible Turbulent Free Jets," Report 81, Department of Aerospace Engineering, Polytechnic Institute of Brooklyn, NY (1966).

48. Trentacoste, N., and P. M. Sforza. "Further Experimental Results for Three Dimensional Free Jets," *J. AIAA* 5:885–891 (1967).
49. Yevdjevich, V. M. "Diffusion of Slot Jets with Finite Orifice Length-Width Ratios," Hydraulics Paper 2, Colorado State University, Fort Collins, CO (1966).
50. Pani, B. S. "Three Dimensional Turbulent Wall Jets," PhD Thesis, Department of Civil Engineering, University of Alberta, Edmonton, Alberta (1972).

CHAPTER 11
SINGLE-FLUID FLOW THROUGH POROUS MEDIA

R. A. Greenkorn
School of Chemical Engineering
Purdue University
West Lafayette, Indiana 47907

CONTENTS

INTRODUCTION

The movement of materials through porous media is of interest in many disciplines: in chemical engineering—adsorption, chromatography, filtration, flow in packed columns, ion exchange, reactor engineering; in petroleum engineering—displacement of oil with gas, water and miscible solvents, including surface-active agent solutions and description of reservoirs; in hydrology—movement of trace pollutants in water systems, recovery of water for drinking and irrigation, salt water encroachment into fresh water reservoirs; in soil physics—movement of water, nutrients and pollutants into plants; in biophysics—life processes such as flow in the lung and the kidney.

The parameters that relate the media and single-fluid flow are porosity, permeability, tortuosity and connectivity. Complex potentials can be used to model single-fluid flow problems as Laplace's equation results in single-fluid-steady flow situations. Also, compressible flow can be modeled as steady flow.

Much of the mathematics and techniques for transient flow in multifluid systems is an extension of the single flow situation. Many multifluid transient situations are modeled as single-fluid flow. Inertial effects—wave propagation—also are modeled with a single fluid model. The topic of non-Newtonian flow in porous media often is introduced as a single-fluid flow problem.

This chapter discusses the fundamentals of single-fluid flow through porous media: it discusses the parameters of porosity, permeability, tortuosity and connectivity, and relates these coefficients to the geometry of porous media.

A porous medium is a solid with holes in it. Usually the number of holes or pores is sufficiently large that a volume average is needed to calculate its pertinent properties. Pores that occupy some definite fraction of the bulk volume form a complex network of voids.

The matrix of a porous medium is the material in which the holes or pores are embedded. The manner in which the holes are embedded, how they are interconnected and the description of their location, size, shape and interconnection characterize the porous medium. The porosity is a quantitative property that describes the fraction of the medium that has voids. When we are concerned with flow, the pores or fraction of the medium that contribute to flow is called effective porosity.

An extremely large array of materials can be classified as porous media. Broadly speaking, porous media are classified as unconsolidated or consolidated; as ordered or random. Examples of unconsolidated media are beach sand, glass beads, catalyst pellets, column packings, soil, gravel and packing, such as charcoal. Examples of consolidated media are most of the naturally occurring rocks, such as sandstones and limestones. In addition, concrete, cement, bricks, paper, cloth, and so forth are man-made consolidated media. Wood can be considered a consolidated medium, as can the human lung. Ordered media include regular packing of various types of materials, such as spheres, column packings and wood. Random media are media without any particular correlating factor. They are hard to find since one can correlate some factor with almost any medium if one looks carefully.

Darcy's law [1] relates the volumetric flowrate, Q, of a fluid flowing linearly through a porous medium directly to the energy loss, inversely to the length of the medium, and proportionally to a factor called the hydraulic conductivity, K.

Darcy's law is expressed as

$$Q = \frac{KA(h_1 - h_2)}{\Delta l} \tag{1}$$

where

$$\Delta h = \Delta z + \frac{\Delta p}{\rho} + \text{constant} \tag{2}$$

Darcy's law is empirical in that it is not derived from first principles; rather it is the result of experimental observation. Thus, although Darcy's law is empirical, DeWiest [2] has heuristically demonstrated that it is the empirical equivalent of the Navier-Stokes equations.

Darcy's law usually is considered valid for creeping flow, where the Reynolds number, as defined for a porous medium, is less than one. The Reynolds number in open conduit flow is the ratio of inertial to viscous forces and is defined in terms of a characteristic length perpendicular to flow for the system. Using four times the hydraulic radius to replace the length perpendicular to flow and correcting the velocity with porosity yields a Reynolds number in the form

$$Re_p = \frac{D_p v_\infty \rho}{\mu(1 - \phi)} \tag{3}$$

Darcy's law is considered valid where $Re_p < 1$.

The hydraulic conductivity, K, defined by Darcy's law (Equation 1), is dependent on the properties of the fluid, as well as on the pore structure of the medium. The hydraulic conductivity is temperature dependent because the properties of the fluid, density and viscosity, are temperature dependent. Hydraulic conductivity can be written more specifically in terms of the intrinsic permeability and the properties of the fluids:

$$K = \frac{k\rho g}{\mu} \tag{4}$$

where k is the intrinsic permeability of the porous medium and is a function only of the pore structure. The intrinsic permeability is not temperature dependent. Darcy's law often is written in differential form, so that in one dimension,

$$\frac{Q}{A} = q = -\frac{k}{\mu}\frac{dp}{dx} \tag{5}$$

The minus sign results from the definition of Δp, which is equal to $p_2 - p_1$, a negative quantity. The term q is called the seepage velocity and is equivalent to the velocity of approach, v_∞, which was used in the definition of the Reynolds number.

Permeability is normally determined using linear flow in the incompressible or compressible form, depending on whether a liquid or gas is used as the flowing fluid. Most often liquid is used as the flowing fluid, because one does not have to correct for compressible effects or slip flow. The volumetric flowrate, Q (or Q_m), is determined at several pressure drops. Q (or Q_m) is plotted versus the average pressure, p_m. The slope of this line will yield the fluid conductivity, K, or, knowing the fluid density and viscosity, the intrinsic permeability, k. For gases, the fluid conductivity depends on pressure so that

$$\tilde{K} = K\left(1 + \frac{b}{p}\right) \tag{6}$$

where b is a parameter dependent on the fluid and the porous medium. Under such circumstances, a straight line results as with a liquid but does not go through the origin; instead, it has a slope of bK and intercept K. This difference between the liquid and gas flow was pointed out by Klinkenberg [3]. The explanation for this phenomenon is that gases do not always stick to the walls of the porous medium. This slip shows up as an apparent dependence of the permeability on pressure.

DESCRIPTION OF POROUS MEDIA

The main interest in describing porous media is to understand and predict the passage of materials into and out of the media. It is possible to break down the description or characterization of porous media in terms of geometric or structural properties of the matrix that affect flow and in terms of the flow properties that describe the matrix from the point of view of the contained fluid. The problem of description of a porous medium is one of describing the geometric or structural properties in some average fashion and relating these average structural properties to the flow properties.

A microscopic description characterizes the structure of the pores. The objective of pore structure analysis is to provide a description that can be related to the macroscopic or bulk flow properties. The bulk properties that we want to relate to pore description are porosity, permeability, tortuosity and connectivity. When one examines different samples of the same medium, such as sandstone, it is apparent that the number of pore sizes, shapes, orientations and interconnections is enormous. Because of this complexity, pore structure description is most often a statistical distribution of apparent pore sizes. This distribution is apparent because to convert measurements to pore sizes we resort to models that provide "average" or model pore sizes. One way to define a pore size distribution is to model the porous medium as a bundle of straight cylindrical capillaries. The diameters of the model capillaries are distributed according to some distribution function.

One method of obtaining a pore-size distribution is from capillary pressure measurements or mercury porosimetry. Brutsaert [4] describes how to obtain a pore size distribution from capillary pressure data. Another method of obtaining a pore size distribution is from photomicrographs of thin sections of a porous medium and the pore sizes are inferred by direct measurement of the pores on a set of such photomicrographs. The approach used to interpret the photomicrograph depends most often on the equipment that is available to interpret them. Dullien and co-workers [5–7] have discussed photomicrographic methods in detail.

Pore structure for unconsolidated media is inferred from the particle size distribution, the geometry of the particles and the manner in which the particles are packed. The theory of packing has been determined for symmetrical shapes, especially spheres. A knowledge of particle size, symmetry and the theory of packing allows one to establish relationships between pore size distributions and particle size distributions [8–10].

A macroscopic description of a porous medium is a description in terms of average or bulk properties at sizes much larger than a single pore. When we reach the point of characterizing a porous medium macroscopically, we must cope with the scale of description. The scale used will depend on the way, and on what size, we wish to model the porous medium. For a macroscopic reading, assume, if possible, that a given medium is "ideal," that is, homogeneous, uniform and isotropic. The term homogeneous normally implies that an average property can be used to replace the entire medium. For example, a single value of permeability can be used for a "homogeneous" reservoir, and this value of permeability will characterize flow in this reservoir. Unfortunately, a medium that is "homogeneous" in this sense for one property may not be "homogeneous" for another.

The term "reservoir description" often means description in the sense of homogeneous, as opposed to heterogeneous, as discussed above. Reservoir description means describing the reservoir at a level at which a property changes enough so that more than a single average must be used to model flow. In this sense, a reservoir composed of a section of coarse gravel and a section of fine sand is heterogeneous, where these two materials are separated and have significantly different permeabilities. Defining the dimensions, locating the area and determining the average properties of the gravel and the sand is reservoir description. This description is satisfactory for reservoir-level problems; however, if one studies mechanisms of fluid flow or a process at a given scale and wants to use these data at a different scale, the effects of nonideal media at a given scale require more specific definitions.

Statistical models often are useful to relate the micro- and macrostructure of a porous medium. Guin et al. [11,12] discuss generalized models for a network of randomly intersecting straight capillaries. They show that the ergodic assumption–that spatial averages coincide with mathematical expectation–is not necessarily satisfied in all models [13–16]. Dullien [17] gives an excellent review of statistical models. It is interesting that MacDonald et al. [18] conclude that network models composed of tubes or channels are satisfactory to predict flowrate and pressure drop.

Slattery [19] describes averages to use in volume-averaging the equations of change for flow in porous media. Equations describing incompressible creeping flow of a non-Newtonian fluid of constant viscosity through a porous medium are the continuity equation,

$$(\underline{\nabla} \cdot \underline{v}) = 0 \tag{7}$$

and the void distribution function,

$$\alpha(\delta) = \begin{cases} 1 \text{ if } \delta \text{ is in the pores} \\ 0 \text{ if } \delta \text{ is in the matrix} \end{cases} \tag{8}$$

Since $\alpha(\delta)$ is unknown, we volume-average the equation of continuity and equation of motion.

Assume for every point in space an average volume that contains the volume of the fluid, V_f. The porosity of this space is

$$\phi = \frac{1}{V} \int_V \alpha(\delta)\, dV \tag{9}$$

Assume a microscopic characteristic length, d, over which there is significant variation in point velocity. Further assume that a macroscopic characteristic length, L, represents the distance over which significant variation in the average velocity, $<v>$, takes place. Let d approximate pore diameter or grain size and L approximate the length of the system. The volume average of a point quantity associated with the fluid is

$$<\psi> = \frac{1}{V}\int_{V} \psi \, dV \tag{10}$$

If we plot $<\psi>$ versus V, starting with V in the solid, as V increases portions of the fluid are contained within V. The volume average, $<\psi>$, passes through several fluctuations until V reaches a size where $<\psi>$ smooths. Figure 1 [20] is a schematic of $<\psi>$ versus V. For values of V larger than at the dashed line on Figure 1, the volume average $<\psi>$ is smooth. However, it is not necessarily constant.

Let l be a characteristic macroscopic length for the average value and $d << l$ and $<<\psi>> = <\psi>$. Then the volume average of $<\psi>$ is

$$<<\psi>> = \frac{1}{V}\int_{V} <\psi> dV \tag{11}$$

which leads to $l << L$. Whitaker [20] shows for the above restrictions that the volume and area averages are essentially equivalent.

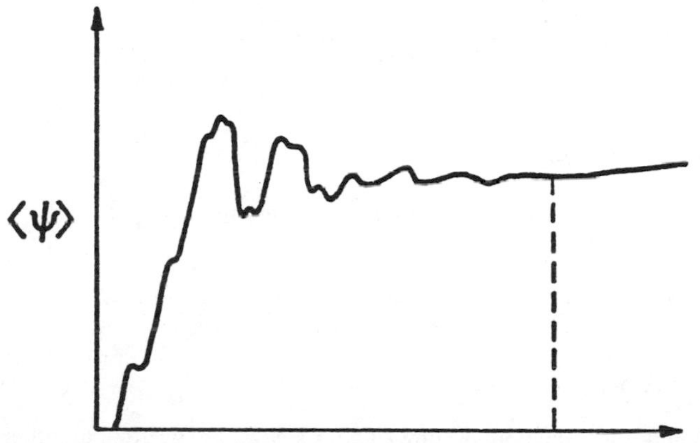

Figure 1. Dependence on averaging volume [20].

Bear [21] uses porosity as a point quantity in a porous medium and representative elementary volume (REV) to characterize a medium macroscopically. This definition is a specific case of the more general discussion of Whitaker [20]. The size of the REV around a point, P, is smaller than the total medium so it can represent flow at P. There must be enough pores to allow statistical averaging. If porosity varies, the maximum length is the characteristic length that indicates the rate of change of porosity. The minimum length is the pore size. The REV is equivalent to the generalized value defined by Whitaker [20] as the dashed line in Figure 1. $<\psi>$ is porosity. Bear [21] replaces the porous medium in a "continuum sense" with a model that gives the correct average porosity at any point in the medium.

The terms heterogeneity, nonuniformity and anisotropy must be defined in the volume-average sense. Because the definitions for nonideal properties are arbitrary, it seems reasonable to define them in direct relation to flow. The definitions of heterogeneity, nonuniformity and anisotropy can be defined at the level of Darcy's law in terms of permeability. Permeability is more sensitive to conductance, mixing and capillary pressure than is porosity.

Greenkorn and Kessler [22] define heterogeneity, nonuniformity and anisotropy in reference to permeability as follows. First, by macroscopic they imply averaging over elemental volumes of radius, ϵ, about a point in the medium where ϵ is large enough that Darcy's law can be applied for appropriate Reynolds numbers. In other words, volumes are large with respect to that of a single pore. Further, ϵ is to be the minimum radius that satisfies such a condition; otherwise, by making ϵ too large, certain nonidealities may be obscured by burying their effects far within the elemental volume. Obviously, for all practical purposes one can remove certain effects by scale.

The definitions of heterogeneity, nonuniformity and anisotropy are based on the probability density distribution of permeability of random macroscopic elemental volumes selected from the medium, the permeability being expressed by the one-dimensional form of Darcy's law.

For a uniform medium, the probability density function for permeability is either a Dirac delta function, δ, or a linear combination of N functions that satisfy the following relation:

$$f(k) = \sum_{i=1}^{N} \xi_i \delta \tag{12}$$

where

$$\sum_{i=1}^{N} \xi_1 = 1 \tag{13}$$

EQUATIONS OF CHANGE

Darcy's law is an empirically determined relation for single fluid flow in a linear system:

$$q = \frac{Q}{A} = -\frac{k}{\mu}\frac{dp}{dx} \tag{14}$$

Slattery [19] raises questions concerning Equation 14.

1. How should the equation of flow—Darcy's law—be written in other geometries and with other boundary conditions?

Usually the general three-dimensional equation for Darcy's law is written heuristically as

$$\underline{q} = -\frac{k}{\mu}\underline{\nabla} p \tag{15}$$

2. How should Darcy's law be written if the medium is anisotropic (oriented)?

An oriented porous medium has distinct directional effect on flow, that is, at a "point" the resistance to flow may be a function of direction. It usually is assumed that permeability is a second-order tensor and Darcy's law is written [15,23,24] as

$$\underline{q} = -\frac{\underline{\underline{k}}}{\mu}\underline{\nabla} p \tag{16}$$

Whitaker [20] poses the problem raised by Slattery [19] in a slightly different man-

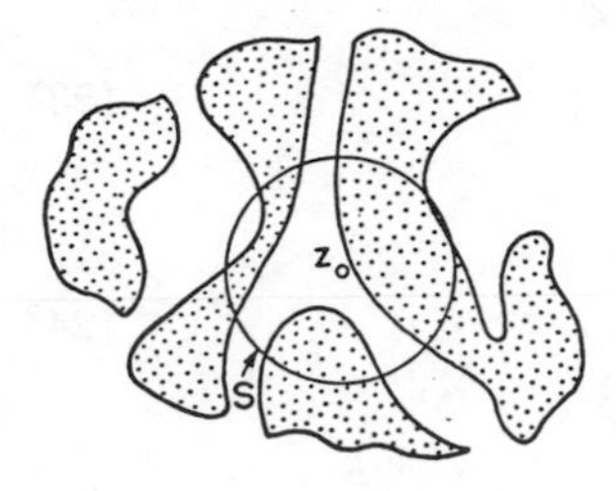

Figure 2. Averaging surface, S, to be associated with every point, z, in the porous medium [19].

ner. For the creep flow of a constant-viscosity single Newtonian fluid in a rigid porous medium, the continuity equation of the fluid is

$$\frac{\partial \rho}{\partial t} + (\underline{\nabla} \cdot \underline{v}) = 0 \tag{17}$$

The equation of motion for the fluid is

$$\frac{\partial \rho \underline{v}}{\partial t} + (\underline{\nabla} \cdot \rho\, \underline{vv}) - (\underline{\nabla} \cdot \underline{\underline{\tau}}) - \rho \underline{F} = 0 \tag{18}$$

where $\underline{\underline{\tau}}$ are the stresses in the fluid and $\underline{F}$ are body forces. With these two equations, a void distribution function is required where

$$\alpha(\delta) = \begin{cases} 1, \text{ if } \delta \text{ is in the fluid} \\ 0, \text{ if } \delta \text{ is in the solid} \end{cases} \tag{19}$$

We cannot determine Equation 19 and we cannot solve Equations 17 and 18. The method posed to overcome the dilemma is to volume-average the equations of change. The volume average of a point quantity, ψ, associated with the fluid is given by Equation 10.

To volume-average the equations of change, we need gradients of averages and averages of gradients. Slattery [25] describes a procedure for the volume average of a gradient,

$$< \underline{\nabla}\, \psi > = \underline{\nabla} < \psi > + \frac{1}{V} \int_{A_i} \psi\, \underline{n}\, dA \tag{20}$$

where $\underline{n}$ is the outward directed normal for V or V_f, and A_i is the area of the solid-fluid interface. The divergence is described by

$$< (\underline{\nabla} \cdot \psi) > = (\underline{\nabla} \cdot < \psi >) + \frac{1}{V} \int_{A_i} (\psi \cdot n)\, dA \tag{21}$$

Consider a point, z, in a porous medium of closed surface, S, and volume, V. $\underline{z}$ is either in the matrix or in a pore or on the boundary, as sketched in Figure 2. Apply Equations 20 and 21 to the equation of continuity, Equation 17, and the equation of motion, Equation 18 for the region enclosed by S, sketched in Figure 2. The mass balance for the fluid inside the closed surface, S, is

$$\int_{V_f} \left[\frac{\partial \rho}{\partial t} + \left(\underline{\nabla} \cdot \rho\, \underline{v} \right) \right] dV = 0 \tag{22}$$

Using Equations 20 and 21,

$$\frac{\partial <\rho>}{\partial t} + (\underline{\nabla} \cdot <\rho\, \underline{v}>) = o \tag{23}$$

Equation 23 simplifies for an incompressible fluid to

$$(\underline{\nabla} \cdot <\underline{v}>) = o \tag{24}$$

The equation of motion for the fluid inside the closed surface, S, is

$$\int_{V_f} \left[\frac{\partial \rho \underline{v}}{\partial t} + (\underline{\nabla} \cdot \rho\, \underline{vv}) - (\underline{\nabla} \cdot \underline{\underline{\tau}}) - \rho \underline{F}\right] dv = 0 \tag{25}$$

Applying Equations 20 and 21 again,

$$\frac{\partial(\rho\, \underline{v})}{\partial t} + (\underline{\nabla} \cdot <\rho\, \underline{vv}>) - (\underline{\nabla} \cdot <\underline{\underline{\tau}}>) - <\rho \underline{F}> - \frac{1}{V}\int_S (\underline{\underline{\tau}} \cdot \underline{n})\, dS = 0 \tag{26}$$

For an incompressible fluid without any inertial effects present and with body forces represented by a potential function, $\underline{F} = -\underline{\nabla}\,\phi$,

$$(\underline{\nabla} \cdot \{<\underline{\underline{\tau}}> - \rho\hat{\phi}\, \underline{\underline{I}}\}) + \frac{1}{V}\int_{S_w} (\{\underline{\underline{\tau}} - \rho\hat{\phi}\underline{\underline{I}}\} \cdot \underline{n})\, dS = o \tag{27}$$

where S_w is the interface between the fluid and the solid.

Slattery [19] shows that the second term of Equation 27 is a function of the difference between the local average fluid velocity, $<\underline{v}>$, and the local average solid velocity, $<\underline{u}>$. Therefore, the force per unit volume an incompressible fluid exerts on an anisotropic (oriented) porous medium is in addition to the hydrostatic force and local pressure. Let $\mathcal{p} = p - p_o + \rho\,\hat{\phi}$. Then Equation 27 can be written as

$$\underline{\nabla}\, \mathcal{p} - \mu\nabla^2 <\underline{v}> + \mu\underline{\underline{k}}^{-1} <\underline{v}> = o \tag{28}$$

The second term in Equation 28 represents viscous drag at the boundaries and was recognized heuristically by Brinkman [26]. Howells [27] and Hinch [28] confirm the validity of Equation 28 by considering slow flow in random arrays of fixed spheres and for suspensions, respectively. Normally, the viscous drag term is neglected and

$$<\underline{v}> = -\frac{\underline{\underline{k}}}{\mu}\, \underline{\nabla}\, p \tag{29}$$

If $<\underline{v}>$ is equal to $\underline{q}$, then Equation 29 is Darcy's law in three dimensions for an anisotropic porous medium. The permeability tensor, $\underline{\underline{k}}$, is symmetrical.

The following are the equations of change for single fluid flow in a porous medium based on the above discussion. The equation of continuity is

$$\phi\frac{\partial \rho}{\partial t} + (\underline{\nabla} \cdot \rho \underline{q}) = o \tag{30}$$

where ϕ is the porosity of the medium, and ρ is the density of the fluid in the pores. The equation of motion—Darcy's law—in three dimensions is

$$\underline{q} = -\frac{\underline{\underline{k}}}{\mu}\, \underline{\nabla}\left(p - \frac{d}{dz}\rho g z\right) \tag{31}$$

where g is the acceleration due to gravity.

If we define a flow potential, Φ, such that

$$\Phi = p + \rho g z \tag{32}$$

and combine Equations 30, 31 and 32, we obtain the equation of condition for single fluid flow in a porous medium:

$$\phi \mu \frac{\partial \rho}{\partial t} = (\underline{\nabla} \cdot \rho \underline{\underline{k}} \nabla \Phi) \tag{33}$$

For flow of an incompressible fluid, ρ is constant and $\partial \rho / \partial t = 0$, so that

$$(\underline{\nabla} \cdot \underline{\underline{k}} \underline{\nabla} \Phi) = 0 \tag{34}$$

If the medium is homogeneous and isotropic, the permeability can be taken outside of the derivative:

$$\nabla^2 \Phi = 0 \tag{35}$$

For compressible liquids,

$$c_l = \frac{1}{\rho} \frac{\partial \rho}{\partial p} \tag{36}$$

If c_l is constant, then

$$\rho \underline{\nabla} p = \frac{1}{c_l} \underline{\nabla} \rho \tag{37}$$

Assuming the effect of gravity is small compared to the pressure, Equation 33 becomes

$$\phi \mu c_l \frac{\partial \rho}{\partial t} = (\underline{\nabla} \cdot \underline{\underline{k}} \underline{\nabla} \rho) \tag{38}$$

For slightly compressible liquids, assume the following equation for density as a function of pressure:

$$\rho = \rho_o e^{c_l (p - p_o)} = \rho_o \left[1 + c_l (p - p_o) + \frac{1}{2!} c_l^2 (p - p_o)^2 + \ldots \right] \tag{39}$$

Assume the higher-order terms of Equation 39 are negligible and differentiate Equation 39 to obtain

$$d\rho = \rho_o c_l dp \tag{40}$$

Substituting Equation 40 into Equation 38 gives

$$\phi \mu c_l \frac{\partial p}{\partial t} = (\underline{\nabla} \cdot \underline{\underline{k}} \underline{\nabla} p) \tag{41}$$

which is the equation of condition for a slightly compressible liquid.

For flow of an ideal gas in a porous medium, substitute the ideal gas equation,

$$\rho = \frac{M}{RT} p \tag{42}$$

into Equation 33 and assume gravity is negligible to obtain

$$\phi\mu \frac{\partial p}{\partial t} = (\underline{\nabla} \cdot \underline{\underline{k}}\, p \underline{\nabla} p) \tag{43}$$

On rearrangement,

$$\phi\mu \frac{\partial p}{\partial t} = (\underline{\nabla} \cdot \underline{\underline{k}}\, \nabla p^2) \tag{44}$$

POROSITY, PERMEABILITY, TORTUOSITY AND CONNECTIVITY OF NONIDEAL MEDIA

Four macroscopic properties of nonideal porous media that may be used to describe single fluid flow will be discussed. (Inversely, single fluid flow may be used to infer these properties.) To calculate flow through a nonideal porous medium, the effects of heterogeneity, nonuniformity and anisotropy must be considered on these macroscopic properties. These four properties are described as follows:

Porosity macroscopically characterizes the effective pore volume of the medium. The porosity is directly related to the size of the pores relative to the matrix. When porosity is substituted we lose the details of the structure.

Permeability is the conductance of the medium defined with direct reference to Darcy's law. The permeability is related to the pore size distribution, since the distribution of the sizes of entrances, exits and lengths of the pore walls makes up the major resistances to flow. The permeability is the single parameter that reflects the conductance of a given pore structure. The permeability and porosity are related because if the porosity is zero the permeability is zero. Although there may be a correlation between porosity and permeability, permeability cannot be predicted from porosity alone as we need additional parameters that contain more information about pore structure. These additional parameters are the next two macroscopic properties.

Tortuosity is the relative average length of a flow path, the average length of the flow paths to the length of the medium. The tortuosity is a macroscopic measure of both the sinuousness of the flow path and the variation in pore size along the flow path. Like porosity, tortuosity correlates with permeability but cannot be used alone to predict permeability, except in some limiting cases.

Connectivity is the manner and number of pore connections. When all the pores are the same size, connectivity is the average number of pores per junction [29]. The connectivity is a macroscopic measure of the number of pores at a junction and, further, the manner of connection, that is, the different size of the pores at the junction. Like porosity and tortuosity, connectivity correlates with permeability but cannot be used alone to predict permeability, except in certain limiting cases.

The real problem lies in the conceptual simplifications that result from replacing the real porous medium with macroscopic parameters that are averages and that relate to some idealized model of the medium. Tortuosity and connectivity represent different features of the pore structure and are useful to interpret macroscopic flow properties, such as permeability, capillary pressure and dispersion.

Kozeny [30] represents a porous medium as an ensemble of channels of various cross sections of the same length and solves the Navier-Stokes equations for all channels passing a cross section normal to the flow to obtain

$$S^2 = \frac{c\phi^3}{k} \tag{45}$$

where the Kozeny constant, c, is a shape factor that takes on different values, depending on the shape of the capillary (c = 0.5 for a circular capillary). S is the specific surface of the channels. For other than circular capillaries, we include a shape factor:

$$r^2 = \frac{ck}{\phi} \tag{46}$$

The specific surface for cylindrical pores is

$$S_A = \frac{n2\pi rL}{n\pi r^2 L} = \frac{2}{r} \tag{47}$$

and

$$S_A{}^2 = \frac{2\phi}{\sqrt{8}\,k} \tag{48}$$

If we replace $2/\sqrt{8}$ with a shape parameter, c, and S_A with a specific surface,

$$S = \phi\, S_A \tag{49}$$

we obtain the Kozeny equation.

The tortuosity, τ, was introduced as a modification to the Kozeny equation to account for the fact that in a real medium the pores are not straight. The length of the most probable flow path is longer than the overall length of the porous medium and

$$S^2 = \frac{c\phi^3}{\tau k} \tag{50}$$

Many authors include a shape factor in the tortuosity. In any event, there is an inferred relationship among permeability, porosity, tortuosity and pore structure.

The connectivity must be related to the permeability because the manner and number of intersections affects the flow resistance. Fatt [29,31] uses connectivity to relate the structure of his model to permeability. Since the tortuosity of his model is one, tortuosity does not enter the problem. It would seem that the Kozeny relation must be modified so that tortuosity includes the geometric effect and a connectivity parameter that itself includes the effect of intersections (and thus shape):

$$S^2 = \frac{a\phi^3}{\tau k} \tag{51}$$

where a is connectivity.

In single fluid flow situations we are concerned with determining properties in a real medium, that is, a medium that has macroscopic heterogeneity, nonuniformity and anisotropy. Although porosity can be a function of position and thus be heterogeneous and vary nonuniformly, as long as porosity is constant in time an average value is all that is required in the equation of condition since

$$\phi \frac{\partial \rho}{\partial t} = -\,(\underline{\nabla} \cdot \rho \underline{q}) \tag{52}$$

Orientation of the pores does not cause a directional effect through porosity as porosity is not a tensor, so anisotropy need not be considered for porosity.

To find the average porosity of a homogeneous, but nonuniform, medium, we determine the correct mean of the distribution of porosity. It has been observed that data on porosity of both natural and artificial media usually are normally distributed [32, 33]. Figure 3 shows the porosity distributions of samples from a heterogeneous sandstone:

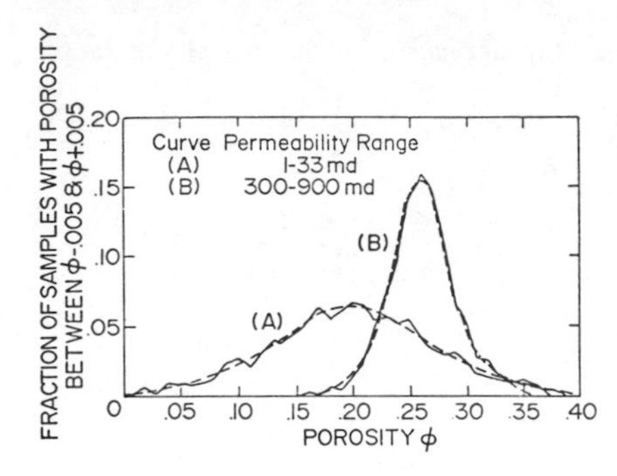

Figure 3. Porosity distributions [34]. From *Flow of Fluids through Porous Media,* by R. F. Collins. Copyright (1961) by Van Nostrand Reinhold Co. Reprinted by permission of the publisher.

there are two distinct distributions with widely differing permeabilities. The two distributions are nonuniform. Also, in Figure 3 in both sets the distribution of the nonuniform porosity is approximately normal. The mean of the normal distribution is

$$<\phi> = \frac{\sum_{i=i}^{n} \phi_i}{n} \tag{53}$$

the arithmetic average. The average porosity of a heterogeneous nonuniform medium, as in Figure 3, is the volume-weighted average of the number average, Equation 53:

$$<<\phi>> = \frac{\sum_{i=i}^{m} V_i <\phi_i>}{\sum_{t=1}^{m} V_i} \tag{54}$$

For the case of Figure 3, m = 2, there are two distinct distributions.

The average nonuniform permeability is a function of position. For a homogeneous but nonuniform medium, the average permeability is the correct mean (first moment) of the permeability distribution function. Permeability for a nonuniform medium is usually skewed, as shown in Figure 4. Most data for nonuniform permeability show permeability to be distributed log-normal [32,35,36]. The correct average for a homogeneous, nonuniform permeability, assuming it is distributed as a log-normal, is the geometric mean:

$$<k> = [\prod_{i=1}^{n} k_i]^{1/n} \tag{55}$$

For flow in heterogeneous media, the average permeability will depend on the arrangement and geometry of the nonuniform elements, each of which will have a differ-

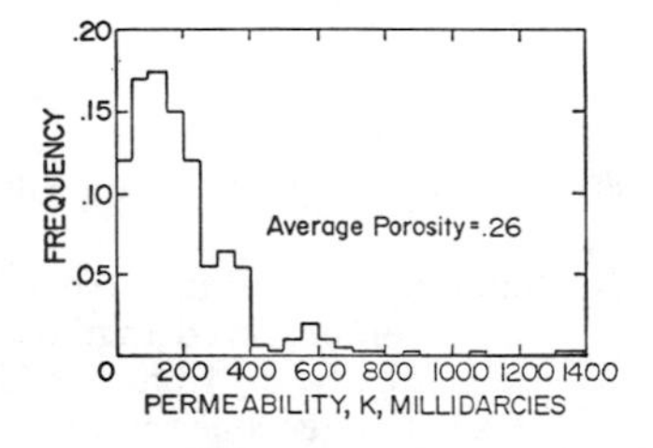

Figure 4. Permeability distribution [32].

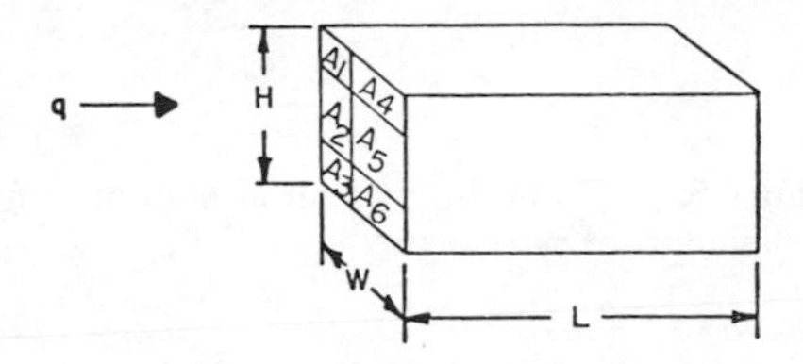

Figure 5. Flow through parallel nonuniform elements of porous media [37].

ent average permeability. Consider flow through a set of nonuniform elements in which the elements are parallel to the flow, as sketched in Figure 5. Because flow is through parallel elements of different constant area, Darcy's law for each element, assuming overall length of each element is equal, is

$$Q_1 = -\frac{A_1 < k_1 > \Delta p}{\mu L} \tag{56}$$

$$Q_2 = -\frac{A_2 < k_2 > \Delta p}{\mu L} \tag{57}$$

$$\vdots$$

The flowrate through the entire system of elements is

$$Q = Q_1 + Q_2 + \ldots \tag{58}$$

Substituting Equations 56 and 57 . . . into 58 and canceling terms,

$$A << k >>_p = A_1 < k_1 > + A_2 < k_2 > + \ldots \tag{59}$$

or

$$<< k >>_p = \frac{A_1 < k_1 > + A_2 < k_2 > + \ldots}{A} \tag{60}$$

Thus, the average permeability for this heterogeneous medium made up of parallel nonuniform elements is the area-weighted average of the average permeability of each of the elements, which, if the permeability of each element is log-normally distributed, are the geometric means. Many reservoirs and soils are composed of heterogeneities that are nonuniform layers, so that only the thickness of the layers varies. So $<< k >>_p$ simplifies to

$$<< k >>_{ph} = \frac{h_1 < k_1 > + h_2 < k_2 > + \ldots}{h} \tag{61}$$

If all the layers have the same thickness,

$$<< k >>_{p\bar{h}} = \frac{\sum_{i=1}^{h} < k_i >}{n} \tag{62}$$

where n is the number of layers.

The other simple situation we can consider is flow through a set of nonuniform elements that are in series, as sketched in Figure 6. In this instance, the volumetric flowrate through each element is equal:

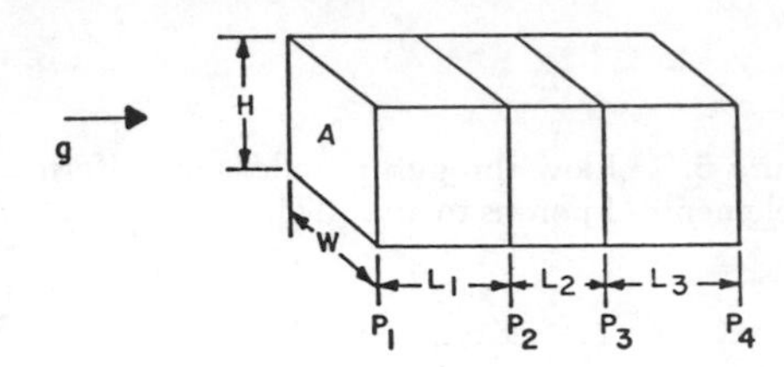

Figure 6. Flow through serial nonuniform elements of porous media [37].

$$Q = -\frac{<k_1> A(p_1 - p_2)}{\mu L_1} \tag{63}$$

$$Q = -\frac{<k_2> A(p_2 - p_3)}{\mu L_2} \tag{64}$$

$$\vdots$$

because

$$(p_1 - p_n) = (p_1 - p_2) + (p_2 - p_3) + \ldots + (p_{n-1} - p_n) \tag{65}$$

$$\frac{L}{<<k>>_s} = \frac{L_1}{<k_1>} + \frac{L_2}{<k_2>} + \ldots \tag{66}$$

and

$$<<k>>_s = \frac{L}{\dfrac{L_1}{<k_1>} + \dfrac{L_2}{<k_2>} + \ldots} \tag{67}$$

If all the elements are of equal length,

$$<<k>>_{sL} = \frac{n}{\sum_{i=1}^{n} <k_i>} \tag{68}$$

where n is the number of elements. The correct average permeability for a heterogeneous series of nonuniform elements is the harmonic mean of the geometric mean permeability of each of the elements.

For a radial system, the area is a function of radius and

$$\frac{Q}{A} = -\frac{<k>}{\mu}\frac{dp}{dr} \tag{69}$$

which leads to

$$<<k>>_r = \frac{\ln (r_3/r_1)}{\dfrac{1}{<k_1>} \ln (r_2/r_1) + \dfrac{1}{<k_2>} \ln (r_3/r_2)} \tag{70}$$

and the correct average for heterogeneous elements that are radially distributed is the harmonic mean weighted by the logarithm of the ratio of the radii of the heterogeneous elements.

For systems of heterogeneous, nonuniform elements that are not simple geometries, the computation of an average permeability for the entire system is difficult [38].

Permeability in general is a tensor, and we determine the elements of the tensor in anisotropic systems. Again, in natural systems we have the dilemma of deciding whether directional effects are ordered heterogeneities taken into account, or whether permeability is really tensorial in that the directional effect is a point property. (Interestingly enough, in most situations either way will give a usable answer). Permeability is a volume-averaged property for a finite, but small, volume of a medium. Anisotropy in natural or artificially packed media may result from particle (or grain) orientation, bedding of different sizes of particles or layering of media of different permeability. It is difficult to say at what level we should treat a directional effect as anisotropy or as an oriented heterogeneity. Assuming the problem of scale is solved, let us look at directional effects represented by the permeability tensor.

In principle, simply use the general form of Darcy's law, Equation 16. The significance of the tensor, $\underline{\underline{k}}$, is that in an anisotropic (oriented) medium, the velocity, $\underline{q}$, and the pressure gradient ,$\underline{\nabla}p$, are not in the same direction.

In an oriented porous medium, the resistance to flow is different, depending on the direction. So, if there is a pressure gradient between two points and a spot of fluid is marked, unless the pressure gradient is parallel to oriented flow paths one would find that the spot would not go from the original point to the point expected, but rather it would drift. This is a result of anisotropy of the porous medium.

We have an additional dilemma in that to determine anisotropy experimentally, finite chunks of porous media must be used and, unfortunately, there are two extremes of possible conditions plus the intermediate possibilities [15,39]. For one extreme, permeability is measured in the direction of pressure gradient. The velocity component parallel to the pressure gradient is

$$q_n = (\underline{n} \cdot \underline{q}) \tag{71}$$

The directional permeability measured is

$$k_n = (\underline{n} \cdot [\underline{\underline{k}} \cdot \underline{n}]) \tag{72}$$

and, in two dimensions,

$$k_n = k_{11}\cos^2\alpha + k_{22}\cos^2\beta \tag{73}$$

where k_{11} and k_{22} are the major and minor values of the permeability ellipse in a plot of $1/\sqrt{k_n}$ vs the angle of measurement. α and β are the angles of the principal axes of the tensor, with the unit vector, $\underline{n}$.

For the other extreme, permeability is measured in the direction of flow. The component of pressure parallel to the flow is

$$(\underline{\nabla}p)_n = (\underline{n} \cdot \underline{\nabla}p) \tag{74}$$

The directional permeability measured is

$$k_n' = \frac{1}{([\underline{n} \cdot \underline{\underline{k}}^{-1}] \cdot \underline{n})} \tag{75}$$

and, in two dimensions,

$$\frac{1}{k_n'} = \frac{\cos^2\alpha}{k_{11}} + \frac{\cos^2\beta}{k_{22}} \tag{76}$$

so the $\sqrt{k_n'}$ vs measured angle plots is an ellipse, where k_{11} and k_{22} are the major and minor values for the tensor.

Both Scheidegger [15] and Marcus [39] discuss the error involved when making either measurement. This error is a maximum at 45° from the principal axis of the permeability tensor. Fortunately, the error is small unless the difference between the values of k_{11} and k_{22} is large.

Parsons [40] makes this point when discussing the work of Marcus [40] and Greenkorn et al. [41]. The maximum variation of measured permeability is a function of anisotropy. The difference between the two extremes is not large if the ratio of $k_{11}:k_{22}$ is small.

As a way of separating the effects of heterogeneity and anisotropy, Greenkorn et al. [41] assumed

$$\underline{\underline{k}} = k_m \underline{\underline{I}} + \underline{\underline{k}}' \tag{77}$$

where k_m is the minor principal axis of the permeability tensor (in two dimensions) and represents the overall (isotropic) permeability at a point. $\underline{\underline{k}}'$ is a tensor added to $k_m \underline{\underline{I}}$ to give the directional value at a point. It represents the directional effect. In two dimensions,

$$\underline{\underline{k}} = k_m \begin{pmatrix} 1 & 0 \\ 0 & 1 \end{pmatrix} + \begin{pmatrix} k_{11} - k_m & k_{12} \\ k_{21} & k_{22} - k_m \end{pmatrix} \tag{78}$$

Greenkorn et al. [41] show that for a sandstone system, k_m correlates with grain size (heterogeneity) and $\underline{\underline{k}}'$ correlates with bedding—the directional effect.

The tortuosity and connectivity are difficult to relate to the nonuniformity and anisotropy of a medium. If we attempt to predict permeability from a pore structure model, then tortuosity is needed to correct the model for pore length and connectivity for intersection shape. For a given model, including nonuniformity and anisotropy, tortuosity is constant (2.25 for the model of Haring and Greenkorn [13], 1 for the model of Fatt [29]). Tortuosity enters the problem significantly through fluid flow, that is, in diffusion in pores and in dispersion when we are concerned with flow of miscible fluids. In mass transfer by diffusion in pores, the tortuosity is used to relate diffusivity and effective diffusivity in the pores [42]:

$$D_{eff} = \frac{D}{\tau} \tag{79}$$

STEADY FLOW-COMPLEX POTENTIALS

Two-dimensional steady flow of an incompressible single fluid in heterogeneous, anisotropic porous media often is described using complex potentials. For steady single fluid flow, where we are concerned mainly with the pressure and velocity, we normally do not differentiate between uniform and nonuniform media, other than determining the correct permeability distribution and its average. For steady flow of an incompressible single fluid, Equations 34 and 35 apply, where $\Phi = p + \rho gz$. To discuss steady flow in terms of a complex potential, we need some definitions used with the flow of ideal fluids. Solution of Equation 34 or 35 in two dimensions models flow hypothetically, as if the matrix were not present. We obtain results superficially imposed on the porous medium modeled.

A streamline is an imaginary line in the fluid that has direction of the local velocity at all points. In steady flow, fluid moves along streamlines. If we put a dye in a fluid flowing in a porous medium, we would see, from a distance, a fuzzy streamline. However, if we investigated the actual fluid movement closely, it is complex because of dispersion as it moves and mixes in the pores. Since a streamline has the direction of velocity at all points, there is no flow across the streamline, only along it.

A streamtube is a tube whose surfaces are made of all streamlines passing through a closed curve. There is no flow through the imaginary wall of the streamtube because there can be no flow across streamlines. For steady flow, a streamtube is fixed in space.

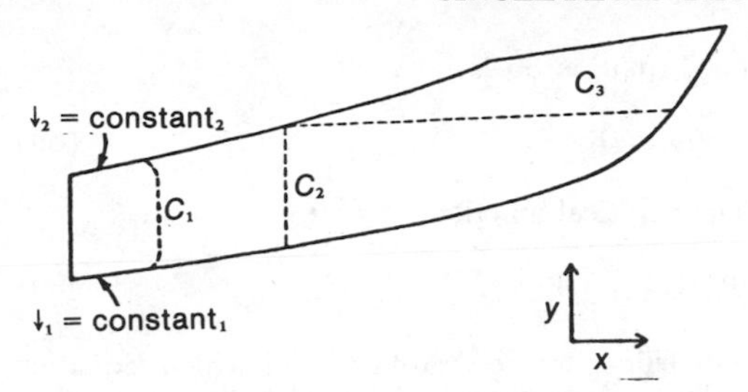

Figure 7. Two-dimensional streamtube [37].

The streamfunction is a quantity defined in such a way that it is constant along a streamline. The two-dimensional streamtube of Figure 7 is constructed of two streamlines represented by the streamfunctions ψ_1 = constant $_1$ and ψ_2 = constant$_2$. Because the flowrate inside a streamtube is constant, we may integrate across any path to evaluate the mean flowrate. For path C_1 of Figure 7,

$$w = \int_{c1} \rho(\underline{v} \cdot \underline{n})\, ds = \int_{c1} \rho v \cos\alpha\, ds \tag{80}$$

where α is the angle between the velocity vector and the outward-directed normal. The value of the streamfunction is defined by

$$w = (\Psi_1 - \Psi_2)\rho \tag{81}$$

For a streamtube,

$$\int \rho d\Psi = \text{constant} \tag{82}$$

Equation 82 is true for any path joining two streamlines, such as C_1, C_2 or C_3 of Figure 7. For path C_2, $\delta s = \delta y$ and $v_x = v \cos\alpha_2$; for path C_3, $\delta s = \delta_x$ and $v_y = v \cos\alpha_3$. Since $\cos\alpha_2 = -1$,

$$\delta\Psi = -v_x \big|_x \delta y \tag{83}$$

And since $\cos\alpha_3 = 1$,

$$\delta\Psi = v_y \big|_y \delta x \tag{84}$$

In the limit,

$$v_x = -\frac{\delta\Psi}{\partial y} \tag{85}$$

$$v_y = \frac{\partial\Psi}{\partial x} \tag{86}$$

A potential is the pressure, or gravity, or their sum, causing flow along a streamline, and is orthogonal to the streamline so that

$$v_x = -\frac{\partial\Phi}{\partial x} \tag{87}$$

and

$$v_y = -\frac{\partial\Phi}{\partial y} \tag{88}$$

For a homogeneous, isotropic medium, Equation 35 is

$$\nabla^2\Phi = 0 \tag{89}$$

The significance of this equation is that an analytical function,

$$w(z) = \Phi(x,y) + i\,\Psi(x,y) \tag{90}$$

represents the streamlines and equipotential lines for flow problems for which Equation 89 is true. Further, this allows conformal transformations to be used to transform flow in complex geometry into simple geometries.

Let us consider single fluid steady flow in a homogeneous, isotropic porous medium in plane radial flow. We might use this to model flow into or out of a well drilled into a reservoir. The flow is sketched schematically in Figure 8 as flow between concentric cylinders, where r_w represents the well radius and r_e a drainage radius. Laplace's equation, Equation 89 in polar coordinates, is

$$\frac{1}{r}\frac{\partial}{\partial r}\left(r\frac{\partial\Phi}{\partial r}\right) + \frac{1}{r^2}\frac{\partial^2\Phi}{\partial\theta^2} = 0 \tag{91}$$

The problem is symmetrical and $\partial^2\Phi/\partial\theta^2 = 0$, so

$$\frac{\partial}{\partial r}\left(r\frac{\partial\Phi}{\partial r}\right) = 0 \tag{92}$$

The stream function, considering the symmetry, is

$$\frac{\partial^2\Psi}{\partial\theta^2} = 0 \tag{93}$$

To solve this flow problem, we look for an analytical function,

$$w(re^{i\theta}) = \Phi(r,\theta) + i\,\Psi(r,\theta) \tag{94}$$

that satisfies the boundary conditions.

Since the flow is radial, we know lines of constant Φ are concentric circles about the origin, and lines of constant Ψ are a family of radiating straight lines. An analytical function that describes this situation is

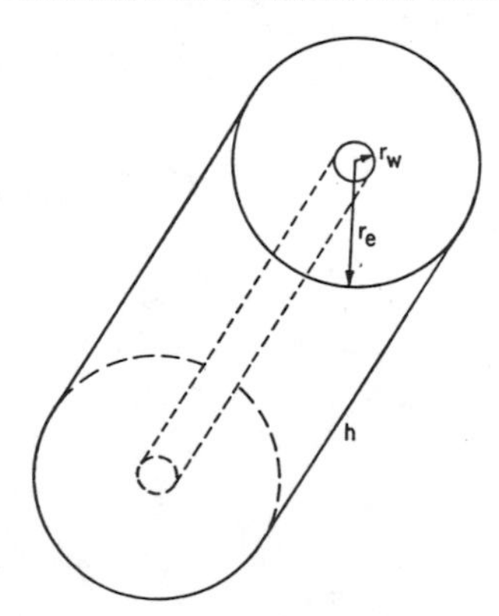

Figure 8. Potential flow between two concentric cylinders [37].

$$w(z) = A \ln Z + B = A \ln r^{i\theta} + B$$

and (95)

$$\Phi = A \ln r + b, \Psi = A\theta + b', B = b + ib'$$

In using complex flow potentials, the problem is to determine an equation for w(z). There are books that contain solutions or different forms of w that give different flow lines and further transformations of these situations. The boundary conditions for C are

$$\Phi\,(r_w) = \Phi_w \tag{96}$$

$$\Phi\,(r_e) = \Phi_e \tag{97}$$

and, therefore,

$$\Phi = \frac{\Phi_e - \Phi_w}{\ln r_e/r_w} \ln \frac{r}{r_w} + \Phi_w \tag{98}$$

The boundary conditions for Ψ are

$$\Psi(o) = 0 \tag{99}$$

$$\Psi(2\,\pi) = -\frac{\rho_w Q_w}{h} \tag{100}$$

and

$$\Psi = -\frac{\rho_w Q_w}{2\pi h}\,\theta \tag{101}$$

The value of A in Equation 95 is determined from both Φ and Ψ, so equating A yields

$$-\rho_w Q_w = 2 \ln \frac{\Phi_e - \Phi_w}{\ln r_e/r_w} \tag{102}$$

The resulting flow field from plotting Equations 98 and 99 is sketched in Figure 9.

Greenkorn et al. [43] used complex potentials to study the effect of the size and shape of a single heterogeneity in single fluid flow in a Hele–Shaw model. A Hele-Shaw model is an analog to two-dimensional flow in porous media. A Hele-Shaw model is constructed by placing two parallel flat plates, usually glass, close together and flowing fluid through the slit between the two plates. The equipotential lines and streamlines are the same for the model and a porous medium if

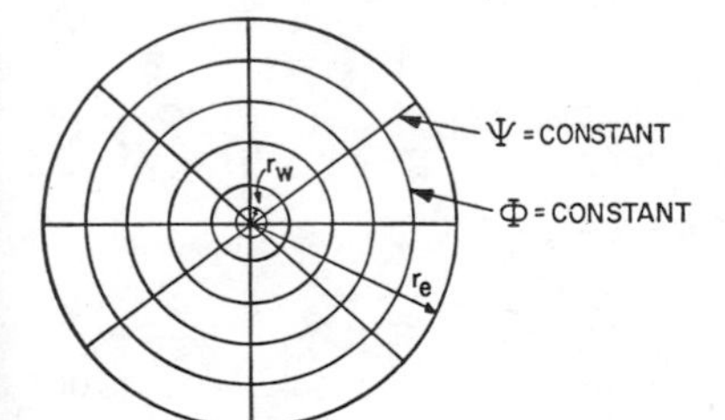

Figure 9. Equipotentials and streamlines for potential flow between concentric cylinders [37].

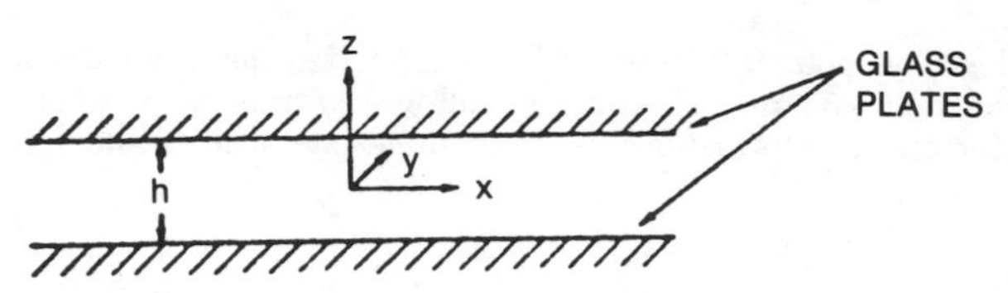

Figure 10. Cross section through Hele-Shaw model [37].

$$k = \frac{h^2}{12} \tag{103}$$

where h is the space between the flat plates, as shown in Figure 10. Experimentally, streamlines are produced by injecting a continuous filament of dye into the flowing fluid. Heterogeneity is introduced into the model by changing the spacing between the plates, etching or creating a partial obstruction.

Greenkorn et al. [43] determined streamlines for single heterogeneities of various shapes in a linear flowfield. Some of their results are shown in Figure 11. As a result of their experiments, they show that heterogeneity is defined approximately for a mathematical model by knowing its size and shape. Further, they found that the actual value of the transmissivity is not required if the transmissivity ratio between the heterogeneity and the flowfield is less than 1/10 or greater than 10. To reach these conclusions, they determined the potential solution to the flow past a circular heterogeneity of infinite permeability (an obstruction) or through a circular heterogeneity of zero permeability (an infinite sink).

The problem of including anisotropy in calculations of steady flow of a single fluid using complex potentials is handled using a coordinate transformation. Consider the equation of condition for steady flow of a single fluid in an anisotropic medium in two dimensions:

$$k_{11}\frac{\partial^2\Phi}{\partial x^2} + k_{22}\frac{\partial^2\Phi}{\partial y^2} = 0 \tag{104}$$

Equation 104 is not Laplace's equation. We can solve the equation numerically or we can apply complex potentials by introducing the transformations

$$\bar{x} = \frac{x}{\sqrt{k_{11}}}$$

and (105)

$$\bar{y} = \frac{y}{\sqrt{k_{22}}}$$

Equation 104 becomes

$$\frac{\partial^2\Phi}{\partial \bar{x}^2} + \frac{\partial^2\Phi}{\partial \bar{y}^2} = 0 \tag{106}$$

Equation 106 is Laplace's equation in the transformed coordinates.

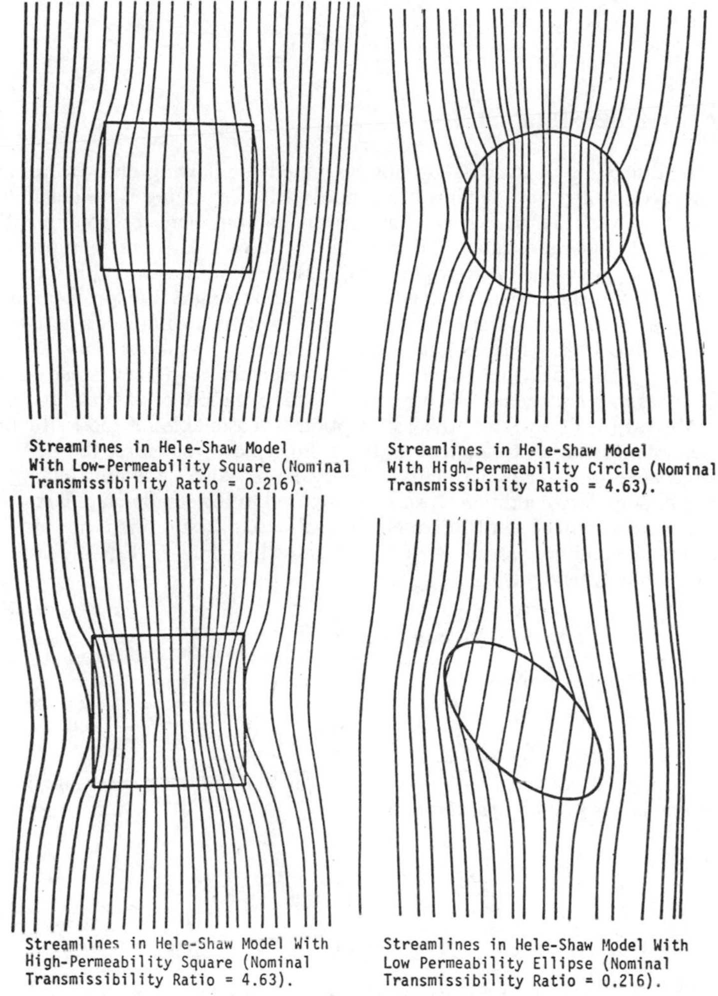

Streamlines in Hele-Shaw Model
With Low-Permeability Square (Nominal
Transmissibility Ratio = 0.216).
Streamlines in Hele-Shaw Model
With High-Permeability Circle (Nominal
Transmissibility Ratio = 4.63).
Streamlines in Hele-Shaw Model With
High-Permeability Square (Nominal
Transmissibility Ratio = 4.63).
Streamlines in Hele-Shaw Model With
Low Permeability Ellipse (Nominal
Transmissibility Ratio = 0.216).

Figure 11. Streamlines in Hele-Shaw model for different permeability conditions [43].

In the transformed system, equipotential lines and streamlines are orthogonal. In the original system, equipotential lines and streamlines are not orthogonal. The angle between the equipotential lines and the streamlines in the original system is given by

$$\cos\theta = \frac{k_{11}\left(\frac{\partial\Phi}{\partial x}\right)^2 + k_{22}\left(\frac{\partial\Phi}{\partial y}\right)^2}{\left|\tilde{q}\right| \left|\underline{i}\frac{\partial\Phi}{\partial x} + \underline{j}\frac{\partial\Phi}{\partial y}\right|} \tag{107}$$

and

$$k_n = \frac{1}{\frac{\cos^2\theta_{11}}{k_{11}} + \frac{\cos^2\theta_{22}}{k_{22}}} \tag{108}$$

The average isotropic permeability is

$$\langle k \rangle = \sqrt{k_{11} k_{22}} \tag{109}$$

STEADY COMPRESSIBLE FLOW

So far, the discussion of steady-state flow assumed the flowing fluid is incompressible. We often are concerned with compressible fluids when studying transient flow, that is, where time is a variable. However, for certain combinations of compressibility and system size for liquids and gases it is possible to have steady-state or pseudo-steady-state compressible flow. Such systems of compressible fluid flow are either in permanent steady state or in a succession of steady states. Time enters the problem in the succession of steady states as a parameter; time is not explicit, so Laplace's equation applies at each successive time. It is assumed that boundary conditions vary as a function of time. The pressure distribution changes immediately to a new steady-state distribution. This assumption means the velocity of propagation of a pressure disturbance is infinite. Muskat [44] shows that this assumption is reasonable for most situations, even though the system is large. Compared to the normal flow velocities, the actual velocity that a disturbance propagates is relatively very large. Also, assuming change in mass flux, change in boundary conditions is small compared to the steady-state flux. This means there is no attenuation of a disturbance, a reasonable assumption for most systems. The equation of condition for flow of a compressible fluid is Equation 38. Because density is not time dependent,

$$(\underset{\sim}{\nabla} \cdot \underset{\approx}{k}\, \underset{\sim}{\nabla} \rho) = 0 \tag{110}$$

If the porous medium is isotropic,

$$\nabla^2 \rho = 0 \tag{111}$$

To use the equation, we need to know density as a function of position. For a linear system with densities ρ_1 and ρ_2 at the boundaries,

$$\rho = \rho_2 + \frac{1}{L}(\rho_1 - \rho_2)x \tag{112}$$

where L is the length of the system. Assume an equation of state, such as

$$\rho = \rho_o e^{c_l p} \tag{113}$$

Substituting Equation 113 into Equation 112 gives the pressure distribution of the system

$$p = \frac{1}{c_l} \ln \left[\frac{1}{L}(e^{c_l p_1} - e^{c_l p_2})x + e^{c_l p_2} \right] \tag{114}$$

The flowrate per unit area is

$$Q = -\frac{k}{\mu c_l L}(p_1 - p_2) \tag{115}$$

For a radial system with the densities at the boundaries of ρ_w and ρ_e,

$$\rho = \rho_w + \frac{(\rho_e - \rho_w)}{\ln r_e/r_w} \ln \frac{r}{r_w} \tag{116}$$

The pressure distribution for the radial system using Equation 113 is

$$p = \frac{1}{c_l} \ln \left\{ (e^{c_l p_e} - e^{c_l p_w}) \frac{\ln r/r_w}{\ln r_e/r_w} + e^{c_l p_w} \right\} \tag{117}$$

The flowrate for the system of thickness, h, is

$$Q = \frac{2\pi kh}{c_l \mu} \frac{(p_e - p_w)}{\ln r_e/r_w} \tag{118}$$

The equation of condition for flow of an ideal gas is Equation 44. For steady state in an isotropic medium,

$$\nabla^2 p^2 = 0 \tag{119}$$

From the ideal gas law, ρ is proportional to p; therefore, in terms of density,

$$\nabla^2 \rho^2 = 0 \tag{120}$$

The pressure distribution for a linear system is

$$p^2 = \frac{1}{L} (p_1{}^2 - p_2{}^2) \, x + p_2{}^2 \tag{121}$$

The flowrate per unit area is

$$Q = -\frac{k\rho}{2\mu L} (p_2{}^2 - p_1{}^2) \tag{122}$$

The pressure distribution for the ideal radial system is

$$p^2 = p_w{}^2 + \frac{(p_e{}^2 - p_w{}^2)}{\ln r_e/r_w} \ln r_e/r_w \tag{123}$$

The flowrate for a radial system of constant thickness, h, is

$$Q = \frac{\pi kh\rho}{\mu} \frac{(p_e{}^2 - p_w{}^2)}{\ln r_e/r_w} \tag{124}$$

To determine the pressure distribution and flowrate for real gases, an appropriate equation of state is used in place of the ideal gas equation.

For systems that are heterogeneous and anisotropic, $\underline{\underline{k}} = \underline{\underline{k}}\,(x,y,z)$. The approach for a purely anisotropic system is to use a transformation to change Equation 110 to Laplace's equation. For heterogeneous systems, the solution for real situations is normally numerical.

TRANSIENT FLOW

Transient flow is described by including time as an independent variable in the equations used to model flow. The equation of condition for a compressible fluid is given by Equation 38. The variation in pressure and flux in a compressible system includes the time variation of density. For slightly compressible systems Equation 41 is used, with pressure the dependent variable. Generally, Equation 38 correctly describes transient flow, but for most situations of practical interest we use Equation 41 and its solutions to model transient flow. For transient flow of highly compressible fluids—gases—we consider variation of density and pressure through appropriate equations of state. It also may be necessary to include the variation of viscosity.

Equations 38 and 41 and their simplified forms and boundary conditions that follow are variations of the diffusion equation. This equation and its solutions have been used extensively in the study of the transport of momentum, heat and mass. Solutions for the equations for physical situations analogous to those discussed here may be found in the scientific literature. See, for example, the books by Crank [45], Carslaw and Jaeger [46] and Bird et al. [47].

For a linear system of semiinfinite length, Equation 41 becomes

$$\frac{\partial p}{\partial t} = \frac{k}{\phi \mu c_l} \frac{\partial^2 p}{\partial x^2} \tag{125}$$

If we let $\alpha = \dfrac{k}{\phi \mu c_l}$, then

$$\frac{\partial p}{\partial t} = \alpha \frac{\partial^2 p}{\partial x^2} \tag{126}$$

Equation 126 is analogous to diffusion of mass in one dimension, conduction of heat in one dimension and transport of momentum in one dimension.

Consider a porous medium occupying the semiinfinite space from $x = 0$ to $x = \infty$. The system is initially at pressure p_0. At time $t = 0$, the pressure is raised to p_1 on the face of the porous medium at $x = 0$, and maintained at p_1 for all times greater than zero. What is the time-dependent pressure, $p(x,t)$? The problem just posed is a classic problem in transport phenomena. By analogy with the solution to the transport of momentum, heat and mass, we write Equation 126 in terms of the dimensionless pressure $p^* = p - p_o/p_i - p_o$. Then the initial conditions and the boundary conditions are

$$\begin{aligned} p^*(x_1 0) &= 0\,, \text{ all } x \\ p^*(0_1 t) &= 1\,, t > 0 \\ p^*(\infty_1 t) &= 0\,, t > 0 \end{aligned} \tag{127}$$

Analogously, we let $\eta = x/\sqrt{4\alpha t}$ and rewrite Equation 126 in terms of p^* and η:

$$\frac{d^2 p^*}{d\eta^2} + 2\eta \frac{dp^*}{d\eta} = 0 \tag{128}$$

with initial and boundary conditions

$$\begin{aligned} p^*\ (0) &= 1 \\ p^*\ (\eta) &= \infty \end{aligned} \tag{129}$$

The solution of Equation 128 with boundary conditions (Equation 129) is

$$p^* = 1 - \frac{2}{\sqrt{\eta}} \int_0^{\eta} e^{-\eta^2}\, d\eta \tag{130}$$

or

$$\frac{p - p_0}{p_i - p_0} = 1 - \operatorname{erf} \frac{x}{\sqrt{4\alpha t}} \tag{131}$$

For an isotropic radial system of thickness h, the equation of condition, Equation 41, is

$$\frac{1}{\alpha} \frac{\partial p}{\partial t} = \frac{\partial^2 p}{\partial r^2} + \frac{1}{r} \frac{\partial p}{\partial r} \tag{132}$$

Solutions to this equation are numerous, especially for reservoir situations. Earlougher [48] writes a generalized solution to Equation 132 in terms of the initial pressure, p_i, for a reservoir of porous rock with a single well with source q as

$$p_i - p(t,r) = 141.2 \frac{qB\mu}{kh} [p_D(t_D, r_D, C_D, \text{geometry}, \ldots) + s] \tag{133}$$

Where B is the formation volume factor, p_D is the dimensionless pressure solution to Equation 132 for appropriate boundary conditions, C_D is a wellbore storage constant, and s is the skin effect. r_D is a dimensionless radius defined by

$$r_D = \frac{r}{r_w} \tag{134}$$

where r_w is the well radius. t_D is the dimensionless time based on wellbore radius

$$t_D = \frac{0.0002637kt}{\phi \mu c_t r_w^2} \tag{135}$$

Sometimes t_{DA}, a dimensionless time based on total drainage area, is used.

$$t_{DA} = \frac{.0002637kt}{\phi \mu c_t A} = t_D \left(\frac{r_w^2}{A}\right) \tag{136}$$

In Equations 135 and 136, c_t is total compressibility (including the small effect of rock compressibility). Figure 12 is a plot of p_D vs t_D/r_D^2 for a single source in an infinite radial system. Figure 13 is a plot of p_D vs t_{DA} for a 4:1 rectangle with well at $x/L = 0.75$ and $y/W = 0.5$.

Let's examine the solution of Equation 132 for the case of an infinite radial system. Assume that at a point the pressure increases or declines (flow into or flow out of) over a line of length, h, at a point in the medium. The initial condition and boundary conditions are similar to those for the linear system of semiinfinite length:

$$\begin{aligned} p(r,0) &= p_o \quad &, \text{all } r \\ q(0,t) &= \text{constant} \quad &, t > o \\ p(\infty,t) &= p_o \quad &, t > o \end{aligned} \tag{137}$$

The solution to Equation 132 with these initial and boundary conditions is

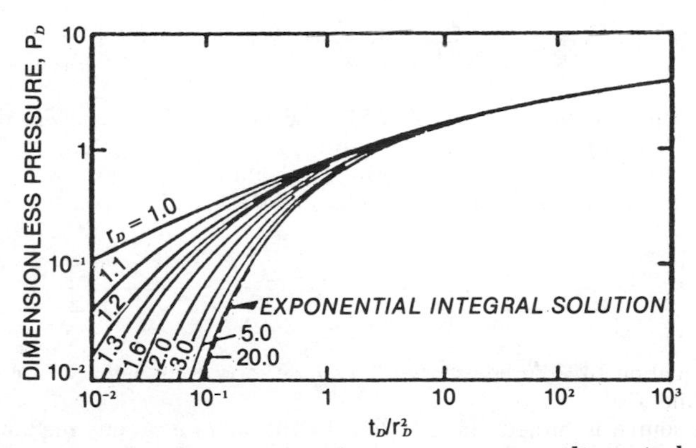

Figure 12. Dimensionless pressure function at various dimensionless distances from a well located in an infinite system [49].

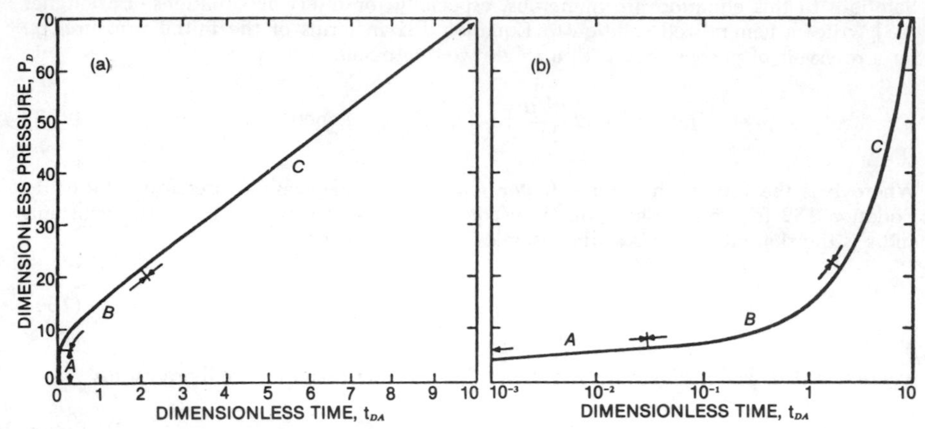

Figure 13. Transient flow regimes: (A) infinite acting; (B) transition; (C) pseudo-steady state. 4:1 rectangle with the well at x/L = 0.75, y/W = 0.5 [50].

$$p = p_o = \frac{q\mu}{4\pi kh}\left[-\text{Ei}\left(-\frac{r^2}{4\alpha t}\right)\right] \tag{138}$$

where the Ei function or exponential integral is defined as

$$-\text{Ei}(-x) = \int_x^{\infty} \frac{e^{-\tau}}{\tau}\, d\tau \tag{139}$$

and is tabulated for values of x. The curve labeled exponential integral in Figure 12 is this solution. This solution, Equation 138, is satisfactory for use with sources, $-q$, and sinks, $+q$, of finite radius, as long as

$$\frac{r^2}{4\alpha t} > 100 \tag{140}$$

so this solution is good for large systems or for short times.

The exponential integral solution is sometimes approximated with a logarithmic expression so that

$$p - p_o = \frac{q\mu}{4\pi kh}\left[\ln\frac{4\alpha t}{r^2} + 0.80907\right] \tag{141}$$

The error in this approximation is about 2%. Since Equation 132 is a linear differential equation, the effect of all point (line) sources and sinks may be superimposed to obtain solutions for multiple sources or sinks in a porous medium, or for successive changes in q (or p) at a given source or sink. Since n sources or sinks with q_i occur at time $> t_i$,

$$p - p_o = \frac{\mu}{4\pi kh}\sum_{i=1}^{n}(q_i - q_{i-1})\left[-\text{Ei}\left(-\frac{r^2}{4\alpha(t - t_i)}\right)\right] \tag{142}$$

As mentioned, Equation 142 can be used with a single source (or sink) where q changes as a function of time.

If at time t_s a source is turned off, the fluid in the infinite porous medium will redistribute itself to establish a uniform pressure. The pressure as a function of r and t is represented by superposition of the source of strength, q, and the existing sink starting at t_s:

$$p - p_o = -\frac{q\mu}{4\pi kh}\left[-\mathrm{Ei}\left(-\frac{r^2}{4\alpha t}\right) + \mathrm{Ei}\left(-\frac{r^2}{4\alpha(t-t_s)}\right)\right] \tag{143}$$

If we use a logarithmic approximation,

$$p - p_o = -\frac{q\mu}{4\pi kh}\ln\frac{t}{t-t_s} \tag{144}$$

Define the period of time the source is turned off as $\Delta t = t - t_s$. Then

$$p - p_o = -\frac{q\mu}{4\pi kh}\ln\frac{t_s + \Delta t}{\Delta t} \tag{145}$$

is valid for large Δt. A plot of p vs $\ln (t_s + \Delta t)/\Delta t$ will yield a straight line of slope $-q\mu/4\pi kh$. If we know q, μ and h, the value of k for the porous medium can be determined.

If the flowrate or pressure of a line source is varied as a function of time, the response at some radius, r, can be determined. Johnson et al. [51] generated a series of pressure pulses at one well in a reservoir and observed the response in an adjacent well to determine the properties of the region between the wells. Figure 14 is a schematic of the resultant pulses and responses showing the time lag and response amplitude. The dimensionless time lag, $t_{DL} = t_L/\Delta t$, can be shown to be approxmately

$$t_{DL} = \frac{r^2}{11.63\alpha\Delta t} \tag{146}$$

The pulse response amplitude is

$$\Delta p_m = \frac{70.6qB}{T}(\bar{p}_m - \bar{p}_{tm}) \tag{147}$$

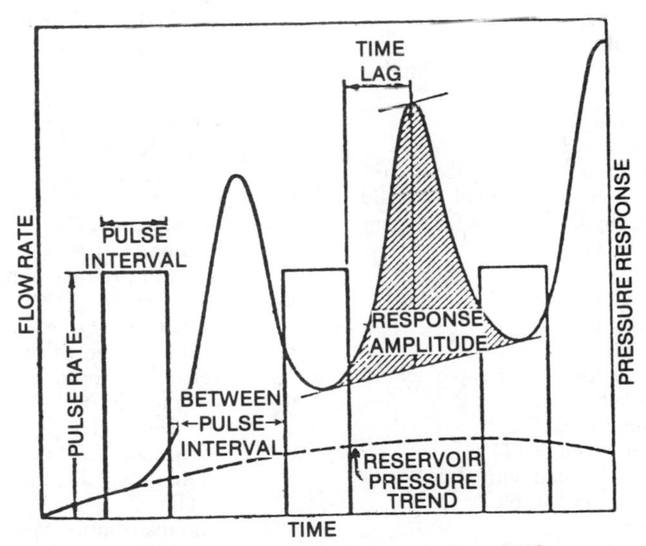

Figure 14. Pulse-test terminology [51].

where $T = kh/\mu$. $\bar{p}_m$ is from the solution of Equation 147 at $\Delta t + t_L$; $\bar{p}_m$ is the pressure response observed at this time. Knowing response amplitude and time lag, we can determine α from Equation 142.

The equation of condition for flow of an ideal gas is given by

$$\frac{\phi\mu}{2p}\frac{\partial p^2}{\partial t} = (\underline{\nabla} \cdot \underline{\nabla}\, p^2) \tag{148}$$

Aronofsky [52] solved Equation 148 numerically for a linear isotropic system. Jenkins and Aronofsky [53] solved Equation 148 numerically for radial flow with a single constant rate source in an infinite radial system.

The effects of anisotropy and heterogeneity can be included. For the infinite two-dimensional radial case, anisotropy is included by writing

$$p - p_o = \frac{q\mu}{4\pi h\sqrt{k_{11}k_{22}}}\left\{-\,\mathrm{Ei}\left[-\,\frac{\phi\mu c_t}{4t}\left(\frac{(x-x')^2}{k_{11}} + \frac{(y-y')^2}{k_{22}}\right)\right]\right\} \tag{149}$$

where k_{11} and k_{22} are the principal values of the permeability tensor when the axes coincide with the coordinates. x and y locate the observation point in the medium. (x' and y') is the location of the source (or sink). The effect of heterogeneity is included in the solution of Equation 41 by keeping $\underline{\underline{k}}$ inside the derivative. In this case, we need to know k as a function of position. Solutions for heterogeneous media are usually numerical.

NOMENCLATURE

A	area (L^2)
a	connectivity, accessibility (1)
B	formation volume factor (1)
b	parameter in slip flow expression for $K(t^2L/M)$
c	shape factor (1)
c_l	fluid compressibility (Lt^2/M)
D	diameter (L)
D_p	particle diameter (L)
d	characteristic microscopic length (L)
d	differential operator (1)
$\underline{F}$	body forces (LM/t^2)
g	acceleration due to gravity (L/t^2)
h	energy per unit weight of fluid (L) (hydraulic head for water)
h	thickness (L)
$\underline{\underline{K}}$	hydraulic conductivity (L/t)
$\tilde{K}$	slip conductivity (L/t)
k	permeability (L^2)
l	length (L)
L	characteristic macroscopic length (L)
l	pore length (L)
M	molecular weight (1)
$\underline{n}$	outward-directed normal
$\mathcal{p}$	$p - p_o + \rho\phi$ (M/t^2L)
p	pressure (M/t^2L)
p_m	arithmetic average pressure, $(p_1 + p_2/2)$ (M/t^2)
Q	volumetric flowrate (L^3/t)
Q_m	volumetric flowrate at average pressure (p_m) (L^3/t)
$\underline{q}$	seepage velocity (L/t)
$\bar{R}$	gas constant (L^2M/RT)
Re_p	Reynolds number based on particle diameter (1)
r	radius (L)
S	specific surface (1)
S_A	surface area of pores (L^2)
S_w	interface (1)
T	temperature (T)
t	time (t)
u	displacement (L)
V	volume (L^3)
V_o	total pure volume of medium (L^3)
V_f	fluid volume (L^3)
v	velocity (L/t)
$\langle v \rangle$	area-averaged velocity (L/t)
v_∞	velocity of approach (L/t)
w	flowrate (M/t)
x	coordinate (L)
y	coordinate (L)
z	coordinate (direction of gravity) (L)
$\alpha(\delta)$	void distribution factor (1)

Δ difference
δ pore diameter (L); Dirac delta
$\underline{\nabla}$ vector partial differential operator (1/L) $\underline{i}(\partial/\partial x) + \underline{j}(\partial/\partial y) + \underline{k}(\partial/\partial z)$
∇^2 $(\partial/\partial x^2) + (\partial/\partial y^2) + (\partial/\partial z^2)$, $(1/L^2)$
μ viscosity (M/Lt)
Φ potential (depends on coordinates)
ϕ porosity (1)
ψ streamfunction (depends on coordinates)
ρ density (M/L^3)
τ fluid stress (M/Lt^2)
τ tortuosity (1)

Overscores

$\hat{}$ identifying mark
< > space average
$-$ time average

Subscripts

D dimensionless quantity
i coordinates
w well

REFERENCES

1. Darcy, H. *Les Fountaines Publiques de la Ville de Dijon* (Paris: Dalmont, 1856).
2. DeWiest, R. J. M. *Geohydrology* (New York: John Wiley & Sons, Inc., 1965).
3. Klinkenberg, L. J. *API Drill. Proc. Pract.* 200 (1941).
4. Brutsaert, W. "Probability Laws for Pore-Size Distributions," *Soil Sci.* 101(2): 85 (1966).
5. Dullien, F. A. L., and V. K. Batra. "Determination of the Structure of Porous Media," *Ind. Eng. Chem.* 62(10):25 (1970).
6. Dullien, F. A. L., and P. N. Mehta. "Particle Size and Pore (Void) Size Determination by Photomicrographic Methods," *Powder Technol.* 7:305 (1971/72).
7. Dullien, F. A. L., and G. K. Dhawan. "Characterization of Pore Structure by a Combination of Quantitative Photomicrography and Mercury Porosimetry," *J. Colloid Interface Sci.* 47(2):337 (1974).
8. Graton, S. C., and H. S. Fraser. "Systematic Packing of Spheres with Particular Reference to Porosity and Permeability," *J. Geol.* 43(8):785 (1935).
9. Mayer, R. P., and R. A. Stowe. "Mercury Porosimetry–Breakthrough Pressure for Penetration between Packed Spheres," *J. Colloid. Sci.* 20:893 (1965).
10. Haughey, D. P., and G. S. G. Beveridge. "Local Voidage Variation in a Randomly Packed Bed of Equal-Sized Spheres," *Chem. Eng. Sci.* 21:905 (1966).
11. Guin, J. A., D. P. Kessler and R. A. Greenkorn. "The Permeability Tensor for Anisotropic Non-uniform Porous Media," *Chem. Eng. Sci.* 26:1475 (1971).
12. Guin, J. A., D. P. Kessler and R. A. Greenkorn. "The Dispersion Tensor in Anisotropic Porous Media," *Ind. Eng. Chem. Fund.* 11(4):477 (1972).
13. Haring, R. E., and R. A. Greenkorn. "A Statistical Model of a Porous Medium with Non-uniform Pores," *AIChE J.* 16(3):477 (1970).
14. Matheron, G. *Eléments Poreus une Theorie des Milieux Poreus* (Paris: Maison, 1967).
15. Scheidegger, A. E. *The Physics of Flow Through Porous Media* (New York: Macmillan Publishing Co., Inc., 1957).
16. Saffman, P. G. "A Theory of Dispersion in a Porous Medium," *Fluid Mech.* 6:21 (1959).

17. Dullien, F. A. L. *Porous Media Fluid Transport and Pore Structure* (New York: Academic Press, Inc., 1979).
18. MacDonald, I. F., M. S. El-Sayad, K. Mow and F. A. L. Dullien. "Flow Through Porous Media–the Ergun Equation Revisited," *Ind. Eng. Chem. Fund.* 18:199 (1979).
19. Slattery, J. C. *Momentum, Energy and Mass Transfer in Continua* (New York: McGraw-Hill Book Co., 1972).
20. Whitaker, S. "Advances in Theory of Fluid Motion in Porous Media," in *Flow Through Porous Media*, R. J. Nunge, Ed. (Washington, DC: American Chemical Society, 1970), p. 31.
21. Bear, J. *Dynamics of Fluids in Porous Media* (New York: Elsevier, North-Holland, Inc., 1972).
22. Greenkorn, R. A., and D. P. Kessler. "Dispersion in Heterogeneous Non-uniform, Anisotropic Porous Media," in *Flow Through Porous Media*, Richard J. Nunge, Ed. (Washington, DC: American Chemical Society, 1970).
23. Ferrandon, J. "Les Lois de l'êcoulement de Filtration," *Genie Civil* 125(2):24 (1948).
24. Scheidegger, A. E. "Directional Permeability of Porous Media to Homogeneous Fluids," *Geofis. Pura. Appl.* 28:75 (1954).
25. Slattery, J. C. "Single-Phase Flow Through Porous Media," *AIChE J.* 15(6):866 (1969).
26. Brinkman, H. C. "A Calculation of the Viscous Force Exerted by a Flowing Fluid on a Dense Swarm of Particles," *Appl. Sci. Res.* A1:27 (1947).
27. Howells, I. D. "Drag Due to the Motion of a Newtonian Fluid Through a Sparse Random Array of Small Fixed Rigid Objects," *J. Fluid Mech.* 64:449 (1974).
28. Hinch, E. J. "An Averaged-Equation Approach to Particle Interactions in a Fluid Suspension," *J. Fluid Mech.* 83:695 (1977).
29. Fatt, I. "The Network Model of Porous Media, I. Capillary Pressure Characteristics," *Pet. Trans. AIME* 207:144 (1956).
30. Kozeny, J. "Ober Kapillare Leitung des Wassers im Boden," *S. Ber. Wiener Akad. Abt. IIa*, 136:271 (1927).
31. Fatt, I. "The Network Model of Porous Media, II. Dynamic Properties of a Single Size Tube Network," *Pet. Trans. AIME* 207:160 (1956).
32. Law, J. "A Statistical Approach to the Interstitial Heterogeneity of Sand Reservoirs," *Pet. Trans. AIME* 155:202 (1944).
33. Bennion, D. W., and J. C. Griffiths. "A Stochastic Model for Predicting Variations in Reservoir Rock Properties," *Pet. Trans. AIME* 237:9 (1966).
34. Collins, R. E. *Flow of Fluids Through Porous Media* (New York: Van Nostrand Reinhold Co., 1961).
35. Csallany, S., and W. C. Walton. "Yields of Shallow Dolomite Wells in Northern Illinois," *Illinois State Water Surv. Dept. Invest.* 26 (1963).
36. Seaber, P. R., and E. F. Hollyday. "Statistical Analysis of Regional Aquifers," paper presented at the annual meeting of the Geological Society of America, San Francisco, 1966.
37. Greenkorn, R. A. *Flow Phenomena in Porous Media* (New York: Marcel Dekker, Inc., 1982).
38. Johnson, C. R., and R. A. Greenkorn. "Comparison of Core Analysis and Drawdown-Test Results from a Water-Bearing Upper Pennsylvanian Sandstone of Central Oklahoma," *Bull. Int. Assoc. Scientific Hydrol.* VII:46 (1962).
39. Marcus, H. "The Permeability of a Sample of an Anisotropic Porous Medium," *J. Geophys. Res.* 67(13):5215 (1962).
40. Parsons, R. W. "Discussion of Directional Permeability of Heterogeneous Anisotropic Porous Media by Greenkorn, Johnson, and Shallenberger," *Pet. Trans. AIME* 231: SPE J. 364 (1964).
41. Greenkorn, R. A., C. R. Johnson and L. K. Shallenberger. "Directional Permeability of Heterogeneous Anisotropic Porous Media," *Pet. Trans. AIME* 231: SPE J. 124 (1964).

42. Satterfield, C. N., and T. K. Sherwood. *The Role of Diffusion in Catalysis* (Reading, MA: Addison-Welsey Publishing Co., Inc., 1963).
43. Greenkorn, R. A., R. E. Haring, H. O. Jahns and L. K. Shallenberger. "Flow in Heterogeneous Hele-Shaw Models," *Pet. Trans. AIME* 231: SPE J. 307 (1964).
44. Muskat, M. *The Flow of Homogeneous Fluids Through Porous Media* (New York: McGraw-Hill Book Co., 1937), Chapt. X.
45. Crank, J. *The Mathematics of Diffusion* (Oxford, England: Oxford University Press, 1956).
46. Carslaw, H. S., and J. C. Jaeger. *Heat Conduction in Solids,* 2nd Ed. (Oxford, England: Oxford University Press, 1959).
47. Bird, R. W., W. E. Stewart and E. N. Lightfoot. *Transport Phenomena* (New York: John Wiley & Sons, Inc., 1960), p. 127.
48. Earlougher, R. C., Jr. *Advances in Well Test Analysis,* Monograph Vol. 5 (New York: Society of Petroleum Eng. of AIME, 1977).
49. Mueller, T. D., and P. A. Witherspoon. "Pressure Interference Effects Within Reservoirs and Aquifers, *Pet. Trans. AIME* 234: SPE J. 471 (1965).
50. Earlougher, R. C., Jr., and H. J. Ramey, Jr. "Interference Analysis in Bounded Systems," *J. Can. Pet. Technol.* (October–December 1973).
51. Johnson, C. R., R. A. Greenkorn and E. G. Woods. "Pulse Testing: a New Method for Describing Reservoir Flow Properties Between Wells," *Pet . Trans. AIME* 237: 1599 (1966).
52. Aronofsky, J. S. "Effect of Gas Slip on Unsteady Flow of Gas Through Porous Media," *J. Appl. Phys.* 25:48 (1954).
53. Jenkins, R., and J. S. Aronofsky. "Unsteady Radial Flow of Gas Through Porous Media," in *Proc. 1st U.S. Nat. Cong. Appl. Mech.,* Chicago, ASME Applied Mech. Div., paper 52-A26 (1951).

CHAPTER 12
DISPERSION MODELS FOR CONTINUOUS FLOW IN CHEMICAL REACTORS

L. T. Fan
Department of Chemical Engineering
Kansas State University
Manhattan, Kansas 66506

CONTENTS

INTRODUCTION

The performance of a continuous-flow chemical reactor is dependent on many factors, including the kinetics of chemical reactions involved and dispersion of reactant molecules or particles induced by the motion of the reacting stream flowing through the reactor. However, it is usually difficult to derive rigorous continuum or statistical mechanical equations governing such motion. At best, the resultant expressions are often cumbersome to solve. More often than not, therefore, we resort to modeling approaches in deriving the governing equations for the performance of a continuous-flow chemical reactor.

The extremely complex process of reactant mixing or dispersion in a continuous-flow chemical reactor consists, in reality, of a multitude of physicochemical phenomena. In evaluating their effect on the performance of the reactor, such phenomena have been classified conveniently into two classes: one is the so-called macromixing, and the other the so-called micromixing. This chapter is concerned with modeling these two manifestations of mixing or dispersion in flow reactors. Both deterministic and stochastic approaches are discussed.

MACROMIXING IN CONTINUOUS-FLOW CHEMICAL REACTORS

Macromixing is the primary aspect of mixing accompanied by gross fluid or particle motion in a flow chemical reactor, induced by the additive effects of convective flow, motion of eddies and molecular diffusion. The macromixing component gives rise to the variation in the residence times experienced by molecules or particles flowing through the reactor. The time elapsed between the entry of a fluid molecule or flowing particle into a reactor and its exit from the reactor, is its residence time. As different fluid elements or particles generally stay in the flow reactor for varying lengths of time, their residence times usually are distributed. The mean of the residence time distribution can be defined as

$$\bar{t} = \frac{V}{q_0} \tag{1}$$

The time, t, can be normalized or made dimensionless by dividing it by the mean residence time, $\bar{t}$, i.e.,

$$\theta = \frac{t}{\bar{t}} \tag{2}$$

The exit age distribution function (E), or the residence time distribution (RTD), function of a fluid or stream of particles flowing through a reactor is defined in such a way that $E(t)dt$ is the fraction of material in the exit stream with ages between t and $t + dt$. The RTD is a characteristic of the flow pattern established in the reactor; it completely defines macromixing. Figure 1 illustrates a typical residence time distribution. The properties of the RTD are

$$\left.\begin{array}{ll} \text{a.} & \int_0^\infty E(t)dt = 1 \text{ or } \int_0^\infty E(\theta)d\theta = 1 \\ \text{b.} & \int_0^\infty tE(t)dt = \bar{t} \text{ or } \int_0^\infty \theta E(\theta)d\theta = 1 \end{array}\right\} \tag{3}$$

Note that $E(t)$, or $E(\theta)$, is naturally nonnegative.

One of the two extremes of the RTD is typified by that of a plug flow reactor (PFR), in which all the fluid elements or particles pass through the reactor together in one

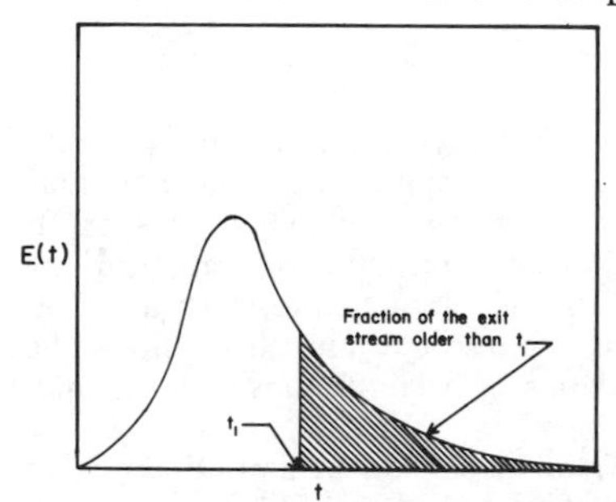

Figure 1. Residence time distribution of a continuous-flow chemical reactor.

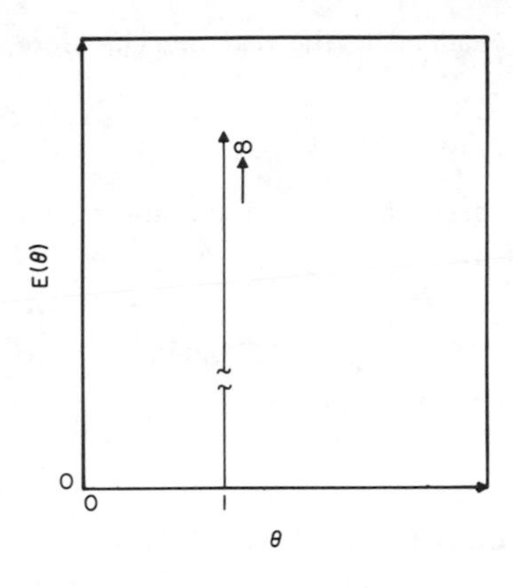

Figure 2. Residence time distribution of a plug-flow reactor.

nominal holding time, as shown in Figure 2. The other is the RTD of a continuous-stirred-tank reactor (CSTR) in which all the fluid elements or particles within the reactor have an equal probability of departing from the reactor at any moment. The RTD of a CSTR is an exponential decay function as shown in Figure 3. The RTD of a PFR can be written as

$$E(\theta) = \delta(\theta - 1) \tag{4}$$

and that of a CSTR as

$$E(\theta) = e^{-\theta} \tag{5}$$

The information contained in the RTD may be expressed by one of the following related distribution functions.

1. Cumulative age distribution, $F(t)$. This is the fraction of all molecules or particles that stay in the reactor for residence times shorter than t.
2. Internal age distribution, $I(\alpha)$. $I(\alpha)d\alpha$ is the fraction of the molecules or particles in the reactor with ages between α and $\alpha + d\alpha$.
3. Internal life expectation distribution, $\psi(\lambda)$. $\psi(\lambda)d\lambda$ is the fraction of the molecules or particles in the reactor with life expectation between λ and $\lambda + d\lambda$.
4. Intensity function, $\Delta(t)$. $\Delta(t)dt$ is the fraction of the molecules or particles inside the reactor having ages between t and t + dt, which will exit in the next dt units of time.

In these definitions, the life expectation, λ, measures the length of time that a fluid element or particle remains in the reactor before it exits, and the internal age, α, is the

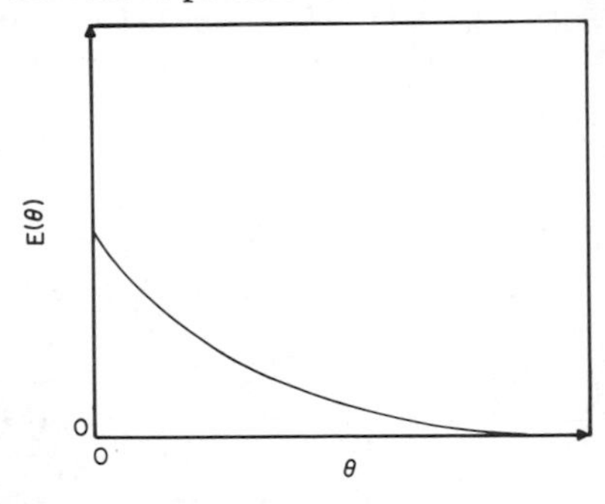

Figure 3. Residence time distribution of a continuous-stirred-tank reactor.

length of time that the element (or molecule) has remained in the reactor. Therefore, we can write

Total residence time = present age + life expectation

The different distribution functions defined in the preceding paragraph are related through the following expressions [1–3]:

$$\left.\begin{array}{llll} \text{a.} & F(t) = \int_0^t E(t)\,dt & \text{b.} & I(\alpha) = \frac{1}{\bar{t}}[1 - F(\alpha)] \\ \text{c.} & \psi(\lambda) = \frac{1}{\bar{t}}[1 - F(\lambda)] & \text{d.} & E(t) = \bar{t}\Delta(t)\, I(t) \end{array}\right\} \quad (6)$$

MACROMIXING MODELS FOR CONTINUOUS-FLOW CHEMICAL REACTORS

As stated in the preceding section, RTD alone is not sufficient to determine the conversion of reactant in a continuous-flow chemical reactor unless the kinetics of its conversion are linear (isothermal and first order). For nonlinear kinetics, however, knowledge of the residence time distribution enables us to predict the limits between which the conversion of the reactant must lie. These limits correspond to two extremes of micromixing–complete segregation and maximum mixedness–presented in the next section. Some of the relatively simple macromixing models are presented in this section.

Continuous-Stirred-Tank Reactor Model

This model (CSTR model) is shown schematically in Figure 4. It is probably the simplest model for a continuous-flow chemical reactor, and its RTD is the exponential decay function given in Figure 3.

Plug-Flow Reactor Model

The plug-flow reactor (PFR) model also is shown in Figure 4, and its RTD is an impulse function, peaking at a time corresponding to the mean residence time, as shown in Figure 2.

CSTR's-in-Series Model

The CSTR's-in-series model is depicted in Figure 5. It is composed of a sequence of completely mixed stirred-tank reactors in series. The reactant enters the first reactor and exits from the last reactor. As described earlier, the RTD of this model can be determined by considering an impulse input of a tracer to the first reactor. A mass balance of the tracer around each reactor gives

$$V_i \frac{dC_i}{dt} + q_0 C_i = q_0 C_{i-1}, \; i = 1, 2, \ldots, n \quad (7)$$

Solving for C_i, $i = 1, 2, \ldots, n$ by means of the Laplace transform technique, we obtain the RTD of the model as

$$E(\theta) = \frac{n^n \theta^{n-1} e^{-n\theta}}{(n-1)!} \quad (8)$$

where

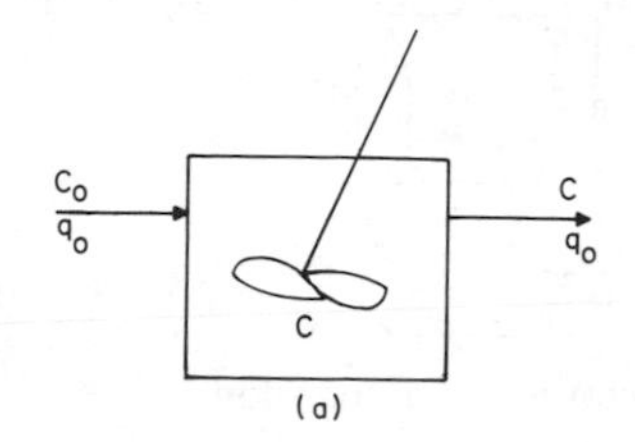

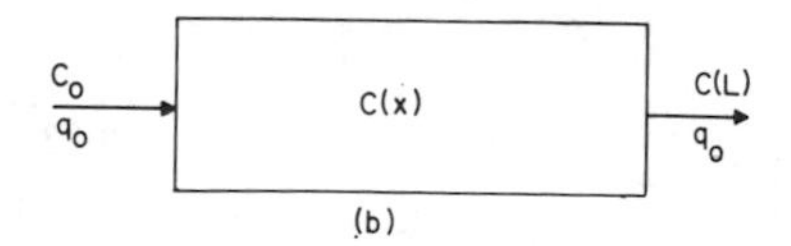

Figure 4. Schematic representations of (a) the CSTR model and (b) PFR model.

$$\bar{t} = \frac{V}{q_0}, \quad \theta = \frac{t}{\bar{t}} \quad \text{and } V = nV_i$$

Notice that V_i is the volume of a single tank in a series of perfectly mixed tanks of equal volume. For a CSTR's-in-series model with unequal volume, the expression for the RTD is not straightforward [3]. As the number of CSTR's in the model increases, the behavior of this model approaches that of the PFR model.

Laminar-Flow Reactor Model

To devise the RTD for this model, we consider the reactor to be cylindrical with a large length-to-diameter ratio [4]. The fluid (or particles flowing through the reactor) is visualized as composed of an infinite number of coaxial annuli, and the fraction of the fluid whose residence time is between t and t + dt is determined.

Consider an annulus of fluid lying between r and r + dr. The velocity profile in a Newtonian laminar flow reactor is

$$v(r) = \frac{(-\Delta P)}{4L\mu}(R^2 - r^2) \tag{9}$$

The residence time in the laminar flow reactor is a function of the radius, r, as the linear velocity, v, is a function of r. The residence time (or the time of passage) at the radial position, r, is (see Figure 6)

$$t = \frac{L}{\dfrac{(-\Delta P)}{4L\mu}(R^2 - r^2)} \tag{10}$$

At the tube axis, the residence time is

$$t_0 = \frac{L}{\dfrac{(-\Delta P)R^2}{4L\mu}} \tag{11}$$

Figure 5. CSTR's-in-series model.

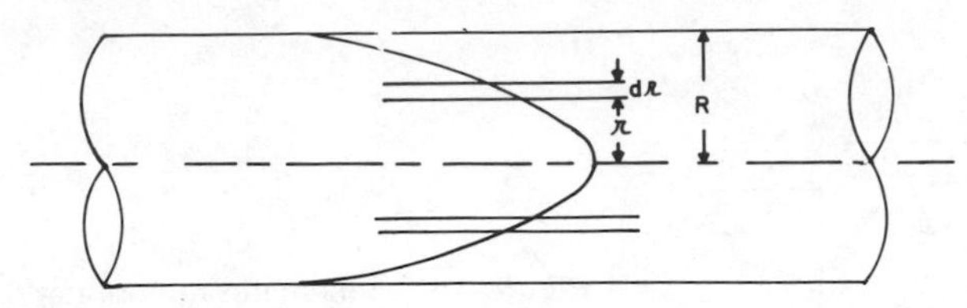

Figure 6. Velocity profile in a tubular reactor with laminar flow.

Dividing this equation by Equation 10, we have

$$\frac{t_O}{t} = \frac{R^2 - r^2}{R^2} \tag{12}$$

Differentiating both sides of this expression and simplifying, we have

$$\frac{R^2 t_O}{t^2}\,dt = 2rdr \tag{13}$$

The volumetric rate of flow through the annulus lying between r and r + dr is

$$dq_r = 2\pi \frac{L}{t}\,rdr \tag{14}$$

Substituting for rdr from Equation 13 and integrating between $t = t_O$ at $r = 0$ and $t \to \infty$ at $r = R$, we obtain

$$q_R = \frac{\pi R^2 L}{2t_O} \tag{15}$$

The residence time distribution or the fraction of the fluid with exit age between t and t + dt is

$$E(t)dt = \frac{dq_r}{q_R} = \frac{2t_O{}^2}{t^3} \text{ or } E(t) = \frac{2t_O{}^2}{t^3} \tag{16}$$

CSTR's-in-Series with Backflow Model

A variation of the CSTR's-in-series model is the CSTR's-in-series with backflow model shown in Figure 7. The RTD for each reactor and for the model can be calculated by considering an impulse input of the tracer to the model. Mass balances for the tracer around reactors 1, $i(1 < i < n)$, and n give rise, respectively, to

$$V_1 \frac{dC_1}{dt} + (q_O + q')C_1 = q_O C_O + q' C_2 \tag{17}$$

$$V_i \frac{dC_i}{dt} + (q_O + q')C_i + q'C_i = (q_O + q')C_{i-1} + q'C_{i+1} \tag{18}$$

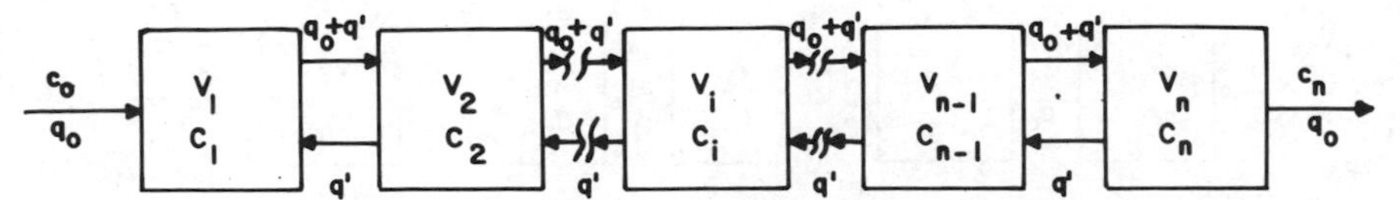

Figure 7. CSTR's-in-series with backflow model.

$$V_n \frac{dC_n}{dt} + (q_o + q')C_n = (q_o + q')C_{n-1} \tag{19}$$

Closed-form analytical solutions for the RTD of this model can be obtained when the number of reactors in the model is relatively small; however, numerical solutions generally are required.

Axial Dispersion Model

One of the most widely used models is the axial dispersion model, in which the diffusion in the direction of flow is superimposed on plug flow. Therefore, it is often called the dispersed plug-flow model [3,5–7]. A mass balance of the tracer around a differential volume between z and (z + dz) gives

$$\frac{\partial c}{\partial t} = E_z \frac{\partial^2 c}{\partial z^2} - \bar{u} \frac{\partial c}{\partial z} \tag{20}$$

E_z is termed the axial dispersion coefficient. It is assumed to be independent of concentration and position, although it can be a function of either one or both. This model is used frequently for turbulent flow of fluids in pipes, flow through packed beds and flow through fluidized beds. In fact, the model has been applied to many other homogeneous and heterogeneous reactors, in which axial symmetry can be assumed and in which the flow behavior closely approximates that of plug flow. Provided that no dispersion takes place outside the reactor proper, mass balances of the tracer around the inlet and exit of the reactor give the following boundary conditions [3,5,8,9]:

$$\left.\begin{aligned} \bar{u}c_{z\to 0^-} &= \bar{u}c_{z\to 0^+} - E_z \left(\frac{\partial c}{\partial z}\right)_{z\to 0^+} \\ \left(\frac{\partial c}{\partial z}\right)_{z=L} &= 0 \end{aligned}\right\} \tag{21}$$

Other simple macromixing models are available [1–3], and an unlimited number of complex macromixing models can be generated by combining in series and/or parallel some of the simple models. Most of the recent monographs and books related to chemical and biochemical reactor analysis and design devote substantial pages to macromixing models [10–14].

Governing or performance equations of continuous-flow reactors are obtained by adding terms representing the consumption of reactants or generation of products to the mass balance equations for the tracer, if the kinetics involved are linear (isothermal first order). If they are nonlinear, micromixing must be taken into account.

MICROMIXING IN CONTINUOUS-FLOW CHEMICAL REACTORS

Knowledge of micromixing provides information regarding the environmental variation experienced by the reacting fluid elements or particles during their passage through the reactor. Micromixing encompasses all aspects of mixing not defined by macromixing or the RTD, and is concerned with the extent of mixing on a molecular level. Danckwerts [15] and Zwietering [16] introduced the concept of a "point" of fluid as a volume of fluid very small in size compared to the overall volume of the reactor, but still sufficiently large to contain many molecules.

Molecules entering the reactor have an age $\alpha = 0$ and various life expectations described by the RTD, $E(\lambda)$. Molecules leaving the reactor, however, have a life expectation $\lambda = 0$ and various exit ages given by the RTD, $E(\alpha)$ or $E(t)$. Notice that a transition takes place in the reactor from a grouping of molecules with an identical age to a grouping of molecules with an identical life expectation. This transition between the molecular groupings is regarded as micromixing.

Models representing the two extremes of micromixing are the so-called complete segregation and maximum mixedness models. In the complete segregation model, mixing occurs as late as possible, and a fluid element entering the reactor breaks into small, but finite, discrete fragments in which the molecules entering together remain together. Under these conditions, each fluid element is totally isolated or segregated from the other points in the system. In the maximum mixedness model, mixing occurs as early as possible, and an incoming fluid element is dispersed on a molecular scale in a time much less than the mean residence time of the system. Any molecule entering a maximum mixedness chemical reactor immediately becomes associated with other molecules with which it eventually will leave the reactor, that is, the molecules of the same life expectation are completely mixed regardless of their ages. These two extreme models are shown in Figure 8.

Danckwerts [15] and Zwietering [16] have proposed a measure for micromixing, the "degree of segregation." The degree of segregation, J, is defined as the ratio of the variance of the ages between the points in the reactor to the variance of the ages of all molecules in the reactor. In other words,

$$J = \frac{\mathrm{var}\ \alpha_p}{\mathrm{var}\ \alpha} = \frac{\frac{1}{V}\int_V (\alpha_p - \bar{\alpha})^2\, dV}{\int_0^\infty (\alpha - \bar{\alpha})^2 I(\alpha)\, d\alpha} \tag{22}$$

Note that regardless of its RTD, the degree of segregation of any completely segregated reactor, J, is one because α_p is identical to α at each point or in each fluid particle. On the other hand, for a maximum mixedness reactor the points mix at the earliest possible moment. Thus, J assumes the minimum possible value, for a given RTD, which is not necessarily zero. It is important to note that a continuous-stirred-tank reactor may have completely segregated or maximum mixedness flow and, correspondingly, it may assume any value of J from zero through one. For any other flow reactor whose RTD is different from that of a CSTR, J cannot reach the value of zero. If the CSTR's-in-series model is used to generate a RTD between the RTD of the CSTR and that of the PFR, the farther the RTD of the model deviates from the RTD of the CSTR, the farther the J under the state of maximum mixedness deviates positively from zero. Any micromixing condition lying between these extremes is considered as partial segregation or

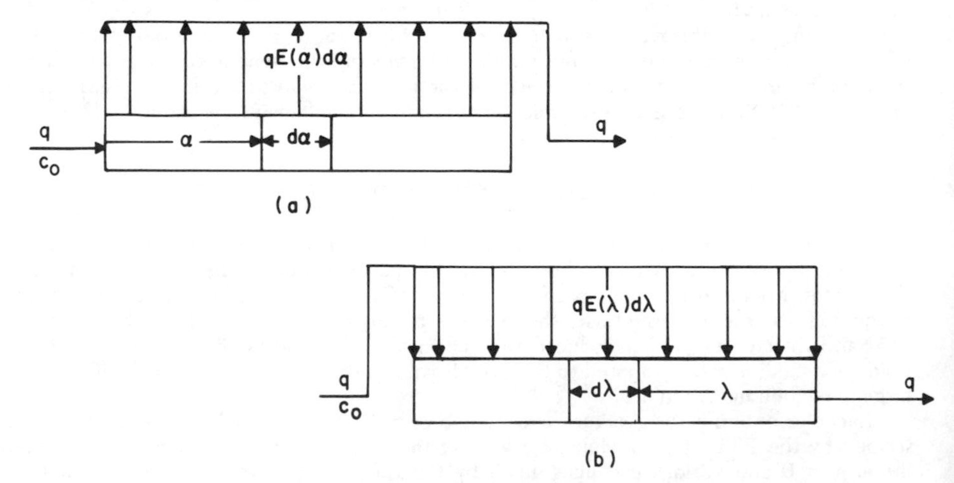

Figure 8. (a) Complete segregation model and (b) maximum mixedness model [16].

incomplete micromixing. For a partially segregated system, the degree of segregation can be calculated if the age distribution function within the points over the entire space is known.

MICROMIXING MODELS FOR CONTINUOUS-FLOW CHEMICAL REACTORS

We shall discuss in this section not only the two extreme models of micromixing, but also other models that describe intermediate cases of micromixing.

Complete Segregation Model

The complete segregation model consists of a plug-flow reactor in which the feed stream enters from one side and the fluid stream leaves through a series of exits (Figure 8a).

A steady-state material balance of the reactant around a differential volume (age-space) of the reactor between α and $\alpha + d\alpha$ leads to

$$\underbrace{q\{1 - F(\alpha)\}\, C(\alpha)}_{\left[\text{input to the differential volume (age-space)}\right]}$$

$$= \underbrace{q\{1 - F(\alpha + d\alpha)\}\, C(\alpha + d\alpha)}_{\left[\text{output from the differential volume (age-space)}\right]} + \underbrace{C(\alpha) qE(\alpha) d\alpha}_{\left[\text{outflow from the side streams}\right]} + \underbrace{r(C)\, q\{1 - F(\alpha)\} d\alpha}_{\left[\text{loss due to reaction}\right]} \tag{23}$$

On simplification, we have

$$\frac{dC}{d\alpha} = -r(C) \tag{24}$$

Equation 24 suggests that each point in a completely segregated chemical reactor behaves like a batch reactor. The average concentration of the reactant over all fluid elements at the exit of the reactor is the expected reactant concentration of all the fluid elements over the entire RTD, that is,

$$C_{seg} = \int_0^\infty C_B(\alpha)\, E(\alpha) d\alpha \tag{25}$$

Maximum Mixedness Model

In the maximum mixedness model, the feedstreams enter the plug-flow reactor at various points and there is only one exit stream, as shown in Figure 8b. The molecules within the reactor are classified according to their life expectation; this eventually leads to the earliest possible mixing of all fluid elements.

For this reactor, a mass balance is written in terms of the life expectation, λ. For a differential volume (age-space) between λ and $\lambda + d\lambda$, we have

$$\underbrace{q\{1 - F(\lambda + d\lambda)\} C(\lambda + d\lambda)}_{\left[\text{input to the differential volume (age-space)}\right]}$$

$$= q\{1 - F(\lambda)\}C(\lambda) + r(C)\, q\{1 - F(\lambda)\}d\lambda - qE(\lambda)C_0 d\lambda \tag{26}$$

[output from the differential volume (age-space)] [loss due to reaction] [input from the side streams]

Simplifying Equation 26, we obtain

$$\frac{dC}{d\lambda} = r(C) - \frac{E(\lambda)}{1 - F(\lambda)}[C_0 - C(\lambda)] \tag{27}$$

with the boundary condition

$$\frac{dC}{d\lambda} = 0, \qquad \lambda \rightarrow \infty \tag{28}$$

The desired concentration at the reactor outlet is C(O).

Consecutive Model

This model, advanced by Weinstein and Adler [17], contains a parameter, age α^*. The part of the reactor having fluid elements younger than α^* is considered to be in the state of complete segregation. The other part of the reactor is assumed to be in the state of maximum mixedness. As these two parts of the flow reactor are assumed to be consecutive, the model is termed the consecutive model (Figure 9a). The fluid elements leaving the first part of the reactor will have a concentration equal to

$$C_{seg} = \int_0^{\alpha^*} C_B(\alpha)\, E(\alpha)d\alpha \tag{29}$$

For the second part of the reactor, we have

$$\frac{dC_{mm}}{d\lambda} = r(C_{mm}) + \frac{E(\lambda)}{1 - F(\lambda)}[C_B(\alpha^*)C_{mm}(\lambda)] \tag{30}$$

with the boundary condition

$$\frac{dC_{mm}}{d\lambda} = 0, \qquad \lambda \rightarrow \infty$$

The exit concentration of the chemical reactor is

$$C_e = \int_0^{\alpha^*} C_B(\alpha)E(\alpha)d\alpha + C_{mm}(\alpha^*) \int_{\alpha^*}^{\infty} E(\alpha)d\alpha \tag{31}$$

Parallel Model

This model is similar to the consecutive model, except that the two parts of the reactor are in parallel, as shown in Figure 9b. The exit concentration of the reactor is

$$C_e = \int_0^{\alpha^*} C_B(\alpha)\, E(\alpha)d\alpha + C_{mm}(O) \int_{\alpha^*}^{\infty} E(\alpha)d\alpha \tag{32}$$

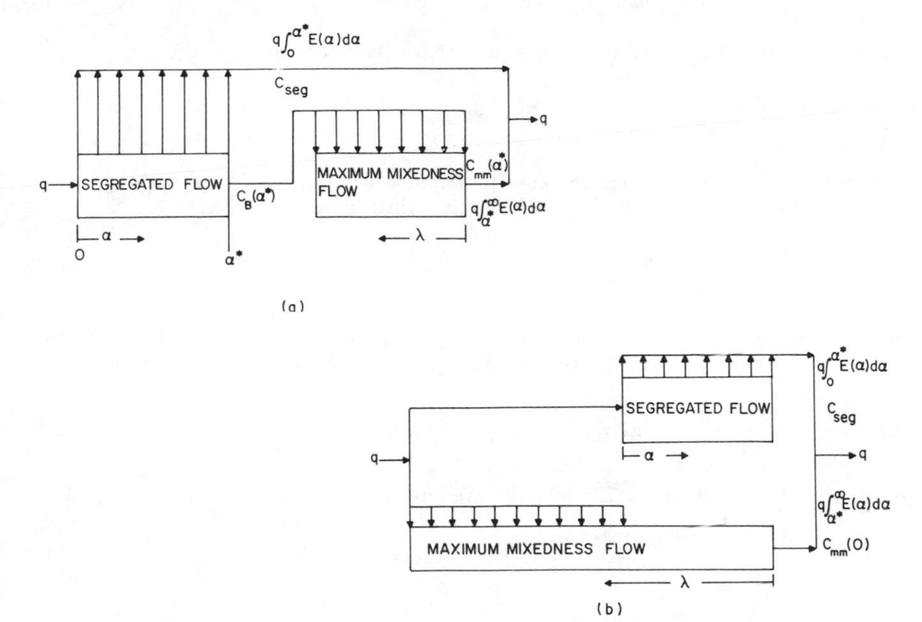

Figure 9. (a) Consecutive model and (b) parallel model [17].

In the limit of $\alpha^* \rightarrow \infty$, both models lead to the completely segregated model, and as $\alpha^* \rightarrow 0$, both models lead to the maximum mixedness model (Figure 8).

Two-Environmental Model

Ng and Rippin [18–20] have proposed a model consisting of two environments. One of these is the entering environment, in which the fluid elements are completely segregated; the other is the leaving environment, which contains molecules in a state of maximum mixedness. The material transfers from the entering environment to the leaving environment at a rate proportional to the amount of material remaining in the entering environment.

Consider that there are m units of material in the entering environment at any time, t. The rate of transfer out of the entering environment is

$$\frac{dm}{dt} = -mR \tag{33}$$

where R is the specific rate of transfer. Integration of Equation 33 leads to

$$m = m_0 e^{-Rt} \tag{34}$$

where m_0 is the value of m at t = 0. From these equations it is apparent that the fraction of material remaining in the entering environment, whose age is α, is given by

$$F_{en}(\alpha) = 1 - e^{-R\alpha} \tag{35}$$

and the fraction of the material in the leaving environment is

$$F_{le}(\alpha) = 1 - e^{-R\alpha} \tag{36}$$

The total volume of material with life expectation between λ and $\lambda + d\lambda$ is

$$Vd\lambda \int_0^\infty g(\alpha, \lambda)d\alpha \tag{37}$$

Zwietering [16] has shown that the joint probability of a fluid molecule with age α and life expectation λ is related to the residence time distribution, $E(t)$, as

$$g(a,\lambda) = \frac{1}{\bar{t}} E(\alpha + \lambda) \tag{38}$$

where $\bar{t}$ is the mean residence time. Thus, the volume fraction of material in the vessel having ages between α and $\alpha + d\alpha$ and life expectations between λ and $\lambda + d\lambda$ is

$$g(\alpha, \lambda)d\alpha d\lambda = \frac{1}{\bar{t}} E(\alpha + \lambda)d\alpha d\lambda \tag{39}$$

Substituting Equation 39 into Equation 37, we have

$$\frac{Vd\lambda}{\bar{t}} \int_0^\infty E(\alpha + \lambda)d\alpha = I_1 \, Vd\lambda \tag{40}$$

where

$$I_1 = \frac{1}{\bar{t}} \int_0^\infty E(\alpha + \lambda)d\alpha \tag{41}$$

The material in the leaving environment with life expectation λ to $\lambda + d\lambda$ is

$$\frac{Vd\lambda}{\bar{t}} \int_0^\infty (1 - e^{-R\alpha})E(\alpha + \lambda)d\alpha = I_2 Vd\lambda \tag{42}$$

where

$$I_2 = \frac{1}{\bar{t}} \int_0^\infty (1 - e^{-R\alpha})E(\alpha + \lambda)d\alpha \tag{43}$$

The volume of material in the entering environment with life expectation λ to $\lambda + d\lambda$ is

$$\frac{Vd\lambda}{\bar{t}} \int_0^\infty e^{-R\alpha}E(\alpha + \lambda)d\alpha \tag{44}$$

Thus, the rate at which material of life expectation λ to $\lambda + d\lambda$ is transferred from the entering to the leaving environment is

$$\frac{RVd\lambda}{\bar{t}} \int_0^\infty e^{-R\alpha}E(\alpha + \lambda)d\alpha = I_3 RVd\lambda \tag{45}$$

where

$$I_3 = \frac{1}{\bar{t}} \int_0^\infty e^{-R\alpha}E(\alpha + \lambda)d\alpha \tag{46}$$

The quantity of material in the entering environment of life expectation λ to $\lambda + d\lambda$ is

$$\frac{Vd\lambda}{\bar{t}} \int_0^\infty C(\alpha)e^{-R\alpha}E(\alpha + \lambda)d\alpha \tag{47}$$

Thus, the rate at which reactant with life expectation between λ and $\lambda + d\lambda$ is transferred from the entering to the leaving environment is

$$\frac{RVd\lambda}{\bar{t}} \int_0^\infty C(\alpha)e^{-R\alpha}E(\alpha + \lambda)d\alpha = I_4C_ORVd\lambda \tag{48}$$

where

$$I_4 = \frac{1}{\bar{t}} \int_0^\infty \frac{C(\alpha)}{C_O} e^{-R\alpha}E(\alpha + \lambda)d\alpha \tag{49}$$

The mean concentration at the reactor exit is obtained by mixing together the material in the leaving environment, with a zero life expectation, and the material remaining in the entering environment, with a zero life expectation. The amount of reactant in the entering environment with zero life expectation is

$$\left[\frac{Vd\lambda}{\bar{t}} \int_0^\infty C(\alpha)e^{-R\alpha}E(\alpha + \lambda)d\alpha\right]_{\lambda=0} = I_4C_OVd\lambda \Big|_{\lambda = 0} \tag{50}$$

The reactant concentration in the leaving environment, C_{le}, is determined as a function of the life expectation by solving the differential equation, which is derived from the mass balance around the differential volume (age-space) of the leaving environment between λ and $\lambda + d\lambda$:

$$C_{le}I_2V \Big|_{\lambda + d\lambda}$$

[input to the differential volume in the leaving environment]

$$= C_{le}I_2V \Big|_{\lambda} + RCI_4Vd\lambda + r(C)I_2Vd\lambda \tag{51}$$

[output from the differential volume in the leaving environment] [transfer from the entering environment] [loss by reaction]

Simplifying Equation 51 yields

$$\frac{I_2C_{le}\Big|_{\lambda + d\lambda} - I_2C_{le}\Big|_{\lambda}}{d\lambda} + RC_OI_4 - r(C_{le})I_2 = 0 \tag{52a}$$

or

$$\frac{d(I_2C_{le})}{d\lambda} = -RC_OI_4 + I_2r(C_{le}) \tag{52b}$$

or

$$I_2\frac{dC_{le}}{d\lambda} = -C_{le}\frac{dI_2}{d\lambda} - C_ORI_4 + I_2r(C_{le}) \tag{52c}$$

The rate of decrease in the amount of material in the entering environment is equal to the rate of increase in the amount of material in the leaving environment; therefore, we have

$$\frac{dI_3}{d\alpha} = \frac{dI_2}{d\lambda} = -I_3R \tag{53}$$

Substituting Equation 53 into Equation 52c, we obtain

$$\frac{dC_{le}}{d\lambda} = r(C_{le}) + \frac{R(I_3C_{le} - I_4C_O)}{I_2} \tag{54}$$

The boundary condition for Equation 54 is

$$\frac{dC_{le}}{d\lambda} = 0, \qquad \lambda \rightarrow \infty \tag{55}$$

If q_O is the volumetric flowrate through the reactor, the total quantity of fluid leaving in any time interval, $d\lambda$, is $q_Od\lambda$. Hence, the mean concentration, $\bar{C}$, at the reactor exit is

$$\bar{C} = \left[\frac{I_4C_OVd\lambda + I_2VC_{le}d\lambda}{q_Od\lambda}\right]\Bigg|_{\lambda = O} \tag{56}$$

$$= \bar{t}\{C_O(I_4)_{\lambda=O} + (I_2C_{le})_{\lambda=O}\}$$

The viscosity of a polymerization mixture often increases significantly during the reaction. Toward the end of reaction, the increase in viscosity enhances the segregation of fluid elements with different residence times in a flow reactor. In effect, the conditions in a flow polymerization reactor may be visualized as changing from maximum mixedness to complete segregation. A reversed two-environmental model can be used to model such a mixing situation. However, from the thermodynamic viewpoint, this model should be regarded as far more empirical than the original two-environmental model [21].

STOCHASTIC MODELING OF A CONTINUOUS-FLOW CHEMICAL REACTOR

The significance of the random or probabilistic nature of the residence time in a continuous flow system was pointed out by Danckwerts [6] as early as 1953. In fact, liberal uses of statistical and stochastic terminologies and concepts have been made in modeling flow chemical reactors in terms of their residence time distributions or corresponding mathematical expressions. Approaches have largely been ad hoc, however, from the standpoint of mathematics or stochastic processes [22,23].

In recent years, substantial progress has been made in analyzing and modeling flow chemical reactors from the purely or rigorously stochastic points of view. A brief survey of such progress is given in this section.

Macromixing Models

Countless numbers of theorems, theories and methodologies of statistics, probability and stochastic processes are at our disposal for developing ways to analyze and model flow reactors. However, one of the simplest ways to model macromixing in flow reactors stochastically is to simulate numerically the movements of fluid elements, molecules or particles. For example, Mann and O'Leary [24] have employed the Monte Carlo simulation technique to generate the RTD of an arbitrarily interconnected complex system. The technique is implemented by simulating the path of a "test particle" as it passes through the system and the time it resides in each region it visits. The specific path is selected according to the probability that the "test particle" takes a certain route at each point where the stream splits. The accuracy of this technique depends on the number of "test particles" used in the simulation and the size of the time interval selected as a step of transition. As the geometric complexity of the system increases, however, this technique becomes increasingly cumbersome and time consuming.

Apparently, the most extensively developed field of stochastic processes is the so-called Markov processes. It has been shown that we can conveniently resort to a variety of theories and methodologies of the Markov processes for mechanistically modeling flow reactors, especially their macromixing characteristics.

A stochastic process is said to be a Markov process if, for any set of n time points, $t_1 < t_2 < \ldots < t_n$, the conditional distribution of random variable $X(t_n)$ for the given values of $X(t_1), X(t_2), \ldots, X(t_{n-1})$ depends only on x_{n-1}, the value of $X(t_{n-1})$. More precisely, for any real numbers, $x_1, x_2, \ldots, x_n$,

$$\begin{aligned} \Pr[X(t_n) \leq x_n \mid X(t_1) = x_1, X(t_2) = x_2, \ldots, X(t_{n-1}) = x_{n-1}] \\ = \Pr[X(t_n) \leq x_n \mid X(t_{n-1}) = x_{n-1}] \end{aligned} \tag{57}$$

Equation 57 means that, given the "present" of the process, the "future" is independent of its "past." Markov processes usually are classified on the basis of the nature of their parameter and state spaces, as outlined below.

- those with discrete parameter and state spaces, i.e., Markov chains;
- those with continuous-parameter and discrete state spaces, i.e., time continuous Markov chains, or simply Markov processes; and
- those with continuous-parameter and state spaces, i.e., diffusion processes.

Obviously, the first step in conducting Markovian or, for that matter, general stochastic analysis and modeling of a flow reactor is identification of the random variable, $X(t)$, the state property considered to be probabilistically varying as time progresses. For example, it can be the position of a fluid element or particle, the number of particles, the discretized velocity or any other attribute of the reactant stream.

Schmalzer and Hoelscher [25] have described the fluid flow through a packed bed in terms of a Markov chain of velocity states. In contrast, Srinivasan and Metha [26] have employed a time continuous Markov chain (Markov process). Although they have not obtained the results in terms of the RTD or equivalent mathematical models, their results will be useful indirectly for simulating the performance of the packed bed when it is used as a flow reactor.

Gibilaro et al. [27] have employed the Markov chain to generate the residence time distribution for a continuous-flow system, consisting of a number of well mixed vessels connected together by any fashion. By resorting to Markov processes, Fan et al. [28] have derived the unsteady-state age distributions for a variety of flow systems. Fan et al. [29] have presented a technique for modeling and simulating totally interconnected stirred tank networks by using the theory of Markov processes (time continuous Markov chains). Raghuraman and Mohan [30] have proposed a time continuous Markov chain model to describe the residence time and contact time distributions of the fluid in a packed bed. Mann and Rubinovitch [31] have employed a Markov chain approach for analyzing particulate processes. Their analysis is based on describing the movement of a single particle as it passes through a flow system.

Nassar et al. [32,33] have shown that both chemical reactions with linear kinetics and with dispersive mixing in several classes of continuous-flow chemical reactors under the unsteady-state and steady-state operations can be analyzed and modeled in a unified fashion by means of Markov processes. Their approach can be applied to both the time homogeneous and time heterogeneous processes. Nauman [34] has analyzed the RTD of the axial dispersion model from the standpoint of Markov processes.

For reactors with nonlinear chemical reactions, however, microscopic or molecular-level diffusion and interaction, i.e., micromixing of reactant molecules and particles, need to be considered.

Micromixing Models

Only a limited number of stochastic models are available to describe the state of micromixing in a flow reactor; they include the models developed by Kattan and Adler [35] and Krambeck et al. [36]. The contribution by McCord [37] appears to provide a rigorous foundation for stochastic analysis and modeling of micromixing in a flow reactor. Nevertheless, much remains to be done in this field.

ACKNOWLEDGMENTS

The author wishes to express his appreciation for the assistance given by Mr. Shun-See Rong and Mr. Jui-Rze Too.

NOMENCLATURE

C_O initial concentration of the reactant or the concentration of the reactant at the inlet of the reactor (mol/cm³)

C concentration of the reactant (mol/cm³)

$\bar{C}$ mean concentration of the reactant at the exit of the reactor (mol/cm³)

C_e concentration of the reactant at the exit of the reactor (mol/cm³)

C_{en} concentration of the reactant in the entering environment (mol/cm³)

C_i concentration of the trace or the reactant in the ith reactor (mol/cm³)

C_{le} concentration of the reactant in the leaving environment (mol/cm³)

C_{mm} concentration of the reactant in the flow from the maximum mixedness portion (mol/cm³)

C_{seg} average concentration of the reactant in the segregation portion (mol/cm³)

$C_B(\alpha)$, $C_B(\lambda)$ concentration of the reactant in a batch reactor at age α and life expectation λ, respectively (mol/cm³)

$E(t)$ residence time or exit age distribution function (sec^{-1})

$E(\theta)$ dimensionless residence time distribution function (—)

$E(\lambda)$ life expectation distribution function (sec^{-1})

$F(t)$ cumulative age distribution function (—)

$F_{en}(\alpha)$ fraction of the material remaining in the entering environment at age α (—)

$F_{le}(\alpha)$ fraction of the material remaining in the leaving environment at age α (—)

$g(\alpha,\lambda)$ joint distribution function of a molecule with age α and life expectation λ (sec^{-1})

$I(\alpha)$ internal age distribution in the reactor (sec^{-1})

$I_p(\alpha)$ internal age distribution within a point p in the reactor (sec^{-1})

J degree of segregation (—)

k specific reaction rate constant (cm³/mol·sec)

L length of the tubular reactor (cm)

m_O value of m at $t = 0$ (—)

m units of materials (—)

n number of stages (—)

$p_{en}(\alpha)$, $p_{le}(\alpha)$ probability of finding the molecule in the entering and leaving environment, respectively (—)

$p_{en}(\lambda)$, $p_{le}(\lambda)$ probability of the fluid element with life expectation λ in entering and leaving environment, respectively (—)

ΔP pressure drop of fluid through the reactor (dyne/cm^2)
q volumetric flowrate of the fluid through the reactor (cm^3/sec)
q' backflow rate (cm^3/sec)
r any radius r (cm)
$r(C)$ rate of consumption of the reactant (mol/cm^3·sec)
R radius of the tube (cm) or the specific rate of transfer of the two-environmental model (sec^{-1})
t_0 residence time at the tube axis (sec)
t time or residence time (sec)
$\bar{t}$ V/q_0, mean residence time (sec)
$\bar{u}$ mean velocity in the axial dispersion model (cm/sec)
V_i volume of the ith reactor (cm^3)
$v(r)$ velocity profile in the tubular reactor (cm/sec)
V total volume of the reactor (cm^3)

Greek Symbols

α internal age of a molecule in the reactor (sec)
$\bar{\alpha}$ mean age of molecules in the reactor (sec)
α^* age parameter in the consecutive model (sec)
$\bar{\alpha}_p$ mean age of molecules within a point p in the reactor (sec)
$\Delta(t)$ intensity function (sec)
θ $t/\bar{t}$, dimensionless time (—)
μ viscosity of the fluid (poise)
λ internal life expectation (sec)
$\psi(\lambda)$ internal life expectation distribution function (sec^{-1})

Subscripts

O initial state or the state of inlet flow
i number stages, $i = 1, 2, \ldots, n$

REFERENCES

1. Smith, J. M. *Chemical Engineering Kinetics*, 2nd ed., (New York: McGraw-Hill Book Co., 1970).
2. Levenspiel, O. *Chemical Reaction Engineering* (New York: John Wiley & Sons, Inc., 1972).
3. Wen, Y., and L. T. Fan. *Models for Flow Systems and Chemical Reactors*, (New York: Marcel Dekker, Inc., 1975).
4. Denbigh, K. G. *Chemical Reactor Theory* (Cambridge, England: Cambridge University Press, 1966).
5. Langmuir, I. "The Velocity of Reactions in Gases Moving through Heated Vessels and the Effect of Convection and Diffusion," *J. Am. Chem. Soc.* 30:1742 (1908).
6. Danckwerts, P. V. "The Effect of Incomplete Mixing on Homogeneous Reaction," *Chem. Eng. Sci.* 2:1 (1953).
7. Fan, L. T., and R. C. Bailie. "Axial Diffusion in Isothermal Tubular Flow Reactors," *Chem. Eng. Sci.* 13:63 (1960).
8. Wehner, J. F., and R. H. Wilhelm. "Boundary Conditions of Flow Reactor," *Chem. Eng. Sci.* 6:89 (1956).
9. Fan, L. T., and Y. K. Ahn. "Critical Evaluation of Boundary Conditions for Tubular Flow Reactors," *Ind. Eng. Chem. Process Des. Dev.* 1:190 (1962).
10. Carberry, J. J. *Chemical and Catalytic Reaction Engineering* (New York: McGraw-Hill Book Co., 1976).
11. Bailey, J. E. D., and F. Ollis. *Biochemical Engineering Fundamentals*, (New York: McGraw-Hill Book Co., 1977).
12. Holland, C. D., and R. G. Anthony. *Fundamentals of Chemical Reaction Engineering*, (Englewood Cliffs, NJ: Prentice-Hall, Inc. 1979).

13. Butt, J. B. *Reaction Kinetics and Reactor Design* (Englewood Cliffs, NJ: Prentice-Hall, Inc., 1980).
14. Satterfield, C. N. *Heterogeneous Catalysis in Practice* (New York: McGraw-Hill Book Co., 1980).
15. Danckwerts, P. V. "Continuous Flow System-Distribution of Residence Times," *Chem. Eng. Sci.* 8:93 (1958).
16. Zwietering, N. "The Degree of Mixing in Continuous Flow Systems," *Chem. Eng. Sci.* 11:1 (1959).
17. Weinstein, H., and R. J. Adler. "Micromixing Effects in Continuous Chemical Reactors," *Chem. Eng. Sci.* 22:65 (1967).
18. Ng, D. Y. C., and D. W. T. Rippin. "The Effect of Incomplete Mixing on Conversion in Homogeneous Reactions," *Proc. 3rd Eur. Symp. of Chem. Eng.*, Amsterdam, Netherlands (1964).
19. Ng, D. Y. C. "The Effect of Incomplete Mixing on Conversion in Homogeneous Reactions," PhD Thesis, Imperial College of Science and Technology, London (1965).
20. Rippin, D. W. T. "Segregation in a Two-Environment Model of a Partially Mixed Chemical Reactor," *Chem. Eng. Sci.* 22:247 (1967).
21. Chen, M. S. K., and L. T. Fan. "Reversed Two-Environment Model for Micromixing in a Continuous Flow Reactor," *Can. J. Chem. Eng.* 49:704 (1971).
22. Kemeny, J. G., and J. L. Snell. *Finite Markov Chains* (New York: D. Van Nostrand Co., 1960).
23. Karlin, S., and H. M. Taylor. *A First Course in Stochastic Processes,* (New York: Academic Press, Inc., 1975).
24. Mann, U., and M. O'Leary. "Modeling of Flow Systems by Stochastic (Monte Carlo) Simulation," *Proc. 2nd World Cong. Chem. Eng.*, Montreal, Quebec, Canada (1981).
25. Schmalzer, D. K., and H. E. Hoelscher. "A Stochastic Model of Packed-Bed Mixing and Mass Transfer," *AIChE J.* 17:104 (1971).
26. Srinivasan, S. K., and K. M. Metha. "A Markov Chain Model for a Packed Bed," *AIChE J.* 18:650 (1972).
27. Gibilaro, L. G., H. W. Dropholler and D. J. Spikins. "Solution of a Mixing Model due to van de Vusse by a Simple Probability Method," *Chem. Eng. Sci.* 22:517 (1967).
28. Fan, L. T., L. S. Fan and R. F. Nassar. "A Stochastic Model of the Unsteady State Age Distribution in a Flow System," *Chem. Eng. Sci.* 34:1172 (1979).
29. Fan, L. T., J. R. Too and R. Nassar, "Stochastic Flow Reactor Modeling: A General Continuous Time Compartmental Model with First Order Reactions," Lecture Note Presented at Holiday and Science on the Rhine, Bad-Honnef, West Germany, Aug. 15–25, 1982.
30. Raghuraman, J., and V. Mohan. "A Markov Chain Model for a Packed Bed," *Chem. Eng. Sci.* 30:549 (1975).
31. Mann, U., and M. Rubinovitch. "New Methodology for Analyzing Particulate Processes," paper presented at the AIChE meeting, New Orleans, LA, November 8–12, 1981.
32. Nassar, R., L. T. Fan, J. R. Too and L. S. Fan. "A Stochastic Treatment of Unimolecular Reactions in an Unsteady State Continuous Flow System," *Chem. Eng. Sci.* 36:1307 (1981).
33. Nassar, R., J. R. Too and L. T. Fan. "Stochastic Modeling of Polymerization in a Continuous Flow Reactor," *J. Appl. Polymer Sci.* 26:3745 (1981).
34. Nauman, E. B. "Residence Time Distributions in Systems Governed by the Dispersion Equation," *Chem. Eng. Sci.* 36:957 (1981).
35. Kattan, A., and R. T. Adler. "A Stochastic Mixing Model for Homogeneous, Turbulent, Tubular Reactors," *AIChE J.* 13:580 (1967).
36. Krambeck, F. J., R. Shinnar and S. Katz. "Stochastic Mixing Models for Chemical Reactors," *Ind. Eng. Chem. Fund.* 6:276 (1967).
37. McCord, J. R. "A Stochastic Analysis of the Extremes of Micromixing in a Non-steady-State Continuous-Flow Chemical Reactor," *Chem. Eng. Sci.* 27:1613 (1972).

CHAPTER 13
OPEN-CHANNEL FLOWS

Nicholas P. Cheremisinoff
Exxon Research & Engineering Company
Florham Park, New Jersey 07932

Paul N. Cheremisinoff
Department of Civil and Environmental Engineering
New Jersey Institute of Technology
Newark, New Jersey 07102

CONTENTS

INTRODUCTION

Preceding chapters in this section have focused largely on closed-conduit flows. Such flows are encountered widely throughout industry, and the principles and phenomena described have laid the foundation for engineering designs for piping distribution systems. Piping systems often are connected to open channels, and it thus becomes important to understand how the transient is transmitted through the open channel. Examples of this are process waste streams from paper pulp mills, chemical plants, oil refineries and breweries, which are transmitted to wastewater treatment stations; the forebay channel, which leads to a hydroelectric installation; and systems in which water is pumped into and out of canals or rivers for process supplies, cooling or consumption and recreational purposes. Included in this subject is the general problem of flood routing in canals and rivers.

This chapter discusses the characteristics of open-channel flows, with emphasis on industrial considerations. The traditional approach to handling design problems involving optimum control of changes of flow from one steady regime, such as a pipe, to another steady regime in a channel, or vice versa, has largely been empirical. Although empirical approaches have proven adequate for many analyses of open-channel flows, more fundamental and theoretical attention must be given this subject to generalize problem solving.

In this chapter, the characteristics of open-channel flows are described and empirical flow expressions summarized. An introduction to numerical methods for analyzing open-channel flows is provided. The chapter concludes with a discussion of some design considerations for weirs and flumes. Although some of this latter information can be found in standard hydraulics textbooks, it is included here for the sake of completeness for the user-oriented reader.

LOCAL AND AVERAGE FLOW VELOCITY

Velocity Distributions

The simplest open-channel flow situation is that of a straight channel having a constant rectangular cross section for a sufficient length upstream. A typical velocity profile as a function of depth is illustrated in Figure 1, in which the point of maximum velocity occurs at some distance down from the free surface. Profile A represents the vertical velocity distribution on the centerline of a rectangular channel in which the depth is one-half the breadth. Curve B shows the vertical distribution of the mean velocity, i.e., each point on the curve is the average velocity along the horizontal across the section at that level.

Secondary circulation is observed in channel flow, as shown in Figure 2. The solid lines in Figure 2, called *isovels,* are drawn through points of equal velocity. The arrows indicate a double-spiral secondary circulation. The number of "cells" of circulation varies with the proportions of the cross section. The relative strength of these circulation cells is sensitive to the presence of bends or disturbances capable of producing transitions. Generally, flow around a bend results in a single spiral (the faster-moving fluid near the free surface moves toward the outside of the bend, attempting to balance its radial acceleration with that of slower-moving layers near the bed). Spiral flow in bends is called "secondary circulation of the first kind." When it occurs in a straight channel it is known as "secondary circulation of the second kind."

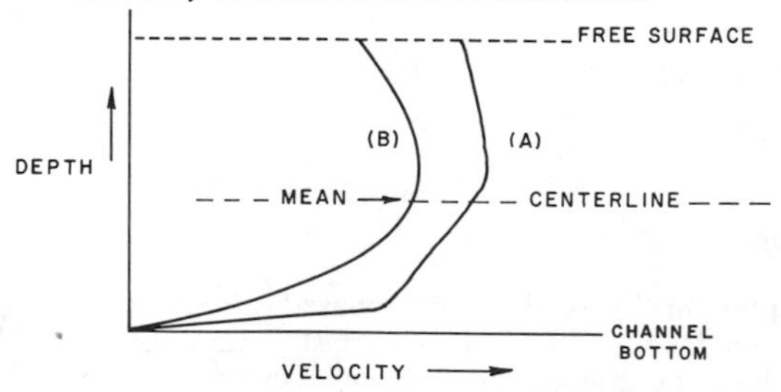

Figure 1. Typical velocity profile for flow in a channel of rectangular cross section.

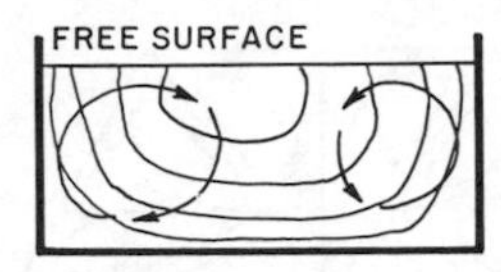

Figure 2. Double-spiral secondary circulation in open-channel flow.

Channel geometry has a dramatic effect on the velocity profile. Figure 3 shows typical velocity profiles for water flow in various noncircular pipes. The profiles shown are for full pipe flow. Figure 4 illustrates secondary flow patterns for these flow geometries. Secondary patterns have been observed to be superimposed over the longitudinal motion of the fluid velocity streams. The effect of secondary flow on the lines of constant velocity in Figure 3 causes these lines to transpose in the direction of secondary flow. This results in the corners of these lines being pushed toward the corner. In the vicinity of the wall they are pushed away from the wall. Unfortunately, no extensive studies have been made of velocity distributions in open channels of irregular geometries.

Empirical Velocity Formulas

From a design standpoint, an average or mean velocity has the most significance because it enables a volumetric or mass flowrate to be computed. Boundary layer theory has shown that the mean velocity for flow in a rectangular channel is at a distance of approximately 37% of the total fluid level above the channel bottom. Velocity measurements obtained at this depth over a number of verticals can be averaged to obtain a mean velocity over the cross section. Another simple approach to obtaining the mean velocity is to average measurements obtained at 0.2 h and 0.8 h, where h is the height of the fluid surface above the channel bottom (referred to as the "bed").

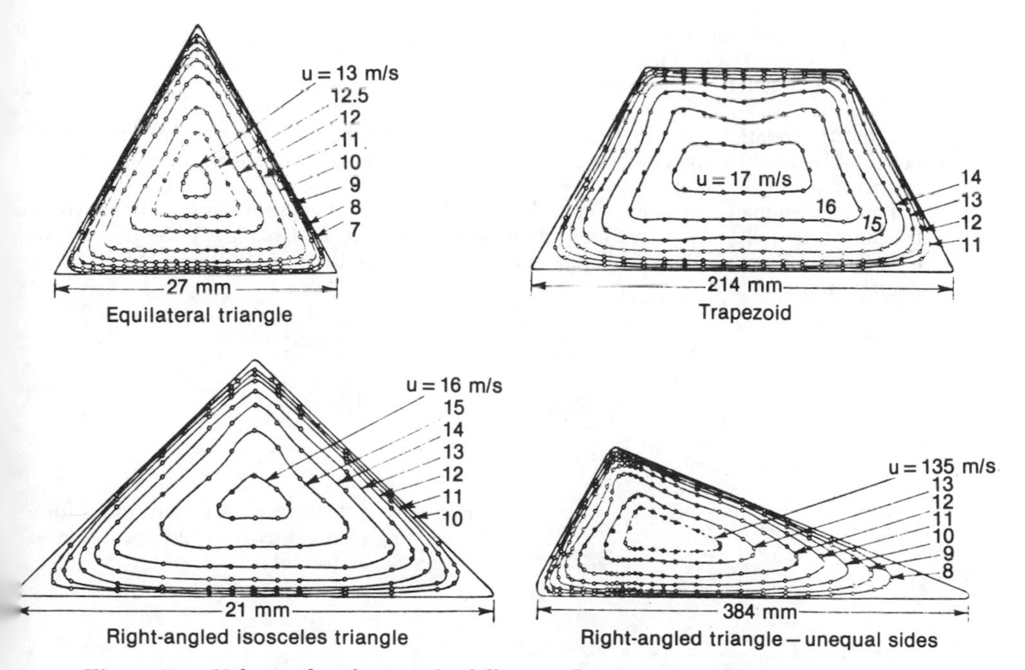

Figure 3. Velocity distributions for full water flow in noncircular pipes [1,2].

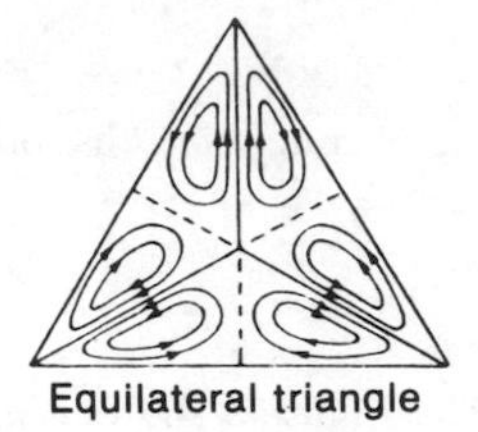

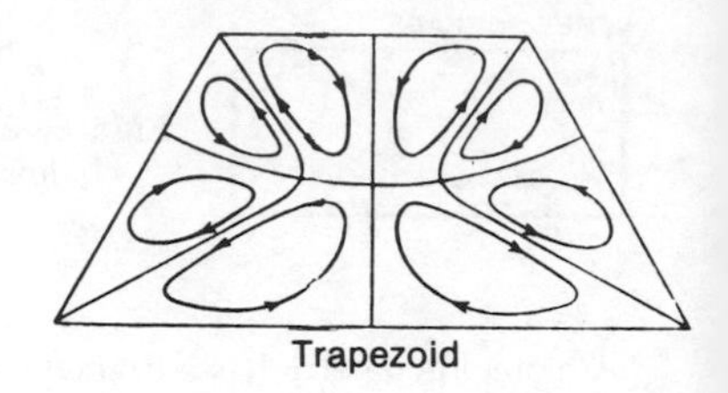

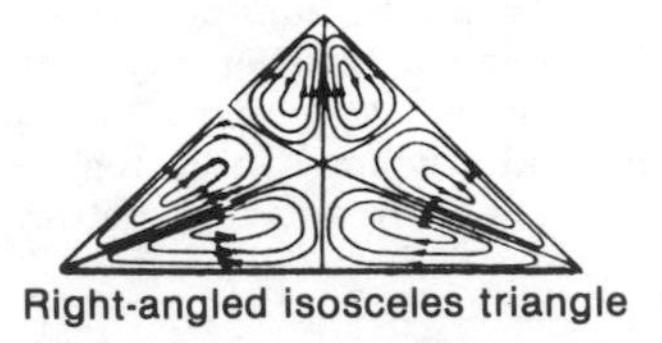

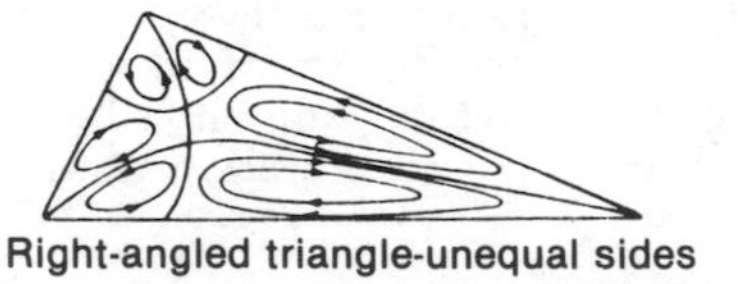

Figure 4. Secondary flow patterns for full water flow in noncircular conduits [1,2].

From an engineering standpoint, flow capacity is of prime importance in open channels. Once a value for the mean fluid velocity has been determined for a given channel cross section, the discharge through that section can be established. Early work to establish engineering formulas was entirely empirical. The first relation developed for open-channel flow was the Chezy formula, the basis of which is the following empirical expression:

$$\text{Flow resistance} = C'pLU^2 \tag{1}$$

where C' = friction coefficient
p = the wetted perimeter of the channel
L = length of the channel

From this relation, the Chezy formula may be developed for the general problem of flow in a rectangular channel, inclined at some angle, θ, from the horizontal. The general system is defined in Figure 5.

For a Newtonian liquid, the slope, S, of the fluid's free surface is $\sin\theta = h/L$, where h denotes the full level over a finite fluid element length, L. Defining W as the weight of a finite volume of fluid (i.e., the specific weight of the fluid) designated between points 1 and 2, then

$$WS = ALWS \tag{2}$$

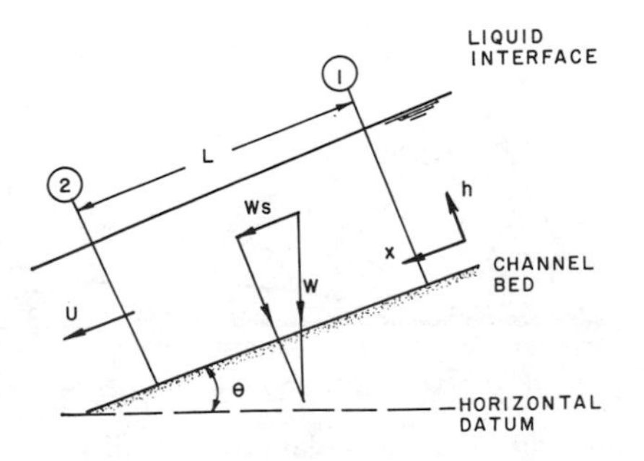

Figure 5. Definition of various terms for a sloping channel used in the derivation of the Chezy formula.

where A is the flow cross section.

WS is actually a force causing flow; hence, equating Equations 1 and 2 and rearranging,

$$U^2 = \frac{W}{K}\frac{A}{p}S \tag{3}$$

Equation 3 is the Chezy formula, which is most often written as

$$U = C\sqrt{h_D S} \tag{4}$$

where h_D is the hydraulic mean depth = A/p, and C is the Chezy constant = $\sqrt{W/K}$.

Chezy's constant can be evaluated from the following formula:

$$C = \frac{23 + \frac{0.00155}{S} + \frac{1}{\epsilon}}{1 + \left(23 + \frac{0.00155}{S}\right)\frac{\epsilon}{\sqrt{h_D}}} \tag{5}$$

where ϵ is a coefficient of roughness for the channel material.

A second empirical formula heavily used for open-channel flows is the Manning equation:

$$U = \frac{1}{\epsilon} h_D{}^{2/3} S^{1/2} \qquad \text{in SI units} \tag{6a}$$

or

$$U = \frac{1.49}{\epsilon} h_D{}^{2/3} S^{1/2} \qquad \text{in British units} \tag{6b}$$

For accuracy, the Manning formula is preferred.

Unlike full pipe flow, the roughness element, ϵ, for open-channel flows is sensitive to the hydraulic state of the channel. Usually, assigning a reliable value for ϵ depends on experience and judgment. Table I gives typical values for ϵ for different straight channel conditions. For nonstraight channels, ϵ values should be increased by 30% [3].

Table I. Range of Typical Values for Roughness Coefficients for Open-Channel Flows [3]

Channel Description	Range of ϵ (in.)
Steel Channels	0.011–0.017
Planned Timber, Joints Flush	0.010–0.014
Sawn Timber, Joints Uneven	0.011–0.015
Concrete	0.011–0.017
Brickwork	0.012–0.018
Excavated Channels (earth)	0.016–0.030
(gravel)	0.022–0.030
(rock cut, smooth)	0.023–0.040
(rock cut, jagged)	0.035–0.060
Natural Channels (clean, regular)	0.025–0.040
(rocky, brushwood, debris)	0.050–0.150
Floodplains (short grass pasture)	0.025–0.035
(brushwood)	0.035–0.070

The simple flow expressions presented in this section form the starting basis for describing open-channel flow and channel discharge. However, the complexity of analysis in describing the characteristics of open-channel flow depends on the nature of the flow itself. There are three general flow cases that are encountered frequently: uniform flows, nonuniform flows and unsteady flows in open channels. Each of these systems, along with the general expressions for estimating volumetric flows, are presented below. Computational methods based on empirical and semiempirical formulas are presented first. A more rigorous treatment of transient steady-state and unsteady flows is given later in this chapter.

UNIFORM FLOWS IN CHANNELS

Volumetric Flowrate

For uniform, fully developed flow in open channels, the Chezy and Manning formulas provide adequate estimates of the mean fluid velocity from which a volumetric discharge rate can be computed from a knowledge of the flow area, A. From the Manning equation, the volumetric discharge rate is

$$Q = \frac{1}{\epsilon} A h_D^{2/3} S^{1/2} \qquad (7)$$

The group $Ah_D^{2/3}$ is named the *section factor* and contains information on the geometry of the cross section. The group $(1/\epsilon)Ah_D^{2/3}$ is referred to as the *conveyance* and contains information on both the channel geometry and the channel's relative roughness. These definitions become useful when long computations are undertaken.

Estimating Depth

For flow in a rectangular cross-sectional channel, the depth of liquid corresponding to uniform flow for a specified discharge and bed slope will be constant at all points. The fluid depth for this special case is referred to as the normal depth, h_o. For nonrectangular channels of uniform flow, normal depth is defined as the maximum depth in the cross section. The normal depth meets the criterion of

$$\frac{dh}{dx} = 0 \qquad (8)$$

For the special case of a wide channel ($b >> h_o$, where b is the channel width), the normal depth can be estimated directly from the Chezy formula:

$$Q = AC\sqrt{h_o S} = bh_o C\sqrt{h_o S} \qquad (9)$$

whence it follows that since the hydraulic mean depth approaches the depth for a wide channel,

$$h_o = \sqrt[3]{\frac{Q^2}{b^2 S C^2}} \qquad (10)$$

Defining q as the discharge per unit width,

$$q = Q/b \qquad (11)$$

then

$$h_o = \sqrt[3]{q^2/SC^2} \tag{12}$$

The Manning formula gives the following relation for wide channels:

$$h_o = \left(\frac{\epsilon q}{S^{1/2}}\right)^{3/5} \tag{13}$$

For the more general case, in which h_o is comparable to channel width b, the hydraulic mean depth (a definition analogous to hydraulic radius in pipe flow) may be used:

$$h_D = \frac{bh_o}{b + 2h_o} \tag{14}$$

Combining this with the Manning relationship,

$$Q = \frac{1}{\epsilon}(bh_o)(h_D)^{3/2}\sqrt{S} \tag{15}$$

Note that if uniform flow does not exist in the channel there may be several possible fluid levels for a specified discharge. In such cases, the solution to Equation 15 becomes one of trial and error.

For nonrectangular cross-sectional channels, the following procedure is useful for determining the normal depth in a channel for uniform flow, with the Manning formula rewritten as

$$\frac{Q}{\sqrt{S}} = \frac{1}{\epsilon} A h_D{}^{2/3} = K' \tag{16}$$

Step 1. Compute values of K′ for a suitable range, where K′ is the conveyance of h (up to the maximum depth in the cross section).
Step 2. Prepare a plot of conveyance K′ versus depth, as shown in Figure 6.
Step 3. Using the plot, read the normal depth for the K′ value corresponding to the specified Q and S values.

The specific channel geometry greatly affects the mean hydraulic depth, which, in turn, affects not only the volumetric flow, but the economics of the design as well. The most economical channel design is based on a geometry that requires the minimum flow cross section for a specified discharge rate. From the Manning formula, this becomes an exercise in maximizing the term A^5/p^2, or since A is often fixed, p is optimized. Obviously, a semicircular channel would provide the most economical or *best hydraulic section;* however, this often poses fabrication problems.

The two most common geometries employed are rectangular and trapezoidal channels. Table II gives formulas for the cross-sectional area of flow, wetted perimeter and the

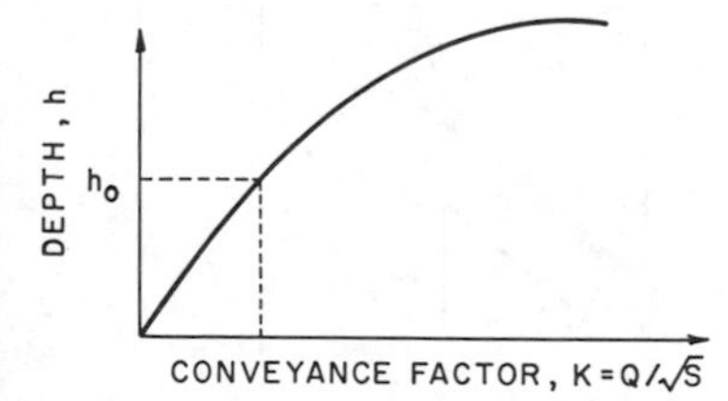

Figure 6. Plot of conveyance vs flow depth to determine normal channel depth from the Manning formula.

Table II. Geometric Definitions for Rectangular and Trapezoidal Channels

Channel Description	Cross-Sectional Area of Flow	Wetted Perimeter	Best Hydraulic Depth
Rectangular Channel 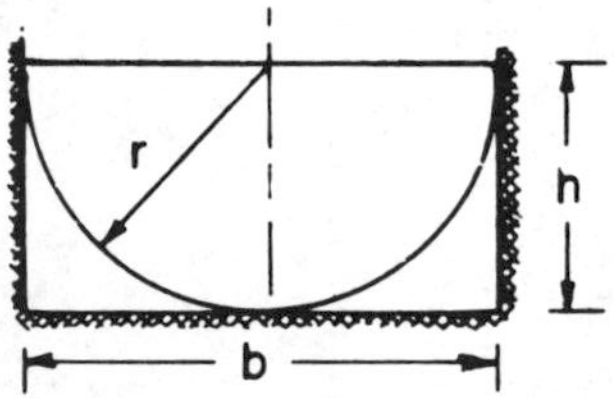	$A = bh$	$p = 2h + b$ $= 2h + \frac{A}{h}$	$h_D = \frac{bh}{2h + b} = \frac{2h^2}{4h}$ $= h/2$
Trapezoidal Channel 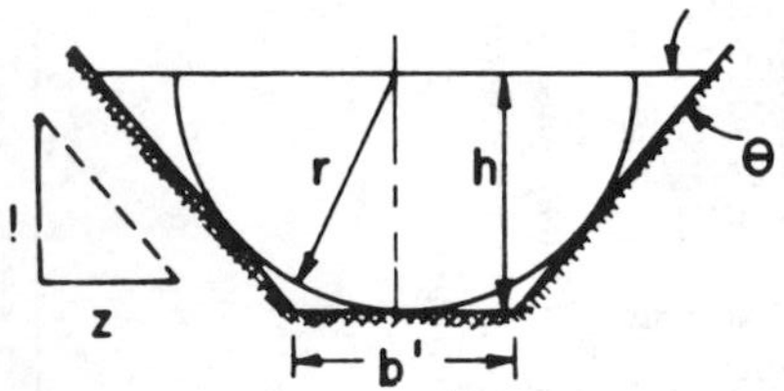	$A = h(zh + b')$ $= h(zh + p - 2h\sqrt{(z^2 + 1)})$	$p = 2h\sqrt{(z^2 + 1)} + b'$ where z is the side slope; if z is a variable and 'A' and 'h' are fixed, then – $p = \frac{A}{h} - zh + 2h\sqrt{(z^2 + 1)}$	$h_D = h/2$; with $\theta = 60°$ for the best hydraulic section

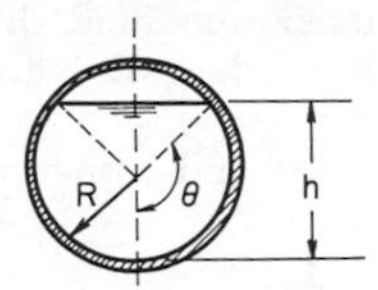

Figure 7. Defines system of partial flow in a circular pipe.

best hydraulic depth for these two geometries. The best hydraulic depth is determined by minimizing p with respect to depth (i.e., the best hydraulic depth exists for $dp/dh = 0$).

Partial Flow in Pipes

Partial pipe flow frequently is encountered in sewer pipes and storm drainage lines. Such a system is illustrated in Figure 7. Note that the angle subtended by the wetter perimeter is 2θ, and the mean fluid velocity can be obtained through the Manning formula. The following relationships follow from Figure 7:

Wetted perimeter:

$$p = 2R\theta \tag{17}$$

Cross-sectional flow area:

$$A_L = R^2\theta - R^2 \frac{\sin 2\theta}{2} \tag{18}$$

Hydraulic mean depth:

$$h_D = A/p = \frac{R}{2} - \frac{R\sin 2\theta}{4\theta} \tag{19}$$

Differentiating the Manning formula (Equation 6), an expression for the maximum discharge rate may be derived as outlined in the following steps:

$$\frac{dQ}{d\theta} = 0$$

whence

$$\frac{d(A^5/p^2)}{d\theta} = 0,$$

$$\frac{5A^4}{p^5} \cdot \frac{dA}{d\theta} - 2\frac{A^5}{p^3} \cdot \frac{dp}{d\theta} = 0$$

which simplifies to

$$5p\frac{dA}{d\theta} - 2A\frac{dp}{d\theta} = 0 \tag{20}$$

Substituting expressions for A and p (Equations 17 and 18) into Equation 20 and combining terms,

$$3\theta - 5\theta\cos 2\theta + \sin 2\theta = 0 \tag{21}$$

Equation 21 becomes satisfied when $\theta = 151°12'$. Hence, the maximum fluid discharge will occur when the water depth is 0.938D (from Equation 19), where D is the pipe diameter.

Combining Equations 7 (Manning expression for volumetric flowrate) and Equation 18 and rearranging terms, a relationship between Q and θ is obtained:

$$Q = \frac{1}{\epsilon}\frac{D^{8/3}}{10.08}\left(\theta - \frac{\sin 2\theta}{2}\right)\left(1 - \frac{\sin 2\theta}{2\theta}\right)^{2/3} S^{1/2} \tag{22}$$

For full pipe flow, h = D and $\theta = \pi$, whence the volumetric discharge rate for full pipe flow becomes

$$Q_f = \frac{\pi D^{8/3} S^{1/2}}{10.08\,\epsilon} = \frac{D^{8/3} S^{1/2}}{3.21\,\epsilon} \tag{23}$$

For the maximum discharge rate, where h = 0.938D,

$$Q_{max} = \frac{\pi D^{8/3} S^{1/2}}{9.38\,\epsilon} = \frac{D^{8/3} S^{1/2}}{2.98\,\epsilon} \tag{24}$$

And for the pipe running half full,

$$Q_{1/2} = \frac{\pi D^{8/3} S^{1/2}}{20.16\,\epsilon} = \frac{D^{8/3} S^{1/2}}{6.41\,\epsilon} \tag{25}$$

The above equations are in English units.

Estimating Holdup for Partial Pipe Flows

The term holdup is used to describe the average cross section of flow or nonflow. The definition given by Equation 18 divided by the full pipe cross section is the liquid holdup for the pipe; however, because it is defined in terms of angle θ, it is in a form that is not readily usable. From a practical standpoint, holdup may be estimated from geometric considerations of the flow and a single measurement of the average fluid depth over the centerline of the pipe. Figure 8 shows the geometric parameters defining holdup for the liquid, R_L (= A_L/A), and nonflowing gas phase, R_G (= A_G/A). The angle γ is now 2θ. Using the standard definition for equivalent diameter (i.e., four times the cross-sectional area of the phase divided by the wetted perimeter), equivalent diameters for the flowing and nonflowing phases are as follows:

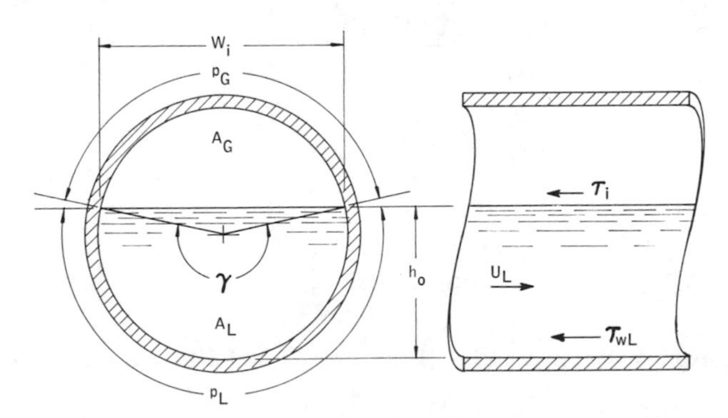

Figure 8. Partial pipe flow system for estimating liquid holdup.

$$D_{EL} = 4A_L/p_L \tag{26}$$

$$D_{EG} = 4A_G/(p_G + w_i) \tag{27}$$

where p_G, p_L are the gas- and liquid-phase wetted perimeters, respectively, and w is the width of the gas-liquid interface subtended by angle γ.

From consideration of the geometry in Figure 8, the following dimensionless parameters are developed:

$$h_o/D = (1 - \cos(\gamma/2))/2 \tag{28}$$

$$\tilde{w}_i = w_i/D = 2[(h_o/D) - (h_o/D)^2]^{1/2} \tag{29}$$

$$R_L = A_L/A = (\gamma - \sin\gamma)/2\pi \tag{30}$$

$$R_G = A_G/A = 1 - R_L \tag{31}$$

$$\begin{aligned}\tilde{p}_L = p_L/p = \gamma/2\pi &= 1 - p_G/p \\ &= 1 - \tilde{p}_G\end{aligned} \tag{32}$$

where

$$p = p_L + p_G$$

$$\tilde{D}_{EL} = D_{EL}/D = 4A_L/(p_L D) = R_L/\tilde{p}_L \tag{33}$$

$$\begin{aligned}\tilde{D}_{EG} &= D_{EG}/D = 4A_G/(p_G + w_i)D \\ &= R_G/[(p_G/p) + (w_i/\pi D)]\end{aligned} \tag{34}$$

The dimensionless geometric parameters are shown plotted against fractional liquid holdup, R_L, in Figures 9 and 10. A plot of h_o/D versus fractional liquid holdup is given in Figure 11.

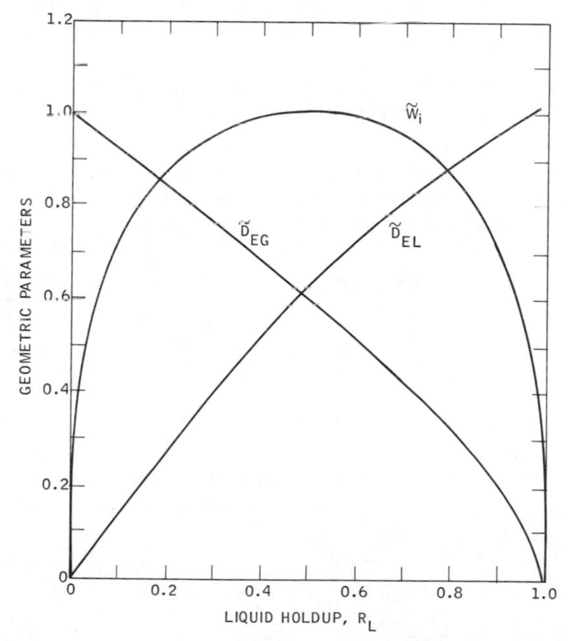

Figure 9. Dimensionless geometric parameters plotted as a function of fractional liquid holdup [4].

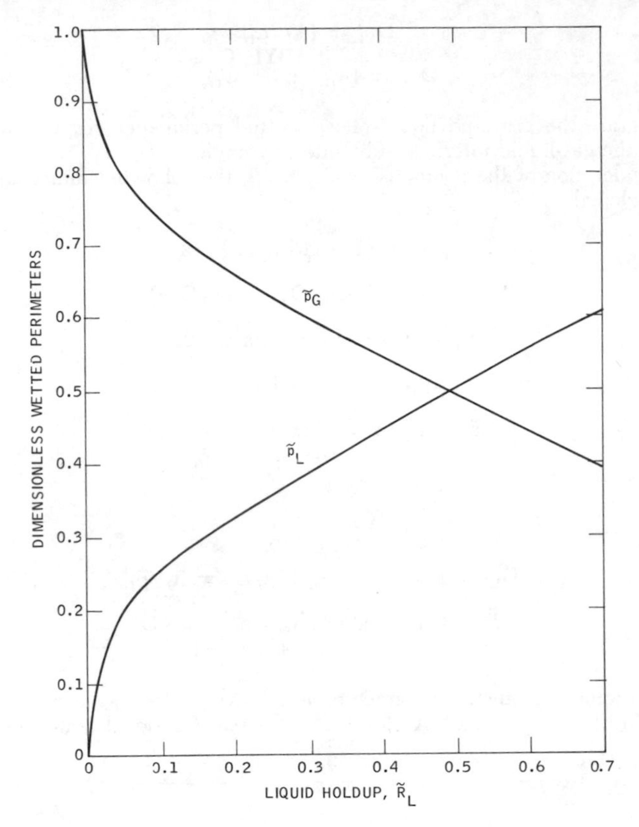

Figure 10. Dimensionless wetted perimeters of flowing and nonflowing phases, plotted as a function of fractional liquid holdup [4].

Figure 11 is the most useful of the three plots in that it provides an estimate of the liquid holdup from a measure of the average flow depth in the pipe. Figures 9 and 10 become useful when estimates of friction factors are needed.

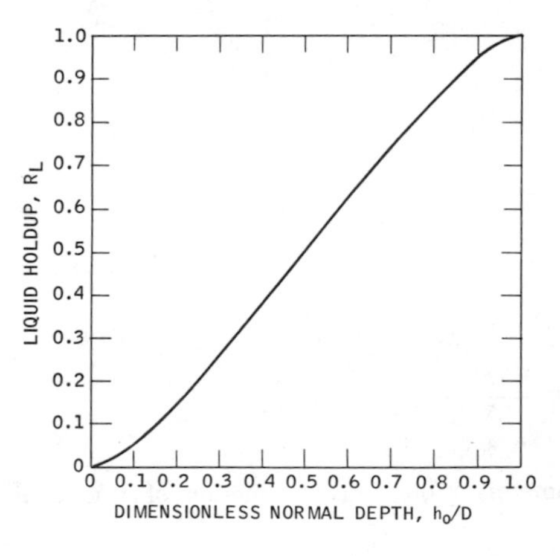

Figure 11. Plot of fractional liquid holdup vs average dimensionless liquid depth.

NONUNIFORM FLOWS: GRADUALLY VARYING

Nonuniform open-channel flow is characterized by a changing fluid level in the direction of flow. This change can occur either gradually or suddenly. This somewhat arbitrary distinction of the rate of level change as a function of distance along the flow axis enables those specific situations that are frequently encountered to be described.

Channel flows undergoing gradual changes normally involve small changes in depths and, consequently, the concern here is with relatively long channels. This type flow is most often encountered in natural channels, such as rivers.

Specific Energy Relation

The system of nonuniform gradually varying flow is illustrated in Figure 12A, from which Bernoulli's equation may be written to obtain the total energy (head) of the flowing fluid:

$$\Delta H = \frac{U^2}{2g} + h + Z \tag{35}$$

Note from Figure 12A that if a length of channel is considered in profile, a line may be drawn above the fluid surface to denote the total energy of flow. This is called the *total energy line*. That is, each point on this line is $U^2/2g$ above the fluid surface. For an ideal fluid, there would be no energy losses and the total energy line would be horizontal. For a real fluid of uniform flow, the line is parallel with the fluid surface and channel bed. In the case of nonuniform flow considered here, the total energy line, the line describing the fluid's free surface profile and the channel bed all may be of different slopes.

Specific energy (also called the *specific head*) refers to the fluid's energy for a cross section. With $Z = 0$ and noting that $U = Q/A = q/h$ for a rectangular channel, Equation 35 can be restated as follows:

$$E = \frac{q^2}{2gh^2} + h \tag{36}$$

where E is the specific energy comprised of the sum of the kinetic energy of the fluid and its depth (Figure 12B). Specific energy is referenced to the horizontal datum. Hence, if the channel bottom is sloped, then the specific energy of a specified discharge may vary along the flow axis. Variations in specific energy may not necessarily be accompanied by changes in the total energy flow. The two are related as follows:

$$H = E + Z \tag{37}$$

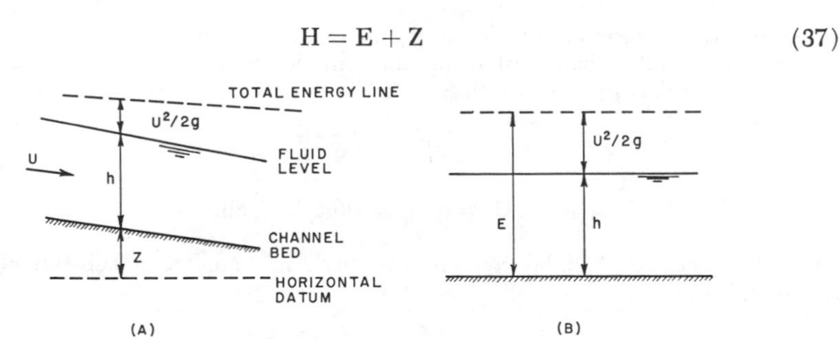

Figure 12. System for nonuniform flow.

For flows in which the velocity distribution is highly nonuniform, a kinetic energy correction term is needed to correct the specific energy:

$$E = \alpha \frac{U^2}{2g} + h \tag{38}$$

The relationship between specific energy and depth is shown in Figure 13. The plot of E versus h (called the specific energy curve) is for a constant discharge per unit channel width, q. At any depth, h, the specific energy is the sum of kinetic energy (curve A) and potential energy (curve B). Note that for a particular value of the specific energy there are two possible depths, h_1 and h_2. The point of minimum specific energy, E_{min}, is called the *critical depth*, h_c. At E_{min}, the Froude number ($Fr = U/\sqrt{gh}$) has a value of unity. The velocity at this point is referred to as the critical velocity. At depths greater than h_c, the mean velocity is referred to as being subcritical, since $U < U_c$ (also called streaming or tranquil flow). Similarly, at depths less than h_c, $U > U_c$ and the flow is referred to as supercritical. Rewriting Equation 38 for a channel of any geometry,

$$E = \frac{Q^2}{2gA^2} + h \tag{39}$$

Differentiating this expression with respect to depth and noting that at $dE/dh = 0$, $h = h_c$, the following is obtained:

$$\left(\frac{Q^2}{gA_c^3}\right)\frac{dA}{dh} = 1 \tag{40}$$

where A_c is the critical flow area. Note that $dA/dh \simeq \Delta A/\Delta h$, and if Δh is taken to be a thin sliver of fluid in the channel flow near the surface, the flow area of a finite fluid element can be approximated as a rectangle such that $\Delta A/\Delta h \simeq b$. Equation 41, the expression for critical flow, becomes

$$\frac{Q^2 b_c}{gA_c^3} = 1 \tag{41}$$

where subscript c refers to the critical condition. Since $Q = U_c A_c$, then Equation 41 can be rearranged to give

$$U_c = \sqrt{\frac{gA_c}{b_c}} \tag{42a}$$

$$U_c = \sqrt{gh_{mc}} \tag{42b}$$

where h_{mc} is the mean critical depth of an irregularly shaped channel.

For a rectangular channel showing constant depth, Equation 39 can be developed in terms of the flow per unit width to give the following relationships:

$$h_c = \sqrt[3]{q^2/g} \tag{43}$$

$$U_c = q/h_c = \sqrt[3]{gq} = \sqrt{gh_c} \tag{44}$$

These two relations can be combined through h_c and, when substituted back into Equation 39, give

$$E_c = \frac{U_c^2}{2g} + \frac{2U_c^2}{2g} \tag{45a}$$

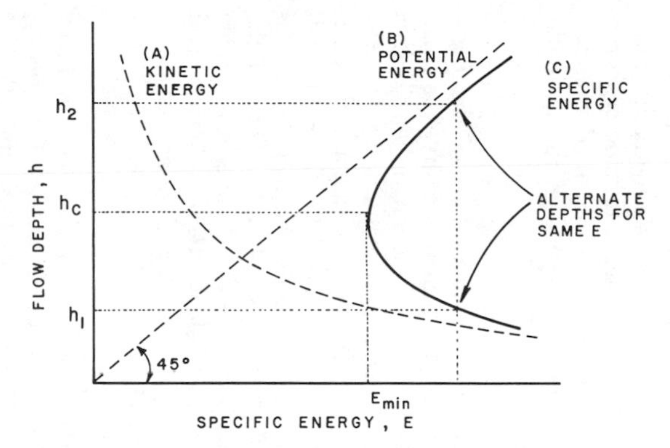

Figure 13. Specific energy curve at constant q for open-channel flow.

or

$$h_c = \frac{2}{3} E_c \tag{45b}$$

This last expression states that the critical flow depth is two-thirds the critical specific energy. The above discussion has been based on the assumption that the discharge in the channel is constant. However, the reverse also may be true, that is, the specific energy may be fixed and q may be a function of depth. In this case, it follows that at the critical depth the discharge is at a maximum. Foregoing the mathematics, it can be shown through a derivation similar to the above that

$$q_{max} = \frac{2}{3} \frac{\sqrt{2g}}{\sqrt{3}} E_c^{3/2} \tag{46}$$

for the critical depth given by Equation 45b.

Surface Profile Classification

Proper classification of the surface profile is essential to analyzing open-channel flows. For water flows, emphasis is placed on establishing the proper free surface slope for a particular flow situation. The surface profile slope often is classified by reference to the critical slope, S_c. If the slope is less than S_c, the flow at the normal depth is referred to as being *mild*. For supercritical flow, $S > S_c$, the slope is said to be *steep*.

The critical slope occurs when the critical and normal depths coincide. Because the flow at the critical depth is uniform, then the bed and total energy line must have the same slope. From the Manning formula, an expression for the critical slope is

$$S_c = \frac{g h_c \epsilon^2}{h_{mc}^{4/3}} \tag{47}$$

Twelve free-surface profiles have been classified to describe gradually varied flow situations [5]. These are summarized in Table III. Examples of each flow situation are included in the table, showing the critical depth line and the normal depth line. Table III can serve as a guide to classifying profiles for different open channel flows.

Table III. Free-Surface Profiles, as Reported by Bakhmeteff [5][a]

Slope Description	Profile Classification	Examples	Comments
Horizontal Slope ($S = 0$)	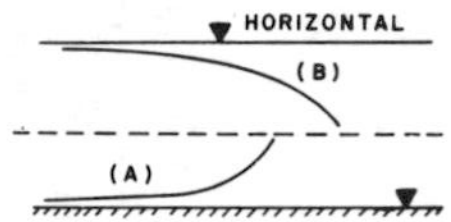 HORIZONTAL (B) (A)	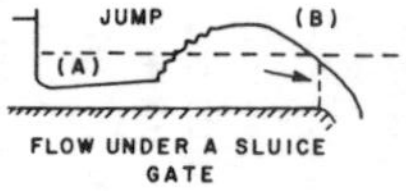 JUMP (B) (A) FLOW UNDER A SLUICE GATE	Here $h_o \to \infty$, so uniform flow cannot exist for $h > h_o, h_c$. For example, profiles A and B are joined by a hydraulic jump. Profile B constitutes a *drawdown* curve; the profile passes through the critical depth again at the end of the hydraulic jump.
Mild Slope ($S < S_C$)	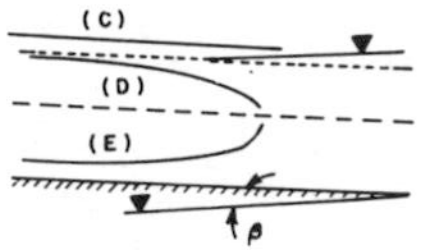 (C) (D) (E) β	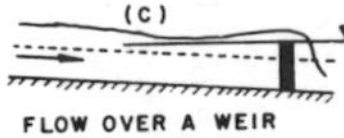 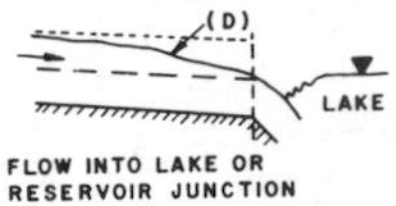 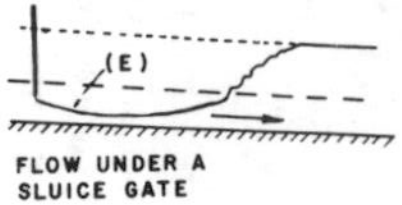 (C) FLOW OVER A WEIR; (D) LAKE FLOW INTO LAKE OR RESERVOIR JUNCTION; (E) FLOW UNDER A SLUICE GATE	Profile C is referred to as the *backwater curve;* commonly observed in channels of changing section or flow over a weir. Curve A is asymptotic both to the normal depth line upstream and the horizontal downstream. The D profile will occur at the junction with a reservoir if the fluid level is below the level of the normal depth in the channel exit. Profile E occurs below steep spillways and sluices. Profiles D and E are generally short compared to C.
Critical Slope ($S = S_C$)	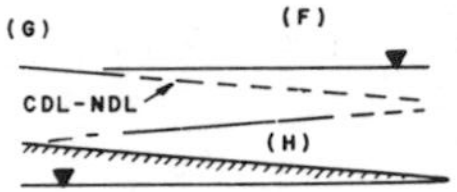 (G) (F) CDL-NDL (H)	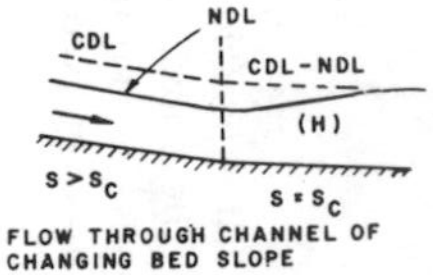 NDL CDL CDL-NDL (H) $S > S_C$ $S = S_C$ FLOW THROUGH CHANNEL OF CHANGING BED SLOPE	Profiles F and H are close to being straight horizontal lines. Theoretically, profile G corresponds to uniform flow at the normal depth. At the critical slope the fluid surface oscillates about h_o for uniform flow. This gives rise to a series of stationary surface waves, CDL, critical depth line.

Steep-Slope
($S > S_C$)

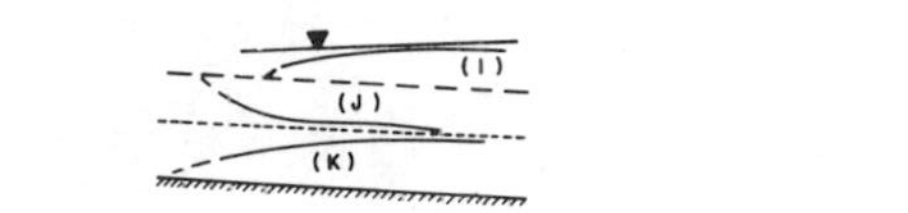

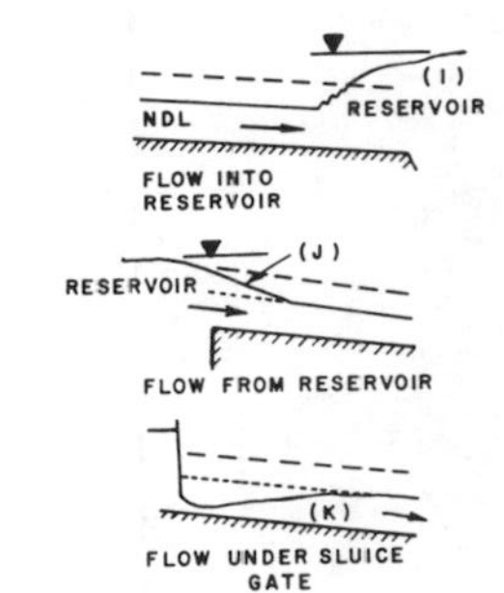

Such profiles are observed in spillways and chutes. Profiles are short and asymptotic downstream. Profile I is normally found downstream from a jump. Profile J is a continuation of a hydraulic drop formed when the channel steepens or when flow exits a reservoir. Profile K is formed below a sluice gate or below the junction with a steeper channel.

Adverse Slope
($S < 0$)

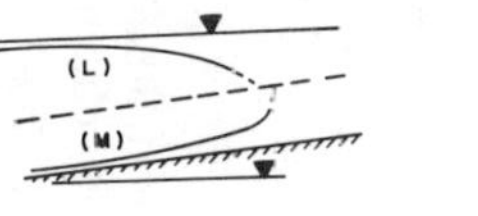

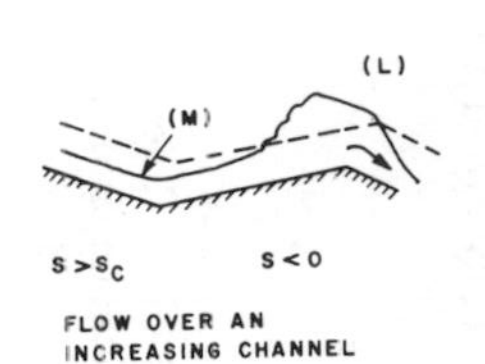

Uniform flow cannot exist in the direction shown. Profiles L and M normally occur together and are joined by a jump on the apron below a spillway. Profile L also may be found in front of profile J (see above) when a steep channel runs out of a reservoir. These profiles are very short.

---

[a] – – – – – Denotes critical depth line (CDL).
------------- Denotes normal depth (NDL).
⟶ Denotes direction of flow.

Calculating Surface Profiles

The equations outlined above enable detailed calculations of the free-surface profile to be made once the profile has been classified properly. The methods outlined by Sellin [3] and Henderson [6] are most suited for gradually varied flows, and the reader is referred to these sources for details. Starting with the total energy head expression for gradually varied nonuniform flow (Equation 35), Sellin [3] applied the following assumptions:

1. If the slope of the channel bed, S, is small, the depth, h, may be considered a normal distance, i.e., $\cos\theta = 1$ (refer to Figure 5).
2. The velocity distribution across the channel is uniform, i.e., the results for uniform flow are applied locally to gradually varied flow.
3. The total energy head, H, and the bed elevation, Z, are considered positive above a horizontal datum.
4. The relationship between change in depth and change in cross-sectional area can be approximated by $dA/dh = b$, where b is the channel width.

The general equation of gradually varied flow is

$$\frac{dh}{dx} = \frac{S - j}{1 - \dfrac{Q^2 b}{gA^3}} \tag{48}$$

where j is the slope of the total energy line.

For a rectangular cross-sectional channel, Equation 48 becomes

$$\frac{dh}{dx} = \frac{S - j}{1 - q^2/gh^3} \tag{49}$$

where $A = bd$
$q = Q/b$

From assumption 2, we can introduce the Manning formula to obtain an expression for j:

$$j = \frac{U^2 \epsilon^2}{{h_{Dm}}^{4/3}} \tag{50}$$

where h_{Dm} is the hydraulic mean depth. Substituting into Equation 49, a workable expression for a rectangular channel is obtained:

$$\frac{dh}{dx} = \frac{S - \dfrac{U^2}{C^2 h_{Dm}}}{1 - U^2/gh} \tag{51}$$

The gradually varied flow expression may be solved to obtain the free-surface profile, that is, the fluid depth as a function of x along the channel. The general expression, Equation 48, can be rearranged to solve for x:

$$x = \int dx = \int \left(\frac{1 - B}{S - d}\right) dh$$

where

$$B = \frac{Q^2 b}{gA^3} \text{ and } d = \frac{\epsilon^2 Q^2}{A^2 {h_{Dm}}^{4/3}} \tag{52}$$

Note that B and d are functions of h (since A, b, h_{Dm} are functions of h). Hence, Equation 52 is actually $x = \int f(h)\,dh$.

If f(h) is a simple function (usually it is not), the integral may be integrated directly.

Simpson's rule may be used to evaluate quadratic functions numerically. In applying Simpson's rule, Equation 52 is treated in the form dx/dh = f(h), whereby values of dx/dh corresponding to a chosen range of h values are first computed. Simpson's rule is then applied to evaluating the integral

$$\Delta x = \int_{h_1}^{h} \left(\frac{dx}{dh}\right) dh \tag{53}$$

whereby Δx is a distance along the channel between specified depths h_1 and h_2 (i.e., Equation 53 gives the area under the curve between the coordinates h_1 and h_2 for the curve dx/dh versus h). A quadratic is fitted to match values at h_o, h_1, h_2, h_3, h_4 and so forth. The number of intervals must be even, and the area under each successive three points is approximated by the area under the quadratic. Hence, this is a second-order method, giving

$$\text{Area under curve} = \frac{h_n - h_o}{3n} [f(h_o) + 4f(h_1) + 2f(h_2) + 4f(h_3) + 2f(h_4) + \ldots + 4f(h_{n-1}) + f(h_n)] \tag{54}$$

Note that the error incurred in the approximation technique depends on the order of the approximation; however, it always decreases as the number of intervals increases. On the other hand, the error accumulation in the summation process in Simpson's rule increases with n. Thus, the relative error must have a minimum for some value of n.

Various numerical techniques for treating this problem are outlined in the literature [6–8]. Although many of these methods may be used via hand computations, they can be tedious and are better adapted to programming on the computer.

NONUNIFORM FLOWS: RAPIDLY VARYING

The distinction between gradually varying and rapidly varying flows is made based on the criterion of streamline curvature. In the former, only small changes in depth are observed over the channel length; hence, energy losses due to channel boundary friction are the primary means of energy dissipation. In the case of rapidly varying flows, energy losses become complicated by the appearance of hydraulic jumps.

A hydraulic jump is a stationary surge or shock wave in which the speed of advance of the wave front (celerity) is equalized by the velocity of flow in the opposite direction. The position of the hydraulic jump will depend on a combination of the upstream and downstream flow conditions. The form or shape, as well as the performance of the jump, will depend on the upstream and downstream depths and the velocity of the approaching flow. The Froude number ($Fr = U_1/\sqrt{gh_1}$) is used to characterize the jump. There are three types of hydraulic jumps: undular, transitional and direct jumps (Figure 14).

The undular jump is described as having a flow that first expands slowly and then oscillates. Hence, this type jump has the form of a smooth initial wave, followed by a series of waves of decreasing amplitude. Energy losses associated with this type jump are small.

A transitional jump produces an increasing rate of energy dissipation, resulting in the first few wave crests breaking, and isolated "reverse rollers" form on the upstream facing the slope of these waves below the crest. Once waves begin to break, any waves formerly downstream of them become lost in surface agitation.

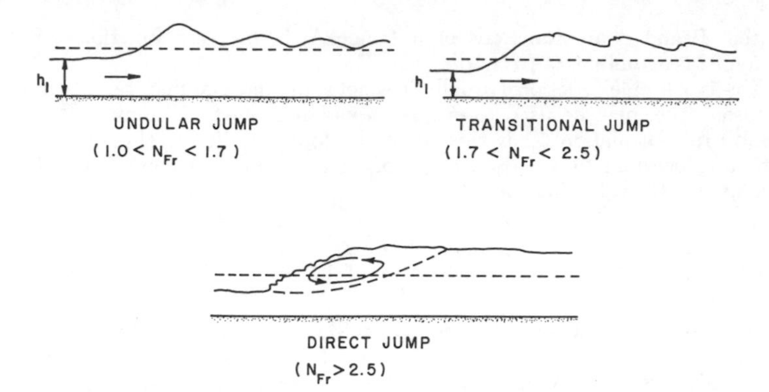

Figure 14. Three types of hydraulic jumps.

Direct hydraulic jumps occur at higher Froude numbers and often are characterized by high turbulence and, consequently, high energy losses. Such jumps consist of a single monoclinical wave covered by an intensely turbulent reverse roller, which normally shows a froth (called "white water") due to air entrainment. Figure 14 gives the range of Froude numbers in which the different hydraulic jumps occur.

Hydraulic jumps can be induced to the flow for a variety of reasons. Basically, it provides a means of dissipating excess energy, which can prove useful in such applications as mixing polyelectrolytes and various chemicals in wastewater treatment, in irrigation and water abstraction schemes by raising downstream water levels, and in handling flood discharge carried from reservoirs by spillways.

An analysis of hydraulic jumps is given by Sellin [3] and Bakhmeteff [5]. Their descriptions of a hydraulic jump are in terms of the *specific force* function, F(h):

$$F(h) = \frac{Q^2}{gA} + AZ \tag{55}$$

The specific force function is shown plotted against depth, along with the specific energy curve in Figure 15. Both relationships exhibit a minimum at the critical depth. Note that for any F(h) value two depths are possible: one depth, h_1, corresponds to a depth less than the critical depth; the other, h_2, corresponds to a depth greater than the critical.

A stable hydraulic jump may exist between two depths, referred to as *conjugate depths,* where the corresponding energies are E_1 and E_2 in Figure 15. The difference between these two energy states, ΔE, corresponds to the specific energy loss associated

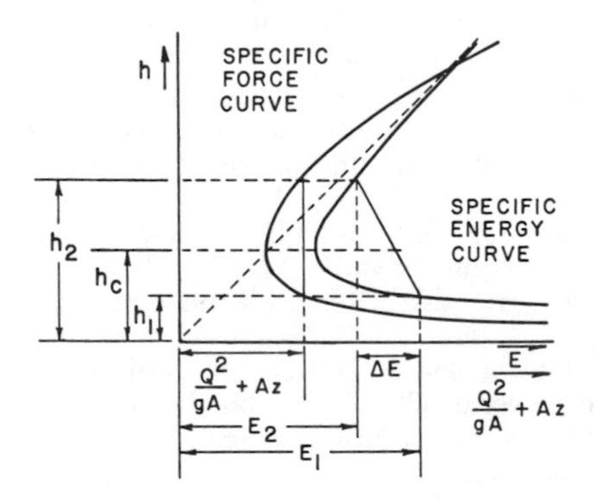

Figure 15. Plots of specific energy and specific force for a hydraulic jump.

with the hydraulic jump. Note that $E_1 > E_2$, hence ΔE is negative, meaning that energy is dissipated in a hydraulic jump, not added.

The following limiting conditions can be observed from Figure 15: as the height of the jump decreases, i.e., $h_2/h_1 \rightarrow 1$, $\Delta E \rightarrow 0$.

For a wide horizontal channel, Sellin [3] gives the following relationship for the height of the jump:

$$\frac{h_2}{h_z} = \tfrac{1}{2}[\sqrt{1 + 8\,Fr^2} - 1] \tag{56}$$

Other important physical characteristics of a hydraulic jump that may be described from momentum considerations and Figure 15 are the energy loss, the length of the jump and the location of the jump along the channel axis.

For a wide channel, the energy loss may be computed from the following expression:

$$\Delta E = \frac{(h_2 - h_1)^3}{4h_1h_2} \tag{57}$$

The length of the jump typically lies between the lmits $S < l/h_2 < 6.5$, where l is the length of the hydraulic jump.

If a stable jump occurs, it will form at a point along the channel axis where suitable conjugate depths exist. The jump's location will depend on whether the normal depth is present. If the normal depth forms one of the limits of the jump, the following calculation procedure may be used to estimate the location of the jump:

Step 1. Compute the conjugate depth from Equation 56 for steep slopes or from a parallel form of that equation for mild slopes.

Step 2. Compute the free surface profile of the varied flow forming the other limit of the jump (refer to Table III for profile descriptions).

Step 3. Locate the jump by selecting the point on that profile at which the depth is the same as the required conjugate depth.

For the case in which the normal depth is not present in the flow, the reader is referred to the procedure outlined by Sellin [3].

UNSTEADY FLOWS

Characteristics of Unsteady Flows

Unsteady or transient flows are characterized on the basis of disturbances at the free surface. There are three general types of disturbances: solitary waves, monoclinical waves, and surges [9–11]. These are illustrated in Figure 16.

Solitary waves (Figure 16A) are a form of gradually varied flow and are characterized by a single smooth profile wave. This type of wave travels along the channel with a constant velocity of propagation (called "celerity"), c. The wave amplitude gradually diminishes as the flow energy is dissipated by friction. An empirical expression for celerity is

$$c = \sqrt{\frac{g}{2}(h_1 + h_2)} \tag{58a}$$

or

$$c \simeq \sqrt{gh} \tag{58b}$$

Equation 58 is accurate if the wave amplitude is comparable to h_1. Solitary waves are generated by the introduction of a single disturbance (in Figure 16A the wave is caused by the movement of a paddle in the flow stream).

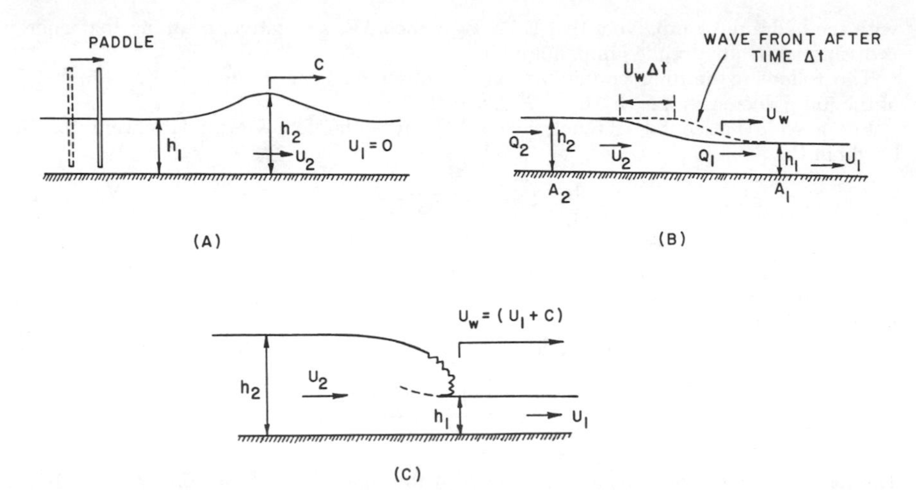

Figure 16. Three types of interfacial phenomena observed in unsteady open-channel flows: (A) paddle generation of a solitary wave for water flow; (B) monoclinical rising wave; (C) channel surge wave.

Monoclinical waves result in an increase in the fluid level at some point after its passage (Figure 16B). Such waves are generated in a channel by gradually increasing the flowrate to some steady value, Q_2.

A surge results when a monoclinical wave undergoes an abrupt change in its surface curvature, causing the slope at its forward end to break down and form a rapidly varying wave profile. The length of a surge wave is short in comparison to the two other types discussed. In addition, the effect of channel friction on its behavior is smaller. Surges may advance either upstream or downstream, depending on how they are initiated. The passage of surges can produce either an increase in depth at a particular point (called a positive surge) or a decrease in depth (called a negative surge). Consequently, there are four types of surges that are classified according to their frontage: a positive surge advancing upstream, a positive surge advancing downstream, a negative surge advancing upstream, and a negative surge advancing downstream. Note that positive surge profiles are relatively stable, whereas negative surges tend to be unstable and quickly disperse. For illustration, we shall analyze a positive surge wave from a theoretical basis.

Positive Surge Waves

A positive surge wave results from a sudden change in flow caused, for example, by a sluice gate closing or opening. Streeter and Wylie [12] analyzed the increase in depth caused by a surge wave in a rectangular channel. By superposition of the celerity (surge velocity), the surge wave is reduced to a steady-state situation, as shown in Figure 17. Neglecting friction, the momentum equation per unit width for a finite volume of flowing fluid may be written as follows:

$$\frac{\Delta P}{2}\,(h_1^2 - h_2^2) = \frac{\Delta P}{g}\,h_1\,(U_1 + c)\,(U_2 + c - U_1 - c) \tag{59}$$

Equation 59 assumes the wall shear stress at the channel bottom is negligible, which is reasonable if the channel is deep.

From continuity,

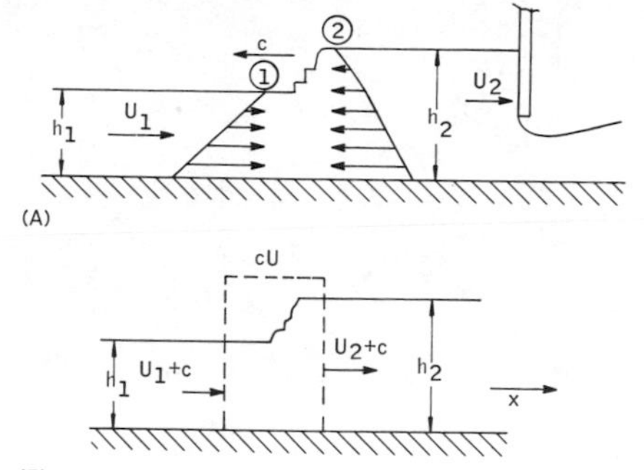

Figure 17. (A) Formation of a positive surge wave by partially closing a sluice gate; (B) by superposition of the celerity (surge velocity), the situation is reduced to a steady-state problem.

$$(U_1 + c)\, h_1 = (U_2 + c)\, h_2 \tag{60}$$

Combining Equations 59 and 60 through U_2, the following is obtained:

$$U_1 + c = \sqrt{gh_1}\left[\frac{h_2}{2h_1}\left(1 + \frac{h_2}{h_1}\right)\right]^{1/2} \tag{61}$$

The above expressions contain five variables: U_1, U_2, c, h_1 and h_2. Hence, if any three are specified, the other two may be computed.

These expressions may be written in a more general form by defining a fluid discharge per unit width of channel, $q = Uh$, and introducing the following dimensionless groups:

$$\tilde{h} = h_2/h_1 \tag{62}$$

$$\tilde{q} = q_2/q_1 \tag{63}$$

$$\tilde{F}_1 = U_1^2/gh_1 \tag{64}$$

The continuity equation then becomes

$$c = \frac{q_1}{h_1}\frac{\tilde{q} - 1}{1 - \tilde{h}} = \sqrt{gF_1h_1}\,\frac{\tilde{q} - 1}{1 - \tilde{h}} \tag{65}$$

An alternate form of this expression is

$$\tilde{c} = c/U_1 = \frac{\tilde{q} - 1}{1 - \tilde{h}} \tag{66}$$

Equation 66 gives a dimensionless surge velocity, which is shown plotted against $\tilde{h}$ for several values of $\tilde{q}$ in Figure 18. From the physics of the problem, values of $\tilde{c}$ for $\tilde{h} \leq 1$ are meaningless. When $\tilde{q} = 1$, $\tilde{c} = 0$, which is the special case of a hydraulic jump.

The momentum expression in dimensionless form becomes

$$(\tilde{h} - 1)\sqrt{\tilde{h}\,(\tilde{h} + 1)} = \sqrt{2F_1}\,(\tilde{h} - \tilde{q}) \tag{67}$$

Equation 67 holds true for unique values of $\tilde{q}$ and $\tilde{h}$ for a given flow situation. A plot of each side of Equation 67 is shown in Figure 19, where $f_1 = (\tilde{h} - 1)\sqrt{\tilde{h}\,(\tilde{h} + 1)}$ and $f_2 = \sqrt{2F_1}\,(\tilde{h} - \tilde{q})$. Function f_1 plots as a positive sloping curve, whereas f_2 is that of a straight line with slope $\sqrt{2F_1}$ and an ordinate intercept of $-\sqrt{2F_1}\,\tilde{q}$. The

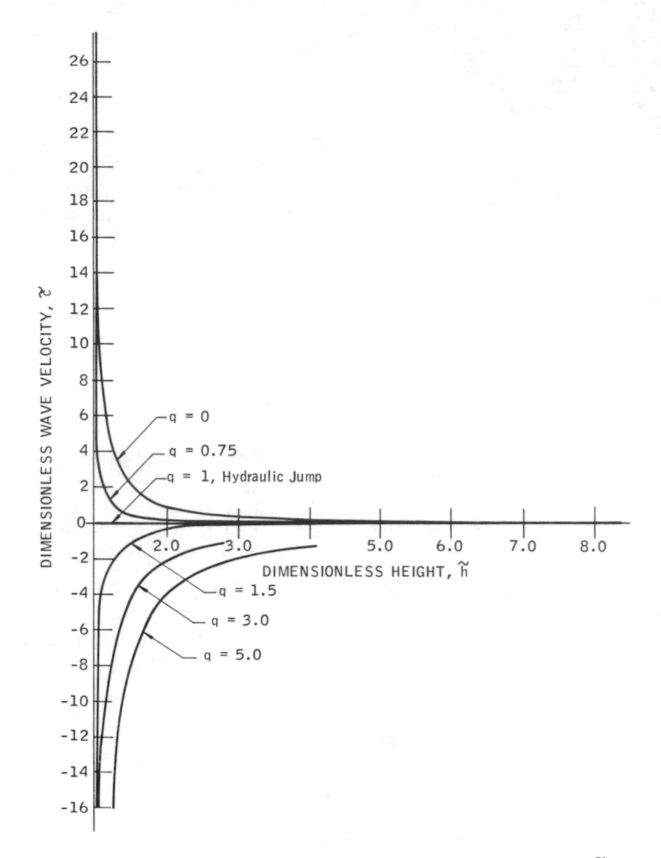

Figure 18. Plot of dimensionless celerity vs $\tilde{h}$.

intersection of these two functions represents a solution to Equation 67. Hence, for a given F_1 and $\tilde{q}$, a value for h can be obtained from the chart by constructing a straight line of slope, $\sqrt{2F_1}$, and ordinate intercept, $-\sqrt{2F_1}\,\tilde{q}$. From the $\tilde{h}$ value, Figure 18 may be used to obtain c. Note from Figure 19 that two solutions of $\tilde{h}$ are possible for $\sqrt{2F_1}\,\tilde{q} < 0$ for a range of F_1. Also, the hydraulic jump occurs at slope $\sqrt{2F_1} = 1$.

Numerical Techniques for Transient Open-Channel Flows

The simple analysis in the preceding subsection illustrates that modeling open-channel flows from first principles is possible. Streeter [13–17] and others [18–22] discuss various numerical techniques based on the method of specified time intervals and the characteristic grid method for solving differential equations for unsteady pipe and channel flows.

A more advanced technique based on Eulerian formulations has emerged in recent years and has been highly successful in analyzing fluid dynamics problems involving free boundaries. Harlow et al. [23,24] developed the first successful computer code based on Eulerian finite-difference methods, and Hirt and his co-workers [25,26] and McMaster et al. [27,28] have improved on the technique. The general approach involves the use of an Eulerian mesh of rectangular cells having variable sizes, as shown in Figure 20. The Navier-Stokes equations are then solved by the method outlined below:

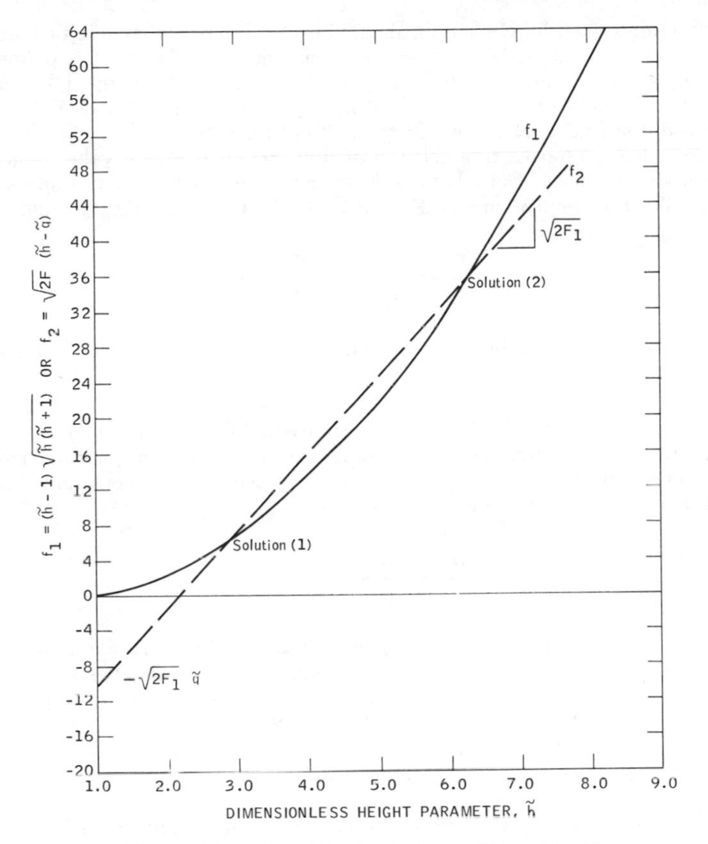

Figure 19. Graphical solution to Equation 67.

$$\frac{\partial u}{\partial t} + u\frac{\partial u}{\partial x} + v\frac{\partial u}{\partial y} = -\frac{1}{\rho}\frac{\partial P}{\partial x} + g_x + \nu\left[\frac{\partial^2 u}{\partial x^2} + \frac{\partial^2 u}{\partial y^2} + \xi\left(\frac{1}{x}\frac{\partial u}{\partial x} - \frac{u}{x^2}\right)\right] \quad (68)$$

$$\frac{\partial v}{\partial t} + u\frac{\partial v}{\partial x} + v\frac{\partial v}{\partial y} = -\frac{1}{\rho}\frac{\partial P}{\partial y} + g_y + \nu\left[\frac{\partial^2 v}{\partial x^2} + \frac{\partial^2 v}{\partial y^2} + \frac{\xi}{x}\frac{\partial v}{\partial x}\right] \quad (69)$$

where the velocity components, u, v, may be in either Cartesian or cylindrical coordinates (selection is based on the value for ξ; $\xi = 0$ for Cartesian and $\xi = 1$ for cylindrical).

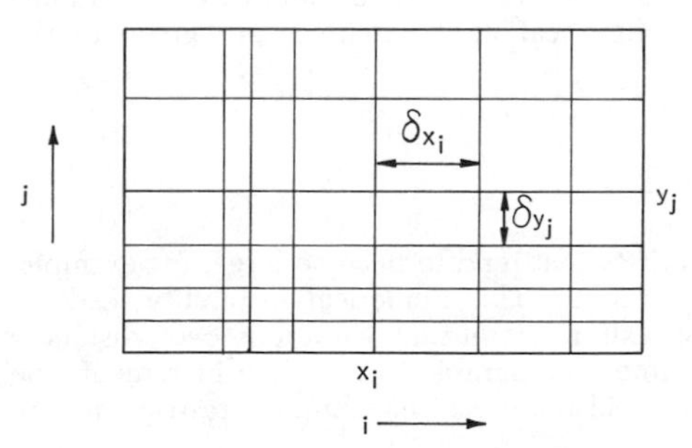

Figure 20. Finite difference mesh with variable rectangular cells.

The method described by Nichols et al. [26] uses a fractional volume of fluid scheme for tracking free boundaries. In Eulerian mesh, a function F (x, y, t) is defined whose value is unity at any point occupied by a fluid and zero elsewhere in the mesh. By averaging over the cells of a computational mesh, the average value of F in a cell becomes equal to the fractional volume of the cell occupied by the fluid. So, for example, a unit value of F corresponds to a cell full of fluid, whereas a zero value indicates that the cell contains no fluid. Cells having F values between zero and unity contain a free surface. The time dependence of F is governed by the following equation:

$$\frac{\partial F}{\partial t} + u\frac{\partial F}{\partial x} + v\frac{\partial F}{\partial y} = 0 \tag{70}$$

The basic procedure for advancing a solution through one increment in time, δt, is as follows:

1. Explicit approximations of the momentum equations (68 and 69) are used to compute an initial estimate for new time-level velocities using initial conditions or previous time-level values for advective, pressure and viscous terms.
2. The continuity equation must be satisfied:

$$\frac{\partial u}{\partial x} + \frac{\partial v}{\partial y} + \xi\frac{u}{x} = 0 \tag{71}$$

   Pressures are iteratively adjusted in each cell, and the corresponding velocity changes for each pressure change are added to the velocities computed in Step 1. An iteration is required because the change in pressure required in a single cell to satisfy Equation 71 will upset the balance in adjacent cells.
3. The F function defining fluid regions must be updated to estimate the new fluid configuration.

By repeating these steps, a solution is advanced through any desired time interval. At each step, suitable boundary conditions must be imposed at all mesh and free-surface boundaries. Details of these steps and boundary conditions are given by Nichols et al. [26].

Using the method outlined above, the problem was analyzed of an undular bore generated when a stream of fluid encounters a dam. The calculation of the free-surface profile and velocity vectors are shown in Figure 21. The solution was obtained by the iterative process, in which the computational mesh was swept row by row, starting at the bottom and working upward. The solution shown represents a two-dimensional one.

Numerical techniques based on Eulerian finite-difference formulation, with pressure and velocity as the primary dependent variables, offer a relatively straightforward approach to analyzing/modeling a variety of open-channel flow situations. In addition, much theory and suitable algorithms have been developed for handling transients in piping systems, thus enabling entire process flow systems to be analyzed. Additional information for those requiring details of numerical fluid dynamics are given in the literature [29–37].

Numerical Instability

Numerical computations often have quantities that tend to become large, for example, high-frequency oscillations in space, time or both. This numerical instability leads to unstable solutions. When computed results exhibit significant variations over distances comparable to a single cell width or over times comparable to the time increment, the accuracy of results must be in suspect. To avoid numerical instabilities, restrictions are normally imposed in defining mesh increments, the time increment and any differencing parameters used.

Accuracy is obtained by specifying mesh increments small enough to resolve expected spatial variations in all dependent variables. Sometimes this is not possible because of limitations in computing time and/or memory requirements. In such cases, care must be used in interpreting the computations. For example, in computing the flow patterns in a pretreatment mixing tank containing a variety of baffle arrangements, it is extremely difficult to resolve thin boundary layers along the confining walls. In some applications, such as computing the flow in a ship canal containing several locks and bed levels, the presence of thin boundary layers is unimportant, and freeslip boundary conditions can be justified as a good approximation.

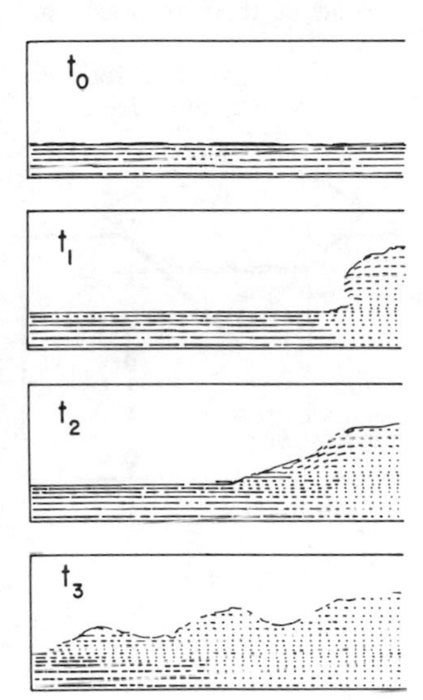

Figure 21. Velocities and free-surface profiles for breaking bore calculation.

After the mesh size has been selected, the choice of the time increment necessary for stability is based on several criteria. One criterion is that the flow cannot move through more than one cell in a single time step (the difference equations normally used assume fluxes only between adjacent cells). Hence, the time increment should satisfy an inequality of the following form:

$$\Delta t < \min \left\{ \frac{\delta x_i}{|u_{i,j}|}, \frac{\delta y_j}{|v_{i,j}|} \right\} \tag{72}$$

The minimum is with respect to each cell in the mesh.

Another important criterion is that the fluid momentum should not diffuse more than roughly one cell in a single time increment. From linear stability analysis it can be shown that

$$\nu \Delta t < \frac{1}{2} \frac{\delta x_i^2 \delta y_j^2}{\delta x_i^2 + \delta y_i^2} \tag{73}$$

where ν is the kinematic viscosity.

Other criteria often applied in computations relate to surface tension and to differencing parameters used in inducing any numerical smoothing.

FLOW MEASUREMENT

Flow measurement is often an integral part of a piping and channel flow network in plant operations. The design of flow monitoring and transfer facilities requires knowledge of flowrates, flow variability and total flow. Proper measuring and instrumentation selection will depend on such factors as cost, type and accessibility of the conduit or channel, available hydraulic head, and fluid characteristics. Once these properties are defined, a suitable measuring system may be chosen. There is a variety of flow measuring devices and methods available for open-channel flow monitoring, many of which are described by the author [38–40]. Table IV summarizes various types of flow measurement systems and gives a relative rating of their accuracy and applicability.

Because many of these systems are described elsewhere, only the working formulas and some design aspects of large channel flow measurement systems are summarized here.

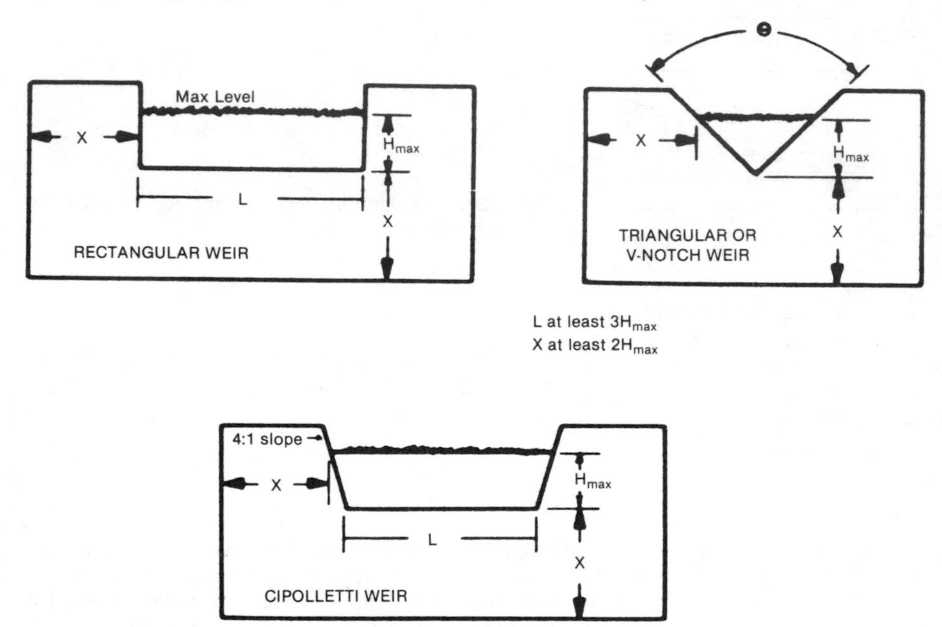

Figure 22. Common types of sharp-crested weirs.

Weirs

Weirs, most often used for measuring waste flows or in canals, represent the simplest and often most economical approach for large flows. The three most common designs are given in Figure 22.

The basis of measurement is the fluid level at a given distance upstream from the weir, which is proportional to the flow. The form of the crest is important for accurate measurements. A problem associated with rectangular weirs for water flow is that the flow will be contracted as it passes over the weir. Hence, the effective width of the weir is smaller than the crest. Cipolletti weirs were designed with sloping sides to compensate for this contraction, which enables the use of the width of the crest for flow calculations.

The design formula for flow over a weir (in English units) is

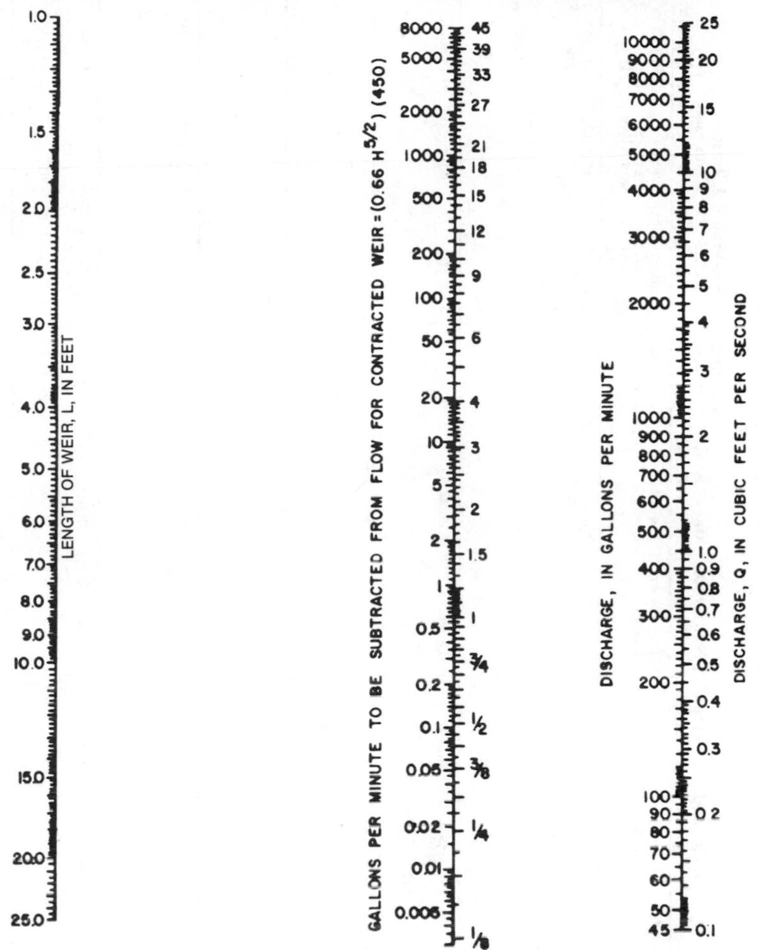

Note: Based on Frances weir formula as follows:

$Q = 3.33\ LH^{3/2}$ (for suppressed weir)

or

$Q = 3.33\ (L - 0.2H)\ H^{3/2}$

$= 3.33\ LH^{3/2} - 0.66\ H^{5/2}$ (for contracted weir with two end contractions)

Where:

Q = discharge, in cubic feet per second.

L = length of weir, in feet.

H = head in feet.

Figure 23. Nomograph for estimating the discharge over rectangular weirs [41].

$$q = C_w^{2/3} \sqrt{2g}\, H^{3/2} \tag{74}$$

where H is the upstream head above the crest, q is the flow per unit width (cfs/ft), and C_w is the nonuniformity coefficient to compensate for the nonuniformity of flow ($C_w \leq 1$).

Permanently installed weirs should be calibrated after installation to determine the proper coefficient.

Rectangular weirs are of the straight or notched variety. The former is called a suppressed weir without end contractions. A notched weir may have one or two end contractions. All engineering weir formulas have as their basis the Francis formula. Figure 23 provides a nomograph of the Francis formula, which can be used for a suppressed weir or a weir with standard end contractions [41]. Note that Figure 23

Table IV. Types and Characteristics of Flow Measurement Devices for All Services

	Rotameter	Diff. Pressure with Orifice, Flow Tube, etc.	Turbine	Magnetic	Weirs and Flumes	Forced Vortex	Nat. Vortex Meter	Mass (Ang. Momentum	Positive Displacement	Ultrasonic Thru-Wall Transducers	Ultrasonic Clamp on Type	Fluidic
Service	Liquids or gases	Liquids or gases	Liquids or gases	Electrically conductive liquids or slurries	Liquids in open channels	Gases	Liquids or gases	Liquids or gases	Liquids or gases	Liquids sonically conductive	Liquids sonically conductive	Liquids or gases
Flow Limits Liquids	0.1 cc/min to 150 gpm	0.1 cc/min to any desired max. flow	1–16,000 gpm	0.1–100,000 gpm: 0.3 fps	50 gpm to any desired maximum flow	–	10–1,400 gpm	2,000–300,000 pph	0.1–3,000 gpm liq.	0.1–100,000 gpm: 0.1 fps	1–100,000 gpm: 0.1 fps	1–1,000 gpm
Gases as air equiv. metered at STP	50 cc/min to 600 scfm	50 cc/min to any desired max. flow	–	–	–	1–2,000 actual ft/min	–	250–60,000 pph	–	–	–	8–8,000 scfh
Scale or Signal Characteristic	Linear or logarithmic	Square root	Linear	Linear	3/2 power or 5/2 power for V-notch weir	Linear	Linear	Linear	Linear	Linear	Linear	Linear
Flow Rangeability Without adjusting the span	10:1	4:1	10:1 to 15:1	10:1	20:1	10:1 to 25:1	8:1 to 15:1	10:1	10:1	25:1	25:1	10:1
With adjusting the span, max. usable minutes	No adjustment	12:1 (with same flow element)	15:1	100:1	100:1	25:1	15:1	–	–	100:1	–	–
Accuracy including dead band, hysteresis, nonlinearity	±2% of max. (uncalibrated) to ±1/2% of rate (calibrated)	±1% of max. (uncalibrated) including transmitter	±1/2% of rate (±1/4% of rate excluding nonlinearity)	±1/2% of rate	Approximately 5% of rate	1% of rate	±1% to ±2% of rate	±1% Rate	±1/2% Rate	±1% Rate	3.5% with good piping	±1% max.
Factory Flow Calibrated	Optional	Not for orifice (optional for other elements)	Always	Always	No	Always	Always	Always	Always	Usually	No	Always
Display Only (Copowered)	Yes, glass or metal tube	Yes	No	No	Yes	No	No	No	Yes totalization	No	No	No
Transmission Types	Analog pneumatic/ Analog electrical	Analog pneumatic/ Analog electrical	Analog electrical/ Digital pulse rate	Analog electrical/ Digital pulse rate	Analog pneumatic/ Analog electrical	Analog electrical/ Digital pulse rate	Analog electrical/ Digital pulse rate	Analog electrical/ Digital pulse rate	Analog electrical/ Digital pulse rate	Analog electrical/ Digital pulse rate	Analog electrical/ Digital pulse rate	Analog electrical/ Digital pulse rate

Electrical Classification	Available as intrinsically safe	Available either as Class 1, Group C & D, Div. I or intrinsically safe	Class 1, Group C & D, Div. I, Class II Group B, C & D, Div. II	Class 1, Group B, C & D, Div. II or Class 1 Group C and D, Div. I with purge	Available as intrinsically safe	Class 1, Group C & D, Div. I	Class 1, Group C & D, Div. I or intrinsically safe	Class 1, Group C & D, Div. I	Class 1, Group C & D, Div. I	Class 1, Group C & D, Div. II	Class 1, Group C & D, Div. II	Class 1, Group D, Div. I
Relative Installed Cost	Local display low; transmitters 2 low 2 medium	Local display medium; transmitters low to medium	Medium	Medium to high	Low to medium	Medium to high (with P&T compensation high)	Low to medium	High	Medium-high	Medium-high	Low	Medium
Viscosity Effect	Good immunity to viscosity, except in very small meters	Fairly good immunity for orifice plate; less immunity for venturi or flow tube	Fairly good immunity	No effect of viscosity on accuracy	Normally not used on viscous liquids	Good immunity to viscosity of glasses	Fair immunity to viscosity. Viscosity can increase the minimum flow	Immune to viscosity	Fair immunity to viscosity	Good immunity on large sizes	Good immunity on large sizes	Viscosity sets lower flow limit
Max. Overall Pressure Loss, in. H_2O (typically)	3–30	10–200	30–120	Negligible if meter is line size	3–15 for Parshall flumes, 10–40 for weirs	10–500	3–150	110–138	140–300 Liquid	Negligible	Negligible	120
Other Considerations	Materials of float fittings Float-bounce limits for gas	Matls. of element and DP sensor freezing Gases, Δp max. must be less than 2% of absolute pressure	Matls. of meter and bearings Pulse voltage Type bearings normally ceramic	Matls. of liner and electrodes Vacuum possibility of damage to Teflon®[a] liners	Matls. of float Freezing possibility Sufficient gradient for free flow conditions	Matls. of meter and O-Ring (316 SS only) Pressure and temperature compensation	Materials of meter (316 SS only) No slurries	Clean fluids only; motor driven parts may require maintenance program	Relatively clean fluids Mech. wear requires periodical proving	Relatively clean liquids Sensitive to low Reynolds numbers; flow profile sensitive	Sensitive to low Reynolds No. and internal surface conditions	Clean low viscosity fluid required; horizontal installation recommended
Effect of Installation Piping	Negligible	Per ASME Stds.	Per ISA RP31-1	Negligible	Straight upstreat conduit required	Straight piping required	Straight piping required	Not critical	Negligible	Critical for single beam; negligible for 4 beam	Critical	Straight piping required
Use on Dirty Service	Fair	Fair	Good	Excellent	Excellent	—	—	—	Fair	Excellent	Excellent	Good
Use for Liquids Containing Vapors	Good	Excellent	Good	Excellent	Excellent	—	Excellent	Excellent	Fair	Fair	Fair	Good
Use for Vapors Containing Condensate	Good	Good-Poor	Excellent	—	—	Excellent	Good	Good	Poor	Not Applicable	Not Applicable	Good
Ease of Changing Capacity	Excellent	Excellent	Good	Excellent	Excellent	Fair	Fair	Fair	Excellent	Good	Good	Excellent

[a]Registered trademark of E.I. du Pont de Nemours and Company, Inc., Wilmington, Delaware.

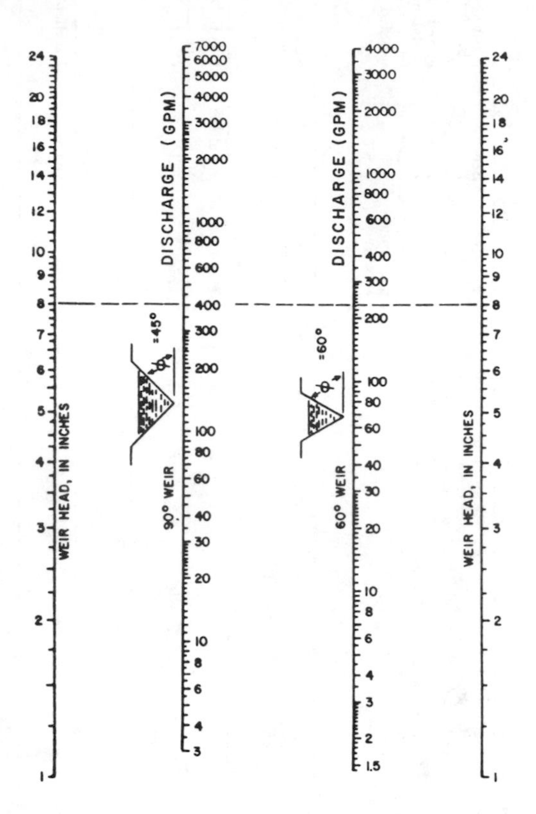

Figure 24. Nomograph for estimating flowrates over 60° and 90° triangular weirs [41].

may be used for large flows only. Discharges with very low heads, which can cause the nappe of the fluid to cling to the weir face, should not be estimated from this nomograph.

For measuring low flows (less than 0.03 m^3/s), a V-notched weir is preferred. A nomograph for estimating flowrates over 60° and 90° triangular weirs is given in Figure 24 [41].

The formulas used for the nomographs assume the fluid velocity of approach is small. When this is not the case, a correction factor, $V^2/2g$, should be added to the head term, where V is the approach velocity. Note also that if the crest height is too high, the flow may occur over the full width of the weir and problems may develop when a vacuum forms under the nappe. For rectangular weirs, the crest height should not exceed 5H. For triangular weirs, Figure 25 provides a rough estimate of minimum discharge without forming a vacuum for heads lower than 0.1 m.

Measurement with Flumes

Parshall flumes represent the most common design for measuring flows in sewers. The basic configuration consists of three parts: a converging section, a throat section and a diverging section. Dimensions and capacities of Parshall flumes are given in Figure 26. Flow measurement is based on the fluid surface in the converging section. The flow level normally is measured back from the crest of the flume at a distance of two-thirds the length of the converging section. Measurements always should be made in a stilling well instead of the flume itself, because sudden flow changes are dampened in stilling wells. Figure 27 estimates flow capacities for Parshall flumes.

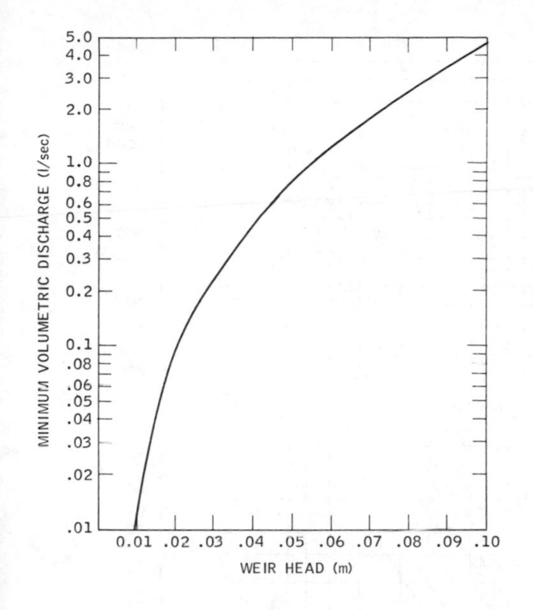

Figure 25. Practical minimum discharge for 90° V-notched weirs [41].

A Palmer-Bowlus flume is an even simpler design, which, for sewer flows, may simply consist of a level section of floor placed into a sewer. The length of the floor is generally sized at the same diameter as the conduit. Figure 28 shows various configurations of Palmer-Bowlus flumes. Flow through a Palmer-Bowlus flume is given by the following formulas (in English units):

$$\frac{Q^2}{g} = \frac{A_c^{\,3}}{b} \tag{75}$$

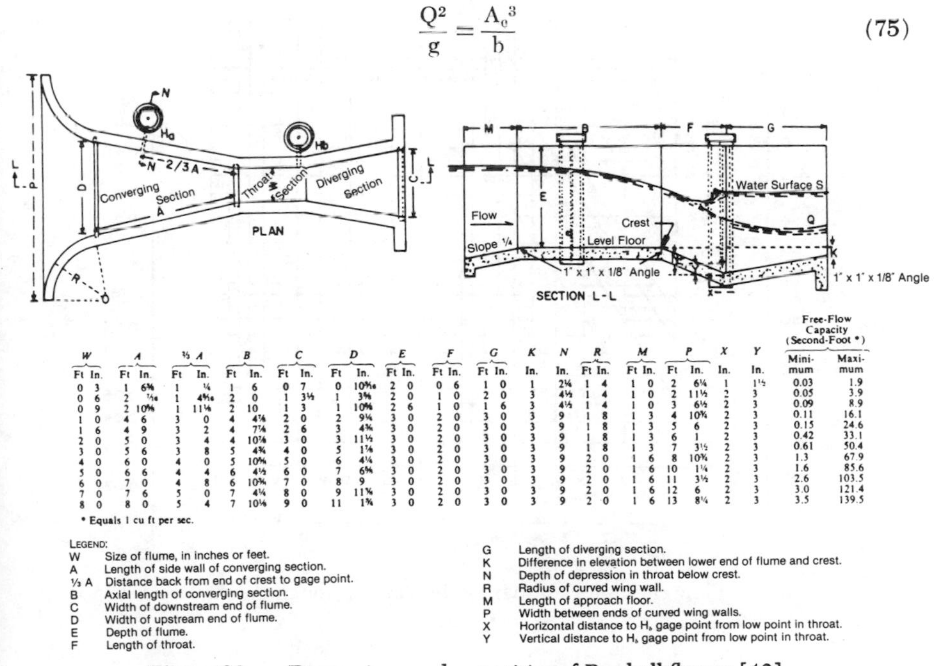

W Ft	W In.	A Ft	A In.	⅔ A Ft	⅔ A In.	B Ft	B In.	C Ft	C In.	D Ft	D In.	E Ft	E In.	F Ft	F In.	G Ft	G In.	K In.	N In.	R Ft	R In.	M Ft	M In.	P Ft	P In.	X In.	Y In.	Free-Flow Capacity (Second-Foot *) Minimum	Maximum
0	3	1	6⅜	1	¼	1	6	0	7	0	10³⁄₁₆	2	0	0	6	1	0	1	2¼	1	4	1	0	2	6¼	1	1½	0.03	1.9
0	6	2	⁷⁄₁₆	1	4⁵⁄₁₆	2	0	1	3½	1	3⅝	2	0	1	0	2	0	3	4½	1	4	1	0	2	11½	2	3	0.05	3.9
0	9	2	10⅝	1	11⅛	2	10	1	3	1	10⅝	2	6	1	0	1	6	3	4½	1	4	1	0	3	6½	2	3	0.09	8.9
1	0	4	6	3	0	4	4⅞	2	0	2	9¼	3	0	2	0	3	0	3	9	1	8	1	3	4	10¾	2	3	0.11	16.1
1	6	4	9	3	2	4	7⅞	2	6	3	4⅜	3	0	2	0	3	0	3	9	1	8	1	3	5	6	2	3	0.15	24.6
2	0	5	0	3	4	4	10⅞	3	0	3	11½	3	0	2	0	3	0	3	9	1	8	1	3	6	1	2	3	0.42	33.1
3	0	5	6	3	8	5	4¾	4	0	5	1⅞	3	0	2	0	3	0	3	9	1	8	1	3	7	3½	2	3	0.61	50.4
4	0	6	0	4	0	5	10⅝	5	0	6	4¼	3	0	2	0	3	0	3	9	2	0	1	6	8	10¾	2	3	1.3	67.9
5	0	6	6	4	4	6	4½	6	0	7	6⅝	3	0	2	0	3	0	3	9	2	0	1	6	10	1¼	2	3	1.6	85.6
6	0	7	0	4	8	6	10⅜	7	0	8	9	3	0	2	0	3	0	3	9	2	0	1	6	11	3½	2	3	2.6	103.5
7	0	7	6	5	0	7	4¼	8	0	9	11⅜	3	0	2	0	3	0	3	9	2	0	1	6	12	6	2	3	3.0	121.4
8	0	8	0	5	4	7	10⅛	9	0	11	1¾	3	0	2	0	3	0	3	9	2	0	1	6	13	8¼	2	3	3.5	139.5

* Equals 1 cu ft per sec.

LEGEND:

W Size of flume, in inches or feet.
A Length of side wall of converging section.
⅓ A Distance back from end of crest to gage point.
B Axial length of converging section.
C Width of downstream end of flume.
D Width of upstream end of flume.
E Depth of flume.
F Length of throat.
G Length of diverging section.
K Difference in elevation between lower end of flume and crest.
N Depth of depression in throat below crest.
R Radius of curved wing wall.
M Length of approach floor.
P Width between ends of curved wing walls.
X Horizontal distance to H_b gage point from low point in throat.
Y Vertical distance to H_b gage point from low point in throat.

Figure 26. Dimensions and capacities of Parshall flumes [42].

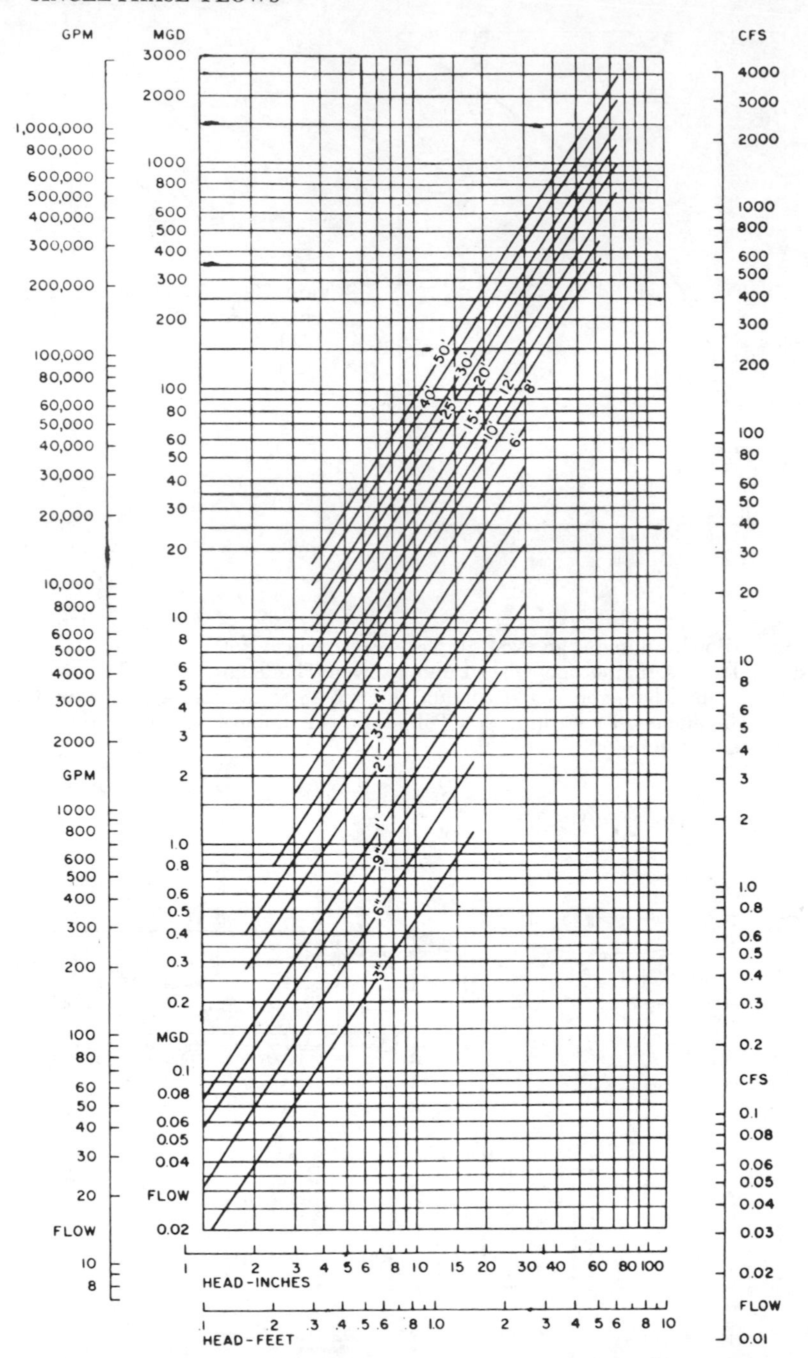

Figure 27. Flow curves for estimating discharge through Parshall flumes [43].

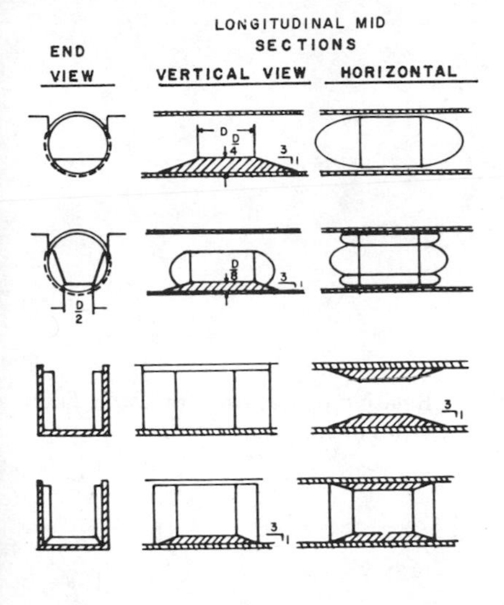

Figure 28. Various configurations of Palmer-Bowlus flumes.

and

$$\frac{U_c^2}{2g} = \frac{A_c}{2b} = \frac{h_c}{2} \tag{76}$$

where A_c = area of critical depth
h_c = critical depth
U_c = critical velocity
b = width of flume

The accuracy of flow measurements with these flumes is comparable to that obtained with a Parshall flume.

Figure 29 provides a summary and approximate rating of the flow measuring systems described in the form of a plot of flow versus head losses. These are approximate head losses for water flows in weirs and flumes.

Estimating Sewer Flows from the Manning Formula

Often, particularly in older plants, slop lines and/or storm drains are not equipped with flowmeters or suitable measuring devices. It occasionally becomes necessary to estimate these flows either for control purposes or to determine an operating range for a flow measuring system to be selected. Order of magnitude estimates of flows can be made by obtaining a measurement of the average depth in the line and using the Manning formula to compute the mean velocity. The flowrate then may be obtained from the continuity equation. This approach has the disadvantage that an estimate of the coefficient of roughness and the slope of the energy grade line are needed.

As an approximation, the energy grade line for full pipe flow can be assumed, and Figure 30 can be used to obtain an estimate of the average flow velocity for a partially filled pipe. Figure 30 relates the ratio of average flow velocity and superficial velocity to dimensionless depth (i.e., ratio of depth to pipe diameter). Also, given in this figure is the relationship for the ratio of volumetric and superficial volumetric flow. Liquid holdup may be obtained from Figure 9.

This approach is only useful for an order of magnitude estimate of the flow. It does, however, provide useful information on ranges of flow, which can help size more accurate instruments for measurement.

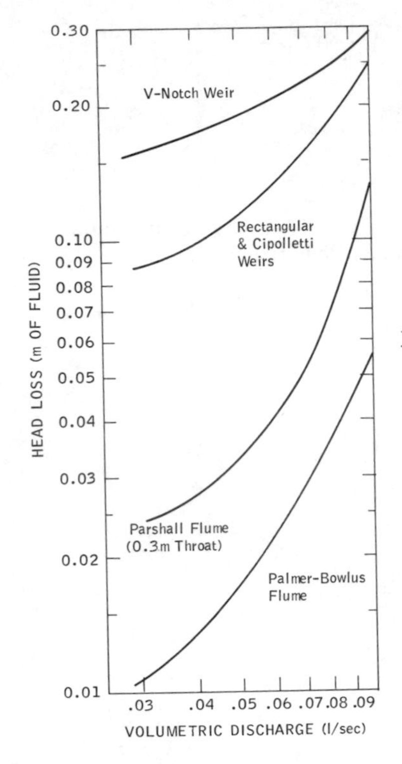

Figure 29. Relative head losses for water flows in different types of weirs and flumes.

CONCLUSIONS

Although the approximations outlined in the latter sections are of little value in analyzing the complexity of partial pipe and open-channel flows, they represent quick techniques for the field engineer in specifying operating ranges and measurement instrumentation. For design purposes and for understanding the complexity of the flow, the most powerful tools presently available are computer codes for the solution of two-dimensional transient fluid flow with free boundaries. The concept of a fractional volume of fluid developed by the Los Alamos Laboratories appears the most straightforward approach for tracking free boundaries for flows in complex channel configurations. These techniques have proven highly useful in structural dynamics for many years. Although it has been customary to employ Lagrangian coordinates as the basis for numerical solution algorithms, Eulerian coordinates appear more suited for fluid dynamics problems. Further effort is needed, however, in extending these techniques to multicomponent fluid-fluid and fluid-solid flows in open channels. Eulerian approximations have the drawback of smoothing variations in flow quantities and, in some treatments, information is lost on the interfaces of discontinuities.

From an operational control standpoint, considerable theory has been applied successfully to handling transients in piping systems to establish and maintain steady motion, for example, when a valve motion ceases. A similar analogy is needed for open-channel flows, where it is necessary to establish an optimum adjustment of levels either at the inlet or discharge to maintain steady or even uniform flow. Because celerity varies both with flow velocity and depth, the problem is more complex than the full pipe flow case. The computer codes mentioned could have immense application in this area. For example, one simple control scheme for controlling rates at a pumping

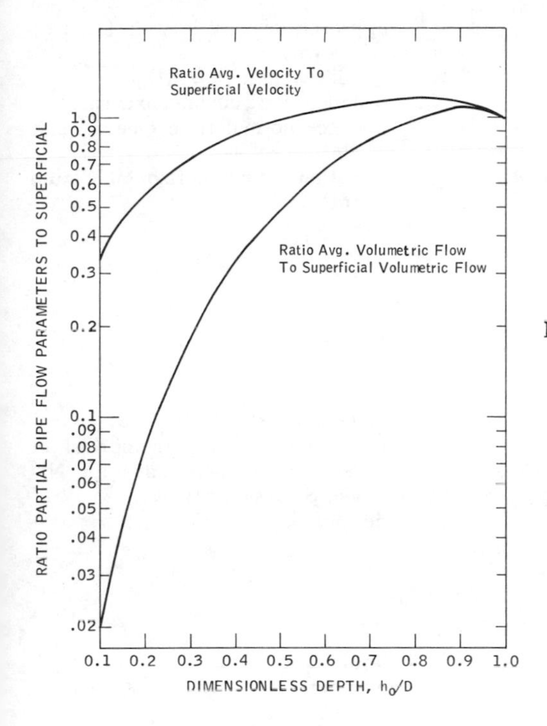

Figure 30. Ratios of average velocity and volumetric flow to superficial conditions [44].

station from a reservoir would be to generate a hydraulic bore by rapidly opening an upstream gate. This action would cause the bore to move downstream of the canal and, on arrival at the pumping station, the pumps could be put into operation. The codes could be used either to check the operation or to actually control it by specifying the rate at which the gate should be opened or computing the time it takes for the bore to reach the pumping station. If a combination of gates is used, both at the inlet and channel discharge, a number of control criteria may be devised and optimized by such codes.

NOMENCLATURE

- A cross section (m^2)
- b channel width (m)
- B function defined in Equation 52, dimensionless
- c wave speed (m/sec)
- C Chezy discharge constant, Equation 5
- C′ flow friction coefficient
- C_w nonuniformity flow coefficient
- d function defined in Equation 52 ($m^{8/3}sec^{-2}$)
- D diameter (m)
- E total energy (m)
- Fr Froude number, dimensionless
- $g_{x,y}$ gravitational acceleration term in x and y directions (m/sec^2)
- h depth (m)
- h_D hydraulic depth (m)
- H total head (m)
- j slope of total energy line, refer to Equation 50 ($m^{8/3}sec^{-2}$)
- K′ conveyance factor ($m^{5/3}$)
- l length of hydraulic jump (m)
- L channel length (m)
- p perimeter (m)
- P pressure (n/m^2)
- q volumetric flow per unit width (m^2/sec)

Q volumetric flowrate (m^3/sec)
R radius (m)
$R_{L,G}$ liquid- and gas-phase fractional holdup
S slope of free surface (m/m)
t time (sec)
u local velocity or velocity component in x direction (m/sec)
U average velocity (m/sec)
v velocity component in y direction (m/sec)
V approach velocity to weir (m/sec)
w width (m)
W specific weight (kg/m^3)
X distance along channel axis (m)
y distance normal from channel axis (m)
Z elevation above horizontal datum (m)

Greek Symbols

α kinetic energy correction term
γ angle (°)
δ small difference or increment
ϵ roughness coefficient (mm or in.)
θ angle (°)
ν kinematic viscosity (m/sec)
ξ parameter denoting cylindrical or rectangular coordinates for Navier-Stokes equations
ρ density (kg/m^3)

Subscripts

C critical condition
D hydraulic
E equivalent
f full pipe flow
G gas
L liquid
max maximum
MC mean critical
1/2 half flow condition

REFERENCES

1. Prandtl, L. *Proc. Int. Cong. Appl. Mech.* 2nd Cong., Zurich (1927), p. 62.
2. Nikuradse, T. *Ing. Arch.* 1:306 (1930).
3. Sellin, R. H. *Flow in Channels* (London, England: MacMillan St. Martins Press, 1969).
4. Cheremisinoff, N. P., and E. J. Davis. "Stratified Turbulent-Turbulent Gas-Liquid Flow," *AIChE J.* 25(no. 1):48–56 (1979).
5. Bakhmeteff, B. A. *Hydraulics of Open Channels,* Engineering Societies Monographs (New York: McGraw-Hill Book Co., 1932).
6. Henderson, F. M. *Open Channel Flows* (London, England: Collier Publishers, 1966).
7. Woodward, S. M., and C. J. Posey. *Hydraulics of Steady Flow in Open Channels* (New York: John Wiley & Sons, Inc., 1941).
8. King, H. W., and E. F. Brater. *Handbook of Hydraulics* 5th ed. (New York: McGraw-Hill Book Co., 1963).
9. Chow, V. T. *Open Channel Hydraulics* (New York: McGraw-Hill Book Co., 1959).
10. Kuiper, E. *Water Resources Development* (London, England: Butterworth Publishers, Inc., 1965).
11. Morris, H. M. *Applied Hydraulics in Engineering* (New York: Ronald Press, 1963).
12. Streeter, V. L., and E. B. Wylie. *Hydraulic Transients* (New York: McGraw-Hill Book Co., 1967).
13. Streeter, V. L. "Waterhammer Analysis with Nonlinear Frictional Resistance," Proc. 1st Aust. Conf. Hydraulics Fluid Mech. (Elmsford, NY: Pergamon Press, Inc., 1963).

14. Streeter, V. L., W. F. Keitzer and D. F. Bohr. "Energy Dissipation in Pulsatile Flow Through Distensible Tapered Vessels," in *Pulsatile Blood Flow*, E. O. Attinger, Ed. (New York: McGraw-Hill Book Co., 1964).
15. Streeter, V. L., W. F. Keitzer and D. F. Bohr. "Pulsatile Pressure and Flow through Distensible Vessels," *Circ. Res.* 13:3–20 (1963).
16. Streeter, V. L., and C. Lai. "Waterhammer Analysis Including Fluid Friction," *J. Hydraul. Div.*, ASCE 88(Hy3):79–112 (1962).
17. Streeter, V. L., and E. B. Wylie. "Resonance in Governed Hydro Piping Systems," *Proc. Int. Symp. Waterhammer Pumped Storage Projects*, ASME, Chicago, IL (1966).
18. Churchill, R. V. *Fourier Series and Boundary Value Problems*, 2nd ed., (New York: McGraw-Hill Book Co., 1963).
19. Lister, M. "The Numerical Solutions of Hyperbolic Partial Differential Equations by the Method of Characteristics," in *Mathematical Methods for Digital Computers*, A. Ralston and H. S. Wilf, Eds. (New York: John Wiley & Sons, Inc., 1960).
20. Marchal, M., G. Flesch and P. Suter. "The Calculation of Waterhammer Problems by Means of the Digital Computer," *Proc. Int. Symp. Waterhammer Pumped Storage Projects*, ASME, Chicago, IL (1966).
21. Paynter, H. M. "Studies in Unsteady Flow," *J. Boston Soc. Civil Eng.* 39:120–165 (1952).
22. Stoker, J. J. "Numerical Solution of Flood Prediction and River Regulation," New York University Institute of Mathematical Sciences, IMM-NYU 200 (1953).
23. Harlow, F. H., and J. E. Welch. "Numerical Calculation of Time-Dependent Viscous Incompressible Flow," *Phys. Fluids* 8:2182 (1965).
24. Welch, J. E., F. H. Harlow, J. P. Shannon and B. J. Daly. "The MAC Method: A Computing Technique for Solving Viscous, Incompressible, Transient Fluid-Flow Problems Involving Free Surfaces," Los Alamos Scientific Laboratory report LA-3425 (1966).
25. Hirt, C. W., B. D. Nichols and N. C. Romero. "SOLA-A Numerical Solution Algorithm for Transient Fluid Flows," Los Alamos Scientific Laboratory report LA-5852 (1975).
26. Nichols, B. D., C. W. Hirt and R. S. Hotchkiss. "SOLA-VOF: A Solution Algorithm for Transient Fluid Flow with Multiple Free Boundaries," Los Alamos Scientific Laboratory reports LA-8355, VC-32 and VC-34 (1980).
27. McMaster, W. H., and E. Y. Gong. "PELE-IC User's Manual," Lawrence Livermore Laboratory report UCRL-52609 (1979).
28. McMaster, W. H., D. F. Quinones, C. S. Landram, D. M. Norris, E. Y. Gong, N. A. Machen and R. E. Nickell. "Applications of the Coupled Fluid-Structure Code PELE-IC to Pressure Suppression Analysis—Annual Report to NRC for 1979," NURER/CR-1179, UCRL-52733 (1980).
29. Nichols, B. D., and C. W. Hirt. "Numerical Simulation of BWR Vent-Clearing Hydrodynamics," *Nucl. Sci. Eng.* 73:196 (1980).
30. Anderson, W. G., P. W. Huber and A. A. Sonin. "Small Seals Modeling of Hydrodynamic Forces in Presssure Suppression Systems," final report, Department of Mechanical Engineering, Massachusetts Institute of Technology, Nuclear Regulatory Commission report NUREG/CR-0003 (1978).
31. Lamb, H. *Hydrodynamics* (New York: Dover Publications, Inc., 1945).
32. Stoker, J. J. *Water Waves* (New York: Interscience Publishers, 1957).
33. Hirt, C. W. "Heuristic Stability Theory for Finite-Difference Equations," *J. Computer Phys.* 2:339 (1968).
34. Noh, W. F., and P. Woodward. "The SLIC (Simple Line Interface Calculation) Method," Lawrence Livermore Laboratory report UCRL-5211 (1976).
35. Kershner, J. D., and C. L. Mader. "2DE: A Two-Dimensional Continuous Eulerian Hydrodynamic Code for Computing Multicomponent Reactive Hydrodynamic Problems," Los Alamos Scientific Laboratory report LA-4846 (1972).

36. Nichols, B. D., and C. W. Hirt. "Methods for Calculating Multi-Dimensional, Transient Free Surface Flows Past Bodies," *Proc. 1st Int. Conf. Num. Ship Hydrodynamics,* Gaithersburg, MD (1976).
37. Chan, R. K. C., and R. L. Street. "SUMMAC—A Numerical Model for Water Waves," Stanford University, Department of Civil Engineering, report 135 (1970).
38. Cheremisinoff, N. P. *Applied Fluid Flow Measurement: Fundamentals and Technology* (New York: Marcel Dekker, Inc., 1979).
39. Cheremisinoff, N. P. *Fluid Flow: Pumps, Pipes and Channels* (Ann Arbor, MI: Ann Arbor Science Publishers, Inc., 1981).
40. Cheremisinoff, N. P. *Process Level Instrumentation and Control* (New York: Marcel Dekker, Inc., 1981).
41. "Manual of Disposal of Refinery Wastes," *Volume of Liquid Wastes* (Washington, DC: American Petroleum Institute, 1969).
42. "Planning and Making Industrial Waste Surveys," Ohio River Valley Water Sanitation Commission (April 1952).
43. *Handbook for Monitoring Industrial Wastewater* (Washington, DC: U.S. Environmental Protection Agency, 1973).
44. "Design and Construction of Sanitary and Storm Sewers," ASCE manuals and reports on engineering practice, No. 37, New York, *WPCF Manual of Practice No. 9* (1969).

SECTION II

GAS-LIQUID FLOWS

CONTENTS

CHAPTER 14
FUNDAMENTALS OF GAS-LIQUID FLOWS

Nicholas P. Cheremisinoff
Exxon Research & Engineering Company
Florham Park, New Jersey 07932

CONTENTS

INTRODUCTION

Two-phase gas-liquid flows have been studied extensively for many years, largely because of their immense frequency of occurrence throughout industry. Applications involving these flows range from straightforward transfer systems, such as pumping, to those involving heat and/or mass transfer, such as heat exchangers, thermosyphon reboilers, scrubbers, cooling towers, bubble columns and nuclear reactors. The intensity of research into this subject, largely aimed at providing theoretical and experimental information that will allow prediction of key parameters for design and operation, is witnessed by a volume of literature exceeding 12,000 published papers.

The immense volume of literature has often produced confusion for the designer because of arbitrary definitions and differentiation assigned condensing systems, evaporating systems, boiling and nonboiling systems, vertical and horizontal flow orientations, and their combinations. Although some of these distinctions are important, some are irrelevant. For example, horizontal and vertical two-phase annular and stratified flows do not necessarily require distinction when the interfacial shear and pressure drop forces are dominant.

The designer's key interests are in the following parameters: percentage liquid holdup, void fraction, frictional pressure losses, rates of heat and mass transfer, percentage liquid entrainment and tube burnout. These parameters are related to various controlled variables such as fluid properties, flowrates, and the purpose, size, configuration and orientation of equipment.

Two approaches have evolved for developing conservation laws of mass, momentum and energy for predicting these parameters. The one dominating design calculation is the traditional engineering approach, which is empirical or semiempirical. Although this approach has produced useful first-generation direct relations, there is little general applicability to design problems, as in contrast to pressure drop or heat transfer correlations of single-phase flows. In fact, for mass transfer rates this direct approach does not seem to establish correlations that can be safely extrapolated outside of the range of variables studied in experimental work. Furthermore, much of the gas-liquid flow information in the open literature is limited to air-water and steam-water systems. Few experimental validations of these correlations have been made based on systems whose physical properties differ greatly.

The second and more recent approach focuses on the development of more sophisticated models based on fundamentals. The fundamental approach starts with local instantaneous relationships and systematically develops time-averaged relations for parameters of interest. These types of treatments have contributed greatly to the basic understanding of the nature of two-phase flows but, unfortunately, have not been widely accepted in design practice.

The purpose of this chapter is twofold: (1) to introduce the subject of two-phase gas-liquid flow, and (2) to acquaint the newcomer with the major parameters of industrial interest. To these ends, basic definitions and a review of some of the more successful correlations in two-phase flow are presented.

The intent of this series is to present the state-of-the-art of the fluid dynamics of complex flows. Without an understanding of the fluid dynamics of the system, it is not possible to rationally design for heat and mass transfer. In this chapter, then, we shall attempt to relate fluid mechanics conclusions to heat and mass transfer processes in a general way. Successive chapters in this section provide in-depth discussion and analysis of flow phenomena.

FLOW REGIMES AND INSTABILITY

The flow regimes encountered in two-phase flows and their prediction are discussed in detail in the next chapter. For an introduction, we shall only present definitions for regimes encountered in horizontal flows. Hewitt and Hall-Taylor [1] have listed the following six regimes for horizontal flows in a pipeline:

1. **Stratified Flow**—Gas and liquid phases are separated into cocurrently flowing phases, with the liquid flowing as a layer along the channel bottom.
2. **Wavy Flow**—This is stratified flow, in which flow instabilities generate a wavy gas-liquid interface, except at very low flowrates of the gas and liquid.
3. **Slug Flow**—Liquid waves tend to bridge the gap between the liquid surface and the channel top, causing the gas phase to move as slugs.
4. **Plug Flow**—Gas bubbles tend to agglomerate and nearly fill the cross section of the channel, moving as asymmetrical bullet-shaped entities.
5. **Bubbly Flow**—The gas tends to distribute as discrete bubbles in the continuous liquid phase, with bubbles rising toward the top of the channel.
6. **Annular Flow**—At very high gas flowrates, the liquid climbs the walls of the tube, forming a ring of nonuniform thickness around the central core of gas. The gas-liquid interface is highly irregular and waves tend to break off, giving rise to dispersed annular flow. At sufficiently high gas flowrates the flow becomes dispersed, during which liquid droplets are distributed in the continuous gas phase.

Figure 1A illustrates the irregular nature of the gas-liquid interface with oscillograph tracings of the wave structure obtained from conductivity probes for air-water flow.

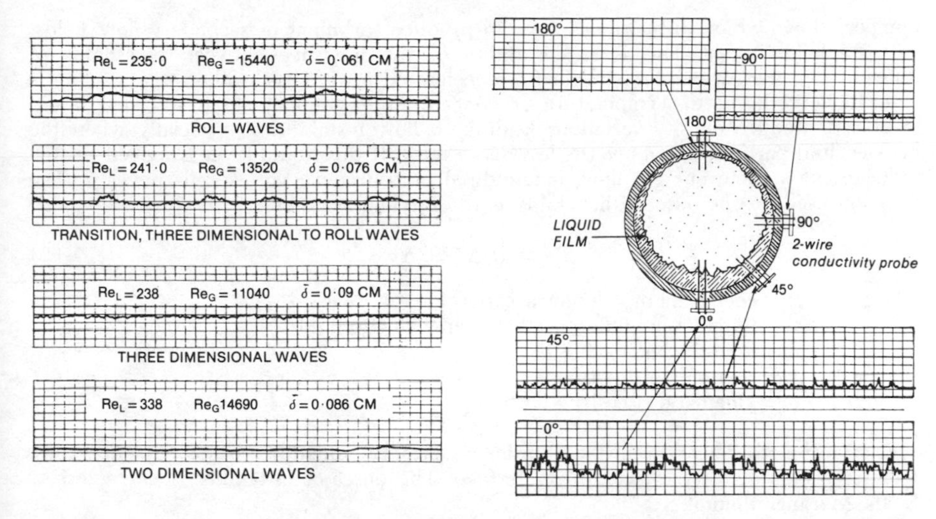

Figure 1. (A) Oscillograph tracings using a conductivity probe of stratified flow two-dimensional waves, three-dimensional waves and roll waves; (B) tracings of interface of horizontal annular flow.

Shown are smooth stratified flow (rarely encountered because of wave action), two-dimensional waves, squalls or three-dimensional waves, and roll waves. Figure 1B shows tracings of the interface for annular flow.

As shown from the tracings, two-dimensional waves are of relatively small amplitude and short wavelength. As the gas flowrate is increased, the two-dimensional wave pattern becomes unstable, giving rise to three-dimensional waves. The formation of this fine structure at the interface gives the appearance of a pebbled or sandy plane. At still higher gas flowrates, long-wavelength, large-amplitude roll waves form and sweep along the top of a base film of flowing liquid. Note in Figure 1 that the fine three-dimensional structure remains at the surface of the roll waves.

Standard flow maps only present broad, general characteristics of gas-liquid flows and do not represent important features of the fine structure. Details of this fine structure are important, however, for developing reliable heat and mass transfer correlations.

Analysis of interfacial instabilities for cocurrent gas-liquid flows were attempted by Ostrach and Koestel [2], Hanratty and Engen [3,4] and Davis [5]. There are five principal types of hydrodynamic instabilities that can lead to the interfacial disturbances described:

1. **Kelvin-Helmholtz.** Instabilities arise because of the interaction between two fluid phases for the case of different stratified heterogeneous fluids in relative motion. If the viscosities of the fluids are neglected, this type of instability is characterized by the Richardson number:

$$J = -\frac{g(dP/dz)}{\rho(dU/dz)^2} \tag{1}$$

where J is the ratio of buoyant force to inertia or shear force. When viscosities are considered, the analysis is considerably more complicated.

2. **Tollmein-Schlichting.** Here, flow undergoes a transition from laminar to turbulent, that is, infinitesimal disturbances become amplified by viscosity, thus changing the turbulence level. The onset of this instability is characterized by a critical Reynolds

number, above which disturbances grow to produce turbulent or secondary flow fields.

3. **Rayleigh-Taylor.** This is an instability of the interface between two fluids of different densities, which are stratified or accelerated toward each other.

4. **Rayleigh-Bernard.** Temperature or concentration gradients in a stratified liquid layer can produce density variations leading to flow instabilities, especially when the heavier fluid particles are above the less dense ones. If this liquid layer is moved by the influence of a cocurrent gas flow, longitudinal or transverse vortex rolls develop. The criterion used for the onset of instability is the Rayleigh number:

$$\mathrm{Ra} = g\beta\Delta T\delta^3/\alpha\nu \tag{2}$$

where β = coefficient of volumetric expansion
ΔT = temperature difference between wall and fluid
δ = fluid thickness
α = thermal diffusivity
ν = kinematic viscosity

5. **Marangoni.** This is a surface tension instability sometimes observed when mass transfer occurs across the gas-liquid interface. The onset of instability is characterized by the Marangoni number:

$$\mathrm{Ma} = -(\partial\sigma/\partial T)_o(\Delta T)\delta/\alpha\mu \tag{3}$$

DESIGN PARAMETERS OF INTEREST

From an engineering viewpoint, analysis of the instabilities of two-phase flows are useful only if they provide insight into data that allow prediction of parameters for design and scaleup. There are several key parameters of importance in the design and operation of equipment. Each is described briefly below:

1. **Mean Phase Content.** This is the fraction by volume or cross-sectional area of a particular flowing phase. For the gas phase, the term *void fraction* is used for the mean phase content. For the liquid phase, the term *holdup* is used. Phase content is important not only for describing heat and mass transfer characteristics, but also in computing inventories of materials from economic and safety standpoints. In nuclear reactor systems containing vapor and liquid mixtures, for example, the void fraction is important in determining the extent of neutron absorption and, hence, reactivity.

2. **Pressure Drop.** Information on mean phase content and pressure drop is required for design/scaleup of equipment, both from the point of view of the fluid dynamics of the system and for the prediction of heat and mass transfer characteristics. If a system is to be designed for a fixed pressure loss, the relationship between pressure drop and the flow velocities establishes the velocity-dependent parameters, such as the heat transfer coefficient and mass or heat flux limitations. For designs in which fixed flows are required, pressure drop determines the power input required for pumping.

3. **Heat Transfer Coefficient.** This parameter is the ratio of heat flux to some appropriate temperature differential. A local heat transfer coefficient is defined in terms of the temperature difference between the apparatus wall and the bulk fluid. An overall coefficient is defined in terms of the temperature difference of the two fluid streams.

4. **Mass Transfer Coefficient.** This parameter is the ratio of mass transfer rate to an appropriate concentration difference. It is analogous to the heat transfer coefficient.

5. **Flux Limitations.** Improper equipment design may result in poor system performance because of limitations in fluxes of mass and heat. Examples of mass flux limitations are critical flow or flooding in countercurrent operations, such as in a reflux condenser or an absorption tower. Heat flux limitations can occur in boiling, where exceeding the limiting heat flux may result in physical damage to equipment due to excessive increases in wall temperature.

Additional parameters of interest are related to stability and control of operations and potential mechanical problems with equipment. Instabilities result in fluctuations in flowrates, pressures and other operating parameters. Linked to stability are problems of transient behavior. Transient conditions almost always plague startup and shutdown and are of greatest concern during emergency operations. The most common transients are pressure surges and water hammer effects. To achieve the goal of proper and safe design of control systems, an understanding of the time-response behavior of multiphase systems is essential.

Finally, many design problems are strongly influenced by mechanical considerations, which are related to flow behavior. Flow instabilities, if not properly avoided or compensated for in design, can lead to poor service factors and equipment failures. Examples are tube vibrations in water-walled furnaces and heat exchangers, as well as excessive erosion and corrosion problems.

EMPIRICAL CORRELATIONS OF HOLDUP, VOID FRACTION AND PRESSURE DROP

The most widely referenced correlation for pressure drop and holdup is that of Lockhart and Martinelli [6]. With the exception of several improvements to this correlation over the years [3,7,10], the Lockhart-Martinelli correlation is still used extensively despite the poor predictions it provides for many flow configurations. The correlation employs information on single-phase flow to relate data in terms of the following two parameters:

$$\phi_{G \text{ or } L} = \left\{\frac{(dP_f/dz)}{(dP_f/dz)_{G \text{ or } L}}\right\}^{1/2} \quad (4)$$

$$X = \left\{\frac{(dP_f/dz)_L}{(dP_f/dz)_G}\right\}^{1/2} \quad (5)$$

$(dP_f/dz)_G$ and $(dP_f/dz)_L$ are the pressure gradients for the gas and liquid phases flowing alone in single-phase flow in the channel, respectively. The correlation distinguishes between laminar (viscous) single-phase flow and turbulent single-phase flow for each phase. The correlation is given in Figure 2. To illustrate the use of Figure 2, ϕ_{Gvt} refers to ϕ_G computed for laminar liquid-phase flow and a turbulent gas flow. Void fraction, R_G, also is shown correlated in Figure 2.

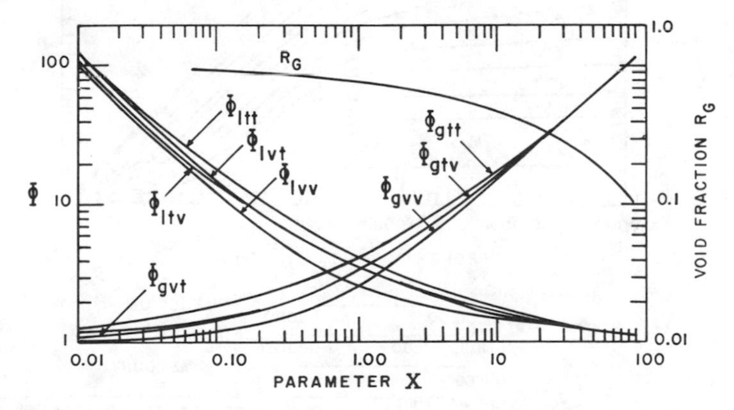

Figure 2. Lockhart-Martinelli correlation for frictional pressure drop and void fraction [6].

Several investigators have compared experimental data with the Lockhart-Martinelli correlation. Chenoweth and Martin [7] compared two-phase frictional pressure drop in 38-mm and 76-mm id. pipes at two different pressures and found disagreement as great as 60% for the larger pipe at 4.8 kPa. Aziz and Govier [11] compared the Lockhart-Martinelli correlation to one developed by Chenoweth and Martin [7] for annular-mist flow of natural gas-water mixtures. The comparison showed that although the Lockhart-Martinelli correlation only predicted pressure drops to within ±65%, it was better than the Chenoweth-Martin method. The correlation provides particularly poor predictions for stratified wavy flows. Overpredictions of pressure drop by more than 100% at high-liquid-volume fractions have been found for flows involving small-amplitude wave conditions.

The Lockhart-Martinelli correlation does not agree well with data at higher pressures. This is due in part to the exclusion of surface tension effects in the correlation. Casagrande [12] proposed a correlation that included surface tension for steam-water flows:

$$-\frac{dP_f}{dz} = \frac{0.43}{d_o^{1.2}}\left(\frac{\sigma}{73}\right)^{0.4}\left(\frac{\mu_L}{0.016}\right)^{0.04}\left(\frac{G^2}{\rho_H}\right)^{0.75} \tag{6}$$

where σ and μ are surface tension and viscosity, respectively, d_o is the pipe diameter, and G and ρ_H are the two-phase mass flux and density, respectively. The units of Equation 6 are given in cgs.

Another correlation, based on considerable experimental data but rather complex in its usage, is that of Baroczy [10]. His correlation related a two-phase flow multiplier, ϕ_{Lo}, to a physical properties index $(\mu_L/\mu_G)^{0.2}/(\rho_L/\rho_G)$:

$$\phi_{Lo} = \left\{\frac{(dP_f/dz)}{(dP_f/dz)_{Lo}}\right\}^{1/2} \tag{7}$$

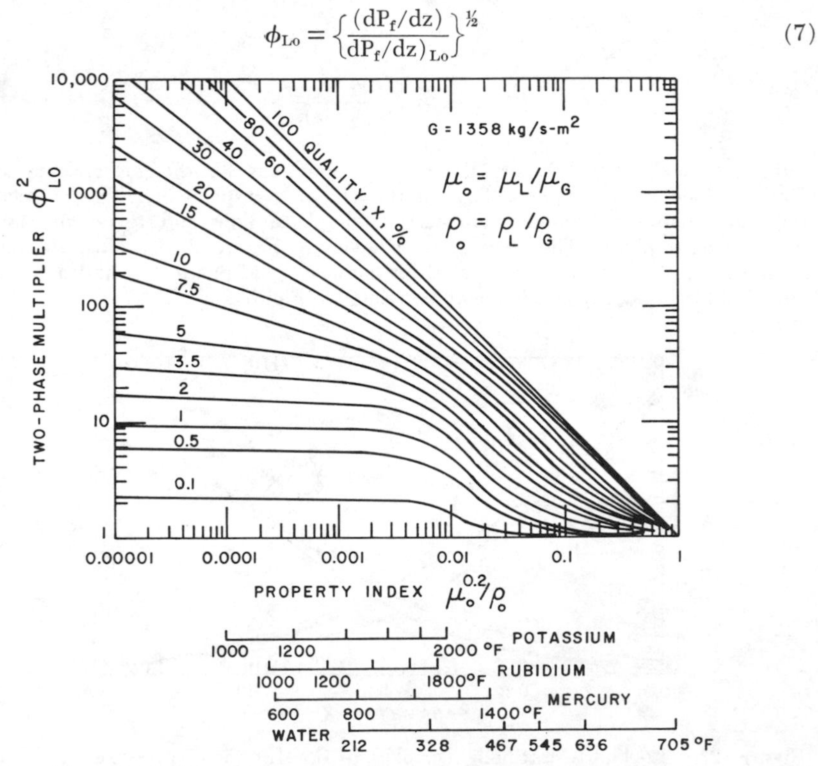

Figure 3. Two-phase frictional pressure drop correlation proposed by Baroczy [10].

The correlation is given in Figure 3 for a specific mass flux, G, of 1358 kg/sec-m². For mass flux conditions different from that of the reference value, a correlation factor, Ω, given in Figure 4 is used to correct ϕ_{Lo}. Note that the pressure drop, $(dp_f/dz)_{Lo}$, is computed for the total flow based on liquid-phase physical properties.

Hughmark [13] and Hoogendoorn [14] proposed correlations for holdup, and Dukler et al. [15] have reviewed and compared them to that of Lockhart and Martinelli. Among the three correlations, Hughmark's was found to agree best with data obtained for the air-water system. Hughmark obtained the following expression relating the volume fraction of gas, R_G, to the Martinelli parameter, X, and a flow parameter, K:

$$\frac{1}{X} = 1 - \frac{\rho_L}{\rho_G}\left(1 - \frac{K}{R_G}\right) \tag{8}$$

The flow parameter, K, was correlated empirically with the Reynolds number, defined as

$$Re = DG(R_L\mu_L + R_G\mu_G) \tag{9}$$

the Froude number,

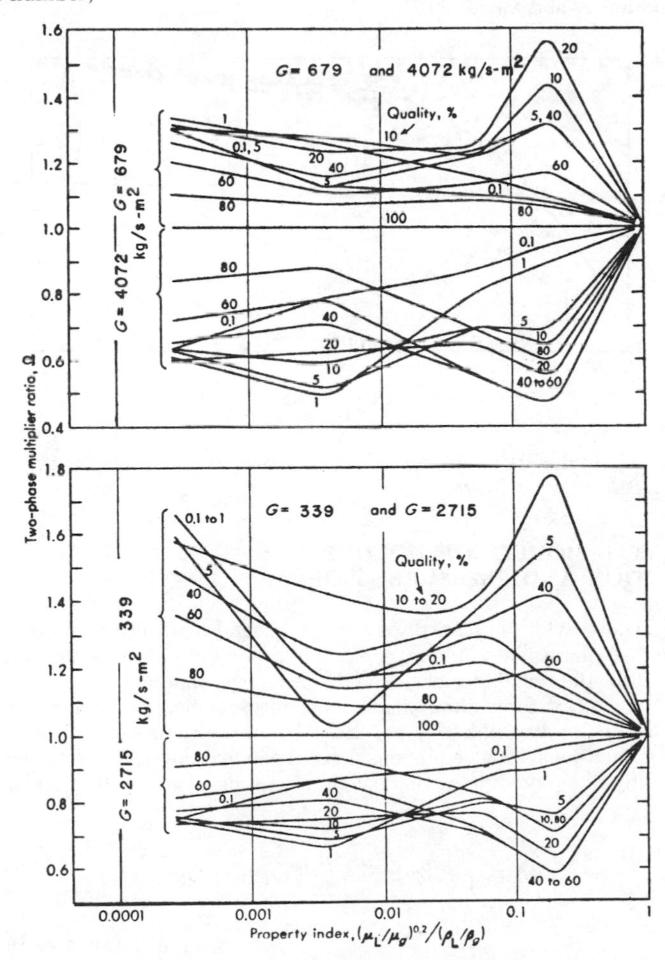

Figure 4. Correction factors for the Baroczy [10] pressure drop correlation.

$$\mathrm{Fr} = \mathrm{U}^2/\mathrm{gD} \tag{10}$$

and the liquid volume that would exist if there were no slip between the phases:

$$\tilde{y_L} = \frac{W_L V_L}{W_L V_L + W_G V_G} \tag{11}$$

The parameters W and V are the mass flowrate and specific volume of the appropriate phase, respectively. Note that U in the Froude number (Equation 10) is the velocity computed, assuming no slip between the phases. G is the total mass velocity. The graphic correlation for parameter K is given in Figure 5.

All the empirical correlations for pressure drop and holdup are intrinsically limited by the assumption of one-dimensional flow. They do not account for the actual velocity profiles or other fluid dynamic characteristics of the flow; hence, they cannot be expected to provide highly accurate predictions of pressure drop and holdup.

Further state-of-the-art reviews of pressure drop and holdup have been written by Delhaye [16], Anderson and Russell [17], Hewitt and Hall-Taylor [1], Hughes et al. [18] and Lahey and Moody [19].

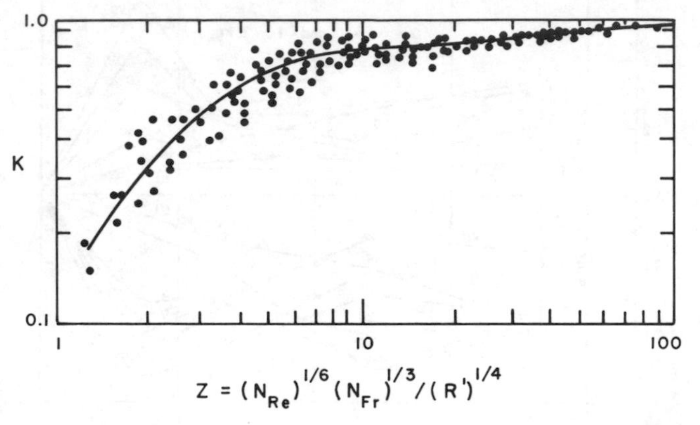

Figure 5. Hughmark's correlation [13] for flow parameter K to obtain holdup.

FUNDAMENTAL MODELS OF HOLDUP, VOID FRACTION AND PRESSURE DROP

Hewitt and Hall-Taylor [1] have discussed the general momentum and energy balance equations for two-phase flows, and part of their presentation is repeated here as background needed for the remainder of this and subsequent chapters.

Although the general three-dimensional continuity, momentum and energy equations apply in principle to two-phase flows, they have often proven intractable. Consequently, one-dimensional flow equations have been employed to describe two-phase flows. Extending the one-dimensional single-phase flow momentum balance to two phases,

$$\int_A \left[P - \left(P + \frac{dP}{dZ}\,\delta z \right) \right] dA = \int_S \tau_w \delta z ds + \int_A \frac{d}{dz}\left(G_L u_L + G_G u_G \right) dA + \int_A \rho g \sin\theta \,\delta z dA \tag{12}$$

For Equation 12, the flow is in the z direction, ρ is the local density weighted on an area basis, and G and u are the local mass flux and velocity, respectively. τ_w is the wall shear stress, and subscripts L and G refer to liquid and gas, respectively.

The energy balance for a unit mass of two-phase fluid flowing through area dA is

$$d[xPV_G + (1+x)PV_L] + [\delta q + \delta F - Pd(xV_G + (1-x)V_L)]$$

$$+ d\left[\frac{1}{2} \times u_G^2 + \frac{1}{2}(1-x)u_L^2\right] + g \sin\theta \, dz = \delta q - \delta W \qquad (13)$$

where
x = gas mass fraction
V = specific volume
δq = added head
δW = work performed
δF = amount of mechanical energy converted to internal energy

For the simple case of horizontal flow, the potential energy term can be dropped, since $\sin\theta = 0$.

Although Equations 12 and 13 adequately describe the system of cocurrent horizontal gas-liquid flow, they are not useful for computations. Most often, information on local mass fluxes, G_L and G_G, local velocities, u_L and u_G, and/or local density, ρ, are not known. Use of a time-averaged mean-phase content can simplify these expressions, where

$$G_L = \rho_L u_L (1 - R_G) \qquad (14a)$$

$$G_G = \rho_G u_G R_G \qquad (14b)$$

$$\rho = R_G \rho_G + (1 - R_G)\rho_L \qquad (14c)$$

where R_G is the time-averaged volume fraction of gas and applies to any point within the cross section of flow. Hence, three of the six variables—ρ, G_L, u_L, G_G, u_G and R_G—are eliminated.

Following Hewitt and Hall-Taylor [1], we shall develop the system equations for horizontal separated flows for the simple case of no entrainment. The following three assumptions may be applied to simplify the governing equations:

1. The liquid and gas phases are separated physically, with the gas phase occupying area $R_G A$ and the liquid phase area being $(1 - R_G)A = R_L A$.
2. Shear stress on the channel wall, τ_w, is independent of peripheral position.
3. Each phase is assumed to have plug flow, that is, density, mass velocity and velocity are constant over the entire region occupied by each phase.

The first assumption is reasonable for stratified flows and even annular, if entrainment rates are low. The second is questionable for flow through irregularly shaped and rectangular conduits [20]; however, it is considered a fair approximation for annular and stratified flows in circular pipes. The last assumption is highly questionable, as shown by Hanratty et al. [3] and Davis [20], but for the purpose of developing this simple model we shall stand by it for now.

The one-dimensional momentum balance (Equation 12) is integrated using these assumptions to give

$$-\frac{dP}{dZ} = \frac{S}{A}\tau_w + \frac{d}{dz}[R_G G_G u_G + (1 - R_G) G_L u_L] \qquad (15)$$

This equation describes horizontal flow in a tube, where S is the tube periphery.

Introducing x as the mass fraction of the gas phase, then the mass flowrate of liquid and the liquid-phase velocity is

$$G_L = G\frac{(1-x)}{(1-R_G)} \tag{16a}$$

$$u_L = G\frac{(1-x)}{\rho_L(1-R_G)} \tag{16b}$$

and the mass flowrate of the gas and the velocity is

$$G_G = G\frac{x}{R_G} \tag{17a}$$

$$u_G = \frac{Gx}{\rho_G R_G} \tag{17b}$$

Hence, the momentum equation becomes

$$-\frac{dP}{dz} = \frac{S}{A}\tau_w + G^2\frac{d}{dz}\left[\frac{x^2}{R_G\rho_G} + \frac{(1-x)^2}{\rho_L(1-R_G)}\right] \tag{18}$$

And with a similar analysis the energy equation becomes

$$-\frac{dP}{dz} = \underbrace{\rho_H\frac{dF}{dz}}_{-dP_F/dz} + \underbrace{\frac{1}{2}\rho_H G^2\frac{d}{dz}\left[\frac{x^3}{R_G\rho_G{}^2} + \frac{(1-x)^3}{(1-R_G)^2\rho_L{}^2}\right]}_{-dP_a/dz} \tag{19}$$

The term ρ_H is the homogeneous density of the two-phase mixture and is defined as

$$\frac{1}{\rho_H} = \frac{x}{\rho_G} + \frac{1-x}{\rho_L} \tag{20}$$

Note that Equation 19 gives the total pressure loss, which is the sum of the frictional pressure gradient, $-dP_F/dz$, and the acceleration pressure gradient, $-dP_a/dz$. The frictional pressure gradient is defined conventionally by the momentum equation:

$$-\frac{dP_F}{dZ} = \frac{S}{A}\tau_w \tag{21}$$

Equations 18 and 19 are useful from an instructive standpoint. It is clear, however, that considerable information is needed to calculate or estimate the acceleration pressure drop, even for the simple case of no entrainment. If experimental data are not available, the empirical correlations described earlier must be used to predict the frictional pressure drop, and Equation 21 can be used to compute wall shear.

Modeling Stratified Flows

The approach of assuming one-dimensional flow has obvious limitations, and discussions may be found in the text books of Hewitt and Hall-Taylor [1], Collier [21], Ishii [22] and Delhaye [16]. Fundamental mathematical models based on iterative or mechanistic approaches have evolved in lieu of this approach. These approaches are based on knowledge of the flow patterns. Three specific models describing the stratified flow case, and that readily lend themselves to design computations, are described below.

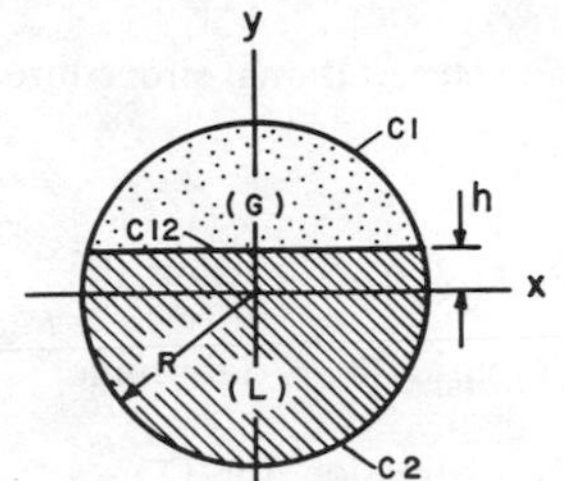

Figure 6. Schematic of flow cross section analyzed by Yu and Sparrow [23].

These models extend over the entire practical range of stratified flows, namely, laminar liquid/laminar gas, laminar liquid/turbulent gas and turbulent liquid/turbulent gas flows.

Laminar/Laminar Flow

Yu and Sparrow [23] treated laminar/laminar two-phase flows analytically and developed closed-form solutions for special limiting cases. Figure 6 defines the coordinates and system geometry for this model. Yu and Sparrow plotted their solutions of the equations of motion as Q_G/Q_G^* versus μ_G/μ_L and Q_L/Q_L^* versus μ_G/μ_L, where subscripts L and G refer to liquid and gas phases, respectively. The parameter Q_G^* is the flow volume passing through the part of the conduit cross section that lies above the line $y = h$ where the dust is filled with gas. Similarly, Q_L^* is the flow volume through the region below $y = h$ where thc liquid phase fills the tube. The following equations were developed in the analysis:

$$\frac{Q_G^*}{Q_{G,\,full}} = 1 - \frac{Q_L^*}{Q_{L,\,full}} = \frac{1}{2} - \frac{2}{3\pi}\left\{\frac{h}{R}\left(1 - \frac{h^2}{R^2}\right)^{3/2} + \frac{3}{2}\frac{h}{R}\left(1 - \frac{h^2}{R^2}\right)^{1/2} + \frac{3}{2}\sin^{-1}\frac{h}{R}\right\} \tag{22a}$$

and

$$Q_{L,\,G,\,full} = \frac{\pi R^4}{8\mu_{L,\,G}}\left(\frac{\Delta P}{L}\right) \tag{22b}$$

These results are shown plotted in Figure 7. Equation 22 and the plots given in Figure 7 form the basis of a design procedure for estimating pressure drop. For

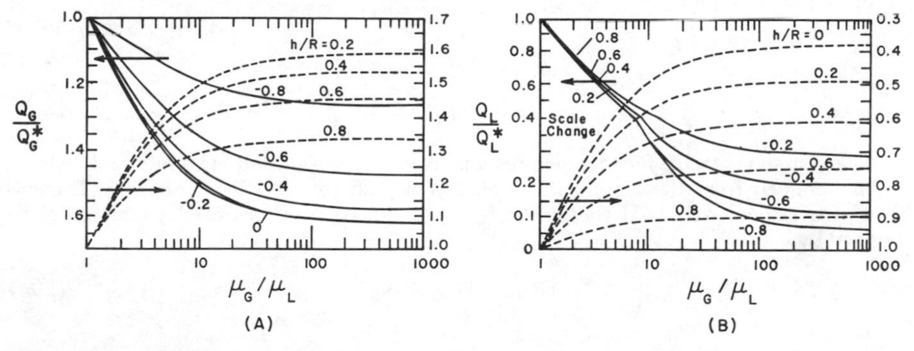

Figure 7. Results of Yu and Sparrow [23] for the flowrate of (A) the gas phase and (B) the liquid phase.

known viscosity ratio μ_G/μ_L and the volumetric flowrates, the computational procedure for estimating ΔP is as follows:

1. Assume a value for h/R.
2. Determine Q_G^* and Q_L^* from information on Q_G, Q_L and μ_G/μ_L using the plots given in Figure 7.
3. Using Equation 22a, compute $Q_G^*/Q_{G,full}$ and $Q_L^*/Q_{L,full}$. Then determine $Q_{G,full}$ and $Q_{L,full}$.
4. Compute $\Delta P/L$ from Equation 22b using the proper expression for $Q_{G,full}$ and $Q_{L,full}$.

If the two values of pressure drop computed in step 4 do not agree, a new value for h/R is chosen and the procedure repeated until convergence is achieved.

Laminar/Turbulent Flow

For laminar liquid/turbulent gas flow, Etchells [24] proposed estimating the frictional pressure drop from single-phase friction factors, f_L and f_G, and equivalent diameters, D_{EL} and D_{EG}.

$$(\Delta P/L)_{tp} = 2f_L\rho_L\bar{u}_L^2/D_{EL} \tag{23}$$

$$(\Delta P/L)_{tp} = 2f_G\rho_G\bar{u}_G^2/D_{EG} \tag{24}$$

where

$$D_{EL} = 4A_L/p_L \tag{25a}$$

$$D_{EG} = 4A_G/(p_G + W_i) \tag{25b}$$

The friction factors used in Equations 23 and 24 are the same as those for single-phase flow, but the appropriate Reynolds numbers are based on actual in-situ average velocities, $\bar{u}_L$ and $\bar{u}_G$, and the appropriate equivalent diameters.

Govier and Aziz [25] developed a similar procedure based on the following force balances:

$$A_G(\Delta P/L)_{tp} = \tau_{wG}p_G + \tau_iW_i \tag{26}$$

$$A_L(\Delta P/L)_{tp} = \tau_{wL}p_L - \tau_iW_i \tag{27}$$

with

$$\tau_{wG} = f_G\rho_G\bar{u}_G^2/2 \tag{28a}$$

$$\tau_{wL} = f_L\rho_L\bar{u}_L^2/2 \tag{28b}$$

The system geometry for these expressions is given in Figure 8. The interfacial shear, τ_i, is obtained from the empirical correlation of Ellis et al. [26]. Following Etchells [24], Govier and Aziz [25] considered f_L and f_G to be functions of Reynolds numbers defined by

$$Re_G = D_{EG}\bar{u}_G/\nu_G \tag{29a}$$

and

$$Re_L = D_{EL}\bar{u}_L/\nu_L \tag{29b}$$

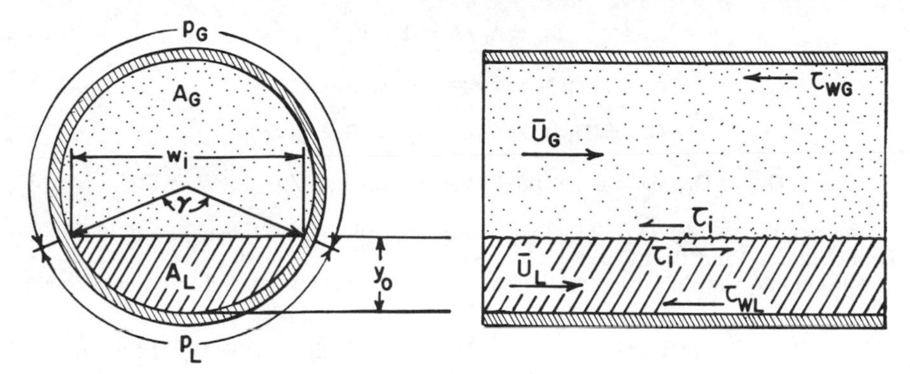

Figure 8. System coordinates and geometry for stratified flow models.

Govier and Aziz's model was modified further by Agrawal et al. [27] using an extended velocity profile to determine the mean value of the liquid flowrate. The mean value of the liquid flowrate used in the model is similar to that which would be "seen" by the pipe wall under single-phase flow conditions, with the same velocity profile in the portion of the pipe occupied by liquid. Agrawal et al. [27] applied this to laminar liquid/turbulent gas flow and to turbulent liquid/turbulent gas flow. In the latter case, they used the one-seventh power law for the turbulent liquid flow instead of the Hagen-Poiseuille equation for the laminar liquid layer.

A less empirical model of laminar liquid/turbulent gas stratified flow was developed by Russell et al. [28]. The gas phase was treated similar to that used by Agrawal et al. by using the Blasius friction factor:

$$f_G = 0.079 \, Re_G^{-0.25} \tag{30}$$

The gas-phase Reynolds number, Re_G, is defined in Equation 29a. The velocity distribution in the liquid phase was determined by an approximate solution of the Navier-Stokes equation. By integrating the velocity distribution, an expression relating the volumetric flowrate, Q_L, to pressure drop, fluid properties and the position of the interface in the pipe was obtained.

As a design procedure, each of the methods of Etchells [24], Govier and Aziz [25], Agrawal et al. [27] and Russell et al. [28] involves an iterative calculation procedure to determine the pressure drop and holdup from information on the gas and liquid flowrates and the fluid physical properties.

Because Russell's method was the most extensively tested, the computational procedure is outlined below:

1. From specified gas and liquid flowrates, physical properties and pipe diameter, and using an appropriate flow map (see the next chapter), determine whether the flow regime is stratified. Compute the superficial Reynolds numbers, $Re_L^* = DQ_L/A\nu_L$ and $Re_G^* = DQ_G/A\nu_G$. If $Re_L^* < 2100$ and $Re_G^* > 2100$, the analysis of Russell et al. [29] applies.

2. Assume y_o/D and determine D_{EL}, D_{EG}, w_i, p_L and p_G. These geometric definitions are defined by the system coordinates shown in Figure 8 and are the same formulations outlined in Chapter 13 of Section I for estimating liquid holdup in partial pipe flows. For convenience, these definitions are restated here:

$$y_o/D = (1 - \cos \gamma/2)/2 \tag{31}$$

$$\tilde{W}_i = W_i/D = 2[(y_o/D) - (y_o/D)^2]^{1/2} \tag{32}$$

$$\tilde{R}_L = A_L/A = (\gamma - \sin \gamma)/2\pi \tag{33}$$

$$R_G = A_G/A = 1 - R_L \tag{34}$$

$$\tilde{p}_L = p_L/p = \gamma/2\pi = 1 - p_G/p = 1 - \tilde{p}_G \tag{35}$$

$$\tilde{D}_{EL} = D_{EL}/D = 4A_L/p_L D = R_L/(p_L/p) \tag{36}$$

$$\tilde{D}_{EG} = D_{EG}/D = 4A_G/(p_G + W_i)D = R_G/[(p_G/p) + (W_i/\pi D)] \tag{37}$$

The parameters $\tilde{W}_i$, $\tilde{D}_{EL}$, $\tilde{D}_{EG}$, $\tilde{p}_G$ and $\tilde{p}_L$ are plotted as functions of the holdup, R_L, in Figures 9 and 10 of Chapter 13, in Section I.

3. Compute the dimensionless liquid volumetric flowrate, Q^*, from the following:

$$Q^* \equiv Q_L\mu_L/8R^4(\Delta P/L)_L = -\frac{C_B}{96}\Big[2\hat{\alpha}^3\beta - 15\hat{\alpha}\beta + (3 + 12\beta^2)\sin^{-1}\hat{\alpha} + \frac{H_G}{D}(6\hat{\alpha} - 2\hat{\alpha}^3 - 6\beta\sin^{-1}\hat{\alpha})\Big] \tag{38}$$

where $\hat{\alpha} = W_i/D$, $\beta = 1 - y_o/R$, $R_G = 4A_G/p_G$ and C_B is unity for $h/D \leq 0.1$. For larger values of h/D, refer to Russell et al. [28].

4. Determine $(\Delta P/L)_L$ from Q^* and the known values of Q_L, μ_L and R using the definition of Q^* in Equation 38.

5. Compute $(\Delta P/L)_G$ from the following:

$$\left(\frac{\Delta P}{L}\right)_G = 2f_G\bar{U}_G^2\rho_G/H_G \tag{39}$$

where $\bar{U}_G = Q_G/A_G$ and

$$f_G = 0.079\,Re_G^{-0.25} \tag{40a}$$

and

$$Re_G = H_G\bar{U}_G/\nu_G \tag{40b}$$

6. If $(\Delta P/L)_G$ is not equal to $(\Delta P/L)_L$, assume a new value of y_o/D and repeat the procedure until convergence is obtained.

Russell et al. [28] compared their model, the Lockhart-Martinelli correlation and the model of Agrawal et al. [27] with air-water and air-glycerine data in tubes of 25.4-mm, 38.1-mm and 50.8-mm I.D. For $y_o/D > 0.30$, the Lockhart-Martinelli and Agrawal et al. methods were found to underpredict the liquid volumetric flowrate, and the Agrawal et al. model is particularly poor for $y_o/D < 0.20$. The model of Russell et al. is in good agreement with the liquid volumetric flowrate and predicts the pressure drop and holdup considerably better than other correlations.

Turbulent/Turbulent Flow

Davis and Cheremisinoff [29,30] developed a model for predicting two-phase pressure drop and liquid holdup for stratified flows based on the theory of turbulent liquid film flow. As shown by Figure 1, under conditions of turbulent/turbulent flow the gas-liquid interface is often highly irregular. Hence, y_o in Equation 31 must be considered to be an average distance from the bottom of the tube to the interface on a vertical plane through the tube centerline.

Taking y as the distance from the tube wall in the direction of the inward normal to the wall, the position of the interface becomes

$$y_i/R = 1 - \cos(\gamma/2)(1 + \tan\theta)^{1/2} \tag{41a}$$

where

$$\theta = \cos^{-1}\left(\frac{R - y_o}{R - y_i}\right) \tag{41b}$$

For a turbulent liquid flow, the momentum flux or shear stress can be expressed by

$$\tau = \rho_L(\nu_L + \epsilon_M')\frac{du}{dy} \tag{42}$$

where the eddy viscosity, ϵ_M', is assumed to be a function of y. The shear stress distribution in the liquid can be expressed by a force balance on a liquid layer extending from $y = 0$ to any arbitrary position, $y = y$:

$$\tau = \frac{R\gamma}{2(R - y)\theta}\tau_{WL} - \frac{(Dy_o - y_o^2)^{1/2} - [(Dy_o - y_o^2) - (Dy_i - y_i^2)]^{1/2}}{(R - y)\theta}\tau_i - \frac{(\Delta P/L)}{4}\left[\frac{R^2(\gamma - \sin\gamma)}{(R - y)\theta} - (R - y)\left(2 - \frac{\sin 2\theta}{\theta}\right)\right] \tag{43}$$

The interfacial shear, τ_i, and the wall shear, τ_{WL}, are related by

$$\tau_i = \tau_{WL}\frac{R\gamma}{W_i} - \left(\frac{\Delta P}{L}\right)\frac{R^2}{2W_i}(\gamma - \sin\gamma) \tag{44}$$

If the holdup is small, i.e., γ is small, the approximation for thin films may be applied to give $\tau \simeq \tau_{WL}$.

This approximation was assumed valid over the entire range of values of the holdup. Assuming that the turbulence characteristics of the liquid phase are similar to those for single-phase flow, the friction velocity definition may be applied along with the usual dimensionless coordinate, y^+, and dimensionless velocity, u^+:

$$u^* = (\tau_{WL}/\rho_L)^{1/2} \tag{45}$$

$$y^+ = u^* y/\nu_L \tag{46}$$

$$u^+ = u/u^* \tag{47}$$

For the region near the wall ($0 \leq y^+ \leq 20$) Deissler's [31] expression for the eddy viscosity is

$$\epsilon_M' = n^2uy[1 - \exp(-n^2uy/\nu_L)] \tag{48}$$

where $n = 0.10$. For the turbulent core ($y^+ \geq 20$) von Karman's [32] equation is

$$\epsilon_M' = \chi^2\frac{(du/dy)^3}{(d^2u/dy^2)^2} \tag{49}$$

where $\chi = 0.36$.

By combining Equations 45 through 49 with Equation 42, the following is obtained:

$$\frac{\tau}{\tau_{WL}} \simeq 1 = \{1 + n^2u^+y^+[1 - \exp(-n^2u^+y^+)]\}\frac{du^+}{dy^+} \tag{50a}$$

for $0 \leq y^+ \leq 20$,

and

$$\frac{\tau}{\tau_{WL}} \simeq 1 = \chi^2 \frac{(du^+/dy^+)^4}{(d^2u^+/dy^{+2})^2} \tag{50b}$$

for $y^+ \geqq 20$.

Equation 50a may be solved numerically to obtain u^+ as a function of y^+, the universal velocity profile, and Equation 49 can be solved analytically by applying the following boundary conditions:

$$\frac{du^+}{dy^+} \to \infty \text{ as } y^+ \to 0 \tag{51a}$$

$$u^+ = u_{20}^+ \text{ at } y^+ = y_{20}^+ = 20 \tag{51b}$$

where u_{20}^+ is obtained from the numerical solution of Equation 50a:

$$u^+ = u_{20}^+ + \frac{1}{2\chi} \ln(y^+/20) \tag{52}$$

The liquid flowrate is obtained by integrating the velocity distribution over the area occupied by the liquid phase. The dimensionless flowrate is given by

$$W^+ = \frac{W_L}{\mu_L R} = 2 \int_0^{\gamma/2} \int_0^{y_1^+} u^+(y^+)(1 - y^+/R^+)dy^+ d\theta \tag{53}$$

where

$$R^+ = Ru^*/\nu_L \tag{54}$$

Equation 53 may be integrated numerically for various values of the parameter R^+. Typical results are given in Figure 9 in the form of a plot of the holdup, R_L, versus W^+ for various values of R^+. Note that the holdup enters through the upper limits on the

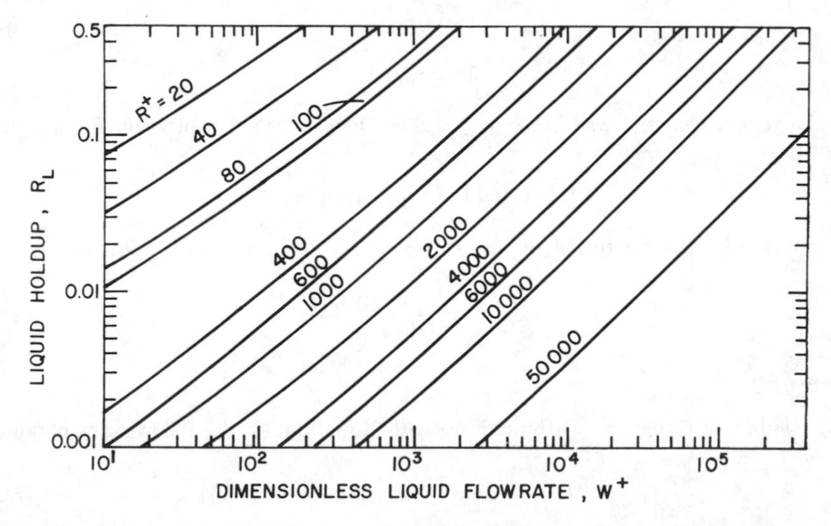

Figure 9. In-situ liquid holdup as a function of dimensionless liquid flowrate and the dimensionless tube size parameter, R^+.

integrals in Equation 53 since $y_i^+ = y_i u^*/\nu_L$ is related to angle γ through Equation 41a and γ, in turn, is related to R_L by Equation 33. Thus, the holdup can be determined from the liquid flowrate using Figure 9, provided R^+ is known. Because R^+ is a function of u^* (Equation 54), information on wall shear is needed.

The liquid-phase wall shear, τ_{WL}, is related to R_L, $\Delta P/L$, τ_{WG} and τ_i through Equations 26, 27, 33 and 34. For the gas-phase wall shear, the definitions proposed by Govier and Aziz (Equation 28) were used where an appropriate single-phase friction factor expression was chosen based on D_{EG} (Equation 25b) and $\overline{U}_G(=Q_G/A_G)$.

To define the interfacial shear, τ_i, in the force balance expressions (Equations 26 and 27) requires a reliable friction factor expression, which, in turn, requires information on the detailed structure of the interface. For stratified flows with three-dimensional and roll waves, the following expressions were derived experimentally [33,34]:

$$\tau_i = 0.0142(\rho_G \overline{U}_G{}^2/2) \tag{55}$$

for three-dimensional waves, and

$$\tau_i = f_i \rho_G (\overline{U}_G - C)^2/2 \tag{56}$$

for roll waves.

C in Equation 56 is the wave velocity and f_i was obtained by fitting data for air-water flows:

$$f_i = 0.080 + 2.00 \times 10^{-5}\,Re_L \tag{57}$$

for $100 \leq Re_L \leq 1700$.

As in the previous models, an iterative procedure was used to determine pressure drop and holdup. The procedure requires an initial estimate of the holdup, R_L, but computations are greatly facilitated by starting with a reliable estimate of R_L.

One approximate method is to use the Lockhart-Martinelli correlation for holdup, which correlates R_L with the Lockhart-Martinelli parameter, X, defined by Equation 5. However, because the Lockhart-Martinelli correlation for R_L is poor for stratified flow, an improved version was obtained by accounting for the geometric parameters of the flow system. Following the method of Dukler and Taitel [35], the following dimensionless expression was obtained:

$$X^2(\tilde{D}_{EL}\tilde{U}_L)^{-m}\,\tilde{U}_L{}^2\,\frac{\tilde{p}_L}{\pi R_L} - (\tilde{D}_{EG}\tilde{U}_G)^{-q}\,\tilde{U}_G{}^2\left(\frac{\tilde{W}_i}{R_G} + \frac{\tilde{p}_G}{\pi R_G} + \frac{\tilde{W}_i}{R_L}\right) = 0 \tag{58}$$

where

$$X^2 = \frac{\dfrac{4C_L}{D}\left(\dfrac{U_L^S D}{\nu_L}\right)^{-m}\dfrac{\rho_L(U_L^S)^2}{2}}{\dfrac{4C_G}{D}\left(\dfrac{U_G^S D}{\nu_G}\right)^{-q}\dfrac{\rho_G(U_G^S)^2}{2}} = \frac{(\Delta P/L)_L^S}{(\Delta P/L)_G^S} \tag{59}$$

and $\tilde{W}_i$, $\tilde{p}_L$, $\tilde{p}_G$, $\tilde{D}_{EL}$ and $\tilde{D}_{EG}$ are given as before and

$$\tilde{U}_L = \overline{U}_L/U_L^S = 1/R_L \tag{60}$$

$$\tilde{U}_G = \overline{U}_G/U_G^S = 1/R_G \tag{61}$$

Hence, all the geometric parameters associated with stratified flow can be evaluated for a unique value of the Lockhart-Martinelli parameter, X, for various values of pipe size, fluid properties and flowrates. Equation 58 was found to be a better approximation than the Lockhart-Martinelli correlation and, therefore, provides a better trial value of R_L to begin the iterative procedure.

The design procedure involves a trial-and-error solution of Equations 26 and 27 to obtain the two-phase pressure drop and holdup. The procedure is as follows:

1. From the gas and liquid flowrates, pipe size and physical properties of the individual phases, calculate the superficial mass velocities, G_G and G_L, and establish whether the flow regime is stratified using an appropriate flow map (refer to the next chapter). Further, ascertain from the literature [29,30,36] whether the interface involves small amplitude waves or roll waves.
2. Calculate a trial value of the holdup, R_L, from the Lockhart-Martinelli parameter, X, based on single-phase flow pressure drops, using Equation 58 or a plot thereof.
3. Determine D_{EL}, D_{EG}, W_i, p_L and p_G for the tube diameter under consideration.
4. Calculate Re_L and Re_G using Equations 29a and 29b to determine whether the flow is turbulent/turbulent. If the liquid-phase flow is laminar ($Re_L < 1000$), the method of Russell et al. [28] is recommended.
5. Calculate $W^+ = W_L/\mu_L R$ from the liquid mass flowrate, W_L. From W^+ and R_L determine R^+, either by interpolating from Figure 9 or from computations.
6. Calculate the friction velocity, u^*, from R^+, and then calculate τ_{WL} from u^* using Equation 45. Evaluate τ_i and τ_{WG} from the appropriate equations.
7. Solve Equations 26 and 27 separately for $(\Delta P/L)_{tp}$. If the two calculated values do not agree to within some specified accuracy, a new value of R_L is assumed and the procedure is repeated starting from step 3.

When convergence is attained, the pressure drop and current value of R_L and τ_{WL} are the predictions of the model.

Cheremisinoff [36] tested the model with extensive data for the air-water system. Figures 10 and 11 show comparisons between measured and predicted values of pressure

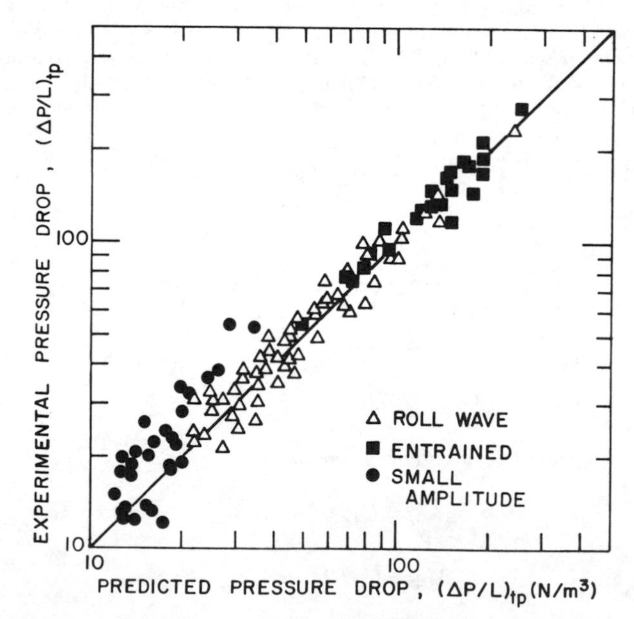

Figure 10. Comparison of predicted and experimental values of pressure drop for the air-water system.

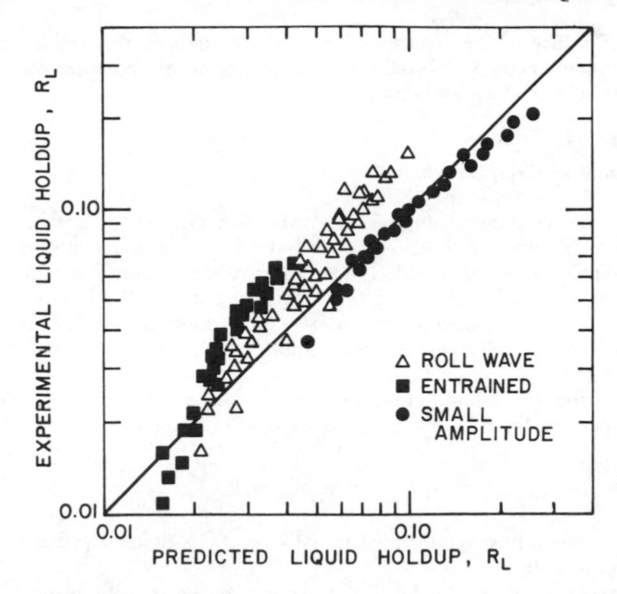

Figure 11. Comparison of predicted and experimental values of holdup for the air-water system.

drop and holdup for stratified wavy flows. The average percentage deviation of the predicted pressure drop and holdup for all runs was about 16%, with the model underpredicting for roll waves.

The predictions by the more fundamental-based models described above are much better than the Lockhart-Martinelli correlations for pressure drop and holdup. In fact, Equation 59, also tested by Cheremisinoff, was found to be superior to the Lockhart-Martinelli method, but not as good as those predictions obtained by using iterative procedures.

PRINCIPLES OF HEAT TRANSFER

There are many industrial operations and pieces of equipment that encounter heat transfer with two-phase flows. Examples include thermosyphon reboilers, cooler-condensers, gas-liquid chemical reactors and a variety of heat exchanger equipment. A large number of these services involve forced-convection heat transfer, but natural circulation also is employed in systems such as vertical-tube reboilers. Often, condensation or evaporation plays a role in the heat transfer process, causing different flow regimes to exist along the length of the pipe or vessel. Proper design of such systems requires information on the heat transfer characteristics for each flow regime.

In many applications, the main resistance to heat transfer is attributed to the liquid film adjacent to the pipe or equipment walls. In this case, the controlling mechanism of heat transfer is either molecular conduction through a laminar liquid layer, or a combination of molecular transport and turbulent transport through a liquid layer. If the wall temperature is sufficiently high to sustain bubble nucleation, then the heat transfer rate is greatly enhanced by turbulence induced by bubble growth and departure from the wall.

Because two-phase flows are characterized by rather complex fluid mechanics, it is often difficult to solve the governing energy equations to obtain the temperature field and heat transfer characteristics. For this reason, there are hundreds of empirical and semitheoretical correlations based on heat and momentum analogies scattered throughout the literature. Because of this overabundance, discussions in this section are limited to describing the phenomena and briefly introducing some of the concepts behind developing analogies between heat and momentum transfer.

Heat and Momentum Transfer Analogies

Consider the simple case of single-phase turbulent flow of an isothermal fluid along a straight length of pipe. If the flow is well developed and incompressible, then the radial local velocity may be divided into three distinct zones as noted earlier—the laminar sublayer, the buffer layer and the turbulent core. In the laminar sublayer, heat transfer is controlled by molecular transport (i.e., conduction). The buffer layer and turbulent core mechanically mix warmer and colder portions of the fluid by eddy diffusion, better known as convection.

We can describe the combination of these two heat transfer mechanisms by an equation analogous to the momentum expression (Equation 42):

$$q = -(k + \rho C_p \epsilon_H) \frac{dT}{dy} \tag{62}$$

where ϵ_H is the eddy diffusivity for heat transfer, and y is defined as before, the distance measured from the wall.

For fully developed flow, the local shear stress is linear with respect to the radius and we obtain $\tau/\tau_W = r/r_W$. A similar relation may be assumed for the heat flux:

$$q/q_W = r/r_W \tag{63}$$

Hence, we can write

$$\frac{\tau_W}{\rho}(1 - y/r_W) = \left(\frac{\mu}{\rho} + \epsilon_M'\right)\frac{du}{dy} \tag{64a}$$

$$\frac{q_W}{\rho C_p}(1 - y/r_W) = -\left(\frac{k}{\rho C_p} + \epsilon_H\right)\frac{du}{dy} \tag{64b}$$

The terms in parentheses on the right-hand side (RHS) of Equation 64a represent the sum of the molecular and eddy diffusivities of momentum. The RHS parenthetical terms in Equation 64b represent the sum of the molecular and eddy diffusivities of heat. These two equations form the starting basis for several important analogies between heat and momentum transfer for single-phase, fully developed flows that have been extended to describe two-phase systems. We shall first describe the analogies for single-phase flow.

The first of these analogies is that of Reynolds for a Prandtl number of unity. Assuming the sum of the thermal diffusivities to equal the sum of the momentum diffusivities and integrating the ratio of Equations 64a and 64b over the limits $0 \leq u \leq \bar{u}$ and $T_W \leq T \leq T_b$, the following expression is obtained:

$$q_W\bar{u}/C_pT_W = T_W - T_b \tag{65}$$

Note that subscript b refers to the fluid bulk properties and the overbar denotes an average. By definition,

$$h \equiv q_W/(T_W - T_b)$$

$$\tau_W = f\rho\bar{u}^2/2$$

Hence,

$$\frac{h}{C_p \bar{u} \rho} = \frac{f}{2} \tag{66}$$

Equation 66 is called the Reynolds analogy and provides good estimates of the heat transfer coefficient for common gases at moderate temperature differences and for Prandtl numbers close to 1.

The second and more advanced analogy is that of von Karman [32,37], in which the resistance to heat transfer is considered to be composed of three parts. For the laminar sublayer, $y^+ \leq 5$, the molecular thermal diffusivity and the kinematic viscosity are accounted for in the governing equations, but the eddy terms are neglected. Hence, an expression for the temperature drop over this region is obtained. For the buffer layer, $5 < y^+ \leq 30$, both molecular and eddy transport terms are included. The following empirical expression for the velocity profile in the buffer layer is used to obtain the eddy viscosity:

$$u^+ = 5 + 5\,ln\,(y^+/5) \tag{67}$$

As before, the eddy viscosity is assumed equal to the eddy thermal diffusivity in the heat flux equation. Integration of the heat flux equation containing both the eddy and molecular thermal diffusivities gives an expression for the temperature drop over the buffer layer. For the turbulent core, $y^+ > 30$, the Reynolds analogy is applied to obtain an expression for the temperature drop associated with this thermal resistance. Summing all three temperature drops provides an expression for the overall temperature difference for the flow. The heat transfer coefficient is obtained by dividing the heat flux, q, by the overall temperature difference:

$$h = \frac{f\bar{u}\rho C_p/2}{1 + 5\sqrt{f/2}\,\{Pr - 1 + ln[(1 + 5\,Pr)/6]\}} \tag{68}$$

Note that Equation 68 reduces to the Reynolds analogy for $Pr = 1$.

The last analogy for single-phase flows is that of Martinelli [38] for uniform heat flux. In the Martinelli analogy, the turbulent core is treated in the following manner: The derivative of the velocity field, $u^+ = 5.5 + 2.5 ln y^+$, is combined with Equation 64a and, by appropriate manipulation, the following final expression is obtained:

$$h = \frac{\bar{u}\rho C_p\sqrt{f/2}}{\frac{T_w - T_b}{T_w - T_c}(5)\left[\,Pr + ln(1 + 5Pr) + 0.5N_{DR} ln\,\frac{Re}{60}\sqrt{\frac{f}{2}}\right]} \tag{69}$$

where T_c is the temperature at the center of the tube, $N_{DR} = \dfrac{\epsilon_H}{\epsilon_H + \dfrac{k}{\rho C_p}}$, and the values of N_{DR} and $\dfrac{T_w - T_b}{T_w - T_c}$ are given in Tables I and II, respectively.

Table I. Tabulated Values of N_{DR} as a Function of Re and Pe

Pe	Re →	10^4	10^5	10^6
10^2	....	0.18	0.098	0.052
10^3	....	0.55	0.45	0.29
10^4	....	0.92	0.83	0.65
10^5	....	0.99	0.985	0.980
10^6	....	1.00	1.00	1.00

Table II. Tabulated Values of $(T_w - T_b)/(T_w - T_c)$ as a Function of Re and Pr

Pr	Re →	10^4	10^5	10^6	10^7
0	....	0.564	0.558	0.553	0.550
10^{-4}	....	0.568	0.560	0.565	0.617
10^{-3}	....	0.570	0.572	0.627	0.728
10^{-2}	....	0.589	0.639	0.738	0.813
10^{-1}	....	0.692	0.761	0.823	0.864
1.0	....	0.865	0.877	0.897	0.912
10	....	0.958	0.962	0.963	0.966
10^2	....	0.992	0.993	0.993	0.994
10^3	....	1.00	1.00	1.00	1.00

It should be noted that a number of modified analogies have been proposed [39–48]; however, they are similar, or all have as a starting basis one of the three analogies noted above.

There have been a number of attempts to apply the momentum-heat transfer analogy of single-phase flow to analyze the turbulent heat transfer in two-phase flow, among them those by Rosson and Myers [49], Dukler [50], Hewitt [51], Altman et al. [52], Kosky and Staub [53], Hughmark [54] and Cheremisinoff and co-workers [55–57].

Dukler [50] developed a model to describe heat transfer for downward cocurrent annular flow, which Hewitt [51] later modified for the case of upward cocurrent annular flow. Again, the key assumption made was that $\epsilon_M' = \epsilon_H = \epsilon$. The value of ϵ was computed from the Deissler expression (Equation 48) for $y^+ < 20$, and for $y^+ > 20$ the von Karman expression based on mixing length, l, was used:

$$\epsilon = k^* \frac{(du/dy)^3}{(d^2u/dy^2)^2} \tag{70a}$$

where

$$l = k^* \frac{(du/dy)}{(d^2u/dy^2)} \tag{70b}$$

From these, the heat flux expression (Equation 62) may be expressed in the following dimensionless forms:

$$1 = \left[\frac{1}{Pr} + n^2u^+y^+(1 - e^{-n^2u^+y^+})\right]\frac{dT^+}{dy^+} \tag{71}$$

for $y^+ \leq 20$, and

$$1 = k\frac{(du^+/dy^+)^3}{(d^2u^+/dy^{+2})^2}\frac{dT^+}{dy^+} \tag{72}$$

for $y^+ > 20$.
where T^+ is defined as follows:

$$T^+ = \frac{C_{PL}\rho_L u^*}{q_W/A_W}(T_W - T) \tag{73}$$

The relationship between u^+ and y^+ is obtained as described previously, and T^+ as a function of y^+ is obtained by numerical integration of Equations 71 and 72. The heat transfer coefficient for the heat flow through the liquid film is defined in terms of the dimensionless temperature difference, T_i^+, across the film. T_i^+ is evaluated at the appropriate value of the dimensionless film thickness, m^+:

$$h = \frac{q_W/A_W}{T_W - T_i} = \frac{C_P \rho_L u^*}{T_i^+} \tag{74}$$

Davis and Cheremisinoff [56,57] extended their fluid mechnics model for stratified gas-liquid flow for turbulent liquid/turbulent gas conditions to predict the heat transfer characteristics of the liquid layer. Using wall shear stress predicted by the method outlined previously and using the geometric definitions of the fluid mechanics model, liquid-phase Nusselt numbers were predicted to within 20% of experimental values for the air-water stratified flow system. In this model, the heat flux for the liquid layer can be written in dimensionless form as

$$\frac{q}{q_w} = \left(\frac{1}{Pr} + \frac{\epsilon_M'}{\nu_L Pr_t}\right)\frac{dT^+}{dy^+} \tag{75}$$

where $Pr_t = \epsilon_M'/\epsilon_H \simeq 1$, and it was assumed that $q/q_w \simeq 1$. Note that q_W is considered to be a constant, i.e., the case applies to constant wall heat flux. Equation 71 was used to obtain the temperature profile over the laminar sublayer and buffer zone, $0 \leqq y^+ \leqq 20$. For $y^+ > 20$, the von Karman expression for ϵ_M' was used to give

$$1 = \chi \frac{(du^+/dy^+)^3}{(d^2u^+/dy^{+2})^2}\frac{dT^+}{dy^+} \tag{76}$$

Equation 76 can be integrated analytically and Equation 71 must be integrated numerically to obtain T^+ as a function of y^+.

The bulk or mixing cup temperature is defined by

$$T_b - T_o = \frac{\int_0^{\gamma/2}\int_0^{y_i} u(T - T_o)(R - y)\,dy\,d\theta}{\int_0^{\gamma/2}\int_0^{y_i} u(R - y)\,dy\,d\theta} \tag{77}$$

where T_o is a reference temperature, taken to be the local wall temperature, T_W, and assumed uniform along the periphery of the tube wetted by the liquid. In dimensionless form this equation is

$$T_b^+ = \frac{\int_0^{\gamma/2}\int_0^{y_i^+} u^+T^+(1 - y^+/R^+)\,dy^+d\theta}{(W^+/2)} \tag{78}$$

where W^+ is defined by Equation 53.

The heat transfer coefficient for the liquid phase may be expressed as follows:

$$h_L = C_{PL}\rho_L u^*/T_b^+ \tag{79}$$

And the Nusselt number for heat transfer to the liquid phase is

$$Nu_L = \frac{h_L y_o}{k_L} = \frac{Pr_L y_o^+}{T_b^+} \tag{80}$$

where $y_o^+ = y_o u^*/\nu_L$.

For the gas phase, it was assumed that the gas could be treated as if it were in single-phase flow in a duct of irregular shape. The heat transfer from the wall to the gas phase was found to correlate reasonably well with the modified Dittus-Boelter equation:

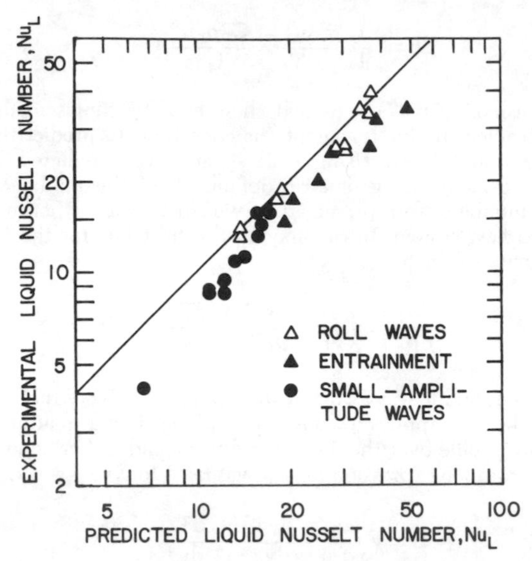

Figure 12. Comparison between measured liquid-phase Nusselt number and predicted, using the method of Davis and Cheremisinoff [56,57].

$$Nu_G = 0.022\, Re_G^{0.8}\, Pr_G^{0.6} \tag{81}$$

where Re_G is defined by Equation 29a and

$$Nu_G = \frac{h_G D_{EG}}{k_G} \tag{82}$$

Figures 12 and 13 compare model predictions to measured liquid- and gas-phase Nusselt numbers for the air-water system, respectively. Note that the predicted values in Figures 12 and 13 are based on predicted values of holdup from the fluid mechanics portion of the model.

Similar success has been noted by other investigators applying single-phase heat and momentum analogies to two-phase flows. Rosson and Myers [49] applied the von Karman analogy to condensation studies in stratified flows. Altman et al. [52] and Kosky and Staub [53] applied the Martinelli analogy to annular two-phase flow.

Evaporation in Two-Phase Flows

Much of the literature on evaporation in two-phase flows is limited to single component evaporation. Many of the early correlations report only average values or overall heat transfer coefficients, with emphasis on annular and annular-mist flows. Correlations for single-component evaporation fall into two general categories: (1) those that correlate the heat transfer coefficient as a function of the Lockhart-Martinelli parameter, X_{tt}, and (2) those based on dimensionless groups.

Correlations of the first type are of the form

$$\frac{h}{h_L} = C\left(\frac{1}{X_{tt}}\right)^{n'} \tag{83}$$

In Equation 83, h_L is the heat transfer coefficient for the liquid portion as if it were in total flow in the pipe. Constants C and n′ have been evaluated for several systems

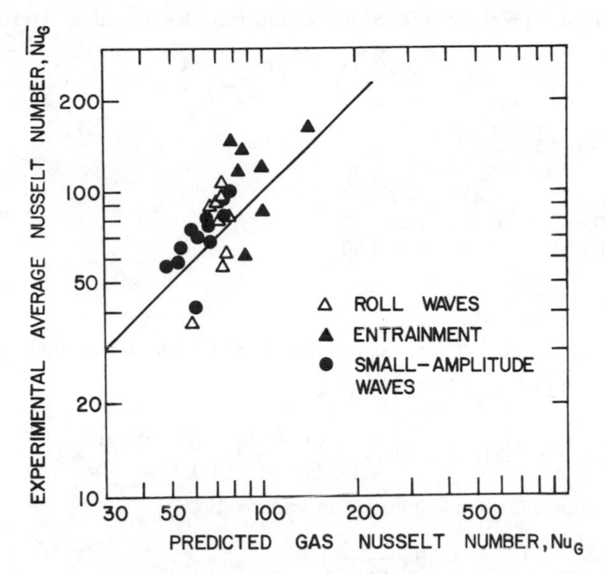

Figure 13. Comparison between measured average gas-phase Nusselt number and predicted, using the method of Davis and Cheremisinoff [56,57].

by a number of investigators. Table III gives typical values for C and n′ reported for different systems. Note that these values are valid in the nucleation-suppressed region and for vertical flows.

Various correlations of the second type are given in Table IV. The correlations listed in Table IV were developed for horizontal flows.

Mechanisms of Flow Boiling

High heat flux systems are of particular interest to nuclear reactor cooling design. When a liquid contacts a wall that has a surface temperature greater than the saturation temperature of the liquid at the pressure of the system, nucleate boiling can occur. There are several types of boiling that merit distinction.

Table III. Values of Constants C and n′ in Equation 83

Investigators	C	n′	Fluid	Flow Orientation	Maximum Exit Quality	h_L Base
Dengler and Adams [58]	3.5	0.5	Water	Upward	70	h_{L_T}
Guerrieri and Talty [59]	3.4	0.45	Pentane, heptane	Upward	12	h_{L_f}
Schrock and Grossman [60]	2.5	0.75	Water	Upward	59	h_{L_T}
Wright [61]	2.721	0.581	Water	Downward	19	h_{L_f}
Somerville [62]	7.55	0.328	n-Butanol	Downward	31	h_{L_f}
Collier et al. [63]	2.167	0.699	Water	Upward	66	h_{L_f}

Table IV. Various Correlations for Single-Component Evaporation (Horizontal Flows)

- $$\frac{h_D}{k_L} = 0.060 \left(\frac{\rho_L}{\rho_G}\right)^{0.28} \left(\frac{DG_T x}{\mu_L}\right)^{0.87} \left(\frac{C_p \mu_L}{k_L}\right)^{0.4}$$
- $$\frac{hD}{k_G} = C \left(\frac{D \rho_G \bar{U}_G}{\mu_G}\right)^{0.8}$$

where $C = 0.84 \quad 0.4 \leq x \leq 0.5$
$C = 0.887 \quad 0.55 \leq x \leq 0.65$
$C = 0.904 \quad 0.70 \leq x \leq 0.80$

- $$\frac{h}{h_L} = 1.5 \left(\frac{1 - x}{1 - R_G}\right)^{0.8}$$

where R_G = steam void fraction calculated from Lockhart-Martinelli correlation
$$h_L = 0.023 \left(\frac{k_L}{D}\right)\left(\frac{C_p \mu_L}{k_L}\right)^{0.4} \left(\frac{DG_T}{\mu_L}\right)^{0.8}$$

- $$\frac{h_D}{k} = \left[4.3 + 5.0 \times 10^{-4} \left(\frac{V_{LG}}{V_L}\right)^{1.64} x \right]\left(\frac{q}{G\lambda}\right)^{0.464} \left(\frac{DG}{\mu_L}\right)^{0.808}$$

where V_{LG} = specific volume increase on vaporization

- $$\frac{h}{h_L} \left(\frac{G_G \lambda A}{q}\right)^{0.1} = 6.59 \left(\frac{1 + x}{1 - x}\right)^{1.16}$$

where $h_L = 0.023 \left(\frac{k_L}{D}\right)\left(\frac{C_p \mu_L}{k_L}\right)^{0.4} \left[\frac{DG_T(1 - x)}{\mu_L}\right]^{0.8} \left(\frac{\mu}{\mu_w}\right)^{0.14}$

Pool boiling. In this case, the bulk liquid does not flow along the heated surface due to forced convection. When boiling is suppressed, the liquid reverts back to a stagnant pool.

Flow boiling. Here the liquid moves relative to the heated surface due to forced convection.

Subcooled boiling. In this case, the bulk liquid temperature is below the saturation temperature. Bubbles that form at the hot wall collapse upon entering the colder bulk liquid.

Saturated boiling. In this case, the bulk liquid is at the saturation temperature. For a flowing system, the vapor formed results in two-phase flow.

Both convective boiling and condensation have been reviewed by Collier [21]. Extensive reviews of nucleate boiling are given by Hewitt and Hall-Taylor [1], Collier [21] and Leppert and Pitts [64]. A partial listing of correlations for two-phase forced convection and nucleate boiling is given in Table V.

One of the most widely used correlations for nucleate boiling is the Rohsenow equation:

$$\frac{C_{PL}\,\Delta T_{sat}}{\lambda_{vap}} = C_{sf} \left\{ \frac{\phi}{\mu_L \lambda_{vap}} \left[\frac{\sigma}{g(\rho_L - \rho_G)} \right]^{1/2} \right\}^{n} Pr_L^{\,m+1} \tag{84}$$

where ϕ = heat flux at the wall
C_{PL} = specific heat of the liquid
μ_L = viscosity of the liquid
σ = surface tension
λ_{vap} = latent heat of vaporization
ΔT_{sat} = temperature difference between the wall and saturation temperature

C_{fs} is a constant that is a function of the liquid-surface combination. Table VI gives values of C_{fs} for various combinations.

Table V. Correlations for Estimating Heat Transfer Coefficients in Forced Convection and Nucleate Boiling Regions

Investigators	Two-Phase System—Fluid	Equation for Two-Phase Forced Convective Region	Criterion for Initiation of Nucleation	Modification for Nucleate Boiling in Liquid Film
Dengler and Adams [58]	2.54 cm × 6.1 m vertical steam-heated tube—water	$\frac{h_{TP}}{h_{fo}} = 3.5\left(\frac{1}{X_{tt}}\right)^{0.5}$	$(\Delta T_{SAT})_{ONB}(°C) = 7.9\, u_f^{0.3}$ where u_f is the local liquid velocity (m/s) given by $u_f = \frac{G(1-x)}{\rho_f(1-R_G)}$	$\frac{h_{TP}}{h_{fo}} = (F)3.5\left(\frac{1}{X_{tt}}\right)^{0.5}$ $F = 0.67\left[(\Delta T_{SAT} - (\Delta T_{SAT})_{ONB})\left(\frac{dp}{dT_{SAT}}\frac{D}{\sigma}\right)_w\right]^{0.1}$
Guerrieri and Talty [57]	1.9 cm × 1.83 m vertical tube—various organics	$\frac{h_{TP}}{h_f} = 3.4\left(\frac{1}{X_{tt}}\right)^{0.45}$	$\frac{r^*}{\delta} > 0.049$ r^* is the radius of an equilibrium bubble corresponding to the wall superheat. δ is the thickness of the laminar film given by $\delta = \frac{10\mu_f}{\rho_f}\left[\frac{4\rho_f}{(dp/dz)D}\right]^{0.5}$	$\frac{h_{TP}}{h_f} = (E)3.4\left(\frac{1}{X_{tt}}\right)^{0.45}$ $E = 0.187\left(\frac{r^*}{\delta}\right)^{-5/9}$
Bennet [65]	Internally heated vertical annulus—water	$\frac{h_{TP}}{h_f} = 0.64\left(\frac{1}{X_{tt}}\right)^{0.74}(\phi)^{0.11}$ ϕ in Btu/hr-ft^2		Used Rohsenow equation (84) for nucleate boiling
Schrock and Grossman [60]	Vertical tubes heated electrically—water	These authors proposed one equation for both regions of heat transfer $\frac{h_{TP}}{h_f} = \gamma\left[\left(\frac{\phi}{Gi_{fg}}\right) + m\left(\frac{1}{X_{tt}}\right)^n\right]$ where $\gamma = 7.39 \times 10^3$, $m = 1.5 \times 10^{-4}$, and $n = 0.66$		

Table VI. Values of C_{fs} for Rohsenow's Equation (84)[a]

Liquid-Surface Combination	C_{sf}
n-Pentane on polished copper	0.0154
n-Pentane on polished nickel	0.0127
Water on polished copper	0.0128
Carbon tetrachloride on polished copper	0.0070
Water on lapped copper	0.0147
n-Pentane on lapped copper	0.0049
n-Pentane on emery-rubbed copper	0.0074
Water on scored copper	0.0068
Water on ground and polished stainless steel	0.0080
Water on Teflon-pitted stainless steel	0.0058
Water on chemically etched stainless steel	0.0133
Water on mechanically polished stainless steel	0.0132

[a]Whenever possible, it is recommended that a pool boiling experiment be carried out to determine the value of C_{sf} applicable to the particular conditions of interest. In the absence of such information, a value of C_{sf} of 0.013 may be used as a first approximation.

In general, a high degree of turbulence is generated by nucleate boiling, which can enhance heat transfer rates. Designs based on nonboiling convective heat transfer tend to be conservative with respect to estimating heat transfer coefficients.

PRINCIPLES OF MASS TRANSFER

In designing for two-phase flow with mass transfer, it is necessary to make distinctions not only with respect to the fluid dynamics of the system, but also taking into account whether chemical reactions occur. Systems involving both chemical reactions and changing flow regimes are exceedingly complex and difficult to design. Danckwerts [66] gives a good introduction to simultaneous mass transfer and chemical reaction in stagnant and well defined gas-liquid systems. Reviews of reactor model development with discussions on two-phase flows may be found in the works of Cichy et al. [67,68] and Shah [69]. In this section, fundamental concepts of mass transfer in two-phase flows are presented.

Perhaps the earliest work, and the most significant from an engineering standpoint, is that of Lewis and Whitman [70], who proposed the film theory of mass transfer describing the absorption of gas into liquid. They postulated that a stagnant film of liquid exists adjacent to the free surface and that the liquid film constitutes the only resistance to mass transfer. Both the gas and liquid are at equilibrium at the interface. The following expression describes the mass flux of gas into liquid:

$$N = k_L (C^* - C_o) \tag{85}$$

k_L is the mean transfer coefficient equal to D/δ, where D is the diffusion coefficient of the dissolved gas, and δ is the film thickness. C^* and C_o are the equilibrium concentration and the bulk concentration in the liquid phase, respectively. When the liquid is in the form of a flowing film, experiments suggest that k_L and δ are not determined uniquely by the fluid dynamics. Under flowing conditions, k_L is found to be approximately proportional to $D^{1/2}$. This early attempt provides qualitative insight into the driving force needed to effect mass transfer.

Later, Higbie [71] proposed the so-called "penetration theory," which assumed that eddies generated in the liquid are exposed at the interface for a short time, where unsteady-state diffusion takes place. The following expression for k_L was obtained from this interpretation:

$$k_L = 2\sqrt{\frac{D}{\pi\theta_c}} \tag{86}$$

where θ_c is the contact time at the gas-liquid interface. Equation 86 assumes that the contact time is the same for all eddies and that diffusion occurs as if it were into a stagnant liquid of infinite depth.

The penetration theory was modified by Danckwerts [72] by assuming that there is a distribution of surface ages resulting from a constant rate of surface renewal. The resulting mass transfer coefficient is

$$k_L = \sqrt{D\Phi} \tag{87}$$

where Φ is the fractional rate of surface renewal. The model was extended to include gas absorption accompanied by chemical reaction.

The principles set forth in the Higbie-Danckwerts models have been extended by several investigators, such as Hanratty [73], Perlmutter [74], Dobbins [75] and Kishinevsky [76,77].

Penetration-surface renewal models have not been readily adapted into design practices, partly because they are difficult to verify experimentally. For well defined flow systems, eddy diffusivity models analogous to those used for momentum and heat transfers have been proposed. In this approach, investigators [78–80] have argued that due to surface tension, the eddy diffusivity increases continuously from zero at the interface to a high value in the bulk liquid. This enables the convective diffusion equation,

$$u\frac{\partial c}{\partial x} = \frac{\partial}{\partial y}\left[(D + \epsilon_D)\frac{\partial c}{\partial y}\right] \tag{88}$$

to be solved to obtain expressions for the mass flux. In the fluid region in which the concentration profile is fully developed, the eddy diffusivity may be expressed as

$$\epsilon_D = a y^n \tag{89}$$

The mass transfer coefficient is redefined as k_c, and approximated by

$$k_c = \frac{\dot{n}}{\pi} a^{1/\dot{n}} D^{\frac{\dot{n}-1}{\dot{n}}} \sin(\pi/\dot{n}) \tag{90}$$

Hence, the mass flux N is

$$N = k_c(C_s - C_b) \tag{91}$$

Constants a and n must be determined experimentally.

Lamourelle et al. [81] tested the eddy diffusivity model by measuring the rates of absorption of helium, hydrogen, oxygen and carbon dioxide into water in a wetted-wall column. Their results are expressed in terms of the dimensionless Sherwood number:

$$Sh = \frac{k_c\delta}{D} = 1.76 \times 10^{-5}\, Re^{1.506}\, Sc^{1/2} \tag{92}$$

Note that δ is the liquid film thickness and Sc is the Schmidt number, $Sc \equiv \nu/D$.

Equation 92 is equivalent to an eddy diffusivity expression proportional to $D^{0.5}$. Hence, no distinction could be made between the eddy diffusivity model and the Danckwerts surface renewal theory.

The experimental investigations of Lamourelle and others suggest that a number of important factors must be accounted for in designing for two-phase flows with mass transfer:

Table VII. Partial Listing of Mass Transfer Correlations

Investigators	Correlations	Flow Regimes
Jagota, Rhodes and Scott [90]	$k_La = 226\ \epsilon_L^{0.48}$, (h^{-1})	Vertical annular
	where $\epsilon_L = (\Delta P/L)_{TP} V_{SL}$	
	V_{SL} = the superficial liquid velocity	
	$k_La = 235(1 - \bar{E})\epsilon_L^{0.50}$,	Vertical annular/mist
	where $\bar{E}$ = the average entrainment	
Lamourelle and Sandall [81]	$k_L = 0.339\ Re^{0.839}D^{0.5}$ (ft/h)	Vertical stratified, no interfacial shear
	where $Re = 4\Gamma/\mu_L$	
	Γ = the flowrate per wetted perimeter	
	D = the pipe diameter	
Jepsen [86]	$k_La = 3.47\ \epsilon^{0.40}D_{AB}^{0.50}\sigma^{0.50}\mu_L^{0.05}d^{-0.68}$, for $\epsilon < 0.05$ atm/s (s^{-1})	Horizontal froth
	where $\epsilon = (\Delta P/L)_{TP}(V_{SG} + V_{SL})$	
	D_{AB} = the diffusion coefficient	
	d = the pipe diameter	
Banerjee, Scott and Rhodes [91]	$k_La = 18.75\ \epsilon^{0.79}D_{AB}^{0.50}\sigma^{0.50}\mu_L^{0.05}$, for $\epsilon > 0.05$ atm/s	Horizontal annular
	$k_La = 1595(\Delta P)^{1.577}$, 18-in. coil (h^{-1}) (psi)	Annular flow in helically coiled tubes
	$k_La = 1312(\Delta P)^{2.010}$, 12-in. coil	
	$k_La = 1132(\Delta P)^{1.913}$, 6-in. coil	
	$a = 6.18\ E^{0.4018}$, 18-in. coil (cm^{-1}) (J/S)	
	$a = 7.9\ E^{0.642}$, 12-in. coil	
	$a = 4.43\ E^{0.557}$, 6-in. coil	
	where $E = Q_L(\Delta P)$, and Q_L is the volumetric liquid flowrate	
Banerjee, Scott and Rhodes [92]	$k_L = 2.93 \times 10^{-3}\ D_{AB}^{0.5}Re^{0.993}$	Falling wavy, turbulent liquid films
	for $2000 < Re < 10{,}000$	
	where D_{AB} is the diffusion coefficient and	
	$Re = 4\Gamma_L/\nu_L$,	
	where Γ_L is the liquid flowrate per wetted perimeter	

Howard and Lightfoot [93]	$$\frac{k_L}{CV_s} = \Lambda \sqrt{\frac{6D}{\pi L V_s}}$$ where C is the total molar density, V_s is the surface velocity, and Λ is the amplification factor which is unity for a smooth surface, and $\Lambda > 1$ for a wavy surface.	Wavy film flow
Vivian and Peaceman [94]	$$k_L = 7.13\, D^{0.5}\Gamma^{0.4}L^{-0.5}$$ where D is the tube diameter, Γ is the mass flowrate per wetted perimeter, and L is the tube length.	Vertical laminar liquid film and turbulent gas
Wales [82]	Graphical correlation	Horizontal annular and dispersed flow
Hughmark [88]	$$\frac{k_G}{V_G - V_L} = \frac{f_G/2}{\phi}\frac{\rho_G}{M}$$ $$\left(k_g \text{ in } \frac{\text{lb-mol}}{\text{h-ft}^2\text{-atm}}\right)$$ $$\frac{k_L}{V_L} = \frac{f_G/2}{\phi}$$ $$(k_L \text{ in ft/h})$$ where $f_G = \frac{(\Delta P/L)_{TP} g_c D_c}{2(V_G - V_L)^2 \rho_G}$, D_c is the core diameter $\phi = \phi(\mathrm{Re}, \mathrm{Pr})$, $\mathrm{Re}_L = \frac{D_c V_L \rho_L}{\mu_L}$, $\mathrm{Re}_G = \frac{D_c (V_G - V_L)\rho_G}{\mu_G}$	Horizontal annular
Anderson et al. [87]	Graphical presentation of data	Horizontal annular
Downing and Knowles [95]	$$k_L = \frac{28 V_L^{0.8} \delta_L^{-1.1}(1 - e^{-100\delta_L})}{1 + 4W^{-1.3}}$$ where W is the river width, and δ is the depth.	Rivers, large bodies of turbulent water

1. **Interfacial disturbances.** Rippling and wave action greatly increase the contact area and lead to local convection, which generally enhances mass transfer at the interface.
2. **Miscellaneous instabilities.** Thermal and concentration variations can arise when chemical species are transported across an interface. This can result in a Marangoni instability, which generally enhances mass transfer rates.
3. **Range of empirical correlations.** Special caution must be exercised when using empirical correlations for design purposes. Because the constants in various mass transfer models depend on the fluid dynamics, the correlations cannot be extrapolated safely beyond the range of conditions for which they were developed.

Part of the uncertainty in using mass transfer coefficients results from the difficulties in estimating the interfacial area, particularly when the flow regime is complex, such as annular mist. It has become convention to express the mass transfer rate per unit volume, aN, as follows:

$$aN = k_L a (C_s - C_o) \tag{93}$$

where a is the transfer area per unit volume. For the most part, it is not possible to predict or measure a and k_L separately; hence, the product $k_L a$ is most often reported. This poses a problem in that both a and k_L are affected by the fluid dynamics of the system, but independently. For example, interfacial waves may increase by 30% for some systems, whereas they can increase k_L by as much as 300% due to convection. Hence, it is not advisable to extrapolate $k_L a$ beyond the range of experiments studied for developing the correlation.

Using penetration theory to calculate the effective interfacial area and penetration contact time, overall mass transfer coefficients may be defined in terms of the individual gas and liquid film coefficients, k_G and k_L. For the gas phase, these coefficients are related by

$$\frac{1}{k_G a} = \frac{H}{k_L a} + \frac{1}{k_G a} \tag{94}$$

where H is the Henry's law constant. Wales [82] studied physical desorption and chemical absorption of carbon dioxide in water in annular and dispersed horizontal flow and rearranged this equation to the following form:

$$\frac{1}{k_G a} = \frac{H}{a[D_a k_r (OH)_o]^{1/2}} + \frac{1}{k_G a} \tag{95}$$

where k_r = reaction rate constant
$(OH)_o$ = initial concentration of the hydroxyl ion
D_a = diffusion coefficient for solute A in the solvent

By fitting the data to Equation 95, values for $k_L a$ and k_G were obtained. Wales also presented data on $k_G a$ and the specific area, a.

Numerous experimental studies in both vertical and horizontal systems have been aimed at developing correlations for the separate cases of liquid-phase- and gas-phase-controlled mass transfer. Vertical systems have been studied most extensively, and introductory discussions may be found in standard textbooks such as those by Treybal [83] and McCabe and Smith [84]. Horizontal systems have received considerably less attention, and only a handful of in-depth studies have produced useful correlations [85–89].

Table VII provides a partial listing of various correlations for mass transfer coefficients and the two-phase flow regimes, for which the correlations are applicable. Most of the correlations only provide estimates of the liquid-phase composite coefficient, $k_L a$. In general, correlations for gas-phase-controlled mass transfer are sparse.

CONCLUSIONS

In this chapter we have highlighted some important concepts for interpreting and modeling complex gas-liquid flows, with emphasis on design-oriented practices. Attempts have been made in recent years to organize the almost overwhelming collection of literature into a workable format for engineering practices. Clearly this is a formidable task, particularly if emphasis is placed on the organization of empirical models.

Fundamental models provide a more organized approach to developing design recommendations. The mechanistic approaches to estimating pressure drop and holdup developed by Dukler, Agrawal et al., Russel et al., Davis and Cheremisinoff, and Yu and Sparrow [24] have shown good results for horizontal stratified flows and appear to have promising extensions to other flow configurations. In addition, these approaches may be coupled with heat transfer analogies. The weakness with these approaches is that few validations have been made with systems other than air-water and, hence, they cannot be recommended as reliable design procedures.

The state of the art for heat transfer and mass transfer in two-phase flows is limited by our knowledge of the fluid mechanics. Heat transfer is in somewhat better shape in that where the flow pattern is well defined, reliable heat transfer correlations are available. Laminar flow condensation, particularly for vertical systems, is probably the most extensively studied, and rigorous theoretical analyses may be found in the literature.

Both turbulent film condensation and evaporation with two-phase flow have numerous empirical correlations, but are poorly organized. In addition, many of the early investigations did not report flow regimes, making it difficult to apply correlations to design.

The area of boiling has been studied extensively, and detailed reviews in numerous textbooks and monographs are available. For an introduction to this subject the reader is referred to Rohsenow et al. [96].

Our understanding of mass transfer in two-phase flows is generally poor. Other than vertical systems, such as wetted wall columns and absorption towers, which have been studied extensively, mass transfer data are sparse, and it is difficult to recommend any of the correlations reported in the literature. Clearly, considerable effort is needed both in developing theory and in experimental studies aimed at design methods for mass transfer. A major reason for our deficiency in this area is the lack of understanding and poor correlations that exist for entrainment, droplet interaction and deposition rates, and liquid film thickness for various flow regimes. That is, a large piece of the mass transfer puzzle may be found with the missing information on the fluid mechanics.

In the chapters to follow, several authors will attempt to present these missing pieces of the puzzle. With an understanding of the fluid mechanics, we can begin to address the heat and mass transfer problems of these complex flows in a rational manner.

NOMENCLATURE

A area (m^2)
a mass transfer area per unit volume (m^{-1})
C constant
C^* equilibrium concentration (kg/m^3)
C_{fs} liquid-surface combination function, Equation 84
C_o bulk concentration (kg/m^3)
C_p specific heat (J/kg-°K)
D diameter (m)
$\mathcal{D}$ diffusivity (m^2/sec)
d_o pipe diameter (m)
F mechanical energy (J/kg)
Fr Froude number, Equation 10
f friction factor
G two-phase mass flux (kg/m^2-hr)
g gravity acceleration (m/sec^2)
H Henry's law constant
h level defined in Figure 6 (m)
h_L heat transfer coefficient (kg/sec^3-°K)
J Richardson number, Equation 1

K Hughmark flow parameter, Equation 8
K_G overall mass transfer coefficient (kg/sec^3-°K)
k fluid thermal conductivity (kg-m/sec^3-°K)
k_c mass transfer coefficient (kg/m^2-sec)
k_L mean mass transfer coefficient (kg/m^2-sec)
k_r reaction rate constant (sec^{-1})
l mixing length (m)
Ma Marangoni number, Equation 3
m^+ dimensionless film thickness
N mass flux (kg/m^2-sec)
Nu Nusselt number
N_{DR} parameter defined in Equation 69
n' exponent in Equation 83
$\dot{n}$ parameter in Equation 90
P pressure, Pa
Pe Peclet number
Pr Prandtl number
p wetted perimeter (m)
Q volume flow (m^3/sec)
$Q_{G,L}^*$ flow volume defined by Yu and Sparrow [24] (m^3/sec)
q heat flux (J/m^2-sec)
δq added head (m)
R tube radius (m)
Ra Rayleigh number, Equation 2
R_L liquid holdup
R_G void fraction
Re, N_{Re} Reynolds number
R^+ dimensionless tube radius parameter, Equation 54
Re^* Reynolds number based on superficial phase velocity
S tube periphery (m)
Sc Schmidt number
Sh Sherwood number
T temperature (°K)
T^* dimensionless temperature, Equation 73
U velocity (m/sec)
u^+ dimensionless velocity
u^* friction velocity, Equation 45 (m/sec)
V specific volume (m^3/kg)
W mass flowrate (kg/hr)
δW work (J/kg)
W^+ dimensionless flowrate
W_i interface width (m)
X Lockhart-Martinelli parameter, Equation 5
x gas mass fraction
y_o distance normal from bottom of pipe wall (m)
y^+ dimensionless distance, Equation 46
Z parameter defined in Figure 5

Greek Symbols

α thermal diffusivity (m^2/sec)
$\hat{\alpha}$ interface width to diameter ratio
β coefficient of volume expansion (m/m-°K)
γ angle (°)
δ film thickness (cm)
ϵ_D eddy diffusivity (m^2/sec)
ϵ_H eddy thermal diffusivity (m^2/sec)
ϵ_M' eddy viscosity of momentum (m^2/sec)
θ angle (°), defined by Equation 41b
θ_c interface contact time (hr)
λ latent heat (J/kg-mol)
μ viscosity (poise)
ν kinematic viscosity (m^2/sec)
ρ density (kg/m^3)
ρ_H two-phase density (kg/m^3)
σ surface tension (dyne/cm)
τ_w wall shear stress (N/m^2)
Φ fractional rate of surface removal
ϕ Lockhart-Martinelli pressure drop parameter, see Equation 4 or wall heat flux (J/m^2-sec)
ϕ_{LO} two-phase flow multiplier, see Equation 7
χ von Karman constant = 0.36
Ω Baroczy correlation factor, see Figure 4

Subscripts

a acceleration
b bulk
EG refers to equivalent diameter of gas phase

EL	refers to equivalent diameter of liquid phase	L	liquid
		o	reference state
f	frictional	sat	saturated
G	gas	vap	vaporization
i	interface	w	wall

REFERENCES

1. Hewitt, G. F., and N. S. Hall-Taylor. *Annular Two-Phase Flow* (Elmsford, NY: Pergamon Press, Inc., 1970).
2. Ostrach, S., and A. Koestel. "Film Instabilities in Two-Phase Flows," *AIChE J.* 11 (2): 294 (1965).
3. Hanratty, T. J., and J. M. Engen. "Interaction between Turbulent Air Stream and a Moving Water Surface," *AIChE J.* 3: 299 (1957).
4. Hanratty, T. J. "Initiation of Roll Waves," *AIChE J.* 7(3): 488 (1961).
5. Davis, E. J. "An Analysis of Liquid Film Flow," *Chem. Eng. Sci.* 20: 265 (1965).
6. Lockhart, R. W., and R. C. Martinelli. "Proposed Correlation of Data for Isothermal Two-Phase, Two-Component Flow in Pipes," *Chem. Eng. Prog.* 45(1): 39 (1949).
7. Chenoweth, J. M., and M. W. Martin. "Turbulent Two-Phase Flow," *Pet. Ref.* 34 (10): 151 (1955).
8. Bertuzzi, A. F., M. R. Tek and F. H. Poettmann. "Simultaneous Flow of Liquid and Gas through Horizontal Pipe," *Pet. TAIME* 207: 17 (1956).
9. Hughmark, G. A. "Pressure Drop in Horizontal and Vertical Cocurrent Gas-Liquid Flow," *Ind. Eng. Chem. Fund.* 2(4): 315 (1963).
10. Baroczy, C. J. "A Systematic Correlation for Two-Phase Pressure Drop," *Chem. Eng. Prog. Symp. Series* 62 No. 64: 232 (1965).
11. Aziz, K., and G. W. Govier. "Horizontal Annular-Mist Flow of Natural Gas-Water Mixtures," *Can. J. Chem. Eng.* 51 (April 1962).
12. Casagrande, I., L. Cravardo and A. Hassid. "Researches on Adiabatic Two-Phase Flow," *Energia Nucl.* 9: 148 (1962).
13. Hughmark, G. A. "Holdup in Gas-Liquid Flow," *Chem. Eng. Prog.* 58(4): 62 (1962).
14. Hoogendoorn, C. J. "Gas-Liquid Flow in Horizontal Pipes," *Chem. Eng. Sci.* 9: 205 (1959).
15. Dukler, A. E., M. Wicks and R. G. Cleveland. "Frictional Pressure Drop in Two-Phase Flow: A Comparison of Existing Correlations for Pressure Loss and Holdup," *AIChE J.* 10(1): 38 (1964).
16. Delhaye, J. M. "Basic Equations for Two-Phase Flow Modelling," in *Two-Phase Flow and Heat Transfer in the Power and Process Industries* (New York: Hemisphere Publishing Corp., 1980).
17. Anderson, R. J., and T. W. F. Russell. "Designing for Two-Phase Flow—Parts 1, 2, 3" *Chem. Eng.* No. 25, 26, 139, 199 (1965); 73 (1): 87 (1966).
18. Hughes, E. D., R. W. Lyczkowski and J. H. McFadden. *An Evaluation of the State-of-the-Art of Two Velocity Two-Phase Flow Models and their Applicability to Nuclear Reactor Transient Analysis,* EPRI Report No. NP143 (1976).
19. Lahey, R. T., and F. J. Moody. *The Thermal-Hydraulics of a Boiling Water Nuclear Reactor,* American Nuclear Society (1977).
20. Davis, E. J., "Interfacial Shear Measurement for Two-Phase Gas-Liquid Flow by Means of Preston Tubes," *Ind. Eng. Chem. Fund.* 8: 153 (1969).
21. Collier, J. G. *Convective Boiling and Condensation* (London: McGraw-Hill Book Co., 1972).
22. Ishii, *Thermal-Fluid Dynamic Theory of Two-Phase Flow,* Eyrolles (1975).
23. Yu, H. S., and E. M. Sparrow. "Stratified Laminar Flow in Ducts of Arbitrary Shape," *AIChE J.* 13: 10 (1967).

24. Etchells, A. W. "Stratified Horizontal Two-Phase Flow in Pipe," PhD Thesis, University of Delaware, Newark, DE (1970).
25. Govier, G. W., and A. Aziz. *The Flow of Complex Mixtures in Pipes* (New York: Van Nostrand Reinhold Co., 1972).
26. Ellis, S. R. M., and B. Gay. "The Parallel Flow of Two Fluid Streams: Interfacial Shear and Fluid-Fluid Interaction," *Trans. Inst. Chem. Eng.* 37: 206 (1959).
27. Agrawal, S. S., G. A. Gregory and G. W. Govier. "An Analysis of Horizontal Stratified Flow in Pipes," *Can. J. Chem. Eng.* 51: 280 (1973).
28. Russell, T. W. F., A. W. Etchells, R. H. Jensen and P. J. Arruda. "Pressure Drop and Holdup in Stratified Gas-Liquid Flow," *AIChE J.* 20: 664 (1974).
29. Davis, E. J., and N. P. Cheremisinoff. "Stratified Two-Phase Flow: Pressure Drop and Holdup," AIChE-DIMP Report 4, Clarkson College of Technology, Potsdam, NY (1977).
30. Cheremisinoff, N. P., and E. J. Davis. "Stratified Turbulent-Turbulent Gas-Liquid Flow," *AIChE J.* 25(1) (1979).
31. Deissler, R. E. "Analytical and Experimental Investigation of Adiabatic Turbulent Flow in Smooth Tubes," *NACA-TN*-2138 (1952).
32. von Karman, T. "The Analogy between Fluid Friction and Heat Transfer," *Chem. Eng.* 148: 210 (1939).
33. Cohen, L. S., and T. J. Hanratty. "Effect of Waves at a Gas-Liquid Interface on a Turbulent Air Flow," *J. Fluid Mech.* 31: 467 (1968).
34. Miya, M., D. Woodmansee and T. J. Hanratty. "A Model for Roll Waves in Gas-Liquid Flow," *Chem. Eng. Sci.* 26: 1915 (1971).
35. Taitel, Y., and A. E. Dukler. "A Model for Predicting Flow Regime Transitions in Horizontal and Near Horizontal Gas-Liquid Flow," *AIChE J.* 22(1): 47–54 (1976).
36. Cheremisinoff, N. P. "An Experimental and Theoretical Investigation of Horizontal Stratified and Annular Two-Phase Flow with Heat Transfer," PhD Thesis, Clarkson College of Technology, Potsdam, NY (1977).
37. von Karman, T. "The Analogy between Fluid Friction and Heat Transfer," *Trans. ASME* 61: 705 (1939).
38. Martinelli, R. C. "Heat Transfer to Molten Metals," *Trans. ASME* 69: 947 (1947).
39. Prandtl, L. "Eine Beziehung Zwischen Wärmeaustauch and Ströhmungswiederstand der Flüssigkeiten," *Phys. Zeit* 10: 1072 (1910).
40. Taylor, G. I. *Brit. Advisory Comm. Aeronaut. Rep. Mem.* 272(31): 423 (1916).
41. Colburn, A. P. "A Method of Correlating Forced Convection Heat Transfer Data and a Comparison with Fluid Friction," *Trans. AIChE* 29: 174 (1933).
42. Hoffmann, E. *Forsch. Gebiete Ing.* 11: 159 (1940).
43. Boelter, L. M. K., R. C. Martinelli and F. Jonassen. "Remarks on the Analogy between Heat Transfer and Momentum Transfer," *Trans. ASME* 63: 447 (1941).
44. Lyon, R. N. "Liquid Metal Heat-Transfer Coefficient," *Chem. Eng. Prog.* 47(2): 75 (1951).
45. Deissler, R. G. "Investigation of Turbulent Flow and Heat Transfer in Smooth Tubes Including the Effects of Variable Properties, *Trans. ASME* 73: 101 (1951).
46. Friend, W. L., and A. B. Metzner. "Turbulent Heat Transfer Inside Tubes and Analogy Among Heat, Mass, and Momentum Transfer," *AIChE J.* 4(4): 393 (1958).
47. Kropholler, H. W., and A. D. Carr. "The Prediction of Heat and Mass Transfer Coefficients for Turbulent Flow in Pipes at All Values of the Prandtl and Schmidt Numbers," *Int. J. Heat Mass Transfer* 5: 1191 (1962).
48. Hughmark, G. A. "Momentum, Heat and Mass Transfer Analogy for Turbulent Flow in Circular Pipes," *Ind. Eng. Chem. Fund.* 8(1): 31 (1969).
49. Rosson, H. F., and J. A. Myers. "Point Values of Condensing Film Coefficients Inside a Horizontal Pipe," *Chem. Eng. Prog. Symp. Series 61* No. 59: 190 (1965).
50. Dukler, A. E. "Fluid Mechanics and Heat Transfer in Falling Film Systems," *Chem. Eng. Prog. Symp. Series 56* No. 30 (1960).
51. Hewitt, G. F. "Analysis of Annular Two-Phase Flow, Application of the Dukler Analysis to Vertical Upward Flow in a Tube," AERE-R 3680, H. 7.S.O. (1961).

52. Altman, M., F. W. Staub and R. H. Norris. "Local Heat Transfer and Pressure Drop for Refrigerant-22 Condensing in Horizontal Tubes," *Chem. Eng. Prog. Symp. Series 56* No. 30: 151 (1960).
53. Kosky, P. G., and F. W. Staub. "Local Condensing Heat Transfer Coefficients in the Annular Flow Regime," *AIChE J.* 17(5): 1037 (1971).
54. Hughmark, G. A. "Heat Transfer in Horizontal Annular Gas-Liquid Flow," *Chem. Eng. Prog.* 59(7): 54 (1963).
55. Davis, E. J., N. P. Cheremisinoff and G. Sambasivan. "Heat and Momentum Transfer Analogies for Horizontal Stratified Two-Phase Flow," AIChE-DIMP Report, Clarkson College of Technology, Potsdam, NY (1976).
56. Davis, E. J., N. P. Cheremisinoff and C. J. Guzy. "Heat Transfer with Stratified Gas-Liquid Flow," *AIChE J.* 25(6) (1979).
57. Davis, E. J., and N. P. Cheremisinoff. "Heat Transfer to Two-Phase Stratified Gas-Liquid Flow," AIChE-DIMP Report 5, Clarkson College of Technology, Potsdam, NY (1977).
58. Dengler, C. E., and J. N. Addams, "Heat Transfer Mechanism for Vaporization of Water in a Vertical Tube," *Chem. Eng. Prog. Symp. Series* 52(95) (1956).
59. Guerrieri, S. A., and R. D. Talty, "A Study of Heat Transfer to Organic Liquids in Single Tube, Natural Circulation Vertical Tube Boilers," *Chem. Eng. Prog. Symp. Series*, Heat Transfer, Louisville, 52(18): 67–77 (1956).
60. Shrock, V. E., and L. M. Grossman, "Forced Convection Boiling Studies," Forced Convection Vaporization Project, USAEC Rept. TID-14632 (1959).
61. Wright, R. M., "Downflow Forced-Convection Boiling of Water in Uniformly Heated Tubes," University of California, Lawrence Radiation Labs., UCRL-9744 (Aug. 1961).
62. Somerville, G. F., "Downflow Boiling of n-Butanol in a Uniformly Heated Tube," University of California, Lawrence Radiation Labs, UCRL-10527 (Oct. 1962).
63. Collier, J. G., P. M. C. Lacey and D. J. Pulling, *Trans. Inst. Chem. Engrs.*, 42 (1964).
64. Leppert, G., and C. C. Pitts. "Boiling," *Adv. Heat Transfer* 1: 185 (1964).
65. Bennett, A. W., J. G. Collier and P. M. C. Lacey, AERE-R3804 (Aug. 1961).
66. Danckwerts, P. V. *Gas-Liquid Reactions* (New York: McGraw-Hill Book Co., 1970).
67. Cichy, P. T., J. S. Ultman and T. W. F. Russell. "Two-Phase Reactor Design-Tubular Reactors," *Ind. Eng. Chem. Fund.* 61(8): 6 (1969).
68. Cichy, P. T., and T. W. F. Russell. "Two-Phase Reactor Design Tubular Reactors," *Ind. Eng. Chem. Fund.* 61: 15 (1969).
69. Shah, Y. T. *Gas-Liquid-Solid Reactor Design* (New York: McGraw-Hill Book Co., 1979).
70. Lewis, W. K., and W. G. Whitman: "Principles of Gas Absorption," *Ind. Eng. Chem. Fund.* 16: 1215 (1924).
71. Higbie, R., C. J. King and C. R. Wilke. "Gas-Liquid Mass Transfer in Cocurrent Froth Flow," *AIChE J.* 11(5): 866 (1965).
72. Danckwerts, P. V. "Gas Absorption Accompanied by Chemical Reaction," *AIChE J.* 1: 456 (1955).
73. Hanratty, T. J. "Turbulent Exchange of Mass and Momentum with a Boundary," *AIChE J.* 2(3): 359 (1965).
74. Perlmutter, D. D. "Surface Renewal Models in Mass Transfer," *Chem. Eng. Sci.* 16: 287 (1961).
75. Dobbins, W. E. "Biological Treatment of Sewage and Industrial Wastes," W. L. McCabe and W. W. Eckenfelder, Eds. (New York: Van Nostrand Reinhold Co., 1956).
76. Kishinevsky, M. K. *J. Appl. Chem. USSR* 28: 881 (1955).
77. Kishinevsky, M. K., and V. T. Serebryansky. "The Mechanism of Mass Transfer at the Gas-Liquid Interface with Vigorous Stirring," *J. Appl. Chem. USSR* 29: 29 (1956).
78. Levich, V. G. *Physicochemical Hydrodynamics* (Englewood Cliffs, NJ: Prentice-Hall, Inc., 1962).

79. Davis, J. T. "The Effects of Surface Films in Damping Eddies at a Free Surface of a Turbulent Liquid," *Proc. Roy. Soc.* A290: 515 (1966).
80. King, C. J. "Turbulent Liquid Phase Mass Transfer at a Free Gas-Liquid Interface," *Ind. Eng. Chem. Fund.* 5: 1 (1966).
81. Lamourelle, A. P., and O. C. Sandall. "Gas Absorption into a Turbulent Liquid," *Chem. Eng. Sci.* 27: 1035 (1972).
82. Wales, C. E. "Physical and Chemical Absorption in Two-Phase Annular and Dispersed Horizontal Flow," *AIChE J.* No. 6: 1166 (1966).
83. Treybal, R. E. *Mass Transfer Operations*, 2nd ed. (New York: McGraw-Hill Book Co., 1968).
84. McCabe, W. L., and J. C. Smith. *Unit Operations of Chemical Engineering*, 3rd ed. (New York: McGraw-Hill Book Co., 1976).
85. Hayduk, W. "Gas Absorption in Horizontal Cocurrent Bubble Flow," PhD Thesis, University of British Columbia, Vancouver, BC, Canada (1964).
86. Jepsen, J. C. "Mass Transfer in Two-Phase Flow in Horizontal Pipelines," *AIChE J.* 16(5): 705 (1970).
87. Anderson, J. D., R. E. Bollinger and D. E. Lamb. "Gas Phase Controlled Mass Transfer in Two-Phase Annular Horizontal Flow," *AIChE J.* No. 5: 640 (1964).
88. Hughmark, G. A. "Mass Transfer in Horizontal Annular Gas-Liquid Flow," *Ind. Eng. Chem. Fund.* 4(3): 361 (1965).
89. Bollinger, R. E. MS Thesis, University of Delaware, Newark, DE (1960).
90. Jagota, A. K., E. Rhodes and D. S. Scott, "Mass Transfer in Upwards Cocurrent Gas-Liquid Annular Flow," *Chem. Eng. J.*, 5(23) (1973).
91. Banerjee, S., D. Scott and E. Rhodes, "Cocurrent Gas-Liquid Flow in Helically Coiled Tubes," *Can. J. Chem. Eng.* 48(5): 542 (1970).
92. Banerjee, S., D. Scott and E. Rhodes, "Mass Transfer to Falling, Wavy, Liquid Films in Turbulent Flow," *Ind. Eng. Chem. Fund.* 7(1): 22 (1968).
93. Howard, D. W., and E. N. Lightfoot, AIChE J. 14: 458 (1968).
94. Vivian, J. E., and D. W. Peaceman, "Liquid-Side Resistance in Gas Absorption," *AIChE J.*, 2(437) (1966).
95. Downing, A. L., and G. Knowles, *International Conference on Water Pollution, Advances in Water Pollution Research, Proceedings Vol. 1*, p. 263 (1964).
96. Rohsenow, W. M., and J. P. Hartnett, Eds. *Handbook of Heat Transfer* (New York: McGraw-Hill Book Co., 1973).

CHAPTER 15
TWO-PHASE FLOW PATTERNS

Joel Weisman
Department of Chemical and Nuclear Engineering
University of Cincinnati
Cincinnati, Ohio 45221

CONTENTS

INTRODUCTION

When a liquid and vapor flow together in a channel, there must be an interface between the phases. The phase boundaries take a variety of configurations, or "flow patterns." The term "flow pattern" is now preferred over the earlier designation of "flow regime," because the term "regime" often is associated with the presence or absence of turbulence. Thus, Lockhart and Martinelli [1] defined four two-phase flow regimes, depending on whether the vapor or liquid flows were or were not turbulent. These regimes were not related directly to the configuration assumed by the liquid and vapor.

The typical flow patterns encountered in cocurrent flow through a horizontal tube are illustrated in Figure 1. At the lowest gas and liquid flowrates, the liquid flows along the bottom of the tube and the gas along the top, which is designated as stratified flow. At higher gas flowrates, waves are found on the gas-liquid interface; this is called "wavy flow."

At higher liquid flowrates than those that occur during stratified flow, we encounter "bubble flow" if the gas flowrate is very low. Here, small bubbles of gas tend to flow in the upper portion of the tube. When the gas flowrate is increased, large bullet-shaped bubbles, called plugs, are formed. In plug flow, these plugs move through the liquid along the upper portion of the tube. An increase in gas flow leads to "slug flow," where frothy slugs of liquid move across the upper region of the tube. Between the slugs there is a wavy layer of liquid at the bottom of the tube.

At high gas velocities "annular flow" appears. Here the liquid flows around the outside of the tube while the gas flows in a central core. Further increases in gas velocities lead to entrainment of some of the liquid as droplets carried along in the central gas core.

At very high mass flowrates we encounter dispersed flow. At low quantities this is seen as a froth of tiny bubbles essentially uniformly distributed in the liquid. At high quantities it is seen as a mist of fine droplets suspended in the vapor.

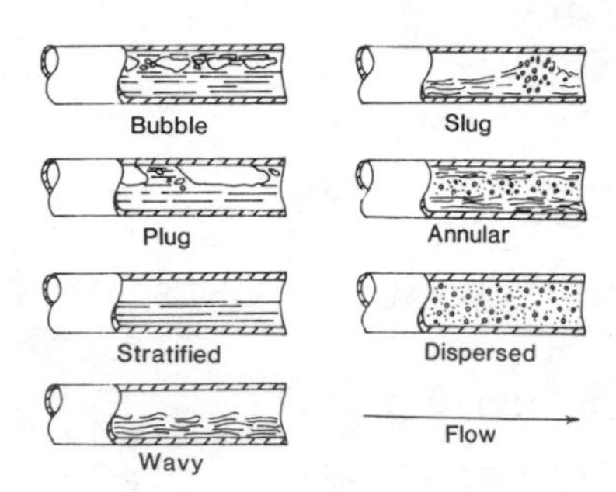

Figure 1. Flow pattern configurations in horizontal flow.

It has been recognized for some time that pressure drop and interphase mass and momentum transfer are influenced significantly by the flow pattern prevailing. More recently, Ngugen and Spedding [2] established the effect of flow patterns on void fraction. Nevertheless, most current computational analyses of two-phase systems tend to ignore the flow pattern. This stems in part from the difficulty of establishing reliable criteria for the transition from one pattern to another. This situation is now somewhat improved, and greater attention to flow pattern effects can be expected.

FLOW PATTERN OBSERVATION AND MAPPING

Determination of flow pattern behavior has been accomplished primarily by visual observation of the flow in transparent channels. The range of visual observations often is enhanced by flash photography or cine photography. However, because flow pattern changes are generally gradual, it is often difficult to determine the exact point at which the transition occurs.

In an effort to eliminate some of the subjectivity inherent in visual observations, several investigators have supplemented or replaced visual observations by other measurements. Hubbard and Dukler [3] used a spectral density analysis of the pressure drop fluctuation between two wall locations a few inches apart. Weisman et al. [4] also used these pressure drop fluctuations but used a set of simple criteria that could be applied to an oscillograph trace. Conductance probes have been used by a number of investigators. More recently, Barnea et al. [5] used a conductance probe with several electrodes to distinguish among flow patterns in horizontal flow.

Jones and Zuber [6] found that a spectral density analysis of X-ray void fraction measurements also could be used for flow pattern determination. The void measurement approach has been extended recently [7] by using X-ray scanning techniques similar to those now being employed in medicine. With this approach, it is possible to obtain the image of the cross section showing the vapor and liquid regions at any desired location.

The experimental flow patterns observations are usually presented in a flow pattern map. The coordinates of such maps are generally superficial mass or linear velocities for the vapor and liquid. A symbol, indicating the flow pattern, is placed at each location where an observation has been made. Figure 2 shows such an experimental flow pattern map for observations made with air and water in a horizontal tube. The shaded regions

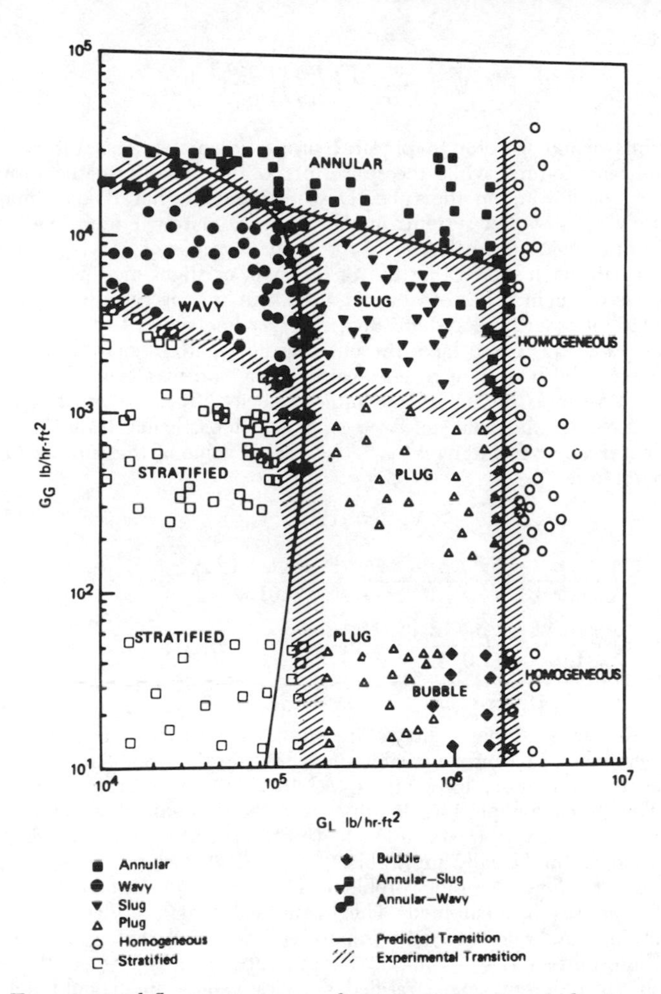

Figure 2. Experimental flow pattern map for air-water system in horizontal flow: line diameter = 2 in. [8].

represent the locations of the flow pattern transitions based on the observations recorded, and the solid lines represent the investigator's prediction of the locations of the flow pattern transitions.

FLOW PATTERN TRANSITIONS IN HORIZONTAL ADIABATIC LINES

Baker [9] was perhaps the earliest investigator to attempt to systematize flow pattern data. His approach to the correlation of flow pattern transition data relied on the use of a flow pattern map in dimensionless coordinates, which were believed to apply to all of the flow pattern transitions. Baker plotted his map in terms of G_G/λ and $G_L \lambda \psi/G_G$, where G_G and G_L are the superficial mass velocities of the vapor and liquid, respectively, and the factors ψ and λ are given by

$$\lambda = \left[\left(\frac{\rho_G}{\rho_a}\right)\left(\frac{\rho_L}{\rho_w}\right)\right]^{1/2}$$

$$\psi = \left(\frac{\sigma_w}{\sigma}\right)\left[\left(\frac{\mu_L}{\mu_w}\right)\left(\frac{\rho_w}{\rho_L}\right)^2\right]^{1/3} \tag{1}$$

The subscripts a and w refer to physical properties of air and water at atmospheric pressure and temperature, while the subscripts L and G indicate the properties of the liquid and vapor flowing in the tube. Despite the fact that Baker's map was based almost entirely on air-water systems in tubes approximately 1 inch in diameter, it was used widely to estimate the behavior of the fluids in tubes of varying sizes. Subsequent experimental data have shown that the Baker map, or slight modifications of it [10], is reliable for air-water in tubes of about 1 inch, but that it furnishes a poor prediction of the behavior of other fluid systems and in larger line sizes.

Mandhane et al. [11] also have presented a flow pattern map for horizontal tubes, which has been used widely in recent years. Their map was in terms of the superficial gas velocity, V_{SG}, and the superficial liquid velocity, V_{SL}. To use the map for fluids other than air-water, the value of V_{SG} at which a particular transition occurs is determined from the map for a given V_{SL}. A corrected value of the superficial gas velocity, V_{SG}', is formed from

$$V_{SG}' = f(x,y)(V_{SG}) \tag{2}$$

where $$x = \left[\left(\frac{\rho_G}{0.0013}\right)^{0.2}\left(\frac{\rho_L}{1.0}\frac{72.4}{\sigma}\right)^{0.25}\left(\frac{\mu_G}{0.018}\right)^2\right]$$

$$y = \left[\left(\frac{\mu_L}{1.0}\right)^{0.2}\left(\frac{\rho_L}{1.0}\frac{72.4}{\sigma}\right)^{0.25}\right]$$

Their map was again based largely, but not entirely, on air-water data. The map relies on the idea that fluid property effects are similar for all transitions and that diameter effects are negligible. Presently available data indicate that this is not the case.

Most recent investigators have attempted to obtain a separate prediction for each individual flow transition pattern. It appears to be recognized generally that the significant parameters vary from transition to transition and that no simple set of dimensionless groupings is universally applicable.

To simplify the flow mapping problem, Hubbard and Dukler [3] proposed that several flow patterns be combined. They proposed that stratified wavy and annular flow be combined and considered as "separated" flow and that plug and slug be considered as "intermittent flow." Choe et al. [8] subsequently argued that annular flow was substantially different from stratified and wavy flow and should not be included in the same category. Choe et al. [8] therefore concluded that the horizontal flow patterns of major interest were separated, intermittent, annular and dispersed. We shall generally follow this classification scheme. With this categorization, there are only four flow pattern transitions that must be considered: onset of annular flow, onset of dispersed flow, separated-intermittent transition, and bubble-intermittent transition.

Many recent investigators have attempted to develop physical models of the various transitions, and encouraging progress has been made in some areas. There appears to be reasonable agreement on the mechanism governing the transition from dispersed to nondispersed flow. When the mixing forces of the turbulent eddies overcome the tendency of the bubbles to rise and coalesce, dispersed flow occurs. Husain and Weisman [12] presented a model of this type and based their development on an analogy with the minimum power required to suspend solid particles in a pipeline. Taitel and Dukler [13] presented a somewhat similar model using a vertical force balance for their derivation. They concluded that this transition should be governed by the dimensionless groups X and T, where X is as defined in Equation 5 and T is given by

$$T = \left(\frac{dp/dx\,|_L^S}{(\rho_L - \rho_G)g}\right)^{1/2} \tag{3}$$

Weisman et al. subsequently compared the Taitel-Dukler correlating approach to data for a variety of fluid systems. They found that the effect of physical property variation was accounted for inadequately and that T was nearly independent of X. The data were well correlated by

$$T[\sigma/(\rho_L - \rho_G)Dg]^{-1/4} = 0.97 \quad (4)$$

Weisman et al. [4] also showed that the correlation resulting from this revision was very similar to that obtained by Husain and Weisman [12].

The transition from separated flow occurs with increasing liquid or gas flow. As the flow increases, the liquid level rises and waves are formed. Taitel and Dukler [13] concluded that further flowrate increases cause wave growth and flow blockage. The flow blockage takes the form of intermittent flow at low gas flowrates and annular flow at high gas flowrates. Based on the conditions for a wave to become unstable and grow, they concluded that the dimensionless groups governing the separated-intermittent transition were

$$V^*_{SG} = V_{SG}\left[\left(\frac{\rho_G}{\rho_L - \rho_G}\right)\left(\frac{1}{Dg}\right)\right]^{1/2} \quad (5)$$

$$X = \left[\frac{(dp/dx\,|^S_L)}{(dp/dx\,|^S_G)}\right]^{1/2}$$

Taitel and Dukler showed that the correlation based on these groups was in agreement with the Mandhane et al. [11] flow map. However, Weisman et al. [4] found that the correlation disagreed sharply with glycerine solution data.

For the separated intermittent-transition, Choe [8] used an approach similar to that developed by Wallis and Dobson [14] to determine the conditions when large-sized waves first reached the top of the tube. He concluded that there should be a relationship between V^*_{SG} and the void fraction, α. They found that they could correlate the then available data by

$$V^*_{SG} = 2.5\exp[-12(1-\alpha)] + 0.03\alpha \quad (6)$$

When this correlation was used with glycerine solution data [4], moderate agreement was obtained if observed values of α were used in Equation 6. However, attempts to use an α calculated from available quality vs α relationships yielded poor results. In view of this difficulty, the use of α was replaced by V_{sG}/V_{sL}. Further, experimental studies of Freon 113 systems (high gas density) showed that the gas density variation predicted by Equation 6 was inaccurate. The correlation finally obtained was

$$V^{1/2}_{SG}/gD = 0.25\,(V_{sG}/V_{sb})^{1.1} \quad (7)$$

Several approaches have been suggested for the intermittent-annular transition. Taitel and Dukler [13] argued that the choice between which of the possible nonseparated flow patterns is reached depends on the equilibrium level of liquid in the duct. According to their hypothesis, if h_L/D is less than 0.5, the waves will deposit the liquid on the top of the tube. They concluded this would occur at a constant value of X. Comparison of available data with this hypothesis has shown poor agreement.

Choe et al. [8] proposed an alternative criterion for the separated-annular transition. In this model, liquid is deposited on the top of the tube by periodic waves or slugs. The liquid deposited drains during the period between the waves or slugs. If the frequency of the waves or slugs is sufficiently high so that drainage is incomplete, then annular flow is said to be established. While the approach appears to fit air-water data, it did not represent the data from air-glycerine solutions.

Wallis [15] has suggested that for vertical lines the annular intermittent transition occurs at a constant value of V^*_{SG}. Weisman et al. [4] observed that this is also approximately true for horizontal flow. However, the approach overemphasizes the effect of gas density, and better agreement was obtained by using the product of the Froude number and the Kutadelaze number. They finally proposed

$$1.9\ (V_{SG}/V_{SL})^{1.8} = (\text{KuFr})^{0.2} \tag{8}$$

and found that this fitted all the available data.

Although many horizontal flow maps omit bubble flow, this flow pattern is found at very low gas flows and moderate liquid flows. The theoretical approaches to the transition from bubbly to intermittent flow have been based on the idea that this transition occurs when the bubbles become so closely packed that many collisions occur. Radovich and Moissis' [16] semitheoretical approach led to the conclusion that this occurs when the void fraction is about 0.3. Taitel et al. [17] used a somewhat similar approach to obtain a relationship between superficial gas and liquid velocities and physical properties. For vertical flow they proposed

$$V_{SL} = 3.0\ V_{SG} - 1.15 \left[\frac{g(\rho_L - \rho_G)\sigma}{\rho_L^2}\right]^{1/4} \tag{9}$$

Weisman and Kang [18] found that this expression represented the actual data rather poorly and obtained an alternative correlation for inclined and vertical lines. Extrapolation of their correlation to horizontal lines yields

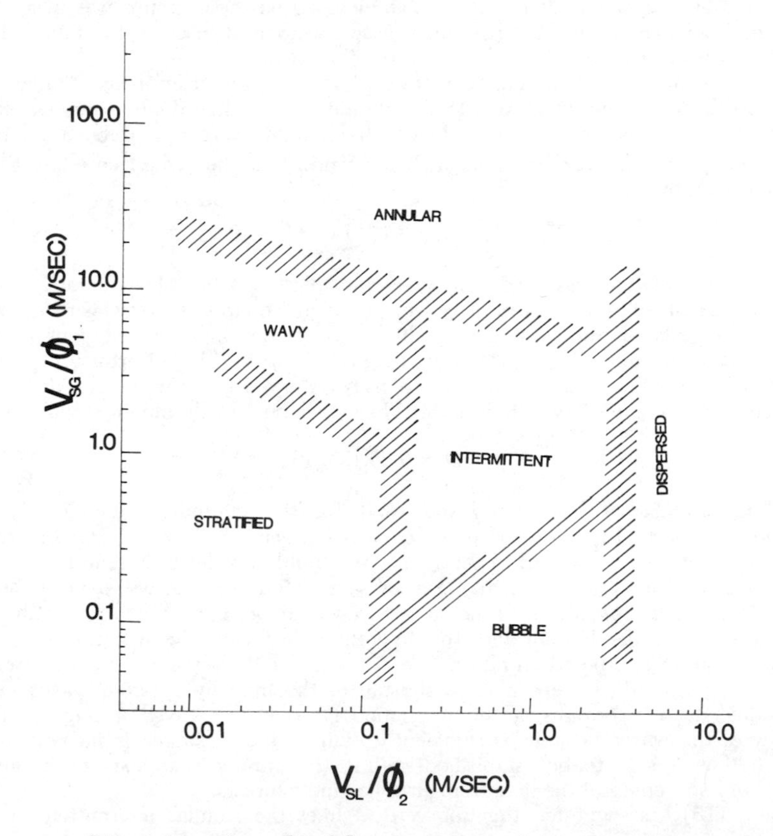

Figure 3. Generalized flow pattern map for horizontal flow.

$$\frac{V_{SG}}{\sqrt{gD}} = 0.15 \left(\frac{V_{SG} + V_{SL}}{\sqrt{gD}}\right)^{0.78} \tag{10}$$

Equation 10 appears to be in agreement with the limited horizontal line data available at these low gas flowrates.

Weisman et al. [4] have devised a generalized flow map that incorporates what they believe are the best transition line correlations. The map is in terms of modified superficial gas and liquid velocities. As shown in Figure 3, the coordinates are V_{sG}/ϕ_1 and V_{SL}/ϕ_2. For water and air in a 1-inch line at room temperature, $\phi_1 = \phi_2 = 1.0$. For other fluids and pipe diameters the values of ϕ_1 and ϕ_2, which are derived from the transition line correlations of Weisman et al. [4] and Weisman and Kang [18], are obtained from Table I.

FLOW PATTERNS IN UPWARDLY INCLINED AND VERTICAL LINES (ADIABATIC CONDITIONS)

A significant change in flow patterns occurs when a line is inclined slightly to the horizontal. The stratified flow pattern disappears and is replaced by intermittent flow. Weisman and Kang [18] suggest that stratified flow will disappear when

$$\sin\theta > D/L \tag{11}$$

where θ is the angle of inclination, and L is the length of line being examined.

Wavy flow is still seen in inclined lines, but the wavy flow region is restricted. As the inclination angle increases, the gas flowrate at which wavy flow begins also increases. At a sufficiently sharp angle of inclination, wavy flow disappears altogether. However, the bubble flow area expands as inclination is increased.

In vertical lines and lines at very sharp angles of inclination, slug flow disappears and the only flow patterns seen are bubbly, plug, churn and annular (Figure 4). Churn flow is a chaotic mixture of large packets of gas and liquid, which seem to have a churning motion. In accordance with the simplification suggested previously, we shall consider churn and plug flow as intermittent flow.

Taitel et al. [17] have suggested a series of models for the various flow pattern transitions. The model for the transition to dispersed flow is again based on the existence of sufficient turbulence to disperse small bubbles and is similar to their model for horizontal flow. Their model for the annular flow transition is based on the idea that annular flow cannot exist unless the gas velocity in the gas core is sufficient to lift entrained droplets. If the gas velocity is insufficient, the droplets are believed to fall back and form a bridge, which leads to intermittent flow. On the basis of this analysis, they concluded that at the annular-intermittent transition,

$$\mathrm{Ku} = 3.1 \tag{12}$$

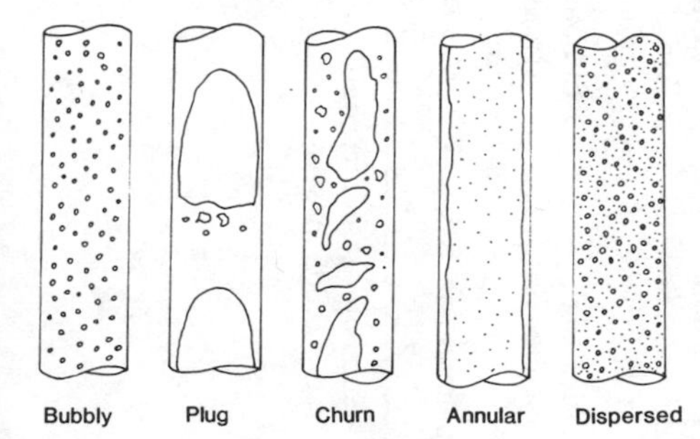

Figure 4. Flow pattern configurations in upwardly inclined and vertical flow.

Table I. Property and Pipe Diameter Corrections to Overall Flow Map[a]

Flow Orientation		ϕ_1	ϕ_2
	Transition-to-dispersed flow	1.0	$\left(\frac{\rho_L}{\rho_{SL}}\right)^{-0.33} \left(\frac{D}{D_S}\right)^{0.16} \left(\frac{\mu_{SL}}{\mu_L}\right)^{0.09} \left(\frac{\sigma}{\sigma_S}\right)^{0.24}$
Horizontal, vertical and inclined flow	Transition-to-annular flow	$\left(\frac{\rho_{SG}}{\rho G}\right)^{0.23} \left(\frac{\Delta\rho}{\Delta\rho_S}\right)^{0.11} \left(\frac{\sigma}{\sigma_S}\right)^{0.11} \left(\frac{D}{D_S}\right)^{0.415}$	1.0
Horizontal and slightly inclined flow	Intermittent-separated transition		$\left(\frac{D}{D_S}\right)^{0.45}$
Horizontal flow	Wavy-stratified transition	$\left(\frac{D_S}{D}\right)^{0.17} \left(\frac{\mu_G}{\mu_{SG}}\right)^{1.55} \left(\frac{\rho_{SG}}{\rho_G}\right)^{1.55} \left(\frac{\Delta\rho}{\Delta\rho_S}\right)^{0.69} \left(\frac{\sigma_S}{\sigma}\right)^{0.69}$	1.0
Vertical and inclined flow	Bubble-intermittent transition	$\left(\frac{D}{D_s}\right)^{n} (1 - 0.65 \cos\theta)$ $n = 0.26e^{0.17(V_{SL}/V_{SL}^{S})}$	1.0
Inclined flow	Wavy-intermittent transition	$1 + 5 \sin\theta$	1.0

[a]S denotes standard conditions. $D_S = 1.0$ in. $= 2.54$ cm, $\rho_{SG} = 0.0013$ kg/1, $\rho_{SL} = 1.0$ kg/1, $\mu_{SL} = 1$ cP, $\sigma_S = 70$ dynes/cm, $V_{SL}^{s} = 1.0$ ft/sec $= 0.305$ m/sec.

The Taitel et al. [17] bubble transition approach was discussed when considering horizontal tubes.

Weisman and Kang [18] noted that at high velocities pipeline orientation should have little effect. They found the dispersed flow transition criterion proposed by Weisman et al. [4] for horizontal flow to be equally applicable to vertical flow. The Taitel et al. [17] approach for this transition yielded similarly good results.

Comparison of the Taitel et al. [17] approach for the annular flow transition with actual data [18] showed reasonable agreement. However, the Weisman et al. [18] approach to this transition in horizontal flow also worked well here.

The annular transition line was unaffected by the angle of inclination. Weisman and Kang [18] did observe that the behavior of the ½-inch line was anomalous. Annular flow began at gas flowrates lower than those predicted.

The bubble-intermittent criterion proposed by Taitel et al. [17] was not found entirely satisfactory. Weisman and Kang [18] found they could correlate all the data for inclined and vertical lines using the gas-phase and total-flow Froude number. They proposed

$$\frac{V_{SG}}{\sqrt{gD}} = 0.45\left(\frac{V_{SG} + V_{SL}}{\sqrt{gD}}\right)^{0.78} (1 - 0.65 \cos \theta) \qquad (13)$$

Figure 5 shows a flow map for vertical and inclined lines, which is essentially the vertical map of Weisman and Kang [18]. Their vertical map has been modified to include a wavy region. Again the values of ϕ_1 and ϕ_2 to be used are to be found in Table I. It should be noted that the value of ϕ_1 reported for the wavy-intermittent boundary is only approximate, as the effects of fluid properties and pipeline diameter

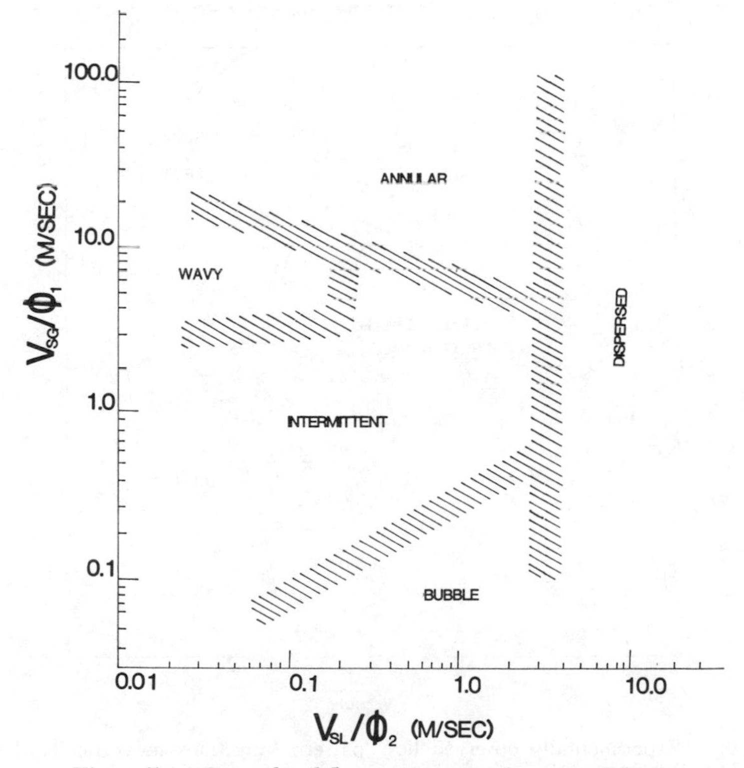

Figure 5. Generalized flow pattern map for vertical flow.

are not yet known. Use of the map for prediction of annular flow transition should be limited to line sizes of 1 inch and larger.

FLOW PATTERNS IN COCURRENT DOWNWARD FLOW

Tests at the University of Cincinnati have shown that as the pipeline is inclined downward from the horizontal, the separated flow region expands at the expense of the intermittent region. When the flow is vertically downward, intermittent flow is restricted to a small region. The separated flow region has become what is often called a "falling film," or "wetted wall," region. The distinction between this region and the annular flow region is nebulous, although at least one set of investigators [19] claims to observe a difference.

Figure 6 shows some of the limited data presently available for downward flow inside round tubes. Considerable scatter is observed, and no definite quantitative correlations or models have yet emerged. By comparison of Figures 3 and 6, it would appear that dispersed flow may begin at only slightly lower values of V_{SG} than predicted by the horizontal map. The bubble-intermittent transition also would appear to be in the vicinity of the predictions of the horizontal map. However, the intermittent-falling film

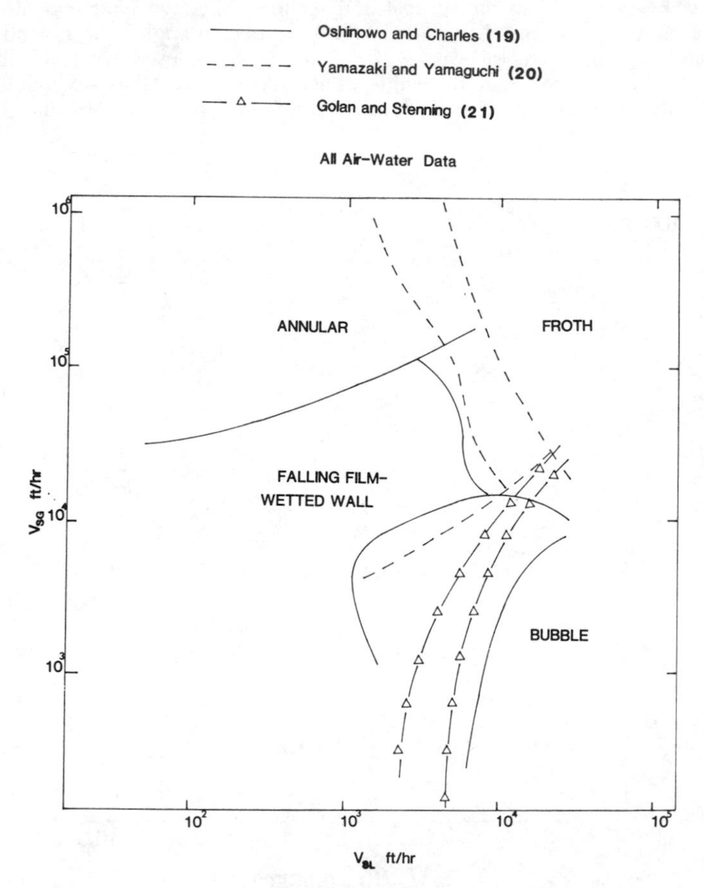

Figure 6. Experimentally observed flow pattern transitions in vertically downward flow [19–21].

transition is unclear. It appears to be very roughly in the position of the slug-plug transition in horizontal flow. Additional work in this area clearly is needed.

FLOW PATTERNS IN NONCIRCULAR DUCTS

Hosler [22] indicates that the flow patterns and transitions observed in rectangular channels are similar to those seen in round tubes. Gibson [23] reports that the flow patterns seen with horizontal air-water flow in an annulus were the same as in a round tube. The location of the flow pattern transitions differed only very slightly from those seen in a round tube.

The presence of screens, grids or orifices in a flow line generally leads to a considerable degree of homogenization in the vicinity of the obstruction. The flow pattern reverts to the normal behavior in a relatively short distance. Similar mixing is seen just downstream of a contraction, and such mixing significantly affects the observed pressure drop [24].

Peterson and Williams [25] have examined flow patterns in a vertical flow of steam-water mixtures in rod bundles. Weisman and Kang [18] compared these results with the vertical flow pattern map shown in Figure 5. The data taken at 27 bar show the same sequence of events as expected: bubble, intermittent and annular flow successively appear with increasing gas flowrate. The transitions between these flow patterns are in the general region of the round tube transitions. At higher pressures, the transitions are again in the regions expected, but intermittent flow has been replaced by froth flow. As noted previously, obstructions present in a line tend to homogenize the flow immediately downstream of the obstruction. Weisman and Kang [18] suggest that the bundles' spacer grids tend to homogenize the flow and that this homogenization is easier at high pressures, where there is less of a density difference between phases.

FLOW IN BENDS AND COILS

Collier [26] notes that a bend or system of bends acts to separate the phases. A bend can cause coalescence of bubbles to form plugs and can separate entrained droplets in annular flow. The curvature of a pipeline induces secondary flow in the fluid in the form of a double spiral.

Boyce et al. [27] report on the behavior of air-water systems in horizontal helical coils. The flow patterns seen are similar to those seen in horizontal flow, except that bubbly flow is absent. In annular flow, it is found that the liquid tends to travel as a film on the inside surfaces of the coils at low flows. At higher flows, liquid is entrained from the flow on the inside surface and is carried across the tube diameter to be deposited farther downstream. The liquid is then returned to the inside surface of the coil via circulation of the liquid film.

FLOW PATTERNS UNDER ADIABATIC CONDITIONS

When boiling occurs in a heated vertical tube, we see subcooled nucleate boiling near the inlet, with bubbles present only at the wall if the entering fluid is subcooled significantly. Further along the tube, when the subcooling has decreased sufficiently so that bubbles detaching from the wall are not immediately condensed, we find bubbles distributed across the stream. At low-to-moderate flows, bubbly flow is succeeded by churn flow as the void fraction is increased. At higher mass flows, bubbly flow is succeeded by plug flow. At still higher void fractions, annular flow is seen. The succession of bubble to intermittent to annular flow is what is to be expected on the basis of adiabatic flow maps.

Quantitative observations of flow pattern transitions in vertical heated tubes have been made by Bergles et al. [28] and Hosler [22], among others. In the experiments of Bergles et al. [28], the inlet subcooling was large and nonequilibrium effects were important, thus making it difficult to compare his results to adiabatic flow maps. In Hosler's [22] experiments the nonequilibrium effects were much less important, and comparisons may be accomplished more readily.

When Hosler's flow maps, which are given in terms of total mass velocity and quality for runs with a pressure of 600–2000 psi, are converted to the form used in Figures 3 and 5, comparisons of the bubble-intermittent and annular transition are possible. It is found that the bubble-intermittent transition lines are in fair agreement with the predictions of Figure 5. In comparing the predictions and observations of the annular-intermittent transition, account should be taken of the observation that in small-sized (½-in. and less) inclined lines the annular transition takes place at lower velocities than indicated by Figure 5. With this modification, fair agreement between prediction and observations is obtained, the predictions being somewhat below the observations.

Flow pattern observations recently have been carried out with water boiling at 1.6–3.5 bar in an annulus with a centrally heated rod. These data showed good agreement between predictions and observations of the bubble-intermittent and intermittent-annular transitions. Predictions were based on Figure 5, but the annular-intermittent transition was taken to be at about 60% of the value of Figure 5 since the equivalent diameter of the passage was approximately ½ inch.

On the basis of the limited information available it would appear that transitions of flow patterns in flowing systems with heat addition are generally similar to those observed in adiabatic flow. Figure 5 may be used as a guide for their prediction. Note that when small lines are used, annular flow appears to occur at lower gas flowrates than indicated by Figure 5. As shown by the data of Bergles et al. [28], estimation of the vapor flowrate must include nonequilibrium effects.

Flow patterns during horizontal tube-side condensation have been investigated most recently by Palen et al. [29]. Their general observations are indicated in Figure 7. At low mass flows, the stream progresses from annular to slug to plug to subcooled liquid. At high mass flows, the stream progresses from annular to wavy to subcooled liquid. These progressions are in qualitative agreement with the horizontal flow map of Figure 3. Palen et al. [29] suggest that at low mass flows, preliminary categorization is accomplished using the dimensionless quantity V^*_{SG} (see Equation 5). They suggest that their transition region (semiannular, semiannular wave) occurs at values of V^*_{SG} between 0.5 and 1.5. We have noted previously that Wallis [15] proposed V^*_{SG} as a tran-

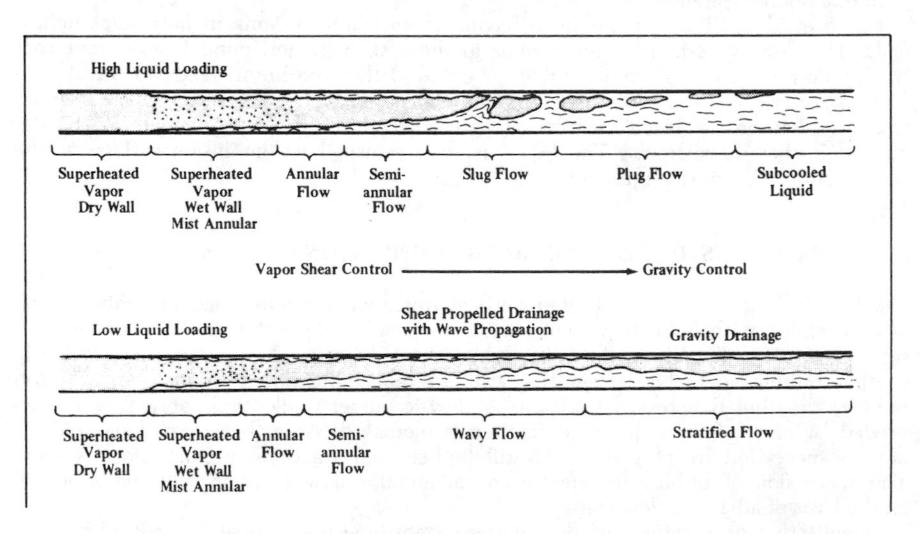

Figure 7. Flow pattern configurations seen in tube-side condensation [29].

sition criterion. Palen et al. state that a proprietary flow map based on V^*_{SG}, or a similar parameter, and a liquid volume fraction provides a full description of the data. Weisman et al. [4] show that a graph of V^*_{SG} vs V_{SG}/V_{SL} provides a fair correlation of these data but is not as good as their final approach. They noted that the transition occurred at V^*_{SG} values between 0.4 and 2.0, which is the same range suggested by Palen et al. [29]. In view of this general agreement, and because the qualitative observations reported are in agreement with Figure 5, it is suggested that Figure 5 be used as a guide for tube-side condenser flow patterns until more definitive information is available.

FLOW PATTERNS DURING SYSTEM TRANSIENTS

Since it is only recently that substantiated progress has been made in prediction of steady-state flow patterns, relatively little work has been directed toward behavior during transients. Perhaps the earliest experimental work was that of Sakaguchi and co-workers [30,31]. They subjected the air-water system to step increases in gas or liquid flowrates. With step increases in water flow, the flow patterns observed during the transient were always the same as those at the final conditions. However, step changes in air flowrates often produced transient flow patterns not seen at the final conditions. For transients ending in the stratified region, wavy flow often could be observed temporarily, while slug flow often could be observed temporarily for transients ending in the wavy region. Sakaguchi et al. concluded that the boundaries of temporary slug and wavy flow could be represented on a plot of V_{SL} vs the final value of V_{SG} as a series of curves with the initial value of V_{SG} as a parameter.

More recently, Prassinos and Liao [32] have compared the flow patterns detected in LOFT loss-of-coolant tests to the steady-state flow pattern maps of Mandhane et al. [11]. Because visual observation was not possible, flow patterns were determined by evaluation of the readings of a three-beam gamma densitometer, which provided densities along three separate chords. They were able to distinguish dispersed flow, intermittent flow and what they considered "stratified flow." However, it seems likely that the so-called "stratified" points may really be a misinterpretation of the data. When annular flow begins, nearly all the liquid is at the bottom of the pipe and only a very thin layer of liquid flows along the top and sides. Such behavior easily could be interpreted as separated flow when only densitometer readings are available. If the "stratified" flow points are taken as annular flow, then the limited data are in moderate agreement with the flow map of Figure 3. However, annular flow appears to begin at a significantly lower flowrate than predicted by the steady-state data.

Taitel et al. [33] have attempted to provide a theoretical framework for prediction of flow patterns during transients based on the earlier steady-state prediction scheme of Taitel and Dukler [13]. The steady-state criteria are expressed in terms of the stable, stratified liquid level, h_L, and the actual gas and liquid velocities, U_G and U_L, which would be found with stable stratified flow. Taitel et al.'s steady-state transition equations then are given by the following:

Stratified to intermittent or annular:

$$U_G \geq \left(1 - \frac{h_L}{D}\right)\left[\frac{(\rho_L - \rho_G)gA_G}{\rho_G A_L}\right]^{1/2} \tag{14}$$

Intermittent to annular:
Equation 14 is satisfied and

$$h_L/D < 0.5 \text{ annular}$$
$$h_L/D > 0.5 \text{ intermittent}$$

Stratified smooth to stratified wavy:

$$U_G \geq \left[\frac{400\nu_L(\rho_L - \rho_G)g}{\rho_G U_L}\right]^{1/2} \tag{15}$$

Intermittent to dispersed bubble:

$$U_L \geq \left[\frac{4A_G}{S_i}\frac{g}{f_L}\left(\frac{\rho_L - \rho_G}{\rho_L}\right)\right]^{1/2} \tag{16}$$

If the foregoing are correct, the transition lines are then functions of U_G, U_L and h_L/D only for any given liquid and gas pair. Taitel et al. solved a set of coupled momentum and continuity equations to determine h_L/D during flow transients. For rapid step changes in gas velocity, with total liquid flow constant, the liquid level and velocity were slow to respond to changes. Thus, in the beginning of the transient a much higher gas velocity, U_G, is present, but U_L and h_L/D are at the original values. This can lead to a flow pattern during the transient that differs from those seen at initial or final conditions. Predictions based on the foregoing agreed with observations of a limited number of air-water tests made with step changes in gas rate originating from the separated and annular regions. When the gas rate is held constant and the liquid flowrate is increased, Taitel et al. predict that the transitions to intermittent and annular flows are delayed (occur at higher flowrates) when the liquid is originally at a subcritical velocity. Comparison of theoretical predictions of the time for the appearance of the first slug to the time experimentally determined showed reasonable agreement.

Although the theoretical predictions of Taitel et al. appear promising, a number of questions remain. The use of the equilibrium liquid level in stratified flow for predictions of behavior at low flowrates where separated and annular flow (with most liquid at bottom of tube) predominate, may be meaningful. However, another approach would appear to be required for prediction of behavior originating in the intermittent or dispersed regions. Further, since the Taitel-Dukler theory has not worked in the steady state for fluids whose properties differ substantially from the air-water system, difficulties may be expected when its extension is used for predictions of transient behavior of these systems.

More recently, Ishi and Mishima [34] have argued that the usual approach of basing flow pattern transition criteria on volumetric gas and liquid flow may not be suitable for rapid transients or entrance regions. They believe that void fraction and interfacial area are more appropriate parameters for such conditions. For vertical flow, they suggest, as did Radovich and Moissis [16], that the bubble-to-intermittent-transition criterion should correspond to the void fraction at which the probability of bubble collisions is large. Hence, they proposed $\alpha \geq 0.3$ as the criterion.

The plug-churn transition is postulated as occurring when the void fraction in the liquid portion reaches the void fraction in the bubble portion. On the basis of their analysis, they concluded that the transition occurred when

$$\alpha \geq 1 - 0.813\left[\frac{0.2\left(1-\sqrt{\frac{\rho_g}{\rho_L}}\right)V_S + 0.35\sqrt{\frac{\Delta\rho g D}{\rho_L}}}{V_S + 0.75\sqrt{\frac{gD\Delta\rho}{\rho_L}}\left(\frac{\Delta\rho\, gD^3}{\rho_L \nu_L{}^2}\right)^{1/18}}\right]^{0.75} \tag{17}$$

Ishi and Meshima [34] argue that the churn-annular transition behaves differently in small- and large-sized tubes. For small-sized tubes, they believe the transition occurs where

$$\left\{V_{SG}\sqrt{\frac{\rho_G}{(\Delta\rho)gD}}\right\} > \alpha - 0.1 \tag{18}$$

However, for large-sized tubes this criterion reverts to superficial gas flowrate, and they propose the criterion to be

$$V_{SG} > \left(\frac{\sigma g \Delta\rho}{\rho_G^2}\right)^{1/4} [\mu_L/(\rho_L\sigma\sqrt{\sigma/\Delta\rho g})^{1/2}]^{-0.2} \tag{19}$$

Ishi and Mishema [34] compared their criteria to the steady-state vertical air-water flow map presented by Govier and Aziz [35] and found moderate agreement. Somewhat better agreement was obtained with the Taitel et al. [13] map. No attempt was made to compare these criteria with the limited experimental data on transient flow patterns, and no firm conclusion can be drawn as to their applicability under transient situations. Additional work in this area is still required.

CURRENT STATUS OF FLOW PATTERN KNOWLEDGE

Sufficient information is now available so that modeling of interphase transfer processes can consider the important effects of flow patterns on transfer rates in some instances. Reasonably reliable flow mapping techniques are now available for adiabatic horizontal, upwardly inclined and vertically upward flow in round tubes. The recommended techniques have been compared to data on a variety of fluids and line sizes. The limited comparisons that have been made indicate that these same techniques may be used for diabatic flow in tubes. The understanding of the behavior in downwardly inclined and vertical downward flow in tubes is less satisfactory. The new work by Barnea et al. [36,37] should help improve this situation.

Care also must be used in extrapolating currently available techniques to geometries other than round tubes. Further studies of rod bundle geometries would be particularly useful.

Our knowledge of behavior during transients is still very limited. Considerably more experimental study is needed before any of the currently proposed approaches may be considered to have been validated.

NOMENCLATURE

A_G, A_L	gas and liquid flow areas, respectively (L^2)
$(dp/dx)_G^S$ $(dp/dx)_L^S$	pressure drop per unit length due to gas or liquid, respectively, flowing alone (F/L^3)
D	pipe diameter (L)
f_L	friction factor for liquid flow
Fr	Froude number, $V_{SG}^2/(gD)$
g	acceleration due to gravity (L/T^2)
g_c	gravitational conversion factor $(M/F)(L/T^2)$
G_g, G_L, G_T	gas, liquid and total mass flowrates, respectively, based on superficial area (M/TL^2)
h_L	height of liquid above bottom of tube (L)
Ku	Kutadelaze number, $V_{SG}\rho_G^{1/2}/[g(\Delta\rho)\sigma]^{1/4}$
L	pipe length (L)
S_i	interfacial perimeter (L)
U_G, U_L	gas and liquid velocities, respectively (L/T)
V_{SG}, V_{SL}	gas and liquid superficial velocities, respectively, based on one phase flowing separately (L/T)
V^*_{SG}	$(\rho_G^{1/2} V_{SG})[gD(\rho_L - \rho_G)]^{1/2}$
x	distance (L)
α	void fraction
$\Delta\rho$	$(\rho_L - \rho_G)$
ν	kinematic viscosity
ν_L	kinematic viscosity of liquid (L^2/T)
μ_G, μ_L	gas and liquid viscosities, respectively (M/LT)
ρ_G, ρ_L	gas and liquid
σ	interfacial tension (F/L)
θ	angle of inclination

REFERENCES

1. Lockhart, R. W., and R. C. Martinelli. *Chem. Eng. Prog.* 45: 39 (1949).
2. Nguyen, V. T., and P. L. Spedding, *Chem. Eng. Sci.* 32: 1003 (1977).
3. Hubbard, N. G., and A. E. Dukler. "The Characterization of Flow Regimes in Horizontal Two-Phase Flow," Heat Transfer and Fluid Mechanics Institute, Stanford University (1966).
4. Weisman, J., D. Duncan, J. Gibson and T. Crawford. *Int. J. Multiphase Flow* 5: 437 (1979).
5. Barnea, D., O. Shoham and Y. Taitel. *Int. J. Multiphase Flow* 6:387 (1980).
6. Jones, O., and N. Zuber. "Statistical Methods for Measurement and Analysis of Two-Phase Flow," *Proc. 1974 Int. Heat Transfer Conf.*, Tokyo, Japan (1974).
7. Schlosser, P., Ohio State University. Personal communication (1980).
8. Choe, W. G., L. Weinberg and J. Weisman. "Observation and Correlation of Flow Pattern Transitions in Horizontal Cocurrent Gas-Liquid Flow," in *Two-Phase Transport and Reactor Safety*, T. N. Veziroglu and S. Kakac, eds. (Washington, DC: Hemisphere Press, 1978).
9. Baker, O. *Oil Gas J.* 53(12):185 (1954).
10. Scott, D. S. *Adv. Chem. Eng.* 4:199 (1963).
11. Mandhane, J. M., G. A. Gregory and K. A. Aziz. *Int. J. Multiphase Flow* 1:537 (1974).
12. Husain, A., and J. Weisman. *AIChE Symp. Series 174* No. 74: 205 (1978).
13. Taitel, Y., and A. E. Dukler. *AIChE J.* 22: 47 (1976).
14. Wallis, G., and J. E. Dobson. *Int. J. Multiphase Flow* 1: 173 (1973).
15. Wallis, G. B. "Vertical Annular Flow–A Simple Theory," paper presented at AIChE annual meeting, Tampa, FL (1968).
16. Radovich, N. A., and R. Moissis. "The Transition from Two-Phase Bubble Flow to Slug Flow," MIT Report 7-7633-22 (1962).
17. Taitel, Y., D. Barnea and A. E. Dukler. *AIChE J.* 26: 345 (1980).
18. Weisman, J., and S. Y. Kang. "Flow Pattern Transitions in Vertical and Upwardly Inclined Lines," *Int. J. Multiphase Flow* 7:27 (1981).
19. Oshinowo, T., and M. E. Charles. *Chem. Eng.* 52: 438 (1974).
20. Yamazaki, Y., and K. Yamaguchu. *J. Nuc. Sci. Tech.* 16:245 (1979).
21. Golan, L. P., and A. H. Stenning. *Proc. Inst. Mech. Eng.* 184:105 (1969–70).
22. Hosler, E. R. *AIChE Symp. Series 64* No. 82: 54 (1968).
23. Gibson, J. "Effect of Liquid Properties, Angle of Inclination, and Pipeline Geometry on Flow Patterns in Cocurrent Liquid-Gas Flow," MS Thesis, University of Cincinnati, Cincinnati, OH (1981).
24. Weisman, J., A. Husain and B. Harshe. "Two-Phase Pressure Drop Across Abrupt Area Changes," in *Two-Phase Flow and Reactor Safety*, T. Veziroglu and S. Kakac, Eds. (Washington, DC: Hemisphere Press, 1978).
25. Peterson, A. C., and C. L. Williams. *Nucl. Sci Eng.* 68: 155 (1978).
26. Collier, J. G. "Convective Boiling and Condensation," (New York: McGraw-Hill Book Co., 1972).
27. Boyce, B. E., J. G. Collier and J. Levy. "Hold Up and Pressure Drop Measurements in Two-Phase Flow of Air-Water Mixture in Helical Coils," paper presented at Int. Symp. on Research in Cocurrent Gas Liquid Flow, Waterloo, Ontario, Canada, September, 1968.
28. Bergles, A. E., R. F. Lopina and M. P. Fiorri. *Trans. ASME, J. Heat Transfer* 89C (1):69 (1967).
29. Palen, J. W., G. Breber and J. Taborek. *Heat Transfer Eng.* 1(2): 47 (1979).
30. Sakagachi, T., K. Akagawa and H. Hamagachi. "Transient Behavior of Air-Water Two-Phase Flow in Horizontal Tubes," ASME paper 73-WA/HT-21 (1973).
31. Sakaguchi, T., K. Akagawa and H. Hamguchi. "Transient Behavior of Flow Patterns for Air-Water in Two-Phase Horizontal Flow," *Theoret. Appl. Mech. (Japan)* 26: 445 (1976).

32. Prasinos, P. G., and C. K. Liao. "An Investigation of Two-Phase Flow Regimes in Loft Piping During Loss-of-Coolant Experiments," Report NUREG/CR-0606, Idaho National Engineering Laboratory (1979).
33. Taitel, Y., N. Lee and A. E. Dukler. *AIChE J.* 24: 920 (1978).
34. Ishi, M., and K. Mishima. "Study of Two-Fluid Model and Interfacial Area," Report NUREG/CR-1823, Argonne National Laboratory (1980).
35. Govier, G. W., and N. Aziz. "The Flow of Complex Mixtures in Pipes," (New York: Van Nostrand Reinhold Co., 1972).
36. Barnea, D., O. Shoham and Y. Taitel. *Chem. Eng. Sci.* 37: 735 (1982).
37. Barnea, D., O. Shoham and Y. Taitel. *Chem. Eng. Sci.* 37: 741 (1982).

CHAPTER 16
HYDRODYNAMICS OF GAS-LIQUID FLOWS

David Azbel and Athanasios I. Liapis
Department of Chemical Engineering
University of Missouri–Rolla
Rolla, Missouri 65401

CONTENTS

INTRODUCTION

The engineering fields dealing with motions of gas-liquid mixtures are very diverse. Numerous operations in the fields of nuclear and chemical engineering are based on the use of gas-liquid mixtures. Related to these types of processes are motion of vapor (gas)-liquid mixtures in tubes and various heat exchange equipment; atomization of liquid by spray nozzles, and other equipment directly contacting gases and liquids; and breakup and entrainment of a liquid by a gas stream.

This chapter does not attempt to cover all the specific solutions available at present in the field of two-phase flows, and it is not intended to be exhaustive either in breadth or depth of coverage. The problems under consideration are associated with a liquid layer through which a gas is blown (is bubbling), a dynamic two-phase—or, more accurately, a two-component—layer. The discrete elements of this layer, i.e., gas-liquid mixture (bubbles and drops), generally change their shape in motion, and often also their mass, as a result of coalescence or breakup of separate bubbles and drops.

Bubbling through a liquid layer is used widely in various industrial processes. These include various types of bubble columns used in chemical technology, the Bessemer process used in metallurgy, the scrubbing of steam in steam boilers, and others. Thus, the types of motion in two-phase flows are considerably more diversified and their laws more complicated than the types of motion and laws of hydrodynamics in homogeneous media.

To simplify the analysis of the hydrodynamics of the bubbling mixtures and to show at which flowrates the viscous and inertial effects appear, two limiting cases of bubbling regimes will be considered first: *rapid bubbling* and *slow bubbling*. The regime of rapid bubbling is applicable in shallow liquid pools (about 3–15 cm deep) and for relatively large superficial gas velocities (more than 0.5 m/sec), in other words, for values of Froude number ($Fr = V_s^2/gh$) greater than unity. This kind of liquid pool is found in the use of sieve and bubble plates in the distillation and absorption processes. On the other hand, the regime of slow bubbling is characterized by a deep pool, often termed a bubble column (with a depth more than 15 cm), and a superficial gas velocity less than 0.2 m/sec which implies a Froude number much less than unity. This regime occurs extensively in gas-liquid reactors, i.e., reactors in which oxidation, hydrogenation and fermentation processes occur.

The flow in the dynamic two-phase mixture consists of numerous discrete elements (bubbles or droplets) enclosed in a continuous fluid (liquid or gas depending on the gas void fraction). It is known [1–5] that formulation of the fundamental equations of hydrodynamics is only possible for single discrete elements of the two-phase flow. In this chapter, where a theoretical analysis of the total two-phase flow is considered, the approach of calculus of variations is adopted as the method for obtaining information on the flow characteristics. By taking account of the geometric characteristics of the system in which the bubbling process takes place—the physical properties of the gas and liquid, the gas and liquid flowrates and the boundary conditions—we can determine the fundamental characteristics of the dynamic two-phase mixture, for example, the distribution of the gas and liquid in the flow.

As an example of how the physical properties of the gas and liquid may affect the dynamics of the flow, consider a gas-liquid flow in which we have surface-active substances present, i.e., some kind of contaminant present at the interface of the bubbles and the liquid. The presence of such a substance can result in a change in the capillary force, as well as in the appearance of additional surface forces, leading to a significant change in the hydrodynamic regime of the bubble process. When surface-active substances are present in the gas-liquid mixture, the bubble surfaces are covered by a monolayer of surface-active molecules, and the energy dissipation in this layer is known to be small compared to the dissipation of energy in the two-phase mixture [6]. The form drag, F_o, of a bubble, considered as a sphere for simplicity, will be

$$F_o \simeq C_D \frac{\pi r_b^2 \rho_f v_r^2}{2} \tag{1}$$

where C_D is the drag coefficient (at large Reynolds numbers, $C_D \cong 0.5 = \text{constant}$). As a result of the flow of the liquid around the bubble, a saturated monolayer of the surface-active substance is accumulated at the rear of the bubble, and this area, s_o (less than $4\pi r_b^2$), of the bubble remains covered by a nondeforming monolayer of the surface-active substance while the bubble rises. During the entire period of motion, this monolayer on the rear of the bubble will cause the liquid velocity on the bubble surface to become zero, just as on the surface of a solid body. Separation of liquid from the bubble surface then occurs, accompanied by a new factor in the form drag [7]. With this physical picture in mind, the ratio of viscous drag, F_D, to form drag, F_o, is

$$\frac{F_D}{F_o} \simeq \frac{12\pi \mu_f v_r r_b}{C_D s_o \rho_f v_r^2} = \frac{12\pi r_b^2}{C_D s_o Re} \tag{2}$$

where $$Re = \frac{\rho_f v_r r_b}{\mu_f} \tag{3}$$

The ratio F_D/F_o will be less than unity when

$$s_o > \frac{12\pi r_b^2}{C_D Re} = \frac{3 s_b}{C_D Re} \tag{4}$$

where s_b is the bubble surface area. From Equation 4 the following inequality is obtained:

$$\frac{s_o}{s_b} > \frac{3}{C_D Re} \tag{5}$$

This condition is satisfied in the case in which the bubble surface, s_o, covered by a monolayer of surface-active substances is more than a fraction 6/Re of the total bubble surface area (when Re >> 6). Further, in this case the problem of the motion of the gas bubble with surface contaminants in a continuous liquid may be reduced to the problem of the motion of a solid sphere in an ideal, infinite liquid [6,8], thus considerably simplifying the analysis.

RAPID BUBBLING WITH IDEAL LIQUID

The gas void fraction, ϕ, defined as the ratio of gas volume to total volume, is a very important characteristic parameter of any two-phase flow and, as such, will be considered first in this section. We will obtain the void fraction as a function of system properties for the case of rapid bubbling of gas in liquid, when the viscosity of the liquid is negligible.

For moderate and large values of the Reynolds number, which are typical of bubbling processes utilized in commercial distillation and absorption equipment, the viscous forces (being small compared to the inertial forces) have little influence on the hydrodynamics of the gas-liquid mixture and, therefore, may be ignored in a study of such flows. Theoretical [5,6] and experimental studies [9–12] concerning the influence of viscosity on the behavior of the two-phase flow in a bubbling process confirm this conclusion, giving validity to our assumption of ideal liquid. Many detailed experimental studies [13–37] on mass bubbling operations (i.e., in shallow pools) have indicated that three apparent groups of parameters influence the gas-liquid flow:

1. geometric characteristics of the flow system,
2. physical properties of the gas and liquid, and
3. dynamic factors.

The height of the liquid in a flow system in which the liquid is static, the diameter of the equipment and the geometry of the gas-distributing device (i.e., the orifices) belong in the first group. To the second group belong the viscosity, density and surface tension of the gas and liquid; and to the third group belong the gas and liquid flowrates.

Let us now examine a two-phase flow, which is sufficiently distant from the walls of the equipment and from the gas distribution devices so that any effects they may have on the flow are eliminated. Then one can take the two-phase flow to be one-dimensional, and at a distance, x, from the gas inlet a differential layer located perpendicular to the direction of motion of the gas stream is considered, with thickness dx. The fraction of the volume of this differential layer occupied by the gas is

$$\phi = \frac{4}{3} \pi r_b^3 n \tag{6}$$

An energy balance of a unit cross section of the differential layer during the bubbling process is given by

$$dE = dE_1 + dE_2 + dE_3 \tag{7}$$

where dE is the total energy of the layer, dE_1 is the potential energy of the liquid, dE_2 is the kinetic energy of the layer, and dE_3 is the energy of surface tension of the bubbles in the layer. Taking into account that buoyancy forces are balanced by the resistance forces, it is assumed that the potential energy of the gas does not change during the motion of a bubble. The potential energy of the liquid per unit cross-sectional area is

$$dE_1 = (1 - \phi)\rho_f gx\, dx \tag{8}$$

The kinetic energy of the layer is comprised of the kinetic energy of the bubbles of gas (equal to $\frac{\rho_g v^2}{2}\frac{4}{3}\pi r_b^3 n\, dx$, where v is the velocity of the gas in the differential layer, and ρ_g is the density of the gas) and the kinetic energy of the liquid carried along by the bubbles. To take both components into account, it is simplest to consider each bubble of gas as a rigid sphere, the mass of which consists of the mass of the gas enclosed on it, plus the mass of a volume of liquid equal to half the volume of the bubble (the so-called "additional mass") [6,8,38]. The total kinetic energy in the layer is then given as

$$dE_2 = \left(\rho_g + \frac{\rho_f}{2}\right)\phi\frac{v^2}{2}\, dx \tag{9}$$

where we have used Equation 6. From the mass conservation of the gas, one can write

$$\phi v = v_s \tag{10}$$

where v_s is the gas velocity in the region of the flow that is free of liquid (i.e., the gas superficial velocity). Then Equation 9 becomes

$$dE_2 = \left(\rho_g + \frac{\rho_f}{2}\right)\frac{v_s^2}{2\phi}\, dx \tag{11}$$

Finally, the energy of surface tension can be calculated using the expression

$$dE_3 = 4\pi r_b^2 \sigma n\, dx = \frac{3\phi}{r_b}\sigma dx \tag{12}$$

where σ is the surface tension coefficient, and we have used Equation 6. By substituting into Equation 7 the expressions for dE, dE_2 and dE_3 from Equations 8, 11 and 12, respectively, dE takes the form

$$dE = \left[(1 - \phi)\rho_f gx + \left(\rho_g + \frac{\rho_f}{2}\right)\frac{v_s^2}{2\phi} + \frac{3\sigma}{r_b}\phi\right] dx \tag{13}$$

The total energy of a two-phase mixture of height x_1 is then

$$E = \int_0^{x_1}\left[(1 - \phi)\rho_f gx + \left(\rho_g + \frac{\rho_f}{2}\right)\frac{v_s^2}{2\phi} + \frac{3\sigma}{r_b}\phi\right] dx \tag{14}$$

It is a fundamental axiom of physical theory that for any system the steady state of the system is that for which the available energy of the system is at a minimum. Hence, the steady-state distribution of the gas void fraction, ϕ, occurs when this energy is at a minimum. Thus, to find ϕ it is necessary to determine the minimum of the integral in Equation 14. This problem [39] can be reduced to that of finding the minimum of the integral E under the condition of invariability of the amount of liquid in the system, i.e., with

$$h = \int (1 - \phi)\, dx = \text{constant} \tag{15}$$

where h is the static liquid height. Equation 15 is a boundary condition on the variation of Equation 14. The method of calculus of variations is now considered by starting with the following function of a function:

$$E = \int f[\phi(x), \phi'(x), x]\, dx \tag{16}$$

where f is some known function of its argument, $\phi(x)$ is an unknown function, and $\phi'(x) \equiv d\phi/dx$. The calculus of variations is a method to obtain a function $\phi(x)$ such that the variation of E is zero:

$$\delta E = 0 \tag{17}$$

which also means that the function E is at a minimum (or possibly a maximum) for that function $\phi(x)$. This is done subject to the constraint that

$$\psi(\phi, x) = 0 \tag{18}$$

where ψ is some known function. Using the method of variations on Equations 16, 17 and 18, it can be shown that we obtain the Euler equation:

$$\frac{\partial f}{\partial \phi} - \frac{d}{dx}\left(\frac{\partial f}{\partial \phi'}\right) + \lambda \frac{\partial \psi}{\partial \phi} - \lambda \frac{d}{dx}\left(\frac{\partial \psi}{\partial \phi'}\right) = 0 \tag{19}$$

where λ is the Lagrange multiplier and is determined by the constraint Equation 18. Equation 19, which is usually differential, but sometimes algebraic, is then used to obtain that function $\phi(x)$, which gives a minimum of E.

When we apply this analysis to Equation 14 with the constraint Equation 15, we find that

$$f(\phi,\phi', x) = (1-\phi)\rho_f g x + \left(\rho_g + \frac{\rho_f}{2}\right)\frac{v_s^2}{2\phi} + \frac{3\sigma}{r_b}\phi \tag{20a}$$

$$\psi(\phi,x) = (1-\phi) \tag{20b}$$

and for the Euler equation we obtain

$$-\rho_f g x - \left(\rho_g + \frac{\rho_f}{2}\right)\frac{v_s^2}{2\phi^2} + \frac{3\sigma}{r_b} - \lambda = 0 \tag{21}$$

We have ignored the variation due to the unknown upper limit, x_1, in the calculus of variation analysis given above. This upper limit is the point at which the two-phase nature of the flow breaks down, in other words, when $\phi = 1$. We next show that it is acceptable to ignore the variation due to x_1.

The problem of the floating limit of an integral is called the problem with natural boundary conditions [39]. If we take

$$E = \int_0^{x_1} f(\phi,\phi',x)\,dx \tag{21a}$$

and we need to find the $\phi(x)$, which gives an extremum of E when $\phi(0) = \phi_o$ is set (the lower limit is fixed) and $\phi(x_1)$ is not set (no condition is made on the upper limit), then the value of δE will be given not only by the variation of f under the integral, but also by the variation at the upper limit. The most general expression of the variation δE can be written as

$$\delta E = \Big[\, f\phi'\delta\phi \,\Big]_o^{x_1} + \int_o^{x_1}\Big[f\phi - \frac{d}{dx}(f\phi')\Big]\,\delta\phi\,dx \tag{21b}$$

where we are not taking system constraints into account. In the case we are now considering, $f = f(\phi,x)$. In other words, we do not have a term $\phi' = d\phi/dx$, $f\phi' \equiv 0$, and the term integrated in brackets is equal to zero, $[f\phi'\delta\phi]_o^{x_1} = 0$. Therefore, the floating limit of x has no effect and Euler's equation is as shown in Equation 19. And, since the derivative ϕ' does not enter into Equation 19, Equation 21 is not differential but algebraic, and one finds

$$\phi(x) = \left[\frac{v_s^2\left(\rho_g + \frac{\rho_f}{2}\right)}{2\left(\frac{3\sigma}{r_b} - \rho_f g x - \lambda\right)} \right]^{1/2} \tag{22}$$

Further, since the derivative $\phi'(x)$ does not appear in Equation 21, we cannot impose any additional conditions on the values of $\phi(x)$ at the limits of the interval 0 to x_1. Therefore, this function may become discontinuous at the ends of the interval. However, from physical considerations it must be the case that the function $\phi(x)$ is continuous at all values of x between the limits and, therefore, can be determined from Equation 22. The value of x_1 now can be determined from the equality

$$1 = \left[\frac{v_s^2\left(\rho_g + \frac{\rho_f}{2}\right)}{2\left(\frac{3\sigma}{r_b} - \rho_f g x_1 - \lambda\right)} \right]^{1/2} \tag{23}$$

The undetermined Lagrange multiplier has to be eliminated. To do this, we use ϕ from Equation 22 in the constraint equation 15:

$$h = \int_0^{x_1} (1 - \phi)\,dx = \int_0^{x_1} \left\{ 1 - \left[\frac{v_s^2\left(\rho_g + \frac{\rho_f}{2}\right)}{2\left(\frac{3\sigma}{r_b} - \rho_f g x - \lambda\right)} \right]^{1/2} \right\} dx \tag{24}$$

By integrating Equation 24 and eliminating λ using Equation 23, we get

$$h = x_1 + \frac{\left(\rho_g + \frac{\rho_f}{2}\right)}{\rho_f}\frac{v_s^2}{g} - \frac{2}{\rho_f g}\left\{ \left[\frac{\left(\rho_g + \frac{\rho_f}{2}\right) v_s^2}{2} \right] \left[\rho_f g x_1 + \frac{\left(\rho_g + \frac{\rho_f}{2}\right) v_s^2}{2} \right] \right\}^{1/2} \tag{25}$$

Let us define the term

$$A = 1 + \frac{2\rho_f g x_1}{\left(\rho_g + \frac{\rho_f}{2}\right) v_s^2} \tag{25a}$$

which means that

$$x_1 = \frac{1}{\rho_f g}\left(\rho_g + \frac{\rho_f}{2}\right) v_s^2 (A - 1) \tag{25b}$$

By substituting this value of x_1 into Equation 25, we obtain, after simplifying,

$$(1/2)A - 1/2 + 1 - A^{1/2} = \frac{\rho_f g h}{\left(\rho_g + \frac{\rho_f}{2}\right) v_s^2} \tag{26}$$

We now define

$$B = \frac{2\rho_f gh}{\left(\rho_g + \frac{\rho_f}{2}\right) v_s^2} \tag{26a}$$

and Equation 26 takes the form

$$A - 2A^{1/2} + 1 - B = 0 \tag{26b}$$

and, solving with respect to $A^{1/2}$,

$$A^{1/2} = 1 \pm B^{1/2} \tag{26c}$$

We now replace A and B and, after some manipulation, we find

$$\left[\rho_f g x_1 + \frac{\left(\rho_g + \frac{\rho_f}{2}\right) v_s^2}{2}\right]^{1/2} = \left[\frac{\left(\rho_g + \frac{\rho_f}{2}\right) v_s^2}{2}\right]^{1/2} \pm \left[\rho_f gh\right]^{1/2} \tag{26d}$$

In the above equation, the positive sign $(+)$ is selected, since when $v_s = 0$, the condition $x_1 = h$ must be fulfilled.

The equation for the height of the dynamic two-phase mixture in terms of the static liquid level is then

$$x_1 = 2\left[\frac{h\left(\rho_g + \frac{\rho_f}{2}\right) v_s^2}{2\rho_f g}\right]^{1/2} + h \tag{26e}$$

Assuming $\rho_g << \rho_f$, we can write this in the form

$$x_1 = \left[\frac{v_s^2 h}{g}\right]^{1/2} + h = h\,(Fr^{1/2} + 1) \tag{27}$$

where Fr is the Froude number. It follows from Equation 27 that the height of the dynamic two-phase mixture varies linearly with the superficial velocity of the gas, v_s. The value of $\phi(x)$ will be determined next. From Equation 23 one can write the following expressions:

$$\lambda = \frac{3\sigma}{r_b} - \rho_f\, g\left(x_1 + \frac{v_s^2}{4g}\right) \tag{27a}$$

where we have again used the inequality $\rho_g << \rho_f$. By substituting the value of x_1 from Equation 27 into Equation 27a, one obtains

$$\lambda = \frac{3\sigma}{r_b} - \rho_f g\left(hFr^{1/2} + h + \frac{v_s^2}{4g}\right) \tag{27b}$$

Taking into consideration that

$$\frac{hv_s^2}{4gh} = \frac{h}{4}\,Fr \tag{27c}$$

we can write

$$\frac{3\sigma}{r_b} - \lambda = \rho_f gh\left(Fr^{1/2} + \frac{Fr}{4} + 1\right) \tag{27d}$$

and using this in Equation 22, the following expression is obtained:

$$\phi(x) = \left[\frac{\rho_f v_s^2}{4\left[\rho_f gh\left(Fr^{1/2} + \frac{Fr}{4} + 1\right) - \rho_f gx\right]}\right]^{1/2} \tag{28}$$

or

$$\phi(x) = \left[\frac{F}{(1 + F^{1/2})^2 - \frac{x}{h}}\right]^{1/2} \tag{28a}$$

where $F = (Fr/4)$.

Thus, we have obtained the local gas void fraction, $\phi(x)$, solely in terms of the Froude number, Fr, and the location, x, for the case of an ideal liquid. Equation 28a is illustrated [40] for a particular flow configuration in Figure 1. Note the close correspondence to the experimental data. All the results we will obtain in this section, where we are assuming ideal-fluid conditions, will be shown to give such close correspondence for tests in equipment that has a typical dimension less than 20 cm in diameter.

It follows that the gas content (or void fraction) does not depend on the form and size of the bubbles of gas, because the bubble radius does not enter into the equation. Consequently, this equation can be used for calculations over a broad range of velocities of the gas.

Let us now determine the average gas content (or void fraction) of the two-phase mixture:

$$\phi_{av} = \frac{\int_0^{x_1} \phi(x)dx}{\int_0^{x_1} dx} = \frac{\int_0^{x_1} \frac{F^{1/2}}{\left[(1 + F^{1/2})^2 - \frac{x}{h}\right]^{1/2}} dx}{x_1}$$

$$= -\frac{2h}{x_1} F^{1/2} \left\{\left[(1 + F^{1/2})^2 - \frac{x_1}{h}\right]^{1/2} - (1 + F^{1/2})\right\} \tag{29}$$

After substitution into Equation 29 of the expression for x_1 from Equation 27 and letting $F = (Fr/4)$, we obtain

$$\phi_{av} = \frac{1}{1 + \frac{1}{Fr^{1/2}}} \tag{30}$$

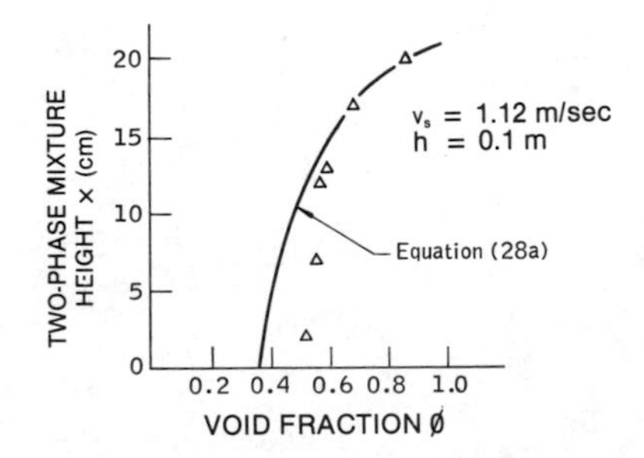

Figure 1. Plot of height vs void fraction.

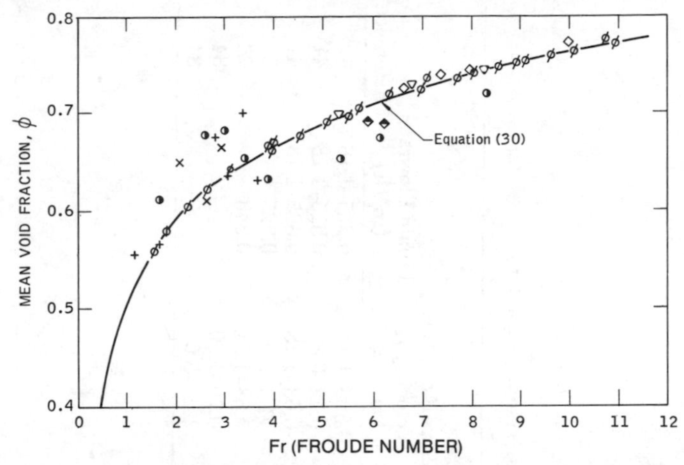

Figure 2. Mean void fraction vs Froude number for air-water flows. The data for this system are given in Table I.

This result shows that the average void fraction in a rapidly bubbling flow with an ideal liquid is given by a very simple function of the Froude number alone. Equation 30 is illustrated in Figure 2 for an air-water system together with data [41] obtained for various orifice and sieve configurations (also see Table I). Clearly, Equation 30 gives very good results, regardless of flow geometries. Figure 3 shows that over a wide range of different gases and liquids (represented by different symbols) [42,43], Equation 30 still holds good for flow systems with a typical dimension less than about 20 cm in diameter (also see Table II).

The average relative density of the gas-liquid mixture (the "specific gravity of the foam") is determined by the expression

$$\psi = \frac{h}{x_1} \tag{31}$$

where ψ is now the average relative density. If we replace x_1 according to Equation 27, we have

$$\psi = \frac{1}{1 + Fr^{1/2}} \tag{32}$$

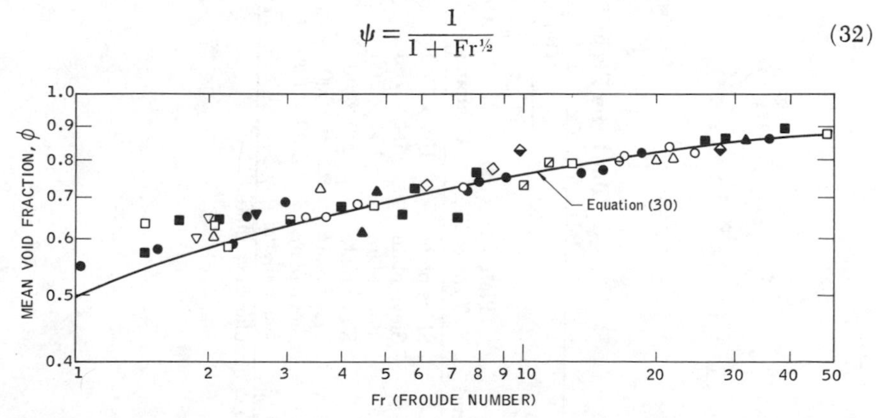

Figure 3. Mean void fraction vs Froude number for various mixtures. Symbols used in this figure are given in Table II.

Table I (for Figure 2). Plate Geometry and Liquid and Gas Flowrates (system: air-water)

Legend	Plate Type	Dimensions, mm: Of Column	Of Orifice, d_o	Pitch, t'	Plate Thickness, δ	Free Area %	Weir Height, $h_{w.h.}$ (mm)	Superficial Gas Velocity, v_s (m/sec)	Liquid Flowrate (m^3/h)	Reference
◑	Sieve plate	104 × 700	3.5	11.0	3.0	8.8	0.0	0.325–1.1	0.325–1.18	41
×	Sieve plate	104 × 700	3.5	11.0	3.0	8.8	27.0	0.42–0.8	0.325–1.18	41
+	Sieve plate	104 × 700	3.5	11.0	3.0	8.8	65.0	0.68–1.18	0.325–1.18	41
◇	Sieve plate	200 × 60	3.0	6.0	5.0	19.2	20.0	1.4–2.0	0.325	34
▽	Sieve plate	Φ [a]200	2.0	7.2	2.0	6.95	15.0	0.68–1.31	0.348	22
⬘	Sieve plate	200 × 600	4.0	6.0	5.0	25.5	20.0	0.70–1.60	–	34
∅	Sieve plate	Φ 95	2.0	5.0	12.3	–	–	1.25–3.0	–	34

[a]Φ ≡ circular sieve plate.

Table II (for Figure 3). Summary of Data[a]

Legend	Liquid	μ_g,[b] cP	μ_l,[c] cP	σ,[d] $\frac{dyn}{cm^2}$	Reference
○	Filter liquid	0.019	2.40	74.30	42,43
●	Water	0.019	1.0	72.0	42,43
□	Water[e]	0.0084	1.0	72.0	42,43
⬓	Isopropanol	0.019	3.26	22.5	42,43
■	Aqueous NH_4OH solution	0.019	1.0	69.0	42,43
▽	Heptane	0.019	0.45	22.0	42,43
▼	26.4 vol% $CuSO_4 \cdot 5H_2O$ solution	0.019	1.09	79.9	42,43
△	16.6 vol% $CuSO_4 \cdot 5H_2O$ solution	0.019	2.19	73.4	42,43
▲	25 vol% Na_2CO_3 solution	0.019	6.12	63.3	42,43
⊖	20 vol% NaCl solution	0.019	1.33	77.1	42,43
⍁	Aqueous glycerine solution	0.019	6.70	74.0	42,43
◇	Aqueous glycerine solution	0.019	5.40	70.3	42,43
⬙	Aqueous glycerine solution	0.019	21.8	68.0	42,43

[a]In all experiments but one, air was used as the gas phase.
[b]$\mu_g \equiv$ viscosity of gas.
[c]$\mu_l \equiv$ viscosity of liquid.
[d]$\sigma \equiv$ surface tension between the gas and the liquid.
[e]Hydrogen was used as the gas phase.

RAPID BUBBLING WITH REAL LIQUID

In this section an expression will be developed for the void fraction ϕ in rapid bubbling by using the same notion of energy minimization, except that we will not allow for energy dissipation due to the viscosity of the liquid and due to liquid turbulence. We again use the energy balance for the differential layer, Equation 7, except that dE_2 now represents the energy dissipation in the layer, while the potential energy, dE_1, of the liquid is given by Equation 8. The expression for the dissipative energy, dE_2, is developed through the use of the model [44] of a uniform bubble distribution in the differential layer, with each bubble being located in the center of a "spherical compartment" formed by adjacent bubbles. Calculation of drag based on this model agrees fairly well with experimental data on liquid-liquid and gas-liquid systems [45–48]. With this model, the energy dissipation of the bubble is obtained by using the following expression [44]:

$$dE_2 = 12\pi\mu_f r_b v^2 \left(\frac{1-\phi^{5/3}}{(1-\phi)^2}\right) \tau n dx \tag{33}$$

where τ is the time scale of turbulent eddies, and the number of bubbles, n, per unit volume is

$$n = \frac{\phi}{\frac{4}{3}\pi r_b{}^3} \tag{34}$$

Combining Equations 33 and 34 results in the following expression for calculating the energy dissipation produced by motion of the bubbles in the layer

$$dE_2 = \frac{9\mu_f v^2 \phi\,(1 - \phi^{5/3})}{r_b{}^2\,(1 - \phi)^2}\,\tau dx \tag{35}$$

For the time scale of the eddies most likely to have an effect on the bubble motion, an expression developed by Azbel [44] is used:

$$\tau = \frac{k r_b{}^2 (1 - \phi)^2}{9\nu_f (1 - \phi^{5/3})} \tag{36}$$

where k is an added mass coefficient and can be taken to be constant. Using Equation 36 and the continuity condition, $v_s = v\phi$, in Equation 35, the expression of dE_2 takes the form

$$dE_2 = k\rho_f \frac{v_s{}^2}{\phi}\,dx \tag{37}$$

The surface tension energy, dE_3, is given in Equation 12. Substituting into Equation 7 the values of dE_1, dE_2 and dE_3, as given by Equations 8, 37 and 12, we obtain

$$dE = \left[(1 - \phi)\rho_f g x + k\rho_f \frac{v_s{}^2}{\phi} + \frac{3\sigma}{r_b}\phi \right] dx \tag{38}$$

Consequently, the total energy of a two-phase layer of height x_1 is

$$E = \int_0^{x_1} \left[(1 - \phi)\rho_f g x + k\rho_f \frac{v_s{}^2}{\phi} + \frac{3\sigma}{r_b}\phi \right] dx \tag{39}$$

As in the previous section, the equilibrium distribution of the gas void fraction, ϕ, is established when this energy is at a minimum. Therefore, to determine ϕ we must find the minimum of the integral (Equation 39).

Following the analysis of the previous section, the Euler equation (with the variation with respect to x_1 being neglected as before) in this case is

$$-\rho_f g x - k\rho_f \frac{v_s{}^2}{\phi^2} + \frac{3\sigma}{r_b} - \lambda = 0 \tag{39a}$$

Hence,

$$\phi(x) = \left[\frac{k\rho_f v_s{}^2}{\dfrac{3\sigma}{r_b} - \lambda - \rho_f g x} \right]^{1/2} \tag{40}$$

With $\phi(x_1) = 1$, where x_1 is the point where the flow is totally gaseous, the value of x_1 may be determined from the above equation:

$$1 = \left[\frac{k\rho_f v_s{}^2}{\dfrac{3\sigma}{r_b} - \lambda - \rho_f g x_1} \right]^{1/2} \tag{41}$$

As before, we can eliminate the undetermined Lagrange multiplier by using the constant Equation 15:

$$h = \int_0^{x_1} (1 - \phi)dx = \int_0^{x_1} \left\{ 1 - \left[\frac{k\rho_f v_s^2}{\frac{3\sigma}{r_b} - \rho_f g x - \lambda} \right]^{1/2} \right\} dx \tag{42}$$

and integrating, while making allowance for Equation 41, we obtain

$$h = x_1 + \frac{kv_s^2}{g} - \frac{2}{\rho_f g} [k\rho_f v_s^2]^{1/2} [\rho_f g x_1 + k\rho_f v_s^2]^{1/2} \tag{43}$$

Solving this equation with respect to x_1, we get an equation suitable for calculating the height of a dynamic two-phase flow, allowing for a real liquid:

$$x_1 = h[2(kFr)^{1/2} + 1] \tag{44}$$

It follows from this equation that, as it is for the ideal liquid, the height of the two-phase flow is a linear function of the superficial gas velocity, v_s. We can now develop the expression for the gas void fraction of the flow. From Equations 41 and 44, it follows that

$$\lambda = \frac{3\sigma}{r_b} - \rho_f g \left[h\ (4kFr)^{1/2} + h + \frac{kv_s^2}{g} \right] \tag{45}$$

and since $(kv_s^2/gh)h = khFr$, we then may write

$$\frac{3\sigma}{r_b} - \lambda = \rho_f gh\ [(4kFr)^{1/2} + 1 + kFr] \tag{46}$$

Equation 46 is used in Equation 40, and the functional form of $\phi(x)$ becomes as follows:

$$\phi(x) = \left[\frac{kFr}{(kFr)^{1/2} + kFr + 1 - \frac{x}{h}} \right] \tag{47}$$

From this equation it follows that the gas content does not depend on the shape and size of the gas bubbles, since the bubble radius does not occur in the equation. It is noteworthy that the analysis of rapid bubbling for a real liquid gives a formula for the void fraction very similar to that for an ideal liquid (Equation 28a). The average gas content, ϕ_{av}, of the flow is given by

$$\phi_{av} = \frac{\int_0^{x_1} \phi(x)dx}{\int_0^{x_1} dx} = \frac{2(kFr)^{1/2}}{1 + 2(kFr)^{1/2}} \tag{48}$$

Equation 48 shows that for a real liquid (in other words, allowing for dissipative forces) we result with a simple formula for the mean void fraction in terms of the Froude number and a constant. Note the similarity between this equation and Equation 30 for an ideal liquid.

The relative density of the two-phase flow is

$$\psi = \frac{h}{x_1} \tag{49}$$

and, on substituting the expression for x_1 from Equation 44, we find

$$\psi = \frac{1}{1 + 2(\text{kFr})^{1/2}} \tag{50}$$

For air-water systems, the constant, k, has been shown to be of the order of unity. Hence, we can write Equations 44, 48 and 50 as

$$x_1 = h\ (2\,\text{Fr}^{1/2} + 1) \tag{51}$$

for the height of the two-phase mixture,

$$\phi_{av} = \frac{2\text{Fr}^{1/2}}{1 + 2\text{Fr}^{1/2}} \tag{52}$$

for the average void fraction, and

$$\psi = \frac{1}{1 + 2\text{Fr}^{1/2}} \tag{53}$$

for the relative density.

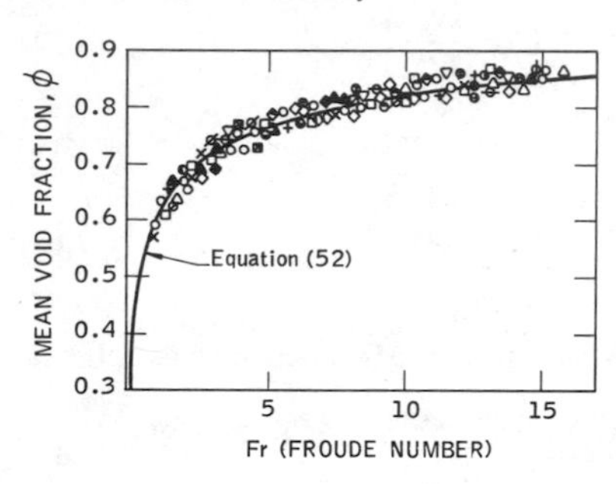

Figure 4. Mean void fraction vs Froude number for the air-water system. Data for the plot are given in Table III.

It follows from Equations 51, 52 and 53 that the main hydrodynamic parameters of the flow in rapid bubbling depend neither on the physical properties of the liquid and gas nor on the geometric characteristics of the gas-distributing devices. They are determined by the ratio of the liquid and gas flowrates (where v_s is a characteristic gas velocity), and are characterized by the ratio between the inertial forces and the gravity forces. In other words, they are functions only of the Froude number. Equations 52 and 53 are illustrated in Figures 4 (also see Table III) and 5 (also see Table IV), respectively, in comparison with experimental data [41,42] for air-water over a wide range of geometric conditions and flowrates. All the data are for flow systems in which the characteristic dimension is greater than 20 cm in diameter. Note the close agreement between experiment and theory.

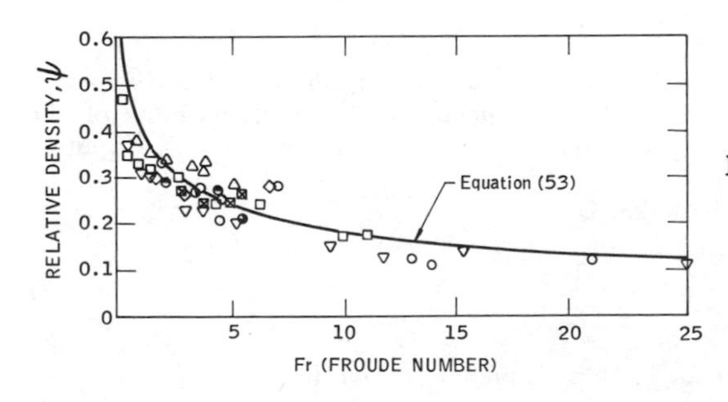

Figure 5. Relative density vs Froude number for the air-water system. Data for the plot are given in Table IV.

Table III. Summary of Data for Figure 4 (sieve plates geometry and liquid loads)

Legend	Orifice Diameter d_o (mm)	Free Area, F_a (%)	Plate Thickness, δ (mm)	Weir Height, $h_{w.h.}$ (mm)	Irrigation Density, L/b[a] (m^3/(hr)(mm))
△	1.6	16.6	1.0	30	0.334
⧅	1.6	16.6	1.0	30	3.70
▽	2.0	4.2	1.0	30	0.334
⊖	2.0	4.2	1.0	30	3.70
○	2.0	32.0	3.0	30	0.55
◆	2.0	32.0	3.0	30	1.20
□	3.0	16.6	1.0	30	2.30
⧗	3.0	16.6	1.0	30	2.00
◇	4.0	16.6	3.0	30	1.80
⊗	4.0	16.6	3.0	30	1.65
▲	2.0	16.6	2.0	5	0.80
◆	2.0	16.6	2.0	5	0.75
●	2.0	16.6	1.0	10	0.62
∅	2.0	16.6	1.0	10	0.45
◆	2.0	16.6	2.0	17	0.68
-⊖-	2.0	16.6	2.0	17	0.92
◑	2.0	16.6	1.0	25	0.73
+	2.0	16.6	1.0	25	0.87
×	2.0	16.6	1.0	35	0.47

[a]L ≡ Liquid flowrate (m^3/hr); b ≡ Length of weir crest (mm).

Table IV. Summary of Data for Figure 5 (column diameter, D = 200 mm; weir height, $h_{w.h.}$ = 32 mm)

Legend	Orifice Diameter, d_o (mm)	Irrigation Density, L/b[a] (m^3/(hr)(cm))	Free Area, Fa (%)
□	0.9	3.024	6.95
△	6.0	3.024	6.95
▽	2.0	3.024	14.60
○	2.0	3.024	22.60
⊖	2.0	3.024	6.95
◇	2.0	87	6.95
◑	2.0	174	6.95
⊠	2.0	348	6.95

[a]L ≡ Liquid flowrates (m^3/hr); b ≡ length of weir crest (cm).

MAIN PARAMETERS OF THE TWO-PHASE FLOW IN THE SLOW BUBBLING REGIME

In the following paragraphs, a study is presented of the main hydrodynamic parameters of the two-phase flow in the *slow bubbling* regime. In this regime, geometric specifications of the gas-distributing devices are known to have no practical influence on the gas void of the two-phase flow [11,49–51].

The gas void fraction, ϕ, by definition is a measure of the amount of gas per unit volume, and we can write a continuity equation for the void fraction:

$$\frac{\partial \phi}{\partial t} + \frac{\partial}{\partial x_i}(u_i \phi) \qquad (54)$$

where u_i ($i = 1, 2, 3$) is the gas velocity vector at vector coordinate, x_i, and the cartesian tensor convention of summation of repeated indices is implied.

Because the turbulence of the two-phase flow, the gas content, ϕ, and velocity, u_i, change their values in space and time, and we can decompose them into two components —the mean component and the fluctuating component—as

$$\phi = \bar{\phi} + \phi' \\ u_i = \bar{u}_i + u_i' \tag{55}$$

where $\bar{\phi}$ is the mean and ϕ' the fluctuation component of the void fraction, and $\bar{u}_i$ is the mean and u_i' the fluctuation component of the velocity vector. The turbulent exchange is defined by

$$\overline{u_i'\phi'} = -D_t \frac{\partial \bar{\phi}}{\partial x_i} \tag{56}$$

where D_t is the coefficient of turbulent diffusion. Substituting the value of $\overline{u_i'\phi'}$ from the above equation into Equation 54 and time-averaging gives

$$\frac{\partial \bar{\phi}}{\partial t} + \frac{\partial}{\partial x_i}(\bar{u}_i \bar{\phi}) = \frac{\partial}{\partial x_i}\left(D_t \frac{\partial \bar{\phi}}{\partial x_i}\right) \tag{57}$$

This modified equation of continuity takes into account gas mass transfer as a result of the interaction of the fluctuating velocity and void fraction.

As a first approximation we may assume that the gas content, ϕ, can be averaged over the flow cross section so that it depends only on the height, x, in the two-phase flow. Then, for a steady-state flow, Equation 57 may be rewritten as

$$\frac{d}{dx}\left(\bar{v}\bar{\phi} - D_t \frac{d\phi}{dx}\right) = 0 \tag{58}$$

where $\bar{v}$ is now the mean gas velocity in the x direction. The expression in parentheses in this equation is obviously a constant, and in a liquid-free zone, $\bar{v} = v_s$ (the superficial gas velocity), and $d\bar{\phi}/dx = 0$ because $\phi = 1$, so that

$$\bar{v}\bar{\phi} - D_t \frac{d\bar{\phi}}{dx} = v_s \tag{59}$$

Taking as a boundary condition

$$\bar{\phi}(x_i) = 1 \tag{60}$$

the solution of Equation 59, assuming as a first approximation that $\bar{v}$ and D_t are independent of the coordinate x, will be

$$\bar{\phi}(x) = \frac{v_s}{\bar{v}} + \left(1 - \frac{v_s}{\bar{v}}\right) \exp\left\{-\frac{\bar{v}}{D_t}(x_1 - x)\right\} \tag{61}$$

It follows from Equation 61 that the gas void fraction has only a weak dependency on the position in the two-phase mixture when the liquid height, h, is sufficiently high ($\sim$1 m) that $\bar{v}/D_t\ (h - x) >> 1$. Thus, this equation implies that if the undisturbed liquid level, $h(< x_1)$, is large, then $\phi \cong$ constant over most of the flow. Local measurements of the gas content [51] confirm this conclusion.

In the lower region [40,50,51] of the two-phase flow, in its initial section, the void fraction is defined by the cross-sectional area of the gas-distributing device, for example, the orifices. The main region of the two-phase flow, which has a fairly constant void fraction, is upstream of this small initial section, and the void fraction increases in the upper part of the main region, approaching unity at the "interface" of the two-phase flow and the free-gas region. The gas void fraction is higher in the upper part because the surface bubbles burst more slowly than the rate of rise of the newly generated bubbles. But for flow systems, where h is large and the superficial gas velocity is small, the influence of the upper part on the space-average gas void fraction of the two-phase flow is not great. However, in flow systems in which h has a moderate value ($5 \text{ cm} < h < 120 \text{ cm}$), the mean void fraction, $\bar{\phi}$, is dependent on the value of h, and there may be a significant effect on the average. The location at which the void fraction starts increasing rapidly to unity is basically determined by the superficial gas velocity. For $v_s \leq 0.2$ m/sec, the transition is at a height of 1–2 cm, and for $v_s > 0.2$ m/sec it may be extended to 30–35 cm.

The energy balance of a differential layer, dx (of unit cross section), for a regime of slow bubbling is given as

$$dE = dE_1 + dE_2 + dE_3 \tag{62}$$

where dE is the total energy of the layer, dE_1 is the potential energy of the liquid, dE_2 is the dissipative energy of the layer, and dE_3 is the energy of surface tension. The expression of dE_1 is given by Equation 8 and that of dE_2 has the form

$$dE_2 = C_D \frac{\rho_f v^2}{2} \pi r_b^2 \, n x dx \tag{63}$$

where C_D is the drag coefficient, v is the local gas velocity, and n is the number of bubbles per unit volume. We have, as before, for gas mass conservation,

$$\phi v = v_s \tag{64}$$

and the fraction of the volume of the differential layer occupied by the gas is

$$\phi = \frac{4}{3} \pi r_b^3 n \tag{65}$$

Taking the above two equations into consideration, Equation 63 becomes

$$dE_2 = \left(\frac{3}{8}\right) \frac{C_D \rho_f v_s^2}{r_b \phi} x dx \tag{66}$$

The expression for the energy of surface tension is given [44] by

$$dE_3 = 4\pi r_b^2 \sigma n dx = \frac{3\sigma}{r_b} \phi dx \tag{67}$$

Substituting the expressions of dE_1, dE_2 and dE_3 from Equations 8, 66 and 67 into Equation 62, the following expression for the total energy of the mixture of height x_1 is obtained:

$$E = \int_0^{x_1} \left[(1 - \phi) \rho_f g x + \frac{3}{8} \frac{C_D \rho_f v_s^2}{r_b \phi} x + \frac{3\sigma}{r_b} \sigma \right] dx \tag{68}$$

The kinetic energy of the two-phase flow is ignored in Equation 68 because its value is usually three orders of magnitude less than the energy of dissipation for slow bubbling.

As in the analysis of rapid bubbling, the steady-state distribution of the gas content, ϕ, sets in when the energy given by Equation 68 is at a minimum [52]. Thus, to find the value of ϕ, it is necessary to determine the minimum of the integral under the condition that the height of the liquid in the equipment is constant (Equation 15). Ignoring variations due to x_1, the Euler's equation in this case assumes the form

$$-\rho_f g x - \frac{3C_D \rho_f v_s^2 x}{8 r_b \phi^2} + \frac{3\sigma}{r_b} - \lambda = 0 \tag{69}$$

where, as in the earlier cases, λ is the Lagrange multiplier. Rearranging, we find

$$\phi = \left[\frac{3C_D \rho_f v_s^2 x}{8 r_b \left(\frac{3\sigma}{r_b} - \rho_f g x - \lambda \right)} \right]^{1/2} \tag{70}$$

The height of the two-phase layer, x_1, can be determined from Equation 15 by substituting into it the value of ϕ from the above equation:

$$h = \int_0^{x_1} \left\{ 1 - \left[\frac{3\, C_D \rho_f v_s^2 x}{8 r_b \left(\frac{3\sigma}{r_b} - \rho_f g x - \lambda \right)} \right]^{1/2} \right\} dx \tag{71}$$

Integrating, we obtain

$$h = x + \frac{v_s}{g} \left(\frac{3C_D}{8 r_b \rho_f} \right)^{1/2} \left[\left(\frac{3\sigma}{r_b} - \lambda \right) x - \rho_f g x^2 \right]^{1/2} - \frac{v_s}{2} \left(\frac{3\sigma}{r_b} - \lambda \right) \left(\frac{3C_D \rho_f}{8 r_b} \right)^{1/2} \frac{1}{(\rho_f g)^{3/2}} \sin^{-1} \left[\frac{2\rho_f g x - \left(\frac{3\sigma}{r_b} - \lambda \right)}{\left(\frac{3\sigma}{r_b} - \lambda \right)} \right] \Bigg|_0^{x_1} \tag{72}$$

From Equation 70, for $\phi(x_1) = 1$, it follows that

$$\frac{3\sigma}{r_b} - \lambda = \rho_f g x_1 + \frac{3C_D \rho_f v_s^2 x_1}{8 r_b} \tag{73}$$

and taking this into account we can rewrite Equation 72 in the form

$$h = x_1 + \frac{3C_D v_s^2}{8 r_b g} x_1 - \left(\frac{3C_D}{8 g r_b} \right)^{1/2} \left(1 + \frac{3C_D v_s^2}{8 r_b g} \right) \frac{v_s x_1}{2} \left\{ \sin^{-1} \left[\frac{1 - \frac{3C_D v_s^2}{8 r_b g}}{1 + \frac{3C_D v_s^2}{8 r_b g}} \right] + \frac{\pi}{2} \right\} \tag{74}$$

Thus, the height of the two-phase mixture is given by

$$x_1 = \frac{h}{(1 + ab)\left\{ 1 - \frac{(ab)^{1/2}}{2} \left[\sin^{-1} \left(\frac{1 - ab}{1 + ab} \right) + \frac{\pi}{2} \right] \right\}} \tag{75}$$

and the average gas void fraction is simply

$$\phi_{av} = \frac{x_1 - h}{x_1} = 1 - \frac{h}{x_1} = ab - \frac{(ab)^{1/2}}{2} (1 + ab) \left[\sin^{-1} \left(\frac{1 - ab}{1 + ab} \right) + \frac{\pi}{2} \right] \tag{76}$$

where $a = \frac{3C_D}{8}$

$$b = \frac{v_s^2}{gr_b} \tag{77}$$

Equation 76 is illustrated in Figure 6 in comparison to experimental data [10,12]. Now, for superficial gas velocities, $v_s < 0.1$ m/sec for instance, in the bubbling process, which usually takes place in chemical reactors, the drag coefficient can be estimated by the following formula [53]:

$$C_D = 0.82 \left(\frac{g\mu_f^4}{\rho_f\sigma^3}\right)^{1/4} Re \tag{78}$$

Equations 75, 76 and 78 can be used to obtain the height and average void fraction, respectively, of the two-phase flow in the slow bubbling regime. The radius of the bubbles is given by the expression

$$r_b = \frac{C_D v_b^2}{4g} \tag{79}$$

where v_b is the velocity of the bubble, and we can use for v_b the velocity of buoyancy [7,44] given by

$$v_b = \frac{2}{3}\left(\frac{4\sigma^2 g}{3s\rho_f\mu_f}\right)^{1/5} \tag{80}$$

The factor s is a shape factor, and when the bubbles are spherical s = 1. The derivations of Equations 79 and 80 are given by Azbel [44].

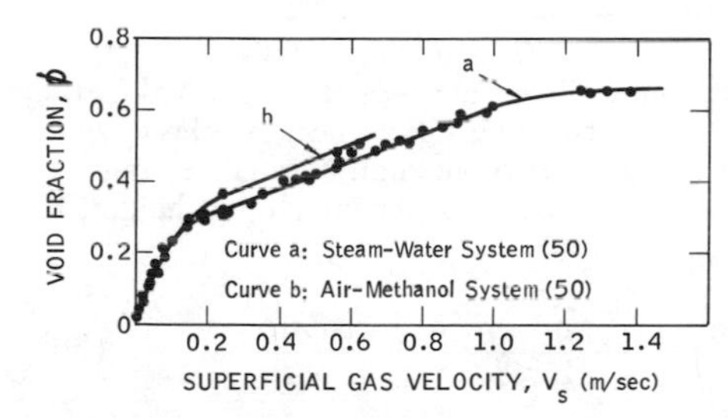

Figure 6. Void fraction vs superficial gas velocity.

MAIN PARAMETERS OF THE TWO-PHASE FLOW WITH DISSIPATION AND INERTIA

The influences of the dissipative and inertial forces have been investigated separately in the preceding sections. We will now consider the general problem of the calculation of the average void fraction when dissipative and inertial forces are present at the same time. For this case, we shall use the equation for the relative motion of a gas bubble in the two-phase flow, whose form is given as follows [44]:

$$g(\rho_f - \rho_g)V - C_D\frac{\rho_f v^2}{2}A_b - (\rho_g + k\rho_f)V\frac{dv}{dt} = 0 \tag{81}$$

where ρ_f is the density of liquid, ρ_g is the density of gas, V is the volume of the bubble, C_D is the drag coefficient, A_b is the cross-sectional area of a bubble, and k is the "apparent additional mass" coefficient (for a spherical bubble $k = 0.5$). Multiplying Equation 81 by v, we obtain an expression for the change of energy of a single bubble during its motion:

$$g(\rho_f - \rho_g)gVv - C_D \frac{\rho_f v^3}{2} A_b - (\rho_g + k\rho_f)V \frac{d}{dt}\frac{v^2}{2} = 0 \tag{82}$$

or, in the integral form,

$$E_b = (\rho_f - \rho_g)gVx - C_D \frac{\rho_f A_b}{2} \int_0^x v^2 dx - \frac{(\rho_g + k\rho_f)Vv^2}{2} \tag{83}$$

where E_b is the energy of a single bubble.

Assuming that the differential layer, dx, contains n bubbles per unit volume, whose volume is a fraction $\phi = 4/3\ \pi r_b^3 n$ of the total volume in the layer, we can express the energy of all the bubbles in a unit cross section of the layer dx as follows:

$$E_b\, ndx = \left[\left(\rho_g + \frac{\rho_f}{2}\right)\frac{v^2}{2}\phi + C_D \frac{\rho_f}{2}\phi \frac{A_b}{V}\int_0^x v^2 dx - g(\rho_f - \rho_g)\phi x\right] dx \tag{84}$$

Note that, as in Equation 11, we have set $k = ½$. If we combine the potential energy of the liquid and the energy of surface tension with the energy of the bubbles, the total energy of the two-phase layer will be

$$E = \int_0^{x_1}\left[(\rho_f - \rho_g)gx(1-\phi) + 1/2\left(\rho_g + \frac{\rho_f}{2}\right)v^2\phi + C_D \frac{3}{4r_b}\phi\frac{\rho_f}{2}\int_0^x v^2 dx + \frac{3\sigma}{r_b}\phi\right] dx \tag{85}$$

We shall ignore the influence of the transition zones (lower and upper) described in the previous section on the average void fraction of the two-phase flow because these zones essentially balance each other out. With this condition of negligibly small change in the void fraction with height, the energy of the two-phase flow per unit liquid height, h, will be

$$\frac{E}{h} = \rho_f gh(1-\phi)\frac{\eta_1^2}{2} + \frac{\rho_f v_s^2 \eta_1}{4\phi} + \frac{3C_D\rho_f v_s^2 \eta_1^2 h}{16 r_b \phi} + \frac{3\sigma\phi\eta_1}{r_b} \tag{86}$$

when $\eta_1 = x_1/h$ (h is the height of the static liquid, and x_1 is the height of the two-phase flow), v_s is the superficial gas velocity, and we have used the relation $v = v_s/\phi$ and the approximation $(\rho_f - \rho_g) \simeq \rho_f$.

Noting that $\phi_{av} \simeq \phi \simeq$ constant by neglecting the transition regions, we find that

$$\eta_1 \simeq \frac{1}{1-\phi}$$

and Equation 86 can be rewritten as follows:

$$\frac{E}{h} = \frac{\rho_f gh}{2(1-\phi)} + \frac{\rho_f v_s^2}{4\phi(1-\phi)} + \frac{3C_D\rho_f v_s^2 h}{16 r_b \phi(1-\phi)^2} + \frac{3\sigma\phi}{r_b(1-\phi)} \tag{87}$$

where, as in the rest of this section, ϕ will denote the average void fraction. The equilibrium distribution of the void fraction, ϕ, is established when the specific energy,

E/h, of the two-phase flow is at a minimum. Consequently, to determine this equilibrium, ϕ, we must vary Equation 87 with respect to ϕ to obtain

$$\frac{\rho_f gh}{2(1-\phi)^2} - \frac{\rho_f v_s^2(1-2\phi)}{4\phi^2(1-\phi)^2} - \frac{3C_D\rho_f v_s^2 h(1-3\phi)}{16r_b\phi^2(1-\phi)^3} + \frac{3\sigma}{r_b(1-\phi)^2} = 0 \tag{88}$$

After algebraic transformations, we can rewrite this equation as

$$\phi^3 - B\phi^2 - 3A\phi + A = 0 \tag{89}$$

where A and B are the dimensionless quantities:

$$A = \frac{(1/2)v_s^2\left(1 + \dfrac{3C_D h}{4r_b}\right)}{\left(gh + \dfrac{6\sigma}{\rho_f r_b}\right)}, \tag{90}$$

and

$$B = 1 - \frac{v_s^2}{\left(gh + \dfrac{6\sigma}{\rho_f r_b}\right)} \tag{91}$$

In many practical applications quantity B approaches unity and, therefore, in this case Equation 89 gives

$$A \simeq \frac{\phi^2(1-\phi)}{1-3\phi} \tag{92}$$

In Figure 7, a plot is given of $A^{1/2}$ against ϕ according to Equation 92. This curve can be used to determine the void fraction, once the value of A is known, for $B \simeq 1$.

Let us now eliminate from the quantities A and B the part that is dependent only on one flow system parameter, h, using Equations 90 and 91:

$$\frac{A}{1\ \ B} = (1/2)\left(1 + \frac{3C_D h}{4\,r_b}\right) = \alpha \tag{93}$$

where α is a function of h.

Taking this into account, we can rewrite Equation 89 in the form

$$\phi^3 - B\phi^2 - 3\alpha(1-B)\phi + (1-B)\alpha = 0 \tag{94}$$

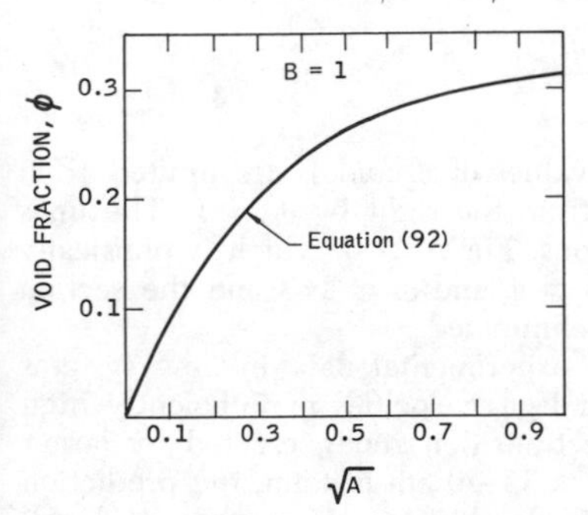

Figure 7. Void fraction vs factor $(A)^{1/2}$.

or

$$B = \frac{\phi^3 - 3\phi\alpha + \alpha}{\phi^2 - 3\phi\alpha + \alpha} \tag{94a}$$

From this equation it follows that for values of ϕ that are the roots of the equation,

$$\phi^2 - 3\alpha\phi + \alpha = 0 \tag{95}$$

$B = -\infty$ (because $0 \leq \phi \leq 1$), which means, as indicated in Equation 91, that for h being constant, the superficial velocity, v_s, is infinite.

Thus, Equation 95 is applicable when we have flows with large superficial velocity, and the roots of this equation will be

$$\phi = \frac{3\alpha}{2}\left[1 \pm \left(1 - \frac{4}{9\alpha}\right)^{1/2}\right] \tag{96}$$

Because ϕ must be less than, or equal to, unity, we have only one value of ϕ from this equation, which is

$$\phi = \frac{3\alpha}{2}\left[1 - \left(1 - \frac{4}{9\alpha}\right)^{1/2}\right] \tag{97}$$

For the above equation, for $\alpha = 1$ (about the smallest possible value of α as is clear from Equation 93 because $h/r_b >> 1$), ϕ achieves a maximum value, equal to $\phi = 0.382$. This, then, is the value of the void fraction in the central zone, which cannot be surpassed either by decreasing the height, h, or by increasing the gas velocity, v_s, beyond a certain (high) value.

Further, for $\alpha \to \infty$, we find

$$\phi \simeq \frac{3\alpha}{2}\left(\frac{2}{9\alpha}\right) = \frac{1}{3} \tag{98}$$

and this is the limit the void fraction reaches when there are no restrictions on the upper limits of gas velocity and static liquid height.

It should be noted that according to experimental data, the void fraction in this center zone actually achieves higher values than we have estimated because of the effect of the two transition zones, which we neglected in our analysis. Nevertheless, for high gas velocities we can see that the analysis gives the result that the void fraction, ϕ, lies in the range 0.3 to 0.4.

Figure 8 gives a graph of quantity B against ϕ with different values of the parameter A, according to Equation 89, in the form

$$B = \phi + \frac{A(1 - 3\phi)}{\phi^2} \tag{99}$$

Thus, we can see in the figure that the possible values of ϕ and B are limited from above by line $B = 1$, from the left by $\phi = 0$ and from the right by $\phi = 1$. The lines $B = \phi$ and $\phi = 1/3$ divide the field into four sections. For $A > 0$, which is physically necessary, the section limited by the lines $B = 1$, $B = \phi$ and $\phi = 1/3$ and the section limited by the lines $B = 0$, $B = \phi$ and $\phi = 1/3$ are eliminated.

Equations 89 through 97 agree fairly well with experimental data in flow systems where the liquid depth, h, is large and the superficial gas velocities in sufficiently high regions of the liquid are small. If the extent of the transition zones, created by larger superficial gas velocities (more than 0.5–0.6 m/sec) is 35–40 cm in total, the prediction of ϕ by these equations will still hold when the liquid depth is large (in excess of 1 m).

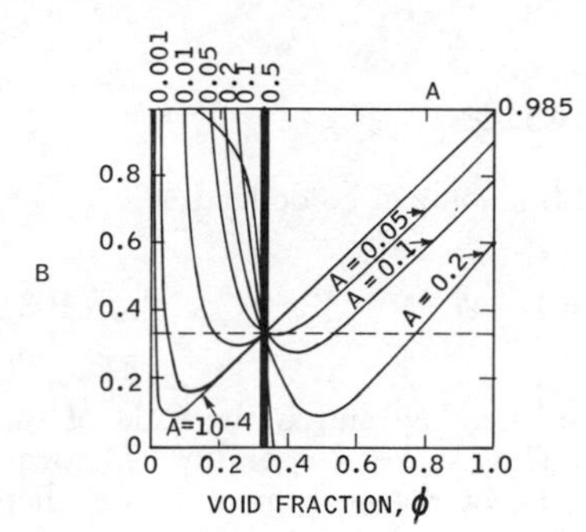

Figure 8. Plot of void fraction vs factors A,B.

The upper limit for using the equations (divergence from experimental data being no more than ± 10%) is for $B \leq 0.985$.

To be able to use this set of equations, we must have a means of determining the bubble radius, r_b, and the drag coefficient, C_D. The bubble radius and drag coefficient have been shown [44,54] to be related in the following form:

$$r_b = \frac{C_D v_b^2}{4g} \tag{100}$$

where v_b is the velocity of the bubble. Consequently,

$$\frac{3}{8}\frac{C_D}{r_b} = \frac{3g}{2v_b^2} \tag{101}$$

and we can use for v_b the velocity of buoyancy given [7] by

$$v_b = \frac{2}{3}\left[\frac{4\sigma^2 g}{3s\rho_f\mu_f}\right]^{1/5} \tag{102}$$

The factor s is a shape factor, and when the bubbles are spherical $s = 1$. It is possible to take other values of s. For example, by taking $s = 4$—the ratio of the major to minor axes of the bubble then being equal to 2—we get better results. Equation 101 then can be used in Equation 93. We still have to determine the factor $6\sigma/\rho_f r_b$ in Equations 90 and 91, and to simplify the calculation, r_b in this term is taken to be 0.2 cm (the influence of this term on the final result being very small). Figure 9 shows experimental data [37] obtained for an air-water system, as well as the curve corresponding to Equation 89. Note the close correspondence between theory and experiment.

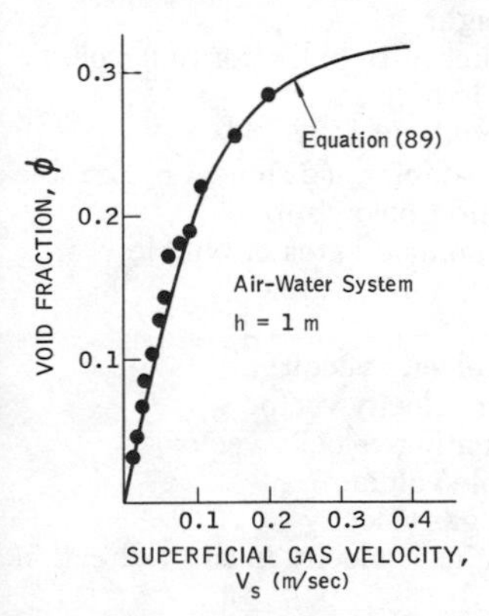

Figure 9. Void fraction vs superficial gas velocity for the air-water system.

EFFECT OF STATIC LIQUID HEIGHT AND EQUIPMENT DIAMETER ON THE VOID FRACTION

In Equation 76 it is seen that the average gas void fraction can be defined as

$$\phi_{av} = \frac{x_1 - h}{x_1} = 1 - \frac{h}{x_1} = 1 - \psi \tag{103}$$

where ψ is the average relative density of the two-phase flow and is the ratio of the density of the gas-liquid mixture to that of the liquid. Thus, if ψ (or x_1 and h) is known, ϕ_{av} can be obtained. However, when the diameter, D, of the equipment is less than 0.2 m and the static liquid level, h, is less than 1 m, we must introduce correction factors into the above equation to obtain

$$\phi_{av} = 1 - k_d k_h \psi \tag{104}$$

where k_d is the correction factor for the equipment diameter, and k_h is the correction factor for the static liquid level.

Analyzing experimental data with the use of similarity theory yields

$$k_d = 1 - \exp\left\{ - 1.1 \left[\frac{v_s}{v_b}\right]^{1/2} \left[\frac{(gd)^{1/2}}{v_s}\right]^{3/4}\right\} \tag{105}$$

and

$$k_h = 1 - \exp\left\{ - 0.405 \left(\frac{v_s}{v_b}\right)^{0.7} \frac{(gh)^{1/2}}{v_s}\right\} \tag{106}$$

where v_b is the rising velocity of a single bubble, given by the formula in [44]

$$v_b = 1.18 \left[\frac{g\sigma(\rho_f - \rho_g)}{\rho_f^2}\right]^{1/4} \tag{107}$$

NOMENCLATURE

- A_b cross-sectional area of bubble
- C_D drag coefficient
- D_t turbulent diffusivity
- d diameter of equipment
- dE total energy of layer
- dE_1 potential energy of layer
- dE_2 kinetic energy of layer
- dE_2 dissipation in layer
- dE_3 surface tension energy
- E total energy of two-phase mixture
- E_b energy of a bubble
- F_D drag force
- F_o form drag
- Fr Froude number
- g gravitational acceleration
- h liquid column height
- k coefficient of apparent additional mass
- k_d correction factor for equipment diameter
- k_h correction factor for equipment height
- n number of bubbles per unit volume
- r_b bubble radius
- Re Reynolds number
- s shape factor (distance of bubble center below surface)
- s_o contaminated area of bubble
- s_b bubble surface area
- t time
- u_i gas velocity vector
- $\overline{u_i}$ mean velocity vector
- u_i' fluctuating velocity vector
- V bubble volume
- v local gas velocity
- $\overline{v}$ mean gas velocity in the x direction

v_b bubble velocity
v_r velocity of the bubble relative to the liquid
v_s superficial gas velocity
x direction of flow (vertical)
x_1 height of two-phase mixture

Greek Symbols

λ Lagrange multiplier
μ_f liquid dynamic viscosity
ν_f liquid kinematic viscosity
ρ_f liquid density
ρ_g gas density
σ surface tension
τ eddy time scale
ϕ gas void fraction
$\bar{\phi}$ mean void fraction
ϕ' fluctuating void fraction
ϕ_{av} average gas void fraction
ψ foam specific gravity

REFERENCES

1. Teletov, S. G. *DAN SSSR* 4 (1945).
2. Teletov, S. G. *Vest. Mosk. Univer. N2 seria mekhaniki,* Moscow (1958).
3. Teletov, S. G. Doctoral Dissertation, ENIN, AN SSSR, Moscow (1948).
4. Teletov, S. G. Candidate's Dissertation. Moscow University, USSR (1938).
5. Levich, V. G. *K Teorii Poverkhnostnykh Yavlenii* (Theory of Surface Phenomena) (Moscow: Izd. Sov. Nauka, 1941).
6. Landau, L. D., and E. M. Lifshitz. *Fluid Mechanics* (Elmsford, NY: Pergamon Press, Inc., 1959).
7. Levich, V. G. *Physico-chemical Hydrodynamics* (Englewood Cliffs, NJ: Prentice-Hall, Inc., 1962).
8. Kochin, N. E., I. A. Kibel and N. N. Roze. *Teoreticheskaya Gidrodinamika* (Moscow: Gostechisdat, 1955).
9. Akselrod, L. S., and V. V. Dilman. 1956, *Zh. Prikl. Khim.* 29(12): 1803 (1956).
10. Kutateladze, S. S., and M. A. Styrikovich. *Hydraulics of Gas-Liquid Systems,* Wright Field, trans. F-TS-9814v (1958).
11. Kasatkin, A. G., Yu. I. Dyinerskii and D. M. Popov. *Khim. Prom.* 7: 482 (1961).
12. Aizenbud, M. B., and V. V. Dilman. *Khim. Prom.* 3: 199 (1961).
13. Akselrod, L. S., and V. V. Dilman. *Zh. Prikl. Kim.* 27: 15 (1954).
14. Chhabra, P. S., and S. P. Mahajan. *Indian Chem. Eng.* 16(2): 16 (1974).
15. Bhaga, D., and M. E. Weber. *Can. J. Chem. Eng.* 50(3): 323 (1972).
16. Bhaga, D., and M. E. Weber. *Can. J. Chem. Eng.* 50(3): 329 (1972).
17. Pruden, B. B., and M. E. Weber. *Can. J. Chem. Eng.* 48: 162 (1970).
18. Pozin, M. E., I. P. Mukhlenov and E. Ya. Tarat. 1957, *Zh. Prinkl. Khim.* 30(1): 45 (1957).
19. Zuber, N., and J. A. Finlay. *J. Heat Transf. Trans. ASME* C 87: 453 (1965).
20. Chekhov, O. S., Candidate's Dissertation, MIKhM, Moscow (1960).
21. Solomakha, G. P., Candidate's Dissertation, MIKhM, Moscow (1957).
22. Artomonov, D. S., Candidate's Dissertation, MIKhM, Moscow (1961).
23. Brown, R. W., A. Gomezplata and J. D. Price. *Chem. Eng. Sci.* 24: 1483 (1969).
24. Ribgy, G. R., and C. E. Capes. *Can. J. Chem. Eng.* 48: 343 (1970).
25. Gomezplata, A., R. E. Munson and J. D. Price. *Can. J. Chem. Eng.* 50: 669 (1972).
26. Gomezplata, A., and P. T. Sung. *Chem. Eng. Sci.* 48: 336 (1970).
27. Koide, K., T. Hirahara and H. Kubota. *Kogaku Koguku* 5(1): 38 (1967).
28. Stepanek, J. *Chem. Eng. Sci.* 25: 751 (1970).
29. Pruden, B. B., W. Hayduk and H. Laudie. *Can. J. Chem. Eng.* 52: 64 (1974).
30. Laudie, H. A. MSc Thesis, Ottawa, Ontario, Canada (1969).
31. Bell, R. L. *AIChE J.* 18(3): 498 (1972).
32. Kuz'minykh, I. N., and G. A. Koval'. *Zh. Prinkl. Khim.* 28(1): 21 (1955).
33. Noskov, A. A., and V. N. Sokolov. *Trudy L.T.I. im. Lensoveta* 39: 110 (1957).

34. Pozin, M. E. *Sbornik: voprosy massoperedachi* 148 (1957).
35. Shepherd, E. B. *Ind. Chemist* 4: 175 (1956).
36. Jackson, R. *Ind. Chemist* 336(16) (1953).
37. Hughes, R. R. *Chem. Eng. Prog.* 51(12): 555 (1955).
38. Loitsyanskii, L. G. *Mekhanika Zhidkosti i Gaza, Gostekhizdat* (1957).
39. Smirnov, V. I. *Kurs Vysshey Matematiki Gostechisdat* 4 (1941).
40. Vinokur, Ya. G., and V. V. Dilman. *Khim. Prom.* 7:619 (1959).
41. Kashnikov, A. M. Candidate's Dissertation, Moscow, D. I. Mendeleev Institute of Chemical Technology, Moscow (1965).
42. Rodionov, A. I., A. M. Kashnikov and V. M. Radikovsky. *Khim. Prom.* 10:17 (1964).
43. Azbel, D. S., and A. N. Zeldin. *Teoreticheskie Osnovy Khimicheskoi Tekhnologii* 5:863 (1971).
44. Azbel, D. S. *Two-Phase Flows in Chemical Engineering* (New York: Cambridge University Press, 1981).
45. Akselrod, L. S., and V. V. Dilman. *Khim. Prom.* 1:8 (1954).
46. Azbel, D. S. *Khim. Prom.* 11:854 (1962).
47. Gal-Or, B. *AIChE J.* 12(3):604 (1966).
48. Marucci, G. *Ind. Eng. Chem. Fund.* 2:224 (1965).
49. Kurbatov, A. V. *Trudy MEI* 2:82 (1953).
50. Sterman, L. S. *Zh. Tekh. Phys.* 26(7):1512 (1956).
51. Sterman, L. S., and A. B. Surnov. *Teploenergetika* 8:39 (1955).
52. Lavrent'ev, M. A., and L. A. Luysternak. *A Course of Variation Calculus* (in Russian) (Moscow: GITTL, 1950).
53. Peebles, F. N., and H. J. Garber. *Chem. Eng. Prog.* 49(2):88 (1953).
54. Azbel, D. S. *Khim. Prom.* 1:43 (1964).

CHAPTER 17
MECHANISMS OF LIQUID ENTRAINMENT

David Azbel and Athanasios I. Liapis
Department of Chemical Engineering
University of Missouri–Rolla
Rolla, Missouri 65401

CONTENTS

INTRODUCTION

This chapter presents the description and analysis of the different mechanisms involved in the phenomenon of liquid entrainment. This phenomenon proceeds under various conditions of interaction of a gas stream with a liquid; liquid droplets are entrained when,

1. a gas bubbles through a liquid layer,
2. a gas tears away from a free liquid surface, and
3. droplets break up whenever a liquid jet as well as single droplets impinge against a liquid surface and/or the walls of processing equipment and are reentrained in a passing gas stream.

The transportation of the entrained droplets may be considered to occur by either of the following two modes: (1) transport of droplets by the gas stream, in which the droplets are considered to be in a state of suspension (the droplets are suspended in the gas stream), or (2) droplet throwing up as a result of dynamic interaction of gas jets.

Formation of liquid droplets of various sizes and their entrainment by a gas stream, as well as partial separation of droplets from a gas on the walls of flow channels, take place in a large amount of engineering processing equipment (i.e., fractionation and absorption columns, steam boilers, nuclear and chemical reactors) when a gas stream interacts with a liquid. This interaction fundamentally affects the changes occurring in the state variables of the gas-liquid mixture and the transfer of momentum, heat and mass from one point of space to another separated from the first by an interface. In many cases, sudden changes in the velocity vector and pressure develop at the phase boundaries, which result in the establishment of conditions such that instabilities may occur.

It is well known that droplet entrainment decreases the efficiency of fractionation and absorption columns, since this phenomenon occurs long before flooding and decreases the coefficient of enrichment, and so it does not permit the gas (vapor) rate in the column to increase. This example indicates the need to analyze the mechanisms of liquid entrainment, the knowledge of which will enable one to choose the appropriate column diameter and space between trays, depending on the gas (vapor) and liquid loads, as well as on the physical properties of the working media. Thus, rational design of equipment in which gas-liquid mixtures are processed requires a knowledge of the basic laws governing the phenomenon of liquid entrainment.

The study reported in this chapter is important because it will show which factors contribute to the maximization (or minimization) of the amount of liquid entrained by the flowing gas.

The complicated nature of the two-phase behavior mitigates against the development of anything approaching a complete analytical description, and thus several different possible mechanisms will be presented. It should be noted that although the following analysis is applied to the case of gas passing through an orifice (or orifices) into liquid, the results obtained are in no way restricted to this flow system.

ENTRAINMENT AT THE LIQUID-FREE SURFACE DURING BUBBLING

In this section, possible mechanisms of entrainment in the form of liquid droplets at the upper (free) surface of the liquid column are considered.

Cellular Foam Condition

When gas bubbles through a liquid layer, the bursting of the shells of the bubbles coming to the surface and the simultaneous formation of droplets result in a considerable decrease in the total surface area of the bubbles. This decrease is usually many times greater than the increase in the total surface area contributed by the newly formed drops; therefore, drop formation and propulsion during bubbling must be due not only to the kinetic energy of the gas, but also to the surface energy released when bubbles burst.

At a moderate rate of bubbling (at a small superficial gas velocity) and with a considerable depth of liquid through which bubbles rise, which is typical of most engineering equipment, the kinetic energy of the gas approaching the surface is relatively small and the surface energy of the bubble shell plays a major role in the total energy balance. Thus, for example, the average velocity of gas rising in liquid under conditions of high-pressure steam boilers ($p \simeq 100$ to 120 atm) does not exceed 0.7 m/sec, and in low-pressure distillation columns ($p \simeq 1$ atm) this velocity reaches a maximum of about 2–3 m/sec. Taking into account that the kinetic energy of a bubble is

$$L_k = \frac{1}{2} V \rho_g v_b^2 \tag{1}$$

and that the surface energy of a bubble is

$$L_\sigma = \pi d_b^2 \sigma \tag{2}$$

one can make a rough estimate of L_k and L_σ for these two cases. Letting, for illustration, the mean bubble diameter be $d_b = 3$ mm in a steam boiler and $d_b = 5$ mm in a distillation column, we get

For a high-pressure boiler:

$$\sigma \simeq 1.75 \times 10^{-2} \text{ N/m}$$

$$L_k \simeq 4.2 \times 10^{-9} \text{ J}$$

$$L_\sigma \simeq 5 \times 10^{-7} \text{ J}$$

For a distillation column:

$$\sigma = 7.28 \times 10^{-2} \text{ N/m}$$

$$L_k \simeq 2 \times 10^{-7} \text{ J}$$

$$L_\sigma \simeq 5.7 \times 10^{-6} \text{ J}$$

Thus, under these conditions, the total kinetic energy of the gas is much less than the released surface energy, and an analysis of the mechanism by which the surface energy affects droplet formation is of clear interest. At low gas flowrates, the gas breaks through the liquid mass at regular intervals in the form of bubbles. When the superficial gas velocity is increased, a cellular foam layer, consisting of tightly packed bubbles, much enlarged and deformed, is developed on the surface of the liquid, and at a given ratio of flowrates of gas and liquid all the liquid may be changed into cellular foam [1].

A possible mechanism for liquid entrainment under foaming condition is suggested by experimental investigations [2] of this phenomenon in the so-called "bubble by bubble regime," where it is observed that at low flowrates of gas through individual orifices the bubbles do not coalesce to form jets when coming to the surface. Under these cellular foam conditions, drops of liquid usually are generated by bursting of the bubble film, and subsequently can be entrained by the gas escaping from the cellular foam layer at the liquid surface.

Bubble Breakup

In this section, a mathematical model is developed to describe entrainment of liquid due to bubble breakup at the free surface of the liquid [3].

When a gas bubble is released in a stagnant liquid, it will rise under the action of the buoyancy force, F_b, to the liquid surface, where it is arrested by the surface tension force, F_σ, exerted by the liquid membrane, with base radius, r_m, protruding out of the liquid surface and forming the top part of the bubble boundary, as shown in Figure 1. It is assumed, for simplicity, that the bubble remains spherical in shape, with radius, r_b, and its center, 0, located at a distance, s, below the liquid surface, under equilibrium

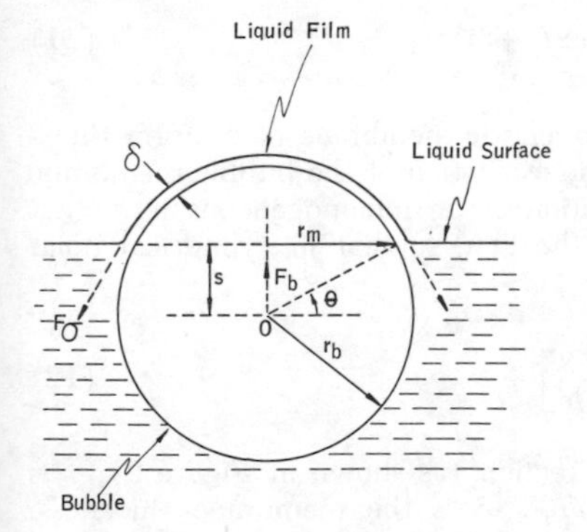

Figure 1. Sketch of equilibrium bubble configuration.

conditions. We further assume that near the edge of the liquid membrane the surface tension force, F_σ, acts in a direction tangential to the boundary of the bubble. Now the buoyancy force, F_b, acting in the vertical direction is

$$F_b = V_d\,(\rho_f - \rho_g)g, \tag{3}$$

where V_d is the displacement volume of the bubble, or the volume of that portion of the bubble that lies below the liquid surface. The vertical component of the surface tension force acting on the edge of the membrane is

$$F_{\sigma_v} = 2\pi r_m \sigma \cos\theta \tag{4}$$

where θ is the angle between the horizontal and the bubble radius ending at the edge of the membrane; r_m is the base radius of the membrane; and σ is the interface surface tension.

For use in Equations 3 and 4, the following relations can be obtained from geometric considerations:

$$V_d = \frac{2}{3}\pi r_b^3 + \pi s r_b^2 - \frac{1}{3}\pi s^3 \tag{5}$$

$$\cos\theta = \left(1 - \frac{s^2}{r_b^2}\right)^{1/2} \tag{6}$$

$$r_m = (r_b^2 - s^2)^{1/2} \tag{7}$$

At equilibrium, $F_b = F_{\sigma_v}$. Therefore, from Equations 3–7 we have

$$\eta^3 - \beta\eta^2 - 3\eta + (\beta - 2) = 0 \tag{8}$$

where

$$\eta = \frac{s}{r_b},\ \beta = \frac{6\sigma}{(\rho_f - \rho_g) g r_b^2} \tag{9}$$

This cubic equation has a root at $\eta = -1$, but the particular root we are concerned with will be given by

$$\eta^2 - (\beta + 1)\,\eta + (\beta - 2) = 0 \tag{10}$$

or

$$\eta = \frac{1}{2}\,[(\beta + 1) - (\beta^2 - 2\beta + 9)^{1/2}] \tag{11}$$

In a manner similar to the case of a thin, planar elastic membrane of uniform thickness subjected to small transverse displacements, the oscillation of the bubble membrane is governed by the wave equation, with the addition of an inhomogeneous term that accounts for the static pressure difference across the film, so that in cylindrical polar coordinates [4] we have

$$\frac{\partial^2 y}{\partial t^2} = a^2\left[\frac{\partial^2 y}{\partial r^2} + \frac{1}{r}\frac{\partial y}{\partial r}\right] + P \tag{12}$$

where $y(r,t)$ is the displacement in the vertical direction (as shown in Figure 2); r is now the cylindrical radial coordinate, $a^2 = 2\sigma/\rho_f\delta$; δ is the membrane thickness;

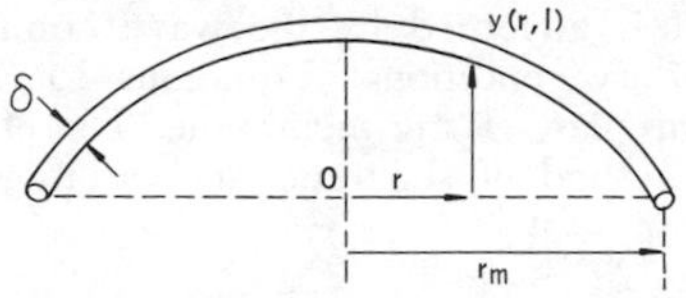

Figure 2. Sketch of film cap in oscillation.

$P = (P_g - P_a)/\rho_f\delta$ is the bubble pressure increment; P_g is the bubble pressure; and P_a is the ambient pressure. This equation is subject to the following boundary conditions:

$$y(r,t) = 0 \tag{12a}$$

$$\frac{\partial y(0,t)}{\partial r} = 0 \tag{12b}$$

We will let the membrane displacement, $y(r,t)$, be made up of two parts: one to be time invariant, y_o, the other to be time dependent, $y'(r,t)$:

$$y(r,t) = y_o(r) + y'(r,t) \tag{13}$$

Introducing Equation 13 into Equations 12–12b, we have for each of the new dependent variables, $y_o(r)$ and $y'(r,t)$, a simpler governing equation. For the steady displacement, $y_o(r)$, we have the governing equation

$$\frac{d^2y_o}{dr^2} + \frac{1}{r}\frac{dy_o}{dr} = -\frac{P}{a^2} \tag{14}$$

subject to the boundary conditions

$$y_o(r_m) = 0 \tag{14a}$$

$$\frac{dy_o(0)}{dr} = 0 \tag{14b}$$

and for the time-dependent displacement, $y'(r,t)$, we then have (from Equations 12 and 14),

$$\frac{\partial^2 y'}{\partial t^2} = a^2 \left[\frac{\partial^2 y'}{\partial r^2} + \frac{1}{r}\frac{\partial y'}{\partial r}\right] \tag{15}$$

subject to the boundary conditions

$$y'(r_m,t) = 0 \tag{15a}$$

$$\frac{\partial y'(0,t)}{\partial r} = 0 \tag{15b}$$

A solution for Equation 14 for the steady displacement $y_o(r)$, satisfying the boundary conditions, then can be obtained readily as

$$y_o(r) = c\left(1 - \frac{r^2}{{r_m}^2}\right) \tag{16}$$

where $c = \frac{P{r_m}^2}{4a^2}$

Now, the time-dependent displacement, $y'(r,t)$, is governed by the wave equation (Equation 15), subject to the homogeneous boundary conditions (Equations 15a and 15b), similar to the case of oscillations of a flat, circular, elastic membrane. Therefore, a solution for $y'(r,t)$ is readily obtained by the method of separation of variables in the form

$$y'(r,t) = W(r)\,G(t) \tag{17}$$

and, substituting this into Equation 15, we have

$$\ddot{G} + (ak)^2 = 0 \qquad k = \text{constant} \tag{18}$$

and $\ddot{G}$ denotes d^2G/dr^2, and

$$W'' + \frac{1}{r}W' + k^2W = 0 \tag{19}$$

where W' and W'' denote dW/dr, d^2w/dr^2.

Introducing the substitution $S = kr$, we obtain from Equation 19

$$\frac{d^2W}{dS^2} + \frac{1}{S}\frac{dW}{dS} + W = 0 \tag{20}$$

This is a form of Bessel's equation, and for our application has the solution

$$W(r) = J_o(S) = J_o(kr) \tag{21}$$

where J_o is the Bessel function of the first kind and order zero.

Using the boundary conditions of Equation 15, we find that

$$J_o(k_nr_m) = 0 \tag{22}$$

where $k_n = \alpha_n/r_m$ and $\alpha_n(n = 1, 2, 3 \ldots)$ are the positive zeros of the Bessel function, J_o. Hence,

$$W_n(r) = J_o\left(\frac{\alpha_n r}{r_m}\right), n = 1, 2, 3 \ldots \tag{23}$$

is the solution of Equation 19, with boundary conditions given by Equations 15a and 15b.

The corresponding solutions of Equation 18 are easily obtained as

$$G_n(t) = A_n \cos(ak_nt) + B_n \sin(ak_nt) \tag{24}$$

where A_n and B_n are constants, giving a solution of the wave equation (Equation 15):

$$y'_n(r,t) = W_n(r)G_n(t) = [A_n\cos(ak_nt) + B_n\sin(ak_nt)]J_o(k_nr) \tag{25}$$

and the corresponding frequencies of oscillation are

$$\Omega_n = \frac{a\alpha_n}{2\pi r_m} = \frac{\alpha_n}{2\pi r_m}\left(\frac{2\sigma}{\rho_f\delta}\right)^{1/2}, n = 1, 2, \ldots \tag{26}$$

the first of which is the fundamental normal mode,

$$\Omega_1 = \frac{\alpha_1}{2\pi r_m}\left(\frac{2\sigma}{\rho_f \delta}\right)^{1/2} \tag{27}$$

where $\alpha_1 \simeq 2.404$.

We thus have a formula specifying the natural frequency of oscillations of the membrane; we now need to obtain a formula for the oscillation frequency of the gas in the bubble. It is well known that for a compressible fluid (the gas), the velocity potential, ϕ, of the motion is described by the wave equation [5,6]:

$$\frac{\partial^2 \phi}{\partial t^2} = c_s \left(\frac{\partial^2 \phi}{\partial r^2} + \frac{2}{r}\frac{\partial \phi}{\partial r}\right) \tag{28}$$

where c_s is the speed of sound in the gas, and r is now the radial distance from the bubble center, in spherical coordinates. Note that in Equation 28 we have assumed spherical symmetry for the flow in the bubble interior. We need boundary and initial conditions to obtain a solution of this equation, and so we consider the gas to be constrained in a rigid sphere (the bubble surface). This is, of course, not the case for the bubble, but our intention here is only to obtain an approximate value for the natural frequencies of the gas motion, and so we take as the boundary condition

$$\left.\frac{\partial \phi}{\partial r}\right|_{r=r_b} = 0 \tag{29}$$

which states that the normal velocity of the gas at the bubble surface is zero. As for the analysis of the membrane motion, we use the method of separation of variables:

$$\phi(r,t) = T(t)\,R(r) \tag{30}$$

which, in conjunction with Equation 18, produces

$$\ddot{T}(t) + \lambda^2 c_s\, T(t) = 0 \tag{31}$$

and

$$R''(r) + \frac{2}{r}R'(r) + \lambda^2 R(r) = 0 \tag{32}$$

where λ is a constant, and $\ddot{T}$, R' and R'' denote differentiations as in Equations 18 and 19, with the boundary condition, Equation 29, being

$$\left.\frac{dR}{dr}\right|_{r=r_b} = 0 \tag{33}$$

The general solution of Equation 29 is of the form

$$R(r) = C_1 \frac{\sin\lambda r}{r} + C_2 \frac{\cos\lambda r}{r} \tag{34}$$

where C_1, and C_2 are constants. Clearly $C_1 = 0$ because otherwise the solution would diverge as r approaches zero, and C_2 can be set equal to unity without loss of generality. Hence, we can write

$$R(r) = \frac{\sin\lambda r}{r} \tag{35}$$

and using this in the boundary condition (Equation 33), we obtain

$$\tan\lambda r_b = \lambda r_b \tag{36}$$

If we let the eigenvalues of Equation 36 be λ_n, n = 1, 2, . . . , then the solution of Equation 31 will be of the form

$$T_n(t) = C_3\cos(\lambda_n c_s t) + C_4\sin(\lambda_n c_s t) \tag{37}$$

and the solution of Equation 28 is of the form

$$\phi_n(r,t) = \frac{\sin\lambda_n r}{r}\left[C_3\cos(\lambda_n c_s t) + C_4\sin(\lambda_n c_s t)\right] \tag{38}$$

where C_3 and C_4 are constants.

The natural frequencies of the gas motion are then

$$\omega_n = \frac{\lambda_n c_s}{2\pi},\ n = 1, 2, \ldots \tag{39}$$

the first of which is the fundamental normal mode:

$$\omega_1 = \frac{\lambda_1 c_s}{2\pi} \tag{40}$$

where $\lambda_1 \simeq \dfrac{4.493}{r_b}$, this being obtained by solving Equation 36.

We can now postulate that the amplification of the membrane oscillations, fed by background acoustic noise in the liquid, is mainly responsible for the rupture, and that the direct coupling of the membrane to the bulk liquid at its rim is generally not sufficient for this feeding, but that the gas bubble cavity plays the central role of an intermediary in the process. When a bubble is stable, coupling via the cavity is weak, with the membrane cavity system oscillating passively in the manner of a "kettledrum." When the natural frequencies of the membrane and the cavity are close, however, this coupling becomes significant, and we propose that the latter condition is achieved at some point during the draining or runoff of the membrane. Since the speed of sound in the bulk liquid far exceeds both that of the gas cavity and the speed of the associated membrane wave, only fundamental modes of motion of the cavity and the membrane can be expected to receive efficient excitation, and then the onset of rupture is interpreted as the state in which this strong coupling allows ambient energy to preferentially enter the membrane oscillation in its fundamental mode. Subsequently, with the amplitude of motion increasing progressively beyond the linear range, secondary pumping, through mode coupling from the fundamental to the higher modes, then becomes possible, and this stage is just prior to the final disintegration of the membrane into droplets.

We thus arrive at the condition for rupture of the membrane by equating the fundamental frequencies of oscillation of the membrane, Ω_1, and of the gas cavity, ω_1, from Equations 27 and 40, respectively, to get

$$\frac{r_m^2}{r_b^2} = \frac{0.5726\sigma}{\rho_f \delta c_s^2} \tag{41}$$

which, by use of Equation 7, can be written as

$$\eta = (1 - \gamma)^{1/2} \tag{42}$$

where $\gamma = \left[\frac{0.5726\sigma}{\rho_f \delta c_s^2}\right]$ and $\eta = \frac{s}{r_b}$, as defined previously.

The combination of Equations 11 and 42 provides a relationship between the bubble size, r_b, and the critical membrane thickness, δ_c, at which the condition of rupture is established:

$$\gamma = \frac{1}{2}\left[(\beta + 1)(\beta^2 - 2\beta + 9)^{1/2} - (\beta^2 + 3)\right] \tag{43}$$

A plot of this in terms of the nondimensional bubble radius,

$$r_b\left[\frac{(\rho_f - \rho_g)g}{6\sigma}\right] = \frac{1}{\beta^{1/2}} \tag{44}$$

and the nondimensional critical film thickness,

$$\delta_c\left[\frac{\rho_f c_s^2}{0.5726\sigma}\right] = \frac{1}{\gamma} \tag{45}$$

is shown in Figure 3.

The mass of liquid entrained due to the bursting of a single gas bubble then can be computed from the total mass of the liquid in the membrane at the onset of rupture:

$$M_b = \rho_f \delta_c A_m \tag{46}$$

where $A_m = 2\pi r_m(r_b - s)$ is the surface area of the membrane. Introducing Equations 11 and 42 into Equation 46, we obtain, after simplification, the liquid entrainment from one gas bubble:

$$M_b = k_1 r_b^2 \left[\frac{\left(1 - \frac{2r_b^2}{k_2} + \frac{9r_b^4}{k_2^2}\right)^{1/2} + \left(\frac{r_b^2}{k_2} - 1\right)}{\left(1 + \frac{3r_b^2}{k_2}\right) - \left(1 - \frac{2r_b^2}{k_2} + \frac{9r_b^4}{k_2^2}\right)^{1/2}}\right]^{1/2} \tag{47}$$

where $k_1 = \frac{1.15\pi\sigma}{c_s^2}$ and $k_2 = \frac{6\sigma}{(\rho_f - \rho_g)g}$.

We are now in a position to compare this theory of liquid entrainment due to bubble rupture with experimental results. A study [2] of droplet formation due to the bubbles collapsing at the gas-liquid interface indicates that droplets of two distinctly different

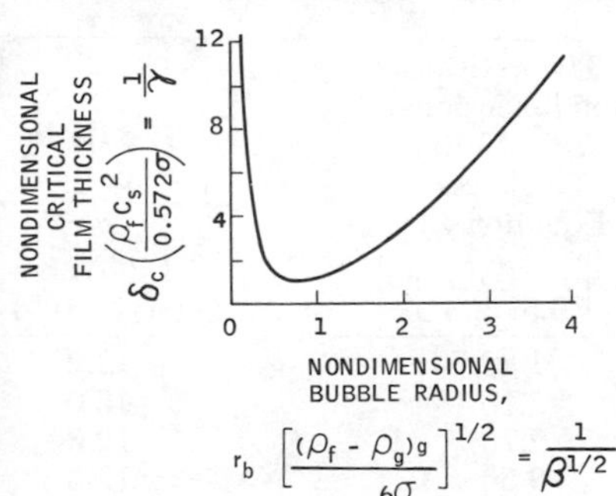

Figure 3. Critical film thickness vs bubble size.

size ranges are generated: small droplets, due to the disintegration of the bubble membranes, and large droplets, which appear only for bubbles of small enough size and are due to the breaking up of a liquid jet induced from the bubble cavity. Owing to a limitation on the measuring technique, droplets smaller than 20 μm, which were expected to show up in large numbers, were not measured; however, the number of droplets from one single bubble of a given size was reported for each of these two distinctly separated droplet size ranges for a number of bubble sizes; the corresponding sizes of droplets were evaluated on the basis of the Sauter mean diameter, $d_{s.m.} = \Sigma d_b^3/\Sigma d_b^2$.

The use of the Sauter mean diameter is probably more justifiable for the larger than for the smaller droplets, since the total surface area per unit mass of the smaller droplets is significantly greater than that of the larger droplets. For these reasons, no quantitative comparison can be made between the theoretical result we have obtained and the experimental results for the small droplet size range, which relates only to the disintegration of the membrane. However, it might not be unreasonable to expect that the mean size of droplets, which are generated as a result of the disintegration of the bubble membrane, will be related to the critical thickness of the membrane at which disintegration first takes place. In fact, the dependence of the experimental mean droplet size for small droplets on bubble size has been found to agree quite well qualitatively with the dependence on the theoretical critical film thickness, as shown in Table I.

A similar experiment [7], over a much wider bubble size range, was conducted in which the size distributions of both the larger and smaller droplets down to a size of 4.9 μm (created by the bursting air bubbles) were measured, yielding results in terms of the liquid mass entrainment per unit mass of air as a function of the bubble size. Knowing the bubble size, we can readily obtain the mass of air in the bubble by first computing the static pressure inside the bubble, with the surface tension effect included, and then the total liquid mass entrainment per air bubble is simply the product of the specific (per unit mass of air) liquid mass entrainment and the mass of air so computed. By this method, the specific mass entrainment data can be compared with the corresponding theoretical results of Equation 47, as shown in Figure 4.

Only results for bubbles with radii greater than 0.25 cm have been included in this comparison, because droplets from bubbles above 0.25 cm radius were found to be produced entirely from the collapse of the bubble dome [8]. Close agreement on liquid entrainment has been found between theory and experiment for bubbles of these larger sizes, with the exception of a much lower measured liquid entrainment for the largest size bubble than that predicted by the theory. This discrepancy could be due partly to shortcomings in the measuring technique employed to determine the size and number distributions of droplets.

Table I. Comparison of Dependence of Experimental Mean Droplet Size for the Small Droplet Size Range and Theoretical Film Thickness on Bubble Size

Bubble Radius, r_b (cm)	Experimental Sauter Mean Diameter of Small Droplets [2], $d_{s.m.}$ ($m \times 10^6$)	Theoretical Nondimensional Critical Film Thickness (Equation 45) $\delta_c\left[\frac{\rho_f c_s^2}{0.5726\sigma}\right]$	The Ratio: $\dfrac{d_{s.m.}}{\delta_c\left[\frac{\rho_f c_s^2}{0.5726\sigma}\right]}$ ($m \times 10^6$)
0.265	22	1.80	12.2
0.233	26	2.16	12.0
0.205	34	2.70	12.6
0.180	45	3.60	12.5
0.156	58	4.70	12.3

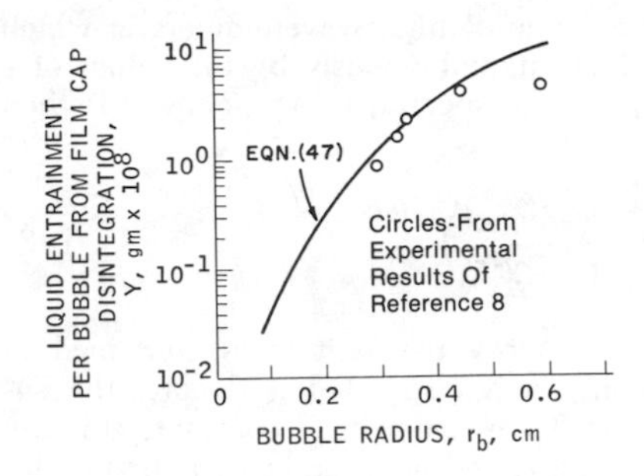

Figure 4. Comparison of theory with experimental data on liquid entrainment from film cap disintegration.

ENTRAINMENT OF TURBULENT EDDIES

During intensive bubbling (if the Froude number $Fr = v_s^2/gh$ is greater than unity), the gas-liquid flow is markedly turbulent. Disordered motions, during which the various parameters, i.e., pressure, velocity and void fraction, undergo random changes in space and time, are generated in this mixture.

In this section we will consider a possible entrainment mechanism in which the flowing gas interacts with the turbulent liquid, such that eddies that are sufficiently small are carried off by the gas. It is assumed that the eddies most likely to be so entrained are those in the "inertial subrange," because eddies smaller than this are strongly dominated by liquid viscosity so are less likely to be separated from the bulk of the liquid. Also, eddies larger than this are considered too large to be carried off by the gas. We first will consider some basic characteristics of turbulent flows.

Turbulization of the bubbling flow results in superposition of motions with decreasing periods or, in other words, of eddies of decreasing size [9]. From experimental data, it is known that turbulence may be represented by a succession of three main stages of eddy sizes, or of wavenumbers ($\kappa = 2\pi n/v$, where n denotes the turbulent frequency and v the time-average of the velocity). As a result of inertia interactions, energy is transmitted from a stage with high wavenumbers, through an intermediate stage, to a stage with low wavenumbers.

It should be noted that turbulent motions differ in systems of different construction and also may depend on the discharge rates of the two separate phases. For this reason, the time-averaged properties of a two-phase (gas-liquid) flow depend on the vertical position in the mixture. Averaged flowrates in the chamber also vary from one lateral point to another (due to the wall effect), so that fluctuating velocities vary from point to point and give rise to an interaction between fluctuating and averaged components of the motion, this interaction being related to the existence of transport effects.

For example, passage of gas through a sieve plate, with regularly arranged orifices and mounted perpendicular to the flow, superimposes a random velocity distribution on the motion in the bubbling mixture. These random motions degenerate as the distance from the plate increases, and seem, therefore, to be not statistically uniform. However, the rate at which the turbulence degenerates is so small that for practical applications this turbulence can be taken to be uniform, and, in this case, the turbulence closely resembles uniform isotropic turbulence, whose characteristics do not depend on the position and direction of the coordinate axes. In practice, the turbulence of a gas-liquid mixture may be developed completely, and the energy spectrum then drops steeply at high frequencies, n (in other words, at large wavenumbers, κ). Neglecting this fact may result in too high a predicted value for the entrainment of liquid from the two-phase mixture.

Although the wavenumber spectrum of an actual turbulent flow is not continuous, it is possible to assign a definite amount of the total energy to each value of the wavenumber. In other words, we can derive the relationship between the wavenumbers in the form of an energy spectrum [10]. According to Kolmogorov's first law [11,12], "at sufficiently high Reynolds numbers there exists a range of high wavenumbers in which the turbulence is in static equilibrium, being defined unambiguously by the values of ϵ (energy dissipation) and ν (kinematic viscosity)." The spectral energy density is then given by the following expression [10]:

$$E(\kappa) = \left(\frac{8\epsilon}{9\gamma}\right)^{2/3} \kappa^{-5/3} \left(1 + \frac{8\nu_f^3\kappa^4}{3\gamma^2\epsilon}\right)^{-4/3} \quad (48)$$

where $E(\kappa)$ denotes the spectral density of the energy per unit mass contained in eddies with wavenumbers in the interval between κ and $(\kappa + d\kappa)$, ϵ denotes the energy dissipation per unit mass and unit time, $\gamma = 0.4$ is a universal constant, κ is the wavenumber ($\kappa = r^{-1}$, where r denotes the eddy size), and ν_f is the kinematic viscosity coefficient.

For high Reynolds numbers, in the subrange of wavenumbers of finite magnitude (inertial subrange) defined by the inequality

$$\kappa_e << \kappa << \kappa_d \quad (49)$$

the effect of dissipation is negligibly small compared with energy transmitted under the action of inertial forces. Also, in this subrange the energy spectrum is determined unambiguously by a single ϵ value and does not depend on ν_f. Here, κ_e denotes the wavenumber defining the range of energy containing eddies:

$$\kappa_e = 1/r_e \quad (50)$$

(r_e is the mean size of energy containing eddies) and κ_d denotes the wavenumber defining the range of marked effect of viscous forces, being equal to the inverse size of the smallest eddy:

$$\kappa_d = \frac{1}{r_d} = [3\gamma^2\epsilon/8\nu_f^3]^{1/4} \quad (51)$$

At wavenumbers within the inertial subrange of the energy spectrum, the viscous term in Equation 48 can be neglected. The equation for the spectral energy density then becomes

$$E(\kappa) = \left(\frac{8}{3.6}\right)^{2/3} \epsilon^{2/3} \kappa^{-5/3} = 1.7\epsilon^{2/3} \kappa^{-5/3} \quad (52)$$

If $N(\kappa)$ denotes the number of eddies per unit mass of liquid and per unit interval of wavenumbers, κ, and

$$e = \rho_f \frac{4}{3} \pi r^3 \frac{v_e^2}{2} \quad (53)$$

denotes the energy of one eddy, then the spectral density, $E(\kappa)$, can be expressed by the number of eddies, $N(\kappa)$, and by the energy of one eddy by

$$E(\kappa) = N(\kappa)e \quad (54)$$

Substituting Equation 52 for $E(\kappa)$ and Equation 53 for e into Equation 54, the following expression is obtained:

$$E(\kappa) = 1.7\epsilon^{2/3} \kappa^{-5/3} = N(\kappa) \frac{2\pi\rho_f v_e^2}{3\kappa^3} \quad (55)$$

Hence, it follows that the number of eddies is

$$N(\kappa) = 0.812 \frac{\epsilon^{2/3} \kappa^{-5/3} \kappa^3}{\rho_f v_e^2} \tag{56}$$

According to Kolmogorov's law for the inertial subrange of the spectrum,

$$v_e^2 = \beta(\epsilon/\kappa)^{2/3} \tag{57}$$

where $\beta = 8.2$.
Equation 56 now can be transformed into

$$N(\kappa) = \frac{0.812}{8.2} \frac{\epsilon^{2/3} \kappa^{-5/3} \kappa^3 \kappa^{2/3}}{\rho_f \epsilon^{2/3}} = 0.1 \frac{\kappa^2}{\rho_f} \tag{58}$$

Let us now assume the following mechanism for the entrainment of liquid from a two-phase flow. The gas intrudes between the liquid eddies (intrusion evidently being possible at the moment the eddy forms). If the gas is capable of lifting the eddy formed, liquid is atomized and entrained from the bulk of the liquid, but if the liquid eddy is too large and cannot be lifted by the gas, it remains in the bulk of the liquid.

We can determine the maximum radius, r_o, of the liquid eddies that can be entrained by the gas from the liquid, and to do so we set the force due to gravity on the liquid eddy equal to the lifting force exerted by the gas:

$$\frac{4}{3}\pi r_o^3 \rho_f g = C_D \rho_g v^2 \pi r_o^2 \tag{59}$$

where v (x) denotes the velocity of the gas in the liquid, and C_D is a drag coefficient.

From the above equation one can obtain

$$r_o = \frac{3}{4} C_D \frac{\rho_g v^2}{\rho_f g} \tag{60}$$

or

$$\kappa_o = \frac{4}{3} \frac{\rho_f g}{C_D \rho_g v^2} \tag{61}$$

Note that for $r_o \leqq r_d$ (or $\kappa_o \geqq \kappa_d$), turbulent entrainment is clearly impossible because r_d denotes the smallest eddy size for which liquid viscosity is not significant. It is obvious that under this condition no turbulence will develop in the two-phase flow because the mechanism of liquid atomization and entrainment from the system is also the mechanism of development of turbulence. So for $r_o \leqq r_d$, the regime is a pure bubbling regime without entrainment of liquid. We introduce the dimensionless ratio

$$\eta = \frac{r_o}{r_d} = \frac{\kappa_d}{\kappa_o} \tag{62}$$

which characterizes the regime of entrainment of liquid from a two-phase mixture during bubbling, so that for $\eta < 1$ we have no liquid entrainment and for $\eta > 1$ we have entrainment.

The mass of liquid entrained, Y, per unit liquid mass and time from the system, is proportional to the number of eddies formed per unit time and capable of being entrained by the gas. The expression for Y is

$$Y = c_4 \int_{\kappa_0}^{\kappa_d} N(\kappa) \frac{4}{3} \frac{\pi \rho_f}{\kappa^3} \frac{1}{\tau(\kappa)} d\kappa \tag{63}$$

where c_4 is a constant of proportionality.
The formation time of an eddy of size r is

$$\tau(\kappa) = \frac{r}{v_e(r)} = \frac{1}{\kappa v_e(\kappa)} \tag{64}$$

and substituting the value of $v_e(\kappa)$ given by Equation 57 we get

$$\tau(\kappa) = 0.35 \frac{1}{\epsilon^{1/3}} \frac{1}{\kappa^{2/3}} \tag{65}$$

Substituting the values of $N(\kappa)$ given by Equation 58 and $\tau(\kappa)$ given by Equation 65 into the entrainment expression yields

$$Y = c_4 \, \epsilon^{1/3} \int_{\kappa_0}^{\kappa_d} \frac{d\kappa}{\kappa^{1/3}} \tag{66}$$

By carrying out the integration, we obtain

$$Y = c \, \epsilon^{1/3} \left(\kappa_d^{2/3} - \kappa_o^{2/3}\right) \tag{67}$$

where c is a constant.
Through the use of Equation 62, the expression of Y given in Equation 66 takes the form

$$Y = c \, \epsilon^{1/3} \kappa_d^{2/3} \left(1 - \eta^{-2/3}\right) \tag{68}$$

Equation 68 is the final result of the analysis of the contribution of turbulent motion to liquid entrainment. If, for any gas-liquid flow system being considered (for instance, two-phase flow in a pipe, gas injection through an orifice), we can find the quantities κ_d, η (or κ_0) and ϵ, then the equation gives us a value for both the mass of liquid per unit liquid mass and time, which is entrained by the gas. In fact, Equations 51 and 61 give κ_d and κ_o in terms of flow system properties; further, in combination with Equation 62 they indicate whether liquid entrainment can occur for that flow system. Equation 62 also defines the region of applicability of Equation 68, namely, $\eta \geqq 1$. Finally, it should be noted that the constant, c, must be determined empirically.

As an example, let us consider a flow system consisting of air being injected into water from below, through an array of orifices. Let the depth of liquid, h, be 2 cm, the superficial gas velocity, v_s, be 25 cm/sec, and the pressure drop in the air during passage through the water, Δp, be 2 cm of water (equivalent to 196 Pa). The kinematic viscosity of water, ν_f, is 1×10^{-6} m^2/sec; the density of water, ρ_f, is 10^3kg/m^3; and the density of air, ρ_g, is 1.2 kg/m^3. The superficial gas velocity, v_s, and the local gas velocity, v, are related by the expression

$$v = \frac{v_s}{\phi} \tag{69}$$

where ϕ is the void fraction, that is, the ratio of the instantaneous volume of the gas to the total volume. The void fraction, ϕ, can be obtained [13,14] from

$$\phi = \frac{2Fr^{1/2}}{1 + 2Fr^{1/2}} \tag{70}$$

By using Equation 70 we obtain $\phi = 0.53$. Therefore, from Equation 69 it is found that

$$v = \frac{25}{0.53} = 47 \text{ cm/sec}$$

The energy dissipation per unit mass and time is given by

$$\epsilon = \frac{v\Delta p}{\rho_f h} = 4.6 \text{ J/(kg)(sec)}$$

We can then find the size, r_d, of the smallest eddy formed in the liquid from Equation 51:

$$r_d = \left(\frac{8\,\nu_f^3}{3\gamma^2\epsilon}\right)^{1/4} = 4.36 \times 10^{-5} \text{ m}$$

On the other hand, the largest eddy capable of being carried off by the airflow is given by Equation 61:

$$r_o = \frac{3}{4} C_D \frac{\rho_g}{\rho_f} \frac{v^2}{g}$$

and the drag coefficient is usually of the order unity, so we can write

$$r_o \simeq \frac{3\rho_g\, v^2}{4\rho_f\, g} = 2.02 \times 10^{-5} \text{ m}$$

Hence, for this example, $r_o < r_d$ and the smallest eddy present in the turbulent liquid is too large to be entrained by the gas, so we have no entrainment.

DROPLET DYNAMICS

In the preceding section, a mechanism for the production of liquid droplets during the bubbling process was discussed. Now the behavior of droplets in the gas flow is considered. After formation, the drops move with a considerable initial velocity, which can cause them to rise (in a practically motionless gas) to a great height. For example, it has been observed [15] that drops were ejected in air to more than 2 m above a bubbling air-water mixture. However, such a height is reached only by individual drops that have the greatest velocity and nearly vertical direction of flight at the instant of their breakoff, and the bulk of the drops are ejected to a considerably lesser height.

For purposes of illustration, we will consider a flow system consisting of an orifice (or orifices) through which gas is being passed into liquid of finite depth. As the gas reaches the liquid-free surface, droplets are entrained by one or more of the mechanisms described earlier in this chapter, and possibly by other mechanisms. We define h_s to be the perpendicular distance from the liquid-free surface to the height at which the liquid entrainment is being evaluated and, obviously, the liquid entrainment measured will be a function of the location, h_s, of the observation point. From experimental data [16] it is apparent that an increase in the height of the gas space, h_s, from 0.28 to 0.7 m decreases the entrainment coefficient, $Y = \Sigma M_f/M_g$ (where ΣM_f is the total mass of entrained droplets per unit time, and M_g is the mass of gas per unit time) from 0.16 to 0.0007, or over two orders of magnitude. In other words, at the height, $h_s = 0.7$ m, less than 0.5% of all the droplets rising from the surface of the liquid to a height of 0.28 m were entrained. It would be of interest to determine whether the droplet concentration becomes zero at some finite value of h_s, or if, for a given superficial gas velocity, v_s, some droplets will always be transported by the gas flow to any value of h_s.

It is useful to consider under what conditions the liquid droplets will tend to rise or settle in the gas space above the liquid level, and at what rate. Whether settling or rising occurs depends on the size, shape, viscosity and density of the drops, and on the density of the gas. As a model, we will consider the dispersion of droplets in gas where they are sufficiently separated that there are no collisions of other interactions between droplets. Let us assume spherical droplets of density, ρ_f, and diameter, d, in a gas of density, ρ_g, and viscosity, μ_g. In this case, the gravitational force causing a droplet to fall $(\rho_f > \rho_g)$ will be

$$F_{grav.} = \frac{\pi d^3}{6}(\rho_f - \rho_g)g \tag{71}$$

Motion of the droplet in gas results in the drag force

$$F_D = C_D \frac{\rho_g v_r^2}{2} \frac{\pi d^2}{4} \tag{72}$$

where C_D is the drag coefficient, and v_r is the velocity of the droplet relative to the gas.

After an initial period of acceleration or deceleration, the droplet attains a uniform velocity (called the terminal settling velocity), under which the drag force just balances the gravitational force, and this settling velocity may then be obtained from

$$\frac{\pi d^3}{6}(\rho_f - \rho_g)g = C_D \frac{1}{2} v_r^2 \frac{\pi d^2}{4} \rho_g \tag{73}$$

and

$$v_r = 1.155 \left(\frac{g(\rho_f - \rho_g)d}{\rho_g C_D}\right)^{1/2} \tag{74}$$

For turbulent flow, $C_D \simeq 0.45$ ($Re = \rho_g v_r d/\mu_g > 800$) and then

$$v_r = 1.72 \left(\frac{g(\rho_f - \rho_g)d}{\rho_g}\right)^{1/2} \tag{75}$$

Equation 75 indicates that the terminal relative velocity of a liquid droplet in the gas increases with droplet size and decreases with gas density. Thus, for gases of high density, or for very small droplets, resistance of the gas begins to play a noticeable role in droplet trajectories. So, for example, for droplets 0.2 mm in diameter flying up vertically at an initial velocity of 2 m/sec in motionless air, the height of ascent decreases from 0.2 m at p = 1 atm (ρ_g = 1.2 kg/m^3) to 10 mm at p = 110 atm (ρ_g = 100 kg/m^3).

Under considerable gas velocities we must take into account the effect of gas velocity on the maximum height of droplet ascent. The presence of a high superficial gas velocity, v_s, increases the total rise height of all droplets, and this circumstance becomes of significance for entrainment when the quantity v_s becomes comparable with the settling velocity of the droplets. When $v_s \geqq v_r$ (where we are considering absolute values), the height of ascent of droplets becomes unlimited, and they will be carried off with the gas stream. Liquid droplets, torn away from the liquid at the initial stage of their trajectory, may rise with a velocity considerably exceeding the upward velocity of the gas. In this case, droplet motion is noticeably slowed by drag of the gas, and is especially pronounced for small droplets and at high gas density. As the upward velocity of the droplets decreases, the drag of the gas decreases rapidly and becomes negative for $v_r \leqq v_s$, so that the gas stream begins to entrain the drops upward.

Thus, we see that if the gas superficial velocity, v_s, exceeds in magnitude the relative settling velocity of the droplets, they will be carried upwards by the gas to an unlimited height (completely transportable droplets). However, if the settling velocity of the droplets is greater than the velocity of the gas, after losing their initial energy the droplets begin to fall with a velocity equal to $v_r - v_s$. The maximum height to which such droplets rise depends on the vertical component of the initial velocity of the drops, on the settling velocity, v_r, and on the upward velocity of the gas, v_s. When $v_r >> v_s$, the rise height of the drops depends only slightly on v_s, and this is a flow of nearly complete ejection. On the other hand, when the quantities v_r and v_s are approximately equal, the height of ascent will be determined almost entirely by the value v_s. Thus, this is a flow of droplet transportation.

We now consider the effect of the walls of the enclosing chamber on droplet motion. Formulas for calculation of liquid entrainment are usually obtained assuming wall-free conditions, when the geometric configuration (cross-sectional area) of the chamber in which the gas-liquid system moves is not taken into account. Evaluation of the effect of the finite dimensions of the chamber (using a suitably defined equivalent diameter) on the ascent velocity of a single droplet is the first step in a solution of the problem of a cloud of droplets in the same chamber. Consider the steady flow of gas-liquid droplet dispersion in a vertical duct of uniform cross section, where the diameter of a drop is comparable with the diameter of the duct (Figure 5). In this case, the drag of the drop is determined by the relative velocity,

$$v_r = v_d - v_s \tag{76}$$

where v_d is the absolute velocity of the drop, and v_s is the superficial velocity of the gas.

Let us first evaluate the gas velocity, v_a, in the annular space between the drop and the wall by considering the equivalence of the volume of the liquid moving up and the volume of gas moving down in the annular space, in a coordinate system moving at the superficial gas velocity:

$$(v_d - v_s)\pi r_d^2 = -(v_a - v_s)\pi(R^2 - r_d^2) \tag{77}$$

or

$$v_d \pi r_d^2 = -v_a \pi (R^2 - r_d^2) + v_s \pi R^2 \tag{77a}$$

where r_d is the droplet radius, and $R = D/2$ is the equivalent radius of the chamber. After manipulation, we obtain

$$\frac{v_d\, r_d^2}{v_s\, R^2} = \frac{-v_a}{v_s}\left(1 - \frac{r_d^2}{R^2}\right) + 1 \tag{77b}$$

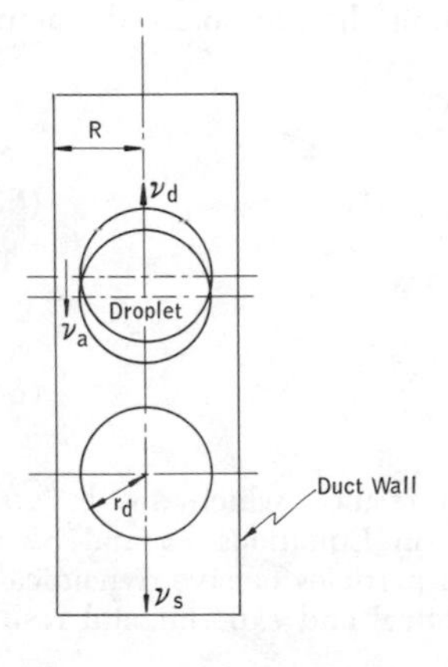

Figure 5. Flow of bubbles in a narrow duct.

and

$$\frac{v_a}{v_s} = \frac{1 - \dfrac{v_d\, r_d^2}{v_s\, R^2}}{1 - \dfrac{r_d^2}{R^2}} \tag{78}$$

The terminal velocity of a single droplet is determined by balancing gravitational and drag forces, and for a droplet in an infinite medium we have, as in Equation 73,

$$\frac{4}{3}\pi r_d^3 \rho_f g = C_D \pi r_d^2 \frac{1}{2} \rho_f (v - v_{d_\infty})^2 \tag{79}$$

or

$$v_{d_\infty} = v - \left(\frac{8 r_d g}{3 C_D}\right)^{1/2} \tag{80}$$

where v_{d_∞} is the droplet absolute velocity in an infinite medium, and for a single drop in a finite medium

$$v_f = v_a - \left(\frac{8 r_d g}{3 C_D}\right)^{1/2} \tag{81}$$

Combining Equations 80 and 81 we obtain

$$v_d = v_{d_\infty} \frac{v_a - \left(\dfrac{8 r_d g}{3 C_D}\right)^{1/2}}{v - \left(\dfrac{8 r_d g}{3 C_D}\right)^{1/2}} \tag{82}$$

By combining Equations 78 and 82 we obtain a formula illustrating the effect of the ratio (r_d/D) on the droplet velocity, and we now consider empirical evidence for such a dependence. According to measurements [15], the minimum size of the equivalent diameter of a chamber in which this wall correction is to be negligible is at least ten (and preferably fifteen) times the diameter of the largest droplet occurring in the two-phase flow. Experiments disclose a more or less substantial influence on the ratio r_d/R (or d/D) on v_d. Thus, for example, the empirical correction factor, K, is given [16] below by which the free ascent velocity should be multiplied to obtain the actual velocity of the droplet:

For laminar flow:

$$K = 1 - \left(\frac{d}{D}\right)^2 \tag{83}$$

For turbulent flow:

$$K = 1 - \left(\frac{d}{D}\right)^{1/5} \tag{84}$$

These equations give reasonable order of magnitude results, which are in good accordance with the elementary calculation given above in Equations 78 and 82. It should be noted also that for small sizes, droplets and solid particles behave dynamically in a very similar fashion, so it is possible to use the theoretical and experimental results obtained for solid spheres in analyzing droplet motion.

For large values of (d/D), we can assume [17] that the surface of the falling sphere near its equator and of the nearby cylindrical container can be approximated by two flat plates, and an analysis based on this assumption leads to the expression

$$K = 0.6\left(1 - \frac{d}{D}\right)^{2.5} \tag{85}$$

This equation is valid when inertial effects are negligible, so under these conditions we have

$$\frac{v_d}{v_{d_\infty}} = K = 0.6\left(1 - \frac{d}{D}\right)^{2.5} \tag{86}$$

Analysis of the wall effect on the rate of fall of solid spheres, assuming negligible inertial effects and a small (d/D) ratio, has been carried out [7,18–20]. The resulting equation for the terminal velocity correction factor is

$$\frac{v_d}{v_{d_\infty}} = K = 1 - 2.105\left(\frac{d}{D}\right) + 2.087\left(\frac{d}{D}\right)^3 \tag{87}$$

A binomial of $[1 - (d/D)]^{2.1}$ leads to

$$K = 1 - 2.1\left(\frac{d}{D}\right) + 1.155\left(\frac{d}{D}\right)^2 - 0.0385\left(\frac{d}{D}\right)^3 \tag{88}$$

and the difference in the two expressions is small in the range over which they may be applied. So for the case of negligible inertia and small (d/D), we have

$$\frac{v_d}{v_{d_\infty}} \simeq \left[1 - \left(\frac{d}{D}\right)\right]^{2.1} \tag{89}$$

All the above correction factors suggest that the correction factor may be expressed in a general form:

$$K = \phi\left(1 - \frac{d}{D}\right)^n \tag{90}$$

In a different approach, if we consider that section of the cylindrical chamber near the equator of the droplet to be approximated by the section of a large sphere, we can use spherical coordinates and, after analysis, one can obtain the following equation for the range of negligible inertial effects and low d/D ratio:

$$K = \frac{1}{1 + \frac{9}{4}\left(\frac{d}{D}\right) + \left(\frac{9}{4}\frac{d}{D}\right)^2} \tag{91}$$

Now, a binomial expansion of $[1 - (d/D)]^{-2.25}$ is

$$1 + 2.25\left(\frac{d}{D}\right) + 3.66\left(\frac{d}{D}\right)^2 + 5.18\left(\frac{d}{D}\right)^3 + \ldots$$

which, in the range of variables over which it is applicable, does not differ greatly from the previous expression. Therefore, for the case of negligible inertia and small (d/D), an alternative formula to Equation 89 is

$$\frac{v_d}{v_{d_\infty}} \simeq \left[1 - \left(\frac{d}{D}\right)\right]^{2.25} \tag{92}$$

It has been shown [21] that for large inertial effects,

$$\frac{v_d}{v_{d_\infty}} = K = \left[1 - \left(\frac{d}{D}\right)^2\right]\left[1 - \frac{1}{2}\left(\frac{d}{D}\right)^2\right]^{1/2} \tag{93}$$

Finally, the equation of momentum has been used [21] to arrive at an equation for the case of (d/D) nearly equal to unity and negligible viscous forces. The result is

$$K = \frac{\left[1 - \left(\frac{d}{D}\right)\right]^2}{\left(\frac{d}{D}\right)} \tag{94}$$

Equations 83–94 furnish formulas that can be used to correct the droplet velocity when allowing for the finite dimensions of the enclosing chamber.

DROPLET STATISTICS

When determining quantitative relationships of the droplet entrainment process with a large number of factors (physical, structural and operational) that affect entrainment, it is advantageous to isolate what seem to be the two principal ones: the superficial velocity of the gas, v_s, and the gas space above the level of the liquid, h_s. To analyze this extremely complex phenomenon, a simplified model is examined in which we assume that entrainment takes place in a unique manner for given values of v_s and h_s. Obviously, the accuracy of such a model would be increased considerably by increasing the number of determining factors in the process, but such an attempt would provide an impractical solution because of its greatly increased complexity.

In the gas space above the liquid, many droplets of different sizes are moving at various velocities, but the time-averaged entrainment behavior of this large number is found to be independent of the individual random features of the flight of the drops, and the entrainment is found to be virtually nonrandom. The stability of this time-averaged entrainment, repeatedly confirmed by experiment [22–25], enables probability methods to be used to establish quantitative relations for predicting entrainment.

Now, due to breakdown of bulk liquid in the bubbling mixture, droplets of various radii are being produced continuously. These droplets have initial escape velocities, w, from the liquid surface, having random values scattered around a certain value, w_o, which depends on the escape velocity of the gas from the bubbling layer. The subsequent motion of each drop is determined by gravity and drag of the gas, as we have outlined in the previous section. We make the following two assumptions, which have experimental support [26]:

1. The random quantities w and r_d are independent of one another.
2. The probability distribution law for the random quantities is close to normal.

Indeed, if it is assumed that the deviation of the individual drop velocities from the mean value, w_o, is due to the effect of the large number of independently acting random factors, each of which alters its velocity a little, then in this case the distribution law for the drop velocities will be close to normal [27].

Based on assumption 1, the probability of an event, M, which means an ejected droplet having a radius in the range r_d to $(r_d + dr_d)$ and an escape velocity in the range w to $(w + dw)$, is given by

$$P\{M\} = \int_{r_d}^{r_d+dr_d} f(r_d)\, dr_d \int_{w}^{w+dw} f(w)dw \tag{95}$$

where f(w) is the probability density distribution of the velocity, w, and $f(r_d)$ is the probability density distribution of the droplet radii.

The total mass of all such drops is then

$$Y = \int_{r_d}^{r_d+dr_d} f(r_d) \left(\frac{4}{3}\pi r_d^3 \rho_f\right) dr_d \int_{w}^{w+dw} f(w)dw \tag{96}$$

where $4/3\ \pi r_d^3 \rho_f$ is the mass of a drop of radius, r_d. The entrainment at height h_s, corresponding to an escape velocity, w, is then

$$Y = \int_{r_{d_{min}}}^{r_{d_{max}}} \frac{4}{3} \pi r_d^3 \rho_f\, f(r_d) dr_d \int_{w}^{w_{max}} f(w)dw \tag{97}$$

It is clear now that the total entrainment of liquid from the surface of a two-phase mixture is

$$Y_o = \int_{r_{d_{min}}}^{r_{d_{max}}} f(r_d) \left(\rho_f \frac{4}{3} \pi r_d^3\right) dr_d \tag{98}$$

and using this, Equation 97 becomes

$$Y(w) = Y_o \int_{w}^{w_{max}} f(w)\, dw \tag{99}$$

The probability density distribution of the drop initial (or escape) velocity, according to assumption 2, is

$$f(w) = \frac{\dfrac{1}{\sigma_w (2\pi)^{1/2}}\, e^{-1/2 \left(\frac{w - w_o}{\sigma_w}\right)^2}}{P(w > 0)} \tag{100}$$

where σ_w is the standard deviation of the velocity, w, and $P(w > 0)$ is the probability that the velocity $w > 0$. The denominator of Equation 100 is a normalization factor, so that

$$\int_0^\infty P(w) = 1 \tag{101}$$

and the probability that w is negative is zero:

$$P(w < 0) = 0 \tag{102}$$

Therefore, the probability that the velocity $w > 0$ is

$$P(w > 0) = \frac{1}{(2\pi)^{1/2}\sigma_w} \int_0^\infty \frac{e^{-1/2 \left(\frac{w - w_o}{\sigma_w}\right)^2}}{P(w > 0)} dw = 1 \tag{103}$$

and Equation 99 takes the form

$$Y(w) = \frac{Y_o}{(2\pi)^{1/2}\sigma_w} \int_{w}^{w_{max}} e^{-1/2 \left(\frac{w - w_o}{\sigma_w}\right)^2} dw \tag{104}$$

We now determine entrainment as a function of the gas space height, h_s; in other words, we replace the limits of integration in Equation 104:

$$Y(h_s) = \frac{Y_o}{(2\pi)^{1/2}\sigma_w} \int_{(2gh_s)^{1/2}}^{(2gh_s)^{1/2}_{max}} e^{-1/2\left(\frac{w - w_o}{\sigma_w}\right)^2} dw \tag{105}$$

$$\simeq \frac{Y_o}{(2\pi)^{1/2}\sigma_w} \int_{(2gh_s)^{1/2}}^{\infty} e^{-1/2\left(\frac{w - w_o}{\sigma_w}\right)^2} dw \tag{105a}$$

and, using the substitution $(w - w_o)/\sigma_w = x$, we obtain the final equation for calculating the entrainment of liquid as

$$Y(h_s) = \frac{Y_o}{(2\pi)^{1/2}\sigma_w} \sigma_w \int_{\frac{(2gh_s)^{1/2} - w_o}{\sigma_w}}^{\infty} e^{-\frac{x^2}{2}} dx = \frac{Y_o}{(2\pi)^{1/2}} \int_{\frac{(2gh_s)^{1/2} - w_o}{\sigma_w}}^{\infty} e^{-\frac{x^2}{2}} dx \tag{106}$$

or

$$Y = Y_o \left\{ \frac{1}{(2\pi)^{1/2}} \int_{\frac{(2gh_s)^{1/2} - w_o}{\sigma_w}}^{o} e^{-\frac{x^2}{2}} dx + \frac{1}{(2\pi)^{1/2}} \int_{o}^{\infty} e^{-\frac{x^2}{2}} dx \right\} \tag{106a}$$

Finally,

$$Y = Y_o \left\{ \frac{1}{2} - \frac{1}{(2\pi)^{1/2}} \int_{o}^{\frac{(2gh_s)^{1/2} - w_o}{\sigma_w}} e^{-\frac{x^2}{2}} dx \right\} + Y_c \tag{107}$$

where Y_c is an entrainment component, which is independent of the spacing height.

To calculate liquid droplet entrainment by Equation 107, it is first necessary to determine the parameters Y_o, w_o, σ_w and Y_c. For this reason, we use experimental entrainment values, Y_e, which have been obtained for various values of the superficial gas velocity, v_s, and gas space height, h_s. For this we formulate the function

$$\psi(h_s, v_s, Y_o, w_o, \sigma_w, Y_c) = \sum_{i=1}^{n} \frac{(Y_i - Y_{e_i})^2}{Y_{e_i}} \tag{108}$$

where Y_i is determined from Equation 107, Y_{e_i} are the experimental values, and n is the number of experimental points. The unknown quantities–Y_o, w_o, σ_w and Y_c–are found from the conditions for a minimum of Equation 108. This can be done by using appropriate optimization methods for multivariable functions [28,29]; these algorithmic methods are usually programmed on a digital computer. For an air-water system with a fixed value of h_s and various values of v_s, we find

$$Y_o = 34.29\, v_s - 9.24 \text{ (kg/m}^2\text{-min)}$$
$$w_o = 0.3342\, v_s - 0.0179 \text{ (m/sec)}$$
$$\sigma_w = 0.1261\, v_s + 0.8997 \text{ (m/sec)}$$
$$Y_c = 6 \times 10^{-4} \text{ (kg/m}^2\text{-min)}$$

Using these values in Equation 107, we obtain

$$Y \simeq (34\,v_s - 9)\left(0.5 - \frac{1}{(2\pi)^{1/2}}\int_0^s e^{-\frac{x^2}{2}}dx\right) + 6 \times 10^{-4}\ (\text{kg/m}^2\text{-min}) \quad (109)$$

where

$$s \simeq \frac{4.42\,(h_s)^{1/2} - 0.334\,v_s + 0.018}{0.126\,v_s + 0.9} \quad (110)$$

and v_s is in m/sec and h_s is in m.

For a given value of s, the probability integral $\frac{1}{(2\pi)^{1/2}}\int_0^s e^{-\frac{x^2}{2}}dx$ can be found from standard math tables, and Equation 109 then gives us a value for the liquid entrainment from an air-water flow solely in terms of the height above the water level, h_s, and the superficial air velocity, v_s. Equations similar to Equation 109 can be obtained from Equation 108 by the same method for any gas-liquid combination (for example, methane-water, air-kerosene, Freon 12-water), if experimental data are available [30]. It has been found that the use of Equation 109 gives predicted results accurate to ±10% with experimental results. The usefulness of Equation 107 is diminished in that experimental data are needed to obtain the final Equation 109, and also the analysis of these data in minimizing Equation 108 is quite laborious.

SIMILARITY ANALYSIS

It is clear from the discussion in earlier sections of this chapter that a total analytical treatment of the phenomenon of liquid entrainment in a two-phase flow is very difficult. Nevertheless, we can approach the problem by considering the equations of motion of the liquid and the gas, then using the method of similarity analysis to obtain dimensionless groups, which provide a convenient way of collecting and interpreting data.

We make the assumption that the gas phase is turbulent, and that the "apparent" turbulent stresses dominate viscous stresses in the gas. It is assumed that the liquid phase is dominated by molecular viscosity, and that the two phases are coupled by the stresses occurring at the interface.

The equation of motion for the (incompressible) liquid is

$$\frac{\partial u_{f_i}}{\partial t} + u_{f_k}\frac{\partial u_{f_i}}{\partial x_k} = -\frac{1}{\rho_f}\frac{\partial p_f}{\partial x_i} + \frac{\mu_f}{\rho_f}\frac{\partial^2 u_{f_i}}{\partial x_k \partial x_k} + g \quad (111)$$

where u_{f_i} is the velocity vector of the liquid; x_i is the cartesian coordinate vector; ρ_f is the liquid density; p_f is the liquid pressure; μ_f is the liquid dynamic viscosity; and g is the acceleration due to gravity. Note that Equation 111 is simply the Navier-Stokes equations written in cartesian tensor notation, with the summation of index notation being used. So,

$$\frac{\partial^2}{\partial x_k \partial x_k} \equiv \frac{\partial^2}{\partial x^2} + \frac{\partial^2}{\partial y^2} + \frac{\partial^2}{\partial z^2}$$

Also, the continuity equation for the liquid is

$$\frac{\partial u_{f_i}}{\partial x_i} = 0 \quad (112)$$

For the gas phase, again assumed incompressible, we have an equation similar to Equation 111, except that the viscous term is replaced by the "apparent" turbulent stress term, and the gravitational force is neglected:

$$\frac{\partial \bar{u}_i}{\partial t} + \bar{u}_k \frac{\partial \bar{u}_i}{\partial x_k} = -\frac{1}{\rho_g}\frac{\partial \bar{p}_g}{\partial x_i} - \frac{\partial \overline{u_i' u_k'}}{\partial x_k} \tag{113}$$

Here, $\bar{u}_i$ is the time-averaged gas velocity vector and u_i' is the fluctuating gas velocity vector (where the gas velocity vector, u_i, is given as $u_i = \bar{u}_i + u_i'$). The term $\bar{p}_g$ is the time-averaged gas pressure.

The continuity equation for the gas is

$$\frac{\partial \bar{u}_i}{\partial x_i} = 0 \tag{114}$$

We now need to couple the liquid and gas equations, and these are coupled by the interface boundary conditions. We have a balance of normal stresses at the interface:

$$-p_f + 2\mu_f \frac{\partial u_{f_3}}{\partial x_3} = -\bar{p}_g - \rho_g \overline{u_3'^2} - \sigma\left(\frac{1}{R_1} + \frac{1}{R_2}\right) \tag{115}$$

Where the index 3 refers to the coordinate normal to the interface and indices 1 and 2 refer to coordinates parallel to the interface; the fundamental radii of curvature of the interface are given by R_1 and R_2; and σ is the surface tension. The second terms on each side of the equation are the normal viscous and turbulent stresses, respectively. For the shear stresses at the interface, we find

$$\mu_f\left(\frac{\partial u_{f_1}}{\partial x_2} + \frac{\partial u_{f_3}}{\partial x_1}\right) = -\rho_g \overline{u_2' u_3'} \tag{116}$$

where the left-hand side of the equation represents liquid viscous shear and the right-hand side represents gas turbulent shear.

Equations 111 through 116 provide a complete description of the two-phase motion, but this group of equations presents major analytical difficulties. Instead of attempting a solution, we will introduce dimensionless functions into the equations, thereby generating the dimensionless groups useful in understanding experimental data. The dimensionless functions are

$$U_{f_i} = \frac{u_{f_i}}{v_s},\ \bar{U}_i = \frac{\bar{u}_i}{v_s},\ U_i' = \frac{u_i'}{v_s}$$

$$P_f = \frac{p_f}{\Delta p},\ \bar{P}_g = \frac{\bar{P}_g}{\Delta p}$$

$$T = \frac{t}{\tau},\ X = \frac{x}{h_s},\ R_{o_i} = \frac{R_i}{h_s}$$

Here, v_s is the gas superficial velocity; Δp is a typical flow system pressure difference (for example, between the walls of a chamber); τ is a characteristic time; and h_s is the distance above the free liquid level.

Using these in the above equations we obtain

$$\frac{v_s}{\tau}\left(\frac{\partial U_{f_i}}{\partial T}\right) + \frac{v_s^2}{h_s}\left(U_{f_k}\frac{\partial U_{f_i}}{\partial X_k}\right) = -\frac{\Delta p}{\rho_f h_s}\left(\frac{\partial p_f}{\partial X_i}\right) + \frac{\mu_f v_s}{\rho_f h_s^2}\left(\frac{\partial^2 U_{f_i}}{\partial X_k \partial X_k}\right) + g \tag{117}$$

$$\frac{v_s}{h_s}\left(\frac{\partial U_{f_i}}{\partial X_i}\right) = 0 \tag{118}$$

$$\frac{v_s}{\tau}\left(\frac{\partial \bar{U}_i}{\partial T}\right) + \frac{v_s^2}{h_s}\left(U_{g_k}\frac{\partial U_i}{\partial X_k}\right) = -\frac{\Delta p}{\rho_g h_s}\left(\frac{\partial \bar{P}_g}{\partial X_i}\right) - \frac{v_s^2}{h_s}\left(\frac{\partial \overline{U_i' U_k'}}{\partial X_k}\right) \tag{119}$$

$$\frac{v_s}{h_s}\left(\frac{\partial \bar{U}_i}{\partial X_i}\right) = 0 \tag{120}$$

$$-\Delta p\ (P_f) + \frac{\mu_f v_s}{h_s}\left(2\frac{\partial U_3}{\partial X_3}\right) = -\Delta p(\bar{P}_g) - \rho_g v_s^2(\overline{U_3'^2}) - \frac{\sigma}{h_s}\left(\frac{1}{R_{o_1}} + \frac{1}{R_{o_3}}\right) \tag{121}$$

In the above equations, all the terms in parentheses are of the order unity, so in any study of the relative influence of pressure, viscosity, surface tension, gravity and time fluctuations we need only consider the coefficients in front of the parentheses. It turns out that for this procedure we only need to use Equations 117 and 121. Hence, we find

$$\frac{v_s}{\tau}(0(1)) + \frac{v_s^2}{h_s}(0(1)) = -\frac{\Delta p}{\rho_f h_s}(0(1)) + \frac{\mu_f v_s}{\rho_f h_s^2}(0(1)) + g \tag{122}$$

or

$$\frac{h_s}{v_s \tau}(0(1)) + (0(1)) = -\frac{\Delta p}{\rho_f v_s^2}(0(1)) + \frac{\mu_f}{\rho_f h_s v_s}(0(1)) + \frac{g h_s}{v_s^2} \tag{122a}$$

and

$$-\Delta p(0(1)) + \frac{\mu_f v_s}{h_s}(0(1)) = -\Delta p(0(1)) - \rho_g v_s^2(0(1)) - \frac{\sigma}{h_s}(0(1)) \tag{123}$$

or

$$\frac{-\Delta p}{\rho_f v_s^2}(0(1)) + \frac{\mu_f}{\rho_g h_s v_s}(0(1)) = -\frac{\Delta p}{\rho_f v_s^2}(0(1)) - \frac{\rho_g}{\rho_f}(0(1)) - \frac{\sigma}{\rho_g h_s v_s^2}(0(1)) \tag{123a}$$

In the above, 0(1) signifies that the term is of the order unity. It is clear from Equations 122a and 123a that we have several important nondimensional groups in the two-phase flow described by the original equations:

$$\left(\frac{\Delta p}{\rho_f v_s^2}, \frac{\rho_f v_s h_s}{\mu_f}, \frac{v_s^2}{g h_s}, \frac{\rho_g v_s^2 h_s}{\sigma}, \frac{v_s \tau}{h_s}, \frac{\rho_g}{\rho_f}\right)$$

We can write these as (Eu, Re, Fr, We, Ho, ρ_g/ρ_f), where

$Eu = \Delta p/\rho_f v_s^2$, the Euler number, ratio of pressure to inertial forces
$Re = \rho_f v_s h_s/\mu_f$, the Reynolds number, ratio of inertial to viscous forces
$Fr = v_s^2/g h_s$, the Froude number, ratio of inertial to gravitational forces
$We = \rho_g v_s^2 h_s/\sigma$, the Weber number, ratio of inertial to surface tension forces
$Ho = v_s \tau/h_s$, the Homochronity number, ratio of spatial to temporal inertial forces

All these are independent except for the Euler number. This is dependent because it expresses the relationship between the pressure and velocity fields. Once the velocity field is given, the pressure field is fully determined, and vice versa. Hence, we can say

$$Eu = Eu\left(Re, Fr, We, Ho, \frac{\rho_g}{\rho_f}\right) \tag{124}$$

This relationship holds not only for the Euler number, but for any dependent quantity, so, for example, for the entrainment coefficient $Y = \Sigma M_f/M_g$, where ΣM_f denotes the total liquid (droplet) entrainment per unit cross-sectional area and time and M_g is the mass flowrate of gas, we will have

$$Y = Y\left(\text{Re, Fr, We, Ho}, \frac{\rho_g}{\rho_f}\right) \tag{125}$$

Experimental determination of this relationship presents serious difficulties, but experience indicates that this set of dimensionless numbers may, to some advantage, be combined into groups.

Thus we can write

$$A = \frac{\text{Fr}^{1/2}\text{We}^{3/2}}{\text{Re}^2} \frac{\rho_g}{(\rho_f - \rho_g)} = \frac{\left(\frac{v_s^2}{gh_s}\right)^{1/2}\left[\frac{(\rho_f - \rho_g)v_s^2h_s}{\sigma}\right]^{3/2} \rho_g}{\left(\frac{v_sh_s}{\nu_f}\right)^2 (\rho_f - \rho_g)} \tag{126}$$

or

$$A = \frac{v_f^2 v_s^2 \rho_g}{g^2\left(\frac{\sigma}{g(\rho_f - \rho_g)}\right)^{3/2} h_s(\rho_f - \rho_g)} \tag{126a}$$

Note that we have replaced ρ_g by $(\rho_f - \rho_g)$ in the Weber number. This is simply an alternative definition of We.

Also,

$$B = \frac{\text{Ho Fr}^{1/2}}{\text{We}^{1/2}} = \frac{\frac{v_s\tau}{h_s}\left(\frac{v_s^2}{gh_s}\right)^{1/2}}{\left[\frac{v_s^2(\rho_f - \rho_g)h_s}{\sigma}\right]^{1/2}} = \frac{v_s\,\tau\,\sigma^{1/2}}{h_s^2[(\rho_f - \rho_g)g]^{1/2}} \tag{127}$$

In most applications, the Homochronity number is approximately equal to unity. Hence,

$$\tau \simeq \frac{h_s}{v_s} \tag{128}$$

and Equation 127 may be rewritten as

$$B = \frac{\left[\frac{\sigma}{g(\rho_f - \rho_g)}\right]^{1/2}}{h_s} \tag{129}$$

The value of factor A in Equation 126a for any system is determined by the magnitudes of gravitational, viscous and surface tension forces during droplet generation, and also the gravitational and inertial forces during droplet motion in the gaseous region. Factor B in Equation 127 is determined by the relative magnitude of gravitational, inertial and surface tension forces alone. The simplest possible expression of Equation 125 in terms of factors A and B is

$$Y = C\,A^m B^n \tag{130}$$

where C, m and n are constants to be determined experimentally. It now remains to be seen whether Equation 130 adequately "condenses" experimental data. Experiments

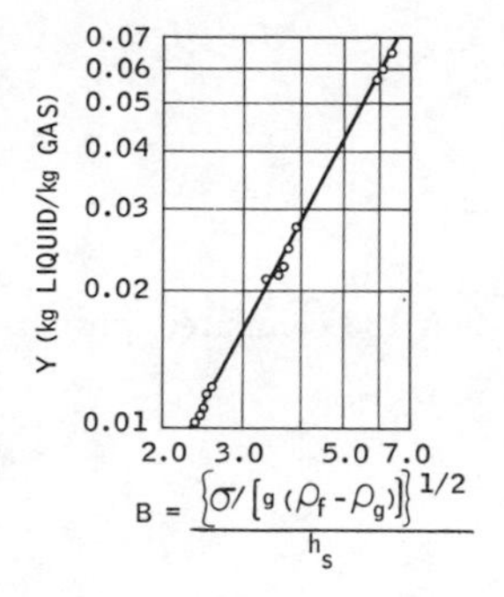

Figure 6. Entrainment coefficient against factor B.

have been conducted [23] using an air-water system to test this equation. The straight line in Figure 6 represents the least-squares fit to the data, and has a slope of 1.8. Hence, constant n is seen to have the value of 1.8. The effect of factor A on Y has also been determined experimentally, as seen in Figure 7 for various liquid-gas flow systems.

The least-squares fitted straight line through the data represents the expression

$$\frac{Y}{B^{1.8}} = f(A) \tag{131}$$

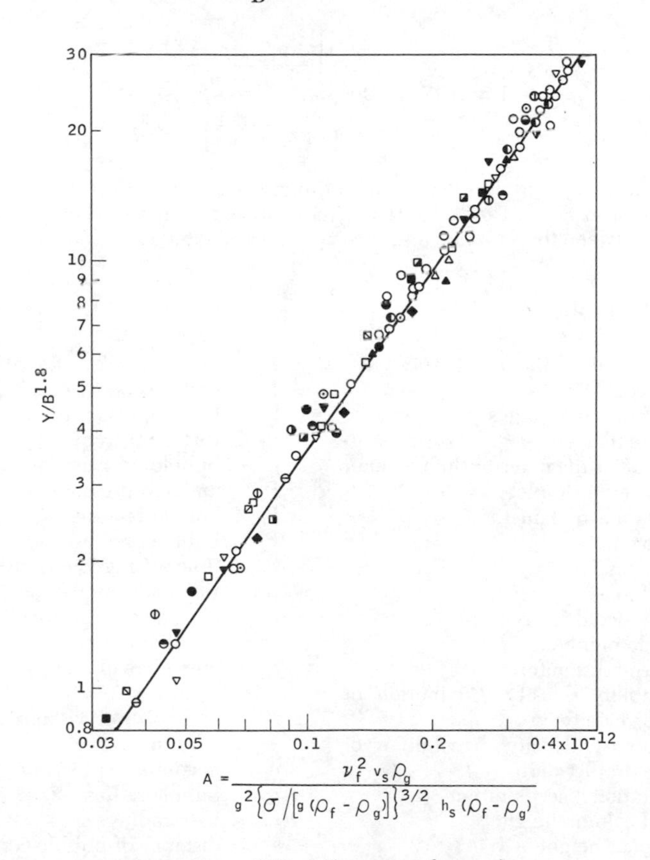

Figure 7. ($YB^{-1.8}$) against factor A.

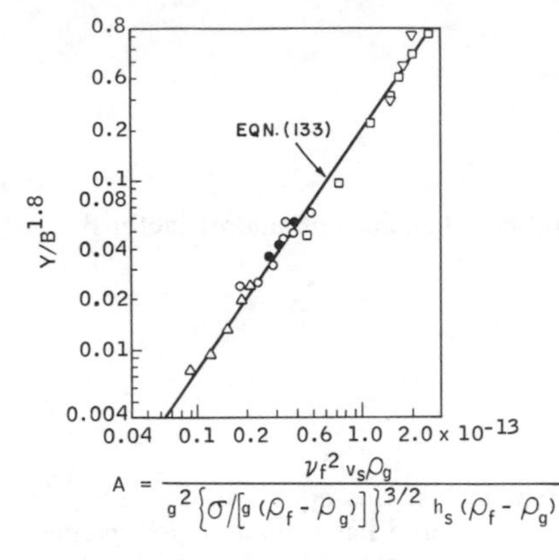

$$A = \frac{\nu_f^2 v_s \rho_g}{g^2 \{\sigma/[g(\rho_f - \rho_g)]\}^{3/2} h_s(\rho_f - \rho_g)}$$

Figure 8. Equation 133 compared with experimental data.

and from this line we find $C = 3.17 \times 10^{13}$, and $m = 1.4$. Hence, Equation 130 is verified and takes the form

$$Y = 3.17 \times 10^{13} \cdot A^{1.4} B^{1.8} \tag{132}$$

or

$$Y = 3.17 \times 10^{13} \frac{\left| \dfrac{\nu_f^2 v_s^2 \rho_g}{g^2(\rho_f - \rho_g)} \right|^{1.4}}{\left| \dfrac{\sigma}{g(\rho_f - \rho_g)} \right|^{1.2} h_s^{3.2}} \tag{133}$$

Equation 133 is compared to data from two-phase systems [13] over a range of physical properties (for instance, Freon $12/H_2O$, Air/CCl_4) in Figure 8. Clearly, there is a good correlation between the equation and the experimental data.

NOMENCLATURE

A_m	surface area of the membrane
C_D	drag coefficient
c_s	speed of sound in gas
d_b	bubble diameter
D	equivalent diameter of the chamber
d	diameter of droplet
$d_{s.m}$	Sauter mean diameter
Eu	Euler number
F_b	buoyancy force
F_D	drag force
F_{grav}	gravitational force
Fr	Froude number
F_σ	surface-tension force
f(w)	probability density distribution of the velocity, w
$f(r_d)$	probability density distribution of the droplet radii
g	gravitational acceleration
h	liquid column height
h_s	gas space height
Ho	Homochronity number
L_k	kinetic energy of a bubble
L_σ	surface energy of a bubble
M_b	liquid mass entrained/bubble
n	turbulent frequency
p	bubble pressure increment
P_a	ambient pressure
P_f	liquid pressure
P_g	bubble pressure
$\bar{P}_g$	time-averaged gas pressure
Re	Reynolds number
r	eddy size; cylindrical radial coordinate
r_e	mean size of energy-containing eddies
r_d	eddy size of viscous effect; droplet radius
r_o	maximum radius of entrained eddies
r_b	bubble radius
r_m	base radius
s	distance of bubble center below surface

t time
U_i velocity vector
$\overline{U}_i$ time-averaged velocity vector
U_i' fluctuating velocity vector
U_{fi} fluid velocity vector
V volume of a bubble
v time average of the velocity
V_d displaced bubble volume
v_b bubble velocity
v_d droplet absolute velocity
$v_{d\infty}$ droplet absolute velocity in infinite medium
v_e small eddy characteristic velocity
v_r velocity of the droplet relative to the gas
v_s superficial gas velocity
We Weber number
w initial escape velocity
w_o mean value of drop velocities
Y total mass of entrained drops/area/time
$\underline{Y}$ entrainment coefficient/area/time
$\overline{Y}_c$ entrainment component independent of height/area/time
Y_o total entrainment/area/time
Y membrane displacement in the vertical direction

Greek Symbols

δ membrane thickness
δ_c critical membrane thickness
ϵ dissipation energy
κ wavenumber
κ_d wavenumber of eddies with viscous effect
κ_e wavenumber of energy-containing eddies
μ_g gas viscosity
μ_f liquid viscosity
ν_f liquid kinematic viscosity
ρ_f liquid density
ρ_g gas density
σ surface tension
ϕ gas void fraction; velocity potential
ω_n gas motion natural frequency

REFERENCES

1. Azbel, D. S. *Khim. Mashinostroenie* 5:14 (1960).
2. Newitt, D. M., N. Dombrowski and F. H. Kneeman. *Trans. Inst. Chem. Eng. (London)* 32:4 (1954).
3. Azbel, D. S., S. L. Lee and T. S. Lee. *Proc. 1978 Int. Seminar on Momentum, Heat and Mass Transfer in Two-Phase Energy and Chemical Systems*, Dubrovnik, Yugoslavia (1979).
4. Kreyszig, E. *Advanced Engineering Mathematics* (New York: John Wiley & Sons, Inc., 1967).
5. Rayleigh, I. W. *The Theory of Sound* (New York: Dover Publications, Inc., 1954).
6. Koshliakov, N. S. *Principal Differential Equations of Mathematical Physics* (Moscow: ONTI, 1936) (in Russian).
7. Wakiya, S. J. *Phys. Soc. Japan* 8:254 (1953).
8. Garner, F. H., S. R. M. Ellis and T. A. Lacey. *Trans. Inst. Chem. Eng. (London)* 32:4 (1954).
9. Hinze, I. O. *Turbulence* (New York: McGraw-Hill Book Co., 1959).
10. Batchelor, G. K. *Theory of Homogeneous Turbulence* (New York: Cambridge University Press, 1970).
11. Kolmogorov, A. N. *Dokl. Akad., Nauk. SSSR* 30:299 (1941).
12. Kolmogorov, A. N. *Dokl. Akad., Nauk SSSR* 31:538 (1941).
13. Azbel, D. S. Doctoral Dissertation (in Russian), D. I. Mendeleev Moscow Chemical-Technological Institute, Moscow (1966).
14. Azbel, D. S. *Two-Phase Flows in Chemical Engineering* (New York: Cambridge University Press, 1981).
15. Blinov, V. I., and Ye. L. Feynberg. *Zh. Tekh. Fiz.* 3:5 (1933).
16. Kutateladze, S. S., and M. A. Styrikovich. *Hydraulics of Gas-Liquid Systems* (Moscow: Gosudazstrennoe Energieticheskoye Izdatelstro, 1958).

17. McNown, J. S., H. M. Lee, M. B. MacPherson and S. M. Engez. *Proc. VII Int. Cong. Appl. Mech. London* (1948).
18. Ladenburg, A. *Ann. Phys.* 23:447 (1970).
19. Faxen, H. *Arkiv. Mat. Astron. Fysik* 19: (13) (1925).
20. Happel, J., and B. J. Byrne. *Ind. Eng. Chem.* 46:1181 (1954).
21. Newton, I. *Mathematical Principles* (Berkeley, CA: University of California Press, 1934), p. 348.
22. Azbel, D. S. *Khim. Mashinostroenie* 6:14 (1960).
23. Azbel, D. S. *Theoretical Found. Chem. Eng.* 1(1):91 (1967).
24. Holbrook, G. E., and E. M. Baker. *Trans. Am. Inst. Chem. Eng.* 30:S.20 (1933–1934).
25. Jones, J. B., and C. Pyle. *Chem. Eng. Prog.* 51:424 (1955).
26. Cumo, M., and G. Farello. Paper presented at the Tenth International Heat Transfer Conference, AIChE-ASME, 1968.
27. Dunin-Barkovskii, I. V., and N. V. Smirnov. *The Theory of Probability and Mathematical Statistics in Engineering* (Moscow: Gostekhizdat, 1955) (in Russian).
28. Azbel, D. S., and A. F. Narazhenko. *Teoreti. Osnovy Khim. Tekhnol.* 1:129 (1967).
29. Beveridge, G. S. G., and R. S. Schecter. *Optimization Theory and Practice* (New York: McGraw-Hill Book Co., 1970).
30. Hunt, Ch. A., D. H. Hanson and C. R. Wilke. *AIChE J.* 1:441 (1955).

CHAPTER 18
GAS-LIQUID FLOW IN PIPELINE SYSTEMS

D. Chisholm
Glasgow College of Technology
Glasgow G4 0BA
Scotland

CONTENTS

INTRODUCTION

This chapter gives methods for predicting pressure gradients during the flow of gas-liquid mixtures through pipeline systems. The word "gas" is used in the sense of a substance either above or below its thermodynamic critical point; in other words, methods for vapor-liquid mixtures are also discussed.

The words "pipeline systems" in the title of the chapter have been used to emphasize that in addition to discussing pressure gradients in pipes, pressure changes over pipe fittings, such as bends and changes of section, also are included.

FLOW PATTERNS

The methods discussed below implicitly use the flow pattern map shown in Figure 1. The base of the figure, the Lockhart-Martinelli [1] parameter, is defined as follows:

$$X^2 = Dp_L/Dp_G \tag{1}$$

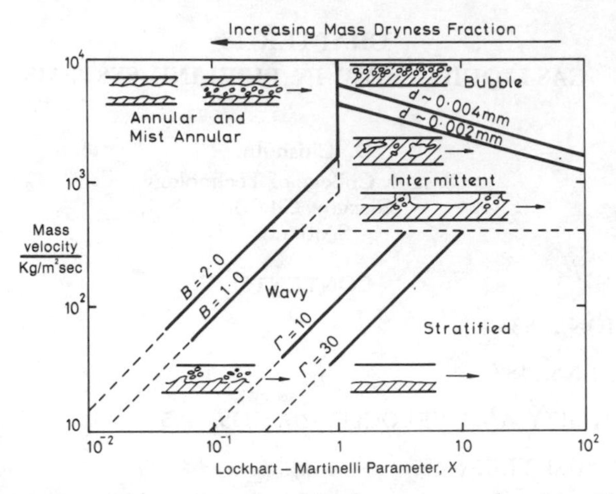

Figure 1. Flow pattern map based on pressure drop transitions.

where Dp_L and Dp_G are, respectively, the gradients if the liquid and gas components flow alone. If the friction factor is represented in the form of the Blasius equation,

$$\lambda = c/Re^n \tag{2}$$

then for friction

$$X = \left(\frac{1 - x}{x}\right)^{(2-n)/2} \left(\frac{V_L}{V_G}\right)^{1/2} \left(\frac{\mu_L}{\mu_G}\right)^{n} \tag{3}$$

Where one is concerned with gradients due to momentum flux changes, or where the friction factor is independent of Reynolds number, Equation 3 reduces to

$$X = \left(\frac{1 - x}{x}\right)\left(\frac{V_L}{V_G}\right)^{1/2} \tag{4}$$

The transition from bubble and intermittent flow to annular and annular mist flow is seen to occur [2,3] at

$$X = 1 \tag{5}$$

It occurs, therefore, when the momentum fluxes of the two components would be the same if they flowed alone. The transition from dispersed bubble to intermittent flow [2] can be approximated by

$$\frac{Dp_{FL} \times V_L}{(V_G - V_L)g} = \frac{1}{X^{0.25}} \tag{6}$$

This particular transition is not used in this chapter; it is given here for completeness.

In the section entitled Friction: Turbulent Flow, the equations for the boundary between stratified and wavy flow [4] are obtained as

$$G = \frac{1226 \times X}{\Gamma + 1} \tag{7}$$

where [5]

$$\Gamma = \left(\frac{V_G}{V_L}\right)^{½} \left(\frac{\mu_G}{\mu_L}\right)^{n/2} \tag{8}$$

The boundary between the wavy and annular/mist patterns [4] is

$$G = 818BX \tag{9}$$

The coefficient B [6] is discussed in the section entitled Friction: Turbulent Flow.

The transition from stratified to intermittent flow is shown as

$$G = 400 \text{ kg/m}^2\text{-sec} \tag{10}$$

In Figure 1, the dotted curves signify regions of greatest uncertainty. The transitions represented by Equations 7–10 were arrived at from noting transitions in the pressure drop characteristics, rather than visual observation of the flow pattern.

MIXTURE DENSITY AND VELOCITY RATIOS

Defining the velocity ratio,

$$K = U_G/U_L \tag{11}$$

enables the mixture specific volume to be expressed as

$$V_m = \frac{xV_G + K(1-x)V_L}{x + K(1-x)} \tag{12}$$

The "homogeneous" specific volume is that in which $K = 1.0$:

$$V_H = V_L\left\{1 + \left(\frac{V_G}{V_L} - 1\right)x\right\} \tag{13}$$

Over a wide range of conditions [7,8] the following formulas can be used for the velocity ratio:

$$X > 1 \quad K_o = \left(\frac{V_H}{V_L}\right)^{½} \tag{14}$$

$$X \leq 1 \quad K_o = \left(\frac{V_G}{V_L}\right)^{¼} \tag{15}$$

Equation 14 reduces to Equation 15 at $X = 1$. The subscript zero is used to signify velocity ratios evaluated using Equations 14 and 15.

Figure 2 compares these equations with experimental data [9]. A more extensive comparison is given by Friedel [10]. Of 14 methods tested there, this was the most satisfactory in density prediction. In fact, in that comparison Equation 14 was used over the whole range of dryness fraction. The use of Equation 15 should further improve comparison with data. It is known [11] that in the annular region, $X < 1$, the velocity ratio is a function of mass velocity; however, as the gradient due to gravitational effects is small in this region, Equation 15 is satisfactory for most purposes.

With decreasing mixture velocity in vertical pipes, the velocity ratio will gradually increase above values given by K_o. Labuntsov's [12] version of drift flux theory [11] can be stated as

$$U_G = U_H + U_{WD} \tag{16}$$

where the weighted drift fluxes, for small bore pipes is

$$U_{WD} = 0.35 W_D \left\{ gd \left(1 - \frac{V_L}{V_G}\right) \right\}^{1/2} \tag{17}$$

and for large bore pipes,

$$U_{WD} = 1.5 W_D \left\{ \sigma g V_L \left(1 - \frac{V_L}{V_G}\right) \right\}^{1/4} \tag{18}$$

The drift flux term is

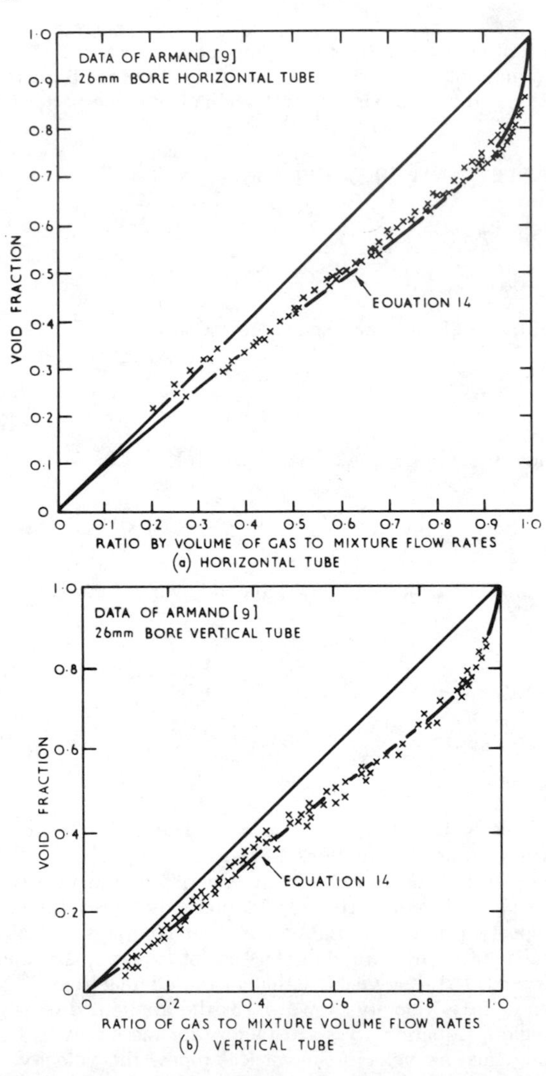

Figure 2. Void fractions for a base ratio of gas to mixture volume flowrates: air-water mixtures (pressure $\simeq$ 1 bar).

$$W_D = 1.4 \left(\frac{V_G}{V_L}\right)^{1/5} \left(1 - \frac{V_L}{V_G}\right)^5 \tag{19}$$

The transitional diameter is obtained by equating Equations 17 and 18. The homogeneous velocity is

$$U_H = GV_H \tag{20}$$

Equation 16 should be used when

$$U_H < (U_G - U_{WD}) \tag{21}$$

where the gas velocity is

$$U_G = G\{xV_G + K_o(1 - x)V_L\} \tag{22}$$

Using Equation 14, the liquid cross section can be expressed as

$$\alpha_L = \frac{1}{\frac{1}{1-x}\left(\frac{V_H}{V_L}\right)^{1/2} + 1 - \left(\frac{V_L}{V_H}\right)^{1/2}} \tag{23}$$

This can be approximated as

$$\alpha_L = (1 - x)\left(\frac{V_L}{V_H}\right)^{1/2} \tag{24}$$

This expression will be used in the section entitled Friction: Turbulent Flow. It is useful in that it can be integrated to give an estimate of the average liquid cross section along a pipe.

THE MOMENTUM FLUX

Using a separated flow model, the momentum flux of the mixture can be expressed as

$$MF = x\frac{M}{A}U_G + (1 - x)\frac{M}{A}U_L$$

$$= G^2\{xV_G + K(1 - x)V_L\}\left(x + \frac{1-x}{K}\right) \tag{25}$$

Define [13] an effective volume, V_e, by the equation

$$MF = G^2V_e \tag{26}$$

Then

$$\frac{V_e}{V_L} = \frac{1}{V_L}\{xV_G + K(1 - x)V_L\}\left(x + \frac{1-x}{K}\right)$$

$$= 1 + \left(\frac{V_G}{V_L} - 1\right)\{Bx(1 - x) + x^2\} \tag{27}$$

where the B coefficient is

$$B = \frac{\frac{1}{K}\left(\frac{V_G}{V_L}\right) + K - 2}{\frac{V_G}{V_L} - 1} \tag{28}$$

Provided $V_G/V_L >> 1$, then, approximately,

$$B = 1/K \tag{29}$$

Alternatively, Equation 27 can be expressed as

$$\frac{V_e}{V_L} = (1 - x)^2\left(1 + \frac{C}{X} + \frac{1}{X^2}\right) \tag{30}$$

where the C coefficient is

$$C = \frac{1}{K}\left(\frac{V_G}{V_L}\right)^{1/2} + K\left(\frac{V_L}{V_G}\right)^{1/2} \tag{31}$$

and the Lockhart-Martinelli parameter is defined by Equation 4.

It should be noted that

$$\frac{V_e}{V_m} = \{x + K(1 - x)\}\left(x + \frac{1 - x}{K}\right) \tag{32}$$

Thus, the effective volume, V_e, and that based on the tube content, V_m, can vary significantly.

Define now an effective mixture velocity

$$U_e = GV_e \tag{33}$$

From Equation 26,

$$MF = U_e^2/V_e \tag{34}$$

and

$$MF = GU_e \tag{35}$$

The momentum flux if the mixture flows as liquid is

$$MF_{LO} = G^2V_L \tag{36}$$

By defining

$$\psi = \frac{\frac{MF}{MF_{LO}} - 1}{\frac{V_G}{V_L} - 1} \tag{37}$$

then from Equations 26, 27, 36 and 37,

$$\psi = Bx(1 - x) + x^2 \tag{38}$$

The group ψ [14] has been referred to as the "normalized two-phase multiplier."

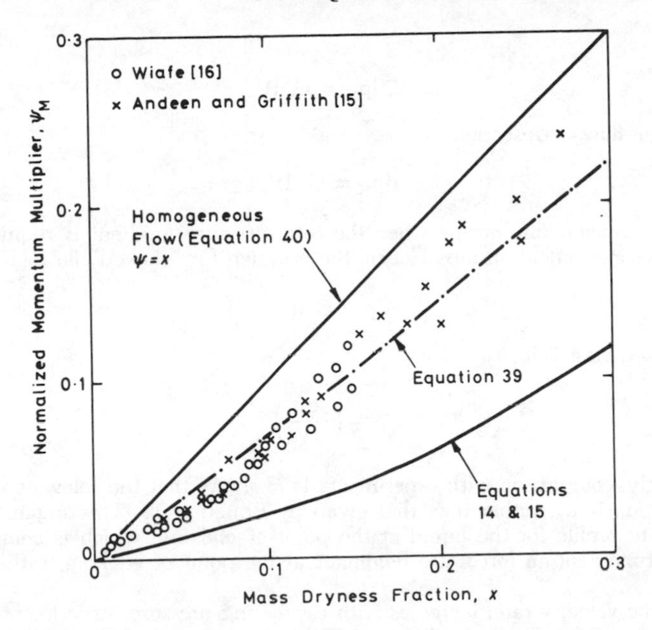

Figure 3. Normalized momentum multiplier, ψ_M, to a base of dryness fraction x: air-water mixtures at atmospheric pressure.

Andeen and Griffiths [15] and Wiafe [16] have measured the momentum flux during two-phase flow. Wiafe's measurements are shown in Figure 3 in the form of a $\psi - x$ plot. In these tests, void fractions were measured and are well represented by K_o; however, the momentum flux measurements were more closely represented by a velocity ratio:

$$K_e = K_o^{0.28} \tag{39}$$

This is illustrated in Figure 3. The subscript e is used to indicate the "effective" velocity ratio in the momentum equation.

For homogeneous conditions $(K = 1)$, Equation 37 reduces to

$$\psi = x \tag{40}$$

This line is also shown in Figure 3.

It is understood that the effective velocity ratio is given by Equation 39, rather than by K_o, due to the large radial variation in liquid velocity. The liquid momentum flux is considerably greater than

$$MF = (1 - x) GU_L \tag{41}$$

THE MOMENTUM GRADIENT AND CHOKED FLOW

The pressure gradient due to momentum changes is, using Equation 35,

$$Adp_M = d(MU_e) \tag{42}$$

Alternatively,

$$dp_M = GdU_e \tag{43}$$

and

$$V_e dp_M = U_e dU_e \quad (44)$$

For a constant-flow cross section,

$$dp_M = G^2 dV_e \quad (45)$$

The maximum flowrate occurs when the total pressure gradient is required to overcome changes in momentum flux. Hence, the criterion for "choked" flow [13] is

$$dp = G^2 dV_e \quad (46)$$

and the choked mass velocity is

$$G_C = \left(\frac{dp}{dV_e}\right)^{1/2} \quad (47)$$

Surprisingly, comparison with experiments [17] shows that the relevant velocity ratio in this equation is K_o, rather than that given by Equation 39. This suggests a uniform radial velocity profile for the liquid at the point of choking, which is compatible with the fact that momentum forces are dominant at the point of choking, rather than wall shear forces.

Because the velocity ratio increases with decreasing pressure, experiments at vacuum are a severe test of a theory in this area. Table I compares prediction, using K_o, with the data of Burnell [18]. At the lowest pressure of 0.2 bar, the phase density ratio is 7520, and K_o, as $X > 1$, is 9.31 from Equation 15. There are only two causes for possible concern regarding this comparison: (1) whether in the light of the work of Henry [19] and Presco et al. [20] the critical pressure is correctly measured; and (2) whether the mixture was, as assumed in the prediction, in thermal equilibrium. The latter seems likely as nonequilibrium would show itself as a requirement for a larger velocity ratio to fit the data.

It is well known that nonequilibrium effects are significant with vapor-liquid mixtures at low dryness fractions. Define [19,21]

$$N = x_N / x \quad (48)$$

where x_N is the actual mass dryness fraction, and x that for equilibrium conditions. Equation 27 can now be written as

Table I. Choking Mass Velocities: Comparison with Data of Burnell [18] for Water

		Mass Velocity ($kg\ m^{-2}\ sec^{-1}$)	
Pressure (Bar)	Dryness Fraction, x	Experiment	Equations 14, 15 and 47
1.02	0.025 8	2010	2136
0.70	0.018 32	1730	1332
0.68	0.046 5	1150	1313
0.60	0.024 4	1530	1414
0.48	0.035 1	1100	1075
0.43	0.040 2	858	941
0.35	0.025 0	930	909
0.34	0.048 0	676	726
0.29	0.056 25	550	596
0.23	0.037 6	564	569
0.20	0.069 1	353	394

$$\frac{V_e}{V_L} = 1 + \left(\frac{V_G}{V_L} - 1\right)\left\{\frac{x_N}{K}(1 - x_N) + x_N{}^2\right\} \tag{49}$$

Nonequilibrium effects, as mentioned above, are significant at low mass dryness fraction. Hence, approximately,

$$\frac{V_e}{V_L} = 1 + \left(\frac{V_G}{V_L} - 1\right)\left\{\frac{N}{K}x(1 - x) + x^2\right\} \tag{50}$$

This, of course, is Equation 27 with

$$B = N/K \tag{51}$$

The problem reduces once more to evaluating the coefficient B.

Fair agreement with experiment is shown in Figure 4 on the following basis:

$$X \leqslant 1 \quad B = 1/K_o \tag{52}$$

$$X > 1 \quad B = (x/x_t)(V_G/V_L)^{1/4} \tag{53}$$

where x_t is the mass dryness fraction corresponding to X = 1. (In Table I, the value of X is approximately unity or less.)

$$x_t = \frac{1}{\left(\frac{V_G}{V_L}\right)^{1/2} + 1} \tag{54}$$

From Equations 53 and 54,

$$X > 1 \quad B = \left\{\left(\frac{V_G}{V_L}\right)^{1/4} + \left(\frac{V_L}{V_G}\right)^{1/4}\right\}x \tag{55}$$

This can also be expressed, where K_o is the value at X = 1, as

$$X > 1 \quad B = \left\{\frac{1}{K_o}\left(\frac{V_G}{V_L}\right)^{1/2} + K_o\left(\frac{V_L}{V_G}\right)^{1/2}\right\}x \tag{56}$$

The continuous curves in Figure 4 are evaluated on this basis. The discontinuity occurs at X = 1 and arises from the change in slope of the B − x function at that point. Both Henry [19] and Zaloudek [22] observed decreases in the ratio $G_c{:}G_{cH}$ in this region with decreasing mass velocity.

The above analysis parallels that of Henry and co-workers [19–21] in assuming B − x is a constant (for given physical properties). Better agreement with experiment is obtained, however, when Equation 53 is replaced by

$$X > 1 \quad B = \frac{1}{K_o}\frac{x_N}{x} \tag{57}$$

where the nonequilibrium mass dryness fraction is

$$x_N = \left(\frac{x}{x_t}\right)^2 x \tag{58}$$

The velocity ratio, K_o, is evaluated with mass dryness fraction, x_N. The uppermost curve at low mass dryness fraction is obtained on this basis.

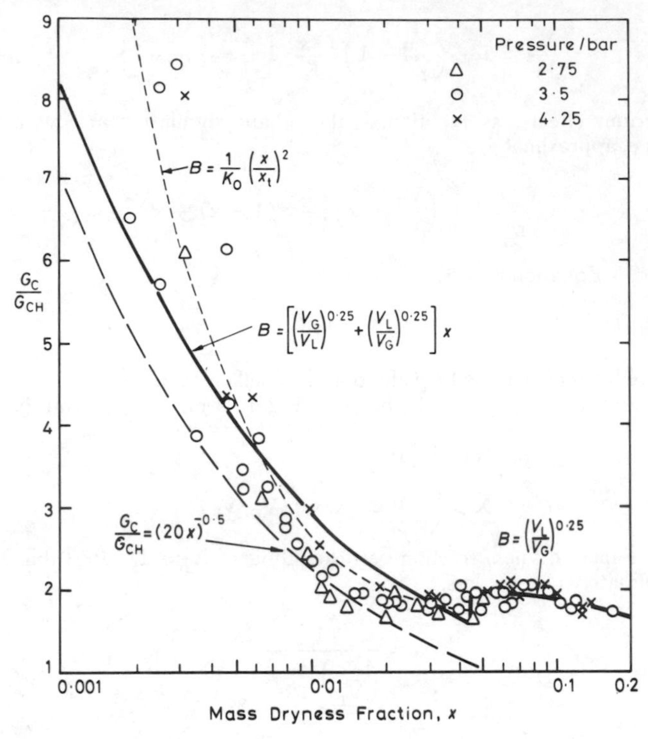

Figure 4. Ratio of choking mass velocity to "homogeneous value" pressure: 2.75–4.25 bar fluid: water [19].

On the basis that at low mass dryness fraction the assumption that

$$dx/dp = 0 \tag{59}$$

gives good agreement with experiments, leads [17] to the conclusion that

$$\frac{B}{x} = -\frac{1}{2np\,\frac{dx}{dp}} \tag{60}$$

where n is the vapor expansion exponent.

Table II compares Equations 55 and 60. Equation 60 would give a curve about 10% below the continuous curve in Figure 4. Also shown in Figure 4 is the curve corresponding to Henry's [19] procedure.

FRICTION: TURBULENT FLOW

In this section, equations are developed for that component of static pressure gradient attributable to wall friction. For single-phase flow,

$$Dp_F = \frac{\lambda G^2 V}{2d} \tag{61}$$

Table II. Comparison of Predictions for B/x by Equations 55 and 60

	B/x	
p bar	$(V_G/V_L)^{1/4} + (V_L/V_G)^{1/4}$	$-\dfrac{1}{2np\,\dfrac{dx}{dp}}$
0.1	11.1	11.55
0.5	7.64	8.78
1.0	6.53	7.59
5.0	4.54	5.07
10.0	3.91	4.10
50.0	2.78	2.07
100.0	2.41	1.23
200.0	2.04	0.22

where D is used to signify a derivative with respect to axial pipe length.
It might be expected that for a two-phase flow,

$$Dp_F = \frac{\lambda G^2 V_e}{2d} \tag{62}$$

To satisfy the all-liquid and all-gas condition, the product λV_e requires the following values:

$$x = 0 \quad \lambda V_e = \lambda_L V_L \tag{63}$$

$$x = 1 \quad \lambda V_e = \lambda_G V_G \tag{64}$$

where λ_L and λ_G are, respectively, the friction factors if the liquid and gas flow alone.
An equation satisfying these conditions is

$$Dp_F = \frac{G^2}{2d}\{xV_G\lambda_G + K_e(1-x)V_L\lambda_L\}\left(x + \frac{1-x}{K_e}\right) \tag{65}$$

where K_e is the "effective" velocity ratio associated with the momentum flux, Equation 39. This equation also can be expressed as

$$\frac{Dp_F}{Dp_{FL}} = \phi_{FL}^2 = 1 + \frac{C}{X} + \frac{1}{X^2} \tag{66}$$

where

$$C = Z + \frac{1}{Z} \tag{67}$$

and

$$Z = \frac{1}{K_e}\left(\frac{\lambda_L}{\lambda_G}\frac{V_G}{V_L}\right)^{1/2} \tag{68}$$

In Equation 66, the opportunity has been taken of defining [1] the two-phase multiplier, ϕ_{FL}^2, the ratio of the friction pressure gradient to that if the liquid flows alone.
Figure 5 shows data [23] for air-water mixtures as a plot of $\phi_{FL}^2 - 1$ to a base of X. Lockhart and Martinelli [1] originally plotted ϕ_{FL} to this base. The data are well represented using Equation 66, with

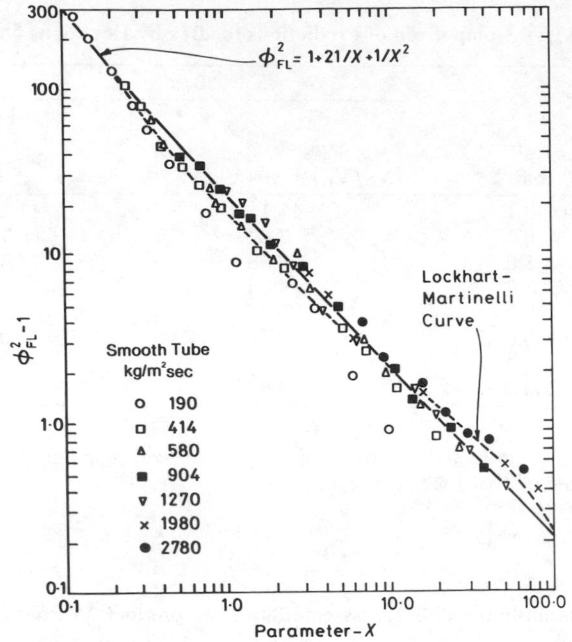

Figure 5. $\phi_{FL}{}^2$ against x: air-water mixtures in a 27-mm bore horizontal tube at atmospheric pressure.

$$C = 21 \tag{69}$$

Table III gives the C coefficients obtained from Equation 68, and the percentage difference in multiplier using Equations 68 and 69. The phase density and viscosity ratios were taken as 850 and 52, respectively, and the Blasius exponent as 0.2. The agreement between prediction and experiments in Table III will give the reader a wrong impression regarding our ability generally to predict friction pressure gradient; the model was chosen to suit the occasion.

When both phases would flow turbulently when flowing alone (turbulent-turbulent flow), at the thermodynamic critical point [14]

$$C = 2^{2-n} - 2 \tag{70}$$

where n is the Blasius exponent.

This is also the approximate solution [24] for stratified flow.

Figure 6 shows C coefficients recommended by Chisholm and Sutherland [5].

If the two-phase multiplier is taken with respect to the pressure gradient if the total mixture flows as liquid, Dp_{FLO}, then Equation 66 can be transformed—exactly if the Blasius exponent n = 0, and with a maximum error of about 2.5%, if n = 0.25—to

Table III. Comparison of Elementary Model for Friction with Empirical Values

X	0.01	0.1	1.0	10	100
C, Equations 67 and 68	32.4	25.13	19.5	20.3	18.2
$\phi_L{}^2$ Based on Estimated C	13 240	353	21.5	3.03	1.18
$\phi_L{}^2$ Based on C = 21	12 100	310	23.0	3.1	1.21
$\phi_L{}^2$ Estimated/$\phi_L{}^2$ Experiment	1.09	1.14	0.93	0.98	0.98

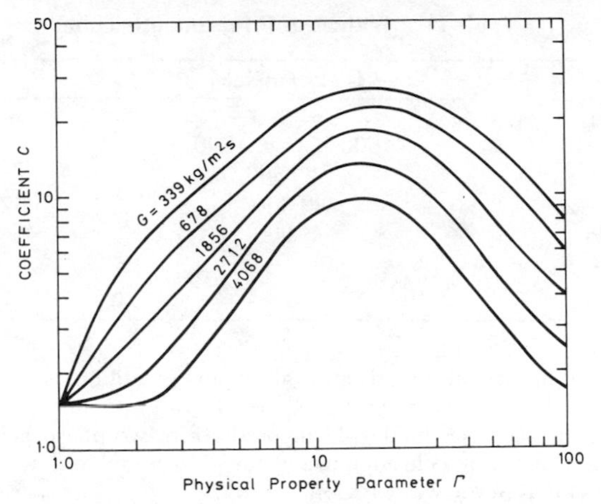

Figure 6. The coefficient C as a function of the mass velocity, G, and the physical property parameter, Γ [5].

$$\frac{Dp_F}{Dp_{FLO}} = \phi_{FLO}^2 = 1 + (\Gamma^2 - 1)\{Bx^{2-n/2}(1-x)^{2-n/2} + x^{2-n}\} \tag{71}$$

where

$$B = \frac{C\Gamma - 2^{2-n} + 2}{\Gamma^2 - 1} \tag{72}$$

and

$$\Gamma^2 = \frac{Dp_{FGO}}{Dp_{FLO}} = \frac{V_G}{V_L}\left(\frac{\mu_G}{\mu_L}\right)^n \tag{73}$$

The pressure gradient, if the total mixture flows as gas, is Dp_{FGO}. Equation 73 is a more general statement of Equation 8.

Where B = 1.0 and n = 0, Equation 72 reduces to homogeneous theory. Equation 61, where the coefficient C is evaluated [25] from

$$C = \left(\frac{V_G}{V_L}\right)^{1/2} + \left(\frac{V_L}{V_G}\right)^{1/2} \tag{74}$$

is commonly referred to as a "pseudo"-homogeneous theory. These two assumptions are identical with a Blasius exponent n = 0.

B coefficients can be obtained [14] from Table IV. These were originally presented for use with both horizontal and vertical flow, although it was appreciated that they would not be applicable to the case of stratified flow.

The B coefficients in Table IV are appropriate for smooth tubes. For rough tubes [26], the following equation is recommended:

$$\frac{B_R}{B_S} = 0.5\left\{1 + \left(\frac{\mu_G}{\mu_L}\right)^2 + 10^{-600\epsilon/D}\right\}^{(0.25-n)/0.25} \tag{75}$$

where B_S is the value from Table IV. This equation will result in a surface roughness only, producing a modest increase in pressure gradient with increasing surface rough-

Table IV. Values of B for Smooth Tubes

Γ	G (kg/m²sec)	B
	$\leqq 500$	4.8
9.5	$500 < G < 1900$	$2400/G$
	$\geqq 1900$	$55/G^{0.5}$
9.5 Γ 28	$\leqq 600$	$520/(\Gamma G^{0.5})$
$\geqq$ 28	> 600	$21/\Gamma$
		$\dfrac{15{,}000}{\Gamma^2\ G^{0.5}}$

ness, except close to the all-liquid and all-vapor conditions. This is in line with experiments.

Recently, progress has been made [4] in predicting two-phase behavior through the transition from stratified flow, through wavy flow, to annular flow (see Figure 1). It has been shown [4] that in the wavy region,

$$\psi = (1 + 0.001223\Gamma G)x^{2-n} \tag{76}$$

where ψ, which already has been defined for the case of momentum multipliers in Equation 36 is

$$\psi = \frac{\dfrac{Dp_F}{Dp_{FLO}} - 1}{\Gamma^2 - 1} \tag{77}$$

Figure 7 gives [4] a comparison between experiments and Equation 76. The experimental points shown [27] are given in Table V; Equation 76 implies Dp/Dp_{FG} is constant for a given G.

Define now a "flow pattern" parameter,

$$e = \frac{\psi - \psi_S}{\psi_A - \psi_S} \tag{78}$$

where ψ_S is the value during stratified flow, and ψ_A that during annular flow. Associated B-coefficients are B_S and B_A. For stratified flow [24],

Table V. Sample of McMillan's Data [27]

Test No.	xM (kg/sec)	(1 − x)M (kg/sec)	M (kg/sec)	Dp/Dp_{FG}
8	0.0633	0.2147	0.2780	2.56
23	0.0947	0.1854	0.2801	3.09
38	0.1240	0.1510	0.2750	2.55
47	0.1495	0.1303	0.2798	3.14
72	0.1773	0.1063	0.2836	3.09
101	0.1912	0.0747	0.2659	3.01
14	0.0603	0.2851	0.3454	3.73
29	0.0907	0.2598	0.3505	3.68
44	0.1034	0.2401	0.3435	4.27
59	0.1840	0.1686	0.3526	3.44
75	0.1753	0.1690	0.3443	3.82
104	0.2050	0.1510	0.3560	3.75

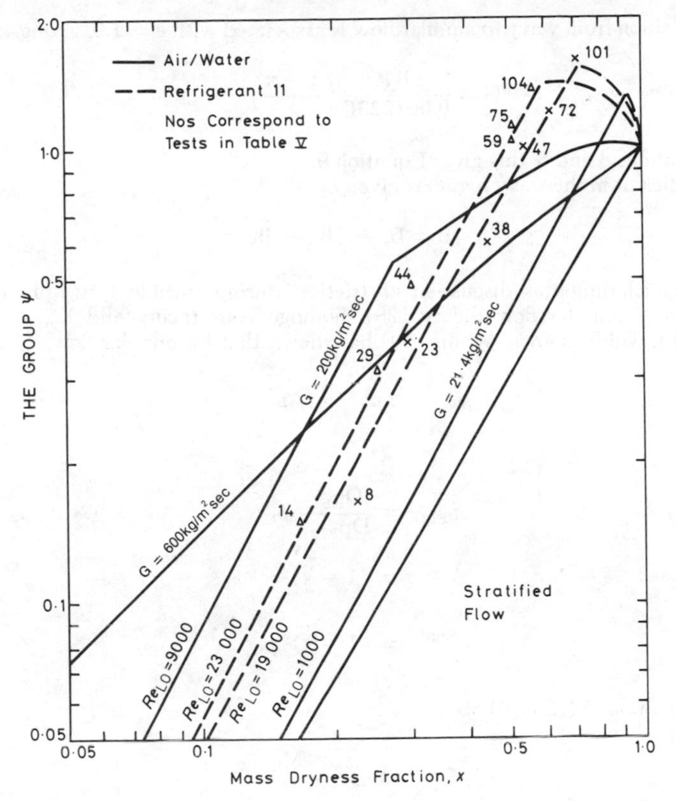

Figure 7. The dimensionless two-phase multiplier group, ψ, to base of mass dryness fraction.

$$B_S = \frac{2^{2-n} - 2}{\Gamma + 1} \tag{79}$$

and B_A is obtained from Table IV.

From Equations 76–78,

$$e = \frac{0.001223\Gamma G}{B_A - B_S}\left(\frac{x}{1-x}\right)^{(2-n)/2} - \frac{B_S}{B_A - B_S} \tag{80}$$

The stratified-wavy transition is associated with e = 0; hence, from Equation 80 at this transition,

$$G = \frac{B_S}{0.001223\Gamma}\left(\frac{1-x}{x}\right)^{(2-n)/2} \tag{81}$$

or using Equations 3 and 8,

$$G = \frac{B_S X}{0.001223} \tag{82}$$

Substituting Equation 79, assuming n = 0.2, gives Equation 7.

The transition from wavy to annular flow is associated with e = 1.0, giving

$$G = \frac{B_A}{0.001223\Gamma}\left(\frac{1-x}{x}\right)^{(2-n)/2} \tag{83}$$

Using Equations 3 and 8, this gives Equation 9.
The B coefficient in the wavy region is given by

$$B = B_S + (B_A - B_S)e \tag{84}$$

Before concluding this discussion of friction during turbulent two-phase flow, the relationship of annular flow theory [28], homogeneous theory and K_o, as defined by Equation 14, will be shown. Wallis [29] has shown that for annular flow, approximately,

$$\phi_{FL}^2 = 1/\alpha_L^2 \tag{85}$$

Also,

$$\begin{aligned}\phi_{FLO}^2 &= \frac{Dp_{FL}}{Dp_{FLO}}\phi_L^2 \\ &= (1-x)^{2-n}\phi_L^2 \\ &\doteq (1-x)^2\phi_L^2\end{aligned} \tag{86}$$

From Equations 24, 85 and 86

$$\phi_{FLO}^2 = \frac{V_H}{V_L} \tag{87}$$

which is the two-phase multiplier, assuming homogeneous theory and a two-phase viscosity corresponding to that of the liquid. Thus, as Equation 24 is derived, assuming K_o (for $X > 1$), then K_o is the velocity ratio that will result [7] in annular flow theory and homogeneous theory giving similar predictions.

FRICTION: LAMINAR FLOW

For the case in which one or the other, or both, components would flow laminarly when flowing alone, the C coefficient values in Table VI have been recommended [30].

Kubie and Oates [31] simulated conditions near the thermodynamic critical point, using two liquids of similar densities but immiscible. For each of the three mechanisms in Table VI it was found that

$$C = 2.0 \tag{88}$$

Table VI. C Coefficients for Flow with Laminar Component

Liquid-Gas	C	m Equation 90
Turbulent-Laminar	12	0.345
Laminar-Turbulent	10	0.31
Laminar-Laminar	5	0.16

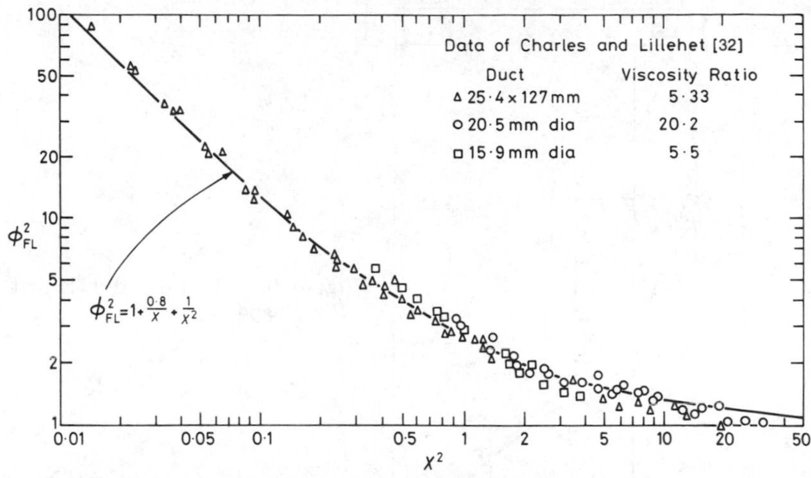

Figure 8. $\phi_{FL}{}^2$ to a base of X^2 for turbulent-laminar flow of two immiscible liquids; multiplier relative turbulent component.

This is surprising. From Equation 70 it would be expected that with laminar-laminar flow, as $n = 1$, the C coefficient should be zero.

In contrast to the findings of Kubie and Oates [31], the data of Charles and Lillelehet [32], for two immiscible liquids in which one component flowed laminarly, is well represented, as shown in Figure 8, by

$$C = 0.8 \tag{89}$$

As an interim measure, the following equation is recommended where one component flows laminarly:

$$C = 2 + \left(\frac{V_G}{V_L}\right)^m \tag{90}$$

The value of m is obtained from Table VI.

BENDS

In discussing the static pressure change in pipe fittings in the following sections, the change attributable to the device generally is considered; downstream effects are included. The pressure change attributable to a bend is specified in Figure 9. Where the bend is in the vertical plane, part of the loss will be due to gravitational effects; this discussion is concerned primarily, however, with the pressure change due to friction and momentum changes. Gravitational effects are estimated using equilibrium values of the velocity ratio, although undoubtedly the bend will influence the velocity ratio also.

For simplicity, therefore, consider a bend in the horizontal plane. For single-phase flow, define a single-phase loss coefficient:

$$k_B = -\frac{\Delta p_B}{2G^2V} \tag{91}$$

It has been shown [33] that Equation 73 is applicable to bends where

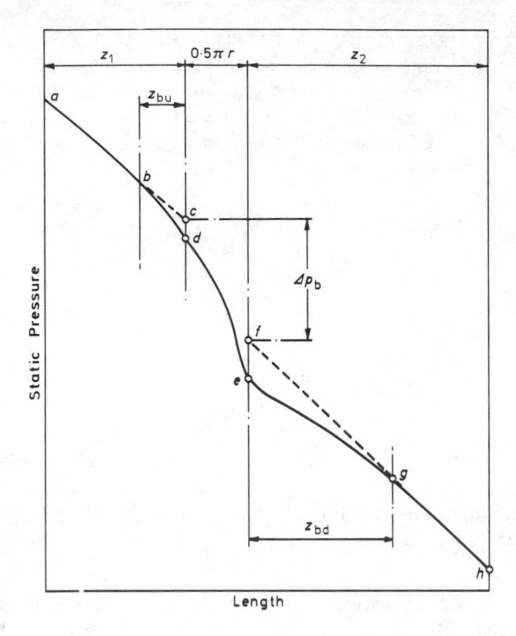

Figure 9. Pressure distribution over 90° bend.

$$\Gamma^2 = \frac{k_{BGO}\, V_G}{k_{BLO}\, V_L} \tag{92}$$

and, for 90° bends,

$$B = 1 + \frac{2.2}{k_{BLO}\left(2 + \dfrac{R}{d}\right)} \tag{93}$$

where R is the radius of the bend and d the pipe bore.

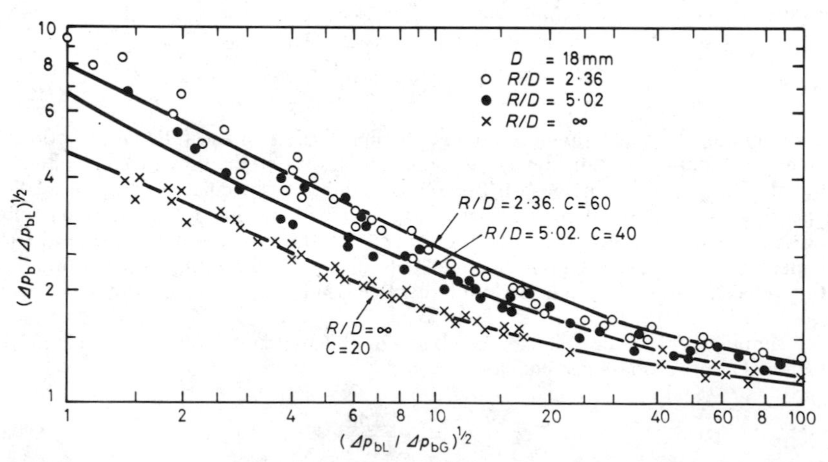

Figure 10. The root of the two-phase multiplier for 90° bends to a base of the Lockhart-Martinelli parameter [33,35].

The form of this equation was derived [33,34] from a model assuming that the major effects of the bend were associated with momentum change within and downstream of the bend. It follows, of course, that Equation 66 is also applicable where, using Equation 72,

$$C = \left\{1 + \frac{2.2}{k_{BLO}\left(2 + \frac{R}{d}\right)}\right\}\left(\Gamma - \frac{1}{\Gamma}\right) + \frac{2^{2-n} - 2}{\Gamma} \quad (94)$$

Except near the thermodynamic critical point, the final term on the right-hand side can be neglected. Figure 10 compares experiments [35] and prediction [33].

For bends other than 90°, the following formula is recommended:

$$B_{B\theta} = 1 + (B_{B90} - 1)\frac{k_{B90}}{k_{B\theta}} \quad (95)$$

CHANGES OF SECTION: INCOMPRESSIBLE FLOW

Considering first the flow to the vena contracta after a sharp inlet (from Equation 44), assuming incompressible flow of the components, no phase change and no change in velocity ratio gives

$$-\Delta p = \frac{U_e^2}{2V_e}(1 - \sigma^2)$$

$$= \frac{G^2 V_e}{2}(1 - \sigma^2) \quad (96)$$

Defining two-phase multipliers now in terms of pressure changes, rather than gradients, leads again [8] to either Equation 66 or 71, where B is given by Equation 29 and C by Equation 31.

Figure 11 gives a comparison with Thom's data [36] for steam-water flow through an orifice, using K_o as the velocity ratio. More extensive comparisons for flow to the vena contracta of sharp-edged orifices are given by Chisholm [8]. As discussed later, K_o also has been found to apply to some of the data for conical nozzles; some of the data, however, and data for round-edged inlet, have an effective velocity ratio to the throat of

$$K = K_o^{0.4} \quad (97)$$

In evaluating the momentum flux at other flow cross sections, it is recommended that the value based on momentum flux measurement be used (Equation 39). Experiments with enlargements [37] suggest a higher velocity ratio; however, the evidence [38] is inconsistent.

In the case of entry to a pipe, downstream of the vena contracta, the pressure change can be expressed as

$$\Delta p_{23} = \frac{MF_2}{C_c} - MF_3 \quad (98)$$

where the subscript 2 refers to the plane of the vena contracta, and 3 to the point of reattachment downstream. Using Equations 24, 25 and 98 gives

$$\Delta p_{23} = \Delta p_{LO23}\left(1 + \left\{\frac{V_G}{V_L} - 1\right\}\{B_{23}x(1 - x) + x\}\right) \quad (99)$$

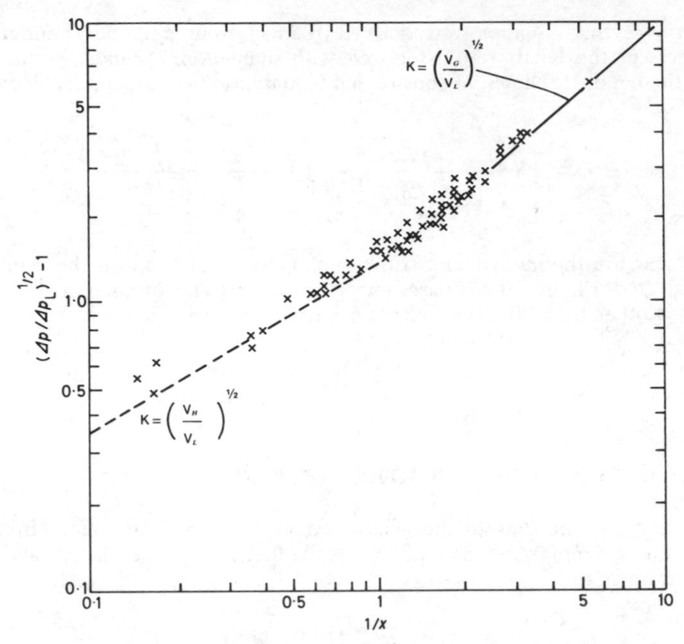

Figure 11. $(\Delta p/\Delta p_L)^{1/2} - 1$ to a base of $1/X$ [36].

where

$$\Delta p_{LO23} = G^2 V_L \left(\frac{1}{C_c} - 1\right) \tag{100}$$

and

$$B_{23} = \frac{\dfrac{B_2}{C_c} - B_3}{\dfrac{1}{C_c} - 1} \tag{101}$$

More generally, it can be shown [34] that for an incompressible flow through a series of fittings,

$$B = \frac{\Sigma\, k \times B/\sigma^2}{\Sigma\, k/\sigma^2} \tag{102}$$

On this basis, the following equations can be developed, all for the case of sharp inlets. Sudden contraction:

$$B = \frac{\left\{\left(\dfrac{1}{C_c\sigma}\right)^2 - 1\right\}\dfrac{1}{K_o} - \dfrac{2}{C_c K_o \sigma^2} + \dfrac{2}{K_o^{0.28}\sigma^2}}{\left(\dfrac{1}{C_c\sigma}\right)^2 - 1 - \dfrac{2}{C_c\sigma^2} + \dfrac{2}{\sigma^2}} \tag{103}$$

Thin plate:

$$B = \frac{\left\{\left(\frac{1}{C_c\sigma}\right)^2 - 1\right\}\frac{1}{K_o} - \frac{2}{C_c\sigma K_o} + \frac{2}{K_o^{0.28}}}{\left(\frac{1}{C_c\sigma}\right)^2 - 1 - \frac{2}{C_c\sigma} + 2} \tag{104}$$

Thick plate:

$$B = \frac{\left\{\left(\frac{1}{C_c\sigma}\right)^2 - 1\right\}\frac{1}{K_o} - \frac{2}{C_c\sigma K_o} + \frac{2}{K_o^{0.28}} - \frac{2}{\sigma K_o^{0.28}} + \frac{2}{K_o^{0.28}}}{\left(\frac{1}{C_c\sigma}\right)^2 - 1 - \frac{2}{C_c\sigma^2} + \frac{2}{\sigma^2} - \frac{2}{\sigma} + 2} \tag{105}$$

It is a simple matter, if laborious, to generate equations of this type for other conditions of interest.

For the sudden contraction, it has been shown experimentally [37,39] that the pressure change is satisfactorily predicted using homogeneous theory. This appears surprising in the light of the prominent position of K_o in Equation 103. However, Table VII, which shows predicted B values, illustrates that even with a velocity ratio as high as 5 at the vena contracta the value of B is 0.82 and greater, depending on the area ratio; that is, within 18% of homogeneous theory.

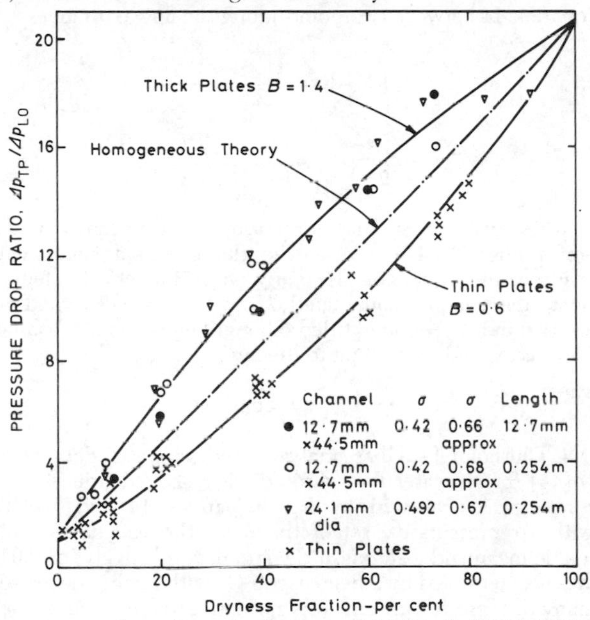

Figure 12. Steam-water flow through plates at 7 MN/m² (1000 lbf/in.²) abs. [34].

Figure 12 compares Equations 104 and 105 with experiments. The agreement is satisfactory, the change in B coefficient from a value well below unity with a thin plate to well above unity with a thick plate being well predicted. Zaloudek's [40] tests indicate that when the length to bore ratio exceeds 2, almost all the pressure recovery expected of a thick plate is obtained. Thin plate characteristics are exhibited where the length to bore ratio is of the order of 0.5 and less.

A gate valve can be treated as either a thick or thin plate, depending either on the seat length to bore ratio, z/d, or the single-phase characteristics. For globe valves, a B coefficient of 2.3 is [34] recommended.

Table VII. B Coefficient for Sudden Contraction as Function K_o (Equation 103)

Area Ratio	0.01			0.1			0.5		
Contraction Coefficient	0.61			0.62			0.69		
K_o	1.1	2	5	1.1	2	5	1.1	2	5
B	1.0	0.96	0.82	1.0	0.98	0.84	1.0	1.18	1.12

Approximate B coefficient values [34] may be summarized as follows:

	z/d	B
Plates/Gate Valves	< 0.5	0.5
	0.5	1.5
Globe Valves		2.3

CHANGES OF SECTION: COMPRESSIBLE FLOW

For the case in which the components are compressible and there is no phase change, integrating Equation 44 between two points along the flow path gives

$$\int_1^2 V_e\,dp = \frac{U_{e2}^2}{2} - \frac{U_{e1}^2}{2}$$

$$= \frac{G_2^2}{2}\left\{V_{e2}^2 - V_{e1}^2\left(\frac{A_2}{A_1}\right)^2\right\} \tag{106}$$

The case in which wall friction and gravitational effects can be ignored is considered. A number of studies [13,41] have solved this equation, mathematically integrating the left-hand side of the expression. In many ways, numerical integration by computer is as convenient; the computational method can be equally valid for flashing flows, which are discussed in the section entitled Changes of Section: Flashing Flow.

Assuming the gas expands according to the law,

$$PV_G^n = \text{constant} \tag{107}$$

where n = 1.4, Chisholm [42] has analyzed the air-water data of Graham [43] and Watson et al. [44] for air-water flow through sharp-edged orifices using Equation 106. The velocity ratio was evaluated by K_o, Equations 14 and 15. Table VIII makes comparison with a representative set of the data; the 150 data points were predicted with an arithmetic mean and a standard deviation, respectively, of 0.015 and 0.063.

Choked flow, as determined by a decrease in G_2 with further decrease in back pressure, occurred in many of these tests; there was no noticeable change in the accuracy of prediction in these cases.

If it is assumed that during the expansion the gas and liquid remain at the same temperature, the following equation for the expansion exponent [45] in Equation 107 is obtained:

$$n = \frac{(1-x)C_{pL} + xC_{pG}}{(1-x)C_{pL} + xC_{vG}} \tag{108}$$

In using n = 1.4 rather than values from this equation, it is being assumed that in flow through an orifice there is insufficient time for heat transfer to take place to maintain gas and liquid at the same temperature.

Table VIII. Sharp-Edged Orifice: 9.53- and 25.4-mm Bore Upstream Bore: 50.8-mm Air-Water Mixture

p_1 (bar)	Δp (bar)	M (kg/sec)	r	x	$\frac{M_{meas}}{M_{cal}}$	r_c	C_c
9.53-mm bore							
6.55	5.74	0.798	0.1237	0.0244	1.034	0.211	0.770
5.79	4.91	0.346	0.1508	0.1243	0.959	0.320	0.795
6.45	5.61	0.692	0.1295	0.0427	1.037	0.259	0.775
6.11	5.27	0.452	0.1380	0.0786	0.929	0.310	0.784
5.55	4.71	0.293	0.1516	0.1563	0.970	0.364	0.789
5.39	4.52	0.224	0.1619	0.2215	0.954	0.371	0.792
5.74	4.90	0.351	0.1474	—	0.971	0.361	0.784
5.48	4.62	0.574	0.1582	0.0445	0.979	0.243	0.783
5.34	4.47	0.594	0.1635	0.0389	0.994	0.325	0.779
4.56	3.63	0.443	0.2044	0.0564	0.976	0.288	0.776
3.04	2.08	0.411	0.3155	0.0338	1.005	—	0.773
5.49	4.53	0.493	0.1752	0.0678	1.027	0.341	0.774
6.43	5.45	0.740	0.1520	0.0398	1.091	0.236	0.778
2.50	1.53	0.264	0.3867	0.0576	0.996	—	0.772
2.34	1.35	0.266	0.4217	0.0508	1.006	—	0.760
25.4-mm bore							
1.65	0.60	2.373	0.6375	0.0070	1.066	—	0.709
1.64	0.56	2.253	0.6597	0.0078	1.038	—	0.692
1.64	0.641	1.354	0.6092	0.0391	0.914	—	0.713
1.62	0.628	1.815	0.6128	0.0198	1.028	—	0.707
1.59	0.59	1.692	0.6277	0.0215	0.990	—	0.704
1.50	0.51	1.508	0.6606	0.0257	1.012	—	0.702
1.49	0.50	1.044	0.6620	0.0507	0.952	—	0.707
1.48	0.47	2.096	0.6822	0.0096	1.179	—	0.687
1.47	0.47	0.757	0.6993	0.0884	0.988	—	0.701
1.46	0.48	0.718	0.6741	0.1011	1.025	—	0.708

With sharp inlets, the contraction coefficient was evaluated [46] from

$$C_c = \frac{xV_G + K(1-x)V_L}{\dfrac{xV_G}{C_G} + \dfrac{K(1-x)V_L}{C_L}} \qquad (109)$$

where C_G and C_L are the contraction coefficients for the gas and liquid components if they flowed alone. For the gas, the equations of Jobson [47] were used [42] in evaluating the contraction coefficient. Values of the estimated contraction coefficients are given in Table VIII; as the values are as high as 0.8, compared with 0.61 for incompressible flow, it will be appreciated that the accuracy of Equation 109 plays a significant part in the accuracy of prediction.

Progress in correlating two-phase compressible flow originally was made [46] by plotting the two-phase multiplier, ϕ_L^2, to a base of the following dimensionless group:

$$Y = \frac{C_L}{C_G}\,\frac{x}{1-x}\left(\frac{V_{G1}}{V_L}\right)^{1/2} F \qquad (110)$$

where for unchoked flow,

$$F = \left(\frac{1-n}{r^{2/n}} \frac{1}{1-r^{(n-1)/n}} \frac{n-1}{n}\right)^{1/2} \tag{111}$$

and for choked flow,

$$F = \left\{\frac{2(1-r)}{n}\right\}^{1/2} \left(\frac{n+1}{2}\right)^{[(n+1)/(n-1)]1/2} \tag{112}$$

The pressure ratio is r, and n the gas expansion exponent.

Figure 13 shows the air-water data for flow through orifices correlated on this basis. Chisholm [46] deduced that the velocity ratio at the vena contracta corresponded to

$$K = \left(\frac{\rho_L}{\rho_{G2}}\right)^{1/2} \frac{2}{Fr^{1/2n}\{C + \sqrt{(C^2-4)}\}} \tag{113}$$

This equation gives velocity ratios of the same magnitude as K_0. The coefficient C is that (5.3) in the equation fitting the data in Figure 13.

The round inlet nozzle data of Graham [43] and Watson et al. [44] can be correlated to a similar accuracy as for the orifice data, using Equation 97 for K.

On the other hand, data for flow to the throat of straight-sided venturis is well represented using K_0. Table IX compares prediction on this basis with the data of Muir and Eichorn [48]. Thus, at present the influence of the nozzle profile on the velocity ratio to the contraction is not fully understood.

In the case of a sudden contraction, for example, the recovery after the vena contracta is again given by Equation 98. With compressible flows, equations of the form of Equation 101 are no longer obtained, however. The pressure differences, however, can be evaluated readily by numerical means, the momentum fluxes being evaluated using Equation 26; the velocity ratio recommendations made for incompressible flow conditions would apply here also.

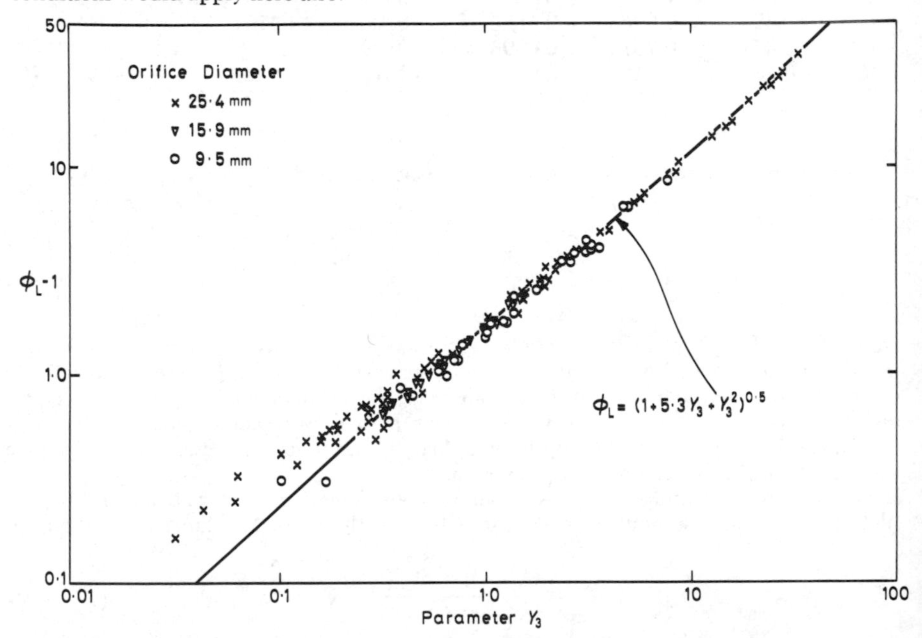

Figure 13. Air-water flow through sharp-edged orifices [46].

Table IX. Air-Water Flow Through Converging Nozzle; Throat 6.35 × 12.7 mm [48]

Initial Pressure (bar)	Throat Pressure (bar)	Mass Dryness Fraction, x	Mass Flowrate (kg/sec)	
			Experiment	Equations 14 and 106
6.34	1.08	0.000 13	2.54	2.50
5.52	1.08	0.000 183	2.29	2.27
3.83	1.12	0.000 295	1.74	1.71
3.50	1.12	0.000 361	1.59	1.58
3.01	1.12	0.000 471	1.38	1.36
2.74	1.12	0.000 577	1.25	1.24
2.43	1.12	0.000 734	1.10	1.08
2.21	1.12	0.009 922	0.964	0.95
1.74	1.12	0.001 54	0.659	0.65
1.50	1.12	0.002 42	0.459	0.47
1.393	1.12	0.003 93	0.377	0.352
1.386	1.12	0.005 48	0.309	0.329

CHANGES OF SECTION: FLASHING FLOW

The expansion of a vapor-liquid mixture at an orifice, for example, results in evaporation of the liquid due to pressure reduction; this is referred to as flashing.

The energy equation can be written as

$$dh + d\left(\frac{xU_G^2}{2} + \frac{1-x}{2}\frac{U_L^2}{2}\right) = 0 \qquad (114)$$

where a horizontal adiabatic path is considered for simplicity. The enthalpy term can be expanded to

$$dh = d(i + pV_H) = di + pdV_h + V_H dp \qquad (115)$$

The internal energy of the fluid, the energy for expansion, heat supply (when it exists) and internal frictional dissipation are related:

$$di + pdV_h = dQ + dF_F = T\,ds \qquad (116)$$

Combining Equations 114–116 gives

$$V_H\,dp + d\left(\frac{xU_G^2}{2} + (1-x)\frac{U_L^2}{2}\right) + dF_F = 0 \qquad (117)$$

This compares with the equation obtained from momentum considerations:

$$V_e\,dp_M = U_e dU_e \qquad (44)$$

The forces between the phases do not appear in this equation, nor do other effects of frictional dissipation. Due to the internal frictional dissipation term in Equation 117, the expansion to the throat of a nozzle, for example, will not be isentropic. Nor from Equation 114 is it enthalpic. The expansion will lie between these two limiting conditions. The mass dryness along an adiabatic flow path can be obtained by integrating Equation 114, solving simultaneously with Equation 44. For water, it can be shown

[49] that there is little difference in the magnitudes of mass dryness fraction evaluated on either the basis of an isentropic or isenthalpic expansion.

In using Equation 44 to evaluate flow at changes of section, it is recommended that the velocity ratio equations already discussed in the previous two sections be used.

In discussing choked flow in the section entitled The Momentum Gradient and Choked Flow, thermodynamic nonequilibrium effects already have been discussed. These also can occur at sudden contractions and, as in the case of choking in a pipe, they can be handled analytically through appropriate selection of the B coefficient.

The subject of thermodynamic nonequilibrium flows is imperfectly understood at present. Effects of the scale of the device are seen clearly in some of the experimental data [44,50], which are not allowed for in current design methods, nor are the effects of noncondensable gas, which, by providing evaporation sites, presumably assist the mixture to reach equilibrium. It has also been shown [51] that nonequilibrium effects arising at inlet can be maintained to approaching z/d of 100.

An approximate procedure for including thermodynamic nonequilibrium effects in nozzles is as follows: Where $x > x_t$, as given in Equation 54, thermodynamic equilibrium is assumed. For lower mass dryness fractions, using the same form of relationship as used in correlating nonequilibrium effects at pipe exit, Equation 58, but now introducing the initial mass dryness fraction, x_o, gives

$$x_N = x_o + \left(\frac{x - x_o}{x_t - x_o}\right)^2 (x - x_o) \tag{118}$$

Using Equations 49, 97, 106 and 118 gives, assuming an isentropic expansion, the predicted curves in Figure 14 for steam-water flow through a de Laval nozzle. Equation 106 was integrated numerically using a digital computer. The maximum difference between experiment and prediction is +23/−21%. The descriptions "stagnation pressure" and "stagnation mass dryness fraction" used in Figure 14 indicate values of these terms with zero kinetic energy.

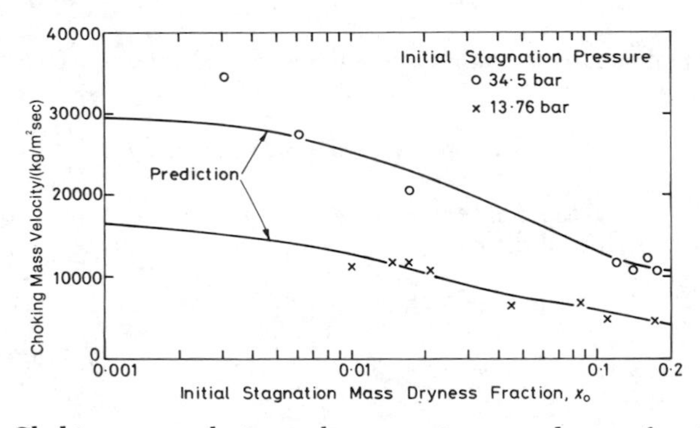

Figure 14. Choking mass velocity to the stagnation mass dryness fraction: steam-water flow through de Laval nozzles.

Pressure recovery downstream of the throat can be handled on the assumption of thermodynamic equilibrium; the procedure is similar to that discussed previously for incompressible mixtures.

SUMMARY AND CONCLUSIONS

The recommended procedures for predicting two-phase pressure gradients involve the evaluation of B coefficients and velocity ratios. Methods for evaluating those terms are summarized in Table X.

Table X. Summary: Values B and K

B or K		Comments
1. Gravitational Component		When
$X > 1 \quad K_o = (V_H/V_L)^{1/2}$	(14)	$U_H < (U_G - U_{WD})$
$X < 1 \quad K_o = (V_G/V_L)^{1/4}$	(15)	use drift flux model
2. Momentum Flux		
$K_o^{0.28}$	(39)	
3. Choking		
$X < 1 \quad B = 1/K_o$	(52)	
$X > 1 \quad B = \frac{1}{K_o}\frac{x_N}{x}$	(57)	x_N: Equation 58
4. Friction		
Turbulent-turbulent		Stratified flow:
B: Table IV		Equations 80–84
Laminar component		
B: Table VI		
5. Bends		
B. Equation 93		
6. Contraction (to throat)		
K_o		Round inlet $K_o^{0.4}$
Evaporating flow		
$X < 1 \quad B = 1/K_o$	(52)	
$X > 1 \quad B = \frac{1}{K_o}\frac{x_N}{x}$	(57)	x_N: Equation 118

NOMENCLATURE

A	cross-sectional flow area
B	coefficient, Equations 27 and 71
B_A	B coefficients for annular flow, Table IV
$B_{B\theta}$	B coefficient for bend of θ degrees
B_R	B coefficient for rough tubes, Equation 75
B_S	B coefficient for stratified flow, Equation 79
C	coefficient, Equations 30 and 66
C_c	contraction coefficient, Equation 98
C_G	contraction coefficient for gas
C_L	contraction coefficient for liquid
C_p	specific heat at constant pressure
C_v	specific heat at constant volume
c	constant in Blasius Equation (2)
D_p	pressure gradient
Dp_F	pressure gradient due to friction
Dp_{FGO}	pressure gradient due to friction, if mixture flows as gas
Dp_{FL}	pressure gradient due to friction, if liquid flows alone
Dp_{FLO}	pressure gradient due to friction, if mixture flows as liquid
Dp_G	pressure gradient due to friction or momentum when gas component flows alone
Dp_L	pressure gradient due to friction or momentum when liquid component flows alone
d	bore
e	flow pattern parameter, Equation 78
F	dimensionless group, Equation 111
F_F	frictional dissipation, Equation 116
G	mass velocity
G_c	mass velocity at choking
g	gravitational acceleration
h	enthalpy (specific), Equation 115
i	internal energy (specific), Equation 116
K	phase velocity ratio, U_G/U_L
K_e	effective velocity ratio in momentum change, Equation 39

K_o	velocity ratio, Equations 14 and 15
k_B	single-phase loss coefficient for bend, Equation 91
k_{BGO}	single-phase loss coefficient for bend, if mixture flows as gas
k_{BLO}	single-phase loss coefficient for bend, if mixture flows as liquid
$k_{B\theta}$	single-phase loss coefficient for bend of θ degrees
M	mass flowrate
MF	momentum flux
MF_{LO}	momentum flux, if mixture flows as liquid
m	exponent, Equation 90
N	nonequilibrium mass fraction ratio, Equation 48
n	gas expansion exponent, Equation 60
n	Blasius exponent, Equation 2
Δp_b	pressure drop over bend
p	pressure
Δp_M	pressure changes associated with momentum changes
Q	heat supply
R	radius of bend
Re	Reynolds number
r	downstream to upstream pressure ratio, Equation 111
r_c	pressure ratio with choking, Table VIII
s	entropy, Equation 116
T	temperature, Equation 116
U_e	effective velocity, Equation 33
U_G	velocity of gas
U_H	homogeneous velocity
U_L	velocity of liquid
U_{WD}	weighted drift flux, Equation 16
V_e	effective volume, Equation 26
V_G	specific volume of gas
V_H	homogeneous specific volume, Equation 13
V_L	specific volume of liquid
V_m	specific volume of mixture, Equation 12
W_D	drift flux term, Equation 19
X	Lockhart-Martinelli parameter, Equation 1
x	mass dryness fraction
x_N	mass dryness fraction under nonequilibrium conditions
x_t	mass dryness fraction, corresponding to $X = 1$, Equation 54
Y	dimensionless group, Equation 110
Z	coefficient defined by Equation 67
z	axial length

Greek Symbols

α_L	liquid cross section
Γ	physical property parameter, Equation 8
ϵ	surface roughness, Equation 75
λ	friction factor
λ_G	friction factor for gas flowing alone
λ_L	friction factor for liquid flowing alone
μ_G	absolute viscosity of gas
μ_L	absolute viscosity of liquid
σ	ratio of downstream to upstream flow cross sections
ϕ_L^2	two-phase multiplier; ratio of two-phase pressure gradient to gradient, if liquid flows alone
ϕ_{FL}^2	two-phase multiplier for friction, Equation 66
ϕ_{FLO}^2	two-phase multiplier for friction, Equation 71
ψ	normalized multiplier, Equation 37
ψ_A	normalized multiplier for annular flow
ψ_S	normalized multiplier for stratified flow

REFERENCES

1. Lockhart, R. W., and R. C. Martinelli. "Proposed Correlation of Data for Isothermal Two-Phase Two-Component Flow in Pipes," *Chem. Eng. Prog.* 45(1):39–48 (1949).
2. Taitel, Y., and A. E. Dukler. "A Model for Predicting Flow Regime Transitions in Horizontal and Near-Horizontal Flow," *AIChE J.* 22:47–55 (1976).
3. Berber, G., J. W. Palen and J. Taborek. "Prediction of Horizontal Tube-Side Condensation of Pure Components Using Flow Regime Criteria," in *Condensation Heat*

Transfer, Proc. 18th Nat. Heat Transfer Conf., San Diego, CA, August 6–8, 1979 (New York: American Society of Mechanical Engineers, 1979), pp. 1–8.

4. Chisholm, D. "The Turbulent Flow of Two-Phase Mixtures in Horizontal Pipes at Low Reynolds Number," *J. Mech. Eng. Sci.* 22(4):199–202 (1980).
5. Chisholm, D., and L. A. Sutherland. "Prediction of Pressure Changes in Pipeline Systems During Two-Phase Flow," paper no. 4, presented at the I.mech.E./I.chem.E. Joint Symposium on Fluid Mechanics and Measurements in Two-Phase Systems, University of Leeds, September 24–25, 1969.
6. Chisholm, D. "Pressure Gradients During the Flow of Evaporating Two-Phase Mixtures," NEL Report No 470, National Engineering Laboratory, East Kilbride, Glasgow, Scotland (1970).
7. Chisholm, D. "Void Fraction During Two-Phase Flow," *J. Mech. Eng. Sci.* 15(3): 235–236 (1973).
8. Chisholm, D. "Two-Phase Flow Through Sharp-Edged Orifices," *J. Mech. Eng. Sci.* 19(3):128–130 (1977).
9. Armand, A. A. "The Flow Mechanism of a Two-Phase Mixture in a Vertical Tube," in *Hydrodynamics and Heat Transfer During Boiling in High Pressure Boilers,* N. A. Styrikovick, Ed. *Akad. Nauk SSSR,* 19–34 (1955) (in Russian). English translation, AEC-tr-4490, Washington, DC, Office of Technical Services, U.S. Department of Commerce.
10. Friedel, L. "Mean Void Fraction and Friction Pressure Drops: Comparison of Some Correlations with Experimental Data," paper A7, presented at the European Two-Phase Flow Group Meeting, Grenoble, France, June 6–9, 1977.
11. Zuber, N., and J. A. Findlay. "Average Volumetric Concentration in Two-Phase Flow Systems," *Trans. ASME J. Heat Transfer* 87(4):453–468 (1965).
12. Labuntsov, D. A., et al. "Vapour Concentration of a Two-Phase Adiabatic Flow in Vertical Ducts," *Thermal Eng.* 15(4):62–67 (1968).
13. Chisholm, D. "Critical Conditions During Flow of Two-Phase Mixtures Through Nozzles," *Proc. Inst. Mech. Eng.* 182(3H):145–151 (1967–68).
14. Chisholm, D. "Pressure Gradients Due to Friction During the Flow of Evaporating Two-Phase Mixtures in Smooth Tubes and Channels," *Int. J. Heat Mass Transfer* 16(2):347–358 (1973).
15. Andeen, G. B., and P. Griffith. "Momentum Flux in Two-Phase Flow," *Trans. ASME J. Heat Transfer* 90(2):211–222 (1968).
16. Wiafe, F. "Two-Phase Flow in Rough Tubes," PhD Thesis, The University of Strathclyde, Glasgow, Scotland (1970).
17. Chisholm, D. "Mass Velocities Under Choked Flow Conditions in Two-Phase Flashing Pipe Flow," *J. Mech. Eng. Sci.* 23(6):309–311 (1981).
18. Burnell, J. G. "Flow of Boiling Water Through Nozzles, Orifices and Pipes," *Engineering, London* 164(4272):572–576 (1947).
19. Henry, R. E. "A Study of One- and Two-Component Two-Phase Critical Flow at Low Qualities," Report No. ANL-7430, Argonne National Laboratory, Argonne, IL (1968).
20. Prisco, M. R., R. E. Henry, M. N. Hutcheson and J. L. Linehan. "Non-equilibrium Critical Discharge of Saturated and Subcooled Liquid Freon-11," *Nucl. Sci. Eng.* 63: 365–375 (1977).
21. Henry, R. E., and H. K. Fauske. "The Two-Phase Critical Flow of One-Component Mixtures in Nozzles, Orifices and Short Tubes," *Trans. ASME, Series C, J. Heat Transfer* 93:179–187 (1971).
22. Zaloudek, F. R. "The Low Pressure Critical Discharge of Steam-Water Mixtures from Pipes," USAEC Report HW-68934 (rev.), Contract (AT45-1)-1350, General Electric Company, Hanford Atomic Products Operation, Richland, WA, Office of Technical Services, U.S. Department of Commerce, Washington, DC (1961).
23. Chisholm, D., and A. D. K. Laird. "Two-Phase Flow in Rough Tubes," *Trans. Am. Soc. Mech. Eng.* 80(2):276–286 (1958).

24. Chisholm, D. "Discussion of Paper by T. Johannsen," *Int. J. Heat Mass Transfer* 16:225–226 (1973).
25. Chisholm, D. "Pressure Gradients During Flow of Incompressible Two-Phase Mixtures Through Pipes, Venturis and Orifice Plates," *Brit. Chem. Eng.* 12(9):1368–1371 (1967).
26. Chisholm, D. "Influence of Pipe Surface Roughness on Friction Pressure Gradient During Two-Phase Flow," *J. Mech. Eng. Sci.* 20(6):353–354 (1978).
27. McMillan, H. K. "A Study of Flow Pattern and Pressure Drop in Horizontal Two-Phase Flow," PhD Thesis, Purdue University (1963).
28. Hewitt, G. F., and N. S. Hall-Taylor. *Annular Two-Phase Flow* (Oxford, England: Pergamon Press, Inc., 1970).
29. Wallis, G. B. *One-Dimensional Two-Phase Flow* (New York: McGraw-Hill Book Co., 1969).
30. Chisholm, D. "A Theoretical Basis for the Lockhart-Martinelli Correlation for Two-Phase Flow," *Int. J. Heat Mass Transfer* 10(12):1767–1778 (1967).
31. Kubie, J., and H. S. Oates. "Aspects of Two-Phase Frictional Pressure Drop in Tubes," *Trans. Inst. Chem. Eng.* 56:205–209 (1978).
32. Charles, M. E., and L. U. Lillelehet. "Correlation of Pressure Gradients for Stratified Laminar-Turbulent Flow of Two Immiscible Liquids," *Can. J. Chem. Eng.* 44(1):47–49 (1966).
33. Chisholm, D. "Two-Phase Pressure Drop in Bends," *Int. J. Multiphase Flow* 6:363–367 (1980).
34. Chisholm, D. "Prediction of Pressure Drop at Pipe Fittings During Two-Phase Flow," *Proc. 13th Int. Inst. Refrig. Cong.* 2:781–789 (1971).
35. Sekoda, G., et al. "Horizontal Two-Phase Air-Water Flow Characteristics in the Disturbed Region Due to 90° Bend," *Trans. Jap. Soc. Mech. Eng.* 35(279):2227–2233 (1969). English translation: NEL 2237 T6914, National Engineering Laboratory, East Kilbride, Glasgow, Scotland (1971).
36. Thom, J. R. S. "The Flow of Steam-Water Mixtures Through Sharp-Edged Orifices," Research Department Report No. 1/62/65, Renfrew, Babcock and Wilcox (1963).
37. Ferrel, J. K., and J. W. McGee. "Two-Phase Flow Through Abrupt Expansions and Contractions," Department of Chemical Engineering, North Carolina State University, Raleigh, NC, TID-23394, Vol. 3 (1966).
38. Janssen, E. "Two-Phase Pressure Loss Across Abrupt Contractions and Expansions, Steam-Water at 600 and 1400 lbf/in^2," in *Proc. Third Int. Heat Transfer Conf.*, Vol. 5 (New York: American Institute of Chemical Engineers, 1966), pp. 13–25.
39. Geiger, G. E., and W. M. Rohrer. "Sudden Contraction Losses in Two-Phase Flow," *J. Heat Transfer* 2:1–9 (1966).
40. Zaloudek, F. R. "The Critical Flow of Hot Water Through Short Tubes," HW-77594, General Electric Co., Hanford Atomic Products Operation, Richland, WA (1963), p. 36.
41. Henry, R. E. "The Compressible Flow of Two-Component, Two-Phase Mixtures in Nozzles and Orifices," Report to NASA TM-X-52729, E-5440, Lewis Research Center, Cleveland, OH, National Aeronautics and Space Administration, Washington, DC (1970).
42. Chisholm, D. "Flow of Compressible Two-Phase Mixtures Through Sharp-Edged Orifices," *J. Mech. Eng. Sci.* 23(1):45–48 (1981).
43. Graham, E. J. "The Flow of Air-Water Mixtures Through Nozzles," NEL Report No. 308, National Engineering Laboratory, East Kilbride, Glasgow, Scotland (1967).
44. Watson, G. G., V. E. Vaughan and M. W. McFarlane. "Two-Phase Pressure Drop with a Sharp-Edged Orifice," NEL Report No. 290, National Engineering Laboratory, East Kilbride, Glasgow (1967).
45. Tangren, R. F., G. H. Dodge and H. S. Seifert. "Compressibility Effects in Two-Phase Flow," *J. Appl. Phys.* 20(7):637–645 (1949).

46. Chisholm, D. "Flow of Compressible Two-Phase Mixtures Through Throttling Devices," *Chem. Process Eng.* 48(12):73–78 (1967).
47. Jobson, D. A. "On the Flow of Compressible Fluids Through Orifices," *Proc. Inst. Mech. Eng.* 169(37):767–776 (1955).
48. Muir, J. F., and R. Eichorn. "Compressible Flow of an Air-Water Mixture through a Vertical Two-Dimensional Converging-Diverging Nozzle," in *Proc. Heat Transfer and Fluid Mech. Inst.* (Stanford, CA: Stanford University Press, 1963), pp. 183–204.
49. Linning, D. L. "The Adiabatic Flow of Evaporating Fluids in Pipes of Uniform Bore," *Proc. Inst. Mech. Eng.* 1B(2):64–75 (1952).
50. Silver, R. S., and J. A. Mitchell. "The Discharge of Saturated Water Through Nozzles," *Trans. N.E. Coast Inst. Eng. Shipb.* 62:51–72, D15–D30 (1945).
51. Henry, R. E. "The Two-Phase Critical Discharge of Initially Saturated or Subcooled Liquid," *Nucl. Sci. Eng.* 41:336–342 (1970).

CHAPTER 19
COCURRENT GAS-LIQUID DOWNFLOW IN PACKED BEDS

R. Gupta
Exxon Research & Engineering Company
Florham Park, New Jersey 07932

CONTENTS

INTRODUCTION

Two-phase gas-liquid flow in packed columns is encountered in many chemical and petroleum processes involving absorption and chemical reactions. The flow of the gas and liquid phases can be either cocurrent or countercurrent. In the cocurrent mode, the gas-liquid flow can be either downward or upward. In the countercurrent mode, only downward flow of liquid with the gas flowing upward is possible. Countercurrent operation with upward liquid flow is not possible because the liquid would flood the column and prevent gas flow. In this chapter, we will discuss the dynamics of cocurrent gas-liquid downflow. Cocurrent upflow and countercurrent operations will be discussed in the next two chapters.

The most common use of packed beds with cocurrent gas-liquid downflow is in petroleum hydroprocessing, in which a liquid or a partially vaporized petroleum fraction is reacted with hydrogen gas in a fixed bed of catalyst particles. The catalyst size is usually in the range of 1 to 3 mm. Common hydroprocessing reactions are hydrocracking of heavy petroleum fractions and hydrodesulfurization of various petroleum fractions. Fixed beds with gas-liquid flow are also used in the chemical industry, although not as extensively as in the petroleum industry. Some examples of their use in the chemical industry are for the hydrogenation of various solvents and for hydrogen peroxide production. Packed columns with cocurrent gas-liquid downflow sometimes are also used for absorption. The mass transfer coefficient in the cocurrent downflow columns is usually higher than that in the conventional countercurrent absorption columns. However, because the average driving force for mass transfer is higher in a countercurrent operation, the use of cocurrent absorption is usually limited to operations requiring less than one theoretical stage or to those in which the driving force for mass transfer is sustained by the depletion of the absorbed species by a reaction.

The cocurrent downflow operation has several advantages. The flow is close to plug flow. In contrast to the countercurrent flow or cocurrent upflow, the packing is not fluidized and is held in place. The throughput is not limited by bed flooding. Also, the pressure drop is usually small unless very small catalyst size is used or the bed is plugged by the deposition of foulants. The major disadvantages of cocurrent downflow are that very small catalyst particles, which are more effective in diffusion-limited reactions, cannot be used from pressure drop considerations. In addition, flow maldistribution and incomplete catalyst utilization may occur unless sufficiently high flowrates are employed. Also, liquid holdup, which is important if homogeneous liquid-phase reactions are involved, is less than in cocurrent upflow and countercurrent flow operations.

The design and operation of packed beds with gas-liquid downflow is affected strongly by such hydrodynamic parameters as flow pattern, holdup of gas and liquid phases, contacting and interaction between various phases, pressure drop, and heat and mass transfer in the bed. The hydrodynamics and transport processes in packed beds with gas-liquid downflow will be discussed in the following paragraphs. Additional information on the subject can be found in review papers by Satterfield [1], Hofmann [2], Goto et al. [3] and Charpentier [4], and in a book on multiphase reactor design by Shah [5].

HYDRODYNAMICS

Flow Regimes

Various flow regimes such as trickle flow, pulsed flow, spray flow and bubble flow can exist in packed beds with cocurrent gas-liquid downflow. Transition between these flow regimes depends on the gas and liquid flowrates and the physical properties of the fluids, as well as the packing.

In trickle flow, the regime most common in commercial hydroprocessing reactors, the gas phase is continuous while the liquid trickles down the packing in the form of discontinuous films and rivulets. The trickle flow regime is obtained at relatively

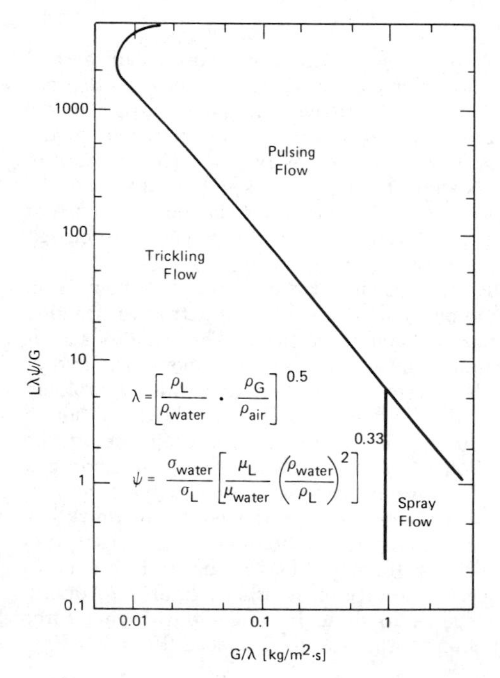

Figure 1a. Flow regime map for nonfoaming liquids [7].

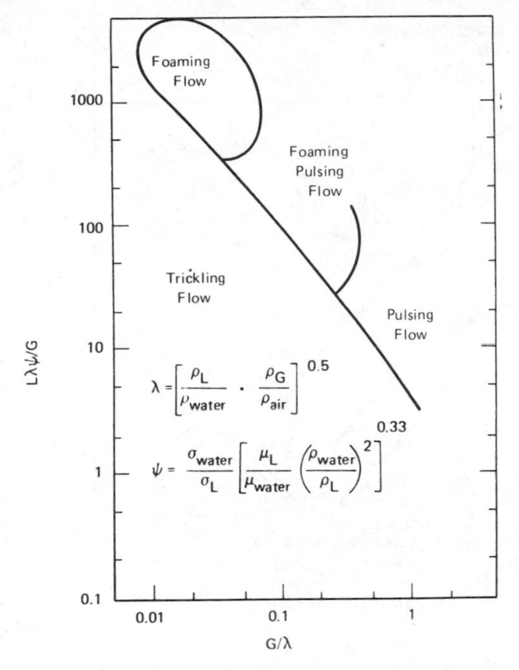

Figure 1b. Flow regime map for foaming liquids [7].

low liquid and gas mass velocities. If the flowrates are adequately raised, the increased shear between the gas and the liquid causes the liquid films to separate from the packing. The separated liquid momentarily blocks the flow channels in the packing but is soon propelled down by the advancing gas. This phenomenon gives rise to the pulsed flow regime in which liquid- and gas-rich zones alternately travel down the bed. If the gas flowrate is raised further, the increased shear between the gas and the liquid causes the liquid to break up in the form of a spray suspended as a mist in a continuous gas phase. For certain flow conditions, yet another flow regime, called the bubble flow regime, is obtained in which the gas flows as a dispersed phase in a continuous liquid phase.

The transition boundaries between the various flow regimes have been analyzed by many investigators [6–10]. The flow regime transitions are dependent on many variables, including the ability of the liquid to foam. In the case of foaming liquids, two additional regimes—foaming and foaming pulsing—are observed. Flow regime maps for foaming and nonfoaming liquids presented by Charpentier and Favier [7] are given in Figures 1a and 1b. Figures 1a and 1b use gas and liquid properties normalized with respect to air and water. Alternate flow maps proposed by Talmor [10], based on the different forces acting on the system, are given in Figures 2a and 2b.

It has been recognized increasingly that the transition between the various flow regimes is a more complex function of the flowrates and the fluids and packing properties than what is apparent from Figures 1 and 2. Thus, the boundaries separating the various flow regimes in Figures 1 and 2 should be considered approximate and may not be taken very rigorously. Chou et al. [9], for example, have shown that in addition to the flowrates and the fluid properties, the packing wettability and the bed void fraction also influence the flow regime transition. In the case of a nonwettable packing, the onset of pulsing takes place at flowrates higher than those predicted from Figures 1 and 2. In beds packed with a lower void fraction, the transition from trickle to pulsed flow takes place at lower flowrates. In a recent publication, Gupta and Koros [11] show that bed packing defects also influence the flow regime transition and that a flow channel in a fixed bed can lead to flow pulsations.

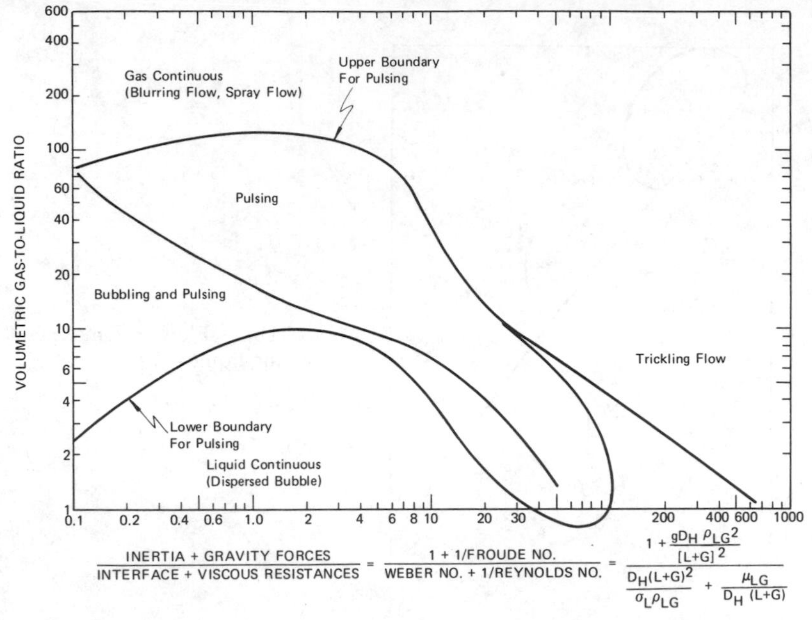

Figure 2a. Flow regime map for nonfoaming liquids [10].

Flow Distribution

Uniform flow distribution is an important objective in the design of fixed beds with gas-liquid downflow. Without uniform distribution, some regions of the bed would be

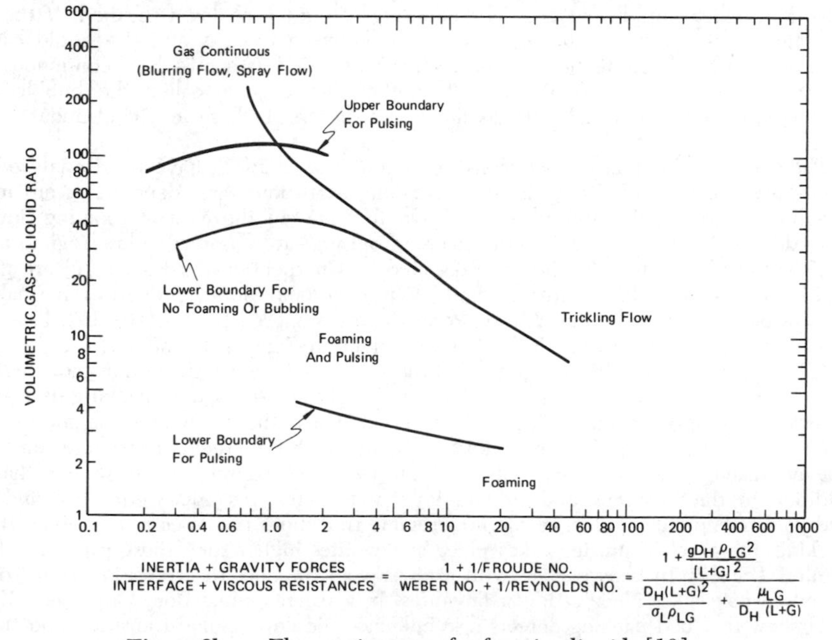

Figure 2b. Flow regime map for foaming liquids [10].

irrigated poorly, catalyst wetting would not be uniform and catalyst utilization may not be complete. In addition, hot spots may result if exothermic reactions are involved.

Liquid distribution in the bed can vary immensely, depending on the gas and liquid flowrates, the type and size of packing, the bed dimensions and the type of flow distributor used for initial flow distributon at the top of the bed. The influence of these variables on flow distribution is not understood completely. However, some general statements can be made from the reported studies [12–14]. Operation at high liquid mass velocity generally gives a more uniform distribution. The minimum liquid mass velocity required for uniform distribution and complete catalyst utilization will be discussed later in this chapter in a section on catalyst wetting. Increasing the bed height and the bed to particle diameter ratio also gives a more uniform distribution. Herskowitz and Smith [12] have recommended that the bed to particle diameter ratio should be greater than eighteen. Some other workers have reported a uniform flow distribution at a lower bed to particle diameter ratio, provided the liquid mass velocity is high.

Particle shape also seems to have an effect on flow distribution. For example, Herskowitz and Smith [12] found that a uniform distribution was achieved at lower ratios of bed to particle diameter for granular particles than for spherical or cylindrical particles. Also, a smaller bed height was necessary to obtain a uniform flow distribution in the case of granular particles. For the trickle flow regime, Herskowitz and Smith [12] have developed a model to predict liquid distribution as a function of the gas and liquid flowrates, the shape and size of particles and the bed dimensions. Sylvester and Pitayagulsarn [13] also have observed that the particle shape influences flow distribution. In their studies, at low gas rates the liquid distribution was fairly uniform in beds packed with pellets, Intalox saddles or raschig rings. However, when the gas flowrate was increased, the liquid flowed somewhat faster at the center of the bed packed with pellets or Intalox saddles, but not in the bed packed with raschig rings.

It is important that the initial flow distribution at the top of the bed be uniform. Using radioactive tracers, Ross [15] related the poor performance of some reactors to inadequate flow distribution at the bed top. A flow distributor tray is necessary in large-diameter beds (all commercial reactors). If a distributor is not used or if the spacing between the distribution tubes in the tray is excessive, then the initial depth of the bed serves primarily to even out the distribution.

If liquid distribution at the bed top is not uniform, the liquid will spread out laterally as it flows down from a point source at the bed surface. The bed depth at which a uniform flow distribution is established can be calculated approximately as [16,17].

$$Z = 2.5\, Y^2\, d_p^{-0.5} \tag{1}$$

where Z (in m) is the bed depth required to even out the distribution, Y (in m) is the diameter of the area over which the flow is to be distributed, and d_p (in m) is the particle diameter. When no distributor is used, Y is equal to the bed diameter. If a distributor is used, Y is equal to the spacing between the distributor tubes.

Even with a flow distributor, some flow maldistribution may occur in the bed because of separation between the gas and the liquid phases. A natural tendency for the phase separation exists because the gas and liquid phases flowing separately through different paths in the bed yield a lower pressure drop than when they share the same paths. A complete phase separation does not occur, however, because the fluids also tend to fill the available space uniformly.

For example, Specchia et al. [14] have observed that at low gas flowrates the liquid distribution across the bed diameter is fairly uniform. As the gas rate is increased, the liquid is expelled from the peripheral area of the bed to the center, while the gas flows preferentially through the peripheral area. At still higher gas rates, the liquid distribution becomes uniform again. Liquid velocity profiles at various gas rates are shown in Figure 3 for a 14.1-cm-diameter column packed with 6-mm glass beads.

The above observations of Specchia et al. can be explained by the phase separation phenomenon described earlier. At very low gas rates, the pressure drop, as well as the

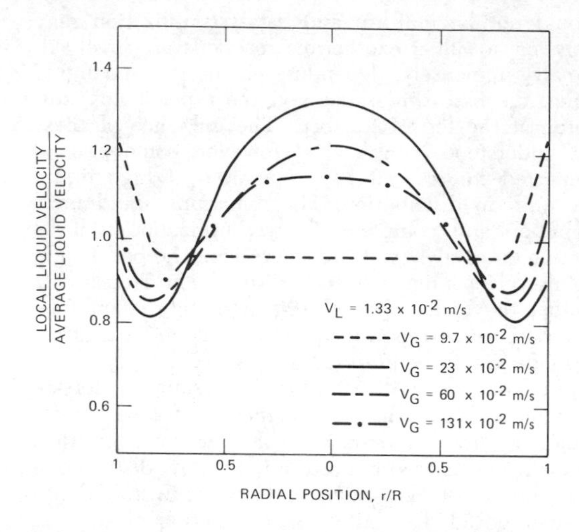

Figure 3. Liquid velocity vs radial position for $V_L = 1.33 \times 10^{-2}$ m/sec at various V_G [14].

driving force for pressure drop reduction by phase separation, are small. Thus, the liquid flows down the bed with a fairly flat profile without undergoing any phase separation. As the gas rate is increased, the increased driving force for phase separation leads to the creation of preferential zones for gas and liquid flows. The reason why the preferential zones for liquid and gas exist at the center and the peripheral area, respectively, is explained as follows. The gravitational force acting on the liquid opposes the radial liquid displacement necessary for creating the preferential flow zones. To create a preferential gas zone of a given area, a smaller liquid displacement is necessary if the liquid is displaced from the wall to the center of the bed, rather than if the liquid displacement is in the other direction. Thus, a preferential gas zone near the

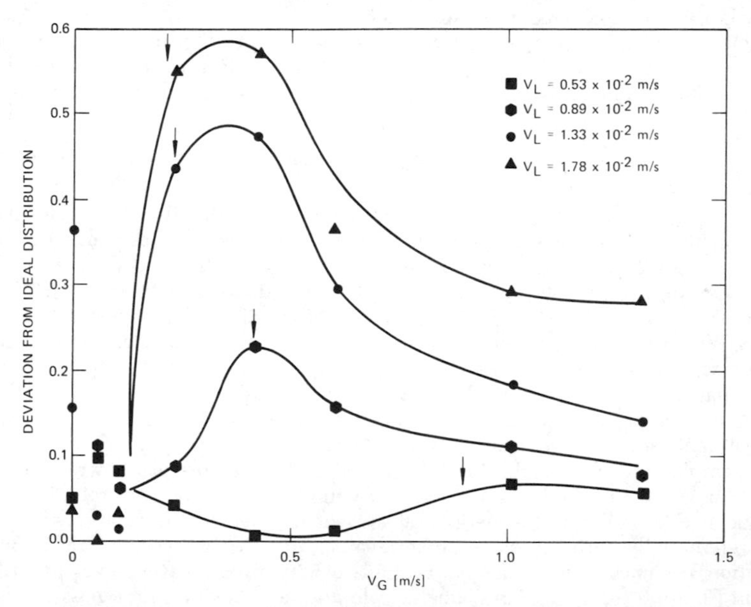

Figure 4. Flow maldistribution as a function of gas velocity at various liquid velocities. The vertical arrows represent the beginning of the pulsing regime [14].

wall and a preferential liquid zone at the center of the bed represent phase separation with a minimum of radial liquid displacement.

As the gas rate is increased even further, the inertial forces become predominant. The inertial forces overcome the tendency of the phases to separate. The result is that the flow distribution becomes uniform again.

By measuring the liquid flowrate into four concentric vessels at the bed bottom, Specchia et al. [14] measured flow maldistribution in the bed at various gas and liquid flowrates. The four concentric vessels did not include a small peripheral area adjacent to the wall to delete the effect of wall on flow. The deviation from ideal flow distribution was calculated as

$$\text{Deviation from ideal flow} = \frac{1}{4}\sum_{i=1}^{4}\left(\frac{V_i - V_{AVG}}{V_{AVG}}\right)^2 \tag{2}$$

where V_i is the liquid velocity in concentric collector i, and V_{AVG} is the liquid velocity under the condition of perfect flow distribution.

The deviation from ideal distribution is shown in Figure 4. The deviation is low at very low and very high gas rates. The deviation from ideal distribution, attributed to phase separation, is high at the intermediate gas rates, particularly at the onset of pulsing flow denoted by vertical arrows in Figure 4.

Very significant phase separation occurs in the pulsed flow regime. In addition to the well demarcated axial phase separation that leads to the gas and liquid pulses, significant phase separation also occurs in the radial direction. For example, Weekman and Myers [8] have observed that the liquid pulses do not bridge the entire bed

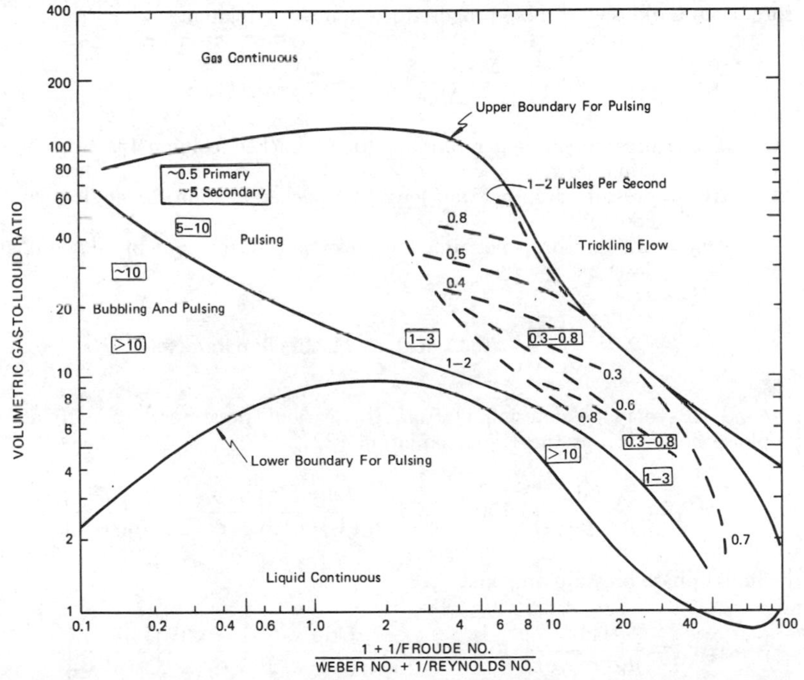

Figure 5. Pulse frequency map for gas-liquid downflow through packed beds; non-foaming liquids. Numbers in boxes represent frequency of pressure drop oscillations (s^{-1}). The dotted lines represent contours of constant frequency in the 0.3 to 0.8 s^{-1} range [10].

diameter but have the shape of a torus. The existence of gas holes within the liquid pulse also has been reported by Beimesch and Kessler [18]. The pulse velocity is lower than the gas velocity but higher than the liquid velocity [6]. The existence of gas holes in the liquid pulse allows the leakage of the fast-moving gas across the pulses.

Another characteristic of pulsed flow is that the frequency of pulsing can vary immensely, depending on the flowrates. Figure 5 [10] shows the observed frequencies at different flowrates.

Pressure Drop

The pressure drop is an important hydrodynamic parameter. This is particularly true for commercial reactors designed for high mass velocity operation to assure uniform flow distribution and intimate solid-liquid contacting.

The pressure drop of cocurrent gas-liquid downflow through a packed column depends on whether the liquid foams. For identical flowrates, the pressure drop of a foaming liquid is much higher than that of a nonfoaming liquid. A number of pressure drop correlations are available in the literature for the foaming, as well as the nonfoaming, liquids. The accuracy of the available correlations for the foaming liquids may be limited and they should be used for approximate calculations only.

A widely used pressure drop correlation for nonfoaming liquids resulted from the work of Larkins et al. [19]. This empirical correlation for packed beds is an extension of the original Lockhart and Martinelli [20] correlation for pressure drop in pipes. The general form of the correlation has been supported by the pressure drop data of several other investigators [8,21]. The correlation is based on obtaining the two-phase pressure drop from the single-phase gas pressure drop and the single-phase liquid pressure drop. Larkins et al. [19] proposed the following relation:

$$\log_{10}\left(\frac{\Delta P_{LG}}{\Delta P_L + \Delta P_G}\right) = \frac{0.416}{(\log_{10} X)^2 + 0.666} \tag{3}$$

where ΔP_{LG} = pressure drop per unit length of packed column for two-phase gas-liquid flow

ΔP_L = pressure drop per unit length of packed column for single-phase liquid flow

ΔP_G = pressure drop per unit length of packed column for single-phase gas flow

$$X = \sqrt{\frac{\Delta P_L}{\Delta P_G}}, \text{ the Lockhart-Martinelli parameter.}$$

ΔP_L and ΔP_G may be determined from the various pressure drop correlations for single-phase flow, such as the Ergun equation [22],

$$\Delta P_L = \frac{Ld_E}{\mu_L(1-\epsilon)}\left[150 + 1.75\frac{Ld_E}{\mu_L(1-\epsilon)}\right]\left[\frac{1-\epsilon}{\epsilon}\right]^3\left[\frac{\mu_L{}^2}{\rho_L d_E 3}\right] \tag{4a}$$

for the liquid-phase pressure drop and

$$\Delta P_G = \frac{Gd_E}{\mu_G(1-\epsilon)}\left[150 + 1.75\frac{Gd_E}{\mu_G(1-\epsilon)}\right]\left[\frac{1-\epsilon}{\epsilon}\right]^3\left[\frac{\mu_G{}^2}{\rho_G d_E 3}\right] \tag{4b}$$

for the gas-phase pressure drop.

Equation 3 is independent of the hydrodynamic flow regime in the bed. Specchia and Baldi [23] have proposed that the hydrodynamic flow regime should be incorporated into the pressure drop correlation. Only two basic hydrodynamic regimes involving poor or high interaction between the gas and the liquid may be considered [23]. The poor interaction regime consists of gas continuous trickling flow that occurs at low liquid and gas flowrates. The high interaction regime occurs at higher flowrates and includes all other flow regimes, such as pulsing and spray flow for the nonfoaming liquids, and foaming, foaming pulsing and spray flow for the foaming liquids.

In the poor interaction regime, the flow of one phase is affected very little by the other phase. For the poor interaction regime, Specchia and Baldi [23] used a model that considers the flow of gas through a bed restricted by the liquid. An Ergun-type pressure drop equation was found to fit the pressure drop data satisfactorily:

$$\Delta P_{LG} = \frac{k_1[1-\epsilon(1-\beta_s-\beta_D)]^2\mu_G V_G}{[\epsilon(1-\beta_s-\beta_D)]^3} + \frac{k_2[1-\epsilon(1-\beta_s-\beta_D)]\rho_G V_G^2}{[\epsilon(1-\beta_s-\beta_D)]^3} \qquad (5)$$

where k_1 and k_2 are constants that depend on the packing shape and size. β_S and β_D are the static and dynamic liquid holdups, respectively. The liquid holdups and correlations for obtaining them are discussed in the next section.

The constants k_1 and k_2 may not be determined from the pressure drop of gas through a dry packing since wetting changes the packing shape. These constants could be determined, however, from the pressure drop of gas through a wetted packing under the conditions of no liquid flow. Without liquid flow, only the static liquid holdup is applicable, and the preceding equation reduces to

$$\Delta P_G\text{, wet packing} = \frac{k_1[1-\epsilon(1-\beta_s)]^2\mu_G V_G}{[\epsilon(1-\beta_s)]^3} + \frac{k_2[1-\epsilon(1-\beta_s)]\rho_G V_G^2}{[\epsilon(1-\beta_s)]^3} \qquad (6)$$

In the high interaction regime, the pressure drop depends on whether the liquid foams. For identical gas and liquid flowrates, the pressure drop of a foaming liquid can be an order of magnitude higher than that of a nonfoaming liquid. Charpentier et al. [24] have proposed that the two-phase pressure drop may be correlated in terms of the energy dissipations for single-phase gas and liquid flows, rather than in terms of the single phase pressure drops. Instead of using a parameter, X, based on single-phase pressure drops (Equation 3), modified parameter, X′, based on energy dissipation may be used. X′ is defined as

$$X' = \sqrt{\frac{\xi_L}{\xi_G}} \qquad (7)$$

where

$$\xi_L = \frac{L}{\epsilon} \qquad (8)$$

and

$$\xi_G = \frac{G}{\epsilon}\left[1 + \frac{\Delta P_G}{\rho_G g}\right] \qquad (9)$$

Midoux et al. [25] have proposed the following pressure drop correlation for foaming liquids in the high interaction regime:

$$\left(\frac{\xi_{LG}}{\xi_L}\right)^{0.5} = 1 + \frac{1}{X'} + \frac{6.55}{(X')^{0.42}} \tag{10}$$
$$0.5 \leq X' \leq 100$$

where

$$\xi_{LG} = \frac{1}{\epsilon g}\left[\frac{L}{\rho_L} + \frac{G}{\rho_G}\right] \Delta P_{LG} + \frac{L+G}{\epsilon} \tag{11}$$

For the nonfoaming liquids in the high interaction regime, Equation 3 continues to be valid. Alternatively, the following correlations proposed by Midoux et al. [25] and Morsi [26] may be used for the nonfoaming liquids in the poor, as well as the high interaction, regime:

$$\left(\frac{\Delta P_{LG}}{\Delta P_L}\right)^{0.5} = 1 + \frac{1}{X} + \frac{1.14}{X^{0.54}} \tag{12}$$

$$0.1 \leq X \leq 80$$

if the liquid viscosity is less than 6 cP, and

$$\left(\frac{\xi_{LG}}{\xi_L}\right)^{0.5} = 1 + \frac{1}{X'} + \frac{7.11}{(X')^{0.55}} \tag{13}$$

$$0.05 \leq X' \leq 100$$

if the liquid viscosity is greater than 6 cP.

For the correct pressure drop correlation to be used, the foaming propensity of the liquid must be known a priori. The foaming propensity of the liquid cannot be predicted a priori on the basis of its static surface tension. For example, cyclohexane and kerosene have practically identical surface tensions, but only kerosene foams [27].

Foaminess is caused by the decreased coalescence of gas bubbles trapped in the liquid. A technique for determining the foaming propensity of a liquid has been devised by Charpentier and co-workers [27–29]. The technique involves injecting pairs of gas bubbles in the liquid and determining the percentage of bubbles that coalesce. Charpentier [27] determined the bubble coalescence in cyclohexane, to which various amounts of gas-oil had been added. For cyclohexane containing up to 6% gas-oil, 100% coalescence was obtained, and such mixtures were found to be nonfoaming. Between 6% and 14% of added gas-oil, the coalescence decreased from 100% to 0%, and the foaming propensity of such mixtures was found to increase drastically. Thus, this bubble coalescence technique shows promise for qualitatively determining whether a liquid would foam. Once the foaming propensity of the liquid is determined, the appropriate pressure drop correlation can be used.

The pressure drop correlations outlined above are empirical or semitheoretical in nature. A correlation based on fundamental theoretical analysis would be highly desirable. However, the complex hydrodynamics of the system is not easily amenable to theoretical analysis, and a number of questions remain unresolved. For example, Kan and Greenfield [30] reported a multiplicity in hydrodynamic states, depending on the flow history in the column. If the gas rate was increased to a higher value and then decreased to its original value, a lower pressure drop was obtained. The higher the value to which the gas rate was increased before returning to the original flowrate, the lower was the obtained pressure drop. Morsi and Charpentier [31] have observed that on returning to the original gas flowrate, a lower pressure drop may not necessarily be obtained. If the liquid rate is high, the obtained pressure drop is higher. This multiplicity in pressure drop is not accounted for in the available correlations.

Another area where more work is desired is the modeling of pressure drop buildup in packed beds subject to fouling. The correlations presented in this chapter are applicable only to clean beds with a constant void fraction. Many hydroprocessing reactors of significant industrial importance yield an increasing pressure drop with time as foulants deposit in the bed. Mathematical modeling of the transient pressure drop buildup as a function of the bed fouling pattern would be very useful.

Liquid Holdup

The liquid holdup affects the pressure drop, the thickness of the liquid film surrounding the packing and the effectiveness with which the packing is wetted. Its effect on the performance of a fixed-bed reactor depends on the nature of the reaction and the mechanism of mass transfer. A high liquid holdup will be advantageous if the liquid-phase reactants control the reaction rate. For example, in the hydrodesulfurization of a heavy petroleum fraction the sulfur compounds are present mainly in the liquid phase, and it is the concentration of these sulfur compounds at the catalyst surface that controls the rate of reaction. A high liquid holdup will improve catalyst wetting, assure intimate contact between the liquid-phase reactants and the catalyst, and increase catalyst utilization. A high liquid holdup also can be advantageous if the reactions take place in the homogeneous liquid phase, as well as on the catalyst surface.

A high liquid holdup can be deleterious if the rate-limiting reactants are present in the vapor phase. For example, in the hydrogenation of an organic solvent, hydrogen could be the rate-limiting reactant. In such a reaction, the hydrogen transport to the catalyst surface may be hindered by the liquid film that surrounds the catalyst particle. In addition, liquid in the catalyst pores could also increase the intraparticle mass transfer resistance.

The total liquid holdup, h_T, is defined as the volume of liquid in a unit volume of the packed column. If the packing is porous, there is an internal holdup, h_{INT}, representing the intraparticle liquid in the pores of the packing. The exterior holdup, h_{EXT}, representing the interparticle liquid in the bed void spaces, is made of two parts. The first part, called the dynamic holdup, h_D, represents the liquid that would drain from the column after the flow to the column is shut off. The second part, called the static holdup, h_S, represents the liquid that would not drain from the column but would be retained on the surface of the particles and in lenses between particles. Thus

$$h_T = h_{INT} + h_{EXT} = h_{INT} + h_D + h_S \quad (14)$$

The static liquid holdup depends on the packing size and the physical properties of the liquid. It has been correlated to the dimensionless Eötvos number [32–34], as shown in Figure 6. Eötvos No. $= \rho_L g\, d_p^2/\sigma_L$ represents the ratio of gravity to surface forces. The correlation line for the porous catalyst in Figure 6 includes the internal holdup. For typical trickle bed operation, the Eötvos number is less than 5. Thus, from Figure 6, the static liquid holdup is about 6% of the reactor volume. The static liquid holdup may not be interpreted as the stagnant or semistagnant liquid in the bed. The static holdup is a constant, whereas the volume of the stagnant liquid varies with the operating variables, such as the liquid mass velocity.

Several correlations are available in the literature for the dynamic liquid holdup. Predictions of several of the available correlations vary considerably. Therefore, experimental data should be used whenever possible. Care should be exercised in using the published correlations because some are for the total holdup, whereas others are for the dynamic holdup. Also, the holdup has been defined differently in many correlations as the liquid volume per unit void volume, β, rather than as the liquid volume per unit column volume, h. β and h are related as

$$h = \epsilon\beta \quad (15)$$

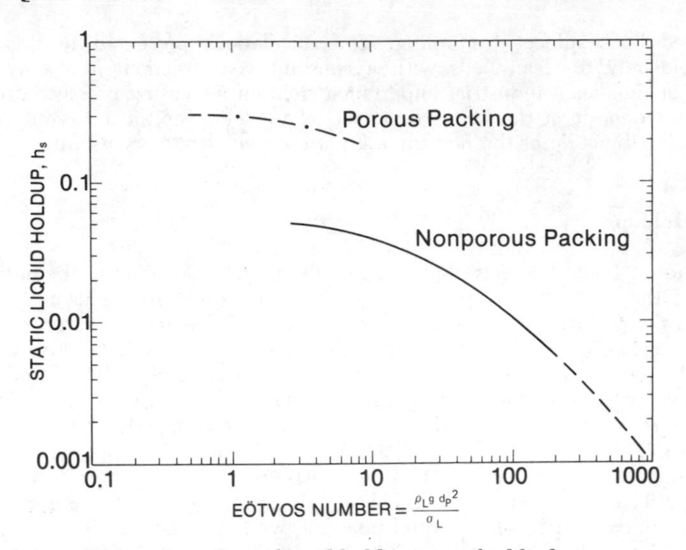

Figure 6. Static liquid holdup in packed beds.

The most recent correlations for liquid holdup result from the work at Midoux et al. [25], Specchia and Baldi [23] and Charpentier et al. [7,27]. For the nonfoaming liquids in all flow regimes and for the foaming liquids in trickle flow, Midoux et al. [25] have proposed the following correlation:

$$h_s + h_D = \epsilon[\beta_s + \beta_D] = \epsilon \frac{0.66\ (X)^{0.81}}{1 + 0.66\ (X)^{0.81}} \tag{16}$$
$$0.1 < X < 80$$

For the foaming liquids in the high interaction regime, the proposed correlation is

$$h_s + h_D = \epsilon[\beta_s + \beta_D] = \epsilon \frac{0.92\ (X')^{0.3}}{1 + 0.92\ (X')^{0.3}} \tag{17}$$

$$0.05 < X' < 100$$

where X and X′ are defined by Equations 3 and 8, respectively.

The holdup correlation proposed by Specchia and Baldi [23] takes the flow regime into account. For the poor interaction regime (trickle flow), their correlation is

$$h_D = \epsilon\beta_D = 3.86\epsilon\ (Re_L)^{0.545}\ (Ga^*)^{-0.42} \left(\frac{a_s d_p}{\epsilon}\right)^{0.65} \tag{18}$$

where

$$Re_L = \frac{V_L\ \rho_L\ d_p}{\mu_L}$$

and Ga* is the modified Galileo number defined as

$$Ga^* = \frac{d_p{}^3 \rho_L (g\rho_L + \Delta P_{LG})}{\mu_L{}^2} \tag{18a}$$

For the high interaction regime, the correlations proposed by Specchia and Baldi [23] are

$$h_D = \epsilon\beta_D = 0.125\epsilon\left(\frac{Z'}{\psi^{1.1}}\right)^{-0.312}\left(\frac{a_s d_p}{\epsilon}\right)^{0.65} \tag{19}$$

for the nonfoaming liquids and

$$h_D = \epsilon\beta_D = 0.0616\epsilon\left(\frac{Z'}{\psi^{1.1}}\right)^{-0.172}\left(\frac{a_s d_p}{\epsilon}\right)^{0.65} \tag{20}$$

for the foaming liquids. In the above two equations, Z' and ψ are defined as

$$Z' = \frac{(Re_G)^{1.167}}{(Re_L)^{0.767}} \tag{21}$$

and

$$\psi = \frac{\sigma_{Water}}{\sigma_L}\left[\frac{\mu_L}{\mu_{Water}}\left(\frac{\rho_{Water}}{\rho_L}\right)^2\right]^{1/3} \tag{22}$$

Finally, the following correlation by Charpentier et al. [7,27] has been claimed to apply to the foaming, as well as to the nonfoaming, liquids in all flow regimes:

$$\log_{10}\left[\left(\frac{h_s + h_D}{\epsilon}\right)\cdot\left(\frac{\mu_{Water}}{\mu_L}\right)^{0.2}\right] \tag{23}$$

$$= -0.363 + 0.168\log_{10}X' - 0.043(\log_{10}X')^2$$

Wetting of Packing

A complete wetting of the packing is desirable to obtain a high surface area for mass transfer in absorption. A high liquid holdup and complete catalyst wetting also are desirable in reactors in which the rate limiting reactants are present in the liquid phase, as mentioned previously. In practice, however, only a fraction of the available surface area is wetted by the liquid. How this wetted fraction changes with the flowrates, the physical properties of the fluids and the physical properties of the packing is understood only qualitatively. In general, increasing the liquid mass velocity improves wetting. Low liquid surface tension and high particle wettability also seem to improve wetting.

For nonporous solids, a number of correlations for wetting effectiveness have been proposed in the literature. These correlations have been summarized by Schwartz et al. [35]. Just as for liquid holdup, the predictions of the available correlations differ significantly and, therefore, experimental data should be used whenever possible. If experimental data are not available, the following correlation by Onda et al. [36] is recommended:

$$\frac{a_w}{a_s} = 1 - \exp\left[-0.225\left(\frac{L}{a_s\mu_L}\right)^{0.43}\right] \tag{24}$$

where a_w and a_s are the wetted and the total areas per unit column volume.

The entire wetted area may not be effective in mass transfer. Some of the solid particles are submerged in stagnant or semistagnant liquid pockets that can exist in a fixed bed. The existence of these stagnant liquid pockets can be particularly significant if the liquid mass velocity is very low. Since there may not be sufficient mass interchange between the stagnant liquid and the flowing liquid, the stagnant liquid tends to get saturated in a short time and becomes ineffective for further mass transfer. Thus, only that part of the wetted area in contact with the moving liquid participates in mass transfer.

The situation is even more complicated in reactors packed with a porous catalyst. Although the external surface area of the catalyst particle may be wetted only partially, the internal pores usually are filled completely with liquid due to the capillary forces. Despite the completely wetted internal surface, the reaction will not proceed very far unless there is mass exchange between the internal liquid and the external liquid. This mass exchange can take place only by way of the liquid film that wets the catalyst external surface.

Satterfield [1] has defined the catalyst contacting effectiveness as the ratio of the apparent kinetic constant obtained in a fixed-bed reactor to the actual kinetic constant. The actual kinetic constant is obtained in a well stirred reactor in which the catalyst pellet is completely immersed in the liquid. Figure 7 [1] shows how the catalyst contacting effectiveness in a packed-bed reactor might change with the liquid mass velocity. At very low liquid mass velocities, the liquid trickles mainly as rivulets through the interstitial openings between the catalyst particles. To envelope the catalyst particle with a liquid film to obtain effective contacting and efficient catalyst utilization, the liquid mass velocity has to be sufficiently high. A minimum liquid mass velocity of 3–8 kg/m^2-sec has been recommended [1,37,38]. The minimum liquid mass velocity for effective catalyst utilization also depends on the reaction rate and the diffusivity of reactants inside the porous catalyst particle. If the reaction rate is faster than the diffusion rate, i.e., the reaction takes place mostly at the exterior surface of the catalyst, higher liquid mass velocities are necessary to continuously bathe the catalyst with a liquid film and thereby replenish reactants at the catalyst surface. On the other hand, if the reaction rate is slow and the reactants at the exterior catalyst surface are not depleted rapidly, a complete liquid film envelope to continuously replenish the reactants is not necessary, and a somewhat lower mass velocity may be acceptable.

Axial and Radial Dispersion

A number of studies on axial mixing or dispersion have been reported in the literature. These studies have been summarized in a review by Ostergaard [39] and in a book by Shah [5]. The studies show that axial dispersion in the gas phase depends on the gas and liquid flowrates and the properties of the fluids and the packing. Axial dispersion in the liquid phase depends on all the above variables, except the gas flowrate.

The axial dispersion coefficient in the liquid phase, as measured by the dimensionless Peclet number, is signficantly higher in trickle flow than in single-phase flow. This is

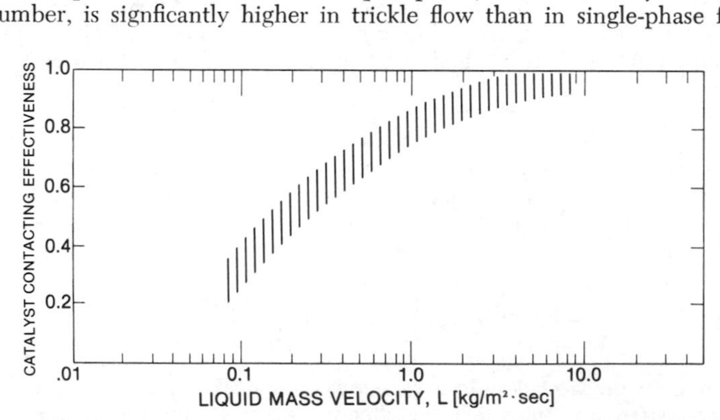

Figure 7. Effect of liquid mass velocity on catalyst contacting effectiveness [1].

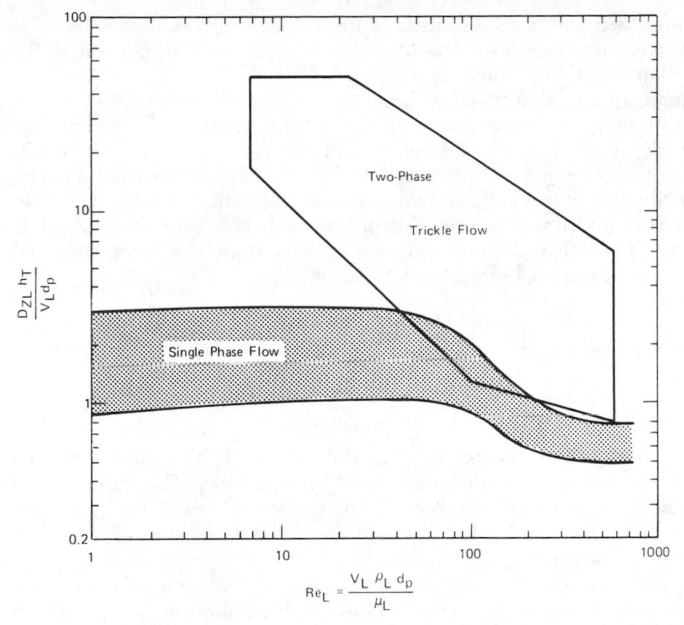

Figure 8. Axial liquid dispersion in single-phase and two-phase trickle flows [37].

shown in Figure 8 [37]. In Figure 8, the reciprocal of the Peclet number, $Pe_{ZL} = V_L d_p / D_{ZL} h_T$, is shown as a function of the liquid Reynolds number. It is only at high Reynolds numbers that the Peclet number for trickle flow approaches the Peclet number for single-phase flow.

From among the numerous correlations for the axial Peclet number reported in the literature, the following correlations by Hochman and Effron [40] are recommended: For the liquid phase,

$$Pe_{ZL} = 0.042\,[Re_L']^{0.5} \tag{25}$$

and for the gas phase,

$$Pe_{ZG} = 1.8\,[Re_G']^{-0.7} \cdot 10^{-0.005 Re_L'} \tag{26}$$

where the liquid and gas Reynolds numbers are defined, respectively, as

$$Re_L' = \frac{V_L \rho_L d_p}{\mu_L (1 - \epsilon)} \tag{27a}$$

and

$$Re_G' = \frac{V_G \rho_G d_p}{\mu_G (1 - \epsilon)} \tag{27b}$$

The axial dispersion coefficients in the above correlation are for a nonporous packing. Schiesser and Lapidus [41] have shown that the liquid-phase dispersion coefficient can be significantly higher for a porous packing. For example, if the dispersion coefficient is determined from residence time distribution using a tracer technique, the response curve will be more dispersed for a porous packing as the tracer is first entrapped and then released gradually from the pores. Thus, the axial dispersion coefficient is affected by the intraparticle and static liquid holdups. Van Swaaij et al. [32] have shown that the intraparticle and static liquid holdups will affect axial dispersion if these holdups are greater than about 10% of the total liquid holdup.

Very significant radial dispersion has been observed under trickle flow conditions. Koros [38] studied the charcoal-catalyzed decomposition of dilute hydrogen peroxide in the presence of nitrogen gas. Nitrogen does not participate in the chemical reaction; its only role was to simulate the two-phase trickle flow. Koros found very significant flow maldistribution in the bed, as evidenced by different flowrates and space velocities in the inner core and the outer annulus of the bed. However, in spite of the different space velocities, identical conversions were obtained in the inner core and the outer annulus, thus indicating a high rate of radial mixing.

TRANSPORT PROCESSES

Mass Transfer

Resistance to mass transfer can exist in the exterior of the solid particle and, in the case of catalytic reactors, in the interior of the particle. The external mass transfer takes place in three steps: (1) transfer from the gas phase to the gas-liquid interface; (2) transfer from the gas-liquid interface to the liquid; and (3) transfer from the liquid to the solid surface.

In trickle bed hydroprocessing reactors, it may be assumed that the gas-phase mass transfer resistance is negligible, i.e., vapor-liquid equilibrium exists between the gas and the gas-liquid interface [1]. This assumption is valid because hydrogen, the gaseous reactant, is present in high fractional concentration in the gas phase and is relatively insoluble in the liquid phase. In absorption from a dilute gas mixture, on the other hand, the gas-phase mass transfer resistance can be significant.

The gas-phase mass transfer coefficient may be calculated approximately using Reiss's correlation [42]:

$$k_{GI}\, a_L = 2.0 + 0.1\, E_G^{0.67} \tag{28}$$

where a_L is the gas-liquid interfacial area (m^2/m^3 column volume), and E_G is an energy dissipation term ($w\ m^{-3}$) for gas flow, defined as

$$E_G = \Delta P_{LG}\, V_G \tag{29}$$

Equation 28 is based on data mostly in the pulsed and spray flow regimes. Data obtained by other investigators [43,44] show that Equation 28 overpredicts the value of $K_{GI}\, a_L$, particularly in the trickle flow regime. Charpentier [4] has suggested that for trickle flow, conservative estimates of $k_{GI}\, a_L$ may be obtained by using mass transfer correlations for countercurrent gas-liquid flow.

The gas-liquid interfacial area, a_L, is usually much larger than the surface area of the packing. The following correlations may be used for determining a_L:
In the trickle flow regime,

$$\frac{a_L}{a_S} = 0.05\left[\frac{\Delta P_{LG}\epsilon}{a_s}\right]^{1.2} \tag{30}$$

when

$$\frac{\Delta P_{LG}\epsilon}{a_s} < 12$$

and in the pulse and spray flow regime,

$$\frac{a_L}{a_s} = 0.25 \left[\frac{\Delta P_{LG} \epsilon}{a_s}\right]^{0.5} \tag{30a}$$

The mass transfer rate from the gas-liquid interface to the liquid, measured by the liquid-phase mass transfer coefficient, $k_{IL}\, a_L$, is affected by the gas and liquid flowrates. The following correlations [1,4,42] are recommended for estimating $k_{IL}\, a_L$:

$$k_{IL}\, a_L = 0.008 \tag{31a}$$

when $E_L < 5$, i.e., for the trickle flow regime with very low gas and liquid flowrates;

$$k_{IL}\, a_L = 0.0011\, E_L \left(\frac{D}{2.4 \times 10^{-9}}\right) \tag{31b}$$

for $5 < E_L < 100$, i.e., for the trickle flow regime; and

$$k_{IL}\, a_L = 0.0173\, E_L{}^{0.5} \left(\frac{D}{2.4 \times 10^{-9}}\right)^{0.5} \tag{31c}$$

for $E_L > 60$–100, i.e., for the pulse or spray flow regimes.

D in Equations 31b and 31c is the gas diffusivity in the liquid under the assumption that the liquid viscosity is close to that of water. Otherwise, a corrected value of D based on $D\mu_L{}^{0.8}$ = constant should be used. E_L in the above correlations is defined as

$$E_L = \Delta P_{LG} \cdot V_L \tag{32}$$

The mass transfer coefficient, k_{LS}, between the liquid and the solid surface is approximately twice as large as in liquid-only flow. The reason for a higher k_{LS} in two-phase flow is the higher liquid velocity within the bed, since a part of the column void volume is occupied by the gas phase.

In trickle flow regime, k_{LS} may be determined from the van Krevelen and Krekels [45] correlation if the solid particles are larger than 3 mm:

$$\frac{\epsilon k_{LS} d_p}{D} = 1.56\, (Re_L)^{0.5}\, (Sc_L)^{0.33} \tag{33a}$$

For a particle diameter smaller than 3 mm, Equation 33a somewhat overestimates the value of k_{LS}. A conservative estimate of k_{LS} may be obtained by using the correlation for liquid-only flow:

$$\frac{\epsilon k_{LS} d_p}{D} = 0.8\, (Re_L)^{0.5}\, (Sc_L)^{0.33} \tag{33b}$$

In the pulse and dispersed bubble flow regimes, Hirose et al. [46] found that k_{LS} is proportional to the reciprocal of the liquid holdup. Using the correlation for liquid holdup, they proposed the following correlation for k_{LS}:

$$\frac{\epsilon k_{LS} d_p}{D} = 20\, a_s{}^{-0.33}\, X^{-0.22}\, (Re_L)^{0.5}\, (Sc_L)^{0.33} \tag{34}$$

The variable X in Equation 34 is the Lockhart-Martinelli parameter defined in Equation 3.

An alternate correlation for k_{LS}, based on the literature survey by Dharwadkar and Sylvester [47], which may be used in all flow regimes, is

$$\frac{k_{LS}d_p}{D} = 1.637 \, (Re_L)^{0.669} \, (Sc_L)^{0.33} \tag{35}$$

The correlations for k_{LS} given by the above equations are based on the assumption that the total surface area of the solids participates in liquid-solid mass transfer. Since only a fraction of the total surface may be wetted, the true mass transfer coefficient is obtained by multiplying k_{LS} from the above equations with a_S/a_W.

Packed beds with cocurrent gas-liquid downflow are used in absorption and in operations involving multiphase chemical reactions. In absorption, only the mass transfer coefficients at the gas-liquid interphase ($k_{GI}\,a_L$ and $k_{IL}\,a_L$) are to be considered. The mass transfer coefficients are generally higher in the cocurrent absorption columns than in the conventional countercurrent absorption columns. However, because the countercurrent columns have a higher average driving force for mass transfer, the use of cocurrent absorption usually is limited to operations requiring less than one theoretical stage [44]. An exception to this occurs in systems in which the driving force for mass transfer is sustained by the depletion of the absorbed species by a chemical reaction. Cocurrent columns with their superior mass transfer coefficients are ideal for such systems.

In systems involving reactions with porous catalyst particles, the intraparticle mass transfer resistance is usually much higher than the mass transfer resistances in the exterior of the catalyst particle. Therefore, the external mass transfer resistances may be ignored in relation to the intraparticle diffusional resistance. If the intraparticle diffusional resistance controls the overall reaction rate, smaller catalyst particles should be considered, provided the smaller particles are acceptable from pressure drop considerations.

The magnitude of the intraparticle diffusional resistance depends on whether the catalyst is completely wet (i.e., the catalyst pores are completely filled with liquid) or partially wet (i.e., some pores are filled with vapors). Vapor-filled pores generally have a lower diffusional resistance. If intraparticle diffusional resistance is the rate-limiting step, the partially wet catalyst may yield a higher reaction rate. For example, Satterfield and Ozel [48], in their study of benzene hydrogenation, found a partially wet catalyst to be more effective than a completely wet catalyst. Liquid-filled pores severely limited the transport of hydrogen into the catalyst and yielded a very slow reaction rate. Dry pores, on the other hand, offered very little resistance to the transport of the reactants. The readily transported hydrogen and benzene vapors reacted on the active catalyst sites, and the reaction progressed at a fast rate.

Dry catalyst can be more effective only when all the reactants are volatile. No reaction can take place on the dry portion of the catalyst if any one of the reactants is nonvolatile. Thus, for nonvolatile reactants, a completely wet catalyst is more effective. In addition to being completely wet, the catalyst should be well irrigated by the nonvolatile reactants. Without good irrigation, depletion of the reactants may take place on the catalyst surface. As mentioned earlier in this chapter, high liquid mass velocities should be employed to obtain good catalyst irrigation and complete wetting.

Heat Transfer

Fixed beds with gas-liquid flow usually are operated adiabatically, and very few data exist on the heat transfer characteristics of these beds. An early paper on heat transfer by Weekman and Myers [49] interpreted the radial heat transfer as due to an overall effective thermal conductivity of the bed (one-parameter model). More recently, Hashimoto et al. [50] and Specchia and Baldi [51] have used a two-parameter model (effective thermal conductivity and a heat transfer resistance in the wall zone) to interpret the radial heat transfer.

Under the assumptions of no temperature difference between the phases at any one point and negligible dispersion of heat in the axial direction, the two-parameter radial heat transfer model is described by the following equations:

$$(LC_{PL} + G\,C_{PG})\frac{\partial T}{\partial z} = k_e\left(\frac{1}{r}\frac{\partial T}{\partial r} + \frac{\partial^2 T}{\partial r^2}\right) \tag{36}$$

$$T = T_i \text{ at } z = 0, r \geqq 0 \tag{36a}$$

$$\frac{\partial T}{\partial r} = 0 \text{ at } z \geqq 0, r = 0 \tag{36b}$$

$$h_w(T_w - T_R) = k_e\frac{\partial T}{\partial r} \text{ at } z \geqq 0, r = R \tag{36c}$$

where C_{PL} and C_{PG} are specific heats of the liquid and the saturated gas, k_e, is the overall effective radial thermal conductivity of the bed with two-phase flow, and h_w is the wall heat transfer coefficient.

The overall effective radial conductivity, k_e, can be considered to be made up of three components:

$$k_e = (k_e)_o + (k_e)_G + (k_e)_L \tag{37}$$

where $(k_e)_o$ is the conductivity of the stagnant bed, i.e., the conductivity of the bed with gas in the void spaces but no flow. $(k_e)_G$ and $(k_e)_L$ are the dynamic contributions due to the radial mixing when the gas and liquid flows are turned on. The following correlations have been proposed based on data for air-water flow through packings of 6-mm glass spheres, 6-mm ceramic rings and 12.3-mm ceramic spheres [51–53]:

$$(k_e)_o = k_G\left[\epsilon + \frac{(1-\epsilon)}{0.67\,\dfrac{k_G}{k_s} + 0.22\,\epsilon^2}\right] \tag{38}$$

$$(k_e)_G = \frac{V_G\rho_G\,C_{PG}\,d_s}{8.65\left[1 + 19.4\left(\dfrac{d_s}{D_{BED}}\right)^2\right]} \tag{39}$$

$$(k_e)_L = \frac{V_L\rho_L\,C_{PL}\,d_p}{0.041\left(\dfrac{V_L\rho_L\,d_p}{(h_D + h_s)\mu_L}\right)^{0.87}} \tag{40a}$$

for the poor interaction regime, and

$$(k_e)_L = \frac{V_L\rho_L\,C_{PL}\,d_p}{338\left(\dfrac{V_L\rho_L\,d_p}{\mu_L}\right)^{0.675}(h_D + h_s)^{0.29}\left(\dfrac{a_s d_p}{\epsilon}\right)^{-2.7}} \tag{40b}$$

for the high interaction regime. $(k_e)_L$ is usually much larger than $(k_e)_o + (k_e)_G$ due to the much higher heat capacity of the liquid.

As for the heat transfer coefficient at the wall, h_w, Specchia and Baldi [51] found it to be much higher than for gas-only flow, and highly dependent on the flow regime. In the poor interaction regime, h_w generally increases with the gas and liquid flowrates and is correlated as

$$\frac{h_w d_s}{k_L} = 0.057\left[\frac{V_L\rho_L d_s}{(h_s + h_D)\mu_L}\right]^{0.89}\left[\frac{C_{PL}\mu_L}{k_L}\right]^{0.33} \tag{41}$$

In the high interaction regime, h_w has a much higher value than in the poor interaction regime. The wall heat transfer is attributed to a liquid film that bathes the wall. Since a complete wall wetting, i.e., saturation of the wall by the liquid film, takes place in the high interaction regime, h_w no longer continues to increase with the flowrates. Specchia and Baldi [51] measured a constant h_w of 2100 w/m^2-°K in the high interaction regime.

NOMENCLATURE

a external area of particle (packing) per unit particle volume (m^{-1})
a_L gas-liquid interfacial area per unit bed volume (m^{-1})
a_s external area of packing per unit bed volume [m^{-1}, $a_s = a(1-\epsilon)$]
a_w wetted packing area per unit bed volume (m^{-1})
C_{PG} specific heat of saturated gas (J/kg-°K)
C_{PL} specific heat of liquid (J/kg-°K)
D diffusivity (m^2/sec)
D_{BED} bed diameter (m)
d_E equivalent particle diameter (m), d_E = 6X particle volume/particle external area
D_H bed hydraulic diameter (m), $D_H = 2\epsilon D_{BED}/[2 + 3(1-\epsilon)(D_{BED}/d_E)]$
d_p nominal particle diameter (m)
d_s diameter of sphere with same external area as the particle (m)
D_{ZG} axial dispersion coefficient for gas (m^2/sec)
D_{ZL} axial dispersion coefficient for liquid (m^2/sec)
E_G energy dissipation term for gas flow defined by Equation 29 (w/m^3)
E_L energy dissipation term for liquid flow defined by Equation 32 (w/m^3)
g acceleration due to gravity (9.8 m/s^2)
G gas mass velocity (kg/m^2-sec), $G = V_G\rho_G$
Ga^* modified Galileo number defined by Equation 18a, dimensionless
h liquid holdup volume per unit bed volume, dimensionless
h_D dynamic liquid holdup volume per unit bed volume, dimensionless
h_{EXT} interparticle liquid holdup volume per unit bed volume, dimensionless
h_{INT} intraparticle liquid holdup volume per unit bed volume, dimensionless
h_s static liquid holdup volume per unit bed volume, dimensionless
h_T total liquid holdup volume per unit bed volume, dimensionless
h_w wall heat transfer coefficient (J/m^2-sec-°K)
k_1, k_2 constants in Equations 5 and 6
k_{GI} mass transfer coefficient for transport from the gas phase to the gas-liquid interface (m/sec)
k_{IL} mass transfer coefficient for transport from the gas-liquid interface to the liquid phase (m/sec)
k_{LS} mass transfer coefficient for transport from the liquid phase to the solid surface (m/sec)
k_e overall effective radial conductivity of the bed (J/m-sec-°K)
$(k_e)_o$ conductivity of stagnant bed (J/m-sec-°K)
$(k_e)_G$ bed conductivity due to gas flow (J/m-sec-°K)
$(k_e)_L$ bed conductivity due to liquid flow (J/m-sec-°K)
k_G gas conductivity (J/m-sec-°K)
k_L liquid conductivity (J/m-sec-°K)
k_s conductivity of packing material (J/m-sec-°K)
L liquid mass velocity (kg/m^2-sec), $L = V_L\rho_L$
Pe_{ZL} Peclet number for liquid dispersion in the axial direction, dimensionless ($Pe_{ZL} = V_L d_p/D_{ZL}h_T$)
Pe_{ZG} Peclet number for gas dispersion in the axial direction, dimensionless {$Pe_{ZG} = V_G d_p/[D_{ZG}(\epsilon - h_D - h_s)]$}
r bed radial position (m)
R bed radius (m)
Re_G gas Reynolds number, dimensionless ($Re_G = V_G\rho_G d_p/\mu_G$)

Re_L liquid Reynolds number, dimensionless ($Re_L = V_L \rho_L d_p / \mu_L$)
Re_G' gas Reynolds number, dimensionless [$Re_G' = Re_G/(1-\epsilon)$]
Re_L' liquid Reynolds number, dimensionless [$Re_L' = Re_L/(1-\epsilon)$]
Sc_L liquid Schmidt number, dimensionless ($Sc_L = \mu_L/\rho_L D$)
T temperature (°K)
T_i temperature at bed inlet (°K)
T_R bed temperature at radius R (°K)
T_w wall temperature (°K)
V_G superficial gas velocity (m/sec)
V_L superficial liquid velocity (m/sec)
X Lockhart-Martinelli parameter defined by Equation 3, dimensionless
X' parameter defined by Equation 7, dimensionless
Y diameter of the area over which the flow is to be distributed (m)
z axial position in the bed (m)
Z' defined by Equation 21, dimensionless

Greek Symbols

β liquid holdup volume per unit volume of bed void space, dimensionless
β_D dynamic liquid holdup volume per unit volume of bed void space, dimensionless
β_S static liquid holdup volume per unit volume of bed void space, dimensionless
β_T total liquid holdup volume per unit volume of bed void space, dimensionless
ϵ bed void fraction, dimensionless
ΔP_G pressure drop per unit length of packed column for single-phase gas flow (N/m³)
ΔP_L pressure drop per unit length of packed column for single-phase liquid flow (N/m³)
ΔP_{LG} pressure drop per unit length of packed column for two-phase gas-liquid flow (N/m³)
ρ_G gas density (kg/m³)
ρ_L liquid density (kg/m³)
ρ_{LG} gas-liquid density (kg/m³)

$$\rho_{LG} = \left[\frac{1}{\rho_L}\frac{L/G}{1+L/G} + \frac{1}{\rho_G}\frac{1}{1+L/G}\right]^{-1}$$

ξ_G defined by Equation 9 (kg/m²-sec)
ξ_L defined by Equation 8 (kg/m²-sec)
ξ_{LG} defined by Equation 11 (kg/m²-sec)
μ_G gas viscosity (kg/m-sec)
μ_L liquid viscosity (kg/m-sec)
μ_{LG} gas-liquid viscosity (kg/m-sec),

$$\mu_{LG} = \left[\mu_L \frac{L/G}{1+L/G} + \mu_G \frac{1}{1+L/G}\right]$$

σ_L liquid surface tension (N/m)
λ defined in Figure 1a, dimensionless
ψ defined by Equation 22, dimensionless

REFERENCES

1. Satterfield, C. N. "Trickle Bed Reactors," *AIChE J.* 21:209 (1975).
2. Hofmann, H. P. "Multiphase Catalytic Packed Bed Reactors," *Catal. Rev. Sci. Eng.* 17:71 (1978).
3. Goto, S., J. Levec and J. M. Smith. "Trickle Bed Oxidation Reactors," *Catal. Rev. Sci. Eng.* 15:187 (1977).
4. Charpentier, J. C. "Recent Progress in Two Phase Gas-Liquid Mass Transfer in Packed Beds," *Chem. Eng. J.* 11:161 (1976).
5. Shah, Y. T. *Gas-Liquid-Solid Reactor Design* (New York: McGraw-Hill Book Co., 1979).
6. Sato, Y., T. Hirose, F. Takahashi, M. Toda and Y. Hashiguchi. "Flow Pattern and Pulsation Properties of Cocurrent Gas-Liquid Downflow in Packed Beds," *J. Chem. Eng. (Japan)* 6:315 (1973).

7. Charpentier, J. C., and M. Favier. "Some Liquid Holdup Experimental Data in Trickle-Bed Reactors for Foaming and Nonfoaming Hydrocarbons, *AIChE J.* 21: 1213 (1975).
8. Weekman, V. W., Jr., and J. E. Myers. "Fluid Flow Characteristics of Cocurrent Gas-Liquid Flow in Packed Beds," *AIChE J.* 10:951 (1964).
9. Chou, T. S., F. L. Worley, Jr. and D. Luss. *Ind. Eng. Chem. Process Des. Dev.* 16: 424 (1977).
10. Talmor, E. "Two Phase Downflow Through Catalyst Beds," *AIChE J.* 23:868 (1977).
11. Gupta, R., and R. M. Koros. "Effect of Packing Imperfections on Pulsed Flow Inception in Fixed Bed Reactors with Gas-Liquid Downflow," *Ind. Eng. Chem. Fund.* 20 (1982).
12. Herskowitz, M., and J. M. Smith. "Liquid Distribution in Trickle Bed Reactors," *AIChE J.* 24:439 (1978).
13. Sylvester, N. D., and P. Pitayagulsarn. "Radial Liquid Distribution in Cocurrent Two Phase Downflow in Packed Beds," *Can. J. Chem. Eng.* 53:599 (1975).
14. Specchia, V., A. Rossini and G. Baldi. "Distribution and Radial Spread of Liquid in Two Phase Concurrent Flows in a Packed Bed," *Ing. Chim. Italy* 10:171 (1974).
15. Ross, L. D. "Performance of Trickle Bed Reactors," *Chem. Eng. Prog.* 61(10):77 (1965).
16. Koros, R. M. "Scale-up Considerations for Mixed Phase Catalytic Reactors," paper presented at the NATO Advanced Study Institute on Multiphase Chemical Reactors, Portugal, 1980.
17. Hoftyzer, P. J. "Liquid Distribution in a Column with Dumped Packing," *Trans. Inst. Chem. Eng.* 42:T109 (1964).
18. Beimesch, W. E., and D. P. Kessler. "Liquid-Gas Distribution Measurements in the Pulsing Regime of Two Phase Cocurrent Flow in Packed Beds," *AIChE J.* 17: 1160 (1971).
19. Larkins, R. P., R. R. White and D. W. Jeffrey. "Two-Phase Cocurrent Flow in Packed Beds," *AIChE J.* 7:231 (1961).
20. Lockhart, R. W., and R. C. Martinelli. "Proposed Correlation of Data for Isothermal Two-Phase, Two Component Flow in Pipes," *Chem. Eng. Prog.* 45:39 (1949).
21. Sato, Y., T. Hirose, F. Takahashi and M. Toda. "Pressure Loss and Liquid Holdup in Packed Bed Reactors with Cocurrent Gas-Liquid Down Flow," *J. Chem. Eng. Japan* 6:147 (1973).
22. Ergun, S. "Fluid Flow Through Packed Columns," *Chem. Eng. Prog.* 48:89 (1952).
23. Specchia, V., and G. Baldi. "Pressure Drop and Liquid Holdup for Two Phase Cocurrent Flow in Packed Beds," *Chem. Eng. Sci.* 32:515 (1977).
24. Charpentier, J. C., C. Prost and P. LeGoff. *Chem. Eng. Sci.* 24:1777 (1969).
25. Midoux, N., M. Favier and J. C. Charpentier. "Flow Pattern, Pressure Loss and Liquid Holdup Data in Gas-Liquid Downflow Packed Beds with Foaming and Nonfoaming Hydrocarbons," *J. Chem. Eng. Japan* 9:350 (1976).
26. Morsi, B. I. Chemical Engineering Thesis, University of Nancy, France (1977).
27. Charpentier, J. C. "Hydrodynamics of Two Phase Flow through Porous Media," in *Chemical Engineering of Gas-Liquid-Solid Catalyst Reactions,* G. A. L'Homme, Ed., Proceedings of an International Symposium held at the University of Liege, Belgium, March, 1978 (1978).
28. Loisier, P. Chemical Engineering Thesis, University of Nancy, France (1975).
29. Lormier, R. Chemical Engineering Thesis, University of Nancy, France (1976).
30. Kan, K. M., and P. F. Greenfield. "Multiple Hydrodynamic States in Cocurrent Two-Phase Downflow Through Packed Beds," *Ind. Eng. Chem. Process Des. Dev.* 17:482 (1978).
31. Morsi, B. I., and J. C. Charpentier. "Hydrodynamics and Gas-Liquid Interfacial Parameters with Organic and Aqueous Liquids in Catalytic and Non-Catalytic Packings in Trickle Bed Reactors," paper presented at the NATO Advanced Study Institute on Mass Transfer with Chemical Reaction in Multiphase Systems, Turkey, 1981.

32. Van Swaaij, W. P. M., J. C. Charpentier and J. Villermaux. "Residence Time Distribution in the Liquid Phase of Trickle Flow in Packed Columns," *Chem. Eng. Sci.* 24:1083 (1969).
33. Mersmann, A. *Chem. Eng. Tech.* 37:672 (1965).
34. Verhoeven, L., and P. van Rompay. *Proc. 7th Eur. Symp. Comp. Appl. Chem. Ind.*, Erlangen (1974), p. 90.
35. Schwartz, J. G., E. Weger and M. P. Dudukovic. "A New Tracer Method for Determination of Liquid-Solid Contacting Efficiency in Trickle-Bed Reactors," *AIChE J.* 22:894 (1976).
36. Onda, K., E. Sada, C. Kido and T. Tanaka. *Kagaku Kogaku* 27:140 (1963).
37. Hofmann, H. "Hydrodynamics, Transport Phenomenon and Mathematical Models in Trickle Bed Reactors," *Int. Chem. Eng.* 17:19 (1977).
38. Koros, R. M. "Catalyst Utilization in Mixed Phase Packed Bed Reactors—Pilot Plant Reactors," *Proc. Fourth Int. Symp. Chem. Reactor Eng.*, Heidelberg, West Germany (1976).
39. Ostergaard, K. "Gas-Liquid-Particle Operations in Chemical Reaction Engineering, *Adv. Chem. Eng.* 7:71 (1968).
40. Hochman, J. M., and E. Effron. "Two Phase Cocurrent Downflow in Packed Beds," *Ind. Eng. Chem. Fund.* 8:63 (1969).
41. Schiesser, W. E., and L. Lapidus. "Further Studies of Fluid Flow and Mass Transfer in Trickle Beds, *AIChE J.* 7:163 (1961).
42. Reiss, L. P. "Cocurrent Gas Liquid Contacting in Packed Columns," *Ind. Eng. Chem. Process Des. Dev.* 6:486 (1967).
43. Shende, B. W., and M. M. Sharma. "Mass Transfer in Packed Columns: Cocurrent Operation," *Chem. Eng. Sci.* 29:1763 (1974).
44. Gianetto, A., V. Specchia and G. Baldi. "Absorption in Packed Towers with Cocurrent Downward High Velocity Flows II. Mass Transfer," *AIChE J.* 19:916 (1973).
45. Van Krevelan, D. W., and J. T. C. Krekels. *Recl. Trav. Chim* 67:512 (1948).
46. Hirose, T., Y. Mori and Y. Sato, "Liquid to Particle Mass Transfer in Fixed Bed Reactor with Cocurrent Gas Liquid Downflow," *J. Chem. Eng. Japan* 9:220 (1976).
47. Dharwadkar, A., and N. D. Sylvester. "Liquid Solid Mass Transfer in Trickle Beds," *AIChE J.* 23:376 (1977).
48. Satterfield, C. N., and F. Ozel. "Direct Solid-Catalyzed Reaction of a Vapor in an Apparently Completely Wetted Trickle Bed Reactor," *AIChE J.* 19:1259 (1973).
49. Weekman, V. W. Jr., and J. E. Myers. "Heat Transfer Characteristics of Concurrent Gas-Liquid Flow in Packed Beds," *AIChE J.* 11:13 (1965).
50. Hashimoto, K., K. Muroyama, K. Fujiyoshi and S. Nagata. "Effective Radial Thermal Conductivity in Cocurrent Flow of a Gas and Liquid Through a Packed Bed," *Int. Chem. Eng.* 16:720 (1976).
51. Specchia, V., and G. Baldi. "Heat Transfer in Trickle Bed Reactors," *Chem. Eng. Commun.* 3:483 (1979).
52. Specchia, V., G. Baldi and S. Sicardi. "Heat Transfer in Packed Bed Reactors with One Phase Flow," *Chem. Eng. Commun.* 4:361 (1980).
53. Kunii, D., and J. M. Smith. "Heat Transfer Characteristics of Porous Rocks," *AIChE J.* 6:71 (1960).

CHAPTER 20
COCURRENT UPFLOW IN FIXED BEDS

Hanns Hofmann
Institut für Technische Chemie I
Universität Erlangen-Nürnberg
Erlangen, Federal Republic of Germany

CONTENTS

INDUSTRIAL APPLICATIONS AND TECHNOLOGICAL ASPECTS

Fixed beds of catalytic (or noncatalytic) solid particles passed by a cocurrent upflow of two fluid phases (usually a gas and a liquid) are used industrially as chemical reactors or absorbers for gas absorption, accompanied by a chemical reaction, e.g., in amination of alcohols; selective hydrogenation of acetylenes, olefins and heteroaromates; ethynylation of formaldehyde; oxidative treatment of waste liquids; and in biotechnology [1]. These so-called packed bubble columns are used as alternatives to trickle beds, which are operated in cocurrent downflow of the phases; neither operation is limited by flooding.

The upflow operation with continuous liquid and dispersed gas phase is preferred when the liquid has to be treated with a small amount of gas or when a large liquid residence time is required, i.e., if the overall reaction rate is low. At low liquid flowrates, the contacting efficiency of the solid, as well as the gas-liquid mass transfer and, thus, the apparent reaction rate, is higher in upflow, than in downflow, operation. Better sweeping of the catalyst and good heat release from the particles sometimes result in less catalyst decay. Further, in packed bubble columns the liquid residence time can be varied easily, and the liquid holdup ($\epsilon_L \leqq 0.4$) and the liquid to solid ratio are rather high. Therefore, the temperature control is improved because of the large heat capacity of the reaction mixture. On the other hand, liquid-phase mixing and pressure drop are larger than in downflow operation, fluidization may be a problem, and the catalyst must be kept in place by suitable mechanical methods to avoid fluidization. Good phase distribution over the whole entrance cross section is a prerequisite for a satisfying operation of industrial packed bubble columns.

Compared to empty bubble columns (with suspended solid particles), gas back-mixing and bubble coalescence are reduced substantially in packed bubble columns, whereas the gas-liquid interface based on void volume can be improved 15–100%. However, based on total column volume, the improvement is less because a substantial part ($\approx$50%) of the column is occupied by solid. Therefore, it is sometimes desirable to use packings with high voidage, like screen packings.

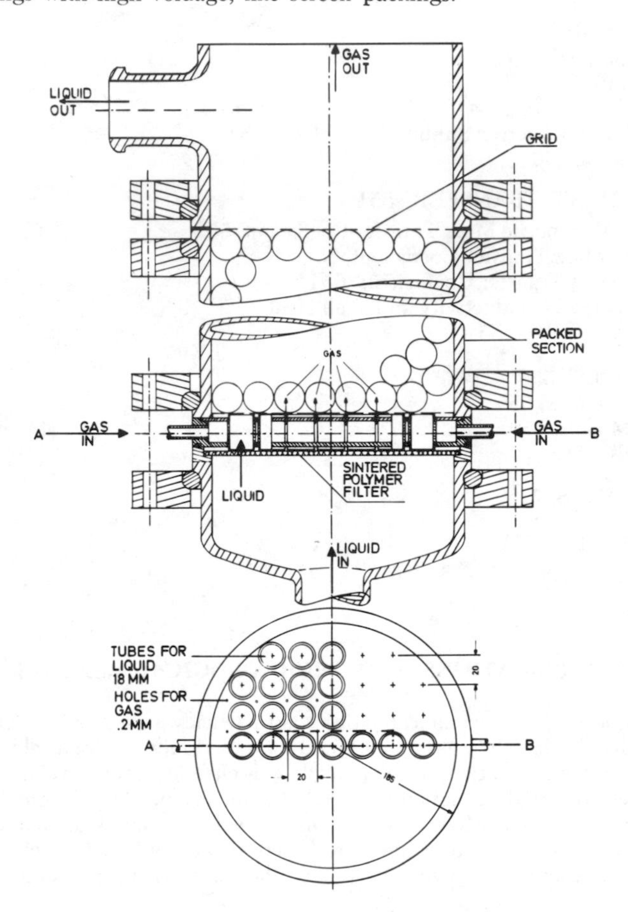

Figure 1. Sketch of a packed bubble column.

Figure 1 shows a sketch of a typical packed bubble column, with its gas and liquid distributors at the bottom, the supporting grid below and the fixing grid at the top of the packing, as well as the gas-liquid disengaging zone above the packing [2]. The usual aspect ratio, H/D, of technical columns is in the range of 5 to 40.

Although some monographs on gas-liquid-solid reactors have been published [3–6] and several symposia on this special subject have been organized [1,7–9], an a priori design has not yet been directly possible until now. Much information is still missing. This chapter summarizes the available information and indicates gaps to be bridged by future research.

FLUID DYNAMICS

Flow regimes, pressure drop, holdup of and dispersion in the fluid phases, bubble size distribution and bubble rise velocity, as well as gas-liquid interfacial area, are important design parameters of packed bubble columns that depend on fluid dynamics. As it is impossible to solve the momentum balance equations for a multiphase system, a rigorous description of fluid dynamics based on first principles is impossible. Nevertheless, much information about the dependence of the abovementioned parameters on the operating variables has been collected, and attempts have been made to correlate the data according to plausible principles.

Flow Regimes

Whereas in single-phase packed beds only two flow regimes exist (laminar and turbulent flow), characterized by a single dimensionless number, the Reynolds number ($Re_{pc} \approx 150$), the fluid dynamics in packed bubble columns is much more complex. Depending on flowrates of the phases, packing geometry and physical properties of the phases, four different flow regimes are observed:

1. Bubble flow. At low gas and any liquid flowrates, it is characterized by insulated gas bubbles, rising in a continuous liquid phase.
2. Distorted bubble flow (surging or shell flow). At medium gas and small liquid flowrates, it is characterized by bubble coalescence and larger bubbles surrounding several particles.
3. Pulse Flow (slug flow). At medium to high gas and liquid flowrates, it is characterized by alternate portions of more and less dense phases passing through the bed.
4. Spray flow (churn flow). At high gas and low liquid flowrates, it is characterized by a continuous gas phase with insulated liquid drops and liquid films over the packing.

As pressure drop, transport phenomena and overall reaction rate depend on flow mode, so-called flowcharts have been constructed to indicate how flow regimes depend on the operating variables [3,7,10–14]. Figure 2 shows such a flowchart according to Shah [3]. It should be noted that most of these diagrams, except those of Ogarkov et al. [14], in which the gas holdup and gas Froude number are used as coordinates, show only the dependence of the flow regimes on gas and liquid throughput ($\dot{G}$ and $\dot{L}$, respectively) and are mainly based on the air-water system. Information about systems

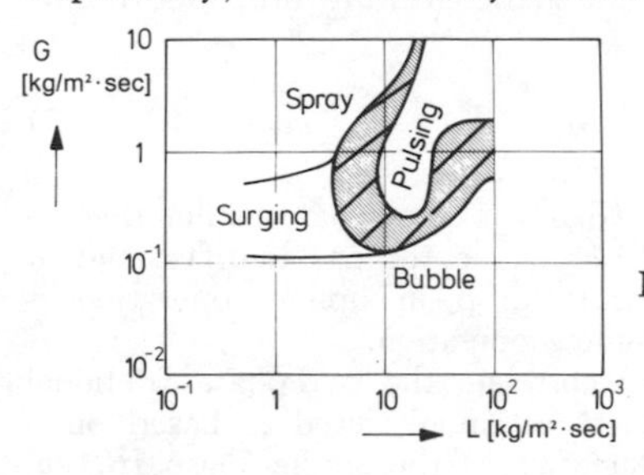

Figure 2. Flowchart for packed bubble columns.

having different physical properties is not available. It is obvious that the physical properties of various systems have to be included in a universal flowchart, e.g., by the dimensionless numbers

$$\lambda = \left[\frac{\rho_G}{\rho_{air}} \cdot \frac{\rho_L}{\rho_{H_2O}}\right]^{1/2} \text{ and } \psi = \frac{\sigma_{H_2O}}{\sigma_L}\left[\frac{\mu_L}{\mu_{H_2O}}\left(\frac{\rho_{H_2O}}{\rho_L}\right)\right]^{1/2} \quad (1a+b)$$

as has been done already in flowcharts for downflow operation. Furthermore, a dimensionless number characterizing the packing geometry, such as

$$\Delta = \frac{a_s\, d_p}{\epsilon} \text{ and/or } \Gamma = d_t/d_p \quad (2a+b)$$

seems to be missing in the flowcharts presented up to now. Recently, Fukushima [15] has tried to describe empirically the flow regimes as dependent on Re_G, Re_L and, also, Γ.

This situation may be an indication that we are still far from a full understanding of the fluid dynamics in multiphase systems. Flowcharts should be based on fluid dynamic considerations of the interaction between the flowing phases, such as in the work of Kirillov et al. 16]. They showed that the pure bubble flow regime can exist only if the spacing between packing elements is greater than the bubble size, leading to a critical radius (similar to an hydraulic radius, r_h),

$$r_c \geqq \left[\frac{\sigma_L}{g(\rho_L - \rho_G)}\right]^{0.5} \quad (3)$$

as boundary of this domain. They also showed that pulsing cannot occur if

$$\frac{1-\beta_G}{\epsilon \cdot \beta_G} \leqq \frac{3\pi\sigma_L}{\rho_L\, u_p^2\, d_t} \quad (4)$$

where u_p is the traveling velocity of a pulse. But more work of this type is needed, even when most of the commercial units are operated only in the bubble or distorted bubble flow regime.

Pressure Drop

Pressure drop is a measure of the energy dissipation in the packed bed and, thus, an important design variable. Figure 3 shows (as an example) how the total pressure drop in packed bubble columns depends on liquid and gas flowrate in a packing of 10 × 20 mm porcelain cylinders [2]. The total two-phase pressure drop is the sum of the frictional pressure drop of the flowing phases and the static pressure drop due to the liquid phase:

$$(\delta_{LG})_t = (\delta_{LG})_f + g \cdot \rho_L \quad (5)$$

In a bubble flow regime at low liquid flowrates, $(\delta_{LG})_f$ is somewhat smaller than in a trickle operation [11] due to the buoyancy force acting on the gas bubbles. But at high flowrates of the phases, the frictional pressure drop in upflow operation is dominant and becomes very similar to that in downflow operation.

There are two different empirical approaches to correlate the two-phase frictional pressure drop and the operating variables. The most commonly used is based on a two-phase friction factor [10], defined in a similar way as the single-phase friction factor:

$$f_{LG} = \frac{(\delta_{LG})_f \, d_{pe}}{2 \, \rho_G \, u_{OG}^2} \tag{6}$$

where d_{pe} is an equivalent particle diameter defined as

$$d_{pe} = 4 \, r_h = \frac{\epsilon}{1-\epsilon} \cdot \frac{V_p}{S_p} = \frac{2}{3} \, d_p \frac{\epsilon}{1-\epsilon} \tag{7}$$

Based on experimental data, this two-phase friction factor has been correlated as

$$\ln f_{LG} = 8.0 - 1.12(\ln Z) - 0.0769(\ln Z)^2 + 0.0152(\ln Z)^3 \qquad 0.3 \leq Z \leq 500 \tag{8}$$

with the two-phase Z factor,

$$Z = (Re_G^{1.167}/Re_L^{0.767}) \, (\mu_{H_2O} / \mu_L)^{0.90} \tag{9}$$

A slightly different correlation for the two-phase friction factor has been used by Larkins [17]:

$$(\delta_{LG})_f = \left(f_L + \frac{1.75}{Re_G^{0.3}} \cdot \frac{u_{OG}}{u_{OL}} \right) \frac{\rho_L \, u_{OL}^2}{2} \frac{a_s}{\epsilon} \tag{10}$$

where f_L is the single-phase friction factor for the liquid,

$$f_L = \frac{40}{Re_L} + \frac{2.26}{Re_L^{0.5}} + 0.46 \tag{11}$$

Both correlations can be recommended for low interaction of gas and liquid phases (bubble and distorted bubble flow).

For the high-interaction regimes (pulse and spray flow), the second approach introduced by Larkins [17] seems to be more appropriate. Here, the total two-phase pressure drop is regarded as enhanced (compared to the sum of the frictional pressure drop of the pure phases) by the factor $(\delta_{LG})_t/(\delta_L + \delta_G)$. This enhancement factor

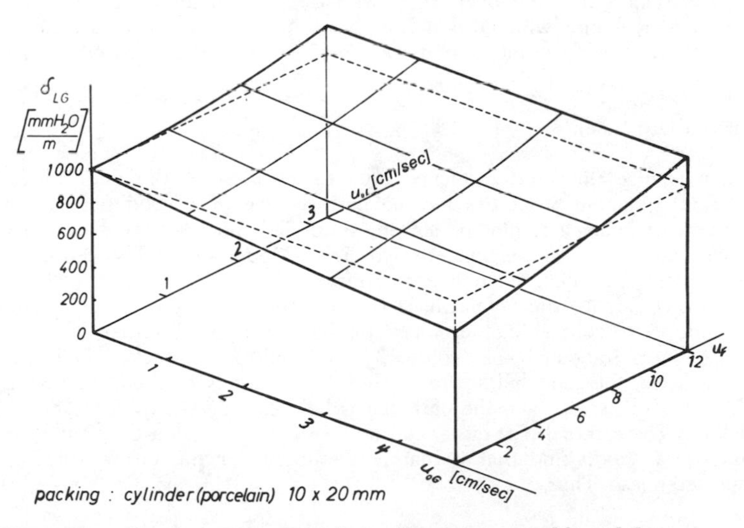

Figure 3. Total pressure drop dependence on gas and liquid flowrate.

has been correlated with the Lockhart-Martinelli parameter, $\chi = (\delta_L/\delta_G)^{0.5}$ (taken for the single-phase flow of the phases at the same flowrates and in the same packing) as

$$\log \frac{(\delta_{LG})_t}{\delta_L + \delta_G} = \frac{0.416}{0.666 + (\log \chi)^2} \qquad 0.05 \leq \chi \leq 30 \tag{12}$$

Colquhoun-Lee and Stepanek [18] recently have correlated the frictional pressure drop in the moderate-to-high flowrate region more fundamentally with the total rate of energy dissipation in the packing, E_L'. They assume that E_L' can be separated into two components, namely the rate of energy dissipation due to the interaction of the liquid with the solid packing, and the energy transferred to the liquid from the gas by virtue of the shear between both phases. Assuming that the first term is just the rate of energy dissipation corresponding to single-phase liquid flow through the bed and that the second component may be expressed in terms of the ratio of the slip velocity to some characteristic length, it follows that

$$(\delta_{LG})_f = \frac{\epsilon}{u_{OL} + u_{OG}} \left\{ E_L' + B \cdot \beta_{Gt} \left[\frac{Q_L}{1 - \beta_{Gt}}\right]^2 \left[1 - \frac{u_{OL}}{u_{OG}}\right]^2 \right\} \times \left\{1 + \beta_{Gt} \left[\frac{\rho_G}{\rho_L}\left(\frac{u_{OG}}{u_{OL}}\right)^3 - 1\right]\right\} \tag{13}$$

where $E_L' = [u_{OL}/(1 - \beta_{Gt})]$. δ_L is the rate of energy dissipation per unit free volume for a single-phase liquid flowing through the bed at the same velocity as in two-phase flow, and B is an empirical factor, being equal to $6.09 \cdot 10^{12}$ and $4.07 \cdot 10^{12}$ for cylinders and spheres, respectively. Equations 8 and 13 have been compared by Ramachandran and Chaudhari [4] and found to give similar results. But to use Equation 13, β_{Gt}, as well as δ_L, have to be known, which might be regarded as a certain disadvantage.

It should be noted that investigations of packings with small particles [19,20], e.g., $d_p \leq 3$ mm, have shown a different behavior of the packing due to a stronger influence of capillary forces; therefore, more work is required using smaller particles.

Furthermore, it is important to know that foaming liquids, viscous liquids and dispersions with a coalescing ability different from the air-water system may lead to other correlations [21]. Moreover, the correlations given above are based mainly on small-diameter columns with good initial-phase distribution, whereas commercial units may exhibit less gas-liquid interaction and, thus, smaller pressure drop [22,23].

Holdup and Dispersion

Mean residence time and residence time distribution of the fluid phases are important fluid dynamic properties for the performance of packed bubble columns. As the momentum balance cannot be solved, simplified models have been developed for multiphase flow (see the section entitled Flow Models) defining certain parameters that characterize both fluid dynamic properties.

The mean residence time of a phase is dependent on one of these parameters, the phase holdup. In general, holdup of a phase is expressed either as fractional bed volume, ϵ, or as fractional void volume, β. In multiphase packed bed reactors with porous particles, the total holdup of a phase, ϵ_t, consists of the part that is held internally in the pores, ϵ_{int}, and the part that is held externally in the interstices between the packing. The external part either can be free flowing, called dynamic holdup, ϵ_d, or stagnant, ϵ_s, a subdivision that is mainly important for the mean residence time of the dispersed phase. Thus,

$$\epsilon_{Gt} = (\epsilon_{Gd} + \epsilon_{Gs}) + \epsilon_{Gint} \tag{14a}$$

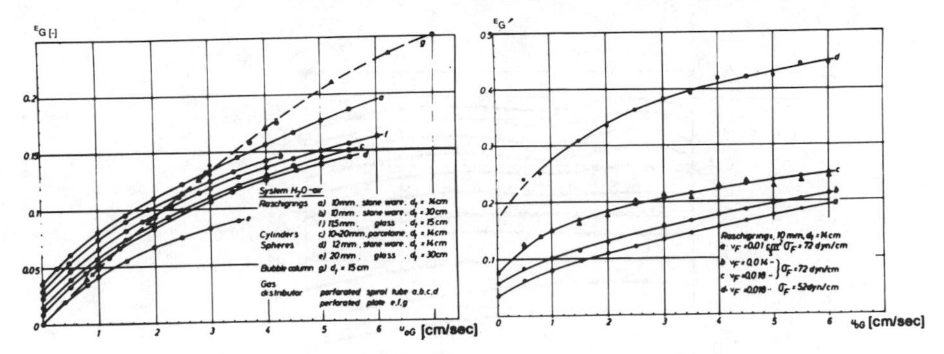

Figure 4. Total average gas holdup in packed bubble columns for $u_{OL} = 0$.

where

$$\epsilon_{Gt} + \epsilon_{Ft} = \epsilon = 1 - \epsilon_s \tag{14b}$$

and

$$\beta_{Gt} = \beta_{Gd} + \beta_{Gs} \tag{14c}$$

where

$$\beta_{Gt} \cdot \epsilon = \epsilon_{G\,ext} \tag{14d}$$

As the amount of the stagnant holdup (which is ineffective in mass transfer [24] and, thus, does not contribute to the conversion) depends on the degree of turbulence (i.e., on the flowrate of the phases), the only meaningful method for its determination is via tracer flow experiments, using appropriate mathematical models (PE or PDE model). These models make allowance for stagnant zones (outside, and probably also inside, of the particles). A residual (instead of a stagnant) and a free-draining (instead of a dynamic) holdup—used by Blass and Kurtz [25]—makes little sense in the performance of packed bubble columns, as they belong to zero flowrates only.

First, data and correlations on the average total gas holdup in packed bubble columns, which can be determined naturally without any model, can be found in the work of Weber [26]. Typical values for $u_{OL} = 0$ are given in Figure 4. ϵ_{Gt} appears to be proportional to $u_{OG}^{0.5}$ [27]. At low gas load, ϵ_{Gt} is higher than in empty bubble columns and it increases with higher liquid viscosity, as well as with lower surface tension, but seems to be independent of the type of gas distributor [2]. The cocurrent liquid velocity should raise the bubble velocity and, hence, decrease the total gas holdup. However, at the same time the increasing turbulence reduces the bubble size and, therefore, the bubble rising velocity, leading finally to a nearly constant total gas holdup for $u_{OL} \leq 1.5$ cm/s (Figure 5).

Based on some theoretical analysis, the average total gas holdup for $u_{OL} = 0$ has been correlated by Zighanshin and Ermakova [28] as

$$\bar{\epsilon}_{Gt} = \frac{1}{1 + 1.3\,(Fr_G)^{0.5}} \tag{15}$$

where $Fr_G = u_{OG}^2/g \cdot dp$ is the Froude number based on gas superficial velocity. Later on, to include physical properties of the system also, this correlation has been extended [29] to give

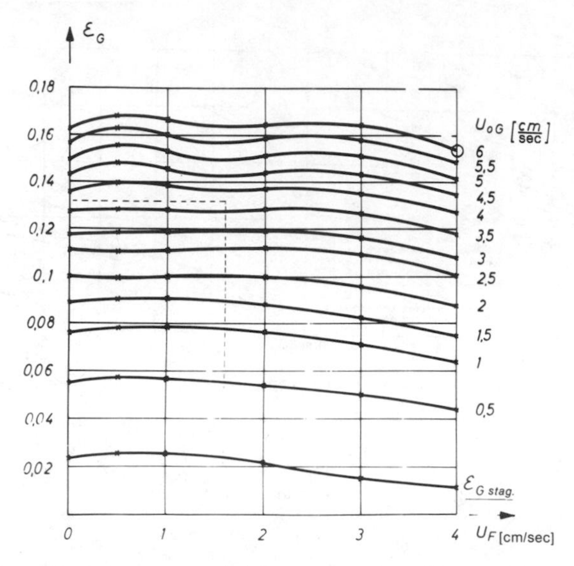

Figure 5. Total average gas holdup as function of gas and liquid velocity.

$$\bar{\epsilon}_{Gt} = \frac{1}{1 + 0.45\ Fr_G{}^{0.66}\ Re_G{}^{0.1}\ We_G{}^{-0.15}} \tag{16}$$

Other scientists [27] also have included a packing number in their correlation:

$$\bar{\epsilon}_{Lt} = 1.47\ Re_L{}^{0.11} \cdot Re_G{}^{-0.14}\ (a_s\ d_p)^{-0.41} \tag{17}$$

or used the Lockhart-Martinelli parameter [12]:

$$\bar{\beta}_{Lt} = 0.6\ a_s{}^{0.33}\chi^{0.16} \tag{18}$$

A theoretical model also was proposed for the average liquid holdup in cocurrent upflow through a packed bed [30]. Assuming that $\epsilon_{Lt} = f(u_{OL}, (\Delta P/\Delta L)_{LG})$, it has been shown that

$$\bar{\epsilon}_{Lt} = u_{OL}\left(\frac{S'}{d_p{}'^{0.5}}\right)\left[\left(\delta_{LG}\right)_t \cdot \frac{1}{\rho_L} - g\right]^{0,5} \tag{19}$$

where $d_p{}'$ is the characteristic packing length, and S′ is a shape factor for the packing. As $(\delta_{LG})_t$ depends on u_{OG}, Equation 19 predicts ϵ_{Lt} to depend on u_{OL}, as well as u_{OG}. But the agreement between theory and experiment is only fair [3]. It has been found [27] that ϵ_{Lt} is higher in a rectangular column than in a circular column of equal cross section.

As a matter of fact, the theoretical understanding of the phenomenon is still rather inadequate, possibly because in reality, a radial [31] as well as an axial (Figure 6) [2,32], holdup profile exists in the bed, whereas up to now only average values for the whole column have been correlated. In this connection, it is interesting to observe [2] that the change of ϵ_{Gt} in large columns due to gas-phase expansion (according to the pressure gradient) leads only to a higher number of bubbles, whereas the bubble diameter remains constant and in the range of the particle diameter (or the capillary constant as defined by Equation 3).

There is only one publication in the literature [24] on the stagnant gas holdup, ϵ_{Gs}. In this study, ϵ_{Gs} was found to be almost independent of u_{OG}, but decreasing with u_L,

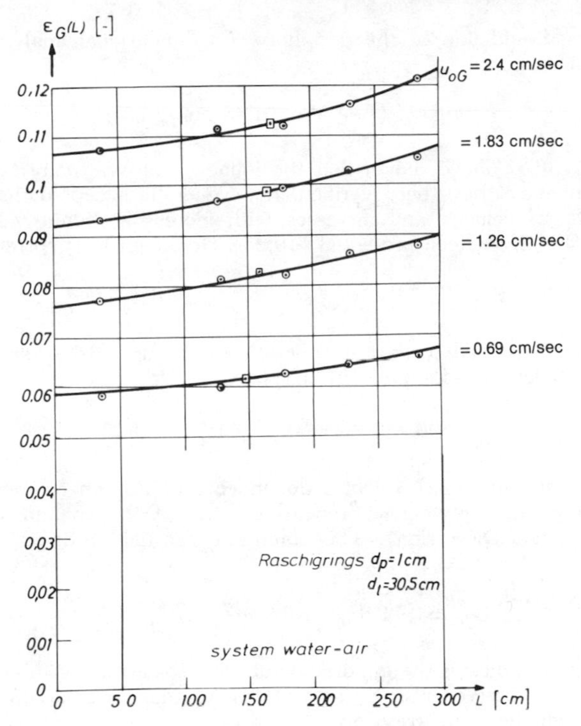

Figure 6. Local gas holdup in packed beds.

approaching zero at the minimum fluidization velocity, u_{Lmf}. Regarding the dependence of ϵ_{Gs} on the physical properties of the system, there are some indications that the Eötves number, $Eö = \rho_L g d_p^2/\sigma_L$, or Bond number, $Bn = \rho_L g/\sigma_L a_s^2$, may be an appropriate correlation parameter [25,33]. But more data are needed to establish a universal correlation.

Dispersion is also of some influence on packed bubble column performance, mainly in large-diameter units and at a high degree of conversion. Therefore, many efforts were made to correlate dispersion, (characterized by an axial particle Bodenstein (or Peclet) number, $Bo = (u d_p/D_a)$ with operating variables. To define Bo, the plug-flow dispersion (PD) model (see the section entitled Flow Models) has been used almost exclusively for the analysis of experimental data.

For gas-phase dispersion, the deterministic PD model seems quite inadequate because of the probabilistic character of the bubble movement. The scattering of data, depicted in Figure 7 [34] (the only one found in literature), may support this statement. Nevertheless, an attempt has been made [34] to correlate the available data as

$$\left(\frac{D_a}{\nu}\right) = Sc_G' = 6.02 \cdot 10^2 \, Re_G^{0.55} \, Re_L^{0.15} \tag{20}$$

But, in general, it is believed that gas-phase dispersion can be neglected in packed bubble columns.

More important is the liquid-phase dispersion, created by the reverse flow of liquid that draws in gas bubbles (small-scale circulation) and by the nonuniform distribution of bubble phase across the bed section (large-scale circulation). It has been shown experimentally [35] that as with empty bubble columns (at least in packed columns of larger diameter), a rising flow of liquid still exists in the center, counterbalanced by a descending liquid flow at the periphery. By dimensional analysis [19] it might be expected that

$$Bo_L = f\,(Re_L, We_L, d_t/d_p, u_{OL}/u_{OG}) \tag{21}$$

Several studies [36,37] have shown that the influence of We_L may be neglected, but the other dependencies have been verified. It is generally accepted that Bo_L decreases with increasing gas velocity and increases with increasing liquid velocity [7]. First correlations for Bo_L have been presented [19,26]. Heilmann [38] correlated his data as

$$Bo_L = 2.13 \cdot 10^{-3}\,\epsilon_{Lt}\,[Re_L/\epsilon_{Gt} \cdot d_p^{3.3}]^{0.736} \tag{22}$$

a correlation that was used successfully for other data also [3], whereas dispersion in beds of small particles (3.8–4.4 mm) [27] has been correlated as

$$Bo_L = 0.128\,Re_L^{0.245}\,Re_G^{-0.16}\,(a_s\,d_p)^{0.53} \tag{23}$$

There are some indications [27,34] of a dependence of Bo_L on column height, as one might expect from the growing ϵ_{Gt} toward the top of the column (see the section entitled Holdup and Dispersion). The column to particle ratio is included in the literature [37].

$$Bo_L = 0.11\,Re_L^{0.45}\,Re_G^{0.47}\,(d_t/d_p)^{-0,31} \tag{24}$$

The last correlation indicates the importance of large-scale mixing in columns having a larger diameter, as observed also in the work of Magnussen et al. [39]. But it might be questionable whether the correlation

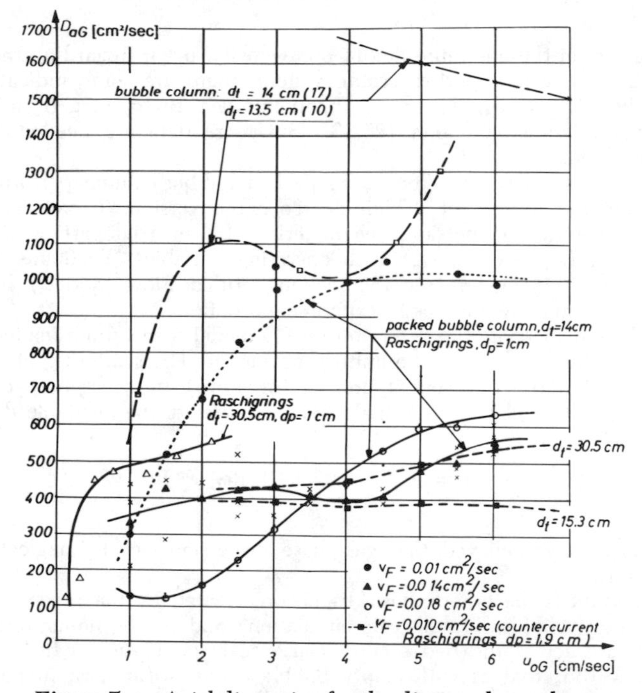

Figure 7. Axial dispersion for the dispersed gas phase.

$$Bo_L = 0.1\, Re_L^{0.2}\, Fr_G^{0.1} + 2.5 \cdot 10^{-4}\, (Re_L^2/Ga_G^{0.4}) \left[1 - 14\frac{\nu_L}{\nu_G} \cdot \frac{u_{OG}}{u_{OL}} \cdot \frac{(1-\beta_G)}{\beta_G} Re_L^{-0,5}\right]^2 \quad (25)$$

given by Kirillov et al. [29] is of real practical value because of its complex structure. $Ga_G = g\, d_p^3/\nu_G^2$ is the Galileo number. Equation 25 exhibits a minimum in the $Bo_L = f(Re_L)$ relationship (e.g., at $Re_L \approx 200$ for $Re_G = 150$ and spheres), which is not observed in other studies.

Bo_L values for the pulsing regime are given by Ohsasa et al. [40], whereas Bo_L values for cocurrent operation are compared to those for countercurrent operation in the work of Heilmann [2] and Takeuchi et al. [41].

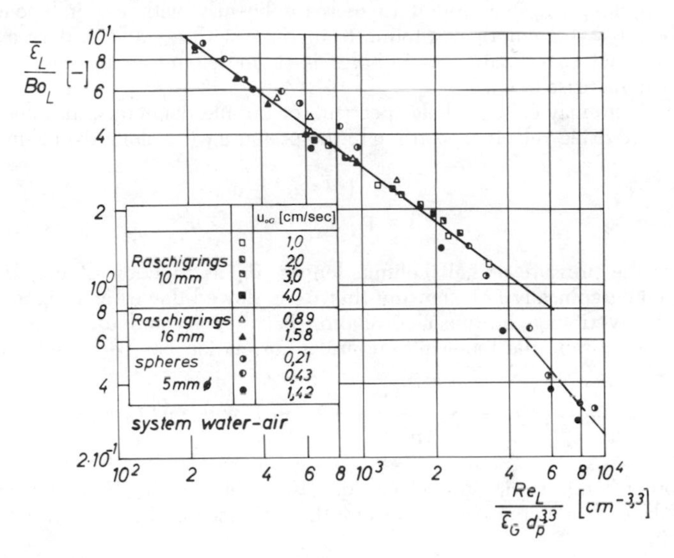

Figure 8. Axial dispersion for the continuous liquid phase.

It might be indicative for the insufficient and to some extent controversial situation that in a recent publication on liquid-phase backmixing [36] the authors have omitted any theory and correlated dispersion data, just as

$$Bo_L = a \cdot u_{OL}^{\,b} \cdot 10^{c \cdot u_{OG}} \quad (26)$$

where a, b and c are empirical parameters, depending on the type of packing and the physical properties of the phases.

It should be noted also that some values have been published [42] of the radial dispersion coefficient, D_r, in cocurrent upflow through packed beds, showing a maximum value in the transient region from bubble to pulse flow (not observed in downflow operation). The order of magnitude of the dispersion coefficient is 0.1–0.4 m^2/h for 6-mm ceramic balls ($0 < \dot{G} < 2500$ kg/m²-h; $11{,}800 < \dot{L} < 94{,}000$ kg/m²-h). No correlation of the data is available.

Bubble Size and Effective Bubble Rise Velocity

A better understanding of fluid dynamics in packed bubble columns requires knowledge about gas bubble size and effective rise velocity of the bubbles, particularly with their dependence on packing type, physical properties of the system and phase loads.

Oshima et al. [24] calculated an average bubble diameter, d_b, according to

$$d_b = 6\,\epsilon_{Gd}/a \tag{27}$$

as a hypothetical value, assuming bubbles to be spheres: a is the specific gas-liquid interface based on unit volume of the reactor. At a particle diameter of $d_p = 0.43$ cm, the bubble diameter has been found to be constant, $d_b \approx 0.4$, almost independent of u_{OL} and ϵ_{Gd}, in accordance with the work of Heilmann [2]. As a consequence, a linear relation between a and ϵ_{Gd} should exist. When $d_p = 0.28$ cm, d_b has a constant value of approximately 0.2 cm, almost independent of ϵ_{Gd}, as long as $u_{OL} < u_{OLmf}$. When $u_{OL} > u_{OLmf}$, d_b was found to increase linearly with ϵ_{Gd} independent of u_L. Finally, when $d_p = 0.1$ cm, d_b was found to increase with ϵ_{Gd} and to decrease with u_{OL}. This suggests that consolidation of bubbles is promoted by an increase in ϵ_{Gd} and is retarded by an increase in u_{OL}.

In reality, there may exist a whole spectrum of bubble diameters, and one has to deal with an average rising velocity according to the pressure p(z) along the column:

$$\bar{u}_G = \frac{1}{1 + p^*/p(z)} \cdot \frac{u_{OG}}{\epsilon_G\,(z)} \tag{28}$$

where p^* is the pressure at half column length. $\bar{u}_G$ as function of u_{OG} has been determined photographically [2], showing that $\bar{u}_G$ is slowed down by a higher viscosity of the liquid, but verifying the linear correlation between $\bar{u}_G$ and u_{OG} given in Equation 28. By data regression, the following empirical correlation has been derived:

$$\bar{u}_G = \left(24.5 - 8\,\frac{\nu_L}{\nu_{H_2O}}\right) + (2.2 + 1.32\,d_p)\,u_{OG}\ [\text{cm/s}] \tag{29}$$

On the other hand, floating-up velocity of gas bubbles in steady-state motion through a granular layer is determined [14] from the balance of drag, deformation and lift forces, leading to

$$\bar{u}_G = \frac{1}{9}\,\frac{g\,r_c^2}{\nu_L}\,\frac{(1-\epsilon)^2}{(1-\epsilon^{5/3})}\left(1 - \frac{3}{2}C \cdot \text{Eö}\right) \tag{30}$$

Here, C is a constant, reflecting the structure of the packing, which must be determined experimentally ($C \approx 0.64$ for glass spheres).

Interfacial Area

The effective interfacial area between gas and liquid—a, when based on total column volume, or a′, when based on dispersion volume—as measured by chemical methods [6], is a very important design variable when mass transfer between both phases is controlling. In packed bubble columns, a ranges from 0.5 to 8.0 cm^2/cm^3 [6,43], varying substantially for different packings. Until now, not enough data have been available to develop a universal correlation for the dependence of effective interfacial area on all variables. Also it is judicious to consider that a′ in bubble flow regime (and with $d_p \geqq 3$ mm) will vary with about the 0.5th power of the superficial gas velocity [44], regardless of packing size and type, column to packing diameter ratio, and liquid velocity or type of gas distributor. This seems understandable in view of the $\epsilon_{Gt} - u_{OL}^{0.5}$ relationship and the constant (close to particle dimension) bubble diameter observed by Oshima et al. [24]. a′ also increases with liquid flowrate and is always a few times higher than the geometrical packing area, a_s, except at very low gas flowrates and very small packing diameter. The upflow values are higher than downflow values for the same liquid and gas flowrates [11,15], but the difference (u_{OG} being equal) diminishes with the increase in liquid flowrate, perhaps because at low liquid loads the gas-liquid interaction in upflow is more intensive than in downflow. There exist empirical correlations for various flow regimes [15] of the following general type:

$$\frac{a \cdot d_p}{(1 - \beta_{Lt})} = \alpha \, Re_L{}^m \cdot Re_G{}^n \, (d_t/d_p)^{-o} \tag{31}$$

where the constants, α, m, n and o depend on the flow regime. A comparison with data from other authors [11,45] supports $\alpha = 2.9 \cdot 10^{-4}$, m = 0.66, n = 0.2 and o = 2.5 for the pulsing flow within ±20%. In the work of Kirillov and Orgakov [46] data on interfacial area from different authors are reported for various systems; these data could be correlated for $Re_G < 500$ (for $Re_L < 200$) by

$$Eö' = \frac{\sigma a^2}{\rho_L g} = 2.66 \cdot 10^{-3} \left(\frac{S_p{}^{0.5}}{V_p{}^{0.33}}\right)^{1.8} Re_L{}^{0.36} \, Re_G{}^{0.45} \tag{32a}$$

and, if $200 < Re_L < 1000$, by

$$Eö' = 2.34 \cdot 10^{-5} \left(\frac{S_p{}^{0.5}}{V_p{}^{0.33}}\right)^{1.8} Re_L{}^{1.21} \, Re_G{}^{0.45} \tag{32b}$$

where S_p and V_p are particle surface and particle volume, respectively. Equations 32a and 32b show that a increases if surface tension decreases, as was found by Mashelkar and Sharma [47].

In another approach, based on Kolmogoroff's theory on isotropic turbulence [48–50], the energy dissipation density is used for data correlation in regimes of high-phase interaction. Specchia et al. [11] used $(-\Delta P/\Delta L) \, \epsilon/a_s$ as a measure for the energy dissipation density, and experimental data could be represented with a relative mean quadratic error of 6.2% by

$$\frac{a}{a_s} = 0.29 \, [(-\Delta P/\Delta L)\epsilon/a_s]^{1.17} + 0.61 \tag{33}$$

All correlations given above concern an average effective gas-liquid interfacial area. But as with ϵ_{Gt}, a must vary with column length. Furthermore, immediately above the gas distributor different bubble diameters may exist, depending on the distributor orifices [51]. Finally, it is well known from empty bubble columns that the coalescence behavior of the gas-liquid system is influenced strongly by electrolytes and organic compounds in the liquid. Therefore, even though this does not seem to be significant in packed bubble columns of small particles [52], all correlations given in this chapter should be used carefully if physical properties of the system deviate drastically from the air-water system.

MASS AND HEAT TRANSFER

In many cases, mass and heat transfer processes are decisive for the performance of packed bubble columns. The different transport steps encountered in the most complex type of reactions are depicted in Figure 9. It is commonly agreed that in packed bubble columns gas film resistance at the gas-liquid interface may be neglected in practical cases. This is certainly not correct for an instantaneous, irreversible reaction of a gaseous reactant in the liquid, where $Ha > 10 \, E_i$ (with E_i the enhancement factor for instantaneous reaction), but it is unlikely that such a reaction will be performed in a packed bubble column.

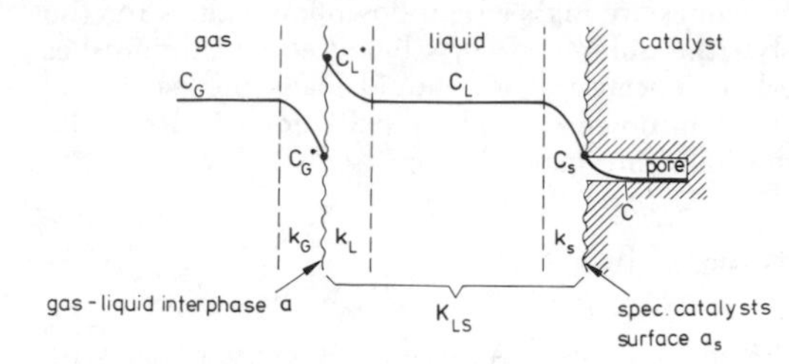

Figure 9. Transport steps in multiphase catalytic reactions.

Liquid-Side Volumetric Mass Transfer Coefficient

Only a few measurements concerning liquid-side volumetric mass transfer coefficient in packed bubble columns have been published [11,20,41,47,52] showing that k_La varies between $0.5 \cdot 10^{-2}$ and $12 \cdot 10^{-2}\ s^{-1}$ [6]. In bubble flow, k_La increases proportionally to $u_{OG}^{0.5}$ (just as effective interfacial area does), suggesting a constant liquid-side mass transfer coefficient, k_L, between $1 - 8 \cdot 10^{-2}$ cm/s for any gas-liquid system [52]. Thus, in this regime energy dissipation mainly creates a new gas-liquid interface, but cannot change microturbulence at the boundary layer. As long as $u_{OL} < 1.5$ cm/s, the volumetric mass transfer coefficient is also independent of u_{OL} (because $\epsilon_{Gt} =$ const.) and, therefore, a constant value of $0.1\ s^{-1}$ for k_La seems to be a conservative assumption for bubble flow. In pulse flow and spray flow regimes, upflow values of k_La are greater by 100% than downflow values because gravitational forces lead to a higher liquid holdup and pressure drop. At very high gas and liquid flowrates, k_La may exceed $1\ s^{-1}$. Both k_L and a increase with u_{OL} and u_{OG} and correlate with the pressure drop [11], as well as with ϵ_{Gd} [24].

An empirical equation for k_La as function of gas and liquid load for various packings is given by Shah [3]. The data also could have been reasonably correlated with Reynolds numbers, as was done by Kirillov and Orgakov [46]:

$$\frac{k_L}{D_L}\left(\frac{\sigma}{\rho_L g}\right)^{0.5} = Sh' = 2.3 \cdot 10^2 \left(\frac{S_p^{0.5}}{V_p^{0.33}}\right) Re_G^{0.18}\, Re_L^{0.33} \qquad (34)$$

or by Fukushima and Kusaka [15] for high-interaction regimes within ±30% error as

$$\frac{k_L \cdot d_p}{D_L} = Sh = 5 \cdot 10^2 \left(\frac{S_p}{d_p^2}\right) Re_G^{0.20} \cdot Re_L^{0.33} \cdot Sc^{0.5} \cdot (d_t/d_p)^{-2.2} \qquad (35)$$

But in this regime energy dissipation seems to be a more universal correlation parameter [11], as shown in Figure 10, where for upflow the correlation

$$\frac{k_L \cdot \epsilon}{u_{OL}} = 7.96 \cdot 10^{-3} \left[(-\Delta P/\Delta L)\frac{g \cdot \epsilon}{a_s\, \rho_L\, u_{OL}^2}\right]^{0.275} - 9.41 \cdot 10^{-3} \qquad (36)$$

is depicted.

It should be noted here also that most experimental values concern air and water or aqueous salt solutions, i.e., systems with different physical properties may behave differently.

Liquid-Solid Mass Transfer Coefficient

The liquid film surrounding the solid particles creates a resistance to the transport of reactants to the surface of the solid particle, which may be the limiting step of the

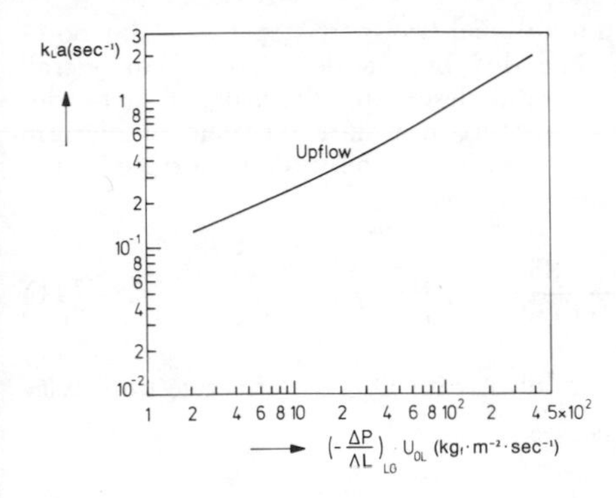

Figure 10. Liquid-side volumetric mass transfer coefficient, k_La, in cocurrent multiphase flow.

whole process. In upflow two-phase operation, mass transfer is always higher than the corresponding values observed in the single-phase flow. The respective volumetric mass transfer coefficient, k_sa_s, for cocurrent packed bubble operation was first measured by Snider and Perona [53] and Kirillov and Kasamanian [54]. In the latter publication, the experimental values have been correlated as

$$Sh = 4.825\ Sc^{0.33}\ Re_L^{0.33} \left(\frac{d_p}{\delta}\right)^{0.33} \left[0.7 + 0.12 \left(\frac{u_{OG}(1-\beta_{Gt})}{u_{OF}\,\beta_{Gt}}\right)^{0.62}\right] \tag{37}$$

where δ is the average thickness of liquid layer on the surface of a pellet, calculated by the empirical formula [43]

$$\left(\frac{\delta}{d_p}\right) = 0.15\ Fr_G^{0.1} \tag{38}$$

It has been shown [55] that the Calderbank and Moo-Young [56] energy dissipation assumption can be extended into the low Re_L regime to correlate mass transfer rates in other than aqueous systems—those used by Snider and Perona [53], for example. The same authors have shown that their data [57], together with other data [18,58,59], could be correlated uniquely by

$$Sh^* \cdot Sc^{-0.33} = f\ (Re_L^*) \tag{39}$$

using the hydraulic diameter, $d_h = \epsilon_L\ d_p/1.5\ (1 - \epsilon_L)$, and the average actual liquid velocity, $u_L = u_{OL}/\epsilon_L$, instead of d_p and u_{OL} in the dimensionless numbers. At low Re_L^* values, where mass transfer rate is mainly dominated by gas turbulence, Sh^* does not depend on Re_L^* but on d_p, whereas in a high Re_L^* regime the values asymptotically approached the correlation for single-phase liquid flow.

Recently [60], new k_sa_s data for bubble flow and distorted bubble flow regime in packed bubble columns have been correlated with the rate of energy dissipation in the liquid E_L per unit volume of the liquid phase as

$$Sh \cdot Sc^{-0.33} = 0.28 \left[\frac{E_L\ (1-\beta_{Gt})\ \rho_L^2\ d_p^4}{\mu_L^3}\right]^{0.218} \tag{40}$$

It seems important to notice that in two-phase flow mass transfer can be increased up to 300% over the single-phase flow values [60], but in a bed packed with small particles, the enhancement due to the accompanying gas flow is not observed [20]. Again, more data for pulsing flow conditions should be available to confirm the correlations given above for this flow regime.

If the solid particles are porous, there is an additional transport step inside the pores of the particle. This phenomenon usually is included by the definition of an overall pore efficiency. In accordance with a simple model describing diffusion, a first-order reaction within a spherical particle and mass transport in a stationary film of uniform thickness surrounding the grain, the degree of utilization of the internal surface of the grain is determined by [61]

$$\eta_o = \frac{3\ \mathrm{Sh}}{\Phi^2}\left[1 + \frac{\mathrm{Sh}}{\Phi\ (\mathrm{cth}\ \Phi) - 1}\right]^{-1} \tag{41}$$

where $\Phi = (d_p/2)(k_1/D_{eff})^{0.5}$ is the Thiele modulus, with D_{eff} the effective diffusivity of the reactant in the catalyst pores.

Solid-Fluid Heat Transfer Coefficients

In the case of chemical reactions performed on the active sites in porous catalytic particles, it can become important how fast the heat transfer is between a gas-liquid mixture and a fixed packing. There is only one publication dealing with this subject [46], in which it is shown that the dependence of the heat transfer coefficient on gas and liquid velocities is of a complex nonmonotonic nature, evidently associated with the different hydrodynamic regimes. Starting with relatively low values in the bubble flow regime, h_p passes through a maximum at the pulsing onset, reaching again very low values in spray flow. To obtain an approximate relationship, it was found convenient to employ the well known Nu $Pe^{-0.33}$ relationship, taking into consideration the two-phase nature of the fluid by mixed properties of the dispersion, such as

$$\bar{c}_p = \lambda_1 c_{p_L} = c_{p_L}\left[1 + \frac{c_{p_G}\,\beta_G}{c_{p_L}(1-\beta_G)}\right]$$
$$\text{and} \quad \bar{u}_o = \lambda_2 u_{OL} = u_{OL} \times \left[1 + \frac{u_{OG}\,(1-\beta_G)}{u_{OL}\,\beta_G}\right] \tag{42}$$

On the basis of these mixed properties, Nu $[Pe_L \cdot \lambda_1 \cdot \lambda_2]^{-0.33}$ could be correlated (for low and high interaction regimes separately) with β_G in a unique way.

Effective Thermal Conductivity and Wall Heat Transfer Coefficient

Heat transfer performance of cocurrently operated multiphase packed beds is characterized in terms of an effective thermal conductivity of the bed, λ_{eff}, and the wall heat transfer coefficient, h_w. Systematic studies on these parameters for packed bubble columns have been published only recently. It has been shown [46] that for large values of H/d_p and not too steep axial temperature gradients in the column the influence of longitudinal heat transfer through the solid particles can be neglected (as it is usually done also with single-phase flow). The effective thermal conductivity in the multiphase upflow system is always larger than in the downflow system. By increasing the gas flowrate at constant liquid flowrate, λ_{eff} in the upflow system increases temporarily and decreases in the bubble flow regime, being constant or gradually increasing in the pulse flow regime, very similar to the radial dispersion coefficient. This indicates the dominant contribution of radial mixing to radial heat transfer rate [42]. The effects of gas and liquid flowrates on the wall heat transfer coefficient are very complicated. The coefficient has a maximum or minimum with increasing liquid flowrate, and its dependence on the gas velocity is reversed at a certain liquid flowrate [62]. Similar observations (minimum values of h_w in certain ranges of u_{OG} and h_w close to those of empty bubble columns) have been reported [63].

FLOW MODELS

As already mentioned, it is impossible to solve the momentum balance equations for a packed bubble column. Therefore, a rigorous description of fluid dynamics based on first principles is impossible. In view of this situation, several attempts have been made to define fluid dynamic parameters relevant for the description of the behavior of multiphase systems via appropriate models.

The most common models, although probably not the most realistic types, are continuum models of the same form as used for single-phase flow through packings. Its probable reason might be the chemical engineers' familiarity with the numerical mathematics needed to solve the model equations. In these models, each phase—gas, liquid and solid—is regarded as a continuum in the system, and differential balance equations are formulated separately [61] for volume elements of each phase.

Stagewise models, in which the packing with its voids is regarded as a series of ideally (or nonideally) mixed cells, which are connected deterministically or stochastically by interstage fluid streams, are closer to reality. In this case, algebraic balance equations are formulated for each cell separately. There are two different approaches: (1) the Deans-Lapidus approach [64], in which all transport, as well as reaction steps, are proceeding in the cells themselves, whereas in the deterministically proposed connecting streams nothing happened; and (2) the Crine percolation approach [65], in which the transformations happen more realistically in bonds, connecting nodes (i.e., cells) that are distributed regularly in the packing in a probabilistic way. In both cases, elements (stages) of the model have the same dimension as packing particles.

A third category of models also has been developed—the so-called zone (or recirculation) models, taking into account the flow inhomogeneities in the packing, with a rising flow in the center and descending flow at the periphery [66]. Here, the dimension of the model element (zone) is in the same order of magnitude as the column dimension.

Continuum Models

The continuum models are based on the differential material (and eventually energy) balances, not only including convective (P) and dispersion (D) terms to characterize flow and mixing in a phase, but, if desired, also exchange terms for transport between the phases, as well as reaction rate terms for consumption or production in the phases.

As shown above in multiphase flow, under certain flow conditions some of the phases cannot be regarded any longer as homogeneous because of stagnant (dead) zones. In this case, separate balance equations for the dynamic and the stagnant part have to be formulated, also including exchange (E) terms between both parts.

If the behavior of the packed bubble column is sufficiently described by balance equations of only one of the phases, the model is called pseudohomogeneous; otherwise, it is called heterogeneous. Furthermore, continuum models are in most cases one dimensional, not taking into account radial variations of the state variables in a cross section. Finally, dynamic models also include the change of state variables by derivatives with respect to time, whereas stationary models belong only to the constant (i.e., time-independent) operation.

Stationary pseudohomogeneous or heterogeneous plug-flow or plug-flow dispersion models are used more or less exclusively in the literature [3–5,20,43,46]. Reasons why more complex models are not in use might be that (1) specific parameter values to be used in the models are not available, or (2) mathematical work for the solution of model equations should be kept small.

A typical set of equations for a stationary one-dimensional, isothermal, heterogeneous plug-flow dispersion model, applied to a catalytic gas-liquid surface reaction on a porous catalyst,

$$A\ (gas) + B\ (liq) \rightarrow C\ (liq)$$

looks like the following [3,67]:
Gas phase:

$$\frac{1}{Bo_G}\frac{d^2y_a}{dz^2} - \frac{dy_a}{dz} - \tau_G\, k_La\, (x_a{}^* - x_a) = 0 \tag{43a}$$

Liquid phase:

$$\frac{1}{Bo_L}\frac{d^2x_a}{dz^2} - \frac{dx_a}{dz} + \tau_L\, k_La\, (x_a{}^* - x_a) - \tau_L\, k_sa_s\, (x_a - x_{as}) = 0 \tag{43b}$$

$$\frac{1}{Bo_L}\frac{d^2x_b}{dz^2} - \frac{dx_b}{dz} - \tau_L\, k_sa_s\, (x_b - x_{bs}) = 0 \tag{43c}$$

Solid phase:

$$k_sa_s\, (x_a - x_{as}) = k_sa_s\, (x_b - x_{bs}) = k_2\, \eta_o\, x_{as}\, x_{bs} \tag{43d}$$

It can be simplified if one of the reactants is in great excess (leading to a pseudo first-order reaction) or if one of the transport steps becomes controlling. The solution of this set of differential equations can be performed via a standard integration routine, such as the Runge-Kutta-method.

Stagewise Models

As an alternative to the continuum model, the stagewise model of a different degree of complexity also is in use [3,68]. In its simplest form (the series of the stirred-tank model), it can be written, e.g., for reactants in the liquid phase, as

$$\left|(x_{io} - x_i) = \frac{d_p\, \epsilon_L}{u_{OL}}\, k_s\, a_s (x_i - x_{is})\right|_{n,i} \qquad \begin{matrix} n = 10, \ldots, 40 \\ i = \ 1, \ldots, N \end{matrix} \tag{44}$$

etc. That is, the convective term dx/dz has been substituted by the difference between entrance and exit concentration of the nth cell in series and an appropriate cell residence time. The various aspects of this model have been discussed by Nishiwaki and Kato [69] and Ham and Coe [70], among others. The solution of this set of algebraic equations demands an appropriate numerical technique. As long as there are no recycle streams between the cells, the solution is simplified, as a stage-by-stage procedure can be used. Otherwise, solution techniques used in the design of stagewise separation units can be used.

Zone Models

An example for a zone model applied on packed bubble columns is given by Kats [66]. In this model, the backmixing of a liquid is described by a recirculation model with a velocity profile and cross transfer between zones with an effective cross transfer coefficient, $D_\perp$. This model has four parameters: an average velocity, u_r, the cross transfer coefficient, $D_\perp$, the column diameter, d_t, and the column length, H. The advantage of this model seems to be that d_t and H account for the scale factor.

Numerical analysis of the model for a parabolic velocity profile leads to the conclusion that for values of $\tau_c/\tau_\perp$ (the ratio of the time constants for the convective to the cross-flow transport) larger than one, i.e., with large H and small d_t, the one-dimensional dispersion model is adequate because the scale of local turbulence is suppressed considerably by the packing. On the other hand, the recirculation in the packed tower is expressed more sharply since cross flow is suppressed. Hence, one should expect a stronger influence of the scale factor, d_t, on axial mixing than in unpacked towers. If, on the other hand, $\tau_c/\tau_\perp < 0.3$, pure convection is observed, whereas for $0.3 < \tau_c/\tau_\perp < 1$ an intermediate region exists. As shown by experiments, in packed bubble columns for low and moderate gas velocities (depending on d_t and H), any of the three cases may be realized, and the dispersion model gives a good description only if $u_{OG} > 0.3$ m/s. The greatest effects on longitudinal dispersion are those of gas velocity and tower diameter.

CONCLUSIONS

This chapter shows the significant progress that has been made toward understanding the fluid dynamic behavior and modeling of cocurrent upflow in fixed beds. This progress, made mainly in the last 5–10 years through intensive research, is reflected in the growing number of publications on packed bubble columns. But at the same time, it became clear that the phenomena are more complex than they first appear. More information is needed about fluid phases that differ from the air-water system, about non-Newtonian and viscous liquids, and about small-diameter packings used often in practice. This information mainly is needed for scaleup from laboratory units to commercial reactor dimensions because both units may behave differently in fluid dynamics. Therefore, results from small laboratory columns should be analyzed carefully as they may not be representative for commercial units. Additional results would be very helpful for a safe design.

NOMENCLATURE

a	specific gas-liquid interface based on column volume
a'	specific gas-liquid interface based on dispersion volume
a_s	specific solid surface based on column volume
Bn	Bond number, $Bn = \rho_L g/\sigma_L a_s^2$
Bo	Bodenstein number, $Bo = u\, d_p/D_a$
c_p	specific heat
d_b	bubble diameter
d_p	particle diameter
d_t	tower diameter
D	molecular diffusion constant
D_a	axial dispersion coefficient
D_L	cross flow coefficient
D_r	radial dispersion coefficient
E, E′	energy dissipation density
E_i	enhancement factor
Eö	Eötves number, $Eö = \rho_L g\, d_p^2/\sigma_L$
Fr	Froude number, $Fr = u_o^2/g \cdot d_p$
f	friction factor
g	gravity constant
$\dot{G}$	mass flowrate for the gas phase
Ga	Galileo number, $Ga = g\, d_p^3/\nu^2$
h_w	wall heat transfer coefficient
h_p	particle heat transfer coefficient
H	column height
Ha	Hatta number, $Ha = [k_1 D/k^2]^{0.5}$
k_1	first-order reaction rate constant
K	overall liquid-phase mass transfer coefficient
k	mass transfer coefficient
$\dot{L}$	mass flowrate for the liquid phase
L	length
n	number of unit cells in series
N	number of reactants in the system
Nu	Nusselt number, $Nu = h_p d_p/\lambda$
p	pressure
$\Delta P/\Delta L$	pressure drop
Pe	Peclet number, $Pe = u d_p c_p \rho/\lambda$
Q	volumetric load
r	radius
Re	Reynolds number, $Re = u_o d_p/\nu$
S	surface
Sc	Schmidt number, $Sc = \nu/D$
Sh	Sherwood number, $Sh = k\, d_p/D$
u	velocity

V volume

We Weber number, $We = \rho u_o^2 d_p/\sigma_L$

x concentration in the liquid phase

y concentration in the gas phase

z dimensionless column length, $0 < z < 1$

Z two-phase Z factor, $Z = Re_G^{1.167}/Re_L^{0.767} \cdot (\mu_{H_2O}/\mu_L)^{0.9}$

Greek Symbols

β fractional holdup based on dispersion volume

Γ tower to packing ratio, $\Gamma = d_t/d_p$

δ liquid film thickness; pressure drop

Δ packing number, $\Delta = a_s d_p/\epsilon$

ϵ fractional holdup based on column volume

η pore efficiency

λ density factor $\lambda = [(\rho_G/\rho_{air})/(\rho_L/\rho_{H_2O})^{0.5}$; thermal conductivity; correction factor defined in Equations 1a and 35, respectively

μ dynamic viscosity

ν kinematic viscosity

ρ density

σ surface tension

τ time constant

Φ Thiele modulus, $\Phi = (d_p/2)(k_1/D_{eff})^{0.5}$

ψ correction factor defined in Equation 1b

χ Lockhart-Martinelli parameter, $\chi = (\delta_L/\delta_G)^{0.5}$

Indices

c critical

d dynamic

f frictional

G gas phase

GL two phase (gas-liquid)

L liquid phase

o superficial, entrance, overall

p particle, pulse

s solid phase, static

t total, tower

mf minimum fluidization

i ith component in the reaction mixture

REFERENCES

1. Germain, A., G. L'Homme and A. Lefebvre. In: *Chemical Engineering* of Gas-Liquid-Solid Reactions, G. A. L'Homme, Ed. (Liège, Belgium: Cebedoc, 1979).
2. Heilmann, W. PhD Thesis, University of Erlangen/Nürnberg (1969).
3. Shah, Y. T. *Gas-Liquid-Solid Reactor Design* (New York: McGraw-Hill Book Co., (1979).
4. Ramachandran, P. A., and R. V. Chaudhari. *Three-Phase Catalytic Reactors* (New York: Gordon & Breach Science Publishers, Inc., 1982).
5. Hofmann, H. "Multiphase Catalytic Packed Bed Reactors," *Catal. Rev. Sci. Eng.* 17:71 (1978).
6. Charpentier, J. C. "Gas-Liquid Absorptions and Reactions," *Adv. Chem. Eng.* 11:9 (1981).
7. Hirose, T. *Proc. Symp. Multiphase Cocurrent Fixed Beds,* Okayama, Japan 1 (1978).
8. Rodrigues, A. E., paper presented at the NATO ASI on Multiphase Chemical Reactors, Vimeiro, Portugal, 1980.
9. Alper, E., paper presented at the NATO ASI on Mass Transfer with Chemical Reaction in Multiphase Systems, Cesme, Turkey, 1981.
10. Turpin, J. L., and R. L. Huntington. *AIChE J.* 13:1196 (1967).
11. Specchia, V. S., S. Sicardi and A. Gianetto. *AIChE J.* 20:646 (1974).

12. Sato, Y., T. Hirose and T. Ida. *Kagaku Kogaku* 38:534 (1974).
13. Charpentier, J. C. *Proc. Chisa '75 Sec G. Destillation and Absorption,* Prague (1975).
14. Ogarkov, B. L., V. A. Kirillov and V. A. Kuzin. *J. Eng. Phys. (USSR)* 31:800 (1976).
15. Fukushima, S., and K. Kusaka. *J. Chem. Eng. Japan* 12:296 (1979).
16. Kirillov, V. A., B. L. Ogarkov and V. G. Voronov. *J. Eng. Phys. (USSR)* 31:402 (1976).
17. Larkins, R., R. White and D. Jeffrey. *AIChE J.* 7:231 (1961).
18. Colquhoun-Lee, I., and J. B. Stepanek. *Trans. Inst. Chem. Eng.* 56:136 (1978).
19. Hofmann, H. *Chem. Eng. Sci.* 14:193 (1961).
20. Goto, S., J. Levec and J. M. Smith. *Ind. Eng. Chem. PDD* 14:473 (1975).
21. Gianetto, A., G. Baldi, V. Specchia and S. Sicardi. *AIChE J.* 24:1087 (1978).
22. Kan, K. M., and P. F. Greenfield. *Ind. Eng. Chem. PDD* 17:482 (1978).
23. Kan, K. M., and P. F. Greenfield. *Ind. Eng. Chem. PDD* 18:740 (1979).
24. Ohshima, S., T. Takematsu, Y. Kuriki, K. Shimada, M. Suzuki and J. Kato. *J. Chem. Eng. Japan* 9:29 (1976).
25. Blass, B., and R. Kurtz. *Verfahrenstechnik* 11:44 (1977).
26. Weber, H. H. PhD Thesis, University of Darmstadt, Darmstadt, West Germany (1961).
27. Stiegl, G. J., and Y. T. Shah. *Ind. Eng. Chem. PDD* 16:37 (1977).
28. Zighanshin, G. K., and A. Ermakova. *Theor. Found. Chem. Eng. (USSR)* 4:594 (1970).
29. Kirillov, V. A., M. A. Kasamanyan and V. A. Kuzin. *Theor. Found. Chem. Eng. (USSR)* 9:870 (1975).
30. Hutton, B. E. T., and L. S. Leung. *Chem. Eng. Sci.* 29:1681 (1974).
31. "Pittsburgh Energy Research Center Quarterly Reports" (1975–1976).
32. Achwal, S. K., and J. B. Stepanek. *Chem. Eng. J.* 12:69 (1976).
33. Charpentier, J. C., C. Prost, W. van Swaaij and P. Le Goff. *Chim. Ind. Génie Chem.* 99:803 (1968).
34. Böxkes, W. PhD Thesis, University of Erlangen/Nürnberg (1973).
35. Kats, M. B. *Theor. Found. Chem. Technol. (USSR)* 12:598 (1978).
36. Achwal, S. K., and J. B. Stepanek. *Can. J. Chem. Eng.* 57:409 (1979).
37. Bezdenezhnykh, A. A., V. I. Taranov and A. P. Orlov. *Theor. Found. Chem. Technol.* (USSR) 5:163 (1971).
38. Heilmann, W. and H. Hofmann. *Proc. 4th Europ. Symp. Chem. React. Eng.*, Brussels, Belgium (1968) p. 169.
39. Magnussen, P. V. Schumacher, G. W. Rotermund and F. Hafner. *Chem. Ing. Technol.* 50:811 (1978).
40. Ohsasa, K., T. Tanahashi and M. Sambuicki. *Kagaku Kogaku, Ronb.* 3:314 (1978).
41. Takeuchi, H., K. Onda and Y. Maeda. *Kagaku Kogaku, Ronb.* 1:267 (1975).
42. Nakamura, M. *Proc. Symp. Multiphase Cocurrent Fixed Beds,* Okayama, Japan (1978), p. C-5.
43. Ermakova, A., V. A. Kirillov, E. F. Stefoglu and G. A. Berg. *Theor. Found. Chem. Technol. (USSR)* 11:688 (1977).
44. Böxkes, W., and H. Hofmann. *Verfahrenstechnik* 9:112 (1975).
45. Sato, Y., T. Hirose, T. Ida and S. Fujiyama. *Preprints–8th Autumn Meeting Soc. Chem. Eng.* Japan, Tokyo (1974) p. 302.
46. Kirillov, V. A., and B. L. Orgakov. *Int. Chem. Eng.* 20:478 (1980).
47. Mashelkar, R. A., and M. M. Sharma. *Trans. Inst. Chem. Eng.* 48:T 162 (1970).
48. Kolmogoroff, A. *Comp. Rend. Acad. Sci. USSR* 30:301 (1941).
49. Batchelor, G. K. *Proc. Cambridge Phil. Soc.* 47:359 (1951).
50. Hinze, J. O. *Turbulence* (New York: McGraw-Hill Book Co., 1958).
51. Voyer, R. D., and A. I. Miller. *Can. J. Chem. Eng.* 46:335 (1968).
52. Saada, M. Y. *Chem. Ind. Génie Chim.* 105:1415 (1972).
53. Snider, J. W., and J. J. Perona. *AIChE J.* 20:1172 (1974).
54. Kirillov, V. A., and M. A. Kasamanian. *Int. Chem. Eng.* 16:538 (1978).

55. Mochizuki, S. *AIChE J.* 24:1138 (1978).
56. Calderbank, P. H., and M. B. Moo-Young. *Chem. Eng. Sci.* 16:39 (1961).
57. Mochizuki, S., and T. Matsui. *Chem. Eng. Sci.* 29:1328 (1974).
58. Nosaka, M., H. Kawasaki and H. Tanaka. *Proc. Meeting Soc. Chem. Eng. Japan,* Hokuriku, Japan (1976) p. A-10.
59. Specchia, V., G. Baldi and A. Gianetto. *Ind. Eng. Chem. PDD* 17:362 (1978).
60. Delaunay, G., A. Storck, A. Laurent and J. C. Charpentier. *Ind. Eng. Chem. PDD* 19:514 (1980).
61. Denbigh, K. G., and J. C. R. Turner. *Chemical Reactor Theory* (London: Cambridge University Press, 1971).
62. Kato, Y., S. Oshima, T. Kago and S. Morooka. *Kagaku Kogaku Ronb.* 4:328 (1978).
63. Oshima, S., M. Suzuki, K. Shimada, T. Takematsu, Y. Kuriki and J. Kato. *Proc. Symp. Multiphases Cocurrent Fixed Beds,* Okayama, Japan (1978) pp. G1–307.
64. Deans, H. A., and L. Lapidus. *AIChE J.* 6:656 (1960).
65. Crine, M. *Chem. Eng. Sci.* 35:51 (1980).
66. Kats, M. B. *Theor. Found. Chem. Technol.* (*USSR*) 12:598 (1978).
67. Parlica, R. T., and J. H. Olson. *Ind. Eng. Chem.* 62(12):45 (1970).
68. Rozenbaum, G. E., P. E. Kunin and L. I. Blyakhman. *Theor. Found. Chem. Technol.* (*USSR*) 5:920 (1971).
69. Nishiwaki, A., and Y. Kato. *Can. J. Chem. Eng.* 52:276 (1974).
70. Ham, A., and H. S. Coe. *Chem. Met. Eng.* 19:663 (1918).

CHAPTER 21
COUNTERCURRENT FLOW THROUGH FIXED-BED REACTORS

N. P. Cheremisinoff
Exxon Research & Engineering Company
Florham Park, New Jersey 07932
P. N. Cheremisinoff
Department of Civil and Environmental Engineering
New Jersey Institute of Technology
Newark, New Jersey 07102

CONTENTS

INTRODUCTION

The concept of countercurrent gas-liquid flows through fixed-bed reactors has been studied extensively for a variety of process systems. Countercurrent flows, in which two fluids enter at different ends of the reactor vessel and pass in opposite directions through the unit, are used extensively in a variety of industrial equipment such as heat exchangers, cooling towers, absorption columns and various chemical reactors. For many systems this type of flow affords the most efficient means of mass or energy transfer between gas and liquid phases, since the largest driving force is realized from the contact of the purest streams.

Because the use of counterflow fixed-bed reactors is considered standard practice in today's industry, relatively few generalized advances have been made in recent years. Emphasis has been placed on the refinement of more effective internals or packings to promote highly efficient and intimate contacting between phases. The concern of this chapter is essentially with packed towers applied to the absorption of gases. The processes of absorption and desorption of gases are important in many chemical plants. The aim of this chapter is to provide the engineer with the methods and data needed for the preliminary design of full-scale towers. Because of our emphasis on design, the fundamental theories of diffusion and analogies among fluid friction, heat transfer and mass transfer are reviewed only briefly. Such discussions are given in standard textbooks such as those by Sherwood and Pigford [1], McCabe and Smith [2] and Treybal [3].

OVERVIEW OF ABSORPTION FUNDAMENTALS

The process of absorption is applied to the removal of one or more selected components from a gas mixture. Perhaps the most extensive application in recent years has been in the control of gaseous pollutant emissions. The process of absorption conventionally refers to the intimate contact of a mixture of gases with a liquid so that part of one or more of the gaseous constituents dissolves in the liquid phase. The contact conventionally is accomplished in a packed column, which, in engineering jargon, is referred to as a wet scrubber. The net effect is the transfer of a soluble gas from the gas phase into the counterflowing liquid phase. The purpose of such gas scrubbing operations might include product recovery or process reaction, as well as pollutant removal, or a combination thereof. For example, after being removed from a contaminated gas stream, the pollutant sulfur dioxide also may be recovered to provide a valuable by-product in the form of sulfuric acid (a common practice in the paper industry).

The necessary condition for gas absorption is the solubility of the gaseous constituent in the absorbing liquid. The rate of mass transfer of the soluble constituents from the gas to the liquid phase is established by the diffusional processes occurring on each side of the gas-liquid interface. For example, consider the process that occurs when a mixture of air and sulfur dioxide is brought into contact with water. Sulfur dioxide is soluble in water and, hence, those molecules that contact the water dissolve immediately. Initially, however, the sulfur dioxide molecules are dispersed throughout the gas phase and can only reach the water surface by diffusing through the air medium, which is substantially insoluble in water. When the sulfur dioxide at the water surface has dissolved, it is distributed throughout the bulk water phase by a diffusional process. Hence, the rate of absorption is dependent on the rate of diffusion in both the gas and liquid phases.

The existence of an equilibrium state plays an important role in the absorption process. The rate at which the gaseous constituent diffuses into an absorbent liquid is a function of the deviation or departure from the equilibrium state. The rate at which equilibrium is established thus depends on the constituent's rate of diffusion through the non-absorbed gas medium and the liquid.

Quantitatively, the rate at which mass is transferred between phases is expressed in terms of a mass transfer coefficient. The mass transfer coefficient equates the quantity of mass being transferred to a conveniently defined driving force. The transfer process ceases on attainment of equilibrium.

The engineering design principles for gas absorption equipment outlined in the following section are based on application of the principles of diffusion, equilibrium and mass transfer. The primary requirement in the design is to bring the gas and liquid phases into intimate contact. That is, the design must provide a large interfacial area and a high intensity of interface renewal, while minimizing resistance and maximizing the driving force at the same time. This contacting of the phases can be achieved in many different types of equipment, the most frequently used being either packed or plate columns. Final selection is based on various engineering and economic criteria that must be met. As an example, if the pressure drop through the column is so large that compression costs become significant, a packed column may be preferable to a plate-type column due to the lower pressure drop.

In most gas absorption applications the gas stream is the processed fluid; hence, its inlet conditions (flowrate, composition and temperature) are usually known. Furthermore, the temperature and composition of the outlet gas are usually specified. The primary design objectives for an absorption column are then the determination of the solvent flowrate and the principal dimensions of the equipment (i.e., column diameter and height) needed to accomplish the operation. These design variables may be evaluated for a selected solvent at a given flowrate in terms of the number of theoretical separation units (stages or plates). These theoretical separation units may be related either to practical units of column height or number of actual plates through the use of empirical correlations.

The general design approach involves several important considerations including the following:

1. Solvent selection.
2. Equilibrium data evaluation.
3. Estimation of operating data (usually consisting of mass and energy balances, in which the energy balance decides whether the absorption process can be considered isothermal or adiabatic).
4. Column selection (the final specification often is based on economic considerations).
5. Column diameter (for packed columns, this is usually based on flooding conditions and, for plate columns, on the optimum gas velocity or the liquid-handling capacity of the plate).
6. Column height or the number of plates. For packed columns, the column height is obtained by multiplying the number of transfer units, obtained from a knowledge of equilibrium and operating data, by the height of a transfer unit. For plate columns, the number of theoretical plates determined from the plot of equilibrium and operating lines is divided by the estimated overall plate efficiency to give the number of actual plates, which, in turn, allows the column height to be estimated from the plate spacing.
7. Pressure drop through the column. For packed columns, correlations dependent on packing type, column operating data and physical properties of the constituents involved are available to estimate the pressure drop through the packing; for plate columns, the pressure drop per plate is obtained and multiplied by the number of plates.

Before examining design criteria in more depth, a brief review of absorption principles is given.

THE TWO-FILM THEORY

Regardless of the equipment configuration employed for absorption, the foremost criterion is that the gas and liquid phases intimately contact each other. The absorption process is explained diagramatically in Figure 1 [4]. The illustration shows thin films of gas and liquid separated by a defined interface. The ordinates are the partial pressure of the soluble gas in the gas phase and the concentration of the dissolved gas in the liquid. (Note that the break at the interface is a reflection of the difference in units used for pressures and concentrations and does not denote an interfacial resistance to diffusion.) Hence, from Figure 1, the driving force across the gas film is $p_g - p_i$ and that across the liquid film is $c_i - c_l$. The rather simplified explanation is referred to as Whitman's two-film theory, and although it is far from reality, it generally provides an adequate basis for designing absorption columns.

Whitman's theory assumes that when the gas and liquid are in contact, the bulk of both fluids undergo turbulent mixing. On either side of the interface, however, thin films of gas and liquid are assumed to exist in streamline flow. Furthermore, it is assumed that the resistances to diffusion of the soluble gas species are entirely in

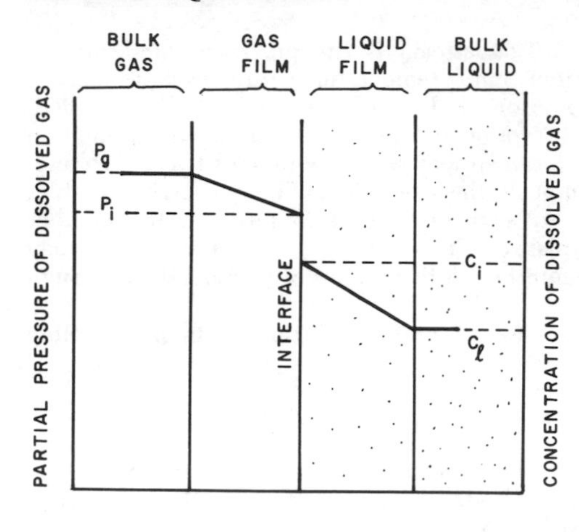

Figure 1. The principle of the two-film theory [4].

these two films and that the main bulk of both fluids, along with the interface itself, present no resistance.

For an infinitesimal portion of interfacial area, dA, at any point in a column, the rate of absorption across the gas film is

$$dW = k_g \, dA \, (p_g - p_i) \tag{1}$$

and, for the liquid film,

$$dW = k_l \, dA \, (c_i - c_l) \tag{2}$$

where k_g and k_l are the gas- and liquid-film absorption coefficients, respectively.

Because conditions at the interface are difficult to evaluate, overall coefficients are defined either on the basis of the gas or of the liquid phase. On the basis of the gas phase, the overall driving force is defined as the difference between the actual partial pressure of the soluble gas, p_g, and the partial pressure, p_l, corresponding to equilibrium with the liquid of concentration, c_l. On the basis of the liquid phase, it is the difference between the concentration, c_g, corresponding to equilibrium with gas of partial pressure, p_g, and the actual concentration of the soluble species, c_l.

Hence, in terms of the overall coefficients, Equations 1 and 2 may be expressed as follows:

$$dW = k_g \, dA \, (p_g - p_l) \tag{3}$$

$$dW = k_l \, dA \, (c_g - c_l) \tag{4}$$

Overall coefficients must be derived from the individual film coefficients. The relationship between overall and film coefficients is developed by interpreting the mass transfer process in a manner analogous to electrical circuitry. That is, we consider the coefficients to be, in effect, conductances. Hence, the diffusional resistance of each film may be interpreted as the reciprocals of the film coefficients and, by the usual laws of additivity of resistances in series, we may write

$$\frac{1}{K_g} = \frac{1}{k_g} + \frac{1}{Hk_l} \tag{5}$$

And for the liquid film,

$$\frac{1}{K_l} = \frac{H}{k_g} + \frac{1}{k_l} \tag{6}$$

The factor H is the solubility coefficient, defined as the ratio of the change in concentration of dissolved gas to the change in equilibrium partial pressure. For ideal solutions, Henry's law applies:

$$C_l = Cp_l \tag{7}$$

where C is a constant. When Henry's law is applicable, H is the constant for an isothermal case (for nonisothermal systems, H varies with concentration). In reality, H is actually nothing more than a unit conversion factor, since allowance must be made when adding the film resistances, i.e., the partial pressures and concentrations are not measured in the same units.

CHARACTERISTICS AND PROPERTIES OF PACKINGS

Classifications

A packed tower essentially consists of a vessel filled to a specified level with some suitable solid material. The liquid is introduced at the top of the tower and flows down through the packing to the bottom. In doing so, it exposes a large surface area to contact the gas phase. Note that the gas may be introduced at either end of the tower or perpendicular to the liquid flow, i.e., the gas and liquid flows may be either cocurrent, countercurrent or operated in cross flow. There are a wide variety of packings available ranging from crushed stone or lumps of coke or quartz to complicated manufactured ceramic and plastic geometries. We shall direct our discussion to common configurations, which include simple solid materials, grids and rings.

Solids such as coke, quartz or crushed stone are the simplest, as well as the earliest, form of packing used in absorbers. They are supplied in nominal size ranges in which their upper limit is $\sqrt{2}$ times the lower. Solid packings are the least expensive and, in general, provide good interfacial contact area because of the tightness of the packing. However, they are limited to handling only low gas throughputs. They are limited, therefore, to towers of large cross sections.

Grid elements are employed as either plain or serrated along their lower edges. The latter are employed to minimize excessive drip which occurs when the spacing between elements exceeds 2.5 cm. Grids are particularly suited for high gas to liquid rate ratios. In addition, they are less susceptible to *choking* than are more complex configurations and, hence, may be utilized with liquids containing suspended solids (one example being gas scrubbing with lime slurry). A major disadvantage with grids in comparison to other standard packings is that they require special arrangements to ensure even liquid distribution in the column.

Various types of ring configurations commercially available include Raschig rings, in which their height is equal to the ring outer diameter; there are also shallow rings of a height less than the outer diameter. Others are Lessing, or partition, rings. In general, Raschig rings have proven most reliable for the spectrum of operating variables, i.e., transfer coefficients, pressure drop, throughput, as well as cost per unit volume. There are two general arrangements for rings in towers, namely, stacked and randomly packed. As a general rule, stacked rings should be 7.6 cm in diameter or larger. With smaller sizes, stacking is often uneconomical, particularly for large towers. Stacked rings involve layering rings such that their open areas overlap. In general, the performance of stacked rings is inferior to that of grids; however, they have the advantage of being widely available in noncorrosive materials.

Rings smaller than 7.6 cm in diameter are difficult, as well as costly, to stack. Hence, they are usually randomly packed in the vessel. This arrangement produces favorably higher transfer coefficients than do the stacked rings, but at the cost of higher pressure drop. Random packings are best suited for use in high-pressure columns for low gas to liquid velocity ratios. These are not recommended for liquid loadings containing solids because of the risk of choking.

Liquid Flow Over Packings

For absorption limited by the gas film, the performance of packings (based on the total dry surface area) drops off as the liquid loading rate is reduced below certain critical values. Under such conditions, the proportion of packing surface wetted by the liquid is reduced. Thus, to ensure the most effective use of packing, the liquid loading must exceed a minimum value referred to as the minimum wetting load (MWL). At liquid rates above the MWL, conditions are eventually attained with stacked rings and grids where the liquid no longer flows over the packing surface as a film. It tends to break away and undergo free fall through the tower until it impinges on a layer of packing farther down the column, i.e., cascading. Cascading can lead to undesirable liquid entrainment by the gas.

In specifying the MWL, it is convenient to express the liquid rate as a wetting rate. For packed towers, this is

$$L = \frac{V_l a}{p} = \frac{V_l}{S} \tag{8}$$

where L is the liquid flowrate per unit periphery (i.e., the wetting rate—m^3/hr-m), V_l is the liquid rate (m^3/hr-m^2), a is the packed cross-section of the tower (m^2), p is the periphery of the packed section (m), and S is the surface area of packing per unit volume (m^2/m^3). Typical values of S along with other properties for different packings are given in Table I.

Morris and Jackson [5] recommend a value of MWL of 0.08 m^3/hr-m for all packings except rings of diameters greater than 7.6 cm and grids of pitch greater than 5.1 cm. For remaining packings, a value of 0.12 m^3/hr-m is recommended. The higher rates minimize the effects of liquid dripping and its resulting entrainment.

As implied from the discussion thus far, a major feature in the design of an absorption column is the required surface area to effect mass transfer. In terms of a design parameter, this translates to the required height of packing for a given range of operating conditions. Equations 3 and 4 may be integrated in terms of their appropriate driving forces to give

$$A = \frac{W}{K_{gm}\Delta p_m} \tag{9}$$

$$A = \frac{W}{K_{lm}\Delta C_m} \tag{10}$$

where Δp_m and ΔC_m are the mean driving forces on the gas and liquid phases, respectively. K_{gm} and K_{lm} are the average overall coefficients on the gas- and liquid-phase bases, respectively.

We shall discuss the evaluation of the mean driving force later. For now, simply note that to determine the height of packing, the surface area per unit volume of packing must be known (Table I):

$$A = Sla \tag{11}$$

where l is the height of the packing. Hence, Equations 10 and 11 may be expressed as follows:

$$l = \frac{W}{K_{gm}Sa\Delta p_m} \tag{12}$$

$$l = \frac{W}{k_{lm}Sa\Delta C_m} \tag{13}$$

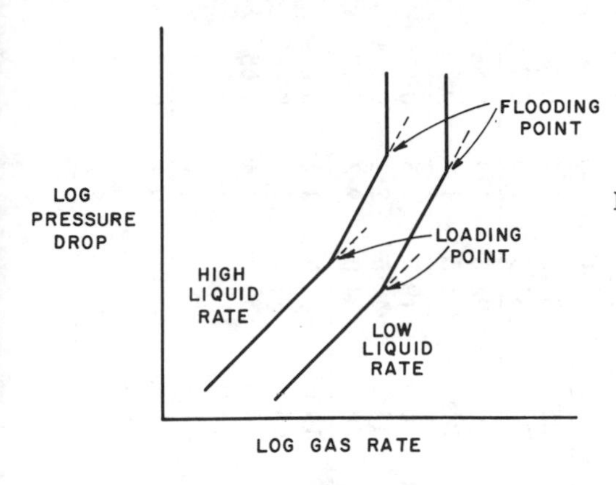

Figure 2. Loading and flooding points on a plot of pressure drop vs gas rate.

Limitations on Gas Velocities and Loadings

If the gas velocity through a randomly packed tower is sufficiently turbulent, then the pressure drop through the tower is observed to be proportional to the square of the gas rate (up to a critical point) for a constant liquid rate. For stacked packings, pressure drop is proportional to the gas rate raised to a power slightly less than 2. At higher gas rates, a critical point eventually is reached at which the pressure drop is proportional to the gas rate raised to a power greater than 2. Further increases in gas rate result in an increase in liquid holdup in the column. This point is referred to as the loading point, and the corresponding gas rate is the loading rate. At gas rates beyond the loading rate, the liquid holdup is sufficient to interrupt the flow of gas and the tower floods. The loading and flooding points are shown on a logarithmic plot of pressure drop versus gas rate in Figure 2. Figure 2 is somewhat idealized, as experiments by the authors have shown that loading points for grids and stacked rings often show a great deal of scatter on such plots. At the flooding condition, gas bubbles through the liquid and the pressure drop through the tower increases infinitely. The gas rate corresponding to this condition is referred to as the flooding rate.

There is no generalized correlation among loading rates, the proportions of packings and the physical properties of gases and liquids. Some of the earlier literature [5–10] report experimental loading rates for the air-water system for different packings. Some of the data reported by Morris and Jackson [5] are shown plotted in Figure 3 as wetting rate, L, versus the gas to liquid ratio at the loading point. The types of packings used for the data given in the plot are summarized in Table II. Note that parameter ϕ on the gas-liquid ratio is a correction factor for the density of air ($\phi = \sqrt{\rho/1.205}$, where ρ is in kg/m^3).

Generalized correlations for flooding rates are given in Figure 4 [11–13], which represents flooding rates for random rings and solids. It consists of a plot of flooding rate as the group YV_{gf} versus the properties of the packing in the form of the ratio XS/ϵ^3. Parameters X and Y are correction factors for fluid properties, defined as follows:

$$X = 2.5\mu_l^{0.2} \tag{14a}$$

$$Y = 28.8\sqrt{\rho/\rho_l} \tag{14b}$$

where μ_l is the viscosity of the liquid in poise, and ρ,ρ_l are the densities of the gas and liquid, respectively, at the tower's operating conditions. Note that ϵ is the voidage for various values of the liquid rate. Typical values of S/ϵ^3 for various packings are given in Table III.

Table I. Properties of Various Packing Materials [5]

Packing												
			Dimensions, in.						Surface Area per Unit Volume, S^b			
Type	Material	Serial No.	Pitch or Diameter	Height	Thickness	Number per Cubic Foot[a]	Number per Cubic Meter[a]	Voidage, ϵ	(ft²/ft³)	(m²/m³)	Gas-Film Packing Factor, R_g	Liquid-Film Packing Factor, R_l[c]
Plain Grids	Metal	1	1	1	1/16	–	–	0·93	25·5	83·5	2·5	0·9[d]
		2	1	2	1/16	–	–	0·93	24·8	81·5	2·0[d]	0·72[d]
	Wood	3	1	1	1/4	–	–	0·75	30·0	98·5	1·8	0·88
		4	1	2	1/4	–	–	0·75	27·0	88·5	1·5	0·70[d]
Serrated Grids	Wood	5	4	4[e]	1/2	–	–	0·89	6·0	19·5	1·5	0·60[d]
		6	2	2[e]	3/8	–	–	0·83	13·0	42·5	2·4	0·72[d]
		7	1½	1½[e]	3/16	–	–	0·89	16·5	54·0	2·3[d]	0·77
Stacked Rings	Stoneware	8	4	4	3/8	27	950	0·73	19·0	62·5	1·4[d]	0·60[d]
		9	3	3	3/8	65	2,300	0·66	25·0	82·0	1·4	0·68[d]
		10	3	3	1/4	65	2,300	0·76	25·0	82·0	1·4[d]	0·68
		11	2	2	1/4	210	7,400	0·67	36·0	118	1·4[d]	0·70[d]
		12	2	2	3/16	210	7,400	0·72	36·0	118	1·4[d]	0·70[d]
Random Rings	Metal	13	2	2	1/16	175	6,180	0·92	30·0	98·5	3·3	0·67
		14	1	1	1/16	1,350	47,600	0·86	59·0	194	3·0[d]	0·90[d]
		15	½	½	1/32	10,500	370,000	0·87	115	377	3·1	0·88

		16	3	3	3/8	52	1,840	0·72	20·0	65·5	2·5	0·60[d]
		17	2	2	1/4	165	5,820	0·74	28·0	92·0	2·7[d]	0·65[d]
		18	2	2	3/16	170	6,000	0·79	29·0	95·0	2·8	0·67
	Stone-ware	19	1½	1½	3/16	400	14,100	0·73	38·0	125	2·7[d]	0·75[d]
		20	1	1	3/32	1,300	46,000	0·80	56·0	184	2·7	0·88
		21	¾	¾	3/32	3,000	106,000	0·74	72·0	236	2·7[d]	0·88[d]
		22	½	½	1/16	10,500	370,000	0·73	115	377	2·7[d]	0·88[d]
		23	2	2	1/4	165	5,820	0·74	28	92·0	2·7[d]	0·65[d]
	Carbon	24	1	1	3/16	1,250	44,000	0·66	52	170	2·7[d]	0·88
		25	½	½	1/16	10,500	370,000	0·73	115	377	2·7[d]	0·88[d]
		26		3		—	—	0·50[f]	15	49·0	2·5[d]	0·45[d]
Solids	Coke	27		1–2		—	—	0·40[f]	35	115	2·1	0·55[d]
		28		1		—	—	0·45[f]	40	131	2·5[d]	0·58
	Quartz	29		2		—	—	0·46	19	62·5	2·7[d]	0·50[d]
		30		½–1¼		—	—	0·40	44	144	2·2[d]	0·55[d]

[a]Values given apply when the diameter of the tower is about 20 times the nominal diameter of the rings. For small towers, the number may be approximately 10% less, and for large towers approximately 10% greater than those shown.

[b]The surface area includes that of the edges. No allowance has been made for surface lost as a result of contact between packing elements. The surface area of spacer bars has been neglected in calculating values of S for grid packings. The figures given can be used to calculate the equivalent diameter of the packing, which is given by $4\epsilon/S$ (ft or m); this quantity is required if it is necessary to determine the Reynolds number.

[c]At the standard wetting rate.

[d]Estimated value.

[e]From top of grid bar to tip of serration.

[f]Does not include the pore space.

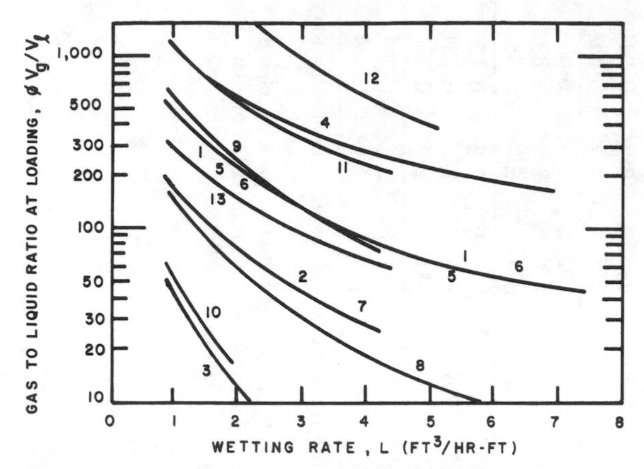

Figure 3. Correlations of wetting rate for the air-water system reported by Morris and Jackson [5]. Curve numbers correspond to the packings listed in Table II.

Pressure Losses Through Packing

Pressure losses may be expressed in terms of the number of velocity heads lost per unit height of packing, i.e., the ratio of pressure drop per unit height of packing, expressed as a head, to the velocity head of the gas. The gas velocity of interest is the average value equivalent to an empty tower, i.e., superficial gas velocity, V_o. The pressure drop through the packing is

$$\Delta P = 7.35 \times 10^{-5} N \rho V_o^2 l \tag{15}$$

where ΔP is in psi, ρ is in kg/m³, V_o is in m/sec, l (the height of packing) is in m, and N is the number of velocity heads lost per unit height of packing.

Table II. Packing Materials Tested for Correlations Given in Figure 3

Packing Material	Packing Dimensions (in.)	Curve No.
Random Metal Rings	2 × 2 × 1/16	1
	1 × 1 × 1/16	2
	1/2 × 1/2 × 1/32	3
Random Stoneware Rings	3 × 3 × 3/8	4
	2 × 2 × 1/4	5
	2 × 2 × 3/16	6
	1 × 1 × 3/32	7
Random Carbon Rings	1 × 1 × 3/16	8
Quartz	2	9
	1/2–1 1/4	10
Plain Grids	1 × 2 × 1/4	11
Serrated Grids	1 1/2 × 1 1/2 × 3/16	12
Stacked Stoneware Rings	2 × 2 × 1/4	13

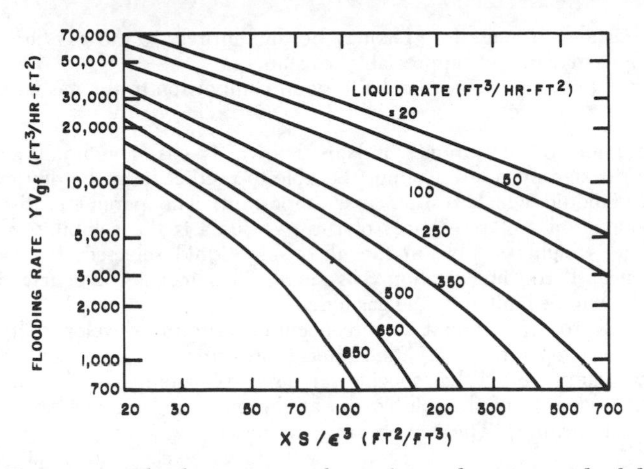

Figure 4. Flooding rate correlation for random rings and solids.

Table III. Typical Values of S/ϵ^3 for Random Rings and Solid Packings

Packing				S/ϵ^3	
	Dimensions (in.)				
Type	Diameter	Height	Thickness	(ft^2/ft^3)	(m^2/m^3)
Random Stoneware Rings	3	3	3/8	54	177
	2	2	1/4	69	226
	2	2	3/16	59	194
	1½	1½	3/16	98	322
	1	1	3/32	109	358
	3/4	3/4	3/32	178	585
	3/4	½	1/16	295	970
Random Carbon Rings	2	2	1/4	69	226
	1	1	3/16	180	590
	½	½	1/16	295	970
Coke		3		120	394
		1–2		550	1,800
		1		440	1,440
Quartz		2		195	640
		½–1¼		690	2,260

DESIGN METHOD FOR PACKED TOWERS

Criteria for Film Resistance

As explained earlier by the two-film model, in all absorption processes the soluble gas must diffuse through the gas film before contacting the gas-liquid interface and, in doing so, must encounter a diffusional resistance. If the gas is not readily soluble in the liquid, an additional resistance arises to the passage of the dissolved gas through the liquid film. If we consider the combined resistances of both films, then three conditions could occur:

1. The resistance to mass transfer may be due entirely to the gas film.
2. Both films may provide appreciable resistances.
3. The liquid film resistance may be large in comparison to the gas film.

Conditions 1 and 3 are extremes and are known as gas- and liquid-film limits, respectively. For design purposes, we must be able to predict from the physical properties of the gas and liquid which resistance is important. The parameter that establishes the relative magnitudes of resistances of the two films is the solubility coefficient, H. For a gas that is highly soluble in the absorbing liquid selected, the value of H is large and, hence, $1/Hk_l$ in Equation 5 is small. This means, of course, that the absorption process will be limited by the gas film.

Morris and Jackson [5] suggest a more accurate criterion developed from similarity theory. For tower operations at atmospheric pressure in the absence of chemical reactions, the quantity ρ/HP has specific ranges corresponding to the three conditions (where ρ_s is the density of the soluble gas at the operating temperature and pressure, and P is the total pressure). These limits are as follows:

1. For $\rho_s/HP < 5 \times 10^{-4}$, the gas-film limit is likely.
2. For $5 \times 10^{-4} < \rho_s/HP < 0.2$, both films are likely to contribute resistance.
3. For $\rho_s/HP > 0.2$, the liquid-film limit is likely.

Packing Selection

Preliminary selection of suitable packing may be made in the early stages of design. In theory, packing should be selected on the basis of economics to ensure that the total annual charges for the tower are minimum. However, selection generally is limited by considerations involving the flow of gases and liquids over packings, as discussed earlier. In atmospheric and low-pressure towers, operation above the loading point is often avoided. The plots given in Figure 3 can be used as a guide to determining for which packings the wetting rate is satisfactory for a specified $\phi V_g/V_l$. Alternatively, if the wetting rate is specified, these plots may be used to ascertain the maximum allowable gas rate for each packing without exceeding the loading rate. After several packings have been evaluated as being suitable for the specified duty, preliminary estimates of the packed height and packing cross section of the resulting towers should be made and compared to each other.

The ratio of tower diameter to packing diameter also becomes an issue in packing selection. Large packings are not suitable for use in towers of small diameters because of the resulting increase in voidage near the walls (potentially giving rise to a preferential flow of gas in this region). Their usage also results in a reduction in available surface area. As a rule of thumb for random packings, the diameter of the tower should not be less than six to eight times the diameter of a single packing element.

Finally, packing should be selected from consideration of the materials being handled. If the fluids are corrosive in nature, then suitable corrosion-resistant packing material must be chosen. If the deposition of solids is anticipated, then the use of open grid packing may be preferable.

In addition to packing selection, the specifications established in a preliminary design are the cross-sectional area of packing and the height of the packing. Both are dependent on the type of packing selected. We shall present a design method for specifying these two features in the immediate discussions to follow. An alternate design method is described in the following section. Note that the methods presented are applicable only to the problem of mass transfer and do not account for the kinetics of chemical reactions. In addition, tower specifications from these methods are for isothermal conditions only.

Calculating Gas- and Liquid-Film Coefficients

The overall absorption coefficients cannot be computed directly from the physical properties and flowrates. It is necessary to evaluate these coefficients for the gas and liquid films separately and apply Equations 5 and 6. For the gas-film coefficient, the following empirical correlation is recommended by Morris and Jackson [5] and Carey and Williamson [14]:

$$k_g = 36.1\, R_g C \nu^{0.75} \frac{P}{(P - p)_{lm}} \cdot \frac{1}{p^{0.25}} \left(\frac{T_r}{T_f}\right)^{0.56} \tag{16}$$

where
k_g = gas film coefficient (kg/hr-m^2-atm)
R_g = gas-film packing factor
C = gas-mixture constant
ν = gas velocity relative to the liquid interface (m/sec)
P = total gas pressure (atm)
$(P - p)_{lm}$ = log-mean partial pressure of insoluble gas in the gas film (atm)
T_r = reference temperature = 293°K
T_f = absolute temperature of the gas film (°K)

Equation 15 is applicable to turbulent gas flow only (for random rings and solids the critical Reynolds number, Re_c, i.e., the transition from laminar to turbulent gas flow, is 100–300; for grids and stacked rings Re_c = 500–1500). The Reynolds number should always be computed as a check for turbulent conditions. It is desirable to avoid laminar flow mainly because of economic considerations and because of the difficulty in maintaining even distribution of the gas and liquid. When calculating Re, use the equivalent diameter of the packing, where equivalent diameter = $4\epsilon/S$.

Values for the gas-film packing factor, R_g, are reported by Morris and Jackson [5], and a few of their values are given in Table I. The following empirical expression may be used to obtain the gas-mixture constant, C:

$$C = \left(\frac{\rho_r}{\mu_r}\right)^{0.25} D_r^{0.5} \tag{17}$$

where ρ_r is the density of the gas mixture in g/cm^3, μ_r is the gas mixture viscosity, in poise, and D_r is the gas diffusion coefficient in cm^2/sec.

The gas velocity relative to the liquid surface is defined as

$$\nu = \nu_g + \nu_l \tag{18}$$

where ν_g is the average velocity of the gas through the packing alone, and ν_l is the liquid surface velocity. Note that

$$\nu_g = \frac{V_g}{\epsilon} \tag{19}$$

where V_g is the gas rate and ϵ is the voidage of the packing. Values of voidage for dry packing are given in Table I.

Note that Equation 17 is only a crude approximation at best, since allowance has not been made for the liquid holdup in the packing (liquid holdup is defined as the product of the liquid-film thickness and the surface area per unit volume of packing). Although a number of liquid-film thickness correlations are reported in the literature, it is best to obtain an empirical correlation from laboratory measurements for the specific packing and absorbing liquid to be used. Various techniques for measuring liquid-film thicknesses based on conductivity and resistivity principles have been described by Cheremisinoff and Davis [15,16].

The log-mean pressure term in Equation 16 can be evaluated as follows:

$$(P - p)_{lm} = \frac{(P - p_i) - (P - p_g)}{\log_e \left(\frac{P - p_i}{P - p_g}\right)} \tag{20}$$

p_g and p_i are the actual and interfacial partial pressures of the soluble gas, respectively. Note that if the gas film controls absorption, then the quantity $P - p_i$ may be approximated by $P - p_l$, where p_l is the partial pressure of the soluble gas in equilibrium with the liquid.

Various correlations are reported in the literature for the liquid-film coefficient in a packed tower. One correlation found to give reasonable predictions for NO_x-water and SO_2-water systems is

$$k_l = R_l \kappa L^n \tag{21}$$

where R_l is the liquid-film packing factor (values given in Table I). The constant κ must be determined experimentally for the system. Exponent n is found to have a value between 0.7 and 1.0. Once the gas- and liquid-film coefficients have been evaluated, the overall coefficients can be computed from Equations 5 and 6.

Note that if Henry's law is applicable, the solubility coefficient, H, is constant for constant temperature conditions. For nonisothermal conditions, H varies with the concentration of the soluble gas in the liquid. This means that the overall coefficients may vary greatly over the tower height.

Tower Cross Section and Height

To estimate tower cross section (Equations 9 and 10) and tower height (Equations 12 and 13), the mean driving force must be evaluated either on the gas- or liquid-phase basis. Appreciation for the magnitude of the driving force is best obtained from constructing a "driving force diagram." These diagrams illustrate the variation of driving force as absorption proceeds.

On the basis of the gas phase, the diagram consists of a plot of concentrations of dissolved gas in the liquid versus partial pressures of the soluble gas, as shown in Figure 5. As shown, two curves are obtained. The *equilibrium curve* represents the change in the equilibrium partial pressure of the soluble gas as the composition of the liquid changes. Note that its shape is established by solubility data; however, corrections must be made for changes in temperature. For most applications it is sufficient to

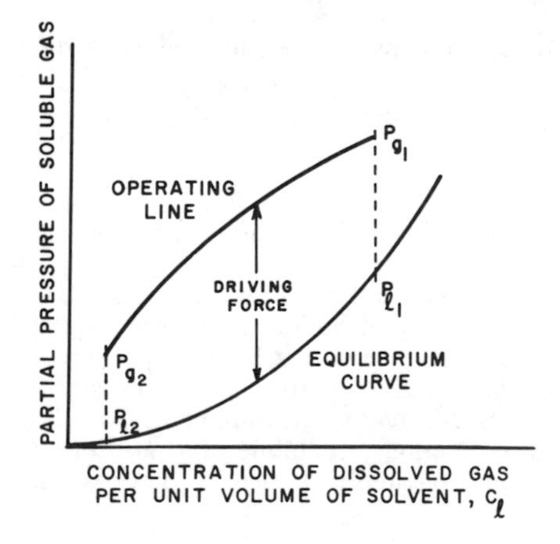

Figure 5. Driving force diagram for the absorption process.

estimate an overall temperature change from a heat balance and assume the temperature is proportional to the amount of gas absorbed.

The operating line in Figure 5 indicates the change in the partial pressure, p_g, of the soluble gas in the gas actually passing through the column, as the composition of the liquid changes due to the absorption process. Its shape is established from a material balance.

The driving force, Δp, at any point in the column is $p_g - p_l$, i.e., the difference between the operating and equilibrium curves corresponding to the appropriate concentration of soluble gas in the liquid. Note, that for absorption to proceed, $p_g - p_l$ must always be positive.

The driving force diagram on the liquid-phase basis differs in that the ordinates of Figure 5 are concentrations of dissolved gas in the liquid. In this case, the operating line reflects the change in the actual concentration, c_l, of dissolved gas in liquid as the liquid composition changes. The equilibrium curve indicates the change in equilibrium concentration, c_g, of dissolved gas as the composition of the liquid changes. The driving force, Δc, at any point in the tower is $c_g - c_l$. Hence, for absorption to proceed, the equilibrium line must lie above the operating line over the entire column, in contrast to the gas-phase basis diagram (Figure 5) when it lies below the operating line. Driving force diagrams on the liquid-phase basis are obviously convenient for use when absorption is liquid-film controlled. When the resistance of both gas and liquid films is significant, the gas-phase basis diagram must be used.

To apply Equations 12 and 13 to obtain the overall height of packing, the mean driving force is defined in terms of the log-mean of the terminal driving forces. For example, on the gas-phase basis, this is

$$\Delta p_{lm} = \frac{\Delta p_1 - \Delta p_2}{\log_e \left(\frac{\Delta p_1}{\Delta p_2}\right)} \tag{22}$$

where Δp_1 and Δp_2 are the terminal driving forces (a similar definition is applied to the liquid-film using concentrations Δc_1 and Δc_2).

An approximate method for determining the mean driving force involves the use of an arithmetic mean of the terminal values. Details are provided in the literature [3,15].

When the average overall coefficient or a mean value of the product of the overall absorption coefficient and driving force cannot be used, integration (either graphically or numerically) of Equations 3 and 4 is required for the rate of absorption. These terms may be rearranged to give the tower cross section:

$$A = Sal = \int_0^W \frac{dW}{K_g(p_g - p_l)} \tag{23a}$$

$$A = Sal = \int_0^W \frac{dW}{K_l(p_g - p_l)} \tag{23b}$$

For graphic integration, a plot of $1/K_g(p_g - p_l)$ or $1/K_l(c_g - c_l)$ versus the rate of soluble gas transfer is prepared and the area under the curve evaluated. From Equations 23a and 23b the tower cross section and height may be computed.

ALTERNATE DESIGN METHOD

An alternate method for sizing towers, and one that has been used in the United States for many years, is the transfer unit method. The procedure provides quick estimates of the required height of packing for a specified duty and thus enables ready comparisons of the merits of different packings. As with the previous method, its use is restricted to nonreacting systems.

When a gas mixture passes through a column, the partial pressure of the soluble gas is reduced from p_{g1} to p_{g2} as absorption proceeds. The rate of absorption may then be expressed as follows:

$$W = V_g \, a\rho_s \left(\frac{p_{g1} - p_{g2}}{P}\right) \tag{24}$$

where ρ_s is the density of the soluble gas.

Also from Equations 12 and 13 we may write

$$W = K_{gm} \, Sal \, \Delta p_m \tag{25}$$

and for the liquid phase,

$$W = K_{lm} \, Sal \, \Delta c_m \tag{26}$$

This last expression assumes that the dissolved gas in the liquid increases from c_{l2} to c_{l1}. Combining Equations 24 and 25 we obtain

$$\frac{p_{g1} - p_{g2}}{\Delta p_{lm}} = \frac{K_{gm} \, Sl}{V_g \, (\rho_s/P)} \tag{27}$$

Similarly, for the liquid film,

$$\frac{c_{l1} - c_{l2}}{\Delta c_{lm}} = \frac{K_{lm} \, Sl}{V_l} \tag{28}$$

The right-hand sides of Equations 27 and 28 represent the ratios of the change in composition (for either the gas phase or liquid phase) to the appropriate mean driving force. These ratios are defined as the gas- or liquid-phase overall transfer units, i.e.,

$$N_{og} = \frac{p_{g1} - p_{g2}}{\Delta p_{lm}} \tag{29}$$

$$N_{ol} = \frac{c_{l1} - c_{l2}}{\Delta c_{lm}} \tag{30}$$

We introduce the definition of the height of an overall transfer unit from Equations 27 and 28 as follows:

$$U_{og} = \frac{V_g \, (\rho_s/P)}{K_{gm} \, S} \tag{31}$$

$$U_{ol} = \frac{V_l}{K_{lm} \, S} \tag{32}$$

where U_{og} and U_{ol} are the heights of the overall transfer unit on both the gas and liquid-phase bases. Since l has dimensions of length in Equations 27 and 28, the left-hand sides of these expressions are dimensionless, and we may write

$$N_{og} = \frac{l}{U_{og}} \tag{33}$$

$$N_{ol} = \frac{l}{U_{ol}} \tag{34}$$

Rearranging Equations 33 and 34 in terms of the packing height, l, and setting the expressions equal to each other, a relationship between N_{og} and N_{ol} is obtained:

$$\frac{N_{ol}}{N_{og}} = \left(\frac{c_{l1} - c_{l2}}{\Delta c_{lm}}\right)\left(\frac{\Delta p_{lm}}{p_{g1} - p_{g2}}\right) = \frac{p_{l1} - p_{l2}}{p_{g1} - p_{g2}} \tag{35}$$

where p_{l1} and p_{l2} are the equilibrium partial pressures of the soluble gas at the entrance and exit of the tower.

Defining the partial pressure change ratio as θ, we may simply write Equation 35 as

$$N_{ol} = \theta\, N_{og} \tag{36}$$

And, similarly,

$$U_{og} = \theta\, U_{ol} \tag{37}$$

An alternate expression for θ may be derived from Equation 24 by noting that $W = V_l\, a\, (c_{l1} - c_{l2})$, to give

$$\theta = \left(\frac{V_g}{V_l}\right)\left(\frac{\rho_s}{HP}\right) \tag{38}$$

As follows from the expression for the overall gas coefficient (Equation 5), we may write the following:

$$U_{og} = \frac{V_g\left(\frac{\rho_s}{P}\right)}{k_g S} + \frac{V_g\left(\frac{\rho_s}{P}\right)}{Hk_l S} \tag{39a}$$

and defining dimensionless groups, we obtain

$$U_{og} = U_g + \theta\, U_l \tag{39b}$$

where U_g and U_l are the heights of transfer units for the gas and liquid films, respectively.

In a similar manner,

$$U_{ol} = \left(\frac{1}{\theta}\right) U_g + U_l \tag{40}$$

Note that the height of a transfer unit for the gas film is defined as

$$U_g = \frac{V_g\left(\frac{\rho_s}{P}\right)}{k_g S} \tag{41}$$

Equation 16 gives an empirical expression for the gas-film coefficient. Assuming that the proportion of soluble gas by volume does not exceed 5–10% and that an average overall coefficient and a log-mean driving force can be used, Equations 41 and 16 can be combined and simplified to the following:

$$U_g = \frac{3600\, \epsilon}{36.1\, R_g C S} \nu_g^{0.25} \left(1 + \frac{\nu_l}{\nu_g}\right)^{-0.75} \tag{42}$$

Table IV. Gas- and Liquid-Film Transfer Unit Heights for Various Packing Materials [5]

Packing						Height of a Transfer Unit			
			Dimensions, in.			Gas Film, U_g		Liquid Film, U_l	
Type	Material	Serial No.	Pitch or Diameter	Height	Thickness	(ft)	(m)	(ft)	(m)
Plain Grids	Metal	1	1	1	1/16	3·2	0·98	1·5	0·46
		2	1	2	1/16	4·0	1·22	1·9	0·58
	Wood	3	1	1	1/4	2·9	0·89	1·55	0·47
		4	1	2	1/4	4·0	1·22	1·95	0·59
Serrated Grids	Wood	5	4	4	1/2	22·2	6·8	2·3	0·70
		6	2	2	3/8	5·9	1·80	1·9	0·58
		7	1½	1½	3/16	5·1	1·55	1·8	0·55
Stacked Rings	Stoneware	8	4	4	3/8	5·9	1·80	2·3	0·70
		9	3	3	3/8	3·7	1·13	2·0	0·61
		10	3	3	1/4	4·5	1·37	2·0	0·61
		11	2	2	1/4	2·4	0·73	1·95	0·59
		12	2	2	3/16	2·5	0·76	1·95	0·59

Random Rings	Metal	13	2	2	1/16	1·5	0·46	2·0	0·61
		14	1	1	1/16	0·75	0·23	1·5	0·46
		15	½	½	1/32	0·31	0·10	1·55	0·47
	Stoneware	16	3	3	⅜	2·6	0·80	2·3	0·70
		17	2	2	¼	1·6	0·49	2·1	0·64
		18	2	2	3/16	1·6	0·49	2.0	0·61
		19	1½	1½	3/16	1·1	0·34	1·8	0·55
		20	1	1	3/32	0·79	0·24	1·55	0·47
		21	¾	¾	3/32	—	—	1·55	0·47
		22	½	½	1/16	0·28	0·09	1·55	0·47
	Carbon	23	2	2	¼	1·7	0·52	2·1	0·64
		24	1	1	3/16	0·7	0·22	1·6	0·49
		25	½	½	1/16	0·28	0·09	1·55	0·47
Solids	Coke	26		3		2·4	0·73	3·0	0·92
		27		1–2		0·81	0·25	2·5	0·77
		28		1		0·61	0·19	2·3	0·70
	Quartz	29		2		1·5	0·46	2·7	0·83
		30		½–1¼		0·51	0·16	2·5	0·77

Note that the relative velocities of the gas and liquid also have been included in this expression from Equation 18. The units in this expression are in cgs.

Note that the term $\epsilon/36.1\ R_gCS$ is a constant that depends only on the packing material for a specific gas mixture. Also, the term ν_l/ν_g is generally small, and Equation 42 can be approximated as

$$U_g \simeq \frac{3600\ \epsilon}{36.1\ R_gCS} \nu_g^{0.25} \tag{43}$$

The height of a transfer unit for the liquid film may be derived in a similar manner through the use of the following definition for U_l and Equation 21:

$$U_l = \frac{V_l}{k_l\ S} \tag{44}$$

Hence, for the liquid film,

$$U_l = \frac{V_l}{R_l \kappa L^n S} \tag{45}$$

or, since $V_l = SL$,

$$U_l = \frac{1}{\kappa R_l} L^{1-n} \tag{46}$$

The term $1/\kappa R_l$ is a constant that depends only on the properties of the packing for a chosen liquid system. The height of packing required for a specific gas-liquid combination can be computed from Equations 33 and 34 for a specified packing.

The height of an overall transfer unit can be computed from Equations 39b and 40 if the partial pressure change, θ, is known. Table IV gives estimated heights of gas- and liquid-film transfer units for various types of packing. The tabulated values can be used as a rough guide comparing the effectiveness of different packings setting the preliminary design.

MALDISTRIBUTION IN PACKED TOWERS

The condition of maldistribution in packed towers exists when the gas to liquid ratio varies across the packing cross section. Maldistribution causes a reduction in tower performance. Hence, considerable attention should be given to specifying and/or designing a proper liquid distribution system.

Although accurate predictions of the magnitude of tower performance reduction occurring during maldistribution are not possible, the important parameters causing poor performance can be evaluated a priori by assuming that a tower in which maldistribution occurs is equivalent to two towers in parallel (that is, each column has half the gas flow between them, but the liquid to each is the same as in the existing unit). It can be shown that the effect of maldistribution on the exit gas composition, in terms of the ratio of the actual performance to that achieved if perfect distribution is obtained, is proportional to the number of overall transfer units, N_{og}, and the partial pressure change ratio, θ, as follows:

1. The exit gas composition increases with the number of overall transfer units on the gas phase basis, N_{og}.
2. For a given number of overall transfer units, the exit concentration is a maximum for values of θ around 0.5.
3. When θ is zero, the exit gas concentration is negligible.

In a similar manner, the effect of maldistribution on the composition of the exit liquid varies with the number of overall transfer units, N_{ol}, and with the concentration change ratio. (Colburn and Pigford [17] give more details.)

It should be emphasized that the effect of maldistribution on tower performance greatly increases when wetting rates fall below the MWL, or if channeling occurs.

NOMENCLATURE

- A interfacial area (m^2)
- a area of cross section of tower (m^2)
- C gas mixture constant
- c_g concentration of soluble gas in equilibrium with gas of partial pressure p_g (kg/m^3)
- c_i concentration of soluble gas at interface (kg/m^3)
- c_l actual concentration of soluble gas in liquid (kg/m^3)
- D absolute gas diffusion coefficient (kg/m^3)
- H solubility coefficient (kg/m^3-atm)
- K_g overall coefficient on gas-phase basis (kg/hr-m^2-atm)
- K_{gm} mean overall coefficient on gas-phase basis (kg/hr-m^2-atm)
- K_l overall coefficient on liquid-phase basis (kg/hr-m^2 or kg/m^3)
- k_g gas-film coefficient (kg/hr-m^2-atm)
- k_l liquid-film coefficient (kg/hr-m^2 or kg/m^3)
- L wetting rate (m^3/hr-m)
- l height of packing (m)
- N number of velocity heads lost per unit height of packing
- N_{og} number of overall transfer units on gas-phase basis
- N_{ol} number of overall transfer units on liquid-phase basis
- n exponent
- P total gas pressure (atm)
- p_g actual partial pressure of soluble gas (atm)
- p_i partial pressure of soluble gas at interface (atm)
- p_l partial pressure of soluble gas in equilibrium with liquid of concentration c_l (atm)
- R_g, R_l gas- and liquid-film packing factors, respectively
- S surface area per unit volume of packing (m^2/m^3)
- T_f absolute temperature of gas film (°K)
- T_r reference temperature (°K)
- U_g, U_l height of transfer unit for gas- and liquid film, respectively (m)
- U_{og}, U_{ol} height of overall transfer unit on a gas-phase basis and a liquid-phase basis, respectively (m)
- V_g gas rate per unit area of cross section (m^3/hr-m^2)
- V_l liquid rate per unit area of cross section (m^3/hr-m^2)
- W rate of absorption (kg/hr)
- Y correction factor for gas and liquid density

Greek Symbols

- ϵ voidage
- θ partial pressure change or concentration change ratio
- κ experimentally determined constant
- μ viscosity (poise)
- ν gas velocity relative to liquid surface (m/sec)
- ν_g gas velocity through packing (m/sec)
- ν_l liquid surface velocity (m/sec)
- ρ density (kg/m^3)

REFERENCES

1. Sherwood, T. K., and R. L Pigford. *Absorption and Extraction*, 2nd ed. (New York: McGraw-Hill Book Co., 1952).

2. McCabe, W. L., and J. C. Smith. *Unit Operations of Chemical Engineering,* 3rd ed. (New York: McGraw-Hill Book Co., 1976).
3. Treybal, R. E. *Mass-Transfer-Operations,* 2nd ed. (New York: McGraw-Hill Book Co., 1968).
4. Whitman, W. G. "The Two-Film Theory of Absorption," *Chem. Met. Eng.* 29 (1923).
5. Morris, G. A., and J. Jackson. *Absorption Towers* (London: Butterworth Publishers, Inc., 1953).
6. Shulman, H. L., et al. *AIChE J.* 1:247, 253, 259 (1955).
7. Shulman, H. L., et al. *AIChE J.* 3:157 (1957).
8. Shulman, H. L., et al. *AIChE J.* 5:280 (1959).
9. Shulman, H. L., et al. *AIChE J.* 6:175, 469 (1960).
10. Shulman, H. L., et al. *AIChE J.* 9:479 (1963).
11. Sherwood, T. K., G. H. Shipley and F. A. L. Holloway. "Flooding Velocities in Packed Towers," *Ind. Eng. Chem.* 30 (1938).
12. Bain, W. W., and O. A. Hougen. "Flooding Velocities in Packed Columns," *Trans. Am. Inst. Chem. Eng.* 40 (1944).
13. Lobo, W. E., et al. "Limiting Capacity of Dumped Tower Packings," *Trans. Am. Inst. Chem. Eng.* 41 (1945).
14. Carey, W. F., and G. J. Williamson. "Gas Cooling and Humidification–Design of Packed Towers from Small-Scale Test," *Proc. Inst. Mech. Eng.* 163 (W. E. P. No. 56.) (1950).
15. Cheremisinoff, N. P., and E. J. Davis. "Stratified Turbulent-Turbulent Gas Liquid Flow," *AIChE J.* 25(1) (1979).
16. Cheremisinoff, N. P., and E. J. Davis. "Pressure Drop and Holdup," *AIChE-DIMP Report 4* (1977).
17. Colburn, A. P., and R. L Pigford. "General Theory of Diffusional Operations," in *Chemical Engineers Handbook,* 3rd ed., J. H. Perry, ed. (New York: McGraw-Hill Book Co., 1950).

CHAPTER 22
HYDRODYNAMICS OF BUBBLE COLUMNS

Y. T. Shah
Department of Chemical and Petroleum Engineering
University of Pittsburgh
Pittsburgh, Pennsylvania 15261

W.-D. Deckwer
FB9/Chemie
Postfach 25-03
D2900 Oldenburg
West Germany

CONTENTS

INTRODUCTION

Bubble column reactors are widely used in chemical process industries for carrying out gas-liquid and gas-liquid-solid reactions such as oxidation, hydrogenation, chlorination, aerobic fermentation, coal liquefaction, etc. Some applications of bubble columns for gas-liquid reactions are outlined in Table I.

Bubble column is a term used to describe a vertical column wherein gas is bubbled through either a moving or stagnant pool of liquid or a liquid-solid slurry. In general, however, the column need not be vertical; it can be horizontal or even coil shaped. This chapter examines the hydrodynamics of vertically sparged bubble columns. The

Table I. Industrial Applications of Bubble Columns

System	Reference
1. Absorption of CO_2 in ammoniated brine for the manufacture of soda ash	Danckwerts and Sharma [1], Kafarov et al. [2]
2. Absorption of CO_2 in aqueous buffer solutions and amines	Danckwerts and Sharma [1], Danckwerts [3]
3. Simultaneous absorption of NH_3 and CO_2	Ramachandran and Sharma [4]
4. Absorption of isobutylene and butenes in aqueous solutions of H_2SO_4	Gehlawat and Sharma [5], Kroper et al. [6], Deckwer et al. [7,8], Deckwer [8]
5. Partial oxidation of ethylene to acetaldehyde	Smidt et al. [9], Smidt et al. [10), Jira et al. [11]
6. Oxidation of ethylene in acetic acid solution	Krekeler and Schmitz [12]
7. Oxidation of acetaldehyde to acetic acid	Sittig [13], Kostylak et al. [14]
8. Oxidation of acetaldehyde to acetic anhydride	Yau [15]
9. Acetic acid manufacture from oxidation of Sec. butanol	Sittig [13]
10. Oxidation of butanes to acetic acid and methyl ethyl ketone	Broich et al. [16], Broich [17], Höfermann [18], Saunby and Kiff [19]
11. Oxidation of toluene to benzoic acid	Kaeding et al. [20]
12. Oxidation of xylene to phthalic acid	Haase [21]
13. Oxidation of cumene to phenol and acetone	Hattori et al. [22], Hagberg and Krupa [23]
14. Oxidation of ethylbenzene to acetophenon	Sittig [13]
15. Wet oxidation of wastewater	Beyrich et al. [24]
16. Alkylation of phenols with isobutylene diluted with inert gas	Gehlawat and Sharma [25]
17. Ethylation and propylation of benzene to ethylbenzene and cumene	*Hydrocarbon Processing* [26], Griesbaum et al. [27]
18. Carbonylation of methanol to acetic acid	von Kutepow et al. [28], Hjortkjaer and Jensen [29]
19. Hydroformylation of olefins to aldehydes and alcohols	Oliver and Booth [30], Weissermei and Arpe [31], Albright [32]
20. Oxychlorination of ethylene to dichloroethane	Friend et al. [33], Ramachandran and Sharma [4], Rassaertz and Witzel [34]
21. Chlorination of aliphatic and aromatic hydrocarbons	Hawkins [35], Sittig [13], van den Berg [36], Rathjen [37], Tan and Ratcliffe [38]
22. Oxydesulfurization of paraffins for the manufacture of paraffin sulfonates	*Hydrocarbon Processing* [39]

discussion is restricted to gas-liquid systems. Various types of bubble columns used in chemical process industries are described by Shah et al. [40].

Bubble columns have found diverse applications in chemical industries because of their simple construction and the absence of moving parts. Due to high degree of backmixing, bubble columns have good heat transfer properties, which make them useful for reactions with high heat generation. Large liquid holdup also makes them useful when carrying out slow reactions that require large liquid holdup. The design and

Table II. Some Characteristics of Industrial Bubble Columns

A. Ranges of Operating Variables	
Size	
Volume	
Chemical process industry	< 200 m^3
Biochemical processes	< 3000 m^3
Wastewater treatment	< 20000 m^3
Diameter	0.2–20 m
Length to diameter ratio	3–10
Volume throughputs	
Superficial gas velocity	< 100 cm/sec
Superficial liquid velocity	< 10 cm/sec
Typical liquid-phase properties	
Viscosities	0.5–100 mPas
Densities	0.6–2 g/cm^3
Surface tension	20–73 dyne/cm
Kind of gas sparger	
Perforated plates, single and multiorifice nozzles, various types of two-phase nozzles	
B. Advantages	
Simple construction	
No moving parts	
Low energy input (only gas compression)	
High liquid circulation rate and, hence, good mixing	
Good heat transfer	
High liquid holdup	
Ability to process corrosive and toxic gases at high temperatures and pressures	
Possibility of reacting gases that form explosive mixtures	
C. Disadvantages	
Complex hydrodynamic flow pattern	
Uncertainties in scaleup	
Short residence time of gas	
Volume demand usually increased by axial mixing	

scaleup of a bubble column are difficult because of the complex flow pattern prevailing in the column. The effects of the complex flow pattern on the mixing and transport properties are understood at present only in empirical terms. Some typical characteristics of industrial bubble columns, their advantages and disadvantages are described briefly in Table II.

This chapter describes the state-of-the-art of the hydrodynamics of gas-liquid bubble columns. The discussion is restricted mainly to the cocurrent upflow system with the practical range of operating variables. An assessment of the existing theories is given, along with their experimental verifications, whenever possible. Some recommendations for future work are also outlined.

FLOW REGIME, BUBBLE DYNAMICS AND PRESSURE DROP

Flow Regime

The phase holdup, mixing and transport characteristics of a bubble column depend significantly on the prevailing flow regime in the column [41]. Govier and Aziz [42] summarized all the flow regimes that can be observed visually for gas-liquid flows through pipes. The liquid and gas flowrates ranged from 0.01 to 3 m/sec and 0.01 to 30 m/sec, respectively. For most bubble column applications in chemical and petroleum industries, the ranges of liquid velocities of 0 to 5 cm/sec and of gas velocities of 1 to about 50 cm/sec are important. For these ranges, the nature of flow depends largely on the gas velocity and column diameter. At low gas velocities (less than approximately 5 cm/sec), ideal bubbly flow (homogeneous flow) is obtained, whereas at high gas velocities, slug flow in small-diameter columns and churn turbulent flow in large-diameter columns are obtained. A flow regime plot is shown in Figure 1 for an air-water system. The determination of the flow regimes and their boundaries often is based on visual observation, which can lead to some inaccuracies. As shown in Figure 1, there is a large transition zone between the flow regimes.

The boundaries and transition zones between flow regimes also depend on the gas sparger design and physicochemical properties of the fluids [40,43]. As shown in Figure 2, the effect of the fluid properties on the flow regime boundaries was reported by Cichy et al. [41], who used the modified Govier coordinate system ($G_L\lambda\psi G_G$ vs G_G/λ), and by Oshinowo and Charles [44], who introduced their own coordinate systems. A more detailed discussion of these flowcharts has been given recently by Miller [45]. Unfortunately, the flow regime plots shown in Figure 2 have not been tested for large-diameter columns, in which the churn-turbulent flow is more likely to occur than the slug flow.

Bubble Dynamics

If the gas velocity is low, the bubble sizes are rather uniform and their rise velocities approximately equal. In this case, the bubbles hardly interact with each other. They rise upward in an uniform manner without significant coalescence, and the flow is called "pseudohomogeneous" or "bubbly." In the bubble flow regime, if the bubble sizes and their rise velocities are constant, a maximum amount of gas can be transported through the column. This gas throughput can be estimated from the flooding point calculations [46,47]. Downflow bubble columns [48] always operate in the bubble flow regime, and their upper limit of operating conditions corresponds with the maximum gas throughput in the bubble flow regime. The amount of gas transportable in the bubble flow regime can be increased considerably by increasing the cocurrent liquid flowrate [48,49].

If the gas flowrate is increased above the flooding point, large bubbles with high buoyancy are formed and the transition to churn-turbulent or heterogeneous flow occurs.

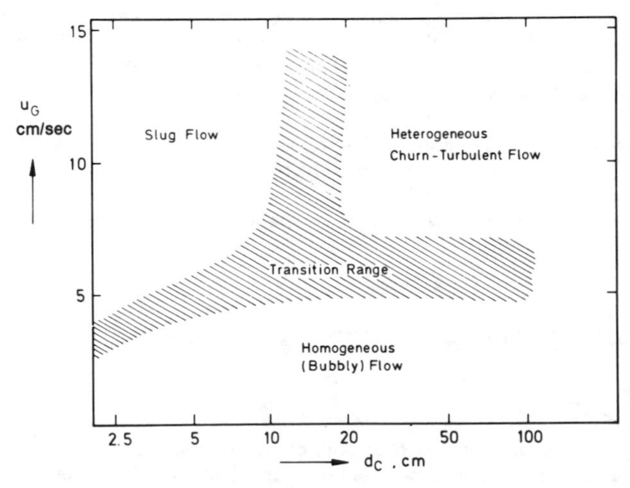

Figure 1. Flow regime plot for bubble columns (air-water system).

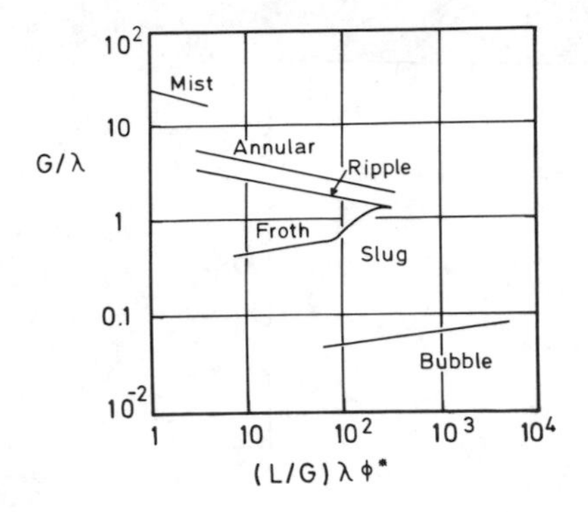

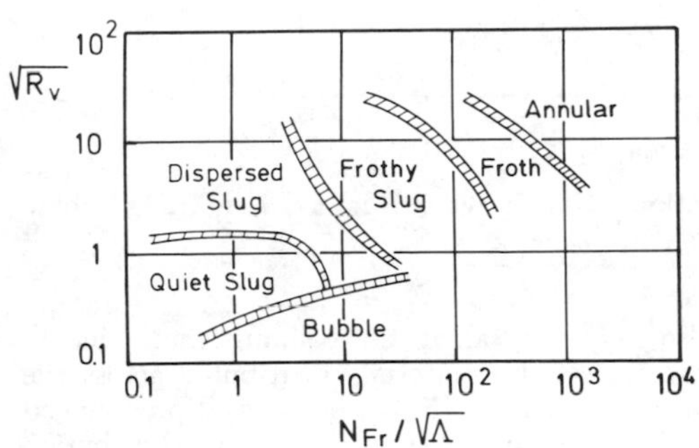

Figure 2. Flow regime charts [41,44].

In this flow regime, which is found in most industrial applications, large bubbles moving predominantly at the centerline in a churn-like motion coexist with small bubbles typically found in the homogeneous flow regime [50,51]. However, with increasing gas velocity, the amount of gas transported through the column in the form of large bubbles increases. At high gas velocities, almost all gas is transported by large bubbles and bubble clusters.

A peculiar situation can be observed in small-diameter bubble columns, in which the large bubbles are stabilized by the wall to form slugs (Figure 3). Slug flow is a commonly observed phenomenon for the flow through ducts, but is not typical for the bubble columns of industrial size. Although bubbles and bubble clusters of about 10 cm in diameter are observable in bubble column, they generally do not form slugs if the column diameter is larger than about 15 cm [45,52]. While the calculations indicate that the maximum stable bubble sizes are of the order of 7 cm [53], in highly viscous Newtonian and non-Newtonian media and in tall bubble columns bubble slugs can be found even when the column diameter is large [54,55]. As shown in Figure 4, for Carboxyl Methyl Cellulose (CMC) solutions with high effective viscosities slug flow can be observed in a 14-cm ID bubble column even at low gas velocities, say $u_G < 3$ cm/sec.

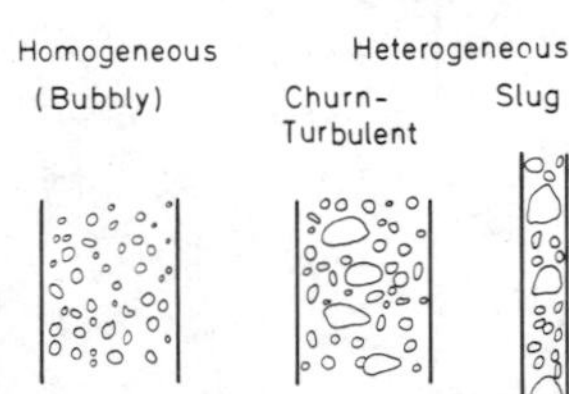

Figure 3. Bubble appearance in different flow regimes in bubble columns.

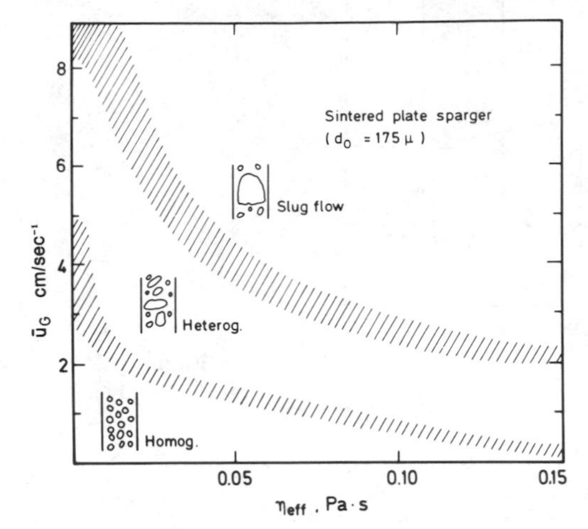

Figure 4. Flow regimes in CMC solutions (0.6–2 wt %) in a 14-cm I.D. bubble column [55].

The gas holdup in bubble columns also may vary along the column length due to coalescence. Only at low gas velocities can the gas be uniformly distributed across the cross-sectional area. Figure 5 presents the results of Hills [56], who showed pronounced radial holdup profiles in bubble columns. This nonuniformity of the bubble distribution plays an important role in hydrodynamic models of the bubble columns.

Beinhauer [50] measured the holdup of the large and small bubbles in a bubble column of 10 cm in diameter using water as the liquid phase and a sintered plate as a gas sparger. These results are shown in Figure 6. Although the holdup of small bubbles is rather large (i.e., about 20% of the reactor volume) for the gas velocities larger than 4 cm/sec, one should note that the large bubbles rise much more quickly as shown in Figure 7. The constant rise velocity above u_G = 80 cm/sec results because the velocity of the slugs is determined mainly by wall friction. Transition from the bubble

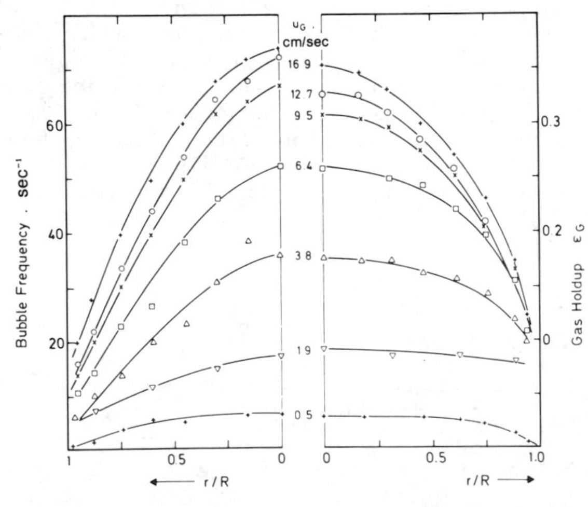

Figure 5. Bubble frequency and gas holdup as a function of the radius [52].

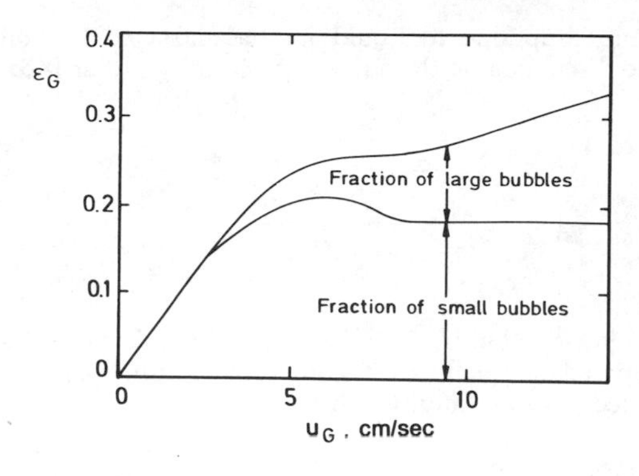

Figure 6. Fractional gas holdup of large and small bubbles [50,51].

flow to heterogeneous flow usually can be recognized easily by the sharp increase of bubble rise velocity, $u_G{}^*$.

Pressure Drop

In vertically sparged reactors, the pressure drop due to friction at reactor walls can be neglected. The pressure drop is composed of the drop exerted by the gas sparger and the hydrostatic head of the liquid. Therefore, the pressure profile within the column is given by

$$P(z) = P_T(1 + \alpha(1 - z)) \tag{1}$$

where z is the dimensionless axial coordinate, P_T the pressure at column top, and α represents the ratio of the hydrostatic head to the pressure at column top:

$$\alpha = \frac{\rho_L(1 - \epsilon_L)gL}{P_T} \tag{2}$$

where L is the dispersion height.

As the gas expands in the column, the axial pressure variation in the pressure should be considered for the calculation of the local gas velocities, provided α is large (say $\alpha > 0.2$). This may be important for the processes that are carried out in tall reactors at normal pressure. Examples are fermentation reactors, where α may be larger than 1, or for the absorption of isobutane in sulfuric acid [57,58]. At elevated pressures, i.e., for small α, the pressure drop can be neglected. Hills [52,56] and Miller [45] pointed

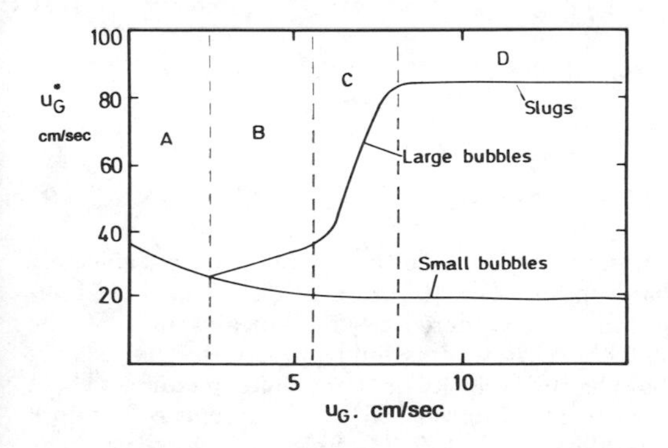

Figure 7. Rise velocities of large and small bubbles [50,51].

out that at high gas velocities pressure drop due to liquid inertia and wall friction may be significant. The most up-to-date discussion on the subject is given by Miller [45].

BUBBLE FLOW MODELS

Ideal Bubbly Flow

Slip Velocity Concept

A widely accepted method for treating the bubble columns at low gas flowrates is the concept of ideal bubbly flow, proposed originally by Lapidus and Elgin [46]. According to this, the slip velocity is defined from continuity and is given by

$$u_s = \frac{u_G}{\epsilon_G} \pm \frac{u_L}{1 - \epsilon_G} \tag{3}$$

and u_s is assumed to be independent of the flowrates of either phase and only a function of the bubble rise velocity, $u_{B\infty}$, and the gas holdup, ϵ_G:

$$u_s = u_{B\infty}\, \phi\, (\epsilon_G) \tag{4}$$

In Equation 3, + refers to countercurrent flow and − refers to cocurrent flow. Various expressions for $\phi\,(\epsilon_G)$ have been proposed that are summarized by Lockett and Kirkpatrick [47] as follows:

Turner [59]:

$$\phi(\epsilon_G) = 1 \tag{5}$$

Davidson and Harrison [60]:

$$\phi(\epsilon_G) = 1/(1 - \epsilon_G) \tag{6}$$

Wallis [61]:

$$\phi(\epsilon_G) = (1 - \epsilon_G)^{n-1} \tag{7}$$

n = 2 for small bubbles   n = 0 for large bubbles

Richardson and Zaki [62]:

$$\phi(\epsilon_G) = (1 - \epsilon_G)^{n-1} \tag{8}$$

where n depends on Re in case of air-water: n = 2.39.

Marrucci [63]:

$$\phi(\epsilon_G) = (1 - \epsilon_G)/(1 - \epsilon_G{}^{5/3}) \tag{9}$$

These expressions are barely distinguishable at low values of ϵ_G but differ significantly at higher values, say $\epsilon_G \geqq 0.15$. Richardson and Zaki [62] and Bridge et al. [64] indicated that the value of n generally can be considered as an indicator of how the individual bubble is affected by its neighbors, as well as fluid properties. The velocity $u_{B\infty}$ in Equation 4 is the terminal bubble rise velocity in an infinite medium. These were measured, for instance, by Haberman and Morton [65], and the entire literature on the subject is well summarized by Clift et al. [53]. Bubble diameters, normally encountered in small-diameter bubble columns, lie between 1 and 10 mm, corresponding to rise velocities of single bubbles ranging from 10 to 35 cm/sec [53,65]. For batch liquid systems, Equations 3 and 4 give

$$u_G = \epsilon_G \, u_{B\infty} \, \phi \, (\epsilon_G) \tag{10}$$

Freedman and Davidson [66] showed that for this case Turner's [59] approach describes the measured data for the air-water and air-electrolyte solutions fairly well up to gas velocities of about 5 cm/sec (corresponding to ϵ_G values of about 0.25). However, for aeration of commercial pink paraffin, where holdup values up to 0.5 could be achieved, the data points closely follow the expression derived by Marrucci [63].

Drift Flux Concept

Wallis [61] introduced the concept of "drift flux" as a means of relating the phase flow velocities and gas holdup. Drift flux, γ_{CD}, is defined as the volumetric flux of gas relative to the surface moving at an average velocity, i.e.,

$$\gamma_{CD} = u_s \epsilon_G \, (1 - \epsilon_G) \tag{11}$$

Combining Equations 4 and 11, we obtain

$$\gamma_{CD} = u_{B\infty} \epsilon_G (1 - \epsilon_G) \phi(\epsilon_G) = u_G (1 - \epsilon_G) \pm u_L \epsilon_G \tag{12}$$

Rice et al. [67] pointed out that the usefulness of the drift flux structure resides in the simplicity of a graphic solution to Equation 12. Thus, as shown in Figure 8 for given values of u_G, u_L (and assuming that $\phi(\epsilon_G)$ also is known), ϵ_G can be determined from the intersection of the linear right-hand side and the usually nonlinear function on the left-hand side of Equation 12.

Using $\phi(\epsilon_G) = 1 - \epsilon_G$, Rice et al. [67] tested the validity of Equation 12 for air-water flow in the presence of a surfactant (terpineol) and found good agreement between theory and experiment. Often, a Richardson-Zaki type relationship gives a good description of measured data [47,49,68]. Using n = 2.39 in the Richardson-Zaki

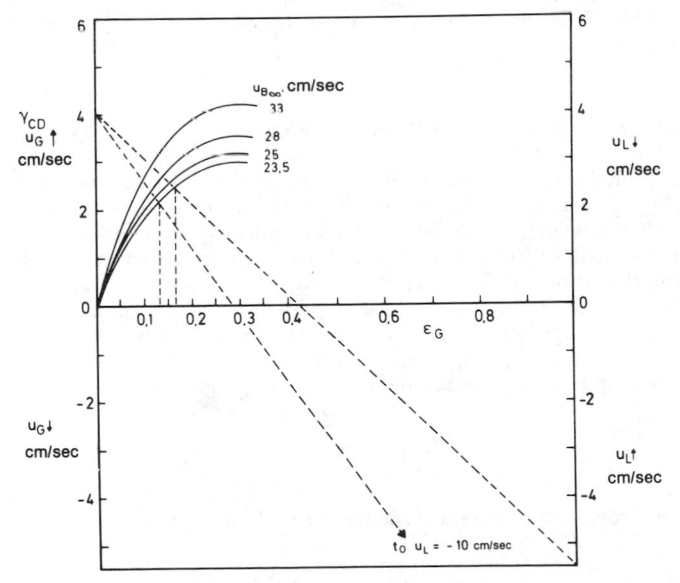

Figure 8. Drift flux vs gas holdup.

relation, Equation 12 is plotted for various values of $u_{B\infty}$ in Figure 8. One can discern from this figure that for $u_{B\infty} = 23.5$ cm/sec and cocurrent upflow of gas and liquid with $u_G = 4$ cm/sec and $u_L = -5.5$ cm/sec, a gas holdup of 0.165 is predicted. If the cocurrent liquid flowrate is increased to $u_L = 10$ cm/sec, ϵ_G reduces to 0.13.

Weiland [49] measured gas holdup in a bubble column with external liquid recycle (airlift reactor) and used cocurrent liquid flowrates up to 36 cm/sec. The agreement between his data and the predictions of the drift flux model generally was good, being better at larger liquid flowrates and smaller gas flowrates. This result is in agreement with the predictions of Freedman and Davidson [66], who analyzed the flow pattern in a bubble column with an inserted draft tube. These authors observed lower gas holdup values with increasing liquid circulation. The drift flux model should be applicable particularly to downflow bubble columns [48], which always are operated in the bubble flow regime.

In a special experimental arrangement with stationary bubble clouds, Lockett and Kirkpatrick [47] found the gas holdup values up to 0.66. Various equations describing the bubble flow regime were tested by these authors, but none of the proposed equations was found to be completely satisfactory. Lockett and Kirkpatrick used the Richardson-Zaki relation with n = 2.39 and introduced an empirical correction factor, which accounted for the bubble deformation. The final equation used to describe the data is

$$\gamma_{CD} = u_{B\infty}\epsilon_G(1 - \epsilon_G)^{2.39}(1 + 2.55\epsilon_G{}^3) \tag{13}$$

High gas holdups achieved by Lockett and Kirkpatrick usually are not observed in the industrial bubble columns. Gas holdup values up to 60% have been reported for cocurrent upflow by Schugerl et al. [69] and Oels et al. [70] for special liquid systems (synthetic fermentation media) and by the use of two-phase nozzle spargers only. These data, however, were not analyzed using the concept of ideal bubbly flow.

Normally, the pseudohomogeneous bubbly flow regime can be maintained for the gas holdup up to about 15–20% only. According to Lockett and Kirkpatrick [47], breakdown of bubbly flow has to be attributed to (1) flooding, (2) liquid circulation, and (3) bubble coalescence, which lead to formation of large bubbles or bubble clusters with higher rise velocities. Furthermore, high interphase mass transfer rate would allow the bubble flow regime to occur only at low ϵ_G [71]. The radial holdup profiles [56] also lead to the breakdown of bubble flow and the onset of liquid recirculation.

Model of Zuber and Findlay: Churn-Turbulent Flow

As pointed out by Govier and Aziz [42], many workers have recognized that the flow of gas-liquid mixtures is affected greatly by the local relative velocity between gas and liquid phases, which is caused most commonly by gravitational forces. In this respect, the treatment of Zuber and Findlay [72] is of particular value. The continuity equations for the dispersed flow systems represent the starting point of this theory, which predicts a relation for the gas holdup that has a broad application in the interpretation of the holdup data. This relation applies to both bubble and churn-turbulent flow regimes and can be applied, at least empirically, to slug and annular flow regimes.

A general holdup relation of Zuber and Findlay [72] is given by

$$\frac{V_{S\alpha}}{\epsilon_\alpha} = V_\alpha = C_O V_M + V_{\alpha j} \tag{14}$$

where ϵ_α is the cross-sectional average fraction of α phase,

$$\epsilon_\alpha = \frac{1}{A}\int_0^A \epsilon_\alpha(r)\,dA \tag{15}$$

and $\epsilon_\alpha(r)$ is the local volume fraction of α phase. V_α is the actual α phase velocity or interstitial velocity, and $V_{S\alpha}$ presents the superficial velocity, i.e., the volume throughput divided by the cross-sectional area. V_M is the superficial velocity of the mixture and is given by

$$V_M = V_{S\alpha} + V_{S\beta} \tag{16}$$

The quantity $V_{\alpha j}$ presents the average drift velocity weighted with regard to the volume fraction:

$$V_{\alpha j} = \frac{\overline{\epsilon_\alpha(r) v_{\alpha j}(r)}}{\epsilon_\alpha} \tag{17}$$

Here, $v_{\alpha j}$ is the local drift velocity or the local velocity of the α phase relative to the local velocity of the mixture $v_M(r)$:

$$v_{\alpha j}(r) = v_\alpha(r) - v_M(r) \tag{18}$$

With consideration of

$$v_M(r) = \epsilon_\alpha(r) v_\alpha(r) + \epsilon_\beta(r) v_\beta(r) \tag{19}$$

and the relative velocity

$$v_r(r) = v_\alpha(r) - v_\beta(r) \tag{20}$$

the local drift velocity $v_{\alpha j}(r)$ can be written as

$$v_{\alpha j}(r) = (1 - \epsilon_\alpha(r))\, v_r(r) \tag{21}$$

The constant C_o in Equation 14 is given by

$$C_O = \frac{\overline{\epsilon_\alpha(r) v_M(r)}}{\epsilon_\alpha V_M} \tag{22}$$

Equation 14 also can be written in the form

$$\frac{V_{S\alpha}}{\epsilon_\alpha V_M} = \frac{C_\alpha}{\epsilon_\alpha} = C_O + \frac{V_{\alpha j}}{V_M} \tag{23}$$

where C_α is the relative volume throughput of α phase:

$$C_\alpha = \frac{V_{S\alpha}}{V_{S\alpha} + V_{S\beta}} \tag{24}$$

For the special case of $V_{\alpha j} = 0$, Equation 23 shows that C_O is the reciprocal to Bankoff's K factor [72], which also is denoted as a distribution coefficient.

In the nomenclature used in context with the bubble column, Equation 14 can be written as

$$u_G{}^* = \frac{u_G}{\epsilon_G} = C_O(u_G + u_L) + u_{Gj} \tag{25}$$

This equation presents a linear relationship between u_G/ϵ_G and $(u_G + u_L)$ and, when plotting u_G/ϵ_G vs $(u_G + u_L)$, the slope is C_O and the intercept is the volume fraction-weighted drift velocity, which, according to Equation 17, is obtained from

$$u_{Gj} = \frac{1}{\epsilon_a}\frac{1}{A}\int_0^A \epsilon_G(r)\, u_{Gj}(r)\, dA \tag{26}$$

or, by consideration of Equations 20 and 21,

$$u_{Gj} = \frac{1}{\epsilon_a}\frac{1}{A}\int_0^A \epsilon_G(r)\,(1 - \epsilon_G(r))\,(u_G(r) - u_L(r))\,dA \tag{27}$$

Zuber and Findlay [72] illustrated the effects of nonuniform flow and holdup distribution parameter, C_O. They assumed that in a vertical circular pipe flow both profiles are axially symmetric, and they introduced profiles of the form

$$\frac{v_M(r)}{v_{MO}} = 1 - \left(\frac{r}{R}\right)^m \tag{28}$$

where v_{MO} is the value of v_M at the centerline and

$$\frac{\epsilon_a(r) - \epsilon_{aW}}{\epsilon_{aO} - \epsilon_{aW}} = 1 - \left(\frac{r}{R}\right)^n \tag{29}$$

Here, the subscript W refers to the wall. With these profiles, it follows from the definition of C_O, i.e., Equation 28, that

$$C_O = 1 + \frac{2}{m + n + 2}\left[1 - \frac{\epsilon_{aW}}{\epsilon_a}\right] \tag{30}$$

when expressed in terms of ϵ_{aW} or

$$C_O = \frac{m + 2}{m + n + 2}\left[1 + \frac{\epsilon_{ao}}{\epsilon_a}\frac{n}{m + 2}\right] \tag{31}$$

when expressed in terms of ϵ_{ao}. It follows from the above equations that

$$C_O = 1$$

if $\epsilon_{aW} = \epsilon_{ao} = \epsilon_a$, regardless of the flow velocity distribution. Usually, the holdup is larger at the centerline than at the wall, $\epsilon_{ao} > \epsilon_{aW}$, and in this case

$$C_O > 1$$

If the holdup and velocity drop linearly from the center to the wall ($m = n = 1$), the value of C_O varies from 1.5 to 1 for fully established profiles (equilibrium conditions). If both holdup and velocity follow a parabolic distribution, C_O ranges between 1.33 to 1.0, depending on the value of $\epsilon_{aW}/\epsilon_{ao}$. For other profiles, the values of the distribution parameters C_O also are given by Zuber and Findlay, who discuss how n and m can be determined from the experimental holdup data.

The local drift flux velocity, v_{aj}, represents the local velocity of the bubble phase with respect to the local volumetric flux density of the mixture. The simplest expression for this velocity is to assume that the bubbles do not interact, i.e., each bubble moves independently and is not affected by the presence of other bubbles. In such a case, the drift velocity is equal to the terminal rise velocity of the bubbles. Hence, the local drift velocity of the gas is given by the following expressions:

$$v_{aj} = 0.35\left(\frac{g\Delta\rho\sigma}{\rho_L^2}\right)^{1/2} \tag{32}$$

for the slug flow regime and

$$v_{aj} = 1.53 \left(\frac{\sigma g \Delta\rho}{\rho_L{}^2}\right)^{1/4} \tag{33}$$

for churn-turbulent flow. Zuber and Findlay [72] show that if Equations 32 and 33 apply, then the average drift velocity weighted with regard to the volume fraction also is given by these equations, i.e.,

$$V_{aj} = \frac{\overline{\epsilon_a(r) v_{aj}(r)}}{\epsilon_a} = v_{aj} = u_{Gj} \tag{34}$$

Therefore, for churn-turbulent flow it follows from Equation 25 that

$$\frac{u_G}{\epsilon_G} = C_O(u_G + u_L) + 1.53\left[\frac{\sigma g \Delta\rho}{\rho_L{}^2}\right]^{1/4} \tag{35}$$

Zuber and Findlay [72] used this equation to analyze approximately 120 data points reported by Petrick [73] for air-water flow in a 14-cm I.D. circular pipe. The result of this analysis is shown in Figure 9. It can be seen that the data points follow a linear relation with respect to the average volumetric flux density of the mixture. The value of C_O is 1.6, which indicates that bubble density is considerably larger at the centerline than at the wall. The value of the intercept corresponds to the weighted mean drift velocity for churn-turbulent flow and is predicted by Equation 32.

When the presence of other bubbles affects the motion of a given bubble, then the drift velocity, v_{aj}, will depend on the number of bubbles present in a given dispersion volume. For this case, Zuber and Findlay [72] proposed relations such as

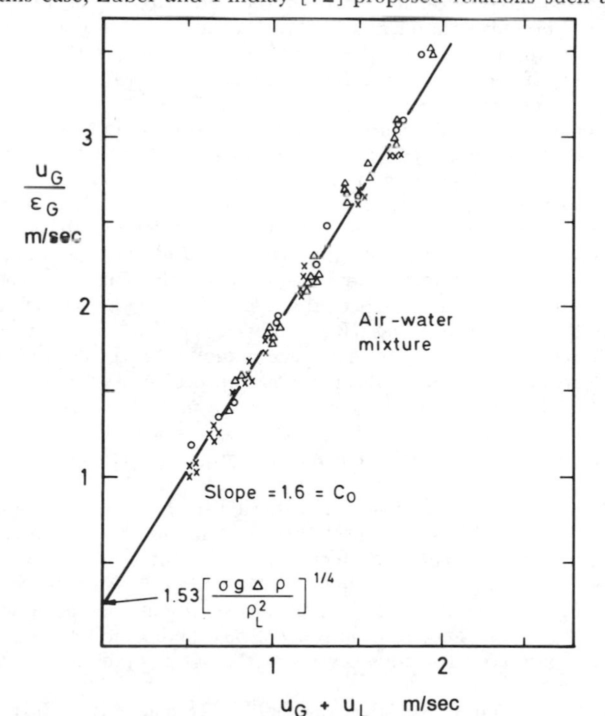

Figure 9. Experimental data of Petrick [73] plotted according to Equation 35 [72].

$$v_{aj} = u_{B\infty} (1 - \epsilon_G)^k \tag{36}$$

where the exponent k, which is a measure of the bubble interference and the terminal rise velocity of a single bubble in an infinite medium, $u_{B\infty}$, depends on the bubble size. It should be noted that Equation 36 is equal to Equation 4 for the slip velocity. This underlines the similarity between the slip velocity concept and the approach of Zuber and Findlay [72].

Hills [52] measured gas holdup in a bubble column of 15-cm I.D. by 10.5-m height at gas velocities of 0.07–3.5 m/sec and liquid velocities of up to 2.7 m/sec. For $u_L >$ 0.3 m/sec, the results were well correlated by the equation

$$\frac{u_G}{\epsilon_G} = 0.24 + 1.35\,(u_G + u_L)^{0.93} \tag{37}$$

which has the same structure as Equation 25. Kara [74] also has correlated gas holdup data in gas-liquid-particle flow by equations similar to Equations 25 and 37, respectively.

Bubble Swarm Dynamics—Coalescence and Breakup of Bubbles

Bubble coalescence and breakup are important phenomena in the design and scaleup of tall and unstaged bubble columns, in which all the gas is distributed from the bottom. Due to coalescence, large bubbles may be generated along the column length, which decreases the interfacial area and, hence, the mass transfer rate. Coalescence predominates at high gas velocities, i.e., for high bubble densities. The transition from bubbly flow to churn-turbulent flow is a result of a coalescence process. Calderbank et al. [75] suspect that coalescence sets in if the ratio of the distance between two bubble centers and the mean Sauter diameter of the bubbles is less than 2. The bubble breakup also may be important in both the bubbly and churn-turbulent flow regimes. As a matter of fact, under certain conditions, breakup of bubbles even may be more dominant than coalescence [76–78].

While the mechanism of coalescence in bubble swarms is not yet well understood, generally it is believed to be dependent on a number of operating parameters. Besides flow velocities, it is the composition of the liquid phase and its physical properties (especially liquid viscosity), which affect coalescence rates. Thus, in highly viscous Newtonian and non-Newtonian media, spherical cap bubbles and bubble slugs can be observed even at very low gas velocities [54,55,75]. In addition, there are some indications that for some systems (e.g., diluted solutions of alcohols [79] and electrolytes [80–81]) initial bubble sizes and, hence, the type of sparger, determine equilibrium bubble sizes and, therefore, coalescence. It is believed that coalescence takes place more readily in pure liquids. In this case, the interface is mobile, and the liquid film between bubbles can be narrowed rapidly to the point of rupture by a stretching process [82–84]. In solutions, as well as for the liquids with trace impurities, film stretching is opposed by interfacial tension gradients, which lead to a lower film-thinning rate. However, in high viscous media, another mechanism might be responsible for the rapid coalescence.

Coalescence in bubble column reactors was studied in the pioneering work of Calderbank et al. [75]. These authors used the light and gamma ray transmission technique to measure the local gas holdup and interfacial area in a rectangular column. From these data, the number of bubbles per unit volume of dispersion, N, and the Sauter diameter of the bubbles, d_s, can be calculated. The measurements were done at low gas velocities, $u_G \leq 5.1$ cm/sec. Therefore, in low viscous media the prevailing flow regime was bubbly flow. Some typical results for the air-water system are shown in Figures 10A and 10B.

It can be seen from Figure 10A that the holdup is constant (0.05), and bubbles of about 3.5 mm are generated at the sieve plate sparger. These bubbles coalesce until

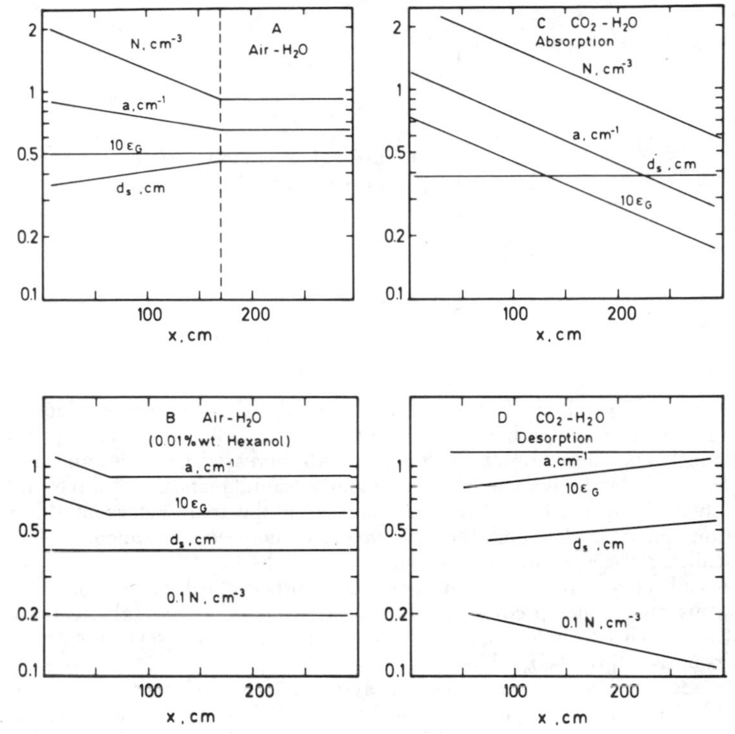

Figure 10. Gas holdup, interfacial area, bubble diameter and bubble density as a function of the column height [75].

their diameter is 5 mm, which occurs at a height of 170 cm above the sparger. If 0.01 wt% of n-hexanol is added to the water, coalescence is suppressed and the equilibrium bubble diameter is reached already after a short distance above the sieve plate (Figure 10B). In the case of absorption of CO_2 in water, the measured and calculated N, d_s data are shown in Figure 10C. Here, the bubble density, holdup and interfacial area decrease rapidly, while the bubble diameter is constant along the length of a column. If CO_2 desorbs from saturated water into air bubbles (Figure 10D), the bubble diameter and holdup increase. For desorption, some coalescence can be observed, but it is less pronounced than for absorption.

Calderbank et al. [75] explained their results on the basis that the bubble rise velocity depends on the bubble diameter up to a value of $d_s = 0.4$ cm and is approximately constant at higher values of d_s. Therefore, in this latter region variation of bubble size may occur without appreciable coalescence (Figure 10D). However, if absorption occurs, the bubble sizes may decrease below 0.4 cm, where the rise velocity depends on d_s. Then rapid coalescence may occur by an "overtaking" mechanism. Thus, if absorption takes place, a dynamic equilibrium is being established whereby the mean bubble diameter remains approximately constant.

Calderbank et al. [75] also studied coalescence in glycerol-water mixtures. In these solutions with higher viscosities, coalescence rates were much more pronounced than the ones in low viscous media. Figure 11 gives the conditions for coalescence within a distance of about 3 m above the gas sparger as a function of the gas holdup and viscosity. If the viscosity is larger than about 70 cP, the gas holdup decreases sharply due to the formation of spherical caps by coalescence. From the visual observation, Cal-

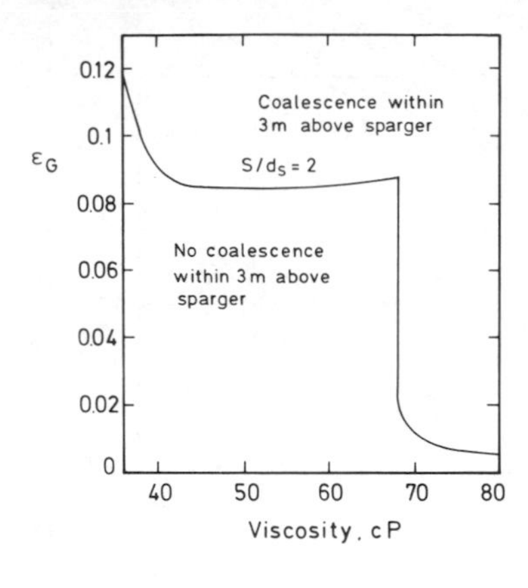

Figure 11. Region of coalescence and non-coalescence [75].

derbank et al. [75] found that the generation of spherical cap bubbles starts with clustering of ellipsoidal bubbles into assemblies, whose shape already approaches that of a spherical cap. These bubble clusters rise with increased velocities and, after a short period of time, the individual bubbles coalesce simultaneously. Bubbles in the path of the spherical cap are overtaken and coalesce at the rear surface of the cap. Thus, the capture cross section and the rise velocity increase continuously, leading to a process called self-accelerating coalescence.

Calderbank et al. [75] have developed a semitheoretical model for coalescence in high viscous media that predicts the height at which the first spherical cap appears. The authors have concluded that the process of coalescence sets in with a formation of clusters of six ellipsoidal bubbles.

Coalescence in air-water and air-electrolyte solutions also was studied by Lee and Ssali [84]. They interpret their data on the basis of a simple model of coalescence, which has been used already by Calderbank et al. [75]. It is assumed that the rate of coalescence (the change in bubble density) is proportional to the bubble density:

$$-\frac{dN}{dt} = f\,N \tag{38}$$

where f represents a coalescence frequency factor. From their experimental results, which refer to low gas velocities ($u_{Go} \leq 2.8$ cm/sec), Lee and Ssali evaluated the frequency factors. The f values obtained were small ($f \leq 0.023$ sec^{-1}), indicating that only modest coalescence had taken place.

Otake et al. [76] were the first to show that in a medium of lower viscosity not only coalescence, but also the breakup of bubbles, represent a common phenomenon. These authors studied the three-dimensional motion of bubbles in bubble swarms using high-speed photography. Figure 12 shows the results of Otake et al. [76] in aqueous glycerol solutions (viscosity 12.5 cP). The coalescence and breakup frequencies were defined as the number of events of coalescence and breakup per unit dispersion volume and time. The frequency of bubble passage is the number of bubbles passing per unit cross-sectional area and time. Figure 12 shows that the coalescence frequency is high immediately after the nozzle sparger. The zone of high coalescence is followed by a region in which bubble breakup achieves a maximum value. At a distance of about 20 cm above the nozzle, coalescence and breakup rates are approximately equal. Therefore, the frequency of bubble passage is approximately constant along the column except for a region of about 10 cm above the nozzle, where the bubble frequency decreases sharply. The effect of liquid properties on the variation of the bubble-passing frequency with the dispersion height is shown in Figure 13. For the case of ethanol, the bubble-passing frequency increases in the lower part of the column, indicating that breakup of bubbles predominates, whereas in aqueous glycerol

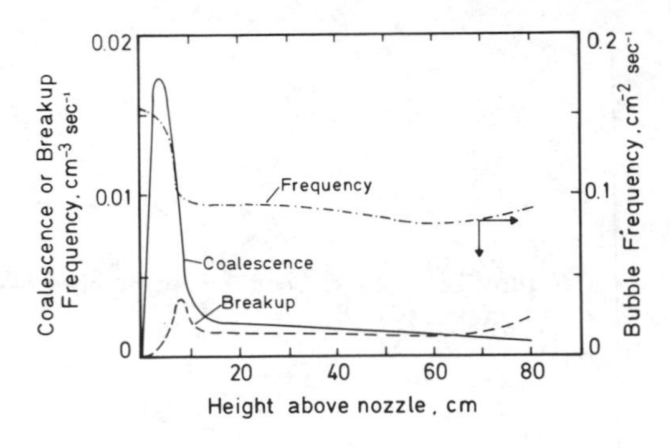

Figure 12. Frequencies of bubble passing, coalescence and breakup, schematically redrawn from Otake et al. [76].

solutions bubble coalescence takes place. For all liquids, the bubble-passing frequency is constant along the column above a height of 20 cm from the sparger. Thus, a dynamic equilibrium between bubble coalescence and breakup is established.

The technique applied by Otake et al. [76] permitted a direct observation of the bubble coalescence and breakup events. Otake et al. found that coalescence takes place when more than about half the projected area of the following bubble is overlapped by that of the leading bubble at a critical distance of about three to four times the bubble diameter. On the other hand, bubble breakup occurs if the overlapping is less than about one-half the projected area of the following bubble. In general, a tendency for the coalescence was found if the leading bubble is larger than the following one. On the other hand, if the leading bubble has a smaller size than the following one, the following one tends to break up.

The occurrence of bubble breakups in bubble swarms also was shown indirectly by Deckwer et al. [77,78]. These authors investigated the absorption of CO_2 (from air-CO_2 mixtures) in water in tall bubble columns under the conditions of high mass transfer rates between gas and liquid phases. CO_2 axial concentration profiles in both phases and the local gas holdup were measured. The bubble diameter determined at 3.5-m and 6-m heights above the sparger were found to be approximately the same for both absorption and desorption experiments. From the experimental data, Deckwer et al. [77,78] could calculate the local bubble density, N (bubbles/cm^3) and the bubble flux, J (bubbles/cm^2-sec), which is equivalent to the bubble-passing frequency. It was assumed that the local variation of J is proportional to the bubble density as

$$\frac{dJ}{dx} = k\,N \tag{39}$$

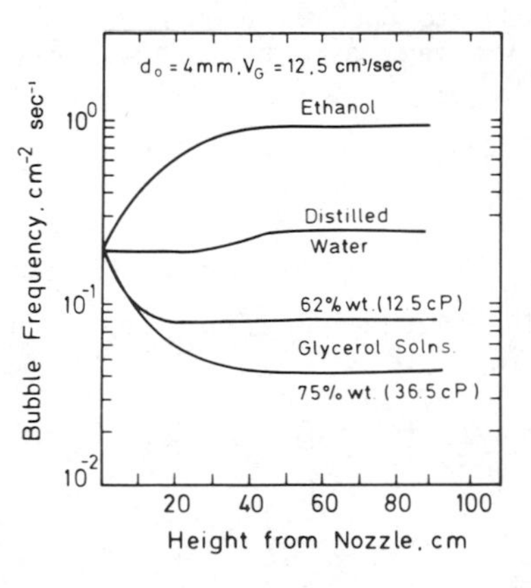

Figure 13. Bubble frequency (flux) for different systems, schematically redrawn from the work of Otake et al. [76].

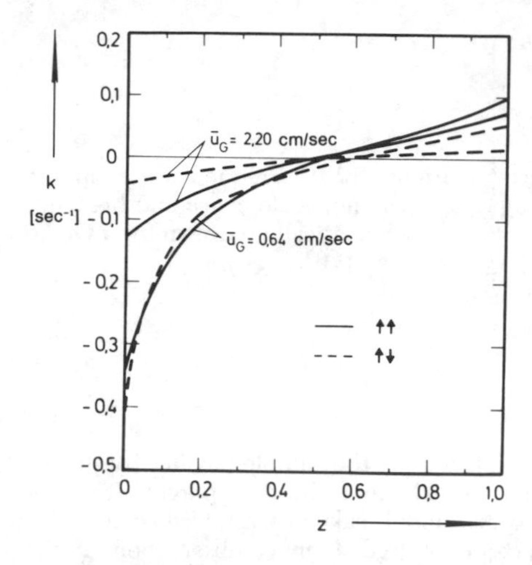

Figure 14. Frequency factor in absorption runs [77,78].

Here, k again can be interpreted as a frequency factor of coalescence and breakup events. Obviously, if $k < 0$ coalescence predominates, while breakup of bubbles prevails if k is positive. For an absorption experiment, a typical data evaluation for the calculation of k is shown in Figure 14. In the lower part of the column, where strong absorption occurred, the bubble density and flux decreased, giving negative k values. In the immediate vicinity of the gas sparger, the coalescence frequency factor was as high as 0.3 sec^{-1}. Comparable values were reported by Calderbank et al. [75] for the absorption experiments. In the upper part of the column, the volumetric gas flowrate increased due to reduced pressure and possible desorption of CO_2 [78]. As the bubble diameter was found to be constant, bubble breakup predominated in the upper part of the column. As shown in Figure 15, for the desorption of CO_2 breakup prevailed along the entire column length. Breakup frequencies were the largest at the bottom for the cocurrent flow and at the top for the countercurrent flow.

Bubble breakup was not observed by Calderbank et al. [75] and Lee and Ssali [84]. However, the observation of Otake et al. [76] clearly demonstrates that both coalescence and breakup of bubbles are common phenomena, even at very low bubble densities. The bubble size distribution generated by the sparger is significantly different from the distribution at a short distance above the sparger, which is a result of simultaneous coalescence and breakup processes. The findings of Deckwer et al. [77,78] confirm the phenomenon of breakup, even at higher gas velocities (up to 5 cm/sec). If mass transfer rates between the phases are high, a dynamic equilibrium between coalescence and breakup gives an approximately constant bubble size along the main part of the column. However, this conclusion is only valid for water as a liquid phase. The

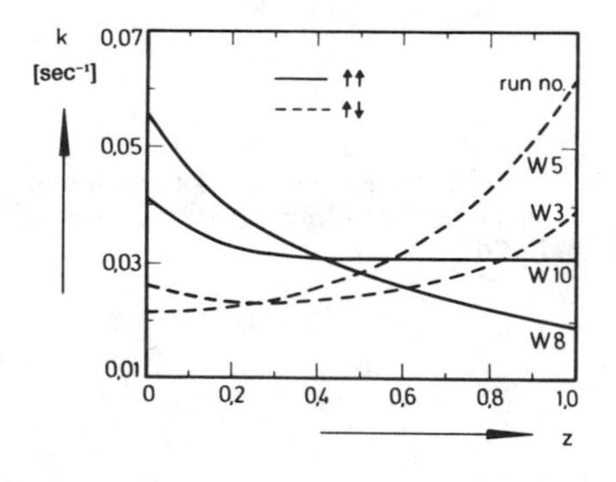

Figure 15. Frequency factor in desorption runs [77].

coalescence-breakup behavior of bubbles in organic liquids and liquid mixtures has not been studied yet. Further experimental and theoretical work are required to determine which other physical properties besides viscosity are responsible for the onset of coalescence and breakup.

Summary

The treatment given here for the ideal bubbly flow regime and the churn-turbulent flow regime is valuable for the analysis and interpretation of experimental data. Both concepts are limited in predictions, however, and apply preferentially at high phase velocities. So far, only very simplified models have been proposed to describe the coalescence phenomena in bubble swarms, and a priori predictions are not possible. Additional studies in this area also should consider the bubble breakup.

MODELS OF FLOW PATTERN IN THE LIQUID PHASE

In the past decade, a number of attempts have been made [67,85–94] to describe the flow pattern in the liquid phase of a bubble column. Moving gas bubbles carry liquid to an extent such that in typical bubble column operations the transport of liquid from the bottom to the top in the center of the column is usually larger than the average volumetric throughput. Therefore, liquid must flow downwards in the annular region near the wall to satisfy the continuity equation for the liquid phase, and this results in the circulation of liquid. The liquid circulation velocity depends on the superficial gas velocity, column diameter, gas holdup, bubble diameters, bubble use velocity and liquid phase viscosity.

For high-viscosity liquids and at low gas velocities, the flow pattern is laminar. Laminar liquid circulation caused by bubble streets and chains has been modeleld by Rietema and Ottengraph [85] and Crabtree and Bridgewater [95]. The analyses of these authors are useful in understanding basic features of the liquid circulation. However, in most bubble columns the flow pattern is essentially turbulent; therefore, the above models will not be considered here. Although the liquid circulation model of Bhavraju et al. [92] is applicable to turbulent conditions, it assumes multiple circulation cells in traverse direction. Such a situation can be expected only in shallow bubble beds, which are used when treating waste water in lagoons, for instance. In bubble columns used in chemical industries, the length to diameter ratio usually is large, say $L/d_C > 5$; therefore, traverse multiple circulation cells cannot be encountered. In addition, this model does not account for the energy dissipation at the gas-liquid interface and is unable to predict the circulation velocity. The details of this model are given by Joshi and Shah [96].

Recirculating Flow Model of Miyauchi and Co-Workers

In this model, the bubble column is assumed to be of an infinite height (i.e., end effects are neglected), in which all the liquid flows vertically and the pressure is constant over the entire cross section [56]. A force balance on an annular element leads to [56,93,94]

$$-\frac{1}{r}\frac{d}{dr}(r\tau) = \frac{dP}{dz} + (1-\epsilon_G)\rho_L g \qquad (40)$$

The shear stress, τ, is defined through the turbulent kinematic viscosity, as molecular kinematic viscosity can be neglected:

$$\tau = -\nu_t \rho_L \frac{du'_L}{dr} \tag{41}$$

Miyauchi and co-workers [93,94] assumed the flow pattern of liquid in a bubble column, as shown in Figure 16. Liquid flow is upwards and turbulent in the innermost region, $a < r < b$; it is downwards and turbulent as well in the annular region, $b < r < c$. In the laminar sublayer, $c < r < d$, the liquid flow is downwards and laminar. To simplify the mathematical treatment, the existence of a buffer layer between the laminar sublayer and the turbulent downward flow is neglected. As the thickness of the sublayer is much smaller than the column diameter, one can assume that the peripheral velocity of the turbulent core, u_W, extends to very nearly $r = R$. Then the boundary condition can be written as

$$u_W = -11.63\sqrt{|\tau_W|/\rho_L} \qquad \text{at } r = R \tag{42}$$

Another boundary condition is

$$\frac{du'_L}{dr} = 0 \qquad \text{at } r = 0 \tag{43}$$

If the turbulent viscosity, ν_t, is regarded as constant within the turbulent core and if the radial gas holdup profile is approximated by

$$\frac{\epsilon_G}{\bar{\epsilon}_G} = \frac{n+2}{n}(1 - \bar{\phi}^n) \tag{44}$$

with $\bar{\phi} = r/R$, then the solution of Equation 40, which satisfies both boundary conditions, is given by

$$u'_L + |u_W| = \frac{R^2}{\nu_t}\left[\left(\frac{\tau_w}{2R\rho_L} + \frac{\bar{\epsilon}_G g}{2n}\right)(1 - \bar{\phi}^2) - \frac{\bar{\epsilon}_G g}{n(n+2)}(1 - \bar{\phi}^{2+n})\right] \tag{45}$$

The solution incorporates unknown quantities $|u_w|$ and τ_w, which can be eliminated by further manipulations with the help of Equation 42.

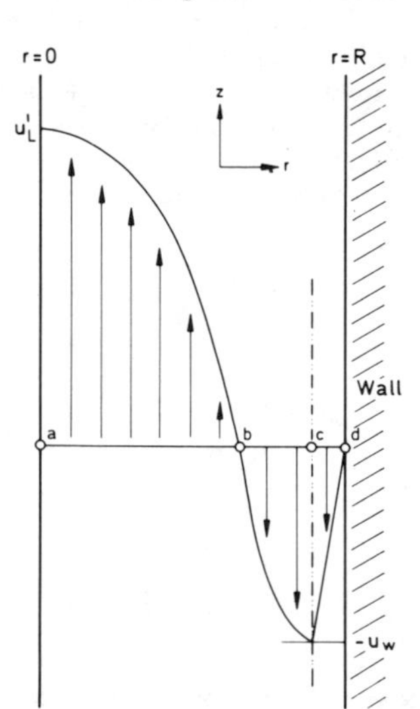

Figure 16. Liquid flow pattern [94].

The complete solution with consideration of a more general gas holdup profile,

$$\epsilon_G(\bar{\phi}) = \epsilon_G(0)\ (1 - C\bar{\phi}^n) \tag{46}$$

which was proposed empirically by Kato et al. [95] and also has been reported by Kojima et al. [97] as

$$u'_L(\bar{\phi}) = u_w\ [\Gamma_1(1 - \bar{\phi}^2) - 1] + \frac{\rho_L g d_c^2}{4\mu_t}[I(\bar{\phi}) - (1 - \bar{\phi}^2)\Gamma_o] \tag{47}$$

where

$$u_w = \frac{269.2\mu_t\Gamma_1}{\rho_L d_c} \times \left[-1 + \sqrt{1 + \frac{(d_c/2)^3\rho_L^2 g[2\Gamma_o - I(1)]}{134.6\ \mu_t^2\Gamma_1^2}}\right] \tag{48}$$

Γ_o, Γ_1, $I(\bar{\phi})$ and $I(1)$ are functions of $\epsilon_G(\bar{\phi})$ and are given by the following expressions

$$\Gamma_o = \epsilon_G(o)\ \frac{(1 - \epsilon_G(o))(n+2)^2[(n+2)(n+4) - 8C] + 8\epsilon_G(o)C[(n+2)^2 - C(n+4)]}{4(n+2)^2[(1 - \epsilon_G(o))(n+2)(n+4) + 8\epsilon_G(o)C]} \tag{49}$$

$$\Gamma_1 = \frac{2(n+4)[(1 - \epsilon_G(o))(n+2) + 2\epsilon_G(0)C]}{(1 - \epsilon_G(o))(n+2)(n+4) + 8\epsilon_G(o)C} \tag{50}$$

$$I(\bar{\phi}) = \epsilon_G(o)\left[\frac{1 - \bar{\phi}^2}{4} - \frac{C(1 - \bar{\phi}^{n+2})}{(n+2)^2}\right] \tag{51}$$

$$I(1) = \frac{\epsilon_G(o)(n + 2 - 2C)}{2(n+2)} \tag{52}$$

The gas holdup in the centerline of the column is given by

$$\epsilon_G(o) = \frac{\bar{\epsilon}_G}{1 - \dfrac{2C}{n+2}} \tag{53}$$

where $\bar{\epsilon}_G$ is the mean holdup value.

From the above equations, the distribution of the liquid-phase velocity can be calculated for any value of the turbulent viscosity, μ_t, provided the radial variation of the gas holdup, i.e., the values of n and C, is known.

The turbulent viscosity is not known a priori. Ueyama and Miyauchi [94] and Kojima et al. [97] determined the value of μ_t by matching experimental and predicted liquid velocity profiles. As suspected by Miyauchi and Shyu [93], the turbulent viscosity is independent of the gas velocity and determined mainly by the column diameter. The data were correlated empirically by

$$\mu_t = 0.0536\ d_c^{1.77} \tag{54}$$

As shown in Figure 17, this correlation includes data from 10- to 550-cm-diameter columns. Substitution of Equation 54 gives the complete velocity profile, which strongly depends on d_C.

Ueyama and Miyauchi [94] have reviewed the literature on the radial variation of the gas holdup. The average values of n and C in Equation 46 were found to be two

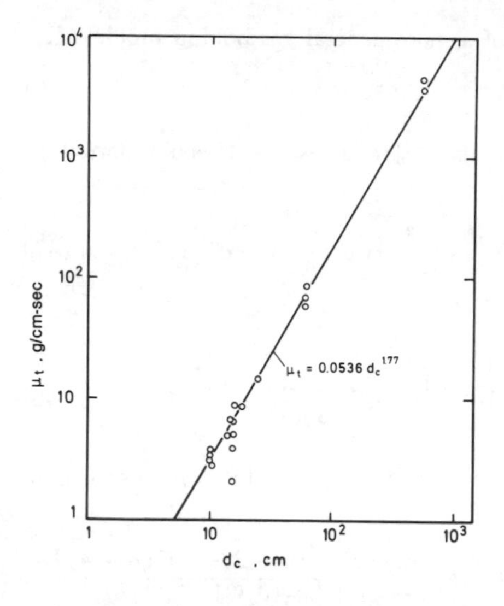

Figure 17. Correlation of turbulent viscosity with column diameter [97].

and one, respectively. Using approximate correlations for the liquid velocity at reactor wall and the mean holdup, Ueyama and Miyauchi [94] have found that

$$\left|\frac{\tau_w}{R\rho_L}\right| << \frac{\bar{\epsilon}_G g}{n} \tag{55}$$

Therefore, Equation 45 simplifies to

$$\frac{u'_L + |u_w|}{u'_{Lo} + |u_w|} = (1 - \bar{\phi}^2)^2 \tag{56}$$

where

$$u'_{Lo} + |u_w| = \frac{d_c^2 g \bar{\epsilon}_G}{32\nu_t} \tag{57}$$

and, with consideration of Equation 54, one obtains

$$u'_{Lo} + |u_w| = 0.583\, g\bar{\epsilon}_G d_c^{0.23}\, \rho_L \tag{58}$$

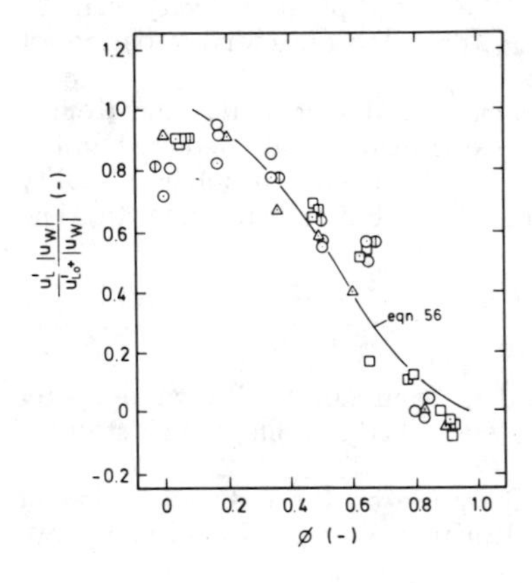

Figure 18. Comparison of experimental data [56, 98,99] with Equation 56 [94].

As shown in Figure 18, good agreement is found between theoretical predictions and the experimental velocity profiles in 13.8- to 60-cm-diameter bubble columns [56,98, 99]. However, Kojima et al. [97] reported that in a 5.5-cm-diameter bubble column, the liquid velocity profile is better fitted using $n = 3$ and $C = 0.7$ in Equation 46.

The hydrodynamic model of Miyauchi and co-workers [93,94] also gives an explicit expression for the mean slip velocity in bubble columns with recirculating turbulent flow as

$$u_s = \frac{u_G}{\bar{\epsilon}_G} - \frac{1}{1 - \bar{\epsilon}_G}\left(\pm u_L + \frac{u'_{Lo} + |u_w|}{n + 4}\right) \tag{59}$$

Equation 59 shows that an apparent interstitial liquid velocity, $(u'_{Lo} + |u_w|)/(n + 4)$, is added to the net liquid flowrate. If there is no recirculation flow, Equation 57 reduces to Equation 3 for the bubble flow regime. The available experimental data are in fair agreement with Equation 59. Ueyama and Miyauchi [94] also have compared bubble rise velocities of single bubbles with u_s calculated from Equation 59 and found only slight differences. They concluded from this result that in the churn-turbulent flow regime bubbles apparently have little interaction, and each bubble rises independently.

The model of Miyauchi and co-workers presents a very fundamental approach on the hydrodynamic behavior of bubble columns in a churn-turbulent flow. Application of this model requires the knowledge of the radial variation of the gas holdup. For the air-water system, the values of n and C in Equations 44 and 46, respectively, are known with sufficient accuracy. However, for other gas-liquid systems, n and C may vary. Therefore, for the estimation of the liquid velocity profile in other systems, additional measurements are needed to clarify the radial gas holdup profiles and only then can the treatment of Miyauchi and co-workers be generalized.

Riquarts [100] neglected the effect of wall friction on the pressure gradient and made the same assumptions as those of Ueyama and Miyauchi. In addition, he assumed that the velocity gradient is zero in the column centerline and obtained the following solution of the basic balance equation, i.e., Equation 40:

$$u'_L(\bar{\phi}) = u'_{Lo} - \frac{\bar{\epsilon}_G g\, d_c^2}{32\nu_t}[1 - (1 - \bar{\phi}^2)^2] \tag{60}$$

where the liquid velocity in the centerline is given by

$$u'_{Lo} = \left(\frac{1 - 0.75\,\bar{\epsilon}_G}{1 - \bar{\epsilon}_G}\right)\frac{\bar{\epsilon}_G\, g\, d_c^2}{48\,\nu_t} \tag{61}$$

Compared to the solution of Kojima et al. [97], i.e., Equations 47 through 52, the calculation of the liquid velocity profile from the above equations is appreciably simpler. As shown below in the section entitled Dispersed Plug-Flow Model and Recirculating Flow, Riquarts [100] also derived an expression that permits calculation of ν_t with the knowledge of the column diameter and gas velocity. It should be noted that the solution of Riquarts does not involve the radial holdup profile but only the mean gas holdup.

The velocity distribution given by Equation 56 and illustrated in Figure 18 apparently correlates the measured data. However, Riquarts [100] pointed out that if u'_L/u'_{Lo} is plotted against $\bar{\phi}$, the agreement between the theory and the experiment is poor, particularly for high values of $\bar{\phi}$. Therefore, Riquarts proposed an approximate correlation for the data on the basis of regression polynomials as

$$\frac{u'_L}{u'_{Lo}} = a_o + a_1\bar{\phi} + a_2\bar{\phi}^2 + a_3\bar{\phi}^3 + a_4\bar{\phi}^4 \tag{62}$$

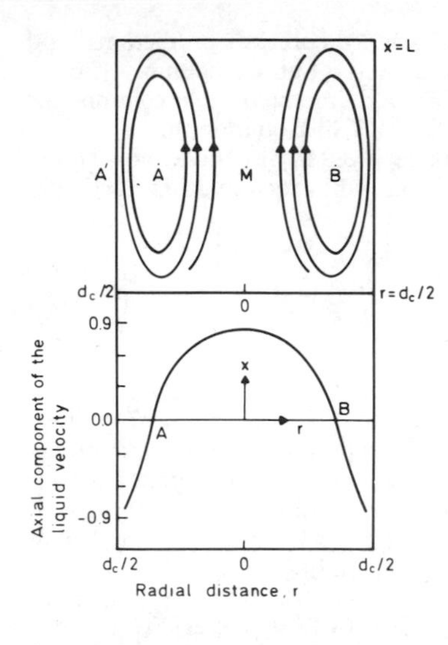

Figure 19. Liquid circulation and velocity profile [90].

The coefficients are determined from the data reported by Hills [56], Pavlov [101] and Ueyama and Miyauchi [94] for 13.8-, 17.2- and 60-cm-diameter bubble columns and are compiled in Table III. Equation 62 describes the measured liquid velocity distributions with striking agreement, and its use appears to be the simplest way to calculate the velocity distribution.

Energy Balance Method

Freedman and Davidson [66] developed a "gulf stream model" for a two-dimensional (rectangular) bubble column based on a pressure balance. In general, the solution of the model equations required a considerable mathematical effort, and the velocity field and the bubble paths predicted therefrom were in good agreement with the experimental observations. However, the subsequent work of Whalley and Davidson [87] and Joshi and Sharma [88–91] showed that the energy balance method is considerably superior to the pressure balance. Therefore, these models are considered below.

Whalley and Davidson Model

Whalley and Davidson [87] applied the energy balance for the calculation of the velocity field in shallow ($L/d_c = 1$) bubble beds. The basis of the energy balance method is to equate (1) the rate of energy input to the column by the gas to (2) the rate of energy dissipation by the fluid motion. The energy input is given approximately by the gas volume flowrate and the pressure drop across the liquid height as

$$E_i = u_G \frac{\pi d_c^2}{4} (\rho_L - \rho_G) L (1 - \bar{\epsilon}_G) g \tag{63}$$

Table III. Coefficients for Calculating u_L'/u_{Lo}' from Equation 62

Range	a_o	a_1	a_2	a_3	a_4
$0 \leq \bar{\phi} \leq 0.7$	1	0	−0.5	0	−3
$0.7 < \bar{\phi} \leq 0.9$	−234.397	1166.379	−2142.483	1723.390	−513.702

The kinetic energy can be neglected. There are several modes of energy dissipation:

1. Energy dissipation in the wakes behind the bubbles is

$$E_1 = \frac{\pi}{4} d_c^2 L \epsilon_G (1 - \bar{\epsilon}_G)(\rho_L - \rho_G) g u_{B\infty} \quad (64)$$

The above equation assumes $u_s = u_{B\infty}$:

2. Viscous energy dissipates along the walls of the contactor.
3. Kinetic energy is dissipated when the liquid decelerates during downward liquid motion.
4. Energy dissipates in the hydraulic jump at the liquid surface.

For the case of axisymmetrical bubble columns, Whalley and Davidson [87] have shown that only the first and fourth modes are important. The rate of energy dissipation in the hydraulic jump then results in the energy balance:

$$E = E_i - E_1 \quad (65)$$

For the calculation of velocity profiles, E also is obtained by an alternate way. For an axisymmetrical case, Lamb [102] has given the following equation for the vorticity, ω, in the inviscid fluids:

$$\omega = -\frac{1}{r}\left[\frac{\partial^2\psi}{\partial x^2} + \frac{\partial^2\psi}{\partial r^2} - \frac{1}{r}\frac{\partial\psi}{\partial r}\right] \quad (66)$$

where the coordinates are explained in Figure 19.
Further, Lamb [102] has shown that the vorticity divided by the radial distance from the axis of symmetry is dependent on the stream function only. The following equation has been used successfully by Whalley and Davidson for the prediction of bubble path in a shallow pool of liquid:

$$\frac{\omega}{r} = K_2\psi \quad (67)$$

The boundary conditions are

$$\psi = 0 \quad \text{at } r = 0 \quad \text{and } r = d_c/2 \quad (68)$$

$$\psi = 0 \quad \text{at } x = 0 \quad \text{and } x = L \quad (69)$$

The procedure for the solution of Equation 66 is given by Whalley and Davidson. The solution of Equation 67 gives the following velocity profile [91]:

$$u_L(z,\bar{\phi}) = \frac{2\omega_o}{K_2 A d_c}\left\{\left[\frac{\pi}{L}\frac{R^1}{\bar{\phi}}\cos(\pi z)\right]^2 + \left[\frac{2}{d_c\bar{\phi}}\frac{dR^1}{d\bar{\phi}}\sin(\pi z)\right]^2\right\}^{1/2} \quad (70)$$

where ω_o is maximum vorticity or the circulation strength.
Whalley and Davidson [87] assumed that all the kinetic energy associated with the upward-moving liquid at the plane AMB (in Figure 19) is dissipated in the hydraulic jump and equals E given by Equation 65. Therefore, one obtains

$$E = \frac{\pi\rho_L(1 - \bar{\epsilon}_G)\psi_o^3}{(d_c/2)^4}\int_0^{2A/d_c}\frac{1}{\bar{\phi}^2}\left(\frac{dR^1}{d\bar{\phi}}\right)^3 d\phi \quad (71)$$

where R^1 is the radial component of the stream function.

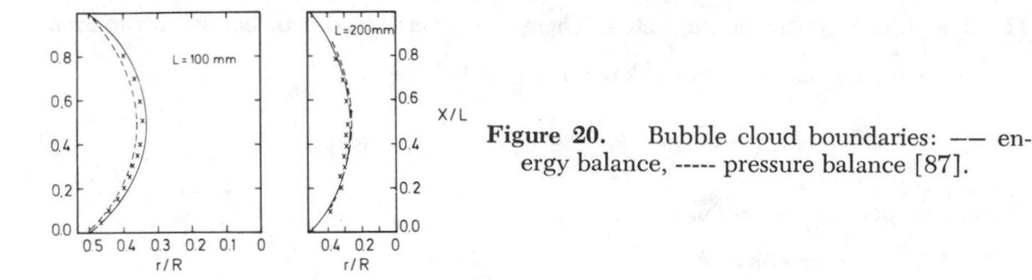

Figure 20. Bubble cloud boundaries: —— energy balance, ----- pressure balance [87].

Whalley and Davidson [87] determined the shape of the bubble cloud and the liquid velocity in the centerline of an axisymmetrical shallow bubble bed ($d_c = 30$ cm, $L/d_c < 1$). The gas was sparged only within a central circular region of 7.5-cm radius. As shown in Figures 20 and 21, the measured bubble cloud shape and the variations of maximum axial superficial liquid velocity with gas velocity are in general agreement with the predictions of both the pressure balance method of Freedman and Davidson and the energy balance method. However, Whalley and Davidson concluded that the energy balance method predicts the bubble cloud shape better than the pressure balance method.

The energy balance model has the definite advantage that for the calculation of the liquid velocity profile the knowledge of the mean gas holdup and the rise velocity of bubbles is sufficient. However, Whalley and Davidson [87] demonstrated the applicability of this method only for a very shallow bubble column, $L/d_c < 1$, for a special sparger design and for low gas velocities, $u_G = 1$ cm/sec. Such conditions are rarely encountered in commercial bubble column reactors.

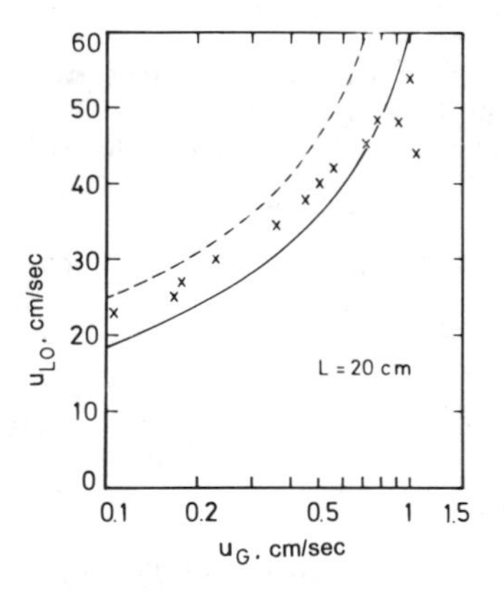

Figure 21. Maximum axial liquid velocity as a function of the gas velocity [87].

Joshi and Sharma Model

Joshi and Sharma [91] pointed out that direct application of the energy balance method to bubble columns with $L/d_c > 4$ suffers from the following drawbacks:

1. For the case of $L/d_c > 1$, the values of liquid velocity at any point predicted by the energy balance method are much higher (by a factor of even 2 to 3) compared to the experimental values reported by Pavlov [101], Pozin et al. [103], Miyauchi and Shyu [93], Hills [56], Ueyama and Miyauchi [94] and Kojima et al. [97].

2. This method predicts that the liquid circulation velocity should be proportional to the cube root of height of dispersion [104]. Consequently, the predicted values of parameters such as ϵ_G, axial dispersion coefficient, bed-wall heat transfer coefficient, the minimum superficial gas velocity for the suspension of solids, u_{cs}, etc., which

are strongly dependent on the liquid circulation velocity, vary with the height of dispersion. However, there is sufficient evidence in the literature to conclude that the above parameters are essentially independent of the liquid height in the range $3 < L/d_c < 12$ [105].

To overcome these difficulties, Joshi and Sharma developed a multiple-cell circulation model. It is viewed that many circulation cells prevail in the axial direction (Figure 22). The height of each circulation cell is adjusted so that the liquid-phase vorticity is minimum. It has been shown that the height of each circulation cell equals the column diameter, regardless of the superficial gas velocity, the height of dispersion and the column diameter. Therefore, Equations 63 and 64 take the following forms:

$$E_i = \frac{\pi}{4} d_c^3 u_G g \rho_L (1 - \bar{\epsilon}_G) \tag{72}$$

$$E_1 = \frac{\pi}{4} d_c^3 \bar{\epsilon}_G (1 - \bar{\epsilon}_G) g \rho_L u_{B\infty} \tag{73}$$

The liquid velocity profile was obtained using the following procedure.

1. For any circulation cell, the kinetic energy associated with the downward-moving liquid at the plane A′A (Figure 22) is not recoverable as the pressure energy and is lost as the turbulent dissipation. This value can be obtained from Equation 71 and by integration at the plane A′A:

$$E = \frac{\pi \rho_L \omega_o^3}{(d_c/2)^4} \int_{2A/d_c}^{1} \frac{1}{\bar{\phi}^2} \left(\frac{dR^1}{d\bar{\phi}}\right)^3 (1 - \epsilon_G)\, d\bar{\phi} \tag{74}$$

where ϵ_G is the local value of the fractional gas holdup and can be obtained from Equation 44. If the radial variation of the gas holdup is neglected, then Equation 74 takes the following form:

$$E = \frac{\pi \rho_L (1 - \bar{\epsilon}_G) \psi_o^3}{(d_c/2)^4} \int_{2A/d_c}^{1} \frac{1}{\bar{\phi}^2} \left(\frac{dR^1}{d\bar{\phi}}\right)^3 d\bar{\phi} \tag{75}$$

Joshi and Sharma [91] have shown that the maximum difference in the results obtained by Equations 74 and 75 is only 5%, and Equation 75 was recommended.

2. The order of magnitude calculation indicated that the turbulent energy dissipation and the viscous drag at the gas-liquid interface are the major modes of energy dissipation. The energy required for the gas dispersion was found to be negligible. Therefore, E in Equation 75 can be obtained from Equations 63 and 64 as

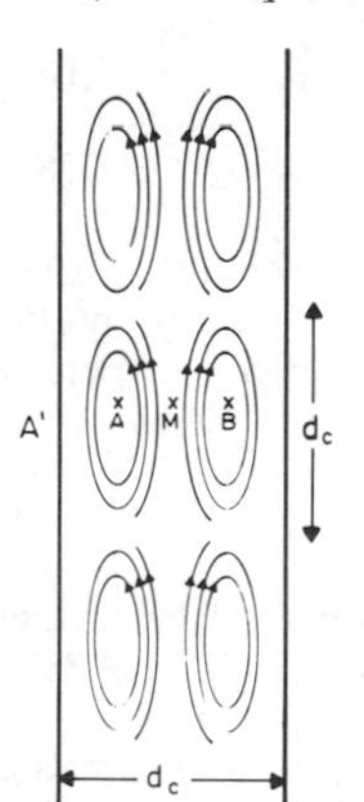

Figure 22. Flow pattern in circulation cell model of Joshi and Sharma [91].

$$E = E_i - E_1 = \frac{\pi}{4} d_c^3 \rho_L g (u_G - \epsilon_G u_{B\infty})(1 - \epsilon_G) \quad (76)$$

Joshi and Sharma [91] have discussed the procedure for the calculation of $dR^1/d\phi$ and the integral in Equation 75. Its value depends on L/d_c, and it works out to be 24.22 for the case of $L/d_c = 1$. Substitution of Equation 75 into 76 gives

$$\psi_o = \frac{d_c^2}{11.55} [g d_c (u_G - \epsilon_G u_{B\infty})]^{1/3} \quad (77)$$

3. The vertical component of the liquid velocity is given by

$$u_z = \frac{1}{r} \frac{\partial \psi}{\partial z} \quad (78)$$

or

$$u_z = \frac{\psi_o}{(d_c/2)^2} \frac{1}{\phi} \frac{dR^1}{d\phi} \sin(\pi z) \quad (79)$$

Equation 79 gives the velocity profile, which was averaged over the height of a circulation cell. Hills [56] and Kojima et al. [97] have measured the velocity profiles in 0.138- and 5.5-m I.D. columns, respectively. The agreement between the experimental and the predicted velocity profiles is as good as in the case of the recirculating flow model of Miyauchi and co-workers described earlier [94].

4. The average liquid circulation velocity was obtained by the integration of Equation 77 from A to $d_c/2$ at the plane A′A and is given by

$$u_c (m/sec) = 1.31 [g d_c (u_G - \epsilon_G u_{B\infty})]^{1/3} \quad (80)$$

If the liquid phase is also continuous, then the above equation takes the following form:

$$u_c = 1.31 \left[g d_c \left(u_G \pm \frac{\epsilon_G u_L}{1 - \epsilon_G} - \epsilon_G u_{B\infty} \right) \right]^{1/3} \quad (81)$$

where the minus and plus signs represent the cocurrent and countercurrent flow patterns, respectively.

The average axial radial components of the liquid velocity are given by the following equations:

$$u_a (m/sec) = 1.18 [g d_c (u_G - \epsilon_G u_{B\infty})]^{1/3} \quad (82)$$

$$u_r (m/sec) = 0.42 [g d_c (u_G - \epsilon_G u_{B\infty})]^{1/3} \quad (83)$$

The velocity distribution in the liquid phase and the average velocities derived therefrom were applied successfully to describe characteristic hydrodynamic properties of bubble columns. Joshi and Sharma [91] have shown that the liquid-phase dispersion reported in the literature can be correlated with a standard deviation of 7% by

$$E_L = 0.31\, d_c u_c \quad (84)$$

According to Equation 80, u_c varies with $d_c^{1/3}$. As a result, E_L varies with $d_c^{4/3}$, which is in close agreement with the correlations reported by Deckwer et al. [106] and Baird and Rice [107].

Joshi et al. [108,109] also have presented a generalized procedure for the estimation of wall-to dispersion heat transfer coefficients in bubble columns. Their analysis indicates that the higher heat transfer coefficients observed in gas-liquid flow, as compared to single-phase flow, can be explained on the basis of increased local liquid velocities caused by the presence of rising gas bubbles. The final correlation has the following form:

$$\frac{h_w d_c}{k} = 0.087 \left[\frac{d_c^{4/3} g^{1/3} (u_G - \epsilon_G u_{B\infty})^{1/3} \rho}{\mu}\right]^{0.8} \left(\frac{C_p \mu}{k}\right)^{1/3} \left(\frac{\mu}{\mu_w}\right)^{0.14} \tag{85}$$

In agreement with the experimental findings, Equation 85 shows that h_w is practically independent of d_c ($h_w \propto d_c^{0.064}$). Heat-transfer coefficients predicted from Equation 85 are in agreement with the experimental data within 30%.

Joshi and Sharma [91] also presented a procedure for the estimation of the gas holdup in bubble columns, which is based on the circulation velocity and the mean bubble rise velocity. The circulation flow pattern itself is caused by the introduction of gas, and the average circulation velocity is dependent on gas holdup. The proposed procedure is an iterative one and rather cumbersome. However, the experimental data for the air-water system can be described within 20–30% error. It is not clear whether the procedure of Joshi and Sharma [92] for estimating ϵ_G can be applied to other systems. Until then, the use of empirical correlations is recommended [40].

The energy balance method of Whalley and Davidson [87] (and its extension to bubble column of industrial size by Joshi and Sharma [91]) is capable not only of describing liquid-flow velocity distribution in bubble columns, but also delivers the average circulation velocity, which can be correlated to the liquid-phase dispersion coefficient and dispersion-to-wall heat transfer coefficient. It is important to note that with the knowledge of gas velocity and the average gas holdup the average circulation velocity can be calculated by means of a simple equation.

Dispersed Plug-Flow Model and Recirculating Flow

The correlation of the liquid-phase dispersion coefficient to the average liquid circulation velocity, as proposed by Joshi and Sharma [91] (i.e., Equation 84), apparently presents a contradiction. The liquid dispersion coefficient is a measure of the irregular stochastic mixing processes. In contrast, the circulation velocity derived from the energy balance is deterministic in nature. This problem is discussed by Riquarts [110] in detail. He has analyzed the governing mass balance equations by taking into account a simplified liquid-velocity distribution, i.e., by dividing the entire column in a central upflow and an annular downflow region. Riquarts finally arrives at a balance equation equivalent to the one-dimensional dispersed plug-flow model, and concludes that the liquid dispersion coefficient is, in fact, a measure of stochastic mixing processes, even when the circulation flow is chiefly responsible for the axial mass transfer. Following the analogy to the dispersion relation in fixed beds, i.e., Pe = 2, Riquarts [110] assumes that the following relation should be valid in bubble columns:

$$E_L = \frac{1}{2} u_{B\infty} d_B \tag{86}$$

As dispersion is due mainly to the motion of large bubbles, the bubble rise velocity can be expressed by

$$u_{B\infty} = 0.71\sqrt{d_B g} \tag{87}$$

If the number of bubbles of diameter d_B present in a longitudinal section of length d_c is n_B, the following relation holds:

$$u_{B\infty} d_B^3 \frac{\pi}{6} = \bar{\epsilon}_G d_c^3 \frac{\pi}{4} \tag{88}$$

Introducing Equations 87 and 88 into Equation 86 and using the resulting expression for the gas holdup [111],

$$\epsilon_G = 0.1 \left(\frac{u_G{}^3}{\nu_L g}\right)^{1/4} \tag{89}$$

it follows that

$$E_L = \frac{0.137}{\sqrt{n_B}} d_c\sqrt{d_c g} \left(\frac{u_G{}^3}{\nu_L g}\right)^{1/8} \tag{90}$$

This equation has the same form as standard empirical correlations and describes the measured data with reasonable accuracy if n_B is taken to be 4. By introducing dimensionless numbers, Equation 90 can be written as [110]

$$Pe = 14.7 \left(\frac{Fr^3}{Re}\right)^{1/8} \tag{91}$$

A derivation of the above equation shows that the liquid-phase dispersion in bubble columns also can be modeled physically solely by a stochastic mixing process, which is the basis of Equation 86. Moreover, Riquarts [100] has shown that both the one-dimensional liquid dispersion coefficient, E_L, and the turbulent viscosity, ν_t, which are responsible for the radial momentum transfer, are interrelated with the centerline liquid velocity as

$$\frac{u'_{Lo} d_c}{E_L} = 3.1 \tag{92}$$

and

$$\frac{u'_{Lo} d_c}{\nu_t} = 19 \tag{93}$$

Therefore, it can be concluded that the apparent discrepancy between the deterministic and stochastic approaches is immaterial. Both physical concepts are equally capable of representing transfer phenomena in the liquid phase of bubble columns and can be regarded as limiting cases of a more general model.

APPLICATION OF KOLMOGOROFF'S THEORY OF ISOTROPIC TURBULENCE

During the past decade, the parameters of Kolmogoroff's theory of isotropic turbulence, i.e., the length and velocity scale of microeddies, have found increasing use for correlating characteristic parameters of the bubble columns. Some examples are summarized in Table IV. Calderbank [112] showed that as predicted by Kolmogoroff's theory, the maximum diameter of a stable bubble is inversely proportional to the energy dissipation rate per unit volume raised to the power 0.4 (i.e., $(E/V)^{0.4}$).

In a series of papers, Nagel et al. [113–116] made extensive use of this result and showed that high interfacial areas in various kinds of gas-liquid contactors depend very strongly on the energy dissipation rate per unit volume. A correlation for the liquid-phase dispersion coefficient derived on the basis of Kolmogoroff's theory has been shown [107] to correlate well with the experimental data for the bubble columns and other multiphase contactors. This fact once again indicates that the mass dispersion in bubble columns should be the result predominantly of a stochastic turbulent mixing

Table IV. Examples of Correlations of Characteristic Hydrodynamic and Transfer Parameters of Bubble Columns by the Application of Kolmogoroff's Theory of Isotropic Turbulence

Parameter	Equation	Reference
Maximum stable bubble diameter (d_{Bmax})	$d_{Bmax} = C \dfrac{\sigma^{0.6}}{(E/V)^{0.4} \rho_L{}^{0.2}}$	Calderbank [112]
Gas-liquid interfacial area (a)	$a = c' \, (E/V)^{0.4} \epsilon_G^n$	Nagel et al. [113–116]
Volumetric gas-liquid mass transfer coefficient (k_La)	$k_La = c'' \, u_G{}^{0.8}$ $Sh = 0.13 \, Re_T{}^{3/4} \, Sc^{1/3}$ where $Re_T = d_B{}^{4/3} \, \rho_L \, (E/V)^{2/7}/H_L$	Kastenek [117] Calderbank and Moo-Young [125]
Liquid-phase dispersion coefficient (E_L)	$E_L = c \, d_c{}^{4/3} \, (g \, u_G)^{1/3}$	Baird and Rice [107]
Liquid-solid (suspended by aeration) mass transfer coefficient (k_s)	$\dfrac{k_s d_s}{D_L} = 2 + 0.545 \left(\dfrac{\nu_L}{D_L}\right)^{1/3} \left(\dfrac{\epsilon d_s{}^4}{\nu_L{}^3}\right)^{0.264}$ $Sh = 2 + 0.545 \, Sc^{1/3} \, Re^{0.264}$	Sänger and Deckwer [118]
Wall-to-dispersion heat transfer coefficient (h)	$\dfrac{h_w}{Lc_p u_G} = 0.1 \left[\dfrac{u_G{}^3}{\nu_L g} \left(\dfrac{\nu_L \rho_L c_p}{k}\right)^2\right]^{1/4}$ $St = 0.1 \, (ReFrPr^2)^{1/4}$	Deckwer [119]

process [100]. It is also common practice to evaluate the liquid-solid mass transfer data by the Kolmogoroff theory [118,120,121].

In general, the mean turbulent flow is superimposed by a spectrum of velocity fluctuations. The energy of the turbulent eddies depends on their size. Figure 23 shows the spectral energy distribution for a single-phase flow reported by Hinze [122]. The wave number, k, is proportional to the frequency of velocity fluctuations, and the inverse of k is roughly proportional to the eddy size.

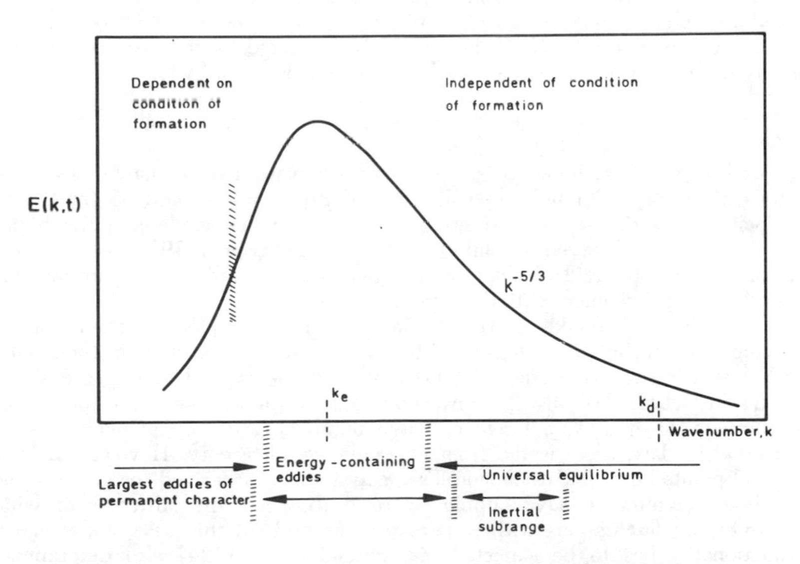

Figure 23. Energy spectrum for single-phase flow [122].

There are large primary eddies, which are formed either by disturbances or during the formation of flow. These macroscale eddies correspond to large-velocity fluctuations of low frequency, and their scale may be of the same order of magnitude as characteristic geometric sizes of the contactors. The large eddies characterized by wave number in the range of k_e contain the major part of energy but contribute only little to the viscous energy dissipation. However, by the diffusive motion of the energetic large eddies, small eddies are generated and, thereby, the kinetic energy is transferred to small scales. In these small eddies, the energy dissipation due to viscous forces becomes most effective. These eddies are characterized by wave numbers in the order of k_d. Thus, on one hand we have the large eddies as an energy reservoir and, on the other hand, a dissipation process in the small eddies. This picture forms the basis of Kolmogoroff's theory of locally isotropic turbulence.

If the Reynolds number of turbulent motion is sufficiently high, Kolmogoroff postulates [122] that there is a wave number range, k_d, at which turbulence is statistically in equilibrium and solely determined by the energy dissipation rate per unit mass, ϵ, and the viscosity, ν. This state of equilibrium is called "universal," as the character of the turbulence is independent of the external conditions. The miscroscale eddies behave locally isotropic, hence do not depend on the motion of the large-scale eddies. While formulating the correlation functions, Kolmogoroff introduced two quantities from the dimensional reasoning, namely, a length scale,

$$\eta = (\nu^3/\epsilon)^{1/4} \tag{94}$$

and a velocity scale,

$$v = (\nu\epsilon)^{1/4} \tag{95}$$

which are characteristic properties of the microscale eddies in which the energy is dissipated. The energy transfer from large to small eddies gives a spectral energy distribution for the inertial subrange, which is given by

$$E\ (k,t) = A'\,\epsilon^{2/3}\,k^{-5/3} \tag{96}$$

The above equation is the Kolmogoroff spectrum law.

In bubble columns with stagnant liquid phase, the energy is introduced by the gas flow (see Equation 63). If all the energy is dissipated by the viscous forces in the small eddies, the energy dissipation rate per unit mass is given by

$$\epsilon = u_G g \tag{97}$$

The usefulness of Kolmogoroff's theory to correlate hydrodynamic and transfer properties in bubble columns is to make use of Equations 94 and 95 for the length and velocity scales of the microeddies. For instance, when explaining the high heat transfer coefficients observed in bubble columns, Deckwer [119] assumed that the mean contact time of fluid elements at the heat transfer area is proportional to η/ν and found striking agreement with the experimental data.

Very recently, Zakrzewski [123] and Zakrzewski et al. [124] carried out careful measurements on turbulence intensity in the bubble columns and, from their data, the spectral energy distribution could be calculated. Some typical results are shown in Figure 24 [123,124]. All distributions show a uniform decline with the increasing wave numbers. Over a wide range of wave numbers, the curves follow a k^{-2} law instead of $k^{-5/3}$ law, as expected from the Kolmogoroff theory. However, Zakrzewski et al. [124] pointed out that Kolmogoroff's spectral law was derived for grid turbulence. In a bubble column, we have bubble swarm turbulence, the structure of which is subject to strong fluctuations with respect to time and, in this case, a decrease with k^{-2} functionality has to be expected. Zakrzewski et al. [124] also determined the energy dissipation spectra and found characteristic differences for different spargers.

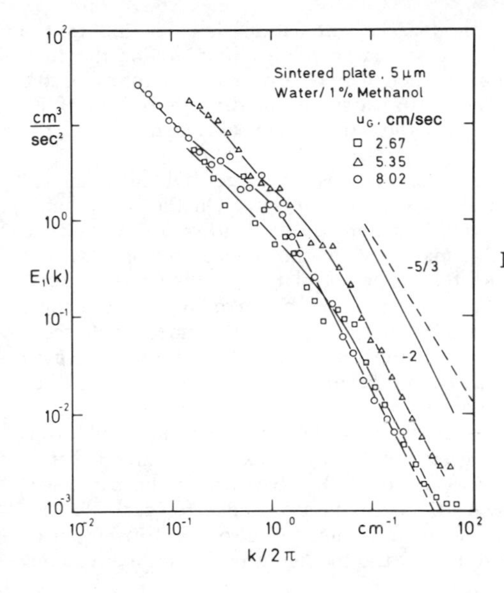

Figure 24. Spectral energy distribution of liquid velocity fluctuation measured in bubble columns.

All spectra do show, however, that as expected from turbulence theory the overwhelming part of the introduced energy is dissipated in the high wave number range. In general, one can conclude, therefore, that in gas-liquid flow the introduced energy is dissipated mainly by a mechanism that is equal or similar to that proposed by Kolmogoroff for the single-phase flow.

OUTLOOK AND RECOMMENDATIONS FOR FUTURE WORK

The slip velocity and the drift flux approaches are helpful in interpreting the experimental data. Both concepts apply only to the bubbly flow regime and, within this regime, the effect of the phase velocities (gas and liquid) on the average holdup can be predicted. The treatment of Zuber and Findlay [72] presents a more general approach and is not restricted to the bubbly flow. So far, these models do not allow a priori predictions of the holdup in bubble columns. In general, their applicability can be recommended at high liquid velocities and moderate gas velocities. Unfortunately, in most industrial bubble columns low liquid velocities and high gas velocities are found.

In the Miyauchi flow model [93,94], the recirculation cell extends from the bottom of the column to the top. For calculating the liquid velocity profile, the radial gas holdup distribution is required. As an important result, the recirculation flow model also delivers an expression for the slip velocity in bubble swarms. The computation of the velocity profile is rather cumbersome; however, by introducing a simplification in the governing balance equation, Riquarts [100] has obtained a simpler expression that involves only the mean gas holdup. Also, the energy balance method, which is the basis of the circulation cell model of Joshi and Sharma [89–91], predicts the liquid velocity profile with the knowledge of mean gas holdup only. The model of Joshi and Sharma seems to be more realistic as it accounts for the radial flow. It is understood that all macroscopic flow models are capable of describing the liquid-phase dispersion phenomenon.

One can anticipate that the mixing energy distribution due to the turbulence in gas-liquid flow is similar to that shown in Figure 23 for single-phase flow. This view is supported by the recent findings of Zakrzewski et al. [124]. One may suspect, therefore, that various complex flow phenomena have their origin in this energy distribution. The large energetic eddies, which probably are a result of the high rise velocities of large bubble and bubble clusters, are responsible for the circulation flow leading to the pronounced liquid velocity profile. Therefore, the large eddies provide for the high overall macromixing in the bubble columns, which can be expressed in terms of the dispersion coefficient. Indeed, Baird and Rice [107] assumed in their derivation of E_L that the diameter of the turbulent eddies has the same order of magnitude as the column diameter.

Large eddies also cause deformation and oscillations of large bubbles and are probably involved in bubble clustering, coalescence and breakup. On the other hand, the small eddies of high frequency, which provide the energy dissipation, are responsible for micromixing phenomena and play a dominant role in the mechanism of heat and mass interphase transfer. Therefore, the heat transfer correlation involving the microscale parameters [11] is more realistic than a correlation based on the macroscopic circulation flow [108,109]. One can expect that more sophisticated measuring techniques will give a deeper insight into the structure and character of turbulence in bubble columns. This knowledge will form a more reliable and accurate basis of our understanding of the bubble column behavior.

More information is desirable on the hydrodynamic behavior around bubbles in bubble swarms. Here, the application of special probes and devices may give a better insight into gas-liquid mass transfer. Our knowledge on the structure of the gas distribution in churn-turbulent flow is still poor. Additional data are needed for gas velocities above 5 cm/sec. It is not only the bubble size distribution, but also the distribution of the rise velocities, which is of major importance in the modeling of the bubble column reactor.

Coalescence and rupture of bubbles are not well understood. It is obvious that new experimental techniques have to be developed to explain these phenomena, which should be studied particularly under conditions of high interphase mass transfer rates. Our knowledge of bubble swarm behavior and bubble dynamics is still incomplete.

All hydrodynamic models proposed so far have been tested largely for the air-water system. More data are needed on traverse and lateral holdup variations and velocity distributions for the organic liquids, liquid mixtures and highly viscous Newtonian and non-Newtonian media.

NOMENCLATURE

a	specific interfacial area
A	cross-sectional area or distance between the column axis and the point of maximum vorticity
A′	a constant
B	point of maximum vorticity
C,C′,C″	constants
C_o	a constant defined by Equation 22
c_p	heat capacity
d_B	mean bubble diameter
d_c	column diameter
d_s	particle diameter
D	diffusion coefficient
E	energy dissipation rate in the hydraulic jump
E(k,t)	energy spectrum
E/V	energy dissipation per unit volume
E_i	energy input rate
E_1	energy dissipation rate in wakes behind the bubbles
E_L	liquid-phase dispersion coefficient
Fr	Froude number, $u_G^2/d_c g$
g	gravitational constant
G	gas mass flowrate (kg/m^2-sec)
h_w	wall-to-dispersion heat transfer coefficient
k	thermal conductivity or wave number
$k_L a$	volumetric mass transfer coefficient
k_s	liquid-solid mass transfer coefficient
L	column length (dispersion height) or liquid mass flowrate (kg/m^{2}sec)
N_{Fr}	two-phase Froude number, V_M^2/gd_c
n	a constant in Equation 42

n_B	number of bubbles present in longitudinal section of height d_c
P	pressure at the column top
P_T	pressure at temperature, T
Pe	Peclet number, $u_G d_c/E_L$
Pr	Prandtl number, $\nu_L \rho c_p/k$
r	radial distance from column axis
R	column radius
R^1	radial component of the dimensionless stream function, $\psi^* = \psi/\psi_o$
R_V	gas to liquid volumetric ratio
Re	Reynolds number, $u_G d_c/\nu_L$
St	Stanton number, $h_w/\rho_L c_p u_G$
u	superficial velocity
u′	interstitial velocity
u	mean superficial velocity
u_a	average axial component of liquid circulation velocity
u_B	terminal rise velocity of single bubbles
u_c	average liquid circulation velocity
u_G^*	mean rise velocity of bubbles in bubble swarm, u_G/ϵ_G, or interstitial velocity of gas
$u_{Gj}(r)$	local drift velocity of gas phase
u_{Gj}	volume fraction weighted-average drift velocity of gas phase
u_r	average radial component of liquid circulation velocity
u_s	slip velocity
V_α	axial interstitial velocity of α phase
$v_{\alpha j}$	weighted-average drift velocity, defined by Equation 17
v	local velocity or velocity scale of microeddies
$v_{\alpha j}(r)$	local drift velocity, defined by Equation 18
V_M	superficial velocity of gas-liquid mixture
$v_M(r)$	local velocity of gas-liquid mixture
v_r	relative velocity, defined by Equation 20
$V_{S\alpha}$	superficial velocity of α phase
x	axial coordinate
z	dimensionless axial coordinate

Indices

A	air
G	gas
L	liquid
o	centerline
W	water or wall
α,β	phase

Greek Symbols

α	pressure parameters, defined by Equation 2
γ_{CD}	drift flux velocity, Equations 11 and 12
ϵ	volume fraction
$\bar{\epsilon}_G$	average gas holdup
ϵ	energy dissipation rate per unit mass
η	length scale of microeddies
λ	density parameter, $(\rho_G^* \rho_L^*)^{1/2}$
Λ	property parameter, $u_L^*/(\rho_L^* \sigma^3)^{1/4}$
μ_{eff}	effective dynamic viscosity
μ_t	turbulent dynamic viscosity
μ_L^*	viscosity ratio, μ_L/μ_W
ν	kinematic viscosity
ν_t	turbulent kinematic viscosity
ρ	density
ρ_G^*	density ratio, ρ_G/ρ_A
ρ_L^*	density ratio, ρ_L/ρ_W
σ^*	surface tension ratio, σ/σ_w
τ	shear stress or bubble residence time
$\bar{\phi}$	dimensionless radial coordinate, r/R
$\phi(\epsilon_G)$	function of gas holdup
ϕ^*	physical properties parameter, $u_L^*/(\rho_L^* \sigma^{*3})^{1/4}$
ψ	stream function
ψ_o	maximum value of stream function
ψ^*	dimensionless stream function
ω	vorticity
ω_o	maximum vorticity or circulation strength

REFERENCES

1. Danckwerts, P. V., and M. M. Sharma. *Chem. Eng. Rev. Series* 2:244 (1966).
2. Kafarov, V. V., V. A. Rentsii and T. Y. Zhuravleva. *J. Appl. Chem. USSR* 49:2724 [1976].
3. Danckwerts, P. V. *Gas-Liquid Reactions* (New York: McGraw-Hill Book Co., 1970).
4. Ramachandran, P. A., and M. M. Sharma. *Trans. Inst. Chem. Eng.* 49:253 (1971).
5. Gehlawat, J. K., and M. M. Sharma. *Chem. Eng. Sci.* 23:1173 (1968).
6. Kröper, H., K. Schlömer and H. M. Weitz. *Hydrocarbon Proc.* 48:195 (1969).
7. Deckwer, W. D., U. Allenbach and H. Bretschneider. *Chem. Eng. Sci.* 32:43 (1977).
8. Deckwer, W.-D. *Chem. Eng. Sci.* 32:51 (1977).
9. Smidt, J., W. Hafner, J. Sedlmeier, R. Sieber, R. Ruttinger and H. Kojer. *Angew. Chemie* 71:176 (1959).
10. Smidt, J., W. Hafner, R. Jira, R. Sieber, J. Sedlmeier and A. Sabel. *Angew Chemie* 74:93 (1962).
11. Jira, R., W. Blau and D. Grimm. *Hydrocarbon Proc.* 97 (March 1976).
12. Krekeler, H., and H. Schmitz. *Chem. Ing. Tech.* 40:785 (1968).
13. Sittig, M. *Organic Chemical Process Encyclopedia* (Park Ridge, NJ: Noyes Development Corporation, 1967).
14. Kostylak, N. G., S. U. L'ov, V. B. Falkowski, A. U. Starkov and N. M. Levina. *J. Appl. Chem. USSR* 35:1939 (1962).
15. Yan, A. V., A. E. Hemielec and A. I. Johnson. Paper presented at the International Symposium on Research in Cocurrent Gas-Liquid Flow, Waterloo, Ontario, September, 1968.
16. Broich, F., H. Höfermann, H. Hunsmann and H. Simmrock. *Erdöl Kohl-Erdgas-Petrochemie* 16:284 (1963).
17. Broich, F. *Chem. Ing. Tech.* 36:417 (1964).
18. Höfermann, H. *Chem. Ing. Tech.* 36:423 (1964).
19. Saunby, J. B., and B. W. Kiff. *Hydrocarbon Proc.* 247 (November 1976).
20. Kaeding, W. W., R. O. Lindblom, R. G. Temple and H. I. Mahon. *Ind. Eng. Chem. Proc. Des. Dev.* 4:97 (1965).
21. Haase, H. *Chem. Ing. Tech.* 44:987 (1972).
22. Hattori, K., Y. Tanaka, H. Suzuki, T. Ikawa and H. Kubota. *J. Chem. Eng. Japan* 3:72 (1970).
23. Hagberg, C. G., and F. X. Krupa. *Proc. 4th Int. Symp. Chem. React. Eng.* Heidelberg, West Germany (1976), p. IX-409.
24. Beyrich, J., W. Gautschi, W. Regenass and W. Wiedmann. *Proc. 12th Symp. Comp. Appl. Chem. Eng.*, Montreux, Switzerland (1979).
25. Gehlawat, J. K., and M. H. Sharma. *J. Appl. Chem.* 20:93 (1970).
26. *Hydrocarbon Proc.* 226 (November 1977).
27. Griesbaum, K., et al. *Ullmanns Enzyklopädie der Technischen Chemie,* Vol. 14 (Weinheim: Verlag Chemie, 1974).
28. von Kutepow, M., W. Himmele and H. Hohenschutz. *Chem. Ing. Tech.* 37:383 (1965).
29. Hjortkjaer, J., and V. W. Jensen. *Ind. Eng. Chem. Proc. Des. Dev.* 15:46 (1976).
30. Oliver, K. L., and F. B. Booth. *Hydrocarbon Proc.* 49(4):112 (1970).
31. Weissermel, K., and H. J. Arpe. *Industrielle Organische Chemie* (Weinheim: Verlag Chemie, 1976).
32. Albright, L. F. *Chem. Eng.* 179 (December 4, 1967).
33. Friend, L., L. Wender and I. C. Yarze. *Adv. Chem. Series* 168 (1968).
34. Rassaertz, H., and D. Witzel. In: *Ullmanns Enzyklopädie der Technischen Chemie,* Vol. 9 (Weinheim: Verlag Chemie, 1974).
35. Hawkins, P. A. *Trans. Inst. Chem. Eng.* 43:T287 (1965).
36. van den Berg, H. Ph.D. Thesis, University of Gröningen, Holland (1973).
37. Rathjen, H. In: *Ullmanns Enzyklopädie der Technischen Chemie,* Vol. 9 (Weinheim: Verlag Chemie, 1974).

38. Tan, P. M., and J. S. Ratcliffe. *Mech. Chem. Eng. Trans.* 10:29 (1974).
39. *Hydrocarbon Proc.* 356 (November 1977).
40. Shah, Y. T., B. G. Kelkar, S. Godbole and W.-D. Deckwer. "Design Parameter Estimations for Bubble Column Reactors," *AIChE J.* 28(3):353 (1982).
41. Cichy, P. T., J. S. Ultman and T. W. F. Russell. *Ind. Eng. Chem.* 61(8):6 (1969).
42. Govier, G. W., and K. Aziz. "The Flow of Complex Mixtures in Pipes" (New York: Van Nostrand Reinhold, 1972).
43. Shah, Y. T., and W.-D. Deckwer. "Fluid-Fluid Reactors," in *Scaleup in the Chemical Process Industries,* A. Bisio and R. Kabel, Eds. (New York: John Wiley & Sons, Inc., in press.
44. Oshinowo, T., and M. E. Charles. *Can. J. Chem. Eng.* 52:25 (1974).
45. Miller, D. N. *Ind. Eng. Chem. Proc. Des. Dev.* 19:371–377 (1980).
46. Lapidus, L., and J. C. Elgin. *AIChE J.* 3:63 (1957).
47. Lockett, M. J., and R. D. Kirkpatrick. *Trans. Inst. Chem. Eng.* 53:267 (1975).
48. Herbrechtsmeier, P., and R. Steiner. *Chem. Ing. Tech.* 52:468 (1980).
49. Weiland, P. Dr.-Ing. Thesis, University of Dortmund, West Germany (1978).
50. Beinhauer, R. Dr.-Ing. Thesis, Technical University of Berlin, West Germany (1971).
51. Kölbel, H. R., Beinhauer and H. Langemann. *Chem. Ing. Tech.* 44:697 (1972).
52. Hills, J. H. *Chem. Eng. J.* 12:89 (1976).
53. Clift, R., J. R. Grace and M. E. Weber. "Bubbles, Drops and Particles," (New York: Academic Press, Inc., 1978).
54. Schumpe, A. Dr.-Ing. Thesis, University of Hannover, West Germany (1981).
55. Schumpe, A., and W.-D. Deckwer. Paper presented at ACS Meeting, Atlanta, GA, April, 1981.
56. Hills, J. H. *Trans. Inst. Chem. Eng.* 52:1 (1974).
57. Deckwer, W.-D. *Chem. Eng. Sci.* 31:309 (1976).
58. Deckwer, W.-D. *Chem. Eng. Sci.* 32:51 (1977).
59. Turner, J. C. R. *Chem. Eng. Sci.* 21:971 (1966).
60. Davidson, J. F., and D. Harrison. *Chem. Eng. Sci.* 21:731 (1966).
61. Wallis, G. B. *One Dimensional Two-Phase Flow* (New York: McGraw-Hill Book Co., 1969).
62. Richardson, J. F., and W. N. Zaki. *Trans. Inst. Chem. Eng.* 32:35 (1954).
63. Marrucci, G. *Ind. Eng. Chem. Fund.* 4:224 (1965).
64. Bridge, A. G., L. Lapidus and J. C. Elgin. *AIChE J.* 10:819 (1964).
65. Haberman, W. L., and R. K. Morton. David Taylor Model Basin Rep. No. 802 (1953).
66. Freedman, W., and J. F. Davidson. *Trans. Inst. Chem. Eng.* 47:T251 (1969).
67. Rice, R. G., J. M. I. Tuppurainen and R. M. Hedge. Paper presented at ACS Annual Meeting, Las Vegas, NV, August, 1980.
68. Deckwer, W.-D., Y. Louisi, A. Zaidi and M. Ralek. Paper presented at AIChE Annual Meeting, San Francisco, CA, November, 1979.
69. Schugerl, K., J. Lucke and U. Oels. *Adv. Biochem. Eng.* 7:1 (1977).
70. Oels, U., J. Lucke, R. Buchholz and K. Schugerl. *Ger. Chem. Eng.* 1:115 (1978).
71. Deckwer, W.-D., I. Adler and A. Zaidi. Preprints of European Symposium, Nüremberg, West Germany, March 28–30, 1977.
72. Zuber, M., and J. A. Findlay. *J. Heat Transfer* (*Trans. ASME*) 453 (November 1965).
73. Petrick, M. Argonne Natl. Lab. Report ANL-6581 (1962).
74. Kara, S. PhD Thesis, University of Pittsburgh, PA (1981).
75. Calderbank, P. H., M. B. Moo-Young and R. Bibbey. *Proc. 3rd Eur. Symp. Chem. React. Eng.,* Amsterdam (1964).
76. Otake, T., S. Tone, K. Nakao and Y. Mitsuhashi. *Chem. Eng. Sci.* 32:377 (1977).
77. Deckwer, W.-D., A. Zaidi and I. Adler. Preprints of European Congress, "Transfer Processes in Particle Systems," Nüremberg, West Germany, March 28–30, 1977.
78. Deckwer, W. D., I. Adler and A. Zaidi. *Can. J. Chem. Eng.* 56:43 (1978).
79. Schügerl, K., J. Lücke and U. Oels. *Adv. Biochem. Eng.* 7:1 (1977).
80. Marucci, G., and L. Nicodemo. *Chem. Eng. Sci.* 22:1257 (1967).

81. Voigt, J., and K. Schügerl. *Chem. Eng. Sci.* 34:1221 (1979).
82. Lee, J. C., and T. D. Hodgson. *Chem. Eng. Sci.* 23:1375 (1968).
83. Marucci, G. *Chem. Eng. Sci.* 24:975 (1969).
84. Lee, J. C., and G. W. K. Ssali. Paper presented at the Joint Meeting on Bubbles and Foams, VTG/VDI-Inst. Chem. Engr., Nüremberg, West Germany, 1971.
85. Rietema, K., and S. P. P. Ottengraph. *Trans. Inst. Chem. Eng.* 48:T54 (1970).
86. Crabtree, J. R., and J. Bridgewater. *Chem. Eng. Sci.* 24:1755 (1969).
87. Whalley, P. B., and J. F. Davidson. *Inst. Chem. Eng. Symp. Series* 38:55 (1974).
88. Joshi, J. B., and M. M. Sharma. *Trans. Inst. Chem. Eng.* 54:42 (1976).
89. Joshi, J. B., and M. M. Sharma. *Can. J. Chem. Eng.* 56:116 (1978).
90. Joshi, J. B., and M. M. Sharma. *Can. J. Chem. Eng.* 57:375 (1979).
91. Joshi, J. B., and M. M. Sharma. *Trans. Inst. Chem. Eng.* 57:244 (1979).
92. Bhavraju, S. M., T. W. F. Russel and H. W. Blanch. *AIChE J.* 24:454 (1978).
93. Miyauchi, T., and C. N. Shyu. *Kagaku Kogaku* 34:958 (1970).
94. Ueyama, K., and T. Miyauchi. *AIChE J.* 25:258 (1979).
95. Kato, Y., M. Nishiwaka and S. Morroka. *Kogaku Kogaku Ronb.* 1:530 (1975).
96. Joshi, J. B., and Y. T. Shah. *Chem. Eng. Commun.* 11:165 (1981).
97. Kojima, E., H. Unno, Y. Sato, T. Chida, H. Imai, K. Endo, I. Inoue, J. Kobayashi and K. M. Yamamoto. *Chem. Eng. Japan* 13:16 (1980).
98. Yamagoshi, T., B.S. Thesis, Department of Chemical Engineering, University of Tokyo, Japan (1969).
99. Ueyama, K., and T. Miyauchi. *Kagaku Kogaku Ronb.* 3:19 (1976).
100. Riquarts, H. P. *Chem. Ing. Tech.,* in press.
101. Paulov, V. P. *Khim. Prom.* 9:698 (1965).
102. Lamb, H. *Hydrodynamics* (New York: Cambridge University Press, 1932).
103. Pozin, L. S., M. E. Aerov and T. A. Bystrova. *Theor. Found. Chem. Eng.* (English translation) 3:71A (1969).
104. Field, R. W., and J. F. Davidson. *Trans. Inst. Chem. Eng.* 58:228 (1980).
105. Mashelkar, R. A. *Brit. Chem. Eng.* 15:1297 (1968).
106. Deckwer, W. D., R. Burckhart and G. Zoll. *Chem. Eng. Sci.* 29:2177 (1974).
107. Baird, M. H. I., and R. G. Rice. *Chem. Eng. J.* 9:171 (1975).
108. Joshi, J. B., M. M. Sharma, Y. T. Shah, C. P. P. Singh, M. Ally and G. E. Klinzing. *Chem. Eng. Commun.* 6:1 (1980).
109. Joshi, J. B., M. M. Sharma and Y. T. Shah. Paper presented at ACS Meeting, Las Vegas, NV, August, 1980.
110. Riquarts, H. P. *Ger. Chem. Eng.* 4:18 (1981).
111. Riquarts, H. P., and Th. Pilhofer. *Verfaherenstechnik (Mainz)* 12:77 (1978).
112. Calderbank, P. H. *Chem. Eng.* CE 209 (1967).
113. Nagel, O., H. Kürten and B. Hegner. *Chem. Ing. Tech.* 45:913 (1973).
114. Nagel, O., and H. Kürten. *Chem. Ing. Tech.* 48:513 (1976).
115. Nagel, O., B. Hegner and H. Kürten. *Chem. Ing. Tech.* 50:934 (1978).
116. Nagel, O., H. Kürten and B. Hegner. In: *Two-Phase Momentum, Heat and Mass Transfer in Chemical Process and Energy Engineering Systems,* Vol. 2, F. Durst, G. V. Tsiklauri and N. H. Afgan, Eds. (Washington, DC: Hemisphere Publishing Corporation, 1979), p. 835.
117. Kastanek, F. *Coll. Czechoslov. Chem. Commun.* 42:2491 (1977).
118. Sanger, P., and W. -D. Deckwer. *Chem. Eng. J.*, in press.
119. Deckwer, W. -D. *Chem. Eng. Sci.* 35:1341 (1980).
120. Levins, D. M., and J. R. Glastonbury. *Chem. Eng. Sci.* 27:537 (1972).
121. Sano, Y., N. Yamaguchi and T. Adachi. *J. Chem. Eng. Japan* 7:255 (1974).
122. Hinze, J. O. *Turbulence* (New York: McGraw-Hill Book Co., 1959).
123. Zakrzewski, W. Dr. rer. nat. Thesis, University of Hannover, West Germany (1980).
124. Zakrzewski, W., J. Lippert, A. Lübbert and K. Schügerl. *Chem. Ing. Tech.* 53:135 (1981).
125. Calderbank, P. M., and M. B. Moo-Young. *Chem. Eng. Sci.* 16:39 (1961).

Section 3

GAS-SOLIDS FLOWS

CONTENTS

CHAPTER 23
RHEOLOGY OF PARTICULATE SOLIDS

F. A. Zenz
Chemical Engineering Department
Manhattan College
New York, New York

CONTENTS

INTRODUCTION

As it becomes more evident, quantitatively as well as qualitatively, that relationships for flow, vapor-liquid equilibria, internal shear and, likely, many other areas, hold as well for particulate solids as for gases and liquids, the "rheology" of particulate solids is becoming an extremely fascinating field. Understanding the relatively macroscopic properties and behavior of particulates should inevitably foster a more mechanistic comprehension of the properties of gases and liquids. Certainly particulates represent greater complexities when one realizes that a molecule of any known compound in the gaseous or liquid state is dimensionally always identical, whereas in the solid state a particle of a given material can differ in dimensional shape or porosity even when of the same size and density.

Handbooks are crammed with the properties of liquids, gases, solutions and mixtures, yet no equally organized compilation exists with regard to particulate solids, despite their reasonable similarity within any given industry. For any product, whether new or old, it should be standard practice to compile its properties methodically in a sequence encompassing at least: (1) composition (chemical components, particle shape, size distribution and moisture content); (2) mass (true and apparent densities and bulk densities); (3) aerodynamics (single particles and bulk solids); (4) two-phase equilibrium (bed and freeboard composition); and (5) deformation (frictional planes and bulk resilience).

COMPOSITION

Chemical Components

Chemical composition is certainly of interest with regard to a material's potential solubility, crystalline or amorphous structure, adsorptive capacity, reactivity, toxicity, explosive or combustible hazard, and similar "handling" or storage aspects. The same properties are commonly recorded for liquids and gases; however, of the four principal descriptive characteristics of particulates, these are otherwise usually the least important.

Particle Shape

Other than the "physical shape," which should be recorded in a photomicrograph, shape becomes more of an aerodynamically related factor. If the particle sizes are such as would overlap in sieve, Coulter Counter, Bahco or sedimentation analyzers, then a shape factor can be derived from a comparison between sieve and Bahco analyses, for example. If the largest size is detectable by the eye and insoluble in some transparent liquid, then its terminal velocity in a deep tank of the fluid can be measured (and fitted to the drag coefficient curve for spheres) to derive the particle's hydro- or aerodynamically equivalent spherical diameter. Comparison with the sieve opening passing the largest particle results in a so-called shape factor, j, which can be assumed to apply equally well to all smaller sizes in the sample:

$$D_p^* = j\, D_p$$

where D_p = measurable physical dimension or diameter
D_p^* = diameter of a hydro- or aerodynamically equivalent sphere
j = shape factor

For perfect spheres, $j = 1$; for irregular particles, offering more surface friction, j is less than unity. It is not unusual to find fly ash in bark or hog boiler flue gases to yield a j of the order of 0.2, or bank sand a j of 0.5. Without a terminal velocity determination, shape factors have been derived with reasonable accuracy from photomicrographs using the dimensional ratio approximation suggested by Heywood [1]:

$$j = \frac{3}{\pi\left[\dfrac{\text{max. length}}{\text{max. width}} \Big/ \dfrac{\text{max. width}}{\text{max. thickness}}\right]}$$

A direct analogy to shape with respect to gases and liquids lies in molecule complexity or type, for example, an organic amine, acid, paraffinic hydrocarbon, ester, aldehyde, alcohol, etc., of equal molecular weight.

Size Distribution

The most pertinent characterization of a particulate material lies in its size analysis, conveniently represented on log-probability coordinates in the form illustrated in Figure 1. A straight line with the illustrated shape (over the major portion of the analysis) represents the skewed probability distribution of a naturally occurring or

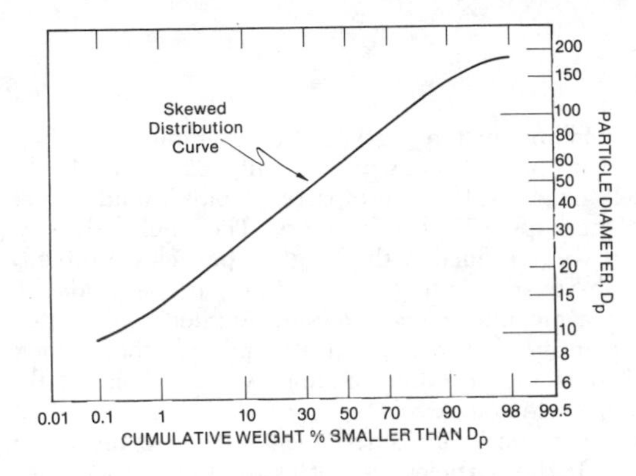

Figure 1. Conventional representation of ground particle-size distributions.

minimal internally abrasive grind. Allowing for the analogy between particle diameter and fluid molecular weight, the average diameter of the distribution becomes the diameter corresponding to about the 33.3% point on the abscissa of Figure 1, which does not differ significantly from the general practice of accepting the 50% (geometric weight mean) point diameter as a representative average.

Moisture Content

It is difficult to find an obvious analogy between the moisture effect on particle flow and a similar effect on gas or liquid flow. At best, one might consider the purging of a gas from a container filled with 0.01-μ spheres of colloidal silica. Although the voids in the container could amount to 99% by volume, the affinity of the gas to the particles and its difficulty in flowing through the interstices would make it nearly impossible to pour a heavy interstitial gas out of the container, for example, unless the colloidal enveloping solids also were removed. Because this mixture could stand at 90% or more to the horizontal, the container cannot readily be emptied of either gas or solids in free flow. If the solids can be imagined to swell in size to occupy more like 30–50% of the container volume, and the interstitial gas to bear a component such as water vapor with a high angle of wetting, then the larger and otherwise free-flowing sand particles would be trapped in a network of fluid film in much the same manner as the gas molecules were trapped in the network of colloidal silica. The higher the fluid's surface tension and the lower the particle's mass, the more difficult it becomes to separate particles bonded by the fluid film between them.

The relative adsorptive capacity of porous solids such as molecular sieves is known to correlate with the pseudosurface tension of gases, as the pore diameters approach molecular dimensions. Therefore, it is not surprising to find fine particles adhering to each other, even in the presence of certain dry surrounding gases, although adsorbed water molecules are predominantly the more common cause of "sticking" or nonfree flowability in conventional solids handling operations. Moisture content is conventionally reported on a weight percent basis, although volume percent would be a more relevant figure; 1% by weight of water in a bed of lead particles would constitute a far greater free-flow deterrent than 1% by weight in a bed of ground cork particles.

MASS

Unlike the constancy of atomic and molecular weights for fluids (and, hence, the constancy of their voidless packed densities), a particle-form solid can exhibit varied weight-volume characteristics.

True and Apparent Particle Densities

True density refers to the material making up the particle. For example, a particle of solid glass ordinarily would have a true particle density of about 162 lb/ft^3. If the particle contained minute fissures, cracks or pores, these intraparticle voids would reduce the density to an "apparent" value of perhaps 150 lb/ft^3 or less. The "bulk" density, obtained by weighing a vessel of given volume filled with the glass particles, probably would be in the neighborhood of 75 lb/ft^3, depending on the interparticle voids. If, after determining the bulk density, the same undisturbed volume is filled with water, mercury or some measurable liquid quantity that will permeate through the particle bed, then this liquid volume subtracted from the total container volume holding the particle mass will give a measure of the particle volume.

The particle volume divided by the previously measured weight then gives either the apparent or the true particle density. If the particles are not fissured or porous, then the result is the true particle density. In such a case, the true particle density would equal the apparent density. On the other hand, if the particles are porous, the result may be somewhat in doubt, depending on the technique employed. If it is suspected that the porosity may be high, then the use of a liquid with low wetting characteristics, such as mercury, would be preferred simply to measure the interparticle voids and, thus, permit determination of the apparent particle density.

Water, or a similar liquid, might penetrate the particle pores so that the displacement thus measured would yield the true particle density. Visual examination of the particles with an appreciation of the possible effects of liquid wetting and capillary action generally will permit evaluation of the probability that a given displacement technique gives a measure of the true or apparent particle density.

Bulk Densities

In determining the simple bulk density, some indication should be given of the degree of bed packing, i.e., whether it is loose or dense. There appear to be three relatively satisfactory standard techniques for handling powders in the determination of bulk density. The most common is simply to feed the powder gently into a graduated container without disturbances that might create compaction effects. The second technique is to take this same container and subject it to tapping or shaking for a sufficient period of time to compact the bed to the minimum porosity. A third technique involves passing air up through the powder bed at a velocity sufficient to fluidize the bed, and then very gradually diminishing the air rate so that the bed will settle in its probably loosest stable packing.

The order of magnitude of differences between loose and compacted bed voidages ranges normally from 5 to 20%. For example, microspheroidal fluid catalytic-cracking (FCC) catalyst consists of an amorphous silica composition containing a small percentage of alumina. The particles are hard, smooth and almost perfect spheres. The particle-size distribution is log-normal, with about a twentyfold variation from the largest to the smallest and a geometric weight mean diameter of about 60 μ. Voidages vary from the loosest bed at 70% to the densest bed at 58%.

Calcium fluoride, as pure chemical reagent powder, consists of regularly shaped and fairly evenly sized cubes or regular octahedra. The average particle size is about 30 μ. The voidage varies from the loosest bed at 42% to the densest bed at 33.5%.

Reagent-quality anhydrous sodium carbonate consists of elongated crystals approximately 50 μ wide $\times$ 150 μ long. The voidage varies from 82% to 68.8%.

Titanium dioxide pigment particles are relatively equidimensional, with a diameter of about 0.5 μ. This material has a strong tendency to pack like a snowball in room air; it varies in voidage from 86.2% to 77.7%.

Refined kaolin clay powder is made up of soft, porous, irregular particles, each of which is an aggregate of the clay mineral crystallites. The aggregate particles have an irregular rounded shape and are 2–10 μ in diameter; the crystallites are 0.3–0.8 μ in diameter. This material also has a strong snowball-packing characteristic and varies in voidage from the loosest, at 93.1%, to the densest, at 88%.

Activated fuller's earth is an acid-treated calcium montmorillonite clay of highly open, spongy structure. The aggregate particles have an internal porosity of the order of 70%; they are irregular in outline and are mostly 5–25 μ in diameter, with a few larger than 50 μ. This material snowballs to a lesser degree. The extremes of the void fractions vary from 85.9% to 75.9%.

Ordinary portland cement consists of irregularly-shaped particles of about 5–15 μ in diameter. Void fractions vary from the loosest bed, 70.6%, to the densest bed, at 52.3%.

Self-rising baking flour consists of regular, disklike particles of about 25 μ in diameter, with very strong snowballing characteristics. Voidage varies from 73.1% to 55.5%.

The more irregular the particle shapes and the wider the particle-size distributions, the more difficult the process of particle reorientation into denser and more compact configurations.

AERODYNAMICS

Single Particles

The two most pronounced aerodynamic characteristics of a solid particle are the minimum velocity required to convey it vertically upward and that required to convey it horizontally.

The terminal or vertical free fall velocity considered to be equal to the minimum velocity required to suspend the particle in mid-air (any increment in velocity will begin to convey the particle) may be regarded as analogous to a fluid's triple point temperature.

Free fall or suspending velocities should be determined experimentally and fitted to the conventional drag coefficient versus Reynolds number correlation, which can be used to determine the particle size corresponding to an aerodynamically equivalent sphere. Comparison with sieve sizes permits determination of a so-called shape factor. Particle size analyses of powders preferably are carried out by techniques that yield the aerodynamically equivalent sphere size, such as Roller, Bahco and Sedigraph.

The so-called saltation velocity, necessary to keep a particle in suspension in horizontal conveyance, is more of a practical than a fundamental characteristic. It relates to the thickness of the laminar film or the velocity gradient at the wall surface, which, if small relative to the particle's narrowest dimension, will not restrain the particle's motion and will allow it to be swept back into the mainstream. A similarly "practical" figure is the threshold velocity, which will pick up a particle already at rest in the bottom of a pipe or channel as fluid velocity is increased.

Bulk Solids

When a fluid such as air is passed upward through a bed of fine particles, it encounters a frictional resistance in flowing through the interstices between the particles. As the flow is increased, a point is finally reached at which the frictional resistance is equal to the bed weight. A further increase in gas velocity will cause the bed either to dilate or bubble, depending on its average particle size and size distribution. The gas velocity and the bed voidage at incipient bubbling are fundamental characteristics related by analogy to the boiling point of a pure liquid or a mixture. Figure 2 qualitatively illustrates the observations relating incipient buoyancy and incipient bubbling.

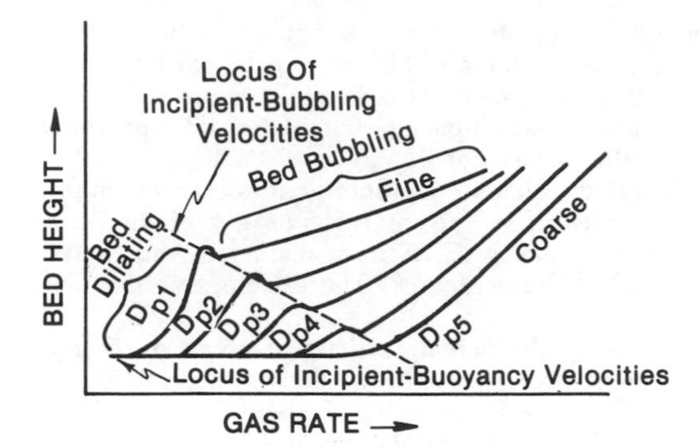

Figure 2. Particle size affects incipient-buoyancy, dilation and incipient-bubbling velocities.

Experimentally, incipient bubbling frequently is difficult to determine visually. A less ambiguous and reproducible procedure involves vigorously fluidizing the bed, instantaneously shutting off the gas flow, recording bed height as a function of time and extrapolating these observations to time zero, as illustrated in Figure 3. The elapsed time on the abscissa of Figure 3 may be less than 10 seconds over the constant-rate bed dilation range, so electronic timing is essential. At superficial velocities incrementally greater than incipient bubbling, a bed of mixed particle sizes begins to segregate. This results in varied distortions in pressure drop and voidage distributions with time and, hence, confusion with regard to the significance and interpretation of the shape of a pressure drop versus velocity relationship.

Bulk solids are frequently referred to as exhibiting a characteristic "deaeration time." This is extremely important in relation to the solids flow characteristics, but also rather ill defined and poorly understood. It can best be regarded as the time period over which the unaerated bulk solids behave like a liquid, either exhibiting a viscosity (measurable with a falling ball or paddle viscosimeter), flushing from a hole in a storage bin (as though under the influence of its own head, e.g., bulk density times bed depth), or being pumpable through a pipeline (with the entire line filled with bulk solids).

No methodical investigation of deaeration time has ever been undertaken, apparently. There are bulk solids that flow like a liquid and are pumpable in bulk, exhibiting a deaeration time that is infinite; however, if shaken, they can be walked on. Others can be pumped like a liquid and exhibit a viscosity. They cannot be vibrated to a load-bearing compaction, yet will not flow through a hole in a vessel. In general, the coarser the particles, the shorter their deaeration time; coarse gravel has a deaeration time of essentially zero. Deaeration time is neither the dilatancy period of Figure 3 nor does it appear to be proportional to the depth of bed and, therefore, related through the incipient bubbling velocity. In relation to the rheology of particulate solids, there is hardly a more directly related, yet less understood or predictable, property.

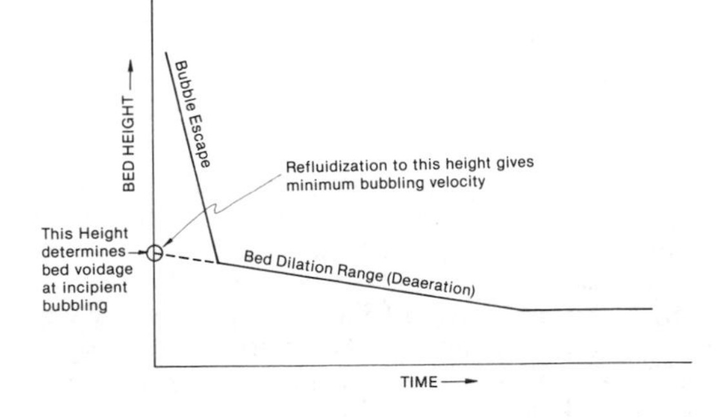

Figure 3. Determination of incipient bubbling conditions.

TWO-PHASE EQUILIBRIUM

Bed and Freeboard Compositions

Evidence of a direct analogy, if not a fundamental equivalence, between liquids and particulate solids appears more than circumstantial when considering the composition and rate of particle carryover from a fluidized bed [2]. As exemplified in Figure 4a, a multicomponent liquid is held at constant temperature above its initial boiling point and the resultant vapor above the splash region is totally condensed and returned. In Figure 4b, a multiparticle-size bulk solid is held at a constant superficial velocity above its incipient bubbling velocity. The resultant equilibrium entrainment existing above transport disengaging height (TDH) is separated totally and returned to the bed.

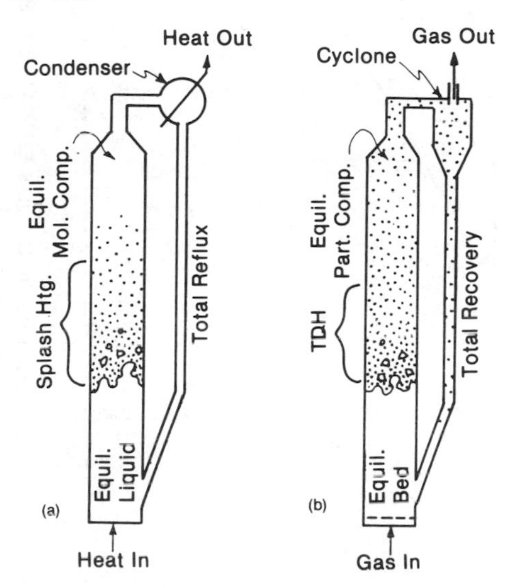

Figure 4. Analogies between molecular and particulate equilibria.

Since, for ideal liquids, concentration in the vapor is proportional to the product of vapor pressure and concentration in the liquid, analogies between temperature and velocity suggest that a basic property akin to vapor pressure should exist for each particle component in a bed as some function of velocity. Since, for a homologous series of hydrocarbons, it can be shown that $\Delta T/\sqrt{\text{mol wt}}$ is essentially constant over equal vapor pressure increments, analogously, entrainment above TDH might be expected to correlate with $\Delta U\sqrt{d}$, where ΔT could be regarded as (T − melting point) and ΔU as $(U - V_t)$. Above TDH, the dimensionless quantity $(U - V_t)/\sqrt{g\,d}$ therefore should correlate with the relative rates of entrainment (or relative volatility) of each component particle size. This term is not necessarily compatible with the bed source, since absolute entrainment (or vapor pressure) must be zero above a non-bubbling bed; U must exceed U_{mb}. Assuming that the total rate of entrainment varies with $(U - U_{mb})/\sqrt{g\,d_{avg}}$ in the same functional relationship as the relative entrainment of each component varies with $(U - V_t)/\sqrt{g\,d}$, it has been shown that the correlation of Figure 5 satisfactorily reproduced entrained compositions and total rates for several systems, explored experimentally in substantial detail. To what degree Figure 5, or any modifications thereof, can be considered universally applicable has yet to be demonstrated.

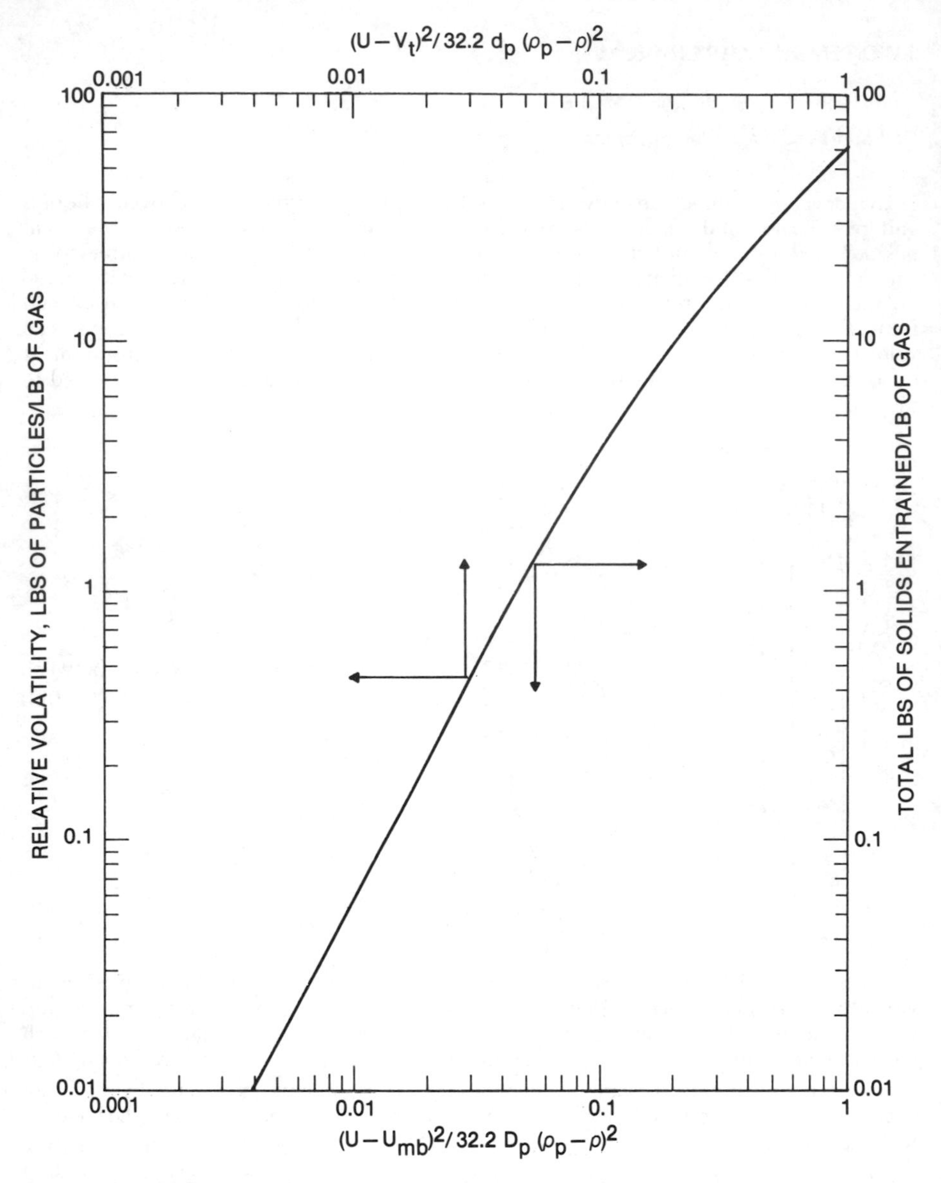

Figure 5. Relative and total steady-state ("equilibrium") particle carryover rates ("vapor pressures").

DEFORMATION

Equilibrium Frictional Planes

When fully drained from a vessel, or poured on a level surface, freely flowing bulk solids exhibit their so-called angle of repose, which represents an equilibrium limit between stationary bulk solids and their surrounding fluid medium. In the course of flowing from the full bin, the actual core of flowing material takes on a conical shape, whose slope (measured from the horizontal) is an average of 27° steeper than the solids' angle of repose. This equilibrium between flowing and stationary bulk solids is referred to as the angle of draw in mining, the wedge of maximum thrust or angle of rupture in soil mechanics, the angle of wall friction in civil engineering, determining the slope of mass flow hoppers, and the angle of internal friction in the field of fluid-particle technology.

It has been shown [3] experimentally and derived theoretically that the dimensionless relationship

$$W = \sqrt{gh\rho(\Delta\rho)/\tan\alpha} \tag{1}$$

applies equally to the flow of gases, liquids and bulk solids. In the case of solids, h represents the width or diameter of the opening; $\Delta\rho$ the difference between bulk density, ρ_B, and surrounding fluid density, ρ; and α the bulk solids' angle of internal friction. For gases flowing over submerged weirs, liquids obeying the Francis weir equation and average bulk solids, α is approximately 70°. Equation 1 therefore can be rearranged in the following form:

$$\frac{W}{\rho} = \frac{1}{\sqrt{\tan 70°}}\sqrt{\frac{gh\Delta\rho}{\rho}}$$

or (2)

$$V = 0.6\sqrt{g\Delta P/\rho}$$

Since α is a measurable characteristic of bulk solids, this raises interesting questions concerning the physical significance of the commonly measured coefficient of 0.6 for liquids and gases; it suggests that α for bulk solids might be analogous to an angle of wetting for liquids.

The angle of slide represents the frictional force between grains and an inclined solid surface. It is generally measured as the angle to which a steel plate supporting the material must be tipped from horizontal before the solids will slide off, owing to the force of gravity. This angular property is primarily of interest in determining whether hoppers or hopper cars handling relatively free-flowing materials can be expected to empty all their contents when discharged, or whether additional shakers will be required. It also determines shallowest permissible hopper bottoms in instances in which elevation must be kept to a minimum. In the academic sense, the angle of slide always must be accompanied by qualifications of the nature of the surface against which it is measured.

Bulk Resilience

A fluid's viscosity is fundamentally a measure of its resilience or deformability. The greater a fluid's viscosity, the less its ability to deform and, therefore, to flow. The smaller and farther apart the fluid molecules, the lower its viscosity; the smaller the solid particles and the greater the bulk solids' voidage, with or without aeration, the lower its viscosity.

Experiments with a paddle viscometer have demonstrated [4] that the point at which further addition of fines is needed to reduce the viscosity of a mixture of two sizes of a particle has a relatively minor effect. This occurs when the concentration of the finer particles has reached the level corresponding to the quantity necessary to coat the surface of each of the large particles with a monolayer [5,6], as illustrated in Figure 6. Extending this to multicomponent systems (in which each particle is present in sufficient num-

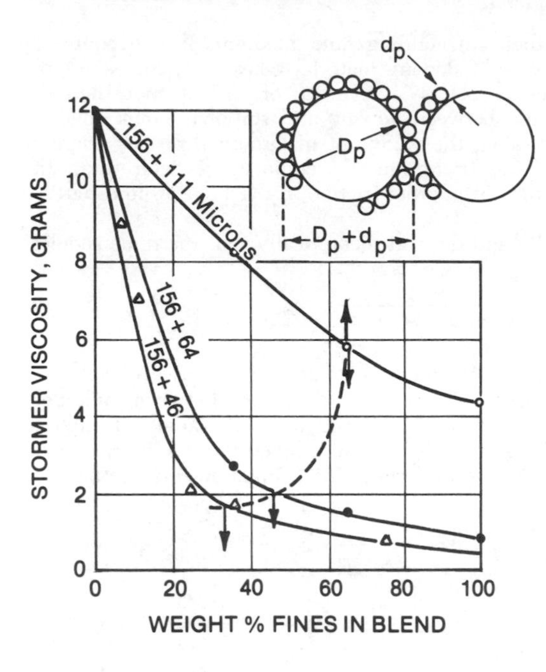

Figure 6. Mixture viscosity and minimum fines for particle "coating."

bers to coat the largest with a monolayer) results in the skewed or log-normal size distribution of Figure 1. This corresponds to that of soils such as sand, which obviously would tend toward minimum internal stress, or viscosity, in the process of their formation.

The limiting case of maximum density of conveying would occur when all solids in a pipe maintain their bulk density and move along like a plastic material extruded through a pipelike die. For a powder to undergo extrusion flow, it must possess a property that might be called "bulk deformability" or, essentially, viscosity. Just as a liquid will allow a denser solid body to fall through it, powders capable of extrusion flow display a similar continuous yield or bulk viscosity. This is the basis of a very simple, but effective, test [7] related to the falling ball viscometer.

A 250-ml cylinder is filled with the powder to be tested. A 1-cm-diameter steel ball is released from just above the surface and if it falls to the bottom, as if through a liquid, the powder is capable of extrusion flow. All finely powdered coal particles less than 44 μ in diameter readily pass this test, even after standing in the test cylinder for weeks. If, however, the charge is tapped and thereby compacted the falling weight will rest on the surface and not penetrate even under pressure.

Some powders, which alone do not pass the falling ball or drop weight tests, can be rendered deformable by the addition of colloidal silica. An equal volume added to an FCC catalyst (a 4% by weight addition) gives it a deformability comparable to that of powdered coal. The dropped weight fell readily and the mixture moved easily in extrusion flow through the test pipe. The volumes of colloidal silica and catalyst were approximately additive so that the density of the mix was almost half that of the original catalyst. Under magnification, the silica was seen to be dispersed as small-diameter agglomerates within a matrix of the catalyst, suggesting that its role was that of interspersed deformable or "soft" particles, analogous to small gas bubbles trapped in a fluidized bed.

Such simple tests offer clues to the role of fine particles in altering the rheological properties of a powder. The significance lies in the identification of powders suitable for a vast number of commercial applications. A prime example involves the charging of powdered coal to blast furnace tuyeres, where extrusion flow would afford essentially equal coal feed to all tuyeres with nearly zero dilution air. The more uniform charging of coal or limestone to fluid-bed gasifiers, combustors and boilers and the design of pressure letdown standpipes are additional examples.

REFERENCES

1. Heywood, H. *J. Imp. Coll. Chem. Eng. Soc.* 4:17 (1948).
2. Gugnoni, R. J., and F. A. Zenz. *Fluidization,* J. R. Grace and J. M. Matsen, Eds. (New York: Plenum Publishing Corp. 1980), p. 501.
3. Zenz, F. A., and F. E. Zenz. *Ind. Eng. Chem. Fund.* 18(4):345 (1979).
4. Matheson, G. L., W. A. Herst and P. H. Holt. *Ind. Eng. Chem.* 4:1099 (1949).
5. Trawinski, H. *Chem. Ing. Tech.* 23:416 (1951).
6. Trawinski, H. *Chem. Ing. Tech.* 25:201 (1953).
7. Rowe, P. N., and F. A. Zenz. In: *Fluidization Technology, II,* D. L. Kcairns, Ed. (New York: McGraw-Hill Book Company, 1975), p. 151.

CHAPTER 24
CONCEPTS AND CRITERIA FOR GAS-SOLIDS FLOW

S. J. Rossetti
Exxon Research & Engineering Company
Florham Park, New Jersey 07932

CONTENTS

INTRODUCTION

Gas-solids flow is a fundamental part of most industrial processes. Whether included in a fluidized bed operation or to convey solids into and between reactors, bins, hoppers and other process vessels, reliable solids flow is essential for smooth process operation. Systems designed with excessively high gas rates are subject to increased energy consumption, solids degradation and erosion. These can result in an economically unattractive operation. Systems designed for gas rates that are too low or solids rates that are too high are subject to erratic operation due to deposition or may be completely inoperable because of blockage. This chapter presents concepts and criteria that can aid in designing a smoothly operating gas-solids flow system or in troubleshooting a problem operation.

Gas-solids flow systems generally are designed as either dilute- or dense-phase systems. Dilute-phase systems are low-pressure systems with low solids concentrations and high gas velocities that result in low-pressure drop per foot of conveying line compared to dense-phase systems. Dense-phase systems operate at higher pressure levels and feature low gas velocities and high solids concentrations to minimize solids degradation and line erosion. Between these regimes is a region that generally is avoided because of unpredictable solids flow and a tendency toward line blockage. There are no quick and generalized methods for predicting when a system will be operating in the dilute- or dense-phase regime. Although the nature of the flow regime depends on complex relations among solids, gas and line characteristics, it is generally safe to consider systems with solids loadings of 5 lb solid/lb gas or less to be dilute-phase systems and those with loadings higher than 50 to be dense-phase systems.

In this chapter some principles of single-particle dynamics will be reviewed, followed by a detailed description of dilute-phase flow and an overview of dense-phase flow systems. Since dilute-phase flow is commonly encountered industrially and is more amenable to laboratory investigation, many generalized design relations are available for such systems. Much less information is available for dense-phase flow, and these systems often must be designed empirically. A knowledge of some basic principles of particle dynamics is helpful in understanding these flow modes.

PARTICLE DYNAMICS

Vertical—Terminal Settling Velocity and "Slip"

In vertical solids transport, gravity causes particles to fall behind an upflowing gas. The difference between the gas and solids velocity is defined as "slip." The amount of slip defines the solids velocity, the flow regime and, ultimately, the pressure drop due to solids holdup and frictional effects. The velocity of a single particle flowing up a vertical line, u_{so}, often is defined as the difference between the gas superficial velocity, v, and the particle's "terminal settling velocity," u_{to}:

$$u_{so} = v - u_{to} \tag{1}$$

The single-particle terminal settling velocity can be determined by Stokes' law [1] for particle Reynolds number, $Re_p = d_p u_{to} \rho_g / \mu$, less than 0.4:

$$u_{to} = \frac{d_p^2(\rho_p - \rho_g)g}{18\mu} \tag{2}$$

At higher Reynolds numbers, expressions referred to as the "intermediate law" and "Newton's law" can be used to calculate the single-particle terminal velocity.
Intermediate law:

$$u_{to} = \left[\frac{4}{225}\frac{(\rho_p - \rho_g)^2 g^2}{\rho_g \mu}\right]^{1/3} d_p \qquad 0.4 < Re_p < 500 \tag{3}$$

Newton's law:

$$u_{to} = \left[\frac{3.1(\rho_p - \rho_g) g d_p}{\rho_g}\right]^{1/2} \qquad 500 < Re_p < 200{,}000 \tag{4}$$

Unfortunately, use of these analytical expressions to determine the terminal settling velocity can be a trial and error procedure. The particle Reynolds number must be determined after computing the terminal settling velocity to determine whether the proper relation was used. If the Reynolds number determined is beyond the range of the correlation, then the terminal velocity must be recalculated using the appropriate expression. Zenz and Othmer [2] have developed a plot that simplifies the calculation and precludes trial and error. As shown in Figure 1, the terminal velocity divided by a system-dependent constant is plotted as a function of particle size divided by another system constant. Both constants are functions of fluid and particle properties so that the complete range of particle Reynolds numbers is covered.

Horizontal—Saltation and Pickup Velocities

The "saltation velocity" is one of the key design indices for horizontal gas-solids flow systems. One definition of the saltation velocity is that it is the minimum velocity in a system containing a horizontal line that will prevent solids deposition on the bottom

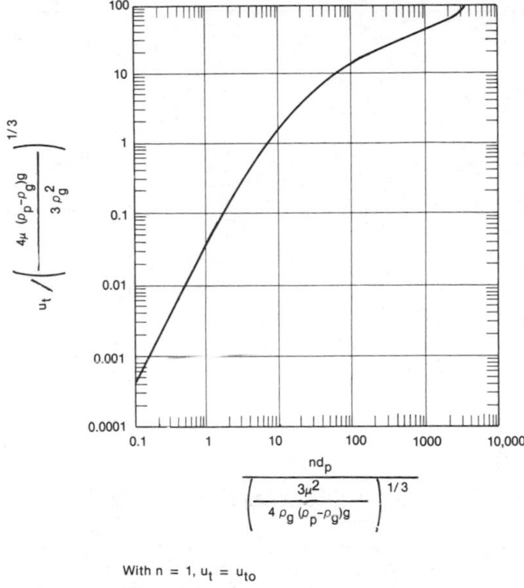

Figure 1. Graphic representation of relation between terminal settling velocity, u_{to} or u_{et}, and particle and gas properties [2].

of the line. Zenz [3] experimentally determined single-particle saltation velocities for a wide variety of spherical- and angular-shaped materials. He was able to correlate the minimum velocity required to transport a particle, without saltation and without obviously rolling or bouncing on the bottom of the tube, with particle drag coefficients, C_D. In this correlating procedure, $\sqrt[3]{Re_p C_D}$ is plotted against $\sqrt[3]{C_D Re_p^2}$. This is effectively a log-plot of the single-particle saltation velocity, V_{so}, divided by a system-dependent constant versus the particle size divided by another system-dependent constant.

The resultant curves exhibit a minimum indicating that for very fine solids, smaller particles can require a higher velocity to prevent saltation than larger particles. Zenz explains this by having smaller particles becoming trapped in an inviscid boundary layer at the wall of the pipe, which prevents resuspension. Larger particles are resuspended more easily since they exceed the size of the boundary layer and are subject to higher-velocity gradients, even at the same line superficial velocities. Zenz also presents data on mixed particle sizes and proposes an empirical method by which the saltation velocity of a mixed particle size distribution could be estimated. In this approach, the saltation velocity is a function both of the size and spread of the particle-size distribution.

The "pickup velocity" concept is closely related to the saltation velocity. The pickup velocity is the fluid velocity required to resuspend a particle initially at rest on the bottom of a line. Figure 2 shows the forces acting on a single particle initially at rest on a surface. By defining each of these forces, the fluid velocity at which a particle will

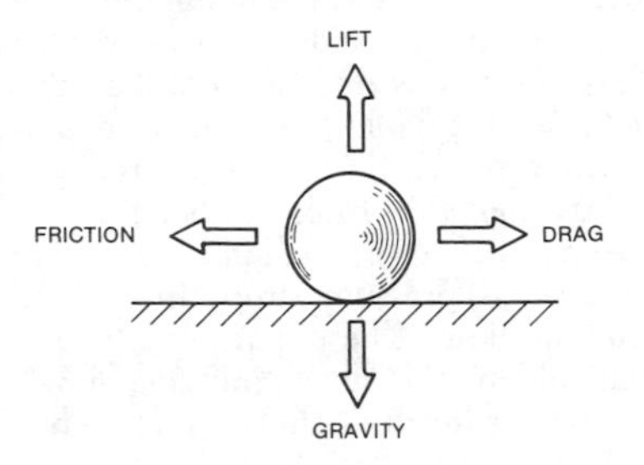

Figure 2. Forces acting on a single particle.

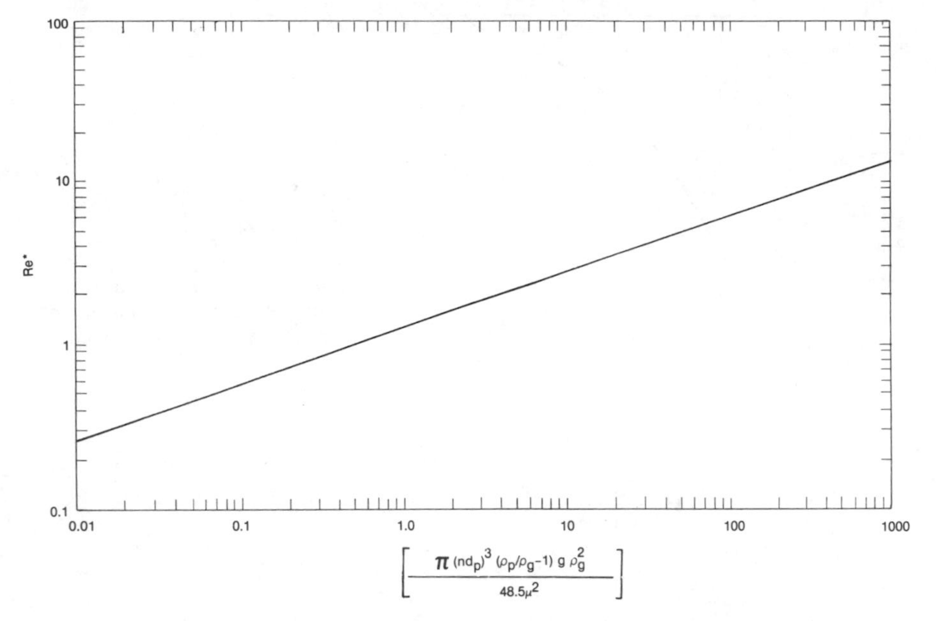

Figure 3. Plot to determine Re*, low range [4].

roll, slide or become suspended can be determined. This approach was developed theoretically and verified experimentally by Halow [4] for materials having a wide range of physical and chemical characteristics. Materials as diverse as Rice Krispies® and lead shot were used in his studies. Halow's results for particulate suspension in turbulent flow are summarized in Figures 3, 4 and 5. The particle pickup velocity can be determined from these figures if the pipe size and physical properties of the solids and gas are known.

DILUTE-PHASE CONVEYING

Vertical Flow—Flow Regimes, Choking, Solids Velocity and Pressure Drop

Three types of flow patterns, illustrated in Figure 6, have been observed in dilute-phase vertical flow [5,6]. At high velocities, solids and gas both flow uniformly up the pipe and frictional pressure drop is significant. As the velocity is decreased, the solids at the wall slow down and, at some point, have no net vertical movement. These particles appear to swirl, float or flutter near the wall. The solids rate is maintained by flow through the central core of the pipe, and essentially all of the solids pressure drop is due to the head of solids in the pipe. When the gas velocity is decreased still further for fine solids, the solids at the wall actually move downward, while the solids rate is maintained by the upward flow of solids in the core of the pipe. In this "annular" regime, the total pressure drop due to solids actually can be less than that from the head of solids in the line. However, flow in this annular regime is quite unstable, and small variations in the gas velocity can result in large differences in pressure drop. In addition, coarse particles cannot make the transition to annular flow. These particles tend to exhibit a dense-phase type of slugging flow, instead of annular flow. Slugging flow results in sharply higher pressure drop, which will catastrophically "choke" lines where limited pressure drop is available for conveying.

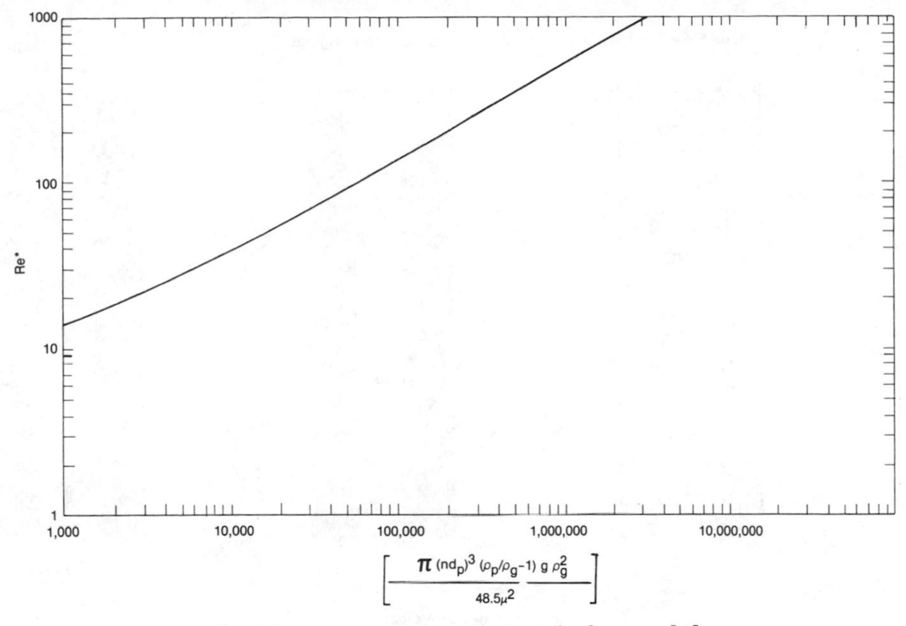

Figure 4. Plot to determine Re*, high range [4].

Yang [7] has proposed a criterion for determining whether choking will occur for a particular gas-solids system. His criterion is that to avoid choking, the dimensionless ratio $u_{to}/(gD)^{1/2}$ must be less than 0.35. For systems in which choking can occur, it is

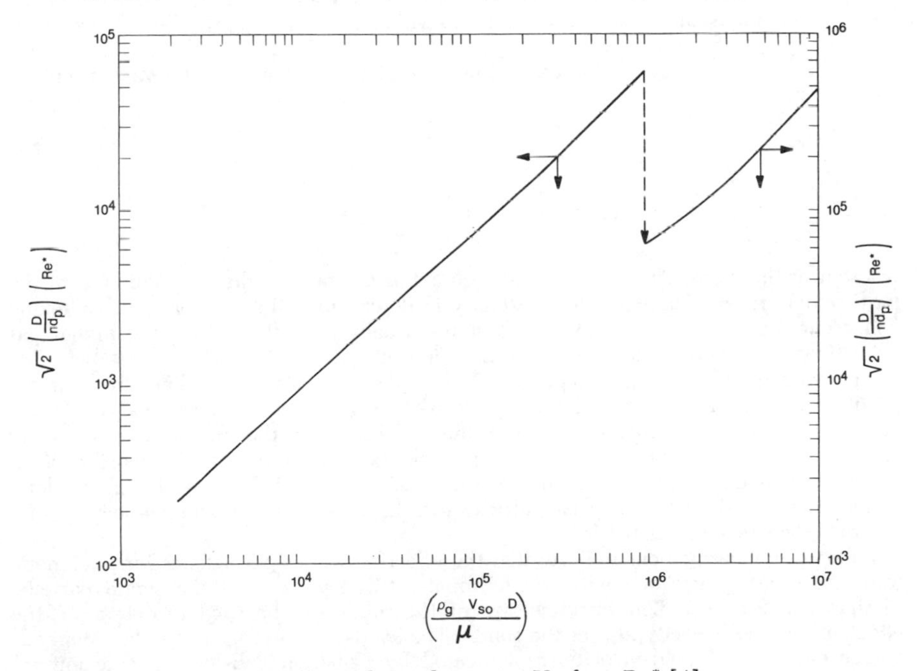

Figure 5. Plot to determine V_{so} from Re* [4].

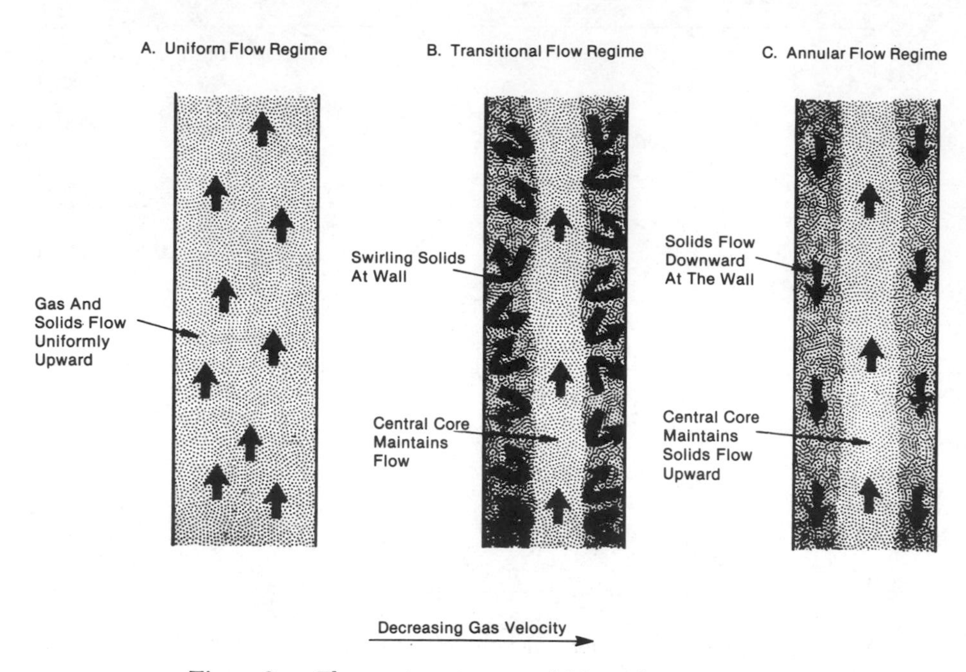

Figure 6. Flow regimes in vertical dilute-phase conveying.

important to be able to determine the critical choking condition. Leung et al. [8] suggested that choking occurs when the voidage in the transport line, ϵ_c, is equal to 0.97. Yang [9] proposes that the gas velocity and voidage at choking can be calculated on the assumption of a solids friction factor of 0.01 at choking. The choking point then would be determined by solving the following equations for ϵ_c and the gas velocity at choking, v_c:

$$u_o = (v_c - u_{to})(1 - \epsilon_c) \tag{5a}$$

$$(\epsilon_c^{-4.7} - 1) = \frac{0.01\,(v_c - u_{to})^2}{2\,g_c D} \tag{5b}$$

where u_o is the superficial solids velocity defined as the ratio of the mass flux, G_s, to the particle density, ρ_p. There is still uncertainty in determining the point at which choking will occur. For this reason, conveying systems usually are designed for the higher gas velocities, where both solids and gas move uniformly up the pipe.

The operating flow regime depends on the solids velocity in the line. In a single-particle system, where there are no particle-particle interactions, the solids velocity in vertical lines can be approximated as the difference between the superficial gas velocity and the terminal settling velocity of the particle (see Equation 1). For more concentrated systems this is not necessarily so. Fine particles can behave as though agglomerated or clustered [10], and coarse particles may have an effective terminal velocity that is actually less than calculated [5].

Laboratory investigations indicate that the effective size of particulates for pneumatic transport can be correlated with the terminal settling velocity of the single particle, as shown in Figure 7. The effective size of the solids can be used to determine the effective terminal velocity, u_{et}, of the solids. The solids velocity, u_s, in the line depends on the effective terminal velocity, as indicated by Equation 6 below for the uniform flow regime. This relation was derived in the manner suggested by Nakamurra and

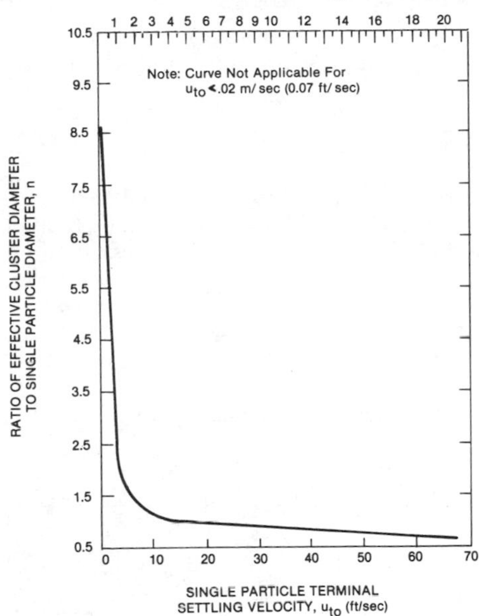

Figure 7. Effective cluster diameter to single-particle diameter ratio as a function of single-particle terminal settling velocity.

Capes [11] through solution of the momentum equations for the gas and the solids using the Richardson-Zaki relation for evaluating interparticle forces and approximating the solids friction factor as a constant divided by the solids velocity:

$$u_s = \frac{v - u_{et}\left[1 - \dfrac{\Delta P_g}{2\rho_p H}\dfrac{g_c}{g}\right]}{1 + \dfrac{Cu_{et}}{4gD}} \tag{6}$$

In this relation, ΔP_g is the frictional pressure drop per unit height of line due to the gas alone, D is the diameter of the line, and C is a system constant. For metal lines, C has been found [12] to have a value of 0.20 m/sec (0.66 ft/sec).

The solids velocity determines both the operating flow regime and the pressure drop due to solids. The total pressure drop per length of vertical line, $\Delta P_v/H$, can be calculated using Equation 7 below, once the solids have been fully accelerated:

$$\frac{\Delta P_v}{H} = \frac{f_g \rho_g v^2}{2g_c D} + \left(\frac{\Delta P_f}{H}\right)_s + \frac{G_s}{u_s}\left(\frac{g}{g_c}\right) \tag{7}$$

total pressure drop = gas frictional pressure drop + solids friction + solids head

In this relation, f_g is the Darcy friction factor (4 × Fanning friction factor), and G_s is the solids mass flowrate per unit of pipe cross-sectional area. The pressure drop due to solids friction $(\Delta P_f/H)_s$ is a function of the solids velocity in the line. This relationship can be expressed by

$$\left(\frac{\Delta P_f}{H}\right)_s = f_s\left[\frac{\rho_{dp}u_s^2}{2g_cD}\right] \tag{8}$$

where f_s is the solids friction factor, and ρ_{dp} is the dispersed-phase density defined by

$$\rho_{dp} = \frac{G_s}{u_s} \tag{9}$$

In the uniform flow regime, the solids friction factor, f_s, has been found to be approximately inversely proportional to the solids velocity [5,12–14]. If the solids friction factor were exactly proportional to the inverse of the solids velocity, the pressure drop due to solids could be expressed by

$$\left(\frac{\Delta P_t}{H}\right)_s = \left(\frac{\Delta P_f}{H}\right)_s + \frac{G_s}{u_s}\left(\frac{g}{g_c}\right) = \frac{Cu_s\rho_{dp}}{2g_cD} + \rho_{dp}\left(\frac{g}{g_c}\right) \tag{10}$$

This suggests the parameter Cu_s/gD for correlating frictional pressure drops.

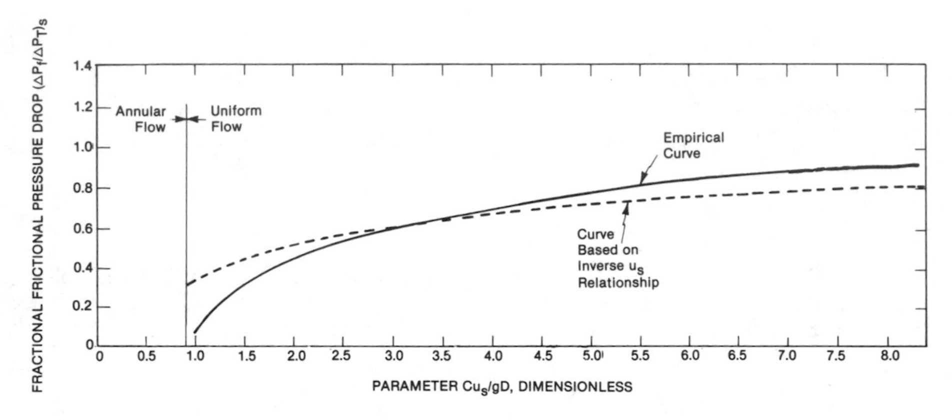

Figure 8. Correlation for vertical frictional pressure drop.

Figure 8 shows a plot of experimentally determined ratios of frictional to total solids pressure drop as a function of the parameter $[Cu_s/gD]$ [12]. The dashed line in the figure shows what would be obtained using the inverse u_s assumption for solids friction factor. The empirical curve shown by the solid line in the figure indicates that the inverse solids velocity assumption is a reasonable approximation in the uniform flow regime. The empirical curve, however, should be used in determining the solids friction pressure drop $(\Delta P_f/H)_s$ in Equation 7. Large deviations from the theoretical basis occur in the annular flow region, $Cu_s/gD < 1$. Pneumatic conveying systems for coarse particulates should not be designed for $Cu_s/gD < 1$ because of the possibility of operating in a slugging flow regime with sharply higher pressure drops that could choke the line. Extreme care should be exercised in operating fine particle systems at conditions in which $Cu_s/gD < 1$. In this regime, small changes in gas flowrate can result in marked pressure drop excursions.

Horizontal Flow—Flow Regimes, Saltation Velocity and Pressure Drop

Somewhat different concepts are involved in horizontal dilute-phase gas-solids flow. Two distinct regimes exist for this conveying mode: flow at gas velocities above and below the saltation velocity. Figure 9 illustrates the two flow regimes. Above the

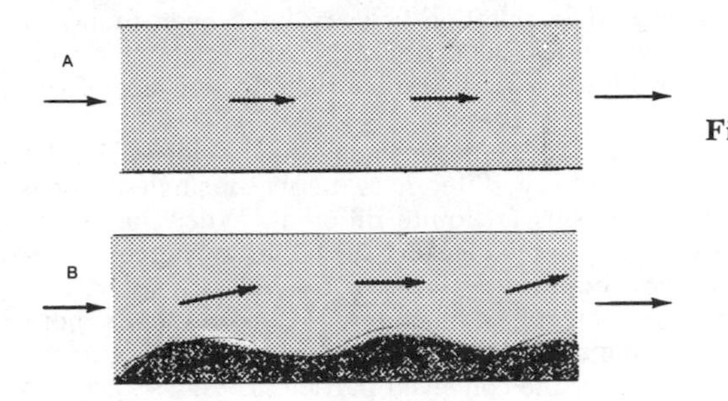

Figure 9. Flow regimes in horizontal dilute-phase conveying. (A) Flow at gas velocities above saltation. (B) Flow at gas velocities below saltation.

saltation velocity, at high gas velocities, the solids flow is dispersed in the gas phase similar to the uniform flow regime in vertical flow. The solids are suspended and move in the same direction as the gas. In this regime, the pressure drop increases with gas velocity. As the gas velocity is decreased, however, its carrying capacity is eventually exceeded and solids begin to deposit at the bottom of the line. The gas velocity at which the deposition occurs is the saltation velocity. At velocities below the saltation velocity, the pressure drop increases with decreasing velocity because of increased solids deposition.

For coarse solids, the saltation phenomenon is dramatic and quickly results in increased pressure drop. For fine solids, the deposition occurs slowly and nonuniformly down the length of the pipe. In this case, the overall pressure drop increases more slowly and is quite erratic due to the unstable nature of the nonuniformly sized and distributed deposits that roll down the pipe like dunes in a desert. These dunes form and travel the length of the pipe, slamming into elbows, bends, fitting or valves at the end of the piping. The pressure drop increases that can occur due to the change in flow regimes are illustrated schematically by Figure 10 [2].

Therefore, the saltation velocity is a key criterion for designing and troubleshooting gas-solids flow systems containing horizontal components. Most systems should be operated at gas velocities about 10–15% above the saltation velocity. This condition avoids solids deposition while decreasing horsepower requirements, erosion and solids degradation in dilute-phase systems. The saltation velocity, V_s, has been found [3,12] to be a linear function of the solids mass flux and to depend on particle and gas properties and the spread of the flowing solids size distribution. One way of expressing this relationship is shown below [12]:

$$V_s = V_{so}\left(1 + \frac{G_s}{k\rho_p}\right) \tag{11}$$

In this relation, V_{so} is an effective single-particle saltation velocity, and $G_s/k\rho_p$ is a factor that quantifies the effect of solids flux on the saltation velocity. Experimentally

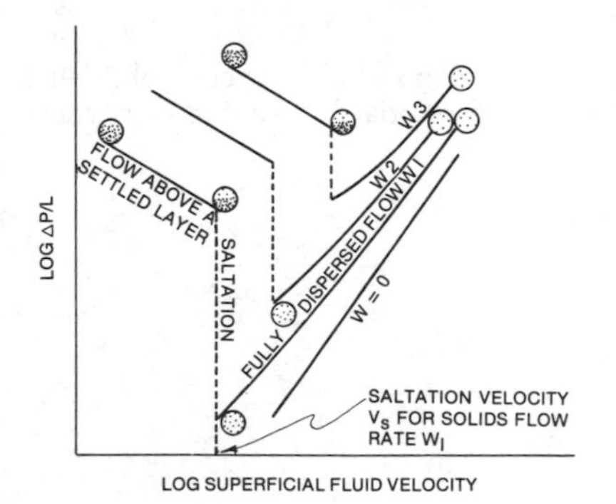

Figure 10. Effect of saltation on pressure drop [2].

determined values of the effective single-particle saltation velocity have been found to be most closely approximated by pickup velocities calculated using the relations developed by Halow [4], together with those in Figure 7 to determine the effective size of the particle.

That the pickup velocity rather than the single-particle saltation velocity is the significant particle property can be explained by differences between single-particle dynamics and dilute-phase flow. The mechanisms are quite different. When solids are being conveyed in a line, there is a dynamic interchange between the flowing particles and deposited solids. In reality, the process equilibrates around the velocity where the gas is no longer capable of picking up deposited solids. Therefore, V_{so} corresponds more closely to the pickup velocity than the single-particle saltation velocity. This pickup velocity depends, in turn, on the effective size of the conveyed particles. Besides knowing V_{so}, Equation 10 indicates that it is also necessary to determine the effect of solids rate on the saltation velocity. This effect is great for solids having a wide particle size distribution and slight for solids with a very narrow size distribution. The effect of the size distribution on saltation velocity has been included in Equation 11 by means of the mass flux factor, k. In the experimental studies previously cited [12], k was found to correlate with the standard deviation of the particle size distribution. The standard deviation, σ, is a convenient indicator of the spread of a particle size distribution. The relationship is presented in Figure 11. These relations provide a means of estimating the saltation velocity from particle and fluid properties and operating conditions.

In horizontal flow there is no pressure drop due to the head of solids in the line, so that the overall equilibrium pressure drop per unit length is obtained by adding the gas and solids frictional ΔP contributions:

$$\frac{\Delta P_H}{L} = \frac{f_g \rho_g v^2}{2 g_c D} + \left(\frac{\Delta P_f}{L}\right)_{SH} \tag{12}$$

Total pressure drop per unit length = gas-only frictional pressure drop per unit length + frictional pressure drop per unit length due to solids

At gas velocities above the saltation velocity, the flow is similar to the uniform flow regime in vertical flow. Therefore, as with vertical flow, the solids friction factor is

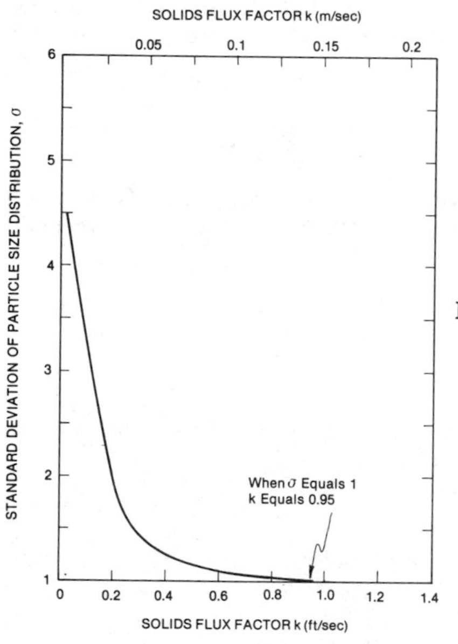

Figure 11. Relationship between solids flux factor, k, and standard deviation of particle size distribution.

approximately inversely proportional to the solids velocity, so that the pressure drop due to solids friction per unit length of horizontal pipe can be expressed as

$$\left(\frac{\Delta P_f}{L}\right)_{SH} = \frac{C_H G_S}{2 g_c D} \tag{13}$$

at gas velocities above the saltation velocity. As in vertical uniform flow, C_H has been found to have a value of 0.20 m/sec (0.66 ft/sec) in metallic pipelines [12]. Very little work has been done to determine pressure drop at gas velocities below saltation. Due to the erratic nature of the flow under these conditions, any method of determining the pressure drop can only provide an overall estimate and cannot account for the fluctuations encountered. At superficial gas velocities below the saltation velocity, the area available for flow is restricted by solids flowing and/or deposited at the bottom of the pipe. The gas above the deposited layer flows at the saltation velocity. Some gas percolates through the deposited layer, causing it to expand. The effective diameter of the line, De, available for flow has been determined for experimental studies carried out at velocities below the saltation velocity [12]. This has been correlated in terms of the calculated diameter of the open line, D_C, in Figure 12. D_C is the equivalent calculated diameter of the unblocked area, assuming that there is no air within the deposited solids, so that

$$D_C = \sqrt{\frac{vD^2}{V_{SO}} - \frac{4\,W_S}{\pi k \rho_p}} \tag{14}$$

where W_s is the solids mass flowrate. With this definition and using Figure 12, the pressure drop at gas velocities below saltation can be estimated from the relation

$$\left(\frac{\Delta P}{L}\right)_H = \frac{\left[0.0056 + \dfrac{0.5}{\left(\rho_g v \left(\dfrac{D}{Dc}\right)^2 De/\mu\right)^{0.32}}\right]\left[\rho_g V_{SO}^2 \left(1 + \dfrac{4\,W_s}{\pi k \rho_p De^2}\right)^2\right]}{2\,g_c\,De} + \frac{2 C_H W_S}{\pi g_c De^3} \tag{15}$$

Although it is possible to operate systems at velocities below the saltation velocity, this operation will tend to be unsteady, especially for fine particles. The operation becomes increasingly unstable the more the saltation velocity exceeds the superficial gas velocity. Operations with De/D less than 0.75 are susceptible to plugging, particularly at elbows, so they are not recommended.

Flow Through Elbows—Flow Models and Pressure Drop

Dilute-phase conveying systems generally include elbows. Elbows are an important source of system pressure drop which have received relatively little attention in the fluid-solids literature. The development presented for elbows in this chapter was confirmed in experimental studies carried out at the Exxon Research & Engineering Company [12]. Solids flow patterns through an elbow are highly dependent on the nature of the solids being conveyed. Fine particles travel at velocities very close to the gas velocity when entering an elbow. They are slowed down in the elbow by sliding around its outer edge and generating a frictional pressure drop. On leaving the elbow, they must be reaccelerated to velocities close to the gas velocity at the expense of gas energy or pressure drop. Coarse or dense materials are slowed down, as indicated in Figure 13, more by bouncing at the point of impact than by sliding around the elbow. There is subsequently less reacceleration energy loss when leaving the elbow.

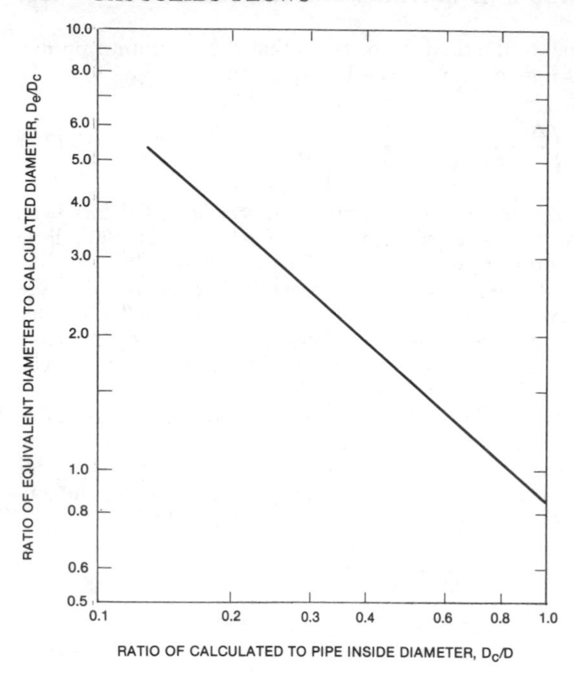

Figure 12. Curve for determining equivalent pipe diameter below saltation.

The pressure drop due to an elbow is the sum of the pressure drop across the elbow itself and the additional pressure drop in the piping downstream of the elbow due to solids reacceleration because of the presence of the elbow in the line. This pressure drop can be determined from the relation

$$\Delta P_e = (k_g + k_{el}\eta)\,\frac{\rho_g v^2}{2\,g_c} \tag{16}$$

where k_g is a coefficient accounting for the gas-only pressure drop, and k_{el} accounts for the frictional pressure drop across the elbow itself and the additional downstream pressure drop resulting from the elbow. The solids concentration effect is accounted for by the solids loading, η, which is the solids to gas mass flow ratio.

Based on studies using a limited number of elbow geometries, Rossetti [12] reports a value of 0.35 for k_g. The solids flow model described above indicates that the pressure drop due to solids flow through the elbow is a function of the particle terminal settling velocity so that the elbow pressure drop also should depend on the terminal velocity. Figure 14 shows the relation developed between the solids elbow coefficient, k_{el}, and the terminal settling velocity of the particles. Use of Figure 14 with Equation 16 enables estimation of elbow pressure drop effects for vertical-horizontal and horizontal-horizontal elbows at gas velocities above saltation in uniform flow regimes. At gas velocities below saltation, the pressure drop can be many times higher than calculated. In the investigation cited, the effect of the elbow radius of curvature to pipe inside diameter

Figure 13. Elbow flow characteristics.

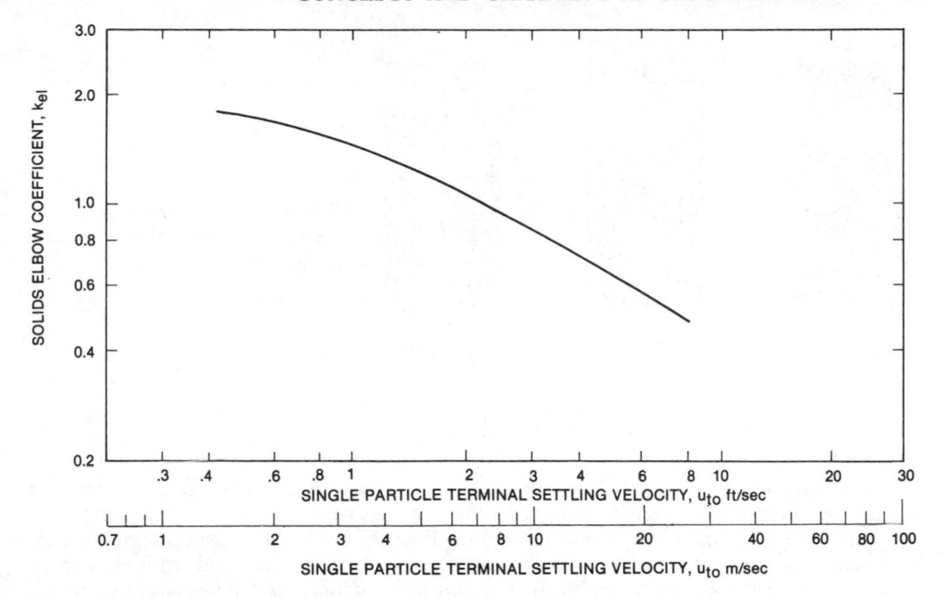

Figure 14. Relation between the solids elbow coefficient and single-particle terminal settling velocity.

ratio was studied for values of 1 and 4.2, and no effect was discernible. It is possible, however, that at higher ratios the values of k_{el} and k_g could change. Additional information or testing with elbows having higher ratios is needed to clarify this issue.

DENSE-PHASE CONVEYING

Vertical Systems—Flow Models and Pressure Drop

Vertical dense-phase conveying is most commonly encountered in standpipes and risers in petroleum and chemical plants. Several types of dense-phase flow modes occur in vertical systems. In addition to the annular flow regime previously described for fine particles, two other flow patterns have been reported—slugging and "extrusion," or "packed-bed," flow.

In slugging or bubble flow, gas bubbles act as pistons between slugs of solids to transport the solids. This is similar to the action observed in a slugging fluidized bed. The size of the bubbles in the line is controlled and determined by the pipe size. As the superficial gas velocity is decreased further, transport occurs in the "extruded," or "packed-bed," mode. In this mode, the solids move very slowly through the pipe as at a density close to the bulk density of the material. There is essentially no relative motion between the particles (as in a packed bed), and the solids behave like a deformable material being extruded in the pipe. Figure 15 schematically illustrates the flow patterns observed for the three dense-phase modes described.

Leung and Wiles [15] suggest the following criteria to estimate the gas velocity for transition to packed-bed flow:

$$0.55\,v - 0.45\,u_o = 0.55\,v_{mf} \qquad (17)$$

where u_o is the superficial solids velocity defined by G_s/ρ_p and v_{mf} is the minimum fluidization velocity of the solids being conveyed.

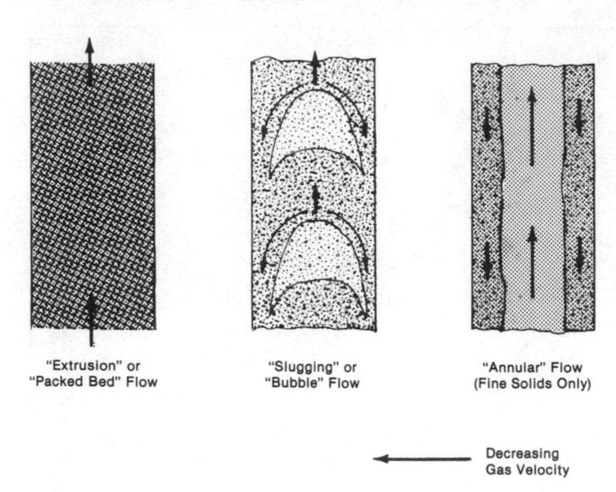

Figure 15. Flow modes in vertical dense-phase conveying.

Nakamurra and Capes [11] have proposed a model that describes the annular flow regime and presents relations to determine the average voidage and, therefore, the pressure drop across a line transporting in annular flow. However, the relations require experimental determination of empirical constants and cannot be used without experimental data on the system being evaluated. No other generalized information or model has been proposed for the annular flow mode of transport.

Slugging flow is encountered more commonly and is amenable to analytical description using bubble theory. The relation below proposed by Matsen [16] can be used to determine the voidage, ϵ, for a slugging system:

$$(1-\epsilon)/(1-\epsilon_{mf}) = \{U_b + W_s/[\rho_p(1-\epsilon_{mf})]\}/(v + U_b - v_{mf} + W_s/\rho_p) \quad (18)$$

In the relation, ϵ_{mf} is the voidage at minimum fluidization and U_b is the bubble rise velocity, which equals $0.35\sqrt{gD}$ in a slugging pipeline. Since frictional losses are not considered to be very significant in slugging systems due to the low velocities involved, pressure drop can be estimated by determining the solids head in the line as indicated below:

$$\Delta P_v = \rho_p(1-\epsilon)\, H \left(\frac{g}{g_c}\right) \quad (19)$$

Vertical extrusion or packed-bed flow is not usually encountered; however, Leung and Wiles [15] have proposed that the pressure drop for such a system may be calculated from a modified Ergun-type equation:

$$\frac{\Delta P d_p \epsilon^3}{H\rho[(v-u_o)\epsilon]^2\,(1-\epsilon)} = 150/Re_{sl} + 1.75 \quad (20)$$

In this relation, Re_{sl} is the particle Reynolds number based on the slip velocity

$$Re_{sl} = \frac{\rho_g\,(v-u_o)\,d_p}{\mu}\left(\frac{\epsilon}{1-\epsilon}\right) \quad (21)$$

and ϵ is presumably the voidage at minimum fluidization. This voidage is frequently within 10% of the voidage calculated from the poured bulk density, using the relation

$$\rho_B = \rho_p\,(1-\epsilon) \quad (22)$$

Horizontal Systems—Features, Advantages and Disadvantages, and Types

Dense-phase conveying systems, which include horizontal components, are used primarily for bulk transport and unloading; weighing, batching and delivery; conveying from storage areas; and conveying with inert gases. In specific instances, however, the advantages offered by a dense-phase conveying system can make it attractive for high-pressure transfers between process vessels and for fines disposal transfers. These types of systems are used in the cement, glass, plastics, chemical, ceramics, food, mining and petroleum industries. Unfortunately, however, dense-phase conveying technology is not as well developed as that for dilute-phase transfers. With the exception of vertical standpipes and risers, where bubble theory has been applied successfully, few tools are available for the design of dense-phase systems. Most designing is left to vendors, who generally test the material to be conveyed in their own facilities. These tests enable the vendor to specify the gas rate, line size and pressure level required to convey a given rate of the material for the desired distance. Various feeding vessel, gas introduction and even piping designs are recommended by different vendors. Little independent information is available to evaluate those recommendations.

Essential features of dense-phase conveying systems include low gas rates and velocities, high rates of solids transfer at low velocities, high pressure drop and material-specific operability problems. Not all materials can be conveyed successfully by all the systems available. Systems normally operate at gas velocities between 5 and 40 ft/sec (well below the saltation velocity) and pressure levels between 15 and 125 psig. Because of these features, horizontal dense-phase systems have certain inherent advantages and disadvantages when compared to dilute-phase operations. Some of these are summarized below (Table I).

In dense-phase systems, solids are either conveyed in the extruded flow mode or in the form of a series of short plugs separated by the conveying gas. Systems are designed to minimize the size of the plugs while maintaining operability. This is done to reduce the required pressure level for conveying since the pressure needed to move a solid plug in a pipe is proportional to the square of the length of the plug. Therefore, it is necessary to allow the pressure to be transmitted throughout the particle mass while reducing the plug length of the solids conveyed. The means by which these plugs are formed differentiates types of horizontal dense-phase systems. Most dense-phase systems use high-pressure gas to form the desired plugs or to transmit the pressure throughout the solids being pushed through the line. This gas may issue from an aeration or fluidization device, which aids the solids feeding, from a small perforated internal tube running the length of the pipe or from external bypass pipes for gas only.

In some systems, a pulsed-air injection device, cutter or air knife is used at the beginning of the pipeline. Dense-phase conveying systems also differ in feed hopper design. The principal differences in feed hopper design are whether the solids are fluidized or aerated by an external gas supply; whether the hopper is pressurized; and

Table I. Horizontal Dense-Phase Systems

Advantages	Disadvantages
Little solids, degradation or attrition.	High-pressure solids feeding systems need to be rugged, expensive and/or complicated.
Minimize line erosion.	Higher-pressure compressors are often more expensive.
Easier gas-solid separation (less and smaller separation equipment at end of system).	Not all materials are capable of being conveyed in all systems.
Generally smaller line sizes required.	No generalized technology is available for predicting dense-phase conveyability or for designing systems.
Possible energy advantage in high-pressure, low-volume operation.	

whether the solids are discharged internally from a fluidized bed or forced from the hopper into an external conveying line. The system chosen depends largely on the characteristics of the solids being conveyed. Table II shows solids characteristics associated with some of the most common types of horizontal dense-phase conveying systems.

Table II. Solids Characteristics Associated with Common Types of Horizontal Dense-Phase Conveying Systems

Solids Characteristics	Feed Hopper Design Features	Type of Solids Discharge	Plug Size Control
Somewhat free flowing, fluidizable, moderate air retention (the higher the air retention the longer the allowable conveying distance)	External fluidization gas	Internal, vertical	None
Somewhat free flowing, high air retention, can handle nonfluidizable fines	No fluidization, pressurized hopper	External	None
Free flowing, fluidizable, moderate air retention	External fluidization gas with pressurized hopper	Internal, vertical	None
Free flowing, fluidizable, moderate air retention	Pressurized	Across venturi throat	None
Cohesive, fluidizable, nonsticky powders, fine powders	Pressurized	External	Internal plug-making pipe
Fluidizable, fragile, abrasive, heavy	Pressurized	External	External compressed air lines
Fluidizable, cohesive, fine powders	Aerated	External	Air cutter
Granular, narrow size distribution, nonfluidizable, low air retention, fragile or abrasive	Nonfluidized, pressurized	External	Intermittent solids flow

NOMENCLATURE

Symbol	Definition	Dimension
C	system frictional constant, vertical flow	(l/t)
C_D	particle drag coefficient	—
C_H	system frictional constant, horizontal flow	(l/t)
D	pipe inside diameter	(l)
D_c	calculated equivalent line diameter with nonexpanded layer at velocities below the saltation velocity in horizontal flow	(l)
De	actual equivalent line diameter at velocities below saltation in horizontal flow	(l)

Symbol	Definition	Dimension
d_p	weight median particle diameter: the 50% point on a log-probability plot of particle size versus cumulative weight percent	(l)
f_g	Darcy friction factor for gas = 4 × Fanning friction factor	–
f_s	Darcy friction factor for solids	–
g	gravitational acceleration	(l/t^2)
g_c	gravitational constant needed in English system of units only = 32.2; in metric units, use 1.0 with no dimensions	$\left(\frac{m}{f}\frac{l}{t^2}\right)$
G_s	solids mass flowrate per unit cross-sectional area	$[m/(t\text{-}l^2)]$
H	vertical height	(l)
k	particle-size distribution factor on saltation velocity	(l/t)
k_{el}	elbow coefficient accounting for both reacceleration downstream and pressure drop by solids across an elbow	
k_g	coefficient accounting only for pressure drop across an elbow due to solids	
L	length of horizontal line	(l)
ΔP_e	pressure drop due to presence of an elbow	(f/l^2)
$(\Delta P_f)_s$	frictional pressure drop due to solids in vertical flow	(f/l^2)
$(\Delta P_f)_{SH}$	frictional pressure drop due to solids in horizontal flow	(f/l^2)
ΔP_g	pressure drop due to gas only	(f/l^2)
$(\Delta P_t)_s$	total pressure drop due to solids in vertical flow	(f/l^2)
ΔP_H	horizontal line pressure drop (excluding acceleration)	(f/l^2)
ΔP_v	total vertical flow pressure drop (excluding acceleration)	(f/l^2)
Re_p	particle Reynolds number based on terminal velocity	–
Re_{se}	particle Reynolds number based on slip velocity	–
U_b	bubble rise velocity	(l/t^2)
u_{et}	solids effective terminal settling velocity	(l/t^2)
u_o	superficial solids velocity	(l/t^2)
u_s	solids velocity in line	(l/t^2)
u_{so}	single-particle velocity	(l/t^2)
u_{to}	single-particle terminal settling velocity	(l/t^2)
v	superficial gas velocity	(l/t^2)
v_c	gas velocity at choking	(l/t^2)
v_{mf}	minimum fluidization velocity	(l/t^2)
V_s	saltation velocity	(l/t^2)
V_{so}	effective single-particle saltation velocity	(l/t)
W_s	solids flowrate	(l/t)
ϵ	voidage in line	–
ϵ_c	voidage at choking	–
ϵ_{mf}	voidage at minimum fluidization	–
η	solids loading ratio in mass of solids per unit mass of gas	–
μ	gas viscosity	$[m/(l\text{-}t)]$
ρ_{dp}	dispersed-phase density	(m/l^3)
ρ_g	gas density	(m/l^3)
ρ_p	particle density	(m/l^3)
σ	standard deviation of particle size distribution	–

Note: Letter symbols in dimension columns have the following dimensions

Dimension/Symbol	l	m	t	f
SI	meters	kilograms	seconds	newton
English	feet	pound mass	seconds	pound force

REFERENCES

1. Stokes, G. G. *Combined Trans.* IX (1851).
2. Zenz, F. A., and D. F. Othmer. *Fluidization and Fluid-Particle Systems* (New York: Van Nostrand Reinhold Co., 1960).
3. Zenz, F. A. "Conveyability of Materials of Mixed Particle Size," *Ind. Eng. Chem. Fund.* 3:1 (February 1964).
4. Halow, J. S. "Incipient Rolling, Sliding and Suspension of Particles in Horizontal and Inclined Turbulent Flow," *Chem. Eng. Sci.* 28:1–12 (1973).
5. Capes, C. E., and K. Nakamurra. "Vertical Pneumatic Conveying: an Experimental Study with Particles in the Intermediate and Turbulent Flow Regimes," *Can. J. Chem. Eng.* 51:31 (February 1973).
6. Van Swaaij, W. P. M., C. Burman and J. W. Van Breugel. "Shear Stresses on the Wall of a Dense Gas-Solids Riser," *Chem. Eng. Sci.* 25:1818 (1970).
7. Yang, W. C. "A Criterion for Fast Fluidization," *Proc. Pneumotransport* 3, *BHRA Fluid Engineering* (1976).
8. Leung, L. S., R. J. Wiles and D. J. Nicklin. "Correlation for Predicting Choking Flow Rates in Vertical Pneumatic Conveying," *Ind. Eng. Chem. Proc. Des. Dev.* 2(10):183 (1971).
9. Yang, W. C. "A Mathematical Definition of Choking Phenomenon and a Mathematical Model for Predicting Choking Velocity and Choking Voidage," *AIChE J.* 21(5):1013 (1975).
10. Arundel, P. A., S. D. Bibb and R. G. Boothroyd. "Dispersed Density Distribution and Extent of Agglomeration in a Polydisperse Fine Particle Suspension Flowing Turbulently Upwards in a Duct," *Powder Technol.* 4:302–312 (1970/71).
11. Nakamurra, K., and C. E. Capes. "Vertical Pneumatic Conveying: a Theoretical Study of Uniform and Annular Flow Models," *Can. J. Chem. Eng.* 51:39 (1973).
12. Rossetti, S. J. "Laboratory Investigations of Dilute Phase Gas-Solids Flow in Vertical and Horizontal Lines and Elbows," (to be published).
13. Reddy, K. V. S., and D. C. T. Pei. "Particle Dynamics in Solids-Gas Flow in a Vertical Pipe," *Ind. Eng. Chem. Fund.* 8(3):490 (1969).
14. Konno, H., and S. Saito. "Pneumatic Conveying of Solids Through Straight Pipes," *J. Chem. Eng. Japan* 2:2 (1969).
15. Leung, L. S., and R. J. Wiles. "A Quantitative Design Procedure for Vertical Pneumatic Conveying Systems," *Ind. Eng. Chem. Proc. Des. Dev.* 15:4 (1976).
16. Matsen, J. M. "Flow of Fluidized Solids and Bubbles in Standpipes and Risers," *Powder Technol.* 7:93–96 (1973).

CHAPTER 25
FLUIDIZING FINE POWDERS

P. Harriott
School of Chemical Engineering
Cornell University
Ithaca, New York 14853

S. Simone
Lagoven, S.A.
Refineria de Amuay
Edo. Falcon, Venezuela

CONTENTS

INTRODUCTION

Fluidization of fine solids with gases is generally of the bubbling type (aggregative, heterogeneous), although some powders may exhibit particulate fluidization (homogeneous, nonbubbling) over a limited range of velocities near the minimum fluidization point. With such powders, the bed expands uniformly as the gas velocity is increased, reaches a maximum height at about the velocity where bubbles are first noticed, and gradually collapses to a minimum height with further increases in velocity. Eventually the bubble flow becomes predominant, and the bed again expands with increasing velocity, as shown in Figure 1.

Solids exhibiting this anomalous behavior fall in Group A of Geldart's [1] classification, and commercial silica-alumina cracking catalyst is a typical example.

With solids that are somewhat coarser or more dense, bubbles start forming at the minimum fluidization point, and the bed expands nonuniformly; such materials are classified as Group B solids. Geldart's classification diagram, shown in Figure 2, also includes Group C powders, which are too fine for normal fluidization because of cohesive forces, and very large or dense particles in Group D, which can form spouted beds.

This chapter deals with the behavior of Group A powders, which show particulate fluidization at low velocities and are somewhat different from Group B solids, even in the bubbling region. There have been many studies of the minimum fluidization velocity, so the subject is reviewed only briefly. The data and correlations for the minimum bubbling velocity are discussed in more details, as there is some uncertainty in the results. The bed expansion during particulate fluidization is compared with predictions from laminar flow theory and from empirical correlations, and the bed expansion at the

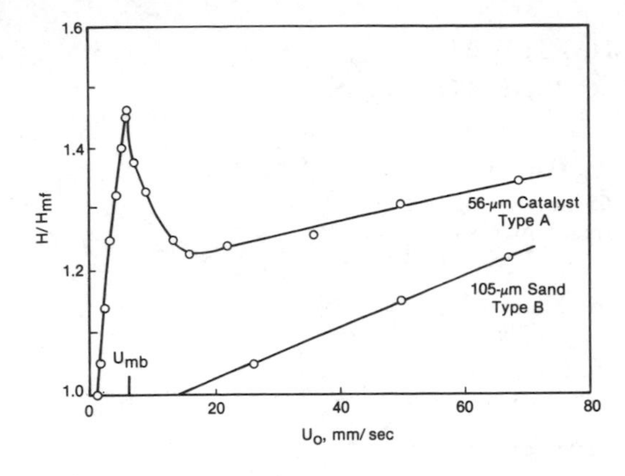

Figure 1. Bed expansion for Type A and Type B solids.

bubble point offers a check of different theories for particulate fluidization. Finally, the characteristics of the dense bed in the bubbling region are discussed; with type A solids, the dense-phase void faction and relative velocity are quite different from values at incipient fluidization, which may affect the performance of the fluidized bed as a reactor or mass transfer device.

MINIMUM FLUIDIZATON VELOCITY

A bed of solids should start to fluidize when the pressure drop across the bed becomes equal to the weight of the bed per unit cross section, allowing for the buoyant force of the fluid. Sometimes the particles start to move and the bed appears fluidized before the pressure drop reaches the critical value, which means that only part of the solid is supported by the fluid. At somewhat higher velocity, complete fluidization is obtained and the pressure drop does equal the weight of bed. To avoid the uncertainty associated with either the onset of fluidization or the velocity at which "complete" fluidizations is reached, the minimum fluidization velocity is generally determined by extrapolating the graph of pressure drop for the fixed bed until it reaches the horizontal line corresponding to the weight of the bed per unit area or the final value of the pressure drop.

Using the Ergun equation [2] for pressure drop per unit length of fixed bed and equating this to the net weight of the bed gives a quadratic equation for the minimum fluidization velocity, U_{mf}:

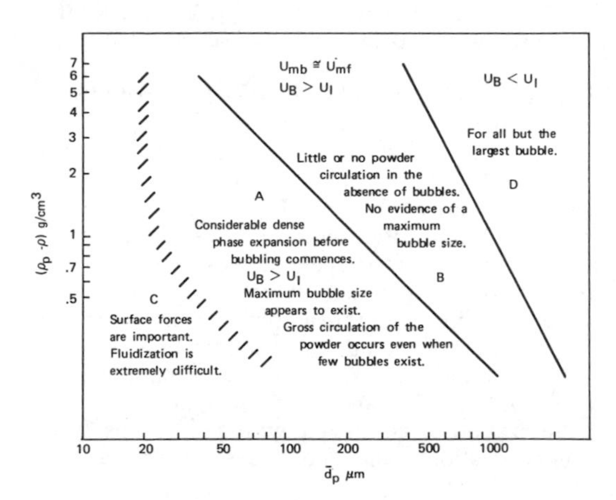

Figure 2. Particle classification proposed by Geldart [1].

$$\frac{\Delta P}{L} = g(1-\epsilon_{mf})(\rho_p-\rho) = \frac{150\mu U_{mf}(1-\epsilon_{mf})^2}{(\phi_s d_p)^2\epsilon_{mf}^3} + \frac{1.75\rho U_{mf}^2(1-\epsilon_{mf})}{\phi_s d_p \epsilon_{mf}^3} \quad (1)$$

For very small particles, only the laminar flow term is important, and a simpler equation can be obtained:

For $N_{Re_p} < 1$,

$$U_{mf} = \frac{g(\rho_p-\rho)(\phi_s d_p)^2 \epsilon_{mf}^3}{150\mu(1-\epsilon_{mf})} \quad (2)$$

One problem in using Equation 1 or Equation 2 to predict U_{mf} is that ϵ_{mf} depends on the particle size, shape and size distribution of the solid, and an error in ϵ_{mf} leads to a much greater percentage error in the value of U_{mf}. Correlations for ϵ_{mf} are not very reliable, and an experimental value is preferable. The value of ϵ_{mf} can be approximated from the bulk density of the solid, although ϵ_{mf} is often slightly greater than the value in a packed bed because the particles tend to pack more tightly when dumped into a bed. The best way to measure ϵ_{mf} is to fluidize the solids briefly and allow the bed to settle slowly to its final density.

When dealing with a particle-size distribution, the surface-volume mean, d_{sv}, should be used for d_p. For nonspherical particles, the product of the nominal size and the sphericity or shape factor, ϕ_s, is used to give the equivalent diameter. However, it is hard to determine ϕ_s from inspection of irregular solids, and it is usually determined from the pressure drop through a fixed bed or from the equation for minimum fluidization velocity. Values of ϕ_s range from about 0.7 for coal or angular sand to over 0.9 for round sand or microspheroidal cracking catalyst.

Several authors have presented correlations for U_{mf} that are based on Equation 2 but do not contain ϵ_{mf} or ϕ_s. For example, the following equation, due to Davidson and Harrison [3], corresponds to Equation 2 with $\epsilon_{mf} = 0.42$ and $\phi_s = 1.0$:

$$U_{mf} = \frac{0.00081\, g(\rho_p-\rho)\, d_p^2}{\mu} \quad (3)$$

A similar equation with a constant 0.001065 was given by Frantz [4].

Many other authors have presented empirical correlations for U_{mf} with exponents for d_p, μ, and $(\rho_p - \rho)$ that differ slightly from the values in Equation 2. A review by Grewal and Saxena [5] shows that Leva's [6] correlation fits published data for U_{mf} about as well as any:

$$U_{mf} = \frac{0.0093 d_p^{1.82} (\rho_p-\rho)^{0.94}}{\mu^{0.88}\rho^{0.06}} \quad (4)$$

The exponent of 1.82 rather than 2.0 for the particle size probably comes partly from the tendency for smaller-sized fractions of most solids to have higher values of ϵ_{mf}. The interparticle forces become more important for small sizes, leading to agglomeration and formations of pockets that would not exist in beds of large particles. The deviation from the expected exponents also may come from inclusion of data a little beyond the laminar range.

MINIMUM BUBBLING VELOCITY

The determination of the bubble point or minimum bubbling velocity, U_{mb}, is subject to the same kind of uncertainty as U_{mf}, if based on visual observations. A few bubbles may be noticed while the bed is still expanding in particulate fluidization, but these may be caused by wall effects or column irregularities, and the velocity might not be reproducible. More consistent results can be obtained from a plot of bed height versus velocity. The simplest definition of U_{mb} is the velocity at the maximum bed height, and this definition is used when taking values of U_{mb} from published graphs. In a few cases [7], workers have observed a broader maximum in height than shown in Figure 1, and U_{mb} has been determined by extending the initial linear portion of the curve until it reaches the maximum height. This gives a lower value of U_{mb} than the definition used here, but the difference is not enough to affect the correlations presented.

If there is no maximum in bed height or if the bed expansion is only about 1% when bubbles are first noticed, it is questionable whether the initial fluidization should be classified as the particulate type; data in this range were omitted when preparing correlations.

Geldart [1] showed that the major factor affecting the bubble point is the particle size, and he proposed a linear correlation between U_{mb} and d_{sv}. However, his data for catalyst particles can be fitted better with a lower slope, with a separate line for the larger particles of diakon polymer. A recent study [8] showed that U_{mb} varied with about the 0.6 power of the particle size for narrow fractions of cracking catalyst and was about the same for samples with a wide size distribution. To check this particle size effect, data from several sources for fluidizing cracking catalyst in air are plotted in Figure 3. The results from different investigators are in quite good agreement and indicate that U_{mb} varies with $d_p^{0.7}$. Some of the data are for spent catalyst or impregnated catalyst, and in the range 800 to 1500 kg/m³ the solid density does not appear to affect the minimum bubbling velocity.

Bubble points for several other types of solids fluidized with air are shown in Figure 4. The values of U_{mb} are generally higher than those for cracking catalyst, but no correlation with other physical properties is evident. There may be effects of particle shape, surface roughness or static charge, which would influence the interparticle forces. Up to almost the bubble point, the bed is known to have some structure, even though fluidized, since the top surface does not stay horizontal when the bed is tilted slightly [7]. The bed may contain weak aggregates of particles and some irregular cavities or channels, which help stabilize the bed to higher velocities before bubbling starts [9]. Particles with a strongly corrugated surface did give somewhat higher values of U_{mb} and ϵ_{bp} than relatively smooth particles [9], but there is a lot of overlap in the data for spheres and angular particles, although angular solids should have stronger interparticle effects.

Some of the runs for diakon fractions support the dependence of U_{mb} on $d_p^{0.7}$, but the results from different laboratories are not in agreement about the magnitude of U_{mb}. The data of Rowe et al. [10] for a commercial silica-based catalyst indicate a large change of U_{mb} with d_{sv}, but this could be an effect of particle size distribution, since the original catalyst had 27% fines ($< 45\mu m$), and the fines content was decreased by elutriation or sieving to get the other samples. Most other workers have found little effect of the size distribution on U_{mb} [8,11,12]. The correlation of U_{mb} proposed by Abrahamsen and Geldart [13] includes the fraction fines in a term $\bar{d}_p$ exp (0.716 F),

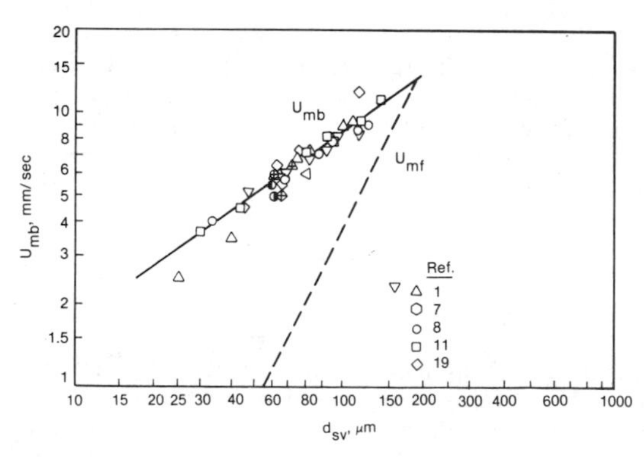

Figure 3. Minimum bubbling velocity for cracking catalyst.

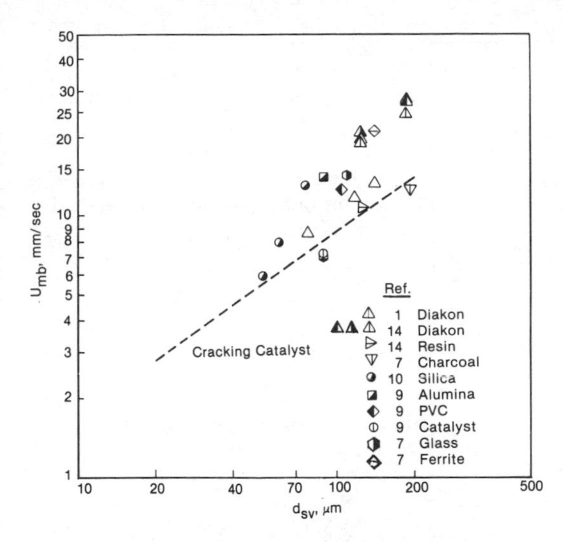

Figure 4. Minimum bubbling velocity for various solids.

but since F is large for small values of $\bar{d}_p$, the data probably could be fitted about as well using $d_p^{0.6}$ or $d_p^{0.7}$ and no term for the fines fraction.

The effect of gas density on U_{mb} was determined by fluidizing solids at high pressures. Goddard and Richardson [14] found that U_{mb} varied with $P^{0.06}$, and the 0.06 exponent was confirmed by Abrahamsen and Geldart [13]. The minimum fluidization velocity is either constant or decreases slightly with increasing pressure, so high-pressure operation gives a greater range of velocity over which particulate fluidization occurs. Crowther and Whitehead [15] found that fine coal particles exhibited particulate fluidization over the entire range of velocities tested at $\rho \geqq 120$ kg/m³, although at lower pressures a transition to bubbling fluidization was observed. Other studies have shown that high-pressure fluidization of fine solids results in smaller gas bubbles in the bubbling region [16,17] and lower dense-phase viscosities [18]. Pressure has little or no effect on bubble size or bed viscosity for particles too large to show any region of particulate fluidization.

Once the effect of gas density on U_{mb} was known, the effect of viscosity could be measured by using different gases. Increasing the viscosity significantly decreases the minimum bubbling velocity, although not as much as it decreases the minimum fluidization velocity. Abrahamsen and Geldart [13] reported that $U_{mb}/\rho^{0.06}$ varied with $(1/\mu)^{0.347}$. In an earlier study, Mutsers and Rietema [7] fluidized cracking catalyst and other solids with four gases and showed that the void fraction at the bubble point, ϵ_{bp}, was appreciably higher at higher viscosity. To compare their data with others, values of U_{mb} were estimated by assuming $\epsilon^3/(1-\epsilon)$ was proportional to the gas velocity. Figure 5 shows that U_{mb} varies with about the -0.45 power of the viscosity. Averaging the results of these two studies gives an exponent of -0.4, and an empirical equation for the minimum bubbling velocity is

$$U_{mb} = a d_{sv}^{0.7} \rho^{0.06} \mu^{-0.4} \tag{5}$$

For silica-alumina cracking catalyst the value of a is 4.23×10^{-3} if d_{sv} is in mm, ρ in kg/m³ and μ in Pa-s. There are not enough data to give values of a for other solids or to develop a more general equation accounting for all solid properties.

The ratio U_{mb}/U_{mf} is a useful index for the range of particulate fluidization. Using Equation 2 for U_{mf} leads to

$$\frac{U_{mb}}{U_{mf}} = b\, d_{sv}^{-1.3} \rho^{0.06} \mu^{0.6} \tag{6}$$

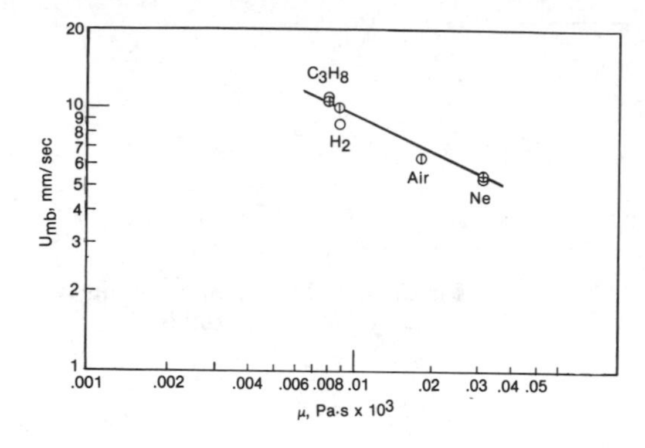

Figure 5. Effect of gas viscosity on bubble point (O corrected to ρ_{air} using $U_{mb} \alpha \rho^{0.06}$).

The range of particulate fluidization is greatest for fine particles, increases slightly at high pressures and is significantly greater when the gas viscosity is high.

BED EXPANSION

The expansion of a bed of solids in the region between the minimum fluidization velocity and the bubble point can be expressed as the change in bed height, bed density or void fraction, and several types of correlations have been proposed. Correlations based on void fraction permit a more direct comparison with theories of particulate fluidization. If flow between the particles is laminar, and if the pressure drop through the slightly expanded bed is assumed to have the same dependence on ϵ as for a fixed bed, the term $\epsilon^3/(1-\epsilon)$ should be directly proportional to the superficial velocity. Rearranging Equation 2 to make U_o the independent variable gives

$$\frac{\epsilon^3}{1-\epsilon} = U_o \frac{150\,\mu}{g(\rho_p - \rho)(\phi_s d_p)^2} \tag{7}$$

Equation 7 comes fairly close to fitting the data for cracking catalyst, as shown in Figure 6. On the log-log plot, the lines are quite straight but have slopes of 0.79 to 1.0, with lower values for the smaller fractions. Similar results were obtained by DeJong and Nomden [11], whose exponents of 1.2 to 1.0 for the group $\epsilon^3/(1-\epsilon)$ are the reciprocals of the slopes of log-log plots such as Figure 6. Because the range of ϵ is not very great, the same data also give straight lines on an arithmetic plot of $\epsilon^3/(1-\epsilon)$

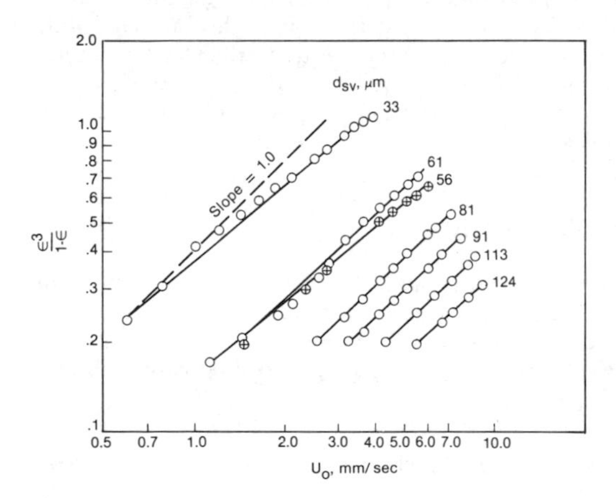

Figure 6. Bed expansion between $U_{m\rho}$ and U_{mb}.

versus U_o, but in most cases the lines have a positive intercept on the y axis, which is consistent with a slope less than 1.0 for the log-log plots.

The slopes that are less than 1.0 indicate less expansion than predicted by Equation 7 or a more permeable bed. The deviations could come from small differences in local porosity due to the bed structure. A bed with some channels or cavities larger than the average has a higher permeability than a bed with uniform channels. Photographs of 100-μm powder fluidized with air at velocities below the bubble point [19] showed cavities and microchannels at the wall as large as twice the particle size, but the size and frequency of cavities in the bed interior is not known. Another reason for slopes less than 1.0 may be that the function $\epsilon^3/(1-\epsilon)$, which comes from the Ergun or Carman-Kozeny models for flow in parallel channels, is not really appropriate for a highly expanded bed. The lowest slopes are found for the smallest size fractions, which have the highest maximum values of ϵ. Perhaps more precise data would show a gradual decrease in slope as ϵ increased.

The bed expansion in particulate fluidization also is frequently correlated with the simple equation used by Richardson and Zaki [20] and others [21]:

$$\frac{U_o}{U_i} = \epsilon^n \tag{8}$$

Based on studies of fluidization of solids with liquids, n was reported to vary from 4.65 for small particles to 2.4 for large particles, and equations were developed to predict n as a function of the particle Reynolds number [20]. The value of U_i was approximately equal to U_t, the terminal velocity for a single particle.

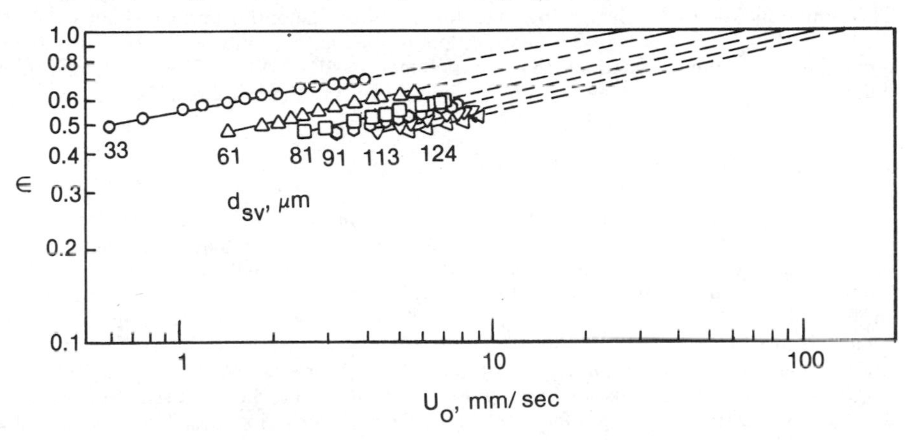

Figure 7. Bed expansion with particulate fluidization.

Data for particulate fluidization of solids with gases can be fitted to Equation 8, but the values of n are generally larger than for liquids at the same Reynolds number. Figure 7 shows data for several fractions of cracking catalyst, and n ranges from 5.6 for the smallest size to 4.3 for the largest size, compared to an expected range of 4.65 to 4.1. The higher than expected values of n are another way of stating that the bed expansion with gases is not quite as great as with liquids. The values of U_i are not equal to U_t, but no particular significance is attached to the exact value of U_i/U_t. Goddard and Richardson [14] found that n for diakon powder fluidized with air at 1 atm was 4.7–5.5, but n approached the values for liquid fluidization as the pressure was increased. With a low-density phenolic resin, n was about 8.5, indicating much less expansion than for liquid fluidization. Even higher values of n (7–20) were reported for particulate fluidization of coal with gases at high pressure [15], and it is not clear whether these high values are caused mainly by the irregular shape of the particles or

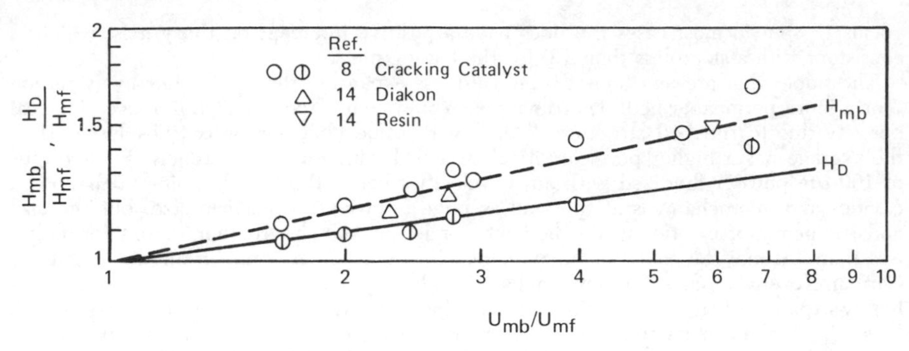

Figure 8. Maximum expansion and dense-phase expansion.

the fact that ϵ was in the range of 0.7 to 0.9. Even if uniform expansion was obtained, the value of n might be expected to increase as ϵ approached 1.0.

If the minimum fluidization velocity and the bubble point are known, the bed expansion at the bubble point can be predicted using the empirical equation of Abrahamsen and Geldart [13]:

$$\frac{H_{mb}}{H_{mf}} = \left(\frac{U_{mb}}{U_{mf}}\right)^{0.22} \tag{9}$$

This equation was established using data for alumina, ballontini and cracking catalyst fluidized with different gases. The correlation is shown in Figure 8, along with some other data for catalyst fractions [8] and polymer particles [14]. The data are fitted quite well by Equation 9, even though the values of n range from 4.3 to 8.5.

THEORIES FOR MAXIMUM BED EXPANSION

The bubble point was defined as the velocity at which the bed height reaches a local maximum, so the bed porosity at the bubble point is also a maximum value. This porosity ϵ_{bp} or ϵ_{max} depends on the ratio U_{mb}/U_{mf} or the velocity range for particulate fluidization, and on the initial porosity, ϵ_{mf}. With these values from measurements or empirical correlations, the value of ϵ_{bp} could be predicted reasonably well, any error coming from uncertainty in the slope of plots such as Figure 6 or Figure 7. However, ϵ_{bp} also can be correlated directly with the physical properties of the system using empirical or theoretical equations as a guide. Several approaches to predicting ϵ_{bp} are discussed by Oltrogge and Kadlec [22], and a simple criterion based on the Froude number is given here.

The bed expansion is assumed to follow the equation for laminar flow up to a critical Froude number, and the density of the gas is neglected relative to the solid density.

From Equation 7,

$$\frac{\epsilon^3_{bp}}{1-\epsilon_{bp}} = \frac{150\mu U_{mb}}{g\, d_p^2\, \rho_p} \tag{10}$$

The Froude number is based on the superficial velocity and particle diameter, and a critical value of about 0.1 is used, as originally suggested by Wilhelm and Kwauk [23] to distinguish particulate and bubbling fluidization:

$$N_{Fr} = \frac{U_o^2}{gd_p} \tag{11}$$

$$\frac{U^2_{mb}}{gd_p} = 0.1 \quad (12)$$

Combining Equations 10 and 12,

$$\frac{\epsilon^3_{bp}}{1-\epsilon_{bp}} = \frac{47.4\,\mu}{g^{1/2}d_p^{3/2}\rho_p} = \frac{47.4}{N_{Ga}^{1/2}} \quad (13)$$

where $N_{Ga} = gd^3{}_p\rho_p{}^2/\mu^2$.

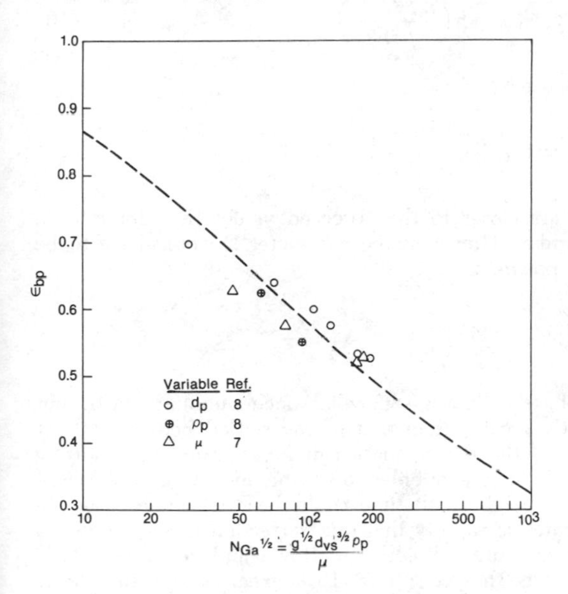

Figure 9. Maximum bed porosity.

Oltrogge and Kadlec [22] showed that Equation 13 gave a reasonable fit to their data and to published results, and similar equations with different constants or more complex functions of ϵ_{bp} did not fit quite as well. However, the data fell in a broad band, and to show how well this equation matches the measured effects of different variables, some recent data are plotted in Figure 9. For each set of data points there is only one variable, such as particle size, solid density or gas viscosity. The data for different fractions of cracking catalyst fall very close to the predicted line, except for the finest fraction, which had a broader size distribution. The two points for unfractioned catalyst with densities of 850 and 1310 kg/m³ fall a little below the line, showing that the effect of density is properly accounted for, but suggesting ϵ_{bp} is lower for a wide size distribution. The data for fluidizing catalyst with different gases (air, He, Ne, Freon®) show mainly the effect of viscosity differences, and the change in ϵ_{bp} with viscosity is significantly less than predicted. The range of gas viscosities encountered in practice is not very great, so the wrong dependence on μ has only theoretical implications. Similarly, the effect of gas density is not included in Equation 13, but there is a significant increase in ϵ_{bp} (0.51 to 0.54) on going from 1 to 14 atm. Increasing pressure has almost no effect on U_{mf}, but it does lead to higher values of U_{mb} and ϵ_{bp} and smaller bubbles in the bubbling region, for reasons that are not yet clear.

An alternate equation for ϵ_{bp} can be obtained using the empirical equation for U_{mb} and the theoretical equations for U_{mf} and bed expansion:

$$\frac{\epsilon^3_{bp}}{1-\epsilon_{bp}} = \frac{\epsilon^3_{mf}}{1-\epsilon_{mf}} \times \frac{U_{mb}}{U_{mf}} = \frac{150\mu}{g(\rho_p-\rho)d_p^2} \times \frac{adp^{0.7}\rho^{0.06}}{\mu^{0.4}} \quad (14)$$

$$\frac{\epsilon^3_{bp}}{(1-\epsilon)_{bp}} = \frac{150a\,\mu^{0.6}\,\rho^{0.06}}{g\,\rho_p\,d_p^{1.3}} \tag{15}$$

The exponents for ρ_p, d_p and ρ are in reasonable agreement with Equation 13, but the difference in the viscosity terms shows the need for refinements in the theory.

The effect of interparticle forces on bed expansion was considered by Mutsers and Rietema [7], who introduced an elasticity coefficient, E, for the powder structure. They predicted the maximum porosity would be a function of a fluidization number, N_F:

$$\epsilon_{bp} = f(N_F) = f\left(\frac{\rho_p{}^3 d_p{}^4 g^2}{\mu^2 E}\right) \tag{16}$$

To compare this with Equation 13, use $N_F{}^{-1/3}$:

$$\epsilon_{bp} = f'\left(\frac{\mu^{2/3}\,E^{1/3}}{g^{2/3} d_p{}^{4/3} \rho_p}\right) \tag{17}$$

The exponents for ρ_p, d_p and μ are closer to the observed values than for criteria expressed in terms of the Galileo number. However, the parameter E would have to be obtained by experiment for different powders.

DENSE-PHASE EXPANSION

With type A solids, the dense phase collapses somewhat once bubbling starts, but the void fraction remains significantly greater than that of the settled bed throughout the bubbling region. The expansion of the dense phase can be measured by shutting off the gas flow, waiting a few seconds for the bubbles to escape and noting the height of the bed. Measurements of the bed height with time could be extrapolated back to zero time, but usually the bed deaerates so rapidly that this correction is not necessary. The dense-phase void fraction was measured directly by Rowe and co-workers [10] using X-ray absorption, and these tests showed up to 10% expansion of the dense phase, in reasonable agreement with expected values.

Some data for the expansion of the dense phase when fluidizing cracking catalyst with air are shown in Figure 8. The change in height of the dense phase is about half that measured for the bed at the bubble point. The data shown are for a porous distributor in a small column, which gave the maximum expansion at the bubble point. With a perforated-plate distributor, the expansion at the bubble point was not as great and the expansion of the dense phase was somewhat less [8]. However, the effect of an imperfect distributor was less noticeable in a tall bed [8,24], and others [25] have reported that the type of distributor does not have a significant effect on the dense-phase properties for a vigorously bubbling bed. The dense-phase expansion was slightly greater in a larger-diameter column [8], perhaps because the bubbles were larger and there was a lower rate of transfer of gas from bubbles to the dense phase.

Significant expansion of the dense phase was found for several other type A solids fluidized with various gases [22]. The values of ϵ_D for the well bubbling region were just below (0.1–0.05) the values of ϵ_{bp}, but the values of ϵ_{mf} were not given and no general correlation for ϵ_D was presented.

The expansion of the dense phase indicates an interstitial velocity appreciably greater than that at U_{mf}. The superficial velocity of the gas relative to the solid can be estimated from the change in void fraction and Equation 2. A 20% expansion of the dense phase means about a twofold increase in relative velocity. The absolute velocity of the gas in the dense phase depends on the solids flow, which is downward over most of the bed and may make the absolute gas velocity zero or negative. Although it might seem worthwhile to develop correlations for predicting the absolute velocity or actual gas flow in the dense phase, in most fluidized-bed reactors the value of U_o is so much larger than U_{mf} that the gas flow in the dense phase can be neglected in kinetic models.

The major significance of the dense phase expansion comes from the resulting changes in fluidity, bubble stability and bubble velocity. The viscosity of a fluidized bed of fine catalyst is much less than for a bed of coarser solids [18,26], and bubbles of a given size rise more rapidly when the bed is composed of fine particles [27]. Probably of more importance is the lower average bubble size found when fluidizing fine particles. Tsutsui and Miyauchi [28] found that beds of catalytic cracking catalysts with mean sizes of 50–60 μm (Type A) had mean bubble sizes of 10 mm, compared to a mean size of about 50 mm for 180-μm alumina particles (Type B). The smaller bubbles mean a greater range of operation without slugging and better mass transfer between bubbles and dense phase, which is critical to the performance of a fluidized reactor. High-pressure fluidization leads to even smaller bubbles in beds of fine powders, but has little effect with coarser solids [29]. The theoretical basis for these effects is still uncertain.

NOMENCLATURE

d_p particle size
d_{sv} surface-mean size
E elasticity coefficient
F fraction of fines
g gravitational constant
H_{mb} bed height at bubble point
H_{mf} bed height at minimum fluidization
L bed length
n exponent in Equation 8
N_F fluidization number
N_{Fr} Froude number
N_{Ga} Galileo number
N_{Re_p} particle Reynolds number
P pressure
ΔP pressure drop across bed
U_o superficial velocity
U_{mf} minimum fluidization velocity
U_{mb} minimum bubbling velocity
U_i velocity for $\epsilon = 1.0$ (extrapolated)
U_t terminal velocity for a single particle

Greek Symbols

ϵ external void faction
ϵ_D void faction of dense phase
ϵ_{mf} void faction at minimum fluidization
ϵ_{bp} void faction at bubble point
ρ fluid density
ρ_p particle density
μ viscosity
ϕ_s sphericity

REFERENCES

1. Geldart, D. *Powder Technol.* 7:285 (1973).
2. Ergun, S. *Chem. Eng. Prog.* 48:89 (1952).
3. Davidson, J. F., and D. Harrison. *Fluidised Particles* (London: Cambridge University Press, 1963).
4. Frantz, J. F. *Chem. Eng. Prog. Symp. Ser.* 62:21 (1966).
5. Grewal, N. S., and S. C. Saxena. *Powder Technol.* 26:229 (1980).
6. Leva, M., T. Shirai and C. Y. Wen. *Gen. Chim.* 75(2):33 (1956).
7. Mutsers, S. M. P., and K. Rietema. *Powder Technol.* 18:233 (1977).
8. Simone, S., and P. Harriott. *Powder Technol.* 26:161 (1980).
9. Donsi, G., and L. Massimilla. *AIChE J.* 19:1104 (1973).
10. Rowe, P. N., L. Santoro and J. G. Yates. *Chem. Eng. Sci.* 33:133 (1978).
11. DeJong, J. A. H., and J. F. Nomden. *Powder Technol.* 9:91 (1974).
12. Richardson, J. F. In: *Fluidization*, J. F. Davidson and D. Harrison, Eds. (New York: Academic Press, Inc., 1971), p. 25.
13. Abrahamsen, A. R., and D. Geldart. *Powder Technol.* 26:35 (1980).

14. Goddard, K., and J. F. Richardson. *Ind. Chem. Eng. Symp. Ser.* 30:126 (1968).
15. Crowther, M. E., and J. C. Whitehead. In: *Fluidization,* J. F. Davidson and D. L. Keairns, Eds. (London: Cambridge University Press, 1978), p. 65.
16. Subzwari, M. P., R. Clift and D. L. Pyle. In: *Fluidization,* J. F. Davidson and D. L. Keairns, Eds. (London: Cambridge University Press, 1978), p. 50.
17. Guedes de Carvalho, J. R. F., and D. Harrison. *Fluidised Combustion, Inst. Fuel Symp. Ser.* 1 (1975).
18. King, D. F., F. R. G. Mitchell and D. Harrison. *Powder Technol.* 28:55 (1981).
19. Massamilla, L., G. Donsi and C. Zucchini. *Chem. Eng. Sci.* 27:2005 (1972).
20. Richardson, J. F., and W. N. Zaki. *Trans. Inst. Chem. Eng.* 32:35 (1954).
21. Lewis, W. K., E. R. Gilliland and W. C. Bauer. *Ind. Eng. Chem.* 44:1104 (1949).
22. Oltrogge, R. D., and R. H. Kadlec. Paper presented at the Detroit AIChE meeting, June, 1973.
23. Wilhelm, R. H., and M. Kwauk. *Chem. Eng. Prog.* 44:201 (1948).
24. Lehmann, J., H. Ritzmann and B. Schugerl. *Proc. Int. Symp. Fluidization,* Toulouse, France (1973), p. 107.
25. Abrahamsen, A. E., and D. Geldart. *Powder Technol.* 26:57 (1980).
26. Matheson, G. L., W. A. Herbst and P. H. Holt. *Ind. Eng. Chem.* 41:1099 (1949).
27. Rowe, P. N. In: *Fluidization,* J. F. Davidson and D. Harrison, Eds. (London: Academic Press, Inc., 1971), p. 145.
28. Tsutsui, T., and T. Miyauchi. *Trans. Int. Chem. Eng.* 20:386 (1980).
29. Guedes de Carvalho, J. R. F. *Chem. Eng. Sci.* 36:413 (1981).

CHAPTER 26
FLOW MODELING CONCEPTS OF FLUIDIZED BEDS

C. Y. Wen and L. H. Chen
Department of Chemical Engineering
West Virginia University
Morgantown, West Virginia 26506

CONTENTS

INTRODUCTION

Fluidized bed technology has been applied widely to the chemical, petroleum and associated industries since their advent during World War II. Fluidized-bed cat crackers considerably aided the Allied victory by producing large quantities of cheap aviation fuel.

Applications of this technology include cracking and reforming of hydrocarbons, Fischer-Tropsch synthesis, coal carbonization and gasification, ore roasting, aviline production, acrylonitrile production, phthalic anhydride production, polyethylene production, calcining, coking, aluminum production, drying, granulation, etc.

In the future, increased application of fluidized-bed gasification and combustion of coal, oil shale and other low-crude fuels may enable the United States to turn away from imported oil and back to indigenous fossil fuels. In addition, as the country's supply of essential raw materials gradually decreases, application of fluidization technology to these fields certainly can be accelerated.

Despite its importance and wide application, the design of the fluidized bed system is, at best, difficult, imprecise and based mainly on experience and know-how; knowledge of the basic phenomenon is still very rudimentary. The flow behavior of fluidized beds is sensitive to scale and operating conditions. Going to commercial-scale units usually provides surprises for which laboratory data are often of little use. The procedure for the scaleup of fluidized-bed reactors has been to build a sequence of large prototype reactors. This procedure is costly and time consuming and has led to the overdesign of reactors.

Therefore, there is critical need for more fundamental information, particularly data on the flow of gas and solids in a large-scale operation of fluidized beds. As more is learned about the scaleup problems of fluidized-bed reactors, less risk will be incurred in new applications. This chapter summarizes the current state-of-the-art on flow models applied to fluidized-bed reactors. Emphasis is made on flow models preferred by the authors. As more information is available, refinement of the models and model parameters

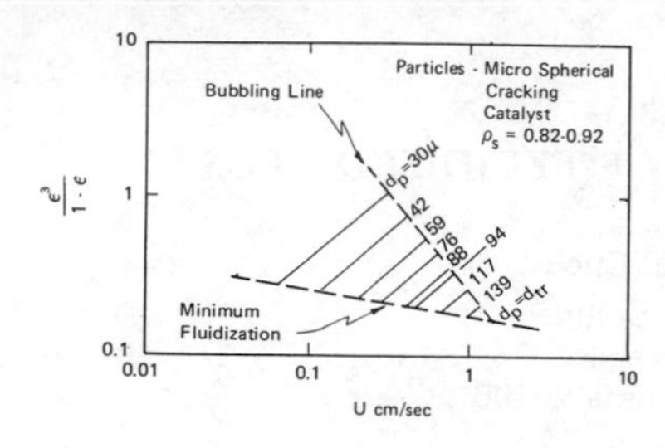

Figure 1. The region of bubble-free operation [1].

to represent the behavior of large fluidized beds must be made. Only through such an iterative approach can design and scaleup of fluidized-bed reactors eventually be achieved on a priori basis.

CRITERIA ON FLUIDIZED-BED FLOW REGIMES

Bubbleless Fluidization

The mode of fluidization depends on the particle size and the gas velocity. When fine particles are fluidized, there is a regime in which the bed expands without formation of bubbles. This bubble-free regime disappears when the gas velocity is increased above the bubbling velocity. For larger particles, the minimum fluidization velocity usually coincides with the bubbling velocity. This is reported by De Jong and Nomden [1] and is shown in Figure 1.

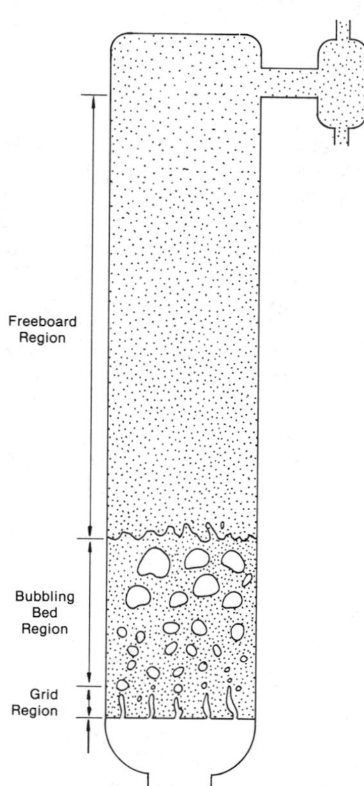

Figure 2. Particle entrainment from a fluidized bed [2].

Bubbling Fluidization

The state of fluidization above the bubbling velocity is called the bubbling regime when the gas velocity is higher than the bubbling velocity. Figure 2 illustrates three distinct zones in the bubbling fluidized-bed regime. Close to the bottom of the bed is the *grid region,* which is represented by vertical or horizontal gas jets and/or small-sized gas bubbles; the type of grid determines the form of the gas voids near the grid. For example, vertical jets or bubbles are produced by perforated plates, horizontal jets by tuyeres and small bubbles by porous plates. Above the grid region is the *bubbling zone.* In this zone, bubbles grow by coalescence and rise to the surface of the bed where they break. The rising bubble voids consist of three important portions, as illustrated in Figure 3:

1. a wake of solids, which is carried upward below the bubble void,
2. a shell or cloud of gas surrounding the bubble void, and
3. a small fraction of solid particles entering the bubble void.

The upward movement of solids in the wakes of bubbles is compensated with a net downward flow of solids in the emulsion phase. This downflow of solids can lead to the surprising result that under certain conditions the gas in the emulsion phase is flowing downward [4,5]. Thus, the gas velocity in the emulsion phase, U_e, can be expressed as

$$U_e = (U_{mf}/\epsilon_{mf}) - U_s \quad (1)$$

where U_s = absolute velocity of solids in the emulsion phase

The solids velocity is obtained by mass balance, i.e.,

$$U_s = U_b \cdot \delta \cdot f_w/[1 - (1 + f_w)] \quad (2)$$

where f_w = ratio of the wake volume to the bubble void volume $\simeq 0.33$
U_b = bubble velocity = $U_{br} + U - U_{mf}$ (U_{br} is shown in Equation 5-ix)
δ = fraction of bed occupied by bubble voids (Equation 5-v)

As bubbles break at the surface of the bed, particles are thrown above the bed surface and entrained by the upward-flowing gas stream. This zone above the bed surface is the

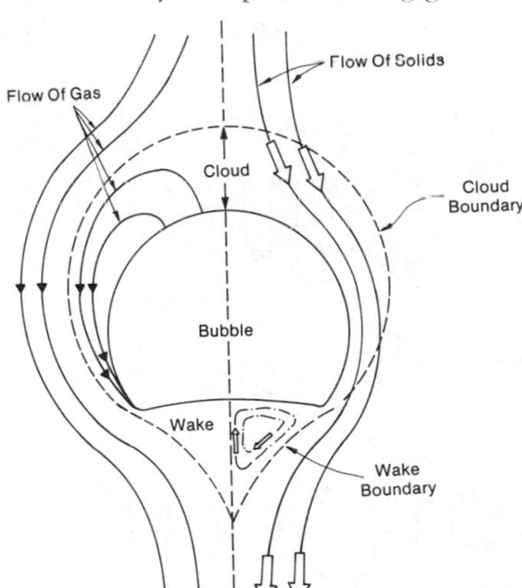

Figure 3. A fast, two-dimensional bubble and its cloud and wake [3].

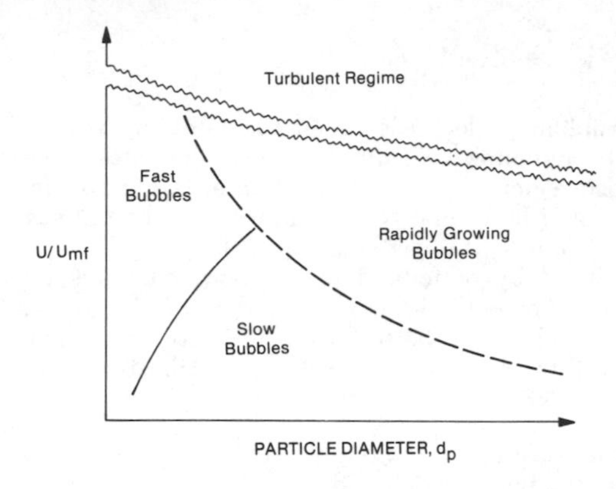

Figure 4. Map of large-particle fluidization regimes [6].

freeboard zone. In this zone, some particles are carried far above the bed surface and elutriated, while others fall back to the bed. This freeboard zone provides an opportunity for the disengagement of particles and additional lean-phase reactions.

The bubbling bed can be further distinguished into three different regimes depending on the particle size and the bubble growth rate, as shown in Figure 4 [6].

With extremely fine particles, gas bubbles rise much faster than the emulsion gas, hence, the designation "fast bubble regime." In this case, the rising bubbles are surrounded by a thin cloud of circulating gas. However, with larger particles, the emulsion gas rises as fast as, or faster than, the bubbles. This regime is the "slow bubble regime." Here the clouds of the bubbles are loosely attached to the bubble and can overlap with other clouds taking up the whole bed. The emulsion gas uses the bubbles as a short cut through the bed. The difference in gas streamlines between a fast and a slow bubble is shown in Figure 5.

The "exploding bubbles" regime occurs when bubbles grow much faster than they rise and when the bed is large enough. The criteria for the exploding bubbles is shown as follows [6]:

$$\frac{d(d_b)}{dt} \geqq U_b \tag{3}$$

where d_b = the bubble diameter
U_b = the bubble rising velocity

These three different bubbling regimes are also depicted in Figure 6.

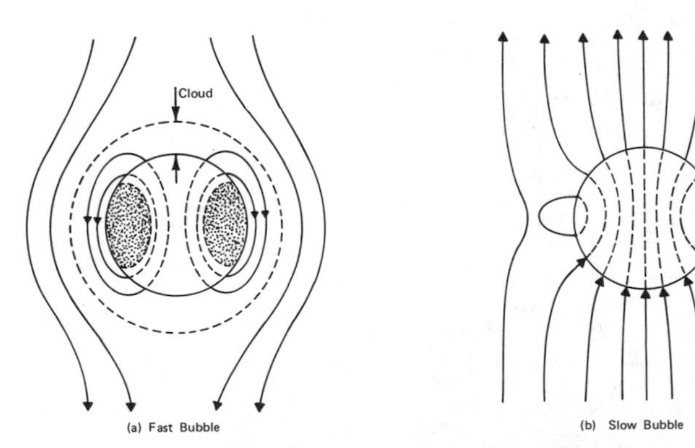

Figure 5. Differences in gas streamlines between a fast bubble and a slow bubble [6].

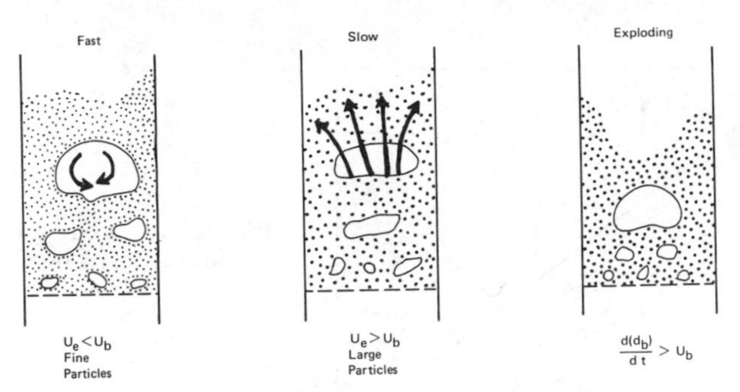

Figure 6. Flow regime for the bubbling bed.

Slugging Fluidization

When the gas velocity is increased and the bubble size approaches the diameter of the bed, the bed is in a state of slug flow regime. Several criteria have been proposed for slug flow; the most comprehensive one is shown as follows [7]:

$$Fr_{ms} = 51.4\ (d_c/L_{mf})^{1.79}\ (\rho_G/\rho_s)^{0.09} + 0.0042\ Ar^{0.41} \tag{4}$$

where
Fr_{ms} = Froude number at minimum slugging point
$= \rho_G\ U_{ms}^2/(g\Delta\rho dp)$
Ar = Archimedes number
$= \rho_s \cdot g \cdot \Delta\rho \cdot dp^3/\mu_G^2$
d_c = bed diameter
L_{mf} = height of settled bed
U_{ms} = superficial velocity at the minimum slugging point
$\Delta\rho = \rho_s - \rho_G$
μ_G = viscosity of gas

The geometry of the bubble or slug can be conveniently related to the particle types grouped by Geldart [8] as a function of size, dp, and relative density, $\rho_s - \rho_G$. Figure 7 shows these four groups, which are summarized in Table I.

Turbulent and Circulating Fluidization

At a higher gas velocity, the bubbling bed becomes a "turbulent" and "circulating" bed, or "fast" bed. Transitions from bubbling to turbulent and turbulent to fast have been summarized by Van Swaaij [9] for small particles and by Staub and Canada [10] for medium and large particles shown in Table II. Figure 8 shows the change of the flow regime as the gas velocity is increased.

FLUIDIZED BED MODELING

Bubbleless Fluidization

This flow regime is characterized by smooth bed expansion without formation of bubbles. Such a phenomenon occurs predominantly for very fine particles and in liquid-phase fluidization. Additional flow of gas beyond the minimum fluidization is accommodated through uniform expansion of the bed by the particles moving farther apart without formation of particle aggregates and gas bubbles. For practical purposes, bubble-

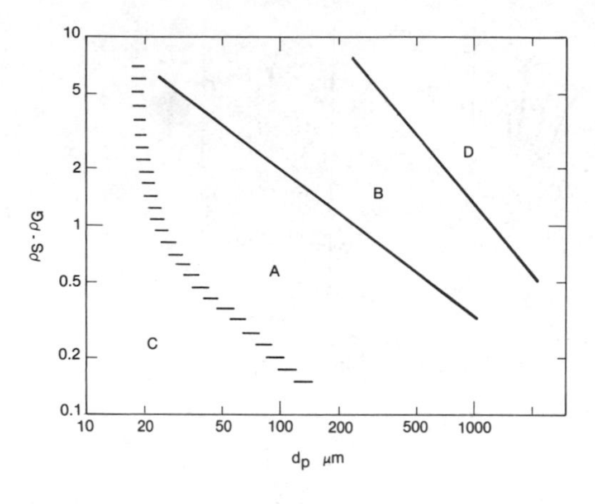

Figure 7. Four groups of particles proposed by Geldart [8].

Table I. Distinguishing Features of Four Groups of Particles (drawn largely from the work of Geldart [8])

Group	C	A	B	D
Distinguishing Word or Phrase	Cohesive	Aeratable	Bubble readily	Spoutable
Example	Flour	Fluid cracking catalyst	Sand	Wheat
Particle Size for $\rho_S - 2.5$ g/cm^3	≤ 20 μm	$20 < \overline{d}_p \leq 90$ μm	90 μm $< \overline{d}_p \leq 650$ μm	> 650 μm
Channeling	Severe	Little	Negligible	Negligible
Spouting	None	None	Shallow beds only	Readily
Collapse Rate	–	Slow	Rapid	Rapid
Expansion	Low because of channeling	High; initially bubble free	Medium	Medium
Bubble Shape	Channels, no bubbles	Flat-base spherical cap	Rounded with small indentation	Rounded
Rheological Character of Dense Phase	High yield stress	Apparent viscosity of order 1 p	Apparent viscosity of order 5 p	Apparent viscosity of order 10 p
Solids Mixing	Very low	High	Medium	Low
Gas Backmixing	Very low	High	Medium	Low
Slugging Mode	Flat-raining plugs	Axisymmetric	Mostly axisymmetric	Mostly wall slugs
Effect of $\overline{d}_p$ (within group) on Hydrodynamics	Unknown	Appreciable	Minor	Unknown
Effect of Particle Size Distribution	Unknown	Appreciable	Negligible	Can cause segregation

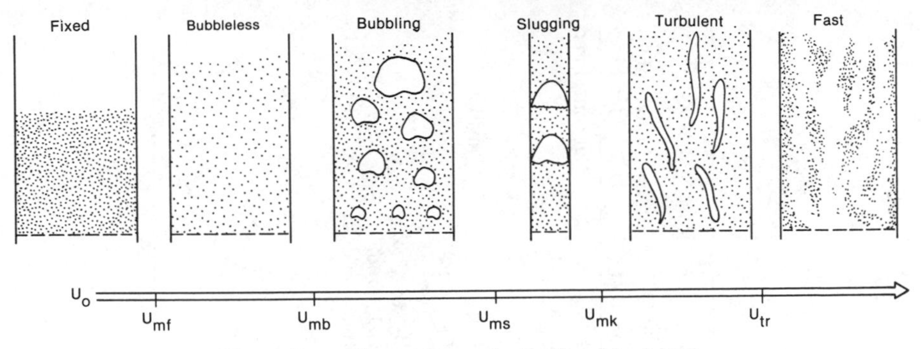

Figure 8. Flow regime of a fluidized bed [61].

less fluidized beds act like expanded packed beds. A homogeneous dispersion model can be used for the modeling of the bed in this regime [11]. The material balance equation for the dispersion model is shown in Equation 39.

Bubbling Fluidization

Bubbling Zone

This section deals with the effect of design and operating variables on bubbling fluidized bed modeling. The emphasis of this section centers on the influence of the bubble zone, with the effects of the grid and freeboard regions treated separately.

A large number of mathematical models, varying in sophistication, have been proposed, critically evaluated and compared by several authors, including Rowe [12], Pyle [13], Calderbank and Toor [14], Horio and Wen [15] and Barreteau et al. [16]. Seventeen models have been classified by Horio and Wen [15] into three major groups, as shown in Tables III and IV.

Level I models are two-phase models, with several adjustable parameters not directly related to bubble behavior. Consequently, these models are not well suited to scaleup and design, except under well defined circumstances.

Level II models attempt to relate bed parameters to bubble size, which is treated either as a constant or adjustable parameter. Extension of the Level II bubbling model to include varying bubble size leads to the Level III bubbling models. These are the

Table II. Fluidization Regimes and Transitions [9,10]

Regime	Criteria
Slug Flow	Equation 4
Free Bubbling	Equation 4 and $U < U_{mk}$
Turbulent Flow	$d_p < 100\ \mu m$, $U \simeq 0.5$ (m/sec) (ambient)
	$d_p < 650\ \mu m$, $U \simeq 2.5$ to 3.0 or $\frac{U}{U_T} \simeq 0.55$ to 0.65
	$d_p \simeq 2600\ \mu m$, $U \simeq 4$ to 5.0 or $\frac{U}{U_T} \simeq 0.30$ to 0.35

Table III. Code ($n_1n_2n_3n_4n_5n_6$) for the Classification of Models [15]

Factors Values of n_i	n_1 Way of Dividing the Phase	n_2 Way of Flow Assignment	n_3 Cloud Volume	n_4 Gas Exchange Coefficient	n_5 Bubble Diameter	n_6 Effect of Jet
1	Three phases B – C – E[a]	$U_{ob} = U_o - U_{mf}$ $U_{oe} + U_{oc} = U_{mf}$ or $U_{oe} = U_{mf}$	Davidson model[b] $R = \frac{\alpha + 2}{\alpha - 1}$ (Three-dimensional)	Two interphases Kunii and Levenspiel model	Constant and adjustable	Not considered
2	Two Phases B – (C + E)	$U_e = U_{mf}/\epsilon_{mf}$ $U_c = U_b/\epsilon_{mf}$	Murray model $R = \frac{\alpha}{\alpha - 1}$	One interphase Orcutt and Davidson model	Constant $d_{be} = \begin{cases} d_b \text{ at } L_f/2 \\ \text{or} \\ d_b \text{ at } L_{mf}/2 \end{cases}$	Considered
3	Two phases (B + C) – E	$U_e = U_{mf}/\epsilon_{mf} - U_s$	$R = 1$	One interphase Partridge and Rowe model	Constant D_{be} from bed expansion data	
4		$U_{ob} = U_o - U_{mf}$ $U_{oc} = 0$ $U_{oe} = 0$		One interphase Empirical correlation by Kobayashi et al.	Varied along the bed axis	
5		$U_{ob} = U_o$ $U_{oc} = 0$ $U_{oe} = 0$		The others		

[a]B = bubble phase; C = cloud phase; E = emulsion phase.
[b]$\alpha = U_b\epsilon_{mf}/U_{mf}$.

Table IV. Classification of Fluidized-Bed Reactor Models [15]

Level	Description	Models	Code $(n_1n_2n_3n_4n_5n_6)^a$
I	Parameters are constant along the bed	Shen and Johnstone [17]	21xxx1
		van Deemter [18]	25xxx1
		Johnstone et al. [19]	31xxx1
	Parameters are not related to the bubble behavior ($n_3 \sim n_5$ cannot be specified)	May [20]	21xxx1
		Kobayashi and Arai [21]	31 (or 5) xxx1
		Muchi [22]	31xxx1
II	Parameters are constant along the bed	Orcutt, Davidson, and Pigford [23]	21x211
		Kunii and Levenspiel [24]	141111
		Fryer and Potter [25]	131111
	Parameters are related to the bubble size, which is adjustable, $n_5 = 1 \sim 3$	—	
III	Parameters are related to the bubble size	Mamuro and Muchi [26]	21x541
		Toor and Calderbank [27]	311241
		Partridge and Rowe [28]	322341
	Bubble size is varied along the bed axis	Kobayashi, et al. [29]	312541
		Kato and Wen [30]	351441
		Mori and Muchi [31]	21x541
	$n_5 = 4$	Fryer and Potter [25]	21x241
		Mori and Wen [32]	351442

[a]The definition of the code is given in Table III. Example: 21xxx1 means $n_1 = 2$ (i.e., the bed is divided into two phases and the cloud phase is included in the emulsion phase); $n_2 = 1$ (i.e., the flow rate in bubble phase is $u_o - u_{mf}$); factors $n_3 \sim n_5$ are not specified (i.e., the parameters are not related to the bubble characteristics); and $n_6 = 1$ (i.e., the effect of jet is not considered).

most realistic from a mechanistic point of view, particularly if they can account for special conditions of the grid region and of the freeboard. Level III models should be used for scaleup and process evaluation, unless simple models can be justified. However, the parameters developed for Level III models so far apply only to Geldart's class B particles and freely bubbling beds. Further development of the model parameters is needed if Level III models are to be applied for other systems.

Simple models—homogeneous and two phase. A number of simple models have been proposed by Horio and Wen [15] to account for general features of the gas-fluidized bed. Figure 9 depicts the models that are either homogeneous (P and M) or two phase (P-P, P-M, C). When used judiciously, these simple models can be reasonable predictors of performance. The material balance equations for the simple model first-order gas-phase reaction are shown as follows:

Homogeneous Model

(a) P Model (plug flow), mass balance over incremental bed height:

$$U \frac{dC}{dL} = kC \tag{5}$$

(b) M model (complete mixing):

$$U(1 - C_{out}) = kL_{mf}\, C_{out} \tag{6}$$

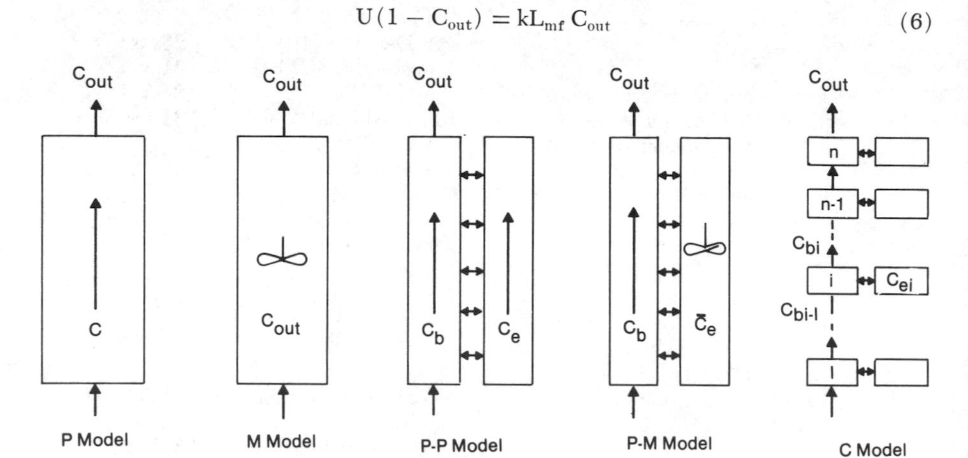

Figure 9. Simple homogeneous and two-phase models—gas-phase reaction.

Two-Phase Model

For the two-phase models, the excess gas above the minimum fluidization velocity flows through the bed in the form of bubbles [33]. The gas flow-through in the emulsion phase, U_e, as shown in Equation 1, could be either upward or downward. Here, the gas flow-through in the emulsion phase, U_e, is assumed to be negligible, i.e., the exit concentration of the bubble phase is a good approximation of the overall exit concentration from the bed. This assumption is reasonable for beds with $U_o/U_{mf} \simeq 6 \sim 11$. However, reaction occurs in both phases, and gas exchange takes place between the phases. General relationships for the two-phase models are:

Bubble phase

$$\begin{matrix}\text{change in concentration} \\ \times \text{ bubble flowrate}\end{matrix} = \begin{matrix}\text{material exchanged} \\ \text{with emulsion}\end{matrix} - \begin{matrix}\text{material reacted} \\ \text{in the bubble phase}\end{matrix} \tag{7}$$

Emulsion phase

$$\begin{matrix}\text{material exchanged} \\ \text{with bubbles}\end{matrix} = \begin{matrix}\text{material reacted in} \\ \text{the emulsion phase}\end{matrix} \tag{8}$$

Thus,

(c) P-P model (plug-flow bubble phase, plug-flow emulsion phase) for the bubble phase:

$$\frac{d\,C_b}{d\xi} = N_M\,(C_e - C_b) - N_R\,\gamma_s \cdot C_b \tag{9}$$

where N_M = number of transfer unit

$= F_{be}\delta\, L_f/U$ (10a)

N_R = number of reaction unit

$= k\, L_{mf}/U$ (10b)

F_{be} = gas-exchange coefficient (bubble to emulsion)

k = first-order reaction rate constant

L_f, L_{mf} = bed height fluidized, at minimum fluidization

U = superficial gas velocity

δ = volume of bubbles/total bed volume

$= (L_f - L_{mf})/L_f$ (10c)

C_b, C_e = concentration in bubble, emulsion phase

γ_s = fraction of particles in bubble phase (including cloud) based on total volume of particles

ξ = dimensionless bed height

$= L/L_f$ (10d)

For the emulsion phase,

$$N_M\,(C_e - C_b) = -N_R\,(1 - \gamma_s)\,C_e \quad (11)$$

(d) M-M model (well mixed bubble phase, well mixed emulsion phase)
For the bubble phase,

$$1 - C_b = N_M\,(C_b - C_e) + N_R\,\gamma_s \cdot C_b \quad (12)$$

For the emulsion phase, Equation 11 applied.

(e) P-M model (plug-flow bubble phase, well mixed emulsion phase).
Assuming $\gamma_s \to 0$, the following equations apply:
For the bubble phase,

$$\frac{d\,C_b}{d\xi} = N_M\,(C_e - C_b) \quad (13)$$

Overall balance,

$$(1 - C_{out}) = N_R \cdot C_e \quad (14)$$

(f) C model (compartments)
The bed is divided into n compartments and the general equations are as follows:
For the bubble phase,

$$C_{bi} - (C_b)_{i-1} = (N_M/n)\,(C_{ei} - C_{bi}) - (N_R/n)\,\gamma_s \cdot C_{bi} \quad (15)$$

For the emulsion phase,

$$(N_M/n)\,(C_{ei} - C_{bi}) = -(N_R/n)\,(1 - \gamma_s)\,C_{ei} \quad (16)$$

The compartment model, C, will approach the bubble-free plug-flow model, P, when the number of compartments is appreciable ($n > 10$). An estimate of the appropriate value for n is obtained by assuming

$$n = L_f/\bar{d}_b \quad (17)$$

where $\bar{d}_b$ is the estimated bubble size at mid-bed height. The value of F_{be} is itself related to $\bar{d}_b$, and recommended correlations for these two parameters are given in Table V. Level III bubbling models recognize the variation of d_b with height, and they are, in effect, sophisticated versions of the compartment model, C. For example, the bubble

Table V. Relationships Used in the Bubble Assemblage Model

Parameter	Relationship	Equation
Bubble Size at Given Level (L)	$d_b = d_{bm} - (d_{bm} - d_{bo}) \exp[-0.3 L'/d_c]$ where $L' = L - L_J$, L_J = jet height (equation 22)	5-i
Maximum Bubble Size	$d_{bm} = 0.652 [A (U - U_{mf})]^{2/5}$	5-ii
Initial Bubble Size	$d_{bo} = 0.00376 (U - U_{mf})^2$ (porous plate)	5-iii
	$d_{bo} = 0.347 [A (U - U_{mf})/n_d]^{2/5}$ (perforated) where n_d = number of perforations	5-iv
Bed Expansion Ratio	$(L_f - L_{mf})/L_{mf} = \delta/(1-\delta) = (U - U_{mf})/[\phi(g \bar{d}_b)^{1/2}]$ where δ = volume of bubbles/total bed volume	5-v
Compartment Height	$\Delta L_n = \dfrac{d_{bn}'}{1 + 0.15(d_{bn}' - d_{bm})/d_c}$	5-vi
Number of Bubbles in Compartment	$N = 6 A (L_f - L_{mf})/(\pi L_f (\Delta L_n)^2)$	5-vii
$\left[\dfrac{\text{Diameter cloud}}{\text{Diameter bubble}}\right]^3$	$r = (U_{br} + 2 U_{mf}/\epsilon_{mf})/U_{br} - U_{mf}/\epsilon_{mf})$ where	5-viii
	$U_{br} = \phi (g \Delta L_n)^{1/2} \geq U_{mf}/\epsilon_{mf}$ with $\phi = \begin{cases} 0.64 & d_c \leq 10 \text{ cm} \\ 0.255 \cdot d_c^{0.4} & 10 \leq d_c \leq 100 \text{ cm} \; [34] \\ 1.6 & d_c \geq 100 \text{ cm} \end{cases}$	5-ix
Volume of Bubble Phase (void + cloud)	$V_{bn} = N \pi (\Delta L_n)^3 \cdot r/6 = V_{bVn} \cdot r$ where V_{bVn} = volume of bubble void	5-xi
Volume of Cloud Phase	$V_{cn} = N \pi (\Delta L_n)^3 (r - 1)/6$	5-xii
Volume of Emulsion Phase	$V_{en} = A \Delta L_n - V_{bn}$	5-xiii
Voidage	$1 - \epsilon = L_{mf}(1 - \epsilon_{mf})/L_f, \quad L \leq L_{mf}$	5-xiv
	$1 - \epsilon = [L_{mf}(1 - \epsilon_{mf})/L_f] - \dfrac{[L_{mf}(1 - \epsilon_{mf})(L - L_{mf})]}{2 L_f (L_f - L_{mf})}$ for $L_{mf} \leq L \leq L_{mf} + 2 (L_f - L_{mf})$	5-xv
Gas Exchange Coefficient	$F_{be}' = F_{be}/r$ and	5-xvi
	$F_{be} = 11/d_b$	5-xvii

assemblage model has variable compartment size equal to the bubble size at a given height.

Application of simple models. Simple models may be used to estimate conversion in gas-fluidized beds. The most useful models are those designated M, P, P-P and P-M. The conditions appropriate to their application are outlined below.

(a) Slow reactions ($k \leq 0.5\ \text{sec}^{-1}$; $N_R < 1.0$)

The predicted conversion is relatively insensitive to the value of N_M or the model used because the system is limited by chemical reaction rather than mass transfer considerations. Almost any model may be used, although the bubbleless P model is probably the most appropriate (the M model may be applied to estimate the lower limit of conversion).

In this situation, emphasis should be placed on obtaining accurate kinetic data, rather than assessing bed hydrodynamics.

(b) Intermediate reactions (nominally $0.5 < k < 5.0\ sec^{-1}$; $1 < N_R < 10$)

Conversion may be limited by chemical reaction or mass transfer between the bubble and emulsion phases. The importance of the bed hydrodynamics requires the use of one of the bubble models. The criteria should be (1) P-P model, if $U/U_{mf} < 6 \sim 11$; or (2) P-M model, if $U/U_{mf} > 6 \sim 11$.

The parameter N_M becomes important and either can be estimated, using Equation 10a, or determined experimentally according to Table II in the work by Van Swaaij [9]. If the interaction between conversion and the dependent variables such as bed diameter, grid design, internals, etc., is to be evaluated with more confidence, it will be necessary to apply a Level III model.

(c) Fast reaction (nominally $k > 5\ sec^{-1}$; $N_R > 10$)

For this type of reaction the conversion again becomes less sensitive to the bubble structure and mass transfer in the bubbling zone. Conversion will be high in the grid region, where gas exchange is rapid, and in the freeboard, where dispersed-phase contact occurs. The P-M model, combined with a grid region model and a freeboard model, is recommended. When using Level I and II models, special considerations should be given to cases in which a very high reactant conversion in the fluidized-bed reactor is required. For example, if more than 90% conversion of the reactant must be achieved, the effects of bed diameter, grid design and bed internals become critical and must be taken into consideration carefully. Additional discussion on this point is presented in a later section.

Bubble assemblage model. A large number of models have been proposed for simulating the performance of fluidized-bed reactors (Table IV). Level III models are the most sophisticated since they allow for variations in bubble size with bed height. One such model, the bubble assemblage model (BAM) [30,35] is described below.

An essential feature of the BAM is that the bed is divided into compartments, the height of which is adjusted to the bubble size at that level. This approach greatly reduces computing time without loss of accuracy. For example, Shaw et al. [36] compared the BAM with the Level II model of Orcutt et al. [23] and the Level III model of Partridge and Rowe [28] and found it to be more accurate with only 10–30% of the computing time.

The BAM depicted in Figure 10 is based on the following assumptions:

1. A fluidized bed may be represented by "n" compartments in series. The height of each compartment is equal to the size of each bubble at the corresponding bed height. Correlations used to calculate the bubble growth and the compartment height are listed in Table V.
2. Each compartment consists of a bubble phase and an emulsion phase; gas within each phase is completely mixed.
3. The void space within the emulsion phase is equal to that of the bed at incipient fluidization.
4. The bubble phase consists of spherical bubbles surrounded by spherical clouds. Table V also gives the relationship between the cloud and bubble diameters, as predicted by Davidson [37].
5. Gas exchange occurs between the two phases. An empirical relationship between the exchange coefficient, F_{be}, and the bubble size is given (Equation xvii) in Table V.
6. The voidage is constant from the bottom to a bed height corresponding to L_{mf}. It then increases linearly to unify at a level corresponding to $L_{mf} + 2\ (L_f - L_{mf})$. Table V details the relationships for voidage computation.

The BAM relies on solving the mass balance equations around each compartment from the bottom to the top of the bed.

The *mass balance for the bubble phase* in the nth compartment gives

$$A\,U\,[(C_b)_{n-1} - (C_b)_n] = [F_{be}'\,V_b\,(C_b - C_e)]_n + (r_b\,V_c)_n \quad (18)$$

where F_{be}' = gas exchange coefficient percent volume of bubbles (see Table V)
r_b = reaction rate in the cloud per unit volume
= $k\,C_b$, for first-order reaction with respect to gas composition
V_b, V_c = total volume of bubble phase (void and cloud) and bubble clouds in the compartment (see Table V)

The *mass balance for the emulsion phase* gives

$$[F_{be}'\,V_b\,(C_b - C_e)]_n = (r_e\,V_e)_n \quad (19)$$

where $r_e = k\,C_e$, for a first-order reaction
V_e = volume of emulsion phase (see Table V)

The BAM is amenable to further modification as new and more reliable correlations become available for such parameters as bubble size, gas exchange coefficient, etc.

Defects of bubbling-bed models. Although bubbling-bed models, such as the bubble assemblage model, can successfully predict experimental conversion and selectivity [36], an inadequacy exists in these models. Fryer and Potter [38] and Chavarie and Grace [39] reported concentration profiles along a fluidized-bed reactor for ozone decomposition experiments. None of the present Level III models can account for the few cases of reported concentration profiles that depict a minimum. Fryer and Potter [38] presented profiles of average axial concentration of ozone from a three-dimensional bed, as shown in Figure 11. The authors attributed the change of concentration profiles to the flow of gas reversing its direction and estimated what they termed the critical velocity, U_{cr}.

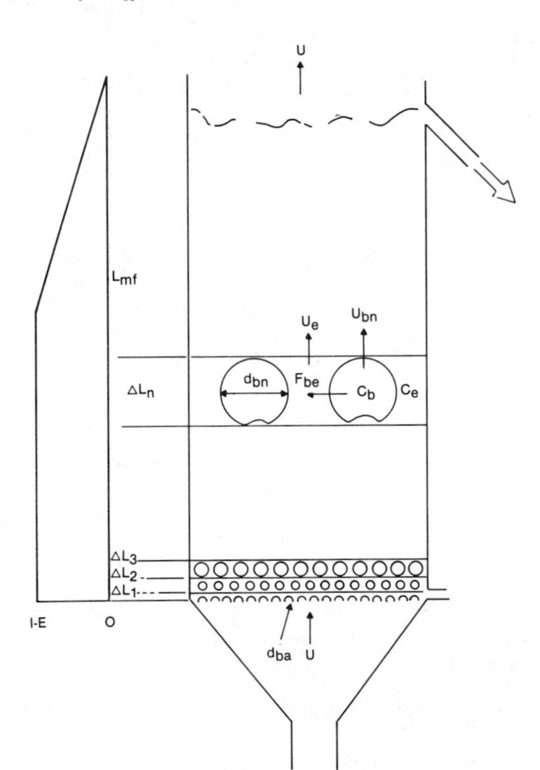

Figure 10. Bubble assemblage model [30] (see Table V for parameter relationships).

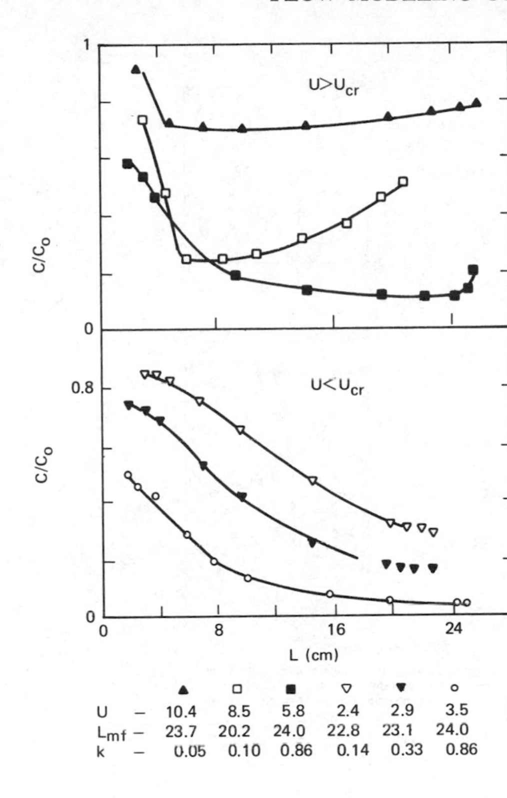

Figure 11. Observed ozone concentration by Fryer and Potter [38] ($d_c = 22.9$ cm, $d_p = 117$ μm).

Here, U_{cr} can be estimated from

$$\frac{U_{cr}}{U_{mf}} = \left(1 + \frac{1}{\epsilon_{mf}\, f_w}\right) [1 - \delta\,(1 + f_w)] \tag{20}$$

where U_{cr} = value of superficial gas velocity, U, to give backmix flow of emulsion gas ($U_e < 0$)

For typical values of ϵ_{mf}, f_w and δ, Equation 20 gives

$$\frac{U_{cr}}{U_{mf}} > 6 \sim 11 \tag{21}$$

The reverse flow of gas in the emulsion phase is caused by the downward flow of solids in a portion of the emulsion phase.

This downflow of solids can lead to the surprising result that under certain conditions the gas in the emulsion phase is also flowing downward [4,5,40–42]. Nguyen et al. [42] and their experiments with a 1.49-m² bed showed that the solid movement is clearly related to the bubble patterns and the gas movement in the particulate (emulsion) phase is related to the solids movement. They observed that the stable bubble pattern in the upper region of the bed leaves a persistent central area relatively bubble free; a pronounced solids downflow in this area is associated with strong gas backmixing, as illustrated in Figure 12.

Therefore, the reverse flow in the emulsion phase must be introduced in the bubbling bed model to describe the concentration profile of a fluidized-bed reactor when flow is above U_{cr}.

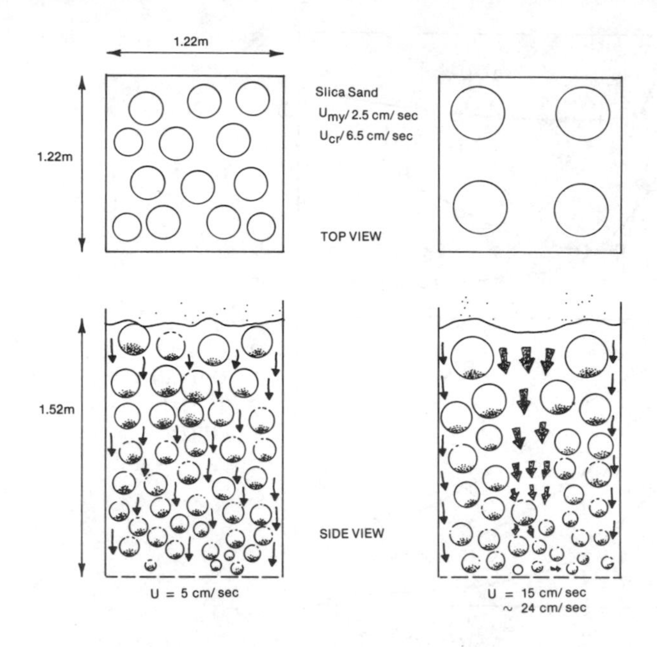

Bubbles rise uniformly at velocity 5 cm/sec. Bubbles tend to rise along the corners at velocity 15–24 cm/sec. At this velocity, the solids circulate downward in the center portion of the bed, dragging some of the gas down into the bed (backmixing the gas).

Figure 12a. Bubble movement and solids movement in a tubeless fluidized bed [42].

Examination of the parameters that could affect the concentration profiles reveals two key parameters that may be inaccurate and may considerably affect the profile. These are (1) the gas exchange rate between the bubble and emulsion phases, and (2) the gas velocity in the emulsion phase. The gas exchange rates near the bottom of the bed could be considerably greater than those found substantially above the distributor because the bubbles are smaller and moving much more slowly near the distributor. This suggests a need for further research on phenomena near the grid region. The gas flowrate in the emulsion phase is estimated from measurements of the so-called visible bubble flowrate, and it is assumed that this rate equals the true gas flowrate through the bubbles. As previously discussed, measurement of visible bubble flowrate shows that the gas flow in the emulsion phase is not constant throughout the bed, as assumed in the two-phase theory and by Fryer and Potter [38]. The concentration profile of the gaseous reactants in the emulsion phase may be affected significantly by the variation in the gas flowrate through the emulsion phase. In addition, there is room for questioning the assumption that the visible bubble flowrate is the true gas flowrate through bubbles.

The Grid Region

It is well accepted that the design of the grid (or gas distributor) is crucial to the success of a fluidized-bed reactor. Experimental evidence [43] indicates that with fast reactions most of the conversion occurs in the lower part of the bed and that the degree of conversion is influenced by the grid design parameters. Porous plate distributors, commonly used in small-scale experiments, have been shown to give increased efficiency compared with perforated plate distributors more typical of a large-scale operation [44].

Gas entering the bed through a grid will emerge as pulsating jets, which detach to yield formation-size bubbles [45,46]. For a given free area of grid, the jets become shorter and the initial bubbles smaller as the grid orifice size is reduced. In the extreme,

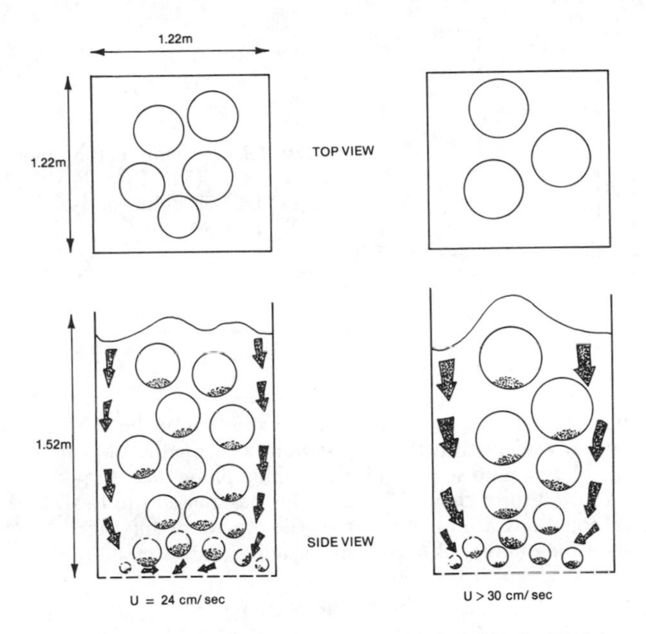
1.22m
1.22m
TOP VIEW
1.52m
SIDE VIEW
U = 24 cm/ sec
U > 30 cm/ sec
As gas velocity is increased to 24 cm/sec, bubbles begin to rise in the center of the bed.
At a velocity greater than 30 cm/ sec, gas bubbles predominantly rise in the center and
solids move downward along the wall.

Figure 12b. Bubble movement and solids movement in a tubeless fluidized bed [42].

a perforated grid plate behaves very much like a porous plate when large numbers of small orifices are used [32]. From this viewpoint, the grid region has two ways of influencing the reactor performance. First, if pulsating gas jets exist, they can be expected to provide a zone of very effective gas exchange between the lean and dense phases. Secondly, the influence of the initial bubble size on coalescence and growth of the bubble in the bubbling zone may be profound.

Few data are available for transfer processes and reactions around grid jets. Values of F_{Je} (the jet to emulsion gas exchange coefficient per unit vessel volume) derived from the data of Behie and Kehoe [47] range from about 18 to 4 (sec^{-1}), which can be compared with values of δF_{be} (bubble to emulsion coefficient per unit vessel volume), typically from 2 to 0.2 (sec^{-1}), assuming a value of $\delta = 0.5$. These limited data suggest that the gas transfer process in the grid region is roughly an order of magnitude faster than that in the bubbling region.

The extent of the grid region can be assumed to be equal to the jet height, L_J, which can be predicted by the correlation of Wen et al. [48]:

$$\frac{L_J}{d_o} = 1150 \left[\frac{(U_J - U_{mfJ})\,\mu_G}{\rho_G\, d_p{}^2 \cdot g}\right]^{0.42} \left(\frac{\rho_G}{\rho_s}\right)\left(\frac{d_p}{d_o}\right)^{0.66} \tag{22}$$

Mori and Wen [32] incorporate the jet height equation into the modified bubble assemblage model to calculate the height of the first compartment, ΔL_1, which is assumed to be well mixed. Thus,

$$\Delta L_1 = L_J + L_c \tag{23}$$

where L_c = critical height above the jets at which the bubble clouds overlap

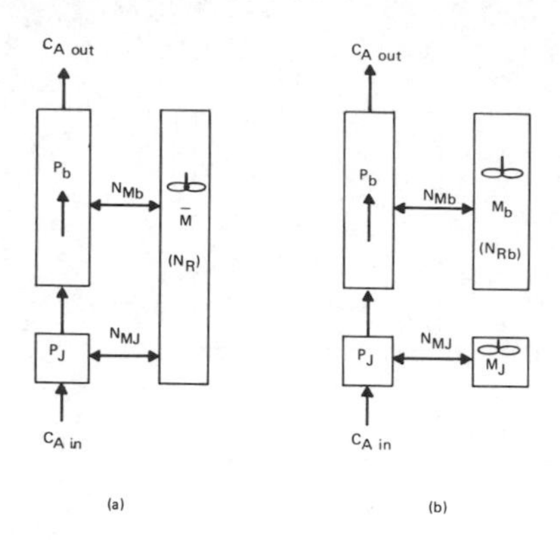

Figure 13. Simple P-M models with grid region: (a) $P_J\overline{M}/P_b\overline{M}$ model ("grid model" of Behie and Kehoe [47]); (b) P_JM_J/P_bM_b model.

Behie and Kehoe [47] have proposed a simple extension of the P-M model to account for the enhanced gas exchange in the grid region. The model, depicted in Figure 13a, treats the jets as a plug flow in series with the bubble phase, also in plug flow; the emulsion phase is well mixed throughout. This model is designated the $P_J\overline{M}/P_b\overline{M}$ model. The material balance equations are identical with the equations for the P-M model, except for the definition of the number of mass transfer units. In this case,

$$\overline{N}_M = N_{Mb} + N_{MJ} \tag{24}$$

where

$$N_{Mb} = \text{mass transfer units in the bubble zone}$$

$$= F_{be} \cdot \delta \cdot (L_f - L_J)/U \tag{25}$$

and

$$N_{MJ} = \text{mass transfer units in the jet zone}$$

$$= F_{Je}\, L_J/U \tag{26}$$

Applying Equation 26 to the experimental data of Behie and Kehoe [47] gives values of N_{MJ} from 4 to 3 transfer units, which may be considerably more than is achieved in the bubbling zone, particularly for shallow beds, high gas flows and large bubble sizes. The extent of conversion in the bed may be limited either by gas exchange or by reaction rate, and it is possible that in the grid region the conversion becomes reaction-rate limited for slow reactions. For fast reactions in the grid region, the conversion should be balanced by the rapid exchange of gas and should only become gas-exchange limited for exceptionally fast reaction rates. The $P_J\overline{M}/P_b\overline{M}$ model of Behie and Kehoe [47] should be suitable for fast reactions, but it may be inadequate for slow reactions since it assumes that gas transferred from the jets has access to the whole emulsion zone, underestimating the effect of reaction rate limitation.

Figure 13b depicts a more appropriate model for both fast and slow reactions [49]. This model, designated P_JM_J/P_bM_b, assumes the jets are in plug flow and transfer gas to a well mixed emulsion bounded by the grid region. The P-M grid model is then connected in series with a P-M bubble zone model (Equations 13 and 14).

N_{MJ}, N_{Mb} are given by Equations 25 and 26, and

$$N_{RJ} = k\, L_J/U \tag{27}$$

$$N_{Rb} = k\,(L_{mf} - L_J)/U \tag{28}$$

The calculations show the significance of the grid region, particularly for fast reactions. Differences between the two models are slight for the fast reactions, but considerable for slow reactions. The P_JM_J/P_bM_b model is recommended for use, although the two models converge for deep beds and low gas rates.

Figure 14 shows concentration versus height profiles calculated from the $P_J\overline{M}/P_b\overline{M}$ model (line a and b) and from the P_JM_J/P_bM_b model. (Note that the freeboard profiles are calculated using the plug-flow freeboard model.)

The conversion in the grid region may be predicted by simple extensions of the P-M model or by simple modifications to the bubble assemblage model. A more precise analysis awaits suitable correlations for F_{Je} or K_{Je}. Consideration of the grid section is particularly important for fast reactions.

The Freeboard Region

As gas bubbles erupt at the bed surface of a fluidized bed, particles are ejected upward. These particles either rise or fall in the freeboard region, depending on the size, density and gas velocity. The particles with terminal velocities greater than gas velocities will reach a certain height within the freeboard before falling back to the bed. However, the particles with terminal velocities lower than gas velocities will be carried out of the bed. The entrainment rate of the particles along the freeboard can be represented by the following equation [2]:

$$F_i = F_{i\infty} + (F_{io} - F_{i\infty}) \exp(-ah) \tag{29}$$

where F_{io}, the entrainment rate of particles at the dense bed surface, can be estimated by the following equations:

$$F_i = F_o \cdot X_i \tag{30}$$

and

$$\frac{F_o}{\rho_G (g\, d_b)^{1/2}} = 3.07 \times 10^{-9} \left(\frac{A}{d_b^2}\right)\left[\frac{\rho_G \cdot d_B \cdot (U - U_{mf})}{\mu_G}\right]^{2.5} \tag{31}$$

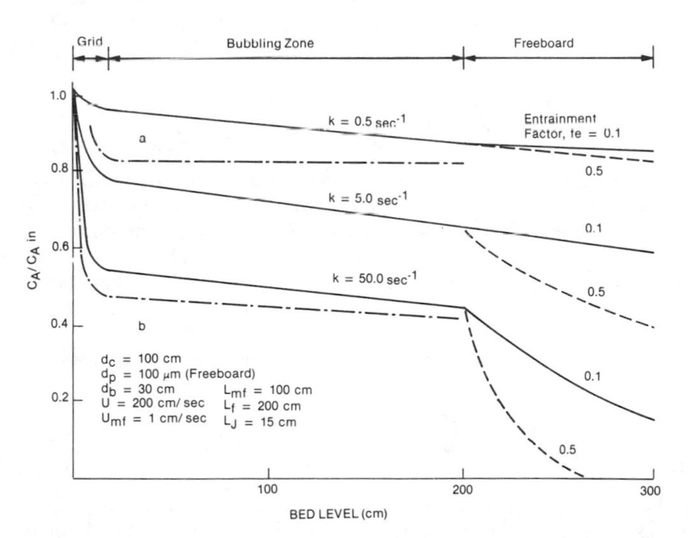

Figure 14. Concentration profiles in gas phase from $P_J\overline{M}/P_b\overline{M}$ model (lines a and b) and from P_JM_J/P_bM_b model with plug-flow freeboard model.

$F_{i\infty}$, the elutriation rate of fines, however, is calculated as follows:

$$F_{i\infty} = E_{i\infty} \cdot X_i \tag{32}$$

and

$$E_{i\infty} = \rho_s (1 - \epsilon_i)(U - U_T) \tag{33}$$

with

$$\epsilon_i = \left[1 + \frac{\lambda (U - U_T)^2}{2 g d_c}\right]^{-\frac{1}{4.7}} \tag{34}$$

$$\frac{\lambda \rho_s}{d_p^2} \left(\frac{\mu_G}{\rho_G}\right)^{2.5} = \begin{cases} 5.17 \, Rep^{-1.5} \, d_c^2, & Rep \leqq 2.38/d_c \\ 12.3 \, Rep^{-2.5} \, d_c, & Rep \geqq 2.38/d_c \end{cases} \tag{35}$$

Here,

$$Rep = \rho_G (U - U_T) \cdot d_p / \mu_G. \tag{36}$$

When bubbles burst at the bed surface, the particles are thrown upward with different initial velocities. This initial solid velocity distribution ranges from 0 to 8 times the bubble velocity at the bed surface, with 2.1 times as the average value [50].

To estimate the extent of the solid-gas reaction, it is necessary to estimate the solid concentration in the freeboard. The concentration of the particles in the freeboard is calculated using the relation between the solid entrainment rate and the solid velocity. Accordingly [2],

$$H_d = \int_0^{F_i} \frac{dF_i}{U_{si}} + \int_0^{F_i'} \frac{dF_i'}{U_{si}'} \tag{37}$$

Here, U_{si} and U_{si}' are the solid ascending and descending velocities, and F_i and F_i' are the ascending and descending flowrates of particles, respectively. The downward flowrate of particles, F_i', is obtained from the material balance of the particles in the freeboard as

$$F_i' = (F_{io} - F_{i\infty}) \exp(-ah) \tag{38}$$

The gas flow pattern for the freeboard is another factor that may affect the concentration profiles of the reactants in the freeboard. Most of the investigators [51–54] used either a plug-flow or complete mixing-flow model. Considering that some degree of back-mixing of the gas occurs in the freeboard region reported by Horio et al. [55], the equal-sized compartment-in-series model or the axial dispersion model for the gas hydrodynamics in the freeboard is therefore adopted for fluidized-bed freeboard modeling [56, 57].

The material balance equation for the axial dispersion flow is as follows:

$$E_\delta \frac{d^2 C_i}{dh^2} - U \frac{d C_i}{dh} + R_i = 0 \tag{39}$$

where R_i is the production rate of species i, and E_δ is the axial dispersion coefficient, which can be estimated from the Peclet number of gas flow in the freeboard region by the following correlations [58]:

$$\frac{1}{N_{Pe}} = \frac{1}{N_{Rep}\,N_{sc}} + \frac{N_{Rep}\,N_{sc}}{192} \quad \text{for } N_{Rep} < 2000$$

$$\frac{1}{N_{Pe}} = \frac{3 \times 10^7}{N_{Rep}^{\,2.1}} + \frac{1.35}{N_{Rep}^{\,0.13}} \quad \text{for } N_{Rep} \geqq 2000 \tag{40}$$

with

$$N_{Rep} = \frac{d_c \cdot U \cdot \rho_G}{\mu_G} \tag{41}$$

$$N_{Pe} = \frac{d_c \cdot U}{E_\delta} \tag{42}$$

and

$$N_{sc} = \frac{\mu_G}{d_c \cdot \rho_G} \tag{43}$$

The boundary conditions used for this case are

$$\text{(a)}\ \ C_i(0-) = C_{io} = C_i(0+) - \frac{E_\delta}{U}\frac{d\,C_i(0+)}{dh} \tag{44}$$

and

$$\text{(b)}\ \ \frac{d\,C_i\,(H)}{dh} \tag{45}$$

For a first-order catalytic reaction, R_i can be represented by the following equation:

$$R_i = -k_{ov} \cdot C_i \tag{46}$$

and for the surface reaction,

$$k_{ov} = \frac{1}{\sum_i Y_i \dfrac{d_{pi}}{6\,h_{mi}(1-\epsilon)} + \dfrac{1}{k_s\,S}} \tag{47}$$

and for the volumetric reaction,

$$k_{ov} = \frac{(1-\epsilon)}{\sum_i Y_i \left(\dfrac{d_p}{6\,D_{eff}}\right)_i + \dfrac{1}{k_t}} \tag{48}$$

Here, D_{eff} is the gas diffusivity through the pores. ϵ, the voidage in the freeboard, can be calculated from the solid concentration, H_d, as follows:

$$\epsilon = 1 - \frac{H_d}{\rho_s} \tag{49}$$

and Y_i is the weight fraction of the particle size, d_{pi}, in the freeboard, while h_{mi} represents the mass transfer coefficient across the gas film. k_t is the intrinsic reaction rate and S is the total surface area.

Slugging Fluidization

Slug flow tends to occur in small-diameter fluidized beds typical of laboratory and pilot-plant reactors. In addition, large commercial-scale reactors can be designed deliberately to operate in this regime to simplify the scaleup. The criteria for slug flow have been given in Equation 4.

A simple model for the case of a catalyzed gas-phase reaction in a slugging bed has been presented by Hovmand and Davidson [59]. This model was extended by Raghuraman and Potter [60] to describe the cyclic variations occurring in a steady-state catalytic system. The development of the model is similar to that of the P-P model, except that the reaction in the bubble phase is ignored and the flow through the emulsion phase is not neglected. Assuming the emulsion phase gas flow is U_{mf} (on a superficial area basis), the following mass balances apply:

Mass balance over bed cross section

$$U_{mf} \frac{d C_e}{d \xi} + (U - U_{mf}) \frac{d C_b}{d \xi} + N_R C_e = 0 \tag{50}$$

Mass balance on a rising slug

$$\frac{d C_b}{d \xi} + N_{MS} (C_b - C_e) = 0 \tag{51}$$

where $N_R = k L_{mf}/U$ (see Equation 10b)
N_{MS} = number of gas exchange transfer units in the slugging bed
= (gas exchange rate per slug volume) $\times$ (L_f/U_s)

The gas exchange rate per slug volume is analogous to the gas exchange coefficient, F_{be}, but it cannot be expected to have the same magnitude. A theoretical approximation of the gas exchange process by Hovmand and Davidson [59] gives

$$N_{MS} = \frac{L_{mf}}{0.35 (g d_c)^{1/2} \cdot d_c m} U_{mf} + \frac{16 \epsilon_{mf} I}{1 + \epsilon_{mf}} \left(\frac{D}{\pi}\right)^{1/2} \left(\frac{g}{d_c}\right)^{1/4} \tag{52}$$

where D = gas diffusivity
I = surface integral over the slug surface
= f (slug length/bed diameter, L_s/d_c) (Table VI gives numerical values of I and an empirical relationship for L_s/d_c)
m = slug shape factor (Table VI)

Turbulent and Circulating (Fast) Fluidized Beds

The high superficial gas velocities and the lack of conventional bubble structure suggest the use of a simple plug-flow (P) model for these regimes. Table VII compares the bed height requirement of circulating (fast) fluidized beds for a range of reaction rates. Careful studies of gas hydrodynamics for turbulent beds and circulating fluidized

Table VI. Parameters to Use in the Slug Flow Model (Equation 52) [59]

$L_S/d_c = 0.3 + 3.9 (U - U_{mf})/(0.35 (g d_c)^{1/2})$ (i)
$m = L_S/d_c - 0.495 (L_S/d_c)^{1/2} + 0.061$ (ii)
L_S/d_c = 0.3, 0.5, 1.0, 2.0, 3.0, 4.0, 5.0
I = 0.13, 0.21, 0.39, 0.71, 0.98, 1.24, 1.48
$L_f = L_{mf} [1 + (U - U_{mf})/(0.35 (g d_c)^{1/2})]$ (iii)

Table VII. Length Required for Fast-Bed Reactors [9][a]

k Packed Bed (sec^{-1})	L (95% conversion), m; No Mass Transfer Limitation	L (95% conversion), m; with Mass Transfer Limitation L_M = 2 m
0.1	500	500
1	50	56
3	17	23
10	5	11
30	1.7	9.8

[a]First-order reaction; U = 5 m/sec; solids holdup = 18 vol. %; plug-flow model.

beds have not been available, and further investigation of these flow regimes is critically needed.

NOMENCLATURE

- a constant in the entrainment equation (1/m)
- A cross-sectional area of the bed (m^2)
- C molar concentration in gas phase (kg-mol/m^3)
- C_A molar concentration of reactant A in the gas phase (kg-mol/m^3)
- C_{AF} gas concentration at top of freeboard region (kg-mol/m^3)
- C_b gas concentration in bubble (kg-mol/m^3)
- C_e gas concentration in emulsion (kg-mol/m^3)
- d_b bubble diameter (m)
- d_{bo} initial bubble diameter (m)
- d_{bm} maximum bubble diameter (m)
- d_c bed diameter (m)
- d_o orifice diameter (m)
- d_p particle diameter (m)
- D_{eff} effective diffusivity through the pores (m^2/sec)
- D gas diffusivity (m^2/sec)
- E_δ axial dispersion coefficient (m^2/sec)
- $E_{i\infty}$ elutriation rate constant (kg/m^2-sec)
- f_e fraction of wake solids ejected
- f_w ratio of wake volume/bubble void volume
- F_i entrainment rate of particle size i (upward) (kg/m^2-sec)
- F_i' entrainment rate of particle size i (downward) (kg/m^2-sec)
- F_{io} entrainment rate of particle size i at the bed surface (kg/m^2-sec)
- F_o total particle entrainment rate at the bed surface (kg/m^2-sec)
- $F_{i\infty}$ elutriation rate of particle size i (kg/m^2-sec)
- F_{be} gas exchange coefficient (bubble to emulsion), based on bubble void volume (m/sec)
- F_{be}' gas exchange coefficient, based on bubble phase (void + cloud) volume (m/sec)
- h freeboard height (m)
- h_{mi} mass transfer coefficient through the gas film (m/sec)
- H_d solid holdup (kg/m^3)
- I surface integral in Equation 52
- k first-order (gas) reaction rate constant per unit volume of settled bed (1/sec)
- k_{ov} overall reaction rate constant (1/sec)
- k_s surface reaction rate constant (1/m^2-sec)
- k_t intrinsic reaction rate constant (1/sec)
- L height from bottom of fluidized bed (m)
- L' height from top of jets (m)
- L_c critical height above the jets at which the bubble clouds overlap (m)
- L_f height of fluidized bed (m)
- L_J height of jets (m)
- L_{mf} height of bed at minimum fluidization (m)
- ΔL_n compartment height in bubble assemblage model (m)
- L_s slug length (m)
- m slug shape factor in Equation 52
- n number of compartments

n_d number of holes in perforated plate
N number of stages
N_M number of gas mass transfer units
N_{Mb} number of transfer units in bubbling zone
N_{MJ} number of transfer units in grid (jet) region
N_{MS} number of transfer units in slugging bed
N_R number of reaction units
N_{Rb} number of reaction units in bubbling zone
R_i production rate of species i (kg-mol/m^3-sec)
r ratio cloud sphere volume/bubble volume
r_b reaction rate in the cloud per unit volume $= kC_b$ (kg-mol/m^3-sec)
r_c radius of unreacted core (m)
r_e reaction rate in the emulsion per unit volume $= kC_b$ (kg-mol/m^3-sec)
S total particle surface area (m^2)
t time (sec)
U superficial gas velocity (m/sec)
U_b bubble velocity (m/sec)
U_{br} rise velocity of isolated bubble (m/sec)
U_{cr} critical gas velocity for gas back-mixing (m/sec)
U_e gas velocity in the emulsion phase (m/sec)
U_{ob} superficial velocity through bubbles (m/sec)
U_{oc} superficial velocity through clouds (m/sec)
U_{oe} superficial velocity through emulsion (m/sec)
U_J jet velocity (m/sec)
U_{mfJ} minimum fluidization velocity through the orifice (m/sec)
U_{si} solid velocity (ascending) (m/sec)
U_{si}' solid velocity (descending) (m/sec)
U_{mb} minimum bubbling velocity (m/sec)
U_{mf} minimum fluidization velocity (m/sec)
U_{ms} superficial velocity at the minimum slugging point (m/sec)
U_s absolute velocity of solids in the emulsion phase (m/sec)
U_T terminal velocity of the particle (m/sec)
U_{mk} superficial velocity at turbulent flow (m/sec)
V_b volume of bubble phase (void + cloud) (m^3)
V_{bV} volume of bubble void (m^3)
V_c volume of cloud phase (m^3)
V_e volume of emulsion phase (m^3)
X_i weight fraction of particle size i in the bed
Y_i weight fraction of particle size i in the freeboard

Greek Symbols

γ_s fraction of particles in bubble phase
δ volume of bubbles/total bed volume
ϵ voidage
ϵ_{mf} voidage at minimum fluidization
ξ dimensionless bed height
μ_G gas viscosity (kg/m-sec)
ρ_G gas density (kg/m^3)
ρ_s solid density (kg/m^3)

REFERENCES

1. DeJong, J. A. H., and F. F. Nomden. *Powder Technol.* 9:91 (1974).
2. Wen, C. Y., and L. H. Chen. "Fluidized Bed Freeboard Phenomena: Entrainment and Elutriation," *AIChE J.* 28(1):117 (1982).
3. Murray, J. D. *Chem. Prog. Symp. Ser.* 62(62):71 (1966).
4. Stephens, G. K., R. J. Sinclair and O. E. Potter. *Powder Technol.* 1:157 (1967).
5. Kunii, D., K. Yoshida and I. Hiraki. *Proc. Int. Symp. on Fluidization* (Amsterdam, Netherlands: Netherlands University Press, 1967), p. 243.
6. Catipovic, N. M., G. N. Govanovic and T. J. Fitzgerald. *AIChE J.* 24:543 (1978).
7. Broadhurst, T. E., and H. A. Becker. *AIChE J.* 21(2):238 (1975).
8. Geldart, D. *Powder Technol.* 7:285 (1973).
9. Van Swaaij, W. P. M. *Chem. Reaction Eng. Rev. ACS Symp. Ser.* 72:193 (1978).

10. Staub, F. W., and G. S. Canada. In *Fluidization* (London: Cambridge University Press, 1978), p. 339.
11. Potter, O. E. In *Fluidization*, J. F. Davidson and D. Harrison, Eds. (London: Academic Press, Inc., 1971).
12. Rowe, P. N. *Proc. Second Int. Symp. Chem. Reaction Eng.* (1972), p. A9.
13. Pyle, D. L. *Adv. Chem. Series* 109:106 (1972).
14. Calderbank, P. H., and F. D. Toor. In *Fluidization*, J. F. Davidson and D. Harrison, Eds. (London: Academic Press, Inc., 1971), p. 383.
15. Horio, M., and C. Y. Wen. *AIChE Symp. Series* 73(161):9 (1977).
16. Barreteau, D., C. Laguerie and H. Angelino. In *Fluidization* (London: Cambridge University Press, 1978), p. 292.
17. Shen, C. Y., and H. F. Johnstone. *AIChE J.* 1:349 (1955).
18. Van Deemter, J. J. *Chem. Eng. Sci.* 13:143 (1961).
19. Johnstone, H. F., J. D. Batchelor and W. Y. Shen. *AIChE J.* 1:318 (1955).
20. May, W. G. *Chem. Eng. Prog.* 55(12):49 (1959).
21. Kobayashi, H., and F. Arai. *Chem. Eng. Tokyo* 29:885 (1965).
22. Muchi, I. *Memories Faculty Eng. Nagoya Univ.* 17:79 (1965).
23. Orcutt, J. C., J. F. Davidson and R. L. Pigford. *Chem. Eng. Prog. Symp. Series No. 38* 58:1 (1962).
24. Kunii, D., and O. Levenspiel. *Ind. Eng. Chem. Fund.* 2:446 (1968).
25. Fryer, C., and O. E. Potter. *Ind. Eng. Chem. Fund.* 11:338 (1972).
26. Mamuro, T., and I. Muchi. *J. Ind. Chem. Tokyo* 68:126 (1965).
27. Toor, F. D., and P. H. Calderbank. *Proc. Int. Symp. on Fluidization* (Amsterdam: Netherlands University Press, 1967), p. 373.
28. Partridge, B. A., and P. N. Rowe. *Trans. Inst. Chem. Eng.* 44:T349 (1966).
29. Kobayashi, H., F. Arai, T. Chiba and Y. Tanaka. *Chem. Eng. Tokyo* 33:274 (1969).
30. Kato, K., and C. Y. Wen. *Chem. Eng. Sci.* 24:1351 (1969).
31. Mori, S., and I. Muchi. *Chem. Eng. Japan* 5:251 (1972).
32. Mori, S., and C. Y. Wen. *AIChE J.* 21:109 (1975).
33. Toomey, R. D., and H. F. Johnstone. *Chem. Eng. Prog.* 48:220 (1952).
34. Werther, J. *Ger. Chem. Eng.* 1:243 (1978).
35. Mori, S., and C. Y. Wen. *Fluidization Technol.* 1:179 (1976).
36. Shaw, I. D., T. W. Hoffman and P. M. Reilly. *AIChE Symp. Series No. 141* 70:41 (1974).
37. Davidson, J. F. *Trans. Inst. Chem. Eng.* 39:230 (1961).
38. Fryer, C., and O. E. Potter. Preprints of the International Fluidization Conference, Pacific Grove, CA (1975), p. III-1.
39. Chavarie, C., and J. R. Grace. *Ind. Eng. Chem. Fund.* 14:75, 79, 86 (1975).
40. Latham, R. L., C. J. Hamilton and O. E. Potter. *Brit. Chem. Eng.* 13:666 (1968).
41. Nguyen, H. V., and O. E. Potter. *Adv. Chem. Series* 133:290 (1974).
42. Nguyen, H. V., A. B. Whitehead and O. E. Potter. *AIChE J.* 23:913 (1977).
43. Cooke, M. J., W. Harris, J. Highley and D. F. Williams. *Proc. Tripartite Chem. Eng. Conf. Symp. on Fluidization* Vol. 1 (1968), p. 14.
44. Gomezplata, A., and W. W. Shuster. *AIChE J.* 6:454 (1960).
45. Zenz, F. A. *Proc. Tripartite Chem. Eng. Conf. Symp. on Fluidization* Vol. 1 (1968), p. 136.
46. Zenz, F. A. *Chem. Eng.* 81 (December 1977).
47. Behie, L. A., and P. Kehoe. *AIChE J.* 19(5):1070 (1973).
48. Wen, C. Y., N. R. Deole and L. H. Chen. "A Study of Jets in a Three-Dimensional Gas Fluidized Bed," *Powder Technol.* 31:175 (1982).
49. Grace, J. R., and H. I. De Lasa. *AIChE J.* 24:364 (1978).
50. George, S. E., and J. R. Grace. *AIChE Symp. Series* 74(176):67 (1978).
51. Miyauchi, T., and S. Furusaki. *AIChE J.* 20(6):1087 (1974).
52. Yates, J. G., and P. N. Rowe. *Trans. Inst. Chem. Eng.* 55:137 (1977).
53. De Lasa, H. I., and J. R. Grace. *AIChE J.* 25(6):984 (1979).
54. Beer, J. M., A. F. Sarofim, P. K. Sharma, T. Z. Chaung and S. S. Sandhu. In: *Fluidization*, 3rd international conference on fluidization (1980), p. 185.

55. Horio, M., A. Taki, Y. S. Hsieh and I. Muchi. In: *Fluidization,* 3rd international conference on fluidization (1980), p. 509.
56. Rajan, R. R., and C. Y. Wen. *AIChE J.* 26(4):642 (1980).
57. Chen, L. H., and C. Y. Wen. "Model of Solid Gas Reaction Phenomena in the Fluidized Bed Freeboard," paper presented at the 74th Annual Meeting of AIChE, New Orleans, LA (1981).
58. Wen, C. Y., and L. T. Fan. *Model for Flow Systems and Chemical Reactors* (New York: Marcel Dekker, Inc., 1975).
59. Hovmand, S., and J. F. Davidson. *Fluidization* (London: Academic Press, Inc. 1971), p. 193.
60. Raghuraman, J., and O. E. Potter. *Chem. Reaction Eng. ACS Symp. Series* 65:400 (1978).
61. Grace, J. R. Personal Communication (1981).

CHAPTER 27
SOLIDS INTERACTION AND TRANSFER BETWEEN FLUIDIZED BEDS

Te-Yu Chen
UOP Process Division
Des Plaines, Illinois 60016

Walter P. Walawender and L. T. Fan
Department of Chemical Engineering
Kansas State University
Manhattan, Kansas 66056

CONTENTS

INTRODUCTION

In industry, there are a number of fluidized bed processes that consist of circuits of two or more fluidized-bed reactors and connecting transfer lines for the circulation of solids between reactors. Examples of such processes are given by Zenz and Othmer [1] and Kunii and Levenspiel [2]. Probably, the best-known example of solids transfer between fluidized beds is the fluid catalytic cracking (FCC) process, a reactor-regenerator process, in which catalyst particles are circulated between the reactor and regenerator. Attaining a successful circulation of solids is a vital step in the design and operation of such a process. The principle of solids circulation, developed for the FCC process, also has been applied elsewhere in the petroleum industry, e.g., fluid hydroforming for reforming naphtha vapor [3], fluid coking for the treatment of heavy oil [4] and sand cracking for the thermal cracking of petroleum feedstocks [5]. Applications also extend outside the petroleum industry. Cox [6] described the drying of air by circulation of silica gel beads between two multistage fluidized beds. Hasegawa et al. [7] described a solid waste gasification system that circulates hot solids and char between two fluidized beds.

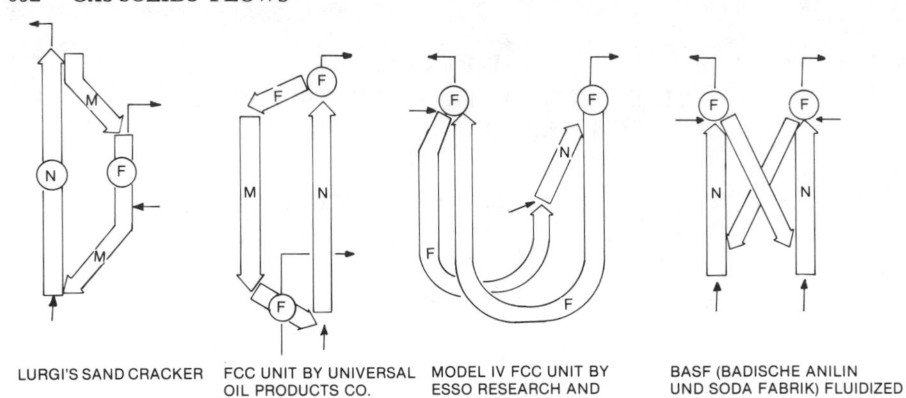

Figure 1. Schematics of various circulation systems [13].

The transport of solids in a fluid-bed circuit can be classified into two types: (1) transport within a fluidized bed itself (or internal circulation), and (2) transport between fluidized beds through transfer lines (or external circulation). Solids transport within a fluidized bed is usually very rapid and can be modeled satisfactorily based on the bubbling phenomenon [8–11] or the slugging phenomenon [12]. In contrast, solids transport in the transfer lines connecting fluidized beds is more complicated than internal circulation because of the different flow regimes that are possible and the different geometric configurations that can be used. Moreover, since internal circulation is generally rapid, external circulation is usually the controlling step in a circulation circuit involving two or more fluidized beds. Therefore, the emphasis of this chapter will be on external circulation. Even with this limited scope, we do not attempt to cover all the work reported in the literature. Instead, some selected works will be reviewed to show that a theoretical principle based on the macroscopic momentum balance can be used to unify these works.

CLASSIFICATION OF TRANSPORT SYSTEMS

The flow regime and the pipe inclination are factors that have a significant influence on external circulation. Hence, they are the basis for classifying transport systems.

Flow Regime

The flow regime can be classified into three types: pneumatic-conveying, fluidized flow and moving-bed flow. Figure 1 depicts the transfer circuits used in four industrial fluidized bed processes [13]. While all the circuits include fluidized flow and pneumatic-conveying, some also include moving-bed flow. These three flow regimes result from different modes of gas-solid contacting. Pneumatic-conveying* (or dilute-phase transport) is the type of flow in which the solids are apparently evenly dispersed and are low in volume concentration (i.e., $< 5\%$). In fluidized flow (or dense-phase transport), solids are still suspended in the gas but have a high volume concentration (i.e., up to approximately 50%). The dividing line between these two regimes is at a solids volume

*Discrepancies exist with respect to the usage of this term in the literature. For example, in the work of Leung and Wiles [14] the same term implies all three flow regimes.

fraction of about 0.1 [2]. In moving-bed flow, solids move en bloc at a voidage corresponding to that of a packed bed, with hardly any relative motion between particles.

Three criteria have been proposed to classify gas-solids flow regimes, Reh [15] suggests the specific force, sf, which is the ratio of the pressure drop in the transfer line to the gravity force of the mixture in the pipe. It is defined as

$$sf = \frac{\Delta P}{L\,(\rho_s - \rho_g)\,g\,\epsilon_s} \tag{1}$$

The value of sf distinguishes the three flow regimes as follows:

1. $sf < 1$, moving-bed flow
2. $sf = 1$, fluidized flow
3. $sf > 1$, pneumatic-conveying

Lapidus and Elgin [16] suggest the slip velocity as a criterion. The slip velocity is defined as

$$v_{sl} = \frac{u_g}{\epsilon_g} - \frac{u_s}{\epsilon_s} = \frac{u_g}{\epsilon_g} - \frac{u_s}{1 - \epsilon_g} \tag{2}$$

The distinction between the three flow regimes is made by comparing the slip velocity with the incipient fluidization velocity, u_{mf}, or the terminal velocity, v_t, of the particles as follows:

1. $\frac{u_{mf}}{\epsilon_{mf}} > v_{sl}$, moving-bed flow
2. $\frac{u_{mf}}{\epsilon_{mf}} \leq v_{sl} < v_t$, fluidized flow
3. $v_{sl} \geq v_t$, pneumatic-conveying

This type of analysis eventually gives rise to the so-called drift-flux model [17].

Leung and Jones [18] also suggest using the slip velocity as part of the criterion. They combined the features of Kojabashian's analysis [19] with the classification proposed by Leung and Jones [20] to give the following criteria:

1. $v_{sl} > \frac{u_{mf}}{\epsilon_{mf}}$ and $\left(\frac{\partial u_g}{\partial \epsilon}\right)_{J_s} > 0$, fluidized flow
2. $v_{sl} > \frac{u_{mf}}{\epsilon_{mf}}$ and $\left(\frac{\partial u_g}{\partial \epsilon}\right)_{J_s} < 0$, pneumatic-conveying
3. $v_{sl} < \frac{u_{mf}}{\epsilon_{mf}}$ and $\epsilon \neq f(v_{sl})$, moving-bed flow
4. $v_{sl} < \frac{u_{mf}}{\epsilon_{mf}}$ and $\epsilon = f(v_{sl})$, transition moving-bed flow

These criteria require an experimental relationship between v_{sl} and ϵ so that the second part of each criterion can be examined. The partial derivatives can be evaluated from Equation 2, noting that $J_s = \epsilon_s u_s$. Leung and Jones [18] give an alternative procedure to use when an experimental relationship is not available.

Methods for estimating the minimum fluidization velocity are available from various sources. Babu et al. [21] reviewed several correlations. Among them, the Kunii and Levenspiel [2] correlation seems to have the most theoretical foundation. A general form of the correlation may be written as [21]

$$K_1\,(Re_{mf})^2 + K_2\,Re_{mf} = Ga \tag{3}$$

where

$$Re_{mf} = \frac{\rho_g \, d_p \, u_{mf}}{\mu_g} \tag{4}$$

$$Ga = \frac{\rho_g d_p^3 \, (\rho_s - \rho_g) \, g}{\mu_g^2} \tag{5}$$

$$K_1 = \frac{1.75}{\psi_s \epsilon_{mf}} \tag{6}$$

$$K_2 = \frac{150 \, (1 - \epsilon_{mf})}{\psi_s^2 \, \epsilon_{mf}^3} \tag{7}$$

When values of ψ_s and ϵ_{mf} are not available, K_1 and K_2 can be considered as empirical parameters.

Pipe Inclination

Because of the important role that gravity plays in gas-solids flow, the angle between the pipe and the gravity field is a critical factor. Consequently, flows in pipes have been classified conveniently into vertical flow, horizontal flow and inclined flow.

Vertical flow represents one extreme of pipe inclination with the pipe parallel to the gravity field. All three flow regimes are possible in vertical pipes. A phenomenon related to the transition in flow regime from pneumatic to fluidized slug flow is choking. Choking is illustrated in Figure 2 [1]. The figure is a logarithmic plot of pressure drop per unit length of pipe versus superficial gas velocity, with the solids mass velocity as the parameter. It shows that, if the superficial gas velocity is reduced at a given solids mass velocity, the solids velocity will decrease and the solids concentration will increase. Choking occurs when the superficial gas velocity is reduced to such an extent that it can no longer support the solids in an evenly dispersed fashion; therefore, the entire suspension collapses, and fluidized slug flow begins.

An extensive review of methods for determining criteria for the existence of choking and for estimating the choking velocity was conducted by Leung and Wiles [14]. Only the methods deemed superior by them are summarized here. Choking does not occur

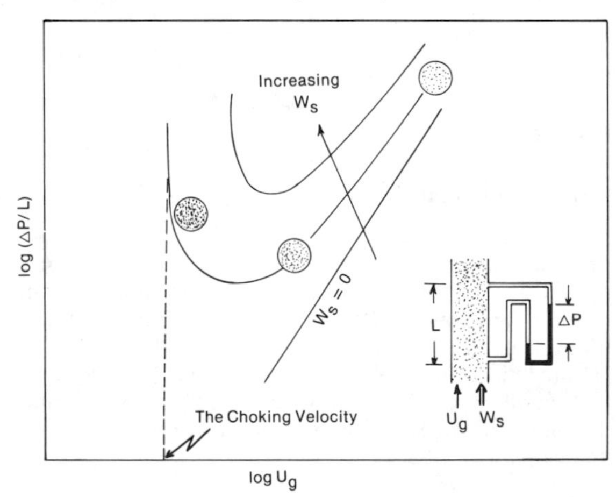

Figure 2. Choking in vertical gas-solids flow [1].

in all gas-solids systems [22,23]. Yang's [24] criterion states that for no choking to occur,

$$\frac{v_t}{\sqrt{gD}} < 0.35 \tag{8}$$

and for choking to occur,

$$\frac{v_t^2}{gD} > 0.12 \tag{9}$$

Yang [25] also has provided a method for estimating the choking velocity, u_c, by simultaneous solution of the following two equations for u_g and u_c:

$$u_c = (u_g - v_t)(1 - \epsilon_c) \tag{10}$$

and

$$\frac{(u_g - v_t)^2}{2gD} = 100\,(\epsilon_c^{-4.7} - 1) \tag{11}$$

Equations 8, 9 and 10 are for uniformly sized particles. In practice, however, the particles are often of mixed sizes. For particles of mixed sizes, two modifications to these equations are possible [14], although their reliability still needs to be justified experimentally. One is to modify the terminal velocity terms in the equations as

$$v_t = \sum_i X_{ti} v_{ti} \tag{12}$$

and the other is to modify the equations as

$$X_{ti} u_c = X_{ti}\,(1 - \epsilon_c)\,(u_g - v_{ti}) \tag{13}$$

and

$$[\sum_i X_{ti}\,(u_g - v_{ti})]^2 = 200\,gD\,(\epsilon_c^{-4.7} - 1) \tag{14}$$

Leung [23] recommends that Yang's equation tentatively be adopted for the prediction of the demarcation between choking and nonchoking systems. Leung [23] also presents a quantitative flow regime diagram in terms of the loading ratio and gas velocity, with the solids flux as a parameter for a solid-air system of the choking type.

Horizontal flow represents another extreme of pipe inclination, with the pipe perpendicular to the gravity field. Figure 3 presents a typical plot of the pressure drop per unit length of pipe versus the gas velocity. When the gas velocity is sufficiently high, the solids are in the pneumatic-conveying flow regime. In this regime, a decrease in the gas velocity results in an increase in the solids concentration and a decrease in the pressure gradient. If the gas velocity is reduced to the saltation velocity, however, sedimentation or salting of solids begins, resulting in an abrupt increase in the pressure gradient. The saltation velocity varies with solid loading.

No theoretical correlation is yet available for the prediction of the saltation velocity, although some methods for calculating the incipient suspension of single particles have been developed theoretically [26]. An empirical correlation of the saltation velocity for uniformly sized particles (both spherical and angular) has been given by Zenz [27]. To estimate the saltation velocity for mixed-sized particles, a very lengthy procedure was suggested by Zenz [27]. The procedure has been summarized by Kunii and Levenspiel [2] and Wen and Galli [28].

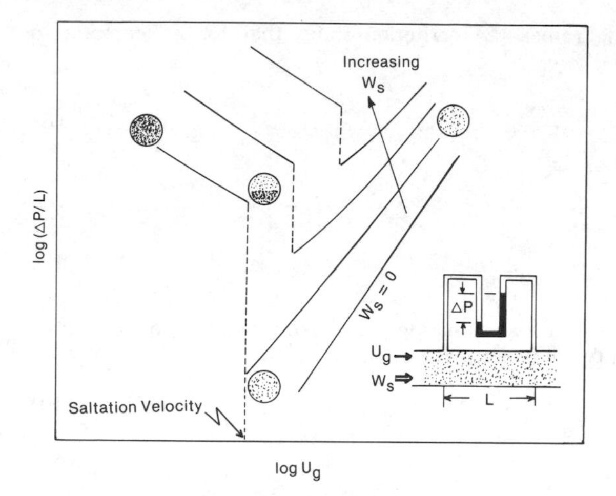

Figure 3. Saltation phenomena in horizontal gas-solids flow [1].

Although the Zenz correlation is accepted widely, Jones and Leung [29] point out that it cannot be used with confidence for very small particles. They reviewed and statistically compared eight correlations for the saltation velocity and concluded that the semitheoretical correlation of Thomas [30] predicts the saltation velocity more accurately than all the other correlations.

Between the two extremes of vertical and horizontal flow is flow in inclined pipes. The flow regimes possible in inclined pipes are closely related to the angle of inclination and the solids properties, such as the angle of repose and the angle of internal friction [1]. Figure 4 presents a typical plot of the pressure gradient versus the gas velocity and shows the influence of pipe inclination on the flow regime transition and

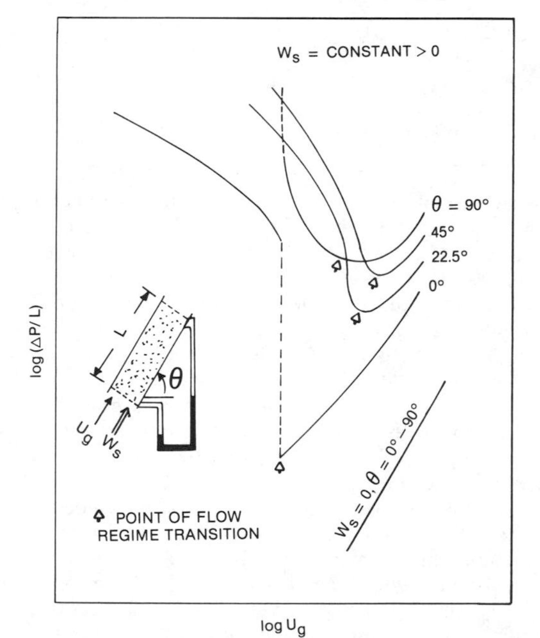

Figure 4. Effect of pipe inclination on cocurrent gas-solids flow [1].

Table I. Possible Combinations of Flow Regimes and Pipe Geometry

Geometry	Flow Regime		
	Moving Bed	Fluidized Bed	Pneumatic
Vertical	Yes	Yes	Yes
Horizontal	No[a]	Yes	Yes
Inclined	Yes[b]	Yes	Yes

[a]The special system of Chari [31] is excluded.
[b]There may be a critical angle of inclination below which the solids stop flowing.

the pressure gradient. Table I summarizes the possible flow regimes in vertical, horizontal and inclined pipes.

THEORETICAL FOUNDATIONS

Theoretical developments in the field of two-phase (gas-solid) flow systems can be divided into two classes—microscopic and macroscopic. In the microscopic approach, the law of conservation of momentum is applied to a "microscopic" volume element through which the phases are flowing, whereas in the macroscopic approach the law of conservation of momentum is applied to a "macroscopic" system, with one entrance and one exit for the flowing phases.

Although several investigators [32–39] have attempted to formulate the governing equations for a microscopic volume in heterogeneous flow systems, their results are not in complete agreement because the models used are essentially intuitive and contain terms whose form must be determined empirically. More rigorous approaches based on the use of local volume averages in the derivation of equations of motion have been attempted by other investigators [40–43]. Although it is difficult to apply the microscopic equations to practical flow systems, they may provide a rational basis for construction of mechanistic models.

The macroscopic momentum balance equation offers another option for describing gas-solids flow systems. It is of particular utility for the study of external circulation as it applies to two-phase flow in a straight pipe. The details of this equation are discussed below.

According to Newton's second law of motion, the rate of momentum change of a system is equal to the rate of momentum influx through the external surface of the system plus the contributions of momentum influx due to external surface forces and external body forces exerted on the system. Thus, we can write [44]

$$\frac{d}{dt}\int_V \rho\bar{v}\, d\bar{V} = -\oint_{A_e} \{\rho\bar{v}(\bar{v}-\bar{v}_e) + P\bar{\bar{U}} + \bar{\tau}\} \cdot d\bar{A} + \sum_i \int_V \rho_i \bar{F}_i dV \tag{15}$$

If the only external force is gravity, we have

$$\bar{F}_i = -g\bar{k} \tag{16}$$

and the last term on the right-hand side of Equation 15 can be expressed as

$$\sum_i \int_V \rho_i \bar{F}_i dV = -gk \int_V \rho dV \tag{17}$$

Now let us consider a two-phase, one-dimensional cocurrent flow with no mass transfer through the side walls, as depicted in Figure 5. Since the inlet (position 1) and outlet (position 2) are fixed,

$$\bar{v}_{1e} = \bar{v}_{2e} = 0 \tag{18}$$

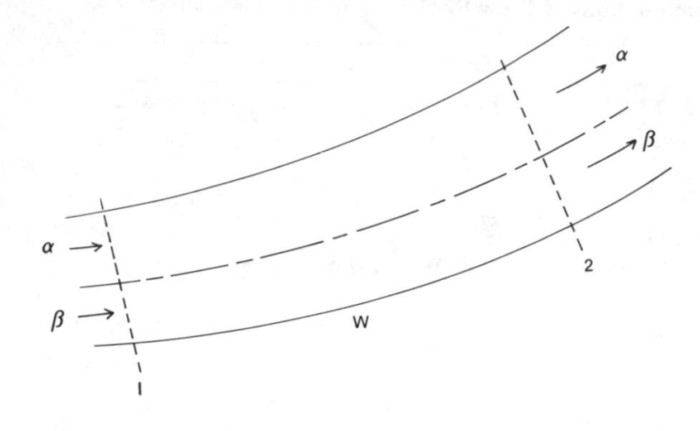

Figure 5. One-dimensional cocurrent flow.

Moreover, since the side wall is impervious, we have at the wall

$$\bar{v}_{ew} = \bar{v}_w \tag{19}$$

The external surface can be divided into three parts, i.e.,

$$A_e = A_1 + A_2 + A_w \tag{20}$$

Equation 15 can then be simplified to

$$\frac{d\bar{H}}{dt} = (W\bar{v})_{\alpha 1} + (W\bar{v})_{\beta 1} - (W\bar{v})_{\alpha 2} - (W\bar{v})_{\beta 2} - (P\epsilon\bar{A})_{\alpha 1} - (P\epsilon\bar{A})_{\beta 1} - (P\epsilon\bar{A})_{\alpha 2} - (P\epsilon\bar{A})_{\beta 2} - \bar{F}_w - g\bar{k}M \tag{21}$$

where the momentum of the material in the system is given by

$$\bar{H} \equiv \int_V \rho\bar{v}dV \tag{22}$$

the mass flowrate of a phase is given by

$$W \equiv \rho v \epsilon A \tag{23}$$

the wall force is given by

$$\bar{F}_w \equiv \int_{A_w} (P\bar{\bar{U}} + \bar{\bar{\tau}}) \cdot d\bar{A} \tag{24}$$

and the mass of material in the system is given by

$$M \equiv \int_V \rho dV \tag{25}$$

For the special case in which the flow conduit is straight, we can define $\bar{l}$ as a unit vector parallel to the pipe and pointing in the direction of flow. If we also assume that the pressure is constant over a cross section, then by taking the dot-product of Equation 21 and $\bar{l}$, we have the following equation:

$$\frac{dH}{dt} = (Wv)_{\alpha 1} + (Wv)_{\beta 1} - (Wv)_{\alpha 2} - (Wv)_{\beta 2} + P_1A_1 - P_2A_2 - F_w - g\cos\phi\, M \tag{26}$$

where

$$H \equiv \bar{l} \cdot \bar{H} \tag{27}$$

$$F_w \equiv \bar{l} \cdot \bar{F}_w \tag{28}$$

and

$$\cos \phi \equiv l \cdot \bar{k} \tag{29}$$

The wall force, F_w, can be further divided into two components: the pressure contribution and the friction force, i.e.,

$$F_w = F_{wp} + F_{w\tau} \tag{30}$$

where

$$F_{wp} = \bar{l} \cdot \int_{A_w} P d\bar{A} \tag{31}$$

and

$$F_{w\tau} \equiv \bar{l} \cdot \int_{A_w} \bar{\bar{\tau}} \cdot d\bar{A} \tag{32}$$

Since $\bar{l}$ and $d\bar{A}$ are perpendicular to each other for a conduit of constant cross section, we have, from Equation 31,

$$F_{wp} = 0 \tag{33}$$

Thus, Equation 26 can be simplified further to

$$\frac{dH}{dt} = (Wv)_{\alpha 1} + (Wv)_{\beta 1} - (Wv)_{\alpha 2} - (Wv)_{\beta 2} - A(P_2 - P_1) - F_{w\tau} - g \cos \phi\, M \tag{34}$$

At steady state, Equation 34 reduces to

$$\begin{aligned} 0 = {} & (Wv)_{\alpha 1} + (Wv)_{\beta 1} - (Wv)_{\alpha 2} - (Wv)_{\beta 2} \\ & - A(P_2 - P_1) - F_{w\tau} - g \cos \phi\, M \end{aligned} \tag{35}$$

Equation 35 shows a balance of the inertial forces (the first four terms on the right-hand side), the pressure force, the friction force from the wall and the force of gravity. It is the theoretical basis of most of the studies reviewed in the following section. The importance of each term varies from one flow regime to another, and the determination of the friction term, $F_{w\tau}$, is usually the focus of most investigations.

SELECTED EXAMPLES

Works selected are presented in four categories: vertical transport, horizontal transport, inclined-pipe flow and others. The fourth category includes works on systems that use combinations of geometry and flow regime and slide valves.

Vertical Transport

Leung and Wiles [14] have reviewed this topic comprehensively. Their review supplies a major part of the following summary.

For pneumatic-conveying, the inertia terms of the gas phase can be neglected compared to those of the solids; thus, after some rearrangement, Equation 35 can be simplified to

$$-\Delta P = \rho_s \epsilon_s v_s^2 + \rho_s \epsilon_s g L + \frac{F_{w\tau}}{A}$$
$$= \rho_s \epsilon_s v_s^2 + \rho_s \epsilon_s g L + \frac{4\tau L}{D} \quad (36)$$

The main problems encountered in the use of this equation are the determination of the solids velocity, v_s, the solids volume fraction, ϵ_s, and the friction loss, τ.

The friction loss usually consists of two components—one due to the gas and the other due to the solids—and may be written as [45,46]

$$\tau = \tau_g + \tau_s \quad (37)$$

where

$$\tau_g = \frac{1}{2} f_g \rho_g v_g^2 \quad (38)$$

and

$$\tau_s = \frac{1}{2} f_s \rho_s \epsilon_s v_s^2 \quad (39)$$

Two different correlations for f_s are recommended by Leung and Wiles [14] for high and low pressures. For high-pressure processes (operated up to 700 psia), the following Knowlton and Bachovchin [47] correlation may be used:

$$f_s = 0.02515 \left(\frac{J_s}{\rho_g v_g}\right)^{0.415} \left(\frac{v_s}{v_g}\right)^{-0.859} - 0.03 \quad (40)$$

where J_s, the solids mass flux, is defined by

$$J_s \equiv \rho_s (1 - \epsilon_g) v_s \quad (41)$$

For normal pressure operations, the following simple, but less accurate, correlation may be used:

$$f_s = 0.05 v_s^{-1} \quad (42)$$

For better accuracy, the Yang [48] correlation may be used:

$$f_s = 0.0206 \frac{\epsilon_s}{4\epsilon_g^3} \left[\epsilon_s \frac{Re_t}{Re_p}\right]^{-0.869} \quad (43)$$

where Re_t and Re_p are defined as

$$Re_p \equiv \frac{d_p (v_g - v_s) \rho_g}{\mu_g} \quad (44)$$

$$Re_t \equiv \frac{d_p v_t \rho_g}{\mu_g} \quad (45)$$

For estimating the solids velocity, v_s, and the voidage, ϵ_g, the equation proposed by Yang [49] may be used in combination with Equation 43:

$$v_s - v_g = \left[\left(1 + \frac{2f_s v_s^2}{D}\right) \frac{4}{3} \frac{(\rho_s - \rho_g) d_p}{\rho_g C_{DS}} \epsilon_g^{4.7}\right]^{0.5} \tag{46}$$

where C_{DS} is the drag coefficient for a single particle in an infinite fluid. With the solids flow rate, W_s, known, Equations 36–46 can be solved simultaneously for ΔP, v_s and ϵ_g.

Note that the above correlation applies to the well developed flow region. Yang [50] proposed design equations for the acceleration region in both vertical and horizontal transport. The procedure is based on a differential momentum balance. The pressure drop is a total of several integrals of pressure differentials. His paper should be consulted for details.

Jones et al. [46] calculated the friction loss due to gas from the standard Fanning friction factor equation. However, the contribution of this term is usually small compared to that due to the solid particles and, hence, can be neglected [45].

For fluidized flow, the gravity term in Equation 35 is the only significant contribution to the pressure drop. Thus, the main task is to determine the solids volume fraction, ϵ_s. To accomplish this, the two types of fluidized flow, i.e., fluidized flow with slugging and without slugging, have to be treated separately. For the former, the solids volume fraction, ϵ_s, is given by the Matsen equation [51]:

$$\frac{\epsilon_s}{\epsilon_{s.mf}} = \frac{v_b + J_s/\rho_s \, \epsilon_{s.mf}}{u_g + v_b - u_{mf} + J_s/\rho_s} \tag{47}$$

where

$$v_b = 0.35\ (gD)^{1/2} \tag{48}$$

at low gas velocity, i.e., $(u_g - u_{mf}) < 20$ cm/sec, and

$$v_b = 0.35\ (2gD)^{1/2} \tag{49}$$

at high gas velocity, i.e., $(u_g - u_{mf}) > 50$ cm/sec.

In the case of fluidized flow without slugging, no reliable model is yet available. As an approximation, a voidage ranging from 0.6 to 0.8 is recommended by Leung and Wiles [14].

For descending moving-bed flow, Yoon and Kunii [52] have found that the Ergun [53] equation for the pressure drop of fluid flow in a fixed bed can be applied to a moving bed, with the following modification:

$$\frac{(-\Delta P) d_p \epsilon_g}{L \rho_g \epsilon_s v_s^2} = \frac{150}{Re_{sl}} + 1.75 \tag{50}$$

where

$$Re_{sl} \equiv \frac{\rho_g d_p v_{sl} \epsilon_g}{\mu_g \epsilon_s} \tag{51}$$

Note that in the modification the gas velocity in the Ergun equation is replaced by the slip velocity. Equation 50 can be combined with Equation 35 to solve for the wall friction, F_{wT}.

Sandy et al. [54], in their study on vertical dense-phase gas-solids transport, used Equation 50 to calculate the pressure drop due to gas-solids friction. It is believed that their system was incipiently fluidized and, hence, Equation 50 is applicable.

Leung et al. [55] used Equations 50 and 51, coupled with a modified orifice equation, to describe solids downflow through a standpipe and slide valve. They used these equations to assess the effect of aeration on the solid mass flowrate and the limit of moving-bed flow.

Horizontal Transport

For horizontal pneumatic-conveying, the gravity term disappears in addition to the gas inertia terms. Therefore, Equation 35 simplifies to

$$-\Delta P = \rho_s \epsilon_s v_s^2 + \frac{4\tau L}{D} \quad (52)$$

With the friction loss from solids defined in Equation 39, the following equations by Yang [48] can be used together with Equation 52 to estimate the pressure drop:

$$f_s = 0.117 \frac{\epsilon_s}{4\epsilon_g^3} \left[\epsilon_s \frac{Re_t}{Re_p} \frac{v_g}{\sqrt{gD}} \right]^{-1.15} \quad (53)$$

and

$$v_t^2 = \left(\frac{2f_s v_s^2}{D}\right)\left(\frac{4}{3}\right)\left[\frac{(\rho_s - \rho_g)d_p}{\rho_g C_{DS}}\right] \epsilon_g^{4.7} \quad (54)$$

where f_s, Re_p, Re_t and C_{DS} are as defined previously. Yang tested his equations with Hinkle's [56] data and found them to be accurate within ± 20%.

Wen and Galli [28] proposed that the total pressure drop, ΔP_t, is given by the sum of the pressure drop due to the drag force of gas on the particles, ΔP_d, and the pressure drop due to the friction of the pipe wall, P_f, or

$$-\Delta P = (-\Delta P_d) + (-\Delta P_f) \quad (55)$$

Here, $(-\Delta P_f)$ is given by the pressure drop when no solid particles are present, and $(-\Delta P_d)$ can be obtained by integrating

$$\frac{dP_d}{dL} = 3C_{DS}\left(\frac{\rho_g^2}{\rho_s - \rho_g}\right)\frac{(v_g - v_s)^2}{d_p}(R)\left(1 - \frac{4\rho_g R}{\rho_s - \rho_g}\right)^{-4.7} \quad (56)$$

where R, the solid to gas input ratio, is defined by

$$R = \frac{4\,W_s}{\pi D^2 \rho_g v_s} \quad (57)$$

Wen and Galli [28] compared their method with the experimental data of Richardson and McLeman [57] and found that their method and the data were in reasonably good agreement. Note that while the Wen and Galli method needs information on the particle velocity, the Yang method requires knowledge of the voidage.

An empirical correlation for the fluidized flow regime in horizontal pipes has been given by Wen and Galli [28]:

$$\left(\frac{-\Delta P}{L\rho_s\epsilon_s}\right)\left(\frac{D}{d_p}\right)^{0.25} = 2.5\,v_s^{0.45} \quad (58)$$

This correlation is supported by the data of Wen and Simons [58], Carney [59] and

Koble et al. [60], as well as of Zenz [1]. Additional empirical correlations by other investigators also are given by Wen and Galli [28].

Inclined-Pipe Flow

In the macroscopic momentum balance (Equation 35), the gravity term is the only one explicitly exhibiting the influence of the angle of inclination. When the flow regime is other than pneumatic-conveying, however, the angle of inclination may change the distribution of the gas flow, as well as the solids flow, in the pipe. In this case, the flow situation becomes very complicated, and the applicability of Equation 35 has to be tested experimentally.

Trees [61] obtained data on the moving-bed flow of solids in inclined pipes connecting two fluidized beds and correlated his data empirically. Because different effects of back pressure on solids flow were observed in different ranges, Trees divided the flow into two regions: the free flow region and the pressure-impeded region. His correlations, which also depend on pipe size, are summarized below:

For pipes greater than 7.6 cm in diameter:

1. Free-flow region with $P_2 > P_1$,
$$J_s = (\text{constant})\ D^{1.4} L^{0.08} \theta^{0.6} (P_2 - P_1)^{-0.034}$$
2. Free-flow region with $P_2 < P_1$,
$$J_s = (\text{constant})\ D^{1.4} L^{0.08} \theta^{0.06} (P_1 - P_2)^{0.034}$$
3. Pressure-impeded region, where P_2 is always greater than P_1,
$$J_s = (\text{constant})\ D^{2} L^{-0.2} \theta^{6} (P_2 - P_1)^{-1.67}$$

For pipes smaller than 7.6 cm in diameter:

1. Free-flow region with $P_2 > P_1$,
$$J_s = (\text{constant})\ L^{0.08} \theta^{2} (P_2 - P_1)^{-0.034}$$
2. Pressure-impeded region with $P_2 > P_1$,
$$J_s = (\text{constant})\ L^{-0.2} \theta^{6} (P_2 - P_1)^{-1.67}$$

Trees' [61] work is a valuable source of experimental data, and his correlations are capable of accurately predicting the solids flowrate. However, the relationships between the solids flowrate and the design and operating parameters of his correlations appear to be unduly complicated.

Chen et al. [62] applied Equation 35 to Trees' data and obtained the following simplified form:

$$A\,\Delta P - F_{WT} + Mg \sin\theta = 0 \tag{59}$$

By combining the above equation and Equation 39, they found the following correlation between f_s and v_s:

$$f_s = 438\ D^{0.691}\ v_s^{-1.78} \tag{60}$$

This single correlation was capable of describing all of Trees' [61] data. The fit of the correlation is illustrated in Figure 6.

Chen et al. [62] also showed that the moving-bed solids flow behavior of Trees' data may be expressed in an alternative manner using the method of Metzner and Reed [63] for non-Newtonian fluids. The shear stress at the wall, τ, and the average shear rate, $\dot{\gamma}$, are defined by the expressions

$$\tau = \frac{DF_{WT}}{4AL} \tag{61}$$

and

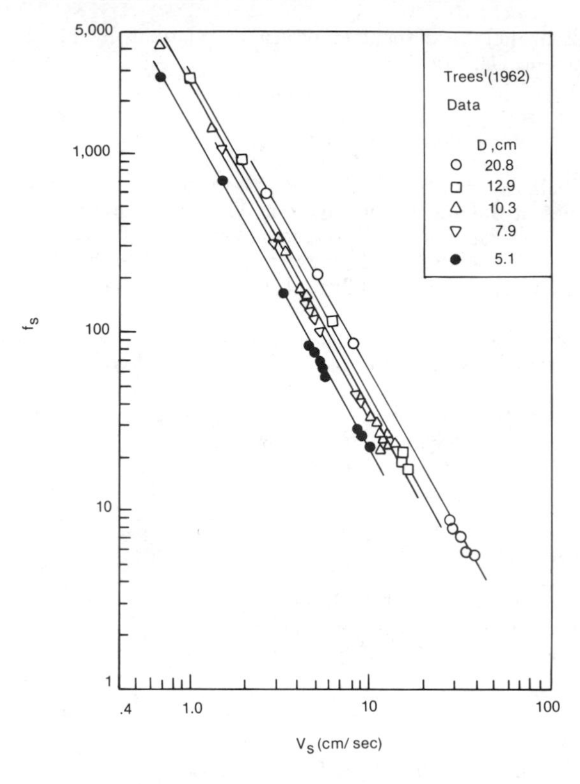

Figure 6. Friction factor vs solids velocity [62].

$$\dot{\gamma} = \frac{8v_s}{D} \tag{62}$$

These expressions have been used for the study of fluidized solids by Botterill and Bessant [64,65].

The wall shear stress and the average shear rate evaluated from Trees' [61] data are plotted in Figure 7, which shows that the flow curve for each pipe diameter is approximately linear and that there is no significant variation in the slope of these lines. Following Metzner and Reed [63],

$$n = \frac{d\, ln\, \tau}{d\, ln\, \dot{\gamma}} \tag{63}$$

and

$$\tau = k\, \dot{\gamma}^n \tag{64}$$

where k and n can be evaluated from Figure 7 by linear regression. The values of k and n are shown in Figure 7. Note that the value of k increases with an increase in pipe diameter.

Metzner and Reed [63] proposed a generalized Reynolds number so that the conventional correlation of the Fanning friction factor for Newtonian fluids in pipe flow could be applied to a large variety of time-independent non-Newtonian fluids. The generalized Reynolds number is defined as

$$Re_{MR} = \frac{D^n\, v_s^{2-n} \rho_b}{m} \tag{65}$$

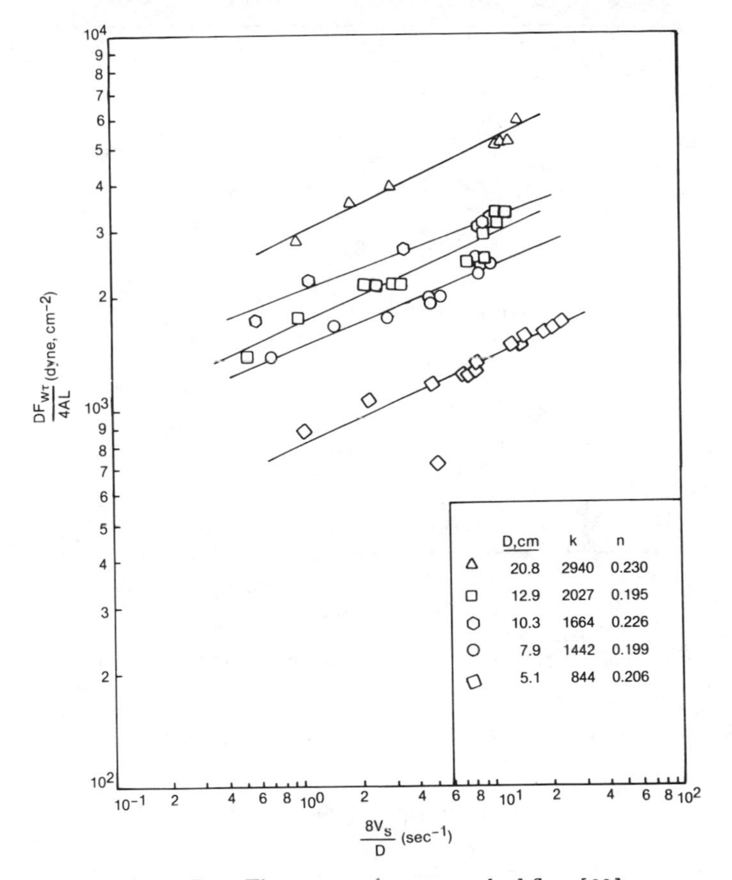

Figure 7. Flow curves for moving-bed flow [62].

where

$$m = k\,8^{n-1} \tag{66}$$

Thus, in the laminar flow region where $Re_{MR} < 2100$, the following relationship holds:

$$f_s = \frac{16}{Re_{MR}} \tag{67}$$

Figure 8 shows good agreement between Equation 67 and Trees' data.

The correlation in Figure 7 shows that for the particular range of operating conditions the friction factor is only a function of the solids flowrate and the pipe diameter. The dependence on the pipe diameter arises from the non-Newtonian behavior of the gas-solids mixture. This dependence can often be found in two-phase flow systems [66].

In contrast to Trees' experiments, which varied the geometric parameters, Chen et al. [67,68] studied moving-bed transfer into a fluidized bed, while changing the fluidized bed operating conditions. The study considered both gas flow and solids flow in the pipe. A dimensional analysis approach was used to formulate a correlation for the rate of gas leakage. (Their paper [67] should be consulted for details.) Here, we summarize their findings on the solids flow properties [68].

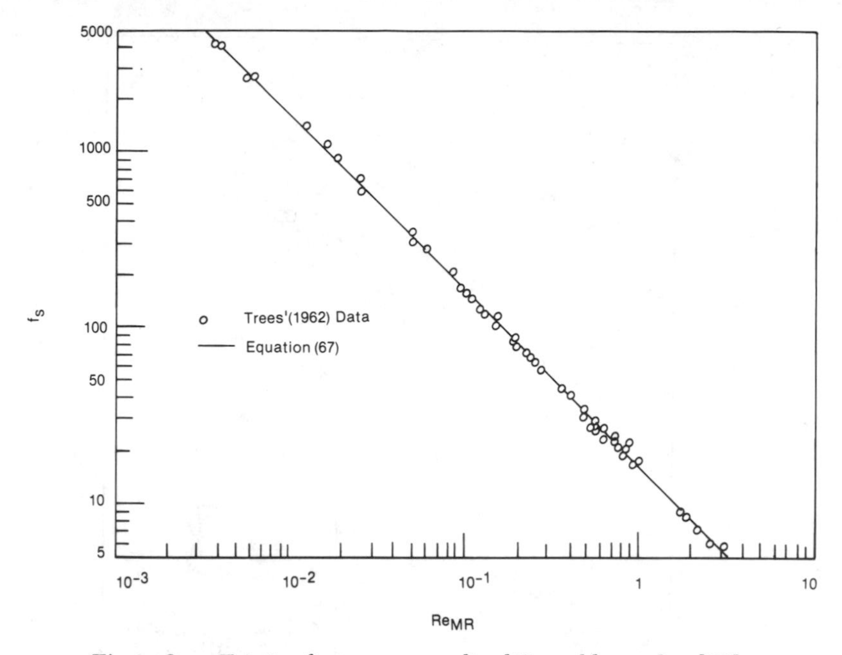

Figure 8. Friction factor vs generalized Reynolds number [62].

Equation 59 was applied in conjunction with Equation 39 to formulate a relation for evaluation of the friction factor. The friction factor was correlated by the following dimensional expression:

$$f_s = 1.314 \times 10^5 D_t^{0.7} d_p^{1.655} u_g^{-0.663} v_s^{-0.553} \tag{68}$$

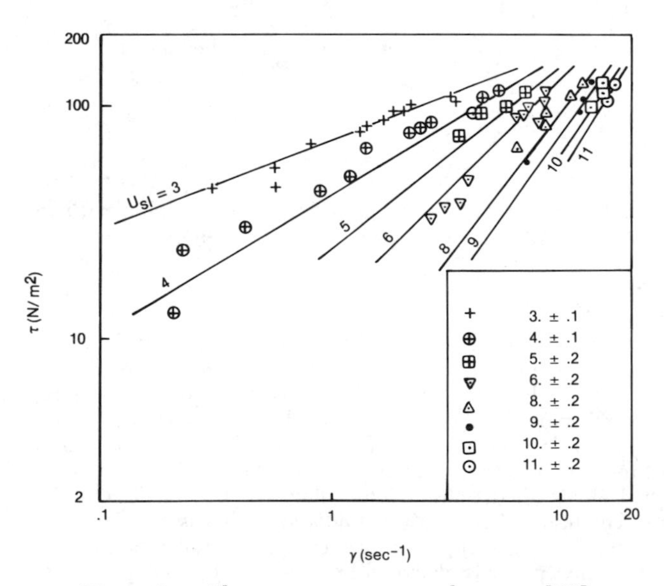

Figure 9. Shear stress vs average shear rate [68].

The exponent of pipe diameter agrees with that obtained from Trees' data. That of the particle velocity differs slightly.

The flow curves offer an alternative method for treating the friction loss. Figure 9 presents a typical plot of the wall shear stress, τ, against the average shear rate, $\dot{\gamma}$, with the superficial slip velocity as a parameter. The data in this figure were obtained with the same distributor, pipe size and particle size. The superficial slip velocity is defined as

$$u_{sl} = u_g + \epsilon_g v_s \tag{69}$$

Data within the specified deviations from the same slip velocity are plotted with the same symbol in Figure 9. The following mechanistic model, as represented by the straight lines in the figure, was fitted to the data

$$\tau = k\dot{\gamma}^n \tag{64}$$

where

$$k = a_1 + \exp(a_2 - a_3 u_{sl}) \tag{70}$$

and

$$n = a_4\{\exp[a_5(u_{sl} - a_7)] - \exp[a_6(u_{sl} - a_7)]\} \tag{71}$$

The parameters a_1 through a_7 were evaluated by means of a nonlinear parameter estimation technique. The resulting relationships between k and n and the superficial slip velocity are shown in Figures 10 and 11.

Figure 12 plots the friction factor, f_s, against the generalized Reynolds number by Metzner and Reed [63], Re_{MR}, for data from different distributors. As shown, the data agree well with the prediction of Equation 67. By analogy, the flow can be considered laminar ($Re_{MR} \leq 2100$), meaning that, as observed, the particles move in the pipe without significant change of their radial positions.

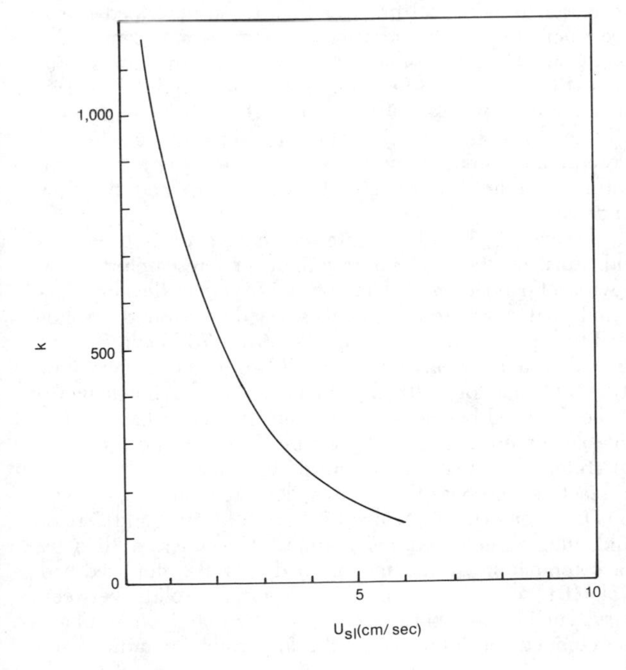

Figure 10. Variation of k with slip velocity [68].

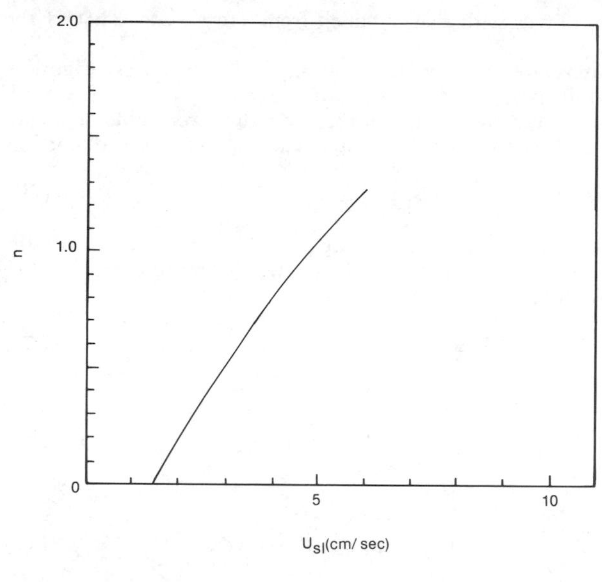

Figure 11. Variation of n with slip velocity [68].

Chen et al. [68] also used this approach to study the effect of k and n on distributor design. Their paper should be consulted for details.

Others

The work of Mason and Arundel [69] includes both vertical and horizontal transfer, namely, the flow in a horizontal duct immediately following a vertical riser. This flow system is more complicated and closer to industrial practice than those described previously. The pressure drop can be calculated by considering the two sections separately.

In practice, slide valves usually are used in connecting standpipes and risers. For fluidized flow across slide valves, the solids transfer rate has been found by several investigators [70–72] to be a function of the pressure drop across the valve. Leung [73] has summarized and combined those results with the Matsen [51] equation for the design of fluidized gas-solids flow in standpipes. Leung and Jones [18] present a comprehensive review of the literature on four flow regimes for standpipes and the flow of gas-solids mixtures through slide valves.

The solids transport inside a fluidized bed with an internal, e.g., a draft tube, has found industrial application and, thus, has been of interest to several researchers. The British Gas Council (e.g., the work of Horsler and Thompson [74]) has developed a reactor of such a type for oil and coal gasification; however, no details on the solids transfer were reported. For fossil-fuel processing, Yang and Keairns [75] have investigated a similar system and termed it a recirculating fluidized-bed reactor. Two flow regimes were reported to exist in their reactor, with the pneumatic-conveying flow regime inside the draft tube and the fluidized-bed flow regime outside. They have found Equations 50 and 51 to be suitable for predicting the pressure drop outside the draft tube, and Equations 36 through 46 to be suitable for that inside the tube.

Two other draft tube-type reactors composed of other flow regimes have been studied. The one investigated by Davidson [76], LaNauze [77] and LaNauze and Davidson [78] consists of incipient and slugging fluidized-bed regimes. The other, studied by Ishida and Shirai [79,80], is a combination of the moving-bed and the fluidized-bed flow regimes. Bachovchin et al. [81] examined pulsed transport of solids between fluidized beds. Their transport system involves both moving-bed flow and dilute-phase transport. They experimentally examined the influence of the important operating vari-

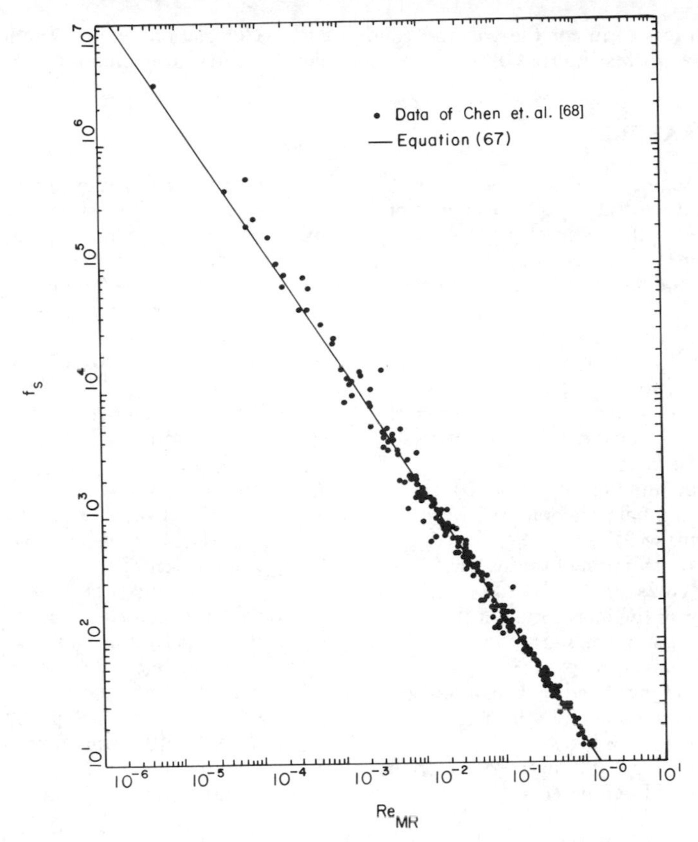

Figure 12. Friction factor vs generalized Reynolds number [68].

ables and applied the macroscopic momentum balance in describing their experimental results.

SUMMARY

The flow regime in a transfer line for solids transfer between fluidized beds may assume one or more of three forms, i.e., pneumatic-conveying, fluidized flow and moving-bed flow. Depending on the purposes of the process, different flow regimes may be desirable in different sections of the solids circulation system. Therefore, it is important to be able to predict the type of flow and, furthermore, to be able to evaluate the pressure drop and solids transfer rate. Characteristic velocities that distinguish these different flow regimes are the incipient fluidization velocity and the choking (or saltation) velocity. Available empirical or semiempirical methods for estimating these velocities have been summarized in this chapter.

The macroscopic momentum balance appears to be the basis of the majority of previous works on solids transfer in fluidized bed processes. The contribution of various terms in this equation varies with different flow regimes. Examples have been provided in this chapter to illustrate these differences.

Because of the complex nature of gas-solids flow systems, theoretical evaluation of the friction loss term in the momentum balance equation is impractical. Therefore, empirical approaches have been used predominantly by workers in the field for evaluating the friction loss term for the gas and solids phases. Correlations of the friction factor, which characterizes the friction loss in various flow regimes, are summarized.

NOMENCLATURE

A	area (cm^2)	$\bar{l}$	unit vector in the direction of flow
$d\bar{A}$	an outwardly directed element of area of the external surface (cm^2)	M	total mass of the system
		m	parameter defined by Equation 66
		n	flow behavior index
C_{DS}	drag coefficient of a single particle	P	pressure (dyne/cm^2)
D	pipe diameter (cm)	ΔP	pressure drop, i.e., $P_2 - P_1$ (dyne/cm^2)
d_p	particle diameter (cm)		
F	external forces (dyne)	R	solid to gas ratio defined by Equation 57
$\bar{F}$	external force vector (dyne)		
$\bar{F}_i$	body force acting on unit mass of the ith constituent (dyne/g)	Re	Reynolds number
		sf	specific force defined by Equation 1
f	friction factor	t	time (sec)
Ga	Galileo number (Equation 5)	$\bar{\bar{U}}$	unit tensor
g	gravitational acceleration (cm/sec^2)	u	superficial velocity (cm/sec)
		u_{sl}	superficial slip velocity (cm/sec)
H	total momentum of the system (g cm/sec)	V	volume (cm^3)
		v	velocity (cm/sec)
$\bar{H}$	vector of the momentum of the system (g cm/sec)	$\bar{v}$	velocity vector (cm/sec)
		v_{sl}	slip velocity (cm/sec)
J	solids mass flux (g/cm^2sec)	v_t	terminal velocity (cm/sec)
k_1	parameter defined by Equation 6	W	mass flowrate (g/sec)
k_2	parameter defined by Equation 7	X_{fi}	volume fraction of particles in the feed with terminal velocity v_{ti}
k	consistency index		
$\bar{k}$	unit vector in the upward direction	X_{ti}	volume fraction of particles in the pipe with terminal velocity v_{ti}
L	length of the pipe (cm)		

Greek Symbols

$\dot{\gamma}$	average shear rate (sec^{-1})	τ	wall shear stress (dyne/cm^2)
ϵ	volume fraction of either phase	$\bar{\bar{\tau}}$	symmetrical stress tensor
θ	angle of inclination to the horizontal	ϕ	angle defined by Equation 17
μ	viscosity (poise)	ψ_s	shape factor of solid particles
ρ	density (g/cm^3)		

Subscripts

b	bubble	s	solid phase
c	quantity at the choking condition	w	quantity at the pipe wall
e	quantity at external surface	α	α phase
g	gas phase	β	β phase
i	ith particle size or component	τ	friction
mf	quantity at the incipient fluidization condition	1	quantity at inlet of the pipe
		2	quantity at outlet of the pipe
p	pressure		

REFERENCES

1. Zenz, F. A., and D. F. Othmer. *Fluidization and Fluid-Particle Systems* (New York: Van Nostrand Reinhold Co., 1960).

2. Kunii, O., and O. Levenspiel. *Fluidization Engineering* (New York: John Wiley & Sons, Inc., 1969).
3. Kraft, W. W., W. Ullrich and W. O'Connor. "The Significance of Details in Fluid Catalytic Cracking Units: Engineering Design, Instrumentation and Operation," in *Fluidization,* D. F. Othmer, Ed. (New York: Van Nostrand Reinhold Co., 1956).
4. Krebs, R. W. "Fluid Coking," in *Fluidization,* D. F. Othmer, Ed. (New York: Van Nostrand Reinhold Co., 1956).
5. Schmalfeld, P. "How Lurgi Improved Sand Cracker," *Hydrocarbon Proc. Pet. Ref.* 42(7):145–148 (1963).
6. Cox, M. "A Fluidized Adsorbent Air-drying Plant," *Trans. Inst. Chem. Eng.* 36:29–42 (1958).
7. Hasegawa, M., J. Fukuka and D. Kunii. "Research and Development of Circulation System Between Fluidized Beds for Application of Gas-Solid Reactions," *Proc. 2nd Pachec '77* (1977), pp. 176–182.
8. Kunii, D., K. Yoshida and O. Levenspiel. "Axial Movement of Solids in Bubbling Fluidized Beds," in *Fluidization, Proc. Tripartite Chemical Engineering Conf.*, Montreal (1968), pp. 79–84.
9. Woollard, I. N. M., and O. E. Potter. "Solids Mixing in Fluidized Beds," *AIChE J.* 14:388–391 (1968).
10. Ishida, M., and C. Y. Wen. "Effect of Solid Mixing on Noncatalytic Solid-Gas Reactions in a Fluidized Bed," *AIChE Symp. Ser.* 69(128):1–7 (1973).
11. Rowe, P. N. "Estimation of Solids Circulation Rate in a Bubbling Fluidized Bed," *Chem. Eng. Sci.* 28:979–980 (1973).
12. Potter, O. E., and W. Thiel. "Solids Mixing in Slugging Fluidized Beds," in *Fluidization Technology,* Vol. II, D. L. Keairns, Ed. (Washington, DC: Hemisphere Publishing Corp., 1976), pp. 185–192.
13. Kunii, D., T. Kunugi, O. Yuzawa and N. Kunii. "Flow Characteristics of Circulation Systems with Two Fluidized Beds," *Int. Chem. Eng.* 14:588–593 (1974).
14. Leung, L. S., and R. J. Wiles. "A Quantitative Design Procedure for Vertical Pneumatic Conveying Systems," *Ind. Eng. Chem. Proc. Des. Dev.* 15:552–557 (1976).
15. Reh, L. "Fluidized Bed Processing," *Chem. Eng. Proc.* 67(2):58–63 (1971).
16. Lapidus, L., and J. C. Elgin. "Mechanics of Vertical-Moving Fluidized Systems," *AIChE J.* 3:62–68 (1957).
17. Wallis, G. B. *One-Dimensional Two-Phase Flow* (New York: McGraw-Hill Book Co., 1969).
18. Leung, L. S., and P. J. Jones. "Flow of Gas-Solid Mixtures in Standpipes, A Review," *Powder Technol.* 20:145–160 (1978).
19. Kojabashian, C. "Properties of Dense-Phase Fluidized Solids in Vertical Downflow," Ph.D. Thesis, Massachusetts Institute of Technology, Cambridge, MA (1958).
20. Leung, L. S., and P. J. Jones. "Coexistence of Fluidized Solids Flow and Packed Bed Flow in Stand Pipes," in *Proc. Int. Fluidization Conf.*, J. F. Davidson, Ed. (London: Cambridge University Press, 1978).
21. Babu, S., B. Shah and A. Talwalkar. "Fluidization Characteristics of Coal Gasification Materials," paper presented at the AIChE 69th annual meeting, Chicago, IL, November 28–December 2, 1976.
22. Kehoe, P. W. K., and J. F. Davidson. "Continuously Slugging Fluidized Beds," in *Proc. Chemica '70 Conf., Inst. Chem. Eng.*, London (1970).
23. Leung, L. S. "Vertical Pneumatic Conveying: a Flow Regime Diagram and Review of Choking Versus Non-Choking Systems," *Powder Technol.* 25:185–190 (1980).
24. Yang, W. C. "A Criterion for Fast Fluidization," in *Proc. Pneumo-transport 3, BHRA Fluid Eng.* (1976).
25. Yang, W. C. "A Mathematical Definition of Choking Phenomenon and a Mathematical Model for Predicting Choking Velocity and Choking Voidage," *AIChE J.* 21:1013–1015 (1975).
26. Halow, J. S. "Incipient Rolling, Sliding, and Suspension of Particles in Horizontal and Inclined Turbulent Flow," *Chem. Eng. Sci.* 28:1–12 (1973).
27. Zenz, F. A. "Conveyability of Materials of Mixed Particle Size," *Ind. Eng. Chem. Fund.* 3:65–75 (1964).

28. Wen, C. Y., and A. F. Galli. "Dilute Phase Systems," in *Fluidization,* J. F. Davidson and D. Harrison, Eds. (New York: Academic Press, Inc., 1971), pp. 677–710.
29. Jones, P. J., and L. S. Leung. "A Comparison of Correlations for Saltation Velocity in Horizontal Pneumatic Conveying," *Ind. Eng. Chem. Proc. Des. Dev.* 17:571–575 (1978).
30. Thomas, D. G. "Transport Characteristics of Suspensions: Part VI. Minimum Transport Velocity for Large Particle Size Suspensions in Round Horizontal Pipes," *AIChE J.* 8:373–378 (1962).
31. Chari, S. S. "Pressure Drop in Horizontal Dense Phase Conveying of Air-Solid Mixtures," *AIChE Symp. Ser.* 67(116):77–84 (1971).
32. van Deemter, J. J., and E. T. van der Laan. "Momentum and Energy Balances for Dispersed Two-Phase Flow," *Appl. Sci. Res.* A10:102–108 (1961).
33. Hinze, J. O. "Momentum and Mechanical-Energy Balance Equation for a Flowing Homogeneous Suspension with Slip Between the Two Phases," *Appl. Sci. Res.* A11: 33–46 (1962).
34. Jackson, R. "The Mechanics of Fluidized Beds," *Trans. Inst. Chem. Eng.* 41:13–28 (1963).
35. Soo, S. L. "Dynamics of Multiphase Flow Systems," *Ind. Eng. Chem. Fund.* 4:426–433 (1965).
36. Murray, J. D. "On the Mathematics of Fluidization. Part I. Fundamental Equations and Wave Propagation," *J. Fluid Mech.* 21:81–87 (1965).
37. Pigford, R. L., and T. Baron. "Hydrodynamic Stability of a Fluidized Bed," *Ind. Eng. Chem. Fund.* 4:81–87 (1965).
38. Eleftheriades, C. M., and M. R. Judd. "The Design of Downcomers Joining Gas-Fluidized Beds in Multiple Stages," *Powder Technol.* 21:217–225 (1978).
39. Arastoopour, H., and D. Gidaspow. "Vertical Counter Current Solids Gas Flow," *Chem. Eng. Sci.* 34:1063–1066 (1979).
40. Anderson, T. B., and R. Jackson. "A Fluid Mechanical Description of Fluidized Beds," *Ind. Eng. Chem. Fund.* 6:527–539 (1967).
41. Slattery, J. C. "General Balance Equations for a Phase Interface," *Ind. Eng. Chem. Fund.* 6:108–115 (1967).
42. Whitaker, S. "Diffusion and Dispersion in Porous Media," *AIChE J.* 13:420–427 (1967).
43. Chen, Y. C. "The Mass and Momentum Equations for Heterogeneous Flow Systems and Their Applications," Ph.D. Thesis, Northwestern University, Evanston, IL (1975).
44. Standart, G. "The Mass, Momentum, and Energy Equations for Heterogeneous Flow Systems," *Chem. Eng. Sci.* 19:227–236 (1964).
45. Stemerding, S. "The Pneumatic Transport of Cracking Catalyst in Vertical Rises," *Chem. Eng. Sci.* 17:599–608 (1962).
46. Jones, J. H., W. G. Braun, T. E. Daurbert and H. D. Allendorg. "Estimation of Pressure Drop from Vertical Pneumatic Transport of Solids," *AIChE J.* 13:608–611 (1967).
47. Knowlton, T. M., and D. M. Bachovchin. "The Determination of Gas-Solids Pressure Drop and Choking Velocity as a Function of Gas Density in a Vertical Pneumatic Conveying Line," in *Fluidization Technology,* Vol. II, D. L. Keairns, Ed. (Washington, DC: Hemisphere Publishing Corp., 1976), pp. 253–282.
48. Yang, W. C. "Correlations for Solid Friction Factors in Vertical and Horizontal Pneumatic Conveyings," *AIChE J.* 20:605–607 (1974).
49. Yang, W. C. "Estimating the Solid Particle Velocity in Vertical Pneumatic Conveying Lines," *Ind. Eng. Chem. Fund.* 12:349–352 (1973).
50. Yang, W. C. "A Unified Theory on Dilute Phase Pneumatic Transport," *J. Powder Bulk Solids Technol.* 1:89–95 (1978).
51. Matsen, J. M. "Flow of Fluidized Solids and Bubbles in Standpipes and Risers," *Powder Technol.* 7:93–96 (1973).
52. Yoon, S. M., D. Kunii. "Gas Flow and Pressure Drop through Moving Beds," *Ind. Eng. Chem. Proc. Des. Dev.* 9:559–565 (1970).

53. Ergun, S. "Fluid Flow through Packed Columns," *Chem. Eng. Proc.* 48:89–94 (1952).
54. Sandy, C. W., T. E. Daubert and J. H. Jones. "Vertical Dense-Phase Gas-Solids Transport," *CEP Symposium Series* 66(105):133–142 (1970).
55. Leung, L. S., P. J. Jones and T. M. Knowlton. "An Analysis of Moving-Bed Flow of Solids Down Standpipes and Slide Valves," *Powder Technol.* 19:7–15 (1978).
56. Hinkle, B. L. "Acceleration of Particles and Pressure Drops Encountered in Horizontal Pneumatic Conveying," Ph.D. Thesis, Georgia Institute of Technology, Atlanta, GA (1953).
57. Richardson, J. F., and M. McLeman. "Solids Velocities and Pressure Gradients in a One-inch Horizontal Pipe," *Trans. Inst. Chem. Eng.* 38:257–266 (1960).
58. Wen, C. Y., and H. P. Simons. "Flow Characteristics in Horizontal Fluidized Solids Transport," *AIChE J.* 5:263–267 (1959).
59. Carney, W. J. "Simplified Correlation for Heating Solid-Air Mixtures Transported from the Fluidized Bed," Ph.D. Thesis, West Virginia University, Morgantown, WV (1954).
60. Koble, R. H., P. R. Jones and W. A. Koehler. "Blending and Conveying of Ceramic Raw Materials by Fluidization," *Am. Ceramic Soc. Bull.* 32:367–372 (1953).
61. Trees, J. "A Practical Investigation on the Flow of Particulate Solids through Sloping Pipes," *Trans. Inst. Chem. Eng.* 40:286–296 (1962).
62. Chen, T. Y., W. P. Walawender and L. T. Fan. "Moving-Bed Solids Flow Between Two Fluidized Beds," *Powder Technol.* 22:89–96 (1979).
63. Metzner, A. B., and J. C. Reed. "Flow of Non-Newtonian Fluids: Correlation of the Laminar, Transition, and Turbulent-Flow Regions," *AIChE J.* 1:434–440 (1955).
64. Botterill, J. S. M., and D. J. Bessant. "The Flow Properties of Fluidized Solids," *Powder Technol.* 8:213–222 (1973).
65. Botterill, J. S. M., and D. J. Bessant. "The Flow Properties of Fluidized Solids," *Powder Technol.* 14:131–137 (1976).
66. Govier, G. W., and K. Aziz. *The Flow of Complex Mixtures in Pipes* (New York: Van Nostrand Reinhold Co., 1972).
67. Chen, T. Y., W. P. Walawender and L. T. Fan. "Moving-Bed Solids Flow Between an Inclined Pipe Leading into a Fluidized Bed: Part I. Gas Leakage and Pressure Drop," *AIChE J.* 26:24–30 (1980).
68. Chen, T. Y., W. P. Walawender and L. T. Fan. "Moving-Bed Solids Flow in an Inclined Pipe Leading into a Fluidized Bed: Part II. The Solids Flow Properties," *AIChE J.* 26:31–36 (1980).
69. Mason, J. S., and P. A. Arundel. "Pressure Measurements in a Flowing Alumina Suspension along a Horizontal Duct Immediately Following a Vertical Riser," *Powder Technol.* 8:261–272 (1973).
70. Massimilla, L., V. Betta and C. D. Rocca. "A Study of Streams of Solids Flowing from Solid-Gas Fluidized Beds," *AIChE J.* 7:502–510 (1961).
71. Jones, D. R. M., and J. F. Davison. "The Flow of Particles from a Fluidized Bed through an Orifice," *Rheol. Acta* 4:180–192 (1965).
72. de Jong, J. A. H., and Q. E. J. J. M. Hoelen. "Cocurrent Gas and Particle Flow during Pneumatic Discharge from a Bunker through an Orifice," *Powder Technol.* 12:201–208 (1975).
73. Leung, L. S. "Design of Fluidized Gas-Solids Flow in Standpipes," *Powder Technol.* 16:1–7 (1977).
74. Horsler, A. G., and B. H. Thompson. "Fluidization in the Development of Gas Making Processes," in *Fluidization, Proc. Tripartite Chemical Engineering Conf.*, Montreal (1968), pp. 59–66.
75. Yang, W. C., and D. L. Keairns. "Recirculating Fluidized-Bed Reactor Data Utilizing a Two-dimensional Cold Model," *AIChE Symp. Ser.* 70(141):27–40 (1974).
76. Davidson, J. F. "Differences Between Large and Small Fluidized Beds," *AIChE Symp. Ser.* 69(128):16–17 (1973).

77. La Nauze, R. D. "A Circulating Fluidized Bed," *Powder Technol.* 15:117–127 (1976).
78. La Nauze, R. D., and J. F. Davidson. "The Flow of Fluidized Solids," in *Fluidization Technology*, Vol. II, D. L. Keairns, Ed. (Washington, DC: Hemisphere Publishing Corp., 1976), pp. 113–124.
79. Ishida, M., and T. Shirai. "Circulation of Solid Particles within the Fluidized Bed with a Draft Tube," *J. Chem. Eng. Japan* 8:477–481 (1975).
80. Ishida, M., and T. Shirai. "Equilibrium Bed Heights when a Fluidized Bed and a Fixed Bed Are Connected through an Opening," *J. Chem. Eng. Japan* 9:249–259 (1976).
81. Bachovchin, D. M., P. R. Mulik, R. A. Newby and D. L. Keairns. "Pulsed Transport of Bulk Solids Between Adjacent Fluidized Beds," *Ind. Eng. Chem. Proc. Des. Dev.* 20:19–26 (1981).

CHAPTER 28
TIME-DEPENDENT REACTIVE MODELS OF FLUIDIZED-BED AND ENTRAINED-FLOW CHEMICAL REACTORS

H. H. Klein, D. E. Dietrich, S. R. Goldman,[*] D. H. Laird, M. F. Scharff and B. Srinivas

JAYCOR
San Diego, California 92138

CONTENTS

INTRODUCTION

Much of today's industrial technology in petroleum refining, petrochemical processing and coal conversion centers around the use of fluidized-bed and entrained-flow reactors, both of which involve the flow of solid and gas mixtures undergoing chemical reactions. To date, it has been at the least a difficult, to at the most an impossible problem to gain

[*]Present address: Los Alamos National Laboratory, Los Alamos, New Mexico.

a detailed understanding of the very complex thermodynamic and transport phenomena occurring in chemical process reactors. This is due primarily to the experimental difficulty of obtaining data in the high-temperature, high-pressure hostile environments in which these phenomena occur, and to the analytical difficulty of quantifying the various competing processes. However, owing to recent advances in numerical processing on high-speed computers, it is now becoming possible to obtain solutions for the very complex chemical-fluid mechanical systems that had previously defied analysis.

Using these new techniques, it is now reasonable to simulate mathematically detailed fluidized-bed and entrained-flow reactor dynamics at moderate computer costs. The usefulness of such models lies in the fact that they can give valuable insights into the crucial mechanisms that control the process of interest. These insights are indispensable for effective design, optimization, scaleup, reactor improvements and hazard analysis.

Over the past several years, JAYCOR has been involved in the development of the FLAME (flow and multiprocess engineering) code for modeling entrained-flow reactors and the FLAG (fluidized-bed agglomerating gasification) code, which models fluidized-bed reactors. The approach adopted in developing these codes has been that of using the fundamental principles of physics and chemistry whenever practical so that the number of assumptions is minimized. For example, the complete set of time-dependent, compressible Navier-Stokes equations is solved for the gas dynamics, rather than simplifying those equations by making ad hoc assumptions for specific cases. For Newtonian fluids, the solution naturally involves flow patterns such as eddies and recirculation. However, it is not always possible to describe real-world processes adequately, or economically, using first principles. In those cases, we use a small number of appropriate empirical correlations selected from published literature, which can be refined or replaced later when better theories or data become available. Other models of fluidized-bed and entrained-flow reactors also have been presented [1–3].

Both the FLAME and FLAG codes are space dependent, two dimensional, axisymmetric and time dependent, with hydrodynamic swirl included as a field variable. Space dependence can give the process engineer information about the spatial variation of the flow field. Time-dependent models were chosen over their steady-state counterparts because the initial value problem is well posed mathematically. Time-dependent codes can calculate steady states when they exist, and are useful for studying situations in which there exists no steady state (such as bubbles in a fluidized bed) and nonsteady hazards and safety issues such as startup, shutdown, transients and operational instabilities.

The codes are modular in architecture, with several modules common to both codes. The major modules include gas flow dynamics with turbulence, gas-phase chemistry, particle chemistry, particle dynamics, particle energetics, radiation and heat transfer. The FLAG code includes two additional modules—particle collisions and agglomeration.

The following sections contain a detailed description of the contents of each of these modules and the methods used in them to obtain a solution. The work described herein was performed under subcontract R-10785-C-22990 for Stearns-Roger, Inc. (DOE contract DE-AC01-80ET14705), and contract DE-AC21-78-ET10329 for the U. S. Department of Energy.

FLUID MECHANICS

Two-Phase Averaging

Entrained-flow and fluidized-bed reactors involve the flow of mutually interacting gases and solids (through drag forces and chemical reactions). A continuum approach can be applied to the gas, where equations of mass, momentum and energy conservation describe the gas flow at each spatial location where gas is present. These are the Navier-Stokes equations. With a single-phase fluid or a lightly loaded two-phase system, the procedure is straightforward, but in a two-phase system where particles and gases are occupying equal amounts of volume, such as a fluidized bed, the representation becomes complicated. Since significant variations in voidage in the bed (bubbles) can influence the gas flow pattern considerably, it is important to include voidage effects in the gas flow description.

The continuum approach already involves an average over length scales that is large compared to a molecular diameter. A method of resolving the high particle loading difficulty is to extend the averaging over length scales large compared to mean particle separation, but small compared to major flow variations. By integrating over the mass, momentum and energy exchange at particle-gas interfaces, the particle graininess virtually disappears, and the system can be viewed as two interpenetrating fluids. The drag and reaction terms then become source and sink terms in the fluid equations and sink and source terms in the corresponding particle equations, thereby conserving total mass, momentum and energy.

The particle-scale average lends itself naturally to our solution technique, in which we subdivide the reactor into a number of computational zones, each containing a small volume of the reactor. The fluid dynamic variables are taken as constant throughout each of these zones and vary slowly from zone to zone. Thus, we take averages of the Navier-Stokes equations over each computational zone. This averaging volume is certainly large compared to the particle spacing in a fluidized bed. In an entrained-flow gasifier with a small fraction of solids, we can estimate the average distance between particles, λ,

$$\lambda = \frac{1}{n_p^{1/3}}$$

$$\lambda = \left(\frac{4\pi}{3\phi}\right)^{1/3} r_p$$

where n_p is the particle number density, r_p is the particle radius, and ϕ is the solid volume fraction. If ϕ is as small as 0.01, $\lambda \simeq 7.5\ r_p$, which is small compared to a zone size. The averaging procedure is then valid for entrained-flow gasifiers also.

The method of integrating the conservation equations over control volumes has been described by Crowe and Smoot [4]. A similar and more mathematically formal method has been presented by Anderson and Jackson [5], who averaged all point variables over a region containing many particles and obtained local mean variables.

We remark that the averaging procedure, in addition to yielding the mean gas-phase properties, also can be used to describe the particles in terms of their average properties. We describe our treatment of the particles in the section entitled Particle Dynamics and Energetics.

Gas-Phase Equations

Gas-Phase Continuity

We wish to derive an equation that describes the change of gas mass inside a fixed control volume that also contains particles. Although the control volume is fixed, the volume occupied by the gas can change by movement of particles across the fixed control volume boundaries or by volume changes of individual reacting particles. Figure 1 shows a surface, S, bounding a control volume, V, containing gas and particles. The gas occupies a portion, V_g, of the control volume and flows through a part, S_g, of the surface. We denote the gas density as ρ, the gas velocity across S_g as v_g, the particle density as ρ_p, and the regression velocity due to chemical reactions at S_p (the surface of the particle) as $\overline{w}$. The conservation of gas mass inside V_g is

$$\frac{d}{dt}\int_{V_g} \rho d^3X = \int_{V_g} \frac{\partial \rho}{\partial t} d^3X - \int_{S_p} \rho_p \overline{w} \cdot \bar{n}_p d^2X \tag{1}$$

where $\bar{n}_p$ is the unit outward normal to the particle surface. The first term on the right-hand side of Equation 1 represents the change of density inside V_g. Using the gas-phase continuity equation at a point and Gauss's theorem to convert a volume integral to a surface integral, this term becomes

$$\int_{V_g} \frac{\partial \rho}{\partial t} d^3X = -\int_{S_g} \rho \bar{v}_g \cdot \bar{n}_g d^2X \tag{2}$$

where $\bar{n}_g$ is the unit outward normal to the surface, S_g. The second term on the right-hand side of Equation 1 is an integral over all particle surfaces and represents the rate of change of V_g as the particles gasify and lose volume. If we define r_M as the average rate at which mass is gained by the gas per unit volume due to the chemical reactions, the second term becomes

$$-\int_{S_p} \rho_p \bar{w} \cdot \bar{n}_p d^2X = r_M \int_V d^3X \tag{3}$$

We define the local void fraction, θ, as the volume occupied by the gas per unit volume of the gas-particle mixture, i.e.,

$$\theta = \frac{\partial V_g}{\partial V}$$

The integrals in Equations 1 and 2 become

$$\frac{d}{dt}\int_V \rho\theta d^3X + \int_S \rho\theta\bar{v}_g \cdot \bar{n}_g d^2X = \int_V r_M d^3X \tag{4}$$

This equation expresses the conservation of gas mass inside a control volume containing a mixture of gas and particles, that is, the rate of change of gas mass inside the control volume equals the rate at which mass is entering or leaving the volume, plus the rate at which mass is produced from chemical reactions with particles.

Equation 2 can be cast in differential form. First we define a function, $S(\bar{r})$, which is constant on the surface of the control volume, and the gradient of S is normal to the control surface. We also define an average over the control volume:

$$\langle A \rangle = \frac{\int_V A\theta d^3X}{\int_V d^3X} \tag{5}$$

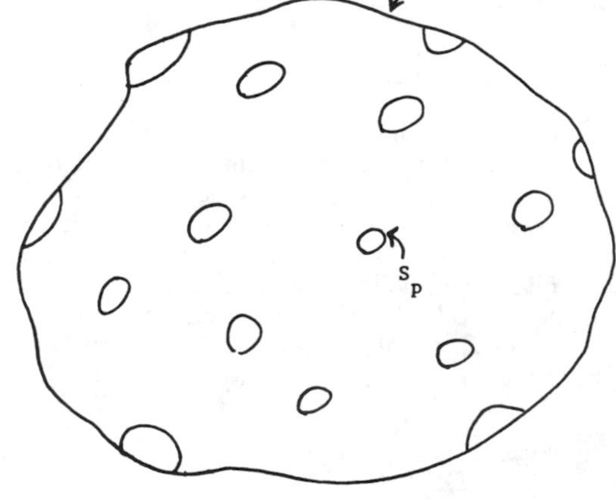

Figure 1. Control volume bounded by surface S. S_p is the surface bounding the particles.

Then Equation 4 becomes

$$\frac{d}{dt} <\rho> \Delta V + \frac{\partial}{\partial S} <\rho v_g \cdot \nabla S> \Delta V = r_M \Delta V \tag{6}$$

where

$$\Delta V = \int_V d^3X$$

By dividing through Equation 6 by ΔV, we arrive at the continuity equation in differential form:

$$\frac{d}{dt} <\rho> + \frac{\partial}{\partial S} <\rho v_g \cdot \nabla S> = r_M \tag{7}$$

Gas-Phase Momentum

In this section, we derive an expression for the rate of change of momentum of a gas inside a control volume containing gas and particles. We follow the procedure of the previous section, in which we presented an expression for the conservation of gas mass inside the control volume.

If $\overline{F}_v$ is the force on the gas per unit volume, the differential equation relating the momentum density at a point to $\overline{F}_v$ is

$$\frac{\partial \rho \overline{v}_g}{\partial t} + \nabla \cdot \rho \overline{v}_g \cdot \overline{v}_g = \overline{F}_v \tag{8}$$

The rate of change of momentum of the gas inside a control volume is given by

$$\frac{d}{dt}\int_{V_g} \rho \overline{v}_g d^3X = \int_{V_g} \frac{\partial}{\partial t} \rho \overline{v}_g d^3X - \int_{S_p} \rho_p (\overline{v}_p + \overline{w})\, \overline{w} \cdot \overline{n}_p d^2X \tag{9}$$

where $\overline{v}_p$ is the particle velocity and $\overline{w}$ is the regression velocity of the particle surface. Combining Equations 8 and 9, we obtain

$$\overline{F} = \int_V \overline{F}_v d^3X = \frac{d}{dt}\int_V \theta \rho \overline{v}_g d^3X + \int_S \theta \rho \overline{v}_g \overline{v}_g \cdot \overline{n}_g d^2X + \int_{S_p} \rho_p (\overline{v}_p + \overline{w})\, \overline{w} \cdot \overline{n}_p d^2X \tag{10}$$

where we have used Gauss's theorem to transform the second term on the right-hand side of Equation 10 to a surface integral. $\overline{F}$ is the total force acting on the fluid element, and the right-hand side of Equation 10 represents the change in momentum of the element. The last term on the right-hand side of Equation 10 can be expressed as

$$\int_{S_p} \rho_p (\overline{v}_p + \overline{w})\, \overline{w} \cdot \overline{n}_p\, d^2X = -\int_V r_M \overline{v}_p d^3X + \int_{S_p} \rho_p \overline{w}\overline{w} \cdot \overline{n}_p\, d^2X \tag{11}$$

The force on the fluid element includes the sum of the pressure, shear stress and gravity:

$$\overline{F} = \int_V - \theta \nabla P d^3X + \int_V \nabla \cdot \overline{\tau} \theta d^3X + \int_V \theta \rho_g \overline{g} d^3X \tag{12}$$

In addition, there are forces on the gas arising from interaction with the particles. These include pressure forces at the particle surfaces and drag forces between gas and particles:

$$\overline{F} = -\int_{S_p} (P - P_S)\, \overline{n}_p d^2X + \int_V \overline{f}_p d^3X \tag{13}$$

where $\overline{f}_p$ is the drag force per unit volume of the mixture, and P_S is the pressure at the particle surface. Assuming a thin boundary layer at the surface of the particles, the pressure change across the layer is

$$P - P_S = -\rho_p \, ww \tag{14}$$

Combining Equations 10–14, the change in momentum of the fluid element is

$$\frac{d}{dt}\int_V \theta\rho\overline{v}_g d^3X = \int_S \theta\rho\overline{v}_g \overline{v}_g \cdot \overline{n} d^2X = -\int_V \theta\nabla P d^3X + \int_V \theta\nabla \cdot \overline{\tau} d^3X + \int_V \theta\rho\overline{v}_g \overline{g} d^3X + \int \overline{f}_p d^3X + \int_V r_M \overline{v}_p d^3X \tag{15}$$

Using the same procedure that converted the continuity integral equation to differential form, the differential form of the momentum equation is

$$\frac{d}{dt} <\rho\overline{v}_g> + \frac{\partial}{\partial S} <\rho\overline{v}_g \overline{v}_g \cdot \nabla S> = -<\nabla P> - <\nabla \cdot \overline{\tau}> - <\rho\overline{g}> + \overline{f}_p + S_{\overline{v}} \tag{16}$$

where $S_{\overline{v}}$ represents that last integral on the right-hand side of Equation 15 and is the average rate of momentum production resulting from particle reactions per unit volume inside the control volume.

Gas-Phase Energy Equation

The first law of thermodynamics states that the change in energy of a fluid system is equal to the rate at which heat is added to the system, minus the work done by the fluid on the surroundings. The energy is the sum of the internal plus kinetic energy. In equation form, the first law is

$$\frac{dE}{dt} = \dot{Q} - \dot{W} \tag{17}$$

where $\dot{Q}$ is the rate at which heat is added and $\dot{W}$ is the work done on the surroundings. The change in energy can be written as

$$\frac{dE}{dt} = \frac{d}{dt}\int_V \theta\rho e d^3X + \int_S \theta\rho e\overline{v}_g \cdot d^2\overline{X} + \int_{S_p} \rho_p e_S \overline{w} \cdot d^2\overline{X} \tag{18}$$

where e is the total energy per unit mass of the fluid, and e_S is evaluated at the particle surface. The second and third terms on the right-hand side represent the amount of energy flowing into the system from the surroundings and the energy change resulting from particle reactions.

The heat transferred from the system is composed of conduction across the surface, conduction from the particles and radiation:

$$\dot{Q} = \int_S \overline{q}\theta \cdot d^2X - \int_{S_p} \overline{q}_p \cdot d^2\overline{X} + \int_V \theta S_r d^3X \tag{19}$$

where $\overline{q}$ is the heat flux through the surface, $\overline{q}_p$ is the heat flux to the particles, and S_r is the radiation energy density flowing into the volume per unit time.

The work done by the gas includes the work done against pressure and shear force, particle drag forces and gravity:

$$\dot{W} = \int_S \theta(P\bar{v}_g \cdot \bar{\tau} \cdot \bar{v}_g)\, d^2\bar{X} + \int_V \bar{f}_p \cdot \bar{v}_p d^3X - \int_V \theta\rho\bar{v}_g \cdot \bar{g} d^3X + \int_{S_p} P_S(\bar{v}_p + \bar{w}) \cdot d^2\bar{X} \tag{20}$$

Writing the total energy per unit mass as

$$e = i + v_g^2/2 \tag{21}$$

where i is the internal energy per unit mass, and collecting terms from Equations 18–20, gives

$$\frac{d}{dt}\int_V \theta\rho(i + v_g^2/2)d^3X + \int_S \theta\rho(h + v_g^2/2)\bar{v}_g \cdot d^2\bar{X}$$
$$= \int_S \theta[\bar{\tau} \cdot \bar{v}_g + (1-\theta)\, P_S\bar{v}_p] \cdot d^2\bar{X} + \int_V \theta[\rho\bar{v}_g \cdot \bar{g} - \bar{v}_p \cdot \bar{f}_p$$
$$- S_r - S_e]\, d^3X - \int_S \theta\bar{q} \cdot d^2X + \int_V r_M\left(h_S + \frac{v_p^2}{2} + \frac{w^2}{2}\right) d^3X \tag{22}$$

In Equation 22, h is the enthalpy $(i + P/\rho)$, h_S is the enthalpy at the surface of the particles, and S_e is the heat transfer per unit volume from the gas to the particles.

If we multiply the momentum equation, Equation 15, by $\bar{v}_g$ and subtract from the total energy equation, we arrive at an equation for the change in internal energy of the fluid:

$$\frac{d}{dt}\int \theta\rho i_g d^3X + \int \theta\rho h\bar{v}_g \cdot d^2\bar{X} = -\int_S [\theta\bar{q} + (1-\theta)\,\rho\bar{v}_p + \theta\bar{\tau} \cdot \bar{v}_g]d^2X$$
$$+ \int_V [\theta\nabla P \cdot \bar{v}_g + (\bar{v}_g - \bar{v}_g) \cdot \bar{f}_p - \bar{v}_p\bar{v}_g r_M + \theta\bar{v}_g\nabla \cdot \tau]d^3X$$
$$+ \int_V \left[-S_r - q_P + r_M\left(h_S + \frac{v_p^2}{2} + \frac{w^2}{2}\right)\right] d^3X \tag{23}$$

We can express the internal energy, i_g, as

$$\rho i_g = P\frac{C_v}{R} = \frac{P}{\gamma - 1} \tag{24}$$

and h as

$$\rho h = \rho C_p T \tag{25}$$

where C_p and C_v are the specific heats at constant pressure and volume, $\gamma = C_p/C_v$, and R is the gas constant divided by the average molecular weight. The energy equation becomes

$$\frac{d}{dt}\int_V \theta P d^3X + \int \theta\gamma P\bar{v} \cdot d^2X = (\gamma - 1)\int \theta\nabla P \cdot \bar{v}_g d^3X + \int S_E(\gamma - 1)d^3X \tag{26}$$

where the last integral in Equation 26 represents the remaining terms in Equation 23.

In the differential form, the energy equation is

$$\frac{d}{dt} <P> + \frac{\partial}{\partial S} <\gamma P \bar{v} \cdot \nabla S> = (\gamma - 1) <v_g \cdot \nabla P> + (\gamma - 1) S_E \tag{27}$$

The purpose of the FLOW module of the FLAG and FLAME codes is to integrate the above gas mass, energy and momentum conservation equations. In the two codes there are sources of gas mass, energy and momentum from gas phase and particle chemical reactions, radiation, conduction (particle to gas) and particle drag. These sources are calculated in various submodules called by the main program.

Turbulence

In regions of the reactor in which the velocity gradients can become too large to be relaxed by molecular viscous effects (e.g., around jet inlets), the fluid can break into large-scale eddies. These eddies couple their energy to smaller and smaller scales, at which point the instability saturates. This turbulence in the fluid leads to an enhanced diffusion and mixing of momentum and energy of both the gas and the particle, and is thus an important part of the physical process in a reactor. At present, we use a first-order turbulence model in both FLAME and FLAG, which is applicable to nearly axial flow in cylindrical geometry. Turbulent flow in such geometry is dominated strongly by the walls, except in regions near an injector, where the flow is developing, and in regions of rapid change in geometry, where flow separation occurs. A strong swirl component can modify the turbulence structure.

Our model of turbulence is based on the Prandtl mixing hypothesis, in which the effects of turbulence are represented by a diffusion coefficient. The turbulence diffusion coefficient, Γ, is given by

$$\Gamma = 0.15 \, \rho l_m \, \overline{v'^2}^{1/2} \tag{28}$$

where ρ is the gas density, l_m is a prespecified mixing length, and $\overline{v'^2}^{1/2}$ is the rms fluctuating velocity about the mean velocity. The level of fluctuation is dependent on the amount of velocity shear:

$$\overline{v'^2}^{1/2} = 0.4 \, l_m \left[\left(\frac{\partial v_x}{\partial y} \right)^2 + \left(\frac{\partial v_y}{\partial x} \right)^2 \right]^{1/2} \tag{29}$$

Summary of Gas-Phase Equations

Using the foregoing generalized equations, we summarize below the gas mass, energy and momentum conservation equations in their component form as solved in the FLOW module. The differential forms of these conservation equations that are derived above represent the conservation of these quantities inside a control volume. To solve for the spatial distribution of mass, momentum and energy, we subdivide the total volume of the reactor into a number of subvolumes, or computational cells. We then calculate the change in average mass, momentum and energy of the gas in each of these cells by solving the equations in their integral forms.

Axial Momentum Equation:

$$\frac{\partial U}{\partial t} = - \frac{\partial}{\partial x} (\alpha U U) - \frac{1}{r} \frac{\partial}{\partial r} (r \alpha V U) - \epsilon \frac{\partial P}{\partial x} - g \rho$$

$$+\frac{\partial}{\partial x}\left(\Gamma\frac{\partial(\alpha U)}{\partial x}\right)+\frac{1}{r}\frac{\partial}{\partial r}\left(\Gamma r\frac{\partial(\alpha U)}{\partial r}\right)+S_U \tag{30}$$

Radial Momentum Equation:

$$\frac{\partial V}{\partial t}=-\frac{\partial}{\partial x}(\alpha UV)-\frac{1}{r}\frac{\partial}{\partial r}(r\alpha VV)+\frac{\alpha WW}{r}-\epsilon\frac{\partial P}{\partial r}$$

$$+\frac{\partial}{\partial x}\left(\Gamma\frac{\partial(\alpha V)}{\partial x}\right)+\frac{\partial}{\partial r}\left(\frac{\Gamma}{r}\frac{\partial(r\alpha V)}{\partial r}\right)+S_V \tag{31}$$

Swirl Momentum Equation:

$$\frac{\partial W}{\partial t}=-\frac{\partial}{\partial x}(\alpha WW)-\frac{1}{r}\frac{\partial}{\partial r}(r\alpha VW)-\frac{\alpha VW}{r}$$

$$+\frac{\partial}{\partial x}\left(\Gamma\frac{\partial(\alpha W)}{\partial x}\right)+\frac{\partial}{\partial r}\left(\frac{\Gamma}{r}\frac{\partial(r\alpha W)}{\partial r}\right)+S_W \tag{32}$$

Mass Continuity:

$$\frac{\partial\rho\epsilon}{\partial t}=-\frac{\partial U}{\partial x}-\frac{1}{r}\frac{\partial(rV)}{\partial r}+r_M \tag{33}$$

Energy Equation:

$$\epsilon\frac{\partial P}{\partial t}=-\gamma R\left[\frac{\partial}{\partial x}(UT)+\frac{1}{r}\frac{\partial}{\partial r}(rVT)\right]+(\gamma-1)\left\{\alpha U\epsilon\frac{\partial P}{\partial x}+\alpha V\epsilon\frac{\partial P}{\partial r}\right.$$

$$+C_P\left[\frac{\partial}{\partial x}\left(\Gamma\frac{\partial T}{\partial x}\right)+\frac{1}{r}\frac{\partial}{\partial r}\left(r\Gamma\frac{\partial T}{\partial r}\right)\right]-\sum_{i=1}^{N}\frac{\partial(h_iC_i)}{\partial t}+S_r$$

$$\left.+S_c+C_PT\,r_M\right\}+\gamma\frac{P}{R}\frac{\partial R}{\partial t}-\gamma P\frac{\partial\epsilon}{\partial t} \tag{34}$$

Equation of State:

$$P\alpha=RT \tag{35}$$

Species Equation:

$$\frac{\partial C_i}{dt}=-\frac{\partial}{\partial x}(UC_i\alpha)-\frac{1}{r}\frac{\partial}{\partial r}(rVC_i\alpha)+S_i+r_i+\left[\frac{\partial}{\partial x}\left(\Gamma\frac{\partial C_i\alpha}{\partial x}\right)\right.$$

$$\left.+\frac{1}{r}\frac{\partial}{\partial r}\left(\Gamma r\frac{\partial C_i\alpha}{\partial r}\right)\right] \tag{36}$$

In writing these equations, we have made the approximation, for example,

$$\langle\rho\rangle\simeq\epsilon\rho$$

where ϵ is the average void fraction in the cell. For ease of notation, we have also dropped the brackets indicating volume integration and have neglected the shear stress terms in comparison to the turbulence diffusion terms.

In these equations, U, V and W are the volume-averaged momentum densities in the x, r and θ directions, α is the inverse of the volume-averaged density, $<\rho>$, and C_i is the volume-averaged concentration of species i. The S terms are specific source terms to these equations.

The term $\sum_{i=1}^{N} \partial(h_i C_i)/\partial t$ on the right-hand side of Equation 34 represents the change in gas energy from homogeneous chemical reactions. The last term in Equation 34 represents the change in pressure in the gas as particles enter and leave the control volume, changing the volume occupied by gas. In an adiabatic system, this represents the conservation of entropy, $P\epsilon^{\gamma} = \text{constant}$.

Computational Procedure

The basic solution procedure of the governing equations is a step-by-step integration with respect to time and space, using a semiimplicit, second-order, accurate, finite-difference scheme over a computational zone network chosen by the user (Figure 2). That is, the interior of the reactor is subdivided into a number of zones to describe the spatial variation of the fluid quantities. Each zone is treated as a well mixed reactor and has its own value of pressure, temperature, density, velocity and chemical composition. The size of the zone can be chosen by the user according to need; hence, they need not be equal to each other. The larger the number of zones, the more accurate the computation will be, but the cost of computing will be higher. Therefore, fine zoning may be used where greater resolution is desired (e.g., around jet inlets), and coarse zoning used in regions where flow and composition variations are not drastic. Implied in the difference approximation is that spatial variation of fluid quantities on scales smaller than a zone size (e.g., fine scale turbulence) cannot be resolved.

Figure 3 shows a computational zone. The pressure, temperature, density, swirl velocity and species concentration are stored at the cell center. The axial and radial mass fluxes are stored on cell boundaries. This staggered grid formulation gives a more accurate difference scheme and leads to a set of difference equations that conserve mass, momentum and energy. In Appendix A of this chapter, we present the equations in their finite-difference forms.

The set of nonlinear equations to be solved has the form

$$\frac{\partial \Phi}{\partial t} + O(\Phi) = S(\Phi) \tag{37}$$

where Φ is the vector representation of the dependent variables, S is the source vector, and O is the operator representing pressure gradients, convective fluxes and diffusive fluxes. A second-order Crank-Nicholson time marching scheme is used to integrate the equations in time:

$$(\Phi^{n+1} - \Phi^{n}) + \frac{\Delta t}{2}[O(\Phi^{n+1}) + O(\Phi^{n})] = \frac{\Delta t}{2}[S(\Phi^{n+1}) + S(\Phi^{n})] \tag{38}$$

The operators, O, are represented by their centered finite-difference form. Recently, Kansa [6] has presented an implicit technique for solution of equations of this type. We have preferred to solve Equation 36 with a modified ICE technique [7] due to its lower computational cost.

The $\nabla \cdot T\bar{U}$ terms in the energy equations represent the propagation of sound waves in the system. Since this term is orders of magnitude larger than most of the other terms in the set of equations, it must be solved implicitly if a reasonable time step, on the order of fluid flow times and much larger than sound propagation times, is to be used.

Other large terms requiring implicitization are the gas-particle drag force, the radiation loss, the gas-particle heat conduction and the pressure gradient terms in the momentum equation. A semiimplicit iteration scheme is used to solve the remaining terms in Equation 38. The solution procedure is as follows.

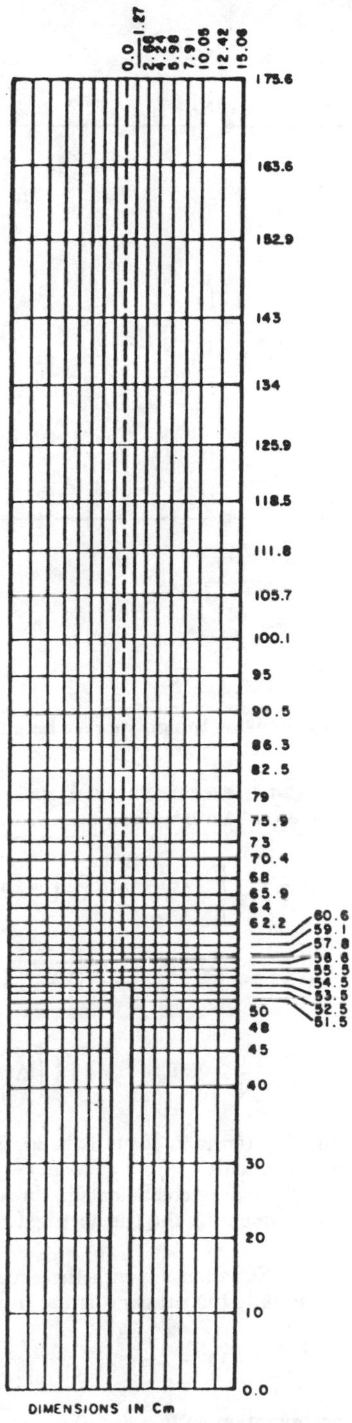

Figure 2. Computational zone network.

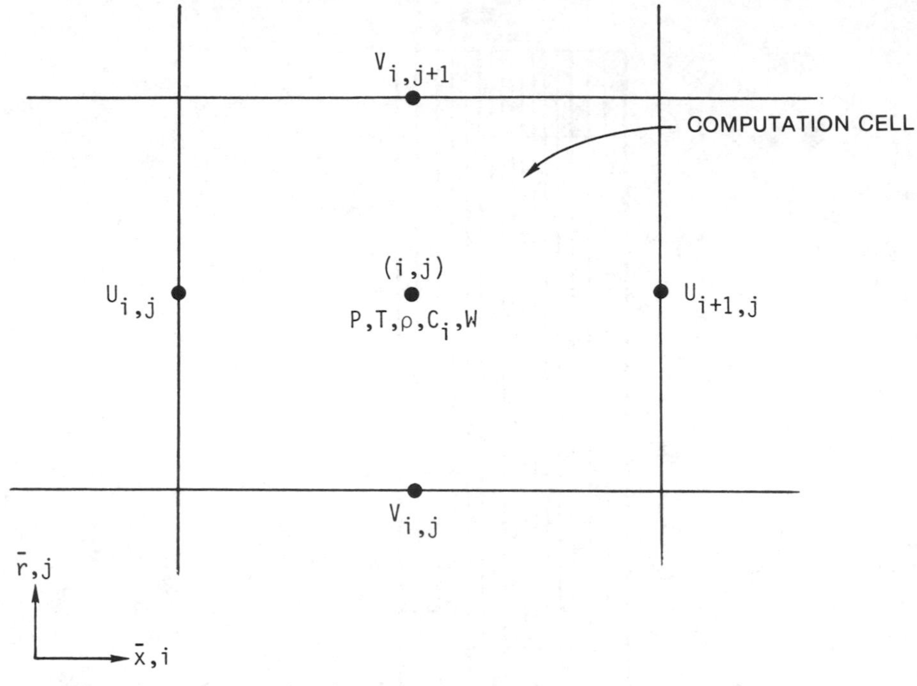

Figure 3. Staggered grid showing the positions where the variables are defined.

At each inner iterate k and time level n the axial and radial momentum Equations 30 and 31, and the energy Equation 34 are written as:

$$U_{k+1}^{n+1} = U_k^* - \left[\epsilon \frac{\partial p_{k+1}^{n+1}}{\partial x} + D_u U_{k+1}^{n+1}\right] \Delta t \tag{39}$$

$$V_{k+1}^{n+1} = V_k^* - \left[\epsilon \frac{2p_{k+1}^{n+1}}{\partial r} + D_v V_{k+1}^{n+1}\right] \Delta t \tag{40}$$

$$\epsilon P_{k+1}^{n+1} + \left[\frac{\partial}{\partial x} \gamma R T_k^{n+1} U_{k+1}^{n+1} + \frac{1}{r}\frac{\partial}{\partial r} r\gamma R T_k^{n+1} V_{k+1}^{n+1}\right] \Delta t = P_k^* - D_p P_{k+1}^{n+1} \Delta t \tag{41}$$

where U^*, V^*, P^* = remaining terms in Equations 30, 31 and 34, which are to be solved explicitly

D_u, D_v = gas-particle drag coefficients in the axial and radial directions

D_pP = term representing the gas heating from conduction and radiation

By replacing U^{n+1} and V^{n+1} in Equation 41 with the expressions given by Equations 39 and 40, an equation involving only the pressure is the result:

$$O\, P_{k+1}^{n+1} = \Psi_k \tag{42}$$

where O = finite difference operation on P

Ψ = remaining explicit terms in the equation

Standard solution techniques can be used to find the solution P of Equation 42. We have used direct [8] and iterative [9] techniques with success.

After the new pressure has been determined, the new axial and radial momenta are computed according to Equations 39 and 40. The change in pressure from the previous iterate is also computed, and if the change arising from the explicit terms in the equations is too large, the procedure is repeated. The iteration is terminated when

$$\frac{|P_{k+1} - P_k|}{|P_k|} < \epsilon \tag{43}$$

where ϵ is a small number.

Usually this requires three to four iterations when the time step is that determined by the fluid CFL condition.

PARTICLE DYNAMICS AND ENERGETICS

Particle Description

In the section entitled Fluid Mechanics, we presented an averaging procedure that describes gas-fluid flow in a mixture of particles. The treatment of the particles in a reacting environment is more difficult because the particles are distributed in size, as well as in chemical conversion. These distributions, as shown in the previous chapter, can affect the gas-fluid dynamics; thus, an averaging over all particles to obtain a single fluid phase can be insufficient for their treatment.

On the other hand, tracking every particle in the reactor is economically out of the question. We compromise and treat the particle phase with multifluids to account for the size and conversion distributions throughout the reactor. Blake et al. [1] have treated the particles as a single fluid and track flow and composition variations.

The gas description we presented was Eulerian, i.e., the gas variables are determined at each point in space as functions of time. Defining the multifluid particle variables at each point in the reactor is costly and unnecessary since there will be certain regions in the reactor containing no particles of a given type. An alternative treatment is termed Lagrangian; this model of the particles circumvents the difficulties of the Eulerian description. From the Lagrangian point of view, a number of particles is represented by a fluid element, which is localized in a region of the reactor and moves through the reactor according to Newton's second law of motion. All particles represented by the fluid element are identical; they have the same size, temperature conversion, etc. The elements do have spatial extent, however, and can interact with each other through viscous drag as they pass through each other. An element also interacts with the gas chemically and dynamically in the region of the reactor through which it moves.

Thus, the particle description we have adopted is a combination fluid dynamic-statistical one in that the particles are treated as a fluid, but there are many localized fluids to account for the size and conversion distribution throughout the reactor. This technique is an extension of the particle-in-cell (PIC) method [10] for describing fluid flow. A particle treatment related to ours is given by Chen and Saxena [11] and Weimer and Clough [12], who used a probability density function to describe the particle size and conversion distributions throughout the reactor.

Particle-Gas Drag Force

As a particle moves in the reactor, its motion is affected by the gas flowing around it, collisions with other particles and chemical reactions at its surface. In this section, we treat the drag force exerted by the particles on the gas and the subsequent slowing down of the gas and acceleration of the particles. In subsequent sections, we describe the other two effects.

Since all the particles in a given fluid element are taken to be identical, they all have the same size and mass and move with the same mean velocity. (There is no dispersion in velocity about this mean velocity in the model at present.) Thus, the motion of the fluid element is the same as that of any particle in the element, and we can track the trajectory of the element by tracking a single particle in the element. This particle thus becomes a representative of all the particles in the fluid element.

The drag force, which the gas and particles exert on each other as they flow through one another, is a complicated function of the relative gas-particle velocity, gas density and viscosity, particle size and the fraction of volume occupied by the particles within the gas. Fortunately, experimental data exist for the various parameter regimes of interest, which allow an empirical determination of the drag force.

If a gas of density ρ_g, occupying a fraction, ϵ, of the local volume with the particles, flows past a group of particles of diameter, d, with relative velocity, u, it exerts a drag force, F_d, on the particles. We define a drag coefficient, C_D, as

$$C_D = \frac{F_d}{\pi/8 \, \rho_g \, \epsilon^2 u^2 d^2} \tag{44}$$

If μ is the kinematic viscosity of the gas, a Reynolds number for the particle is given by

$$R_e = \frac{d \, \epsilon \, u \, \rho_g}{\mu} \tag{45}$$

Figure 4 is taken from Zenz [13], who has collected data from experiments with various values of voidage, particle size and Reynolds number. Zenz compares $(R_e/C_D)^{1/3}$ versus $(C_D/R_e^2)^{1/3}$ for various void fractions. We note that for the $\epsilon = 1$ curve with Reynolds number less than unity, Stokes' law is recovered, i.e., $F_d = 3\pi/2 \, \mu du$.

From these curves we can obtain the local drag forces acting on the particles. First, we recast the curves and plot C_D versus R_e for various void fractions. These curves are shown in Figure 5. We have derived analytical fits to each curve representing a given void fraction.

The actual value of the drag force is determined in the following manner. The gas velocity, temperature and density are calculated in the fluid dynamics section of the code. A local gas viscosity is calculated, and a Reynolds number is determined based on the particle size, the local void fraction and the relative gas-particle velocity as given in Equation 45. From this number, the drag coefficient is obtained from the fit of Figure 5. Finally, the drag force is calculated from Equation 44.

Once the drag force is determined, the position and velocity of the fluid element are updated by integrating the equation of motion of the representative particle for the time step:

$$m_p \frac{d\bar{v}_p}{dt} = m_p \bar{g} + D_p (\bar{v}_g - \bar{v}_p) \tag{46}$$

$$\frac{d\bar{X}_p}{dt} = \bar{v}_p \tag{47}$$

where $\bar{X}_p$ is the position of the particle, $\bar{v}_p$ its velocity, m_p its mass, $\bar{v}_g$ the gas velocity, and D_p the drag slowing-down frequency. The second term on the right-hand side of Equation 46 represents the drag force. This procedure is repeated for each fluid element.

Once the new positions and velocities of the particles have been updated, terms such as void fractions, drag forces and heating of the gas, which are required for the next fluid dynamic time step, can be calculated. The particle fluid element is taken to be localized within a computational zone. If the element contains N_e particles of diameter d_e, the void fraction at cell i,j, with volume $\Delta V_{i,j}$, is

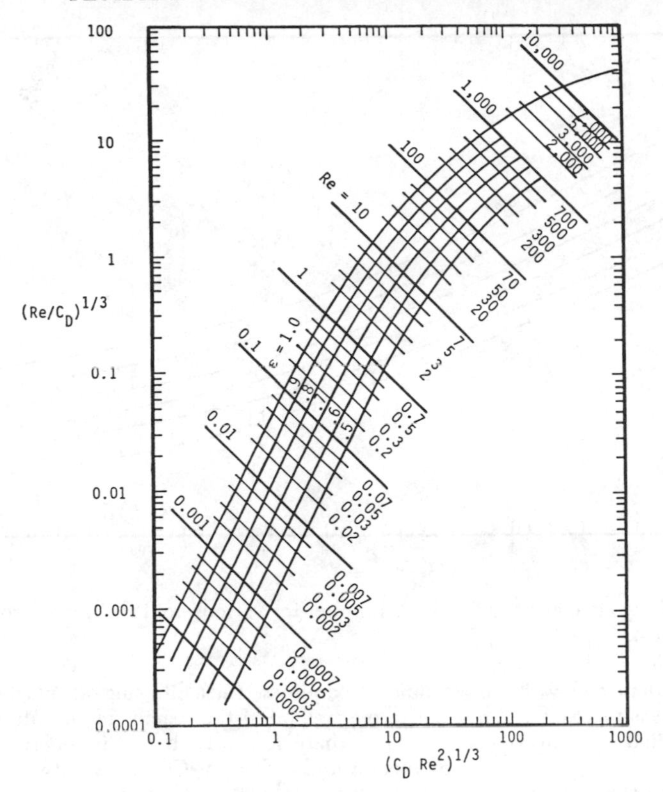

Figure 4. Curves given by Zenz [12] from which one can calculate the particle drag force for various Reynolds numbers and void fractions.

$$\epsilon_{i,j} = 1 - \frac{\pi}{6\,\Delta V_{i,j}} \sum_e N_e\, d_e^3 \tag{48}$$

where the sum is over the elements in the cell.

The drag force per unit volume on the gas by the particle is obtained once again by summing over all the elements in a cell:

$$\overline{F}_{i,j} = -\frac{1}{\Delta V_{i,j}} \sum_e N_e\, D_e\, (\overline{v}_g - \overline{v}_e) \tag{49}$$

where $(\overline{v}_g - \overline{v}_e)$ is the relative gas-element velocity.

For energy to be conserved in the gas-particle drag interaction, a term representing an increase in thermal energy of the gas must be added to the gas energy conservation equation. This frictional heating term is given by

$$H_{i,j} = \frac{1}{\Delta V_{i,j}} \sum_e D_e\, (\overline{v}_g - \overline{v}_e) \cdot (\overline{v}_g - \overline{v}_e) \tag{50}$$

Interparticle Collisions

In a heavily loaded fluid particle system, such as a fluidized bed, particles colliding is the dominant mechanism for heat and mass mixing throughout the bed. Particles

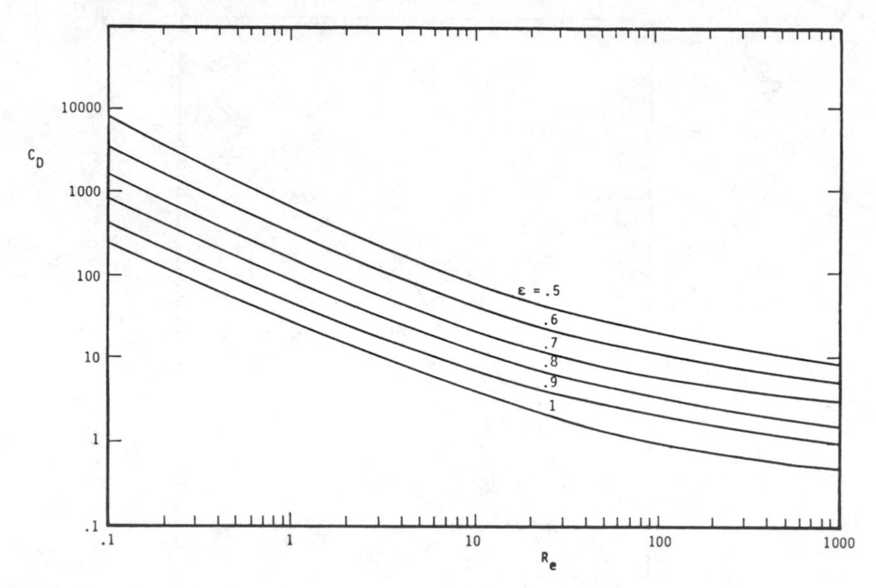

Figure 5. Drag coefficient vs Reynolds number for various void fractions (cross plotted from Figure 4).

collide at random and walk across fluid streamlines, thus diffusing about the system. A particle that is reacting and heating in one region of the reactor might, after a series of collisions, find itself at a cooler region, where it can exchange its energy with the surrounding gas and particles by conduction and radiation. Collisions between particles also provide resistance to compression and shear, which manifest themselves as effective shear and bulk viscosities of the bed.

As noted earlier, our description of the particles was a hybrid statistical-fluid one, in which a number of fluid elements represent the total particle population. With enough fluid elements, the distribution of particle sizes and conversions can be represented adequately with this model. Since the collision process is purely random, a natural extension of the particle model is to include collisions between fluid elements giving rise to a random force on the elements, in addition to the other forces acting on them. It is necessary to find the average collision rate between all the particles in two interpenetrating fluid elements and the angle of scatter after the collision. This procedure is described below.

Consider two clouds of particles moving through each other, as shown in Figure 6, one of which is assumed to be composed of particles A and the other of particles B. Their centers of gravity are denoted by O_1 and O_2. We assume that each cloud of particles moves monoenergetically, with no dispersion of its velocity.

There can be a number of collisions between the two classes of particles, and the velocities after the collisions are in a number of directions, since any one collision

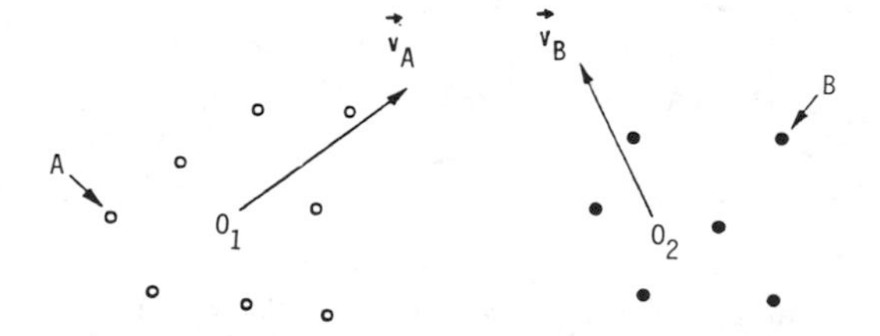

Figure 6. Collisons between two monoenergetic clouds of particles.

can range from head-on to tangential. We must perform an average over all these collisions and scatter the particles appropriately. To do this, we continue the description of a single particle representing a cloud of particles that we developed earlier.

If we represent the particles in cloud A by a single particle, which moves through cloud B, particles A can collide at random with any of the particles in cloud B. The probability that A will suffer a collision before moving a distance, S, through B is

$$P(s) = \int_0^S \frac{dS'}{\lambda(S')} \exp\left[- \int_0^{S'} dS''/\lambda(S'') \right] \tag{51}$$

where λ is the mean free path between collisions,

$$\lambda^{-1} = n_B \, \pi (r_A + r_B)^2 \tag{52}$$

n_B is the number density of cloud B, and r_A and r_B are the radii of particles A and B. So particle A (and most of the particles represented by A) probably will suffer a collision before traveling a few mean free paths through B. In a fluidized bed, λ is roughly a particle diameter. By generating a set of random numbers distributed according to Equation 51, we determine that particle A has suffered a collision a distance S through cloud B, where S is the solution to the equation

$$\chi = \int_0^S dS'/\lambda \tag{53}$$

and χ is from the set of random numbers.

It can be shown that the random numbers, χ, distributed according to Equation 51, can be generated from the random number distribution, ξ, uniform in the range $0 < \xi < 1$, by solving

$$\chi = -ln\xi \tag{54}$$

Computational Procedure

At each time step, we calculate the distance ΔS, which each fluid element has traversed during the step. The total mean free path at the time step for each element is also determined by summing Equation 52 for all the elements that overlap it. Thus, if elements A and B are interpenetrating, with velocities $\overline{v}_A$ and $\overline{v}_B$, the cross section that particles A see in colliding with particles B is

$$\sigma_{A,B} = \pi (r_A + r_B)^2 \frac{|\overline{v}_A - \overline{v}_B|}{|\overline{v}_A|} \tag{55}$$

where r_A and r_B are the radii of particles A and B. The mean free path for A colliding with B is

$$\lambda_{A,B}^{-1} = n_B \, \sigma_{A,B} \tag{56}$$

The total mean free path for element A is obtained by summing over all overlapping elements:

$$\lambda_A^{-1} = \sum_B \lambda_{A,B}^{-1} \tag{57}$$

Once ΔS and λ^{-1} are known, the integral in Equation 53 is incremented and a test is made to determine whether

$$\sum_{i} \Delta S_i/\lambda_i \ldots \geqq \chi \tag{58}$$

where the summation is over time steps and χ has been specified previously according to Equation 54.

If the inequality in Equation 58 is satisfied for an element, the element is said to suffer a collision. Its velocity and that of its colliding partner are then changed, based on the laws of conservation of mass and energy. If there is more than one element overlapping A, a choice is made (based on the sizes of the relative mean free paths) as to which of the other elements collides with A.

The collision process need not be elastic, in which case some of the kinetic energy of the particles is lost during the collision and is transformed into internal energy. The amount of kinetic energy conversion is characterized by a restitution coefficient for the particles, which is the fraction of energy absorbed. This coefficient is a function of particle temperature and composition, and can be obtained from the literature for specific types of particles.

Particle Energetics

In a two-phase reactive system, the particles undergo chemical reactions with and exchange heat with the surrounding gas by conduction and, also, radiation. These energy exchange processes serve to equilibrate the heat throughout the system. This exchange of energy will affect the fluid dynamics and must be included for a complete representation of reactive two-phase flow.

The internal energy of a particle can change as a result of chemical reactions with gases at the particle's surface, convective heat flow across the particle-gas boundary layer, and energy exchange with the radiation field surrounding the particle. Before writing an equation representing the change in particle internal energy, we list a set of assumptions made in deriving it:

1. The temperature is uniform throughout the particle volume. That is, the heat transfer within the particle's surface is much faster than between the particle and the gas.
2. All reactions within the particles and at the particle-gas boundary layer are assumed to take place at temperature T_p.
3. The heat capacity of the particle-gas boundary layer is negligible compared to the particle heat capacity.
4. Heat conduction across the particle-gas boundary layer has the form

   $$k\, N_u\, \pi\, d(T_g - T_p)$$

   where k is the heat transfer coefficient of the gas, N_u the Nusselt number, d the particle diameter, T_g the gas temperature, and T_p the particle temperature.
5. Radiation energy loss across the particle-gas boundary layer has the form

   $$\frac{\pi}{4} d^2 \epsilon_p (U - 4\sigma T_p{}^4)$$

   where ϵ_p is the emissivity of the particle, U the energy flux of the radiation field across the particle's surface, and σ Stefan-Boltzman's constant. This implies that the particle is a gray body in radiative equilibrium at temperature T_p with the radiation field. The radiation field is taken to be isotropic.

Using these assumptions, we write the particle internal energy equation as

$$\frac{\partial}{\partial t} m_p C_p T_p = -\sum_g r_g(T_p)\, H_g(T_g) + k\, N_u\, \pi\, d(T_g - T_p)$$
$$+ \frac{\pi}{4} d^2\, \epsilon_p [U(\vec{r}) - 4\sigma\, T_p^4] \tag{59}$$

where C_p is the specific heat of the particle, m_p the particle mass, r_g the rate of production of gas component g at and in the particle, and H_g is the specific enthalpy of gas component g. The first term on the right-hand side of Equation 59 represents the heat released to the particle in the gas-particle chemical reaction process.

Since the change in mass of the particle during the reaction process is known, Equation 59 can be transformed into an equation representing the change in the temperature of the particle:

$$\frac{\partial T_p}{\partial t} = \frac{1}{m_p C_p} \Big\{ -\sum_g r_g(T_p)\, H_g - r_p\, H_p + k\, N_u d(T_g - T_p)$$
$$+ \frac{\pi}{4} d^2\, \epsilon_p\, [U(\vec{r}) - 4\sigma\, T_p^4] \Big\} \tag{60}$$

Equation 60 describes the change in temperature of a single particle. As we have shown, each particle that we follow computationally actually represents many thousands of real particles in a single fluid element. To describe the coupling between gas and particles, the chemical reaction term and the heat conduction term must be summed over all particles in the fluid element. The sums for all elements in a computational cell are added together, and the resulting term, divided by the volume of the cell in which the element resides, is the heating (or cooling) rate per unit volume of the gas resulting from gas-particle reactions. This heating rate is employed in the fluid dynamics section of the codes to determine the change in gas temperature arising from gas-particle interaction.

MODELING OF CHEMICAL REACTIONS AND SPECIES TRANSPORT IN GAS-PARTICULATE SYSTEMS

A reactor containing gases and solids, such as a fluidized bed, usually operates at elevated temperatures and often at elevated pressures, where chemical reactions occur both in the gas phase and in the solid particles. The specific rates, or kinetics, of these reactions are a significant factor in determining reactor behavior. In addition, gas-solid interphase transport of mass, momentum and energy is coupled with the intrinsic reaction kinetics. The exact nature and mathematical representation of these rate phenomena will depend on the specific reaction system under consideration. In this section, we discuss JAYCOR's approach to modeling the kinetics and transport for the gasification of coal in such reactors as entrained-flow and fluidized-bed systems. The following discussions center around the reaction model and the computational procedures for determining the reaction rates. The latter part of this section discusses the heating from chemical reactions.

Chemical Reaction Model

Consider the mass balance of the ith gaseous species in the reactor:

$$\frac{\partial C_i}{\partial t} = \nabla \cdot \Gamma \nabla C_i - \nabla \cdot \bar{v} C_i + S_i \tag{61}$$

where C_i is the concentration of the ith species (kg-mol/m³), Γ is the turbulence diffusion coefficient (m²/sec), $\bar{v}$ is the fluid velocity (m/sec), and S_i is the source term due to the chemical reactions (kg-mol/m^3-sec). In the numerical procedure, the reactor is divided into computational zones, and each zone is assumed to be well mixed with respect to species concentration. Equation 61 applies for each zone, and the diffusive and convective fluxes are evaluated at the boundaries of each cell. Then S_i is the net rate of production of ith species per unit void volume of the cell (kg-mol/m^3-sec). In this section we are concerned with the reaction model that will determine the term S_i. Coal devolatilization, subsequent char combustion and gasification and the gas-phase reactions constitute this term.

Reaction Rates

When coal is introduced into the reactor, devolatilization takes place almost immediately. This step is extremely fast compared to the slower char reactions. Although there are various reaction mechanisms and rate expressions describing this devolatilization process, we do not consider this reaction in such detail for the present application. Here we will assume that the devolatilization reactions can be represented as coal → char + gaseous products and take place instantaneously.

Coal is assumed to be made up of $C_\alpha H_\beta O_\gamma N_\delta A$, where A represents ash. The solid product char is assumed to be base carbon, C, embedded in an ash matrix, A. So the hydrogen, oxygen and nitrogen present in coal are driven out as gaseous products, which will be assumed to be H_2O, CO, CO_2, N_2 and CH_4. Hence, the heat and mass effects due to this reaction are added at the point (or cell) at which coal is introduced.

With devolatilization taking place instantaneously, the reactor contains only char/ash particles. For this study, we will assume that the gas phase (ambient) contains only O_2, CO, CO_2, H_2, H_2O, N_2 and CH_4. Under these conditions, the following reactions will be assumed to take place:

R1. $C + 1/2\,O_2 \rightarrow CO$
R2. $C + H_2O \rightarrow CO + H_2$
R3. $C + CO_2 \rightarrow 2\,CO$
} Heterogeneous

R4. $CO + 1/2\,O_2 \rightarrow CO_2$
R5. $H_2 + 1/2\,O_2 \rightarrow H_2O$
R6. $CO + H_2O \rightleftarrows CO_2 + H_2$
R7. $CH_4 + 2\,O_2 \rightarrow CO_2 + 2\,H_2O$
} Homogeneous

In this study, hydrogasification ($C + 2\,H_2 \rightarrow CH_4$) is neglected because it is negligibly slow compared to other char reactions.

As temperatures of interest, hydrogen oxidation (R5) is extremely fast and can be assumed to go to completion instantaneously. Hence, H_2 and O_2 cannot coexist. Also, at high temperatures the water-gas shift reaction (R6) is essentially at equilibrium, so if H_2 cannot coexist with O_2, CO cannot coexist with O_2. That is, by combining reactions R5 and R6, reaction R4 proceeds instantaneously. Combustion of methane (R7) also is assumed to be instantaneous.

The numerical procedure deals with computational zones, which are assumed to be chemically well mixed. At each instant of time, appropriate sets of equations will be used, depending on whether the zone contains oxygen. If oxygen is present in a zone, it is consumed first. Then the other reactions proceed.

We have used the following set of reaction rates to describe the seven chemical reactions:

R1. $C + 1/2\,O_2 \rightarrow CO$ [14] (62)
$r_1 = k_1\,C_{O_2}$ [kg-mol-C/m^2(external area)-sec]
$k_1 = 3.007 \times 10^5 \exp(-17966/T)$ (m/sec)

The steam gasification and CO_2 gasification rate constants are adjusted to fit the reaction rate data of Western Kentucky char:

R2. $C + H_2O \rightarrow CO + H_2$ (63)
$r_2 = k_2 C_{H_2O}$ [kg-mol H_2O/m^2(external area)-sec]
$k_2 = 95.5 \exp(-17594.4/T)$ (m/sec)

R3. $C + CO_2 \rightarrow 2\,CO$ (64)
$r_3 = k_3 C_{CO_2}$ [kg-mol CO_2/m^2(external area)-sec]
$k_3 = 6.25 \times 10^5 \exp(-29491/T)$ (m/sec)

R6. Water-gas shift equilibrium: $CO + H_2O \rightleftarrows CO_2 + H_2$ [15] (65)
$\log_{10}K = -1.6945 + 1855.6/T$ (T in °K)

Particle Model

A char particle is assumed to be spherical and surrounded by a gas film characterized by mass and heat transfer coefficients. All heterogeneous reactions are lumped at the external surface of the particle. For simplicity, the dependence of reactions rates on the amount of carbon present in the char particle is accounted for by a shrinking-core approximation given by the function $\phi = (1 - X)^{2/3}$, where X is the carbon conversion of the particle defined as the ratio of the amount of carbon gasified to the original amount of carbon. So the mass balance of the ith species at the core surface is given by the steady-state approximation

$$-k_g(C_i - C_{i,s}) = \sum_j \alpha_{i,j}\, r_j\, \phi \tag{66}$$

where r_j = specific rate of reaction j (kg-mol/m²-sec)
$\alpha_{i,j}$ = stoichiometric coefficient of ith species, jth reaction
k_g = mass transfer coefficient (m/sec)
$C_{i,s}$ = concentration of ith species at particle surface (kg-mol/m³)
C_i = concentration of ith species in the bulk (kg-mol/m³)

The mass transfer coefficient is, from the Sherwood number,

$$Sh = \frac{2k_g R_C}{\mathscr{D}} = 2 + 0.654\, Re^{1/2}\, S_c^{1/3} \tag{67}$$

where $\mathscr{D}$ is the molecular diffusion coefficient, Re the Reynolds number and S_c the Schmidt number.

The rate of change of particle mass is given by

$$\frac{dM}{dt} = -12 \times 4\pi R_c^2 \sum_j r_j\phi \tag{68}$$

and the rate of conversion by

$$\frac{dX}{dt} = \frac{36}{\rho_O W_b R_c} \sum_j r_j\phi \tag{69}$$

Computational Procedure

Case 1: Concentration of $O_2 > 0$ in a Computational Zone

Since H_2 and CO cannot coexist with oxygen, the gas phase in the combustion zone contains only O_2, CO_2, H_2O, N_2 and CH_4. The reactions that take place are

R1. $C + 1/2\,O_2 \rightarrow CO$
R2. $C + H_2O \rightarrow CO + H_2$

R3.	$C + CO_2 \rightarrow 2\,CO$	
R4.	$CO + 1/2\,O_2 \rightarrow CO_2$	(instantaneous)
R5.	$H_2 + 1/2\,O_2 \rightarrow H_2O$	(instantaneous)
R6.	$CO + H_2O \rightleftarrows CO_2 + H_2$	(equilibrium)
R7.	$CH_4 + 2\,O_2 \rightarrow CO_2 + 2\,H_2O$	(instantaneous)

Particle Equations

Since R4 occurs instantaneously, combining R1 and R4 yields

$$CO + O_2 \rightarrow CO_2 \text{ at } r_1 \tag{70a}$$

Similarly, combining R2, R4 and R5 yields

$$C + O_2 \rightarrow CO_2 \text{ at } r_2 \tag{70b}$$

and combining R3 and R4 yields

$$C + O_2 \rightarrow CO_2 \text{ at } r_3 \tag{70c}$$

Hence, the net reaction at the particle in the combustion zone is

$$C + O_2 \rightarrow CO_2 \text{ at rate } (r_1 + r_2 + r_3) \tag{70d}$$

So we need to write mass balances only for O_2 and CO_2 at the particle surface:

$$O_2\text{:}\ k_g(C_{O_2} - C_{O_2,s}) = (r_1 + r_2 + r_3)\phi \tag{71a}$$

$$CO_2\text{:}\ k_g(C_{CO_2} - C_{CO_2,s}) = -(r_1 + r_2 + r_3)\phi \tag{71b}$$

i.e.,

$$k_g(C_{O_2} - C_{O_2,s}) = (k_1 C_{O_2,s} + k_2 C_{H_2O,s} + k_3 C_{CO_2,s})\phi \tag{72a}$$

and

$$k_g(C_{CO_2} - C_{CO_2,s}) = -(k_1 C_{O_2,s} + k_2 C_{H_2O,s} + k_3 C_{CO_2,s})\phi \tag{72b}$$

and since there is no net production or consumption of H_2O,

$$H_2O\text{:}\ C_{H_2O,2} = C_{H_2O} \tag{73}$$

By linear combinations of 72a and 72b, we have

$$C_{CO_2,s} = C_{O_2} + C_{CO_2} - C_{O_2,s} \tag{74}$$

Therefore,

$$C_{O_2,s} = \frac{k_g C_{O_2} - \phi k_2 C_{H_2O} + \phi k_1 (C_{O_2} + C_{CO_2})}{(k_g + \phi k_1 - \phi k_3)} \tag{75}$$

Thus, the rate of production of O_2 and CO_2 for each particle is

$$O_2: \; -(k_1C_{O_2,s} + k_2C_{H_2O} + k_3C_{CO_2,s})\phi\, 4\pi R_c^2 \qquad \text{(kg-mol/sec)} \tag{76a}$$

$$CO_2: \; +(k_1C_{O_2,s} + k_2C_{H_2O} + k_3C_{CO_2,s})\phi\, r\pi R_c^2 \qquad \text{(kg-mol/sec)} \tag{76b}$$

and the rate of change of conversion is

$$\frac{dX}{dt} = \frac{36}{\rho_O W_b R_c}(1-X)^{2/3}(k_1C_{O_2,s} + k_2C_{H_2O} + k_3C_{CO_2,s}) \tag{77}$$

Although H_2 and CO do not exist in a combustion zone, there could be an influx of H_2 and CO into the combustion zone by diffusion and convection. Incoming H_2 and CO will react, according to R5 and R6, instantaneously; hence, these effects should be accounted for. The rate at which R4 takes place is N_{CO}, where N_{CO} is the inflow rate of CO per unit cell void volume (kg-mol/m³-sec). Similarly, R5 and R7 take place at N_{H_2} and N_{CH_4}, the inflow rates of hydrogen and methane per unit cell void volume (kg-mol/m³-sec). Hence, the net rate of production of species due to homogeneous reactions is

$$\begin{aligned} &H_2O: \; N_{H_2} + 2N_{CH_4} && \text{(kg-mol/m}^3\text{-sec)} \\ &CO_2: \; N_{CO} + N_{CH_4} && \text{(kg-mol/m}^3\text{-sec)} \\ &CH_4: \; N_{CH_4} && \text{(kg-mol/m}^3\text{-sec)} \\ &O_2: \; -1/2\,(N_{H_2} + N_{CO} + 4N_{CH_4}) && \text{(kg-mol/m}^3\text{-sec)} \end{aligned} \tag{78}$$

Therefore, S_i's for the combustion zone are given by

$$S_{O_2} = \sum_{k=1}^{NP_1} -[(k_1C_{O_2,s} + k_2C_{H_2O} + k_3C_{CO_2,s})\phi\, r\pi R_c^2]_k - \frac{1}{2}(N_{H_2} + N_{CO} + 4N_{CH_4}) \tag{79a}$$

$$S_{H_2O} = N_{H_2} + 2N_{CH_4} \tag{79b}$$

$$S_{CO_2} = \sum_{k=1}^{NP_1} + [(k_1C_{O_2,s} + k_2C_{H_2O} + k_3C_{CO_2,s})\phi\, 4\pi R_c^2]_k + N_{CO} + N_{CH_4} \tag{79c}$$

$$S_{N_2} = 0 \tag{79d}$$

$$S_{CH_4} = 0 \tag{79e}$$

Note that if all the oxygen within the cell is consumed within a time step, the cell becomes a gasification zone in the next time step.

Case 2: Concentration of $O_2 \equiv 0$ in a Computational Zone

With the gasification zone, the gas phase contains only CO, CO_2, H_2, H_2O, N_2 and CH_4, and no O_2. The reactions that are taking place are

R2. $C + H_2O \rightarrow CO + H_2$
R3. $C + CO_2 \rightarrow 2\,CO$
R4. $CO + H_2O \rightleftarrows CO_2 + H_2$ (equilibrium)

If there is an inflow of O_2 into the cell, reactions R4, R5 and R7 take place instantaneously.

Particle Equations

$$H_2O: \quad -k_g(C_{H_2O} - C_{H_2O,s}) = -\phi k_2 C_{H_2O,s} \tag{80a}$$

$$CO_2: \quad -k_g(C_{CO_2} - C_{CO_2,s}) = -\phi k_3 C_{CO_2,s} \tag{80b}$$

$$CO: \quad -k_g(C_{CO} - C_{CO,s}) = -\phi(2k_3 C_{CO,s} + k_2 C_{H_2O,s}) \tag{80c}$$

$$H_2: \quad -k_g(C_{H_2} - C_{H_2,s}) = -\phi k_2 C_{H_2O,s} \tag{80d}$$

Therefore,

$$C_{H_2O,s} = \frac{k_h\, C_{H_2O}}{(k_g + \phi k_2)} \tag{81}$$

$$C_{CO_2,s} = \frac{k_g\, C_{CO_2}}{(k_g + \phi k_3)} \tag{82}$$

Particle Conversion

$$\frac{dX}{dt} = \frac{36}{\rho_O W_b R_c}(1 - X)^{2/3}\,(k_2 C_{H_2O,s} + k_3 C_{CO_2,s}) \tag{83}$$

The rate of production of ith species due to reaction of each particle is

$$H_2O: \quad -\phi k_2 C_{H_2,s}\, 4\pi R_c^2 \qquad \text{(kg-mol/sec)} \tag{84a}$$

$$CO_2: \quad -\phi k_3 C_{CO_2,s}\, 4\pi R_c^2 \qquad \text{(kg-mol/sec)} \tag{84b}$$

$$CO: \quad \phi(k_2 C_{H_2O,s} + 2k_3 C_{CO_2,s})\, 4\pi R_c^2 \qquad \text{(kg-mol/sec)} \tag{84c}$$

$$H_2: \quad \phi k_2 C_{H_2,s}\, 4\pi R_c^2 \qquad \text{(kg-mol/sec)} \tag{84d}$$

O_2 coming into the gasification zone reacts instantaneously with H_2 present in the cell at a rate determined by the oxygen inflow rate, N_{O_2} (kg-mol/sec-m^3). At the same time, the water-gas shift reaction occurs. The net rate of production of the ith species due to homogeneous reaction is given by

$$H_2O: \quad 2N_{O_2} - r_6 \tag{85a}$$

$$H_2: \quad -2N_{O_2} + r_6 \tag{85b}$$

$$CO_2: \quad r_6 \tag{85c}$$

$$CO: \quad -r_6 \tag{85d}$$

Therefore, the source term S_i is given by

$$H_2O: \quad S_{H_2O} = \sum_{k-1}^{NP_i} -(\phi 4\pi R_c^2\, k_2 C_{H_2O,s})_k + 2N_{O_2} - r_6 \tag{86a}$$

$$CO_2: \quad S_{CO_2} = \sum_{k=1}^{NP_i} -(\phi 4\pi R_c^2\, k_3 C_{CO_2,s})_k + r_6 \tag{86b}$$

$$CO:\ S_{CO} = \sum_{k=1}^{NP_1} [\phi 4\pi R_c^2 (k_2 C_{H_2O,s} + 2k_3 C_{CO_2,s})]_k - r_6 \tag{86c}$$

$$H_2:\ S_{H_2} = \sum_{k=1}^{NP_1} (\phi 4\pi R_c^2 C_{H_2O,s})_k - 2N_{O_2} + r_6 \tag{86d}$$

Since R6, the water-gas shift reaction, is in equilibrium, the rate term r_6 must be eliminated in the species mass balance (Equation 61) after S_i has been substituted for by the above relations. And the loss of an equation is to be replaced by the equilibrium condition

$$\frac{C_{CO_2}\, C_{H_2}}{C_{H_2}\, C_{CO}} = K \tag{87}$$

where

$$\log_{10} K = -\,1.6945 + 1855.6/T_g$$

and T_g is the gas temperature.

Heating Produced by the Chemical Reactions

A certain amount of heat is released (or absorbed) as a result of the heterogeneous and gas-phase chemical reactions. In the section entitled Particle Dynamics and Energetics, we discussed particle heating as a result of heterogeneous reactions. Our model applies all the heat released to the particle, and this energy is coupled back to the gas through conduction and particle radiation. The energy supplied to the gas resulting from homogeneous reactions is the change in species concentration times the enthalpy before and after the reaction. The total amount of gas heating per unit volume of the gas due to homogeneous reactions is

$$H = \sum_i \Delta\, (C_i H_i)$$

where the sum is over all species and H_i is the enthalpy per kmol, including heat of formation of species i. This heating is calculated for each computational cell at each time step and is used in the fluid-dynamics section of the codes.

RADIATIVE HEAT TRANSFER

Radiation Model

In a chemically reactive system with regions of high temperature, e.g., where combustion is taking place, significant amounts of thermal radiation are emitted. This radiation is absorbed in cooler regions of the reactor, including the walls and other internal cold surfaces, and is thus a major contributor to heat transfer in a burning reactor. In addition to transfering heat throughout the reactor, the radiation mechanism serves to quench any thermal runaway of burning gases or particles since the emission rate scales as the fourth power of the temperature. Because it can affect reactor behavior, radiation transport is an important element to be included in reactive flow modeling. In this section, we describe the radiation model used in the FLAME and FLAG codes.

When a body is heated, the atoms and molecules composing the body are excited to higher electronic, vibrational and rotational energy levels. Radiation occurs when the molecules fall back to their original lower energy level. This radiation is electromagnetic and is composed of waves of all frequencies, but with relative amplitudes that depend on the temperature of the emitting body. The body not only emits radiation, but absorbs and scatters radiation emitted by surrounding bodies. In the absence of outside heat sources, the temperature of the body changes until it comes into a state of thermal equilibrium with the radiation field, at which point it emits and absorbs radiation at the same rate. What is necessary, then, is to integrate over all the radiation sources and sinks in the reactor and determine the radiation field level. Once this level is known, the amount of absorption can be found. The emission of energy of the gas or particles is taken to be blackbody, i.e., proportional to the temperature to the fourth power.

Radiation Transport Equation and the Diffusion Approximation

To find the energy exchange by radiation, the radiation field must be determined at all points in space. In this section we derive an equation that describes the transport of radiation through a scattering medium. This derivation parallels that for the transport of neutrons [16].

Let $N(\vec{r}, \vec{\Omega}, E, t)$ be the number of photons in the group at $\vec{r}$ and at time t per unit volume, having energies in dE about E and traveling in the direction $\vec{\Omega}$ with a spread $d\vec{\Omega}$. Let $\lambda_{AS}(E)$ be the mean free path that a photon travels before suffering an absorption or scattering collision, and c the photon velocity. Traveling with this group, the number of photons lost from the group by scattering and absorption in a time Δt is

$$(Nc/\lambda_{AS}) \tag{88}$$

Other photons can be scattered into the bundle. If P is the probability that a photon with energy E′ and direction $\vec{\Omega}'$ is scattered into E and $\vec{\Omega}$, the number scattered into the group is

$$[\int\int P(\vec{r}; \vec{\Omega}', E' \rightarrow \Omega, E)\, N'\, c/\lambda'\, d\vec{\Omega}'\, dE'] \tag{89}$$

Sources also can emit radiation into the group. We denote this source as

$$Q(\vec{r}, \vec{\Omega}, E, t) \tag{90}$$

The total change in N to an observer traveling with the packet is then

$$\frac{dN}{dt} + cN/\lambda_{AS} = \int\int cP\, N'/\lambda'\, d\vec{\Omega}'\, dE' + Q \tag{91}$$

At a fixed point,

$$\frac{dN}{dt} = \frac{\partial N}{\partial t} + c\vec{\Omega} \cdot \nabla N$$

The radiation settles to a steady state instantaneously, compared with the other processes of interest. Therefore, we drop the time derivative and arrive at the radiation transport equation for determining the radiation level at a point

$$c\Omega \cdot \nabla N + cN/\lambda_{AS} = \int\int PcN'/\lambda_{AS}'\, d\vec{\Omega}'\, dE' + Q \tag{92}$$

This equation also can be rewritten in terms of the radiation intensity $U(\vec{r}, \vec{\Omega}, E, t)$, which is the amount of energy per unit time transported across an area, $d\vec{A}$, which is in the direction $\vec{\Omega}$.

Equation 92 is an integrodifferential equation, which is very difficult to solve analytically or numerically. However, if the mean free path for scattering is small compared to any large-scale changes in the radiation field, and if the scattering is isotropic, Equation 92 can be simplified. This is accomplished by expressing the angular dependence of the flux as a sum over Legendre polynomials and retaining only the first two terms of the expansion. This is justified if the mean free path is relatively small and the scattering is isotropic. The result is a diffusion equation for the total energy flux, $\mathcal{U}$, crossing an area.

Equation 92 as written in terms of U is:

$$\vec{\Omega} \cdot \nabla U = Q - U/\lambda_{AS} \quad (93)$$

where the adsorption term has been approximated by U/λ_{AS}. Integrating Equation 93 over $d\vec{\Omega}$ results in:

$$\nabla \cdot I = S - \mathcal{U}/\lambda_{AS} \quad (94)$$

where

$$\vec{I} = \int U \vec{\Omega}\, d\vec{\Omega} \quad (95)$$

$$S = \int Q d\vec{\Omega} \quad (96)$$

$$\mathcal{U} = \int U d\vec{\Omega} \quad (97)$$

If the radiation field is isotropic, multiplying Equation 93 by $\vec{\Omega}$ and integrating over $d\vec{r}$ results in:

$$\tfrac{1}{3}\nabla \mathcal{U} = -\vec{I}/\lambda_{AS} \quad (98)$$

Combining Equations 94 and 98 gives the diffusion equation for $\mathcal{U}$:

$$-\nabla \cdot \frac{\lambda_{AS}}{3} \nabla \mathcal{U} + \mathcal{U}/\lambda_{AS} = S \quad (99)$$

(The reader interested in further details of the derivation of Equation 99 from Equation 92 is referred to the work of Bell and Glasstone [16].)

Equation 99 is the equation we solve to determine the radiation field in FLAME and FLAG. As we pointed out, certain approximations were made in its derivation. We can estimate the range of its validity.

If n_p is the number density of particles in the reactor, ϕ the fraction of volume they occupy and r_p their radius, the mean free path for radiation to travel before being absorbed is

$$\lambda = \frac{1}{n_p \pi r_p^2} \quad (100a)$$

But $\phi = n_p\, 4/3\, \pi r_p^3$. Thus,

$$\lambda = \frac{4r_p}{3\phi} \qquad (100b)$$

If the solid volume fraction is 0.5 (fluidized-bed reactor), the mean free path is 8/3 r_p, which is quite small. Even if the volume fraction is 0.01 (entrained-flow reactor), the mean free path is 133 r_p, which is also small compared to the dimensions of the reactor. Therefore, the diffusion approximation is justified for these reactors.

Solution Procedure

We solve Equation 99 for the radiation field $\mathscr{U}$ in each computational cell. The radiation source per unit volume is obtained in each cell by summing the individual source terms for all particles and gas in the cell. The source for a single particle is

$$S_p = 4\epsilon_p \sigma\, T_p{}^4 n_p \pi r_p{}^2 \qquad (101)$$

where ϵ_p is its emissivity, σ is Stefan-Boltzman's constant, n_p is the number density of real particles it represents, and r_p is the particle radius. The source for the gas is

$$S_g = 4K_g\, \sigma\, T_g{}^4 \qquad (102)$$

where K_g is the reciprocal absorption length for the gas.

The diffusion coefficient in Equation 99 is the sum of K_g and the reciprocal mean free paths for all the particles in the cell.

Once the equilibrium radiation field has been obtained, the amount of particles and gas heating or cooling can be calculated. Coupling in the particle energy balance has been discussed in the section entitled Particle Dynamics and Energetics. The gas heating per unit volume in each cell is

$$H = K_g\, (U - 4\sigma\, T_g{}^4) \qquad (103)$$

APPLICATION OF THE FLAME AND FLAG CODES

To show the range of applicability of the FLAME and FLAG codes and their potential usefulness to the process engineer, we present results from a reactive flow calculation by FLAME and a nonreactive calculation of jet penetration into a fluidized bed by FLAG.

FLAME Calculation of the BI-GAS Gasifier

Figure 7 is a schematic drawing of the BI-GAS two-stage, entrained-flow gasifier, and Figure 8 shows the zoning used in representing its interior axisymmetrically for the FLAME code. Using this zoning, the entire gasifier internal flow was computed, from the slag tap hole at the bottom through the product stream exit at the top. Subsequent figures depict those portions of the flow in the lower (combustion) stage, the interstage region and the lower portion of the upper (gasification) stage.

Figure 9 is a series of snapshots from a computer movie showing particle locations and temperature contours for early, nonsteady-state conditions in the lower portion of the gasifier. Figure 10 shows the temperature contours at near steady-state conditions. Figures 11, 12 and 13 show concentration contours of H_2, CO and CO_2 at steady state. Figure 14 shows mole percentages of all gas species on the centerline of the reactor as a function of height in the gasifier.

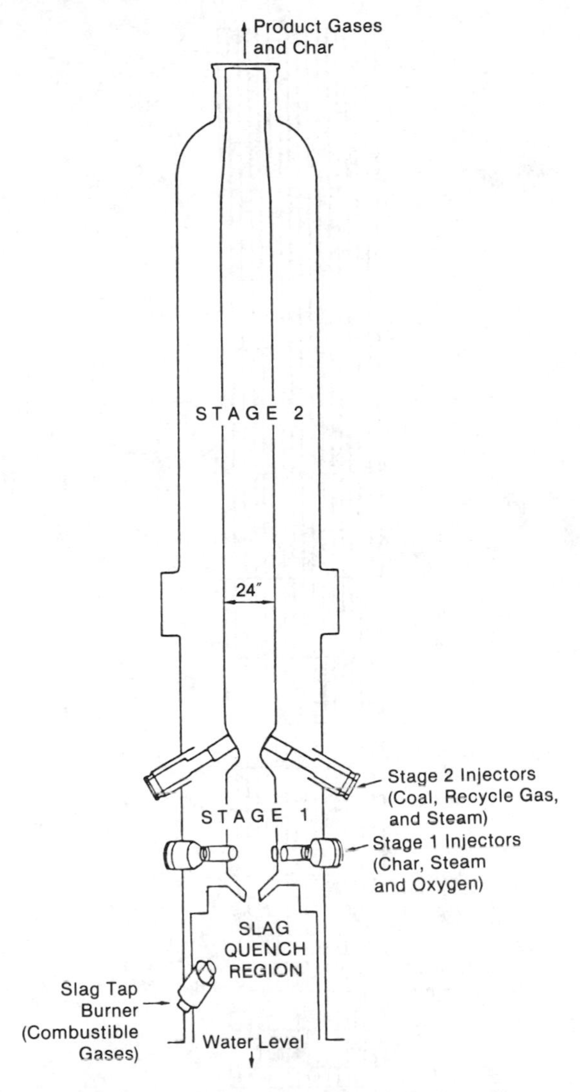

Figure 7. Schematic of the BI-GAS gasifier.

FLAG Calculation of Jet Penetration into a Fluidized Bed

Knowlton and Hirsan [17] at the Institute of Gas Technology (IGT) have performed exhaustive experiments to determine the effect of system pressure on jet penetration in fluidized beds. Their experimental equipment consisted of a semicircular fluidization column 12 inches in diameter. The column was fitted with a plexiglass face plate, which enabled visual observation of the gas-solid flow patterns in the bed. A semicircular, one-inch-diameter nozzle attached to the plexiglass face plate was the source of the gas jet. Nitrogen was used as the fluidizing and jet gas.

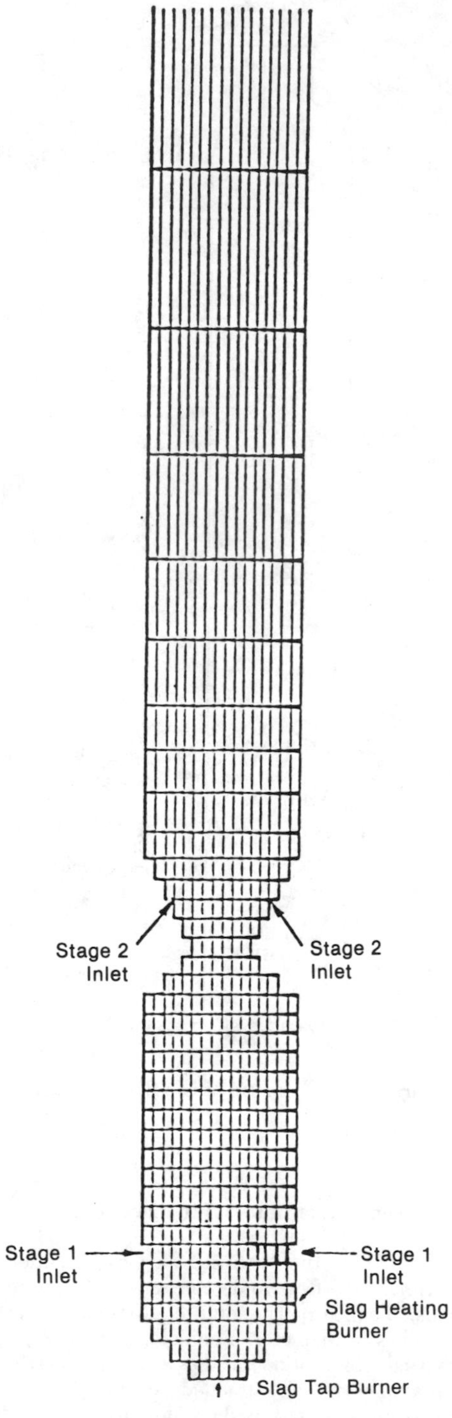

Figure 8. Computational zones for modeling the BI-GAS gasifier.

Experiments were conducted for three different materials over a wide range of pressure and jet velocities. Since the penetration of the gas jet into the fluidized bed fluctuated in length over a wide range, Knowlton and Hirsan identified three different penetration lengths: L_{MIN}, L_{MAX} and L_B. It was found that a jet consisted of a gas-solid-dilute phase (much more dilute than the gas emulsions phase), followed by bubbles emanating from the top of the jet. The minimum and maximum lengths between which the dilute phase fluctuated were labeled as L_{MIN} and L_{MAX}. At L_{MIN} the jet had the shape of a torch or flame, while at L_{MAX} it appeared as a series of coalescing bubbles with

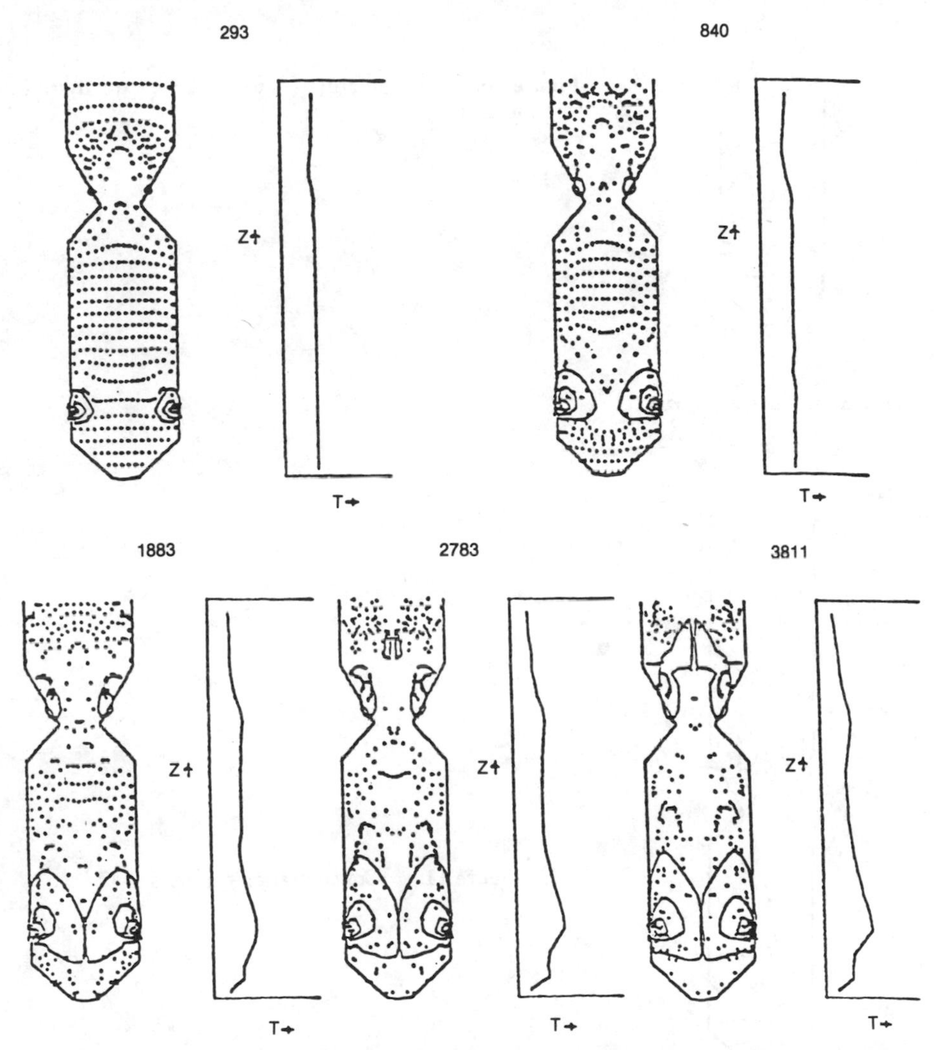

Figure 9. Selected snapshots from a computer movie of a 2-D test calculation. Times are shown in msec (293 = 0.293 sec, 3811 = 3.811 sec). Dots are tracer particle positions and solid lines are temperature contours. Centerline temperatures are plotted to the right of each contour plot.

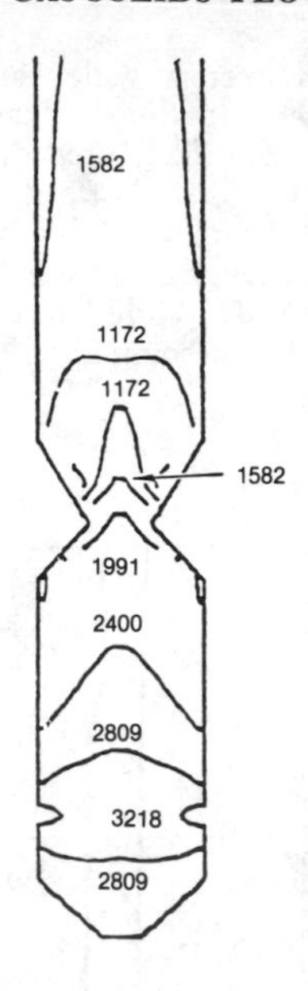

Figure 10. Temperature contours at near steady-state conditions.

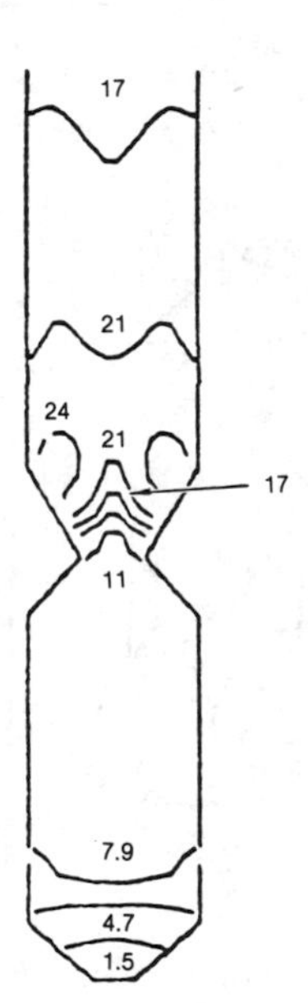

Figure 11. Concentration contours of H_2.

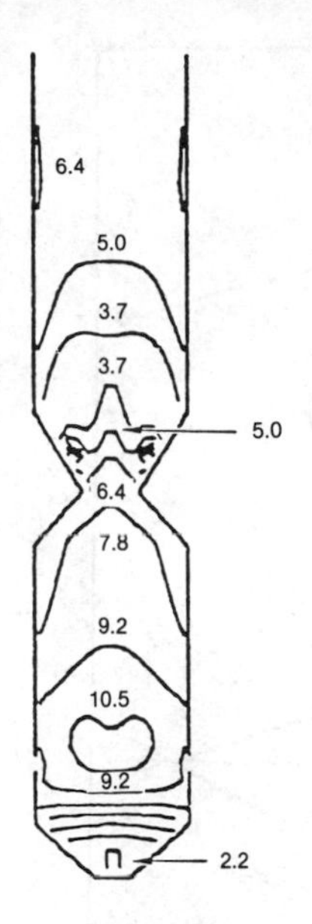

C0 MOLE PERCENT AT T = 17.82 SEC

Figure 12. Concentration contours of CO.

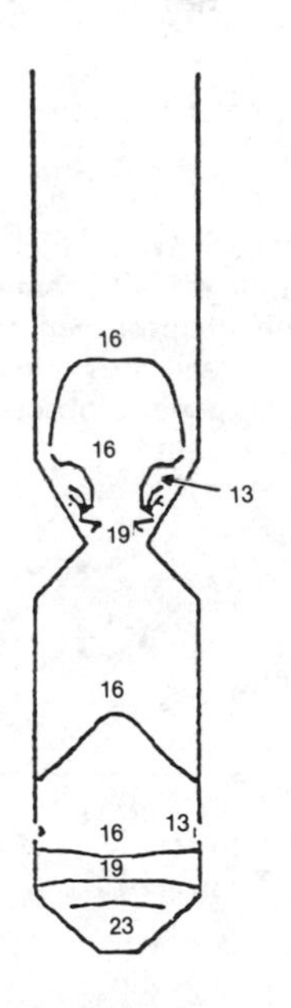

C02 MOLE PERCENT AT T = 17.82 SEC

Figure 13. Concentration contours of CO_2.

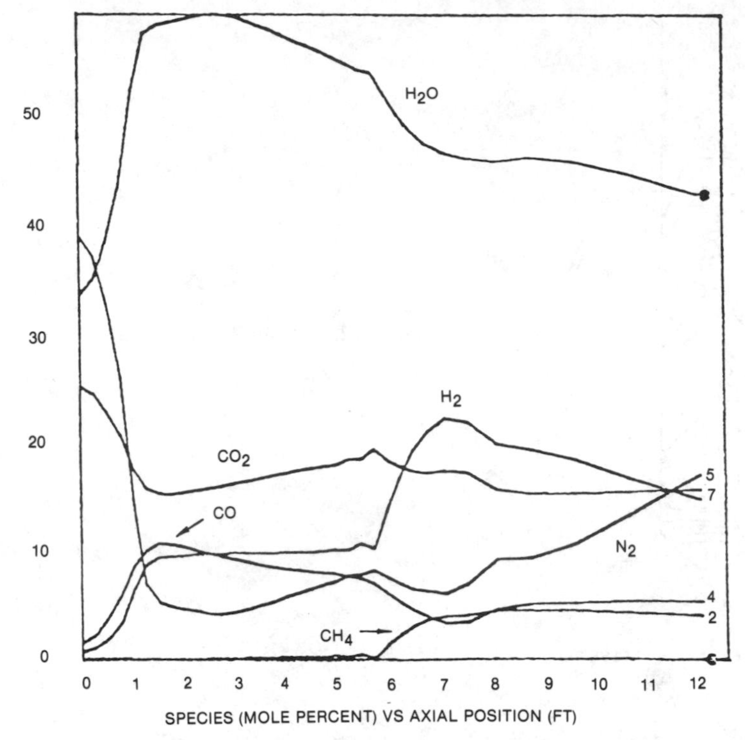

Figure 14. Mole percentages of gas species on the centerline of a reactor.

periodic necks. In some cases in which L_{MAX} did not penetrate the bed completely, it was found that a spout formed at the top of the bed, indicating that the jet momentum penetrated beyond L_{MAX}. Consequently, L_B was defined as the deepest penetration of the jet bubbles into the bed before losing their momentum, which was identified as the point at which the bubbles drifted from the vertical path. The general observation made in the study was that all L_{MIN}, L_{MAX} and L_B increased with pressure.

For the present calculations, we have chosen the following cases:

Material: Sand particle density = 164 lb/ft³
− 20 + 60 mesh
Particle diameter = 0.001335 ft

Pressure	V_{cf}
1. 4.195 atm (424.7 kPa)	0.713 ft/sec (21.74 cm/sec)
2. 21.15 atm (2142.5 kPa)	0.416 ft/sec (12.67 cm/sec)
3. 34.64 atm (3509.2 kPa)	0.363 ft/sec (11.07 cm/sec)

Here, V_{cf} is the bottom grid fluidizing gas velocity and is equal to the value that Knowlton and Hirsan determine to be the "complete fluidization" velocity.

These conditions were chosen to supply a fairly wide range of pressures for comparison of jet penetration lengths and bubble shape and flow behavior. The grid system that was used to model the experiment is shown in Figure 15.

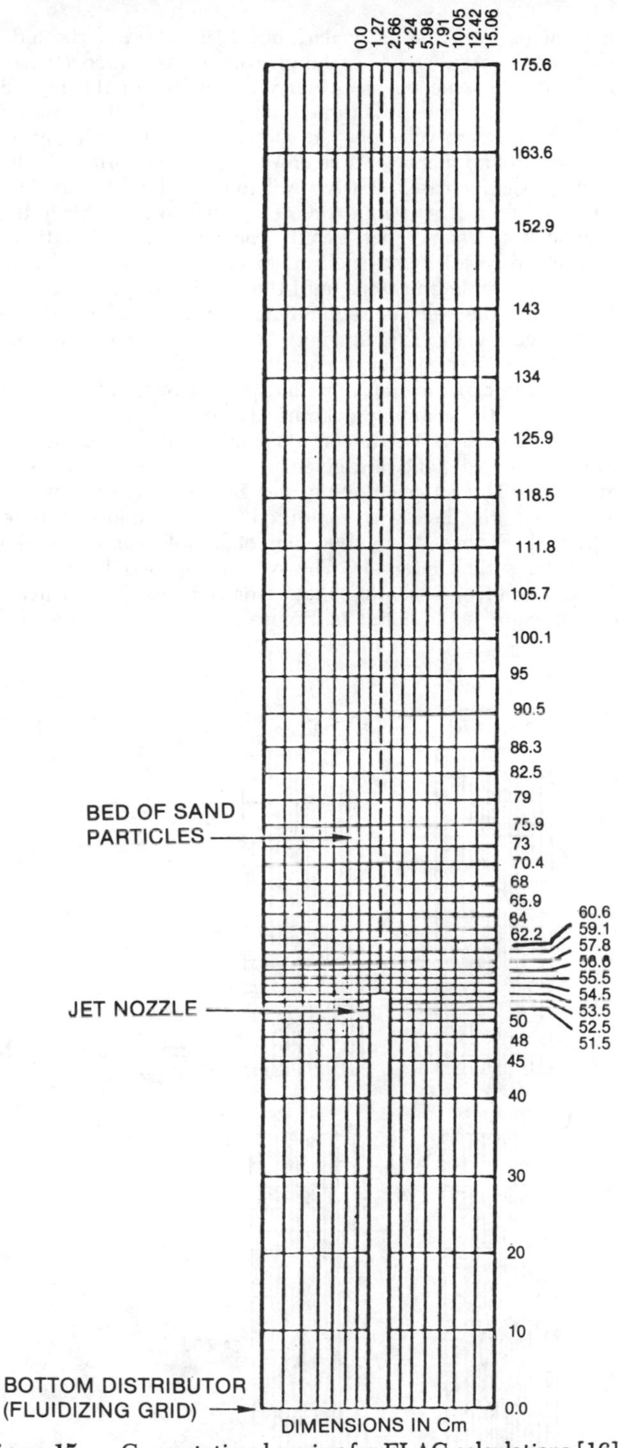

Figure 15. Computational zoning for FLAG calculations [16].

In these experiments [16], the particulate bed is fluidized at the indicated complete fluidization velocity, V_{cf}. At that point, the primary jet is turned on, and the jet penetration and bubbling phenomenon are observed visually. In the equivalent numerical experiments, the fluidizing gas was turned on (into a bed of particles distributed throughout the bed with zero velocities) at the experimental value of V_{cf}.

In each case, it was found that this value of V_{cf} did not uniformly fluidize the particles. So the grid velocity was increased slowly until the net flux of particles across a plane perpendicular to the bed axis was approximately equal to zero. Since this is a transient problem, it is difficult to characterize exactly the point of fluidization. Our approach was to increase the grid velocity to such an extent that the net flux of particles was upward, and then to decrease it in steps until the net flux of particles was approximately zero. It was found that the velocity required to fluidize the bed is about 10% higher than the V_{cf} indicated by the experiments. The difference can be attributed to the empiricism in the Zenz's drag coefficient correlation [12] being used in this calculation, and to the fact that the experimental apparatus is a semicircular cylinder. The FLAG calculations are done in the complete cylindrical geometry.

The experimental parameters are shown in Table I, and the corresponding parameters used in our numerical experiments are given in Table II. In each case, the central jet was turned on slowly. The jet penetration and bubble evolution were recorded. The voidage distribution in the bed was computed as a function of time, and voidage contours were plotted. Figures 16–18 show sets of bubble (or voidage) plots for Cases I, II and III at various time intervals. The region enclosed by these contours is approximately 90–95% void, representing the jet or bubbles. The penetration (L_B/dO) as a function of pressure is presented in Figure 19. The calculated and experimental

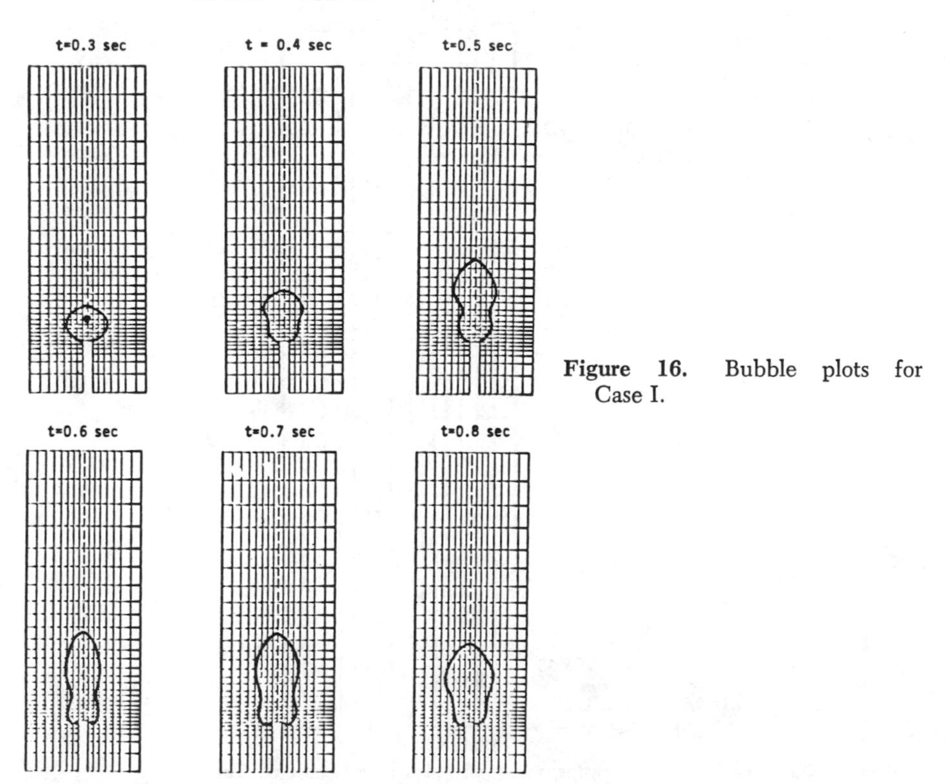

Figure 16. Bubble plots for Case I.

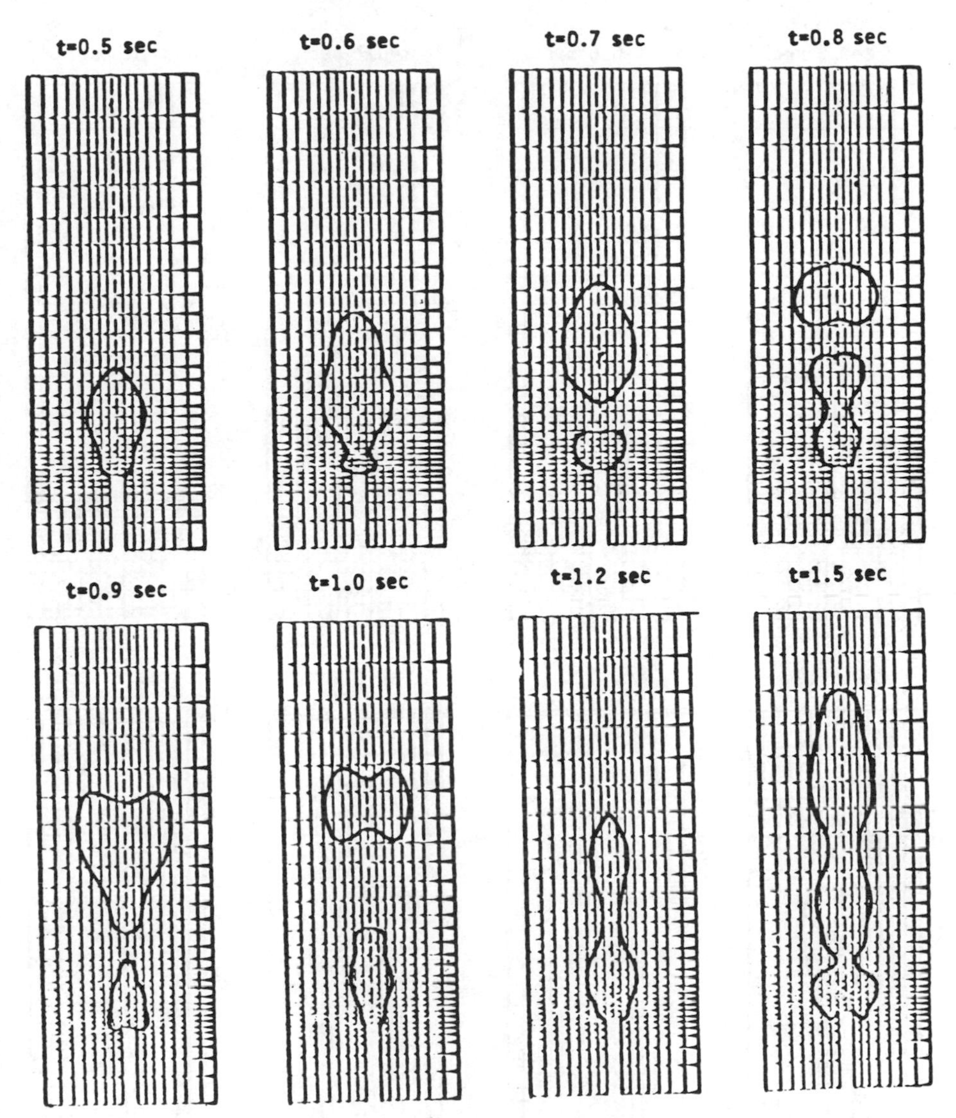

Figure 17. Bubble plots for Case II.

values of L_B/dO are presented in Table III, and the agreement seems to be good. Consistently, the penetrations predicted by these experiments are lower than the ones calculated, and this, we believe, is due to factors such as the jet in the experiment being semicircular and the presence of a plexiglass face plate, which could have caused some hindrance. The shapes of the bubbles computed were compared with the snapshots taken during the experiments, and the agreement is excellent.

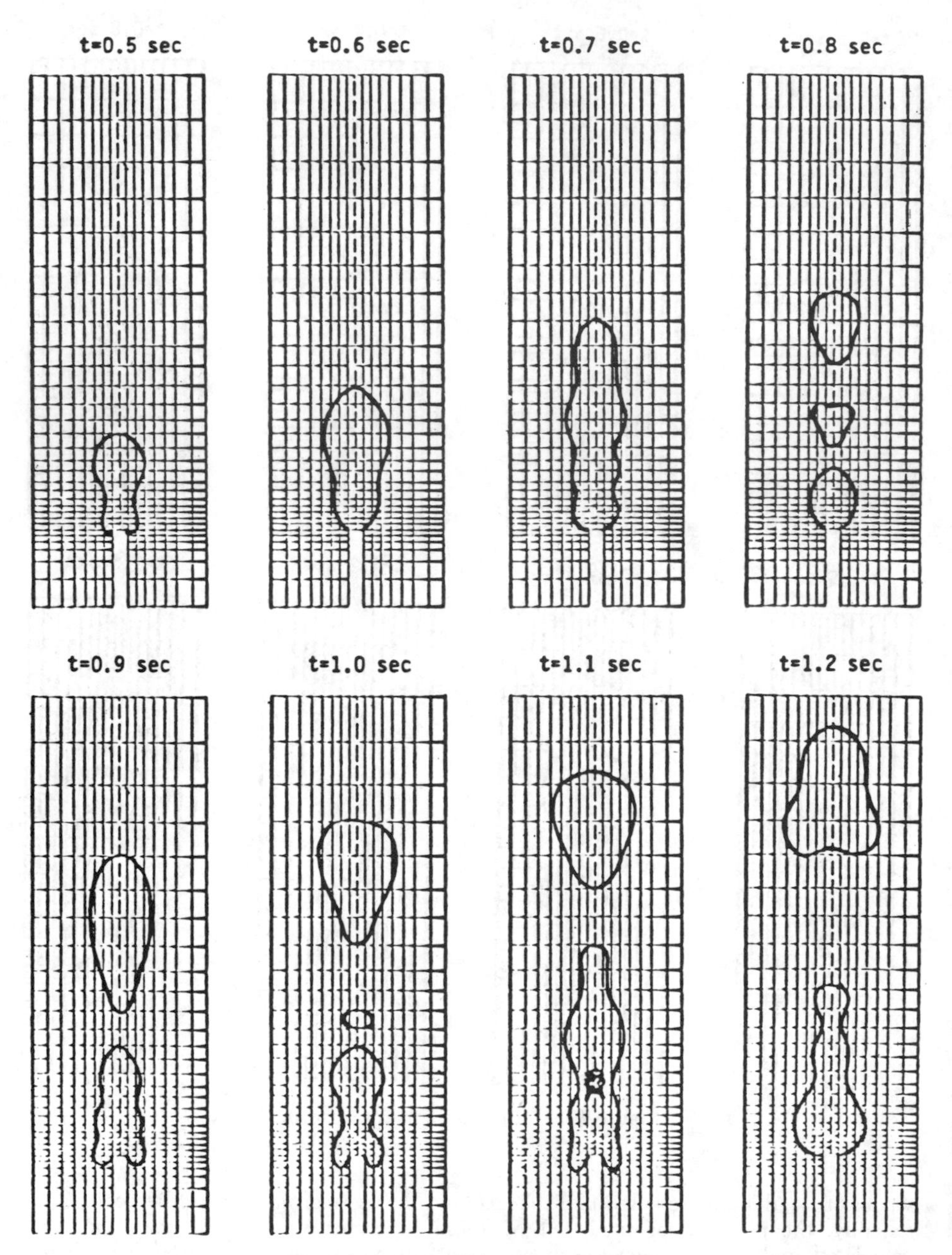

Figure 18. Bubble plots for Case III.

In conclusion, the jet penetration, the shape of the jet and the bubbles formed due to jet action that are predicted by the FLAG code are in good agreement with those observed in the experiments.

Table I. Experimental Conditions

	Pressure		V_{cf}		V_O	
	(atm)	(kPa)	(ft/sec)	(cm/sec)	(ft/sec)	(cm/sec)
Case I	4.195	424.7	0.713	21.74	25	762
Case II	21.15	2142.5	0.416	12.67	Same	
Case III	34.64	3509.2	0.363	11.07	Same	
Particles	Sand, −20 + 60 mesh, density = 164.4 lb/ft³ (2629 kg/m³)					

Table II. Calculation Conditions

	Pressure		V_{cf}		V_O	
	(atm)	(kPa)	(ft/sec)	(cm/sec)	(ft/sec)	(cm/sec)
Case I	4.195	424.7	0.82	25	25	762
Case II	21.15	2142.5	0.45	13.7	Same	
Case III	34.64	3509.2	0.318	11.6	Same	
Particles	133.5×10^{-6} ft (438 μm), density = 164.4 lb/ft³ (2629 kg/m³)					

Table III. Calculation and Experimental Values of L_B/dO

	L_B/dO Calculated	L_B/dO Experimental
Case I	7.68-8.92	7
Case II	22.95	20
Case III	29.33	29

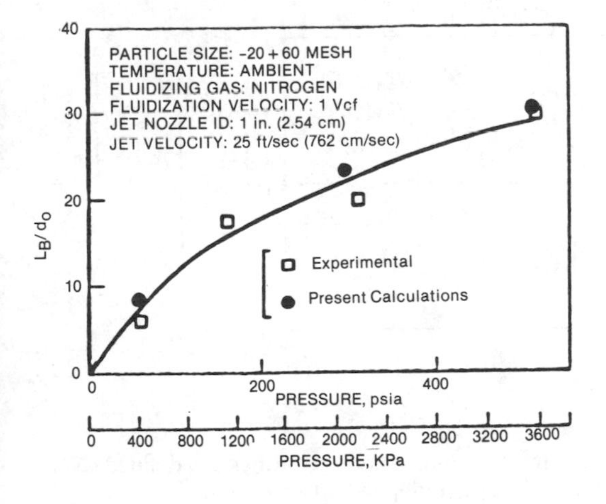

Figure 19. Jet penetration length vs gas pressure.

APPENDIX A—FINITE-DIFFERENCE FORMS OF GAS EQUATIONS

This section contains the finite-difference forms of the gas equations that were derived in the section entitled Fluid Mechanics. These forms are the ones solved by FLAME and FLAG. We also indicate the manner of time and spatial averaging used in both codes.

We pointed out in the Fluid Mechanics section that the variables for which we solve are averages of microscopic quantities over macroscopic volumes. The volumes over which we average form the computational grid network. Figure A-1 shows a grid and the location in which each quantity is defined. For example, the cell vortex is denoted by indices i and j; then the quantity $\alpha_{1+1/2,j+1/2}$ is an average inside the cell formed by (i,j), (i + 1,j) (i,j + 1) and (i + 1,j + 1). Similar volumes are formed for quantities defined at the cell faces.

We first give each equation in differential form and then its difference form.

Axial Momentum Equation

$$\underset{\substack{\text{time rate}\\ \text{of change}}}{\frac{\partial U}{\partial t}} + \underset{\substack{\text{axial}\\ \text{advection}}}{\frac{\partial \alpha UU}{\partial x}} + \underset{\substack{\text{radial}\\ \text{advection}}}{\frac{1}{r}\frac{\partial r\alpha VU}{\partial r}} - \underset{\substack{\text{axial}\\ \text{diffusion}}}{\frac{\partial}{\partial x}\Gamma\frac{\partial \alpha U}{\partial x}} - \underset{\substack{\text{radial}\\ \text{diffusion}}}{\frac{1}{r}\frac{\partial}{\partial r} r\Gamma\frac{\partial \alpha U}{\partial r}}$$

$$= \underset{\substack{\text{pressure}\\ \text{force}}}{-\epsilon\frac{\partial P}{\partial x}} \underset{\substack{\text{gravity}\\ \text{force}}}{- g/\alpha} \underset{\substack{\text{drag force}\\ \text{against particles}}}{- \sum_p K_p(\alpha U - N_p v_p)} + \underset{\substack{\text{remaining}\\ \text{forces}}}{S_U}$$

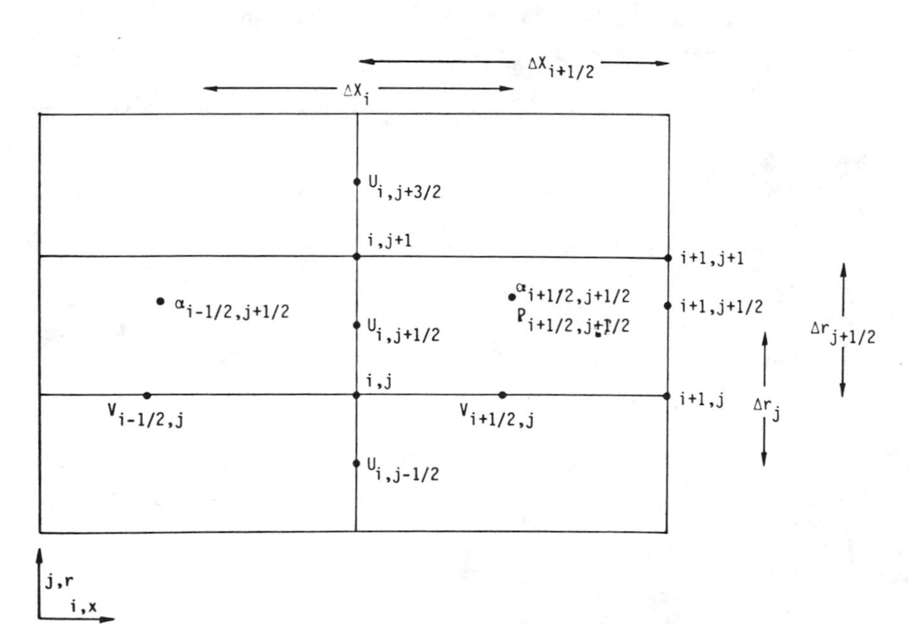

Figure A-1. Computational zones showing locations where quantities are defined. Note that r_i and Δr_j are independent of i, and Δ_{Xi} is independent of r.

where U = volume-averaged axial momentum density $\equiv <\rho v_g> = \epsilon\rho v_g$
ϵ = volume-averaged void fraction
ρ = gas density
v_g = axial gas velocity
α = inverse of gas density $\equiv \frac{1}{<\rho>} = \frac{1}{\epsilon\rho}$
x = axial distance
r = radial distance
V = volume-averaged radial momentum density $\rightarrow \epsilon\rho V$
Γ = gas density times diffusion coefficient
P = volume-averaged gas pressure divided by void fraction $\equiv \frac{<\rho>}{\epsilon}$
g = gravitational acceleration
K_p = drag coefficient
N_p = number of real particles represented by representative particle, p
S_U = axial force per unit volume
v_p = velocity of representative particle p

First we define

1. Time average of a quantity, A, at inner iterate $k + 1$ and time $(n + 1/2)\ \Delta t$:

$$\bar{A} \equiv \frac{[A^{n+1}(k) + A^n]}{2}$$

2. Space average of A in x direction at grid point (i,j):

$$<A>_{ij} \equiv \frac{(A_{i-1/2,j} + A_{i+1/2,j})}{2}$$

3. Space average of A in y direction at grid point (i,j):

$$\{A\}_{i,j} \equiv \frac{(A_{i,j+1/2} + A_{i,j-1/2})}{2}$$

The finite-difference form of the axial momentum equation for advancing $U^{n+1}_{i,j+1/2}$ in time is

Time Rate of Change:

$$\frac{\partial U}{\partial t} \rightarrow \frac{U^{n+1}_{i,j+1/2}\ (k+1) - U^{n}_{i,j+1/2}}{\Delta t}$$

Axial Diffusion:

$$\frac{\partial}{\partial x}(\alpha UU) \rightarrow \frac{[<<\bar{\alpha}>\bar{U}>_{i+1/2,j+1/2}<U>_{i+1/2,j+1/2} - <<\alpha>\bar{U}>_{i-1/2,j+1/2}<\bar{U}>_{i-1/2,j+1/2}]}{\Delta x_i}$$

Radial Advection:

$$\frac{1}{r}\frac{\partial}{\partial r} r(\alpha VU) \rightarrow \frac{(r_{j+1}<\{\bar{\alpha}\}\bar{V}>_{i,j+1}\{\bar{U}\}_{i,j+1} - r_j<\{\bar{\alpha}\}\bar{V}>_{i,j}\{\bar{U}\}_{i,j})}{r_{j+1/2}\ \Delta r_{j+1/2}}$$

Axial Diffusion:

$$\frac{\partial \Gamma}{\partial x}\frac{\partial \alpha U}{\partial x} \rightarrow \frac{1}{\Delta x_i}\Big[\Gamma_{i+1/2,j+1/2}\frac{(<\bar{\alpha}>_{i+1,j+1/2}\overline{U}_{i+1,j+1/2} - <\bar{\alpha}>_{i,j+1/2}\overline{U}_{i,j+1/2})}{\Delta x_{i+1/2}} - \Gamma_{i-1/2,j+1/2}\frac{(<\bar{\alpha}>_{i,j+1/2}\overline{U}_{i,j+1/2} - <\bar{\alpha}>_{i-1,j+1/2}\overline{U}_{i-1,j+1/2})}{\Delta x_{i-1/2}}\Big]$$

Radial Diffusion:

$$\frac{1}{r}\frac{\partial}{\partial r} r\Gamma \frac{\partial \alpha U}{\partial r} \rightarrow \frac{1}{r_{j+1/2}\,\Delta r_{j+1/2}}\Big[r_{j+1}\Gamma_{i,j+1}\frac{(<\bar{\alpha}>_{i,j+3/2}\overline{U}_{i,j+3/2} - <\bar{\alpha}>_{i,j+1/2}\overline{U}_{i,j+1/2})}{\Delta r_{j+1}} - r_j\Gamma_{ij}\frac{(<\bar{\alpha}>_{i,j+1/2}\overline{U}_{i,j+1/2} - <\bar{\alpha}>_{i,j-1/2}\overline{U}_{i,j-1/2})}{\Delta r_j}\Big]$$

Pressure Force:

$$\epsilon\frac{\partial P}{\partial x} \rightarrow <\epsilon>_{i,j+1/2}\frac{[P^{n}_{i+1/2,j+1/2} - P^{n}_{i-1/2,j+1/2} + P^{n+1}_{i+1/2,j+1/2}(k+1) - P^{n+1}_{i-1/2,j+1/2}(k+1)]}{2\Delta x_i}$$

Note that we solve for $P^{n+1}(k+1)$ implicitly here.

Gravity Force:

$$g/\alpha \rightarrow g < \overline{\frac{1}{\alpha}} >_{i,j+1/2}$$

Drag Force:

$$\sum_p K_p\,(\alpha U - N_p\,v_p) \rightarrow \sum_p K_{p_{i,j+1/2}}\,[<\bar{\alpha}>_{i,j+1/2}\,U^{n+1}_{i,j+1/2}(k+1) - (N_p\,v_p)]$$

Note that we solve for $U^{n+1}(k+1)$ implicitly here.

Radial Momentum Equation

$$\underset{\text{time rate of change}}{\frac{\partial V}{\partial t}} + \underset{\text{axial direction}}{\frac{\partial \alpha UV}{\partial x}} + \underset{\text{radial direction}}{\frac{1}{r}\frac{\partial r\alpha VV}{\partial r}} - \underset{\text{axial diffusion}}{\frac{\partial}{\partial x}\Gamma\frac{\partial \alpha V}{\partial x}} - \underset{\text{radial diffusion}}{\frac{\partial}{\partial r}\frac{\Gamma}{r}\frac{\partial \alpha Vr}{\partial r}} + \underset{\text{centrifugal force}}{\frac{\alpha WW}{r}}$$

$$= \underset{\text{pressure force}}{-\epsilon\frac{\partial P}{\partial r}} - \underset{\text{axial diffusion}}{\sum_p K_p\,(\alpha V - N_p v_p)} + \underset{\text{remaining forces}}{S_V}$$

where
V = volume-averaged radial momentum density $\equiv <\rho v> = \epsilon\rho v$
W = volume-averaged swirl momentum density $\equiv <\rho w> = \epsilon\rho w$
v = radial gas velocity
w = swirl gas velocity
v_p = radial velocity of particle p
S_V = radial force per unit volume

The finite difference of the radial momentum equation is solved for $V^{n+1}_{i+1/2,j}$.

Time Rate of Change:

$$\frac{\partial V}{\partial t} \rightarrow \frac{V^{n+1}_{i+1/2,j}(k+1) - V^{n}_{i+1/2,j}}{\Delta t}$$

Axial Advection:

$$\frac{\partial \alpha UV}{\partial x} \rightarrow \frac{[\{<\overline{\alpha}>\overline{U}\}_{i+1,j}<\overline{V}>_{i+1,j} - \{<\overline{\alpha}>\overline{U}\}_{i,j}<\overline{V}>_{i,j}]}{\Delta x_{i+1/2}}$$

Radial Advection:

$$\frac{1}{r}\frac{\partial r\alpha VV}{\partial r} \rightarrow \frac{[r_{j+1/2}\{\{\overline{\alpha}\}\overline{V}\}_{i+1/2,j+1/2}\{\overline{V}\}_{i+1/2,j+1/2} - r_{j-1/2}\{\{\overline{\alpha}\}\overline{V}\}_{i+1/2,j-1/2}\{\overline{V}\}_{i+1,2j-1/2}]}{r_j\,\Delta r_j}$$

Axial Diffusion:

$$\frac{\partial}{\partial x}\Gamma\frac{\partial \alpha V}{\partial x} \rightarrow \frac{1}{\Delta x_{i+1/2}}\left[\frac{[\Gamma_{i+1,j}(\{\overline{\alpha}\}_{i+3/2,j}\overline{V}_{i+3/2,j} - \{\overline{\alpha}\}_{i+1/2,j}\overline{V}_{i+1/2,j})]}{\Delta x_{i+1}} - \frac{\Gamma_{i,j}(\{\overline{\alpha}\}_{i+1/2,j}\overline{V}_{i+1/2,j} - \{\overline{\alpha}\}_{i-1/2,j}\overline{V}_{i-1/2,j})}{\Delta x_i}\right]$$

Radial Diffusion:

$$\frac{1}{r}\frac{\partial}{\partial r}\frac{\Gamma}{r}\frac{\partial \alpha Vr}{\partial r} \rightarrow \frac{1}{\Delta r_j}$$

$$\left[\frac{\Gamma_{i+1/2,j+1/2}}{r_{j+1/2}}\frac{(\{\overline{\alpha}\}_{i+1/2,j+1}\overline{V}_{i+1/2,j+1}r_{j+1} - \{\overline{\alpha}\}_{i+1/2,j}\overline{V}_{i+1/2,j}r_j}{\Delta r_{j+1/2}} - \frac{\Gamma_{i+1/2,j-1/2}}{r_{j-1/2}}\frac{(\{\overline{\alpha}\}_{i+1/2,j}\overline{V}_{i+1/2,j}r_j - \{\overline{\alpha}\}_{i+1/2,j-1}\overline{V}_{i+1/2,j-1}r_{j-1})}{\Delta r_{j-1/2}}\right]$$

Centrifugal Force:

$$\frac{\alpha WW}{r} \rightarrow \frac{\{\overline{\alpha}\overline{W}\}_{i+1/2,j}\{\overline{W}\}_{i+1/2,j}}{r_j}$$

Pressure Force:

$$\epsilon\frac{\partial P}{\partial r} \rightarrow \{\epsilon\}_{i+1/2,j}$$

$$\frac{(P^n_{i+1/2,j+1/2} - P^n_{i+1/2,j-1/2} + P^{n+1}_{i+1/2,j+1/2}(k+1) - P^{n+1}_{i+1/2,j-1/2}(k+1)}{2\Delta r_j}$$

Note that we solve for P^{n+1} term implicitly here.

Particle Drag:

$$\sum_p K_p(\alpha V - N_p v_p) \rightarrow \sum_p K_{p_{i+1/2,j}}[\{\overline{\alpha}\}_{i+1/2,j}V^{n+1}_{i+1/2,j}(k+1) - (N_p v_p)_{i+1/2,j}]$$

Note that we solve for V^{n+1} implicitly here.

Momentum Source:

$$S_V \rightarrow S_{V_{i+1/2,j}}$$

Swirl Momentum Equation

$$\underset{\substack{\text{time rate}\\\text{of change}}}{\frac{\partial W}{\partial t}} + \underset{\substack{\text{axial}\\\text{advection}}}{\frac{\partial \alpha UW}{\partial x}} + \underset{\substack{\text{radial}\\\text{advection}}}{\frac{1}{r}\frac{\partial r\alpha VW}{\partial r}} - \underset{\substack{\text{axial}\\\text{diffusion}}}{\frac{\partial}{\partial x}\Gamma\frac{\partial \alpha W}{\partial x}} - \underset{\substack{\text{radial}\\\text{diffusion}}}{\frac{\partial}{\partial r}\frac{\Gamma}{r}\frac{\partial \alpha W}{\partial r}} - \underset{\substack{\text{Coriolis}\\\text{force}}}{\frac{\alpha WW}{r}}$$

$$= -\underset{\substack{\text{drag against}\\\text{particles}}}{\sum_p K_p(\alpha W - N_p w_p)} - \underset{\substack{\text{remaining}\\\text{forces}}}{S_W}$$

where w_p = particle swirl velocity
S_W = swirl force per unit volume

Time Rate of Change:

$$\frac{\partial W}{\partial t} \rightarrow \frac{W^{n+1}_{i+1/2,j+1/2}(k+1) - W^{n}_{i+1/2,j+1/2}}{\Delta t}$$

Axial Advection:

$$\frac{\partial \alpha UW}{\partial x} \rightarrow \frac{[\langle\bar{\alpha}\rangle_{i+1,j+1/2}\bar{U}_{i+1,j+1/2}\langle\overline{W}\rangle_{i+1,j+1/2} - \langle\bar{\alpha}\rangle_{i,j+1/2}\langle\bar{U}\rangle_{i,j+1/2}\langle\overline{W}\rangle_{i,j+1/2}]}{\Delta x_{i+1/2}}$$

Radial Advection:

$$\frac{1}{r}\frac{\partial r\alpha VW}{\partial r} \rightarrow \frac{[r_{j+1}\{\bar{\alpha}\}_{i+1/2,j+1}\bar{V}_{i+1/2,j+1}\{\overline{W}\}_{i+1/2,j+1} - r_j\{\bar{\alpha}\}_{i+1/2,j}\bar{V}_{i+1/2,j}\{\overline{W}\}_{i+1/2,j}]}{r_{j+1/2}\,\Delta r_{j+1/2}}$$

Axial Diffusion:

$$\frac{\partial}{\partial x}\Gamma\frac{\partial \alpha W}{\partial x} \rightarrow \frac{1}{\Delta x_{i+1/2}}\left[\Gamma_{i+1,j+1/2}\frac{(\bar{\alpha}_{i+3/2,j+1/2}\overline{W}_{i+3/2,j+1/2} - \bar{\alpha}_{i+1/2,j+1/2}\overline{W}_{i+1/2,j+1/2})}{\Delta x_{i+1}} - \Gamma_{i,j+1/2}\frac{(\bar{\alpha}_{i+1/2,j+1/2}\overline{W}_{i+1/2,j+1/2} - \bar{\alpha}_{i-1/2,j+1/2}\overline{W}_{i-1/2,j+1/2})}{\Delta x_i}\right]$$

Radial Diffusion:

$$\frac{\partial}{\partial r}\frac{\Gamma}{r}\frac{\partial \alpha W r}{\partial r} \rightarrow \frac{1}{\Delta r_{j+1/2}}$$

$$\left[\frac{\Gamma_{i+1/2,j+1}}{r_{j+1}}\frac{(\bar{\alpha}_{i+1/2,j+3/2}\overline{W}_{i+1/2,j+3/2}r_{j+1/2} - \bar{\alpha}_{i+1/2,j+1/2}\overline{W}_{i+1/2,j+1/2}r_{j-1/2})}{\Delta r_{j+1}} - \frac{\Gamma_j}{r_j}\frac{(\bar{\alpha}_{i+1/2,j+1/2}\overline{W}_{i+1/2,j+1/2}r_{j+1/2} - \bar{\alpha}_{i+1/2,j-1/2}\overline{W}_{i+1/2,j-1/2}r_{j-1/2})}{\Delta r_j}\right]$$

Coriolis Force:

$$\frac{\alpha V W}{r} \rightarrow \frac{\{\{\bar{\alpha}\}\overline{V}\}_{i+1/2,j+1/2}\overline{W}_{i+1/2,j+1/2}}{r_{j+1/2}}$$

Momentum Force:

$$S_W \rightarrow S_{W_{i+1/2,j+1/2}}$$

Particle Drag:

$$\Sigma K_p(\alpha W - N_p w_p) \rightarrow \sum_p K_{p_{i+1/2,j+1/2}}[\bar{\alpha}_{i+1/2,j+1/2}\overline{W}_{i+1/2,j+1/2} - (N_p w_p)_{i+1/2,j+1/2}]$$

Mass Continuity Equation

$$\underset{\text{time rate of change}}{\frac{\partial\langle\rho'\rangle}{\partial t}} + \underset{\text{axial advection}}{\frac{\partial U}{\partial x}} + \underset{\text{radial advection}}{\frac{1}{r}\frac{\partial rV}{\partial r}} = \underset{\text{mass source}}{r_M}$$

where $\langle\rho'\rangle$ = volume-averaged gas density $\equiv \epsilon\rho$

Time Rate of Change:

$$\frac{\partial\langle\rho'\rangle}{\partial t} \rightarrow \frac{\langle\rho'\rangle^{n+1}_{i+1/2,j+1/2}\ (k+1) - \langle\rho'\rangle^{n}_{i+1/2,j+1/2}}{\partial t}$$

Axial Advection:

$$\frac{\partial U}{\partial x} \rightarrow \frac{(\overline{U}_{i+1,j+1/2} - \overline{U}_{i,j+1/2})}{\Delta x_{i+1/2}}$$

Radial Advection:

$$\frac{1}{r}\frac{\partial rV}{\partial r} \rightarrow \frac{(r_{j+1}\overline{V}_{i+1/2,j+1} - r_j\overline{V}_{i+1/2,j})}{r_{j+1/2}\,\Delta r_{j+1/2}}$$

Mass Source:

$$r_M \rightarrow r_{M_{i+1/2,j+1/2}}$$

Energy Equation

$$\frac{\partial \delta P}{\partial t} + \frac{\gamma R}{\epsilon}\left[\frac{\partial U(T_B+\delta T)}{\partial x} + \frac{1}{r}\frac{\partial rV(T_B+\delta T)}{\partial r}\right] - \frac{R}{\epsilon}\frac{\partial}{\partial x}\Gamma\frac{\partial T}{\partial x} - \frac{R}{\epsilon r}\frac{\partial}{\partial r} r\Gamma\frac{\partial T}{\partial r}$$

time rate of change; axial advection; radial advection; axial diffusion; radial diffusion

$$= \frac{\partial S_p}{\partial P}\delta P + (\gamma - 1)\,\alpha U\frac{\partial P}{\partial x} + (\gamma-1)\,2V\frac{\partial P}{\partial r} - \gamma\frac{P}{\epsilon}\frac{d\epsilon}{dt} + S_p(P_B)$$

implicit heat source; axial work; radial work; work by particles in reducing available gas volume; all other heating sources

where
P_B = base pressure ≡ constant
δP = difference between total pressure and a base state pressure $\equiv P - P_B$
T = gas temperature $\equiv P\epsilon/(R\langle\rho\rangle)$
T_B = base state temperature ≡ constant
$\delta T = T - T_B$
$S_p(P_B)$ = heat source as a function of the base pressure
$\frac{\partial S_p}{\partial P}$ = change of heat source with pressure

Time Rate of Change:

$$\frac{\partial \delta P}{\partial t} \rightarrow \frac{\delta P^{n+1}_{i+1/2,j+1/2}(k+1) - \delta P^{n}_{i+1/2,j+1/2}}{\Delta t}$$

Axial Advection:

$$\frac{1}{\epsilon}\frac{\partial U(T_B+\delta T)}{\partial x} \rightarrow \frac{1}{\epsilon_{i+1/2,j+1/2}\,\Delta x_{i+1/2}}$$

$$\left[\frac{1}{2}\langle T_B\rangle_{i+1,j+1/2}(U^n_{i+1,j+1/2} + U^{n+1}_{i+1,j+1/2}(k+1)) + \langle\delta\overline{T}\rangle_{i+1,j+1/2}\overline{U}_{i+1,j+1/2}\right.$$

$$\left. - \frac{1}{2}\langle T_B\rangle_{i,j+1/2}(U^n_{i,j+1/2} + U^{n+1}_{i,j+1/2}) - \langle\delta\overline{T}\rangle_{i,j+1/2}\overline{U}_{i,j+1/2}\right]$$

The $U^{n+1}(k+1)$ terms are solved implicitly here.

Radial Advection:

$$\frac{1}{\epsilon r}\frac{\partial rV(T_B+\delta T)}{\partial r}\;\frac{1}{\epsilon_{i+1/2,j+1/2}\,\Delta r_{j+1/2} r_{j+1/2}}\left[\frac{1}{2}\{T_B\}_{i+1/2,j+1}\right.$$

$$(V^n_{i+1/2,j+1} + V^{n+1}_{i+1/2,j+1}(k+1)) + \{\delta\overline{T}\}_{i+1/2,j+1}\overline{V}_{i+1/2,j+1} - \frac{1}{2}\{T_B\}_{i+1/2,j}$$

$$\left.(V^n_{i+1/2,j} + V^{n+1}_{i+1/2,j+1}(k+1)) - \{\delta\overline{T}\}_{i+1/2,j}\overline{V}_{i+1/2,j}\right]$$

The $V^{n+1}(k+1)$ terms are determined implicitly here.

Axial Diffusion:

$$\frac{1}{\epsilon}\frac{\partial\Gamma}{\partial x}\frac{\partial T}{\partial x} \rightarrow \frac{1}{\epsilon_{i+1/2,j+1/2}\,\Delta x_{i+1/2}}\Gamma_{i+1,j+1/2}\frac{(\overline{T}_{i+3/2,j+1/2}-\overline{T}_{i+1/2,j+1/2})}{\Delta x_{i+1}} - \Gamma_{i,j+1/2}\frac{(\overline{T}_{i+1/2,j+1/2}-\overline{T}_{i-1/2,j+1/2})}{\Delta x_i}$$

Radial Diffusion:

$$\frac{1}{\epsilon r}\frac{\partial r\Gamma}{\partial r}\frac{\partial T}{\partial r} \rightarrow \frac{1}{\epsilon_{i+1/2,j+1/2}r_{j+1/2}\Delta r_{j+1/2}}\, r_{j+1}\Gamma_{i+1/2,j+1} \frac{(\overline{T}_{i+1/2,j+3/2}-\overline{T}_{i+1/2,j+1/2})}{\Delta r_{j+1}} - r_j\Gamma_{i+1/2,j}\frac{(\overline{T}_{i+1/2,j+1/2}-\overline{T}_{i+1/2,j-1/2})}{\Delta r_j}$$

Axial Work:

$$\alpha U\,\frac{\partial P}{\partial x} \rightarrow \frac{1}{2}\langle\overline{\alpha}\rangle_{i+1,j+1/2}\overline{U}_{i+1,j+1/2}\frac{(\delta\overline{P}_{i+3/2,j+1/2}-\delta\overline{P}_{i+1/2,j+1/2})}{\Delta x_{i+1}} + \frac{1}{2}\langle\overline{\alpha}\rangle_{i,j+1/2}\overline{U}_{i,j+1/2}\frac{(\delta\overline{P}_{i+1/2,j+1/2}-\delta\overline{P}_{i-1/2,j+1/2})}{\Delta x_i}$$

Radial Work:

$$\alpha V\,\frac{\partial P}{\partial r} \rightarrow \frac{1}{2}\{\overline{\alpha}\}_{i+1/2,j+1}\overline{V}_{i+1/2,j+1}\frac{(\delta\overline{P}_{i+1/2,j+3/2}-\delta\overline{P}_{i+1/2,j+1/2})}{\Delta r_{j+1}} + \frac{1}{2}\{\overline{\alpha}\}_{i+1/2,j}\overline{V}_{i+1/2,j}\frac{(\delta\overline{P}_{i+1/2,j+1/2}-\delta\overline{P}_{i+1/2,j-1/2})}{\Delta r_j}$$

Work in Reducing Available Volume:

$$\frac{P}{\epsilon}\frac{\partial\epsilon}{\partial t} \rightarrow \frac{(\delta\overline{P}_{i+1/2}+P_B)}{\overline{\epsilon}_{i+1/2,j+1/2}}\frac{(\epsilon^{n+1}_{i+1/2,j+1/2}-\epsilon^{n}_{i+1/2,j+1/2})}{\Delta t}$$

Implicit Heating:

$$\frac{\partial S_p}{\partial P}\delta P \rightarrow \frac{\partial S_p}{\partial P_{i+1/2,j+1/2}}\,\delta P^{\,n+1}_{\,i+1/2,j+1/2}(k+1)$$

The $\delta P^{n+1}(k+1)$ term is solved implicitly here.

Explicit Heating:

$$S_p \rightarrow S_{p_{i+1/2,\,j+1/2}}$$

NOMENCLATURE

c speed of light (m-sec^{-1})
C_D drag coefficient
C_i concentration of ith gas species (kmol-m^{-3})

$C_{i,s}$ concentration of ith gas species at particle surface (kmol-m^{-3})

C_p specific heat at constant pressure (J-kg^{-1}-°K^{-1}), particle specific heat (J-kg^{-1}-°K^{-1})

C_v specific heat at constant volume (J-kg^{-1}-°K^{-1})

d particle diameter (m)

d_e particle diameter in element e (m)

$d\vec{A}$ incremental area (m^2)

D_p particle drag slowing-down frequency (kg-sec^{-1})

$\mathscr{D}$ molecular diffusion coefficient (m^2-sec^{-1})

e specific energy of the gas (J-m^{-1})

e_S specific energy at the particle's surface (J-m^{-1})

E energy in the gas (J), energy of the photon (J)

$\bar{f}_p$ particle drag force (N)

$\bar{F}$ force on the gas (N)

F_d particle drag force (N)

$F_{i,j}$ drag force per unit volume on gas at i,j (N-m^{-3})

$\bar{F}_v$ force per unit volume on the gas (N-m^{-3})

g gravity (N-m^{-2})

$\bar{g}$ gravity (N-kg^{-1})

h specific enthalpy (J-kg^{-1})

h_i specific enthalpy of species i (J-kmol^{-1})

H heat released per unit volume of gas from homogeneous reactions (J-m^{-3}), gas heating rate by radiation (W-m^{-3})

H_g specific enthalpy of gas component g (J-kg^{-1})

H_i specific enthalpy of gas species i (J-kmol^{-1})

$H_{i,j}$ frictional heating of gas at cell i,j (W)

i specific internal energy (J-kg^{-1})

I identity matrix

J Jacobian for implicit solution procedure

k heat transfer coefficient (J-m^{-1}-°K^{-1}), iteration number

k_g mass transfer coefficient (m-sec^{-1})

k_j gas constant of jth chemical reaction

K equilibrium constant of water-gas shift reaction

K_g reciprocal absorption length of radiation in the gas (m^{-1})

l_m turbulence mixing length (m)

m_p particle mass (kg)

M particle mass (kg)

n number density of particles in an element (m^{-3}), time step number

$\bar{n}$ unit outward normal vector

n_p particle number density (m^{-3})

N number of photons per unit volume, with energy, E, traveling in direction $\vec{\Omega}$, number of gas species

N_e number of particles in an element

NP_i number of particles in cell i

N_u Nusselt number

O difference operator

p pressure (N-m^{-2})

P probability of scattering from $(E', \vec{\Omega}') \rightarrow (E, \vec{\Omega})$, collision probability

$\bar{q}$ heat flux (W-m^{-2})

q_p heat flux from particle to gas (W-m^{-2})

Q source of radiation (sec^{-1})

Q heating rate (W)

r radial space coordinate (m)

$\bar{r}$ position vector (M)

$\vec{r}$ position vector (m)

r_A radius of particle type A (m)

r_g rate of production of component g (kg-sec^{-1})

r_i heterogeneous chemical reaction rate of ith chemical species (kmol-m^{-3}-sec^{-1})

r_j specific rate of jth reaction (kmol-m^{-2}-sec^{-1})

r_M mass source from particles (kg-m^{-3}-sec^{-1})

r_p particle radius (m)

R gas constant (J-kmol^{-1}-°K^{-1})

R_c particle radius (m)

Re particle Reynolds number

S distance to a collision (m), control surface, source for difference equations

S_c energy to gas from particles by conduction (W-m^{-3}), Schmidt number

S_g gas emission rate (W-m^{-3})

S_i net rate of production of ith gas species due to chemical reactions (kmol-m^{-3}-sec^{-1})

S_p particle emission rate (W)

S_r radiation heating rate of gas (W-m^{-3})

S_u external momentum source to gas from particles in x direction (N-m^{-3})

S_v external momentum source to gas from particles in r direction ($N\text{-}m^{-3}$)
$S_{\bar{v}}$ total external momentum source to gas from particles ($N\text{-}m^{-3}$)
S_w external momentum source to gas from particles in θ direction ($N\text{-}m^{-3}$)
t time (sec)
T temperature (°K)
T_g gas temperature (°K)
T_p particle temperature (°K)
u total radiant energy flux crossing a surface ($W\text{-}m^{-2}$)
U gas mass flux in x direction ($kg\text{-}m^{-2}\text{-}sec^{-1}$), radiation intensity ($W\text{-}m^{-2}$) fluid velocity ($m\text{-}sec^{-1}$)
$\mathcal{U}$ radiant energy flux ($W\text{-}m^{-2}$)
$\bar{v}$ velocity ($m\text{-}sec^{-1}$)
$\bar{v}'$ fluctuating velocity ($m\text{-}sec^{-1}$)
$\bar{v}_e$ element velocity ($m\text{-}sec^{-1}$)
$\bar{v}_g$ gas velocity ($m\text{-}sec^{-1}$)
$\bar{v}_p$ particle velocity ($m\text{-}sec^{-1}$)
V gas mass flux in r direction ($kg\text{-}m^{-2}\text{-}sec^{-1}$), control volume
$\bar{w}$ particle regression velocity ($m\text{-}sec^{-1}$)
W gas mass flux in θ direction ($kg\text{-}m^{-2}\text{-}sec^{-1}$)
$\dot{W}$ work done by gas on its surroundings (W)
W_b original particle carbon content by weight
x axial space coordinate (m)
X carbon conversion
$\bar{X}_p$ particle position (m)
α inverse of the average gas density, $1/\langle\rho\rangle$ ($m^3\text{-}kg^{-1}$)
$\alpha_{i,j}$ stoichiometric coefficient, ith species and jth reaction
Γ turbulent viscosity ($kg\text{-}m^{-1}\text{-}sec^{-1}$), turbulence diffusion coefficient ($m^2\text{-}sec^{-1}$)
γ ratio of C_p and C_p
Δt time step (sec)
ΔV volume of a cell (m^3)
$\Delta V_{i,j}$ volume of cell i,j (m^3)
λ mean free path between collisions (m), average interparticle distance (m)
λ_{AS} absorption and scattering mean free path (m)
$\vec{\Omega}$ unit vector
σ Stefan-Boltzman's constant ($W\text{-}m^{-2}\text{-}°K^{-4}$)
θ local void fraction
Φ vector of dependent variables
ϕ function defined in the text
ρ gas density ($kg\text{-}m^{-3}$)
ρ_g gas density ($kg\text{-}m^{-3}$)
ρ_O original particle density ($kg\text{-}m^{-3}$)
$\bar{\tau}$ shear stress tensor in gas ($N\text{-}m^{-2}$)
χ uniform random number distribution
ξ random number distributed
ϵ void fraction
ϵ_p emissivity of the particle
μ viscosity ($N\text{-}sec\text{-}m^{-2}$)

REFERENCES

1. Blake, T. R., et al. "Computer Modelling of Coal Gasification Reactors," ERDA/DOE annual reports (June 1975–September 1979).
2. Prichett, J. W., T. R. Blake and S. K. Garg. "A Numerical Model of Fluidized Beds," In: *Fluidization: Application to Coal Conversion Processes, AIChE Symp. Series* 74 (1978).
3. Smoot, L. D., and D. T. Pratt, eds. *Pulverized Coal Combustion and Gasification* (New York: Plenum Publishing Corp., 1979).
4. Crowe, C. T., and L. D. Smoot. "Multicomponent Conversion Equations," in *Pulverized Coal Combustion and Gasification*, L. D. Smoot and D. T. Pratt, Eds. (New York: Plenum Publishing Corp., 1979).
5. Anderson, T. B., and R. Jackson. "A Fluid Mechanical Description of Fluidized Beds," *Ind. Eng. Chem.* 6:527 (1967).
6. Kansa, E. J. "An Algorithm for Multidimensional Combusting Flow Problems," *J. Comp. Phys.* 42:152 (1981).
7. Harlow, F. H., and A. A. Amsen. "A Numerical Fluid Dynamics Calculations Method for All Flow Speeds," *J. Comp. Phys.* 8:197 (1971).

8. Dietrich, D. E., B. E. McDonald and A. Warn-Varnas. "Optimal Block Implicit Relaxation," *J. Comp. Phys.* 18:421 (1975).
9. Kershaw, D. S. "The Incomplete Cholesky-Conjugate Gradient Method for Iterative Solution of Systems of Linear Equations," *J. Comp. Phys.* 26:24 (1978).
10. Evans, M. E., and F. H. Harlow. "The Particle-in-Cell Method for Hydrodynamic Calculations," Los Alamos Scientific Laboratory report LA-2139 (1959).
11. Chen, T. P., and S. C. Saxena. "A Mechanistic Model Applicable to Coal Combustion in Fluidized Beds," *AIChE Symp. Series* 74(176):149 (1978).
12. Weimer, A. W., and D. E. Clough. "Modeling of Char Particle Size/Conversion Distributions in a Fluidized-Bed Gasifier: Non-Isothermal Effects," *Powder Technol.* 27:85 (1980).
13. Zenz, F. A. "Fluidized-Bed Reactors: Design Scale-Up Problem Areas," *AIChE Today Series* 2-2 (1974).
14. Field, M. A., D. W. Gill, B. B. Morgan and P. G. W. Hawksley. *Combustion of Pulverized Coal* (Banbury, England: BCURA, Cheney & Sons, Ltd., 1967).
15. von Fredersdorff, C. G., and M. A. Elliott. "Coal Gasification," in *Chemistry of Coal Utilization*, H. H. Lowry, Ed., Supl. Vol. (New York: John Wiley and Sons, Inc., 1963).
16. Bell, G. I., and S. Glasstone. *Nuclear Reactor Theory* (New York: Van Nostrand Reinhold Co., 1970).
17. Knowlton, T. M., and I. Hirsan. In: *Fluidization*, J. R. Grace and J. M. Matsen, Eds. (New York: Plenum Publishing Corp., 1980).

CHAPTER 29
PNEUMATIC TRANSPORT

S. L. Soo
Department of Mechanical and Industrial Engineering
University of Illinois at Urbana-Champaign
Urbana, Illinois 61801

CONTENTS

INTRODUCTION

Since the late nineteenth century, pneumatic conveying has found application in various industries. Table I is a chronology of innovations in devices and applications based on a survey by Zandi [1]. Many different materials have been transported by pneumatic pipelines. Table II is an outline of materials transported and is based principally on the same survey. Although pneumatic means air action, other gases may be used. There are basically two groups of concerns in the design of a pneumatic pipeline [2].

1. Safety, health and automation are the concerns and motivations of short-distance conveyance of sensitive or inert materials in granular form.
2. Economy becomes a concern when dealing with transport over a distance, such as in pneumatic mine hoists.

Table I. Chronology of the Pneumatic Pipeline

Year	Innovator	Subject
1667	Denis Papin	Double pneumatic pump—Royal Society of London
1810	George Medhurst	Letter conveyer
1886	B. F. Sturtevant	Dust and waste removal
	A. K. Williams	Grain handling
1888	Frederic Eliot Duckham	Grain handling by suction system
1919	Alonzo G. Kinyon	Fuller-Kinyon pump positive-pressure blower, fluidization, long-distance transport
1924	J. Gasterstadt	Experimental research, paper VDI
1942	Contractor of Grand Coulee Dame	Cement transport in 355-mm pipe Over 2.3 km
1944	E.C.I. Air-Flyte Corp.	Laundry and solid waste system
1971	Radmark Engineering Co.	Mine hoist for 75-mm coal
1971	Walt Disney World	500-mm transport line for trash

At this point, most of the applications belong to group 1, in which we have sufficient experience to identify some of the limitations in detail and can account for them in design. Applications in group 2, except in a few isolated examples now, are largely in the future. Limitations in addition to those mentioned in group 1 also have been identified.

In a closed pneumatic system, the environment and hazards are readily controlled; automation via such a system reduces the number of personnel in hazardous areas and such a reduction further decreases the chance for human error. The savings achieved by these features pay back the investment rather rapidly. Often the power consumption is not a controlling factor.

The solid to gas mass flow ratio tends to be low, usually less than 1 kg solid/kg air (dilute suspension) for transport over short distances for reliability; economy is usually gained in labor savings. A long distance of several kilometers may call for transmission of a suspension of large mass ratio up to 10 kg solid/kg air, but rarely exceeds 10% by volume of solid. In such a case, one strives for economy of operation [3].

MINIMUM TRANSPORT VELOCITY

Of all the parameters to be chosen in designing a dilute gas-solid suspension system, the most fundamental is a minimum permissible gas velocity without settling of solids. The aim is to maintain a steady flow condition. A flowrate below this value leads to unstable operation and possibility of stoppage. However, operation of a system at a

Table II. Materials Transported

Materials
Fine powders, flour, chemicals
Nuclear fuel pellets in reprocessing
Granular materials, grains, explosives
Metal parts, ball bearings
Solid wastes, wood chips, plant foliage
Food, oysters, clams
Rocks, coal (up to 8 cm in size)
Metal plates (up to 15 cm long)
Whole fish
Whole chicken up to 2 kg
Laundry, sheets

velocity significantly above this velocity leads to unnecessarily high power consumption. In either of these two extreme cases the system will be operating uneconomically.

The importance of accurately predicting the minimum air flowrate as a fundamental design parameter is seen from the following:

1. Blower power requirements increase approximately as the cube of the gas velocity.
2. Pipe erosion increases significantly as gas velocity increases.
3. Particle attrition occurs at high velocity impact.

Zenz's correlation [4] is applicable to large-scale coal-conveying installations of 400-mm and 300-mm pipe diameter and coal sizes ranging from 200 mesh to less than 20 mm. His correlation is summarized as follows for $\beta > 10$:

$$\alpha = C_1 \beta^s \tag{1}$$

where $C_1 \simeq 0.90$ for spherical particles and 0.5 for angular particles, $S \simeq 0.45$, and

$$\beta = d_p/\Delta = 3[(C_D \, Re_p^2)^{1/2}] \tag{2}$$

where

$$\Delta = [(3\nu^2/4g)/(\rho^* - 1)]^{1/3} \tag{3}$$

and

$$\alpha = V_s/6.27 \, \omega \, D_m^{0.5} \tag{4}$$

where

$$\omega = [(4g\nu/3)(\rho^* - 1)]^{1/3} \tag{5}$$

and

$$Re_p = d_p \, V_g/\nu \tag{6}$$

ρ^* is the material density ratio, $\bar{\rho}_p/\bar{\rho}_g$, of particle to gas, d_p is the particle diameter, and D_m is the pipe diameter in m. Other groups are dimensionless. ν is the kinematic viscosity of the gas, g is the gravitational constant, and V_s is the minimum suspension velocity of a single particle. For the minimum transport velocity, V_c, in m/sec, an approximate relation of Zenz's may be rewritten as

$$[\dot{m}_p/(\pi/4) \, D^2 \, \bar{\rho}_p] = 0.21 \, k \, S^{1.5} \, (V_c/V_s - 1) = \dot{m}^* \, V_c \, \rho_g/\rho_p \tag{7}$$

where a suspension parameter, k, may be introduced to account for additional effects in addition to those mentioned already. By comparison to measurements of Konchesky and co-workers [5,6], one gets $k \geqq 0.81$ for 1.6-mm coal and $k \geqq 7.084$ for 38-mm coal, while the experimental data of Radmark [7] and Ball and Tweedy [8] yield $k \geqq 2.733$ to 5.551 for 75-mm coal.

These results show the level of the safety factor for the relations of Zenz for a critical design. A k value of 10 appears useful for a critical economic design for long-distance transport because of reduced end effects. For short distances of hundreds of meters, use of the minimum transport velocity is often not the controlling design parameter for an economic system.

Minimum transport velocity in vertical transport tends to be lower than in horizontal transport [6,9], but the risk of stoppage precludes utilization of such a minor difference. When there is a size distribution, settling of the largest particles should be the principal concern. The predicament of an overly critical design could be for a vertical section to start acting on a fluidized bed that becomes dense with time and for a horizontal pipe to develop dunes that grow in size. Fundamental relations of an inclined pipe were treated by Soo and Tung [10], but no specific experimental results are available.

When dealing with fines below 10 μm in size, the minimum transport velocity often is determined by the magnitude of turbulent diffusion necessary to overcome electrostatic deposition [10] (see the section entitled Electrostatic Effects and Hazards).

In designing a pneumatic system especially for handling hazardous materials, a higher speed than minimum suspension velocity is usually chosen to avoid stoppage and its related danger. In doing so, we have to live with the fines generated by impact and provide safety measures.

FRICTION FACTOR

An extensive systematic effort to correlate the pressure drop in pipe flow with suspension has been made by Pfeffer et al. [11]. Because of wide data scatter, no general correlation is possible; however, some trends have been noted. Their data are based on an extensive survey, to which one may add other results [5,12].

For isothermal pipe flow of a gaseous suspension in the system with pipe diameter D, length L, elevation H at one end and total flow of solids and air, m_p and m_a, respectively, the pressure drop, dP, over a length, dx, is given by

$$dP = -4\, f_m (dx/D)[\bar{\rho}_g (V_g^2/2)] - [d(\rho_m V_g^2)] - \rho_m\, g\, dH \tag{8}$$

where f_m is the friction factor of the mixture of gas and solids, V_g is the gas velocity, ρ_m is the density of mixture, $\rho_m = \rho_g + \rho_p$, g is the gravitational acceleration, ρ_g, in kg/m, and gH is the rise in elevation over dx. In Equation 8, the first term on the right-hand side is pressure drop due to friction, the second term is acceleration and the third term is the gravity effect due to change in elevation. Note that mass flow, G_g, is given by

$$\rho_{g1} V_{g1} = \rho_{g2} V_{g2} = \rho_g V_g = \dot{m}_g/(\pi/4)\, D^2 = G_g \tag{9}$$

Subscripts 1 and 2 denote inlet and outlet, respectively.

The friction factor of turbulent pipe flow of a simple fluid, e.g., air, in a smooth pipe is given by

$$f_g = 0.046/Re^{0.2} \tag{10}$$

where the Reynolds number, Re, is given by

$$Re = D\, \bar{\rho}_g V_g/\mu_g = D\, G_g/\mu_g \tag{11}$$

where μ_g is the viscosity of gas.

For small changes in pressure or density of the gas phase, Equation 8 can be integrated as an incompressible fluid [13,14]:

$$P_1 - P_2 = 4f_m (L/D)(G_g^2/2\bar{\rho}_g) + (1+\dot{m}^*)\rho_g (V_{g2}^2 - V_{g1}^2) + (1+\dot{m}^*)\, \rho_g H_g \tag{12}$$

for small ϕ, $P \simeq \bar{P}$ of the gas; $\bar{P} = (P_1 + P_2)/2$. For large changes of pressure, P_1 to P_2, of the gas, the effect of compressibility is accounted for via integration to

$$\begin{aligned}[1 - (P_2^2/P_1^2)] = {} & 4f_m (L/D)(\rho_{g1}^2 V_{g1}^2 R_g T/P_1^2) \\ & + (1+\dot{m}^*)(2G_g^2 R_g T/P_1^2)\, ln\, (P_1/P_2) \\ & + (1+\dot{m}^*)(2\bar{P}^2 g H/R_g T P_1^2)\end{aligned} \tag{13}$$

Konchesky's results appear to be adequately correlated by the relation proposed by Dogin and Lebedev [15], according to

$$f_m = f_g + A(d_p/D)^{0.1}\, Re^{0.4}\, Fr^{-0.5}\, (\bar{\rho}_p/\bar{\rho}_g)\, \dot{m}^* \tag{14}$$

where Fr is the Froude number,

$$Fr = V_g^2/g\,D \tag{15}$$

which accounts for the gravity effect in horizontal pipe flow. A is a parameter depending on the roughness of the pipe. Based on the data of Konchesky,

$$A \simeq 2 \cdot 10^{-7} \tag{16}$$

instead of $10^{-6} < A < 2.10^{-6}$ proposed by Dogin and Lebedev [15]. $A = 2.10^{-6}$ seems to correlate the data on coal dust.

Another correlation [16] to account for the effect of mass flow of solid and density ratio of phases can be expressed in the form

$$f_m = f_g + (\pi/8)\, \dot{m}^*\, (\bar{\rho}_p/\bar{\rho}_g)^{1/2}\, \psi \tag{17}$$

where ψ is a function of Re having a value below 10^{-5} for Re > 35,000; however, calculations from Konchesky's data give $\psi \simeq 10^{-4}$.

Pfeffer et al. [11], coding various sources of data, suggested

$$f_m = f_g(1 + \dot{m}^*)^{0.3} \tag{18}$$

without regard to other factors. This correlation tends to give an optimistic estimate of pressure drop, and we may treat it as the lower bound of f_m.

HEAT TRANSFER AND HUMIDITY CONTROL

Heat transfer to pipes conveying a suspension is important, especially when the pipe is installed outdoors. Proper insulation and humidity control of the carrier gas might be needed to ensure satisfactory operation at various weather conditions (humidity control, frost prevention and preservation of material properties). Heat transfer and humidity need to be checked carefully to prevent plugging by condensation. The data are plentiful but also anomalous [11].

All refer to loading, that is, the ratio of the rates of throughput of particles to gas. This is at least one source of anomaly besides m^* because modification of heat transfer rate of a gas by the added particles is influenced by the fluid-particle interaction and electrostatic charge on the particles [17]. As a matter of logical choice, some authors based their correlations on a weighted specific heat ratio:

$$C^* = [(c_p\, \rho_p\, V_p)/(c_g\, \rho_g\, V_g)] = m^*(c_p/c_g)(V_p/V_g) \tag{19}$$

Denoting h_m as the heat transfer coefficient of the suspension and h as that due to gas alone, various correlations have been presented. Wachtell et al. [18] suggested correlation with

$$h_m/h = 16.9\, (Re)^{-0.3}\, (1 + C^*)^{0.45} \tag{20}$$

which appears to cover a relatively wide range of data. Their studies used tube diameters of 8 mm and 22 mm and $\dot{m} = 0$–90.

A case in which an application is seen in a pellet packaging system is shown in Figure 1. In winter, pellets may be delivered at outdoor temperatures. If they are taken into

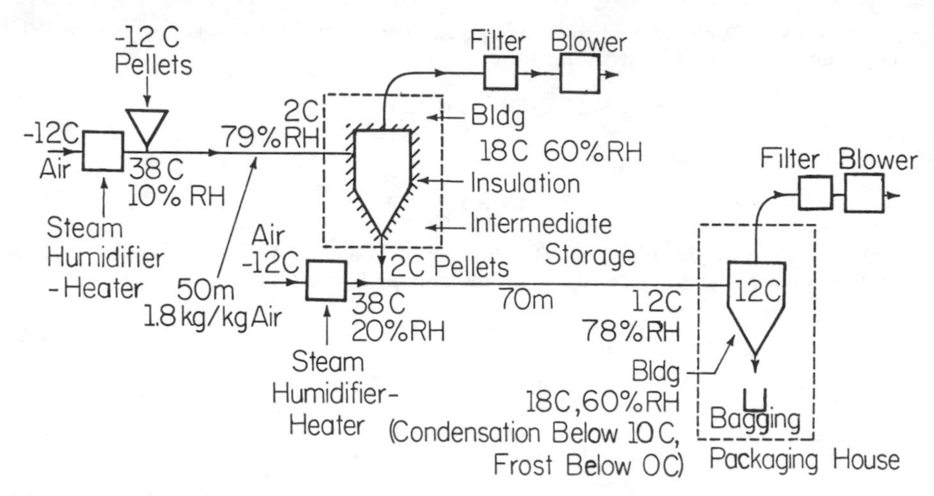

Figure 1. Humidity and temperature control during winter operation to prevent condensation and preserve weighing accuracy.

the packaging operation kept at room conditions of 18° C and 60% relative humidity, frost will form on the pellets, thus affecting the weighing operation. The design chosen, together with hazard isolation and humidity control of static electricity, is shown in Figure 1. Not only is humidity controlled, but pellets are heated by the conveying air in two stages: one into the intermediate storage, another into the packaging house. Of course, the pipes are insulated from outdoor temperatures. No frost or dew was then formed on the pellets.

The same pipe insulation is often desirable in summer operation.

ESSENTIAL COMPONENTS

Basic components of a steady-flow pneumatic transport system include bins from which particles are fed, the pipeline, bends and branches, eventual collection devices and blowers. An unsteady flow system may include blowtanks and blowing at high pressures with compressors charging the tanks. This section highlights some of the essential fractures and cautions in design.

Feed Valves

The conveyance air velocity is often chosen to be double that of minimum transport velocity to make sure there is no stoppage from the point of view of safety or process continuity.

The case that such a specification might not be enough is illustrated in Figure 2, which is quite common for transferring solid particles from a bin into a conveying line, using a rotary valve. The situation may happen to be that the air velocity in the small pipe in Figure 2a is sufficient to transport the solid, but with the enlargement at the tee, the air velocity would be greatly reduced locally and clogging may occur. A better proportion is as shown in Figure 2b, in which the tee will not affect the flow velocity greatly.

The humidity of air in such a system has other effects. In one case of double jeopardy, the situation in Figure 2a combined with high humidity in air in the pipe and the hygroscopic powder material became soggy at room temperature. This not only enhanced plugging, but the rotary valve became stuck and twisted off the worm gear

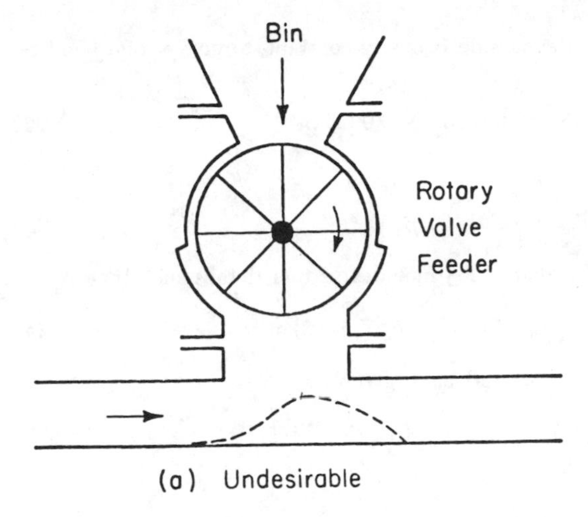

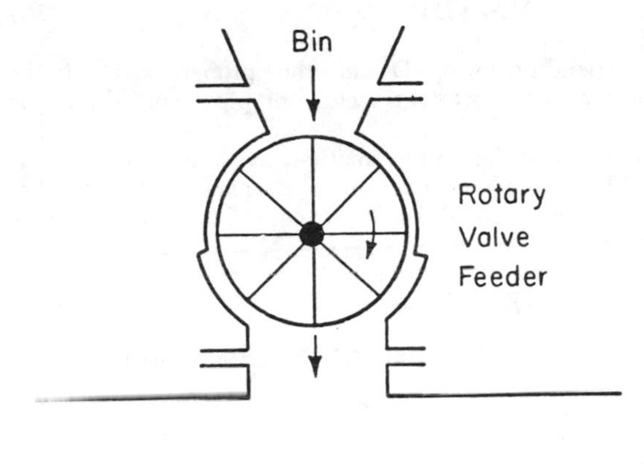

Figure 2. Proportion of feed connection to pipeline.

drive shaft. The solution was to raise the air temperature by 20° C in that particular case to keep the system dry.

Bends and Branching

Pipe bends and branching provide pneumatic-conveying systems with a large degree of flexibility by allowing routing and distribution. Limited experimental results of Mason and Smith [19] suggest a correlation of bend resistance number:

$$\psi_{bp} = 0.028\,\dot{m}^* - 0.0025\,\dot{m}^{*2} \tag{21}$$

for bends of a ratio of curvature $D_b/D \simeq 20$ (where D_b is twice the radius of the bend for a Reynolds number of nearly 10^5). D typically has values of 51 mm and, for alumina particles, 15 to 70 μm. The loss on the air side is nearly constant, to give a total bend resistance number of

$$\psi_{bf} = 0.025 + 0.027\,\dot{m}^* - 0.0025\,\dot{m}^{*2} \quad (22)$$

The pressure drop is given by

$$\Delta P_{bf} = \psi_{bf}(1/2)\,\bar{\rho}_g\,V_g^2 \quad (23)$$

The correlation by Ikemori and Munakata [20] includes further details such that

$$\Delta P_{bf} = f_{zb}(L_b/D)\,\dot{m}^*\,[(\bar{\rho}_g\,V_g^2)/2] \quad (24)$$

where f_{zb} is the friction factor for a horizontal band given by

$$f_{zb} = f_{pb}(V_{ps}/V_g) + [4(V_{pt}/V_g)(V_{ps}/V_g)]/(D_b/D) \quad (25)$$

where f_{pb} is given approximately by

$$f_{pb} = K_p/(V_g^2/g\,D)^{1/2} \quad (26)$$

where K_p is an empirical constant depending on d_p, D and other properties, V_{ps} is the fully developed mean particle velocity in the straight section of pipe, and V_{pt} is the terminal velocity of the particles.

Branched piping is used widely for distribution in pneumatic-conveying systems. However, very little has been quantified beyond a recent report of Harman et al. [21]. Applicable systems are in the forms of a tee and a manifold. In the case of a tee, the pressure loss in branching is given by

$$\Delta P_{Br} = \zeta_1[(\bar{\rho}_g\,V_g^2)/2] \quad (27)$$

where ζ_1 is called Morikawa's branching coefficient [22]. The trend is seen that for an air suspension (subscript a)

$$\xi_{ca} = -0.1 \text{ for } 0 \leq n \leq 0.4$$
$$\xi_{ca} = 0.8\,n - 0.4 \text{ for } 0.4 \leq n \leq 1$$

where n is the ratio of lateral flow from the branch point to header flow into the branch point. The total branching factor, ξ_c, is given by

$$\xi_c = \alpha' + \xi_{ca} \quad (28)$$

where α' is the solid component of Morikawa's correlation and is given by

$$\alpha' = 0.264 - 1.0825\,n\,\dot{m}^* \quad (29)$$

For variable flow, minimum transport velocity in each branch needs to be checked to avoid deposition on the one hand and to make sure that there is ample available power for the throughput on the other.

Another form of branching is for delivering into one branch at a time, such as in using rubberized pitch valves in packaging systems. In that case, the branches serve as turns from a tee.

Still another form of branching is the feeding from several bins intermittently into a pipeline by valving. The volume of the transition piece between the valve and the pipeline should not be so large that plugging may occur because the air velocity slows down; this tends to happen especially when the valve is in a closed position. In general, branches should be made with smooth tees only.

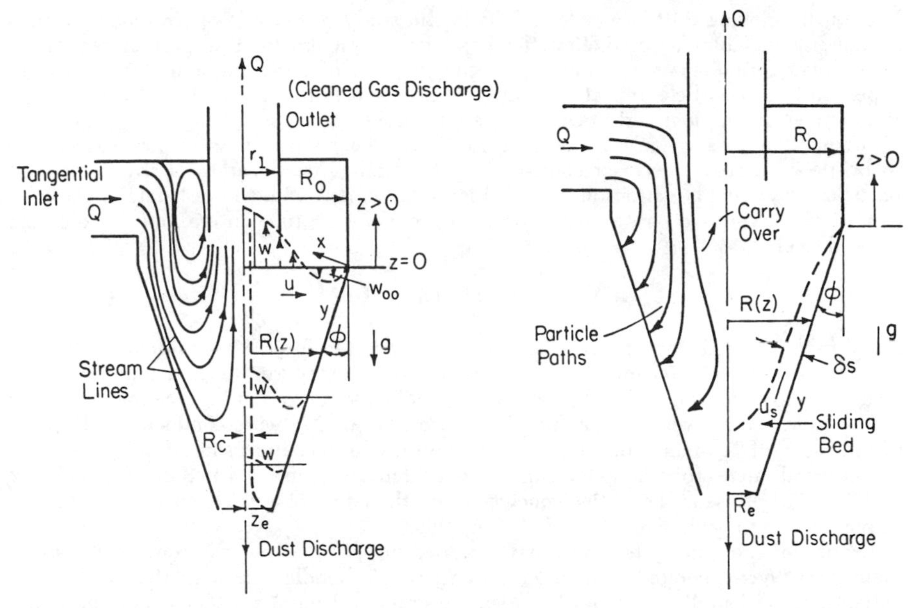

Figure 3. Conditions inside a cyclone separator and coordinates [23]. Left: fluid velocity distribution; right: particle path and sliding bed.

Cyclone Collectors

It was shown via a rigorous integration procedure that the collection efficiency, η_c, of a cyclone separator with a steep cone is given by [23]

$$\eta_c = 1 - \exp\left[-\sigma\left(\frac{2\pi\, C^2\, L}{F\, Q_g R^2} + 8\pi\,\frac{\rho_{p1}\, q^2\, R^2\, L}{\epsilon_o\, m^2\, F\, Q_g} + 2\pi\, E_o\,\frac{q}{m}\,\frac{RL}{FQ_g}\right)\right]. \tag{30}$$

where σ is the sticking probability of the particles to the wall or the layer of collected particles, C is the maximum vorticity of the system, L is the length and R is the maximum radius of the cyclone, Q_g is the volume flowrate, $F = 3\pi\,\mu\, d/m$ is the inverse relaxation time of particle-fluid momentum transfer for particle mass m, size d and fluid viscosity μ, ρ_{p1} is the density of the particle cloud per unit volume of the mixture, ϵ_o is the permittivity of free space, q is the charge per particle, and E_o is the applied electric field. The first term in parentheses is the parameter for collection by centrifugal force exerted by the vortex; the second term is that by space charge of the particles; and the third term is that by the applied electric field. The magnitude of σ depends on contacting surfaces and depositing forces. It was shown that for continuous removal of particles at the bottom of the collector cone a high value of sticking probability is often not desirable because it would hamper continuous removal of collected particles. However, this is not a concern when a cyclone separator is used for collecting aerosol samples and sticking probability near 1 will be desirable. Consistency of results is achieved via controlled test conditions. The flow relations are shown in Figure 3.

EROSION AND ATTRITION

Erosion by impact of particles on parts of the pneumatic piping system causes concerns of wear in the system and contamination of the conveyed solid by the eroded material.

Erosion by moving dust is a special field in the study of wear [24]. Insufficient information is available for predicting the behavior of pneumatic transport systems, and even fewer formulations are suitable for the purpose of design. For a straight run of pipe, erosion by particle impact is small in comparison to that at bends and turns where the number of collisions and force of impact are large.

Among materials, nylon is susceptible to ductile wear while glass is more susceptible to brittle wear than steel. Correlation of ductile and brittle wear by a stream of dust particles impacting at various angles [25] is in agreement with the test data of Arundel et al. [26], giving wear rate in volume removed per unit time in cm^3/min of alumina on mild steel as

$$W = 4 \times 10^{-8} (\dot{m}^*)^{0.5} V^{3.3} \tag{31}$$

for V in m/sec and impact angle between 20° and 30° for a 4.8-mm-diameter nozzle giving a velocity of 300 m/sec. These results from short-term (1 min), high-velocity tests have to be extended by theoretical correlations when designing systems operating below 50 m/sec gas velocity. Tests by Mills and Mason [27] were conducted with pipes of mild steel of 140-mm bend radius and 50-mm inside diameter using 70-μm sand. The mass eroded ranges from 20 g/ton conveyed at 26 m/sec at $\dot{m}^* = 4$ to 3 g/ton at $\dot{m}^* = 2$ at 25 m/sec with sand. Definitive conclusions on the effect of particle shape and concentration remain to be drawn.

Erosion of the pipe system also may produce fine dust. However, concern is caused mainly by fines generated from hazardous materials handled through the system. Attrition in grain handling systems degrades the grain and increases the waste, deposition of fines and dust explosion hazard.

Attrition of material by impact has not been documented extensively. Mills and Mason [28] reported degradation of quartz particles in various pipe materials (steel, nylon and fiberglass) and the degree of fragmentation of quartz particles on steel. Available data on attrition are meager. Certain facts are demonstrated by experimental results on particle degradation, particularly at pipe bends. The fines generated tend to give a bimodal distribution. In the case of quartz particles on steel, for an initial mean size of 180 μm the product of fragmentation distribution was around 40 μm.

The extent of attrition depends both on the number of impacts a particle undergoes as it proceeds through the pipe system and the compressive force acting on a particle during each impact. Hence, the major cause of attrition is the impact at pipe bends. The pressure at the point of contact during a collision may crush the particle locally. The maximum local compressive stress at the point of impact is given by [29]

$$P = K_A E_p (\bar{\rho}_p U_p^2 \sin\theta / E_p)^{1/5} (1 - \nu_p^2)^{-4/5} \tag{32}$$

for the case in which the modulus of elasticity of the wall, E_w, is much greater than that of the particle material, E_p, ν_p is the Poisson ratio, and θ is the angle of impact between the velocity vector and the wall. The proportionality constant, K_A, in Equation 32 is around 0.17 and depends on details of shape, air resistance, etc. The above relation to velocity is verified in part by the experimental results of Goodwin et al. [30], for single impact of quartz particles of 125–150 μm on 11% chrome steel targets. Their results indicate a threshold velocity of 15 m/sec for breakage of the particles. This quantity, when substituted into Equation 32 for a modulus of elasticity of 7×10^{10} Pa for quartz, gives a compressive strength of 10^9 Pa, which is perhaps three times larger than that from testing a large specimen because of sharp corners of the test particles. If this value of P or P_{th} for the threshold compressive stress is expressed in terms of $(P - P_{th})/P_{th}$ and is plotted versus fraction fragmented, a straight line is obtained:

$$\text{fraction fragmented} = 0.27(P - P_{th})/P_{th} \tag{33}$$

for velocities between 15 m/sec and 300 m/sec. Other results show saturation at large velocities for large particles and lack of fragmentation for small particles below 20 m for a target of steel and 60 μm for a target of nylon. This is because of significant slowing down of fine particles by air near the target surface. Other materials, such as TNT, depending on their shape, have a fracture strength of 3×10^7 Pa.

Depending on materials and flow velocities, attrition of particles in a recycled test system tends to reach an asymptotic size, giving the experimental limit [31].

ELECTROSTATIC EFFECTS AND HAZARDS

The major hazard in pneumatic conveyance of powder and bulk materials arises from electrostatic effects [2]. The problems encountered may arise from fines generated by attrition or handling of fine powders. Interruption of a process may occur by electrostatic deposition even when the powders are nonreactive. The danger of explosion exists when handling detonatable explosives or combustible powders such as flour in air. The latter often has been attributed erroneously to spontaneous combustion.

Particle Charge

Charging of solid particles by surface contact has been studied extensively [31]. However, prediction is still uncertain because charging is influenced strongly by the nature of contaminants on contacting surfaces. Because particles are, in general, nonspherical, a lower limit of saturation charge is often the breakdown electric field, E_b, at a sharp point of an irregular shaped particle of radius of curvature a_s, which is given by

$$E_b = q/4\pi \epsilon_o a_s^2 \tag{34}$$

where ϵ_o is the permittivity of free space. The charge to mass ratio of such a particle is thus limited to, for nominal radius a_p,

$$q/m = 3(\epsilon_o E_b/a_p)(a_s/a_p)^2 \tag{35}$$

For a breakdown field of, say, 5 kV/cm in air for a 10-μm particle, the limiting charge will be 10^{-3} C/kg for $a_s/a_p \simeq 1/50$. The breakdown electric field in air is strongly influenced by its humidity.

Measurements on air suspensions of polystyrene particles (25–250 μm) in brass and nylon pipes (22-mm diameter up to 45 m long) with mass flow ratios of up to 3 were made by Cole et al. [32]. The nylon pipe was covered externally by aluminum foil and current to ground to pipe sections of various lengths were measured at air flow velocities of up to 100 m/sec. Saturation of charging was indicated by the relations of current to ground versus pipe length to give the following charge to mass ratios at particle sizes ranging from 25 to 250 μm:

- 25 μm, 5.613×10^{-5} C/kg
- 125 μm, 9.225×10^{-6} C/kg
- 250 μm, 3.456×10^{-6} C/kg

Charging by contact and separation includes separation from long duration of contact for particles in a container. Many solid particles become charged when poured from containers of polyethylene, glass, metals or cardboard. Charge elimination needs to be considered at unloading points of a processing facility. The above relations of particle charging by surface contact and the upper limit of particle charge to mass ratio confirms that fine particles tend to be a greater problem than large particles.

In the case of the example at the end of the section entitled Erosion and Attrition, it is seen that for 2.2 kg/m³ of particles originally charged to 10^{-5} C/kg conveyed at a velocity of 30 m/sec, fragmentation of 3% to 40 μm with 10^{-3} C/kg will increase the volumetric charge density to 8.7×10^{-5} C/m³, or four times the original value. Hence, attention should be given to the expected amount of fines to be generated in the prototype when conducting model tests.

Electrostatic Deposition

Fines are susceptible to electrostatic deposition on the walls and components of a conveying system. Electrostatic deposition in a pipe begins to occur when the volume fraction solid, ϕ, in a suspension is such that the space charge force begins to form a wall layer with volume fraction equal to that of a sliding or packed bed or, according to Soo and Tung [10],

$$\phi > 2/\bar{\alpha}$$

where the characteristic number $\bar{\alpha}$ correlated space charge effect and particle diffusion are given by

$$\bar{\alpha} = \bar{\rho}_p (q/m)^2 R^2/4\epsilon_o D_p F \tag{36}$$

and D_p is the particle diffusivity produced by the fluid turbulence or other excitations, F is the inverse relaxation time for momentum transfer, $F = 9\bar{\mu}/2a^2 \bar{\rho}_p$ for Re_p based on relative motion less than 1, and $F = 0.33(\bar{\rho}/\bar{\rho}_p)\ (U - U_p)/d$ for $Re_p > 100$ [33]. The thickness of the deposited layer, δ_s, increases with time t according to

$$d\,\delta_s/dt = \rho_{pR}\, u_{pR}/\rho_{ps} \tag{37}$$

where ρ_{pR} and u_{pR} are the particle cloud density and depositing velocity at the wall, and ρ_{ps} is the density of the sliding bed of the same solid. u_{pR} is given by

$$u_{pR} = [\rho_p (q/m)^2 R/2\epsilon_o F] - D_p\, \partial\phi/\partial r|_R \tag{38}$$

The last term gives the velocity of reentrainment by diffusion. This steady deposition may continue until the pipe is plugged or may include periodic bed movement in the form of dunes with repeated pileup and blowing away.

Where electrostatic discharges from layers of powder are shown [34] to be rather uncertain, a bootstrap mechanism was suggested. A 1-mm-thick layer may constitute a significant hazard [35].

Voltage and Discharge in Piping and Components

The minimum concentration and the threshold energy for the initiation of an explosion by electrostatic discharges are given in Table III, collected from various sources and for various materials [29]. We note that grain dust in air is more hazardous than TNT in its susceptibility to initiation by electrostatic discharges. Static electricity has been shown to be the major cause of explosion in grain storages because of the charge generated in grain dust during handling [35]. The scaling relation is such that pipe diameters less than 1 cm are completely safe from discharge [35,36].

Particle charges after passing over a length of pipe may achieve 10^{-3} C/kg for 40-μm particles and 2×10^{-6} C/kg for 2-mm particles [23,37]. This is true for coal, polyethylene and a number of explosives such as TNT chips and pellets of smokeless powder. When charges of those magnitudes are delivered into a cyclone or bins for the collection of particles, and if the cyclone interior becomes insulated from ground by a layer of deposit of the material handled, the cyclone will act as a van de Graaff generator. The space charge voltage of the particle cloud is approximately

$$V = \rho_p (q/m)\, R^2/3\epsilon_o \tag{39}$$

An overall check of various hoppers and cyclones in a typical pneumatic-conveying system gave voltages ranging from 9 to 40 kV when parts become ungrounded. Capacitances of various plant components are given by Gibson [38].

Table III. Initiation by Electrostatic Discharge [29,35]

Material	Size	Lower Limit of Concentration (kg/m³)	Threshold Energy (J)
In Air			
Flour, Maize	<50 μm	0.059	<0.1–0.2
Flour, Soya	<50 μm	0.066	<0.1–0.2
Cornstarch		0.045	0.04
Powdered Sugar		0.05	0.03
Aspirin Dust		0.05	0.025
Hexamine Dust		0.4	0.13–0.0013
Coal	<44 μm	0.74	
Aluminum Flakes		0.045	0.01
Sulfur		0.035	0.015
Cellulon Acetate		0.04	0.015
Cellulon Acetate			0.015
Sawdust			0.015
Air Not Needed			
TNT Flake	5 mm		0.5
TNT Dust	40 μm	0.05	0.075
Composition B Dust	40 μm	0.065	0.024–0.075
Pellet, Smokeless Powder	4 mm		>5
Dust, Smokeless Powder	<40 μm		0.075

Current to Ground

When $V_w - V_p$ is finite, collisions of a particle cloud continue to produce charge transfer even when the particle charge does not increase. For a pipe section of length L and radius R, the current to ground is

$$i = \dot{m}_p (Q/m) (\langle v_p^2 \rangle^{1/2} / U_p) (L/2R) \propto U_p^{1.6} \tag{40}$$

where $\dot{m}_p$ is the flowrate of particles in kg/sec. For 16-μm coal particles, at a charge transfer per impact of $Q = 10^{-16}$ C, $L/2R = 8$, and at a total flow of 0.3 kg/sec $i_G = 4.3 \times 10^{-6}$ amp. This is not a large current, but loss of ground will lead to rapid buildup of voltage to, say, 10 kV. The proportionality in Equation 40 agrees with the experimental results of Masuda et al. [39].

It is readily shown that the voltage buildup for finite ground resistance, R, is given by the relaxation mechanism [40]:

$$V = Q/C = R\, i_G (1 - e^{-t/RC}) \tag{41}$$

for capacitance, C, of the pipe section. This confirms the speculation that a resistance to ground of 10^6 Ω might be tolerated [34] and, in the above example, 4 V may be reached.

CHARGE REDUCTION AND SAFETY MEASURES

The primary safety procedure toward reducing the deposition of material is proper grounding at every flange of the piping to reduce the charging and chance of arcover within the pipe. Painstaking design for grounding all components in the system is vital, including all values. When rubber pinch valves are used, the rubber should be metallized and properly grounded electrically.

When handling nonreactive material such as polyethylene, deposition often occurs in cyclones and bins, which prevents proper unloading. This difficulty can be eased by vibrators attached to the apparatus or by using nuclear static eliminators [41].

Humidity Control

When handling combustible or explosive materials, a nuclear static eliminator should not be used because of the danger of nuclear contamination of a wide area if explosion occurs. The available means is humidity control. Even though the mechanism of static elimination by humidity is not clearly known, measurements have shown charge reduction by one-half at 15% relative humidity and to zero at 60% relative humidity at room conditions [41]. Alteration of particle charge over a range can be affected by humidity control. Figure 4 gives the distribution of charge to mass ratio of 83-μm glass particles at various relative humidities correlated [29] from the data of Turner and Balasubramanian [42]. The design in Figure 1 also illustrates the control of humidity.

Use of humidity for charge elimination must be checked against condensation and frost formation. The heat transfer relation should therefore be checked. In the case of explosives, final collection of fines must be done in a wet collector.

Dilution

Humidity control is not suitable for a hygroscopic material such as flour, which is known to be an explosion hazard. In this case, grounding and dilution are left as the

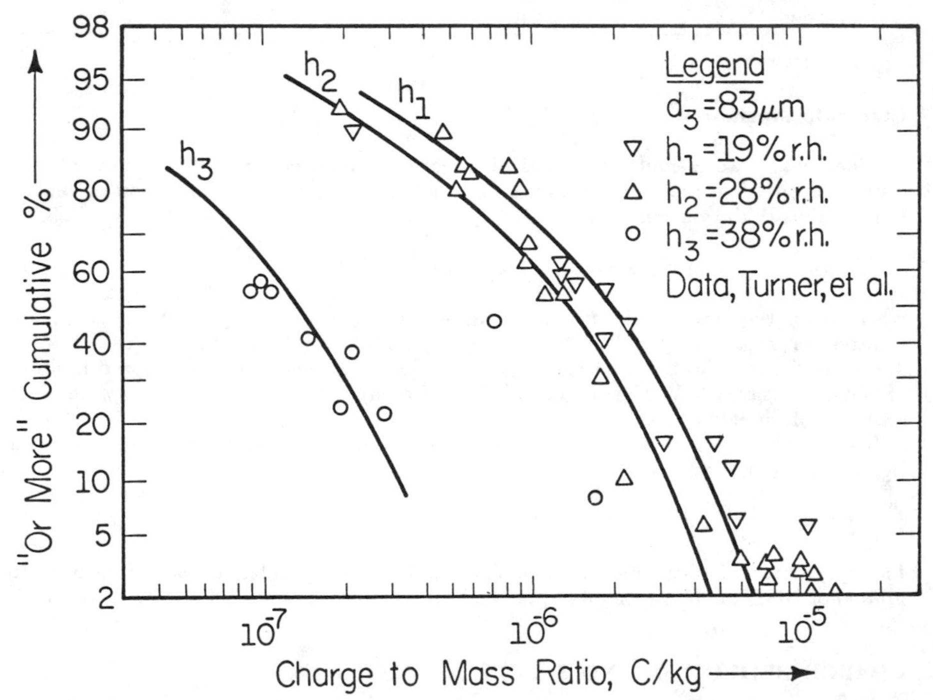

Figure 4. Cumulative percentage of particles with charge to mass ratio equal to or greater than a certain value.

basic means of charge control. A low mass flow ratio far below 1 should be maintained in the dry sections with collection of fines in a wet collector.

Control by dilution means using a dilute suspension and increased pumping power via a large pipe or a high velocity. High velocity above 30 m/sec is desired for most materials of a few mm size to prevent stoppage. This is often double the minimum suspension velocity. For handling explosives, 100% additional blower power will be desirable to ensure that no plugging will occur. Positive displacement-type blowers are preferred to centrifugal type to avoid the surge characteristics of the latter.

Other Prevention and Maintenance

When handling hazardous substances, prevention of fusing from one storage point to the other can be achieved by alternate durations of operation. Where extreme caution is justified, intermittent feeding may be adopted to prevent fusing within the same line.

In almost all cases, accumulation of fines below 40 μm may cause an explosion hazard. This concern often justifies thorough periodic washing of the whole system with water. Hazard control also may include flame detectors and deluge. Bins and cyclone separators are to be equipped with spring-loaded doors in addition to rupture discs on pipes and smaller components.

Coal dust used in power plants is normally conveyed with preheated air at a mass ratio of one, which is usually far below the oxygen concentration to sustain burning. A hazardous condition exists when a sufficient concentration of volatile hydrocarbon from coal accumulates in the system.

In a pneumatic conveying system, safety needs often dictate the design instead of optimum power consumption.

Model Testing

Prior to building a pneumatic system for handling hazardous materials, it is often desirable to test dummy particles in a scaled-down model [36].

In a case in which a prototype was modeled for particle density and velocity, a 19-mm-diameter pipe system was used to model and predict the performance of a 152-mm-diameter system. For particles, velocity and mass flow ratio chosen for the prototype, the flow velocity of the model, which was chosen to be 10 m/sec for large particles of 3 mm in diameter, is 30 m/sec in the prototype. For similarity in drag force, hollow glass particles are available at the specified mass per unit volume of particles. Next, the charge-to-mass ratio is determined. For an ungrounded section, we note that when the model gives 1.9 kV, we could expect 61 kV to be reached in the prototype when its q/m is 10^{-5} C/kg. The prediction and scaling can be complicated further by the presence of fines. If the above model tests were made with particles, velocities and mass flow ratios similar to those in the prototype, the voltage reached in the model would be 4.2 kV, but the distribution due to momentum transfer would not be similar.

LONG-DISTANCE TRANSPORT

Concerns of energy efficiency arise when materials are to be transferred in dry form over a distance. The case in Table I of transporting cement over 2.3 km is an example. Possibilities in the near future may include transport of coal over a distance of up to 10 km. This permits storage of coal away from congested areas and transfer to a power plant, for instance, after preliminary crushing to 6 mm or larger. Use also can be seen in rapid transshipment by barges or ships or gathering from mines (Table I on mine hoists). For the latter applications, pipes would be easier to use than a conveyor belt system. All these possibilities improve the overall energy or material supply options.

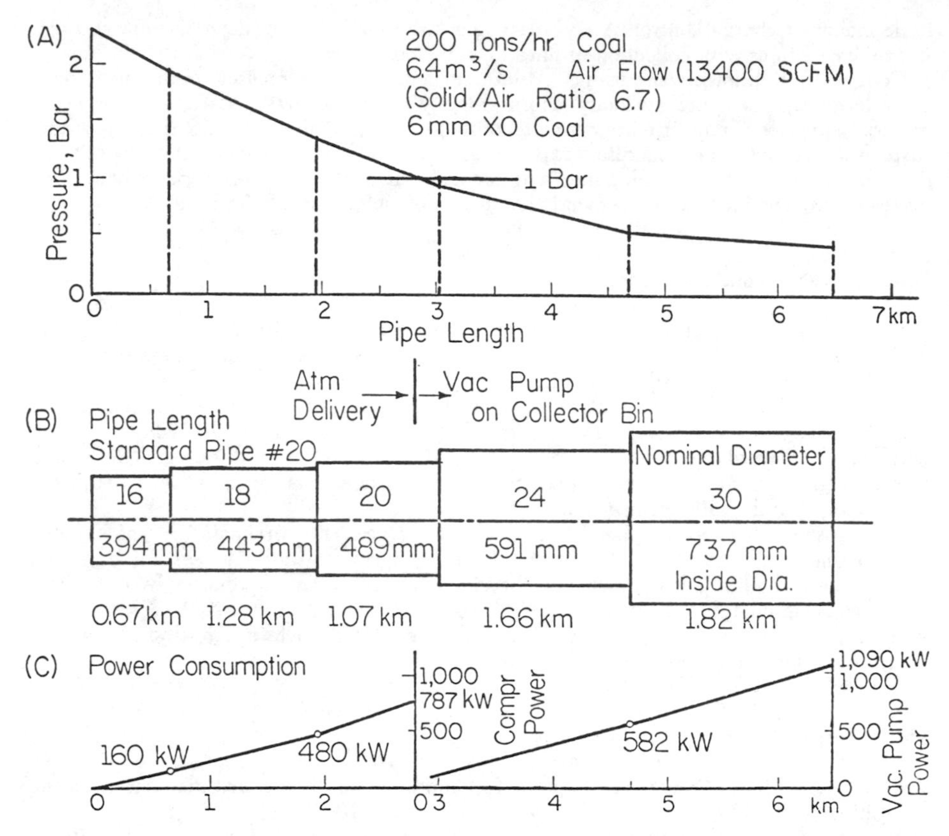

Figure 5. Dimensions and power requirements of a 200-ton/hr pneumatic pipeline.

When transporting over a distance, a pneumatic system differs significantly from a hydraulic transfer system because of compressibility of air. For a uniform pipe size, the flow velocity will increase greatly as the pressure is lowered. Hence, when transporting over a distance the economical design is one in which as the pressure lowers the pipe diameter is changed to the next larger standard-sized pipe, so that minimum transport velocity is maintained at the inlet of each section of pipe. This measure is called telescoping.

Table IV. Pipelines for 6 mm × O Coal (Figure 5)

Total Length (km)	Inlet Pressure Bar	Pipe (km)	Description (Figure 5)	Compressor Power (kW)	Vacuum Power (kW)
1	1.45	1	(from 1.95 km)	250	Atmospheric delivery
2	1.86	1.14 0.86	(from 0.81 km) (from 1.95 km)	500	Atmospheric delivery
2.5	2.31	2.5	(from 0)	700	Atmospheric delivery
3	2.31	3	(from 0)	790	100
4	2.31	4	(from 0)	790	400
5	2.31	5	(from 0)	790	660
6.5	2.31	6.5	(from 0)	790	1090

As an example, the basic dimensions of a telescoped pneumatic pipeline for transferring 6 mm × 0 size coal over a distance to 6.5 km at the rate of 200 ton/hr is shown in Figure 5. Figure 5 presents basic information on (A) the pressure variation over distances along the pipeline; (B) the length of each branch of standard-sized pipes of Schedule 20 of various nominal diameters (in.) and inside diameters (mm); and (C) the power consumption of the compressor (80% efficiency) for delivery at above atmosphere pressure and of the vacuum pump (75% efficiency) for delivery at below atmosphere pressure. The design constraint here is that the rotary airlock for the inlet hopper is limited to 1.36 bar pressure difference, while the cyclone at the delivery point is limited by the pressure of 0.32 bar that can be produced by a positive displacement blower, still using airlock for delivery below atmosphere pressure.

The design in Figure 5 is applicable to various distances of transport according to the selections in Table IV.

For discharge at a higher elevation than the inlet point, for this present design 1 m of elevation is equivalent to 1.6 m of length. For larger-sized coal, such as 50 mm × 0, a preference is seen [3] for using the blow tank system instead of the push-pull system in the above.

Transportation of solids over distance on a continuing basis represents a relatively new potential application of pneumatic pipelines.

NOMENCLATURE

Letter Symbols

A	pipe cross-sectional area (m^2)
a	radius (m)
a_p	particle radius (m)
C	capacitance (Farad)
C	vorticity of fluid motion (m^2/sec)
c	specific heat (J/kg-°K)
D_p	particle diffusivity (m^2/sec)
D	pipe diameter (m)
d_p	particle diameter (m)
E	electric field (V/m)
F	inverse relaxation time for fluid-particle momentum transfer (sec^{-1})
G	mass flow (kg/m^2-sec)
g	gravitational constant (W/m^2-°K)
H	elevation or height (m)
h	heat transfer coefficient (W/m^2-°K)
i_G	current to ground (amp)
L	length or pipe length (m)
m	particle mass (kg)
$\dot{m}$	flowrate (kg/sec)
P	static pressure or stress (N/m^2)
Q	charge transfer per impact (C)
Q_g	volume flowrate (m^3/sec)
q	electrostatic charge per particle (C)
R	gas constant (J/kg-°K)
R	pipe radius or container radius (m)
R	resistance (ohm)
t	time (sec)
u_{pR}	depositing velocity (m/sec)
U_p	mean velocity of particles (m/sec)
V	electric potential (V)
V_s	minimum suspension velocity for a single particle (m/sec)
V	nominal velocity ($\dot{m}/\rho A$) (m/sec)
W	volume removed by wear per unit time (m^3/sec)
W_D	volume removed per unit area per unit time (m^3/m^2-sec)
x	length along pipe (m)
ΔP	pressure drop (N/m^2)
$\langle v_p^2 \rangle^{1/2}$	root mean squared random velocity of particles (m/sec)

Dimensionless Group

A	correlation constant as defined in Equation 14
C_1	constant as defined in Equation 1
C^*	heat capacity ratio
C_D	drag coefficient, $4gd_p(\bar{\rho}_p - \bar{\rho}_g)/3\rho_g V_t^2$
f	friction coefficient

F_r Froude number (V^2/gD)
K_A proportionality constant for particle motion in bend
k suspension parameter as defined in Equation 7
K_p empirical constant for particle motion in bend
m^* mass ratio of solid to gas (ρ_p/ρ_g)
$\dot{m}^*$ mass flow ratio of solid to gas ($\dot{m}_p/\dot{m}_g$)
n ratio of lateral flow from branch point to header flow
Re pipe Reynolds number (VD/ν)
S exponential correlation constant as defined in Equation 24
ρ^* density ratio ($\bar{\rho}_p/\bar{\rho}_g$)

Greek

α parameter as defined in Equation 1 or electrostatic parameter
$\bar{\alpha}$ characteristic number relating space charges and particle motion
α' solid component correlation in Equation 29
β $d/\Delta = 3(C_D\, Re_p^2)^{1/3}$; parameter defines spread of sizes of particles
Δ $[3\nu^2/4g(\rho^* - 1)]^{1/3}$ (m)
ϵ_D energy to remove a given volume of material (J/m^3)
ϵ_0 permittivity of free space, 8.854×10^{-12} f/m, c^2/J-m, C/V,m, s/Ω,m
μ viscosity (kg/m-sec)
ν kinematic viscosity (m^2/sec), or Poisson ratio
ν Poisson ratio
ξ branching coefficient
ρ density (kg/m^3)
$\bar{\rho}$ density of material (kg/m^3)
δ_s thickness of deposited layer of solids (m)
ϕ volume fraction solid
ψ correlation constant as defined in Equation 17
ω $[(4/3)\, g\nu(\rho^* - 1)]^{1/3}$ (m/sec)

Subscripts

a air
b packed or sliding bed or bend
Br branch
c minimum transport condition or branching factor
g gas
m mixture or in unit of meter
p particle
R at radius R or wall
s sharp point
th theoretical
w wall
1 inlet
2 outlet

REFERENCES

1. Zandi, I. "Transport of Road Solid Commodities via Freight Pipeline," report no. DOT-TST-76 T-36, University of Pennsylvania, to Department of Transportation, Washington, DC, Vol. II (1976).
2. Soo, S. L. *J. Powder Bulk Solids Technol.* 4(2/3):33–43 (1980).
3. Soo, S. L., and S. T. Leung. "Transportation and Distribution of Coal via Pneumatic Pipelines," *Proc. Int. Powder and Bulk Solids Handling and Processing*, Rosemont, IL, (1977), pp. 28–34.
4. Zenz, F. A. *Ind. Eng. Fund.* 3(1):65–75 (1964).
5. Konchesky, J. L., and T. G. Craig. "Air and Power Requirements for the Vacuum Transport of Crushed Coal in Horizontal Pipelines," *Pneumotransport* 2:paper B4 (1973).

6. Konchesky, J. L., T. J. George and J. G. Craig. *Trans. ASME J. Eng. Ind.* 94-100; 101–106 (1975).
7. Radmark, Engineering Division of Radar Pneumatics, Ltd. Data of the Radmark Pneumatic Conveying Systems (1976).
8. Ball, D. G., and D. H. Tweedy. "Pneumatic Hoist," Radmark Engineering, Portland, OR.
9. Govier, G. W., and K. Aziz. *The Flow of Complex Mixtures in Pipes* (New York: Van Nostrand Reinhold Company, 1972).
10. Soo, S. L., and S. K. Tung. *J. Powder Technol.* 6(5):283–294 (1972).
11. Pfeffer, R., S. Rossetti and S. Lieblein. *NASA TN* D-3603 (1966).
12. Sproson, J. C., W. A. Gray and J. Haynes. "Pneumatic Conveying Coal," *Pneumotransport* 2:paper B2 (1973).
13. Rose, H. E., and R. A. Duckworth. *The Engineer* 227(5703):392–396, 430–433, 590 (1969).
14. Duckworth, R. A. "Pressure Gradient and Velocity Correlation and their Application to Design," *Pneumotransport* I:paper R2 (1971).
15. Dogin, M. E., and V. P. Lebedev. *Ind. Chem. Eng.* (*USSR*) 2(1):64–67 (1962).
16. Rose, H. E., and H. E. Barnacle. *The Engineer* 898–901, 939–941 (1957).
17. Soo, S. L., and G. T. Trezek. *Ind. Eng. Chem. Fund.* 5:388 (1966).
18. Wachtell, G. P., J. P. Waggener and W. H. Steigelman. "Evaluation of Gas-Graphite Suspension in Vertical Channels," report no. NYO-9672, AEC (1961).
19. Mason, J. S., and B. U. Smith. "Pressure Drop and Flow Behavior for the Pneumatic Transport of Fine Particles around 90 Degree Bends," *Pneumotransport* 2:paper A3 (1973).
20. Ikemori, K., and H. Munakata. "A New Method of Expressing Pressure Drop in Horizontal Pipe Bend in Pneumatic Transport of Solids," *Pneumotransport* 2:paper A3 (1973).
21. Harman, P. D., R. A. Bajura and T. Kubo. "Analytical Study of Manifold Flow Distribution Systems for the Pneumatic Transport of Pulverized Coal," report PHH/RAD/RK-78-1, Department of Mechanical Engineering, West Virginia University, Morgantown, WV (1978).
22. Morikawa, T. *Trans. JSME* 42:361, 2787–2794 (1976).
23. Soo, S. L. *Particle Technol. 1973, Proc. 1st Int. Conf. in Particle Technol.* IIT Research Institute (1973), pp. 9–16.
24. Tilly, B. P. *Wear* 14:63 (1969).
25. Soo, S. L., *Powder Technol.* 17:163–259 (1977).
26. Arundel, P. A., I. A. Taylor and W. Dean. "The Rapid Erosion of Various Pipe Wall Materials by a Stream of Abrasive Alumine Particles," *Pneumotransport* 2:paper E1-1-15 (1973).
27. Mills, D., and J. S. Mason. *Powder Technol.* 17:35–73 (1977).
28. Mills, D., and J. S. Mason. "The Effect of Pipe Bends and Conveying Length upon Particle Degradation in Pneumatic Conveying Systems," *Proc. 2nd Int. Powder and Bulk Solids Handling and Processing*, Rosemont, IL (1978).
29. Soo, S. L. *J. Pipelines* 1(1):57–68 (1981).
30. Goodwin, J. E., W. Sage and G. P. Tilly. *Proc. Inst. Mech. Eng.* 184:1 (1969–70).
31. Cheng, L., and S. L. Soo. *J. Appl. Phys.* 41(2):585–591 (1970).
32. Cole, B. N., M. R. Baum and F. R. Mobbs. *Proc. Inst. Mech. Eng.* 184(3C):79–85 (1969–70).
33. Soo, S. L. *Fluid Dynamics of Multiphase Systems* (Blaisdell Publishing Co., 1967).
34. Gibson, N. Paper presented at the Conference on Electrostatic Hazards, Electrostatic Society of America and AIChE, October 1979.
35. Williams, G. M. Paper presented at the Conference on Electrostatic Hazards in the Storage and Handling of Powders and Liquids, Electrostatic Society of America and AIChE, October, 1979.
36. Soo, S. L. *Particle Technology*, The Institution of Chemical Engineers Symp. Series, ND63, D1/D/1-13 (1981).

37. Soo, S. L. *Proc., Pneumotransport 1 Conf.*, BHRA, Cambridge, England, R1, (1971), pp. 1–20.
38. Gibson, N. *Industrial Safety Handbook,* W. Handley, Ed. (New York: McGraw-Hill Book Co., 1977), pp. 132–146.
39. Masuda, H., T. Komatsu, N. Mitsui and K. Iinoya. *J. Electrostatics* 2:341–350 (1976).
40. Bright, A. W. *J. Electrostatics* 4:131–141 (1978).
41. Loeb, L. B. *Static Electrification* (New York: Springer-Verlag, 1958).
42. Turner, G. A., and M. Balasubramanian. *J. Electrostatics* 2:85–89 (1976).

CHAPTER 30
DESIGN OF GAS-SOLIDS INJECTORS

Matthias Bohnet
Institute for Process Technology
Technical University of Braunschweig
Braunschweig, Federal Republic of Germany

CONTENTS

INTRODUCTION

Gas-solids injectors often are used in industry as feeding devices for bulk solids into pipelines, e.g., in pneumatic conveying. With injectors it is possible to feed granulated or pulverized solids into pipelines under pressure, without any additional valves. It is of considerable advantage to use feeders without moving parts rather than appliances with rotating parts, such as rotary valves and screw feeders. However, apart from the advantages there are also some disadvantages that must be taken into consideration.

Injectors have a large energy consumption, often exceeding the energy required for the actual conveying of the solid material. To benefit from the latest progress in the design of pneumatic-conveying systems, it is essential to improve the performance of gas-solids injectors. At present, there is no reliable design procedure for this type of feeder due to a shortage of theoretical and experimental data.

To obtain satisfactory performance and high efficiency in transforming kinetic energy into static pressure energy, a detailed hydrodynamic calculation and correct dimensioning are essential. Consequently, the influence of several important parameters, such as geometry, location of nozzle, mixing section and diffuser, properties of the solids, mass flowrate of the solids, gas flowrate and driving pressure, should be taken into consideration.

EQUATION OF MOTION FOR THE FLOW OF GAS-SOLIDS MIXTURES

One of the most important parameters for the dimensioning of injectors is the velocity of the solid particles. For pipe flow, the velocity can be calculated from a force balance. Relating the drag force of solid particles to the mass of solid material and introducing the terminal velocity of the particles, w_S, as a measure of the drag force, leads to an equation suggested by Barth [1]:

$$K_W = m_P g \left(\frac{w - w_P}{w_S}\right)^{2-n} \tag{1}$$

In the Stokes regime ($Re_P < 1$), the exponent is $n = 1$; in the Newtonian regime ($10^3 < Re_P < 2 \cdot 10^5$), $n = 0$. For most technical calculations it is sufficiently accurate to approximate the intermediate range with $n = 0.5$. In this case, the exponent of the Stokes regime should be used for $Re_P < 4$ and the exponent of the Newtonian regime for $Re_P > 652$. Looking at the drag force coefficient for single spheres, the following equations should be used:

$$C_w = 24/Re_P \qquad Re_P < 4 \tag{2}$$

$$C_w = 12/(Re_P)^{1/2} \qquad 4 < Re_P < 652 \tag{3}$$

$$C_w = 0.47 \qquad Re_P > 652 \tag{4}$$

Re_P is defined as

$$Re_P = \frac{(w - w_P) d_P \rho}{\eta} \tag{5}$$

For the terminal velocity of the solid particles, the following is obtained:

$$w_S = \sqrt{\frac{4}{3} \cdot \frac{g}{C_w} \frac{\rho_P - \rho}{\rho} d_P} \tag{6}$$

Assuming that the friction between the particles and the wall is proportional to the inertial forces of the particles, the friction force can be obtained [2] as

$$K_R = m_P \frac{w_P^2 \lambda_S^*}{2d} \tag{7}$$

where λ_S^* is the coefficient of friction. It depends on the wall and particle properties. Average values are $\lambda_s^* = 0.002 - 0.005$ [3]. The resistance due to gravity is proportional to the solid mass and can be expressed as follows:

$$K_S = m_P \, g \, \beta \tag{8}$$

For horizontal conveying, an average β value of 0.4 can be used for the flow conditions in an injector. The force of acceleration can be calculated from

$$K_B = m_P \frac{dw_P}{dt} \tag{9}$$

With the force balance

$$K_W - K_R - K_S - K_B = 0 \tag{10}$$

and introducing $w_P = dl/dt$, the equation of motion is as follows:

$$\left(\frac{w - w_P}{w_S}\right)^{2-n} - \frac{w_P{}^2 \lambda_S{}^*}{2gd} - \beta - \frac{1}{g}\frac{w_P \cdot dw_P}{dl} = 0 \tag{11}$$

Introducing the following dimensionless numbers,

$$Fr = \frac{w^2}{gd}$$

$$Fr^* = \frac{w_S{}^{2-n} \cdot w^n}{gd}$$

$$L^* = \frac{lg}{w_S{}^{2-n} \cdot w^n}$$

$$W_P{}^* = \frac{w_P}{w} \tag{12}$$

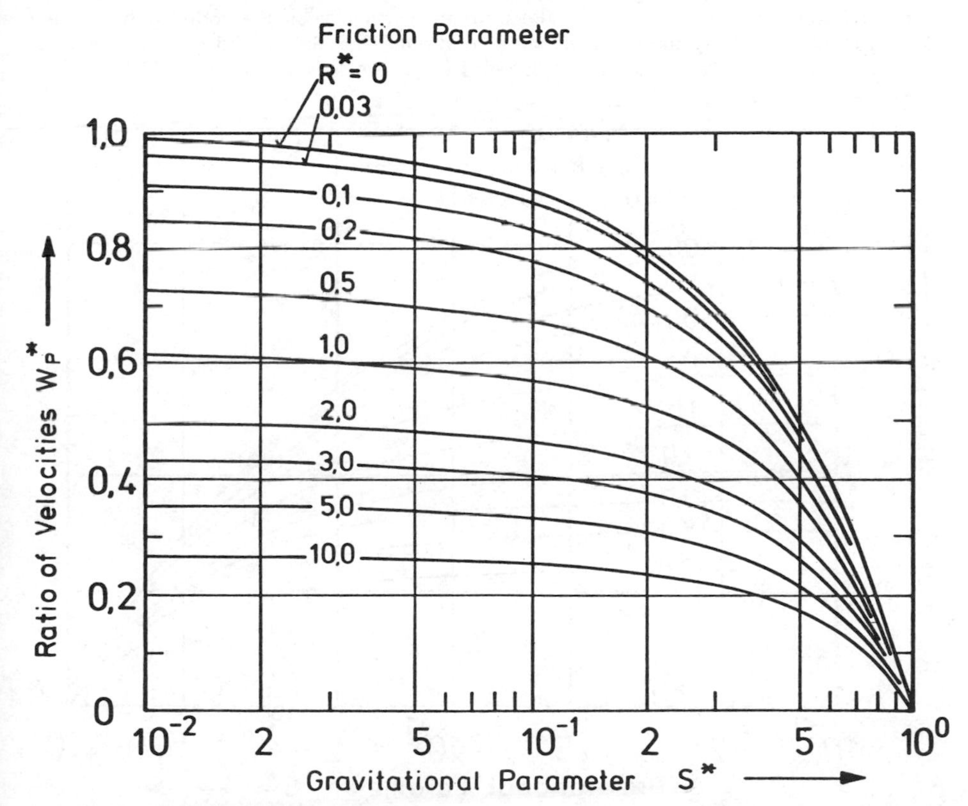

Figure 1. Dimensionless particle velocity as a function of gravitational parameter at different friction parameters, and n = 1.

Equation 11 transforms to

$$dL^* = \frac{W_P^* \cdot dW_P^*}{(1 - W_P^*)^{2-n} - Fr^* \frac{\lambda_S^*}{2} W_P^{*2} - \frac{Fr^*}{Fr}\beta} \tag{13}$$

Defining

$$R^* = Fr^* \frac{\lambda_S^*}{2} \tag{14}$$

as a wall friction parameter and

$$S^* = \frac{Fr^*}{Fr}\beta \tag{15}$$

as a gravitational parameter, a general form of the equation of motion is obtained:

$$dL^* = \frac{W_P^* \, dW_P^*}{(1 - W_P^*)^{2-n} - R^* W_P^{*2} - S^*} \tag{16}$$

At constant gas velocity, Equation 16 can be integrated easily. For other cases, a numerical solution can be obtained. The ultimate particle velocity produced by a gas flow of constant velocity under steady-state conditions, i.e., at the end of the acceleration period, may be read from Figures 1 and 2 for two resistance laws. Neglecting the

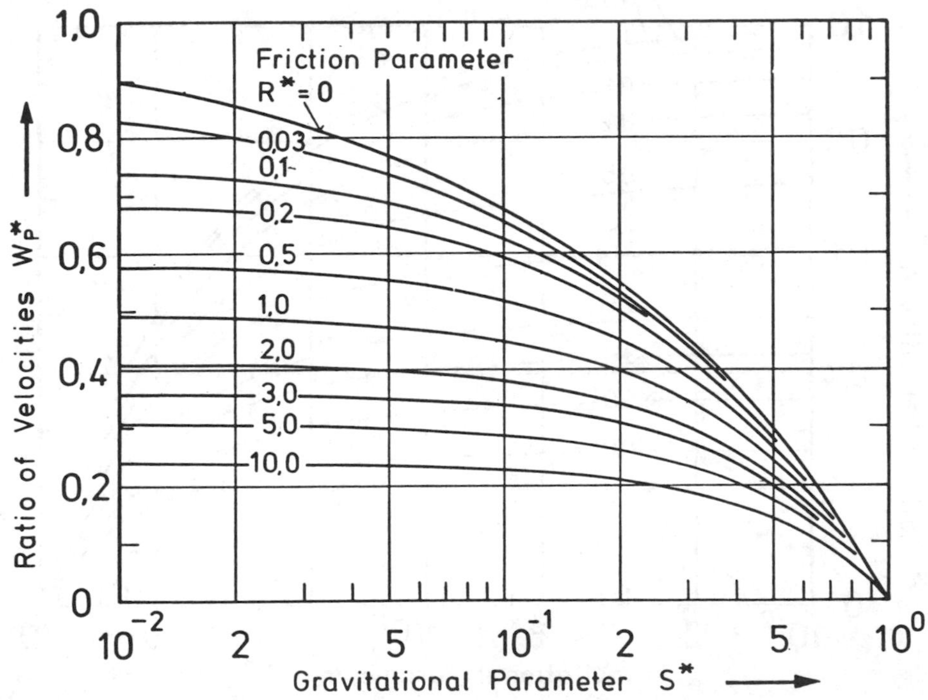

Figure 2. Dimensionless particle velocity as a function of gravitational parameter at different friction parameters, and n = 0.

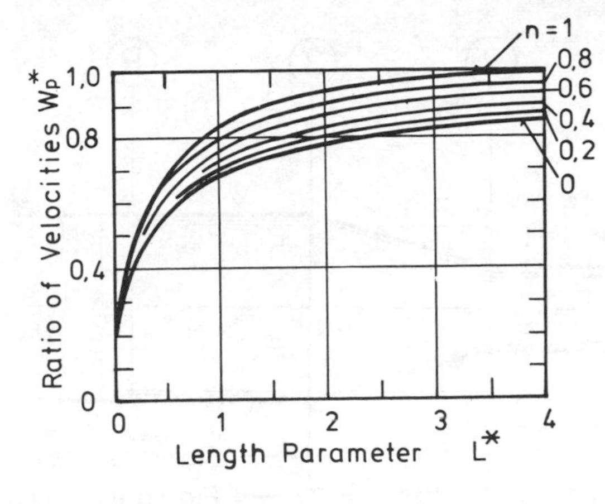

Figure 3. Acceleration of solid particles in a horizontal gas stream, neglecting friction and gravitational forces for different resistance laws.

wall friction and gravitational influence, the particle velocity dependent on the acceleration path may be read from Figure 3.

FLOW MECHANISMS IN GAS-SOLIDS INJECTORS

Following a proposal of Bohnet and Wagenknecht [4], the injector is divided into four sections, as shown in Figure 4. The first section, Ⓣ - ①, includes the hopper and the driving nozzle. In this area, the solids are accelerated by the driving jet. Depending on the pressure level, gas may be sucked in from the atmosphere or blown out. The second section, ① - ②, consists of the mixing tube and the third, ② - ③, consists of the diffuser. In section ③ - Ⓡ, the solids are accelerated or retarded in the adjoining conveying pipe to their final velocity.

Hopper

The static head required to achieve a certain velocity of the driving jet can be calculated for short nozzles, neglecting drag forces, as follows:

$$P_V = P_T + \frac{\rho}{2} w_T^2 [1 - (d_T/d_V)^4] \tag{17}$$

In the case of an open hopper, the flowrate of gas sucked in from the atmosphere or blown out through the hopper is approximately

$$\dot{M}_A = F_A \sqrt{2\rho(P_T - P_U)} \tag{18}$$

The solids that are fed into the hopper by means of a dosing device are accelerated by the driving gas jet. Owing to the high solids concentration in the hopper and high turbulence of the driving gas jet, it is very difficult to calculate exactly the decrease of gas velocity between the nozzle and the beginning of the mixing section because the expansion angle of the driving jet is unknown. According to a model representation, it is therefore assumed that the solids are accelerated in this region by a gas jet of average velocity:

$$w_m = (w_T + w_M)/2 \tag{19}$$

defined as an average value of driving jet velocity and mixing tube gas velocity.

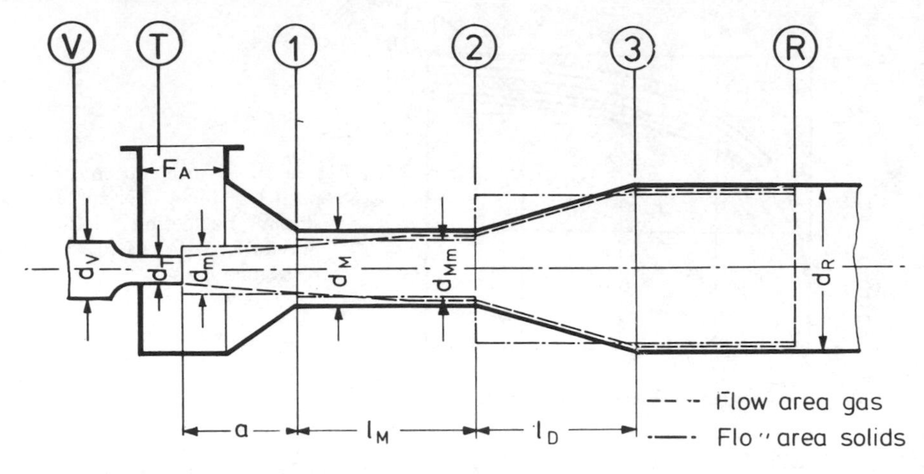

Figure 4. Schematic of injector.

With $\dot{M} = \dot{M}_T + \dot{M}_A$ and $\dot{M}_m = (\dot{M}_T + \dot{M})/2$, it follows that

$$w_m = \dot{M}_m/\rho_m F_m$$

whereby

$$\rho_m = (\rho_T + \rho_1)/2$$

The average jet diameter, d_m, can now be calculated from $F_m = \pi d^2_m/4$:

$$d_m = d_T d_M \sqrt{\frac{2\left(\frac{\dot{M}_T}{\dot{M}} + 1\right)}{\frac{\dot{M}_T}{\dot{M}}\left(1 + \frac{\rho_1}{\rho_T}\right) d_M{}^2 + \left(1 + \frac{\rho_T}{\rho_1}\right) d_T{}^2}} \tag{20}$$

The increase in static pressure between the levels Ⓣ and ① is the result of the decrease of the driving gas velocity from w_T to w_m as a consequence of the expansion of the gas jet. Due to the feeding of the solids, the transformation of kinetic energy of the gas stream into pressure energy is accompanied by considerable loss in energy. It is therefore assumed that the transformation of the gas flow energy occurs with an efficiency η_T. It should be taken into consideration that, in addition to the driving gas jet, extra gas may be sucked in from the atmosphere or blown out through the open hopper, depending on the pressure conditions. Further, it should be noted that to obtain the pressure difference between the levels Ⓣ and ① it is necessary to subtract from the calculated pressure recovery the pressure losses required for the acceleration of the sucked gas and the solids. It follows that

$$P_1 - P_T = \frac{\rho_T}{2} w_T{}^2 \left\{\eta_T \cdot \frac{1}{2}\left(1 + \frac{\rho_1}{\rho_T}\right)\left[1 - \left(\frac{\dot{M}/\dot{M}_T + 1}{1 + (\rho_1/\rho_T)}\right)^2 \left(\frac{d_T}{d_M}\right)^4\right] - \left[\left(\frac{\dot{M}}{\dot{M}_T}\right)^2 - 1\right]\frac{2}{1 + (\rho_1/\rho_T)}\left(\frac{d_T}{d_M}\right)^4 - \mu\frac{w_{P_1}}{w_T}\left(\frac{\dot{M}}{\dot{M}_T} + 1\right)\left(\frac{d_T}{d_M}\right)^2\right\} \tag{21}$$

If gas is blown out ($\dot{M}/\dot{M}_T < 1$), the second term of Equation 21 has to be neglected.

Mixing Tube

If the solids have not yet attained their final velocity at the entrance of the mixing tube, they will still be accelerated in the mixing section. In conventionally designed injectors, the expansion of the driving jet is normally not finished at the entrance of the mixing pipe. In section ① - ②, therefore, a considerable rise of static pressure can be observed. In calculating the particle velocity in this section, it is assumed that the particles are accelerated in a jet of constant gas velocity. The mean velocity of this gas stream is defined as

$$w_{M_m} = (w_{m_1} + w_M)/2 \tag{22a}$$

whereby

$$w_{m_1} = \frac{\dot{M}}{F_m \rho_1} \tag{22b}$$

Hence, it follows that

$$d_{M_m} = 2 d_m d_M \sqrt{\frac{1}{\left(1 + \rho_2/\rho_1\right)\left(\frac{\rho_1}{\rho_2} d_m^2 + d_M^2\right)}} \tag{23}$$

Calculating the transformation of pressure in the mixing tube, it is not allowed to neglect the friction losses. These are calculated as usual for gas-solids flow [2]. If the expansion of the driving jet is not yet finished at the entrance of the mixing tube, part of the kinetic energy is transformed into pressure energy due to the deceleration of the jet. Transformation losses are taken into account by applying an efficiency, η_M.

For the total transformation of the pressure energy in this area, it can be written

$$P_2 - P_1 = \eta_M \frac{\rho_{M_m}}{2} w_M^2 \left[\left(\frac{\rho_2}{\rho_1}\right)^2 \left(\frac{d_M}{d_m}\right)^4 - 1\right] - \mu \rho_{M_m} w_{M_m} (w_{P_2} - w_{P_1}) - (\lambda_G + \mu \lambda_S) \frac{\rho_{M_m}}{2} w_{M_m}^2 \frac{l_M}{d_M} \tag{24}$$

The first term of Equation 24 describes the pressure recovery due to expansion of the gas flow, during which the gas velocity decreases from w_m to w_M. Acceleration of the particles is reflected in the second term and friction losses of the gas and solids in the third term. Relating all the losses to velocity w_M and introducing $\rho_{Mm} = (\rho_1 + \rho_2)/2$, Equation 24 can be written as

$$P_2 - P_1 = \frac{\rho_1}{2} w_M^2 \left\{\eta_M \left[\left(\frac{\rho_2}{\rho_1}\right)^2 \left(\frac{d_M}{d_m}\right)^4 - 1\right]\left(\frac{1 + (\rho_2/\rho_1)}{2}\right) - 2\mu \left(\frac{w_{P_2} - w_{P_1}}{w_M}\right)\left(\frac{d_M}{d_{M_m}}\right)^2 \left(\frac{\rho_2}{\rho_1}\right) - (\lambda_G + \mu \lambda_S) \frac{l_M}{d_M} \left(\frac{d_M}{d_{M_m}}\right)^4 \left(\frac{\rho_2}{\rho_1}\right)^2 \cdot \frac{2}{1 + (\rho_2/\rho_1)}\right\} \tag{25}$$

Diffuser

Particle velocity changes only slightly in the diffuser because of the very short flight path. The pressure recovery following the deceleration of solids occurs mostly in the cylindrical tube behind the diffuser. Measurements of Morimoto et al. [5] show that the retardation path is five times longer than the acceleration path. The change in pressure in the diffuser between levels ② and ③ therefore results mainly from the decreasing gas velocity and, to a lesser extent, from the retardation of the solids. Including the pressure recovery due to the decrease of particle velocity in the cylindrical tube behind the diffuser and introducing as average gas density $(\rho_2 + \rho_3)/2$, the following is obtained:

$$P_R - P_2 = \frac{\rho_R}{2} w_R^2 \left\{ \eta_D \frac{(1 + \rho_2/\rho_3)}{2} \left(\frac{\rho_R}{\rho_3}\right) \left[\left(\frac{\rho_3}{\rho_2}\right)^2 \left(\frac{d_R}{d_M}\right)^4 - 1 \right] + \eta_S 2\mu \frac{w_{P_R} - w_{P_2}}{w_R} \right\} \quad (26)$$

If acceleration of the particles is not completed on leaving the mixing tube, there will be further losses in the diffuser and the adjoining tube. The steady-state velocity of the particles behind the injector can be calculated by means of Equation 16.

SIMPLIFIED CALCULATION

For numerous technical applications, injectors with closed hoppers are used. In this special case, no gas is sucked in from atmosphere or blown out, and it is $\dot{M}_T/\dot{M} = 1$. Further, injectors are used mostly for applications that require only small pressure changes. In this case, the gas density remains nearly constant. It is useful, therefore, to first estimate the injector dimensions for this special case. Then, in a second step, a detailed calculation would be meaningful.

Using only apparatus diameters, the following can be written:

$$\frac{d_T}{d_m} = \sqrt{\frac{1}{2}\left[1 + \left(\frac{d_T}{d_M}\right)^2\right]} \quad (27)$$

$$\frac{d_M}{d_m} = \sqrt{\frac{1}{2}\left[1 + \left(\frac{d_M}{d_T}\right)^2\right]} \quad (28)$$

$$\frac{d_M}{d_{Mm}} = \sqrt{\frac{1}{4}\left[3 + \left(\frac{d_M}{d_T}\right)^2\right]} \quad (29)$$

Relating all gas velocities to the driving jet velocity, w_T, the equations for calculating the pressure changes in the different injector regions are as follows:

Nozzle:

$$\frac{P_V - P_T}{\frac{\rho}{2} w_T^2} = 1 - \left(\frac{d_T}{d_V}\right)^4 \quad (30)$$

Hopper:

$$\frac{P_1 - P_T}{\frac{\rho}{2} w_T^2} = \eta_{To} \left\{1 - \frac{1}{4}\left[1 + \left(\frac{d_T}{d_M}\right)^2\right]^2\right\} - \mu \frac{w_{P_1}}{w_T} \cdot \left[1 + \left(\frac{d_T}{d_M}\right)^2\right] \quad (31)$$

Mixing tube:

$$\frac{P_2 - P_1}{\frac{\rho}{2} w_T^2} = \eta_{Mo} \left\{ \frac{1}{4} \left[\left(\frac{d_T}{d_M} \right)^2 + 1 \right]^2 - \left(\frac{d_T}{d_M} \right)^4 \right\} - \mu \frac{w_{P_2} - w_{P_1}}{w_T} \cdot \frac{1}{2} \left[3 \left(\frac{d_T}{d_M} \right)^2 + 1 \right]$$

$$- (\lambda_G - \mu\lambda_S) \frac{l_M}{d_M} \cdot \frac{1}{16} \left[3 \left(\frac{d_T}{d_M} \right)^2 + 1 \right]^2 \qquad (32)$$

Diffuser:

$$\frac{P_R - P_2}{\frac{\rho}{2} w_T^2} = \eta_D \left(\frac{d_T}{d_M} \right)^4 \left[1 - \left(\frac{d_M}{d_R} \right)^4 \right] + \eta_S \cdot 2\mu \frac{w_{P_2} - w_{P_R}}{w_T} \left(\frac{d_T}{d_M} \right)^2 \left(\frac{d_M}{d_R} \right)^2 \qquad (33)$$

It should be noted that the efficiencies η_{To} and η_{Mo} indicate the efficiency values for the case $\dot{M}_T/\dot{M} = 1$.

SOLIDS VELOCITY AND STATIC PRESSURE

Pressure distribution, as well as particle and gas velocities for two series of measurements, are presented in Figure 5. Although the gas velocities in the hopper region, mixing tube and diffuser are very different, a continuous increase in the solid velocity is observed. A substantial rise in pressure occurs in the driving jet region and continues in

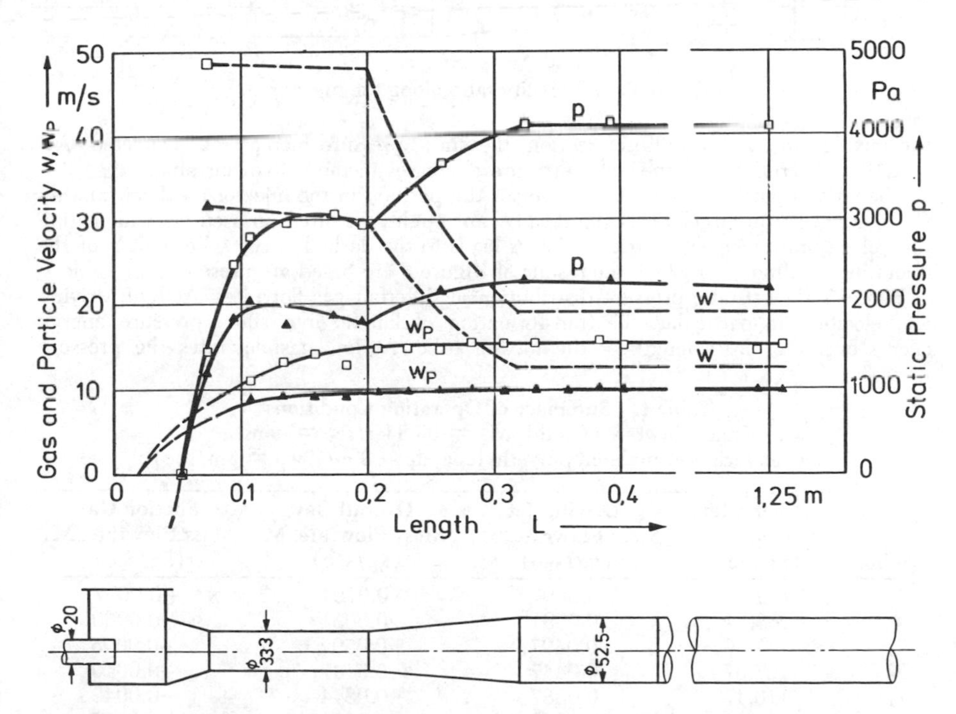

Figure 5. Gas and solids velocities and static pressure in the injector: □ driving jet velocity, 141.8 m/sec; △ driving jet velocity, 105.0 m/sec.

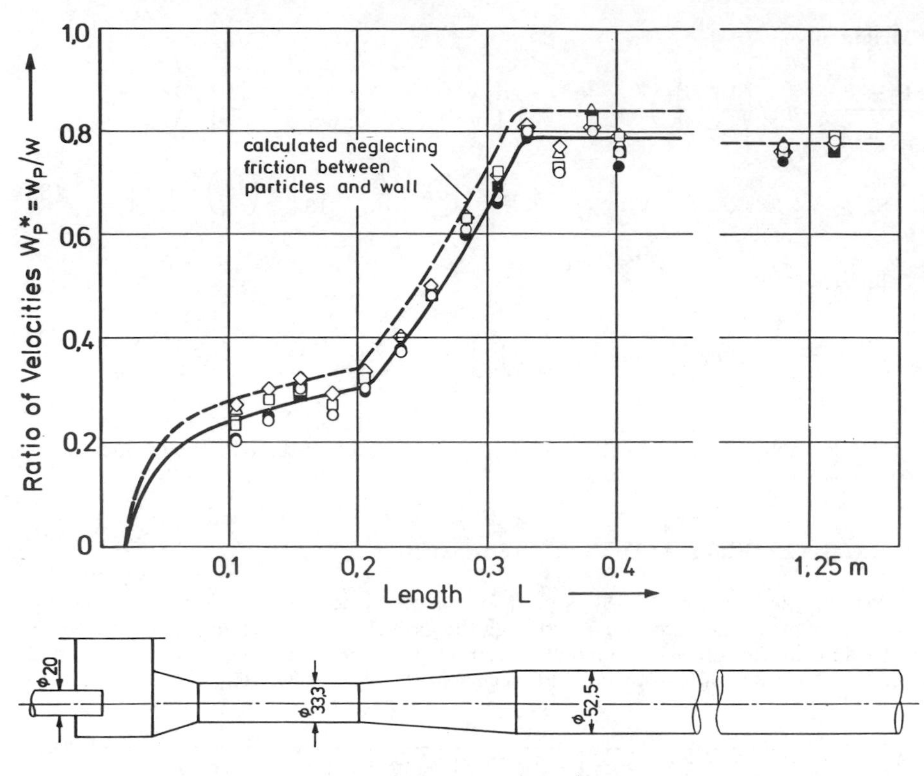

Figure 6. Velocity ratio along the injector.

the mixing tube. In the diffuser section, the static pressure increases as expected. According to friction losses, the static pressure decreases in the conveying pipe.

The ratio of particle to gas velocity versus the position in the injector has been plotted in Figure 6. It can be seen that the results can roughly be fitted on a curve. Calculating the velocity ratio neglecting friction losses leads to the dashed curve. The details of the operating conditions on which the results of Figure 6 are based are presented in Table I.

Figure 7 shows static pressure distribution at different gas flowrates. At high driving jet velocities, in particular, the transformation of kinetic into static pressure energy occurs over the entire length of the mixing tube. At low gas flowrates the pressure

Table I. Summary of Operating Conditions (mass flowrate of solids $\dot{M}_P$ = 0.08 kg/sec = const.; test material: granulated polyethylene, $d_P \approx$ 3 mm, w_S = 8m/sec)

Symbol	Driving Jet Velocity, w_T (m/sec)	Driving Jet Mass Flowrate, $\dot{M}_T$ (kg/sec)	Overall Gas Mass Flowrate, $\dot{M}$ (kg/sec)	Suction Gas Mass Flowrate, $\dot{M}_A$ (kg/sec)
○	199.04	0.0697	0.0794	+0.0097
●	168.74	0.0591	0.0651	+0.0060
□	141.80	0.0497	0.0502	+0.0005
■	127.67	0.0447	0.0437	−0.0010
△	110.47	0.0387	0.0344	−0.0043
▲	105.02	0.0368	0.0321	−0.0047
◇	99.19	0.0347	0.0300	−0.0047

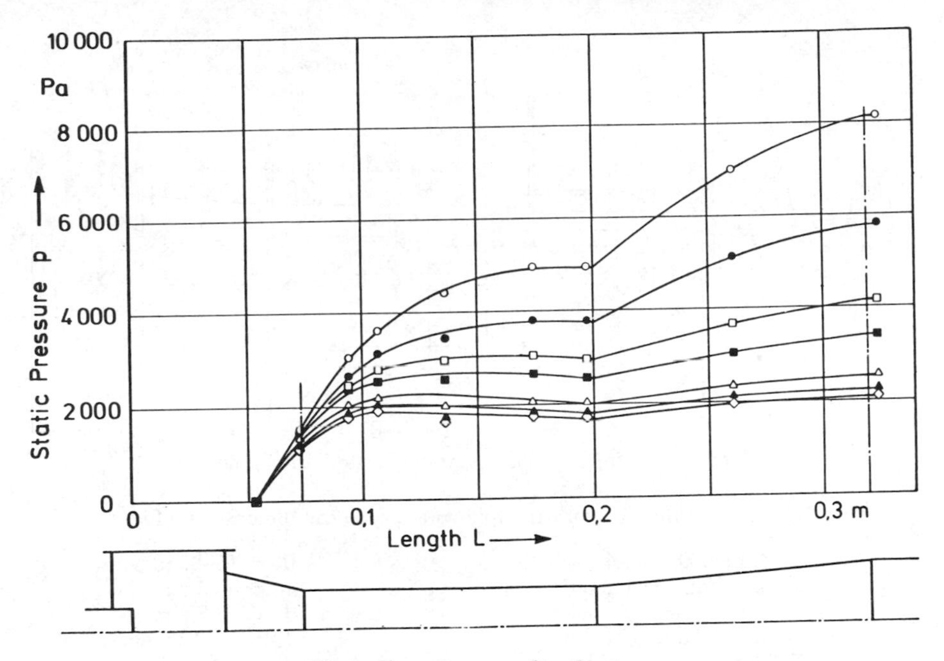

Figure 7. Pressure distribution.

recovery can be observed only in the inlet region of the mixing tube because later on the static pressure falls due to friction and acceleration losses, which exceed the pressure recovery. Pressure distribution in the diffuser corresponds to expectations; however, it is remarkable that the transformation of pressure related to gas flowrate appears to be practically free of losses.

With the results of measurements as shown in Figures 5, 6 and 7, the transformation efficiencies of energy can be calculated for all regions of the injector.

EFFICIENCIES

To determine the efficiencies η_T, η_M, η_D and η_S, the results of experiments done by Wagenknecht [6] will be used. The solids used in the experiments had the physical data listed in Table II.

Table II. Physical Data of Solids

Solid	Mean Particle Diameter, d_{P50} (mm)	Particle Density, ρ_P (kg/m^3)	Particle Terminal Velocity, w_S (m/sec)
Quartz Sand I	1.60	2700	10.00
Polyethylene	3.00	918	8.00
Styropor®	1.10	1050	4.50
Quartz Sand II	0.38	2700	3.20
Oat Hulls			2.50
Quartz Sand III	0.16	2700	1.10
Cement	0.04	3100	0.15

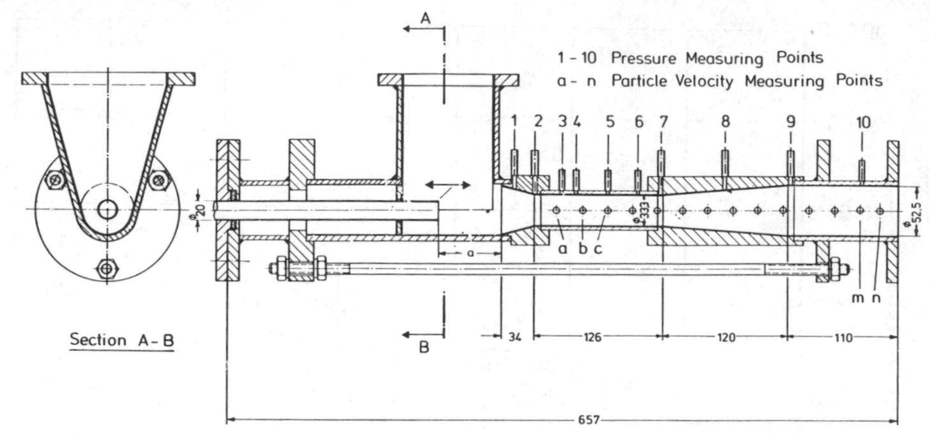

Figure 8. Injector with centrally located driving nozzle.

Wagenknecht used an injector with the following geometric dimensions (Figure 8):

Jet Nozzle Diameter	$d_T = 15–30$ mm
Mixing Tube Diameter	$d_M = 33.3$ mm
Tube Diameter	$d_R = 52.5$ mm
Mixing Tube Length	$l_M = 30–126$ mm
Diffuser Length	$l_D = 120$ mm
Distance	$a = 0–80$ mm

Figure 9 shows, by the way of example, measured efficiencies η_T and η_M as functions of the Reynolds numbers Re_T and Re_M, defined as

$$Re_T = w_m \, d_m \, \rho/\eta \tag{34}$$

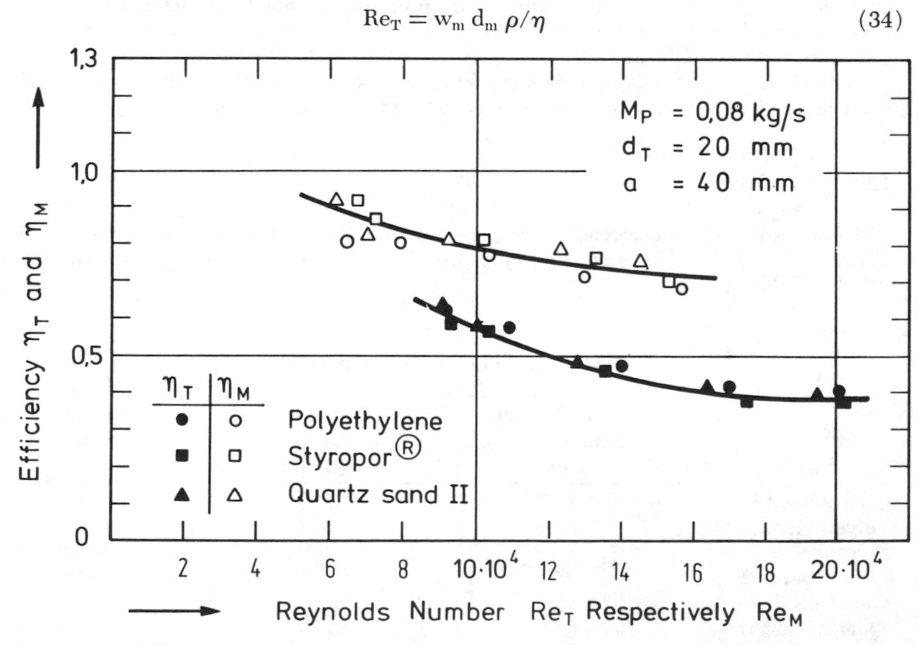

Figure 9. Efficiency of energy transformation in the hopper area and the mixing tube.

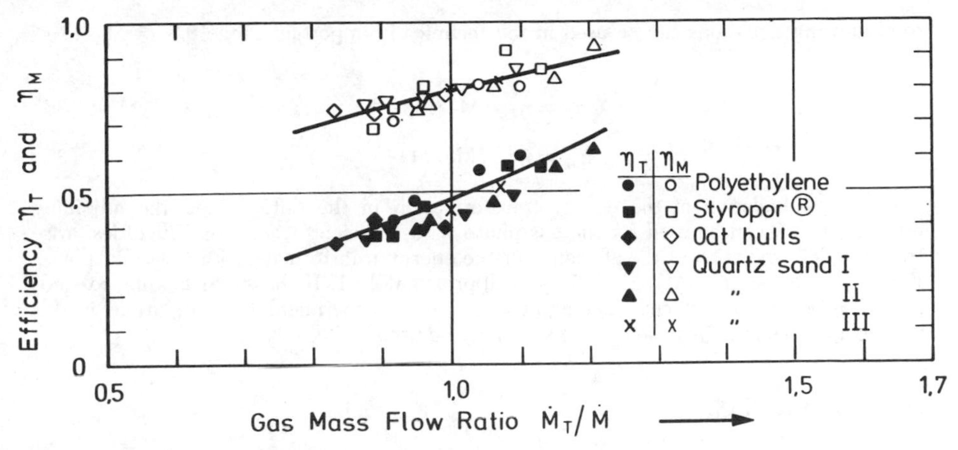

Figure 10. Efficiency of energy transformation in the hopper area and the mixing tube as a function of the gas mass flow ratio.

$$Re_M = w_M\, d_M\, \rho/\eta \tag{35}$$

Using an injector with an open hopper, a considerable amount of gas may be sucked in or blown out. The calculated efficiencies therefore depend strongly on the gas mass flow ratio, $\dot{M}_T/\dot{M}$, as shown in Figure 10. To evaluate the curves in Figure 10, one must consider that the increase of the efficiencies with the gas mass flow ratio is because in the case of flow ratios below 1 gas is sucked in and has to be accelerated. This leads analytically to a reduction of the efficiency, and vice versa.

For the representation of the efficiency as a function of the Reynolds number, it seems to be more reasonable to reduce the efficiency values to the case $\dot{M}_T/\dot{M} = 1$. This describes also the very important case of an injector with a closed hopper. Figure 11 shows the efficiencies η_{T0} and η_{M0}, which are calculated for $\dot{M}_T/\dot{M} = 1$ using the experimental data of Figures 9 and 10.

Using the data of Figure 11, it is no problem to calculate also the efficiencies for cases in which gas is sucked in or blown out through an open hopper. In these cases,

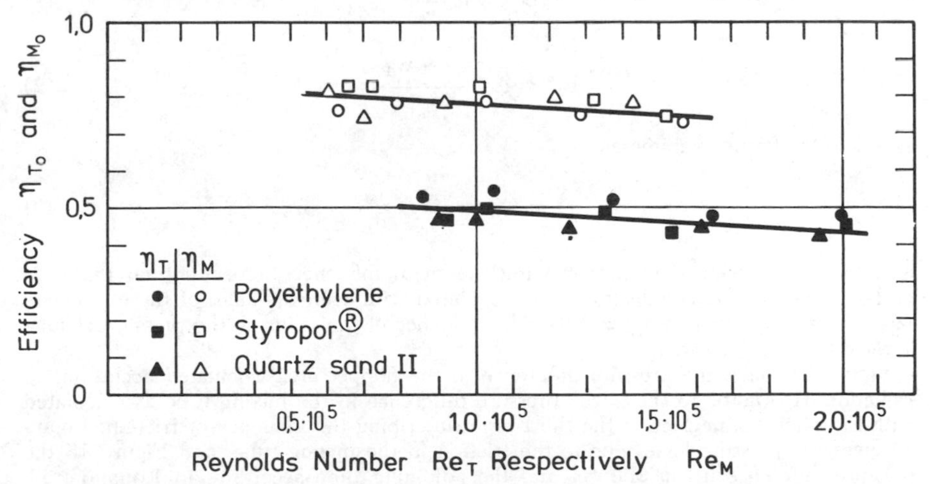

Figure 11. Efficiency of energy transformation in the hopper area and the mixing tube for $\dot{M}_T/\dot{M} = 1$.

the following equations can be used in the technically important range $0.8 \leq \dot{M}_T/\dot{M} \leq 1.2$:

$$\eta_T = \eta_{To} (\dot{M}_T/\dot{M})^{5/3} \tag{36}$$

$$\eta_M = \eta_{Mo} (\dot{M}_T/\dot{M})^{2/3} \tag{37}$$

For the calculation of the energy transformation in the diffuser and the adjoining tube, $\eta_D \approx 1.0$ can be used for the gas phase with sufficient accuracy. Difficulties arise from the determination of the efficiency of the energy transformation for the solid phase. For solids loading $\mu = \dot{M}_P/\dot{M} < 5$, η_S is approximately 1. If the solids loading exceeds $\mu = 5$, η_S has to be reduced. As long as no results of experimental investigations in this regime are available, $\eta_S = 1/(\mu - 4)^{-1/3}$ may be used.

DESIGN DIAGRAMS

Based on the equations presented in the section entitled Simplified Calculation, dimensionless design diagrams can be drawn. Introducing the dimensionless pressure differences,

$$p_1^* - p_T^* = \frac{P_1 - P_T}{\frac{\rho}{2} w_T^2} \tag{38}$$

$$p_2^* - p_1^* = (p_2^* - p_1^*)_M - (p_2^* - p_1^*)_R = \frac{P_2 - P_1}{\frac{\rho}{2} w_T^2} \tag{39}$$

$$p_R^* - p_2^* = \frac{P_R - P_2}{\frac{\rho}{2} w_T^2} \tag{40}$$

and the acceleration parameters,

$$b_1^* = \mu \frac{w_{P_1}}{w_T} \tag{41}$$

$$b_2^* = \mu \frac{w_{P_2} - w_{P_1}}{w_T} \tag{42}$$

as well as the friction parameter,

$$r^* = (\lambda_G + \mu\lambda_S) \frac{l_M}{d_M} \tag{43}$$

lead to the following diagrams showing the main influences dependent on the ratio of diameters, d_T/d_M. The diagrams are calculated with average values of the efficiencies ($\eta_{To} = 0.45$; $\eta_{Mo} = 0.78$; $\eta_D = 1.0$). The influence of the solids on the pressure change in the diffuser is neglected.

Figure 12 shows the pressure difference in the hopper area calculated according to Equation 31. Figure 13 shows the pressure difference in the mixing tube as calculated with Equation 32, neglecting the third term describing the influence of friction. Figure 14 gives the pressure losses due to the friction in the mixing tube, and Figure 15 the pressure difference in the diffuser and the adjoining tube according to Equation 33. Using these equations for the calculation of injectors, one must ensure that the assumptions made will be considered.

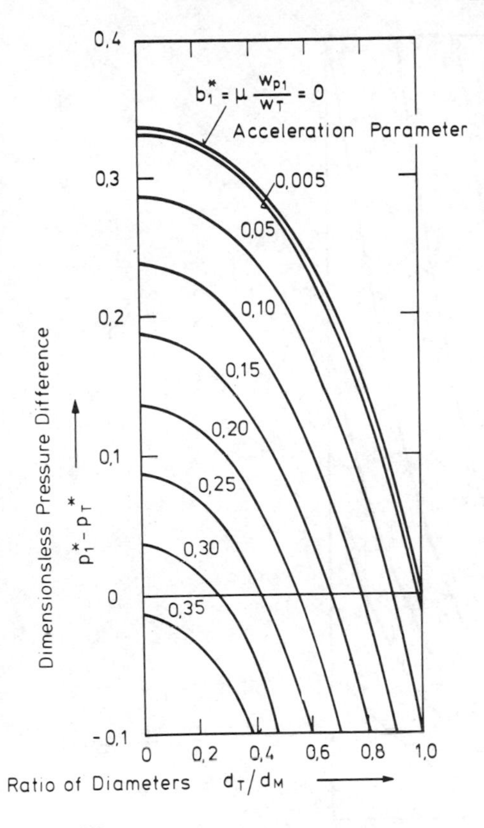

Figure 12. Dimensionless pressure difference in the hopper area for a closed hopper.

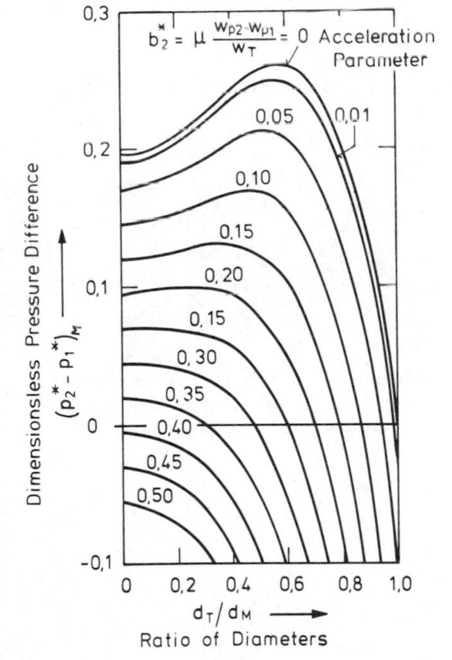

Figure 13. Dimensionless pressure difference in the mixing tube, neglecting friction forces.

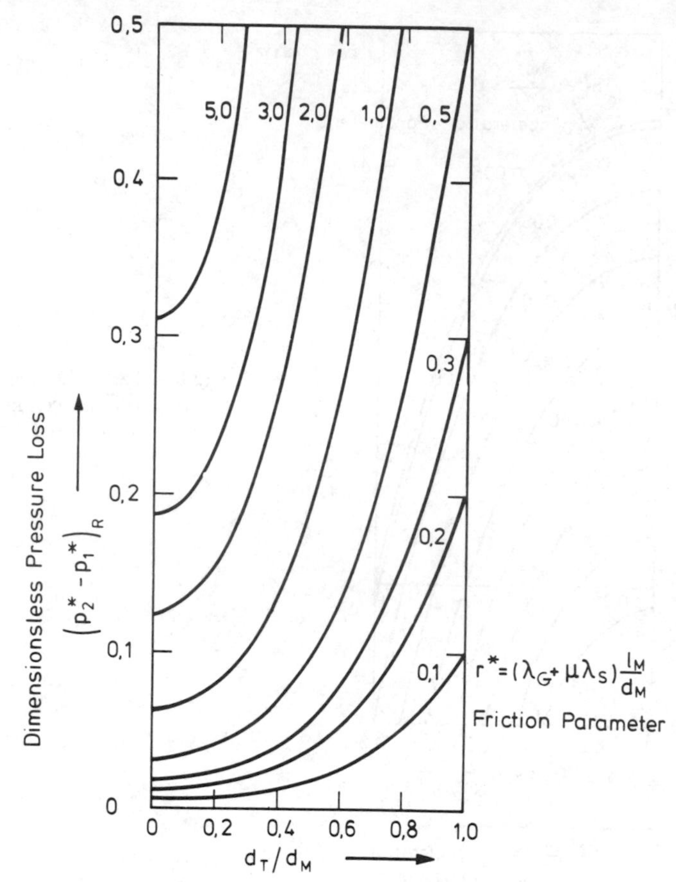

Figure 14. Dimensionless pressure loss in the mixing tube due to solids and gas friction.

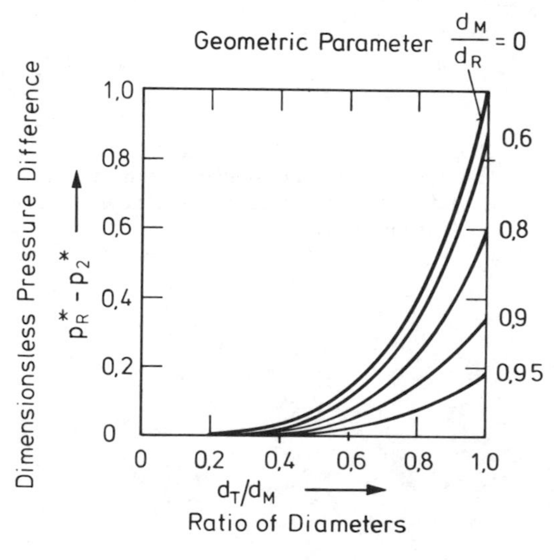

Figure 15. Dimensionless pressure difference in the diffuser, neglecting the influence of the solids.

COMPARISON OF THEORY TO EXPERIMENT

Figures 16 and 17 show measured particle velocities and static pressures for the injector described in Figure 8, fed with a constant solids mass flow for different driving jet mass flowrates. The flow conditions are presented in Table III. As usual for gas-solids flow in pipes, the loss coefficients for the gas phase and the solids phase are calculated according to Bohnet [2]:

$$\lambda_G = 0.3164/\mathrm{Re}^{0.25} \tag{44}$$

$$\lambda_S = \frac{w_P}{w}\lambda_S^* + \frac{2\beta}{\frac{w_P}{w}\mathrm{Fr}} \tag{45}$$

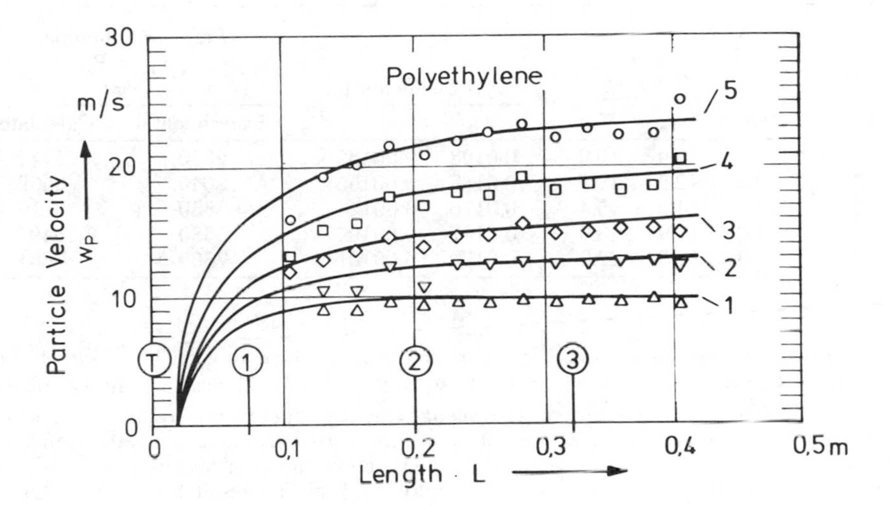

Figure 16. Particle velocity for different gas velocities.

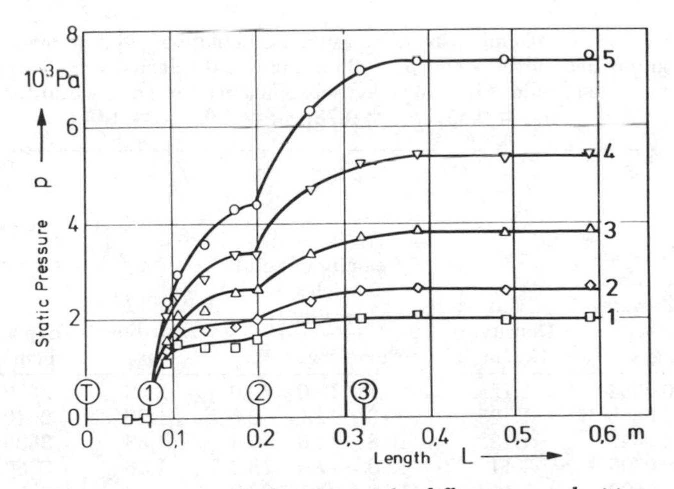

Figure 17. Pressure distribution for different gas velocities.

Table III. Comparison Between Theory and Experiment
(granulated polyethylene, $d_P \approx 3$ mm; terminal velocity $w_S \approx 8$ m/sec; mass flowrate of solids $\dot{M}_P = 0.08$ kg/sec; efficiencies of energy transformation: $\eta_{To} = 0.45$; $\eta_{Mo} = 0.78$; $\eta_D = 1.0$; $\eta_S = 1.0$)

Run	Driving Jet Mass Flowrate, $\dot{M}_T$ (kg/sec)	Gas Mass Flow Ratio $\dot{M}_T/\dot{M}$	Solids Loading (μ)	Gas Velocity, w (m/sec) w_T	w_M	w_R
1	0.0354	1.10	2.48	101.3	33.0	13.1
2	0.0408	1.04	2.04	116.8	40.0	15.9
3	0.0492	0.95	1.54	140.8	52.8	20.7
4	0.0590	0.92	1.25	168.9	65.1	25.2
5	0.0692	0.89	1.03	198.0	78.7	30.0

Run	Velocity of Solid Particles, w_P (m/sec) w_{P1}	w_{P2}	w_{P_R}	Loss Coefficients λ_G	λ_S	Pressure Difference, $P_R - P_T$ (Pa) Experimental	Calculated
1	7.9	10.0	10.0	0.0198	0.00183	2050	1741
2	9.3	12.2	12.6	0.0189	0.00155	2640	2502
3	10.8	14.6	15.4	0.0176	0.00123	3830	3710
4	13.0	17.9	18.9	0.0167	0.00109	5380	5497
5	16.0	21.5	22.9	0.0159	0.00101	7360	7651

To calculate the coefficient λ_S for the mixing pipe, an average particle velocity has to be used. In this case it is $w_P = (w_{P_1} + w_{P_2})/2$. Table III shows that the calculated pressure difference is in sufficient agreement with the experimental results. Neglecting the influence of sucked in or blown out gas, a comparison is made using the simplified calculation and the design diagrams. Table IV shows the results. The calculated and measured pressure differences deviate for ±20%, a sufficient result for the first step in designing gas-solids injectors.

Table IV. Comparison of Simplified Calculation and Experiment
(granulated polyethylene, $d_P \approx 3$ mm; terminal velocity $w_S \approx 8$ m/sec; mass flowrate of solids $\dot{M}_P = 0.08$ kg/sec; efficiencies of energy transformation: $\eta_{To} = 0.45$; $\eta_{Mo} = 0.78$; $\eta_D = 1.0$; $\eta_S = 1.0$

Run	Driving Jet Mass Flowrate, $\dot{M}_T$ (kg/sec)	Gas Density, ρ (kg/m³)	Velocity of Solid Particles, w_P (m/sec) w_{P1}	w_{P2}	w_{P_R}	Solids Loading (μ)	Pressure Difference, $P_R - P_T$ (Pa) Experimental	Calculated (simplified)
1	0.0354	1.12	7.9	10.0	10.0	2.26	2050	1678
2	0.0408	1.125	9.3	12.2	12.6	1.96	2640	2400
3	0.0492	1.13	10.8	14.6	15.4	1.63	3830	3991
4	0.0590	1.14	13.0	17.9	18.9	1.36	5380	6198
5	0.0692	1.145	16.0	21.5	22.9	1.16	7360	8982

DESIGN EXAMPLE

In a pneumatic-conveying system, $\dot{M}_P$ = 500 kg/hr of granulated solids with a terminal velocity w_S = 8 m/sec have to be transported in a pipe with a diameter of d_R = 100 mm. The mass flowrate of the gas is $\dot{M}$ = 750 kg/hr, and the needed pressure for the conveying is 4000 Pa. An injector with the following dimensions is available: d_T = 40 mm, d_M = 65 mm, d_R = 100 mm, l_M = 300 mm, and a = 75 mm.

The friction coefficients are $\lambda_S^* = 0.0035$, $\beta = 0.4$ and the energy transformation efficiencies $\eta_{To} = 0.45$, $\eta_{Mo} = 0.78$, $\eta_D = 1$, $\eta_S = 1$. For an injector with a closed hopper and a gas density of $\rho = 1.2$ kg/m³, the velocity of the solid particles is calculated using Equation 16. With the velocity ratios $w_{P1}/w_T = 0.095$, $w_{P2}/w_T = 0.154$ and $w_{PR}/w_T = 0.152$, it follows for the acceleration and the friction parameters that: $b_1^* = 0.063$, $b_2^* = 0.0391$ and $r^* = 0.0722$. The dimensionless pressure differences are then $p_1^* - p_T^* = 0.1491$, $p_2^* - p_1^* = 0.1965$, $p_R^* - p_2^* = 0.1182$ and $p_R^* - p_T^* = 0.4638$. The pressure at the end of the diffuser is 5307 Pa and sufficient for the problems discussed.

Looking at the gas and particle velocity at the beginning of the conveying pipe, it is seen that the particle velocity with 21.2 m/sec exceeds remarkably the steady-state particle velocity of 16.8 m/sec. This means that a considerable amount of energy is wasted for the overacceleration of the solid material. To avoid this overacceleration, an injector with a larger diameter for the mixing tube and the driving jet nozzle should be used.

Central nozzle and long mixing tube

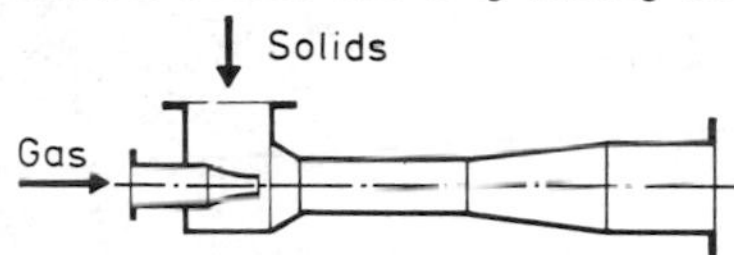

Central nozzle and short mixing tube

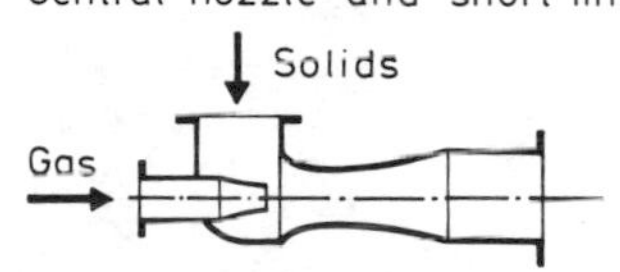

Central solids feed and ring nozzle

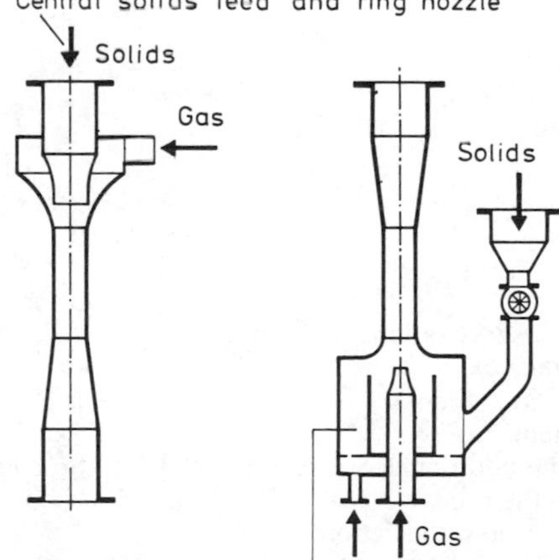

Central nozzle and fluidized-bed feed of solids

Figure 18. Different designs of injectors.

Using the design diagrams, it is finally found that an injector with a driving jet nozzle diameter of $d_T = 42.5$ mm and a mixing tube diameter of $d_M = 69$ mm lead to a particle velocity at the end of the diffuser of $w_{P2} = 18.9$ m/sec and a total pressure difference of $P_R - P_T = 4102$ Pa. As a result of the enlargement of the driving jet nozzle and the mixing tube diameter, the static pressure required to achieve the driving jet velocity drops by 22%.

DIFFERENT INJECTOR DESIGNS

Figure 18 shows the most important types of gas-solids injectors. For all types of injectors it is important to choose the length of the mixing tube so that at the end of the mixing tube the steady-state velocity of the particles in the conveying pipe is reached. The most advantageous aerodynamic solution is an injector with central solids feed and a ring nozzle. Whether this system can be used depends on the free-flowing behavior of the solids and the local conditions. For feeding high amounts of solid material, the fluidized-bed feeder is a well-proven device.

NOMENCLATURE

a	distance between driving jet and mixing tube (m)
d_V, d_T, d_M, d_R, d	geometric diameter (m)
d_m, d_{Mm}	average diameter (m)
d_P	particle diameter (m)
F_A	cross-sectional area of the hopper (m^2)
Fr, Fr^*	Froude number
g	acceleration of gravity (m/sec^2)
K_W, K_R, K_S, K_B	forces (N)
l_M, l_D	length (m)
L^*	parameter of length
$\dot{M}_A, \dot{M}_T, \dot{M}$	mass flowrate of gas (kg/sec)
M_P	mass flowrate of solids (kg/sec)
m_P	mass of solids (kg)
P_V, P_T, P_A, P_R, P_U	pressure (Pa)
p_V^*, p_T^*, p_R^*	dimensionless pressure
R^*	friction parameter
Re	Reynolds number
S^*	gravitational parameter
w_T, w_M, w_R	gas velocity (m/sec)
w_S	terminal velocity (m/sec)
w_P	particle velocity (m/sec)
W_P^*	velocity ratio

Greek Symbols

β	coefficient, considering the gravitational influence
η	dynamic viscosity (kg/m-sec)
$\eta_T, \eta_{To}, \eta_M, \eta_{Mo}, \eta_D, \eta_S$	efficiency of energy transformation
λ_G	tube friction loss coefficient
λ_S	loss coefficient due to the additional pressure needed for the conveying of the solids in the tube
λ_S^*	loss coefficient due to particle wall friction
$\mu = \dot{M}_P/\dot{M}$	solids loading
ρ	density (kg/m^3)

REFERENCES

1. Barth, W. "Strömungsvorgänge beim Transport von Festteilchen und Flüssigkeitsteilchen in Gasen mit besonderer Berücksichtigung der Vorgänge bei pneumatischer Förderung," *Chemie-Ing.-Tech.* 30(3):171–180 (1958).
2. Bohnet, M. "Experimentelle und theoretische Untersuchungen über das Absetzen, das Aufwirbeln und den Transport feiner Staubteilchen in pneumatischen Förderleitungen," *VDI-Forschungsheft* 507, (1965).
3. Muschelknautz, E. "Theoretische und experimentelle Untersuchungen über die Druckverluste pneumatischer Förderleitungen unter besonderer Berücksichtigung des Einflusses von Gutreibung und Gutgewicht," *VDI-Forschungsheft* 476 (1959).
4. Bohnet, M., and U. Wagenknecht. "Investigations on Flow Conditions in Gas/Solid-Injectors," *Ger. Chem. Eng.* 1(5):298–304 (1978).
5. Morimoto, T., A. Yamamoto, T. Nakao and Y. Morikawa. "On the Behavior of Air-Solids Mixture in a Pipeline for Pneumatic Conveyance with a Single or Double T-Branches," *Bull. JSME* 20(143):600–606 (1977).
6. Wagenknecht, U. "Untersuchung der Strömungsverhältnisse und des Druckverlaufes in Gas/Festsoff-Injektoren," Dissertation, Technical University of Braunschweig, West Germany (1981).

CHAPTER 31

FLOW OF SOLIDS IN BUNKERS

J. Bridgwater
Department of Chemical Engineering
University of Birmingham
Edgbaston, Birmingham, England

A. M. Scott
Koninklijke/Shell-Laboratorium, Amsterdam
Amsterdam, The Netherlands

CONTENTS

INTRODUCTION

Solids are frequently one of the feedstocks or products in a chemical process. Even in processes that have fluid feeds and fluid products it is common to find a solid being used, e.g., as a catalyst or to effect drying. Chemical engineering has developed around the processing of oil, with particular emphasis being given to the mechanics of gases and liquids. With such fluids it sometimes has been possible to develop useful methods of analysis. In the flow of solids, knowledge has not advanced so rapidly, partly because of inherent difficulties in the subject and partly due to lack of fashion. Development of technology using solids and a general requirement for tighter process design and operation is warranting the increasing attention of the chemical engineer to understanding properly the factors controlling the processing of solids.

While here we focus on the flow of solids in bunkers—the immediate relevance being to packed beds, moving beds and solids storage vessels—note that the principles and phenomena are equally relevant to a wide range of processing equipment, including items such as standpipes, the continuous or emulsion phase in fluidized beds, spouted beds, dense-phase conveyors and the stirring of solids with blades (e.g., the ribbon blender).

Matters of interest to the process designer include the forces exerted on the equipment by the particles and also the forces on the particles throughout a system. The latter may be particularly important, controlling factors such as the voidage of the packing and the flow of a fluid through it under an applied fluid pressure gradient. It will also affect the tendency of the particles to be crushed or ground away.

Although the study of solids flow in bunkers and process equipment has much in common with soil mechanics, the stresses of significance in the former are generally lower, being typically 1–100 kPa. While peak stresses in bunkers are of obvious importance, the behavior and flowrates are controlled by conditions at the hopper outlet, where the stresses tend to be near zero. The characterization of solids under such conditions is not easy.

The discussion here opens with a brief summary of the elements of force analysis in solids and methods of testing. This is illustrated by particular reference to the design of hoppers and is followed by a brief discussion of the effects of inertia, interstitial fluid and packing non-uniformity.

MOTION OF POWDERS

Unlike liquids, solids can sustain shear stresses and the stress in a solid thus depends on the orientation of the plane on which it acts. Correct analysis of these stresses is a prerequisite for the successful solution of problems involving the statics or dynamics of solids.

Initially, the methods of analyzing solids stresses and motion will be discussed without considering the role of interstitial fluid; this will be dealt with formally below. For now, it is sufficient to remark [1] that fluid effects, be they due to buoyancy or a pressure gradient due to fluid motion, can be separated from the effects due to the forces between the particles and that the latter can be analyzed independently. This approach has been shown to be sound. The discussion that follows is thus directly relevant to a particulate bulk solid having within its pores an interstitial fluid of density much less than the solid and, furthermore, with the fluid moving with respect to the solid at a rate sufficiently low to ensure that fluid pressure gradients have little effect on the solid stresses.

Before passing to principally two-dimensional arguments, it must be noted that in the steady state any three-dimensional state of stress (including both normal and shear stress components) is equivalent to a set of normal stresses only acting on a set of mutually perpendicular planes, the normal stresses being termed principal stresses. These planes, on which no shear stress acts, are termed principal planes. If we denote the normal stresses on the principal planes by σ_1, σ_2, σ_3, then the state of stress is described most fundamentally in terms of the three invariants of stress:

$$I_1 = \sigma_1 + \sigma_2 + \sigma_3 \tag{1}$$

$$I_2 = \sigma_1\sigma_2 + \sigma_2\sigma_3 + \sigma_3\sigma_1 \tag{2}$$

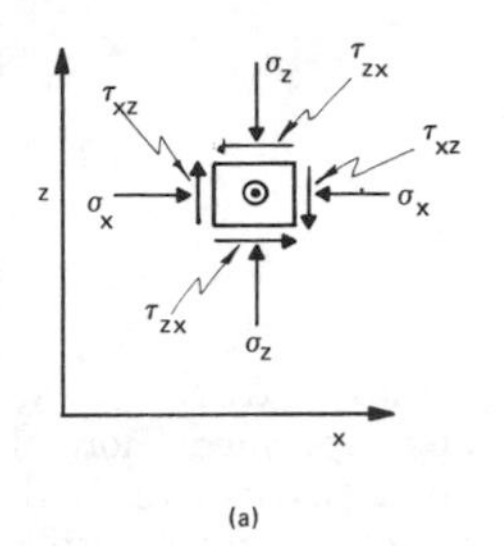

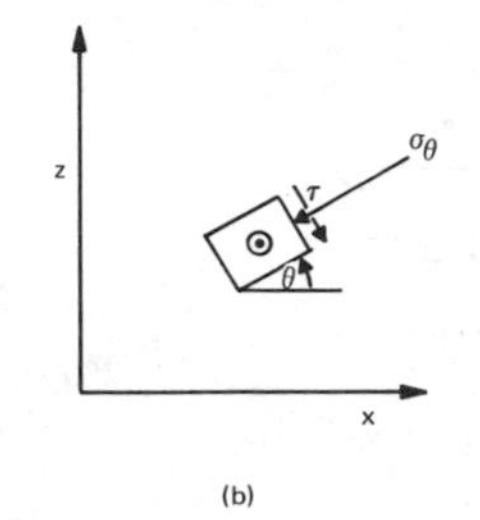

Figure 1. (a) Forces on an element of particulate solid; (b) Rotation of an element through angle θ about the point 0.

and

$$I_3 = \sigma_1 \sigma_2 \sigma_3 \tag{3}$$

In practical terms, however, measurement in a reliable fashion of the properties of a powder or particulate solid is difficult enough in two dimensions, let alone three.

For problems in two dimensions, the states of stress on planes of various orientations are related in a simple manner by force balances. Figure 1a shows the stresses acting on an element, σ_x and σ_z, denoting the normal compressive stresses on the element. For the shear stresses, τ, the first subscript denotes the plane on which the stress acts and the second the direction in which it acts. Figure 1b shows the orientation of a plane on which the stresses are sought. Figure 2 shows the locus of stresses mapped out by varying θ, the point (σ_θ, τ) describing a circle with its center on the $\tau = 0$ axis. This formalism follows by application of a force balance to the element. The convention is followed here that normal stress is positive when it is compressive. Counterclockwise shear stresses are positive, and θ is positive when measured counterclockwise. The stresses τ_{zx} and τ_{xz} are thus complementary, i.e., $\tau_{xz} = -\tau_{zx}$, an identity established by taking moments about the center of the element. Major and minor principal planes, which can be found in Figure 2, have normal stresses σ_1 and σ_3, respectively, acting on them.

Similar arguments apply to linear strain, $\hat{\epsilon}$, and shear strain, γ. The locus of $(\hat{\epsilon}, \gamma/2)$ varies with θ, describing a circle with its center on the axis $\gamma/2 = 0$. Amplification of such points may be found by referring to soil mechanics textbooks such as those by Harr [2], Atkinson and Bransby [1] or Scott [3].

Note that two-dimensional methods ignore the role of the intermediate principal stress, σ_2, which enters into the stress invariants I_1, I_2 and I_3. Most approaches overlook this problem, and the topic remains incompletely resolved in academic approaches. For practical problems, σ_2 is sometimes found to be irrelevant or may, by physical argument, be set equal to σ_1 or σ_3. Although such procedures may appear arbitrary to the purist, usable methods have been developed.

If strain is imposed on a powder sample under a constant normal stress (Figure 3), it is usually found that the shear stress first increases and then decreases to an asymptotic value (route I). If the shear stress is removed and subsequently, reimposed, τ rises to its limiting value, the peak in shear stress being virtually eliminated (route II). Route

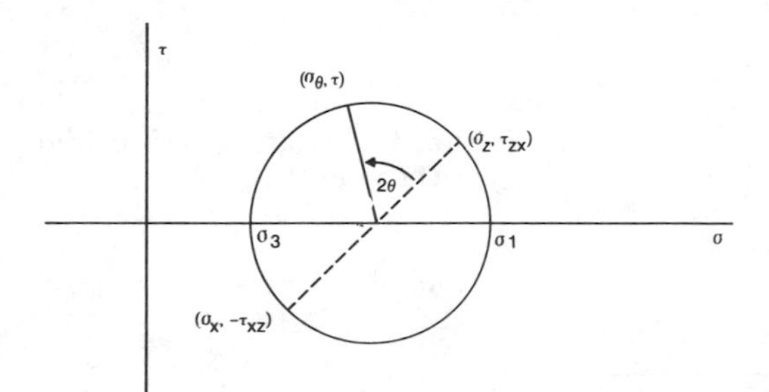

Figure 2. Mohr's circle construction. The relationship between normal stress, σ, shear stress, τ, and angle of rotation, θ.

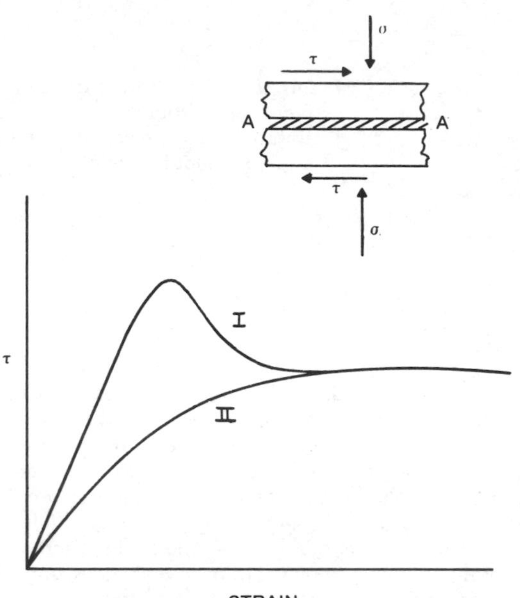

Figure 3. Shear stress, τ, vs strain under specified normal stress, σ, for a particulate solid. Route I is for an overconsolidated material and route II is for a material very loosely packed. AA denotes the slipping region (slip zone or failure zone) that is found for the overconsolidated state, the most usual one.

II is also taken by very loosely packed material. When a constant shear stress is reached, the particulate solid has developed a structure in the region of high strain, the failure zone, which permits indefinite slip without change in volume. In the testing of solids and in analyzing the flow in equipment, the importance of the specific volume of material in the regions of high strain, or regions in which high strain is to be developed, is of central significance.

If a gently settled particulate solid is compressed under the action of a normal stress, it is found that the specific volume is directly proportional to the logarithm of the stress. If the sample is then unloaded, the specific volume is proportional again to the logarithm of stress, although the relationship is different. This idealized behavior is illustrated in Figure 4, which shows a solid being consolidated, unloaded and reloaded. The reloading consolidation process initially is reversible (not absolutely true in practice) and then irreversible when the normal consolidation line is reached. Material lying to the left of the normal consolidation line is termed overconsolidated. At low stresses the logarithmic relationship must fail since it would predict an infinite specific volume

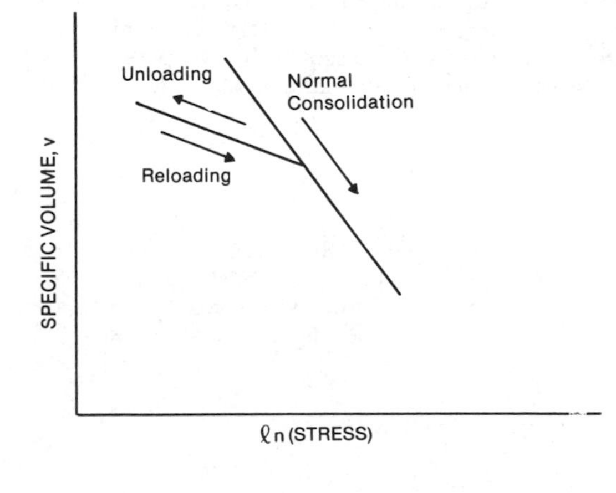

Figure 4. Loading and unloading of a sample.

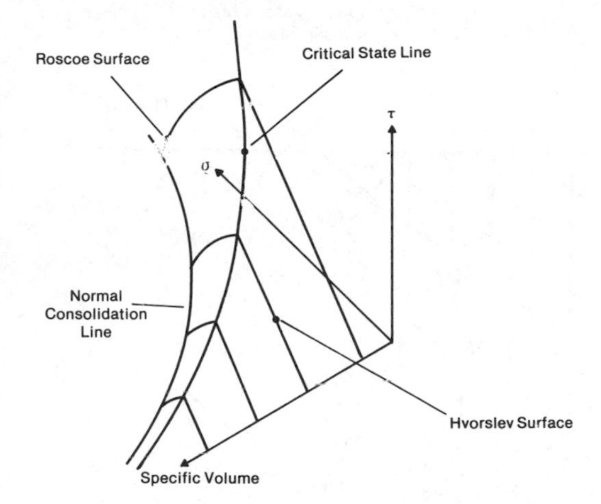

Figure 5. Specific volume, v, plotted schematically against normal stress and shear stress. Illustration of the Hvorslev surface, the Roscoe surface and the critical state line.

at zero normal stress. Since a particulate solid is necessarily dominated by the particle contacts, the relationship must alter with the specific volume attaining a specified value at low stresses. This matter remains incompletely resolved experimentally and theoretically, imposing limits on the reliability of results for some problems.

As has been mentioned, solids deform elastically at small applied shear stresses. At some limiting value, unlimited strain takes place. Thereafter, the shear stress assumes a steady value, which may be lower than that required for initial failure. For a given specific volume, the locus of (σ, τ) values for initial failure is referred to as the yield locus.

More complicated models of behavior have been developed. The limits to which a particulate solid may be consolidated before failure occurs may be delineated by the Hvorslev failure surface which, combined with another surface called the Roscoe surface, has been used to describe the failure of sands, clays and particulate solids used in the chemical industry. This is illustrated schematically in Figure 5 which, for conceptual simplicity, is presented as a plot of normal stress, shear stress and specific volume; invariants of the stress tensor are needed strictly. Most of the applications in the process industries have large strains and small stresses, in contrast to soil mechanics applications with small strains and large stresses where the purpose is to prevent excessive movement. In most instances, solids will be overconsolidated and will suffer initial failure by intersection of the stress path with the Hvorslev surface. The ultimate stress and voidage state will be located on the critical state line. Once a system reaches a point on this line, indefinite strain can occur without further changes in the external conditions.

If the solid is able to sustain tension due to internal cohesive forces, the Hvorslev surface will cut the plane $\sigma = 0$ with $\tau > 0$. The critical state concepts have been elaborated by Atkinson and Bransby [1], Brown and Richards [4], Williams and Birks [5] and Heertjes et al. [6]. We shall look at the Jenike testing procedure in the light of this approach.

LIMITING FAILURE ANALYSES

Slip Plane

One of the most powerful general methods of analysis stems from the classic work of Coulomb, who assumed a linear relationship between limiting values of σ and τ, i.e., a linear yield locus. He further suggested that when failure takes place, slipping surfaces develop and prediction is possible by assuming these to be planar. This is illustrated by a blade moving into a body of particulate solid (Figure 6). There is one

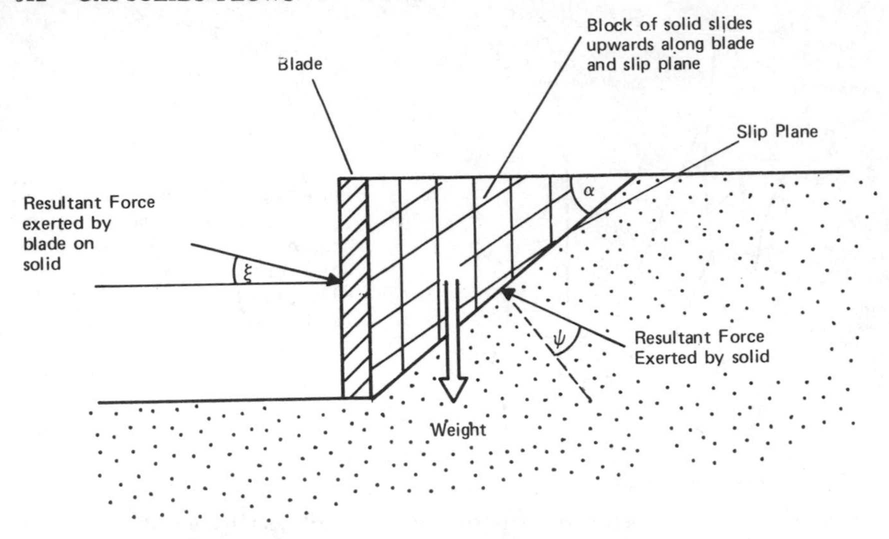

Figure 6. Movement of a blade into a bed of solids.

particular inclination of the slip plane, α, which will give the lowest force exerted by the blade, and this will determine the weakest plane in the material. Other shapes of failure surface can be considered, the ability to change the surface shape leading to lower values of the force on the blade.

Conversely, if the blade is moved gradually away from the solid, the block of solid slides down the slip surface and the blade, reversing the inclinations (ψ, ξ) of the resultant forces to the respective plane normals. The solid then develops a slip plane, the material failing in the manner that makes the force exerted on the blade as large as possible.

This general approach does not employ a force balance on an element. Rather, the analysis is conducted in terms of the overall forces exerted on the body of solid, evading questions of the distribution of stresses along the surfaces. Harr [2], in his fifth chapter, develops the ideas in a soil mechanics context. Hancock and Nedderman [7] apply such ideas to the design of bunkers, but these have not formed the basis of a general design method.

If the blade is advancing into the solid, the state of failure is termed passive, the blade seeking to provide horizontal compression. If the blade is moving away from the solid, the state of failure is termed active, the blade extending the solid horizontally. Here we see one of the key aspects of solids failure, the existence of two possible modes of failure. The selection of the appropriate mode depends on physical understanding of the motion during failure.

Limiting Equilibrium

Consider a body of particulate solid with a horizontal surface (Figure 7a). The vertical normal stress exerted on a horizontal plane, σ_z, is given by

$$\sigma_z = \rho g z \tag{4}$$

where ρ is the bulk density of the solid, and g is the acceleration due to gravity. Let us suppose that the solid has been deposited by settling under gravity, the shear stresses on the element illustrated being zero. The state point for the top surface of the element (ρgz, 0) is indicated in Figure 7b.

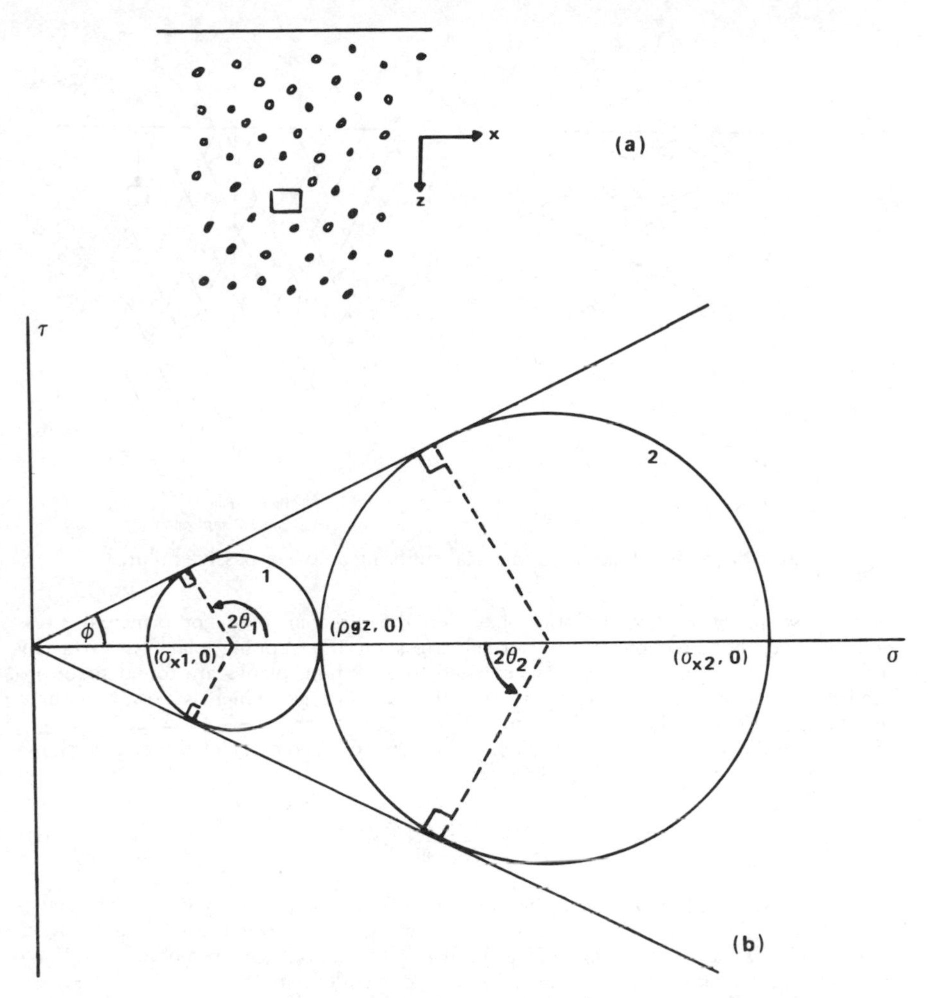

Figure 7. (a) Element distance, z, below the free surface of the material; (b) active and passive failure of a particulate solid at a distance, z, below free surface.

Figure 7b shows a failure envelope defined by

$$|\tau| = \sigma \tan \phi \tag{5}$$

where ϕ is the angle of internal friction.

We now seek to find the stress on a vertical plane, σ_x, and will consider that the solid has been deformed, say by horizontal movement of a retaining wall, so that it is in a state of failure. In this event there exists a plane at the point lying on the failure envelope. It is found that there are two possible circles, 1 and 2, with two possible values of σ_x, denoted by σ_{x1} and σ_{x2}. σ_x will equal σ_{x1} if the material has reached its failure state by horizontal extension, the state of stress being active in accordance with the discussion above. In complementary fashion, σ_x will equal σ_{x2} if the material has reached its failure state by horizontal compression, the state of stress then being passive.

If the whole of the material is in a state of active failure, failure is occurring on planes inclined at an angle θ_1 to the plane on which the stresses are $(\rho gz, 0)$, the horizontal

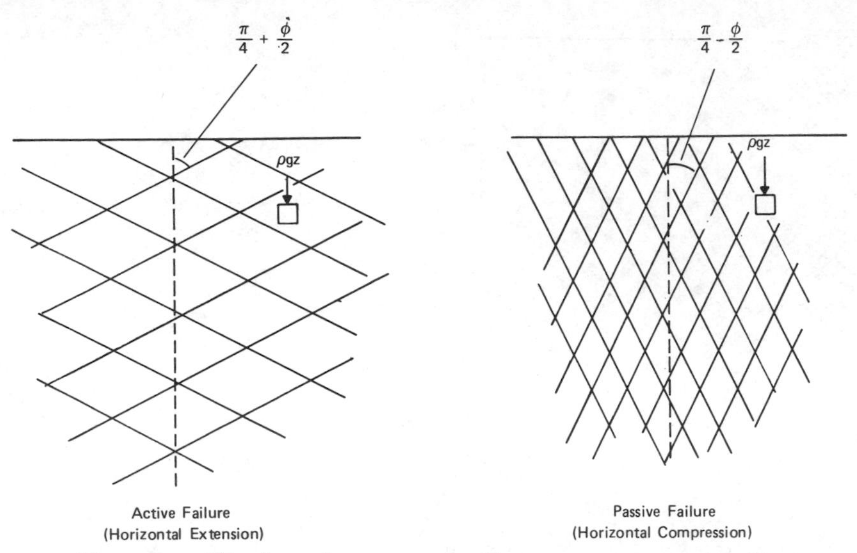

Figure 8. Slip planes for a material suffering active or passive failure.

plane. By simple geometry, the angle between the direction of major principal stress (ρgz, 0) and the direction of the normal stress on the slipping plane is given by $\pm(\pi/4 + \phi/2)$. If passive failure is analyzed, the slipping planes are found to be inclined at angles of $\pm(\pi/4 - \phi/2)$ to the direction of ρgz, which is now the minor principal stress. This is illustrated in Figure 8.

If the cohesion of the material is taken to be zero, the geometry of the circles shows that the ratio of major to minor principal stress is

$$\frac{\sigma_1}{\sigma_3} = \frac{1 + \sin\phi}{1 - \sin\phi} \tag{6}$$

General methods of analysis may be developed from this assuming that a material is in a state of failure everywhere, the extension first having been made by Sokolovski. Such techniques have been exploited by Jenike in his hopper design method and have been followed by other workers, e.g., Horne and Nedderman [8].

The analysis proceeds as follows. For convenience, we choose to rotate the set of axes shown in Figure 1a so that the z axis points downwards in the direction of gravity. Force balances on an element of material (Figure 9) in vertical and horizontal directions give

$$\frac{\partial\sigma_z}{\partial z} + \frac{\partial\tau_{zx}}{\partial x} = \rho g$$

$$\frac{\partial\sigma_x}{\partial x} + \frac{\partial\tau_{zx}}{\partial z} = 0 \tag{7}$$

The Mohr-Coulomb failure criterion is assumed. This is expressed as

$$\sigma_x = \sigma_o\,(1 + \sin\phi\cos 2\theta)$$

$$\sigma_y = \sigma_o\,(1 - \sin\phi\cos 2\theta)$$

$$\tau_{xz} = -\tau_{zx} = \sigma_o \sin\phi \sin 2\theta \tag{8}$$

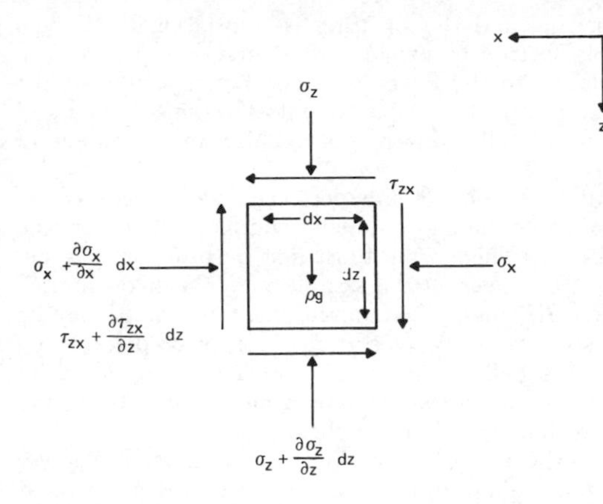

Figure 9. Stresses acting on an element of material. (For clarity, complementary shear stresses are not fully specified.)

where σ_o is the mean stress, i.e., $\sigma_o = \frac{1}{2} (\sigma_1 + \sigma_3)$, and θ is an angular coordinate, measured counterclockwise from the x axis. Equation 8 may be substituted into Equation 7 to give a set of two partial differential equations. These reduce to the two ordinary differential equations:

$$d\sigma_o + 2\sigma_o \tan \phi \, d\theta = dz + \tan \phi \, dx$$

$$d\sigma_o - 2\sigma_o \tan \phi \, d\theta = dz - \tan \phi \, dx \qquad (9)$$

in two directions, which are referred to as the characteristic directions and are given by

$$\frac{dz}{dx} = \tan (\theta \pm \beta) \qquad (10)$$

where

$$2\beta = \pi/2 - \phi \qquad (11)$$

Details of the solution for two-dimensional bins and hoppers are given by Horne and Nedderman [8–10]. The weakness of such an approach is that we know that material, generally being overconsolidated, develops a few slip planes, while the bulk of the material moves as blocks that behave in an elastic manner. This behavior is controlled by the loading and unloading characteristics illustrated in Figure 4.

If failure is occurring against a wall, we have then the alternative or additional constraint that on the plane coincident with the wall

$$|\tau| = \sigma \tan \phi' + a \qquad (12)$$

where ϕ' is the external angle of friction between the material and the wall and 'a' denotes the adhesion. For a non-adherent material, a = 0.

MEASUREMENT OF FLOW PROPERTIES

The Jenike Shear Tester

A tester commonly used to measure flow properties under conditions appropriate to bunker design is the Jenike shear apparatus [11]. In this device, the yield locus is obtained by making a number of measurements of the normal and shear stresses acting on a given plane at failure.

The cell itself (Figure 10) is a split ring and the shearing is translational. A given normal force is applied and the sample is free to expand or contract in the vertical direction. The upper ring is moved relative to the lower at a constant speed, and the applied shear force is measured by a transducer. The shear zone itself is lens-shaped and does not take up the whole volume of the cell [12] so that reliable measurement of strain is not feasible.

As we mentioned in the section entitled Motion of Powder, each yield locus corresponds to a given initial value of the solids voidage (or specific volume). To determine a point on the yield locus, the material in the shear zone must first be brought to a defined, reproducible voidage. This is done by shearing the sample under a given normal stress until a constant shear stress is reached. Under such conditions, the sample may be assumed to be in the critical state. The stresses are derived from the measured forces by dividing by the cross-sectional area of the cell. Because the travel of the cell is inherently limited, special filling and pre-shearing measures are usually taken to ensure that the critical state is achieved within a short translational shear distance [13].

When the sample has been brought to the critical state, shearing is stopped, a lower normal stress applied and shearing recommenced until the applied shear stress has passed through a maximum or become steady. This maximum shear stress, τ, and the corresponding normal stress, σ, give a point on the yield locus that is characterized by the initial critical state. Additional points on the yield locus are generated in the same way. For each point the cell is refilled, preferably with fresh material, and the shear zone is brought to the critical state. Usually a number of yield loci corresponding to a number of initial critical states are determined.

If solids spend time at rest in a bunker, the strength is usually found to increase with time. This effect can be simulated by first bringing the shear zone to the critical state and then applying a normal stress equal to the principal consolidating stress during shearing. This is allowed to remain for the processing time in question, after which the sample is sheared under a lower normal stress to give a point on the yield locus for that consolidation time. As the split ring cell itself can be removed from the apparatus, a number of samples can undergo time consolidation simultaneously.

A typical yield locus is shown in Figure 11. Although the locus may extend into regions of tensile strength, only the compressive part can be determined satisfactorily in the Jenike shear tester. The dashed portion of the yield locus may not be found by this technique. The unconfined yield stress, f_c, is determined by drawing the Mohr's circle, which passes through $\sigma = 0$ and just touches the yield locus. The major principal stress acting during the critical state consolidation, σ_1, is found by drawing a critical Mohr's

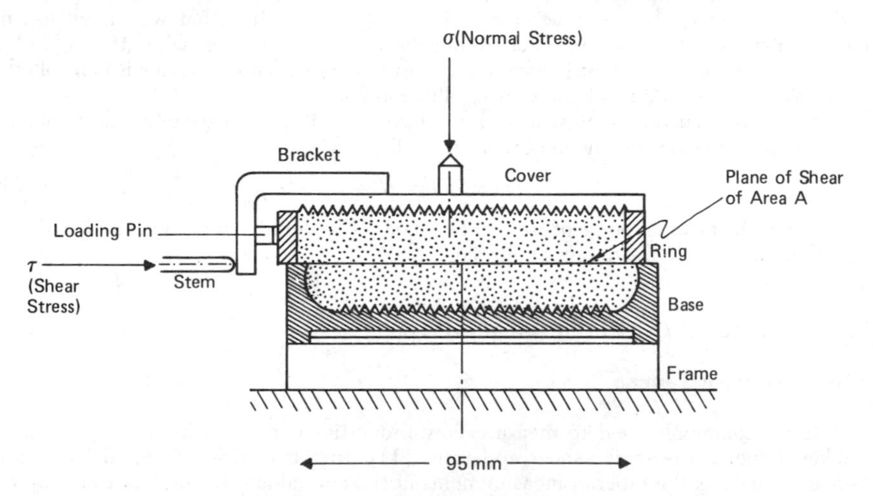

Figure 10. The Jenike shear cell.

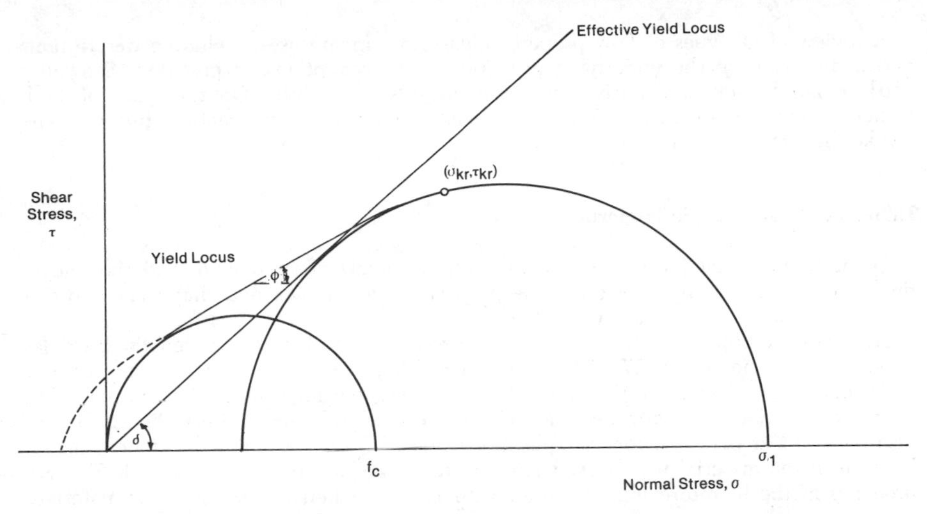

Figure 11. Derivation of principal consolidating stress, σ_1, unconfined yield stress, f_c, effective angle of internal friction, δ, and angle of internal friction, ϕ, from yield locus.

circle, which passes through (σ_{kr},τ_{kr}), the coordinates of the point representing the initial consolidation procedure in the cell. Several constructions are possible; that recommended [11] is a circle through (σ_{kr},τ_{kr}), which is tangential to the yield locus. The assumptions underlying this construction are discussed by Jenike [11] and Schwedes [12].

Other parameters are also found from the yield locus. ϕ (Figure 11) is the angle of internal friction. Experimentally, it is found that the envelope of critical Mohr's circles for various initial critical states is often a straight line. The slope of this is $\tan\delta$ (Figure 11), where δ is defined as the effective angle of internal friction.

From a number of yield loci determinations, the unconfined yield stress can be determined as a function of the major principal consolidating stress. Such a plot was called the flow function (FF) of the solid by Jenike [11] and forms a characterization of the cohesive properties of the material.

In the design of mass flow bunkers (described later) it is also necessary to know the angle of friction between the solids and the bunker wall material. This can be determined by replacing the lower half of the cell by a plate of the relevant material and measuring a number of (σ, τ) points.

Other Flow Property Testers

There are a number of other flow property testers, some of which are intended to be alternatives to the Jenike apparatus, and others that are more suitable for fundamental research studies on particulate materials.

In the first category, rotational cells are often used. These are generally similar in conception to the Jenike apparatus, but the shearing motion is rotational rather than translational. Examples are the annular shear cell [14] and the circular type of apparatus [15]. Rotational cells have the disadvantage that the strain is dependent on radial position. However the unlimited shear allows the critical state to be obtained more easily and with more certainty than in the Jenike apparatus.

A number of devices control more closely the shear plane and the applied stresses than in the Jenike apparatus but are more exacting and time consuming to use, thus making them more suitable for research purposes. These include the simple shear cell and the biaxial and triaxial apparatuses. The latter two are less suitable for low stresses.

A review of all types of flow property measuring apparatuses, including descriptions and a discussion of the underlying principles and assumptions, is given by Schwedes [16]. A number of comparative measurements also are given. On the basis of both principles and measurements, Schwedes recommended that for practical purposes the Jenike shear apparatus should be used.

Influence of Microscopic Properties

Quantitative prediction of the effective internal angle of friction, δ, and the unconfined yield stress, f_c, from basic particle properties such as size and shape and the distribution of these variables is not possible.

However, description of the yield locus in terms of the forces between the particles is possible to some extent [17]. Forces considered include capillary forces in the case of moist particles and van der Waals' forces, which become important with fine particles.

If the material shape deviates markedly from spherical, the particles being fibrous, needle-like or as platelets, for example, the value of δ will be markedly dependent on how much the material was packed and the direction in which it is sheared. The approaches in the literature must be used with caution when processing such materials.

BUNKER DESIGN

Introduction to Bunker Design

Requirements for Bunker Storage

A bunker is a piece of process equipment into which particulate solids are charged, stored and then discharged via an opening in the lower part. In general a bunker will consist of an upper, parallel-sided part, the bin, and a lower, convergent part, the hopper (Figure 12). In most cases a bunker will have a plane or an axis of symmetry, and a single outlet will be located at the bottom of the hopper.

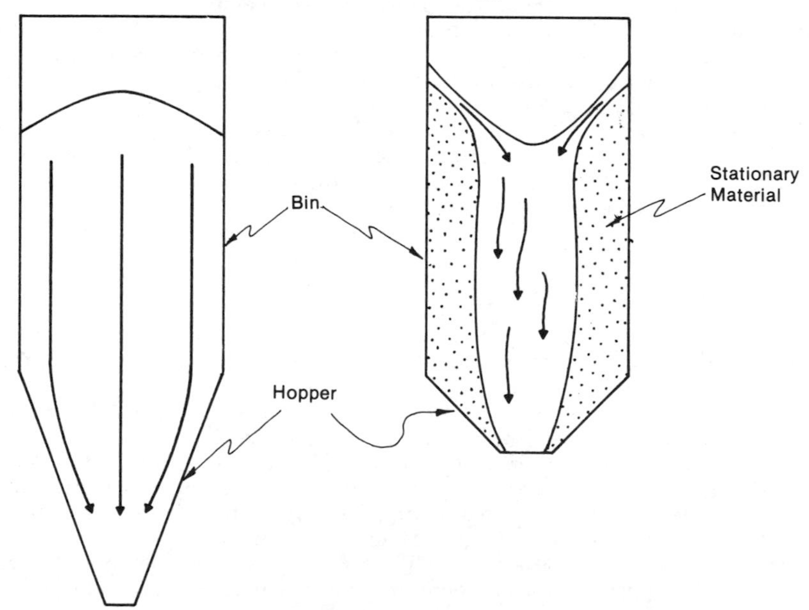

Figure 12. Flow patterns in mass flow (left) and funnel flow (right).

A general requirement for bunkers is that solids should discharge from the bunker as and when desired at a given discharge rate. There may be further requirements as, for example, an approximately uniform residence time distribution if the solids degrade with time. Additionally, of course, the bunker must be sufficiently strong to withstand the forces acting on it without recourse to gross overdesign.

In the following, flow in bunkers is discussed and design methods to ensure that process requirements are met are presented. The forces exerted on the bunker walls by the solids are discussed also, but the structural design of the vessel to enable it to withstand these forces is not considered further.

Flow Patterns in Bunkers

If the walls of a bunker hopper are sufficiently smooth and steep, the solids will fail against the walls (Figure 12). Thus, the solids are all in motion and, in general, the velocity profile is approximately uniform, particularly in the bin. This flow pattern is referred to as "mass flow." If the walls are not sufficiently smooth, and steep, solids flow occurs in a narrow core above the outlet, surrounded by a stagnant zone of solids. Material flowing from the core is replenished from above (Figure 12). This type of flow is known as "funnel (or core) flow." In some cases, the core (or funnel) may expand to meet the walls of a tall bin section, while in others the funnel may extend to the top surface. The terms were introduced by Jenike [11].

Jenike [11,13] stresses a number of advantages of the mass flow discharge pattern:

1. Steady, uniform discharge at a constant bulk density is obtained almost independently of the height of fill.
2. There are no dead regions. First-in, first-out flow is thereby achieved, avoiding prolonged consolidation and/or degradation of part of the material.
3. Segregation of particles by size or density is minimized.

Disadvantages include the following:

1. The steep hopper section requires a higher constructional framework.
2. Particle–wall movement may lead to wall erosion.

Obstructions to Flow

A common practical occurrence is that no discharge takes place on opening a bunker discharge gate. This occurs if the solids develop sufficient strength to resist the forces causing them to flow. There are many ways in which flow might be inhibited, but in practice [11] only two are met. These are (Figure 13) arching (or bridging), in which the solids form a stable arch across the hopper, and piping (or ratholing), in which a stable empty channel is formed above the outlet. The latter occurs only with funnel flow bunkers, whereas the former may occur with either mass or funnel flow types.

Arching also may be caused by mechanical interlocking (Figure 13) as opposed to cohesive forces. However, such interlocking does not generally occur when the outlet dimension exceeds ten times the particle dimension [13].

As is discussed in the following sections, arching and piping may be avoided if the outlet dimension is sufficiently large.

Bunker Design

From the foregoing, it can be seen that designing a bunker for process requirements consists of selecting a desired flow pattern and, if mass flow is chosen, determining what steepness and smoothness of the hopper wall are required. Then an outlet dimension must be chosen that is sufficiently large to give the required discharge rate and to avoid obstructions to flow. For the structural design the forces acting on the walls must be evaluated.

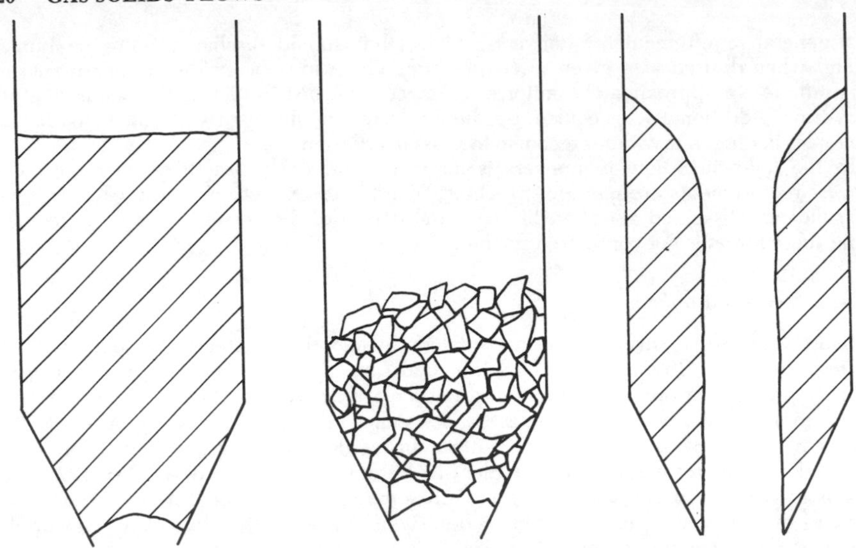

Figure 13. Obstructions to flow: cohesive arching (left), mechanical interlocking (center) and piping (right).

The first requisite for design is information on the stresses acting throughout the material in a bunker, especially those close to the outlet where obstructions to flow are likely to occur. This may then be combined with information cn the strength of the material stored as a function of consolidation, such as may be obtained from a shear cell test (see the section entitled Measurement of Flow Properties) to determine the minimum outlet dimension and, if required, the wall slope for mass flow.

There are a number of models describing stresses in bunkers. In the following paragraphs, the model of Walker [18] is discussed. However, the most complete and generally accepted bunker design procedure is that of Jenike [11,13], which is examined in some detail. Following this, additional information on forces acting on bunker walls is discussed followed by models for the prediction of discharge rate.

Stresses Acting in Particulate Solids in Bunkers

Stresses in the Bin

Janssen [19] developed a model for stresses in a solid contained within parallel walls. A force balance over an element of height, dz (Figure 14), gives

$$-\frac{d\sigma_z}{dz} + \rho g = \frac{S}{A}\tau_w \qquad (13)$$

where
σ_z = the vertical stress in the solids
τ_w = the wall shear stress
ρ = the solids bulk density
S = the perimeter of the element
A = the cross-sectional area of the element

In Equation 13 it is assumed that σ_z is constant over the cross-section. Janssen further assumed that

$$\tau_w = \sigma_w \tan\phi' \qquad (14)$$

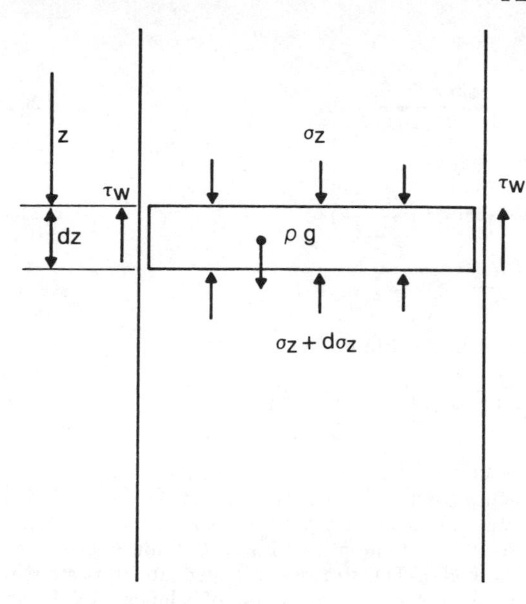

Figure 14. Stresses acting on an element in a cylindrical or parallel-sided channel.

where σ_w is the normal stress on the wall, and ϕ' is the wall friction angle. He also assumed that

$$\sigma_w = K \sigma_z \tag{15}$$

where K is a constant.

With these assumptions and the boundary condition,

$$\sigma_z = 0, \quad z = 0$$

Equation 13 integrates to

$$\sigma_z = \frac{\rho g A}{K S \tan \phi'} \left[1 - \exp\left(\frac{-K S z \tan \phi'}{A} \right) \right] \tag{16}$$

Janssen suggested that K should be evaluated experimentally. However, a simple approximation for K is to assume that the ratio of σ_w/σ_z is similar to the ratio of the principal stresses appropriate in an active stress field (see the section entitled Limiting Failure Analyses), giving

$$K = \frac{1 - \sin \delta}{1 + \sin \delta} \tag{17}$$

Walker's [18] treatment is similar but takes into account that the vertical stress at the wall is not a principal stress. Walker assumes it to be given by

$$(\sigma_z)_w = D^* \sigma_z \tag{18}$$

where σ_z is the average vertical stress over the bin cross-section and D^* is termed a distribution factor, which can in principle be obtained by radial integration of σ_z. The Mohr's circle for the stress at the walls is shown in Figure 15. From this it is found that

$$\tau_w = B^* (\sigma_z)_w \tag{19}$$

where

$$B^{*} = \frac{\sin \epsilon \sin \delta}{1 - \cos \epsilon \sin \delta} \tag{20}$$

and

$$\epsilon = \phi' + \arcsin\left(\frac{\sin \phi'}{\sin \delta}\right) \tag{21}$$

$$\left(\frac{\sin \phi'}{\sin \delta} > \pi/2\right)$$

Integrating the force balance over the element (Figure 14) then gives

$$\overline{\sigma}_z = \frac{\rho\, g\, A}{B^{*}\, D^{*}\, S}\left[1 - \exp\left(\frac{-\,B^{*}\, D^{*}\, Sz}{A}\right)\right] \tag{22}$$

The distribution factor, D^{*}, may be calculated from equilibrium considerations and is slightly less than unity until ϕ' approaches δ when D^{*} drops to about 0.6 [18]. It will be noted that the Mohr's circle (Figure 15) assumes a solid with no cohesion. However, during flow at the critical state (see the section entitled Limiting Failure Analyses) cohesive solids also will conform to this model. The stresses predicted are thus stresses during flow. However, as inertia terms do not occur in the force balance, the forces should not change significantly on stopping the flow.

Stresses in the Hopper

Walker [18] calculated stresses in a hopper in a similar manner to those in a bin. The following assumptions were made:

1. There is mass flow, giving slip along the walls. The wall stresses are given by the higher intersection of the wall yield locus with the Mohr's circle (i.e., there is a passive stress field, the material being compressed horizontally).
2. The material is everywhere in the critical state appropriate to the local stresses.
3. The distribution factor, D^{*}, is taken as independent of position.
4. The major and minor principal stresses lie in a plane normal to the wall.

The differential slice considered is shown in Figure 16 and the stress conditions at the wall are given in Figure 17. From this it is derived that the vertical shear stress at the wall, $(\tau_z)_w$, is given by

$$(\tau_z)_w = B^{**}\,(\sigma_z)_w \tag{23}$$

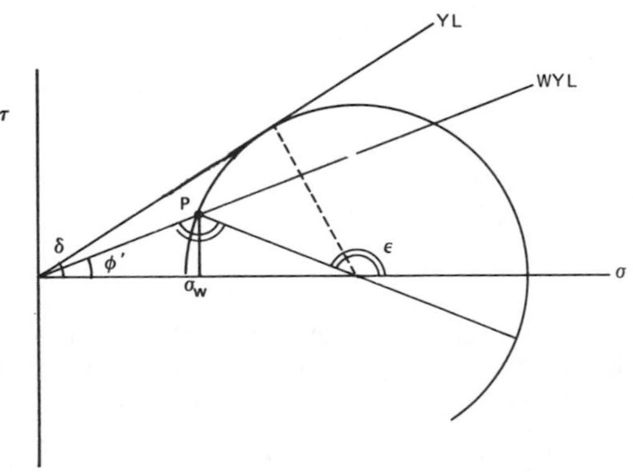

Figure 15. Mohr's circle for stress at a vertical wall [18]. P represents conditions at the wall.

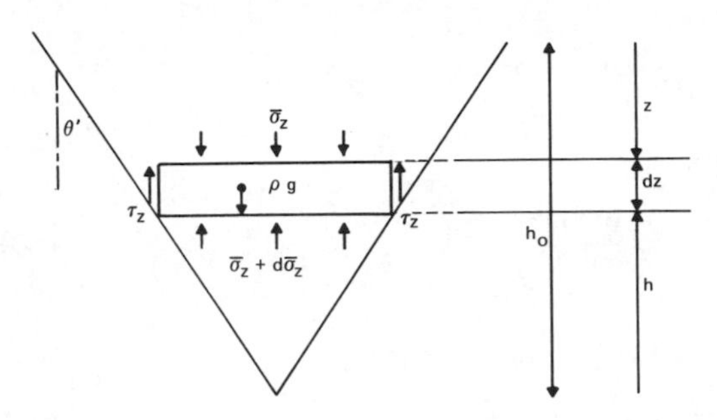

Figure 16. Differential slice in hopper.

where

$$B^{**} = \frac{\sin \delta \sin 2(\theta' + \beta)}{1 - \sin \delta \cos 2(\theta' + \beta)} \tag{24}$$

and

$$\beta = \tfrac{1}{2}\left[\phi' + \arcsin\left(\frac{\sin \phi'}{\sin \delta}\right)\right] \tag{25}$$

The vertical normal stress at the wall $(\sigma_z)_w$ is again related to the average vertical normal stress, $\bar{\sigma}_z$, by

$$(\sigma_z)_w = D^* \bar{\sigma}_z \tag{26}$$

Taking a vertical force balance over the element of Figure 16 and integrating with boundary condition

$$\bar{\sigma}_z = 0, \quad h = h_o \tag{27}$$

for a hopper with a horizontal upper free surface gives

$$\bar{\sigma}_z = \frac{\rho g h}{C - 1}\left[1 - \left(\frac{h}{h_o}\right)^{C-1}\right] \tag{28}$$

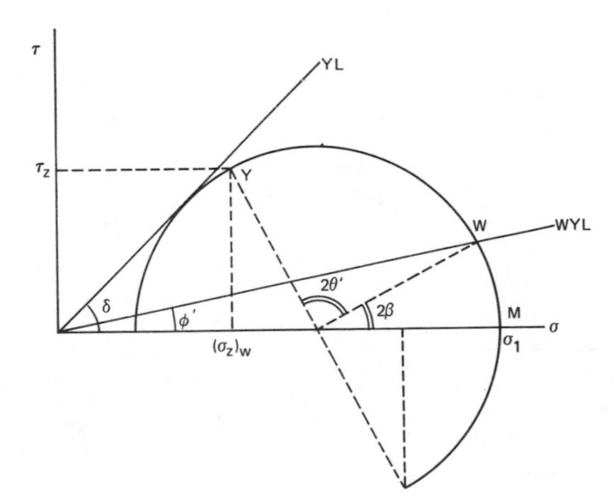

Figure 17. Mohr's circle for stress at a hopper wall [18]. M represents the major principal plane, i.e., that on which σ_1 acts; W represents conditions at the wall plane; and Y represents the vertical plane adjacent to the wall.

where

$$C = \frac{2\,B^{**}D^{*}}{\tan\theta'} \tag{29}$$

for a cone, or

$$C = \frac{B^{**}D^{*}}{\tan\theta'} \tag{30}$$

for a two-dimensional wedge.

If there is a surcharge, e.g., a loading from the bin section,

$$\bar{\sigma}_z = \sigma_{zo}, \quad h = h_o \tag{31}$$

and integration then gives

$$\bar{\sigma}_z = \frac{\rho gh}{C-1}\left[1 - \left(\frac{h}{h_o}\right)^{C-1}\right] + \sigma_{zo}\left(\frac{h}{h_o}\right)^{C} \tag{32}$$

Walker advances both experimental and theoretical arguments to suggest that there is little error in assuming the distribution factor, D^*, to be unity.

Equation 28 suggests that the vertical stress increases with depth to some maximum value and then decreases to a value of zero at the apex (Figure 18). The stresses in any other direction can be related to the vertical stress via the Mohr's circle and thus would show a similar profile. In particular, the major principal stress, σ_1, is given by

$$\frac{\sigma_1}{\bar{\sigma}_z} = \frac{1 + \sin\delta}{1 - \sin\delta\cos 2(\theta' + \beta)} \tag{33}$$

For practical mass flow, C (Equation 29) is rather larger than unity. Thus, toward the hopper apex the vertical stress becomes

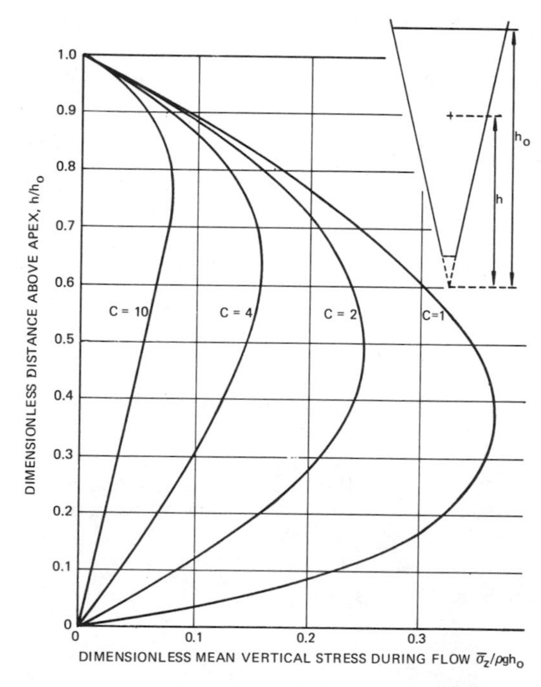

Figure 18. Mean vertical stress in a hopper during flow as a function of height, from Equation 28 [18].

$$\overline{\sigma}_z = \frac{\rho\, gh}{C - 1} \tag{34}$$

and the major principal stress becomes

$$\sigma_1 = \frac{\rho\, gh}{C - 1}\left(\frac{1 + \sin\delta}{1 - \sin\delta\cos 2(\theta' + \beta)}\right) \tag{35}$$

Equation 35 shows that the stresses toward the apex of the hopper depend only on the properties of the solid, the hopper angle and the distance above the apex, and that these are independent of the size and scale of the bunker and the height of fill. This is an important result as it allows obstructions to flow (which nearly always occur close to the outlet) to be analyzed independently of bunker geometry and size away from the outlet.

The Jenike Bunker Design Procedure

Introduction

There are a number of analyses similar to Walker's [18] that lead to broadly similar conclusions [20,26]. Some authors, for example, Walker [18] and Arnold and McLean [25,26], combine an analysis of the stability of arches with the stress field prediction to derive a flow/no-flow criterion for bunker discharge. However, the most comprehensive and widely accepted bunker design procedure is that of Jenike [11,13]. The method includes a number of separate aspects, namely:

1. Determination of the flow properties of the solids by shear cell testing (see the section entitled Measurement of Flow Properties) which is necessary for every solid.
2. Solution of the equations for the stress and velocity fields in the bunker, in particular those in the region close to the outlet.
3. Analysis of the stability of obstructions to flow.
4. Determination of the outlet dimension required to avoid obstructions and the wall slope required for mass flow for the particular solid tested.

The second and third points above, which lead to general graphic solutions, are considered first, after which practical design is examined.

Stress and Velocity Fields

The starting point for Jenike's stress analysis is a force balance on a Cartesian element similar to that discussed in the section entitled Limiting Failure Analyses (Equations 7 and following). It is assumed that the material is in failure everywhere and a third equation is thus given by the equation for the Mohr's circle touching the effective yield locus, i.e., the relationship between the stresses when the material is in steady-state flow. The density is assumed to be a known function of the major principal stress. This gives a set of four equations (two partial differential, two algebraic) for the four unknowns—σ_x, σ_z, τ and ρ. If axisymmetric flow is considered instead of two-dimensional flow, the circumferential stress is set equal to the major principal stress in the radial plane, thereby giving a set of five equations for five unknowns. Inertia terms are taken to be small and do not appear in the stress equations.

The velocity field is given by the equation of continuity and the assumption of the principle of isotropy, which state that the directions of the principal stress and the principal rate of strain coincide. Assuming that stresses are known, two partial differential equations may be solved for the two unknown components of velocity.

The partial differential equations for both the stress and velocity fields are hyperbolic and thus can be solved numerically by the method of characteristics [27]. Numerical solutions are given by Jenike [11].

However, a more important result is Jenike's proof that toward the apex of a converging channel, either two or three dimensional, the stress field reduces to a particularly simple form, called the radial stress field. In such a field the stress is proportional to the distance from the apex, r, and a geometric parameter, s, which is a function of the angular coordinate, θ. In particular, the principal stress, σ_1, is given by

$$\sigma_1 = r\,\rho\,g\,(1 + \sin\delta)\,s\,(\theta) \tag{36}$$

These stresses are assumed to remain constant when the flow stops, and the strength developed by given solids thus depends on the local value of σ_1 and the time at rest.

Solutions of the radial stress field are not given in terms of σ_1, however, but in terms of the ratio $\sigma_1/\overline{\sigma}_1$, which is called the flow factor (ff) by Jenike. σ_1 is the major principal stress at a given point in the hopper during flow or during storage with the outlet closed. $\overline{\sigma}_1$ (the standard Jenike notation) is the major principal stress that would act in an arch at the same point in the bunker. It is shown in the section entitled Bunker Design that $\overline{\sigma}_1$ is determined primarily by the span of the arch and the solids density. Thus, both σ_1 and $\overline{\sigma}_1$ are linear with distance from the apex (provided the radial stress field still holds), and the ratio $\sigma_1/\overline{\sigma}_1$ is a constant for a given material in a given hopper. As the material is being consolidated by stress, σ_1, while the arch is tending to be broken by stress, $\overline{\sigma}_1$, the lower the ratio the less the chance of arching and the better the flowability of the hopper.

Jenike [11,13] gives graphical solutions to the radial stress field for mass flow in which contours of the flow factor are plotted on axes of hopper wall inclination, θ', and wall friction angle, ϕ', for a given internal friction angle, δ. Examples for $\delta = 50°$ are given in Figures 19 and 20 for plane and axisymmetric flow, respectively. As friction angles may depend on stresses, some iteration to find the exact value of ff may be required. For funnel flow there is one less independent parameter, and contours of the no piping flow factor for all δ are given in one graph (Figure 21).

It will be observed that for axisymmetric mass flow the area in which contours occur is much more limited than in plane flow. In both cases, the limits on ϕ' and θ' are those that allow convergence to the radial stress field in a channel where there is a slip at the walls. Combinations of ϕ' and θ' outside these limits give rise to funnel flow or no flow.

Determination of the flow factor concludes one stage of the design procedure.

Obstructions to Flow

As was pointed out previously, the most important obstructions to flow are bridges or arches and, in funnel flow bunkers only, pipes or ratholes. If these obstructions are not stable, it is found empirically that others will not be stable either.

The forces acting in an arch depend on its span. Jenike [11] considers a smooth arch of arbitrary constant vertical thickness. The stress acting along the arch is a major principal stress. At the walls it has value $\overline{\sigma}_1$. It is assumed that no vertical forces from the material above are transferred to the arch. For vertical equilibrium we have (Figure 22)

$$2\,(1 + m)\,\overline{\sigma}_1 \sin\omega \cos\omega = B\,\rho\,g \tag{37}$$

where $m = 0$ for plane geometry or $m = 1$ for axisymmetric geometry. The arch will be stable only if

$$\overline{\sigma}_1 < f_c$$

where f_c is the unconfined yield strength of the solids. Hence, an arch will fail if B, the minor dimension of the outlet, is given by

$$B > \left[\frac{2\,(1+m)\,f_c}{\rho\,g} \sin\omega \cos\omega \right]_{max} \tag{38}$$

that is, noting that the maximum value of 2 sin ω cos ω is unity, if

$$B > \frac{(1+m)\,f_c}{\rho\,g} \tag{39}$$

In a later, more rigorous analysis of arch stability, Jenike and Leser [28] show that for an arch to fail,

$$B > \frac{H\,(\theta')\,f_c}{\rho\,g} \tag{40}$$

where θ' is the inclination of the hopper wall to the vertical. Function $H(\theta')$ is given graphically. At $\theta' = 0$, $H(\theta')$ is 1 for plane flow and 2 for axisymmetric flow, thereby making Equation 40 identical to Equation 39. $H(\theta')$, however, increases slowly with θ'.

The analysis of pipe stability [11] is more complicated but leads to a similar result, namely that a pipe will fail if

$$B > \frac{G\,(\phi)\,f_c}{\rho\,g} \tag{41}$$

where $G(\phi)$ is a graphically specified function of ϕ, the angle of internal friction of the solids.

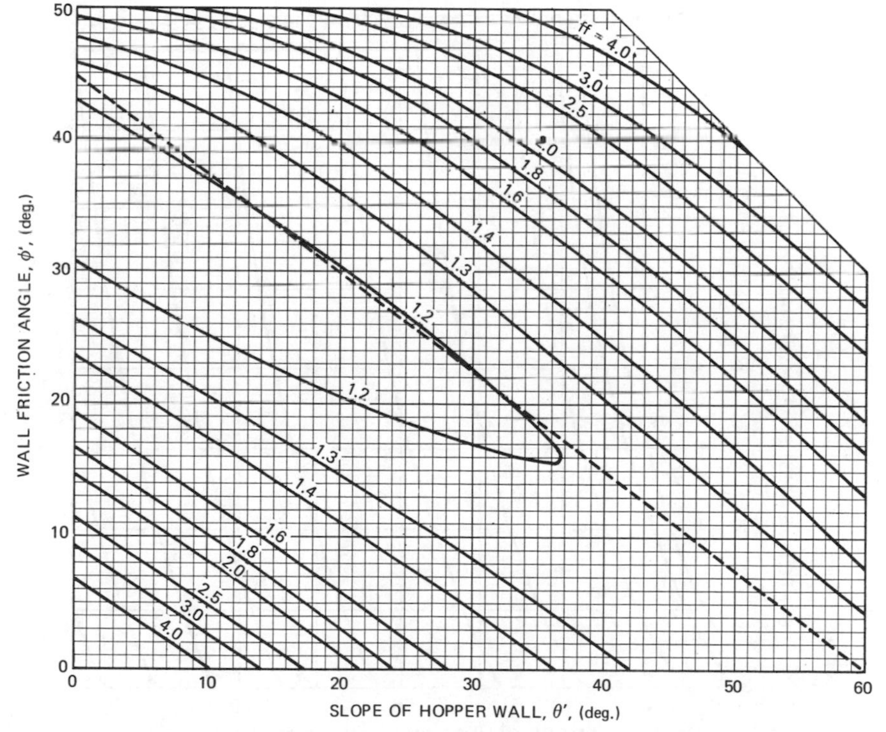

Figure 19. Flow factor (ff) contours for mass flow in symmetrical plane flow hoppers at $\delta = 50°$ [13]. Dashed line gives empirically recommended limit for design.

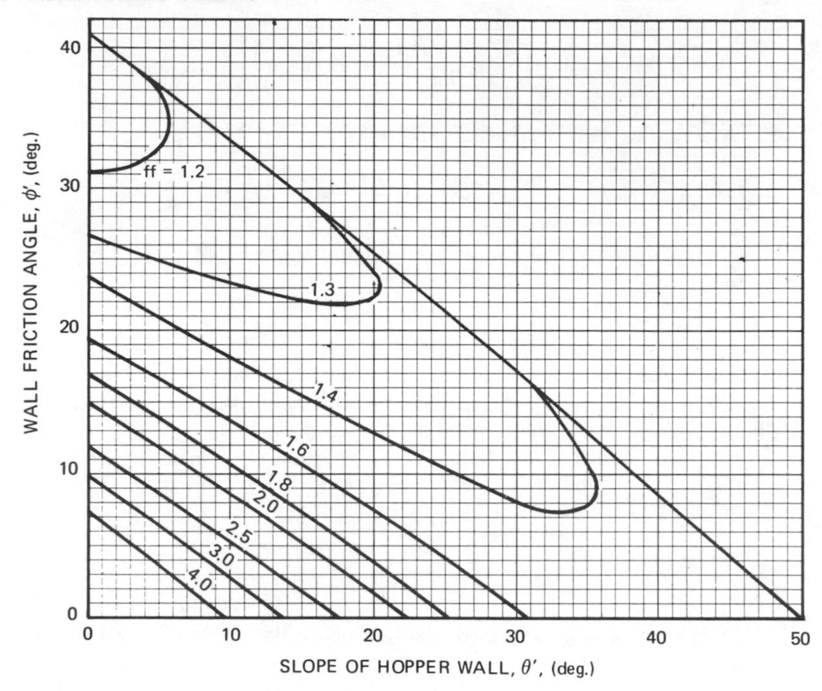

Figure 20. Flow factor (ff) contours for mass flow in conical hoppers at $\delta = 50°$ [13].

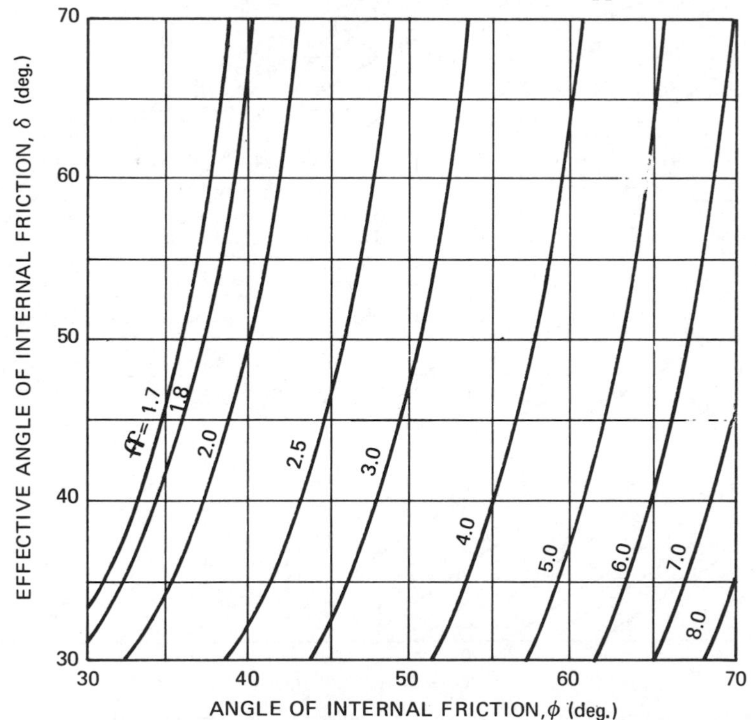

Figure 21. No piping flow factors for funnel flow [13].

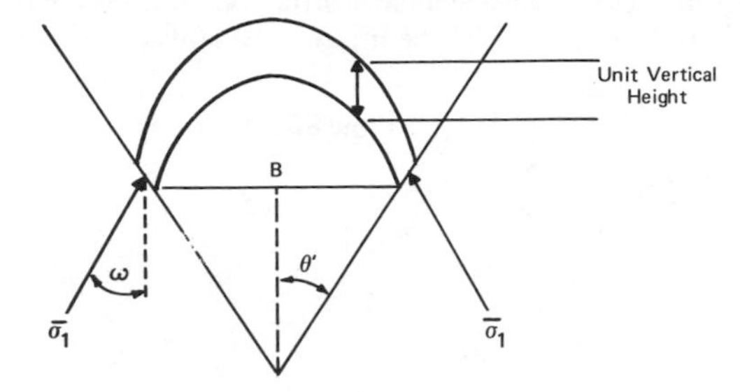

Figure 22. Arch of unit thickness in a hopper.

Practical Design According to the Jenike Method

The first requisite for bunker design according to the Jenike method is to have flow property data for the solids, such as can be obtained from shear cell tests (see the section entitled Measurement of Flow Properties). If a mass flow bunker is required, permissible hopper wall angles can be found from the flow factor plots (e.g., Figures 19 and 20). In the case of axisymmetric flow, a safety margin of 3–5° with respect to the boundary (Figures 19 and 20) should be allowed. To achieve plane flow, a bunker should have a slot outlet, with a length to width ratio greater than 3 [13]. The side wall slope should be chosen such that the flow factor is close to the locus of minimum flow factors (Figure 19). One reason for this large safety margin with respect to the predicted area within which mass flow is possible is that as the distance above a slot outlet increases in a practical bunker, the deviations from true plane flow become progressively greater. If the walls at the end of a slot outlet are to slope instead of being vertical, they should not exceed the inclination appropriate for axisymmetric mass flow. In addition to the wall slope, the flow factor for that solid in the hopper selected is also read from the chart.

In analyzing obstructions to flow, it has been remarked already that an arch will fail if the principal stress acting in the arch, $\bar{\sigma}_1$, exceeds the unconfined yield strength of the solids, f_c. To determine whether arching will occur in a given situation, both the principal stress acting in an arch, $\bar{\sigma}_1$, and the unconfined yield stress of the solids, f_c, are plotted against the principal consolidating stress, σ_1. In other words, the flow factor ($ff = \sigma_1/\bar{\sigma}_1$) and the flow function ($FF = \sigma_1/f_c$) are plotted together (Figure 23). If $\bar{\sigma}_1 > f_c$ (Figure 23a), the stress acting in the arch is sufficient to overcome its strength and flow takes place. If $\bar{\sigma}_1 < f_c$ (Figure 23b), the stress acting in the arch is insufficient to overcome the strength and the arch will be stable. Where $\bar{\sigma}_1 = f_c$, i.e., at the intersection of the flow factor and the Flow Function, the arch is critical. The stress acting in this critical arch is $\bar{\sigma}_{1,critical}$ (Figure 23c), and from this value the critical outlet dimension, above which arching will not occur, $B_{critical}$, can be determined from Equation 40:

$$B_{critical} = \frac{H(\theta')\,\bar{\sigma}_{1\,critical}}{\rho\, g} \tag{42}$$

This dimension is the diameter of a circular outlet in axisymmetric flow or the width of a slot in plane flow.

In the case of funnel flow, the analysis proceeds similarly but there are no constraints on hopper wall inclination. As the principal obstruction to flow considered is a circular pipe (this requires a larger outlet than that merely to prevent bridging), only the case of axisymmetric flow is relevant. The no-piping flow factor (Figure 21) is used, and the minimum outlet dimension found is the diameter of a circular outlet or the diagonal of a rectangle. If such a rectangular outlet is used, it should be checked that the minor dimension is still large enough to prevent two-dimensional arching.

A fuller description of the procedure and several worked examples are given by Jenike [13] and also by Arnold et al. [29].

The existence of a minimum outlet dimension above which obstructions to flow will not be stable depends on the solids flow function (FF) being convex upwards or at

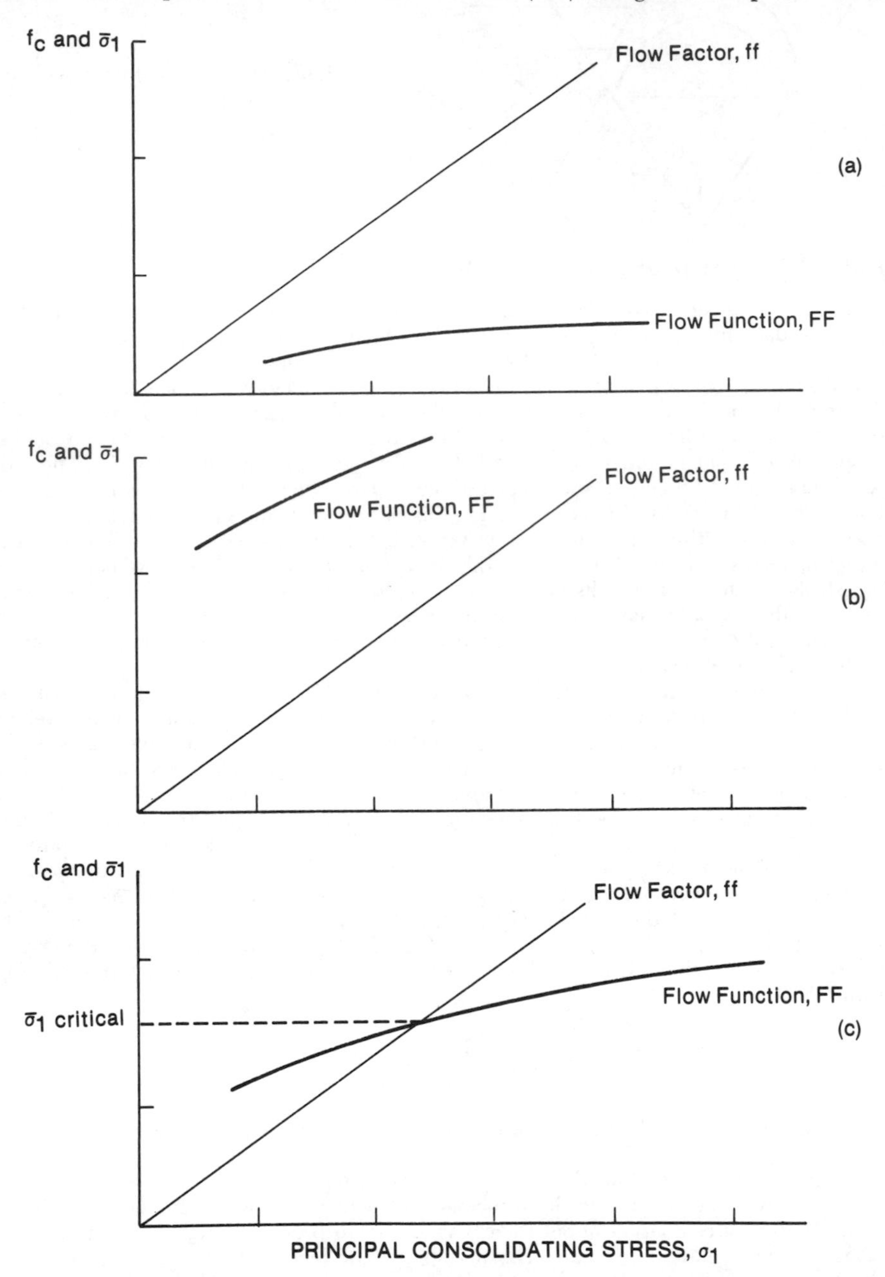

Figure 23. Plots of flow factor and flow function on same axes to determine critical outlet opening: (a) ff above FF, no arching; (b) ff below FF, arching at all stresses; (c) intersection, giving critical value of $\bar{\sigma}_1$ (corresponding to critical outlet) for no arching.

least being linear with a positive intercept on the f_c axis (Figure 23b). This is empirically found to be the case for most cohesive particulate materials.

It should be noted that the stress distributions on which the procedure is based assume flow of the solids to have taken place. In a later edition (November 1976) of his 1964 work, Jenike [13] bases the funnel flow design procedure only on a model of Johanson [30] for the stress distribution after filling before flow has taken place. For large bunkers in particular, this leads to very much larger minimum outlet dimensions than the procedure outlined above.

In mass flow, Johanson [30] suggests that the stress distribution after filling is less likely to lead to arching than that which develops on flow taking place.

Experimental Checks on Jenike Design Method

A number of workers [31–36] have compared the predictions of the Jenike bunker design method, based on shear cell testing of the solids, with observations of the flow pattern and/or the critical outlet diameter for arching in practice. Only Wright [34] used a plane flow hopper. All the authors were interested primarily in mass flow. Some [32,35] reported that the predicted minimum outlet exceeded the observed by more than a factor of 2, while others [33,34,36] reported good agreement between predictions and observations. None of the authors reported arching over diameters larger than the predicted minimum. In axisymmetric flow, the transition to funnel flow occurred at about the predicted angle. The recommended safety factor of 3–5° [13] was sufficient to cover the scatter. However, in plane flow Wright [34] found an overdesign on a wall slope of 5–10°. This is perhaps not surprising in view of the wide theoretical range of slopes for plane mass flow (Figure 19).

Stresses on Bunker Walls

Introduction

The procedures for the design of bunkers for flow outlined in the previous sections use predictions of the stresses acting in the solids to determine the stability of obstructions to flow. The same predictions can be used to determine the stress acting on the walls of a bunker. Jenike [11,13], for example, derives the stress on the wall from the radial stress field and plots $\sigma_w/\rho g B'$ as a function of δ, ϕ' and θ', where B' is the width of flow channel.

However, care should be exercised in applying such models when estimating the required strength of the bunker shell, as they may estimate average or lower bound stresses, while the bunker shell must be stable under the largest loading that can occur.

In the following paragraphs, experimental measurements of forces acting on bunker walls are discussed. Stress prediction methods that may be suitable for shell design purposes are then presented.

Experimental Measurements of Forces Acting on Bunker Walls

In general, the stress acting on a bunker wall will differ depending on whether the bunker has just been filled, has just started discharging or has achieved steady-state flow. On filling, some slight vertical compression of the solids takes place. This leads to the major principal stress having a large vertical component, even in the hopper. If it is assumed that a fully active stress field (see the section entitled Limiting Failure Analyses) prevails everywhere, the orientation of the major principal stress is as shown in Figure 24. When flow takes place, the material in the hopper is compressed horizontally and the major principal stress becomes largely horizontal, resulting in a passive stress field (see the section mentioned above) and a decrease in stress toward the apex. However, in the bin of a mass flow bunker, the material moves as a plug and the stress field remains active (Figure 24).

On starting discharge after filling, the material must dilate in order to flow. Thus, a boundary forms between moving regions, where there is a passive stress field, and stationary regions, where there is an active stress field, and this boundary travels upward from the outlet. At this boundary the wall stress changes from the active value, with its steady increase toward the apex, to the passive value, with a decrease toward the apex (compare Equation 32). To maintain equilibrium, a higher stress must be generated above the boundary, and this is concentrated at the boundary as this is the only region where the deformation accompanying such an increase in stress occurs. Thus, on starting flow, a stress peak (referred to as a switch stress) travels up the hopper. In mass flow bunkers this peak becomes trapped at the bin/hopper transition as the material in the bin remains in the active state.

An alternative way of explaining the switch stress is to consider the vertical stress in the material, which remains unchanged in those regions above the dilation boundary. Immediately above the boundary, the vertical stress on the center line is the major principal stress, and the horizontal stress is thus a minor principal stress and a factor K_{active} smaller, where $K_{active} = (1 - \sin \delta)/(1 + \sin \delta)$. Immediately below the boundary the vertical stress is unchanged, but the horizontal stress is now the major principal stress and is thus a factor $1/K_{active}^2$ greater than immediately above the boundary.

A number of workers have determined the stresses acting on bunker walls. Usually they have used small pressure transducers mounted flush with the wall [36–42]. These studies show that during flow the force on the transducer is not steady but varies with time (Figure 25). The peak stress may be a factor of 1 to 2 times the time-averaged value [36,37]. Fluctuation frequencies of 15–85 Hz were reported by Richards [36], with

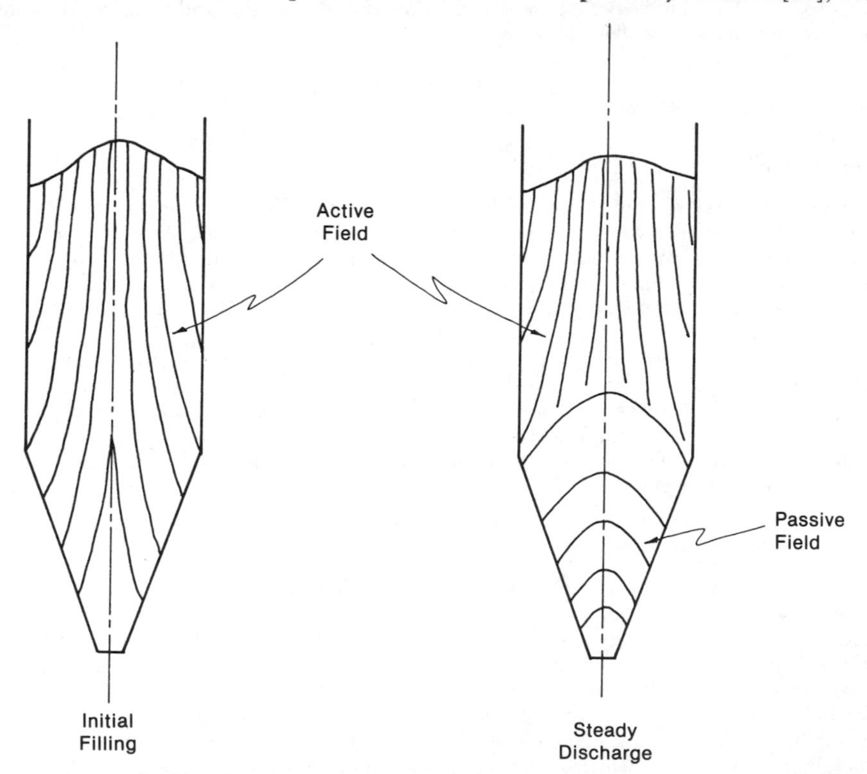

Figure 24. Lines of action of major principal stress after initial filling (active field assumed) and during steady discharge (active field assumed in bin, passive field in hopper).

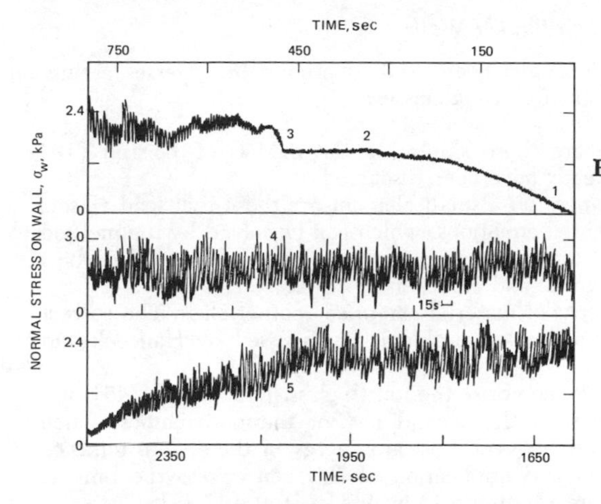

Figure 25. Wall stress in bin of 0.3-m-diameter bunker as a function of time: 1, start filling; 2, stop filling; 3, start flow; 4, steady flow with refilling; 5, discharge [39].

a dependence on the discharge rate but none on the particle size. However, the force traces of Richards [36] and Blair-Fish and Bransby [38] also show longer periods of several seconds during which the stress, although fluctuating, remains considerably above the time-averaged value.

In addition to this variation with time, there also may be large variations in the mean value over short distances. Jenike and co-workers [43–45] and van Zanten and Mooij [40] have shown that irregularities in the bunker wall, which may be as small as 1–2 mm in thickness, cause local stresses that are much higher than those found elsewhere.

Studies such as the above also confirm the existence of a switch stress at the bin-hopper transition in steadily flowing mass flow bunkers. A typical profile of mean measured wall stress over a mass flow bunker is shown in Figure 26.

Studies on funnel flow bunkers have been more limited. It might be expected that high stresses, corresponding to the switch stress, would be found where the funnel meets the wall. These generally have not been recorded, however [41,42,46].

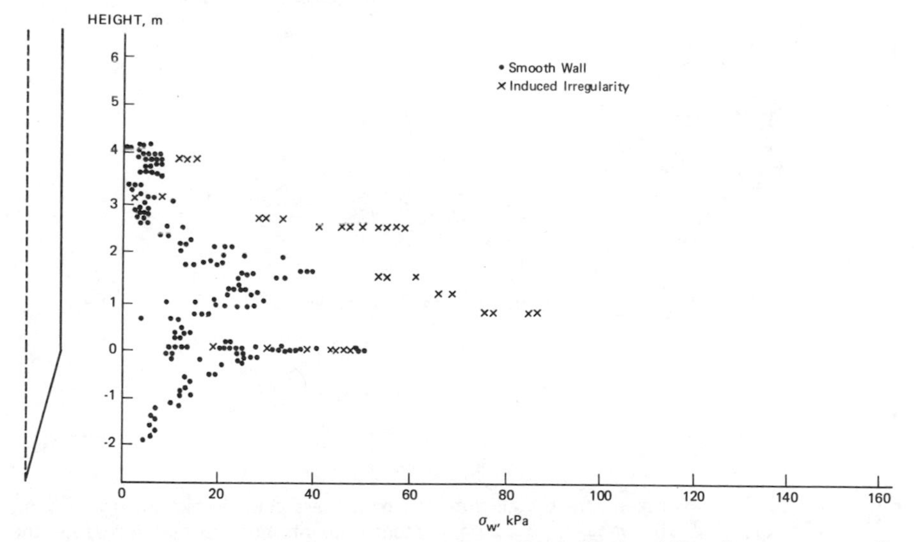

Figure 26. Time-averaged stresses acting at the wall of a 1.5-m-diameter mass flow bunker with sand [40].

Models for the Prediction of Stress on Bunker Walls

A number of different approaches have been used to predict the stresses acting on the wall of a bunker. They fall into the following classes:

1. "Differential slice" models are those similar to the models of Janssen [19] and Walker [18], which already have been discussed.
2. Models based on force balances on a small element are those that lead to sets of hyperbolic partial differential equations, which can be solved by the method of characteristics [27]. Rigorous numerical solutions for two-dimensional bins and hoppers are given by Horne and Nedderman [8–10].
3. Models based on sliding blocks of material comprise another class. The correct application of this method to bunker problems is discussed by Hancock and Nedderman [7].
4. Models based on strain energy comprise the fourth class. Jenike et al. [45] suggested that in accordance with the second law of thermodynamics, when solids flow from a bunker the recoverable strain energy of the system tends to a minimum. The evaluation and application of this concept to the bins of mass and funnel flow bunkers are discussed by Jenike et al. [47,48].

None of the above models predicts the rapid fluctuation of stresses that is observed in practice, although many commentators suggest physical reasons, such as changes in the boundary between active and passive regions. Jenike's strain energy model is claimed to represent an upper bound to the stresses observed.

In general, the differential slice models are fairly easy to apply and achieve a reasonable degree of accuracy. A model of this type, which can be applied to predict both initial and steady flow stresses in axisymmetrical bins and hoppers, is that of Walters [21,22]. His approach is similar to that of Walker [18], but he analyzes the distribution factor theoretically and shows that different values are appropriate in active and passive stress fields. In addition, he performs his force balance in the hopper over a frustrum

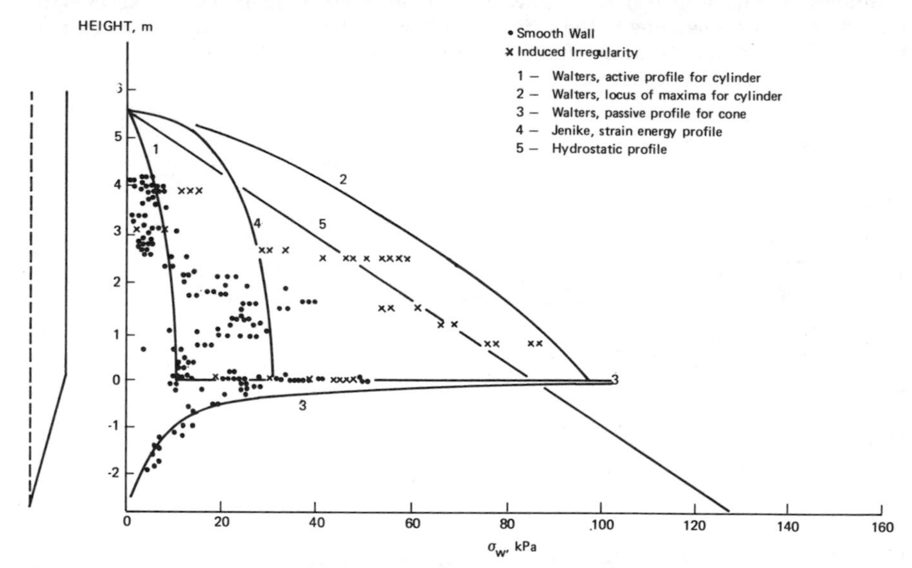

Figure 27. Time-averaged stresses at the wall of a mass flow bunker: $D = 1.5$ m, $\delta = 37°$, $\phi'_{cone} = 18°$, $\phi'_{cylinder} = 23.5°$ (measurements and profiles 4,5 from the work of van Zanten and Mooij [40]).

of a cone rather than over a cylindrical section, as used by Walker. The initial filling stress is then predicted by Walters, assuming an active stress field in both bin and hopper. The switch stress, which moves up to the transition on starting flow, is predicted by assuming that the stress field in the hopper is active above the switch and passive below it. In steady flow, Walters assumes an active stress field in the bin and a passive one in the hopper. The switch stress at the transition and the stress distribution over the whole hopper are predicted from the passive hopper formula with a surcharge equal to the (active) vertical stress at the transition. Walters assumes that peak stresses in the bin are associated with a transition from an active to a passive stress field within the bin. On this basis, the envelope of peak stresses is calculated.

Walters' model is compared with the data of van Zanten and Mooij [40] in Figures 27 and 28, which show the time-averaged stress and the highest instantaneous stress found, respectively. Other models also are shown for wall stresses evaluated by these authors.

In funnel flow, the flow pattern is less well defined and, hence, the prediction of wall stresses is more difficult. It is probably not realistic to expect to predict more than an envelope of maximum stresses. Jenike et al. [45] suggest that in short bunkers with a bin height to diameter ratio of less than two the funnel is unlikely to reach the wall, and stresses should thus conform to the initial filling pattern. In taller funnel flow bunkers, they recommend determining the envelope of maximum stresses by the same methods used for mass flow.

Standard Codes

A bunker shell must be stable with respect to bursting under the horizontal stresses imposed on it. It also must not buckle under the total vertical load transferred to it through the wall shear stresses. There are a number of national codes that take both these aspects into consideration. A widely used one is the German standard DIN 1055 [49]. However, this has some deficiencies [50] and is currently being revised. Extensive discussions of various codes and comparisons with experimental measurements are given in the proceedings of a recent conference organized by the Powder Advisory Centre [51].

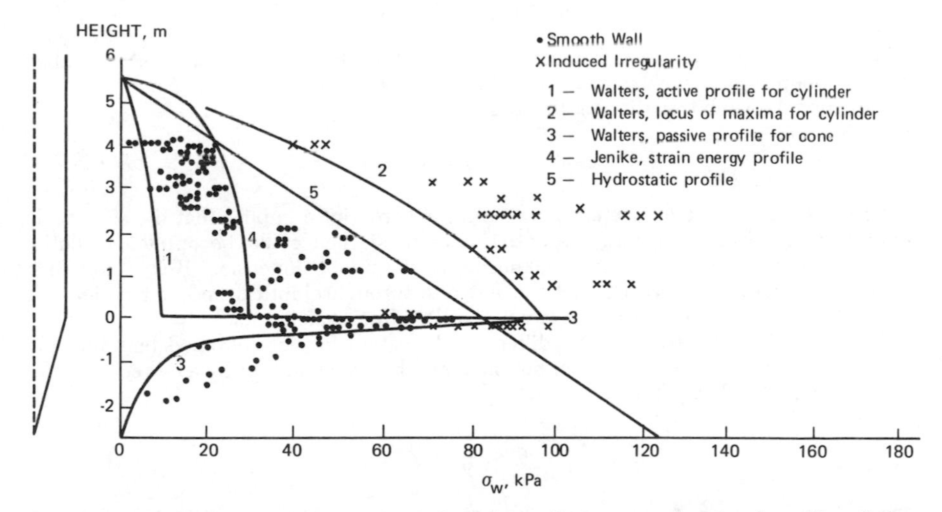

Figure 28. Maximum stresses recorded at the wall of a mass flow bunker: D = 1.5 m, $\delta = 37°$, $\phi'_{\text{cone}} = 18°$, $\phi'_{\text{cylinder}} = 23.5°$ (measurements and profiles 4,5 from the work of van Zanten and Mooij [40]).

INERTIA EFFECTS: RATE OF DISCHARGE FROM BUNKERS

Introduction

Studies on the rate of discharge from bunkers have shown that the rate is substantially independent of the head of material in the bunker and the size of the bunker. As the stresses near the outlet show a similar independence (see the previous section), this is perhaps to be expected. Furthermore, the rate of discharge of coarse granular materials through a given outlet appears to depend only weakly on the material properties, but for cohesive materials or fine powders there is a substantial dependence. In the following, coarse materials are discussed first, after which the effects of cohesion and the presence of fines are examined.

Discharge Rate of Coarse Granular Materials

Experimental Correlations

Assuming the independence of flowrate from head noted previously, the mass flowrate, $\dot{m}$, is a function of the bulk density of the material, ρ, the outlet dimension, D_o, and gravitational acceleration, g. Dimensional analysis then gives

$$\dot{m} = k \rho g^{1/2} D_o^{5/2} \tag{43}$$

where k is a constant.
Many early workers, such as Franklin and Johanson [52], reported correlations of this form, but with a higher exponent for D_o. Brown and Richards [53] found that in flat-bottomed bins there was a statistically empty annulus next to the perimeter of the discharge opening and that the thickness of this annulus was related to the particle size and shape. Beverloo et al. [54] used this concept to correlate their own and previous data for circular orifices in flat-bottomed bins in the form

$$\dot{m} = 0.58 \rho g^{1/2} (D_o - k' d)^{5/2} \tag{44}$$

where d is the mean particle diameter, and k′ is a constant. The constant k′ was about 1.4 for spherical particles but slightly higher for sand.

Equation 44 can be rewritten in the form

$$\dot{m} = 0.74 \rho A_e \sqrt{g D_e} \tag{45}$$

where A is the outlet cross-sectional area and the subscript e implies that the statistically empty perimeter is taken into account. Equation 45 then could be applied to outlets in flat-bottomed bins of any shape, taking D_e as the minor dimension. Brown and Richards [55] suggested, however, that the flowrate through elliptical and slot outlets was about 20% less than that through a circular outlet of equivalent area.

Harmens [56] derived a slightly different correlation for flat-bottomed bins but suggested that the influence of a hopper bottom could be accounted for by a correction factor of $(\tan \delta \tan \theta')^{-0.35}$, provided

$$\theta' < \frac{\pi}{2} - \delta$$

Taking Equation 45 as appropriate for flat-bottomed bins, this gives the following expression for hopper bottoms:

$$\dot{m} = 0.74 (\tan \delta \tan \theta')^{-0.35} \rho A_e \sqrt{g D_e} \tag{46}$$

Harmens suggested that his correlations applied to particles larger than 1 mm. Other workers have suggested a slightly lower limit, but there is general agreement that below the limit mass flowrate decreases rapidly with particle size.

Theoretical Studies

A number of attempts have been made to introduce inertia terms into force balance models of the type discussed in the previous section and to solve these for the discharge velocity. The approach of Savage [57] is discussed below. An approach which involves decoupling the stress and velocity fields, solving the stress field and then, using the principle of isotropy, solving the velocity fields as was done by Jenike [11] does not lead to the discharge velocity because, as Jenike himself points out, the velocity fields determined in this way are not unique.

Savage [57] assumes uniform flow through a straight-walled converging channel (Figure 29). Polar coordinates with origin 0 are used. The wall friction is set equal to zero so that there is no variation in the θ direction. A steep mass flow channel is assumed so that $\cos \theta' \simeq 1$. For a conical channel (Savage gives the plane flow case as well), the equation of motion in the r direction becomes

$$-\frac{d\sigma_r}{dr} + 2\left(\frac{\sigma_\theta - \sigma_r}{r}\right) - \rho g = \rho\, u \frac{du}{dr} \tag{47}$$

where σ_r and σ_θ are the stresses in the radial and tangential directions, respectively, and u is the velocity in the r direction.

From continuity,

$$u = \frac{A^*}{r^2} \tag{48}$$

where A^* is a constant.

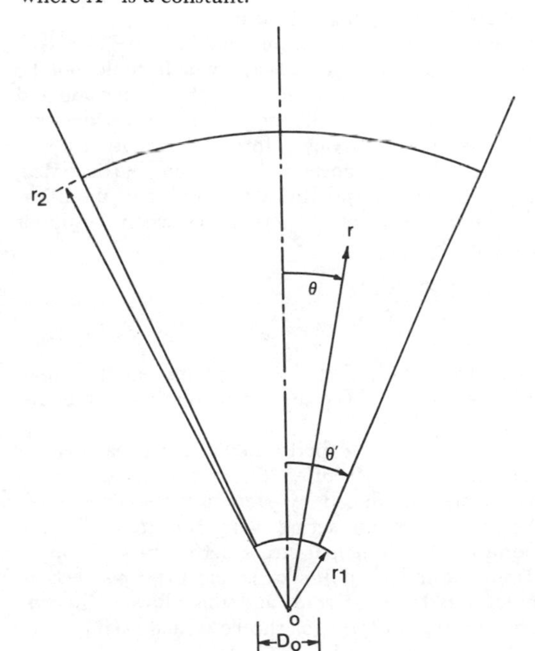

Figure 29. Coordinate system for flow in a converging channel [57].

The stress field in the hopper is assumed to be passive so that

$$\sigma_\theta = K\,\sigma_r \tag{49}$$

where $K = (1 + \sin\delta)/(1 - \sin\delta)$. Substituting Equations 48 and 49 into Equation 47 and integrating with the boundary conditions,

$$\sigma_r(r_1) = \sigma_r(r_2) = 0 \tag{50}$$

gives an analytical expression for the velocity u. At the outlet this gives

$$u_1 = -\left[\frac{(1+K)\,g\,D_o}{2\,[2(K-1)-1]\sin\theta'}\left(\frac{1-(r_1/r_2)^{2(K-1)-1}}{1-(r_1/r_2)^{2(K+1)}}\right)\right]^{1/2} \tag{51}$$

For a substantial head, i.e., $r_2 >> r_1$, this reduces to

$$u_1 \doteqdot \left[\frac{(K+1)\,g\,D_o}{2\,[2K-3]\sin\theta'}\right]^{1/2} \tag{52}$$

which shows the velocity to be independent of head. The mass flowrate, $\dot{m}$, is then

$$\dot{m} = \left[\frac{K+1}{2(2K-3)\sin\theta'}\right]^{1/2}\frac{\pi}{4}\,\rho\, g^{1/2}\, D_o^{2.5} \tag{53}$$

which is of the same form as the empirical equations given earlier. Savage points out that the empty annulus concept could be introduced by replacing D_o by $(D_o - k'\,d)$. Even so, the predictions of Equation 53 are about 50% in excess of observations. This is attributed to the neglect of wall friction and the approximation for $\cos\theta'$.

Davidson and Nedderman [58] start from the same assumptions as Savage [57] and deduce the same equation for mass flowrate. In addition, they examined the consequences of reducing the internal friction in the material by partial fluidization.

Williams [59] relaxed the restriction on flow being radial and allowed a finite wall friction to act. This leads to a set of partial differential equations, which could not be solved analytically. However, analytical solutions were obtained along the center line and along a wall, and from these solutions upper and lower bounds on the mass flowrate were derived. The upper-bound solution reduces to Savage's formula, Equation 53, if Savage's restriction that $\cos\theta' = 1$ is introduced. To account for a finite particle size, Williams introduced the effective diameter of Brown and Richards. With this modification, observed flowrates of large particles from a range of hoppers fell between the upper and lower bounds, which differed by about 20%.

Discharge Rate of Cohesive Materials

With cohesive materials, the discharge rate (assuming the minimum outlet dimension is exceeded) may be substantially below that of coarse non-cohesive materials. There are only a few quantitative studies.

The major contribution is that of Johanson [31]. He performed a force balance on an arch of constant thickness at a mass flow hopper outlet. Stress transmission from above was ignored, but cohesive forces were introduced by assuming the unconfined yield stress f_c of the material to be the principal stress acting along the arch. The appropriate value of f_c is that corresponding to the principal stress acting in the hopper at that point, and this can be found from the model of Jenike in the previous section. The acceleration of the arch at the outlet was taken as zero, and this allowed the discharge rate to be found. The final expression was in terms of the flow factor (ff),

$$\dot{m} = \frac{1}{2\tan^{1/2}\theta'}\left(1 - \frac{ff}{ff_a}\right)^{1/2}\frac{\pi}{4}\,\rho\, g^{1/2}\, D_o^{5/2} \tag{54}$$

for a circular outlet, or

$$\dot{m} = \frac{1}{\sqrt{2}\tan^{1/2}\theta'}\left(1 - \frac{ff}{ff_a}\right)^{1/2} \rho L B (gB)^{1/2} \quad (55)$$

for a slot of width B and length L.

The parameter ff is then the hopper flow factor, as defined in the previous section, while ff_a is the ratio of the unconfined yield stress of the solids to the principal consolidating stress (i.e., f_c/σ_1) evaluated at the value of σ_1 acting at the hopper outlet. This value of σ_1 can be estimated from charts in the work of Jenike [11].

Discharge Rate of Fine Particles

It has been remarked already that below a particle size of around 1 mm, the discharge rate decreases with particle size. The explanation usually advanced is that to flow out the particles must dilate, and this dilation gives rise to a region of underpressure. With large particles, the permeability is large and the magnitude of the underpressure is small. With small particles, however, the permeability is small and the underpressure is significant, causing a considerable reduction in flowrate compared with coarse material. Reductions to only 10% of the rate found for coarse materials are not uncommon.

Prediction of this effect in terms of easily measurable powder flow properties is difficult, partly because it is difficult to relate the voidage to the stress. Crewdson et al. [60] and Spink and Nedderman [61] have examined this problem experimentally and theoretically for axisymmetric and plane flow, respectively. The principal features of the observations were well reproduced, but quantitative agreement was only fair. Holland et al. [62] predicted the discharge rate in terms of the pressure gradient existing at the outlet, but in general this will not be known.

FLUID FLOW

Introduction

The significance of fluid flow on the discharge rate of finer particles has just been noted. If the particles in a bed of particulate solid are immersed in a stationary fluid of identical true density, the particles would be neutrally buoyant and the stresses carried by the particles would be zero. If ρ_t denotes the true density of the particles and ρ_l the interstitial fluid density, then for the consideration of stresses borne by the solids we replace the term ρ in the equations by $(\rho_t - \rho_l)(1 - \epsilon_o)$, where ϵ_o is the bed voidage. More generally, if fluid is flowing, the normal stress borne by the solids is given by the difference between the total stress on a section and the pressure of the interstitial fluid. This is termed the effective stress.

The separation of the total stress exerted on a unit cross-section of the system into an effective solid stress and an interstitial fluid pressure is termed the Terzaghi assumption; there is substantial evidence that this concept is sound [1]. It provides a powerful fundamental method of attacking problems of simultaneous flow of solids and fluids.

Interaction of Fluid Flow and Solid Flow

The influence of fluid flow on solids motion is significant, the effect of the pressure gradient arising from material dilation due to failure in the region of the solids outlet having been noted already. Another example is the control of solids flowrates by addition of air via pads adjacent to a hopper outlet. A further problem, examined recently by Murfitt and Bransby [63], is that of deaeration of a fine powder in a bunker and the transient development of solid stresses.

The extent to which the pressure gradient in the interstitial fluid differs from the static pressure gradient is one of the factors affecting the flowrate of the fluid. The permeability of a packing to flow may be measured experimentally or may be predicted by correlations such as the Ergun equation. The pressure gradient driving the fluid through the interstices is given by

$$\frac{dp}{dy} = 150\,\frac{(1-\epsilon_o)^2}{\epsilon_o^{\,3}}\,\frac{\mu\,U_o}{(\phi_s\,d)^2} + 1.75\,\frac{(1-\epsilon_o)}{\epsilon_o^{\,3}}\,\frac{\rho_l\,U_o^{\,2}}{\phi_s\,d} \tag{56}$$

where ϵ_o is the voidage, μ is the fluid viscosity, U_o is the superficial fluid velocity, d_p is the particle diameter, and ϕ_s is a shape factor defined by

$$\phi_s = \frac{\text{surface area of sphere of equal volume to the particle}}{\text{surface area of particle}} \tag{57}$$

If there is a range of particle sizes, the surface area mean diameter, $\bar{d}$, is probably the most suitable where

$$\bar{d} = \left[\sum_n \frac{X_n}{d_n}\right]^{-1} \tag{58}$$

where X_n is the mass fraction of solids of diameter d_n [64]. A recent critique of the Ergun in relation to other equations has been given by MacDonald et al. [65].

The spreading of a powder due to fluid flow through the surface can be analyzed [66]. The theory relies on a "sliding block" argument. Consider unit area of a slope of the cohesionless powder inclined at an angle to the horizontal θ and let the fluid above the powder be at a constant pressure (Figure 30). Since the pressure gradient in the direction of the slope is zero, the fluid must pass normally through the surface. Let the pressure gradient be dp/dy. The normal stress, σ, borne by the solid a distance, y, from the free surface, is

$$(1-\epsilon_o)(\rho_t-\rho_l)yg\cos\theta - y(dp/dy)$$

If dp/dy is increased to the limiting value $(dp/dy)_l$ so that the slope is about to slip, the limiting shear stress, $\sigma\tan\phi$, where ϕ denotes an angle of friction, is equal to the component of the gravitational force down the plane and, hence,

$$(1-\epsilon_o)(\rho_t-\rho_l)g\sin\theta = \tan\phi\,\{(1-\epsilon_o)(\rho_t-\rho_l)g\cos\theta - (dp/dy)_l\} \tag{59}$$

or

$$\tan\theta = \tan\phi\left[1 - \frac{1}{(1-\epsilon_o)(\rho_t-\rho_l)g\cos\theta}\left(\frac{dp}{dy}\right)_l\right] \tag{60}$$

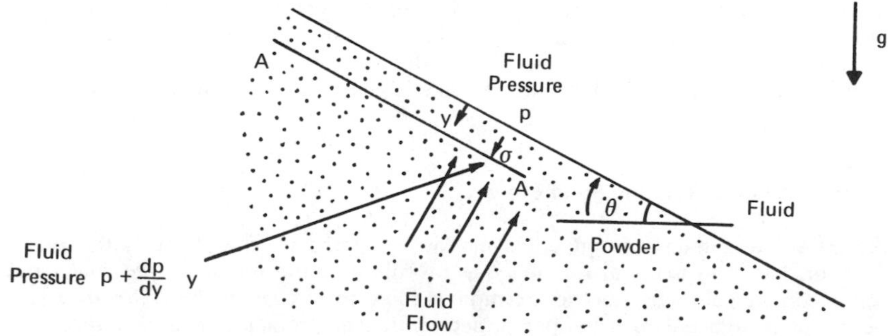

Figure 30. Fluid flow through an inclined flat slope ($0 \leq \theta \leq \pi/2$).

Now the pressure gradient $(dp/dy)_{mf}$ that would produce the minimum fluidization velocity, measured in the conventional manner with a level bed, is given by $(1-\epsilon_o)/(\rho_t-\rho_l)g$, and, hence,

$$\tan\theta = \tan\phi\left(1-\frac{(dp/dy)_l}{(dp/dy)_{mf}\cos\theta}\right) \tag{61}$$

If the Reynolds number of the flow in the pores of the powder is low enough for Darcy's law to apply, then,

$$\tan\theta = \tan\phi\left(1-\frac{U_l}{U_{mf}\cos\theta}\right) \tag{62}$$

where U_l,U_{mf} denote the limiting superficial velocity and the minimum fluidization velocity, respectively.

Influence of Packing Heterogeneity

The Ergun equation indicates that a small change in voidage can have a marked effect on fluid flowrate. It is found that the uniform packing of the solids into a vessel is a matter of some skill, and a great deal of practical know-how resides in techniques of packing catalytic packed bed reactors, for example. There is a change in packing structure near to the wall but this extends only about five particle diameters into the solid. Potentially much more serious variations in specific volumes stem from the rate and speed of introduction of particles. If the rate is low, each particle can settle into a position of greatest stability on its own, whereas this is inhibited by particle interaction. The specific volume can be influenced by loading and unloading if this occurs in a systematic manner, for example, in a vessel that is subject to pressure or temperature variations, where consolidation and increased stress will arise due to fatigue or creep. This can crush particles, rupture the vessel or lead to internal failure within the solid material. Rational design for such problems remains to be worked out but progress here should be possible.

Variation in bed permeability to fluid flow also can arise due to particle segregation in the bunker. This may occur within the bulk material due to the development of slip zones leading to interparticle percolation, as studied by Bridgwater et al. [67]. This is the drainage of small particles through larger ones in the high strain regions. Such a phenomenon is likely only to affect the microstructure in a bunker and will be of less importance than the phenomenon of free-surface segregation which, although well known for many years, has recently been subject to close scrutiny by Drahun and Bridgwater [68]. This is segregation due to the falling of material onto a surface and cascading down the free surface. Segregation in a bunker may be affected by segregation in the feed material, the free-fall distance onto the surface, particle size, particle density and, possibly particle shape.

The practical consequences of such work may be illustrated by considering the filling and emptying of two bunkers, one of mass flow design and the other with core flow (Figure 31). On filling a bunker with large free fall height, smaller particles bounce to the vessel periphery. On the other hand, if laid gently on the surface, small particles percolate down through the layers of sliding particles and congregate in the center. During discharge, the mass flow hopper achieves a reasonably uniform downflow and the various fractions are discharged at the same rate, even though there will be radial inhomogeneity. With core flow, however, the contents at the bunker center are discharged first. The discharging core is then replenished by surface layers cascading into the vessel center. Core-flow bunkers have static regions that further promote segregation.

Some of the complications due to fluid flow maldistribution that can arise due to variation in bed voidage have been given by Craven [69] and Szekely et al. [64]. There can be marked effect on reactor yield, selectivity and stability.

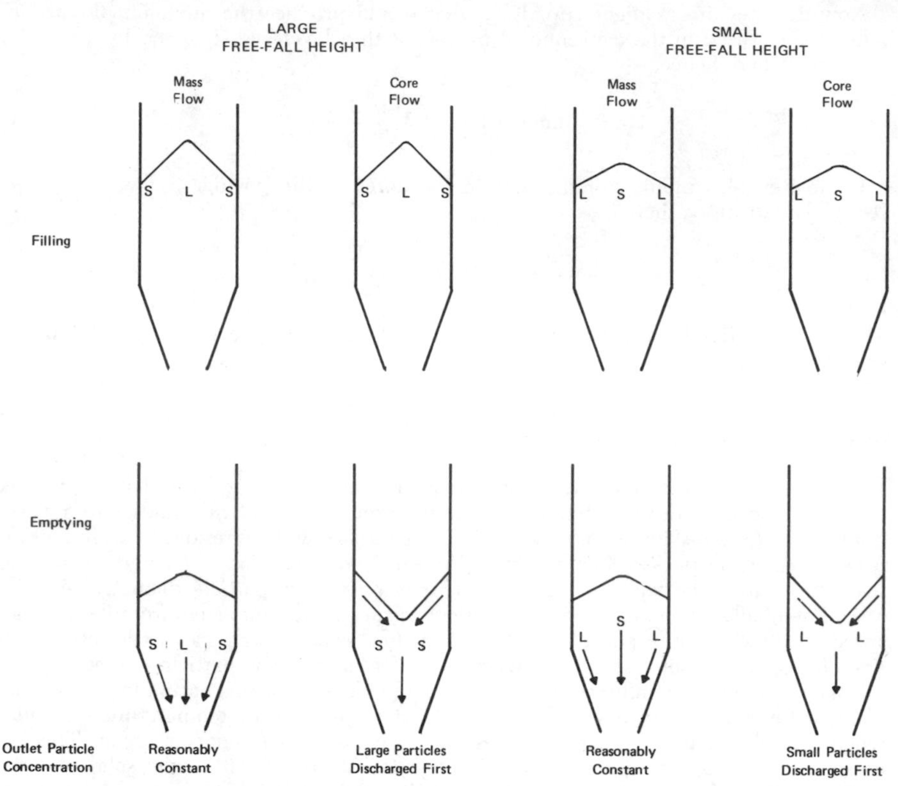

Figure 31. Influence of bunker design and filling method on particle residence times. L denotes regions with predominantly large particles, S denotes regions with predominantly small particles.

CONCLUSIONS

It is possible to separate the effects of solid and interstitial fluid in bunker flow problems. Successful analytical methods are available for stress fields in two-dimensional problems. For three-dimensional problems, practical solutions for the stress field often can be obtained by a pseudo-two-dimensional approach, although such methods sometimes lack rigor. Such methods have been particularly successful in predicting whether solids will discharge from a bunker and what the flow pattern will be. Solution of the combined stress and velocity fields is possible only in a few simple cases. For most practical design problems, this approach rapidly becomes too complex and one must resort to simpler approaches, such as deriving a stress field by ignoring inertia terms and then deducing compatible velocity fields.

All these approaches treat the solids as a continuum. As long as this description is maintained, the failure properties of the material can be well described by the Hvorslev-Roscoe surfaces, for example. However, prediction of such failure criteria based on measurements of particulate properties such as size and density is hardly possible.

The particulate properties also influence the structure of the continuum. Although there is good qualitative understanding of how particulate properties may do this, by segregation among particles of various sizes, for example, it is not possible to take into account such inhomogeneities in the continuum model. For problems of this nature, an

understanding of continuum behavior should be coupled with a practical knowledge of how individual particles behave.

NOMENCLATURE

a	adhesion (kPa)
A	cross-sectional area (m^2)
A_e	effective area of outlet (m^2)
B	minor dimension of outlet (m)
B′	width or diameter in hopper (m)
B^*	constant given by Equation 20
B^{**}	constant given by Equation 24
C	constant given by Equations 29 or 30
D	diameter (m)
D_o	diameter of circular outlet (m)
D_e	effective diameter of circular outlet (m)
D^*	distribution factor, i.e., ratio of vertical stress at wall to average vertical stress
d	particle size (m)
f_c	unconfined yield stress (kPa)
ff	flow factor of hopper, $\sigma_1/\bar{\sigma}_1$
FF	flow function of a solid, σ_1/f_c
g	gravitational acceleration (m/sec^2)
h	height above apex of hopper (m)
h_o	total height of hopper, measured from apex (m)
I_1,I_2,I_3	invariants of stress tensor (kPa, kPa^2, kPa^3
k	constant in Equation 43
k′	constant in Equation 44
K	ratio of horizontal to vertical normal stresses
L	length of slot (m)
m	constant = 0 for plane hopper = 1 for conical hopper
$\dot{m}$	mass flowrate (kg/sec)
p	fluid pressure (kPa)
r	radial coordinate (m)
r_1	radial position of outlet (m)
r_2	radial position of top surface (m)
S	perimeter (m)
u	radial velocity component (m/sec)
u_1	radial velocity at outlet (m/sec)
U_o	superficial velocity of interstitial fluid (m/sec)
U_l	limiting superficial velocity for slope stability (m/sec)
U_{mf}	minimum fluidization velocity (m/sec)
v	specific volume (m^3/kg)
x	horizontal coordinate (m)
y	distance from free surface (m)
z	vertical coordinate (m)
α	inclination of slip plane to horizontal
β	angle between major principal stress and normal to hopper wall
γ	shear strain
δ	effective angle of internal friction
ϵ	angle between major principal stress and normal to bin wall
$\hat{\epsilon}$	linear strain
ϵ_o	voidage of a bed of particles
ψ	angle between resultant force and normal to slip plane
μ	fluid viscosity (ksecPa)
θ	angular coordinate
θ'	inclination of hopper wall to vertical
ρ	bulk density of solids (kg/m^3)
ρ_l	density of interstitial fluid (kg/m^3)
ρ_t	true density of particles (kg/m^3)
σ	normal stress (kPa)
σ_o	$(\sigma_1 + \sigma_3)/2$ (kPa)
σ_1	major principal stress (kPa)
σ_2	intermediate principal stress (kPa)
σ_3	minor principal stress (kPa)
$\bar{\sigma}_1$	major principal stress acting in an arch (kPa)
$\bar{\sigma}_{1\,critical}$	major principal stress acting in critical arch, i.e., one that is just stable (kPa)
(σ_{kr},τ_{kr})	coordinates of consolidation point in shear cell testing (kPa,kPa)
σ_w	normal stress on a wall (kPa)
σ_z	normal stress acting in vertical direction (kPa)
$\bar{\sigma}_z$	average value of normal vertical stress over a cross section (kPa)
σ_{zo}	surcharge stress on top surface (kPa)
$(\sigma_z)_w$	vertical normal stress at the wall (kPa)
τ	shear stress (kPa)
τ_w	shear stress acting a wall (kPa)

$(\tau_z)_w$ vertical shear stress at the wall (kPa)
ϕ angle of internal friction
ϕ' wall friction angle
ϕ_s particle shape factor, Equation 57
ξ inclination of resultant force to normal to a blade
ω inclination of $\bar{\sigma}_1$ to vertical

REFERENCES

1. Atkinson, J. H., and P. L. Bransby. *The Mechanics of Soils, an Introduction to Critical State Soil Mechanics* (New York: McGraw-Hill Book Co., 1978).
2. Harr, M. E. *Foundations of Theoretical Soil Mechanics* (New York: McGraw-Hill Book Co., 1966).
3. Scott, R. F. *Principles of Soil Mechanics* (Reading, MA: Addison-Wesley Publishing Co., Inc., 1963).
4. Brown, R. L., and J. C. Richards. *Principles of Powder Mechanics* (Elmsford, NY: Pergamon Press, Inc., 1966).
5. Williams, J. C., and A. H. Birks. "The Comparison of the Failure Measurements of Powders with Theory," *Powder Technol.* 1:199–206 (1967).
6. Heertjes, P. M., G. K. Khoe and D. Kuster. "A Condition Diagram for Some Noncohesive Round Glass Particles," *Powder Technol.* 21:63–71 (1978).
7. Hancock, A. W., and R. M. Nedderman. "Prediction of Stresses on Vertical Bunker Walls," *Trans. Inst. Chem. Eng.* 52:170–179 (1974).
8. Horne, R. M., and R. M. Nedderman. "Analysis of the Stress Distribution in Two-Dimensional Bins by the Method of Characteristics," *Powder Technol.* 14:93–102 (1976).
9. Horne, R. M., and R. M. Nedderman. "An Analysis of Switch Stresses in Two-Dimensional Bunkers," *Powder Technol.* 19:235–241 (1978).
10. Horne, R. M., and R. M. Nedderman. "Stress Distribution in Hoppers, *Powder Technol.* 19:243–254 (1978).
11. Jenike, A. W. "Gravity Flow of Bulk Solids," *Utah Eng. Expt. Station Bull.* 108 (1961).
12. Schwedes, J. "Measurement of Powder Properties for Hopper Design," *J. Eng. Ind. Trans. ASME Ser. B.* 95:55–59 (1973).
13. Jenike, A. W. "Storage and Flow of Solids," *Utah Eng. Expt. Station Bull.* 123 (1964).
14. Carr, J. F. and D. M. Walker. "The Annular Shear Cell for Granular Materials," *Powder Technol.* 1:369–373 (1967–68).
15. Peschl, I. A. S. Z., and H. Colijn. "New Rotational Shear Cell Testing Technique," *J. Powder Bulk Solids Technol.* 1:55–60 (1977).
16. Schwedes, J. "Vergleichende Betrachtungen zum Einsatz von Schergeräten zur Messung von Schüttguteigenschaften," in *Proc. "Partec '79", European Conference on Transfer Processes in Particle Systems*, Nuremberg (1979), pp. 278–300.
17. Molerus, O. "Theory of Yield of Cohesive Powders," *Powder Technol.* 12:259–275 (1975).
18. Walker, D. M. "An Approximate Theory for Pressures and Arching in Hoppers," *Chem. Eng. Sci.* 21:975–997 (1966).
19. Janssen, H. A. "Versuche über Getreidedruck in Silozellen," *VDI Zeitschrift* 39:1045–1049 (1895).
20. Aoki R., and H. Tsunakawa. "The Pressure in a Granular Material at the Walls of Bins and Hoppers," *J. Chem. Eng. Japan* 2:126–129 (1969).
21. Walters, J. K. "A Theoretical Analysis of Stresses in Axially-Symmetric Hoppers and Bunkers," *Chem. Eng. Sci.* 28:779–789 (1973).
22. Walters, J. K. "A Theoretical Analysis of Stresses in Silos with Vertical Walls," *Chem. Eng. Sci.* 28:13–21 (1973).
23. Enstad, G. "On the Theory of Arching in Mass Flow Hoppers," *Chem. Eng. Sci.* 30:1273–1283 (1975).

24. Enstad, G. "A Note on the Stresses and Dome Formation in Axially Symmetric Mass Flow Hoppers," *Chem. Eng. Sci.* 32:337–339 (1977).
25. Arnold, P. C., and A. G. McLean. "An Analytical Solution for the Stress Function at the Wall of a Converging Channel," *Powder Technol.* 13:255–260 (1976).
26. Arnold, P. C., and A. G. McLean. "Improved Analytical Flow Factors for Mass Flow Hoppers," *Powder Technol.* 15:279–281 (1976).
27. Sokolovski, V. V. "Statics of Granular Media" (Oxford, England: Pergamon Press, Ltd., 1965).
28. Jenike, A. W., and T. Leser. "A Flow-No Flow Criterion in the Gravity Flow of Powders in Converging Channels," paper presented at the Fourth International Congress on Rheology, Brown University (1963).
29. Arnold, P. C., A. G. McLean and A. W. Roberts. "Bulk Solids: Storage, Flow and Handling," Tunra Ltd., University of Newcastle, New South Wales, Australia (1978).
30. Johanson, J. R. "Effect of Initial Pressures on the Flowability of Bins," *J. Eng. Ind. Trans. ASME Ser. B.* 91:395–400 (1969).
31. Johanson, J. R. "Stress and Velocity Fields in the Gravity Flow of Bulk Solids," *J. Appl. Mech. Trans. ASME Ser. E.* 86:499–506 (1964).
32. Walker, D. M. "A Basis for Bunker Design," *Powder Technol.* 1:228–236 (1967).
33. Buick, K. "Schertests im Hinblick auf die industrielle Auslegung von Bunkern," *Chem.-Ing.-Tech.* 48:261 (1976).
34. Wright, H. "An Evaluation of the Jenike Bunker Design Method," *J. Eng. Ind. Trans. ASME Ser. B.* 95:48–54 (1973).
35. Eckhoff, R. K., and P. G. Leversen. "A Further Contribution to the Evaluation of the Jenike Method for the Design of Mass Flow Hoppers," *Powder Technol.* 10: 51–58 (1974).
36. Richards, P. C. "Bunker Design Part I: Bunker Outlet Design and Initial Measurement of Wall Pressures," *J. Eng. Ind. Trans. ASME Ser. B.* 99:809–813 (1977).
37. Walker, D. M., and M. H. Blanchard. "Pressures in Experimental Coal Hoppers," *Chem. Eng. Sci.* 22:1713–1745 (1967).
38. Blair-Fish, P. M., and P. L. Bransby. "Flow Patterns and Wall Stresses in a Mass-Flow Bunker," *J. Eng. Ind. Trans. ASME Ser. B* 95:17–20 (1973).
39. Smid, J. "Pressure of Granular Material on Wall of Model Silo," *Collect. Czech. Chem. Commun.* 40:2424–2437 (1975).
40. van Zanten, D. C., and A. Mooij. "Bunker Design Part 2: Wall Pressures in Mass Flow," *J. Eng. Ind. Trans. ASME Ser. B.* 99:814–818 (1977).
41. van Zanten, D. C., P. C. Richards and A. Mooij. "Bunker Design Part 3: Wall Pressure and Flow Patterns in Funnel Flow," *J. Eng. Ind. Trans. ASME Ser. B.* 99:819–823 (1977).
42. Blight, G. E. and D. Midgley "Pressure Measured in a 20-m Diameter Coal Load-out Bin," *J. Powder Bulk Solids Technol.* 5:21–31 (1981).
43. Jenike, A. W., and J. R. Johanson. "Bin Loads," *J. Struct. Div. Trans. ASCE* 95 (ST4):1011–1041 (1968).
44. Jenike, A. W., and J. R. Johanson. "On the Theory of Bin Loads," *J. Eng. Ind. Trans. ASME Ser. B* 91:339–344 (1969).
45. Jenike, A. W., J. R. Johanson and J. W. Carson. "Bin Loads—Part 2: Concepts," *J. Eng. Ind. Trans. ASME, Ser. B* 95(1):1–5 (1973).
46. Deutsch, G. P., and L. C. Schmidt. "Pressures on Silo Walls," *J. Eng. Ind. Trans. ASME Ser. B* 91:450–459 (1969).
47. Jenike, A. W., J. R. Johanson and J. W. Carson. "Bin Loads—Part 3: Mass Flow Bins," *J. Eng. Ind. Trans. ASME, Ser. B* 95(1):6–12 (1973).
48. Jenike, A. W., J. R. Johanson and J. W. Carson. "Bin Loads—Part 4: Funnel Flow," *J. Eng. Ind. Trans. ASME Ser. B* 95(1):13–16 (1973).
49. "Design Loads for Buildings. Part 6: Loads in Silo Bins," *DIN* 1055 (1964).
50. Pieper, K. E., and F. Wenzel. "Aktuelle Fragen des Entwurfs, der Belastung, der Berechnung und der Bauausführung von Silozellen, *Beton u. Stahlbetonbau* 73: 192–199 (1978).

51. Powder Advisory Centre. Papers presented at the International Conference on the Design of Silos for Strength and Flow, University of Lancaster, UK September 2–4, 1980.
52. Franklin, F. C., and L. N. Johanson. "Flow Rate of Granular Material Through a Circular Orifice, *Chem. Eng. Sci.* 4:119–129 (1955).
53. Brown, R. L., and J. C. Richards. "Profile of Flow of Granules through Apertures," *Trans. Inst. Chem. Eng.* 38:243–256 (1960).
54. Beverloo, W. A., H. A. Leniger and J. van de Velde. "The Flow of Granular Solids Through Orifices," *Chem. Eng. Sci.* 15:260–269 (1961).
55. Brown, R. L., and J. C. Richards. "Exploratory Study of the Flow of Granules Through Apertures," *Trans. Inst. Chem. Eng.* 37:108–119 (1959).
56. Harmens, A. "Flow of Granular Material Through Horizontal Apertures," *Chem. Eng. Sci.* 18:297–306 (1963).
57. Savage, S. B. "The Mass Flow of Granular Materials Derived from Coupled Velocity-Stress Fields," *Brit. J. Appl. Phys.* 16:1885–1888 (1965).
58. Davidson, J. F., and R. M. Nedderman. "The Hour Glass Theory of Hopper Flow," *Trans. Inst. Chem. Eng.* 51:29–35 (1973).
59. Williams, J. C. "The Rate of Discharge of Coarse Granular Materials from Conical Mass Flow Hoppers," *Chem. Eng. Sci.* 32:247–255 (1977).
60. Crewdson, B. J., A. L. Ormond and R. M. Nedderman. "Air-Impeded Discharge of Fine Particles from a Hopper," *Powder Technol.* 16:197–207 (1977).
61. Spink, C. D., and R. M. Nedderman. "Gravity Discharge Rate of Fine Particles from a Hopper," *Powder Technol.* 21:245–261 (1978).
62. Holland, J., J. E. P. Miles, C. Schofield and C. A. Shook. "Fluid Drag Effects in the Discharge of Particles from Hoppers," *Trans. Inst. Chem. Eng.* 47:154–159 (1969).
63. Murfitt, P. G., and P. L. Bransby. "Deaeration of Powders in Hoppers," *Powder Technol.* 24:149–162 (1980).
64. Szekely, J., J. W. Evans and H. Y. Sohn. *Gas-Solid Reactions* (New York: Academic Press, Inc., 1976).
65. MacDonald, I. F., M. S. El-Sayed, F. A. L. Dullien and K. Mow, "Flow through Porous Media–the Ergun Equation Revisited," *Ind. Eng. Chem. Fund.* 18:195–207 (1979).
66. Bridgwater, J. "The Influence of Fluid Flow on Slope Stability," *Powder Technol.* 11:199–201 (1975).
67. Bridgwater, J., M. H. Cooke and A. M. Scott. "Interparticle Percolation: Equipment Development and Mean Percolation Velocities," *Trans. Inst. Chem. Eng.* 56:157–167 (1978).
68. Drahun, J. A., and J. Bridgwater. "Free Surface Segregation," *Inst. Chem. Eng. Symp. Ser.* 65:S4/Q/1–14 (1980).
69. Craven, P. "The Effect of Gas Maldistribution on Catalyst Performance," *Brit. Chem. Eng.* 15:918 (1970).

CHAPTER 32
DESIGN METHODS FOR AEROCYCLONES AND HYDROCYCLONES

Matthias Bohnet
Institute for Process Technology
Technical University of Braunschweig
Braunschweig, Federal Republic of Germany

CONTENTS

INTRODUCTION

In 1886 O. M. Morse of the Knickerbocker Company registered for a patent for a dust sampler and received the first patent letter for a cyclone separator. Five years later, Bretney received a patent for a hydrocyclone [1]. Despite its successful use in industry for the past 100 years and investigations by scientists such as Prandtl and Barth, the proper design of cyclones is still difficult to achieve.

Figure 1 shows the principal setup of cyclone separators. The solid-fluid mixture is introduced tangentially or axially into a cylindrical bin with a mostly conical lower section. The torque of the fluid stream is produced either by tangential inlet or by guide vanes, which are located at the circumference of the outer cyclone wall. Due to the rotational flow within the separation space, centrifugal forces act on the solid particles and throw them against the cyclone wall. For aerocyclones, the particles slide down the wall into a collection hopper or are carried out with a liquid stream from the hydrocyclone. The cleaned fluid leaves the cyclone via a cylindrical fluid outlet pipe, sometimes called a vortex finder.

THEORY OF SEPARATION

Total Separation Efficiency

To evaluate the efficiency of cyclone separators, separation efficiencies η_A and η_T have to be determined. For aerocyclones it is

$$\eta_A = 1 - \frac{\dot{M}_{p_i}}{\dot{M}_{p_e}} \tag{1}$$

In contrast to aerocyclones, in which the total gas stream leaves the cyclone through the clean gas outlet, part of the liquid leaves the hydrocyclone, together with the separated solids, through the apex. For the separation efficiency of hydrocyclones, the proposal of Kelsall [2] leads to the following relationship:

$$\eta_T = \left[\left(1 - \frac{\dot{M}_{p_i}}{\dot{M}_{p_e}}\right) - \left(1 - \frac{\dot{V}_i}{\dot{V}_e}\right)\right]\frac{\dot{V}_e}{\dot{V}_i} \tag{2}$$

Flow Conditions

A simplified model is used to calculate the complex flow conditions within the cyclone. According to a method proposed by Barth [3], the cyclone separator is divided into two areas. The first area includes the flow conditions from the inlet cross section e-e to a cylindrical surface i-i, which is formed from the circumference of the fluid outlet pipe and the height, h (Figure 2). The second area considers the flow conditions behind the cylindrical surface until the end of the fluid outlet at the outlet cross section o-o. If there were potential flow within the cyclone, the tangential velocity would rise from the outer radius, r_a, to the outlet pipe radius, r_i, following the relation $u \cdot r = \text{const.}$ In reality, the velocity changes in a way characterized by curve b in Figure 3. It can be seen that the highest tangential velocity appears at the outlet radius, r_i. Particles within the rotational flow of a cyclone are subject to different centrifugal forces. The centrifugal force acting on a spherical particle is

$$Z = \frac{\pi}{6} d_p^3 (\rho_p - \rho) \frac{u^2}{r} \tag{3}$$

The centrifugal force tries to transport the particle outwards toward the cyclone wall, but because the fluid has to leave through the outlet pipe and, in hydrocyclones, partially

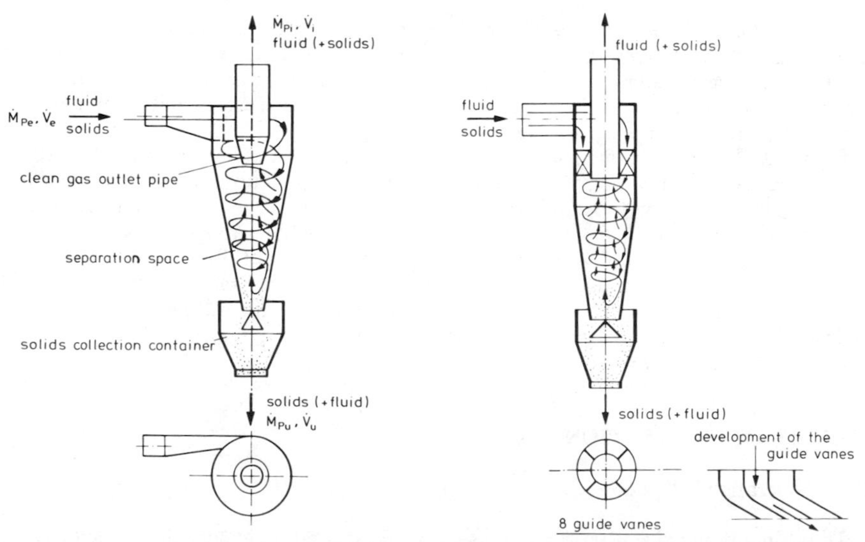

Figure 1. Principal setup of cyclone separators with tangential or axial torque generation.

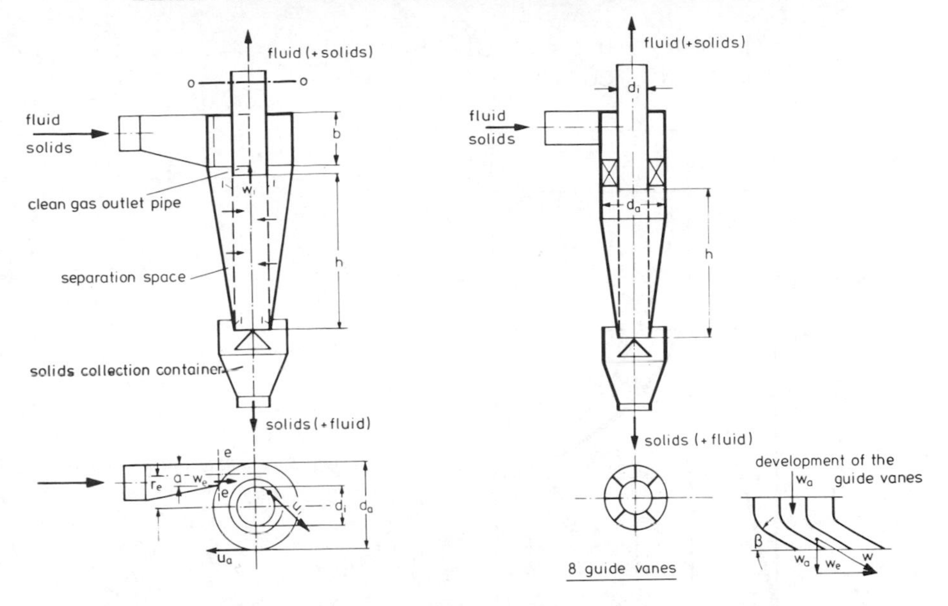

Figure 2. Cyclone separator with tangential or axial torque generation.

through the apex, the fluid flows through the cylindrical surface i-i, causing a flow resistance on the particle contrary to the centrifugal force:

$$W = c_w \frac{\pi}{4} d_p^2 \frac{\rho}{2} w_r^2 \tag{4}$$

The separation efficiency obviously depends on the centrifugal acceleration. Therefore, the locus with the highest centrifugal acceleration is important for the following considerations. The highest tangential velocity occurs at the outlet pipe radius, r_i (Figure 3). The centrifugal acceleration there is u_i^2/r_i. The mean radial velocity with which the fluid flows through the cylindrical surface i-i toward the fluid outlet and the apex, respectively, is

$$w_{r_i} = \frac{\dot{V}_e}{2\,\pi\, r_i\, h} \tag{5}$$

Critical Particle Size

The particle size, for which the centrifugal force and the flow resistance caused by the radial velocity are balanced, is decisive for the separation. All particles with a diameter larger than the so-called critical particle size, d_p^*, will be transported outward and separated. All particles with a diameter smaller than the critical particle size will be carried out through the outlet pipe with the fluid. From the force balance Z = W, it follows for the critical particle size that

$$d_p^* = \frac{3}{4} c_w \left(\frac{\rho}{\rho_p - \rho}\right) \left(\frac{w_{r_i}}{u_i}\right)^2 r_i \tag{6}$$

In practice, only the separation of small particles creates difficulties. Their flow resistance is described by the Stokes resistance law, for which the resistance coefficient is

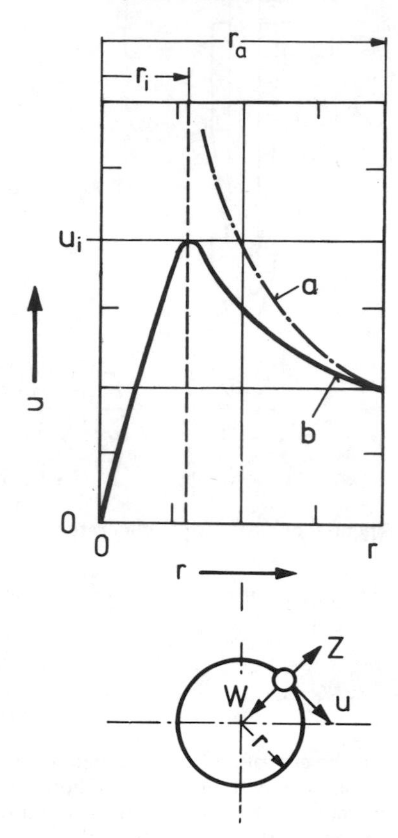

Figure 3. Tangential velocity in cyclones.

$$c_w = \frac{24}{Re_p} = 24 \frac{\eta}{w_{r_i}\, d_p\, \rho} \tag{7}$$

Introducing this value into Equation 6, it follows that

$$\begin{aligned} d_p{}^* &= \left(18 \frac{\eta}{\rho_p - \rho} \frac{w_{r_i}}{u_i} r_i\right)^{1/2} = 3\left(\frac{\eta}{(\rho_p - \rho)} \cdot \frac{\dot{V}_e}{\pi\, h\, u_i{}^2}\right)^{1/2} \\ &= 3\left(\frac{u_i}{w_e}\right)^{-1}\left(\frac{h}{r_i}\right)^{-1/2}\left(\frac{\eta}{\rho_p - \rho}\right)^{1/2}\left(\frac{F_e}{F_i}\right)^{1/2}\left(\frac{r_i}{w_e}\right)^{1/2} \end{aligned} \tag{8}$$

For practical calculations, it is almost always more advantageous to introduce the particle terminal velocity as the characteristic figure rather than the particle diameter because it influences decisively the movement of the particles within a fluid flow. For the region of validity of the Stokes law, the interrelation between the critical terminal velocity, $w_s{}^*$, and the critical particle diameter, $d_p{}^*$, is

$$d_p{}^* = \left[\frac{18\, \eta\, w_s{}^*}{(\rho_p - \rho) g}\right]^{1/2} \tag{9}$$

The critical terminal velocity follows from Equations 8 and 9:

$$w_s^* = \frac{\dot{V}_e\, g}{2\,\pi\, h\, u_i^2} = \left(\frac{u_i}{w_e}\right)^{-2} \left(\frac{h}{r_i}\right)^{-1} \left(\frac{F_e}{F_i}\right)\left(\frac{r_i\, g}{2\, w_e}\right) \tag{10}$$

For calculating the critical terminal velocity, w_s^*, the tangential velocity, u_i, has to be known. Following a proposal of Barth [3], the tangential velocity is calculated based on the assumption that all friction losses within the cyclone occur at a friction surface that is proportional to a cylinder with a mean radius $r_i\sqrt{r_a/r_i}$ and a height h. This model representation is based on the hypothesis that the fluid flows free of losses to this friction surface. At this surface it undergoes a velocity loss, Δu, and flows on without additional losses. From a momentum balance the following equation for the calculation of u_i is gained:

$$\frac{u_i}{w_e} = \frac{1}{\alpha\,\dfrac{r_i}{r_e} + \dfrac{F_i}{F_e}\dfrac{h}{r_i}\,\lambda} \tag{11}$$

Unknown thus far is the correction factor for the contraction of the fluid at the cyclone inlet, α, and the friction factor, λ. For the friction factor, an approximate calculation can be made with $\lambda = 0.02$. If this accuracy is not sufficient, the results of Muchelknautz and Krambrock [4] may be used.

Fractional Separation Efficiency

With the tangential velocity, u_i, on the outlet radius, r_i, the terminal velocity, w_s^*, belonging to the critical particle diameter, d_p^*, can be calculated. The separation behavior of cyclones then can be predicted if the so-called fractional separation efficiency curve is known. This curve shows which fraction of a defined particle size fraction is separated by the cyclone. Figure 4 shows the fractional separation efficiency curve. The real shape of the curve gained by experiments can deviate markedly from the one calculated theoretically. As many investigations show, the fractional separation efficiency, $\eta_F = 0.5$, is not always found at the critical particle diameter, d_p^*. The actual shape of the fractional separation efficiency curve depends essentially on the geometric dimensions of the cyclone and, for hydrocyclones, also on the ratio of the fluid streams, V_i/V_e, as shown in Figure 5. A steep curve indicates a sharp separation within the cyclone.

TORQUE GENERATION

For calculating the tangential velocity, u_i, at the outlet pipe radius, r_i, the correction factor, α, for the contraction at a tangential inlet must be known. The contraction has the result that the actual inlet velocity of the fluid is higher than the velocity calculated at the entrance cross section, F_e. Figure 6 shows measured correction factors [6] for different shapes of the fluid inlet. For a spiral inlet, no contraction is observed, which means that $\alpha = 1$. For slit inlets, the contraction and, therefore, the correction factor, α, depend essentially on the width of the slit. The entrance radii for the flow for spiral inlets are

$$r_e = r_a + a/2 \tag{12}$$

and for slit inlets are

$$r_e = r_a - a/2 \tag{13}$$

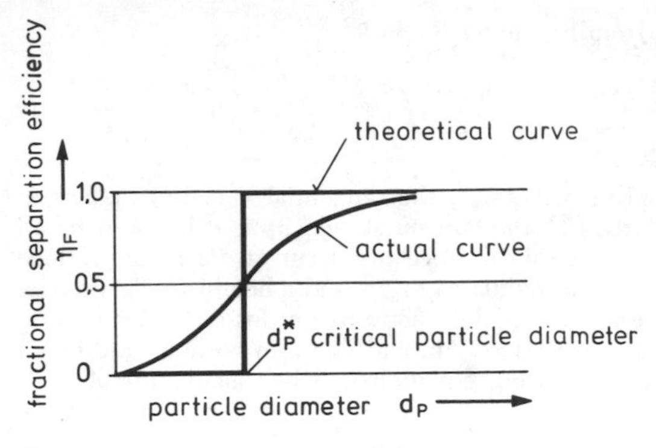

Figure 4. Fractional separation efficiency.

In axial cyclones, the torque is generated by guide vanes, which are placed at the circumference of the cyclone (see Figure 2). The tangential velocity, w_e, depends on the axial velocity, w_a, and on the blade angle, β:

$$w_e = \frac{w_a}{\tan \beta} = \frac{\dot{V}_e}{\frac{\pi}{4}\,(d_a{}^2 - d_L{}^2)\tan \beta} \tag{14}$$

For the guide vanes arrangement shown in Figure 2, the following values have to be introduced into Equation 11:

$$r_e = \left[\frac{1}{18}(d_a{}^2 - d_L{}^2)\right]^{1/2} \tag{15}$$

and

$$F_e = \frac{\pi}{4}\,(d_a{}^2 - d_L{}^2)\tan \beta \tag{16}$$

For plain blades, α is approximately 1. The calculation of the tangential velocity is then similar to the calculation of tangential cyclones.

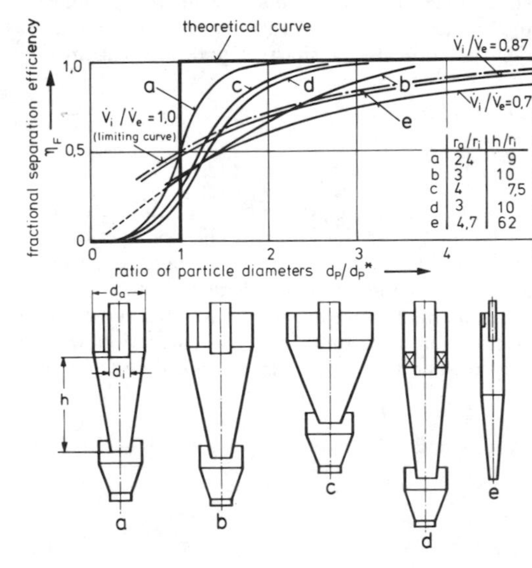

Figure 5. Fractional separation efficiency curves for different aero- and hydrocyclones [5].

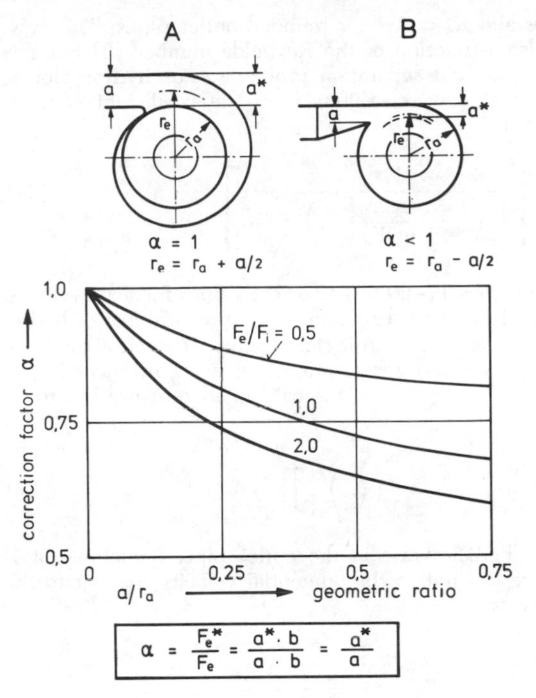

Figure 6. Correction factor, α, for different shapes of the fluid inlet.

PRESSURE LOSS

For calculating the pressure loss, the loss is divided into two parts. The first part, Δp_e, is due to the inlet losses and the losses in the cyclone due to friction within the area e-e and the cylinder surface i-i. The second part, Δp_i, is due to the flow losses in passing through the outlet pipe. It is

$$\Delta p = \Delta p_e + \Delta p_i = (\epsilon_e + \epsilon_i)\,\frac{\rho}{2}\,u_i^2 = \epsilon\,\frac{\rho}{2}\,u_i^2 \tag{17}$$

The pressure loss depends on the geometry of the cyclone and the fluid mass flowrate. For aerocyclones, the pressure loss coefficients are

$$\epsilon_e = \frac{r_i}{r_a}\left[\left(1 + \frac{u_i}{w_e}\,\frac{F_i}{F_e}\,\frac{h}{r_i}\,\lambda\right)^{-2} - 1\right] \tag{18}$$

and

$$\epsilon_i = \left[K\left(\frac{u_i}{w_e}\,\frac{F_i}{F_e}\right)^{-2/3} + 1\right] \tag{19}$$

In this equation, K = 4.4 has to be introduced for sharp-edged outlet pipes and K = 3.4 for rounded-outlet pipes. Equation 19 can be used for velocity ratios $(u_i/w_e)\cdot(F_i/F_e) \geqq 1$ only. For values $(u_i/w_e)\cdot(F_i/F_e) < 1$, the following equation should be used:

$$\epsilon_i = \left\{K\left(\frac{u_i}{w_e}\,\frac{F_i}{F_e}\right)^{-2/3} + 1 + K_o\left(\frac{u_i}{w_e}\,\frac{F_i}{F_e}\right)^{-1}\left[\left(\frac{u_i}{w_e}\,\frac{F_i}{F_e}\right)^{-1} - 1\right]\right\} \tag{20}$$

with $K_o = 2.0$ for sharp-edged pipes and $K_o = 1.1$ for rounded-outlet pipes. The pressure loss coefficients are in reality also a function of the Reynolds number [3] but this influence can be neglected for most practical separation problems. For hydrocyclones, the pressure loss coefficient for the inlet flow, ϵ_e, differs from the coefficient of aerocyclones as calculated by Equation 18. It is

$$\epsilon_e = \left(\frac{r_a}{r_i}\right)^{-1/2}\left[\frac{1}{\frac{r_a}{r_i}\left(1 - \frac{u_i}{w_e}\frac{F_i}{F_e}\frac{h}{r_i}\lambda\right)^2} - 1\right] \tag{21}$$

The pressure loss calculated with Equations 17–20 gives correct values for aerocyclones. The use of Equations 17, 19, 20 and 21 for the calculation of the pressure loss of hydrocyclones leads to values for separators operated with a closed apex. In such a case, the hydrocyclone is only streamed through, without a separation; however, an opened apex is necessary for a separation effect. For this real case, the calculated pressure has to be reduced accordingly:

$$\Delta p = \left\{\epsilon - \left(\frac{u_i}{w_e}\frac{F_i}{F_e}\right)^{-2}\left[1 - \left(\frac{\dot{V}_i}{\dot{V}_e}\right)^2\right]\right\}\frac{\rho}{2}u_i^2 \tag{22}$$

In this equation, $\dot{V}_i$ is the actual fluid flowrate in the outlet pipe. Sometimes it is more convenient to relate the pressure loss not to the tangential velocity, u_i, but to the inlet velocity, w_e, or the outlet velocity, w_i:

$$\Delta p = \xi_e \frac{\rho}{2} w_e^2$$

with

$$\xi_e = \epsilon\left(\frac{u_i}{w_e}\right)^2$$

and

$$\xi_e = \xi_{ee} + \xi_{ei} \tag{23}$$

$$\Delta p = \xi_i \frac{\rho}{2} w_i^2$$

with

$$\xi_i = \epsilon\left(\frac{u_i}{w_e}\right)^2\left(\frac{F_i}{F_e}\right)^2$$

and

$$\xi_i = \xi_{ie} + \xi_{ii}$$

(24)

The pressure loss coefficient of Equation 24 should be used only for aerocyclones, for which $\dot{V}_e = \dot{V}_i$.

To reduce the pressure loss of cyclone separators, several measures can be taken to transfer part of the high kinetic energy of the outlet fluid flow into static pressure energy.

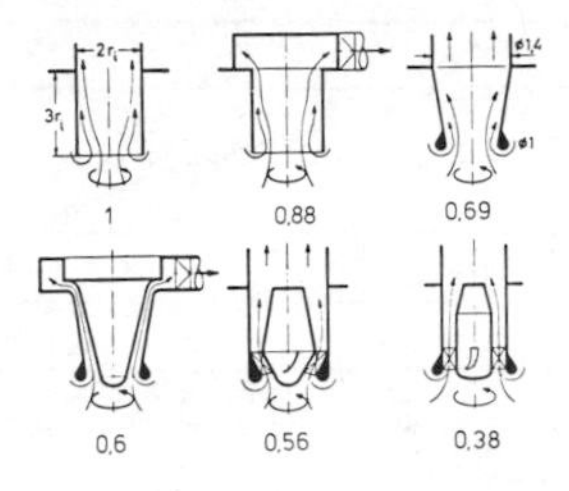

Figure 7. Ratio of pressure loss coefficients for different fluid outlet pipe constructions [8].

A first, but very effective, measure is the use of conical outlet pipes, which reduce the axial velocity. For the partial recovery of torque energy, outlet spirals may be used. Also, the use of guide vanes may lead to a remarkable pressure recovery [7]. A comparison of the efficiencies of different constructions is shown in Figure 7. The values are always related to the pressure loss coefficient of a cylindrical outlet pipe. Evaluating measures to change kinetic into pressure energy, one must consider that—for fine dust particles and aerocyclones mainly—there is a risk of the small outlet pipe cross section being blocked by dust deposits.

SEPARATION BEHAVIOR

Considerations always have been based on the assumption that the dispersed phase does not influence the fluid flow within the cyclone. This is valid for aerocyclones only up to a solids loading of approximately 10 g/m³. If the solids loading is higher than the named value, which occurs very often in cyclones used in process technology, the calculation of aerocyclones becomes more difficult. It is well known from the theory of the pneumatic transport of solids that the gas stream is not able to transport unlimited amounts of solids floating in the gas. Part of the solid material moves in the conveying pipe as a stream at the bottom of the pipe [9]. A similar behavior is observed in aerocyclones. If the loading with dispersed phase exceeds a critical value, part of the solids are separated as a stream immediately after entering the cyclone separator. With the coarse particles, fine particles also are transported to the cyclone wall, although they should not be separated due to their terminal velocity. Knowing the critical solids loading at which the unfractionated separation of the dispersed phase starts, the total separation efficiency of aerocyclones can be calculated. The calculation of the separation efficiency based on the fractional efficiency curve is then done only for that part of the solids that is not separated as a stream entering the cyclone. The critical solids loading can be calculated using the following equation [10]:

$$\mu_c \approx 10^{-1}\,\lambda\left(\alpha\frac{F_e}{F_i}\frac{r_a}{r_e}\right)^{1/2}\left(\frac{h}{r_i}\right)\left(\frac{r_a}{r_i}\right)^{3/2}\left(\frac{r_a}{r_i}-1\right)^{-1}\left(\frac{u_\cdot}{w_e}\frac{F_i}{F_e}\right)^{3/2}\left(\frac{w_s^*}{w_s}\right) \qquad (25)$$

In Equation 25, w_s characterizes the terminal velocity of the mean particle size, d_{p50}, of the dispersed solids entering the cyclone separator with the gas stream.

The tendency of fine particles to agglomerate is of great importance in the separation behavior of gas-solids separators. If solids that are able to form agglomerates have to be separated, one must consider that the shear field of the rotating fluid does not destroy the agglomerates. For such cases, it is appropriate to use aerocyclones with a large diameter ratio, d_a/d_i. For adhesive dust particles, cylindrical, polished cyclone separators are advantageous.

Due to the small density difference between solids and liquids, a similar critical solids loading is not observed in hydrocyclones, but the reciprocal interaction of the solid particles influences the separation procedure. If the inlet volume concentration of the solids exceeds 10^{-4} m³/m³, the total separation efficiency, calculated from the fractional

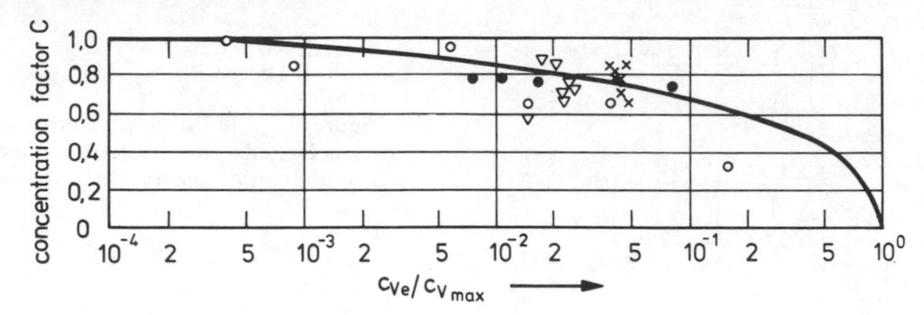

Figure 8. Concentration factor, depending on the concentration ratio [11].

separation efficiency curve, has to be diminished. In a first approximation, the diminution can be calculated by the following empirical equation [5]:

$$\eta_{T_{eff}} = 0.5 \left(ln \frac{c_{vmax}}{c_{v_e}} \right)^{1/3} \cdot \eta_T = C \, \eta_T \tag{26}$$

which is a result of the measured curve shown in Figure 8. c_{vmax} indicates the highest solids volume concentration at which the suspension is still flowing.

OPTIMAL CYCLONE DESIGN

The most important factors in designing cyclone separators are the critical particle size, d_p^*, and the critical terminal velocity, w_s^*, respectively. These figures determine which

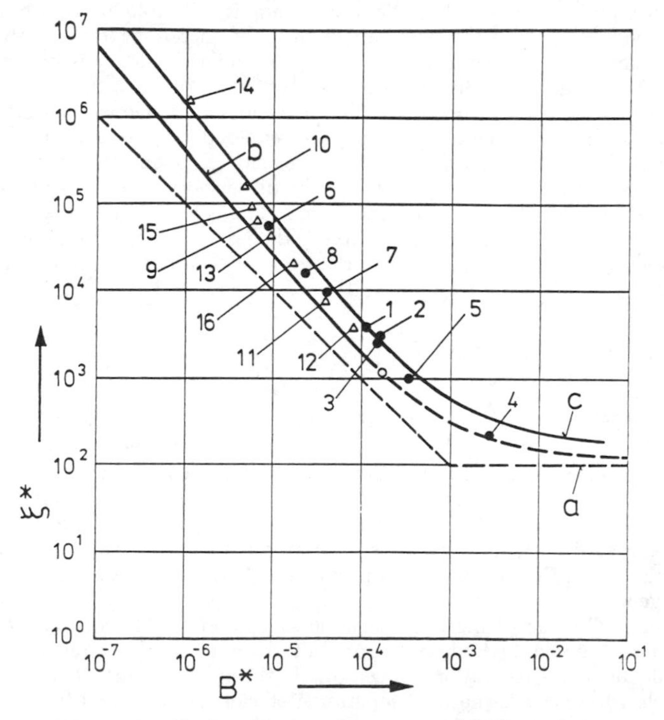

Figure 9. Design diagram for aero- and hydrocyclones.

particle size will be separated and to what extent. As the critical terminal velocity, w_s^*, is inversely proportional to the square of the tangential velocity, u_i, it is theoretically possible to reduce the critical terminal velocity for a given fluid throughput, if only the inlet velocity and, along with it, the tangential velocity, are increased. But this is a fallacy because for a given fluid throughput caused by the unavoidable reduction of the cyclone size, not only does a rise of the inlet velocity occur, but also a rise of the radial velocity and, therefore, also a rise of the flow resistance of the particles. The increase of the centrifugal force is compensated therefore by the increasing flow resistance.

Further, the pressure loss rises with increasing fluid velocity but the pressure loss influences essentially the operation costs of the cyclone. An optimal design requires that not only the critical particle size, but also the pressure loss, be taken into consideration. The definition of the critical particle size also determines the size of the cyclone separator. The latter influences the investment costs.

Cyclones can be optimized either by the needed area or the needed space. When optimizing cyclones on the basis of the needed space, it is useful to relate the fluid throughput to a suitable characteristic cyclone dimensional size. For the characteristic size chosen,

$$r_a^* = \sqrt[3]{r_a^2 h} \tag{27}$$

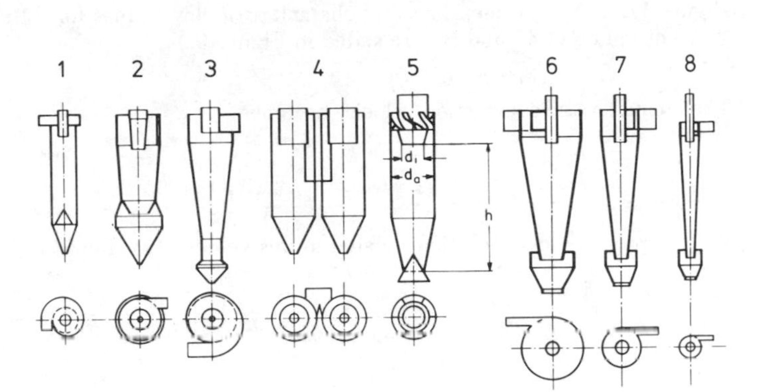

aerocyclone		1	2	3	4	5**	6	7	8
geometric dimensions	r_i m	0,450	0,770	0,260	0,635	0,045	0,050	0,050	0,050
	r_a m	1,12	2,00	0,588	1,00	0,075	0,250	0,150	0,150
	h m	7,35	6,70	2,41	3,40	0,650	0,900	0,900	1,80
	r_a/r_i –	2,49	2,60	2,26	1,58	1,67	5,0	3,0	3,0
	r_e/r_i –	3,73	1,95	3,08	1,14	1,44	4,48	2,48	2,48
	h/r_i –	16,3	8,70	9,27	5,35	14,4	18,0	18,0	36,0
	F_e/F_i –	1,97	1,10	1,00	1,25	0,852	1,00	1,00	1,00
	a/r_a –	1,00	0,482	0,723	0,553	-	0,205	0,341	0,341
	α –	1,00	0,720	1,00	0,700	1,00	0,840	0,790	0,790
	λ –	0,02	0,02	0,02	0,02	0,02	0,02	0,02	0,02
velocity ratio (equation 11)	u_i/w_e –	2,30	1,89	1,96	1,43	0,94	1,83	1,47	0,963
vortex finder coefficients	K –	4,4	4,4	3,4	4,4	4,4	3,4	3,4	3,4
	K_0 –	–	–	–	–	–	–	–	1,1
pressure loss coefficient (equation 17)	ε –	5,61	4,50	3,83	5,22	5,76	4,78	4,81	7,74
volume based pressure loss coefficient (equation 29)	ξ^* –	$3{,}62 \cdot 10^3$	$3{,}03 \cdot 10^3$	$2{,}51 \cdot 10^3$	$2{,}15 \cdot 10^2$	$1{,}03 \cdot 10^3$	$5{,}53 \cdot 10^4$	$9{,}16 \cdot 10^3$	$1{,}60 \cdot 10^4$
volume based separation number (equation 30)	B^* –	$1{,}11 \cdot 10^{-4}$	$1{,}65 \cdot 10^{-4}$	$1{,}48 \cdot 10^{-4}$	$2{,}69 \cdot 10^{-3}$	$3{,}33 \cdot 10^{-4}$	$9{,}22 \cdot 10^{-6}$	$3{,}97 \cdot 10^{-5}$	$2{,}31 \cdot 10^{-5}$

*blade angel $\beta = 31°$

Figure 10. Main dimensions and characteristic flow values for different aerocyclones. (The values of ξ^* and B^* are stated in Figure 9.)

hydrocyclone			9	10	11	12	13	14	15	16
geometric dimensions	r_i	m	0,01	0,01	0,01	0,01	0,01	0,00255	0,0053	0,0075
	r_a	m	0,05	0,05	0,025	0,025	0,05	0,025	0,025	0,025
	h	m	0,33	1,0	0,33	0,15	0,33	0,332	0,329	0,304
	r_a/r_i	–	5,0	5,0	2,5	2,5	5,0	9,81	4,71	3,33
	r_e/r_i	–	4,49	4,49	1,99	1,99	4,28	9,29	4,25	2,79
	h/r_i	–	33	100	33	15	33	130	62	40,5
	F_e/F_i	–	1	1	1	1	2	1,02	0,93	1,05
	a/r_a	–	0,2	0,2	0,4	0,4	0,289	0,104	0,200	0,328
	α	–	0,855	0,855	0,766	0,766	0,731	0,909	0,864	0,786
	λ	–	0,02	0,02	0,02	0,02	0,02	0,02	0,02	0,02
velocity ratio	u_i/w_e	–	1,18	0,457	0,957	1,46	2,00	0,378	0,651	0,950
vortex finder coefficients	K	–	3,4	3,4	3,4	3,4	3,4	3,4	3,4	3,4
	K_0	–	-	1,1	1,1	-	-	1,1	1,1	1,1
pressure loss coefficient	ε	–	5,42	21,26	5,79	3,81	4,73	36,81	11,14	6,52
volume based pressure loss coefficient	ξ^*	–	$5{,}84 \cdot 10^{4}$	$1{,}51 \cdot 10^{5}$	$6{,}46 \cdot 10^{3}$	$3{,}46 \cdot 10^{3}$	$3{,}66 \cdot 10^{4}$	$1{,}47 \cdot 10^{6}$	$8{,}35 \cdot 10^{4}$	$1{,}83 \cdot 10^{4}$
volume based separation number	B^*	–	$6{,}59 \cdot 10^{-6}$	$4{,}79 \cdot 10^{-6}$	$4{,}01 \cdot 10^{-5}$	$8{,}34 \cdot 10^{-5}$	$9{,}18 \cdot 10^{-6}$	$1{,}12 \cdot 10^{-6}$	$5{,}98 \cdot 10^{-6}$	$1{,}68 \cdot 10^{-5}$

Figure 11. Main dimensions and characteristic flow values for different hydrocyclones. (The values of ξ^* and B^* are stated in Figure 9.)

The characteristic comparison velocity becomes

$$W_{St}^{*} = \frac{\dot{V}_e}{\pi \, r_a^{*2}} \tag{28}$$

For the pressure loss coefficient related to this velocity it is found that

$$\xi^{*} = \frac{\Delta p}{\frac{\rho}{2} w_{St}^{*2}} = \epsilon \left(\frac{u_i}{w_e}\frac{F_i}{F_e}\right)^2 \left[\left(\frac{r_a}{r_i}\right)^2 \left(\frac{h}{r_i}\right)\right]^{4/3} \tag{29}$$

The critical terminal velocity, w_s^*, and the dimension sizes of the cyclone are combined following a proposal of Barth and Leineweber [12] in the separation number, B^*:

$$B^{*} = \frac{w_s^{*} \cdot w_{St}^{*}}{2\, r_a^{*}\, g} = \left[2 \frac{r_a}{r_i}\frac{h}{r_i}\frac{u_i}{w_e}\frac{F_i}{F_e}\right]^{-2} \tag{30}$$

Figure 9 shows the relation between the volume-based pressure loss coefficient, ξ^*, and the volume-based separation number, B^*. The dashed curve, a, was given by Barth and Leineweber [12] as the limiting curve for a rotating centrifugal windsifter and in the horizontal part for the axial cell without torque. The theoretically optimized cyclone types are found on curve b. These cyclones are relatively long for aerocyclones. For industrial use, the aerocyclones are normally shorter than calculated using curve b. It is proposed that aero- and hydrocyclones be designed in such a way that the operating and design figures are described by points in the neighborhood of curve b and should not exceed values of curve c.

To illustrate, Figures 9, 10 and 11 show the geometric dimensions, the velocity ratio, u_i/w_e, and the pressure loss coefficient, ϵ, for eight aerocyclones and for eight hydrocyclones. The corresponding volume-based pressure loss coefficients, ξ^*, and separation numbers, B^*, are included in the tables and shown in the diagram. For evaluating the hydrodynamic quality of hydrocyclones, they have to be calculated for the case of a closed apex. All data given in Figure 11 are calculated for the hypothetical operation of hydrocyclones with $\dot{V}_i = \dot{V}_e$.

NOMENCLATURE

B^* volume-based separation number
F cross section (m^2)
$\dot{M}$, $\dot{M}_p$ mass flowrate of fluids or solids, respectively (kg/sec)
Re Reynolds number
$\dot{V}_e$ volume flowrate of the suspension (m^3/sec)
W flow resistance, N
Z centrifugal force, N
a width of the inlet cross section (m)
b height of the inlet cross section (m)
c_v volume concentration of the solids in the suspension (m^3/m^3)
c_w drag coefficient
d diameter (m)
d_L diameter of the guide vanes holder (m)
d_p^* diameter of the critical particle size (m)
g acceleration of gravity (m/sec^2)
h clearance between cyclone apex and outlet gas pipe (m)
Δp pressure loss (N/m^2)
r radius (m)
r_a outer radius of the cyclone (m)
r_e entrance radius (m)
r_i gas outlet pipe radius (m)
u tangential velocity (m/sec)
w_a axial velocity (m/sec)
w_e inlet velocity (m/sec)
w_i outlet velocity (m/sec)
w_r radial velocity (m/sec)
w_s terminal velocity (m/sec)
w_s^* critical terminal velocity (m/sec)
w_{St}^* characteristic fluid velocity (m/sec)
α correction factor for the contraction
β blade angle
ϵ pressure loss coefficient
η dynamic viscosity (Pa-s)
η_A, η_T separation efficiency
η_F fractional separation efficiency
λ friction factor
μ_c limiting solids loading
ξ_e pressure loss coefficient
ξ_i pressure loss coefficient
ξ^* volume-based pressure loss coefficient
ρ fluid density (kg/m^3)
ρ_p solids density (kg/m^3)

REFERENCES

1. Bradley, D. *The Hydrocyclone* (Oxford, England: Pergamon Press, Inc., 1965).
2. Kelsall, D. F. "A Further Study of the Hydraulic Cyclone," *Chem. Eng. Sci.* 2:254–272 (1953).
3. Barth, W. "Berechnung und Auslegung von Zyklonabscheidern auf Grund neuerer Untersuchungen," *Brennstoff-Wärme-Kraft* 8(4):1–9 (1956).
4. Muschelknautz, E., and W. Krambrock. "Aerodynamische Beiwerte des Zyklonabscheiders auf Grund neuer und verbesserter Messungen," *Chem. -Ing. Tech.* 42 (5):247–255 (1970).
5. Bohnet, M. "Kriterien für die Auswahl von Zyklonen und Hydrozyklonen," *Chem. Ind.* XXXIII:91–95 (1981).
6. Muschelknautz, E., and K. Brunner. "Untersuchungen an Zyklonen," *Chem. -Ing. Tech.* 39(9/10):531–538 (1967).
7. Schiele, O. "Möglichkeiten zur Wiedergewinnung der Drallenergie von Zyklonabscheidern," *VDI-Tagungsheft* 3:20–22 (1954).
8. Muschelknautz, E. "Theorie der Fliehkraftabscheider mit besonderer Berücksichtigung hoher Temperaturen und Drücke," *VDI-Bericht* 363:49–60 (1980).
9. Bohnet, M. "Experimentelle und theoretische Untersuchungen über das Absetzen, das Aufwirbeln und den Transport feiner Staubteilchen in pneumatischen Förderleitungen," *VDI-Forschungsheft* 509 (1965).
10. Muschelknautz, E. "Auslegung von Zyklonabscheidern in der technischen Praxis." *Staub.* 30(5):187–195 (1970).

11. Bohnet, M. "Neuere Untersuchungen über die Trennwirkung und den Druckverlust von Hydrozyklonen," *Verfahrenstechnik* 3(9):376–381 (1969).
12. Barth, W., and L. Leineweber. "Beurteilung und Auslegung von Zyklonabscheidern," *Staub* 24(2):41–55 (1964).

SECTION IV

LIQUID-SOLIDS FLOWS

CONTENTS

CHAPTER 33
RHEOLOGY OF SUSPENSIONS

D. T. Y. Kao
Department of Civil Engineering
University of Kentucky
Lexington, Kentucky 40506

CONTENTS

INTRODUCTION

The mechanics of liquid-solid suspensions is regarded as a branch of science of great complexity. Depending on the components involved and their fundamental properties, fluid-solid suspensions often exhibit flow behavior that is governed by a new set of properties, which lead to entirely different responses when subjected to stress or motion.

Rheology deals with the determination of a relationship involving deformation and flow of matter. The relationships between the rate of deformation (rate of shear) and corresponding development of shear stress are referred to as the rheological behavior (rheological property) of the fluid.

Fundamental Properties of Fluids

There are many basic fluid properties that are of fundamental importance, including mass density, weight density, specific volume, surface tension, viscosity and specific gravity. Among these properties, viscosity requires the greatest attention in the study of fluid flow. The viscous nature of a liquid is a direct derivative of the molecular attraction that offers resistance to shear and, consequently, to the resulting flow or deformation induced by the shear.

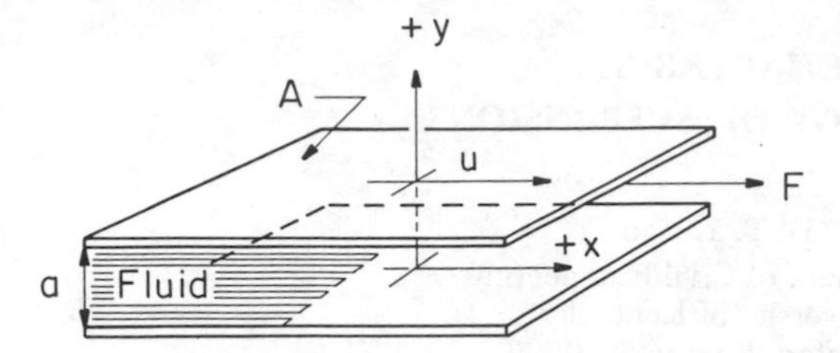

Figure 1. Viscous fluid motion between two parallel plates.

Considering two parallel plates of area A, as shown in Figure 1, with the small space, a, between the plates being filled with viscous fluid, a force, F, must be applied to initiate and maintain the top plate in motion at a constant speed, u, in the given direction, x. The fluid in the space will deform linearly at an angular rate, $du/dy = u/a$. The force, F, is known to be directly proportional to the magnitude of rate of angular deformation and the area of the plate. By introducing a constant of proportionality, μ, an equation for Newton's law of viscosity is formed to give

$$\tau = \frac{F}{A} = \mu \frac{du}{dy} \tag{1}$$

Where τ is the shear stress and μ is the dynamic or absolute viscosity of the fluid having a unit $N\text{-}m^{-2}$ and $N\text{-}sec\text{-}m^{-2}$ ($lb\text{-}ft^{-2}$ and $lb\text{-}sec\text{-}ft^{-2}$), respectively. In a shear stress vs rate of shear (du/dy, rate of angular deformation) plot using an arithmetic coordinate system, μ represents the slope of a straight line. A fluid that behaves in accordance with Newton's law of viscosity is referred to as a Newtonian fluid. Otherwise, it is referred to as a non-Newtonian fluid.

Definition of Terms:

Kinematic or relative viscosity. This is the ratio of viscous effect to the mass density of the fluid, $\mu/\rho = \nu$, expressed in $m^2\text{-}sec^{-1}$ ($ft^2\text{-}sec^{-1}$).

Apparent Viscosity. This is the slope of a straight line connecting the origin and a given point on a shear stress vs rate of shear (flow) curve of a non-Newtonian fluid. Mathematically, it is defined as

$$\mu_a = \frac{\tau}{\left(\frac{du}{dy}\right)} \tag{2}$$

Differential Viscosity. This is the slope of a tangent line at a given point on a shear stress vs shear rate curve of a non-Newtonian fluid. Mathematically, it can be expressed as

$$\mu_d = \frac{d\tau}{d\left(\frac{du}{dy}\right)} \tag{3}$$

Rheogram. This is a graphic representation of the functional relationship between the shear stress, τ, and rate of shear (rate of angular deformation), du/dy.

Pseudohomogeneous. This refers to a manner of spurious homogeneity.

Pure Viscous Fluids. These are fluids that on removal of shearing force do not recover from any deformation they may have undergone under its action.

Viscoelastic Fluids. These are fluids that on removal of the shearing force do recover partially from the deformation they have undergone during the shearing action. These fluids exhibit rheological behavior of combined properties of pure viscous fluid and elastic solid.

Time Independent. Rheological behavior that does not vary with time.

Time Dependent. Of those pure viscous fluids, rheological behavior that does vary with time.

Heterohomogeneous. This refers to heterogeneous dispersion of some coarser solids particles in a homogeneous suspension of a liquid-solids mixture of the same solids particles.

CLASSIFICATION OF LIQUID-SOLIDS MIXTURES

Depending on the physical properties of the components involved, a liquid-fine solids particle mixture may exist in truly homogeneous form or appear in spurious homogeneity. Under such conditions, the mixture flow will behave as a single-phase fluid.

If the suspended component involves relatively coarser solids particles, the power of the suspending component may not be strong enough to provide a uniform dispersion of the former throughout the fluid regime. Instead, higher solids concentration will be observed in its lower portion. A suspension left in such a manner is a heterogeneous one.

Flow of a heterogeneous suspension will behave as a multiphase system in that the rheological behavior of the liquid phase will predominate. The degree of uniformity to which the solids are dispersed is affected by the size consistency of solids and the velocity of the flow. To a lesser extent, the dispersion phenomenon also is influenced by the breadth of the space in which the fluid is contained (diameter of pipe in a closed conduit flow, for example). (Detailed discussion of these subjects is provided in other chapters of this volume.)

The rheological behavior of single-phase suspensions can become rather complex. Based on their rheological behavior, these fluids can be classified as Newtonian or non-Newtonian. Those fluids that strictly follow Newton's Law of viscosity in their behavior, as expressed in Equation 1, are classified as the former. For all others, non-Newtonian classification is assigned.

A fluid also can behave as a purely viscous substance, which does not have the ability to recover its deformation imposed by the action of a shearing force, even after the latter is removed. Those that have the ability to partially recover the exerted deformation on removal of the shear stress are called viscoelastic fluids.

The behavior of non-Newtonian pure viscous fluids sometimes can vary with time. If so, they are further classified as time-dependent fluids. Thixotropic and rheopectic fluids belong to this classification.

Time-independent purely viscous non-Newtonian fluids constitute the most important and commonly encountered suspensions in engineering systems. Based on the different rheological behaviors, they can be classified further as Bingham plastic, pseudoplastic or dilatant fluids, as shown in Figure 2.

A Bingham plastic fluid has the simplest rheological behavior of the three because after an initial stress level (yield stress) is reached the behavior of the fluid follows that of a Newtonian fluid in terms of linear shear stress and rate of shear relationship. The apparent viscosity, μ_a, for a Bingham fluid declines as the rate of shear, du/dy, increases, while the differential viscosity, μ_d, remains constant.

A pseudoplastic fluid exhibits a curvilinear shear stress vs rate of shear relationship with a gradually decreasing apparent viscosity, as well as differential viscosity in the direction of increasing rate of shear. For a simple pseudoplastic fluid, the flow curve passes the origin of the rheogram, indicating the immediate occurrence of angular deformation as a result of initial shear stress application. If no resulting angular deformation takes place until a finite value of shear stress, τ_y, is applied, the fluid is referred to as yield pseudoplastic.

Contrary to pseudoplastic behavior, a dilatant fluid demonstrates increasingly apparent and differential shear stresses as the rate of shear increases. Such a unique behavior was best explained originally by Osborne Reynolds. He attributed this peculiar behavior

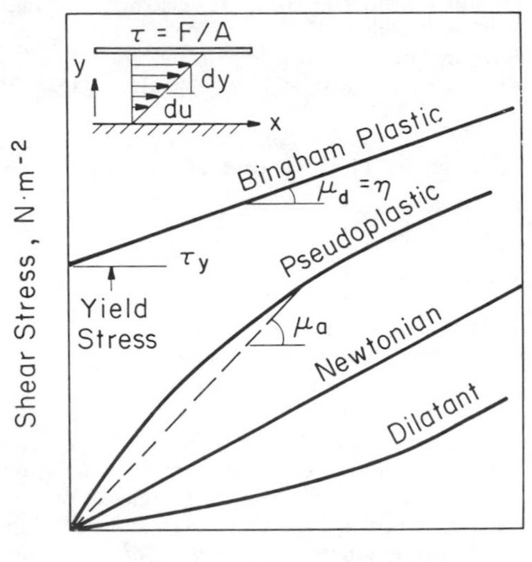

Figure 2. Typical flow curve for Newtonian and non-Newtonian fluids.

to be a result of structural transformation of the suspension from being at rest to being in motion.

When at rest, Reynolds assumed the suspension to have a structure consisting of densely packed solids particles (perhaps maximum possible), with small amounts of fluid merely sufficient to fill the voids. As the motion first begins, the fluid provides lubrication that makes the passage of one particle over another subject to relatively small friction resistance. Once the ordered array of the maximum particle packing is disturbed, there will not be sufficient liquid to completely fill the increased void volume to enable smooth particle passage. As a result, direct particle to particle interaction takes place, causing the shear stress to increase proportionately more than the rate of shear.

Time-dependent non-Newtonian fluids have similar rheological properties to those of pseudoplastic or yield-pseudoplastic fluids, except that their behavior is dependent on the duration of shear. If shear stress decreases with time at a given rate of shear, the fluid is called thixotropic. On the other hand, if a fluid behaves opposite to that of thixotropic, it is called a rheopectic fluid.

The chart shown in Figure 3 provides an overall view of the classification scheme of complex mixtures. Included is the schematic drawing depicting the degree of solids particle dispersion (regime delineation) as a function of solids particle size and mean flow velocity when the mixture behaves as a multiphase system.

Time-Independent Viscous Fluids

As shown in Figure 3, there are six different types of fluids belonging to this classification:

1. Newtonian
2. Bingham
3. Simple Pseudoplastic
4. Yield Pseudoplastic
5. Simple Dilatant
6. Yield Dilatant

These fluids share two common rheological properties: purely viscous and time independent. They can be either single-phase true homogeneous fluids or two-phase fine

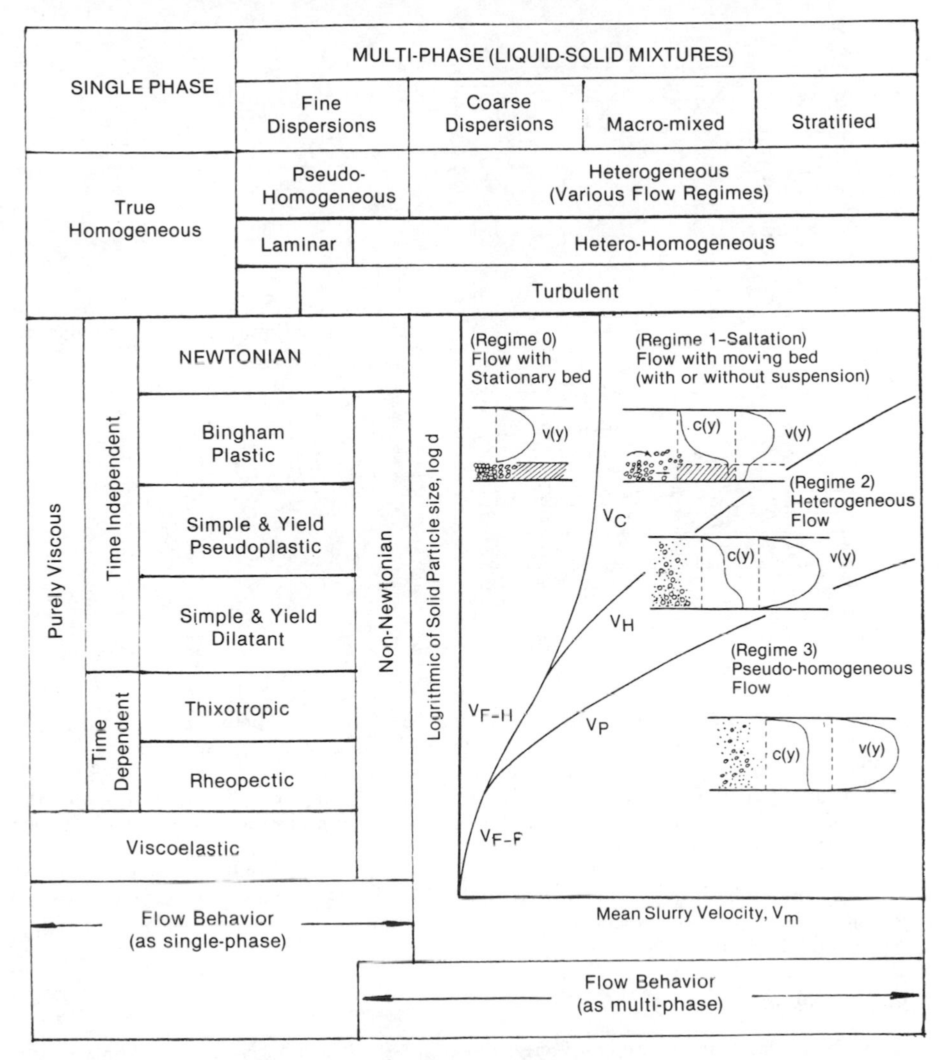

Figure 3. Classification of fluid suspensions.

dispersions of liquid-solids mixtures existing in pseudohomogeneous form. Two-phase pseudohomogeneous suspensions remain stable even in the absence of turbulence.

The rheological behavior of these fluids differs from one to the other as evidenced by the manifestation of different shear stress vs rate of shear relationships (as described in the previous section). In this section, various modeling techniques of these fluids will be presented.

Newtonian Fluid

All single-phase liquids with simple (nonelongated) molecules, solutions of low-molecular-weight (nonpolymeric) materials and pseudohomogeneous suspension of spherical particles in simple liquids are found to have Newtonian behavior. Newtonian

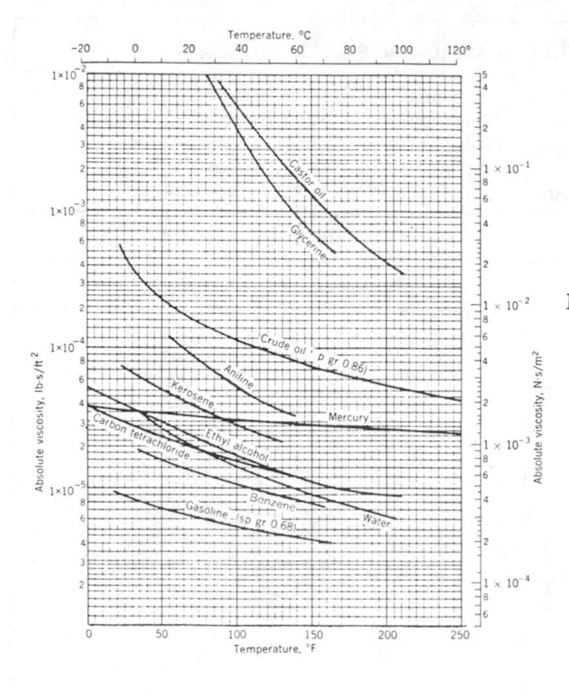

Figure 4. Viscosity of some common liquids as a function of temperature.

fluids are those that exhibit a direct (linear) proportionality between shear stress, τ, and the rate of shear, du/dy. This can be expressed mathematically by Equation 1.

The absolute (dynamic) viscosity, μ, needs to be determined experimentally using an instrument commonly referred to as a viscometer. A more detailed description of the instrument and techniques for determining this property and other rheological behavior of liquids will be presented in a later section.

For a liquid, the value of μ decreases as the temperature of the fluid increases. This is generally true regardless of whether the fluid has Newtonian or non-Newtonian behavior. A plot of the absolute viscosity of some common liquids as they vary with temperature is presented in Figure 4.

Bingham Plastic Fluids

Bingham plastic fluids are considered the simplest of all non-Newtonian fluids in the sense that the relationship between shear stress and shear rate differs from that of a Newtonian fluid only because the linear flow curve does not pass through the origin of the rheogram. Instead, it must have an application of a finite value of shear stress before the fluid deformation can begin. Once the finite shear stress, τ_y, which is referred to as yield stress, is reached and the initial movement has started, the fluid behaves in the same manner as a Newtonian fluid, with the shear stress increasing linearly with the rate of shear, as shown in Figure 2. Mathematical modeling of such rheological behavior can be expressed as

$$\tau - \tau_y = \eta \left(\frac{du}{dy}\right) \tag{4}$$

In the above equation, η is the coefficient of rigidity or plastic viscosity that represents the slope of the straight line portion of the flow curve. The coefficient of rigidity, η, is the same as the differential viscosity, μ_d, and has a constant value for a given

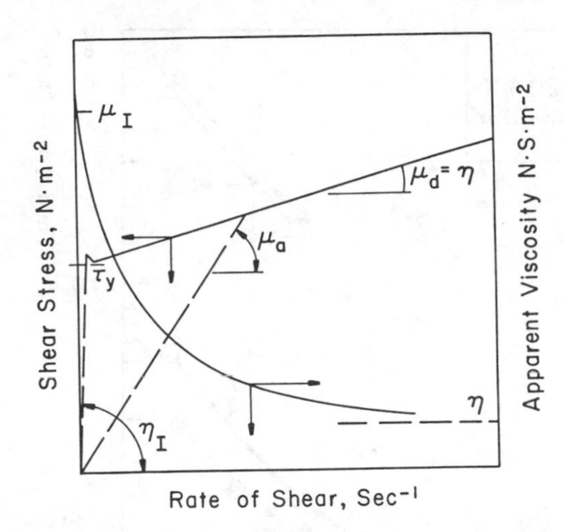

Figure 5. Typical flow curve for a Bingham plastic fluid.

Bingham plastic fluid. By definition, the apparent viscosity, μ_a, for a Bingham plastic can be written as

$$\mu_a = \frac{\tau_y}{\left(\frac{du}{dy}\right)} + \eta \tag{5}$$

Equation 5 indicates that theoretically the apparent viscosity, μ_a, should have an initial value of infinity and decreases with increasing rate of shear, du/dy, asymptotically until the value of the coefficient of rigidity is reached. Actually, du/dy = 0 only when $\tau = 0$, and deformation of fluid does begin immediately as a shearing force is applied to the fluid. However, the magnitude of the initial deformation is too small to be easily detectable using the currently available rheometer. Nonetheless, such a behavior is mechanically sound and can be predicted as shown by the dashed line in Figure 5.

Within this shear stress range, $\tau - \tau_y$, the fluid can be regarded as an elastic substance. The slope of the initial portion of the rheological behavior curve, η_I, gives the initial apparent viscosity of the fluid and has a physical meaning analogous to that of the modulus of elasticity.

Examples of fluids that have been identified to approximate Bingham plastic behavior are water-galena suspensions, suspensions of chalk, sewage sludge, a fine coal-water mixture (Figure 6) [1,2] and, for a household pseudohomogeneous liquid-solids suspension—toothpaste.

Simple and Yield Pseudoplastic Fluids

Most non-Newtonian fluids fall into this classification. A pseudoplastic fluid displays, on arithmetic coordinates, a concave downward flow curve toward the axis for the shear rate. Both the apparent viscosity, μ_a, and differential viscosity, μ_d, decrease with increasing rate of shear, as shown in Figure 7. Shown also in Figure 7 is the typical rheogram for a yield pseudoplastic fluid.

Many different mathematical models have been proposed by various researchers attempting to describe the rheological behavior of pseudoplastic fluids. In terms of an explicit expression for apparent viscosity, this includes the following equations:
Ostwald-deWaele (power law):

$$\mu_a = K\left(\frac{du}{dy}\right)^{n-1} \tag{6}$$

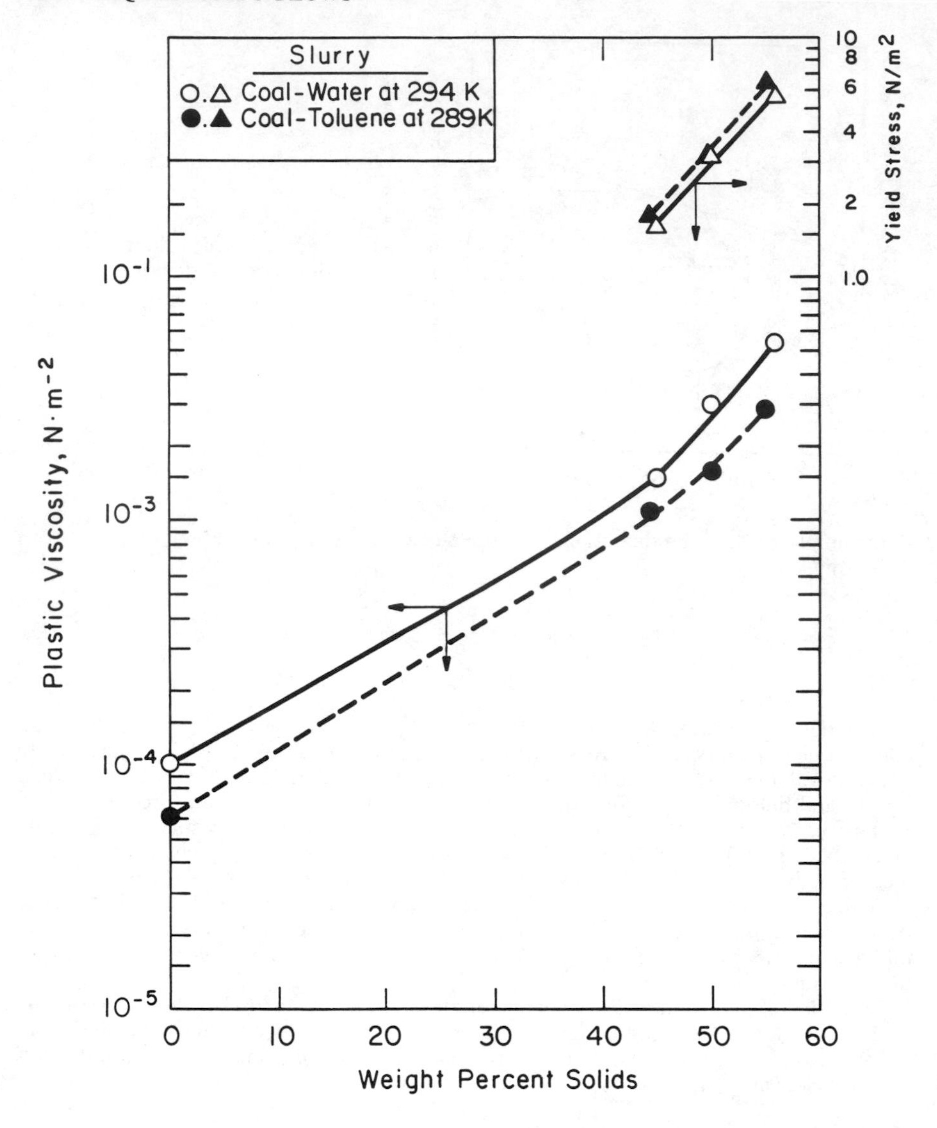

Figure 6. Bingham plastic behavior of coal-liquid suspensions [2].

Prandtl-Eyring:

$$\mu_a = A\ \mathrm{Sinh}^{-1}\left(\frac{du/dy}{B}\right) \Big/ \left(\frac{du}{dy}\right) \tag{7}$$

Ellis:

$$\mu_a = \frac{1}{\left[\frac{1}{\mu_d} + \left(\frac{1}{K}\right)^{\frac{1}{n}} \tau^{\alpha - 1}\right]} \tag{8}$$

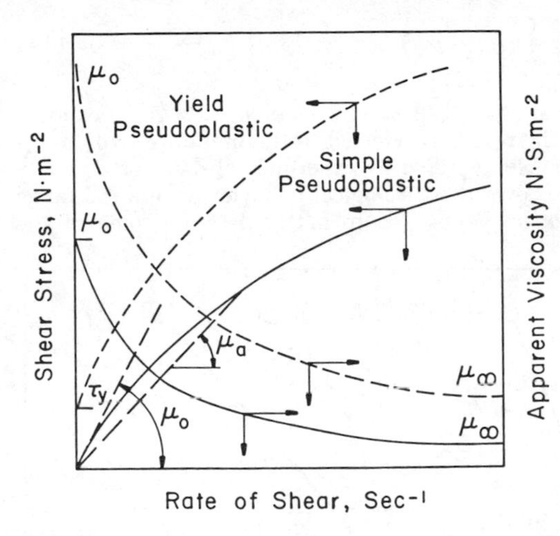

Figure 7. Typical flow curves for simple and yield pseudoplastic fluids.

Reiner-Philippoff:

$$\mu_a = \mu_\infty + \frac{\mu_o + \mu_\infty}{1 + \left(\frac{\tau}{\tau_s}\right)^2} \tag{9}$$

Sisko:

$$\mu_a = a + b\left(\frac{du}{dy}\right)^{c-1} \tag{10}$$

Cross:

$$\mu_a = \mu_\infty + \frac{\mu_o - \mu_\infty}{1 + 2\left(\frac{du}{dy}\right)^{2/3}} \tag{11}$$

Meter:

$$\mu_a = \mu_\infty + \frac{\mu_o + \mu_\infty}{1 + \left(\frac{\tau}{\tau_m}\right)^{\alpha - 1}} \tag{12}$$

Herschel-Bulkley (yield-pseudoplastic):

$$\mu_a = \frac{\tau_y}{\left(\frac{du}{dy}\right)} + K\left(\frac{du}{dy}\right)^{n-1} \tag{13}$$

Some of the notations used in the above equations are defined in Figure 7. The coefficients K, A, a and b are generally referred to as consistency indices, which characterize the consistency or thickness of a fluid and are analogous to the viscosity of a Newtonian fluid. The exponents, n, α and c are the flow behavior indices that characterize the extent of the deviation of a fluid from Newtonian fluid behavior. μ_o and μ_∞ represent the apparent viscosity at zero and infinite (very large) rate of shear and are constants related to the shear stability of the structure. The mean shear stress, τ_m, is defined as the shear stress at a point where the apparent viscosity

$$\mu_a = \frac{1}{2}(\mu_o + \mu_\infty) \tag{14}$$

and can be approximated by the value at $\mu_a = 1/2\ \mu_o$ in case of $\mu_o >> \mu_\infty$. For more detailed treatments of these equations, the reader is referred to the literature [3,4].

Although each of these equations has been verified by experimental data for a variety of pseudoplastic fluids, none was able to cover the rheological behavior for the full range of pseudoplastics. Instead, one equation may be used to better describe some of the

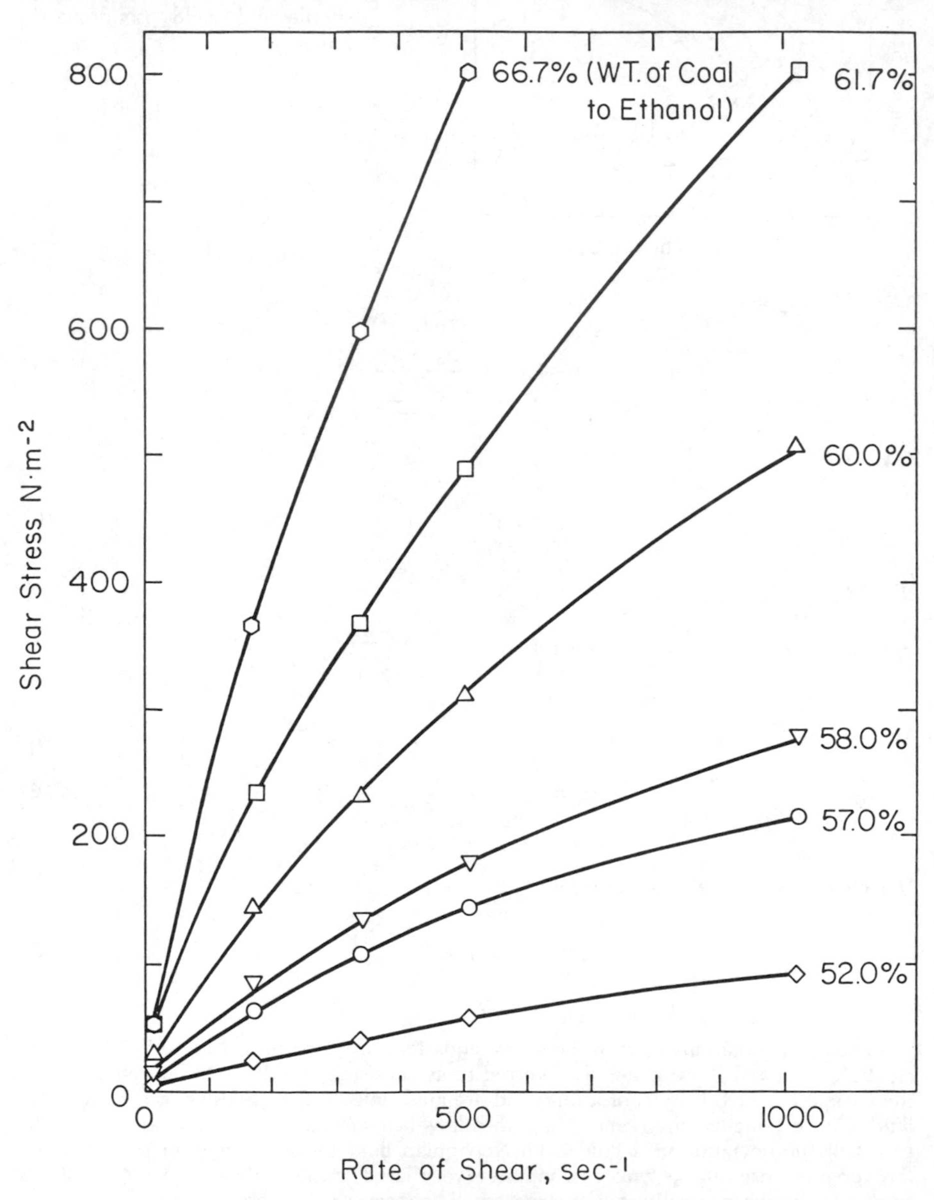

Figure 8. Pseudoplastic behavior of coal-ethanol suspensions [5].

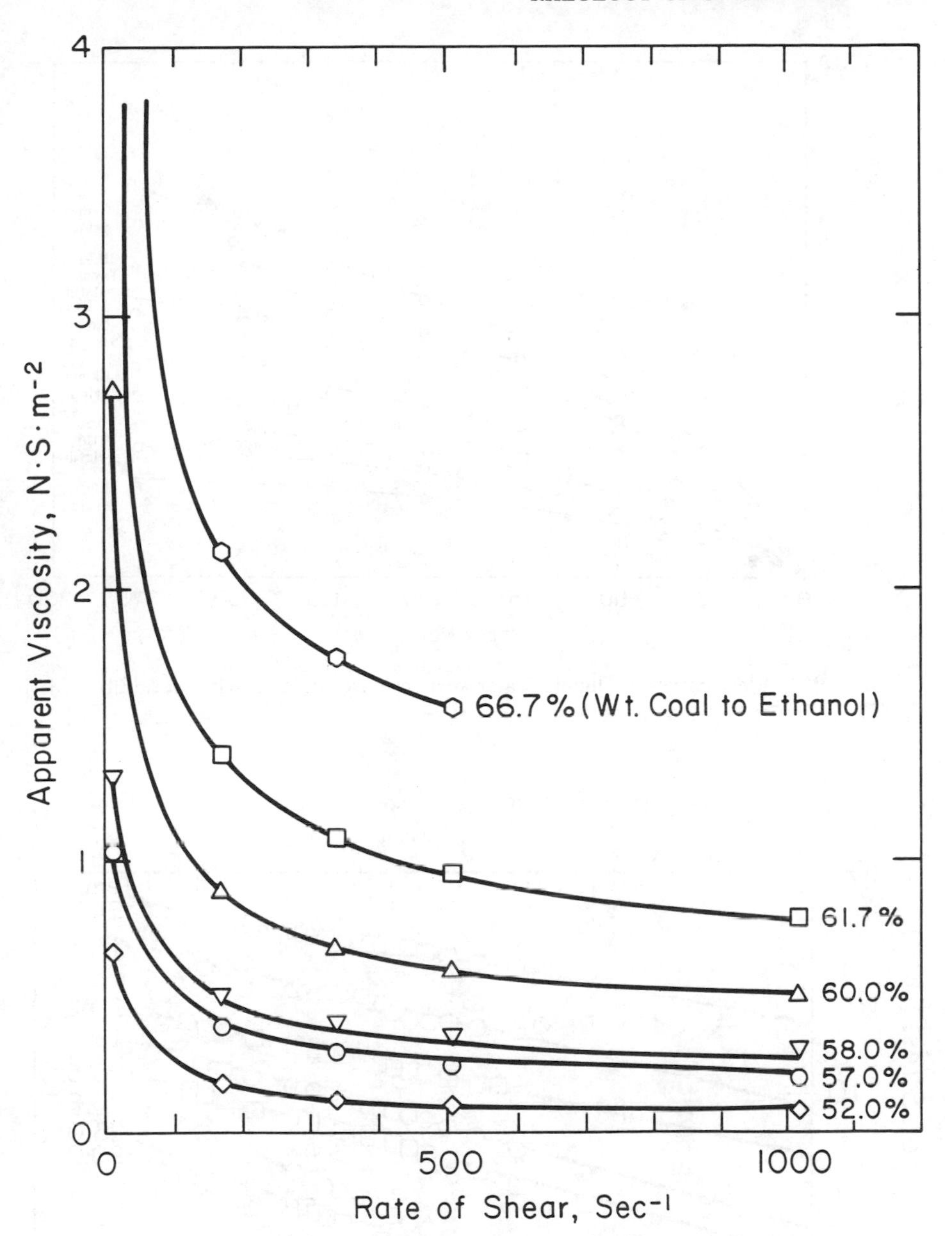

Figure 9. Apparent viscosity of coal-ethanol suspensions [5].

pseudoplastic fluids than the others while being unsuitable for use in some other fluids. The selection therefore must be left to individual engineers and researchers.

Among these equations, however, the power law was found to be most acceptable for use in describing most of the common pseudoplastic fluids. It is also the simplest to use and is widely applied for engineering calculations of pseudoplastic fluid flow.

The power law equation is expressed as

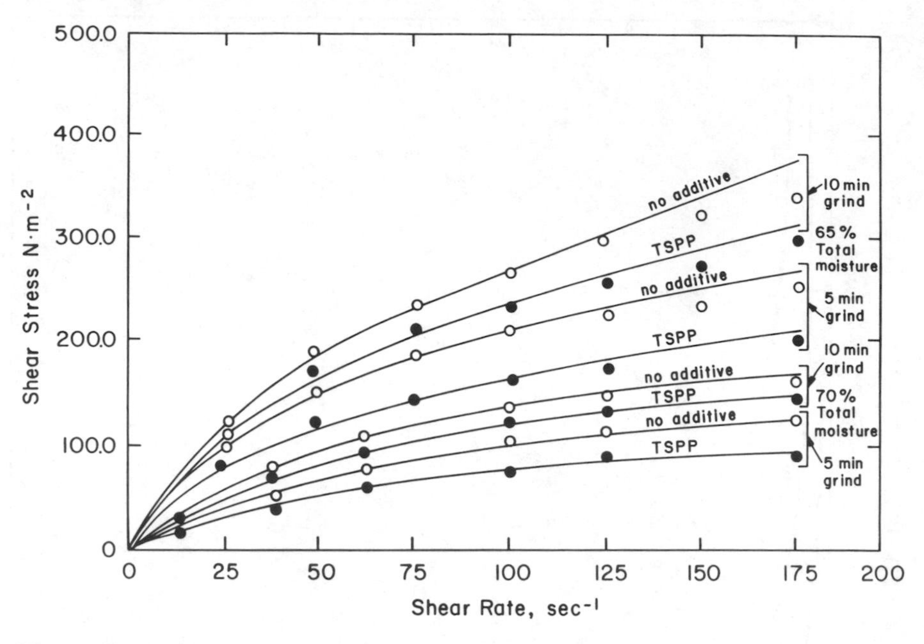

Figure 10. Flow curves for lignite-water suspensions with and without additives [6].

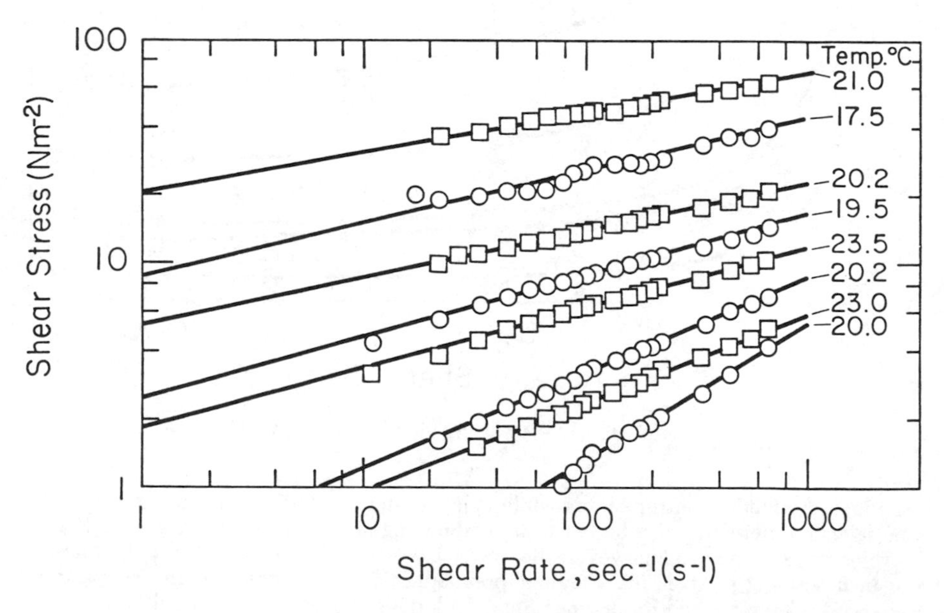

Figure 11. Logarithmic plot of pseudoplastic metalliferous mud flow curves [7].

$$\tau = K\left(\frac{du}{dy}\right)^n \tag{15}$$

Since the flow behavior index, n, has values less than unity, a logarithmic plot of Equation 15 will appear as a straight line having a slope less than 45°. The consistency index, K, has the value of shear stress, τ, at the point at which the straight line intersects the cycle representing the rate of shear, du/dy = 1.

Examples of materials that exhibit pseudoplastic behavior include polymeric solutions or melts (such as cellulose acetate, rubbers and napalm), paper pulps, paints, detergent slurries and some other liquid suspensions of fine inert unsoluted solids, such as coal-ethanol (Figures 8 and 9) [5], coal-methanol mixtures, fine coal or lignite in water (Figure 10) [6] and metalliferous mud (Figure 11) [7].

Simple and Yield Dilatant Fluids

The mechanism that may have caused the fluid dilatancy has been presented in an earlier section based on the Reynolds hypothesis. In a more trivial way, the dilatant fluid is defined as a fluid for which a shear stress vs rate of shear plot is concave toward the stress, as shown in Figure 12.

It is commonly recognized that dilatancy can be clearly observed only in suspensions of high concentration. However, the work of Metzner and Whitlock [8] indicated that the above condition appears to be necessary but may not be sufficient for observation of dilatancy. The behavior is likely to occur when high shear stress also is encountered. To illustrate the point, they suggested a general shear stress-shear rate relationship for concentrated suspensions, as shown in Figure 13.

The flow curve for the concentrated suspension in Figure 13 indicated a Newtonian range behavior at low shear rate. As the rate of shear increases, the corresponding shear stress increases at a lesser rate and demonstrates pseudoplastic behavior. Further increases in shear rate are likely to lead to the observation of dilatancy.

Using suspensions of water and titanium dioxide mixtures in their experimental work, they obtained results as shown in Figure 14, which clearly demonstrated the effect of concentration on the degree of dilatancy. Similar investigations were made by Govier et al. [9] using concentrated water-galena suspensions with mean particle sizes of 20–30 μ. Their results are shown in Figure 15.

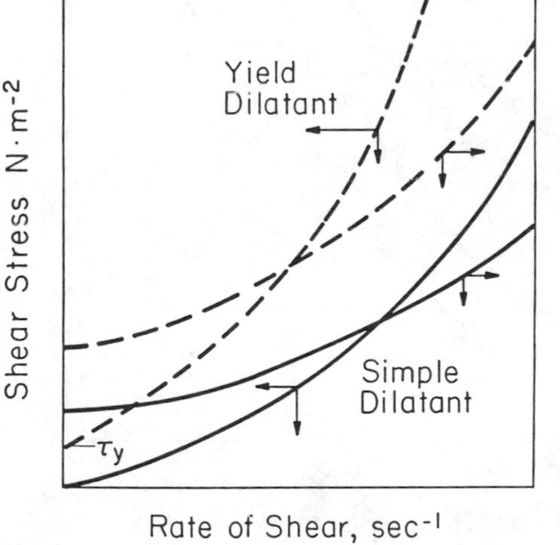

Figure 12. Flow curves for simple and yield dilatant fluids.

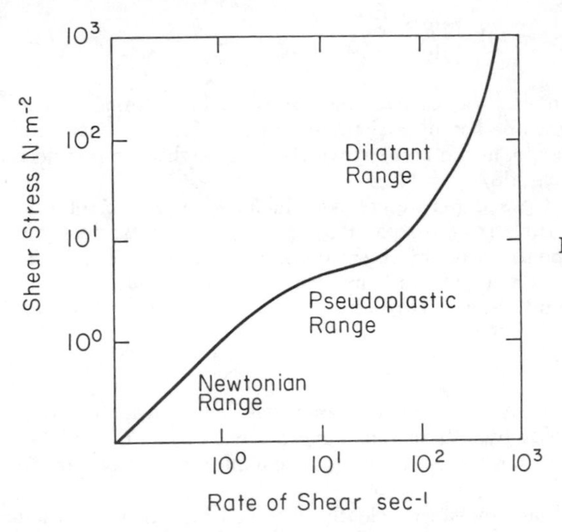

Figure 13. Ranges of flow behavior of concentrated suspensions [8].

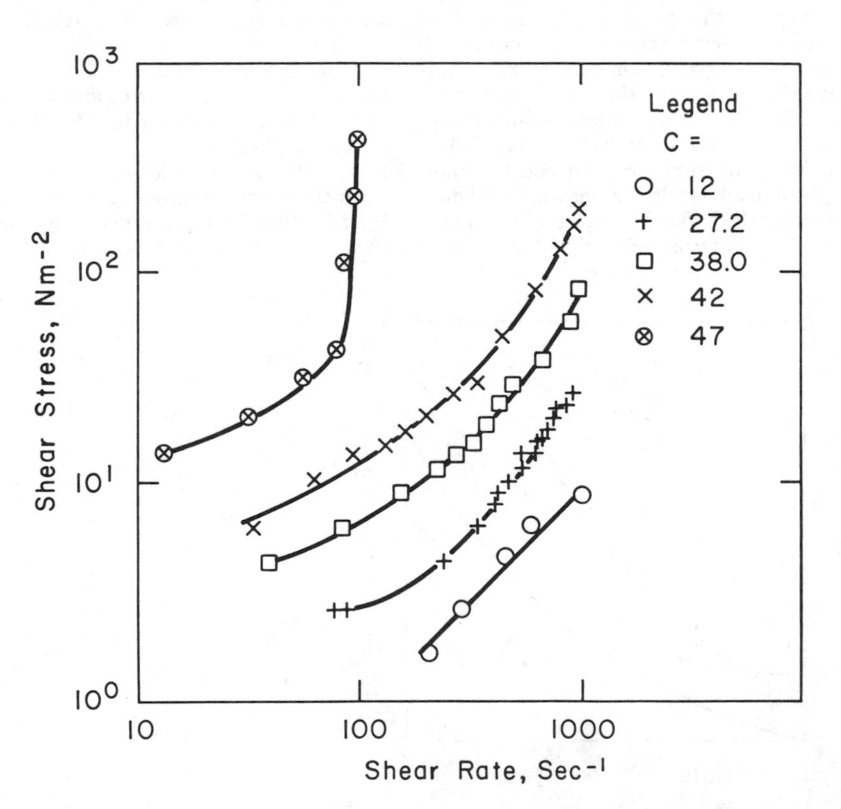

Figure 14. Effect of volume concentration of solids on dilatant behavior of suspensions [8].

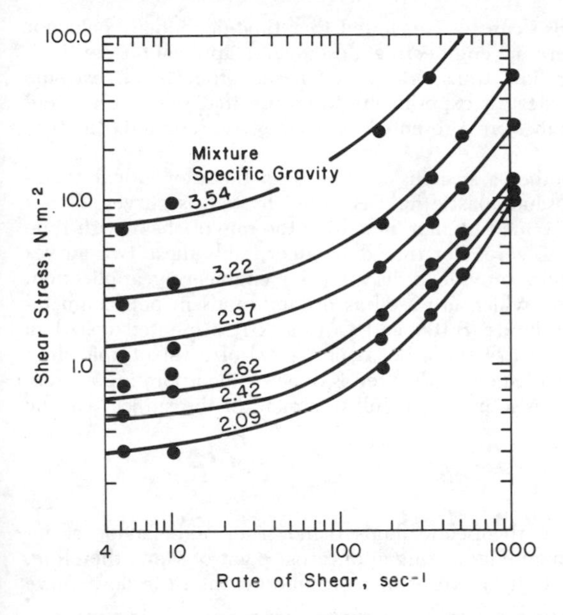

Figure 15. Dilatant behavior of concentrated water-galena suspensions [9].

Time-Dependent Viscous Fluids

In a liquid suspension of nonspherical solids particles, the resulting viscous fluid possesses a given structural configuration, and solids particles are likely to have near random orientation. Both the structural configuration and the particle orientation are sensitive to shear. As a result of flow, realignment of particles takes place. The gradual change of the structural configuration and solids particle orientation will result in a corresponding variation of the rheological behavior of the fluid and, thus, a time-dependent shear stress vs shear rate relationship.

Accepting particle realignment as a cause of pseudoplastic and dilatant behavior and assuming that no irreversible change in either suspended particles or the suspending liquid will occur, it is logical to expect that such time-dependent phenomena will not continue indefinitely. Instead, an equilibrium condition will arrive and the rheological behavior once again will become independent of time.

For the equilibrium condition to result in improved alignment of particles, reduced resistance to flow will be observed, manifesting itself in decreased shear stress with time at a constant rate of shear. Such materials are classified as thixotropic fluids.

Analogous to that change of the void ratio that leads to dilatant behavior and, in addition, the occurrence of particle agglomeration causing limited irreversible structural change, some materials may demonstrate an increase in shear stress with time at a given rate of shear. This can be a result of strengthening the structural reformation property on its course to reach a new equilibrium condition. The behavior is then referred to as rheopexy, and materials demonstrating such behavior are defined as rheopectic fluids.

Thixotropic Fluids

A great many fluids consisting of liquid-solids suspensions can be classified into this category. However, most will require only a very short transition time before reaching equilibrium. Under such circumstances, the fluids are readily regarded as pseudoplastic instead.

Based on these observations, it also can be concluded that time-dependent behavior usually does not cause great concern in engineering design and application for it is unlikely to occur in relatively long flow transport lines. On the other hand, extreme caution should be given to bench-scale test experiments to ensure that the results were not recorded during a transition but were presented as if they represented the true equilibrium flow behavior.

One quick way to determine whether a suspension may be a time-dependent thixotropic fluid or a time-independent pseudoplastic fluid is to use hysteresis curves in that the shear stress-shear rate curve is obtained by first increasing the rate of shear with time and then decreasing gradually back to zero. For time-dependent fluids these two curves do not coincide; however, these results provide little value for engineering application.

Water and bentonite clay mixtures, which are used as drilling muds in petroleum industries, are examples of thixotropic fluids. Ritter and Govier [10] presented test data of crude oil from Alberta and found it to have a behavior typical of a thixotropic fluid (Figure 16). Mathematical modeling of such fluids often is done by incorporating a time-dependent variable into the pseudoplastic model. A full treatment of the subject can be found in the literature [3,4].

Rheopectic Fluids

As described in an earlier section, rheopectic fluids demonstrate a behavior of increasing shear stress with time. From an engineering application point of view, therefore, a shear stress vs time plot at constant shear rate is of more interest than the flow curve

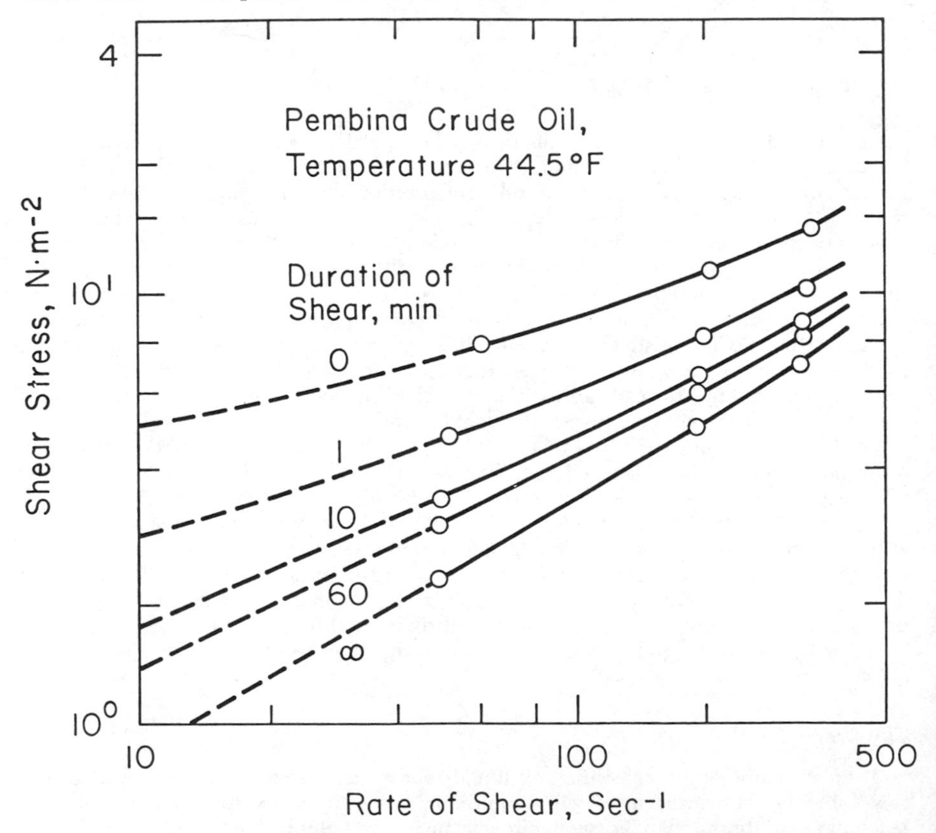

Figure 16. Thixotropic behavior of Pembina crude oil [10].

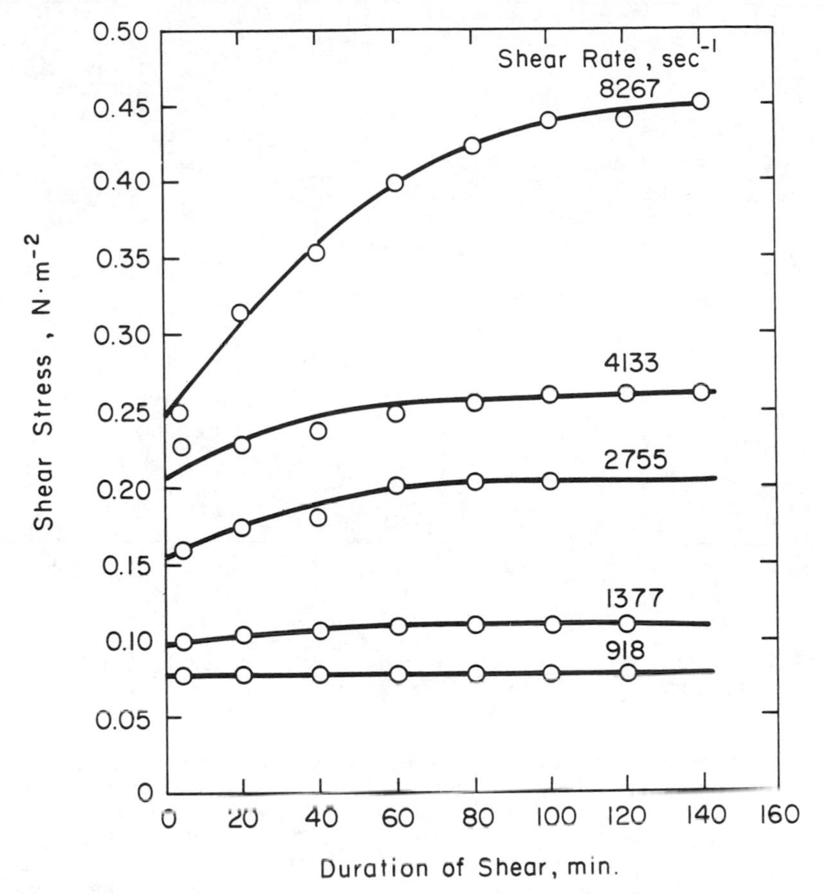

Figure 17. Rheopathic behavior of 2000 molecular-weight-saturated polyester [11].

plotting the shear stress vs shear rate relationship. The former provides information on the possible maximum equilibrium shear stress for a designated flow condition.

Rheopectic behavior occurs less commonly in liquid-solids mixtures. Studying the structural buildup phenomena, Steg and Katz [11] used 2000 molecular-weight-saturated polyester as testing fluid and found it to be highly rheopectic at high shear rate ($du/dy = 8267\ sec^{-1}$) (Figure 17). On the other hand, the fluid appears time independent at lower shear rate ($du/dy < 1377\ sec^{-1}$). The degree of time dependency between these observed limiting values was found to be directly proportional to the magnitude of the shear rate.

Viscoelastic Fluids

Viscoelastic fluids exhibit the effect of partial elastic recovery of the resulting deformation on removal of the shear stress, which caused the deformation. The viscous properties of a viscoelastic fluid may be non-Newtonian time independent or time dependent. All pseudohomogeneous liquid-solids mixtures demonstrate some degree of viscoelastic behavior at low shear rate range near the initial motion period.

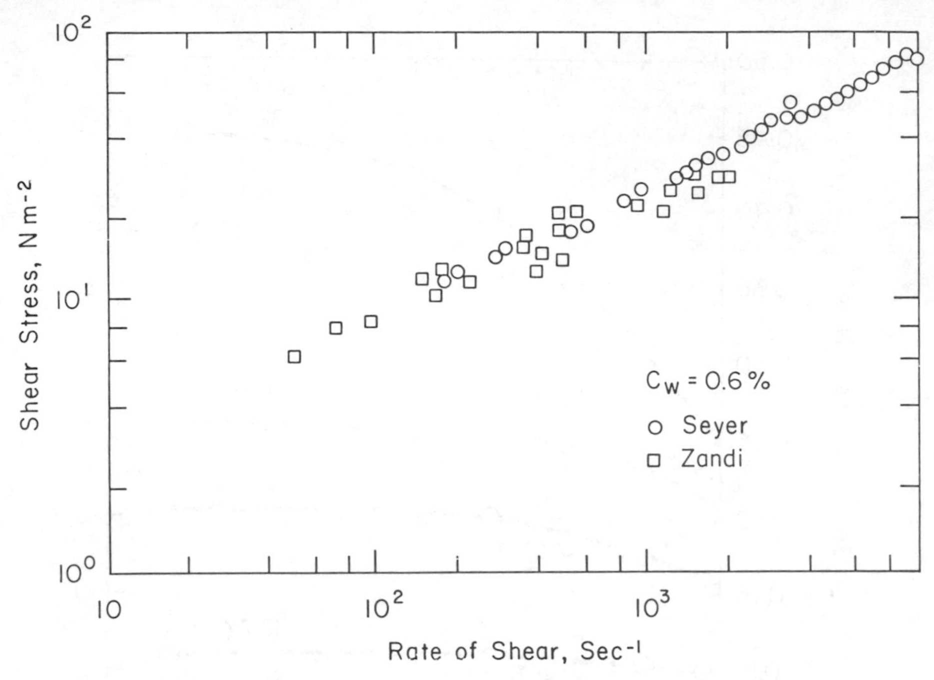

Figure 18. Rheogram of a viscoelastic fluid consisting of a water solution of high-molecular-weight chain polymer, ET-597 [12,13].

In terms of engineering application, viscoelastic behavior was not considered important except in a few special cases, including sudden change in flowrate, such as during startup and shutdown periods. Sudden change of flow direction, such as flow near entrance of a pipe from a reservoir, also will help to illustrate the importance of viscoelastic behavior of a fluid. The influence of the viscoelastic property also will be felt in a rapid oscillatory flow system. Special considerations are usually given to the viscoelastic effect during a flow process involving extrusion or very high local shear rate situations.

The effect of viscoelasticity on the pressure drop of a turbulent flow and flow near the entrance to a pipe system were observed by Seyer [12] and Zandi and Razaghi [13]. Using a water solution of a high-molecular-weight chain polymer (ET-597, product of Dow Chemical), they observed the fluid to exhibit strong viscoelastic behavior. A rheogram for 0.6% ET-597 solution, as shown in Figure 18, is plotted on their experimental results.

Although the viscoelastic effect normally is considered unimportant for steady flow, its effect is also rather insignificant for transient flow over a long process time or during oscillating flow over a long period. However, high-molecular-weight polymeric solutions do have a considerable effect on turbulent flow in a pipe system.

DETERMINATION OF FLOW RHEOLOGY

Measurement of viscosity is one of the major tasks involved in the work of determining the rheological properties of a fluid. The objective is usually accomplished in an indirect manner. Direct laboratory measurements are taken to record simultaneously the shear stress (torque or pressure drop) and rate of shear (at a given rate of angular deformation). By dividing the latter into the former, a value for viscosity or apparent viscosity is obtained for Newtonian or non-Newtonian fluids, respectively.

Measurement Techniques

Three major types of instruments making use of two different basic flow principles are commonly used to conduct these experiments. They are:

1. Hagen-Poiseuille flow: capillary viscometer;
2. couette flow: concentric cylinder viscometer; and
3. couette flow: cone-and-plate viscometer.

A capillary viscometer (Figure 19a) has the advantage of simplicity for it can be constructed with no moving parts involved in the meter itself. The two major components are the reservoir and a capillary tube. Pressure head differences are measured along the capillary tube at two or three different points of known distances apart. Recordings of the flowrate and the pressure drop give information for rheological property computation. The disadvantage of the system is its inability to produce a long running time using the same sample without resulting in some change of the flow parameter. It is definitely not very suitable for time-dependent non-Newtonian fluid flow measurements.

Couette flow refers to flow between two parallel plates induced by the motion of one or both plates. A practical way of producing such a flow system is to employ two concentric cylinders with one moving and the other stationary (Figure 19b). The fluid-filled annular space between the two cylinders constitutes a continuous couette flow system. Measurements of the rotational speed of the moving cylinder and the resulting torque applied to the stationary cylinder provide data for rheological property determination of the fluid involved. The advantage of this system lies in its ability to process a broad range of data during a relatively long time period using the same sample. To minimize the end effect, a fixed circular plate is inserted near the bottom between the moving and stationary cylinders, as shown in Figure 19b.

The cone-and-plate viscometer consists mainly of an inverted cone and a circular plate (Figure 19c). To produce couette flow in the small space between the cone and the plate, the angle of the cone is normally made less than 3°. This viscometer, like the concentric cylinder system, is suitable for use in repeated measurements on a single sample of Newtonian or non-Newtonian fluids at low to moderate shear rates. Measurements of rotational speed of the circular disk and the exerted torque on the inverted cone permit the computation of shear stress and rate of shear and, in turn, determination of the rheological behavior of the fluid. The advantages of the system are the absence of the secondary current and ease in data interpretation.

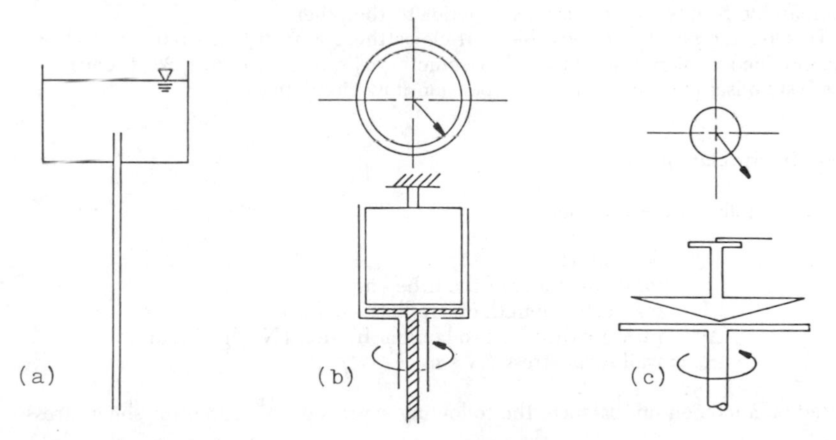

Figure 19. Schematics of capillary tube, concentric cylinder and cone-and-plate viscometers.

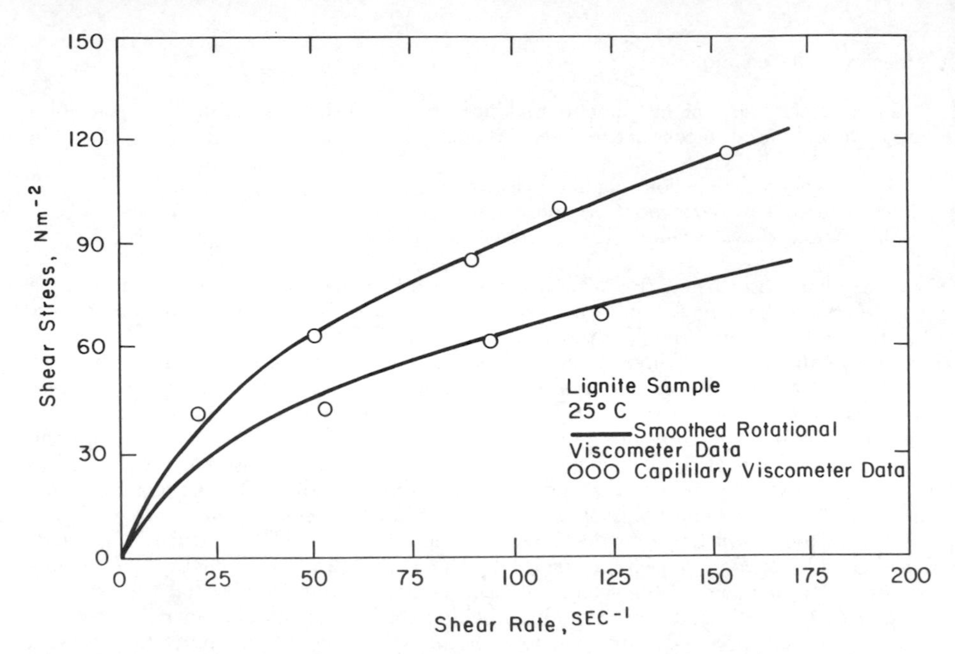

Figure 20. Comparison of rotational and capillary viscometer data of lignite-water suspensions [6].

Since viscosity varies with temperature of the fluid, all measuring instruments as described above are usually equipped with a constant-temperature bath. Should the concern of particle settling in a liquid-solids suspension arise, a small agitator also may be integrated into the system to ensure homogeneous solids dispersion.

Comparison of results obtained from capillary (Hagen-Poiseuille flow) or rotational viscometer (couette flow) were made by Sadler and Senapati [6] using lignite-water mixtures and by Pouska and Link [1] using a water suspension of mineral particles, as shown in Figures 20 and 21, respectively. It can be seen from these results and also as indicated based on past experiences of others that the two basic types of viscometer produce comparable results. Neither seems superior to the other.

To solve the problem of possible particle settling, Ferrini and Ercolani [14] presented the modified version of a rotational viscometer, as shown in Figure 22. Incorporated into the system is a pumping unit for suspension flow circulation.

Data Interpretation

(a) Capillary tube viscometer:

Letting Q = flowrate ($m^3 \cdot T^{-1}$)
D = inside diameter of the tube (m)
L = test section length of capillary (m)
Δp = pressure drop due to laminar friction ($N \cdot m^{-2}$), and
τ_w = wall shear stress ($N \cdot m^{-2}$)

Based on a momentum balance, the following expression is written for shear stress:

$$\tau_w = \frac{D\Delta p}{4L} \tag{16}$$

The rate of shear, $-du/dr$, at the capillary wall is determined from the general equation of Rabinowitsch [15] and Mooney [16] as

$$\left(-\frac{du}{dr}\right)_w = \left(\frac{3n' + 1}{4n'}\right)\frac{8V}{D} \tag{17}$$

where

$$n' = \frac{d\left(ln\,\frac{D\Delta p}{4L}\right)}{d\left(ln\,\frac{8V}{D}\right)} \tag{18}$$

is the slope of the logarithmic plot of $D\Delta p/4L$ vs $8V/D$. The minus sign indicates that r is measured from the center toward the tube wall. A logarithmic plot of $D\Delta p/4L$ vs $8V/D$, as shown in Figure 21, gives the flow curve of the fluid relating the shear stress to the rate of shear.

(b) Concentric Cylinder:

Letting D_1 = diameter of inner cylinder (m)
D_2 = diameter of outer cylinder (m)
θ = angle of torque spring deflection (rad)
K = spring constant ($N\text{-}m\text{-}rad^{-1}$)
h = height of stationary cylinder (m)
N = speed of rotating cylinder (sec^{-1})
τ_{D_1} = shear stress on stationary cylinder

we can express the shear stress in terms of the system parameters, also based on a momentum balance, as

$$\tau_{D_1} = \frac{2K\theta}{\pi D_1^2 h} \tag{19}$$

and the rate of shear as

$$\frac{du}{dr} = \frac{4\pi N}{1 - S^2}\,F_{KM} \tag{20}$$

where $S = D_2/D_1$, and F_{KM} is a correction factor proposed by Krieger and Maron [17] relating the rate of shear and the deviation of system geometry from plane parallel plates:

$$F_{KM} = 1 + \frac{S^2 - 1}{2S^2}\left(1 + \frac{2}{3}\,ln\,S^2\right)\left(\frac{1}{n''} - 1\right) + \frac{S^2 - 1}{6S^2}\,ln\left[\left(\frac{1}{n''} - 1\right)^2 + \frac{d\left(\frac{1}{n''} - 1\right)}{d\log(K\theta)}\right] \tag{21}$$

where $n'' = d\,ln(K\theta)/d\,ln N$ is the slope of the logarithmic plot of the torque, $K\theta$, vs the rotational speed, N.

(c) Cone-and-plate:

Letting D = diameter of the base of the cone (m)
ϕ = angle between the cone and the plate (rad)
N = speed of rotation (sec^{-1})
θ = angle of torque spring deflection (rad)
K = spring constant ($N\text{-}m\text{-}rad^{-1}$)

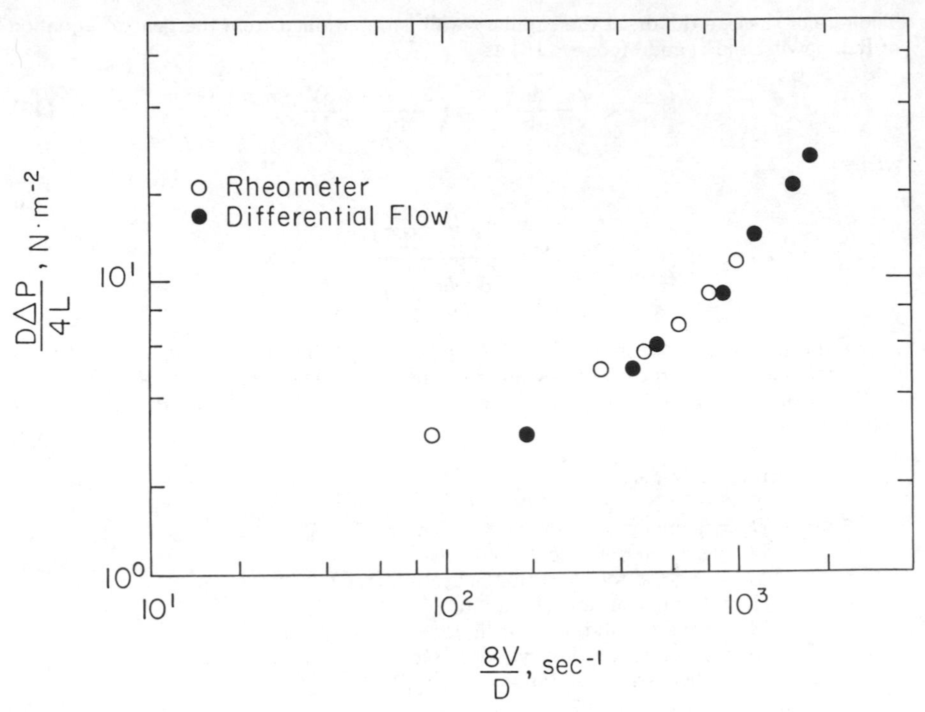

Figure 21. Comparison of rheometer and differential (pipe flow) viscometer data of mineral-water suspensions [1].

and carrying out integration over the surface under the projection of the cone, we have

$$\tau = \frac{3K\theta}{2\pi\left(\frac{D}{2}\right)^3} \tag{22}$$

For the rate of shear evaluation, we have

$$\frac{du}{dr} = \frac{2\pi rN}{r \tan \phi} \cong \frac{2\pi N}{\phi} \tag{23}$$

as $\tan \phi \cong \phi$ for small angle ϕ.

Criteria for Preliminary Liquid-Solids Mixture Classification

Assuming Durand's [18] results using sand in water, they can be extended to fluid suspensions of fine solids in general. Duckworth [19] suggested the use of particle settling velocity Reynolds number, Re_p, as a criterion for suspended flow classification, in which the fluid is as follows (Table I):

where

$$Re_p = \frac{\rho V_0 d}{\mu} \tag{24}$$

For spherically shaped particles the terminal settling velocity, V_0, can be written as

$$V_0 = \frac{4dg\left(\frac{\rho_p}{\rho} - 1\right)}{3\,C_D} \tag{25}$$

where ρ_p and ρ are mass densities of particle and liquid, respectively, and d is the particle diameter. Duckworth further noted that the above equation may be transformed into dimensionless form as

Table I. Range of Particle Settling Velocity Reynolds Numbers for Different Fluids

Non-Newtonian	$Re_p < 0.02$
Transition Region	$0.02 < Re_p < 0.1$
Non-Settling Newtonian	$0.1 < Re_p < 0.8$
Settling Suspension	$0.8 < Re_p < 525$
Settling with Saltation	$525 < Re_p$

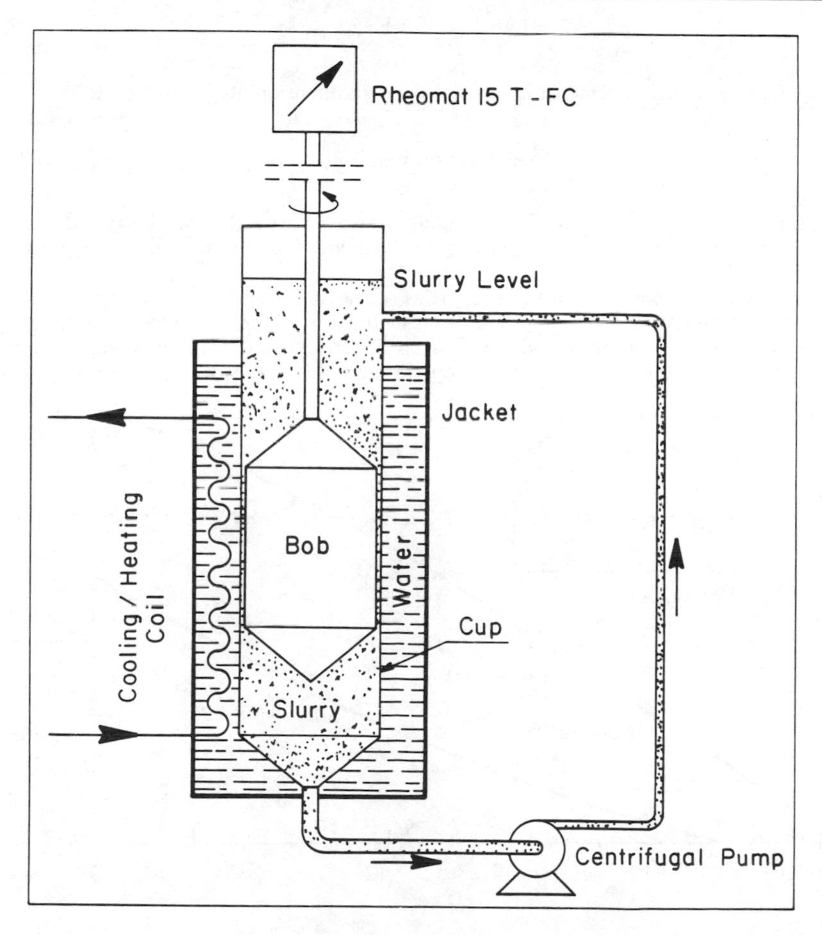

Figure 22. Modified rotational viscometer [14].

$$\frac{d}{\Delta} = (Re_p \cdot C_D)^{1/3} \qquad (26)$$

with

$$\Delta = \left[3\nu^2/4g\left(\frac{\rho_p}{\rho} - 1\right)\right]^{1/3} \qquad (27)$$

where ν is the kinematic viscosity of the fluid. The drag coefficient, C_D, of a submerged particle is known to be a function of the particle Reynolds number, Re_p. Therefore, Equation 26 can be regarded as

$$\frac{d}{\Delta} = \text{funct.}(Re_p) \qquad (28)$$

Thus, for a given particle shape, such as a sphere, a curve relating d/Δ and Re can be plotted as shown in Figure 23.

In a similar manner, letting

$$V' = \left[4g\left(\frac{\rho_p}{\rho} - 1\right)\nu/3\right]^{1/3} \qquad (29)$$

Equation 25 may be transformed into an alternative dimensionless form such that

$$\frac{V_0}{V'} = (Re_p/C_D)^{1/3} \qquad (30)$$

Taking the same steps as used for Equation 28, a single curve also can be plotted for a given particle shape in terms of V_0/V' vs Re_p. Figure 23 shows such a curve for a spherical particle.

In practical applications, when the properties of the solids and the fluid are specified, the values for Δ and V' can be computed using Equations 27 and 29. Therefore, locating the point representing the magnitude of d/Δ on Figure 23, the particle Reynolds number is determined. Using this Reynolds number and the curve of V_0/V' vs Re_p, one

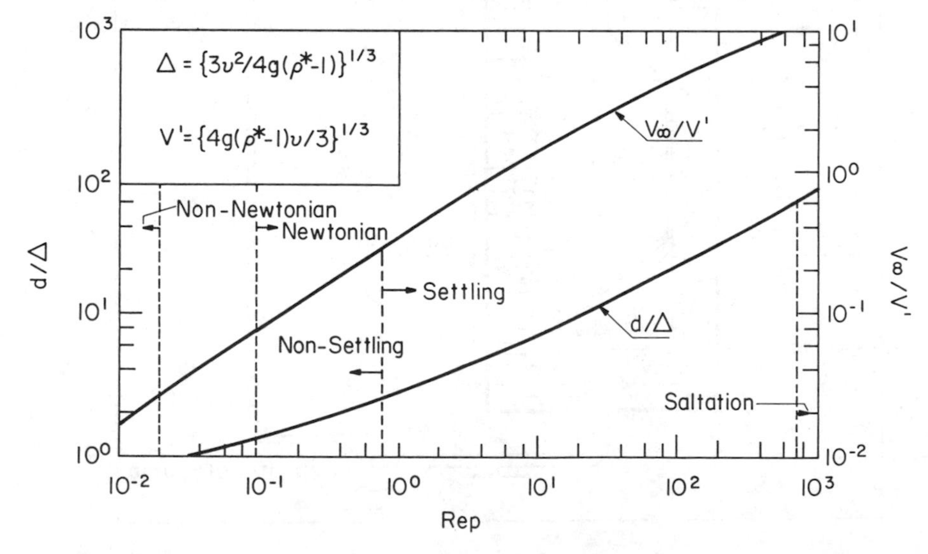

Figure 23. Dimensionless curves for preliminary mixture classification [19].

can read off directly from the second curve a value for V_0/V', from which the settling velocity, V_0, may be calculated.

For the purpose of flow classification, the proposed criteria using the particle Reynolds number as listed in Table I can be used. These criteria also are depicted in Figure 23.

It should be noted, however, that the settling velocity, V_0, obtained using a d/Δ vs Re_p curve is for a single spherical particle falling in an infinite fluid. Hence, the particle Reynolds, Re_p, determined in such a way can be used only to describe the motion of a single particle through an infinite fluid region.

When the effect of concentration is taken into consideration, the magnitude of the settling velocity of a particle is somewhat reduced. By applying Equation 31 below, the hindered settling velocity, V_{oh}, of the particle may be computed as a function of volumetric concentration, C:

$$\frac{V_{oh}}{V_0} = e^{-5.9\,C} \tag{31}$$

With the hindered settling velocity, a new particle Reynolds number, $Re_{ph} = \rho\, V_{oh}\, d/\mu$, can be computed and used to obtain an alternative classification. This procedure can be used consecutively until no significant change in V_0 is noticed. The preliminary classification of the suspension is thus accomplished.

Viscosity of Dilute Liquid-Solids Suspensions

For a Newtonian liquid-solids suspension with uniformly suspended noninteractive spherical particles, the viscosity of the mixture is determined by the viscosity of the carrier liquid and the concentration and size distribution of the solids.

To determine the viscosity of a suspension, various equations were proposed. Einstein [20,21] proposed

$$\mu_m = \mu(1 + 2.5C) \tag{32}$$

for low volumetric concentration, C, and

$$\mu_m = \mu(1 + 4.5C) \tag{33}$$

for high C values.

Extended from these equations, Thomas [22] proposed the empirical equation

$$\mu_m = \mu[1 + 2.5C + 10.05C^2 + 0.00273e^{16.6C}] \tag{34}$$

which offers a reasonably good prediction over the range of all available data.

EFFECT OF RHEOLOGY ON MECHANICS OF SUSPENSION FLOW

Mechanics of Flow of Bingham Plastic Fluids in Pipes

As expressed in Equation 4, the relationship between the shear stress and shear rate for Bingham fluids is

$$\tau - \tau_y = \eta\left(\frac{du}{dy}\right) \tag{4}$$

Buckingham [23] integrated this rheological equation for isothermal flow of a Bingham plastic fluid in pipes and obtained:

$$\frac{8V}{D} = \frac{\tau_w}{\mu}\left[1 - \frac{4}{3}\frac{\tau_y}{\tau_w} + \frac{1}{3}\left(\frac{\tau_y}{\tau_w}\right)^4\right] \tag{35}$$

where V is the mean velocity, D is diameter of pipe, τ_w is the wall shear, η is the coefficient of rigidity, and τ_y is the yield stress.

By applying the relationship between the wall shear stress and pressure drop, the Buckingham equation becomes:

$$\frac{8V}{D} = \left(\frac{D\Delta p}{4L}\right)\frac{1}{\eta}\left[1 - \frac{4}{3}\tau_y\left(\frac{4L}{D\Delta p}\right) + \frac{\tau_y{}^4}{3}\left(\frac{4L}{D\Delta p}\right)\right] \tag{36}$$

Knowing that the pressure drop, Δp, for Bingham plastics varies with the system parameters similar to that for Newtonian fluid flow, and

$$p = \text{funct}\ (D, L, V, \rho, \mu, \tau_y) \tag{37}$$

The resulting nondimensional equation produced by using dimensional analysis procedure gives

$$\frac{\Delta p}{\rho V^2} = \text{funct}\left[\frac{DV\rho}{\eta}, \frac{\tau_y D^2 \rho}{\eta^2}\right] \tag{38}$$

The left-hand side of Equation 38 may be replaced by Fanning's friction coefficient, f.

The foregoing derivations indicate that the pressure drop computation equations, such as the Hagen-Poiseuille and Darcy-Weisbach equations for Newtonian fluid flow, also may be used for flow systems of Bingham plastics. However, Equation 38 also suggests that the Fanning's friction coefficient for Bingham plastics is a function of two dimensionless groups for a given flow system, instead of one.

The first dimensionless group is the well known Reynolds number, Re. The second one has been termed as the Hedstrom number, He, in honor of the man who first proposed its use.

A plot of the Fanning friction coefficient vs the Reynolds number, Re, for constant Hedstrom number, He, is shown in Figure 24 and was originally proposed by Hanks

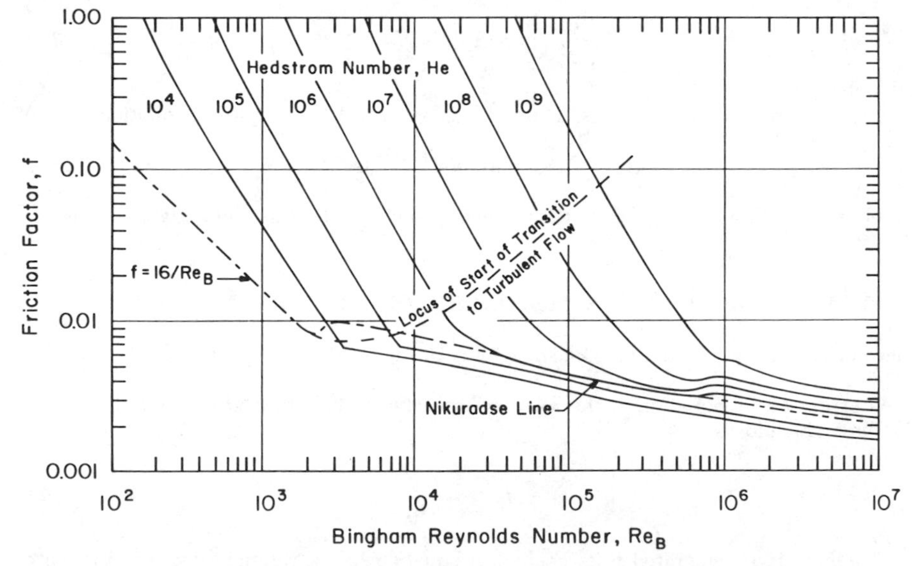

Figure 24. Friction factor correlation for Bingham plastic fluids [24].

Table II. Rheological Properties of Different Suspensions [26,28]

	Yield Stress		Coefficient of Rigidity	
	(Nm^{-2})	($lbf\text{-}ft^{-2}$)	($N\text{-}sec\text{-}m^{-2}$)	($lbf\text{-}sec\text{-}ft^{-2}$)
20% Fly Ash in Water	0.2	4.2×10^{-3}	0.003	0.063×10^{-3}
Chalk Slurry in Water				
41% by weight	4.0	84.0×10^{-3}	0.003	0.063×10^{-3}
55% by weight	23.0	480×10^{-3}	0.02	0.42×10^{-3}
Coal, in Oil of Kinematic Viscosity of 100 cSt, Mean Size 115 μm				
50% by weight	7.5	157×10^{-3}	3.6	75×10^{-3}
60% by weight	21.0	440×10^{-3}	7.5	157×10^{-3}
Coal in Water Mean Size 150 μm				
55% by weight		100×10^{-3}	0.03	0.63×10^{-3}
65% by weight		$610–810 \times 10^{-3}$	1.02–1.05	$2.5–3.1 \times 10^{-3}$

and Dadia [24]. Rheological properties of several liquid-solids suspensions that exhibit characteristics of Bingham plastics are listed in the Table II.

Metzner-Reed Generalized Approach To Power Law Fluid Flow

The empirical power law equation used to describe the shear stress vs rate of shear relationship of pseudoplastic fluids is

$$\tau = K\left(\frac{du}{dy}\right)^{n} \tag{15}$$

where K is the consistency index, and n is the flow-behavior index. This equation is strictly for laminar flow applications. Much work done based on this relationship was found to be inapplicable to flow outside the laminar region or in regions in which the onset of turbulence occurs.

Based on the Rabinowitsch-Mooney equation, Metzner and Reed [25] developed a generalized friction factor correlation to give

$$\left(\frac{du}{dy}\right)_{w} = \left(\frac{1+3n'}{4n'}\right)\frac{8V}{D} \tag{39}$$

where

$$n' = \frac{d\left(\text{Ln}\,\frac{D\Delta p}{4L}\right)}{d\left(\text{Ln}\,\frac{8V}{D}\right)} \tag{40}$$

For a given value of 8V/D, the slope of the tangent line gives the value of n′, and the intersection of the line with the axis of 8V/D = 1 yields a value of K′. Therefore, it is permissible to write

$$\frac{D\Delta p}{4L} = \tau_o = K'\left(\frac{8V}{D}\right)^{n'} \tag{41}$$

where

$$n' = n \tag{42}$$

and

$$K' = \left(\frac{1 + 3n}{4n}\right)^{n} K \tag{43}$$

Combining the above equations and the definition of Fanning's friction coefficient and knowing that f = 16/Re, Metzner and Reed derived a generalized Reynolds number that is equally applicable to both Newtonian and non-Newtonian fluids:

$$Re_{MR} = \frac{D^{n'} V^{2-n'} \rho}{K' 8^{n'-1}} \tag{44}$$

They carried out a semitheoretical analysis to determine the relationship between the friction coefficient, f, and the generalized Metzner-Reed Reynolds number and obtained

$$\frac{1}{\sqrt{f}} = A_n \log [Re_{MR} f^{(1-n'/2)}] + C_n' \tag{45}$$

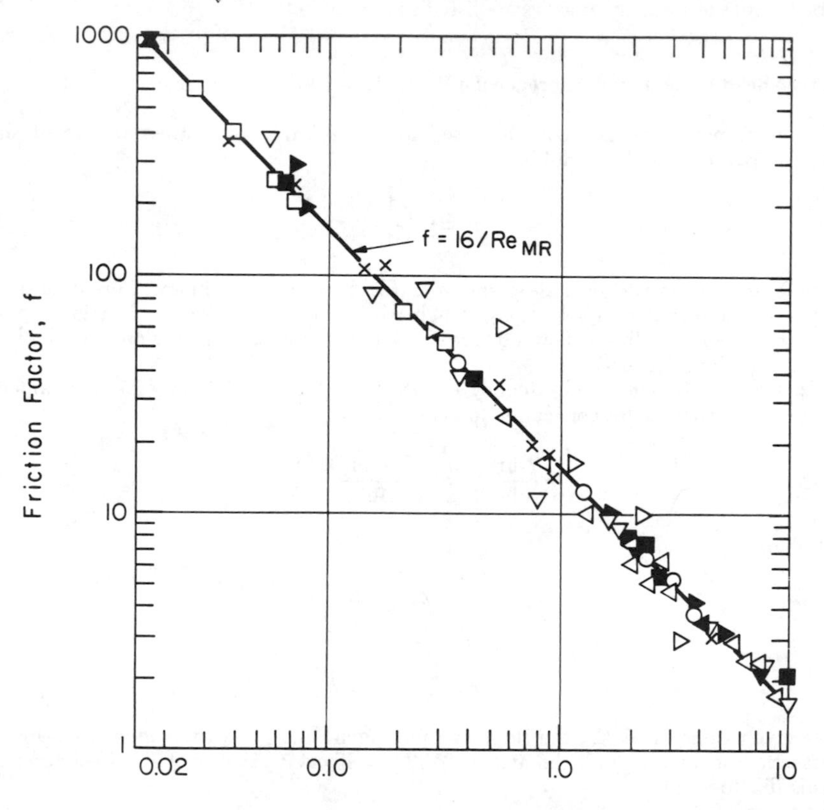

Figure 25. Friction factor for laminar flow of non-Newtonian fluids [26].

where

$$C_n' = A_n \log \left[\frac{1}{8} \left(\frac{2 + 6n'}{n'} \right)^{n'} \right] + C_n \tag{46}$$

The values of A_n and C_n' were determined empirically to be $4/n'^{0.75}$ and $-0.4/n'^{1.2}$, respectively. By substitution, Equation 45 becomes

$$\frac{1}{\sqrt{f}} = \frac{4}{n'^{0.75}} \log \left[Re_{MR}\, f^{(1 - n'/2)} \right] - \frac{0.4}{n'^{1.2}} \tag{47}$$

Table III. Some Published Rheological Data for Materials Following the Logarithmic Law [26,28]

Composition	n′	K′ × 10³
23% Illinois Yellow Clay in Water	0.23	5.4
0.7% Carboxy-methyl-cellulose (CMC) in Water	0.72	2.1
1.5% CMC in Water	0.55	11.2
3% CMC in Water	0.57	35
33% Lime Water	0.17	5.4
10% Napalm in Kerosene	0.52	14
4% Paper Pulp in Water	0.58	79
54% Cement Rock in Water	0.15	1.8
19% Mississippi Clay in Water	0.02	0.4
14% Clay in Water	0.35	0.28
25% Clay in Water	0.19	1.2
40% Clay in Water	0.13	11.7
23% Lime in Water	0.18	5.9

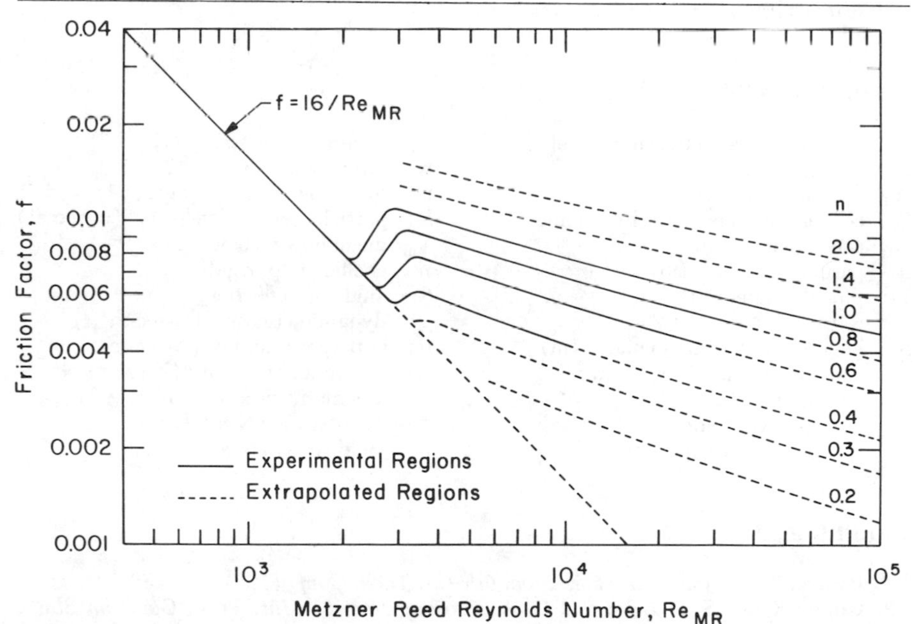

Figure 26. Extended friction factor vs Metzner-Reed Reynolds number relationship [27].

This relationship is plotted as shown in Figures 25 [26] and 26 [27]. A list of some published rheological data for materials following the power law relationship is given in Table III.

It should be stressed that the Metzner-Reed generalized approach is applicable to non-Newtonian fluid of all types, as well as Newtonian fluids. For n' (or n) value being unity, a Newtonian fluid is encountered. Values of n' (or n) less than unity indicate pseudoplastic or Bingham plastic behavior. With values of n' (or n) greater than unity, dilatant behavior is characterized.

Furthermore, because the variation of the flow-behavior index, n', with $D\Delta p/4L$ is very small, this approach may be extended beyond the laminar region. A trial and error procedure is recommended by Metzner [26] for the f value determination. If n' is a function of wall shear, τ_w (a nonlinear logarithmic plot between $D\Delta p/4L$ and $8V/D$), it is necessary to evaluate the appropriate n' at the point of interest for the problem at hand.

CONCLUSIONS

Rheology of suspensions has received a great deal of attention among researchers. Many published works on this subject can be found in various sources. Materials presented in this chapter are intended to provide an overview of the different treatments of various fluid systems and to be used as a readily available reference. Emphasis has been placed on physical understanding of the fluid systems and their respective behaviors, rather than detailed mathematical derivations of the equations. For a more detailed discussion of the analytical aspect of the subjects covered herein, the readers are referred to the original work, as listed in the bibliography.

The classification chart as described in Figure 3 included the regime delineation scheme for suspensions of heterogeneous and heterohomogeneous solids distributions. This is intended to provide a preview of the materials that are treated in detail in other publications listed in the suggested readings section.

NOMENCLATURE

a	small distance between parallel plates (m)
A	area of plate (m^2)
C	concentration of solids by volume
C_D	coefficient of drag
d	particle diameter (m)
D	pipe diameter (m)
F	force (N)
h	height of stationary cylinder (m)
K	consistency index ($N\text{-}sec\text{-}m^{-2}$)
L	length (m)
n	flow behavior index
N	speed of rotation (sec^{-1})
Δp	differential pressure ($N\text{-}m^{-2}$)
r	radius (m)
u	local flow velocity ($m\text{-}sec^{-1}$)
V_o	particle settling velocity ($m\text{-}sec^{-1}$)
V_m	mean flow velocity ($m\text{-}sec^{-1}$)
η	coefficient of rigidity
θ	angle of deflection
μ	dynamic viscosity ($N\text{-}sec\text{-}m^{-2}$)
μ_a	apparent viscosity ($N\text{-}sec\text{-}m^{-2}$)
μ_d	differential viscosity ($N\text{-}sec\text{-}m^{-2}$)
ν	kinematic viscosity ($m^2\text{-}sec^{-1}$)
τ	shear stress ($N\text{-}m^{-2}$)
τ_y	yield stress ($N\text{-}m^{-2}$)

REFERENCES

1. Pouska, G. A., and J. M. Link. *Proc. 6th Int. Tech. Conf.* (1981), p. 138.
2. Gupta, K. K., S. P. Babu and G. A. Pouska. *Proc. 5th Int. Tech. Conf. on Slurry Transport* (1980), p. 229.
3. Skelland, A. H. P. *Non-Newtonian Flow and Heat Transfer* (New York: John Wiley & Sons, Inc., 1967).

4. Govier, G. W., and K. Aziz. *The Flow of Complex Mixtures in Pipes* (New York: Van Nostrand Reinhold Co., 1972).
5. Davis, P. K., C. B. Muchmore, R. J. Missavage and G. N. Coleman. *J. Pipelines* 1:139 (1981).
6. Sadler, L. Y., and K. Senapati. *J. Pipelines* 1:197 (1981).
7. Kazanskij, I., and H. J. Mathias. *HydroTransport* 5:C3-35 (1978).
8. Metzner, A. B., and M. Whitlock. *Trans. Soc. Rheol.* II:239 (1958).
9. Govier, G. W., C. A. Shook and E. O. Lilge. *Trans. Can. Inst. Mining Met.* 60:147 (1957).
10. Ritter, R. A., and G. W. Govier. *Can. J. Chem. Eng.* 48:505 (1970).
11. Steg, I., and D. Katz. *J. Appl. Polymer Sci.* 9:3177 (1965).
12. Seyer, F. A. M.Ch. Eng. Thesis, University of Delaware (1965).
13. Zandi, I., and A. R. Razaghi. *ASCE, Proc. Env. Conf.* (1967).
14. Ferrini, F., and D. Ercolani. *Proc. 3rd Int. Tech. Conf.* (1978), p. 195.
15. Rabinowitsch, B. *Z. Physik. Chem. Ser. A.*, 145:1 (1929).
16. Mooney, M. *J. Rheology* 2:210 (1931).
17. Krieger, I. M., and S. H. J. Maron. *Appl. Phys.* 25:72 (1954).
18. Durand, R. *Houille Blanche* 8:124 (1953).
19. Duckworth, R. A. *Hydrotransport* 3 (1974).
20. Einstein, A. *Ann. Phys.* 19:289 (1905).
21. Einstein, A. *Ann. Phys.* 34:591 (1911).
22. Thomas, B. W. *Ind. Eng. Chem.* 45:87A (1953).
23. Buckingham, E. *ASTM Proc.* 21:1154 (1921).
24. Hanks, R. W., and B. H. Dadia. *AIChE J.* 17:554 (1971).
25. Metzner, A. B., and J. C. Reed. *AIChE J.* 1:434 (1955).
26. Metzner, A. B. *Advances in Chemical Engineering*, Vol. 1 (New York: Academic Press, Inc., 1955).
27. Dodge, D. W., and A. B. Metzner. *AIChE J.* 5:189 (1959).
28. Metzner, A. B. *Ind. Eng. Chem.* 49:1429 (1957).

SUGGESTED READINGS

Bain, A. G., and S. T. Bonnington. *The Hydraulic Transport of Solids in Pipes*, (Elmsford, NY: Pergamon Press, Inc., 1970).

Chen, D. C. H. *Hydrotransport-2* c3-21 (1972).

Cheremisinoff, N. P. *Fluid Flow–Pumps, Pipes and Channels*, (Ann Arbor, MI: Ann Arbor Science Publishers, Inc., 1981).

Cross, M. M. *J. Colloid Sci.* 20:417 (1965).

Frankel, N. A., and A. Acrivos. *Chem. Eng. Sci.* 22:847, (1967).

Hedstrom, B. O. A. *Ind. Eng. Chem.* 44:651 (1951).

Herschel, W. H., and R. Bulkley. *ASTM Proc.* 26:621 (1926).

Kao, D. T. Y. "Hydraulic Transport of Solids in Pipes," University of Kentucky (1976).

Masauyama, T., and T. Kawashima. *Bull. JSME* 22:48 (1979).

Meter, D. M., and R. B. Bird. *AIChE J.* 19:878 (1964).

Philippoff, W. *Kolloid-z.* 71:1 (1935).

Prandtl, L. Z. *Angew. Math. Mech.* 8:85 (1928).

Sisko, A. W. *Ind. Eng. Chem.* 50:1789 (1958).

Wasp, E. J., J. P. Kenny and R. L. Gandhi. Solid-Liquid Flow Slurry Pipeline Transportation," *Trans. Tech. Publications* (1977).

Eying, H. J. *J. Chem. Phys.* 4:283 (1936).

CHAPTER 34
MOTION OF SOLID PARTICLES IN A LIQUID MEDIUM

David Azbel and Athanasios I. Liapis
Department of Chemical Engineering
University of Missouri-Rolla
Rolla, Missouri 65401

CONTENTS

INTRODUCTION

A large number of operations in chemical technology involve the simultaneous flow of liquid and solid particles. The discipline of liquid-solid particle flows encompasses a tremendous range of practical problems; however, we will deal here only with dilute liquid-solid particle systems in which the particles are far enough removed from one another that we may consider the motion of individual particles independently of the motion of the other particles in the system of interest.

The motion of solid particles in dilute liquid-solid particle systems is often of primary concern when dealing with problems of mass transfer, heat transfer and kinetics in chemical technology. Therefore, liquid-solid particle mixtures have been studied extensively, both experimentally and analytically [1–20]. However, our knowledge of such flows is limited because most of the useful correlations are of empirical origin and have application only to the specific systems from which they were developed. Also, sometimes these correlations result in limited accuracy when applied to other systems. Thus, we will attempt to characterize dispersion of a low concentration of solid particles in a liquid flow by means of the volume fraction of solid particles, and also the superficial velocities, v_{s_f} and v_s, of the liquid and solid phases, respectively, in an attempt to develop formulas having wider applicability.

AVERAGE PARTICLE FLOW VELOCITY AND AVERAGE FRACTION OF PARTICLES IN A LIQUID FLOW

This section describes some basic terms that will be used to describe liquid-solid particle flows. The average velocity of solid particles in a liquid stream or, in other words, the superficial particle velocity, v_s, when velocities of the two phases are low and the duct is large enough (so that there are no wall effects), may be defined if we imagine some control plane to pass through the flow perpendicular to the direction of the time-averaged flow velocity vector. Then we write the volumetric flowrate of solid particles per unit area of this control plane as

$$v_s = \frac{Q}{A} = \frac{1}{A\Delta\tau}\int_{\Delta\tau}\int_{A} v_p \, dA d\tau \tag{1}$$

where Q (m^3/sec) is the volumetric flowrate of particles, A (m^2) is the flow system cross-sectional area, $\Delta\tau$ (sec) is a period of time substantially greater than the quantity 1/f (where f (sec^{-1}) is the frequency of movement of solid particles through the cross section in question), and v_p is the actual velocity of the particles. At a given instant in time, the particles do not occupy the whole cross section of the duct and, in general, a part of every cross section of the duct is occupied by the liquid. Therefore, Equation 1 must be written in the form of a sum of integrals:

$$v_s = \frac{1}{A\Delta\tau}\int_{\Delta\tau}\left(\sum_i \int_{A_i} v_p \, dA\right) d\tau \tag{2a}$$

where i is the number of particle formations, or groups, in a given cross section at a given instant of time. The ratio

$$\phi = \frac{1}{A}\sum_i A_i \tag{2b}$$

is defined as the instantaneous value of that fraction of the section of the dispersed flow occupied by particles (this section being of thickness dn, where n is normal to the surface A). The true average velocity of particles in the liquid flow is then

$$\overline{v}_p = \frac{1}{\Delta\tau \sum_i A_i}\int_{\Delta\tau}\left(\sum_i \int_{A_i} v_p \, dA\right) d\tau \tag{3}$$

and this equation, with Equation 2, indicates that the true average velocity and superficial velocity of the particles are related by

$$\overline{v}_p = \frac{v_s}{\phi} \tag{4}$$

The ratio of the total mass flowrate of the phases to the product of the cross-sectional area of the channel and the density of the liquid is usually designated as the flow velocity of the two-component flow:

$$v = \frac{\rho_p v_s + \rho_f v_{s_f}}{\rho_f} \tag{5}$$

where ρ_p and ρ_f are the particle and fluid densities, respectively, and v_{s_f} is the liquid superficial velocity. It provides a measure of the velocity of a liquid flow with mass flowrate equal to that of the mixture.

The average relative velocity of the solid phase is defined by

$$\overline{v}_r = \overline{v}_f - \overline{v}_p = \frac{v_{s_f}}{1-\phi} - \frac{v_s}{\phi} \tag{6}$$

and, hence,

$$\overline{v}_r \phi^2 - (\overline{v}_r - v_{s_f} - v_s)\phi - v_s = 0 \tag{7}$$

Solving this equation for ϕ, taking into account that when $v_s = 0$ we have $\phi = 0$, we obtain

$$\phi = \frac{\overline{v}_r - v_{s_f} - v_s}{2\overline{v}_r} - \left[\left(\frac{\overline{v}_r - v_{s_f} - v_s}{2\overline{v}_r}\right)^2 + \frac{v_s}{\overline{v}_r}\right]^{1/2} \tag{8}$$

When the average relative velocity of the particles, $\overline{v}_r$, is equal to zero, we see from Equation 7 that the volumetric flow fraction, defined below, becomes equal to ϕ:

$$\beta \equiv \frac{v_s}{v_{s_f} + v_s} = \phi \tag{9}$$

and when the particle relative velocity is positive, the true fraction of the particles in the two-phase flow is less than the flow fraction, β.

The inequality of the averaged phase velocities, which is a measure of the "slip" of the particles relative to the liquid flow, changes the energy losses in the two-phase system, as compared to the case in which the relative phase velocity is zero. The mass flowrate of the mixture per unit area is

$$M = \rho_p v_s + \rho_f v_{s_f} \tag{10}$$

and introducing the concept of average density of the mixture we have

$$\rho = \frac{M}{v_s + v_{s_f}} \tag{11a}$$

or

$$\rho = (1-\phi)\rho_f + \phi\rho_p \tag{11b}$$

leading to

$$\rho = \rho_f - (\rho_f - \rho_p)\phi \tag{11c}$$

Note that ϕ is then the equivalent, for solid particles, of the void fraction, ϕ, used in describing gas-liquid flows. The relative density of the mixture is then given by

$$\Psi = \frac{\rho}{\rho_f} = 1 - \frac{\rho_f - \rho_p}{\rho_f}\phi \tag{12}$$

STEADY MOTION OF SOLID PARTICLES IN LAMINAR VISCOUS FLOW

When a spherical particle moves through a viscous liquid in which the ratio of inertial to viscous forces is very small (the Reynolds number $Re = 2\rho_f v_r r_p/\mu_f << 1$), the inertial forces may be neglected. Viscosity is then the controlling factor [21], and the resistance of a liquid to the motion of a particle is expressed by the well known Stokes' law:

$$F_D = 6\pi\mu_f r_p v_r \tag{13}$$

where μ_f is the liquid dynamic viscosity, r_p is the particle radius, and v_r is the particle relative velocity. In the derivation of Equation 13, the inertia terms were omitted from the equations of motion so that the equation may be considered only as a first approximation, and a second approximation, which partly takes into account inertial forces, was obtained by Oseen [22] as

$$F_D = 6\pi\mu_f r_p v_r \left(1 + \frac{3}{16}\,Re\right) \tag{14}$$

and, somewhat more accurately, by Langhaar [23]:

$$F_D = 6\pi\mu_f r_p v_r \left(1 + \frac{3}{16}\,Re - \frac{19}{1280}\,Re^2 + \ldots\right) \tag{15}$$

In this "creeping flow" ($Re << 1$), the resistance is thus proportional to the particle velocity. However, the region of large Reynolds numbers, where inertial forces cannot be ignored, is characterized by the resistance varying as v_r^n, where n steadily increases with Re. So, in general, this nonlinear dependency of resistance has to be expressed in two variables (size and velocity of the particle) instead of one. Introduction of the dimensionless quantity

$$C_D = \frac{F_D}{\frac{1}{2}\,\rho_f v_r^2 \pi r_p^2} \tag{16}$$

called the drag coefficient, which is a unique function of the Reynolds number, enables this complex problem to be simplified considerably. Where Stokes' law applies, the drag coefficient becomes

$$C_D = \frac{24}{Re} \tag{17}$$

while Oseen's formula is expressed by

$$C_D = \frac{24}{Re} + 4.5 \tag{18}$$

A large number of empirical formulas relating C_D and Re have been obtained [24], and the most successful from the standpoint of simplicity and accuracy seems to be the one given by Klyachko [25]:

$$C_D = \frac{24}{Re} + \frac{4}{Re^{1/3}} \tag{19}$$

in the range $3 < Re < 500$.

NONUNIFORM MOTION OF SOLID PARTICLES

The motion of a particle with constant velocity, which was examined in the previous section, must be regarded as ideal because the velocity is always changing, both in magnitude and direction. The nonuniform motion of solid particles is certainly more complex than motion at constant velocity, and the relevant differential equations can be solved in comparatively few cases. In most cases, one must resort to numerical methods. Only a few of the more important cases of nonuniform motion will be discussed here, assuming for simplicity that the particles are spherical.

The differential equation for nonsteady rectilinear motion of a particle in a liquid has been obtained and modified by several authors [22,26–32]. Neglecting external forces, the final form of the equation for the motion of a spherical particle in a turbulent flow, with nonzero mean velocity, is usually obtained as

$$\frac{\pi d_p^3 \rho_f}{6}\frac{dv_p}{dt} = \frac{\pi d_p^3 \rho_f}{6}\frac{dv_f}{dt} + \frac{1}{2}\frac{\pi d_p^3 \rho_f}{6}\left(\frac{dv_f}{dt} - \frac{dv_p}{dt}\right) + 3\pi\nu_f\rho_f\, d_p(v_f - v_p) + \frac{3}{2}\, d_p^2\, \rho_f \pi^{1/2}\nu_f \int_{t_0}^{t}\left(\frac{dv_f}{dx} - \frac{dv_p}{dx}\right)\frac{dx}{(t-x)^{1/2}} \tag{20}$$

where d_p is the diameter of the particle, ρ_p is the particle density, ρ_f is the density of the liquid, v_p is the velocity of a particle, v_f is the absolute velocity of a liquid, ν_f is the liquid kinematic viscosity, and t is the time.

The first term on the right-hand side of the equation is the surface force acting on the particle resulting from the pressure variation on the surface. This variation, in turn, is a function of the pressure field in the entraining liquid. The second term describes the inertia of the added mass (entrained with the relative motion of the particle in the entraining liquid) and is equivalent to an increase in mass of the spherical particle equal to half the mass of the liquid displaced. The third term describes the viscous force due to the relative motion of the particle in liquid, and the fourth is the resistance due to the energy expended in setting the liquid itself in motion, the so-called "history" term.

Equation 20 is valid when the following relations hold [33,34]:

$$\frac{d_p^2}{\nu_f}\frac{\partial v_f}{\partial x} << 1 \tag{21}$$

$$\frac{v_f}{d_p^2 \dfrac{\partial^2 v_f}{\partial x^2}} >> 1 \tag{22}$$

where x is a linear dimension. Following an approach described elsewhere [34], we write Equation 20 in the form

$$\frac{dv_p}{dt} + av_p + c\int_{t_0}^{t}\frac{dv_p/dx}{(t-x)^{1/2}}\,dx = av_f + b\frac{dv_f}{dt} + c\int_{t_0}^{t}\frac{dv_f/dx}{(t-x)^{1/2}}\,dx \tag{23}$$

where

$$a = \frac{18\nu_f}{[(\rho_p/\rho_f) + 0.5]\, d_p^2} \tag{24}$$

$$b = \frac{3}{2[\rho_p/\rho_f + 0.5]} \tag{25}$$

and

$$c = \frac{9\nu_f^{1/2}}{[(\rho_p/\rho_f) + 0.5]\,\pi^{1/2}} \tag{26}$$

Now expressing v_f and v_p in terms of Fourier integrals, we obtain

$$v_f = \int_0^\infty (\zeta \cos\omega t + \lambda \sin\omega t)\,d\omega \tag{27a}$$

$$v_p = \int_0^\infty (\sigma \cos\omega t + \psi \sin\omega t)\,d\omega \tag{27b}$$

where ζ, λ, σ and ψ are Fourier amplitudes and ω is the frequency. And substituting Equations 27 into Equation 23 yields, after straightforward manipulation,

$$\sigma = (1 + f_1)\,\zeta + f_2\lambda \tag{28}$$

and

$$\psi = -f_2\zeta + (1 + f_1)\lambda \tag{29}$$

in which

$$f_1 = \frac{\omega[\omega + c(\pi\omega/2)^{1/2}]\,(b-1)}{[a + c(\pi\omega/2)^{1/2}]^2 + [\omega + c(\pi\omega/2)^{1/2}]^2} \tag{30}$$

and

$$f_2 = \frac{\omega[a + c(\pi\omega/2)^{1/2}](b-1)}{[a + c(\pi\omega/2)^{1/2}]^2 + [\omega + c(\pi\omega/2)^{1/2}]^2} \tag{31}$$

In developing the theory to deal with the influence of turbulence on the motion of a particle in a very dilute flow, certain assumptions and modifications of Equation 20 are necessary. For example, we may neglect the integral (or "history") term in Equation 20, we may modify it by neglecting the added mass and the integral, or we may ignore the pressure gradient effect of the fluid acceleration, in addition to neglecting the added mass and the integral. In fact, deviation of the particle motion from the liquid motion has been investigated [35] for these three types of approximations, as follows:

Type I: When ignoring the integral "history" term in Equation 23,

$$a = \frac{18\nu_f}{[(\rho_p/\rho_f) + 0.5]\,d_p^2}, \quad b = \frac{3}{2[(\rho_p/\rho_f) + 0.5]}, \quad c = 0$$

Type II: When ignoring the integral and added mass terms,

$$a = \frac{18\nu_f}{d_p^2(\rho_p/\rho_f)}, \quad b = \frac{1}{\rho_p/\rho_f}, \quad c = 0$$

Type III: When, in addition to the conditions of Type II, the pressure gradient effect is neglected:

$$a = \frac{18\nu_f}{d_p^2(\rho_p/\rho_f)}, \quad b = c = 0$$

We now introduce the concepts of "amplitude ratio," and "phase angle," between the particle and liquid motions and, expressing the velocity as a Fourier integral, we have

$$v_p = \int_0^\infty \eta[\zeta\cos(\omega t + \beta) + \lambda\sin(\omega t + \beta)]\, d\omega \tag{32}$$

where the amplitude ratio is given by

$$\eta = [(1 + f_1)^2 + f_2^2]^{1/2} \tag{33}$$

and the phase angle is given by

$$\beta = \tan^{-1}\left\{\frac{f_2}{1 + f_1}\right\} \tag{34}$$

The above formulas allow the various approximations to be compared by contrasting the respective phase angles and amplitude ratios, and, for convenience, f_1 and f_2 can be expressed in dimensionless form for the general case, so that we have

$$f_1 = \frac{\left[1 + \frac{9}{2^{1/2}(s + 1/2)} N_s\right]\left[\frac{1 - s}{s + 1/2}\right]}{\frac{81}{(s + 1/2)^2}\left[2N_s^2 + \frac{N_s}{2^{1/2}}\right]^2 + \left[1 + \frac{9}{2^{1/2}(s + 1/2)} N_s\right]^2} \tag{35}$$

$$f_2 = \frac{\frac{9(1 - s)}{(s + 1/2)^2}\left[2N_s^2 + \frac{N_s}{2^{1/2}}\right]}{\frac{81}{(s + 1/2)^2}\left[2N_s^2 + \frac{N_s}{2^{1/2}}\right]^2 + \left[1 + \frac{9N_s}{2^{1/2}(s + 1/2)}\right]^2} \tag{36}$$

where

$$s = \rho_p/\rho_f \tag{37}$$

and

$$N_s = \left[\frac{\nu_f}{\omega d_p^2}\right]^{1/2} \tag{38}$$

where N_s is the Stokes number.

Hence, for the Type I approximation, we find

$$f_1 = \frac{(1 - s)/(s + 1/2)}{\left[\frac{18N_s^2}{s + 1/2}\right]^2 + 1} \tag{39a}$$

and

$$f_2 = \frac{\frac{18N_s^2}{(s + 1/2)^2}(1 - s)}{\left[\frac{18N_s^2}{s + 1/2}\right]^2 + 1} \tag{39b}$$

For the Type II approximation,

$$f_1 = \frac{(1-s)/s}{(18N_s^2/s)^2 + 1} \qquad (40a)$$

and

$$f_2 = \frac{18N_s^2(1-s)/s^2}{(18N_s^2/s)^2 + 1} \qquad (40b)$$

and for the Type III approximation,

$$f_1 = \frac{-1}{(18N_s^2/s)^2 + 1} \qquad (41a)$$

and

$$f_2 = \frac{-18N_s^2/s}{(18N_s^2/s)^2 + 1} \qquad (41b)$$

The importance of added mass, the integral (or "history") term and the pressure gradient due to fluid acceleration in their effects on particle motion in turbulent fluid, is indicated by analyzing the approximations of Equation 20 given above. Only for

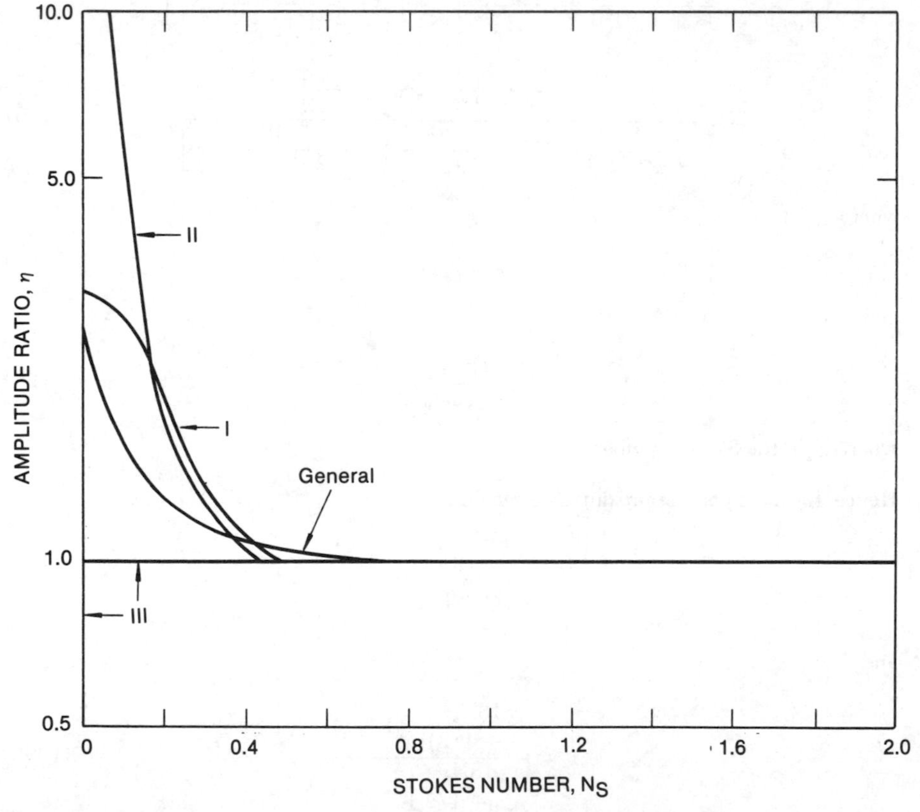

Figure 1a. Amplitude ratio of discrete particles in a turbulent fluid; density ratio is 8.6×10^{-5}.

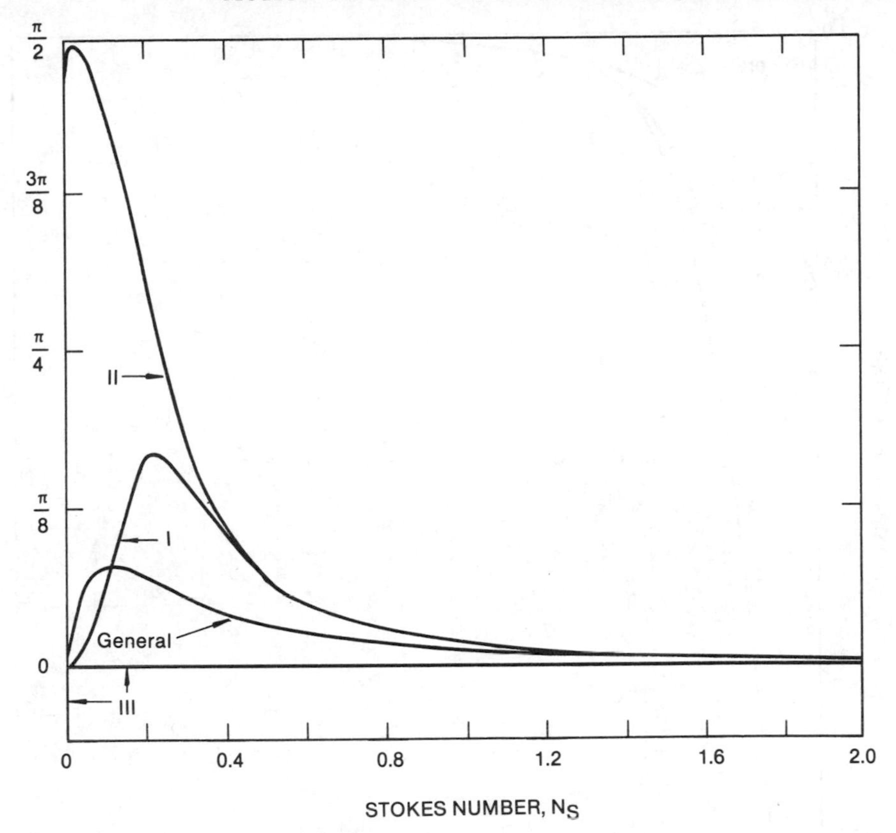

Figure 1b. Phase angle diagram for discrete particles in a turbulent fluid; density ratio is 8.6×10^{-5}.

high-density ratios and small particles were these terms found to be unimportant, and for liquid-solid particle systems, the greatest deviations from the general theory occur within the frequency range of general interest, so other approaches are needed.

However, computations of amplitude ratio and phase angle for various density ratios based on the general case (Equations 20, 35 and 36) and the three types (described by Equations 39–41 have been made [35], and are shown in Figures 1–3 for density ratios of 8.6×10^{-5}, 2.65 and 1000, respectively. It is seen that the approximations are good for high Stokes numbers (low frequencies), although at higher frequencies a considerable error arises for Type I, neglecting the integral or "history" term, the difference between this and the general case being greater than that between it and Type II, neglecting additional mass. For Type III, neglecting the pressure term, we have for the small density ratio case a negative phase angle, which is an incorrect prediction. However, these curves may be of some use in indicating the circumstances in which each approximation may be made.

EFFECT OF SYSTEM WALLS ON PARTICLE VELOCITIES IN A DILUTE SUSPENSION

The preceding section does not consider the possible effects on the liquid-solid particle flow of the finite dimensions of the flow system. It is obvious that there will be, in gen-

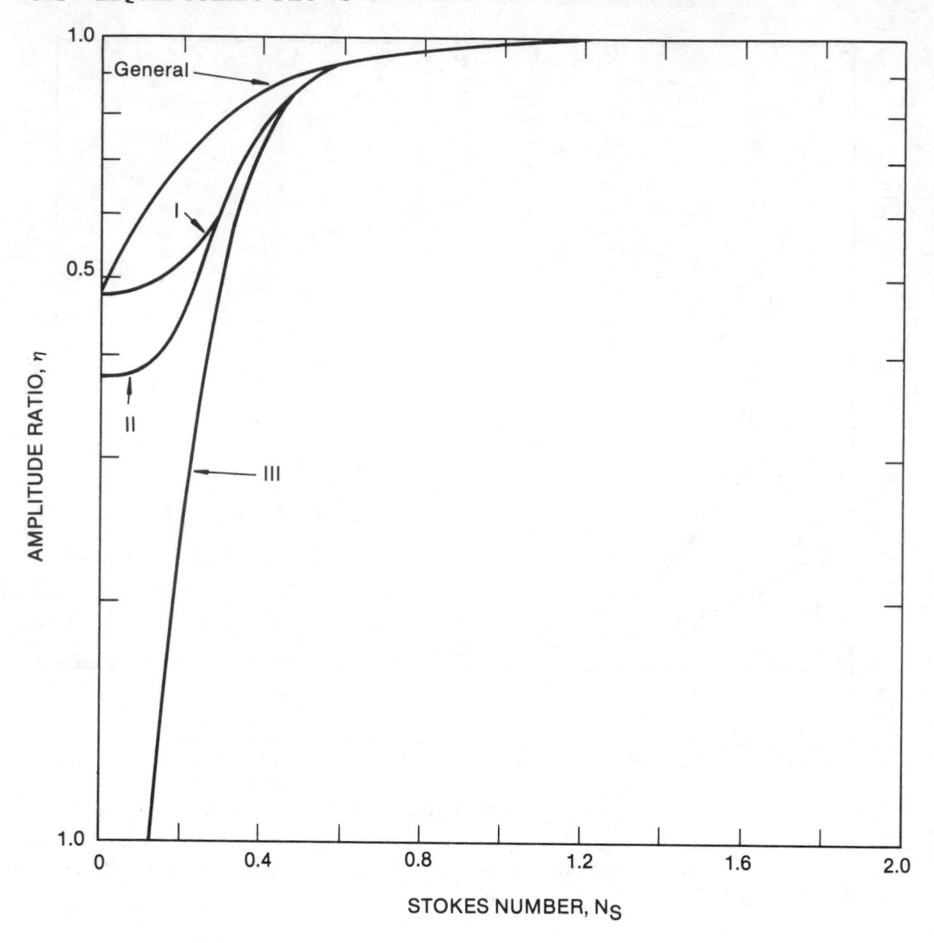

Figure 2a. Amplitude ratio of discrete particles in a turbulent fluid; density ratio is 2.65.

eral, a nontrivial modification to the flow due to this so-called "wall effect." In fact, for a dilute suspension of solid particles in a liquid the effect is the same as that on a gas-solid flow described in the chapter entitled Mechanisms of Liquid Entrainment. We refer the reader to the second section of that chapter, in which this wall effect is discussed. The conclusions drawn there are found to be applicable to liquid-solid particle flow.

BASIC EQUATIONS FOR THE MOTION OF SOLID PARTICLES IN A TURBULENT FLOW

In studying the dynamics of solid particles in a turbulent flow, it is possible to apply precisely the same procedure used when developing the momentum and energy equations for a single liquid. It also is possible to consider the effect of turbulence by applying the well known Reynolds procedure [34]. The main physical assumption needed to realize these possibilities is that the size of the suspended particles be small in comparison with the characteristic scale of turbulence because this provision makes it possible to treat the particles as an admixture distributed continuously in the entraining liquid.

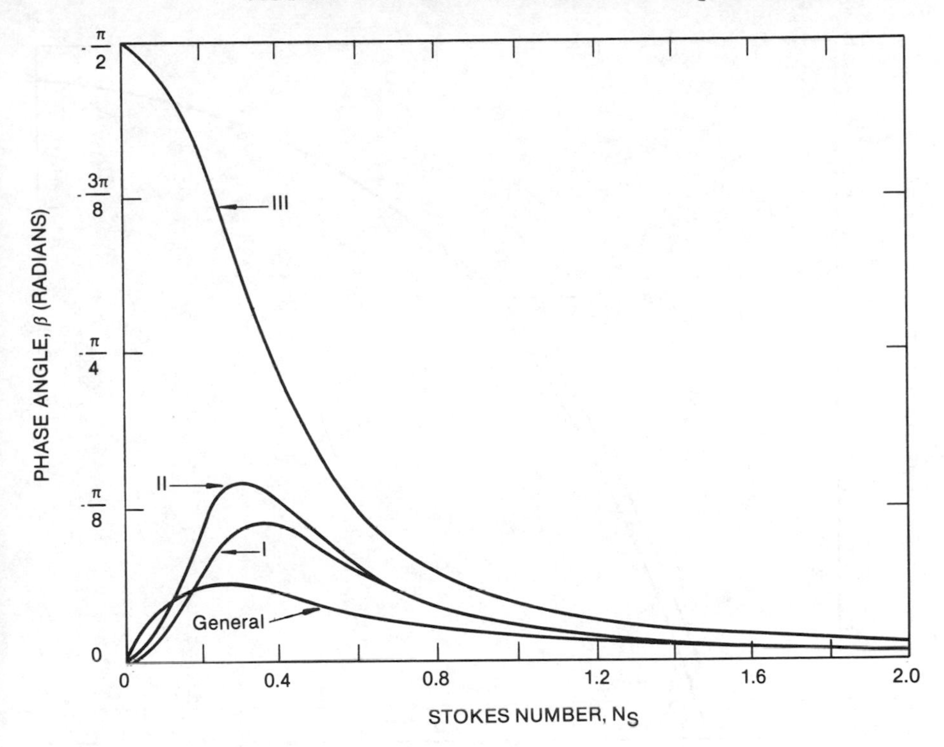

Figure 2b. Phase angle diagram for discrete particles in a turbulent fluid; density ratio is 2.65.

The motion of the entraining liquid and suspended solid particles may be regarded as the flow of a homogeneous medium if we write the mass and momentum equations for the solid and liquid components separately and then add them to derive the corresponding equations for the mixture as a whole. We can introduce the notions of the density and velocity of the mixture and, for an elementary volume of space, the continuity equation is given as

$$\frac{\partial \rho}{\partial t} + \frac{\partial (\rho u_i)}{\partial x_i} = 0 \tag{42}$$

where i is the Einstein summation index, indicating a summation over all the values of the repeated index, and where

$$\rho = \rho_f(1 - \phi) + \rho_p\phi = \rho_f + (\rho_p - \rho_f)\phi \tag{43}$$

is the mixture density and

$$u_i = \frac{\rho_f(1 - \phi)u_{f_i} + \rho_p\phi u_{p_i}}{\rho} \tag{44}$$

is the mixture velocity vector.
Here, ϕ is the solids content of the mixture (by volume), u_{f_i} is the average liquid velocity vector in the interparticle space, u_{p_i} is the average particle velocity vector, and x_i is the position vector.

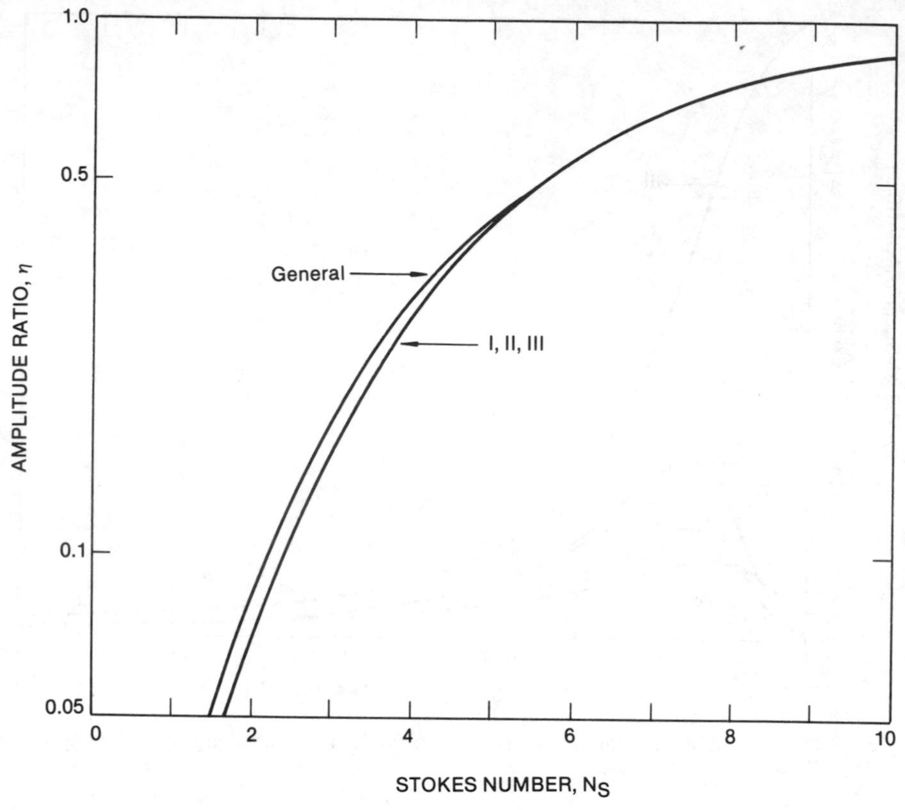

Figure 3a. Amplitude ratio for discrete particles in a turbulent fluid; density ratio is 1,000.

The liquid and the solid particles are incompressible, so one may eliminate the densities ρ_f and ρ_p as constant factors from the mass conservation equations and obtain conservation equations for the liquid and solid particles, as follows:

$$\frac{\partial}{\partial t}(1-\phi)+\frac{\partial}{\partial x_i}(1-\phi)\,u_{f_i}=0 \tag{45}$$

$$\frac{\partial \phi}{\partial t}+\frac{\partial}{\partial x_i}\left(\phi u_{p_i}\right)=0 \tag{46}$$

If we add Equations 45 and 46 we obtain another mass conservation or continuity equation for the liquid-solid particles mixture:

$$\frac{\partial}{\partial x_i}\left[(1-\phi)u_{f_i}+\phi u_{p_i}\right]=0 \tag{47}$$

The momentum equations for the liquid and the solid particles may be determined by making appropriate allowances for stresses in the liquid phase due to the motion of the solid particles and the reciprocal stresses on the particles due to liquid motion. These equations may be written [32] as

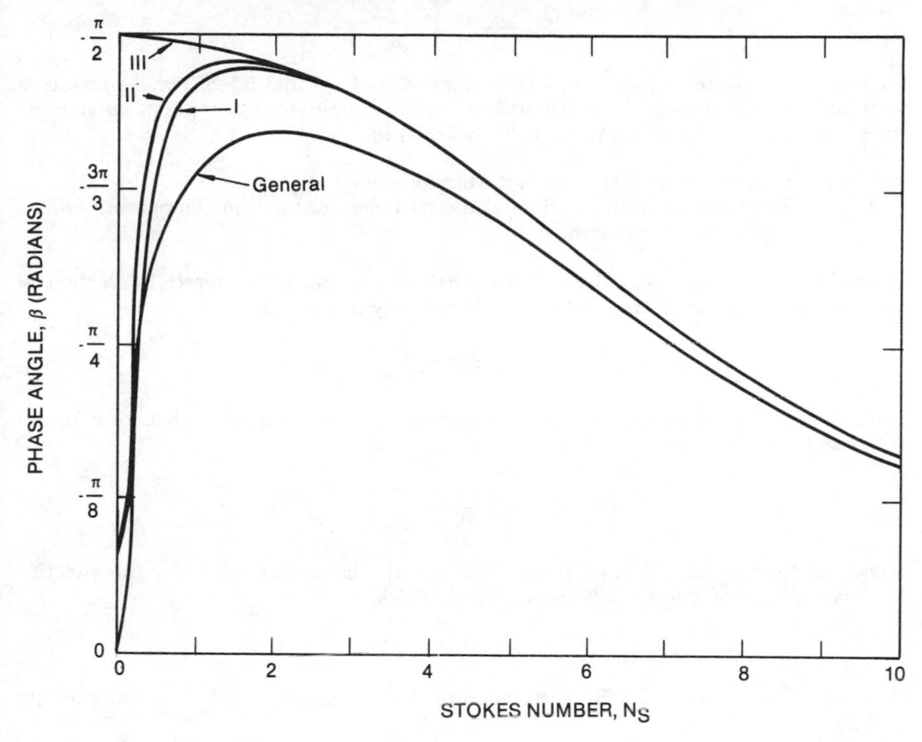

Figure 3b. Phase angle diagram for discrete particles in a turbulent fluid; density ratio is 1,000.

$$\frac{\partial}{\partial t}\left[\rho_f(1-\phi)u_{f_i}\right]+\frac{\partial}{\partial x_j}\left[\rho_f(1-\phi)u_{f_i}u_{f_j}\right]=$$
$$\rho(1-\phi)g_i-(1-\phi)\frac{\partial p}{\partial x_i}+\frac{\partial \tau_{ij}^{(f)}}{\partial x_j}-F_i \tag{48}$$

and

$$\frac{\partial}{\partial t}\left(\rho_p u_{p_i}\right)+\frac{\partial}{\partial x_j}\left(\rho_p\phi u_{p_i}u_{p_j}\right)=\rho_p\phi g_i-\phi\frac{\partial p}{\partial x_i}+\frac{\partial \tau_{ij}^{(p)}}{\partial x_j}+F_i \tag{49}$$

where g_i is the acceleration due to gravity, p is the pressure, $\tau_{ij}^{(f)}$ is a tensor representing viscous stresses in the liquid, $\tau_{ij}^{(p)}$ is a stress tensor resulting from the direct interaction of solid particles, and F_i is the interaction force per unit volume between the liquid and solid particles.

Adding Equations 48 and 49 gives a momentum equation for the mixture:

$$\frac{\partial(\rho u_i)}{\partial t}+\frac{\partial M_{ij}}{\partial x_j}=\rho g_i-\frac{\partial p}{\partial x_i}+\frac{\partial \tau_{ij}}{\partial x_j} \tag{50}$$

where the tensors for momentum flux and shear stresses [36,37] are, respectively,

$$M_{ij}=\rho_f(1-\phi)u_{f_i}u_{f_j}+\phi\,\rho_p u_{p_i}u_{p_j} \tag{51}$$

$$\tau_{ij} = \tau_{ij}^{(f)} + \tau_{ij}^{(p)} \tag{52}$$

We now have a system of equations (Equations 42–44, 47 and 50–52) that provides a basis for further analysis of the problem. Two further simplifying assumptions are now required to close this system of equations. They are as follows:

1. The concentration of particles in the liquid is small, $\phi << 1$.
2. The accelerations of the liquid and the particles are small in comparison with the acceleration due to gravity.

Since the concentration and sizes of the particles are small, the direct interaction of particles may be ignored. Or, in other words, we may assume that

$$\tau_{ij}^{(p)} << \tau_{ij}^{(f)}$$

so the tensor of shear stresses, τ_{ij}, may be written in the same form as that for a homogeneous viscous, incompressible liquid:

$$\tau_{ij} = \mu\left(\frac{\partial u_i}{\partial x_j} + \frac{\partial u_j}{\partial x_i}\right) \tag{53}$$

where, for the mixture viscosity, μ, one may use the Einstein formula for the effective dynamic viscosity of a liquid with suspended particles:

$$\mu = \mu_f(1 + C\phi) \tag{54}$$

Here, μ_f is the viscosity of the corresponding homogeneous liquid, and the constant $C \simeq 2.5$.

In addition, in a cartesian coordinate system let the x_3 axis be directed vertically upward so that

$$g_1 = g_2 = 0, \quad g_3 = -g \tag{55}$$

If we have the x_3 axis forming a small angle, θ, with the vertical, then

$$g_1 = g \sin\theta, \quad g_2 = 0, \quad g_3 = -g\cos\theta \simeq -g \tag{56}$$

As we stated in our assumptions, the size of the particles is small, and the substantial or total accelerations of the particles are small in comparison with that due to gravity. Therefore, the components of the particle velocities and those of the liquid in planes normal to the x_3 axis must coincide, and the components u_{p_3} and u_{f_3} must differ by the value of the gravitational settling velocity of the particles (their terminal velocity, u_r). This means that

$$u_{p_i} = u_{f_i} - u_r\delta_{ij} \tag{57}$$

where δ_{ij} is the Kronecker delta function ($\delta_{ij} = 1, i = j$; $\delta_{ij} = 0, i \neq j$).

This assumption means that the additional inertia force on a particle in its accelerated relative motion is neglected, which makes it possible to obtain equations of motion for the liquid-solids mixture in which the terminal velocity of the particles appears in place of the interaction forces between the suspended particles and the liquid.

Equation 57 then yields the following relationships for the components of the total mass velocity and momentum flux:

$$u_i = u_{f_i} - \frac{\rho_p}{\rho}\phi u_r\delta_{i3} = u_{p_i} + \frac{\rho_f}{\rho}(1-\phi)u_r\delta_{i3} \tag{58}$$

$$M_{ij} = \rho_f(1-\phi)u_{f_i}u_{f_j} + \rho_p\phi u_{p_i}u_{p_j} = \rho u_i u_j + \frac{\rho_f\rho_p\phi}{\rho}(1-\phi)u_r^2\delta_{i3}\delta_{j3} \tag{59}$$

When we use Equations 58 and 59 in the momentum equation (Equation 50), and the continuity equation (Equation 47), we find

$$\frac{\partial(\rho u_i)}{\partial t} + \frac{\partial(\rho u_i u_j)}{\partial x_j} = \rho g_i - \frac{\partial p}{\partial x_i} + \frac{\partial \tau_{ij}}{\partial x_j} - \rho_f\rho_p\frac{\partial}{\partial x_3}\left[\frac{\phi(1-\phi)u_r^2}{\rho}\right]\delta_{i3} \tag{60}$$

and

$$\frac{\partial u_i}{\partial x_i} = -(\rho_p - \rho_f)\frac{\partial}{\partial x_3}\left[\frac{\phi(1-\phi)u_r}{\rho}\right] \tag{61}$$

Equations 57–61 and Equation 42 now form a closed system of five equations corresponding to five unknowns. Because we are considering the case of a relatively small suspension concentration, we can use the conditions

$$\phi << 1, \quad \Delta\rho\phi << 1, \quad \Delta\rho = \left(\frac{\rho_p - \rho_f}{\rho_f}\right) \tag{62}$$

and the terminal velocity, u_r, of the particles may as an approximation be regarded as a constant quantity equal to the uniform settling velocity of an individual particle in an unbounded liquid. Consequently, Equation 61 may be written as

$$\frac{\partial u_i}{\partial x_i} = -\Delta\rho u_r\frac{\partial\phi}{\partial x_3} \tag{63}$$

and the equation of mass conservation will thus take the form of a "balance" equation for the suspended particles:

$$\frac{\partial\phi}{\partial t} + u_i\frac{\partial\phi}{\partial x_i} = u_r\frac{\partial\phi}{\partial x_3} \tag{64}$$

Hence, the right-hand side of Equation 61 may be disregarded in all cases except that of the suspension balance calculation, so the equation may be written as

$$\frac{\partial u_i}{\partial x_i} = 0 \tag{65}$$

The momentum equation also may be simplified into the following:

$$\frac{\partial u_i}{\partial t} + \frac{\partial}{\partial x_j}(u_i u_j) = (1+\Delta\rho\phi)g_i - \frac{1}{\rho_f}\frac{\partial p}{\partial x_i} + \frac{1}{\rho_f}\frac{\partial\tau_{ij}}{\partial x_j} - \frac{\partial(u_i u_j)}{\partial x_j} \tag{66}$$

This analysis shows that the term $\Delta\rho\phi$ in the coefficient of g_i is dynamically quite essential when applying this equation to the n_3 ($i = 3$) direction because in this case $g_3 = -g$ is a large factor. Thus, the equations of momentum (66) and continuity (65) are identical with analogous equations for the motion of liquids of nonhomogeneous density.

The closed system of basic equations we have derived above and which are available for determination of u_i, ϕ and p, provides the possibility of analyzing the dynamics of solid particles in a turbulent liquid flow, and we will now consider a particle suspension in a flow in a horizontal duct, the suspension being caused by turbulent fluctuations of the flow velocity. In this case, the flow characteristic quantities may be represented as the sum of their mean and fluctuating quantities:

$$u_i = \bar{u}_i + u_i'; \quad p = \bar{p} + p'; \quad \rho = \bar{\rho} + \rho' \tag{67}$$

The fluctuation of the mixture density, ρ, is entirely the result of the fluctuation of concentration, ϕ (both components of the mixture being incompressible). Hence, from Equation 43 we obtain

$$\rho' = (\rho_p - \rho_f)\phi' \tag{68}$$

By averaging, we obtain the following equations:
(a) For "volume conservation" (Equation 61):

$$\frac{\partial \bar{u}_i}{\partial x_i} = -\Delta\rho u_r \frac{\partial \bar{\phi}}{\partial x_3} \tag{69}$$

(b) For mass conservation (Equation 42 with Equations 68 and 69):

$$\frac{\partial \bar{\phi}}{dt} + \bar{u}_i \frac{\partial \bar{\phi}}{\partial x_i} + \frac{\partial \overline{\phi' u'_i}}{\partial x_i} = u_r \frac{\partial \bar{\phi}}{\partial x_3} \tag{70}$$

(c) For momentum (Equation 66):

$$\frac{\partial \bar{u}_i}{\partial t} + \frac{\partial \bar{u}_i \bar{u}_j}{\partial x_j} = (1 + \overline{\Delta\rho\phi}) g_i - \frac{1}{\rho_f}\frac{\partial \bar{p}}{\partial x_i} + \frac{1}{\rho_f} + \frac{\partial \bar{\tau}_{ij}}{\partial x_j} - \frac{\partial \overline{u_i' u_j'}}{\partial x_j} \tag{71}$$

As the suspension balance has been calculated already, the averaged equation of volume conservation, Equation 69, takes the simple form

$$\frac{\partial \bar{u}_i}{\partial x_i} = 0 \tag{72}$$

To study turbulent flow, these averaged equations of motion must be supplemented with a turbulent energy equation, and this may be obtained by multiplying the momentum equations with their corresponding velocity component, u_i, and summing up the equations obtained in this manner over the index, i. Applying this method to Equation 71 and using the continuity equation (72), we obtain, after some manipulation, the following equation for the kinetic energy balance for the mean flow:

$$\frac{\partial E}{\partial t} + \bar{u}_i \frac{\partial E}{\partial x_i} = \underbrace{-\frac{\partial}{\partial x_i}\left(\bar{u}_i \frac{\bar{p}}{\rho_f}\right)}_{\text{I}} + \underbrace{\frac{\partial}{\partial x_i}\left[\bar{u}_j\left(\frac{\bar{\tau}_{ij}}{\rho_f} - \overline{u_i' u_j'}\right)\right]}_{\text{II}} + \underbrace{\bar{u}_i g_i}_{\text{III}} \underbrace{- \bar{\epsilon}}_{\text{IV}} + \underbrace{\overline{u_i u_j}\frac{\partial \bar{u}_i}{\partial x_j}}_{\text{V}} \tag{73}$$

with

$$E = (1/2)\overline{u_i u_j}, \quad \bar{\epsilon} = \frac{1}{\rho_f}\bar{\tau}_{ij}\frac{\partial \bar{u}_i}{\partial x_j} \tag{74}$$

The right-hand side of Equation 73 represents the following factors affecting the loss of mean flow energy per unit mass and time:

I. work performed against the pressure field,
II. work performed against viscous and turbulent stresses,
III. work against gravity force,
IV. dissipation of mean flow energy under the action of viscosity, and
V. transition of mean flow energy into that of turbulence (in other words, the generation of turbulent energy).

Applying the same operation to Equations 66 and 68, we obtain an equation for the turbulent kinetic energy time dependence in a liquid flow transporting solid particles:

$$\frac{\partial e}{\partial t} + \bar{u}_i \frac{\partial e}{\partial x_i} = \overbrace{\frac{\partial D_i}{\partial x_i}}^{I} + \overbrace{\frac{\partial}{\partial x_i}\left(\frac{\overline{u_j'\tau_{ij}'}}{\rho_f}\right)}^{II} \overbrace{- \epsilon_t}^{III} + \overbrace{B}^{IV} \overbrace{- \overline{u_i'u_j'}\frac{\partial \bar{u}_i}{\partial x_j}}^{V} \tag{75}$$

$$e = (1/2)\overline{u_i'u_i'}, \quad D_i = -u_i\left(\frac{\rho'}{\rho_f} + \frac{\overline{u_j'u_j'}}{2}\right), \quad \epsilon_t = \frac{1}{\rho_f}\left(\overline{\tau_{ij}'\frac{\partial u_i}{\partial x_j}}\right)$$

$$B = \Delta\rho\, g_i\, \overline{\phi' u_i'} \simeq -\Delta\rho g\,(\overline{\phi' u_3'}), \left(\Delta\rho = \frac{\rho_p - \rho_f}{\rho_f}\right)$$

When we compare the energy balance equation (73) for the mean flow with the analogous equation for a suspensionless liquid, we realize that their structures are identical, the difference being a sign change for term V. Also, in the turbulent energy balance equation (75), all the terms (except IV) are present in the analogous equation for a suspensionless liquid. The terms on the right-hand side of Equation 75 signify

I. turbulent diffusion of "total" turbulent energy,
II. fluctuation work of viscous stresses in turbulent motion,
III. dissipation of turbulent energy into heat,
IV. consumption of turbulent fluctuation energy in suspending the solid particles, in other words, the work of suspension, and
V. transition of the energy of mean flow into that of turbulence, in other words, the generation of turbulent energy.

The equations of energy balance (73 and 75) are found to be identical with analogous equations for a density-stratified homogeneous liquid. In both cases, factor B, which may be written in the form

$$B = \frac{g}{\rho}\overline{\rho' u_3'} \simeq -\Delta\rho g \overline{\phi' u_3'} \tag{76}$$

can be interpreted as work spent in overcoming the buoyancy force arising from the turbulent displacement of liquid elements.

In fully developed turbulence, the stresses due to viscosity are negligibly small in comparison with Reynolds turbulent stresses. Further, the first term in Equation 75, which is the contribution of pressure fluctuations to the diffusional flux of turbulent energy, is very small and may be neglected. For the same reason, the second term, the work due to viscous stress fluctuations, may be neglected. This last assumption becomes valid when we compare this term with the main part of the first term (where this part may be interpreted as the work of the Reynolds stresses in the field of turbulent fluctuations).

If we now add the equation of turbulent energy balance to the system of basic equations of the problem and make the simplifications referred to above, we obtain

$$\frac{\partial u_i}{\partial t} + \bar{u}_j \frac{\partial \bar{u}_i}{\partial x_j} = (1 + \overline{\Delta\rho\phi}) g_i - \frac{1}{\rho_f}\frac{\partial \bar{p}}{\partial x_i} - \frac{\partial \overline{u_i'u_j'}}{\partial x_j}$$

$$\frac{\partial \bar{u}_j}{\partial x_j} = 0$$

$$\frac{\partial \bar{\phi}}{\partial t} + \bar{u}_i \frac{\partial \bar{\phi}}{\partial x_i} = u_r \frac{\partial \bar{\phi}}{\partial x_3} - \frac{\partial \overline{\phi' u'_1}}{\partial x_i}$$

$$\frac{\partial e}{\partial t} + u_i \frac{\partial e}{\partial x_i} = -\frac{1}{2}\frac{\partial}{\partial x_i}\overline{u_i u_j' u_j} - \overline{u_i u_j}\frac{\partial \bar{u}_i}{\partial x_j} - \Delta\rho g \phi' u_3' - \epsilon_t \tag{77}$$

This system of equations (77) shows that the influence of the particles on the dynamics of the entraining flow will be negligibly small where the flow work done to suspend the particles is sufficiently small or, in other words, when term B of the equation for the turbulent energy balance is small in comparison with the other terms of this equation. Note that in this case the term $\Delta\rho\bar{\phi}$ in the equation of momentum flux is negligibly small. When this is the case, the system of equations (77) breaks up into a set of independent equations: the first, second and fourth constitute the equations of motion, and the third is that of suspension balance.

Using these equations (77), we follow an analysis [38–40] for the special case of steady, plane, uniform flow in an open channel with small slope of angle θ and with horizontal and depth h. In this case, the system of basic equations may be considerably simplified because all mean flow characteristics depend on coordinate x_3 only.

The equation of volume conservation (69) then may be rewritten as

$$\frac{d\bar{u}_3}{dx_3} = -\Delta\rho u_r \frac{d\bar{\phi}}{dx_3} \tag{78}$$

and, after integration, we find

$$u_3 = -\Delta\rho u_r \bar{\phi} \tag{79}$$

This equation indicates that the velocity of mean flow of the mixture has a small component in the negative n_3 direction, which is due to the overall mean flux of each substance (which in the present case is zero in the x_3 direction) includes fluctuating components. In other words,

$$\overline{\phi u_{p_3}} = \bar{\phi}\bar{u}_{p_3} + \overline{\phi' u_{p_3}'} = 0, \quad \overline{(1-\phi)u_{f_3}} = (1-\bar{\phi})\bar{u}_{f_3} + \overline{\phi' u'_{f_3}} = 0 \tag{80}$$

It may also be shown that

$$\bar{u}_{f_3} = \bar{\phi} u_r, \quad \bar{u}_{p_3} = -(1-\bar{\phi})u_r \tag{81}$$

Since the velocity component, u_3, is small, the equation of mass balance (54) may be written in the approximate form:

$$\frac{d\,\overline{\phi' u_3'}}{dx_3} = u_r \frac{d\bar{\phi}}{dx_3} - u_3 \frac{d\bar{\phi}}{dx_3} \tag{82}$$

If we substitute Equations 80 into this equation, then the last term turns out to be a small quantity of second order (compared to $\bar{\phi}$) and, therefore, is negligible. And, integrating, yields

$$\overline{\phi' u_3'} = u_r \bar{\phi} \tag{83}$$

This is the classic equation of suspension balance (mass conservation) for a uniform flow.

By neglecting terms of small magnitude, we may write the momentum equations for the n_1 and n_3 directions as

$$\frac{d\overline{u_1' u_3'}}{dx_3} = g\sin\theta, \quad \frac{1}{\rho_f}\frac{d\bar{p}}{dx_3} = -(1+\Delta\rho\bar{\phi})g + \frac{d(\overline{u_3'^2})}{dx_3} \tag{84}$$

Also, by using Equation 83, the expression for the work of suspension, Equation 76 may be written in the form

$$B = -\Delta\rho g \overline{\phi' u_3'} = -\Delta\rho g u_r \bar{\phi} \tag{85}$$

Further, if we again ignore terms of small magnitude, then the equation of energy balance may be written as

$$\frac{d}{dx_3}\overline{u_3'u_i'u_i'} - \overline{u_1'u_3'}\frac{d\bar{u}_1}{dx_3} - \Delta\rho g u_r\bar{\phi} - \epsilon_t = 0 \tag{86}$$

The first momentum equation (84) can be integrated, the constant of integration being determined from the condition that the shear stress is zero on the free surface, and we obtain the usual expression for the distribution of turbulent shear stress, at any position, n_3:

$$-\overline{u_1'u_3'} = (v^*)^2\left(1 - \frac{x_3}{h}\right), \quad (v^* = (gh\sin\theta)^{1/2}) \tag{87}$$

Here, v^* is, by convention, the "shear velocity." Using this in Equation 86, we find

$$v^{*2}\left(1 - \frac{x_3}{h}\right)\frac{d\bar{u}_1}{dx_3} + \frac{d}{dx_3}\overline{u_3'u_i'u_i'} - \Delta\rho g u_r\bar{\phi} - \epsilon_t = 0 \tag{88}$$

and these equations, together with the equation of suspension balance,

$$\overline{\phi u_3'} = u_r\bar{\phi} \tag{89}$$

comprise the basic system of equations for the solution of the problem. However, this system is not complete because we need to simplify the turbulent energy equation. To achieve this, we ignore the second term, expressing the diffusional transfer of turbulent energy on the grounds that its influence is small for most of the depth of the flow [38–40]. We can express the correlations in Equations 87 and 89 in the usual manner, through coefficients of turbulent exchange:

$$\overline{u_1'u_3'} = -k_1\frac{d\bar{u}_1}{dx_3}, \quad \overline{\phi' u_3'} = k_2\frac{d\bar{\phi}}{dx_3} \tag{90}$$

For determination of the coefficients k_1, k_2 of turbulent exchange, as well as the dissipation, ϵ_t, of turbulent energy, we use Kolmogoroff's hypothesis, according to which the coefficients of exchange and dissipation are uniquely determined by the magnitude of turbulent energy, and by some system linear value, l, which may be regarded as a scale of turbulence or, for example, the Prandtl mixing length. Hence, we are led to

$$k_1 = le^{1/2}, \quad k_2 = ale^{1/2}, \quad \epsilon_t = cl^{-1}e^{3/2} \tag{91}$$

where a and c are nondimensional constant parameters. Substituting these expressions into the above equations, we obtain

$$le^{1/2}\frac{d\bar{u}_1}{dx_3} = v^{*2}\left(1 - \frac{x_3}{h}\right) \tag{92}$$

$$ale^{1/2}\frac{d\bar{\phi}}{dx_3} + u_r\bar{\phi} = 0 \tag{93}$$

$$v^{*2}\left(1 - \frac{x_3}{h}\right)\frac{d\bar{u}_3}{dx_3} = \Delta\rho g u_r\bar{\phi} + c\frac{\phi^{3/2}}{l} \tag{94}$$

This new system of equations will be closed if the dependence of the linear parameter,

l, on coordinate x_3 (or on the unknown quantities) is known. Using Equation 93 in Equation 85 we obtain

$$B = a\Delta\rho g l e^{1/2} \frac{d\overline{\phi}}{dx_3} \tag{95}$$

and using this we find from the equation of turbulent energy (94) an explicit expression for the turbulent energy:

$$e = \frac{1}{c^{1/2}} v^{*2} \left(1 - \frac{x_3}{h}\right) (1 - a\,k)^{1/2} \tag{96}$$

where

$$k = -\Delta\rho g \frac{d\overline{\phi}/dx_3}{(d\overline{u}_1/dx_3)^2}, \left(\Delta\rho = \frac{\rho_p - \rho_f}{\rho_p}\right)$$

The nondimensional factor, k, is analogous to the Richardson number occurring in the theory of stratified flows. This conclusion should not be surprising because the method outlined above does, in fact, describe mathematically the flow of a discrete two-component flow as if it were a continuum with variable (but continuously distributed) density. If suspended particles are absent, then $k = 0$ and Equation 96 gives

$$e = \frac{1}{c^{1/2}} v^{*2} \left(1 - \frac{x_3}{h}\right) \tag{97}$$

Comparison of this equation and Equation 96 shows that the presence of suspended particles in the flow leads to a reduction of turbulent energy, this reduction increasing with k. In the present case, k gives the ratio of work produced by the flow for the suspending of particles in the flow to the overall expenditure of turbulent energy. And, if the work of suspension is small in comparison with the dissipative term in the turbulent energy equation, we may apply the usual diffusion theory for suspensions, which neglects the reactive influence of suspended particles on the flow dynamics. Thus, from the theory [38] we have outlined above, we conclude that the diffusion theory is a reasonable approximation if

$$k << 1$$

It should be noted that the work of suspension is proportional to the particle terminal velocity and to the particle concentration, so that k is not small when only the terminal velocity is small.

It remains to define the linear parameter, l, and, again following the analysis given elsewhere [38–40], we use a modified form of the well known von Karman formula, giving

$$l = -K \frac{d\overline{u}_1}{dx_3} \left(\frac{d^2 u_1}{dx_3{}^2}\right)^{-1} f(ak) \tag{98}$$

Here, K is the so-called von Karman "constant" and $f(ak)$ is a function that must decrease with increasing k (or increase of the work of suspension), and become unity at $k = 0$.

After some analysis, these considerations lead us to conclude that the character of the flow depends not only on k, but also on another nondimensional parameter, ω, which occurs in various versions of the diffusion theory:

$$\omega = \frac{u_r}{a\,Kv^*} \simeq \frac{u_r}{Kv^*} \tag{99}$$

In this equation, u_r is the terminal velocity of the particles, K is the von Karman "constant," and v^* is the dynamic (shear) velocity.

The character of the flow appears to differ considerably for $\omega < 1$ and $\omega > 1$. For $\omega > 1$ (when the flow velocity is low and the particles are large), the transport of particles occurs mainly near the bottom surface of the flow. In the upper region of the flow, the influence of suspended particles on the dynamics of the transporting flow is small; however, in the region near the bottom, the theory (based on the assumption of small concentration $\overline{\phi}$) is inapplicable and must be replaced by a more general and exact theory. Far from the bottom region, $k \to 0$, and the distribution of suspended particles asymptotically approaches the distribution obtained from the diffusion theory.

For $\omega < 1$ (when the flow velocity is high and the particles are small), the transport of particles occurs within the main body of the flow. The particle distribution at some distance from the bottom surface asymptotically approaches a certain finite self-similar distribution, ($k \to$ const) for which, in the case of very large depth, $h \to \infty$, the concentration $\overline{\phi}$ is inversely proportional to x_3 and the velocity parallel to the surface is

$$\overline{u}_1 = \frac{1}{ku_r} \ln x_3 + \text{const} \tag{100}$$

Under these conditions, a certain limiting saturation occurs for the suspension flow, and for concentrations beyond that point the theory becomes inapplicable in the region near the bottom.

As early as 1944, investigations of the sediment-transporting flow in an open channel discovered that the presence of suspended sand particles in a flow in an open channel caused a decrease of the so-called von Karman universal constant and a reduction of the resistance coefficient in comparison with the corresponding flow of a homogeneous fluid. It was concluded that the particles acted to dampen the turbulence. These experimental results have been confirmed by many subsequent experiments. The theory of turbulent particle or sediment-transporting flow outlined above provides a theoretical explanation of the causes of this reduction of both the von Karman "constant" and the resistance coefficient. We relate the latter phenomenon to the reduction of velocity fluctuations in any flow transporting suspended particles.

MICROPARTICLES SUSPENDED IN A TURBULENT FLOW

A turbulent flow is considered to be composed of three main ranges of eddy size (or wave number)—small, medium and large—in which, as a result of inertia interactions, energy is transmitted from higher wave numbers ($\kappa = 2\pi n/v$, where n denotes the turbulence frequency and v the time-averaged velocity) to lower wave numbers. Turbulent motions differ in apparatus of different construction, and, in general, the motion occurring depends on the position in the flow. Although the wave number spectrum of any actual turbulent flow is not continuous, it is possible to assign a definite amount of the total energy to certain ranges of wave numbers. In other words, it is possible to derive a relationship between wave numbers in the form of an energy spectrum.

According to Kolmogoroff's first law [41–43], "at sufficiently high Reynolds numbers there exists a range of high wave numbers in which the turbulence is in static equilibrium, being defined unambiguously by the values of ϵ (energy dissipation) and ν (kinematic viscosity)." Let us consider the theory of uniform, isotropic turbulence developed by Kolmogoroff, this obviously being a theory for a flow field whose characteristics do not depend on the position or direction of the coordinate axes [34,41–44] because this theory has become well developed and because it fairly closely approximates flow regimes occurring in commercial equipment.

The largest eddy scale corresponds to the most rapid eddy motions, and the order of magnitude of the velocities of these eddies is given by the characteristic velocity of the system, say U, the maximum mean flow velocity. So, for example, the turbulent flow in a pipe has a largest eddy scale approximately equal to the pipe diameter, and the eddy velocity will approximate the variation of the average velocity over the pipe diameter. Or, in this case, it will be equal to the maximum velocity, which occurs in the center of the pipe [45].

Eddies of any scale can be expressed in terms of a Reynolds number defined as

$$\mathrm{Re}_\lambda = \frac{v_e \lambda}{\nu_f} \tag{101}$$

where v_e is the velocity of eddies of scale λ, and ν_f is the coefficient of kinematic viscosity. Now, for some $\lambda = \lambda_o$, the Reynolds number $\mathrm{Re}_\lambda \simeq 1$, and the eddy motion is accompanied by energy dissipation. Using dimensional considerations [46], it can be shown that for $\lambda > \lambda_o$ the eddy velocity may be written as

$$v_e = (\epsilon\lambda)^{1/3}; \lambda > \lambda_o, \tag{102}$$

where ϵ is the energy dissipation per unit mass. And, when $\lambda \simeq \lambda_o$, by definition (from Equation 101),

$$v_{e_o} = \frac{\nu_f}{\lambda_o} \tag{103}$$

The acceleration is

$$a_e \simeq \left(\frac{\epsilon^3}{\nu_f}\right)^{1/4} ; \lambda > \lambda_o \tag{104}$$

Next, for $\lambda < \lambda_o$ the motion acquires a quasiviscous character, interaction between separate eddies ceases and eddy frequencies become independent of scale. Then the eddy velocity can be derived from the following relationship:

$$v_e \simeq \frac{\lambda}{\tau} ; \lambda < \lambda_o \tag{105}$$

where τ is the eddy period, and the acceleration term assumes the form

$$a_e \simeq \left(\frac{v_{e_o}}{\lambda_o}\right)^2 \lambda ; \lambda < \lambda_o \tag{106}$$

For eddies of scale, $\lambda > \lambda_o$, there corresponds a period, τ, of the motion that is dependent on the scale, λ, and the turbulent dissipation, ϵ, only, because for this range the influence of viscosity may be ignored (Kolmogoroff's first law) and τ is given by

$$\tau = \left(\frac{\lambda^2}{\epsilon}\right)^{1/3} ; \lambda > \lambda_o \tag{107}$$

By comparison, for eddies of scale ($\lambda \simeq \lambda_o$), τ is given by

$$\tau_o \simeq \frac{\lambda_o}{v_e} \simeq \left(\frac{\nu_f}{\epsilon}\right)^{1/2} \tag{108}$$

With the expressions given above, let us now consider the motion of particles in a homogeneous isotropic turbulent flow. We will use a theory developed elsewhere in which we consider the solid particles and liquid as one fluid, and in using such an approach we note that the problem of calculating liquid pressure on the surfaces of the particles no longer arises, this being an internal system constraint [21].

For the motion of a particle in a homogeneous isotropic turbulent flow, let the liquid surrounding the particle have an acceleration, a_f. In this case, we would like to determine what the acceleration, a_p, of the particle will be. Neglecting gravitational forces and perturbation of the flow caused by the particle itself, all the elements of the liquid in the neighborhood of the particle have identical accelerations, a_f, and to this acceleration there corresponds a pressure drop in the direction of the acceleration vector given [47] by

$$\frac{\partial p}{\partial x} = \rho_p a_f \tag{109}$$

As a result of this pressure drop, on a particle with volume V located in the liquid there acts a force $V\rho_f a_f$ in the direction of the acceleration vector, but since the density, ρ_p, of the particle is not equal to the density, ρ_f, of the liquid, the acceleration, a_p, of the particle is also not equal to the acceleration of the liquid, and the particle thus moves relative to the liquid with an acceleration equal to $(a_p - a_f)$. For relative motion of a particle in the liquid with a nonconstant velocity, the effect exerted on the particle by the liquid is given by an additional term, $\rho_f V_a$, added to the mass of the particle [21]. Here, V_a can be envisioned as a volume of liquid considered attached to the particle and that moves with the particle. This additional term means that the total influence of liquid motion on the particle can be represented simply as an increase in its mass and, consequently, as an increase of the total system inertia. So, as a result of the relative motion of the particle in the liquid, a resistance is set up that is opposite to the driving force and proportional to the "added mass," $\rho_f V_a$, that is, equal to $\rho_f(a_p - a_f)V_a$. Then, the equation of relative motion of a particle in a turbulent flow can be represented in the form

$$\rho_p V a_p = \rho_f[a_f V - (a_p - a_f)V_a] \tag{110}$$

From Equation 110 we can find the particle acceleration as a function of the total liquid acceleration:

$$a_p = a_f \frac{(V + V_a)}{\left(V\frac{\rho_p}{\rho_f} + V_a\right)} \tag{111}$$

Taking account of the "added mass" effect thus makes a noticeable contribution to the value of a_p and, for example, for particles approaching a spherical form, the added mass is equal to half the mass of the displaced liquid [48].

The motion of particles that are considerably smaller than the scale of turbulence is determined by the small-scale eddies, and the maximum value for the acceleration of these liquid elements is given [46] by Equation 104. It follows from Equation 111 that for $\rho_p > \rho_f$ the acceleration, a_p, of the particle will be less than the acceleration, a_f, of the flow. Therefore, we can write

$$a_p = a_e - a_r \tag{112}$$

where we have used the relative acceleration of the motion in the form

$$a_r \simeq \frac{v_r}{\tau_p}, \tag{113}$$

where v_r is the particle relative velocity, and τ_p is the period of particle motion given [45] by

$$\tau_p \simeq \frac{\lambda_o}{v_r} \tag{114}$$

Note that for the smaller turbulent eddies we have [46]

$$\lambda_o \simeq \left(\frac{\nu_f^3}{\epsilon}\right)^{1/4} \tag{115}$$

Using Equations 104, 113, 114 and 115 in Equation 112, we obtain

$$a_p \simeq \left(\frac{\epsilon^3}{\nu_f}\right)^{1/4} - \frac{v_r^2}{\left(\frac{\nu_f^3}{\epsilon}\right)^{1/4}} \tag{116}$$

and using this equation for the particle acceleration, a_p, and Equation 104 for the acceleration, a_f, of the liquid, we find that

$$v_r \simeq (\epsilon\nu_f)^{1/4}\left(1 - \frac{V + V_a}{V\frac{\rho_p}{\rho_f} + V_a}\right)^{1/2} \tag{117}$$

Clearly, for $\rho_p >> \rho_f$, in Equation 117 the factor

$$1 - \left(\frac{V + V_a}{V\frac{\rho_p}{\rho_f} + V_a}\right)$$

approaches unity and, in this case,

$$v_r \simeq (\epsilon\nu_f)^{1/4} \tag{118}$$

Taking into account that the velocity of small-scale eddies [21] is given by Equation 103 with λ_o given by Equation 115, we find

$$v_{e_o} \simeq (\epsilon\nu_f)^{1/4} \tag{119}$$

Consequently, when the density of the particles is significantly greater than the density of the liquid, $\rho_p >> \rho_f$, the average value of the relative velocity of a particle (given by Equation 118) is equal to the velocity of eddies of scale λ_o (Equation 119) and is seen to be uniquely determined by the viscosity of the liquid and the energy dissipation per unit mass for constant-density fluids.

In actual industrial processes, the turbulence created in the liquid by use of agitators, for example, or that occurring in a bubbling apparatus, usually has an extremely anisotropic character, this being determined by large-scale eddies. Therefore, it is of interest to investigate to what degree Equation 117 is valid for the case of anisotropic turbulence. It is well known [46] that large-scale eddies, whose size (or scale) is determined by a characteristic system dimension (for example, the diameter of an agitator or the height of a two-phase mixture), contain most of the energy of the flow, this energy being obtained directly from the system generating the turbulence. However, transfer of energy to small-scale eddies is accompanied by increased disorder of the motion and by a diminishing of the effect of the large-scale eddies on the motion. In addition, the anisotropy of the large-scale eddies has no effect on eddies of the order of the internal scale of the turbulence. Therefore, application of the rigorous condition of homogeneity of the turbulence to all the turbulence scales is not required, because the motion of these small-scale eddies can be assumed to take place as in purely homogeneous and isotropic turbulence.

By means of the theory of dimensional analysis, the energy dissipated per unit mass and time is given by

$$\epsilon \simeq \frac{U^3}{l} \tag{120}$$

where U is some average velocity of the flow, and l is a characteristic length scale (equal to the size of largest eddies). Substituting into Equation 117 the value of ϵ from Equation 120, we determine the relative velocity of a microparticle suspended in the turbulent flow to be

$$v_r = \text{const.} \left(1 - \frac{V + V_a}{V \dfrac{\rho_p}{\rho_f} + V_a}\right)^{1/2} \left(\frac{\nu_f U^3}{l}\right)^{1/4} \tag{121}$$

Equation 121 shows that the relative velocity, v_r, of a particle increases with the mean velocity, U, of the flow in the form

$$v_r \propto U^{3/4} \tag{122}$$

Equation 121 may be used to calculate the particle relative velocity in a tubular apparatus because in this application it is possible to measure the average flow velocity. The tube diameter can be used as the characteristic dimension, and the equation has been correlated with experimental data [49] for a liquid droplets-air system; the flow conditions were inferred from the system pressure and from assumed saturation conditions, as shown in Table I.

Droplet diameters for each run were measured, and the corresponding velocities, v_d, calculated by a distance/time correlation. The drops for each run were of relatively uniform size and were assumed to be moving in one dimension and at constant velocity, i.e., with zero absolute acceleration.

The vapor velocity was inferred from a vertical force balance of drag against gravity:

$$F_D = F_g \tag{123a}$$

or

$$\frac{1}{8} C_D \pi d^2 \rho_g (v_g - v_d)^2 = \frac{1}{6} \pi d^3 \rho_f g \tag{123b}$$

where C_D is the drag coefficient, assumed to have the value 0.45, v_g is the vapor velocity, and d is the drop diameter. This reduces to

$$v_g = v_d + \left(\frac{4d\rho_f g}{3C_D \rho_g}\right)^{1/2} \tag{123c}$$

To compare the theory with experimental data, we can use Equation 121 if the size of droplets is actually less than the size of small-scale eddies. The size of small-scale eddies is given by

$$r_d = \left(\frac{8\nu_f^3}{3\gamma^2 \epsilon}\right)^{1/4} \tag{124}$$

where $\gamma = 0.4$ is a universal constant, and ϵ is the energy dissipation, obtained from Equation 120 in the form

$$\epsilon = \frac{v_g^3}{l} \tag{125}$$

where l is the height of the duct ($l = 4.88$ m). According to the experimental data, $\nu_f \simeq 0.025 m^2/sec$ and $v_g \simeq 10.0$ m/sec, so from Equation 124 we find that the typical

Table I. Experimental Data for Computing the Particle Relative Velocity

Run No.	p ($\times 10^5$Pa)	σ ($\times 10^{-3}$N/m)	ρ_f (kg/m³)	ρ_g (kg/m³)	Flood Rate (cm/sec)
1	1.73	5.79	955.08	0.99	5.08
2	1.24	6.00	936.68	0.73	2.03
3	1.03	6.11	967.56	0.61	1.52
4	1.03	6.11	967.56	0.61	1.52
5	3.78	5.24	933.91	2.07	14.99
6	1.30	5.97	962.55	0.76	2.54

size (or radius) of small-scale eddies is $r_d \simeq 3.3$ cm, which is much greater than any droplet sizes occurring in these experiments and, therefore, we may use Equation 121 to compare the theory with experimental data. Experimental [50] and calculated data for particle relative velocities indicate that the average variance with experimental data is 3.8%, which indicates that the theory is a good model for the flow.

MOTION OF MICROPARTICLES IN THE INTENSIVE BUBBLING REGIME

When gas is injected into a liquid, intensive motion of bubbles can be observed under certain flow conditions, and this motion causes a significant mixing of particle-carrying liquid. It is usually the practice [51] to distinguish three principal bubbling regimes (unsteady motion, steady motion and the regime of gas sprays and liquid jets), the transition from one regime to another being determined by gas and liquid distribution devices. Investigation of the steady regime is of practical importance, as well as being the regime most accessible to analysis.

When there is intensive bubbling, it is best to consider the gas-liquid mixture from the point of view of homogeneous isotropic turbulence [50] because it is known that fully developed turbulence is in fact created as a result of the liquid agitation by the moving bubbles. We make the assumption that all the gas kinetic energy is transferred to the liquid-particle mixture, eventually being dissipated by the turbulent motion. Then the pressure drop in turbulent flow is determined [46] by

$$\Delta p \simeq \rho_f U^2. \tag{126}$$

Thus, from Equations 121 and 126, we obtain

$$v_r = \text{const.} \left(\frac{\nu_f}{l}\right)^{1/4} \left(1 - \frac{V + V_a}{V \dfrac{\rho_p}{\rho_f} + V_a}\right)^{1/2} \left(\frac{\Delta p}{\rho_f}\right)^{3/8} \tag{127}$$

where l could be the depth of the bubbling mixture, for example.

MOTION OF MICROPARTICLES IN A BUBBLE APPARATUS WITH AGITATORS

It is known that the energy, E, consumed during liquid mixing in a baffled apparatus can be calculated by the following [52] equation:

$$\frac{E}{\rho_f n^3 D_m^5} = \text{const.} \left(\frac{\rho_f n D_m^2}{\mu_f}\right)^{\alpha} \left(\frac{n^2 D_m}{g}\right)^{\beta} \left(\frac{T}{D_m}\right)^{\lambda} \tag{128}$$

where $Fr = n^2D_m/g$ is the centrifugal Froude number, $Re = \rho_f n D_m^2/\mu_f$ is the centrifugal Reynolds number (and μ_f is the liquid dynamic viscosity), $\Gamma = T/D_m$ is a geometric similarity parameter, n is the revolution rate of the mixer where D_m is the diameter of the mixer, and T is the diameter of the chamber. Gas is dispersed uniformly in the whole liquid volume when intensively mixed and, because bubble sizes are small in comparison with the size of the apparatus and mixer, the gas-liquid system may be considered a homogeneous mixture with a uniform average density. For this case, the energy consumption is equivalent to that occurring in homogeneous liquid as described in the previous section, except that the "liquid" density is equal to the average density of the mixture. In other words,

$$\rho_{av} = \rho_g\phi + \rho_f(1 - \phi) \tag{129}$$

where ϕ (the gas void fraction) may be expressed [53] as a product of the relative gas flowrate, q, and its residence time, τ, in the system:

$$\phi = q\tau = \frac{qx_1}{v_b} \tag{130}$$

where x_1 is the two-phase mixture height, and v_b is the average rise velocity of a bubble.

The velocity for a bubble whose size is greater than the scale of internal turbulence is independent of the viscosity of the liquid, and the bubble motion is determined by the "auto-modeling" zone of the turbulent regime, so we find [54] that

$$v_b = 1.76\left[\frac{d_b g(\rho_f - \rho_g)}{\rho_f}\right] \tag{131}$$

where d_b is the bubble diameter. The gas content in an apparatus with a mixer, once we have calculated the bubble diameter on the basis of homogeneous turbulence, can then be shown to be of the form [55]

$$\phi = \text{const.}\,\frac{q^3T^{0.4}\rho_g^{0.3}E^{0.2}}{\sigma^{0.3}\rho_f^{0.2}g^{0.5}} \tag{132}$$

where σ is the surface tension. Combining Equations 129 and 132, we obtain a value for ρ_{av}, which can be substituted into Equation 128 in place of the term ρ_f.

Experimental coefficients for Equation 128 have been obtained that describe power consumption during mixing with a turbine-type mixer in a gas-liquid system, leading to

$$K_E = 1.08Re^{0.1}Fr^{-0.03}We^{-0.12}K_v^{-0.2}s_g\Gamma^{-0.4} \tag{133}$$

where $We = \rho n^2 D_m/\sigma$ is the centrifugal Weber number, $s_g = \rho_g/\rho_f$ is a dimensionless density, and $K_v = U/n$ is a distribution coefficient.

This equation is not convenient for calculation of energy consumed by liquid mixing for liquids with properties close to water, but it is possible to use the following simplified formula [36]:

$$E \sim 0.9\,\frac{n^3D_m^{5.22}}{T^{0.4}q^{0.2}} \tag{134}$$

We can express the dissipation energy, ϵ, for $Re > 10^4$ in terms of the process parameters that characterize the turbulent regime in an apparatus with mixers as [54]

$$\epsilon = \frac{E}{Q_l} \sim \frac{n^3D_m^5}{T^2x_1q^{0.2}} \tag{135}$$

Taking into account that

$$\epsilon = \frac{U^3}{l} \tag{136}$$

and, in this application, that

$$l \simeq D_m \tag{137}$$

Equation 135 gives

$$U = \text{const.}\,\frac{nD_m^{\,2}}{(T^2x_1q^{0.2})^{1/3}}, \quad Re > 10^4 \tag{138}$$

Equations 137 and 138 conform well with experimental data for systems of various geometries over a broad range of hydrodynamic parameters [37].

So, substituting values l and U from Equations 137 and 138 in Equation 121 gives us an expression for the relative velocity of microparticles in turbulent flow in a bubble apparatus with a mixer:

$$v_r = \text{const.}\left[1 - \frac{V + V_a}{V\dfrac{\rho_p}{\rho_f} + V_a}\right]^{1/2}\left[\frac{n^3D_m^{\,5}\nu_f}{T^2x_1q^{0.2}}\right]^{1/4} \tag{139}$$

MOTION OF A MACROPARTICLE SUSPENDED IN A TURBULENT FLOW

Let us now consider the motion of a solid particle whose size is significantly greater than the size of small-scale eddies, noting that this will include taking into account the unsteady velocity field. In this case, the motion of a particle is determined by the large-scale eddies, and the liquid acceleration depends on ϵ and λ in the form [46]

$$a_e \simeq \left(\frac{\epsilon^2}{\lambda}\right)^{1/3} \tag{140}$$

where a_e is now the acceleration of large-scale eddies, and λ is the scale of these eddies.

For $\rho_p \neq \rho_f$, the particle acceleration, a_p, will be less than the liquid flow acceleration, which we take to be equal to the acceleration, a_e, of large-scale eddies, so that we will have

$$a_p = a_e - a_r \tag{141}$$

where a_r is the relative acceleration. The equation of relative motion of a particle is similar to Equation 110 and we can determine the acceleration term, a_p, from this equation. The relative acceleration is given by Equation 113 as

$$a_r = \frac{v_r}{\tau} \tag{142}$$

where the period of particle motion [46] is

$$\tau = \frac{r_p^{\,2}}{\nu_f} \tag{143}$$

So Equation 141 becomes

$$a_p = \left(\frac{\epsilon^2}{\lambda}\right)^{1/3} - \frac{v_r \nu_f}{r_p^2} \tag{144}$$

From dimensional theory, we can determine the eddy period [46] as

$$\tau = \left(\frac{\lambda^2}{\epsilon}\right)^{1/3} \tag{145}$$

and, taking this into account, with Equation 143 we now can substitute into Equation 111 the liquid acceleration a_e, from Equation 140 and particle acceleration, a_p, from Equation 144, and write the relative velocity as

$$v_r = \left(\frac{\epsilon}{\nu_f}\right)^{1/2} r_p \left(\frac{V\frac{\rho_p}{\rho_f} - V}{V\frac{\rho_p}{\rho_f} + V_a}\right) \tag{146}$$

In reality, when there is a system of discrete particles in a turbulent flow, each particle is influenced by the particles in its immediate neighborhood, as well as those more distant. So the pattern of the motion of the discrete particles is to a large measure determined by their concentration [53]. For determination of this effect of constrained motion, we will use the "cell" model. Let us suppose each particle is in the center of a sphere formed by adjacent particles. With this assumption, the problem of constrained motion may be reduced to the problem of liquid motion between two concentric spheres, the inner one being the particle and the outer one the surrounding particles.

The influence of particles in the turbulent energy spectrum, $E(\kappa)$, has been investigated qualitatively [53], and it seems that the addition of particles does not change the spectrum in the range of $\kappa << \kappa_d \sim 1/r_d$ (where κ_d and r_d are the wavenumber and eddy size, respectively, at which viscous effects first appear significant) but has the effect of extinguishing eddies for $\kappa >> \kappa_d$. To simplify the analysis of the problem, we thus can seriously consider the short wavelength range of $E(\kappa)$, starting from some characteristic wavenumber, κ_0, to be eliminated and considering that, in general, for $\kappa < \kappa_0$ the spectrum $E(\kappa)$ is not changed by adding particles. In fact, if the particle concentration is sufficiently large, eddies smaller than the particles are extinguished completely.

As a first approximation, at a sufficiently high concentration of particles, when the radius of the "outer sphere" mentioned above becomes comparable to about twice the diameter of a particle, the motion of liquid inside the sphere may be considered as laminar. Further, under conditions of high particle concentration and because the particle size is less than the eddy scale, λ, the cluster of neighboring particles will be involved in the same motion as the central particle, being in the same eddy. Consequently, the central particle and "spherical shell," composed of neighboring particles, are moving with equal velocities. In other words, we may consider the outer shell to be stationary with respect to the central particle and, in this case we can write expressions for radial and tangential components of the liquid velocity, relative to the sphere center [53]:

$$v_R = \left(\frac{A}{r^3} + \frac{B}{r} + C + Dr^3\right)\cos\theta - v_r\cos\theta \tag{147}$$

$$v_\theta = \left(\frac{A}{2r^3} - \frac{B}{2r} - C - 2Dr^3\right)\sin\theta + v_r\sin\theta \tag{148}$$

where

$$A = \frac{-v_r r_m^3}{2 - 3\gamma + 3\gamma^5 - 2\gamma^6}$$
$$B = \frac{v_r r_m (2\gamma^5 + 3)}{2 - 3\gamma + 3\gamma^5 - 2\gamma^6}$$
$$C = \frac{-v_r (2\gamma^6 + 3\gamma)}{2 - 3\gamma + 3\gamma^5 - 2\gamma^6} \tag{149}$$
$$D = \frac{v_r \gamma^3}{(2 - 3\gamma + 3\gamma^5 - 2\gamma^6) r_m^2}$$

and $\gamma = r_m/r_p$, where r_m is the radius of the outer "shell," and r_p is the particle radius. Here, r and θ are the radial and angular coordinates, with velocity components v_R and v_θ, respectively.

From physical considerations, for the case in which the Schmidt number Sc >> 1, Equations 147 and 148 may be reduced to the form [37]

$$v_\theta = v_r \Psi(\gamma) \frac{\eta}{r_m} \sin\theta \tag{150}$$

$$v_R = -v_r \Psi(\gamma) \left(\frac{\eta}{r_m}\right)^2 \cos\theta \tag{151}$$

where $\eta = (r - r_m)$.

Thus, the constraint conditions we have introduced for the motion may be represented by a separate factor, $\Psi(\gamma)$, so that

$$v_r \sim v_{r_o} \Psi(\gamma) \tag{152}$$

and

$$\Psi(\gamma) = \frac{3 - \frac{9}{2}\gamma + \frac{9}{2}\gamma^3 - 3\gamma^6}{3 + 2\gamma^5} \tag{153}$$

The relative velocity of a particle in this constrained motion then may be obtained from Equation 146 as

$$v_r = \left(\frac{\epsilon}{\nu_f}\right)^{1/2} r_p \left(\frac{\rho_p - \rho_f}{\rho_p + 0.5\rho_f}\right) \Psi(\gamma) \tag{154}$$

It follows from Equation 154 that the relative velocity of a particle in the constrained motion is determined by the particle size, the energy dissipation and volume content of the solid phase (the latter due to γ). Now, with the energy dissipation being $\epsilon \simeq U^3/l$, and particle relative velocity being $v_r \sim U^{3/2}$, using Equations 138 and 128 (which determine the flow patterns) the relative velocity of a particle is obtained as

$$v_r = \text{const.} \left(\frac{\rho_p - \rho_f}{\rho_p + 0.5\rho_f}\right) \Psi(\gamma) \frac{U^{3/2} r_p}{(l\nu_f)^{1/2}} \tag{155}$$

for a tubular apparatus,

$$v_r = \text{const.} \left(\frac{\rho_p - \rho_f}{\rho_p + 0.5\rho_f}\right) \Psi(\gamma) \Delta P_{st}^{3/4} \frac{r_p}{(x_1\nu_f)^{1/2}\rho_f} \tag{156}$$

for a bubbling apparatus, and

$$v_r = \text{const.} \left(\frac{\rho_p - \rho_f}{\rho_p + 0.5\rho_f}\right)\left(\frac{E^3 D_m{}^5}{T^2 x_1}\right)\frac{r_p}{(\nu_f \rho_f)^{1/2}} \tag{157}$$

for a bubbling apparatus with a mixer.

NOMENCLATURE

- A flow system cross-sectional area
- a_e eddy acceleration
- a_f fluid acceleration
- a_p particle acceleration
- a_r relative acceleration
- C_D drag coefficient
- D_m diameter of mixer
- d drop diameter
- d_b bubble diameter
- d_p particle diameter
- F_D drag force
- F_g gravitational force
- Fr Froude number
- F_i interaction force per unit volume between liquid and solids particles
- g gravitational acceleration
- K von Karman constant
- K_v distribution coefficient
- k work of suspension
- k_1,k_2 coefficients of turbulent exchange
- l characteristic length scale
- M mass flowrate of mixture/area
- N_s Stokes number
- n turbulent frequency; revolution rate
- p pressure
- $\overline{p}$ mean pressure
- p' fluctuating pressure
- Q volumetric flowrate of particles
- q relative gas flowrate
- Re Reynolds number
- Re_λ Reynolds number defined by Equation 101
- r_d size of small-scale eddies
- r_p particle radius
- s_g dimensionless density
- t time
- T diameter of chamber
- u_{fi} fluid velocity vector
- u_i mixture velocity vector
- $\overline{u}_i$ mean velocity vector of mixture
- u_i' fluctuating velocity vector of mixture
- u_{pi} particle velocity vector
- u_r particle terminal velocity
- U maximum mean flow velocity
- V particle volume
- V_a liquid volume equivalent to additional mass
- v time-averaged velocity
- v^* shear velocity
- v_d droplet absolute velocity
- v_b bubble velocity
- v_e small eddy characteristic velocity
- v_f liquid velocity
- v_g vapor velocity
- v_p particle velocity
- $\overline{v}_p$ average particle velocity
- v_s solid superficial velocity
- v_{sf} liquid superficial velocity
- v_r relative velocity of particle
- $\overline{v}_r$ average relative velocity of particles
- We Weber number
- x_i position vector
- x_1 height of two-phase mixture

Greek Symbols

- β phase angle
- Δp pressure drop
- ϵ energy dissipation/mass/time
- ϵ_t dissipated turbulent energy
- η amplitude ratio
- κ wavenumber
- κ_d wavenumber of eddies with viscous effect
- λ microscale of turbulent eddies
- μ mixture dynamic viscosity
- μ_f liquid dynamic viscosity
- ν kinematic viscosity
- ν_f liquid kinematic viscosity
- ρ mixture density
- ρ_f liquid density
- ρ_p particle density

$\bar{\rho}$ mean mixture density
ρ' fluctuating mixture density
σ surface tension
τ eddy period
$\tau_{ij}^{(f)}$ tensor representing viscous stresses in the liquid
$\tau_{ij}^{(p)}$ stress tensor resulting from the direct interaction of solids particles
τ_p period of particle motion
ϕ solid content of mixture
$\bar{\phi}$ mean solid content of mixture
Ψ relative density of the mixture

REFERENCES

1. Leva, M. *Fluidization* (New York: McGraw-Hill Book Co., 1959).
2. Zenz, F. A., and D. F. Othmer. *Fluidization and Fluid Particle Systems* (New York: Van Nostrand Reinhold Co., 1960).
3. Orr, C., Jr. *Particulate Technology* (New York: Macmillan Publishing Company, Inc., 1966).
4. Soo, S. L. *Fluid Dynamics of Multiphase Systems* (Waltham, MA: Blaisdell Publishing Co., 1967).
5. Brodkey, R. S. *The Phenomena of Fluid Motions* (Reading, MA: Addison-Wesley Publishing Company, Inc., 1967).
6. Wallis, G. B. *One-Dimensional Two-Phase Flow* (New York: McGraw-Hill Book Co., 1969).
7. Zenz, F. A. *Pet. Ref.* 36(4):173 (1957).
8. Zenz, F. A. *Pet. Ref.* 36(5):261 (1957).
9. Zenz, F. A. *Pet. Ref.* 36(6):133 (1957).
10. Zenz, F. A. *Pet. Ref.* 36(7):175 (1957).
11. Zenz, F. A. *Pet. Ref.* 36(8):147 (1957).
12. Zenz, F. A. *Pet. Ref.* 36(9):305 (1957).
13. Zenz, F. A. *Pet. Ref.* 36(10):162 (1957).
14. Zenz, F. A. *Pet. Ref.* 36(11):321 (1957).
15. Frantz, J. F. *Chem. Eng.* 69:61 (September 17, 1962).
16. Frantz, J. F. *Chem. Eng.* 69:89 (October 1, 1962).
17. Frantz, J. F. *Chem. Eng.* 69:103 (October 29, 1962).
18. Julian, F. M., and A. E. Dukler. *AIChE J.* 11:853 (1965).
19. Goveir, G. W., and K. Aziz. *The Flow of Complex Mixtures in Pipes* (New York: Van Nostrand Reinhold Company, 1972).
20. Shook, C. A., and S. M. Daniel. *Can. J. Chem. Eng.* 47:196 (1969).
21. Lamb, H. *Hydrodynamics* (New York: Dover Publications, Inc., 1945).
22. Oseen, C. *Neuere Methoden und Ergebnisse in der Hydrodynamik* (Leipzig, West Germany, 1927).
23. Langhaar, H. L. *Dimensional Analysis and Theory of Models* (New York: John Wiley & Sons, Inc., 1951).
24. Handl. *Exp. Phys.* 4(4):13 (1932).
25. Klyachko, L. *Otopleneie Vent.* No. 4 (1934).
26. Basset, A. B. *A Treatise on Hydrodynamics,* Vol. 2 (Cambridge, MA: Deighton, Bell and Co., 1888).
27. Boussinesq, J. *Theorie Analitique de Chaleur,* Vol. 2 (Paris, 1903), p. 204.
28. Lumley, J. "Some Problems Connected with the Motion of Small Particles in Turbulent Fluid," PhD Thesis, Johns Hopkins University, Baltimore, MD (1957).
29. Soo, S. L. *Chem. Eng. Sci.* 5:57 (1956).
30. Friedlander, S. K. *AIChE J.* 3:381 (1957).
31. Liu, V. C. *J. Meteorol.* 13:399 (1956).
32. Tchen, C. M. "Mean Values and Correlation Problems Connected with the Motion of Small Particles Suspended in a Turbulent Fluid," PhD Thesis, Delft, The Netherlands (1947).
33. Corrsin, S., and J. Lumley. *Appl. Sci. Res.* A6:114 (1956).
34. Hinze, J. O. *Turbulence* (New York: McGraw-Hill Book Co., 1959).

35. Hjelmfelt, A. T., Jr., and L. F. Mockros. *Appl. Sci. Res.* 16:149–161 (1966).
36. Braginski, L. N., and L. S. Pavlushenko. *Sbornik Prot. Khim. Tekhnol.* (1965).
37. Schwartzberg, H. G., and R. E. Treyball. Ind. Eng. Chem. Fund. 7(1):6 (1968).
38. Barenblatt, G. I. *Zh. Prik. Mat. i Mech.* 17:3 (1953).
39. Barenblatt, G. I. *Zh. Prik. Mat. i Mech.* 19:1 (1955).
40. Barenblatt, G. I. *Vest. Mosk. Univ.* 8 (1955).
41. Kolmogoroff, A. N. *DAN SSSR* 30:229 (1941).
42. Kolmogoroff, A. N. *DAN SSSR* 31:538 (1941).
43. Kolmogoroff, A. N. *DAN SSSR* 32:19 (1941).
44. Batchelor, G. K. *The Theory of Homogeneous Turbulence* (New York: Cambridge University Press, 1953).
45. Levich, V. G. *Physicochemical Hydrodynamics* (Englewood Cliffs, NJ: Prentice-Hall, Inc., 1962).
46. Landau, L., and E. Lifshitz. *Fluid Mechanics* (London: Pergamon Press, Inc., 1959).
47. Prandtl, L., and O. G. Tietijens. *Hydro- and Aerodynamics* (New York: Dover Publications, Inc., 1957).
48. Kochin, N. E., A. I. Kibel and H. V. Roze. *Teoret. Gidrod.* GTTI, (1955) (in Russian).
49. Smith, T. A. "Heat Transfer and Carryover of Low Pressure Water in a Heated Vertical Tube," MA Thesis, MIT, Cambridge, MA (1976).
50. Azbel, D. S., and A. F. Narozhenko. *Teoretich. Oxnov. Khim. Teknol.* 3:508 (1969).
51. Ramm, V. M. *Absorption of Gases* (Moscow: Chimia, 1966).
52. Kasatkin, A. G. *Osnov. Prot. i App. Khim. Tekhnol.* Goskhimizdat (1966).
53. Ruckenstein, E. *Chem. Eng. Sci.* 19:131 (1964).
54. Piterskikh, P. P., and E. R. Valashek. *Khim. Prom.* 35:1 (1956).
55. Azbel, D. S. *Two-Phase Flows in Chemical Engineering* (New York: Cambridge University Press, 1981).

[illegible]

University Press, [illegible]

[illegible]

University Press, [illegible]

CHAPTER 35
PIPELINE FLOW OF COARSE-PARTICLE SLURRIES

C. A. Shook
Department of Chemistry and Chemical Engineering
University of Saskatchewan
Saskatoon, Saskatchewan SN7 0W0
Canada

CONTENTS

INTRODUCTION

As in the case of fine particles, the dominant problems in the design of systems for handling coarse particles concern predicting pressure drops and limit-deposit velocities. The first successful generalization of experimental headloss results was the Durand-Condolios correlation [1]:

$$\phi = 81\,\Psi^{3/2} \tag{1}$$

where ϕ represents the dimensionless excess frictional headloss,

$$\phi = (i - i_f)/C_{vd}\, i_f \tag{2}$$

and Ψ^{-1} is a form of the Froude number,

$$\Psi^{-1} = V^2 \sqrt{C_D}\,/gD(S_s - 1) \tag{3}$$

Proposed originally for monosize particles, some variation in particle size can be accommodated by the weighted mean drag coefficient, C_{DM}:

$$1/C_{DM} = \Sigma(x/C_D)_i \tag{4}$$

for the fractions that comprise the mixture.

The first recognition of the inherently different character of coarse-particle slurries was given by Newitt et al. [2]. A separate equation was derived by considering the coulombic friction between particles and the pipe wall:

$$i = i_f + 0.8\,(S_s - 1)C_{vd} \tag{5a}$$

or

$$\phi = 0.4\Psi\sqrt{C_D}/C_f \tag{5b}$$

This expression was shown to be in reasonable agreement with results obtained in a 25-mm pipeline when the particle settling tendency was high. The theoretical basis was improved substantially in the derivation of Wilson and co-workers [3], and subsequent studies have allowed this approach to be extended to mixtures of coarse and fine particles [4]. The following presentation is, with minor modifications, a statement of the Wilson model and considers the flow to be represented by two hypothetical layers.

THE WILSON MODEL

The axial force balances for the two layers can be written in terms of the quantities shown in Figure 1:

$$0 = -\frac{dP}{dz} - \rho_1 g\frac{dh}{dz} - \frac{(\tau_1 S_1 + \tau_{12} S_{12})}{A_1} \tag{6}$$

$$0 = -\frac{dP}{dz} - \rho_2 g\frac{dh}{dz} + \frac{(\tau_{12} S_{12} - \tau_2 S_2)}{A_2} \tag{7}$$

Since the upper layer is pseudohomogeneous, the wall stress, τ_1, can be estimated from the fluid relationship:

$$\tau_1 = C_{f_1} V_1^2 \rho_1/2 \tag{8}$$

with

$$C_{f_1} = C_{f_1}(D_{e_1} V_1 \rho_1/\mu_1, \epsilon/D_1) \tag{9a}$$

Actually, for convenience one can use the mean mixture velocity, V, and the diameter of the whole pipe to simplify the calculation procedure, i.e.,

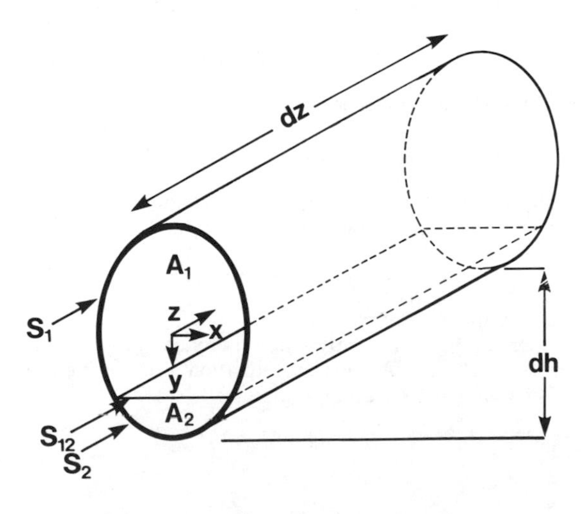

Figure 1. Element of pipe length.

$$C_{f_1} = C_f = C_f(DV\rho_1/\mu_1, \epsilon/D) \tag{9b}$$

The interfacial stress at the boundary is calculated (following Wilson) by a modification of the rough-boundary expression of Colebrook:

$$\tau_{12} = C_{f_{12}} (V_1 - V_2)^2 \rho_1/2 \tag{10}$$

where

$$\sqrt{2}/\sqrt{C_{f_{12}}} = 4 \log_{10}(D/d_2) + 3.36 \tag{11}$$

and d_2 is the diameter of the coarse particles that comprise the lower layer. According to the study of Televantos et al. [5], the magnitude of the interfacial friction factor is not of critical importance in the model.

The boundary stress, τ_2, in Equation 7 is the result of particle wall and fluid wall effects. Since particles are in point contact with the wall, the particle wall stress that will be computed is the mean value for the whole boundary, S_2. This means that the mean fluid wall stress adds to the particle wall stress to produce the net effect,

$$\tau_2 S_2 = \tau_{f_2} S_2 + \tau_{s_2} S_2 \tag{12}$$

where τ_{f_2} is a fluid stress estimated from the pseudohomogeneous portions of the lower layer:

$$\tau_{f_2} = C_{f_2} V_2^2 \rho_1/2 \tag{13}$$

although in practice C_f can be used for C_{f_2}. The boundary stress, τ_{s_2}, is presumed to be due to the normal stress, σ_{rs} (Figure 2):

$$\tau_{s_2} S_2 = 2R \int_{\theta=\pi/2-\beta}^{\pi/2} \sigma_{rs}\, \eta_s \, d\theta \tag{14}$$

where η_s is the coefficient of kinematic friction between the particles and the pipe wall. To obtain σ_{rs}, we consider the equilibrium of forces in the lower layer for the vertical (y) direction. Momentum equations can be written for the fluid, the suspended particles and the contact load particles. Since the flow is macroscopically steady, using the notation of Wallis [6]

$$0 = -\frac{\partial P}{\partial y} + \rho_f g + f_{fsy} \tag{15}$$

$$0 = -\frac{\partial P}{\partial y} + \rho_s g + f_{sfy} \tag{16}$$

$$0 = -\frac{\partial P}{\partial y} + \rho_s g + f_{swy} \tag{17}$$

Equations 15 and 16 reflect that for the fluid and suspended particles there is no wall-caused force in the y direction because there is no net motion in this direction. The interfacial forces of the fluid and these particles must cancel in accordance with Equation 18:

$$C_1 f_{sfy} + (1 - C_1 - C_2) f_{fsy} = 0 \tag{18}$$

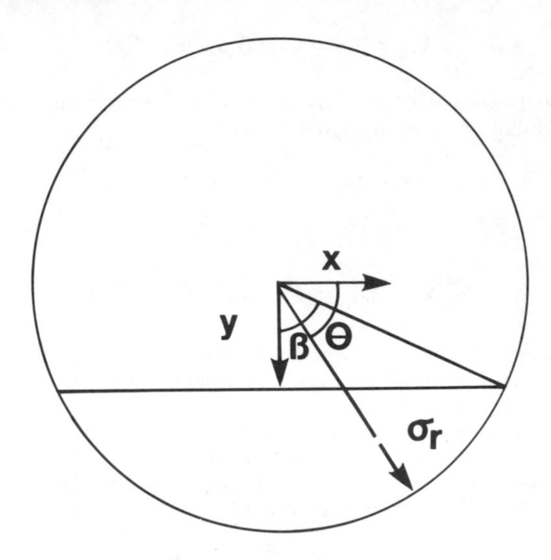

Figure 2. Cross section of pipe.

where C_1 is the concentration of suspended particles, and C_2 is the contact load concentration. The sum is the limiting concentration, C_{max}, taken as 0.60.

The term f_{sw} is the interparticle force per unit volume, which ultimately produces the normal stress, σ_r. Eliminating the pressure gradient from Equation 17 with Equations 15, 16 and 18 yields the expression for the interparticle force per unit volume:

$$f_{swy} = -(1 - C_1 - C_2)(\rho_s - \rho_f)\, g/(1 - C_2)$$

This force can be written in terms of an interparticle stress tensor, whose components are considered to be distributed over the faces of an elemental control volume (Figure 3):

$$f_{swy} = \frac{-1}{C_2}\left[\frac{\partial \sigma_y}{\partial y} + \frac{\partial \tau_{xy}}{\partial x} + \frac{\partial \tau_{zy}}{\partial z}\right]_s \tag{19}$$

In one-dimensional shearing experiments conducted at high concentrations, Bagnold [7] showed that shear stresses were proportional to normal stresses, as, for example,

$$\tau_{yz} = \sigma_y \tan \phi \tag{20}$$

where ϕ is the "angle of internal friction" for the sheared particle mass. A generalization of Bagnold's relationship has been proposed by Savage [8]. However, for the two-dimensional shearing that occurs within the lower layer, further experimental studies are required before σ_r can be predicted with complete confidence. As an approximation in the present state of knowledge, C_2 is taken as constant in the lower layer, the shear stress terms in Equation 19 are neglected and σ_r is taken as σ_y at the boundary. With these assumptions, Equation 14 becomes

$$\tau_{s2}S_2 = 2R\,\eta_s \int_{\pi/2-\beta}^{\pi/2} \int_{R\cos\beta}^{R\sin\theta} (\rho_s - \rho_f)\, g\, \frac{C_2(1 - C_2 - C_1)}{(1 - C_2)}\, dy\, d\theta + 2\beta R \tau_{12}\eta_s/\tan\phi$$

$$= 2R\eta_s\,[R(\rho_s - \rho_f)\, g\, (\sin\beta - \beta\cos\beta)\, C_2(1 - C_2 - C_1)/(1 - C_2) + \beta\tau_{12}/\tan\phi] \tag{21}$$

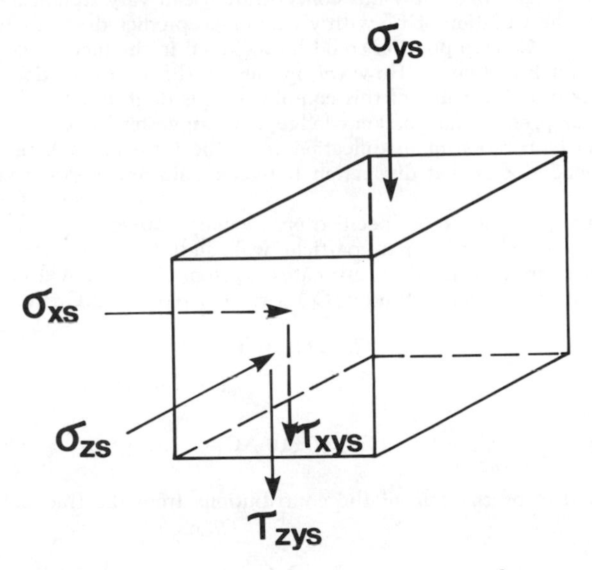

Figure 3. Normal and shearing stresses in y direction.

As the model was developed originally by Wilson for unsheared lower layers of constant composition, it has been customary to take tan ϕ as its value in typical static granular beds (0.577).

APPLICATION OF THE MODEL

Equations 1 and 2 now can be solved in combination with the mass balances, provided values of all the parameters are known. If slip within the lower layer is neglected, the mean velocity, V, is a convenient variable for design defined by

$$V = Q/\pi R^2$$

or

$$V = V_1(\pi - \beta + \sin\beta \cos\beta) + V_2(\beta - \sin\beta \cos\beta) \tag{22}$$

The mean in situ volumetric concentration, C_t, can be calculated from the concentrations and velocities in this case as

$$C_t = C_1(\pi - \beta + \sin\beta \cos\beta) + (C_1 + C_2)(\beta - \sin\beta \cos\beta) \tag{23}$$

The delivered or discharge concentration is obtained from the volumetric flowrate of solids as

$$C_{vd}Q = \pi R^2 [C_1 V_1(\pi - \beta + \sin\beta \cos\beta) + (C_1 + C_2) V_2(\beta - \sin\beta \cos\beta)] \tag{24}$$

When the particles are coarse and when the density difference is high, slip is likely in the lower layers. This can be estimated for the case in which C_1 and C_2 are constant in the manner shown by Televantos et al. [5].

It will be shown later that the solids concentration can vary significantly in the lower layer. However, the variation of C_2 with y cannot be predicted satisfactorily. In principle, the variation of C_2 with position could be included in the model and would produce a different form for Equation 21. However, in view of the approximations that were used to obtain σ_r, the practical value of this complication is doubtful.

Thus, with our present state of knowledge it is suggested that the two-layer model be used without elaboration or modification from the form given here. A consequence of this recommendation is that distinction between fluid and particle velocities is also of limited value.

The remaining quantities to be specified or calculated are η_s, C_1 and C_2. Presumably, values of η_s can be obtained from particle wall sliding friction studies. Wilson has suggested a procedure that, in effect, estimates C_2 from C_{vd}. This will be described now.

The mean contact load concentration, C_c, is related to C_2 and C_1 by

$$C_1 = C_{vd} - C_c$$

and

$$C_2 = 0.60 - C_1 \tag{25}$$

and is considered to be the sum of the contributions from the fractions that comprise the mixture

$$C_c = \Sigma\, C_{ci} \tag{26}$$

For each fraction, the value of C_{ci} is considered to depend on the settling velocity, $V_{\infty i}$, the particle diameter, d_i, the pipe diameter, D, and the friction velocity, u_*:

$$C_{ci} = x_i\, C_{vd}\, [0.6V_{\infty i} \exp\,(45\, di/D)/u_*]^2 \tag{27}$$

An expression such as Equation 27 is essential if the two-layer model is to be used. Although it is likely to be modified in the light of future research, Equation 27 is sufficiently reliable for use in generalizing experimental measurements. The recommended procedure is then as follows:

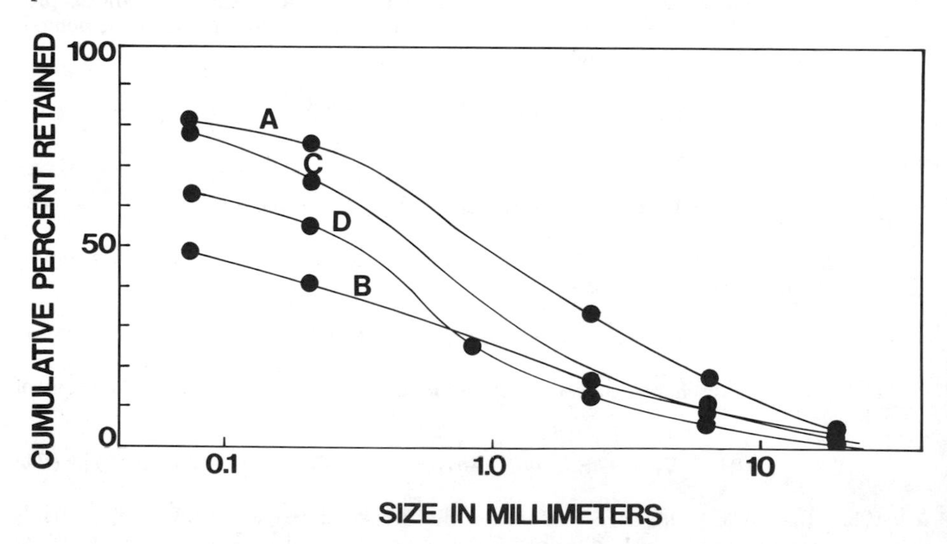

Figure 4. Coal size distributions.

1. When experimental data are available, these should be used to determine η_s, using Equation 27.
2. When no data are available, a value of η_s should be assumed.

Actually, because of the tentative status of Equation 27, it may be more convenient to replace C_{vd} by the mean spatial concentration, C_{vt}.

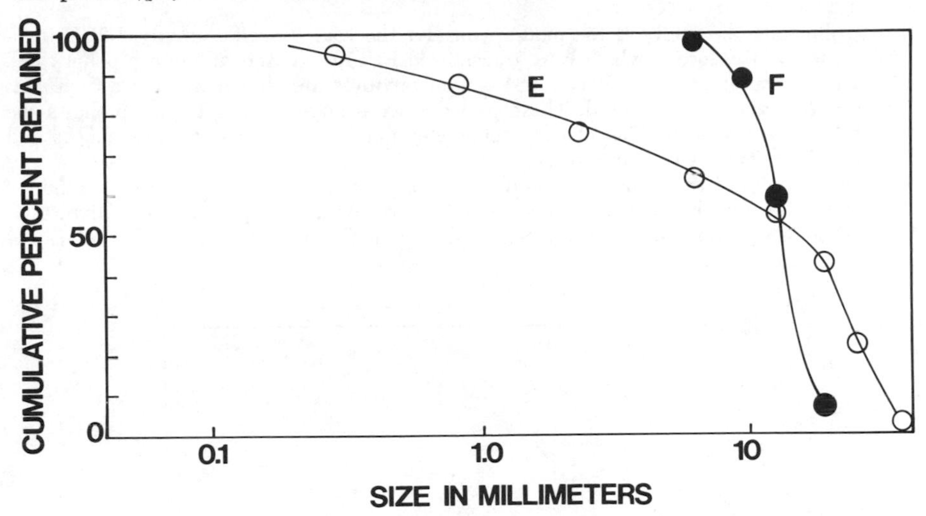

Figure 5. Gravel size distributions.

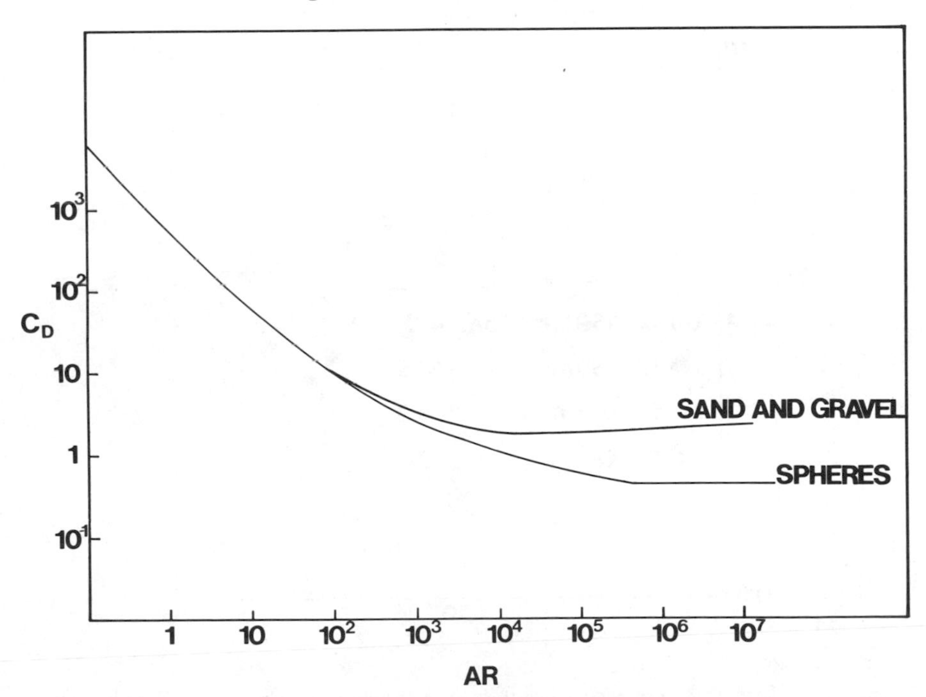

Figure 6. Gravel drag coefficients.

The input variables for the model having been defined, the pressure gradient, V_1, V_2 and β then can be obtained. Since C_c depends on the friction velocity, iteration is required. However, convergence is rapid and a stable algorithm is easily obtained [5].

COMPARISON OF MODEL PREDICTIONS TO EXPERIMENTS

To illustrate the utility of the model, consider the coal (specific gravity 1.307) mixtures shown in Figure 4, which were transported in 159-mm and 495-mm pipelines [9]. Coarser, denser (specific gravity 2.65) gravel mixtures are shown in Figure 5; in this case, a 263-mm pipe was used. These particles were angular, and Figure 6 shows the drag coefficient as a function of the Archimedes number. The coal particle shapes and drag coefficients were similar to these.

Figure 7 shows the limitations of the ϕ vs ψ^{-1} empirical correlation technique by comparing results obtained with the coal at C_t approximately equal to 0.45. Mixtures C and D show a reasonable correlation, indicating that the weighting procedure of Equa-

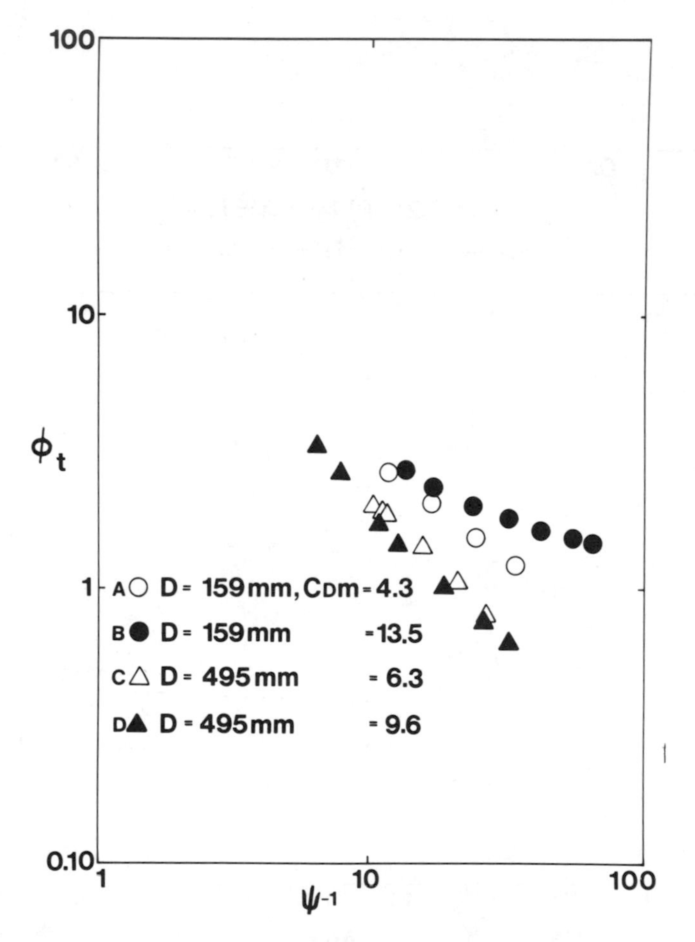

Figure 7. Dimensionless excess headloss as a function of a modified Froude number for the coal of Figure 4, with in situ concentrations of 45% by volume.

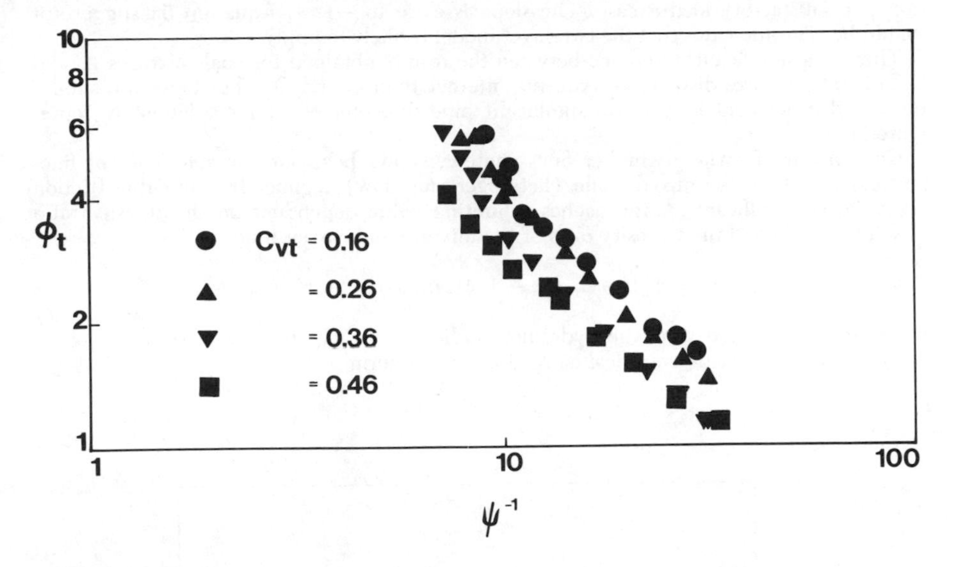

Figure 8. Dimensionless excess headlosses as functions of a modified Froude number for the coal of Figure 4, with size distributions bounded by A and C.

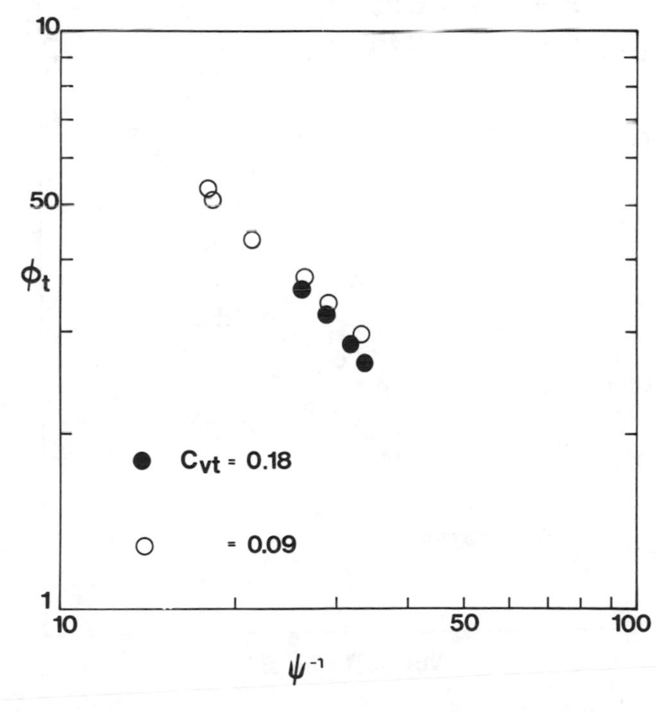

Figure 9. Dimensionless excess headloss as a function of a modified Froude number for gravel F of Figure 5.

tion 4 is satisfactory in this case. The slope is close to −1, as Equation 5a suggests it should be. This indicates that the two-layer model is likely to apply.

There is a significant difference between the results obtained for coal mixtures A and C, although the size distributions are no different than C and D. The slopes are similar but the displacement indicates a significant pipe diameter effect not adequately represented in ψ.

Coal mixture B, with its higher fines content, shows behavior characteristics of fine-particle slurries in the mixed-mode (heterogeneous flow) regime. If coulombic friction ceases to be significant, ϕ approaches a limiting value dependent on the density ratio of solids to fluid and the viscosity ratio of the mixture and carrier fluid:

$$\phi_t = \{[1 + C_{vt}(S_s - 1)](\nu m/\nu_f)^{-n} - 1\}/C_{vt}$$

where ϕ_t is computed from the hydraulic gradient using C_{vt} instead of C_{vd}.

Although mixed-mode empirical correlations of the form

$$\phi_t = (S_s - 1) + B\,\psi^{-n}$$

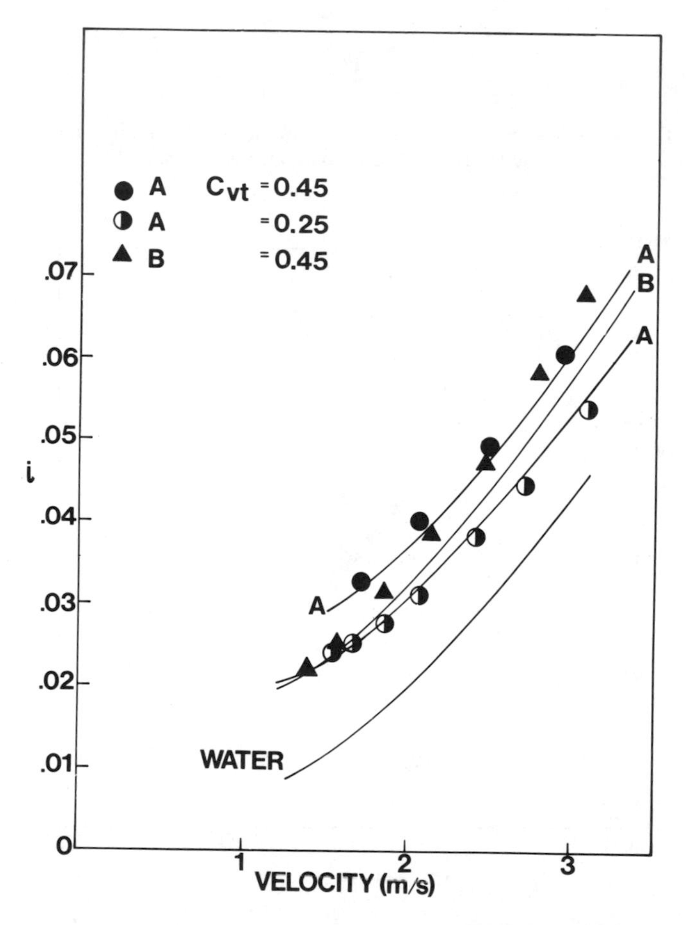

Figure 10. Headlosses as functions of velocity for the coal of Figure 4 in a 159-mm pipeline. The solid lines are computed from the two-layer model, with $\eta_s = 0.32$.

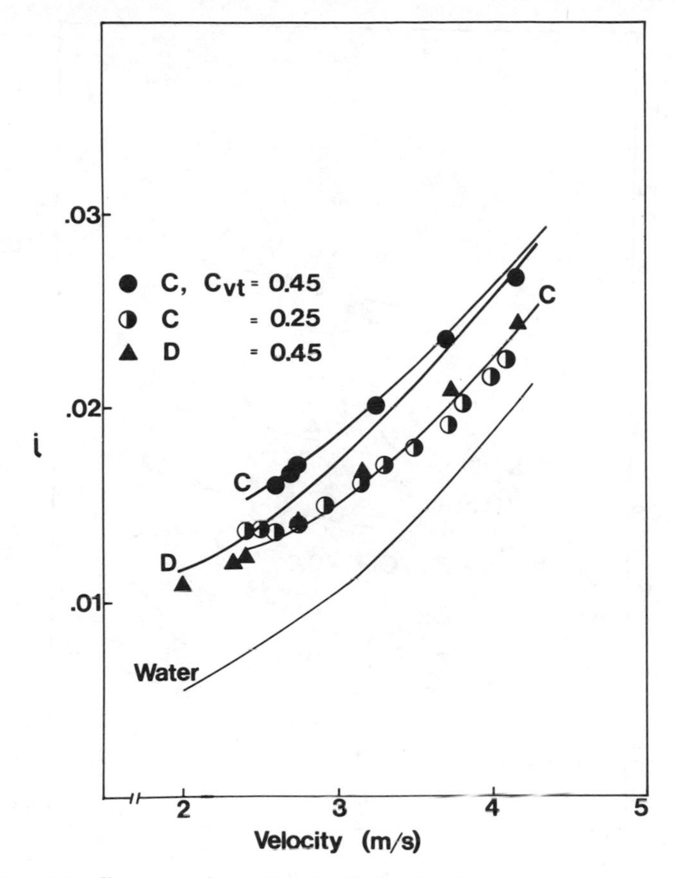

Figure 11. Headloss as a function of velocity for the coal of Figure 4 in a 495-mm pipeline. The solid lines are computed from the two-layer model, with $\eta_s = 0.32$.

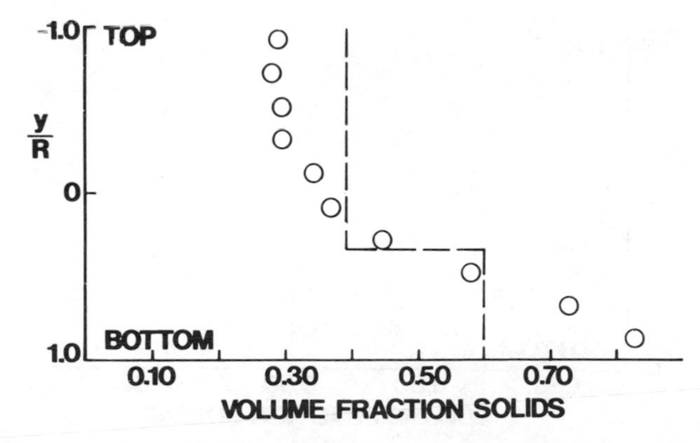

Figure 12. Chord-average solids concentration measurements compared with model predictions (dashed lines) for the coal of Figure 4.

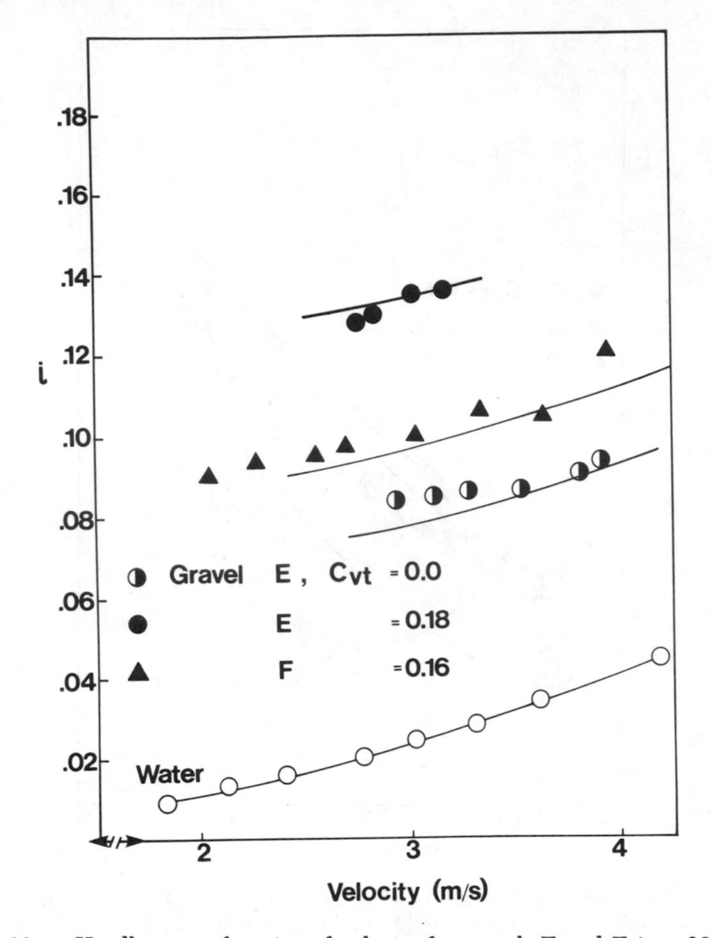

Figure 13. Headloss as a function of velocity for gravels E and F in a 263-mm pipeline. The solid lines are computed from the model, with $\eta_s = 0.22$.

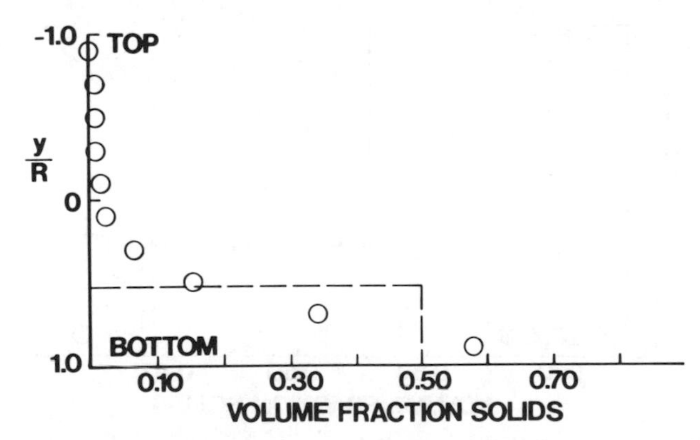

Figure 14. Chord-averaged concentration measurements compared with model predictions (dashed lines) for gravel F.

are useful for particular slurries, there is no universal correlation of this type, as Figure 7 has shown. Systematic deviations in a ϕ_t vs ψ presentation also will arise from solids concentration effects for a given slurry. Figure 8 shows experimental results obtained with this coal in a 263-mm pipe at various concentrations and size distributions bounded by A and C in Figure 6.

The systematic displacements of Figure 8 are difficult to interpret because the effect of small errors in C_t is most significant at low concentrations. However, the two-layer model suggests that the major effect of C_t differences arises from slurry viscosity effects, which become important at high concentrations.

This displacement was probably part of the reason why an exponent of 3/2 was obtained in earlier ϕ - ψ correlations. A second reason for the higher empirical exponent is that the negative slope of ϕ vs ψ tends to increase somewhat after a stationary layer of solids forms in the pipe. A third cause is that C_t exceeds C_{vd} significantly at low velocities.

Figure 9 shows the experimental results for gravel mixture F in Figure 5. Again, the slope is near −1, but there is a substantial displacement from the coal results of Figure 7, confirming that there is no ϕ vs ψ correlation for headlosses. The reasonable agreement at the two concentrations is probably because slurry viscosity changes were minor for these slurries.

The utility of the Wilson two-layer model is illustrated in Figures 10 and 11, which present the experimental coal headlosses and the model predictions. The value of η_s used in the latter was 0.32. This is rather lower than expected and suggests that C_1 may be underestimated by Equations 23, 25, 26 and 27. Evidence for this is given by the chord-average concentration distribution obtained by gamma ray absorption and shown in Figure 12. Nevertheless, the constant value of η_s is reasonably successful in representing the data in the low-velocity region, which is of greatest practical importance.

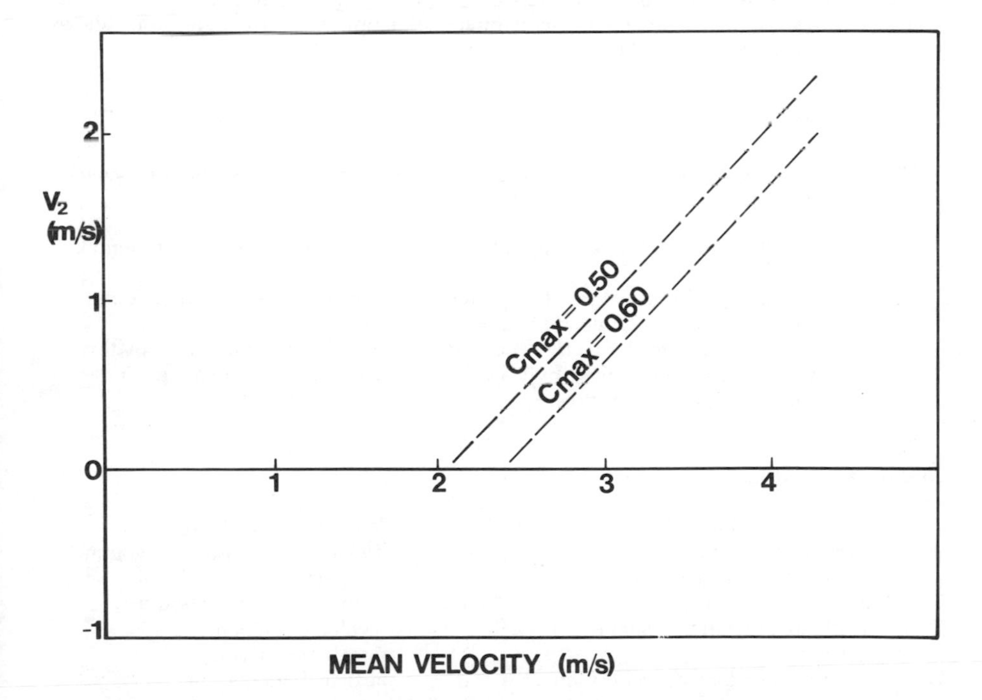

Figure 15. Effect of assumed C_{max} value on lower layer velocity for gravel F at C_{vt} = 0.08.

Figure 13 shows the corresponding comparison of predicted and observed headlosses for gravel mixtures E and F from Figure 5. Here, η_s was 0.22, which is even lower than for the coal. The gamma ray absorption results of Figure 14 show that fluid lift forces were significant in this case [10]. In this case, the particle settling velocity is very high compared to u_* so that lift from turbulent velocity fluctuations is probably not very important. Magnus forces arising from particle rotation are possibly the source of this fluid lift effect.

The low values of η_s (which compensate for the C_1 predictions) imply that as long as Equations 23–27 are used, η_s must be regarded as a correlation parameter. Thus, direct determination in a kinematic friction experiment is not recommended.

The common feature of Figures 12 and 14 is the near-linear variation of concentration in the bottom half of the pipe. The value of C_2 used in the solution has only a minor effect on the headloss prediction: it affects the location of the interface and the velocity, V_2. Figure 15 shows the effect of the choice of C_2 on V_2 for particle mixture F at $C_{vt} = 0.18$. It would be reasonable to assume a higher value of C_2 with a broad particle size distribution, which, in fact, has been done in Figures 10, 11 and 13. However, the value of the two-layer model is that it gives some information concerning the approach of the deposit condition because the model predicts V_2. Figures 13 and 15 indicate that the model is not yet sufficiently developed to predict the limit-deposit velocity. However, it is superior to empirical correlations for coarse-particle slurries [11]. Used with discretion, it may be a useful vehicle for scaleup of experimental results. For the coal particles of Figure 3, limit-deposit velocities gave values of F_L, defined by

$$F_L = V_c/\sqrt{2gD\,(S_s - 1)}$$

of the order of 1.3 in pipes of 158–495 mm in diameter. In the 263-mm pipe, F_L was approximately 0.8 for mixture E and 0.84 for mixture F. This difference between coal and gravel results demonstrates the inadequacy of empirical correlations and shows why testing is required for satisfactory system design.

CONCLUSIONS

The following major points can be concluded based on the model presented for coarse-particle slurry flows:

1. The Wilson two-layer model is satisfactory for generalizing experimental results and for scaleup.
2. It is recommended that values of the effective coefficient of kinematic friction, η_s, be inferred from experimental headloss results.
3. With further research, the method of predicting C_1 may be improved. In this case, η_s then can be obtained by a kinematic friction measurement that does not necessarily involve pipe flow.

NOMENCLATURE

A	cross-sectional area (m^2)
Ar	Archimedes number $4\rho_f(\rho_s - \rho_f)gd^3/3\mu_f^2$
B	empirical coefficient
C	solids concentration, volume fraction
C_c	mean contact load concentration
C_D	drag coefficient for single-particle settling at terminal velocity $4\,gd(S_s - 1)/3V_\infty^2$
C_f	Fanning friction factor
C_v,C_{vd}	discharge concentration, volume fraction
C_{vt}	mean in situ concentration, volume fraction
D	pipe diameter (m)
d	particle diameter (m)
f	force per unit volume (N/m^3)
g	gravitational acceleration (m/sec^2)

h elevation above datum (m)
i hydraulic gradient (m fluid/m pipe)
i_f hydraulic gradient of carrier fluid at the given velocity (m/m)
P pressure (N/m^2)
R pipe radius (m)
S partial pipe diameter (m)
S_s density ratio of solids to fluid
u_* friction velocity $(\tau_w/\rho)^{1/2}$ (m/sec)
V area-mean velocity (m/sec)
V_c limit-deposit velocity (m/sec)
V_∞ terminal falling velocity of particle in the carrier fluid (m/sec)
x mass fraction solids of a given size
x,y,z coordinate directions

Greek Symbols

β angular coordinate defining bed depth (Figure 2) (rad)
ϵ pipe roughness (m)
η_s coefficient of particle-wall kinematic friction
θ angular coordinate of a point on the pipe wall (Figure 2) (rad)
μ viscosity (kg/m-sec)
ρ density (kg/m^3)
σ normal stress (N/m^2)
τ shear stress (N/m^2)
τ_w mean wall shear stress (N/m^2)
ϕ dimensionless excess headloss, Equation 2
ϕ_t ϕ computed using C_{vt} instead of C_{vd}
$\tan\phi$ ratio of shearing stress to normal stress within particle mass
ψ correlating parameter

Subscripts

d discharge value
f fluid
m mixture
s solid particles
t in situ value
w pipe wall
1 upper layer
2 lower layer
12 interface

REFERENCES

1. Durand, R. *La Houille Blanche* 6:384, 609 (1951).
2. Newitt, D. M., J. F. Richardson, M. Abbott and R. B. Turtle. *Trans. Inst. Chem. Eng.* 33:93 (1955).
3. Wilson, K. C., M. Streat and R. A. Bantin. *Proc. 2nd Int. Conf. on Hydrotransport of Solids in Pipes,* BHRA Fluid Engineering, Cranfield UK, Paper B1 (1972).
4. Wilson, K. C. *Proc. 4th Int. Conf. on Hydrotransport of Solids in Pipes,* BHRA Fluid Engineering, Cranfield, UK, Paper A1 (1976).
5. Televantos, Y., C. A. Shook, A. J. Carleton and M. Streat, *Can. J. Chem. Eng.* 57: 255 (1979).
6. Wallis, G. B. *One-Dimensional Two Phase Flow* (New York: McGraw-Hill Book Co., 1969).
7. Bagnold, R. A. *Phil. Trans. Roy. Soc. A* 249:235 (1956).
8. Savage, S. B. *J. Fluid Mech.* 92:53 (1979).
9. Shook, C. A., R. Gillies, D. B. Haas, W. H. W. Husband and M. Small. Paper presented at the Symposium on Coal Utilization, 64th Conference, Chemical Institute of Canada, 1981.
10. Shook, C. A., R. Gillies, D. B. Haas, W. H. W. Husband and M. Small. *Can. J. Chem. Eng.* (1981).
11. Oroskar, A. R., and R. M. Turian. *AIChE J.* 26:550 (1980).

CHAPTER 36
DESIGN ASPECTS OF SLURRY PIPELINES

R. L. Gandhi and P. E. Snoek
Bechtel Petroleum, Inc.
San Francisco, California 94119

CONTENTS

INTRODUCTION

Slurry pipelines have been utilized for many years. Among the earliest installations were municipal storm and sewage pipelines and hydraulic dredge systems. The classification covers a wide range of systems, including relatively small and short in-plant process lines, as well as long and massive cross-country mineral transport systems.

Simply speaking, a slurry pipeline can be defined as any system transporting solids in a liquid carrier. The carrier is usually water, although many other fluids could be used, including crude oil and various petroleum products. As alluded to earlier, the solids carried by the first systems were mostly waste products, which were transported to a point of disposal. It was only in the 1950s that the technology was developed sufficiently to allow this means of transport to be used for the mineral industry. During the decade from 1950 to 1960, systems were constructed in Trinidad, the United States and France to transport limestone, coal and gilsonite.

Since that period, a large number of other installations have been added. These systems transport iron concentrate, phosphate, copper concentrate, various tailings and kaolin, in addition to the minerals previously named. One measure of the rapid development in this technology is the growth in system capacity, as expressed in million ton-km (MT-km). The largest system installed during the 1950s was the Consolidation Coal pipeline in Ohio, which had a capacity of 209 MT-km. With the Black Mesa system in Arizona, which began operation in 1970, the capacity was nearly 12 times greater. New systems under design, such as the Energy Transportation System (ETSI) coal pipeline, will have capacities nearly 300 times greater than the Ohio installation.

In spite of the wide range in system size, type of solid transported, etc., there is a common design basis for all slurry systems. However, for purposes of classification, two major groups can be considered. The first is what is known as "long-distance slurry pipelines." These systems cover the transport of relatively fine solids at carefully controlled concentrations and relatively low velocities. The second group can be designated as "coarse-slurry pipelines." In this case, the relatively large solids are often transported under a wider range of concentrations and at significantly higher velocities.

Both systems include the same major components, although the specific equipment utilized is normally quite different for each of the two cases. First there is a collection vessel to accept the slurry from the upstream processing plant. Then there is a pumping installation to provide sufficient energy to transport the required amount of solids to the next downstream pump station or terminal. The third component is the actual pipeline itself, including its internal lining and external coating, if required. Also, there must be a means of controlling the system that includes sufficient instrumentation to measure the process variables. Finally, there are the terminal facilities that may again include, in the case of a mineral pipeline, a collection vessel to feed a processing plant.

LONG-DISTANCE PIPELINES

Design Criteria

Long-distance slurry pipelines are designed to last for many years without replacement or rotation of the pipeline. To achieve long life, the abrasion of the pipe needs to be controlled. Abrasion of the pipe wall can be controlled either by using an abrasion-resistant lining or by selecting a size consist, slurry concentration and flow velocity combination that avoids heterogeneous flow. Abrasion-resistant pipes are several times more expensive than unlined steel pipe, and the friction losses are also higher when solids are transported in a heterogeneous flow regime. Therefore, long-distance pipelines generally are designed to operate either in a homogeneous or compound flow regime.

The design of a slurry pipeline system requires the establishment of the following variables:

- concentration range
- velocity range
- pipe diameter
- friction losses

All these variables are governed by the size consist and specific gravity of the solid particles. To maintain nearly homogeneous flow, a significant fraction of particles should

Table I. Properties of Solids Transported Over Long Distance by Slurry Pipeline

Material	Maximum Particle Size (μm)	Percentage Finer than 44 μm	Solids Specific Gravity	Slurry Concentration (wt %)
Coal	2400	20	1.4	50
Limestone	600	70	2.7	70
Iron Concentrate	150	80	5.0	60
Copper Concentrate	200	80	4.0	60
Phosphate Concentrate	200	40	3.0	60

be smaller than 44 μm (325 Tyler mesh) in size. The maximum particle size and fraction of particles smaller than 44 μm used in the design of some of the transported materials are given in Table I.

The following criteria may be used for estimating the maximum allowable particle size [1]:

$$\mathrm{Log}(C/C_a) = -1.8\ W/BKU^* \qquad (1)$$

$$\geqq 0.8$$

where C = volume fraction solids at 0.08 D from the pipe top
C_A = average volume fraction solids in slurry
W = settling velocity of maximum size particle in slurry (m/sec)
B = ratio of mass transfer to momentum transfer coefficient
K = von Karman constant
U^* = friction velocity = $V\sqrt{f/2}$ (m/sec)
V = velocity of flow (m/sec)
f = Darcy-Weisbach friction factor

The settling velocity of particles decreases with increased solids concentration. Therefore, according to Equation 1, the C/C_A value for a given size particle will increase with the slurry concentration. The criteria given by Equation 1 can be used, therefore, to determine the minimum concentration suitable for pipelining.

For a given size of pipe, the specific energy consumption decreases as the solids concentration increases up to a certain value [2], as shown in Figure 1. Beyond a certain maximum concentration, the specific energy consumption starts increasing with slurry concentration.

Specific energy consumption is defined as the energy required to move 1 ton of solids over a distance of 1 km in a horizontal pipe (kWh/T-km). The maximum slurry concentration can be estimated by plotting specific energy consumption as a function of concentration and selecting the maximum concentration limit as the concentration corresponding to the minimum specific energy consumption value.

The maximum concentration limit also can be estimated by plotting slurry viscosity as a function of solids concentration, as shown in Figure 2. Beyond a certain concentration, the increase in slurry viscosity for a unit change in concentration becomes quite high. The concentration at which this change occurs can be used as the maximum limit [3].

The velocity of flow should be at least 0.15–0.3 m/sec (0.5–1.0 ft/sec) above the estimated deposition velocity or laminar turbulent transition velocity, whichever is larger. The maximum velocity of flow should be less than about 3 m/sec (10 ft/sec) to avoid abrasion in unlined steel pipe.

The pipe diameter required to transport solids at a given throughput rate can be determined using the known values of slurry concentration and velocity of flow. The following example illustrates the method of selecting pipe diameter.

Example 1. Determine the pipe diameter required to transport 10 million ton/yr of coal, assuming that the deposition velocity is 1.4 m/sec and the slurry concentration is 45% by volume. Assume that the pipeline availability is 95%. The specific gravity of coal is 1.4.

$$\text{The hourly coal throughput} = \frac{10 \times 10^6 \text{ T/y}}{365 \frac{\text{d}}{\text{y}} \times 24 \frac{\text{h}}{\text{d}} \times 0.95}$$

$$= 1202 \text{ T/h}$$

$$\text{Coal flowrate} = 1202 \frac{\text{T}}{\text{h}} \times \frac{\text{m}^3}{1.4 \text{ T}}$$

$$= 858 \text{ m}^3/\text{h}$$

$$\text{Slurry flowrate} = 858 \frac{\text{m}^3 \text{ coal}}{\text{h}} \times \frac{\text{m}^3 \text{ slurry}}{0.45 \text{ m}^3 \text{ coal}}$$

$$= 1907 \text{ m}^3/\text{h} = 0.53 \text{ m}^3/\text{sec}$$

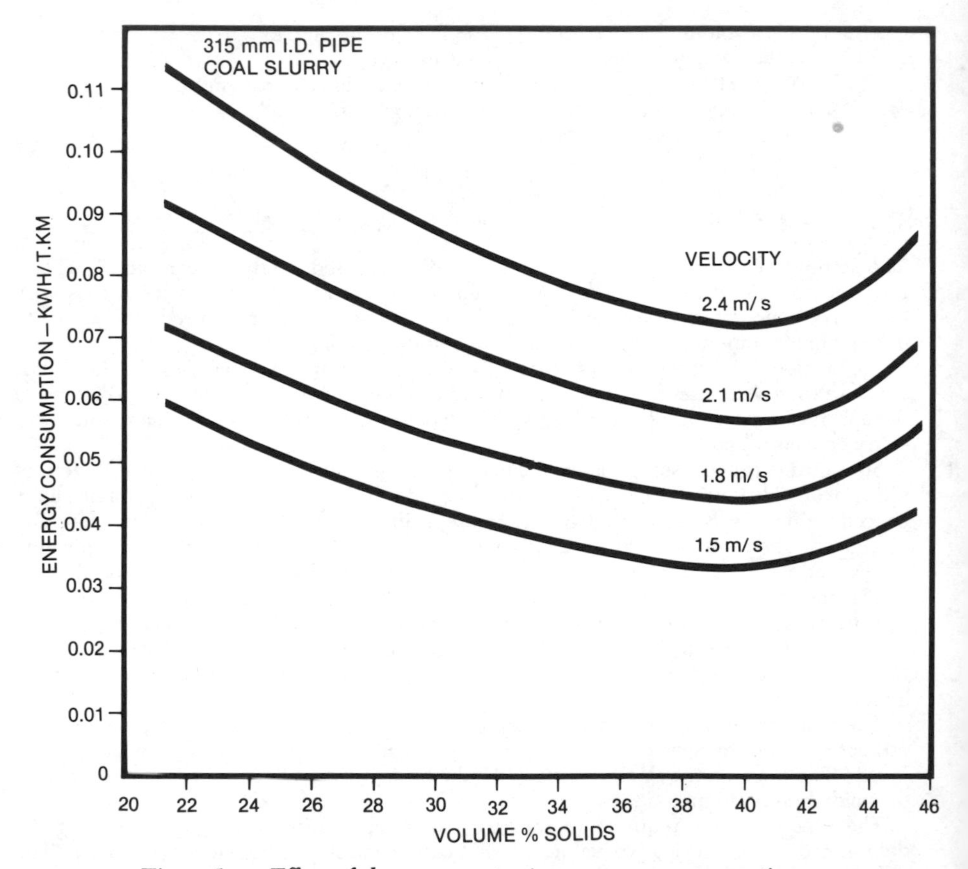

Figure 1. Effect of slurry concentration on energy consumption.

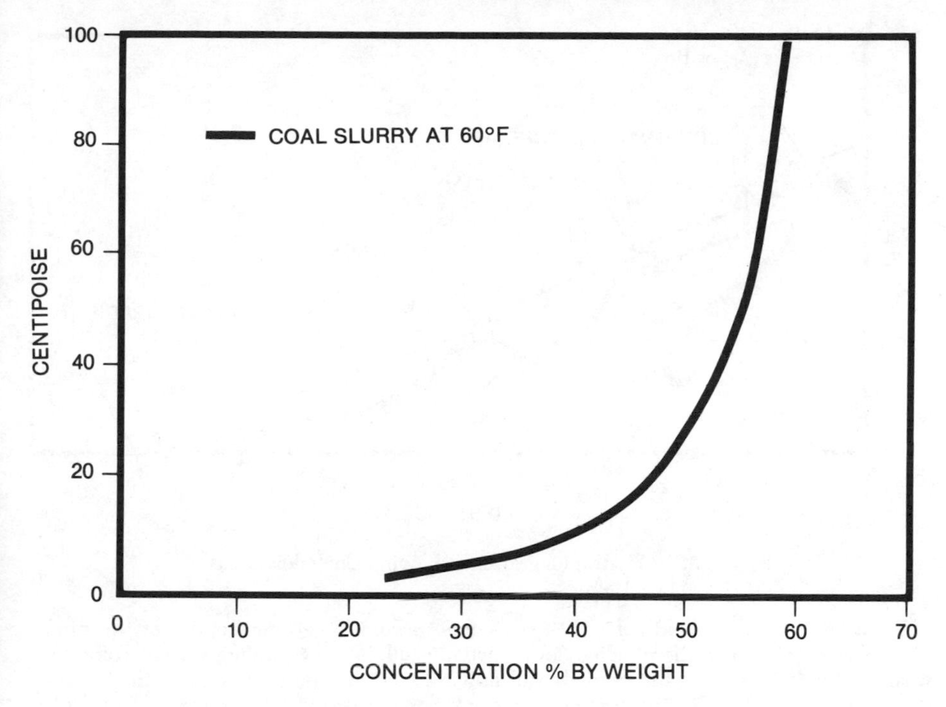

Figure 2. Effect of solids concentration on slurry concentration.

$$\text{Let design velocity} = 1.7 \text{ m/sec}$$

$$\text{Pipe internal area required} = \frac{0.53 \text{ m}^3\text{/sec}}{1.7 \text{ m/sec}} = 0.312 \text{ m}^2$$

$$\text{Pipe inside diameter} = 0.630 \text{ m}$$

The nearest available commercial pipe size is 0.610 m (24-in.). Assuming a 6.35-mm (0.25-in.) wall thickness, the inside diameter of pipe becomes 0.597 m and the velocity of flow becomes 1.89 m/sec. The friction losses can be estimated by the methods described in Chapter 2.

Hydraulic Design

To develop the hydraulic design of the pipeline, it is necessary to prepare a pipeline profile. The pipeline profile can be established from a topographical map of the area traversed by the pipeline. The total pumping pressure requirement is calculated from the following equation:

$$P = \Delta P\, L + E_T - E_o \tag{2}$$

where P = pumping pressure (m of slurry)
ΔP = friction loss (m slurry/km)
L = pipeline length (km)
E_T = elevation of pipe outlet at the terminal end of pipe (m)
E_o = elevation of pipe at the starting point (m)

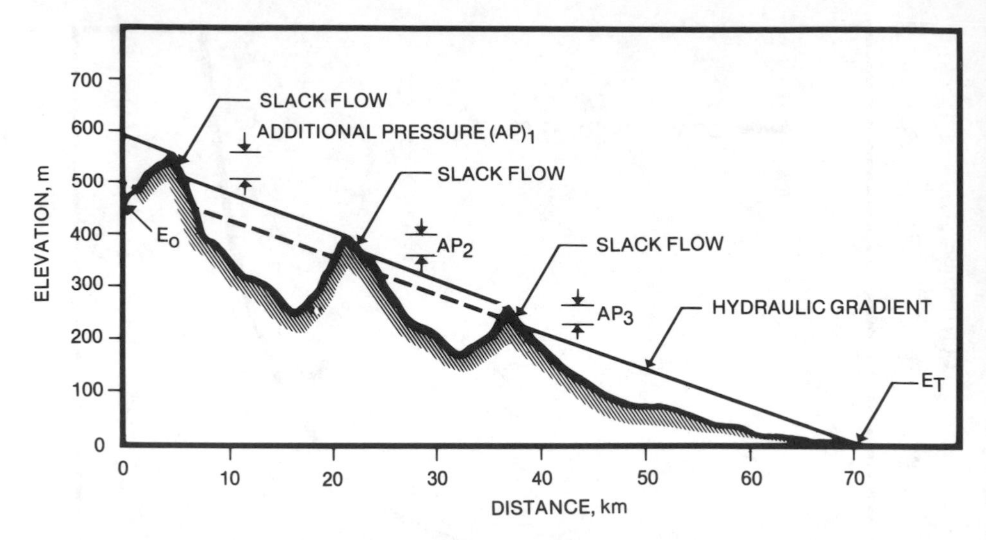

Figure 3. Hydraulic gradient showing slack flow areas.

Equation 2 does not hold true when slack flow occurs in certain sections of the pipe. Slack flow occurs when the pipeline flows partially full. The hydraulic gradient coincides with the liquid surface in the partially full flowing section. Figure 3 is a sketch showing the hydraulic gradient in the presence of slack flow. The additional pumping pressure required due to slack flow near intervening peaks also is shown in this figure. The problems associated with slack flow and methods of controlling slack flow are considered in a later section. Depending on the magnitude of the pumping pressure, P, and the pipe diameter, more than one pump station may be desirable. The optimum number of pump stations can be determined by balancing the savings in cost of pipe steel against the increase in cost of additional pump stations.

The pipe steel requirement is determined by the internal pressure at various points along the pipeline and the allowable stress in the pipe. It should include an allowance in wall thickness for the anticipated corrosion rate inside the pipeline.

Selection of Pumps

Pumps used in slurry pipelines generally fall into two categories: reciprocating positive displacement (PD) pumps and centrifugal pumps. An important consideration is that pumping equipment is available in discrete unit sizes, according to inlet and outlet pipe size, maximum pressure capability, flow capability and horsepower rating. Other considerations include initial capital investment, operating cost and the degree of skill required to operate and maintain the equipment. For example, although centrifugal pumps have lower capital costs than piston or plunger pumps, they have higher power costs because of their lower mechanical-hydraulic efficiencies.

In cases in which only a single pump station is needed, the head requirement and maximum particle size that the system must handle may be sufficient to decide the kind of pump to be used. Figure 4 shows the general areas in which major pump types are suitable [4]. As the figure shows, there are conditions for which only one type of pump is applicable, and there are gray areas where more than one pump type can be used.

In long-distance pipelines, the number of pump stations and the required lift per station is prescribed by the designer. Generally this is decided from the results of a comprehensive economic analysis in which the cost of steel pipe is balanced against

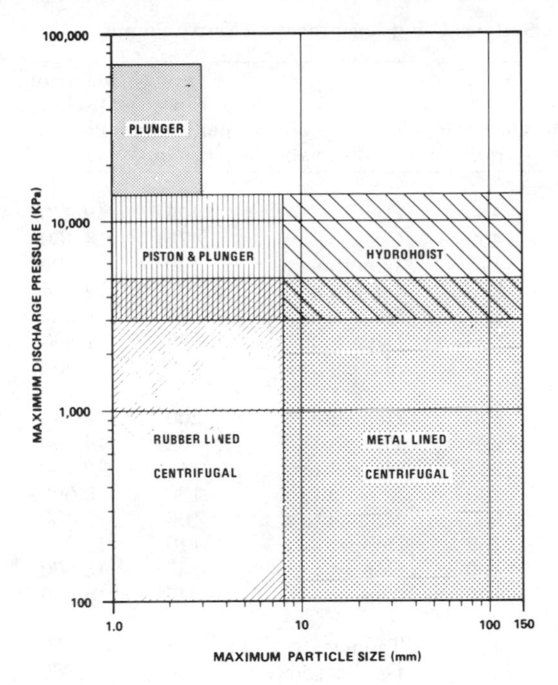

Figure 4. Slurry pump capacity.

the capital and operating cost of the pump stations for various pipe diameters, solids concentrations, pipeline routes, etc. Although there has been a tendency to select systems with relatively few pump stations and high discharge pressures, each case must be considered carefully. It is quite possible that project conditions could warrant a pipeline system having many stations with relatively low to moderate heads.

Reciprocating Pumps

Reciprocating pumps generally are selected when high discharge pressures are required. As shown in Table II, many of the long-distance slurry pipelines constructed to date have used this type of equipment. Reciprocating pumps are available in several different variations. Flow and pressure capabilities of each type are listed in Table III. The two major types of reciprocating pumps found in slurry service are piston and plunger units.

Reciprocating pumps have several desirable features:

1. The flowrate of the pump is independent of system pressure.
2. They can meet any reasonable system discharge pressure requirement. Pumps capable of producing 15,900 kPa (2300 psi) pressure have been used in magnetite pipelines. Units capable of discharge pressures over 34,500 kPa (5000 psi) are available.
3. The overall efficiency of the pump, including drive train, is relatively high—on the order of 85%.
4. Pipeline flowrate can be determined without the use of a flowmeter.

The independence of flow with respect to system pressure and the high overall pump efficiency are very important advantages of reciprocating pumps.

The following disadvantages are associated with this type of unit:

Table II. Selected Pipeline Systems Using Reciprocating Slurry Pumps

System	Length (km)	Diameter (mm)	Material	Flow per Pump (m^3/hr)	Maximum Discharge Pressure (kPa)
Piston Pumps—Double Acting, Duplex					
Consolidation	194	250	Coal	125	8,270
Black Mesa	490	450	Coal	475	14,480
Calaveras	31	175	Limestone	190	8,790
Rugby	103	250	Limestone	65	14,480
Trinidad	11	200	Limestone	70	6,620
Columbia	10	200	Limestone	175	1,380
Japan	68	300	Slime	180	2,930
Plunger Pumps					
Samarco	396	500	Hematite	200	14,480
Valep	120	225	Phosphate	132	14,900
Savage River	95	225	Magnetite	88	13,790
Sierra Grande	36	200	Magnetite	130	9,660
Las Truchas	31	250	Magnetite	190	8,480
Bougainville	31	150	Copper	110	13,790
West Irian	124	100	Copper	30	15,860
Pinto Valley	20	100	Copper	45	15,860
Mars Pumps					
English China Clay	4	106	Tailings	74	3,450
Vaal Reefs, RSA	—	—	Gold ore fines	130	7,860
Oppu Mining	8	27	Milled ore	75	2,940
Hokuroku	72	300	Mill tailings	175	4,900
Diaphragm Pumps					
Cuban American	—	—	Iron ore	—	4,140
West Germany	—	—	Bauxite	162	13,180

1. The maximum flowrate per pump is limited to less than about 890 m^3/hr (3900 gpm). Furthermore, this capacity is only available at relatively low discharge pressures. Therefore, a large number of pumps operating in parallel are needed to handle the high flowrates and working pressures found in large long-distance systems. For example, seven 930-kW (1250 hp) units are used at each pump station of the Samarco system.
2. Initial capital costs and maintenance costs are usually high. Skilled labor is required for operation and maintenance.
3. Variable-speed drives are needed to vary flowrates.
4. The flow through the pump is pulsating, which requires greater attention to station piping design to avoid vibration and fatigue problems.
5. The maximum size of particles that can be pumped is restricted by the check valve seal requirements.

Table III. Flow and Pressure Capabilities of Reciprocating Pumps

Type of Pump	Maximum Flow, m^3/hr (gpm)	Maximum Pressure, kPa (psi)	Maximum Particle Size (mm)
Piston	890(3,900)	15,000(2,200)	8
Plunger	225(990)	41,000(6,000)	8
Mars	204(898)	7,860(1,149)	1
Diaphragm	120(530)	20,500(3,000)	4

For material with a maximum particle size of less than 3 mm (0.1-in.) and with discharge pressures up to about 13,000 kPa (2000 psi), either piston or plunger pumps can be used. Slurries having a maximum particle size of 2.5 mm (0.1-in.) to 6 mm (0.25-in.) also may be handled with these types of pumps if special design pump valves are used.

The decision to use piston or plunger pumps usually is based on the results of a Miller abrasivity test. Material with a Miller number below 30 can be handled using piston pumps, and material with a Miller number above 60 should be pumped with plunger units. Between these values, the choice of what type of pump to use is based on other considerations:

1. Piston pumps can be of double-acting design so that about twice the flowrate can be obtained for the same physical pump size.
2. Plunger pumps are more adaptable to flushing and lubrication. A flushed stuffing box can prolong parts life. However, a flush fluid free of solids must be provided, and some dilution of the slurry will result.

Plunger Pumps

Because abrasive slurries can greatly reduce the life of the pistons and cylinder liners of conventional piston units, plunger pumps are often used. This type of pump maintains a clear liquid barrier between the plunger and packing by means of a flushing system, as shown in Figure 5. Major wear is limited to the plunger, valves and packing.

Plunger pumps should be used at pressures in the range of 13,800 kPa (2000 psi) or higher and with slurries of maximum particle size less than 2.5 mm (0.1-in.). At high pressures, the piston rod for a double-acting piston pump becomes so large that the double-action features of the pump provide only a small increase in throughput over a plunger pump of similar size.

Life of Reciprocating Pump Parts

Slurry valves are the single most important fluid end wear part. With all four types of reciprocating pumps, the valves are exposed to wear from the slurry. To prevent rapid erosion, the flow velocity through the valve area should not exceed about 3.7 m/sec (12 ft/sec). To extend valve life, an elastomer is provided at the sealing surface. Polyurethane has proved to be a successful elastomer; however, it cannot be used at temperatures above 90°C. A metal-to-metal seat is required at higher temperatures. A ball valve is useful in this situation since it offers a multitude of seating surfaces. Up to this time, ball valves have been used in smaller pump units only, such as some of the

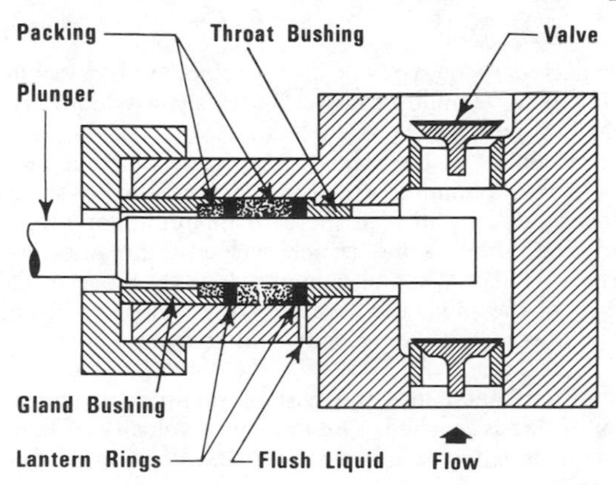

Figure 5. Plunger pump fluid end.

Mars pumps. The pumps installed on the Sierra Grande magnetite line also have ball valves, but parts life has not yet been established.

The life of wear parts in a pump depends on the discharge pressure, pump strokes per minute, materials of construction and slurry characteristics (particularly slurry abrasivity). Parts life for valves reported in Table IV are generally indicative of a moderate-discharge pressure application.

An important consideration for installations in developing nations is whether the consumable pump parts can be manufactured locally or must be imported. Although the situation will vary greatly from country to country, it is becoming increasingly common to find that these parts, including liners and valves, can be made in local machine shops and foundries. The most likely items to be imported are pistons and plunger packing and polyurethane valve parts.

Selection of Valves

There are many manufacturers and types of valves offered for low-pressure slurry service. These designs are backed by years of experience in the mining and dredging industries. Selection of valves for use in positive displacement slurry pump stations introduces some new parameters besides those that must be considered in selecting low-pressure valves. These include vibration of the piping and packing of solids into the void spaces in the valves.

Generally speaking, in high-pressure duty, where a full opening is not required, lubricated plug valves have been applied. Depending on the service, the wetted parts of these valves should be hardfaced to minimize abrasive wear during the opening or closing cycles. Gate or ball valves have been used where a full round opening is required. Both types introduce the problem of removing solids from the body cavity after the valve has been operated in slurry. Again, it may be necessary to hardface some wetted parts to extend the valve life.

Slurry valves will require regular lubrication and maintenance to retain their usefulness. An important feature in selecting valves is the ease of disassembly for maintenance. Some ball valves, for instance, have to be sent to the factory to renew the seats or seals. Some maintenance problems may simply be a matter of solids finding their way into the valve workings, with the solution simply being disassembly and cleanup of the valve. Some slurries (such as limestone) may have some cementation properties. Therefore, it is good practice to regularly lubricate and exercise valves that otherwise are used infrequently.

Corrosion-Erosion Control

In a slurry pipeline, metal loss is expected to be a result of corrosion, with possible erosion of the corrosion products taking place simultaneously. Under some conditions, mechanical abrasion also will play a part in producing the metal loss.

Abrasive wear is governed by the size consist of the solids, slurry concentration and velocity of flow. In a slurry pipeline, these parameters are interdependent to some extent. For example, use of large-sized solids requires an increase in minimum transportation velocity. It has been found that above some critical velocity the abrasive wear increases with the cube of slurry velocity. Wear also increases with the size of solid particles. Thus, by reducing the size of the solids the abrasive wear can be reduced substantially due to the combination of lower required velocity and reduction in wear due to smaller particle size.

The minimum required velocity, abrasion metal loss and cost of pumping decrease with particle size until a certain minimum size is reached. The minimum velocity of flow and the pumping cost increase with a further decrease in particle size because of the resulting increase in slurry viscosity.

Table IV. Typical Wear Life of Reciprocating Pump Components

Description	Expendable Parts Life, hr			
	Piston	Plunger	Mars	Diaphragm
Piston Rod	3000	–	–	–
Plunger Sleeve	–	700	–	–
Piston Liner	4000	–	–	–
Packing	6000	425	–	–
Diaphragm	–	–	–	8000
Valves				
-Seat			300	5000
-Bodies	1100	500	270	
-Inserts			205	

The additional cost of grinding and dewatering of fine solids should be compared with the savings in pumping cost and abrasive wear before selecting the size to which the material should be ground. Grinding often is required for ore benefication. From experience it has been found that the metal loss due to abrasion is significant if the velocity of flow is more than about 3 m/sec (10 ft/sec). For long-distance slurry pipelines, velocities in the range of 1.2 to 2.0 m/sec (4 to 6 ft/sec) result in an optimum design from the standpoint of economics. Thus, when possible a particle size should be selected so that the slurry is suspended nearly homogeneously at velocities in the 1.2 to 2.0 m/sec range.

Corrosion can be controlled by passivating either the anodic or the cathodic reactions at the pipe wall. Elimination of dissolved oxygen and the adjustment of slurry pH can reduce the corrosion rate substantially [5].

Slurry pipelines normally are designed to have a useful life of more than 10 years. In determining the design wall thickness of the pipeline, a corrosion allowance is added to the wall thickness requirement determined for the maximum working pressure. The corrosion metal loss is minimized using various types of inhibitors. Based on these data, a corrosion inhibitor is selected that is environmentally acceptable and economically justifiable. The actual dosage of inhibitor and the corrosion allowance are determined from an economic tradeoff analysis.

Slack Flow and Its Control

Slack flow occurs when the available static head between the discharge point of the pipeline and a given point in the pipeline exceeds the friction loss between those points at a given flowrate. It usually occurs just downstream of peaks where the hydraulic gradient intersects the ground profile.

In slack flow areas, the velocity of flow is governed by the pipe slope. The velocity in the slack flow is higher than that in the fully packed pipe section. Erosion takes place in the slack flow area due to the increase in flow velocity, as well as due to possible cavitation at the point where the pipeline changes from slack flow to a packed flow condition [6].

Due to the higher density of slurry compared to that of water, slack flow can occur when water is displaced by slurry. For example, Derammelaere and Chapman [7] have described the development of slack flow in the Samarco iron concentrate pipeline during operation of the pipeline in batch mode.

Slack flow can be avoided by using a smaller-diameter pipe section or by the use of orifice chokes. The velocity of flow in the smaller-diameter pipe should not exceed about 3 m/sec; otherwise, erosion may occur. Either orifice chokes or a combination of orifice chokes and smaller-diameter pipe may be used to achieve flexibility and economy. On the Samarco iron concentrate pipeline, five orifice stages or loops are installed in series. Each stage dissipates 100 m of head. Isolation valves are provided on each stage so that one or more stages can be isolated from the pipeline to obtain variations in choke pressure. The ceramic orifice beans used on the Samarco pipeline have shown no wear after 4000 hours of operation [7].

Storage

It is very seldom that the flowrate of solids in a slurry pipeline matches its rate of production at the mine site or its rate of utilization at the terminal facility. To provide a means by which the pipeline can operate efficiently by transporting solids in a nearly continuous fashion, storage facilities are required at both ends of the line. The type and amount of storage are determined by the specific system operating parameters, as well as by the characteristics of the material being transported.

Tanks

For live slurry storage at the head and tail ends of the system, agitated tanks typically are utilized. The number and size to be used are determined by comparing the plant and pipeline availabilities, in addition to normal engineering criteria, such as allowable soil loadings and an economic analysis of available sizes. If the pipeline is to be operated in batch mode for the first few years of operation, due to low throughput requirements the size of the tanks may be determined by the minimum batch length chosen for the system. This, in turn, is a function of pipeline length.

Agitated storage also may be required at intermediate pump stations. The purpose would be to facilitate pipeline section reconnection during restart operations as well as to hold slurry that has been flushed from station piping during pump changeout.

The maximum size of agitated storage tanks constructed to date is the 125-ft-diameter × 87-ft-high coal slurry tanks of the type installed at the Mohave Power Plant in Nevada at the terminus of the Black Mesa pipeline. Each of these tanks holds 7.8 million gallons, which satisfies a 48-hour peak coal demand for one 750-MW generating unit. A centerline-mounted, top-entering, multiple-impeller agitator is supported from a steel structure above the tank. Each agitator is driven by five 100-hp motors operating through gear reducers to pinions on a bull gear.

Ponds

For storage of high volumes of solids over intermediate- to long-term periods, storage ponds are utilized. A prime advantage of this method of storage is that no energy is required to maintain suspension of the solids.

Slurry storage ponds can be classified into two basic types: semiactive and dead storage. Semiactive ponds are equipped for recovery and reslurrying on short notice. Depending on design, it may be necessary to remove all the stored material before refilling the pond.

Dead storage ponds indicate an infrequent filling and removal, such as for "force majeure" conditions. The range in sophistication can vary from a simple impoundment type, such as the emergency storage ponds used at the Black Mesa pump stations, to lined basins with an underdrain system.

The major consideration in pond design is recovery of a uniform solids size consist. Segregation of the solids can occur during the filling or recovery operations. The resulting coarse- and fine-slurry slugs can be very difficult to handle in the downstream processes. Three major recovery methods have been utilized: mechanical, dredge and the Marconaflo technique.

Mechanical recovery uses conventional earth-moving equipment to remove the settled solids. Prior to the start of recovery operations, the bulk of the free water must be removed from the pond by evaporation, natural drainage, sand points or an underdrain system.

Dredge recovery requires the maintenance of a water layer over the settled solids to float the dredge. The recovery dredge may be maintained onsite or moved in for the recovery operation, depending on whether semiactive or dead storage is required.

The Marconaflo technique [8] was developed originally to remove settled slurries from the holds of ore ships. A high-pressure water jet on a rotating head undercuts and reslurries the solids and the material flows into an underdrain system for pumping away. Eight Marcona ponds have been installed at the Mohave Power Plant for the semiactive storage of coal, with each pond holding up to 75,000 tons of solids [9]. The ponds are dish shaped: 436 ft in diameter and 40 ft deep at their centers. A structural steel tower with a monorail and a 15-ton hoist is located at the center of each pond to lower and raise a portable reclaim unit.

Instrumentation

The instrumentation used and the major parameters measured and controlled are as follows.

1. **Pressure.** Both the suction and discharge pressures at a pump station are measured and controlled. Suction pressure is controlled to prevent cavitation. The discharge pressure is controlled so as not to exceed the maximum allowable pressure for the pump unit or pipeline system. Bourdon-type pressure gauges with an isolation diagram generally are used.

2. **Density.** The pumping head requirement varies with the changes in slurry density. If the slurry density is not controlled within the pipeline design limits, the pipeline system can be overpressurized or the flow velocity could drop below the deposition critical velocity.

Slurry density normally is measured by radiation gauges. Density meters may be needed at each pump station in a multiple-station system. The density meters normally are installed on the suction side of a pump station.

Density is controlled by adding dilution water to a high-density slurry. If the density is too low, the slurry can be returned back to the preparation plant for concentrating.

3. **Solids Particle Size.** The maximum particle size, as well as the amount of fines in a slurry, are controlled within design limits. Increase in the maximum particle size results in an increase in the deposition velocity. A decrease in the amount of fines decreases the ability of the slurry to suspend coarse fractions. This could lead to higher friction losses as well as a higher deposition velocity, which could produce unstable pipeline operation and may lead to plug formation. Pipe erosion also could occur if larger-sized particles are dragged along the pipe bottom due to a decrease in fines content or an increase in the top size of the particles.

Automatic particle size measuring devices are available but have not been justified on a cost-benefit basis. Once a grinding plant is stabilized, the solid particle size changes very slowly. Samples of slurry produced by the grinding plant periodically are taken and screen analyzed to monitor the solids size consist.

4. **Slurry pH.** For concentrate systems, pipeline corrosion control often is achieved by maintaining the slurry pH within a specified range. Suitable inline devices for automatically registering slurry pH are avaliable but are unproven in commercial operating slurry pipelines. In most cases, a laboratory reading of the pH from a hand-collected sample of slurry is used to measure pH.

5. **Pipeline Corrosion Monitoring.** Three methods have been used for monitoring corrosion and erosion. The first is the use of metal sample probes inserted into the line. Readings are made periodically, and any sudden changes in the corrosion rate can serve as a signal that immediate action must be taken to determine and correct the cause of the increased corrosion. The second procedure is to install preweighed test spools in the line. At periodic intervals (usually 6 months to 2 years), they are removed, cleaned

and reweighed. Although this procedure is comparatively awkward, particularly for a large-diameter pipeline, the use of precise weighing techniques will ensure that pipe material losses are being measured accurately.

The third and most recent procedure is the use of an ultrasonic wall thickness gauge. This method allows many locations to be checked quickly with a portable instrument. However, care must be taken to mark accurately the pipe circumference to ensure that all future readings are made at the same place.

6. **Flowrate.** Because slurries are often abrasive by nature, normal inline flow measuring devices, such as orifice plates and pitot tubes, are not applicable. For primary slurry flowmetering, magnetic flowmeters are considered the best choice. However, these meters are not useful with oil slurries. Ultrasonic flowmeters may be required in applications in which magnetic flowmeters are unsuitable. However, these meters are not accurate and reliable.

7. **Slurry Tank Level Measurement.** A number of different systems have been used for recording the level in slurry storage tanks. Sonar devices were used initially in the 1960s on several pipeline systems, with generally unsatisfactory results. Since then, other types of sonar devices have been developed but are still commercially unproven.

Conductance probes for level alarms and a pressure sensor mounted in the wall of the tank are used to obtain a rough indication of level of slurry in the tank. This procedure has proven satisfactory, although the accuracy of the level measurement with a pressure sensor is affected by changes in slurry density. A way to compensate for this effect is available but not commonly used. This can be achieved by measuring density of the slurry in the tank and correcting for density changes.

Process Alarms and Safety Devices

Because of the mechanical vibrations and pressure pulsations induced by a positive displacement-type pump system, the location and mounting of process alarms and safety devices are critical [10,11]. They cannot be mounted on the actual pipeline because numerous false alarms and/or shutdowns will occur that will result in a completely unreliable parameter monitoring system. For the individual pump units, the following are monitored:

- Pump vibration
- Bearing temperatures
- Pump speed
- Cooling water temperature
- Lube oil temperature
- Number of pump strokes

For each pump station, the following parameters are monitored:

- Suction pressure
- Discharge pressure
- Slurry density
- Slurry flowrate
- Integrated mass flow through station

Pump Station Control

Slurry pipeline systems range from single-station low-pressure centrifugal pump installations to multistation high-pressure reciprocating pump systems. In all cases, the basic requirement for successful slurry pumping is to maintain pipeline flow above a minimum operating velocity. The minimum operating velocity is set at a desired margin of safety above the critical velocity. The critical velocity is determined, in turn, by the solids screen analysis, solids density and concentration, as well as the specific system characteristics—pipe diameter, slurry temperature, etc.

In a positive displacement system, the flow is controlled by varying the pump speed. This can be accomplished by the use of a fluid coupling, eddy current coupling, rotating dc and scr varispeed drive or a hydraulic clutch system. Depending on the system flow and discharge pressure requirements, seven or more pump units operating in parallel may be installed. For optimum system efficiency, most of the pumps should be operated at their maximum design speed. Since a pipeline operates in a tight line condition to minimize pipeline wear, a fast control response is needed to maintain this condition.

Flowrate is the primary control parameter for the first pump station. The controller is set to maintain flowrate above the minimum operating flowrate. An alarm is produced when the flowrate drops below the minimum permissible flowrate. In a pump station using positive displacement pumps, the flowrate is determined by the speed of the pumps. The pump speed is controlled by the flow controller, except when the pump station maximum discharge pressure limit is exceeded. The maximum discharge pressure limit is generally set at about 10% below the maximum allowable discharge piping pressure. Once the maximum discharge pressure limit is reached, the pump speed is allowed to vary so as not to exceed this limit. At the same time, a high pump station discharge pressure alarm is produced so that the pipeline operator is made aware of potential problems.

Booster pump stations in the pipeline are controlled to maintain a given suction pressure. A suction pressure controller is used to vary the speed of the mainline pumps to maintain the required suction pressure. However, when the maximum discharge pressure for that pump station is reached, the pump speed is controlled by the discharge

Table V. Worldwide Slurry Pipeline Systems

Material	System/Location	Throughput Capacity (mmty)	Length (km)	Start of Operations
Coal	Consolidation Coal	1.2	174	1957[a]
	Russia	4.0	11	1966
	France (Merlebach)	1.5	9	1954
	Black Mesa	5.0	439	1970
Limestone	Columbia	0.9	9	1963
	Trinidad	0.5	9	1959
	Rugby cement	1.5	92	1964
	Calaveras	1.4	27	1971
Magnetite	Savage River	2.3	85	1967
	Waipipi	1.0	6 and 3	1971
	Pena Colorada	1.6	48	1974
	Las Truchas	1.4	27	1975
	Sierra Grande	1.9	32	1977
Hematite	Samarco	12.0	398	1978
Copper Concentrate	Bougainville	1.0	27	1972
	West Irian	0.3	111	1973
	Pinto Valley	0.4	18	1974
Phosphate	Valep	2.0	113	1979
Gilsonite	Utah	0.4	116	1957[a]
Gold Tailings	South Africa	1.0	34	N/A[b]
Nickel Tailings	Australia (Western Mining)	0.1	7	1970
Copper Tailings	Japan	0.5	71	1968
Kaolin	Georgia	0.6	25	N/A[b]

[a]Not currently operating.
[b]Several systems in existence.

pressure controller so that the maximum discharge pressure limit is not exceeded. As for the initial pump station, a high-discharge pressure alarm is produced for the pipeline operator's attention when the maximum discharge pressure limit is reached or exceeded.

Today, more than 20 long-distance slurry systems are in operation throughout the world. As listed in Table V, these systems transport a wide range of solids over distances up to 439 km. The total operating experience for these systems exceeds 150 years. At present, the Samaro hematite pipeline has the largest capacity—12 million metric tons per year (mmty). However, pipelines for transporting coal are being designed that will be capable of transporting throughputs exceeding 30 mmty.

COARSE-SLURRY PIPELINES

Design Considerations

Coarse solids can be transported by slurry pipelines over short distances. Transportation of tailings from mineral processing plants, sand and gravel from dredging operations, and concrete pumping and coarse coal from the mine face to a preparation plant are typical examples of coarse-slurry systems. In many applications, the size consist and slurry concentration cannot be confined within narrow limits. The maximum solids concentration and flowrate are the only major variables that can be controlled.

The solids normally are transported in heterogeneous or compound flow regimes. In a heterogeneous flow regime, the in situ concentration can be significantly higher than the input or delivered solids concentration. The maximum input concentration is generally limited to less than 40% solids by volume.

Many designers prefer to conduct loop tests for establishing hydraulic design criteria. The following limitations of loop tests should be recognized during interpretation of the loop test data:

1. Attrition of solids can change the hydraulic behavior of a slurry. Attrition mainly occurs due to repeated passage of the same solids through a pump.
2. Inventory of the fastest-settling particles may be limited. The deposition velocity, as well as the friction loss at and below deposition velocity, may be reduced due to settling out of these particles.

The minimum operating velocity should be at least 0.3 m/sec above the estimated deposition velocity.

Pumps

As mentioned previously, reciprocating pumps are not suitable when the maximum particle size exceeds 6 mm (0.25 in.). For a coarse solids pipeline, the following types of pumps may be considered:

- Centrifugal
- Hydrohoist
- Jet pumps
- Air lift pumps

Advantages and limitations of these pumps are discussed in the following sections.

Centrifugal Pumps

Centrifugal pumps are used extensively for pumping slurries under relatively low pressure. A list of some large-capacity installations is shown in Table VI. The main advantages of these pumps are as follows:

Table VI. Selected Slurry Pipeline Installations Using Centrifugal Pumps

System	Material	Pipe Length (km)	Flow (m^3/hr)	Pipe Diameter (m)	Maximum Discharge Pressure (KPa)	No. of Pumps in Series
Waipipi–New Zealand	Iron Sand	3.3	300	1465	4720	6
Kudremukh–India	Magnetite	67.0	400	1115	3100	5
Shirasu–Japan	Ash	6.8	600	4100	2300	3

1. High flowrates can be achieved with a single unit at a relatively low initial cost. Pumps capable of 5700 m^3/hr (25,000 gpm) are available.
2. Very few moving and wearing parts are involved.
3. They are simple to operate as well as to maintain.
4. There is no practical restriction on the maximum size of solids that can be handled.
5. The flow through the pump is pulse free.
6. They require little space.
7. Valves are not required for the operation of the pump.

Centrifugal pumps have the following shortcomings:

1. The maximum discharge pressure is limited to less than 900 kPa (130 psi) for a single-stage pump. With several pumps in series, a maximum pressure of about 5200 kPa (750 psi) can be achieved.
2. The flowrate of the pump is governed by the system pressure.
3. Attrition of friable solids may occur due to the high velocity of flow through the pump. Attrition can become important if the slurry has to pass through a number of pumps.
4. Seal water is required for good packing life. The seal water dilutes the slurry. The amount of dilution could become significant when the slurry passes through a number of pumps.
5. Centrifugal slurry pumps are made robust because of the abrasion of parts coming in contact with slurry, resulting in a low pump efficiency.

Characteristics of Centrifugal Pumps

A typical flow discharge head characteristic curve for a centrifugal slurry pump is shown in Figure 6. It shows the slurry flow decreasing as the discharge head increases. If an increase in pressure head results from the growth of a stationary bed of solids in the pipeline, the flowrate will be reduced further, which will intensify the problem. A variable-speed drive may be used to maintain a constant flowrate under varying discharge head conditions.

As shown in Figure 6, the flow through a centrifugal pump is defined by the point of intersection of the pump characteristic curve and the system head curve. A centrifugal pump should be selected so that it operates at its best efficiency point. Operating a pump above or below the best efficiency point gives rise to accelerated wear in the pump, as well as increased power requirements.

The performance of a centrifugal pump is affected by the presence of solids. Compared to water service, the pump head (expressed in feet of slurry) and efficiency may increase or decrease, depending on whether the characteristics of the slurry cause the mixture to behave as a Newtonian fluid or as a Bingham plastic [12]. Most researchers [13–18] have reported a reduction in head, as well as in the efficiency of the pump, when handling slurry. The reduction in the head and efficiency increases with the size, as well as with the concentration of the solids.

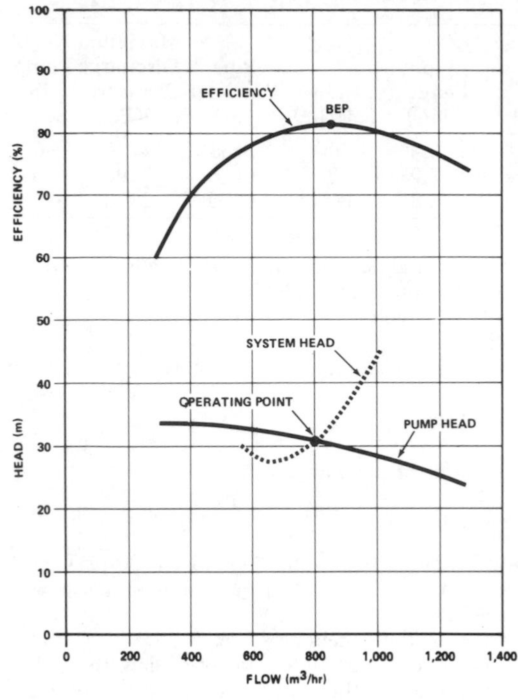

Figure 6. Centrifugal pump characteristics.

The following equation was developed by Sellegren [17] for determining the head reduction in pumping various types of ore:

$$\frac{H_s}{H_w} = 1 - 0.32\, C_W^{0.7}(S-1)^{0.7}/C_D^{0.25} \tag{3}$$

where H_s = head (m of slurry)
H_w = head (m of water)
C_w = fractional weight of solids in slurry
S = specific gravity of solids
C_D = average drag coefficient of solids

Depending on the extent of the reduction in pump discharge head, the pressure produced by the pump in handling slurry (expressed in Pascals) could be greater than, equal to or less than the pressure produced by the pump with clear water. For the case of a fine mineral slurry with a relatively high specific gravity, the pump discharge head at a given speed and flowrate will be greater than the clear water discharge head (expressed in Pascals). This could lead to unstable operation when the pump is used to flush a pipeline full of slurry, unless a variable-speed drive or a high-pressure water pump is used to provide the additional pressure required.

In the case of coarse solids, the pump characteristic curve (expressed in Pascals for slurry) could be lower than the clear water curve. Obviously, in this case the problem of a reduction in flowrate when switching to water will not occur.

Construction and Parts Life

End-suction, single-stage centrifugal pumps are limited by the number and size of the bolts that can be used to hold the two halves of the pump case together. This limits

Table VII. Life of Centrifugal Pump Parts

Part	Life (hr)
Impeller	400–8000
Casing	400–8000
Packing	1000–2000

the case pressure to a range of 3500 kPa (490 psi) to 5200 kPa (750 psi). It should be noted that pumps made by different manufacturers with similar flowrate capability may have different case pressure ratings. Common case pressure ratings are 400, 1000, 1400, 1900, 2400 and 3400 kPa.

The centrifugal pump parts exposed to wear from slurry are the casing, impeller and gland seal. Mechanical gland seals have not proven to be effective in slurry installations. A seal that incorporates a water flush to keep solids from entering the gland is used in most cases.

To obtain good service, the casing and impeller should be lined with abrasion-resistant material. Both rubber-lined and "Ni-hard" (metal-lined) units are used widely. The size of the solids to be pumped determines the type of pump to be selected. Rubber-lined pumps are generally used with particles up to about 10 mm (0.39 in.), and "Ni-hard" pumps can be used for relatively coarse solids. Because particles smaller than 6 mm (0.24 in.) usually do not have enough energy on impact with the "Ni hard" pump parts to keep the exposed part surfaces work hardened, rubber-lined pump parts usually have longer service life with fine materials. Coal is one material for which "Ni-hard"-lined pumps have shown better parts life than rubber-lined units. The life of each centrifugal pump part is given in Table VII.

To obtain good pump parts life, it is good practice to limit the impeller tip speed to less than 1200 m/min (4000 ft/min), or about 25 m (80 ft) of lift per stage. The pump parts life on units running faster than this speed drops in proportion to the square of the impeller tip speed. A careful economic analysis should take into account the reduced parts life that results from high pump impeller tip speed. Otherwise, a low initial cost pump may be selected that may result in a relatively high total pumping cost over the life of the project.

Hydrohoist

The limitation of maximum particle size for positive displacement pumps and the limitation of pump discharge pressure for centrifugal pumps can be eliminated by using a hydrohoist system. The hydrohoist consists of one or more chambers that can be filled with solids, either in dry form (lockhopper) or as a slurry. Once the chamber is filled with solids, water under high pressure is admitted to the chamber to push the solids into the pipeline. A major application is in vertical transportation of coarse ore from deep mines. Some of the hydrohoist system installations are listed in Table VIII. The advantages of these devices are as follows:

1. Large-sized solids can be pumped under high pressure without their having to pass through a number of centrifugal pumps. The attrition of solids is thereby reduced.
2. A high-pressure water pump is used that has relatively high efficiency compared to a centrifugal slurry pump.
3. The number of moving parts exposed to slurry is considerably less than with reciprocating units.
4. The life of slurry check valves is extended because of the reduced frequency of operation compared to reciprocating pumps; normally, the valves open and close in the presence of water.

Table VIII. Selected Installations Using the Hydrohoist System

Place	Length, m		Pipe Diameter (mm)	Material	Maximum Pressure (kPa)
	Vertical	Horizontal			
Hungary	214	550	150	Coal	36,000
Hungary	69	8,012	200	Coal	46,000
Hungary	15	6,000	200	Fly ash	10,000
Japan	—	—	—	Bauxite	3,430
Australia	—	—	—	Bauxite	5,490
Japan	—	—	—	Sulfide ore	5,390
Japan	488	25,000	200	Coal	7,350

5. A larger capacity can be obtained with a hydrohoist than with a reciprocating pump operating at the same discharge pressure.

The disadvantages of the hydrohoist system are as follows:

1. A sophisticated control system is needed to open and close the slurry valves.
2. The flowrate of this device depends on the system discharge pressure because a high-pressure centrifugal water pump is used.
3. Because of the complex control system, skilled operating and maintenance labor is required.

The various types of hydrohoist devices available differ mainly in the method of feeding solids and the country of origin. Similar systems include lockhoppers [19], pipe feeders [20], hydrohoist [21,22], high-pressure feeders [23] and the recirculatory slurry pump system [24].

Jet Pumps

A jet pump is used to transport solids over short distances [25]. For this device, part of the energy from a high-pressure jet of water is transferred to the solids in a mixing tube before the solids are discharged into the pipeline. The equipment has the following advantages:

1. Construction is simple.
2. There are no moving parts exposed to wear from slurry.
3. Investment is low.
4. It can handle large volumes of slurry without any particle-size limitation.
5. There is no particle attrition.

The disadvantages of this type of equipment are as follows:

1. It has a very low efficiency.
2. It has a limited discharge pressure capability.

Air Lift Pumps

This type of pump is used for vertical lifting of solids and fluid. It is a simple device with no moving parts exposed to slurry wear. The driving force results from buoyancy caused by the introduction of air at a certain height in a pipe that is submerged in water.

The higher-density water adjacent to the air water-solids mixture in the pipe results in the solids being driven upward.

These devices are used in dredging and underground mining operations [26]. Air lift pumps also have been considered for deep sea mining of manganese nodules.

Abrasion Control

Pipe abrasion is a major concern in coarse solids pipelines. In temporary systems the pipeline may be replaced periodically. For example, the phosphate mining industry in Florida replaces steel pipes after about two years of operation [27]. Use of special steel pipes, nonferrous pipes and steel pipes lined with abrasion-resistant linings can increase significantly the life of pipe carrying coarse solids.

Alloy Steel

Use of 3.2% chrome steel cladding inside a steel pipe was investigated by Takada et al. [28]. They report a wear rate reduction of 75% for this material compared to plain carbon steel with a silica sand slurry. The cost of cladded steel pipe was indicated to be about 20% higher than carbon steel, and field weldability would be difficult.

In another approach, the hardness of the inside surface of the pipe is increased by heating it to 1000°C and then quenching it with 20°C water. The result is that about half the inside wall thickness of the pipe is transformed from a relatively soft pearlite-ferrite mixture to abrasion-resistant martensite with a hardness of between 55 and 60 Rockwell C.

Nonferrous Pipes

For low-pressure applications, a number of nonferrous materials can be considered for transporting coarse solids. Some of the most common materials are listed in Table

Table IX. Lined Steel Pipes

		Pipe Availability				
Pipe	General Abrasion Resistance	Maximum Pipe Length	Maximum Pipe Diameter	Lining Thickness	Joints[a]	Miscellaneous
Rubber-Lined Steel	Good to very good	60 ft[b]	[c]	0.125–0.5 in.	Flanged	Natural rubber, normally superior to synthetic
Polyurethane-Lined Steel	Very good	60 ft	24 in.	0.187–0.375 in.	Flanged	Few slurry pipeline applications
Polyethylene-Lined Steel	Good	[d]	48 in.	0.125–2.0 in.	Flanged	Unproven in slurry service

[a]Victaulic couplings are also utilized for low-pressure service.
[b]Not in all pipe diameters.
[c]Diameters greater than 48 in. available.
[d]Up to several thousand feet for pulled-in-place systems.

IX. The advantages and limitations of various materials are discussed by Snoek and Carney [29].

Internally Lined Steel Pipe

The inherent pressure limitation of the nonferrous pipe materials can be overcome by using them as internal linings in a steel pipe. The main features of the three major types of linings are summarized in Table IX. Other types of internal linings are used also. These include basalt lining, castable ionomer and ceramic tiles. These types of internally lined steel pipes cannot be welded. They require the use of relatively expensive mechanical joints.

Tierling and Baron [30] have described a method of in situ internal lining of welded steel pipe. A plastic pipe is pulled inside a steel pipe and the annular space between the two pipes grouted with cement. This method reduces considerably the number of mechanical joints in the pipe.

NOMENCLATURE

B	ratio of mass transfer to momentum transfer coefficient
C	volume fraction solids at 0.08 D from the pipe top
C_A	volume fraction solids in the slurry
C_D	drag coefficient
C_W	fractional weight of solids in slurry
D	pipe inside diameter (m)
E_o	elevation at the pipeline inlet (m)
E_T	elevation at the pipeline terminal (m)
H_S	head (m of slurry)
H_W	head (m of water)
K	von Karman constant
L	pipeline length (m)
P	pumping pressure (m of slurry)
S	specific gravity of solids
U^*	friction velocity $= V\sqrt{f/2}$ (m/sec)
V	flow velocity (m/sec)
W	settling velocity (m/sec)
f	Darcy-Weisbach friction factor
ΔP	friction loss (m/m)

REFERENCES

1. Wasp, E. J., J. P. Kenny and R. L. Gandhi. "Solid Liquid Flow-Slurry Pipeline Transportation," *Trans. Tech. Publications* (1977).
2. Gandhi, R. L., P. E. Snoek and M. D. Weston. "Design Considerations for Slurry Pipeline Systems," *Proc. 3rd Int. Tech. Conf. on Slurry Transportation,* Las Vegas, Nevada, March 29–31 (1978).
3. Wasp, E. J., T. L. Thompson and T. C. Aude. "Initial Economic Evaluation of Slurry Pipeline Systems," *Transportation Eng. J., Proc. ASCE* 97(TE2):271–279 (1971).
4. Gandhi, R. L., P. E. Snoek and J. C. Carney. "An Evaluation of Slurry Pumps," in *Proc. 5th Int. Tech. Conf. on Slurry Transportation,* Lake Tahoe, Nevada, March 26–28 (1980).
5. Gandhi, R. L., B. L. Ricks and T. C. Aude. "Control of Corrosion-Erosion in Slurry Pipelines," paper presented at the 1st International Conference on Internal and External Protection of Pipes, BHRA Fluid Engineering, Cranfield, UK, Paper G-2, September, 1972.
6. Thompson, T. L., and T. C. Aude. "Slurry Pipeline Design and Operation Pitfalls to Avoid," paper presented at the joint Petroleum Mechanical Engineering and Pressure Vessels and Piping Conference, ASME, Mexico City, Mexico, September 19–24, 1976.
7. Derammelaere, R. H., and J. P. Chapman. "Slack Flow in the World's Largest Iron Concentrate Slurry Pipeline," in *Proc. 4th Int. Tech. Conf. on Slurry Transportation,* Las Vegas, Nevada, March 28–30 (1979).

8. Lutjen, G. P. "Marconaflo—the System and the Concept," *Eng. Mining J.* 67–75 (May 1970).
9. Cobb, D. B., C. O. Giles, J. D. Hornbuckle and F. O. Leavitt. "Coal Slurry Storage and Reclaim Facility for Mohave Generating Station," in *Proc. 3rd Int. Tech. Conf. on Slurry Transportation*, Las Vegas, Nevada, March 29–31 (1978).
10. Buckwalter, R. K., and J. C. Carney. "Instrument Design Consideration for Slurry Pipelines Utilizing Positive Displacement Pump," paper presented at HYDRO-TRANSPORT-4, BHRA Fluid Engineering, Cranfield, UK, May 18–21, 1976.
11. Hill, R. A., and J. D. Pitts. "Control Aspects of Slurry Pipelines," paper presented at the IEEE conference, Houston, TX, September 10, 1980.
12. Stepanoff, A. J. "Pumping Solid-Liquid Mixtures," reprint of Paper No. 63-WA-102, ASME Publication (1963).
13. Wilson, C. "The Design Aspects of Centrifugal pumps for Abrasive Slurries," paper presented at HYDROTRANSPORT-2, BHRA Fluid Engineering, Cranfield, UK, Paper H2, September, 1972.
14. Wiedenroth, W. "The Influence of Sand and Gravel on the Characteristics of Centrifugal Pumps. Some Aspects of Wear in Hydraulic Transportation Installations," paper presented at HYDROTRANSPORT-1, BHRA Fluid Engineering, Cranfield, UK, Paper 1, May, 1970.
15. Vocadlo, J. J., J. K. Koo and A. J. Prang. "Performance of Centrifugal Pumps in Slurry Service," paper presented at HYDROTRANSPORT-3, BHRA Fluid Engineering, Cranfield, UK, Paper J2, May, 1974.
16. Cave, I. "Effect of Suspended Solids on the Performance of Centrifugal Pumps," paper presented at HYDROTRANSPORT-4, BHRA Fluid Engineering, Cranfield, UK, Paper H3, May, 1976.
17. Sellegren, A. "Performance of a Centrifugal Pump when Pumping Ores and Industrial Minerals," paper presented at HYDROTRANSPORT-6, BHRA Fluid Engineering, Cranfield, UK, Paper G1, September, 1979.
18. Burgess, K. E., and J. A. Reizes. "The Effect of Sizing, Specific Gravity and Concentration on the Performance of Centrifugal Slurry Pumps," *Inst. Mech. Eng.* 190(36) (1976).
19. Dierks, H. A., and H. B. Link. "Developing a Lock-Hopper Feeder for Hydraulic Hoisting of Coal," Bureau of Mines Report of Investigation 6347 (1964).
20. Kocsanyi, L. "High Pressure Hydraulic Transport of Coal and Other Mining Products by Means of Pipe-Feeders without Contamination," paper presented at HYDROTRANSPORT 3, BHRA Fluid Engineering, Cranfield, UK, Paper A1, May, 1974.
22. Sakamoto, M., K. Uchida and Y. Kamino. "Vertical Type Hydrohoist for Hydraulic Transportation of Fine Slurry," paper presented at HYDROTRANSPORT 6, BHRA Fluid Engineering, Cranfield, UK, Paper F1, September, 1979.
23. Funk, E. D., M. D. Barrett and D. W. Hunter. "Pilot Experiences with Run of Mine Coal Injection and Pipelining, paper presented at HYDROTRANSPORT 5, BHRA Fluid Engineering, Cranfield, UK, Paper F3, May, 1978.
24. Huso, M. A. "Recirculatory Slurry Pump System," in *Proc. 3rd Int. Conf. on Slurry Transportation,* Las Vegas, Nevada, March 29–31 (1978).
25. Debreczeni, E., I. Tarjan and T. Meggyes. "Hydraulic Transport Systems in the Mining Industry Using Jet Slurry Pumps," paper presented at HYDROTRANS-PORT 6, BHRA Fluid Engineering, Cranfield, UK, Paper G3, September, 1979.
26. Weber, M., and Y. Dedegil. "Transport of Solids According to the Air Lift Principle," paper presented at HYDROTRANSPORT 4, BHRA Fluid Engineering, Cranfield, UK, Paper H1, May, 1976.
27. Faddick, R. R., and O. D. Staman. "Pipeline Transportation of Phosphate Slurries—A Survey," *Colorado School of Mines Res. Inst. Min. Ind. Bull.* V20 (6) (1977).
28. Takada, I., A. Shiga and S. Yamaguchi. "Mechanical Properties, and Field Weldability of 3.2 percent Cr Steel Clad ERW Pipe for Slurry Transportation," Research Laboratories, Kawasaki Steel Corporation, Chiba, Japan.

29. Snoek, P. E., and J. C. Carney. "Pipeline Material Selection for Transport of Abrasive Tailings," in *Proc. 6th Int. Tech. Conf. on Slurry Transportation,* Las Vegas, Nevada, March 24–27 (1981).
30. Tierling, K., and J. Baron. "Multi-Wall System Curbs Sour Gas Corrosion," *Pipeline Ind.* 41–44 (September 1981).

CHAPTER 37
DEEP BED FILTRATION

Chi Tien
Department of Chemical Engineering and Materials Science
Syracuse University
Syracuse, New York 13210

CONTENTS

INTRODUCTION

In the classical treatise on chemical engineering by Walker et al. [1], filtration is described as a process in which solids present in a liquid slurry are separated from the liquid by forcing the slurry through a supporting mesh or cloth. The end products of the operation are a clear liquid (filtrate) and solids retained at the surface of the supporting mesh (filter cake). Such a process (often referred to as cake filtration) is effective for slurries with relatively high solids concentration. It is this type of filtration that is commonly described in chemical engineering textbooks.

There is another type of filtration process, the objective of which is to remove particulate matters from dilute liquid suspensions (solids concentrations in the range of 100 ppm). Solids separation (for clarification) is effected by passing the suspension through a granular porous medium of substantial thickness, leading to the retention of particulates throughout the bed. This is known as deep bed filtration (or depth filtration).

A typical example of deep bed filtration is the use of sand filters for water treatment. This is perhaps one of the oldest engineering practices, dating back more than a century. In more recent years, the application of deep bed filtration has extended to other situations as well, including waste treatment, removal of solids present in feedstocks to catalytic beds and purification of molten aluminum to ensure the quality of aluminum ingot. Furthermore, in many separation processes (such as membrane and fixed bed adsorption processes), the use of deep bed filters to pretreat the feed streams often is required.

Paradoxically, in spite of its long history of varied applications, the design of deep bed filtration systems remains largely empirical and, in many instances, based on specifications promulgated by government agencies. There is a significant lack of basic knowledge in every aspect of the problem. In terms of scientific understanding, deep bed filtration is a relatively unexplored area.

In more recent years there has been resurgent interest in the study of deep bed filtration. A substantial number of these investigations were conducted on a fairly fundamental level. The results of these studies have yielded an outline of a plausible theory that incorporates the essential features of deep bed filtration. The purpose of this chapter is to present an overview of this effort, to discuss the success as well as the failures experienced so far and to speculate on the direction of future studies. This chapter is not intended to be an exhaustive literature survey on deep bed filtration. Rather, it is the intention of the author to make the presentation as complete and self-sufficient as possible. However, for a more complete literature survey, the reader may wish to refer to some of the earlier review articles [2–4].

PHENOMENOLOGICAL DESCRIPTIONS

Deep bed filtration may be defined as a process in which a fluid suspension is passed through a filter composed of granular or fibrous substances. As the suspension flows through the filter bed, particulate matter is transported from the suspension to the surface of the filter grain (or fiber) and becomes attached to the surfaces. The accumulation of particulate matter within the bed results in a continuous change of the structure of the granular medium. Consequently, the particle collecting ability of the medium and the pressure drop across the bed necessary to maintain a constant flowrate vary with time. This time-dependent behavior is the main feature of deep bed filtration.

The fundamental equations describing the macroscopic behavior of deep bed filters are the conservation equation and the rate equation. For an axial flow filter with a constant cross-sectional area, the conservation equation can be written as

$$u\left(\frac{\partial c}{\partial z}\right) + \frac{\partial(\sigma + \epsilon c)}{\partial t} = 0 \tag{1}$$

The independent variables are the axial distance, z, and time, t. The dependent variables are the particle concentration in the suspension, c, and the specific deposit of particles in the bed, σ. The assumptions involved are one-dimensional plug flow and negligible axial dispersion of particulate matter. The porosity of the bed, ϵ, changes with time. If the particle deposits are composed largely of particle aggregates, ϵ and σ can be related by the simple expression

$$\epsilon = \epsilon_0 - \frac{\sigma}{1 - \epsilon_d} \tag{2}$$

where ϵ_d is the porosity of deposits. The use of Equation 2 requires caution. This is especially true of the deposit morphology of the blocking mode (see the discussion in the section entitled Prediction of Dynamic Behavior of Deep Bed Filters).

Similar to all fixed bed processes, it is more convenient to employ a corrected time variable, θ, defined as

$$\theta = t - \int_0^z \frac{\epsilon dz}{u} \tag{3}$$

In terms of z and θ, Equation 1 becomes

$$u\left(\frac{\partial c}{\partial z}\right) + \frac{\partial \sigma}{\partial \theta} = 0 \tag{4}$$

The term $\partial\sigma/\partial\theta$ represents the rate of filtration i.e., volume of particles retained per unit volume of bed per unit of time. In general, it can be written as

$$\left(\frac{\partial \sigma}{\partial \theta}\right) = G[\underline{\gamma}, c, \sigma] \tag{5}$$

where $\underline{\gamma}$ is the parameter vector characteristic of the filtration process. The dependence of the filtration rate on σ is a direct consequence of the change in media structure due to particle deposition. It is obvious that the rate of filtration depends on the number of particles present in the suspension.

The expression of Equation 5 is only qualitatively useful. To be quantitatively useful, specific functional forms must be assigned to G. The simplest expression advanced by Iwasaki [5], based on slow sand filter data and subsequently applied to rapid sand filters [6], is

$$\frac{\partial \sigma}{\partial \theta} = u \lambda c \tag{6}$$

where λ is the filter coefficient. Furthermore, as shown in later sections, λ is directly related to the unit collector efficiency of a filter bed. As such, λ is independent of c and is only a function of the state of the filter media. Accordingly, one may write

$$\lambda = \lambda_o F_1(\underline{\alpha}, \sigma) \tag{7}$$

where λ_o is the clean filter coefficient.

Equations 4, 6 and 7, together with the appropriate initial and boundary conditions, give a complete description of the particle distributions (both in the suspension and within the bed). Specifically, they provide a means for estimating the history of the effluent concentration for a given set of conditions. This, of course, assumes that values of λ_o and the functional form of F_1 are known.

For a clean filter bed, the pressure gradient flowrate relationship is governed by the Carman-Kozeny equation, namely

$$-\left(\frac{\partial p}{\partial z}\right)_o = \frac{150}{d_g^2} u \frac{(1-\epsilon_o)^2}{\epsilon_o^2} \tag{8}$$

To account for the change in pressure gradient due to filter clogging, the following expression may be used:

$$\frac{\left(\frac{\partial p}{\partial z}\right)}{\left(\frac{\partial p}{\partial z}\right)_o} = F_2(\underline{\beta}, \sigma) \tag{9}$$

and the pressure drop across the filter bed can be obtained directly from the integration of Equation 9, or

$$\Delta p = p - p_o = \left(\frac{\partial p}{\partial z}\right)_o \int_o^z F_2[\underline{\beta}, \sigma(z'\theta)]\, dz' \tag{10}$$

Equations 1, 4, 6 and 7 therefore give a complete description of the dynamic behavior of deep bed filtration for a given set of operating and system variables. The parameters presented in these equations are λ_o, $(\partial p/\partial z)_o$, $F_1(\gamma, \sigma)$ and $F_2(\underline{\beta}, \sigma)$. As stated previously, the initial pressure gradient can be found according to the Carman-Kozeny equation (8). The clean filter coefficient, λ_o, has the dimension of $(\text{length})^{-1}$ and, typically, the order of magnitude 10^{-2} to 10^{-1} cm^{-1}. Both λ_o and the two functions, i.e., F_1 and F_2, can be determined experimentally or theoretically. These determinations will form the major part of this chapter, as shown below.

THEORETICAL ANALYSIS OF DEPOSITION IN FILTER BEDS

The phenomenological equations presented in the preceding section provide a description of the dynamic behavior of deep bed filtration, provided the values of λ_o and the functional forms of F_1 and F_2 are known. On the other hand, a phenomenological formulation does not provide any detailed insight into the physical phenomena of filtration. Analyses on a more microscopic level are required to formulate a comprehensive theory that can be used to predict the various parameters present in the phenomenological equations.

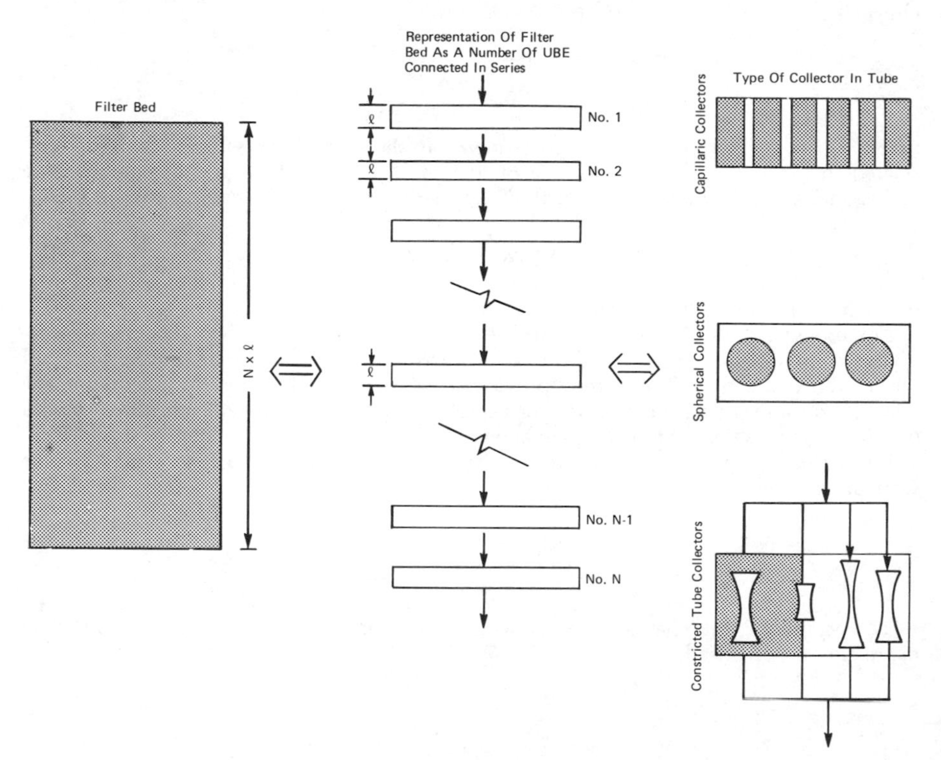

Figure 1. Representation of filter beds as an assembly of collectors.

The starting point of such a formulation is to view a filter bed as an assembly of particle collectors. This is a natural assumption because the major function of deep bed filters is to remove fine or colloidal particles from fluids. In the terminology of Payatakes et al. [7], a deep bed filter is composed of a series of unit bed elements (UBE) of thickness l, each of which is composed, in turn, of a number of geometrically similar collectors (unit cells) that may or may not be the same size. This is shown schematically in Figure 1.

To complete the description it is necessary to specify the geometry, size and size distribution of the unit cells, as well as the flow field around (or within) the unit cell. In other words, the selection of a particular porous media model is required. Once the model is chosen, the specification of the collector geometry, size, size distribution and flow field follows.

The selection of the porous media model for the representation of filter beds is by no means unique. However, there are certain practical limitations. The large number of models available generally can be classified into two categories; external flow models and internal flow models, the difference being whether the emphasis is on the filter grains or the pores of the granular media. In most of these cases, the flow field is obtained by solving the linearized Navier-Stokes equation applied to the particle geometry chosen for the collector.

External Flow Models

The prominent aspect of the geometry of these models is the collector (or filter grain) itself, rather than the pore. The fluid is visualized to flow over the collector. Although in certain cases the effects of neighboring grains is partially accounted for, their presence generally cannot be considered explicitly.

Most of the external flow models consider the filter grain to be of spherical geometry. The collector radius, a_c, is given as

$$a_c = \langle d_g \rangle / 2 \tag{11}$$

where $\langle d_g \rangle$ is the average grain diameter. The unit bed element thickness, l, is the length that contains one filter grain on average. Accordingly,

$$\pi a_c^2 \, l(1-\epsilon) = \frac{4}{3} \pi \, a_c^3 \tag{12}$$

or

$$l = \frac{4a_c}{3(1-\epsilon)} \tag{13}$$

The flow field around the spherical collector is assumed to be axisymmetrical and can be represented by the stream function, ψ. The velocity components V_r and V_θ are given as

$$V_r = \frac{-1}{r^2 \sin\theta} \frac{\partial \psi}{\partial \theta} \tag{14a}$$

$$V_\theta = \frac{1}{r \sin\theta} \frac{\partial \psi}{\partial r} \tag{14b}$$

Isolated Sphere Model

In this case, each filter grain is regarded as a single sphere, and the individual collectors (spheres) are assumed to be independent of each other. The fluid that flows over the collector therefore can be assumed to be infinite in extent. If the flow is creeping, the flow field is given by the Stokes solution, or

$$\psi = \frac{1}{4} V a_c^2 \sin^2 \theta \left[2 \left(\frac{r}{a_c}\right)^2 - 3 \frac{r}{a_c} + \frac{a_c}{r} \right] \tag{15}$$

where V is the superficial velocity in the filter. In fact, it is assumed that the flow field of the fluid is not influenced by the presence of particles in the fluid phase. This is true only if the suspension is dilute.

Brinkman's Model

Brinkman's original formulation [8] considered a spherical grain imbedded in a porous mass. The flow through this mass is described by a combination of the Navier-Stokes equation (without the inertial terms) and Darcy's equation. The stream function is given as

$$\psi = \psi_\infty \left(\frac{a_c}{r}\right)^3 \left[\left(\frac{r}{a_c}\right)^3 - 1 \right] + \left(\frac{3}{p_1^2}\right) \cdot \left[\left(1 + p_1 \frac{r}{a_c}\right) \exp\left(p_1 - p_1 \frac{r}{a_c}\right) - (1 + p_1) \right] \tag{16}$$

where

$$\psi_\infty = \left(\frac{1}{2}\right) V a_c^2 \left(\frac{r}{a_c}\right)^2 \sin^2\theta \tag{17}$$

$$p_1 = \frac{\left[9 + 3 \left(\frac{8}{1-\epsilon} - 3 \right)^{1/2} \right]}{\left[\frac{4}{1-\epsilon} - 6 \right]} \tag{18}$$

Happel's Model

According to Happel's model, a filter bed is assumed to be comprised of spherical collectors (radius a_c), each of which is surrounded by a fluid envelope (radius b). The stream function, ψ, is given as [9]

$$\psi = \left(\frac{V}{2}\right) \sin^2\theta \, a_c^2 \left[K_1 \left(\frac{a_c}{r}\right) + K_2 \left(\frac{r}{a_c}\right) + K_3 \left(\frac{r}{a_c}\right)^2 + K_4 \left(\frac{r}{a_c}\right)^3 \right] \tag{19}$$

for $a_c < r < b$

$$\left.\begin{aligned} K_1 &= \frac{1}{w} \\ K_2 &= \frac{(3 + 2p'^2)}{w} \\ K_3 &= \frac{p'(3 + 2p'^5)}{w} \\ K_4 &= -\frac{p'^5}{w} \\ w &= 2 - 3p' + 3p^5 - 2p'^6 \end{aligned}\right\} \tag{20}$$

and

$$p' = (1 - \epsilon)^{1/3} = \frac{b}{a_c} \tag{21}$$

Near the collector surface, ψ can be approximated to be

$$\psi = \frac{3}{4} A_s V (r - a_c)^2 \sin^2\theta \tag{22}$$

where

$$A_s = \frac{2(1 - p'^5)}{w} \tag{23}$$

This approximate expression also has been used in the calculation of λ_o [10]. The results were presented as those based on Happel's model. In fact, Equation 22 differs significantly from Equation 16 for large distances from the collector surface.

Internal Flow Models

In contrast to the various external flow models discussed above, the internal flow models focus their attention on the pores of the filter medium. This approach has the distinct advantage that the transport processes in a porous medium often are controlled by the geometry of the pores and the flow field within the pores. For example, the convergent-divergent flow behavior that is characteristic of flow through porous media is difficult to incorporate in an external flow model. For internal flow models, the surfaces of the pores act as particle collectors. Two examples are given below.

Capillaric Model

This is the simplest and also the oldest porous media model. A porous medium is viewed as a bundle of capillaries. To adapt the capillaric model to deep bed filtration, a unit bed element is assumed to have N (per unit bed cross-sectional area) capillaries with radius a_c and length l. The model parameters–N, a_c and l–are related to the macroscopic properties of the filter bed by the following expressions

$$N\, a_c^2 = \epsilon \tag{24}$$

$$\frac{\epsilon a_c^2}{8} = k \tag{25}$$

$$l = \langle d_g \rangle \tag{26}$$

where k is the permeability of the bed.

The flow within the capillaries may be assumed to be Poiseuillean, i.e.,

$$V = 2V_i \left[1 - \left(\frac{r}{a_c}\right)^2\right] \tag{27}$$

where V_i, the average velocity in the pore, is given as

$$V_i = \frac{V}{N\, \pi\, a_c^2} \tag{28}$$

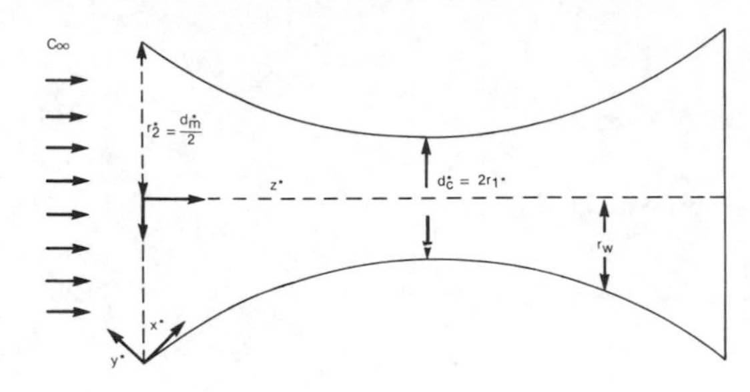

Figure 2. Representation of particle deposition in a constricted tube.

Constricted Tube Model

The individual collectors in this case are assumed to be unit cells in the shape of three parameters (Figure 2): d_{m_i}, the maximum diameter, d_{c_i}, the constriction diameter, and h_i, its height. Three different geometries have been used to describe the wall of the constricted tube: parabolic [11], hyperbolic [12] and sinusoidal [13]. The unit cells in the original formulation of Payatakes et al. [11] are assumed to follow a certain size distribution, which can be determined experimentally. As a simplification they may be assumed to be of uniform size. Similarly, different flow fields within the unit cells have been presented [12,14–16]. Table I lists formulas for the estimation of the various model parameters.

Regardless of the specific models used to describe the unit cells (collectors), the effectiveness of any unit bed element in removing suspended matter can be written in terms of the particle concentrations of the inlet and outlet streams to that unit bed element, or

$$\eta = \frac{(c_{i-1} - c_i)}{c_{i-1}} \tag{29}$$

On the other hand, if one applies Equations 4 and 6 to the unit bed element, assuming that λ is constant (for any instant) over the thickness of the unit bed element, one has

Table I. Parameters of a Constricted Tube Model (case of uniform unit cell size)

1. Number of constricted tubes per unit area (N_c):

$$N_c^* = N_c \cdot d_g^2 = \frac{6 \cdot \epsilon^{1/3}(1 - S_w)^{1/3}(1 - \epsilon)^{2/3}}{\pi}$$

2. Length of unit bed element:

$$l^* = \frac{l}{d_g} = \left[\frac{\pi}{6}\frac{1}{(1 - \epsilon)}\right]^{1/3}$$

3. Constriction diameter (d_c[a]):

$$d_c^* = \frac{d_c}{d_g} = 0.35 = 2\,r_1^* \text{ (approximately)}$$

4. Height of the unit cell, h:

$$h = d_g$$

5. Maximum diameter of the unit cell, d_m:

$$d_m^* = \frac{d_m^*}{d_g} = \left[\frac{\epsilon(1 - S_{wi})}{(1 - \epsilon)}\right]^{1/3} = 2\,r_2^*$$

[a] d_c and S_{wi} can be determined from capillary pressure-saturation data.

$$n\frac{c_{i-1}}{c_i} = \lambda l \tag{30}$$

Comparing Equations 29 and 30, one has

$$\lambda = \frac{1}{l}\, n\frac{1}{1-\eta}$$

or

$$\lambda \simeq \frac{\eta}{l} \quad \text{if} \quad \eta << 1 \tag{31}$$

The collection efficiency of the unit bed element, η, represents the average collection efficiency of the unit cells. For the special case in which all the unit cells are identical in size, the collection efficiencies of the individual collectors are the same as η. Accordingly, one has

$$F_1(\underline{\alpha}, \sigma) = \frac{\lambda}{\lambda_o} = \frac{\eta}{\eta_o} \tag{32}$$

On the other hand, the pressure drop associated with fluid flow through porous media can be considered as the cumulated drag forces acting on the surface of the media network. If a unit bed element is considered to be an assembly of identical unit cells, one can write

$$F_2(\underline{\beta}, \sigma) = \frac{\left(\frac{\partial p}{\partial z}\right)}{\left(\frac{\partial p}{\partial z}\right)_o} = \frac{F_D}{F_{D_o}} \tag{33}$$

where F_{D_o} is the drag force acting on a clean unit cell, and F_D is the drag force acting on the unit cell with a number of particles deposited on it. This development is intended to demonstrate at least a plausible and rational way for studying deep bed filtration.

Equations 4, 6, 7 and 10 provide a description of the dynamic behavior of deep bed filtration. Equations 23, 24 and 25 provide the link between the parameters of the phenomenological equations and physically significant quantities when filtration is viewed on a more microscopic level. In principle, it also points to the future direction of research if a fundamental understanding of deep bed filtration is to be achieved.

TRAJECTORY CALCULATION OF CLEAN COLLECTORS

With the representation of a filter bed as an assembly of collectors (unit cells) of specified geometry, deep bed filtration can be viewed in terms of the study of particle deposition from suspension flowing past these collectors. Specifically, one is interested in the estimation of the collection efficiency of these unit cells and the drag force acting on these unit cells corresponding to different stages of particle retention. Although present knowledge of deep bed filtration has not enabled sufficiently accurate predictions of all these quantities, the estimation of the initial collection efficiency (i.e., collection efficiency for clean unit cells) from trajectory calculation has been found to be practical and useful. This will be discussed in the following sections.

The principle of trajectory calculation is to estimate the collection efficiency of unit cells from the location of particle trajectories, especially the so-called limiting trajectory. Implicit in its use is the assumption of certain criteria that determine the outcome once

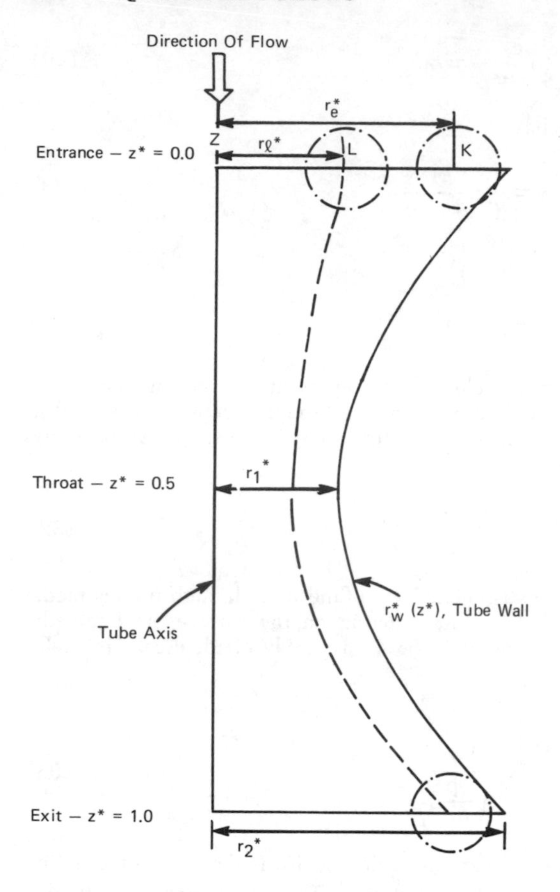

Figure 3. Schematic representation of limiting trajectory.

a particle makes contact with a collector (unit cell). This is demonstrated in Figure 3, in which a constricted tube collector is considered. Assume that particle K enters the tube at a radial distance of $(r_e)_i$ and makes contact with the tube at its exit (point N). Furthermore, if one assumes that a particle becomes deposited automatically once it makes contact with the collector (this is a good assumption for liquid suspension in deep bed filters), it is obvious that all particles that enter the tube with a radial distance greater than $(r_e^*)_i$ will be collected. If one assumes that the particle concentration distribution at the inlet is uniform, the collection efficiency of the unit cell, defined as the fraction of particles entering the unit cell, is given as

$$\eta_o = \frac{\psi^*\left[\frac{1}{2}, (r_e^*)_i\right] - \psi\left[\frac{1}{2}, (r_o^*)_i\right]}{\psi^*\left[\frac{1}{2}, 0\right] - \left[\frac{1}{2}, (r_o^*)_i\right]} \tag{34}$$

where ψ^* is the dimensionless stream function for the flow inside the constricted tube [11]. This trajectory (indicated by the dotted line in Figure 3) is the limiting one.

A detailed discourse on the determination of particle trajectories (or limiting trajectories, in particular) is beyond the scope of this chapter. Only a brief account will be presented. To initiate trajectory calculations, the following procedure has to be followed:

1. Specify the geometry size and size distribution of collectors.
2. Specify the flow field around or within the collectors.
3. Specify the forces acting on the collector.

Steps 1 and 2 are tantamount to the selection of a particular porous media model to characterize the filter bed. As for Step 3, it is obvious that the results of trajectory calculation depend on the kind of forces considered, as well as the expressions used to describe these forces. It is also worth noting that since trajectory calculation yields deterministic results, the inclusion of Brownian diffusion force, which is stochastic, becomes impossible. Consequently, the use of trajectory calculations to estimate the collection efficiency (or filter coefficient) can be made only from relatively large particles ($d_p \geqq 2\ \mu m$), with negligible Brownian diffusion effect.

The use of trajectory calculation for the estimation of the clean filter coefficient, λ_o, represents one of the major achievements in the study of deep bed filtration during the last decade. A summary of the various trajectory calculation studies is given in Table II. The results of these studies are summarized on the following pages.

Table II. Summary of Results on Trajectory Calculations for Deep Bed Filtration

Model	Investigator	Remarks
Single-Sphere Model	Yao et al. [17] and Yao [18]	Surface interaction not included; drag correction neglected.
Single-Sphere Model	Rajagopalan and Tien [19] and Rajagopalan [20]	Surface interaction with retardation effect for London force. Drag correction considered.
Capillaric Model	Payatakes et al. [21]	Surface interaction with retardation effect for London force. Drag correction considered.
Capillaric Model	Hung and Tien [22]	Surface interaction and drag correction included; nonvanishing fluid velocity across collecting surface.
Sphere-in-Cell Model (Happel's model)	Spielman and Fitzpatrick [10] and Fitzpatrick [23]	Approximate fluid velocity expression valid for small particles. No retardation effect.
Sphere-in-Cell Model (Brinkman model)	Payatakes et al. [21]	Surface interaction and drag correction included.
Sphere-in-Cell Model (Happel's model)	Rajagopalan and Tien [19]	Surface interaction with retardation effect for London force and drag correction included.
Constricted Tube Model	Payatakes et al. [11]	Surface interaction with retardation effect for London force and drag correction included. Constricted nature of flow channels considered.

1. The particle trajectories are obtained from the integration of the appropriate equations of particle motion. Regardless of the geometry used for the collection, these equations are first order because of the omission of particle inertia. Furthermore, because of its stochastic nature, the Brownian diffusion force cannot be included.
2. Many dimensionless groups appear in these equations of particle motion, a listing of which is given in Table III. That the filter coefficient (or collection efficiency) depends on a large number of dimensionless groups underscores the difficulty of establishing a general correlation based on experimental data alone.
3. Generally speaking, the determination of the limiting trajectory can be made only through the numerical integration of the equations of particle motion. Consequently, it is not possible to obtain closed-form expressions for filter coefficients. However, in two cases, collection efficiencies obtained from the numerical calculation with the use of the isolated sphere model and the Happel cell model [19,25] have been correlated empirically with relevant parameters, thus overcoming one of the major objections about the use of trajectory calculation for the estimation of the filter coefficient, namely, that the method is tedious and time consuming. The expression with Happel's cell model is

$$\eta_o = 0.72\ A_s N_{Lo}^{1/8} N_R^{15/8} + 2.4 \times 10^{-3} A_s N_G^{1.2} N_R^{-0.4} \tag{35}$$

for $N_R < 0.09$ and favorable surface interaction.
4. As expected, the calculated filter coefficient values are found to vary with the collector model. However, with the exception of the capillaric model, which gives significantly lower values of λ_o [21], calculated values of λ_o, according to different collector geometries, are comparable in magnitude and are of the same order of magnitude with experimental data. This is shown in Table IV.
5. The above statement should not be construed to mean that there is no difference among the various collector models. Subtle and significant differences

Table III. Relevant Dimensionless Groups of Deep Bed Filtration and Ranges of Values

Group	Definition	Ranges of Values
N_{DL}	κa_p *Do*uble *La*yer group; dimensionless double layer thickness	10 to 10^5
N_{E1}	$\nu\kappa(\zeta_c^2 + \zeta_p^2)12\pi\mu V$ *E*lectrokinetic group *1*	10^{-1} to 10^3
N_{E2}	$2\zeta_c\zeta_p/(\zeta_c^2 + \zeta_p^2)$ *E*lectrokinetic group *2*; Electrokinetic "parity" group	-1 to $+1$
N_G	$2\, a_p^2\, (\rho_p - \rho_f) g/9\mu V$ *G*ravity group; dimensionless settling velocity of the particle	10^{-2} to 10^0
N_{Lo}	$H/9\pi\mu a_p^2 V$ *Lo*ndon group	10^{-5} to 10^{-2}
N_{Pe}	Vd_c/D_{BM} *Pe*clet number	10^6 to 10^8
N_R	a_p/a_c *R*elative size group; so-called "aspect ratio"	10^{-4} to 10^{-1}
N_{Rtd}	$2\pi a_p/\lambda_e$ *R*e*t*ar*d*ation group; accounts for electromagnetic retardation of the London force	10^0 to 10^3

Table IV. Comparison of Experimental Data [26,27] and Calculated λ_o based on Happel's and Constricted Tube Models [25] ($\lambda \times 10^2$, cm^{-1})

	$d_p = 2.75\ \mu m$			$d_p = 4.5\ \mu m$			$d_p = 9.0\ \mu m$		
Run No.	Experimental Data	Payatakes [24]	Rajagopalan [20][a]	Experimental Data	Payatakes [24]	Rajagopalan [20][a]	Experimental Data	Payatakes [24]	Rajagopalan [20][a]
I	6.0	2.4	2.4	7.6	3.2	3.4	8.8	7.1	9.6
II	8.1	3.1	2.6	11.0	4.6	3.9	14	10.5	10.4
III	11.0	4.1	2.7	15.0	6.2	4.1	16.5	14.2	10.4
IV	3.1	1.8	2.2	4.6	2.5	3.7	6.4	5.0	9.2
V	3.1	1.2	1.8	4.4	1.7	3.3	5.6	3.7	9.1
VI	4.5	2.1	1.5	5.8	3.3	2.2	6.3	7.7	6.0
VII	3.9	1.9	1.1	4.4	3.1	1.5	4.6	8.2	3.5
VIII	2.7	1.7	0.7	3.9	3.1	1	5.3	8.2	2.3

[a]The corresponding values based on Spielman and Fitzpatrick [10] are given by Payatakes et al. [24] and are markedly higher, especially for large d_p. For example, for d_p = 9 μm they are about four to ten times higher than those based on the rigorous application of the sphere-in-cell model. See also the work of Rajagopalan and Tien [25].

indeed exist. For example, according to the single-sphere model [19], η_o (or λ_o) decreases monotonically with the increase of N_R, whereas the results according to Happel's model exhibit a minimum in the η vs N_R curve [25].

6. In general, the strong dependence of η_o on the various predicted surface interaction groups is consistent with experimental observations. Only in the case of favorable surface interaction is the agreement between theory and experiment observed (see Table IV). The problem concerning the role of surface roughness on the collector in particle attachment remains poorly understood. This will be discussed in later sections.

DEPOSITION OF SUBMICRON PARTICLES

Transport of submicron particles is caused largely by Brownian diffusion. The Brownian diffusion force cannot be included in the equations of particle motion because of its stochastic nature, as mentioned earlier. The problem, however, can be treated as that of diffusive mass transfer. Depending on the nature of surface interaction between the Brownian particles and the collector, the boundary conditions at the collector surface can be taken to be either zero concentration or a first-order chemical reaction. The former situation corresponds to the case of favorable surface interaction. Based on previous results of mass transfer in packed beds [28,29], the collection efficiency, η_{BM_o}, is found to be

$$\eta_{BM_o} = 4\ A_s^{1/3}\ N_{pe}^{-2/3} \tag{36}$$

On the other hand, if the surface interactions are unfavorable, namely, the surface interaction forces between the particle and the collector is repulsive, the rate of particle deposition is retarded. Several investigators [30–32] have shown that the presence of a repulsive force can be considered to be equivalent to the presence of a first-order chemical reaction, with the submicron particles as reacting species. Rajagopalan and Karis [32], in particular, have shown based on their numerical results, that a correction factor can be introduced to the expression of Equation 36 to account for the effect of the repulsive force. Their results are expressed as

$$\eta_{BM_o} = 4\,A_s^{1/3}\,N_{pe}^{-2/3}\,f(\S) \tag{37}$$

$$\S = 0.71\,A_s^{-1/3}\,N_{pe}^{2/3}\,k_r^* \tag{38}$$

$$k_r^* = \frac{D_{BM}}{V}\int_o^\infty \left[\frac{e^{\phi(\delta)}}{R(\delta)} - 1\right] d\delta \tag{39}$$

where ϕ is the surface interaction force potential. $R(\delta)$ is a function of position that accounts for the hydrodynamic retardation effect. Similar expressions have been developed recently based on the constricted tube model. The results differ somewhat from those of Equations 38 and 39, but only marginally.

EFFECT OF SURFACE INTERACTION ON DEPOSITION

Under conditions commonly encountered in liquid deep bed filtration, when particles move sufficiently close to filter grains a significant manifestation of surface forces is expected. The surface forces involved are principally of two kinds: the molecular dispersion force (commonly known as London van der Waals force) and the double ionic layer interaction force. The force potential of the power is given as

$$\phi_{LO} = -\frac{H}{6}\left[\frac{2(\delta^+ + 1)}{\delta^+(\delta^+ + 2)} - n\left(\frac{\delta^+ + 2}{\delta^+}\right)\right] \tag{40}$$

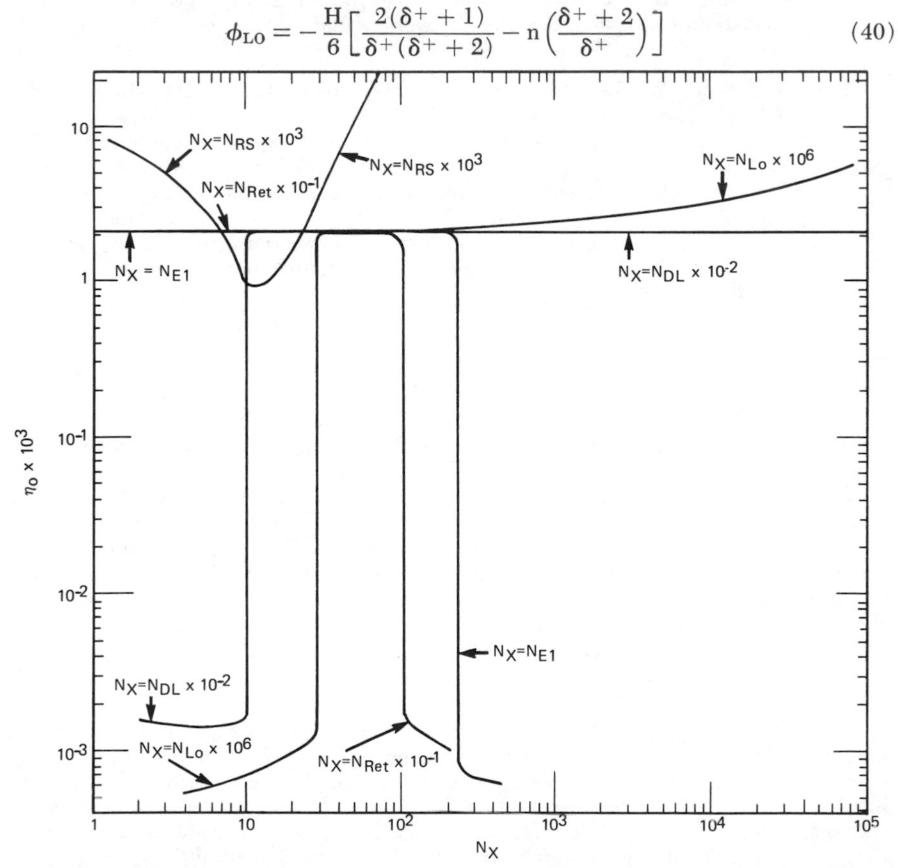

Figure 4. Effect of various surface interaction groups on the unit collection efficiency, η_o.

where δ^+ is the dimensionless separation distance defined as δ/a_p, and H is the Hamaker constant. Equation 40 is derived for the case of sphere flat plate, which applies to deep bed filtration if $N_R << 10^{-1}$.

The force potential of the double ionic layer interaction force, according to the DLVO theory, is

$$\phi_{DL} = \epsilon a_p \left(\frac{\delta_p^2 + \delta_c^2}{4}\right)\left[\frac{2\delta_p\delta_c}{\delta_p^2 + \delta_c^2} \, ln \, \frac{1 + e^{\kappa\delta}}{1 - e^{-\kappa\delta}} + ln(1 - e^{-2\kappa\delta})\right] \qquad (41)$$

Equation 41 is also for the plate-sphere case, with the added assumption that the surface potentials of the particle and the collector are the same.

The effect of surface interaction on the deposition of submicron particles has been discussed before. The result is given by Equations 37 and 38, with k_r^* evaluated from Equation 39. The surface interaction potential that appears under the integral sign on the right side of Equation 39 is the algebraic sum of ϕ_{LO}, i.e., Equation 40, and ϕ_{DL}, i.e., Equation 41. The discussions given below will be restricted therefore to the case of non-Brownian particles (i.e. particles with radius greater than 1 μm).

Some of the trajectory calculation studies cited in Table II included both the double layer force and the London force [11,19,25]. These results indicate that with the presence of a repulsive force barrier near the collector surface particle deposition diminishes to nil. This conclusion holds true for a number of collector models (single-sphere model, Happel's model and the constricted tube model). As an example, the results based on the constricted tube model are shown in Figure 4. In Figure 4, curves of η_o vs N_x are

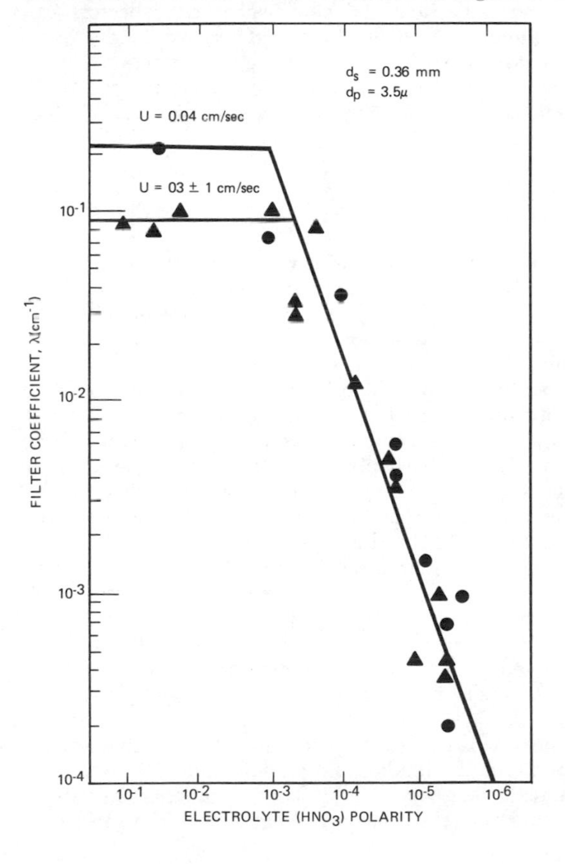

Figure 5. Effect of surface interaction due to pH change of filter coefficient [23].

shown where N_x denotes various surface interaction parameters. The results indicate that the value of η_0 undergoes a step function change to virtually zero at certain threshold values of N_x corresponding to the presence of a net repulsion interaction force between the particle and the collector. The limited experimental data of Spielman and Fitzpatrick [10] seem to offer some qualitative endorsement of this conclusion (Figure 5).

On the other hand, attempts to validate the conclusion based on trajectory calculations with single collector experiments have not been successful. A number of investigators [33,34] measured the rate of particle deposition from suspensions flowing past spherical collectors. With the adjustment of the electrolyte concentration in the suspension and by the use of surface coating techniques for the collector, the nature of the surface interaction between particles in the suspension and the collector can be made to alter from favorable to unfavorable. For the case of favorable interaction, the experimental results were found to agree with trajectory calculations within the experimental consistency. However, the catastrophic decline predicted from trajectory calculation with the presence of repulsive force barriers was not observed. While some decrease in deposition was displayed in certain instances, there was hardly any difference between the favorable and unfavorable surface interaction cases.

A different approach for considering the effect of surface interaction on particle deposition was advanced by Spielman and Cukor [35]. It was argued that deposition does not necessarily require direct contact between particle and collector. Particles become collected if, through a balance of forces, they are held stationary in the proximity of the collector. A delineation scheme that describes the nature of deposition in terms of the surface interactions between particle and single-spherical collectors can be established. This approach subsequently was generalized and applied to the interpretation of single-collector experimental data [36]. By applying this criterion, agreement between theory and experiment was improved, but only in a marginal sense. An understanding of the surface interactions remains elusive. This will be discussed further in later sections.

EFFECT OF SURFACE ROUGHNESS

In the discussion on trajectory calculations given above, it was assumed that particles and collectors have smooth surfaces. It is on this basis that the surface interaction force potentials were formulated (i.e., Equations 40 and 41). Although it is generally recognized that filter grains and particles present in suspensions do not normally have smooth surfaces, the effect of surface roughness has hardly been studied.

The importance of surface roughness on particle deposition can be viewed from two directions. First, it is obvious that expressions such as Equations 40 and 41 are not valid if the magnitude of surface roughness is comparable to, or even greater than, the ranges of the surface interaction forces. Secondly, particles escape collection if (1) they are repelled from the collector, or (2) they fail to become attached to the collector after they have made contact. Surface roughness, in conjunction with surface interactions, is expected to play an important role in the second situation.

In an interesting study, Gimbel [37,38] considered the incorporation of the effect of surface roughness in particle deposition. Gimbel observed that his data on filter coefficients differed systematically from those based on trajectory calculation. Specifically, for small particles the predicted filter coefficient is less than the experimental value. For the large particles, the reverse is observed. This is shown in Figure 6.

The main arguments advanced by Gimbel can be summarized as follows:

1. Because of the surface roughness, it is necessary to consider the effective size of a spherical collector to be the sum of its radius and a small value, R_K, where R_K is related to the characteristic height of the surface roughness. This effective size will be used in considering the transport of particles to the surface of the collector.

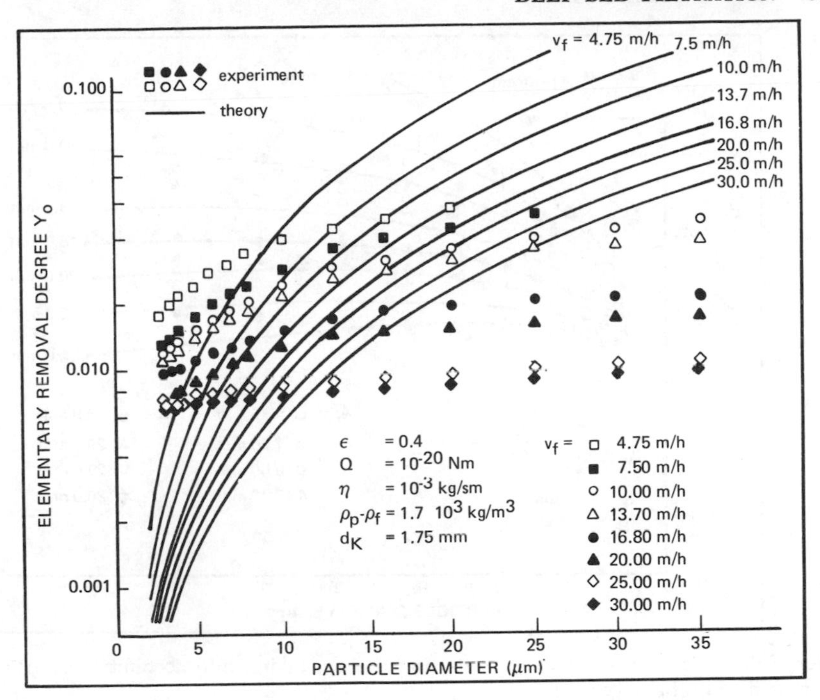

Figure 6. Comparison between Gimbel's data [37] and the prediction of Spielman and Fitzpatrick [10].

2. To account for the situation once a particle makes contact with the collector, the particle can be considered to be placed in a shear flow field and the forces and torque acting on the particle can be viewed as shown in Figure 7. Depending on the magnitude of the hydrodynamic drag force, torque and adhesion force, as well as the characteristic height of the surface roughness, R_K, a particle may become attached or roll away. Since the adhesion forces are determined on a statistical basis (i.e., percent of particles removed for a given applied force), the probability that a requisite force will be manifested can be taken as the probability of particle attachment. The actual filter coefficient is taken to be the product of the filter coefficient determined from trajectory calculation multiplied by the attachment probability.

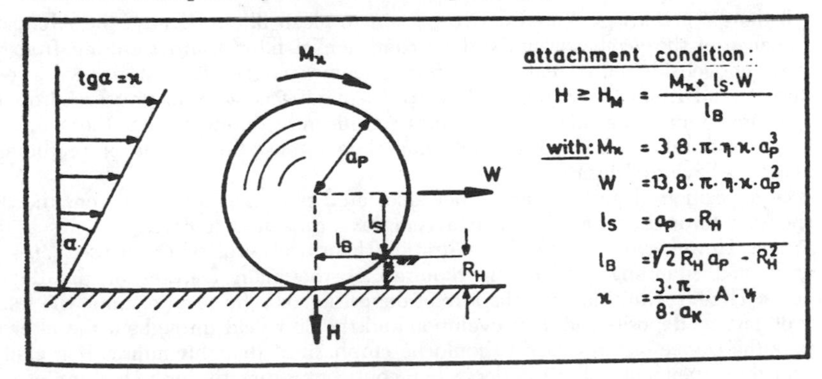

Figure 7. Attachment model proposed by Gimbel [38].

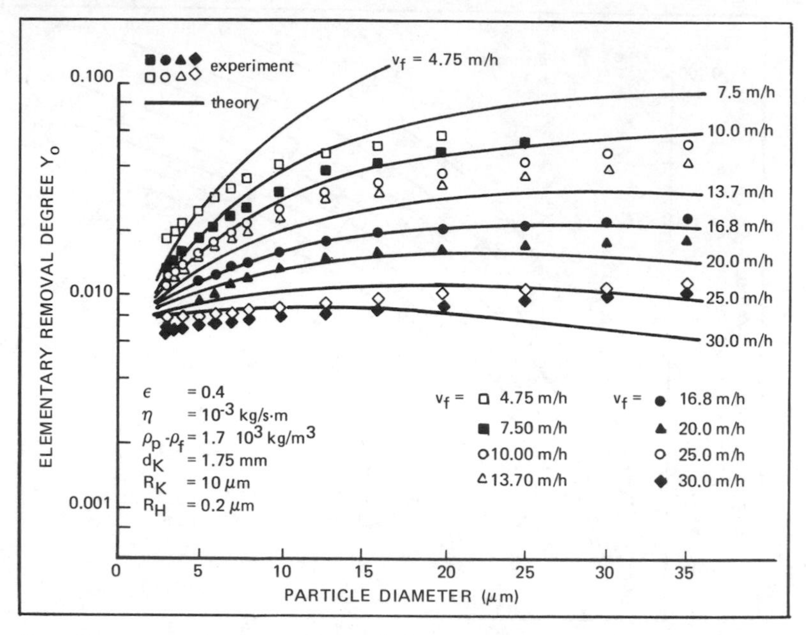

Figure 8. Comparison between data and theory, taking into account the particle attachment probability [37].

It can be seen that the effect of (1) is to increase the magnitude of the predicted filter coefficient because of the use of a greater collector radius. The effect of (2) is to reduce the magnitude of the filter coefficient. The former is more pronounced for smaller particles, while the latter is more important for large particles. With these corrections, the prediction and experiment become much closer, as shown in Figure 8.

The analysis given by Gimbel was based on an extremely simplified model. Nevertheless, it provides the first indication of how to incorporate surface roughness in the consideration of particle deposition. It also shows, in a correct way, the necessity of combining surface geometry and surface forces in the consideration of particle attachment.

SIMULATION MODEL OF PARTICLE DEPOSITION

All the discussions presented above pertain to clean filters because the effect due to the change of the geometry and surface condition of filter grains resulting from deposition is not considered. In reality, as a filter becomes clogged its performance in particle collection and its hydrodynamic behavior (as manifested by the pressure drop necessary to maintain given flowrate) becomes significantly different from those of a clean filter. A comprehensive deep bed filtration theory must be capable of predicting the transient behavior of filtration processes.

The deposition of particles and other associated phenomena arising from the flow of suspension through porous media is a complex problem that perhaps defies exact description. However, for a deep bed filtration theory to be of practical use, some quantitative understanding certainly is required of the main features of the deposition process. These features include the rate of deposition as a function of time, the morphology of particle deposits and their evolution and the flow field throughout the filter media during the course of filtration. It should be emphasized that this information cannot be obtained independently. Particle deposition contributes directly to the buildup of particle deposits and their morphology. The flow field is influenced directly by the presence of particle deposits and the rate of deposition over time depends strongly on the flow field around the collector.

A direct method of obtaining all information is the creation of a deposition process through computer experiments. If properly constructed, such computer experiments or simulation models would provide detailed information on the dynamics of the deposition process. This approach has been pursued by a number of investigators [39–42], principally for the study of deposition on fibers. The same approach can be readily adapted to liquid filtration in deep beds. A version of this simulation model based on the use of the constricted tube model is given below to illustrate the main feature of this approach.

The basic set of hypotheses used in describing particle deposition was first postulated by Tien et al. [39] and subsequently expanded by Wang et al. [40]. Two basic concepts were introduced:

1. **Shadow Effect.** When a particle becomes deposited on a collector, the part of the collector surface immediately adjacent to the deposited particle no longer will be accessible to particles for deposition. The shadow effect is a direct result of the finiteness of particles. A diagrammatic representation is shown in Figure 9.

   There are two important consequences associated with the creation of shadow areas by deposited particles. First, it implies that particle deposition occurs in a discrete manner. Secondly, particles that ordinarily would have been deposited within the shadow area will now be deposited on the deposited particle, leading to the formation of particle aggregates of various sizes and geometries.
2. **Singular and Random Behavior.** Although the particle concentration of a suspension is uniform macroscopically, the locations of the individual particles in the fluid at any instant are stochastic. For example, consider that particles approach the collector at a long distance from the collector. In general, the particles will be positioned at irregular intervals. Consequently, if one considers an imaginary control surface on a plane normal to the direction of the main flow upstream from the collector, the passage of the particle across the control surface takes place one particle at a time (namely singularly), and their posi-

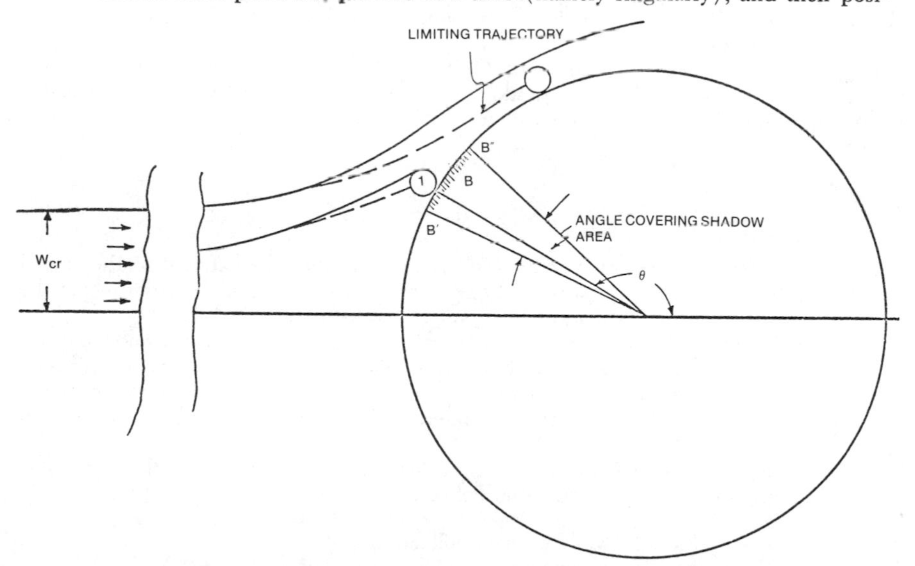

Figure 9. Illustration of the shadow effect.

tions are distributed randomly throughout the control surface (i.e., random behavior).

With the use of the constricted tube model, filtration in granular media can be considered as the deposition of particles on the surface of the constricted tube from suspension flowing through the tube. The possibility of collection vs escape of each and every particle entering the tube can be determined by following the particles' trajectories. In the event that a particle is collected, its position of deposition is recorded and becomes part of the inventory. The procedure for the simulation can be divided into three steps, described briefly as follows:

1. **Assignment of Initial Positions of Particles at Tube Inlet.** The basic assumption used in assigning initial positions for entering particles is that particles enter the tube one at a time and are distributed randomly at the inlet. For convenience, the initial position can be given in terms of the cylindrical polar coordinates (r_{in}, ϕ_{in}). Let r_o be the radial distance, beyond which no particle can be placed.

$$2\, r_o = d_m - 2a_p \tag{42}$$

where a_p is the particle radius.

For any position (r, ϕ), $o < r < r_o$ and $o < \phi < 2\pi$, the probability of a particle (in terms of its center) being within the area element (r, d, ϕ, dr) at the inlet is proportional to the suspension volumetric flowrate through the area element. Thus, one has

$$\text{Probability } (o < r_{in} < r_o) = 1 \tag{43}$$

$$\text{Probability } (o < \phi < 2\pi) = 1 \tag{44}$$

$$G_1(r) = \text{Probability } (r_{in} < r) = \frac{\psi(o, r) - \psi(o, o)}{\psi(o, r_o) - \psi(o, o)} \tag{45}$$

$$G_2(\phi) = \text{Probability } (\phi_{in} < \phi) = \frac{\phi}{2\pi} \tag{46}$$

where $\psi(z, r)$ is the stream function corresponding to the flow within the constricted tube of specified wall geometry.

From the cumulative density functions, $G_1(f)$ and $G_2(\phi)$, of random variables, r_{in} and ϕ_{in}, respectively, one can find the random variables in terms of the two uncorrelated random numbers, R_1 and R_2, distributing uniformly over $(o, 1)$ as follows:

$$r_{in} = G_1^{-1}(R_1) \tag{47}$$

$$\phi_{in} = G_2^{-1}(R_2) = 2\pi\, R_2 \tag{48}$$

A standard random number generator (GGUBF) from the International Mathematical and Statistical Library (IMSL) was used to generate sets of R_1 and R_2 from which the initial positions of the succeeding entering particles were determined.

2. **Determination of Particle Trajectory.** Once the initial position of any entry particle is given, its trajectory is known deterministically, assuming that the Brownian diffusion effect is negligible. The particle trajectory can be obtained from the integration of the equations of particle motion with the knowledge of the flow field within the constricted tube.

3. **Determination of the Outcome of Entering Particles.** There are three possible outcomes for any particle entering the unit collector, namely:

(a) deposition on the tube wall (primary collection),
(b) deposition on an already deposited particle (secondary collection), and
(c) escape out of the unit collector.

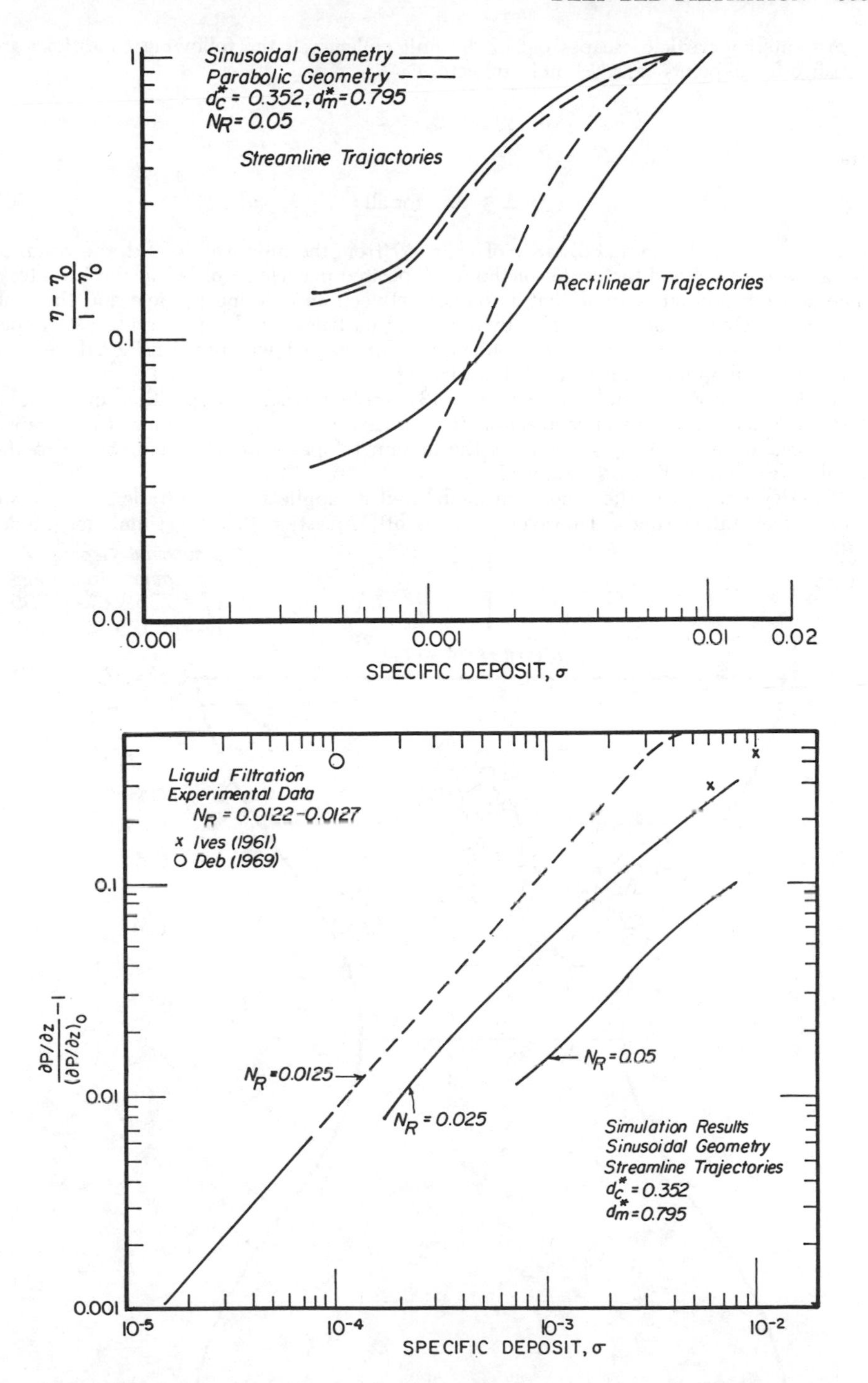

Figure 10. Increase in collection efficiency and pressure drop as predicted from a simulation model [43].

An entering particle escapes out of the unit collector if the following conditions are satisfied for all points (z, r, ϕ) on its trajectory:

$$y(z, r, \phi) > 2a_p \tag{49}$$

and

$$d_i > d_p \quad \text{for all i} \tag{50}$$

where y denotes the normal distance of (z, r, ϕ) from the tube wall, and d_i the distance between (z, r, ϕ) and the center of the ith deposited particle recorded in the inventory. The above conditions assume that a contact between an entering particle and the wall of unit collector (or one of the already deposited particles) leads automatically to deposition. The possibility of particles bouncing off on impact was not considered because it is not an important factor in liquid filtration.

With the approach outlined, Pendse [43] studied particle deposition in granular filter beds under a variety of conditions. His results were expressed in terms of the change in collection efficiency, $\eta/\eta_o - 1$, vs the amount of particles deposited. Some of the results are shown in Figures 10 and 11.

The development of the simulation model and its application to filtration studies are still in their infant stages. However, there is little question that the simulation model

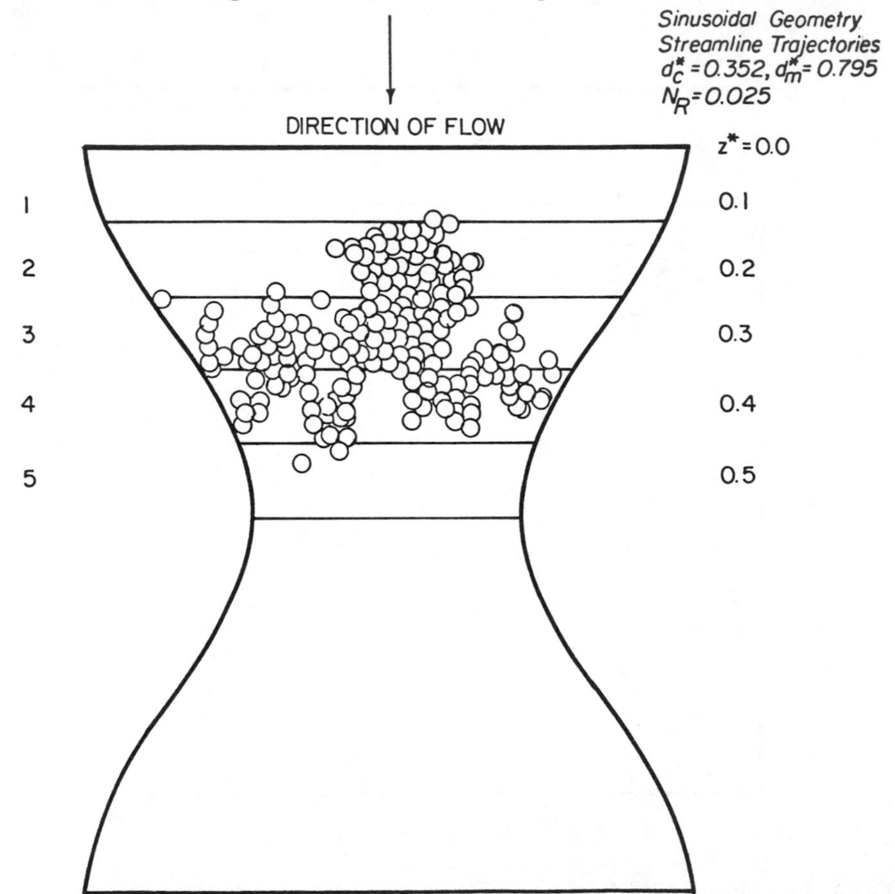

Figure 11. Morphology of particle deposits predicted from a simulation model [43].

offers significant possibilities in obtaining detailed information on particle deposition. This approach, together with certain experimental observations obtained with the use of two-dimensional filters, is potentially capable of revealing many of the salient features of particle deposition that are indispensable if a comprehensive filtration theory is to be developed.

PREDICTION OF DYNAMIC BEHAVIOR OF DEEP BED FILTERS

The phenomenological equations describing the dynamic behavior of deep bed filtration were discussed in the section entitled Phenomenological Descriptions. They are:

$$u \frac{\partial c}{\partial z} + \frac{\partial \sigma}{\partial \theta} = 0 \tag{4}$$

$$\frac{\partial \sigma}{\partial \theta} = u\lambda \cdot c \tag{6}$$

$$\lambda = \lambda_o \, F_1(\underline{\alpha}, \sigma) \tag{7}$$

$$p = p - p_o = \left(\frac{\partial p}{\partial z}\right)_o \int_o^z F_2[\underline{\beta}, \sigma(z' \, \theta)] \, dz' \tag{10}$$

$$c = c_o(\theta) \quad \text{at} \quad z = 0 \tag{51}$$

$$\left.\begin{array}{l} \sigma = \sigma_i(z) \\ c = c_i(z) \end{array}\right\} \; z > 0, \theta \leq 0 \tag{52}$$

In most cases of practical interest, $\sigma_i = c_i = 0$.

The prediction of the dynamic behavior of deep bed filtration therefore involves two problems:

1. development of efficient logarithms for the integration of the above system of equations, and
2. prediction of the values of λ_o and $(\partial p/\partial z)_o$ and the functional forms of F_1 and F_2.

These will be discussed separately.

Integration of the Filter Equation

The major problem involved here is the integration of Equations 4, 6 and 7 with appropriate initial and boundary conditions. The integration of Equation 10 is straightforward once the functional form of F_2 is known.

Equations 4, 6 and 7 comprise the form characteristic of fixed bed processes, such as adsorption, cross-flow heat exchanger, etc. Thus, a large number of algorithms for the solution of the semilinear hyperbolic equations can be applied readily. However, because of the form of the rate expression (Equation 6), this system of equations can be reduced to a pair of ordinary differential equations, as was done by Herzig et al. [44]. The results are

$$\frac{\partial \sigma_o}{d\theta} = c_o \, u \, \lambda_o \, F_1(\underline{\alpha}, \sigma_i) \tag{53}$$

with

$$\sigma_o = 0 \quad \text{at} \quad \theta = 0 \tag{54}$$

and

$$\frac{d\sigma}{dz} = -\lambda_o F_1(\underline{\alpha}, \sigma)\,\sigma \tag{55}$$

with

$$\sigma = \sigma_o \quad \text{at} \quad z = 0 \tag{56}$$

Also,

$$\frac{c}{c_o} = \frac{\sigma}{\sigma_o} \tag{57}$$

where subscript o denotes the inlet conditions. The integration of Equation 53 with the initial condition of Equation 54 can be made first, which gives the expression of the history of the specific deposit at the inlet. Once σ_o is known, it can be used as the initial condition for the integration of Equation 55. The effluent quality (c) can be found readily from Equation 56 once c_o, σ and σ_o are known.

Values of λ_o and $(\partial p/\partial z)_o$

As stated before, $(\partial p/\partial z)_o$, the initial (or clean filter) pressure gradient, can be estimated using the Carman-Kozeny equation (Equation 8). The clean filter coefficient, λ_o, can be obtained either experimentally or from trajectory calculation (Equation 35). In the case of Brownian particles, Equations 36 or 37 should be used, depending on the nature of the surface interaction.

Functional Forms of F_1 and F_2

In essence, the functions F_1 and F_2 account for the effects of particle deposition on filter performance. The simulation model discussed in the section entitled Effect of Surface Roughness, in principle, is capable of obtaining the values of F_1 and F_2. For example, the data shown in Figure 10 provide the values of F_1 for a given set of conditions; however, because of the many simplifying assumptions used in the simulation model, F_1 (or F_2) predicted from the model is not of sufficient accuracy to be of practical use [45].

At present, there are two ways of obtaining information concerning F_1 and F_2. One may obtain the necessary information directly from experiments. This is the approach used by most investigators. The filter coefficient and various degrees of deposition can be determined from experimental data, and the ratio λ/λ_o yields values of F_1. Similarly, one can obtain values of F_2 by comparing the pressure profiles of filter beds at different stages of deposition. A summary of the various functional forms established by this model is shown in Table V.

The difficulty with this approach is the lack of generality of the functions derived. Equally important, the accuracy of the data on which F_1 (or F_2) was established, was limited. In addition, most of the functional expressions listed in Table V were evaluated from data in an ad hoc manner. An optimization procedure was proposed [58] for data treatment. However, because of the relatively poor quality of this type of data in general, the improvement was found to be only marginal.

Alternatively, one may establish F_1 (or F_2) by assuming the morphology of particle deposits. Tien et al. [58] have shown that for a given deposit morphology relationships can be established between the extent of particle deposition and variations of filter

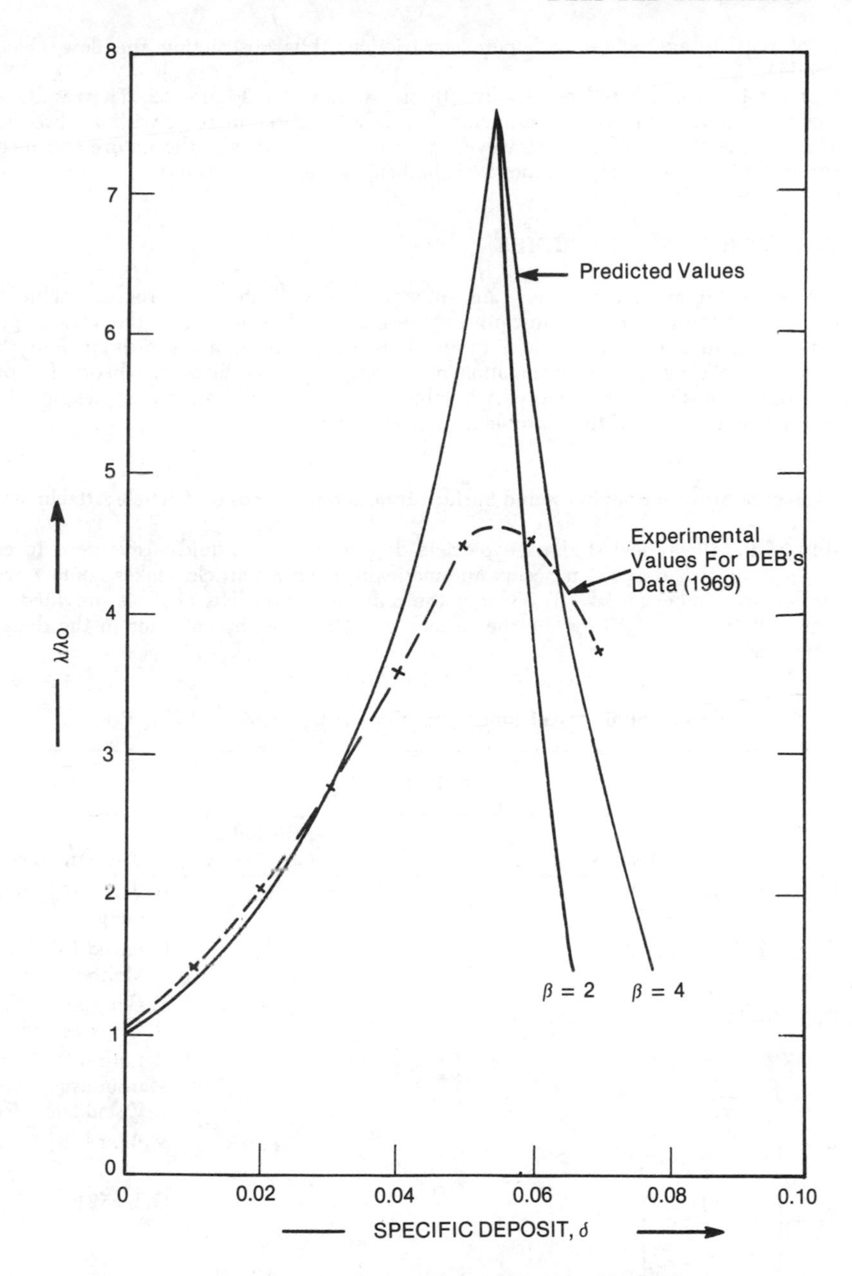

Figure 12. Prediction of the variation of λ_0 [58].

media structure. This, in turn, can be used to obtain expressions of the filter coefficient and the pressure gradient as functions of the extent of deposition (i.e., σ). Specifically, these authors assumed that during the initial stage of filtration particle deposition is in the form of smooth coatings outside filter grains (smooth coating model). As the extent of particle deposition increases, particle deposition is manifested mainly in the lodgement of particle aggregates with pore constriction, thus obstructing the flow (blocking mode).

A prediction of F_1, based on this hypothesis, is shown in Figure 12. The results are in qualitative agreement with experiments that show an initial increase with a subsequent decrease in the filter coefficient. However, the basic question as to the nature and mechanism of the transition from one mode to another was left unanswered.

TOPICS FOR FUTURE STUDIES

The presentations given above are intended to indicate the progress achieved, mainly during the last decade, toward a more basic understanding of the various phenomena associated with deep bed filtration. It is also clear from these discussions that in spite of the progress the formulation of a comprehensive filtration theory for predictive purposes is not yet a reality. A number of major problems remain unexplored. A brief discussion of some of these problems is given below.

The Effect of Surface Roughness and Surface Interaction Forces on Particle Attachment

Most of the theoretical studies on particle deposition from liquid suspensions to collectors assume that deposition occurs automatically once a particle makes contact with the collector. Gimbel's work [37,38] was the only exception. His analysis provided the first quantitative attempt to relate the surface conditions of the collector in the deposition process.

Table V. Summary of Functional Forms: $F_1(\underline{\alpha},\sigma)$ and $F_2(\underline{\beta},\sigma)$

$$\frac{\lambda}{\lambda_o} = F_1(\underline{\alpha},\sigma)$$

Expression	Adjustable Parameters	Investigators
1. $F_1 = 1 + k_1\sigma,\ k_1 > 0$	k_1	Iwasaki [5], Stein [46]
2. $F_1 = 1 - k_2\sigma,\ k_2 > 0$	k_2	Ornatski [47], Mehter [48]
3. $F_1 = 1 - \dfrac{\sigma}{\epsilon_o}$	–	Shekhtman [49], Heertjes and Lerk [50]
4. $F_1 = 1 - \dfrac{\sigma}{\sigma_{ult}}$	–	Maroudas and Eisenklam [51][a]
5. $F_1 = \left(\dfrac{1}{1+\nu\sigma}\right)^n \quad \nu > 0,\ n > 0$	ν, n	Mehter [48]
6. $F_1 = \left[\dfrac{\phi(\sigma)/\phi_o}{\epsilon_o - \dfrac{\sigma}{1-\epsilon_d}}\right]^A$	A	Deb [52][b]
7. $F_1 = \left(1 + \dfrac{\beta'\sigma}{\epsilon_o}\right)^a \left(1 - \dfrac{\sigma}{\epsilon_o}\right)^b;\ \beta' > 0$	β, a, b	Mackrle [53]
8. $F_1 = 1 + c'\sigma - \dfrac{\phi'\sigma^2}{\epsilon_o - \sigma};\ c' > 0,\ \phi' > 0$	c', ϕ'	Ives [6]
9. $F_1 = \left(1 + \dfrac{\beta'\sigma}{\epsilon_o}\right)^a \left(1 - \dfrac{\sigma}{\epsilon_o}\right)^b \left(1 - \dfrac{\sigma}{\sigma_{ult}}\right)^c;\ \beta' > 0$	β, a, b, c	Ives [54]

$$\frac{(\partial p/\partial z)}{(\partial p/\partial z)_o} = F_2\,(\underline{\beta},\sigma)$$

Expression	Adjustable Parameters	Investigators
1. $F_2 = 1 + k_3\sigma;\ k_3 > 0$	k_3	Mehter [48]
2. $F_2 = 1 + k_4\dfrac{\sigma}{t_o};\ k_4 > 0$	k_4	Mints [55]
3. $F_2 = \left(\dfrac{1}{1-\delta\sigma}\right)^m,\ \delta > 0, m > 0$	δ, m	Mehter [48]
4. $F_2 = \left(1 - \dfrac{2\sigma}{\tilde{\beta}}\right)^{-2};\ \tilde{\beta} > 0$	–	Maroudas and Eisenklam [51][a]
5. $F_2 = \{1 + G\,(1 - 10^{-k\sigma/(1-\epsilon_d)})\}\cdot\left\{\dfrac{\epsilon_o}{\epsilon_o - \dfrac{\sigma}{1-\epsilon d}}\right\}^3$	G, k	Deb [52][b]
6. $F_2 = \left(1 + \dfrac{\beta'\sigma}{\epsilon_o}\right)^{y'}\left(1 - \dfrac{\sigma}{\epsilon_o}\right)^{z'};\ B > 0, y' > 0,\ z' >> 0, c',\phi',p' > 0$	y', z', β'	Ives [54]
7. $F_2 = 1 + p'\left\{(\lambda_o + \phi'\epsilon_o)\sigma + \left(\dfrac{c'+\phi'}{2}\right)\sigma^2 + \phi'\,\epsilon_o^2\, ln\left(\dfrac{\epsilon_o - \sigma}{\epsilon_o}\right)\right\}$	c', ϕ', p'	Ives [56][c]
8. $F_2 = \left(\dfrac{\epsilon_o}{\epsilon_o - \sigma}\right)^3\left(\dfrac{1-\epsilon_o+\sigma}{1-\epsilon_o}\right)^2\left\{\sqrt{\left(\dfrac{\sigma}{3(1-\epsilon_o)} + \dfrac{1}{4}\right)} + \dfrac{\sigma}{3(1-\epsilon_o)} + \dfrac{1}{2}\right\}$	–	Camp [57][d]

[a]Expression derived on the basis of blocking mode of particle deposition, where β is the blocking factor.

[b]Expression derived on the basis of smooth coating mode, taking into account the contact with neighboring grains, where $\phi\ (\sigma)$ is the shape factor of filter grains, and ϕ_o is its value at the clean bed stage.

[c]p' is determined on the basis of the following assumption: $\Delta(\partial p/\partial z)/\Delta\sigma = p'\ (\partial p/\partial z)_o\,\lambda_o\,F_1\,(\sigma;c',\phi',\beta')$

[d]Expression derived on the basis of smooth coating mode of particle deposition, assuming no contact with neighboring grains.

Gimbel's work admittedly represents a most simplified version of the deposition process. However, it is not difficult to generalize his concept. The physical model shown in Figure 7 considers a single protrusion from the collector surface. In reality, surface roughness is manifested both by protrusions and depressions with various depths. If one assumes that these geometric obstacles follow a random distribution, one can easily simulate the particle movement in the following manner. Consider a particle being transported to the surface of the collector. The combined effect of the adhesion force between the particle and the collector, the local surface condition of the collector at the contact point and the hydrodynamic force and torque acting on the particle determine whether the particle will attach to the collector at the point of contact. This process continues as the particle rolls along the surface and encounters a number of protrusions and depressions (Figure 13). An overall attachment probability can be calculated readily. Even if the quantitative results may not be directly applicable to the calculation of deep bed filters, the analysis is expected to yield qualitative information identifying

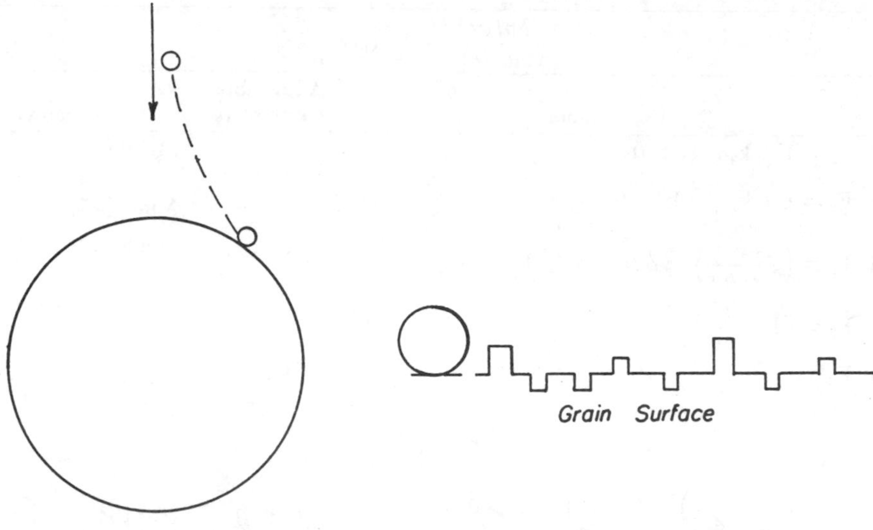

Figure 13. Proposed model of particle attachment.

those relevant parameters of particle attachment that will be useful in data correlation. The validation of the analysis can be made by comparing the results of analysis with the type of adhesion measurements reported by Kneilmann [59] and Johannisson [60].

Particle Reentrainment During Filtration

The importance of particle reentrainment and its effect on effluent quality has long been recognized. Mints' expression [55] for the rate of filtration,

$$\frac{\partial \sigma}{\partial \theta} = \lambda uc - \alpha\sigma \tag{58}$$

was formulated to include the reentrainment effect by introducing the second term on the right side of Equation 58. It should be noted that reentrainment and particle adhesion refer to similar, but different, phenomena. With particle adhesion, one is concerned with whether a particle in contact with a collector would be collected. Particle reentrainment, on the other hand, pertains to the breakaway of once-deposited particles from the collector to the suspension.

A basic approach that can be applied to the study of particle reentrainment is to examine the various forces and torques acting on deposited particles. In principle, this can be done if the geometry of particle deposits is known and one has the means of estimating the various forces acting on particle deposits. However the problem is perhaps too complex to be treated analytically, at least for now.

A more direct method of observing the occurrence of particle reentrainment (as well as other aspects of the deposition process) was devised by Payatakes et al. [61]. These investigators employed the use of a two-dimensional filter composed of an array of constricted tubes. The occurrence was observed of the progressive blockade of constriction and subsequent reopening of some of these under various conditions. The problem remaining to be resolved is to draw an inference from this large amount of seemingly disparate data so that the data can be incorporated into the description of the filtration process.

Experimental Data on the Transient Behavior of Deep Bed Filters

As a fixed bed process, deep bed filtration is necessarily unsteady in nature. During the course of filtration, both the effluent quality and the pressure gradient across the bed change with time. As stated previously, this time-dependent behavior is the main feature of deep bed filtration processes.

Paradoxically, in spite of its obvious importance there are scarcely any existing data on the transient behavior of deep bed filters. Much experimental work has been done with deep bed filtration. However, the conditions under which the measurements were made and the systems used usually are not well defined. The data of these studies undoubtedly are useful for a variety of purposes but are neither suitable as a basis of comparison with theories nor can they be used to formulate new theories.

Experimental data needed to advance the basic understanding must be of the following kind:

1. The experimental system must be simple and well defined. As a starting point, the filter grain should be uniform in size with defined surface conditions. The suspension used should be monodispersed.
2. The experimental conditions should be well controlled. For example, the effect of flocculation should be reduced to a minimum, if not completely eliminated.
3. The complete dynamic behavior should be recorded, namely, both the effluent quality and the pressure profile should be recorded as functions of time.

The accumulation of this type of data is necessarily time consuming and expensive because of the high cost of monodispersed particles. On the other hand, the potential value of this type of data cannot be overemphasized. They provide the very basis for the validation and improvement of existing theories, as well as for the formulation of new ones.

ACKNOWLEDGMENT

This study was conducted under grant No. CPE 7905962, from the National Science Foundation.

NOMENCLATURE

A_s	defined by Equation 23
a_c	collector radius
a_p	particle diameter
b	radius of Happel's outer cell
c	particle concentration
c_i	particle concentration leaving the ith unit bed element
D_{BM}	Brownian diffusivity
d_c	collector diameter
d_g	grain diameter
F_D	drag force acting on a unit collector
F_{D_o}	drag force acting on a clean unit collector
F_1	defined as λ/λ_o
F_2	defined as $(\partial p/\partial z)/(\partial p/\partial z)_o$
$G[\underline{\gamma}\, c\, \sigma]$	general expression of filtration rate
G_1, G_2	probability functions
H	Hamaker constant
K_1, K_2, K_3, K_4	constants defined by Equation 20
k	permeability
k^*	pseudo-rate constant defined by Equation 39
l	unit bed element thickness
N	number of constricted tubes per unit cross section
$N_{DL}, N_{E1}, N_{E2}, N_G, N_{LO}, N_{pe}, N_R, N_{Rtd}$	dimensionless groups defined in Table III
p	pressure

p_o	inlet pressure
p'	defined by Equation 21
p_1	defined by Equation 18
$R(\delta)$	factor to account for hydrodynamic retardation effect (see Equation 39)
r	radial distance
t	time
u	superficial velocity
V_r, V_o	fluid velocity components around the collector
V	approach velocity
V_i	average velocity in the pore
w	defined by Equation 20
y	distance from collector surface
z	axial distance

Greek Symbols

$\underline{\alpha}, \underline{\beta}, \underline{\gamma}$	vectors of relevant variables
δ	separation distance between particle and collector
δ^+	dimensionless values of δ defined as δ / a_p
ϵ	porosity of bed or dielectric constant
ϵ_d	porosity of deposit
ϵ_o	initial bed porosity
ζ_c	zeta potential of collector
ζ_p	zeta potential of particle
η	unit collector efficiency
η_o	clean unit collector efficiency
θ	corrected time defined by Equation 3
κ	Debye-Huckel reciprocal thickness
λ	filter coefficient
λ_o	clean filter coefficient
λ_e	wavelength of electron oscillation, $\lambda_e = 1000$ Å
μ	viscosity
ρ_f	fluid density
ρ_p	particle density
σ	specific deposit
σ_o	specific deposit at $z = 0$
θ	surface interaction potential
ψ	stream function
ψ_∞	defined by Equation 17

REFERENCES

1. Walker, W. H., W. K. Lewis, W. H. McAdams and E. R. Gilliland. *Principles of Chemical Engineering*, 3rd ed. (New York: McGraw-Hill Book Co., 1937), p. 327.
2. Tien, C., and A. C. Payatakes. *AIChE J.* 25:737 (1979).
3. Rajagopalan, R., and C. Tien. "The Theory of Deep Bed Filtration," in *Progress in Filtration and Separation*, Vol. 1 (New York: North-Holland, Inc., 1979), pp. 179–270.
4. Spielman, L. A. "Particle Capture from Low Speed Laminar Flow," *Ann. Rev. Fluid Mech.* 9:297 (1977).
5. Iwasaki, T. *J. Am. Water Works Assoc.* 29:1591 (1937).
6. Ives, K. J. *Proc. Inst. Civil Eng. (London)* 16:189 (1960).
7. Payatakes, A. C., C. Tien and R. M. Turian. *AIChE J.* 19:58 (1973).
8. Brinkman, H. C. *Appl. Sci. Res.* A1:27, 86 (1947).
9. Happel, J. *AIChE J.* 4:197 (1958).
10. Spielman, L. A., and J. A. Fitzpatrick. *J. Colloid Interface Sci.* 42:607 (1973).
11. Payatakes, A. C., C. Tien and R. M. Turian. *AIChE J.* 20:889 (1974).
12. Venkatesan, M., and R. Rajagopalan. *AIChE J.* 26:694 (1980).
13. Fedkiw, P., and J. Newman. *AIChE J.* 23:255 (1977).
14. Payatakes, A. C., C. Tien and R. M. Turian. *AIChE J.* 19:67 (1973).
15. Neira, M. A., and A. C. Payatakes. *AIChE J.* 24:43 (1978).
16. Chow, J. C. F., and K. Soda. *Phys. of Fluids* 15:1700 (1972).
17. Yao, K-M., M. T. Habibian and C. R. O'Melia. *Environ. Sci. Technol.* 5:1105 (1971).
18. Yao, K-M. "Influence of Suspended Particle Size on the Transport Aspect of Water Filtration," PhD Thesis, University of North Carolina, Chapel Hill, NC (1968).
19. Rajagopalan, R., and C. Tien. *Can. J. Chem. Eng.* 55:246 (1977).

20. Rajagopalan, R. "Stochastic Modelling and Experimental Analysis of Particle Transport in Water Filtration," PhD Thesis, Syracuse University, Syracuse, NY (1974).
21. Payatakes, A. C., R. Rajagopalan and C. Tien. *Can. J. Chem. Eng.* 52:727 (1974).
22. Hung, C. C., and C. Tien. *Desalination* 18:173 (1976).
23. Fitzpatrick, J. A. "Mechanisms of Particle Capture in Water Filtration," PhD Thesis, Harvard University, Cambridge, MA (1972).
24. Payatakes, A. C. "A New Model for Granular Porous Media–Application to Filtration through Packed Beds," PhD Thesis, Syracuse University, Syracuse, New York (1973).
25. Rajagopalan, R., and C. Tien. *AIChE J.* 22:523 (1976).
26. Ison, C. R. "Dilute Suspensions in Filtration," PhD Thesis, University of London, London, England (1967).
27. Ison, C. R., and K. J. Ives. *Chem. Eng. Sci.* 24:717 (1969).
28. Ruckenstein, E. *Chem. Eng. Sci.* 19:131 (1964).
29. Pfeffer, R., and J. Happel. *AIChE J.* 10:605 (1974).
30. Spielman, L. A., and S. K. Friedlander. *J. Colloid Interface Sci.* 44:22 (1974).
31. Ruckenstein, E., and D. C. Prieve. *J. Chem. Soc. (London) Farad. Trans.* 69(11): 1522 (1973).
32. Rajagopalan, R., and T. E. Karis. *AIChE Symp. Ser. No. 190* 75:75 (1979).
33. Rajagopalan, R., and C. Tien. *Can. J. Chem. Eng.* 55:256 (1967).
34. Onorato, F. J., and C. Tien. *Chem. Eng. Commun.*
35. Spielman, L. A., and P. M. Cukor. *J. Colloid Interface Sci.* 43:51 (1973).
36. Onorato, F. J. "The Effect of Surface Interactions on Particle Deposition in Aqueous Media–Single Collector Study," PhD Thesis, Syracuse University, Syracuse, NY (1979).
37. Gimbel, R. D. "Untersuchungen zur Partikeltabscheidung in Schnellfiltern," Dr. -Ing. Dissertation, Karlsruhe, West Germany (1978).
38. Gimbel, R. D., and H. Southeimer. "Recent Results on Particle Deposition in Sand Filters," in *Deposition and Filtration of Particles from Gases and Liquids, Soc. Chem. Ind. (London)* (1978).
39. Tien, C., C. S. Wang and D. T. Barot. *Science* 196:985 (1977).
40. Wang, C. S., M. Beizaie and C. Tien. *AIChE J.* 23:879 (1977).
41. Kanaoka, C., H. Emi and T. Myajyo. *Kaguku Koguku* 4:538 (1978).
42. Nielsen, K. A., and J. A. Hill. *AIChE J.* 26:678 (1980).
43. Pendse, H. "A Study of Certain Problems Concerning Deep Bed Filtration," PhD Thesis, Syracuse University, Syracuse, NY (1979).
44. Herzig, J. P., D. N. Leclerc and P. LeGoff. *Ind. Eng. Chem.* 62:8 (1970).
45. Payatakes, A. C., R. M. Turian and C. Tien. *Proc. 2nd World Cong. on Water Resources* 241 (1975).
46. Stein, P. C. "A Study of the Theory of Rapid Filtration of Water through Sand," DSc Thesis, Massachusetts Institute of Technology, Cambridge, MA (1940).
47. Ornatski, N. V., E. M. Sergeev and Y. M. Shekhtman. "Investigations of the Process of Clogging of Sands," University of Moscow (1975).
48. Mehter, A. A. "Filtration in Deep Beds of Granular Activated Carbon," MSc Thesis, Syracuse University, Syracuse, NY (1970).
49. Shekhtman, Y. M. "Filtration of Suspensions of Low Concentration," Institute of Mechanics of USSR Academy of Science (1961).
50. Heertjes, P. N., and C. F. Lerk. *Trans. Inst. Chem. Eng.* 45:T138–T145 (1967).
51. Maroudas, E., and P. Eisenklam. *Chem. Eng. Sci.* 20:867 (1965).
52. Deb, A. K. *J. San. Eng. Div., ASCE* 95:399 (1969).
53. Mackrle, V., Q. Dracka and J. Svec. "Hydrodynamics of Disposal of Low Level Radioactive Waste in Soil," International Atomic Energy Agency, Contract Report no. 98, Czechoslovakia Academy of Science, Institute of Hydrodynamics, Prague (1965)
54. Ives, K. J. "Theory of Filtration," Special Subject No. 7, International Water Supply Congress, Vienna, Austria, 1969.
55. Mints, D. M. *Dokl, Akad Naak, S.S.S.R.* 78:315 (1951).

56. Ives, K. J. *J. San. Eng. Div., ASCE* 87:23 (1961).
57. Camp, T. R. *J. San. Eng. Div., ASCE* 90:3 (1964).
58. Tien, C., R. M. Turian and H. Pendse. *AIChE J.* 25:385 (1979).
59. Kneilmann, R. Dipl. -Ing. Thesis, Karlsruhe, West Germany (1977).
60. Johannisson, R. Dipl. -Ing. Thesis, Karlsruhe, West Germany (1979).
61. Payatakes, A. C., H. Y. Park and J. Petrie. *Chem. Eng. Sci.* 36:1319 (1981).

SECTION V

SPECIAL TOPICS: FLOW PHENOMENA

CONTENTS

CHAPTER 38
HYDRODYNAMICS OF LIQUID DROPS IN IMMISCIBLE LIQUIDS

J. R. Grace
Department of Chemical Engineering
University of British Columbia
Vancouver, British Columbia V6T 1W5
Canada

CONTENTS

DROP FORMATION

Formation Due to Flow Through an Orifice

Drops usually are formed in liquid-liquid contacting equipment by dispersing one liquid into the other through a series of orifices or nozzles [1]. To maximize the interfacial area, it is desirable to form small droplets and to keep adjacent orifices from feeding liquid into the same drop. It is therefore of interest to consider drop formation at a single orifice. Formation in mechanically agitated systems will not be considered in this section because drop dispersal occurs due to turbulent field breakup mechanisms, which will be treated later in this chapter.

Consider an orifice of an inside diameter, d_{or}, that opens in the direction of drop motion (upward for $\rho_d < \rho_c$ and downward for $\rho_d > \rho_c$). Liquid is being dispersed into a second liquid, which forms a stagnant continuous phase. At very low flowrates, the volume, V_D, of drops forming at the orifice is obtained from a balance of interfacial tension and gravity forces as

$$V_D = \lambda_H \pi d_{or} \sigma / g \Delta \rho \tag{1}$$

where λ_H, called the Harkins correction factor, accounts for the volume of a residual drop that is left clinging to the orifice when the principal drop detaches. This correction factor can be calculated from the following empirical equations:

$$\lambda_H = 1/[0.92878 + 0.87638\, d^* - 0.261\, (d^*)^2] \qquad (0.6 < d^* < 2.4) \tag{2a}$$

or

$$\lambda_H = 1.000 - 0.66023 d^* + 0.33936\, (d^*)^2 \qquad (0 \leq d^* \leq 0.6) \tag{2b}$$

where $d^* = d_{or}/V_D^{1/3}$. Since λ in Equation 1 is a function of V_D, Equations 1 and 2 must be solved iteratively. If formation is occurring at the end of a projecting tube that is preferentially wetted by the dispersed phase, the outer tube diameter should be used in place of d_{or} in Equations 1 and 2. Note that the drop volume is independent of the dispersed-phase flowrate, Q, through the orifice for flowrates that are small enough that Equation 1 is valid.

A number of equations are available for predicting the size of drops formed at somewhat higher flowrates. Some of these equations are entirely empirical and are not recommended outside the range of conditions for which they were obtained. Others are based on mechanistic considerations whereby one balances forces acting on the forming drop. The equation of Scheele and Meister [2] is recommended, giving

$$V_D = \frac{\lambda_H}{g \Delta \rho} \left\{ \pi d_{or} \sigma + \frac{13 \mu d_{or} Q}{V_D^{2/3}} - \frac{16 \rho_d Q^2}{3 \pi d_{or}^2} + \frac{9}{2} [g d_{or}^2 Q^2 \sigma \rho_d \Delta \rho]^{1/3} \right\} \tag{3}$$

For low Q, where interfacial tension is dominant, Equation 3 reduces to Equation 1, which balances gravity and interfacial tension forces. The other terms on the right-hand side arise from drag, momentum and volume added during the process of drop detachment. The 16/3 coefficient in the second to last term assumes a parabolic velocity profile of the liquid reaching the orifice and should be replaced by 4 if the velocity profile there is uniform. Humphrey et al. [3] have used tracers to measure the local velocity distribution and calculate pressure profiles inside drops of carbon tetrachloride, chlorobenzene and tributyrin forming in water.

Equation 3 is only valid when drop formation occurs right at the orifice itself. Above a flowrate given by

$$Q_{jet} = c\, [\sigma d_{or}^3\, (1 - d^*/1.24)/\rho_d]^{1/2} \tag{4}$$

a cylindrical jet forms from the orifice. Here, c = 1.36 or 1.57, depending on whether the velocity profile at the orifice is parabolic or uniform. For $Q > Q_{jet}$, drop formation results from breakup of the jets due to growth of instabilities. Theories that can be used to predict drop sizes and/or jet lengths have been presented by Meister and Scheele [4,5]. The jet first increases in length with increasing Q as axisymmetrical disturbances are observed to grow by the well known mechanism observed by Rayleigh [6]. However, the jet length passes through a maximum at some flowrate, Q_{max}, as asymmetrical sinuous disturbances begin to occur, causing forming drops to be ejected laterally, out of the path of the approaching jet. The flowrate, Q_{max}, has been correlated [7] based on the experimental results of Christiansen and Hixson [8] by

$$Q_{max} = 2.11\, d_j^2 \left\{ \frac{\sigma / d_j}{0.514 \rho_d + 0.472 \rho_c} \right\}^{1/2} \tag{5}$$

where d_j is the jet diameter, which can be calculated from

$$d_j = d_{or}\,[1 + 0.485\,Eo_{or}]^{-1} \qquad (Eo_{or} = g\Delta\rho d_{or}^2/\sigma \leq 0.616) \qquad (6a)$$

or

$$d_j = d_{or}\,[0.12 + 1.51\,Eo_{or}^{0.5}]^{-1} \qquad (Eo_{or} = g\Delta\rho d_{or}^2/\sigma > 0.616) \qquad (6b)$$

For $Q = Q_{max}$, the resulting drop size is quite uniform, and the interfacial area also is maximized approximately.

At some higher flowrate, denoted Q_{at}, atomization of the jet occurs at the orifice itself. In this case, the dispersed liquid forms a host of droplets of nonuniform size. Although this leads to a large interfacial area between the phases, the resultant drops tend to be noncirculating and difficult to separate, so that it is generally undesirable to operate at such high flowrates in liquid-liquid systems. These different flow regimes are shown in Figure 1 for a typical system in which heptane was injected into water through a 1.6-mm-diameter orifice.

Hozawa and Tadaki [9] have presented a method of calculating drop sizes for $Q_{jet} < Q < Q_{at}$ based on experimental results for systems with relatively low viscosities, with water as the continuous phase and five nozzles ranging from 1 to 2.6 mm in diameter. The results are normalized with respect to the conditions corresponding to Q_{max}.

Other factors that influence drop formation at an orifice are the orifice shape and orientation, flow or agitation of the continuous phase, mass transfer and the presence

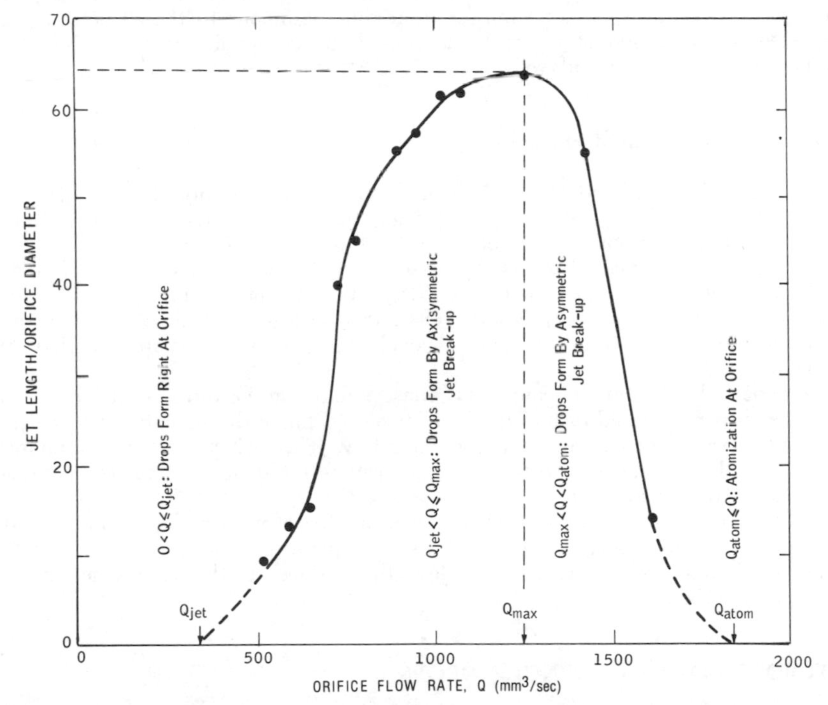

Figure 1. Jet development and regimes of drop formation for injection of heptane into water through an orifice of a diameter of 1.6 mm. The experimental results are from Meister and Scheele [5]. The viscosity and density of the dispersed phase were 3.9×10^{-4} kg/ms and 683 kg/m^3, respectively.

of fixed packings in the column. The influence of orifice geometry on the formation of gas bubbles has been reviewed by Kumar and Kuloor [10], and it is expected that the conclusions should be applicable also to liquid drops. Deviations from the performance of a circular orifice of the same area are most prominent at low Q, where some of the perimeter for noncircular orifices is ineffective. Inclined orifices differ from horizontal ones chiefly in having only a component of the interfacial tension force and the momentum of the incoming liquid acting in the vertical direction. A model for a vertically oriented orifice (horizontal liquid injection) is presented by Kumar and Kuloor [10]. Flow of the continuous phase in the vertical direction affects the drag on the forming drop and, hence, may decrease or increase the size of drop produced, depending on whether the flow is in the same or opposite direction to that of the dispersed liquid. Lateral flows tend to produce somewhat larger drops by lessening the updraught effect from previously formed drops.

Drop formation in packed columns has been studied extensively by Gayler and Pratt [11]. The volume-surface (Sauter) mean diameter of drops leaving the packing was correlated as a function of the physical properties of the liquids, volume fraction of the packing and flowrate of the dispersed phase [11]. The packing can play a considerable role in determining the dispersion characteristics. If the flow of the dispersed liquid exceeds a critical value, flooding of the packing occurs, and further increases in flow simply lead to the formation of a layer of this liquid at the surface of the packing.

Most studies of drop formation have been for mutually saturated systems so that mass transfer was not a factor. However, in practical cases mass transfer is usually important, especially during the formation stage, when the dispersed phase is freshly exposed to the continuous phase. This may lead to significant changes in drop formation [12]. Generally speaking, transfer from the continuous to the dispersed phase is expected to cause premature jet breakup and, hence, smaller drops, while transfer in the opposite direction leads to larger drops [13].

Other Drop Formation Mechanisms

In steel-making and other metal refining operations, two liquid layers (one being a slag layer) are often present. Bubbles of gas traveling across the slag-metal interface tend to carry with them a layer of the heavier liquid, which then peels off, leaving behind a series of small droplets dispersed in the upper lighter liquid. If a gas jet impinges downward on the top liquid with sufficient momentum, a hollow is formed exposing the lower (heavier) liquid. When this occurs, emulsification of the heavier liquid in the lighter one occurs again. Szekely [14] provides a more complete review of these phenomena.

Secondary droplets of the continuous-phase liquid may be introduced into the larger drops of dispersed liquid or into the bulk dispersed phase during interdrop coalescence or coalescence of drops at the phase boundary. This arises from the asymmetrical rupture of the film during coalescence and can result in loss of efficiency in liquid-liquid contacting equipment.

In the experimental study of single drops, a number of workers have employed simple "dumping cup" methods, whereby a beaker or hemispherical cup is filled with the dispersed liquid and inverted just below the surface of the continuous phase for $\rho_d > \rho_c$, or at the bottom for $\rho_d < \rho_c$.

HYDRODYNAMICS OF SINGLE DROPS

Shape Regimes

Consider a liquid drop rising or falling freely under gravity in a second liquid. The two liquids are immiscible or only sparingly soluble in one another. It is then convenient to consider five dimensionless groups that govern the motion of the drop:

1. Reynolds number:

$$Re = \frac{\rho_c d_e U_D}{\mu_c} \tag{7}$$

2. Eötvös number:

$$Eo = \frac{\Delta\rho g d_e^2}{\sigma} \tag{8}$$

3. M-group:

$$M = \frac{g\mu_c^4 \Delta\rho}{\rho_c^2\, \sigma^3} \tag{9}$$

4. Viscosity ratio:

$$\kappa = \mu_d/\mu_c \tag{10}$$

5. Density ratio:

$$\gamma = \rho_d/\rho_c \tag{11}$$

where the subscripts c and d denote the continuous and dispersed phases, respectively, $\Delta\rho = |\rho_c - \rho_d|$, d_e is the volume-equivalent sphere diameter ($=[6V_D/\pi]^{1/3}$) and U_D and V_D are the drop steady rising or falling velocity and volume, respectively. Other dimensionless groups are also in common usage, but these all may be formed from the five groups defined above. The most common of these alternate groups are

6. P-group:

$$P = \frac{\rho_c^2 \sigma^3}{g\mu_c^4 \Delta\rho} = M^{-1} \tag{12}$$

7. Drag coefficient:

$$C_D = \frac{4}{3}\frac{\Delta\rho}{\rho_c}\frac{g d_e}{U_D^2} = \frac{4}{3}\frac{Eo^{1.5}}{Re^2 M^{0.5}} \tag{13}$$

8. Weber number:

$$We = \frac{\rho_c d_e U_D^2}{\sigma} = \frac{Re^2 M^{0.5}}{Eo^{0.5}} \tag{14}$$

For any particular liquid-liquid combination, M, κ and γ are all constant, leaving Re to vary as a function of Eo. By plotting terminal velocity data from the literature as Re vs Eo for constant values of M for a host of liquid-liquid pairs, Grace et al. [15] were able to cross-plot and obtain a "regime plot," as shown in Figure 2. Both κ and γ play minor roles and are not included. Re and, hence, U_D, can be estimated rapidly from this diagram, although more accurate methods of calculating terminal velocities are available and will be presented below. Many of the data on which Figure 2 is based were derived from gas bubbles rising in liquids, so this diagram is also applicable to gas-in-liquid systems.

Drops rising or falling freely in liquids generally are considered to belong to one of three different shape regimes. Boundaries of these regimes appear as heavy lines in Figure 2. Broadly speaking, these regimes may be characterized as follows:

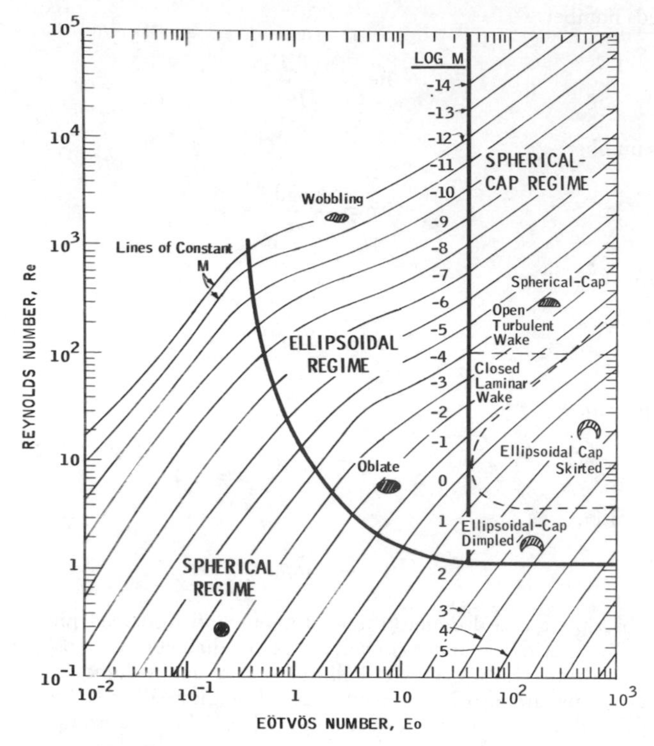

Figure 2. Regime plot for liquid drops and gas bubbles in free motion through an immiscible liquid. Curves are for constant values of M. Boundaries between the major regimes are shown as heavy black lines. Boundaries of subregimes are shown as dashed lines.

1. **Spherical regime.** Drops are spherical, or nearly so, at low values of either Eo or Re, i.e., when either interfacial tension or viscous forces are dominant. This will occur in all systems if the drop is sufficiently small. The maximum value of Eo at which drops still may be considered spherical increases with the viscosity of the continuous phase, i.e., as M increases.
2. **Ellipsoidal regime.** Somewhat larger drops rise or fall with a shape that, on a time mean basis, can be approximated as an oblate ellipsoid of revolution. The instantaneous shape may depart quite radically from that of a spheroid, however, especially for drops that undergo wobbling and shape dilations. Moreover, the eccentricity of the fore and aft sections may not be the same, with either section being more flattened than the other. As shown in Figure 2, drops in low M systems show an extensive ellipsoidal regime. As M increases, the extent of the regime shrinks, and the regime is bypassed altogether for $M > 10^2$.
3. **Spherical-cap regime.** Large drops ($Eo > 40$ for all $M \leq 10^2$ and even larger Eo for $M > 10^2$) assume a shape whose leading edge can be approximated as a segment of a sphere or an oblate spheroid. The aft section may be indented or flat. For some conditions, a thin annular "skirt" of dispersed fluid may trail behind the drop [16]. For simplicity, and to be consistent with common nomenclature, this entire range of shapes will be referred to as spherical cap.

Various subregimes of these three major shape regimes have been mapped by Wairegi [17] and Bhaga and Weber [18]. Some of these subregimes are indicated on Figure 2, with the boundaries plotted as dashed lines. Because Wairegi dealt with both liquid-liquid and gas-liquid systems, whereas Bhaga and Weber studied only air bubbles in liquids, preference has been given to the former study in cases in which the proposed subregimes are at variance. In any case, all the boundaries shown in Figure 2 are somewhat arbitrary because the transitions tend to be gradual, rather than sharp. Moreover, as discussed below, the degree of purity of the liquid-liquid system from surface-active contaminants (not included in the diagram) can play a significant role, causing some variation in the position of the constant-M lines and in the regime and subregime boundaries. The experimental data on which Figure 2 is based are for systems in which exceptional measures have neither been taken to eliminate all contaminants nor have surfactants been purposely added. Therefore, the diagram corresponds to conditions normally encountered in practice.

Spherical Drops at Low Reynolds Numbers

At low Re and Re/κ, creeping flow can be assumed in both the dispersed and continuous fluid phases. The analytical solution was obtained independently by Hadamard [19] and Rybczynski [20] for a spherical drop, free of surface-active contamination and moving steadily at its terminal velocity. The boundary conditions [21] are uniform stream flow in the $-z$ direction as $r \to \infty$, no flow across the drop surface, $r = R$, and continuity of tangential stress and tangential velocity at $r = R$. The resulting Stokes stream functions for the dispersed and continuous phases relative to the drop are then

$$\psi_d = \frac{U_D r^2 \sin^2\theta}{4(\kappa + 1)} \left[1 - \frac{r^2}{R^2}\right] \tag{15}$$

and

$$\psi_c = \frac{-U_D r^2 \sin^2\theta}{2} \left[1 - \frac{R(3\kappa + 2)}{2r(\kappa + 1)} + \frac{R^3 \kappa}{2r^3(\kappa + 1)}\right] \tag{16}$$

The vorticity at $r = R$ is given by

$$\zeta_R = \frac{U_D \sin\theta}{2R} \left[\frac{\kappa + 1}{3\kappa + 2}\right] \tag{17}$$

The terminal (rising or falling) velocity is obtained by balancing the net weight minus buoyancy force against the total drag force. The result is

$$U_D = \frac{2\Delta\rho g R^2}{3\mu_c} \left[\frac{\kappa + 1}{3\kappa + 2}\right] \tag{18}$$

Alternatively, the drag coefficient can be expressed as

$$C_D = \frac{8}{Re} \left[\frac{3\kappa + 2}{\kappa + 1}\right] \tag{19}$$

In the limit of $\kappa \to \infty$, Equations 18 and 19 reduce to the well known Stokes [22] results for creeping flow past a solid sphere:

$$U_{DS} = \frac{2}{9} \frac{\Delta\rho g R^2}{\mu_c} \tag{20}$$

and

$$C_{DS} = 24/Re \tag{21}$$

In the opposite extreme, $\kappa \rightarrow 0$, the analysis gives the results for a freely circulating gas bubble in a liquid or for a liquid drop whose viscosity is much less than that of the continuous phase. The Hadamard-Rybczynski analysis has been generalized by Hetsroni and co-workers [23,24] to cases in which the velocity field at infinity, instead of being uniform, is an arbitrary solution to the creeping flow equations.

In practice, surface-active contaminants very often cause the terminal velocity of spherical drops to follow the Stokes law relationship, Equation 20, rather than the Hadamard-Rybczynski result, Equation 18. This results from damping of the internal motion of the dispersed-phase liquid due to accumulation of surface-active agents at the interface [25]. Whereas Equation 15 signifies an internal Hill's spherical vortex with fore and aft symmetry, observation by Savic [26] of drops subject to some contamination shows that the eye of the vortex motion shifts forward, while a "stagnant cap" forms over the rear surface. This happens because hydrodynamic forces tend to sweep the surfactant preferentially to the rear, hence establishing a tangential gradient of surfactant concentration along the surface. This, in turn, results in an effective surface tangential stress, which retards the surface motion [27]. As the front of the drop still may be swept clear of surfactant, the forward part of the drop may continue to circulate, in which case the value of U_D will lie between those predicted from Equations 18 and 20.

Stream functions and an expression for the terminal velocity for drops subject to surface tensions gradients have been presented in terms of Gegenbauer polynomials by Levan and Newman [28]. The analysis confirms that surfactant tends to accumulate at the rear and that the terminal velocity is reduced toward the Stokes value as surfactant accumulation increases. Various approaches used to predict the effect of cap angle and Eötvös number on the ratio U_D/U_{DS} have been reviewed by Clift et al. [21]. Even small cap angles can cause significant drop retardation. The results confirm that the smaller the drop, the more likely that surface contamination will completely suppress internal circulation and cause U_D to approach U_{DS}. An approximate criterion of Bond and Newton [29] proposes that internal circulation only occurs and, hence, the drop is only unimpeded, for $Eo > 4$. While this criterion is greatly oversimplified, it is a useful first approximation for water and many aqueous solutions in which a certain degree of surface contamination is inevitable. In practice, the transition from fully circulating to fully stagnant conditions occurs over a range of drop sizes. However, the type of contaminants and their concentration are seldom well enough known to allow more accurate estimates to be made.

Although the creeping flow solutions (Equations 15 and 16) were derived without imposition of a boundary condition requiring that normal stresses be balanced at $r = R$, it can be shown that this condition is also satisfied provided that the creeping flow assumptions are valid in both phases and that σ is constant around the entire drop surface. Removing the creeping flow restriction for the inner fluid has relatively little effect on this conclusion [30]. Surface contamination also has been shown to have little influence on drop shape [31]. Hence, a sufficient condition for drops to be essentially spherical in practice is that $Re < 1$, and this is consistent with Figure 2.

Spherical Drops at High Reynolds Numbers

As shown in Figure 2, drops also can be spherical at $Re > 1$, provided that Eo is sufficiently small. For such cases analytical solutions are unavailable, and detailed knowledge of the motion for pure systems is provided by numerical solutions of the equations of motion for both inner and outer fluids, subject to appropriate boundary conditions. While most solutions have been obtained for the special cases of $\kappa \rightarrow 0$, $\gamma \rightarrow 0$ (bubbles in liquids) and $\kappa = 55$, $\gamma = 790$ (water drops in air), some solutions also have been obtained for γ of order unity and $\kappa = 1$ [32]; $\kappa = 0.1, 0.27, 0.3, 0.55,$

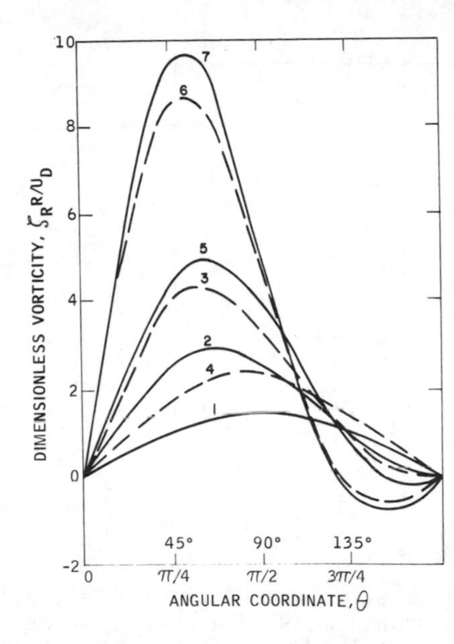

Figure 3. Dimensionless surface vorticities for spherical drops at different values of Re and κ, as derived numerically by Rivkind et al. [35].

Curve	Re	κ
1	0.5	1
2	20	1
3	20	∞
4	100	0
5	100	1
6	100	10
7	100	∞

0.71 and 1.4 [33]; $\kappa = 1, 3, 5, 10, 100$ and 1000 [34]; and $\kappa = 0.33, 1, 3$ and 10 [35]. Dimensionless surface vorticities for a number of these cases are plotted in Figure 3, and these curves illustrate a number of important features of the flow. At low Re ($\mathrm{Re} = 0.5$, $\kappa = 1$), the surface vorticity (curve 1) is nearly symmetrical about $0 = 90°$ and similar to the prediction of the Hadamard-Rybczynski solution, i.e., Equation 17. With increasing Re (compare curves 1, 2 and 5), the eye of the internal vortex moves forward, the maximum vorticity increases and flow separation occurs at the rear of the drop. At constant Re (compare curves 4–7), increasing κ leads to increased vorticity generation, greater asymmetry in the flow about the equator and earlier boundary layer separation.

Rivkind and Ryskin proposed an expression for the drag coefficient:

$$C_D = \frac{1}{\kappa + 1}\left[\kappa\left(\frac{24}{\mathrm{Re}} + \frac{4}{\mathrm{Re}^{1/3}}\right) + \frac{14.9}{\mathrm{Re}^{0.78}}\right] \tag{22}$$

Predictions from this equation appear in Figure 4. In addition to giving a reasonable approximation to the standard drag curve [21] for rigid spheres as $\kappa \to \infty$ and reducing to an equation for gas bubbles as $\kappa \to 0$, Equation 22 provides a good fit to numerical results for freely circulating drops at intermediate κ. As shown in Figure 4, experimental results for pure systems are also reasonably well fitted by the above expression. Abdel-Alim and Hamielec [33] correlated their own numerical data by means of another empirical equation:

$$C_D = \frac{26.5}{\mathrm{Re}^{0.74}}\left[\frac{(1.3 + \kappa)^2 - 0.5}{(1.3 + \kappa)(2 + \kappa)}\right] \tag{23}$$

and showed this expression to be in good agreement with limited experimental data. A third expression for C_D has been obtained from a boundary layer analysis by Harper and Moore [37] to terms of order $\mathrm{Re}^{-3/2}$. The terminal velocity for undeformed drops in pure systems can be estimated from any of these C_D expressions, provided that $\mathrm{Re} > 1$. Some experimental data for relatively pure systems are also shown in Figure 4.

Surface-active contaminants result in damping of internal motion and retardation of a drop's terminal rising or settling velocity outside the creeping flow range, just as for

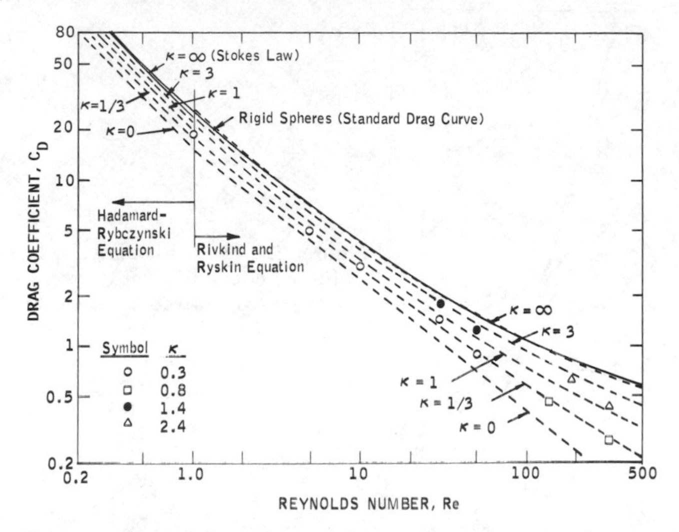

Figure 4. Drag coefficients for spherical drops in pure systems. Curves for Re < 1 are from the Hadamard-Rybczynski equation (Equation 19). For Re > 1, curves are derived from the equation of Rivkind and Ryskin [34] (Equation 22). Also shown is the standard drag curve for rigid spheres as given by Clift et al. [21] and some experimental data from Abdel-Alim and Hamielec [33] and Winnikow and Chao [36].

Re → 0. Accurate experimental data for spherical drops with known concentrations of surfactants are generally unavailable. The results of Edge and Grant [38] and Winnikow and Chao [36] indicate that the effect on U_D can be even larger than at low Re.

Magarvey and Bishop [39] have published an excellent series of photographs showing the complex evolution of wake structure over an extended Re range of spherical drops.

Ellipsoidal Drops

As a first approximation, Figure 2 can be used to estimate drop terminal velocities, as well as shape regimes. For drops in the ellipsoidal regime, more accurate predictions can be obtained from a correlation developed by Grace et al. [15] based on the form of the correlation adopted by Hu and Kintner [40] as extended by Johnson and Braida [41]. The resulting correlation, based on nearly 1500 data points from 20 different experimental studies in which stringent measures to eliminate all surfactants were not in force, is valid for $M < 10^{-3}$, Eo < 40 and Re > 0.1. The terminal velocity is correlated by

$$U_D = \frac{\mu}{\rho d_e} M^{-0.149}(J - 0.857) \tag{24}$$

where

$$J = 0.94H^{0.757} \quad \text{for } 2 < H \leq 59.3 \tag{25a}$$

or

$$J = 3.42H^{0.441} \quad \text{for } H > 59.3 \tag{25b}$$

with

$$H = \frac{4}{3} \mathrm{EoM}^{-0.149} (\mu_c/0.0009)^{-0.14} \tag{26}$$

Here, the continuous-phase viscosity, μ_c, is expressed in SI units (kg/ms). The gradient discontinuity at H = 59.3 corresponds approximately to the transition between non-oscillating and oscillating drops. These equations may be used for gas bubbles as well as for liquid drops in a quiescent liquid continuous phase. Where the continuous phase is in turbulent motion, the terminal velocity can be lowered substantially [42].

For small deformations and high M systems, the deformation of drops can be predicted by a matched asymptotic expansion technique developed by Taylor and Acrivos [43], Pan and Acrivos [30] and Brignell [44]. For small We, the deformation is predicted to result in spheroidal shapes which, for cases of physical significance, are oblate. For low M (higher Re) deformation there are few theoretical results, although Wairegi [17] had some success in predicting shapes based on an approximation whereby both the front and rear portions were assumed to be semispheroids of different eccentricity subject to a balance of hydrostatic, hydrodynamic and interfacial tension forces.

An empirical correlation for the time-mean aspect ratio (E = maximum height/maximum horizontal dimension) of drops and bubbles in normally contaminated systems appears in Figure 5. For $M \leq 10^{-6}$ and Eo < 40, the aspect ratio is given by

$$\overline{E} = 1/[0.163\,\mathrm{Eo}^{0.757} + 1] \tag{27}$$

For higher M systems, there is less deformation at a given value of Eo. Aspect ratios for drops in the spherical and spherical-cap regimes, as well as in the ellipsoidal regime, appear in Figure 5.

For contaminated drops, the onset of an attached wake tends to occur in the Re range 5 to 20, the lower value applying to cases in which significant deformation from the spherical has taken place already and the higher value to systems in which the drop is still nearly spherical at Re = 20 [21]. Measurements of wake dimensions and shedding frequencies for deformed drops have been reported by Hendrix et al. [45] and Yeheskel and Kehat [46].

Shedding of elements from the wake commonly results in secondary motion as drops rise or fall through a second liquid. This secondary motion may involve rocking from side to side, zig-zag motion, spiral trajectories and/or shape dilations (usually referred to as oscillations). For drops subject to contamination, the onset of wake shedding and, therefore, of secondary motion tends to be at Re $\simeq$ 200. Since secondary motion leads to increased drag, the maximum frequency observed in the U_D vs d_e curve also tends to occur at about this Reynolds number. The natural frequency of the fundamental

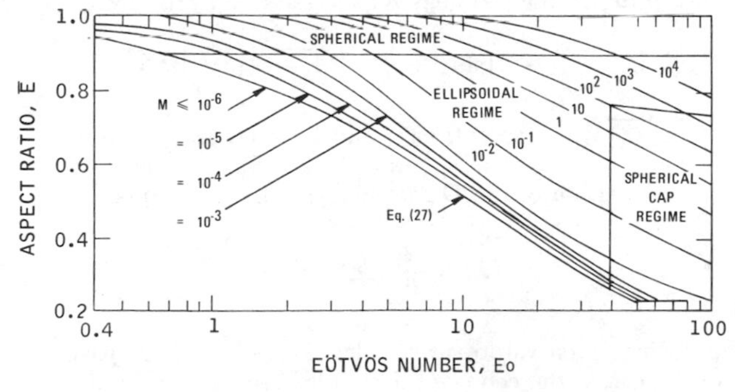

Figure 5. Time-mean aspect ratio of drops and bubbles in contaminated systems.

mode for stationary drops subject to oscillations of small amplitude in the absence of viscous effects is given [47] by

$$f_o = (48\sigma/[\pi^2 d_e^3(2\rho_c + 3\rho_d)])^{1/2} \quad (28)$$

This expression has been modified to account for viscous effects, translation, surface contamination and large amplitude. In practice, observed oscillation frequencies for freely rising or falling drops tend to be somewhat smaller than given by Equation 28, typically by 10–20% for contaminated systems. The amplitude of the oscillations tends to vary erratically. Correlation of the amplitude has been attempted by Schroeder and Kintner [48]. For low M systems, the oscillations may become so prominent that the drops are said to exhibit "random wobbling." Oscillations are important in promoting rapid mass transfer in liquid-liquid systems, as well as in augmenting drag.

The previous results in this section have all referred to cases in which some surface-active impurities have been present. If stringent measures are taken to ensure that all impurities are absent, pronounced changes in behavior may be observed. These changes can be even more dramatic than those for the spherical drops described in the previous sections, especially for low κ systems. Useful experimental results have been obtained by Winnikow and Chao [36] and Edge and Grant [38]. Compared with impure systems, drops in very pure systems exhibit:

1. increased internal circulation,
2. a delay in the onset of wake formation and wake shedding,
3. in view of 2, a delay in the onset of secondary motion, including shape oscillations,
4. smaller wakes at corresponding values of d_e or Re,
5. in view of 1–4 above, significant increases in terminal velocity, and
6. in view of 5, increased deformation from the spherical, i.e., reductions in the time-mean aspect ratio, $\overline{E}$.

Empirical correlations for estimating both the aspect ratio and the increase in terminal velocity due to complete purity of pure drops and bubbles in the ellipsoidal regime have been presented by Clift et al. [21].

Spherical-Cap or Ellipsoidal-Cap Regime

This regime corresponds to rather large drops. As shown by Figure 2, Eo must be greater than 40 for most systems; for high M (very viscous) systems, still higher values of Eo are required. As a result, this regime is not encountered very often in practice. Its analog, the spherical-cap or ellipsoidal-cap gas bubble, has received considerably greater attention because of its more frequent appearance and application.

For ellipsoidal-cap drops, the terminal velocity is given [16] by

$$U_D = (1/e^3)[\sin^{-1}e - e\sqrt{1 - e^2}]\sqrt{gb\Delta\rho/\rho} \quad (29)$$

where $e = \sqrt{1 - b^2/a^2}$ is the eccentricity, and b and a are the vertical and horizontal semiaxes, respectively, of the oblate spheroidal cap. For a spherical cap ($e \rightarrow 0$), this reduces to a modified version of the Davies and Taylor [49] equation:

$$U_D = \frac{2}{3}\sqrt{gR\Delta\rho/\rho}$$

where R is the radius of curvature of the leading edge. The wake angle, measured from the nose to the rim at the center of a circular arc, which fits the front of the drop over its forward portion, has been fitted [21] by the expression

$$\theta_w = 50° + 190° \exp\{-0.62\ Re^{0.4}\} \tag{31}$$

For $We/Re = \mu_c U_D/\sigma > 2.3$ and $Re > 4$, "skirts" of the dispersed fluid have been found to trail behind the drops. Various configurations have been observed, ranging from smooth steady skirts to those that grow with time or exhibit other unsteady motion, such as shedding of small droplets [16]. The skirt thickness can be estimated from

$$\delta = \sqrt{6\mu_d U_D/g\Delta\rho} \tag{32}$$

and is typically of the order of 1 mm for liquid drops. The occurrence of skirts appears to have little effect on the forward part of the drop or on its terminal velocity.

The wakes of ellipsoidal-cap fluid particles are closed and steady for $Re < 110$, with the wake volume given [18] by

$$V_W/V_D = 0.037\ Re^{1.4} \tag{33}$$

For $Re > 110$, the wake is open and subject to unsteady shedding.

WALL EFFECTS

Results given in the previous sections all have been for cases in which the volume-equivalent sphere drop diameter, d_e, is much less than the diameter, D, of the containing vessel. For cases in which the ratio d_e/D is appreciable, the presence of the walls can cause significant changes in behavior. The most noticeable changes are generally an elongation in the vertical direction and a retardation in terminal velocity. Other common changes are a reduction in wake volume and changes in external and internal flow patterns.

Retardation in terminal velocities is more marked at low Re than at high Re. For small drops subject to surface contamination, the interface acts as if it were rigid, and at $d_e/D < 0.3$, the standard corrections for rigid spheres can be used to predict the change in U_D. Thus, for example, the terminal velocity, U_D, of a spherical drop of diameter, d, on the axis of a cylindrical column of diameter, D, may be related to that of the same drop in an infinitely large container, $U_{D\infty}$, by the following relationship [50]:

$$U_D/U_{D\infty} = [(D - d)/(D - 0.475d)]^4 \tag{34}$$

Somewhat smaller corrections are needed if the drop is pure or large enough to be fully circulating. Haberman and Sayre [51] obtained an expression for a drop settling along the centerline of stagnant fluid in a cylindrical tube:

$$\frac{U_D}{U_{D\infty}} = \frac{1 - 0.7017k_1\lambda + 2.0865k_2\lambda^3 + 0.5689k_3\lambda^5 - 0.72603k_4\lambda^6}{1 + 2.2757\lambda^5(1-\kappa)/(2+3\kappa)} \tag{35}$$

where
$$k_1 = (2+3\kappa)/(1+\kappa)$$
$$k_2 = \kappa/(1+\kappa)$$
$$k_3 = (2-3\kappa)/(1+\kappa)$$
$$k_4 = (1-\kappa)/(1+\kappa) \tag{36}$$

and

$$\lambda = d/D \tag{37}$$

For a spherical drop traveling at a velocity, U, parallel to the axis of the tube, the drop center being a distance, b [< (D − d)/2], from the axis, with a parabolic flow of the continuous phase far upstream, Hetsroni et al. [23] derived the net drag force in the axial direction as

$$F_D = \pi\mu_c d\left(\frac{3\kappa+2}{\kappa+1}\right)\left[\left\{U_o(1-\beta^2) - U\right\}\left\{1 + \frac{\lambda}{3}\left(\frac{3\kappa+2}{\kappa+1}\right)f(\beta) + \frac{\lambda^2}{9}\left(\frac{3\kappa+2}{\kappa+1}\right)^2 f^2(\beta)\right\} - \left(\frac{2\kappa}{3\kappa+2}\right)U_o\lambda^2 + O(\lambda^3)\right] \quad (38)$$

Here, U_o is the centerline velocity far upstream, $\beta = 2b/D$, and values of the function $f(\beta)$ from Greenstein and Happel [52] are reproduced in Table I. For small deformations, the degree of deformation can be estimated from the normal stress distribution on the drop. Some lateral migration also is predicted by this analysis, this migration being outward (toward the wall), if U_o is in the same direction as the drop terminal velocity, or inward, if U_o opposes the terminal velocity. Experimental agreement with the predictions of this analysis are quite good for Re < 1 and $\lambda \lesssim 0.3$.

For drops mostly in the ellipsoidal regime with Re > 200, Eo < 40 and $d_e/D \leq 0.6$, Clift et al. [21] correlated data available in the literature by means of the following relationship:

$$U_D/U_{D\infty} = [(D^2 - d_e^2)/D^2]^{3/2} \quad (39)$$

For large drops and bubbles with Eo > 40, retardation is negligible for $d_e/D < 0.125$, while for $0.125 \leq d_e/D \leq 0.6$ the equation of Wallis [53],

$$U_D/U_{D\infty} = 1.13 \exp\{-d_e/D\} \quad (40)$$

is recommended. For columns of noncircular cross section, the hydraulic diameter (4 × cross-sectional area ÷ wetted perimeter) should be used in place of D in all expressions above.

For $d_e > 0.6$ D, the influence of the containing wall becomes predominant and the drop becomes a "slug." As for the spherical-cap regime (see above), the great majority of work on this regime has been for gas bubbles rather than for liquid drops, but the two types of systems appear to be analogous, provided that dimensionless groups are used with care [54,55]. The terminal velocity of a single slug then can be determined by the graphic correlation of White and Beardmore [56], as shown in generalized form in Figure 6. In this diagram, a Froude number group, $Fr_D = \sqrt{\rho/\Delta\rho}\, U_D/\sqrt{gD}$, is plotted as a function of $Eo_D = \Delta\rho g D^2/\sigma$ for constant values of M. Clift et al. [21] discuss special cases, as well as more complex geometries, e.g., a slug rising in an annular channel or in a tube that is inclined to the vertical.

Table I. Values of Function f(β) Required for Use of Equation 38, as calculated by Greenstein and Happel [52]

β	$f(\beta)$	β	$f(\beta)$	β	$f(\beta)$
0.00	2.10444	0.30	2.05687	0.65	2.28060
0.05	2.10270	0.35	2.04800	0.70	2.45850
0.10	2.09758	0.40	2.04388	0.75	2.742
0.15	2.08962	0.45	2.04819	0.80	3.20
0.20	2.07937	0.50	2.06557	0.85	3.96
0.25	2.06801	0.55	2.10274	0.90	5.30
		0.60	2.16980		

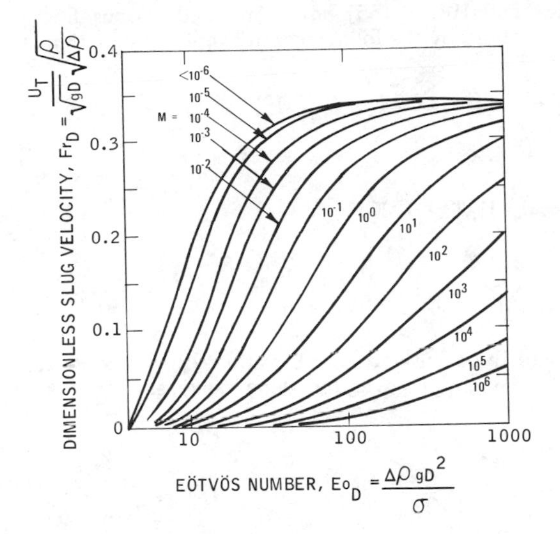

Figure 6. Generalized correlation based on White and Beardmore [56] for a rise velocity of liquid or gaseous slugs.

NON-NEWTONIAN FLUIDS

In earlier sections, we have considered only Newtonian fluids, characterized by a linear relationship between stress and rate of strain. In practice, many fluids, especially polymer solutions, polymer melts and emulsions or suspensions, are nonlinear, and many of these also show time dependency. These factors complicate greatly the study of drops in non-Newtonian fluids. Analysis is more difficult due to the increased complexity and nonlinearity of the governing equations. Experimental studies are also harder due to the added difficulty in characterizing fully the properties of the fluids. Most experiments have been carried out with polymer solutions, which have a tendency to be both elastic and surface active, but rheological properties reported usually have been limited to power law consistency factors and indices. Many of the reported findings have been only qualitative in character, while experiments have been carried out in small columns in which wall effects have been an added factor.

The most noteworthy changes that occur with non-Newtonian fluids are in the shapes of the drops. These changes are more marked if the continuous phase, rather than the dispersed phase, is non-Newtonian. Whereas drops in Newtonian liquids pass from spherical through oblate to spherical cap (see above), drops in rheologically complex media may show prolate shapes [57], and tear-drop forms (with a cusped trailing surface) are not uncommon, especially in viscoelastic media (e.g., see the work of Wilkinson [58]).

Very large drops (e.g., $d_e > 16$ mm) tend to have a rising velocity and shape in non-Newtonian systems much as in Newtonian systems because viscous effects are unimportant for such cases [59,60]. However, it has been shown both experimentally and theoretically that small drops in non-Newtonian media can behave quite differently from their counterparts in Newtonian systems, especially if the continuous phase shows time-dependent behavior. Some cases in which non-Newtonian properties lead to appreciable qualitative and quantitative differences have been reviewed by Leal [61]. Breakup of drops in shear fields of viscoelastic fluids and formation of drops in power law liquids are considered by Flumerfelt [62] and Kumar and Kuloor [10], respectively.

Terminal velocity data for drops in non-Newtonian fluids have been reported by Mhatre and Kintner [63], Fararoui and Kintner [59], Mohan et al. [64] and Kawase and Hirose [65]. For power law fluids, it is common to replace the viscosity in forming Re and K (see Equations 7 and 10) by $KU_D{}^{n-1}/d_e{}^{n-1}$, where K is the consistency, and n is the power law index. Kawase and Hirose [65] have correlated viscous flow of a Newtonian drop through a power law fluid by the following semiempirical equation:

$$C_D = \frac{8}{Re_{PL}} \frac{3\kappa_{PL} + 2}{\kappa_{PL} + 1} \{2.1 - 17.6\,(n - 0.75)^2\} \tag{41}$$

where

$$Re_{PL} = d_e^n\, U_D^{2-n} \rho_c / K \tag{42}$$

and

$$\kappa_{PL} = \mu_d U_D^{1-n} / K d_e^{1-n} \tag{43}$$

This equation reduces to the Hadamard-Rybczynski result, Equation 19, for $n \rightarrow 1$, where $K \rightarrow \mu_c$. Equation 41 is expected to apply over the same range as Equation 19, i.e., for $Re_{PL} < 0.3$.

DROP BREAKUP

Drops in Stagnant Media

Perturbations at the surface between a denser liquid and a less dense one underneath will grow in amplitude if the wavelength of the disturbance exceeds $2\pi\sqrt{\sigma/g\Delta\rho}$. As the disturbance grows, forming a spike or indentation on the surface of a drop, the disturbance tends to be swept around toward the equator. Drop splitting occurs if the disturbance grows rapidly enough relative to the time required for it to be carried to the side. This mechanism of drop splitting has led to a method for predicting maximum stable drop sizes [66]. For liquid-liquid systems of low viscosity, instability is reached when $Eo \approx 16$, i.e.,

$$(d_e)_{max} \approx 4\sqrt{\sigma/g\Delta\rho} \tag{44}$$

Larger maxima are predicted and observed when viscous effects are significant. Some typical experimental and predicted values of $(d_e)_{max}$ [21] appear in Table II. These sizes determine the largest size of drops that can be observed because rupture will occur by this Taylor instability mechanism in the absence of other mechanisms (shear and turbulence), which are discussed below.

Breakup Due to Shear

Rumscheidt and Mason [67] carried out a thorough series of experimental and theoretical studies on the rotation, deformation and breakup of drops in shear fields and hyperbolic flows. For an initially spherical drop of diameter, d, subject to an impulsively started velocity gradient, G, they predicted that breakup will occur for

$$G \geq \frac{\sigma}{d\mu_c} \frac{1 + \kappa}{1 + 19\kappa/16} \tag{45}$$

Experimental results were in good agreement with this relationship. Torza et al. [68] found that rupture did not occur for $\kappa > 3$. In addition, they found that the mode of breakup and the number of resultant daughter drops depend on the rate, dG/dt, at which shear is applied. Splitting due to shear was found to be most likely for $0.3 \leq \kappa \leq 0.9$.

Table II. Experimental and Predicted Maximum Stable Diameters for Several Liquid-Liquid Systems

Dispersed Phase	Continuous Phase	Maximum Stable Diameter: Experimental (mm)	Maximum Stable Diameter: Predicted (mm)
Carbon Tetrachloride	Water	10.4	13.4
Nitrobenzene	Water	15.4	15.6
Chlorobenzene	Water	31.6	30.0
Bromoform	Water	5.6	6.9
Diphenyl Ether	Water	32.5	31.7
Carbon Tetrachloride	31% aqueous sucrose	135	135

Turbulent Breakup

Drops are subjected to turbulent motion in agitated liquid-liquid dispersions and in turbulent pipe flows. The earliest work on drop stability in turbulent flows was by Kolmogoroff [69], who proposed that breakup is due to relatively small eddies, large eddies merely transporting the drops from one region to another. Equating the dynamic pressure difference caused by eddies to the interfacial tension forces leads to the prediction of a maximum stable diameter, given by

$$d_{max} = k\,(8\sigma/\rho_c)^{0.6}\epsilon^{-0.4} \tag{46}$$

where k is a constant, and ϵ is the turbulent energy dissipation per unit mass. Hinze [70] postulated that splitting would occur if the local shear stress, τ, were sufficient to overbalance the interfacial tension and internal viscous forces, i.e., if

$$\tau d_e > \sigma + \mu_d\sqrt{\tau/\rho_d} \tag{47}$$

This mechanism has been used successfully by Hughmark [71] to explain the breakup of drops in turbulent pipe flow. Coulaloglou and Tavlarides [72] have assumed that drop deformation is due to turbulent pressure fluctuations and that drop rupture will occur if the kinetic energy transmitted to a drop by turbulent eddies is greater than the surface energy of the drop. To apply this concept, the turbulent flow field was taken to be locally isotropic, with the drop size within the inertial subrange of eddies. With these assumptions, a breakage frequency function was obtained as follows:

$$\text{Fraction of drops breaking/time} \propto d_e^{-2/3}\,\epsilon^{1/3}\exp\{-k\sigma/\rho_d\epsilon^{2/3}\,d_e^{5/3}\} \tag{48}$$

where k is a constant, and ϵ is the local energy dissipation per unit mass. The size distribution of droplets formed by breakage was assumed to be normally distributed. Coupled with a suitable expression for drop coalescence, this mechanistic model has led to good predictions of droplet size distributions in agitated systems [73]. This approach is preferred to the earlier models as it more realistically predicts a distribution of drop sizes, rather than a single discrete maximum.

DROP-DROP INTERACTIONS

Long-Range and Medium-Range Interactions

Analytical solutions have been obtained by a number of workers for low Reynolds number motion of two fluid spheres along their line of centers. A solution that is general, except in the sense that the particles are constrained to remain spherical, is presented by Rushton and Davies [74]. This reduces to a number of earlier solutions as special cases, including those for rigid spheres and gas bubbles, as κ for both drops approaches ∞ or 0, respectively. Two equal spherical drops in creeping flow under gravity in an unbounded field are predicted to maintain their separation distance as they fall or rise.

A solution, also constrained to spherical drops, has been obtained by the method of reflections by Hetsroni and Haber [75] for two fluid spheres moving in creeping motion at right angles to their line of centers in an unbounded arbitrary velocity field. More general motions can be obtained by vector summation of the cases in which the motions are parallel and perpendicular to the line of centers. As an example, Hetsroni and Haber [75] obtained estimates of the drag force and terminal settling velocities of two droplets falling in an unbounded quiescent fluid.

There appear to be no solutions available for interacting spherical drops in liquid-liquid systems at higher Reynolds numbers. Results for gas bubbles under these conditions [76] suggest that two drops traveling along their line of centers should experience less drag than when in isolation.

The interaction of ellipsoidal drops is complicated by secondary motion effects. This may, for example, upset the vertical alignment of pairs of drops. In general, however, the velocity of a trailing drop is increased significantly by the presence of a leading drop, while the leading one is influenced very little by its follower [77]. On the other hand, drops side by side (in a horizontal row) travel more slowly than corresponding single drops [78]. The same general behavior is observed for pairs of ellipsoidal-cap drops [17]. For both cases, it is possible to apply a superposition principle whereby each drop moves with a velocity that is the vector sum of the velocity it would have in isolation and the velocity that would be caused by the other drop at the position of the nose of the drop if it were absent. Clearly, this superposition principle can be extended readily to multiple drop systems.

Film Drainage and Drop Coalescence

When two drops collide they may coalesce or rebound. Coalescence is preceded by a period of film drainage in which the film of continuous-phase liquid separating the two drops thins further. The film ruptures when it becomes sufficiently thin, the thickness at rupture typically being of the order of 100–1000 Å [79]. Rupture may lead to complete coalescence or to partial coalescence only. Rebounding tends to occur when film drainage is slow.

There has been considerable study of the coalescence between a drop and a plane interface related to phase separation in liquid-liquid contactors. An extensive review has been given by Hartland and Hartley [80]. The factors influencing coalescence are the same, for the most part, as those for interdrop coalescence, a case that has received less attention in the literature.

For two deformable drops of diameters d_{e1} and d_{e2}, subject to a force, F, pushing the drops together, the time for the film to thin from a thickness δ_o to a rupture thickness δ_r is given [72] by

$$t = \frac{3\mu_c F}{16\pi\sigma^2} \frac{d_{e1} d_{e2}}{(d_{e1} + d_{e2})} \left\{ \frac{1}{\delta_r^2} - \frac{1}{\delta_o^2} \right\} \tag{49}$$

Some additional factors, beyond those that appear in Equation 49, that affect the drainage time in practice, are as follows:

1. Mobility of the interface. Drainage is fastest for a mobile interface, so that internal circulation tends to promote drainage while film thinning is retarded by traces of surface-active agents.

2. Electrical fields can be used to promote rapid film thinning and coalescence [81].
3. Electrical double layers tend to retard film drainage.
4. Mass transfer affects coalescence time through the generation of local interfacial tension gradients and interfacial turbulence. Transfer out of the drops generally accelerates coalescence, whereas transfer into the dispersed phase causes slower drainage, thereby stabilizing the drops [82].
5. Presence of a third component. A third component, such as a solid phase, tends to promote rapid coalescence, especially if wetted by the dispersed phase [83].
6. Oscillations, vibrations (including ultrasonics) and turbulence are factors.
7. Geometric constraints, such as surrounding drops in a closely packed dispersion [84], also are factors.

NOMENCLATURE

a,b	semiaxes of ellipsoid of revolution
b	distance from containing vessel axis to center of drop
C_D	drag coefficient as defined in Equation 13
C_{DS}	drag coefficient for sphere in uniform Stokes flow
c	constant
D	diameter of cylindrical column
d	spherical drop diameter
d^*	dimensionless orifice diameter, $d_{or}/V_D^{1/3}$
d_e	volume-equivalent drop diameter
$(d_e)_{max}$	value of d_e corresponding to maximum stable drop size
d_j	jet diameter
d_{or}	orifice diameter
$\overline{E}$	mean aspect (maximum height/maximum width) ratio
Eo	Eötvös number as defined in Equation 8
Eo_D	Eötvös number based on D, i.e., $g\Delta\rho D^2/\sigma$
Eo_{or}	Eötvös number based on d_{or}, i.e., $g\Delta\rho d_{or}^2/\sigma$
e	eccentricity, $= \sqrt{1 - b^2/a^2}$
Fr_D	Froude number group, $= \sqrt{\rho/\Delta\rho}\, U_D/\sqrt{gD}$
$f(\beta)$	function given in Table I
f_o	natural frequency
G	velocity gradient in shear field
g	acceleration of gravity
H,J	dimensionless groups defined by Equations 24 and 26
K	consistency factor for power law fluid
k	constant
M	dimensionless property group defined by Equation 9
n	power law index
P	reciprocal of M, Equation 12
Q	volumetric flowrate through orifice
Q_{at}	minimum value of Q required to cause jet atomization
Q_{jet}	minimum value of Q required for jetting to occur
Q_{max}	value of Q corresponding to maximum jet length
R	radius of spherical drop
Re	Reynolds number as defined in Equation 7
Re_{PL}	Reynolds number for power law continuous phase, Equation 42
r	radial coordinate
t	time required for film drainage
U_D	terminal velocity of drop
U_{DS}	Stokes terminal velocity
$U_{D\infty}$	terminal velocity in an unbounded quiescent liquid
V_D	drop volume
V_W	wake volume
We	Weber number as defined in Equation 14

Greek Symbols

β	dimensionless drop position, $2b/D$
γ	density ratio, ρ_d/ρ_c
$\Delta\rho$	absolute density difference, $\lvert\rho_c - \rho_d\rvert$
δ	skirt thickness

δ_o initial film thickness
δ_r film thickness at rupture
ϵ turbulent energy dissipation per unit mass liquid
ζ_R vorticity at spherical drop surface, $r = R$
θ angular polar coordinate
κ viscosity ratio, μ_d/μ_c
κ_{PL} viscosity ratio with power law continuous phase, Equation 43
λ dimensionless drop size, d/D or d_e/D
λ_H Harkins correction factor
μ_c, μ_d viscosity of continuous dispersed phase
ρ_c, ρ_d density of continuous dispersed phase
σ interfacial tension
τ shear stress
ψ_c, ψ_d stream function for continuous dispersed phase

REFERENCES

1. Cavers, S. D. In: *Handbook of Solvent Extraction*, T. C. Lo, M. H. I. Baird and C. Hanson, Eds. (New York: John Wiley & Sons, Inc., 1982).
2. Scheele, G. F., and B. J. Meister. "Drop Formation at Low Velocities in Liquid-Liquid Systems," *AIChE J.* 14:9–19 (1968).
3. Humphrey, J. A. C., R. L. Hummel and J. W. Smith. "Experimental Study of the Internal Fluid Dynamics of Forming Drops," *Can. J. Chem. Eng.* 52:449–456 (1974).
4. Meister, B. J., and G. F. Scheele. "Generalized Solution of the Tomotika Stability Analysis for a Cylindrical Jet," *AIChE J.* 13:682–688 (1967).
5. Meister, B. J., and G. F. Scheele. "Prediction of Jet Length in Immiscible Liquid Systems," *AIChE J.* 15:689–699 (1969).
6. Rayleigh, Lord. "On the Instability of Liquid Jets," *Proc. London Math. Soc.* 10: 4–13 (1878).
7. Treybal, R. E. *Liquid Extraction*, 2nd ed. (New York: McGraw-Hill Book Co., 1963).
8. Christiansen, R. M., and A. N. Hixson. "Breakup of a Liquid Jet in a Denser Liquid," *Ind. Eng. Chem.* 49:1017–1024 (1957).
9. Hozawa, M., and T. Tadaki. "Drop Formation from Single Nozzles at High Nozzle Velocities in Liquid-Liquid Systems," *Kazaku Kozaku* 37:827–833 (1973).
10. Kumar, R., and N. R. Kuloor. "The Formation of Bubbles and Drops," *Adv. Chem. Eng.* 8:256–368 (1970).
11. Gayler, R., and H. R. C. Pratt. "Liquid-Liquid Extraction: Further Studies of Droplet Behaviour in Packed Columns," *Trans. Inst. Chem. Eng.* 31:69–77 (1953).
12. Jackson, R. "The Formation and Coalescence of Drops and Bubbles in Liquids," *Chem. Eng.* 178:CE107–118 (May 1964).
13. Sawistowski, H. In: *Recent Advances in Liquid-Liquid Extraction*, C. Hanson, Ed. (Oxford, England: Pergamon Press, Inc., 1971), pp. 293–366.
14. Szekely, J. *Fluid Flow Phenomena in Metals Processing* (New York: Academic Press, Inc., 1979).
15. Grace, J. R., T. Wairegi and T. H. Nguyen. "Shapes and Velocities of Single Drops and Bubbles Moving Freely through Immiscible Liquids," *Trans. Inst. Chem. Eng.* 54:167–173 (1976).
16. Wairegi, T., and J. R. Grace. "The Behaviour of Large Drops in Immiscible Liquids," *Int. J. Multiphase Flow* 3:67–77 (1976).
17. Wairegi, T. "The Mechanics of Large Drops and Bubbles Moving through Extended Liquid Media," PhD Thesis, McGill University, Montreal, Quebec (1974).
18. Bhaga, D., and M. E. Weber. "Bubbles in Viscous Liquids: Shapes, Wakes and Velocities," *J. Fluid Mech.* 105:61–85 (1981).
19. Hadamard, J. S. "Mouvement permenant lent d'une sphère liquide et visqueuse dans un liquide visqueux," *Comp. Rend. Aca. Sci.* 152:1735–1738 (1911).
20. Rybczynski, E. "Über die fortschreitende Bewegung einer flüssigen Kugel in einem zähen Medium," *Bull. Int. Acad. Sci. Cracovie* (*Ser. A*) 40-46 (1911).

21. Clift, R., J. R. Grace and M. E. Weber. *Bubbles, Drops and Particles* (New York: Academic Press, Inc., 1978).
22. Stokes, G. G. "On the Effect of the Internal Friction of Fluids on the Motion of Pendulums," *Trans. Camb. Phil. Soc.* 9:8–27 (1851).
23. Hetsroni, G., and S. Haber. "The Flow in and around a Droplet or Bubble Submerged in an Unbounded Arbitrary Velocity Field," *Rheol. Acta* 9:488–496 (1970).
24. Hetsroni, G., S. Haber and E. Wacholder. "The Flow Fields in and around a Droplet Moving Axially within a Tube," *J. Fluid Mech.* 41:689–705 (1970).
25. Levich, V. G. *Physiochemical Hydrodynamics* (Englewood Cliffs, NJ: Prentice-Hall, Inc., 1962).
26. Savic, P. "Circulation and Distortion of Liquid Drops Falling Through a Viscous Medium," NRC lab. report MT-22 (1953).
27. Scriven, L. E. "Dynamics of a Fluid Interface," *Chem. Eng. Sci.* 12:98–108 (1960).
28. Levan, M. D., and J. Newman. "The Effect of Surfactant on the Terminal and Interfacial Velocities of a Bubble or Drop," *AIChE J.* 22:695–701 (1976).
29. Bond, W. N., and D. A. Newton. "Bubbles, Drops and Stokes' Law," *Phil. Mag.* 5: 794–800 (1928).
30. Pan, F. Y., and A. Acrivos. "Shape of a Drop or Bubble at Low Reynolds Number," *Ind. Eng. Chem. Fund.* 7:227–232 (1968).
31. Wasserman, M. L., and J. C. Slattery. "Creeping Flow Past a Fluid Globule When a Trace of Surfactant Is Present," *AIChE J.* 15:533–547 (1969).
32. Rivkind, V. Y., G. M. Ryskin and G. A. Fishbein. "The Motion of a Spherical Drop in the Flow of a Viscous Fluid," *Fluid Mech. Sov. Res.* 1:142–151 (1972).
33. Abdel-Alim, A. H., and A. E. Hamielec. "A Theoretical and Experimental Investigation of the Effect of Internal Circulation on the Drag of Spherical Droplets Falling at Terminal Velocity in Liquid Media," *Ind. Eng. Chem. Fund.* 14:308–312 (1975).
34. Rivkind, V. Y., and G. M. Ryskin. "Flow Structure in Motion of a Spherical Drop in a Fluid Medium at Intermediate Reynolds Number," *Fluid Dynamics* 1:5–12 (1976).
35. Rivkind, V. Y., G. M. Ryskin and G. A. Fishbein. "Flow Around a Spherical Drop at Intermediate Reynolds Numbers," *Appl. Math. Mech.* 40:687–691 (1976).
36. Winnikow, S., and B. T. Chao. "Droplet Motion in Purified Systems," *Phys. Fluids* 9:50–61 (1966).
37. Harper, J. F., and D. W. Moore. "The Motion of a Spherical Liquid Drop at High Reynolds Number," *J. Fluid Mech.* 32:367–391 (1968).
38. Edge, R. M., and C. D. Grant. "The Motion of Drops in Water Contaminated with a Surface-Active Agent," *Chem. Eng. Sci.* 27:1709–1721 (1972).
39. Magarvey, R. H., and R. L. Bishop. "Transition Ranges for Three-Dimensional Wakes," *Can. J. Phys.* 39:1418 (1961).
40. Hu, S., and R. C. Kintner. "The Fall of Single Liquid Drops Through Water," *AIChE J.* 1:42–50 (1955).
41. Johnson, A. I., and L. Braida. "The Velocity of Fall of Circulating and Oscillating Liquid Drops Through Quiescent Liquid Phases," *Can. J. Chem. Eng.* 35:165–172 (1957).
42. Kubie, J. "Settling Velocity of Droplets in Turbulent Flows," *Chem. Eng. Sci.* 35: 1787–1793 (1980).
43. Taylor, T. D., and A. Acrivos. "On the Deformation and Drag of a Falling Viscous Drop at Low Reynolds Number," *J. Fluid Mech.* 18:466–476 (1964).
44. Brignell, A. S. "The Deformation of a Liquid Drop at Small Reynolds Number," *Quart. J. Mech. Appl. Math.* 26:99–107 (1973).
45. Hendrix, C. D., S. B. Dave and H. G. Johnson. "Translation of Continuous Phase in the Wakes of Single Rising Drops," *AIChE J.* 13:1072–1077 (1967).
46. Yeheskel, J., and E. Kehat. "The Size and Rate of Shedding of Wakes of Single Drops Rising in a Continuous Medium," *Chem. Eng. Sci.* 26:1223–1233 (1971).
47. Lamb, H. *Hydrodynamics*, 6th ed. (New York: Cambridge University Press, 1932).

48. Schroeder, R. R., and R. C. Kintner. "Oscillations of Drops Falling in a Liquid Field," *AIChE J.* 11:5–8 (1965).
49. Davies, R. M., and G. I. Taylor. "The Mechanics of Large Bubbles Rising Through Extended Liquids and Through Liquids in Tubes," *Proc. Royl. Soc.* A200:375–390 (1950).
50. Francis, A. W. "Wall Effect in a Falling Ball Method for Viscometry," *Physics* 4: 403–406 (1933).
51. Haberman, W. L., and R. M. Sayre. "Motion of Rigid and Fluid Spheres in Stationary and Moving Liquids Inside Cylindrical Tubes," *David Taylor Model Basin Report* 1143 (1958).
52. Greenstein, T., and J. Happel. "Theoretical Study of the Slow Motion of a Sphere and a Fluid in a Cylindrical Tube," *J. Fluid Mech.* 34:705–710 (1968).
53. Wallis, G. B. *One-Dimensional Two-Phase Flow* (New York: McGraw-Hill Book Co., 1969).
54. Harmathy, T. Z. "Velocity of Large Drops and Bubbles in Media of Infinite or Restricted Extent," *AIChE J.* 6:281–288 (1960).
55. Rader, D. W., A. T. Bourgoyne and R. H. Ward. "Factors Affecting Bubble Rise Velocity of Gas Kicks," *J. Petrol. Technol.* 27:571–584 (1975).
56. White, E. T., and R. H. Beardmore. "The Velocity of Rise of Single Cylindrical Air Bubbles Through Liquids Contained in Vertical Tubes," *Chem. Eng. Sci.* 17:351–361 (1962).
57. Shirotsuka, T., and Y. Kawase. "Drop Shape in Non-Newtonian Fluid Systems," *J. Chem. Eng. Jap.* 8:336–338 (1975).
58. Wilkinson, W. L. "Tailing of Drops Falling Through Viscoelastic Liquids," *Nature Phys. Sci.* 240:44 (November 13, 1972).
59. Fararoui, A., and R. C. Kintner. "Flow and Shape of Drops in Non-Newtonian Fluids," *Trans. Soc. Rheol.* 5:369–380 (1961).
60. Marrucci, G., G. Apuzzo and G. Astarita. "Motion of Liquid Drops in Non-Newtonian Systems," *AIChE J.* 16:538–541 (1970).
61. Leal, L. G. "The Motion of Small Particles in Non-Newtonian Fluids," *J. Non-Newtonian Fluid Mech.* 5:33–78 (1979).
62. Flumerfelt, R. W. "Drop Breakup in Simple Shear Fields of Viscoelastic Fluids," *Ind. Eng. Chem. Fund.* 11:312–318 (1972).
63. Mhatre, M. U., and R. C. Kintner. "Fall of Liquid Drops Through Pseudoplastic Liquids," *Ind. Eng. Chem.* 51:865–867 (1959).
64. Mohan, V., R. Nagarajan and D. Venkateswarlu. "Fall of Drops in Non-Newtonian Media," *Can. J. Chem. Eng.* 50:37–40 (1972).
65. Kawase, Y., and Y. Hirose. "Motion of Drops in Non-Newtonian Fluid Systems at Low Reynolds Number," *J. Chem. Eng. Japan.* 10:68–70 (1977).
66. Grace, J. R., T. Wairegi and J. Brophy. "Break-Up of Drops and Bubbles in Stagnant Media," *Can. J. Chem. Eng.* 56:3–8 (1978).
67. Rumscheidt, F. D., and S. G. Mason. "Deformation and Burst of Fluid Drops in Shear and Hyperbolic Flow," *J. Colloid Interface Sci.* 16:238–261 (1961).
68. Torza, S., R. G. Cox and S. G. Mason. "Transient and Steady Deformation and Burst of Liquid Drops," *J. Colloid Interface Sci.* 38:395–411 (1972).
69. Kolmogoroff, A. N. *Dokl. Akad. Nauk. SSSR* 66:825–828 (1949).
70. Hinze, J. O. "Fundamentals of the Hydrodynamic Mechanism of Splitting in Dispersion Processes," *AIChE J.* 1:289–295 (1955).
71. Hughmark, G. A. "Drop Breakup in Turbulent Pipe Flow," *AIChE J.* 17:1000 (1971).
72. Coulaloglou, C. A., and L. L. Tavlarides. "Description of Interaction Processes in Agitated Liquid-Liquid Dispersions," *Chem. Eng. Sci.* 32:1289–1297 (1977).
73. Cruz-Pinto, J. J. C., and W. J. Korchinsky. "Drop Breakage in Countercurrent Flow Liquid-Liquid Extraction Columns," *Chem. Eng. Sci.* 36:687–694 (1981).
74. Rushton, E., and G. A. Davies. "The Slow Motion of Two Spherical Particles Along Their Line of Centres," *Int. J. Multiphase Flow* 4:357–381 (1978).

75. Hetsroni, G., and S. Haber. "Low Reynolds Number Motion of Two Drops Submerged in an Unbounded Arbitrary Velocity Field," *Int. J. Multiphase Flow* 4:1–17 (1978).
76. Harper, J. F. "On Bubbles Rising in Line at Large Reynolds Numbers," *J. Fluid Mech.* 41:751–758 (1970).
77. Zabel, T., C. Hanson and J. Ingham. "The Influence of System Purity, Drop Separation and Heat Transfer on the Terminal Velocity of Falling Drops in Liquid-Liquid Systems," *Trans. Inst. Chem. Eng.* 51:162–164 (1973).
78. Raghavendra, N. M., and M. N. Rao. "Studies of Momentum Transfer in the Case of Assemblages of Liquid Drops Falling in a Stagnant Column of an Immiscible Liquid," *Indian J. Technol.* 3:303–307 (1965).
79. Scheele, G. F., and D. E. Leng. "An Experimental Study of Factors Which Promote Coalescence of Two Colliding Drops Suspended in Water," *Chem. Eng. Sci.* 26: 1867–1879 (1971).
80. Hartland, S., and R. W. Hartley. *Axisymmetric Fluid-Liquid Interfaces* (Amsterdam: Elsevier Publishing Co., 1976).
81. Brown, A. H., and C. Hanson. "The Effect of Oscillating Electric Fields on the Coalescence of Liquid Drops," *Chem. Eng. Sci.* 23:841–848 (1968).
82. Groothuis, H., and F. J. Zuiderweg. "Influence of Mass Transfer on Coalescence of Drops," *Chem. Eng. Sci.* 12:288–289 (1960).
83. Charles, G. E., and S. G. Mason. "The Coalescence of Liquid Drops with Flat Liquid/Liquid Interfaces," *J. Colloid. Sci.* 15:236–267 (1960).
84. Vohra, D. K., and S. Hartland. "Effect of Geometrical Arrangement and Interdrop Forces on Coalescence Time," *Can. J. Chem. Eng.* 59:438–449 (1981).

CHAPTER 39

FLOW OF IMMISCIBLE LIQUIDS THROUGH POROUS MEDIA

F. G. McCaffery
Occidental Research Corporation
Irvine, California 92713

J. P. Batycky
Petroleum Recovery Institute
Calgary, Alberta T2L 2A6
Canada

CONTENTS

INTRODUCTION

A proper understanding of the flow of immiscible liquids and gas-liquid systems in porous media is particularly important in the areas of petroleum reservoir engineering and groundwater hydrology. Immiscible displacement is the major mechanism involved in recovering oil from a reservoir, the most common example being the production of oil in response to waterflooding. Quantification of the flow characteristics of the reservoir rock is essential to the proper design of a recovery strategy and correct interpretation of the history of oil production.

As our petroleum resources become depleted and the application of more sophisticated processes is considered to improve oil recovery efficiency beyond levels attainable by straight waterflooding, it becomes necessary to have a high level of understanding of the fluid flow behavior for particular reservoir systems. This is required to decrease the uncertainty in predictions of increased oil recovery.

When conducting laboratory studies of displacement processes for a particular reservoir, one must attempt to duplicate the conditions that exist in the reservoir. This often means that the reservoir fluids should be returned to the formation temperature and the reservoir rock subjected to the proper overburden pressure. With all immiscible displacement tests, the reservoir rock samples, or cores, should possess the same wettability and initial fluid saturations as in the reservoir. Without this correspondence, which can be difficult to achieve with certainty, any displacement data determined in laboratory tests are less likely to be useful for providing reliable predictions of field behavior.

A major problem associated with laboratory displacement tests concerns the size of the available core samples, typically 3.5 cm in diameter and 7 cm in length, and whether such minute specimens are representative of hectares of reservoir rock lying unsampled in the formation between well locations. Certain assumptions of continuity and homogeneity within specific layers of the reservoir usually must be made that simplify problems (but not uncertainties) in extrapolating core displacement test results to help the field situation. Core plugs subjected to displacement tests also must be homogeneous within certain bounds so that a unique set of mathematical relationships can be ascribed to the fluid flow properties of the rock.

A qualitative description of homogeneity is approached by attempting a microscopic description of a volume segment of the porous medium. The system is composed of a solid substrate that surrounds voids or pores that are generally interconnected through pore throats. In a petroleum reservoir, the solid surfaces usually are composed of a variety of mineral assemblages assumed to be at thermodynamic equilibrium with the reservoir fluids when they are at the initial saturation conditions. The pore throats and bodies are of various sizes, randomly connected so that the rock is heterogeneous microscopically. However, by expanding the volume being considered, a sufficient sampling of each variable is included to obtain a distribution function identical to that of the entire sample. When the size of this minimum volume is small in comparison to the total rock volume, the rock is then defined as homogeneous. This minimum volume, defined by Bear [1] as the representative volume element, or by Slattery [2] as contained by the closed average surface, then can be viewed conceptually as a continuum composed of the rock and the void that is homogeneously occupied by the resident fluids.

As summarized by Bear, the actual heterogeneous porous medium "is replaced by a fictitious continuum; a structureless substance, to any point of which we can assign kinematic and dynamic variables and parameters." Mathematical descriptions then can be developed using smooth, continuous functions of each relevant variable. This concept has been adapted to computer simulation of reservoir and laboratory displacement where block sizes are chosen to suit computational stability and speed. The convenience of this approach unfortunately has simplified our comprehension and treatment of the very complex processes that control oil displacement during waterflooding and the application of enhanced oil recovery.

DISPLACEMENT CONCEPTS

Multiphase immiscible flow is described through the application of Darcy's law for each flowing fluid. For linear flow of phases w (wetting) and n (nonwetting),

$$q_w = -\frac{k_{rw}KA}{\mu_w}\frac{\partial P_w}{\partial x} \tag{1}$$

$$q_n = -\frac{k_{rn}KA}{\mu_n}\frac{\partial P_n}{\partial x} \tag{2}$$

where q_w and q_n = the flowrates of each phase
k_{rw} and k_{rn} = the relative permeabilities
K = the absolute permeability
A = the cross-sectional area of the flow
P_w and P_n = the pressures in each phase
μ_w and μ_n = the viscosities
x = the displacement direction

The capillary pressure, ΔP_c, is defined as

$$\Delta P_c = P_n - P_w \tag{3}$$

The concepts of relative permeability and capillary pressure are fundamental to the study of immiscible fluid flow in porous media. For information on this subject as it applies to oil recovery, in addition to what is presented in this brief chapter, the reader is referred to the work of Muskat [3], Collins [4], Richardson [5] and Craig [6].

Pore Geometry and Capillary Pressure

To understand fluid distribution and, consequently, fluid displacement behavior, one must appreciate the solid surface geometry and the manner by which pore spaces interconnect within a porous medium. Two idealized approaches that characterize early attempts in this direction involved the use of packings of equalized spheres and bundles of parallel capillary tubes. The capillary tube model cannot depict the occurrence of pore bodies interconnected through pore throats, however, and cannot provide proper representation of fluid saturations and their distribution that are typical for porous rock.

Equal-sized spheres can provide precise definition of pore body-throat geometries for any selected packing geometry. Fluid distributions then can be computed readily by assuming that spherical geometries apply. Fluid movements probably can be followed by solving the Navier-Stokes equations on a pore-by-pore basis. However, since most porous rock is created by natural sorting and grading phenomena, the real value of such an idealized treatment becomes questionable. The simplified models also ignore the secondary geological processes of diagenetic cementation, dissolution and authigenic deposition of materials.

The best that any idealized model can hope to accomplish is to enable simplistic interpretations and generalizations of more realistic behavior. A capillary network with pore bodies interconnected by capillary pore throats has been used by Wardlaw and Taylor [7] to describe the process of injection and withdrawal of a nonwetting fluid (mercury) to better understand the phenomenon. Morrow [8] and Batycky and Singhal [9] employed a cubic packing of spheres to quantify the mobilization requirements for trapped oil ganglia. Ng and Payatakes [10] and Payatakes et al. [11] studied the behavior of populations of ganglia being mobilized, broken up and stranded again. These efforts clearly were directed toward describing specific phenomena, and a more general approach would be pursued and developed more eagerly if it were not possible to obtain an interpretation of pore geometry from standard capillary pressure measurements.

Capillary pressure experiments most appropriate for the definition of pore geometry are best performed with a pair of fluids in which one is strongly nonwetting. In this case there can be little or no ambiguity in the interpretation of the results. For most purposes, an ideal system is considered to be mercury-vacuum or low pressure air. Mercury is a nonwetting fluid under the subcritical pressures and temperatures that can be expected to exist, so that the gas phase occupies the interior of the pore spaces and contacts the solid surfaces with a near-zero contact angle. In the course of such experiments, both the pressure and volume of the mercury contained in a core are monitored to obtain plots of pressure versus saturation during sequential mercury injection and ejection. Once the values of pressure have been obtained, pore sizes can be estimated at any mercury pressure using Laplace's equation:

$$\Delta P_c = \sigma\left(\frac{1}{r_1} + \frac{1}{r_2}\right) \tag{4}$$

where r_1 and r_2 are mutually orthogonal radii of curvature, and σ is the fluid-fluid interfacial tension (for mercury-air in this case). For pore shapes that result in nearly spherical interfaces, $r_1 = r_2 = r$, and

$$\Delta P_c = \frac{2\sigma}{r} \tag{5a}$$

whereas for sheet pores, where $r_1 = r >> r_2$,

$$\Delta P_c = \frac{\sigma}{r} \tag{5b}$$

It is reasonable to assume that for lower pressures and in more permeable rock samples Equation 5a is the more appropriate for computing r, unless microscopic examination has indicated otherwise. For low-permeability dolomite rock and in the zones of microporosity in sandstone caused by extensive crystal growth, sheet pores commonly occur [7].

A typical mercury porosimetry scan is presented in Figure 1. After the threshold capillary pressure, ΔP_c^*, is exceeded during primary injection, D_1, mercury begins to invade the larger pore throats existing near the exposed surfaces. As pressures are increased, smaller pore throats, as well as interconnected larger pores, begin to be invaded during this primary drainage cycle. The last pores to be invaded at S_{I1} are those interconnected by the smallest throats and those portions of the pore spaces containing microporosity. An average throat size, computed using Equation 5a, could be that corresponding to the point of inflection on the primary drainage curve. If the distribution is bi- or multimodal, then additional inflection points would occur. As might be expected, a reasonable correlation will exist between absolute permeability and the average value of pore throat radius. For instance, with 27 sandstone and carbonate samples, Wardlaw and Cassan [12] obtained correlation coefficient values of 67 and 71 between the computed radius, r, at $S_{Hg} = 0.5$, and each of K and $\sqrt{K/\phi}$. Here, K is the permeability and ϕ is the porosity. The slope of the primary drainage curve also can be used to derive a distribution of pore throat radii and, in simplistic terms, the smaller the slope, the more homogeneous the core about the mean radius.

The slope of the mercury withdrawal or imbibition curve to a maximum saturation, S_{w1}, as zero pressure is approached provides for further interpretation of the pore geometry (Figure 1). The withdrawal step (I), which is a reflection of the pressure drops during the relaxation of surface curvatures, typically displays an inverse threshold pressure before the nonwetting phase saturation begins to decrease. Then, withdrawal occurs to a limit of S_{w1}. When compared with the maximum mercury saturation, the movable mercury is indicative of pore body-pore throat radii contrasts, as well as the number or interconnectivity of pore bodies. Mercury withdrawal is characterized by the progressive isolation of more and more of the pore space as pressures are reduced. The less the microscopic scale heterogeneity, the more efficient the recovery during imbibition and the less trapping of nonwetting phase. Hysteresis between the secondary drainage, D_2, and the preceding imbibition (I) curves, which may be dependent to some degree on contact angle hysteresis, in addition to the pore throat-pore body size contrasts and pore accessibility, has not been related quantitatively to heterogeneity to date.

During primary drainage it is possible to initiate scanning imbibition curves at intermediate pressures, as is shown by the withdrawal at S_{I2}, rather than injecting

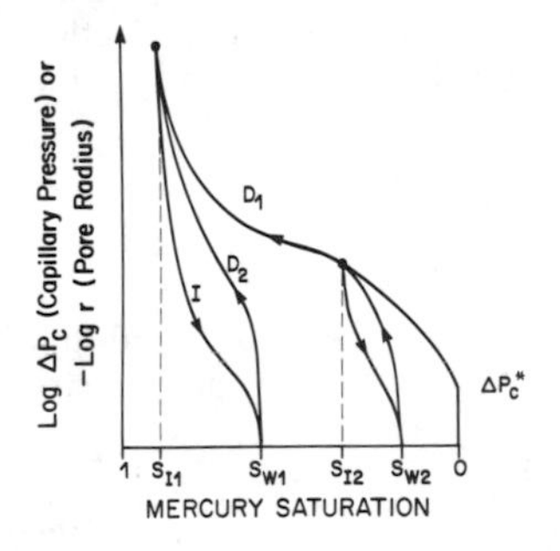

Figure 1. Mercury-air capillary pressure curves.

to the maximum attainable pressure of the apparatus. Recoveries then can be obtained as a function of maximum injection pressure. However, Wardlaw and Taylor [7] have shown that by simply subtracting the secondary drainage or reinjection curve, D_2, from the initial drainage relation, D_1, it is possible to determine the dependence of residual saturation on the initial (maximum) nonwetting phase saturation.

The foregoing discussion assumes that mercury is perfectly nonwetting and that all phenomena are related to pore geometry effects. As, under strongly wetting conditions, capillary effects are greater than viscous displacement effects at normal displacement rates, much of the data obtained, including the dependence of residual mercury saturation on maximum injected saturation, can be extended to at least qualitatively show the dependence on residual oil contents for an oil-water system in the same porous medium under strongly water-wet conditions.

A comment must be made with respect to heterogeneity. In an Indiana limestone system investigated by Wardlaw and Taylor [7], where sample bulk volumes were varied from 1.7 to 25.2 ml, the mercury injection data of each were similar. Consequently, Bear's [1] representative volume element would be less than 1.7 ml for this rock sample. It must be pointed out also that samples tested by this technique are normally selected for good homogeneity and intergranular porosity, and for the absence of visible moldic porosity. Therefore, extrapolation of the pore geometry and displacement properties derived from the mercury tests to larger systems, such as may be used in oil-water core displacement tests, also should be approached with caution.

An alternative procedure for defining pore geometry, which has received less use than capillary pressure methods, is based on the analysis of stable miscible displacement tests on cores. The core displacement results are usually interpreted according to the dispersion-capacitance model [13]. Two types of pore space are considered in this model—one with fluid flowing at the average velocity and the other in which the fluid is stagnant. Material movement between the two types is considered to be due to mass transfer. The effluent concentration behavior following a change in the injected fluid composition is used to determine the values of each of three parameters:

1. the fraction of pore space contributing to flow, f,
2. the dispersion coefficient in the flowing fraction, K_D, and
3. the mass transfer rate constant, K_M.

The roles of each of the three parameters can be interpreted by examining the computer-simulated concentration gradients of the flowing fluid, c, at the exit of a linear core following a step change in injected concentration at the core inlet from 0 to c_o (Figure 2(a)) [14]. If there is no dispersion, $K_D = 0$, and a step change in outlet concentration occurs that is identical to that imposed at the inlet. As the dispersion coefficient increases, breakthrough of the injected component occurs earlier. However, when there is no stagnant fraction ($f = 1$), the effluent concentration (c/c_o) is always 0.50 at one pore volume of injection, and the concentration gradient is mirror symmetrical about this injection volume.

The effect of decreasing the flowing fraction, f, to thereby increase the stagnant fraction, is shown in Figure 2(b). As f is decreased from 1, the effluent concentration curve moves to the left and the flowing fraction is the value of pore volume corresponding to $(c/c_o) = 0.50$. Since $K_M = 0$, the symmetry of the effluent profile is retained. When, as would be expected, there is mass transfer between the stagnant and flowing fractions, the effluent profiles become skewed with increasing values of mass transfer rate, K_M, as given in Figure 2(c). As K_M is increasing, asymmetrical tailing increases, and profile shifts occur to the right at the $(c/c_o) = 0.50$ value of injected pore volume. Then, as K_M—which is dependent on viscosities, diffusion coefficient and the occurrence of eddies—increases further, the behavior begins to approach that corresponding to larger values of f and smaller values of K_M, as given in the profile corresponding to K_{M4}. Since there can be overlapping behavior, determination of the three parameters may not be unique, and interpretation using "best fit" parameters may be somewhat sub-

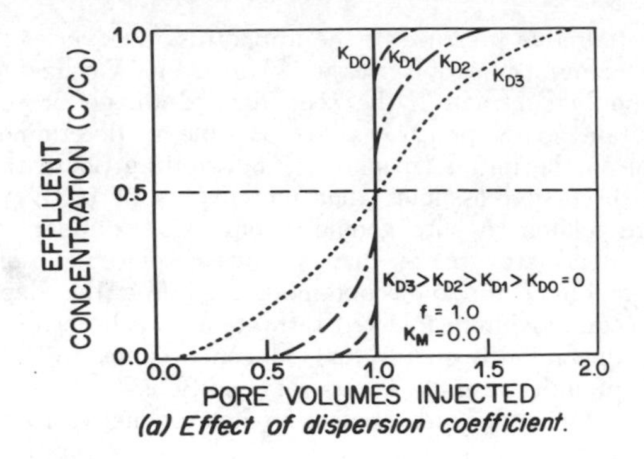

(a) Effect of dispersion coefficient.

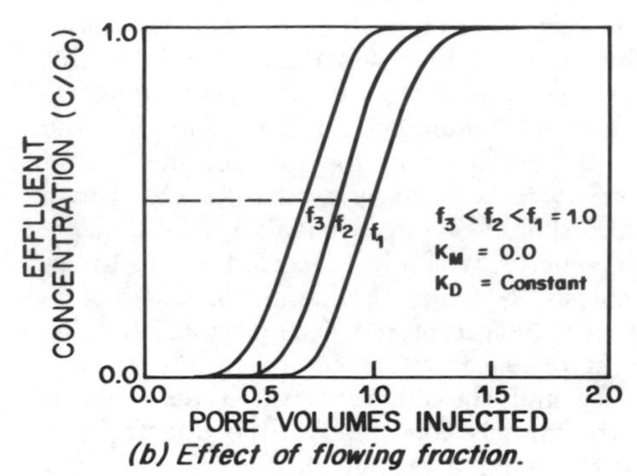

(b) Effect of flowing fraction.

Figure 2. Computed effects of dispersion coefficient, flowing fraction and mass transfer rate constants on miscible displacement [14].

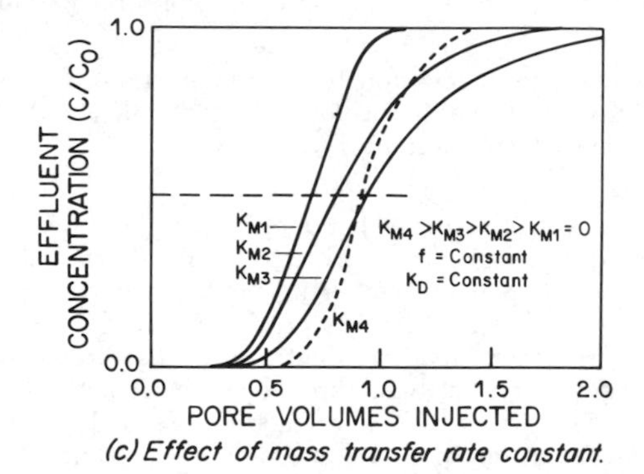

(c) Effect of mass transfer rate constant.

jective. However, this probably would occur only at high rates not appropriate to displacement conditions in an oil reservoir.

These parameters, besides depending on the pore geometry, are also affected by the fluid properties, the displacement rate and, when immiscible fluids are present, the fluid saturations and wetting behavior. An example of efficient gas-gas displacement in fairly homogeneous Berea sandstone is shown in Figure 3(a) [15]. As the velocity is increased, the effluent curve is sharper and the recovery more efficient. When water was present in the core at a 24.5% pore volume (PV) saturation, the resulting concentration profiles (not illustrated) were similar, and it could be concluded that the presence of a wetting phase did not alter the pore topology and geometry. The flowing fraction approached unity for this core and, as such, no stagnant portion was evident.

A series of heterogeneous carbonate cores displaying dual- and multiple-porosity types were subjected to a similar set of displacements [15]. Results obtained on car-

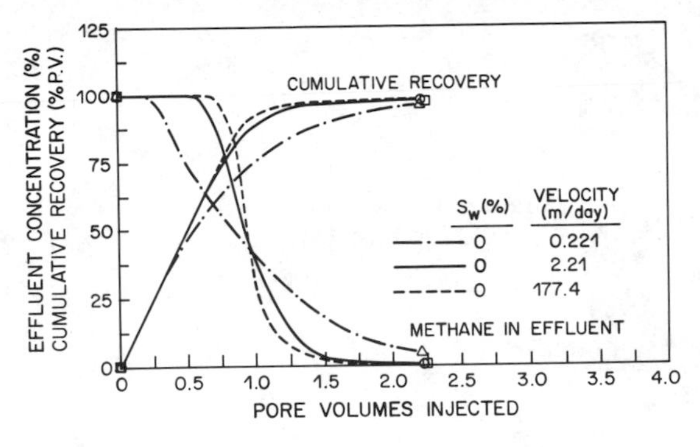

Figure 3(a). Miscible displacement of methane with nitrogen in a dry sandstone core [15].

bonate core having a fracture-matrix pore geometry are shown in Figure 3(b). In the dry core, i.e., at $S_w = 0$, the recovery became less efficient with increasing velocity, a behavior that was opposite to the Berea sandstone results. This result indicates that mass transfer, which is time dependent, is more important in the carbonate core and that the pores are more poorly interconnected than in the Berea sandstone. The flowing fraction was found to fall from 0.94 to 0.74 with increasing rate, which points out possible limitations in the dispersion-capacitance model. It also can be noted from Figure 3(b) that the presence of an immobile water saturation in the core significantly impaired recovery efficiency. The flowing fraction was found to be only 0.52 PV in this case.

The influence of pore geometry on flow behavior has yet to be studied in detail. Quantification of parameters is possible with the dispersion-capacitance or other similar model, and the results obtained relate to actual flow conditions. We believe that the wider use of such approaches for both single- and multiphase flow shows good potential for providing new and useful insight into macroscopic displacement phenomena of practical importance. Dispersion during two-phase flow has been examined by Raimondi and Torcaso [16], who studied the distribution of the oil phase during displacement by water. Recently, Delshad et al. [17] utilized tracers and multiphase flow interpretations of the dispersion equation to monitor their water and surfactant flooding experiments. Our understanding of flow in porous media could improve significantly through the wider application of such approaches.

Wettability

The wettability of porous media is important in a wide variety of practical situations. In petroleum recovery, it has become well recognized that wetting preferences displayed by oil and water for surfaces of the reservoir rock can strongly influence the

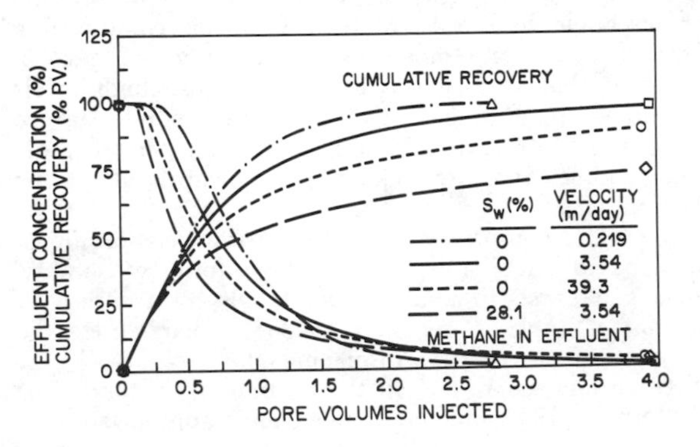

Figure 3(b). Miscible displacement of methane with nitrogen in a fractured carbonate core [15].

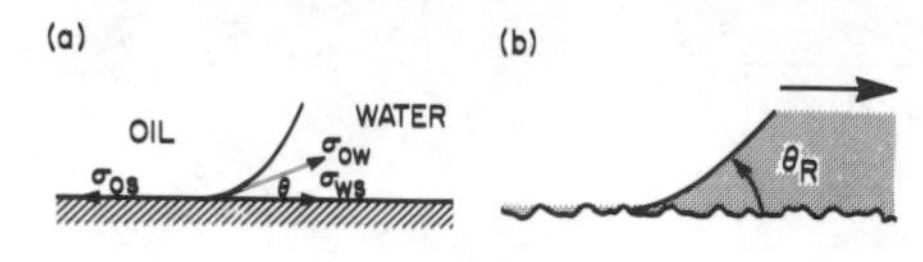

Figure 4. Contact angles for oil-water-solid systems.

oil recovery efficiency achieved in a waterflood. Reservoir wetting behavior is determined by the interaction of several factors. The surfaces that define the complex shapes of the pore space in reservoir rocks are generally heterogeneous in mineral composition and exhibit varying degrees of roughness. The wetting behavior of these surfaces is determined by solid-fluid interactions, which may be influenced strongly by trace amounts of polar compounds adsorbed on the rock.

Originally it was widely thought that most oil reservoirs are preferentially water-wet because water was always the phase initially contained in the porous medium. Over the past 15 years, however, considerable laboratory and field evidence has become available that shows that a preponderance of reservoirs have no distinct wetting preference for either oil or water. This led to use of the terms "intermediate" or "mixed" wettability.

The degree of wetting of a solid by a liquid-fluid pair can be defined by the contact angle exhibited on the solid. Over 170 years ago, Young proposed that the contact angle results from the mechanical equilibrium of three interfacial tensions, as illustrated in Figure 4(a) for a plane solid surface with oil and water. Equating the horizontal components of these tensions,

$$\sigma_{os} = \sigma_{ws} + \sigma_{ow} \cos \theta \tag{6}$$

where s, o and w denote solid, oil and water, respectively.
Young's equation has been shown to represent a condition of thermodynamic equilibrium at a three-phase boundary involving an incompressible solid and two immiscible liquids, or one liquid and a vapor phase [18].

In most practical situations, contact angles exhibit two limiting values rather than a single equilibrium one. Their magnitude depends on the direction of prior movement of the liquid-fluid or oil-water interface, as shown for the receding contact angle, θ_R, in Figure 4(b) and the advancing angle in Figure 4(c). Surface roughness [19] and phenomena associated with the adsorption of surface-active materials from solution [20] are principal causes of contact angle hysteresis. Morrow [19] has conducted a comprehensive study of the effect of surface roughness on contact angle. Figure 5, based on the work of Morrow, shows that surface roughness can cause high levels of hysteresis over most of the contact angle range. Liquids having smooth surface or intrinsic contact angles of up to about 50° can exhibit near-zero receding contact angles on rough surfaces, while roughening can cause contact angles of up to 20° to be reduced to zero under advancing conditions.

Morrow and McCaffery [21] prepared a comprehensive report on the relationship of contact angles to displacement behavior in porous media of uniform wettability. Treiber et al. [22] reported contact angle tests to assess the wettability of a variety of reservoirs. The water-advancing contact angle is generally taken to characterize the reservoir wettability for the most important case of water displacing oil.

Another factor influencing the wetting condition is the saturation history. The advancing contact angle data of Morrow [19] would be applied most appropriately to

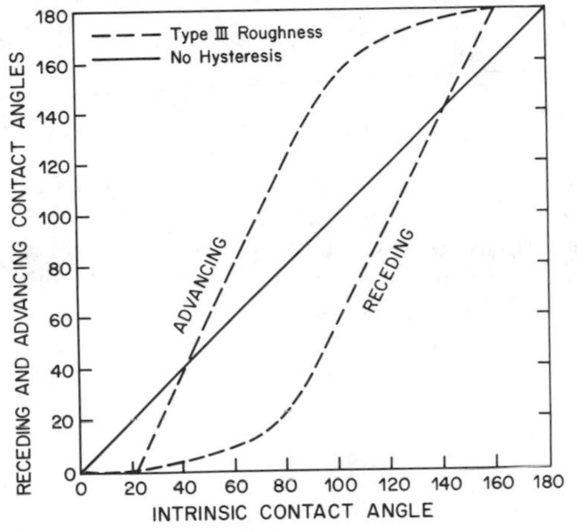

Figure 5. Effect of surface roughness on contact angle [19].

simple surface behavior, as in Figure 4(c), on an initial imbibition or exposure to a wetting fluid. If a roughened surface has been contacted previously by a wetting fluid and the movement of the three-phase line corresponds to a secondary imbibition, the resulting contact angle shown in Figure 4(d) is much lower. The actual contact angle will be between the advancing value suggested from Figure 5 and zero degrees due to trapped or immobile droplets of the advancing fluid that already reside in the microporosity of the surface. The formation of such a composite surface is affected by the relative sizes of the rock pores and the surface heterogeneities. Batycky [23] has discussed receding and advancing behavior for composite surfaces as a function of roughness, as well as the manner in which wetting phase contact angles are reduced compared to those measured on a smooth surface.

The rate and extent of imbibition are also commonly used to measure rock wetting preference [24]. In this context, imbibition, sometimes called spontaneous imbibition, refers to a process in which a wetting fluid replaces a nonwetting fluid in a porous medium by capillary forces alone, similar to an ink blotter soaking up ink and expelling air. The rate of imbibition depends on other factors besides wettability. These include permeability, pore geometry, fluid viscosities, saturation history and interfacial tension [5,21].

Amott [25] proposed combining spontaneous imbibition and forced displacement tests to determine wettability indices, based on ratios of the displacement volumes determined from the following four measurements: (1) spontaneous imbibition when water displaces oil, (2) forced displacement of water by oil, (3) spontaneous displacement of oil by water, and (4) forced displacement of oil by water. This method has been recommended by Raza et al. [26] and Craig [6]. The areas under capillary pressure drainage and imbibition curves also have been used to define a wettability scale [27]. Numerous other techniques of characterizing wettability have been suggested, most of which are referenced in the previously noted review articles.

Figure 6 [28] shows capillary pressure curves for systems of differing wettability. The schematic curves pictured in Figure 6(a), which generally relate to oil as the nonwetting phase and water as the wetting phase, are representative of a strongly and uniformly wetted rock. The features of these curves are similar to those for mercury-air shown in Figure 1, except that oil-water capillary pressure data are obtained using a restored-state method [4–6] or centrifugation [27]. The minimum saturation (S_{wi}) reached in Figure 6(a) generally corresponds to the irreducible wetting phase saturation, while the maximum wetting phase saturation reached at zero capillary pressure following imbibition defines the residual nonwetting phase saturation by difference.

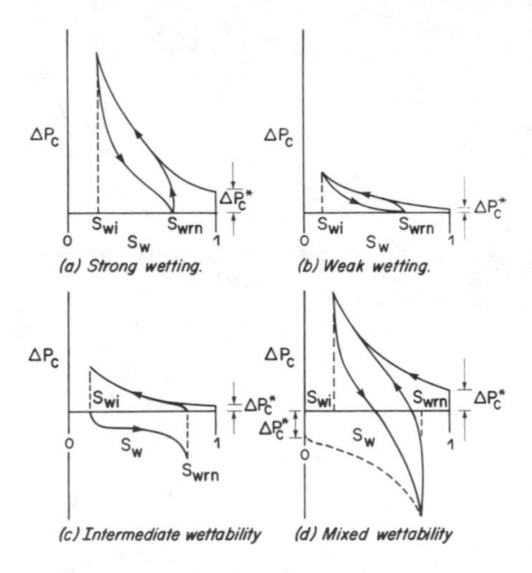

Figure 6. Capillary pressure characteristics for systems of differing wettability [28].

If the surface properties are such that wetting preferences are not as strong, but the wettability is still uniform in the rock, similar behavior occurs but at lower capillary pressures. This is illustrated for the weakly wetting case pictured in Figure 6(b). Here, the threshold value and other capillary pressures are lower than for the system represented in Figure 6(a), and endpoint saturations also may differ. The work of Morrow and McCaffery [21] should be examined for more definitive data on the effect of minor changes in wettability on capillary pressure-saturation relations.

From their detailed studies of the displacement characteristics of synthetic porous media having a well-defined and uniform wettability, Morrow and McCaffery [21] identified three distinct classes of wetting behavior that were described as wetted, intermediately wetted and nonwetted. Intermediate wettability, which covered a significant range of wetting situations, was characterized by displacements that were never spontaneous for either fluid phase. The fluid phase entering the porous medium always has a convex curvature corresponding to the nonwetting phase, with the large pores tending to fill first. Thus, all displacements correspond strictly to a drainage situation, with the displaced fluid always behaving as the wetting phase. The gross change in fluid distribution with increasing and decreasing saturation of a given phase causes significant hysteresis in capillary pressure (Figure 6(c)) for intermediately wetted systems.

The concept of uniform wettability is generally recognized as a simplification and idealization. Reservoir materials often are thought to possess various forms of mixed wettability, one possible example of which is pictured in Figure 6(d). As with the intermediate wettability system (Figure 6(c)), the subscripts w and n have been retained for descriptive purposes, even though definition of the wetting and nonwetting phases now depends on saturation and saturation history and the concept loses meaning.

Starting primary injection at $S_w = 1$ in Figure 6(d), a threshold capillary pressure again is indicated before entry of phase n can occur. On entry of phase n, the reduction in phase w saturation corresponding to increases in ΔP_c is similar to that dictated for a uniformly wetted system. However, after a minimum value of S_w has been attained and reinjection of phase w is initiated, there are differences in the resulting capillary pressure relation. As phase n is withdrawn and phase w saturation increased, some isolation of segments of phase n results and the capillary pressure falls. Once a value of $\Delta P_c =$ 0 has been obtained, further removal of phase n occurs because there still is continuity within the phase, and ΔP_c becomes increasingly negative. This means that according to Equation 3, surfaces that were concave toward phase n for $\Delta P_c > 0$ have become

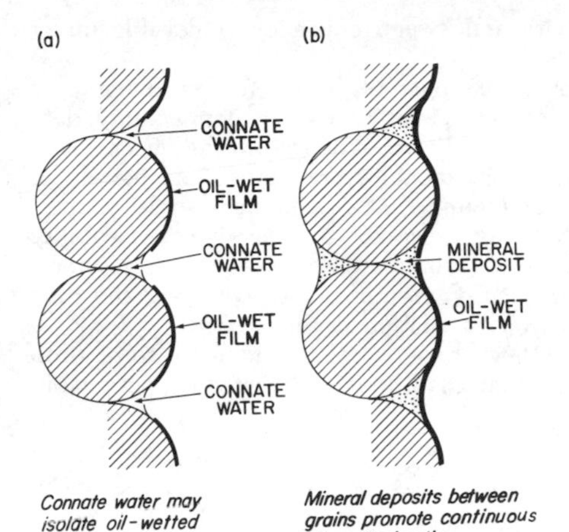

Figure 7. Conceptual representation of mixed-wettability systems [29].

convex with the injection and increasing saturation of phase w. Another injection of phase n and withdrawal of phase w can lead to the hysteresis pictured.

In theory, the process could have been initiated by first injecting phase w into a phase n saturated rock after overcoming a threshold capillary pressure to phase w. The dotted line pictured in Figure 6(d) would result. If, rather than displaying an entirely heterogeneous or "salt-and-pepper" type of mixed wettability, which might correspond to high levels of capillary pressure, there exist generally continuous wetting paths for each phase through most of the pores, then both the threshold and maximum pressures will be very low. As the likelihood of isolation of elements of each fluid then would be reduced, low values of S_{wi} and high values of S_{wrn} also are likely to occur.

Salathiel [29] has presented the principal discussion of mixed-wettability oil reservoir systems that provide high recovery efficiencies when gravity drainage or segregation are effective. Figure 7(a) shows oil-wetted surfaces separated by pendular rings of water initially in the reservoir. The situation depicted in Figure 7(b), on the other hand, allows for development of a continuous path of oil-wetting that causes an oil permeability to persist to low oil saturations. Salathiel postulates that the flow of oil occurs by surface drainage or rivulets over strongly oil-wetted pore surfaces, forming continuous oil-wet paths extending through the pore structure. Field performance studies and reservoir core test results provide increasing evidence for the occurrence of oil reservoirs with such mixed wettability characteristics.

Reservoir wettability is now recognized as an important factor in regulating immiscible displacement behavior and the amount and distribution of residual hydrocarbon. Initial and transient wetting characteristics of reservoirs also can influence the success of enhanced oil recovery processes. There are a number of possible causes for certain oil reservoirs to be other than water-wet and, as a rule, the wettability of a given reservoir is extremely hard to determine. Nevertheless, ample evidence of nonwater-wetness in oil reservoirs has been presented in the literature [22,29].

Relative Permeability

The description that has evolved of the flow of two or more immiscible fluid phases within porous media is based on applying Darcy's law, or the concept of permeability, to each phase separately. The analytical description of reservoir performance requires the knowledge of relative permeabilities as one of the variables in the equations of fluid motion. In most cases, the assignment of relative permeability-saturation characteristics for a particular system essentially will fix the answers that are determined from a reservoir model study. Relative permeability information finds direct use in the prediction of the oil recovery, fluid injectivity and sweep efficiency achievable under different fluid injection schemes.

Considerable effort has been devoted over the past 30 years to improving core handling and laboratory testing procedures so that reliable relative permeability data might be obtained on reservoir rock samples. In addition, much study has been given to theoretical approaches for predicting relative permeability relations. The petroleum literature reveals many diverse opinions about factors that can influence relative permeability and about the best method of obtaining representative relative permeability data. Even after reliable results are obtained on individual core samples, reservoir engineers are faced with the often uncertain task of grouping and averaging the data before the data can be used as input for reservoir simulation studies.

During two-phase flow, the flowrate of either phase is related to a pressure drop gradient by means of Darcy's law, as given in Equations 1 and 2. The effective permeability, k_{ei}, of a given phase is defined as

$$k_{ei} = K \cdot k_{ri} \tag{7}$$

where subscript i can denote either the wetting (w) or nonwetting (n) phase. Effective permeability, k_e, is defined as the permeability to a fluid when the porous medium is only partially saturated with that fluid; relative permeability, k_r, is the ratio of the effective permeability to some base value, such as the absolute permeability, K. The sum of the effective permeabilities is always less than the absolute or single-phase permeability due to mutual interference between fluids flowing through an interconnected pore system.

The effective permeabilities to each phase depend on the saturation and saturation history of the phases. During single-phase flow, when only one mobile phase is present, the effective permeability to that phase is a maximum and the flowrate of the injected phase is equivalent to the total flowrate. The effective permeability to the immobile phase is zero so that its flowrate is also zero.

At any point when two phases flow simultaneously and, hence, the flowrates of the phases obey Equations 1 and 2, a difference in pressure between the phases usually exists. This capillary pressure is determined by local curvatures of the fluid-fluid interfaces. When fluids flow within a core and saturation changes occur along it, as in an unsteady-state displacement, the capillary pressure and effective permeabilities, or Darcy's law relationships, apply simultaneously.

As with capillary pressure, the relative permeability properties of porous media reflect the composite effects of pore geometry, wettability, fluid distribution and saturation history [6]. Morgan and Gordon [30] have demonstrated the influence of pore geometry on the water-oil relative permeability properties of water-wet sandstones. Detailed attention also has been given to the effect of wettability and saturation history on the two-phase relative permeability properties of fairly homogeneous, intergranular porous media [31,32]. Less effort has been devoted to the complex problem of obtaining valid relative permeabilities for heterogeneous porous media, such as that containing appreciable amounts of solution porosity and fractures [33].

Both steady- and unsteady-state methods are available for the measurement of relative permeability-saturation characteristics of core samples. Loomis and Crowell [34] have presented a comprehensive report on a variety of test methods. The usual two-phase steady-state method involves the coinjection of immiscible fluids into one end of a cylindrical core sample at constant flowrates. The pressure drop across the sample obtained under steady-state conditions is used to define the relative permeability of each phase for the independently determined saturation of the core. By varying the injection ratio of the fluids in a stepwise fashion, a relative permeability-saturation relationship is determined for the system that applies to the particular saturation and saturation history of the core.

An alternative, and more common, procedure is to derive relative permeability-saturation properties from unsteady-state displacement test results [6,35]. In the usual test, water is injected into an oil-containing core while monitoring the pressure drop and oil recovery response. This data then are used for the calculation of relative permeability properties.

In the past, most core displacement tests of the type outlined have been performed at flowrates greatly exceeding those encountered in an oil reservoir. The objective was to negate the effect of capillary forces on the flooding behavior of the short core samples at slow displacement rates, which could cause a spreading of the displacement front and interference from the well known capillary end effect, which gave a buildup of the wetting phase at the end of the core.

It is being recognized increasingly that high flooding velocities in core tests can give oil recovery and relative permeability information that is nonrepresentative of reservoir behavior. Recent work of Batycky et al. [28] has concentrated on the interpretation of unsteady-state tests performed at reservoir rates, of the order of one foot per day. At low displacement rates, it is possible to reduce potential problems due to fines migration or emulsion formation, as well as to permit a realistic representation of time-dependent interfacial or wettability phenomena that can occur during water flooding or chemical flooding. In an extension of previous work by the Petroleum Recovery Institute [33], measured recovery and pressure drop behavior are matched using a least-squares procedure to derive relative permeability relations that give an optimum fit of the data. This approach, which takes capillary forces into account, also has been utilized to provide information on the wetting and capillary pressure characteristics of the system.

There is a voluminous amount of literature on how best to handle and test reservoir cores to preserve or restore their native wettability condition. The reason for this interest is illustrated in Figures 8(a) and 8(b) [31], which show the effect of wettability on oil-water relative permeability and displacement efficiency. Figure 8(b) shows that the highest oil recovery efficiency at water breakthrough is characteristic of the most water-wet system, having a zero degree contact angle. However, very oil-wet systems may provide the greatest oil recovery at high volumes of water throughput [29]. Figure 8(a) also indicates extension of relative permeability data to high water saturations (high oil recovery levels) for the most oil-wet case. However, gravity drainage and segregation effects have to be significant in an oil-wet reservoir system for the high oil recovery levels to be realized in practice.

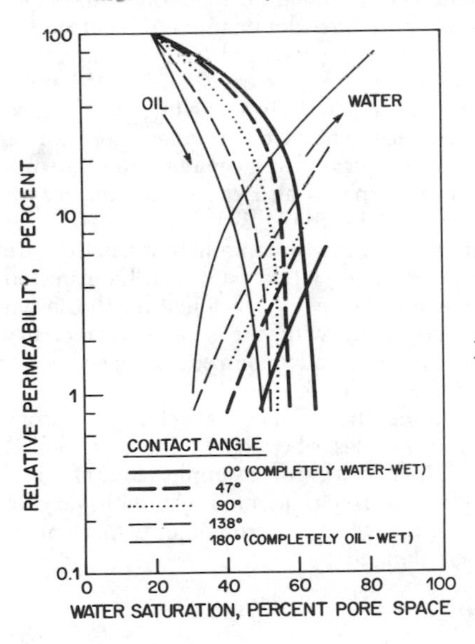

Figure 8(a). Relative permeabilities for the range of wetting conditions [31].

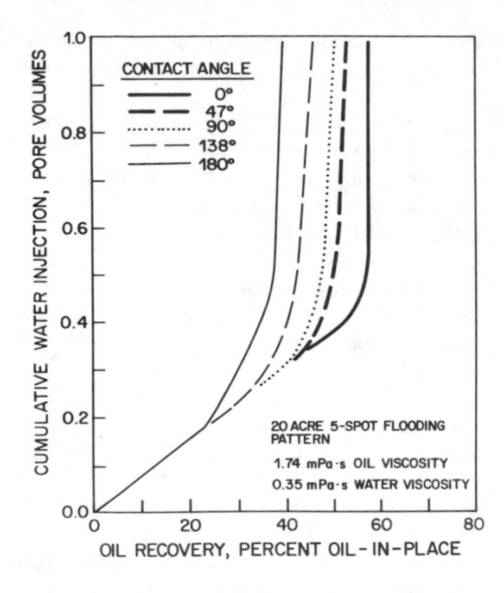

Figure 8(b). Effect of wettability on waterflood performance calculations [31].

RESIDUAL SATURATION

The residual saturation of principal concern is the pore volume fraction of oil left trapped in the porous medium following displacement by water. The question of predicting the amounts of residual oil following waterflooding in a reservoir has received much study because of its obvious economic importance. Past work in this area suggests that the role of wettability in the ultimate oil recovery is still poorly understood.

For water-wet systems, there is a fairly good understanding of the roles of pore geometry and saturation history on residual oil saturations, which is consistent with the mercury porosimetry behavior described previously. In strongly water-wet systems, the residual oil level obtained by water displacement is similar to that following spontaneous water imbibition because the capillary forces are dominant and control the recovery. Perhaps the best interpretation of the process is presented by an examination of the "doublet model," most recently discussed by Stegemeier [36]. The doublet model indicates that the capillary forces existing across each oil-water interface, as expressed by Equation 5, overwhelmingly favor displacement in the smallest pores. This leads to bypassing and trapping in the larger ones unless displacement rates are unrealistically large. Snapoff of oil droplets within pore spaces also can occur, the extent depending on pore body-throat contrasts, as described by Roof [37].

Consequently, as pore body-pore throat contrasts and microscopic heterogeneities are increased, recovery efficiencies during linear displacements within cores can be expected to decline. Figure 9 [36] indicates the dependence of recovery efficiency on the initial oil saturation for a variety of rock types. In agreement with the schematic mercury capillary pressure curves of Figure 1, the higher the initial oil (nonwetting phase) saturation, the greater the subsequent residual oil saturation level.

Residual oil saturations can be reduced by altering the balance between the viscous displacing forces and the capillary trapping forces, as described by Moore and Slobod [38], Taber et al. [39] and Melrose and Brandner [40], for example. Melrose and Brandner have provided a set of correlations (Figure 10) that, for rocks with varying levels of heterogeneity, give the fractional oil recovery (microscopic displacement efficiency) as a function of the capillary number, defined by

$$N_{cap} = \frac{V\mu}{\phi\sigma} \tag{8}$$

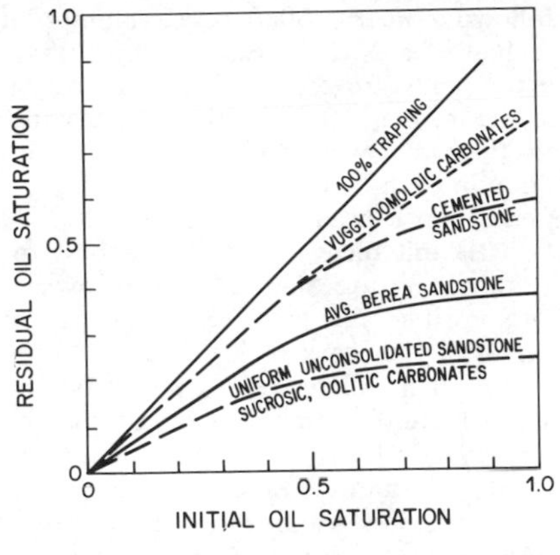

Figure 9. Effect of initial oil saturation on residual oil for different types of porous media [36].

where V is the superficial velocity (flowrate/area, m/sec), μ is the displacing fluid viscosity (mPa-sec), ϕ is the porosity (fraction), and σ is the oil-water interfacial tension (mN/m). Also included in Figure 10 is a curve summarizing similar data obtained by Batycky and McCaffery [41] for a sandpack with a narrow pore size distribution.

At normal reservoir rates and for water displacing oil, recovery efficiencies correspond to the data at the left side of Figure 10. As mentioned earlier, residual oil levels result from differing drive-water velocities through the interconnected pore paths that lead to isolation and trapping of the oil. Reduction of the residuals then requires increases in the capillary number by several orders of magnitude. Since it is impractical to significantly alter reservoir flow velocities or drive-water viscosity, the only parameter that can be manipulated reasonably is the interfacial tension which, for a typical sandstone geometry, must be reduced to $1 \mu N/m$ or lower. This reduction can be achieved either by approaching miscibility conditions with carbon dioxide or hydrocarbon gases [42] or through the use of surfactant materials [43,44].

At the other extreme of wetting, i.e., strongly oil-wet behavior, the recovery during water injection is characterized by early water breakthrough. With continued water injection, the oil remains connected and coats the pore linings as illustrated in Figure 7(b). The oil is still exposed to the hydrostatic pressure drop due to the water flow, so that high water cuts with persistent oil production occur. In some cases, the wetting oil phase residuals can be substantially lower than typical nonwetting residuals. This concept seems to be supported by data of Salathiel [29], who obtained

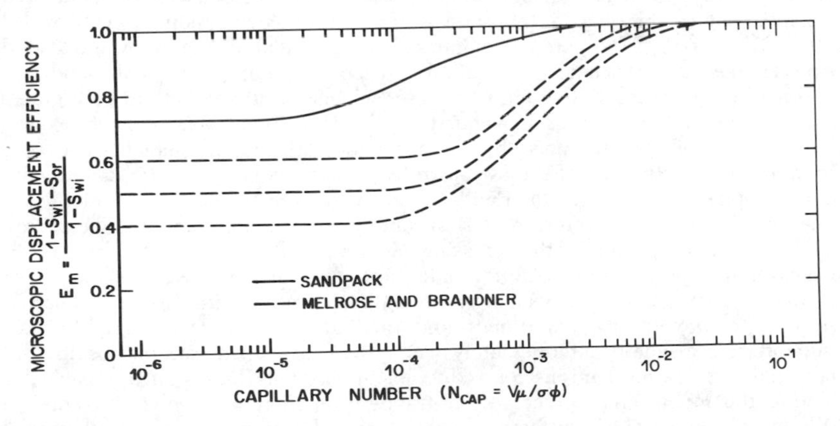

Figure 10. Relationships between microscopic displacement efficiency and capillary number [40,41].

residual oil saturations of less than 10% PV following nearly 5000 pore volumes of water injection, and also by Slobod [45], who achieved a 15% residual wetting phase level after the injection of 3000 pore volumes of the nonwetting phase. An examination of the work of Owens and Archer [31] (also see Figure 8(a)), shows that the residual oil saturation was 18% for the oil-wet case, while the water-wet residual was 32%, following more than 50 pore volumes of water injection in each case.

Contrasting these results are the waterflood performance predictions (Figure 8(b)), which show that for one pore volume of injection the oil recovery would be best for the water-wet system. Where it is reported in the literature that oil-wet recoveries are poorer than those for water-wet systems, it is possible that insufficient volumes of water were injected. Examples of this are contained in the work of Lefebvre du Prey [46], who stated that four pore volumes were injected in all tests in a study on the effects of wettability on recovery. Even though the highest ultimate recovery may be attainable for an oil-wet system, the high recovery levels may not be reached in practice because of economic limits on the producing water to oil ratio.

The foregoing discussion was necessary to better assess various data concerning the effects of other wettability conditions on residual oil saturations. In many cases, residual levels are higher for intermediately wet systems than for water-wet systems. This is supported in the work of Owens and Archer [31] (Figure 8(a)), Lefebvre du Prey [46] and Singhal et al. [47]. Lefebvre du Prey used uniformly wetted sintered media composed of alumina, stainless steel or TFE plastics and various immiscible fluid-fluid pairs. Singhal et al. used unconsolidated mixtures of TFE and glass beads of varying compositions, thus simulating "salt-and-pepper" or heterogeneous, wettability. Owens and Archer established different levels of wettability in consolidated sandstone, supported by contact angle measurements, by varying the concentration of a surfactant in the oil phase.

Salathiel [29], on the other hand, showed that the residuals he obtained with what was described as a mixed-wettability system were lower than for a water-wet system. Mixed wettability was achieved by contacting sandstone cores with a selected East Texas crude oil, which would deposit a film on the rock surfaces while at a connate (initial) water saturation before being displaced by a mineral oil. Mixed wettability was described to be a condition in which continuous wetting paths existed for each phase in all pore spaces contacted by both phases. Whether these results would reflect an intermediately wet or an oil-wet condition, the influence of geometry and oil-film formation pictured in Figure 7(b) gives credence to the following interpretation of Batycky [48], which extends Salathiels' concepts.

If, as shown in Figure 11, the pore geometry is such that any contact angle movement of the trapped droplet of oil can be accomplished by the continuous movement of the three-phase contact lines along the solid surface, then low residual levels will be achieved. By contrast, if the porous medium is less consolidated, then movement of the three-phase lines is restricted. For example, when examining the small sand grain at the rear of the ganglion in the left and center schematics at the bottom of Figure 11 it may be seen that continuous movement of the three-phase line cannot occur. A condition of instability determined by the maximum advancing contact angle resists further movement in the force field, as illustrated in the center schematic. One of three things can happen: (1) the ganglion will be detached from the sand grain at the rear, (2) the ganglion can be severed, or (3) it simply can remain trapped. The increase in mobilization requirements with increasing oil wetness for a packing of spheres has been measured experimentally and the adhesion phenomenon observed [48]. From these interpretations, it can be seen why ultimate recoveries with oil-wet systems are better than those with water-wet ones and qualitatively under what conditions of wetting and pore geometry intermediately wet systems can yield variable results.

Understanding the mechanisms for reduction in residuals has been approached by considering the mobilization process of a trapped ganglion to involve movement of only the leading and trailing surfaces of a geometrically complex trapped oil ganglion [9,40]. A force balance using Laplace's equation applied to each surface can predict the critical pressure drop necessary to overcome the capillary trapping forces, so as to

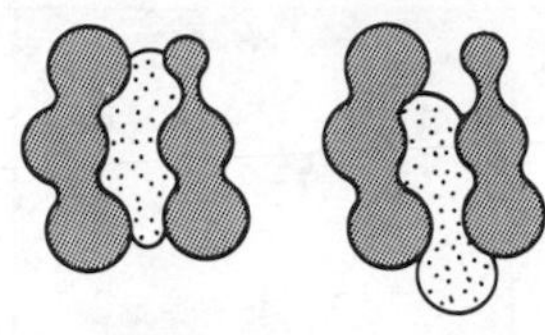

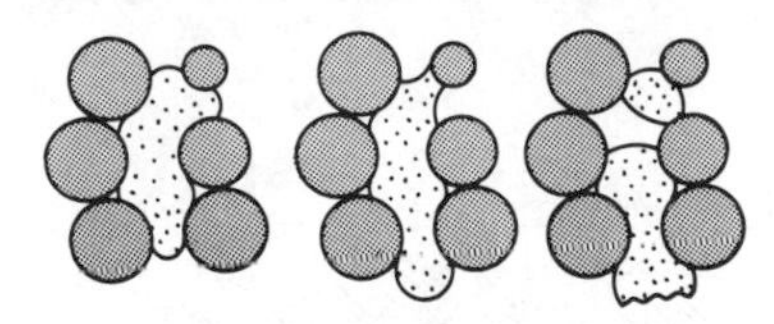

Figure 11. Mechanisms of residual oil mobilization [48].

attain mobilization [40]. If the leading edge is of minimum radius, r_c, and the trailing edge is a maximum size, r_t, then for a spherical system

$$\Delta P_{crit} = 2\sigma\left(\frac{1}{r_c} - \frac{1}{r_t}\right) \tag{9}$$

where ΔP_{crit} is applied by gravity and viscous forces. The role of interfacial tension reduction in reducing the value of ΔP_{crit} is clearly indicated.

Unfortunately, Equation 9 can be applied rigorously to water-wet systems only because as contact angles increase the ganglia begin to wet portions of the solid surfaces. Depending on the geometry, a detachment of a ganglion will add to the value of ΔP_{crit}. In an experimental study [23,48] using an idealized sphere pack, it was found that ΔP_{crit} increased with contact angle in a manner inconsistent with the roughness concepts of Morrow [19]. Smooth surface mobilization requirements were found to be higher than those for roughened spheres due to the formation of composite surfaces, which lowered contact angle. Also, the observation of detachment preceding mobilization indicated the mechanics of the process to be similar to those in Figure 4(b), where the force balance used in arriving at Young's equation needs to be evaluated normal to the surface.

In his comprehensive displacement study with uniformly wetted systems. Lefebvre du Prey [46], examined the reductions of residuals over the entire wetting range by varying the capillary number. While he determined that wetting phase residuals were higher than those of nonwetting phase residuals, as mentioned previously, he also found, consistent with other studies, that residuals in the intermediately wet system fell between both extremes. For all the systems examined, it was shown that the residual levels were reduced as capillary number was increased. Thus, reductions in interfacial tensions tend to reduce residuals or improve recovery levels for all wetting conditions.

SWEEP EFFICIENCY

The immiscible displacement efficiency achieved on a macroscopic scale has an important bearing on oil recovery levels, in addition to the microscopic displacement behavior. In waterflooding, unrecovered oil will occur as residual oil in waterswept regions of the reservoir and in parts of the reservoir that have been completely bypassed by the injected water phase. The greatest proportion of the unrecovered oil often resides in the unflushed zones [49]. The appropriately named volumetric sweep efficiency

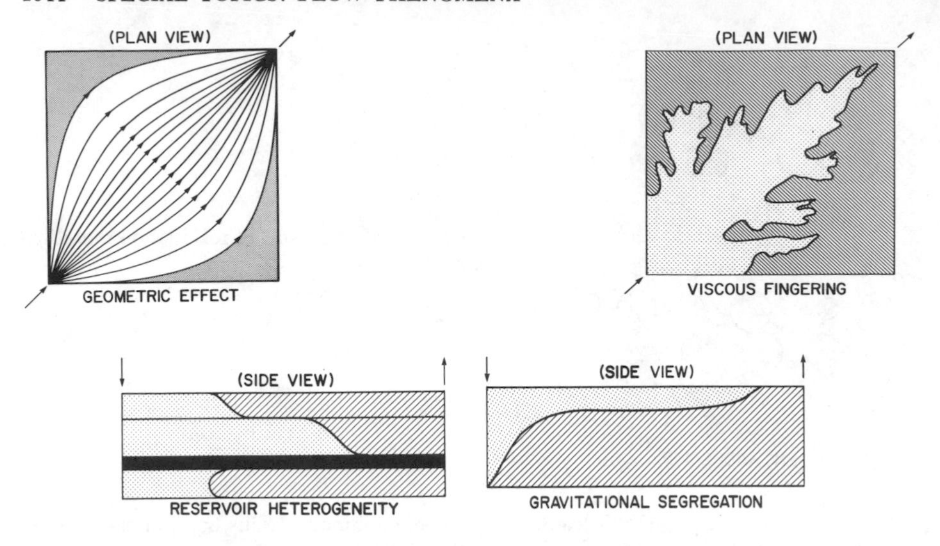

Figure 12. Problems in effective reservoir contact [50].

or conformance is commonly separated into the consideration of both areal and vertical sweep efficiencies.

Oil reservoirs generally must be flooded in a horizontal fashion using injection and production wells arranged in a pattern. A variety of conditions hinder injection fluids from effectively contacting the entire reservoir, as illustrated in Figure 12 from a recent paper by Welch [50]. The areal sweep efficiency is at its optimum when the mobility (generally expressed as the endpoint relative permeability divided by viscosity) of the displacing phase is less than, or equal to, that of the phase being displaced. The upper left-hand side of Figure 12 shows how the geometry of a displacement pattern will influence the amount of contact that can be obtained, even for favorable mobility ratios. The lines within the swept zone depict the paths of the flowing fluids, with velocities that are inversely related to their lengths. It is easy to realize that injected water would break through at the producing well before a complete sweep of the pattern is achieved.

When the ratio of the mobilities of the displacing to the displaced phases is greater than one, the displacement can be unstable and rate dependent. In this case, "fingers" of the less viscous injected fluid develop that lead to its early breakthrough with a resulting poor sweep efficiency (Figure 12). The onset of instability during two-phase immiscible displacement in homogeneous porous media has been studied recently by Peters and Flock [51], who report on an extension of the stability theory developed by Chuoke et al. [52]. Gas displacing oil and water displacing highly viscous oil in a horizontal flooding mode are situations in which severe bypassing due to viscous fingering would seriously affect the recovery efficiency, even for ideally homogeneous reservoirs. Peters and Flock present a dimensionless number and its critical value for predicting the onset of instability during immiscible displacement in porous media. Further experiments are needed to test the general applicability of this number (which includes terms for viscous and interfacial forces) to a variety of porous media and different sized systems.

Reservoir heterogeneity and density differences between the oil and displacing fluid also can lead to poor sweep efficiencies. Examples of poor vertical sweep efficiency due to these factors are shown schematically at the bottom of Figure 12. A significant proportion of the world's hydrocarbon reserves is contained in highly heterogeneous reservoirs containing natural fractures. As would be expected, extremely poor sweep efficiencies can characterize such dual-porosity systems if the recovery process depends on horizontal displacement.

The understanding of reservoir heterogeneities occurring on a macroscopic scale and the successful prediction of their effects on oil recovery behavior continue to be principal challenges for reservoir engineers. Sophisticated logging tools, pressure interference test methods and tracer studies now are used in conjunction with computer simulation of the reservoir behavior to better describe the system and to identify optimum recovery strategies in many of our more geologically complex reservoirs.

DIRECTIONS OF RESEARCH

Multiphase fluid flow in porous media is of great economic importance. Basic investigations of the subject as it relates to oil recovery began in the 1930s, and soil scientists were studying related questions much earlier. The rapid escalation of the value of oil and natural gas over the last several years has made them even more important commodities to find and optimally produce. Fundamental to the latter activity is further research into how the hydrocarbons traverse and are best transported through porous media to points of production.

The evolving commercialization of various enhanced oil recovery methods will continue to provide the principal incentive for extensive new experimental and theoretical research on miscible and immiscible fluid flow through porous media. Enhanced oil recovery methods usually are taken to mean those processes used to recover additional oil compared to standard waterflooding in light oil reservoirs, as well as recovery procedures such as thermal methods, as may be applied at any time in the life of heavy oil reservoirs. Along with the natural geology of a reservoir, three forces—viscous, gravitational and capillary—influence the flow of fluids through porous media and, therefore, oil production rates and recovery levels. The control or modification of these forces in a beneficial manner by the introduction of heat, chemicals (surfactants) or miscible gases (solvents) forms the basis of most enhanced oil recovery methods. A few billion dollars probably have been spent already on basic research and pilot testing of various enhanced oil recovery processes. Doscher [53], in a recent review of enhanced oil recovery, estimates past expenditures of at least $250–$500 million on the research and development of chemical flooding processes.

The following paragraphs provide a brief additional comment on the direction of some ongoing research that is believed to be important to improving our understanding of the flow of immiscible fluids in porous media. As noted earlier, our current macroscopic description of multiphase flow is based on assigning relative permeability and capillary pressure characteristics to the system. These properties depend on saturation history, as well as on the absolute value of saturation. The nature and causes of hysteresis in relative permeability relations have yet to be fully understood, despite years of theoretical study and experimentation. The accuracy of performance predictions for many enhanced oil recovery processes depends strongly on a good knowledge of hysteresis in the flow properties. Two examples are steam stimulation and tertiary oil recovery by carbon dioxide flooding. In each case, the porous medium can encounter both increasing and decreasing oil saturations over time, for which drainage and imbibition conditions apply, respectively.

In our opinion, more extensive use of numerical simulation of laboratory core test results [28,33] can help to better define and understand complete hysteresis effects in two- and three-phase immiscible flow situations. The saturation-pulse phenomena observed by Gladfelter and Gupta [54] during steady-state relative permeability tests also should be examined in greater detail because its occurrence raises some questions about the validity of this relative permeability measurement method.

The industry as a whole is tending to conduct more immiscible displacement tests at near-reservoir rates, and with native reservoir core and fluids. This approach recognizes the need to keep capillary, as well as viscous, forces realistically in play for the laboratory situation. Such is often required for valid flow property results, particularly when the rock system does not exhibit a strong wetting preference for either the water or oil phases. There also is an increasing awareness by reservoir engineers and their supporting researchers that rock mineralogy and brine chemistry can have an important bearing on flow behavior. The chemical makeup of typical reservoir oil-brine-rock systems is extremely complex. Disruptions to the chemical equilibrium due to the injection of brine of a different chemistry, for example, can cause significant permeability reductions and wettability changes in reservoir rock. For this reason alone, it is now considered extremely important to conduct tests at simulated reservoir conditions of temperature and pressure, along with the actual fluids whenever possible. For unconsolidated sand systems, there is an additional problem of preserving the in situ particle arrangement and being able to restore reservoir stress conditions for accurate laboratory displacement testing.

Another important area of research that is receiving renewed attention deals with the complex problem of scaling enhanced oil recovery processes in the laboratory. Both physical and numerical modeling approaches are utilized, and both have certain acknowledged limitations. Parsons and Jones [55] have suggested that multiple-fluid injection processes such as micellar-polymer flooding can be linearly scaled to a good approximation. Doscher and El-Arabi [56] recently reported on scaled physical model studies of oil recovery by carbon dioxide injection. These tests revealed the domination of the high mobility of the carbon dioxide on the recovery process and suggested that an optimum slug size exists for achieving the most efficient recovery.

The abovementioned current research activities are but a few involving multiphase fluid flow in porous media. Petroleum recovery is very much an activity requiring multidisciplinary lines of attack. Engineers and earth scientists having a variety of backgrounds, along with chemists, physicists and mathematicians, all have contributed to the art and science of petroleum recovery, an activity that will continue to receive much attention in the foreseeable future.

REFERENCES

1. Bear, J. *Dynamics of Fluids in Porous Media,* (New York: Elsevier North-Holland, Inc., 1972).
2. Slattery, J. C. *Momentum, Energy, and Mass Transfer in Continua,* (New York: McGraw-Hill, 1972).
3. Muskat, M. *Physical Principles of Oil Production* (New York: McGraw-Hill Book Co., 1949).
4. Collins, R. E. *Flow of Fluids through Porous Materials* (New York: Van Nostrand Reinhold Co., 1961).
5. Richardson, J. G. "Flow Through Porous Media," in *Handbook of Fluid Dynamics,* V. L. Streeter, Ed. (New York: McGraw-Hill Book Co., 1961).
6. Craig, F. F., Jr. *The Reservoir Engineering Aspects of Waterflooding,* Monograph Series 3, Society of Petroleum Engineers of AIME, Dallas, TX (1971).
7. Wardlaw, N. C., and R. P. Taylor. *Bull. Can. Pet. Geol.* 24:2 (June 1976).
8. Morrow, N. R. *J. Can. Pet. Technol.* 18:3 (July–September 1979).
9. Batycky, J. P., and A. K. Singhal. *Research Report RR-35,* Petroleum Recovery Institute, Calgary (1977).
10. Ng. K. M., and A. C. Payatakes. *AIChE J.* 26:3 (May 1980).
11. Payatakes, A. C., K. M. Ng and R. W. Flumerfelt. *AIChE J.* 26:3 (May 1980).
12. Wardlaw, N. C., and J. P. Cassan. *Bull. Can. Pet. Geol.* 27:2 (June 1979).
13. Coats, K. H., and B. D. Smith. *Soc. Pet. Eng. J.* 8:73 (1964).
14. Batycky, J. P., B. B. Maini and D. B. Fisher. Paper SPE 9233, presented at 55th Annual Fall Meeting of SPE, Dallas, TX, September, 1980.
15. Batycky, J. P., M. I. Mirkin, G. J. Besserer and C. H. Jackson. *Can. J. Pet. Technol.* 104 (January–March 1981).
16. Raimondi, P., and M. A. Torcaso. *Soc. Pet. Eng. J.* 8:49 (1964).

17. Delshad, M., D. G. MacAlister, G. A. Pope and B. A. Rouse. Paper SPE 10201, presented at 56th Annual Fall Meeting of SPE, San Antonio, TX, October 1981.
18. Johnson, R. E. *J. Phys. Chem.* 63:1655 (1959).
19. Morrow, N. R. *J. Can. Pet. Technol.* 14:42 (October–December, 1975).
20. McCaffery, F. G., and N. Mungan. *J. Can. Pet. Technol.* 9:185 (1970).
21. Morrow, N. R., and F. G. McCaffery. "Displacement Studies in Uniformly Wetted Porous Media," in *Wetting, Spreading and Adhesion*, J. F. Padday, Ed. (New York: Academic Press, Inc., 1978).
22. Treiber, L. E., D. L. Archer and W. W. Owens. *Soc. Pet. Eng. J.* 12:531 (1972).
23. Batycky, J. P. In: *Surface Phenomena in Enhanced Oil Recovery*, D. O. Shah, Ed. (New York: Plenum Publishing Corp., 1981).
24. Bobek, J. E., C. C. Mattax and M. O. Denekas. *Trans. AIME* 213:155 (1958).
25. Amott, E. *Trans. AIME* 216:156 (1959).
26. Raza, S. H., L. E. Treiber and D. L. Archer. *Prod. Monthly* 32(4):2 (1968).
27. Donaldson, E. C., R. D. Thomas and P. B. Lorenz. *Soc. Pet. Eng. J.* 9:13 (1969).
28. Batycky, J. P., F. G. McCaffery, P. K. Hodgins and D. B. Fisher. *Soc. Pet. Eng. J.* 21:296 (1981).
29. Salathiel, R. A. *J. Pet. Technol.* 25:1216 (1973).
30. Morgan, J. T., and G. T. Gordon. *J. Pet. Technol.* 1199 (October 1970).
31. Owens, W. W., and D. L. Archer. *J. Pet. Technol.* 873 (July 1971).
32. McCaffery, F. G., and D. W. Bennion. *J. Can. Pet. Technol.* 42 (October–December 1974).
33. Sigmund, P. M., and F. G. McCaffery. *Soc. Pet. Eng. J.* 19:15 (1979).
34. Loomis, A. G., and D. C. Crowell. *Bull. 599*, U.S. Bureau of Mines (1962).
35. Jones, S. C., and W. O. Roszelle. *J. Pet. Technol.* 30:807 (1978).
36. Stegemeier, G. L. In: *Improved Oil Recovery by Surfactant and Polymer Flooding*, D. O. Shah and R. S. Schechter, Eds. (New York: Academic Press, Inc., 1977).
37. Roof, J. G. *Soc. Pet. Eng. J.* 10 (March 1970).
38. Moore, T. F., and R. L. Slobod. *Prod. Monthly* 20:10 (August 1956).
39. Taber, J. J., J. C. Kirby and F. V. Schroeder. Preprint of 71st National Meeting of AIChE, Dallas, TX (1972).
40. Melrose, J. C., and C. F. Brandner. *J. Can. Pet. Technol.* 13:4 (October–December 1974).
41. Batycky, J. P., and F. G. McCaffery. Paper No. 78-29-26, presented at 29th Annual Technical Meeting of Petroleum Society of CIM, Calgary, Alberta, June 13–16, 1978.
42. Rosman, A., and E. Zana. Paper No. SPE 6723, presented at 52nd Annual Fall Technical Conference and Exhibition of SPE of AIME, Denver, CO, October 9–12, 1977.
43. Morgan, J. C., R. S. Schechter and W. H. Wade. In: *Improved Oil Recovery by Surfactant and Polymer Flooding*, D. O. Shah and R. S. Schechter, Eds. (New York: Academic Press, Inc., 1977).
44. Reed, R. L., and R. N. Healy. In: *Improved Oil Recovery by Surfactant and Polymer Flooding*, D. O. Shah and R. S. Schechter, Eds. (New York: Academic Press, Inc., 1977).
45. Slobod, R. L. *Pet. Trans. AIME* 189 (1950).
46. Lefebvre du Prey, E. J. *Soc. Pet. Eng. J.* 13:1 (February 1973).
47. Singhal, A. K., D. P. Mukherjee and W. H. Somerton. Preprint of 26th Annual Meeting of Petroleum Society of CIM, Banff, Alberta (1975).
48. Batycky, J. P. *Proc. I.E.A. Workshop on Enhanced Oil Recovery*, Bartlesville, OK (1980).
49. Richardson, J. G. *Oil Gas J.* 75(35):235 (1977).
50. Welch, L. W. Paper presented at 4th International Petroleum Seminar, Nice, France, March 4–11, 1981.
51. Peters, E., and D. L. Flock. *Soc. Pet. Eng. J.* 21 (1981).
52. Chuoke, R. L., P. van Meurs and C. van der Poel. *Trans. AIME* 216:188 (1959).

53. Doscher, T. M. *Am. Scientist* 69:193 (March–April 1981).
54. Gladfelter, R. E., and S. P. Gupta. *Soc. Pet. Eng. J.* 20:508 (1980).
55. Parsons, R. W., and S. C. Jones. *Soc. Pet. Eng. J.* 17:11 (1977).
56. Doscher, T. M., and M. El-Arabi. *Paper* SPE/DOE 9787, presented at Second SPE/DOE Joint Symposium on Enhanced Oil Recovery, Tulsa, OK, April, 1981.

CHAPTER 40
HYDRODYNAMICS OF LIQUID-LIQUID SPRAY COLUMNS

L. Steiner and S. Hartland
Swiss Federal Institute of Technology
Department of Industrial and Engineering Chemistry
Zürich, Switzerland

CONTENTS

INTRODUCTION

The spray column is one of the simplest and oldest types of equipment used to contact two immiscible liquids in countercurrent flow. It consists of an empty vertical tube with expanded sections at either end. A distributor in one of the sections disperses one of the liquids into drops. These drops then rise or fall against the flow of the continuous phase, collecting at the other end of the column and finally coalescing to form a layer of clear liquid, which is withdrawn from the column. The aim is to bring two immiscible liquids into close contact with a large interfacial area to facilitate mass or heat transfer. In contrast to packed or plate columns, there is no additional agitation to modify the drop size and affect the drop velocity once the drops are formed. In addition, the continuous phase flows freely through the column so that backflows and circulation can develop easily.

Spray columns are simple and cheap to run and can handle very large liquid loads. However, due to internal mixing their extraction efficiency is low, and they cannot be used for separations in which many theoretical stages are required. On the other hand, they are very suitable for direct heat transfer between large amounts of immiscible liquids, especially in the desalination of water and production of fertilizers, and also in the experimental study of drop swarms and mass transfer between liquid phases. They represent a good starting point for the development of models for use in more sophisticated extraction columns and enable experimental techniques to be tested under defined conditions. Serious research on spray column extraction was begun some 30 years ago by Elgin [1] and Geankoplis [2] and is still being pursued. Many publications are available on this subject. In this chapter, only the hydrodynamics will be discussed, with mass and heat transfer between drops and a surrounding liquid being discussed only when it affects the behavior of the drops.

Many topics, such as drop formation, movement and coalescence, are not specific for spray columns but are of importance in many other types of equipment. However, other topics, such as axial mixing, are more specific to conditions pertaining to spray columns, and the available information is often conflicting and confusing. A comprehensive treatment is unfortunately not possible here; for each topic considered, equations must be selected and tested on experimental data collected by various authors. Very few relevant equations have a sound theoretical basis, and the only way to check the available empirical equations is against actual spray column data. Unfortunately, these usually were determined for columns of relatively small diameter, up to, say, 0.5 m. As the performance of spray columns is sensitive to the internal circulation of the continuous phase, many phenomena are dependent on the column diameter. Even drop formation on a distributor plate can be affected. Therefore, parameters can only be estimated, and experimental verification is usually necessary for design purposes.

EXAMPLE OF SPRAY COLUMN CONSTRUCTION

Figure 1 shows the layout of a general purpose pilot-plant size spray column [3]. The column is constructed of technical glass and is 0.1 m in diameter, the cylindrical part being 3.4 m high. The drops of lighter dispersed liquid are formed on a distributor in the lower section. The conical section between the expanded and cylindrical parts of the column serves as a funnel, which has been described by Elgin [1] and Minard and Johnson [4]. This compresses the drops into a smaller volume to achieve denser dispersions in the column. The funnel-shaped distributor is made of stainless steel and is covered with a Teflon plate of 100 mm in diameter, having 109 stainless steel nozzles of 1 mm inner diameter extending about 10 mm above the upper surface of the plate to prevent the coalescence of neighboring drops during formation. The upper ends of the nozzles were ground into sharp edges with an apex angle of 30° so that very uniform drops could be produced in quantities sufficiently large for operation with densely packed drops and holdups of dispersed phase above 60%.

In the upper part of the column, the cylindrical section was extended well into the upper section by means of a metallic riser. Grooves were milled into the upper part of the riser to ensure regular distribution of the incoming continuous phase along the column periphery. The introduction of this phase must be performed carefully to avoid additional circulation, which can extend over a considerable part of the column height. One possibility is to introduce the continuous phase through a set of tubes distributed evenly over the cross section of the column. The construction with the cylindrical riser and introduction of the continuous phase at the periphery is simple and suitable for columns of pilot-plant dimensions. Another advantage of this arrangement is the possibility of influencing the coalescence area for the drops at the main interface in the column head, thus enabling us to work with dense dispersions (referred to later in this chapter).

The section in the upper part of the column is sufficiently large to allow complete coalescence of the drops coming from the column. Both the upper and lower sections are

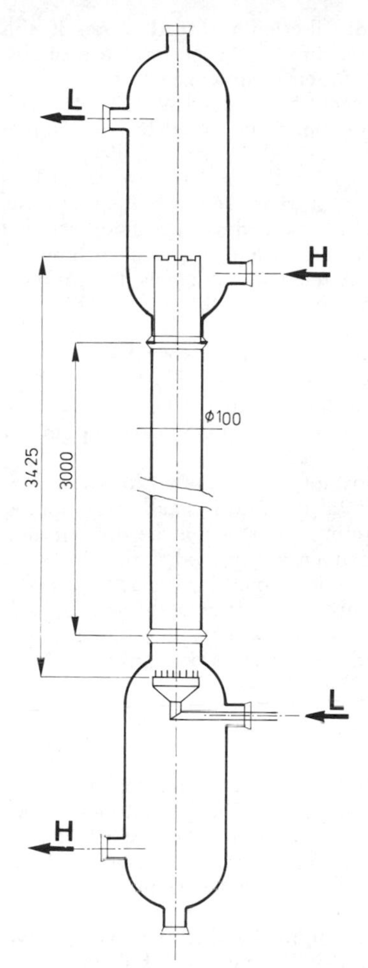

Figure 1. Example of a spray column construction: H = heavy phase; L = light phase.

0.25 m in diameter, with lengths of approximately 1 m. The interface position is regulated by an overflow valve. The flowrates of both phases are indicated by flowmeters and controlled by automatic devices. Four tanks, each 400 *l* in volume, enable continuous, steady-state operation to be carried out for at least 2 hours. Further details about this column have been given by Ugarcic [3].

DROP FORMATION ON DISTRIBUTORS

The necessary condition for the performance of any spray column is that one liquid is distributed in the form of drops into the second liquid. This is usually done by a distributor, which can be a perforated plate or a set of nozzles. Drop formation on an orifice is of primary importance in the hydrodynamics of spray columns. This question has been studied frequently and numerous papers appear in the literature. However, up to now the prediction of drop size has been possible only under special circumstances (the most important having been given by Hayworth and Treybal [5]), in that reproducible results can only be obtained with sharp-edged nozzles so designed that the dispersed phase cannot spread over, or wet, the outer surface. In all other cases, the drops are not uniform and their size distribution is broad. Even if this condition is met for each nozzle the problem still remains of the performance of distributors with multiple closely spaced orifices generally used in spray columns.

Qualitatively, drop formation on a single orifice can be described as follows: With an extremely slow supply of dispersed phase, the formation is governed by static forces only, the buoyancy and gravitational forces being balanced by the capillary forces. This kind of formation, described some 60 years ago by Harkins and Brown [6], is plainly of little importance in actual work. However, it represents the limiting case, and the principles enunciated by these authors have only been developed and extended by their successors. The drop size grows with increasing feedrate as additional liquid is supplied after the drop starts the detaching process. A maximum is reached eventually, and with further increase of the feedrate a jet begins to form at the orifice. This is of a more or less cylindrical shape and disintegrates into drops at some distance from the orifice. The drops formed by jet disintegration are generally smaller than those formed directly at the orifice at lower rates. A further increase in feedrate leads to an enlargement of the jet, and when the critical velocity in the orifice is reached the drops become irregular and the jet disintegrates into drop clouds. Velocities of this magnitude are of no further interest for spray column processes.

With distributors provided with orifices in close proximity to each other, the drops are first formed at a small number of orifices when the total flowrate is small. The number increases with the feedrate until all orifices are operating, after which the drop formation is governed by laws similar to those for single openings, provided the drops do not touch each other during formation. The detachment of one drop triggers the detachment of neighboring drops so that formation in clusters can frequently be observed. With jet formation below the critical velocity, the drops are usually quite uniform and relatively small, thus representing the preferable condition for practical application.

Formation of Single Drops on Orifices

This problem has been studied frequently, and there are at least five relatively well founded equations in the literature. Each author claims that the previous equations are not applicable under conditions different from the experimental set used to derive the corresponding formula, and then presents a new one. As in all cases involving empirical coefficients, checking can be done only by comparing the results with more data.

Hayworth and Treybal [5] were among the first to improve the Harkins and Brown equations for static formation. The final form, expressed in SI units, is as follows:

$$V_D + 0.0411\, V_D^{2/3} \left(\frac{\rho_d v_n^2}{\Delta\rho}\right) = 0.21 \left(\frac{\sigma d_n}{\Delta\rho}\right) + 5.035 \left(\frac{d_n^{0.747} v_n^{0.36} \mu_c^{0.186}}{\Delta\rho}\right) \quad (1)$$

As in most other cases, this equation gives the drop volume implicitly so that numerical solution is necessary. To facilitate manual evaluation, a graphic solution is given by the original authors.

Null and Johnson [7] criticized this equation and found deviations of up to 300%. They derived their own formula but this is even more complicated and does not improve the fit, as shown later by Scheele and Meister [8]. These authors also developed their own correlation, which may be written in the following form:

$$V_D = F \left[\frac{\pi \sigma d_n}{\Delta\rho\, g} + \frac{5\mu_c \pi d_n^3 v_n}{\Delta\rho\, d_D\, g} - \frac{\pi \rho_d d_n v_n^2}{3\Delta\rho\, g} + 4.50 \left(\frac{\pi^2 \rho_d \sigma d_n^6 v_n^2}{\Delta\rho\, g^2}\right)^{1/3}\right] \quad (2)$$

It will be shown that Equation 2 is better than Equation 1, although a more recent study by de Chazal and Ryan [9] proved that the values predicted by it were sometimes too high. The alternative equation is as follows:

$$V_D = \frac{\pi \sigma d_n}{\Delta\rho\, g}\left[F + 0.824 \frac{\Delta\rho\, d_n\, g\, v_n\, V_d^{1/3}}{\sigma\, v_D} - 0.428 \frac{\rho_d\, d_n\, v_n^2}{\sigma}(1+\beta)\right] \quad (3a)$$

where

$$\beta = 0 \quad \text{if} \quad \sqrt{\frac{\rho_d\, d_n\, v_n^2}{2\sigma}} \leq 1.07 - 0.75\sqrt{\frac{\Delta\rho\, d_n\, g}{4\,\sigma}} \quad (3b)$$

Otherwise,

$$\beta = 0.286\sqrt{\frac{\Delta\rho\, d_n^2\, g}{4\sigma}} \quad (3c)$$

The drop velocity, v_D, in the second term in the bracket may be calculated from Hu and Kintner's formula (given later). Development did not cease with de Chazal and Ryan's formula, and further equations have been introduced since. One of the more recent ones, which has the advantage of an explicit expression of the drop volume, was published by Kagan et al. [10]:

$$V_D = \frac{\pi \sigma d_n F}{\Delta\rho\, g}\left(1 + 2.39 \frac{d_n}{\sqrt{\frac{8\,\sigma}{\Delta\rho\, g}}} We^{1/3} - 0.485\, We\right) \quad (4)$$

where

$$We = \frac{(\rho_c + \rho_d)\, d_n\, v_n^2}{4\,\sigma} \quad (5)$$

The term F in Equations 2–4 is the Harkins and Brown correction factor which, for drops formed under static conditions, gives the relative volume of the detached drop in relation to all liquid present above the nozzle at the time of drop detachment; its value was approximated by Hayworth and Treybal [5] by 0.655. More precise calculation is possible by using a diagram published by Scheele and Meister, which, in turn, can be approximated by the following equation:

$$F = 0.6 + 0.4 \exp\left(-2 d_n^3 \sqrt{\frac{\Delta\rho\, g}{\pi\, \sigma\, d_n}}\right) \quad (6)$$

Equations 1–4 are compared with each other in Figures 2 and 3 for high and low interfacial tensions, respectively, using nozzles of 1 and 2 mm in diameter. One can see that with high interfacial tension the results are in reasonable agreement so that all the equations may be used for moderate nozzle velocities. With low interfacial tensions Equation 1 fails completely, while the rest show good agreement. This proves that Equation 1 is not reliable and should not be used.

Of the remaining three, Equation 2 actually gives higher results, as indicated by de Chazal and Ryan. Equation 4 of Kagan is close to Equation 3 and has the advantage of simpler calculation, as the drop volume is given explicitly. With Equation 3 numerical solution is necessary. Moreover, this equation requires the knowledge of the drop velocity, which must be supplied from other empirical equations.

An attempt was made to digitalize the plots published by Scheele and Meister [8] and de Chazal and Ryan [9] in an effort to obtain sets of experimental data for checking Equations 1–4. The results were considerably scattered and, predictably, the authors'

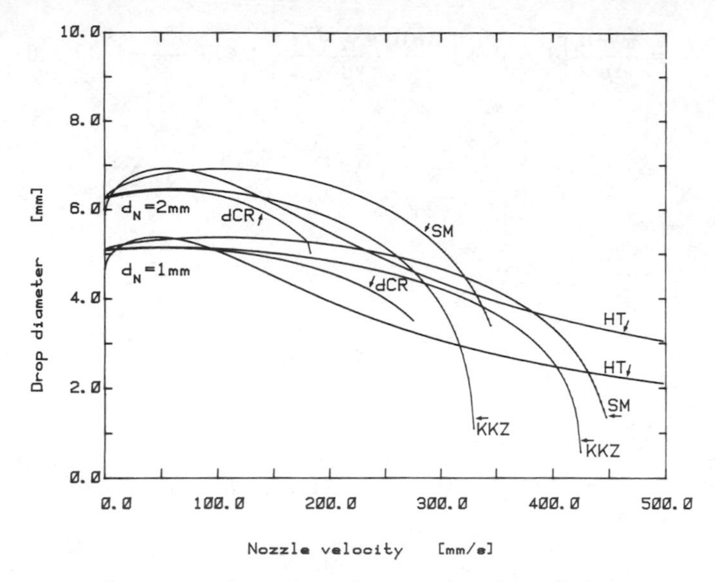

Figure 2. Drop formation on an orifice. The system is toluene in water: HT—Hayworth and Treybal [5]; SM—Scheele and Meister [8]; dCR—de Chazal and Ryan [9]; and KKZ—Kagan et al. [10].

data were correlated better by their own equations. The overall accuracy lay mostly within the range of 20%, which shows that none of the equations is entirely satisfactory.

Jetting

As already stated, drop formation directly on an orifice is only possible at low nozzle velocities. With higher velocities, a jet forms above the orifice and the drops are formed by its disintegration. The jet length increases with nozzle velocity up to a maximum, then decreases until it disappears completely so that very small drops are formed again directly on the orifice. This latter phenomenon is known as atomization and is of no importance in the performance of spray columns as such drops are too small and their production is energy consuming. The most important region of drop formation is that between the onset of jetting and the maximum jet length, described by Skelland and Johnson [11] as the critical condition. The drops are very regular and their interfacial area large. A mathematical description of jetting phenomena has been attempted repeatedly. The literature in this field is extensive, beginning with the work of Rayleigh [12] in 1892. A survey is given in Skelland and Johnson's work [11] that represents the latest significant paper.

From a practical point of view, it is necessary to determine the nozzle velocity when jetting starts ("jetting velocity"), the velocity with maximum jet length ("critical velocity") and the diameter of the drops formed from the jet.

Scheele and Meister [13] give a relationship for the jetting velocity as follows:

$$v_j = \left[\frac{2\,\sigma}{\rho_d\, d_n}\left(1 - \frac{d_n}{d_D}\right)\right]^{1/2} \tag{7}$$

where d_D is the diameter of the drop that would form without jetting. A similar result can be determined if Equation 3b is used, rewritten in the following form:

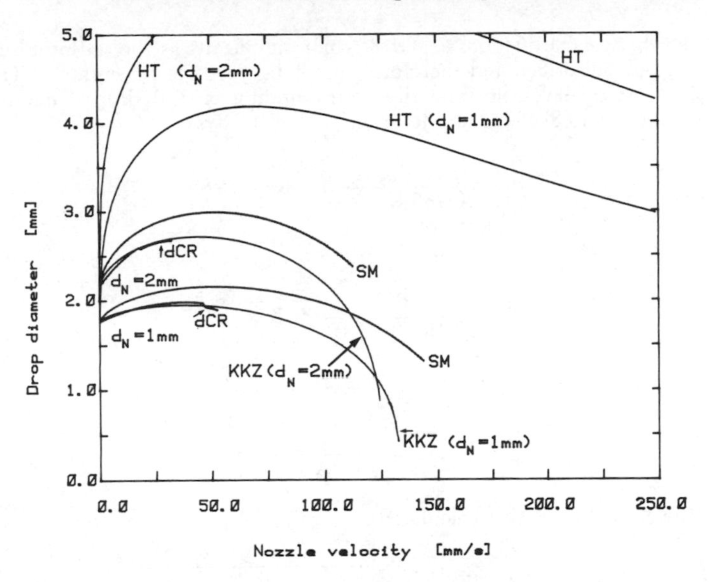

Figure 3. Drop formation on an orifice. The system is butylalcohol in water.

$$v_j = \sqrt{\frac{2\,\sigma}{\rho_d\, d_n}} \left(1.07 - 0.75 \sqrt{\frac{\Delta\rho\, d_n^2\, g}{4\,\sigma}} \right) \tag{8}$$

Equations 7 and 8 are compared in Figure 4 for systems with high and low interfacial tensions. It can be seen that jetting starts much sooner in the latter case and is strongly dependent on nozzle diameter. Equation 8 is recommended for general use since no implicit evaluation of the drop diameter is necessary. (In Equation 7, the drop diameter was evaluated from Equation 2, which originated from the same authors as Equation 7.)

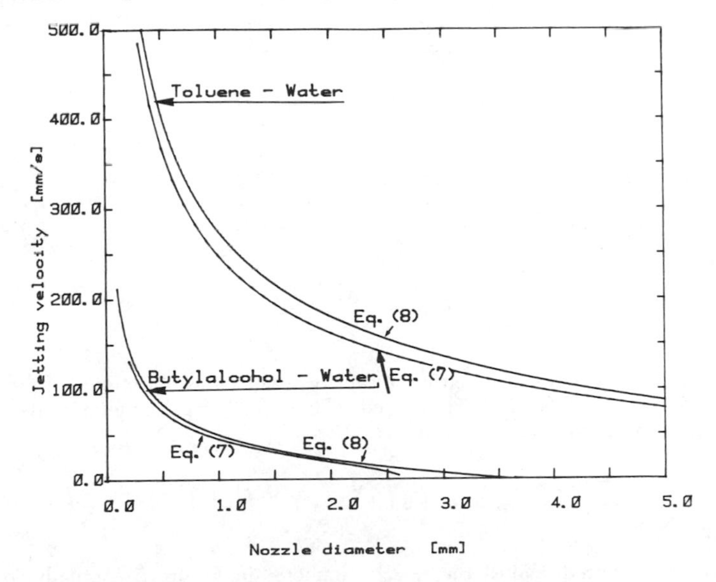

Figure 4. Jetting velocity correlations.

The deviation between Equations 7 and 8 is not significant as the jetting also depends on the quality of the orifice and therefore cannot be predicted accurately. The critical conditions, i.e., critical velocity and the corresponding critical drop diameter, can be calculated according to Skelland and Johnson [11] as follows:

$$d_{jc} = \frac{d_n}{0.485\,K^2 + 1} \text{ for } K < 0.785 \tag{9a}$$

or

$$d_{jc} = \frac{d_n}{1.51\,K + 0.12} \text{ for } K \geqq 0.785 \tag{9b}$$

where

$$K = \frac{d_n}{\sqrt{\sigma/\Delta\rho\, g}} \tag{9c}$$

From this, the critical velocity is calculated:

$$v_{jc} = 2.69\,\frac{d_{jc}^{\,2}}{d_n^{\,2}}\sqrt{\frac{\sigma}{d_{jc}(0.514\,\rho_d + 0.472\,\rho_c)}} \tag{10}$$

At this critical velocity it was found experimentally that

$$d_{DC} = K_1\, d_{jc} \tag{11}$$

where K_1 lay between 1.7 and 2.6, depending on the physical properties of the system. (With low interfacial tensions, higher values were found.) Alternatively, the following formulas given by Treybal [14] may be used:

$$d_{DC} = \frac{2.07\, d_n}{0.485\,\text{Eö} + 1} \qquad \text{for Eö} < 0.615 \tag{12a}$$

and

$$d_{DC} = \frac{2.07\, d_n}{1.51\,\text{Eö}^{1/2} + 0.12} \qquad \text{for Eö} \geqq 0.615 \tag{12b}$$

where Eö is the Eötvös number, defined as

$$\text{Eö} = \frac{\Delta\rho\, g\, d_n^{\,2}}{\sigma} \tag{13}$$

To bridge the gap between the jetting and critical velocities, an empirical equation was suggested by Horvath et al. [15] that correlated the drop diameter with the above-mentioned variables d_{jc} and v_{jc}, as follows:

$$d_D = d_{jc}\left(2.06\,\frac{v_{jc}}{v_n} - 1.47\, ln\,\frac{v_{jc}}{v_n}\right) \tag{14}$$

Equation 10 is plotted against the nozzle diameter in Figure 5. Actually, distributors should be designed so that the nozzle velocities are close to the critical velocity.

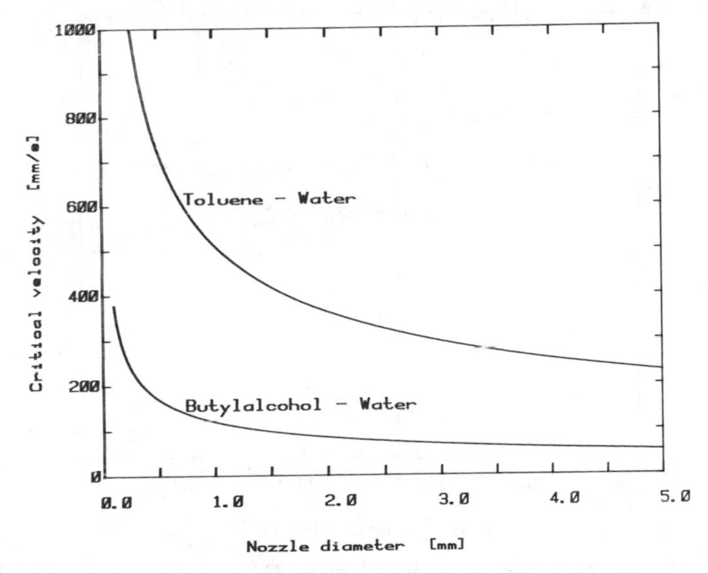

Figure 5. Critical jetting velocity from Equation 10.

Drop Formation on Multiple Nozzles

All rules discussed to this point were developed originally for single orifices only. However, several investigators have shown that they also can be used for multiple openings, provided the growing drops do not touch each other. If the drops are formed on perforated plates, the seeping of the continuous phase through the openings should be prevented by choosing a sufficiently high nozzle velocity. Mersmann [16] reports that the minimum Weber number (defined with nozzle velocity and the density of the dispersed phase, i.e., $We = v_n^2 d_n \rho_d / \sigma$) should be greater than 2 to prevent seeping and secure drop formation on all openings.

Miller and Pilhofer [17] experimentally investigated the formation of drops on large perforated plates using toluene dispersed in water. They found that before the onset of jetting the drops were generally smaller than those formed on single orifices. They explained this phenomenon by showing that with small loadings some orifices do not operate. The openings in which drops actually were formed operated under conditions close to the jetting velocity. When all openings operated, the drop size decreased until it reached a value predicted for critical formation. Above the critical velocity the drop size started to grow again. Some of the typical data have been digitalized from the published graphs and are shown in Figure 6, together with some of the equations discussed. One sees that Equation 4, (as representative of Equations 1–4) actually gives higher values than measured. The deviation was compensated for mostly by modifying the Harkins and Brown factor, F, to the value of 0.625, as recommended by de Chazal and Ryan. Equation 12 for the critical drop velocity and Equation 14 for the intermediate region are reasonably applicable.

A similar comparison was made with data published by Horvath [18] for *o*-xylene and water. This author used multiple nozzles with tube lengths of approximately 10 mm arranged in a triangular pitch of 7 mm. It may be seen from Figure 7 that the situation is similar to that in Figure 6. Also here the drops formed at low velocities are smaller than those predicted by Equations 1–4, and the deviations were compensated for by reducing the value of the Harkins and Brown factor to the same value as in the previous figure. The jetting velocity is now higher due to the smaller diameter of the

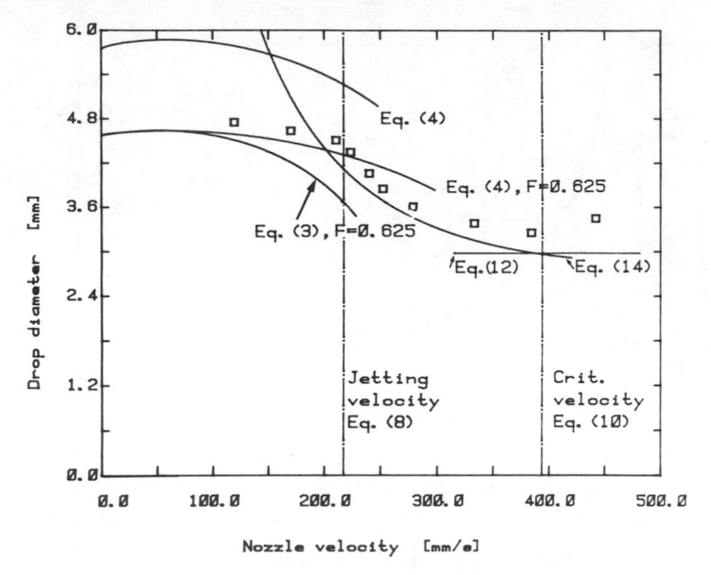

Figure 6. Drop formation on a perforated plate. Data are from Miller and Pilhofer [17]; the system is toluene in water, $d_n = 1.5$ mm.

nozzles. This time it is the equation of de Chazal and Ryan (Equation 3) that describes the data up to the jetting velocity.

The change in the Harkins and Brown factor demonstrated on the data obtained on multiple openings possibly can be explained by disturbances connected with the detachment of neighboring drops because in this case the continuous liquid is not perfectly calm and the drops detach earlier than they would on single nozzles. More

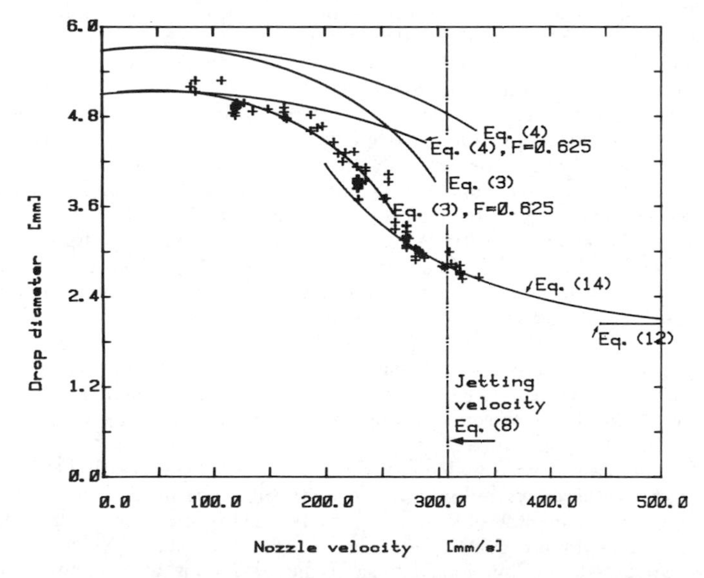

Figure 7. Drop formation on multiple nozzles. Data are from Horvath [18]; the system is *o*-xylene in water, $d_n = 1$ mm.

experimental work involving the evaluation of the number of openings actually forming drops is necessary before this question can be answered accurately.

Drop Size Distribution

If the drops are formed on distributors with multiple openings, they are never perfectly uniform. The sizes of individual drops are scattered about some mean value and the distribution may be more or less broad. (With the exception of the static drop formation at extremely low velocities, even drops formed on a single nozzle are not perfectly uniform.) To characterize a drop swarm with drops of different dimensions the definition of a mean value is necessary. In many publications the Sauter diameter is used for this purpose, although its application is justified only in mass transfer processes. By definition, it is the diameter of a sphere with the same volume to surface ratio as the drop swarm in question, which can be calculated from the following formula:

$$d_{32} = \frac{6 \Sigma V_D}{\Sigma S_D} \tag{15}$$

where V_D and S_D are the drop volume and the drop surface, respectively. (More about the use of the Sauter diameter and other possibilities of characterizing a drop swarm is given by Mugele and Evans [19].) Practically all existing publications on spray columns assume that the actual drop field behaves like a monodispersed one, with the drop size equal to the mean diameter.

Compared with other column types, the properties of the distributor are essential for the performance of a spray column. The drops usually remain as they were formed during their stay in the column and are not affected by obstacles (such as in packed or plate columns), turbulence (which is very low) or interdrop coalescence (which must be avoided by system selection). For this reason, distributions of curious shapes are sometimes found that cannot be reproduced by simple mathematical functions. Horvath [18] reports that with low nozzle velocities the drops were very uniform (standard deviation less than 0.02 mm). Increasing the nozzle velocities resulted in a two-peak distribution, which persisted until the critical velocity was reached, after which regular drops were again produced. This can be seen in Figure 8, the explanation being that at initially low nozzle velocities the drops are produced by a single drop formation mechanism and therefore are uniform and large. At higher velocities, additional openings start to operate and eventually some drops are formed by the jetting mechanism. This corresponds to the appearance of the second peak on the distribution diagram, the height of which increases as more openings start operating in the jetting region. The larger drops disappear only when the nozzle velocity reaches the critical value given in Equation 10. From this point on, the drops recover their uniformity and the distribution becomes narrow. Distributions with two distinctive peaks also were reported by Miller and Pilhofer [17].

Mathematical description of the drop size distribution is rather difficult. For narrow distributions with one peak only, the Gaussian normal distribution may be used [18]. Alternatively, the functions discussed by Mugele and Evans [19], such as the

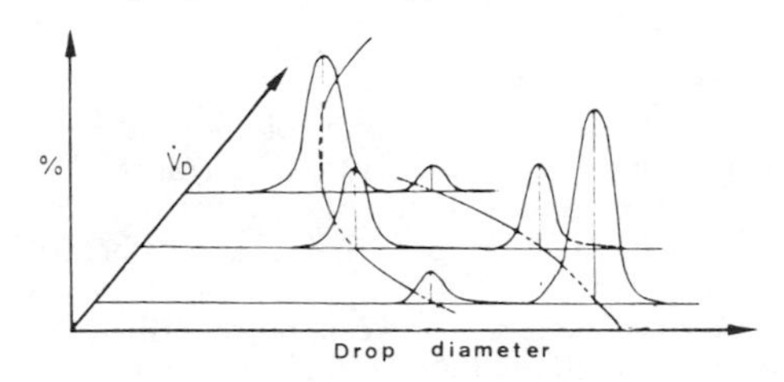

Figure 8. Effect of dispersed-phase flowrate on drop size distribution [18].

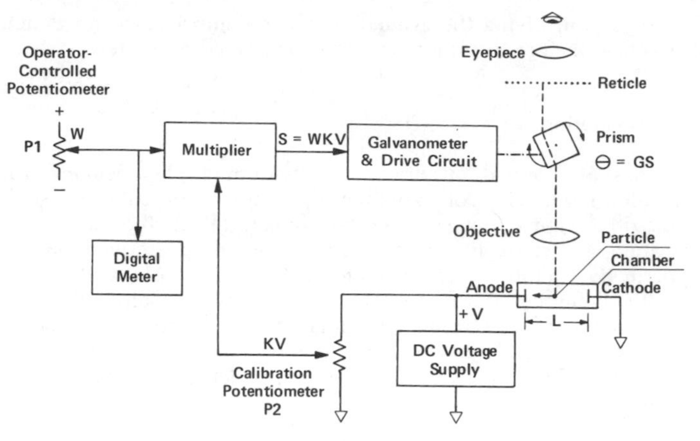

Figure 8a.

logarithmic-normal or upper limit, may be tried. The logarithmic-normal distribution was used by the authors of this chapter and proved to be in reasonable agreement with experimental data, while containing only two parameters. As no better results were obtained with distribution using more parameters, it may be recommended. Principally, it is assumed that the parameter $ln\ (d_D/\bar{d})$ is normally distributed, so that

$$\frac{dn}{d_D} = \frac{\delta}{\sqrt{\pi}\, d_D} \exp\left[-\delta^2\left(ln\frac{d_D}{d}\right)^2\right] \tag{16}$$

where δ and d are constants that characterize the distribution.

For distributions with two maxima, a sum of two functions can be used. However, in this case four parameters must be determined but this is only possible if the peaks are distinctly separated. A simple and often sufficiently accurate method is to characterize the swarm by drops of two different sizes corresponding to the two peaks. Various conditions may be simulated by varying the proportion of these two kinds of drops.

BEHAVIOR OF DROP SWARMS

Having been formed on a distributor, the drops ascend or fall through a layer of a continuous liquid until they reach the other end of the column. During this time, the mass or heat transfering process takes place for which spray columns are operated. The velocity of the drop movement determines the holdup of the dispersed phase inside the column, which, in turn, affects the interfacial area, axial mixing, residence time and flooding limits. This section therefore represents the essential part of the hydrodynamic description of the spray column.

Rise or Fall of Individual Drops

Most of the following equations have as a parameter the velocity of movement of single drops in a large amount of quiescent liquid. Since the laws that control the movement of a fluid particle in a liquid are rather complicated because of the unstable shape of fluid drops, empirical equations have been developed that are satisfactory for practical purposes.

Very small drops behave like rigid spheres, and their velocity can be described by Stokes' law in the laminar region and by corresponding drag coefficient correlations in the turbulent region. Larger drops may develop an internal circulation, and very large drops tend to change their shape during ascent as they travel on zig-zag paths. They may even split or coalesce. A comprehensive description of these phenomena is given by Slater [20], who also provides an extensive bibliography. A generalized equation derived originally by Yilmaz for the motion of rigid particles is given by Pilhofer and Mewes [21] in the following form:

$$v_D = \frac{\mu_c}{\rho_c\, d_D} \frac{|Ar|}{\left(324 + 23.1\,|Ar|^{0.54} - \dfrac{1.51(10^{-5})\,|Ar|^{5/3}}{1 + 7.07(10^{-7})\,|Ar|} + 0.368\,|Ar|\right)^{1/2}} \tag{17}$$

This equation reduces to Stokes' law at very low values of the Reynolds number and is applicable well into the turbulent region. For actual drops, empirical equations of several authors are available, the most frequently used being those of Hu and Kintner [22], which can be rewritten in the following form:

$$v_D = \frac{\mu_c\, P^{0.15}}{\rho_c\, d_D}(0.798\, Y^{0.784} - 0.75) \text{ for } 2 \leq Y \leq 70, \tag{18}$$

and

$$v_D = \frac{\mu_c\, P^{0.15}}{\rho_c\, d_D}(3.701\, Y^{0.422} - 0.75) \text{ for } Y > 70. \tag{19}$$

The groups are defined as follows:

$$P = \frac{\rho_c^2\, \sigma^3}{\mu_c^4\, \Delta\rho\, g} \qquad Y = \frac{4\, \Delta\rho\, d_D^2\, g}{3\, \sigma} P^{0.15}$$

Alternatively the equations of Klee and Treybal [23] may be used which, according to the authors, are applicable in a broader region of physical properties:

$$v_D = 3.042\, \rho_c^{-0.45}\, \Delta\rho^{0.58}\, \mu_c^{-0.11}\, d_D^{0.7} \tag{20}$$

and

$$v_D = 4.96\, \rho_c^{-0.55}\, \Delta\rho^{0.28}\, \mu_c^{0.1}\, \sigma^{0.18} \tag{21}$$

Equation 20 is used for small drop diameters up to the value when the result from both equations is identical. From this value onward, Equation 21 is used. The transition diameter is given as the solution to these two equations, which may be written as follows:

$$d_{DC} = 1.772\, \rho_c^{-0.14}\, \Delta\rho^{-0.43}\, \mu_c^{0.3}\, \sigma^{0.24} \tag{22}$$

In Figures 9 and 10 all these equations are compared for toluene and butylalcohol dispersed in water. It can be seen that they give very similar results; however, there is some deviation in the region of very small drops (below 1 mm) where Equation 17 probably would be considered most reliable.

The transition from circulating to oscillating drops depends on the purity of the system. When surfactants are present, the drop develops a "rigid" skin, which delays transition and thus enables the drop to follow the laws described by Equations 18 or 20 longer than predicted. Thus the transition point can only be approximate, and considerable deviations may be found in practice. Therefore, it is not possible to say

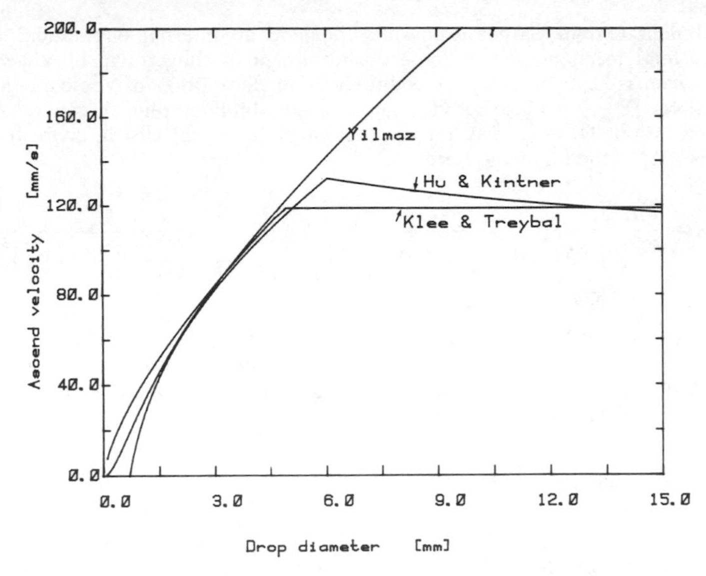

Figure 9. Rise velocity of single drops. The system is toluene in water.

which equations are to be preferred. In the following calculations, the velocity will be calculated from Equation 17 from zero to the point where this equation coincides with Equation 18; from this point upward, Equations 18 and 19 with the prescribed transition Y = 70 are used. (Equation 18 fails with very small diameters, giving a negative result.)

Velocity in Drop Swarms

Drops in spray columns do not move independently. They collect into more or less dense swarms and are contained by the column wall. The continuous phase flows in a more or less opposite direction to their movement. All these phenomena cause deviations in the drop velocity which must be taken into account. To describe the rather complex conditions met in drop swarms, the relative velocity is defined as the sum of the mean velocities of the drops and the velocity of the dispersed phase, so that

$$v_r = \frac{\dot{V}_c}{A(1-\epsilon)} + \frac{\dot{V}_d}{A\epsilon} \tag{23}$$

where ϵ is the holdup of the dispersed phase. This relative velocity is then usually correlated to different parameters, and the holdup is calculated from it numerically. A review of earlier works was made by Slater [20], so only a brief introduction is given here. There is a group of equations for relative velocity based on Stokes' law for the sedimentation of solid particles. Introducing some simplifications, the following equation was derived by Thornton [24] for spray columns:

$$v_r = v_x\,(1-\epsilon) \tag{24}$$

This was extended by Steinour [25] to

$$v_r = v_x\,(1-\epsilon)\exp(-4.19\,\epsilon) \tag{25}$$

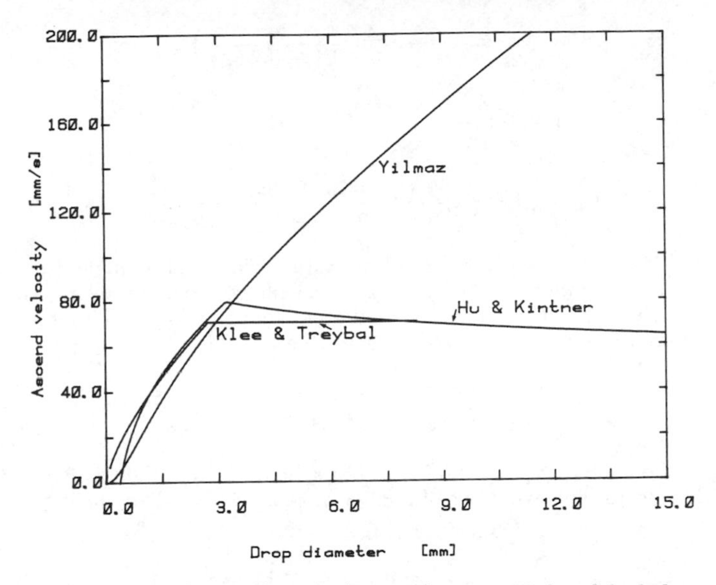

Figure 10. Rise velocity of single drops. The system is butylalcohol in water.

and by Misek [26] to

$$v_r = v_x (1 - \epsilon) \exp[-(4.19 - z)\epsilon] \tag{26}$$

where the constant, z, should express the influence of the coalescence along the column height. An exact relationship is hardly possible, however, so that this parameter may be considered as an empirical constant, characteristic for a given system and column arrangement.

Another approach begins with relationships for the pressure drop in packed beds, starting from the equation derived by Andersson [27], who introduced the tortuosity and cross section factors. His final formula is as follows:

$$\frac{\Delta p}{L} = 36\, zq^2 \frac{(1-\epsilon)^2}{\epsilon^3} \frac{\mu_c v_D}{d_D^2} + 6\, C_i\, q^3 \frac{1-\epsilon}{\epsilon^3} \frac{\rho_c v_D^2}{d_D} \tag{27}$$

It was found that zq^2 and q^3 were functions of the holdup only; C_i was related to the drag coefficient. Ferrarini [28] related this equation to the holdup of the dispersed phase, using the assumption that under hydrostatic conditions the pressure drop over unit length may be written as

$$\frac{\Delta p}{L} = \epsilon\, \Delta\rho\, g \tag{28}$$

Finally, Pilhofer [29] improved the definition of the correcting factors, obtaining equations in the following form:

$$\mathrm{Ar} = \frac{36\, zq^2 \epsilon}{(1-\epsilon)^2} \mathrm{Re} + \frac{6\, C_i\, q^3}{(1-\epsilon)} \mathrm{Re}^2 \tag{29a}$$

where

$$zq^2 = \frac{1-\epsilon}{\epsilon} \exp\left(\frac{\epsilon}{0.4 - 0.244\,\epsilon}\right) \tag{29b}$$

$$q^3 = 5\left(\frac{\epsilon}{1-\epsilon}\right)^{0.45} \left[1 - 0.31\left(\frac{\mu_c}{\mu_d}\right)^{0.39}\right] \tag{29c}$$

$$C_i = \frac{1}{8}\left(\xi - \frac{24}{Re_D}\right) = \frac{1}{6}\frac{\Delta\rho\, d_D\, g}{\rho_c\, v_D^2} - \frac{1}{3}\frac{\mu_c}{\rho_c\, d_D\, v_D} \tag{29d}$$

In Equation 29d, the term C_i is evaluated using the velocity of a single drop in a large amount of quiescent liquid, v_D, which may be calculated from the relationships shown in the previous paragraph. (Pilhofer used the equations of Hu and Kintner.) Solving Equation 29 for Re, the following equation can be obtained:

$$Re = \frac{3\, zq^2\, \epsilon}{(1-\epsilon)q^3\, C_i}\left[\left(\frac{(1-\epsilon)^3 Ar\, q^3\, C_i}{54(zq^2)^2\, \epsilon^2} + 1\right)^{1/2} - 1\right] \tag{30}$$

According to the author, this equation is valid for holdups from 0.06 to 0.55. Replacing Equation 29b by the following one, Equation 30 may be used for holdups from 0.55 to 0.74:

$$zq^2 = 2.2\frac{1-\epsilon}{\epsilon} \exp\left(\frac{0.44\,\epsilon}{1 - 0.61\,\epsilon}\right) \tag{31}$$

Kumar et al. [30] recently published a simple equation for the relative velocity in the following form:

$$v_r = \left[\frac{k\,\Delta\rho\, d_D\, g}{\rho_c}\left(\frac{1-\epsilon}{1+\epsilon^{1/3}}\right)^{n}\right]^{1/2} \tag{32}$$

This is a modification of the method published originally by Zenz [31] and extended later by Barnea and Mizrahi [32].

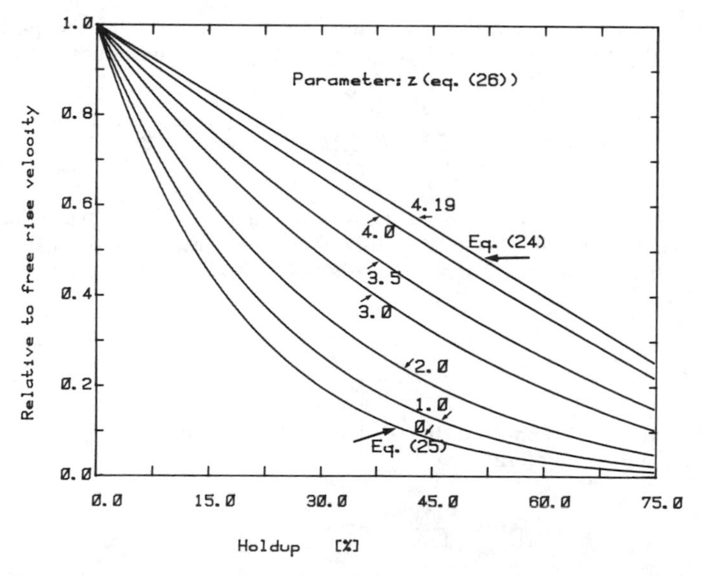

Figure 11. Equation 26 [26], with Equations 24 and 25 as its limits.

Equation 32 was obtained by proper definition of the drag coefficient and the mean viscosity of the dispersion, as well as calculation of the coefficients k and n from extensive sets of experimental data of different authors. According to Kumar, Equation 32 is applicable for Reynolds numbers from 7 to 2450 and holdups from 0.01 to 0.75, with the following values of the constants:

$$k = 2.725$$
$$n = 1.834$$

As the relative velocity is of fundamental importance to all spray column hydrodynamics, Equations 26, 30 and 32 were checked against several sets of experimental data found in the literature, the data sources being those shown in Table I. In all cases, the continuous phase was water. Ferrarini [28], Garwin and Smith [33] and Loutaty et al. [34] carried out their measurements while heat transfer was taking place so that the physical properties of the phases changed slightly along the column height.

To check and compare Equations 24–26, they were plotted in Figure 11 against the holdup. One sees that Equations 24 and 25 are limits of Equation 26, which, according to the value of the parameter z, covers a rather large area of the figure. The parameters v_x and z may be obtained from the experimental data, plotting the value of $\log(v_r/(1-\epsilon))$ against the holdup, ϵ. This was done in Figures 12 and 13 for the data mentioned, selecting fractions with drop diameters in the narrow ranges.

Table I. Holdup and Drop Diameter Sizes for Various Systems

Author	Dispersed Phase	Holdup Range (%)	Drop Diameter Range (mm)
Horvath [18]	*o*-Xylene	0–15	2.6–5.2
Ugarcic [3]	*o*-Xylene	3–70	2.0–2.8
Garwin and Smith [33]	Benzene	4–74	5.8–8.0
Loutaty et al. [34]	Kerosene	3–75	2.8–4.7
Ferrarini [28]	Spindle oil	4–25	5.3–8.8

With the exception of Horvath's and Ugarcic's data at low holdups, almost horizontal straight lines were obtained. This indicates that the parameter 4.19–z should be of zero value and that of Equations 24–26, Equation 24 has the best chance of correlating the spray column data. In fact, this could have been expected as all the authors worked under conditions with no coalescence inside the column. The deviations from the straight lines in the region of very low holdups perhaps may be explained by the bulk circulation of the continuous phase inside the columns, which causes drop velocities well above those of single drops in a quiescent liquid.

The values of v_x shown in Table II and Figure 14 were obtained by drawing the best horizontal lines through the data. The comparison to free rise velocities of single drops calculated from Equations 18 and 19 show that they are not identical and that considerable deviations may be found. The data of Garwin and Smith are not involved in the figure as the velocity, v_x, calculated from the data was over 0.4 m/sec. This indicates that the simple equation (24) can be used only in connection with a characteristic velocity, v_x, which should be calculated from experimental data obtained with the same system and arrangement.

An approximation by the free rise velocity as calculated from Equations 18 or 19 may fail completely, as with Garwin and Smith's data, or cause considerable deviations. Consequently, Equations 24 or 26 should be used only for interpolation. If a few data exist and can be plotted as in Figures 12 and 13, so that a straight line is obtained, interpolation can be used safely to obtain other holdups. However, the characteristic velocity, v_x, and the parameter, z (in cases in which no horizontal line results) should

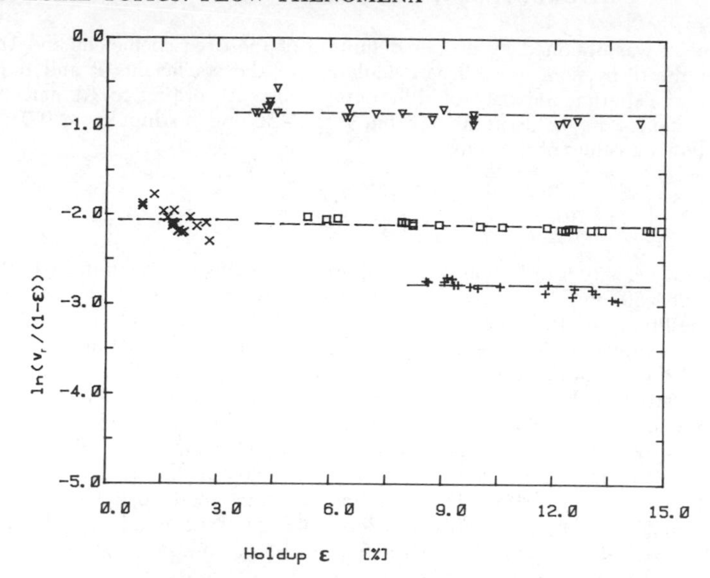

Figure 12. Evaluation of the parameters in Equation 26 with best horizontal lines ($z = 4.19$): + Horvath [18], d = 3±0.2 mm; □ Ferrarini [28], d = 5±0.5 mm; x Horvath [18], d = 5±0.2 mm; ▽ Garwin and Smith [33], d = 8±0.5 mm.

be obtained experimentally. The characteristic velocity need not be identical with the free rise velocity of single drops, as may be seen from Table II.

The results obtained from Equations 30 and 32 are plotted in Figures 15 and 16, respectively. There is a considerable difference between these equations, especially in the region of low holdups. Pilhofer's equation (30) predicts a steep initial increase

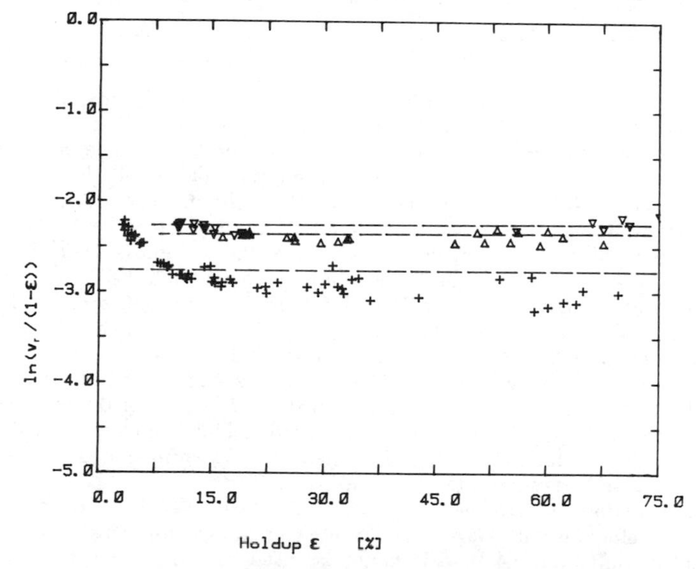

Figure 13. Evaluation of the parameters in Equation 26 with best horizontal lines ($z = 4.19$): + Ugarcic [3], d = 2.4±0.2 mm; △ Loutaty et al. [34], d = 3.0±0.5 mm; and ▽ Loutaty et al. [34], d = 4.0±0.5 mm.

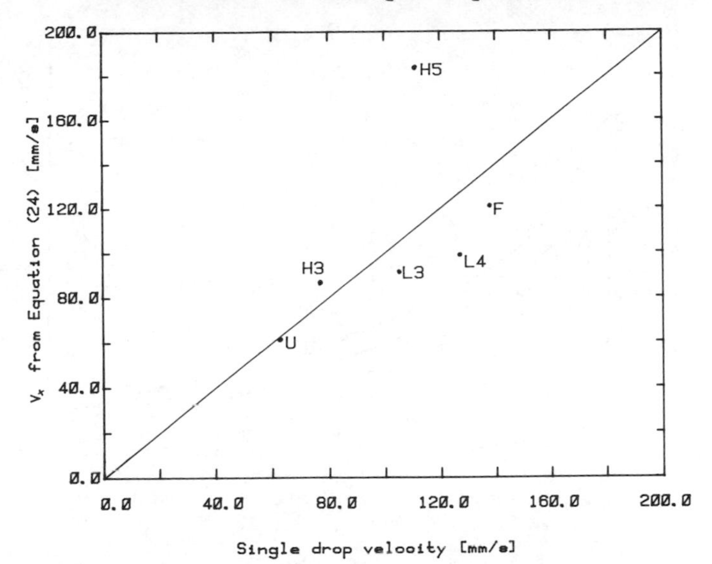

Figure 14. Comparison of best fitting velocity, v_x, with free rise velocity calculated from Equations 18 and 19: H3–Horvath [18], $d_D = 3$ mm; H5–Horvath [18], $d_D = 5$ mm; L3–Loutaty et al. [34], $d_D = 3$ mm; L4–Loutaty et al. [34], $d_D = 4$ mm; F–Ferrarini [28], $d_D = 6$ mm; U–Ugarcic [3], $d_D = 2.4$ mm.

of the relative velocity with holdup, and its extrapolation (outside its region of validity) to holdups under 4% results in velocities well above the free rise velocity of single drops. This may appear wrong, but high velocities in this region were found by Horvath et al. [15] and explained by circulation in the continuous phase. It is improbable, however, that Equation 30 explains these phenomena correctly. The jump on the right-hand side of Figure 15 was caused by transition from Equation 29b to Equation 31 for the expression of the term zq^2. As no such jump can be shown experimentally, the equations should be modified to give smooth curves.

A further difference between Equation 30 and 32 is shown in Figures 15 and 16 for the region of large drops (6 and 8 mm). While in Equation 30 the result is practically identical, due to the application of Hu and Kintner's equations for the free rise velocity, there is a considerable difference in the relative velocity calculated from Equation 32. As both equations are semitheoretical at best, the choice of applicability between them can be done only by evaluating experimental data. Kumar et al. [30] used practically all available data to evaluate the constants in their equation and claim a

Table II. Characteristic Velocities Predicted by Governing Equations

Data Source	$\bar{d}$ (mm)	Equation 26 v_x	Equation 26 z	Equation 24 v_x	Equations 18 and 19 v_D
Horvath [18]	3	0.087	0.84	0.062	0.077
	5	0.184	−16.0	0.126	0.111
Ugarcic [3]	2.4	0.076	3.26	0.062	0.063
Garwin and Smith [33]	8	0.486	2.52	0.424	0.124
Loutaty et al. [34]	3	0.092	4.17	0.091	0.105
	4	0.099	4.29	0.102	0.127
Ferrarini [28]	6	0.139	− 2.85	0.140	0.138

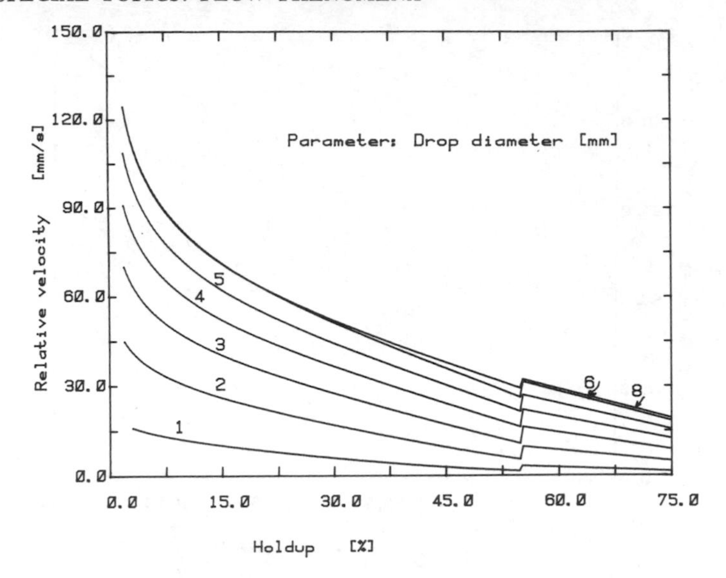

Figure 15. Relative velocity calculated from Equation 30. The system is toluene in water.

much better agreement than any other equation. However, most of the data used were obtained under similar conditions, so no clear influence of the physical properties or drop diameter can be found. An attempt has been made to sort the data at least according to drop diameter to show fitting in different regions. Using the same data as listed earlier, the results are shown in Figure 17. Starting with holdups of some 15%, Equation 32 seems better than Equation 30. Both Ugarcic's [3] and Loutaty et

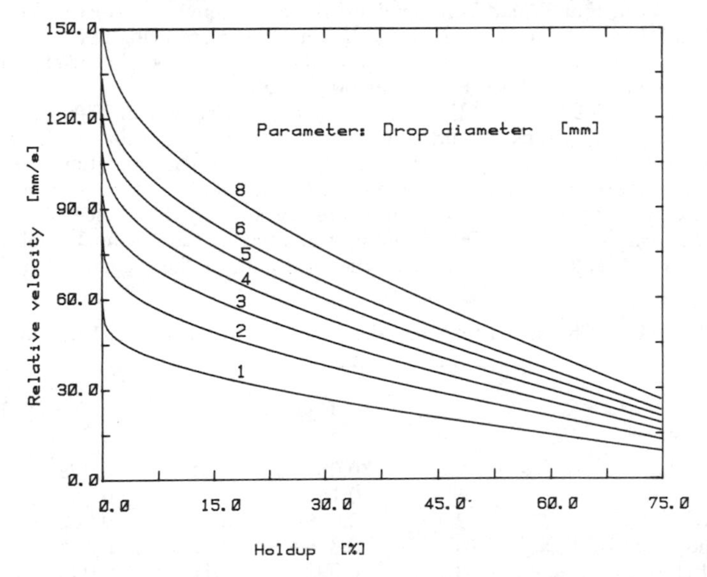

Figure 16. Relative velocity calculated from Equation 32. The system is toluene in water.

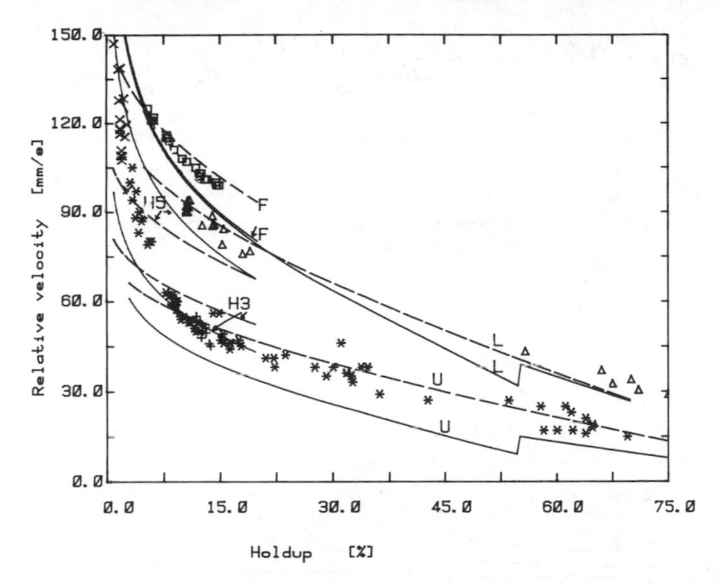

Figure 17. Direct comparison of Equations 30 and 32 with experimental data. Thick lines: Equation 30; broken lines: Equation 32: + Horvath [18], d = 3 mm (H3); x Horvath [18], d = 5 mm (H5); * Ugarcic [3], d = 2.4 mm (U); △ Loutaty et al. [34], d = 4 mm (L); □ Ferrarini [28], d = 6 mm (F).

al.'s [34] data are correlated reasonably well. Ferrarini's data fit well to Equation 32, even in the region of lower holdups. Garwin and Smith's data do not fit at all, as the relative velocity is over 0.4 m/sec in this case. The data of Horvath [18] and Ugarcic [3] for the region of low holdups are better reproduced by Equation 30, probably due to the reported circulation of the continuous phase. To summarize, neither equation gives exact results, and considerable deviation must be expected if they are applied to a new system. Unfortunately, there is no substitute for them at present, but they give at least an estimation of the behavior of drop swarms. Equation 32 has an additional advantage in its simplicity.

Holdup of the Dispersed Phase

Holdup is the fraction of the total volume occupied by the dispersed phase. It is a very important parameter that determines the interfacial area between the phases and the flooding conditions of a column (see later sections). It may be calculated from the relative velocity discussed in the previous paragraph by using some numerical method for a common solution of Equation 23 and some of the equations discussed for the relative velocity (Equations 24, 30 or 32). This can be seen in Figure 18, using Equation 32. It has been shown experimentally that each of the two solutions has a physical meaning corresponding to loose and dense dispersions, respectively. More about this is given by Letan and Kehat [35] and Loutaty et al. [34]. The numerical calculation of holdup, which is a parameter of primary importance, is rather tedious as additional error is introduced.

To speed up the procedure, a graphic solution based on Equation 32, giving both the relative velocity and holdup in dependence on two dimensionless parameters, is shown in Figures 19 and 20. Another equation, giving the holdup directly as a function of the operational parameters, was published by Mersmann [16]. Rewritten in the symbols used here, his equation is as follows:

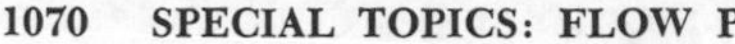

$$\epsilon(1-\epsilon)^2 = 0.14\,\frac{\dot{V}_d}{A}\left(\frac{\rho_c{}^2}{\sigma\,\Delta\rho\,g}\right)^{1/4}\left(\frac{\sigma^3\,\rho_c{}^2}{\mu_c{}^4\,\Delta\rho\,g}\left(\frac{\rho_c}{\rho_d}\right)^{2/3}\right)^{1/24}\left(\frac{\rho_c}{\Delta\rho}\right)^{1/3} \tag{33}$$

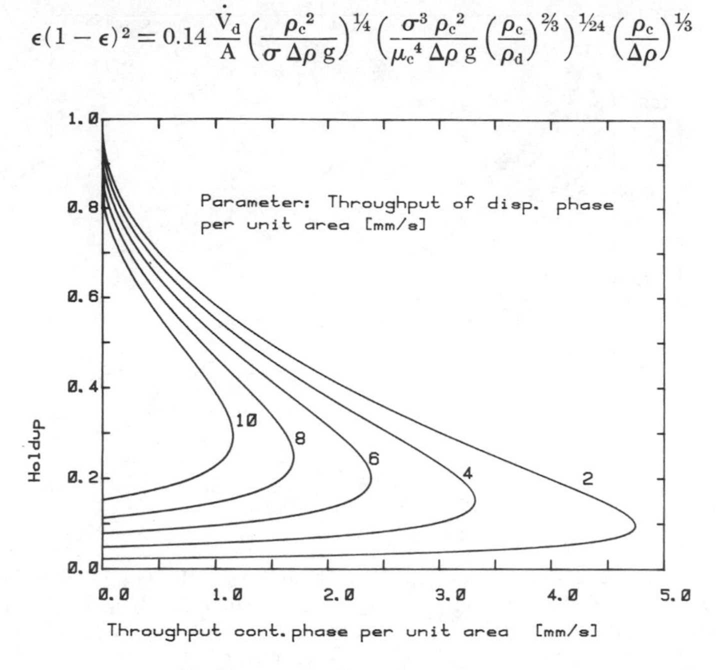

Figure 18. Variation of holdup with phase throughputs per unit area. The system is toluene in water. Drop diameter is 3 mm. The relative velocity is from Equation 32.

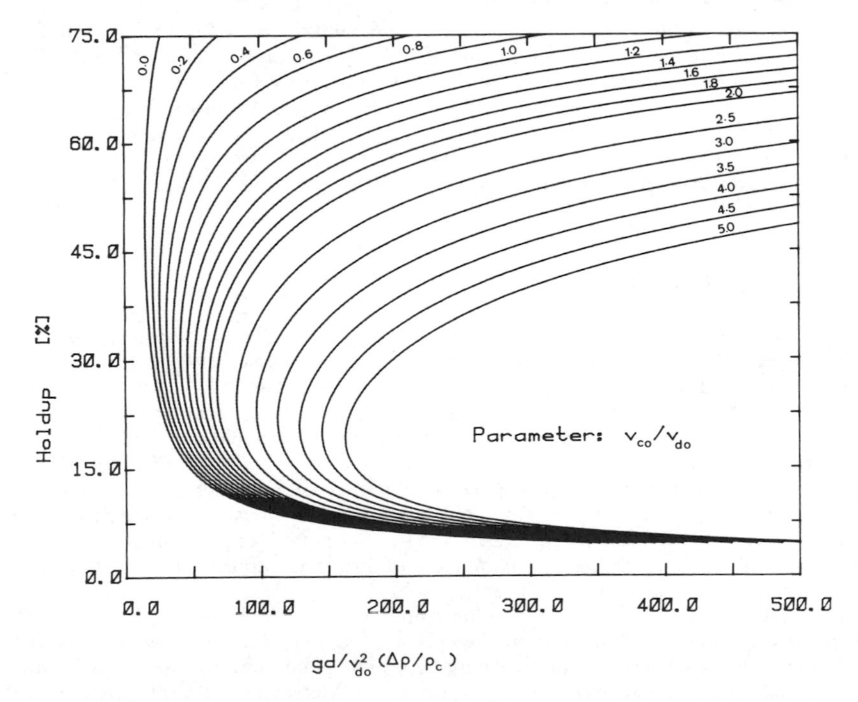

Figure 19. Equation 32 in dimensionless form; calculation of holdup.

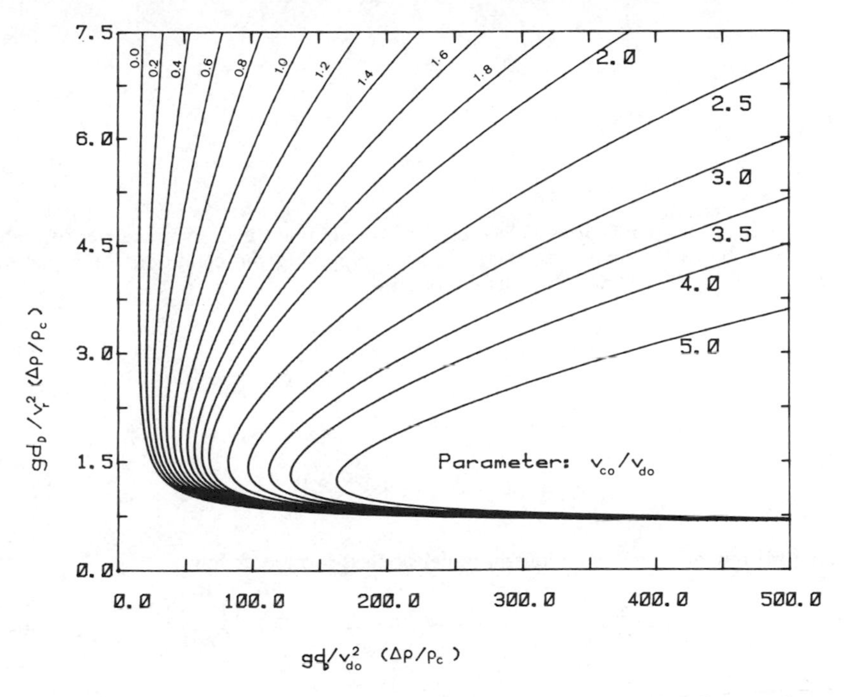

Figure 20. Equation 32 in dimensionless form; calculation of relative velocity.

In this way, a function of the holdup on the left-hand side should be directly proportional to the empty tube velocity of the dispersed phase and independent of the velocity of the continuous phase. This contradicts Equation 23, in which the dependence on this velocity is considerable unless the holdup is low, as demonstrated in Figure 18. Therefore, Equation 33 can only apply in the region in which the curves in Figure 18 run more or less parallel to the horizontal axis.

In comparison with the experimental data from Table I, this was roughly confirmed, and deviations were found in all cases as soon as the holdup exceeded 0.1. Therefore, the equation has no advantage over indirect calculation of holdup from the relative velocity.

FLOODING

If the flowrate of one of the phases is increased constantly while the flowrate of the second phase remains constant, the holdup generally will increase. This is the situation shown in Figure 18, where the lower part of the curves applies. The point is reached eventually at which the curve bends upward and begins to return. This limit is called the flooding point and indicates a maximum possible loading under given circumstances. In practice, it is not possible to go over this point and transfer smoothly to the upper part of the curves. However, appropriate organization allows single points from the upper part of the curves to be obtained if dense dispersion is stabilized in the column. The flooding point is an important threshold, as the mass or heat transfer is high if the column operates close to the flooding capacity. The prediction of the flooding rates is, therefore, an important factor in column design.

Thornton [24] analyzed the conditions for column flooding and stated that one of the following relationships for the flooding point must be true:

$$\frac{d\,v_{co}}{d\,\epsilon} = 0 \tag{34a}$$

or

$$\frac{d\,v_{do}}{d\,\epsilon} = 0 \tag{34b}$$

where v_{co} and v_{do} are empty tube velocities that are obtained if the volumetric throughput is divided by the column cross-sectional area. Applying the conditions in Equation 34 to Equation 23, the following results are obtained:

$$v_{fc} = \epsilon^2 \left(v_r - (1 - \epsilon) \frac{d\,v_r}{d\,\epsilon} \right) \qquad [v_{co}] \tag{35a}$$

or

$$v_{fd} = (1 - \epsilon)^2 \left(v_r + \epsilon \frac{d\,v_r}{d\,\epsilon} \right) \qquad [v_{do}] \tag{35b}$$

Pilhofer [36] rewrote these equations in dimensionless form as follows:

$$v_{fd} \left(\frac{\rho_c^2}{\mu_c\,\Delta\rho\,g} \right)^{1/3} = Ar^{-1/3}\,\epsilon^2 \left(Re_r - (1 - \epsilon) \frac{d\,Re_r}{d\,\epsilon} \right) \tag{36a}$$

and

$$v_{fc} \left(\frac{\rho_c^2}{\mu_c\,\Delta\rho\,g} \right)^{1/3} = Ar^{-1/3}\,(1 - \epsilon)^2 \left(Re_r + \epsilon \frac{d\,Re_r}{d\,\epsilon} \right) \tag{36b}$$

In this way, a physical properties group was added to make the velocity dimensionless and the Archimedes number appear on the right-hand side. The relative velocity and the Reynolds number based on it can be calculated from all the formulas discussed in the previous paragraph. Here, Equation 32 will be used because it is simpler than Equation 30 and gives better results with higher holdups, which are of greater importance in flooding calculations. Recognizing that $Re^2 = Ar.Fr.\ (\rho_c/\Delta\rho)$, and that Equation 32 gives $Fr(\rho_c/\Delta\rho)$ as a function of holdup, Equation 36 may be rewritten as

$$v_{fd} \left(\frac{\rho_c^2}{\mu_c\,\Delta\rho\,g} \right)^{1/3} = Ar^{1/6}\,\epsilon^2 \left[\left(Fr_r \frac{\rho_c}{\Delta\rho} \right)^{1/2} - \frac{1 - \epsilon}{2 \left(Fr_r \frac{\rho_c}{\Delta\rho} \right)^{1/2}} \frac{d}{d\,\epsilon} \left(Fr_r \frac{\rho_c}{\Delta\rho} \right) \right] \tag{37a}$$

and

$$v_{fc} \left(\frac{\rho_c^2}{\mu_c\,\Delta\rho\,g} \right)^{1/3} = Ar^{1/6}\,(1 - \epsilon)^2 \left[\left(Fr_r \frac{\rho_c}{\Delta\rho} \right)^{1/2} + \frac{\epsilon}{2 \left(Fr_r \frac{\rho_c}{\Delta\rho} \right)^{1/2}} \frac{d}{d\,\epsilon} \left(Fr_r \frac{\rho_c}{\Delta\rho} \right) \right] \tag{37b}$$

where

$$Fr_r \frac{\rho_c}{\Delta\rho} = 2.725 \left(\frac{1 - \epsilon}{1 + \epsilon^{1/3}} \right)^{1.834}$$

and

$$\frac{d}{d\epsilon}\left(\mathrm{Fr_r}\frac{\rho_c}{\Delta\rho}\right) = 4.998\left(\frac{1-\epsilon}{1+\epsilon^{1/3}}\right)^{0.834}\left(-\frac{3+\epsilon^{-2/3}}{3(1+\epsilon^{1/3})^2}\right) \tag{38}$$

These equations enable a prediction of flooding conditions.

For rapid work, a complete flooding diagram has been prepared and is shown in Figure 21. It is similar to that published by Pilhofer, the deviations being caused by the use of Equation 32 instead of 30 for the relative velocity. It can be seen that after selecting the Archimedes number, which contains the drop diameter, and one of the phase velocities, the other velocity at flooding point and the corresponding holdup are obtained.

It was not easy to find flooding data in the literature to check the diagram shown in Figure 21. The existing sets are mostly over 15 years old, and the definition given of flooding was not always the same. However, three sets eventually were applied, adding the missing values by reading data from published graphs.

It appears that the data of Minard and Johnson [4] are up to 10% lower than predicted. Those of Loutaty and Vignes [37] are some 20% higher. Garwin and Smith's data [33] do not correlate at all, giving values approximately 300% higher than predicted by the diagram. It is possible that the formation of dense dispersion was mistaken for flooding in this case, a view supported by Loutaty and Vignes [37]. It generally can be expected that flooding is predictable with about the same accuracy as for the relative velocity discussed in the previous section. The prediction would be better for "reasonable" systems, with intermediate interfacial tensions and viscosities similar to that of water and a sufficiently large density difference. Systems of this kind were used almost exclusively in the experimental work, so not much can be said about more extreme conditions. As the definition of flooding and its experimental determination are not standardized and, moreover, as the result is dependent on the construction of the actual column, experimental work remains the most reliable way to determine the limiting flowrates.

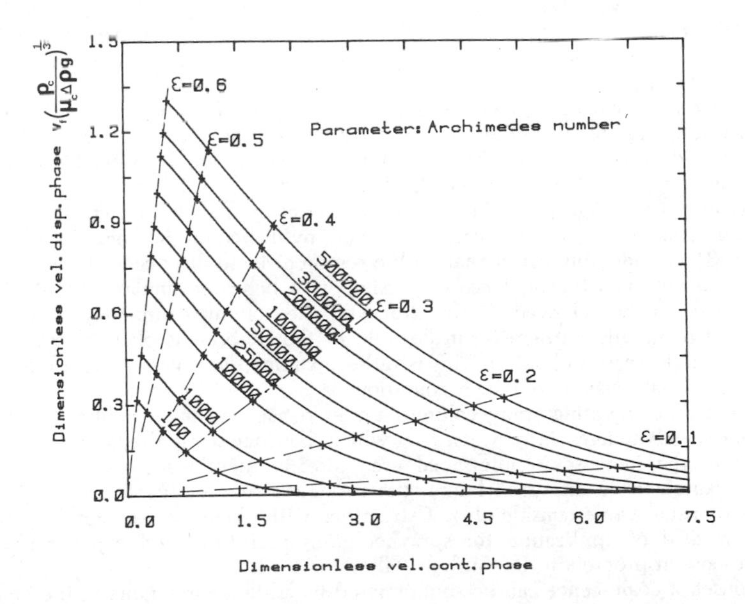

Figure 21. Generalized flooding diagram; relative velocity from Equation 32.

OPERATION WITH DENSE PACKING OF DROPS

It has been mentioned already that two values of holdup can be achieved with every pair of phase flowrates. The dense dispersion corresponding to the upper branch of the curves shown in Figure 18 can be obtained in practice if the coalescence rate of the drops in the column head opposite to the distributor is reduced. This can be achieved by reducing the coalescence area by bringing the interface into the conical part of the head, or sufficiently close to the riser if the construction described in the second section of this chapter is used. In this case, the dense drop packing starts to move into the column, maintaining a sharp interface between loosely and densely packed regions, and eventually fills the entire column up to the distributor. Fine control of the interface allows the boundary between loose and dense dispersions in any position inside the column to be stabilized.

This kind of operation has been investigated by many authors [37–40], who all agree that it is possible to operate with densely packed drops for very long time periods and that this kind of operation has several advantages over the common operation with loose packing. The interfacial area is thus three to five times higher, and backmixing is reduced considerably. These statements have been confirmed by Steiner et al. [41], who found that by positioning the main interface several millimeters above the rim of the riser (Figure 1) the dense dispersion can be controlled so that its lower end remains constant some 30–50 mm above the distributor. In this way, uniform conditions over the entire height of the column were produced without interdrop coalescence taking place. If the lower boundary of the densely packed dispersion is so close to the distributor that the jets above the distributor nozzles extend into it, coalescence takes place and the drops are not uniform. Once coalescence starts, it continues throughout the column height and the operation eventually can break down.

In actual fact, the danger of coalescence is the main limitation in operating with densely packed dispersions. In many transfer processes, the presence of the shared component causes coalescence, even with a loose dispersion. So if acetone is transferred from toluene into water, for example, drops coalesce a short distance above the distributor. With mass transfer in the opposite direction, the drops are stabilized and no problems arise. Examples of coalescing and noncoalescing drops can be seen in Figure 22. It has been stated already that systems that tend to coalesce should not be treated in spray columns as there is no possibility to reform the coalesced drops along the column height.

Not much is known about the dense packing operation of spray columns. Most authors limit themselves to holdup studies and have shown that the behavior shown in Figure 18 can be reproduced experimentally. This indicates that a densely packed dispersion can be compared to a fluidized bed, which also expands if the flowrate of the continuous phase is increased. Figure 17 shows that both sets of data with values in the dense packing region are correlated reasonably well by Equation 32 and, in the case of Ugarcic [3], considerably better than in the region of loose dispersion. In our laboratory it was even observed that a dense dispersion with behavior similar to that of a fixed particle bed can be achieved. If the drops fill the column completely they have nowhere to expand, and an increase in flowrate of the continuous phase does not, in this case, cause a decrease in holdup. This is rather extreme, however, as with most systems coalescence would start under such conditions.

It seems that operating spray columns under dense packing conditions could bring considerable advantages in processes in which noncoalescing drops are involved. In addition to the increase in interfacial area already mentioned, and the elimination of backmixing (to be discussed later), the residence time of the dispersed phase also exceeds its usual value considerably. Extractions with chemical reaction may therefore be a new field of application for spray columns in future, and experiments of this kind are now in progress in the authors' laboratory.

The onset of coalescence can be counteracted by adding surfactants to the continuous phase, thus decreasing the drop size and increasing holdup. Experiments of this kind

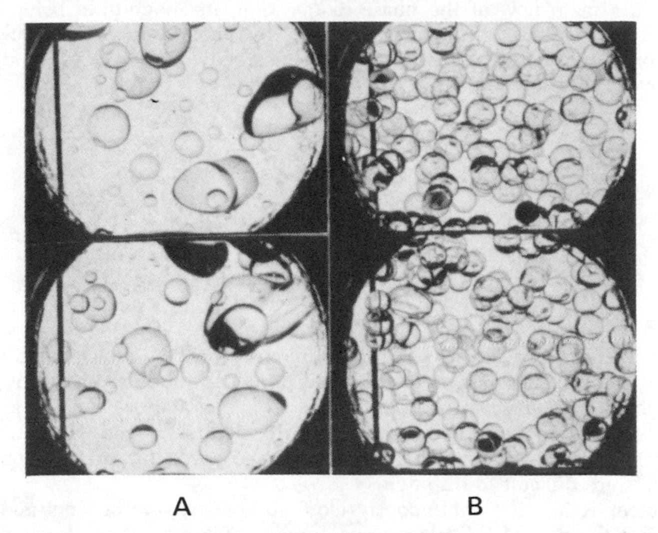

Figure 22. Examples of coalescing (A) and noncoalescing (B) drops in a spray column [18].

have been described by Dunn et al. [42] and Lackme [40]. A practical application of this method and the influence of surfactants on mass and heat transfer rates under dense dispersion conditions have not yet been studied.

AXIAL MIXING

Spray columns are designed to facilitate countercurrent contact between two liquid phases. Ideally, each phase should move in plug flow in its predetermined direction while contacting the other phase. This is impossible to achieve in practice, however, and the flow of the phases is far from ideal. If a short impulse of tracer material is injected into one of the phases, it moves along the column but its shape becomes considerably distorted. In fact, the axial mixing indicated by the distortion can become sufficiently great to control mass or heat transfer in a spray column. Internal mixing was observed in 1950 by Geankoplis and Hixon [2] and explained as axial mixing (backmixing) by Newman [43] and as bulk circulation by Gier and Hougen [44]. Miyauchi and Vermeulen [45] applied the dispersion model to this process. A review of the earlier works on axial mixing is given by Anderson and Pratt [46] and Henton and Cavers [47].

A considerable amount of literature on the phenomenon of mixing has since been accumulated; however, the mechanism responsible has not yet been resolved. It is now clear that axial mixing may be of different kinds and caused by various influences; it is dependent on physical properties, flowrates of both phases, column dimensions, drop size distribution, holdup and the transfer processes.

Up to the present, no comprehensive theory has been developed to describe axial mixing in spray columns. To correlate experimental data and quantitatively describe the mixing phenomena, backmixing is frequently expressed in terms of axial dispersion superimposed on plug flow. Such models do not consider the mechanism of the process but combine all deviations from the plug flow into single-dispersion coefficients, E_c and E_d, which then may be correlated against the different operating parameters.

A detailed description of the dispersion model is given in textbooks, such as that by Levenspiel [48]. Principally, it is assumed that an additional flux exists in the opposite direction to the main flow of the phase in question, its magnitude being proportional to the negative gradient of the concentration by analogy with molecular diffusion. The governing differential equation for the unsteady-state process and with mass transfer between the phases can be written as follows:

$$\frac{\partial x}{\partial t} = v_c \frac{\partial x}{\partial h} + E_c \frac{\partial^2 x}{\partial h^2} - \frac{k_c a}{1 - \epsilon} (x - x^*) \quad (39a)$$

or, for the dispersed phase,

$$\frac{\partial y}{\partial t} = - v_d \frac{\partial y}{\partial h} + E_d \frac{\partial^2 y}{\partial h^2} + \frac{k_c a}{\epsilon} (x - x^*) \quad (39b)$$

The dispersion coefficient, E, has the same dimensions as the molecular diffusivity, but its magnitude depends not only on the physical properties but also on the hydrodynamics of the system. Many special solutions of the above equation and its application in liquid-liquid extraction have been given by Mecklenburgh and Hartland [49]. The applicability of the above model is not generally accepted for spray columns, but the following facts are difficult to dispute:

1. If a tracer is injected continuously close to the exit of the continuous phase, it spreads against the flow of this phase. Eventually, a steady state is reached with tracer distributed over the entire column. Its concentration decreases with the distance from the point of injection, and if the logarithm of its concentration is plotted against distance, straight lines in agreement with Equation 39 are obtained. This was reported by Henton and Cavers [47] and Horvath et al. [15]. In fact, after working with spray columns for many years we have hardly seen a case in which this does not apply.
2. It is possible to reproduce concentration profiles measured experimentally on spray columns by means of the dispersion model, the mean deviations of the experimental points being better than 5%.
3. Most of the special mechanistic models can be converted to the dispersion model, giving some expression for the dispersion coefficient.
4. Due to the complicated character of the actual flows inside the spray column, an accurate description is usually impossible. Use of the dispersion model is a reasonable compromise to characterize the deviations from the plug flow at relatively low cost and in a way comparable to other types of extraction columns, where the application of the dispersion model is accepted.

To make the equations dimensionless, the dispersion coefficient usually is expressed in terms of the Péclet number, defined generally as

$$Pe = vL/E$$

where v is the velocity, and L is some characteristic length. It follows from Equation 39 that v should be the actual phase velocity and L the length of the column. However, many other definitions can be found in the literature, such as empty tube or slip velocities, and column or drop diameter instead of length. It is necessary, therefore, to distinguish these definitions clearly as they lead to large deviations in the magnitude of the Péclet number.

Continuous Phase

Backmixing in the continuous phase is much more important than that in the dispersed phase. Mixon et al. [50] found that under special circumstances it may even control the transfer performance of a spray column, i.e., determining the number of transfer units independently of the actual transfer rate. To show the magnitude of this effect, the data measured in the authors' laboratory (on one system and column

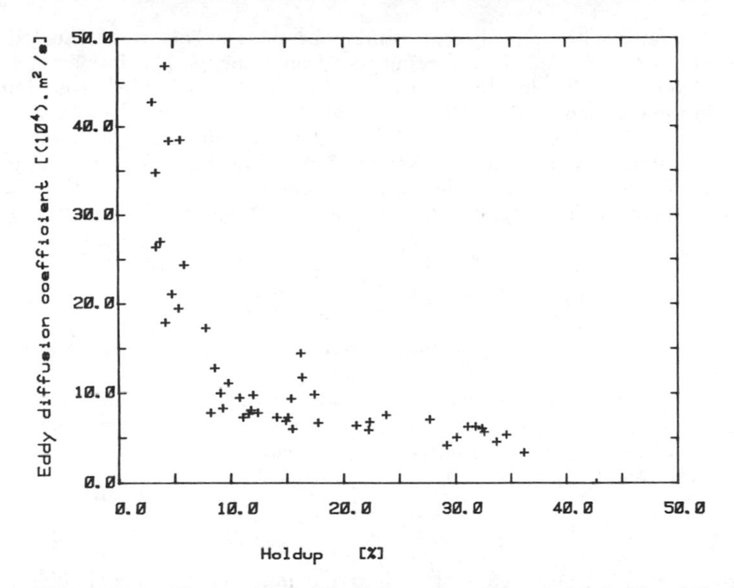

Figure 23. Variation of eddy diffusion coefficient with holdup [3].

with uniform drop size) are shown in Figure 23. A strong dependence on the holdup of the dispersed phase is observed. With no dispersed phase present, the backmixing is caused by the velocity distribution in the continuous phase only. As the Reynolds number for this phase is typically above 2000, turbulent flow may be expected with a relatively flat velocity profile. Therefore, the axial mixing is low and for the situation in Figure 23 gives values of the dispersion coefficient between 5 and 10 ($\times 10^{-4}$) m^2/sec.

This is also true for very low dispersed phase holdups (under 0.01). On increasing the throughput of the dispersed phase, the continuous phase dispersion coefficient jumps to very high levels, caused by the bulk circulation of the continuous phase induced by the upward movement of the drops. At small holdups (under approximately 0.06) the drops move up in the middle part of the column together with a core of continuous liquid, which returns to the bottom along the column walls. This frequently has been observed and described in the literature [46,51].

In addition to the bulk circulation, the drops carry wakes of the continuous phase with them as they move upward. These wakes may persist without change for long distances and cause considerable backmixing, but not with the same degree of intensity as the bulk circulation. They are considered to be mainly responsible for mixing in the region of medium holdups after the bulk circulation has been dampened by the presence of too many drops (over $\epsilon = 0.06$ in Figure 23). In the region of holdups above 0.2, the drops come so close together that no wakes can develop behind them, so the amount of backmixed liquid decreases with increasing holdup, which is contrary to the predictions of the wake theory. This situation can best be described by the same means as for backmixing in packed beds, the drops forming a more or less rigid net of channels through which the continuous phase flows. In dense dispersions with holdups above 0.5 backmixing is virtually nonexistent, as found by Ugarcic [3]. Some of the mechanisms will now be described in detail.

Bulk Circulation

This is the most powerful mechanism for mixing in the continuous phase of spray columns. It arises because the drops concentrate in the middle of the column, where they rise (or fall) with the adjacent continuous phase. This was observed experimentally using dye injection or measuring the drop velocity. The former method was used by Wijffels and Rietema [51] and the latter by Horvath [18], who found that the drops in the core move with velocities up to 50% higher than the free rise velocity of corresponding single drops. It is rather difficult to describe these phenomena quantitatively. An attempt was made by Wijffels and Rietema [51], who applied the momentum balance separately to the core containing the drops and the annulus of pure continuous phase. The final equation can be rewritten in the following form:

$$\frac{E_c}{v_r\, d_D} = \frac{u^{*2}}{24\, n\, \epsilon_k} \frac{D}{d_D} + 2\, \epsilon_k \left(1 + \frac{u^{*2}}{3}\right) + \frac{n}{8} \frac{D}{d_D} \tag{40}$$

The actual influence of the circulation is given by the first term on the right-hand side. Unfortunately, parameters such as turbulent viscosity, n, core porosity, ϵ_k, and dimensionless circulation velocity, u^*, are introduced that are difficult to obtain, even if all the simplifications used by the authors are maintained. It seems, therefore, that much more work in the field is needed before a really good expression for circulation behavior can be found.

Wakes

Hendrix et al. [52] observed that a drop moving in continuous liquid is accompanied by a volume of continuous phase, which follows the drop and from which material may possibly be exchanged with the surrounding liquid. Kehat and Letan [53] and Yeheskel and Kehat [54] studied this behavior, both for single drops and drop assemblies, and suggested that this mechanism controls the performance of spray columns. However, for the wakes to play an important part the bulk circulation must be small, as it provides mixing by an order of magnitude greater than the wakes attached to the drops. This was discussed by Anderson and Pratt [46], who recently used a photochromic dye to illustrate the phenomenon. Several regions of wake behavior were observed, starting with low Reynolds numbers ($Re < 200$), where the wakes were stable and did not share material with the surrounding liquid, up to a turbulent region with an alternating vortex street behind each drop. Yeheskel and Kehat found that in liquid fluidized beds the volume of the wakes was between 0.2 and 2.4 times the drop volume, decreasing with increasing holdup of the dispersed phase.

The typical shedding height, h_{sh}, was about 0.15 m, which means that the wake volume is exchanged with the surrounding liquid in this height. Paules and Perrut [55] have shown that the wake model can be expressed in terms of the dispersion model, the Péclet number being related to the relative wake size and shedding height as follows:

$$Pe = \frac{v_{co}}{v_{co} + M\, v_{do}} \frac{L}{M\, h_{sh}} \tag{41}$$

By evaluation of the experimental data, they obtained relative wake volumes between 0.2 and 0.8; the shedding height was found to be between 0.08 and 0.4 m, the latter parameter being independent of holdup. Using these limits in Equation 41 gives, for a 3-m-high column and phase ratio, $v_c/v_d = 1.5$, a range of Péclet numbers between 10 and 180, which is much greater than the range shown in Figure 23. Both shedding rate and drop volume therefore should be well known for a given situation if the model is to be used for practical purposes. The amount of data available for evaluation is small and mostly obtained on columns of pilot plant dimensions. The prediction of dispersion coefficients on the basis of wake parameters is hardly possible at present. Extensive experimenting on large spray columns would be necessary to determine when wakes are important and to calculate the shedding rates and wake volumes for a broad range of parameters.

Axial Dispersion in Liquid Fluidized Beds

A considerable amount of experimental material exists in the literature about mixing in fluidized beds with the solid dispersed phase. Krishnaswamy et al. [56] collected these data and derived a simple equation for the dispersion coefficient, which can be rewritten as follows:

$$E_c = \frac{v_c L}{2}\left(1 - \frac{0.74}{(1-\epsilon)^{0.25}}\right)^2 \tag{42}$$

It may be assumed that in the absence of bulk circulation and with medium to high holdup of the dispersed phase the flow patterns in a spray column are similar to those in a liquid fluidized bed. Equation 42 therefore may express the axial mixing under these circumstances.

A test on Ugarcic's data [3] (Figure 24) shows that results are actually comparable, Equation 42 giving rather high values in the region of holdups above 0.15. As expected, the data in the region of possible bulk circulation are not correlated so that the equation should be applied only for denser dispersions or columns with diameters below 50 mm where, according to Henton and Cavers [47], no bulk circulation develops. A similar analogy between flow in a spray column and through a layer of packing was investigated by Henton et al. [57], who correlated the Péclet number based on particle diameter and relative velocity (Equation 23) against the holdup; however, the equation obtained did not work on the sets of data used here.

An equation of Zheleznyak and Landau [58] also may be included in this group. It relates the dispersion coefficient to the holdup, a specially defined Reynolds number and the relative velocity, the final form being as follows:

$$E_c = 6.5 \frac{\mu_c}{\rho_c} Re_h^{0.987} \epsilon^{0.814} \bar{\mu}^{3.89} \tag{43a}$$

where

$$Re_h = \frac{\rho_c v_r}{\mu_c} \frac{D(1-\epsilon)}{1.5 \frac{\epsilon D}{d_p} + 1} \tag{43b}$$

and

$$\bar{\mu} = \frac{\mu_c + \mu_d}{\frac{2}{3}\mu_c + \mu_d} \tag{43c}$$

This equation was tested earlier [15] for different data sources and reasonable agreement was obtained. The comparison with Ugarcic's data is shown in Figure 25. Also here, the result is slightly higher in the region of medium holdup and the mixing due to circulation is not well reproduced. The equation is therefore an alternative to Equation 42.

Influence of Column Diameter

This very important factor has hardly been explored, as experimenting with full-scale columns is beyond the limits of the average university research facility. Further, industrial data are not usually made available for publication. It is assumed that axial mixing in large columns is considerably more extensive than in smaller columns, due mostly to bulk circulation. Assuming the dispersion coefficient is proportional to the column diameter, raised to some exponent ($E \sim D^n$), the data of several authors may be compared. Wijffels and Rietema [51] show that the exponent increases from 1 to 2 as the column diameter increases. Henton et al. [57] give the exponent as 2.62.

If the dependence on column diameter is calculated from Equation 43, the exponent would be close to 0.5. However, Equation 43 is not intended for use with circulatory flow, and this small influence of diameter therefore should be interpreted as an increase in axial mixing with a homogeneous distribution of drops. Circulation increases strongly with column diameter but does not lead to perfect mixing, even in columns of several meters in diameter, as it has been reported that at least two theoretical stages are available in such cases. As the applicability of spray columns depends largely on axial mixing, further investigation of its effect is of considerable importance.

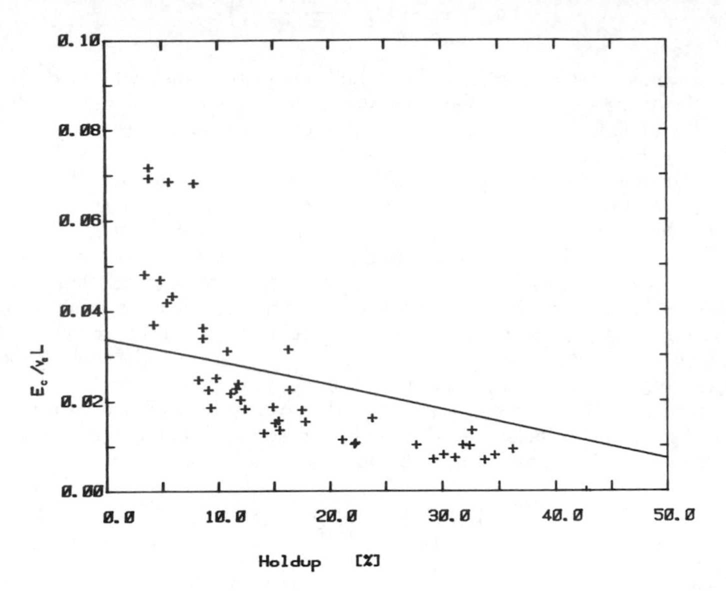

Figure 24. Comparison of Equation 42 with experimental data [3].

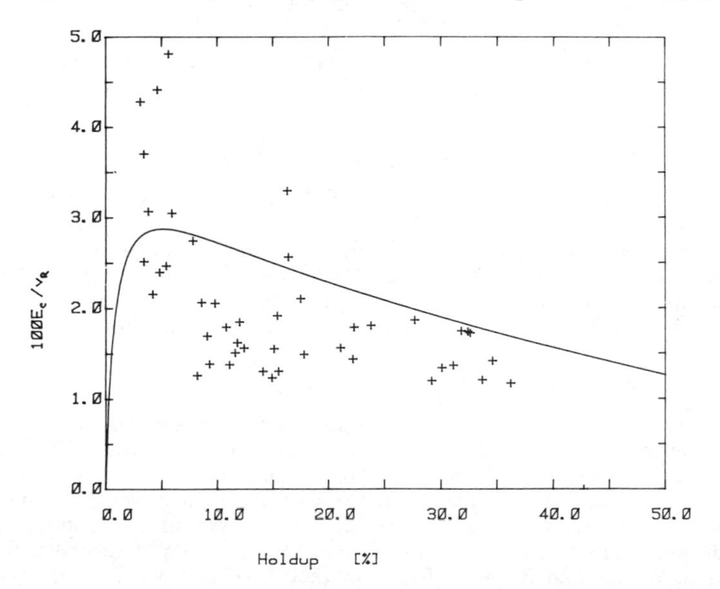

Figure 25. Comparison of Equation 43 with experimental data [3].

Dispersed Phase

If a tracer is injected continuously into the continuous phase close to its exit, it spreads upstream and some tracer may be found over the entire column. On the other hand, if the marked drops are introduced close to the dispersed-phase exit, they are not usually found in positions upstream of the point of injection. This indicates that there is no real "backmixing" in the dispersed phase in spray columns with actual movement of the drops in a direction opposite to the main stream. However, if a short impulse is injected in an upstream position it becomes distorted during its stay in the column, indicating a residence time distribution for particles forming the dispersed phase.

Few publications dealing with the dispersed phase are to be found in the literature. Letan and Kehat [59] were among the first to measure it and found practically no residence time distribution if no coalescence took place; they claimed that under normal conditions the drops are in plug flow. However, Steiner et al. [60] measured axial mixing, even without coalescence, and found the apparent dispersion coefficient was inversely proportional to the holdup. Finally, Laddha et al. [61] showed that the residence time distribution depended on the drop size distribution, this later being confirmed by our own investigations. At lower holdups it was even possible to predict the residence time distribution from the drop size spectrum by using free rise velocities for the drops.

Backward movement of the drops is not usual. The residence time distribution may be simulated from the distribution of free rise velocities due to the different drop diameters. This is shown in Figure 26, which gives the shape of an initially rectangular impulse at two downstream positions in the column. It can be seen that the observed and predicted drop size distribution shapes agree well in the main part of the curves, the tailing of experimental values probably being caused by the presence of very small drops which remain invisible behind the larger ones and therefore were not included in the photographic evaluation. We further showed [62] that it was possible to simulate the mass transfer profiles in a spray column using models that assumed plug flow in the dispersed phase. It can be assumed, therefore, that axial mixing in the dispersed phase is of minor importance in most spray columns. In contrast to the continuous phase, it should not be very dependent on column diameter.

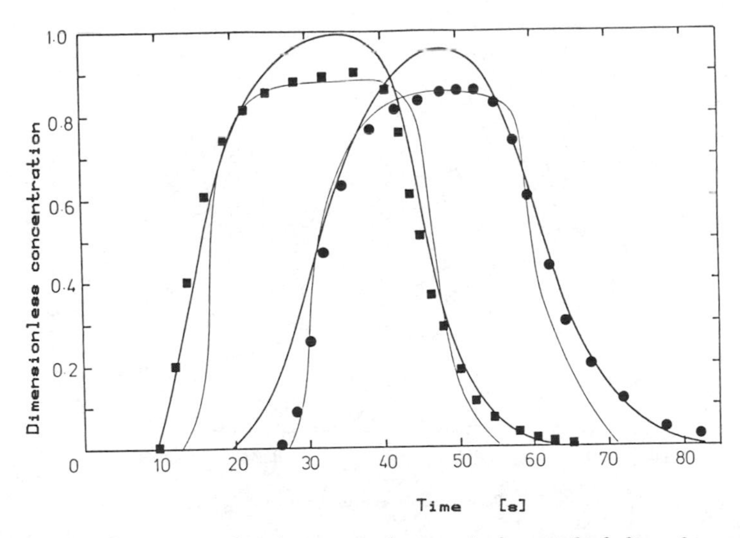

Figure 26. Residence time distribution in dispersed phase. Thick line: best approximation by dispersion model; thin line: approximation by different drop velocities [60].

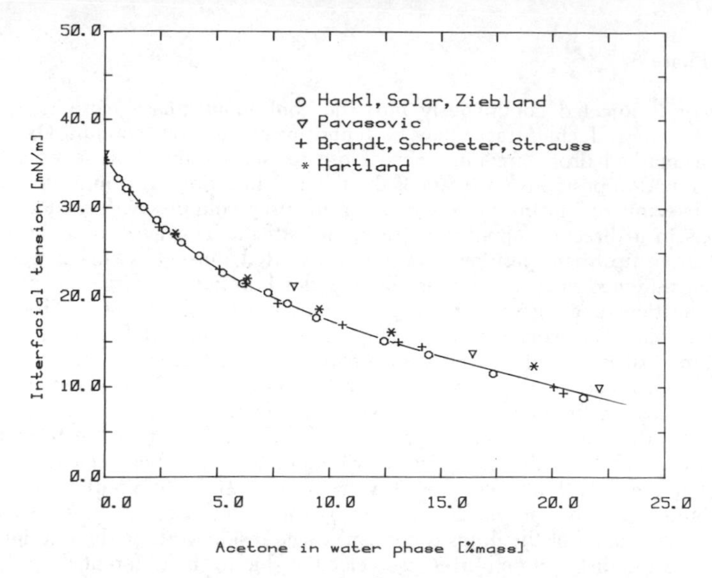

Figure 27a. Interfacial tension. The system is toluene/acetone/water [63].

COALESCENCE

Coalescence can occur only when two drops touch. The collision frequency will increase with the dispersed phase holdup and as the drop size distribution becomes broader because the droplet rise velocities will then differ. However, the countercurrent flow of continuous phase between the drops tends to keep them apart, and the random motion of the drops decreases as the holdup increases. Steiner et al. [41] have operated

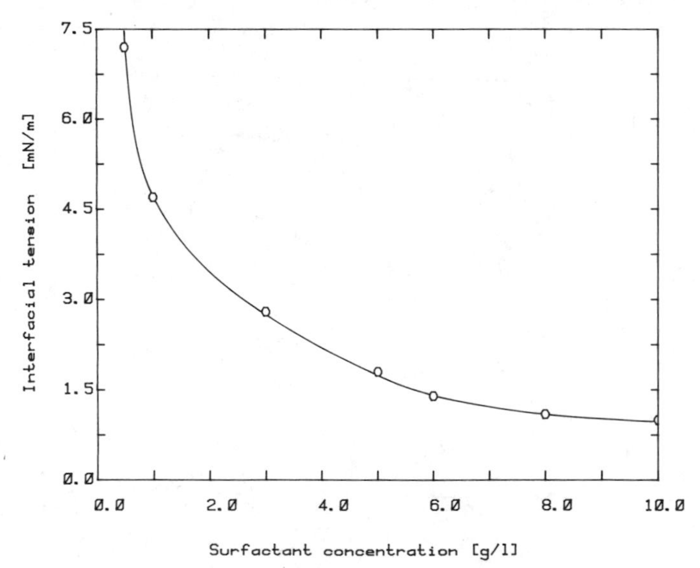

Figure 27b. Effect of Marlophene 87 concentration on the interfacial tension between water and paraffin oil at 22°C [69].

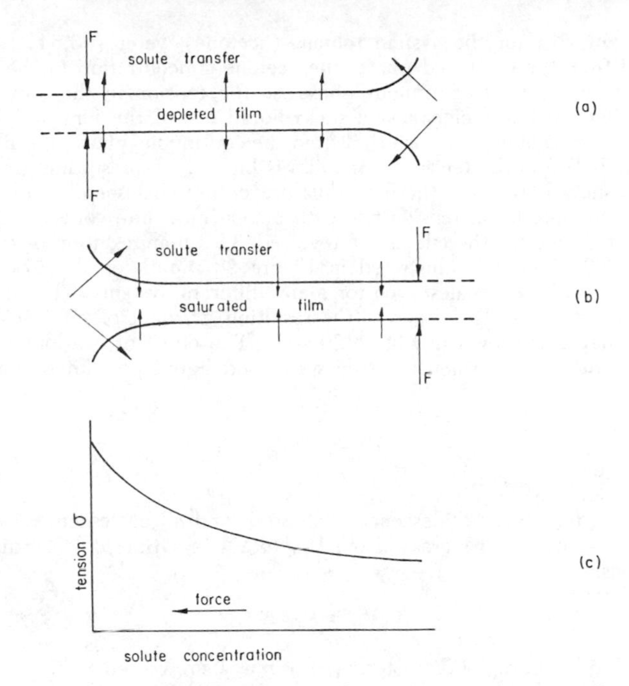

Figure 28. Origin of interfacial tension gradients due to solute transfer: (a) transfer from film; (b) transfer to film; and (c) decrease in interfacial tension with concentration.

spray columns for the *o*-xylene-water system with acetone transferred from water to xylene. These studies employed holdups as high as 70% without coalescence and at throughputs up to 20 m^3/m^2h. The *o*-xylene represented the dispersed phase and water the continuous.

As shown in Figure 22, the most important factor is the direction of mass transfer and its effect on the interfacial tension gradient in the draining film formed between the drops in contact. Normally, the interfacial tension falls with solute concentration, as

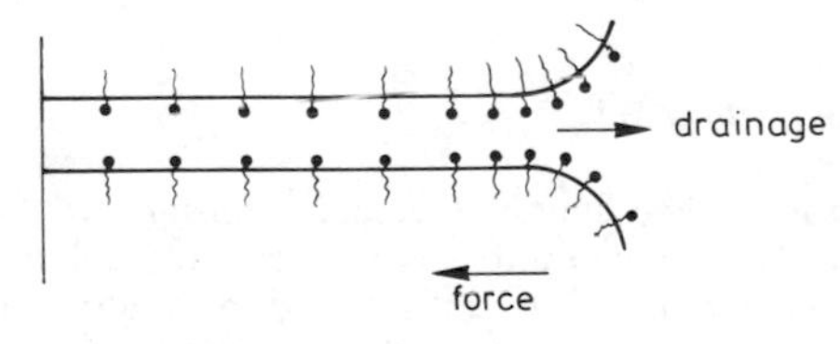

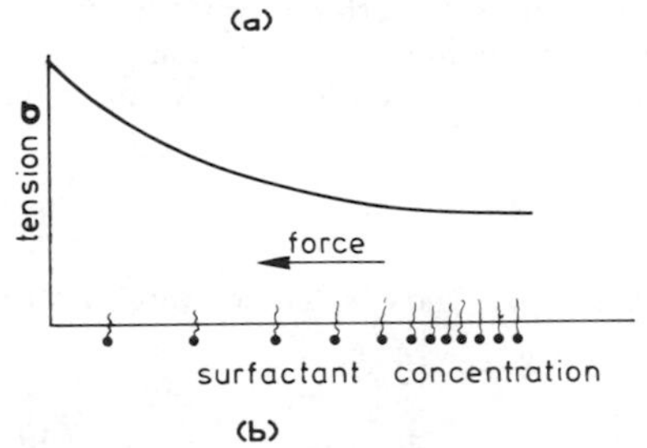

Figure 29. Origin of interfacial tension gradients due to surfactants: (a) effect of drainage on surfactant concentration in interface of draining film; (b) variation of surface tension with surfactant concentration.

shown in Figure 27a for the system toluene-(acetone)-water [63,64]. When acetone is transferred from the dispersed phase, the acetone concentration in the draining film is higher than in the bulk continuous phase, so the tension in the film is lower. The resulting gradient of interfacial tension sucks liquid out of the film and leads to rapid coalescence. When acetone is transferred from the continuous phase, the film is depleted relative to the bulk and the tension in the film is higher. The resulting gradient of interfacial tension sucks liquid into the film, thus preventing coalescence and stabilizing the drops. Should the interfacial tension rise with increasing solute concentration, as occurs with some inorganic salts, the effects are reversed. This manifestation of the well known Marangoni [65,66] effect is illustrated in Figure 28. Smith et al. [67] tabulated the effect of mass transfer on coalescence for many different systems. The presence of surfactants usually stabilizes the drops as the resulting gradient of interfacial tension opposes the drainage, as shown in Figure 29 [68]. The effect of Marlophene 87 concentration on the interfacial tension between water and liquid paraffin is shown in Figure 27b [69].

Dilute Dispersions

If the time, τ, for binary coalescence is constant and all coalescences occur together, the mean drop volume will increase from V_{Do} to V_D in a time, t, according to the following relationship:

$$V_D = V_{Do} 2^{t/\tau}$$

As $V_D = \pi d_D{}^3/6$ for a drop of diameter, d_D, this may be rewritten as

$$d_D = d_{Do} \exp[(ln 2) t/3\tau] \tag{44}$$

For a dispersed phase throughput, $\dot{V}_d$, and cross-sectional area, A, the coalescence time, τ, is given by

$$\tau = lA\epsilon/\dot{V}_d \tag{45}$$

if the drops coalesce together after traversing a column length, l.
Similarly, the time, t, for the drops to traverse a height z is given by

$$t = zA\epsilon/\dot{V}_d \tag{46}$$

so Equation 44 becomes

$$d_D = d_{Do} \exp[(ln 2) z/3\tau] \tag{47}$$

when the dispersed phase holdup, ϵ, and coalescence length, l, remain constant as the drop size increases, as shown in Figure 30a. The same result is obtained if the column is split into a series of perfectly mixed stages, each of length l, as illustrated in Figure 30b.

However, the drop motion in a spray column approximates better to plug flow than perfect mixing and drops coalesce continuously throughout the column. Binary coalescence is a random process, and the coalescence time, τ, represents the mean of a range of values. In a unit column volume there are

$$N = 6\epsilon/\pi d_D^3 \tag{48}$$

drops of diameter, d_D, so the rate of coalescence is N/τ. And, as half of these drops disappear, we may write

$$-\frac{dN}{dt} = \frac{N}{2\tau} \tag{49}$$

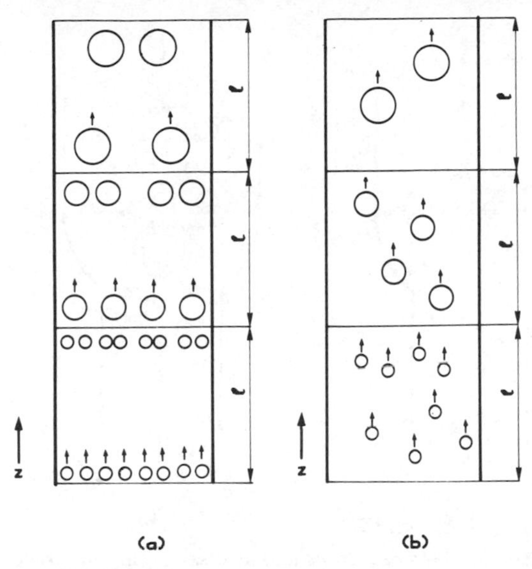

Figure 30. Illustration of coalescence height, l: (a) drops in plug flow; (b) perfectly mixed stages.

If there are N_0 drops at time zero, this integrates to

$$ln\frac{N_0}{N} = \frac{t}{2\tau} \quad (50)$$

Using Equation 44 with constant ϵ, this becomes

$$d_D = d_{Do}\exp(t/6\tau) \quad (51)$$

or, in terms of coalescence length, l, and distance, z,

$$d_D = d_{Do}\exp(z/6l) \quad (52)$$

These equations are similar to Equations 44 and 47, in which the numerical constant 6 replaces $3/ln2$ ($= 4.3$). The coalescence frequency, $f = (1/N)(-dN/dt)$, is related to the coalescence time, τ, through Equation 49 by $f = 1/2\tau$. Combining Equations 48 and 49 gives $f = (3/d_D)(dd_D/dt)$, which was used by Howard [70] to measure the coalescence frequency in stirred tanks in which the intensity of agitation suddenly was reduced.

Densely Packed Dispersions

The drops leave the extraction section with a diameter, d_{D1}, and collect together to form a closely packed dispersion of average holdup, ϵ_b, in which they coalesce with each other and finally with their homophase before leaving the column, as shown in Figure 31a [71]. If the height of the closely packed dispersion is H, its cross-sectional area is A and the binary coalescence time, τ_b, the drop size, d_{Di}, at the disengaging interface is given by combining the appropriate forms of Equations 46 and 51:

$$d_{Di} = d_{D1}\exp(HA_b\epsilon_b/6V_d\tau_b) \quad (53)$$

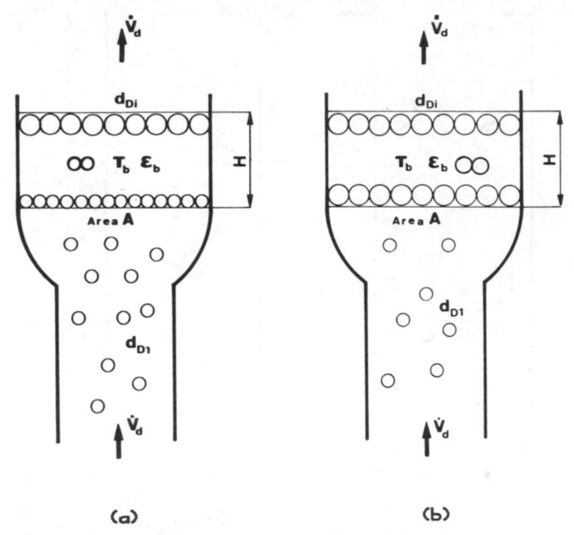

Figure 31. Coalescence between drops and at disengaging interface in densely packed dispersions: (a) drops in plug flow; (b) perfect mixing.

The number of drops at the disengaging interface of area A_i, where the holdup is ϵ_i, is $4A_i\epsilon_i\gamma_i/\pi d_{Di}^2$; in this expression, γ_i is a shape factor, which allows for the nonsphericity of the drops. Dividing by the coalescence time at the interface τ_i and multiplying by the drop volume $\pi d_{Di}^3/6$ gives the volume rate of coalescence, so at the steady state

$$\dot{V}_d/A_i = (2/3)\epsilon_i\gamma_i\, d_{Di}/\tau_i \tag{54}$$

Substituting for d_D from Equation 53 and rearranging gives the dispersion thickness as

$$H = \frac{6\dot{V}_d\tau_b}{A_b\,\epsilon_b}\, ln\left(\frac{3V_d\tau_i}{2\,A_i\epsilon_i\gamma_i d_{D1}}\right) \tag{55}$$

as derived by Hartland and Vohra [72]. In this equation, both ϵ_i and γ_i are close to, but less than, 1. The values of A_b and A_i usually will be equal.

Conditions in a spray column are relatively calm [73], so in the above analysis it is assumed that the drops move through the densely packed dispersion in plug flow. However, if the flow is turbulent and the densely packed dispersion is thoroughly mixed, as shown in Figure 31b, the drop diameter at the interface in Equation 54 is given by combining the relevant forms of Equations 44 and 46, so the dispersion height is given by

$$H = \frac{3\,\dot{V}_d\tau_b}{(ln2)\,A_b\epsilon_b}\, ln\left(\frac{3\,V_d\tau_j}{2\,A_i\epsilon_i\gamma_i d_{D1}}\right) \tag{56}$$

which only differs from the previous equation in the numerical constant.

Empirical Models

To determine H, the values of τ_i, τ_b and τ must be known. (The latter enables d_{D1} to be found from d_{Do}, using Equations 47 or 51.) The values of τ_i and τ_b may themselves be functions of the operating parameters. They are strongly dependent on the geometric configuration of the drops and the flow conditions, as well as on the physical properties of the system.

For example, the binary coalescence time, τ_b, is affected by the degree of turbulence in the dispersion, which we will assume increases with the dispersed phase throughput per unit area, A:

$$\tau_b = k_b(\dot{V}_d/A)^p \tag{57}$$

where p is a dimensionless index, and k_b is a dimensional constant. Equations 55 and 56 then can be rewritten as

$$H = c_b(\dot{V}_d/A)^{p+1}\, ln(c_i\, \dot{V}_d/A) \tag{58}$$

where $c_b = 6k_b/\epsilon_b$ or $3\ k_b/\epsilon_b\ ln2$, respectively, and $c_i = 3\tau_i/2\epsilon_i\gamma_i d_{D1}$. When $p > 1$, this equation is similar to the empirical result of Ryan et al. [74]:

$$H = c(\dot{V}_d/A)^q \tag{59}$$

where c is a constant for a given system, and q usually lies between 2 and 3 for relatively calm dispersions, but may be much higher for turbulent dispersions.

In addition, the coalescence time of a drop increases with the area of the draining film and decreases as the force pressing on it increases [75]. One therefore would expect τ_i to increase with drop size, d_{Di}, and the decrease with dispersion height, H, as shown in Figure 32a. As a simplification, we may write

$$\tau_i = k_i d_{Di}^n/H^m \tag{60}$$

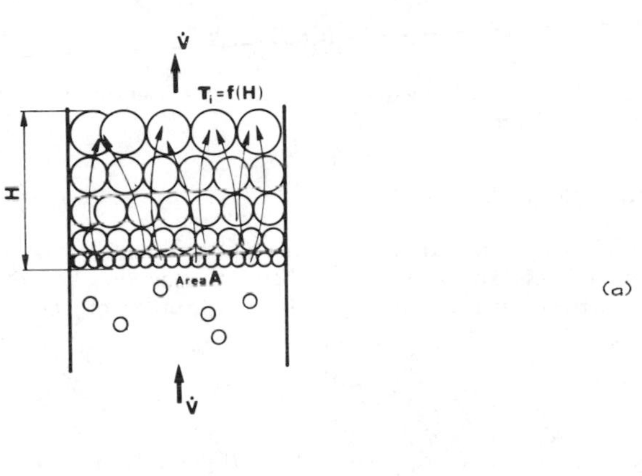

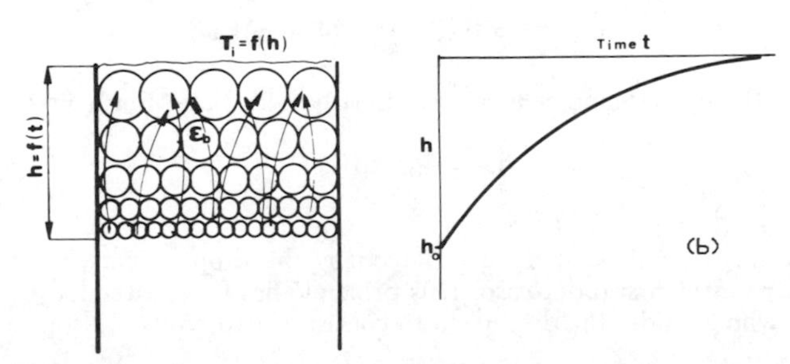

Figure 32. Transmission of gravitational forces: (a) steady-state dispersion of height, H; (b) decaying dispersion in which height, h, decreases with time, t.

where n and m are dimensionless indices and k_i is a dimensional constant. When n = 1, substitution into Equation 54 with $\epsilon_i \simeq \gamma_i \simeq 1$ and $A_b = A_i = A$ gives

$$H = \left(\frac{3}{2} k_i \frac{\dot{V}_d}{A}\right)^{1/m} \tag{61}$$

which is again similar to the empirical equation (59) when $1/3 < m < 1/2$.

The value of k_i can be found by allowing the dispersion to decay, as shown in Figure 32b, and following its height, h, as a function of time, t. With the current assumptions, we may then write

$$-\epsilon_b \frac{dh}{dt} = \frac{2}{3} \frac{h^m}{k_i} \tag{62}$$

because the volume rate of coalescence given by Equation 54 is now equal to the rate of decrease in volume of the dispersion. Integrating this equation with $h = h_o$ at $t = 0$ gives

$$h_o^{1-m} - h^{1-m} = \frac{2t}{3\epsilon_b k_i} \tag{63}$$

so a plot of h^{1-m} versus t thus yields a straight line of slope $1/k_i$, if we assume $\epsilon_b \simeq 2/3$. When $h << h_o$, the variation of h/h_o with $ln t$ is approximately linear with slope $-1/(1-m)$. Furthermore, if the dispersion completely disappears after a time, T, the values of m and k_i are related by

$$(1-m) \ln h_o = \ln(2T/3\epsilon_i k_i) \tag{64}$$

For isobutanol dispersed in water, the value of k_i is about 4.5 $s.mm^{-1/2}$ (with $m = 1/2$) for an initial drop size of 0.8 mm.

Analogy with Chemical Reaction

Because meaningful coalescence times are difficult to obtain, coalescence has been considered as a rate process analogous to a chemical reaction. For example, let us suppose the drop growth in binary coalescence is an nth order process:

$$\frac{dd_D}{dt} = k\, d_D^n \tag{65}$$

Integrating with boundary condition $d_D = d_{Do}$ at $t = 0$ gives, for $n \neq 1$,

$$d_D^{1-n} = d_{Do}^{1-n} + (1+n)kt \tag{66}$$

so when n = 0 the drop diameter increases linearly with time. For a first-order process,

$$d_D = d_{Do} \exp(kt) \tag{67}$$

which is Equation 51, with $k = 1/6\tau$.

A similar result is obtained if d_D is replaced by the drop volume $V_D = \pi d_D^3/6$, in which $k = 1/2\tau$ for the first-order case. This principle has been extended by Stönner and Weisner [76], who consider the dispersion to consist of two kinds of drops:

1. a fraction θ, which reacts in the bulk of the dispersion, and
2. a fraction $1-\theta$, which coalesces at the disengaging interface.

Both reactions are assumed to be first order with rate constants k_1 and k_2. At the steady state, the fraction θ decreases from θ_o at the inlet so the fraction $1 - \theta$ increases until the volume rate of coalescence at the disengaging interface $k_2A_i(1 - \theta)$ balances the volume feedrate of dispersed phase $\dot{V}_d$, so that

$$\dot{V}_d = k_2 A \epsilon_i (1 - \theta) \tag{68}$$

When the dispersion of volume $V = AH$ is perfectly mixed, as shown in Figure 33a, the feedrate of interreacting drops, $\dot{V}_d\theta_o$, must equal the rate at which they disappear in the dispersion bulk, $k_1 V \epsilon_b \theta$, so

$$\dot{V}_d \theta_o = k_1 V \epsilon_b \theta \tag{69}$$

Eliminating θ between these two equations gives the variation in H with $\dot{V}_d/A$ as

$$\frac{\theta_o}{H} = \frac{\epsilon_b k_1 A}{V_d} - \frac{\epsilon_b k_1}{\epsilon_i k_2} \tag{70}$$

The constants $\epsilon_b k_1/\theta_o$ and $\epsilon_i k_2$ can be obtained from the slope and intercept of this linear equation if the variation in dispersion height, H, with throughput $\dot{V}_d/A$ is known from subsidiary experiments. However, since spray column conditions are relatively calm, the dispersion is not likely to be perfectly mixed. When the drops move in plug flow, as shown in Figure 33b, the change in θ with distance, z, is given by

$$-\dot{V}_d \frac{d\theta}{dz} = k_1 A \epsilon_b \theta \tag{71}$$

so at the disengaging interface

$$\theta_i = \theta_o \exp(-l_1 A \epsilon_b H/\dot{V}_d) \tag{72}$$

Setting this in Equation 68 with $\theta = \theta_i$ gives

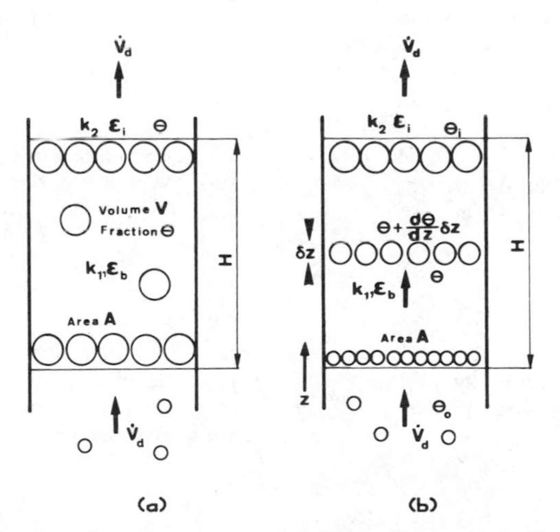

Figure 33. Interreaction of fraction, θ, of drops in dispersion bulk with rate constant, k_1, and of fraction, $1 - \theta$, at disengaging interface with rate constant, k_2: (a) perfectly mixed dispersion; (b) drops in plug flow.

$$\epsilon_b k_1 H \frac{A}{\dot{V}_d} = ln\ \theta_o - ln\left(1 - \frac{\dot{V}_d}{\epsilon_i k_2 A}\right) \tag{73}$$

which reduces to

$$H = \frac{1}{\epsilon_b \epsilon_i k_1 k_2}\left(\frac{\dot{V}_d}{A}\right)^2 \tag{74}$$

when $\theta_o \simeq 1$ and $\dot{V}_d/\epsilon_i k_2 A << 1$ (corresponding to $\theta_i \simeq 1$, so $k_1 \epsilon_b HA/\dot{V}_d << 1$). This equation is then similar to the empirical equation (59). It should apply when $\dot{V}_d/A$ is small or when $k_1\epsilon_b$ is small and $k_2\epsilon_i$ is large. For the system kerosene-water used by Barnea and Mizrahi [77], the values of $k_1\epsilon_b$ and $k_2\epsilon_i$ determined from Equation 70 with $\theta_o = 1$ are 0.004 sec^{-1} and 5.84 mm/sec, respectively, so $1/\epsilon_b\epsilon_i\ k_1k_2 \simeq 4\ sec^2/mm$.

NOMENCLATURE

- A — area (m^2)
- a — interfacial area per unit volume (m^2/m^3)
- c — dimensional constant
- D — column diameter (m)
- d — diameter (m)
- d_1 — drop diameter entering densely packed dispersion (m)
- d_{32} — Sauter mean diameter (m)
- E — dispersion coefficient (m^2/sec)
- F — Harkins-Brown correction factor
- f — coalescence frequency (sec^{-1})
- g — gravitational acceleration (m/sec^2)
- H — steady-state dispersion height (m)
- h — height (m)
- K — mass transfer coefficient (m/sec)
- k — dimensional constant
- k_1,k_2 — rate constants (sec^{-1},m/sec)
- L — column length (m)
- l — coalescence length (m)
- M — relative wake volume
- m — index
- n,p,q — indices
- Δp — pressure difference (kg/m-sec^2)
- S — surface area (m^2)
- T — time (sec)
- t — time (sec)
- N — number of drops per unit volume (m^{-3})
- V — volume (m^3)
- $\dot{V}$ — volume throughput (m^3/sec)
- v — velocity (m/sec)
- x,y — concentrations in continuous and dispersed phases
- z — distance measured from entry of dispersed phase (m)

Dimensionless Groups

- Ar — Archimedes number, $\Delta\rho\ d^3g/\mu^2$
- Eö — Eötvös number, $\Delta\rho\ gd^2/\sigma$
- Fr — Froude number, v^2/gd
- Re — Reynolds number, $\rho\ dv/\mu$
- We — Weber number, $\rho\ dv^2/\sigma$

Greek Symbols

- γ — shape factor
- $\Delta\rho$ — density difference (kg/m^3)
- ϵ — holdup of dispersed phase
- θ — fraction of drops
- μ — viscosity (kg/m-sec)
- ρ — density (kg/m^3)
- σ — interfacial tension (mN/m)
- τ — coalescence time (sec)

Subscripts

- b — binary coalescence
- C — critical
- c — continuous
- D — drop

d	dispersed	n	nozzle
f	flooding	o	specific value
i	disengaging interface	r	relative
j	jetting	x	characteristic velocity

Superscripts

*	equilibrium value

REFERENCES

1. Elgin, J. C., and H. C. Foust. *Ind. Eng. Chem.* 42:1127 (1950).
2. Geankoplis, C. J., and A. N. Hixon. *Ind. Eng. Chem.* 42:1141 (1950).
3. Ugarcic, M. Thesis No. 6825, ETH Zürich, Switzerland (1981).
4. Minard, G. W., and A. I. Johnson. *Chem. Eng. Prog.* 48:62 (1952).
5. Hayworth, C. B., and R. E. Treybal. *Ind. Eng. Chem.* 42:1174 (1950).
6. Harkins, W. D., and F. E. Brown. *J. Am. Chem. Soc.* 41:499 (1919).
7. Null, H. R., and H. F. Johnson. *AIChE J.* 4:273 (1958).
8. Scheele, G. F., and B. J. Meister. *AIChE J.* 14:9 (1968).
9. de Chazal, L. E. M., and J. T. Ryan. *AIChE J.* 17:1226 (1971).
10. Kagan, S. Z., Y. N. Kovalev and A. P. Zakharychev. *Theor. Osn. Chem. Eng.* (*Moscow*) 7:514 (1973).
11. Skelland, A. H. P., and K. R. Johnson. *Can. J. Chem. Eng.* 52:732 (1974).
12. Rayleigh, Lord. *Phil. Mag.* 34:177 (1892).
13. Scheele, G. F., and B. J. Meister. *AIChE J.* 14:15 (1968).
14. Treybal, R. E. *Liquid Extraction* (New York: McGraw-Hill Book Co., 1963).
15. Horvath, M., L. Steiner and S. Hartland. *Can. J. Chem. Eng.* 56:9 (1978).
16. Mersmann, A. *Chem. Ing. Tech.* 49:679 (1977).
17. Miller, H. D., and T. Pilhofer. *Chem. Ing. Tech.* 48: MS 421/76 (1976).
18. Horvath, M. Thesis No. 5774, ETH Zürich, Switzerland (1976).
19. Mugele, R. A., and H. D. Evans. *Ind. Eng. Chem.* 43:1317 (1951).
20. Slater, M. J. *Chem. Ing. Tech.* 46: MS 116/74 (1974).
21. Pilhofer, T., and D. Mewes. *Siebboden-Extraktionskolonnen* (Weinheim, West Germany: Verlag Chemie, 1979).
22. Hu, S., and R. C. Kintner. *AIChE J.* 1:42 (1955).
23. Klee, A. J., and R. E. Treybal. *AIChE J.* 2:444 (1956).
24. Thornton, J. D. *Chem. Eng. Sci.* 5:201 (1956).
25. Steinour, H. *Ind. Eng. Chem.* 36:618 (1944).
26. Misek, T. *Coll. Czech. Chem. Comm.* 28:1613 (1963).
27. Andersson, K. E. B. *Chem. Eng. Sci.* 15:276 (1961).
28. Ferrarini, R. *VDI-Forschungsheft* 551 (1972).
29. Pilhofer, T. *Chem. Ing. Tech.* 46: MS 133/74 (1974).
30. Kumar, A., D. K. Vohra and S. Hartland. *Can. J. Chem. Eng.* 58:154 (1980).
31. Zenz, F. A. *Pet. Ref.* 36:147 (1957).
32. Barnea, E., and J. Mizrahi. *Can. J. Chem. Eng.* 53:461 (1975).
33. Garwin, L., and B. D. Smith. *Chem. Eng. Prog.* 49:591 (1953).
34. Loutaty, R., A. Vignes and P. LeGoff. *Chem. Eng. Sci.* 24:1795 (1969).
35. Letan, R., and E. Kehat. *AIChE J.* 13:443 (1967).
36. Pilhofer, T. *Chem. Ing. Tech.* 48: MS 331, 76 (1976).
37. Loutaty, R., and A. Vignes. *Chim. Ind.* 101:231 (1969).
38. Kehat, E., and R. Letan. *Ind. Eng. Chem., Proc. Des. Dev.* 7:385 (1968).
39. Geskowich, E. J. *Ind. Eng. Chem., Proc. Des. Dev.* 8:593 (1969).
40. Lackme, C. *AIChE Symp. Ser. No. 138,* 70:57 (1974).
41. Steiner, L., M. Ugarcic and S. Hartland. Paper presented at the CHISA Congress, Prague, 1978.

42. Dunn, I., L. Lapidus and J. C. Elgin. *AIChE J.* 11:158 (1965).
43. Newman, M. L. *Ind. Eng. Chem.* 44:2457 (1952).
44. Gier, T. E., and J. O. Hougen. *Ind. Eng. Chem.* 45:1362 (1953).
45. Miyauchi, T., and T. Vermeulen. *Ind. Eng. Chem. Fund.* 2:113 (1963).
46. Anderson, W. J., and H. R. C. Pratt. *Chem. Eng. Sci.* 33:995 (1978).
47. Henton, J. E., and S. D. Cavers. *Ind. Eng. Chem. Fund.* 9:384 (1970).
48. Levenspiel, O. *Chemical Reaction Engineering* (New York: John Wiley & Sons, Inc., 1962).
49. Mecklenburgh, J. C., and S. Hartland. *The Theory of Backmixing* (New York: John Wiley & Sons, Inc., 1975).
50. Mixon, F. O., D. R. Whitaker and J. C. Orcutt. *AIChE J.* 13:21 (1967).
51. Wijffels, J. B., and K. Rietema. *Trans. Inst. Chem. Eng.* 50:224 (1972).
52. Hendrix, C. D., S. B. Dave and H. F. Johnson. *AIChE J.* 13:1072 (1967).
53. Kehat, E., and R. Letan. *AIChE J.* 17:984 (1971).
54. Yeheskel, J., and E. Kehat. *Chem. Eng. Sci.* 26:2037 (1971).
55. Paules, B., and M. Perrut. *Proc. ISEC* (Lyon, France) 997 (1974).
56. Krishnaswamy, P. R., R. Ganapathy and L. W. Shelmit. *Can. J. Chem. Eng.* 56:550 (1978).
57. Henton, J. E., L. W. Fish and S. D. Cavers. *Ind. Eng. Chem. Fund.* 12:365 (1973).
58. Zheleznyak, A. S., and A. M. Landau. *Theor. Osn. Ch. Techn. Moscow* 7:577 (1973).
59. Letan, R., and E. Kehat. *AIChE J.* 15:4 (1969).
60. Steiner, L., M. Stemper and S. Hartland. *Chem. Ing. Tech.* 50: MS 585, 78 (1978).
61. Laddha, G. S., T. R. Krishnan, S. Viswanathan, S. Wedaiyan, T. E. Degaleesan and H. E. Hoelscher. *AIChE J.* 22:456 (1976).
62. Steiner, L., M. Horvath and S. Hartland. *Proc. ISEC* (1977), CIM spec. Vol. 1979, 21:366.
63. Misek, T. "Recommended Systems for Liquid Extraction Studies," European Federation of Chem. Engineers, Institute of Chemical Engineers, London (1978).
64. Hartland, S., and L. Steiner. "Experience with Liquid/Liquid Test Systems in Extraction Columns," paper presented at the NATO Conference, Cesme, Turkey, 1981.
65. Marangoni, C. *Nuovo Cimento* 2:239 (1871).
66. Marangoni, C. *Nuovo Cimento* 3:97 (1878).
67. Smith, A. R., J. E. Caswell, P. P. Larson and S. D. Cavers. *Can. J. Chem. Eng.* 150 (August 1963).
68. Hartland, S. *Chimia* 27:238 (1973).
69. Hartland, S. Paper presented at the Fourteenth International Conference on Surface & Colloid Science, Jerusalem, 1981.
70. Howard, W. J. *AIChE J.* 13:1007 (1967).
71. Hartland, S. *Bunsengesellschaft Phys. Chem.* 85:857 (1981).
72. Hartland, S., and D. K. Vohra. *Chem. Ing. Tech.* 50:673 (1978).
73. Hartland, S. "The Coalescence of Liquid Drops," British University Film Council, 1970.
74. Ryan, A. D., F. L. Daley and R. S. Lowrie. *Chem. Eng. Prog.* 55:70 (1959).
75. Hartland, S., and D. K. Vohra. *Can. J. Chem. Eng.* 59:438 (1981).
76. Stönner, H. M., and P. Weisner. Paper presented at the NIM Symposium, Johannesburg, South Africa, February, 1980.
77. Barnea, E., and J. Mizrahi. *Trans. Inst. Chem. Eng.* 53:83 (1975).

CHAPTER 41

CHARGE GENERATION AND TRANSPORT DURING FLOW OF LOW-CONDUCTIVITY FLUIDS

J. P. Wagner
Exxon Research & Engineering Company
Florham Park, New Jersey 07932

CONTENTS

INTRODUCTION

This chapter on special flow topics presents a fundamental treatment of electrostatic phenomena involving primarily low dielectric constant or insulating liquids in flow situations and, to a lesser extent, processes involving solids and gases. Following an overview of general electrostatic phenomena, a separate section on ionization behavior in low-conductivity liquids is included to acquaint the reader with the chemistry necessary for the basic development of the charge conservation equations.

The equations are formulated on the basis of transport of a flux of uni-univalent ions by conduction, convection and diffusion. They are finally expressed in terms of a time-dependent equation for charge density, rather than in terms of an equation formulated in terms of a streaming or zeta potential. This approach permits one to use heat or mass transfer correlations to describe charge transfer processes. The physical significance of the important dimensionless groups that arise in various charging processes are discussed in the sections entitled Similitude Principles, Boundary Conditions for Charged Interface and Dimensional Analysis. Theoretical solutions to the charge transport equations are illustrated for selected processes.

Four examples involving turbulent charging in capillaries, flow through porous media, charge relaxation and laminar charging in capillaries are analyzed in detail. The successful application of the charge transport equations or dimensional analysis is validated by an extensive body of experimental data for turbulent flow and flow through porous media.

Static hazards and accepted control methods are discussed because of their industrial importance. Finally, experimental methods and conductivity data analysis are highlighted in the Appendix.

An important objective of this chapter is to show the applicability of the charge transport equations or modeling techniques to different processes occurring in dielectric fluids. Possible areas to extend these approaches include separation processes for coal-derived liquids, immiscible liquid-liquid systems and other systems that can be treated as dielectric fluids.

GENERAL ELECTROSTATIC PHENOMENA

Electrostatic or charge separation phenomena can occur wherever two dissimilar phases are brought into contact as a result of migration of some constituent of the phases up to, or across, the interface. Charging mechanisms differ considerably among solids, liquids and gases. Major differences also exist within single classes of matter as in metals, semiconductors, and insulators. Harper [1] has examined how solid surfaces become charged, and provided reasons for movement and termination of the charge carriers. A synopsis on surface charging [1] is given below:

1. **Metals.** Either a surplus or deficit of electrons can move through a metal; however, at equilibrium the charge is confined to the surface. There is no internal field.
2. **Semiconductors.** (**Relatively High Conductivity**) These can concentrate charge into a relatively thin surface layer because of a high concentration of impurity centers. Electrical transfer can take place on contact with a metal or a semiconductor of a different energy level. Charging is now confined to the surfaces. (**Lower Conductivity**) The charged surface layer becomes thicker and extends farther into the interior because of reduced concentration of impurity centers. A longer time is required for the charge to propagate deeper into the material.

Table I. Illustration of a Triboelectric Series

+
Asbestos
Glass
Mica
Wool
Cat fur
Lead
Silk
Aluminum
Paper
Cotton
Wood, Iron
Sealing wax
Ebonite
Nickel, Copper, Silver, Brass
Sulfur
Platinum, Mercury
India rubber
−

3. **Insulators.** (**Electrophobic**) Practically does not charge at all. (**Electrophilic**) Charges freely. The charge carriers can only be ions.

Charging can occur through contact followed by separation or through frictional charging. Frictional charging is illustrated by rubbing of two dissimilar solids (one of which is an insulator), which leads to one becoming charged electropositively and the other electronegatively. The so-called triboelectric series may be obtained for various solid materials. An example of a series containing both insulators and metallic conductors is given in Table I. The materials at the top of the column are positive with respect to the materials lower in the list. Since some of these materials are hygroscopic and are thus affected by humidity and surface ionic impurities, and because the type of contact influences all materials [1], differences in this listing are encountered frequently. Shinohara et al. [2] developed a triboelectric series for p-substituted styrene polymers and other vinyl polymers (Tables IIA and IIB). The authors found that the electron-releasing polymers are situated on the positive side of the series, while electron-attracting polymers are located on the negative side. Strong differences in surface potentials also were noted for charging of the samples with iron powder or nickel powder.

Table IIA. Triboelectric Series Highlights for Selected Styrene and Vinyl Plastics [2]—Triboelectric Series of p-Substituted Styrene Polymers

Polymer	Triboelectric Series	E_{max}(V) For Iron Powder	E_{max}(V) For Nickel Powder
polydimethylaminostyrene	+	+180	+110
polyaminostyrene	↑	+140	− 10
polymethoxystyrene		−210	−152
polymethylstyrene		−170	−200
polystyrene		−200	−190
polychlorostyrene	↓	−220	−178
polynitrostyrene	−	−240	−181

Table IIB. Triboelectric Series Highlights for Selected Styrene and Vinyl Plastics—Relation Between Charging Sign and Chemical Structure

	Chemical Structure	Charging Sign
poly-2-vinyl pyridine	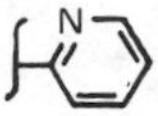 (2-pyridyl ring, N)	+
poly-4-vinyl pyridine	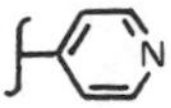 (4-pyridyl ring, N)	+
polyvinylacetate	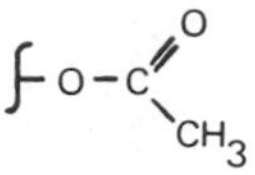 $-O-C(=O)-CH_3$	+
polyvinylchloride	$-Cl$	−

Triboelectric effects were studied for various production fabrics by Frederick [3] (Table III) for the purpose of developing fabric-type filters for solids removal. Criteria for the most effective filters were developed for very fine agglomerating and non-agglomerating particles. To use the information contained in Table III, information is required about the charging tendencies of the dust to be filtered. An experimental setup [2,4] may be used to measure charging tendencies of dusts. This is an area that is not well understood and requires future research efforts.

Table IV illustrates the reasons for movement and termination of charge carriers. Little is understood about charging by material transfer considered under bulk matter transfer. Selected authors [5–7] have used the material transfer approach to explain solids charging phenomena. Following charge termination, or when equilibrium is established, there will be an excess of charge of one sign in one phase and an excess of charge of the opposite sign in another phase or at the interface. This results in various charge separation phenomena (Table V), some unwanted, others of commercial applicability. The details of various technologically important charging processes are provided in the monograph by Moore [8].

Some of the aspects of the examples in Table V, such as the thermocouple junction potential, polarographic analysis using the dropping mercury electrode, electrophoresis, electroosmosis and the electrolytic cell have been studied extensively and are reasonably well understood. Other processes dealing with streaming potential associated with flow past surfaces of relatively high-conductivity liquids are well understood. Examples include flows of aqueous solutions [9,10] conducting organic liquids (hydrocarbons) in turbulent flow through capillaries [11–13] and particulate flows through electrostatic precipitators [14]. Charging processes dealing with solids, e.g., electrification of materials by rubbing [1,2,7,15,16], charge generation and accumulation on synthetic fabrics [17,18], and coalescence of two- [19–26] and three-phase emulsions [27–33] are poorly understood. Basic differences in charging behavior can occur for these examples, depending primarily on large differences in electrical conductivity for low dielectric constant and high dielectric constant media. Table VIA gives some typical values for different materials. (Methods to determine the dielectric constant and the specific conductivity are discussed in Appendix A.)

A simple example will serve to illustrate pronounced differences due to conductivity. For example, consider the distance to which a charged layer on a plane surface extends into a quiescent liquid for which the potential decays to 1/e of its value at the surface. The thickness of this layer is called the Debye double layer (see the sections entitled Development of the Charge Conservation Equations and Illustrations of Selected Solutions of the Charge Transport Equations for details) given by

$$\delta = (D\tau)^{1/2} \tag{1}$$

where D = ionic diffusion coefficient (cm^2/sec)
τ = relaxation time (sec)

Table III. Triboelectric Series for Some Production Fabrics [3]

Positive	
+25	
	Wool felt
+20	
+15	Glass, filament, heat cleaned and silicone treated
	Glass, spun, heat cleaned and silicone treated
	Wool, woven felt, T-2®
+10	Nylon 66, spun
	Nylon 66, spun, heat set
	Nylon 6, spun
	Cotton sateen
+5	Orlon® 81, filament
	Orlon 42, needled fabrics
	Arnel®, filament
	Dacron®, filament
	Dacron, filament, silicone treated
0	Dacron, filament, M-31
	Dacron, combination filament and spun
	Creslan®, spun; Azoton®, spun
	Verel®, regular, spun; Orlon 81, spun (55200)
	Dynel®, spun
−5	Orlon 81, spun
	Orlon 42, spun
	Dacron, needled
−10	Dacron, spun; Orlon 81, spun (79475)
	Dacron, spun and heat set
	Polypropylene 01, filament
	Orlon 39B, spun
−15	Fibravyl, spun
	Darvan®, needled
	Kodel®
−20	Polyethylene B, filament and spun
Negative	

Table IV. Movement and Termination of Charge Carriers [1]

Charge Carrier	Movement	Termination
Electrons (metals, semiconductors, *not* insulators)	Going to lower energy levels	Fermi levels at same height
	or	
	Thermal electromagnetic force (emf)	Hot spot cooled off or Capacitance fully charged
Ions	Going to lower energy levels	Leveling complete
	or	or
	Diffusion down concentration gradient	Shortage of ions back emf
	or	
	Electrolytic emf	Capacitance fully charged
Bulk Matter	Adhesion of parts of opposing surfaces	Surfaces separated
	or	
	Particulate contamination transferred mechanically	Surfaces separated

Table V. Examples of Charge Separation Phenomena

Nature of Interface	
Solid-Solid	Static electrification by rubbing, thermocouple, movement of grain dusts through pneumatic systems, discharge of grain from storage silos, paper printing, charge accumulation on synthetic fabrics, minerals beneficiation, fabric filtration for solids removal
Liquid-Liquid	EMF developed at liquid junction, dropping mercury electrode, tanker cleaning with water jets, dielectrophoresis or electrostatic coalescence
Gas-Liquid,	Gas bubbling through organic liquids
Liquid-Solid	Electrolytic cell, electrophoresis, electroosmosis, electrification by spraying, fabric spinning through spinneret heads, flow of liquids through pipelines and filters, liquid metals (mercury) through capillaries, blood flow past prosthetic surfaces, reactor fouling, electrofiltration, cross-flow electrofiltration, ink jet printing
Gas-Solid	Ionized gases or plasmas flowing past surfaces, electrostatic precipitation, charging in gas fluidized beds
Liquid-Liquid-Solid	Electrostatic coalescence of three-phase emulsions
Free Surface	Unconfined flames or ionized gases

given by

$$\tau = \epsilon\epsilon_0/k \qquad (2)$$

where ϵ = the dielectric constant, dimensionless

ϵ_0 = the permittivity of free space equal to 8.85×10^{-14} sec/Ω-cm

k = the specific conductivity equal to $2CF^2D/R_gT, \Omega^{-1}\ cm^{-1}$

C = total ionic concentration (mol/cm³)

R_g = gas constant $\left(8.32 \frac{V - coul}{°K\text{-}mol}\right)$

T = temperature (°K)

Values of δ for $D = 1 \times 10^{-5}$ cm²/sec for a typical high-conductivity fluid such as a saline solution of $k = 0.1\ \Omega^{-1}\ cm^{-1}$ ($\epsilon = 80$) and a toluene solution for $k = 1 \times 10^{-15}$ $\Omega^{-1}\ cm^{-1}$ ($\epsilon \simeq 2$) are equal to 3×10^{-7} cm and 0.04 cm, respectively. (Table VIB

Table VIA. Specific Conductivity and Dielectric Constant for Several Liquids and Solids

Specific Conductivity ($k - \Omega^{-1}cm^{-1}$)	Dielectric Constant (dimensionless)
Ethanol $1.5 \times 10^{-7} - 1.35 \times 10^{-9}$	25.7
Benzene $7.6 \times 10^{-8} - < 1 \times 10^{-18}$	2.3
Toluene $< 1 \times 10^{-14}$	2.4
Hexane 1×10^{-18}	1.9
Nylon 2.5×10^{-15}	3.5
Polyethylene $10^{-14} - 10^{-18}$	2.3
Polytetrafluoroethylene $< 10^{-15}$	2.0
Polystyrene Molding $10^{-17} - 10^{-19}$	2.5
Water (absolute purity) 10^{-7}	80.4

Table. VIB. Relaxation Time and Double-Layer Thickness for a Range of Specific Conductivities ($D = 1 \times 10^{-5}$ cm²/sec, $\epsilon = 2$)

Specific Conductivity, κ (Ω^{-1} cm^{-1})	Relaxation Time, τ (sec)	Double-Layer Thickness, δ (cm)
1×10^{-11}	1.77×10^{-2}	4.21×10^{-4}
1×10^{-13}	1.77	4.21×10^{-3}
1×10^{-15}	1.77×10^{2}	4.21×10^{-2}
1×10^{-18}	1.77×10^{5}	1.33

gives δ for a range of conductivities). These double layer thicknesses cover effects that are microscopic to macroscopic in behavior. This chapter emphasizes the effects of macroscopically large double layers on different fluid-mechanical and time-dependent charging processes.

IONIZATION BEHAVIOR IN LOW-CONDUCTIVITY LIQUIDS

To have a clear picture of the mechanism of charge transport in flowing fluids it is necessary to discuss the behavior of conducting species in low-dielectric constant fluids. Since the physical chemistry of electrolytic solutions is well documented in various monographs [34,35], only pertinent aspects will be presented here.

The principal conduction mechanism in dielectric liquids of practical interest is ionization of weak electrolytic solutes. When these materials are added intentionally for static suppression [34–37], they are known as antistatic additions. However, ionizable species are often present in hydrocarbon liquids as impurities. For highly purified liquids having conductivities below above 10^{-16} Ω^{-1} cm^{-1}, the presence of dissolved gases and electrons in the hydrocarbon structure can influence conduction profoundly [38,39] (Tables VII and VIII). These two types of conduction mechanisms are not considered in this chapter.

Antistatic additives are usually classified into three principal categories [40]:

1. **Anionic Compounds.** The cations are usually alkalai metals or alkaline earth metals. Structurally, these compounds may be expressed by the formula $(R - A^-)\ M^+$, where M^+ is the metal cation, R is a hydrocarbon that may be partially or fully fluorinated, and A^- may be a sulfonate, phosphate, carboxyate or dithiocarbamate.
2. **Cationic Compounds.** Principal compounds are quaternary ammonium or phosphonium salts in which the quaternary group also may form part of a ring, e.g., imidazoline. Structurally, they may be represented by X^- $(B^+ -$

Table VII. Effect of Sample Purity and Oxygen on Electric Conductance of Cyclohexane [38]

	Specific Conductivity $(\Omega\ \text{cm})^{-1}$, 25°C		
	Original Sample	Purified Sample	
Condition	Oxygen Present	Oxygen Present	No Oxygen
After 1 hr	5×15^{-15}	5×10^{-16}	4×10^{-17}
18 hr	3×10^{-16}	8×10^{-17}	7×10^{-18}
30 hr	5×10^{-17}	2×10^{-17}	$> 4 \times 10^{-18}$
	Activation energy (eV)		
	0.15 0.16 (no oxygen)	0.16	0.16

Table VIII. Electric Conductance of Unsaturated Hydrocarbons [38]

Hydrocarbon Type	Specific Conductivity $(\Omega\ cm)^{-1}$, 25°C	Activation Energy (eV)	
		Conductivity	Viscosity
Cyclohexane	$< 10^{-16}$	0.16	0.1
Cyclohexene	1.5×10^{-15}	0.42	0.07
Cyclohexadiene-1,3	6.5×10^{-15}	0.41	0.07
Cyclohexadiene-1,4	5×10^{-15}	0.42	0.07
Benzene [39]	1.1×10^{-14}	0.41	0.08

$R - B^{+}), X^{-}$ or $(R - B^{+})\ X^{-}$ where X^{-} is typically a halogen or $ROO_2 - SO^{-}$ and R is a hydrocarbon or an acyl group $R - \dot{C} = 0$.

3. **Nonionic compounds.** Principal compounds include polyethylene glycol esters or ethers, fatty acid esters or ethanolamides, mono- and diglycerides and hydroxyethylated fatty amines.

The principal function of an antistatic agent is to increase the electrical conductivity of the liquid from typical values of around $1 \times 10^{-14}\ \Omega^{-1}\ cm^{-1}$ (Table VIA) to values above around $1 \times 10^{-11}\ \Omega^{-1}\ cm^{-1}$. At conductivities of $1 \times 10^{-11}\ \Omega^{-1}\ cm^{-1}$ and above, free charges recombine in a relatively short time period and may be removed readily by conduction to ground (see the section entitled Dimensional Analysis). (The functions of antistats for solid polymers have been discussed by Finck [40] and will not be considered here.) Another essential feature of antistats for liquids is that they have a low vapor pressure to minimize evaporation losses. For oils, a combination of antistatic activity with lubrication is often desirable [41]. Other factors include permanence, toxicity, color, odor, flammability and other very specific and less important factors. These factors are fulfilled by literally hundreds to thousands of chemically different surfactants, many of which are proprietary or covered by patents.

The behavior of very dilute solutions of the above additives dissolved in low dielectric-constant liquids may be explained by Bjerrum's theory of ionic association [34]. In very dilute solutions, nearly the entire ionic solute exists as associated ion pairs, which do not contribute to the conductivity of the liquid; therefore, low values of equivalent conductivity are expected. Furthermore, the association theory predicts that singly charged ions predominate overwhelmingly in solution. Therefore, the development of the equations in the following section is based on transport of univalent ions. However, for sufficiently high conductivities, triple ions such as ABA^{+} or BAB^{-}, or even unchanged quadrupoles, such as ABAB, can exist in solution. This is discussed in greater detail in Appendix B.

DEVELOPMENT OF THE CHARGE TRANSPORT EQUATIONS

A knowledge of the principles of physics covered in courses on electricity and magnetism and the presence of uni-univalent ions will be assumed in the development of the charge transport equation. Following Gavis [42–45], we assume transport of a flux of charge by diffusion, conduction and convection as follows:

$$j_{+} = -D_{+}F\nabla c_{+} - \frac{D_{+}F^{2}c_{+}}{R_{g}T}\nabla\psi + \vec{v}Fc_{+} \tag{3}$$

$$j_{-} = D_{-}F\nabla c_{-} - \frac{D_{-}F^{2}c_{-}}{R_{g}T}\nabla\psi - \vec{v}Fc_{-} \tag{4}$$

Here, the flux of charge associated with positive ions resulting from diffusion is directionally opposite to its concentration gradient. The positive and negative fluxes due to convection are also opposite one another, as denoted by the positive sign in Equation 3 and the negative sign in Equation 4. The total flux of charge transported is given by

$$j = j_+ + j_- \tag{5}$$

On substitution of Equations 3 and 4 into 5, one obtains

$$j = -D\nabla q - \left(\frac{DF^2}{R_gT}\right) C\nabla\psi + \vec{v}\,q - \left(\frac{D'F}{R_gT}\right) q\nabla\psi - D'F\nabla C \tag{6}$$

Here,

$$q = F(C_+ - C_-) \tag{7}$$

is the net charge density

$$C = C_+ + C_- \tag{8}$$

is the total ion concentration

$$D = (D_+ + D_-)/2 \tag{9}$$

is the mean diffusivity of the ions, and

$$D' = (D_+ - D_-)/2 \tag{10}$$

Applying continuity of charge,

$$\frac{\partial q}{\partial t} = -\nabla \cdot j \tag{11}$$

to Equation 6, one obtains

$$\frac{\partial q}{\partial t} = D\nabla^2 q + \frac{DF^2}{R_gT} C\nabla^2\psi + \frac{DF^2}{R_gT}\nabla\psi \cdot \nabla C - \vec{v} \cdot \nabla q + q\nabla \cdot \vec{v} + \frac{D'F}{R_gT}\nabla\psi \cdot \nabla q + FD'\nabla^2 C + \frac{D'F^2}{R_gT} q^2 \tag{12}$$

Equation 12 can be written in terms of q by eliminating the ψ and C dependencies. One assumes that incompressible flow, i.e., $\nabla \cdot \vec{v} = 0$, uses Equations 7 and 8, the assumption of ionic equilibrium:

$$C_+C_- = C_o^2/4 \tag{13}$$

and Poisson's equation

$$\nabla^2\psi = -q/\epsilon\epsilon_o \tag{14}$$

to arrive at [43]

$$\frac{\partial q}{\partial t} = \left[D + \frac{D'q}{(q^2 + F^2C_o^2)^{1/2}}\right]\nabla^2 q - \frac{Fq}{R_gT\epsilon\epsilon_o}\left[D(q^2 + F^2C_o^2)^{1/2} + D'q\right] - \vec{v} \cdot \nabla q + \frac{F}{R_gT}\left[\frac{Dq}{(q^2 + F^2C_o^2)^{1/2}} + D'\right]\nabla q \cdot \nabla\psi + \frac{D'F^2C_o^2}{(q^2 + F^2C_o^2)^{3/2}}(\nabla q)^2 \tag{15}$$

Equation 15 may be written in dimensionless form following introduction of the following dimensionless variables:

$$q_r \equiv q/FC^o \qquad v_r \equiv v/U \qquad V_r \equiv LV$$
$$t_r \equiv t/\tau_o \qquad \psi_r \equiv \psi(\epsilon\epsilon_o/FC^oL^2) \qquad V_r^2 \equiv L^2\nabla^2$$

where L and U are length and velocity characteristics of a particular system, and τ_o is the relaxation time of the original solution before transport has occurred, given by

$$\tau_o = \epsilon\epsilon_o/\kappa_o = \epsilon\epsilon_o/(DF^2C^o/R_gT) \tag{16}$$

$$\frac{\partial q_r}{\partial t_r} = \left(\frac{\delta_1}{L}\right)^2\left[1 + \frac{D^* qr}{(q_r^2+1)^{1/2}}\right]\nabla_r^2 q_r - [(q_r^2+1)^{1/2} + D^* q_r]q_r - \left(\frac{\delta_2}{L}\right)\vec{v_r}\cdot\nabla_r q_r + \left[\frac{q_r}{(q_r^2+1)^{1/2}} + D^*\right]\nabla_r q_r \cdot \nabla_r \psi_r + \frac{D^*(\nabla_r q_r)^2}{(q_r^2+1)^{1/2}} \tag{17}$$

Here,

$$\delta_1 = (D\tau_o)^{1/2} \tag{18}$$

$$\delta_2 = U\tau_o \tag{19}$$

$$D^* = (D_+ - D_-)/(D_+ + D_-) \tag{20}$$

The physical significance of the variables used here will be discussed in the next section.

Equation 15 or 17 cannot be solved as written even for the simplest geometry and boundary conditions. Therefore, Gavis [43] has presented several useful approximate forms, which may be used as starting equations. These are given below.

Simplified Forms of the Conservation Equations:

1. Equal Ionic Diffusivities: $D_+ = D_-$, $D^* = 0$.
   Equations 15 and 17 simplify to

$$\frac{\partial q}{\partial t} = D\nabla^2 q - \left(\frac{FDq}{R_gT\epsilon\epsilon_o}\right)(q^2 + F^2C^{o2})^{1/2} - \vec{v}\cdot\nabla q + \left(\frac{FDq}{R_gT}\right)\frac{\nabla q \cdot \nabla\psi}{(q^2 + F^2C^{o2})^{1/2}} \tag{21}$$

and

$$\frac{\partial q_r}{\partial t_r} = \left(\frac{\delta_1}{L}\right)^2 \nabla_r^2 q_r - q_r(q_r^2+1)^{1/2} - \left(\frac{\delta_2}{L}\right)\vec{v_r}\cdot\nabla_r q_r + \frac{q_r\nabla_r q_r \cdot \nabla_r\psi_r}{(q_r^2+1)^{1/2}} \tag{22}$$

2. Large Charge Density: $q/FC^o >> 1$. Equations 21 and 22 become

$$\partial q/\partial t = D_\pm \nabla^2 q - (FD_\pm/R_gT\epsilon\epsilon_o)q^2 - \vec{v}\cdot\nabla q + (D_\pm F/R_gT)\nabla q \cdot \nabla\psi \tag{23}$$

in which D_+ or D_- is to be used, depending on whether q is positive or negative.

$$\partial q_r/\partial t_r = (\delta_1/L)^2(1 \pm D^*)\nabla_r{}^2 q_r - (1 \pm D^*)q_r{}^2 - (\delta_2/L)\,\vec{v}_r \cdot \nabla_r q_r + (1 \pm D^*)V_r q_r \cdot \nabla_r \psi_r \qquad (24)$$

3. Small Charge Density: $q/FC^o << 1$. Equations 23 and 24 reduce to the linear equations

$$\partial q/\partial t = D\nabla^2 q - q/\tau_o - \vec{v} \cdot \nabla q \qquad (25)$$

and

$$\partial q_r/\partial t_r = (\delta_1/L)^2 \nabla_r{}^2 q_r - q_r - (\delta_2/L)\,\vec{v}_r \cdot \nabla_r q_r \qquad (26)$$

The applicability of the simplified forms of the charge transport equations will be considered in later sections of this chapter, following a general discussion of similitude principles and the nature of the boundary conditions.

SIMILITUDE PRINCIPLES

It is important to consider the physical significance of the various terms in the generalized equation of charge transport—Equation 12. For the approximation $D_+ = D_-$, or $D' = 0$ (ionic diffusivities in extreme cases can vary up to 100% [46]) and incompressible flow, Equation 12 may be simplified to

$$\frac{\partial q}{\partial t} = D\nabla^2 q - q/\tau - \vec{v} \cdot \nabla q + \frac{K}{C}\nabla\psi \cdot \nabla C \qquad (27)$$

where

$$K = \frac{DF^2C}{R_gT}$$

and τ is given by Equation 2. Equation 27 may be further simplified by making an order of magnitude comparison on the four remaining right-hand terms in a standard manner:

$$\frac{\text{Diffusive}}{\text{Convective}} \simeq \frac{Dq/L^2}{Uq/L} = D/UL = \frac{1}{ReSc_m}$$

$$\frac{\text{Diffusive}}{\text{Conductive}} \simeq \frac{Dq/L^2}{q/\tau} = \frac{D\tau}{L^2} = (\delta/L)^2$$

$$\frac{\text{Convective}}{\text{Conductive}} \simeq \frac{Uq/L}{q/\tau} = \frac{U\tau}{L} = ReSc_m(\delta/L)^2$$

$$\frac{\text{Conductive}}{\text{Product of gradients}} \simeq \frac{kq/\epsilon\epsilon_o}{k\psi/L^2} = \frac{q/\epsilon\epsilon_o}{\psi/L^2} = 1$$

where U is the characteristic flow velocity, L is the characteristic length, and $\psi/L^2 \simeq q/\epsilon\epsilon_o$ from Poisson's equation.

Here Re is the Reynolds number, $Re = \rho LU/\mu$, Sc_m is the mass transfer Schmidt number, $Sc_m = \nu/D$ and $\delta = (D\tau)^{1/2}$ is the Debye double layer thickness. The product of the mass transfer Schmidt number, Sc_m, times $(\delta/L)^2$ may be interpreted as an electrokinetic Schmidt number [47,48]; in addition, $ReSc_m(\delta/L)^2$ is an electrokinetic Péclet number.

An alternative way of carrying out this type of order of magnitude analysis is to first nondimensionalize Equation 27. Defining

$$t_r \equiv t/\tau_o \qquad q_r \equiv q/FC^o \qquad \nabla_r^2 \equiv L^2\nabla^2$$

$$\vec{v}_r \equiv \vec{v}/U \qquad \nabla_r \equiv L\nabla \qquad \psi_r \equiv \psi/\psi_o$$

$$\tau_r \equiv \tau/\tau_o \qquad C_r \equiv C/C_o$$

where τ_o, C_o and ψ_o are values of relaxation time, ionic concentration and potential at the start of the transport process, and substituting into Equation 27 one obtains

$$\frac{\partial q_r}{\partial t_r} = \left(\frac{D\tau_o}{L^2}\right)\nabla_r^2 q_r - q_r/\tau_r - \left(\frac{U\tau_o}{L}\right)\vec{v}_r \cdot \nabla q_r + \left(\frac{DF}{R_g T}\right)\left(\frac{\tau_o\psi_o}{L^2}\right)\nabla_r\psi_r \cdot \nabla_r C_r \qquad (28)$$

Using Poisson's equation for an order of magnitude estimate of ψ_o/L^2 and substituting for k_o and τ_o, the coefficient of the last term in Equation 28 becomes unity. For cases of practical interest, such as in turbulent flow through capillaries [11–13] and flow through microporous filters [47–49], neglect of the last term in comparison to the conduction term, even though both coefficients are unity, does not significantly offset the analysis. In these cases, the diffusive and conductive terms are orders of magnitude larger than unity. As an approximation, therefore, one may generally neglect the latter term in analyzing a practical charging process.

An important consideration at this time is to determine the requirements for similar charging in geometrically similar systems. This was considered earlier by Wagner [47] and the present approach is based on this study. The practical consequences are evident here because if similar charging can be obtained in a laboratory setup then one need not resort to full-scale measurements on various processes. In other words, electrostatic charging occurring in large-scale equipment may be studied conveniently on a scaled-down version of some predetermined value in the laboratory. Alternatively, improved confidence in scaling up a laboratory charging process to commercial scale would be anticipated.

For illustrative purposes we will select the steady-state form of Equation 25 and, imposing the requirement of similitude to obtain Equations 29 and 30,

$$D\nabla^2 q - q/\tau - \vec{v} \cdot \nabla q = 0 \qquad (29)$$

$$D'\nabla'^2 q' - q'/\tau' - \vec{v} \cdot \nabla' q' = 0 \qquad (30)$$

where $q' = \lambda_q q$
$v' = \lambda_v v$
$D' = \lambda_D D$
$\tau' = \lambda_\tau \tau$
$L' = \lambda_L L$

Substituting for the primed members into Equation 30, we obtain

$$\lambda_q\lambda_D\lambda_L{}^2 D\nabla^2 q - \frac{\lambda_q}{\lambda_\tau} q/\tau - \lambda_v\lambda_q\lambda_L \vec{v} \cdot \nabla q = 0 \qquad (31)$$

Now the following equalities are necessary for Equation 31 to equal Equation 29:

$$\lambda_D\lambda_q\lambda_L{}^2 = 1 \quad \text{or} \quad q'\frac{D'}{L'^2} = qD/L^2 \qquad (a)$$

$$\lambda_q/\lambda_\tau = 1 \quad \text{or} \quad q'/\tau' = q/\tau \tag{b}$$

$$\lambda_v\lambda_q\lambda_L = 1 \quad \text{or} \quad \frac{v'q'}{L'} = \frac{vq}{L} \tag{c}$$

These three requirements may be stated that similarity will be preserved only if the volumetric charge transport rates by diffusion, conduction and convection are identical for the system and the model. This condition is difficult to meet in general. A further simplification of this requirement results for constant relaxation time, i.e., constant electrical conductivity, because the dielectric constant is a constant for all charging processes of general interest [11–13,48,49]; thus, a and c become

$$D'/L'^2 = D/L^2 \tag{a}$$

and

$$v'/L' = v/L \tag{c}$$

Multiplying a and c by the constant τ and rewriting the equalities in familiar engineering terms,

$$\delta'/L' = \delta/L \tag{a}$$

and

$$Re'Sc'_m(\delta'/L')^2 = ReSc_m(\delta/L)^2 \tag{c}$$

An example is provided below to illustrate the meaning of the requirements a and c.

As a starting point we will determine the requirements for similar charging in a model of scale factor 1/10 actual size, i.e., $L' = (1/10)L$, and assume constant D, ρ, μ and ϵ. Condition a reduces to

$$1/\left\{k'\left(\frac{1}{100}\right)L^2\right\} = 1/kL^2$$

when Equation 2 is substituted for the relaxation time. Thus, k′ must be increased by a factor of 100 for validity of the preceding equality. This condition may be met for slightly conducting hydrocarbon fluids by adding an antistatic agent, as discussed previously. Substituting condition a in condition c we get

$$Re'Sc_m' = Re\ Sc_m \tag{c}$$

As the Schmidt number generally will remain unchanged, we are left with equality of the Reynolds number. This reduces to

$$v'L' = vL \tag{c}$$

for $\rho' = \rho$ and $\mu' = \mu$. Since $L' = 1/10L$, equality of the Reynolds number requires that one increase the velocity for the model tenfold to maintain a fixed Reynolds number for the one-tenth size scale model.

For the above example with constant τ, D, ρ, μ and ϵ, charge transport similarity for a model of a larger system of similar geometry may be preserved by maintaining:

- equality of the ratio of the Debye double layer thickness, δ, to characteristic length, i.e., δ/L = constant, by increasing the conductivity through addition of an antistatic additive; and

- a fixed mass transfer Péclet number, $Pe = ReSc_m$ = constant, by increasing the characteristic flow velocity.

It should be emphasized that conditions a and c resulted from the equality of charge density $q' = q$ when $\tau' = \tau$ from condition b. This is strictly true only in rare cases because q itself is a function of many variables (see the following section). Thus, in small test models it is apparent that only partial similarity can be preserved. Alternatively, errors may result if small-scale test data are used to predict charging behavior in large-scale systems.

Gibson and Lloyd [50] were the first to show experimentally on a large-scale test setup that the theoretical derivations of Koszman and Gavis [11–13], Gibbings and Hignett [51], Hignett [52] and Gavis [43,44] for turbulent flow through capillaries and small tubes required further modifications before they could be used to predict accurately electrification in large-scale systems.

There is one additional requirement concerning charging similarity: an examination of the nature of the boundary conditions. That is, complete similarity (in addition to the requirements outlined here and the requirements of geometric similarity) also requires similar boundary conditions. The nature of the boundary conditions at liquid-solid interfaces is considered in detail in the following section.

BOUNDARY CONDITIONS FOR CHARGED INTERFACE

Wagner [47] has outlined an approach to determine the functional form of the boundary conditions for charging processes that could be expressed in terms of a Nusselt number. This approach will be followed here to illustrate further the similarities between charge transfer and classical mass or heat transfer.

The amount of charge crossing an interface may be given by analogy with classical heat and mass transfer theory as follows:

$$Q' = hAF\Delta C = hA\Delta q \tag{32}$$

where $\Delta C = c_+ - c_-$ the concentration driving force (eq/cc)
Q' = rate of charge transfer across interface (coul/sec)
h = charge transfer coefficient (cm/sec)

and A and F are standard quantities. Strictly speaking, Equation 32 is the definition of the charge transfer coefficient. Let us limit ourselves to flow in conduits and charge transfer occurring within the control volume sketched in Figure 1. Further, if we limit ourselves to charging at the control volume inlet (i.e., Plane 1), although we could specify conditions at Plane 2 or a mean value, the charge crossing the interface is given by

$$Q' = h\pi dL\,(q_{b1} - q_{o1})$$

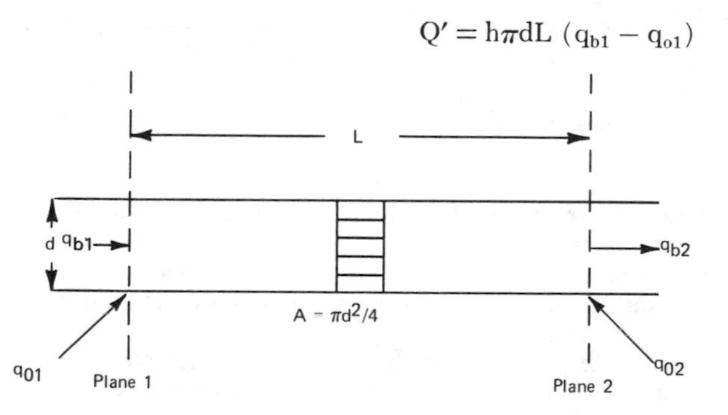

Figure 1. Sketch of control volume for charge transfer.

Equating Q′ to the total charge flow from the fluid to the tube wall over the length, L, we obtain

$$h\pi dL\,(q_{b1} - q_{o1}) = \int_0^L \int_0^{2\pi} -D\left.\frac{\partial q}{\partial r}\right|_{r=d/2} \left(\frac{d}{2}\right) d\theta dz$$

Defining the dimensionless quantities,

$$q_r \equiv \frac{q - q_{o1}}{q_{b1} - q_{o1}} \quad z_r \equiv z/d \; r_r \equiv r/d$$

and inserting into the above equation gives

$$h/D = \frac{1}{2\pi L} \int_0^{L/d} \int_0^{2\pi} -\left(\frac{\partial q}{\partial r_r}\right)_{r_r=1/2} d\theta dz$$

or, in terms of the Nusselt number,

$$Nu = \frac{hd}{D} = \frac{1}{2\pi L/d} \int_0^{L/d} \int_0^{2\pi} -\left(\frac{\partial q_r}{\partial r_r}\right)_{r_r=1/2} d\theta dz$$

Now the solution of the charge transport equation may be denoted symbolically as

$$q_r = q_r\left(r_r, \theta_r\, z_r, \frac{\delta}{L}, \frac{U\tau}{L}, \psi_r\right)$$

The Nusselt number is, therefore, a function of

$$Nu = Nu\left(\frac{\delta}{L}, \frac{U\tau}{L}, \psi_r, \frac{L}{d}\right)$$

Or, if we rewrite $U\tau/L$ as $ReSc_m(\delta/L)^2$, then Nu is given symbolically by

$$Nu = Nu\,(Re, Sc, \delta/L, \psi_r, L/d)$$

or as

$$Nu = ARe^{\alpha}Sc^{\beta}(\delta/L)^{\sigma}\psi_r{}^{\phi}(L/d)^{\omega} \tag{33}$$

where A is a constant, and α–ω are the powers to which the dimensionless groups are raised.

For processes dealing with fully developed flows, the entrance region correction term, $(L/d)^{\omega}$, is unimportant. For cases in which ψ is close to its reference value, ψ_o, then $\psi_r{}^{\phi}$ is unimportant and Equation 33 reduces to the following simplified form:

$$Nu = ARe^{\alpha}Sc^{\beta}(\delta/L)^{\sigma} \tag{34}$$

Correlations for the Nusselt number for charge transport processes are not available in the published literature. Thus, if one is to use a mass or heat transfer correlation to describe charge transfer, how should the added term, δ/L (the ratio of the Debye double layer thickness to characteristic length) be accounted for? In their initial studies, which introduced the use of mass transfer Nusselt number to represent charge transfer in turbulent flow through capillaries, Koszman and Gavis [11–13] did not consider δ/L in the Nusselt number. Since experimental values of δ/L were less than approximately 0.1 and larger than about 0.001, the $(\delta/L)^{\sigma}$ term is unimportant for

$0.1 \lesssim \sigma \lesssim 0.125$ and $\sigma \simeq 0.1$ for values of δ/L equal to 0.1 and 0.001, respectively. Since the Schmidt number for the Nusselt number is to the one-quarter power (see the section entitled Turbulent Charge Generation in Capillaries and Pipes) and considering the usual amount of scatter involved in their electrostatic measurements, omission of the $(\delta/L)^{\sigma}$ term is not immediately evident. However, if one takes $\beta = 1$ and $\sigma = 2$ in Equation 34 and rewrites $Sc_m(\delta/L)^2$ as

$$Sc_m(\delta/L)^2 = \nu/(L^2/\tau) = \nu/(L^2k/\epsilon\epsilon_o)$$

then L^2/τ or $L^2/(\epsilon\epsilon_o/k)$ have the units of $(\text{length})^2/\text{time}$, which are standard units for diffusivity. Thus, this nondimensional number is an electrokinetic analog of the mass transfer Schmidt number. As a first approximation, one should therefore substitute $Sc_e = Sc_m(\delta/L)^2$ for Sc_m or for the Prandtl number, Pr, in mass or heat transfer correlations for the Nusselt number and carry over these correlations to describe charge transfer in geometrically similar systems.

Additional evidence in support of the above arguments is given in later sections of this chapter. However, a brief synopsis is given here for emphasis:

1. Koszman and Gavis's [11–13] dimensionless number, $G = d^2/Re^{7/4}\tau\nu$, is, in terms of the present study, equal to (see Figure 6)

$$d^2/(Re^{7/4}\tau\nu) = \frac{1}{Re^{7/4}\nu/(d^2/\tau)} = \frac{1}{Re^{7/4}Sc_m(\delta/d)^2}$$

   where L, the characteristic length for a capillary, is the tube diameter, d. The charging data in Figure 6 would be plotted more appropriately against $Re^{7/4}Sc_m(\delta/d)^2$ or $Re^{7/4}\tau\nu/d^2$, i.e., against $1/G$.
2. Wagner and Gavis's [47–49] dimensionless number, $v_p\tau_o/d_p$, for flow through microporous media is equal to $Re_{d_p}Sc_m(\delta/d_p)^2$, where the d_p denotes the pore diameter.

   Both dimensionless numbers $Re^{7/4}Sc(\delta/d_p)^2$ and $Re_{d_p}Sc(\delta/d_p)^2$ may be interpreted as electrokinetic Péclet numbers, Pe_e, in the context of this discussion. They also may be written $Re^{7/4}Sc_e$ and $Re_{d_p}Sc_e$, respectively.

DIMENSIONAL ANALYSIS

Two approaches to dimensional analysis that start with the ion transport equations have appeared in the literature [42–44,51]. Gavis's approach [42–44] was considered earlier. The second approach was developed by Gibbings and Hignett [51], who considered electrical and thermal effects, as well as the nature of the boundary.

A different approach based on the well known Buckingham π theorem was utilized by Wagner [47–48] to successfully correlate charging data for the problem of charge generation in flow through microporous media. A detailed discussion of this approach is given in the section entitled Flow Through Porous Media.

ILLUSTRATIONS OF SELECTED SOLUTIONS OF THE CHARGE TRANSPORT EQUATION

We will select the linearized form of the charge transport equation, Equation 25, valid for low charge density, $q/FC^o << 1$, to illustrate the applicability of this equation. Three different examples will be treated here covering steady-state to transient relaxation processes. In certain cases, closed-form solutions to the linearized charge transport equation may be obtained readily from their heat and mass transfer analogs given in the monographs of Carslaw and Jaeger [53] and Crank [54].

The Equilibrium Double Layer in Cylindrical Geometry

For nonflow conditions, steady state and contact with a solid surface exhibiting cylindrical geometry, Equation 25 becomes

$$\frac{D}{r}\frac{d}{dr}\left(r\frac{dq}{dr}\right) + D\frac{d^2q}{dz^2} - q/\tau_o = 0 \tag{35}$$

Substituting the dimensionless variables,

$$q_r \equiv q/FC^o \qquad r_r \equiv r/d \qquad z_r \equiv z/L$$

into Equation 35, where FC^o is the total initial charge density, d is the diameter, and L is the length, we get

$$\left(\frac{D\tau_o}{d^2}\right)\frac{1}{r_r}\frac{d}{dr_r}\left(r_r\frac{dq_r}{dr_r}\right) + \left(\frac{D\tau_o}{L^2}\right)\frac{d^2q_r}{dz_r^{\,2}} - q_r = 0 \tag{36}$$

For length to diameter ratios greater than one, $L/d > 1$, axial diffusion of charge is small with respect to radial diffusion and, therefore, the second term may be neglected. The solution to Equation 36 subject to symmetry about the tube axis and $q_r|_{r_r = ½} = q_{wr}$ is

$$q_r/q_{wr} = I_o(r/\delta)/I_o(\delta/r)^{-1} \tag{37}$$

where I_o is the modified Bessel function of zero order. Equation 37 may be rewritten in terms of the average charge density given by

$$\bar{q} = \frac{\int_o^a qrdr}{\int_o^a rdr} \tag{38}$$

as follows:

$$q/\bar{q} = \frac{d}{4\delta}\left\{\frac{I_o\left(\frac{r/a}{\delta/a}\right)}{I_1(\delta a)^{-1}}\right\} \tag{39}$$

where I_1 is the modified Bessel function of first order. A plot of $q/\bar{q}$ vs r/a for three different values of δ/a shown in Figure 2 shows that the charge distribution is uniform for $\delta/a \gtrsim 5$. Also, for values as low as $\delta/a = 1$, the charge distribution may be considered uniform for $r/a \lesssim 0.7$, i.e., in the central core of the tube.

A physical interpretation of the above transport process is first that ions of one sign will adsorb preferentially on the solid surface leaving an excess of ions of the opposite sign in the adjacent fluid. These excess ions will now be transported by conduction toward the surface that is oppositely charged. At equilibrium, conductive-induced migration is counterbalanced by diffusive transport in the opposite direction so that there is no net flux of charge crossing the interface.

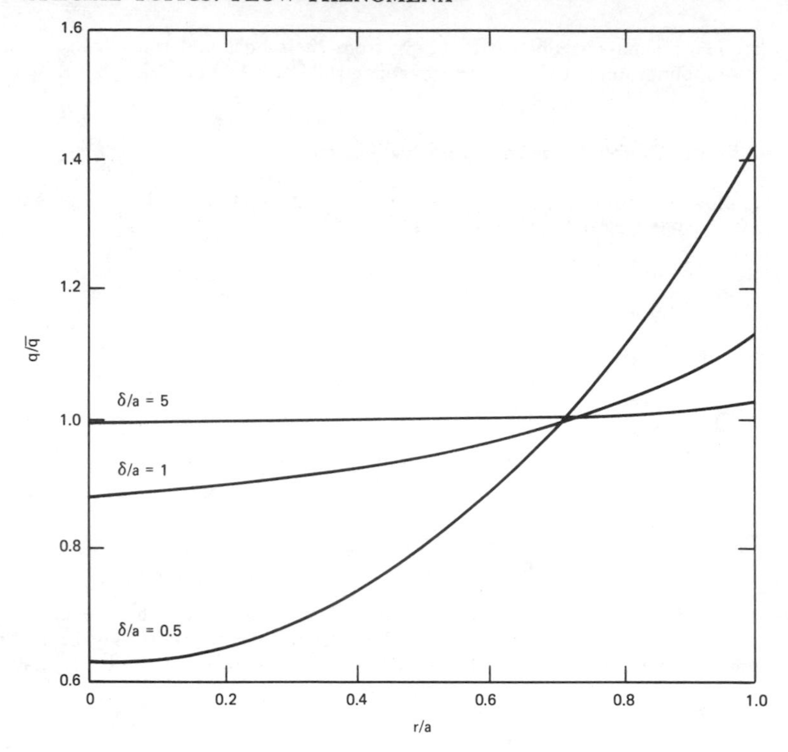

Figure 2. Charge distribution within a tube at different values of δ/a from Equation 39 [55].

Current Applied Potential Relationship Under Influence of Ionization and Ion Recombination Effects

The current applied potential relationship under ionization and ion recombination effects has been developed by Gavis [56] for ionic transport by diffusion and conduction. His numerical solutions to the transport equations in show that for:

1. low applied potential differences—the current is proportional to the potential difference, although the conduction process is generally nonohmic;
2. large ionization rates—ohmic conduction is approached; and
3. high applied potential differences—current reaches a maximum value limited by diffusion.

This approach is an extension of Thomson and Thomson's work [57] in which they ignored the effects of diffusion and assumed a uniform field through most of the fluid. The results differ from those given by Thompson and Thompson in that the limiting currents are smaller for equivalent values of a dimensionless group and that conduction is nonohmic at all applied potentials but only approaches ohmic behavior for large values of the same dimensionless group. Because of the importance of this study, the development of the important assumptions and equations will be illustrated below. This approach then can be extended to different geometries or ionization conditions when required.

The one-dimensional steady-state forms of Equations 3 and 4, ignoring convection, are

$$j_{\pm} = \mp DF \frac{dC_{\pm}}{dx} - \frac{DF^2}{R_gT} C_{\pm} \frac{d\psi}{dx} \tag{40}$$

Applying the divergence theorem,

$$\frac{dj_{\pm}}{dx} = \pm Fr^* \tag{41}$$

where r^* is the net rate of ion production per unit volume on the positive and negative fluxes, to Equation 40 gives

$$\mp DF \frac{d^2c}{dx^2} - \frac{DF^2}{R_gT} C_{\pm} \frac{d^2\psi}{dx^2} - \frac{DF^2}{R_gT} \frac{dC_{\pm}}{dx} \frac{d\psi}{dx} = \pm Fr^* \tag{42}$$

On adding and subtracting the above equation and using $q = F(C_+ - C_-)$, one obtains

$$\frac{d^2q}{dx^2} + \frac{F^2C}{R_gT} C \frac{d^2\psi}{dx^2} + \frac{F^2}{R_gT} \frac{dC}{dx} \frac{d\psi}{dx} = 0 \tag{43}$$

$$D \frac{d^2c}{dx^2} + \frac{D}{R_gT} q \frac{d^2\psi}{dx^2} + \frac{D}{R_gT} \frac{dq}{dx} \frac{d\psi}{dx} = -2r^* \tag{44}$$

Eliminating $d^2\psi/dx^2$ using Poisson's equation (Equation 14) and using

$$j = j_+ + j_- = -D \frac{dq}{dx} - \frac{DF^2}{R_gT} C \frac{d\psi}{dx} \tag{45}$$

to eliminate the potential gradient, Equations 43 and 44 become

$$\frac{d^2q}{dx^2} - \frac{F^2}{R_gT\epsilon\epsilon_o} Cq - \frac{1}{Dc}\left(j + D\frac{dq}{dx}\right)\frac{dc}{dx} = 0 \tag{46}$$

$$D \frac{d^2C}{dx^2} - \frac{D}{R_gT\epsilon\epsilon_o} q^2 - \frac{1}{F^2C}\left(j + D\frac{dq}{dx}\right)\frac{dq}{dx} = -2r^* \tag{47}$$

Gavis assumes that the rate of production of ions per volume, r^*, is a constant and that the recombination rate of ions, r_2, is second order:

$$r_2 = -k_2C_+C_- \tag{48}$$

where k_2 is the recombination rate coefficient.

At equilibrium, the rate of ionization, r_1, and recombination are equal.

$$C_+ = C_- = C^o/2$$

where

$$K_{eq} = (C_+C_-)_{eq} = k_1/k_2 \tag{49}$$

so that

$$r_1 = k_1(C_+C_-)_{eq} \tag{50}$$

and

$$r^* = r_1 + r_2 \tag{51}$$

Gavis shows that r^* is given by

$$r^* = \frac{k_2}{4}\{C^{o2} + (q/F)^2 - C^2\} \tag{52}$$

Four boundary conditions must be specified to solve Equations 46 and 47. The first two are fairly straightforward and express the conditions about the origin:

$$q_o = 0 \tag{53}$$

$$C_o = C^o \tag{54}$$

The third boundary condition imposes the physical requirement that ions of one sign discharge at the surface of each electrode. For negative ions discharging at the positive electrode taken at $x = L$, one obtains

$$j = j_-|_L \tag{55}$$

$$0 = j_+|_L \tag{56}$$

Evaluating these fluxes using Equation 45 and rearranging, Gavis obtains

$$FC_L = \frac{q_L\left(D\left.\frac{dq}{dx}\right|_L + j\right)}{\left(DF\left.\frac{dc}{dx}\right|_L - j\right)} \tag{57}$$

The last boundary condition is based on the physical requirement that the current must be equal to the net rate of ionization throughout the fluid:

$$j = j_-|_L = -F\int_{-L}^{L} r^* dx \tag{58}$$

Substituting Equation 52 into Equation 58 and for symmetry about the origin, we obtain

$$j = -\frac{1}{2}Fk_2\int_o^L [C^{o2} + (q/F)^2 - C^2]\,dx \tag{59}$$

Equations 46 and 47 and the boundary conditions (Equations 53, 54, 57 and 59) may be written as Equations 60–65:

$$\frac{d^2q_r}{dx_r^2} - \alpha^2 C_r q_r - \frac{1}{C_r}\left(j_r + \frac{dq_r}{dx_r}\right)\frac{dc_r}{dx_r} = 0 \tag{60}$$

$$\frac{d^2C_r}{dx_r^2} - \alpha^2 q_r^2 - \frac{1}{C_r}\left(j_r + \frac{dq_r}{dx_r}\right)\frac{dq_r}{dx_r} + \beta(1 + q_r^2 - C_r^2) = 0 \tag{61}$$

$$q_{or} = 0 \tag{62}$$

$$C_{or} = 1 \tag{63}$$

$$C_{1r} = q_{1r} \frac{\left(\left.\frac{dq_r}{dx_r}\right|_{x_r = 1} + j_r\right)}{\left(\left.\frac{dC_r}{dx_r}\right|_{x_r = 1} - j_r\right)} \quad (64)$$

$$j_r = \beta \int_0^1 (1 + q_r^2 - C_r^2)\, dx_r \quad (65)$$

where $q_r \equiv q/FC^o$

$C_r \equiv c/C^o$

$X_r \equiv x/L$

$\psi_r \equiv \psi/(R_gT/F)$

$V_r = V/(R_gT/F)$

$j_r \equiv j/(DFC^o/L)$

and

$$\alpha^2 \equiv L^2/\left(\frac{R_gT\epsilon\epsilon_o}{F^2C^o}\right) = (L/\delta)^2 \quad (66)$$

$$\beta = k_2C^o/(2D/L^2) \quad (67)$$

Equation 66 is the square of the ratio of L to the Debye double layer thickness, and Equation 67 expresses the ratio of the ionization rate to the characteristic diffusion rate.

One remaining equation is used by Gavis to determine the total potential drop across the cell, taking into account the transport potential and the concentration overpotential:

$$\Delta V = \Delta\psi_{\text{transport}} + \Delta V_{\text{o.p.}} \quad (68)$$

$\Delta\psi_{\text{transport}}$ for a half cell may be written from Equation 45 as

$$\Delta\psi = \frac{-R_gT}{DF^2} \int_0^L \left(j + D\frac{dq}{dx}\right) \frac{dx}{C} \quad (69)$$

The concentration overpotential term arises because the discharge rate exceeds the rate at which ions can be supplied to the fluid adjacent to the electrode. The limiting current occurs when the concentration reaches zero. The equilibrium Nernst equation given by

$$\Delta V_{\text{o.p.}} = \frac{RT}{F} \ln \frac{C_+|_o}{C_+|_L} = \frac{R_gT}{F} \ln \frac{C^o}{C_L - (q_L/F)} \quad (70)$$

may be used for the overpotential. Thus, on combining Equations 69 and 70 and expressing the result in nondimensional form, Gavis gives

$$\Delta V_r = -\int_0^1 \left(\frac{dq_r}{dx_r} + j_r\right) \frac{dx_r}{C_r} - \ln (C_{1r} - q_{1r}) \quad (71)$$

Numerical solutions are given for q_r, C_r, $C_\pm$ vs x_r for a range of α^2 and β^2. For the purpose of this discussion, the dimensionless plot of $-j_r$ vs ΔV_r in Figure 3 shows the

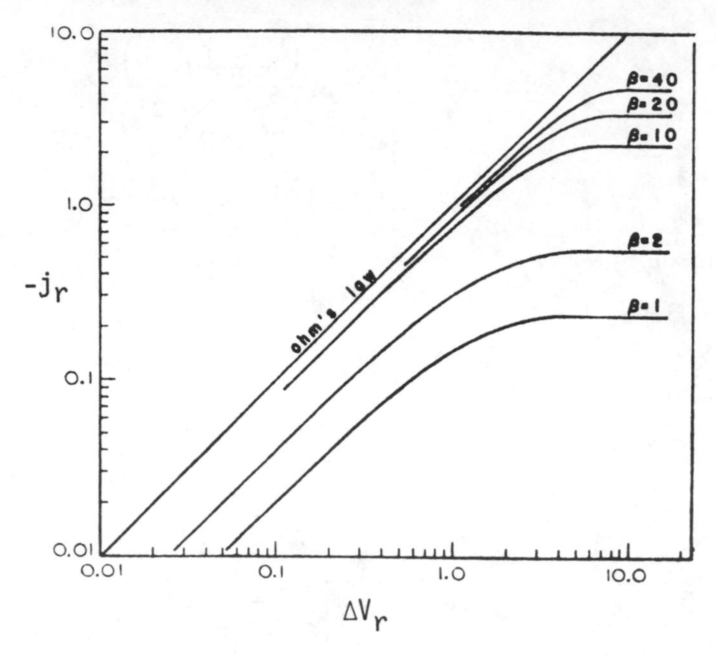

Figure 3. Plots of $-j_r$ as a function of ΔV_r for different values of α^2 and β. $\alpha^2 = 1$ for the curves labeled $\beta = 1,2,10,20$; $\alpha^2 = 4$ for the curves labeled $\beta = 20$, 40; for $\beta = 20$, the curves for $\alpha^2 = 1$, and $\alpha^2 = 4$ are indistinguishable at this scale [56].

effect diffusion of charge can play on introducing deviations from Ohm's law. Note that the limiting currents are smaller for given values of β, in contrast to the approach of Thomson and Thomson, who have ignored diffusion, i.e., assumed large β. Also, the transport of charge is nonohmic at all potential differences and only becomes ohmic for large values of the ratio of the ionization rate to characteristic diffusion rate.

In summary, the effect of diffusion superimposed on conduction leads to a current-applied potential relationship that is nonohmic for small ionization rates or small electrode separations, i.e., small β. The limiting currents are also smaller than those given by ohmic behavior.

Charge Relaxation in a Spherical Container

The rate of discharge of an initial distribution of charge in a spherical container will be used to show the effect that geometry and diffusion can play on dissipating charge to ground. Equation 25 for spherical geometry and quiescent conditions becomes

$$\frac{\partial q}{\partial t} = \frac{D}{r}\frac{\partial}{\partial r}\left(r\frac{\partial q}{\partial r}\right) - q/\tau \tag{72}$$

Solution to Equation 72 that is symmetrical about the origin and subject to the initial conditions

$$q|_{t=0,r} = q_o \tag{73}$$

and

$$q|_{t,r=a} = 0 \tag{74}$$

is from the work of Gavis [58]:

$$q/q_o = \frac{2}{\pi(r/\delta)} \sum_{n=1}^{\infty} \frac{(-1)^n \exp[-(1 + n^2\pi^2\delta^2/a^2)t/\tau] \sin n\pi r/a}{n} \quad (75)$$

For very large values of δ/a, as in a very pure fluid or in small-scale handling equipment, Equation 75 reduces to

$$q/q_o = \frac{2}{\pi(r/a)} \sum_{n=1}^{\infty} \frac{(-1)^n \exp \quad -(n^2\pi^2\delta^2/a^2)t/\tau] \sin n\pi r/a}{n} \quad (76)$$

The total rate of charge relaxation is given by averaging q over a spherical element of volume to give

$$Q/Q_o = \frac{6}{\pi^2} \sum_{n=1}^{\infty} \frac{1}{n^2} \exp[-(n^2\pi^2\delta^2/a^2)t/\tau] \quad (77)$$

Another limiting case is for small values of δ/a, which denotes high-conductivity fluids or flows through large-scale equipment. Equation 77 for this condition reduces to the standard form for ohmic conduction:

$$Q/Q_o = \exp - t/\tau \quad (78)$$

These calculations show that (1) diffusion generally should be considered in small-scale relaxation studies; and (2) relaxation of charge by diffusion is generally unimportant in large commercial fuel handling equipment.

TURBULENT CHARGE GENERATION IN CAPILLARIES AND PIPES

A comprehensive early treatment of this problem is given in the monograph by Klinkenberg and Van der Minne [36]. Extensive work in this area was carried out in the 1950s and early 1960s because of the buildup of incendiary charges accompanying the fueling of jet aircraft and turbulent transport of purified hydrocarbons into storage tanks. A discussion of the ignition hazards from static charging and methods for static elimination will be covered in the section entitled Conditions for a Hazardous Electrostatic Environment. The emphasis in this section is on the charge transport mechanism and solution to Equation 25 applicable to turbulent flow.

Earlier, Koszman and Gavis [11,55] presented a detailed review of existing literature. Prior to their study, theories were based on the classical double layer approach and were lacking in the following key points:

1. Classical theories do not provide the proper experimentally observed charge generation rate (i.e., the measured current) vs conductivity/flow dependence that exhibits a maximum, which is dependent on both conductivity and flow velocity (Figure 4). The proper length dependence also is not provided. Furthermore, the streaming potential is assumed to be a fundamental parameter; however, it was observed earlier to have a functional dependence on the Reynolds number.
2. Modification of the classical theory led either to an equation that gave a parametric dependence for the generated current that was contrary to observation or led to a reasonable prediction of the current vs length-velocity dependence but required introduction of a wall current of unspecified origin.

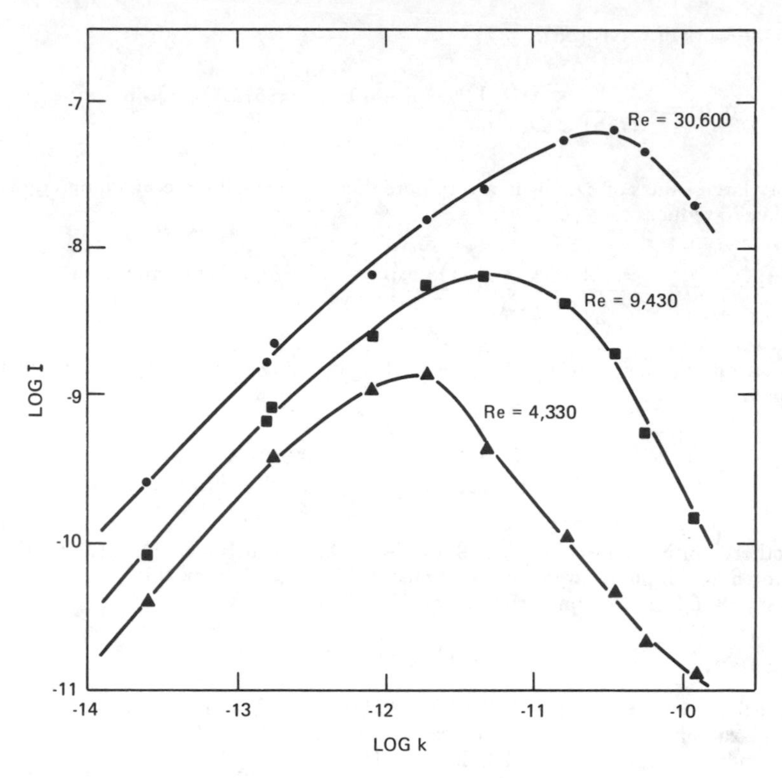

Figure 4. Charging current dependence on conductivity for turbulent flow [55] through platinum capillary —L/D = 168.

These defects provide justification for the charge density viewpoint utilized here.

Koszman and Gavis [12,13] view the charging process as illustrated in Figure 5. The rate-limiting process at the tube wall was assumed to be due to concentration polarization (i.e., ohmic and activation polarizations are small with respect to concentration polarization). Here, the thickness of the layer near the wall, where the concentration changes from c_o in the liquid bulk to c_s at the surface, is known as the diffusion layer thickness, λ. By analogy with heat and mass transfer,

$$\lambda = d/Nu_m \tag{79}$$

where d is the tube diameter, and Nu_m is the Nusselt number for mass transfer. For fluids of high Schmidt number, Deissler's [58] correlation, based on the assumption of a constant flux boundary condition for turbulent flow through tubes, gives for Nu_m

$$Nu_m = 0.022 Re^{7/8} Sc^{1/4} \tag{80}$$

Now, in the absence of an electric field, the surface current flux, j_s, was taken to be

$$j_s = -\frac{DF}{\lambda n_+}(1 - c_s/c_o) = \text{Const.} \tag{81}$$

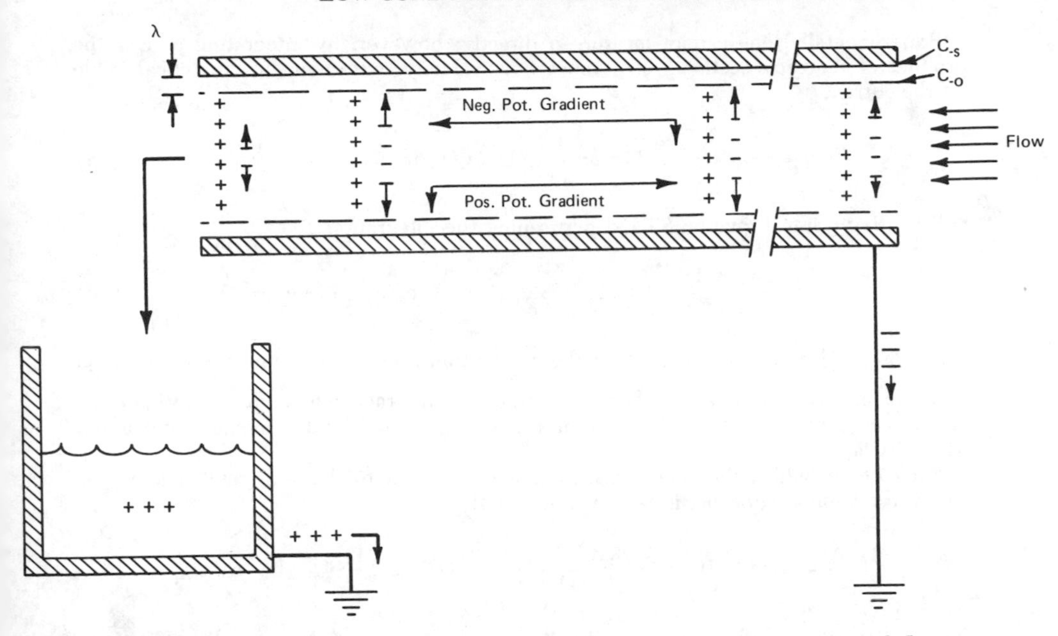

Figure 5. Schematic diagram of generation of the charging current by liquid flowing through a tube [12,13].

where n_+ is the transference number for positive ions. This was equated to the gradient at the tube wall (a rigorous derivation of the boundary conditions is given by Gavis [42,43]) such that

$$D\frac{dq}{dr}\bigg|_{r=a} = \frac{DF}{\lambda n_+}(1 - c_s/c_o) - \frac{KR_gT}{2n_+Fd}Nu_m\,(1 - c_s/c_o) \tag{82}$$

The remaining boundary conditions were

$$q|_{o,r} = 0 \tag{83}$$

i.e., the fluid enters the tube inlet uncharged, and symmetry about the tube axis,

$$\frac{\partial q}{\partial r}\bigg|_{r=o} = 0 \tag{84}$$

for $x \geqq 0$.

The linearized form of the charge transport equation (Equation 25 or 26) subject to above boundary conditions (Equations 82–84) requires a numerical approach for solution as D and v are nonsimple functions of r. A closed-form solution was given for the extreme case of D and v independent of r as

$$q = \frac{F\,(1 - c_s/c_o)}{\lambda n_+}\,\delta\left\{\frac{I_o\left(\dfrac{r/a}{\delta/a}\right)}{I_1\left(\dfrac{1}{\delta/a}\right)} - \frac{2a}{\delta}\sum_{n=o}^{\infty}\frac{J_o\left(\dfrac{\alpha_n r}{a}\right)\exp - \left(\dfrac{1}{v\tau} + \dfrac{\alpha^2{}_n D}{a^2 v}\right)x}{(\alpha^2{}_n - a^2/\alpha^2)\,J_o(\alpha_n)}\right\}$$

where J_o is a Bessel function of the first kind, and I_o, I_1 are the modified Bessel functions of zero and first order.

Experimentally, one cannot measure q directly; however, by integrating q over the product of the cross-sectional area times flow velocity, we obtain an analytical value for the current

$$I = 2\pi \int_{o}^{a} rq(L,r)v(r)dr \tag{85}$$

Substituting q into Equation 85 and performing the integration gives

$$I = \frac{\pi\epsilon\epsilon_o R_g T\bar{v}}{2n_+F}\,(0.022Re^{7/8}Sc_m{}^{1/4})\,(1 - c_s/c_o)\,(1 - e^{-L/\bar{v}\tau}) \tag{86}$$

Equation 86 also may be obtained for the assumption that the initial charge distribution is approximately uniform across the tube. This occurs in practice when $\delta \gtrsim d$ (Figure 2), i.e., for very low-conductivity fluids and in regions unaffected by development of charged boundary layers.

For cases in which the charge distribution is confined to the wall region, and thus within the laminar subzone, it can be shown that

$$I = 4.42 \times 10^{-4}\left\{\frac{\pi KR_gT}{n_+Fd^2D}\,(D\tau)^{3/2}Re^{21/8}Sc_m{}^{1/4}\nu\right\}(1 - c_s/c_o) \tag{87}$$

In practice, this confined charge distribution is approximated by high-conductivity hydrocarbon fluids.

For cases not represented by Equations 86 and 87, an inspectional analysis of the governing equations was performed to deduce the functional relationship given by

$$I/\frac{(1 - e^{-L/\bar{v}\tau})}{I_\infty} = G\left\{\frac{d^2}{Re^{7/4}\tau\nu},\ Re,\ Sc_m\right\} \tag{88}$$

where $G \rightarrow 1$ when $d^2/Re^{7/4}\tau\nu$ is small; $G \rightarrow 0.0395\ (d^2/Re^{7/4}\tau\nu)^{-1/2}Re^{-1/8}Sc_m{}^{-1/2}$ when $d^2/Re^{7/4}\tau\nu$ is large, and where I is the limiting current when $L/\bar{v}\tau >> 1$. The experimental data given in Figure 6, in which

$$\log \frac{I}{\epsilon\epsilon_o\,\bar{v}^{15/8}d^{7/8}(1 - e^{-L/\bar{v}\tau})}$$

is plotted vs $d^2/Re^{7/4}\tau\nu$, substantiated the general approach of Koszman and Gavis over a wide range of system parameters. However, the high-conductivity approximate solution, Equation 87, as given by lines D and E in Figure 6, is a rough approximation of the observed behavior for $d^2/Re^{7/4}\tau\nu \gtrsim 10^{-5}$. The shift in the plotted ordinate from a value of around zero for stainless steel tubes to a value of around 0.7 for platinum tubes was suggested by Koszman and Gavis as being due to surface roughness not accounted for in their theory. This suggestion does not appear to be valid in light of Gibson and Lloyd's study [50], which addressed the surface roughness postulate. They compared internal surfaces of 0.32- and 2.88-cm-diameter tubes for roughness and found the former to be rougher than the latter. However, the experimental value of $I_\infty/\bar{v}^{15/8}d^{7/8}$ obtained with the latter was greater than that obtained for the 0.32-cm tube, whose value for $I_\infty/\bar{v}^{15/8}d^{7/8}$ was close to the value predicted by Koszman and Gavis's equation.

One can speculate on a number of other reasons for the differences noted between the platinum and the stainless steel tubes. Certainly the unusual electromechanical and/or catalytic activity of platinum [59] vs other metals, such as those present in stainless steel, might be expected to enter the picture. However, the suggested reason(s)

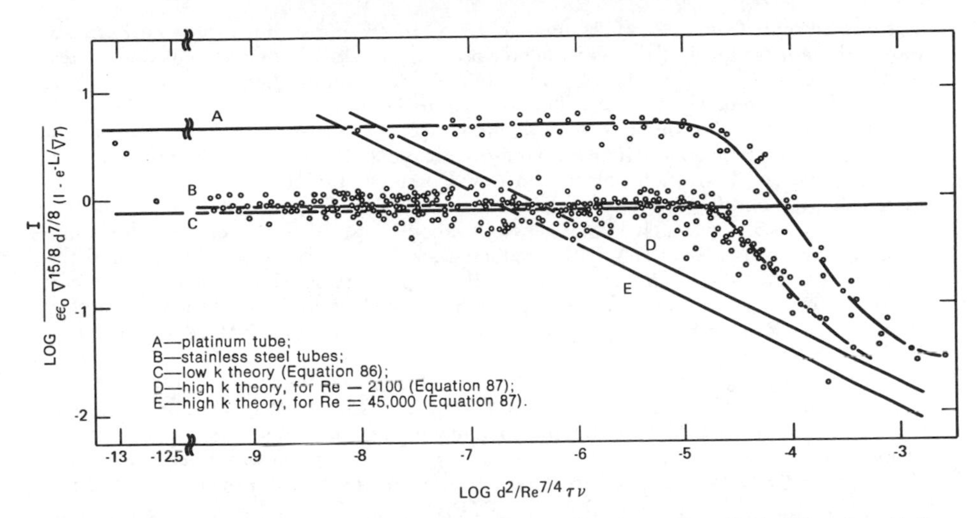

Figure 6. Plot of log $I/\epsilon\epsilon_0 v^{15/8} d^{7/8}(1 - e^{-L/v\tau})$ against log $d^2/Re^{7/8}\tau\nu$ for all the data taken with platinum and stainless steel tubes [12,13].

must preserve the overall features of the Koszman and Gavis approach since the A and B curves in Figure 6 are merely shifted in the values of the ordinate. The simplest way to view this is to say that the assumption $(1 - C_s/C_o) \simeq 1$, i.e., $C_s/C_o << 1$, may not be valid. This may be accounted for by arbitrarily setting $1 - C_s/C_o = 1$ for stainless steel; then a value of $1 - C_s/C_o$ can be determined easily for platinum that will transpose curve A to curve B in Figure 6. Some caution should be exercised in applying the data in Figure 6 to all systems for the following reasons:

1. The plotting variable for the ordinate is not dimensionless. Compare with Equation 86 and note the missing Sc_m number and kinematic viscosity term from the Reynolds number. Thus,

$$Sc_m{}^{1/4} \frac{1}{\nu^{7/8}} = \frac{1}{D^{1/4}\nu^{5/8}}$$

is the omitted term.
Note also that $0.022\pi R_g T/2n_+ F$ was treated as a constant because all tests were carried out at ambient conditions. Substitution of $Sc_e = Sc_m(\delta/d)^2$ for the mass transfer Sc_m shifts the plot shown in Figure 6 upward. For example, the data for the platinum tube—curve A from the range $10^{-5} \lesssim d^2/Re^{7/4}\tau\nu \lesssim 10^{-3}$—becomes around 1.6 as a high and 0 as a low. Since no functional advantage is gained in replotting these data, Figure 6 is left as it was in the original article.

2. For large-scale systems, the empirical correlation of Gibson and Lloyd [50] for infinitely long pipes is

$$I_\infty = K_5 d^{-1.8} v^{1.45 + 0.01\, d}\, e^{0.4 \log d \log v}$$

where $K_5 = (2.42 \pm 0.1) \times 10^{-11}$ should be used to calculate the current.

The abscissa $d^2/Re^{7/4}\tau\nu$, which is equivalent to $1/Re^{7/4}Sc_m(\delta/d)^2$, is left as it was published in the original articles [12,13,55]. The scale will be inverted if $Pe_e = Re^{7/4}Sc_m(\delta/d)^2$ is used as the abscissa instead of $d^2/Re^{7/4}\tau\nu$, i.e., Pe_e will range from around 15 to 5×10^{12}. Reference to the original article [55] shows that the last three data points for the platinum tube ranging from $-12.5 \lesssim \log d^2/Re^{7/4}\tau\nu \lesssim -13$ had measured currents from 5×10^{-14} amps to 2.7×10^{-13} amps. Since these currents are at around the noise level of the electrometer used in this study, they are not believed to be of high precision. Thus, the current obtained from using Figure 6 also would be expected to be erroneous even if a fluid similar to heptane were employed. The conductivity of heptane for this test was reported as $3.7 \times 10^{-17} \Omega^{-1} cm^{-1}$. This is in a range where different conduction mechanisms are possible, as discussed previously in the section entitled Ionization Behavior in Low-Conductivity Liquids.

Note that the onset of the tails in curves A, B and C in Figure 6 start at around log $d^2/Re^{7/4} \tau \nu = -5$, for which k was generally about $5 \times 10^{-12} \Omega^{-1} cm^{-1}$. This value of conductivity coincides closely with conductivities that are close to values around the equivalent conductivity minimum of Figure 2B in Appendix B. Thus, it is not surprising that these curves tail off since one no longer has univalent ions in solution but rather complex triple ions and uncharged quadrupole-type equilibrium. This is discussed in detail in Appendix B.

A recent study dealing with electrification of toluene and toluene/water mixtures in pipelines containing ball valves is given by Gibson and Lloyd [60]. For concentrations of water greater than 0.1–0.2% volume in toluene, they found a marked increase in charging over toluene alone. They attribute this primarily to flow through the valve and not pipeline flow. For this experimental setup (27 m of 4-in. i.d. pipe upstream of the valve containing polytetrafluoroethylene (PTFE) seats followed by a further 2-m length of similar diameter pipe), they recommend that the linear flow velocity should not exceed 1–3 m/sec in the pipe to minimize charging. The possibility of generating dangerous levels of charge can occur especially if the constriction is located at the entrance to a storage vessel. Further study in this area is needed because Gibson and Lloyd point out that analytical procedures are unavailable to predict electrification in constrictions.

FLOW THROUGH POROUS MEDIA

The study by Bruinzeel et al. [61] appears to be one of the first published reports illustrating the potentially serious nature of charging in filters during low-temperature refueling of aircraft. High electric field strengths (5000 V/cm and up)* were measured in the vapor space of an aircraft tip tank, while discharges were observed during normal refueling rates. The following test highlights were obtained using glass fiber and pleated paper filters,† each with water separator elements:

1. There seems to be little difference in the charging behavior of new and old type elements, as well as no gross differences in charging between the two filter types.
2. Simultaneous flow of nitrogen and fuel through a filter does not affect the charging characteristics.

A study by the Coordinating Research Council (CRC) [62] using the filtration equipment supplying J. F. Kennedy Airport in New York showed the following:

1. Fuel characteristics have the major effect on filter charging characteristics.
2. Teflon-coated screen-type separators charge less than pleated paper separators.

Other investigators who have examined charging in filters at some length include Wagner and Gavis [47–49], Leonard and Carhart [63], Masuda and Schon [64] and

*Homogeneous values exceeding around 4000 V/cm produce local values as high as 30,000 V/cm, the breakdown potential for air ionization.

†Here a filter is one specific type of porous material. The latter may be defined as a solid that contains a large number of holes (often referred to as pores) usually interconnected and distributed randomly throughout its volume.

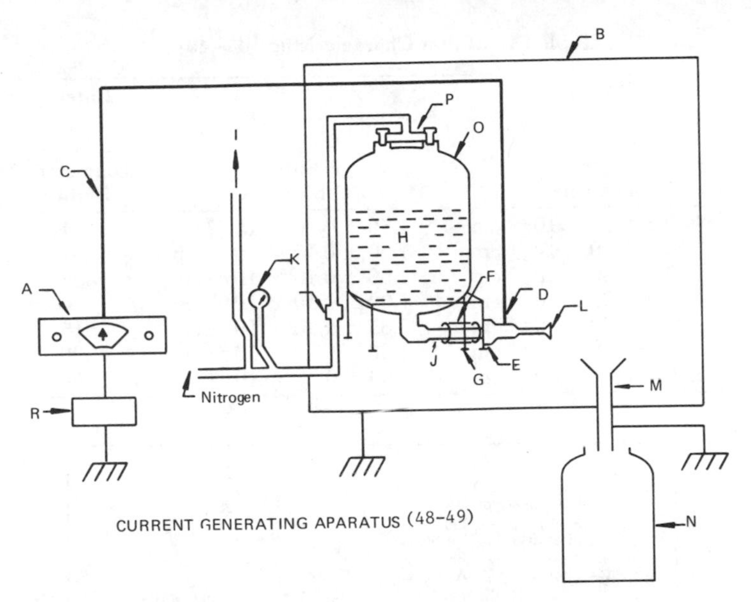

A—Keithley #610-B electrometer
B—Faraday cage
C—Amphenol #140/U coaxial cable
D—Teflon-insulated banana plug
E—Millipore microanalysis filter holder
F—Teflon approach tube: 7/32 in. I.D., with adjustable clamps leading to 1/2 in. O.D. copper pipe nipple
G—Teflon legs
H—A.S.A. #1 solution dissolved in purified heptane
I—Relief valve
J—Teflon separation forced fitted over copper tubing
K—Pressure gauge
L—Threaded Teflon plug
M—Conical copper funnel
N—Storage vessel
O—Stainless steel reservoir, 40-*l* capacity, with a steel ring welded to its top
P—Steel cover over a rubber gasket held in place by hexagonal bolts threaded into the steel ring on top of the reservoir
Q—Recorder

Figure 7. Current generating apparatus [48,49].

Huber and Sonin [65]. More recent studies have investigated electroosmotic flow and pressure distribution in perforated membranes [66] and sintered glass beads [67].

The emphasis in this section is on charge generation studies because experimental data have been correlated successfully over a wide range of variables. In addition, the charge density viewpoint will be detailed for reasons pointed out in the preceding section on turbulent flow through capillaries and pipes.

The study by Wagner and Gavis [47–49] involved the laboratory setup sketched in Figure 7 to measure the rate of charge generation, i.e., the streaming current through a single-filter type (cellulose triacetate Millipore® filters) over a wide range of filter characteristics (Table IX) and flow conditions (Table X). Current-velocity-conductivity data are shown in Figure 8. Because a suitable boundary condition at the wall of the filter similar to Equation 82 for turbulent flow is unavailable for filters of microporous structure, i.e., filters with characteristic pore diameters in the micrometer size range, solution to the charge conservation equation will continue to be a major unknown. In this situation it is recommended to first seek a dimensionless correlation of the experimental data and then, as an illustration of the validity of the equations, to show that under a given set of conditions the experimental data can be predicted. Two approaches are available for data correlation [42,47,48]. The former used Equation 21 minus the term containing the product of the gradients of charge density and potential supplemented by judicious reasoning to reduce the data given in Figure 8 to the dimensionless plot of Figure 9. Because this approach is difficult to conceive realistically, the systematic approach based on the Buckingham π theorem will be used to illustrate this method of data correlation originally published by Wagner [47].

The matrix notation approach detailed in the monograph by Langhaar [68] with the measured current, I_m, selected as the dependent variable in the π analysis is adopted

Table IX. Filter Characteristics [47–49]

Filter Type	Pore Diameter	Filter Thickness	Filter Thickness to Pore Diameter Ratio	Porosity (%)
AA	$8.00 \pm 0.50 \times 10^{-5}$ cm	$1.50 \pm 0.10 \times 10^{-2}$ cm	187	82
DA	$6.50 \pm 0.30 \times 10^{-5}$ cm	$1.50 \pm 0.10 \times 10^{-2}$ cm	231	81
HA	$4.50 \pm 0.20 \times 10^{-5}$ cm	$1.50 \pm 0.10 \times 10^{-2}$ cm	333	79
TW	$4.50 \pm 0.20 \times 10^{-5}$ cm	$1.00 \pm 0.10 \times 10^{-2}$ cm	450	43
GS	$2.20 \pm 0.20 \times 10^{-5}$ cm	$1.35 \pm 0.10 \times 10^{-2}$ cm	615	75
VC	$1.00 \pm 0.08 \times 10^{-5}$ cm	$1.30 \pm 0.10 \times 10^{-2}$ cm	1300	74
VM	$5.00 \pm 0.03 \times 10^{-6}$ cm	$1.30 \pm 0.10 \times 10^{-2}$ cm	3200	72

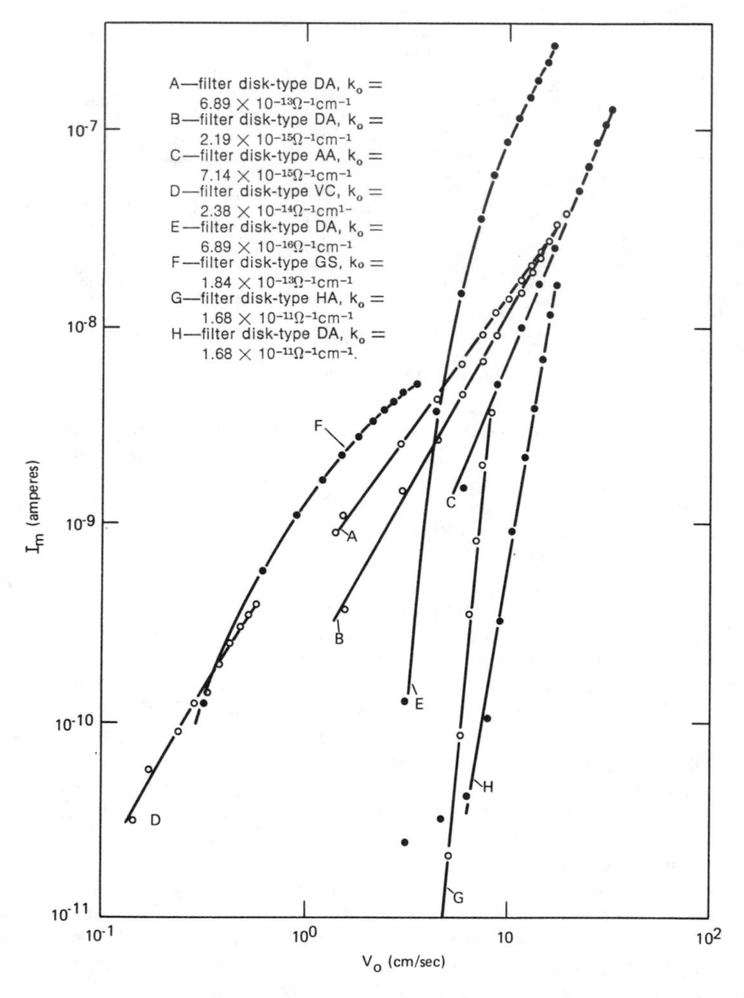

Figure 8. Plot of measured current against velocity for several representative experiments [49].

Table X. Flow Characteristics of Millipore Filters [47–49]

Filter Type	Pressure Drop (psig)	Average Pore Velocity (cm/sec)	Friction Factor, Dimensionless	Reynolds Number, Dimensionless	Friction Factor times Reynolds Number, Dimensionless
AA	5	3.42	2.32×10^{2}	4.81×10^{-2}	11.16
AA	60	3.79×10^{1}	2.27×10^{1}	5.32×10^{-1}	12.07
DA	5	1.65	8.10×10^{2}	1.90×10^{-2}	15.4
DA	60	1.96×10^{1}	6.84×10^{1}	2.26×10^{-1}	15.5
HA	5	7.80×10^{-1}	2.48×10^{3}	6.20×10^{-3}	15.4
HA	60	9.11	2.20×10^{2}	7.20×10^{-2}	15.9
TW	5	8.03×10^{-1}	3.53×10^{2}	3.46×10^{-3}	1.22
TW	60	9.26	3.19×10^{1}	3.98×10^{-2}	1.27
GS	5	3.28×10^{-1}	7.73×10^{3}	2.36×10^{-3}	18.20
GS	60	3.93	6.47×10^{2}	2.83×10^{-2}	18.31
VC	5	4.89×10^{-2}	1.67×10^{5}	9.21×10^{-5}	14.3
VC	60	6.01×10^{-1}	1.14×10^{4}	1.13×10^{-3}	12.9
VM	60	5.31×10^{-1}	1.76×10^{4}	4.67×10^{-4}	8.12
VM	5	1.49×10^{-2}	8.75×10^{5}	1.31×10^{-5}	11.5

along with MLTQ (mass, length, time, charge) system of dimensions. Inspection of Equations 2, 12 and 14 and reference to the work of Langhaar [68] allows one to cast the measured current in the matrix notation (temperature and the ionic diffusivity are treated as constants) given in Table XI.

Following Langhaar's suggested approach, the matrix in Table XI can be reduced to the following dimensionless groups:

$$\pi_{1p} = \frac{I_m \tau_o / d_p^3}{FC^o} \tag{89}$$

$$\pi_{2p} = v_p \tau_o / d_p = Re_{d_p} Sc_m (\delta / d_p)^2 \tag{90}$$

$$\pi_{3p} = d_p^2 / \nu \tau_o \tag{91}$$

$$\pi_{4p} = L_p / d_p \tag{92}$$

$$\pi_{5p} = A_v / d_p^2 \tag{93}$$

According to the π theorem, a functional relationship such as

$$\pi_{1p} = \Phi f\{\pi_{2p}, \pi_{3p}, \pi_{4p}, \pi_{5p}\}$$

exists between the π_{1p}'s, where Φ is an arbitrary constant. It is often possible to express f (above) as a product of the dimensionless groups, each raised to an arbitrary power so that

Table XI. Charge Transport Matrix for Flow Through Microporous Membranes [47,48]

	K_1 I_m	K_2 v_p	K_3 ρ	K_4 L_p	K_5 A_v	K_6 d_p	K_7 FC^o	K_8 τ_o	K_9 μ
M	0	0	1	0	0	0	0	0	1
L	0	1	−3	1	2	1	−3	0	−1
T	−1	−1	0	0	0	0	0	1	−1
Q	1	0	0	0	0	0	1	0	0

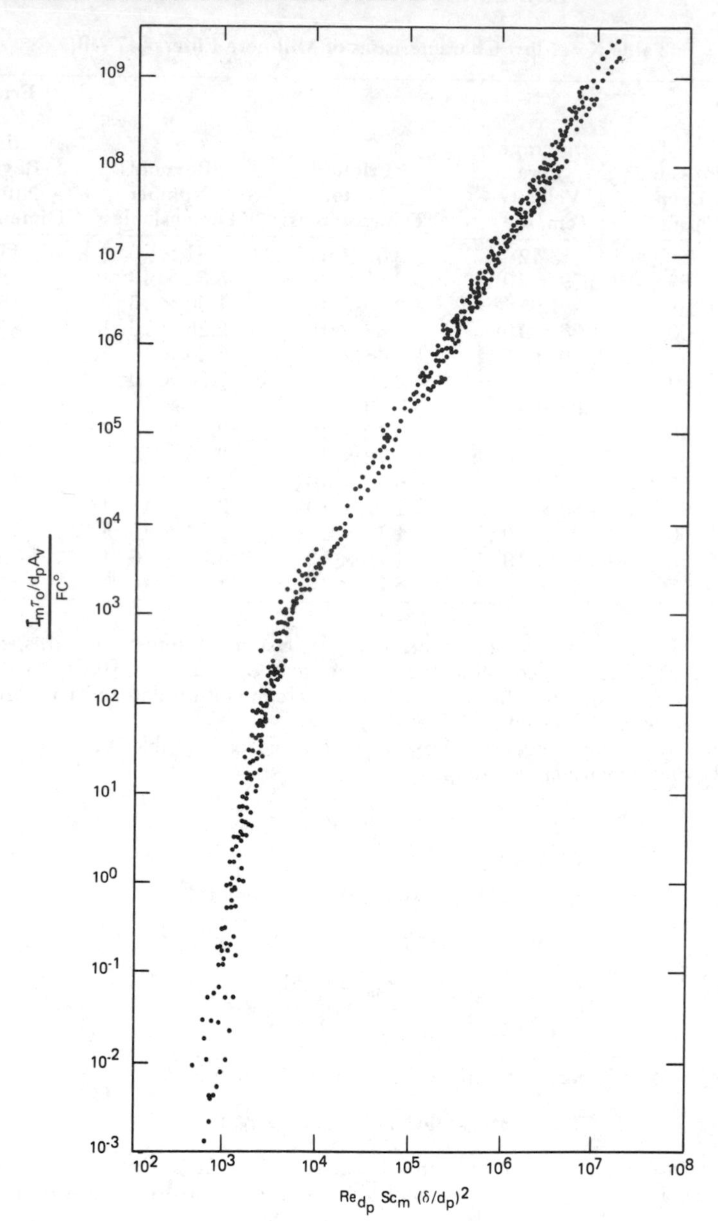

Figure 9. Dimensionless charging current data for flow of hydrocarbon liquid through filters [47–49].

$$\pi_{1p} = \Phi_1 \pi_{2p}{}^{b2} \pi_{3p}{}^{b3} \pi_{4p}{}^{b4} \pi_{5p}{}^{b5}$$

where Φ_1 is another constant.

Returning to Figure 9, we see that $\frac{I_m \tau_o / d_p A_v}{FC^o}$ is merely π_{1p}/π_{5p} and $v_p \tau_o / d_p$ is π_{2p}.

It appears that only three of the five π groups were necessary to correlate the data; π_{3p} and π_{4p} do not appear explicitly. However, π_{3p} appears implicitly in π_{2p} when the latter is written as $Re_{d_p}(\nu\tau_o/d_p^2)$. Total absence of π_{4p} means that the current is independent of L_p/d_p. This follows readily since L_p/d_p is very large (Table IX) and, thus, hydrodynamic entrance region effects are unimportant.

There are two ranges of π_{2p} for which different charging behavior is exhibited in Figure 9. For the range $5 \times 10^3 < \pi_{2p} < 2.5 \times 10^7$,*

$$\frac{I_m\tau_o/d_pA_v}{FC^o} = 2 \times 10^{-4}\{Re_{d_p} Sc_m(\delta/d_p)^2\}^{0.75} \tag{94}$$

while for the range $6 \times 10^2 < \pi_{2p} < 5 \times 10^3$ the current dependence is more complicated and cannot be described as simply.

Each of the π_{1p} to π_{5p} groups has a physical significance that may be interpreted as follows:

1. π_{1p} is the ratio of a dynamic charge density to the original total ionic concentration expressed in charge units.
2. π_{2p} is the ratio of a characteristic electrical length, which may be termed the charging length, to the pore diameter.
3. π_{3p} is the square of the ratio of the pore diameter to another characteristic electrical length.
4. π_{4p} is the ratio of effective pore length to pore diameter.
5. π_{5p} is the ratio of effective filter void area to the square of the pore diameter.

There are alternate ways of interpreting the groups. For example:

1. Writing π_{1p} as $I_m/(d_pFC^o/\tau_o)$, one may interpret the group $I_m/(d_p^3FC^o/\tau_o)$ as the ratio of the measured electrical current to a current of ions at their original concentration.
2. Writing π_{2p} as $v_p/(d_p/\tau_o)$ and multiplying the numerator and denominator by q, one may interpret the group $v_pq/(d_p/\tau_o)q$ as the ratio of charge transported by convection to charge transported by conduction.
3. Writing π_{3p} as $\nu/(d_p^2/\tau_o)$, one may interpret it as the ratio of the diffusivity of momentum to an electrokinetic diffusivity. It is, therefore, an electrokinetic analog of the mass transfer Schmidt number.

Other dimensionless groups will result if one selects different independent variables for the k's in the charge transport matrix. These new groups, however, were not as applicable as the ones illustrated here. Consequently, they are not given.

A theoretical explanation for the charging process is given in Appendix C because of uncertainties in the nature of the boundary condition at the filter wall. However, it is emphasized here that the equations under limiting conditions will reduce to the experimental results given by Equation 94. Although this approach provides indirect verification of the solution of the charge transport equation subject to appropriate boundary conditions for a model of a microporous filter, an unequivocal confirmation requires a suitable boundary condition at the filter wall. Future efforts would be required to determine this unknown.

Leonard and Carhart's study [63] on various filter materials (fiberglass-bonded and baked, nylon, Kel-F, Dacron and ordinary glass wool) appears to be in general agreement with previously observed electrostatic behavior. For example, their data show the typical current increase with conductivity, the passing through a maximum and subsequent decrease with increasing conductivity. Because of unknown filter characteristics such as surface area, effective pore diameter, etc., their data could not be cast

*Current was always negative except when erratic behavior due to filter clogging occurred; positive currents also were obtained in the latter.

into the dimensionless parameters plotted in Figure 9 to check out the adequacy of this correlation for other filter types.

Leonard and Carhart's observation that the sign of the charge on a fuel may be controlled by controlling the nature of the filter surface led to the suggestion of three new approaches for practical control of static electricity. One of their approaches was tested successfully; they divided fuel flow into two separate streams, each containing filters (one that acquired a positive charge, the other negative) and carefully controlled the flow through each, thus leading to virtually no net charge on a receiving tank. This novel approach seems to offer considerable potential for practical control of static electricity in liquids. It is recognized that it might be difficult to have precise flow control under all field conditions; however, even here it might be used to augment existing standard control techniques involving the use of additives to increase electrical conductivity, charge relaxation techniques and corona-type static charge reducers.

Masuda and Schon [64] have investigated the space charge distribution in n-heptane after passage through a metal filter using the experimental apparatus in Figure 10. Here, the fluid is driven through the metal filter (particle size, 0.1–0.2 mm; average pore diameter, 20 mm; porosity, 36–38%; thickness, 3 mm) with pressurized nitrogen and through a second perforated electrode into a reservoir from which a leakage current to ground is measured on an electrometer. The electrical conductivity was varied from $5 \times 10^{-15}\ \Omega^{-1}\ cm^{-1}$ to $5 \times 10^{-12}\ \Omega^{-1}\ cm^{-1}$ with the addition of the antistat Shell A.S.A.#3 (a mixture of equal parts of chromium dialkylsalicylate, calcium didecylsulfosuccinate, copolymer of lauryl methacrylate and methylvinylpyridine as a 50% solution in a hydrocarbon solvent [69]). The distance, h, between the two electrodes was varied, thus permitting one to investigate both conductivity and spatial effects on the measured current. Representative data are illustrated in Figure 11, where the ratio of the total charge density to excess charge density is plotted vs streaming velocity for several different conductivities. Some scatter in their data is noted. The most probable cause for this scatter is due to relaxation of charge in the storage reservoir that is related only indirectly to the current generated from flow through the filter. In Wagner and Gavis's experiments, the second electrode in Figure 10 was used as a support for their Millipore filters so that $h = 0$ for all their studies. Only under this condition is there similarity between these two different studies. An attempt to cast Masuda and Schon's data into a form given by Equation 94 was therefore not undertaken.

Huber and Sonin [65] investigated the generation of streaming potentials and currents during flow of n-heptane doped with Shell A.S.A.#3 through Millipore filters type VC and PH using the test cell given in Figure 12. Filter type VC also was studied by Wagner and Gavis and has the geometric and flow characteristics given in Tables IX and X. Type PH has a pore diameter three times larger than VC, i.e., $3 \times 10^{-5} \pm 0.2$ cm, but a similar thickness and porosity $-L = 1.5 \times 10^{-4} \pm 0.1$ cm, $P = 0.77$. With the test cell in Figure 12, the electrodes could be moved up to 5 cm on either side of the filter. Conductivity variations examined were over a limited range from $6.9 \times 10^{-12} \Omega^{-1} cm^{-1}$ to $4.6 \times 10^{-11} \Omega^{-1} cm^{-1}$ corresponding to relaxation times of around

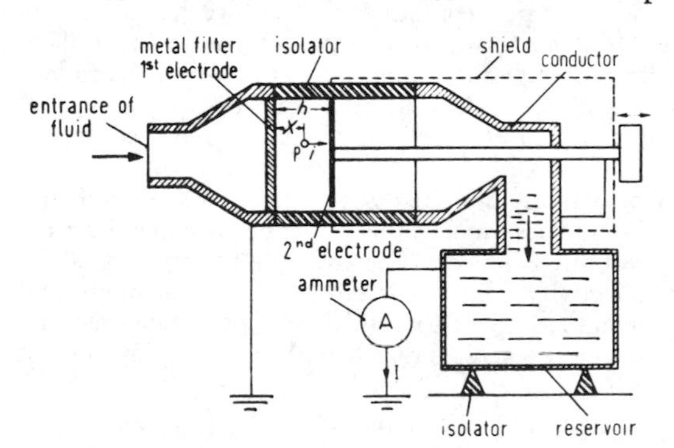

Figure 10. Sketch of apparatus for measurement of space-charge distribution in liquids after passage through filters [64].

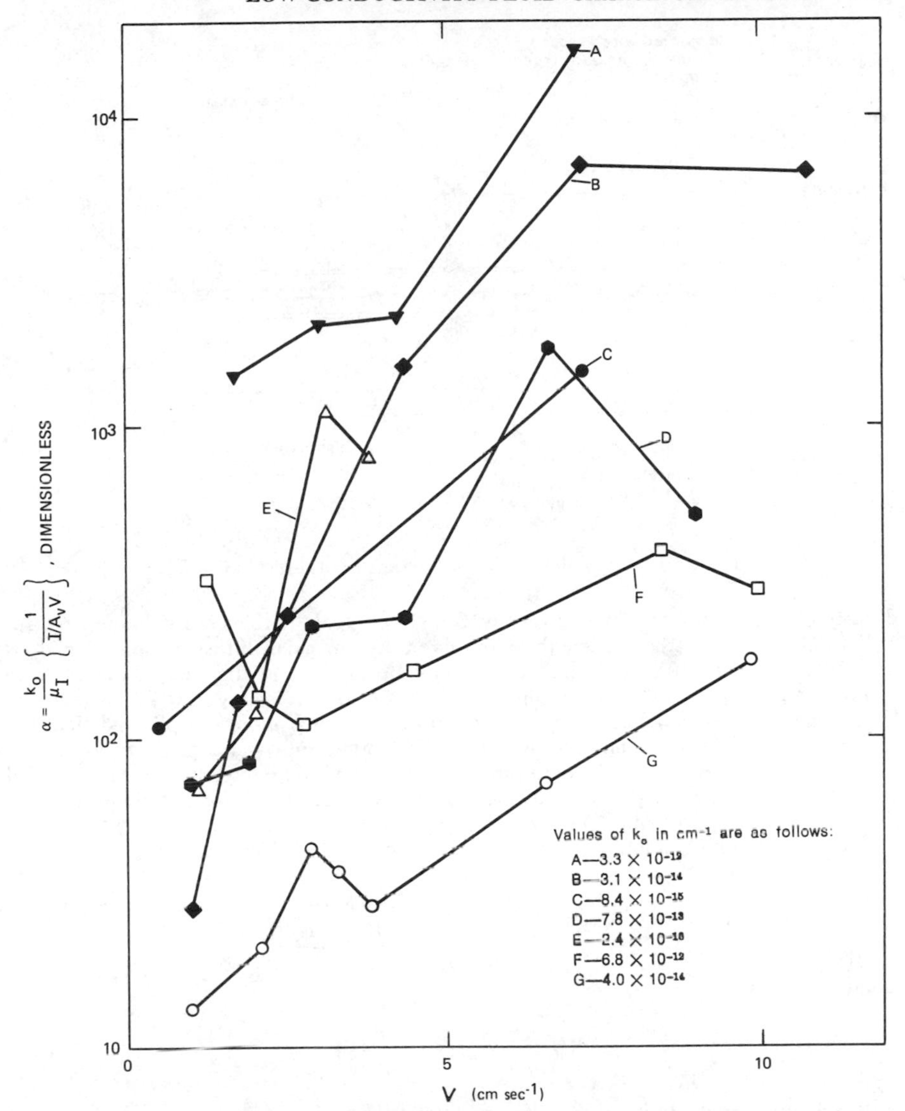

Figure 11. Ratio of the total charge density, k_o/u_I, to the excess charge density, I/A_v, as a function of the streaming velocity, v, with rest conductivity, k_o, as a parameter (assumed mobility, u_I, is $10^{-4} cm^2 V^{-1} sec^{-1}$) [64].

0.025 sec to 0.0035 sec, respectively. For these relaxation times, one generally is not concerned with dangerous charging situations. However, the authors point out that the main purpose of their work was to try to establish a model for the charging mechanism in flow through filters. Their proposed model is relatively simple, as stated by the authors, and resembles those used for flow of aqueous saline solutions through charged membranes [70–72]. Although the authors start with the equations for transport of flux of charge by diffusion, conduction and convection, as given in the section entitled Development of the Charge Transport Equations, they have introduced a serious limitation into their model by ignoring radial diffusion of charge. From the section entitled Similitude Principles, an order of magnitude estimate of the ratio of the

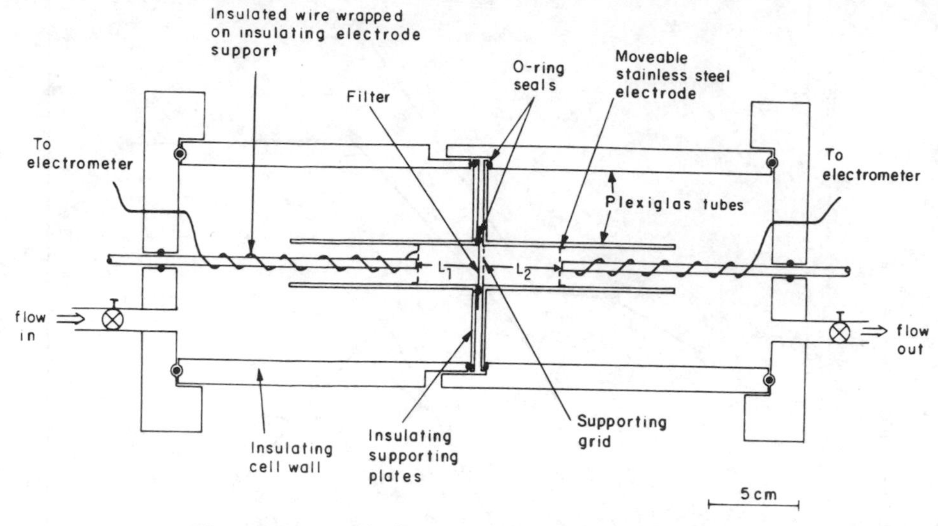

Figure 12. Scale diagram of experimental test cell [65].

diffusive to conductive terms is given by $D\tau/L$, where L is the pore diameter, d_p, for radial diffusion, and L is the length of the filter, L_f, for axial diffusion. Since $D\tau/d_p$ is the order of 10^4 to 10^5, radial diffusion cannot be ignored in their development. One can ignore $D\tau/L_f$, as Huber and Sonin have done, because it is only of $0(10^{-2} - 10^{-3})$. When one includes radial diffusion, the equations are much more complicated than the Huber and Sonin formulation, but also contain an indeterminate boundary condition, as pointed out in Appendix C. The data presented by them look interesting when compared with their predicted current-voltage relation. A brief accounting of highlights of their study is, therefore, presented below.

The current-voltage relation for the setup in Figure 12 is given by

$$I/A_v = -\frac{z_i q_f v_o - K_{ef}\Delta\psi/L_f}{1 + (K_{ef}/k_o)\left\{\dfrac{L_1 + L_2}{L_f}\right\}} \tag{95}$$

where

$$K_{ef} = \frac{PK_p}{t_o^{\,2}} + \frac{(1-P)K_s}{t_o^{\,2}} \tag{96}$$

is the effective conductivity of the filter as a whole and

where k_o, K_s = the fluid and solid conductivities, respectively
K_p = the conductivity of the fluid inside a pore
P = the porosity of filter
q_f = a fixed charge density
L_1, L_2 = upstream and downstream distances from the filter to the measuring electrode (Figure 12)
L_f = the thickness of filter
L_p = the pore length
$\Delta\psi$ = the downstream potential with respect to upstream potential difference
t_o = tortuosity, L_p/L_f
v_o = superficial velocity
z_i = sign of charge

Under certain simplifying assumptions, this equation can be written as

$$I/A_v = \frac{-z_i f(\gamma) L_f}{L_1 + L_2} \frac{k_o v_o}{\mu_I P} - k_o \frac{\Delta\psi}{L_1 + L_2} \tag{97}$$

where

$$\gamma = q_{f+}\mu_I/k_o$$

and

$$f(\gamma) = \frac{\gamma[\gamma + (1+\gamma^2)^{1/2}]}{1 + \gamma[\gamma + (1+\gamma^2)^{1/2}]} \tag{98}$$

This function becomes $f(\gamma) \simeq \gamma$ for $\gamma < 0.3$ and $f(\gamma) \simeq 1$ when $\gamma > 0.3$. Data given in Figure 13 show a linear dependence of streaming current on superficial velocity as given by Equation 97, because filter length to porosity variations, L/P, are negligible with respect to scatter in the data. Similar behavior was observed for the streaming potential dependence on flow velocity for the two electrodes open circuited.

One should exercise caution in applying Equation 97 to situations outside the range of the variables where experimental confirmation of Equation 97 has been validated. For example, in commercial fuel filtering operations there are no upstream or downstream electrodes. One is, however, interested in knowing the rate of charge generation, i.e., the streaming current in the filter and/or the potential buildup in a receiving tank. One cannot use Equation 97 to provide a reliable estimate of the charge generation rate (here L_1 and L_2 would equal zero) because a linear dependence on flow velocity is predicted. However, the measured value given by the experimental data in Figure 9, which holds for up to twelve orders of magnitude in a reduced dimensionless current and six orders of magnitude for the electrokinetic Péclet number, is far from linearly dependent on the flow velocity. As pointed out earlier by Huber and Sonin [65], it is impossible to make a direct comparison of their work with the Wagner and Gavis study [47–49]. As of this writing, this statement is still valid.

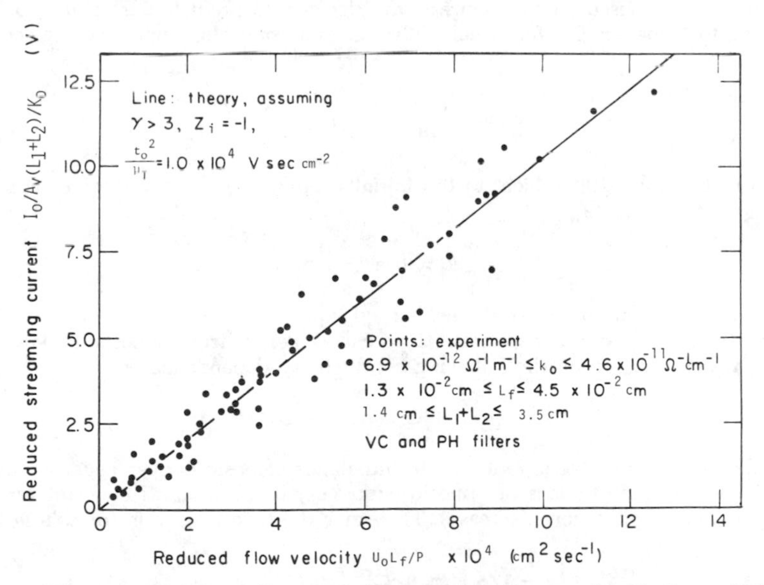

Figure 13. Summary of streaming current data compared with theory [65].

CHARGE RELAXATION IN LOW-CONDUCTIVITY HYDROCARBONS

The occurrence of potentially dangerous charge levels downstream of filters or commercial filter-separator units has been mentioned in the preceding section. One method to decrease charging using antistatic additives through an increase in the fluid conductivity already has been emphasized appropriately. The antistatic approach to charge control is, however, not always applicable. Bustin et al. [73] cite an early example of such a case in jet fuel manufacture, where absorbent treating was used to remove trace levels of surfactants to provide conductivities less than $1 \times 10^{-14} \Omega^{-1} cm^{-1}$ [73]. One would not purposely add an antistat to these fuels where prior efforts to remove these types of materials already had been expended. An important approach commonly practiced to reduce charge downstream of filter separators is to employ a 30-sec holdup in a charge relaxation chamber [73]. These chambers are enlarged conduits or vessels connected downstream of the filter separator that provide increased holdup from a significant reduction in linear velocity and an increase in chamber length. Under these conditions, charge can relax to ground in the chamber. Investigators other than Bustin, Koszman and Tobye, who have addressed the charge relaxation problem in general terms, include Vellenga and Klinkenberg [74] and Gavis [75,76].

Prior to the work of Bustin et al., it was well known that charged liquids relax exponentially according to ohmic theory (see Equations 78 and 101). These investigators apparently were the first to point out that highly charged fluids do not relax exponentially, but much more rapidly according to a hyperbolic relationship. For the simplifying assumption that the fuel is charged positively, the hyperbolic relationship is given by [73]

$$q_+ = q_{o+}/(1 + \mu_I q_{o+} t/\epsilon\epsilon_o) \tag{99}$$

where q_+ = the density of positive charge
q_{o+} = the initial charge density

The experimental data in Figure 14 illustrate the applicability of the hyperbolic equation to predict the proper charge decay rate for a fluid of $1 \times 10^{-16} \Omega^{-1} cm^{-1}$ conductivity. Superimposed on this figure is the decay rate predicted by ohmic behavior. According to Equation 25, for which diffusion and convection may be neglected, we obtain

$$\frac{\partial q_r}{\partial t_r} = -q_r \tag{100}$$

Solution to Equation 100 subject to the initial condition $q_r|_{t_r=o} = 1$ gives the ohmic form plotted in Figure 14:

$$q = q_o e^{-t/\tau_o} \tag{101}$$

where τ_o is the relaxation time of the uncharged liquid.

Vellenga and Klinkenberg propose a so-called generalized hyperbolic law of relaxation, which is based on the following conductivity dependence:

$$k = q\mu_{I+} + C_-(\mu_{I+} - \mu_{I-})F \tag{102}$$

This is derived from the assumption of an imbalance of positive over negative ions and that the concentration of ions of opposite sign (negative) remains constant while the concentration of positive ions decreases. They give the following form for charge decay:

$$1/q = 1/q_o e^{t/\tau_-} + \frac{\mu_{I+}\tau_-}{\epsilon\epsilon_o}\{1 - e^{t/\tau_-}\} \tag{103}$$

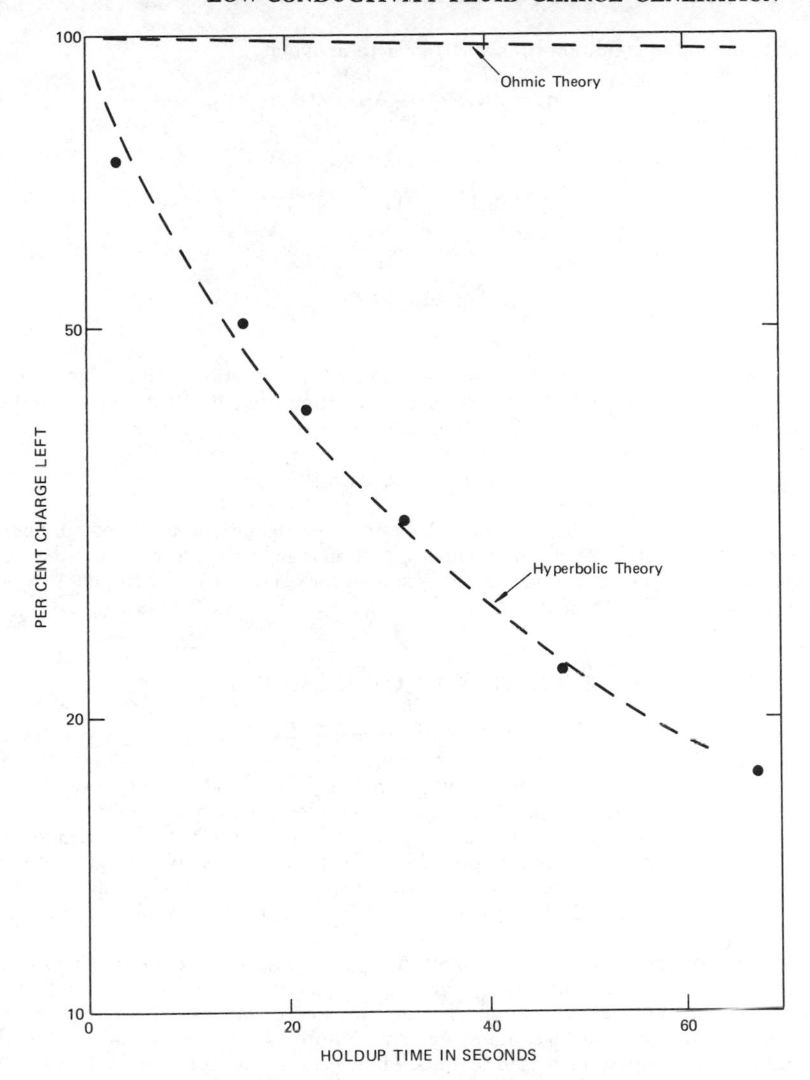

Figure 14. Relaxation of $10^{16}\Omega^{-1}\text{cm}^{-1}$ fuel [73].

where

$$\tau_- = \epsilon\epsilon_0 / C_-(\mu_{I+} - \mu_{I-})F \tag{104}$$

For conditions that $\mu_{I+} = -\mu_{I-}$ and small values of C_-, Equation 103 approaches the hyperbolic equation given by Bustin et al. Although this follows mathematically, it should be pointed out that the assumption that the negative ion concentration (the limiting species) remains constant while the positive ion concentration (the excess species) decreases is opposite to that expected.

Gavis [75] uses Equation 22 written as

$$\frac{dq_r}{dt_r} = -q_r(q_r^2 + 1)^{1/2} \tag{105}$$

subject to the initial condition $q_r(t_r = 0) = 1$ to arrive at

$$q_r = 2A(e^{t/\tau_o} - A^2e^{-t/\tau_o})^{-1} \tag{106}$$

where

$$A = q_{ro}[1 + (q_{ro}^2 + 1)^{1/2}]^{-1}$$

and

$$q_{ro} = q_o/FC^o \tag{107}$$

This differs from Equation 99 given by Bustin et al. as Equation 105 contains the term $(q_{ro}^2 + 1)^{1/2}$, while the right-hand side of the equation used by Bustin et al. contained q_{ro}^2. However, when $q_{ro}^2 >> 1$, Equation 105 for $q_{ro}(t_r = o) = 1$ gives

$$q_r = q_{ro}[1 + q_{ro}t/\tau_o]^{-1} \tag{108}$$

which is identical to Equation 99 when the standard definitions used for conductivity in terms of ionic mobility, relaxation time and similar notation are used for the charge density. Equation 105 for $q_r << 1$ is easily solved for $q_r(t_r = o) = 1$ to give the ohmic form $q = q_oe^{-t/\tau_o}$ (Equation 101).

LAMINAR CHARGE GENERATION IN CAPILLARIES

A recent search of the technical literature reveals no additional studies besides those of Wagner [47] involving laminar charging of low-conductivity fluids in single capillaries. Lack of interest in this flow regime is understandable because fluids are nearly always transported in pipelines in the turbulent regime for economic reasons. In addition, since the charging rates are much less than in the turbulent regime, there is generally little concern about approaching dangerous levels of charging for most fluid conductivities of practical interest. However, from the fundamental viewpoint expressed in this work, such data are important in developing an understanding of the charge transport process.

The data of Wagner (Figure 15) show the typical current increase-maximum value-decrease, with increasing conductivity for both turbulent flow and flow through porous media. This similarity seems to imply that at least one of the earlier approaches might readily predict or correlate these data. An expenditure of some effort along these lines was fruitless. One important reason for lack of success seems to be implicit in the data in Figure 15 for $k \lesssim 3 \times 10^{-13}\Omega^{-1}cm^{-1}$. Here, $I \alpha v^{0.5}$ suggests a possible entrance region problem because the Nusselt number for mass transfer over a flat plate has the same velocity dependence, i.e., $Nu = CRe^{1/2}Sc_m^{1/3}$ [76]. This idea is explored in some detail in the following paragraphs.

Following Levich [77], one may divide the capillary into three principal regions:

1. Region 1—velocity and charge profiles developing.
2. Region 2—velocity profile developed, charge profile still developing, or vice versa.
3. Region 3—both velocity and charge profiles are fully developed as, for example, in an infinitely long tube.

Tube curvature may be ignored in regions in which velocity boundary layer and charge boundary layer thickness are small with respect to the tube diameter. This may now

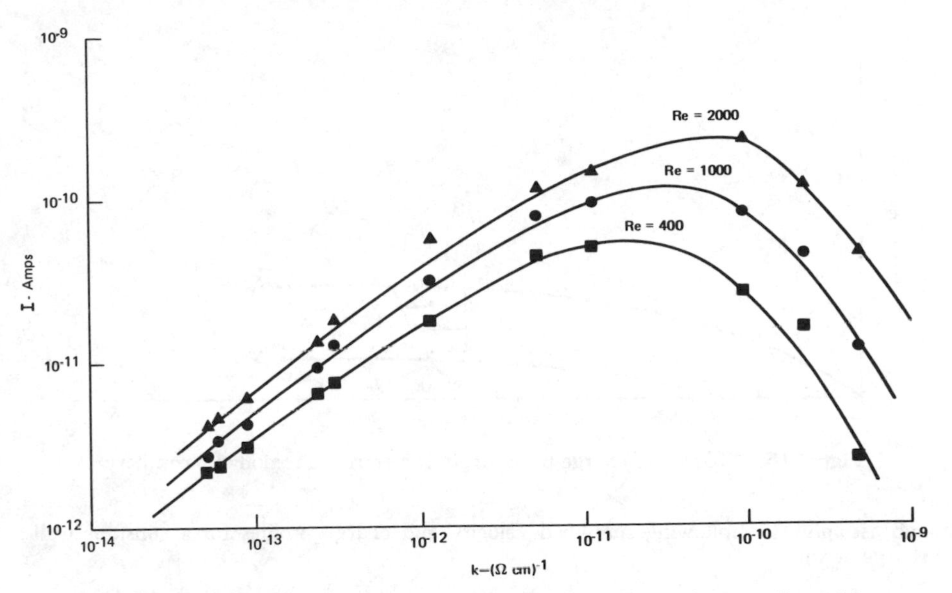

Figure 15. Charging current dependence on conductivity for laminar flow of heptane through capillary, L/d = 600 [47].

be treated as charge transport on a flat plate, as sketched in Figure 16, where δ_v is the velocity boundary layer thickness, U is the free stream flow velocity, Δ is the thickness of the charge boundary layer, q_∞ is the free stream charge density, y is the distance off plate, and x is distance along the plate. (The case $\Delta > \delta_v$ also can be treated; however, it is somewhat more difficult.) Writing Equation 25 in boundary layer form,

$$u\frac{\partial q}{\partial x} + v\frac{\partial q}{\partial y} = D\frac{\partial^2 q}{\partial y^2} - q/\tau \tag{109}$$

and integrating from 0 to Δ, subject to boundary conditions,

$$v = 0 \text{ at } y = 0$$

$$\begin{cases} v = -\int_0^\Delta \frac{\partial u}{\partial x}\,dy & \text{at } y = \Delta \\ q = q_o \end{cases}$$

we obtain

$$\frac{\partial}{\partial x}\int_0^\Delta u(q - q_\infty)dy = -D\left.\frac{\partial q}{\partial y}\right|_0 - \frac{1}{\tau}\int_0^\Delta q dy \tag{110}$$

Defining the local charge transfer coefficient by

$$h = -\frac{D}{q_w - q_\infty}\left(\frac{\partial q}{\partial y}\right)_{y=0} \tag{111}$$

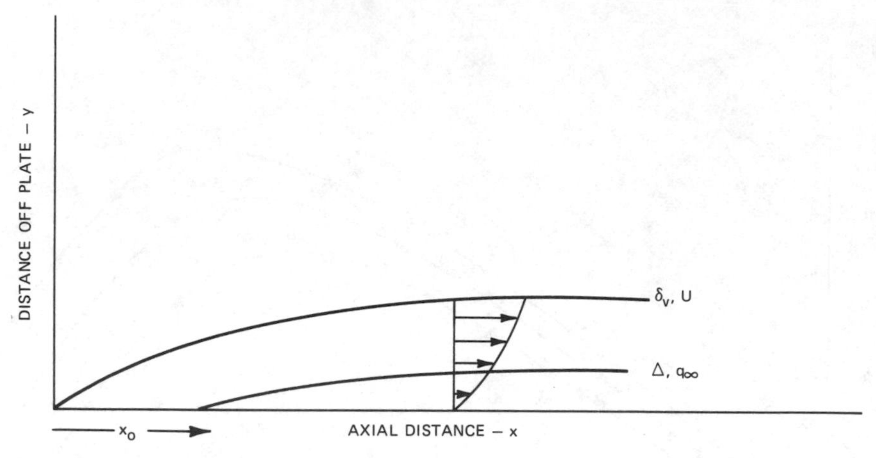

Figure 16. Sketch of charge transport in the entrance region of a capillary.

and assuming the following simplified velocity and charge profiles for a constant wall charge density,

$$u/U = 1.5y/\delta_v - 1/2(y/\delta_v)^3 \tag{112}$$

and

$$\frac{q - q_\infty}{q_w - q_\infty} = 1 - 1.5y/\Delta + \frac{1}{2}(y/\Delta)^3 \tag{113}$$

it can be shown from Equation 109 for high Schmidt number fluids that

$$\frac{\Delta}{\delta_v} = \frac{4}{5}\left\{\frac{10}{3}\frac{1}{U\tau}\left(\frac{q_\infty}{q_\infty - q_w}\right) - \frac{3}{8}\right\}\left(x - \left(\frac{x_o}{x}\right)^{1/4} x_o\right) \tag{114}$$

For our case, the starting length $x_o = 0$, $x = L$ and q_w may be neglected with respect to q_∞, assuming that the electrochemical discharge process at the surface is rapid. Furthermore, if we assume that the analysis is valid only if $\Delta/\delta_v \lesssim 1/10$, then for overall validity

$$U\tau/L \gtrsim 50/3$$

Substituting typical values of $\tau = 1$ sec and $U = 200$ cm/sec into the inequality, we obtain

$$L \lesssim 12 \text{ cm}$$

Since the capillary length was 9.3 cm, the charge transport entrance region lasts the entire length of the capillary, even though the velocity profile is fully developed for $L/D \simeq 150$, i.e., $L = 2.25$ cm.

This simplified analysis shows that laminar charge transport in capillaries occurs through a boundary layer-type mechanism. Further study is needed to develop a satisfactory model to predict the behavior illustrated in Figure 16 or to develop a correlation for the data in Figure 15.

CONDITIONS FOR A HAZARDOUS ELECTROSTATIC ENVIRONMENT

The conditions necessary for a hazardous electrostatic environment vary considerably, depending on whether the material under consideration is a gas, liquid or solid. Although the hazard of greatest concern is production of an explosion, processes leading to spark ignition and subsequent flame propagation are also important.

At least the following conditions appear necessary for electrostatically induced discharges.

1. **Generation and accumulation of electric charge.** Many factors influence the former (see previous sections), whereas charge accumulation is the net difference between generation rate and dissipation to ground or another object, as, for example, in refueling operations, where the other object could be a dip stick.
2. **Accumulation of maximum charge an object can sustain in its environment, followed by self-excitation of the environment or induced breakdown.** The discharge process is classified broadly into two categories of increasing intensity: corona or spark discharge types. Factors influencing the breakdown process are known to include nature of charged surface, geometry (needle points vs hemispherically capped cylinders) and distance of external objects from charged surface, dry air environment (moist or high relative humidity, or inert gas (He or Ar inhibit breakdown)) and others. This is an area in which it is rather difficult to make useful generalizations. Therefore, we will limit discussion to specific accepted approaches following a brief discussion on conditions necessary for an explosive environment.

Conditions necessary for explosions in gaseous fuels and liquids in air are:

1. **Gases.** Fuel concentrations must be within the upper and lower explosion limits, and the electrostatic discharge must be sufficiently energetic to produce ignition.
2. **Liquids.** The conditions of 1 apply for liquids at temperatures corresponding to their flash points. For liquids present in aerosol form, i.e., mists, it is well known that flammable mist-vapor oxidant mixtures can be formed at temperatures below the flash point [78].

For suspended dusts, the review by Hughes [79] shows an even more complex picture. Here, the dust concentration must fall within certain limits and a portion of the cloud must be raised to its ignition temperature. The following factors influence ignition and subsequent combustion of the dust cloud:

- physical characteristics of the particles,
- temperature necessary for propagation, and
- factors of flame propagation through the cloud (inert particles, moisture, and O_2 concentration).

For electrostatic discharge to induce ignition of the dust cloud, the energy dissipated during discharge would have to heat a portion of the cloud to its ignition temperature.

Some electrostatic values useful in hazard assessment are given in Table XII. The minimum ignition energy in A is defined as the total energy transferred from a capacitor in a discharge [81], where C is capacitance, pF, V is potential, V, and Q is charge, coul. The appropriateness of applying this steady-state equation to realistic discharges such as those occurring from a charged fuel surface has been questioned by Leonard and Carhart [82,83] and others. Parameters that may affect E include relative humidity, surface area, physical and chemical characteristics of materials, electrode geometry and dimensions, and fixed or moving objects. The usefulness of the minimum ignition energy

Table XII. Some Electrostatic Values Useful in Hazard Assessment

A. Minimum ignition energy:

0.1 mJ vapors in air
1 mJ mists in air
2.5–5000+ mJ dusts in air

$$E = CV^2/2 \times 10^{-9} = QV/2 \times 10^{-9} \text{ mJ}$$

B. Typical values of capacitance in pF, potential in V:
Human being, C = 100–400; dry carpet or auto seat, 5,000 V
Automobile, C = 500; traveling over pavement, 10,000 V
Belts over pulleys, C = variable; 20,000–25,000 V

C. Percentage of theoretical maximum charge density:

$$\left(2.65 \times 10^{-9} \frac{\text{coul}}{\text{cm}^2}\right)$$

Sliding and rolling contact, up to 8%
Dispersion of dusts, 1–10%
Pneumatic transport of solids and sheets pressing together, up to 60%
Close machining, up to around 100%

D. Particle size vs minimum ignition energy [80]

Particle Size (μm)	Minimum Ignition Energy (mJ)
710–1680	$> 5{,}000$
355– 710	200–500
180– 354	50–250
105– 179	< 10
53– 104	< 10
< 5	< 10

concept follows from a simple example. Let us assume that a human is charged to 5000 V by walking over a dry carpet and take his capacitance as 200 pF; then we obtain E = 2.5 mJ. This is of sufficient energy to ignite gaseous mixtures, such as diethyl ether-air, over approximately 3–8%, and various mists in a more restricted concentration range [81,84].

The magnitude of static hazards also may be assessed in terms of charge per given area (coul/cm^2). The theoretical maximum charge density (area basis), i.e., the value that produces a field strength equal to the breakdown strength of air in a homogeneous field, is 2.65×10^{-9} coul/cm^2. Gibson and Lloyd [85] found that if polyethylene is charged to $1.1\text{–}2.2 \times 10^{-9}$ coul/cm^2, depending on relative humidity, surface area and electrode characteristics, the discharge energy is sufficient to ignite a number of flammable vapor-air mixtures.

Accepted methods for control of static electricity [86,87] include

1. **Grounding.** Grounding is useful for conductors and semiconductors; however, for very good insulators, i.e., materials of low conductivity and, thus, high relaxation time (Table VIA), grounding cannot be particularly effective. Use of one or a combination of the following methods is necessary for insulators.
2. **Humidification.** Increasing the relative humidity (RH) is particularly helpful for static suppression on various solids through increasing surface conductivity and volumetric conductivity, if the material absorbs water (is hygroscopic). For gaseous fuel-air mixtures, humidification also is helpful because the minimum ignition energy increases with % RH. However, for liquids, water leads to pronounced charging by settling mechanisms [88] and, consequently, humidification would not appear applicable to many processes involving liquids.

3. **Increasing electrical conductivity.** One can add antistatic additives, such as those outlined in the section entitled Ionization Behavior in Low-Conductivity Liquids, to hydrocarbon liquids or use aerosol sprays on solid surfaces.
4. **Relaxation techniques.** This was mentioned previously in the section entitled Charge Relaxation in Low-Conductivity Hydrocarbons.
5. **Ionization.** Use of radioactive materials, corona discharge and inductive devices (such as grounded combs, bristles and tinsel bars) also is helpful in providing a conductive path to ground.

Other techniques come to mind; however, they are usually very specific or often employ combinations of the above. Control of static at the manufacturing level is evidenced by the availability of various types of conducting plastics and carpeting (usually containing metallic fibers woven into the fabric).

In the area of control, the introduction of new techniques is a very difficult task. One recent technique, applicable to refueling operations, is the result of the work of Leonard and Carhart [63]. Bright et al. [89] describe an application of this technique, which injects neutralizing charge to suppress static generation during refueling. This injection device has been tested successfully at a full-scale fueling facility.

A review of selected literature emphasizing important static hazards is given in the following section.

Recent Examples of Commercial Static Electrification Hazards

Static electrification problems in supertankers were highlighted in December 1969 when three tankers, each being cleaned with high-velocity water jets, were involved in major explosions [90–92]. The explosions all occurred in rectangular water tanks of 25,000 m^3 in volume (l, w, h = 50 m, 20 m, 25 m) following cleaning with a water cannon operating at a 40-m/sec flow velocity. During impaction of the jet with the surface, a spray of particles with a wide size distribution is produced. During a one-hour washing cycle, the larger droplets settle while the finer particles remain in suspension. Early test results showed that the equilibrium field was as high as 400 kV/m, while spark discharges were known to occur when the field at the surface of the liquid exceeded 500 kV/m.

Although the mechanisms for charge generation and transport are not well understood for this complex two-phase flow situation, methods to minimize dangerous charging include the following [90]:

1. Maintain the initial filling velocity at less than 1 m/sec until the lower structure of the tank is covered.
2. Restrict the maximum filling velocity so that the electric field normal to the free liquid surface should not exceed 500 kV/m.
3. Use only cold clean seawater and do not recirculate the already charged wash water.
4. Install inert gas systems to ensure that the tanker atmosphere is completely outside the explosive range.

The hazards associated with handling, storage and accompanying filtration of powders are discussed in the literature [80,93,94]. Because of the very low conductivities of many polymeric powders—$10^{-16}-10^{-18}\ \Omega^{-1}\ cm^{-1}$ [80]—charge dissipation to ground is very slow, thus permitting high charge accumulations. In addition, it is difficult to suppress static since antistatic additives are not commercially available for powders [93]. Another interesting example of the hazards associated with powders is given by Roberts and Hughes [94], where electrostatic charging in punctured aerosol cans was studied. Certain aerosol formulations were found to produce spark energies as high as 43 mJ, capable of igniting even suspended dusts (Table XII).

APPENDIX A–EXPERIMENTAL METHODS

Conductivity and Dielectric Constant

A number of different methods are available for determination of the conductivity and dielectric constant of low-conductivity fluids. Differences in the precision of the measurements are attributed to the sophistication of the instrumentation used in the measurement and experimental technique. Cleaning of the apparatus in a reproducible manner is especially important to avoid introducing ionic contaminants, which can significantly affect the measurements. As shown in Figure 1B, Appendix B, very minute concentrations of certain ionic additives can produce orders of magnitude increases in conductivity. The dielectric constant, however, is rather insensitive to small and even moderate amounts of soluble impurities, and its value for a pure liquid* may be used for charge generation in hydrocarbon fluids. For two-phase systems, such as hydrocarbon-water dispersions or emulsions, variation in the dielectric constant can be appreciable. It is recommended, therefore, that one obtain an experimentally measured value for these systems. In absence of the necessary instrumentation, Bruggeman's formula given by [96]

$$\frac{\bar{\epsilon} - \epsilon_2}{\epsilon_1 - \epsilon_2}\left(\frac{\epsilon_1}{\bar{\epsilon}}\right)^{1/3} = 1 - \Phi \qquad (1A)$$

where $\bar{\epsilon}$ = dielectric constant of the mixture
ϵ_1 = dielectric constant of water
ϵ_2 = dielectric constant of hydrocarbon
Φ = volume fraction of dispersed phase

may be used to provide an estimate of the limiting dielectric constant at high frequencies. Although Hanai [97] has shown that Bruggeman's formula predicts the observed $\bar{\epsilon}$ vs Φ dependence for water-in-oil emulsions, errors will result for systems of significantly different interfacial tensions from the water-oil system,† or ionic strengths. For suspensions in potassium chloride solutions containing high concentrations of colloidal particles, dielectric constants greater than 2000 have been reported [98] at low frequencies. Thus, for three-phase systems, e.g., water of high ionic strength/hydrocarbon/solids dispersions or colloidal systems, experimental measurements are required to provide reliable dielectric constant data for charge transport studies.

The so-called condenser or Ball method is described in the literature [36,69]. It is intended for use in refinery laboratories only. The apparatus is illustrated in Figure 1A. In this setup, the conductivity is calculated from the time required to reduce a potential difference across a condenser to a certain fraction of its original value. Calibrated low-leakage capacitors with ratings of 100, 500, 1000, 5000, 20,000 and 50,000 pF, a high-value resistor of about $3 \times 10^{11}\Omega$, a battery or series of batteries with a rating of about 125 V, and an electrostatic voltmeter with a relaxation time of not less than 10 minutes and a measuring range of 0 to 150 V are the principal pieces of equipment required. Shielded coaxial cable is required to connect the high-voltage terminal of the voltmeter to the sphere of the condenser.

To calibrate the apparatus, the resistor of $3 \times 10^{11}\Omega$ and a capacitor of about 100 pF are first connected between the terminals of the voltmeter. The voltage is applied for 15 minutes or as long as required to reach a constant discharge time. One then records the times for:

1. a voltage drop from 100 to 80 V, t_1; and

*An extensive tabulation of dielectric constants of pure liquids is given by Maryott and Smith [95].

†Interfacial tensions range estimated at 25 to 35 dyne/cm.

2. obtaining the same voltage drop after disconnecting the capacitor, t_2.

The total capacitance of the system—empty condenser plus coaxial cable plus voltmeter—is given by:

$$C_{total} = \left(\frac{t_2}{t_1 - t_2}\right) C_a \tag{2A}$$

where

$$C_a = 100 \text{ pF}$$

Conductivity in $1 \times 10^{-12} \Omega^{-1} m^{-1}$ units may be calculated from either of two equations given by:

- high conductivity through a voltage drop from 100 to 80 V:

$$K = \left\{\frac{C_{ball}(\epsilon - 1) + C_{total} + C_{ad}}{C_{ball}}\right\} \times \frac{\epsilon_o}{t} \log_e \left(\frac{100}{80} \times 10^{12}\right) \tag{3A}$$

- low conductivity from a voltage drop from 100 to 98 V:

$$K = \frac{C_{ball}(\epsilon - 1) + C_{total}}{C_{ball}} \times \frac{\epsilon_o}{t} \log_e \left(\frac{100}{98} \times 10^{12}\right) \tag{4A}$$

Here, C_{ball} is calculated to be 1.6 pF. The effective capacity in the hydrocarbon is larger by $1.6(\epsilon - 1)$, and is, therefore, added to the other values of capacitance. The capacity of any added capacitors in pF is denoted by C_{ad}.

For accurate work, the following procedure is recommended [69]. Connect the battery with the high-voltage terminal of the voltmeter and, at the same time, start

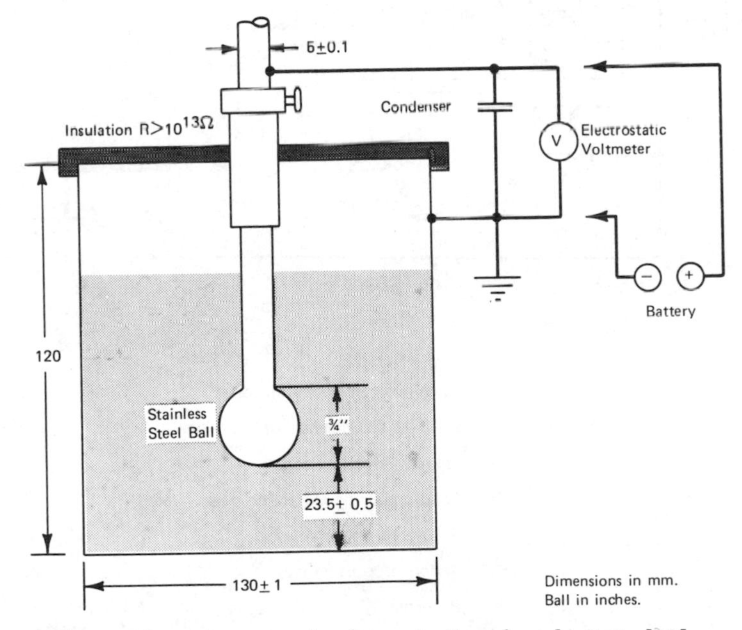

Figure 1A. Apparatus for determination of conductivity [69].

a stopwatch (60-second sweep). After 10 seconds, remove the lead and allow the stopwatch to continue running. With a second stopwatch (3-second sweep), measure the time required for the voltage to decay from 100 V to 80 V. Reconnect the lead. Record the time indicated by the second stopwatch. Repeat the measurement when the first stopwatch indicates a time of 80 seconds and when it indicates a time of 150 seconds. Hence, the three measurements are made within 3 minutes. Empty the container immediately. Take the average of the three discharge times. Repeat the procedure outlined above until consistent results are obtained.

Another method that uses fixed voltage dc to effect a potential drop across a measuring cell that gives rise to a current flow to ground is described in the literature [49,55]. Although conductivities reproducible to around 10% were reported for relatively pure hydrocarbons, this technique is not recommended for two- and three-phase emulsions where dc fields could lead to appreciable errors.

A schematic diagram of a suitable apparatus using the above technique is given in Figure 2A. The principal parts of this setup are:

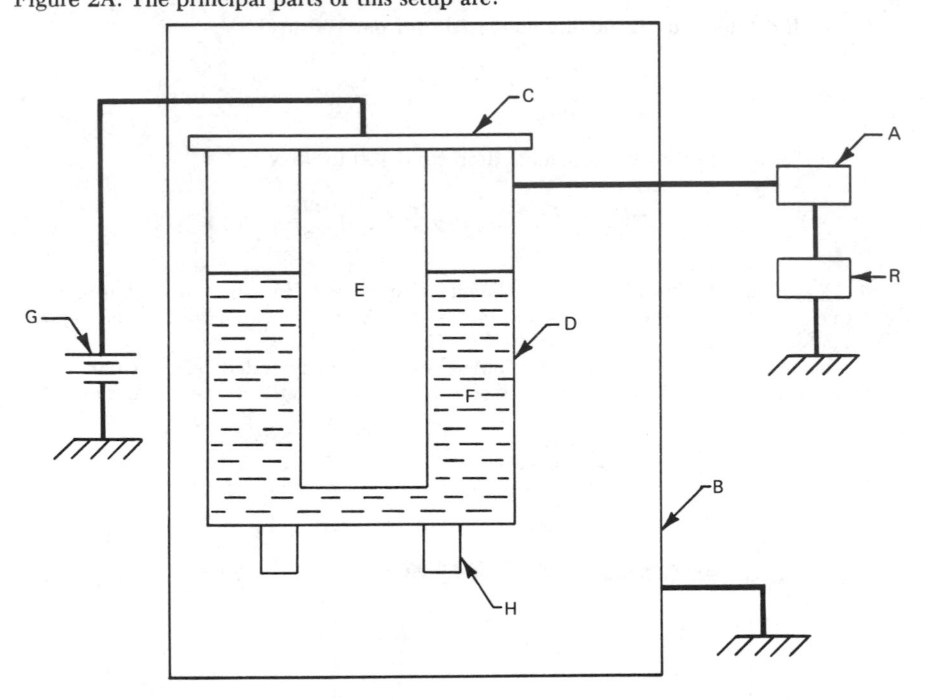

A) Electrometer

B) Faraday cage

C) Machined circular teflon disc

D) Stainless steel beaker or container

E) Stainless steel cylinder

F) Solution for conductivity measurement

G) Regulated power supply

H) Teflon supporting legs

R) Recorder

Figure 2A. Steady voltage conductivity cell.

- an electrometer or picoammeter for measurement of current flow;
- a concentric cylinder conductivity cell, the inner electrode being made from stainless steel rod and the outer electrode being a stainless steel beaker or container, with the interelectrode spacing fixed by a groove machined into the Teflon top plate; and
- a regulated power supply (0–500 V) or battery with a voltage rating of at least 200 V.

Shielded Teflon-insulated coaxial cable is recommended for the entire setup.

To use the above setup for conductivity determinations one must know the cell constant. This may be determined by using a minimum of three KC1 solutions of specific resistivities selected from the range $1 \times 10^3 \Omega$ cm to $1 \times 10^6 \Omega$ cm and measuring the overall cell resistance on a suitable ac conductance bridge. The bridge should be operated at a frequency of at least 1000 Hz to minimize polarization effects at the electrodes. Ohm's law given by

$$R = \rho_R (l/A)_{cell} \tag{5A}$$

with R obtained from the conductance bridge measurement for a known ρ_R of KC1 (starting with the highest-resistivity fluid) is used to determine the cell constant $(l/A)_{cell}$. This calibration method requires that the ohmic current be large with respect to the capacitive current [36]. This may be checked for the equivalent circuit given in Figure 3A.

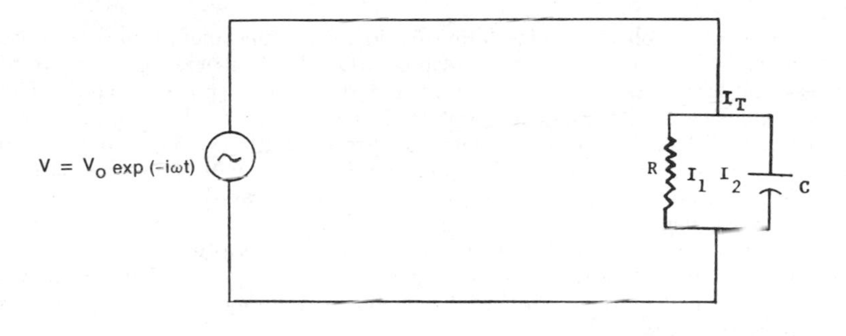

$$Z = R + \frac{1}{2\pi f C}$$

$$I_T = V/Z$$

Figure 3A. Circuit schematic for ac conductivity cell calibration.

where the total circuit impedance, Z, is

$$Z = R + \frac{1}{\omega C} \tag{6A}$$

and the junction current, I_T, is given by

$$I_T = V\left(\frac{1}{R} + \omega C\right) \tag{7A}$$

The cell capacitance may be approximated by the capacitance for two flat plates:

$$C = \epsilon\epsilon_o A/l \tag{8A}$$

For the saline solutions and frequencies recommended,

$$1/\rho_R >> \omega\epsilon\epsilon_o \qquad (9A)$$

This criterion is easily met in practice.

Conductivity ($k = 1/\rho_R$) may be calculated from Ohm's law using the zero-time intercept of current obtained from a plot of the log of current, I, vs time and the known cell constant. Plots of I vs t that cannot be approximated by a straight line should be redone following appropriate cleaning of the conductivity cell [49].

Several different types and models of inductance-type bridges are available commercially that will measure a wide range of conductance, resistance, capacitance or inductance [99]. A model capable of covering a range of conductance from about $1 \times 10^{-15}\Omega^{-1}$ to $1 \times 10^{-7}\Omega^{-1}$ is recommended to cover the conductivity range of $1 \times 10^{-16}\Omega^{-1}cm^{-1}$ to $1 \times 10^{-7}\Omega^{-1}cm^{-1}$ of interest in research applications. The conductivity range may be extended by using conductivity cells having cell constants of 0.01 cm^{-1} to 10 cm^{-1}.

Methods for measuring the dielectric constants and resistivities of insulating liquids and solids are detailed in ASTM standards [100–104]. The dielectric constant is determined by measuring capacitance with a suitable bridge and test cell containing the material of interest, C_m, and dividing by the capacitance of the same cell containing air or a vacuum as the dielectric.

$$\epsilon = C_m/C_{air} \qquad (10A)$$

Different test cells applicable to insulating liquids, waxes and metastable solids are commercially available from Rutherford Research [105]. The cells may be used to measure dielectric constant, conductivity/resistivity and dissipation (power) factor. The temperature range covered is from ambient to 250°C (500°F). The range may be extended below ambient if the dew point is suppressed with dry air to prevent condensation. The cell electrodes are made from stainless steel with polytetrafluoroethylene (TFE) insulators. The heated models have integral aluminum block ovens with heaters, thermostat and thermometer well, as required. Three terminal models permit direct measurement of the air capacitance when the cells are empty. The various cells are designed to operate over a wide frequency range–0.001 Hz to 1 MHz and above, with specific units up to 100 MHz. The limiting factor in the upper frequency range is typically bridge design and cable.

The procedures, equipment and instrumentation outlined above provide various methods to determine the important electrical properties of conductivity and dielectric constant. Other approaches and instrumentation also may be used to determine these properties. For precise work, especially when dealing with emulsions or suspensions, three-terminal type cells and an appropriate precision bridge with conductance and capacitance measurement capability are needed.

Flow Measurement Techniques

As detailed in the text, whenever low-conductivity fluids flow past surfaces free electric charges appear in the liquid. Charges of opposite sign also occur on the surface. If the liquid is caused to flow into a storage tank, these free charges may accumulate to sufficiently high levels, possibly leading to spark discharge.

Based on this phenomenon, different measurement techniques have been devised:

1. Insulate the flow surface from ground by a solid dielectric, such as Teflon, and monitor the current generated as a function of flow and geometric properties using a high impedance input picoammeter connected in series between the solid surface and ground. (Figure 7 in the text illustrates this technique.)

2. Ground the tube and measure the voltage buildup on a receiving tank using an electrostatic voltmeter or multipurpose electrometer with an appropriate voltage range (0 to 10,000 V being typical) or lower range coupled with attenuation voltage probes (10 times −1000 times step-down).
3. Use a picoammeter to measure the current flow from a receiving tank to ground (Figure 10).

Although each of these techniques has its merits, depending on the charging process under study, the former technique is recommended to study streaming current because of its simplicity and direct approach. The latter two techniques are indirect approaches to generally determine an upstream phenomenon as based on a downstream measurement. These methods can lead to error since charge relaxation and recombination are occurring in the bulk fluid simultaneously with the current flow or potential drop.

Zeta Potential Concept for Aqueous Systems

Zeta potential measurements have been used for many years as the standard technique for determining colloidal stability in aqueous media. An illustration of the zeta potential of a particle is given in Figure 4A [106,107]. The net charge acquired is generally always electronegative in aqueous media. Adjacent to the charged particle is a layer of stationary positive charges; next is a diffuse layer of positive and negative ions. The zeta potential is the difference between the charge on the rigid layer, i.e., the so-called Stern layer at the plane of shear and the bulk of the suspending liquid. Naturally occurring colloids suspended in distilled or tap water generally have zeta potentials in the range of −15 to −30 mV. Stability characteristics based on the standard Helmholtz-Smoluchowski equation given by

$$\zeta = 4\pi M_e \mu / D \tag{11A}$$

where M_e is the measured electrophoritic mobility (cm/sec)/(V/cm), for anionically dispersed systems are summarized in Table IA [107]. So that the units in Equation 11A are consistent, one should multiply the equation by $\{8.85 \times 10^{-11}$ coul^2/dyne-cm$^2\}$ $1/\epsilon_0^2$ where ϵ_0 is the permittivity of free space given in the nomenclature. Methods to achieve stability are outlined in Figure 5A. These include:

- adsorption of an anionic or polyelectrolyte to effect strong mutual repulsion, the range of −45 to −70 mV generally being preferred;
- adsorption of a hydrophilic protective colloid on a hydrophobic colloid, in which the affinity for water must exceed the attractive forces of the particles; and
- adsoprtion of a long-chain nonionic polymer to create steric hindrance.

Table IA. Stability Characteristics [107]

Stability Characteristics	Average Zeta Potential (mV)
Maximum Agglomeration and Precipitation	0– +3
Range of Strong Agglomeration and Precipitation	+5– −5
Threshold of Agglomeration	−10– −15
Threshold of Delicate Dispersion	−16– −30
Moderate Stability	−31– −40
Fairly Good Stability	−41– −60
Very Good Stability	−61– −80
Extremely Good Stability	−81––100

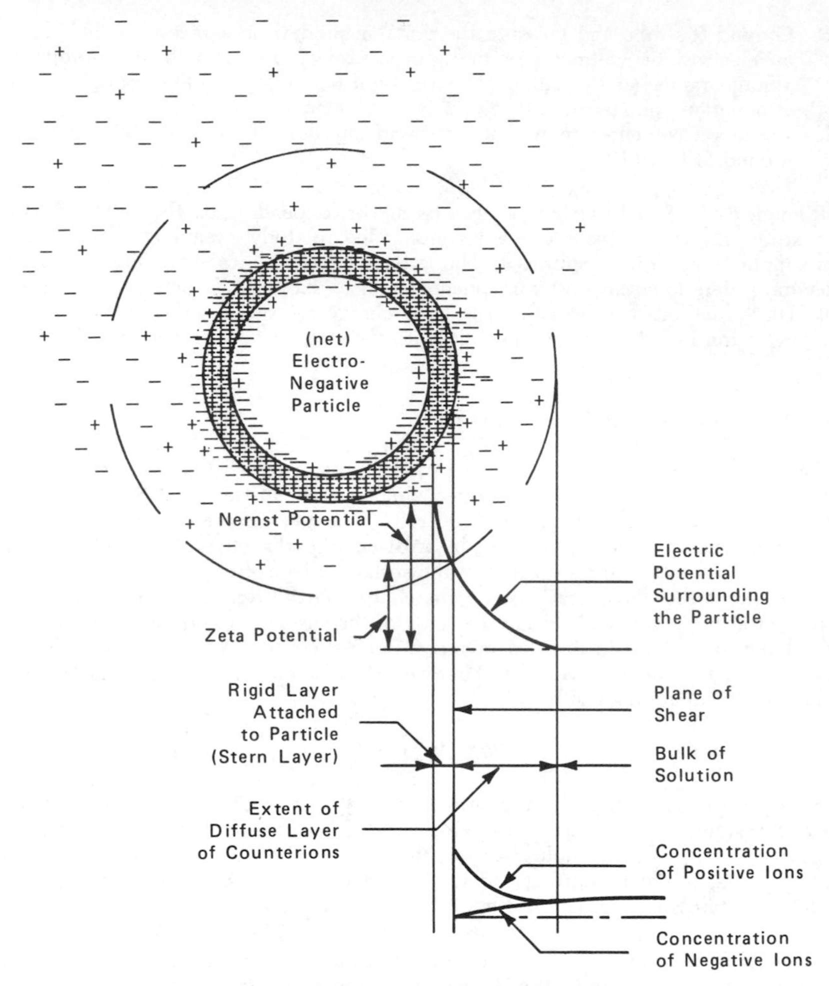

Figure 4A. Concept of the zeta potential [106,107].

Coagulation or agglomeration is the opposite effect of the stability characteristics considered above. Table IA shows that maximum agglomeration occurs in the range of 0 to +3 mV for the zeta potential. Different methods to produce agglomeration or precipitation are detailed by Riddick [106]. They are summarized as follows:

1. Add a strong cationic electrolyte such as aluminum sulfate to lower the zeta potential to zero. Combine a cationic electrolyte with an appropriate alkali to form a hydrous oxide.
2. Add a lyophobic colloid to bind the water in the system.
3. Add a long- or branched-chain anionic or nonionic polyelectrolyte to produce agglomeration.

Measurement Techniques

The conventional method for determining the zeta potential involves measuring the electrophoretic mobility and use of Equation 11A or charts to obtain ζ. A stereomicro-

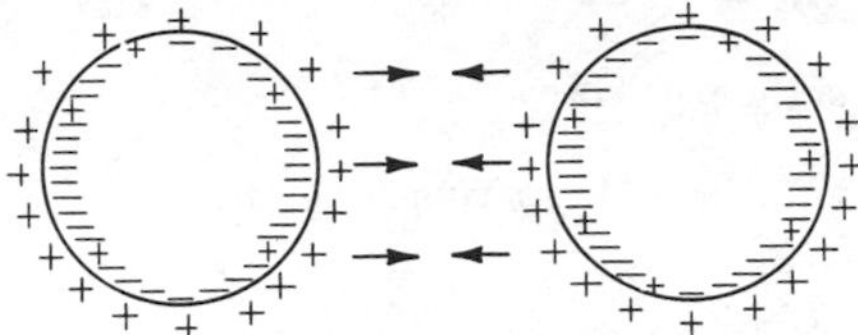

METHOD 1: Mutual repulsion due to high Zeta Potential

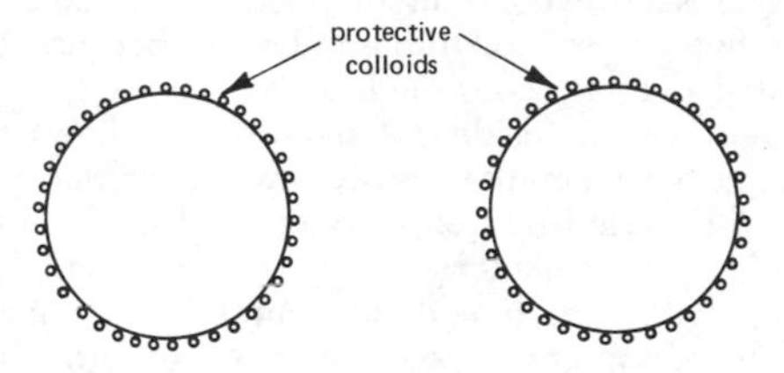

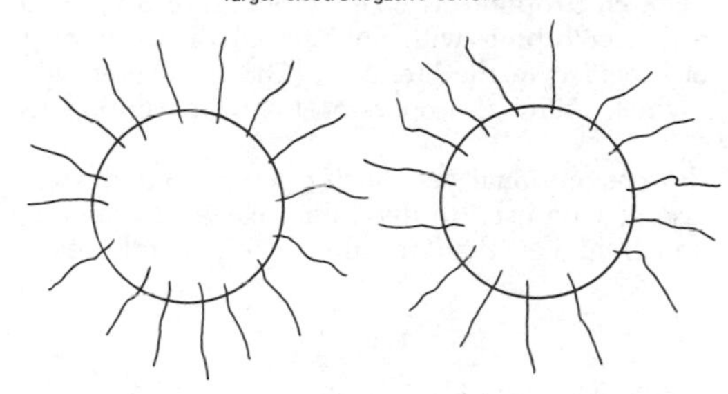

METHOD 3: Steric hindrance due to adsorption of an oriented nonionic polyelectrolyte

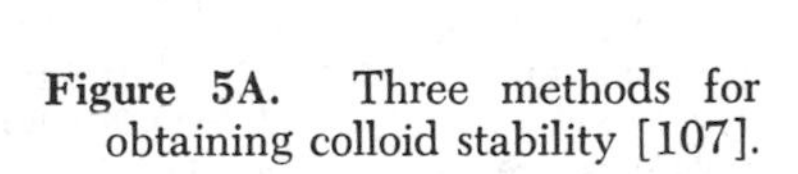

Figure 5A. Three methods for obtaining colloid stability [107].

scope is used to follow the colloid at the line of zero endosmotic flow (Figure 6A) under an imposed potential (0–300 V) in a specially designed electrophoretic cell (Figure 7A). The Zeta Meter Catalogue [107] contains different cell types to obtain this measurement. The time required to track the colloidal particle over a given distance is determined on a multiscale ocular micrometer. Using this information one can calculate the electrophoretic mobility:

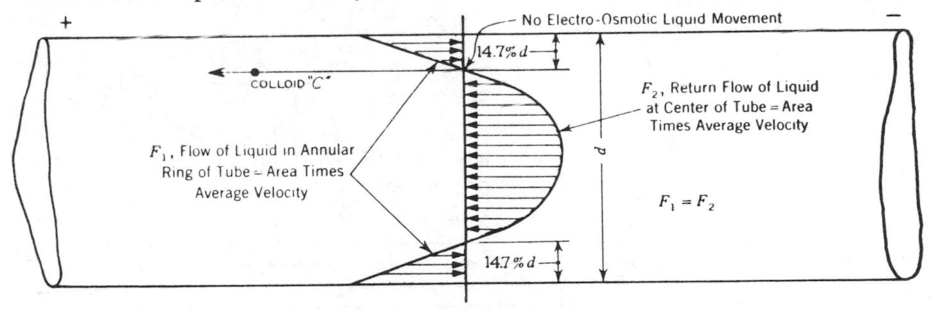

Figure 6A. Plane of zero endosmotic flow [107].

$$M_e = v_e/(V/L) = v_e/E_f \tag{12A}$$

where v_e = the electrophoretic velocity of the particle
V = imposed potential
L = effective chamber length or interelectrode spacing
E_f = electric field strength

Charts are also available to determine the zeta potential, ζ, including a correction factor for time of relaxation.

For low-conductivity hydrocarbons of a dielectric constant of about 2, it is necessary to increase the applied voltage to as high as several thousand volts because the tracking time is attenuated by a factor of 40× over aqueous media.

A new technique based on rotating a cube prism inside the microscope allows the observer to measure the mobility of many particles simultaneously, instead of one particle at a time [108,109]. The rate of rotation is varied by the operator until the average particle motion caused by the applied field is cancelled by the prism rotation. This technique permits rapid generation of data with few time-dependent errors as many particles are viewed simultaneously. A fully automated system capable of providing turbidity, settling velocity, particle size and electrokinetic histograms is illustrated by the block diagram given in Figure 8A. The electrophoresis chamber (Figure 9A) used with this setup is fabricated from pure quartz tubing with an internal diameter of 1 mm. The details of the electrodes are not specified in the literature [108,109]; however, they are flow-through type electrodes, which allow the operator to make continuous readings.

It should be pointed out that both the conventional zeta meter setups and devices based on the rotating prism technique also can be used to determine electrical conductivity. However, equipment costs are much higher than for the experimental setups shown in Figures 1A and 2A.

Zeta Potential Determinations in Low-Conductivity Liquids

The early literature on the stability of colloidal particles in nonaqueous media appears to be limited [110,111]. Soyenkoff [110] reported that stability is not related to particle charge for colloidal graphite particles dispersed in various hydrocarbons. Hendrick et al. [111] have shown for coal particles in oil that both positive and negative charges

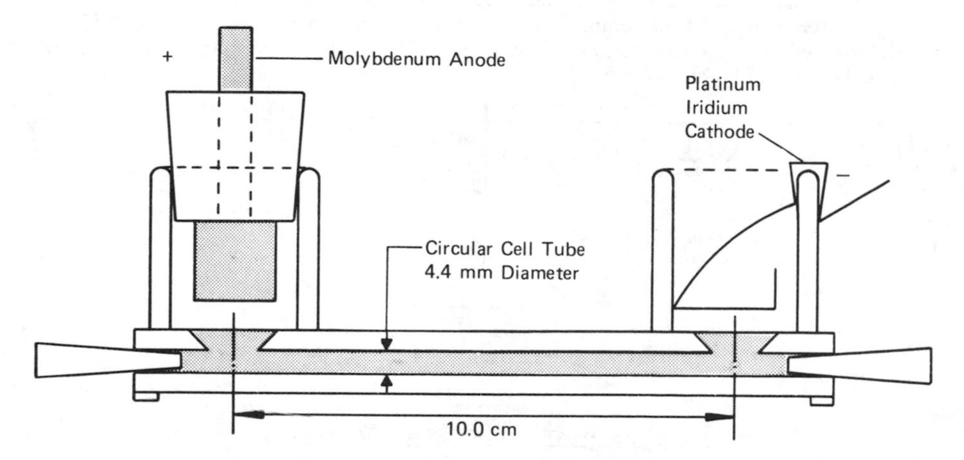

Figure 7A. A cell for electrophoretic measurements [107].

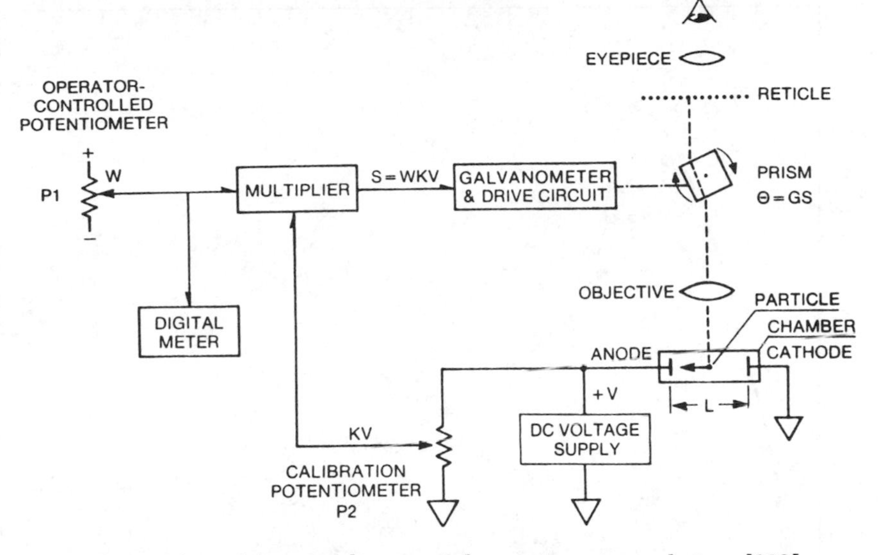

Figure 8A. Schematic diagram of the rotating prism technique [108].

can exist on different areas of the same particle. In addition, the particle charge was concentration dependent:

1. At low coal concentrations the particles were negatively charged because they all migrated to the positive electrode under an imposed field.
2. At high concentrations the particles went to both electrodes simultaneously.

The addition of surface-active agents, such as Aerosol OT and lecithin, prevented agglomeration, even though all the coal particles could be deposited on electrodes.

Other early investigators who have studied the behavior of particles in nonaqueous media include Van der Minne and Hermanie [112,113], McGown et al. [114], Parfitt and Willis [115], Lewis and Parfitt [116] and Kuo and Osterle [117]. The work of Lewis and Parfitt [116] dealing with dispersions of graphitized carbon black in n-heptane containing different concentrations of Aerosol OT surfactant is particularly informative. First, the carbon black particles were found to be negatively charged in all solutions

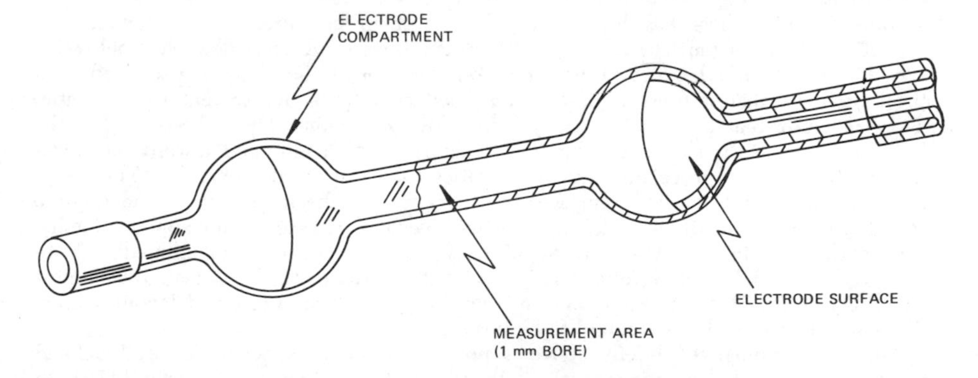

Figure 9A. Microelectrophoresis chamber [109].

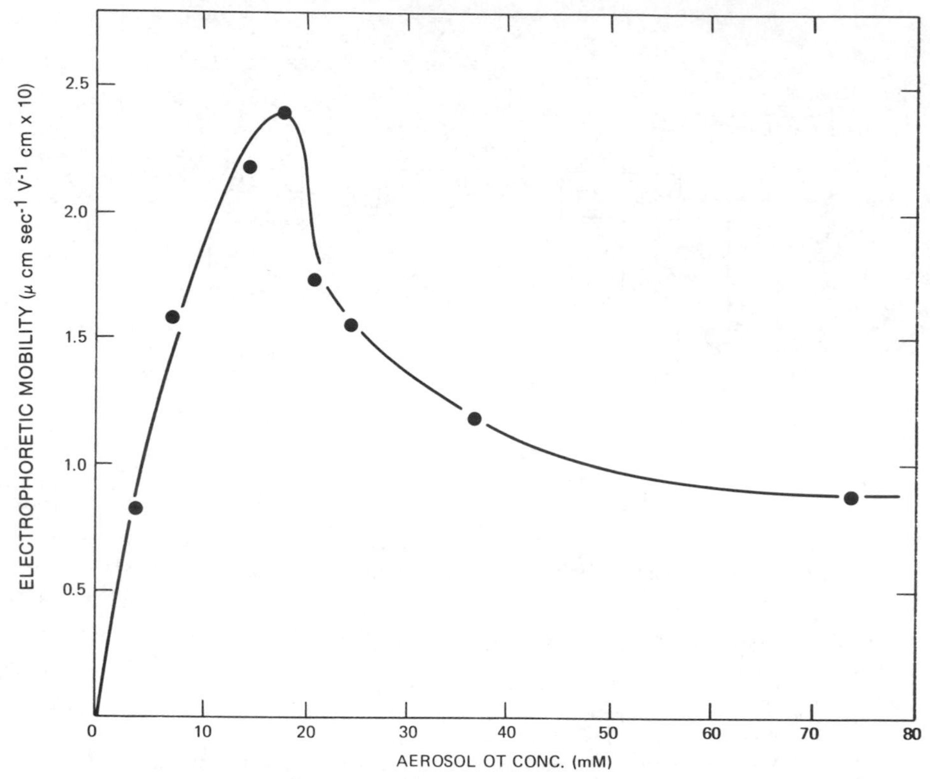

Figure 10A. Electrophoretic mobilities for Sterling MTG in solutions of Aerosol OT in n-heptane [116].

and had the typical maximum in electrophoretic mobility vs concentration behavior—Figure 10A observed for other nonaqueous systems [113,114]. They also found qualitative agreement between the observed stability and predictions based on the Derjaguin-Landau-Verwey-Overbeek (DLVO) [118] theory of colloid stability. However, because they had to adjust values of a constant in the theory by over two orders of magnitude, they suggest that further refinements in the theory are needed.

More recently there has been renewed interest in measurements of the electrophoretic mobility of particles and separation techniques using cross-flow electrofiltration [28–33]. Henry and Jacques [119] report on the sign of the charges associated with the particulate matter present in solvent refined coal for a 3:1 solvent to coal ratio. They conclude that the particles carry a net positive charge. They observed that the larger particles existed as long chain-like structures, which migrated toward the negative electrode with appreciably higher velocities than the smaller particles. Their estimated mobility of 1×10^{-5} (cm/sec)/(V/cm) was of the same order of magnitude reported by Van der Minne and Hermanie with carbon particles in nonaqueous media. For tetralin containing α-Al_2O_3 ($d_p = 0.3$ to 2 μm) and H-coal liquid diluted with 40 wt % xylene and 5.5 wt % tetralin dispersed with Aerosol OT, Lee et al. [32] report a mobility of 5×10^{-6}(cm/sec)/(V/cm). This is only half of Henry and Jacques value. The particle charge also was found to be positive.

This area summarized briefly above is not well understood, even for well defined laboratory colloidal suspensions, such as those studied by Lewis and Parfitt [116]. In dealing with coal liquids that can contain particles ranging from 0.1 to 100 μm at loadings of up to 10 wt %, the problem of obtaining reliable mobility information is even more difficult.

An expanded scope of research activity of an experimental and theoretical nature is clearly needed to obtain a better understanding of stability characteristics in nonaqueous media.

APPENDIX B—CONDUCTIVITY-CONCENTRATION BEHAVIOR OF SHELL ANTISTATIC ADDITION IN HEPTANE

An example based on unpublished data of the author [120] will be used to illustrate the conductivity vs concentration behavior for an antistatic additive in heptane.

For Shell A.S.A.#1* (a mixture of divalent calcium and trivalent chromium ionic solution in a light hydrocarbon) dissolved in heptane (purified to around $1 \times 10^{-15}\Omega^{-1}$ cm^{-1}), the conductivity vs concentration behavior is given in Figure 1B. The concentration of additive is plotted as grams of metal present in A.S.A. #1 per liter of heptane. These data may be replotted as equivalent conductivity, $\Lambda = k/c$, vs the square root of the concentration, c (the total concentration of additive in g-eq/cm^3 is assumed to be twice the concentration of metal in g-eq/cm^3) as shown in Figure 2B. Because the equivalent conductivities to the left of the equivalent conductivity minimum

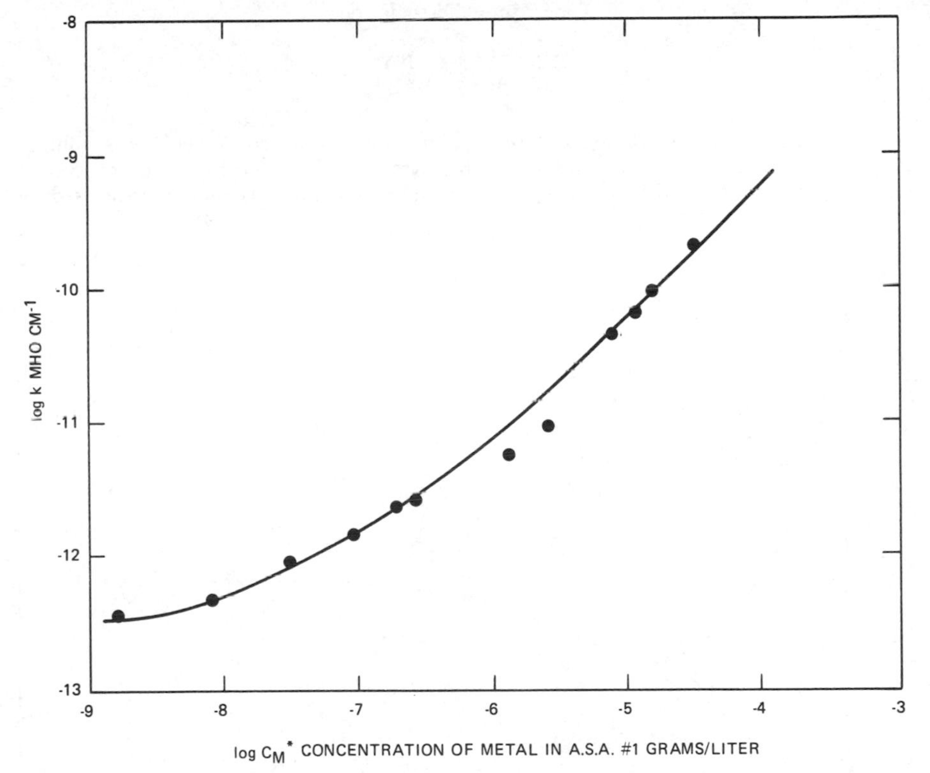

Figure 1B. Conductivity vs concentration behavior for Shell A.S.A. #1 dissolved in heptane [120].

*An early antistatic additive used in the 1960 time period for static suppression in hydrocarbon fuels. Shell A.S.A. #3 is presently available [69].

range from around 5×10^{-5} mho cm^2eq^{-1} to 4×10^{-4} mho cm^2eq^{-1}, ionic dissociations of type 1 below lie strongly to the left:

Type 1 Ionic Dissociative

$$AB \rightleftarrows A^+ + B^-$$

Alternatively, a very small fraction of the antistat exists as dissociated singly charged species. An estimate of the number of dissociated ions to the number of undissociated ion pairs may be obtained from the ratio of equivalent conductivity at low, but finite, concentration to equivalent conductivity at finite dilution, Λ_∞, where the antistat is assumed to be completely dissociated. Conway [121] gives Λ_∞ for fluids of dielectric constant, ϵ, equal to 2 as 100–150 $cm^2/coul\ \Omega$. Thus, the ratio of dissociated ions to undissociated ion pairs is $3 - 5 \times 10^{-6}$.

Further ionic association may occur with an increase in concentration near the minimum, causing triple ion formation. Triple ion formation may be represented by equilibria of type 2 [34] given below.

Type 2 Ionic Dissociation

$$AB + A^+ \rightleftarrows ABA^+$$
$$AB + B^- \rightleftarrows BAB^-$$

Near the minimum, equilibria of types 1 and 2 may both occur. The two equilibria may be analyzed by the theory of Fuoss and Kraus [37] according to the procedure outlined by Robinson and Stokes [34]. The equivalent conductivity may be expressed by

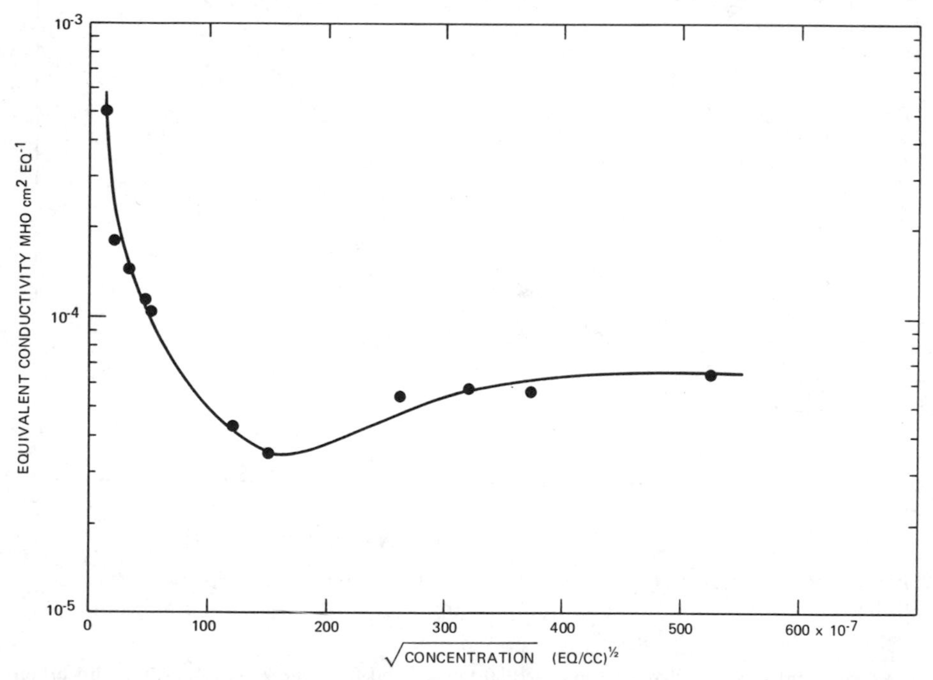

Figure 2B. Equivalent conductivity vs square root of concentration [120].

$$\Lambda = \alpha\Lambda^\circ + \alpha_T\Lambda^\circ_T \tag{1B}$$

or

$$\Lambda = \sqrt{\frac{K_{eq}}{C}}\,\Lambda^\circ + \sqrt{\frac{K_{eq}C}{K_{eq}{}^T}}\,\Lambda^\circ_T \tag{2B}$$

where K_{eq} = equilibrium constant of type 1
$K_{eq}{}^T$ = equilibrium constant of type 2
α = fraction of univalent ions dissociated
α_T = fraction of triple ions dissociated
Λ° = equivalent conductivity of a univalent ion at infinite dilution
Λ°_T = equivalent conductivity of a triple ion at infinite dilution

Equation 2B is of the form

$$\Lambda = aC^{-1/2} + bC^{1/2} \tag{3B}$$

where a and b are constants. Thus, a plot of $\Lambda\sqrt{C}$ vs C should be a straight line. The minimum equivalent conductivity is obtained from $d\Lambda/dC = 0$. This gives the following values for the minimum equivalent conductivity and the concentration evaluated at the minimum:

$$\Lambda_{min} = 2\sqrt{ab} \tag{4B}$$

$$C_{min} = a/b \tag{5B}$$

The minimum equivalent conductivity also may be expressed by [34]

$$\Lambda = 2\alpha_{min}\Lambda^\circ = 2\alpha_{Tmin}\Lambda^\circ_T \tag{6B}$$

A plot of the equivalent conductivity times the square root of concentration vs concentration is shown in Figure 3B. The straight-line behavior justifies using the equivalent conductivity concentration relationship given in Equation 3B. The slope and intercept of this curve are 1.2 cm^2 mho $eq^{-3/2}$ and 6.0×10^{-10} $cm^{+1/2}$ mho $eq^{-1/2}$, respectively. From the intercept and using $\Lambda^\circ = 150$ mho cm^2eq^{-1}, $K_{eq} = 4 - 6 \times 10^{-12}$ eq/cm^3. This low value also confirms that equilibria of type 1 lie far to the left. From Equations 4B and 5B, the values of Λ_{min} and C_{min} are 5.4×10^{-5} mho cm^2 eq^{-1} and 4.9×10^{-10} eq/cm^3. The experimental values of Λ_{min} and C_{min} obtained from Figure 2B are 3.5×10^{-5} mho cm^2 eq^{-1} and 2.9×10^{-10} eq/cm^3.

The agreement between the experimental and theoretical values of Λ_{min} and C_{min} is good, considering the approximations made:

1. Twice the concentration of metal in eq/cm^3 was used to represent the concentration of ionic additive in eq/cm^3.
2. A single un-ionized species undergoing equilibria of types 1 and 2 to represent the complete equilibria occurring in A.S.A. #1-heptane solutions was used despite the presence of another dissociable substance and possible cross equilibria that might occur.

One may conclude that Bjerrum's theory [34], coupled with that of Fuoss and Kraus, is able to explain ionization in A.S.A. #1-heptane solutions. Furthermore, the divalent and trivalent salts present in the A.S.A. #1 additive exist overwhelmingly as univalently charged ions in solution.

Figure 2B now may be divided into three distinct regions. Dissociations of type 1 are responsible for the observed conductivity in the range from infinite dilution down

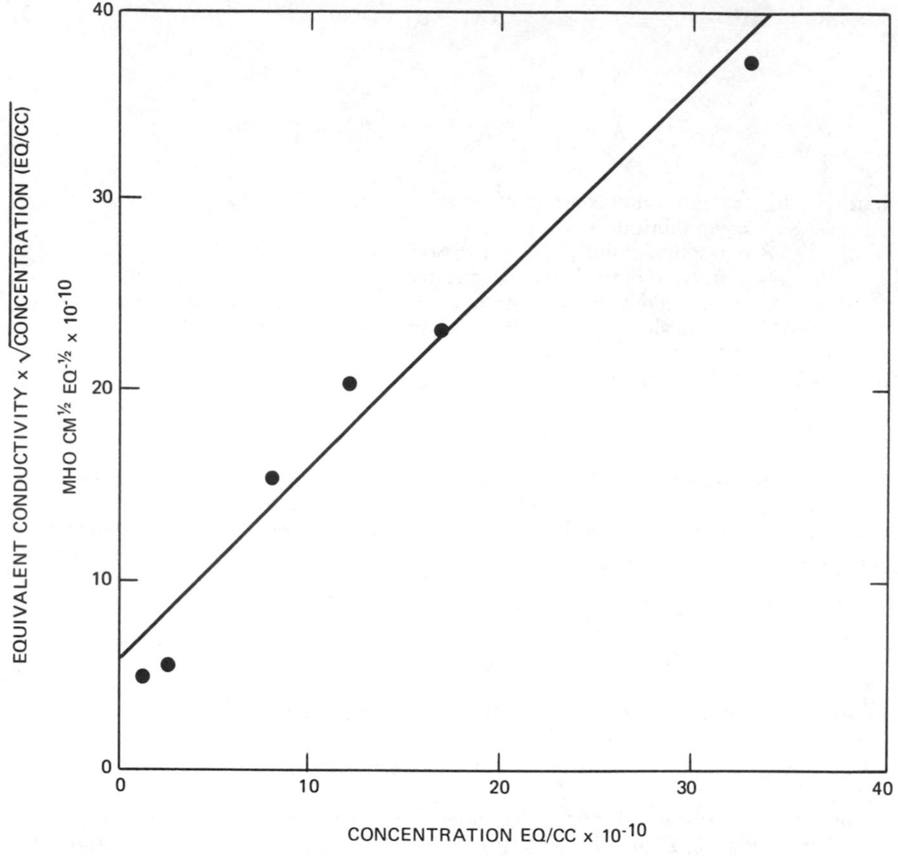

Figure 3B. Equivalent conductivity times the square root of concentration vs concentration [120].

to the minimum; in this range, the equivalent conductivity decreases with increasing concentration. Close to the minimum, triple ions form as dissociations of type 2 take place. At the minimum, the equivalent conductivity is caused equally by single and triple ions, as given by Equation 6B. For concentrations to the right of the minimum, further ionic associations may occur [34] causing formation of uncharged quadrupoles $A^+B^-A^+B^-$; in this range, the equivalent conductivity increases slightly and levels off with increasing concentration.

Because the various antistatic additives ionize to different extents, even in fluids of $\epsilon \simeq 2$, it is important to know the conductivity vs concentration behavior for the additive and solution of interest. Methods to determine the electrical conductivity and dielectric constant are outlined in Appendix A. Because many technologically important and scientifically challenging charge transport processes occur in the region where univalent ions predominate, the total conductivity may be expressed by

$$k = \sum_{i=1}^{n} k_i = \sum_{i=1}^{n} C_i \mu_i F \tag{7B}$$

where C_i = concentration of the ith species

μ_i = ionic mobility of i th species, essentially a constant equal to 5×10^{-4} cm^2/V-sec

F = 96,500 coul/g-eq; the Faraday number

Using the electrokinetic definition of ionic mobility,

$$\mu_i = \frac{D_i F}{R_g T} \tag{8B}$$

Equation 7B may be rewritten as

$$k = (C_+ D_+ + C_- D_-) F^2/R_g T = C F^2 D/R_g T \tag{9B}$$

for the assumption of electroneutrality, i.e., $C_+ = C_- = C/2$, and D is equal to the mean ionic diffusivity. For multicomponent fluids, as μ_i is essentially constant the ionic concentration controls the overall conductivity, assuming the individual fluid volumes are roughly of equivalent volume percentages. Alternatively, the conductivity of a mixture is expected to be close to that of the most conductive components. These approximations are useful in developing the charge transport equations in the section entitled Development of the Charge Conservation Equations. An understanding of the mechanisms of ionic dissolution is also helpful in interpreting deviations from theoretical predictions obtained through solving simplified forms of these equations. A brief discussion of this point was given in the sections entitled Turbulent Charge Generation in Capillaries and Pipes and Flow through Porous Media.

APPENDIX C–ANALYSIS OF CHARGE GENERATION IN FILTERS (MICROPOROUS MEDIA)

The type of porous material considered here is a solid that contains a large number of interconnected holes (often referred to as pores) distributed throughout its volume. The Millipore filters, in particular, are porous media with pore diameters that are very small with respect to other characteristic lengths associated with the charge transport problem. Hence, they may be termed microporous media. Fluids in their motion through porous media may travel distances many times greater than the thicknesses of the media if the pore paths are extremely tortuous and/or interconnected. It is extremely difficult, if not impossible, to describe the fine features or microstructure of the flow field in terms of physically measurable parameters. For this study, an overall macroscopic approach, valid when one deals with regions in space containing a large number of pores, will be considered.

The equation of motion that describes flow in porous media and accounts for distortion of velocity profiles near containing walls [122] is

$$\nabla p = \frac{\mu}{\kappa} \vec{v}_o + \mu \nabla^2 \vec{v}_o + \rho \vec{g}. \tag{1C}$$

Equation 1A minus the viscous term $\mu \nabla^2 \vec{v}_o$ is the familiar Darcy law* for flow through porous media. The velocity, v_o, is the superficial velocity that is related to the pore velocity, v_p, by

$$v_p = v_o/P = \frac{Q_m}{A_v}$$

where Q_m = measured volumetric flowrate (cm³/sec)
$A_v = PA_c$ = void area (cm²)
A_c = cross-sectional area (cm²)

*Comprehensive discussions of Darcy's law, its applicability and limitations, are given in monographs by Scheidegger [123] and Collins [124].

and P is the porosity given by

$$\%P = V_p/V_t \times 100\%$$

where V_p is the pore volume, and V_t is the total volume (solids plus voids). The permeability, κ, may be interpreted as a measure of the ease of flow through a porous medium (high values indicate ease of flow, while low values indicate the contrary). By analogy with flow through single tubes, one may write the Reynolds number and friction factor (a measure of resistance to flow) as

$$Re_{d_p} = \rho v_p d_p/\mu,$$

$$f = \frac{d_p \Delta p}{\rho L_f}\left(\frac{A_v}{Q_m}\right)^2$$

where $\Delta \rho$ = pressure drop imposed over the filter thickness, L_f

If we now apply Equation 1C to the porous medium sketched in Figure 1C and assume that the medium exhibits overall cylindrical geometry, then for constant κ and gravity-free flow we obtain

$$\frac{d^2 v_o}{dr^2} + \frac{1}{r}\frac{dv_o}{dr} - (1/\kappa)v_o = \frac{\Delta p}{\mu L} \tag{2C}$$

The solution to Equation 2C, subject to no slip at r = a and with v_o remaining finite for all values of r, is

$$v_o = \frac{\kappa \Delta p}{\mu}\left\{1 - \frac{I_o(r/\sqrt{\kappa})}{I_o(a/\sqrt{\kappa})}\right\} \tag{3C}$$

Averaging v_o over the cross-sectional area, we obtain

$$v_o = \frac{\int_o^a v_o r dr}{\int_o^a r dr} = \frac{\kappa \Delta p}{\mu L_f}\left\{1 - \frac{2\sqrt{\kappa}}{a}\frac{I_1(a/\sqrt{\kappa})}{I_0(a/\sqrt{\kappa})}\right\} \tag{4C}$$

Equation 4C minus the bracketed terms is Darcy's law. By plotting $1 - I_o(r/\sqrt{\kappa})/I_o(a/\sqrt{\kappa})$ vs r/a for a range of values, $a/\sqrt{\kappa}$, we see that $I_o(r\sqrt{\kappa})/I_o(a\sqrt{\kappa})$ approaches zero for $a = 10\sqrt{\kappa}$. Thus, for $a/\sqrt{\kappa} > 10$, Darcy's law is applicable for flow through porous filters. The validity of $a/\sqrt{\kappa} > 10$ is readily confirmed by using Darcy's law to calculate the permeability:

$$\kappa = \frac{Q_m \mu}{A_v(\Delta p/L_f)} \tag{5C}$$

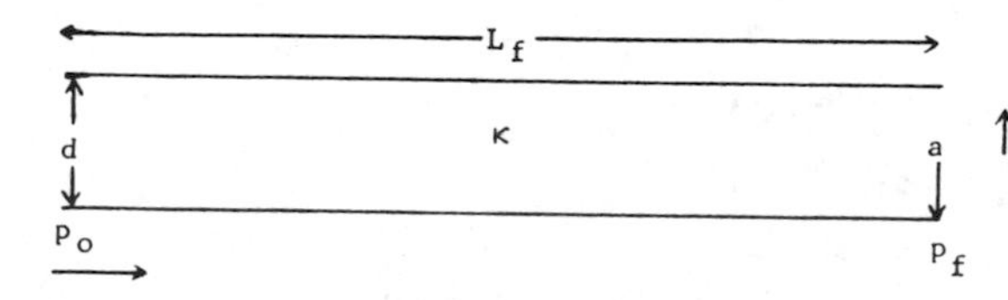

Figure 1C. Simplified model of a porous filter bundle of noninterconnecting capillaries.

Values of κ for the filters in Table IX ranged from 1.1×10^{-12} cm^2 to 4.2×10^{-10} cm^2. Since a = 1.25 cm, and $a/\sqrt{\kappa} >> 10$, the above approach is valid. Strictly speaking, L_f in Equation 5C should be L_{eff}, i.e., the effective filter thickness that accounts for the actual fluid flow path through the porous filter. This is usually accounted for in an empirical manner by the introduction of a correction factor called the tortuosity, t_o, into a geometric model of the porous medium from which κ can be determined. For the simplified model of a bundle of straight, noninterconnecting cylindrical tubes of uniform pore diameter, the tortuosity is usually introduced as t_o^2 to give [123]

$$\kappa = \frac{Pd_p^2}{32t_o^2}. \tag{6C}$$

By comparison of values of κ from Equation 6C, values of t_o were obtained ranging from $0.6 \lesssim t_o \lesssim 2.3$, with a few values ranging from 0.86 to 1.33. Thus, the Millipore filters under consideration here are rather representative of an ideal porous medium.

For the simplified model of the porous medium given above, it can be shown [47,48] (for the assumption that the thickness of the charge transfer layer is small with respect to the pore diameter) that Equation 22 may be written as

$$\frac{D\tau_o}{d_p^2}\frac{dq_r}{dr_r}\bigg|_{ctl} - (q_r^2+1)^{1/2}q_r - \left(\frac{v_p\tau_o}{d_p}\right)\frac{dq_r}{dx_r} = 0 \tag{7C}$$

$q_r = q/FC^o$, $r_r = r/d_p$
$x_r = x/d_p$, and ctl = charge transfer layer*

The boundary condition at the ctl can be shown to be [48]

$$(D\tau_o/d_p^2)\frac{dq_r}{dr_r} = \frac{h\tau_o}{d_p}(q_r^2+1)^{1/2} \tag{8C}$$

The usual entry condition for a fluid entering the filter uncharged is

$$q_r\big|_{x_r=0} = 0 \tag{9C}$$

Solution of Equation 7A, subject to Equations 8A and 9A, gives

$$q_r = A - \frac{1+A^2}{\sinh\left\{(1+A^2)^{1/2}\dfrac{x}{v_p\tau_o} + B\right\} + A}, \tag{10C}$$

where $A = h\tau_o/d_p$ and $B = \sinh^{-1}(1/A)$. To apply Equation 10C, the functional form of the charge transfer coefficient must be determined. This will be accomplished by dimensional analysis.

The dimensional analysis approach given in the section entitled Flow Through Porous Media is applicable here; however, we wish to show that similar results can be obtained using a slightly different approach. In the general case, one expects the charge transfer coefficient to be a function of the same independent parameters as the measured current. However, by analogy with mass transfer coefficients, which are assumed to be independent of concentration differences, h may be taken independent of FC^o (coul/cm^3). In addition, since h should depend on interfacial parameters, not bulk parameters, it must also be independent of A_v. Thus, h may be written as a function

*The boundary condition is written in terms of a charge transfer coefficient, h. We assume, therefore, that there is a region very close to the surface where q is a function of r, i.e., the charge transfer layer. Its extent of penetration into the liquid is the charge transfer layer thickness.

of v_p, L_p, ν, τ_o and d_p. Carrying out the analysis similar to that in the section mentioned above, we obtain the following dimensionless groups:

$$\pi_{1h} = h\tau_o/d_p \qquad (11C)$$

$$\pi_{2h} = v_p\tau_o/d_p = Re_{d_p}\, Sc_m(\delta/d_p)^2 \qquad (12C)$$

$$\pi_{3h} = L_p/d_p \qquad (13C)$$

$$\pi_{4h} = \nu\tau_o/d_p^2 = Sc_m(\delta/d_p)^2 \qquad (14C)$$

According to the π theorem,

$$h\tau_o/d_p = \Phi\, (v_p\tau_o/d_p)^\alpha (L_p/d_p)^\beta (\nu\tau_o/d_p^2)^\sigma \qquad (15C)$$

where ϕ is a constant of proportionality and α, β and σ are constants. If we write the left-hand side of Equation 15C as $(hd_p/D)(D\tau_o/d_p^2)$ and use the right-hand side of Equations 12C and 14C, then hd_p/D, which is a Nusselt number, may be written as

$$Nu_e = hd_p/D = \Phi' Re_{d_p}^{\alpha'} Sc_m^{\beta'} (\delta/d_p)^{\sigma'} (L_p/d_p)^{\omega'} \qquad (16C)$$

Note the similarity of Equations 16C and 34. As Equation 10C contains $h\tau_o/d_p$ terms, Equation 15C is preferred to Equation 16C for this work.

To compare Equation 10C with Equation 94, we substitute

$$v_p = Q_m/A_v \qquad (17C)$$

and

$$q_m = I_m/Q_m$$

(q_m is the mean charge density obtained from the measured current for a given volumetric flowrate) into Equation 94 and obtain

$$q_m/FC^o = 2 \times 10^{-4}\, (v_p\tau_o/d_p)^{0.75} \qquad (18C)$$

Now the only way Equation 10C can equal Equation 18C is for $\{1 + (h\tau_o/d_p)^2\}^{1/2} L_p/v_p\tau_o >> 1$. For this case, Equation 10C reduces to

$$q/FC_o = \Phi\, (v_p\tau_o/d_p)^\alpha (L_p/d_p)^\beta (\nu\tau_o/d_p^2)^\sigma$$

which is identical to Equation 18C for $\Phi = 2 \times 10^{-4}$, $\alpha = 0.75$, $\beta = \tau = 0$. However, it is not possible to show that $\{1 + (h\tau_o/d_p)^2\}^{1/2} \times L_p/v_p\tau_o >> 1$ by direct substitution, as L_p, the pore length, is indeterminate and as h was determined by means of this approximation.

Since the analysis was based on the assumption that the thickness of the charge transfer layer is small with respect to the pore diameter, this condition should be verified from the data. By analogy with ordinary mass transfer,

$$d_{ctl} = d_p/Nu_e \qquad (19C)$$

A suitable Nusselt number is available in $h\tau_o/d_p$ (compare Equations 15C and 18C):

$$Nu_e = h\tau_o/d_p = 2 \times 10^{-4} \left(\frac{v_p\tau_o}{d_p}\right)^{0.75} = 2 \times 10^{-4}\, Re_{d_p}^{\;0.75}\, Sc_e^{0.75} \qquad (20C)$$

where Sc_e was used to replace $\nu\tau_o/d_p^2$.

Table IC. Nusselt Number and Charge Transfer Layer Thickness for Typical Values of $Re_{d_p}\ Sc_m(\delta/d_p)^2$

$Re_{d_p}\ Sc_m(\delta/d_p)^2$	Nu_m	d_{ctl}/d_p
10^7	36.5	0.03
10^6	6.34	0.16
10^5	1.13	0.89
10^4	0.50	2.00
5×10^3	0.12	8.35

Values of Nu_m from Equation 20C and d_{ctl} from Equation 19C are listed as functions of $Re_{d_p}\ Sc_m(\delta/d_p)^2$ in Table 1C with Sc_m set equal to Sc_e [i.e., $(\delta/d_p)^2$ taken as 1].

Although d_{ctl}/d_p is properly small when $Re_{d_p}Sc_m(\delta/d_p)^2 \gtrsim 10^6$, it is apparently too large for smaller values of $Re_{d_p}Sc_m(\delta/d_p)^2$. From Figure 9, the actual limit of applicability of the development is $Re_{d_p}Sc_m(\delta/d_p)^2 = 5 \times 10^3$. Although one can offer many reasons for roughly a thirtyfold reduction of d_{ctl}/d_p at $Re_{d_p}Sc_m(\delta/d_p)^2 = 5 \times 10^3$, it is difficult to provide firm support for these assertions. Certainly d_{ctl} as calculated here really represents an average over the cross-sectional area of the filter and the actual pore length and, consequently, should not be compared with d_p.

Besides the difficulties noted above, there is one additional factor not immediately obvious from the development for conductivities around $10^{-13}\Omega^{-1}cm^{-1}$ to $10^{-14}\Omega^{-1}cm^{-1}$ corresponding to the data in Figure 9 for $Re_{d_p}Sc_m(\delta/d_p)^2 \gtrsim 10^5$. Here, the total number of ions contained in a given pore is so low that continuum theory would not be expected to be valid. This does not appear to be a serious drawback because one can examine regions in space that contain a larger number of ions, similar to the restriction of the fluid flow equations to regions in space that are large with respect to a characteristic dimension.

Values of $Re_{d_p}Sc_m(\delta/d_p)^2 \gtrsim 5 \times 10^3$ were obtained for conductivities close to the equivalent conductivity minimum of Figure 2B. Thus, because of triple ion formation, one also would expect differences in charging behavior.

NOMENCLATURE

- a tube radius (cm)
- A area (cm^2)
- A_c cross-sectional area (cm^2)
- A_v void area (cm^2)
- $C_\pm$ concentration of positive or negative ions (mol/cm^3)
- C_o concentration of ions in bulk liquids (mol/cm^3)
- C_m^* concentration of metallic ions (g/l)
- C_s concentration of ions at surface (mol/cm^3)
- C^o total ionic concentration (mol/cm^3)
- C capacitance (pF)
- d diameter of capillary or pipe (cm)
- d_{ctl} charge transfer layer thickness (cm)
- d_p pore diameter (cm)
- $D_\pm$ diffusivity of positive or negative ions (cm^2/sec)
- D' imbalance of diffusivities, $(D_+ - D_-)/2$ (cm^2/sec)
- D mean diffusivity of positive and negative ions, $(D_+ + D_-)/2$ (cm^2/sec)
- D^* diffusivity ratio, $(D_+ - D_-)/(D_+ + D_-)$
- E minimum ignition energy (mJ)
- E_f electric field strength (V/cm)
- f friction factor
- F Faraday number (96,500 coul/mol)
- g acceleration of gravity (980 cm/sec^2)

G — $d^2/Re^{7/4}\tau\nu$ $1/Re^{7/4}Sc_m(\delta/d)^2$, dimensionless number
h — charge transfer coefficient (cm/sec)
I — current (A)
I_d — current density (A/cm^2)
I_m — measured current (A)
I_∞ — current for infinitely long tubes (A)
I_o — modified Bessel function of first kind, zero order
I_1 — modified Bessel function of first kind, first order
$j\pm$ — flux of charge of positive or negative ions (coul/cm^2-sec)
j — total flux of charge (coul/cm^2-sec)
J_o — Bessel function of zero order
k — electrical conductivity, $\Omega^{-1}cm^{-1}$
k_1 — ionization rate coefficient (cm^3/mol)
K_{ef} — effective conductivity of a filter ($\Omega^{-1}cm^{-1}$)
k_2 — recombination rate coefficient (cm^3/mol)
K_{eq} — equilibrium constant of type 1 (g-eq/cm^3)
$K_{eq}{}^T$ — equilibrium constant of type 2 (g-eq/cm^3)
K_p — conductivity inside a pore ($\Omega^{-1}cm^{-1}$)
K_s — conductivity of a solid ($\Omega^{-1}cm^{-1}$)
L — characteristic length, length of capillary or pipe, interelectrode sparing (cm)
L_p — pore length (cm)
L_f — filter thickness (cm)
M_e — electrophoretic mobility (cm/sec)/(V/cm)
n_+ — transfer number of positive ion
Nu_m — Nusselt number for mass transfer
Nu_e — Nusselt number for charge transfer
p — pressure (dynes/cm^2)
P — porosity
Pe_m — Péclet number for mass transfer
Pe_e — Péclet number for charge transfer
q — volumetric charge density (coul/cm^3)
q_m — measured charge density (coul/cm^3)
q_w — charge density at wall (coul/cm^3)
q — free stream charge density (coul/cm^3)
q_+ — charge density of positive ions (coul/cm^3)
Q — total charge (coul)
Q' — rate of charge transfer (coul/sec)
Q_A — charge per area (coul/cm^2)
Q_m — measured volumetric flowrate (cm^3/sec)
r — radial coordinate (cm)
r^* — net rate of production of ions per volume (mol/cm^{3}sec)
r_1 — ionization rate per volume (mol/cm^{3}sec)
r_2 — ion recombination rate per unit volume (mol/cm^{3}sec)
R_g — gas constant, 8.32 (V − coul)/(°K-mol)
R — ohmic resistance, Ω
Re — Reynolds number
Sc_e — Schmidt number for charge transfer
Sc_m — Schmidt number for charge transfer
t — time (sec)
t_o — tortuosity
T — temperature (°K)
u — flow velocity (cm/sec)
U — characteristic flow velocity (cm/sec)
v — flow velocity (cm/sec)
v_e — electrophoretic velocity of a particle (cm/sec)
v_o — superficial velocity (cm/sec)
v_p — pore velocity (cm/sec)
V — potential (V)
V_p — pore or void volume (cm^3)
V_t — total volume (solids plus voids) (cm^3)
x — axial distance (cm)
x_o — starting length (cm)
y — distance off plate (cm)
z — distance in Z direction (cm)
z_i — sign of charge
Z — equivalent circuit impedance, Ω

Superscripts

— — averaged quantity refers to model condition
$\rightarrow$ — vectoral quantity

Subscripts

r reduced or dimensionless variable
s solid
o initial value of a parameter
eq equilibrium
op overpotential
+,− refer to properties of positive and negative ions

Greek Symbols

δ Debye double layer thickness $(D\tau)^{1/2}$ (cm)
δ_1 Debye double layer thickness $(D\tau_o)^{1/2}$ (cm)
δ_2 Debye double layer thickness $(v\tau_o)$ (cm)
Δ thickness of charge boundary layer (cm)
Δ (property) difference
δ_v velocity boundary layer thickness (cm)
θ angular coordinate
ϵ dielectric constant
ϵ_1 dielectric constant of water phase
ϵ_2 dielectric constant of hydrocarbon phase
ϵ_0 permittivity of free space, 8.85×10^{-14}sec/Ω-cm
ζ zeta potential (V)
κ permeability (cm^2)
λ diffusion layer thickness (cm)
Λ° equivalent conductivity of uni-valent ion at infinite dilution (cm$^2\Omega$ g-eq)
Λ°_T equivalent conductivity of uni-valent ion at infinite dilution (cm$^2/\Omega$ g-eq)
ρ_R specific resistivity, Ω-cm
μ viscosity (g/cm-sec)
μ_I ionic mobility (cm^2/V-sec)
ρ density (g/cm^3)
ν kinematic viscosity (cm^2/sec)
τ relaxation time (sec)
ψ potential (V)
ω angular frequency, 2π times natural frequency (radians/sec)

REFERENCES

1. Harper, R. W. "How Do Solid Surfaces Become Charged," *Proc. Conf. Static Electrification* (1967), pp. 3–10.
2. Shinohara, I., F. Yamamoto, H. Anzai and S. Endo. "Chemical Structure and Electrostatic Properties of Polymers," *J. Electrostatics* 2:99–110 (1976).
3. Frederick, E. R. "Some Aspects of Electrostatic Charges in Fabric Filtration," *J. Air Poll. Control Assoc.* 24(12):1164–1168 (1974).
4. Davies, D. K. "The Generation and Dissipation of Static Charge on Dielectrics in a Vacuum," *Proc. Conf. Static Electrification* (1967), pp. 29–36.
5. Levy, J. B., J. H. Wakelin, W. J. Kauzmann and J. H. Dillon. *Text. Res. J.* 28:897 (1958).
6. Kunkel, W. B. *J. Appl. Phys.* 21:820 (1950).
7. Medley, J. A. *Brit. J. Appl. Phys.* (*Suppl.* 2) 4:S28 (1953).
8. Moore, A. D. *Electrostatics and Its Applications* (New York: John Wiley & Sons, Inc., 1973).
9. Davies, J. T., and E. K. Rideal. *Interfacial Phenomena* (New York: Academic Press, Inc., 1963), pp. 149–335.
10. Kruyt, H. R. *Colloid Science* (Amsterdam: Elsevier Publishing Co., 1952).
11. Gavis, J., and I. Koszman. *J. Colloid Sci.* 16:375 (1961).
12. Koszman, I., and J. Gavis. *J. Chem. Eng. Sci.* 17:1013 (1962).
13. Koszman, I., and J. Gavis. *J. Chem. Eng. Sci.* 17:1023 (1962).
14. Moore, A. D. *Electrostatics and Its Applications* (New York: John Wiley & Sons, Inc., 1973), pp. 180–220.
15. Ward, G. R. *Am. Dyestuff Rep.* 44:220 (1955).

16. McLean, H. T. *Am. Dyestuff Rep.* 44:485 (1955).
17. Steiges, F. H. "Evaluating Antistatic Finishes," *Text. Res. J.* 27(9):721–733 (1958).
18. Seaborne, J., J. P. Wagner and N. R. S. Hollies. "Static Charges in Fire Retarded Cottons," Contract No. 01-4078 for Cotton, Inc., Raleigh, NC, Gillette Research Institute, Rockville, MD (1978).
19. Waterman, L. C. *Electrical Coalescers, CEP* 61(10):51–57 (1965).
20. "Carbonization Research Report 27," The British Carbonization Research Association. Chesterfield, Derbyshire (1976).
21. Moore, A. D. *Electrostatics and Its Applications* (New York: John Wiley & Sons, Inc., 1973), pp. 336–376.
22. Charles, G. E., and S. G. Manson. "The Mechanism of Coalescence of Liquid Drops at Liquid/Liquid Interfaces," *J. Colloid Sci.* 15:105–122 (1960).
23. Charles, G. E., and S. G. Manson. "The Coalescence of Liquid Drops with Flat Liquid/Liquid Interfaces," *J. Colloid Sci.* 15:236–267 (1960).
24. Sadek, S. E., and C. D. Hendricks. "Electrical Coalescence of Water Droplets in Low-Conductivity Oils," *Ind. Eng. Chem. Fund.* 13(2):139–142 (1974).
25. Bailes, P. J. "Solvent Extraction in an Electrostatic Field," paper presented at the AIChE 88th National Meeting, Solvent Extraction Research Session, Part II, 1980.
26. Bailes, P. J., and M. Kalbasi. "Charge Measurements and Leakage for Single Drops in a Liquid/Liquid System," *J. Electrostatics* 10:81–88 (1981).
27. Moulik, S. P., F. C. Cooper and M. Bier. "Forced Flow Electrophoretic Filtration of Clay Suspensions," *J. Colloid Interface Sci.* 24:427–432 (1967).
28. Sprute, R. H., and D. J. Kelsh. "Dewatering and Densification of Coal Wastes by Direct Current–Laboratory Tests, U.S. Bureau of Mines, RI8197," U.S. Department of Interior (1976).
29. "Electrostatic Separation of Solids from Liquids," *Filtration Separation* 140–143 (March/April 1977).
30. Henry, J. D., Jr., L. F. Lawler and C. H. A. Kuo. "A Solid/Liquid Separation Process Based on Cross Flow and Electrofiltration," *AIChE J.* 23(6):851–859 (1977).
31. Rodgers, B. R. "Separation of Micron Sized Particles from Coal Liquids: Verification of Surface Charge," *AIChE Symp. Series, Recent Advances in Separation Techniques II, No. 192* 76:68–73 (1980).
32. Lee, C. H., D. Gidaspow and D. T. Wasan. "Cross-Flow Electrofilter for Nonaqueous Slurries," *Ind. Eng. Chem. Fund.* 19(2):166–175 (1980).
33. Chowdiah, P., D. T. Wasan and D. Gidaspow. "Electrokinetic Phenomena in the Filtration of Colloidal Particles Suspended in Nonaqueous Media, Symposium on Interfacial Phenomena in Phase Separation Processes," paper presented at the AIChE Annual Meeting, Chicago, IL, November 10–20, 1980).
34. Robinson, R. A., and R. H. Stokes. *Electrolytic Solutions*, 2nd ed. (London: Butterworth Publishers, Inc., 1959).
35. Harned, H. S., and B. B. Owen. *The Physical Chemistry of Electrolytic Solutions* (New York: Van Nostrand Reinhold Co., 1943).
36. Klinkenberg, A., and H. L. Van der Minne. *Electrostatics in the Petroleum Industry* (Amsterdam: Elsevier Publishing Co., 1958).
37. Fuoss, R. M., and C. A. Kraus. *J. Am. Chem. Soc.* 55:2387 (1933).
38. Forster, E. O. *J. Chem. Phys.* 40(1):91–95 (1964).
39. Forster, E. O. *J. Chem. Phys.* 37:1021 (1962).
40. Finck, H. W. "Antistatic Agents for Plastics," *Kunststoffe* 64:6–8 (February 1974).
41. Likhterov, S. D., V. P. Lapin and G. I. Shov. *Khim. Tekh. Topliv Masel* 12:41–42 (December 1975).
42. Gavis, J. *Chem. Eng. Ssi.* 19:237 (1964).
43. Gavis, J. *Chem. Eng. Sci.* 22:359 (1967).
44. Gavis, J. *Chem. Eng. Sci.* 22:365 (1967).
45. Gavis, J. *Chem. Eng. Sci.* 24:451 (1969).

46. Adamczewski, I. *Ionization, Conductivity and Breakdown in Dielectric Liquids* (London: Taylor and Frances, 1969), p. 208.
47. Wagner, J. P. *Fire Res. Abst. Rev.* 15(3):189 (1973).
48. Wagner, J. P. "Charge Generation During Flow of a Hydrocarbon Liquid through Micro-Porous Media," PhD Thesis, Johns Hopkins University, Department of Chemical Engineering (1966).
49. Gavis, J., and J. P. Wagner. *Chem. Eng. Sci.* 23:381 (1968).
50. Gibson, N., and F. C. Lloyd. "Electrification of Toluene in Pipeline Flow," *Proc. Conf. Static Electrification*, London (1967), pp. 89–97.
51. Gibbings, J. C., and E. T. Hignett. *Electrochem. Acta.* 11:815 (1966).
52. Hignett, E. T. PhD Thesis, University of Liverpool, Liverpool, England (1963).
53. Carslaw, H. S., and J. C. Jaeger. *Conduction of Heat in Solids* (London: Oxford University Press, 1959).
54. Crank, J. *The Mathematics of Diffusion* (London: Oxford University Press, 1956).
55. Koszman, I. PhD Thesis, Johns Hopkins University, Department of Chemical Engineering (1961).
56. Gavis, J. *J. Chem. Soc. Faraday Trans. II* 71:1115 (1975).
57. Thomson, J. J., and G. P. Thomson. *Conduction of Electricity through Gases, Vol. 1* (London: Cambridge University Press, 1928), pp. 43–47.
58. Deissler, R. G. *NACA Rep.* 1210 (1966).
59. Liebhafsky, H. A., and E. J. Cairns. *Fuel Cells and Fuel Batteries* (New York: John Wiley & Sons, Inc., 1968).
60. Gibson, N., and F. C. Lloyd. *J. Electrostatics* 1:339 (1975).
61. Bruinzeel, C., C. Luttik, S. J. Vellenga and L. Gardner. "A Study of Electrostatic Charge Generation During Low Temperature Refueling of Aircraft," Aeronautical Report LR-387, NTC No. 7794, National Research Council of Canada, Ottawa (1963).
62. "Electrostatic Charging Characteristics of Jet Fuel Filtration Equipment," CRC Report No. 455, Coordinating Research Council, Inc., New York (1973).
63. Leonard, J. T., and H. W. Carhart. "Effect of Conductivity on Charge Generation in Hydrocarbon Fuels Flowing through Fiber Glass Filters," *J. Colloid Interface Sci.* 32(3) (March 1970).
64. Masuda, S., and G. Schon. *Static Electrification*, The Institute of Physics and Physical Society Conference Series No. 4, London (1967), p. 112.
65. Huber, P. W., and A. A. Sonin. Fluid Mechanics Laboratory, Publication No. 74-4 (July 1974).
66. Kobayashi, K., M. Iwata, Y. Hosoda and H. Yukawa. *J. Chem. Eng. Japan* 12(6): 466 (1979).
67. Kobayashi, K., M. Hakoda, Y. Hosoda, M. Iwata and H. Yukawa. *J. Chem. Eng. Japan* 12(6):492 (1979).
68. Langhaar, H. L. *Dimensional Analysis and Theory of Models* (New York: John Wiley & Sons, Inc., 1951).
69. "Safety in Fuel Handling with ASA-3," *Shell Tech. Bull. ICSX* 69:5 (1969).
70. Schlogel, R. Z. *Phyzik. Chem. Neue Folge* 1:73 (1955).
71. Gross, R. J., and J. F. Osterle. *J. Chem. Phys.* 49:228 (1968).
72. Jacazio, G., R. F. Probstein, A. A. Sonin and D. Yung. *J. Phys. Chem.* 76:4015 (1972).
73. Bustin, W. M., I. Koszman and I. T. Tobye. *Div. Ref.* 44(III):548 (1964).
74. Vellenga, S. J., and A. Klinkenberg. *Chem. Eng. Sci.* 20:923 (1965).
75. Gavis, J. *Chem. Eng. Sci.* 22:633 (1967).
76. Knudsen, J. G., and D. L. Katz. *Fluid Dynamics and Heat Transfer* (New York: McGraw-Hill Book Co., 1958), p. 481.
77. Levich, V. G. Physiochemical Hydrodynamics (Englewood Cliffs, NJ: Prentice-Hall, Inc., 1962).
78. "Fire Hazards in Oxygen Enriched Atmospheres," NFPA No. 53M National Fire Protection Association, Boston, MA (1969), pp. 36–38.
79. Hughes, J. W. "Dust Explosion," *Fire Res. Abst. Rev.* 12(1):20 (1970).

80. Gibson, N. *Proc. Filtration Soc.* 382 (July/August 1979).
81. Lewis, B., and G. von Elbe. *Combustion, Flames and Explosion of Gases* (New York: Academic Press, Inc., 1961).
82. Leonard, J. T., and H. W. Carhart. "Electrical Discharges from a Fuel Surface," *Proc. Conf. Static Electrification,* London (1967), pp. 100–111.
83. Leonard, J. T., and H. W. Carhart. "Static Electricity Measurements during Refueling Loading," NRL Report No. 7203, Naval Research Laboratory, Washington, DC (January 5, 1971).
84. Zabetakis, M. G. *Bulletin 627* U. S. Bureau of Mines (1965).
85. Gibson, N., and F. C. Lloyd. *Brit. J. Appl. Phys.* 16:1619 (1965).
86. Factory Mutual Engineering Corporation. *Handbook of Industrial Loss Prevention* (New York: McGraw-Hill Book Co., 1967).
87. Eichel, F. G. *Chem. Eng.* 153 (March 13, 1967).
88. Klinkenberg, A. *Static Electrification,* The Institute of Physics and Physical Society, Conference Series No. 4 (May 1967), p. 63.
89. Bright, A. W., I. F. Parker and I. C. Haig. *IEEE Trans. Ind. Appl.* 1A-15(1):109 (1979).
90. Makin, B. *Phys. Technol.* 109 (May 1975).
91. Bright, A. W., and J. F. Hughes. *J. Electrostatics* 1:37 (1975).
92. Chubb, J. N. *J. Electrostatics* 1:61 (1975).
93. Hughes, J. F., and A. W. Bright. *IEEE Trans. Ind. Appl.* TA-15(1):100 (January/February 1979).
94. Roberts, J. M. C., and J. F. Hughes. *IEEE Trans. Ind. Appl.* 1A-15(1):104 (January/February 1979).
95. Maryott, A. A., and E. R. Smith. *National Bureau of Standards* Circular 514 (August 10, 1951).
96. Bruggeman, D. A. G. *Ann. Physik* 5(24):636 (1935).
97. Hanai, T. *Kolloid-Z.* 177:57 (1961).
98. Schwan, H. P., G. Schwarz, J. Maczuk and H. Pauly. *J. Phys. Chem.* 66:2626 (1962).
99. *General Radio Catalogue,* latest edition, Concord, MA.
100. ASTM D1169-74.
101. ASTM D924-65.
102. ASTM D150-74.
103. ASTM D257-76.
104. ASTM D1531-62.
105. Dialectric Constant and Loss Measurement Test Cells and Accessories Brochure, Rutherford Research Products Co., Rutherford, NJ (1982).
106. Riddick, T. M. *Control of Colloid Stability through Zeta Potential,* Zeta Meter, Inc., NY (1968).
107. Zeta Meter Catalogue No. 8, Zeta Meter, Inc., NY.
108. Goetz, P. J., and J. G. Penniman, Jr. "A New Technique for Microelectrophoretic Measurements," 49th National Colloid Symposium, Clarkson College, Potsdam, NY (1975).
109. Goetz, P. J., and J. G. Penniman, Jr. "A Microprocessor-Based Electrophoresis Sensor/Controller for Optimizing the Electrokinetics of Papermaking," paper presented at the International Seminar on Paper Mill Chemistry, Amsterdam, Netherlands, September 11–13, 1977.
110. Soyenkoff, B. *J. Phys. Chem.* 35:2993 (1931).
111. Hendrick, J. E., A. C. Andrews and J. B. Sutherland. *Ind. Eng. Chem.* 33(8):1055 (1941).
112. Van der Minne, J. L., and P. H. J. Hermanie. *J. Colloid Sci.* 7:600 (1952).
113. Van der Minne, J. L., and P. H. J. Hermanie. *J. Colloid Sci.* 8:38 (1953).
114. McGown, D. N. L., G. D. Parfitt and E. Willis. "Stability of Non-Aqueous Dispersions, I. The Relationship between Surface Potential in Hydrocarbon Media," *J. Colloid Sci.* 20:650–664 (1965).

115. Parfitt, G. D., and E. Willis. "Stability of Non-Aqueous Dispersions, II. Graphon in Solutions of Alkyl-Benzene in n-Heptane," *J. Colloid. Interface Sci.* 22:100–106 (1966).
116. Lewis, K. E., and G. D. Parfitt. "Stability of Non-Aqueous Dispersions, III. Rate of Coagulation of Sterling MTG in Aerosol OT + n-Heptane Solutions," *Trans. Faraday Soc.* 62:1652–1661 (1966).
117. Kuo, S., and F. Osterle. "High Field Electrophoresis in Low Conductivity Liquids," *J. Colloid Interface Sci.* 25:421–428 (1967).
118. Verwey, E., and J. Th. G. Oberbeek. *Theory of the Stability of Lyophobic Colloids* (Amsterdam: Elsevier Publishing Co., 1948).
119. Henry, J. D., Jr., and M. T. Jacques. *AIChE J.* 23(4):607 (1977).
120. Wagner, J. P. Unpublished results (1964).
121. Conway, B. E. *Electrochemical Data* (Amsterdam: Elsevier Publishing Co., 1952).
122. Brinkman, H. C. *Appl. Sci. Res.* A1:27–34, 81–86 (1947).
123. Scheidegger, A. E. *The Physics of Flow through Porous Media* (New York: Macmillan Publishing Co., Inc., 1960), p. 114.
124. Collins, R. E. *Flow of Fluids through Porous Materials* (New York: Van Nostrand Reinhold Co., Inc., 1961), pp. 51, 770.

CHAPTER 42
HYDRODYNAMICS OF THREE-PHASE FLUIDIZATION

Norman Epstein
Department of Chemical Engineering
University of British Columbia
Vancouver, British Columbia V6T 1W5
Canada

CONTENTS

INTRODUCTION

The term "three-phase fluidization" has both a broad connotation and a specific denotation. Here it is used in its special sense, which can be relegated to a particular category within the general classification scheme shown in Figure 1 [1]. The three phases involved are usually gas, liquid and particulate solids, although at least two studies have been performed in which the gas phase is replaced by a second liquid immiscible with the first [2,3]. As in the case of fixed-bed operation, both cocurrent and countercurrent gas-liquid flows are permissible. For each of these flows there exists both a bubble flow regime, in which the liquid is the continuous phase and the gas is dispersed, and a trickle flow regime, in which the gas forms a continuous phase and the liquid is more or less dispersed.

Trickle flow in the countercurrent mode is exemplified by the "turbulent bed," "mobile bed" or "fluidized packing" contactor [4], in which low-density spheres are fluidized by an upward current of gas and irrigated by a downward flow of liquid. This contactor is called a "turbulent contact absorber" when it is used for gas absorption and/or dust removal. In the cocurrent mode, referred to in general as "gas-liquid fluidization" and in which both fluid phases move upward, trickle flow is represented by the gas-continuous region of Figure 2 [5]. It is noteworthy that the transition from bubble to slug to gas-continuous flow in this figure corresponds approximately to the transition from particulate to aggregative fluidization.

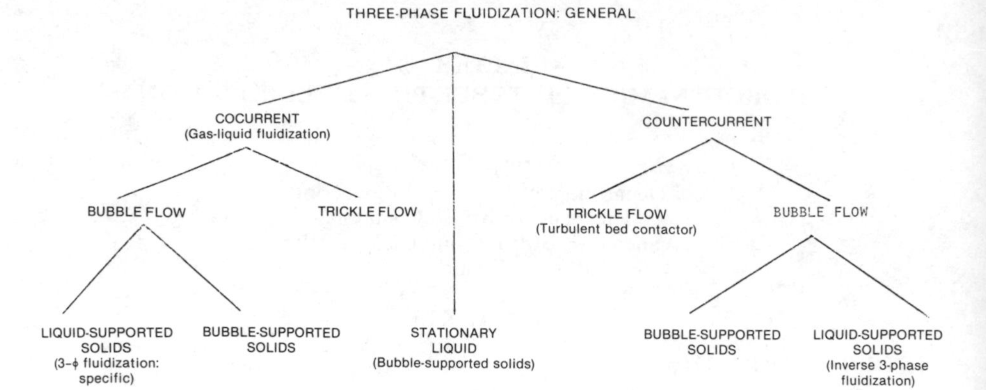

Figure 1. Classification of three-phase fluidized beds.

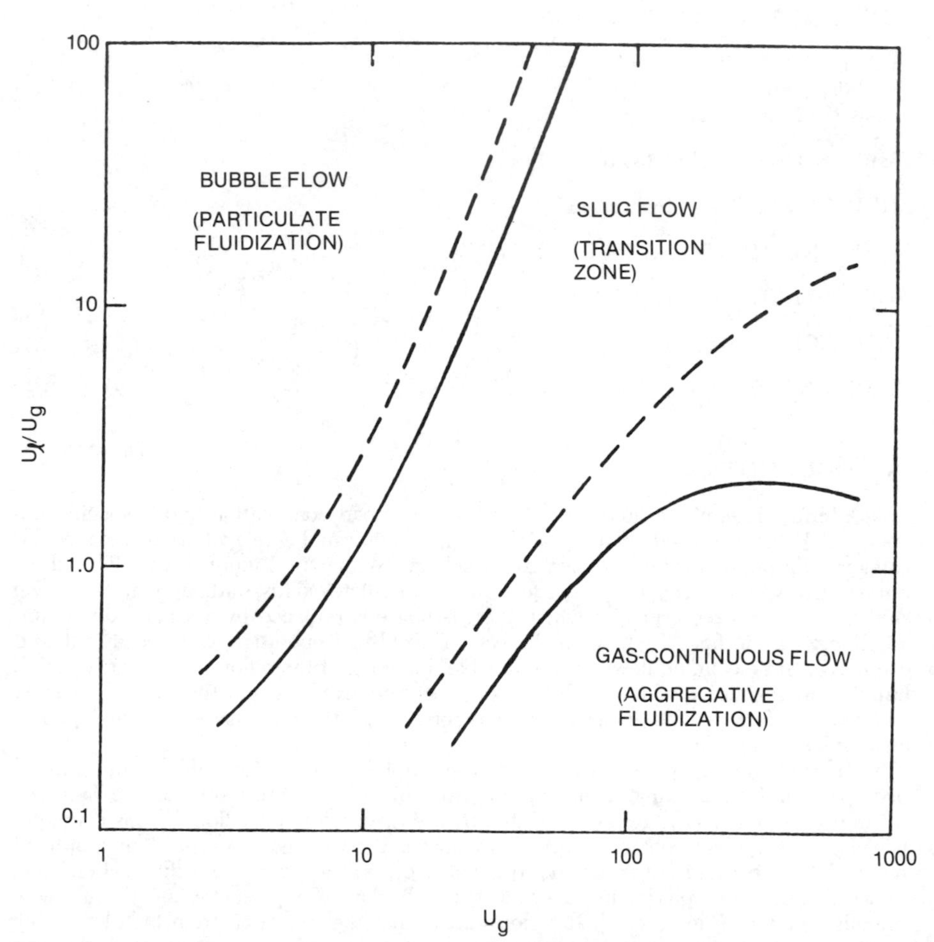

Figure 2. Flow regimes in, and types of, gas-liquid fluidization [5]. Solid lines separate flow regimes, dashed lines separate types of fluidization.

Bubble flow, whether cocurrent or countercurrent, is conveniently subdivided into two modes, namely *liquid-supported solids,* in which the liquid exceeds the minimum liquid-fluidization velocity, and *bubble-supported solids,* in which the liquid is below its minimum fluidization velocity or even stationary [6] and serves mainly to transmit to the solids the momentum and potential energy of the gas bubbles [7], thus suspending the solids. Volpicelli and Massimilla [8], who refer to these two modes as "three-phase liquid fluidization" and "three-phase bubble fluidization," respectively, have found that the much greater gas-liquid flow ratios possible with the latter make it the more practical mode for some processes. Gas throughputs can be increased further by resorting to gas-continuous flow.

Countercurrent bubble flow with liquid-supported solids, which can be effected by downward liquid fluidization of buoyant particles and upward flow of gas, has been called "inverse three-phase fluidization" [9] and only recently has its hydrodynamics received some attention [10]. In contrast, the corresponding cocurrent mode—three-phase liquid fluidization—has been the object of much interest and investigation, and is the primary subject of this chapter. It is to this mode mainly that the appellation three-phase fluidization will apply. Where occasionally the term is used in a more general sense, it will be obvious from the context.

It should be noted also that the term "fluidized bed," as conventionally used, implies a swarm of solid particles with an interface, sharp or fuzzy, between the bed region and contiguous downstream fluid. Such a bed must be distinguished on the one hand from a fixed packed bed, in which the particles are supported at least in part by the vessel rather than entirely by the drag exerted on them by the upward moving fluid, and on the other hand from a transport reactor [11], in which the solid particles are carried away by the moving fluid, leaving no solids-free or dilute-solids freeboard zone in the column. Where the transporting fluid is a liquid, the reactor becomes known as a "slurry reactor." For a three-phase slurry reactor [12], in which the liquid velocity is sufficiently high and the particles sufficiently small and/or low in density to be transported out of the column by the ascending liquid, this involves exceeding the upper liquid velocity limit of an unconstrained three-phase fluidized bed, while unconstrained fixed packed-bed operation involves the use of a liquid velocity below that for minimum fluidization. (In the case of a "bubble column slurry reactor" [13], there is often no net flow of liquid; however, in contrast to a stationary-liquid three phase bubble fluidized bed [6], with its heavier and/or denser particles and its limited height, the liquid is able to support fine/light particles over its whole depth by means of the momentum imparted to it by the gas bubbles.) If expansion of the fluidized bed is constrained by the use of an upper porous grid, then semifluidization [14] results, which then lowers the upper liquid velocity and voidage limit for unconstrained three-phase fluidization.

PRESSURE DROP

For three-phase fluidization, the *total* axial pressure gradient at any bed level is simply the bed weight per unit volume at that level:

$$-\frac{dP}{dz} = (\epsilon_s\rho_s + \epsilon_l\rho_l + \epsilon_g\rho_g)\,g \tag{1}$$

where the individual phase holdups are interrelated as

$$\epsilon_s + \epsilon_l + \epsilon_g = 1 \tag{2}$$

The total pressure drop across a bed of height H is then given by

$$-\Delta P = g\int_0^H (\epsilon_s\rho_s + \epsilon_l\rho_l + \epsilon_g\rho_g)\,dz$$

in which $\epsilon_g \rho_g$ usually can be neglected relative to the other terms to which it is added. When the liquid is the continuous phase, the *dynamic* pressure gradient, as measured by differential manometry, is the total pressure gradient corrected for the hydrostatic head of liquid:

$$-\frac{dp}{dz} = -\frac{dP}{dz} - \rho_l g \tag{3}$$

Substituting Equations 1 and 2 into Equation 3,

$$-\frac{dp}{dz} = [\epsilon_s(\epsilon_s - \rho_l) - \epsilon_g(\rho_l - \rho_g)]g \tag{4}$$

When $\epsilon_g = 0$, $\epsilon_s = 1 - \epsilon_l$ and Equation 4 reduces to the familiar relationship for liquid fluidization:

$$-\frac{dp}{dz} = \epsilon_s(\rho_s - \rho_l)g = (1 - \epsilon_l)(\rho_s - \rho_l)g \tag{5}$$

The *frictional* pressure gradient is the total pressure gradient corrected for the hydrostatic head of a two-phase fluid:

$$-\frac{dp_f}{dz} = -\frac{dP}{dz} - \rho_f g \tag{6}$$

where ρ_f is the composite fluid density given by

$$\rho_f = \frac{\epsilon_l \rho_l + \epsilon_g \rho_g}{\epsilon_l + \epsilon_g} = \frac{\epsilon_l \rho_l + \epsilon_g \rho_g}{1 - \epsilon_s} \tag{7}$$

Substituting Equation 7 into Equation 6,

$$-\frac{dp_f}{dz} = -\frac{dP}{dz} - \frac{\epsilon_l \rho_l + \epsilon_g \rho_g}{1 - \epsilon_s} \cdot g \tag{8}$$

The integral form of Equation 8, neglecting $\epsilon_g \rho_g$ relative to $\epsilon_l \rho_l$, was recently used by Lee and Al-Dabbagh [15] as their expression for pressure drop. Substituting Equations 1 and 7 into 8,

$$-\frac{dp_f}{dz} = \epsilon_s(\rho_s - \rho_f)g = (1 - \epsilon)(\rho_s - \rho_f)g \tag{9}$$

which has the same form as Equation 5, to which it reduces identically in the absence of gas.

Equations 1, 4 and 9 assume that (1) the buoyed weight of the solid particles is supported by the upward fluid drag on these particles; (2) the flow of gas causes negligible additional losses by friction; and (3) wall friction also may be neglected. Justification of these assumptions by experimental verification of Equation 1 has been provided by Ermakova et al. [16], who reported some deviation from this equation as gas flow increased and liquid flow decreased. More recently, Dhanuka and Stepanek [17] found a maximum deviation of only 4% between measured pressure gradients and those predicted by Equation 1. In addition, radial pressure gradients are commonly assumed to be negligible relative to axial gradients, so that the pressure drop experienced by each fluid phase is essentially the same.

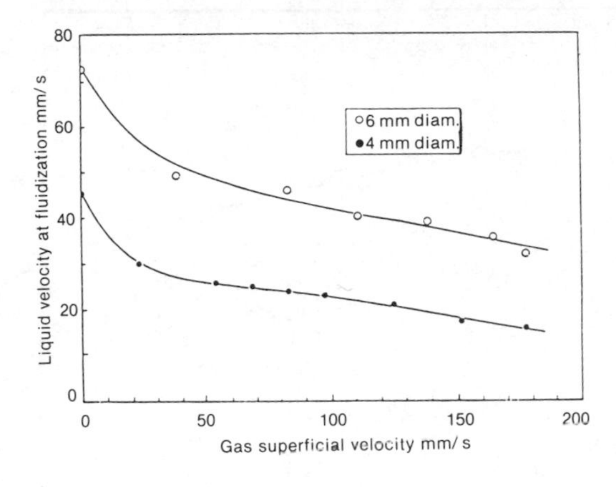

Figure 3. Minimum superficial liquid fluidization velocity at various gas velocities for air-water fluidization of two sizes of glass beads [15,18].

MINIMUM FLUIDIZATION

The frictional pressure gradient at minimum fluidization is given by Equation 9, with $\epsilon = \epsilon_{mf}$, as recently verified by Lee and Al-Dabbagh [15], who also showed that the upward liquid superficial velocity required to initiate fluidization in the presence of an upward gas flow is lower than in its absence. The latter result, illustrated by Figure 3 [15,18], is incorporated in one of the two empirical correlations for minimum fluidization velocity presented by Begovich and Watson [19].

The critical gas flow required to suspend solids in a stationary liquid has been studied by Roy et al. [6] and Narayanan et al. [7].

PHASE HOLDUPS

Solids

The solids holdup, ϵ_s, is defined as the volume fraction of the fluidized bed occupied by particulate solids and, therefore, is given at any bed level by

$$\epsilon_s = \frac{dm/dz}{\rho_s A} \tag{10A}$$

and for the bed as a whole by

$$\epsilon_s = \frac{M}{\rho_s AH} \tag{10B}$$

assuming a univalued solids density, ρ_s, and a column of constant cross-sectional area, A. Where the upper bed surface is defined distinctly, as occurs when relatively coarse and/or dense solids are used, then H can be measured by direct visual observation through a transparent column. For finer and/or lighter solids (e.g., glass beads smaller than 1 mm fluidized by air and water), the upper bed surface becomes increasingly diffuse at higher fluid flowrates, primarily due to particle entrainment, but also due to stratification of solids by size (where a significant particle size variation exists). As shown in Figure 4, a reproducible value of H then can be obtained at the intersection of two straight lines, one of positive slope representing the pressure drop profile in the constant-voidage region of the three-phase bed, which terminates at S, and the other of negative slope representing the pressure drop profile in the solids-free two-phase

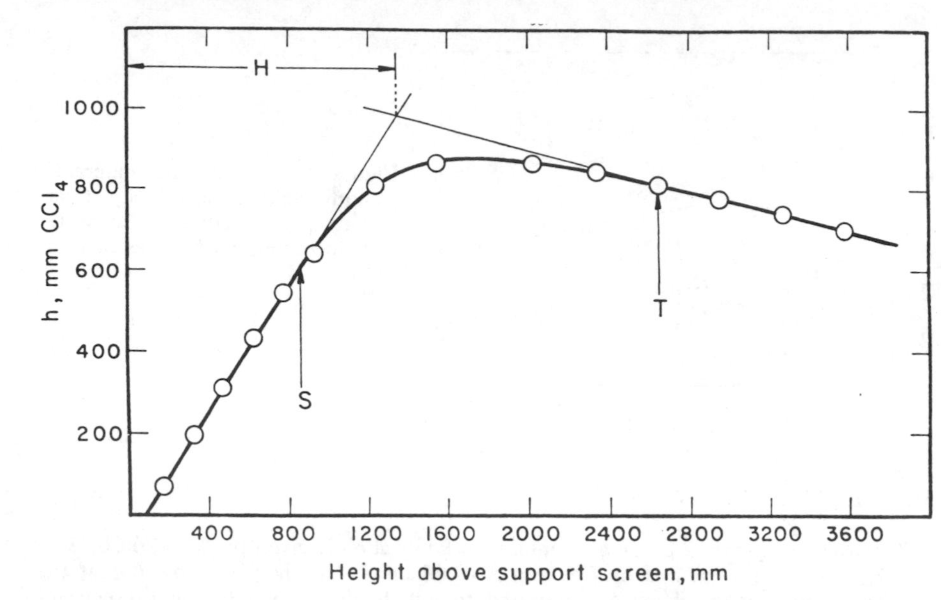

Figure 4. Typical axial pressure drop profile in gas-liquid fluidization [20]. The bottom or reference pressure tap was 95 mm above the support screen.

region above the bed, which starts at T. This value of H represents remarkably well the height that the same fluidized bed would have if the entire solids were distributed uniformly at a concentration which occurs in the actual bed only below S [19,21].

A procedure for measuring ϵ_s locally, thus avoiding the use of Equation 10B and the necessity for estimating H, has been presented by Begovich and Watson [19,22]. The method depends on a local measurement of ϵ_l by an electroconductivity technique (similar to that used for overall measurement of ϵ_l [17]) and of local pressure gradient via a pressure profile. Equations 1 and 2 then are solved simultaneously for ϵ_g and ϵ_s. For a known solids mass, M, a check on the local values of ϵ_s can be obtained by integrating Equation 10A as follows:

$$\int_0^\infty \epsilon_s dz = \frac{1}{\rho_s A}\int_0^M dm = \frac{M}{\rho_s A} \qquad (10C)$$

Under certain circumstances, the introduction of gas to a liquid-fluidized bed or the increase of gas velocity to a gas-liquid fluidized bed [23] results in a contraction of the bed. This counterintuitive result occurs because some of the liquid that otherwise would give support to the solid particles is diverted to the solids-deficient wakes behind the gas bubbles. Where the wake flux is large relative to the remaining liquid flux, as in the case of small and/or light particles in viscous liquids, the resulting contraction effect usually overrides the expansion caused by the presence of the gas bubbles. And where the bubble wake flux is relatively small, as for large and/or heavy particles in nonviscous liquids, the expansion effect tends to predominate. For glass beads in water as the continuous phase, the transition particle size is about 3 mm when the dispersed fluid is a gas [24], and somewhat smaller when the dispersed fluid is either kerosene or toluene [2]. Other three-phase systems show other transition sizes [25].

This behavior for wettable solids can be rationalized most consistently by the "generalized wake model" of Bhatia and Epstein [21]. In this model, as in other wake models, the three-phase fluidized bed is divided into a liquid-fluidized region, a gas-bubble region and a bubble-wake region (Figure 5), so that

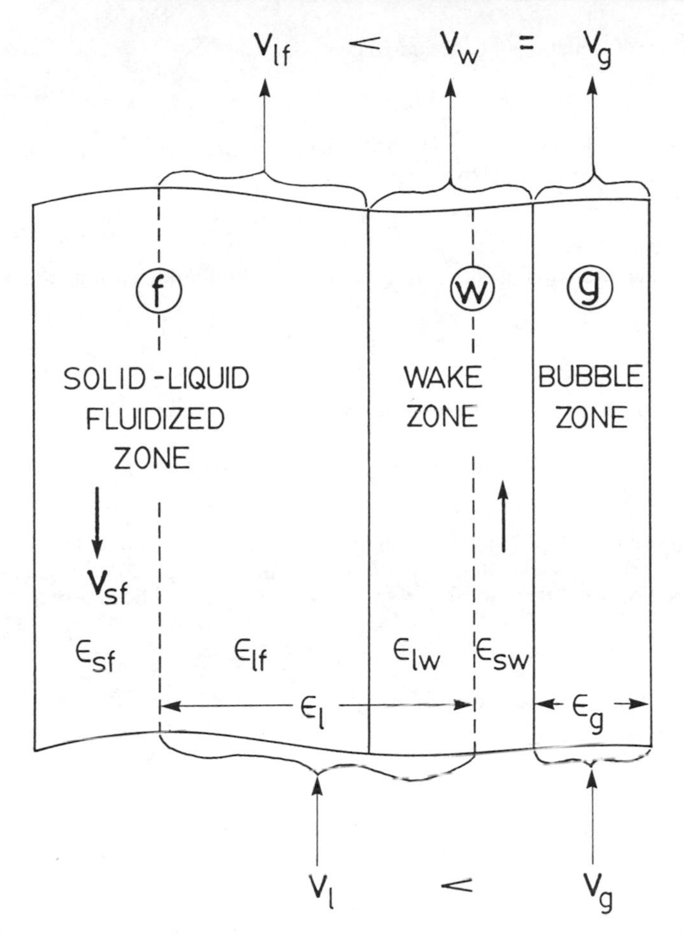

Figure 5. Three-zone model.

$$\epsilon_f + \epsilon_g + \epsilon_w = 1 \tag{11}$$

It also is assumed that the bubbles and their periodically shedding and reforming wakes travel at the same velocity. Whereas other wake models assume either that the wake region is composed of liquid free of solids [26] or that this region has a solids holdup equal to that of the liquid-fluidized region [27], the generalized wake model assumes neither extreme but instead introduces a new parameter, x, the solids partition coefficient, defined by

$$x = \frac{\epsilon_{sw}}{\epsilon_{sf}} = \frac{\epsilon_{sw}}{1 - \epsilon_{lf}} = \frac{1 - \epsilon_{lw}}{\epsilon_{sf}} \tag{12}$$

In general, x lies between zero (solids-free wakes) and unity (wakes solids holdup same as in liquid-fluidized region). The overall bed voidage, ϵ, is the sum of the total gas holdup, ϵ_g, and the total liquid holdup, ϵ_l, the latter being divided between the wake region and the liquid-fluidized region:

$$\epsilon = \epsilon_g + \epsilon_w \epsilon_{lw} + \epsilon_f \epsilon_{lf} \tag{13}$$

Combination of Equations 11, 12 and 13 gives

$$\epsilon = \epsilon_g + \epsilon_w(1 - x) + (1 - \epsilon_g - \epsilon_w + \epsilon_w x)\epsilon_{lf} \tag{14}$$

The net upward flow of liquid, which divides itself between the liquid-fluidized region and the wake region, is given by

$$U_l = U_{lf}\epsilon_f + v_g\epsilon_w\epsilon_{lw} \tag{15}$$

Therefore, the interstitial velocity of liquid in the liquid-fluidized region is given by

$$v_{lf} = \frac{U_{lf}}{\epsilon_{lf}} = \frac{U_l - v_g\epsilon_w(1 - \epsilon_{sw})}{\epsilon_f\epsilon_{lf}} \tag{16}$$

Substituting for ϵ_f from Equation 11 and for ϵ_{sw} from Equation 12,

$$v_{lf} = \frac{U_l - v_g\epsilon_w(1 - x\epsilon_{sf})}{(1 - \epsilon_g - \epsilon_w)\epsilon_{lf}} \tag{17}$$

The net upward flow of solids in the bubble-wakes, which eventually are freed of their solids during the wake-shedding in the freeboard, i.e., in zone ST of Figure 4 [20], must be balanced by a net downward flow of solids in the liquid-fluidized region. Hence,

$$v_g\epsilon_w\epsilon_{sw} = -\,v_{sf}\epsilon_f\epsilon_{sf} \tag{18}$$

Substituting for ϵ_f from Equation 11 and for ϵ_{sw} from Equation 12,

$$v_{sf} = -\frac{v_g\epsilon_w x}{1 - \epsilon_g - \epsilon_w} \tag{19}$$

The liquid velocity in the liquid-fluidized region relative to the solids in this region is given by

$$v_{ls} = v_{lf} - v_{sf} \tag{20}$$

Substituting for v_{lf} from Equation 17, for v_{sf} from Equation 19, and utilizing Equation 12,

$$v_{ls} = \frac{U_l - v_g\epsilon_w(1 - x)}{\epsilon_{lf}(1 - \epsilon_g - \epsilon_w)} \tag{21}$$

An alternative expression for this relative velocity in the liquid-fluidized region is given by the Richardson-Zaki [28] equation for particulate fluidization:

$$v_{ls} = V_1\epsilon_{lf}^{\,n-1} \tag{22}$$

where n is the slope, and V_1 is the intercept at $\epsilon = 1$ when U_l in the absence of gas is plotted against ϵ on log-log coordinates. The parameters n and V_1 may be determined experimentally from expansion measurements on the liquid-fluidized bed, or they may be estimated for a given liquid-solid system and column diameter from the empirical equations of Richardson and Zaki [28] or of Neuzil and Hrdina [29]. Eliminating v_{ls} between Equations 21 and 22 and solving for ϵ_{lf},

$$\epsilon_{lf} = \left\{\frac{U_l - v_g\epsilon_w(1 - x)}{V_1(1 - \epsilon_g - \epsilon_w)}\right\}^{1/n} \tag{23}$$

Substituting this expression for ϵ_{lf} into Equation 14, along with $v_g = U_g/\epsilon_g$, then

$$\epsilon = 1 - \epsilon_s = \epsilon_g + \epsilon_w(1-x) + (1 - \epsilon_g - \epsilon_w + \epsilon_w x) \left\{\frac{U_l - U_g\epsilon_w(1-x)/\epsilon_g}{V_1(1-\epsilon_g-\epsilon_w)}\right\}^{1/n} \tag{24}$$

To use Equation 24 to predict the solids holdup, ϵ_s, for a given gas-liquid-solid system (which fixes n and V_1) and given values of U_l and U_g, requires a knowledge of ϵ_g, $\epsilon_w(=k\epsilon_g)$ and x. The gas holdup, ϵ_g, will be discussed in the next section. The wake/bubble volume ratio, $k(=\epsilon_w/\epsilon_g)$, is best estimated by a procedure that treats each bubble wake as the sphere-completing volume of a spherical-cap bubble in a viscous medium, due allowance being made for hydrodynamic interaction between bubbles [30]. A much less reliable empirical equation [21,30], which can be used iteratively to obtain a rough estimate of k, is

$$k = \left(0.61 + \frac{0.037}{\epsilon_g + 0.013}\right)\epsilon^3 \tag{25}$$

Finally, the solids partition coefficient, x, can be estimated from a correlation developed by El-Temtamy and Epstein [30] for wettable solids:

$$x = 1 - 0.877\frac{V_1}{v_{gl}} \quad [0 < V_1/v_{gl} < 1.14] \tag{26A}$$

$$x = 0 \quad [V_1/v_{gl} > 1.14] \tag{26B}$$

where the velocity of the gas relative to the liquid is given by

$$v_{gl} = \frac{U_g}{\epsilon_g} - \frac{U_l}{\epsilon_l} \tag{27}$$

Equation 26 shows that x increases as d_p decreases (which decreases V_1), as μ_l increases (which again decreases V_1), as σ increases (which increases v_{gl} by decreasing ϵ_g), as U_l decreases (which increases v_{gl}) and as U_g increases (which again increases v_{gl}).

To predict whether a bed of wettable solids will expand or contract for an incremental addition of gas requires differentiation of Equation 24 with respect to U_g to determine whether $d\epsilon/dU_g$ is positive (expansion) or negative (contraction) for the given conditions. Applied to the initial introduction of gas to a liquid-fluidized bed, i.e., for the condition $U_g \rightarrow 0$, $\epsilon_g \rightarrow 0$, this is best done by writing $\epsilon_w = k\epsilon_g$ in Equation 24 and treating both k and x as constants in the differentiation. This procedure predicts bed expansions and contractions in conformity with experiment [31]. Expansions normally occur in beds of large and/or heavy particles, for which $k \sim 0$; contractions usually occur in beds of small and/or light particles for which $k >> 0$, unless conditions (e.g., high-viscosity liquid, very fine particles) are such that x is relatively large, e.g., close to unity. Nonwettable solids have a greater tendency to expand with addition of gas [32, 33], due presumably to their greater propensity for penetrating the gas bubbles and thereby falling into the bubble-wakes, thus increasing the value of x above what is given by Equation 26.

Gas (or Liquid)

If the solids holdup is known, e.g., from Equation 10B, then a pressure gradient measurement allows simultaneous solution of Equations 1 and 2 for ϵ_g and ϵ_l. A point bed density measurement [34] can be used instead of pressure gradient. One simple, but elegant, method for arriving at ϵ_g, employed by El-Temtamy [35], involves measuring pressure gradient for the three-phase bed and for the corresponding two-phase solid-liquid bed at the same value of H (and therefore of ϵ_s) using differential manometry. The three-phase pressure gradient, given by Equation 4, is subtracted from the two-phase gradient, given by Equation 5, and the result is $\epsilon_g(\rho_l - \rho_g)g$, from which ϵ_g can be determined. Liquid holdup is, of course, $1 - \epsilon_g - \epsilon_s$.

A more direct method of measuring ϵ_g is to simply isolate a representative portion of the test section by simultaneously shutting two quick-closing valves and measuring the fraction of the isolated volume occupied by the gas. Care must be taken not to include the two-phase gas-liquid volume above the three-phase bed in the isolated section. Other methods of measuring ϵ_g that also require no knowledge of ϵ_s are the use of electroresistivity (or impedance) probes [36,37], which yield local values of ϵ_g; the use of a γ-emitting gas tracer in conjunction with dynamic analysis [38]; and the measurement of average bubble velocity, v_b, relative to the column walls, whence $\epsilon_g = U_g/V_b$ on the assumption that V_b is equal to the average interstitial gas velocity, v_g. Alternatively, measurements of ϵ_l may be obtained by the electroconductivity technique [17,22] referred to in the previous section; by the use of a γ-emitting liquid tracer in conjunction with dynamic analysis [38]; and by γ-ray transmission measurements [15].

Several empirical correlations for either ϵ_g or ϵ_l as a function of fluid fluxes and of particle and fluid properties are given in the literature. Unfortunately, each is limited to a relatively narrow range of conditions. An attempt by Begovich and Watson [19] to broaden the range of applicability has resulted in the following empirical equation, based on a statistical correlation of 913 data points, many from the literature:

$$\epsilon_g = \frac{(1.612 \pm 0.023) U_g^{0.720 \pm 0.028} d_p^{0.168 \pm 0.061}}{D_c^{0.125 \pm 0.088}} \tag{28}$$

with d_p and D_c in m, and V_g in m/sec. Equation 28 is apparently for wettable solids and shows no effect of fluid properties, but other investigators [39–41] of gas-liquid and gas-slurry column reactors have found that

$$\frac{\epsilon_g}{(1 - \epsilon_g)^4} \propto \frac{\rho_l^{1/8}}{\sigma^{1/8} \nu_l^{1/6}} \tag{29}$$

Because of the preponderance of water data underlying Equation 28, when applying this equation to predict ϵ_g of other liquids, corrections for variation of liquid properties from those of water at room temperature should be made by means of Equation 29. For nonwettable solids, gas holdup decreases even more than does solids holdup relative to similar wettable solids under the same conditions [33].

Equation 28 also shows no effect of liquid flux on ϵ_g, although Begovich and Watson [19] observed that for the liquid-solids fluidization systems they studied, typically involving water and 4.6-mm glass beads, bed expansion always occurred when introducing air into the bed (in line with the predictions of the previous section). They also found that increasing the liquid velocity while holding all other conditions constant "resulted in a slight decrease in the overall gas holdup." These observations for air-water fluidization of relatively large glass beads are similar to those of Østergaard [38], who also found the reverse effects for the air-water system using smaller (e.g., < 3 mm) glass beads, i.e., bed contraction on introducing gas and an increase in gas holdup on raising the liquid flux for a fixed gas flux. Both sets of observations support the generalization that ϵ_g for relatively small particles is lower than ϵ_g for the corresponding (same U_l and U_g) solids-free system, while the reverse is the case for relatively large particles. Further, differences between three-phase and two-phase (gas-liquid) systems in this respect (and others) are attenuated as the liquid flux is increased, i.e., as the bed becomes more dilute in solids. These phenomena are related directly to the bubble characteristics discussed below.

BUBBLE CHARACTERISTICS

The uniform-bubbling, churn-turbulent and slug-flow regimes common in gas-liquid flow also occur in three-phase fluidization, the transition from uniform bubbling to churn turbulence commonly occurring when U_g/U_l exceeds about 2.5 [42]. However, the most striking effects of introducing fluidized solids to an upward-flowing two-phase gas-liquid system are the enhanced bubble growth, which occurs in beds of fine particles, and the increased bubble splitting, which takes place in beds of coarse particles, especially at low bed expansions [43]. As the bubble rise velocity increases with bubble size, the gas holdup ($= U_g/V_b$) is reduced by the bubble growth and enhanced by the bubble splitting. Gas-liquid specific interfacial areas also are reduced and enhanced, respectively. Table I summarizes some of the differences between the two types of behavior. For glass beads or sand fluidized by air and water, the transition particle diameter is about 3 mm. For denser particles it can be expected that the transition diameter would be smaller.

The bubble growth effect has been attributed to the fact that when the solid particles are much smaller than the bubbles, the latter experience the liquid-fluidized bed as a pseudohomogeneous medium of density and viscosity greater than the density and viscosity of the pure liquid. The increased viscosity, in particular, would be expected to increase the bubble coalescence rate by the mechanism of bubble entrapment in the wake of a preceding bubble [44].

The bubble disintegration phenomenon has been explained [45] as arising from Rayleigh-Taylor instability of the bubble roofs, which are caused by the higher liquid velocities necessary to fluidize large, rather than small, particles. This generates turbulence in the liquid phase of the former and liquid fingers that project through the gas bubble roofs. According to this argument, viscosity would tend to stabilize the bubbles and thereby increase their size, a result actually found experimentally by Kim et al. [46]. A different mechanism of bubble breakup, based on particle penetration of the bubble roof, has been proposed by Lee et al. [47], who derived from simple arguments and verified photographically for a few air-water fluidization runs that bubbles break up when

$$\frac{\rho_s V_b^2 d_p}{\sigma} \gtrsim 3 \tag{30}$$

Despite the controversy surrounding the mechanism on which Equation 30 is based [1,45,46], it nevertheless can be taken as a rough empirical criterion for bubble breakup, at least for nonviscous liquids.

The similarity between the value of the particle diameter for transition from a bed-contracting to a bed-expanding system on the introduction of gas, to that for transition from a bubble-coalescing to a bubble-disintegrating system (about 3 mm for air-water-

Table I. Bubble Behavior Relative to Solids-Free Column

Small d_p	Large d_p
Bubbles coalesce	Bubbles disintegrate
d_b increases	d_b decreases
V_b increases	V_b decreases
ϵ_g decreases	ϵ_g increases
$k(=\epsilon_w/\epsilon_g)$ increases	$k(=\epsilon_w/\epsilon_g)$ decreases
Liquid mixing increases	Liquid mixing decreases
k_l increases	k_l decreases
a decreases	a increases
$k_l a$ decreases	$k_l a$ increases

sand), requires explanation. The bed contraction-expansion effect probably is a result, rather than a cause, of the bubble growth-breakup phenomenon. The large, fast-rising bubbles, relatively large wakes and small liquid velocities associated with the small particles of a bubble-coalescing system result in very high bubble-wake fluxes relative to that of the solids-fluidizing liquid, a situation that results in bed contraction, unless x by Equation 26A is very high due to the high viscosity of the liquid [31]; conversely, the small, slow-moving bubbles, relatively small wakes and large liquid velocities associated with the large particles of a bubble-disintegrating system result in relatively low wake fluxes, which gives rise to bed expansion [21,24].

Bubble diameters, which change from the order of centimeters to that of millimeters in transition from a bubble coalescence to a bubble disintegration regime, can be estimated by the following expression relating surface tension forces to the opposing turbulent shear forces [47,48]:

$$d_b = \frac{C\sigma^{0.6}}{P_v^{0.4}\rho_l^{0.2}} \tag{31}$$

where the dimensionless constant, C, varies with the properties of the particles, and P_v, the power dissipation per unit volume of liquid in the bed (corrected for single-phase dissipation of the liquid), is given by [18]

$$P_v = \frac{[(\rho_s\epsilon_s + \rho_l\epsilon_l + \rho_g\epsilon_g)(U_l + U_g) - \rho_l U_l]g}{\epsilon_l} \tag{32}$$

Velocities of single gas bubbles in liquid-fluidized beds [49,50] are best correlated with bubble velocities in pure liquids by expressing the bubble velocity relative to the fluidizing liquid rather than relative to the bed or column walls [49,51]. Then, the classic Davies-Taylor [52] equation can be used to predict the isolated bubble rise velocity, U_{bo}, assuming conditions conducive to the occurrence of spherical-cap bubbles [51]:

$$U_{bo} = \frac{2}{3}(gR)^{1/2} \tag{33}$$

where the radius of curvature, R, of the spherical cap bubbles is related to the bubble diameter, d_e, of an equal volume sphere as

$$R = \frac{d_e}{[4 - 6\cos(\theta/2) + 2\cos^3(\theta/2)]^{1/3}} \tag{34}$$

and the included angle, θ, can be obtained from the kinematic viscosity of the liquid-fluidized bed in the manner of Henricksen and Østergaard [50]. Further work on relating isolated gas bubble rise velocity, $U_{bo} = V_{bo} - U_l$, to the rise velocity of many bubbles, $U_b = V_b - U_l/\epsilon_l$, in three-phase beds is required before the drift flux approach used so successfully in two-phase gas-liquid flows [53] can be applied with confidence in three-phase fluidization [42].

MIXING WITHIN PHASES

The degree of axial mixing in the gas phase is relatively low [38], but in the liquid it is dependent on the bubble characteristics discussed above. Thus, for bubble-coalescing systems axial mixing is augmented, while for bubble-disintegrating systems it is diminished, relative to a solids-free system at the same gas and liquid flowrates [54], presumably because the larger faster-moving bubbles stir the liquid more, while the smaller, slower-moving bubbles stir it less, than the bubbles in the solids-free system. The same trends seem to apply to the degree of radial mixing in the liquid phase, but the radial dispersion coefficients are an order of magnitude smaller than the corresponding axial dispersion coefficients [55]. As in the case of solids-free gas-liquid and liquid-liquid systems, the axial diffusivity for the continuous liquid phase in gas-liquid-solid systems increases sharply with bed diameter [54,56], the most recent generalized correlation [57] being

$$D_l = 0.29(U_l + V_c)D_c \tag{35}$$

where the average liquid circulation velocity, V_c, arising from energy dissipation at the gas-liquid and solid-liquid interfaces is given by

$$V_c = 1.31\left\{gD_c\left[U_g + U_l - \frac{\rho_l U_l}{\epsilon_s\rho_s + \epsilon_l\rho_l} - \epsilon_s\left(\frac{\rho_s}{\epsilon_s\rho_s + \epsilon_l\rho_l} - 1\right)U_{lf} - \epsilon_g U_{bo}\right]\right\}^{1/3} \tag{36}$$

The axially dispersed plug flow model with a single value of the dispersion coefficient to characterize the whole bed has been used most commonly to describe liquid mixing in three-phase fluidized beds [56]. However, as in the case of two-phase fluidization, the grid region displays different continuous-phase mixing behavior than the rest of the bed; therefore, a two-zone model is being advocated for both bubble columns and three-phase fluidized beds [58]. In the bulk of the bed, the axial mixing is contributed mainly by wake formation and shedding behind the gas bubbles. Where these wakes carry appreciable solids, as in the case of fine particles ($d_p < 1$ mm), the same mechanism contributes to solids mixing, and for $d_p < 0.1$ mm (where we are more likely to have slurry transport than fluidization) there is very little difference between the axial diffusivity for the solids and that of the liquid [59,60], at least in laboratory-sized columns ($D_c > 13$ mm). For larger particles and columns, Kato et al. [60] have recommended that the liquid axial diffusivity be divided by the empirical factor, $1 + 0.009\ \mathrm{ReFr}^{-0.4}$, to yield the solids axial diffusivity.

MASS AND HEAT TRANSFER

The difference between bubble-coalescing and bubble-disintegrating systems is manifested most dramatically in the experimental data of Østergaard and Suchozebrski [61] on volumetric liquid-phase mass transfer coefficients, $k_l a$, for gas-liquid mass transfer (Figure 6). These data, especially for the larger particles, are in substantial agreement with those of many subsequent studies [62–65], most recently with those of Dhanuka and Stepanek [66], although not with those of Østergaard [54]. Where techniques have been used to separate k_l and a [54,63,66], a wider spread between the various data has been reported. A conservative value of k_l may be obtained from the Calderbank [67] empirical equation for bubbles rising in a liquid:

$$k_l = 0.31\left\{\frac{(\rho_s - \rho_g)gD^2}{\mu_l}\right\}^{1/3} \tag{37}$$

A more realistic, but more complex, empirical equation containing both fluid velocities as variables has been presented by Fukushima [68], who also proposes a generalized equation for the gas-liquid interfacial area per unit volume of bed, a, which is related to the average bubble diameter, d_b, by

$$a = \frac{6\epsilon_g}{d_b} \tag{38}$$

Solid-liquid mass transfer in gas-liquid fluidized beds and slurry columns can be correlated either by a relative velocity approach or by application of Kolmogoroff's theory of

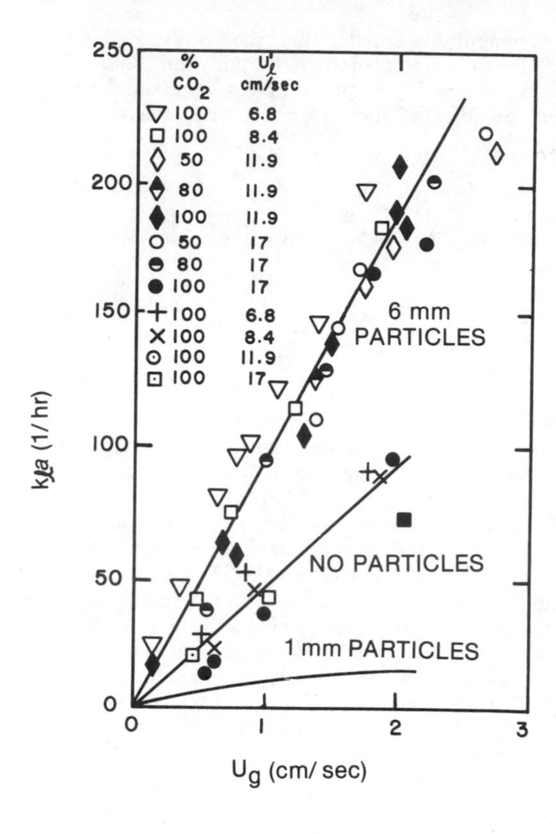

Figure 6. Volumetric absorption coefficients for 1-mm particles (bubble-coalescing system), two-phase gas-liquid flow and 6-mm particles (bubble-disintegrating system) at different liquid velocities [61].

local isotropic turbulence. Shah [12] describes both methods and recommends the latter. His recommendation is probably valid for slurry columns, where relative velocity between liquid and solids may be difficult to determine accurately, but for three-phase fluidized beds the relative velocity method is more straightforward and easier to apply.

Although there are no reported studies of gas-liquid or particle-liquid heat transfer in three-phase fluidized beds, the analogy between heat and mass transfer can be used to generate credible values of heat transfer coefficients where the corresponding mass transfer coefficients have been determined.

Two early studies [69,70] of wall-to-bed heat transfer showed that, at low bed expansions, the heat transfer coefficient for a liquid-fluidized bed was significantly increased by the introduction of gas bubbles, presumably due to the stirring effect of the latter [71]. As the gas rate was increased, the heat transfer coefficient eventually reached a maximum value. Similar trends were observed by the much more broadly based studies of Armstrong, Baker and Bergougnou [72,73], who measured heat transfer from a centrally located cylinder to their two- and three-phase beds. Generally they found that values of surface-to-bed heat transfer coefficients for three-phase beds exceeded those for the corresponding two-phase gas-liquid and liquid-solid systems, which, in turn, exceeded those for pure liquids. Two recent investigations [74,75] have attempted to correlate the data on wall-to-bed heat transfer by means of nondimensional groups. Kato et al. [74] recommend the following equation for the wall-to-bed heat transfer coefficient, h_w, in three-phase fluidized beds:

$$\frac{h_w d_p}{k_l'} \cdot \frac{\epsilon_l}{1-\epsilon_l} = 0.044 \left\{\frac{d_p U_l \rho_l C_{pl}}{(1-\epsilon_l) k_l'}\right\}^{0.78} + 2.0 \left(\frac{U_g^2}{g d_p}\right) \tag{39}$$

In the absence of any gas flow, Equation 39 also gives good correlation of the available wall-to-bed heat transfer data for two-phase, liquid-solid fluidized beds. Convective heat transfer or, for that matter, any other hydrodynamic correlations in the literature for gas-liquid fluidization of liquid-supported solids that do not meet this simple criterion (i.e., the correlation collapses to the case for liquid fluidization when the gas flowrate becomes zero) should be looked on warily by the reader.

ACKNOWLEDGMENT

The author is indebted to the Natural Sciences and Engineering Research Council of Canada for continuing financial support in the form of grants-in-aid of research.

NOMENCLATURE

- A cross-sectional area of column, m^2
- a gas-liquid specific interfacial area, m^{-1}
- C constant in Equation 31, dimensionless
- C_{pl} specific heat capacity of liquid, $J\,kg^{-1}K^{-1}$
- D molecular diffusivity of gas solute in liquid solvent, m^2sec^{-1}
- D_c column diameter, m
- D_l liquid-phase axial dispersion coefficient, m^2sec^{-1}
- d_b mean bubble diameter, m
- d_e diameter of sphere having same volume as bubble, m
- d_p particle diameter, m
- Fr gas-phase Froude number $= U_g^2/gD_c$, dimensionless
- g acceleration of gravity, $msec^{-2}$
- H bed height, m
- h vertical differential manometer reading, m
- h_w wall-to-bed heat transfer coefficient, $W\,m^{-2}K^{-1}$
- k wake volume/bubble volume $= \epsilon_w/\epsilon_g$, dimensionless
- k_l liquid side gas-liquid mass transfer coefficient, $msec^{-1}$
- k_l' thermal conductivity of liquid, $W\,m^{-1}K^{-1}$
- M total solids mass in bed, kg
- m solids mass, kg
- n Richardson-Zaki index for liquid fluidization, dimensionless
- P total pressure, Nm^{-2}
- P_v power dissipation per unit volume of liquid in bed, Wm^{-3}
- p dynamic pressure, Nm^{-2}
- p_f frictional pressure, Nm^{-2}
- R radius of curvature of spherical-cap bubbles, m
- Re terminal particle Reynolds number $= d_pV_1/\nu_l$, dimensionless
- U_b bubble rise velocity relative to liquid, $msec^{-1}$
- U_{bo} isolated bubble rise velocity relative to liquid, $msec^{-1}$
- U_g superficial gas velocity, $msec^{-1}$
- U_l superficial liquid velocity, $msec^{-1}$
- U_{lf} superficial liquid velocity in liquid-fluidized zone, $msec^{-1}$
- V_b bubble velocity relative to walls, $msec^{-1}$
- V_{bo} isolated bubble velocity relative to walls, $msec^{-1}$
- V_c average liquid circulation velocity, $msec^{-1}$
- V_1 Richardson-Zaki intercept for liquid fluidization = terminal velocity of single particle in liquid velocity field of fluidization column
- v_g interstitial gas velocity, $msec^{-1}$
- v_{gl} velocity of gas relative to liquid, $msec^{-1}$
- v_l interstitial liquid velocity, $msec^{-1}$
- v_{lf} interstitial liquid velocity in liquid-fluidized zone, $msec^{-1}$
- v_{ls} velocity of liquid relative to solids in liquid-fluidized zone, $msec^{-1}$
- v_{sf} solids velocity in liquid-fluidized zone, $msec^{-1}$
- x solids partition coefficient, dimensionless
- z distance in upward direction, m

Greek Symbols

- ϵ bed voidage $= \epsilon_g + \epsilon_l$, dimensionless
- ϵ_f volume fraction of liquid-fluidized zone

ϵ_g gas volume fraction
ϵ_l liquid volume fraction
ϵ_{lf} liquid holdup in liquid-fluidized zone
ϵ_{lw} liquid holdup in wake zone
ϵ_{mf} bed voidage at minimum fluidization
ϵ_s solids volume fraction
ϵ_{sf} solids holdup in liquid-fluidized zone $= 1 - \epsilon_{lf}$
ϵ_{sw} solids holdup in wake zone $= 1 - \epsilon_{lw}$
ϵ_w wake volume fraction
θ included angle of spherical-cap bubble
μ_l absolute viscosity of liquid, $kg\ m^{-1}sec^{-1}$
ν_l kinematic viscosity of liquid $= \mu_l/\rho_l$, m^2sec^{-1}
ρ_f composite fluid density, $kg\ m^{-3}$
ρ_g gas density, $kg\ m^{-3}$
ρ_l liquid density, $kg\ m^{-3}$
ρ_s solids density, $kg\ m^{-3}$
σ gas-liquid surface tension, Nm^{-1}

REFERENCES

1. Epstein, N. *Can. J. Chem. Eng.* 59:649 (1981).
2. Dakshinamurty, P., K. Veerabhadra Rao and A. B. Venkatarao. *Ind. Eng. Chem. Proc. Des. Dev.* 18:638 (1979).
3. Roszak, J., and R. Gawronski. *Chem. Eng. J.* 17:101 (1979).
4. O'Neill, B. K., D. J. Nicklin and L. S. Leung. In: *Fluidization and its Applications,* H. Angelino et al., Eds. (Toulouse: Cepadues-Editions, 1974), p. 365.
5. Mukherjee, R. N., P. Bhattacharya and D. K. Taraphdar. In: *Fluidization and its Applications,* H Angelino et al., Eds. (Toulouse: Cepadues-Editions, 1974), p. 372.
6. Roy, N. K., D. K. Guha and M. N. Rao. *Chem. Eng. Sci.* 19:215 (1964).
7. Narayanan, S., V. K. Bhatia and D. K. Guha. *Can. J. Chem. Eng.* 47:360 (1969).
8. Volpicelli, G., and L. Massimilla. *Chem. Eng. Sci.* 25:1361 (1970).
9. Page, R. E. Ph.D. Thesis, Appendix D, University of Cambridge, Cambridge, England (1970).
10. Fan, L.-S., K. Muroyama and S. H. Chern. "Hydrodynamics of Inverse Fluidization in Liquid-Solid and Gas-Liquid-Solid Systems," *Chem. Eng. J.* in press.
11. Pruden, B. B., and M. E. Weber. *Can. J. Chem. Eng.* 48:162 (1970).
12. Shah, Y. T. *Gas-Liquid-Solid Reactor Design* (New York: McGraw-Hill Book Co., 1979).
13. Farkas, E. J., and P. F. Leblond. *Can. J. Chem. Eng.* 47:215 (1969).
14. Fan, L. T., Y. C. Yang and C. Y. Wen. *AIChE J.* 5:407 (1959).
15. Lee, J. C., and N. Al-Dabbagh. *Fluidization,* Proc. 2nd Eng. Found. Conf., J. F. Davidson and D. L. Keairns, Eds. (New York: Cambridge University Press, 1978), p. 184.
16. Ermakova, A., G. K. Ziganshin and M. G. Slinko. *Theor. Found. Chem. Eng.* 4:95 (1970).
17. Dhanuka, V. R., and J. B. Stepanek. *Fluidization,* Proc. 2nd Eng. Found. Conf., J. F. Davidson and D. L. Keairns, Eds. (New York: Cambridge University Press, 1978), p. 179.
18. Lee, J. C., and P. S. Buckley. In: *Biological Fluidized Bed Treatment of Water and Wastewater,* P. F. Cooper and B. Atkinson, Eds. (Ellis Horwood Ltd., 1981), Chapter 4.
19. Begovich, J. M., and J. S. Watson. *Fluidization,* Proc. 2nd Eng. Found. Conf., J. F. Davidson and D. L. Keairns, Eds. (New York: Cambridge University Press, 1978), p. 190.
20. El-Temtamy, S. A., and N. Epstein. In: *Fluidization,* J. R. Grace and J. M. Matsen, Eds. (New York: Plenum Publishing Corp., 1980), p. 519.
21. Bhatia, V. K., and N. Epstein. *Fluidization and its Applications,* H. Angelino et al., Eds. (Toulouse: Cepadues-Editions, 1974), p. 372.
22. Begovich, J. M., and J. S. Watson. *AIChE J.* 24:351 (1978).
23. Michelsen, M. L., and K. Østergaard. *Chem. Eng. J.* 1:37 (1970).
24. Epstein, N. *Can. J. Chem. Eng.* 54:259 (1976).
25. Soung, W. Y. *Ind. Eng. Chem. Proc. Des. Dev.* 17:33 (1978).

26. Stewart, P. S. B., and J. F. Davidson. *Chem. Eng. Sci.* 19:319 (1964).
27. Østergaard, K. *Chem. Eng. Sci.* 20:165 (1965).
28. Richardson, J. F., and W. N. Zaki. *Trans. Inst. Chem. Eng.* 32:35 (1954).
29. Neuzil, L., and M. Hrdina. *Coll. Czech. Chem. Commun.* 30:752 (1965).
30. El-Temtamy, S. A., and N. Epstein. *Int. J. Multiphase Flow* 4:19 (1978).
31. El-Temtamy, S. A., and N. Epstein. *Can. J. Chem. Eng.* 57:520 (1979).
32. Bhatia, V. K., K. A. Evans and N. Epstein. *Ind. Eng. Chem. Proc. Des. Dev.* 11: 151 (1972).
33. Armstrong, E. R., C. G. J. Baker and M. A. Bergougnou. *Fluidization Technology,* D. L. Keairns, Ed., Vol. 1 (Washington, DC: Hemisphere Publishing Corp., 1976), p. 405.
34. Blum, D. B., and J. J. Toman. *AIChE Symp. Series* 73(161):115 (1977).
35. El-Temtamy, S. A. Ph.D. Thesis, Cairo University, Cairo, Egypt (1976).
36. Rigby, G. R., G. P. Van Blockland, W. H. Park and C. E. Capes. *Chem. Eng. Sci.* 25:1729 (1970).
37. Darton, R. C., and D. Harrison. *Inst. Chem. Eng. Symp. Ser.* No. 38, paper B1 (1974).
38. Østergaard, K. *Studies of Gas-Liquid Fluidisation* (Danish Technical Press, 1969).
39. Ying, D. H., E. N. Givens and R. F. Weimer. *Ind. Eng. Chem. Proc. Des. Dev.* 19:635 (1980).
40. Akita, K., and F. Yoshida. *Ind. Eng. Chem. Proc. Des. Dev.* 12:76 (1973).
41. Shah, Y. T., B. G. Kelkar, S. P. Godbole and W.-D. Deckwer. *AIChE J.* 28:353 (1982).
42. Darton, R. C., and D. Harrison. *Chem. Eng. Sci.* 30:581 (1975).
43. Kim, S. D., C. G. J. Baker and M. A. Bergougnou. *Can. J. Chem. Eng.* 53:134 (1975).
44. Østergaard, K. *AIChE Symp. Ser.* 69(128):28 (1973).
45. Henricksen, H. K., and K. Østergaard. *Chem. Eng. Sci.* 29:626 (1974).
46. Kim, S. D., C. G. J. Baker and M. A. Bergougnou. *Chem. Eng. Sci.* 32:1299 (1977).
47. Lee, J. C., A. J. Sherrard and P. S. Buckley. *Fluidization and its Applications,* H. Angelino et al., Eds. (Toulouse: Cepadues-Editions, 1974), p. 407.
48. Calderbank, P. H. "Gas Absorption from Bubbles," *Chem. Eng.* CE 209 (1967).
49. Darton, R. C., and D. Harrison. *Trans. Inst. Chem. Eng.* 52:301 (1974).
50. Henricksen, H. K., and K. Østergaard. *Chem. Eng. J.* 7:141 (1974).
51. El-Temtamy, S. A., and N. Epstein. *Chem. Eng. J.* 19:153 (1980).
52. Davies, R. M., and G. I. Taylor. *Proc. Roy. Soc. London Ser. A,* 200:375 (1950).
53. Wallis, G. B. *One-Dimensional Two-Phase Flow* (New York: McGraw-Hill Book Co., 1969).
54. Østergaard, K. *AIChE Symp. Ser.* 74(176):82 (1978).
55. El-Temtamy, S. A., Y. O. El-Sharnoubi and M. M. El-Halwagi. *Chem. Eng. J.* 18(2): 161 (1979).
56. El-Temtamy, S. A., Y. O. El-Sharnoubi and M. M. El-Halwagi. *Chem. Eng. J.* 18(2): 151 (1979).
57. Joshi, J. B. *Trans. Inst. Chem. Eng.* 58:155 (1980).
58. Alvarez-Cuenca, M., M. A. Nerenberg, C. G. J. Baker and M. A. Bergougnou. "Mass Transfer Models for Bubble Columns and Three-Phase Fluidized Beds," Paper No. 8-3, 29th Can. Chem. Eng. Conf., Sarnia, Ontario, October, 1979.
59. Imafuku, K., T. Y. Wang, K. Koide and H. Kubota. *J. Chem. Eng. Japan* 1:153 (1968).
60. Kato, Y., A. Nishiwaki, T. Fukuda and S. Tanaka. *J. Chem. Eng. Japan* 5:112 (1972).
61. Østergaard, K., and W. Suchozebrski. *Proc. 4th Europ. Symp. Chem. Reaction Eng.* (Elmsford, NY: Pergamon Press, Inc., 1971).
62. Østergaard, K., and P. Fosbøl. *Chem. Eng. J.* 3:105 (1972).
63. Lee, J. C., and H. Worthington. *Inst. Chem. Eng. Symp. Ser.* No. 38, paper B2 (1974).

64. Dakshinamurty, P., C. Chiranveji, V. Subrahmanyam and P. Kameswar Rao. *Fluidization and its Applications*, H. Angelino et al., Eds. (Toulouse: Cepadues-Editions, 1974), p. 429.
65. Dakshinamurty, P., and K. Veerabhadra Rao. *Indian J. Technol.* 14:9 (1976).
66. Dhanuka, V. R., and J. B. Stepanek. *AIChE J.* 26:1029 (1980).
67. Calderbank, P. H. *Trans. Inst. Chem. Eng.* 37:173 (1959).
68. Fukushima, S. *J. Chem. Eng. Japan* 12:489 (1979).
69. Østergaard, K. In: *Fluidisation* (London: Society of Chemical Industry 1964), p. 58.
70. Viswanathan, S., A. S. Kakar and P. S. Murti. *Chem. Eng. Sci.* 20:903 (1965).
71. Østergaard, K. In: *Fluidization*, J. F. Davidson and D. Harrison, Eds. (New York: Academic Press, Inc., 1978).
72. Armstrong, E. R., C. G. J. Baker and M. A. Bergougnou. *Fluidization Technology*, D. L. Keairns, Ed., Vol. I (Washington, DC: Hemisphere Publishing Corp., 1976), p. 453.
73. Baker, C. G. J., E. R. Armstrong and M. A. Bergougnou. *Powder Technol.* 21:195 (1978).
74. Kato, Y., K. Uchida, T. Kago and S. Morooka. *Powder Technol.* 28:173 (1981).
75. Muroyama, K., M. Fukuma and A. Yasunishi. "Heat Transfer in Three-Phase Fluidized Beds," paper presented at Session 36, 20th National Heat Transfer Conference, preprint by Heat Transfer and Energy Conversion Div., AIChE, August, 1981.

INDEX